Airline Transport Pilot TEST PREP

STUDY & PREPARE

For the Aircraft Dispatcher and ATP Part 121, 135, Airplane, and Helicopter FAA Knowledge Exams

READER TIP:
The FAA Knowledge Exam Questions can change throughout the year. Stay current with test changes; sign up for ASA's free email update service at **www.asa2fly.com/testupdate**

Aviation Supplies & Academics, Inc.
Newcastle, Washington

Airline Transport Pilot Test Prep
2013 Edition

Aviation Supplies & Academics, Inc.
7005 132nd Place SE
Newcastle, Washington 98059-3153
425.235.1500
www.asa2fly.com

FAA Questions herein are from United States government sources and contain current information as of: June 2012

None of the material in this publication supersedes any documents, procedures or regulations issued by the Federal Aviation Administration.

ASA assumes no responsibility for any errors or omissions. Neither is any liability assumed for damages resulting from the use of the information contained herein.

Important: This Test Prep should be sold with and used in conjunction with *Computer Testing Supplement for Airline Transport Pilot and Aircraft Dispatcher* (FAA-CT-8080-7C).

ASA reprints the FAA test figures and legends contained within this government document, and it is also sold separately and available from aviation retailers nationwide. Order #ASA-CT-8080-7C.

ASA-TP-ATP-13
ISBN 1-56027-919-2
978-1-56027-919-8

Printed in the United States of America

2013 2012 5 4 3 2 1

Stay informed of aviation industry happenings

Website www.asa2fly.com
Updates www.asa2fly.com/testupdate
Blog www.learntoflyblog.com
Twitter www.twitter.com/asa2fly
Facebook www.facebook.com/asa2fly

About the Contributors

Charles L. Robertson
Associate Professor, UND Aerospace
University of North Dakota

Charles Robertson as flight instructor, associate professor and manager of training at UND Aerospace, contributes a vital and substantial combination of pilot and educator to ASA's reviewing team. After graduating with education degrees from Florida State University in 1967, and Ball State University in 1975, he began his USAF career as Chief of avionics branch, 58th Military Airlift Squadron, and went on to flight instruction, training for aircraft systems, and airport managing, while gaining many thousands of hours flying international passenger and cargo, aerial refueling and airlift missions. As Division Chief in 1988, Robertson directed the Strategic Air Command's "Alpha Alert Force," coordinating daily flight training operations. He holds the CFI Airplane Land, Multi-Engine, Single-Engine and Instrument, the ATP Airplane Land and Multi-Engine, Commercial Pilot, Advanced and Instrument Ground Instructor licenses.

Jackie Spanitz
Director of Curriculum Development
Aviation Supplies & Academics, Inc.

Jackie Spanitz earned a bachelor of science degree with Western Michigan University (WMU), in Aviation Technology and Operations—Pilot option. In her masters program at Embry-Riddle Aeronautical University, she earned a degree in Aeronautical Science, specializing in Management. As Director of Curriculum Development for ASA, Jackie oversees new and existing product development, ranging from textbooks and flight computers to flight simulation software products, and integration of these products into new and existing curricula. She provides technical support, research for product development, and project management. Jackie holds pilot and instructor certificates and is the author of *Guide to the Flight Review, Private Pilot Syllabus, Instrument Rating Syllabus*, and *Commercial Pilot Syllabus*. Jackie is the technical editor for ASA's Test Prep series.

Tina Anderson
Associate Professor and Assistant Chair of Academics, UND Aerospace
University of North Dakota

Tina Anderson holds Airline Transport Pilot and Flight Instructor certificates and has airline experience in the DC-9 and DHC-8 aircraft. She has a bachelor of science degree in Aeronautical Studies, and a master of science degree in Aviation from the University of North Dakota. Tina is involved with Women in Aviation International and the Airline Pilots Association.

About ASA: Aviation Supplies & Academics, Inc. (ASA) is an industry leader in the development and sale of aviation supplies and publications for pilots, flight instructors, flight engineers, air traffic controllers, flight attendants, and aviation maintenance technicians. We manufacture and publish more than 300 products for the aviation industry. Aviators are invited to call 1-800-ASA-2-FLY for a free copy of our catalog. Visit ASA on the web: **www.asa2fly.com**

Contents

Continued

Preface

Welcome to ASA's Test Prep Series. ASA's test books have been helping pilots prepare for the FAA Knowledge Tests since 1984 with great success. We are confident that with proper use of this book, you will score very well on any of the Airline Transport Pilot tests.

Begin your studies with a classroom or home-study ground school course, which will involve reading a comprehensive textbook (see the FAA Knowledge Exam References list on page x). Conclude your studies with this Test Prep or comparable software. Read the question, select your choice for the correct answer, then read the explanation. Use the Learning Statement Codes and references that conclude each explanation to identify additional resources if you need further study of a subject.

The FAA Airline Transport Pilot questions have been arranged into chapters based on subject matter. Topical study, in which similar material is covered under a common subject heading, promotes better understanding, aids recall, and thus provides a more efficient study guide. Study and place emphasis on those questions most likely to be included in your test (identified by the aircraft and test category above each question). For example: a pilot preparing for the ATP Part 121 test would focus on the questions marked "ALL" and "121"; a pilot preparing for the ATP Part 135 test would focus on the questions marked "ALL" and "135"; a pilot preparing for the ATP Part 135 Rotorcraft test would focus on the questions marked "ALL" and "RTC"; and candidates for the Dispatcher certificate would focus on the questions marked "ALL" and "DSP."

It is important to answer every question assigned on your FAA Knowledge Test. If in their ongoing review, the FAA authors decide a question has no correct answer, is no longer applicable, or is otherwise defective, your answer will be marked correct no matter which one you chose. However, you will not be given the automatic credit unless you have marked an answer. Unlike some other exams you may have taken, there is no penalty for "guessing" in this instance.

The FAA exams are "closed tests" which means the exact database of questions is not available to the public. The question and answer choices in this book provide the largest sampling of representative FAA questions available and they are derived from history and experience with the FAA testing process. You might see similar although not exactly the same questions on your official FAA exam. Answer stems may be rearranged from the A, B, C order you see in this book. Therefore, be careful to fully understand the intent of each question and corresponding answer while studying, rather than memorize the A, B, C answer. You may be asked a question that has unfamiliar wording; studying and understanding the information in this book and the associated references will give you the tools to answer all types of questions with confidence.

If your study leads you to question an answer choice, we recommend you seek the assistance of a local instructor. We welcome your questions, recommendations or concerns:

Aviation Supplies & Academics, Inc.
7005 132nd Place SE — Voice: 425.235.1500 — Fax: 425.235.0128
Newcastle, WA 98059-3153 — Email: cfi@asa2fly.com — Website: www.asa2fly.com

The FAA appreciates testing experience feedback. You can contact the branch responsible for the FAA Knowledge Exams at:

Federal Aviation Administration
AFS-630, Airman Testing Standards Branch
PO Box 25082
Oklahoma City, OK 73125 — Email: afs630comments@faa.gov

Updates and Practice Tests

Free Test Updates for the One-Year Life Cycle of Test Prep Books

The FAA rolls out new tests as needed throughout the year. The FAA Knowledge Exams are "closed tests" which means the exact database of questions is not available to the public. ASA combines years of experience with expertise in working with the tests to prepare the most comprehensive test preparation materials available in the industry.

You can feel confident you will be prepared for your FAA Knowledge Exam by using the ASA Test Preps. ASA publishes test books each June and keeps abreast of changes to the tests. These changes are then posted on the ASA website as a Test Update.

Visit the ASA website before taking your test to be certain you have the most current information. While there, sign up for ASA's free email Update service. We will then send you an email notification if there is a change to the test you are preparing for so you can review the Update for revised and/or new test information.

www.asa2fly.com/testupdate

We invite your feedback. After you take your official FAA exam, let us know how you did. Were you prepared? Did the ASA products meet your needs and exceed your expectations? We want to continue to improve these products to ensure applicants are prepared, and become safe aviators. Send feedback to: **cfi@asa2fly.com**

Description of the Tests

All test questions are the objective, multiple-choice type, with three answer choices. Each question can be answered by the selection of a single response. The answer to some questions depend on the response to a previous question to calculate the correct answer.

The FAA Knowledge Exams are designed to test your knowledge in many subject areas. If you are pursuing an airline transport pilot certificate or added rating, you should review the appropriate sections of 14 CFR Part 61 for the specific knowledge areas on each test. Those taking the ATP or "ATP—Airline Pilot (Part 121)" exam will be tested on Part 121 as one of the knowledge areas. Those taking the ATA or "ATP—Airline Pilot (Part 135)" exam will be tested on Part 135 as one of the knowledge areas. None of the other knowledge areas are specified as being for Part 121 or Part 135, so their associated questions may be used on any of the tests.

An applicant for an aircraft dispatcher certificate should review the appropriate sections of 14 CFR Part 65 for the specific knowledge areas on the test. The applicant will be tested on Part 121 as one of the knowledge areas. If Part 135 commuter operators (as defined in DOT Part 298) are required to have aircraft dispatchers in the future, Part 135 questions will be added to the test. The aircraft dispatcher applicant is not required to have the flying skills of an airline transport pilot but is expected to have the same knowledge.

For the most efficient and effective study program, begin by reading the book cover to cover. Study **all** *the questions first,* **then** *refer to the following table, placing emphasis on those questions most likely to be included on your test (identified by the aircraft category above each question number).*

Test Code	Test Name	Test Prep Study	Number of Questions	Min. Age	Allotted Time (hrs)
ADX	Aircraft Dispatcher	ALL, DSP	80	21	3.0
ATP	ATP—Airplane (Part 121)	ALL, 121	80	21	3.0
ATA	ATP—Airplane (Part 135)	ALL, 135	80	21	3.0
ATH	ATP—Helicopter (Part 135)	ALL, RTC	80	21	3.0
ARA	ATP—Airplane–Added Rating (Part 135)*	ALL, 135	50	21	2.5
ARH	ATP—Helicopter–Added Rating (Part 135)*	ALL, RTC	50	21	2.5
ACP	ATP Canadian Conversion**	ALL, 121	40	21	2.0

*** Expert Study Tip**
If it's been more than 24 months since you took the initial ATP FAA Knowledge Exam, we recommend that you prepare for the Add-On test using the "ATP, Rotorcraft, Part 135" test. This will better prepare you for the non-rotorcraft (or "ALL") questions that may be included on your add-on test.

**This test focuses on U.S. regulations, procedures and operations, not airplane know-how.

A score of 70 percent must be attained to successfully pass each test.

Note: All applicants transitioning from ATP airplane and/or helicopter need to take the additional knowledge test. For example, an applicant adding a helicopter rating to an existing ATP–Airplane certificate ***will*** need to take the 50-question add-on test.

Knowledge Test Eligibility Requirements

If you are pursuing an airline transport or aircraft dispatcher certificate, you should review Title 14 of the Code of Federal Regulations (14 CFR) Part 61, §61.23 "Medical Certificates: Requirement and Duration," 14 CFR §61.35 "Knowledge Test: Prerequisites and Passing Grades," and 14 CFR Part 61 (ATP) or Part 65 (ADX) for certificate requirements.

Process for Taking a Knowledge Test

The FAA has designated two holders of airman knowledge testing (AKT) organization designation authorization (ODA). These two AKT-ODAs sponsor hundreds of knowledge testing center locations. The testing centers offer a full range of airman knowledge tests including: Aircraft Dispatcher, Airline Transport Pilot, Aviation Maintenance Technician, Commercial Pilot, Flight Engineer, Flight Instructor, Flight Navigator, Ground Instructor, Inspection Authorization, Instrument Rating, Parachute Rigger, Private Pilot, Recreational Pilot, Sport Pilot, and Military Competence. Contact information for the AKT-ODA holders is provided at the end of this section.

The first step in taking a knowledge test is the registration process. You may either call the testing centers' 1-800 numbers or simply take the test on a walk-in basis. If you choose to use the 1-800 number to register, you will need to select a testing center, schedule a test date, and make financial arrangements for test payment. You may register for tests several weeks in advance, and you may cancel your appointment according to the AKT-ODA holder's cancellation policy. If you do not follow the AKT-ODA holder's cancellation policies, you could be subject to a cancellation fee.

The next step in taking a knowledge test is providing proper identification. Although no prior authorization is necessary, except in the case of failure (see "Acceptable Forms of Authorization" below), proper identification is required to take any airline transport pilot, aircraft dispatcher, or flight navigator knowledge test. Testing center personnel will not begin the test until your identification is verified. For U.S. citizens, an acceptable form of photo I.D. includes, but is not limited to: driver's license, government-issued I.D. card, passport, alien residency card, and/or military I.D. card. For non-U.S. citizens, a passport is required, along with one or more of the following forms of photo I.D.: driver's license, government-issued I.D. card, and/or military I.D. card.

Proper identification contains your photograph, signature, date of birth (must show that you will meet the age requirement for the certificate sought before the expiration date of the Airman Test Report), and actual residential address, if different from your mailing address.

Acceptable Forms of Authorization

1. Requires ***no*** instructor endorsements or other form of written authorization.
2. Failed, passing or expired Airman Knowledge Test Report, provided the applicant still has the ***original*** test report in his/her possession. (*See* Retesting explanation.)

Test-Taking Tips

Prior to launching the actual test, the AKT-ODA holder's testing software will provide you with an opportunity to practice navigating through the test. This practice (or tutorial) session may include a "sample" question(s). These sample questions have no relation to the content of the test, but are meant to familiarize you with the look and feel of the system screens, including selecting an answer, marking a question for later review, time remaining for the test, and other features of the testing software.

Follow these time-proven tips, which will help you develop a skillful, smooth approach to test-taking:

1. Be careful to fully understand the intent of each question and corresponding answer while studying, rather than memorize the A, B, C answer choice—answer stems may appear in a different order than you studied.
2. Take with you to the testing center proof of eligibility for this certificate, photo I.D., the testing fee, calculator, flight computer (ASA's E6-B, Micro E6-B which has the high-speed wind correction slide, or CX-2 Pathfinder), plotter, magnifying glass, and a sharp pointer, such as a safety pin.
3. Your first action when you sit down should be to write on the scratch paper the weight and balance and any other formulas and information you can remember from your study. Remember, some of the formulas may be on your E6-B.
4. Answer each question in accordance with the latest regulations and guidance publications.
5. Read each question carefully before looking at the possible answers. You should clearly understand the problem before attempting to solve it.
6. After formulating an answer, determine which answer choice corresponds the closest with your answer. The answer chosen should completely resolve the problem.
7. From the answer choices given, it may appear that there is more than one possible answer. However, there is only one answer that is correct and complete. The other answers are either incomplete, erroneous, or represent popular misconceptions.
8. If a certain question is difficult for you, it is best to mark it for REVIEW and proceed to the other questions. After you answer the less difficult questions, return to those which you marked for review and answer them. Be sure to untag these questions once you've answered them. The review marking procedure will be explained to you prior to starting the test. Although the computer should alert you to unanswered questions, make sure every question has an answer recorded. This procedure will enable you to use the available time to the maximum advantage.
9. Perform each math calculation twice to confirm your answer. If adding or subtracting a column of numbers, reverse your direction the second time to reduce the possibility of errors.
10. When solving a calculation problem, select the answer nearest to your solution. The problem has been checked with various types of calculators; therefore, if you have solved it correctly, your answer will be closer to the correct answer than any of the other choices.
11. Remember that information is provided in the FAA Legends and FAA Figures.
12. Remember to answer every question, even the ones with no completely correct answer, to ensure the FAA gives you credit for a bad question.
13. Take your time and be thorough but relaxed. Take a minute off every half-hour or so to relax the brain and the body. Get a drink of water halfway through the test.
14. Your test will be graded immediately upon completion. You will be allowed 10 minutes to review any questions you missed. You will see the question only; you will not see the answer choices or your selected response. This allows you to review the missed areas with an instructor prior to taking the Practical exam.

Test Reports

Your test will be graded immediately upon completion. You will be allowed 10 minutes to review any questions you missed. You will see the question only; you will not see the answer choices or your selected response. This allows you to review the missed areas with an instructor prior to taking the Practical exam. After this review period you will receive your Airman Test Report, with the testing center's embossed seal, which reflects your score.

Validity of Airman Test Reports

For an Airman Test Report to be valid, it must be dated within the 24-calendar month period preceding the month you complete the practical test. If the Airman Test Report expires before completion of the practical test, you must retake the knowledge test.

The 24-month limitation does not apply if you:

1. are employed as a flight crewmember by a certificate holder under 14 CFR Parts 121, 125, or 135 at the time of the practical test and have satisfactorily accomplished that operator's approved:
   a. pilot in command aircraft qualification training program that is appropriate to the certificate and rating sought; and
   b. qualification training requirements appropriate to the certificate and ratings sought; or
2. are employed as a flight crewmember in scheduled U.S. military air transport operations at the time of the practical test, and have accomplished the pilot in command aircraft qualification training program that is appropriate to the certificate and rating sought.

Test Reports and Learning Statement Codes

The Airman Test Report lists the learning statement codes for questions answered incorrectly. The total number of learning statement codes shown on the Airman Test Report is not necessarily an indication of the total number of questions answered incorrectly. Study these knowledge areas to improve your understanding of the subject matter. See the *Learning Statement Code/Question Number Cross-Reference* in the back of this book for a complete list of which questions apply to each learning statement code.

Your instructor is required to provide instruction on each of these knowledge areas listed on your Airman Test Report and to complete an endorsement of this instruction. The Airman Test Report must be presented to the examiner prior to taking the practical test. During the oral portion of the practical test, the examiner is required to evaluate the noted areas of deficiency.

Should you require a duplicate Airman Test Report due to loss or destruction of the original, send a signed request accompanied by a check or money order for $1 payable to the FAA. Your request should be sent to the Federal Aviation Administration, Airmen Certification Branch, AFS-760, P.O. Box 25082, Oklahoma City, OK 73125.

Airman Knowledge Testing Sites

The following is a list of the airman knowledge testing (AKT) organization designation authorization (ODA) holders authorized to give FAA knowledge tests. This list should be helpful in case you choose to register for a test or simply want more information. The latest listing of computer testing center locations is available on the FAA website at **http://www.faa.gov/pilots/testing**, under "Knowledge Test Centers" select "Center List" and a PDF will download automatically.

Computer Assisted Testing Service (CATS)
777 Mariners Island Blvd., Suite 200
San Mateo, CA 94404
Applicant inquiry and test registration: 1-800-947-4228
From outside the U.S.: (650) 259-8550

PSI/LaserGrade Computer Testing
16821 S.E. McGillivray, Suite 201
Vancouver, WA 98683
Applicant inquiry and test registration: 1-800-211-2753
From outside the U.S.: (360) 896-9111

Use of Test Aids and Materials

Airman knowledge tests require applicants to analyze the relationship between variables needed to solve aviation problems, in addition to testing for accuracy of a mathematical calculation. The intent is that all applicants are tested on concepts rather than rote calculation ability. It is permissible to use certain calculating devices when taking airman knowledge tests, provided they are used within the following guidelines. The term "calculating devices" is interchangeable with such items as calculators, computers, or any similar devices designed for aviation-related activities.

Guidelines for Use of Test Aids and Materials

The applicant may use test aids and materials within the guidelines listed below, if actual test questions or answers are not revealed.

1. Applicants may use test aids, such as scales, straightedges, protractors, plotters, navigation computers, log sheets, and all models of aviation-oriented calculating devices that are directly related to the test. In addition, applicants may use any test materials provided with the test.
2. Manufacturer's permanently inscribed instructions on the front and back of such aids listed in 1(a), e.g., formulas, conversions, regulations, signals, weather data, holding pattern diagrams, frequencies, weight and balance formulas, and air traffic control procedures are permissible.
3. The test proctor may provide calculating devices to applicants and deny them use of their personal calculating devices if the applicant's device does not have a screen that indicates all memory has been erased. The test proctor must be able to determine the calculating device's erasure capability. The use of calculating devices incorporating permanent or continuous type memory circuits without erasure capability is prohibited.
4. The use of magnetic cards, magnetic tapes, modules, computer chips, or any other device upon which prewritten programs or information related to the test can be stored and retrieved is prohibited. Printouts of data will be surrendered at the completion of the test if the calculating device used incorporates this design feature.
5. The use of any booklet or manual containing instructions related to the use of the applicant's calculating device is not permitted.
6. Dictionaries are not allowed in the testing area.
7. The test proctor makes the final determination relating to test materials and personal possessions that the applicant may take into the testing area.

Testing Procedures For Applicants Requesting Special Accommodations

If you are an applicant with a learning or reading disability, you may request approval from the local FSDO or FAA International Field Office (IFO) to take an airman knowledge test, using the special accommodations procedures outlined in the most current version of FAA Order 8080.6 "Conduct of Airman Knowledge Tests."

Prior to approval of any option, the FSDO or IFO Aviation Safety Inspector must advise you of the regulatory certification requirement of being able to read, write, speak, and understand the English language.

Retesting Procedures

ADX

Retests do not require a 30-day waiting period if the applicant presents a signed statement from an airman holding a certificate and rating sought by the applicant. This statement must certify that the airman has given the applicant additional instruction in each of the subjects failed, and that the airman considers the applicant ready for retesting. Requires a 30-day waiting period for retesting if the applicant presents a failed test report without a signed statement.

Applicants taking retests ***after previous failure*** are required to submit the applicable score report indicating failure to the testing center prior to retesting. The original failed test report shall be retained by the proctor and attached to the applicable sign-in/out log. The latest test taken will reflect the official score.

All Airline Transport Pilot Tests

Applicants retesting ***after failure*** are required to submit the applicable score report indicating failure, along with an endorsement from an authorized instructor who gave the applicant the additional training, and certifying the applicant is competent to pass the test. The original failed test report presented as authorization shall be retained by the proctor and attached to the applicable sign-in/out log. The latest test taken will reflect the official score.

ADX and All Airline Transport Pilot Tests

Applicants retesting ***in an attempt to achieve a higher passing score*** may retake the same test for a better grade after 30 days. The latest test taken will reflect the official score. Applicants are required to submit the ***original*** applicable score report indicating previous passing score to the testing center prior to testing. Testing center personnel must collect and destroy this report prior to issuing the new test report.

Cheating or Other Unauthorized Conduct

Computer testing centers must follow strict security procedures to avoid test compromise. These procedures are established by the FAA and are covered in FAA Order 8080.6, Conduct of Airman Knowledge Tests. The FAA has directed testing centers to terminate a test at any time a test proctor suspects a cheating incident has occurred. An FAA investigation will then be conducted. If the investigation determines that cheating or unauthorized conduct has occurred, then any airman certificate or rating that you hold may be revoked, and you will be prohibited for 1 year from applying for or taking any test for a certificate or rating under 14 CFR Part 61.

Eligibility Requirements for the Airline Transport Pilot Certificate

To be eligible for an Airline Transport Pilot Certificate, a person must:

1. Be at least 23 years old.
2. Be of good moral character.
3. Read, write and understand English, and speak it without impediment that would interfere with radio conversation.
4. Have a current Third-Class Medical Certificate.
5. Pass a knowledge examination on the appropriate subjects with a score of at least 70 percent.
6. Pass an oral and flight check on the subjects and maneuvers in the Airline Transport Pilot and Type Rating Practical Test Standards (#ASA-8081-5).
7. Have a Commercial Pilot Certificate or foreign or military equivalent.
8. For an Airplane rating have:
   a. 250 hours PIC flight time in an airplane including 100 hours cross-country and 25 hours at night
   b. 1,500 hours pilot flight time including:
      i. 500 hours cross-country flight time
      ii. 100 hours night time; and
      iii. 75 hours actual or simulated instrument time, 50 hours of which were in actual flight

   (*See* 14 CFR Part 61 for alternate ways to qualify time.)
9. For a rotorcraft category and helicopter class rating, have 1,200 hours pilot time including:
   a. 500 hours cross-country flight time
   b. 100 hours night time, at least 15 hours in helicopters
   c. 200 hours in helicopters including 75 hours PIC time
   d. 75 hours of actual or simulated instrument time with at least 50 hours in flight and 25 hours PIC time in helicopters.

Knowledge Exam References

The FAA references the following documents to write the FAA Knowledge Exam questions. You should be familiar with the latest revision for all of these as part of your ground school studies, which you should complete before starting test preparation:

FAA-G-8082-1 *Airline Transport Pilot, Aircraft Dispatcher and Flight Navigator Test Guide*

ANA *Aerodynamics for Naval Aviators*

CUG *Aeronautical Chart User's Guide*

Aeronautical Information Manual (AIM)

FAA-H-8083-25 *Pilot's Handbook of Aeronautical Knowledge*

FAA-H-8083-3 *Airplane Flying Handbook*, or FAA-H-8083-21 *Helicopter Flying Handbook*

FAA-H-8083-6 *Advanced Avionics Handbook*

FAA-H-8083-15 *Instrument Flying Handbook*

FAA-H-8083-1 *Aircraft Weight and Balance Handbook*

FAA-H-8083-2 *Risk Management Handbook*

FAA-H-8261-1 *Instrument Procedures Handbook*

FAA-S-8081-5 *ATP Practical Test Standards*

AC 00-6 *Aviation Weather*

AC 00-24 *Thunderstorms*

AC 00-30 *Atmospheric Turbulence Avoidance*

AC 00-45 *Aviation Weather Services*

AC 00-54 *Pilot Wind Shear Guide*

AC 20-117 *Hazards Following Ground Deicing & Ground Operations in Conditions Conducive to Aircraft Icing*

AC 91-6 *Water, Slush and Snow on the Runway*

AC 91-43 *Unreliable Airspeed Indication*

AC 91-51 *Effect of Icing on Aircraft Control and Airplane Deice and Anti-Ice Systems*

AC 91-74 *Pilot Guide Flight in Icing Conditions*

AC 135-17 *Pilot Guide-Small Aircraft Ground Deicing*

AC 120-51 *Crew Resource Management Training*

AC 120-58 *Pilot Guide for Large Aircraft Ground Deicing*

14 CFR Part 1, 25, 61, 63, 71, 91, 97, 119, 121, 125, 135

49 CFR Part 830

Airport/Facility Directory (A/FD)

IFR Enroute High Altitude Chart

IFR Enroute Low Altitude Chart

STAR—Standard Terminal Arrival

U.S. Terminal Procedures

Visit the ASA website for these and many more titles and pilot supplies for your aviation endeavors: **www.asa2fly.com**

ASA Test Prep Layout

The sample FAA questions have been sorted into chapters according to subject matter. Within each chapter, the questions have been further classified and all similar questions grouped together with a concise discussion of the material covered in each group. This discussion material of "Chapter text" is printed in a larger font and spans the entire width of the page. Immediately following the sample FAA Question is ASA's Explanation in *italics*. The last line of the Explanation contains the Learning Statement Code and further reference (if applicable). *See* the EXAMPLE below.

Figures referenced by the Chapter text only are numbered with the appropriate chapter number, i.e., "Figure 1-1" is Chapter 1's first chapter-text figure.

Some Questions refer to Figures or Legends immediately following the question number, i.e., "**8201.** (Refer to Figure 14.)." These are FAA Figures and Legends which can be found in the separate booklet: *Computer Testing Supplement* (CT-8080-XX). This supplement is bundled with the Test Prep and is the exact material you will have access to when you take your computerized test. We provide it separately, so you will become accustomed to referring to the FAA Figures and Legends as you would during the test.

Figures referenced by the Explanation and pertinent to the understanding of that particular question are labeled by their corresponding Question number. For example: the caption "Questions 8245 and 8248" means the figure accompanies the Explanations for both Question 8245 and 8248.

Answers to each question are found at the bottom of each page.

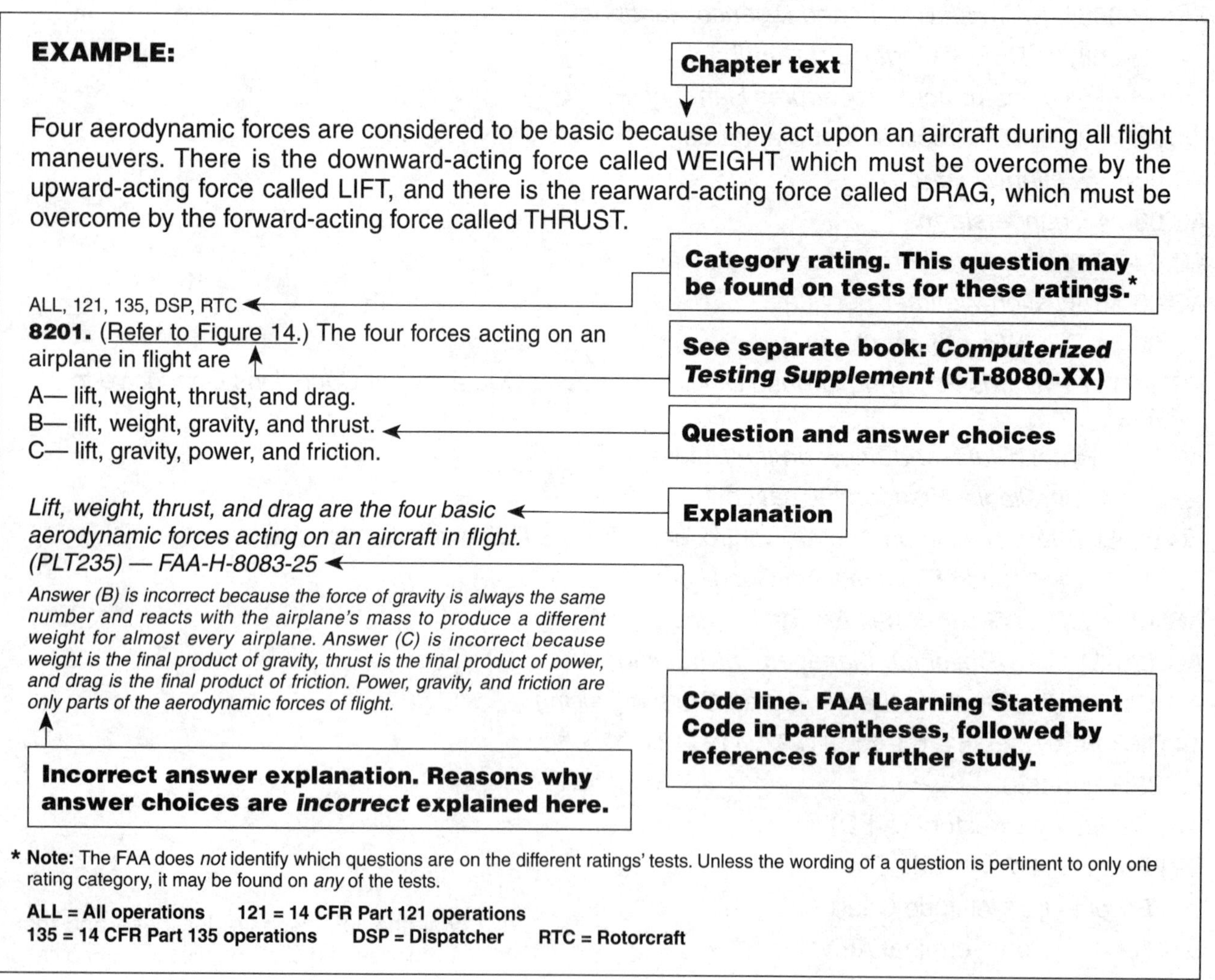

EXAMPLE:

Four aerodynamic forces are considered to be basic because they act upon an aircraft during all flight maneuvers. There is the downward-acting force called WEIGHT which must be overcome by the upward-acting force called LIFT, and there is the rearward-acting force called DRAG, which must be overcome by the forward-acting force called THRUST.

ALL, 121, 135, DSP, RTC

8201. (Refer to Figure 14.) The four forces acting on an airplane in flight are

A— lift, weight, thrust, and drag.
B— lift, weight, gravity, and thrust.
C— lift, gravity, power, and friction.

Lift, weight, thrust, and drag are the four basic aerodynamic forces acting on an aircraft in flight. (PLT235) — FAA-H-8083-25

Answer (B) is incorrect because the force of gravity is always the same number and reacts with the airplane's mass to produce a different weight for almost every airplane. Answer (C) is incorrect because weight is the final product of gravity, thrust is the final product of power, and drag is the final product of friction. Power, gravity, and friction are only parts of the aerodynamic forces of flight.

* **Note:** The FAA does *not* identify which questions are on the different ratings' tests. Unless the wording of a question is pertinent to only one rating category, it may be found on *any* of the tests.

ALL = All operations 121 = 14 CFR Part 121 operations
135 = 14 CFR Part 135 operations DSP = Dispatcher RTC = Rotorcraft

Chapter 1
Regulations

Applicable Regulations

Although "FAR" is used as the acronym for "Federal Aviation Regulations," and found throughout the regulations themselves and hundreds of other publications, the FAA is now actively discouraging its use. "FAR" also means "Federal Acquisition Regulations." To eliminate any possible confusion, the FAA is now citing the federal aviation regulations with reference to Title 14 of the Code of Federal Regulations. For example, "FAR Part 91.3" is now referenced as "14 CFR Part 91 Section 3." The regulations change frequently; answer all questions in compliance with the most current regulations.

Three different Federal Aviation Regulation Parts can apply to operations of aircraft covered by this chapter: Parts 91, 121, and 135. Part 91 encompasses the general operation and flight rules for all aircraft operating within the United States. Often the rules of Part 121 or 135 supplement or even supersede Part 91. When an airplane is not operated for compensation, only the Part 91 rules apply. For the test, assume Part 121 or 135 rules apply unless the question specifically states otherwise.

Part 121 applies to air carriers (airlines) engaged in interstate or overseas air transportation. Carriers which operate under Part 121 engage in **common carriage**. This means that they offer their services to the public and receive compensation for those services.

Part 121 operators are subdivided into three categories. Carriers authorized to conduct scheduled operations within the 48 contiguous states are **domestic air carriers**. **Flag air carriers** conduct scheduled operations inside and outside the 48 contiguous states. A **supplemental carrier** conducts its operations anywhere that its operations specifications permit but only on a nonscheduled basis. There is a fourth category, **commercial operators of large aircraft**, but they must comply with the rules covering supplemental carriers and the distinction is unimportant to this discussion.

Part 135 applies to air taxi operators. These operators are subdivided into two categories, commuter and on-demand operations.

Other parts of the regulations apply as well. Part 61 governs certification of pilots and flight instructors. Part 67 covers the issuing and standards for medical certificates. Part 1 contains definitions and abbreviations.

The ATP Certificate

The pilot-in-command of an air carrier flight must hold an Airline Transport Pilot (ATP) certificate with the appropriate type rating. The co-pilot on an air carrier flight that requires only two pilots need only hold a Commercial Pilot certificate (with an Instrument rating) with the appropriate category and class ratings.

The pilot-in-command of a large aircraft (gross weight over 12,500 pounds) or of a turbojet powered airplane must have a type rating from that aircraft issued under 14 CFR Part 61.

Any type rating(s) on the pilot certificate of an applicant who successfully completes an ATP checkride will be included on the ATP Certificate with the privileges and limitations of the ATP Certificate, provided the applicant passes the checkride in the same category and class of aircraft for which the applicant holds the type rating(s). However, if a type rating for that category and class of aircraft on the superseded pilot certificate is limited to VFR, that limitation will be carried forward to the person's ATP Certificate level.

An ATP certificate holder may give instruction in "air transportation service" in aircraft for which he/she holds category, class and type ratings as an ATP. An ATP may not instruct more than 8 hours a day and not more than 36 hours in any 7-day period.

If a person's pilot or medical certificate is lost or destroyed he/she can request the FAA to send a FAX confirming that they were issued. This FAX can be used as a temporary replacement for the certificates for up to 60 days.

If a pilot certificate holder is convicted of driving under the influence of alcohol or drugs, the pilot must report that conviction to the FAA, Civil Aviation Security Division within 60 days. Failure to do so is grounds for suspending or revoking any pilot or flight instructor certificates held by that person.

A **crewmember** is a person assigned to duty in the aircraft during flight. This includes pilots, flight engineers, navigators, flight attendants or anyone else assigned to duty in the airplane. A **flight crewmember** is a pilot, flight engineer or flight navigator assigned to duty in the aircraft during flight.

No one may serve as a pilot on an air carrier after that person has reached his/her 65th birthday. Note that this rule applies to any pilot position in the aircraft, but it does not apply to other flight crew positions such as flight engineer or navigator. This is known as the "Age 65 Rule."

To exercise **ATP privileges** (such as pilot-in-command of an air carrier flight) a pilot must hold a First-Class Medical Certificate issued within the preceding 6 calendar months. To exercise commercial pilot privileges (e.g., co-pilot on a two-pilot air carrier flight) a pilot must hold either a First- or Second-Class Medical Certificate issued within the preceding 12 calendar months. For example, a First-Class Certificate issued anytime in February would be good for ATP privileges through August 31 of the same year, and good through the last day of February the next year for commercial pilot privileges.

A prerequisite for taking a practical test requires the applicant hold at least a current Third-Class Medical Certificate, if a medical certificate is required. If the practical test is scheduled in an aircraft, the applicant is required to have the Third-Class Medical Certificate. The applicant is not required to hold a medical certificate when taking a test or check for a certificate, rating, or authorization conducted in a flight simulator or flight training device.

ALL
9350. Unless otherwise authorized, when is the pilot-in-command required to hold a type rating?

A—When operating an aircraft that is certificated for more than one pilot.
B—When operating an aircraft having a gross weight of more than 12,500 pounds.
C—When operating a multiengine aircraft having a gross weight of more than 6,000 pounds.

A person must hold a type rating to act as pilot-in-command of a large aircraft (over 12,500 pounds gross takeoff weight), or of a turbojet-powered airplane. (PLT443) — 14 CFR §61.31

Answer (A) is incorrect because an aircraft requiring more than one pilot does not constitute the need for a type rating. Answer (C) is incorrect because it does not matter if the aircraft is single-engine or multi-engine, and the aircraft must weigh over 12,500 lbs., not 6,000.

ALL
9328. A commercial pilot has a type rating in a B-727 and B-737. A flight test is completed in a B-747 for the Airline Transport Pilot Certificate. What pilot privileges may be exercised regarding these airplanes?

A—Commercial – B-737; ATP – B-727 and B-747.
B—ATP – B-747; Commercial – B-727 and B-737.
C—ATP – B-747, B-727, and B-737.

Any type rating(s) on the pilot certificate of an applicant who successfully completes an ATP checkride will be included on the ATP Certificate with the privileges and limitations of the ATP Certificate, provided the applicant passes the checkride in the same category and class of aircraft for which the applicant holds the type rating(s). However, if a type rating for that category and class of aircraft on the superseded pilot certificate is limited to VFR, that limitation shall be carried forward to the person's ATP Certificate level. (PLT443)—14 CFR §61.157

ALL
9329. A commercial pilot has DC-3 and DC-9 type ratings. A flight test is completed for an Airline Transport Pilot Certificate in a B-727. What pilot privileges may be exercised?

A—ATP – B-727 and DC-3; Commercial – DC-9.
B—ATP – B-727 only; Commercial – DC-9 and DC 3.
C—ATP – B-727, DC-3, and DC-9.

Any type rating(s) on the pilot certificate of an applicant who successfully completes an ATP checkride will be included on the ATP Certificate with the privileges and limitations of the ATP Certificate, provided the applicant passes the checkride in the same category and class of aircraft for which the applicant holds the type rating(s). However, if a type rating for that category and class of

Answers

9350 [B] 9328 [C] 9329 [C]

aircraft on the superseded pilot certificate is limited to VFR, that limitation shall be carried forward to the person's ATP Certificate level. (PLT442) — 14 CFR §61.157

ALL

9330. In a 24-hour consecutive period, what is the maximum time, excluding briefing and debriefing, that an airline transport pilot may instruct other pilots in air transportation service?

A—6 hours.
B—8 hours.
C—10 hours.

An airline transport pilot may instruct other pilots in air transportation service in aircraft of the category, class and type for which he/she is rated. However, the ATP may not instruct for more than 8 hours in one day nor more than 36 hours in any 7-day period. (PLT460) — 14 CFR §61.167

ALL

9331. The flight instruction of other pilots in air transportation service by an airline transport pilot is restricted to

A—30 hours in any 7-consecutive-day period.
B—7 hours in any 24-consecutive-hour period.
C—36 hours in any 7-consecutive-day period.

The ATP may not instruct for more than 8 hours in one day nor more than 36 hours in any 7-day period. (PLT460) — 14 CFR §61.167

ALL

9351. When a facsimile replacement is received for an airman's medical certificate, for what maximum time is this document valid?

A—30 days.
B—60 days.
C—90 days.

A person who has lost an Airman's Certificate or a Medical Certificate, or both, may obtain a FAX from the FAA confirming that it was issued. The FAX may be carried as temporary certificate(s) for a period not to exceed 60 days. (PLT447) — 14 CFR §61.29

ALL

9332. How soon after the conviction for driving while intoxicated by alcohol or drugs shall it be reported to the FAA, Civil Aviation Security Division?

A—No later than 30 working days after the motor vehicle action.
B—No later than 60 days after the motor vehicle action.
C—Required to be reported upon renewal of medical certificate.

Each person holding a certificate issued under this part shall provide a written report of each motor vehicle action to the FAA, Civil Aviation Security Division, no later than 60 days after the motor vehicle action. (PLT463) — 14 CFR §61.15

ALL

9325. Which is a definition of the term "crewmember"?

A—Only a pilot, flight engineer, or flight navigator assigned to duty in an aircraft during flight time.
B—A person assigned to perform duty in an aircraft during flight time.
C—Any person assigned to duty in an aircraft during flight except a pilot or flight engineer.

"Crewmember" means a person assigned to perform duty in an aircraft during flight time. (PLT395) — 14 CFR §1.1

Answer (A) is incorrect because "crewmember" pertains to anyone assigned duty in the aircraft during flight. Answer (C) is also incorrect because "crewmember" also includes the pilot and flight engineer.

ALL

9349. When a type rating is to be added to an airline transport pilot certificate, and the practical test is scheduled in an approved flight simulator and an aircraft, the applicant is

A—required to have a least a current third-class medical certificate.
B—required to have a current first-class medical certificate.
C—not required to hold a medical certificate.

A prerequisite for taking a practical test requires that the applicant hold at least a current third-class medical certificate, if a medical certificate is required. In this case, since part of the practical test is scheduled in an aircraft, the applicant is required to have at least a current third-class medical certificate. (PLT427) — 14 CFR §61.39

Answers

9330 [B]	9331 [C]	9351 [B]	9332 [B]	9325 [B]	9349 [A]

ALL
9335. An applicant who is taking a practical test for a type rating to be added to a commercial pilot certificate, in an approved simulator, is

A—required to have a first-class medical certificate.
B—required to have a second-class medical certificate.
C—not required to have a medical certificate.

A prerequisite for taking a practical test requires that the applicant hold at least a current third-class medical certificate, if a medical certificate is required. The applicant is not required to hold a medical certificate when taking a test or check for a certificate, rating, or authorization conducted in a flight simulator or flight training device. In this case, since the practical test is scheduled in an approved flight simulator, the applicant is not required to have a medical certificate. (PLT427) — 14 CFR §61.39 and §61.23

ALL
9333. An applicant who is scheduled for a practical test for an airline transport pilot certificate, in an approved flight simulator, is

A—required to have at least a current third-class medical certificate.
B—not required to have a medical certificate.
C—required to have a first-class medical certificate.

A prerequisite for taking a practical test requires that the applicant hold at least a current third-class medical certificate, if a medical certificate is required. The applicant is not required to hold a medical certificate when taking a test or check for a certificate, rating, or authorization conducted in a flight simulator or flight training device. In this case, since the practical test is scheduled in an approved flight simulator, the applicant is not required to have a medical certificate. (PLT427) — 14 CFR §61.39 and §61.23

ALL
9343. When a type rating is to be added to an airline transport pilot certificate, and the practical test is scheduled in an approved flight training device and/or approved flight simulator, the applicant is

A—required to have at least a third-class medical certificate.
B—is not required to have a medical certificate.
C—required to have a first-class medical certificate.

A prerequisite for taking a practical test requires that the applicant hold at least a current third-class medical certificate, if a medical certificate is required. The applicant is not required to hold a medical certificate when taking a test or check for a certificate, rating, or authorization conducted in a flight simulator or flight training device. In this case, since the practical test is scheduled in an approved flight training device and/or approved flight simulator, the applicant is not required to have a medical certificate. (PLT427) — 14 CFR §61.39 and §61.23

ALL
9340. An applicant who is scheduled for a practical test for an airline transport pilot certificate, in an aircraft, needs

A—a first-class medical certificate.
B—at least a current third-class medical certificate.
C—a second-class medical certificate.

A prerequisite for taking a practical test requires that the applicant hold at least a current third-class medical certificate, if a medical certificate is required. In this case, since the practical test is scheduled in an aircraft, the applicant is required to have at least a current third-class medical certificate. (PLT427) — 14 CFR §61.39

121, DSP
8191. The "age 65 rule" of 14 CFR Part 121 applies to

A—any required pilot crewmember.
B—any flight crewmember.
C—the pilot in command only.

No person may serve as a pilot on an airplane engaged in operations under 14 CFR Part 121 if that person has reached his/her 65th birthday. (PLT443) — 14 CFR §121.383

Answer (B) is incorrect because the "age 65" rule excludes flight engineers and navigators. Answer (C) is incorrect because the "age 65" rule applies to every pilot crewmember.

Answers

9335 [C] 9333 [B] 9343 [B] 9340 [B] 8191 [A]

Flight Engineer Requirements

Many air carrier aircraft have a **flight engineer** as a required flight crewmember. All older airplanes that have a maximum takeoff weight of more than 80,000 pounds must have a flight engineer. On aircraft types certified after 1963, the aircraft's "type certificate" states whether or not a flight engineer is required.

On each flight that requires a flight engineer, at least one other member of the flight crew must be qualified to provide emergency performance of the flight engineer's duties if he/she becomes ill or incapacitated. Either pilot can fulfill the function and they need not hold a Flight Engineer Certificate to be "qualified."

121, DSP

8189. Under which condition is a flight engineer required as a flight crewmember in 14 CFR Part 121 operations?

A—If the airplane is being flown on proving flights, with revenue cargo aboard.
B—If the airplane is powered by more than two turbine engines.
C—If required by the airplane's type certificate.

No certificate holder may operate an airplane for which a type certificate was issued before January 2, 1964, having a maximum certificated takeoff weight of more than 80,000 pounds without a flight crewmember holding a current Flight Engineer Certificate. For each airplane type certificated after January 1, 1964, the requirement for a flight engineer is determined under the type certification requirements of 14 CFR §25.1523. (PLT409) — 14 CFR §121.387

Answer (A) is incorrect because the type certificate is the determining factor for a flight engineer. Answer (B) is incorrect because the type certificate is the determining factor for a flight engineer.

121, DSP

8190. When the need for a flight engineer is determined by aircraft weight, what is the takeoff weight that requires a flight engineer?

A—80,000 pounds.
B—more than 80,000 pounds.
C—300,000 pounds.

No certificate holder may operate an airplane for which a type certificate was issued before January 2, 1964, having a maximum certificated takeoff weight of more than 80,000 pounds without a flight crewmember holding a current Flight Engineer Certificate. (PLT440) — 14 CFR §121.387

121, DSP

8212. An air carrier uses an airplane that is certified for operation with a flightcrew of two pilots and one flight engineer. In case the flight engineer becomes incapacitated,

A—at least one other flight crewmember must be qualified to perform the flight engineer duties.
B—one crewmember must be qualified to perform the duties of the flight engineer.
C—one pilot must be qualified and have a flight engineer certificate to perform the flight engineer duties.

On each flight requiring a flight engineer at least one flight crewmember, other than the flight engineer, must be qualified to provide emergency performance of the flight engineer's functions for the safe completion of the flight if the flight engineer becomes ill or is otherwise incapacitated. A pilot need not hold a Flight Engineer's Certificate to perform the flight engineer's functions in such a situation. (PLT440) — 14 CFR §121.385

121, DSP

8213. When a flight engineer is a required crewmember on a flight, it is necessary for

A—one pilot to hold a flight engineer certificate and be qualified to perform the flight engineer duties in an emergency.
B—the flight engineer to be properly certificated and qualified, but also at least one other flight crewmember must be qualified and certified to perform flight engineer duties.
C—at least one other flight crewmember to be qualified to perform flight engineer duties, but a certificate is not required.

On each flight requiring a flight engineer at least one flight crewmember, other than the flight engineer, must be qualified to provide emergency performance of the flight engineer's functions for the safe completion of the flight if the flight engineer becomes ill or is otherwise incapacitated. A pilot need not hold a Flight Engineer's Certificate to perform the flight engineer's functions in such a situation. (PLT440) — 14 CFR §121.385

Answers

8189 [C]	8190 [B]	8212 [A]	8213 [C]

121, DSP

8188. If a flight engineer becomes incapacitated during flight, who may perform the flight engineer's duties?

A—The second in command only.
B—Any flight crewmember, if qualified.
C—Either pilot, if they have a flight engineer certificate.

On each flight requiring a flight engineer at least one flight crewmember, other than the flight engineer, must be qualified to provide emergency performance of the flight engineer's functions for the safe completion of the flight if the flight engineer becomes ill or is otherwise incapacitated. A pilot need not hold a Flight Engineer's Certificate to perform the flight engineer's functions in such a situation. (PLT440) — 14 CFR §121.385

Flight Attendants

One or more **flight attendants** are required on each passenger carrying airplane that has more than nine passenger seats. The number of flight attendants is determined by the number of installed passenger seats—not by the actual number of passengers on board.

One flight attendant is required on airplanes that can seat from 10 through 50 passengers. Two flight attendants are required on airplanes having a seating capacity from 51 through 100 seats. After that, an additional flight attendant is required for each unit (or partial unit) of 50 seats above 100. For example, three flight attendants are required on airplanes having from 101 through 150 seats, and four flight attendants must be on aircraft with 151 through 200 seats.

121, DSP

8192. An airplane has seats for 149 passengers and eight crewmembers. What is the minimum number of flight attendants required with 97 passengers aboard?

A—Four.
B—Three.
C—Two.

For airplanes having a seating capacity of more than 100 passengers, each certificate holder shall provide at least two flight attendants plus one additional flight attendant for a unit (or partial unit) of 50 passenger seats above a seating capacity of 100 passengers. The number of flight attendants is determined by the number of installed passenger seats (not by the actual number of passengers on board). For an airplane with a seating capacity of 149 passengers, three flight attendants are required. (PLT389) — 14 CFR §121.391

121, DSP

8193. When an air carrier airplane with a seating capacity of 187 has 137 passengers on board, what is the minimum number of flight attendants required?

A—Five.
B—Four.
C—Three.

For airplanes having a seating capacity of more than 100 passengers, each certificate holder shall provide at least two flight attendants plus one additional flight attendant for a unit (or partial unit) of 50 passenger seats above a seating capacity of 100 passengers. The number of flight attendants is determined by the number of installed passenger seats (not by the actual number of passengers on board). For an airplane with a seating capacity of 187 passengers, four flight attendants are required. (PLT389) — 14 CFR §121.391

Answers

8188 [B] 8192 [B] 8193 [B]

121, DSP

8201. What is the minimum number of flight attendants required on an airplane having a passenger seating capacity of 188 with only 117 passengers aboard?

A—Five.
B—Four.
C—Three.

For airplanes having a seating capacity of more than 100 passengers, each certificate holder shall provide at least two flight attendants plus one additional flight attendant for a unit (or partial unit) of 50 passenger seats above a seating capacity of 100 passengers. The number of flight attendants is determined by the number of installed passenger seats (not by the actual number of passengers on board). For an airplane with a seating capacity of 188 passengers, four flight attendants are required. (PLT389) — 14 CFR §121.391

121, DSP

8202. What is the minimum number of flight attendants required on an airplane with a passenger seating capacity of 333 when 296 passengers are aboard?

A—Seven.
B—Six.
C—Five.

For airplanes having a seating capacity of more than 100 passengers, each certificate holder shall provide at least two flight attendants plus one additional flight attendant for a unit (or partial unit) of 50 passenger seats above a seating capacity of 100 passengers. The number of flight attendants is determined by the number of installed passenger seats (not by the actual number of passengers on board). For an airplane with a seating capacity of 333 passengers, seven flight attendants are required. (PLT389) — 14 CFR §121.391

Experience and Training Requirements

For these definitions of training, aircraft are divided into two "groups." **Group I** aircraft are propeller driven. Turbojet aircraft are **Group II. Initial training** is the training required for crewmembers and dispatchers who have not qualified and served in the same capacity (i.e., flight engineer, co-pilot, pilot-in-command) on another aircraft of the same group. **Transition training** is the training required for crewmembers or dispatchers who have qualified and served in the same capacity on another aircraft of the same group. **Upgrade training** is the training required for crewmembers who have qualified and served as second-in-command or flight engineer on a particular airplane type (e.g., Boeing 727) before they can serve as pilot-in-command or second-in-command, respectively, on that airplane. **Differences training** is the training required for crewmembers or dispatchers who have qualified and served on a particular type of airplane before they can serve in the same capacity on a variation of that airplane. For example, a crewmember who is qualified on a Boeing 727-100 would need differences training to serve on a Boeing 727-200.

The pilot-in-command (PIC) of an air carrier flight must have had a proficiency check within the preceding 12 calendar months. In addition, within the preceding 6 calendar months the pilot-in-command must have either passed a proficiency check or completed an approved simulator training course. Pilots other than the PIC must have either passed a proficiency check or completed "line oriented" simulator training within the last 24 calendar months. In addition, the co-pilot must have had a **proficiency check** or any other kind of simulator training within the last 12 calendar months.

The pilot-in-command of an air carrier flight must have completed a **line check** in one of the aircraft types he/she is qualified to fly within the preceding 12 calendar months. If the PIC is qualified in more than one type aircraft, a line check in any of them satisfies this requirement.

Recurrent training and **checkrides** are always due during a calendar month rather than by a certain date. In addition, if recurrent training or a check is taken during, before, or after the month, it is considered to have been taken during the month it was due. For example, if a crewmember had a check due in December, he/she could take it November, December or January and it would be considered as having been done in December. Also, January would be considered a "grace month" in that the crewmember could fly, even though he/she had technically gone beyond the due date of the check.

Continued

Answers

8201 [B] 8202 [A]

Every pilot on an air carrier flight must have made at least 3 takeoffs and landings in the type of airplane flown within the preceding 90 days. If a pilot doesn't meet these requirements, he/she must re-establish the recency of experience by making 3 takeoffs and 3 landings under the supervision of a check airman. These takeoffs and landings must meet the following:

- At least 1 takeoff must be made with a simulated failure of the most critical engine.
- At least 1 landing must be made from an ILS approach to the lowest ILS minimums authorized for the certificate holder.
- At least 1 landing must be made to a full stop.

Air Carriers' Operations Specifications are usually written so that the instrument experience requirements of 14 CFR Part 61 do not apply to their pilots. This test asks four questions on the Part 61 requirements: 9333, 9339, 9342, 9344.

The pilot-in-command of an airplane who has less than one hundred hours in the aircraft type has higher than published landing minimums at the destination airport. Such a pilot-in-command must add 100 feet to the published DH or MDA and add 1/2-mile (or 2,400 feet RVR) to the required visibility. If a flight diverts to an alternate airport, the pilot-in-command may use the published minimums for the approach there, but in no event may the landing minimums be less than 300 and 1. If a pilot has at least 100 hours PIC in another aircraft under Part 121 operations, he/she may reduce the current restriction by 1 hour for each landing, up to 50 hours maximum.

A Category II Instrument Approach is an ILS approach with a published minimum visibility of less than 1,800 RVR but equal to or greater than 1,200 RVR. Most CAT II approaches have published decision heights of 150 and 100 feet HAT. To fly a published CAT II approach, the aircraft must meet certain equipment and maintenance requirements and the pilots must be trained and qualified. Part 61 sets forth requirements for pilot qualification and an Air Carrier's Operations Specifications may modify or replace those requirements. The test limits its questions to Part 61 rules. To qualify for CAT II approach authorization, a pilot must take a CAT II checkride. To be eligible for the checkride he/she must meet all recent experience requirements of Part 61 and have certain recent experience with regard to ILS approaches. Within the previous 6 months the pilot must have made at least 6 ILS approaches down to minimums (CAT I minimums are OK). At least 3 of the approaches must have been hand flown. The other 3 may have been flown using an approach coupler. When issued an original CAT II certification, a pilot is restricted to a DH of 150 feet and a minimum RVR of 1,600. This restriction is lifted when the pilot logs 3 CAT II approaches to the 150-foot DH within the previous 6 months.

An aircraft dispatcher must have spent at least five hours observing flight deck operations within the preceding 12 calendar months. The dispatcher must have done this for at least one of the types for each group he/she is to dispatch.

ALL

9339. A pilot, acting as second-in-command, successfully completes the instrument competency check specified in 14 CFR Part 61. How long does this pilot remain current if no further IFR flights are made?

A—12 months.
B—90 days.
C—6 months.

No pilot may act as pilot-in-command under IFR unless he/she has, within the preceding 6 calendar months in the aircraft category for the instrument privileges sought, logged at least 6 instrument approaches, performed holding procedures, and intercepted and tracked courses through the use of navigation systems, or passed an instrument competency check in the category of aircraft involved. (PLT442) — 14 CFR §61.57

Answer (A) is incorrect because, upon completion of an instrument competency check, a pilot will remain current for 6 months. Answer (B) is incorrect because ninety days defines the 3 takeoffs and landings experience required to carry passengers.

Answers

9339 [C]

ALL

9344. To satisfy the minimum required instrument experience for IFR operations, a pilot must accomplish during the past 6 months at least

A—six instrument approaches, holding, intercepting and tracking courses through the use of navigation systems in an approved flight training device/simulator or in the category of aircraft to be flown.
B—six instrument approaches, three of which must be in the same category and class of aircraft to be flown, plus holding, intercepting and tracking courses in any aircraft.
C—six instrument approaches and 6 hours of instrument time, three of which may be in a glider.

No pilot may act as pilot-in-command under IFR unless he/she has, within the preceding 6 calendar months in the aircraft category for the instrument approaches, performed holding procedures, and intercepted and tracked courses through the use of navigation systems. (PLT442) — 14 CFR §61.57

ALL

9342. What instrument flight time may be logged by a second-in-command of an aircraft requiring two pilots?

A—All of the time the second-in-command is controlling the airplane solely by reference to flight instruments.
B—One-half the time the flight is on an IFR flight plan.
C—One-half the time the airplane is in actual IFR conditions.

A pilot may log as instrument flight time only that time during which he/she operates the aircraft solely by reference to the instruments, under actual or simulated instrument flight conditions. (PLT409) — 14 CFR §61.51

Answers (B) and (C) are incorrect because only when the pilot is flying in actual or simulated instrument flying conditions and is the sole manipulator of the controls may he/she log instrument flight time.

ALL

9334. What recent experience is required to be eligible for the practical test for the original issue of a Category II authorization?

A—Within the previous 6 months, six ILS approaches flown manually to the Category I DH.
B—Within the previous 12 calendar months, six ILS approaches flown by use of an approach coupler to the Category I or Category II DH.
C—Within the previous 6 months, six ILS approaches, three of which may be flown to the Category I DH by use of an approach coupler.

To be eligible for Category II authorization, a pilot must have made at least 6 ILS approaches since the beginning of the 6th month before the test. These approaches must be under actual or simulated instrument flight conditions down to the minimum landing altitude for the ILS approach in the type aircraft in which the flight test is to be conducted. However, the approaches need not be conducted down to the decision heights authorized for Category II operations. At least 3 of these approaches must have been conducted manually, without the use of an approach coupler. (PLT442) — 14 CFR §61.67

Answer (A) is incorrect because only 3 of the approaches must be flown manually to Category I DH. Answer (B) is incorrect because the 6 ILS approaches must be flown within the preceding 6 calendar months and 3 of the approaches must be flown without an approach coupler.

ALL

9345. To be eligible for the practical test for the renewal of a Category II authorization, what recent instrument approach experience is required?

A—Within the previous 6 months, six ILS approaches, three of which may be flown to the Category I DH by use of an approach coupler.
B—Within the previous 6 months, six ILS approaches flown by use of an approach coupler to the Category I DH.
C—Within the previous 12 calendar months, three ILS approaches flown by use of an approach coupler to the Category II DH.

To be eligible for Category II authorization, a pilot must have made at least 6 ILS approaches since the beginning of the 6th month before the test. These approaches must be under actual or simulated instrument flight conditions down to the minimum landing altitude for the ILS approach in the type aircraft in which the flight test is to be conducted. However, the approaches need not be conducted down to the decision heights authorized for Category II operations. At least 3 of these approaches must have been conducted manually, without the use of an approach coupler. (PLT442) — 14 CFR §61.67

Answer (B) is incorrect because only 3 of the 6 approaches may be flown using an approach coupler. Answer (C) is incorrect because the requirement is for a total of 6 approaches, only 3 of which may be flown by the use of an approach coupler. The approaches are not required to be flown down to Category II DH. Also, they must have been flown within the preceding 6 calendar months.

Answers

9344 [A] 9342 [A] 9334 [C] 9345 [A]

ALL

9346. When may a Category II ILS limitation be removed?

A—When three Cat II ILS approaches have been completed to a 150-foot decision height and landing.
B—When six ILS approaches to Category II minimums and landing have been completed in the past 6 months.
C—120 days after issue or renewal.

Upon original issue, a Category II authorization contains a limitation for Category II operations of 1,600 feet RVR and a 150-foot decision height. This limitation is removed when the holder shows that since the beginning of the 6th preceding month he/she has made 3 Category II ILS approaches to a landing under actual or simulated instrument conditions with a 150-foot decision height. (PLT407) — 14 CFR §61.13

ALL

9347. A Category II ILS pilot authorization, when originally issued, is normally limited to

A—Category II operations not less than 1600 RVR and a 150-foot DH.
B—pilots who have completed an FAA-approved Category II training program.
C—Category II operations not less than 1200 RVR and a 100-foot DH.

Upon original issue, a Category II authorization contains a limitation for Category II operations of 1,600 feet RVR and a 150-foot decision height. This limitation is removed when the holder shows that since the beginning of the 6th preceding month he/she has made 3 Category II ILS approaches to a landing under actual or simulated instrument conditions with a 150-foot decision height. (PLT407) — 14 CFR §61.13

Answer (B) is incorrect because all pilots must undergo FAA-approved training for a Category II authorization. The initial limitation is to RVR and DH for 6 months. Answer (C) is incorrect because a 1,200 RVR and a 100-foot DH are the Category II minimums after the initial limitation is removed by the pilot completing 3 ILS approaches to a 150-foot DH in the preceding 6 months.

ALL

9348. What is the lowest decision height for which a Category II applicant can be certified during the original issuance of the authorization?

A—100 feet AGL.
B—150 feet AGL.
C—200 feet AGL.

Upon original issue, a Category II authorization contains a limitation for Category II operations of 1,600 feet RVR and a 150-foot decision height. (PLT420) — 14 CFR §61.13

Answer (A) is incorrect because a 100-foot DH is allowed only after completion of 3 Category II ILS approaches to a 150-foot DH. Answer (C) is incorrect because 200 feet is the standard Category I ILS DH.

121, DSP

8215. The training required by flight crewmembers who have not qualified and served in the same capacity on another airplane of the same group (e.g., turbojet powered) is

A—upgrade training.
B—transition training.
C—initial training.

Initial training is the training required for crewmembers and dispatchers who have not qualified and served in the same capacity on another airplane of the same group. (PLT407) — 14 CFR §121.400

Answer (A) is incorrect because upgrade training is required of a flight engineer or second-in-command when he/she trains for the next higher position in a particular airplane type. Answer (B) is incorrect because transition training is the training required for crewmembers and dispatchers who have qualified and served in the same capacity on another airplane of the same group.

121, DSP

8216. A crewmember who has served as second-in-command on a particular type airplane (e.g., B-727-100), may serve as pilot-in-command upon completing which training program?

A—Upgrade training.
B—Recurrent training.
C—Initial training.

Upgrade training is the training required for crewmembers who have qualified and served as second-in-command or flight engineer on a particular airplane type, before they serve as pilot-in-command or second-in-command respectively, on that airplane. (PLT407) — 14 CFR §121.400

Answer (B) is incorrect because recurrent training is a periodic requirement of crewmembers who are qualified in their positions. Answer (C) is incorrect because initial training is the first training received for crewmembers who have not previously qualified and served in the same airplane group (e.g., turboprop or turbojet).

Answers

9346 [A]	9347 [A]	9348 [B]	8215 [C]	8216 [A]

121, DSP
8217. The training required for crewmembers or dispatchers who have been qualified and served in the same capacity on other airplanes of the same group is

A—difference training.
B—transition training.
C—upgrade training.

Transition training is the training required for crewmembers and dispatchers who have qualified and served in the same capacity on another airplane of the same group. (PLT407) — 14 CFR §121.400

Answer (A) is incorrect because difference training is required of a crewmember who is qualified on a particular type of airplane prior to becoming qualified in a variation of that same type. Answer (C) is incorrect because upgrade training is required of a crewmember who is qualified in a particular type of airplane and then desires to advance to the next higher position in that airplane, e.g., from copilot to pilot.

121, DSP
8205. A pilot in command must complete a proficiency check or simulator training within the preceding

A—6 calendar months.
B—12 calendar months.
C—24 calendar months.

For a person to serve as pilot-in-command he/she must have completed a proficiency check within the preceding 12 calendar months and, in addition, within the preceding 6 calendar months, either a proficiency check or an approved simulator training course. (PLT407) — 14 CFR §121.441

Answer (B) is incorrect because a proficiency check is mandatory within the preceding 12 months. Additionally, a proficiency check or simulator training is required within the preceding 6 months. Answer (C) is incorrect because a 24-month time frame applies to pilots other than the pilot-in-command.

121, DSP
8207. A pilot flight crewmember, other than pilot in command, must have received a proficiency check or line-oriented simulator training within the preceding

A—6 calendar months.
B—12 calendar months.
C—24 calendar months.

Pilots other than the pilot-in-command must have completed either a proficiency check or a line-oriented simulator training course within the preceding 24 calendar months. (PLT407) — 14 CFR §121.441

Answer (A) is incorrect because 6 months is the requirement for pilot-in-command to complete a proficiency check or simulator training. Answer (B) is incorrect because 12 months is the requirement for pilots other than pilot-in-command to receive a proficiency check or "any other kind of simulation training" (not necessarily line-oriented simulator training).

121, DSP
8210. What are the line check requirements for the pilot in command for a domestic air carrier?

A—The line check is required every 12 calendar months in one of the types of airplanes to be flown.
B—The line check is required only when the pilot is scheduled to fly into special areas and airports.
C—The line check is required every 12 months in each type aircraft in which the pilot may fly.

No certificate holder may use any person nor may any person serve as pilot-in-command of an airplane unless, within the preceding 12 calendar months that person has passed a line check in which he/she satisfactorily performs the duties and responsibilities of a pilot-in-command in one of the types of airplanes to be flown. (PLT442) — 14 CFR §121.440

121, DSP
8214. If a flight crewmember completes a required annual flight check in December 2010 and the required annual recurrent flight check in January 2012, the latter check is considered to have been taken in

A—November 2010.
B—December 2011.
C—January 2011.

Whenever a crewmember or aircraft dispatcher who is required to take recurrent training, a flight check, or a competency check, takes the check or completes the training in the calendar month before or after the month in which that training or check is required, he/she is considered to have taken or completed it in the calendar month in which it was required. (PLT449) — 14 CFR §121.401

Answers

8217 [B] 8205 [A] 8207 [C] 8210 [A] 8214 [B]

121, DSP

8208. Which is one of the requirements that must be met by a required pilot flight crewmember in re-establishing recency of experience?

A—At least one landing must be made with a simulated failure of the most critical engine.
B—At least one ILS approach to the lowest ILS minimums authorized for the certificate holder and a landing from that approach.
C—At least three landings must be made to a complete stop.

When a pilot has not made 3 takeoffs and landings within the preceding 90 days, the pilot must make at least 3 takeoffs and landings in the type of airplane in which that person is to serve or in an advanced simulator. These takeoffs and landings must include:

1. *At least 1 takeoff with a simulated failure of the most critical powerplant;*
2. *At least 1 landing from an ILS approach to the lowest ILS minimum authorized for the certificate holder; and*
3. *At least 1 landing to a full stop.*

(PLT442) — 14 CFR §121.439

Answer (A) is incorrect because at least 1 takeoff is required with a simulated failure of the most critical powerplant. Answer (C) is incorrect because only 1 landing to a complete stop is required.

121, DSP

8209. What is one of the requirements that must be met by an airline pilot to re-establish recency of experience?

A—At least one landing must be made from a circling approach.
B—At least one full stop landing must be made.
C—At least one precision approach must be made to the lowest minimums authorized for the certificate holder.

When a pilot has not made 3 takeoffs and landings within the preceding 90 days, the pilot must make at least 3 takeoffs and landings in the type of airplane in which that pilot is to serve, or in an advanced simulator. These takeoffs and landings must include:

1. *At least 1 takeoff with a simulated failure of the most critical powerplant;*
2. *At least 1 landing from an ILS approach to the lowest ILS minimum authorized for the certificate holder; and*
3. *At least 1 landing to a full stop.*

(PLT442) — 14 CFR §121.439

Answers (A) and (C) are incorrect because the only instrument approach required is an ILS approach to the lowest minimums authorized for the certificate holder.

121, DSP

8289. When a pilot's flight time consists of 80 hours' pilot in command in a particular type airplane, how does this affect the minimums for the destination airport?

A—Has no effect on destination but alternate minimums are no less than 300 and 1.
B—Minimums are decreased by 100 feet and 1/2 mile.
C—Minimums are increased by 100 feet and 1/2 mile.

If the pilot-in-command has not served 100 hours as pilot-in-command in operations under Part 121 in the type of airplane he/she is operating, the MDA or DH and visibility landing minimums in the certificate holder's operations specifications for regular, provisional, or refueling airports are increased by 100 feet and 1/2 mile (or the RVR equivalent). (PLT443) — 14 CFR §121.652

121, DSP

9586. (Refer to Figures 115, 116, 117, 118, 118A, 118B, and 118C.) At ARLIN Intersection, PTL 130 is notified that the Phoenix Sky Harbor Airport is closed. PTL 130 is told to proceed to Tucson. PTL 130 is operating under 14 CFR Part 121. The PIC on PTL 130 has less than 100 hours as PIC in the B-727 (approach category C). What are the PICs minimums for the VOR RWY 11L approach at Tucson Intl Airport?

A—2,860-1/2.
B—2,900-1.
C—2,960-1.

If the pilot-in-command has not served 100 hours as PIC under Part 121 operations in the airplane type, the MDA or DH visibility minimums are increased by 100 feet and 1/2 mile above the published minimums. If a flight goes to an alternate airport, the minimums do not have to be raised by 100-1/2, but they can't be less than 300-1. In this case, 36 feet had to be added to the MDA to comply with the 300-1 rule. (PLT083) — 14 CFR §121.652

Answers

8208 [B] 8209 [B] 8289 [C] 9586 [B]

121, DSP

9589. (Refer to Figure 118A.) Determine the 14 CFR Part 121 landing minimums for the LOC BC RWY 26L approach at Phoenix Sky Harbor Intl.

PIC time 94 hours
Airplane V_{SO} maximum certificated weight 105 knots
V_{REF} approach speed 140 knots
DME NOTAMed OTS

A—1,800/1-3/4.
B—1,900/2-1/4.
C—1,900/2-1/2.

Compute the approach category by multiplying 1.3 times the V_{SO} at the airplane's maximum weight. See FAA Legend 7. In this case 105 x 1.3 = 136.5 knots, which is category C. The category C minimums for the LOC BC RWY 26L without Haden is 1800 and 1-3/4. If the pilot-in-command of the airplane has not served 100 hours as PIC in Part 121 operations in the airplane type, the MDA or DH visibility minimums are increased by 100 feet and 1/2 mile above the published minimums. OTS stands for out of service. (PLT083) — 14 CFR §121.652

121, DSP

9646. (Refer to Figures 190, 195, 195A, 196 and 196A.) The PIC of PIL 10 has 87.5 hours and 26 landings as PIC in the B-767, while operating under Part 121. The PIC has 1,876 hours and 298 landings, as PIC in the L-1011 while operating under Part 121. What are the minimums for the ILS/DME RWY 35R approach at DEN for the PIC?

A—5567/18.
B—5667/42.
C—5631/20.

If the pilot-in-command has not served 100 hours as PIC under Part 121 operations in the airplane type, the MDA or DH visibility minimums are increased by 100 feet and 1/2 mile above the published minimums. However, if the pilot has at least 100 hours of PIC time in another airplane under Part 121 operations, he/she may reduce the current restriction by one hour for each landing. If the total of landings and hours in the current airplane add up to 100 or more, the PIC may fly approaches to the published minimums. The maximum allowable reduction is 50 hours. (PLT443) — 14 CFR §121.652

121, DSP

9663. (Refer to Figure 206.) The PIC of PTL 55 has 75 hours and 30 landings as PIC in the B-747, while operating under Part 121. The PIC has 759 hours and 312 landings, as PIC, in the B-767 while operating under Part 121. What are the minimums for the ILS RWY 19L approach at SFO, for the PIC?

A—308/64.
B—208/40.
C—308-1.

If the pilot-in-command has not served 100 hours as PIC under Part 121 operations in the airplane type, the MDA or DH visibility minimums are increased by 100 feet and 1/2 mile above the published minimums. However, if the pilot has at least 100 hours of PIC time in another airplane under Part 121 operations, he/she may reduce the current restriction by one hour for each landing. If the total of landings and hours in the current airplane add up to 100 or more, the PIC may fly approaches to the published minimums. The maximum allowable reduction is 50 hours. (PLT443) — 14 CFR §121.652

Answer (A) is incorrect because the PIC is not required to increase landing weather minimums by the 100 feet and 1/2-mile visibility because he/she substituted one hour for each of the 30 landings made in the 747 while operating under Part 121 (since the pilot also has at least 100 hours as PIC in another type airplane). In this case, the pilot had 759 landings in the B-767 while operating under Part 121. Answer (C) is incorrect because the pilot does not need to adjust the landing minimums (also the visibility adjustment is incorrect).

121, DSP

9680. (Refer to Figures 214 and 182A.) The PIC of TNA 90 has 49 hours and 102 landings as PIC in the MD90 while operating under Part 121. The PIC also has 959 hours and 246 landings, as PIC, in the B-727 while operating under Part 121. What are the minimums for the ILS RWY 9R approach at PHL, for this PIC?

A—221/18.
B—321/42.
C—321/24.

If the pilot-in-command has not served 100 hours as PIC under Part 121 operations in the airplane type, the MDA or DH visibility minimums are increased by 100 feet and 1/2 mile above the published minimums. However, if the pilot has at least 100 hours of PIC time in another airplane under Part 121 operations, he/she may reduce the current restriction by one hour for each landing. If the

Continued

Answers

9589 [B]	9646 [A]	9663 [B]	9680 [B]

total of landings and hours in the current airplane add up to 100 or more, the PIC may fly approaches to the published minimums. The maximum allowable reduction is 50 hours. (PLT443) — 14 CFR §121.652

Answer (A) is incorrect because the PIC is required to increase landing minimums by 100 feet and 1/2-mile visibility because he/she can substitute one hour for each landing (not to exceed 50 hours). Answer (C) is incorrect because the visibility adjustment is incorrect. The published visibility is 18 (1,800 RVR) and should be increased by 1/2 mile (2,400 RVR = 1/2 mile).

121, DSP

9684. (Refer to Figures 214 and 182A.) The PIC on TNA 90 (CAT C aircraft operated under 14 CFR Part 121) has not flown 100 hours as PIC in the MD90. What are the minimums while flying the ILS RWY 9R to land 09R at PHL?

A—321/30.
B—321/36.
C—321/42

If the pilot-in-command has not served 100 hours as PIC under Part 121 operations in the airplane type, the MDA or DH visibility minimums are increased by 100 feet and 1/2 mile above the published minimums. However, if the pilot has at least 100 hours of PIC time in another airplane under Part 121 operations, he/she may reduce the current restriction by one hour for each landing. If the total of landings and hours in the current airplane add up to 100 or more, the PIC may fly approaches to the published minimums. The maximum allowable reduction is 50 hours. (PLT443) — 14 CFR §121.652

Answers (A) and (B) are incorrect because the minimums must be increased by 100 feet and 1/2-mile visibility. The published visibility is 18 (1,800 RVR) and should be increased by 1/2 mile (2,400 RVR = 1/2 mile).

121, DSP

9685. (Refer to Figures 214, 183, and 183A.) The weather at PHL goes below the PICs minimums and TNA 90 (a CAT C aircraft operating under 14 CFR Part 121) diverts to the alternate ACY. Upon arrival at ACY, TNA 90 is cleared for an ILS RWY 13 approach. The PIC has less than 100 hours of PIC time in the MD 90. What are the landing minimums?

A—376/18.
B—376/50.
C—376/42.

If the pilot-in-command with less than 100 hours in type diverts to an alternate airport, then the flight may use the published minimums at the alternate, or 300-1, whichever is higher. (PLT443) — 14 CFR §121.652

121, DSP

9687. (Refer to Figures 214, 183, and 183A.) The PIC, of TNA 90, has 75 hours as PIC of this type airplane. The MD 90 is a Category C aircraft. What is the lowest ceiling/visibility that may be forecast to use ACY as an alternate on the flight plan, and what is the lowest visibility that may exist at ACY prior to the final approach segment to continue the ILS RWY 13 approach?

A—600-2 and 2 miles.
B—600-2 and 50 RVR.
C—700-2 and 18 RVR.

Non-standard alternate minimums for ILS RWY 13 only apply to category D and E aircraft, so only 600-2 is needed to file ACY as an alternate. If a PIC with less than 100 hours in type diverts to an alternate airport, then 100 and 1/2 addition does not apply. The flight may use the published minimums at the alternate or 300-1, whichever is higher. (PLT443) — 14 CFR §121.652

Answer (A) is incorrect because the visibility minimums are wrong. Answer (C) is incorrect because the alternate minimums for ILS RWY 13 approach were for category D and E, thus the alternate minimums for the approach should be 600-2. Additionally, the minima is incorrect due to the PIC's low time in the airplane (less than 100 hours as PIC).

121, DSP

8285. Category II ILS operations below 1600 RVR and a 150-foot DH may be approved after the pilot in command has

A—logged 90 hours' flight time, 10 takeoffs and landings in make and model airplane and three Category II ILS approaches in actual or simulated IFR conditions with 150-foot DH since the beginning of the sixth preceding month, in operations under 14 CFR parts 91 and 121.
B—made at least six Category II approaches in actual IFR conditions with 100-foot DH within the preceding 12 calendar months.
C—logged 100 hours' flight time in make and model airplane under 14 CFR part 121 and three Category II ILS approaches in actual or simulated IFR conditions with 150-foot DH since the beginning of the sixth preceding month.

Answers

9684 [C] 9685 [B] 9687 [B] 8285 [C]

If the pilot-in-command of an airplane has not served 100 hours as pilot-in-command in operations under 14 CFR Part 121 in the type of airplane he/she is operating, the MDA or DH and visibility landing minimums in the certificate holder's operations specifications for regular, provisional, or refueling airports are increased by 100 feet and 1/2 mile (or the RVR equivalent). In addition, CAT II minimums and the sliding scale do not apply. Upon original issue, a Category II authorization contains a limitation for Category II operations of 1,600 feet RVR and a 150-foot decision height. This limitation is removed when the holder shows that since the beginning of the 6th preceding month he/she has made 3 Category II ILS approaches to a landing under actual or simulated instrument conditions with a 150-foot decision height. (PLT444) — 14 CFR §121.652 and §61.13

DSP

8230. To remain current as an aircraft dispatcher, a person must, in addition to other requirements,

A—within the preceding 12 calendar months, spend 2.5 hours observing flight deck operations, plus two additional takeoff and landings, in one of the types of airplanes in each group he/she is to dispatch.

B—within the preceding 12 calendar months, spend at least 5 hours observing flight deck operations in one of the types of airplanes in each group he/she is to dispatch.

C—within the preceding 12 calendar months, spend at least 5 hours observing flight deck operations in each type of airplane, in each group that he/she is to dispatch.

No domestic or flag air carrier may use any person as an aircraft dispatcher unless, within the preceding 12 calendar months, he/she has satisfactorily completed operating familiarization consisting of at least 5 hours observing operations from the flight deck under 14 CFR Part 121 in one of the types of airplanes in each group he/she is to dispatch. (PLT450) — 14 CFR §121.463

Part 135 Flight Crew Requirements

135

8082. What are the minimum certificate and rating requirements for the pilot in command of a multiengine airplane being operated by a commuter air carrier?

A—Airline transport pilot; airplane category; multiengine class.

B—Commercial pilot; airplane category; multiengine class; instrument rating; airplane type rating, if required.

C—Airline transport pilot; airplane category; multiengine class; airplane type rating, if required.

No certificate holder may use a person, nor may any person serve, as pilot-in-command in passenger-carrying operations of a turbojet airplane, or an airplane having a passenger seating configuration, excluding any crewmember seat, of 10 seats or more, or a multi-engine airplane being operated by "commuter operations," unless that person holds an Airline Transport Pilot Certificate with appropriate category and class ratings and, if required, an appropriate type rating for that aircraft. (PLT443) — 14 CFR §135.243

135

8083. What are the minimum certificate and rating requirements for the pilot-in-command of a multiengine airplane in commuter air carrier service under IFR?

A—Airline transport pilot of any category; multiengine class rating.

B—Airline transport pilot; airplane category; multiengine class rating; airplane type rating, if required.

C—Commercial pilot; airplane category; multiengine class and instrument rating.

No certificate holder may use a person, nor may any person serve, as pilot-in-command in passenger-carrying operations of a turbojet airplane, or an airplane having a passenger seating configuration, excluding any crewmember seat, of 10 seats or more, or a multi-engine airplane being operated by "commuter operations," unless that person holds an Airline Transport Pilot Certificate with appropriate category and class ratings and, if required, an appropriate type rating for that aircraft. (PLT443) — 14 CFR §135.243

Answers

8230 [B] 8082 [C] 8083 [B]

135
8094. Which takeoff computation must not exceed the length of the runway plus the length of the stopway for a turbine-engine-powered small transport category airplane?

A—Takeoff distance.
B—Acceleration-stop distance.
C—Acceleration-climb distance.

The accelerate-stop distance, as defined in 14 CFR §25.109, must not exceed the length of the runway plus the length of any stopway. (PLT456) — 14 CFR §135.379 and §135.397

135
8100. A person is assigned as pilot in command to fly both single-engine and multiengine airplanes and has passed the initial instrument proficiency check in a multiengine airplane. Which requirement applies regarding each succeeding instrument check?

A—The instrument check must be taken every 6 calendar months in both a single-engine and a multiengine airplane.
B—The instrument check must be taken alternately in single-engine and multiengine airplanes every 6 calendar months.
C—The instrument check may be taken in either a single-engine or multiengine airplane if taken at intervals of 6 calendar months.

No certificate holder may use a pilot, nor may any person serve as a pilot-in-command of an aircraft under IFR unless, since the beginning of the 6th calendar month before that service, that pilot has passed an instrument proficiency check given by the FAA or authorized check pilot. If the pilot-in-command is assigned to both single-engine aircraft and multi-engine aircraft, that pilot must initially take the instrument proficiency check in a multi-engine aircraft and each succeeding check alternately in single-engine and multi-engine aircraft. (PLT442) — 14 CFR §135.297

135
8103. A person is acting as pilot in command of a multiengine, turboprop-powered airplane operated in passenger-carrying service by a commuter air carrier. If eight takeoffs and landings are accomplished in that make and basic model, which additional pilot-in-command experience meets the requirement for designation as pilot in command?

A—7 hours, and two takeoffs and landing.
B—10 hours, and three takeoffs and landings.
C—10 hours, and two takeoffs and one landings.

No certificate holder may use any person, nor may any person serve, as a pilot-in-command of an aircraft operated by "commuter operations" in passenger-carrying operations, unless that person has completed, on that make and basic model aircraft and in that crewmember position the following operating experience:

1. *Aircraft, single engine—10 hours;*
2. *Aircraft, multi-engine, reciprocating engine-powered—15 hours;*
3. *Aircraft, multi-engine, turbine engine-powered—20 hours; or*
4. *Airplane, turbojet-powered—25 hours.*

The hours of operating experience may be reduced to not less than 50% of the hours required above by the substitution of one additional takeoff and landing for each hour of flight. (PLT407) — 14 CFR §135.244

135
8107. What are the minimum certificate and rating requirements for the pilot in command of a turbojet airplane with two engines being operated by a Commuter Air Carrier (as defined in part 298)?

A—Airline transport pilot; airplane category; multiengine class rating; airplane type rating, if required.
B—Airline transport pilot of any category; multiengine class rating; airplane type rating.
C—Commercial pilot; airplane category; multiengine class rating; instrument rating; airplane type rating.

No certificate holder may use a person, nor may any person serve, as pilot-in-command in passenger-carrying operations of a turbojet airplane, or an airplane having a passenger seating configuration, excluding any crewmember seat, of 10 seats or more, or a multi-engine airplane being operated by "commuter operations" unless that person holds an Airline Transport Pilot Certificate with appropriate category and class ratings and, if required, an appropriate type rating for that aircraft. (PLT443) — 14 CFR §135.243

Answers

8094 [B] 8100 [B] 8103 [B] 8107 [A]

135

8108. A person is acting as pilot in command of a multi-engine, reciprocating engine powered airplane operated in passenger-carrying service by a commuter air carrier. If five takeoffs and landings have been accomplished in that make and basic model, which additional pilot-in-command experience meets the requirement for designation as the pilot in command?

A— Two takeoffs and landings, and 8 hours.
B— Five takeoffs and landings, and 5 hours.
C— Three takeoffs and landings, and 7 hours.

No certificate holder may use any person, nor may any person serve, as a pilot-in-command of an aircraft operated by "commuter operations" in passenger-carrying operations, unless that person has completed, on that make and basic model aircraft and in that crewmember position the following operating experience:

1. *Aircraft, single engine—10 hours;*
2. *Aircraft, multi-engine, reciprocating engine-powered—15 hours;*
3. *Aircraft, multi-engine, turbine engine-powered—20 hours; or*
4. *Airplane, turbojet-powered — 25 hours.*

The hours of operating experience may be reduced to not less than 50% of the hours required above by the substitution of one additional takeoff and landing for each hour of flight. (PLT407) — 14 CFR §135.244

135

8109. A person is acting as pilot in command of a turbojet powered airplane operated in passenger-carrying service by a commuter air carrier. If 10 takeoffs and landings have been accomplished in that make and basic model, which additional pilot-in-command experience meets the requirement for designation as pilot in command?

A— 10 hours.
B— 15 hours.
C— 10 hours, and five takeoffs and landings.

No certificate holder may use any person, nor may any person serve, as a pilot-in-command of an aircraft operated by "commuter operations" in passenger-carrying operations, unless that person has completed, on that make and basic model aircraft and in that crewmember position the following operating experience:

1. *Aircraft, single engine—10 hours;*
2. *Aircraft, multi-engine, reciprocating engine-powered—15 hours;*
3. *Aircraft, multi-engine, turbine engine-powered—20 hours; or*
4. *Airplane, turbojet-powered—25 hours.*

The hours of operating experience may be reduced to not less than 50% of the hours required above by the substitution of one additional takeoff and landing for each hour of flight. (PLT407) — 14 CFR §135.244

135

8110. A pilot's experience includes 8 hours in a particular make and basic model multiengine, turboprop airplane while acting as pilot-in-command. Which additional pilot-in-command experience meets the requirements for designation as pilot in command of that airplane when operated by a commuter air carrier in passenger-carrying service?

A— Twelve takeoffs and landings.
B— Five takeoffs and landings, and 2 hours.
C— Ten takeoffs and landings, and 2 hours.

No certificate holder may use any person, nor may any person serve, as a pilot-in-command of an aircraft operated by "commuter operations" in passenger-carrying operations, unless that person has completed, on that make and basic model aircraft and in that crewmember position the following operating experience:

1. *Aircraft, single engine—10 hours;*
2. *Aircraft, multi-engine, reciprocating engine-powered—15 hours;*
3. *Aircraft, multi-engine, turbine engine-powered—20 hours; or*
4. *Airplane, turbojet-powered—25 hours.*

The hours of operating experience may be reduced to not less than 50% of the hours required above by the substitution of one additional takeoff and landing for each hour of flight. (PLT407) — 14 CFR §135.244

Answers

8108 [A] 8109 [B] 8110 [C]

135

8111. A person is acting as pilot in command of a single-engine airplane operated in passenger-carrying service by a commuter air carrier. If six takeoffs and landings have been accomplished in that make and basic model, which additional pilot-in-command experience meets the requirement for designation as pilot in command?

A—4 hours
B—5 hours
C—6 hours

No certificate holder may use any person, nor may any person serve, as a pilot-in-command of an aircraft operated by "commuter operations" in passenger-carrying operations, unless that person has completed, on that make and basic model aircraft and in that crewmember position the following operating experience:

1. *Aircraft, single engine—10 hours;*
2. *Aircraft, multi-engine, reciprocating engine-powered—15 hours;*
3. *Aircraft, multi-engine, turbine engine-powered—20 hours; or*
4. *Airplane, turbojet-powered—25 hours.*

The hours of operating experience may be reduced to not less than 50% of the hours required above by the substitution of one additional takeoff and landing for each hour of flight. (PLT407) — 14 CFR §135.244

135

9618. (Refer to Figures 173 and 173A.) The PIC of PTZ 70 has less than 100 hours of PIC time in the BE 1900. Due to BUF weather being 100 feet, 1/4 mile in blowing snow, which is below landing minimums, the PIC requested and received clearance to SYR, the filed alternate. Under Part 135, what are the PICs minimums at SYR for the ILS RWY 10?

A—671/40.
B—771/64.
C—800/2.

If the PIC of a turbine-powered airplane does not have 100 hours of PIC time in type, he/she must raise the approach minimums by 100-1/2 above the published minimums. The pilot does not have to raise the approach minimums above the published alternate minimums. Note: 2,400 RVR is equal to 1/2 mile visibility. (PLT407) — 14 CFR §135.225

135

9632. (Refer to Figure 182A.) The PIC on EAB 90 has not flown 100 hours as PIC in the BE 1900 (CAT B aircraft). What are the minimums for the PIC when flying the ILS RWY 09R, at PHL?

A—321/42.
B—221/18.
C—321/36.

If the PIC of a turbine-powered airplane does not have 100 hours of PIC time in type, he/she must raise the approach minimums by 100-1/2 above the published minimums. The pilot does not have to raise the approach minimums above the published alternate minimums. Note: 2,400 RVR is equal to 1/2 mile visibility. (PLT407) — 14 CFR §135.225

135

9633. (Refer to Figure 182A.) The PIC of EAB 90 has 89 hours and 29 landings as PIC in the BE 1900, while operating under Part 135. The PIC has 1,234 hours and 579 landings as PIC in the DC-3 while operating under Part 135. What are the minimums for the ILS RWY 9R approach at PHL, for this PIC?

A—221/18.
B—321/24.
C—321/42.

If the PIC of a turbine-powered airplane does not have 100 hours of PIC time in type, he/she must raise the approach minimums by 100-1/2 above the published minimums. The pilot does not have to raise the approach minimums above the published alternate minimums. Note: 2,400 RVR is equal to 1/2 mile visibility. (PLT443) — 14 CFR §135.225

Answers

8111 [B]	9618 [B]	9632 [A]	9633 [C]

135

9634. (Refer to Figures 183 and 183A.) When the weather at PHL goes below the PICs minimums, the flight diverts to ACY. Upon arrival at ACY, EAB 90 is cleared for the ILS RWY 13 approach. The PIC has 89 hours of PIC time in the BE 1900. What are the PICs minimums?

A—700-2.
B—276/18.
C—376/42.

If the PIC of a turbine-powered airplane does not have 100 hours of PIC time in type, he/she must raise the approach minimums by 100-1/2 above the published minimums. The pilot does not have to raise the approach minimums above the published alternate minimums. Note: 2,400 RVR is equal to 1/2 mile visibility. (PLT443) — 14 CFR §135.225

135, RTC

8018. Which person, other than the second in command, may the pilot in command permit to manipulate the flight controls?

A—A member of the National Transportation Safety Board who holds a pilot certificate appropriate for the aircraft.
B—An authorized FAA safety representative who is qualified in the aircraft, and is checking flight operations.
C—A pilot employed by an engineering firm who is authorized by the certificate holder to conduct flight tests.

No pilot-in-command may allow any person to manipulate the controls of an aircraft during flight unless that person is:

1. *A pilot employed by the certificate holder and qualified in the aircraft; or*
2. *An authorized safety representative of the Administrator who has permission of the pilot-in-command, is qualified in the aircraft, and is checking flight operations.*

(PLT444) — 14 CFR §135.115

135, RTC

8026. A flight attendant crewmember is required on aircraft having a passenger seating configuration, excluding any pilot seat, of

A—15 or more.
B—19 or more.
C—20 or more

No certificate holder may operate an aircraft that has a passenger seating configuration, excluding any pilot seat, of more than 19 unless there is a flight attendant crewmember on board the aircraft. (PLT440) — 14 CFR §135.107

135, RTC

8027. Before each takeoff, the pilot in command of an aircraft carrying passengers shall ensure that all passengers have been orally briefed on the

A—location of normal and emergency exits, oxygen masks, and life preservers.
B—use of safety belts, location and operation of fire extinguishers, and smoking.
C—use of seatbelts, smoking, and location and use of survival equipment.

Before each takeoff the pilot-in-command shall ensure that all passengers have been orally briefed on:

1. *Smoking;*
2. *Use of seatbelts;*
3. *The placement of seat backs in an upright position before takeoff and landing;*
4. *Location and means of opening the passenger entry door and emergency exits;*
5. *Location of survival equipment;*
6. *If the flight involves extended overwater operation, ditching procedures and the use of required flotation equipment;*
7. *If the flight involves operations above 12,000 feet MSL, the normal and emergency use of oxygen; and*
8. *Location and operation of fire extinguishers.*

(PLT384) — 14 CFR §135.117

Answers

9634 [C] 8018 [B] 8026 [C] 8027 [B]

135, RTC

8028. Before takeoff, the pilot in command of an aircraft carrying passengers shall ensure that all passengers have been orally briefed on the normal and emergency use of oxygen

A—if the flight involves operations above 12,000 feet MSL.
B—regardless of the altitude at which the flight will operate.
C—if the flight involves operations at or above 12,000 feet MSL for more than 30 minutes.

Before each takeoff the pilot-in-command shall ensure that all passengers have been orally briefed on the normal and emergency use of oxygen if the flight involves operations above 12,000 feet MSL. (PLT438) — 14 CFR §135.117

135, RTC

8029. The oral before flight briefing required on passenger-carrying aircraft shall be

A—supplemented by an actual demonstration of emergency exit door operation by a crewmember.
B—presented by the pilot in command or another flight crewmember, as a crewmember demonstrates the operation of the emergency equipment.
C—conducted by a crewmember or the pilot in command and supplemented by printed cards for the use of each passenger.

The required oral briefing must be given by the pilot-in-command or other crewmember. It must be supplemented by printed cards which must be carried in the aircraft in locations convenient for the use of each passenger. (PLT384) — 14 CFR §135.117

135, RTC

8034. A commuter air carrier certificate holder plans to assign a pilot as pilot in command of an aircraft having eight passenger seats to be used in passenger-carrying operations. Which experience requirement must that pilot meet if the aircraft is to be flown with an operative approved autopilot and no second in command?

A—100 hours as pilot in command in the category, class, and type.
B—50 hours and 10 landings as pilot in command in the make and model.
C—100 hours as pilot in command in the make and model.

When using an autopilot in lieu of a second-in-command in commuter airline passenger-carrying operations, the pilot-in-command must have at least 100 hours of PIC time in the make and model of aircraft to be flown. (PLT407) — 14 CFR §135.105

135, RTC

8035. Which is a condition that must be met by a commuter air carrier certificate holder to have an aircraft approved for operation with an autopilot system and no second in command?

A—The passenger seating configuration is 10 or more, including any pilot seat.
B—The autopilot system is capable of operating the controls to maintain flight and to maneuver the aircraft about the three axes.
C—The operation is restricted to VFR or VFR over-the-top.

The autopilot used in lieu of a second-in-command must be capable of operating the aircraft controls to maintain flight and maneuver it about the three axes. (PLT443) — 14 CFR §135.105

135, RTC

8036. An autopilot may not be used in place of a second in command in any aircraft

A—being operated in commuter air carrier service.
B—having a passenger seating configuration, excluding any pilot's seat, of 10 seats or more.
C—having a total seating capacity of 10 or more seats and being operated in commuter air service.

Unless two pilots are required by 14 CFR for operations under VFR, a person may operate an aircraft without a second-in-command, if it is equipped with an operative approved autopilot system and the use of that system is authorized by appropriate operations specifications. No certificate holder may operate an aircraft without a second-in-command if that aircraft has a passenger seating configuration, excluding any pilot seat, of 10 seats or more. (PLT443) — 14 CFR §135.99 and §135.105

Answers

8028 [A] 8029 [C] 8034 [C] 8035 [B] 8036 [B]

135, RTC
8044. What is the minimum passenger seating configuration that requires a second in command?

A— 15 seats.
B— 12 seats.
C— 10 seats.

No certificate holder may operate an aircraft without a second-in-command if that aircraft has a passenger seating configuration, excluding any pilot seat, of 10 seats or more. (PLT443) — 14 CFR §135.99

135, RTC
8076. When is a pilot not required to keep the shoulder harness fastened during takeoff and landing while at a pilot station?

A— When operating an aircraft having a passenger seating configuration, excluding any pilot seat, of 10 seats or less.
B— When the pilot cannot perform the required duties with the shoulder harness fastened.
C— When serving as pilot in command or second in command of an aircraft having a total seating capacity of eight seats or less.

Each flight crewmember occupying a station equipped with a shoulder harness must fasten the shoulder harness during takeoff and landing, except that the shoulder harness may be unfastened if the crewmember is unable to perform required duties with the shoulder harness fastened. (PLT464) — 14 CFR §135.171

135, RTC
8095. To serve as pilot in command in an IFR operation, a person must have passed a line check

A— consisting of a flight over the route to be flown, with at least three instrument approaches at representative airports, within the past 12 calendar months, in one type of aircraft which that pilot is to fly.
B— within the past 12 months, which include a portion of a civil airway and one instrument approach at one representative airport, in one of the types of aircraft which that pilot is to fly.
C— since the beginning of the 12th month before that service, which included at least one flight over a civil airway, or approved off-airway route, or any portion of either, in one type of aircraft which that pilot is to fly.

No certificate holder may use a pilot, nor may any person serve, as a pilot-in-command of a flight unless, since the beginning of the 12th calendar month before that service, that pilot has passed a flight check (line check) in one of the types of aircraft that pilot is to fly. The flight check shall:

1. *Be given by an approved check pilot or by the FAA;*
2. *Consist of at least one flight over one route segment; and*
3. *Include takeoffs and landings at one or more representative airports;*
4. *For a pilot authorized for IFR operations, at least one flight shall be flown over a civil airway, an approved off-airway route, or a portion of either of them.*

(PLT442) — 14 CFR §135.299

135, RTC
8096. What are the minimum requirements for the line check required of each pilot in command authorized for IFR air taxi operations? The line check shall be given over

A— one route segment in each type of airplane the pilot is to fly and includes takeoffs and landings at one or more representative airports.
B— a civil airway or an approved off-airway route, or a portion of either of them, in one type of airplane the pilot is to fly and includes takeoffs and landings at one or more representative airports.
C— a civil airway or an approved off-airway route in each make and model airplane the pilot is to fly and includes takeoffs and landings at one or more representative airports.

No certificate holder may use a pilot, nor may any person serve, as a pilot-in-command of a flight unless, since the beginning of the 12th calendar month before that service, that pilot has passed a flight check (line check) in one of the types of aircraft that pilot is to fly. The flight check shall:

1. *Be given by an approved check pilot or by the FAA;*
2. *Consist of at least one flight over one route segment; and*
3. *Include takeoffs and landings at one or more representative airports;*
4. *For a pilot authorized for IFR operations, at least one flight shall be flown over a civil airway, an approved off-airway route, or a portion of either of them.*

(PLT442) — 14 CFR §135.299

Answers

8044 [C] 8076 [B] 8095 [C] 8096 [B]

135, RTC
8097. No certificate holder may use a person as pilot in command unless that person has passed a line check

A—since the beginning of the 12th month before serving as pilot in command.
B—since the beginning of the 6th month before serving as pilot in command.
C—within the past 6 months.

No certificate holder may use a pilot, nor may any person serve, as a pilot-in-command of a flight unless, since the beginning of the 12th calendar month before that service, that pilot has passed a flight check (line check) in one of the types of aircraft that pilot is to fly. (PLT442) — 14 CFR §135.299

135, RTC
8098. A person may act as pilot in command of both type A and type B aircraft under IFR, if an instrument proficiency check has been passed in

A—either type A or B since the beginning of the 12th month before time to serve.
B—type A since the beginning of the 12th month, and in type B since the beginning of the 6th month before time to serve.
C—type A since the beginning of the 12th month, and in type B since the beginning of the 24th month before time to serve.

No certificate holder may use a pilot, nor may any person serve, as a pilot-in-command of an aircraft under IFR unless, since the beginning of the 6th calendar month before that service, that pilot has passed an instrument proficiency check given by the FAA or authorized check pilot. If the pilot-in-command is assigned to pilot more than one type of aircraft, that pilot must take the instrument proficiency check for each type of aircraft to which that pilot is assigned in rotation, but not more than one flight check in each period. (PLT442) — 14 CFR §135.297

135, RTC
8099. A pilot in command is authorized to use an autopilot system in place of a second in command. During the instrument proficiency check, that person is required to demonstrate (without a second in command) the ability to

A—comply with complex ATC instructions with, but not without, the autopilot.
B—properly conduct air-ground communications with, but not without, the autopilot.
C—properly conduct instrument operations competently both with, and without, the autopilot.

If the pilot-in-command is authorized to use an autopilot system in place of a second-in-command, that pilot must show during the required instrument proficiency check, that the pilot is able both with and without using the autopilot to:

1. *Conduct instrument operations competently; and*
2. *Properly conduct air-ground communications and comply with complex air traffic control instructions.*

(PLT442) — 14 CFR §135.297

135, RTC
8101. A person may not serve as pilot in command in an IFR operation unless that person has passed an

A—aircraft competency, an instrument proficiency, and autopilot check within the previous 6 calendar months prior to the date to serve.
B—instrument proficiency check in the airplane in which to serve, or in an approved aircraft simulator, within the previous 12 calendar months.
C—instrument proficiency check under actual or simulated IFR conditions, since the beginning of the 6th calendar month prior to the date to serve.

No certificate holder may use a pilot, nor may any person serve as a pilot-in-command of an aircraft under IFR unless, since the beginning of the 6th calendar month before that service, that pilot has passed an instrument proficiency check given by the FAA or authorized check pilot. If the pilot-in-command is assigned to both single-engine aircraft and multi-engine aircraft, that pilot must initially take the instrument proficiency check in a multi-engine aircraft and each succeeding check alternately in single-engine and multi-engine aircraft. (PLT442) — 14 CFR §135.297

Answers

8097 [A]　　8098 [B]　　8099 [C]　　8101 [C]

135, RTC
8102. A pilot in command who is authorized to use an autopilot system, in place of a second in command, may take the autopilot check

A—concurrently with the instrument proficiency check, but at 12 month intervals.
B—in any aircraft appropriately equipped, providing the check is taken at 6 month intervals.
C—concurrently with the competency check, providing the check is taken at 12 month intervals.

If the pilot-in-command is authorized to use an autopilot system in place of a second-in-command, that pilot must show during the required instrument proficiency check, that the pilot is able both with and without using the autopilot to:

1. *Conduct instrument operations competently; and*
2. *Properly conduct air-ground communications and comply with complex air traffic control instructions.*

(PLT424) — 14 CFR §135.297

135, RTC
8104. Pilot flight time limitations under 14 CFR Part 135 are based

A—on the flight time accumulated in any commercial flying.
B—solely on flight time accumulated in air taxi operations.
C—solely on flight time accumulated during commercial flying, in the last 30 day and/or 12 month period.

Pilot flight time limitations are based on the flight time accumulated under 14 CFR Part 135 and any other commercial flying time. (PLT409) — 14 CFR §135.265

135, RTC
8105. No person may serve, as second in command of an aircraft (under part 135), unless they hold a commercial pilot certificate with the appropriate category, class rating and an instrument rating. For flight under IFR, that person must have accomplished within the last 6 months, the recent instrument requirements of

A—using the navigation systems for interception and tracking of courses, 6 instrument low approaches and holding.
B—using the navigation systems to intercept and track 3 inbound/3 outbound courses, 6 holding patterns and 6 instrument approaches.
C—holding procedures, using the navigation systems for intercepting and tracking courses, and 6 instrument approaches.

To act as second-in-command under IFR, a person must meet the recent instrument experience requirements of 14 CFR Part 61. These requirements are: in the last 6 months, the pilot must have logged 6 instrument approaches, performed holding procedures, and intercepted and tracked courses through the use of navigation systems. (PLT442) — 14 CFR §135.245

135, RTC
8106. With regard to flight crewmember duties, which operations are considered to be in the "critical phase of flight"?

A—All ground operations involving taxi, takeoff, landing, and all other operations conducted below 10,000 feet MSL, including cruise flight.
B—Descent, approach, landing, and taxi operations, irrespective of altitudes MSL.
C—All ground operations involving taxi, takeoff, landing, and all other operations conducted below 10,000 feet, excluding cruise flight.

For the purpose of this section, critical phases of flight include all ground operations involving taxi, takeoff and landing, and all other flight operations conducted below 10,000 feet, except cruise flight. (PLT029) — 14 CFR §135.100

135, RTC
8113. Other than in cruise flight, below what altitude are non-safety related cockpit activities by flight crewmembers prohibited?

A—12,000 feet.
B—10,000 feet.
C—8,000 feet.

No certificate holder shall require, nor may any flight crewmember perform, any duties during a critical phase of flight except those duties required for the safe operation of the aircraft. For purposes of this section, critical phases of flight include all ground operations involving taxi, takeoff and landing, and all other flight operations conducted below 10,000 feet, except cruise flight. (PLT440) — 14 CFR §135.100

Answers

8102 [A] 8104 [A] 8105 [C] 8106 [C] 8113 [B]

Flight Crew Duty Time Limits

The time limits in this section count all commercial flying done by the crewmember in any flight crew position, not just the time flown with the air carrier. Besides the limits on flight time, there are required periods of rest based on the amount of flying done within a 24-hour period. There is also a requirement that a flight crewmember be given at least 24 consecutive hours of rest in any seven consecutive day period. A person cannot be assigned to any ground or flight duties during required rest periods. The term "**deadhead**" is used to describe the transportation of crewmembers by the air carrier to or from their flight assignments when that transportation is not local in character. Time spent in deadhead air transportation cannot be considered as part of a required rest period.

On **flag operations** with a flight crew consisting of two pilots, the following flight time limits apply.

- 1,000 hours during any 12 calendar months
- 100 hours during any calendar month
- 32 hours in any 7 consecutive days
- 8 hours in any 24 without a rest.

On flag operations with a flight crew consisting of two pilots and one additional flight crewmember (e.g., flight engineer) the following flight time limits apply.

- 1,000 hours during any 12 calendar months
- 300 hours during any 90 consecutive days
- 120 hours during any 30 consecutive days
- 12 hours during any 24 consecutive hours

On **supplemental operations with one pilot** the following flight time limits apply.

- 1,000 hours during any calendar year
- 100 hours during any 30 consecutive days
- 8 hours in any 24 without a rest.

No pilot of a supplemental carrier may be on flight deck duty for more than 8 hours in any 24 consecutive hours. If three pilots are assigned to a flight, the crew can be aloft no more than 12 hours in any 24 consecutive hours.

No domestic or flag carrier may schedule a dispatcher to be on duty for more than 10 consecutive hours. If a dispatcher is scheduled for more than 10 hours of duty in 24 consecutive hours, he/she must be given at least 8 hours of rest at or before the end of 10 consecutive hours of duty. A dispatcher must be relieved of all duty with the carrier for at least 24 consecutive hours in any 7 consecutive days.

121, DSP

8227. How does deadhead transportation, going to or from a duty assignment, affect the computation of flight time limits for air carrier flight crewmembers? It is

A—considered part of the rest period if the flightcrew includes more than two pilots.

B—considered part of the rest period for flight engineers and navigators.

C—not considered to be part of a rest period.

Time spent in deadhead transportation to or from duty assignment is not considered part of a rest period. (PLT409) — 14 CFR §121.471, §121.491 and §121.519

Answer (A) is incorrect because deadhead transportation does not count for part of the required rest period. Answer (B) is incorrect because flight engineers and navigators are defined as flight crewmembers. The same rest period requirements apply to them as to pilot and copilot.

Answers

8227 [C]

121, DSP
8228. Duty and rest period rules for domestic air carrier operations require that a flight crewmember

A—not be assigned to any duty with the air carrier during any required rest period.
B—not be on duty aloft for more than 100 hours in any 30-day period.
C—be relieved of all duty for at least 24 hours during any 7 consecutive days.

No domestic air carrier may assign any flight crewmember to any duty with the air carrier during any required rest period. (PLT409) — 14 CFR §121.471

121, DSP
8220. The maximum flight time in 24 consecutive hours that a flag air carrier may schedule a pilot in a two-pilot crew without a rest period is

A—8 hours.
B—10 hours.
C—12 hours.

A flag air carrier may schedule a pilot to fly in an airplane that has a crew of one or two pilots for 8 hours or less during any 24 consecutive hours without a rest period. (PLT409) — 14 CFR §121.481

Answer (B) is incorrect because 10 hours in any consecutive 24 hours is not a specified maximum in 14 CFR Part 121. Answer (C) is incorrect because 12 hours is the limit for a crew consisting of two pilots and one additional crewmember.

121, DSP
8221. The maximum number of hours a pilot may fly in 7 consecutive days as the pilot in command in a two-pilot crew for a flag air carrier is

A—35 hours.
B—32 hours.
C—30 hours.

No pilot may fly more than 32 hours during any 7 consecutive days when flying for a flag air carrier in a one- or two-pilot crew. (PLT409) — 14 CFR §121.481

121, DSP
8219. A flag air carrier may schedule a pilot to fly in an airplane, having two pilots and one additional flight crewmember, for no more than

A—8 hours during any 12 consecutive hours.
B—10 hours during any 12 consecutive hours.
C—12 hours during any 24 consecutive hours.

No flag air carrier may schedule a pilot to fly, in an airplane that has a crew of two pilots and at least one additional flight crewmember, for a total of more than 12 hours during any 24 consecutive hours. (PLT409) — 14 CFR §121.483

121, DSP
8222. The maximum number of hours that a supplemental air carrier pilot may fly, as a crewmember, in a commercial operation, in any 30 consecutive days is

A—100 hours.
B—120 hours.
C—300 hours.

No pilot may fly as a crewmember on a supplemental air carrier more than 100 hours during any 30 consecutive days. (PLT409) — 14 CFR §121.503

121, DSP
8223. A supplemental air carrier may schedule a pilot, on a three-pilot crew, for flight deck duty during any 24-consecutive-hour period for not more than

A—6 hours.
B—8 hours.
C—10 hours.

No supplemental carrier or commercial operator may schedule a pilot for flight deck duty in an airplane that has a crew of three pilots for more than 8 hours in any 24 consecutive hours. (PLT409) — 14 CFR §121.507

121, DSP
9714. Which is the maximum number of hours that a supplemental air carrier airman may be aloft in any 30 consecutive days, as a member of a flight crew that consists of two pilots and at least one additional flight crewmember?

A—100 hours.
B—120 hours.
C—300 hours.

No airman may be aloft as a flight crewmember, of a crew consisting of two pilots and one additional airman, more than 120 hours during any 30 consecutive days. (PLT409) — 14 CFR §121.521

Answers

8228 [A]	8220 [A]	8221 [B]	8219 [C]	8222 [A]	8223 [B]
9714 [B]					

DSP

8211. Normally, a dispatcher for domestic or flag operations should be scheduled for no more than

A—8 hours of service in any 24 consecutive hours.
B—10 hours of duty in any 24 consecutive hours.
C—10 consecutive hours of duty.

Except in cases where circumstances or emergency conditions beyond the control of the air carrier require otherwise, no domestic or flag carrier may schedule a dispatcher for more than 10 consecutive hours of duty. (PLT450) — 14 CFR §121.465

DSP

8224. The flight time limitations established for flight crewmembers include

A—only commercial flying in any flight crewmember position in which 14 CFR Part 121 operations are conducted.
B—all flight time, except military, in any flight crewmember position.
C—all commercial flying in any flight crewmember position.

Flight time limitations are calculated on all commercial flying done by a pilot in any flight crew position. (PLT409) — 14 CFR §121.471, §121.489, and §121.517

Answer (A) is incorrect because the flight time limitations include all commercial flying in any flight crewmember position. Answer (B) is incorrect because flight time limitations apply only to commercial flying operations.

DSP

8229. If a domestic or flag air carrier schedules a dispatcher for 13 hours of duty in a 24-consecutive-hour period, what action is required?

A—The dispatcher should be given a rest period of 24 hours at the end of the 13 hours.
B—The dispatcher should refuse to be on duty 13 hours as 121.465(1) limits the duty period to 10 consecutive hours.
C—The dispatcher should be given a rest period of at least 8 hours at or before the completion of 10 hours of duty.

If a dispatcher is scheduled for more than 10 hours of duty in 24 consecutive hours, the carrier shall provide him/her a rest period of at least 8 hours at or before the end of 10 hours of duty. (PLT409) — 14 CFR §121.465

DSP

8231. An aircraft dispatcher shall receive at least 24 consecutive hours of rest during

A—every 7 consecutive days.
B—any 7 consecutive days or the equivalent thereof within any calendar month.
C—each calendar week.

Each dispatcher must be relieved of all duty with the air carrier for at least 24 consecutive hours during any 7 consecutive days, or the equivalent thereof, within any month. (PLT450) — 14 CFR §121.465

DSP

8238. The maximum number of consecutive hours of duty that an aircraft dispatcher may be scheduled is

A—12 hours.
B—10 hours.
C—8 hours.

Except in cases where circumstances or emergency conditions beyond the control of the air carrier require otherwise, no domestic or flag carrier may schedule a dispatcher for more than 10 consecutive hours of duty. (PLT450) — 14 CFR §121.465

Answers

8211 [C]	8224 [C]	8229 [C]	8231 [B]	8238 [B]

Dispatching and Flight Release

Operational control with respect to a flight, means the exercise of authority over initiating, conducting or terminating a flight.

The air carrier or commercial operator is responsible for operational control. The pilot-in-command and the director of operations are jointly responsible for the initiation, continuation, diversion, and termination of flight in compliance with regulations and the company's operations specifications. The pilot-in-command is responsible for the preflight planning and the operation of the flight.

Each flag and domestic flight must have a **dispatch release** on board. The dispatch release can be in any form but must contain the following information.

- The identification number of the aircraft
- The trip number
- The departure, destination, intermediate and alternate airports
- The type of operation (IFR or VFR)
- The minimum fuel supply
- The latest weather reports and forecasts for the complete flight (may be attached to the release rather than be part of it)

The aircraft dispatcher must provide the pilot-in-command with all available current reports or information on airport conditions and irregularities of navigation facilities that may affect the safety of flight. The aircraft dispatcher must provide the pilot-in-command with all available weather reports and forecasts of weather phenomena that may affect the safety of flight including adverse weather. The aircraft dispatcher must update this information during a flight.

When a domestic flight lands at an intermediate airport named in its original dispatch release and departs again within 1 hour, it does not need a new dispatch release. If it remains on the ground for more than 1 hour, a redispatch release must be issued.

When a flag flight lands at an intermediate airport named in its original dispatch release and departs again within 6 hours, it does not need a new dispatch release. If it remains on the ground for more than 6 hours, a redispatch is required.

The pilot-in-command of a flight shall carry in the airplane to its destination:

- A copy of the completed load manifest
- A copy of the dispatch release
- A copy of the flight plan.

The air carrier must keep copies of these documents for at least 3 months.

Each supplemental carrier or commercial operator flight must have a **flight release** on board. The flight release can be in any form but must contain the following information:

- The company or organization name
- Make, model and registration number of the aircraft used
- The flight or trip number and the date of the flight
- The name of each flight crewmember, flight attendant and the pilot designated as pilot-in-command
- The departure, destination, intermediate and alternate airports and route

Continued

- The type of operation (e.g., IFR or VFR)
- The minimum fuel supply
- The latest weather reports and forecasts for the complete flight (may be attached to the release rather than be part of it)

Before beginning a flight, the pilot-in-command must obtain all available current reports or information on airport conditions and irregularities of navigation facilities that may affect the safety of the flight. During a flight, the pilot-in-command must obtain any additional available information of meteorological conditions and irregularities of facilities and services that may affect the safety of the flight.

A provisional airport is defined as an airport approved by the Administrator for use by a certificate holder for the purpose of providing service to a community when the regular airport used by the certificate holder is not available. A person who is not authorized to conduct direct air carrier operations, but who is authorized by the Administrator to conduct operations as a U.S. commercial operator, will be issued an Operating Certificate. Each certificate holder conducting domestic, flag, or commuter operations must obtain operations specifications containing, among many other provisions, the kinds of operations authorized.

Extended-range twin-engine operational performance standards (ETOPS) is a rating accompanied by a time limit (such as 180-minute ETOPS) that allows twin-engine civil transport aircraft to fly over oceans and deserts provided that the aircraft is never more than 180 minutes away from a suitable airfield. An ETOPS "entry point" is the first point on an ETOPS route at which the airplane is farther than a distance of 60 minutes flying time, with one engine inoperative, from an emergency or diversion airport that is adequate for an airplane with two engines.

When filing an alternate using the 180-minute ETOPS rule, the alternate airport must have rescue and fire fighting services (RFFS) that meet ICAO Category 4 standard or higher. If filing an alternate using the beyond-180-minute ETOPS rule, the alternate must have RFFS that meet the ICAO Category 4 standard or higher, and the aircraft must remain within the ETOPS authorized diversion time from an adequate airport that has RFFS equal to ICAO Category 7 or higher.

ALL

9326. "Operational control" of a flight refer to

A—the specific duties of any required crewmember.
B—exercising authority over initiating, conducting, or terminating a flight.
C—exercising the privileges of pilot-in-command of an aircraft.

"Operational Control," with respect to flight, means the exercise of authority over initiating, conducting or terminating a flight. (PLT432) — 14 CFR §1.1

Answer (A) is incorrect because "crewmember" refers to any person assigned to perform duty in an aircraft during flight time, which includes cabin crew as well as cockpit crew. Answer (C) is incorrect because "pilot-in-command" refers to the pilot responsible for the operation and safety of an aircraft during flight time, which does not include the initiation of a flight.

ALL

8003. Which document specifically authorizes a person to operate an aircraft in a particular geographic area?

A—Operations Specifications.
B—Operating Certificate.
C—Dispatch Release.

Each certificate holder conducting domestic, flag, or commuter operations must obtain operations specifications containing authorization and limitations for routes and areas of operations. (PLT389) — 14 CFR §119.49

Answers

9326 [B]   8003 [A]

ALL

9745. No person may operate a U.S. registered civil aircraft

A—for which an AFM or RFM is required by part 21 section 21.5 unless there is a current, approved operator's manual available.
B—for which an AFM or RFM is required by part 21 section 21.5 unless there is a current, approved AFM or RFM available.
C—for which an AFM or RFM is required by part 21 section 21.5 unless there is a current, approved AFM or RFM available or the manual specified in part 135 section 135.19(b).

No person may operate a U.S.-registered civil aircraft for which an Airplane or Rotorcraft Flight Manual is required by §21.5 unless there is available in the aircraft a current, approved Airplane or Rotorcraft Flight Manual or the manual provided for in §121.141(b). (PLT373) — 14 CFR §91.9

ALL

8429. An airport approved by the Administrator for use by an air carrier certificate holder for the purpose of providing service to a community when the regular airport is not available is a/an:

A—destination airport.
B—provisional airport.
C—alternate airport.

A provisional airport is defined as an airport approved by the Administrator for use by a certificate holder for the purpose of providing service to a community when the regular airport used by the certificate holder is not available. (PLT395) — 14 CFR §119.3

Answer (A) is incorrect because the destination airport is the term used to describe the primary airport of intended landing. Answer (C) is incorrect because the alternate airport is generally defined as an airport at which an aircraft may land if a landing at the intended airport becomes inadvisable.

ALL

8430. A provisional airport is an airport approved by the Administrator for use by an air carrier certificate holder for the purpose of

A—obtaining provisions and fuel when unable, due to winds, to proceed direct to the regular airport.
B—having the aircraft catered (foods, beverages, or supplies).
C—providing service to a community when the regular airport is unavailable.

A provisional airport is defined as an airport approved by the Administrator for use by a certificate holder for the purpose of providing service to a community when the regular airport used by the certificate holder is not available. (PLT389) — 14 CFR §119.3

ALL

8767. A person who is not authorized to conduct direct air carrier operations, but who is authorized by the Administrator to conduct operations as a U.S. commercial operator, will be issued

A—an Air Carrier Certificate.
B—a Supplemental Air Carrier Certificate.
C—an Operating Certificate.

A person who is not authorized to conduct direct air carrier operations, but who is authorized by the Administrator to conduct operations as a U.S. commercial operator, will be issued an Operating Certificate. (PLT389) — 14 CFR §119.5

Answer (A) is incorrect because a person authorized by the Administrator to conduct operations as a direct air carrier is issued an Air Carrier Certificate. Answer (B) is incorrect because wherever in the Federal Aviation Regulations the term "supplemental air carrier operating certificate" appears, it shall be deemed to mean an "Air Carrier Operating Certificate."

Answers

9745 [B]	8429 [B]	8430 [C]	8767 [C]

ALL

8768. The kinds of operation that a certificate holder is authorized to conduct are specified in the

A—certificate holder's operations specifications.
B—application submitted for an Air Carrier or Operating Certificate, by the applicant.
C—Air Carrier Certificate or Operating Certificate.

Each certificate holder conducting domestic, flag, or commuter operations must obtain operations specifications containing, among many other provisions, the kinds of operations authorized. (PLT389) — 14 CFR §119.49

Answers (B) and (C) are incorrect because the operations specifications are continually updated and amended relative to the operator's needs and not contained in the original application or on the certificate itself.

ALL

9782. All 14 CFR Part 139 airports must report

A—accident and incident data annually.
B—noise complaint statistics for each departure procedure or runway.
C—declared distances for each runway.

All 14 CFR Part 139 airports must report 12 consecutive calendar months for each accident or incident in movement areas and safety areas involving an air carrier aircraft and/or ground vehicle. (PLT078) — 14 CFR §139.301

121, DSP

8243. The persons jointly responsible for the initiation, continuation, diversion, and termination of a supplemental air carrier or commercial operator flight are the

A—pilot in command and chief pilot.
B—pilot in command and director of operations.
C—pilot in command and the flight follower.

For operations of supplemental air carriers or commercial operators, the pilot-in-command and the director of operations are jointly responsible for the initiation, continuation, diversion, and termination of a flight. (PLT444) — 14 CFR §121.537

121, DSP

8290. Which information must be contained in, or attached to, the dispatch release for a flag air carrier flight?

A—Type of operation (e.g., IFR, VFR), trip number.
B—Total fuel supply and minimum fuel required on board the airplane.
C—Passenger manifest, company or organization name, and cargo weight.

The dispatch release of a flag or domestic air carrier may be in any form but must contain at least the following information concerning the flight:

1. *Identification number of the aircraft;*
2. *Trip number;*
3. *Departure airport, intermediate stops, destination airports, and alternate airports;*
4. *A statement of the type of operation (IFR, VFR);*
5. *Minimum fuel supply.*

(PLT455) — 14 CFR §121.687

Answers (B) and (C) are incorrect because fuel on board, a passenger list, and cargo weights are found in the load manifest. Although separate items, both the dispatch release and the load manifest are required to be carried on the flight.

121, DSP

8292. What information must be contained in, or attached to, the dispatch release for a domestic air carrier flight?

A—Departure airport, intermediate stops, destinations, alternate airports, and trip number.
B—Names of all passengers on board and minimum fuel supply.
C—Cargo load, weight and balance data, and identification number of the aircraft.

The dispatch release of a flag or domestic air carrier may be in any form but must contain at least the following information concerning the flight:

1. *Identification number of the aircraft;*
2. *Trip number;*
3. *Departure airport, intermediate stops, destination airports, and alternate airports;*
4. *A statement of the type of operation (IFR, VFR);*
5. *Minimum fuel supply.*

(PLT400) — 14 CFR §121.687

Answers (B) and (C) are incorrect because the passenger names, cargo load, and weight and balance data are part of the required load manifest. A copy of the load manifest must also be carried on the flight. The load manifest is not part of the dispatch release.

Answers

8768 [A] 9782 [A] 8243 [B] 8290 [A] 8292 [A]

121, DSP

8293. What information must be included on a domestic air carrier dispatch release?

A—Evidence that the airplane is loaded according to schedule, and a statement of the type of operation.
B—Minimum fuel supply and trip number.
C—Company or organization name and identification number of the aircraft.

The dispatch release of a flag or domestic air carrier may be in any form but must contain at least the following information concerning the flight:

1. *Identification number of the aircraft;*
2. *Trip number;*
3. *Departure airport, intermediate stops, destination airports, and alternate airports;*
4. *A statement of the type of operation (IFR, VFR);*
5. *Minimum fuel supply.*

(PLT412) — 14 CFR §121.687

Answer (A) is incorrect because the proper loading of the airplane is documented in the load manifest. Answer (C) is incorrect because the company or organization name is not required on the dispatch release.

121, DSP

8294. A dispatch release for a flag or domestic air carrier must contain or have attached to it

A—minimum fuel supply and weather information for the complete flight.
B—trip number and weight and balance data.
C—weather information for the complete flight and a crew list.

The dispatch release must contain, or have attached to it, weather reports, available weather forecasts, or a combination thereof, for the destination airport, intermediate stops, and alternate airports, that are the latest available at the time the release is signed by the pilot-in-command and dispatcher. It may include any additional available weather reports or forecasts that the pilot-in-command or the aircraft dispatcher considers necessary or desirable. (PLT412) — 14 CFR §121.687

121, DSP

8280. By regulation, who shall provide the pilot in command of a domestic or flag air carrier airplane information concerning weather, and irregularities of facilities and services?

A—The aircraft dispatcher.
B—Air route traffic control center.
C—Director of operations.

The aircraft dispatcher for a flag or domestic flight shall provide the pilot-in-command all available reports or information on airport conditions and irregularities of navigation facilities that may affect safety of the flight. (PLT398) — 14 CFR §121.601

Answer (B) is incorrect because air route traffic control center may have information concerning irregularities of facilities and service, but it is not the proper source of that information. That information should be provided by the aircraft dispatcher. Answer (C) is incorrect because the director of operations (who may also be the general manager) is an administrative person, responsible for the day-to-day operations and not usually involved in specific flight operations.

121, DSP

8283. Where can the pilot of a flag air carrier airplane find the latest FDC NOTAMs?

A—Any company dispatch facility.
B—Notices To Airmen publication.
C—Airport/Facility Directory.

The Aircraft Dispatcher for a flag or domestic flight shall provide the pilot-in-command all available reports or information on airport conditions and irregularities of navigation facilities that may affect safety of the flight. Since FDC NOTAMs are regulatory in nature and apply to instrument approach procedures and enroute charts, they would have to be available. (PLT323) — 14 CFR §121.601

121, DSP

8284. Who is responsible, by regulation, for briefing a domestic or flag air carrier pilot in command on all available weather information?

A—Company meteorologist.
B—Aircraft dispatcher.
C—Director of operations.

Before the beginning of a flag or domestic flight, the aircraft dispatcher shall provide the pilot-in-command with all available weather reports and forecasts of weather phenomena that may affect the safety of flight. (PLT398) — 14 CFR §121.601

Answers

8293 [B]	8294 [A]	8280 [A]	8283 [A]	8284 [B]

121, DSP

8232. A domestic air carrier flight has a delay while on the ground, at an intermediate airport. How long before a redispatch release is required?

A—Not more than 1 hour.
B—Not more than 2 hours.
C—More than 6 hours.

Except when a domestic air carrier airplane lands at an intermediate airport specified in the original dispatch release and remains there for not more than 1 hour, no person may start a flight unless an aircraft dispatcher specifically authorizes that flight. (PLT452) — 14 CFR §121.593

Answer (B) is incorrect because domestic air carriers may remain at an intermediate stop for 1 hour before a redispatch release is required. Answer (C) is incorrect because flag, supplemental, and commercial operators may remain at an intermediate stop up to 6 hours before a redispatch release is required.

121, DSP

8260. A domestic air carrier airplane lands at an intermediate airport at 1815Z. The latest time it may depart without a specific authorization from an aircraft dispatcher is

A—1945Z.
B—1915Z.
C—1845Z.

Except when a domestic air carrier airplane lands at an intermediate airport specified in the original dispatch release and remains there for not more than 1 hour, no person may start a flight unless an aircraft dispatcher specifically authorizes that flight. (PLT398) — 14 CFR §121.593

121, DSP

8259. A flag air carrier flight lands at an intermediate airport at 1805Z. The latest time that it may depart without being redispatched is

A—2005Z.
B—1905Z.
C—0005Z.

No person may continue a flag air carrier flight from an intermediate airport without redispatch if the airplane has been on the ground more than 6 hours. (PLT398) — 14 CFR §121.595

121, DSP

8266. When a flag air carrier airplane lands at an intermediate airport at 1822Z, what is the latest time it may continue a flight without receiving a redispatch authorization?

A—1922Z.
B—1952Z.
C—0022Z.

No person may continue a flag air carrier flight from an intermediate airport without redispatch if the airplane has been on the ground more than 6 hours. (PLT398) — 14 CFR §121.595

121, DSP

8267. If a flag air carrier flight lands at an intermediate airport at 1845Z, and experiences a delay, what is the latest time it may depart for the next airport without a redispatch release?

A—1945Z.
B—2015Z.
C—0045Z.

No person may continue a flag air carrier flight from an intermediate airport without redispatch if the airplane has been on the ground more than 6 hours. (PLT398) — 14 CFR §121.595

121, DSP

8226. What information must the pilot in command of a supplemental air carrier flight or commercial operator carry to the destination airport?

A—Cargo and passenger distribution information.
B—Copy of the flight plan.
C—Names of all crewmembers and designated pilot in command.

The pilot-in-command shall carry in the airplane to its destination: load manifest, flight release, airworthiness release, pilot route certification, and flight plan. (PLT400) — 14 CFR §121.687

Answer (A) is incorrect because this information is only part of the load manifest. Answer (C) is incorrect because this is only one element of the flight release which is required on board.

Answers

8232 [A]	8260 [B]	8259 [C]	8266 [C]	8267 [C]	8226 [B]

121, DSP

8286. Which documents are required to be carried aboard each domestic air carrier flight?

A—Load manifest (or information from it) and flight release.
B—Dispatch release and weight and balance release.
C—Dispatch release, load manifest (or information from it), and flight plan.

The pilot-in-command of a domestic or flag air carrier flight shall carry in the airplane to its destination:

1. *A copy of the completed load manifest;*
2. *A copy of the dispatch release; and*
3. *A copy of the flight plan.*

(PLT400) — 14 CFR §121.695

121, DSP

8288. A domestic or flag air carrier shall keep copies of the flight plans, dispatch releases, and load manifests for at least

A—3 months.
B—6 months.
C—30 days.

The air carrier shall keep copies of the flight plans, dispatch releases, and load manifests for at least 3 months. (PLT453) — 14 CFR §121.695

121, DSP

8296. Which documents are required to be carried aboard each flag air carrier flight?

A—Dispatch release, flight plan, and weight and balance release.
B—Load manifest, flight plan, and flight release.
C—Dispatch release, load manifest, and flight plan.

The pilot-in-command of a domestic or flag air carrier flight shall carry in the airplane to its destination:

1. *A copy of the completed load manifest;*
2. *A copy of the dispatch release; and*
3. *A copy of the flight plan.*

(PLT400) — 14 CFR §121.695

Answer (A) is incorrect because a dispatch release is required but there is no required document called a weight and balance release. Answer (B) is incorrect because a flight release is used by supplemental air carriers and commercial operators.

121, DSP

8287. How long shall a supplemental air carrier or commercial operator retain a record of the load manifest, airworthiness release, pilot route certification, flight release, and flight plan?

A—1 month.
B—3 months.
C—12 months.

A supplemental air carrier must retain a copy of each load manifest, flight release and flight plan at its principal operations base for at least 3 months. (PLT453) — 14 CFR §121.697

121, DSP

8291. The certificated air carrier and operators who must attach to, or include on, the flight release form the name of each flight crewmember, flight attendant, and designated pilot in command are

A—supplemental and commercial.
B—supplemental and domestic.
C—flag and commercial.

Supplemental air carrier and commercial operators must attach to, or include on, the flight release form, containing at least the following information concerning each flight:

1. *Company or organization name;*
2. *Make, model and registration number of the aircraft being used;*
3. *Flight or trip number and the date of the flight;*
4. *Name of each flight crewmember, flight attendant, and pilot designated as pilot-in-command;*
5. *Departure airport, destination airports, alternate airports, and route;*
6. *Minimum fuel supply; and*
7. *A statement of the type of operation (IFR, VFR).*

(PLT455) — 14 CFR §121.689

Answers (B) and (C) are incorrect because domestic and flag carriers, unlike supplemental and commercial operators, utilize a dispatch release. Commercial operators and supplemental carriers utilize a flight release. A flight release contains the crew names, but a dispatch release does not.

Answers

8286 [C]	8288 [A]	8296 [C]	8287 [B]	8291 [A]

121, DSP

8295. The information required in the flight release for supplemental air carriers and commercial operators that is not required in the dispatch release for flag and domestic air carriers is the

A—weather reports and forecasts.
B—names of all crewmembers.
C—minimum fuel supply.

The flight release of a supplemental air carrier or commercial operator may be in any form but must contain at least the following information concerning each flight:

1. *Company or organization name;*
2. *Make, model and registration number of the aircraft being used;*
3. *Flight or trip number and the date of the flight;*
4. *Name of each flight crewmember, flight attendant, and pilot designated as pilot-in-command;*
5. *Departure airport, destination airports, alternate airports, and route;*
6. *Minimum fuel supply; and*
7. *A statement of the type of operation (IFR, VFR).*

The dispatch release of a flag or domestic air carrier may be in any form but must contain at least the following information concerning the flight:

1. *Identification number of the aircraft;*
2. *Trip number;*
3. *Departure airport, intermediate stops, destination airports, and alternate airports;*
4. *A statement of the type of operation (IFR, VFR);*
5. *Minimum fuel supply.*

(PLT412) — 14 CFR §121.689

Answers (A) and (C) are incorrect because weather reports and forecasts and minimum fuel supply information are required in the flight release for supplemental and commercial operators and in the dispatch release for flag and domestic air carriers.

121, DSP

9746. Before an ETOPS flight may commence, an ETOPS

A—preflight check must be conducted by a certified A&P and signed off in the logbook.
B—pre-departure service check must be certified by a PDSC Signatory Person.
C—pre-departure check must be signed off by an A&P or the PIC for the flight.

An appropriately-trained, ETOPS-qualified maintenance person must accomplish and certify by signature ETOPS specific tasks. Before an ETOPS flight may commence, an ETOPS pre-departure service check (PDSC) Signatory Person, who has been authorized by the certificate holder, must certify by signature, that the ETOPS PDSC has been completed. (PLT425) — 14 CFR §121.374

121, DSP

9746-1. An ETOPS entry point means

A—the first entry point on the route of flight of an ETOPS flight using one-engine-inoperative cruise speed that is more than 60 minutes from an adequate airport for airplanes having two engines.
B—the first entry point on the route of flight of an ETOPS flight using one-engine-inoperative cruise speed that is more than 200 minutes from an adequate airport for airplanes having more than two engines.
C—the first entry point on the route of flight of an ETOPS flight using one-engine-inoperative cruise speed that is more than 90 minutes from an adequate airport for airplanes having two engines.

"ETOPS entry point" means the first point on the route of an ETOPS flight that is (1) more than 60 minutes from an adequate airport for airplanes with two engines, and (2) more than 180 minutes from an adequate airport for passenger-carrying airplanes with more than two engines. This is determined using a one-engine-inoperative cruise speed under standard conditions in still air. (PLT425) — 14 CFR §121.7

Answers

8295 [B] 9746 [B] 9746-1 [A]

121, DSP

9746-2. For flight planning, a Designated ETOPS Alternate Airport

A—for ETOPS up to 180 minutes, must have RFFS equivalent to that specified by ICAO category 4, unless the airport's RFFS can be augmented by local fire fighting assets within 30 minutes.
B—for ETOPS up to 180 minutes, must have RFFS equivalent to that specified by ICAO category 3, unless the airport's RFFS can be augmented by local fire fighting assets within 45 minutes.
C—for ETOPS up to 180 minutes, must have RFFS equivalent to that specified by ICAO category 4, unless the airport's RFFS can be augmented by local fire fighting assets within 45 minutes.

For ETOPS up to 180 minutes, each designated ETOPS alternate airport must have RFFS equivalent to that specified by ICAO as Category 4 or higher. If the equipment and personnel required are not immediately available at an airport, the certificate holder may still list the airport on the dispatch or flight release if the airport's RFFS can be augmented from local fire fighting assets. A 30-minute response time for augmentation is adequate if the local assets can be notified while the diverting airplane is en route. (PLT398) — 14 CFR §121.106

ALL

9761. What is considered "north polar"?

A—north of 60° N latitude.
B—north of 68° N latitude.
C—north of 78° N latitude.

As an example, operations in the "North Polar Area" and "South Polar Area" require a specific passenger recovery plan for each diversion airport. North Polar Area means the entire area north of 78° N latitude. (PLT425) — 14 CFR §121.7

ALL

9762. What is considered "south polar"?

A—south of 60° S latitude.
B—south of 68° S latitude.
C—south of 78° S latitude.

As an example, operations in the "North Polar Area" and "South Polar Area" require a specific passenger recovery plan for each diversion airport. South Polar Area means the entire area south of 60° S latitude. (PLT425) — 14 CFR §121.7

121, DSP

8281. Who is responsible for obtaining information on all current airport conditions, weather, and irregularities of navigation facilities for a supplemental air carrier flight?

A—Aircraft dispatcher.
B—Director of operations or flight follower.
C—Pilot in command.

Before beginning a flight, each pilot-in-command of a supplemental air carrier or commercial operator flight shall obtain all available current reports or information on airport conditions and irregularities or navigation facilities that may affect the safety of the flight. (PLT444) — 14 CFR §121.603

Answer (A) is incorrect because an aircraft dispatcher is responsible for briefing a flag or domestic (not supplemental) air carrier pilot. Answer (B) is incorrect because the director of operations (who may also be the general manager) is an administrative person, responsible for the day-to-day operations and not usually involved in specific flight operations.

121, DSP

8282. During a supplemental air carrier flight, who is responsible for obtaining information on meteorological conditions?

A—Aircraft dispatcher.
B—Pilot in command.
C—Director of operations or flight follower.

During a flight, the pilot-in-command of a supplemental air carrier or commercial operator flight shall obtain any additional available information of meteorological conditions and irregularities of facilities and services that may affect the safety of the flight. (PLT444) — 14 CFR §121.603

Answer (A) is incorrect because an aircraft dispatcher is responsible for obtaining weather information for a flag or domestic air carrier flight. Answer (C) is incorrect because the director of operations (who may also be the general manager) or flight follower is an administrative person, responsible for day-to-day operations and not usually involved in specific flight operations.

Answers

9746-2 [A] 9761 [C] 9762 [A] 8281 [C] 8282 [B]

Fuel Requirements

All **domestic flights** must have enough fuel to:

1. Fly to the airport to which the flight was dispatched;
2. Thereafter, fly to and land at the most distant alternate airport (if an alternate is required) and
3. Thereafter, fly for 45 minutes at normal cruising fuel consumption.

(The fuel required for a **flag flight** landing in the 48 contiguous states or the District of Columbia is the same as for domestic flights.)

(The fuel requirements for **reciprocating-powered supplemental or commercial operations** landing in the contiguous 48 states is the same as for domestic operations.)

If an **alternate is not required** or the flight is being made to a remote airport where **no alternate is available**, the fuel requirements are:

1. Enough fuel to fly to the destination, and then;
2. Fly for two hours at normal cruising fuel consumption.

A **turbojet supplemental flight** (with an alternate available) landing outside the 48 contiguous states must have fuel to:

1. Fly to the destination, then
2. Fly 10% of the total time required to fly to the destination, then
3. Fly to the alternate, then
4. Fly for 30 minutes at holding speed at 1,500 feet above the alternate.

Propeller driven flag flights must have enough fuel to:

1. Fly to the airport to which the flight was dispatched;
2. Thereafter, fly to and land at the most distant alternate; and
3. Thereafter, fly for 30 minutes plus 15% of the total flying time to the destination and the alternate at normal cruising fuel consumption; or fly for 90 minutes, whichever is less.

If an **alternate is not required** or the flight is being made to a remote airport where **no alternate is available**, the fuel requirements for **reciprocating engine powered flights** are:

1. Enough fuel to fly to the destination, and then;
2. Fly for 3 hours at normal cruising fuel consumption.

121, DSP

8268. The reserve fuel supply for a domestic air carrier flight is

A—30 minutes plus 15 percent at normal fuel consumption in addition to the fuel required to the alternate airport.
B—45 minutes at normal fuel consumption in addition to the fuel required to fly to and land at the most distant alternate airport.
C—45 minutes at normal fuel consumption in addition to the fuel required to the alternate airport.

For domestic operations, no person may dispatch or takeoff an airplane unless it has enough fuel to:

1. *Fly to the airport to which it was dispatched;*
2. *Thereafter, to fly to and land at the most distant alternate airport (if an alternate is required); and*
3. *Thereafter, to fly for 45 minutes at normal cruising fuel consumption.*

(PLT413) — 14 CFR §121.639

121, DSP

8269. The minimum amount (planned) of fuel to be aboard a flag air carrier turbojet airplane on a flight within the 48 contiguous United States, after reaching the most distant alternate airport, should be

A—45 minutes at normal cruising fuel consumption.
B—2 hours at normal cruising fuel consumption.
C—enough fuel to return to the destination airport or to fly for 90 minutes at normal cruising fuel consumption, whichever is less.

A turbine-engined flag air carrier operation within the 48 contiguous United States and the District of Columbia may use the fuel requirements of a domestic air carrier. For domestic operations, no person may dispatch or takeoff in an airplane unless it has enough fuel to:

1. *Fly to the airport to which it was dispatched;*
2. *Thereafter, to fly to and land at the most distant alternate airport (if an alternate is required); and*
3. *Thereafter, to fly for 45 minutes at normal cruising fuel consumption.*

(PLT413) — 14 CFR §121.639

Answer (B) is incorrect because 2 hours normal cruising fuel is required at the destination airport when an alternate is not specified and the flight is conducted outside the 48 contiguous United States. Answer (C) is incorrect because there is no provision for return to the destination airport in calculating fuel requirements.

121, DSP

8271. For a flag air carrier flight to be released to an island airport for which an alternate airport is not available, a turbojet-powered airplane must have enough fuel to fly to that airport and thereafter to fly

A—at least 2 hours at normal cruising fuel consumption.
B—for 3 hours at normal cruising fuel consumption.
C—back to the departure airport.

No person may dispatch a turbojet-powered airplane to an airport for which no alternate is available unless it has enough fuel, considering wind and other weather conditions, to fly to that airport and thereafter to fly for at least 2 hours at normal cruising fuel consumption. (PLT413) — 14 CFR §121.645

121, DSP

8272. An alternate airport is not required for a supplemental or commercial air carrier, turbojet-powered airplane on an IFR flight outside the 48 contiguous United States, if enough fuel

A—is aboard to fly to the destination at normal cruise speed and thereafter at least 2 hours at normal holding speed.
B—is aboard the airplane to fly to the destination and then to fly for at least 2 more hours at normal cruising fuel consumption.
C—to fly over the destination for 30 minutes at holding airspeed at 1,500 feet AGL is carried aboard the airplane.

No person may dispatch a turbojet-powered airplane to an airport for which no alternate is available unless it has enough fuel, considering wind and other weather conditions, to fly to that airport and thereafter to fly for at least 2 hours at normal cruising fuel consumption. (PLT413) — 14 CFR §121.645

Answers

8268 [B] 8269 [A] 8271 [A] 8272 [B]

121, DSP

8276. A turbine-engine-powered flag air carrier airplane is released to an airport which has no available alternate. What is the required fuel reserve?

A—2 hours at normal cruise speed in a no wind condition fuel consumption.
B—2 hours at normal cruise fuel consumption.
C—30 minutes, plus 10 percent of the total flight time.

No person may dispatch a turbojet-powered airplane to an airport for which no alternate is available unless it has enough fuel, considering wind and other weather conditions, to fly to that airport and thereafter to fly for at least 2 hours at normal cruising fuel consumption. (PLT413) — 14 CFR §121.645

121, DSP

8273. The fuel reserve required for a turbine-engine-powered (other than turbopropeller) supplemental air carrier airplane upon arrival over the most distant alternate airport outside the 48 contiguous United States is

A—30 minutes at holding speed, at 1,500 feet over the airport.
B—30 minutes, over the airport, at 1,500 feet, at cruising speed.
C—2 hours at the normal cruising fuel consumption rate.

For any flag air carrier, supplemental air carrier, or commercial operator operation outside the 48 contiguous United States or District of Columbia, no person may release for flight or takeoff a turbine engine-powered airplane (other than a turbopropeller-powered airplane) unless, considering wind and other weather conditions expected, it has enough fuel:

1. *To fly to and land at the airport to which it was released;*
2. *After that, to fly for a period of 10% of the total time required to fly from the airport of departure to and land at, the airport to which it was released;*
3. *After that, to fly to and land at the most distant alternate airport specified in the flight release, if an alternate is required; and*
4. *After that, to fly for 30 minutes at holding speed at 1,500 feet above the alternate airport (or destination airport if no alternate is required) under standard temperature conditions.*

(PLT413) — 14 CFR §121.645

121, DSP

8270. What is the fuel reserve requirement for a commercially operated reciprocating-engine-powered airplane flying within the 48 contiguous United States upon arrival at the most distant alternate airport specified in the flight release? Enough fuel to fly

A—30 minutes plus 15 percent of total time required to fly at normal cruising consumption to the alternate.
B—to fly for 90 minutes at normal cruising fuel consumption.
C—45 minutes at normal cruising fuel consumption.

No person may release for flight or takeoff a nonturbine or turbopropeller-powered airplane unless, considering the wind and other weather conditions expected, it has enough fuel to:

1. *Fly to the airport to which it was released;*
2. *Thereafter, to fly to and land at the most distant alternate airport specified in the flight release; and*
3. *Thereafter, to fly for 45 minutes at normal cruising fuel consumption.*

(PLT413) — 14 CFR §121.643

121, DSP

8277. The fuel reserve required for a reciprocating-engine-powered supplemental air carrier airplane upon arrival at the most distant alternate airport during a flight in the 48 contiguous United States is

A—45 minutes at normal cruising fuel consumption.
B—the fuel required to fly to the alternate, plus 10 percent.
C—3 hours at normal cruising fuel consumption.

No person may release for flight or takeoff a nonturbine or turbopropeller-powered airplane unless, considering the wind and other weather conditions expected, it has enough fuel to:

1. *Fly to the airport to which it was released;*
2. *Thereafter, to fly to and land at the most distant alternate airport specified in the flight release; and*
3. *Thereafter, to fly for 45 minutes at normal cruising fuel consumption.*

(PLT413) — 14 CFR §121.643

Answers

8276 [B] 8273 [A] 8270 [C] 8277 [A]

121, DSP

8274. Upon arriving at the most distant airport, what is the fuel reserve requirement for a turbopropeller flag air carrier airplane?

A—90 minutes at holding altitude and speed fuel consumption or 30 minutes plus 15 percent of cruise fuel consumption, whichever is less.
B—45 minutes at holding altitude.
C—30 minutes plus 15 percent of the total time required, or 90 minutes at normal cruise, whichever is less.

No person may dispatch or takeoff in a flag air carrier nonturbine or turbopropeller-powered airplane unless, considering the wind and other weather conditions expected, it has enough fuel:

1. *To fly to and land at the airport to which it is dispatched;*
2. *Thereafter, to fly to and land at the most distant alternate airport specified in the dispatch release; and*
3. *Thereafter to fly for 30 minutes plus 15% of numbers 1 and 2 above, or to fly for 90 minutes at normal cruising fuel consumption, whichever is less.*

(PLT413) — 14 CFR §121.641

121, DSP

8275. The fuel reserve required, for a turbopropeller supplemental air carrier airplane upon the arrival at a destination airport for which an alternate airport is not specified, is

A—3 hours at normal consumption, no wind condition.
B—3 hours at normal cruising fuel consumption.
C—2 hours at normal cruising fuel consumption.

No supplemental air carrier or commercial operator may release a nonturbine or turbopropeller-powered airplane to an airport for which no alternate is specified unless it has enough fuel, considering wind and weather conditions expected, to fly to that airport and thereafter to fly for 3 hours at normal cruising fuel consumption. (PLT413) — 14 CFR §121.643

Answers

8274 [C] 8275 [B]

Carriage of Passengers and Cargo

Before takeoff all the passengers must be briefed on:

- Smoking,
- the location of emergency exits,
- the use of seatbelts,
- the location and use of any required means of emergency flotation.

After the seatbelt sign has been turned off in flight, the passengers must be briefed to keep their seatbelts fastened while seated. In addition to the required briefings, passengers must be provided with printed cards that contain diagrams of and methods of operating the emergency exits and the use of other emergency equipment. Before flight is conducted above FL250, a crewmember must instruct the passengers on the necessity of using oxygen in the event of cabin depressurization, and must point out to them the location and demonstrate the use of the oxygen dispensing equipment.

Each passenger two years old and older must have their own seat or berth and approved seatbelt. During takeoff and landing, all passengers must be in their seat with their seatbelts fastened. A child under two may be held by an adult. During the enroute portion of a flight, two passengers may share a seatbelt while seated in a multiple lounge or divan seat.

There are certain persons who have to be admitted to the flight deck in flight (such as crewmembers, FAA inspectors, etc.) and certain others who may be admitted (e.g., deadheading crew), but the pilot-in-command has emergency authority to exclude any person from the flight deck in the interest of safety. In what is commonly known as the "sterile cockpit rule," crewmembers are required to refrain from nonessential activities during critical phases of flight. As defined in the regulation, critical phases of flight are all ground operations involving taxi, takeoff, and landing, and all other flight operations below 10,000 feet except cruise flight. Nonessential activities include such activities as eating, reading a newspaper, or chatting.

Law enforcement officers may carry firearms on board an air carrier flight if their duties so require. Except in an emergency, the carrier should be given at least one hour prior notice that a person carrying a deadly weapon is going to be on the flight. If a passenger is carrying a firearm in their checked baggage, the weapon must be unloaded and the bag locked. The passenger must retain the key to the bag. The bag must be stowed in a portion of the aircraft that is inaccessible to both the passenger and to crewmembers in flight.

Prisoners are sometimes carried on air carrier flights. The prisoners are always escorted and no more than one prisoner who is classified as "maximum risk" can be allowed on the aircraft. Certain rules apply to the carriage of prisoners. These include:

- The prisoner and escort must be boarded before all other passengers and must stay on board until all other passengers have deplaned.
- The prisoner and escort must sit in the most rearward passenger seats and the escort must sit between the prisoner and the aisle.
- The carrier may serve the prisoner and the escort food and beverages, but neither of them may be served alcohol.

If a person who appears to be intoxicated creates a disturbance on a flight, a report of the incident must be made to the Administrator (the FAA) within 5 days.

Certain passengers may be carried on an all-cargo flight without the carrier having to comply with all the passenger-carrying rules. Passengers carried on an all-cargo flight must have a seat with an approved seatbelt in the cargo compartment. They must have access to the pilot compartment or to an exit. The pilot-in-command must be able to notify them when they must have their seatbelt fastened and when smoking is prohibited. They must receive an emergency briefing from a crewmember prior to takeoff. The pilot-in-command may authorize the passenger to be admitted to the flight crew compartment.

Cargo (including carry-on baggage) may be carried in the passenger compartment of an aircraft if certain conditions are met. If the cargo is carried in an approved cargo bin, it can be located anywhere in the passenger compartment. The bin:

- Must withstand the load factor required of passenger seats multiplied by 1.15
- May not be installed in a position that restricts access to or use of any required emergency exit, or of the aisle in the passenger cabin
- Must be completely enclosed and made of material that is at least flame resistant

If the cargo is not placed in an approved cargo bin it must be located aft of a bulkhead or divider (i.e., not aft of a passenger) and it must meet certain other requirements. These include:

- It must be properly secured by a safety belt or other tie down.
- It must be packaged or covered in a manner so as to avoid injury to occupants of the passenger cabin.
- It must not impose an excessive load on the floor or seat structures of the aircraft.
- Its location must not restrict access to or use of the aisle, any regular exit or any required emergency exit.
- Its location must not obscure any passenger's view of the "seatbelt," "no smoking" or required "exit" signs unless an auxiliary sign is installed.

Each person who has duties concerning the handling or carriage of dangerous articles or magnetized materials must have completed a training course within the preceding 12 calendar months.

ALL

8131. A certificate holder is notified that a person specifically authorized to carry a deadly weapon is to be aboard an aircraft. Except in an emergency, how long before loading that flight should the air carrier be notified?

A—Notification is not required, if the certificate holder has a security coordinator.
B—A minimum of 1 hour.
C—A minimum of 2 hours.

The certificate holder, except in an emergency, must be given at least 1 hour notice when an authorized person intends to have a weapon accessible in flight. (PLT498) — 49 CFR §1544.219

Answers

8131 [B]

ALL

8137. When a passenger notifies the certificate holder prior to checking baggage that an unloaded weapon is in the baggage, what action is required by regulation regarding this baggage?

A—The baggage may be carried in the flightcrew compartment, provided the baggage remains locked, and the key is given to the pilot in command.
B—The baggage must remain locked and carried in an area that is inaccessible to the passenger, and only the passenger retains the key.
C—The baggage must remain locked and stored where it would be inaccessible, and custody of the key shall remain with a designated crewmember.

No certificate holder may knowingly permit any person to transport any unloaded firearm in checked baggage unless the baggage in which it is carried is locked and only the passenger checking the baggage retains the key or combination. The baggage containing the firearm must be carried in an area, other than the flight crew compartment, that is inaccessible to passengers. (PLT498) — 49 CFR §1544.203(f)

Answers (A) and (C) are incorrect because the baggage containing the unloaded firearm will be carried in the baggage area, and only the passenger checking the baggage retains the key.

ALL

9763. What is meant by "sterile cockpit"?

A—All preflight checks are complete and the aircraft is ready for engine starting.
B—Crewmembers refrain from nonessential activities during critical phases of flight.
C—Crewmembers are seated and buckled at their required stations.

Commonly known as the "sterile cockpit rule," 14 CFR §121.542 requires flight crewmembers to refrain from nonessential activities during critical phases of flight. As defined in the regulation, critical phases of flight are all ground operations involving taxi, takeoff, and landing, and all other flight operations below 10,000 feet except cruise flight. Nonessential activities include such activities as eating, reading a newspaper, or chatting. (PLT498) — 14 CFR §121.542

ALL

8132. When a person in the custody of law enforcement personnel is scheduled on a flight, what procedures are required regarding boarding of this person and the escort?

A—They shall be boarded before all other passengers board, and deplaned after all the other passengers have left the aircraft.
B—They shall be boarded after all other passengers board, and deplaned before all the other passengers leave the aircraft.
C—They shall board and depart before the other passengers.

When a person in custody of law enforcement is to be carried on a flight, the prisoner and escort must be boarded before any other passengers and deplaned after all other passengers have deplaned. (PLT325) — 49 CFR §1544.221(f)(1)

ALL

8136. Which applies to the carriage of a person in the custody of law enforcement personnel?

A—The air carrier is not allowed to serve beverages to the person in custody or the law enforcement escort.
B—No more than one person considered to be in the maximum risk category may be carried on a flight, and that person must have at least two armed law enforcement escorts.
C—The person in custody must be seated between the escort and the aisle.

No more than one passenger, of whom the certificate holder has been notified as being in a maximum risk category, can be carried on an airplane. (PLT325) — 49 CFR §1544.221(c)(2), (d)(3)

121, DSP

8225. Which passenger announcement(s) must be made after each takeoff?

A—Keep safety belts fastened while seated and no smoking in the aircraft lavatories.
B—Passengers should keep seat belts fastened while seated.
C—How to use the passenger oxygen system and that there is a $1,000 fine for tampering with a smoke detector.

Answers

8137 [B]	9763 [B]	8132 [A]	8136 [B]	8225 [B]

After each takeoff, immediately before or after turning the seatbelt sign off, an announcement shall be made that passengers should keep their seatbelts fastened, while seated, even when the seatbelt sign is off. (PLT384) — 14 CFR §121.571

Answer (A) is incorrect because the requirement is that each passenger shall be briefed on when, where, and under what conditions smoking is prohibited. "No smoking in the aircraft lavatories" is a required briefing prior to all takeoffs. Answer (C) is incorrect because the emergency use of the passenger oxygen system is a required briefing prior to all takeoffs to pressurized flight above FL250.

121, DSP

8181. A passenger briefing by a crewmember shall be given, instructing passengers on the necessity of using oxygen in the event of cabin depressurization, prior to flights conducted above

A—FL 200.
B—FL 240.
C—FL 250.

Before flight is conducted above FL250, a crewmember shall instruct the passengers on the necessity of using oxygen in the event of cabin depressurization, and shall point out to them the location and demonstrate the use of the oxygen dispensing equipment. (PLT438) — 14 CFR §121.333

121, DSP

8153. When may two persons share one approved safety belt in a lounge seat?

A—When one is an adult and one is a child under 3 years of age.
B—Only during the en route flight.
C—During all operations except the takeoff and landing portion of a flight.

No person may operate an airplane unless there are available during the takeoff, enroute flight, and landing an approved seatbelt for separate use by each person on board the airplane who has reached his/her second birthday, except that two persons occupying a berth may share one approved seatbelt and two persons occupying a multiple lounge or divan seat may share one approved seatbelt during en route flight only. (PLT465) — 14 CFR §121.311

Answer (A) is incorrect because the regulations do not specify an age of persons sharing a seatbelt on a lounge seat. Sharing a seatbelt in a lounge seat can only be done during the enroute portion of the flight. Answer (C) is incorrect because two persons may share one seatbelt in a lounge seat only during the enroute portion of the flight, which excludes taxi and takeoff as well as landing.

121, DSP

8244. The pilot in command has emergency authority to exclude any and all persons from admittance to the flight deck

A—except a FAA inspector doing enroute checks.
B—in the interest of safety.
C—except persons who have authorization from the certificate holder and the FAA or NTSB.

The pilot-in-command has the emergency authority to exclude anyone from the flight deck in the interest of safety. (PLT444) — 14 CFR §121.547

Answers (A) and (C) are incorrect because persons who have specific authorization of the certificate holder and FAA inspectors may be admitted to the flight deck except when excluded in an emergency.

121, DSP

8233. If an intoxicated person creates a disturbance aboard an air carrier aircraft, the certificate holder must submit a report, concerning the incident, to the Administrator within

A—7 days.
B—5 days.
C—48 hours.

If an intoxicated person causes an incident on the aircraft the certificate holder shall, within 5 days, report that incident to the Administrator. (PLT366) — 14 CFR §121.575

121, DSP

8234. When carrying a passenger aboard an all-cargo aircraft, which of the following applies?

A—The passenger must have access to a seat in the pilot compartment.
B—The pilot in command may authorize the passenger to be admitted to the crew compartment.
C—Crew-type oxygen must be provided for the passenger.

When a passenger is allowed on an all-cargo flight, the pilot-in-command may authorize admittance to the flight deck. (PLT444) — 14 CFR §121.583

Answer (A) is incorrect because the seat does not have to be on the flight deck, but there must be an approved seat with an approved seatbelt for each person. Answer (C) is incorrect because crew-type oxygen is not required for passengers. It is only required that the person be briefed on the use of oxygen and emergency oxygen equipment.

Answers

8181 [C] 8153 [B] 8244 [B] 8233 [B] 8234 [B]

121, DSP

8139. What requirement must be met regarding cargo that is carried anywhere in the passenger compartment of an air carrier airplane?

A—The bin in which the cargo is carried may not be installed in a position that restricts access to, or use of, any exit.
B—The bin in which the cargo is carried may not be installed in a position that restricts access to, or use of, any aisle in the passenger compartment.
C—The container or bin in which the cargo is carried must be made of material which is at least flash resistant.

Cargo may be carried anywhere in the passenger compartment if it is carried in an approved cargo bin. The bin must meet the following requirements:

1. *The bin must be able to withstand the load factors and emergency landing conditions applicable to the passenger seats of the airplane in which it is installed, multiplied by a factor of 1.15;*
2. *The cargo bin may not be installed in a position that restricts access to or use of any required emergency exit, or of the aisle in the passenger compartment;*
3. *The bin must be fully enclosed and made of material that is at least flame resistant.*

(PLT385) — 14 CFR §121.285

Answer (A) is incorrect because the bin may not be installed in a position that restricts access to or use of any required emergency exit. Answer (C) is incorrect because the bin must be fully enclosed and made of material that is at least flame resistant.

121, DSP

8175. Which restriction applies to a cargo bin in a passenger compartment? The bin

A—may have an open top if it is placed in front of the passengers and the cargo is secured by a cargo net.
B—must withstand the load factor required of passenger seats, multiplied by 1.15, using the combined weight of the bin and the maximum weight of the cargo that may be carried in the bin.
C—must be constructed of flame retardant material and fully enclosed.

Cargo may be carried anywhere in the passenger compartment if it is carried in an approved cargo bin. The bin must meet the following requirements:

1. *The bin must be able to withstand the load factors and emergency landing conditions applicable to the passenger seats of the airplane in which it is installed, multiplied by a factor of 1.15;*
2. *The cargo bin may not be installed in a position that restricts access to or use of any required emergency exit, or of the aisle in the passenger compartment;*
3. *The bin must be fully enclosed and made of material that is at least flame resistant.*

(PLT385) — 14 CFR §121.285

Answers (A) and (C) are incorrect because the cargo bin must be fully enclosed, and be constructed of materials that are at least flame resistant.

121, DSP

8138. What restrictions must be observed regarding the carrying of cargo in the passenger compartment of an airplane operated under 14 CFR Part 121?

A—All cargo must be separated from the passengers by a partition capable of withstanding certain load stresses.
B—All cargo must be carried in a suitable flame resistant bin and the bin must be secured to the floor structure of the airplane.
C—Cargo may be carried aft of a divider if properly secured by a safety belt or other tiedown having enough strength to eliminate the possibility of shifting.

Cargo may be carried aft of a bulkhead or divider in any passenger compartment provided the cargo is restrained to required load factors, and it is properly secured by a safety belt or other tiedown having enough strength to eliminate the possibility of shifting under all normally anticipated flight and ground conditions. (PLT385) — 14 CFR §121.285

Answers (A) and (B) are incorrect because cargo may be carried in the passenger compartment if it is properly covered and secured so as not to be a hazard.

Answers

8139 [B]   8175 [B]   8138 [C]

Part 135 Carriage of Passengers and Cargo Requirements

135, RTC

8007. Where must a certificate holder keep copies of completed load manifests and for what period of time?

A—1 month at its principal operations base, or at a location approved by the Administrator.
B—30 days at its principal operations base, or another location used by it and approved by the Administrator.
C—30 days, at the flight's destination.

The certificate holder shall keep copies of completed load manifests for at least 30 days at its principal operations base, or at another location used by it and approved by the Administrator. (PLT400) — 14 CFR §135.63

135, RTC

8008. Which is NOT a required item on the load manifest?

A—List of passenger names and the weight of each.
B—Aircraft registration number or flight number.
C—Identification of crewmembers and their crew position.

The load manifest must be prepared before each takeoff and must include:

1. *The number of passengers;*
2. *The total weight of the loaded aircraft;*
3. *The maximum allowable takeoff weight for that flight;*
4. *The center of gravity limits;*
5. *The center of gravity of the loaded aircraft;*
6. *The registration number of the aircraft or flight number;*
7. *The origin and destination; and*
8. *Identification of crewmembers and their crew position assignments.*

(PLT440) — 14 CFR §135.63

135, RTC

8009. Who is responsible for the preparation of a required load manifest?

A—PIC or the Dispatcher.
B—Company official designated by the Administrator.
C—The certificate holder.

For multi-engine aircraft, each certificate holder is responsible for the preparation and accuracy of a load manifest. (PLT440) — 14 CFR §135.63

135, RTC

8032. Which restriction must be observed regarding the carrying of cargo in the passenger compartment?

A—It is packaged or covered to avoid possible injury to occupants.
B—All cargo must be carried in a suitable bin and secured to a passenger seat or the floor structure of the aircraft.
C—Cargo carried in passenger seats must be forward of all passengers.

No person may carry cargo, including carry-on baggage, in or on any aircraft unless one of the three following criteria is met:

1. *It is carried in an approved cargo rack, bin or compartment;*
2. *It is secured by approved means; or*
3. *If number 1 or 2 is not met, then all of the following are met:*
    a. *For cargo, it is properly secured by a safety belt or other tie-down having enough strength to eliminate the possibility of shifting under all normally anticipated flight and ground conditions, or for carry-on baggage, it is restrained so as to prevent its movement during air turbulence;*
    b. *It is packaged or covered to avoid possible injury to occupants;*
    c. *It does not impose any load on seats or on the floor structure that exceeds the load limitation for those components;*
    d. *It is not located in a position that obstructs the access to, or use of, any required emergency or regular exit, or the use of the aisle between the crew and passenger compartment, or located in a position that obscures any passenger's view of the "seatbelt" sign, "no smoking" sign or any required exit sign;*
    e. *It is not carried directly above seated occupants.*

(PLT385) — 14 CFR §135.87

Answers

8007 [B] 8008 [A] 8009 [C] 8032 [A]

135, RTC

9720. A person whose duties include the handling or carriage of dangerous articles and/or magnetized materials must have satisfactorily completed an approved training program established by the certificate holder within the previous

A—6 calendar months.
B—12 calendar months.
C—24 calendar months.

No certificate holder may use any person or perform, and no person may perform, any assigned duties and responsibilities for the handling or carriage of hazardous materials unless within the preceding 12 calendar months that person has satisfactorily completed initial or recurrent training in an appropriate training program established by the certificate holder. (PLT407) — 14 CFR, SFAR 99

135, RTC

8039. In a cargo-only operation, cargo must be loaded

A—so that it does not obstruct the aisle between the crew and cargo compartments.
B—in such a manner that at least one emergency or regular exit is available to all occupants.
C—in such a manner that at least one emergency or regular exit is available to all crewmembers, if an emergency occurs.

For cargo-only operations, the cargo must be loaded so at least one emergency or regular exit is available to provide all occupants of the aircraft a means of unobstructed exit from the aircraft if an emergency occurs. (PLT385) — 14 CFR §135.87

135, RTC

8040. Which is a requirement governing the carriage of cargo, on a scheduled passenger flight?

A—Cargo must be carried in an approved rack, bin, or compartment.
B—Cargo not stowed in an approved bin must be secured by a safety belt or approved tiedown device.
C—All cargo carried in the passenger compartment must be packaged and stowed ahead of the foremost seated passenger.

No person may carry cargo, including carry-on baggage, in or on any aircraft unless one of the three following criteria is met:

1. *It is carried in an approved cargo rack, bin or compartment;*
2. *It is secured by approved means; or*
3. *If number 1 or 2 is not met, then all of the following are met:*
   a. *For cargo, it is properly secured by a safety belt or other tie-down having enough strength to eliminate the possibility of shifting under all normally anticipated flight and ground conditions, or for carry-on baggage, it is restrained so as to prevent its movement during air turbulence;*
   b. *It is packaged or covered to avoid possible injury to occupants;*
   c. *It does not impose any load on seats or on the floor structure that exceeds the load limitation for those components;*
   d. *It is not located in a position that obstructs the access to, or use of, any required emergency or regular exit, or the use of the aisle between the crew and passenger compartment, or located in a position that obscures any passenger's view of the "seatbelt" sign, "no smoking" sign or any required exit sign;*
   e. *It is not carried directly above seated occupants.*

(PLT385) — 14 CFR §135.87

Answers

9720 [B] 8039 [B] 8040 [B]

135, RTC

8041. Which is a requirement governing the carriage of carry-on baggage?

A—All carry-on baggage must be restrained so that its movement is prevented during air turbulence.
B—Carry-on baggage must be stowed under the seat in front of the owner.
C—Pieces of carry-on baggage weighing more than 10 pounds must be carried in an approved rack or bin.

No person may carry cargo, including carry-on baggage, in or on any aircraft unless one of the three following criteria is met:

1. *It is carried in an approved cargo rack, bin or compartment;*
2. *It is secured by approved means; or*
3. *If number 1 or 2 is not met, then all of the following are met:*
   a. *For cargo, it is properly secured by a safety belt or other tie-down having enough strength to eliminate the possibility of shifting under all normally anticipated flight and ground conditions, or for carry-on baggage, it is restrained so as to prevent its movement during air turbulence;*
   b. *It is packaged or covered to avoid possible injury to occupants;*
   c. *It does not impose any load on seats or on the floor structure that exceeds the load limitation for those components;*
   d. *It is not located in a position that obstructs the access to, or use of, any required emergency or regular exit, or the use of the aisle between the crew and passenger compartment, or located in a position that obscures any passenger's view of the "seatbelt" sign, "no smoking" sign or any required exit sign;*
   e. *It is not carried directly above seated occupants.*

(PLT385) — 14 CFR §135.87

135, RTC

8042. If carry-on baggage or cargo is carried in the passenger compartment, it must be

A—stowed ahead of the foremost seated passengers and secured by approved means.
B—placed in an approved rack, bin, or compartment installed in the aircraft.
C—so located that it does not obstruct the access to, or the use of, any required emergency or regular exit.

No person may carry cargo, including carry-on baggage, in or on any aircraft unless one of the three following criteria is met:

1. *It is carried in an approved cargo rack, bin or compartment;*
2. *It is secured by approved means; or*
3. *If number 1 or 2 is not met, then all of the following are met:*
   a. *For cargo, it is properly secured by a safety belt or other tie-down having enough strength to eliminate the possibility of shifting under all normally anticipated flight and ground conditions, or for carry-on baggage, it is restrained so as to prevent its movement during air turbulence;*
   b. *It is packaged or covered to avoid possible injury to occupants;*
   c. *It does not impose any load on seats or on the floor structure that exceeds the load limitation for those components;*
   d. *It is not located in a position that obstructs the access to, or use of, any required emergency or regular exit, or the use of the aisle between the crew and passenger compartment, or located in a position that obscures any passenger's view of the "seatbelt" sign, "no smoking" sign or any required exit sign;*
   e. *It is not carried directly above seated occupants.*

(PLT385) — 14 CFR §135.87

135, RTC

8043. The load manifest must be prepared prior to each takeoff for

A—any aircraft with a passenger seating capacity of 10 seats or more.
B—any aircraft with more than one engine.
C—all helicopters and large aircraft operated by a commuter air carrier.

For multi-engine aircraft, each certificate holder is responsible for the preparation and accuracy of a load manifest. (PLT440) — 14 CFR §135.63

Answers

8041 [A] 8042 [C] 8043 [B]

Emergency Equipment and Operations

Certain emergency equipment must be carried on every air carrier airplane. This equipment includes fire extinguishers, megaphones, first aid kits and a crash ax. All this equipment must:

- Be inspected regularly.
- Be readily accessible to the crew and, for items carried in the passenger cabin, to the passengers.
- Be clearly identified and marked with its method of operation (this applies to any containers in which the equipment is carried).

Only one crash ax is required on the airplane and must be carried on the flight deck. At least one hand fire extinguisher must be carried on the flight deck. The number of extinguishers carried in the cabin is determined by the number of installed passenger seats. The following table applies.

Minimum Number of Hand Fire Extinguishers in the Passenger Cabin:

Passenger Seating Capacity	Extinguishers Required
6 through 30	1
31 through 60	2
61 through 200	3
201 through 300	4
301 through 400	5
401 through 500	6
501 through 600	7
601 or more	8

The number of megaphones carried on the airplane is determined by the number of installed passenger seats. On airplanes with a seating capacity of 60 through 99 passengers, one megaphone must be carried in the most rearward location in the passenger cabin that is readily accessible to a normal flight attendant seat. On airplanes with a seating capacity of 100 or more seats, one megaphone must be carried at the rear of the cabin and another megaphone must be carried at the front of the cabin.

Passenger carrying airplanes must have an emergency exit light system. This system must be operable manually from both the flight crew station and from a point in the passenger compartment readily accessible to a flight attendant. When the system is armed it must come on automatically with the interruption of the airplane's normal electrical power. The exit lights must be armed or turned on during taxiing, takeoff and landing. Every emergency exit (other than an over wing exit) that is more than 6 feet from the ground must have a means of assisting occupants to the ground in the event of an emergency evacuation. The most common means of complying with this requirement is an inflatable slide that deploys automatically when the door is opened. If such an automatic escape slide is installed, it must be armed during taxi, takeoff and landing. If any required emergency exit for passengers is located in other than the passenger compartment (such as the flight deck), the door separating the compartments must be latched open during takeoff and landing.

A public address system and a separate crewmember interphone system must be installed on all airplanes with a seating capacity of more than 19 seats.

Each crewmember on a flight must have a flashlight in good working order readily available.

When operating at flight altitudes above 10,000 feet there must be enough oxygen for all crewmembers for the entire flight at those altitudes, and in no event less than a 2-hour supply.

When operating at flight altitudes above FL250 each flight crewmember on flight deck duty must have an oxygen mask, within immediate reach, so designed that it can be rapidly placed on his/her face. This is commonly referred to as a "**quick-donning**" oxygen mask. To meet the requirements, regulations require that the mask be designed so that it can be put on the user's face within 5 seconds. If, while operating above FL250, one pilot leaves his/her station, the other pilot must put on his/her oxygen mask.

Above FL410 one pilot must wear his/her mask at all times. Notice that the rule applies only to the pilots. Above FL250 the flight engineer need only have a quick-donning mask readily available. *Note:* For Part 135 operations one pilot must wear the oxygen mask above FL350.

The oxygen requirements for passengers vary with the type of aircraft, but oxygen must be provided to all passengers for the entire time the cabin altitude is above 15,000 feet.

Passengers on turbine powered airplanes must be supplied oxygen according to the following schedule.

- For flights at cabin pressure altitudes above 10,000 feet, up to and including 14,000 feet, there must be enough oxygen to supply 10% of the passengers for any time at those altitudes in excess of 30 minutes.
- For flights at cabin pressure altitudes above 14,000 feet, up to and including 15,000 feet, there must be enough oxygen for 30% of the passengers for the entire time at those altitudes.
- For flights at cabin pressure altitudes above 15,000 feet there must be enough oxygen for all the passengers for the entire time of flight at those altitudes.

The amount of oxygen carried for passengers in the event of loss of pressurization varies depending on the ability of the airplane to make an emergency descent. If the aircraft can make a descent to 14,000 feet within 4 minutes it may carry less oxygen than would otherwise be required.

A certain amount of **first aid oxygen** must be carried for passengers on flights that operate above FL250. The amount of oxygen is determined by the actual number of passengers but in no case may there be less than 2 oxygen dispensing units.

On extended over-water flights (more than 50 nautical miles from the shoreline) the airplane must have a life preserver for each occupant of the aircraft, and enough life rafts to accommodate all the occupants. This equipment must be easily accessible in the event of a ditching.

- Each life raft and each life vest must be equipped with a survivor locator light.
- A survival kit, appropriate for the route flown, must be attached to each life raft.
- There must be at least one portable emergency radio transmitter carried on the airplane.

When flag or supplemental carriers or commercial operators fly over uninhabited terrain, the following survival equipment must be carried on the airplane:

- Suitable pyrotechnic signaling devices.
- A survival-type emergency locator transmitter.
- Enough survival kits, appropriate for the route flown, for all the occupants of the airplane.

In an emergency situation that requires immediate decision and action, the pilot-in-command may take any action that he/she considers necessary under the circumstances. In such a case the PIC may deviate from prescribed procedures and methods, weather minimums and regulations to the extent required in the interest of safety. In an emergency situation arising during flight that requires immediate decision and action by an aircraft dispatcher, the dispatcher must advise the pilot-in-command of the emergency, shall ascertain the decision of the pilot-in-command and shall have that decision recorded. If the dispatcher cannot communicate with the pilot, he/she shall declare an emergency and take any action he/she considers necessary under the circumstances.

Each certificate holder (airline) must, for each type and model of airplane, assign to each category of crewmember, as appropriate, the necessary functions to be performed in an emergency or in a situation requiring emergency evacuation. The certificate holder must describe those duties in its manual.

Crewmembers must receive emergency training annually on several subjects. Besides the training they must perform emergency drills in:

- The operation of emergency exits;
- Hand fire extinguishers;
- The emergency oxygen system and protective breathing equipment;
- Donning, inflation and use of individual flotation equipment; and
- Ditching.

Crewmembers who serve above 25,000 feet must receive instruction in hypoxia, respiration and decompression. Crewmembers must actually operate certain emergency equipment in their recurrent training at least once every 24 months.

The pilot-in-command must make a report to the appropriate ground radio station of the stoppage of an engine's rotation in flight (due either to failure or intentional shutdown) as soon as practicable and must keep that station informed of the progress of the flight. As a general rule, when an engine fails or is shutdown, the pilot-in-command must land the aircraft at the nearest suitable airport, time-wise, at which a safe landing can be made. There is an exception to the rule for airplanes with 3 or more engines. If only 1 engine has failed, the pilot-in-command may elect to continue to a more distant airport (possibly the original destination) if this is considered as safe as landing at the nearest suitable airport.

The certificate holder must provide a **cockpit check procedure** (checklist) for each type of aircraft it operates. The procedures must include each item necessary for flight crewmembers to check for safety before starting engines, taking-off or landing, and in engine and systems emergencies. The procedures must be designed so that a flight crewmember will not need to rely on memory for items to be checked. The flight crew must use the approved check procedure.

Whenever a pilot-in-command or dispatcher exercises emergency authority, he/she shall keep the appropriate ATC facility and dispatch centers fully informed of the progress of the flight. The person declaring the emergency shall send a written report of any deviation through the air carrier's operations manager to the Administrator (FAA). A dispatcher must send this report within 10 days after the date of the emergency. A pilot-in-command must send the report within 10 days after returning to his/her home base.

When ATC gives priority to an aircraft in an emergency, the chief of the ATC facility involved may ask the pilot-in-command to submit a report. If asked, the pilot-in-command must submit a detailed written report to the ATC facility manager within 48 hours. This is required whether or not there was a deviation from regulations.

ALL

9625. (Refer to Legend 15 and Figure 177.) Lewiston – Nez Perce Co. is a 14 CFR Part 139 airport. What is the minimum number of aircraft rescue and fire fighting vehicles, and the type and amount of fire fighting agents that the airport should have?

A—Two vehicles and 600 pounds dry chemical (DC) or Halon 1211 or 500 pounds of DC plus 100 gallons of water.
B—One vehicle and 500 pounds of dry chemical (DC) or Halon 1211 or 450 pounds DC plus 100 gallons of water.
C—One vehicle and 500 pounds of dry chemical (DC) or Halon 1211 or 350 pounds DC and 1,000 gallons of water.

Using FAA Figure 177, the second line of the Lewiston entry indicates it is an ARFF (Aircraft Rescue and Fire Fighting) Index A airport. FAA Legend 15 indicates that an index A airport must have at least one vehicle with either 500 pounds of dry chemical or Halon 1211, or 450 pounds of dry chemical plus 100 gallons of water. (PLT078) — Airport/Facility Directory

ALL

9636. (Refer to Legend 15.) Newport News/Williamsburg Intl is a 14 CFR Part 139 airport. The Airport/Facility Directory contains the following entry: ARFF Index A. What is the minimum number of aircraft rescue and fire fighting vehicles, and the type and amount of fire fighting agents that the airport should have?

A—Two vehicles and 600 pounds dry chemical (DC) or Halon 1211 or 500 pounds of DC plus 100 gallons of water.
B—One vehicle and 500 pounds of dry chemical (DC) or Halon 1211 or 450 pounds DC plus 100 gallons of water.
C—One vehicle and 500 pounds of dry chemical (DC) or Halon 1211 or 350 pounds DC and 1,000 gallons of water.

FAA Legend 15 indicates that an index A airport must have at least one vehicle with either 500 pounds of dry chemical or Halon 1211, or 450 pounds of dry chemical plus 100 gallons of water. (PLT143) — Airport/Facility Directory

ALL

9668. (Refer to Legend 15 and Figure 185A.) McCarran Intl (LAS) is a 14 CFR Part 139 airport. What is the minimum number of aircraft rescue and fire fighting vehicles and the type and amount of fire fighting agents that the airport should have?

A—Three vehicles and 500 pounds of dry chemical (DC) or Halon 1211 or 450 pounds DC and 4,000 gallons of water.
B—Two vehicles and 600 pounds dry chemical (DC) or Halon 1211 or 500 pounds of DC plus 4,000 gallons of water.
C—Three vehicles and 500 pounds of dry chemical (DC) or Halon 1211 or 450 pounds DC plus 3,000 gallons of water.

The second line of the Airport/Facility Directory for McCarran indicates that it is an index D airport. FAA Legend 15 indicates that an index D airport must have at least three vehicles with either 500 pounds of dry chemical or Halon 1211, or 450 pounds of dry chemical plus 4,000 gallons of water. (PLT143) — Airport/Facility Directory

ALL

9669. (Refer to Legend 15 and Figure 205.) San Francisco Intl (SFO) is a 14 CFR Part 139 airport. What is the minimum number of aircraft rescue and fire fighting vehicles, and the type and amount of fire fighting agents that the airport should have?

A—Three vehicles and 500 pounds of dry chemical (DC) or Halon 1211 or 450 pounds DC and 4,000 gallons of water.
B—Three vehicles and 500 pounds of dry chemical (DC) or Halon 1211 or 500 pounds of DC plus 5,000 gallons of water.
C—Three vehicles and 500 pounds of dry chemical (DC) or Halon 1211 or 450 pounds DC plus 6,000 gallons of water.

The second line of the Airport/Facility Directory for San Francisco indicates that it is an index E airport. FAA Legend 15 indicates that an index E airport must have at least three vehicles with either 500 pounds of dry chemical or Halon 1211, or 450 pounds of dry chemical plus 6,000 gallons of water. (PLT143) — Airport/Facility Directory

Answers

9625 [B]   9636 [B]   9668 [A]   9669 [C]

ALL

9690. (Refer to Legend 15 and Figure 215.) Windsor Locks/Bradley Intl, is an 14 CFR Part 139 airport. What minimum number of aircraft rescue and fire-fighting vehicles, and what type and amount of fire-fighting agents are the airport required to have?

A—Two vehicles and 600 pounds dry chemical (DC), or Halon 1211 or 500 pounds of DC plus 4,000 gallons of water.
B—Three vehicles and 500 pounds of dry chemical (DC), or Halon 1211 or 450 pounds DC plus 3,000 gallons of water.
C—Three vehicles and 500 pounds of dry chemical (DC), or Halon 1211 or 450 pounds DC and 4,000 gallons of water.

The Airport/Facility Directory indicates the Windsor Locks/Bradley Intl is an index D airport. FAA Legend 15 indicates that an index D airport must have at least three vehicles with either 500 pounds of dry chemical or Halon 1211, or 450 pounds of dry chemical plus 4,000 gallons of water. (PLT143) — Airport/Facility Directory

ALL

9379. During an emergency, a pilot-in-command does not deviate from a 14 CFR rule but is given priority by ATC. To whom or under what condition is the pilot required to submit a written report?

A—To the manager of the General Aviation District Office within 10 days.
B—To the manager of the facility in control within 10 days.
C—Upon request by ATC, submit a written report within 48 hours to the ATC manager.

Each pilot-in-command who (though not deviating from a rule) is given priority by ATC in an emergency, shall, if requested by ATC, submit a detailed report of that emergency within 48 hours to the chief of that ATC facility. (PLT383) — 14 CFR §91.123

ALL

9388. When may ATC request a detailed report on an emergency even though a rule has not been violated?

A—When priority has been given.
B—Anytime an emergency occurs.
C—When the emergency occurs in controlled airspace.

Each pilot-in-command who (though not deviating from a rule) is given priority by ATC in an emergency, shall, if requested by ATC, submit a detailed report of that emergency within 48 hours to the chief of that ATC facility. (PLT044) — 14 CFR §91.123

Answer (B) is incorrect because a pilot may deviate from a regulation in order to meet an emergency, as long as ATC is notified immediately. A detailed report is usually not required if ATC priority was not given. Answer (C) is incorrect because, regardless of the type of airspace in which it occurs, only when priority has been given may a detailed report be requested by ATC.

121, DSP

8177. Which requirement applies to emergency equipment (fire extinguishers, megaphones, first-aid kits, and crash ax) installed in an air carrier airplane?

A—All emergency equipment, must be readily accessible to the passengers.
B—Emergency equipment cannot be located in a compartment or area where it is not immediately visible to a flight attendant in the passenger compartment.
C—Emergency equipment must be clearly identified and clearly marked to indicate its method of operation.

Each item of required emergency equipment must be clearly identified and clearly marked to indicate its method of operation. (PLT404) — 14 CFR §121.309

Answers (A) and (B) are incorrect because the requirement is that the emergency equipment be "readily accessible" to the crew.

121, DSP

8176. Which factor determines the minimum number of hand fire extinguishers required for flight under 14 CFR Part 121?

A—Number of passengers and crewmembers aboard.
B—Number of passenger cabin occupants.
C—Airplane passenger seating accommodations.

The minimum number of hand fire extinguishers carried on an air carrier flight is determined by the seating capacity of the airplane. (PLT408) — 14 CFR §121.309

Answers (A) and (B) are incorrect because passenger capacity, not actual passenger count, determines the number of extinguishers required.

Answers

9690 [C]	9379 [C]	9388 [A]	8177 [C]	8176 [C]

121, DSP

8160. Where should the portable battery-powered megaphone be located if only one is required on a passenger-carrying airplane?

A—The most forward location in the passenger cabin.
B—In the cabin near the over-the-wing emergency exit.
C—The most rearward location in the passenger cabin.

One megaphone must be installed on each airplane with a seating capacity of more than 60 and less than 100 passengers, at the most rearward location in the passenger cabin where it would be readily accessible to a normal flight attendant seat. (PLT462) — 14 CFR §121.309

121, DSP

8161. How many portable battery-powered megaphones are required on an air carrier airplane with a seating capacity of 100 passengers on a trip segment when 45 passengers are carried?

A—Two; one at the forward end, and the other at the most rearward location in the passenger cabin.
B—Two; one at the most rearward and one in the center of the passenger cabin.
C—Two; one located near or accessible to the flightcrew, and one located near the center of the passenger cabin.

Two megaphones are required in the passenger cabin of each airplane with a seating capacity of more than 99 passengers, one installed at the forward end and the other at the rearward location where it would be readily accessible to a normal flight attendant seat. (PLT462) — 14 CFR §121.309

121, DSP

8162. How many portable battery-powered megaphones are required on an air carrier airplane with a seating capacity of 150 passengers on a trip segment when 75 passengers are carried?

A—Two; one located near or accessible to the flightcrew, and one located near the center of the passenger cabin.
B—Two; one at the most rearward and one in the center of the passenger cabin.
C—Two; one at the forward end, and the other at the most rearward location of the passenger cabin.

Two megaphones are required in the passenger cabin of each airplane with a seating capacity of more than 99 passengers, one installed at the forward end and the other at the rearward location where it would be readily accessible to a normal flight attendant seat. (PLT462) — 14 CFR §121.309

121, DSP

8144. The emergency lights on a passenger-carrying airplane must be armed or turned on during

A—taxiing, takeoff, cruise, and landing.
B—taxiing, takeoff, and landing.
C—takeoff, cruise, and landing.

Each emergency exit light must be armed or turned on during taxiing, takeoff, and landing. (PLT404) — 14 CFR §121.310

Answers (A) and (C) are incorrect because the emergency lights are not required to be armed or turned on during cruise.

121, DSP

8159. Federal Aviation Regulations require that interior emergency lights must

A—operate automatically when subjected to a negative G load.
B—be operable manually from the flightcrew station and a point in the passenger compartment.
C—be armed or turned on during taxiing and all flight operations.

The emergency exit light system must be operable from both the flight crew station and from a point in the passenger compartment that is readily accessible to a normal fight attendant seat. Each emergency exit light must be armed or turned on during taxiing, takeoff, and landing. (PLT404) — 14 CFR §121.310

Answer (A) is incorrect because interior emergency lights must operate automatically with the interruption of the airplane's normal electrical power. Answer (C) is incorrect because the interior emergency light system must only be armed during taxi, takeoff, and landing portions of the flight.

Answers

8160 [C] 8161 [A] 8162 [C] 8144 [B] 8159 [B]

121, DSP

8157. If a passenger-carrying landplane is required to have an automatic deploying escape slide system, when must this system be armed?

A—For taxi, takeoff, and landing.
B—Only for takeoff and landing.
C—During taxi, takeoff, landing, and after ditching.

Each passenger-carrying landplane with an emergency exit (other than over-the-wing) that is more than 6 feet from the ground must have an approved means to assist the occupants in descending to the ground. An assisting means that deploys automatically must be armed during taxi, takeoffs, and landings. (PLT404) — 14 CFR §121.310

121, DSP

8158. If there is a required emergency exit located in the flightcrew compartment, the door which separates the compartment from the passenger cabin must be

A—unlocked during takeoff and landing.
B—locked at all times, except during any emergency declared by the pilot in command.
C—latched open during takeoff and landing.

If it is necessary to pass through a doorway separating the passenger cabin from other areas to reach a required emergency exit from any passenger seat, the door must have means to latch it open, and the door must be latched open during each takeoff and landing. (PLT459) — 14 CFR §121.310

Answers (A) and (B) are incorrect because the door must always be latched open during takeoff and landing.

121, DSP

8178. A crewmember interphone system is required on which airplane?

A—A large airplane.
B—A turbojet airplane.
C—An airplane with more than 19 passenger seats.

No person may operate an airplane with a seating capacity of more than 19 passengers unless the airplane is equipped with a crewmember interphone system. (PLT462) — 14 CFR §121.319

Answers (A) and (B) are incorrect because the crewmember interphone system requirement is based upon the number of seats.

121, DSP

8179. An air carrier airplane must have an operating public address system if it

A—has a seating capacity of 19 passengers.
B—has a seating capacity for more than 19 passengers.
C—weighs more than 12,500 pounds.

No person may operate an airplane with a seating capacity of more than 19 passengers unless the airplane is equipped with an operating public address system. (PLT462) — 14 CFR §121.318

121, DSP

8235. Each crewmember shall have readily available for individual use on each flight a

A—key to the flight deck door.
B—certificate holder's manual.
C—flashlight in good working order.

Each crewmember shall, on each flight, have readily available for use a flashlight that is in good working order. (PLT405) — 14 CFR §121.549

121, DSP

8173. How much supplemental oxygen for emergency descent must a pressurized turbine-powered air transport airplane carry for each flight crewmember on flight deck duty when operating at flight altitudes above 10,000 feet?

A—A minimum of 2-hours' supply.
B—Sufficient for the duration of the flight above 8,000 feet cabin pressure altitude.
C—Sufficient for the duration of the flight at 10,000 feet flight altitude, not to exceed 1 hour and 50 minutes.

When operating at flight altitudes above 10,000 feet, the certificate holder shall supply enough oxygen for each crewmember for the entire flight at those altitudes and not less than a 2-hour supply for each flight crewmember on flight deck duty. (PLT438) — 14 CFR §121.331 and §121.333

Answers

8157 [A]	8158 [C]	8178 [C]	8179 [B]	8235 [C]	8173 [A]

121, DSP

8183. Each air carrier flight deck crewmember on flight deck duty must be provided with an oxygen mask that can be rapidly placed on his face when operating at flight altitudes

A—of FL 260.
B—of FL 250.
C—above FL 250.

When operating at flight altitudes above flight level 250, each flight crewmember on flight deck duty must be provided with an oxygen mask so designed that it can be rapidly placed on his/her face from its ready position, properly secured, sealed, and supplying oxygen within 5 seconds; and so designed that after being placed on the face it does not prevent immediate communication between the flight crewmember and other crewmembers over the airplane intercom system. When not being used at flight altitudes above flight level 250, the mask must be kept ready for use and within immediate reach. (PLT438) — 14 CFR §121.333

121, DSP

8184. A flight crewmember must be able to don and use a quick-donning oxygen mask within

A—5 seconds.
B—10 seconds.
C—15 seconds.

When operating at flight altitudes above flight level 250, each flight crewmember on flight deck duty must be provided with an oxygen mask so designed that it can be rapidly placed on his/her face from its ready position, properly secured, sealed, and supplying oxygen within 5 seconds; and so designed that after being placed on the face it does not prevent immediate communication between the flight crewmember and other crewmembers over the airplane intercom system. When not being used at flight altitudes above flight level 250, the mask must be kept ready for use and within immediate reach. (PLT438) — 14 CFR §121.333

121, DSP

8155. If either pilot of an air carrier airplane leaves the duty station while flying at FL 410, the other pilot

A—and the flight engineer shall put on their oxygen masks and breathe oxygen.
B—shall put on the oxygen mask and breathe oxygen.
C—must have a quick-donning type oxygen mask available.

If for any reason at any time it is necessary for one pilot to leave the station at the controls of the airplane when operating at flight altitudes above flight level 250, the remaining pilot at the controls shall put on and use his/her oxygen mask until the other pilot has returned to the duty station. (PLT440) — 14 CFR §121.333

Answer (A) is incorrect because only the remaining pilot is required to put on and use an oxygen mask. Answer (C) is incorrect because the remaining pilot must put on an oxygen mask, not just have a quick-donning type available.

121, DSP

8156. If a turbine-engine-powered, pressurized airplane is not equipped with quick-donning oxygen masks, what is the maximum flight altitude authorized without one pilot wearing and using an oxygen mask?

A—FL 200.
B—FL 300.
C—FL 250.

When operating at flight altitudes above flight level 250, one pilot at the controls of the airplane shall at all times wear and use an oxygen mask secured, sealed, and supplying oxygen, except that the one pilot need not wear and use an oxygen mask while at or below flight level 410 if each flight crewmember on flight deck duty has a quick-donning type oxygen mask. (PLT438) — 14 CFR §121.333

121, DSP

8187. What is the highest flight level that operations may be conducted without the pilot at the controls wearing and using an oxygen mask, while the other pilot is away from the duty station?

A—FL 240.
B—FL 250.
C—Above FL 250.

If for any reason at any time it is necessary for one pilot to leave the station at the controls of the airplane when operating at flight altitudes above flight level 250, the remaining pilot at the controls shall put on and use his/her oxygen mask until the other pilot has returned to the duty station. (PLT438) — 14 CFR §121.333

Answers

8183 [C]	8184 [A]	8155 [B]	8156 [C]	8187 [B]

121, DSP

8174. What is the passenger oxygen supply requirement for a flight, in a turbine-powered aircraft, with a cabin pressure altitude in excess of 15,000 feet? Enough oxygen for

A—each passengers for the entire flight above 15,000 feet cabin altitude.
B—30 percent of the passengers.
C—10 percent of the passengers for 30 minutes.

For flights at cabin pressure altitudes above 15,000 feet, the certificate holder must provide enough oxygen for each passenger carried during the entire flight at those altitudes. (PLT438) — 14 CFR §121.327, §121.329

121, DSP

8186. For flights above which cabin altitude must oxygen be provided for all passengers during the entire flight at those altitudes?

A—15,000 feet.
B—16,000 feet.
C—14,000 feet.

For flights at cabin pressure altitudes above 15,000 feet, the certificate holder must provide enough oxygen for each passenger carried during the entire flight at those altitudes. (PLT438) — 14 CFR §121.327, §121.329

121, DSP

8185. For a 2-hour flight in a reciprocating engine-powered airplane at a cabin pressure altitude of 12,000 feet, how much supplemental oxygen for sustenance must be provided? Enough oxygen for

A—30 minutes for 10 percent of the passengers.
B—10 percent of the passengers for 1.5 hours.
C—each passenger for 30 minutes.

For flight in reciprocating-engine-powered airplanes, at cabin pressure altitudes above 8,000 feet, up to and including 14,000 feet, each certificate holder shall provide enough oxygen for 30 minutes for 10 percent of the passengers. (PLT438) — 14 CFR §121.327

121, DSP

8182. The supplemental oxygen requirements for passengers when a flight is operated at FL 250 is dependent upon the airplane's ability to make an emergency descent to a flight altitude of

A—10,000 feet within 4 minutes.
B—14,000 feet within 4 minutes.
C—12,000 feet within 4 minutes or at a minimum rate of 2,500 ft/min, whichever is quicker.

The supplemental oxygen requirements for passengers on pressurized aircraft is dependent upon the ability of the aircraft to descend to 14,000 feet within 4 minutes in the event of a loss of pressurization. (PLT438) — 14 CFR §121.333

121, DSP

8180. What is the minimum number of acceptable oxygen-dispensing units for first-aid treatment of occupants who might require undiluted oxygen for physiological reasons?

A—Two.
B—Four.
C—Three.

There must be an appropriate number of oxygen dispensing units for first aid treatment of passengers, but in no case less than 2. (PLT438) — 14 CFR §121.333

121, DSP

8164. Which emergency equipment is required for a flag air carrier flight between John F. Kennedy International Airport and London, England?

A—A life preserver equipped with an approved survivor locator light or other flotation device for the full seating capacity of the airplane.
B—An appropriately equipped survival kit attached to each required liferaft.
C—A self-buoyant, water resistant, portable survival-type emergency locator transmitter for each required liferaft.

No person may operate an airplane in extended over-water operations without having on the airplane the following equipment:

1. *A life preserver equipped with an approved survivor locator light for each occupant of the airplane;*

Answers

8174 [A]	8186 [A]	8185 [A]	8182 [B]	8180 [A]	8164 [B]

2. *Enough life rafts (each equipped with an approved survivor locator light) to accommodate the occupants of the airplane;*
3. *At least one pyrotechnic signaling device for each life raft;*
4. *One survival-type emergency locator transmitter;*
5. *A survival kit, appropriately equipped for the route to be flown, must be attached to each life raft.*

(PLT404) — 14 CFR §121.339

Answer (A) is incorrect because a life preserver or other flotation device for each occupant is required. The requirement is not based upon seating capacity. Answer (C) is incorrect because only one survival type emergency locator transmitter is required to be carried in the aircraft, not one for each life raft.

121, DSP

8166. Each large aircraft operating over water must have a life preserver for each

A—aircraft occupant.
B—seat on the aircraft.
C—passenger seat, plus 10 percent.

No person may operate an airplane in extended overwater operations without having on the airplane the following equipment:

1. *A life preserver equipped with an approved survivor locator light for each occupant of the airplane;*
2. *Enough life rafts (each equipped with an approved survivor locator light) to accommodate the occupants of the airplane;*
3. *At least one pyrotechnic signaling device for each life raft;*
4. *One survival-type emergency locator transmitter;*
5. *A survival kit, appropriately equipped for the route to be flown, must be attached to each life raft.*

(PLT417) — 14 CFR §121.339

Answers (B) and (C) are incorrect because unlike some regulations that are based upon the number of seats in the aircraft, the number of life preservers required is based on the number of occupants for a particular flight.

121, DSP

8169. Life preservers required for overwater operations are stored

A—within easy reach of each passenger.
B—under each occupant seat.
C—within easy reach of each seated occupant.

The required life rafts, life preservers, and survival-type emergency locator transmitter must be easily accessible in the event of a ditching without appreciable time for preparatory procedures. (PLT417) — 14 CFR §121.339

121, DSP

8167. For a flight over uninhabited terrain, an airplane operated by a flag or supplemental air carrier must carry enough appropriately equipped survival kits for

A—all of the passengers, plus 10 percent.
B—all aircraft occupants.
C—all passenger seats.

Unless it has the following equipment, no flag or supplemental carrier or commercial operator may conduct an operation over an uninhabited area:

1. *Suitable pyrotechnic signaling devices;*
2. *A survival-type emergency locator transmitter; and*
3. *Enough survival kits, appropriately equipped for the route to be flown, for the number of occupants of the airplane.*

(PLT404) — 14 CFR §121.353

121, DSP

8168. When a supplemental air carrier is operating over an uninhabited area, how many appropriately equipped survival kits are required aboard the aircraft?

A—One for each passenger seat.
B—One for each passenger, plus 10 percent.
C—One for each occupant of the aircraft.

Unless it has the following equipment, no flag or supplemental carrier or commercial operator may conduct an operation over an uninhabited area:

1. *Suitable pyrotechnic signaling devices;*
2. *A survival-type emergency locator transmitter; and*
3. *Enough survival kits, appropriately equipped for the route to be flown, for the number of occupants of the airplane.*

(PLT404) — 14 CFR §121.353

Answers

8166 [A] 8169 [C] 8167 [B] 8168 [C]

121, DSP

8170. An airplane operated by a supplemental air carrier flying over uninhabited terrain must carry which emergency equipment?

A—Survival kit for each passenger.
B—Suitable pyrotechnic signaling devices.
C—Colored smoke flares and a signal mirror.

Unless it has the following equipment, no flag or supplemental carrier or commercial operator may conduct an operation over an uninhabited area:

1. *Suitable pyrotechnic signaling devices;*
2. *A survival-type emergency locator transmitter; and*
3. *Enough survival kits, appropriately equipped for the route to be flown, for the number of occupants of the airplane.*

(PLT404) — 14 CFR §121.353

121, DSP

8171. An airplane operated by a commercial operator flying over uninhabited terrain must carry which emergency equipment?

A—A signal mirror and colored smoke flares.
B—Survival kit for each passenger.
C—An approved survival-type emergency locator transmitter.

Unless it has the following equipment, no flag or supplemental carrier or commercial operator may conduct an operation over an uninhabited area:

1. *Suitable pyrotechnic signaling devices;*
2. *A survival-type emergency locator transmitter; and*
3. *Enough survival kits, appropriately equipped for the route to be flown, for the number of occupants of the airplane.*

(PLT402) — 14 CFR §121.353

121, DSP

8172. An airplane operated by a flag air carrier operator flying over uninhabited terrain must carry which emergency equipment?

A—Suitable pyrotechnic signaling devices.
B—Colored smoke flares and a signal mirror.
C—Survival kit for each passenger.

Unless it has the following equipment, no flag or supplemental carrier or commercial operator may conduct an operation over an uninhabited area:

1. *Suitable pyrotechnic signaling devices;*
2. *A survival-type emergency locator transmitter; and*
3. *Enough survival kits, appropriately equipped for the route to be flown, for the number of occupants of the airplane.*

(PLT404) — 14 CFR §121.353

121, DSP

8245. If an aircraft dispatcher cannot communicate with the pilot of an air carrier flight during an emergency, the aircraft dispatcher should

A—take any action considered necessary under the circumstances.
B—comply with the company's lost aircraft plan.
C—phone the ARTCC where the flight is located and ask for a phone patch with the flight.

If the aircraft dispatcher cannot communicate with the pilot, he/she shall declare an emergency and take any action considered necessary under the circumstances. (PLT403) — 14 CFR §121.557

121, DSP

8198. Which document includes descriptions of the required crewmember functions to be performed in the event of an emergency?

A—Airplane Flight Manual.
B—Certificate holder's manual.
C—Pilot's Emergency Procedures Handbook.

Each certificate holder shall, for each type and model of airplane, assign to each category of required crewmember, as appropriate, the necessary functions to be performed in an emergency or a situation requiring emergency evacuation. The certificate holder shall describe in its manual the functions of each category of required crewmember. (PLT436) — 14 CFR §121.397

Answer (A) is incorrect because the Airplane Flight Manual may contain emergency procedures as a convenience, but they are not required by 14 CFR §121.141. Answer (C) is incorrect because an "Emergency Procedures Handbook" does not exist.

Answers

8170 [B] 8171 [C] 8172 [A] 8245 [A] 8198 [B]

121, DSP

8200. The required crewmember functions that are to be performed in the event of an emergency shall be assigned by the

A—pilot in command.
B—air carrier's chief pilot.
C—certificate holder.

Each certificate holder shall, for each type and model of airplane, assign to each category of required crewmember, as appropriate, the necessary functions to be performed in an emergency or a situation requiring emergency evacuation. The certificate holder shall describe in its manual the functions of each category of required crewmember. (PLT374) — 14 CFR §121.397

Answer (A) is incorrect because, although the pilot-in-command may assign duties as necessary during an emergency, the required crewmember functions shall be assigned and described in the certificate holder's manual. Answer (B) is incorrect because the chief pilot does not have the authority to assign crewmember functions that are to be performed in the event of an emergency. Those functions shall be described in the certificate holder's manual.

121, DSP

8204. The air carrier must give instruction on such subjects as respiration, hypoxia, and decompression to crewmembers serving on pressurized airplanes operated above

A—FL 180.
B—FL 200.
C—FL 250.

Crewmembers who serve in operations above 25,000 feet must receive instruction in respiration, hypoxia, and decompression. (PLT460) — 14 CFR §121.417

121, DSP

8218. How often must a crewmember actually operate the airplane emergency equipment, after initial training? Once every

A—6 calendar months.
B—12 calendar months.
C—24 calendar months.

Emergency drill requirements must be accomplished during initial training and once each 24 calendar months during recurrent training. (PLT407) — 14 CFR §121.417

121, DSP

8236. If an engine's rotation is stopped in flight, the pilot in command must report it, as soon as practicable, to the

A—appropriate ground radio station.
B—nearest FAA district office.
C—operations manager (or director of operations).

The pilot-in-command shall report each stoppage of engine rotation in flight to the appropriate ground radio station as soon as practicable and shall keep that station fully informed of the progress of the flight. (PLT366) — 14 CFR §121.565

121, DSP

8237. If it becomes necessary to shut down one engine on a domestic air carrier three-engine turbojet airplane, the pilot in command

A—must land at the nearest suitable airport, in point of time, at which a safe landing can be made.
B—may continue to the planned destination if approved by the company aircraft dispatcher.
C—may continue to the planned destination if this is considered as safe as landing at the nearest suitable airport.

If not more than one engine of an airplane that has three or more engines fails or its rotation is stopped, the pilot-in-command may proceed to an airport that he/she selects if, after considering the following, he/she decides that proceeding to that airport is as safe as landing at the nearest suitable airport. (PLT406) — 14 CFR §121.565

121, DSP

8241. What action shall the pilot in command take if it becomes necessary to shut down one of the two engines on an air carrier airplane?

A—Land at the airport which the pilot considers to be as safe as the nearest suitable airport in point of time.
B—Land at the nearest suitable airport in point of time at which a safe landing can be made.
C—Land at the nearest airport, including military, that has a crash and rescue unit.

Whenever an engine of an airplane fails or whenever the rotation of an engine is stopped to prevent possible damage, the pilot-in-command shall land the airplane at the nearest suitable airport, time-wise, at which a safe landing can be made. Note: There are no exceptions to this rule for two-engine airplanes. (PLT223) — 14 CFR §121.565

Answers

8200 [C] 8204 [C] 8218 [C] 8236 [A] 8237 [C] 8241 [B]

121, DSP

8163. In the event of an engine emergency, the use of a cockpit check procedure by the flightcrew is

A—encouraged; it helps to ensure that all items on the procedure are accomplished.
B—required by regulations to prevent reliance upon memorized procedures.
C—required by the FAA as a doublecheck after the memorized procedure has been accomplished.

Each certificate holder shall provide an approved cockpit check procedure for each type of aircraft. The approved procedures must include each item necessary for flight crewmembers to check for safety before starting engines, taking off, or landing, and in engine and systems emergencies. The procedures must be designed so that a flight crewmember will not need to rely upon memory for items to be checked. (PLT404) — 14 CFR §121.315

121, DSP

8240. When the pilot in command is responsible for a deviation during an emergency, the pilot should submit a written report within

A—10 days after the deviation.
B—10 days after returning home.
C—10 days after returning to home base.

A pilot-in-command declaring an emergency shall send a written report of any deviation, through the air carrier's director of operations, to the Administrator within 10 days after returning to the home base. (PLT403) — 14 CFR §121.557

121, DSP

8246. Who is required to submit a written report on a deviation that occurs during an emergency?

A—Pilot in command.
B—Dispatcher.
C—Person who declares the emergency.

The person declaring the emergency shall send a written report of any deviation, through the air carrier's director of operations, to the Administrator within 10 days. (PLT366) — 14 CFR §121.557

DSP

8239. An aircraft dispatcher declares an emergency for a flight and a deviation results. A written report shall be sent through the air carrier's operations manager by the

A—dispatcher to the FAA Administrator within 10 days of the event.
B—certificate holder to the FAA Administrator within 10 days of the event.
C—pilot in command to the FAA Administrator within 10 days of the event.

An aircraft dispatcher declaring an emergency shall send a written report of any deviation, through the air carrier's director of operations, to the Administrator within 10 days after the date of the emergency. (PLT394) — 14 CFR §121.557

Part 135 Oxygen Requirements

135

8020. Which is a requirement for flightcrew use of oxygen masks in a pressurized cabin airplane?

A—Both pilots at the controls shall use oxygen masks above FL 350.
B—At altitudes above 25,000 feet MSL, if one pilot leaves the pilot duty station, the remaining pilot at the controls shall use an oxygen mask.
C—At altitudes above FL 250, one of the two pilots at the controls shall use an oxygen mask continuously.

One pilot of a pressurized aircraft must wear an oxygen mask any time the aircraft is flown above 35,000 feet MSL. In addition, one pilot must wear an oxygen mask above a flight altitude of 25,000 feet MSL if the other pilot leaves the duty station. (PLT438) — 14 CFR §135.89

Answers

8163 [B] 8240 [C] 8246 [C] 8239 [A] 8020 [B]

135

8022. Which is a requirement for pilot use of oxygen in a pressurized airplane?

A—The pilot at the controls shall use oxygen continuously any time the cabin pressure altitude is more than 12,000 feet MSL.
B—At FL 250 and above, each pilot shall have an approved quick-donning oxygen mask.
C—At FL 250 and above, the pilot at the controls must have an approved oxygen mask any time the other pilot is away from the duty station.

Each pilot of an unpressurized aircraft shall use oxygen continuously when flying:

1. *At altitudes above 10,000 feet through 12,000 feet MSL for that part of the flight at those altitudes that is more than 30 minutes duration; and*
2. *Above 12,000 feet MSL.*

Whenever a pressurized aircraft is operated with the cabin pressure altitude more than 10,000 feet MSL, each pilot shall comply with the rules for unpressurized aircraft.

Whenever a pressurized airplane is operated above 25,000 feet MSL flight altitude both pilots must have a "quick-donning"-type oxygen mask.

One pilot of a pressurized aircraft must wear an oxygen mask any time the aircraft is flown above 35,000 feet MSL. In addition, one pilot must wear an oxygen mask above a flight altitude of 25,000 feet MSL if the other pilot leaves the duty station. (PLT438) — 14 CFR §135.89

Answer (B) is incorrect because the regulation states "above 25,000 feet MSL." Answer (C) is incorrect because above 25,000 feet MSL, the pilot at the controls must wear an approved oxygen mask any time the other pilot is away from the duty station.

135

8055. The two pilot stations of a pressurized aircraft are equipped with approved quick-donning oxygen masks. What is the maximum altitude authorized if one pilot is not wearing an oxygen mask and breathing oxygen?

A—41,000 feet MSL.
B—35,000 feet MSL.
C—25,000 feet MSL.

One pilot of a pressurized aircraft must wear an oxygen mask any time the aircraft is flown above 35,000 feet MSL. (PLT438) — 14 CFR §135.89

135

8056. At altitudes above 10,000 feet through 12,000 feet MSL, each pilot of an unpressurized airplane must use supplemental oxygen for that part of the flight that is of a duration of more than

A—20 minutes.
B—30 minutes.
C—45 minutes.

Each pilot of an unpressurized aircraft shall use oxygen continuously when flying:

1. *At altitudes above 10,000 feet through 12,000 feet MSL for the part of the flight, at those altitudes, that is more than 30 minutes duration; and*
2. *Above 12,000 feet MSL.*

(PLT438) — 14 CFR §135.89

135

8072. A pressurized airplane being operated at FL 330 can descend safely to 15,000 feet MSL in 3.5 minutes. What oxygen supply must be carried for all occupants other than the pilots?

A—60 minutes.
B—45 minutes.
C—30 minutes.

No person may operate a pressurized aircraft above 15,000 feet MSL unless it is equipped to supply oxygen to each occupant, other than the pilots, for 1 hour. This is reduced to a 30-minute supply if the aircraft, at all times during flight above 15,000 feet MSL, can safely descend to 15,000 feet within 4 minutes. (PLT438) — 14 CFR §135.157

135

8073. At what altitude, in an unpressurized airplane, must all passengers be supplied oxygen?

A—Above 12,000 feet MSL.
B—Above 14,000 feet MSL.
C—Above 15,000 feet MSL.

In unpressurized aircraft, at altitudes above 10,000 feet MSL through 15,000 feet MSL, oxygen must be available for 10% of the occupants, other than the pilots, for the part of the flight, at those altitudes, in excess of 30-minute duration. Above 15,000 feet MSL, oxygen must be available to all occupants, other than the pilots. (PLT438) — 14 CFR §135.157

Answers

8022 [A]	8055 [B]	8056 [B]	8072 [C]	8073 [C]

135

8074. Between what altitudes must oxygen be available to at least 10 percent of the occupants, in an unpressurized airplane, other than the pilots?

A—Above 12,000 feet through 16,000 feet MSL, for any time period.
B—Above 10,000 feet through 15,000 feet MSL, if flight at those altitudes is of more than a 30-minute duration.
C—10,000 feet to 15,000 feet MSL, if flight at those altitudes is of more than a 30-minute duration.

In unpressurized aircraft, at altitudes above 10,000 feet MSL through 15,000 feet MSL, oxygen must be available for 10% of the occupants, other than the pilots, for that part of the flight at those altitudes in excess of 30-minute duration. Above 15,000 feet MSL, oxygen must be available to all occupants. (PLT438) — 14 CFR §135.157

135

8080. The oxygen requirements for occupants of a pressurized airplane operated at altitudes above FL 250 is dependent upon the airplane's ability to descend safely to an altitude of

A—10,000 feet MSL in 4 minutes.
B—12,000 feet MSL at a minimum rate of 2,500 ft/min.
C—15,000 feet MSL in 4 minutes.

No person may operate a pressurized aircraft above 15,000 feet MSL unless it is equipped to supply oxygen to each occupant, other than the pilots, for 1 hour. This is reduced to a 30-minute supply if the aircraft, at all times during flight above 15,000 feet MSL, can safely descend to 15,000 feet within 4 minutes. (PLT438) — 14 CFR §135.157

135, RTC

8021. Above which altitude/flight level must at least one of the two pilots, at the controls of a pressurized aircraft (with quick-donning masks) wear a secured and sealed oxygen mask?

A—FL 300.
B—FL 350.
C—FL 250.

One pilot of a pressurized aircraft must wear an oxygen mask any time the aircraft is flown above 35,000 feet MSL. (PLT438) — 14 CFR §135.89

135, RTC

8023. Which is a pilot requirement for oxygen?

A—Each pilot of a pressurized aircraft operating at FL 180 and above shall have an approved quick-donning type oxygen mask.
B—On pressurized aircraft requiring a flightcrew of two pilots, both shall continuously wear oxygen masks whenever the cabin pressure altitude exceeds 12,000 feet MSL.
C—On unpressurized aircraft, flying above 12,000 feet MSL, pilots shall use oxygen continuously.

Each pilot of an unpressurized aircraft shall use oxygen continuously when flying:

1. *At altitudes above 10,000 feet through 12,000 feet MSL for that part of the flight at those altitudes that is more than 30 minutes duration; and*
2. *Above 12,000 feet MSL.*

(PLT438) — 14 CFR §135.89

Answer (A) is incorrect because quick-donning type oxygen masks are required above 25,000 feet MSL. Answer (B) is incorrect because both pilots should continuously use oxygen masks when the cabin pressure altitude is more than 10,000 feet MSL.

135, RTC

8024. Which requirement applies when oxygen is stored in liquid form?

A—Smoking is not permitted within 50 feet of stored liquid oxygen.
B—Liquefied oxygen is a hazardous material and must be kept in an isolated storage facility.
C—The equipment used to store liquid oxygen must be covered in the certificate holder's approved maintenance program.

When the oxygen is stored in the form of a liquid, the equipment must have been under the certificate holder's approved maintenance program since its purchase new, or since the storage container was last purged. (PLT438) — 14 CFR §135.91

Answers

8074 [B]	8080 [C]	8021 [B]	8023 [C]	8024 [C]

135, RTC

8025. Which is a condition that must be met when a person is administered medical oxygen in flight?

A—The distance between a person using medical oxygen and any electrical unit must not be less than 5 feet.
B—A person using oxygen equipment must be seated to avoid restricting access to, or use of, any required exit.
C—A person being administered oxygen must be monitored by equipment that displays and records pulse and respiration.

Oxygen equipment must be stowed, and each person using the equipment must be seated, so as not to restrict access to or use of any required emergency or regular exit, or of the aisle in the passenger compartment. (PLT438) — 14 CFR §135.91

135, RTC

8030. Which is a requirement regarding the carriage and operation of oxygen equipment for medical use by passengers?

A—No person may smoke within 10 feet of oxygen storage and dispensing equipment.
B—When oxygen equipment is used for the medical treatment of a patient, the rules pertaining to emergency exit access are waived.
C—No person may connect oxygen bottles or any other ancillary equipment until all passengers are aboard the aircraft and seated.

No person may smoke within 10 feet of oxygen-dispensing equipment. (PLT438) — 14 CFR §135.91

135, RTC

8031. If a certificate holder deviates from the provisions of regulations which pertain to medical use of oxygen by passengers, a complete report of the incident shall be sent to the FAA within

A—7 working days.
B—10 working days.
C—10 days of the deviation.

Each certificate holder who deviates from the provisions of the regulations pertaining to use of medical oxygen by passengers, must send a report of the deviation to the FAA Flight Standards District Office within 10 days excluding Saturdays, Sundays, and federal holidays. (PLT438) — 14 CFR §135.91

135, RTC

8081. An unpressurized aircraft with 20 occupants other than the pilots will be cruising at 14,000 feet MSL for 25 minutes. For how many, if any, of these occupants must there be an oxygen supply?

A—Five.
B—Two.
C—None.

In unpressurized aircraft, at altitudes above 10,000 feet MSL through 15,000 feet MSL, oxygen must be available for 10% of the occupants, other than the pilots, for the part of the flight, at those altitudes, in excess of 30-minutes duration. Above 15,000 feet MSL, oxygen must be available to all occupants, other than the pilots. (PLT438) — 14 CFR §135.157

RTC

9576. (Refer to Figures 113 and 114.) When must the pilots use oxygen, under 14 CFR Part 135, on this flight from 0O2 to LAX?

A—Starting 30 minutes after climbing through 10,000 feet until descending below 10,000 feet.
B—30 minutes after level off until descending below 10,000 feet.
C—All the time they are above 10,000 feet MSL.

Each pilot of an unpressurized aircraft shall use oxygen continuously when flying—

1. *At altitudes above 10,000 feet through 12,000 feet MSL for the part of the flight at those altitudes that is more than 30 minutes duration; and*
2. *Above 12,000 feet MSL.*

(PLT438) — 14 CFR §135.89

Answers

8025 [B]	8030 [A]	8031 [B]	8081 [C]	9576 [A]

RTC

9577. (Refer to Figures 113 and 114.) What are the passenger oxygen requirements, under 14 CFR Part 135, on this flight from 0O2 to LAX?

A—Starting 30 minutes after level off, 10 percent of the aircraft occupants must be supplied oxygen until descending below 10,000 feet.
B—When above 10,000 through 15,000 feet, oxygen must be supplied to at least 10 percent of the aircraft occupants.
C—Starting 30 minutes after climbing through 10,000 feet, 10 percent of the aircraft occupants, except the pilots, must be supplied oxygen until descending below 10,000 feet.

No person may operate an unpressurized aircraft at altitudes prescribed in this section unless it is equipped with enough oxygen dispensers and oxygen to supply the pilots under §135.89(a) and to supply when flying—

1. *At altitudes above 10,000 feet through 15,000 MSL, oxygen to at least 10% of the occupants of the aircraft, other than pilots, for the part of the flight at those altitudes that is more than 30 minutes duration; and*
2. *Above 15,000 feet MSL, oxygen to each occupant of the aircraft other than the pilots.*

(PLT438) — 14 CFR §135.157

RTC

9605. (Refer to Figures 162, 163, 164, 165, and 167.) When are the pilots required to use oxygen on this flight from RYN to AEG?

A—Upon climbing through 12,000 feet on ascent, until passing through 12,000 feet on descent.
B—Starting 30 minutes after level off until descending below 10,000 feet.
C—Starting 30 minutes after climbing through 10,000 feet until descending below 10,000 feet.

Each pilot of an unpressurized aircraft shall use oxygen continuously when flying—

1. *At altitudes above 10,000 feet through 12,000 feet MSL for the part of the flight at those altitudes that is more than 30 minutes duration; and*
2. *Above 12,000 feet MSL.*

(PLT438) — 14 CFR §135.89

RTC

9606. (Refer to Figures 162, 163, 164, 165, and 167.) What are the passenger oxygen requirements on this flight, from Tucson/Ryan to Albuquerque/Double Eagle II?

A—Starting 30 minutes after climbing through 10,000 feet, 10 percent of the aircraft occupants must be supplied oxygen until descending below 10,000 feet.
B—Starting 30 minutes after climbing through 10,000 feet, 10 percent of the aircraft occupants, except pilots, must be supplied oxygen until descending below 10,000 feet.
C—When above 10,000 feet through 15,000 feet, oxygen must be supplied to at least 10 percent of the aircraft occupants, including the pilots.

No person may operate an unpressurized aircraft at altitudes prescribed in this section unless it is equipped with enough oxygen dispensers and oxygen to supply the pilots under §135.89(a) and to supply when flying—

1. *At altitudes above 10,000 feet through 15,000 MSL, oxygen to at least 10% of the occupants of the aircraft, other than pilots, for the part of the flight at those altitudes that is more than 30 minutes duration; and*
2. *Above 15,000 feet MSL, oxygen to each occupant of the aircraft other than the pilots.*

(PLT438) — 14 CFR §135.157

RTC

9637. (Refer to Figures 184, 186, 187, 188, and 188A.) When are the pilots required to use oxygen on this 14 CFR Part 135 flight from LAS to PVU?

A—Starting 30 minutes after climbing through 10,000 feet until descending below 10,000 feet.
B—Upon climbing through 12,000 feet on ascent, until passing through 12,000 feet on descent.
C—Starting 30 minutes after takeoff until descending below 10,000 feet.

Each pilot of an unpressurized aircraft shall use oxygen continuously when flying—

1. *At altitudes above 10,000 feet through 12,000 feet MSL for the part of the flight at those altitudes that is more than 30 minutes duration; and*
2. *Above 12,000 feet MSL.*

(PLT078) — 14 CFR §135.89

Answers

9577 [C] 9605 [C] 9606 [B] 9637 [A]

RTC

9638. (Refer to Figures 184, 186, 187, 188, and 188A.) What are the passenger oxygen requirements on this 14 CFR Part 135 flight from Las Vegas to Provo?

A—When above 10,000 feet through 15,000 feet, oxygen must be supplied to at least 10 percent of the aircraft occupants, including the pilots.
B—Starting 30 minutes after climbing through 10,000 feet, 10 percent of the aircraft occupants until reaching cruise at 15,000 feet then all occupants must be supplied oxygen until descending below 15,000 feet, then 10 percent down to 10,000 feet.
C—Starting 30 minutes after climbing through 10,000 feet, 10 percent of the aircraft occupants, except pilots, must be supplied oxygen until descending below 10,000 feet.

No person may operate an unpressurized aircraft at altitudes prescribed in this section unless it is equipped with enough oxygen dispensers and oxygen to supply the pilots under §135.89(a) and to supply when flying—

1. *At altitudes above 10,000 feet through 15,000 MSL, oxygen to at least 10% of the occupants of the aircraft, other than pilots, for the part of the flight at those altitudes that is more than 30 minutes duration; and*
2. *Above 15,000 feet MSL, oxygen to each occupant of the aircraft other than the pilots.*

(PLT438) — 14 CFR §135.157

RTC

9650. (Refer to Figures 197 and 199.) When are the pilots required to use oxygen on this flight from EGE to SLC?

A—Starting 30 minutes after level off until descending below 10,000 feet.
B—Starting 30 minutes after climbing through 10,000 feet until descending below 10,000 feet.
C—Upon climbing through 12,000 feet on ascent.

Each pilot of an unpressurized aircraft shall use oxygen continuously when flying—

1. *At altitudes above 10,000 feet through 12,000 feet MSL for the part of the flight at those altitudes that is more than 30 minutes duration; and*
2. *Above 12,000 feet MSL.*

(PLT438) — 14 CFR §135.89

RTC

9651. (Refer to Figures 197 and 199.) What are the passenger oxygen requirements on this flight from Eagle County Regional (EGE) to Salt Lake City Intl?

A—Starting 30 minutes after climbing through 10,000 feet, 10 percent of the aircraft occupants must be supplied oxygen until descending below 10,000 feet.
B—Starting 30 minutes after climbing through 10,000 feet, 10 percent of the aircraft occupants, except pilots, must be supplied oxygen until descending below 10,000 feet.
C—When above 10,000 feet through 15,000 feet, oxygen must be supplied to at least 10 percent of the aircraft occupants, including the pilots.

No person may operate an unpressurized aircraft at altitudes prescribed in this section unless it is equipped with enough oxygen dispensers and oxygen to supply the pilots under §135.89(a) and to supply when flying—

1. *At altitudes above 10,000 feet through 15,000 MSL, oxygen to at least 10% of the occupants of the aircraft, other than pilots, for the part of the flight at those altitudes that is more than 30 minutes duration; and*
2. *Above 15,000 feet MSL, oxygen to each occupant of the aircraft other than the pilots.*

(PLT438) — 14 CFR §135.157

Answers

9638 [C]   9650 [C]   9651 [B]

National Transportation Safety Board (NTSB)

Aircraft accident means an occurrence associated with the operation of an aircraft that takes place between the time any person boards the aircraft with the intention of flight, and the time all such persons have disembarked, and in which any person suffers death or serious injury, or in which the aircraft receives substantial damage.

Serious injury means any injury that:

- Requires hospitalization for more than 48 hours commencing within 7 days from the date the injury was received.
- Results in fracture of any bone (except simple fractures of fingers, toes or nose).
- Causes severe hemorrhages, nerve, muscle or tendon damage.
- Involves any internal organ.
- Involves second or third degree burns or any burns affecting more than 5% of the body surface.

Substantial damage means damage or failure that adversely affects the structural strength, performance or flight characteristics of the aircraft and that would normally require major repair or replacement of the affected component. Damage not considered substantial for accident reporting purposes are as follows: engine failure or damage limited to an engine if only one engine fails or is damaged, bent fairings or cowling, dented skin, small punctured holes in the skin or fabric, ground damage to rotor or propeller blades, and damage to the landing gear, wheels, tires, flaps, engine accessories, brakes or wing tips.

The operator of an aircraft must immediately notify the nearest National Transportation Safety Board field office if any of the following occur:

- Flight control system malfunction
- An aircraft accident
- Inability of any required flight crewmember to perform his normal flight duties as the result of injury or illness
- Failure of structural components of a turbine engine excluding compressor and turbine blades and vanes
- Inflight fire
- Aircraft collide in flight
- Damage to property, other than the aircraft, estimate to exceed $25,000 for repair or fair market value in the event of total loss whichever is less
- Certain incidents on large, multi-engine airplanes
- An aircraft is overdue and is believed to have been involved in an accident

The operator of an aircraft must submit a written report of an aircraft accident within 10 days of the accident. The operator of an overdue aircraft must submit a written report within 7 days if the aircraft is still missing. The operator of an aircraft that was involved in an incident requiring immediate notification of the NTSB must submit a written report of the incident only if requested to do so by the NTSB.

ALL

8317. What period of time must a person be hospitalized before an injury may be defined by the NTSB as a "serious injury"?

A—72 hours; commencing within 10 days after date of injury.
B—48 hours; commencing within 7 days after date of the injury.
C—10 days, with no other extenuating circumstances.

"Serious injury" means any injury which requires hospitalization for more than 48 hours, commencing within 7 days from the date the injury was received. (PLT366) — NTSB §830.2

ALL

8319. Which of the following constitutes "substantial damage" according to NTSB Part 830?

A—Ground damage to landing gear, wheels, or tires.
B—Damage to wingtips (or rotor blades, in the case of a helicopter).
C—Failure of a component which would adversely affect the performance, and which would require replacement.

"Substantial damage" is defined as damage or failure which would adversely affect the structural strength, performance, or flight characteristics of the aircraft which would normally require major repair or replacement of the damaged component. (PLT395) — NTSB §830.2

Answer (A) is incorrect because ground damage to landing gear, wheels, or tires is not considered "substantial damage" for the purpose of NTSB Part 830. Answer (B) is incorrect because damage to wing tips (or rotorblades, in the case of a helicopter) is not considered "substantial damage" for the purpose of NTSB Part 830.

ALL

8320. Which of the following meets the requirements of a "serious injury" as defined by the NTSB?

A—A simple fracture of the nose or other extremity.
B—An injury which caused severe tendon damage.
C—First-degree burns over 5 percent of the body.

"Serious injury" includes severe tendon damage and second or third degree burns covering more than five percent of the body. (PLT395) — NTSB §830.2

Answer (A) is incorrect because simple fractures, such as of the finger, toe, or nose, are not considered a serious injury. Answer (C) is incorrect because only second and third degree burns or first degree burns over more than 5% of the body are defined as a serious injury. (First degree burns are less serious than second and third degree burns.)

ALL

8318. Within what time period should the nearest NTSB field office be notified when an aircraft is involved in an accident which results in substantial damage?

A—Immediately.
B—7 calendar days.
C—10 days.

The operator of an aircraft shall immediately, and by the most expeditious means available, notify the nearest NTSB field office when an aircraft accident occurs. (PLT366) — NTSB §830.5

ALL

8321. Which incident requires an immediate notification to NTSB?

A—Aircraft colliding on the ground.
B—Flight control system malfunction.
C—Damage to property, other than the aircraft, estimated to exceed $10,000.

The NTSB lists a flight control malfunction or failure as an incident requiring immediate notification to the field office. (PLT416) — NTSB §830.5

ALL

8322. Within how many days must the operator of an aircraft involved in an accident file a report to the NTSB?

A—3 days.
B—7 days.
C—10 days.

The NTSB requires a report to be filed within 10 days of the accident. (PLT366) — NTSB §830.15

ALL

8323. When is an operator of an aircraft, which has been involved in an incident, required to submit a report to the nearest field office of the NTSB?

A—Within 7 days.
B—Within 10 days.
C—Only if requested to do so by the NTSB.

An aircraft involved in an incident is required to file a report only on request from the NTSB. (PLT366) — NTSB §830.15

Answers

8317 [B] 8319 [C] 8320 [B] 8318 [A] 8321 [B] 8322 [C]
8323 [C]

Part 135 Regulations

135
8053. What aircraft operating under 14 CFR Part 135 are required to have a third gyroscopic bank-and-pitch indicator installed?

A—All airplanes that are turbojet powered.
B—All multiengine airplanes that require a two pilot flightcrew.
C—All turbine powered aircraft having a passenger seating capacity of 30 seats or more.

A third gyroscopic pitch-and-bank indicator is required on all turbojet-powered airplanes. (PLT405) — 14 CFR §135.149

135
8054. In airplanes where a third gyroscopic bank-and-pitch indicator is required, that instrument must

A—continue reliable operation for at least 30 minutes after the output of the airplane's electrical generating system falls below an optimum level.
B—be operable by a selector switch which may be actuated from either pilot station.
C—continue reliable operation for a minimum of 30 minutes after total failure of the electrical generating system.

A third gyroscopic pitch-and-bank indicator is required on all turbojet-powered airplanes. This indicator must be able to continue reliable operation for at least 30 minutes after the failure of the aircraft's electrical generating system. (PLT405) — 14 CFR §135.149

135
8069. In which airplanes is a Class A TAWS required?

A—All airplanes having a passenger seating configuration, excluding any pilot seat, of 10 seats or more.
B—Turbine-powered airplanes having a passenger seating configuration, excluding any pilot seat, of 10 seats or more.
C—Turbine-powered aircraft having a passenger seating configuration, including any pilot seat, of 10 seats or more.

No person may operate a turbine-powered airplane having a passenger seating configuration, excluding any pilot seat, of 10 seats or more unless it is equipped with a terrain awareness system (TAWS). (PLT139) — 14 CFR §135.153

135
8075. Which airplanes must have a shoulder harness installed at each flight crewmember station?

A—All airplanes used in commuter air service, having a passenger seating configuration of 9, excluding any pilot seat.
B—All airplanes operating under 14 CFR Part 135, having a seating configuration for 10 persons.
C—All turbojet-powered airplanes.

No person may operate a turbojet aircraft or an aircraft having a passenger seating configuration, excluding any pilot seat, of 10 seats or more unless it is equipped with an approved shoulder harness installed for each flight crewmember station. (PLT464) — 14 CFR §135.171

135
8165. What emergency equipment is required for extended overwater operations?

A—A portable survival emergency locator transmitter for each liferaft.
B—A pyrotechnic signaling device for each life preserver.
C—A life preserver equipped with a survivor locator light, for each person on the airplane.

No person may operate an aircraft in extended overwater operations unless it carries an approved life preserver (easily accessible to each seated occupant) equipped with an approved survivor locator light for each occupant of the aircraft, and enough approved life rafts of a rated capacity and buoyancy to accommodate the occupants of the aircraft. An approved survival-type emergency locator transmitter must be attached to one of the life rafts. (PLT404) — 14 CFR §135.167

Answer (A) is incorrect because only one survival emergency locator transmitter is required to be carried on the airplane. Answer (B) is incorrect because one pyrotechnic signaling device is required for each life raft.

Answers

8053 [A] 8054 [C] 8069 [B] 8075 [C] 8165 [C]

135
8088. If the weather forecasts do not require the listing of an alternate airport on an IFR flight, the airplane must carry sufficient fuel to fly to the destination airport and

A—make one missed approach and thereafter have a 45-minute reserve at normal cruising speed.
B—fly thereafter for 45 minutes at normal cruising speed.
C—fly for 45 minutes thereafter at normal cruise climb speed.

No person may operate an aircraft in IFR conditions unless it carries enough fuel (considering weather reports and forecasts) to:

1. *Complete the flight to the first airport of intended landing;*
2. *Fly from that airport to the alternate airport (if one is required); and*
3. *Fly after that for 45 minutes at normal cruising speed.*

(PLT413) — 14 CFR §135.223

135
8089. If the weather forecasts require the listing of an alternate airport on an IFR flight, the airplane must carry enough fuel to fly to the first airport of intended landing, then to the alternate, and fly thereafter for a minimum of

A—45 minutes at normal holding speed.
B—45 minutes at normal cruise speed and then complete an approach and landing.
C—45 minutes at normal cruise speed.

No person may operate an aircraft in IFR conditions unless it carries enough fuel (considering weather reports and forecasts) to:

1. *Complete the flight to the first airport of intended landing;*
2. *Fly from that airport to the alternate airport (if one is required); and*
3. *Fly after that for 45 minutes at normal cruising speed.*

(PLT413) — 14 CFR §135.223

135
8115. When computing the takeoff data for reciprocating powered airplanes, what is the percentage of the reported headwind component that may be applied to the "still air" data?

A—Not more than 150 percent.
B—Not more than 100 percent.
C—Not more than 50 percent.

When computing takeoff data not more than 50% of the reported headwind component may be taken into account. (PLT011) — 14 CFR §135.389

135
8116. When computing takeoff data, what is the percentage of the effective tailwind component which may be applied to the "still air" data?

A—Not less than 150 percent.
B—Not less than 100 percent.
C—Not more than 50 percent.

When computing takeoff data not less than 150% of the reported tailwind component may be taken into account. (PLT011) — 14 CFR §135.389

135
8050. Which performance requirement applies to passenger-carrying land airplanes being operated over water?

A—Multiengine airplanes must be able to climb, with the critical engine inoperative, at least 50 ft/min at 1,500 feet above the surface.
B—Single-engine airplanes must be operated at an altitude that will allow them to reach land in case of engine failure.
C—Multiengine airplanes must be able to climb, with the critical engine inoperative, at least 100 ft/min at 1,000 feet above the surface.

No person may operate a land aircraft carrying passengers over water unless it is operated at an altitude that allows it to reach land in the case of engine failure. (PLT437) — 14 CFR §135.183

Answers

8088 [B] 8089 [C] 8115 [C] 8116 [A] 8050 [B]

135
8051. What performance is required of a multiengine airplane with the critical engine inoperative, while carrying passengers for hire in IFR weather conditions?

A—Climb at least 100 ft/min at the highest MEA of the route to be flown or 5,000 feet MSL, whichever is higher.
B—Climb at least 50 ft/min at the MEA's of the route to be flown or 5,000 feet AGL, whichever is higher.
C—Climb at least 50 ft/min at the MEA's of the route to be flown or 5,000 feet MSL, whichever is higher.

No person may operate a multi-engine airplane carrying passengers Over-The-Top or in IFR conditions at a weight that will not allow it to climb, with the critical engine inoperative, at least 50 feet a minute when operating at the MEAs of the route to be flown or 5,000 feet MSL, whichever is higher. (PLT223) — 14 CFR §135.181

135
8792. The crewmember interphone system on a large turbojet-powered airplane provides a means of two-way communications between ground personnel and at least one of two flight crewmembers in the pilot compartment, when the aircraft is on the ground. The interphone station for use by ground personnel must be located so that those using the system from that station

A—are always visible, from within the airplane.
B—are able to avoid the intake areas of the engines.
C—may avoid visible detection from within the airplane.

The interphone system station for use by ground personnel must be so located that personnel using the system may avoid visible detection from within the airplane. (PLT462) — 14 CFR §135.150

135
8831. For which of these aircraft may part of the "clearway" distance, for a particular runway, be considered in computing the takeoff distance?

A—Passenger-carrying transport aircraft.
B—Turbine-engine-powered transport airplanes, certificated after September 30, 1958.
C—U.S. certified transport airplane, certificated before August 26, 1957.

"Clearway" may be used in computing the takeoff distance of turbine-engine-powered airplanes certificated after September 30, 1958. (PLT456) — 14 CFR §1.1

135
8832. What requirement must be met regarding cargo that is carried anywhere in the passenger compartment of a commuter air carrier airplane?

A—Cargo may not be carried anywhere in the rear of the passenger compartment.
B—The bin in which the cargo is carried may not be installed in a position that restricts access to, or use of the aisle between the crew and the passenger compartment.
C—The container or bin in which the cargo is carried must be made of material which is at least flash resistant.

No person may carry cargo, including carry-on baggage in an aircraft unless it is in an approved cargo rack, bin, or compartment, and it does not obstruct access to, or use of, the aisle between the passenger and crew compartment. (PLT385) — 14 CFR §135.87

135
8833. Information recorded during normal operation of a cockpit voice recorder in a multiengine turbine powered airplane

A—may all be erased or otherwise obliterated except for the last 30 minutes.
B—may all be erased or otherwise obliterated except for the last 30 minutes prior to landing.
C—may all be erased, prior to each flight, unless the NTSB has requested that it be kept for 60 days.

Information recorded more than 30 minutes earlier may be erased or obliterated. (PLT388) — 14 CFR §135.151

Answers

8051 [C]	8792 [C]	8831 [B]	8832 [B]	8833 [A]

135
8842. An airplane, operated by a commuter air carrier, flying in extended overwater operations must carry enough approved liferafts of a rated capacity and buoyancy to accommodate the occupants of the aircraft. Each liferaft must be equipped with

A—one approved pyrotechnic signaling device.
B—colored smoke flares and a signal mirror.
C—one fishing kit for each person the raft is rated to carry.

Every aircraft flown in extended overwater operations must carry enough appropriately equipped life rafts to accommodate the occupants of the aircraft. Each raft must have an approved pyrotechnic signaling device (either smoke or flare type flare). (PLT082) — 14 CFR §135.167

Answer (B) is incorrect because the survival kit is not required to have colored smoke flares. Answer (C) is incorrect because the survival kit is only required to have one fishing kit per liferaft, not one per person.

135, RTC
8001. A certificate holder must have "exclusive use" of

A—at least one aircraft that meets the requirements of each kind of operation authorized in the Operations Specifications.
B—at least one aircraft that meets the requirements of at least one kind of operation authorized in the certificate holder's Operations Specifications.
C—at least one aircraft that meets the requirements of the specific operations authorized in the certificate holder's Operations Specifications.

Each certificate holder must have the exclusive use of at least one aircraft that meets the requirements for at least one kind of operation authorized in the certificate holder's operations specifications. (PLT454) — 14 CFR §135.25

135, RTC
8005. Where is the certificate holder required to list the name and title of each person authorized to exercise operational control for a particular flight?

A—Operations Specifications.
B—Attached to the load manifest.
C—Certificate holder's manual.

Each certificate holder is responsible for operational control and shall list in the manual the name and title of each person authorized to exercise operational control. (PLT282) — 14 CFR §135.77

135, RTC
8010. An aircraft being operated outside of the United States, over a foreign country, by a 14 CFR part 135 operator must comply with

A—the International Civil Aviation Organization (ICAO), Annex 3, Rules of the Air.
B—regulations of the foreign country.
C—rules of the U.S. State Department and the foreign country.

Each person operating an aircraft under Part 135 while operating outside the United States, shall comply with Annex 2, Rules of the Air, to the Convention of International Civil Aviation or the regulations of any foreign country, whichever applies. (PLT392) — 14 CFR §135.3

135, RTC
8011. Who is responsible for keeping copies of the certificate holder's manual up to date with approved changes or additions?

A—Each of the certificate holder's employees who are furnished a manual.
B—An employee designated by the certificate holder.
C—A representative of the certificate holder approved by the Administrator.

Each employee of the certificate holder to whom a manual (or appropriate portions of it) is furnished shall keep it up to date with changes and additions furnished to them. (PLT282) — 14 CFR §135.21

Answers

8842 [A] 8001 [B] 8005 [C] 8010 [B] 8011 [A]

135, RTC

9807. No person may operate a U.S. registered civil aircraft

A—for which an AFM or RFM is required by part 21 section 21.5 unless there is a current, approved operator's manual available.

B—for which an AFM or RFM is required by part 21 section 21.5 unless there is a current, approved AFM or RFM available.

C—for which an AFM or RFM is required by part 21 section 21.5 unless there is a current, approved AFM or RFM available or the manual specified in part 135 section 135.19(b).

Per 14 CFR §21.5, with each airplane or rotorcraft not type certificated with an Airplane or Rotorcraft Flight Manual and having no flight time before March 1, 1979, the holder of a type certificate (including amended or supplemental type certificates) or the licensee of a type certificate must make available to the owner at the time of delivery of the aircraft a current approved Airplane or Rotorcraft Flight Manual. (PLT373) — 14 CFR §21.5

135, RTC

8013. What is the lowest altitude above the terrain that an autopilot may be used during en route operations, if the Airplane Flight Manual specifies a malfunction under cruise conditions?

A—1,000 feet.

B—500 feet.

C—100 feet.

Except for approaches, no person may use an autopilot at an altitude above the terrain which is less than 500 feet or less than twice the maximum altitude loss specified in the approved Aircraft Flight Manual or equivalent for a malfunction of the autopilot, whichever is higher. (PLT424) — 14 CFR §135.93

135, RTC

8033. Who may be allowed to carry a deadly weapon on board an aircraft operated under 14 CFR Part 135?

A—Official bodyguards attached to foreign legations.

B—Crewmembers and/or others authorized by the certificate holder.

C—Employees of a municipality or a state, or of the United States.

No person may carry a deadly weapon on a Part 135 flight except for:

1. *Officials or employees of a municipality or a state or of the United States, who are authorized to carry arms; or*
2. *Crewmembers and other persons authorized by the certificate holder to carry arms.*

(PLT440) — 14 CFR §135.119

135, RTC

8038. Which person may be carried aboard an aircraft without complying with the passenger-carrying requirements of 14 CFR Part 135?

A—An individual who is necessary for the safe handling of hazardous material on the aircraft.

B—A representative of the Administrator, traveling to attend a meeting.

C—A member of the United States diplomatic corps on an official courier mission.

The following persons may be carried on an aircraft without complying with the passenger-carrying rules of Part 135:

1. *A crewmember or other employee of the certificate holder;*
2. *A person necessary for the safe handling of animals on the aircraft;*
3. *A person necessary for the safe handling of hazardous materials;*
4. *A person performing duty as a security or honor guard accompanying a shipment made by or under the authority of the U.S. Government;*
5. *A military courier or a military route supervisor carried by a military cargo contract air carrier or commercial operator;*
6. *An authorized representative of the Administrator conducting an enroute inspection; or*
7. *A person, authorized by the Administrator, who is performing a duty connected with a cargo operation of the certificate holder.*

(PLT385) — 14 CFR §135.85

Answers

9807 [B] 8013 [B] 8033 [B] 8038 [A]

135, RTC

8004. If previous arrangements have not been made by the operator, where can the procedures for servicing the aircraft be found?

A—Certificate holder's maintenance manual.
B—Certificate holder's manual.
C—Pilot's Handbook.

The certificate holder's manual must contain procedures to be followed by the pilot-in-command to obtain maintenance, preventative maintenance, and servicing of the aircraft at a place where previous arrangements have not been made by the operator, when the pilot is authorized to so act for the operator. (PLT282) — 14 CFR §135.23

135, RTC

8006. Who is directly responsible for determining the status of each mechanical irregularity previously entered in the aircraft maintenance log?

A—Aircraft dispatcher.
B—Line maintenance supervisor.
C—The next pilot in command.

Before each flight, the pilot-in-command shall determine, if the pilot does not already know, the status of each irregularity entered in the maintenance log at the end of the preceding flight. (PLT374) — 14 CFR §135.65

135, RTC

8012. What document contains procedures that explain how the required return-to-service conditions have been met?

A—Maintenance manual.
B—Pilot's Handbook.
C—Certificate holder's manual.

The certificate holder's manual must include procedures for ensuring that the pilot-in-command knows that required airworthiness inspections have been made and that the aircraft has been returned to service in compliance with applicable maintenance requirements. (PLT375) — 14 CFR §135.23

135, RTC

8019. Procedures for keeping copies of the aircraft maintenance log in the aircraft and available to appropriate personnel shall be set forth in

A—the certificate holder's manual.
B—the maintenance procedures handbook.
C—the Operations Specifications.

Each certificate holder shall establish a procedure for keeping copies of the aircraft maintenance log in the aircraft for access by appropriate personnel and shall include that procedure in the manual. (PLT282) — 14 CFR §135.65

135, RTC

8093. If a certificate holder makes arrangements for another person to perform aircraft maintenance, that maintenance shall be performed in accordance with the

A—certificate holder's manual and 14 CFR Parts 43, 91, and 135.
B—provisions of a contract prepared by a certificate holder and approved by the supervising FAA district office.
C—provisions and standards as outlined in the certificate holder's manual.

The certificate holder shall ensure that any maintenance, preventative maintenance, or alteration that is performed by another person is performed under the certificate holder's manual and regulations. (PLT282) — 14 CFR §135.413

135, RTC

8112. Who is responsible for submitting a Mechanical Reliability Report?

A—Each certificate holder.
B—Director of maintenance at the facility that discovers the reportable condition.
C—Chief inspector at the facility where the condition is found.

The certificate holder is responsible for submitting required mechanical reliability reports. (PLT443) — 14 CFR §135.415

Answers

8004 [B] 8006 [C] 8012 [C] 8019 [A] 8093 [A] 8112 [A]

135, RTC

8014. The maximum altitude loss specified for malfunction of a certain autopilot under cruise conditions is 50 feet. What is the lowest altitude this autopilot may be used en route?

A—500 feet AGL.
B—550 feet AGL.
C—600 feet AGL.

Except for approaches, no person may use an autopilot at an altitude above the terrain which is less than 500 feet or less than twice the maximum altitude loss specified in the approved Aircraft Flight Manual or equivalent for a malfunction of the autopilot, whichever is higher. (PLT424) — 14 CFR §135.93

135, RTC

8015. The maximum altitude loss for a particular malfunctioning autopilot under approach conditions is 55 feet. If the TDZE is 571 feet and the MDA is 1,100 feet, to which minimum altitude may you use this autopilot?

A—626 feet MSL.
B—990 feet MSL.
C—1,050 feet MSL.

When using an instrument approach facility other than ILS, no person may use an autopilot at an altitude above the terrain that is less than 50 feet below the approved minimum descent altitude for that procedure, or less than twice the maximum loss specified in the approved Airplane Flight Manual or equivalent for malfunction of the autopilot under approach conditions, whichever is higher. (PLT424) — 14 CFR §135.93

135, RTC

8016. The maximum altitude loss for a malfunctioning autopilot with an approach coupler is 40 feet. To which minimum altitude may the autopilot be used during an ILS approach in less than basic VFR conditions?

A—40 feet AGL.
B—50 feet AGL.
C—80 feet AGL.

For ILS approaches, when reported weather is less than VFR, no person may use an autopilot with an approach coupler at an altitude that is less than 50 feet above the terrain, or the maximum altitude loss specified in the approved Airplane Flight Manual or equivalent, for the malfunction of the autopilot with an approach coupler, whichever is higher. (PLT424) — 14 CFR §135.93

135, RTC

8017. The maximum altitude loss for a malfunctioning autopilot without an approach coupler is 45 feet. If the MDA is 1,620 feet MSL and the TDZE is 1,294 feet, to which minimum altitude may you use the autopilot?

A—1,510 feet MSL.
B—1,339 feet MSL.
C—1,570 feet MSL.

When using an instrument approach facility other than ILS, no person may use an autopilot at an altitude above the terrain that is less than 50 feet below the approved minimum descent altitude for that procedure, or less than twice the maximum loss specified in the approved Airplane Flight Manual or equivalent for malfunction of the autopilot under approach conditions, whichever is higher. (PLT424) — 14 CFR §135.93

135, RTC

8037. The altitude loss for a particular malfunctioning autopilot with an approach coupler is 60 feet. If the reported weather is below basic VFR minimums and an ILS approach using the approach coupler is to be used, what minimum altitude may be used?

A—50 feet AGL.
B—55 feet AGL.
C—60 feet AGL.

For ILS approaches, when reported weather is less than VFR, no person may use an autopilot with an approach coupler at an altitude that is less than 50 feet above the terrain, or the maximum altitude loss specified in the approved Airplane Flight Manual or equivalent, for the malfunction of the autopilot with an approach coupler, whichever is higher. (PLT424) — 14 CFR §135.93

135, RTC

8045. During which time period must a required voice recorder of a passenger-carrying airplane be continuously operated?

A—From the beginning of taxi to the end of the landing roll.
B—From engine start at departure airport to engine shutdown at landing airport.
C—From the use of the checklist before the flight to completion of the final check at the end of the flight.

No person may operate a multi-engine, turbine-powered airplane or rotorcraft having a passenger seating con-

Answers

8014 [A]	8015 [C]	8016 [B]	8017 [C]	8037 [C]	8045 [C]

figuration of 20 or more seats unless it is equipped with an approved cockpit voice recorder that:

1. *Is installed in compliance with Part 23, 25, 27 or 29 as applicable to Part 135; and*
2. *Is operated continuously from the use of the checklist before the flight to completion of the final check at the end of the flight.*

(PLT405) — 14 CFR §135.151

135, RTC

8046. An approved cockpit voice recorder is required equipment in

A—large turbine-powered airplanes having a maximum passenger capacity of 20 or more seats.
B—multiengine, turbine-powered airplanes having a passenger seating configuration of 20 or more seats.
C—all aircraft operated in commuter air carrier service having a passenger seating configuration of 20 seats or more.

No person may operate a multi-engine, turbine-powered airplane or rotorcraft having a passenger seating configuration of 20 or more seats unless it is equipped with an approved cockpit voice recorder. (PLT405) — 14 CFR §135.151

135, RTC

8047. Information recorded during normal operations by a required cockpit voice recorder

A—must be erased or obliterated, except for the last 30 minutes.
B—may be erased or obliterated, only once each flight.
C—may be erased or obliterated, except the last 30 minutes.

No person may operate a multi-engine, turbine-powered airplane or rotorcraft having a passenger seating configuration of 20 or more seats unless it is equipped with an approved cockpit voice recorder that:

1. *Is installed in compliance with Part 23, 25, 27 or 29 as applicable to Part 135; and*
2. *Is operated continuously from the use of the checklist before the flight to completion of the final check at the end of the flight.*

In complying with this section, information recorded more than 30 minutes earlier may be erased or otherwise obliterated. (PLT388) — 14 CFR §135.151

135, RTC

8048. Which aircraft must be equipped with an approved public address and crewmember interphone system?

A—All turbine-engine-powered aircraft having a seating configuration of more than 19 seats.
B—Aircraft having a passenger seating configuration, excluding any pilot seat, of more than 19 seats.
C—Multiengine aircraft having a passenger seating configuration of 10 seats or more.

No person may operate an aircraft having a passenger seating configuration, excluding any pilot seat, of more than 19 unless an approved public address and crew interphone system is installed. (PLT462) — 14 CFR §135.150

135, RTC

8052. To operate an aircraft with certain equipment inoperative under the provisions of a minimum equipment list, what document authorizing it must be issued to the certificate holder?

A—Letter of Authorization from the Regional Airworthiness Office authorizing such an operation.
B—Operations specifications issued by the FAA district office having certification responsibility.
C—Letter of Authorization issued by the FAA district office having certification responsibility.

No person may takeoff with inoperable instruments or equipment installed unless the following conditions are met:

1. *An approved Minimum Equipment List exists for that aircraft.*
2. *The certificate-holding district office has issued the certificate holder operations specifications authorizing operations in accordance with an approved Minimum Equipment List. The flight crew shall have direct access at all times prior to flight to all of the information contained in the approved Minimum Equipment List through printed or other means approved by the Administrator in the certificate holders operations specifications. An approved Minimum Equipment List, as authorized by the operations specifications, constitutes an approved change to the type design without requiring recertification*

(PLT428) — 14 CFR §135.179

Answers

8046 [B] 8047 [C] 8048 [B] 8052 [B]

135, RTC
8058. When a crash ax is required equipment on an aircraft, where should it be located?

A—In the flight crew compartment.
B—At a location inaccessible to the passengers during normal operations.
C—At a location accessible to both the crew and passengers during normal operations.

No person may operate an aircraft having a passenger seating configuration, excluding any pilot seat, of more than 19 seats unless it is equipped with a crash ax carried that is accessible to the crew but inaccessible to passengers during normal operations. (PLT404) — 14 CFR §135.177

135, RTC
8059. How many, if any, approved first aid kits are required on an aircraft having a passenger seating configuration of 20 seats and a passenger load of 14?

A—None.
B—One.
C—Two.

No person may operate an aircraft having a passenger seating configuration, excluding any pilot seat, of more than 19 seats unless it is equipped with one approved first aid kit for the treatment of injuries likely to occur in flight or in a minor accident. (PLT404) — 14 CFR §135.177

135, RTC
8060. An aircraft has a passenger seating configuration of 19 seats, excluding any pilot seats. How many, if any, approved first aid kits are required?

A—One.
B—Two.
C—None.

No person may operate an aircraft having a passenger seating configuration, excluding any pilot seat, of more than 19 seats unless it is equipped with one approved first aid kit for the treatment of injuries likely to occur in flight or in a minor accident. (PLT404) — 14 CFR §135.177

135, RTC
8061. Airborne weather radar equipment must be installed in large transport category aircraft, in the conterminous 48 United States,

A—that are engaged in passenger-carrying operations.
B—that are engaged in either cargo or passenger-carrying operations.
C—and be fully operational, although weather forecasts indicate no hazardous conditions.

No person may operate a large, transport category aircraft in passenger-carrying operations unless approved airborne weather radar equipment is installed in the aircraft. (PLT367) — 14 CFR §135.175

135, RTC
8062. In which aircraft, or under what conditions, is airborne thunderstorm detection equipment required?

A—Large multiengine turbine-powered aircraft having a passenger seating configuration of 19 seats or more being operated by a commuter air carrier.
B—Any aircraft having a passenger seating configuration of 19 seats or more that is engaged in passenger-carrying operations under IFR or at night.
C—Small aircraft having a passenger seating configuration of 10 seats or more, excluding any pilot seat, that are engaged in passenger-carrying operations.

No person may operate an aircraft that has a passenger seating configuration, excluding any pilot seat, of 10 seats or more in passenger-carrying operations unless the aircraft is equipped with either approved thunderstorm detection equipment or approved airborne weather radar equipment. (PLT367) — 14 CFR §135.173

135, RTC
8070. When a ground proximity warning system is required under 14 CFR Part 135, it must

A—convey warnings of any deviation below glide slope and of excessive closure rate with the terrain.
B—convey warnings for excessive closure rates with the terrain but not for deviation from an ILS glide slope.
C—alert the pilot by an audible and visual warning signals when deviation above or below glide slope occurs.

Answers

8058 [B]	8059 [B]	8060 [C]	8061 [A]	8062 [C]	8070 [A]

An approved ground proximity warning system must convey warnings of excessive closure rates with the terrain and any deviations below glide slope by visual and audible means. (PLT139) — 14 CFR §135.153

135, RTC

8071. When a ground proximity warning system is required, it must

A—apply corrective control pressure when deviation below glide slope occurs.
B—incorporate a means of alerting the pilot when a system malfunction occurs.
C—incorporate a backup feature that activates automatically upon total failure of the aircraft's electrical generating system.

An approved ground proximity warning system must convey warnings of excessive closure rates with the terrain and any deviations below glide slope by visual and audible means. It must also incorporate a means of alerting the pilot when a malfunction occurs. (PLT139) — 14 CFR §135.153

135, RTC

8077. Which group of aircraft must have a shoulder harness installed at each flight crewmember station?

A—Aircraft having a passenger seating configuration, excluding any pilot seat, of 10 seats or more.
B—All passenger-carrying aircraft operating under 14 CFR Part 135, having a seating configuration for 10 persons.
C—Large aircraft being operated in commuter air service, having a passenger seating configuration of 9, excluding any pilot seat.

No person may operate a turbojet aircraft or an aircraft having a passenger seating configuration, excluding any pilot seat, of 10 seats or more unless it is equipped with an approved shoulder harness installed for each flight crewmember station. (PLT464) — 14 CFR §135.171

135, RTC

8078. Which is a requirement for life preservers during extended overwater operations? Each life preserver must be equipped with

A—a dye marker.
B—an approved survivor locator light.
C—one flashlight having at least two size "D" cells or equivalent.

No person may operate an aircraft in extended over-water operations unless it carries an approved life preserver equipped with an approved survivor locator light for each occupant of the aircraft. (PLT437) — 14 CFR §135.167

135, RTC

8079. In addition to fully-equipped liferafts and life preservers, what emergency equipment must be provided during extended overwater operations?

A—One water resistant, self-buoyant, portable survival-type emergency radio transmitter for each liferaft.
B—Each aircraft must have at least one liferaft, equipped with a survival-type emergency locator transmitter.
C—One pyrotechnic signaling device for each aircraft.

No person may operate an aircraft in extended over-water operations unless there is attached to one of the required life rafts, a survival-type emergency locator transmitter. (PLT437) — 14 CFR §135.167

135, RTC

8057. A pilot may make an IFR departure from an airport that does not have an approved standard instrument approach procedure if

A—there is a departure alternate within 60 minutes and the weather there is above landing minimums.
B—the Administrator has issued Operations Specifications to the certificate holder approving the procedure.
C—the departure airport is within 30 minutes flying time of another airport that has an approved standard instrument approach procedure.

The Administrator may issue operations specifications to the certificate holder to allow it to depart at an airport that does not have an approved standard instrument approach procedure when the Administrator determines that it is necessary to make an IFR departure from that airport and that the proposed operations can be conducted safely. (PLT459) — 14 CFR §135.215

Answers

8071 [B] 8077 [A] 8078 [B] 8079 [B] 8057 [B]

135, RTC

8063. Assuming the required ceiling exists, an alternate for the destination airport is not required if, for at least 1 hour before and after the ETA, the forecast visibility is at least

A—5 miles, or 3 miles more than the lowest applicable visibility minimums for the instrument approach procedure to be used, whichever is greater.
B—3 miles, or 2 miles more than the lowest applicable visibility minimums for the instrument approach procedure to be used, whichever is greater.
C—3 nautical miles, or 2 nautical miles more than the lowest applicable visibility minimums for the approach procedure to be used, which ever is greater.

An alternate airport need not be designated if the ceiling criteria is met and the visibility is forecast to be at least 3 miles or 2 miles more than the lowest applicable visibility minimums, whichever is the greater, for the instrument approach procedure to be used at the destination airport. (PLT379) — 14 CFR §135.223

135, RTC

8064. A pilot may not designate an airport as an alternate unless the weather reports, or forecasts, or any combination of them indicate that it will be at or above alternate airport landing minimum at the

A—time of departure.
B—estimated time of arrival, plus or minus 1 hour.
C—estimated time of arrival.

No person may designate an alternate airport unless the weather reports or forecasts, or any combination of them, indicate that the weather conditions will be at or above authorized alternate airport landing minimums for that airport at the estimated time of arrival. (E04) — 14 CFR §135.221

135, RTC

8065. A takeoff may not be made from an airport that is below the authorized IFR landing minimums unless

A—there is an alternate airport with the required IFR landing minimums within 60 minutes flying time, at normal cruising speed in still air.
B—the departure airport is forecast to have the required IFR landing minimums within 1 hour.
C—there is an alternate airport with the required IFR landing minimums within 60 minutes flying time, at normal cruising speed in still air with one engine inoperative.

No person may takeoff an aircraft under IFR from an airport where weather conditions are at or above takeoff minimums, but are below landing minimums, unless there is an alternate airport within one hour's flying time (at normal cruising speed in still air) of the airport of departure. (PLT459) — 14 CFR §135.217

135, RTC

8066. A pilot may not begin an IFR operation unless the next airport of intended landing is forecast to be at or above authorized IFR landing minimums at

A—the estimated time of arrival, ±1 hour.
B—the estimated time of arrival.
C—the estimated time of arrival, ±30 minutes.

No person may takeoff an aircraft under IFR or begin an IFR or over-the-top operation unless the latest weather reports or forecasts, or any combination of them, indicate that weather conditions at the estimated time of arrival at the next airport of intended landing will be at or above authorized IFR landing minimums. (PLT459) — 14 CFR §135.219

135, RTC

8068. Which condition must be met to conduct IFR operations from an airport that is not at the location where weather observations are made?

A—An "Authorization Letter" permitting the procedure must be issued by the FAA district office charged with the overall inspection of the certificate holder.
B—A "Letter of Waiver" authorizing the procedure must be issued by the Administrator, after an Investigation by the U.S. National Weather Service and the FSDO which find the standard of safety to be satisfactory.
C—The Administrator must issue Operations Specifications that permit the procedure.

The Administrator may issue operations specifications to the certificate holder to allow it to depart at an airport that does not have an approved standard instrument approach procedure when the Administrator determines that it is necessary to make an IFR departure from that airport and that the proposed operations can be conducted safely. (PLT282) — 14 CFR §135.215

Answers

8063 [B]	8064 [C]	8065 [A]	8066 [B]	8068 [C]

135, RTC

8084. Which is an operational requirement concerning ice, snow, or frost on structural surfaces?

A—A takeoff may be made with ice, snow, or frost adhering to the wings or stabilizing or control surfaces, but polished smooth, if the anti-icing and deicing equipment is operating.
B—If snow, ice, or frost is adhering to the airplane's lift or control surfaces, but polished smooth, a takeoff may be made.
C—A takeoff may not be made if ice or snow is adhering to the wings or stabilizing or control surfaces.

No pilot may takeoff in an aircraft that has snow or ice adhering to the wings, stabilizing, or control surfaces. (PLT493) — 14 CFR §135.227

135, RTC

8085. Which is one required condition for a pilot to take off under IFR with less-than-standard takeoff minimums at an airport where a straight-in instrument approach procedure is authorized and there is an approved weather reporting source?

A—The pilot must have at least 100 hours as pilot in command in the type airplane to be flown.
B—The certificate holder has been approved for such operation and the visibility at the time of takeoff must be at least RVR 16.
C—Wind direction and velocity must be such that a straight-in approach can be made to the runway served by the procedure.

At airports where straight-in instrument approach procedures are authorized, a pilot may takeoff in an aircraft under IFR when the weather conditions are equal to or better than the lowest straight-in landing minimums if:

1. *The wind direction and velocity at the time of takeoff are such that a straight-in instrument approach can be made to the runway served by instrument approach;*
2. *The associated ground facilities upon which the landing minimums are predicated and the related airborne equipment are in normal operation; and*
3. *The certificate holder has been approved for such operations.*

(PLT459) — 14 CFR §135.225

135, RTC

8086. After passing the final approach fix on a VOR approach, a weather report is received indicating the visibility is below prescribed minimums. In this situation, the pilot

A—may continue the approach and land, if at the MDA, the actual weather conditions are at least equal to the minimums prescribed for the procedure.
B—may continue the approach and land regardless of the visibility observed at the MDA, if prior to beginning the approach, the visibility was reported at or above minimums.
C—should leveloff and continue to fly the approach to the MAP, and execute the missed approach.

If a pilot has begun the final approach segment of a VOR, NDB or comparable approach procedure and has passed the final approach fix when he/she receives a weather report indicating below minimum conditions, he/she may continue the approach and, if upon reaching the MDA finds the weather at least equal to the prescribed minimums, may land. (PLT379) — 14 CFR §135.225

135, RTC

8087. An alternate for a destination airport (circling not authorized) is not required if, for at least 1 hour before and after the ETA, the required visibility exists, and the forecast ceiling is at least

A—1,500 feet above the lowest published minimum, or 2,000 feet above the airport elevation, whichever is higher.
B—1,500 feet above the lowest MDA or 2,000 feet above the runway touchdown zone elevation, whichever is higher.
C—1,000 feet above the lowest published minimum, or 1,500 feet above the airport elevation, whichever is higher.

An alternate airport need not be designated if the required visibility criteria exists and the ceiling is forecast to be at least 1,500 feet above the lowest circling approach MDA. If no circling approach is authorized the ceiling must be forecast to be 1,500 feet above the lowest published minimum, or 2,000 feet above the airport elevation, whichever is higher. (PLT380) — 14 CFR §135.223

Answers

8084 [C] 8085 [C] 8086 [A] 8087 [A]

135, RTC

8090. At a military airport, a pilot may not take off under IFR unless the reported weather conditions indicate that the

A—visibility is at least 1 mile.
B—ceiling is at least 500 feet and the visibility is 1 mile or more.
C—airport has landing minimums.

Each pilot making an IFR takeoff or approach and landing at a military or foreign airport shall comply with applicable instrument approach procedures and weather minimums prescribed by the authority having jurisdiction over that airport. In addition, no pilot may takeoff at that airport when the visibility is less than one mile. (PLT459) — 14 CFR §135.225

135, RTC

8091. A pilot may not take off under IFR at a foreign airport unless the visibility is

A—1/2 mile or more above landing minimums.
B—1 mile or more and the ceiling is 500 feet or more.
C—at least 1 mile.

Each pilot making an IFR takeoff or approach and landing at a military or foreign airport shall comply with applicable instrument approach procedures and weather minimums prescribed by the authority having jurisdiction over that airport. In addition, no pilot may takeoff at that airport when the visibility is less than one mile. (PLT459) — 14 CFR §135.225

135, RTC

8092. An instrument approach procedure to an airport may not be initiated unless the latest weather report issued by an authorized weather reporting facility indicates that weather conditions

A—are at or above the circling minimums for the runway the pilot intends to use.
B—are at or above the authorized IFR landing minimums for that procedure.
C—exceed the straight-in minimums for all nonprecision approaches.

No pilot may begin an instrument approach procedure to an airport unless the latest weather report issued by that weather reporting facility indicates that weather conditions are at or above the authorized IFR landing minimums for that airport. (PLT420) — 14 CFR §135.225

135, RTC

8114. What is the minimum ceiling and visibility for an airplane to operate under VFR in Class G airspace?

A—2,000-foot ceiling; 1-mile visibility.
B—2,000-foot ceiling; 1-mile flight visibility.
C—1,000-foot ceiling; 2-miles flight visibility.

No person may operate an airplane under VFR in uncontrolled airspace when the ceiling is less than 1,000 feet unless flight visibility is at least 2 miles. (PLT163) — 14 CFR §135.205

135, RTC

8807. Which document would constitute an approved change to the type design without requiring a recertification?

A—An approved Minimum Equipment List.
B—The Operations Specifications as approved by the Administrator.
C—A special flight permit.

An approved Minimum Equipment List, as authorized by the operations specifications, constitutes an approved change to the type design without requiring recertification. (PLT428) — 14 CFR §135.179

135, RTC

8808. No person may operate an aircraft under 14 CFR Part 135, carrying passengers under VFR at night, unless

A—each flight crewmember has a flashlight having at least two size "D" batteries or the equivalent.
B—it is equipped with a flashlight having at least two size "D" cell or the equivalent.
C—each crewmember has a flashlight having at least two size "D" cells and a spare bulb.

No person may operate an aircraft carrying passengers under VFR at night unless it is equipped with a flashlight having at least two size "D" cells or equivalent. (PLT405) — 14 CFR §135.159

Answers

8090 [A]	8091 [C]	8092 [B]	8114 [C]	8807 [A]	8808 [B]

135, RTC

8809. For operations during the period beginning 1 hour after sunset and ending 1 hour before sunrise (as published in the Air Almanac), no certificate holder may use any person, nor may any person serve, as pilot in command of an aircraft carrying passengers unless that person has made three takeoffs and three landings, within the preceding 90 days,

A—as the sole manipulator of the flight controls in an aircraft of the same category and class and, if a type rating is required, of the same type in which that person is to serve.
B—as pilot in command of an aircraft of the same category and class and, if a type rating is required, of the same type in which that person is to serve.
C—as the sole manipulator of the flight controls in an aircraft of the same type in which that person is to serve.

No person may serve as pilot-in-command of an aircraft carrying passengers unless, within the preceding 90 days, that person has, for operation during the period beginning 1 hour after sunset and ending 1 hour before sunrise (as published in the air almanac), made 3 takeoffs and 3 landings as the sole manipulator of the flight controls in an aircraft of the same category and class and, if a type rating is required, of the same type in which the person is to serve. (PLT442) — 14 CFR §135.247

135, RTC

8813. An employee who performs safety-sensitive functions, for a certificate holder, who has actual knowledge of an accident involving an aircraft for which he or she performed a safety-sensitive function at or near the time of the accident shall not use alcohol

A—until 4 hours after the accident.
B—within 8 hours of the accident.
C—until given a release by the NTSB or FAA.

No covered employee who has actual knowledge of an accident involving an aircraft for which he or she has performed a safety-sensitive function at or near the time of the accident shall use alcohol for 8 hours following the accident. (PLT463) — 14 CFR §135.253

135, RTC

8814. What is the maximum number of hours that a pilot may fly in 7 consecutive days as a pilot in commercial flying and as a pilot for a commuter air carrier?

A—32 hours.
B—34 hours.
C—35 hours.

No certificate holder may schedule any flight crewmember for flight in scheduled operations if that crewmember's total time in commercial flying will exceed:

1. 1,200 hours in any calendar year.

2. 120 hours in any calendar month.

3. 34 hours in any seven consecutive days.

(PLT409) — 14 CFR §135.265

135, RTC

8815. What is the maximum number of hours that a commuter air carrier may schedule a flight crewmember to fly in scheduled operations and other commercial flying in any calendar month?

A—100.
B—110.
C—120.

No certificate holder may schedule any flight crewmember for flight in scheduled operations if that crewmember's total time in commercial flying will exceed:

1. 1,200 hours in any calendar year.

2. 120 hours in any calendar month.

3. 34 hours in any seven consecutive days.

(PLT409) — 14 CFR §135.265

135, RTC

8819. The pilot in command may deviate from 14 CFR Part 135 during an emergency involving the safety of persons or property only

A—after ATC is notified of the emergency and the extent of deviation required.
B—to the extent required to meet that emergency.
C—if required to, by the emergency cockpit checklist.

In an emergency involving the safety of persons or property, the pilot-in-command may deviate from the rules of 14 CFR Part 135 to the extent required to meet that emergency. (PLT444) — 14 CFR §135.19

Answers

8809 [A]	8813 [B]	8814 [B]	8815 [C]	8819 [B]

135, RTC

8820. The training required for flight crewmembers who have not qualified and served in the same capacity on an aircraft is

A—upgrade training.
B—transition training.
C—initial training.

Initial training is the term used for the training required for crewmembers who have not qualified and served in the same capacity on an aircraft. (PLT407) — 14 CFR §135.321

135, RTC

8821. A crewmember who has served as second in command on a particular aircraft type (e.g., BE-1900), may serve as pilot in command upon completing which training program?

A—Upgrade training.
B—Transition training.
C—Initial training.

Upgrade training is the training required of crewmembers who have qualified and served as second-in-command on a particular aircraft before they serve as pilot-in-command of that aircraft. (PLT407) — 14 CFR §135.321

135, RTC

8827. The training required for crewmembers who have been qualified and served in the same capacity on another aircraft is

A—difference training.
B—transition training.
C—upgrade training.

Transition training is the training required of crewmembers who have qualified and served in the same capacity on another aircraft. (PLT407) — 14 CFR §135.321

135, RTC

8828. The certificate holder must give instruction on such subjects as respiration, hypoxia, gas expansion, and decompression to crewmembers who serve in operations above

A—FL 180.
B—FL 200.
C—FL 250.

Crewmembers who serve in operations above 25,000 feet must receive instruction in respiration, hypoxia, duration of consciousness without supplemental oxygen at altitude, gas expansion, gas bubble formation and physical phenomena and incidents of decompression. (PLT460) — 14 CFR §135.331

135, RTC

8829. The air carrier must give instruction on such subjects as gas bubble formation, hypoxia, decompression, and length of consciousness without supplemental oxygen at altitude to crewmembers serving on aircraft operated above

A—FL 250.
B—FL 200.
C—FL 180.

Crewmembers who serve in operations above 25,000 feet must receive instruction in respiration, hypoxia, duration of consciousness without supplemental oxygen at altitude, gas expansion, gas bubble formation and physical phenomena and incidents of decompression. (PLT407) — 14 CFR §135.331

135, RTC

8830. What is one of the requirements that must be met by a pilot in command to re-establish recency of experience?

A—At least one full stop landing must be made from a circling approach.
B—Three takeoffs and landings must be made as the sole manipulator of the controls, in the type, if a type rating is required, if not in the same category and class aircraft that the person is to serve.
C—At least one nonprecision approach must be made to the lowest minimums authorized for the certificate holder.

No person may serve as pilot-in-command of an aircraft carrying passengers unless, within the preceding 90 days, that person has made 3 takeoffs and 3 landings as the sole manipulator of the flight controls in an aircraft of the same category and class and, if a type rating is required, of the same type in which the person is to serve. (PLT442) — 14 CFR §135.247

Answers

8820 [C] 8821 [A] 8827 [B] 8828 [C] 8829 [A] 8830 [B]

135, RTC
8834. Federal Aviation Regulations require that interior emergency lights, on aircraft having a passenger seating configuration of 20 to

A—operate automatically when subjected to a negative G load.
B—be operable manually from the flight crew station and a point in the passenger compartment.
C—be armed or turned on during taxiing and all flight operations.

Emergency exit lights must be operable manually from the flight crew station and from a station in the passenger compartment that is readily accessible to a normal flight attendant seat. (PLT404) — 14 CFR §135.178

Answer (A) is incorrect because the lights must operate automatically either with loss of normal electrical power or when an emergency assist means is activated, depending on the aircraft certification. Answer (C) is incorrect because the lights must be armed or turned on during taxi, takeoff and landing but not necessarily during all other flight operations.

135, RTC
8838. What emergency equipment is required for extended overwater operations?

A—A portable survival emergency locator transmitter for each life raft.
B—A pyrotechnic signaling device for each life preserver.
C—A life preserver equipped with a survivor locator light, for each person on the airplane.

Every aircraft flown in extended overwater operations must carry an approved life preserver for every occupant of the aircraft. This life preserver must be equipped with an approved survivor locator light. A life preserver must be readily accessible to each seated occupant. In addition, there must be enough appropriately equipped life rafts to accommodate all the occupants of the aircraft. One of the life rafts must have a survival type emergency locator transmitter. (PLT437) — 14 CFR §135.167

135, RTC
8840. Each aircraft being operated in extended overwater operations, must have a life preserver for each

A—aircraft occupant.
B—seat on the aircraft.
C—passenger seat, plus 10 percent.

Every aircraft flown in extended overwater operations must carry an approved life preserver for every occupant of the aircraft. (PLT437) — 14 CFR §135.167

135, RTC
8841. Life preservers required for extended overwater operations are stored

A—within easy reach of each passenger.
B—under each occupant seat.
C—within easy access of each seated occupant.

Every aircraft flown in extended overwater operations must carry an approved life preserver for every occupant of the aircraft. A life preserver must be readily accessible to each seated occupant. (PLT437) — 14 CFR §135.167

135, RTC
8843. No person may takeoff an aircraft under IFR from an airport that has takeoff weather minimums but that is below landing minimums unless there is an alternate airport within

A—1 hour at normal indicated airspeed of the departure airport.
B—1 hour at normal cruise speed in still air of the departure airport.
C—1 hour at normal cruise speed in still air with one engine operating.

No person may takeoff an aircraft under IFR from an airport where weather conditions are at or above takeoff minimums, but are below authorized IFR landing minimums unless there is an alternate airport within 1 hour's flying time (in still air) of the airport of departure. (PLT459) — 14 CFR §135.217

Answers

8834 [B] 8838 [C] 8840 [A] 8841 [C] 8843 [B]

Helicopter Regulations

RTC

8002. What minimum rest period must be provided for a pilot assigned to Helicopter Hospital Emergency Medical Evacuation Service (HEMES) who has been on duty for a 47 hour period?

A—16 consecutive hours.
B—14 consecutive hours.
C—12 consecutive hours.

Each pilot must be given a rest period upon completion of the HEMES assignment and prior to being assigned any further duty with the certificate holder of at least 12 consecutive hours for an assignment of less than 48 hours, and at least 16 consecutive hours for an assignment of more than 48 hours. (PLT409) — 14 CFR §135.271

RTC

9043. What is a helicopter pilot's responsibility when cleared to "air taxi" on the airport?

A—Taxi direct to destination as quickly as possible.
B—Taxi at hover altitude using taxiways.
C—Taxi below 100 feet AGL avoiding other aircraft and personnel.

Air taxi is the preferred method for helicopter ground movements on airports. Unless otherwise requested or instructed, pilots are expected to remain below 100 feet AGL. Helicopters should avoid overflight of other aircraft, vehicles, and personnel during air taxi operations. (PLT112) — AIM ¶4-3-17

RTC

9336. What minimum instrument experience in the past 6 calendar months meets the second-in-command requirement to maintain IFR currency in a helicopter?

A—6 hours in actual IFR conditions or 3 hours actual and 3 hours simulated IFR in a helicopter plus six instrument approaches.
B—Holding procedures, intercepting and tracking courses using the navigation equipment, six instrument approaches logged in actual or simulated IFR in a helicopter, simulator or a flight training device.
C—6 hours of actual or simulated time in a helicopter of the same type, plus six instrument approaches.

For flight under IFR, the second-in-command must meet the recent instrument requirements of Part 61: No pilot may act as pilot-in-command under IFR unless the pilot has performed and logged, within the past 6 calendar months, at least 6 instrument approaches, holding procedures, and intercepting and tracking courses through the use of navigation systems in the appropriate category of aircraft for the instrument privileges sought. (PLT442) — 14 CFR §135.245 and §61.57

RTC

9337. What minimum conditions are necessary for the instrument approaches required for second-in-command IFR currency in a helicopter?

A—Six must be performed and logged under actual or simulated instrument conditions in a rotorcraft.
B—Six must be performed and logged under actual or simulated instrument conditions; three must be in a rotorcraft, three may be in an airplane or an approved flight simulator.
C—All must be made in a rotorcraft category of aircraft, or approved simulator, or flight training device and logged while under actual or simulated IFR conditions.

For flight under IFR, the second-in-command must meet the recent instrument requirements of Part 61: No pilot may act as pilot-in-command under IFR unless the pilot has performed and logged, within the past 6 calendar months, at least 6 instrument approaches, holding procedures, and intercepting and tracking courses through the use of navigation systems in the appropriate category of aircraft for the instrument privileges sought. (PLT442) — 14 CFR §135.245 and §61.57

Answers

8002 [C]	9043 [C]	9336 [B]	9337 [C]

RTC

9338. Within the past 6 months, a pilot has accomplished:

Two approaches in a helicopter.
Two approaches in an airplane.
Two approaches in a glider.

What additional instrument experience must the pilot obtain prior to acting as second in command (under 14 CFR part 135) on an IFR flight?

A—Four approaches in an aircraft, approved training device, flight simulator (that is representative of the aircraft category), holding, intercepting and tracking courses using the navigation systems.
B—Passes an instrument proficiency check in any category aircraft, approved simulator or training device.
C—Holding, intercepting and tracking courses (using the navigation systems) in an aircraft, approved simulator or approved flight training device.

For flight under IFR, the second-in-command must meet the recent instrument requirements of Part 61: No pilot may act as pilot-in-command under IFR unless the pilot has performed and logged, within the past 6 calendar months, at least 6 instrument approaches, holding procedures, and intercepting and tracking courses through the use of navigation systems in the appropriate category of aircraft for the instrument privileges sought. (PLT442) — 14 CFR §135.245 and §61.57

RTC

9341. Within the past 6 months, a pilot has accomplished:

Two approaches and intercepting, tracking courses using the navigation systems in a helicopter.

Two approaches, missed approaches and holding in an approved airplane flight simulator.

Two approaches and holding in an approved rotorcraft flight training device.

What additional instrument experience, if any, must the pilot perform to act as second in command (under 14 CFR part 135) on an IFR helicopter flight?

A—None.
B—Two approaches in a rotorcraft category aircraft.
C—Two approaches in either a helicopter or an airplane.

For flight under IFR, the second-in-command must meet the recent instrument requirements of Part 61: No pilot may act as pilot-in-command under IFR unless the pilot has performed and logged, within the past 6 calendar months, at least 6 instrument approaches, holding procedures, and intercepting and tracking courses through the use of navigation systems in the appropriate category of aircraft for the instrument privileges sought. (PLT442) — 14 CFR §135.245, §61.57

RTC

9366. Unless otherwise prescribed, what is the rule regarding altitude and course to be maintained by a helicopter during an off-airways IFR flight over non-mountainous terrain?

A—1,000 feet above the highest obstacle within 4 nautical miles of course.
B—2,000 feet above the highest obstacle within 5 statute miles of course.
C—1,500 feet above the highest obstacle within a horizontal distance of 3 statute miles of course.

In the case of operations over areas that are not designated as mountainous areas; no person may operate an aircraft under IFR below an altitude of 1,000 feet above the highest obstacle within a horizontal distance of 4 nautical miles from the course to be flown. (PLT430) — 14 CFR §91.177

RTC

9367. Unless otherwise prescribed, what is the rule regarding altitude and course to be maintained by a helicopter during an IFR off-airways flight over mountainous terrain?

A—1,000 feet above the highest obstacle within a horizontal distance of 5 statute miles of course.
B—2,500 feet above the highest obstacle within a horizontal distance of 3 nautical miles of course.
C—2,000 feet above the highest obstacle within 5 statute miles of course.

In the case of operations over areas designated as mountainous; no person may operate an aircraft under IFR below an altitude of 2,000 feet above the highest obstacle within a horizontal distance of 5 statute miles from the course to be flown. (PLT430) — 14 CFR §91.177

Answers

9338 [A] 9341 [B] 9366 [A] 9367 [C]

RTC
9371. According to 14 CFR Part 91, when takeoff minimums are not prescribed for a civil airport, what are the takeoff minimums under IFR for a multiengine helicopter?

A—1 SM visibility.
B—1/2 SM visibility.
C—1200 RVR.

If takeoff minimums are not prescribed under Part 97, the takeoff minimums under IFR for helicopters are 1/2 statute mile visibility. (PLT459) — 14 CFR §91.175

RTC
9372. According to 14 CFR Part 91, when takeoff minimums are not prescribed for a civil airport, what are the takeoff minimums under IFR for a single-engine helicopter?

A—1/2 SM visibility.
B—1 SM visibility.
C—1200 RVR.

If takeoff minimums are not prescribed under Part 97, the takeoff minimums under IFR for helicopters are 1/2 statute mile visibility. (PLT459) — 14 CFR §91.175

RTC
9373. What minimum altitude should a helicopter maintain while en route?

A—Over congested areas such as towns, no lower than 1,000 feet over the highest obstacle within a horizontal radius of 2,000 feet of the helicopter.
B—That specifically prescribed by the air carrier for the operation.
C—That prescribed by the Administrator.

Each person operating a helicopter shall comply with routes or altitudes specifically prescribed for helicopters by the Administrator. (PLT430) — 14 CFR §91.119

RTC
9414. In addition to a two-way radio capable of communicating with ATC on appropriate frequencies, which equipment is the helicopter required to have to operate within Class B airspace? (Letter of agreement not applicable.)

A—A VOR or TACAN receiver.
B—DME, a VOR or TACAN receiver, and an appropriate transponder beacon.
C—An appropriate ATC transponder.

An operable ATC transponder is required to operate all aircraft in Class B airspace except for helicopters operated at or below 1,000 feet AGL under the terms of a letter of agreement. (PLT405) — 14 CFR §91.131 and §91.215

RTC
9415. Which of the following is a transponder requirement for helicopter operations?

A—Helicopters with a certified gross weight of more than 12,500 pounds that are engaged in commercial operations are required to be equipped with operable ATC transponders.
B—Helicopters may be operated at or below 1,000 feet AGL within Class B airspace without an operable ATC transponder.
C—Operable ATC transponders are required when operating helicopters within Class D airspace at night under special VFR.

An operable ATC transponder is required to operate all aircraft in Class B airspace except as authorized by ATC. (PLT405) — 14 CFR §91.215

RTC
8975. Which of the following are required for a helicopter ILS approach with a decision height lower than 200 feet HAT?

A—Special aircrew training and aircraft certification.
B—Both a marker beacon and a radio altimeter.
C—ATP helicopter certificate and CAT II certification.

Approaches with a HAT below 200 feet are annotated with the note: "Special Aircraft & Aircraft Certification Required" since the FAA must approve the helicopter and its avionics, and the flight crew must have the required experience, training, and checking. (PLT356) — FAA-H-8261-1

Answers

9371 [B]	9372 [A]	9373 [C]	9414 [C]	9415 [B]	8975 [A]

Chapter 2
Equipment, Navigation and Facilities

Inoperative Equipment

A certificate holder's manual must contain enroute flight, navigation and communication procedures, including procedures for the dispatch, release or continuance of a flight if a required piece of equipment becomes inoperative.

When any required instrument or equipment in an aircraft is inoperative, the airplane cannot be flown unless that aircraft's **Minimum Equipment List (MEL)** allows such a flight.

The pilot-in-command of an aircraft operating IFR in controlled airspace shall report to ATC immediately any malfunction of navigational, approach or communications equipment that occurs in flight. The report must include:

- Aircraft identification;
- Equipment affected;
- Degree to which the capability of the aircraft to operate IFR in the ATC system is impaired; and
- Nature and extent of assistance desired from ATC.

ALL

9407. An approved minimum equipment list or FAA Letter of Authorization allows certain instruments or equipment

A—to be inoperative prior to beginning a flight in an aircraft if prescribed procedures are followed.
B—to be inoperative anytime with no other documentation required or procedures to be followed.
C—to be inoperative for a one-time ferry flight of a large airplane to a maintenance base without further documentation from the operator or FAA with passengers on board.

The Minimum Equipment List and the letter of authorization constitute a supplemental type certificate for the aircraft. The approved Minimum Equipment List must provide for the operation of the aircraft with the instruments and equipment in an inoperable condition (PLT405) — 14 CFR §91.213

ALL

9380. What action is necessary when a partial loss of ILS receiver capability occurs while operating in controlled airspace under IFR?

A—Continue as cleared and file a written report to the Administrator if requested.
B—If the aircraft is equipped with other radios suitable for executing an instrument approach, no further action is necessary.
C—Report the malfunction immediately to ATC.

The pilot-in-command of an aircraft operating IFR in controlled airspace shall report to ATC as soon as practical any malfunction of navigational, approach or communication equipment that occurs in flight. (PLT356) — 14 CFR §91.187

Answer (A) is incorrect because any malfunction of approach equipment must be reported in flight, not by a written report. Answer (B) is incorrect because, although another type of instrument approach may be executed if permission is granted by ATC, any malfunction of approach equipment should be reported.

ALL

9381. What action should be taken if one of the two VHF radios fail while IFR in controlled airspace?

A—Notify ATC immediately.
B—Squawk 7600.
C—Monitor the VOR receiver.

The pilot-in-command of an aircraft operating IFR in controlled airspace shall report to ATC as soon as practical any malfunction of navigational, approach or communication equipment that occurs in flight. (PLT162) — 14 CFR §91.187

Answer (B) is incorrect because, although you have experienced a communications failure, it is only a partial one. You still have one operational VHF radio and all other radios are working normally, so a squawk of 7600 is not needed. Answer (C) is incorrect because you still have an operable VHF radio for communication, so monitoring of a NAVAID is not needed. The only pilot action required is notification to ATC of the problem.

Answers

9407 [A]	9380 [C]	9381 [A]

ALL

9386. While flying IFR in controlled airspace, if one of the two VOR receivers fails, which course of action should the pilot-in-command follow?

A—No call is required if one of the two VOR receivers is operating properly.
B—Advise ATC immediately.
C—Notify the dispatcher via company frequency.

The pilot-in-command of an aircraft operating IFR in controlled airspace shall report to ATC as soon as practical any malfunction of navigational, approach or communication equipment that occurs in flight. (PLT406) — 14 CFR §91.187

Answer (A) is incorrect because any malfunction of a navigational radio should be reported, no matter how slightly it may affect the conduct of the flight. Answer (C) is incorrect because, although this may be a common practice among the air carriers, the regulations require notification to ATC of the malfunction.

ALL

9387. While flying in controlled airspace under IFR, the ADF fails. What action is required?

A—Descend below Class A airspace.
B—Advise dispatch via company frequency.
C—Notify ATC immediately.

The pilot-in-command of an aircraft operating IFR in controlled airspace shall report to ATC as soon as practical any malfunction of navigational, approach or communication equipment that occurs in flight. (PLT406) — 14 CFR §91.187

Answer (A) is incorrect because controlled airspace exists far below positive control airspace (base of 18,000 feet MSL), and any loss of a navigational aid should be reported to ATC. Answer (B) is incorrect because, although this may be a common practice among the air carriers, the regulations require notification to ATC of the malfunction.

121, DSP

8278. If a required instrument on a multiengine airplane becomes inoperative, which document dictates whether the flight may continue en route?

A—A Master Minimum Equipment List for the airplane.
B—Original dispatch release.
C—Certificate holder's manual.

Each certificate holder's manual must contain enroute flight, navigation, and communication procedures for the dispatch, release or continuance of flight if any item of equipment required for the particular type of operation becomes inoperative or unserviceable en route. (PLT426) — 14 CFR §121.135

Pitot-Static Instruments

Modern jet transports usually have three pitot-static systems. There are separate systems for the captain's and co-pilot's instruments plus an auxiliary system that provides a backup for either of the two primary systems. The instruments that require static pressure input are **airspeed, Mach, altitude** and **vertical speed indicators**. In addition, the airspeed and Mach indicators need a source of pitot pressure. Besides the flight instruments, static pressure input is required for the Mach warning, autopilot, flight director, flight recorder and cabin differential pressure. Pitot input is required for all those systems except for cabin differential pressure. The usual source for these non-flight instruments is the auxiliary pitot-static system. *See* Figure 2-1.

Altimeters compare the sea level pressure setting in their window with the outside air pressure sensed through the static system. The difference is displayed as the altitude above sea level. Part of the preflight check is to verify the accuracy of the altimeters. An altimeter should be considered questionable if the indicated altitude varies by more the 75 feet from a known field elevation.

The altimeter setting used by pilots is always the station pressure of the reporting station corrected to sea level. **Station pressure** is the actual pressure at field elevation.

True altitude is the actual height of the aircraft above sea level. This is the same as indicated altitude when standard temperatures exist. When the temperature is warmer than standard, true altitude is higher than indicated altitude. When the temperature is colder than standard day conditions, just the opposite is true. Corrected altitude (approximately true altitude) can be calculated but it is neither practical nor useful

Answers

9386 [B] 9387 [C] 8278 [C]

to do so in most situations. When setting an altimeter, a pilot should just use the appropriate altimeter setting and disregard the effects of nonstandard atmospheric pressures and temperatures.

Pressure altitude is the altitude indicated when the altimeter is set to standard sea level pressure of 29.92" Hg. Density altitude is used in aircraft performance computations. It is pressure altitude corrected for nonstandard temperatures. If the temperature is warmer than standard, density altitude will be higher than pressure altitude.

The local altimeter setting is used when flying below FL180 and the altimeter is 31.00" Hg or less. Special procedures apply when the local pressure is more than 31.00" Hg because most altimeters cannot be set higher than that. In the United States, all altimeters are set to 29.92" Hg when climbing through FL180. Caution: outside the United States the transition altitude is often something other than FL180.

A common reason for altimeter errors is incorrect setting of the altimeter. If the setting in the altimeter is higher than the actual sea level pressure, the altimeter will read higher than the actual altitude. If the setting is too low, the altimeter will read lower than it really is. As a rough rule of thumb, the magnitude of the error is about 1,000 feet for each 1" Hg that the altimeter is off. For example, if the altimeter is set to 29.92" Hg, but the real sea level pressure is 30.57" Hg, the altimeter will read about 650 feet lower than the actual airplane's altitude (30.57 – 29.92 = .65" Hg = 650 feet). In this example, the airplane would be 650 feet higher than the indicated altitude.

Continued

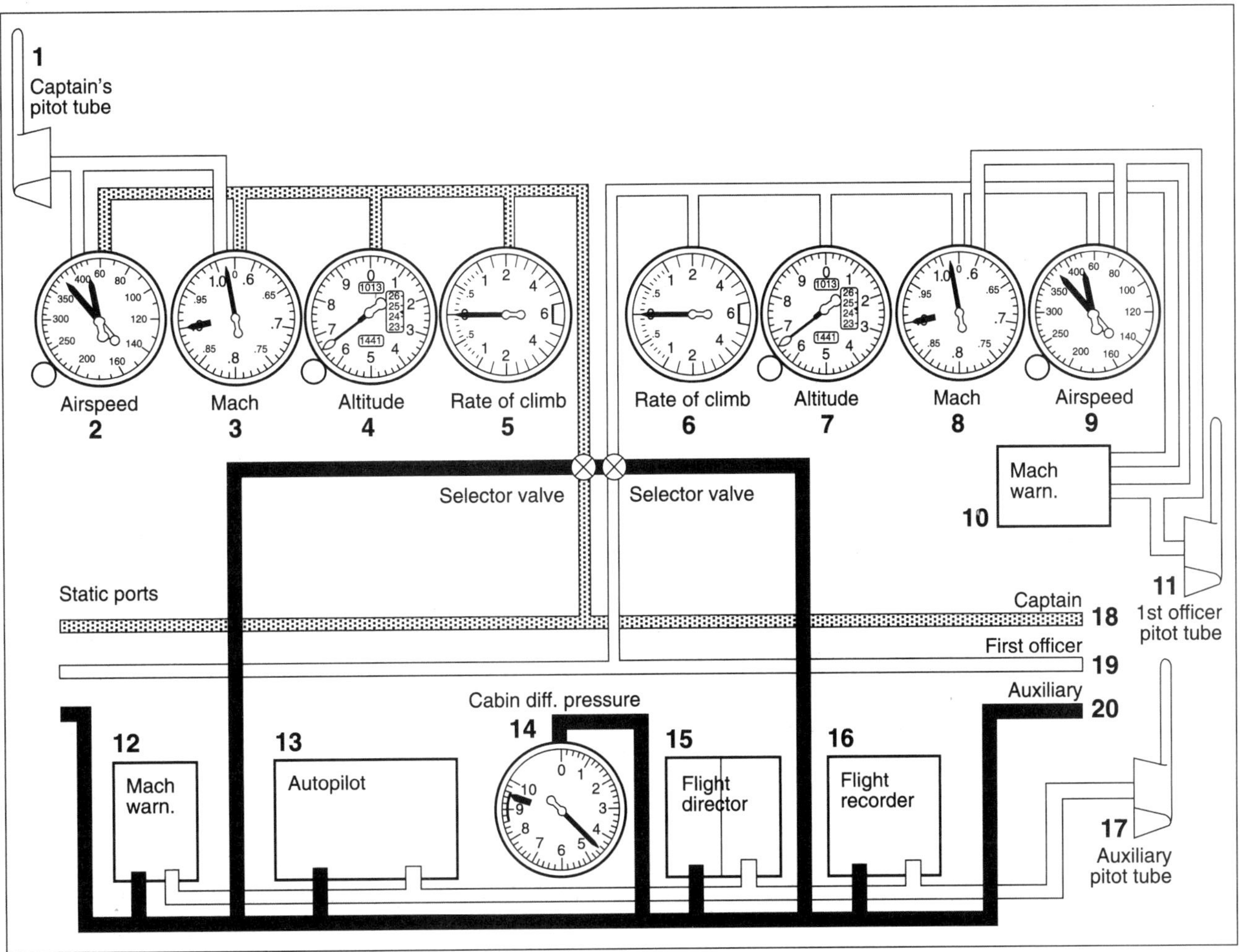

Figure 2-1. Typical pitot-static system

The airspeed indicators compare pitot pressure with static pressure and display the difference as **indicated airspeed**. This indicated airspeed equals the aircraft's actual speed through the air (True Airspeed) only under standard day conditions at sea level. Under almost all flight conditions, true airspeed will be higher than indicated airspeed because of the lower ambient pressures at altitude.

The Machmeter displays aircraft speed as a percentage of the speed of sound. For example, an aircraft cruising at a Mach number of .82 is flying at 82% of the speed of sound. The Machmeter works in a manner similar to the airspeed indicator in that it compares pitot and static pressure, but these inputs are corrected by an altimeter mechanism.

If a pitot tube becomes blocked, the airspeed and Mach indicators will read inaccurately. If pressure is trapped in the pitot line, the airspeed will read inaccurately high as the aircraft climbs, low as it descends, and will be unresponsive to changes in airspeed. The airspeed indicator acts as an altimeter because only the static pressure changes. This situation occurs in icing conditions if both the ram air inlet and the drain hole of the pitot tube become completely blocked by ice.

If the pitot tube is blocked but the static port and the pitot drain hole remain open, the indicated airspeed will drop to zero. The drain pitot tube drain hole allows the pressure in the pitot line to drop to atmospheric and therefore there is no differential between the static and pitot pressures.

Pitot tubes and static ports are electrically heated to prevent ice formations that could interfere with proper operation of the systems. They are required to have "power on" indicator lights to show proper operation. In addition, many aircraft have an ammeter that shows the actual current flow to the pitot and static ports.

Since the magnetic compass is the only direction-seeking instrument in most airplanes, the pilot must be able to turn the airplane to a magnetic compass heading and maintain this heading. It is influenced by magnetic dip which causes northerly turning error and acceleration/deceleration error. When northerly turning error occurs, the compass will lag behind the actual aircraft heading while turning through headings in the northern half of the compass rose, and lead the aircraft's actual heading in the southern half. The error is most pronounced when turning through north or south, and is approximately equal in degrees to the latitude.

The acceleration/deceleration error is most pronounced on headings of east and west. When accelerating, the compass indicates a turn toward the north, and when decelerating it indicates a turn toward the south. The acronym **ANDS** is a good memory aid:

A accelerate

N north

D decelerate

S south

No errors are apparent while on east or west headings, when turning either north or south.

ALL

9174. Which pressure is defined as station pressure?

A—Altimeter setting.
B—Actual pressure at field elevation.
C—Station barometric pressure reduced to sea level.

The pressure measured at a station or airport is "station pressure" or the actual pressure at field elevation. (PLT166) — AC 00-6A, Chapter 3

Answer (A) is incorrect because altimeter setting is the value to which the scale of a pressure altimeter is adjusted to read field elevation. Answer (C) is incorrect because station barometric pressure reduced to sea level is a method to readily compare station pressures between stations at different altitudes.

ALL

9164. What is corrected altitude (approximate true altitude)?

A—Pressure altitude corrected for instrument error.
B—Indicated altitude corrected for temperature variation from standard.
C—Density altitude corrected for temperature variation from standard.

True altitude is indicated altitude corrected for the fact that nonstandard temperatures will result in nonstandard pressure lapse rates. (PLT023) — AC 00-6A, Chapter 3

Answer (A) is incorrect because pressure altitude corrected for instrument error is a nonexistent concept. Answer (C) is incorrect because density altitude is pressure altitude corrected for temperature variation from standard. Density altitude is a final figure and not subject to additional adjustments.

ALL

9099. When setting the altimeter, pilots should disregard

A—effects of nonstandard atmospheric temperatures and pressures.
B—corrections for static pressure systems.
C—corrections for instrument error.

Pilots should disregard the effect of nonstandard atmospheric temperatures and pressures except that low temperatures and pressures need to be considered for terrain clearance purposes. (PLT166) — AIM ¶7-2-2

Answers (B) and (C) are incorrect because altimeters are subject to instrument errors and to errors in the static pressure system. A pilot should set the current reported altimeter setting on the altimeter setting scale. The altimeter should read within 75 feet of field elevation. If not, it is questionable and should be evaluated by a repair station.

ALL

9173. If the ambient temperature is colder than standard at FL310, what is the relationship between true altitude and pressure altitude?

A—They are both the same, 31,000 feet.
B—True altitude is lower than 31,000 feet.
C—Pressure altitude is lower than true altitude.

True altitude is indicated altitude corrected for the fact that nonstandard temperatures will result in nonstandard pressure lapse rates. In warm air, you fly at a true altitude higher than indicated. In cold air, you fly at a true altitude lower than indicated. Pressure altitude is the altitude indicated when the altimeter is set to the standard sea level pressure (29.92" Hg). In the United States, altimeters are always set to 29.92" Hg at and above 18,000 feet. This question assumes the difference between the pressure altitude and the indicated altitude (local altimeter setting) is not significant enough to reverse the effects of the temperature. (PLT023) — AC 00-6A, Chapter 3

Answer (A) is incorrect because both true and pressure altitude would be the same at FL310 if the ambient air temperature was standard. Answer (C) is incorrect because pressure altitude would be lower than true altitude in warmer than standard air temperature.

ALL

9172. If the ambient temperature is warmer than standard at FL350, what is the density altitude compared to pressure altitude?

A—Lower than pressure altitude.
B—Higher than pressure altitude.
C—Impossible to determine without information on possible inversion layers at lower altitudes.

Pressure altitude is the altitude indicated when the altimeter is set to the standard sea level pressure (29.92" Hg). Density altitude is pressure altitude corrected for nonstandard temperature. A warmer than standard temperature will result in a density altitude higher than the pressure altitude. (PLT023) — AC 00-6A, Chapter 3

Answer (A) is incorrect because density altitude is higher when air temperature is warmer than standard. Answer (C) is incorrect because density altitude is pressure altitude corrected for nonstandard temperatures. Pressure altitude is based on a standard pressure atmosphere at a particular altitude, and inversion layers at lower levels have no effect on pressure altitude.

Answers

9174 [B] 9164 [B] 9099 [A] 9173 [B] 9172 [B]

ALL

9163. En route at FL270, the altimeter is set correctly. On descent, a pilot fails to set the local altimeter setting of 30.57. If the field elevation is 650 feet, and the altimeter is functioning properly, what will it indicate upon landing?

A—585 feet.
B—1,300 feet.
C—Sea level.

One inch of Hg pressure is equal to about 1,000 feet of altitude. In the United States, altimeters are always set to 29.92" Hg at and above 18,000 feet. If the altimeter is not reset when descending into an area with a local altimeter setting of 30.57" Hg, an error of 650 feet will result (30.57 – 29.92 = .65 = 650 feet). If the altimeter is set lower than the actual setting, it will read lower than the actual altitude. (PLT166) — AC 00-6A, Chapter 3

Answer (A) is incorrect because 585 feet is the result of subtracting 65 feet rather than subtracting 650 feet. Answer (B) is incorrect because 1,300 feet is the result of adding 650 feet rather than subtracting 650 feet.

ALL

9080. During an en route descent in a fixed-thrust and fixed-pitch attitude configuration, both the ram air input and drain hole of the pitot system become completely blocked by ice. What airspeed indication can be expected?

A—Increase in indicated airspeed.
B—Decrease in indicated airspeed.
C—Indicated airspeed remains at the value prior to icing.

If both the ram air input and the drain hole are blocked, the pressure trapped in the pitot line cannot change and the airspeed indicator may react as an altimeter. The airspeed will not change in level flight even when actual airspeed is varied by large power changes. During a climb the airspeed indication will increase. During a descent the airspeed indication will decrease. (PLT128) — AC 91-43

Answer (A) is incorrect because indicated airspeed will decrease in a descent. Answer (C) is incorrect because indicated airspeed will remain at the same value during level flight.

ALL

9081. What can a pilot expect if the pitot system ram air input and drain hole are blocked by ice?

A—The airspeed indicator may act as an altimeter.
B—The airspeed indicator will show a decrease with an increase in altitude.
C—No airspeed indicator change will occur during climbs or descents.

If both the ram air input and the drain hole are blocked, the pressure trapped in the pitot line cannot change and the airspeed indicator may react as an altimeter. The airspeed will not change in level flight even when actual airspeed is varied by large power changes. During a climb the airspeed indication will increase. During a descent the airspeed indication will decrease. (PLT337) — AC 91-43

Answer (B) is incorrect because the airspeed indicator will show an increase (not decrease) with an increase in altitude. Answer (C) is incorrect because differential pressure between the pitot tube and static air source changes, and so does indicated airspeed.

ALL

9082. If both the ram air input and drain hole of the pitot system are blocked by ice, what airspeed indication can be expected?

A—No variation of indicated airspeed in level flight if large power changes are made.
B—Decrease of indicated airspeed during a climb.
C—Constant indicated airspeed during a descent.

If both the ram air input and the drain hole are blocked, the pressure trapped in the pitot line cannot change and the airspeed indicator may react as an altimeter. The airspeed will not change in level flight even when actual airspeed is varied by large power changes. During a climb the airspeed indication will increase. During a descent the airspeed indication will decrease. (PLT337) — AC 91-43

Answer (B) is incorrect because, during a climb, it will indicate an increase due to the stronger differential pressure in the blocked pitot tube relative to the static vents. Answer (C) is incorrect because indicated airspeed would change with changes in altitude.

ALL

9222. How will the airspeed indicator react if the ram air input to the pitot head is blocked by ice, but the drain hole and static port are not?

A—Indication will drop to zero.
B—Indication will rise to the top of the scale.
C—Indication will remain constant but will increase in a climb.

If the pitot tube becomes blocked but pressure is not trapped in the pitot lines, the indicated airspeed will drop to zero since the pitot pressure will be approximately equal to the static pressure. (PLT337) — AC 00-6A, Chapter 10

Answer (B) is incorrect because the airspeed indication will drop if only the ram air input is blocked. Answer (C) is incorrect because the pressure in the airspeed line will vent out through the hole and the indication will drop to zero.

Answers

9163 [C]	9080 [B]	9081 [A]	9082 [A]	9222 [A]

Electronic Flight Instruments

Electronic flight instrument systems integrate many individual instruments into a single presentation called a primary flight display (PFD). Flight instrument presentations on a PFD differ from conventional instrumentation not only in format, but sometimes in location as well. For example, the attitude indicator on the PFD is often larger than conventional round-dial presentations of an artificial horizon. Airspeed and altitude indications are presented on vertical tape displays that appear on the left and right sides of the primary flight display. The vertical speed indicator is depicted using conventional analog presentation. Turn coordination is shown using a segmented triangle near the top of the attitude indicator. The rate-of-turn indicator appears as a curved line display at the top of the heading/navigation instrument in the lower half of the PFD.

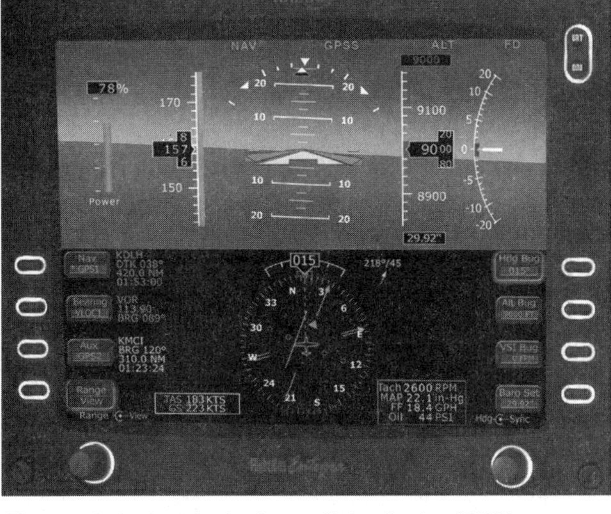

Figure 2-2. A typical primary flight display (PFD)

ALL

8206. (*See* Figure 241.) You see the indication in the figure on your PFD, but your standby indicator reads 120 knots and the power is set for 120-knot cruise in level flight. You decide the

A—pitot tube may be plugged with ice or a bug.
B—standby indicator is defective because there is no red 'X' on the speed tape display.
C—airspeed means attitude is incorrect.

The airspeed indicator on the PFD is indicating a TAS of 64 knots. If this instrument had failed, the numbers would be replaced with a large red X. The stand-by airspeed indicator reading 120 knots suggests this instrument is working fine. The line coming out of the pitot tube splits to feed multiple instruments. The most likely culprit is a bug or ice blockage occurring past the split, in the line that feeds the Air Data Computer (ADC) for the PFD. This would allow the stand-by gauge to work properly, but cause the ASI on the PFD to give a false indication. True Airspeed is calculated in the ADC by correcting CAS with OAT probe data, so this explains why the TAS is correspondingly low. The pitot lines need to be cleared; applying pitot heat may or may not help at this point. (PLT524) — FAA-H-8083-6

Answer (B) is incorrect because you cannot assume the standby is failed if you have cruise power and level attitude; the red Xs appear on the speed tape when the ADC fails or when one of the pressure transducers fail. Answer (C) is incorrect because an attitude instrument savvy pilot would discern attitude correctness by cross referencing other instruments and hearing the pitch of the engine would decide that power and a level attitude must be an indicator problem and have nothing to do with attitude correctness.

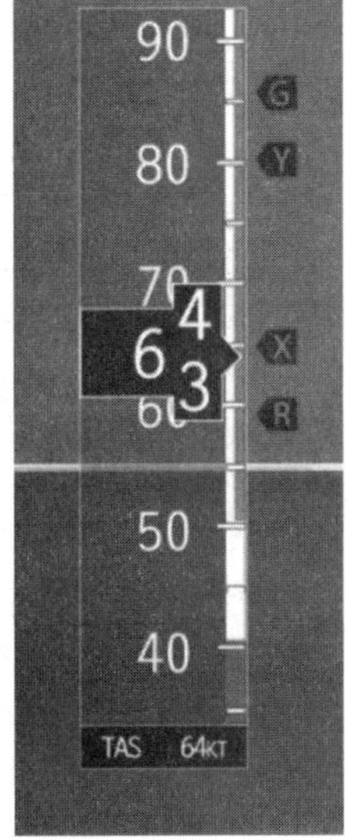

Figure 241

Answers
8206 [A]

ALL

9769. Automated flight decks or cockpits

A—enhance basic pilot flight skills.
B—decrease the workload in terminal areas.
C—often create much larger pilot errors than traditional cockpits.

Advanced avionics were designed to increase safety as well as the utility of the aircraft, particularly during increased workload phases, such as in the terminal areas. (PLT524) — FAA-H-8083-6

Answer (A) is incorrect because automation has been shown to erode some flying skills when they are not kept proficient. Answer (C) is incorrect because while automation can make some errors more evident and hide others, it does not result in larger pilot errors than traditional cockpits.

Safety of Flight Equipment

Airborne weather radar is used to detect and avoid areas of heavy precipitation such as thunderstorms. With few exceptions, all air carrier aircraft must be equipped with an approved airborne weather radar unit. The radar must be in satisfactory operating condition prior to dispatch on an IFR or night VFR flight if thunderstorms (or other hazardous weather) that could be detected by the radar are forecast along the intended route of flight. An aircraft may be dispatched with an inoperative radar unit if one of two conditions is met:

- The flight will be able to remain in day VFR flight conditions, or
- Hazardous weather is not forecast.

An air carrier's operations manual must contain procedures for the flight crew to follow if the weather radar fails in flight.

A ground proximity warning system (GPWS) must be installed on all large turbine powered airplanes. The GPWS gives aural and visual warnings when an aircraft too close to the terrain is in an improper configuration for landing, or when it deviates below glide slope on an ILS approach.

TCAS I (Traffic Alert and Collision Avoidance System) provides proximity warning only, to assist the pilot in the visual acquisition of intruder aircraft. No recommended avoidance maneuvers are provided nor authorized as a result of a TCAS I warning. TCAS II provides traffic advisories (TAs) and resolution advisories (RAs). Resolution advisories provide recommended maneuvers in a vertical direction to avoid conflicting traffic. TCAS does not alter or diminish the pilot's basic authority and responsibility to ensure safe flight. After the conflict, return to the ATC clearance in effect. If a deviation occurs, contact ATC as soon as practical.

Cockpit voice recorders are required on large turbine engine powered airplanes and large four engine reciprocating powered airplanes. The recorder must operate from before the start of the before starting checklist to the completion of the secure cockpit checklist. Although the recorder runs for the entire flight, only the most recent 30 minutes of information need be retained on the recorder tape.

An approved flight recorder must be installed on all airplanes certified for operations above 25,000 feet and on all turbine powered airplanes. What the flight recorder must record varies from airplane to airplane, but at a minimum it must record:

- Time
- Altitude
- Airspeed
- Vertical acceleration
- Heading
- Time of each radio transmission to or from ATC.

Answers

9769 [B]

An air carrier must keep the flight recorder data until an aircraft has been operated at least 25 hours after the data was removed. However, 1 hour of the oldest recorded data may be erased to test the flight recorder.

The cockpit voice and flight recorder data can be used to identify malfunctions and irregularities with the aircraft and in carrying out investigations under NTSB Part 830. It cannot be used by the FAA for enforcement purposes. If an incident occurs which would require the immediate notification of the NTSB, the data must be kept by the operator for at least 60 days.

ALL

9410. Information obtained from flight data and cockpit voice recorders shall be used only for determining

A—who was responsible for any accident or incident.
B—evidence for use in civil penalty or certificate action.
C—possible causes of accidents or incidents.

Information obtained from flight data and cockpit voice recorders is used to assist in determining the cause of accidents or occurrences in connection with investigation under NTSB Part 830. The Administrator does not use the cockpit voice recorder record in any civil penalty or certificate action. (PLT388) — 14 CFR §91.609

Answer (A) is incorrect because flight data or cockpit voice recorders are only used to determine possible causes of accidents or incidents. Answer (B) is incorrect because flight data or cockpit voice recorders may not be used for any civil penalty or certificate action.

ALL

9356. For what purpose may cockpit voice recorders and flight data recorders NOT be used?

A—Determining causes of accidents and occurrences under investigation by the NTSB.
B—Determining any certificate action, or civil penalty, arising out of an accident or occurrence.
C—Identifying procedures that may have been conducive to any accident, or occurrence resulting in investigation under NTSB Part 830.

Information obtained from flight data and cockpit voice recorders is used to assist in determining the cause of accidents or occurrences in connection with investigation under NTSB Part 830. The Administrator does not use the cockpit voice recorder record in any civil penalty or certificate action. (PLT388) — 14 CFR §91.609

Answer (A) is incorrect because cockpit voice recorders and flight data recorders are used to determine causes of accidents or occurrences. Answer (C) is incorrect because flight data recorders and cockpit voice recorders are used to identify any procedures, malfunction, or failure that may have contributed to an accident or occurrence.

ALL

9357. How long is cockpit voice recorder and flight recorder data kept, in the event of an accident or occurrence resulting in terminating the flight?

A—60 days.
B—90 days.
C—30 days.

In the event of an accident or occurrence requiring immediate notification to NTSB Part 830, and that results in the termination of a flight, any operator who has installed approved flight recorders and approved cockpit voice recorders shall keep the recorded information for at least 60 days. (PLT388) — 14 CFR §91.609

121, 135

9428. Each pilot who deviates from an ATC clearance in response to a TCAS II, resolution advisory (RA) is expected to

A—maintain the course and altitude resulting from the deviation, as ATC has radar contact.
B—request ATC clearance for the deviation.
C—notify ATC of the deviation as soon as practicable.

Each pilot who deviates from an ATC clearance in response to a TCAS II RA shall notify ATC of that deviation as soon as practicable and expeditiously return to the current ATC clearance when the traffic conflict is resolved. (PLT195) — AIM ¶4-4-16

Answers

9410 [C] 9356 [B] 9357 [A] 9428 [C]

121, 135, RTC

9425. TCAS I provides

A—traffic and resolution advisories.
B—proximity warning.
C—recommended maneuvers to avoid conflicting traffic.

TCAS I provides proximity warning only, to assist the pilot in the visual acquisition of intruder aircraft. No recommended avoidance maneuvers are provided nor authorized as a result of a TCAS I warning. (PLT195) — AIM ¶4-4-16

Answer (A) is incorrect because traffic and resolution advisories are provided by TCAS II. Answer (C) is incorrect because no recommended avoidance maneuvers are provided nor authorized as a result of a TCAS I warning.

121, 135, RTC

9426. TCAS II provides

A—traffic and resolution advisories.
B—proximity warning.
C—maneuvers in all directions to avoid the conflicting traffic.

TCAS II provides traffic advisories (TAs) and resolution advisories (RAs). (PLT195) — AIM ¶4-4-16

Answer (B) is incorrect because TCAS I provides proximity warning only. Answer (C) is incorrect because resolution advisories provide recommended maneuvers in a vertical direction only to avoid conflicting traffic.

121, 135, RTC

9750. With no traffic identified by TCAS, you

A—can rest assured that no other aircraft are in the area.
B—must continually scan for other traffic in visual conditions.
C—must scan only for hot air balloons.

Traffic data systems are designed to enhance "see and avoid" capabilities. Do not use traffic data systems as a substitute for visual scanning and acquisition of surrounding traffic. (PLT524) – FAA-H-8083-6

Answer (A) is incorrect because TCAS can fail or be affected by power spikes, weather and other onboard aircraft disturbances. Answer (C) is incorrect because TCAS data is supplemental to your traffic awareness for all aircraft.

121, 135, RTC

9427. Each pilot who deviates from an ATC clearance in response to a TCAS advisory is expected to notify ATC and

A—maintain the course and altitude resulting from the deviation, as ATC has radar contact.
B—request a new ATC clearance.
C—expeditiously return to the ATC clearance in effect prior to the advisory, after the conflict is resolved.

Each pilot who deviates from an ATC clearance in response to a TCAS II RA shall notify ATC of that deviation as soon as practicable and expeditiously return to the current ATC clearance when the traffic conflict is resolved. (PLT195) — AIM ¶4-4-16

121, DSP

8150. If an air carrier airplane's airborne radar is inoperative and thunderstorms are forecast along the proposed route of flight, an airplane may be dispatched only

A—when able to climb and descend VFR and maintain VFR/OT en route.
B—in VFR conditions.
C—in day VFR conditions.

No person may dispatch an airplane under IFR or night VFR conditions when current weather reports indicate that thunderstorms, or other potentially hazardous weather conditions that can be detected with airborne weather radar, may reasonably be expected along the route to be flown, unless the weather radar is in satisfactory operating condition. (PLT469) — 14 CFR §121.357

121, DSP

8151. An air carrier airplane's airborne radar must be in satisfactory operating condition prior to dispatch, if the flight will be

A—conducted under VFR conditions at night with scattered thunderstorms reported en route.
B—carrying passengers, but not if it is "all cargo."
C—conducted IFR, and ATC is able to radar vector the flight around areas of weather.

No person may dispatch an airplane under IFR or night VFR conditions when current weather reports indicate that thunderstorms, or other potentially hazardous weather conditions that can be detected with airborne weather radar, may reasonably be expected along the route to be flown, unless the weather radar is in satisfactory operating condition. (PLT469) — 14 CFR §121.357

Answer (B) is incorrect because there is no difference between "all cargo" and "passenger" air carrier operations. The airborne radar

Answers

9425 [B]	9426 [A]	9750 [B]	9427 [C]	8150 [C]	8151 [A]

must be operational prior to dispatch into an area of expected thunderstorms. Answer (C) is incorrect because airborne radar needs to be in operating condition for IFR or night VFR conditions, regardless of ATC's ability to vector the flight around the areas of weather.

121, DSP

8148. What action should be taken by the pilot in command of a transport category airplane if the airborne weather radar becomes inoperative en route on an IFR flight for which weather reports indicate possible thunderstorms?

A—Request radar vectors from ATC to the nearest suitable airport and land.
B—Proceed in accordance with the approved instructions and procedures specified in the operations manual for such an event.
C—Return to the departure airport if the thunderstorms have not been encountered, and there is enough fuel remaining.

No person may dispatch an airplane under IFR or night VFR conditions when current weather reports indicate that thunderstorms, or other potentially hazardous weather conditions that can be detected with airborne weather radar, may reasonably be expected along the route to be flown, unless the weather radar is in satisfactory operating condition. If the airborne radar becomes inoperative en route, the airplane must be operated in accordance with the approved instructions and procedures specified in the operations manual for such an event. (PLT469) — 14 CFR §121.357

Answer (A) is incorrect because radar vectors to land at the nearest suitable airport are not required when airborne weather detection radar malfunctions. Radar vectors to avoid weather would be a wise request. Answer (C) is incorrect because return to the departure airport upon malfunction of airborne weather detection radar would be the correct action if it were the procedure specified in the air carrier's operations manual. However, it is not required by regulation.

121, DSP

8154. Which airplanes are required to be equipped with a ground proximity warning glide slope deviation alerting system?

A—All turbine powered airplanes.
B—Passenger-carrying turbine-powered airplanes only.
C—Large turbine-powered airplanes only.

No person may operate a turbine-powered airplane unless it is equipped with a ground proximity warning/glide slope deviation alerting system. (PLT139) — 14 CFR §121.360

121, DSP

8140. Information recorded during normal operation of a cockpit voice recorder in a large pressurized airplane with four reciprocating engines

A—may all be erased or otherwise obliterated except for the last 30 minutes.
B—may be erased or otherwise obliterated except for the last 30 minutes prior to landing.
C—may all be erased, as the voice recorder is not required on an aircraft with reciprocating engines.

When a cockpit voice recorder is required on an airplane, it must be operated continuously from the start of the use of the checklist (before starting engines for the purpose of flight), to completion of the final checklist at the termination of flight. Information recorded more than 30 minutes earlier may be erased or otherwise obliterated. (PLT405) — 14 CFR §121.359

Answer (B) is incorrect because there is no requirement for information to be retained for 30 minutes after landing. However, under some circumstances involving an accident or occurrence, the certificate holder may be required to retain the information up to 60 days. Answer (C) is incorrect because a cockpit voice recorder is required in large pressurized airplanes with four reciprocating engines.

121, DSP

8141. Which rule applies to the use of the cockpit voice recorder erasure feature?

A—All recorded information may be erased, except for the last 30 minutes prior to landing.
B—Any information more than 30 minutes old may be erased.
C—All recorded information may be erased, unless the NTSB needs to be notified of an occurrence.

When a cockpit voice recorder is required on an airplane, it must be operated continuously from the start of the use of the checklist (before starting engines for the purpose of flight), to completion of the final checklist at the termination of flight. Information recorded more than 30 minutes earlier may be erased or otherwise obliterated. (PLT388) — 14 CFR §121.359

Answer (A) is incorrect because the requirement is that any information more than 30 minutes old may be erased. Answer (C) is incorrect because the requirement is to retain any information that was recorded within the last 30 minutes.

Answers

8148 [B] 8154 [A] 8140 [A] 8141 [B]

121, DSP

8143. A cockpit voice recorder must be operated

A—from the start of the before starting engine checklist to completion of final checklist upon termination of flight.
B—from the start of the before starting engine checklist to completion of checklist prior to engine shutdown.
C—when starting to taxi for takeoff to the engine shutdown checklist after termination of the flight.

When a cockpit voice recorder is required on an airplane, it must be operated continuously from the start of the use of the checklist (before starting engines for the purpose of flight), to completion of the final checklist at the termination of flight. Information recorded more than 30 minutes earlier may be erased or otherwise obliterated. (PLT388) — 14 CFR §121.359

121, DSP

8142. For the purpose of testing the flight recorder system,

A—a minimum of 1 hour of the oldest recorded data must be erased to get a valid test.
B—a total of 1 hour of the oldest recorded data accumulated at the time of testing may be erased.
C—a total of no more than 1 hour of recorded data may be erased.

A total of 1 hour of recorded data may be erased for the purpose of testing a flight recorder or flight recorder system. Any erasure must be of the oldest recorded data accumulated at the time of testing. (PLT388) — 14 CFR §121.343

Answer (A) is incorrect because a maximum of 1 hour of data may be erased for testing. Answer (C) is incorrect because a total of no more than 1 hour of flight recorder data may be erased, but it must be 1 hour of the oldest recorded data.

Communications

Each flag and domestic operator must have a two-way radio system that, under normal conditions, allows reliable and rapid communications between its aircraft and the appropriate dispatch office. For operations within the 48 contiguous states, this system must be independent of any operated by the U.S. government.

One source of current weather information in flight is the **Enroute Flight Advisory Service** (Flight Watch). This service is available nationwide from Flight Service Stations on frequency 122.0 MHz. Flight Watch is limited to weather information only. It is not to be used to open or close flight plans or for any ATC function.

ALL

9258. What type service should normally be expected from an En Route Flight Advisory Service?

A—Weather advisories pertinent to the type of flight, intended route of flight, and altitude.
B—Severe weather information, changes in flight plans, and receipt of position reports.
C—Radar vectors for traffic separation, route weather advisories, and altimeter settings.

The Enroute Flight Advisory Service (Flight Watch) is a weather service on a common frequency of 122.0 MHz from selected FSS's. This service is dedicated specifically to providing weather information to enroute pilots and taking and disseminating pilot reports. (PLT515) — AC 00-45

Answer (B) is incorrect because flight plan changes are services provided through Flight Service Stations, not by EFAS. Answer (C) is incorrect because radar vectors and altimeter settings are services that can be provided by ATC, not EFAS.

ALL

9261. Below FL 180, en route weather advisories should be obtained from an FSS on

A—122.1 MHz.
B—122.0 MHz.
C—123.6 MHz.

The Enroute Flight Advisory Service (Flight Watch) is a weather service on a common frequency of 122.0 MHz from selected FSS's. This service is dedicated specifically to providing weather information to enroute pilots and taking and disseminating pilot reports. (PLT515) — AC 00-45

Answer (A) is incorrect because 122.1 MHz is normally a remote frequency used by FSS at a NAVAID site. A pilot will transmit on 122.1 MHz and the FSS receives on this frequency; the FSS transmits back to the pilot on the NAVAID frequency. Answer (C) is incorrect because 123.6 MHz is the CTAF frequency used by an FSS located on an airport without an operating control tower.

Answers

8143 [A] 8142 [B] 9258 [A] 9261 [B]

ALL

9702. The Federal Aviation Administration's Flight Information Services Data Link (FISDL) is designed to provide data on a common frequency to flight crews from

A—17,500 feet AGL down to 5,000 feet MSL.
B—17,500 feet MSL down to 5,000 feet AGL.
C—5,000 feet MSL to 17,500 feet MSL.

Aeronautical weather and operational information may be displayed in the cockpit through the use of FISDL, and is designed to provide coverage throughout the continental U.S. from 5,000 feet AGL to 17,500 feet MSL, except in those areas where this is unfeasible due to mountainous terrain. (PLT515) — AIM ¶7-1-11

ALL

9712-1. The Federal Aviation Administration's Flight Information Service Data Link (FISDL) provides the following products:

A—METARS, SIGMETS, PIREP's, and AIRMETS.
B—SPECIS, SIGMETS, NOTAM's, and AIRMETS.
C—Convective SIGMETS, PIREPS, AWW's, and NOTAMs.

FAA FISDL provides, free of charge, the following basic products: METARs, SPECIs, TAFs and their amendments, SIGMETs, Convective SIGMETs, AIRMETs, PIREPs and, AWWs issued by the FAA or NWS. (PLT515) — AIM ¶7-1-11

ALL

9712-2. The Federal Aviation Administration's Flight Information Service Data Link (FISDL) products, such as ground radar precipitation maps,

A—may be used instead of the aircraft radar.
B—are not appropriate for finding a path through a weather hazard area.
C—may be used to find a path through a weather hazard area.

FISDL products, such as ground-based radar precipitation maps, are not appropriate for use in tactical severe weather avoidance, such as negotiating a path through a weather hazard area (an area where a pilot cannot reliably divert around hazardous weather, such as a broken line of thunderstorms). FISDL supports strategic weather decision making such as route selection to avoid a weather hazard area in its entirety. The misuse of information beyond its applicability may place the pilot and his/her aircraft in great jeopardy. In addition, FISDL should never be used in lieu of an individual preflight weather and flight planning briefing. (PLT515) — AIM ¶7-1-11

121, DSP

8135. Who must the crew of a domestic or flag air carrier airplane be able to communicate with, under normal conditions, along the entire route (in either direction) of flight?

A—ARINC.
B—Any FSS.
C—Appropriate dispatch office.

Each domestic and flag air carrier must show that a two-way air/ground radio communications system is available at points that will ensure reliable and rapid communications, under normal operating conditions over the entire route (either direct or via approved point to point circuits) between each airplane and the appropriate dispatch office, and between each airplane and the appropriate air traffic control unit. (PLT390) — 14 CFR §121.99

Answer (A) is incorrect because the aircraft must be able to communicate directly with the air carrier dispatch office, not just ARINC. ARINC is a commercial message company which subscribers may use to relay messages, telephone calls, etc. Answer (B) is incorrect because regulations require that the company communications system be independent of any system operated by the FAA or any other third party.

ALL

9783. When should transponders be activated on the ground during taxiing?

A—Only when ATC specifically requests that the transponder to be activated.
B—Any time the airport is operating under IFR.
C—All the time when at an airport with ASDE-X.

If operating at an airport with Airport Surface Detection Equipment - Model X (ASDE-X), transponders should be transmitting "on" with altitude reporting continuously while moving on the airport surface if so equipped. (PLT149) — AIM ¶4-1-20

Answers

9702 [B]	9712-1 [A]	9712-2 [B]	8135 [C]	9783 [C]

ALL

9784. When taxiing on an airport with ASDE-X, you should

A—operate the transponder only when the airport is under IFR or at night during your taxi.

B—operate the transponder with altitude reporting all of the time during taxiing.

C—be ready to activate the transponder upon ATC request while taxing.

If operating at an airport with Airport Surface Detection Equipment - Model X (ASDE-X), transponders should be transmitting "on" with altitude reporting continuously while moving on the airport surface if so equipped. (PLT149) — AIM ¶4-1-20

Navigation Equipment

When an aircraft is flown IFR or VFR Over-the-Top it must have a dual installation of the navigation radios required to fly that route. This means that an aircraft flying Victor airways or jet routes must have two operable VOR systems. Only one ILS system and one marker beacon system is required under Part 121.

When an aircraft is navigating over routes using low frequency, ADF or Radio Range, it only needs one receiver for those NAVAIDs, if it is also equipped with two VOR receivers. If that is the case, the VOR stations must be located such that the aircraft could complete the flight to a suitable airport and make an instrument approach if the low frequency system fails. The airplane must also be fueled to allow for such a failure.

Whenever a different VOR station is tuned, the pilot must listen to the Morse code identification. This will ensure that the correct frequency has been tuned and that a usable signal is available. Occasionally, when a VOR station is undergoing routine maintenance, it will broadcast a signal that is not reliable enough to use for navigation. This condition is indicated in one of two ways. Either the coded ident will be turned off or the ident will be changed to the letters T - E - S - T. Other than the identifier, the station may appear to be broadcasting a normal signal.

To be flown IFR, an aircraft must have had its VORs checked within the past 30 days. The pilots may check the accuracy of the VORs in one of several ways.

The VORs may be checked using a VOT facility on an airport. The VOT broadcasts the 360° radial and so the CDI needle should center either on a setting of 360° with a FROM indication or on 180° with a TO indication. A deviation of ±4° is acceptable for a VOT check.

If a VOT is not available, a VOR checkpoint may be used instead. The aircraft must be moved to the checkpoint and the designated radial set in the CDI course. The acceptable variation for a ground check is ±4°. For an airborne check the allowable variation is ±6°.

If no VOT or VOR check point is available, the VORs may be checked against each other. This is called a "dual VOR check." Tune the VORs to the same station and check the difference in indicated bearing. If they are within 4° of each other, the check is satisfactory. This check can be performed on the ground or in the air.

The person making a VOR check must make an entry in the aircraft log or other record. A proper entry includes the date, place and bearing error. The checker must sign the entry. Besides the VOR check, the altimeter system and the transponder must have been checked within the last 24 calendar months (14 CFR §91.411 and §91.413).

Whenever VOR receivers are required on board an aircraft operating within the United States, it must also have at least one DME receiver on board as well. *Note:* 14 CFR §91.205 requires a DME only if the

Answers

9784 [B]

aircraft is operated above FL240. 14 CFR §121.349 makes the DME required equipment for all air carrier aircraft operating in the U.S. If the DME fails in flight, the pilot must inform ATC as soon as possible.

DME indicates the actual distance from the station to the receiving aircraft in nautical miles. That is different from the horizontal distance because the aircraft is always higher than the DME ground station and altitude is included in the slant range. As a practical matter, the difference between the horizontal distance and the "slant range" is insignificant at distances of more than 10 miles from the station. There is a considerable error close to the station when the aircraft is at high altitudes. In such a situation, almost all of the slant range distance is vertical. When an aircraft passes over a DME station, the distance indicated at station passage is the altitude of the aircraft above the station in nautical miles. For example, if an airplane flew over a VORTAC site 12,000 feet above the station, the DME would indicate 2.0 NM.

A multi-function display (MFD) presents information drawn from a variety of aircraft information systems. The moving map function uses the MFD to provide a pictorial view of the present position of the aircraft, the route programmed into the flight management system, the surrounding airspace, and geographical features. The MFD and moving map can help you maintain the "big picture" and awareness of potential landing sites.

ALL

9019. What would be the identification when a VORTAC is undergoing routine maintenance and is considered unreliable?

A—A test signal, "TESTING," is sent every 30 seconds.
B—Identifier is preceded by "M" and an intermittent "OFF" flag would appear.
C—The identifier would be removed.

During periods of routine or emergency maintenance, coded identification (or code and voice, where applicable) is removed from certain FAA NAVAIDs. During periods of maintenance, VHF ranges may radiate a T-E-S-T code. (PLT300) — AIM ¶1-1-3

Answer (A) is incorrect because a facility may send a T-E-S-T code (not "TESTING") during periods of maintenance. Answer (B) is incorrect because an identifier preceded by "M" designates an identification group for the Microwave Landing System (MLS), a system no longer in operation.

ALL

9020. Which indication may be received when a VOR is undergoing maintenance and is considered unreliable?

A—Coded identification T-E-S-T.
B—Identifier is preceded by "M" and an intermittent "OFF" flag might appear.
C—An automatic voice recording stating the VOR is out-of-service for maintenance.

During periods of routine or emergency maintenance, coded identification (or code and voice, where applicable) is removed from certain FAA NAVAIDs. During periods of maintenance, VHF ranges may radiate a T-E-S-T code. (PLT300) — AIM ¶1-1-3

Answer (B) is incorrect because an identifier preceded by "M" designates an identification group for the Microwave Landing System (MLS), a system no longer in operation. Answer (C) is incorrect because this is used to identify a station and it is removed when the VOR is undergoing maintenance and is considered unreliable.

ALL

9375. What is the maximum permissible variation between the two bearing indicators on a dual VOR system when checking one VOR against the other?

A—4° on the ground and in flight.
B—6° on the ground and in flight.
C—6° in flight and 4° on the ground.

If a dual system VOR (units independent of each other except for the antenna) is installed in the aircraft, the person checking the equipment may check one system against the other. The maximum permissible variation between the two indicated bearings is 4°. (PLT508) — 14 CFR §91.171

Answer (B) is incorrect because 6° is the maximum permissible bearing error when checking a single VOR system against a published radial while in the air, not when checking a dual VOR system. Answer (C) is incorrect because 6° is the maximum permissible bearing error when checking a single VOR system while in the air. Regardless of whether you are on the ground or airborne, the maximum permissible bearing error is only 4° when using a cross-check between dual VORs.

Answers

9019 [C] 9020 [A] 9375 [A]

ALL

9405. During a VOT check of the VOR equipment, the course deviation indicator centers on 356° with the TO/FROM reading FROM. This VOR equipment may

A—be used if 4° is entered on a correction card and subtracted from all VOR courses.
B—be used during IFR flights, since the error is within limits.
C—not be used during IFR flights, since the TO/FROM should read TO.

With the course deviation indicator (CDI) centered, the omni-bearing selector should read 0° (±4°) with the TO/FROM indicator showing FROM or 180° (±4°) with the TO/FROM indicator showing TO. (PLT508) — 14 CFR §91.171

Answer (A) is incorrect because 4° is the maximum permissible bearing error for a VOT check, and no correction card exists for VORs. VORs are either within or not within acceptable limits. Answer (C) is incorrect because a "TO" reading would be indicated if the omni-bearing selector were selected to 180°, not 0°.

ALL

9406. If an airborne checkpoint is used to check the VOR system for IFR operations, the maximum bearing error permissible is

A—plus or minus 6°.
B—plus 6° or minus 4°.
C—plus or minus 4°.

If neither a VOT nor a designated ground checkpoint is available, a pilot may use a designated airborne checkpoint for the VOR check. The maximum permissible bearing error is +6°. (PLT508) — 14 CFR §91.171

Answer (B) is incorrect because the maximum bearing error is ±6. Answer (C) is incorrect because ±4° is the maximum permissible bearing error when using a VOT check or a radio repair facility.

ALL

9376. Which entry shall be recorded by the person performing a VOR operational check?

A—Frequency, radial and facility used, and bearing error.
B—Flight hours and number of days since last check, and bearing error.
C—Date, place, bearing error, and signature.

Each person making the VOR operational check required by regulations shall enter the date, place, bearing error and sign the aircraft log or other record. (PLT508) — 14 CFR §91.171

Answer (A) is incorrect because the frequency and radial used are not required entry items. Answer (B) is incorrect because flight hours and number of days since last check are not required entry items.

ALL

9404. What record shall be made by the pilot performing a VOR operational check?

A—The date, frequency of VOR or VOT, number of hours flown since last check, and signature in the aircraft log.
B—The date, place, bearing error, and signature in the aircraft log or other record.
C—The date, approval or disapproval, tach reading, and signature in the aircraft log or other permanent record.

Each person making the VOR operational check required by regulations shall enter the date, place, bearing error and sign the aircraft log or other record. (PLT508) — 14 CFR §91.171

Answer (A) is incorrect because neither the frequency nor number of hours flown since the last check need to be entered in the log or record. Answer (C) is incorrect because neither the tach reading nor approval or disapproval need to be entered in the record of a VOR operational check.

ALL

9377. Which checks and inspections of flight instruments or instrument systems must be accomplished before an aircraft can be flown under IFR?

A—VOR within 30 days and altimeter systems and transponder within 24 calendar months.
B—ELT test within 30 days, altimeter systems within 12 calendar months, and transponder within 24 calendar months.
C—Airspeed indicator within 24 calendar months, altimeter system within 24 calendar months, and transponder within 12 calendar months.

No person may operate an aircraft under IFR using the VOR system of radio navigation unless the VOR equipment of that aircraft has been operationally checked within the preceding 30 days. No person may operate an airplane in controlled airspace under IFR unless, within the preceding 24 calendar months, each static pressure system, each altimeter instrument, and each automatic pressure altitude reporting system has been tested and inspected. No person may use an ATC transponder required by regulations unless, within the preceding 24 calendar months, it has been tested and inspected. (PLT508) — 14 CFR §91.171, §91.411, §91.413

Answer (B) is incorrect because ELTs do not have to be tested every 30 days, and the altimeter must be checked along with transponder every 24 calendar months (not 12 months). Answer (C) is incorrect because the airspeed indicator is part of the pitot-static system which must be inspected every 24 calendar months and the transponder which must be inspected every 24 calendar months.

Answers

9405 [B] 9406 [A] 9376 [C] 9404 [B] 9377 [A]

ALL

9408. When is DME or suitable RNAV required for an instrument flight?

A—At or above 24,000 feet MSL if VOR navigational equipment is required.
B—In terminal radar service areas.
C—Above 12,500 feet MSL.

If VOR navigational equipment is required, no person may operate a U.S.-registered civil aircraft within the 50 states and District of Columbia, at or above 24,000 feet MSL, unless that aircraft is equipped with approved distance measuring equipment (DME) or a suitable RNAV system. (PLT429) — 14 CFR §91.205

ALL

9023. What DME indications should a pilot observe when directly over a VORTAC site at 12,000 feet?

A—0 DME miles.
B—2 DME miles.
C—2.3 DME miles.

Distance information displayed on DME equipment is slant range from the station in nautical miles. 12,000 feet directly over a VORTAC is almost exactly 2 NM. (PLT202) — FAA-H-8083-15

Answer (A) is incorrect because the DME would indicate 0 DME miles if the DME were sitting on top of the VORTAC site. Answer (C) is incorrect because 2.3 DME miles would be indicated if the airplane were at 13,800 feet (6,000 x 2.3) above the VORTAC site.

ALL

9024. Where does the DME indicator have the greatest error between the ground distance and displayed distance to the VORTAC?

A—High altitudes close to the VORTAC.
B—Low altitudes close to the VORTAC.
C—Low altitudes far from the VORTAC.

Distance information displayed on DME equipment is slant range from the station in nautical miles. The greatest difference between displayed distance and ground distance will occur at high altitudes close to the VORTAC. (PLT202) — FAA-H-8083-15

Answer (B) is incorrect because at low altitudes close to the VORTAC, the slant-range error is less than at high altitudes close to the VORTAC. Answer (C) is incorrect because the slant-range error is at its smallest at low altitudes far from the VORTAC.

ALL

9570. (Refer to Figure 112.) While arcing left on the IAH 10 DME Arc, the pilot experiences a left crosswind component. Where should the bearing pointer be referenced relative to the 90° (wingtip) position to maintain the 10 DME range?

A—On the left wingtip reference.
B—Behind the left wingtip reference.
C—Ahead of the left wingtip reference.

If an aircraft was flying a perfect 10 DME arc to the left in no wind conditions, the RMI bearing would remain on the left wing-tip reference mark indicating that the VOR was exactly 90° to the left of the aircraft's heading. With a left crosswind, the pilot would have to turn the aircraft toward the wind to compensate for the drift to the right. That would place the bearing to the VOR less than 90°, and the bearing pointer would be ahead of the wing-tip reference. (PLT202) — Fly the Wing

121, DSP

8145. When an air carrier flight is operated under IFR or over-the-top on "victor airways," which navigation equipment is required to be installed in duplicate?

A—VOR.
B—ADF.
C—VOR and DME.

No person may operate IFR or Over-the-Top unless the airplane is equipped with the radio equipment necessary for the route, and is able to satisfactorily receive radio navigational signals from all primary en route and approach navigational facilities intended for use, by either of two independent systems. (PLT322) — 14 CFR §121.349

121, DSP

8195. An air carrier operates a flight in VFR over-the-top conditions. What radio navigation equipment is required to be a dual installation?

A—VOR.
B—VOR and ILS.
C—VOR and DME.

No person may operate IFR or Over-the-Top unless the airplane is equipped with the radio equipment necessary for the route and is able to satisfactorily receive radio navigational signals from all primary en route and approach navigational facilities intended for use, by either of two independent systems. (PLT429) — 14 CFR §121.349

Answers

9408 [A]	9023 [B]	9024 [A]	9570 [C]	8145 [A]	8195 [A]

121, DSP

8149. If an air carrier airplane is flying IFR using a single ADF navigation receiver and the ADF equipment fails, the flight must be able to

A—proceed safely to a suitable airport using VOR aids and complete an instrument approach by use of the remaining airplane radio system.
B—continue to the destination airport by means of dead reckoning navigation.
C—proceed to a suitable airport using VOR aids, complete an instrument approach and land.

In the case of IFR operation over routes in which navigation is based on low-frequency radio range or automatic direction finding, only one low-frequency radio range or ADF receiver need be installed if the airplane is equipped with two VOR receivers, and VOR navigational aids are so located and the airplane is fueled so that, in the case of failure of the low-frequency radio range or ADF receiver, the flight may proceed safely to a suitable airport by means of VOR aids and complete an instrument approach by use of the remaining airplane radio system. (PLT429) — 14 CFR §121.349

121, DSP

8147. When a pilot plans a flight using NDB NAVAIDs, which rule applies?

A—The airplane must have sufficient fuel to proceed, by means of one other independent navigation system, to a suitable airport and complete an instrument approach by use of the remaining airplane radio system.
B—The pilot must be able to return to the departure airport using other navigation radios anywhere along the route with 150% of the forecast headwinds.
C—The airplane must have sufficient fuel to proceed, by means of VOR NAVAIDS, to a suitable airport and land anywhere along the route with 150% of the forecast headwinds.

In the case of IFR operation over routes in which navigation is based on low-frequency radio range or automatic direction finding, only one low-frequency radio range or ADF receiver need be installed if the airplane is equipped with two VOR receivers, and VOR navigational aids are so located and the airplane is fueled so that, in the case of failure of the low-frequency radio range or ADF receiver, the flight may proceed safely to a suitable airport by means of VOR aids and complete an instrument approach by use of the remaining airplane radio system. (PLT322) — 14 CFR §121.349

121, DSP

8146. When must an air carrier airplane be DME/suitable RNAV system equipped?

A—In Class E airspace for all IFR or VFR on Top operations.
B—Whenever VOR navigation equipment is required.
C—For flights at or above FL 180.

Whenever VOR navigational receivers are required by regulation, at least one approved distance measuring equipment (DME) unit or suitable RNAV system capable of receiving and indicating distance information from VORTAC facilities must be installed on each airplane when operated in the 50 states and the District of Columbia. (PLT405) — 14 CFR §121.349

Answer (A) is incorrect because DME is only required if VOR equipment is required and not only in Class E airspace. Answer (C) is incorrect because DME is only required if VOR receivers are required.

121, DSP

8152. While on an IFR flight in controlled airspace, the failure of which unit will precipitate an immediate report to ATC?

A—One engine, on a multiengine aircraft.
B—Airborne radar.
C—DME.

If the distance measuring equipment (DME) becomes inoperative enroute, the pilot shall notify ATC of that failure as soon as it occurs. (PLT429) — 14 CFR §121.349

Answers

8149 [A] 8147 [A] 8146 [B] 8152 [C]

ALL

9751. (*See* Figure 242.) The moving map below reflects a loss of

A—position information.
B—the AHRS.
C—the ADC.

Failure indications on the moving map can be quite subtle. The MFD in Figure 242 reflects a loss of position information, indicated by the removal of the aircraft symbol, compass labels, and other subtle differences. (PLT524) — FAA-H-8083-6

Answers (B) and (C) are incorrect because an AHRS or ADC failure would be depicted by red X's on the PFD.

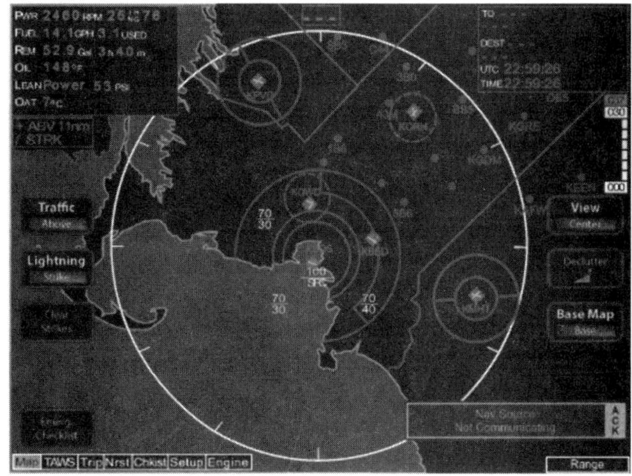

Figure 242

Horizontal Situation Indicator (HSI)

The **Horizontal Situation Indicator (HSI)** is a combination of two instruments: the heading indicator and the VOR. *See* Figure 2-3.

The aircraft heading displayed on the rotating azimuth card under the upper lubber line in Figure 2-2 is 330°. The course-indicating arrowhead that is shown is set to 300°. The tail of the course-indicating arrow indicates the reciprocal, or 120°.

The course deviation bar operates with a VOR/LOC navigation receiver to indicate either left or right deviations from the course that is selected with the course-indicating arrow. It moves left or right to indicate deviation from the centerline in the same manner that the angular movement of a conventional VOR/LOC needle indicates deviation from course.

The desired course is selected by rotating the course-indicating arrow in relation to the azimuth card by means of the course set knob. This gives the pilot a pictorial presentation. The fixed aircraft symbol and the course deviation bar display the aircraft relative to the selected course as though the pilot was above the aircraft looking down.

The TO/FROM indicator is a triangular-shaped pointer. When this indicator points to the head of the course arrow, it indicates that the course selected, and if properly intercepted and flown, will take the aircraft TO the selected facility, and vice versa.

The glide slope deviation pointer indicates the relationship of the aircraft to the glide slope. When the pointer is below the center position, the aircraft is above the glide slope and an increased rate of descent is required.

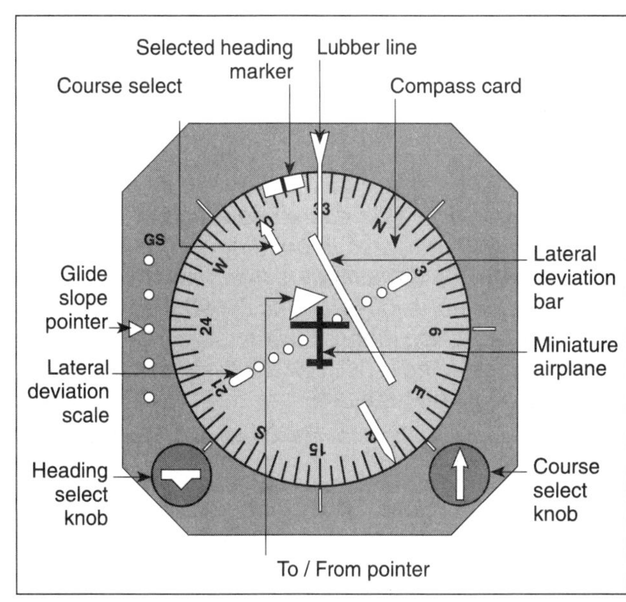

Figure 2-3. Horizontal Situation Indicator (HSI)

Answers

9751 [A]

To orient where the aircraft is in relation to the facility, first determine which radial is selected (look at the arrowhead). Next, determine whether the aircraft is flying to or away from the station (look at the TO/FROM indicator) to find which hemisphere the aircraft is in. Next, determine how far from the selected course the aircraft is (look at the deviation bar) to find which quadrant the aircraft is in. Last, consider the aircraft heading (under the lubber line) to determine the aircraft's position within the quadrant.

Aircraft displacement from course is approximately 200 feet per dot per nautical mile. For example, at 30 NM from the station, 1-dot deflection indicates approximately 1 NM displacement of the aircraft from the course centerline. Therefore, a 2.5-dot deflection at 60 NM would mean the aircraft is approximately 5 NM from the course centerline.

ALL

8999. (Refer to Figures 142 and 143.) To which aircraft position does HSI presentation "D" correspond?

A—4.
B—15.
C—17.

HSI Indicator "D" has a course selection of 180°, and the TO/FROM indicator is pointing to the tail of the course arrow. So the aircraft is flying away FROM the station, and is south of R-270 and R-090. The CDI bar is deflected left, which means the aircraft is west of R-180. The aircraft heading is 180°, which describes position 17. (PLT355) — FAA-H-8083-15

Answer (A) is incorrect because position 4 is to the north of the 270/090 radials, which would require a TO indication. Answer (B) is incorrect because the course deflection bar on position 15 would have a centered deflection bar and a heading of 360°.

ALL

9000. (Refer to Figures 142 and 143.) To which aircraft position does HSI presentation "E" correspond?

A—5.
B—6.
C—15.

HSI Indicator "E" has a course selection of 360°, and the TO/FROM indicator is pointing to the tail of the course arrow. So the aircraft is flying away FROM the station, and is north of R-270 and R-090. The CDI bar is deflected left, which means the aircraft is east of R-180. The aircraft heading is 360°, which describes position 6. (PLT355) — FAA-H-8083-15

Answer (A) is incorrect because position 5 would have a centered deflection bar and a heading of 180°. Answer (C) is incorrect because position 15 is to the south of the R-270 and R-090, which would require a TO indication, and the deflection bar would be centered.

ALL

9001. (Refer to Figures 142 and 143.) To which aircraft position does HSI presentation "F" correspond?

A—10.
B—14.
C—16.

HSI Indicator "F" has a course selection of 180°, and the TO/FROM indicator is pointing to the tail of the course arrow. So the aircraft is flying away FROM the station, and is south of R-270 and R-090. The CDI bar is centered, which means the aircraft is on R-180. The aircraft heading is 045°, which describes position 16. (PLT355) — FAA-H-8083-15

Answer (A) is incorrect because position 10 is north of R-270 and R-090 and east of R-360 and R-180, which would require a TO indication and a right course deflection. Answer (B) is incorrect because position 14 is to the east of R-180, which would require a right course deflection.

ALL

9002. (Refer to Figures 142 and 143.) To which aircraft position does HSI presentation "A" correspond?

A—1.
B—8.
C—11.

HSI Indicator "A" has a course selection of 090°, and the TO/FROM indicator is pointing to the head of the course arrow. So the aircraft is flying TO the station, and is west of R-180 and R-000. The CDI bar is deflected right, which means the aircraft is north of R-270. The aircraft heading is 205°, which describes position 1. (PLT355) — FAA-H-8083-15

Answer (B) is incorrect because position 8 is to the right of R-360 and R-180, which would require a FROM indication. Answer (C) is incorrect because airplane 11 is to the right of R-360 and R-180 and is south of R-270 and R-090, which would require a FROM indication and a left deviation indication.

Answers

8999 [C] 9000 [B] 9001 [C] 9002 [A]

ALL

9003. (Refer to Figures 142 and 143.) To which aircraft position does HSI presentation "B" correspond?

A—9.
B—13.
C—19.

HSI Indicator "B" has a course selection of 270°, and the TO/FROM indicator is pointing to the tail of the course arrow. So the aircraft is flying away FROM the station, and is west of R-180 and R-000. The CDI bar is deflected right, which means the aircraft is south of R-270. The aircraft heading is 135°, which describes position 19. (PLT355) — FAA-H-8083-15

Answer (A) is incorrect because position 9 would require a left course deflection bar indication and a TO indication. Answer (B) is incorrect because position 13 is to the right of R-360 and R-180 and would require a TO indication.

ALL

9004. (Refer to Figures 142 and 143.) To which aircraft position does HSI presentation "C" correspond?

A—6.
B—7.
C—12.

HSI Indicator "C" has a course selection of 360°, and the TO/FROM indicator is pointing to the head of the course arrow. So the aircraft is flying TO the station, and is south of R-270 and R-090. The CDI bar is deflected left, which means the aircraft is east of R-180. The aircraft heading is 310°, which describes position 12. (PLT355) — FAA-H-8083-15

Answer (A) is incorrect because position 6 has a heading of 360° and is north of R-270 and R-090, which would require a FROM indication. Answer (B) is incorrect because position 7 is north of R-270 and R-090 radials, which would require a FROM indication.

ALL

8984. (Refer to Figure 139.) What is the lateral displacement of the aircraft in nautical miles from the radial selected on the No. 1 NAV?

A—5.0 NM.
B—7.5 NM.
C—10.0 NM.

Aircraft displacement from course is approximately 200 feet per dot per nautical mile. For example, at 30 NM from the station, 1-dot deflection indicates approximately 1 NM displacement of the aircraft from the course centerline. Therefore, a 2.5-dot deflection at 60 NM would mean the aircraft is approximately 5 NM from the course centerline. (PLT276) — FAA-H-8083-15

Answer (B) is incorrect because 7.5 NM would be indicated by a displacement of almost 4 dots. Answer (C) is incorrect because 10.0 NM would be indicated by a full deflection.

ALL

8985. (Refer to Figure 139.) On which radial is the aircraft as indicated by the No. 1 NAV?

A—R-175.
B—R-165.
C—R-345.

The No. 1 Nav has a course selection of 350°, and the TO/FROM indicator is pointing to the tail of the course arrow. So the aircraft is flying away FROM the station, and is in the north hemisphere. The CDI bar is deflected right, which means the aircraft is in the northwestern quadrant. The aircraft heading is 140°. The only answer choice in the northwest is R-345. (PLT276) — FAA-H-8083-15

Answer (A) is incorrect because R-175 would require a TO indicator. Answer (B) is incorrect because R-165 would require a TO indicator and a left deflection.

ALL

8986. (Refer to Figure 139.) Which OBS selection on the No. 1 NAV would center the CDI and change the ambiguity indication to a TO?

A—175.
B—165.
C—345.

The No. 1 Nav has a course selection of 350°, and the TO/FROM indicator is pointing to the tail of the course arrow. So the aircraft is flying away FROM the station, and is in the north hemisphere. The CDI bar is deflected 2.5° right and the aircraft heading is 140°, which would put the aircraft on R-345. To center the CDI and change the ambiguity indication to a TO, rotate the OBS to 165° (the reciprocal of R-345). (PLT276) — FAA-H-8083-15

Answers (A) and (C) are incorrect because the aircraft is currently on R-345.

Answers

9003 [C]	9004 [C]	8984 [A]	8985 [C]	8986 [B]

ALL

8987. (Refer to Figure 139.) What is the lateral displacement in degrees from the desired radial on the No. 2 NAV?

A—1°.
B—2°.
C—4°.

Full scale deflection is 10°, so each dot represents 2°. The CDI is displaced two dots. Therefore, the lateral displacement is: 2 dots x 2°/dot = 4°. (PLT276) — FAA-H-8083-15

Answer (A) is incorrect because a 1° lateral displacement would be indicated by a 1/2-dot displacement of the CDI. Answer (B) is incorrect because a 2° lateral displacement would be indicated by a 1-dot displacement of the CDI.

ALL

8988. (Refer to Figure 139.) Which OBS selection on the No. 2 NAV would center the CDI?

A—174.
B—166.
C—335.

Full scale deflection is 10°, so each dot represents 2°. The CDI is displaced two dots (4°). The OBS is set at 170° with a FROM indication, and the aircraft is 4° to the right of course (or on R-174). Simply rotating the OBS to 174° would center the CDI. (PLT276) — FAA-H-8083-15

Answer (B) is incorrect because a right deflection would indicate R-166. Answer (C) is incorrect because the TO-FROM indicator is on FROM, not TO.

ALL

8989. (Refer to Figure 139.) Which OBS selection on the No. 2 NAV would center the CDI and change the ambiguity indication to a TO?

A—166.
B—346.
C—354.

Rotating the OBS to the reciprocal of 170° (350°) under the course arrow will cause the ambiguity indication to change to a TO. The CDI indicates that the aircraft is 4° to the left of course. Rotating the OBS to 354° will center the CDI. (PLT276) — FAA-H-8083-15

Answer (A) is incorrect because a right deflection would currently mean the airplane is on R-166. To change the ambiguity indicator to a TO, 180° must be added to the current radial. Answer (B) is incorrect because a right deflection would mean the airplane is currently on R-166.

ALL

8990. (Refer to Figures 140 and 141.) To which aircraft position(s) does HSI presentation "A" correspond?

A—9 and 6.
B—9 only.
C—6 only.

HSI Indicator "A" is set up with the head of the arrow pointing to 270° (normal sensing). The Course Deviation Indicator is centered; therefore, the aircraft is on the extended centerline of runway #9 and #27. With a heading of 360, Indicator "A" represents an aircraft at position #6 or #9. See the figure below. (PLT355) — FAA-H-8083-15

Answers (B) and (C) are incorrect because the indication will be the same on both the front course and the back course.

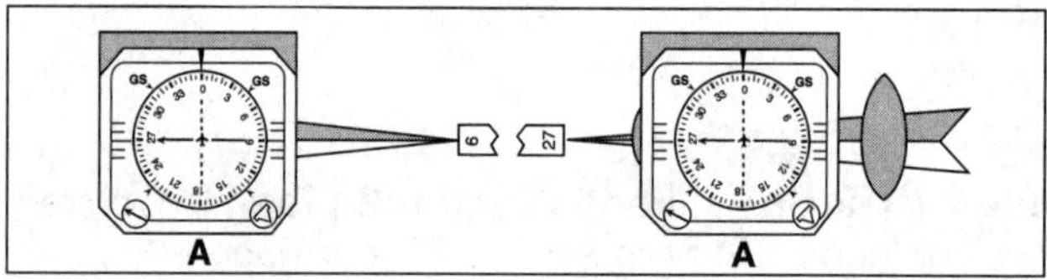

Question 8990

ALL

8991. (Refer to Figures 140 and 141.) To which aircraft position(s) does HSI presentation "B" correspond?

A—11.
B—5 and 13.
C—7 and 11.

HSI Indicator "B" is set up with the head of the arrow pointing to 090° (reverse sensing). The CDI indication is deflected right, which means the aircraft is actually to the south of the extended centerline. Indicator "B" then, with the aircraft flying on a heading of 090°, could be at position #13 and #5. Remember that the local receiver does not know where you are in relationship to the antenna site. See the figure below. (PLT355) — FAA-H-8083-15

Answer (A) is incorrect because position 11 has a 270° heading. Answer (C) is incorrect because positions 7 and 11 have 270° headings.

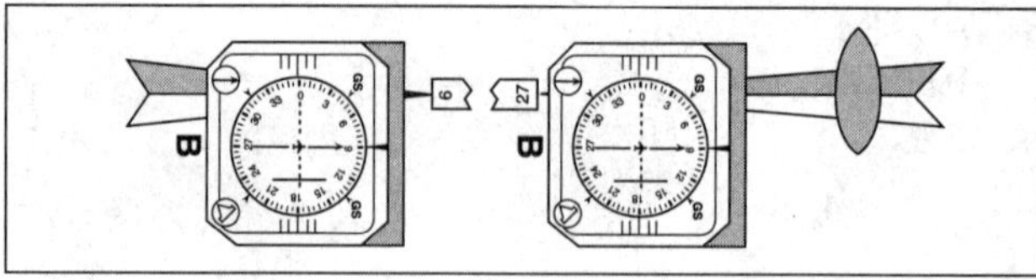

Question 8991

Answers

8987 [C]	8988 [A]	8989 [C]	8990 [A]	8991 [B]

ALL

8992. (Refer to Figures 140 and 141.) To which aircraft position does HSI presentation "C" correspond?

A—9.
B—4.
C—12.

HSI Indicator "C" is set up with the head of the arrow pointing to 090° (reverse sensing). With the CDI centered, the aircraft is on the extended centerline. With a heading of 090°, position #12 is the only one which would have that indication. See the figure below. (PLT355) — FAA-H-8083-15

Answer (A) is incorrect because position 9 has a 360° heading. Answer (B) is incorrect because position 4 has a 270° heading.

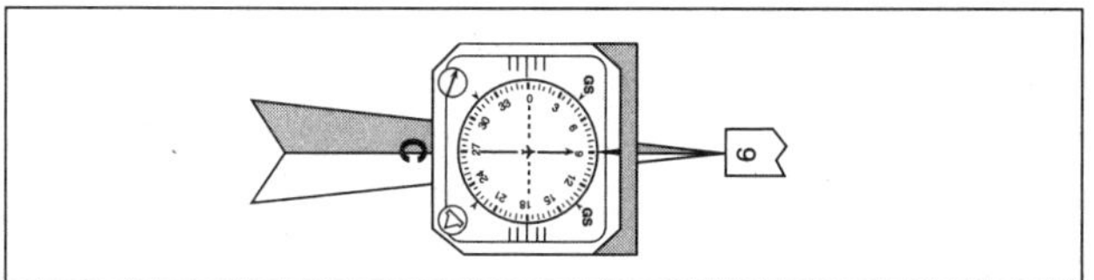

Question 8992

ALL

8993. (Refer to Figures 140 and 141.) To which aircraft position does HSI presentation "D" correspond?

A—1.
B—10.
C—2.

HSI Indicator "D" is set up with the head of the arrow pointing to 090° (reverse sensing). The CDI is deflected right, which means the aircraft is to the south of course. On a heading of 310°, position #2 is the only choice. See the figure below. (PLT355) — FAA-H-8083-15

Answer (A) is incorrect because position 1 is on a 225° heading and is north of the localizer. Answer (B) is incorrect because position 10 has a 135° heading, and is north of the localizer.

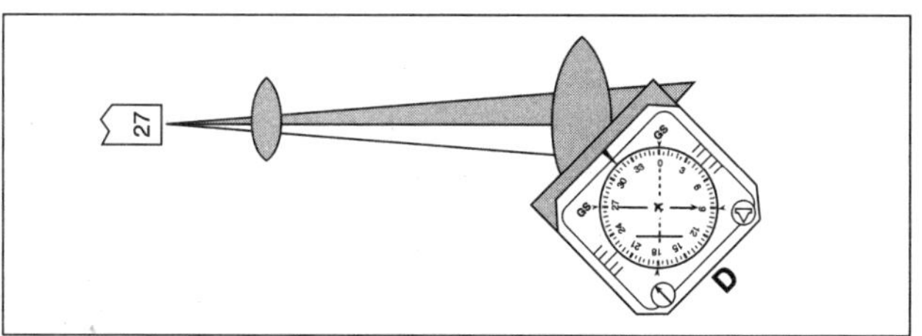

Question 8993

ALL

8994. (Refer to Figures 140 and 141.) To which aircraft position(s) does HSI presentation "E" correspond?

A—8 only.
B—8 and 3.
C—3 only.

HSI Indicator "E" is set up with the head of the arrow pointing to 090° (reverse sensing). With the CDI deflected right, the aircraft is to the south of the extended centerline. On a heading of 045°, position #8 or #3 are the only answers. See the figure below. (PLT355) — FAA-H-8083-15

Answers (A) and (C) are incorrect because both positions 8 and 3 have a 045° heading and are south of the localizer.

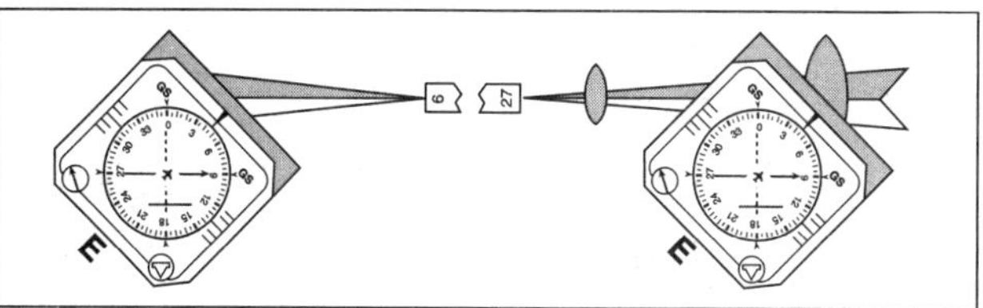

Question 8994

ALL

8995. (Refer to Figures 140 and 141.) To which aircraft position does HSI presentation "F" correspond?

A—4.
B—11.
C—5.

HSI Indicator "F" is set up with the head of the arrow pointing to 270° (normal sensing). The CDI is centered; therefore, the aircraft is on the extended centerline of runway #9 and #27. With a heading of 270°, Indicator "F" represents an aircraft at position #4. See the figure below. (PLT355) — FAA-H-8083-15

Answer (B) is incorrect because position 11 has a left CDI deflection. Answer (C) is incorrect because position 5 has a 090° heading. It also should have a right deflection because it is south of the localizer.

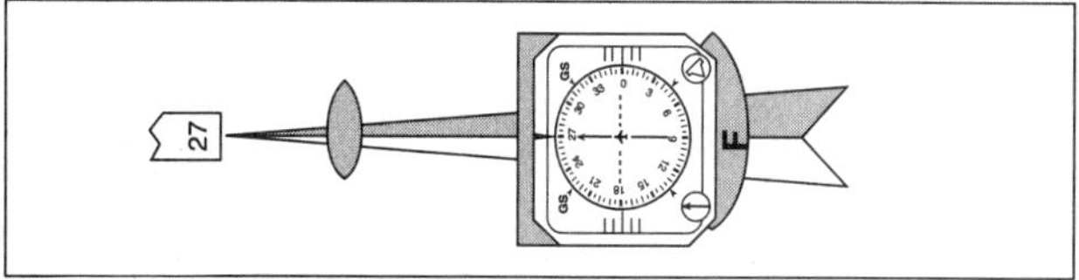

Question 8995

Answers

8992 [C] 8993 [C] 8994 [B] 8995 [A]

ALL

8996. (Refer to Figures 140 and 141.) To which aircraft position(s) does HSI presentation "G" correspond?

A—7 only.
B—7 and 11.
C—5 and 13.

HSI Indicator "G" is set up with the head of the arrow pointing to 270° (normal sensing). The CDI is deflected left; therefore, the aircraft is right of the extended centerline of runway #9 and #27. With a heading of 270°, Indicator "G" represents an aircraft at positions #7 or #11. See the figure below. (PLT355) — FAA-H-8083-15

Answer (A) is incorrect because position 11 is also north of the localizer with a 270° heading. Answer (C) is incorrect because positions 5 and 13 have a 190° heading and are south of the localizer.

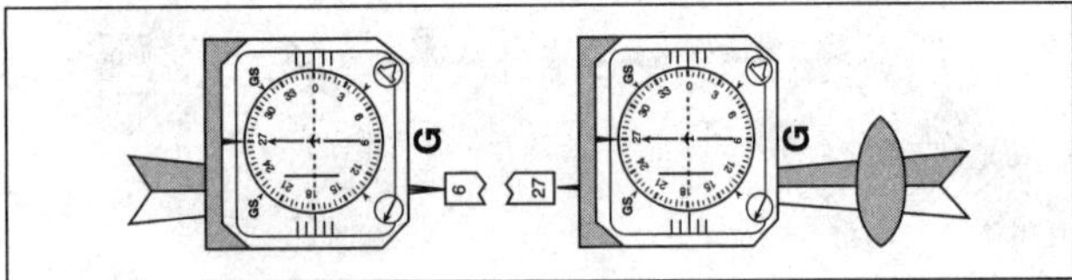

Question 8996

ALL

8997. (Refer to Figures 140 and 141.) To which aircraft position does HSI presentation "H" correspond?

A—8.
B—1.
C—2.

HSI Indicator "H" is set up with the head of the arrow pointing to 270° (normal sensing). The CDI is deflected left; therefore, the aircraft is right of the extended centerline of runway #9 and #27. With a heading of 215°, Indicator "H" represents an aircraft at position #1. See the figure below. (PLT355) — FAA-H-8083-15

Answer (A) is incorrect because position 8 has a heading of 045° and is located south of the localizer. Answer (C) is incorrect because position 2 has a heading of 315° and is located south of the localizer.

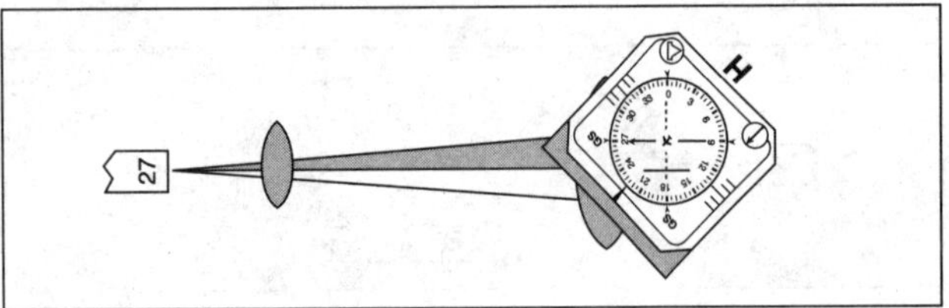

Question 8997

ALL

8998. (Refer to Figures 140 and 141.) To which aircraft position does HSI presentation "I" correspond?

A—4.
B—12.
C—11.

HSI Indicator "I" is set up with the head of the arrow pointing to 090° (reverse sensing). The CDI is deflected left; therefore, the aircraft is north of the extended centerline of runway #9 and #27. With a heading of 270°, Indicator "I" represents an aircraft at positions #7 or #11. See the figure below. (PLT355) — FAA-H-8083-15

Answer (A) is incorrect because position 4 is on the localizer, and has a centered CDI. Answer (B) is incorrect because position 12 is heading 090° and is on the localizer.

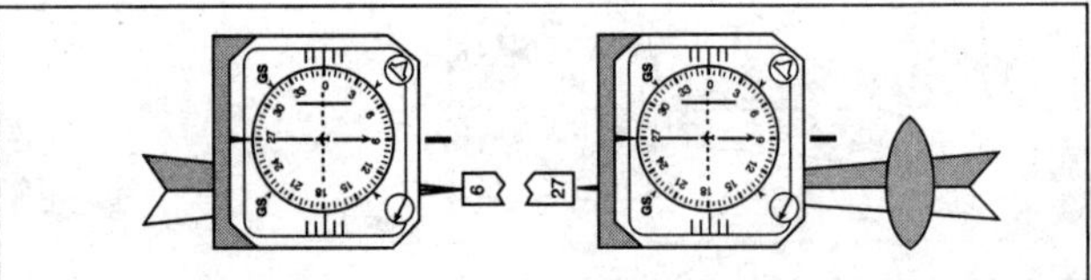

Question 8998

Answers

8996 [B]
8997 [B]
8998 [C]

Radio Magnetic Indicator (RMI)

The compass card shows the aircraft heading at all times under the lubber line. The two needles show the bearings TO and FROM the number 1 and number 2 VORs. The thin needle is usually the number 1 VOR and the double bar needle shows number 2. Often, one or both needles can be selected to display ADF bearing information. The head of each needle shows the magnetic bearing to the station and the tail shows the bearing from (radial). In Figure 2-4, the number 1 needle shows a bearing of 150° TO the station (330° radial) and the number 2 needle shows 255° TO the station (075° radial). *See* Figure 2-4.

To orient where the aircraft is in relation to the facility, first determine which radial is selected to find which quadrant you are in (look at the tail of the needle; if you are trying to orient yourself relative to the VOR, make sure you are using the VOR needle). Next, consider the aircraft heading (under the lubber line) to determine the aircraft's position within the quadrant.

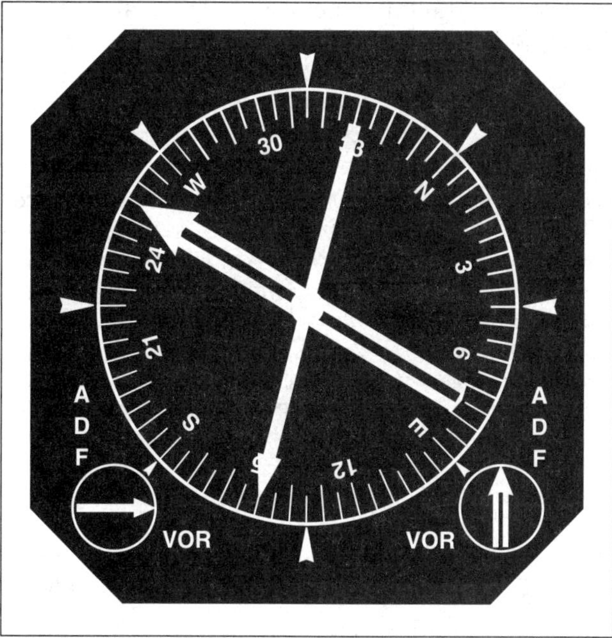

Figure 2-4. Radio Magnetic Indicator (RMI)

ALL

8868. (Refer to Figure 125.) Which RMI illustration indicates the aircraft to be flying outbound on the magnetic bearing of 235° FROM the station? (Wind 050° at 20 knots.)

A—2.
B—3.
C—4.

The magnetic heading of the aircraft is always directly under the index at the top of the instrument. The bearing pointer displays bearings TO the selected station, and the tail displays bearings FROM the station. RMI 3 depicts the magnetic bearing of 235° FROM the station since that is where the tail is pointing. The aircraft heading is also 235° indicating it is tracking outbound on the 235° radial. Since the wind is blowing from 050° at 20 knots (which is a direct tailwind) a wind correction angle would not be required. (PLT091) — FAA-H-8083-15

Answer (A) is incorrect because RMI 2 indicates outbound on the 055° radial from the station. Answer (C) is incorrect because RMI 4 indicates a large wind correction to the right to compensate for a strong crosswind (which does not exist in this question).

ALL

8869. (Refer to Figure 125.) What is the magnetic bearing TO the station as indicated by illustration 4?

A—285°.
B—055°.
C—235°.

The bearing pointer displays bearing TO the selected station. In RMI 4, the needle is pointing to 055°, which is the magnetic bearing TO the station. (PLT091) — FAA-H-8083-15

Answer (A) is incorrect because 285° is the magnetic heading in RMI 4. Answer (C) is incorrect because 235° is the radial from the station that the airplane is crossing in RMI 4.

Answers

8868 [B] 8869 [B]

ALL

8870. (Refer to Figure 125.) Which RMI illustration indicates the aircraft is southwest of the station and moving closer TO the station?

A—1.
B—2.
C—3.

An aircraft southwest of the station and moving closer TO the station would have both the heading and the bearing pointer indicating northeast. This describes RMI 1 which has a heading, and a magnetic bearing TO the station of 055°. (PLT091) — FAA-H-8083-15

Answer (B) is incorrect because RMI 2 indicates northeast of the VOR and flying away (northeast). Answer (C) is incorrect because RMI 3 indicates flying away (to the southwest) from the station.

ALL

8871. (Refer to Figure 125.) Which RMI illustration indicates the aircraft is located on the 055° radial of the station and heading away from the station?

A—1.
B—2.
C—3.

The radial, or magnetic bearing FROM the station, is determined from the tail of the bearing pointer. The aircraft is located on the radial when the aircraft heading matches the radial. RMI 2 is indicating 055° with the tail of the needle, and the aircraft located on that radial with a 055° heading. (PLT091) — FAA-H-8083-15

Answer (A) is incorrect because RMI 1 indicates the airplane is on the 235° radial flying toward the station. Answer (C) is incorrect because RMI 3 indicates the airplane is flying away from the station on the 235° radial.

Long Range Navigation Systems

When an air carrier operates on routes outside of the 48 contiguous states where the aircraft's position cannot be reliably fixed for more than one hour, special rules apply. The aircraft must either be equipped with a "specialized means of navigation" (INS or Doppler Radar), or one of the flight crewmembers must have a current flight navigator certificate. The FAA may also require either a navigator or the specialized navigation on routes which meet the one hour rule if they feel it's necessary. All routes that require either the navigator or specialized means of navigation must be listed in the air carrier's operations specifications.

Certain routes over the North Atlantic Ocean between North America and Europe require better than normal standards of navigation. Appendix C of 14 CFR Part 91 defines these routes and the required navigation standards. The Administrator (the FAA) has the authority to grant a deviation from the navigation standards of Appendix C if an operator requests one.

Inertial Navigation System (INS) is the primary system used by air carriers for over-water navigation. Prior to flight, the pilots enter the present latitude and longitude of the aircraft and the fixes that make up the desired route. The INS constantly updates its position by signals from self contained gyros and accelerometers. The unit then computes the direction and distance to the next fix and displays this information on the aircraft's navigational instruments. The system is completely self-contained and neither needs nor uses signals from any outside navigational source. If the INS gets input of the aircraft's heading and airspeed, it can compute and display the wind and any drift angle. When INS is used as the navigation system, the aircraft must have either two INS units or one INS and Doppler Radar unit (14 CFR §121.355).

Answers

8870 [A] 8871 [B]

ALL

9352. Which publication includes information on operations in the North Atlantic (NAT) Minimum Navigation Performance Specifications Airspace?

A—14 CFR Part 121.
B—ICAO Annex 1, Chapter 2.
C—14 CFR Part 91.

Appendix C of 14 CFR Part 91 establishes performance standards for navigation in the North Atlantic (NAT) airspace. (PLT393) — 14 CFR §91.705

Answer (A) is incorrect because 14 CFR Part 121 provides rules in the United States regarding certification and operations: domestic, flag, and supplemental air carriers and commercial operators of large aircraft. Answer (B) is incorrect because ICAO Annex 1, Chapter 2 concerns licenses and ratings for pilots.

ALL

9353. How may an aircraft operate in North Atlantic (NAT) Minimum Navigation Performance Specifications Airspace with less than the minimum navigation capability required by 14 CFR Part 91, Appendix C?

A—By operating under VFR conditions only.
B—By requesting a deviation from the Administrator.
C—By operating only between 2400Z and 0600Z.

Appendix C of 14 CFR Part 91 establishes performance standards for navigation in the North Atlantic (NAT) airspace. The Administrator authorizes deviations from the requirements of Appendix C. (PLT393) — 14 CFR §91.703

Answer (A) is incorrect because NAT flights, with or without MNPS, may be conducted in IFR weather conditions as well as VFR. Answer (C) is incorrect because NAT flights, with or without MNPS, do not have time restrictions.

ALL

9025. What type navigation system is Inertial Navigation System (INS)? A navigation computer which provides position

A—from information by compass, airspeed, and an input of wind and variation data.
B—from radar-type sensors that measure ground speed and drift angles.
C—by signals from self-contained gyros and accelerometers.

INS is a totally self-contained navigation system, comprised of gyros, accelerometers, and a navigation computer, which provides aircraft position and navigation information in response to signals resulting from inertial effects on system components, and does not require information from external references. (PLT279) — AIM ¶1-1-15

Answer (A) is incorrect because a Doppler radar uses the airplane's compass as a directional reference. It uses radar to detect and measure ground speed and drift angles. Answer (B) is incorrect because a Doppler radar (not INS) provides position information from radar-type sensors that measure ground speed and drift angles.

121, DSP

8196. Routes that require a flight navigator are listed in the

A—Airplane Flight Manual.
B—International Flight Information Manual.
C—Air Carrier's Operations Specifications.

Operations where a flight navigator, special navigation equipment, or both are required, are specified in the operations specifications of the air carrier or commercial operator. (PLT389) — 14 CFR §121.389

121, DSP

8197. Where is a list maintained for routes that require special navigation equipment?

A—Air Carrier's Operations Specifications.
B—International Flight Information Manual.
C—Airplane Flight Manual.

Operations where a flight navigator, special navigation equipment, or both are required, are specified in the operations specifications of the air carrier or commercial operator. (PLT389) — 14 CFR §121.389

Answers (B) and (C) are incorrect because while the International Notices to Airmen and the International Aeronautical Information Manual may contain information on the location and operation of the flight navigation equipment, the air carrier's operations specifications determine the routes in which a flight navigator is required.

Answers

9352 [C] 9353 [B] 9025 [C] 8196 [C] 8197 [A]

121, DSP

8199. A flight navigator or a specialized means of navigation is required aboard an air carrier airplane operated outside the 48 contiguous United States and District of Columbia when

A—operations are conducted IFR or VFR on Top.
B—operations are conducted over water more than 50 miles from shore.
C—the airplane's position cannot be reliably fixed for a period of more than 1 hour.

No certificate holder may operate an airplane outside the 48 contiguous states and the District of Columbia, when its position cannot be reliably fixed for a period of more than one hour, without a flight crewmember who holds a current flight navigator certificate, or unless the aircraft is equipped with an approved specialized means of navigation. (PLT374) — 14 CFR §121.389

Answer (A) is incorrect because whether IFR or VFR-On-Top, the requirement applies if the airplane's position cannot be reliably fixed for more than 1 hour. Answer (B) is incorrect because the requirement applies over water or land if the airplane's position cannot be reliably fixed for more than 1 hour.

121, DSP

8203. An air carrier that elects to use an Inertial Navigational System (INS) must meet which equipment requirement prior to takeoff on a proposed flight?

A—The INS system must consist of two operative INS units.
B—Only one INS is required to be operative, if a Doppler Radar is substituted for the other INS.
C—A dual VORTAC/ILS system may be substituted for an inoperative INS.

If a certificate holder elects to use an Inertial Navigation System (INS) it must be at least a dual system. At least 2 systems must be operational at takeoff. The dual system may consist of either 2 INS units, or 1 INS unit and 1 Doppler radar unit. (PLT429) — 14 CFR §121.355, Part 121 Appendix G

121, DSP

8194. Which equipment requirement must be met by an air carrier that elects to use a dual Inertial Navigation System (INS) on a proposed flight?

A—The dual system must consist of two operative INS units.
B—A dual VORTAC/ILS system may be substituted for an inoperative INS.
C—Only one INS is required to be operative, if a Doppler Radar is substituted for the other INS.

If a certificate holder elects to use an Inertial Navigation System (INS) it must be at least a dual system. At least 2 systems must be operational at takeoff. The dual system may consist of either 2 INS units, or 1 INS unit and 1 Doppler radar unit. (PLT405) — 14 CFR §121.355, Part 121 Appendix G

Answers

8199 [C] 8203 [B] 8194 [C]

Approach Systems

The primary instrument approach system in the United States is the **Instrument Landing System (ILS)**. The system can be divided operationally into three parts: guidance, range and visual information. If any of the elements is unusable, the approach minimums may be raised or the approach may not be authorized at all.

The guidance information consists of the localizer for horizontal guidance and the glide slope for vertical guidance. The localizer operates on one of 40 frequencies from 108.10 MHz to 111.95 MHz. The glide slope operates on one of 40 paired UHF frequencies. The Morse code identifier of the localizer is the letter "I" (• •) followed by three other letters unique to that facility. The portion of the localizer used for the ILS approach is called the front course. The portion of the localizer extending from the far end of the runway is called the back course. The back course may be used for missed approach procedures or for a back course approach if one is published.

Range information is usually provided by 75 MHz marker beacons or, occasionally, by DME. There are four types of marker beacons associated with ILS approaches—the outer marker, the middle marker, the inner marker and the back course marker. Flying over any marker beacon will result in both visual and aural indications. The outer marker is identified by a blue light and continuous dashes in Morse code at a rate of 2 per second. The middle marker is indicated by a flashing amber light and alternating dots and dashes at a rate of 2 per second. The inner marker flashes the white light and sounds continuous dots at 6 per second. The back course marker will also flash the white light and sound a series of 2-dot combinations. *See* Figure 2-5 on the next page.

Often, an ADF facility (called a compass locator) is associated with an ILS approach. Usually it is located at the outer marker, but occasionally it is co-located with the middle marker. An outer compass locator is identified with the first 2 letters of the localizer identification group. A middle compass locator is identified by the last 2 letters of the localizer.

If a middle marker is out of service, the middle compass locator or PAR radar can be substituted. The middle marker being inoperative does not affect minimums during a Category I ILS approach.

The visual information portion of the ILS consists of approach lights, touchdown and centerline lights and runway lights.

The localizer is very narrow. In fact a full scale deflection (CDI moving from the center to full scale left or right) is only about 700 feet at the runway threshold.

Different aircraft will require different rates of descent to stay on glide slope. A good rule of thumb is that the vertical speed in feet per minute will be equal to about five times the ground speed in knots. For example, an aircraft with an approach speed of 140 knots will require a descent rate of about 700 feet per minute (140 x 5 = 700).

The lowest approach minimums that can be used for a normal (Category I) ILS approach are a DH of 200 feet and 1,800 feet RVR. A Category II ILS approach will have minimums as low as a DH of 100 feet and a visibility requirement of 1,200 feet RVR. The approach has to be approved for Category II minimums. In addition to suitable localizer, glide slope and marker beacons, the approach must have certain additional equipment working on the landing runway. This equipment includes an approach light system, High Intensity Runway Lights (HIRL), Touchdown Zone Lights (TDZL), Runway Centerline Lights (CL) and Runway Visual Range (RVR). Radar, VASI and Runway End Identifier Lights (REIL) are not required components of a Category II approach system. To descend below the DH from a Category II approach the pilot must be able to see one of the following:

- The runway threshold;
- The threshold markings;
- The threshold lights;
- The touchdown zone or the touchdown zone markings;
- The touchdown zone lights; or
- The approach light system, except that a pilot may not descend below 100 feet above the touchdown zone unless the red terminating bars or the red side row bars are distinctly visible and identifiable.

The Simplified Directional Facility (SDF) and the Localizer-type Directional Air (LDA) are approach systems that give a localizer-type indication to the pilot, but with some significant differences. The LDA is essentially a localizer, but it is not aligned within 3° of the runway as a localizer must be. The localizer can be any width from 3° to 6° wide. If the LDA is within 30°, straight-in minimums will be published for it; if not, only circling minimums will be published. The SDF may or may not be aligned with the runway. The main difference between it and a localizer is that its width is fixed at either 6° or 12°.

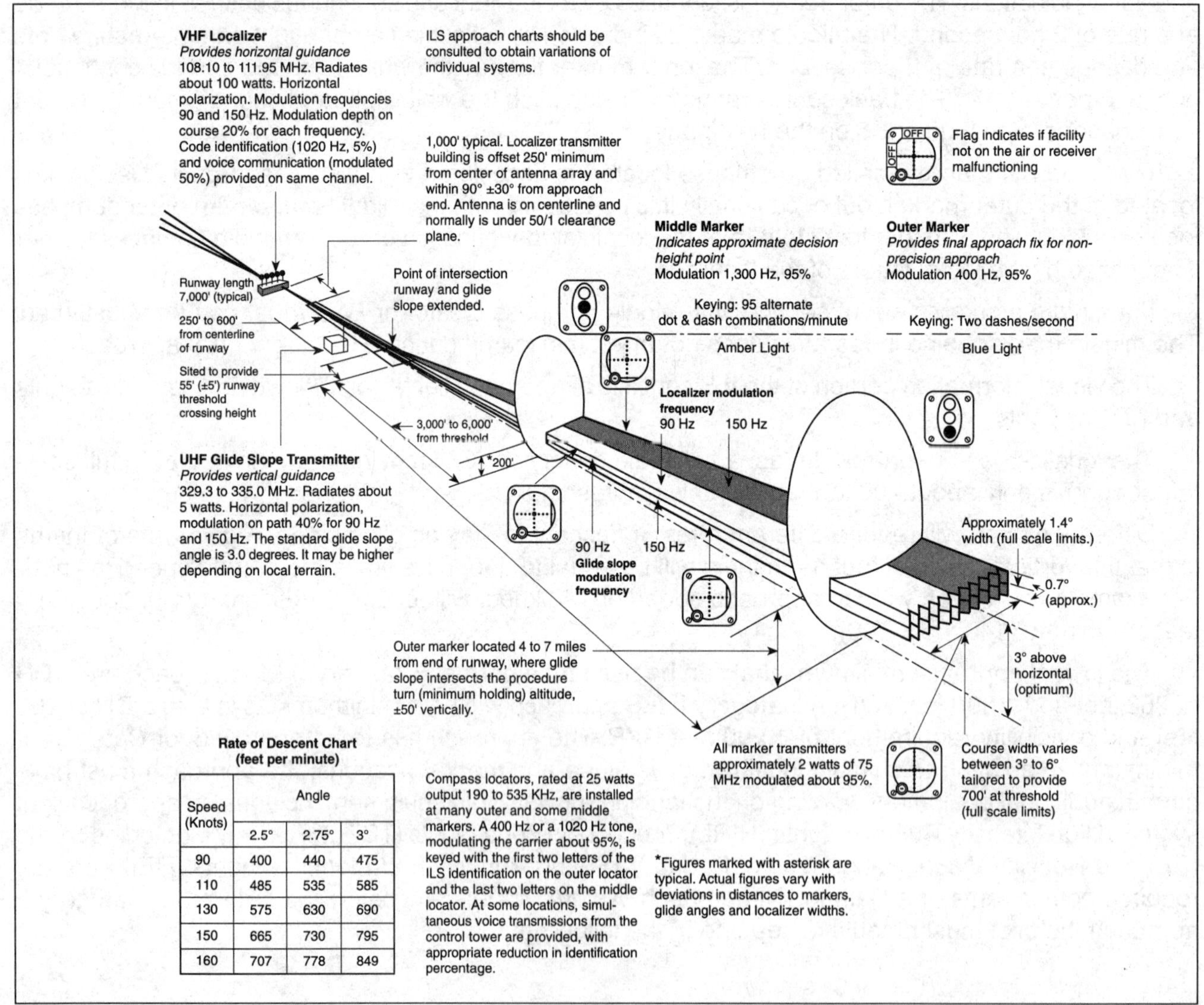

Speed (Knots)	Angle		
	2.5°	2.75°	3°
90	400	440	475
110	485	535	585
130	575	630	690
150	665	730	795
160	707	778	849

Figure 2-5. Instrument Landing System (ILS)

ALL

8961. Within what frequency range does the localizer transmitter of the ILS operate?

A—108.10 to 118.10 MHz.
B—108.10 to 111.95 MHz.
C—108.10 to 117.95 MHz.

The localizer transmitter operates on one of 40 ILS channels within the frequency range of 108.10 to 111.95 MHz. (PLT358) — AIM ¶1-1-9

Answer (A) is incorrect because communications frequencies are above 117.95 MHz. Answer (C) is incorrect because 108.10 to 117.95 MHz is the frequency band in which VORs operate.

ALL

8966. What functions are provided by ILS?

A—Azimuth, distance, and vertical angle.
B—Azimuth, range, and vertical angle.
C—Guidance, range, and visual information.

The ILS system may be divided into three functional parts:

1. *Guidance information—localizer, glide slope;*
2. *Range information—marker beacon, DME; and*
3. *Visual information—approach lights, touchdown and centerline lights, runway lights.*

(PLT356) — AIM ¶1-1-9

Answer (A) is incorrect because azimuth and distance information are provided by a TACAN. Answer (B) is incorrect because a localizer/DME approach provides azimuth and range information.

ALL

8958. What aural and visual indications should be observed over an ILS inner marker?

A—Continuous dots at the rate of six per second.
B—Continuous dashes at the rate of two per second.
C—Alternate dots and dashes at the rate of two per second.

The code and light identifications of marker beacons are as follows:

Marker	***Code***	***Light***
OM	— — —	*BLUE*
MM	• — • —	*AMBER*
IM	• • • •	*WHITE*
BC	• • • •	*WHITE*

(PLT356) — AIM ¶1-1-9

Answer (B) is incorrect because continuous dashes at the rate of two per second indicate the ILS outer marker. Answer (C) is incorrect because alternate dots and dashes at the rate of two per second indicate the ILS middle marker.

ALL

8959. What aural and visual indications should be observed over an ILS middle marker?

A—Continuous dots at the rate of six per second, identified as a high pitch tone.
B—Continuous dashes at the rate of two per second, identified as a low-pitched tone.
C—Alternate dots and dashes identified as a low-pitched tone.

The code and light identifications of marker beacons are as follows:

Marker	***Code***	***Light***
OM	— — —	*BLUE*
MM	• — • —	*AMBER*
IM	• • • •	*WHITE*
BC	• • • •	*WHITE*

(PLT277) — AIM ¶1-1-9

Answer (A) is incorrect because continuous dots at the rate of six per second indicate an ILS inner marker. Answer (B) is incorrect because continuous dashes at the rate of two per second indicate an ILS outer marker.

ALL

8960. What aural and visual indications should be observed over an ILS outer marker?

A—Continuous dots at the rate of six per second.
B—Continuous dashes at the rate of two per second.
C—Alternate dots and dashes at the rate of two per second.

The code and light identifications of marker beacons are as follows:

Marker	***Code***	***Light***
OM	— — —	*BLUE*
MM	• — • —	*AMBER*
IM	• • • •	*WHITE*
BC	• • • •	*WHITE*

(PLT277) — AIM ¶1-1-9

Answer (A) is incorrect because continuous dots at the rate of six per second indicate an ILS inner marker. Answer (C) is incorrect because alternating dots and dashes at the rate of two per second indicate an ILS middle marker.

Answers

8961 [B]	8966 [C]	8958 [A]	8959 [C]	8960 [B]

ALL

8962. If installed, what aural and visual indications should be observed over the ILS back course marker?

A—A series of two dot combinations, and a white marker beacon light.
B—Continuous dashes at the rate of one per second, and a white marker beacon light.
C—A series of two dash combinations, and a white marker beacon light.

The code and light identifications of marker beacons are as follows:

Marker	***Code***	***Light***
OM	— — —	*BLUE*
MM	• — • —	*AMBER*
IM	• • • •	*WHITE*
BC	• • • •	*WHITE*

(PLT277) — AIM ¶1-1-9

Answer (B) is incorrect because this is not a marker indication of any kind, but it most closely resembles an ILS outer marker. Answer (C) is incorrect because this is not marker indication of any kind, but it most closely resembles an ILS middle marker.

ALL

8956. Which component associated with the ILS is identified by the last two letters of the localizer group?

A—Inner marker.
B—Middle compass locator.
C—Outer compass locator.

Compass locators transmit two-letter identification groups. The outer locator transmits the first two letters of the localizer identification group, and the middle locator transmits the last two letters of the localizer identification group. (PLT356) — AIM ¶1-1-9

Answer (A) is incorrect because a simple marker beacon is not identified by letters; only compass locators are so identified. Answer (C) is incorrect because an outer compass locator is identified by the first two letters of the localizer identification group.

ALL

8957. Which component associated with the ILS is identified by the first two letters of the localizer identification group?

A—Inner marker.
B—Middle compass locator.
C—Outer compass locator.

Compass locators transmit two-letter identification groups. The outer locator transmits the first two letters of the localizer identification group, and the middle locator transmits the last two letters of the localizer identification group. (PLT356) — AIM ¶1-1-9

Answer (A) is incorrect because marker beacons are not identified by letters; only compass locators. Answer (B) is incorrect because a middle compass locator is identified by the last two letters of the localizer identification group.

ALL

9403. Which facility may be substituted for the middle marker during a Category I ILS approach?

A—VOR/DME FIX.
B—Surveillance radar.
C—Compass locator.

A compass locator or precision radar may be substituted for the outer or middle marker. (PLT356)—14 CFR §91.175

Answer (A) is incorrect because VOR/DME may be substituted for the outer marker only. Answer (B) is incorrect because surveillance radar may be substituted for the outer marker only.

ALL

8970. If the middle marker for a Category I ILS approach is inoperative,

A—the RVR required to begin the approach in increased by 20%.
B—the DA/DH is increased by 50 feet.
C—the inoperative middle marker has no effect on straight-in minimums.

The middle marker being inoperative does not affect minimums. (PLT277) — 14 CFR §91.175

Answers

8962 [A]   8956 [B]   8957 [C]   9403 [C]   8970 [C]

ALL

8968. When is the course deviation indicator (CDI) considered to have a full-scale deflection?

A—When the CDI deflects from full-scale left to full-scale right, or vice versa.
B—When the CDI deflects from the center of the scale to full-scale left or right.
C—When the CDI deflects from half-scale left to half-scale right, or vice versa.

Full-scale deflection is 5 dots to either side of center. (PLT276) — FAA-H-8083-15

Answers (A) and (C) are incorrect because when the CDI deflects full-scale left to full-scale right (or vice versa), this represents 2 full-scale deflections.

ALL

8969. Which "rule-of-thumb" may be used to approximate the rate of descent required for a 3° glidepath?

A—5 times groundspeed in knots.
B—8 times groundspeed in knots.
C—10 times groundspeed in knots.

The descent rate in feet per minute required to maintain a 3° glide slope is roughly five times your ground speed in knots. (PLT170) — FAA-P-8740-48

Answer (B) is incorrect because 8 times the ground speed in knots would result in a 5% glide slope. Answer (C) is incorrect because 10 times the ground speed in knots would result in a 6% glide slope.

ALL

9749. The rate of descent for a 3.5° angle of descent glidescope is

A—740 ft/min at 105 knots groundspeed.
B—740 ft/min at 120 knots airspeed.
C—740 ft/min at 120 knots groundspeed.

Reference Legend 9. Follow 3.5° to right to 740 ft/min and go up to find the ground speed of 120. (PLT045) — AIM ¶1-1-20

ALL

8971. (Refer to Figures 135 and 138.) Which displacement from the localizer and glide slope at the 1.9 NM point is indicated?

A—710 feet to the left of the localizer centerline and 140 feet below the glide slope.
B—710 feet to the right of the localizer centerline and 140 feet above the glide slope.
C—430 feet to the right of the localizer centerline and 28 feet above the glide slope.

With the CDI two dots to the left, we are to the right of course, and according to FAA Figure 138, at the 1.9 NM location 710 feet to the right of course. The glide slope indication is two dots above the glide slope and at the 1.9 NM mark on FAA Figure 138, that would put you 140 feet above the glide slope. (PLT141) — FAA-H-8083-15

Answer (A) is incorrect because the aircraft is to the right of the localizer and above the glide slope. Answer (C) is incorrect because the 430 feet and 28 feet deviations are at 1,300 feet.

ALL

8972. (Refer to Figures 136 and 138.) Which displacement from the localizer centerline and glide slope at the 1,300-foot point from the runway is indicated?

A—21 feet below the glide slope and approximately 320 feet to the right of the runway centerline.
B—28 feet above the glide slope and approximately 250 feet to the left of the runway centerline.
C—21 feet above the glide slope and approximately 320 feet to the left of the runway centerline.

FAA Figure 136 shows the aircraft is displaced a dot and a half left of course and a dot and a half above the glide slope, eliminating Answer A. Interpolating at the 1,300-foot point, the aircraft is 21 feet above the glide slope and 322 feet left of course. (PLT049) — FAA-H-8083-15

Answer (A) is incorrect because the airplane is above the glide slope, and to the left of the localizer. Answer (B) is incorrect because 28 feet is 2 dots, and 250 feet is less than 1-1/4 dots.

Answers

8968 [B] 8969 [A] 9749 [C] 8971 [B] 8972 [C]

ALL

8973. (Refer to Figures 137 and 138.) Which displacement from the localizer and glide slope at the outer marker is indicated?

A—1,550 feet to the left of the localizer centerline and 210 feet below the glide slope.
B—1,550 feet to the right of the localizer centerline and 210 feet above the glide slope.
C—775 feet to the left of the localizer centerline and 420 feet below the glide slope.

FAA Figure 137 shows the aircraft to be 1 dot below the glide slope and 2 dots left of the localizer, eliminating answer choice B. FAA Figure 138 shows the displacement from the glide slope to be 210 feet and from localizer, 1,550 feet at the outer marker (OM). (PLT276) — FAA-H-8083-15

Answer (B) is incorrect because the airplane is to the left of the localizer, and below the glide slope. Answer (C) is incorrect because at the OM 775 feet is one dot on the localizer, not the glide slope, and 420 feet is two dots on the glide slope.

ALL

8963. The lowest ILS Category II minimums are

A—DH 50 feet and RVR 1,200 feet.
B—DH 100 feet and RVR 1,200 feet.
C—DH 150 feet and RVR 1,500 feet.

The lowest authorized ILS minimums with all required ground and airborne systems components operative, are:

Category I—Decision Height (DH) 200 feet and Runway Visual Range (RVR) 2,400 feet (with touchdown zone and runway centerline lighting, RVR 1,800 Category A, B, C; RVR 2,000 Category D),

Category II—DH 100 feet and RVR 1,200 feet, and

Category IIIA—RVR 700 feet.

(PLT356) — AIM ¶1-1-9

Answer (A) is incorrect because a DH of 50 feet is for Category III operations. Answer (C) is incorrect because a DH of 150 feet is for a pilot's initial Category II authorization (for the initial 6-month period) only and is not the lowest DH for Category II operations.

ALL

9411. Which ground components are required to be operative for a Category II approach in addition to LOC, glide slope, marker beacons, and approach lights?

A—Radar, VOR, ADF, taxiway lead-off lights and RVR.
B—RCLS and REIL.
C—All of the required ground components.

No person may operate a civil aircraft in a Category II or Category III operation unless each ground component required for that operation and the related airborne equipment is installed and operating. (PLT420) — 14 CFR §91.189

Answer (A) is incorrect because radar is not a required Category II ILS ground component. Answer (B) is incorrect because runway end identifier lights (REIL) are used to provide rapid identification of the approach component for Category II ILS.

ALL

9412. When may a pilot descend below 100 feet above the touchdown zone elevation during a Category II ILS instrument approach when only the approach lights are visible?

A—After passing the visual descent point (VDP).
B—When the RVR is 1,600 feet or more.
C—When the red terminal bar of the approach light systems are in sight.

A pilot may descend below the DH on a Category II approach using the approach light system as the sole visual reference. However, the pilot may not descend below 100 feet above touchdown zone elevation (TDZE) using the approach lights as a reference unless the red terminating bars or the red side row bars are also distinctly visible and identifiable. (PLT356) — 14 CFR §91.189

Answer (A) is incorrect because a VDP is not used in conjunction with Category II ILS instrument approaches. Answer (B) is incorrect because, although 1,600 feet may be the required inflight visibility, in order to descend below 100 feet above the touchdown zone elevation based on the approach lights also requires sighting of the red terminating bars.

Answers

8973 [A] 8963 [B] 9411 [C] 9412 [C]

ALL

9413. In addition to the localizer, glide slope, marker beacons, approach lighting, and HIRL, which ground components are required to be operative for a Category II instrument approach to a DH below 150 feet AGL?

A—RCLS and REIL.
B—Radar, VOR, ADF, runway exit lights, and RVR.
C—Each required ground component.

In addition to localizer, glide slope, marker beacons, and approach light system a Category II ILS must have high-intensity runway lights (HIRL), runway centerline lights (RCLS), touchdown zone lights (TDZL), and runway visual range (RVR). (PLT420) — 14 CFR §91.189

Answer (A) is incorrect because runway end identifier lights (REIL) are used to provide rapid identification of the approach end of a runway. Answer (B) is incorrect because radar is not a required ground component for Category II ILS operations.

ALL

8967. How does the LDA differ from an ILS LOC?

A—LDA. 6° or 12° wide, ILS – 3° to 6°.
B—LDA. offset from runway plus 3°, ILS – aligned with runway.
C—LDA. 15° usable off course indications, ILS – 35°.

The LDA is not aligned with the runway. (PLT356) — AIM ¶1-1-9

Answer (A) is incorrect because an SDF (not LDA) is fixed at either 6° or 12° wide. Answer (C) is incorrect because the usable off-course indications are limited to 35° for both types of approaches within 10 NM.

ALL

8965. How does the SDF differ from an ILS LOC?

A—SDF – 6° or 12° wide, ILS – 3° to 6°.
B—SDF – offset from runway plus 4°, ILS – aligned with runway.
C—SDF – 15° usable off course indications, ILS – 35°.

The SDF signal is fixed at either 6° or 12° as necessary to provide maximum flyability and optimum course quality. (PLT361) — AIM ¶1-1-10

Answer (B) is incorrect because an SDF may or may not be aligned with the centerline of the runway. Answer (C) is incorrect because the usable off course indications are limited to 35° for both types of approaches.

ALL

9794. (Refer to Figure 251). You are cleared to HNL and plan to use the RNAV (RNP) RWY 26L approach. Assuming you have received the training, you

A—should be prepared to program the FMS/GPS with the radio frequency to fly this approach.
B—can use the GPS and radio frequency communications to fly this approach to minimums.
C—must know ahead of time whether or not your FMS/GPS has GPS and radius-to-fix capability.

Some RNP approaches have a curved path, also called a radius-to-fix (RF) leg. Since not all aircraft have the capability to fly these arcs, pilots are responsible for knowing ahead of time whether or not they can conduct an RNP approach with an arc. (PLT354) — AIM ¶5-4-18

RTC

9795. (Refer to Figure 253.) You are cleared to LXV in your helicopter and expect to be given the GPS RWY 16 approach. Your helicopter is equipped with an IFR certified WAAS GPS. Your approach minimums will be

A—11,360' MDA and 3/4 mi.
B—11,360' MDA and 1-1/4 mi.
C—11,360' MDA and 6,600 RVR, or 1-1/2 mi.

Helicopters flying conventional (non-copter) SIAP's may reduce the visibility minima to not less than one half the published Category A landing visibility minima, or 1/4 statute mile visibility/1200 RVR, whichever is greater unless the procedure is annotated with "Visibility Reduction by Helicopters NA." (PLT354) — AIM ¶10-1-2

ALL

9796. You arrive at the initial fix for the LPV approach into XYZ. The preflight briefer issued you an unreliable advisory on the approach before you took off. Your avionics indicates good signals and full GPS service is available. You

A—know you can fly the approach down to LPV minimums.
B—cannot use that approach because of the advisory from FSS.
C—must revert to another approach system such as VOR.

Upon commencing an approach at locations with a "WAAS UNRELIABLE" NOTAM, if the WAAS avionics indicate LNAV/VNAV or LPV service is available, then vertical guidance may be used to complete the approach using the displayed level of service. (PLT354) — AIM ¶1-1-20

Answers

9413 [C]	8967 [B]	8965 [A]	9794 [C]	9795 [A]	9796 [B]

GPS

The Global Positioning System (GPS) is a satellite-based radio navigational, positioning, and time transfer system. The GPS receiver verifies the integrity (usability) of the signals received from the GPS satellites through receiver autonomous integrity monitoring (RAIM) to determine if a satellite is providing corrupted information. Without RAIM capability, the pilot has no assurance of the accuracy of the GPS position. If RAIM is not available, another type of navigation and approach system must be used, another destination selected, or the trip delayed until RAIM is predicted to be available on arrival. The authorization to use GPS to fly instrument approaches is limited to U.S. airspace. The use of GPS in any other airspace must be expressly authorized by the appropriate sovereign authority.

If a visual descent point (VDP) is published, it will not be included in the sequence of waypoints. Pilots are expected to use normal piloting techniques for beginning the visual descent. The database may not contain all of the transitions or departures from all runways and some GPS receivers do not contain DPs in the database. The GPS receiver must be set to terminal (±1 NM) course deviation indicator (CDI) sensitivity and the navigation routes contained in the database in order to fly published IFR charted departures and DPs. Terminal RAIM should be automatically provided by the receiver. Terminal RAIM for departure may not be available unless the waypoints are part of the active flight plan rather than proceeding direct to the first destination. Overriding an automatically selected sensitivity during an approach will cancel the approach mode annunciation. The RAIM and CDI sensitivity will not ramp down, and the pilot should not descend to MDA, but fly to the MAWP and execute a missed approach.

It is necessary that helicopter procedures be flown at 70 knots or less since helicopter departure procedures and missed approaches use a 20:1 obstacle clearance surface (OCS), which is double the fixed-wing OCS, and turning areas are based on this speed as well.

The pilot must be thoroughly familiar with the activation procedure for the particular GPS receiver installed in the aircraft and must initiate appropriate action after the missed approach waypoint (MAWP). Activating the missed approach prior to the MAWP will cause CDI sensitivity to immediately change to terminal (±1 NM) sensitivity and the receiver will continue to navigate to the MAWP. The receiver will not sequence past the MAWP. Turns should not begin prior to the MAWP. A GPS missed approach requires pilot action to sequence the receiver past the MAWP to the missed approach portion of the procedure. If the missed approach is not activated, the GPS receiver will display an extension of the inbound final approach course and the ATD will increase from the MAWP until it is manually sequenced after crossing the MAWP.

Any required alternate airport must have an approved instrument approach procedure other than GPS, which is anticipated to be operational and available at the estimated time of arrival and which the aircraft is equipped to fly. Missed approach routings in which the first track is via a course rather than direct to the next waypoint require additional action by the pilot to set the course. Being familiar with all of the inputs required is especially critical during this phase of flight.

ALL

9429. If Receiver Autonomous Integrity Monitoring (RAIM) is not available when setting up for GPS approach, the pilot should

A—continue to the MAP and hold until the satellites are recaptured.
B—proceed as cleared to the IAF and hold until satellite reception is satisfactory.
C—select another type of approach using another type of navigation aid.

If RAIM is not available, another type of navigation and approach system must be used, another destination selected, or the trip delayed until RAIM is predicted to be available on arrival. (PLT354) — AIM ¶1-1-19

ALL

9430. Without Receiver Autonomous Integrity Monitoring (RAIM) capability, the accuracy of the GPS derived

A—altitude information should not be relied upon to determine aircraft altitude.
B—position is not affected.
C—velocity information should be relied upon to determine aircraft groundspeed.

The GPS receiver verifies the integrity (usability) of the signals received from the GPS constellation through RAIM, to determine if a satellite is providing corrupted information. Without RAIM capability, the pilot has no assurance of the accuracy of the GPS position. (PLT354) — AIM ¶1-1-19

ALL

9431. Overriding an automatically selected sensitivity during a GPS approach will

A—cancel the approach mode annunciation.
B—require flying point-to-point on the approach to comply with the published approach procedure.
C—have no affect if the approach is flown manually.

Overriding an automatically selected sensitivity during an approach will cancel the approach mode annunciation. The RAIM and CDI sensitivity will not ramp down, and the pilot should not descend to MDA, but fly to the MAWP and execute a missed approach. (PLT354) — AIM ¶1-1-19

ALL

9432. If a visual descent point (VDP) is published on a GPS approach, it

A—will be coded in the waypoint sequence and identified using ATD.
B—will not be included in the sequence of waypoints.
C—must be included in the normal waypoints.

If a visual descent point (VDP) is published, it will not be included in the sequence of waypoints. Pilots are expected to use normal piloting techniques for beginning the visual descent. (PLT354) — AIM ¶1-1-19

ALL

9722. GPS instrument approach operations, outside the United States, must be authorized by

A—the FAA-approved aircraft flight manual (AFM) or flight manual supplement.
B—a sovereign country or governmental unit.
C—the FAA Administrator only.

The authorization to use GPS to fly instrument approaches is provided by a sovereign country or governmental unit. (PLT354) — AIM ¶1-1-19

ALL

9723. Authorization to conduct any GPS operation under IFR requires that

A—the equipment be approved in accordance with TSO C-115a.
B—the pilot review appropriate weather, aircraft flight manual (AFM), and operation of the particular GPS receiver.
C—air carrier and commercial operators must meet the appropriate provisions of their approved operations specifications.

The GPS operation must be conducted in accordance with the FAA-approved aircraft flight manual (AFM) or flight manual supplement. Flight crewmembers must be thoroughly familiar with the particular GPS equipment installed in the aircraft, the receiver operation manual, and the AFM or flight manual supplement. Air carrier and commercial operators must meet the appropriate provisions of their approved operations specifications. (PLT354) — AIM 1-1-19

Answer (A) is incorrect because the equipment must be approved in accordance with TSO C-129, not TSO C-115a. Answer (B) is incorrect because while the pilot is responsible for reviewing the weather before any flight, this requirement is not specific to GPS operations.

Answers

9429 [C] 9430 [A] 9431 [A] 9432 [B] 9722 [B] 9723 [C]

ALL

9742. A pilot employed by an air carrier and/or commercial operator may conduct GPS/WAAS instrument approaches

A—if they are not prohibited by the FAA-approved aircraft flight manual and the flight manual supplement.
B—only if approved in their air carrier/commercial operator operations specifications.
C—only if the pilot was evaluated on GPS/WAAS approach procedures during their most recent proficiency check.

Air carrier and commercial operators must meet the appropriate provisions of their approved operations specifications. (PLT420) — AIM ¶1-1-20

ALL

9724. Authorization to conduct any GPS operation under IFR requires that

A—the pilot review appropriate weather, aircraft flight manual (AFM), and operation of the particular GPS receiver.
B—air carrier and commercial operators must meet the appropriate provisions of their approved operations specifications.
C—the equipment be approved in accordance with TSO C-115a.

Properly certified GPS equipment may be used as a supplemental means of IFR navigation for domestic enroute, terminal operations, and certain instrument approach procedures (IAPs). This approval permits the use of GPS in a manner that is consistent with current navigation requirements as well as approved air carrier operations specifications. (PLT389) — AIM ¶1-1-19

Answer (A) is incorrect because while the pilot is responsible for reviewing the weather before any flight, this requirement is not specific to GPS operation. Answer (C) is incorrect because the equipment must be approved in accordance with TSO C-129, not TSO C-115a.

ALL

9725. When using GPS for navigation and instrument approaches, a required alternate airport must have

A—an approved instrument approach procedure, besides GPS, that is expected to be operational and available at the ETA.
B—a GPS approach that is expected to be operational and available at the ETA.
C—authorization to fly approaches under IFR using GPS avionics.

Use of a GPS for IFR requires that the avionics necessary to receive all of the ground based facilities appropriate for the route to the destination airport and any required alternate airport must be installed and operational. (PLT354) — AIM ¶1-1-19

Answer (B) is incorrect because the operational nature of GPS is not facility dependent. Answer (C) is incorrect because the GPS equipment, not the pilot-in-command, is authorized for use under IFR.

ALL

9727. A GPS missed approach requires that the pilot take action to sequence the receiver

A—over the MAWP.
B—after the MAWP.
C—just prior to the MAWP.

The pilot must be thoroughly familiar with the activation procedure for the particular GPS receiver installed in the aircraft and must initiate appropriate action after the MAWP. Activating the missed approach prior to the MAWP will cause CDI sensitivity to immediately change to terminal (±1 NM) sensitivity and the receiver will continue to navigate to the MAWP. The receiver will not sequence past the MAWP. Turns should not begin prior to the MAWP. (PLT354) — AIM ¶1-1-19

ALL

9728. If the missed approach is not activated, the GPS receiver will display

A—an extension of the outbound final approach course, and the ATD will increase from the MAWP.
B—an extension of the outbound final approach course.
C—an extension of the inbound final approach course.

A GPS missed approach requires pilot action to sequence the receiver past the MAWP to the missed approach portion of the procedure. If the missed approach is not activated, the GPS receiver will display an extension of the inbound final approach course and the ATD will increase from the MAWP until it is manually sequenced after crossing the MAWP. (PLT354) — AIM ¶1-1-19

Answers

9742 [B]	9724 [B]	9725 [A]	9727 [B]	9728 [C]

ALL

9739. "Unreliable," as indicated in the following GPS NOTAMS: SFO 12/051 SFO WAAS LNAV/VNAV AND LPV MNM UNRELBL WEF0512182025-0512182049 means

A—within the time parameters of the NOTAM, the predicted level of service will not support LPV approaches.
B—satellite signals are currently unavailable to support LPV and LNAV/VNAV approaches.
C—within the time parameters of the NOTAM, the predicted level of service will not support RNAV and MLS approaches.

The term "unreliable" is used in conjunction with GPS NOTAMs; the term is an advisory to pilots indicating the expected level of service may not be available. GPS operation may be NOTAMed UNRELIABLE due to testing or anomalies. (PLT354) — AIM ¶1-1-19

ALL

9743. What does "UNREL" indicate in the following GPS and WAAS NOTAM: BOS WAAS LPV AND LNAV/VNAV MNM UNREL WEF 0305231700 -0305231815?

A—Satellite signals are currently unavailable to support LPV and LNAV/VNAV approaches to the Boston airport.
B—The predicted level of service, within the time parameters of the NOTAM, may not support LPV approaches.
C—The predicted level of service, within the time parameters of the NOTAM, will not support LNAV/VNAV and MLS approaches.

The term UNRELIABLE is used in conjunction with GPS and WAAS NOTAMs for flight planning purposes. The term UNRELIABLE is an advisory to pilots indicating the expected level of WAAS service (LNAV/VNAV, LPV) may not be available. (PLT354) — AIM ¶1-1-20

Answer (A) is incorrect because UNREL indicates the expected level of WAAS service merely ***might not*** *be available, and this states that it is* ***definitely*** *unavailable. Answer (C) is incorrect because MLS approaches are not included in the UNREL advisory.*

ALL

9729. If flying a published GPS departure,

A—the data base will contain all of the transition or departures from all runways.
B—and if RAIM is available, manual intervention by the pilot should not be required.
C—the GPS receiver must be set to terminal course deviation indicator sensitivity.

The GPS receiver must be set to terminal (±1 NM) course deviation indicator (CDI) sensitivity and the navigation routes contained in the data base in order to fly published IFR charted departures and SIDs. Terminal RAIM should be automatically provided by the receiver. Terminal RAIM for departure may not be available unless the waypoints are part of the active flight plan rather than proceeding direct to the first destination. (PLT354) — AIM ¶1-1-19

Answer (A) is incorrect because the data base may ***not*** *contain all of the transitions or departures from all runways and some GPS receivers do not contain SIDs in the data base. Answer (B) is incorrect because certain segments of a SID may require some manual intervention by the pilot, especially when radar vectored to a course or required to intercept a specific course to a waypoint.*

ALL

9729-1. To use a substitute means of guidance on departure procedures, pilots of aircraft with RNAV systems using DME/DME/IRU without GPS input must

A—ensure their aircraft navigation system position is confirmed within 1,000 feet at the start point of takeoff roll.
B—ensure their aircraft navigation system position is confirmed within 2,000 feet of the initialization point.
C—ensure their aircraft navigation system position is confirmed within 1,000 feet of pushback.

For RNAV 1 DPs and STARs, pilots of aircraft without GPS using DME/DME/IRU, must ensure the aircraft navigation system position is confirmed within 1,000 feet at the start point of take-off roll. (PLT354) — AIM ¶5-5-16

ALL

9730. Missed approach routing in which the first track is via a course rather than direct to the next waypoint requires

A—that the GPS receiver be sequenced to the missed approach portion of the procedure.
B—manual intervention by the pilot, but will not be required, if RAIM is available.
C—additional action by the operator to set the course.

Missed approach routings in which the first track is via a course rather than direct to the next waypoint require additional action by the pilot to set the course. Being familiar with all of the inputs required is especially critical during this phase of flight. (PLT354) — AIM ¶1-1-19

Answer (A) is incorrect because a GPS missed approach requires pilot action to sequence the receiver, and routing in which the first track is via a course requires additional action by the operator. Answer (B) is incorrect because manual intervention for GPS missed approach routing is not dependent upon RAIM availability.

Answers

9739 [A] 9743 [B] 9729 [C] 9729-1 [A] 9730 [C]

RTC

9721. Obstacles in most areas where "Copter GPS" instrument approaches are needed, require the approach speed must be limited to

A—80 knots on initial and final segments.
B—60 knots on all segments except the missed approach.
C—70 knots on final and missed approach segments.

As long as the obstacle environment permits, helicopter approaches can be flown at a speed of 70 knots from the initial approach waypoint until reaching the missed approach holding waypoint. It is necessary that helicopter procedures be flown at 70 knots or less since helicopter departure procedures and missed approaches use a 20:1 obstacle clearance surface. (PLT382) — AIM ¶1-1-19

RTC

9726. The maximum speed and obstacle clearance surface (OCS) that a "Copter GPS" standard instrument departure (SID) or departure procedure (DP) is based upon is

A—70 knots and 20:1 OCS.
B—70 knots and 10:1 OCS.
C—60 knots and 20:1 OCS.

As long as the obstacle environment permits, helicopter approaches can be flown at a speed of 70 knots from the initial approach waypoint until reaching the missed approach holding waypoint. It is necessary that helicopter procedures be flown at 70 knots or less since helicopter departure procedures and missed approaches use a 20:1 obstacle clearance surface. (PLT354) — AIM ¶1-1-19

Airport Lighting and Marking

A rotating beacon not only aids in locating an airport at night or in low visibility, it can also help to identify which airport is seen. Civilian airports have a beacon that alternately flashes green and white. A military airport has the same green and white beacon but the white beam is split to give a dual flash of white. A lighted heliport has a green, yellow and white beacon.

FAA Figure 129 shows the basic marking and lighting for a runway with a nonprecision approach. The threshold is marked with 4 stripes on either side of the centerline. 1,000 feet from the threshold, a broad "fixed distance" marker is painted on either side of the centerline (A). The runway lights are white for the entire length of the runway (as are the centerline lights if installed). The threshold is lit with red lights.

FAA Figure 130 shows the somewhat more elaborate ICAO markings for a nonprecision runway. In addition to the fixed distance marker, there are stripes painted on the runway every 500 feet to a distance of 3,000 feet from the threshold. This runway has either High Intensity Runway Lights (HIRL) or Medium Intensity Runway Lights (MIRL) installed. These lights are amber rather than white in the areas within 2,000 feet of the threshold. This gives the pilot a "caution zone" on landing rollout.

FAA Figure 131 shows the lighting and marking for a precision instrument runway. The runway striping has been modified to make it easier to tell exactly how much runway remains. The stripes are still at 500 foot intervals for the 3,000 feet from the threshold. The HIRL or MIRL turns amber for the 2,000 feet closest to the threshold. The centerline lighting has alternating red and white lights from 3,000 feet to 1,000 feet to go, and has all red lights in the 1,000 feet closest to the threshold.

In addition to the markings discussed above, some runways have distance remaining markers. These are simply signs showing the remaining runway in thousands of feet.

Taxi leadoff lights associated with runway centerline lights are green and yellow alternating lights, curving from the centerline of the runway to a point on the exit.

Some runways have Runway End Identifier Lights (REIL) installed at the threshold. These are synchronized flashing lights (usually strobes) placed laterally at either side of the runway threshold. Their purpose is to facilitate identification of a runway surrounded by numerous other lighting systems.

Answers

9721 [C]   9726 [A]

LAHSO is an acronym for "Land And Hold Short Operations." These operations include landing and holding short of an intersecting runway, an intersecting taxiway, or some other designated point on a runway other than an intersecting runway or taxiway. At controlled airports, ATC may clear a pilot to land and hold short. The pilot-in-command has the final authority to accept or decline any land and hold short clearance. The safety and operation of the aircraft remain the responsibility of the pilot. To conduct LAHSO, pilots should become familiar with all available information concerning LAHSO at their destination airport. Pilots should have, readily available, the published Available Landing Distance (ALD) and runway slope information for all LAHSO runway combinations at each airport of intended landing. Additionally, knowledge about landing performance data permits the pilot to readily determine that the ALD for the assigned runway is sufficient for safe LAHSO. If, for any reason, such as difficulty in discerning the location of a LAHSO intersection, wind conditions, aircraft condition, etc., the pilot elects to request to land on the full length of the runway, to land on another runway, or to decline LAHSO, a pilot is expected to promptly inform ATC, ideally even before the clearance is issued. A LAHSO clearance, once accepted, must be adhered to, just as any other ATC clearance, unless an amended clearance is obtained or an emergency occurs. However, a LAHSO clearance does not preclude a rejected landing. The airport markings, signage, and lighting associated with LAHSO consist of a three-part system of yellow hold-short markings, red and white signage and, in certain cases, in-pavement lighting.

ALL

8905. How can a pilot identify a military airport at night?

A—Green, yellow, and white beacon light.
B—White and red beacon light with dual flash of the white.
C—Green and white beacon light with dual flash of the white.

Military airport beacons flash alternately white and green, but are differentiated from civil beacons by a dual peaked (two quick) white flashes between the green flashes. (PLT141) — AIM ¶2-1-9

Answer (A) is incorrect because a sequential green, yellow, and white beacon light identifies a lighted civilian heliport. Answer (B) is incorrect because no type of airfield is marked by a beacon with a red and white light with dual flash of the white.

ALL

8906. How can a pilot identify a lighted heliport at night?

A—Green, yellow, and white beacon light.
B—White and red beacon light with dual flash of the white.
C—Green and white beacon light with dual flash of the white.

A rotating beacon flashing green, yellow and white identifies a lighted heliport. (PLT141) — AIM ¶2-1-9

Answer (B) is incorrect because no type of airfield is marked by a white and red beacon with a dual flash of white. Answer (C) is incorrect because a green and white beacon light with a dual flash of the white identifies a military airfield.

ALL

9421. Holding position signs have

A—white inscriptions on a red background.
B—red inscriptions on a white background.
C—yellow inscriptions on a red background.

Holding position signs are mandatory instruction signs, and mandatory instruction signs have a red background with a white inscription. (PLT141) — AIM ¶2-3-8

ALL

9422. Airport information signs, used to provide destination or information, have

A—yellow inscriptions on a black background.
B—white inscriptions on a black background.
C—black inscriptions on a yellow background.

Information signs have a yellow background with a black inscription. They are used to provide the pilot with information on such things as areas that cannot be seen from the control tower, applicable radio frequencies, and noise abatement procedures. (PLT141) — AIM ¶2-3-12

Answers

8905 [C]	8906 [A]	9421 [A]	9422 [C]

ALL

9735. (Refer to Figure 223.) The "runway hold position" sign denotes

A—an area protected for an aircraft approaching a runway.
B—an entrance to runway from a taxiway.
C—intersecting runways.

Runway holding position signs are located at the holding position on taxiways that intersect a runway or on runways that intersect other runways. (PLT141) — AIM ¶2-3-8

ALL

9735-1. (Refer to Figure 228.) What is the purpose of the runway/runway hold position sign?

A—Denotes entrance to runway from a taxiway.
B—Denotes area protected for an aircraft approaching or departing a runway.
C—Denotes intersecting runways.

Mandatory instruction signs are used to denote an entrance to a runway or critical area and areas where an aircraft is prohibited from entering. The runway holding position sign is located at the holding position on taxiways that intersect a runway or on runways that intersect other runways. (PLT141) — AIM ¶2-3-8

ALL

9735-2. (Refer to Figure 225.) What is the purpose of No Entry sign?

A—Identifies paved area where aircraft are prohibited from entering.
B—Identifies area that does not continue beyond intersection.
C—Identifies the exit boundary for the runway protected area.

The no entry sign prohibits an aircraft from entering an area. Typically, this sign would be located on a taxiway intended to be used in only one direction or at the intersection of vehicle roadways with runways, taxiways or aprons where the roadway may be mistaken as a taxiway or other aircraft movement surface. (PLT141) — AIM ¶2-3-8

Answer (B) is incorrect because this is the purpose of a hold position sign. Answer (C) is incorrect because this is the purpose of the runway boundary sign.

ALL

9735-3. (Refer to Figure 226.) What does the outbound destination sign identify?

A—Identifies entrance to the runway from a taxiway.
B—Identifies runway on which an aircraft is located.
C—Identifies direction to take-off runways.

Outbound destination signs define directions to takeoff runways. (PLT141) — AIM ¶2-3-11

Answer (A) is incorrect because this is a runway marking. Answer (B) is incorrect because this is a runway location sign.

ALL

8901. What is the advantage of HIRL or MIRL on an IFR runway as compared to a VFR runway?

A—Lights are closer together and easily distinguished from surrounding lights.
B—Amber lights replace white on the last 2,000 feet of runway for a caution zone.
C—Alternate red and white lights replace the white on the last 3,000 feet of runway for a caution zone.

Runway edge lights (HIRL or MIRL) are white, except on instrument runways, amber replaces white on the last 2,000 feet or half the runway length, whichever is less, to form a caution zone for landing. (PLT148) — AIM ¶2-1-4

Answer (A) is incorrect because MIRL and HIRL are runway edge light systems, and are not spaced closer together on instrument runways. Answer (C) is incorrect because alternate red and white runway centerline lights are on the last 3,000 feet of a runway to the last 1,000 feet of runway (the last 1,000 feet of runway centerline lights are marked by red lights).

ALL

8902. Identify touchdown zone lighting (TDZL).

A—Two rows of transverse light bars disposed symmetrically about the runway centerline.
B—Flush centerline lights spaced at 50-foot intervals extending through the touchdown zone.
C—Alternate white and green centerline lights extending from 75 feet from the threshold through the touchdown zone.

Touchdown Zone Lighting (TDZL) consists of two rows of transverse light bars disposed symmetrically about the runway centerline in the runway touchdown zone. (PLT148) — AIM ¶2-1-5

Answer (B) is incorrect because flush centerline lights spaced at 50-foot intervals extending the length of the runway, including the touchdown zone, are runway centerline lighting. Answer (C) is incorrect because runway centerline lights extend from 75 feet from the threshold through the touchdown zone and are white, not alternating white and green.

Answers

9735 [C]	9735-1 [C]	9735-2 [A]	9735-3 [C]	8901 [B]	8902 [A]

ALL

8903. Identify runway remaining lighting on centerline lighting systems.

A—Amber lights from 3,000 feet to 1,000 feet, then alternate red and white lights to the end.
B—Alternate red and white lights from 3,000 feet to 1,000 feet, then red lights to the end.
C—Alternate red and white lights from 3,000 feet to the end of the runway.

Centerline lighting systems consist of alternating red and white lights from 3,000 feet remaining to the 1,000-foot point, and all red lights for the last 1,000 feet of the runway. (PLT141) — AIM ¶2-1-5

Answer (A) is incorrect because alternate red and white lights are from 3,000 feet to 1,000 feet, then red lights to the end. Answer (C) is incorrect because runway remaining lighting alternates red and white lights from 3,000 feet to 1,000 feet, and red lights from 1,000 feet to the end of the runway.

ALL

8904. Identify taxi leadoff lights associated with the centerline lighting system.

A—Alternate green and yellow lights curving from the centerline of the runway to the centerline of the taxiway.
B—Alternate green and yellow lights curving from the centerline of the runway to the edge of the taxiway.
C—Alternate green and yellow lights curving from the centerline of the runway to a point on the exit.

Taxiway leadoff lights extend from the runway centerline to a point on an exit taxiway to expedite movement of aircraft from the runway. These lights alternate green and yellow from the runway centerline to the runway holding position or the ILS critical area, as appropriate. (PLT141) — AIM ¶2-1-5

ALL

8907. Identify the runway distance remaining markers.

A—Signs with increments of 1,000 feet distance remaining.
B—Red markers laterally placed across the runway at 3,000 feet from the end.
C—Yellow marker laterally placed across the runway with signs on the side denoting distance to end.

Runway distance remaining markers are signs located along the sides of a runway to indicate the remaining runway distance in increments of 1,000 feet. (PLT141) — AIM ¶2-3-3

Answers (B) and (C) are incorrect because distance remaining markers are along the side of the runway and are black and white.

ALL

8922. (Refer to Figure 129.) What is the runway distance remaining at "A" for a daytime takeoff on runway 9?

A—1,000 feet.
B—1,500 feet.
C—2,000 feet.

The fixed distance marker is located 1,000 feet from the threshold (in this case the end of the runway). (PLT141) — AIM ¶2-3-3

Answer (B) is incorrect because the 1,500 feet distance of runway remaining is not marked on an FAA nonprecision runway. Answer (C) is incorrect because 2,000 feet is the distance remaining on an ICAO (not FAA) nonprecision instrument runway where the runway edge lights are amber.

ALL

8923. (Refer to Figure 130.) What is the runway distance remaining at "A" for a nighttime takeoff on runway 9?

A—1,000 feet.
B—2,000 feet.
C—2,500 feet.

According to the key, a half-shaded circle indicates yellow lights. Each fixed-distance marker marks off 500 feet. Four markers from the departure end to the first yellow light measures 2,000 feet remaining. (PLT141) — AIM ¶2-3-3

Answers (A) and (C) are incorrect because the beginning of yellow runway edge lights on an ICAO nonprecision instrument runway indicates 2,000 feet of runway remaining.

ALL

8924. (Refer to Figure 130.) What is the runway distance remaining at "B" for a daytime takeoff on runway 9?

A—2,000 feet.
B—2,500 feet.
C—3,000 feet.

Each fixed-distance marker measures 500 feet. From the departure end, it is 3,000 feet. (PLT141) — AIM ¶2-3-3

Answer (A) is incorrect because 2,000 feet would be the third distance marker encountered. At night, 2,000 feet is identified by the beginning of amber runway edge lights. Answer (B) is incorrect because 2,500 feet would be the second distance marker encountered.

Answers

8903 [B]	8904 [C]	8907 [A]	8922 [A]	8923 [B]	8924 [C]

ALL

8925. (Refer to Figure 130.) What is the runway distance remaining at "C" for a daytime takeoff on runway 9?

A—2,500 feet.
B—2,000 feet.
C—1,500 feet.

Since each fixed-distance marker in this problem represents 500 feet, the distance to "C" from the departure end is 2,000 feet. (PLT141) — AIM ¶2-3-3

Answer (A) is incorrect because 2,500 feet would be the second distance marker encountered. Answer (C) is incorrect because 1,500 feet would be the fourth distance marker encountered.

ALL

8926. (Refer to Figure 130.) What is the runway distance remaining at "D" for a daytime takeoff on runway 9?

A—500 feet.
B—1,000 feet.
C—1,500 feet.

The fixed-distance marker at "D" corresponds to 1,000 feet remaining. (PLT141) — AIM ¶2-3-3

Answer (A) is incorrect because 500 feet would be the sixth distance marker encountered. Answer (C) is incorrect because 1,500 feet would be the fourth distance marker encountered.

ALL

8927. (Refer to Figure 131.) What is the runway distance remaining at "E" for a daytime takeoff on runway 9?

A—1,500 feet.
B—2,000 feet.
C—2,500 feet.

Each fixed-distance marker marks off 500 feet. "E" is 2,000 feet from the departure end. (PLT141) — AIM ¶2-3-3

Answer (A) is incorrect because 1,500 feet would be the second pair of double markers encountered on either side of the centerline. Answer (C) is incorrect because 2,500 feet would be the second pair of single markers encountered on either side of the centerline.

ALL

8928. (Refer to Figure 131.) What is the runway distance remaining at "A" for a nighttime takeoff on runway 9?

A—2,000 feet.
B—3,000 feet.
C—3,500 feet.

This question and figure reference remaining runway lighting/centerline lighting systems in the final 2,000 feet as viewed from the takeoff or approach position. Alternate red and white lights are seen from the 3,000-foot points to the 1,000-foot points, and all red lights are seen for the last 1,000 feet of the runway. (PLT141) — AIM ¶2-1-5

Answer (A) is incorrect because 2,000 feet is marked by the beginning of amber runway edge lights. Answer (C) is incorrect because the runway remaining lights begin alternating between red and white at 3,000 feet of remaining runway.

ALL

8929. (Refer to Figure 131.) What is the runway distance remaining at "D" for a daytime takeoff on runway 9?

A—3,000 feet.
B—2,500 feet.
C—1,500 feet.

Each fixed-distance marker marks off 500 feet. "D" corresponds to 3,000 feet. (PLT141) — AIM ¶2-3-3

Answer (B) is incorrect because 2,500 feet would be the second pair of single markers encountered on either side of the centerline. Answer (C) is incorrect because 1,500 feet would be the second pair of double markers encountered on either side of the centerline.

ALL

8930. (Refer to Figure 131.) What is the runway distance remaining at "B" for a nighttime takeoff on runway 9?

A—1,000 feet.
B—2,000 feet.
C—2,500 feet.

The runway edge lights are white, except on instrument runways. There, amber replaces white on the last 2,000 feet or half the runway length, whichever is less, to form a caution zone for landings. (PLT141) — AIM ¶2-1-4

Answer (A) is incorrect because 1,000 feet would be indicated by the start of red centerline lighting. Answer (C) is incorrect because the beginning of yellow runway edge lights on an instrument runway indicates 2,000 feet of remaining runway.

Answers

8925 [B]	8926 [B]	8927 [B]	8928 [B]	8929 [A]	8930 [B]

ALL

8931. (Refer to Figure 131.) What is the runway distance remaining at "F" for a daytime takeoff on runway 9?

A—2,000 feet.
B—1,500 feet.
C—1,000 feet.

Each fixed-distance marker marks off 500 feet. "F" corresponds to the 1,000-foot fixed-distance marker. (PLT141) — AIM ¶2-3-3

Answer (A) is incorrect because 2,000 feet would be the first pair of double markers encountered on either side of the centerline. At night it is marked by the beginning of amber runway edge lights. Answer (B) is incorrect because 1,500 feet would be the second pair of double markers encountered on either side of the centerline.

ALL

8932. (Refer to Figure 131.) What is the runway distance remaining at "C" for a nighttime takeoff on runway 9?

A—1,000 feet.
B—1,500 feet.
C—1,800 feet.

All red lights along the runway centerline correspond to the last 1,000 feet of runway. (PLT141) — AIM ¶2-1-5

Answer (B) is incorrect because the start of red lights on centerline lighting indicates 1,000 feet of remaining runway. Answer (C) is incorrect because the start of red lights on centerline lighting indicates 1,000 feet of remaining runway.

ALL

8914. What is the purpose of REIL?

A—Identification of a runway surrounded by a preponderance of other lighting.
B—Identification of the touchdown zone to prevent landing short.
C—Establish visual descent guidance information during an approach.

Runway End Identifier Lights (REIL) are effective for:

1. *Identification of a runway surrounded by numerous other lighting systems,*
2. *Identification of a runway which lacks contrast with surrounding terrain, or*
3. *Identification of a runway during reduced visibility.*

(PLT145) — AIM ¶2-1-3

Answer (B) is incorrect because the touchdown zone is identified by in-runway lighting of two rows of transverse light bars on either side of the runway centerline from 100 feet to 3,000 feet from the landing threshold. Answer (C) is incorrect because a VASI (not REIL) assists in providing visual descent guidance information during an approach.

ALL

8915. Identify REIL.

A—Amber lights for the first 2,000 feet of runway.
B—Green lights at the threshold and red lights at far end of runway.
C—Synchronized flashing lights laterally at each side of the runway threshold.

The REIL system consists of a pair of synchronized flashing lights located laterally on each side of the runway threshold. (PLT145) — AIM ¶2-1-3

Answer (A) is incorrect because amber lights are used on the last 2,000 feet of runway edge lights to form a caution zone on instrument runways. Answer (B) is incorrect because green lights at the threshold mark the runway edge for landing aircraft and red lights at the far end mark the runway edge to a departing or landing aircraft.

ALL

9731. Land and Hold Short Operations (LAHSO) include landing and holding short:

A—of an intersecting taxiway only.
B—of some designated point on the runway.
C—only of an intersecting runway or taxiway.

Land And Hold Short Operations (LAHSO) include landing and holding short of an intersecting runway, an intersecting taxiway, or some other designated point on a runway other than an intersecting runway or taxiway. (PLT140) — AIM ¶4-3-11

ALL

9732. A Land and Hold Short Operations (LAHSO) clearance, that the pilot accepts:

A—must be adhered to.
B—does not preclude a rejected landing.
C—precludes a rejected landing.

A LAHSO clearance, once accepted, must be adhered to unless an amended clearance is obtained or an emergency occurs. However, a LAHSO clearance does not preclude a rejected landing. (PLT140) — AIM ¶4-3-11

Answers

8931 [C]	8932 [A]	8914 [A]	8915 [C]	9731 [B]	9732 [B]

ALL

9733. In conducting Land and Hold Short Operations (LAHSO), the pilot should have readily available:

A—the published Available Landing Distance (ALD), landing performance of the aircraft, and slope of all LAHSO combinations at the destination airport.
B—the published runway length and slope for all LAHSO combinations at the airport of intended landing.
C—the landing performance of the aircraft, published Available Landing Distance (ALD) for all LAHSO combinations at the airport of intended landing, plus the forecast winds.

To conduct LAHSO, pilots should become familiar with all available information concerning LAHSO at their destination airport. Pilots should have, readily available, the published Available Landing Distance (ALD) and runway slope information for all LAHSO runway combinations at each airport of intended landing. Additionally, knowledge about landing performance data permits the pilot to readily determine that the ALD for the assigned runway is sufficient for safe LAHSO. (PLT140) — AIM ¶4-3-11

ALL

9734. The airport markings, signage and lighting associated with Land and Hold Short (LAHSO) consists of:

A—yellow hold-short markings, red and white signage, and in-pavement lights.
B—red and white signage, yellow hold-short markings, and at some airports, in-pavement lights.
C—red and black signage, in-pavement lights, and yellow hold-short markings.

The airport markings, signage, and lighting associated with LAHSO consist of a three-part system of yellow hold-short markings, red and white signage and, in certain cases, in-pavement lighting. (PLT140) — AIM ¶4-3-11

ALL

9416-1. (Refer to Figure 224.) The ILS critical area markings denote

A—where you are clear of the runway.
B—where you must be to start your ILS procedure.
C—where you are clear of the ILS critical area.

The ILS critical area sign is located adjacent to the ILS holding position marking on the pavement and can be seen by pilots leaving the critical area. The sign is intended to provide pilots with another visual cue which they can use as a guide in deciding when they are clear of the ILS critical area. (PLT141) — AIM ¶2-3-9

ALL

9416-2. The ILS critical area sign indicates

A—where aircraft are prohibited.
B—the edge of the ILS critical area.
C—the exit boundary.

The ILS critical area sign is located adjacent to the ILS holding position marking on the pavement and can be seen by pilots leaving the critical area. The sign is intended to provide pilots with another visual cue to use as a guide in deciding when they are clear of the ILS critical area. (PLT141) — AIM ¶2-3-9

ALL

9423-1. (Refer to Figure 227.) The "taxiway ending" marker

A—identifies area where aircraft are prohibited.
B—indicates taxiway does not continue.
C—provides general taxiing direction to named taxiway.

Taxiway ending marker is an airport sign indicating the taxiway does not continue. (PLT141) — AIM ¶2-3-11

Answer (A) is incorrect because this is the purpose of a no entry sign. Answer (C) is incorrect because this is the purpose of direction signs.

121, 135, RTC

9423. Hold line markings at the intersection of taxiways and runways consist of four lines (two solid and two dashed) that extend across the width of the taxiway. These lines are

A—white in color and the dashed lines are nearest the runway.
B—yellow in color and the dashed lines are nearest the runway.
C—yellow in color and the solid lines are nearest the runway.

Holding position markings for taxiway/runway intersections consist of four yellow lines—two solid and two dashed. The solid lines are always on the same side where the aircraft is to hold. (PLT141) — AIM ¶2-3-5

Answers

9733 [A]	9734 [B]	9416-1 [C]	9416-2 [B]	9423-1 [B]	9423 [B]

121, 135, RTC

9436. (Refer to Figure 156.) This sign, which is visible to the pilot on the runway, indicates

A—a point at which the pilot should contact ground control without being instructed by the tower.
B—a point at which the aircraft will be clear of the runway.
C—the point at which the emergency arresting gear is stretched across the runway.

The runway boundary sign has a yellow background with a black inscription with a graphic depicting the pavement holding position. This sign, which faces the runway and is visible to the pilot exiting the runway, is located adjacent to the holding position marking on the pavement. The sign is intended to provide pilots with another visual cue which they can use as a guide in deciding when they are "clear of the runway." (PLT141) — AIM ¶2-3-9

121, 135, RTC

9417. You have just landed at JFK and the tower tells you to call ground control when clear of the runway. You are considered clear of the runway when

A—the aft end of the aircraft is even with the taxiway location sign.
B—the flight deck area of the aircraft is even with the hold line.
C—all parts of the aircraft have crossed the hold line.

An aircraft is not "clear of the runway" until all parts have crossed the applicable holding position marking. (PLT141) — AIM ¶2-3-5

ALL

9764. Taxiway Centerline Lead-Off Lights are color coded to warn pilots that

A—they are within the runway environment or run-up danger critical area.
B—they are within the runway environment or ILS/MLS critical area.
C—they are within the taxiway end environment or ILS/MLS critical area.

Taxiway centerline lead-off lights provide visual guidance to persons exiting the runway. They are color-coded to warn pilots and vehicle drivers that they are within the runway environment or ILS/MLS critical area, whichever is more restrictive. Alternate green and yellow lights are installed, beginning with green, from the runway centerline to one centerline light position beyond the runway holding position or ILS/MLS critical area holding position. (PLT141) — AIM ¶2-1-5

ALL

9772. Taxi lead-off lights associated with the centerline lighting system

A—alternate green and yellow lights curving from the centerline of the runway to the centerline of the taxiway.
B—alternate green and yellow lights curving from the centerline of the runway to the beginning of the taxiway.
C—alternate green and yellow lights curving from the centerline of the runway to the edge of the taxiway.

Taxiway centerline lead-off lights provide visual guidance to persons exiting the runway. They are color-coded to warn pilots and vehicle drivers that they are within the runway environment or ILS critical area, whichever is more restrictive. Alternate green and yellow lights are installed, beginning with green, from the runway centerline to one centerline light position beyond the runway holding position or ILS critical area holding position. (PLT141) — AIM ¶2-3-9

ALL

9785. THL is the acronym for

A—Takeoff hold lights.
B—Taxi holding lights.
C—Terminal holding lights.

The Takeoff Hold Lights (THL) system is composed of in-pavement, unidirectional fixtures in a double longitudinal row aligned either side of the runway centerline lighting. Fixtures are focused toward the arrival end of the runway at the "position and hold" point, and they extend for 1,500 feet in front of the holding aircraft. Illuminated red lights provide a signal, to an aircraft in position for takeoff or rolling, indicating that it is unsafe to takeoff because the runway is occupied or about to be occupied by another aircraft or ground vehicle. Two aircraft, or a surface vehicle and an aircraft, are required for the lights to illuminate. The departing aircraft must be in position for takeoff or beginning takeoff roll. Another aircraft or a surface vehicle must be on or about to cross the runway. (PLT141) — AIM ¶2-1-6

Answers

9436 [B] 9417 [C] 9764 [B] 9772 [C] 9785 [A]

ALL

9786. REL is the acronym for

A—Runway exit lights.
B—Runway entrance lights.
C—Ramp entry lights.

The Runway Entrance Lights (REL) system is composed of flush mounted, in-pavement, unidirectional fixtures that are parallel to and focused along the taxiway centerline and directed toward the pilot at the hold line. A specific array of REL lights include the first light at the hold line followed by a series of evenly spaced lights to the runway edge; and one additional light at the runway centerline in line with the last two lights before the runway edge. When activated, these red lights indicate that there is high-speed traffic on the runway or there is an aircraft on final approach within the activation area. (PLT141) — AIM ¶2-1-6

ALL

9787. (Refer to Figure 241). Hot Spots are depicted on airport diagrams as

A—squares or rectangles around "HS" and a number.
B—circles or polygons around "HS" and a number.
C—triangles or blocks filled with "HS" and a number.

Runway hotspots (some FAA Regions refer to them as high alert areas) are locations on particular airports that historically have hazardous intersections. Hotspots are depicted on some airport charts as circled areas. (PLT149) — FAA-H-8261-1

121, 135, RTC

9437. (Refer to Figure 157.) This is an example of

A—an ILS Critical Area Holding Position Sign.
B—a Runway Boundary Sign.
C—an ILS Critical Area Boundary Sign.

This sign has a yellow background with a black inscription with a graphic depicting the ILS pavement holding position marking. This is located adjacent to the ILS holding position marking on the pavement and can be seen by the pilots leaving the critical area. (PLT141) — AIM ¶2-3-9

121, 135, RTC

9416. When instructed by ATC to "Hold short of a runway (ILS critical area, etc.)," the pilot should stop

A—with the nose gear on the hold line.
B—so that no part of the aircraft extends beyond the hold line.
C—so the flight deck area of the aircraft is even with the hold line.

When the ILS critical area is being protected the pilot should stop so no part of the aircraft extends beyond the holding position marking. (PLT141) — AIM ¶2-3-5

ALL

9798. When you see this pavement marking from the cockpit, you

A—can taxi past this point at your own risk.
B—must hold short until "cleared" to taxi onto or past the runway.
C—may not cross the line until ATC allows you to "enter" or "cross" by instruction.

This question will likely include an onscreen image depicting runway hold position markings. These markings indicate where an aircraft is supposed to stop when approaching a runway. ATC will not use the word "cleared" in conjunction with authorization for aircraft to taxi. (PLT141) — AIM ¶2-3-5, 4-3-18

ALL

9799. The sign shown is an example of

A—a mandatory instruction sign.
B—runway heading notification signage.
C—an airport directional sign.

This question will likely include an onscreen image depicting a runway hold position sign. This is a mandatory instruction sign, used to hold an aircraft on a taxiway located in the approach or departure area for a runway so the aircraft does not interfere with operations on that runway. (PLT141) — AIM ¶2-3-8

Answers

9786 [B]	9787 [B]	9437 [C]	9416 [B]	9798 [C]	9799 [A]

Approach Lighting

An airplane approaching to land on a runway served by a **Visual Approach Slope Indicator (VASI)** must remain on or above the glide slope (except for normal bracketing) until a lower altitude is necessary for a safe landing.

A VASI gives the pilot a visual glide slope to follow when landing on certain runways. A VASI glide slope is normally about 3° (the same as an ILS) and the aim point is about 1,000 feet down the runway from the threshold. The angle and aim point of the VASI can be adjusted as necessary to accommodate the runway conditions. If a pilot of a high performance airplane is flying a VASI with a glide slope steeper than 3.5°, he/she should be aware that a longer than normal roll-out may result from the flare maneuver required by the steep angle.

Many runways used by air carrier aircraft have a three-bar VASI system to accommodate aircraft with a high cockpit such as Boeing 747 or DC-10. These aircraft need a glide slope that has an aim point further down the runway to ensure adequate clearance for the landing gear at the runway threshold. The pilot of such an airplane must use the two upwind lights (middle and far bars) for glide slope information.

The **Precision Approach Path Indicator (PAPI)** approach light system consists of a row of four lights perpendicular to the runway. Each light can be either red or white depending on the aircraft's position relative to the glide slope. The glide slope indications of a PAPI are as follows:

- High—4 white lights
- Slightly high—1 red, 3 white lights
- On glidepath—2 red, 2 white lights
- Slightly low—1 white, 3 red lights
- Low—4 red lights

A tri-color VASI consists of one light projector with three colors; red, green and amber. A "high" indication is amber, an "on glide slope" is green and a "low" is red. A tri-color VASI can be seen at a distance of 1/2 to 1 mile in daylight and up to 5 miles at night.

Pulsating visual approach slope indicators normally consist of a single light unit projecting a two-color visual approach path. The below glidepath indication is normally pulsating red and the above glidepath indication is normally pulsating white. The "on glide slope" indication for one type of system is a steady white light, while for another type it is an alternating red and white.

ALL

9378. A pilot approaching to land a turbine-powered aircraft on a runway served by a VASI shall

A—not use the VASI unless a clearance for a VASI approach is received.

B—use the VASI only when weather conditions are below basic VFR.

C—maintain an altitude at or above the glide slope until a lower altitude is necessary for a safe landing.

An airplane approaching to land on a runway served by a visual approach slope indicator (VASI), shall maintain an altitude at or above the glide slope until a lower altitude is necessary for a safe landing. (PLT147) — 14 CFR §91.129

Answer (A) is incorrect because a VASI should be used at all times when available, and is not considered an instrument approach. Answer (B) is incorrect because a VASI should be used at all times in both VFR and when transitioning out of IFR weather.

Answers

9378 [C]

ALL

8912. A pilot of a high-performance airplane should be aware that flying a steeper-than-normal VASI glide slope angle may result in

A—a hard landing.
B—increased landing rollout.
C—landing short of the runway threshold.

Although normal VASI glidepath angles are 3°, angles at some locations may be as high as 4.5° to give proper obstacle clearance. Pilots of high performance aircraft are cautioned that use of VASI angles in excess of 3.5° may cause an increase in runway length required for landing and rollout. (PLT147) — AIM ¶2-1-2

Answer (A) is incorrect because flying a steeper-than-normal VASI may result in an increased landing rollout in a high-performance airplane. Answer (C) is incorrect because a landing short of the runway threshold would be a result of flying a lower-than-normal VASI glide slope angle.

ALL

8911. What is the advantage of a three-bar VASI?

A—Pilots have a choice of glide angles.
B—A normal glide angle is afforded both high and low cockpit aircraft.
C—The three-bar VASI is much more visible and can be used at a greater height.

Three-bar VASI installations provide two visual glidepaths. The lower glidepath is provided by the near and middle bars and is normally set at 3° while the upper glidepath, provided by the middle and far bars is normally 1/4° higher. This higher glidepath is intended for use only by high cockpit aircraft to provide a sufficient threshold crossing height. (PLT147) — AIM ¶2-1-2

Answer (A) is incorrect because the three-bar VASI provides a glide slope for high cockpit aircraft, not a choice of glide angles for pilots. Answer (C) is incorrect because both the two- and three-bar VASI are visible from 3-5 miles during the day and up to 20 miles or more at night, and the three-bar VASI does not provide use at a greater height.

ALL

8913. The higher glide slope of the three-bar VASI is intended for use by

A—high performance aircraft.
B—helicopters.
C—high cockpit aircraft.

Three-bar VASI installations provide two visual glidepaths. The lower glidepath is provided by the near and middle bars and is normally set at 3° while the upper glidepath, provided by the middle and far bars is normally 1/4° higher. This higher glidepath is intended for use only by high cockpit aircraft to provide a sufficient threshold crossing height. (PLT147) — AIM ¶2-1-2

Answer (A) is incorrect because the higher glide slope of a three-bar VASI is for use only by high cockpit aircraft, which may or may not be high performance aircraft. Answer (B) is incorrect because the higher glide slope of a three-bar VASI is for use only by high cockpit aircraft, not specifically for use by helicopters.

ALL

8921. What does the Precision Approach Path Indicator (PAPI) consist of?

A—Row of four lights parallel to the runway; red, white, and green.
B—Row of four lights perpendicular to the runway; red and white.
C—One light projector with two colors; red and white.

The Precision Approach Path Indicator (PAPI) uses light units similar to the VASI but are installed in a single row of either two- or four-light units. (PLT147) — AIM ¶2-1-2

Answer (A) is incorrect because PAPI has a row of four lights perpendicular to the runway, and projects red and white light. Answer (C) is incorrect because PAPI consists of a row of four light projectors emitting red or white light.

ALL

8908. What are the indications of Precision Approach Path Indicator (PAPI)?

A—High – white, on glidepath – red and white; low – red.
B—High – white, on glidepath – green; low – red.
C—High – white and green, on glidepath – green; low – red.

The Precision Approach Path Indicator (PAPI) uses light units similar to the VASI but are installed in a single row of either two or four light units:

High .... 4 white lights
Slightly high .... 1 red, 3 white lights
On glide path .... 2 red, 2 white lights
Slightly low .... 1 white, 3 red lights
Low .... 4 red lights

(PLT147) — AIM ¶2-1-2

Answer (B) is incorrect because the on glidepath indication of PAPI is both red and white lights. Answer (C) is incorrect because above the glidepath indication of PAPI is all white, on glidepath is two red and two white, and below glidepath is all red.

Answers

8912 [B] 8911 [B] 8913 [C] 8921 [B] 8908 [A]

ALL

8916. What does the tri-color VASI consist of?

A—Three light bars; red, green, and amber.
B—One light projector with three colors; red, green, and amber.
C—Three glide slopes, each a different color; red, green, and amber.

Tri-color visual approach slope indicators normally consist of a single light unit projecting a three-color visual approach path into the final approach area of the runway upon which the indicator is installed. The below glidepath indication is red, the above glidepath indication is amber, and the on glide slope indication is green. (PLT147) — AIM ¶2-1-2

Answer (A) is incorrect because a tri-color VASI consists of a single light projector. Answer (C) is incorrect because a tri-color VASI projects only one glide slope. Below the glide slope is red, above the glide slope is amber, and on the glide slope is green.

ALL

8917. Which color on a tri-color VASI is a "high" indication?

A—Red.
B—Amber.
C—Green.

Tri-color visual approach slope indicators normally consist of a single light unit projecting a three-color visual approach path into the final approach area of the runway upon which the indicator is installed. The below glidepath indication is red, the above glidepath indication is amber, and the on glide slope indication is green. (PLT147) — AIM ¶2-1-2

Answer (A) is incorrect because a red color on a tri-color VASI indicates below the glidepath. Answer (C) is incorrect because a green color on a tri-color VASI indicates on the glidepath.

ALL

8918. Which color on a tri-color VASI is an "on course" indication?

A—Red.
B—Amber.
C—Green.

Tri-color visual approach slope indicators normally consist of a single light unit projecting a three-color visual approach path into the final approach area of the runway upon which the indicator is installed. The below glidepath indication is red, the above glidepath indication is amber, and the on glide slope indication is green. (PLT147) — AIM ¶2-1-2

Answer (A) is incorrect because a red color on a tri-color VASI indicates below the glidepath. Answer (B) is incorrect because an amber color on a tri-color VASI indicates above the glidepath.

ALL

8919. Which color on a tri-color VASI is a "low" indication?

A—Red.
B—Amber.
C—Green.

Tri-color visual approach slope indicators normally consist of a single light unit projecting a three-color visual approach path into the final approach area of the runway upon which the indicator is installed. The below glidepath indication is red, the above glidepath indication is amber, and the on glide slope indication is green. (PLT147) — AIM ¶2-1-2

Answer (B) is incorrect because an amber color on a tri-color VASI is an indication of above the glidepath. Answer (C) is incorrect because a green color on a tri-color VASI is an indication of on the glidepath.

ALL

8920. What is the normal range of the tri-color VASI at night?

A—5 miles.
B—10 miles.
C—15 miles.

Tri-color visual approach slope indicators have a useful range of approximately 1/2 to 1 mile during the day and up to 5 miles at night depending upon visibility conditions. (PLT147) — AIM ¶2-1-2

Answers

8916 [B]	8917 [B]	8918 [C]	8919 [A]	8920 [A]

ALL

8909. What does the pulsating VASI consist of?

A—Three-light system, two pulsing and one steady.
B—Two-light projectors, one pulsing and one steady.
C—One-light projector, pulsing white when above glide slope or red when more than slightly below glide slope, steady white when on glide slope, steady red for slightly below glide path.

Pulsating visual approach slope indicators normally consist of a single light unit projecting a two-color visual approach path into the final approach area of the runway upon which the indicator is installed. The below glidepath indication is normally pulsating red, and the above glidepath indication is normally pulsating white. The on glidepath indication for one type of system is a steady white light, while for another type system, the on glidepath indication consists of an alternating red and white. (PLT147) — AIM ¶2-1-2

Answer (A) is incorrect because the pulsating VASI is a two-light system, in which below glidepath is pulsating red, above glidepath is pulsating white, and on glidepath is a steady white light. Answer (B) is incorrect because the pulsating VASI is a single light projecting unit emitting a two-color visual approach path into the final approach area of the runway.

ALL

8910. What are the indications of the pulsating VASI?

A—High – pulsing white, on glidepath – green, low – pulsing red.
B—High – pulsing white, on glidepath – steady white, slightly below glide slope steady red, low – pulsing red.
C—High – pulsing white, on course and on glidepath – steady white, off course but on glidepath – pulsing white and red; low – pulsing red.

Pulsating visual approach slope indicators normally consist of a single light unit projecting a two-color visual approach path into the final approach area of the runway upon which the indicator is installed. The below glidepath indication is normally pulsating red, and the above glidepath indication is normally pulsating white. The on glidepath indication for one type of system is a steady white light, while for another type system the on glidepath indication consists of an alternating red and white. (PLT147) — AIM ¶2-1-2

Answer (A) is incorrect because the on glidepath indication of a pulsating VASI is either a pulsing red and white or steady white, not green. Answer (C) is incorrect because the pulsating VASI only provides glidepath indications, not lateral, or course, indications. Above glidepath is pulsing white, on glidepath is pulsing red and white or steady white, and below glidepath is pulsing red.

Answers

8909 [C] 8910 [B]

Chapter 3
Aerodynamics

Lift and Drag

The four forces acting on an aircraft in flight are **lift, weight, thrust** and **drag**. Weight always acts vertically toward the center of the earth. Lift acts perpendicular to the relative wind (not always vertically). Thrust and drag act opposite each other and parallel to the relative wind.

Lift is produced by air flowing over the curved wing surfaces. The air flowing over the upper surface of the wing is deflected further than that flowing across the lower surface and therefore is accelerated. **Bernoulli's Principle** states that when a gas is accelerated, its pressure decreases. Thus the pressure on the upper wing surface is lower than that on the lower surface and lift is produced.

Angle of attack is the angle between the relative wind and chord line of wing. At zero angle of attack, the pressure on the upper surface of the wing is still less than atmospheric, but the wing is producing minimum lift. As the angle of attack is increased, the lift developed by the wing increases proportionately. This is true until the angle of attack exceeds a critical value, when the air flowing over the top of the wing breaks up into a turbulent flow and the wing stalls.

Angle of attack and indicated airspeed determine the total lift. An increase in either indicated airspeed or angle of attack increases total lift (up to the stalling angle of attack) and a decrease in either decreases total lift. To maintain the same total lift (i.e., maintain level flight), a pilot has to change the angle of attack anytime indicated airspeed is changed. For example, as indicated airspeed is decreased, the angle of attack must be increased to compensate for the loss of lift. The relationship between indicated airspeed and lift for a given angle of attack involves the law of squares. If the angle of attack does not change, total lift varies with the square of the indicated airspeed. For example, if the airspeed doubles, the lift will increase by four times.

Indicated airspeed can be thought of as having two elements—the actual speed of the airplane through the air (true airspeed) and the density of the air. As altitude increases, air density decreases. To maintain the same indicated airspeed at altitude an aircraft must fly at a higher true airspeed. To produce the same amount of lift at altitude, a higher true airspeed is required for a given angle of attack.

A wing will always stall at the same angle of attack. The load factor, weight and density altitude will cause the stalling true airspeed to vary, but the stall angle of attack will always be the same.

A curve comparing total drag to parasite and induced drag reveals an airspeed at which drag is at a minimum value. At higher airspeeds, total drag increases because of increasing parasite drag. At lower airspeeds, induced drag increases which increases the total drag. Since the lift stays constant (equal to weight), the low point on the curve is the airspeed that produces the best lift to drag (L/D) ratio. This point is referred to as **L/D_{MAX}**. *See* Figure 3-1.

A change in weight changes the L/D curve. The amount of parasite drag is mainly a function of indicated airspeed. The amount of induced drag is a function of angle of attack. When an aircraft's weight is increased, any indicated airspeed will require a higher angle of attack to produce the required lift. This means that induced drag will increase with increases in weight while there will be little change in parasite drag.

When an airplane is within about one wingspan of the ground, the flow of air around the wingtips is inhibited by the close proximity of the ground. This **ground effect** reduces induced drag (and therefore total drag) and increases lift. As an airplane flies out of ground effect on takeoff, the increased induced drag will require a higher angle of attack.

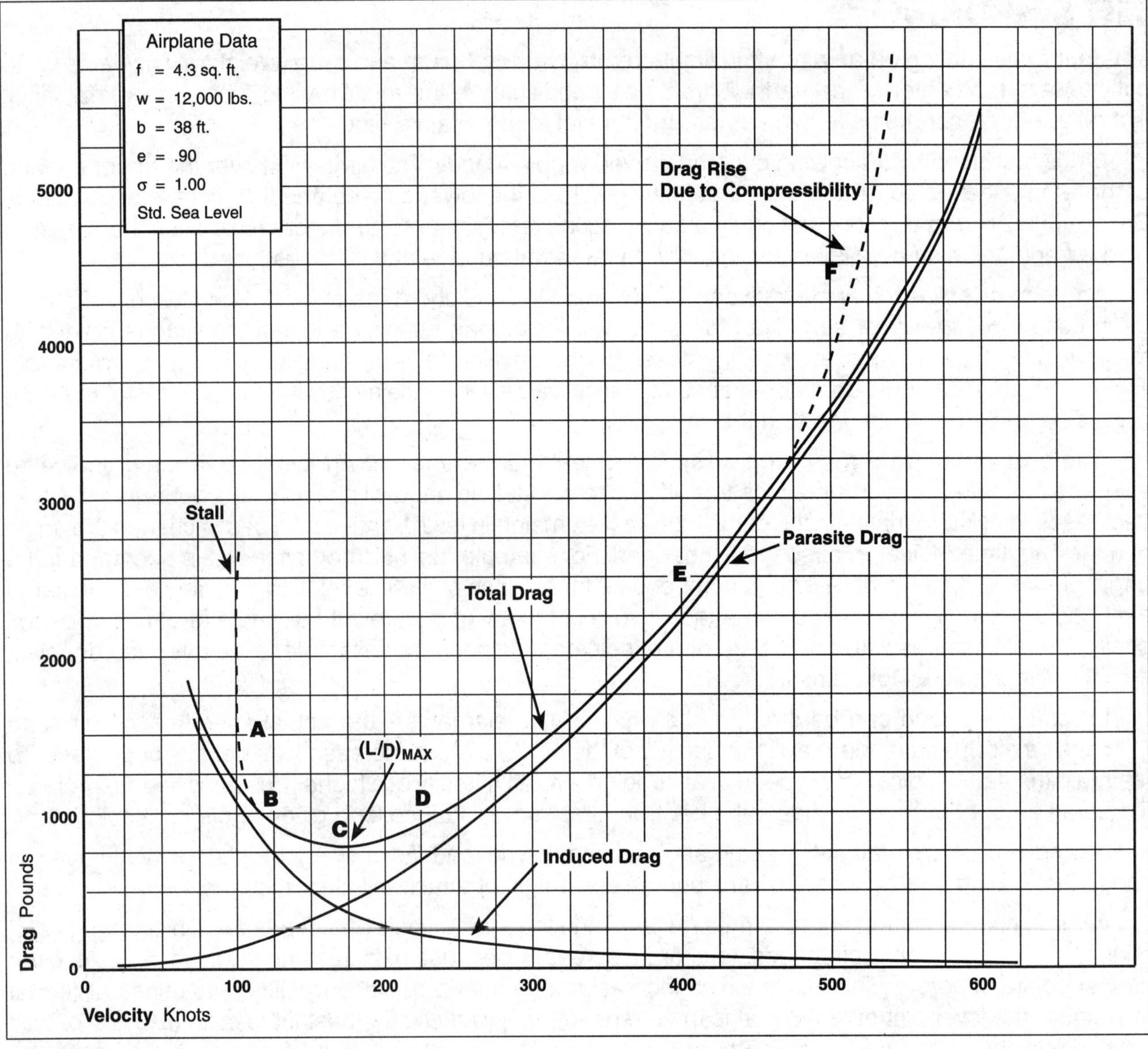

Figure 3-1. Typical drag curves

ALL

8377. What will be the ratio between airspeed and lift if the angle of attack and other factors remain constant and airspeed is doubled? Lift will be

A—the same.
B—two times greater.
C—four times greater.

Lift is proportional to the square of the airplane's velocity. For example, an airplane traveling at 200 knots has four times the lift as the same airplane traveling at 100 knots if the angle of attack and other factors remain constant. (PLT242) — FAA-H-8083-25

Answer (A) is incorrect because lift is proportional to the square of the airplane's velocity, it is not constant. Answer (B) is incorrect because, as airspeed is doubled, the lift will be four times greater.

Answers

8377 [C]

ALL

8378. What true airspeed and angle of attack should be used to generate the same amount of lift as altitude is increased?

A—The same true airspeed and angle of attack.
B—A higher true airspeed for any given angle of attack.
C—A lower true airspeed and higher angle of attack.

If the density factor is decreased and the total lift must equal the total weight to remain in flight, it follows that one of the other factors must be increased. The factors usually increased are the airspeed or the angle of attack, because these can be controlled by the pilot. (H300) — FAA-H-8083-25

Answer (A) is incorrect because true airspeed must be increased (not remain the same) as altitude increases. Answer (C) is incorrect because true airspeed must increase (not decrease) for any given angle of attack.

ALL

8348. What affects indicated stall speed?

A—Weight, load factor, and power.
B—Load factor, angle of attack, and power.
C—Angle of attack, weight, and air density.

An airplane will always stall at the same angle of attack. The indicated airspeed at which the stalling angle of attack is reached will vary with weight, load factor, and (to an extent) power setting. (PLT477) — FAA-H-8083-25

Answers (B) and (C) are incorrect because indicated stall speed is not affected by the angle of attack or air density.

ALL

9808. The stall speed of an airplane

A—is constant regardless of weight or airfoil configuration.
B—is affected by weight, and bank angle.
C—is not affected by dynamic pressures and lift coefficient.

Airplanes stall at the same angle of attack regardless of weight, dynamic pressure, bank angle, etc. The stall speed of the aircraft will be affected by weight, bank angle, and other factors since the product of dynamic pressure, wing area, and lift coefficient must produce the required lift. (PLT477) — ANA

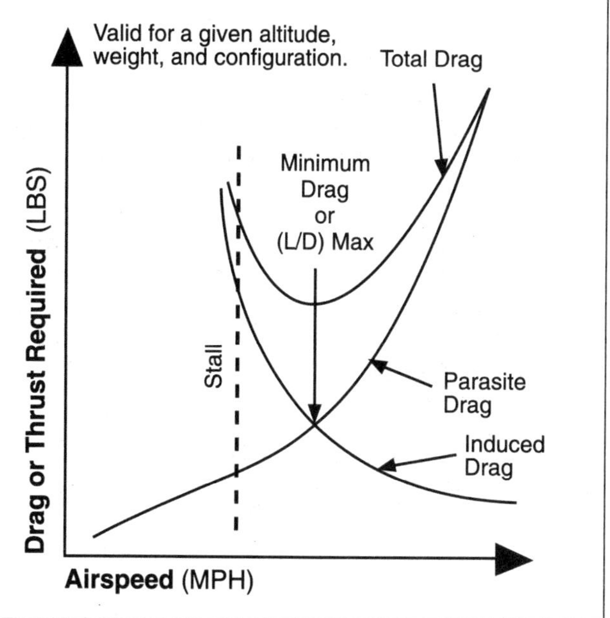

Question 8346

ALL

8346. What is the effect on total drag of an aircraft if the airspeed decreases in level flight below that speed for maximum L/D?

A—Drag increases because of increased induced drag.
B—Drag increases because of increased parasite drag.
C—Drag decreases because of lower induced drag.

Note in the following figure that the airspeed at which minimum drag occurs is the same airspeed at which the maximum lift/drag ratio (L/D) takes place. At speeds below maximum L/D, any decrease in airspeed will result in an increase in total drag due to the increase in induced drag. (PLT303) — FAA-H-8083-25

Answer (B) is incorrect because parasite drag varies directly (not inversely) with airspeed. Answer (C) is incorrect because drag increases (not decreases) from any speed other than that for maximum L/D.

Answers

8378 [B]	8348 [A]	9808 [B]	8346 [A]

ALL

8397. What is the relationship between induced and parasite drag when the gross weight is increased?

A—Parasite drag increases more than induced drag.
B—Induced drag increases more than parasite drag.
C—Both parasite and induced drag are equally increased.

Lift is required to counteract the aircraft's weight. If weight is increased, lift must be also. To increase lift, angle of attack must be increased, so induced drag also increases. Parasite drag is due to form and friction drag, so there would be little or no change in parasite drag. (PLT015) — ANA

Answer (A) is incorrect because parasite drag increases less (not more) than induced drag as airplane weight increases. Answer (C) is incorrect because induced drag increases more than parasite drag with increases in airplane gross weight.

ALL

9767. How does V_S (KTAS) speed vary with altitude?

A—Remains the same at all altitudes.
B—Varies directly with altitude.
C—Varies inversely with altitude.

True airspeed (KTAS) is based on the density of the air, which is affected by pressure, temperature, and humidity—together, these determine air density. While flying at a constant indicated airspeed, an increase in density altitude will indicate that the air has become less dense, and the true airspeed as well as ground speed will increase. (PLT124) — FAA-H-8083-25

ALL

8368. What is the reason for variations in geometric pitch along a propeller or rotor blade?

A—It permits a relatively constant angle of attack along its length when in cruising flight.
B—It prevents the portion of the blade near the hub or root from stalling during cruising flight.
C—It permits a relatively constant angle of incidence along its length when in cruising flight.

"Twisting," or variations in the geometric pitch of the blades permits the propeller to operate with a relatively constant angle of attack along its length when in cruising flight. (PLT214) — FAA-H-8083-25

ALL

8375. What flight condition should be expected when an aircraft leaves ground effect?

A—An increase in induced drag requiring a higher angle of attack.
B—A decrease in parasite drag permitting a lower angle of attack.
C—An increase in dynamic stability.

An airplane leaving ground effect will require an increase in angle of attack to maintain the same lift coefficient, experience an increase in induced drag and thrust required, experience a decrease in stability with a nose-up change in pitch, and produce a reduction in static source pressure and increase in indicated airspeed. (PLT131) — FAA-H-8083-25

Answer (B) is incorrect because, at slow airspeeds when taking off, induced (not parasite) drag predominates. Answer (C) is incorrect because, when leaving ground effect, expect a decrease in stability and a nose-up change in moment.

121, 135, DSP

8382. By changing the angle of attack of a wing, the pilot can control the airplane's

A—lift, gross weight, and drag.
B—lift, airspeed, and drag.
C—lift and airspeed, but not drag.

By changing the angle of attack, the pilot can control lift, airspeed, and drag. Even the total load supported in flight by the wing may be modified by variations in angle of attack. (PLT004) — FAA-H-8083-3

Answer (A) is incorrect because angle of attack cannot control the airplane's gross weight. Answer (C) is incorrect because the pilot can control the amount of induced drag by changing the angle of attack.

121, 135, DSP

8399. At which speed will increasing the pitch attitude cause an airplane to climb?

A—Low speed.
B—High speed.
C—Any speed.

When operating at speeds below L/D_{MAX}, an increase in pitch or decrease in speed causes total drag to increase, thus causing a descent with a fixed power setting. When operating at speeds above L/D_{MAX} and pitch is increased (or airspeed is decreased), total drag will decrease, thus causing a climb with a fixed power setting. (PLT303) — ANA

Answers (A) and (C) are incorrect because below L/D_{MAX}, performance decreases with increases in pitch.

Answers

8397 [B]	9767 [B]	8368 [A]	8375 [A]	8382 [B]	8399 [B]

121, 135, DSP

8379. How can an airplane produce the same lift in ground effect as when out of ground effect?

A—The same angle of attack.
B—A lower angle of attack.
C—A higher angle of attack.

The reduction of the wing-tip vortices due to ground effect alters the spanwise lift distribution and reduces the induced angle of attack and induced drag. Therefore, the wing will require a lower angle of attack in ground effect to produce the same lift coefficient. (PLT131) — FAA-H-8083-3

Answer (A) is incorrect because, if the same angle of attack is maintained, an increase in lift coefficient will result. Answer (C) is incorrect because a lower angle of attack is required to produce the same lift in ground effect.

Critical Engine and V_{MC}

Because of "P-Factor" on most propeller-driven airplanes, the loss of one particular engine at high angles of attack would be more detrimental to performance than the loss of the other. One of the engines has its thrust line closer to the aircraft centerline (*see* Figure 3-2). The loss of this engine would more adversely affect the performance and handling of the aircraft; therefore this is the "critical engine."

For unsupercharged engines, V_{MC} decreases as altitude is increased. Stalls should never be practiced with one engine inoperative because of the potential for loss of control. Engine out approaches and landings should be made the same as normal approaches and landings.

Banking at least 5° into the good engine ensures the airplane will be controllable at any speed above the certificated V_{MC}, that the airplane will be in a minimum drag configuration for best climb performance, and that the stall characteristics will not be degraded. Engine out flight with the ball centered is never correct.

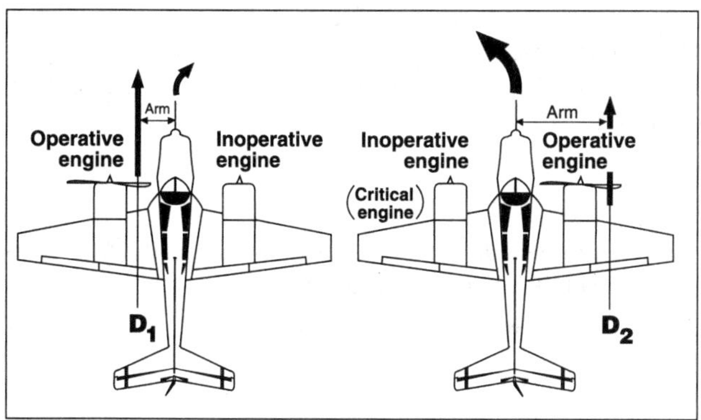

Figure 3-2

Banking at least 5° into the good engine ensures that the airplane will be controllable at any speed above the certificated V_{MC}, that the airplane will be in a minimum drag configuration for best climb performance, and that the stall characteristics will not be degraded. Engine out flight with the ball centered is never correct. (PLT223) — FAA-H-8083-3

121, 135

8357. In a light, twin-engine airplane with one engine inoperative, when is it acceptable to allow the ball of a slip-skid indicator to be deflected outside the reference lines?

A—While maneuvering at minimum controllable airspeed or less to avoid overbanking.
B—When operating at any airspeed of V_{MC} or greater with only enough deflection to zero the side slip.
C—When practicing imminent stalls in a banked attitude of over 60°.

121, 135

8358. What is the safest and most efficient takeoff and initial climb procedure in a light, twin-engine airplane? Accelerate to

A—best engine-out, rate-of-climb airspeed while on the ground, then lift off and climb at that speed.
B—V_{MC}, then lift off at that speed and climb at maximum angle-of-climb airspeed.
C—an airspeed slightly above V_{MC}, then lift off and climb at the best rate-of-climb airspeed.

Lift-off should be made at no less than V_{MC} + 5. After lift-off, the airplane should be allowed to accelerate to the all-engine best-rate-of-climb speed V_Y, and then the climb maintained at this speed with takeoff power until a safe maneuvering altitude is attained. (PLT459) — FAA-H-8083-3

Answers

8379 [B] 8357 [B] 8358 [C]

121, 135

8360. What performance should a pilot of a light, twin-engine airplane be able to maintain at V_{MC}?

A—Heading.
B—Heading and altitude.
C—Heading, altitude, and ability to climb 50 ft/min.

V_{MC} can be defined as the minimum airspeed at which the airplane is controllable when the critical engine is suddenly made inoperative, and the remaining engine is producing takeoff power. This does not mean that the airplane must be able to climb or even hold altitude. It only means that a heading can be maintained. (PLT208) — FAA-H-8083-3

121, 135

8364. What does the blue radial line on the airspeed indicator of a light, twin-engine airplane represent?

A—Maximum single-engine rate of climb.
B—Maximum single-engine angle of climb.
C—Minimum controllable airspeed for single-engine operation.

The airspeed indicator in a twin-engine airplane is marked with a red radial line at the minimum controllable airspeed with the critical engine inoperative (V_{MC}), and a blue radial line at the best rate of climb airspeed with one engine inoperative (V_{YSE}). (PLT132) — FAA-H-8083-3

121, 135, DSP

8359. What procedure is recommended for an engine-out approach and landing?

A—The flightpath and procedures should be almost identical to a normal approach and landing.
B—The altitude and airspeed should be considerably higher than normal throughout the approach.
C—A normal approach, except do not extend the landing gear or flaps until over the runway threshold.

Essentially, an engine-out approach and landing is the same as a normal approach and landing. (PLT223) — FAA-H-8083-3

121, 135, DSP

8361. Which engine is the "critical" engine of a twin-engine airplane?

A—The one with the center of thrust closest to the centerline of the fuselage.
B—The one designated by the manufacturer which develops most usable thrust.
C—The one with the center of thrust farthest from the centerline of the fuselage.

The critical engine is defined as the engine whose failure would most adversely affect performance or handling. Because of "P-factor," most propeller-driven airplanes do not develop symmetrical thrust at high angles of attack. If the engine with the thrust line closest to the airplane centerline fails, the resulting yawing moment will be greater than if the other engine had failed. (PLT347) — FAA-H-8083-3

121, 135, DSP

8362. What effect, if any, does altitude have on V_{MC} for an airplane with unsupercharged engines?

A—None.
B—Increases with altitude.
C—Decreases with altitude.

For an airplane without supercharged engines, V_{MC} decreases as altitude is increased. Consequently, directional control can be maintained at a lower airspeed than at sea level. (PLT314) — FAA-H-8083-3

121, 135, DSP

8363. Under what condition should stalls never be practiced in a twin-engine airplane?

A—With one engine inoperative.
B—With climb power on.
C—With full flaps and gear extended.

With full power applied to the operative engine, as the airspeed drops below V_{MC}, the airplane tends to roll as well as yaw into the inoperative engine. This tendency becomes greater as the airspeed is further reduced. Since this tendency must be counteracted by aileron control, the yaw condition is aggravated by aileron roll (the "down" aileron creates more drag than the "up" aileron). If a stall occurs in this condition, a violent roll into the dead engine may be experienced. (PLT459) — FAA-H-8083-3

Answers

8360 [A] 8364 [A] 8359 [A] 8361 [A] 8362 [C] 8363 [A]

Maneuvering Flight

In a **turn**, centrifugal force is counterbalanced by a portion of the lift of the wing. The horizontal component of lift turns the airplane and the vertical component of lift opposes gravity. When the pilot rolls the airplane into a turn he must increase the total lift of the wing so that the vertical component is equal to the airplane's weight. This is done by increasing the angle of attack. If no compensation is made for the loss of vertical component of lift in a turn, the aircraft will develop a sink rate.

Load factor is the ratio of the weight supported by the wings to the actual weight of the aircraft. For example, if an aircraft with a gross weight of 2,000 pounds were subjected to a total load of 6,000 pounds in flight, the load factor would be 3 Gs. On the ground or in unaccelerated flight the load factor is one. Conditions which can increase the load factor are vertical gusts (turbulence) and level turns. In a **level turn**, the load factor is dependent only on the angle of bank. Airspeed, turn rate or aircraft weight have no effect on load factor.

Rate of turn is the number of degrees per second at which the aircraft turns. The rate of turn is dependent on both the aircraft's airspeed and its angle of bank. To increase the rate of turn, the pilot must increase the angle of bank or decrease the airspeed or both. The rate of turn will decrease if the bank angle is decreased or if the airspeed is increased. The **radius of turn** is also dependent on both the bank angle and the airspeed. If angle of bank is increased or airspeed is decreased, the radius of turn will decrease. If bank angle is shallowed or if airspeed is increased, the radius of turn will increase.

ALL

8349. If no corrective action is taken by the pilot as angle of bank is increased, how is the vertical component of lift and sink rate affected?

A—Lift increases and the sink rate increases.
B—Lift decreases and the sink rate decreases.
C—Lift decreases and the sink rate increases.

When an airplane is banked, its lift can be broken into two vectors, a vertical component of lift and a horizontal component. If the airplane is to maintain altitude in the turn, the vertical component of lift must be equal to the aircraft's weight. This means that total lift must be increased. Lift can be increased either by increasing airspeed or by increasing angle of attack. If the vertical component of lift is less than the aircraft's weight, the airplane will descend. (PLT348) — FAA-H-8083-3

Answer (A) is incorrect because lift will decrease, not increase. Answer (B) is incorrect because the sink rate increases as the lift decreases.

ALL

8350. Why must the angle of attack be increased during a turn to maintain altitude?

A—Compensate for loss of vertical component of lift.
B—Increase the horizontal component of lift equal to the vertical component.
C—Compensate for increase in drag.

When an airplane is banked, its lift can be broken into two vectors, a vertical component of lift and a horizontal component. If the airplane is to maintain altitude in the turn, the vertical component of lift must be equal to the aircraft's weight. This means that total lift must be increased. Lift can be increased either by increasing airspeed or by increasing angle of attack. If the vertical component of lift is less than the aircraft's weight, the airplane will descend. (PLT348) — FAA-H-8083-3

Answer (B) is incorrect because angle of attack is increased in order to increase the vertical component of lift to equal weight. Answer (C) is incorrect because additional thrust (power) is used to compensate for increase in drag.

ALL

8347. What is load factor?

A—Lift multiplied by the total weight.
B—Lift subtracted from the total weight.
C—Lift divided by the total weight.

Load factor is the ratio of the total load supported by the airplane's wings to the actual weight of the airplane and its contents, or the actual load supported by the wings' lift divided by the total weight of the airplane. (PLT310) — FAA-H-8083-25

Answers

8349 [C] 8350 [A] 8347 [C]

ALL

9740. During a skidding turn to the right, what is the relationship between the component of lift and centrifugal force?

A—Centrifugal force is less than the horizontal lift component and the load factor is increased.
B—Centrifugal force is greater than the horizontal lift component.
C—Centrifugal force and the horizontal lift component are equal, and the load factor is decreased.

A skidding turn results from excess centrifugal force over the horizontal lift component, pulling the aircraft toward the outside of the turn. As centrifugal force increases, the load factor also increases. (PLT234) — FAA-H-8083-15

Answer (A) is incorrect because a slipping turn will occur if centrifugal force is less than horizontal lift. Answer (C) is incorrect because centrifugal force and horizontal lift are equal in a coordinated turn and load factor will increase.

ALL

8354. If an aircraft with a gross weight of 2,000 pounds were subjected to a total load of 6,000 pounds in flight, the load factor would be

A—2 Gs.
B—3 Gs.
C—9 Gs.

Load factor is the ratio of the total load supported by the airplane's wings to the actual weight of the airplane and its contents, or the actual load supported by the wings divided by the total weight of the airplane:

6,000 pounds ÷ 2,000 pounds = 3 Gs

(PLT018) — FAA-H-8083-25

ALL

8353. Upon which factor does wing loading during a level coordinated turn in smooth air depend?

A—Rate of turn.
B—Angle of bank.
C—True airspeed.

Load factor is independent of airspeed and dependent on angle of bank; therefore, with a constant bank angle, load factor is not affected. There is no change in centrifugal force for any given bank — the load factor remains the same. (PLT248) — FAA-H-8083-25

Answer (A) is incorrect because in a coordinated turn, the rate of turn does not have any impact on the load factor — it is determined wholly by the angle of bank. Answer (C) is incorrect because true airspeed has no impact on the load factor.

ALL

8396. For a given angle of bank, the load factor imposed on both the aircraft and pilot in a coordinated constant-altitude turn

A—is directly related to the airplane's gross weight.
B—varies with the rate of turn.
C—is constant.

For any given angle of bank, the load factor remains constant. (PLT309) — FAA-H-8083-25

ALL

8351. How can the pilot increase the rate of turn and decrease the radius at the same time?

A—Steepen the bank and increase airspeed.
B—Steepen the bank and decrease airspeed.
C—Shallow the bank and increase airspeed.

Any increase in the angle of bank will increase the rate of turn and decrease the radius of turn. Turn radius will decrease with decreasing airspeed. (PLT348) — FAA-H-8083-3

ALL

8352. What is the relationship of the rate of turn with the radius of turn with a constant angle of bank but increasing airspeed?

A—Rate will decrease and radius will increase.
B—Rate will increase and radius will decrease.
C—Rate and radius will increase.

For any given angle of bank, the rate of turn varies with the airspeed. If the angle of bank is held constant and the airspeed is increased, the rate of turn will decrease. The radius of turn will vary with airspeed. As airspeed is increased the radius will also increase. (PLT248) — FAA-H-8083-3

Answer (B) is incorrect because to maintain a constant angle of bank while increasing airspeed, the radius of turn will increase. Answer (C) is incorrect because to maintain a constant angle of bank while increasing airspeed, the rate of turn will decrease.

Answers

9740 [B]	8354 [B]	8353 [B]	8396 [C]	8351 [B]	8352 [A]

ALL

8345. What effect does an increase in airspeed have on a coordinated turn while maintaining a constant angle of bank and altitude?

A—The rate of turn will decrease resulting in a decreased load factor.
B—The rate of turn will increase resulting in an increased load factor.
C—The rate of turn will decrease resulting in no changes in load factor.

For any given angle of bank, the rate of turn varies with the airspeed. A constant bank angle does not change the load factor. If the angle of bank is held constant and the airspeed is increased, the rate of turn will decrease. The radius of turn will vary with airspeed. As airspeed is increased the radius will also increase. (PLT248) — FAA-H-8083-25

Answer (A) is incorrect because, at a constant bank angle, the higher airspeed will decrease the rate of turn to compensate for added centrifugal force, allowing the load factor to remain the same. Answer (B) is incorrect because, for any bank angle, the rate of turn varies with the airspeed; the higher the speed, the slower the rate of turn.

Stability

Static stability describes the initial reaction of an aircraft after it has been disturbed from equilibrium in one or more of its axes of rotation. If the aircraft has an initial tendency to return to its original attitude of equilibrium, it has **positive static stability**. When it continues to diverge, it exhibits **negative static** stability. If an aircraft tends to remain in its new, disturbed state, it has **neutral static stability**. Most airplanes have positive static stability in pitch and yaw, and are close to neutrally stable in roll.

When an aircraft exhibits positive static stability in one of its axes, the term "dynamic stability" describes the long term tendency of the aircraft. When an aircraft is disturbed from equilibrium and then tries to return, it will invariably overshoot the original attitude and then pitch back. This results in a series of oscillations. If the oscillations become smaller with time, the aircraft has positive dynamic stability. If the aircraft diverges further away from its original attitude with each oscillation, it has negative dynamic stability.

The entire design of an aircraft contributes to its **stability** (or lack of it) in each of its axes of rotation. However, the vertical tail is the primary source of direction stability (yaw), and the horizontal tail is the primary source of pitch stability. The **center of gravity (CG)** location also affects stability. If the CG is toward its rearward limit, the aircraft will be less stable in both roll and pitch. As the CG is moved forward, the stability improves. Even though an airplane will be less stable with a rearward CG, it will have some desirable aerodynamic characteristics due to reduced aerodynamic loading of horizontal tail surface. This type of an airplane will have a slightly lower stall speed and will cruise faster for a given power setting.

ALL

8365. Identify the type stability if the aircraft attitude remains in the new position after the controls have been neutralized.

A—Negative longitudinal static stability.
B—Neutral longitudinal dynamic stability.
C—Neutral longitudinal static stability.

Neutral static stability is the initial tendency of an airplane to remain in a new position after its equilibrium has been disturbed. (PLT236) — FAA-H-8083-25

Answer (A) is incorrect because a negative longitudinal static stability means the airplane would tend to move even further from the original position. Answer (B) is incorrect because, with neutral longitudinal dynamic stability, the airplane would continue to oscillate without a tendency to increase or decrease.

ALL

8372. Identify the type stability if the aircraft attitude tends to move farther from its original position after the controls have been neutralized.

A—Negative static stability.
B—Positive static stability.
C—Negative dynamic stability.

Negative static stability is the initial tendency of the airplane to continue away from the original state of equilibrium after being disturbed. (PLT213) — FAA-H-8083-25

Answers

8345 [C]   8365 [C]   8372 [A]

ALL

8373. Identify the type stability if the aircraft attitude tends to return to its original position after the controls have been neutralized.

A—Positive dynamic stability.
B—Positive static stability.
C—Neutral dynamic stability.

Positive static stability is the initial tendency of the airplane to return to the original state of equilibrium after being disturbed. (PLT236) — FAA-H-8083-25

Answer (A) is incorrect because positive dynamic stability refers to oscillations being dampened or decreasing. Answer (C) is incorrect because neutral dynamic stability refers to oscillations continuing without a tendency to increase or decrease.

ALL

8366. What is a characteristic of longitudinal instability?

A—Pitch oscillations becoming progressively greater.
B—Bank oscillations becoming progressively greater.
C—Aircraft constantly tries to pitch down.

A longitudinally unstable airplane has a tendency to dive or climb progressively into a very steep dive or climb, or even a stall. (PLT213) — FAA-H-8083-25

Answer (B) is incorrect because longitudinal stability refers to pitch (not bank) oscillations. Answer (C) is incorrect because this is not considered a stability problem. Stability is the reaction of the airplane when its equilibrium is disturbed.

ALL

8367. Describe dynamic longitudinal stability.

A—Motion about the longitudinal axis.
B—Motion about the lateral axis.
C—Motion about the vertical axis.

Longitudinal stability is the quality which makes an airplane stable about its lateral axis. (PLT236) — FAA-H-8083-25

Answer (A) is incorrect because motion about the airplane's longitudinal axis is lateral (not longitudinal) stability. Answer (C) is incorrect because motion about the vertical axis is directional stability.

ALL

8376. What characteristic should exist if an airplane is loaded to the rear of its CG range?

A—Sluggish in aileron control.
B—Sluggish in rudder control.
C—Unstable about the lateral axis.

If an airplane is loaded too far rearward it may not dampen out a vertical displacement of the nose. Instead, when the nose is momentarily pulled up, it may alternately climb and dive becoming steeper with each oscillation. (PLT236) — FAA-H-8083-25

Answer (A) is incorrect because an aft location of the CG has a greater effect on the longitudinal stability, not the lateral (aileron) controllability. Answer (B) is incorrect because an aft CG has a greater effect on the longitudinal stability, not vertical (rudder) controllability.

121, 135, DSP

8380. What are some characteristics of an airplane loaded with the CG at the aft limit?

A—Lowest stall speed, highest cruise speed, and least stability.
B—Highest stall speed, highest cruise speed, and least stability.
C—Lowest stall speed, lowest cruise speed, and highest stability.

An airplane loaded with the CG at its aft limit will have a lower stall speed, higher cruise speed, but be less stable. (PLT240) — FAA-H-8083-3

Answer (B) is incorrect because an aft CG would cause the airplane to have the lowest stall speed. Answer (C) is incorrect because an aft CG would cause an airplane to have the highest cruise speed and the least stability.

Answers

8373 [B] 8366 [A] 8367 [B] 8376 [C] 8380 [A]

High Speed Flight

Mach number is the ratio of the true airspeed to the speed of sound (TAS ÷ Speed of Sound). For example, an aircraft cruising at Mach .80 is flying at 80% of the speed of sound. The speed of sound is Mach 1.0.

A large increase in drag occurs when the air flow around the aircraft exceeds the speed of sound (Mach 1.0). Because lift is generated by accelerating air across the upper surface of the wing, local air flow velocities will reach sonic speeds while the aircraft Mach number is still considerably below the speed of sound. With respect to **Mach cruise control**, flight speeds can be divided into three regimes—subsonic, transonic and supersonic. The **subsonic regime** can be considered to occur at aircraft Mach numbers where all the local air flow is less than the speed of sound. The **transonic range** is where some but not all the local air flow velocities are Mach 1.0 or above. In **supersonic** flight, all the air flow around the aircraft exceeds Mach 1.0. The exact Mach numbers will vary with each aircraft type but as a very rough rule of thumb the subsonic regime occurs below Mach .75, the transonic regime between Mach .75 and Mach 1.20, and the supersonic regime over Mach 1.20.

A limiting speed for a subsonic transport aircraft is its critical Mach number (M_{CRIT}). That is the speed at which airflow over the wing first reaches, but does not exceed, the speed of sound. At M_{CRIT} there may be sonic but no supersonic flow.

When an airplane exceeds its critical Mach number, a shock wave forms on the wing surface that can cause a phenomenon known as shock stall. If this shock stall occurs symmetrically at the wing roots, the loss of lift and loss of downwash on the tail will cause the aircraft to pitch down or "tuck under." This tendency is further aggravated in sweptwing aircraft because the center of pressure moves aft as the wing roots shock stall. If the wing tips of a sweptwing airplane shock stall first, the wing's center of pressure would move inward and forward causing a pitch up motion. *See* Figure 3-3.

The less airflow is accelerated across the wing, the higher the critical Mach number (i.e., the maximum flow velocity is closer to the aircraft's Mach number). Two ways of increasing M_{CRIT} in jet transport designs are to give the wing a lower camber and increase wing sweep. A thin airfoil section (lower camber) causes less airflow acceleration. The sweptwing design has the effect of creating a thin airfoil section by inducing a spanwise flow, thus increasing the effective chord length. *See* Figure 3-4 (next page).

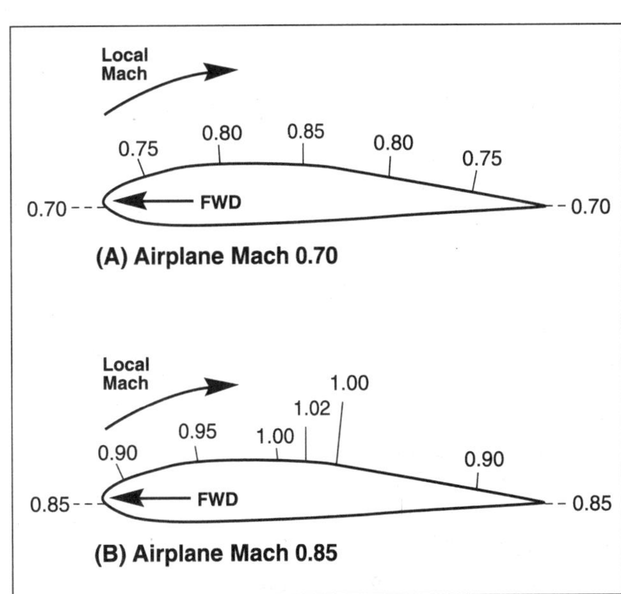

Figure 3-3. Local airstream Mach numbers

Although a sweptwing design gives an airplane a higher critical Mach number (and therefore a higher maximum cruise speed), it results in some undesirable flight characteristics. One of these is a reduced maximum coefficient of lift. This requires that sweptwing airplanes extensively employ high lift devices, such as slats and slotted flaps, to get acceptably low takeoff and landing speeds. The purpose of high lift devices such as flaps, slats and slots is to increase lift at low airspeeds and to delay stall to a higher angle of attack.

Another disadvantage of the sweptwing design is the tendency, at low airspeeds, for the wing tips to stall first. This results in loss of aileron control early in the stall, and in very little aerodynamic buffet on the tail surfaces.

Continued

Dutch roll tendency is typical of sweptwing designs. If such an airplane yaws, the advancing wing is at a higher angle of attack and presents a greater span to the airstream than the retreating wing. This causes the aircraft to roll in the direction of the initial yaw and simultaneously to reverse its direction of yaw. When the yaw reverses, the airplane then reverses its direction of roll and yaw again. This roll-yaw coupling is usually damped out by the vertical stabilizer. But at high speeds and in turbulence, this may not be adequate, so most aircraft are also equipped with a yaw damper to help counteract any Dutch roll tendency.

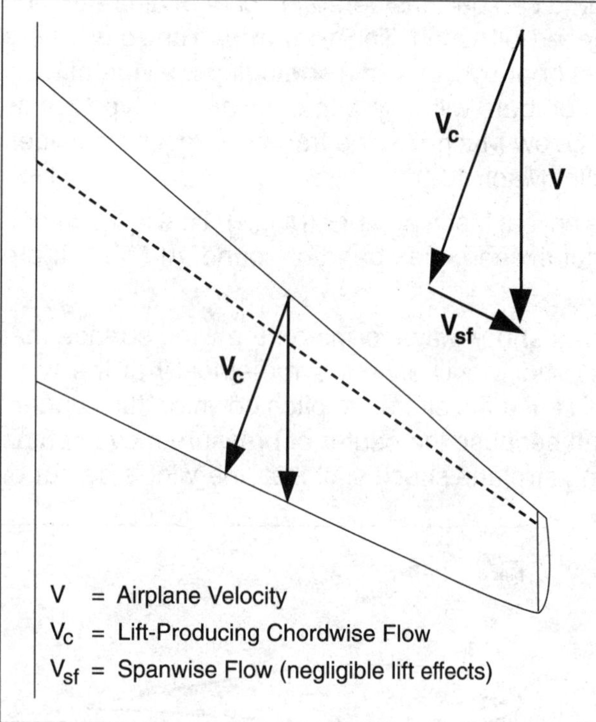

Figure 3-4. Effect of wing sweep on M_{CRIT}

121, 135, DSP
8387. Within what Mach range does transonic flight regimes usually occur?

A—.50 to .75 Mach.
B—.75 to 1.20 Mach.
C—1.20 to 2.50 Mach.

Flight regimes are defined as follows:

Subsonic – Mach numbers below 0.75
Transonic – Mach numbers from 0.75 to 1.20
Supersonic – Mach numbers from 1.20 to 5.00
Hypersonic – Mach numbers above 5.00.

(PLT032) — ANA

Answer (A) is incorrect because .50 to .75 Mach would be subsonic flight. Answer (C) is incorrect because 1.20 to 2.50 Mach would be supersonic flight.

121, 135, DSP
8390. At what Mach range does the subsonic flight range normally occur?

A—Below .75 Mach.
B—From .75 to 1.20 Mach.
C—From 1.20 to 2.50 Mach.

Flight regimes are defined as follows:

Subsonic – Mach numbers below 0.75
Transonic – Mach numbers from 0.75 to 1.20
Supersonic – Mach numbers from 1.20 to 5.00
Hypersonic – Mach numbers above 5.00.

(PLT214) — ANA

Answer (B) is incorrect because .75 to 1.20 Mach would be transonic flight. Answer (C) is incorrect because 1.20 to 2.50 Mach would be supersonic flight.

121, 135, DSP
8388. What is the highest speed possible without supersonic flow over the wing?

A—Initial buffet speed.
B—Critical Mach number.
C—Transonic index.

The highest speed possible without supersonic flow is called the Critical Mach Number. (PLT214) — ANA

121, 135, DSP
8389. What is the free stream Mach number which produces first evidence of local sonic flow?

A—Supersonic Mach number.
B—Transonic Mach number.
C—Critical Mach number.

The highest speed possible without supersonic flow is called the Critical Mach Number. (PLT214) — ANA

Answers

8387 [B]	8390 [A]	8388 [B]	8389 [C]

121, 135, DSP
8392. What is the result of a shock-induced separation of airflow occurring symmetrically near the wing root of a sweptwing aircraft?

A—A high-speed stall and sudden pitchup.
B—A severe moment or "Mach tuck."
C—Severe porpoising.

If the shock-induced separation occurs symmetrically near the wing root, there is an accompanying loss of lift. A decrease in downwash on the horizontal tail will create a diving moment and the aircraft will "tuck under." (PLT214) — ANA

Answer (A) is incorrect because there is a sudden pitch down when a shock-induced separation of airflow occurs symmetrically near the wing root of a sweptwing aircraft. Answer (C) is incorrect because there is a diving moment when a shock-induced separation of airflow occurs symmetrically near the wing root of a sweptwing aircraft.

121, 135, DSP
8395. What is the movement of the center of pressure when the wingtips of a sweptwing airplane are shock-stalled first?

A—Inward and aft.
B—Inward and forward.
C—Outward and forward.

Shock formation at the wing tip first moves the center of pressure forward and inboard and the resulting climbing moment and tail downwash can contribute to "pitch up." (PLT214) — ANA

Answer (A) is incorrect because when the wing tips are shock-stalled first, the center of pressure moves forward. Answer (C) is incorrect because when the wing tips are shock-stalled first, the center of pressure moves inward.

121, 135, DSP
8391. What is the principal advantage of a sweepback design wing over a straightwing design?

A—The critical Mach number will increase significantly.
B—Sweepback will increase changes in the magnitude of force coefficients due to compressibility.
C—Sweepback will accelerate the onset of compressibility effect.

One of the most important advantages of sweep is an increase in critical Mach number, force divergence Mach number, and the Mach number at which the drag rise will peak. In other words, the sweep will delay the onset of compressibility effects. (PLT214) — ANA

121, 135, DSP
8393. What is one disadvantage of a sweptwing design?

A—The wing root stalls prior to the wingtip section.
B—The wingtip section stalls prior to the wing root.
C—Severe pitchdown moment when the center of pressure shifts forward.

When sweepback is combined with taper there is an extremely powerful tendency for the wing tip to stall first. (PLT214) — ANA

121, 135, DSP
9803. Swept wings causes a significant

A—increase in effectiveness of flaps.
B—reduction in effectiveness of flaps.
C—flap actuation reliability issue.

Thin airfoil sections with sweepback impose distinct limitations on the effectiveness of flaps. (PLT266) — ANA

121, 135, DSP
8394. What is the condition that may occur when gusts cause a sweptwing-type airplane to roll in one direction while yawing in the other?

A—Mach buffet.
B—Wingover.
C—Dutch roll.

The tendency of sweptwing airplanes to yaw in one direction and to roll in the other is called "Dutch roll." (PLT214) — ANA

Answers

8392 [B]	8395 [B]	8391 [A]	8393 [B]	9803 [B]	8394 [C]

Primary Flight Controls

Because of the high air loads, it is very difficult to move the flight control surfaces of jet aircraft with just mechanical and aerodynamic forces. So flight controls are usually moved by hydraulic actuators. Flight controls are divided into **primary flight controls** and **secondary** or **auxiliary flight controls.** The primary flight controls are those that maneuver the aircraft in roll, pitch and yaw. These include the ailerons, elevator and rudder. Secondary (or auxiliary) flight controls include tabs, trailing-edge flaps, leading-edge flaps, spoilers and slats. *See* Figure 3-5.

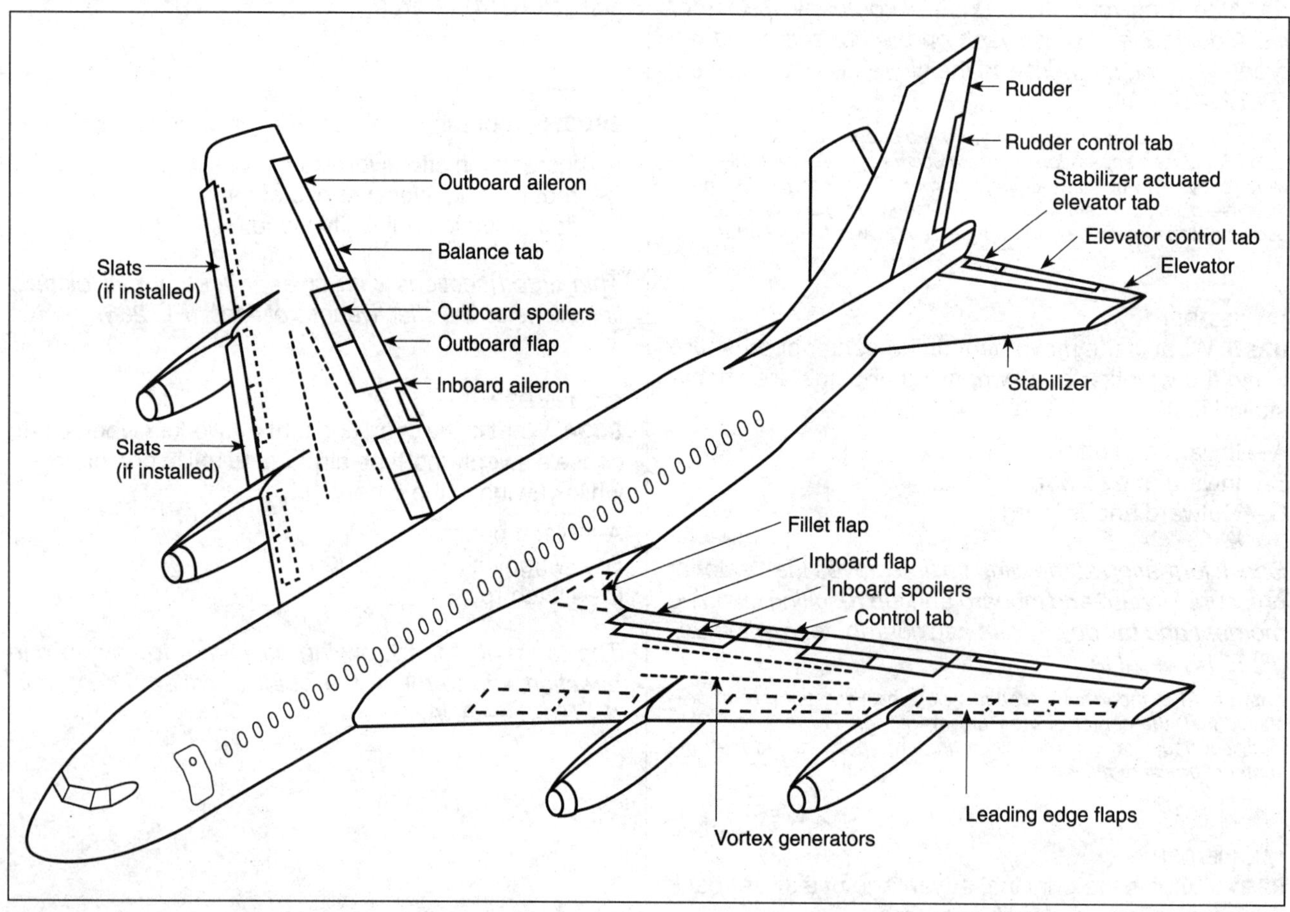

Figure 3-5. Typical transport aircraft flight controls

Roll control of most jet aircraft is accomplished by ailerons and flight spoilers. The exact mix of controls is determined by the aircraft's flight regime. In low speed flight all control surfaces operate to provide the desired roll control. As the aircraft moves into higher speed operations, control surface movement is reduced to provide approximately the same roll response to a given input through a wide range of speeds.

Many aircraft have two sets of ailerons—inboard and outboard. The inboard ailerons operate in all flight regimes. The outboard ailerons work only when the wing flaps are extended and are automatically locked out when flaps are retracted. This allows good roll response in low speed flight with the flaps extended and prevents excessive roll and wing bending at high speeds when the flaps are retracted.

Spoilers increase drag and reduce lift on the wing. If raised on only one wing, they aid roll control by causing that wing to drop. If the spoilers raise symmetrically in flight, the aircraft can either be slowed in level flight or can descend rapidly without an increase in airspeed. When the spoilers rise on the ground at high speeds, they destroy the wing's lift which puts more of the aircraft's weight on its wheels which in turn makes the brakes more effective.

Often aircraft have both flight and ground spoilers. The flight spoilers are available both in flight and on the ground. However, the ground spoilers can only be raised when the weight of the aircraft is on the landing gear. When the spoilers deploy on the ground, they decrease lift and make the brakes more effective. In flight, a ground-sensing switch on the landing gear prevents deployment of the ground spoilers.

Vortex generators are small (an inch or so high) aerodynamic surfaces located in different places on different airplanes. They prevent undesirable airflow separation from the surface by mixing the boundary airflow with the high energy airflow just above the surface. When located on the upper surface of a wing, the vortex generators prevent shock-induced separation from the wing as the aircraft approaches its critical Mach number. This increases aileron effectiveness at high speeds.

121, 135, DSP

8326. Which of the following is considered a primary flight control?

A—Slats.
B—Elevator.
C—Dorsal fin.

The primary group of flight control surfaces consists of ailerons, elevators, and rudders. (PLT346) — FAA-H-8083-25

Answer (A) is incorrect because slats are high-lift devices, not a flight control device. Answer (C) is incorrect because a dorsal fin is not a primary flight control, but is used to provide directional stability.

121, 135, DSP

8327. Which of the following is considered an auxiliary flight control?

A—Ruddervator.
B—Upper rudder.
C—Leading-edge flaps.

Auxiliary wing flight surfaces include trailing edge flaps, leading edge flaps, speed brakes, spoilers, and leading edge slats. (PLT473) — FAA-H-8083-25

Answer (A) is incorrect because a ruddervator is a primary flight control surface that incorporates both a rudder and elevator into one surface. Answer (B) is incorrect because upper rudders (found on the B-727) are stand-by rudders, which are used in the event of a hydraulic system failure.

121, 135, DSP

8343. Which of the following are considered primary flight controls?

A—Tabs.
B—Flaps.
C—Outboard ailerons.

The primary group of flight control surfaces consists of ailerons, elevators, and rudders. (PLT346) — FAA-H-8083-25

121, 135, DSP

8324. When are inboard ailerons normally used?

A—Low-speed flight only.
B—High-speed flight only.
C—Low-speed and high-speed flight.

During low-speed flight, all lateral control surfaces operate to provide maximum stability. This includes all four ailerons, flaps, and spoilers. (PLT346) — FAA-H-8083-25

Answers (A) and (B) are incorrect because the inboard ailerons are used during both high- and low-speed flight.

121, 135, DSP

8325. When are outboard ailerons normally used?

A—Low-speed flight only.
B—High-speed flight only.
C—Low-speed and high-speed flight.

At high speeds, flaps are retracted and the outboard ailerons are locked out of the aileron control system. (PLT346) — FAA-H-8083-25

Answers (B) and (C) are incorrect because the outboard ailerons are locked when airspeed is increased and flaps are raised.

121, 135, DSP

8342. Why do some airplanes equipped with inboard/outboard ailerons use the outboards for slow flight only?

A—Increased surface area provides greater controllability with flap extension.
B—Aerodynamic loads on the outboard ailerons tend to twist the wingtips at high speeds.
C—Locking out the outboard ailerons in high-speed flight provides variable flight control feel.

Aerodynamic loads on the outboard ailerons tend to twist the wing tips at high speeds. This results in deformation great enough to nullify the effect of aileron deflection and create rolling moments opposite to the direction commanded. Because of this, outboard ailerons are used for slow flight only. (PLT346) — FAA-H-8083-25

Answers

8326 [B]	8327 [C]	8343 [C]	8324 [C]	8325 [A]	8342 [B]

121, 135, DSP

8332. What is a purpose of flight spoilers?

A—Increase the camber of the wing.
B—Reduce lift without decreasing airspeed.
C—Direct airflow over the top of the wing at high angles of attack.

The purpose of the spoilers is to disturb the smooth airflow across the top of the wing thereby creating an increased amount of drag and a reduced amount of lift. (PLT473) — FAA-H-8083-25

Answer (A) is incorrect because flaps (not spoilers) increase the camber of the wing. Answer (C) is incorrect because slots and slats direct airflow over the top of the wing at high angles of attack.

121, 135, DSP

8333. For which purpose may flight spoilers be used?

A—Reduce the wings' lift upon landing.
B—Increase the rate of descent without increasing aerodynamic drag.
C—Aid in longitudinal balance when rolling an airplane into a turn.

An additional purpose or use for flight spoilers is to reduce lift when the aircraft lands. (PLT473) — FAA-H-8083-25

Answer (B) is incorrect because spoilers will increase the aerodynamic drag. Answer (C) is incorrect because trim devices (not spoilers) aid in balancing forces on an aircraft about the three axes.

121, 135, DSP

8336. Which is a purpose of ground spoilers?

A—Reduce the wings' lift upon landing.
B—Aid in rolling an airplane into a turn.
C—Increase the rate of descent without gaining airspeed.

Ground spoilers are speed brakes extended on the ground or landing roll to kill lift and keep the aircraft from flying again after touchdown. (PLT473) — FAA-H-8083-25

Answer (B) is incorrect because an aid in rolling an airplane into a turn is a flight spoiler. Answer (C) is incorrect because increasing the rate of descent without gaining airspeed is the purpose of a flight spoiler.

121, 135, DSP

9793. Upon landing, spoilers

A—decrease directional stability on the landing rollout.
B—function by increasing tire to ground friction.
C—should be extended after the thrust reversers have been deployed.

Spoilers should be deployed immediately after touchdown because they are most effective at high speed. The spoilers increase wheel loading by as much as 200 percent in the landing flap configuration. This increases the tire ground friction force making the maximum tire braking and cornering forces available. (PLT170) — FAA-H-8083-3A

121, 135, DSP

8341. Which is a purpose of wing-mounted vortex generators?

A—Delays the onset of drag divergence at high speeds and aids in maintaining aileron effectiveness at high speeds.
B—Increase the onset of drag divergence and aid in aileron effectiveness at low speeds.
C—Breaks the airflow over the wing so the stall will progress from the root out to the tip of the wing.

"Vortex generators" are used to delay or prevent shock wave-induced boundary layer separation encountered in transonic flight. Vortex generators create a vortex which mixes the boundary airflow with the high energy airflow just above the surface. This produces higher surface velocities and increases the energy of the boundary layer. Thus, a stronger shock wave will be necessary to produce airflow separation. (PLT266) — FAA-H-8083-25

Answer (B) is incorrect because vortex generators are most effective at high speeds and the increased drag that they produce is not their primary function. Answer (C) is incorrect because a stall strip breaks the airflow over the wing so the stall will progress from the root out to the tip of the wing.

121, 135, DSP

8356. Airflow separation over the wing can be delayed by using vortex generators

A—directing high pressure air over the top of the wing or flap through slots and making the wing surface smooth.
B—directing a suction over the top of the wing or flap through slots and making the wing surface smooth.
C—making the wing surface rough and/or directing high pressure air over the top of the wing or flap through slots.

Vortex generators prevent undesirable airflow separation from the surface by mixing the boundary airflow with the high energy airflow just above the surface. The vortex generators mix the turbulent outer layers of the boundary layers with the slow moving laminar lower layers thus reenergizing them. (PLT266) — FAA-H-8083-25

Answers

8332 [B] 8333 [A] 8336 [A] 9793 [B] 8341 [A] 8356 [C]

ALL

9759. If the boundary layer separates

A—drag is decreased.
B—the wing is about to stall and stop producing lift.
C—ice will sublimate and not freeze.

The boundary layer gives any object an "effective" shape that is usually slightly different from the physical shape. The boundary layer may also separate from the body, thus creating an effective shape much different from the physical shape of the object. This change in the physical shape of the boundary layer causes a dramatic decrease in lift and an increase in drag. When this happens, the airfoil has stalled. (PLT266) — FAA-H-8083-25

Tabs

Flight control surfaces are sometimes equipped with **servo tabs**. These tabs are on the trailing edge of the control surface and are mechanically linked to move opposite the direction of the surface. If the tab moves up, the surface moves down. This "servo" movement moves the control surface. *See* Figure 3-6.

One method of modifying the downward tail load through changes in airspeed and configuration is by using **trim tabs**. Trim tabs are moved by a separate trim control from the cockpit. Movement of the trim tab (like the servo tab) is opposite that of the primary control surface.

Anti-servo tabs move in the same direction as the primary control surface. This means that as the control surface deflects, the aerodynamic load is increased by movement of the anti-servo tab. This helps to prevent the control surface from moving to a full deflection. It also makes a hydraulically-boosted flight control more aerodynamically effective than it would otherwise be.

Some jet aircraft have **control tabs** for use in the event of loss of all hydraulic pressure. Movement of the control wheel moves the control tab which causes the aerodynamic movement of the control surface. The control tab is used only during manual reversion; that is, with the loss of hydraulic pressure. They work the same as a servo tab but only in the manual mode.

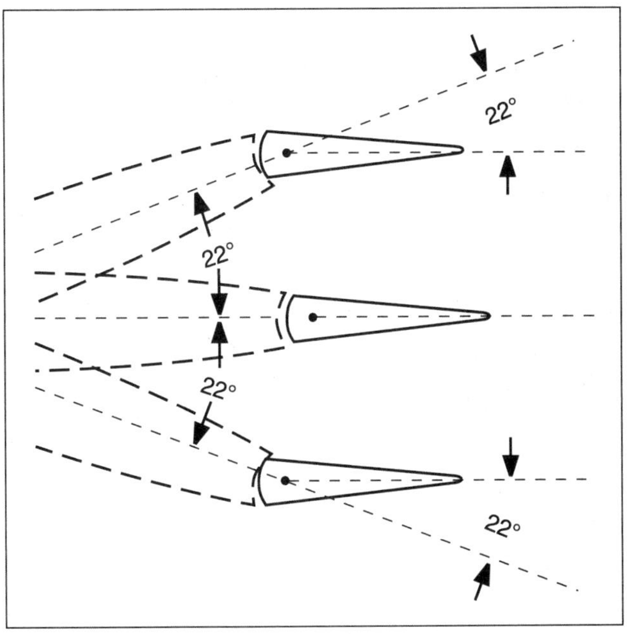

Figure 3-6

121, 135, DSP

8330. What is the purpose of a servo tab?

A—Move the flight controls in the event of manual reversion.
B—Reduce control forces by deflecting in the proper direction to move a primary flight control.
C—Prevent a control surface from moving to a full-deflection position due to aerodynamic forces.

The servo tab moves in response to the cockpit control. The force of the airflow on the servo tab then moves the primary control surface. (PLT473) — FAA-H-8083-25

Answer (A) is incorrect because, in the event of "manual reversion" on some transport category aircraft, the control tabs would move the flight controls. Answer (C) is incorrect because the purpose of the anti-servo tab is to preclude full deflection of control surfaces.

Answers

9759 [B] 8330 [B]

121, 135, DSP

8338. Which direction from the primary control surface does a servo tab move?

A—Same direction.
B—Opposite direction.
C—Remains fixed for all positions.

The servo tab attached to the flight control moves in the opposite direction, to assist in moving and holding the flight control by way of the airflow against it. (PLT473) — FAA-H-8083-25

Answer (A) is incorrect because an anti-servo tab, as found on the trailing edge of stabilators, moves in the same direction as the stabilator to provide a feel to the pilot control pressures. Answer (C) is incorrect because servo tabs move in response to the pilot's control movements.

121, 135, DSP

8339. Which direction from the primary control surface does an elevator adjustable trim tab move when the control surface is moved?

A—Same direction.
B—Opposite direction.
C—Remains fixed for all positions.

Trim tabs remain fixed for all positions of primary control surface movement until mechanically adjusted from the cockpit. A trim tab is hinged to its parent primary control surface but is operated by an independent control. (PLT473) — FAA-H-8083-25

Answer (A) is incorrect because once adjusted, trim tabs remain fixed to the primary control surface as the primary control surface is moved. Answer (B) is incorrect because a tab having a linkage designed to move in the opposite direction from the main control surface is called a balance tab.

121, 135, DSP

8340. What is the purpose of an elevator trim tab?

A—Provide horizontal balance as airspeed is increased to allow hands-off flight.
B—Adjust the speed tail load for different airspeeds in flight allowing neutral control forces.
C—Modify the downward tail load for various airspeeds in flight eliminating flight-control pressures.

The air flowing downward behind the trailing edge of the wing strikes the upper surface of the horizontal stabilizer, creating a downward tail force. The use of the trim tab will allow the pilot to reduce the hinge moment to zero and trim the control forces to zero for a given flight condition. (PLT473) — FAA-H-8083-25

Answer (A) is incorrect because the elevator trim tab permits "hands-off" flight at any airspeed, not only when the airspeed is increasing, and provides longitudinal balance. Answer (B) is incorrect because the elevator trim tab adjusts the downward tail load for various airspeeds in flight allowing neutral control forces.

121, 135, DSP

8329. What is the purpose of an anti-servo tab?

A—Move the flight controls in the event of manual reversion.
B—Reduce control forces by deflecting in the proper direction to move a primary flight control.
C—Prevent a control surface from moving to a full-deflection position due to aerodynamic forces.

Anti-servo tabs add resistance and increase control forces required as the surface moves toward its limit to avoid overcontrolling and offset deflection caused by aerodynamic forces rather than control input. (PLT473) — FAA-H-8083-25

Answer (A) is incorrect because in the event of "manual reversion" the control tabs (not servo or anti-servo tabs) would move the flight controls. Answer (B) is incorrect because the anti-servo usually will increase the pressure required to deflect the control surfaces (the anti-servo tab moves in the same direction as the control surface).

121, 135, DSP

8337. Which direction from the primary control surface does an anti-servo tab move?

A—Same direction.
B—Opposite direction.
C—Remains fixed for all positions.

Anti-servo tabs move in the same direction as the primary control surface. (PLT346) — FAA-H-8083-25

Answer (B) is incorrect because a servo tab moves in the opposite direction from the primary control surface. Answer (C) is incorrect because trim tabs remain fixed during control inputs.

Answers

8338 [B]	8339 [C]	8340 [C]	8329 [C]	8337 [A]

121, 135, DSP

8328. What is the purpose of a control tab?

A—Move the flight controls in the event of manual reversion.
B—Reduce control forces by deflecting in the proper direction to move a primary flight control.
C—Prevent a control surface from moving to a full-deflection position due to aerodynamic forces.

The flight controls of large jet airplanes are usually hydraulically powered. They are equipped with control tabs in the event of a total hydraulic failure. When this happens, the control tab can still be moved with control wheel input and the tab displacement creates an aerodynamic force which moves the control surface. (PLT473) — FAA-H-8083-25

Answer (B) is incorrect because servo tabs reduce control forces by deflecting primary flight controls in the proper direction. Answer (C) is incorrect because this is the purpose and function of an anti-servo tab.

High-Lift Devices

Sweptwing jet aircraft are equipped with a number of high-lift devices. These include leading edge flaps, slots or slats, and trailing edge flaps. The primary purpose of high-lift devices (flaps, slots, slats, etc.) is to increase the **maximum coefficient of lift** (CL_{MAX}) of the airplane and reduce the stall speed. The takeoff and landing speeds are consequently reduced.

The two most common types of **leading-edge devices** are **slats** and **Krueger flaps**. The Krueger flap extends from the leading edge of the wing, increasing its camber. The slat also extends from the wing's leading edge but it creates a gap or slot. This slot allows high energy from under the wing to flow over the top of the wing that delays stall to a higher angle of attack than would otherwise occur. It is common to find Krueger flaps and slats on the same wing.

121, 135, DSP

8384. The primary purpose of high-lift devices is to increase the

A—L/D_{MAX}.
B—lift at low speeds.
C—drag and reduce airspeed.

The primary purpose of high-lift devices (flaps, slots, slats, etc.) is to increase the CL_{MAX} of the airplane and to reduce the stall speed. The takeoff and landing speeds are consequently reduced. (PLT266) — ANA

Answer (A) is incorrect because increasing the lift component is an objective of high-lift devices which increase the ratio of L/D. The primary purpose of high-lift devices is to increase lift at low speeds. Answer (C) is incorrect because increasing the drag to reduce airspeed is the function of spoilers, not high-lift devices.

121, 135, DSP

8331. Which is a purpose of leading-edge flaps?

A—Increase the camber of the wing.
B—Reduce lift without increasing airspeed.
C—Direct airflow over the top of the wing at high angles of attack.

The leading-edge flap extends in a downward direction to increase the camber or total curve of the wing's shape. (PLT473) — ANA

Answer (B) is incorrect because leading-edge flaps increase the maximum lift coefficient at higher angles of attack. Answer (C) is incorrect because a slot will direct airflow over the top of the wing at high angles of attack.

121, 135, DSP

8385. What is the primary function of the leading edge flaps in landing configuration during the flare before touchdown?

A—Prevent flow separation.
B—Decrease rate of sink.
C—Increase profile drag.

The primary purpose of high-lift devices (flaps, slots, slats, etc.) is to increase the CL_{MAX} of the airplane and to reduce the stall speed. The takeoff and landing speeds are consequently reduced. (PLT266) — ANA

Answer (B) is incorrect because, to decrease the rate of sink, the coefficient of lift (CL) must increase. This is done by using leading-edge lift devices, slats, flaps, and other devices with the correct power setting. Answer (C) is incorrect because spoilers increase profile drag and are usually deployed after touchdown to reduce lift.

Answers

8328 [A] 8384 [B] 8331 [A] 8385 [A]

121, 135, DSP

8334. Which is a purpose of leading-edge slats on high-performance wings?

A—Increase lift at relative slow speeds.
B—Improve aileron control during low angles of attack.
C—Direct air from the high pressure area under the leading edge along the top of the wing.

The primary purpose of high-lift devices (flaps, slots, slats, etc.) is to increase the CL_{MAX} of the airplane and reduce the stall speed. The takeoff and landing speeds are consequently reduced. (PLT473) — ANA

Answer (A) is incorrect because lift is increased at relatively slow speeds. Answer (B) is incorrect because leading-edge slats are used at high angles of attack.

121, 135, DSP

8386. What effect does the leading edge slot in the wing have on performance?

A—Decreases profile drag.
B—Changes the stalling angle of attack to a higher angle.
C—Decelerates the upper surface boundary layer air.

The slot delays stall to a higher angle of attack. (PLT266) — ANA

Answer (A) is incorrect because, at low angles of attack, there is little or no profile drag increase. At high angles of attack, the slot delays the stall characteristics of the wing. Answer (C) is incorrect because the leading-edge slot actually increases airflow on the upper wing surface to allow higher angles of attack.

121, 135, DSP

9765. What is a difference between the fowler flap system and split flap system?

A—Fowler flaps produce the greatest change in pitching moment.
B—Fowler flaps produce more drag.
C—Split flaps cause the greatest change in twisting loads.

Fowler flaps slide out and downward from the trailing edge of the wing. When lowered, they increase the wing area as well as the wing camber. The Fowler flap is characterized by the largest increase in CL_{MAX} with the least changes in drag. The Fowler flap also creates the greatest change in pitching moment. Split flaps consist of a hinged plate that deflects downward from the lower surface of the wing and produce the least change in the pitching moments of a wing when it is lowered. The deflection of a flap causes large nose-down moments which create significant twisting loads on the structure and pitching moments that must be controlled with the horizontal tail. Unfortunately, the flap types producing the greatest increases in CL_{MAX} usually cause the greatest twisting moments. The Fowler flap causes the greatest change in twisting moment while the split flap causes the least. (PLT266) — ANA

121, 135, DSP

9766. On which type of wing are flaps most effective?

A—Thin wing.
B—Thick wing.
C—Sweptback wing.

The effectiveness of flaps on a wing configuration depends on many different factors, of which an important one is the amount of the wing area affected by the flaps. Since a certain amount of the span is reserved for ailerons, the actual wing maximum lift properties will be less than that resulting from the flapped two-dimensional section. If the basic wing has a low thickness, any type of flap will be less effective than on a wing of greater thickness. Sweepback of the wing can also cause a significant reduction in the effectiveness of flaps. (PLT266) — ANA

121, 135, DSP

9771. When compared to plain flaps, split flaps

A—produce more lift with less drag.
B—produce only slightly more lift, but much more drag.
C—enhance takeoff performance in high density conditions.

The split flap produces a slightly greater change in CL_{MAX} than the plain flap. However, a much larger change in drag results from the substantial and turbulent wake produced by this type of flap; although greater drag may be advantageous, for example, when steeper landing approaches over obstacles are required. (PLT266) – ANA

Answers

8334 [C] 8386 [B] 9765 [A] 9766 [B] 9771 [B]

Helicopter Aerodynamics

RTC
8355. What is the ratio between the total load supported by the rotor disc and the gross weight of a helicopter in flight?

A—Power loading.
B—Load factor.
C—Aspect ratio.

The load factor is the actual load on the rotor blades at any time, divided by the gross weight (or apparent gross weight; i.e., when the helicopter is in a bank, the apparent gross weight increases). (PLT310) — FAA-H-8083-21

RTC
8402. How should the pilot execute a pinnacle-type approach to a rooftop heliport in conditions of high wind and turbulence?

A—Steeper-than-normal approach, maintaining the desired angle of descent with collective.
B—Normal approach, maintaining a slower-than-normal rate of descent with cyclic.
C—Shallow approach, maintaining a constant line of descent with cyclic.

High winds can cause severe turbulence and downdrafts on the leeward side of rooftop helipads. Under these conditions, a steeper-than-normal approach to avoid downdrafts is desired. Angle of descent is maintained with collective and rate of closure (airspeed) is controlled with cyclic. (PLT170) — FAA-H-8083-21

RTC
8403. How should a quick stop be initiated?

A—Raise collective pitch.
B—Apply aft cyclic.
C—Decrease RPM while raising collective pitch.

The deceleration (or slowing) is initiated by applying aft cyclic to reduce forward speed and simultaneously lowering the collective to counteract any ballooning or climbing tendency. (PLT170) — FAA-H-8083-21

RTC
8404. How does V_{NE} speed vary with altitude?

A—Varies directly with altitude.
B—Remains the same at all altitudes.
C—Varies inversely with altitude.

V_{NE} decreases as altitude increases in most helicopters. Thinner air at higher altitudes means that the rotor blades must operate at higher angles of attack to produce the same amount of lift as they would at lower altitudes. Also, the airspeed indicator reads lower than the TAS at higher altitudes. (PLT124) — FAA-H-8083-21

RTC
8405. What limits the high airspeed potential of a helicopter?

A—Harmonic resonance.
B—Retreating blade stall.
C—Rotor RPM limitations.

The airflow over the retreating blade decreases and the airflow over the advancing blade increases in forward flight. To correct for the resulting dissymmetry of lift, the retreating blade must operate at increasingly higher angles of attack as the forward speed increases, until the retreating blade will stall at some high forward airspeed. (PLT124) — FAA-H-8083-21

RTC
8406. What corrective action can a pilot take to recover from settling with power?

A—Increase forward speed and raise collective pitch.
B—Decrease forward speed and partially raise collective pitch.
C—Increase forward speed and partially lower collective pitch.

By increasing forward speed and/or (if possible) partially lowering collective pitch, the conditions necessary for settling with power are reduced or eliminated. (PLT208) — FAA-H-8083-21

Answers

8355 [B]	8402 [A]	8403 [B]	8404 [C]	8405 [B]	8406 [C]

RTC

8408. The lift differential that exists between the advancing main rotor blade and the retreating main rotor blade is known as

A—Coriolis effect.
B—dissymmetry of lift.
C—translating tendency.

Dissymmetry of lift is created by horizontal flight or by wind during hovering flight, and is the difference in lift that exists between the advancing blade of the rotor disc and the retreating blade of the rotor disc. (PLT470) — FAA-H-8083-21

RTC

8409. During a hover, a helicopter tends to drift in the direction of tail rotor thrust. What is this movement called?

A—Translating tendency.
B—Transverse flow effect.
C—Gyroscopic precession.

The entire helicopter has a tendency to move in the direction of tail rotor thrust when hovering, which is often referred to as "drift or translating tendency." (PLT268) — FAA-H-8083-21

RTC

8410. What is the purpose of the lead-lag (drag) hinge in a three-bladed, fully articulated helicopter rotor system?

A—Offset lateral instability during autorotation.
B—Compensate for Coriolis effect.
C—Provide geometric balance.

When a rotor blade of a three-bladed rotor system flaps upward, the center mass of that blade moves closer to the axis of rotation and blade acceleration takes place. When the blade flaps downward, its center of mass moves further from the axis of rotation and blade deceleration (or slowing) occurs. This increase and decrease of blade velocity in the plane of rotation due to mass movement is known as Coriolis effect. The acceleration and deceleration actions (leading and lagging) are absorbed by dampers or the blade structure itself (hinges) in a three-bladed system. (PLT470) — FAA-H-8083-21

RTC

8411. During an autorotation (collective pitch full down), what is an increase in rotor RPM associated with?

A—An increase in airflow through the rotor system.
B—A decrease in airflow through the rotor system.
C—A decrease in airspeed.

During an autorotation, the flow of air is upward through the rotor. The portion of the blade that produces the forces causing the rotor to turn in autorotation (approximately 25 to 70% of the radius outward from the center) is the driving region. An increase in the aerodynamic forces along the driving region (increase in the airflow through the rotor) tends to speed up the blade rotation. (PLT470) — FAA-H-8083-21

RTC

8412. What corrective action can a pilot take to prevent a retreating blade stall at its onset?

A—Reduce collective pitch and increase rotor RPM.
B—Increase collective pitch and increase rotor RPM.
C—Reduce collective pitch and decrease rotor RPM.

At the onset of blade stall vibration, the pilot should reduce collective pitch, increase rotor RPM, reduce forward airspeed and minimize maneuvering. (PLT470) — FAA-H-8083-21

RTC

8413. Which is a major warning of approaching retreating blade stall?

A—High frequency vibration.
B—Tendency to roll opposite the stalled side of the rotor.
C—Pitchup of the nose.

The major warnings of approaching retreating blade stall are: low-frequency vibration equal to the number of blades per revolution of the main rotor system, pitchup of the nose and tendency for the helicopter to roll towards the stalled (retreating blade) side of the rotor system. (PLT470) — FAA-H-8083-21

Answers

8408 [B]	8409 [A]	8410 [B]	8411 [A]	8412 [A]	8413 [C]

RTC

8417. How does high density altitude affect helicopter performance?

A—Engine and rotor efficiency are increased.
B—Engine and rotor efficiency are reduced.
C—Engine efficiency is reduced, but rotor efficiency is increased.

High elevations, high temperatures, and high moisture content (relative humidity) all contribute to high density altitude, which lessens helicopter performance. The thinner air at high density altitudes reduces the amount of lift of the rotor blades, and unsupercharged engines produce less power. (PLT124) — FAA-H-8083-21

RTC

8418. How is helicopter climb performance most adversely affected?

A—Higher-than-standard temperature and high relative humidity.
B—Lower-than-standard temperature and high relative humidity.
C—Higher-than-standard temperature and low relative humidity.

High elevations, high temperatures, and high moisture content (relative humidity) all contribute to high density altitude, which lessens helicopter performance. The thinner air at high density altitudes reduces the amount of lift of the rotor blades, and unsupercharged engines produce less power. (PLT124) — FAA-H-8083-21

RTC

8420. What causes Coriolis effect?

A—Differential thrust of rotor blades.
B—Changing angle of attack of blades during rotation.
C—Shift in center of mass of flapping blade.

When a rotor blade of a three-bladed rotor system flaps upward, the center mass of that blade moves closer to the axis of rotation and blade acceleration takes place. When the blade flaps downward, its center of mass moves further from the axis of rotation and blade deceleration (or slowing) occurs. This increase and decrease of blade velocity in the plane of rotation due to mass movement is known as Coriolis effect. (PLT197) — FAA-H-8083-21

RTC

8421. Why are the rotor blades more efficient when operating in ground effect?

A—Induced drag is reduced.
B—Induced angle of attack is increased.
C—Downwash velocity is accelerated.

When a helicopter is operated near the surface, the downwash velocity of the rotor blades cannot be fully developed. The reduction in downwash velocity causes the induced angle of attack of each rotor blade to be reduced, which causes the induced drag to be less. (PLT237) — FAA-H-8083-21

RTC

8422. What result does a level turn have on the total lift required and load factor with a constant airspeed?

A—Lift required remains constant and the load factor increases.
B—Lift required increases and the load factor decreases.
C—Both total lift force and load factor increase.

When a helicopter is placed in a bank, the resultant lifting force acts more horizontally and less vertically. To maintain a level turn, the resultant lifting force (total lift force) must be increased. When a helicopter assumes a curved flight path, centrifugal force causes additional stresses (load factor) to be imposed. (PLT248) — FAA-H-8083-21

RTC

8423. What causes a helicopter to turn?

A—Centrifugal force.
B—Horizontal component of lift.
C—Greater angle of attack of rotor blades on upward side of the rotor disc.

When a helicopter is placed in a bank, the rotor disc is tilted sideward causing the horizontal component of lift to be increased. The increased horizontal lift component pulls the helicopter from its straight course. (PLT248) — FAA-H-8083-21

Answers

8417 [B] 8418 [A] 8420 [C] 8421 [A] 8422 [C] 8423 [B]

RTC
8424. What is the primary purpose of the tail rotor system?

A—Maintain heading during forward flight.
B—Act as a rudder to assist in coordinated turns.
C—Counteract the torque effect of the main rotor.

As the main rotor of a helicopter turns in one direction, the fuselage tends to rotate in the opposite direction (Newton's Third Law of Motion: For every action there is an equal and opposite reaction). This tendency to rotate is called torque. The tail rotor is used to produce thrust to counteract the torque effect of the main rotor. (PLT470) — FAA-H-8083-21

RTC
8425. Under what condition would it be necessary to cause the tail rotor to direct thrust to the left on an American-made helicopter?

A—To maintain heading with a left crosswind.
B—To counteract the drag of the transmission during autorotation.
C—To execute hovering turns to the right.

The capability for tail rotors to produce thrust to the left (negative pitch angle) is necessary because during autorotation, the drag of the transmission (with no torque effect present) tends to yaw the nose to the left, in the same direction that the main rotor is turning. (PLT470) — FAA-H-8083-21

RTC
9318. Which statement describes the term "V_{TOSS}"?

A—The takeoff safety speed in a turbine-engine powered transport category airplane.
B—The takeoff safety speed in a Category A helicopter.
C—The takeoff stall speed in the takeoff configuration in a turbopropeller powered airplane.

V_{TOSS} means takeoff safety speed for Category A helicopters. (PLT466) — 14 CFR §1.2

Answers

8424 [C]	8425 [B]	9318 [B]

Chapter 4
Performance

Engine Performance

Note applicable to Chapters 4 and 5: The 135 exam focuses on the BE-1900 and the 121 exam focuses on the B-727, B-737, and DC-9.

There are three types of engines in use on modern airplanes: **reciprocating engine, turboprop engine** and **turbojet engine**. The type of engine selected for a particular airplane design depends primarily on the speed range of the aircraft. The reciprocating engine is most efficient for aircraft with cruising speeds below 250 MPH, the turboprop works best in the 250 MPH to 450 MPH range and the turbojet engine is most efficient above 450 MPH.

Manifold pressure (MAP) is a measurement of the power output of a reciprocating engine. It is basically the pressure in the engine's air inlet system. In a normally-aspirated (unsupercharged) engine, the MAP will drop as the aircraft climbs to altitude. This severely limits a piston-powered airplane's altitude capability.

Most piston-powered airplanes flown by air carriers are turbocharged. On this type of engine, exhaust gas from the engine is used as a power source for a compressor that in turn raises the MAP at any given altitude. The flow of exhaust gas to the turbocharger is controlled by a device called a **waste gate**.

Turbocharging allows an aircraft to fly at much higher altitudes than it would be able to with normally-aspirated engines. The term **critical altitude** is used to describe the effect of turbocharging on the aircraft's performance. The critical altitude of a turbocharged reciprocating engine is the highest altitude at which a desired manifold pressure can be maintained.

The pilots of reciprocating-engine-powered aircraft must be very careful to observe the published limits on manifold pressure and engine RPM. In particular, high RPM and low MAP can produce severe wear, fatigue and damage.

Both turboprops and turbojet engines are types of **gas turbine engines**. All gas turbine engines consist of an air inlet section, a compressor section, the combustion section, the turbine section and the exhaust. Air enters the inlet at roughly ambient temperature and pressure. As it passes through the compressor the pressure increases and so does the temperature due to the heat of compression. **Bleed air** is tapped off the compressor for such accessories as air conditioning and thermal anti-icing.

The section connecting the compressor and the combustion sections is called the diffuser. In the diffuser, the cross sectional area of the engine increases. This allows the air stream from the compressor to slow and its pressure to increase. In fact, the highest pressure in the engine is attained at this point.

Next, the air enters the combustion section where it is mixed with fuel and the mixture is ignited. Note that there is no need for an ignition system that operates continuously (such as the spark plugs in a piston engine) because the uninterrupted flow of fuel and air will sustain combustion after an initial "light off." The combustion of the fuel-air mixture causes a great increase in volume and because there is higher pressure at the diffuser, the gas exits through the turbine section. The temperature of the gas rises rapidly as it passes from the front to the rear of the combustion section. It reaches its highest point in the engine at the turbine inlet. The maximum turbine inlet temperature is a major limitation on turbojet performance, and without cooling, it could easily reach up to 4,000°F, far beyond the limits of the materials used in the turbine section. To keep the temperature down to an acceptable 1,100° to 1,500°F, surplus cooling air from the compressor is mixed aft of the burners.

The purpose of the turbine(s) is to drive the compressor(s) and they are connected by a drive shaft. Since the turbines take energy from the gas, both the temperature and pressure drop.

The gases exit the turbine section at very high velocity into the tailpipe. The tailpipe is shaped so that the gas is accelerated even more, reaching maximum velocity as it exits into the atmosphere. *See* Figure 4-1.

Continued

Combinations of slow airspeed and high engine RPM can cause a phenomenon in turbine engines called **compressor stall**. This occurs when the angle of attack of the engine's compressor blades becomes excessive and they stall. If a transient stall condition exists, the pilot will hear an intermittent "bang" as backfires and flow reversals in the compressor take place. If the transient condition develops into a steady state stall, the pilot will hear a loud roar and experience severe engine vibrations. The steady state compressor stall has the most potential for severe engine damage, which can occur literally within seconds of the onset of the stall.

If a compressor stall occurs in flight, the pilot should reduce fuel flow, reduce the aircraft's angle of attack and increase airspeed.

The turboprop is a turbine engine that drives a conventional propeller. It can develop much more power per pound than can a piston engine and is more fuel efficient than the turbojet engine. Compared to a turbojet engine, it is limited to slower speeds and lower altitudes (25,000 feet to the tropopause). The term **equivalent shaft horsepower (ESHP)** is used to describe the total engine output. This term combines its output in shaft horsepower (used to drive the propeller) and the jet thrust it develops.

As the density altitude is increased, engine performance will decrease. When the air becomes less dense, there is not as much oxygen available for combustion and the potential thrust output is decreased accordingly. Density altitude is increased by increasing the pressure altitude or by increasing the ambient temperature. Relative humidity will also affect engine performance. Reciprocating engines in particular will experience a significant loss of BHP (Brake Horsepower). Turbine engines are not affected as much by high humidity and will experience very little loss of thrust.

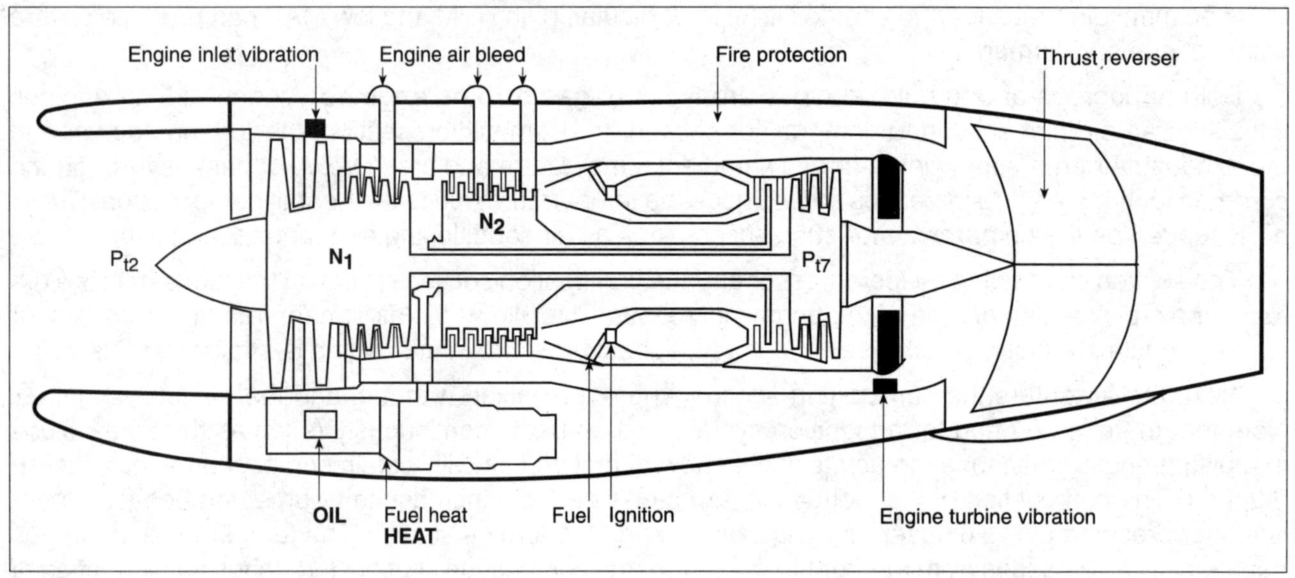

Figure 4-1. Turbojet engine

ALL
9072. Where is the critical altitude of a supercharged-reciprocating engine?

A—The highest altitude at which a desired manifold pressure can be obtained.
B—Highest altitude where the mixture can be leaned to best power ratio.
C—The altitude at which maximum allowable BMEP can be obtained.

The critical altitude of a supercharged reciprocating engine is the highest altitude at which a desired MAP can be maintained. (PLT343) — FAA-H-8083-25

Answer (B) is incorrect because critical altitude is the highest altitude at which a manifold pressure can be obtained. Answer (C) is incorrect because BMEP is pressure representing the mean gas load on the piston during the power stroke.

Answers
9072 [A]

ALL

9073. What is controlled by the waste gate of a turbo-charged-reciprocating engine?

A—Supercharger gear ratio.
B—Exhaust gas discharge.
C—Throttle opening.

A turbocharger drives exhaust gas from the engine. The waste gate controls the flow of the exhaust gas through the turbocharger's turbine. (PLT343) — FAA-H-8083-25

Answer (A) is incorrect because supercharger gear ratio is not controlled by the waste gate. Answer (C) is incorrect because the throttle opening sets the desired manifold pressure.

ALL

9068. Under normal operating conditions, which combination of MAP and RPM produce the most severe wear, fatigue, and damage to high performance reciprocating engines?

A—High RPM and low MAP.
B—Low RPM and high MAP.
C—High RPM and high MAP.

The most severe rate of wear and fatigue damage occurs at high RPM and low MAP. (PLT365) — FAA-H-8083-25

Answer (B) is incorrect because while low RPM and high MAP produce severe wear to high performance reciprocating engines, the most damage is done by high RPM and low manifold pressure. Answer (C) is incorrect because a high RPM and a low MAP produce the most severe wear to high performance reciprocating engines.

ALL

9058. Which place in the turbojet engine is subjected to the highest temperature?

A—Compressor discharge.
B—Fuel spray nozzles.
C—Turbine inlet.

The highest temperatures in any turbine engine will occur at the turbine inlet. This TIT (Turbine Inlet Temperature) is usually the limiting factor in the engine operation. (PLT499) — FAA-H-8083-25

ALL

9060. The most important restriction to the operation of turbojet or turboprop engines is

A—limiting compressor speed.
B—limiting exhaust gas temperature.
C—limiting torque.

The highest temperatures in any turbine engine will occur at the turbine inlet. This TIT (Turbine Inlet Temperature) is usually the limiting factor in the engine operation. In many engines, TIT is measured indirectly as EGT (Exhaust Gas Temperature). (PLT499) — FAA-H-8083-25

Answer (A) is incorrect because the turbine section is the most critical element of the turbojet engine. Temperature control is more restrictive than compressor speed, which may operate above 10,000 RPM continuously. Answer (C) is incorrect because torque is a performance measure used on turbopropeller airplanes, but not generally applicable to turbojet engines. The most important restriction is temperature, even though in cooler weather a torque limitation may be reached before the temperature limitation in a turbopropeller airplane.

ALL

9064. What characterizes a transient compressor stall?

A—Loud, steady roar accompanied by heavy shuddering.
B—Sudden loss of thrust accompanied by a loud whine.
C—Intermittent "bang," as backfires and flow reversals take place.

If a compressor stall is transient and intermittent, the indication will be an intermittent "bang" as backfire and flow reversal take place. If the stall develops and becomes steady, strong vibration and a loud roar develop from the continuous flow reversal. The possibility of damage is immediate from a steady stall. Recovery must be accomplished quickly by reducing throttle setting, lowering the airplane angle of attack, and increasing airspeed. (PLT343) — FAA-H-8083-25

Answer (A) is incorrect because this describes a developed and steady stall. Answer (B) is incorrect because a transient stall is characterized by an intermittent "bang."

Answers

9073 [B]	9068 [A]	9058 [C]	9060 [B]	9064 [C]

ALL

9768. What prevents turbine engines from developing compressor stalls?

A—Deice valves-fuel heat.
B—TKS system.
C—Compressor bleed valves.

At a low RPM, sudden full power application will tend to overfuel the engine resulting in possible compressor surge, excessive turbine temperatures, compressor stall, and/or flameout. To prevent this, various limiters such as compressor bleed valves are contained in the system and serve to restrict the engine until it is at an RPM at which it can respond to a rapid acceleration demand without distress. (PLT499) — FAA-H-8083-3

ALL

9065. What indicates that a compressor stall has developed and become steady?

A—Strong vibrations and loud roar.
B—Occasional loud "bang" and flow reversal.
C—Complete loss of power with severe reduction in airspeed.

If a compressor stall is transient and intermittent, the indication will be an intermittent "bang" as backfire and flow reversal take place. If the stall develops and becomes steady, strong vibration and a loud roar develop from the continuous flow reversal. The possibility of damage is immediate from a steady stall. Recovery must be accomplished quickly by reducing throttle setting, lowering the airplane angle of attack, and increasing airspeed. (PLT343) — FAA-H-8083-25

Answer (B) is incorrect because this describes an indication of a transient stall. Answer (C) is incorrect because a compressor stall will not cause a complete loss of power.

ALL

9066. Which type of compressor stall has the greatest potential for severe engine damage?

A—Intermittent "backfire" stall.
B—Transient "backfire" stall.
C—Steady, continuous flow reversal stall.

If a compressor stall is transient and intermittent, the indication will be an intermittent "bang" as backfire and flow reversal take place. If the stall develops and becomes steady, strong vibration and a loud roar develop from the continuous flow reversal. The possibility of damage is immediate from a steady stall. Recovery must be accomplished quickly by reducing throttle setting, lowering the airplane angle of attack, and increasing airspeed. (PLT343) — FAA-H-8083-25

ALL

8974. Which part(s) in the turbojet engine is subjected to the high temperatures and severe centrifugal forces?

A—Turbine wheel(s).
B—Turbine vanes.
C—Compressor rotor(s) or impeller(s).

The turbine wheels are found at the back of the turbine section, in the area of very high temperatures and high centrifugal forces. Very hot, high pressure gases enter the turbine section from the combustor. The function of the gas generator's turbine wheels is to transfer the energy from the hot, high pressure gases to drive the shaft which is connected to the compressor wheel at the front of the engine. This in turn compresses air into the combustor where fuel is added and ignited. During normal operations, the turbine wheel rotates at many thousands of RPM. (PLT499) — FAA-H-8083-25

Answer (B) is incorrect because although turbine vanes (inlet guide vanes) are exposed to higher temperatures, they are stationary and thus are not subject to centrifugal forces. Answer (C) is incorrect because turbine wheels, or disks, with their attached blades, are the most highly stressed components on a turbojet engine.

ALL

9067. What recovery would be appropriate in the event of compressor stall?

A—Reduce the throttle and then rapidly advance the throttle to decrease the angle of attack on the compressor blades, creating more airflow.
B—Reduce the throttle and then slowly advance the throttle again and decrease the aircraft's angle of attack.
C—Advance the throttle slowly to increase airflow and decrease the angle of attack on one or more compressor blades.

If a compressor stall is transient and intermittent, the indication will be an intermittent "bang" as backfire and flow reversal take place. If the stall develops and becomes steady, strong vibration and a loud roar develop from the continuous flow reversal. The possibility of damage is immediate from a steady stall. Recovery must be accomplished quickly by reducing throttle setting, lowering the airplane angle of attack, and increasing airspeed. (PLT343) — FAA-H-8083-25

Answers

9768 [C]	9065 [A]	9066 [C]	8974 [A]	9067 [B]

ALL
9070. Equivalent shaft horsepower (ESHP) of a turboprop engine is a measure of

A—turbine inlet temperature.
B—shaft horsepower and jet thrust.
C—propeller thrust only.

Turboprop engines get 15 to 25% of their total thrust output from jet exhaust. ESHP (Equivalent Shaft Horsepower) is the term used to describe the shaft horsepower applied to the propeller plus this jet thrust. (PLT500) — FAA-H-8083-25

ALL
9071. Minimum specific fuel consumption of the turboprop engine is normally available in which altitude range?

A—10,000 feet to 25,000 feet.
B—25,000 feet to the tropopause.
C—The tropopause to 45,000 feet.

Minimum specific fuel consumption of a turboprop engine will be obtained in an altitude range of 25,000 to 35,000 feet. The tropopause will be in the neighborhood of 35,000 feet depending on the season and latitude. (PLT130) — FAA-H-8083-25

ALL
9059. What effect would a change in ambient temperature or air density have on gas-turbine-engine performance?

A—As air density decreases, thrust increases.
B—As temperature increases, thrust increases.
C—As temperature increases, thrust decreases.

Turbine engine thrust varies directly with air density. As air density decreases, so does thrust. An increase in temperature will decrease air density. (PLT127) — FAA-H-8083-25

ALL
9061. As outside air pressure decreases, thrust output will

A—increase due to greater efficiency of jet aircraft in thin air.
B—remain the same since compression of inlet air will compensate for any decrease in air pressure.
C—decrease due to higher density altitude.

Thrust output decreases with increasing density altitude. Decreasing air pressure increases density altitude. (PLT127) — FAA-H-8083-25

ALL
9062. What effect will an increase in altitude have upon the available equivalent shaft horsepower (ESHP) of a turboprop engine?

A—Lower air density and engine mass flow will cause a decrease in power.
B—Higher propeller efficiency will cause an increase in usable power (ESHP) and thrust.
C—Power will remain the same but propeller efficiency will decrease.

As altitude is increased, the ESHP of a turboprop engine will decrease due to lower engine mass flow and decreased propeller efficiency. (PLT127) — FAA-H-8083-25

ALL
9063. What effect, if any, does high ambient temperature have upon the thrust output of a turbine engine?

A—Thrust will be reduced due to the decrease in air density.
B—Thrust will remain the same, but turbine temperature will be higher.
C—Thrust will be higher because more heat energy is extracted from the hotter air.

Turbine engine thrust varies directly with air density. As air density decreases, so does thrust. An increase in temperature will decrease air density. (PLT127) — FAA-H-8083-25

ALL
9069. What effect does high relative humidity have upon the maximum power output of modern aircraft engines?

A—Neither turbojet nor reciprocating engines are affected.
B—Reciprocating engines will experience a significant loss of BHP.
C—Turbojet engines will experience a significant loss of thrust.

While turbojet engines are almost unaffected by high relative humidity, reciprocating engines will experience a significant loss of BHP (Brake Horsepower). (PLT365) — FAA-H-8083-25

Answer (A) is incorrect because both reciprocating and turbojet engines are affected by high relative humidity to some degree. Answer (C) is incorrect because turbojet engines will have a negligible loss of thrust.

Answers

9070 [B] 9071 [B] 9059 [C] 9061 [C] 9062 [A] 9063 [A]
9069 [B]

Helicopter Systems

RTC
8407. Which type rotor system is more susceptible to ground resonance?

A—Fully articulated rotor system.
B—Semi-rigid rotor system.
C—Rigid rotor system.

Due to the lead/lag of the blades in a fully articulated rotor system, a shock from a landing gear striking the surface can be transmitted through the fuselage to the rotor, forcing the blades straddling the contact point closer together and unbalancing the rotor system. This can cause a pendulum-like oscillation which will increase rapidly unless immediate corrective action is taken. (PLT470) — FAA-H-8083-21

RTC
9781. Ground resonance occurs when

A—a fully articulated rotor system is unbalanced.
B—a semi-rigid rotor system is out of balance.
C—a pilot lands with over inflated tires.

Ground resonance is an aerodynamic phenomenon associated with fully articulated rotor systems. It develops when the rotor blades move out of phase with each other and cause the rotor disc to become unbalanced. (PLT470) — FAA-H-8083-21

RTC
8414. What type frequency vibration is associated with a defective transmission?

A—Low frequency only.
B—Medium or low frequency.
C—High or medium frequency.

High-frequency vibrations (2,000 cycles per minute or higher) are associated with the engine in most helicopters. Any bearings in the transmission that go bad will result in vibrations with frequencies directly related to the speed of the engine. (PLT472) — FAA-H-8083-21

RTC
8415. What type frequency vibration is associated with the main rotor system?

A—Low frequency.
B—Medium frequency.
C—High frequency.

Abnormal vibrations in the low-frequency range (100 to 400 cycles per minute) are always associated with the main rotor system, and will be somewhat related to the rotor RPM and the number of blades of the main rotor. (PLT472) — FAA-H-8083-21

RTC
9800. A medium or higher frequency vibration mainly present in the anti-torque pedals is

A—usually traceable to engine cooling fan assembly.
B—probably caused by the tail rotor.
C—to be expected and accepted as normal.

Medium-frequency vibrations are a result of trouble with the tail rotor in most helicopters. Improper rigging, imbalance, defective blades, or bad bearings in the tail rotor are all sources of these vibrations. If the vibration occurs only during turns, the trouble may be caused by insufficient tail rotor flapping action. (PLT472) — FAA-H-8083-21

RTC
8416. What type frequency vibration is indicative of a defective tail rotor system?

A—Low and medium frequency.
B—Medium and high frequency.
C—Low and high frequency.

Medium frequency vibrations (1,000–2,000 cycles per minute) and high frequency vibrations (2,000 cycles per minute or higher) are normally associated with out-of-balance components that rotate at a high RPM, such as the tail rotor, engine, cooling fans, and components of the drive train, including transmissions, drive shafts, bearings, pulleys, and belts. (PLT472) — FAA-H-8083-21

Answers

8407 [A] 9781 [A] 8414 [C] 8415 [A] 9800 [B] 8416 [B]

RTC

8426. What is the primary purpose of the free-wheeling unit?

A—To provide speed reduction between the engine, main rotor system, and tail rotor system.
B—To provide disengagement of the engine from the rotor system for autorotation purposes.
C—To transmit engine power to the main rotor, tail rotor, generator/alternator, and other accessories.

The freewheeling coupling provides for autorotative capabilities by disconnecting the rotor system from the engine when the engine stops or slows below the RPM of the rotor system. This allows the transmission to be driven by the main rotor. The tail rotor will continue to be turned by the transmission. (PLT470) — FAA-H-8083-21

RTC

8427. The main rotor blades of a fully articulated rotor system can

A—flap, drag, and feather collectively.
B—flap, drag, and feather independently of each other.
C—flap and drag individually, but can only feather collectively.

Fully articulated rotor systems usually have three or more rotor blades, each of which is attached to the rotor hub by a horizontal hinge (flapping), a vertical hinge (drag), and can be feathered (rotated about their spanwise axis) independently of the other blades. (PLT470) — FAA-H-8083-21

RTC

8428. The main rotor blades of a semi-rigid system can

A—flap and feather as a unit.
B—flap, drag, and feather independently.
C—flap and drag individually, but can only feather collectively.

In a semirigid (two-bladed) rotor system, the blades are rigidly interconnected to the hub and flap as a unit (when one blade flaps up, the other blade flaps down an equal amount). The swash plate changes the pitch angle in each blade (feathers) an equal amount. (PLT470) — FAA-H-8083-21

Takeoff Performance Terminology

Clearway—a plane beyond the end of a runway which does not contain obstructions and can be considered when calculating takeoff performance of turbine-powered transport category airplanes. The first segment of the takeoff of a turbine-powered airplane is considered complete when it reaches a height of 35 feet above the runway and has achieved V_2 speed. Clearway may be used for the climb to 35 feet.

Stopway—an area designated for use in decelerating an aborted takeoff. It cannot be used as a part of the takeoff distance but can be considered as part of the accelerate-stop distance. *See* Figure 4-2.

Regulation requires that a transport category airplane's takeoff weight be such that, if at any time during the takeoff run the critical engine fails, the airplane can either be stopped on the runway and stopway remaining, or that it can safely continue the takeoff. This means that a **maximum takeoff weight** must be computed for each takeoff. Factors which determine the maximum takeoff weight for an airplane include runway length, wind, flap position, runway braking action, pressure altitude and temperature.

In addition to the runway-limited takeoff weight, each takeoff requires a computation of a climb-limited takeoff weight that will guarantee acceptable climb performance after takeoff with an engine inoperative. The climb-limited takeoff weight is determined by flap position, pressure altitude and temperature.

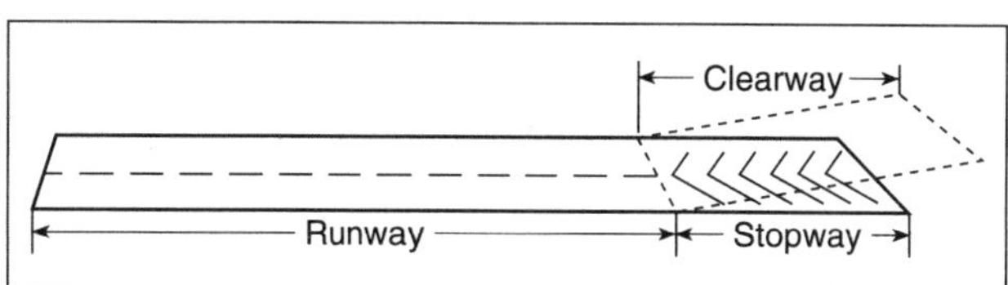

Figure 4-2. Takeoff runway definitions

Continued

Answers

8426 [B] 8427 [B] 8428 [A]

When the runway-limited and climb-limited takeoff weights are determined, they are compared to the maximum structural takeoff weight. The lowest of the three weights is the limit that must be observed for that takeoff. If the airplane's actual weight is at or below the lowest of the three limits, adequate takeoff performance is ensured. If the actual weight is above any of the limits a takeoff cannot be made until the weight is reduced or one or more limiting factors (runway, flap setting, etc.) is changed to raise the limiting weight.

After the maximum takeoff weight is computed and it is determined that the airplane's actual weight is within limits, then V_1, V_R and V_2 are computed. These takeoff speed limits are contained in performance charts and tables of the airplane flight manual, and are observed on the captain's airspeed indicator. By definition they are indicated airspeeds. *See* Figure 4-3.

V_1 (Takeoff Decision Speed) is the speed during the takeoff at which the airplane can experience a failure of the critical engine and the pilot can abort the takeoff and come to a full safe stop on the runway and stopway remaining, or the pilot can continue the takeoff safely. If an engine fails at a speed less than V_1, the pilot must abort; if the failure occurs at a speed above V_1 he/she must continue the takeoff.

V_R (Rotation Speed) is the IAS at which the aircraft is rotated to its takeoff attitude with or without an engine failure. V_R is at or just above V_1.

V_2 (Takeoff Safety Speed) ensures that the airplane can maintain an acceptable climb gradient with the critical engine inoperative.

V_{MU} (Minimum Unstick Speed) is the minimum speed at which the airplane may be flown off the runway without a tail strike. This speed is determined by manufacturer's tests and establishes minimum V_1 and V_R speeds. The flight crew does not normally compute the V_{MU} speed separately.

V_1 is computed using the actual airplane gross weight, flap setting, pressure altitude and temperature. Raising the pressure altitude, temperature or gross weight will all increase the computed V_1 speed. Lowering any of those variables will lower the V_1 speed.

A wind will change the takeoff distance. A headwind will decrease it and a tailwind will increase it. While a headwind or tailwind component does affect the runway limited takeoff weight, it usually has no direct effect on the computed V_1 speed. The performance tables for a few airplanes include a small correction to V_1 for very strong winds. For those airplanes, a headwind will increase V_1 and a tailwind will decrease it.

A runway slope has the same effect on takeoff performance as a wind. A runway which slopes uphill will increase the takeoff distance for an airplane and a downslope will decrease it. A significant slope may require an adjustment in the V_1 speed. An upslope will require an increase in V_1 and a downslope will require a decrease.

If there is slush on the runway or if the antiskid system is inoperative, the stopping performance of the airplane is degraded. This requires that any aborted takeoff be started at a lower speed and with more runway and stopway remaining. This means that both the runway-limited takeoff weight and the V_1 used for takeoff be lower than normal.

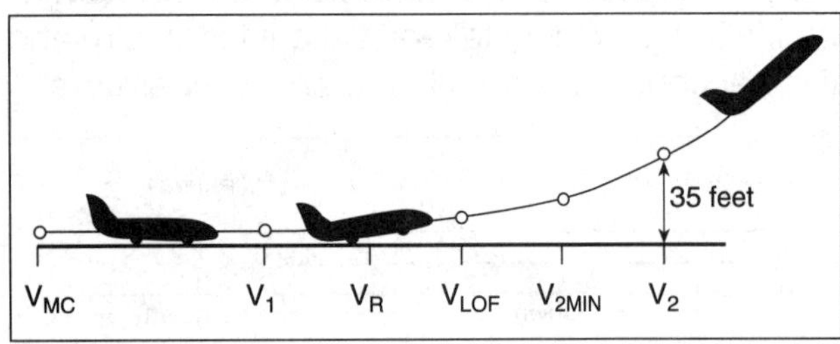

Figure 4-3. Takeoff speeds

121, 135, DSP

9324. What is the name of an area beyond the end of a runway which does not contain obstructions and can be considered when calculating takeoff performance of turbine-powered aircraft?

A—Clearway.
B—Stopway.
C—Obstruction clearance plane.

"Clearway" means, for turbine-powered airplanes, an area beyond the end of the runway, centrally located about the extended centerline and under the control of the airport authorities. Clearway distance may be used in the calculation of takeoff distance. (PLT395) — 14 CFR §1.1

Answer (B) is incorrect because a stopway is an area beyond the takeoff runway, not any less wide than the runway, centered upon the extended centerline of the runway, and able to support the airplane during an aborted takeoff. Answer (C) is incorrect because an obstruction clearance plane is not defined in 14 CFR Part 1.

121, 135, DSP

9327. What is an area identified by the term "stopway"?

A—An area, at least the same width as the runway, capable of supporting an airplane during a normal takeoff.
B—An area designated for use in decelerating an aborted takeoff.
C—An area, not as wide as the runway, capable of supporting an airplane during a normal takeoff.

"Stopway" means an area beyond the takeoff runway, able to support the airplane, for use in decelerating the airplane during an aborted takeoff. (PLT395) — 14 CFR §1.1

Answer (A) is incorrect because it describes the nonlanding portion of a runway behind a displaced threshold, which may be suitable for taxiing, landing rollout, and takeoff of aircraft. Answer (C) is incorrect because it would describe an area that exists before a displaced threshold.

121, 135, DSP

8134. For which of these aircraft is the "clearway" for a particular runway considered in computing takeoff weight limitations?

A—Those passenger-carrying transport aircraft certificated between August 26, 1957 and August 30, 1959.
B—Turbine-engine-powered transport airplanes certificated after September 30, 1958.
C—U.S. certified air carrier airplanes certificated after August 29, 1959.

"Clearway" may be considered when determining the takeoff distance of turbine-engine-powered transport category airplane certificated after August 29, 1959. (PLT034) — 14 CFR §121.189 and §1.1

Answers (A) and (C) are incorrect because the passenger-carrying transport aircraft and air carrier airplane would have to be turbine-engine-powered in order to include the clearway in determining runway length for takeoff purposes.

121, 135, DSP

9317. Which is a definition of V_2 speed?

A—Takeoff decision speed.
B—Takeoff safety speed.
C—Minimum takeoff speed.

V_2 means takeoff safety speed. (PLT466) — 14 CFR §1.2

Answer (A) is incorrect because V_1 is takeoff decision speed. Answer (C) is incorrect because minimum takeoff speed doesn't describe a defined speed.

121, 135, DSP

9319. What is the correct symbol for minimum unstick speed?

A—V_{MU}.
B—V_{MD}.
C—V_{FC}.

V_{MU} means minimum unstick speed. (PLT466) — 14 CFR §1.2

Answer (B) is incorrect because V_{MD} is not a concept that is defined in 14 CFR §1.2. Answer (C) is incorrect because V_{FC}/M_{FC} is maximum speed for stability characteristics.

121, 135, DSP

8774. The maximum speed during takeoff that the pilot may abort the takeoff and stop the airplane within the accelerate-stop distance is

A—V_2.
B—V_{EF}.
C—V_1.

The takeoff decision speed, V_1, is the calibrated airspeed on the ground at which, as a result of engine failure or other reasons, the pilot is assumed to have made a decision to continue or discontinue the takeoff. (PLT506) — 14 CFR §1.2

Answer (A) is incorrect because V_2 is the takeoff safety speed. Answer (B) is incorrect because V_{EF} is the calibrated airspeed at which the critical engine is assumed to fail.

Answers

9324 [A]	9327 [B]	8134 [B]	9317 [B]	9319 [A]	8774 [C]

121, 135, DSP

8775. The minimum speed during takeoff, following a failure of the critical engine at V_{EF}, at which the pilot may continue the takeoff and achieve the required height above the takeoff surface within the takeoff distance is indicated by symbol

A—V_{2min}.
B—V_1.
C—V_{LOF}.

The takeoff decision speed, V_1, is the calibrated airspeed on the ground at which, as a result of engine failure or other reasons, the pilot is assumed to have made a decision to continue or discontinue the takeoff. V_1 is also the speed at which the airplane can be rotated for takeoff and shown to be adequate to safely continue the takeoff, using normal piloting skill, when the critical engine is suddenly made inoperative. (PLT466) — 14 CFR §1.2

Answer (A) is incorrect because V_{2min} is the minimum takeoff safety speed. Answer (C) is incorrect because V_{LOF} is the liftoff speed.

121, 135, DSP

8780. The symbol for the speed at which the critical engine is assumed to fail during takeoff is

A—V_2.
B—V_1.
C—V_{EF}.

V_{EF} is the calibrated airspeed at which the critical engine is assumed to fail. V_{EF} must be selected by the applicant but must not be less than 1.05 V_{MC} or, at the option of the applicant, not less than V_{MCG}. (PLT466) — 14 CFR §23.51

Answer (A) is incorrect because V_2 is the takeoff safety speed. Answer (B) is incorrect because V_1 is takeoff decision speed.

121, 135, DSP

9076. Which performance factor decreases as airplane gross weight increases, for a given runway?

A—Critical engine failure speed.
B—Rotation speed.
C—Accelerate-stop distance.

Critical Engine Failure Speed (also called V_1) increases with an increase in weight, resulting in a decrease in performance. (PLT011) — ANA

121, 135, DSP

9085. Which condition has the effect of reducing critical engine failure speed?

A—Slush on the runway or inoperative antiskid.
B—Low gross weight.
C—High density altitude.

Critical Engine Failure Speed is an obsolete term for V_1 which is now called Takeoff Decision Speed. Lowering the airplane's weight will always have the effect of decreasing V_1, while increasing the density altitude will have the effect of increasing V_1. However, inoperative antiskid or slush on the runway will cause a reduction in the maximum allowable takeoff weight which has the effect of lowering V_1. (PLT347) — ANA

121, 135, DSP

9083. What effect does an uphill runway slope have upon takeoff performance?

A—Increases takeoff distance.
B—Decreases takeoff speed.
C—Decreases takeoff distance.

An uphill runway will have the effect of decreasing an airplane's rate of acceleration during the takeoff roll thus causing it to reach its takeoff speeds (V_1 and V_R) further down the runway than would otherwise be the case. An uphill runway will also necessitate an increased V_1 speed in some airplanes. (PLT129) — ANA

121, 135, DSP

9075. Which condition reduces the required runway for takeoff?

A—Higher-than-recommended airspeed before rotation.
B—Lower-than-standard air density.
C—Increased headwind component.

A headwind, in effect, gives an airplane part of its airspeed prior to starting the takeoff roll. This allows the airplane to reach its takeoff speed after a shorter takeoff roll than in no wind conditions. High rotation speeds and lower air density (high density altitude) both have the effect of increasing takeoff distance. (PLT134) — ANA

Answers (A) and (B) are incorrect because higher-than-recommended airspeed before rotation and lower-than-standard air density would increase the required runway for takeoff.

Answers

8775 [B]	8780 [C]	9076 [A]	9085 [A]	9083 [A]	9075 [C]

ALL

9797. You are rolling out after touchdown and decide you really need to abort your landing, and takeoff. Your airplane is at 116 knots and your engines have spooled down to 71% idle. You need a V_2 of 142 to safely lift off and climb. The airplane will require 6 seconds to accelerate after the engines spool up to takeoff thrust, which requires 4 seconds. How much runway will you require for a safe landing abort from your decision point? (Use an average of 129 knots ground speed.)

A—1,738 feet.
B—2,178 feet.
C—3,601 feet.

At a ground speed of 129 knots, assuming it takes 10 seconds to takeoff, the distance required from the decision point would be calculated with an E6B by placing 129 knots on the A scale over the "36" index on the B scale. The distance required is read on the A scale opposite of 10 seconds on the B scale. (PLT011) — E6B

121, 135, DSP

9801. One typical takeoff error is

A—delayed rotation, which may extend the climb distance.
B—premature rotation, which may increase takeoff distance.
C—extended rotation, which may degrade acceleration.

In training it is common for the pilot to overshoot V_R and then overshoot V_2 because the pilot not flying will call for rotation at, or just past V_R. The reaction of the pilot flying is to visually verify V_R and then rotate. The airplane then leaves the ground at or above V_2. The excess airspeed may be of little concern on a normal takeoff, but a delayed rotation can be critical when runway length or obstacle clearance is limited. (PLT134) — FAA-H-8083-3

121, 135, DSP

9802. Excessive takeoff speeds may result in approximately a

A—4% takeoff distance increase for each 1% of additional takeoff speed.
B—1% takeoff distance increase for each 2% of additional takeoff speed.
C—2% takeoff distance increase for each 1% of additional takeoff speed.

An excessive airspeed at takeoff may improve the initial rate of climb and "feel" of the airplane but will produce an undesirable increase in takeoff distance. Assuming that the acceleration is essentially unaffected, the takeoff distance varies as the square of the takeoff velocity. Thus, a 10% increase excess airspeed would increase the takeoff distance 21%. (PLT134) — ANA

Answers

9797 [C] 9801 [A] 9802 [C]

Calculating "V" Speeds

The table in FAA Figure 82 is used in several problems to determine the pressure altitude from the indicated altitude and the local altimeter setting. The table uses the local altimeter setting to indicate the proper correction to field elevation. For example, assume the local altimeter setting is 29.36" Hg. Enter the table in the left-hand column labeled "QNH IN. HG." and then find the range of altimeter settings that contains 29.36" Hg. Read the correction to elevation in the center column. In this case, add 500 feet to the field elevation to determine the pressure altitude. If the altimeter setting is given in millibars, enter the table in the right-hand column.

B-727 "V" Speeds

V_1, V_R, and V_2 for the B-727 are determined by using the table at the right side of FAA Figure 83. Using Operating Conditions G-1 (FAA Figure 81), follow these steps for determining the "V" speeds. Enter the table at the top left in the row appropriate for the pressure altitude and go across until you come to a column containing a temperature range which includes the given value. In this case, enter in the row labeled 1 to 3 (pressure altitude = 1,550 feet) and go to the first column which contains the temperature of +23°F (be sure to use the Fahrenheit or Celsius ranges as appropriate). Go down the first column until in the row appropriate for a flap setting of 15° and a gross weight of 140,000 pounds. The V_1 and V_R speeds are 122 knots and the V_2 speed is 137 knots. No further adjustments or corrections are required. Notice that on the B-727, V_1 and V_R are always the same speed.

B-737 "V" Speeds

Using Operating Conditions R-1 (FAA Figure 53), follow the steps for determining the "V" speeds (*See* FAA Figure 55). Enter the table at the top left in the row appropriate for the pressure altitude and go across until in a column containing a temperature range which includes the given value. In this case, enter in the row labeled -1 to 1 (pressure altitude = 500 feet, refer to FAA Figure 54) and go to the first column which contains the temperature of +50°F (be sure to use the Fahrenheit or Celsius ranges as appropriate). Go down the first column until in the row appropriate for a flap setting of 15° and a gross weight of 90,000 pounds. The V_1 speed is 120 knots, the V_R speed is 121 knots and the V_2 speed is 128 knots. There are two possible adjustments to make to the V_1 speed only. They are noted at the bottom of the table.

DC-9 "V" Speeds

The first step for calculating V_1 and V_R is to find the basic speeds in the table at the top of FAA Figure 47. Using the conditions from Operating Conditions A-1 from FAA Figure 45, the "V" speeds for a weight of 75,000 pounds are: V_1 = 120.5, and V_R = 123.5 knots. Next, a series of corrections must be applied for pressure altitude, ambient temperature, runway slope, wind component, and engine and wing ice protection. There are table values for all these corrections at the bottom of FAA Figure 47 except for pressure altitude and ambient temperature.

The first step in the altitude correction is to use the table in FAA Figure 46 to determine the pressure altitude. Using the altimeter setting from Operating Conditions A-1 (29.40" Hg), the table shows a correction of +500 feet. The pressure altitude then is 3,000 feet (2,500 + 500).

Next, enter the graphs as shown in FAA Figure 47. Draw two vertical lines representing the pressure altitude of 3,000 feet. Next, draw a horizontal line through the ambient temperature (+10°F) to intersect each of the vertical lines. In this case, the lines meet in the "zero" correction area for both V_1 and V_R.

Notice that the correction is marked by bands. For example, if the lines crossed anywhere in the highest shaded band, the correction would be +1 knot.

Next, set up a table similar to the one below to apply any necessary corrections:

	V_1	V_R
Table Value	120.5	123.5
Pressure Alt & Temp	0	0
Slope (+1%)	+ 1.5	+ .9
10 HW	+ .3	0
Ice Protection	+ .8	+ .8
Corrected Speeds	123.1	125.2

121, DSP

8717. (Refer to Figures 81, 82, and 83.) What is the takeoff safety speed for Operating Conditions G-1?

A—122 knots.
B—137 knots.
C—139 knots.

Pressure altitude = 1,550 feet
V_1 / V_R = 122
V_2 = 137

Note: V_2 is defined as Takeoff Safety Speed.

(PLT011) — FAA-H-8083-25

121, DSP

8718. (Refer to Figures 81, 82, and 83.) What is the rotation speed for Operating Conditions G-2?

A—150 knots.
B—154 knots.
C—155 knots.

Pressure altitude = 1,900 feet
V_1 / V_R = 155
V_2 = 167

Note: V_R is defined as Rotation Speed.

(PLT011) — FAA-H-8083-25

121, DSP

8719. (Refer to Figures 81, 82, and 83.) What are V_1, V_R, and V_2 speeds for Operating Conditions G-3?

A—134, 134, and 145 knots.
B—134, 139, and 145 knots.
C—132, 132, and 145 knots.

Pressure altitude = 4,150 feet
V_1 / V_R = 134
V_2 = 145

(PLT012) — FAA-H-8083-25

121, DSP

8720. (Refer to Figures 81, 82, and 83.) What are V_1 and V_2 speeds for Operating Conditions G-4?

A—133 and 145 knots.
B—127 and 141 knots.
C—132 and 146 knots.

Pressure altitude = 3,150 feet
V_1 / V_R = 132
V_2 = 146

(PLT012) — FAA-H-8083-25

Answers

8717 [B] 8718 [C] 8719 [A] 8720 [C]

121, DSP

8721. (Refer to Figures 81, 82, and 83.) What are rotation and V_2 bug speeds for Operating Conditions G-5?

A—120 and 134 knots.
B—119 and 135 knots.
C—135 and 135 knots.

Pressure altitude = 2,550 feet
$V_1 / V_R = 119$
$V_2 = 135$

Note: V_R is defined as Rotation Speed.

(PLT012) — FAA-H-8083-25

121, DSP

8618. (Refer to Figures 53, 54, and 55.) What is the takeoff safety speed for Operating Conditions R-1?

A—128 knots.
B—121 knots.
C—133 knots.

Pressure altitude = 500 feet
$V_1 = 120 + 1$ *(slope)* $= 121$
$V_R = 121$
$V_2 = 128$

Note: V_2 is defined as Takeoff Safety Speed.

(PLT011) — FAA-H-8083-25

Answer (B) is incorrect because 121 knots is the rotation (V_R) speed. Answer (C) is incorrect because 133 knots is the V_2 speed using 5 flaps.

121, DSP

8619. (Refer to Figures 53, 54, and 55.) What is the rotation speed for Operating Conditions R-2?

A—147 knots.
B—152 knots.
C—146 knots.

Pressure altitude = 3,500 feet
$V_1 = 144 - 1$ *(Wind)* $- 1$ *(Slope)* $= 142$
$V_R = 146$
$V_2 = 150$

Note: V_R is defined as Rotation Speed.

(PLT011) — FAA-H-8083-25

121, DSP

8620. (Refer to Figures 53, 54, and 55.) What are V_1, V_R, and V_2 speeds for Operating Conditions R-3?

A—143, 143, and 147 knots.
B—138, 138, and 142 knots.
C—136, 138, and 143 knots.

Pressure altitude = 1,450 feet
$V_1 = 136 + 1$ *(Wind)* $+ 1$ *(Slope)* $= 138$
$V_R = 138$
$V_2 = 142$

(PLT011) — FAA-H-8083-25

121, DSP

8621. (Refer to Figures 53, 54, and 55.) What are critical engine failure and takeoff safety speeds for Operating Conditions R-4?

A—131 and 133 knots.
B—123 and 134 knots.
C—122 and 130 knots.

Pressure altitude = 1,900 feet
$V_1 = 127 - 2$ *(Wind)* $- 2$ *(Slope)* $= 123$
$V_R = 129$
$V_2 = 134$

Note: Critical Engine Failure Speed is an obsolete term for V_1. V_2 is Takeoff Safety Speed.

(PLT012) — FAA-H-8083-25

121, DSP

8622. (Refer to Figures 53, 54, and 55.) What are rotation and V_2 bug speeds for Operating Conditions R-5?

A—138 and 143 knots.
B—136 and 138 knots.
C—134 and 141 knots.

Pressure altitude = -150 feet
$V_R = 138$
$V_2 = 143$

(PLT012) — FAA-H-8083-25

Answers

8721 [B] | 8618 [A] | 8619 [C] | 8620 [B] | 8621 [B] | 8622 [A]

121, DSP

8583. (Refer to Figures 45, 46, and 47.) What are V_1 and V_R speeds for Operating Conditions A-1?

A—V_1 123.1 knots; V_R 125.2 knots.
B—V_1 120.5 knots; V_R 123.5 knots.
C—V_1 122.3 knots; V_R 124.1 knots.

Pressure altitude = 3,000 feet

	V_1	V_R
Table value	*120.5*	*123.5*
Pressure alt & temp	*0*	*0*
Slope (+1%)	*+ 1.5*	*+ .9*
10 HW	*+ .3*	*0*
Ice protection	*+ .8*	*+ .8*
Corrected speeds	*123.1*	*125.2*

(PLT011) — FAA-H-8083-25

121, DSP

8584. (Refer to Figures 45, 46, and 47.) What are V_1 and V_R speeds for Operating Conditions A-2?

A—V_1 129.7 knots; V_R 134.0 knots.
B—V_1 127.2 knots; V_R 133.2 knots.
C—V_1 127.4 knots; V_R 133.6 knots.

Pressure altitude = 100 feet

	V_1	V_R
Table value	*129.50*	*134.00*
Pressure alt & temp	*+ 1.00*	*+ 1.00*
Slope (-1.5%)	*– 2.25*	*– 1.35*
10 TW	*– .80*	*0*
Ice protection	*—*	*—*
Corrected speeds	*127.45*	*133.65*

(PLT011) — FAA-H-8083-25

Answer (A) is incorrect because 134.0 knots is V_R unadjusted. Answer (B) is incorrect because V_1 and V_R are more than 127.2 and 133.2, respectively.

121, DSP

8585. (Refer to Figures 45, 46, and 47.) What are V_1 and V_R speeds for Operating Conditions A-3?

A—V_1 136.8 knots; V_R 141.8 knots.
B—V_1 134.8 knots; V_R 139.0 knots.
C—V_1 133.5 knots; V_R 141.0 knots.

Pressure altitude = 4,000 feet

	V_1	V_R
Table value	*133.50*	*139.00*
Pressure alt & temp	*+ 2.00*	*+ 2.00*
Slope (+0%)	*—*	*—*
15 HW	*+ .45*	*—*
Ice protection	*+ .80*	*+ .80*
Corrected speeds	*136.75*	*141.80*

(PLT011) — FAA-H-8083-25

121, DSP

8586. (Refer to Figures 45, 46, and 47.) What are V_1 and V_R speeds for Operating Conditions A-4?

A—V_1 128.0 knots; V_R 130.5 knots.
B—V_1 129.9 knots; V_R 133.4 knots.
C—V_1 128.6 knots; V_R 131.1 knots.

Pressure altitude = 5,600 feet

	V_1	V_R
Table value	*125.00*	*129.00*
Pressure alt & temp	*+ 3.00*	*+ 3.00*
Slope (+1.5%)	*+ 2.25*	*+ 1.35*
5 TW	*– .40*	*0*
Ice protection	*—*	*—*
Corrected speeds	*129.85*	*133.35*

Note: No correction is required if engine ice protection only is used. See remarks on table at the top of FAA Figure 47.

(PLT011) — FAA-H-8083-25

Answers

8583 [A] 8584 [C] 8585 [A] 8586 [B]

121, DSP

8587. (Refer to Figures 45, 46, and 47.) What are V_1 and V_R speeds for Operating Conditions A-5?

A—V_1 110.4 knots; V_R 110.9 knots.
B—V_1 109.6 knots; V_R 112.7 knots.
C—V_1 106.4 knots; V_R 106.4 knots.

Pressure altitude = 1,500 feet

	V_1	V_R
Table value	*110.0*	*112.5*
Pressure alt & temp	*+ 2.0*	*+ 2.0*
Slope (-2%)	*– 3.0*	*– 1.8*
20 HW	*+ .6*	*0*
Ice protection	*—*	*—*
Corrected speeds	*109.6*	*112.7*

(PLT011) — FAA-H-8083-25

121, DSP

8642-4. (Refer to Figures 237 and 238.) Given the following conditions, what are the takeoff V speeds?

Weight .... 170,000 lb
Flaps .... 10°
Temperature (OAT) .... 25°C
Field pressure altitude .... 427 ft.
Runway slope 0%
Wind (kts) headwind .... 8 kts
Runway condition .... wet runway

For V_R more than or equal to .1 V_R, round up V_R to the next value (example: 140 + .1 = 141)

A—V_1 133 kts, V_R 140 kts, V_2 145 kts.
B—V_1 140 kts, V_R 140 kts, V_2 145 kts.
C—V_1 138 kts, V_R 141 kts, V_2 145 kts.

1. *Start on FAA Figure 238 since a wet runway is specified.*
2. *Enter the Takeoff speeds chart at 170 (1,000 pounds) and proceed to the Flaps 10 column for V_1 – 133, V_R – 139, V_2 – 145.*
3. *Enter the adjustments table at 27°C and interpolate in each column for pressure altitude of 500 feet.*
4. *For V_1, round down to 133 knots, for V_R round up to 140 knots, for V_2 the adjustment is 0 so it remains 145.*

(PLT011) — FAA-H-8083-25

Calculating Takeoff Power

B-727 Takeoff EPR

EPR (Engine Pressure Ratio) is the thrust indication used on many turbojet aircraft. Basically, it is the ratio of the engine exhaust pressure to the intake pressure. For example, if the exhaust pressure is exactly twice the intake pressure, the EPR is 2.00.

The EPR setting for maximum takeoff thrust will vary with altitude and temperature. In addition, reductions in EPR have to be made when bleed air from the compressor section is used for air conditioning, engine anti-ice and internal regulation of the engine.

The table at the top of FAA Figure 83 is used to determine Takeoff EPR. Enter the table with temperature and pressure altitude. To determine pressure altitude, use FAA Figure 82.

Using the data from Operating Conditions G-1 (FAA Figure 81), the field elevation is 1,050 feet and the altitude correction is +500 feet, which results in a pressure altitude of 1,550 feet. Since there is no listing for this altitude in the table in FAA Figure 83, it is necessary to interpolate. At 1,000 feet and an OAT of 23°F, the EPR for Engines 1 and 3 is 2.15 and for Engine 2 it is 2.16. At 2,000 feet the corresponding values are 2.21 for Engines 1 and 3, and 2.22 for Engine 2. Since the pressure altitude in Condition G-1

Answers

8587 [B] 8642-4 [A]

is roughly halfway between 1,000 and 2,000, the EPR setting should be halfway between as well. The EPR for Engines 1 and 3 is 2.18 and for Engine 2 it is 2.19.

The EPR values in the table are for a given bleed air configuration only. This configuration is noted at the upper right-hand corner of the table: "EPR Bleed Corrections, Eng 1 & 3 on, Eng 2 off." Normally, 1 & 3 are on, and 2 is off—anything different will require a correction. If this is not the configuration in use, corrections are listed at the bottom left-hand corner of the table. The "air conditioning" correction applies only to Engines 1 and 3 and only if the air conditioning is off. The "engine anti-ice on" correction applies only to Engine 2.

The last of the possible corrections is for 6th stage bleed air. This applies only if both of two conditions are met. The 6th stage bleed must be on and the OAT must be 10°C (50°F) or warmer. If these conditions are met, a value of .05 is subtracted from the engine number 2 EPR.

The best method of solving this type of problem is to set up a table which accounts for all the possible corrections. Using the data from Operating Conditions G-1:

	Eng. 1 & 3	**Eng. 2**
Table Value	2.18	2.19
A/C Off	+ .04	—
Eng. AI On	—	– .03
6th Stage Bleed	—	—
Takeoff EPR	2.22	2.16

B-737 Takeoff EPR

The Takeoff EPR table at the top of FAA Figure 55 is similar to the B-727 takeoff EPR. In the table of FAA Figure 55, two EPR values are found: one for temperature and one for altitude (be sure to use the table in FAA Figure 54 to determine the pressure altitude). The lower of the two is the takeoff EPR. For example if the temperature is 50°F at a pressure altitude of 500 feet, the temperature-limited EPR is 2.04 and the altitude-limited EPR is 2.035. (The altitude-limited EPR is 2.01 from sea level up to 1,000 feet.) The only possible correction would be for if the air conditioning bleeds are off.

121, DSP

8712. (Refer to Figures 81, 82, and 83.) What is the max takeoff EPR for Operating Conditions G-1?

A—Engines 1 and 3, 2.22; engine 2, 2.16.
B—Engines 1 and 3, 2.22; engine 2, 2.21.
C—Engines 1 and 3, 2.15; engine 2, 2.09.

Pressure altitude = 1,550 feet

	Eng. 1 & 3	***Eng. 2***
Table value	*2.18*	*2.19*
A/C off	*+ .04*	*—*
Eng. AI on	*—*	*– .03*
6th stage bleed	*—*	*—*
Takeoff EPR	*2.22*	*2.16*

(PLT007) — FAA-H-8083-25

121, DSP

8713. (Refer to Figures 81, 82, and 83.) What is the max takeoff EPR for Operating Conditions G-2?

A—Engines 1 and 3, 2.15; engine 2, 2.16.
B—Engines 1 and 3, 2.18; engine 2, 2.13.
C—Engines 1 and 3, 2.14; engine 2, 2.11.

Pressure altitude = 1,900 feet

	Eng. 1 & 3	***Eng. 2***
Table value	*2.14*	*2.16*
A/C off	*—*	*—*
Eng. AI on	*—*	*—*
6th stage bleed	*—*	*– .05*
Takeoff EPR	*2.14*	*2.11*

(PLT007) — FAA-H-8083-25

Answers

8712 [A]   8713 [C]

121, DSP

8714. (Refer to Figures 81, 82, and 83.) What is the max takeoff EPR for Operating Conditions G-3?

A—Engines 1 and 3, 2.08; engine 2, 2.05.
B—Engines 1 and 3, 2.14; engine 2, 2.10.
C—Engines 1 and 3, 2.18; engine 2, 2.07.

Pressure altitude = 4,150 feet

	Eng. 1 & 3	***Eng. 2***
Table value	*2.14*	*2.15*
A/C off	—	—
Eng. AI on	—	—
6th stage bleed	—	*–.05*
Takeoff EPR	*2.14*	*2.10*

(PLT007) — FAA-H-8083-25

121, DSP

8715. (Refer to Figures 81, 82, and 83.) What is the max takeoff EPR for Operating Conditions G-4?

A—Engines 1 and 3, 2.23; engine 2, 2.21.
B—Engines 1 and 3, 2.26; engine 2, 2.25.
C—Engines 1 and 3, 2.24; engine 2, 2.24.

Pressure altitude = 3,150 feet

	Eng. 1 & 3	***Eng. 2***
Table value	*2.23*	*2.24*
A/C off	—	—
Eng. AI on	—	*–.03*
6th stage bleed	—	—
Takeoff EPR	*2.23*	*2.21*

(PLT007) — FAA-H-8083-25

121, DSP

8716. (Refer to Figures 81, 82, and 83.) What is the max takeoff EPR for Operating Conditions G-5?

A—Engines 1 and 3, 2.27; engine 2, 2.18.
B—Engines 1 and 3, 2.16; engine 2, 2.14.
C—Engines 1 and 3, 2.23; engine 2, 2.22.

Pressure altitude = 2,550 feet

	Eng. 1 & 3	***Eng. 2***
Table value	*2.23*	*2.26*
A/C off	—	—
Eng. AI on	—	*–.03*
6th stage bleed	—	—
Takeoff EPR	*2.23*	*2.23*

(PLT007) — FAA-H-8083-25

121, DSP

8613. (Refer to Figures 53, 54, and 55.) What is the takeoff EPR for Operating Conditions R-1?

A—2.04.
B—2.01.
C—2.035.

Pressure altitude = 500 feet
Altitude-limited EPR = 2.035
Temperature-limited EPR = 2.04
(PLT007) — FAA-H-8083-25

121, DSP

8614. (Refer to Figures 53, 54, and 55.) What is the takeoff EPR for Operating Conditions R-2?

A—2.19.
B—2.18.
C—2.16.

Pressure altitude = 3,500 feet
Altitude-limited EPR = 2.19
Temperature-limited EPR = 2.19

(PLT011) — FAA-H-8083-25

121, DSP

8615. (Refer to Figures 53, 54, and 55.) What is the takeoff EPR for Operating Conditions R-3?

A—2.01.
B—2.083.
C—2.04.

Pressure altitude = 1,450 feet
Altitude-limited EPR = 2.083
Temperature-limited EPR = 2.01
Add .03 for air conditioning off.

(PLT007) — FAA-H-8083-25

121, DSP

8616. (Refer to Figures 53, 54, and 55.) What is the takeoff EPR for Operating Conditions R-4?

A—2.06.
B—2.105.
C—2.11.

Pressure altitude = 1,900 feet
Altitude-limited EPR = 2.105
Temperature-limited EPR = 2.11

(PLT007) — FAA-H-8083-25

Answers

8714 [B]	8715 [A]	8716 [C]	8613 [C]	8614 [A]	8615 [C]
8616 [B]					

121, DSP
8617. (Refer to Figures 53, 54, and 55.) What is the takeoff EPR for Operating Conditions R-5?

A—1.98.
B—1.95.
C—1.96.

Pressure altitude = -150 feet
Altitude-limited EPR = 2.003
Temperature-limited EPR = 1.95

Add .03 for air conditioning off.

(PLT007) — FAA-H-8083-25

Climb Performance

The **best rate-of-climb speed** for any airplane is the speed at which there is the greatest difference between the power required for level flight and the power available from the engines. The L/D_{MAX} speed for any airplane is the one that requires the least power for level flight since it is the lowest drag speed. Because the power output of prop-driven airplanes is relatively constant at all speeds, L/D_{MAX} is the best rate-of-climb speed for them.

Turbojet engines produce more power as the aircraft speed increases. Even though drag increases at speeds above L/D_{MAX}, the engine's power output increases even more so that the maximum difference between power required and power available is achieved at a higher airspeed. For a turbojet, the best rate-of-climb speed is faster than L/D_{MAX}.

DC-9 Performance Tables

The tables in FAA Figures 49 and 50 allow you to determine the time, fuel and distance required for a climb to cruising altitude after takeoff. The table in FAA Figure 49 is used if a "high speed" climb is planned, and the table in FAA Figure 50 is used for a "long range" climb. Each of the tables is broken into four "sub-tables" representing different initial weights. Assume a climb on the long range schedule from sea level to 34,000 feet with an initial weight of 84,000 pounds. The climb will require 17.1 minutes, 2,570 pounds of fuel, and cover a distance of 109.7 NM.

A headwind or tailwind component in the climb will change the distance flown. Assume that there is an average 20-knot headwind in the climb described above. The first step is to compute the average "no wind" ground speed (GS). A distance of 109.7 NM flown in 17.1 minutes works out to a GS of 384.9 knots. A headwind component of 20 knots will reduce this GS to 364.9 knots. The distance flown in 17.1 minutes at 364.9 knots is 104 NM.

Note: Using a CX-2 computer, select "Dist Flown" from the menu and enter TIME and GS. No other adjustments or corrections are required for these tables.

B-737 Climb Performance Tables

The tables in FAA Figures 57 and 58 allow you to determine the time and fuel required for a climb to cruising altitude after takeoff. The table in FAA Figure 57 is for ISA temperatures, and the table in FAA Figure 58 is for ISA +10°C. Each intersection of Brake Release Weight and Cruise Altitude has a box with four numbers. These are the time, the fuel, the distance and the TAS required to climb from a sea level airport to cruise altitude in calm wind conditions. For example, with a brake release weight of 110,000 pounds, a climb to 33,000 feet in ISA +10°C conditions will require 26 minutes, 4,100 pounds of fuel and cover a distance of 154 NM.

A headwind or tailwind component in the climb will change the distance flown. Assume that there is an average 20-knot headwind in the climb described above. The first step is to compute the average

Answers

8617 [A]

"no wind" GS. A distance of 154 NM flown in 26 minutes works out to a GS of 355.4 knots. A headwind component of 20 knots will reduce this GS to 335.4 knots. The distance flown in 26 minutes at 335.4 knots is 145.3 NM.

Note: Using a CX-2 computer, select "Dist Flown" from the menu and enter TIME and GS. Do not use the TAS from the table as that will result in an inaccurate answer.

Departure from an airport that is significantly above sea level will reduce the fuel required for the climb. Notice that departure from a 2,000-foot airport will reduce the climb fuel by 100 pounds, however the effect on time and distance flown is negligible.

B-737 Climb and Cruise Power Tables

The Max Climb & Max Continuous EPR Table at the top of FAA Figure 60 is similar to the one discussed in Takeoff EPR. In this table two EPR values are found—one for temperature and one for altitude. The lower of the two is the maximum climb/continuous EPR. For example, if the temperature is +10°C at a pressure altitude of 10,000 feet, the temperature-limited EPR is 2.04 and the altitude-limited EPR is 2.30. (The altitude-limited EPR is 2.30 from 5,660 feet and up.) The max EPR is 2.04.

The Max Cruise EPR Table supplies one EPR value for a given TAT (Total Air Temperature) in one of two altitude ranges. The correction tables are similar to ones used previously and apply to both tables.

121, 135, DSP

8400. At what speed, with reference to L/D_{MAX}, does maximum rate-of-climb for a jet airplane occur?

A—A speed greater than that for L/D_{MAX}.
B—A speed equal to that for L/D_{MAX}.
C—A speed less than that for L/D_{MAX}.

An airplane's best rate-of-climb is achieved at the airspeed where there is the maximum difference between the power available from the engines and the power required for level flight. The L/D_{MAX} airspeed will require the minimum power for level flight since drag is at its minimum. However, turbojet engines produce more power at high speed than at lower speeds and so the speed with the greatest difference between power required and power available is higher than L/D_{MAX}. (PLT303) — ANA

Answer (B) is incorrect because a speed equal to that of L/D_{MAX} is the maximum rate-of-climb for a propeller airplane. Answer (C) is incorrect because the maximum rate-of-climb is at a speed greater than L/D_{MAX}.

121, DSP

8593. (Refer to Figures 48, 49, and 50.) What is the ground distance covered during en route climb for Operating Conditions W-1?

A—104.0 NM.
B—99.2 NM.
C—109.7 NM.

Long range climb:

No wind time = 17.1 minutes
No wind distance = 109.7 NM
No wind GS = 384.9 knots
Wind adjusted GS = 364.9 knots
Wind adjusted distance = 104 NM

(PLT004) — FAA-H-8083-25

121, DSP

8594. (Refer to Figures 48, 49, and 50.) What is the ground distance covered during en route climb for Operating Conditions W-2?

A—85.8 NM.
B—87.8 NM.
C—79.4 NM.

High speed climb:

No wind time = 12.8 minutes
No wind distance = 85.8 NM
No wind GS = 402.2 knots
Wind adjusted GS = 372.2 knots
Wind adjusted distance = 79.4 NM
(PLT004) — FAA-H-8083-25

Answers

8400 [A] 8593 [A] 8594 [C]

121, DSP

8595. (Refer to Figures 48, 49, and 50.) What is the ground distance covered during en route climb for Operating Conditions W-3?

A—86.4 NM.
B—84.2 NM.
C—85.1 NM.

Long range climb:

No wind time = 13.3 minutes
No wind distance = 84.2 NM
No wind GS = 379.8 knots
Wind adjusted GS = 389.8 knots
Wind adjusted distance = 86.4 NM

(PLT004) — FAA-H-8083-25

121, DSP

8596. (Refer to Figures 48, 49, and 50.) What is the ground distance covered during en route climb for Operating Conditions W-4?

A—58.4 NM.
B—61.4 NM.
C—60.3 NM.

High speed climb:

No wind time = 9.1 minutes
No wind distance = 58.4 NM
No wind GS = 385.1 knots
Wind adjusted GS = 405.1 knots
Wind adjusted distance = 61.4 NM

(PLT004) — FAA-H-8083-25

121, DSP

8597. (Refer to Figures 48, 49, and 50.) What is the ground distance covered during en route climb for Operating Conditions W-5?

A—68.0 NM.
B—73.9 NM.
C—66.4 NM.

High speed climb:

No wind time = 11.3 minutes
No wind distance = 73.9 NM
No wind GS = 392.4 knots
Wind adjusted GS = 352.4 knots
Wind adjusted distance = 66.4 NM

(PLT004) — FAA-H-8083-25

121, DSP

8598. (Refer to Figures 48, 49, and 50.) What is the aircraft weight at the top of climb for Operating Conditions W-1?

A—81,600 pounds.
B—81,400 pounds.
C—81,550 pounds.

Long range climb:

Initial weight	*84,000*
Burn	*– 2,570*
TOC weight	*81,430*

(PLT004) — FAA-H-8083-25

121, DSP

8599. (Refer to Figures 48, 49, and 50.) What is the aircraft weight at the top of climb for Operating Conditions W-2?

A—82,775 pounds.
B—83,650 pounds.
C—83,800 pounds.

High speed climb:

Initial weight	*86,000*
Burn	*– 2,225*
TOC weight	*83,775*

(PLT004) — FAA-H-8083-25

121, DSP

8600. (Refer to Figures 48, 49, and 50.) What is the aircraft weight at the top of climb for Operating Conditions W-3?

A—75,750 pounds.
B—75,900 pounds.
C—76,100 pounds.

Long range climb:

Initial weight	*78,000*
Burn	*– 2,102*
TOC weight	*75,898*

(PLT004) — FAA-H-8083-25

Answers

8595 [A]	8596 [B]	8597 [C]	8598 [B]	8599 [C]	8600 [B]

121, DSP

8601. (Refer to Figures 48, 49, and 50.) What is the aircraft weight at the top of climb for Operating Conditions W-4?

A—86,150 pounds.
B—86,260 pounds.
C—86,450 pounds.

High speed climb:

Initial weight	*88,000*
Burn	*– 1,738*
TOC weight	*86,262*

(PLT004) — FAA-H-8083-25

121, DSP

8602. (Refer to Figures 48, 49, and 50.) What is the aircraft weight at the top of climb for Operating Conditions W-5?

A—89,900 pounds.
B—90,000 pounds.
C—90,100 pounds.

Long-range climb:

Initial weight	*92,000*
Burn	*– 2,079*
TOC weight	*89,921*

(PLT004) — FAA-H-8083-25

121, DSP

8628. (Refer to Figures 56, 57, and 58.) What is the ground distance covered during en route climb for Operating Conditions V-1?

A—145 NM.
B—137 NM.
C—134 NM.

No wind time = 26 minutes
No wind distance = 154 NM
No wind GS = 355.4 knots
Wind adjusted GS = 335.4 knots
Wind adjusted distance = 145.3 NM

(PLT004) — FAA-H-8083-25

121, DSP

8629. (Refer to Figures 56, 57, and 58.) What is the ground distance covered during en route climb for Operating Conditions V-2?

A—84 NM.
B—65 NM.
C—69 NM.

No wind time = 13 minutes
No wind distance = 65 NM
No wind GS = 300 knots
Wind adjusted GS = 320 knots
Wind adjusted distance = 69.3 NM

(PLT004) — FAA-H-8083-25

121, DSP

8630. (Refer to Figures 56, 57, and 58.) What is the ground distance covered during en route climb for Operating Conditions V-3?

A—95 NM.
B—79 NM.
C—57 NM.

No wind time = 16 minutes
No wind distance = 87 NM
No wind GS = 326.3 knots
Wind adjusted GS = 296.3 knots
Wind adjusted distance = 79 NM

(PLT004) — FAA-H-8083-25

121, DSP

8631. (Refer to Figures 56, 57, and 58.) What is the ground distance covered during en route climb for Operating Conditions V-4?

A—63 NM.
B—53 NM.
C—65 NM.

No wind time = 13 minutes
No wind distance = 61 NM
No wind GS = 281.5 knots
Wind adjusted GS = 291.5 knots
Wind adjusted distance = 63.2 NM

(PLT004) — FAA-H-8083-25

Answers

8601 [B] 8602 [A] 8628 [A] 8629 [C] 8630 [B] 8631 [A]

121, DSP

8632. (Refer to Figures 56, 57, and 58.) What is the ground distance covered during en route climb for Operating Conditions V-5?

A—70 NM.
B—52 NM.
C—61 NM.

No wind time = 13 minutes
No wind distance = 70 NM
No wind GS = 323.1 knots
Wind adjusted GS = 283.1 knots
Wind adjusted distance = 61.3 NM

(PLT004) — FAA-H-8083-25

121, DSP

8633. (Refer to Figures 56, 57, and 58.) How much fuel is burned during en route climb for Operating Conditions V-1?

A—4,100 pounds.
B—3,600 pounds.
C—4,000 pounds.

Fuel to climb from sea level = 4,100 lbs
Correction factor = -100 lbs

(PLT012) — FAA-H-8083-25

121, DSP

8634. (Refer to Figures 56, 57, and 58.) How much fuel is burned during en route climb for Operating Conditions V-2?

A—2,250 pounds.
B—2,600 pounds.
C—2,400 pounds.

Fuel to climb from sea level = 2,400 lbs
Correction factor = -150 lbs

(PLT012) — FAA-H-8083-25

121, DSP

8635. (Refer to Figures 56, 57, and 58.) What is the aircraft weight at the top of climb for Operating Conditions V-3?

A—82,100 pounds.
B—82,500 pounds.
C—82,200 pounds.

Fuel to climb from sea level = 2,600 lbs
Correction factor = -100 lbs
Final weight = 85,000 – 2,500 = 82,500 lbs

(PLT004) — FAA-H-8083-25

121, DSP

8636. (Refer to Figures 56, 57, and 58.) What is the aircraft weight at the top of climb for Operating Conditions V-4?

A—102,900 pounds.
B—102,600 pounds.
C—103,100 pounds.

Fuel to climb from sea level = 2,400 lbs
Correction factor = -300 lbs
Final weight = 105,000 – 2,100 = 102,900 lbs

(PLT004) — FAA-H-8083-25

121, DSP

8637. (Refer to Figures 56, 57, and 58.) What is the aircraft weight at the top of climb for Operating Conditions V-5?

A—73,000 pounds.
B—72,900 pounds.
C—72,800 pounds.

Fuel to climb from sea level = 2,100 lbs
Correction factor = -100 lbs
Final weight = 75,000 – 2,000 = 73,000 lbs

(PLT004) — FAA-H-8083-25

Answers

8632 [C] 8633 [C] 8634 [A] 8635 [B] 8636 [A] 8637 [A]

121, DSP

8638. (Refer to Figures 59 and 60.) What is the max climb EPR for Operating Conditions T-1?

A—1.82.
B—1.96.
C—2.04.

Altitude-limited EPR = 2.30
TAT-limited EPR = 1.90
Correction for engine anti-ice = -.08
Max climb EPR setting = 1.82

(PLT007) — FAA-H-8083-25

Answer (B) is incorrect because 1.96 is the max continuous EPR. Answer (C) is incorrect because 2.04 is the max continuous EPR without the bleed air correction for the engine anti-ice ON.

121, DSP

8639. (Refer to Figures 59 and 60.) What is the max continuous EPR for Operating Conditions T-2?

A—2.10.
B—1.99.
C—2.02.

Altitude-limited EPR = 2.20
TAT-limited EPR = 2.10
Engine and wing anti-ice correction = -.12
Air conditioning off correction = +.04
Max continuous EPR setting = 2.02

(PLT007) — FAA-H-8083-25

121, DSP

8640. (Refer to Figures 59 and 60.) What is the max cruise EPR for Operating Conditions T-3?

A—2.11.
B—2.02.
C—1.90.

Max cruise EPR = 2.02
Engine and wing anti-ice correction = -.12
Max cruise EPR setting = 1.90

(PLT007) — FAA-H-8083-25

121, DSP

8641. (Refer to Figures 59 and 60.) What is the max climb EPR for Operating Conditions T-4?

A—2.20.
B—2.07.
C—2.06.

Altitude-limited EPR = 2.30
TAT-limited EPR = 2.20
Engine and wing anti-ice correction = -.14
Max climb EPR setting = 2.06

(PLT007) — FAA-H-8083-25

121, DSP

8642-1. (Refer to Figures 59 and 60.) What is the max continuous EPR for Operating Conditions T-5?

A—2.00.
B—2.04.
C—1.96.

Altitude-limited EPR = 2.30
TAT-limited EPR = 2.00
Air conditioning off correction = +.04
Max continuous EPR setting = 2.04

(PLT007) — FAA-H-8083-25

121, DSP

8642-2. (Refer to Figure 231.) Given the following conditions, what is the takeoff climb limit?

Airport OAT.. 38°C
Airport Pressure Altitude .... 14 ft.
Flaps...15°
Engine Bleed for packs.. On
Anti-ice.. Off

A—136,000 lb.
B—137,500 lb.
C—139,000 lb.

1. *Enter FAA Figure 231 at 38° OAT and proceed up to the 0 ft pressure altitude (this is closest to 14 feet).*
2. *From the point of intersection on the pressure altitude line, draw a line horizontally to the Climb Limit Brake Release Weight of 136,000 pounds.*

(PLT085) — FAA-H-8083-25

Answers

8638 [A]	8639 [C]	8640 [C]	8641 [C]	8642-1 [B]	8642-2 [A]

121, DSP

8642-3. (Refer to Figures 235 and 236.) Given the following conditions, what is the maximum Slush/Standing Water takeoff weight?

Dry field/obstacle limit weight .... 180,000 lb.
Slush/standing water depth .... .25 inches
Temperature (OAT) .... 30°C
Field pressure altitude .... 5431 ft
Field length available .... 9000 ft
No reverse thrust

A— 130,850 lb.
B— 147,550 lb.
C— 139,850 lb.

1. *Enter FAA Figure 235 which is the no reverse thrust chart with an Obstacle Limit Weight of 180 (1,000 pounds).*
2. *Proceed across to the .25 inches water depth column and interpolate to find an adjustment of -40.03 (between 5,000 and 10,000).*
3. *Subtract 40,030 from 180,000 to get 139,970 pounds.*
4. *Determine ISA temperature in order to calculate field length in next step.*
   ISA (°C) = 15 – ((FL/10) x 2)
   15 – (5.431 x 2) = 4.138°C
5. *Go down to the V_1 (MCG) chart and figure out the adjusted field length based on Note 2 and 30°C:*
   Note 2: Adjust field length available by -150 feet for every 5°C above
   30°C – 4.138°C = 25.862°C
   25.862°C / 5°C = 5.2
   5.2 x (-150) = -780
   9,000 – 780 = 8,220 feet.
6. *Enter the Limit Weight table at 8,200 feet and proceed across to the .25 inches column. Interpolating between 5,000 and 10,000 gives you 138.445.*
7. *Enter the Limit Weight table again at 8,600 feet to find the interpolated adjustment to use between 5,000 and 10,000 feet: 159.91.*
8. *Interpolate between the 8,200 feet adjustment (138.445) and the 8,600 adjustment (159.91) for a field length of 8,220 feet (from step 4):*
   159.91 – 138.445 = 21.465
   21.465/400 = 0.053662
   0.053662 x 20 = 1.07325
   138.445 + 1.07325= 139.5182
   139.5182 x 1000 = 139,518
9. *Compare 139,970 pounds to 139,518 pounds as per note 4 and the lesser of the weights is 139,518 pounds.*

(PLT069) — FAA-H-8083-25

Answers

8642-3 [C]

Cruise Performance

The maximum range speed for an aircraft is determined by its L/D curve. Propeller-driven airplanes will achieve best range performance if they are flown at the speed that yields L/D_{MAX}. In turbojet aircraft, a somewhat more complex relationship between lift and drag determines best range. Turbojets always have a best range speed higher than L/D_{MAX}.

A headwind or tailwind will affect the miles per unit of fuel burned. If an airplane is operating at its best-range airspeed and encounters a headwind, it should speed up to minimize the time in the adverse wind. By the same token, an airplane with a tailwind can slow down and let the wind maintain its ground speed with a lower fuel flow. The exact amount of airspeed change that is useful varies with airplane type.

Turbojet engines have a strong preference for operations at high altitudes and airspeeds. Both lower temperatures and higher altitudes increase engine efficiency by requiring a lower fuel flow for a given thrust. Besides increased engine efficiency, lift and drag both decrease at higher altitudes, so less thrust is required.

Turbine engines are much more efficient when operated at the upper end of their RPM range. Generally, the optimum cruise altitude for a turbojet airplane is the highest at which it is possible to maintain the optimum aerodynamic conditions (best angle of attack) at maximum continuous power. The optimum altitude is determined mainly by the aircraft's gross weight at the beginning of cruise.

As an aircraft burns fuel and becomes lighter, the optimum cruise altitude slowly increases and the speed that yields the optimum cruise performance slowly decreases. Since it is seldom practical to change speed and altitude constantly, it is common procedure to maintain a constant Mach cruise at a flight level close to optimum. As fuel is burned, thrust is reduced to maintain the constant Mach number.

121, 135, DSP

8383. What performance is characteristic of flight at maximum L/D in a propeller-driven airplane?

A—Maximum range and distance glide.
B—Best angle of climb.
C—Maximum endurance.

Maximum range and glide distance is achieved at L/D_{MAX}. (PLT242) — FAA-H-8083-3

Answer (B) is incorrect because best angle of climb is at a high angle of attack with both high lift and high drag coefficients, which would not result in a maximum L/D ratio. Answer (C) is incorrect because maximum endurance would be obtained at the point of minimum power required, since this would require the lowest fuel flow to keep the airplane in steady, level flight. This is not at maximum L/D.

121, 135, DSP

8401. At what speed, with reference to L/D_{MAX}, does maximum range for a jet airplane occur?

A—A speed less than that for L/D_{MAX}.
B—A speed equal to that for L/D_{MAX}.
C—A speed greater than that for L/D_{MAX}.

Maximum range is obtained at the aerodynamic condition which produces a maximum proportion between the square root of the lift coefficient and the drag coefficient. It occurs where the proportion between velocity and thrust required is greatest. This point is located by a straight line from the origin tangent to the curve, and is consequently at a higher airspeed than L/D_{MAX}. (PLT303) — ANA

Answer (A) is incorrect because a speed greater than L/D_{MAX} will obtain maximum range for a jet airplane. Answer (B) is incorrect because a speed equal to that of L/D_{MAX} is a jet airplane's maximum endurance, not range.

Answers

8383 [A] 8401 [C]

121, 135, DSP

8398. What should a pilot do to maintain "best range" airplane performance when a tailwind is encountered?

A—Increase speed.
B—Maintain speed.
C—Decrease speed.

While it is only necessary to consider wind velocity effect on cruise speed at wind velocities that exceed 25 percent of the zero wind cruise speed, generally you should increase cruise speed with a headwind and decrease cruise speed with a tailwind. (PLT303) — FAA-H-8083-25

121, 135, DSP

9078. Which procedure produces the minimum fuel consumption for a given leg of the cruise flight?

A—Increase speed for a headwind.
B—Increase speed for a tailwind.
C—Increase altitude for a headwind, decrease altitude for a tailwind.

When flying into a headwind the airspeed should be increased above that used for maximum range in calm winds. Airspeed should be decreased for a tailwind. (PLT015) — ANA

121, 135, DSP

8381. Which maximum range factor decreases as weight decreases?

A—Angle of attack.
B—Altitude.
C—Airspeed.

As fuel is consumed and the airplane's weight decreases, the optimum airspeed and power setting may decrease, or the optimum altitude may increase. The optimum angle of attack does not change with changes in weight. (PLT006) — FAA-H-8083-3

Answer (A) is incorrect because the factors of maximum range are weight, altitude, and aerodynamic configuration of the airplane, not angle of attack. Answer (B) is incorrect because maximum range altitude may increase with a decrease in weight.

121, 135, DSP

9077. Maximum range performance of a turbojet aircraft is obtained by which procedure as aircraft weight reduces?

A—Increasing speed or altitude.
B—Increasing altitude or decreasing speed.
C—Increasing speed or decreasing altitude.

As a turbojet-powered airplane burns fuel, its maximum range of profile can be maintained by increasing the cruise altitude to improve the specific fuel consumption of the engines and by decreasing airspeed to maintain the optimum L/D ratio. (PLT015) — ANA

Answer (A) is incorrect because, as weight decreases, the optimum speed decreases, or altitude increases. Answer (C) is incorrect because, as weight decreases, speed decreases, or altitude increases.

Answers

8398 [C]	9078 [A]	8381 [C]	9077 [B]

Landing Considerations

V_S—stalling speed or the minimum steady flight speed at which the airplane is controllable.

V_{S0}—stalling speed or the minimum steady flight speed in the landing configuration.

V_{REF}—reference speed. It is normally 1.3 x V_{S0}.

Even with all the aircraft's high lift devices extended, a typical air carrier airplane has a high approach speed and a long landing roll. An airplane is normally flown at 1.3 times the V_{S0} speed for the aircraft's weight. Of course, 1.3 times V_{S0} is an indicated airspeed and the ground speed will vary depending on wind, altitude and temperature. A high temperature or high altitude approach will increase an aircraft's ground speed for any given approach speed.

Once an airplane has touched down on a runway there are 3 ways of slowing it to a stop: aerodynamic braking, use of the wheel brakes, and reverse thrust.

The typical technique for stopping an aircraft on a normal landing is to apply reverse thrust (or prop reverse) immediately upon touchdown. This takes maximum advantage of reverse thrust when it is most effective and it saves wear on the wheel brakes, which heat up very rapidly at high ground speeds. Shortly after touchdown, the spoilers are deployed. This reduces lift and increases drag. As the aircraft slows, the main wheel brakes are applied to bring it down to taxiing speed. The brakes are most effective when lift has been reduced (by spoilers and low airspeed) and more of the aircraft's weight is carried by the landing gear.

Water on a runway will increase the landing rollout because the reduced coefficient of friction makes the wheel brakes less effective. This is particularly true at high ground speeds.

A very dangerous possibility when landing on a wet runway is **hydroplaning**. When hydroplaning occurs, the wheel brakes are almost totally ineffective. This not only greatly increases the landing rollout, but also introduces the possibility of losing directional control on sliding off the side of the runway. There are three types of hydroplaning.

Dynamic hydroplaning occurs when a tire rolls through standing water, forms a bow wave, and then rolls up on top of the wave, losing all contact with the runway. The minimum speed at which dynamic hydroplaning can start is related to tire pressure. As a rule of thumb, dynamic hydroplaning will start at speeds of greater than nine times the square root of the tire pressure in pounds per square inch. The practical application is that your nose wheel can hydroplane at a lower speed than the mains because of its lower pressure. Once dynamic hydroplaning has started, it can continue to much lower speeds.

Viscous hydroplaning occurs when there is a thin film of water covering a smooth surface such as a painted or rubber-coated portion of the runway. Viscous hydroplaning can occur at much lower speeds than dynamic hydroplaning.

Reverted rubber hydroplaning occurs during a locked wheel skid. Water trapped between the tire and the runway is heated by friction, and the tire rides along a pocket of steam.

When landing on a water-covered runway, fly the approach as close to "on speed" as possible. Landing at a higher than recommended speed will greatly increase the potential for hydroplaning. After touchdown, use aerodynamic braking and reverse thrust to maximum possible extent, saving the use of wheel brakes until the speed is low enough to minimize the possibility of hydroplaning.

Regulations (14 CFR §121.195) require that when a turbojet aircraft is dispatched to an airport where the runways are forecast to be wet or slippery, the effective length of the landing runway must be 115% of what is required under dry conditions. Since runways cannot be lengthened, the effect of this rule is to lower the maximum allowable landing weight of aircraft on wet runways for dispatch purposes.

121, 135, DSP

9323. Which is the correct symbol for the stalling speed or the minimum steady flight speed at which the airplane is controllable?

A—V_{S0}.
B—V_S.
C—V_{S1}.

V_S means the stalling speed or the minimum steady flight speed at which the airplane is controllable. (PLT466) — 14 CFR §1.2

Answer (A) is incorrect because V_{S0} is the stalling speed or the minimum steady flight speed in the landing configuration. Answer (C) is incorrect because V_{S1} is the stalling speed or the minimum steady flight speed in a specific configuration.

121, 135, DSP

9322. Which is the correct symbol for the minimum steady-flight speed or stalling speed in the landing configuration?

A—V_S.
B—V_{S1}.
C—V_{S0}.

V_{S0} means the stalling speed or the minimum steady flight speed in the landing configuration. (PLT466) — 14 CFR §1.2

Answer (A) is incorrect because V_S is the stalling speed or the minimum steady flight speed at which the airplane is controllable. Answer (B) is incorrect because V_{S1} is the stalling speed or the minimum steady flight speed in a specific configuration.

121, 135, DSP

8374. What effect does landing at high elevation airports have on groundspeed with comparable conditions relative to temperature, wind, and airplane weight?

A—Higher than at low elevation.
B—Lower than at low elevation.
C—The same as at low elevation.

An airplane at altitude will land at the same indicated airspeed as at sea level but, because of the reduced air density, the true airspeed will be greater. Given the same wind conditions, this will also make the ground speed higher than at sea level. (PLT124) — FAA-H-8083-25

Answer (B) is incorrect because at high elevation there is reduced air density, and thus TAS will increase. As TAS increases, ground speed will increase. Answer (C) is incorrect because, under comparable conditions, TAS will increase and cause the ground speed to be higher, at higher elevation.

121, 135, DSP

9074. How should thrust reversers be applied to reduce landing distance for turbojet aircraft?

A—Immediately after ground contact.
B—Immediately prior to touchdown.
C—After applying maximum wheel braking.

Thrust reversers are most effective at high speeds and should be deployed immediately after touchdown. (PLT170) — FAA-H-8083-25

121, 135, DSP

9079. How should reverse thrust propellers be used during landing for maximum effectiveness in stopping?

A—Gradually increase reverse power to maximum as rollout speed decreases.
B—Use maximum reverse power as soon as possible after touchdown.
C—Select reverse-pitch after landing and use idle power setting of the engines.

Reverse thrust is most effective at high airspeeds. It should be used as soon as possible after touchdown. (PLT244) — FAA-H-8083-25

121, 135, DSP

9084. Under which condition during the landing roll are the main wheel brakes at maximum effectiveness?

A—When wing lift has been reduced.
B—At high groundspeeds.
C—When the wheels are locked and skidding.

Wheel brakes are at maximum effectiveness when the weight of the airplane is used to hold the tires in contact with the runway and the rate of wheel deceleration (or slowing) is just below that which would induce a skid. To place the maximum weight on the tires it is necessary to reduce lift as soon as possible after touchdown by lowering the nose wheel to the runway and deploying wing spoilers. Wheel brakes become more effective as an airplane decelerates (or slows down) because of loss of residual lift as the airspeed decreases. (PLT170) — FAA-H-8083-25

Answer (B) is incorrect because, at high ground speeds, the lift is greater and the normal force on the wheels is small, thus the braking friction force is small. Answer (C) is incorrect because, when the wheels are locked and skidding, the braking friction force is small.

Answers

9323 [B]	9322 [C]	8374 [A]	9074 [A]	9079 [B]	9084 [A]

121, 135, DSP

8935. At what minimum speed (rounded off) could dynamic hydroplaning occur on main tires having a pressure of 121 PSI?

A—90 knots.
B—96 knots.
C—110 knots.

Dynamic hydroplaning occurs when there is standing water on the runway surface. Water about 1/10th of an inch deep acts to lift the tire off the runway. The minimum speed at which dynamic hydroplaning occurs has been determined to be about 9 times the square root of the tire pressure in pounds per square inch:

Square root of 121 = 11
11 x 9 = 99

(PLT144) — FAA-H-8083-3

121, 135, DSP

8936. At what minimum speed will dynamic hydroplaning begin if a tire has an air pressure of 70 PSI?

A—85 knots.
B—80 knots.
C—75 knots.

Dynamic hydroplaning occurs when there is standing water on the runway surface. Water about 1/10th of an inch deep acts to lift the tire off the runway. The minimum speed at which dynamic hydroplaning occurs has been determined to be about 9 times the square root of the tire pressure in pounds per square inch:

Square root of 70 = 8.37
8.37 x 9 = 75.3

(PLT144) — FAA-H-8083-3

Answer (A) is incorrect because hydroplaning would occur at 85 knots with a tire pressure of 95 PSI. Answer (B) is incorrect because hydroplaning would occur at 80 knots with a tire pressure of 84 PSI.

121, 135, DSP

8933. A definition of the term "viscous hydroplaning" is where

A—the airplane rides on standing water.
B—a film of moisture covers the painted or rubber-coated portion of the runway.
C—the tires of the airplane are actually riding on a mixture of steam and melted rubber.

Either the moisture or the originally slick surface could cause problems, and the combination is especially dangerous. (PLT144) — FAA-H-8083-3

Answer (A) is incorrect because dynamic hydroplaning occurs when the airplane tires ride on standing water. Answer (C) is incorrect because reverted rubber hydroplaning occurs when the tires of the airplane are actually riding on a mixture of steam and melted rubber.

121, 135, DSP

8938. Compared to dynamic hydroplaning, at what speed does viscous hydroplaning occur when landing on a smooth, wet runway?

A—At approximately 2.0 times the speed that dynamic hydroplaning occurs.
B—At a lower speed than dynamic hydroplaning.
C—At the same speed as dynamic hydroplaning.

Viscous hydroplaning occurs due to the viscous properties of water. In this type, a thin film of fluid (not more than 1/1,000 of an inch in depth) cannot be penetrated by the tire and the tire rolls on top of the film. This can occur at a much lower speed than dynamic hydroplaning but requires a smooth acting surface. (PLT144) — FAA-H-8083-3

121, 135, DSP

8934. Which term describes the hydroplaning which occurs when an airplane's tire is effectively held off a smooth runway surface by steam generated by friction?

A—Reverted rubber hydroplaning.
B—Dynamic hydroplaning.
C—Viscous hydroplaning.

This would typically occur if excessive braking kept a wheel from rotating. (PLT144) — FAA-H-8083-3

Answer (B) is incorrect because dynamic hydroplaning occurs when there is standing water or slush on the runway which forms a wedge that lifts the tire away from contact with the runway surface. Answer (C) is incorrect because viscous hydroplaning occurs on a thin film of water on a smooth (e.g., painted or rubber-coated) runway surface.

Answers

8935 [B] 8936 [C] 8933 [B] 8938 [B] 8934 [A]

121, 135, DSP

8937. What is the best method of speed reduction if hydroplaning is experienced on landing?

A—Apply full main wheel braking only.
B—Apply nosewheel and main wheel braking alternately and abruptly.
C—Apply aerodynamic braking to the fullest advantage.

Since occurrence of dynamic hydroplaning is related to speed, it is prudent to slow the aircraft with spoilers, reverse thrust, etc., as much as possible prior to applying the brakes. (PLT144) — FAA-H-8083-3

Answer (A) is incorrect because applying full main wheel braking may increase or compound the problems associated with hydroplaning. If any brakes are used, a pumping or modulating motion like an antiskid system can be used. Aerodynamic braking is recommended. Answer (B) is incorrect because abrupt use of either the nose wheel or main wheel brakes will lock the wheels and compound the problem.

121, 135, DSP

8939. What effect, if any, will landing at a higher-than-recommended touchdown speed have on hydroplaning?

A—No effect on hydroplaning, but increases landing roll.
B—Reduces hydroplaning potential if heavy braking is applied.
C—Increases hydroplaning potential regardless of braking.

Hydroplaning is most likely to occur during conditions of standing water or slush on a runway with a smooth textured surface. The higher the aircraft speed, the more likely it is to hydroplane. (PLT144) — FAA-H-8083-3

121, DSP

8133. What effective runway length is required for a turbojet-powered airplane at the destination airport if the runways are forecast to be wet or slippery at the ETA?

A—70 percent of the actual runway available, from a height of 50 feet over the threshold.
B—115 percent of the runway length required for a dry runway.
C—115 percent of the runway length required for a wet runway.

No person may takeoff in a turbojet-powered airplane when the appropriate weather reports and forecasts, or combination thereof, indicate that the runways at the destination airport may be wet or slippery at the estimated time of arrival unless the effective runway length at the destination airport is at least 115% of the runway length required for a landing on a dry runway. (PLT144) — 14 CFR §121.195

Answer (A) is incorrect because 70% is the requirement for the turbopropeller aircraft. Answer (C) is incorrect because the effective runway length is based on a dry runway.

Answers

8937 [C] 8939 [C] 8133 [B]

Landing Performance Tables and Graphs

The graphs in FAA Figures 88 through 92 are used to compare landing distances under various conditions. All the graphs are used in the same manner; enter at the bottom with the gross weight, draw a vertical line to the appropriate diagonal line, and from there draw a horizontal line to the landing distance.

Question 8746 is typical of the questions that use these graphs. It asks for a comparison of a 40°-flap landing on a wet runway using three different weights and stopping techniques. Line 1 is the distance using brakes and spoilers at 122,500 pounds. Line 2 is the distance using brakes and reversers at 124,000 pounds. Line 3 is the distance using brakes, spoilers, and reversers at a weight of 131,000 pounds. The graph shows that Answer C will have the shortest stopping distance (about 3,200 feet).

The graphs in FAA Figures 92 and 93 are used to determine the thrust required to maintain a given airspeed with various flap configurations with the landing gear both up and down. The graph in Figure 93 is for a 110,000-pound aircraft, while the graph in Figure 92 is for a 140,000-pound aircraft. Notice that the solid curved lines represent level flight and that the two dashed lines represent a 3° glide slope.

The graph works in a manner similar to those in the previous questions. Enter the graph at the bottom with the airspeed, draw a vertical line to the appropriate curved line and then draw horizontally to the required thrust.

The two dashed lines represent an approach condition, and the V_{REF} speed is marked with a line and circle on each of the curves. For example, V_{REF} for a 140,000-pound airplane with 40° flaps is 123 knots, and with 30° flaps it is 127 knots.

Some problems require the computation of headwind, tailwind and crosswind components. The wind direction, wind speed and course must be known. Most often these problems are part of a takeoff or landing computation. If this is the case, the course corresponds to the runway number (i.e., Rwy 35 = 350° magnetic), and the winds are references to magnetic north and in knots. This type of computation can be accomplished on the CX-2 computer or by use of the Wind Component Graph in FAA Figure 74. This computation cannot be done on an E6-B computer and usually the scales on a CR-type computer do not give an answer of sufficient accuracy.

Determine the headwind and crosswind components for an aircraft landing on runway 35 with the winds from 300° at 20 knots. Using the CX-2 computer: Select "X/H-Wind" from the menu, then enter the wind direction, wind speed and runway number at the prompts. The crosswind is -15 knots (a negative crosswind is from the left, a positive is from the right). The headwind component is -13 knots (a negative number indicates a headwind, a positive indicates a tailwind).

The Landing Speed Table in FAA Figure 75 is used to determine the aircraft's approach speed (V_{REF}) at various weights and flap settings. For example, the 100,000-pound airplane using a 30° flap setting has a V_{REF} of 135 knots. This speed is adjusted by adding one half of any headwind component as noted at the bottom of the table. Gust corrections are not required on this test. If the headwind component were 13 knots, the correction would be 7 knots (always round up) and the corrected V_{REF} would be 143 knots.

The Go-Around EPR Table at the top of FAA Figure 75 is similar to the one discussed for takeoff EPR. In this table there are two EPR values; one for temperature and one for altitude. The lower of the two is the go-around EPR. For example, if the temperature is 15°C TAT at a pressure altitude of 500 feet, the temperature-limited EPR is 2.00. The altitude-limited EPR is 2.01 (you need to interpolate between 2.04 and 1.98). There are possible corrections for air conditioning OFF and for engine and wing anti-ice ON. Notice that there are three different temperature scales.

ALL

9791. Approaching the runway 1° below glidepath can add how many feet to the landing distance?

A—250 feet.
B—500 feet.
C—1,000 feet.

On final approach, at a constant airspeed, the glidepath angle and rate of descent is controlled with pitch attitude and elevator. The optimum glidepath angle is 2.5° to 3° whether or not an electronic glidepath reference is being used. On visual approaches, pilots may have a tendency to make flat approaches. A flat approach, however, will increase landing distance and should be avoided. For example, an approach angle of 2° instead of a recommended 3° will add 500 feet to landing distance. (PLT170) — FAA-H-8083-3A

ALL

9792. Arriving over the runway 10 knots over V_{REF} would add approximately how many feet to the dry landing distance?

A—800 feet.
B—1,700 feet.
C—2,800 feet.

Excess approach speed carried through the threshold window and onto the runway will increase the minimum stopping distance required by 20–30 feet per knot of excess speed for a dry runway. Worse yet, the excess speed will increase the chances of an extended flare, which will increase the distance to touchdown by approximately 250 feet for each excess knot in speed. (PLT170) — FAA-H-8083-3A

121, DSP

8742. (Refer to Figures 88 and 89.) Which conditions will result in the shortest landing distance at a weight of 132,500 pounds?

A—Dry runway using brakes and reversers.
B—Dry runway using brakes and spoilers.
C—Wet runway using brakes, spoilers and reversers.

Dry runway using brakes and reversers = 2,900 feet
Dry runway using brakes and spoilers = 2,750 feet
Wet runway using brakes, spoilers and reversers = 3,100 feet

(PLT008) — FAA-H-8083-25

121, DSP

8743. (Refer to Figure 88.) How much longer is the dry runway landing distance using brakes only compared to using brakes and reversers at 114,000 pounds gross weight?

A—1,150 feet.
B—500 feet.
C—300 feet.

Brakes only distance = 2,900 feet
Brakes and reversers distance = 2,600 feet

(PLT008) — FAA-H-8083-25

121, DSP

8744. (Refer to Figure 88.) How many feet will remain after landing on a 7,200-foot dry runway with spoilers inoperative at 118,000 pounds gross weight?

A—4,200 feet.
B—4,500 feet.
C—4,750 feet.

Landing distance using brakes and reversers = 2,700 feet
7,200 – 2,700 = 4,500 feet

(PLT132) — FAA-H-8083-25

Answer (A) is incorrect because 4,200 feet of runway would remain if brakes only were used. Answer (C) is incorrect because 4,750 feet of runway would remain if brakes and spoilers were operative and used.

121, DSP

8745. (Refer to Figure 88.) What is the maximum landing weight which will permit stopping 2,000 feet short of the end of a 5,400-foot dry runway with reversers and spoilers inoperative?

A—117,500 pounds.
B—136,500 pounds.
C—139,500 pounds.

A landing weight of 136,900 pounds will require a distance of about 3,400 feet. (PLT008) — FAA-H-8083-25

Answer (A) is incorrect because a landing weight of 117,500 pounds would require a landing distance less than 3,400 feet. Answer (C) is incorrect because a landing weight of 139,500 pounds would require a landing distance greater than 3,400 feet.

Answers

9791 [B]	9792 [C]	8742 [B]	8743 [C]	8744 [B]	8745 [B]

121, DSP

8746. (Refer to Figure 89.) Which of the following configurations will result in the shortest landing distance over a 50-foot obstacle to a wet runway?

A—Brakes and spoilers at 122,500 pounds gross weight.
B—Brakes and reversers at 124,000 pounds gross weight.
C—Brakes, spoilers, and reversers at 131,000 pounds gross weight.

Using brakes, spoilers, and reversers at 131,000 pounds gross weight will result in the shortest landing distance of 3,100 feet. (PLT008) — FAA-H-8083-25

121, DSP

8747. (Refer to Figure 89.) How many feet will remain after landing on a 6,000-foot wet runway with reversers inoperative at 122,000 pounds gross weight?

A—2,200 feet.
B—2,750 feet.
C—3,150 feet.

Landing distance using brakes and spoilers = 3,250 feet
6,000 – 3,250 = 2,750 feet
(PLT008) — FAA-H-8083-25

121, DSP

8748. (Refer to Figure 90.) Which configuration will result in a landing distance of 5,900 feet over a 50 foot obstacle to an icy runway?

A—Use of three reversers at 131,000 pounds gross weight.
B—Use of brakes and spoilers at 125,000 pounds gross weight.
C—Use of three reversers at 133,000 pounds gross weight.

A line drawn across from the 5,900-foot distance on the graph passes through the 3 reversers line at 133,000 pounds gross weight. (PLT008) — FAA-H-8083-25

121, DSP

8749. (Refer to Figure 90.) What is the transition distance when landing on an icy runway at a gross weight of 134,000 pounds?

A—400 feet.
B—950 feet.
C—1,350 feet.

Air distance = 950 feet
Air and transition distance = 1,350 feet
1,350 – 950 = 400 feet
(PLT008) — FAA-H-8083-25

121, DSP

8750. (Refer to Figure 90.) What is the maximum landing weight which will permit stopping 700 feet short of the end of a 5,200-foot icy runway?

A—124,000 pounds.
B—137,000 pounds.
C—108,000 pounds.

Using brakes, spoilers and reversers, all three weights will allow a landing with at least 500 feet of runway remaining. (PLT008) — FAA-H-8083-25

121, DSP

8751. (Refer to Figure 90.) What is the landing distance on an icy runway with reversers inoperative at a landing weight of 125,000 pounds?

A—4,500 feet.
B—4,750 feet.
C—5,800 feet.

Proceeding up the 125,000-pound gross weight line to the brakes and spoilers line results in a landing distance of 5,800 feet. (PLT008) — FAA-H-8083-25

121, DSP

8752. (Refer to Figure 91.) How much will landing distance be reduced by using 15° of flaps rather than 0° flaps at a landing weight of 119,000 pounds?

A—500 feet.
B—800 feet.
C—2,700 feet.

A 0° flap landing will require about 3,500 feet. A 15° flap landing will require about 2,700 feet. (PLT008) — FAA-H-8083-25

Answers

8746 [C] 8747 [B] 8748 [C] 8749 [A] 8750 [B] 8751 [C]
8752 [B]

121, DSP

8753. (Refer to Figure 91.) What is the ground roll when landing with 15° of flaps at a landing weight of 122,000 pounds?

A—1,750 feet.
B—2,200 feet.
C—2,750 feet.

Landing distance	*2,750*
Touchdown point	*– 1,000*
Ground roll	*1,750*

(PLT008) — FAA-H-8083-25

121, DSP

8754. (Refer to Figures 91 and 92.) What approach speed and ground roll will be needed when landing at a weight of 140,000 pounds if flaps are not used?

A—138 knots and 3,900 feet.
B—153 knots and 2,900 feet.
C—183 knots and 2,900 feet.

Landing distance	*3,900*
Touchdown point	*– 1,000*
Ground roll	*2,900*

The diagonal line for 0° flaps in FAA Figure 91 states that the V_{REF} speed is the 40° flaps V_{REF} speed plus 60 knots. In FAA Figure 92, at 140,000 pounds, the 40° flap V_{REF} is 123 knots, so the 0° V_{REF} is 183 knots (123 + 60). (PLT008) — FAA-H-8083-25

121, DSP

8755. (Refer to Figure 91.) How much more runway will be used to land with 0° flaps rather than 15° of flaps at a landing weight of 126,000 pounds?

A—900 feet.
B—1,800 feet.
C—2,700 feet.

A 0° flap landing will require about 3,600 feet. A 15° flap landing will require about 2,700 feet. (PLT008) — FAA-H-8083-25

121, DSP

8756. (Refer to Figures 91 and 92.) What approach speed and landing distance will be needed when landing at a weight of 140,000 pounds with 15° of flaps?

A—123 knots and 3,050 feet.
B—138 knots and 3,050 feet.
C—153 knots and 2,050 feet.

Landing distance = 3,050 feet

The diagonal line for 15° flaps on FAA Figure 91 states that the V_{REF} speed is the 40° flaps V_{REF} speed plus 15 knots. In FAA Figure 92, at 140,000 pounds, the 40° flap V_{REF} is 123 knots, so the 15° V_{REF} is 138 knots (123 + 15). (PLT008) — FAA-H-8083-25

121, DSP

8757. (Refer to Figure 92.) What is the maximum charted indicated airspeed while maintaining a 3° glide slope at a weight of 140,000 pounds?

A—127 knots.
B—149 knots.
C—156 knots.

The 3° glide slope curve produces the greatest charted airspeed of 156 knots. (PLT008) — FAA-H-8083-25

121, DSP

8758. (Refer to Figure 92.) What is the thrust required to maintain a 3° glide slope at 140,000 pounds, with gear down, flaps 30°, and an airspeed of V_{REF} +30 knots?

A—13,300 pounds.
B—16,200 pounds.
C—17,700 pounds.

V_{REF} + 30 = 156 knots
Required thrust is 16,200 pounds.

(PLT008) — FAA-H-8083-25

121, DSP

8759. (Refer to Figure 92.) What thrust is required to maintain level flight at 140,000 pounds, with gear up, flaps 25°, and an airspeed of 172 knots?

A—13,700 pounds.
B—18,600 pounds.
C—22,000 pounds.

A gross weight of 140,000 pounds, 172 knots, and 25° flaps will result in a required thrust of 18,600 pounds. (PLT008) — FAA-H-8083-25

Answers

8753 [A]	8754 [C]	8755 [A]	8756 [B]	8757 [C]	8758 [B]
8759 [B]					

121, DSP

8760. (Refer to Figure 92.) What thrust is required to maintain level flight at 140,000 pounds, with gear down, flaps 25°, and an airspeed of 162 knots?

A—17,400 pounds.
B—19,500 pounds.
C—22,200 pounds.

A gross weight of 140,000 pounds, 162 knots, and 25° flaps will result in a required thrust of 19,500 pounds. (PLT008) — FAA-H-8083-25

121, DSP

8761. (Refer to Figure 92.) What thrust is required to maintain level flight at 140,000 pounds, with gear down, flaps 25°, and an airspeed of 145 knots?

A—16,500 pounds.
B—18,100 pounds.
C—18,500 pounds.

A gross weight of 140,000 pounds, 145 knots, and 25° flaps will result in a required thrust of 18,100 pounds. (PLT008) — FAA-H-8083-25

121, DSP

8762. (Refer to Figure 92.) What is the change of total drag for a 140,000-pound airplane when configuration is changed from flaps 30°, gear down, to flaps 0°, gear up, at a constant airspeed of 160 knots?

A—13,500 pounds.
B—13,300 pounds.
C—15,300 pounds.

In level flight, total thrust and total drag are equal.

Drag with 30° flaps, gear down =	*23,800*	*pounds*
Drag with 0° flaps, gear up =	*– 10,300*	*pounds*
Change in drag (thrust) =	*13,500*	*pounds*

(PLT008) — FAA-H-8083-25

121, DSP

8763. (Refer to Figure 93.) What is the maximum charted indicated airspeed while maintaining a 3° glide slope at a weight of 110,000 pounds?

A—136 knots.
B—132 knots.
C—139 knots.

The maximum speed shown for a 30° flap configuration on a 3° glide slope is 136 knots. (PLT007) — FAA-H-8083-25

121, DSP

8764. (Refer to Figure 93.) What is the thrust required to maintain a 3° glide slope at 110,000 pounds, with gear down, flaps 30°, and an airspeed of V_{REF} +20 knots?

A—9,800 pounds.
B—11,200 pounds.
C—17,000 pounds.

V_{REF} + 20 = 131 knots

(PLT008) — FAA-H-8083-25

121, DSP

8765. (Refer to Figure 93.) What thrust is required to maintain level flight at 110,000 pounds, with gear down, flaps 40°, and an airspeed of 118 knots?

A—17,000 pounds.
B—20,800 pounds.
C—22,300 pounds.

A gross weight of 110,000 pounds, 118 knots, and 40° flaps will result in a required thrust of 20,800 pounds. (PLT008) — FAA-H-8083-25

121, DSP

8766. (Refer to Figure 93.) What thrust is required to maintain level flight at 110,000 pounds, with gear up, flaps 25°, and an airspeed of 152 knots?

A—14,500 pounds.
B—15,900 pounds.
C—16,700 pounds.

A gross weight of 110,000 pounds, 152 knots, and 25° flaps will result in a required thrust of 14,500 pounds. (PLT008) — FAA-H-8083-25

121, DSP

8692. (Refer to Figures 73, 74, and 75.) What is V_{REF} for Operating Conditions L-1?

A—143 knots.
B—144 knots.
C—145 knots.

Table value	*135 knots*
Wind correction	*+ 7 knots*
V_{REF}	*142 knots*

(PLT008) — FAA-H-8083-25

Answers

8760 [B]	8761 [B]	8762 [A]	8763 [A]	8764 [B]	8765 [B]
8766 [A]	8692 [A]				

121, DSP

8693. (Refer to Figures 73, 74, and 75.) What is the reference speed for Operating Conditions L-2?

A—140 knots.
B—145 knots.
C—148 knots.

Table value 140 knots
Wind correction + 5 knots
V_{REF} 145 knots

(PLT012) — FAA-H-8083-25

121, DSP

8694. (Refer to Figures 73, 74, and 75.) What is V_{REF} +20 for Operating Conditions L-3?

A—151 knots.
B—169 knots.
C—149 knots.

Table value 141 knots
Wind correction + 8 knots
V_{REF} 149 knots
+ 20
169 knots

(PLT012) — FAA-H-8083-25

121, DSP

8695. (Refer to Figures 73, 74, and 75.) What is V_{REF} +10 for Operating Conditions L-4?

A—152 knots.
B—138 knots.
C—148 knots.

Table value 134 knots
Wind correction + 3 knots
V_{REF} 137 knots
+ 10
147 knots

(PLT012) — FAA-H-8083-25

121, DSP

8696. (Refer to Figures 73, 74, and 75.) What is the maneuvering speed for Operating Conditions L-5?

A—124 knots.
B—137 knots.
C—130 knots.

Table value 124 knots
Wind correction + 6 knots
V_{REF} 130 knots

(PLT002) — FAA-H-8083-25

121, DSP

8687. (Refer to Figures 73 and 75.) What is the go-around EPR for Operating Conditions L-1?

A—2.01 EPR.
B—2.03 EPR.
C—2.04 EPR.

Temperature scale = TAT
Temperature-limited EPR = 2.00
Altitude-limited EPR = 2.04, 1.98 (need to interpolate to find 2.01)
Air conditioning correction = +.03

(PLT007) — FAA-H-8083-25

121, DSP

8688. (Refer to Figures 73 and 75.) What is the go-around EPR for Operating Conditions L-2?

A—2.115 EPR.
B—2.10 EPR.
C—2.06 EPR.

Temperature scale = OAT°F
Temperature-limited EPR = 2.10
Altitude-limited EPR = 2.15
Anti-ice correction = -.04

(PLT007) — FAA-H-8083-25

121, DSP

8689. (Refer to Figures 73 and 75.) What is the go-around EPR for Operating Conditions L-3?

A—2.06 EPR.
B—2.07 EPR.
C—2.09 EPR.

Temperature scale = OAT°C
Temperature-limited EPR = 2.13
Altitude-limited EPR = 2.09, 2.15 (need to interpolate, finding 2.12)
Anti-ice correction = -.06

(PLT007) — FAA-H-8083-25

Answers

8693 [B] 8694 [B] 8695 [C] 8696 [C] 8687 [B] 8688 [C]
8689 [A]

121, DSP

8690. (Refer to Figures 73 and 75.) What is the go-around EPR for Operating Conditions L-4?

A—2.056 EPR.
B—2.12 EPR.
C—2.096 EPR.

Temperature scale = TAT
Temperature-limited EPR = 2.16
Altitude-limited EPR = 2.09, 2.15 (need to interpolate to find 2.096)
Anti-ice correction = -.04

(PLT007) — FAA-H-8083-25

121, DSP

8691. (Refer to Figures 73 and 75.) What is the go-around EPR for Operating Conditions L-5?

A—2.00 EPR.
B—2.04 EPR.
C—2.05 EPR.

Temperature scale = OAT°F
Temperature-limited EPR = 2.00
Altitude-limited EPR = 2.04

(PLT007) — FAA-H-8083-25

Miscellaneous Performance

V_C—design cruising speed.

V_{MO}/M_{MO}—maximum operating limit speed.

The Boeing 727 Holding Table in FAA Figure 85 and the Boeing 737 Holding Table in FAA Figure 69 show the holding EPR, indicated airspeed and fuel flow per engine for various weights and altitudes. You will need to interpolate for conditions between listed weights and altitudes.

Assume a Boeing 727 is holding at 24,000 feet at a weight of 195,000 pounds. What is the EPR, IAS and fuel flow required? This problem will require interpolation both between altitudes and weights. Let's do the steps for EPR first:

1. Since the weight of 195,000 pounds is exactly halfway between two table listed weights, it is easiest to start by determining the required EPR at each of the two nearest altitudes. At 25,000 feet, the EPR for a 200,000-pound airplane is 1.85, and for a 190,000-pound airplane it is 1.81. Add those two values and divide by 2. The EPR is 1.83. The same calculation for 20,000 feet yields an EPR of 1.675.
2. Since the altitude given (24,000) is not halfway between the two table values it is necessary to calculate the amount the EPR changes per 1,000 feet of altitude. Determine the difference between the EPR values at each altitude (1.83 – 1.675 = .155), then divide that number by 5 (.155 ÷ 5 = .031).
3. The EPR variation per 1,000 feet is .031, and the EPR decreases as altitude decreases. To determine the EPR for 24,000 feet, subtract .031 from the EPR for 25,000 feet (1.83 – .031 = 1.799).

Similar interpolations give a holding speed of 264 knots and a fuel flow per engine of 3,508 pounds per hour.

Some questions ask "what is the fuel burned in a particular time?" This is a fuel burn problem identical to those covered in the flight logs section. Remember that the fuel flow rates are given per engine. The Boeing 727 (FAA Figure 85) is a three-engine airplane and the Boeing 737 (FAA Figure 69) is a twin-engine airplane.

An encounter with strong turbulence can result in structural damage to an aircraft, or inadvertent stall. The sudden changes in wind direction and speed can result in very rapid changes in an aircraft's angle of attack. A sudden increase in angle of attack will cause the airplane to accelerate upward, increasing both the load factor and the stalling speed.

Answers

8690 [A] 8691 [A]

For any combination of weight and altitude there will be a recommended "rough air" speed that provides the best protection from stalls and from the possibility of overstressing the aircraft. When clear air turbulence has been reported in the area, a pilot should slow to the rough air speed upon encountering the first ripple of turbulence.

In severe turbulence, it may be impossible to maintain a constant airspeed or altitude. If this happens, the pilot should set the power to that which would maintain the desired airspeed and maintain a level flight attitude, accepting large variations in airspeed and altitude.

Due to the inaccuracy of the EPR gauges in turbulent air, some aircraft use an N_1 power setting to maintain thrust. The Turbulent Air Penetration Table in FAA Figure 64 shows the N_1 power setting for various weights and altitudes. For example, the power setting for a 110,000-pound airplane at 30,000 feet is 82.4%.

This RPM may have to be adjusted for temperature. For example, at 30,000 feet, the RPM must be changed 1.6% for every 10°C deviation from ISA (add for temperatures above ISA and subtract for temperatures below ISA). The ISA TAT (Total Air Temperature) is listed for each altitude. Assume an actual TAT of -8°C. This is 15° warmer than standard and would require adding 2.4% to the table value.

The tables in FAA Figure 87 are used to compute the time, fuel and distance required to descend from cruise altitude. There are four different tables representing different speed schedules in the descent. The table at the top left is used if planning to descend at Mach .80 until intercepting an IAS of 250 knots. The table at the bottom right is for a descent at Mach .80 until intercepting 350 knots and holding that until 10,000 feet and then slowing to 250 knots. Be sure to use the correct table for the planned schedule. For example: Using a descent schedule of .80 M/250 KIAS, if descending from FL370 at a weight of 130,000 pounds, the required time is 26 minutes, the fuel burn is 1,570 pounds and the required distance in air miles (NAM) is 125.5.

ALL

9321. Which is the correct symbol for design cruising speed?

A—V_C.
B—V_S.
C—V_A.

V_C—design cruising speed.

(PLT132) — 14 CFR §1.2

Answer (B) is incorrect because V_S is the stalling speed or minimum steady flight speed at which the airplane is controllable. Answer (C) is incorrect because V_A is maneuvering speed.

ALL

8344. How can turbulent air cause an increase in stalling speed of an airfoil?

A—An abrupt change in relative wind.
B—A decrease in angle of attack.
C—Sudden decrease in load factor.

When an airplane flying at a high speed with a low angle of attack suddenly encounters a vertical current of air moving upward, the relative wind changes in an upward direction as it meets the airfoil. This increases the angle of attack. A downward gust would have the effect of decreasing the angle of attack. (PLT245) — FAA-H-8083-25

Answer (B) is incorrect because a decrease in angle of attack would decrease the possibility of a stall. Answer (C) is incorrect because a sudden decrease in load factor would decrease the stalling speed.

ALL

9129. If severe turbulence is encountered, which procedure is recommended?

A—Maintain a constant altitude.
B—Maintain a constant attitude.
C—Maintain constant airspeed and altitude.

In severe turbulence, the airspeed indicator is inaccurate; therefore the pilot should set power for the recommended rough air speed and then maintain a level flight attitude, accepting variations in indicated airspeed and altitude. (PLT501) — AC 00-30

Answers (A) and (C) are incorrect because severe turbulence causes large variations in both indicated airspeed and altitude. Any attempt to maintain constant airspeed and altitude may overstress the aircraft.

Answers

9321 [A] 8344 [A] 9129 [B]

121, 135, DSP

9320. Which speed symbol indicates the maximum operating limit speed for an airplane?

A—V_{LE}.
B—V_{MO}/M_{MO}.
C—V_{LO}/M_{LO}.

V_{MO}/M_{MO}—maximum operating limit speed.

(PLT466) — 14 CFR §1.2

Answer (A) is incorrect because V_{LE} is maximum landing gear extended speed. Answer (C) is incorrect because V_{LO}/M_{LO} is the maximum speed for operating the landing gear.

121, DSP

8668. (Refer to Figures 68 and 69.) What are the recommended IAS and EPR settings for holding under Operating Conditions O-1?

A—219 knots and 1.83 EPR.
B—223 knots and 2.01 EPR.
C—217 knots and 1.81 EPR.

Interpolation is required for FL310 and 102,000 pounds. At FL350, the EPR value for 102,000 is 1.97. At FL300, the EPR value for 102,000 is 1.77. Interpolation for FL310 results in an EPR value of 1.81. At FL350 the IAS for 102,000 is 219 knots. At FL300 the IAS for 102,000 is 217 knots. Interpolating for FL310 results in an IAS value of 217 knots. (PLT007) — FAA-H-8083-25

121, DSP

8669. (Refer to Figures 68 and 69.) What are the recommended IAS and EPR settings for holding under Operating Conditions O-2?

A—210 knots and 1.57 EPR.
B—210 knots and 1.51 EPR.
C—210 knots and 1.45 EPR.

Interpolation is required for FL230 and 93,000 pounds. At FL250, the EPR value for 93,000 is 1.56. At FL200, the EPR value for 93,000 is 1.44. Interpolation for FL230 results in an EPR value of 1.51. At FL250, the IAS for 93,000 is 210 knots. At FL200, the IAS for 93,000 is 210 knots. Interpolating for FL230 results in an IAS value of 210 knots. (PLT007) — FAA-H-8083-25

121, DSP

8670. (Refer to Figures 68 and 69.) What are the recommended IAS and EPR settings for holding under Operating Conditions O-3?

A—217 knots and 1.50 EPR.
B—215 knots and 1.44 EPR.
C—216 knots and 1.40 EPR.

Interpolation is required for FL170 and 104,000 pounds. At FL200, the EPR value for 104,000 is 1.50. At FL150, the EPR value for 104,000 is 1.40. Interpolation for FL170 results in an EPR value of 1.44. At FL200, the IAS for 104,000 is 216 knots. At FL150, the IAS for 104,000 is 215 knots. Interpolating for FL170 results in an IAS value of 215 knots. (PLT007) — FAA-H-8083-25

121, DSP

8671. (Refer to Figures 68 and 69.) What are the recommended IAS and EPR settings for holding under Operating Conditions O-4?

A—223 knots and 1.33 EPR.
B—225 knots and 1.33 EPR.
C—220 knots and 1.28 EPR.

Interpolation is required for FL080 and 113,000 pounds. At FL100, the EPR value for 113,000 is 1.35. At FL050, the EPR value for 113,000 is 1.29. Interpolation for FL080 results in an EPR value of 1.33. At FL100, the IAS for 113,000 is 223 knots. At FL050, the IAS for 113,000 is 222 knots. Interpolating for FL080 results in an IAS value of 223 knots. (PLT007) — FAA-H-8083-25

121, DSP

8672. (Refer to Figures 68 and 69.) What are the recommended IAS and EPR settings for holding under Operating Conditions O-5?

A—219 knots and 1.28 EPR.
B—214 knots and 1.26 EPR.
C—218 knots and 1.27 EPR.

Interpolation is required for FL040 and 109,000 pounds. At FL050, the EPR value for 109,000 is 1.28. At FL015, the EPR value for 109,000 is 1.24. Interpolation for FL040 results in an EPR value of 1.27. At FL050, the IAS for 109,000 is 218 knots. At FL015, the IAS for 109,000 is 218 knots. Interpolating for FL040 results in an IAS value of 218 knots. (PLT007) — FAA-H-8083-25

Answers

9320 [B]	8668 [C]	8669 [B]	8670 [B]	8671 [A]	8672 [C]

121, DSP

8673. (Refer to Figures 68 and 69.) What is the approximate fuel consumed when holding under Operating Conditions O-1?

A—1,625 pounds.
B—1,950 pounds.
C—2,440 pounds.

Compute the fuel flow for the holding time as follows:

Fuel flow per engine = 2,434 x 2 = 4,868 ÷ 60 x 20 = 1,623

(PLT012) — FAA-H-8083-25

121, DSP

8674. (Refer to Figures 68 and 69.) What is the approximate fuel consumed when holding under Operating Conditions O-2?

A—2,250 pounds.
B—2,500 pounds.
C—3,000 pounds.

Compute the fuel used for the holding time as follows:

Fuel flow per engine = 2,248 x 2 = 4,496 ÷ 60 x 40 = 2,997

(PLT012) — FAA-H-8083-25

121, DSP

8675. (Refer to Figures 68 and 69.) What is the approximate fuel consumed when holding under Operating Conditions O-3?

A—2,940 pounds.
B—2,520 pounds.
C—3,250 pounds.

Compute the fuel used for the holding time as follows:

Fuel flow per engine = 2,518 x 2 = 5,036 ÷ 60 x 35 = 2,938

(PLT012) — FAA-H-8083-25

121, DSP

8676. (Refer to Figures 68 and 69.) What is the approximate fuel consumed when holding under Operating Conditions O-4?

A—2,870 pounds.
B—2,230 pounds.
C—1,440 pounds.

Compute the fuel used for the holding time as follows:

Fuel flow per engine = 2,866 x 2 = 5,732 ÷ 60 x 15 = 1,433

(PLT012) — FAA-H-8083-25

121, DSP

8677. (Refer to Figures 68 and 69.) What is the approximate fuel consumed when holding under Operating Conditions O-5?

A—2,950 pounds.
B—2,870 pounds.
C—2,400 pounds.

Compute the fuel used for the holding time as follows:

Fuel flow per engine = 2,873 x 2 = 5,746 ÷ 60 x 25 = 2,394

(PLT012) — FAA-H-8083-25

121, DSP

8727. (Refer to Figures 84 and 85.) What are the recommended IAS and EPR settings for holding under Operating Conditions H-1?

A—264 knots and 1.80 EPR.
B—259 knots and 1.73 EPR.
C—261 knots and 1.81 EPR.

Interpolation is required for FL240 and 195,000 pounds. At FL250, the EPR value for 195,000 is 1.83. At FL200, the EPR value for 195,000 is 1.68. Interpolation for FL240 results in an EPR value of 1.80. At FL250, the IAS for 195,000 is 264.5 knots. At FL200, the IAS for 195,000 is 261 knots. Interpolating for FL240 results in an IAS value of 264 knots. (PLT007) — FAA-H-8083-25

Answers

8673 [A]	8674 [C]	8675 [A]	8676 [C]	8677 [C]	8727 [A]

121, DSP

8728. (Refer to Figures 84 and 85.) What are the recommended IAS and EPR settings for holding under Operating Conditions H-2?

A—257 knots and 1.60 EPR.
B—258 knots and 1.66 EPR.
C—253 knots and 1.57 EPR.

Interpolation is required for FL170 and 185,000 pounds. At FL200, the EPR value for 185,000 is 1.64. At FL150, the EPR value for 185,000 is 1.52. Interpolation for FL170 results in an EPR value of 1.57. At FL200, the IAS for 185,000 is 254.5 knots. At FL150, the IAS for 185,000 is 252.5 knots. Interpolating for FL170 results in an IAS value of 253 knots. (PLT007) — FAA-H-8083-25

121, DSP

8729. (Refer to Figures 84 and 85.) What are the recommended IAS and EPR settings for holding under Operating Conditions H-3?

A—226 knots and 1.30 EPR.
B—230 knots and 1.31 EPR.
C—234 knots and 1.32 EPR.

Interpolation is required for FL080 and 155,000 pounds. At FL100, the EPR value for 155,000 is 1.34. At FL050, the EPR value for 155,000 is 1.27. Interpolation for FL080 results in an EPR value of 1.31. At FL100, the IAS for 155,000 is 230 knots. At FL050, the IAS for 155,000 is 229 knots. Interpolating for FL080 results in an IAS value of 230 knots. (PLT007) — FAA-H-8083-25

121, DSP

8730. (Refer to Figures 84 and 85.) What are the recommended IAS and EPR settings for holding under Operating Conditions H-4?

A—219 knots and 1.44 EPR.
B—216 knots and 1.42 EPR.
C—220 knots and l.63 EPR.

Interpolation is required for FL180 and 135,000 pounds. At FL200, the EPR value for 135,000 is 1.46. At FL150, the EPR value for 135,000 is 1.36. Interpolation for FL180 results in an EPR value of 1.42. At FL200, the IAS for 135,000 is 216 knots. At FL150, the IAS for 135,000 is 215 knots. Interpolating for FL180 results in an IAS value of 216 knots. (PLT007) — FAA-H-8083-25

121, DSP

8731. (Refer to Figures 84 and 85.) What are the recommended IAS and EPR settings for holding under Operating Conditions H-5?

A—245 knots and 1.65 EPR.
B—237 knots and 1.61 EPR.
C—249 knots and 1.67 EPR.

Interpolation is required for FL220 and 175,000 pounds. At FL250, the EPR value for 175,000 is 1.75. At FL200, the EPR value for 175,000 is 1.61. Interpolation for FL220 results in an EPR value of 1.67. At FL250, the IAS for 175,000 is 249.5 knots. At FL200, the IAS for 175,000 is 247.5 knots. Interpolating for FL220 results in an IAS value of 249 knots. (PLT007) — FAA-H-8083-25

121, DSP

8732. (Refer to Figures 84 and 85.) What is the approximate fuel consumed when holding under Operating Conditions H-1?

A—3,500 pounds.
B—4,680 pounds.
C—2,630 pounds.

Compute the fuel used for the holding time as follows:

Fuel flow per engine = 3,508 x 3 = 10,524 ÷ 60 x 15 = 2,631

(PLT012) — FAA-H-8083-25

121, DSP

8733. (Refer to Figures 84 and 85.) What is the approximate fuel consumed when holding under Operating Conditions H-2?

A—5,100 pounds.
B—3,400 pounds.
C—5,250 pounds.

Compute the fuel used for the holding time as follows:

Fuel flow per engine = 3,398 x 3 = 10,194 ÷ 60 x 30 = 5,097

(PLT012) — FAA-H-8083-25

Answers

8728 [C] 8729 [B] 8730 [B] 8731 [C] 8732 [C] 8733 [A]

121, DSP

8734. (Refer to Figures 84 and 85.) What is the approximate fuel consumed when holding under Operating Conditions H-3?

A—3,090 pounds.
B—6,950 pounds.
C—6,680 pounds.

Compute the fuel used for the holding time as follows:

Fuel flow per engine = 3,089 x 3 = 9,267 ÷ 60 x 45 = 6,950

(PLT012) — FAA-H-8083-25

121, DSP

8735. (Refer to Figures 84 and 85.) What is the approximate fuel consumed when holding under Operating Conditions H-4?

A—3,190 pounds.
B—3,050 pounds.
C—2,550 pounds.

Compute the fuel used for the holding time as follows:

Fuel flow per engine = 2,552 x 3 = 7,656 ÷ 60 x 25 = 3,190

(PLT012) — FAA-H-8083-25

121, DSP

8736. (Refer to Figures 84 and 85.) What is the approximate fuel consumed when holding under Operating Conditions H-5?

A—3,170 pounds.
B—7,380 pounds.
C—5,540 pounds.

Compute the fuel used for the holding time as follows:

Fuel flow per engine = 3,165 x 3 = 9,495 ÷ 60 x 35 = 5,539

(PLT012) — FAA-H-8083-25

121, DSP

9128. What action is appropriate when encountering the first ripple of reported clear air turbulence (CAT)?

A—Extend flaps to decrease wing loading.
B—Extend gear to provide more drag and increase stability.
C—Adjust airspeed to that recommended for rough air.

In an area where significant clear air turbulence has been reported or is forecast, the pilot should adjust the speed to fly at the recommended rough air speed on encountering the first ripple, since the intensity of such turbulence may build up rapidly. (PLT501) — AC 00-30

Answer (A) is incorrect because use of flaps increases the camber of the wing and angle of attack, but does not decrease the amount of wing loading. Answer (B) is incorrect because extending the gear would increase the drag, but would not change the stability of the airplane.

121, DSP

8653. (Refer to Figures 63 and 64.) What is the turbulent air penetration N_1 power setting for Operating Conditions Q-1?

A—82.4 percent.
B—84.0 percent.
C—84.8 percent.

TAT = ISA +15°

Table value	*82.4%*
Temp. correction	*+ 2.4%*
Corrected N_1	*84.8%*

(PLT020) — FAA-H-8083-25

121, DSP

8654. (Refer to Figures 63 and 64.) What is the turbulent air penetration N_1 power setting for Operating Conditions Q-2?

A—78.2 percent.
B—75.2 percent.
C—76.7 percent.

TAT = ISA -10°

Table value	*76.7%*
Temp. correction	*– 1.5%*
Corrected N_1	*75.2%*

(PLT020) — FAA-H-8083-25

Answers

8734 [B]	8735 [A]	8736 [C]	9128 [C]	8653 [C]	8654 [B]

121, DSP

8655. (Refer to Figures 63 and 64.) What is the turbulent air penetration N_1 power setting for Operating Conditions Q-3?

A—77.8 percent.
B—82.6 percent.
C—84.2 percent.

TAT = ISA +20°

Table value	*81.0%*
Temp. correction	*+ 3.2%*
Corrected N_1	*84.2%*

(PLT020) — FAA-H-8083-25

121, DSP

8656. (Refer to Figures 63 and 64.) What is the turbulent air penetration N_1 power setting for Operating Conditions Q-4?

A—76.8 percent.
B—75.4 percent.
C—74.0 percent.

TAT = ISA +10°

Table value	*75.4%*
Temp. correction	*+ 1.4%*
Corrected N_1	*76.8%*

(PLT020) — FAA-H-8083-25

121, DSP

8657. (Refer to Figures 63 and 64.) What is the turbulent air penetration N_1 power setting for Operating Conditions Q-5?

A—70.9 percent.
B—72.9 percent.
C—71.6 percent.

TAT = ISA -15°

Table value	*72.90%*
Temp. correction	*– 1.95%*
Corrected N_1	*70.95%*

(PLT020) — FAA-H-8083-25

121, DSP

8737. (Refer to Figures 86 and 87.) What are descent time and distance under Operating Conditions S-1?

A—24 minutes, 118 NAM.
B—26 minutes, 125 NAM.
C—25 minutes, 118 NAM.

Using the descent schedule of .80/250, if descending from FL370 at a weight of 130,000 pounds, the required time is 26 minutes, the fuel burn is 1,570, and the required distance in air miles (NAM) is 125.5. (PLT012) — FAA-H-8083-25

Answer (A) is incorrect because 24 minutes is the time to descend from FL330. Answer (C) is incorrect because 25 minutes is the time to descend from FL350.

121, DSP

8738. (Refer to Figures 86 and 87.) What are descent fuel and distance under Operating Conditions S-2?

A—1,440 pounds, 104 NAM.
B—1,500 pounds, 118 NAM.
C—1,400 pounds, 98 NAM.

Using the descent schedule of .80/280, if descending from FL350 at a weight of 150,000 pounds, the fuel burn is 1,500, and the required distance in air miles (NAM) is 118. (PLT012) — FAA-H-8083-25

121, DSP

8739. (Refer to Figures 86 and 87.) What are descent fuel and distance under Operating Conditions S-3?

A—1,490 pounds, 118 NAM.
B—1,440 pounds, 110 NAM.
C—1,550 pounds, 127 NAM.

Using the descent schedule of .80/320, if descending from FL410 at a weight of 135,000 pounds, the fuel burn is 1,490, and the required distance in air miles (NAM) is 118. (PLT012) — FAA-H-8083-25

121, DSP

8740. (Refer to Figures 86 and 87.) What are descent time and distance under Operating Conditions S-4?

A—22 minutes, 110 NAM.
B—21 minutes, 113 NAM.
C—24 minutes, 129 NAM.

Using the descent schedule of .85/350, if descending from FL390 at a weight of 155,000 pounds, the required time is 21 minutes, and the required distance in air miles (NAM) is 113. (PLT004) — FAA-H-8083-25

Answers

8655 [C]	8656 [A]	8657 [A]	8737 [B]	8738 [B]	8739 [A]
8740 [B]					

121, DSP

8741. (Refer to Figures 86 and 87.) What are descent fuel and distance under Operating Conditions S-5?

A—1,420 pounds, 97 NAM.
B—1,440 pounds, 102 NAM.
C—1,390 pounds, 92 NAM.

Using the descent schedule of .80/320, if descending from FL330 at a weight of 125,000 pounds, the fuel burn is 1,420, and the required distance in air miles (NAM) is 97. (PLT004) — FAA-H-8083-25

Engine-Out Procedures

V_{MC}—minimum control speed with the critical engine inoperative.

V_{XSE}—best single engine angle-of-climb speed.

V_{YSE}—best single engine rate-of-climb speed.

When an engine fails in flight, the effect on aircraft performance is drastic. For example, the loss of one engine on a two-engine aircraft will result in a loss of climb performance in excess of 50%. Climb performance is determined by the amount of power available in excess of that required for level flight. The one remaining engine must provide all of the power required for level flight. It may be able to develop little or no excess power that would allow for a climb.

When an engine fails in cruise flight, the pilot should slow the aircraft to its best single-engine rate-of-climb speed (V_{YSE}) and apply maximum continuous power on the remaining engine. The airplane may or may not be able to climb. If it cannot climb at the present altitude, at least it will descend at the minimum possible rate of sink and level off at its maximum engine-out altitude. It may be necessary to dump fuel to improve the altitude capability of the aircraft.

A multi-engine airplane should never be flown below its minimum control speed (V_{MC}). If it is below V_{MC} and an engine failure occurs, it may be impossible to maintain directional control with the other engine operating at full power. V_{MC} will vary with the aircraft's center of gravity location. V_{MC} will be highest with the CG at its most rearward-allowed position.

A three- or four-engine turbine-powered airplane, used by an air carrier, may be ferried to a maintenance base with one engine inoperative if certain requirements are met. These requirements include:

- The airplane model must have been test flown to show that such an operation is safe.
- The operator's approved flight manual must contain performance data for such an operation.
- The operating weight of the aircraft must be limited to the minimum required for flight plus any required reserve fuel.
- Takeoffs are usually limited to dry runways.
- The computed takeoff performance must be within acceptable limits (this will vary depending on the type of aircraft).
- The initial climb cannot be over thickly-populated areas.
- Only required flight crewmembers may be on the aircraft.
- Weather conditions at the takeoff and destination airports must be VFR.

The table in FAA Figure 70 shows the time required to dump to a given fuel weight. The column at the left edge of the table is the initial fuel weight and the row at the top of the table is the ending fuel

Answers

8741 [A]

weight. The intersection of any of those weights is the time required to dump from the beginning weight to the ending weight.

Question 8678 asks for the dump time required to reach an aircraft weight of 144,500 pounds given an initial aircraft weight of 180,500 pounds and a Zero Fuel Weight of 125,500 pounds.

1. Determine the initial fuel weight by subtracting the Zero Fuel Weight from the initial aircraft weight:

   180,500 – 125,500 = 55,000 pounds.

2. Determine the ending fuel load by subtracting the Zero Fuel Weight from the ending aircraft weight:

   144,500 – 125,500 = 19,000 pounds.

3. From the table, determine the fuel dump time to go from 55,000 pounds to 19,000 pounds. Interpolate as necessary:

   Dump time = 15.25 minutes

The tables in FAA Figure 72 are used to determine the maximum altitude a Boeing 737 can maintain with one of its engines inoperative. The engine bleed configuration determines which of the three tables is to be used. The upper table is for all anti-ice off, the middle is for engine anti-ice only, and the lower is for when both engine and wing anti-ice are in use. Once the correct table has been selected, find the aircraft weight at the left-hand side and follow that row across to the appropriate ISA temperature. The only possible adjustment to the table value for the level-off altitude is covered by the note at the bottom of the page. The note allows an increase of 800 feet in the level-off altitude if the air conditioning is off and the aircraft is below 17,000 feet.

Assume the aircraft weighs 100,000 pounds at the time of its engine failure, and the engine anti-ice is on. If the temperature is ISA, the level-off altitude is 19,400 feet. Even if the air conditioning were off, it would have no effect since the level-off is above 17,000 feet.

121, 135, DSP

8369. If an engine failure occurs at an altitude above single-engine ceiling, what airspeed should be maintained?

A—V_{MC}.
B—V_{YSE}.
C—V_{XSE}.

If an airplane is not capable of maintaining altitude with an engine inoperative under existing circumstances, the airspeed should be maintained within ±5 knots of the engine-out best rate-of-climb speed (V_{YSE}), in order to conserve altitude as long as possible to reach a suitable landing area. (PLT208) — FAA-H-8083-3

121, 135, DSP

8370. What is the resulting performance loss when one engine on a twin-engine airplane fails?

A—Reduction of cruise airspeed by 50 percent.
B—Reduction of climb by 50 percent or more.
C—Reduction of all performance by 50 percent.

When one engine fails on a light twin, performance is not really halved, but is actually reduced by 80% or more. The performance loss is greater than 50% because an airplane's climb performance is a function of the thrust horsepower, which is in excess of that required for level flight. (PLT223) — FAA-H-8083-3

Answer (A) is incorrect because the power loss affects climb capability much more than it does cruise speed. Answer (C) is incorrect because climb capability is significantly (more than 50%) reduced.

121, 135, DSP

8371. Under what condition is V_{MC} the highest?

A—Gross weight is at the maximum allowable value.
B—CG is at the most rearward allowable position.
C—CG is at the most forward allowable position.

V_{MC} is greater when the center of gravity is at the most rearward-allowed position. (PLT466) — FAA-H-8083-3

Answer (A) is incorrect because the location of the weight (i.e., CG) is more critical than the amount of weight. Answer (C) is incorrect because a forward CG increases rudder effectiveness and reduces V_{MC}.

Answers

8369 [B] 8370 [B] 8371 [B]

121, 135, DSP

9355. Which operational requirement must be observed by a commercial operator when ferrying a large, three-engine, turbojet-powered airplane from one facility to another to repair an inoperative engine?

A—The computed takeoff distance to reach V_1 must not exceed 70 percent of the effective runway length.
B—The existing and forecast weather for departure, en route, and approach must be VFR.
C—No passengers may be carried.

A commercial operator of large aircraft may conduct a ferry flight of a four-engine airplane or a turbine-engine-powered, three-engine airplane with one engine inoperative, to a base for the purpose of repairing the engine. Several restrictions apply to such flights. These include:

1. *The Airplane Flight Manual must include procedures and performance data which allow for the safe operation of such a flight.*
2. *The initial climb cannot be over thickly-populated areas.*
3. *Weather conditions at the takeoff and destination airports must be VFR.*
4. *Only required flight crewmembers may be on board the aircraft.*

(PLT367) — 14 CFR §91.611

Answer (A) is incorrect because runway length allowing V_1 in less than 70% of the runway is not required for ferry flights with one engine inoperative. Answer (B) is incorrect because the weather conditions must be VFR only for takeoff and landing.

121, 135, DSP

9358. A commercial operator plans to ferry a large, four-engine, reciprocating-engine-powered airplane from one facility to another to repair an inoperative engine. Which is an operational requirement for the three-engine flight?

A—The gross weight at takeoff may not exceed 75 percent of the maximum certificated gross weight.
B—Weather conditions at the takeoff and destination airports must be VFR.
C—The computed takeoff distance to reach V_1 must not exceed 70 percent of the effective runway length.

A commercial operator of large aircraft may conduct a ferry flight of a four-engine airplane or a turbine-engine-powered, three-engine airplane with one engine inoperative, to a base for the purpose of repairing the engine. Several restrictions apply to such flights. These include:

1. *The Airplane Flight Manual must include procedures and performance data which allow for the safe operation of such a flight.*
2. *The initial climb cannot be over thickly-populated areas.*
3. *Weather conditions at the takeoff and destination airports must be VFR.*
4. *Only required flight crewmembers may be on board the aircraft.*

(PLT367) — 14 CFR §91.611

121, 135, DSP

9359. Which operational requirement must be observed when ferrying an air carrier airplane when one of its three turbine engines is inoperative?

A—The weather conditions at takeoff and destination must be VFR.
B—The flight cannot be conducted between official sunset and official sunrise.
C—Weather conditions must exceed the basic VFR minimums for the entire route, including takeoff and landing.

A commercial operator of large aircraft may conduct a ferry flight of a four-engine airplane or a turbine-engine-powered, three-engine airplane with one engine inoperative, to a base for the purpose of repairing the engine. Several restrictions apply to such flights. These include:

1. *The Airplane Flight Manual must include procedures and performance data which allow for the safe operation of such a flight.*
2. *The initial climb cannot be over thickly-populated areas.*
3. *Weather conditions at the takeoff and destination airports must be VFR.*
4. *Only required flight crewmembers may be on board the aircraft.*

(PLT367) — 14 CFR §91.611

Answer (B) is incorrect because a ferry flight may be conducted after sunset and before sunrise as long as the takeoff and destination airports are VFR. Answer (C) is incorrect because the weather conditions must be VFR only for takeoff and landing.

Answers

9355 [C] 9358 [B] 9359 [A]

121, 135, DSP

9360. Which operational requirement must be observed when ferrying a large, turbine-engine-powered airplane when one of its engines is inoperative?

A—The weather conditions at takeoff and destination must be VFR.
B—Weather conditions must exceed the basic VFR minimums for the entire route, including takeoff and landing.
C—The flight cannot be conducted between official sunset and sunrise.

A commercial operator of large aircraft may conduct a ferry flight of a four-engine airplane or a turbine-engine-powered, three-engine airplane with one engine inoperative, to a base for the purpose of repairing the engine. Several restrictions apply to such flights. These include:

1. *The Airplane Flight Manual must include procedures and performance data which allow for the safe operation of such a flight.*
2. *The initial climb cannot be over thickly-populated areas.*
3. *Weather conditions at the takeoff and destination airports must be VFR.*
4. *Only required flight crewmembers may be on board the aircraft.*

(PLT367) — 14 CFR §91.611

Answer (B) is incorrect because the weather conditions must be VFR only for takeoff and landing. Answer (C) is incorrect because a ferry flight may be conducted after sunset and before sunrise as long as the takeoff and destination airports are VFR.

121, 135, DSP

9361. When a turbine-engine-powered airplane is to be ferried to another base for repair of an inoperative engine, which operational requirement must be observed?

A—Only the required flight crewmembers may be on board the airplane.
B—The existing and forecast weather for departure, en route, and approach must be VFR.
C—No passengers except authorized maintenance personnel may be carried.

A commercial operator of large aircraft may conduct a ferry flight of a four-engine airplane or a turbine-engine-powered, three-engine airplane with one engine inoperative, to a base for the purpose of repairing the engine. Several restrictions apply to such flights. These include:

1. *The Airplane Flight Manual must include procedures and performance data which allow for the safe operation of such a flight.*
2. *The initial climb cannot be over thickly-populated areas.*
3. *Weather conditions at the takeoff and destination airports must be VFR.*
4. *Only required flight crewmembers may be on board the aircraft.*

(PLT367) — 14 CFR §91.611

Answer (B) is incorrect because the weather conditions must be VFR only for takeoff and landing. Answer (C) is incorrect because only the required flight crewmembers may be aboard.

121, DSP

8678. (Refer to Figure 70.) How many minutes of dump time is required to reach a weight of 144,500 pounds?

Initial weight .... 180,500 lb
Zero fuel weight .... 125,500 lb

A—13 minutes.
B—15 minutes.
C—16 minutes.

1. *Initial weight* 180,500
   Zero fuel wt. – 125,500
   Initial fuel wt. 55,000
2. *Ending weight* 144,500
   Zero fuel wt. – 125,500
   Ending fuel wt. 19,000
3. *Dump time = 15.25 minutes*

(PLT016) — FAA-H-8083-25

121, DSP

8679. (Refer to Figure 70.) How many minutes of dump time is required to reduce fuel load to 25,000 pounds?

Initial weight .... 179,500 lb
Zero fuel weight .... 136,500 lb

A—10 minutes.
B—9 minutes.
C—8 minutes.

1. *Initial weight* 179,500
   Zero fuel wt. – 136,500
   Initial fuel wt. 43,000
2. *Ending fuel wt.* 25,000
3. *Dump time = 7.6 minutes*

(PLT016) — FAA-H-8083-25

Answers

9360 [A] 9361 [A] 8678 [B] 8679 [C]

121, DSP

8680. (Refer to Figure 70.) How many minutes of dump time is required to reach a weight of 151,500 pounds?

Initial weight.. 181,500 lb
Zero fuel weight.. 126,000 lb

A—15 minutes.
B—14 minutes.
C—13 minutes.

1. *Initial weight* *181,500*
   Zero fuel wt. *– 126,000*
   Initial fuel wt. *55,500*
2. *Ending weight* *151,500*
   Zero fuel wt. *– 126,000*
   Ending fuel wt. *25,500*
3. *Dump time = 12.875 minutes*

(PLT016) — FAA-H-8083-25

121, DSP

8681. (Refer to Figure 70.) How many minutes of dump time is required to reduce fuel load to 16,000 pounds (at 2,350 lbs/min)?

Initial weight.. 175,500 lb
Zero fuel weight.. 138,000 lb

A—9 minutes.
B—10 minutes.
C—8 minutes.

1. *Initial weight* *175,500*
   Zero fuel wt. *– 138,000*
   Initial fuel wt. *37,500*
2. *Ending fuel wt.* *16,000*
3. *Dump time = 8.8125 minutes*

(PLT016) — FAA-H-8083-25

121, DSP

8682. (Refer to Figures 71 and 72.) What is the approximate level-off pressure altitude after drift-down under Operating Conditions D-1?

A—19,400 feet.
B—18,000 feet.
C—20,200 feet.

Assume the aircraft weighs 100,000 pounds at the time of its engine failure and the engine anti-ice is on. If the temperature is ISA, the level-off altitude is 19,400 feet. (PLT004) — FAA-H-8083-25

121, DSP

8683. (Refer to Figures 71 and 72.) What is the approximate level-off pressure altitude after drift-down under Operating Conditions D-2?

A—14,700 feet.
B—17,500 feet.
C—18,300 feet.

Assume the aircraft weighs 110,000 pounds at the time of its engine failure. If the temperature is ISA +10°C, the level-off altitude is 17,500 feet. (PLT004) — FAA-H-8083-25

121, DSP

8684. (Refer to Figures 71 and 72.) What is the approximate level-off pressure altitude after drift-down under Operating Conditions D-3?

A—22,200 feet.
B—19,800 feet.
C—21,600 feet.

Assume the aircraft weighs 90,000 pounds at the time of its engine failure. If the temperature is ISA -10°C, the level-off altitude is 21,600 feet. (PLT004) — FAA-H-8083-25

121, DSP

8685. (Refer to Figures 71 and 72.) What is the approximate level-off pressure altitude after drift-down under Operating Conditions D-4?

A—27,900 feet.
B—22,200 feet.
C—24,400 feet.

Assume the aircraft weighs 80,000 pounds at the time of its engine failure. If the temperature is ISA -10°C, the level-off altitude is 24,400 feet. (PLT004) — FAA-H-8083-25

121, DSP

8686. (Refer to Figures 71 and 72.) What is the approximate level-off pressure altitude after drift-down under Operating Conditions D-5?

A—8,800 feet.
B—9,600 feet.
C—13,000 feet.

Assume the aircraft weighs 120,000 pounds at the time of its engine failure. If the temperature is ISA +20°C, the level-off altitude is 8,800 feet. When engine bleed-air for air conditioning is off below 17,000 feet, increase level-off altitude by 800 feet. Therefore, the level-off altitude is 9,600 feet (8,800 + 800). (PLT004) — FAA-H-8083-25

Answers

8680 [C] 8681 [A] 8682 [A] 8683 [B] 8684 [C] 8685 [C]
8686 [B]

Beech 1900 Aircraft Performance

135, DSP

8459. (Refer to Figure 12.) Given the following conditions, what is the minimum torque for takeoff?

Pressure altitude.. 9,000 ft
Temperature (OAT) .... +3°C
Ice vanes .... Extended

A—3,100 foot-pound.
B—3,040 foot-pound.
C—3,180 foot-pound.

The graph on the left of FAA Figure 12 is used if the ice vanes are extended and the graph on the right is used when the ice vanes are retracted. Enter the left-hand graph at +3°C, proceed up to 9,000 feet, then over to the left to read torque of 3,100 foot-pounds. (PLT169) — FAA-H-8083-25

135, DSP

8460. (Refer to Figure 12.) Given the following conditions, what is the minimum torque for takeoff?

Pressure altitude.. 7,500 ft
Temperature (OAT) .... +35°C
Ice vanes .... Retracted

A—2,820 foot-pound.
B—2,880 foot-pound.
C—2,780 foot-pound.

The graph on the left of FAA Figure 12 is used if the ice vanes are extended and the graph on the right is used when the ice vanes are retracted. Enter the right-hand graph at +35°C, proceed up to 7,500 feet, then over to the left to read torque of 2,820 foot-pounds. (PLT169) — FAA-H-8083-25

135, DSP

8461. Refer to Figure 12.) Given the following conditions, what is the minimum torque for takeoff?

Pressure altitude.. 7,500 ft
Temperature (OAT) .... +9°C
Ice vanes .... Extended

A—3,200 foot-pound.
B—3,160 foot-pound.
C—3,300 foot-pound.

The graph on the left of FAA Figure 12 is used if the ice vanes are extended and the graph on the right is used when the ice vanes are retracted. Enter the left-hand graph at +9°C, proceed up to 7,500 feet, then over to the left to read torque of 3,160 foot-pounds. (PLT169) — FAA-H-8083-25

135, DSP

8462. (Refer to Figure 12.) Given the following conditions, what is the minimum torque for takeoff?

Pressure altitude.. 3,500 ft
Temperature (OAT) .... +43°C
Ice vanes .... Retracted

A—3,000 foot-pound.
B—3,050 foot-pound.
C—3,110 foot-pound.

The graph on the left of FAA Figure 12 is used if the ice vanes are extended and the graph on the right is used when the ice vanes are retracted. Enter the right-hand graph at +43°C, proceed up to 3,500 feet, then over to the left to read torque of 3,050 foot-pounds. (PLT169) — FAA-H-8083-25

135, DSP

8463. (Refer to Figure 12.) Given the following conditions, what is the minimum torque for takeoff?

Pressure altitude.. 5,500 ft
Temperature (OAT) .... +29°C
Ice vanes .... Retracted

A—2,950 foot-pound.
B—3,100 foot-pound.
C—3,200 foot-pound.

The graph on the left of FAA Figure 12 is used if the ice vanes are extended and the graph on the right is used when the ice vanes are retracted. Enter the right-hand graph at +29°C, proceed up to 5,500 feet, then over to the left to read torque of 3,200 foot-pounds. (PLT169) — FAA-H-8083-25

Answers

8459 [A] 8460 [A] 8461 [B] 8462 [B] 8463 [C]

135, DSP

8464. (Refer to Figure 13.) Given the following conditions, what is the takeoff distance over a 50-foot obstacle?

Pressure altitude.. Sea Level
Temperature (OAT)..+12°C
Weight .... 16,000 lb
Wind component..16 kts HW
Ice vanes ....Retracted

A—1,750 feet.
B—2,800 feet.
C—2,550 feet.

1. *Enter FAA Figure 13 at the bottom left-hand side at +12°C OAT and proceed upward to the line representing sea level (SL) pressure altitude.*
2. *From the point of intersection on the "pressure altitude" lines, draw a horizontal line to the reference line, then parallel the line which represents an aircraft weight of 16,000 pounds.*
3. *From the point of intersection in the "weight" portion of the graph, draw a horizontal line to the next reference line (wind component), then parallel the line until intersecting with the 16-knot headwind.*
4. *From the point of intersection on the "wind" portion of the graph draw a horizontal line to the next reference line (0-foot obstacle height), then parallel the line required to clear a 50-foot obstacle.*
5. *The distance required is 2,550 feet.*

(PLT011) — FAA-H-8083-25

135, DSP

8465. (Refer to Figure 13.) Given the following conditions, what is the takeoff ground roll and V_1 speed?

Pressure altitude... 4,000 ft
Temperature (OAT) .... 0°C
Weight .... 15,500 lb
Wind component.. 10 kts TW
Ice vanes .... Extended

A—2,900 feet, 106 knots.
B—4,250 feet, 102 knots.
C—2,700 feet, 107 knots.

1. *Enter FAA Figure 13 at the bottom left-hand side at +5°C OAT (adding 5°C to the actual OAT since the ice vanes are extended) and proceed upward to the line representing 4,000 feet pressure altitude.*
2. *From the point of intersection on the "pressure altitude" lines, draw a horizontal line to the reference line, then parallel the line which represents an aircraft weight of 15,500 pounds.*
3. *From the point of intersection in the "weight" portion of the graph, draw a horizontal line to the next reference line (wind component), then parallel the line until intersecting with the 10-knot tailwind.*
4. *From the point of intersection on the "wind" portion of the graph continue to the edge of the graph to find the ground roll required: 2,900 feet.*
5. *The V_1 speed is found using the table in the upper right corner of FAA Figure 13. Interpolate between 16,000 and 14,000 to find 106 knots for 15,500 pounds.*

(PLT011) — FAA-H-8083-25

135, DSP

8466. (Refer to Figure 13.) Given the following conditions, what is the takeoff distance over a 50-foot obstacle?

Pressure altitude... 2,000 ft
Temperature (OAT) ....+15°C
Weight .... 16,600 lb
Wind component..Calm
Ice vanes ....Retracted

A—3,400 feet.
B—3,700 feet.
C—4,200 feet.

1. *Enter FAA Figure 13 at the bottom left-hand side at +15°C OAT and proceed upward to the line representing 2,000 feet pressure altitude.*
2. *From the point of intersection on the "pressure altitudes" lines, draw a horizontal line to the reference line, then parallel the line which represents an aircraft weight of 16,600 pounds.*
3. *Continue past the second reference line to the third reference line (since the wind is calm).*
4. *From the third reference line (0-foot obstacle height), parallel the line required to clear a 50-foot obstacle.*
5. *The distance required is 3,700 feet.*

(PLT011) — FAA-H-8083-25

Answers

8464 [C] 8465 [A] 8466 [B]

135, DSP

8467. (Refer to Figure 13.) Given the following conditions, what is the takeoff ground roll and V_1 speed?

Pressure altitude.. 3,000 ft
Temperature (OAT) ....-10°C
Weight .... 15,000 lb
Wind component.. 8 kts TW
Ice vanes .... Extended

A—2,200 feet, 105 knots.
B—2,000 feet, 113 knots.
C—1,900 feet, 103 knots.

1. *Enter FAA Figure 13 at the bottom left-hand side at -5°C OAT (adding 5°C to the actual OAT since the ice vanes are extended) and proceed upward to the line representing 3,000 feet pressure altitude.*
2. *From the point of intersection on the "pressure altitude" lines, draw a horizontal line to the reference line, then parallel the line which represents an aircraft weight of 15,000 pounds.*
3. *From the point of intersection in the "weight" portion of the graph, draw a horizontal line to the next reference line (wind component), then parallel the line until intersecting with the 8-knot tailwind.*
4. *From the point of intersection on the "wind" portion of the graph continue to the edge of the graph to find the ground roll required: 2,200 feet.*
5. *The V_1 speed is found using the table in the upper right corner of FAA Figure 13. Interpolate between 16,000 and 14,000 to find 105 knots for 15,000 pounds.*

(PLT011) — FAA-H-8083-25

135, DSP

8468. (Refer to Figure 13.) Given the following conditions, what is the takeoff distance over a 50 foot obstacle?

Pressure altitude.. 6,000 ft
Temperature (OAT) ....+35°C
Weight .... 14,500 lb
Wind component...10 kts HW
Ice vanes ....Retracted

A—4,150 feet.
B—4,550 feet.
C—2,600 feet.

1. *Enter FAA Figure 13 at the bottom left-hand side at +35°C OAT and proceed upward to the line representing 6,000 feet pressure altitude.*
2. *From the point of intersection on the "pressure altitude" lines, draw a horizontal line to the reference line, then parallel the line which represents an aircraft weight of 14,500 pounds.*
3. *From the point of intersection in the "weight" portion of the graph, draw a horizontal line to the next reference line (wind component), then parallel the line until intersecting with the 10-knot headwind.*
4. *From the point of intersection on the "wind" portion of the graph draw a horizontal line to the next reference line (0-foot obstacle height), then parallel the line required to clear a 50-foot obstacle.*
5. *The distance required is 4,150 feet.*

(PLT011) — FAA-H-8083-25

135, DSP

8469. (Refer to Figure 14.) Given the following conditions, what is the accelerate-stop field length?

Pressure altitude.. 5,000 ft
Temperature (OAT) ....+20°C
Weight .... 15,000 lb
Wind component...10 kts HW
Ice vanes ....Retracted

A—6,300 feet.
B—4,700 feet.
C—4,300 feet.

1. *Enter FAA Figure 14 at the bottom left-hand side at +20°C OAT and proceed upward to the line representing 5,000 feet pressure altitude.*
2. *From the point of intersection on the "pressure altitude" lines, draw a horizontal line to the reference line, then parallel the line which represents an aircraft weight of 15,000 pounds.*
3. *From the point of intersection in the "weight" portion of the graph, draw a horizontal line to the next reference line (wind component), then parallel the line until intersecting with the 10-knot headwind.*
4. *Proceed from the wind line to the edge of the chart to read the accelerate/stop distance of 4,300 feet.*

(PLT011) — FAA-H-8083-25

Answers

8467 [A] 8468 [A] 8469 [C]

135, DSP

8470. (Refer to Figure 14.) Given the following conditions, what is the accelerate-stop field length?

Pressure altitude.. 2,000 ft
Temperature (OAT) .... -15°C
Weight .... 16,000 lb
Wind component.. 5 kts TW
Ice vanes .... Extended

A—3,750 feet.
B—4,600 feet.
C—4,250 feet.

1. *Enter FAA Figure 14 at the bottom left-hand side at -12°C OAT (adding 3°C to the actual OAT for extended ice vanes) and proceed upward to the line representing 2,000 feet pressure altitude.*
2. *From the point of intersection on the "pressure altitude" lines, draw a horizontal line to the reference line, then parallel the line which represents an aircraft weight of 16,000 pounds.*
3. *From the point of intersection in the "weight" portion of the graph, draw a horizontal line to the next reference line (wind component), then parallel the line until intersecting with the 5-knot tailwind.*
4. *Proceed from the wind line to the edge of the chart to read the accelerate/stop distance of 4,250 feet.*

(PLT011) — FAA-H-8083-25

135, DSP

8471. (Refer to Figure 14.) Given the following conditions, what is the accelerate-stop field length?

Pressure altitude.. 6,000 ft
Temperature (OAT) .... +10°C
Weight .... 16,600 lb
Wind component.. 15 kts HW
Ice vanes .... Retracted

A—4,950 feet.
B—4,800 feet.
C—5,300 feet.

1. *Enter FAA Figure 14 at the bottom left-hand side at +10°C OAT and proceed upward to the line representing 6,000 feet pressure altitude.*
2. *From the point of intersection on the "pressure altitude" lines, draw a horizontal line to the reference line, then parallel the line which represents an aircraft weight of 16,600 pounds.*
3. *From the point of intersection in the "weight" portion of the graph, draw a horizontal line to the next reference line (wind component), then parallel the line until intersecting with the 15-knot headwind.*
4. *Proceed from the wind line to the edge of the chart to read the accelerate/stop distance of 4,950 feet.*

(PLT011) — FAA-H-8083-25

135, DSP

8472. (Refer to Figure 14.) Given the following conditions, what is the accelerate-stop field length?

Pressure altitude.. 8,000 ft
Temperature (OAT) .... -5°C
Weight .... 14,000 lb
Wind component.. 4 kts TW
Ice vanes .... Extended

A—4,500 feet.
B—4,800 feet.
C—5,300 feet.

1. *Enter FAA Figure 14 at the bottom left-hand side at -2°C OAT (adding 3°C to the actual OAT for extended ice vanes) and proceed upward to the line representing 8,000 feet pressure altitude.*
2. *From the point of intersection on the "pressure altitude" lines, draw a horizontal line to the reference line, then parallel the line which represents an aircraft weight of 14,000 pounds.*
3. *From the point of intersection in the "weight" portion of the graph, draw a horizontal line to the next reference line (wind component), then parallel the line until intersecting with the 4-knot tailwind.*
4. *Proceed from the wind line to the edge of the chart to read the accelerate/stop distance of 4,800 feet.*

(PLT011) — FAA-H-8083-25

Answers

8470 [C] 8471 [A] 8472 [B]

135, DSP

8473. (Refer to Figure 14.) Given the following conditions, what is the accelerate-stop field length?

Pressure altitude.. Sea Level
Temperature (OAT) .... +30°C
Weight .... 13,500 lb
Wind component..14 kts HW
Ice vanes ....Retracted

A—2,500 feet.
B—2,850 feet.
C—3,050 feet.

1. *Enter FAA Figure 14 at the bottom left-hand side at +30°C OAT and proceed upward to the line representing sea level (SL) pressure altitude.*
2. *From the point of intersection on the "pressure altitude" lines, draw a horizontal line to the reference line, then parallel the line which represents an aircraft weight of 13,500 pounds.*
3. *From the point of intersection in the "weight" portion of the graph, draw a horizontal line to the next reference line (wind component), then parallel the line until intersecting with the 14-knot headwind.*
4. *Proceed from the wind line to the edge of the chart to read the accelerate/stop distance of 3,050 feet.*

(PLT011) — FAA-H-8083-25

135, DSP

8474. (Refer to Figures 15, 16, and 17.) What is the two-engine rate of climb after takeoff in climb configuration for Operating Conditions BE-21?

A—1,350 ft/min.
B—2,450 ft/min.
C—2,300 ft/min.

1. *Enter FAA Figure 16 at the bottom left-hand side at +10°C OAT and proceed upward to the line representing 2,000 feet pressure altitude.*
2. *From the point of intersection on the "pressure altitude" lines, draw a horizontal line to the reference line, then parallel the line which represents an aircraft weight of 16,600 pounds.*
3. *From the point of intersection in the "weight" portion of the graph, draw a horizontal line to the edge of the chart to read 2,300 fpm.*

(PLT004) — FAA-H-8083-25

135, DSP

8475. (Refer to Figures 15, 16, and 17.) What is the single-engine climb gradient after takeoff in climb configuration for Operating Conditions BE-22?

A—6.8 percent gradient.
B—7.5 percent gradient.
C—5.6 percent gradient.

1. *Enter FAA Figure 17 at the bottom left-hand side at 0°C OAT and proceed upward to the line representing 1,000 feet pressure altitude.*
2. *From the point of intersection on the "pressure altitude" lines, draw a horizontal line to the reference line, then parallel the line which represents an aircraft weight of 14,000 pounds.*
3. *From the point of intersection in the "weight" portion of the graph, draw a horizontal line to the edge of the chart to read 870 fpm.*
4. *Adjust the rate of climb for the extended ice vanes, as indicated at the top of the graph. The single-engine rate of climb with ice vanes extended is 755 fpm (870–115).*
5. *Find 755 fpm at the right edge of the chart and proceed right to the reference line and determine a climb gradient of 5.6%.*

(PLT004) — FAA-H-8083-25

135, DSP

8476. (Refer to Figures 15, 16, and 17.) What is the two-engine rate of climb after takeoff in climb configuration for Operating Conditions BE-23?

A—1,500 ft/min.
B—2,600 ft/min.
C—2,490 ft/min.

1. *Enter FAA Figure 16 at the bottom left-hand side at +20°C OAT and proceed upward to the line representing 3,000 feet pressure altitude.*
2. *From the point of intersection on the "pressure altitude" lines, draw a horizontal line to the reference line, then parallel the line which represents an aircraft weight of 15,000 pounds.*
3. *From the point of intersection in the "weight" portion of the graph, draw a horizontal line to the edge of the chart to read 2,600 fpm.*

(PLT004) — FAA-H-8083-25

Answers

8473 [C] 8474 [C] 8475 [C] 8476 [B]

135, DSP

8477. (Refer to Figures 15, 16, and 17.) What is the two-engine rate of climb after takeoff in climb configuration for Operating Conditions BE-24?

A—2,100 ft/min.
B—2,400 ft/min.
C—1,500 ft/min.

1. *Enter FAA Figure 16 at the bottom left-hand side at +25°C OAT and proceed upward to the line representing 4,000 feet pressure altitude.*
2. *From the point of intersection on the "pressure altitude" lines, draw a horizontal line to the reference line, then parallel the line which represents an aircraft weight of 16,000 pounds.*
3. *From the point of intersection in the "weight" portion of the graph, draw a horizontal line to the edge of the chart to read 2,100 fpm.*

(PLT004) — FAA-H-8083-25

135, DSP

8478. (Refer to Figures 15, 16, and 17.) What is the single-engine rate of climb after takeoff in climb configuration for Operating Conditions BE-25?

A—385 ft/min.
B—780 ft/min.
C—665 ft/min.

1. *Enter FAA Figure 17 at the bottom left-hand side at -10°C OAT and proceed upward to the line representing 5,000 feet pressure altitude.*
2. *From the point of intersection on the "pressure altitude" lines, draw a horizontal line to the reference line, then parallel the line which represents an aircraft weight of 14,000 pounds.*
3. *From the point of intersection in the "weight" portion of the graph, draw a horizontal line to the edge of the chart to read 780 fpm.*
4. *Adjust the rate of climb for the extended ice vanes, as indicated at the top of the graph. The single-engine rate-of-climb with ice vanes extended is 665 fpm (780 – 115).*

(PLT004) — FAA-H-8083-25

135, DSP

8479. (Refer to Figures 15 and 18.) What are the time, fuel, and distance from the start of climb to cruise altitude for Operating Conditions BE-21?

A—10.0 minutes; 290 pounds; 35 NM.
B—10.0 minutes; 165 pounds; 30 NM.
C—11.5 minutes; 165 pounds; 30 NM.

1. *Enter FAA Figure 18 at the bottom left-hand side at -20°C OAT and proceed upward to the line representing 16,000-foot cruise altitude.*
2. *From the point of intersection on the "pressure altitude" lines, draw a horizontal line to the line which represents an aircraft weight of 16,600 pounds.*
3. *From the point of intersection in the "weight" portion of the graph, draw a line down to the bottom of the chart to read the time to climb of 11.5 minutes.*
4. *Continue down to the fuel to climb line to read 190 pounds fuel burned.*
5. *Continue down to the distance-to-climb line to read 32 NM.*
6. *Repeat steps 1 through 5 using an OAT of +10°C OAT and 2,000 feet pressure altitude, and find the time to climb 1.5 minutes, fuel burn 25 pounds, and distance 2 NM.*
7. *Subtract the results of the airport altitudes from the cruise altitude results. Therefore, from 2,000 feet to 16,000 feet, it will take 10 minutes (11.5 – 1.5), 165 pounds of fuel (190 – 25), and 30 NM (32 – 2).*

(PLT012) — FAA-H-8083-25

Answers

8477 [A]   8478 [C]   8479 [B]

135, DSP

8480. (Refer to Figures 15 and 18.) What are the time, fuel, and distance from the start of climb to cruise altitude for Operating Conditions BE-22?

A—12.0 minutes; 220 pounds; 40 NM.
B—11.0 minutes; 185 pounds; 37 NM.
C—10.5 minutes; 175 pounds; 32 NM.

1. *Enter FAA Figure 18 at the bottom left-hand side at -15°C OAT (accounting for the extended ice vanes) and proceed upward to the line representing 18,000-foot cruise altitude.*
2. *From the point of intersection on the "pressure altitude" lines, draw a horizontal line to the line which represents an aircraft weight of 14,000 pounds.*
3. *From the point of intersection in the "weight" portion of the graph, draw a line down to the bottom of the chart to read the time to climb of 12 minutes.*
4. *Continue down to the fuel-to-climb line to read 200 pounds fuel burned.*
5. *Continue down to the distance-to-climb line to read 37.5 NM.*
6. *Repeat steps 1 through 5 using an OAT of +10°C OAT and 1,000 feet pressure altitude, and find the time to climb 1 minute, fuel burn 15 pounds, and distance 0.5 NM.*
7. *Subtract the results of the airport altitudes from the cruise altitude results. Therefore, from 1,000 feet to 18,000 feet, it will take 11 minutes (12 – 1), 185 pounds of fuel (200 – 15), and 37 NM (37.5 – 0.5).*

(PLT012) — FAA-H-8083-25

135, DSP

8481. (Refer to Figures 15 and 18.) What are the time, fuel, and distance from the start of climb to cruise altitude for Operating Conditions BE-23?

A—13.0 minutes; 180 pounds; 35 NM.
B—14.0 minutes; 210 pounds; 40 NM.
C—15.0 minutes; 240 pounds; 46 NM.

1. *Enter FAA Figure 18 at the bottom left-hand side at ISA and proceed upward to the line representing 20,000-foot cruise altitude.*
2. *From the point of intersection on the "pressure altitude" lines, draw a horizontal line to the line which represents an aircraft weight of 15,000 pounds.*
3. *From the point of intersection in the "weight" portion of the graph, draw a line down to the bottom of the chart to read the time to climb of 15 minutes.*
4. *Continue down to the fuel-to-climb line to read 235 pounds fuel burned.*
5. *Continue down to the distance-to-climb line to read 44 NM.*
6. *Repeat steps 1 through 5 using an OAT of +20°C OAT and 3,000 feet pressure altitude, and find the time to climb 1 minutes, fuel burn 25 pounds, and distance 4 NM.*
7. *Subtract the results of the airport altitudes from the cruise altitude results. Therefore, from 3,000 feet to 20,000 feet, it will take 14 minutes (15–1), 210 pounds of fuel (235–25), and 40 NM (44–4).*

(PLT012) — FAA-H-8083-25

135, DSP

8482. (Refer to Figures 15 and 18.) What are the time, fuel, and distance from the start of climb to cruise altitude for Operating Conditions BE-24?

A—12.0 minutes; 220 pounds; 45 NM.
B—9.0 minutes; 185 pounds; 38 NM.
C—10.0 minutes; 170 pounds; 30 NM.

1. *Enter FAA Figure 18 at the bottom left-hand side at 0°C OAT and proceed upward to the line representing 14,000-foot cruise altitude.*
2. *From the point of intersection on the "pressure altitude" lines, draw a horizontal line to the line which represents an aircraft weight of 16,000 pounds.*
3. *From the point of intersection in the "weight" portion of the graph, draw a line down to the bottom of the chart to read the time to climb of 13 minutes.*
4. *Continue down to the fuel-to-climb line to read 210 pounds fuel burned.*
5. *Continue down to the distance-to-climb line to read 38 NM.*
6. *Repeat steps 1 through 5 using an OAT of +25°C OAT and 4,000 feet pressure altitude, and find the time to climb 3 minutes, fuel burn 60 pounds, and distance 8 NM.*
7. *Subtract the results of the airport altitudes from the cruise altitude results. Therefore, from 4,000 feet to 14,000 feet, it will take 10 minutes (13–3), 150 pounds of fuel (210–60), and 30 NM (38–8).*

(PLT012) — FAA-H-8083-25

Answers

8480 [B]　　8481 [B]　　8482 [C]

135, DSP

8483. (Refer to Figures 15 and 18.) What are the time, fuel, and distance from the start of climb to cruise altitude for Operating Conditions BE-25?

A—11.5 minutes; 170 pounds; 31 NM.
B—8.0 minutes; 270 pounds; 28 NM.
C—12.5 minutes; 195 pounds; 38 NM.

1. *Enter FAA Figure 18 at the bottom left-hand side at -30°C OAT (accounting for the extended ice vanes) and proceed upward to the line representing 22,000-foot cruise altitude.*
2. *From the point of intersection on the "pressure altitude" lines, draw a horizontal line to the line which represents an aircraft weight of 14,000 pounds.*
3. *From the point of intersection in the "weight" portion of the graph, draw a line down to the bottom of the chart to read the time to climb of 14.5 minutes.*
4. *Continue down to the fuel-to-climb line to read 235 pounds fuel burned.*
5. *Continue down to the distance-to-climb line to read 44 NM.*
6. *Repeat steps 1 through 5 using an OAT of 0°C OAT and 5,000 feet pressure altitude, and find the time to climb 2 minutes, fuel burn 40 pounds, and distance 6 NM.*
7. *Subtract the results of the airport altitudes from the cruise altitude results. Therefore, from 5,000 feet to 22,000 feet, it will take 12.5 minutes (14.5 – 2), 195 pounds of fuel (235 – 40), and 38 NM (44 – 6).*

(PLT012) — FAA-H-8083-25

135, DSP

8484. (Refer to Figures 19 and 20.) At what altitude is the service ceiling with one engine inoperative for Operating Conditions BE-26?

A—13,000 feet.
B—14,200 feet.
C—13,600 feet.

1. *Enter FAA Figure 20 at the bottom left-hand side (for bleed air on) at -8°C OAT and proceed upward to the line representing 15,500 pounds.*
2. *From the point of intersection in the "weight" portion of the graph, draw a line to the edge of the chart to determine a pressure altitude service ceiling of 13,000 feet with one engine inoperative.*

(PLT065) — FAA-H-8083-25

135, DSP

8485. (Refer to Figures 19 and 20.) Which statement is true regarding performance with one engine inoperative for Operating Conditions BE-27?

A—Climb rate at the MEA is more than 50 ft/min.
B—Service ceiling is below the MEA.
C—Bleed air OFF improves service ceiling by 3,000 feet.

1. *Enter FAA Figure 20 at the bottom left-hand side (for bleed air on) at +30°C OAT and proceed upward to the line representing 16,600 pounds.*
2. *From the point of intersection in the "weight" portion of the graph, draw a line to the edge of the chart to determine a pressure altitude service ceiling of 5,000 feet with one engine inoperative.*
3. *Therefore, the 5,000-foot service ceiling is below the MEA of 5,500 feet.*

(PLT065) — FAA-H-8083-25

135, DSP

8486. (Refer to Figures 19 and 20.) At what altitude is the service ceiling with one engine inoperative for Operating Conditions BE-28?

A—1,500 feet above the MEA.
B—10,400 feet.
C—11,800 feet.

1. *Enter FAA Figure 20 at the bottom right-hand side (for bleed air off) at +5°C OAT and proceed upward to the line representing 16,000 pounds.*
2. *From the point of intersection in the "weight" portion of the graph, draw a line to the edge of the chart to determine a pressure altitude service ceiling of 11,800 feet with one engine inoperative.*
3. *11,800 feet is 2,800 feet above the 9,000-foot MEA.*

(PLT065) — FAA-H-8083-25

Answers

8483 [C] 8484 [A] 8485 [B] 8486 [C]

135, DSP

8487. (Refer to Figures 19 and 20.) Which statement is true regarding performance with one engine inoperative for Operating Conditions BE-29?

A—Service ceiling is more than 100 feet above the MEA.
B—Bleed air must be OFF to obtain a rate of climb of 50 ft/min at the MEA.
C—Climb is not possible at the MEA.

1. *Enter FAA Figure 20 at the bottom left-hand side (for bleed air on) at +18°C OAT and proceed upward to the line representing 16,300 pounds.*
2. *From the point of intersection in the "weight" portion of the graph, draw a line to the edge of the chart to determine a pressure altitude service ceiling of 7,700 feet with one engine inoperative.*
3. *Therefore, the 7,700-foot service ceiling is above the MEA of 7,000 feet by more than 100 feet.*

(PLT065) — FAA-H-8083-25

135, DSP

8488. (Refer to Figures 19 and 20.) At what altitude is the service ceiling with one engine inoperative for Operating Conditions BE-30?

A—9,600 feet.
B—13,200 feet.
C—2,100 feet above the MEA.

1. *Enter FAA Figure 20 at the bottom right-hand side (for bleed air off) at +22°C OAT and proceed upward to the line representing 14,500 pounds.*
2. *From the point of intersection in the "weight" portion of the graph, draw a line to the edge of the chart to determine a pressure altitude service ceiling of 11,600 feet with one engine inoperative.*
3. *Therefore, the 11,600-foot service ceiling is above the MEA of 9,500 feet.*

(PLT065) — FAA-H-8083-25

135

8117. (Refer to Figure 1.) What is the maximum landing distance that may be used by a turbopropeller-powered, small transport category airplane to land on Rwy 24 (dry) at the alternate airport?

A—5,490 feet.
B—6,210 feet.
C—6,405 feet.

The maximum landing percentages for turboprop small transport category airplanes are 60% at destination and 70% at alternate of effective runway length (actual runway minus shaded obstruction clearance portion). Compute for runway 24 as follows:

$$\begin{array}{r} 10{,}350 \\ -\ 1{,}200 \\ \hline 9{,}150 \end{array} \times .7 = 6{,}405$$

(PLT008) — 14 CFR §135.385, §135.387, §135.397

135

8118. (Refer to Figure 1.) What is the maximum landing distance that may be used by a reciprocating-engine-powered, small transport category airplane to land on Rwy 24 (dry) at the destination airport?

A—5,490 feet.
B—6,210 feet.
C—6,405 feet.

The maximum landing percentages for reciprocating transport category airplanes are 60% at destination and 70% at alternate of effective runway length (actual runway minus shaded obstruction clearance portion). Compute for runway 24 as follows:

$$\begin{array}{r} 10{,}350 \\ -\ 1{,}200 \\ \hline 9{,}150 \end{array} \times .6 = 5{,}490$$

(PLT008) — 14 CFR §135.375, §135.377, §135.397

Answers

8487 [A] 8488 [C] 8117 [C] 8118 [A]

135
8119. (Refer to Figure 1.) What is the maximum landing distance that may be used by a turbopropeller-powered, small transport category airplane to land on Rwy 6 (dry) at the alternate airport?

A—5,460 feet.
B—6,210 feet.
C—6,370 feet.

The maximum landing percentages for turboprop small transport category airplanes are 60% at destination and 70% at alternate of effective runway length (actual runway minus shaded obstruction clearance portion). Compute for runway 6 as follows:

10,350
– 1,250
9,100 x .7 = 6,370

(PLT008) — 14 CFR §135.385, §135.387, §135.395

135
8120. (Refer to Figure 1.) What is the maximum landing distance that may be used by a reciprocating-engine-powered, small transport category airplane to land on Rwy 6 (dry) at the destination airport?

A—5,460 feet.
B—6,210 feet.
C—6,370 feet.

The maximum landing percentages for reciprocating transport category airplanes are 60% at destination and 70% at alternate of effective runway length (actual runway minus shaded obstruction clearance portion). Compute for runway 6 as follows:

10,350
– 1,250
9,100 x .6 = 5,460

(PLT008) — 14 CFR §135.375, §135.377, §135.397

135
8121. (Refer to Figure 1.) What is the maximum landing distance that may be used by a turbine-engine-powered, small transport category airplane to land on Rwy 24 (dry) at the destination airport?

A—5,460 feet.
B—5,490 feet.
C—6,210 feet.

The maximum landing percentages for turbine-engine-powered small transport category airplanes are 60% at destination and 70% at alternate of effective runway length (actual runway minus shaded obstruction clearance portion). Compute for runway 24 as follows:

10,350
– 1,200
9,150 x .6 = 5,490

(PLT008) — 14 CFR §135.385, §135.387, §135.395

135
8122. (Refer to Figure 1.) What is the maximum landing distance that may be used by a turbine-engine-powered, small transport category airplane to land on Rwy 6 (wet) at the destination airport?

A—5,460 feet.
B—9,100 feet.
C—6,279 feet.

The maximum landing percentages for turbine-engine-powered small transport category airplanes are 60% at destination and 70% at alternate of effective runway length (actual runway minus shaded obstruction clearance portion). Compute for runway 6 as follows:

10,350
– 1,250
9,100 x .6 = 5,460

14 CFR §135.385 is misleading here because you must increase actual landing distance by 115% to find the effective runway length wet, but the question asks for maximum landing distance, not effective landing distance. (PLT008) — 14 CFR §§135.385, 135.387, 135.395

Answers

8119 [C]	8120 [A]	8121 [B]	8122 [A]

135

8123. (Refer to Figure 2.) What is the maximum landing distance that may be used by a turbopropeller-powered, small transport category airplane to land on Rwy 19 (dry) at the destination airport?

A—6,020 feet.
B—5,820 feet.
C—5,160 feet.

The maximum landing percentages for turboprop small transport category airplanes are 60% at destination and 70% at alternate of effective runway length (actual runway minus shaded obstruction clearance portion). Compute for runway 19 as follows:

$$\begin{array}{r} 9{,}700 \\ -\,1{,}100 \\ \hline 8{,}600 \end{array} \times .6 = 5{,}160$$

(PLT008)—14 CFR §135.377, 135.385, 135.387, 135.397

135

8124. (Refer to Figure 2.) What is the maximum landing distance that may be used by a reciprocating-engine-powered, small transport category airplane to land on Rwy 1 (dry) at the destination airport?

A—5,010 feet.
B—5,820 feet.
C—5,845 feet.

The maximum landing percentages for reciprocating transport category airplanes are 60% at destination and 70% at alternate of effective runway length (actual runway minus shaded obstruction clearance portion). Compute for runway 1 as follows:

$$\begin{array}{r} 9{,}700 \\ -\,1{,}350 \\ \hline 8{,}350 \end{array} \times .6 = 5{,}010$$

(PLT008) — 14 CFR §135.375, §135.377, §135.395

135

8125. (Refer to Figure 2.) What is the maximum landing distance that may be used by a turbine-engine-powered, small transport category airplane to land on Rwy 1 (dry) at the destination airport?

A—5,010 feet.
B—5,820 feet.
C—5,845 feet.

The maximum landing percentages for turbine-engine-powered small transport category airplanes are 60% at destination and 70% at alternate of effective runway length (actual runway minus shaded obstruction clearance portion). Compute for runway 1 as follows:

$$\begin{array}{r} 9{,}700 \\ -\,1{,}350 \\ \hline 8{,}350 \end{array} \times .6 = 5{,}010$$

(PLT008) — 14 CFR §135.385, §135.387, §135.395

135

8126. (Refer to Figure 2.) What is the maximum landing distance that may be used by a turbine-engine-powered, small transport category airplane to land on Rwy 19 (dry) at the destination airport?

A—5,160 feet.
B—5,820 feet.
C—6,020 feet.

The maximum landing percentages for turbine-engine-powered small transport category airplanes are 60% at destination and 70% at alternate of effective runway length (actual runway minus shaded obstruction clearance portion). Compute for runway 19 as follows:

$$\begin{array}{r} 9{,}700 \\ -\,1{,}100 \\ \hline 8{,}600 \end{array} \times .6 = 5{,}160$$

(PLT008) — 14 CFR §135.385, §135.387, §135.395

Answers

8123 [C] 8124 [A] 8125 [A] 8126 [A]

135

8127. (Refer to Figure 2.) May a small transport category, turbine-engine-powered airplane that has a computed landing distance of 5,500 feet use one or both of the runways depicted in the illustration at the destination airport?

A—Neither Rwy 1 nor Rwy 19 may be used if dry conditions exist.
B—Only Rwy 19 may be used provided dry conditions exist.
C—Rwy 1 or Rwy 19 may be used whether conditions are wet or dry.

The maximum landing percentages for turbine-engine-powered small transport category airplanes are 60% at destination and 70% at alternate of effective runway length (actual runway minus shaded obstruction clearance portion). Compute as follows:

1. *Runway 1*

   9,700
   – 1,350
   8,350 x .6 = 5,010

2. *Runway 19*

   9,700
   – 1,100
   8,600 x .6 = 5,160

3. *Computed landing distance is given as 5,500. Since computed landing distance exceeds both runway landing distances even in dry conditions, neither runway 1 nor 19 may be used if dry conditions exist.*

(PLT456) — 14 CFR §135.385

135

8128. (Refer to Figure 2.) May a small transport category, turboprop airplane that has a computed landing distance of 6,000 feet use either or both runways depicted in the illustration at the destination airport?

A—Only Rwy 19 may be used if dry conditions exist.
B—Neither Rwy 1 nor Rwy 19 may be used under any conditions.
C—Either Rwy 1 or Rwy 19 may be used whether conditions are wet or dry.

The maximum landing percentages for turboprop small transport category airplanes are 60% at destination and 70% at alternate of effective runway length (actual runway minus shaded obstruction clearance portion). Compute as follows:

1. *Runway 1*

   9,700
   – 1,350
   8,350 x .6 = 5,010

2. *Runway 19*

   9,700
   – 1,100
   8,600 x .6 = 5,160

3. *Computed landing distance is given as 6,000; therefore neither runway may be used.*

(PLT456) — 14 CFR §135.385, §135.387, §135.395

135

8129. (Refer to Figure 2.) What is the maximum landing distance that may be used for a non-transport category, turbopropeller-driven airplane to land on Rwy 1 (dry) at the alternate airport?

A—5,010 feet.
B—5,845 feet.
C—6,020 feet.

The maximum landing percentages for turboprop small transport category airplanes are 60% at destination and 70% at alternate of effective runway length (actual runway minus shaded obstruction clearance portion). Compute as follows:

9,700
– 1,350
8,350 x .7 = 5,845

(PLT008) — 14 CFR §135.385, §135.387, §135.395

Answers

8127 [A] 8128 [B] 8129 [B]

135

8130. (Refer to Figure 2.) Which condition meets 14 CFR Part 135 operational requirements for a small, transport category, turboprop airplane to land at the destination airport that has the runway environment given in the illustration?

A—The airport may be listed as the destination airport if the landing distance does not exceed 5,160 feet for Rwy 19.

B—The airport may NOT be listed as the destination airport if the landing distance exceeds 5,100 feet for Rwy 19.

C—The airport may be listed as the destination airport if the landing distance does not exceed 5,350 feet for either runway, wet or dry conditions.

The maximum landing percentages for turboprop small transport category airplanes are 60% at destination and 70% at alternate of effective runway length (actual runway minus shaded obstruction clearance portion). Compute as follows:

1. *Runway 1*

   9,700
   – 1,350
   8,350 x .6 = 5,010

2. *Runway 19*

   9,700
   – 1,100
   8,600 x .6 = 5,160

3. *The best answer is to not exceed 5,160 feet.*

(PLT456) — 14 CFR §135.385

Beech 1900 Cruise Tables

The tables in FAA Figures 21 through 25 are used to determine cruise, fuel flow and TAS for various temperatures, weights and altitudes. Use FAA Figure 23 in ISA +10°C conditions, FAA Figure 24 in ISA, and FAA Figure 25 in ISA -10°C.

For example, Operating Conditions BE-31 (FAA Figure 21) specify a weight of 15,000 pounds at a pressure altitude of 22,000 feet and an OAT of -19°C. The first step is to decide which of the three tables to use. Altitudes are listed in the left-hand column of each table, beside each altitude IOAT (Indicated OAT) and the OAT which corresponds to the ISA temperature for the table. All temperatures in these problems are OAT rather than IOAT. Figure 23 (ISA +10°C) shows an OAT at 22,000 feet of -19°C which is the given value. There is no column of data for a 15,000-pound aircraft, but interpolation of the data for 16,000 and 14,000 pounds yields a TAS of 228 knots and a total fuel flow of 633 pounds per hour.

Questions 8489 through 8493 ask for the flight time at cruise, using the data from these tables. The first step is to compute the ground speed using the table TAS, the given winds and true course.

Since the true course is given, it is not necessary to convert the wind direction to magnetic. Given a TAS of 228 knots and the winds and course in Operating Conditions BE-31, the GS is 216.0 knots. The next step is to solve for leg time. At a GS of 216.0 knots, it will take 1 hour 17 minutes 47 seconds to fly 280 NM.

Questions 8494 through 8498 ask for the total fuel burn using the data from these tables. The first step is to compute the leg time as described above. Next, using the total fuel flow from the table, compute the fuel burn. With a fuel flow of 633 pounds per hour and a cruise time of 1 hour 17 minutes, the fuel burned is 812.3 pounds.

Answers

8130 [A]

135, DSP

8489. (Refer to Figures 21, 22, 23, 24, and 25.) What is the en route time of the cruise leg for Operating Conditions BE-31?

A—1 hour 11 minutes.
B—1 hour 17 minutes.
C—1 hour 19 minutes.

Temperature = ISA +10°C
TAS = 228 knots
GS = 216.0 knots
Time = 1 hour 17 minutes 47 seconds

(PLT012) — FAA-H-8083-25

135, DSP

8490. (Refer to Figures 21, 22, 23, 24, and 25.) What is the en route time of the cruise leg for Operating Conditions BE-32?

A—1 hour 13 minutes.
B—1 hour 15 minutes.
C—1 hour 20 minutes.

Temperature = -19°C
TAS = 252 knots
GS = 261.8 knots
Time = 1 hour 13 minutes 20 seconds

(PLT012) — FAA-H-8083-25

135, DSP

8491. (Refer to Figures 21, 22, 23, 24, and 25.) What is the en route time of the cruise leg for Operating Conditions BE-33?

A—1 hour 50 minutes.
B—1 hour 36 minutes.
C—1 hour 46 minutes.

Temperature = ISA -10°C
TAS = 256.5 knots
GS = 225.2 knots
Time = 1 hour 46 minutes 33 seconds

(PLT012) — FAA-H-8083-25

135, DSP

8492. (Refer to Figures 21, 22, 23, 24, and 25.) What is the en route time of the cruise leg for Operating Conditions BE-34?

A—1 hour 7 minutes.
B—1 hour 2 minutes.
C—1 hour 12 minutes.

Temperature = ISA
TAS = 228 knots
GS = 208 knots
Time = 1 hour 06 minutes 15 seconds

(PLT012) — FAA-H-8083-25

135, DSP

8493. (Refer to Figures 21, 22, 23, 24, and 25.) What is the en route time of the cruise leg for Operating Conditions BE-35?

A—1 hour 6 minutes.
B—1 hour 8 minutes.
C—1 hour 10 minutes.

Temperature = ISA +10°C
TAS = 253 knots
GS = 252.4 knots
Time = 1 hour 11 minutes 19 seconds

(PLT012) — FAA-H-8083-25

135, DSP

8494. (Refer to Figures 21, 22, 23, 24, and 25.) What is the fuel consumption during the cruise leg for Operating Conditions BE-31?

A—812 pounds.
B—749 pounds.
C—870 pounds.

Temperature = ISA +10°C
Time = 1 hour 17 minutes (see Question 8489)
Fuel flow = 633
Fuel burn = 812.3 pounds

(PLT012) — FAA-H-8083-25

Answers

8489 [B] 8490 [A] 8491 [C] 8492 [A] 8493 [C] 8494 [A]

135, DSP

8495. (Refer to Figures 21, 22, 23, 24, and 25.) What is the fuel consumption during the cruise leg for Operating Conditions BE-32?

A—1,028 pounds.
B—896 pounds.
C—977 pounds.

Temperature = ISA
Time = 1 hour 13 minutes (see Question 8490)
Fuel flow = 803
Fuel burn = 977.0 pounds

(PLT012) — FAA-H-8083-25

135, DSP

8496. (Refer to Figures 21, 22, 23, 24, and 25.) What is the fuel consumption during the cruise leg for Operating Conditions BE-33?

A—1,165 pounds.
B—1,373 pounds.
C—976 pounds.

Temperature = ISA -10°C
Time = 1 hour 46 minutes (see Question 8491)
Fuel flow = 777
Fuel burn = 1,372.7 pounds

(PLT012) — FAA-H-8083-25

135, DSP

8497. (Refer to Figures 21, 22, 23, 24, and 25.) What is the fuel consumption during the cruise leg for Operating Conditions BE-34?

A—668 pounds.
B—718 pounds.
C—737 pounds.

Temperature = ISA
Time = 1 hour 06 minutes (see Question 8492)
Fuel flow = 653
Fuel burn = 718.3 pounds

(PLT012) — FAA-H-8083-25

135, DSP

8498. (Refer to Figures 21, 23, 24, and 25.) What is the fuel consumption during the cruise leg for Operating Conditions BE-35?

A—900 pounds.
B—1,030 pounds.
C—954 pounds.

Temperature = ISA +10°C
Time = 1 hour 10 minutes (see Question 8493)
Fuel flow = 818
Fuel burn = 954.3 pounds

(PLT012) — FAA-H-8083-25

135, DSP

8499. (Refer to Figure 26.) What are the time and distance to descend from 18,000 feet to 2,500 feet?

A—10.3 minutes, 39 NM.
B—9.8 minutes, 33 NM.
C—10.0 minutes, 36 NM.

Enter FAA Figure 26 from initial pressure altitude on the left side of the graph. Proceed to heavy reference line, then drop to minutes, pounds and/or distance scales as needed. Repeat this process with the final altitude.

	Time	***NM***
Descent to SL	*12.0*	*45*
Field alt.	*– 1.5*	*– 5*
Descent	*10.5*	*40*

(PLT012) — FAA-H-8083-25

135, DSP

8500. (Refer to Figure 26.) What are the distance and fuel consumption to descend from 22,000 feet to 4,500 feet?

A—44 NAM, 117 pounds.
B—48 NAM, 112 pounds.
C—56 NAM, 125 pounds.

Enter FAA Figure 26 from initial pressure altitude on the left side of the graph. Proceed to heavy reference line, then drop to minutes, pounds and/or distance scales as needed. Repeat this process with the final altitude.

	NM	***Fuel***
Descent to SL	*58*	*142*
Field alt.	*– 10*	*– 30*
Descent	*48*	*112*

(PLT012) — FAA-H-8083-25

Answers

8495 [C] 8496 [B] 8497 [B] 8498 [C] 8499 [A] 8500 [B]

135, DSP

8501. (Refer to Figure 26.) What are the time and distance to descend from 16,500 feet to 3,500 feet?

A—9.3 minutes, 37 NAM.
B—9.1 minutes, 35 NAM.
C—8.7 minutes, 33 NAM.

Enter FAA Figure 26 from initial pressure altitude on the left side of the graph. Proceed to heavy reference line, then drop to minutes, pounds and/or distance scales as needed. Repeat this process with the final altitude.

	Time	***NM***
Descent to SL	*11.1*	*41*
Field alt.	*– 2.4*	*– 8*
Descent	*8.7*	*33*

(PLT045) — FAA-H-8083-25

135, DSP

8502. (Refer to Figure 26.) What are the distance and fuel consumption to descend from 13,500 feet to 1,500 feet?

A—30 NAM, 87 pounds.
B—29 NAM, 80 pounds.
C—38 NAM, 100 pounds.

Enter FAA Figure 26 from initial pressure altitude on the left side of the graph. Proceed to heavy reference line, then drop to minutes, pounds and/or distance scales as needed. Repeat this process with the final altitude.

	NM	***Fuel***
Descent to SL	*32.5*	*90*
Field alt.	*– 3.5*	*– 10*
Descent	*29*	*80*

(PLT045) — FAA-H-8083-25

135, DSP

8503. (Refer to Figure 26.) What are the time and distance to descend from 23,000 feet to 600 feet with an average 15-knot headwind?

A—14.2 minutes, 50 NAM.
B—14.6 minutes, 56 NAM.
C—14.9 minutes, 59 NAM.

Enter FAA Figure 26 from initial pressure altitude on the left side of the graph. Proceed to heavy reference line, then drop to minutes, pounds and/or distance scales as needed. Repeat this process with the final altitude.

	Time	***NM***
Descent to SL	*15.4*	*62*
Field alt.	*– .5*	*– 2*
Descent	*14.9*	*60*

Note: The answers are in NM so no wind correction is necessary.

(PLT045) — FAA-H-8083-25

135, DSP

8504. (Refer to Figures 27 and 28.) What is the landing distance over a 50-foot obstacle for Operating Conditions B-36?

A—1,900 feet.
B—1,625 feet.
C—950 feet.

1. *Enter FAA Figure 28 at the bottom left-hand side at +30°C OAT and proceed upward to the line representing sea level (SL) pressure altitude.*
2. *From the point of intersection on the "pressure altitude" lines, draw a horizontal line to the reference line, then parallel the line which represents an aircraft weight of 16,000 pounds.*
3. *From the point of intersection in the "weight" portion of the graph, draw a horizontal line to the next reference line (wind component), then parallel the line until intersecting with the 20-knot headwind.*
4. *From the point of intersection on the "wind" portion of the graph draw a horizontal line to the next reference line (0-foot obstacle height), then parallel the line required to clear a 50-foot obstacle.*
5. *The distance required is 1,900 feet.*

(PLT008) — FAA-H-8083-25

Answers

8501 [C] 8502 [B] 8503 [C] 8504 [A]

135, DSP

8505. (Refer to Figures 27 and 28.) What are the approach speed and ground roll when landing under Operating Conditions B-36?

A—113 knots and 950 feet.
B—113 knots and 1,950 feet.
C—112 knots and 900 feet.

1. *Enter FAA Figure 28 at the bottom left-hand side at +30°C OAT and proceed upward to the line representing sea level (SL) pressure altitude.*
2. *From the point of intersection on the "pressure altitude" lines, draw a horizontal line to the reference line, then parallel the line which represents an aircraft weight of 16,000 pounds.*
3. *From the point of intersection in the "weight" portion of the graph, draw a horizontal line to the next reference line (wind component), then parallel the line until intersecting with the 20-knot headwind.*
4. *From the point of intersection on the "wind" portion of the graph draw a horizontal line to the edge of the chart to find a ground roll of 950 feet.*
5. *The approach speed is found using the table at the top of figure. Interpolate between 16,100 and 14,000 to find the approach speed for 16,000 pounds at 113 knots.*

(PLT008) — FAA-H-8083-25

135, DSP

8506. (Refer to Figures 27 and 28.) What is the remaining runway length when stopped after landing over a 50-foot obstacle for Operating Conditions B-37?

A—2,500 feet.
B—2,000 feet.
C—2,600 feet.

1. *Enter FAA Figure 28 at the bottom left-hand side at +16°C OAT and proceed upward to the line representing 1,000 feet pressure altitude.*
2. *From the point of intersection on the "pressure altitude" lines, draw a horizontal line to the reference line, then parallel the line which represents an aircraft weight of 14,500 pounds.*
3. *From the point of intersection in the "weight" portion of the graph, draw a horizontal line to the next reference line (wind component), then parallel the line until intersecting with the 10-knot tailwind.*
4. *From the point of intersection on the "wind" portion of the graph draw a horizontal line to the next reference line (0-foot obstacle height), then parallel the line required to clear a 50-foot obstacle.*
5. *The distance required is 2,500 feet.*
6. *The remaining runway length when stopped is the runway distance minus the distance required, which is 2,000 feet (4,500 – 2,500).*

(PLT008) — FAA-H-8083-25

135, DSP

8507. (Refer to Figures 27 and 28.) What are the approach speed and ground roll when landing under Operating Conditions B-37?

A—108 knots and 1,400 feet.
B—109 knots and 900 feet.
C—107 knots and 1,350 feet.

1. *Enter FAA Figure 28 at the bottom left-hand side at +16°C OAT and proceed upward to the line representing 1,000 feet pressure altitude.*
2. *From the point of intersection on the "pressure altitude" lines, draw a horizontal line to the reference line, then parallel the line which represents an aircraft weight of 14,500 pounds.*
3. *From the point of intersection in the "weight" portion of the graph, draw a horizontal line to the next reference line (wind component), then parallel the line until intersecting with the 10-knot tailwind.*
4. *From the point of intersection on the "wind" portion of the graph draw a horizontal line to the edge of the chart to find a ground roll of 1,400 feet.*
5. *The approach speed is found using the table at the top of figure. Interpolate between 16,100 and 14,000 to find the approach speed for 14,500 pounds at 108 knots.*

(PLT008) — FAA-H-8083-25

Answers

8505 [A] 8506 [B] 8507 [A]

135, DSP

8508. (Refer to Figures 27 and 28.) What is the landing distance over a 50-foot obstacle for Operating Conditions B-38?

A—1,850 feet.
B—1,700 feet.
C—1,800 feet.

1. Enter FAA Figure 28 at the bottom left-hand side at 0°C OAT and proceed upward to the line representing 2,000 feet pressure altitude.

2. From the point of intersection on the "pressure altitude" lines, draw a horizontal line to the reference line, then parallel the line which represents an aircraft weight of 13,500 pounds.

3. From the point of intersection in the "weight" portion of the graph, draw a horizontal line to the next reference line (wind component), then parallel the line until intersecting with the 15-knot headwind.

4. From the point of intersection on the "wind" portion of the graph draw a horizontal line to the next reference line (0-foot obstacle height), then parallel the line required to clear a 50-foot obstacle.

5. The distance required is 1,700 feet.

(PLT008) — FAA-H-8083-25

135, DSP

8509. (Refer to Figures 27 and 28.) What is the total runway used when touchdown is at the 1,000 foot marker for Operating Conditions B-38?

A—2,000 feet.
B—1,700 feet.
C—1,800 feet.

1. Enter FAA Figure 28 at the bottom left-hand side at 0°C OAT and proceed upward to the line representing 2,000 feet pressure altitude.

2. From the point of intersection on the "pressure altitude" lines, draw a horizontal line to the reference line, then parallel the line which represents an aircraft weight of 13,500 pounds.

3. From the point of intersection in the "weight" portion of the graph, draw a horizontal line to the next reference line (wind component), then parallel the line until intersecting with the 15-knot headwind.

4. From the point of intersection on the "wind" portion of the graph draw a horizontal line to the edge of the chart to find the ground roll distance is 800 feet.

5. The total runway used is the distance from the threshold of the runway to the touchdown point, plus the ground roll distance, which is 1,800 feet (1,000 + 800).

(PLT008) — FAA-H-8083-25

135, DSP

8510. (Refer to Figures 27 and 28.) What is the remaining runway length when stopped after landing over a 50-foot obstacle for Operating Conditions B-39?

A—2,300 feet.
B—2,400 feet.
C—2,500 feet.

1. Enter FAA Figure 28 at the bottom left-hand side at +20°C OAT and proceed upward to the line representing 4,000 feet pressure altitude.

2. From the point of intersection on the "pressure altitude" lines, draw a horizontal line to the reference line, then parallel the line which represents an aircraft weight of 15,000 pounds.

3. From the point of intersection in the "weight" portion of the graph, draw a horizontal line to the next reference line (wind component), then parallel the line until intersecting with the 5-knot tailwind.

4. From the point of intersection on the "wind" portion of the graph draw a horizontal line to the next reference line (0-foot obstacle height), then parallel the line required to clear a 50-foot obstacle, which is 2,500 feet.

5. The remaining runway length when stopped is the runway length minus the landing distance, which is 2,500 feet (5,000 – 2,500).

(PLT008) — FAA-H-8083-25

Answers

8508 [B]   8509 [C]   8510 [C]

135, DSP

8511. (Refer to Figures 27 and 28.) What are the approach speed and ground roll when landing under Operating Conditions B-39?

A—111 knots and 1,550 feet.
B—110 knots and 1,400 feet.
C—109 knots and 1,300 feet.

1. *Enter FAA Figure 28 at the bottom left-hand side at +20°C OAT and proceed upward to the line representing 4,000 feet pressure altitude.*
2. *From the point of intersection on the "pressure altitude" lines, draw a horizontal line to the reference line, then parallel the line which represents an aircraft weight of 15,000 pounds.*
3. *From the point of intersection in the "weight" portion of the graph, draw a horizontal line to the next reference line (wind component), then parallel the line until intersecting with the 5-knot tailwind.*
4. *From the point of intersection on the "wind" portion of the graph draw a horizontal line to the edge of the chart to find a ground roll of 1,400 feet.*
5. *The approach speed is found using the table at the top of figure. Interpolate between 16,100 and 14,000 to find the approach speed for 15,000 pounds at 110 knots.*

(PLT008) — FAA-H-8083-25

135, DSP

8512. (Refer to Figures 27 and 28.) What is the landing distance over a 50-foot obstacle for Operating Conditions B-40?

A—1,500 feet.
B—1,750 feet.
C—1,650 feet.

1. *Enter FAA Figure 28 at the ISA line and proceed to the line representing 5,000 feet pressure altitude.*
2. *From the point of intersection on the "pressure altitude" lines, draw a horizontal line to the reference line, then parallel the line which represents an aircraft weight of 12,500 pounds.*
3. *From the point of intersection in the "weight" portion of the graph, draw a horizontal line to the next reference line (wind component), then parallel the line until intersecting with the 25-knot headwind.*
4. *From the point of intersection on the "wind" portion of the graph draw a horizontal line to the next reference line (0-foot obstacle height), then parallel the line required to clear a 50-foot obstacle, which is 1,650 feet.*

(PLT008) — FAA-H-8083-25

Helicopter Performance

RTC

8533. (Refer to Figure 36.) Given the following conditions, what is the maximum allowable measured gas temperature (MGT) during the power assurance check?

Engine torque .... 57 percent
Pressure altitude .... 2,500 ft
Temperature (OAT) .... +5°C

A—810°C.
B—815°C.
C—828°C.

Follow the example in the inset, upper left corner of FAA Figure 36. Begin at 57% torque, draw a line with a straight-edge parallel to the bold line through pressure altitude. From the point where your line intersects with 2,500 feet, draw a second line through the OAT lines. From the intersection of that line and +5°C OAT, draw a perpendicular line to maximum allowable MGT. Read the correct answer of 828°C. (PLT009) — FAA-H-8083-21

RTC

8534. (Refer to Figure 36.) Given the following conditions, what is the maximum allowable measured gas temperature (MGT) during the power assurance check?

Engine torque .... 49 percent
Pressure altitude .... 5,500 ft
Temperature (OAT) .... +25°C

A—870°C.
B—855°C.
C—880°C.

Follow the example in the inset, upper left corner FAA Figure 36. Begin at 49% torque, draw a line with a straight-edge parallel to the bold line through pressure altitude. From the point where your line intersects with 5,500 feet, draw a second line through the OAT lines. From the intersection of that line and +25°C OAT, draw a perpendicular line to maximum allowable MGT. Read the correct answer of 870°C. (PLT009) — FAA-H-8083-21

Answers

8511 [B] 8512 [C] 8533 [C] 8534 [A]

RTC

8535. (Refer to Figure 36.) Given the following conditions, what is the maximum allowable measured gas temperature (MGT) during the power assurance check?

Engine torque .... 54 percent
Pressure altitude.. 500 ft
Temperature (OAT) .... +25°C

A—840°C.
B—830°C.
C—820°C.

Follow the example in the inset, upper left corner FAA Figure 36. Begin at 54% torque, draw a line with a straight-edge parallel to the bold line through pressure altitude. From the point where your line intersects with 500 feet, draw a second line through the OAT lines. From the intersection of that line and +25°C OAT, draw a perpendicular line to maximum allowable MGT. Read the correct answer of 840°C. (PLT009) — FAA-H-8083-21

RTC

8536. (Refer to Figure 36.) Given the following conditions, what is the maximum allowable measured gas temperature (MGT) during the power assurance check?

Engine torque .... 43 percent
Pressure altitude... 9,000 ft
Temperature (OAT) .... -15°C

A—782°C.
B—768°C.
C—750°C.

Follow the example in the inset, upper left corner FAA Figure 36. Begin at 43% torque, draw a line with a straight-edge parallel to the bold line through pressure altitude. From the point where your line intersects with 9,000 feet, draw a second line through the OAT lines. From the intersection of that line and -15°C OAT, draw a perpendicular line to maximum allowable MGT. Read the correct answer of 768°C. (PLT009) — FAA-H-8083-21

RTC

8537. (Refer to Figure 36.) Given the following conditions, what is the maximum allowable measured gas temperature (MGT) during the power assurance check?

Engine torque .... 52 percent
Pressure altitude... 1,500 ft
Temperature (OAT) .... +35°C

A—880°C.
B—865°C.
C—872°C.

Follow the example in the inset, upper left corner FAA Figure 36. Begin at 52% torque, draw a line with a straight-edge parallel to the bold line through pressure altitude. From the point where your line intersects with 1,500 feet, draw a second line through the OAT lines. From the intersection of that line and +35°C OAT, draw a perpendicular line to maximum allowable MGT. Read the correct answer of 865°C. (PLT009) — FAA-H-8083-21

RTC

8538. (Refer to Figure 37.) What is the maximum gross weight for hovering in ground effect at 3,000 feet pressure altitude and +25°C?

A—17,300 pounds.
B—14,700 pounds.
C—16,600 pounds.

To determine the maximum gross weight, begin in the lower left corner of FAA Figure 37 at +25°C OAT. With a straight-edge, draw a line vertically to intersect 3,000 feet pressure altitude. Draw a second perpendicular line from that point to the right to intersect +25°C OAT. Draw a third line from this intersection vertically to gross weight. Read the correct answer of 17,300 pounds. (PLT048) — FAA-H-8083-21

RTC

8539. (Refer to Figure 37.) What is the maximum gross weight for hovering in ground effect at 6,000 feet pressure altitude and +15°C?

A—17,200 pounds.
B—16,600 pounds.
C—14,200 pounds.

To determine the maximum gross weight, begin in the lower left corner of FAA Figure 37 at +15°C OAT. With a straight-edge, draw a line vertically to intersect 6,000 feet pressure altitude. Draw a second perpendicular line from that point to the right to intersect +15°C OAT. Draw a third line from this intersection vertically to gross weight. Read the correct answer of 16,600 pounds. (PLT048) — FAA-H-8083-21

Answers

8535 [A] 8536 [B] 8537 [B] 8538 [A] 8539 [B]

RTC
8540. (Refer to Figure 37.) What is the maximum gross weight for hovering in ground effect at 7,000 feet pressure altitude and +35°C?

A—13,500 pounds.
B—14,700 pounds.
C—12,100 pounds.

To determine the maximum gross weight, begin in the lower left corner of FAA Figure 37 at +35°C OAT. With a straight-edge, draw a line vertically to intersect 7,000 feet pressure altitude. Draw a second perpendicular line from that point to the right to intersect +35°C OAT. Draw a third line from this intersection vertically to gross weight. Read the correct answer of 13,500 pounds. (PLT048) — FAA-H-8083-21

RTC
8541. (Refer to Figure 37.) What is the maximum gross weight for hovering in ground effect at 4,500 feet pressure altitude and +20°C?

A—14,500 pounds.
B—16,500 pounds.
C—17,000 pounds.

To determine the maximum gross weight, begin in the lower left corner of FAA Figure 37 at +20°C OAT. With a straight-edge, draw a line vertically to intersect 4,500 feet pressure altitude. Draw a second perpendicular line from that point to the right to intersect +20°C OAT. Draw a third line from this intersection vertically to gross weight. Read the correct answer of 17,000 pounds. (PLT048) — FAA-H-8083-21

RTC
8542. (Refer to Figure 37.) What is the maximum gross weight for hovering in ground effect at 2,500 feet pressure altitude and +35°C?

A—16,200 pounds.
B—16,600 pounds.
C—14,600 pounds.

To determine the maximum gross weight, begin in the lower left corner of FAA Figure 37 at +35°C OAT. With a straight-edge, draw a line vertically to intersect 2,500 feet pressure altitude. Draw a second perpendicular line from that point to the right to intersect +35°C OAT. Draw a third line from this intersection vertically to gross weight. Read the correct answer of 16,200 pounds. (PLT048) — FAA-H-8083-21

RTC
8543. (Refer to Figure 38.) What is the maximum gross weight for hovering out of ground effect at 3,000 feet pressure altitude and +30°C?

A—17,500 pounds.
B—14,300 pounds.
C—13,400 pounds.

To determine the maximum gross weight, begin in the lower left corner of FAA Figure 38 at +30°C OAT. With a straight-edge, draw a line vertically to intersect 3,000 feet pressure altitude. Draw a second perpendicular line from that point to the right to intersect +30°C OAT. Draw a third line from this intersection vertically to gross weight. Read the correct answer of 14,300 pounds. (PLT048) — FAA-H-8083-21

RTC
8544. (Refer to Figure 38.) What is the maximum gross weight for hovering out of ground effect at 6,000 feet pressure altitude and +15°C?

A—16,800 pounds.
B—13,500 pounds.
C—14,400 pounds.

To determine the maximum gross weight, begin in the lower left corner of FAA Figure 38 at +15°C OAT. With a straight-edge, draw a line vertically to intersect 6,000 feet pressure altitude. Draw a second perpendicular line from that point to the right to intersect +15°C OAT. Draw a third line from this intersection vertically to gross weight. Read the correct answer of 14,400 pounds. (PLT048) — FAA-H-8083-21

RTC
8545. (Refer to Figure 38.) What is the maximum gross weight for hovering out of ground effect at 7,000 feet pressure altitude and +35°C?

A—14,000 pounds.
B—11,600 pounds.
C—12,500 pounds.

To determine the maximum gross weight, begin in the lower left corner of FAA Figure 38 at +35°C OAT. With a straight-edge, draw a line vertically to intersect 7,000 feet pressure altitude. Draw a second perpendicular line from that point to the right to intersect +35°C OAT. Draw a third line from this intersection vertically to gross weight. Read the correct answer of 11,600 pounds. (PLT048) — FAA-H-8083-21

Answers

8540 [A] 8541 [C] 8542 [A] 8543 [B] 8544 [C] 8545 [B]

RTC

8546. (Refer to Figure 38.) What is the maximum gross weight for hovering out of ground effect at 4,500 feet pressure altitude and +20°C?

A—14,500 pounds.
B—14,000 pounds.
C—17,000 pounds.

To determine the maximum gross weight, begin in the lower left corner of FAA Figure 38 at +20°C OAT. With a straight-edge, draw a line vertically to intersect 4,500 feet pressure altitude. Draw a second perpendicular line from that point to the right to intersect +20°C OAT. Draw a third line from this intersection vertically to gross weight. Read the correct answer of 14,500 pounds. (PLT048) — FAA-H-8083-21

RTC

8547. (Refer to Figure 38.) What is the maximum gross weight for hovering out of ground effect at 2,500 feet pressure altitude and +30°C?

A—17,400 pounds.
B—15,000 pounds.
C—14,500 pounds.

To determine the maximum gross weight, begin in the lower left corner of FAA Figure 38 at +30°C OAT. With a straight-edge, draw a line vertically to intersect 2,500 feet pressure altitude. Draw a second perpendicular line from that point to the right to intersect +30°C OAT. Draw a third line from this intersection vertically to gross weight. Read the correct answer of 14,500 pounds. (PLT048) — FAA-H-8083-21

RTC

8548. (Refer to Figure 39.) What is the takeoff distance over a 50-foot obstacle?

Pressure altitude...3,500 ft
Temperature (OAT) ....+20°C
Gross weight... 15,000 lb

A—1,070 feet.
B—1,020 feet.
C—1,100 feet.

To determine the takeoff distance, begin at +20°C OAT in the lower left corner of FAA Figure 39. Draw a vertical line upward to intersect 3,500 feet pressure altitude. From that point, draw a perpendicular line to 15,000 pounds gross weight. From this point, draw a vertical line downward to takeoff distance. Read the distance of 1,070 feet. (PLT011) — FAA-H-8083-21

RTC

8549. (Refer to Figure 39.) What is the takeoff distance over a 50-foot obstacle?

Pressure altitude... 5,000 ft
Temperature (OAT) ....-10°C
Gross weight... 11,000 lb

A—1,000 feet.
B—920 feet.
C—870 feet.

To determine the takeoff distance, begin at -10°C OAT in the lower left corner of FAA Figure 39. Draw a vertical line upward to intersect 5,000 feet pressure altitude. From that point, draw a perpendicular line to 11,000 pounds gross weight. From this point, draw a vertical line downward to takeoff distance. Read the distance of 870 feet. (PLT011) — FAA-H-8083-21

RTC

8550. (Refer to Figure 39.) What is the takeoff distance over a 50-foot obstacle?

Pressure altitude... 6,500 ft
Temperature (OAT) .... 0°C
Gross weight.. 13,500 lb

A—1,500 feet.
B—1,050 feet.
C—1,100 feet.

To determine the takeoff distance, begin at +0°C OAT in the lower left corner of FAA Figure 39. Draw a vertical line upward to intersect 6,500 feet pressure altitude. From that point, draw a perpendicular line to 13,500 pounds gross weight. From this point, draw a vertical line downward to takeoff distance. Read the distance of 1,050 feet. (PLT011) — FAA-H-8083-21

Answers

8546 [A] 8547 [C] 8548 [A] 8549 [C] 8550 [B]

RTC

8551. (Refer to Figure 39.) What is the takeoff distance over a 50-foot obstacle?

Pressure altitude.. 9,000 ft
Temperature (OAT)... +20°C
Gross weight.. 15,000 lb

A—1,300 feet.
B—1,350 feet.
C—1,250 feet.

To determine the takeoff distance, begin at +20°C OAT in the lower left corner of FAA Figure 39. Draw a vertical line upward to intersect 9,000 feet pressure altitude. From that point, draw a perpendicular line to 15,000 pounds gross weight. From this point, draw a vertical line downward to takeoff distance. Read the distance of 1,350 feet. (PLT011) — FAA-H-8083-21

RTC

8552. (Refer to Figure 39.) What is the takeoff distance over a 50-foot obstacle?

Pressure altitude.. -1,000 ft
Temperature (OAT)... +25°C
Gross weight.. 14,000 lb

A—1,000 feet.
B—900 feet.
C—950 feet.

To determine the takeoff distance, begin at +25°C OAT in the lower left corner of FAA Figure 39. Draw a vertical line upward to intersect -1,000 feet pressure altitude. From that point, draw a perpendicular line to 14,000 pounds gross weight. From this point, draw a vertical line downward to takeoff distance. Read the distance of 900 feet. (PLT011) — FAA-H-8083-21

RTC

8553. (Refer to Figure 40.) What is the climb performance with both engines operating?

Pressure altitude.. 9,500 ft
Temperature (OAT)... -5°C
Heater.. ON

A—925 ft/min.
B—600 ft/min.
C—335 ft/min.

1. *Enter FAA Figure 40 at the left side with 9,500 pressure altitude and draw a horizontal line to the curved line representing -5°C temperature (interpolate between temperatures as necessary).*
2. *From that point of intersection, draw a vertical line to the bottom of the graph and read the rate of climb of 915 fpm.*
3. *With the heater on above 1,800 feet, the rate of climb is 315 fpm less, therefore our rate of climb is 600 fpm.*

(PLT004) — FAA-H-8083-21

RTC

8554. (Refer to Figure 40.) What is the climb performance with both engines operating?

Pressure altitude.. 7,500 ft
Temperature (OAT)... +5°C
Heater.. ON

A—905 ft/min.
B—765 ft/min.
C—1,080 ft/min.

1. *Enter FAA Figure 40 at the left side with 7,500 pressure altitude and draw a horizontal line to the curved line representing +5°C temperature (interpolate between temperatures as necessary).*
2. *From that point of intersection, draw a vertical line to the bottom of the graph and read the rate of climb of 1,080 fpm.*
3. *With the heater on above 1,800 feet, the rate of climb is 315 fpm less; therefore, our rate of climb is 765 fpm.*

(PLT004) — FAA-H-8083-21

Answers

8551 [B] 8552 [B] 8553 [B] 8554 [B]

RTC
8555. (Refer to Figure 40.) What is the climb performance with both engines operating?

Pressure altitude.. 6,500 ft
Temperature (OAT)... +25°C
Heater... OFF

A—285 ft/min.
B—600 ft/min.
C—400 ft/min.

1. *Enter FAA Figure 40 at the left side with 6,500 pressure altitude and draw a horizontal line to the curved line representing +25°C temperature (interpolate between temperatures as necessary).*
2. *From that point of intersection, draw a vertical line to the bottom of the graph and read the rate of climb of 600 fpm.*

(PLT004) — FAA-H-8083-21

RTC
8556. (Refer to Figure 40.) What is the climb performance with both engines operating?

Pressure altitude.. 11,500 ft
Temperature (OAT).. -15°C
Heater... ON

A—645 ft/min.
B—375 ft/min.
C—330 ft/min.

1. *Enter FAA Figure 40 at the left side with 11,500 pressure altitude and draw a horizontal line to the curved line representing -15°C temperature (interpolate between temperatures as necessary).*
2. *From that point of intersection, draw a vertical line to the bottom of the graph and read the rate of climb of 645 fpm.*
3. *With the heater on above 1,800 feet, the rate of climb is 315 fpm less; therefore, the rate of climb is 330 fpm.*

(PLT004) — FAA-H-8083-21

RTC
8557. (Refer to Figure 40.) What is the climb performance with both engines operating?

Pressure altitude.. 3,500 ft
Temperature (OAT).. -10°C
Heater... ON

A—985 ft/min.
B—1,300 ft/min.
C—1,360 ft/min.

1. *Enter FAA Figure 40 at the left side with 3,500 pressure altitude and draw a horizontal line to the curved line representing -10°C temperature (interpolate between temperatures as necessary).*
2. *From that point of intersection, draw a vertical line to the bottom of the graph and read the rate of climb of 1,300 fpm.*
3. *With the heater on above 1,800 feet, the rate of climb is 315 fpm less; therefore, our rate of climb is 985 fpm.*

(PLT004) — FAA-H-8083-21

RTC
8558. (Refer to Figure 41.) What is the single-engine climb or descent performance?

Pressure altitude.. 7,500 ft
Temperature (OAT)... 0°C

A—80 ft/min descent.
B—10 ft/min climb.
C—50 ft/min climb.

1. *Enter FAA Figure 41 at the left side with 7,500 pressure altitude and draw a horizontal line to the curved line representing 0°C temperature (interpolate between temperatures as necessary).*
2. *From that point of intersection, draw a vertical line to the bottom of the graph and read the rate of descent of 80 fpm.*

(PLT004) — FAA-H-8083-21

Answers

8555 [B] 8556 [C] 8557 [A] 8558 [A]

RTC

8559. (Refer to Figure 41.) Given the following, what is the single-engine climb or descent performance?

Pressure altitude..3,000 ft
Temperature (OAT)..+35°C

A—150 ft/min descent.
B—350 ft/min climb.
C—100 ft/min descent.

1. *Enter FAA Figure 41 at the left side with 3,000 pressure altitude and draw a horizontal line to the curved line representing +35°C temperature (interpolate between temperatures as necessary).*
2. *From that point of intersection, draw a vertical line to the bottom of the graph and read the rate of descent of 100 fpm.*

(PLT004) — FAA-H-8083-21

RTC

8560. (Refer to Figure 41.) Given the following, what is the single-engine climb or descent performance?

Pressure altitude..4,700 ft
Temperature (OAT)..+20°C

A—420 ft/min climb.
B—60 ft/min climb.
C—60 ft/min descent.

1. *Enter FAA Figure 41 at the left side with 4,700 pressure altitude and draw a horizontal line to the curved line representing +20°C temperature (interpolate between temperatures as necessary).*
2. *From that point of intersection, draw a vertical line to the bottom of the graph and read the rate of climb of 60 fpm.*

(PLT004) — FAA-H-8083-21

RTC

8561. (Refer to Figure 41.) Given the following, what is the single-engine climb or descent performance?

Pressure altitude..9,500 ft
Temperature (OAT)..-10°C

A—600 ft/min descent.
B—840 ft/min descent.
C—280 ft/min descent.

1. *Enter FAA Figure 41 at the left side with 9,500 pressure altitude and draw a horizontal line to the curved line representing -10°C temperature (interpolate between temperatures as necessary).*
2. *From that point of intersection, draw a vertical line to the bottom of the graph and read the rate of descent of 280 fpm.*

(PLT004) — FAA-H-8083-21

RTC

8562. (Refer to Figure 41.) Given the following, what is the single-engine climb or descent performance?

Pressure altitude..1,500 ft
Temperature (OAT)..+45°C

A—100 ft/min descent.
B—360 ft/min climb.
C—200 ft/min descent.

1. *Enter FAA Figure 41 at the left side with 1,500 pressure altitude and draw a horizontal line to the curved line representing +45°C temperature (interpolate between temperatures as necessary).*
2. *From that point of intersection, draw a vertical line to the bottom of the graph and read the rate of descent of 100 fpm.*

(PLT004) — FAA-H-8083-21

RTC

8563. (Refer to Figure 42.) Given the following, what is the airspeed limit (V_{NE})?

Gross weight..16,500 lb
Pressure altitude..5,000 ft
Temperature (OAT)..-15°C

A—128 KIAS.
B—133 KIAS.
C—126 KIAS.

1. *Use the right-middle table for an aircraft weighing 16,500 pounds.*
2. *Start at an OAT of -15°C and interpolate between 6,000 (126) and 4,000 (131) feet to find V_{NE} for 5,000.*
3. *V_{NE} for 5,000 feet at -15°C is 128.5 knots.*

(PLT002) — FAA-H-8083-21

Answers

8559 [C] 8560 [B] 8561 [C] 8562 [A] 8563 [A]

RTC

8564. (Refer to Figure 42.) What is the airspeed limit (V_{NE})?

Gross weight .... 17,500 lb
Pressure altitude .... 4,000 ft
Temperature (OAT) .... +10°C

A—114 KIAS.
B—120 KIAS.
C—130 KIAS.

1. *Use the right-bottom table for an aircraft weighing 17,500 pounds.*
2. *Interpolate between 0°C and 20°C for 4,000 feet to find V_{NE} for +10°C.*
3. *V_{NE} for 5,000 feet at +10°C is 120 knots.*

(PLT002) — FAA-H-8083-21

RTC

8565. (Refer to Figure 42.) What is the airspeed limit (V_{NE})?

Gross weight .... 15,000 lb
Pressure altitude .... 6,000 ft
Temperature (OAT) .... 0°C

A—135 KIAS.
B—127 KIAS.
C—143 KIAS.

1. *Use the table for 14,500 pounds and 16,500 to interpolate for an aircraft weighing 15,000 pounds.*
2. *Start at an OAT of 0°C and find 6,000 for both 14,500 pounds (138) and 16,500 pounds (122) feet to find V_{NE} for 15,000 pounds.*
3. *V_{NE} for 15,000 feet at 0°C is 135 knots.*

(PLT002) — FAA-H-8083-21

RTC

8566. (Refer to Figure 42.) What is the airspeed limit (V_{NE})?

Gross weight .... 14,000 lb
Pressure altitude .... 8,000 ft
Temperature (OAT) .... -15°C

A—121 KIAS.
B—123 KIAS.
C—113 KIAS.

1. *Use the table for 13,500 pounds and 14,500 to interpolate for an aircraft weighing 14,000 pounds.*
2. *Start at an OAT of -15°C and find 8,000 for both 13,500 pounds (121) and 14,500 pounds (121) feet to find V_{NE} for 14,000 pounds.*
3. *V_{NE} for 14,000 feet at -15°C is 121 knots.*

(PLT002) — FAA-H-8083-21

RTC

8567. (Refer to Figure 42.) What is the airspeed limit (V_{NE})?

Gross weight .... 12,500 lb
Pressure altitude .... 14,000 ft
Temperature (OAT) .... -20°C

A—99 KIAS.
B—108 KIAS.
C—103 KIAS.

1. *Use the table for an aircraft weighing 12,500 pounds.*
2. *Interpolate between -15°C and -25°C for 14,000 feet to find V_{NE} for -20°C.*
3. *V_{NE} for 14,000 feet at -20°C is 103 knots.*

(PLT002) — FAA-H-8083-21

RTC

8568. (Refer to Figure 43.) What is the single-engine landing distance over a 50-foot obstacle?

Gross weight .... 12,000 lb
Pressure altitude .... 3,500 ft
Temperature (OAT) .... +30°C

A—850 feet.
B—900 feet.
C—1,000 feet.

1. *Enter FAA Figure 43 on the left side at the bottom and draw a vertical line from +30°C OAT to the 3,500 feet pressure altitude (interpolate as necessary).*
2. *From that point of intersection, draw a horizontal line to the diagonal representing 12,000 pounds, and then a vertical line from there down to the landing distance of 1,000 feet.*

(PLT011) — FAA-H-8083-21

Answers

8564 [B]	8565 [A]	8566 [A]	8567 [C]	8568 [C]

RTC

8569. (Refer to Figure 43.) What is the single-engine landing distance over a 50-foot obstacle?

Gross weight.. 16,500 lb
Pressure altitude... 5,500 ft
Temperature (OAT) .... -10°C

A—1,700 feet.
B—1,550 feet.
C—1,600 feet.

1. *Enter FAA Figure 43 on the left side at the bottom and draw a vertical line from -10°C OAT to the 5,500 feet pressure altitude (interpolate as necessary).*
2. *From that point of intersection, draw a horizontal line to the diagonal representing 16,500 pounds, and then a vertical line from there down to the landing distance of 1,550 feet.*

(PLT011) — FAA-H-8083-21

RTC

8570. (Refer to Figure 43.) What is the single-engine landing distance over a 50-foot obstacle?

Gross weight.. 15,000 lb
Pressure altitude... 8,000 ft
Temperature (OAT) .... +20°C

A—1,900 feet.
B—1,800 feet.
C—2,000 feet.

1. *Enter FAA Figure 43 on the left side at the bottom and draw a vertical line from +20°C OAT to the 8,000 feet pressure altitude (interpolate as necessary).*
2. *From that point of intersection, draw a horizontal line to the diagonal representing 15,000 pounds, and then a vertical line from there down to the landing distance of 1,900 feet.*

(PLT011) — FAA-H-8083-21

RTC

8571. (Refer to Figure 43.) What is the single-engine landing distance over a 50-foot obstacle?

Gross weight.. 14,000 lb
Pressure altitude... 1,000 ft
Temperature (OAT) .... +10°C

A—650 feet.
B—920 feet.
C—800 feet.

1. *Enter FAA Figure 43 on the left side at the bottom and draw a vertical line from +10°C OAT to the 1,000 feet pressure altitude (interpolate as necessary).*
2. *From that point of intersection, draw a horizontal line to the diagonal representing 14,000 pounds, and then a vertical line from there down to the landing distance of 920 feet.*

(PLT011) — FAA-H-8083-21

RTC

8572. (Refer to Figure 43.) What is the single-engine landing distance over a 50-foot obstacle?

Gross weight.. 17,000 lb
Pressure altitude... 4,000 ft
Temperature (OAT) .... +40°C

A—1,850 feet.
B—2,200 feet.
C—2,000 feet.

1. *Enter FAA Figure 43 on the left side at the bottom and draw a vertical line from +40°C OAT to the 4,000 feet pressure altitude (interpolate as necessary).*
2. *From that point of intersection, draw a horizontal line to the diagonal representing 17,000 pounds, and then a vertical line from there down to the landing distance of 2,000 feet.*

(PLT011) — FAA-H-8083-21

Answers

8569 [B] 8570 [A] 8571 [B] 8572 [C]

RTC

8822. (Refer to Figures 113 and 114.) What TAS would be required to arrive at POM VORTAC 52 minutes after passing DAG VORTAC?

A—114 knots.
B—117 knots.
C—120 knots.

Problem may be solved using E6-B or CX-2 flight computer:

Distance = 80 NM
GS = 92.3 knots
Wind = 290/36, variation = 15° East (found in Figure 113A)
Course = 214°
TAS = 114.2 knots

(PLT012) — FAA-H-8083-15

RTC

8823. (Refer to Figures 113 and 114.) What TAS would be required to arrive at POM VORTAC 1 hour after passing DAG VORTAC?

A—102 knots.
B—105 knots.
C—108 knots.

Problem may be solved using E6-B or CX-2 flight computer:

Distance = 80 NM
GS = 80 knots
Wind = 290/36, variation = 15° East (found in Figure 113A)
Course = 214°
TAS = 102.4 knots

(PLT012) — FAA-H-8083-15

Flight Planning Graphs and Tables

Aircraft manufacturers publish flight planning graphs or tables that enable the flight crews to quickly estimate the time and fuel required to fly certain trips. These tables or graphs allow adjustments for aircraft weight, wind, altitude, cruise speed and other variables.

The graph in FAA Figure 61 is used to determine the time and fuel required for a planned flight. It allows for trip distance, wind, flight altitude, landing weight, and temperature. This example shows the solution for Operating Condition X-1.

Start at the bottom of the graph in FAA Figure 62 with the trip distance (2,000 NM) and draw a vertical line to the reference line representing a zero knot wind component. Condition X-1 states that there is a 50-knot tailwind component. From the reference line follow the curved diagonal line back to the -50-knot line. This will result in an equivalent trip distance of 1,800 NM. From there, draw a vertical line through both sets of diagonal lines representing cruise altitude.

To determine the trip time, find the point of intersection between the vertical line and the diagonal in the upper set which represents the cruise altitude (27,000 feet). From there draw a horizontal line to the reference line representing 0° deviation from ISA. Since the temperature in Condition X-1 is ISA +10°, from the reference line parallel the diagonals down and to the left to the ISA +10° line. From there draw a horizontal line to the trip time (3 hours, 55 minutes).

To determine the trip fuel, find the point of intersection (on the lower set of altitude diagonals) between the vertical line drawn earlier and the diagonal line used for the cruise altitude. From there, draw a horizontal line, the reference line representing a landing weight of 65,000 pounds. Since the landing weight in Condition X-1 is 70,000 pounds, parallel the diagonal lines up and to the right until meeting the vertical 70,000-pound line. From there, draw a horizontal line to the trip fuel (26,000 pounds).

The table in FAA Figure 67 shows the time and fuel required for trips of various distances. For example, a trip of 340 NM requires 55 minutes and consumes 5,550 pounds of fuel.

Continued

Answers

8822 [A] 8823 [A]

A wind will change both the time and fuel required for a given flight. The notes at the bottom of the table explain the correction factors. The formulas are:

Change in time = Time x Wind Component ÷ TAS

and

Change in Fuel = Fuel x Wind Component ÷ TAS

Assume a 25-knot tailwind for the conditions above. The formulas are:

Change in Time = 55 min. x (-25) ÷ 438 = -3.1 minutes

and

Change in Fuel = 5,550 lb. x (-25) ÷ 438 = -316.8 pounds

The corrected time is 51.9 minutes (55 – 3.1), and the corrected fuel is 5,233.2 pounds (5,550 – 316.8). Notice that a tailwind is entered as a negative number because it reduces the required time and fuel. A headwind should be entered as a positive number since it will increase the required time and fuel.

The table in FAA Figure 52 is used to plan for possible diversion to an alternate airport. For example, in Condition L-1 (FAA Figure 51) the alternate is 110 NAM away. It will take 29 minutes to fly there and make an approach. It will also require 3,400 pounds of fuel, which includes an additional 15 minutes of holding fuel.

The conditions specify that there is 15 minutes holding at the alternate. Since Note 2 in FAA Figure 52 specifically states that 15 minutes holding is not included in the time, then we must add the 15 minutes to the time listed in the alternate planning chart. Wind is not a factor since both figures deal in Nautical Air Miles (NAM).

29 minutes + 15 minutes = 44 minutes

Landing weight way be calculated by subtracting fuel burn from gross weight as follows:

85,000 lbs – 3,400 lbs = 81,600 lbs landing weight

121, DSP

8643. (Refer to Figures 61 and 62.) What is the trip time for Operating Conditions X-1?

A—4 hours 5 minutes.
B—4 hours 15 minutes.
C—4 hours.

1. *Enter the bottom of FAA Figure 62 at 2,000 and move up to the reference line.*
2. *From the reference line, follow the diagonal line representing a 50-knot tailwind.*
3. *Move up to the line representing a pressure altitude of 27,000 feet and move left to the reference line.*
4. *From the reference line, follow the diagonal line representing ISA +10°C and continue to the edge of the chart.*
5. *Determine a trip time of 4 hours.*

(PLT012) — FAA-H-8083-25

121, DSP

8644. (Refer to Figures 61 and 62.) What is the trip time for Operating Conditions X-2?

A—5 hours 5 minutes.
B—6 hours 15 minutes.
C—5 hours 55 minutes.

1. *Enter the bottom of FAA Figure 62 at 2,400 and move up to the reference line.*
2. *From the reference line, follow the diagonal line representing a 50-knot headwind.*
3. *Move up to the line representing a pressure altitude of 35,000 feet and move left to edge of the chart.*
4. *Determine a trip time of 6 hours 15 minutes.*

(PLT012) — FAA-H-8083-25

Answers

8643 [C] 8644 [B]

121, DSP

8645. (Refer to Figures 61 and 62.) What is the trip time for Operating Conditions X-3?

A—4 hours 15 minutes.
B—3 hours 40 minutes.
C—4 hours.

1. *Enter the bottom of FAA Figure 62 at 1,800 and move up to the reference line.*
2. *From the reference line, follow the diagonal line representing a 20-knot headwind.*
3. *Move up to the line representing a pressure altitude of 20,000 feet and move left to the reference line.*
4. *From the reference line, follow the diagonal line representing ISA +20°C and continue to the edge of the chart.*
5. *Determine a trip time of 4 hours.*

(PLT012) — FAA-H-8083-25

121, DSP

8646. (Refer to Figures 61 and 62.) What is the trip time for Operating Conditions X-4?

A—6 hours 50 minutes.
B—5 hours 45 minutes.
C—5 hours 30 minutes.

1. *Enter the bottom of FAA Figure 62 at 2,800 and move up to the reference line.*
2. *From the reference line, follow the diagonal line representing a 50-knot tailwind.*
3. *Move up to the line representing a pressure altitude of 29,000 feet and move left to the reference line.*
4. *From the reference line, follow the diagonal line representing ISA -10°C and continue to the edge of the chart.*
5. *Determine a trip time of 5 hours 45 minutes.*

(PLT012) — FAA-H-8083-25

121, DSP

8647. (Refer to Figures 61 and 62.) What is the trip time for Operating Conditions X-5?

A—2 hours 55 minutes.
B—3 hours 10 minutes.
C—2 hours 59 minutes.

1. *Enter the bottom of FAA Figure 62 at 1,200 and move up to the reference line.*
2. *From the reference line, follow the diagonal line representing a 30-knot headwind.*
3. *Move up to the line representing a pressure altitude of 37,000 feet and move left to the reference line.*
4. *From the reference line, follow the diagonal line representing ISA +10°C and continue to the edge of the chart.*
5. *Determine a trip time of 2 hours 55 minutes.*

(PLT012) — FAA-H-8083-25

121, DSP

8648. (Refer to Figures 61 and 62.) What is the trip fuel for Operating Conditions X-1?

A—25,000 pounds.
B—26,000 pounds.
C—24,000 pounds.

1. *Enter the bottom of FAA Figure 62 at 2,000 and move up to the reference line.*
2. *From the reference line, follow the diagonal line representing a 50-knot tailwind.*
3. *Move up to the line representing a pressure altitude of 27,000 feet and move right to the reference line.*
4. *From the reference line, follow the diagonal line representing 70,000 pounds and continue to the edge of the chart.*
5. *Determine a trip fuel of 26,000 pounds.*

(PLT012) — FAA-H-8083-25

Answers

8645 [C] 8646 [B] 8647 [A] 8648 [B]

121, DSP

8649. (Refer to Figures 61 and 62.) What is the trip fuel for Operating Conditions X-2?

A—33,000 pounds.
B—28,000 pounds.
C—35,000 pounds.

1. *Enter the bottom of FAA Figure 62 at 2,400 and move up to the reference line.*
2. *From the reference line, follow the diagonal line representing a 50-knot headwind.*
3. *Move up to the line representing a pressure altitude of 35,000 feet and move right to the reference line.*
4. *From the reference line, follow the diagonal line representing 75,000 pounds and continue to the edge of the chart.*
5. *Determine a trip fuel of 35,000 pounds.*

(PLT012) — FAA-H-8083-25

121, DSP

8650. (Refer to Figures 61 and 62.) What is the trip fuel for Operating Conditions X-3?

A—36,000 pounds.
B—34,500 pounds.
C—33,000 pounds.

1. *Enter the bottom of FAA Figure 62 at 1,800 and move up to the reference line.*
2. *From the reference line, follow the diagonal line representing a 20-knot headwind.*
3. *Move up to the line representing a pressure altitude of 20,000 feet and move right to the reference line.*
4. *From the reference line, follow the diagonal line representing 75,000 pounds and continue to the edge of the chart.*
5. *Determine a trip fuel of 34,500 pounds.*

(PLT012) — FAA-H-8083-25

121, DSP

8651. (Refer to Figures 61 and 62.) What is the trip fuel for Operating Conditions X-4?

A—33,000 pounds.
B—31,500 pounds.
C—34,000 pounds.

1. *Enter the bottom of FAA Figure 62 at 2,800 and move up to the reference line.*
2. *From the reference line, follow the diagonal line representing a 50-knot tailwind.*
3. *Move up to the line representing a pressure altitude of 29,000 feet and move right to the reference line.*
4. *From the reference line, follow the diagonal line representing 65,000 pounds and continue to the edge of the chart.*
5. *Determine a trip fuel of 33,000 pounds.*

(PLT012) — FAA-H-8083-25

121, DSP

8652. (Refer to Figures 61 and 62.) What is the trip fuel for Operating Conditions X-5?

A—15,000 pounds.
B—20,000 pounds.
C—19,000 pounds.

1. *Enter the bottom of FAA Figure 62 at 1,200 and move up to the reference line.*
2. *From the reference line, follow the diagonal line representing a 30-knot headwind.*
3. *Move up to the line representing a pressure altitude of 37,000 feet and move right to the reference line.*
4. *From the reference line, follow the diagonal line representing 90,000 pounds and continue to the edge of the chart.*
5. *Determine a trip fuel of 19,000 pounds.*

(PLT012) — FAA-H-8083-25

Answers

8649 [C] 8650 [B] 8651 [A] 8652 [C]

121, 135, DSP

8658. (Refer to Figures 66 and 67.) What is the trip time corrected for wind under Operating Conditions Z-1?

A—58.1 minutes.
B—51.9 minutes.
C—54.7 minutes.

Change in time = Time x Wind Component ÷ TAS
55 x (-25) ÷ 438 = -3.1 minutes

Therefore, the trip time corrected for wind is 51.9 minutes (55 - 3.1).

(PLT012) — FAA-H-8083-25

121, 135, DSP

8659. (Refer to Figures 66 and 67.) What is the trip time corrected for wind under Operating Conditions Z-2?

A—1 hour 35 minutes.
B—1 hour 52 minutes.
C—1 hour 46 minutes.

Table time = 96 minutes
Change in time = 96 minutes x 45 ÷ 433 = +10 minutes
Trip time = 96 min + 10 min = 106 minutes = 1 hour 46 minutes

(PLT012) — FAA-H-8083-25

121, 135, DSP

8660. (Refer to Figures 66 and 67.) What is the trip time corrected for wind under Operating Conditions Z-3?

A—2 hours 9 minutes.
B—1 hour 59 minutes.
C—1 hour 52 minutes.

Table time = 129 minutes
Change in time = 129 min x (-35) ÷ 433 = -10.4 minutes
Trip time = 129 min – 10.4 min = 118.6 minutes = 1 hour 59 minutes

(PLT012) — FAA-H-8083-25

121, 135, DSP

8661. (Refer to Figures 66 and 67.) What is the trip time corrected for wind under Operating Conditions Z-4?

A—48.3 minutes.
B—50.7 minutes.
C—51.3 minutes.

Table time = 48 minutes
Change in time = 48 min x 25 ÷ 443 = 2.7 minutes
Trip time = 48 min + 2.7 min = 50.7 minutes

(PLT012) — FAA-H-8083-25

121, 135, DSP

8662. (Refer to Figures 66 and 67.) What is the trip time corrected for wind under Operating Conditions Z-5?

A—1 hour 11 minutes.
B—56 minutes.
C—62 minutes.

Table time = 62 minutes
Change in time = 62 min x 60 ÷ 433 = 8.6 minutes
Trip time = 62 min + 8.6 min = 70.6 minute = 1 hour 11 minutes

(PLT012) — FAA-H-8083-25

121, 135, DSP

8663. (Refer to Figures 66 and 67.) What is the estimated fuel consumption for Operating Conditions Z-1?

A—5,230 pounds.
B—5,970 pounds.
C—5,550 pounds.

Table fuel = 5,550 lbs
Change in fuel = 5,500 x (-25) ÷ 438 = -316.8 lbs
Trip fuel = 5,500 lbs – 316.8 lbs = 5,233.2 lbs

(PLT012) — FAA-H-8083-25

121, 135, DSP

8664. (Refer to Figures 66 and 67.) What is the estimated fuel consumption for Operating Conditions Z-2?

A—10,270 pounds.
B—9,660 pounds.
C—10,165 pounds.

Table fuel = 9,300 lbs
Change in fuel = 9,300 x 45 ÷ 433 = 967 lbs
Trip fuel = 9,300 lbs + 967 lbs = 10,267 lbs

(PLT012) — FAA-H-8083-25

Answers

8658 [B] 8659 [C] 8660 [B] 8661 [B] 8662 [A] 8663 [A]
8664 [A]

121, 135, DSP

8665. (Refer to Figures 66 and 67.) What is the estimated fuel consumption for Operating Conditions Z-3?

A—12,300 pounds.
B—11,300 pounds.
C—13,990 pounds.

Table fuel = 12,300 lbs
Change in fuel = 12,300 x (-35) ÷ 433 = -994 lbs
Trip fuel = 12,300 lbs – 994 lbs = 11,306 lbs

(PLT012) — FAA-H-8083-25

121, 135, DSP

8666. (Refer to Figures 66 and 67.) What is the estimated fuel consumption for Operating Conditions Z-4?

A—4,950 pounds.
B—5,380 pounds.
C—5,230 pounds.

Table fuel = 4,950 lbs
Change in fuel = 4,950 x 25 ÷ 443 = 279 lbs
Trip fuel = 4,950 lbs + 279 lbs = 5,229 lbs

(PLT012) — FAA-H-8083-25

121, 135, DSP

8667. (Refer to Figures 66 and 67.) What is the estimated fuel consumption for Operating Conditions Z-5?

A—6,250 pounds.
B—5,380 pounds.
C—7,120 pounds.

Table fuel = 6,250 lbs
Change in fuel = 6,250 x 60 ÷ 433 = 866 lbs
Trip fuel = 6,250 lbs + 866 lbs = 7,116 lbs

(PLT012) — FAA-H-8083-25

121, 135, DSP

8603. (Refer to Figures 51 and 52.) What is the total time from starting to the alternate through completing the approach for Operating Conditions L-1?

A—30 minutes.
B—44 minutes.
C—29 minutes.

29	*minutes*	*En route*
+15	*minutes*	*Holding*
44	*minutes*	*Total*

Note: *Distance is provided as NM so you should disregard the wind component provided in the operating conditions.*

(PLT012) — FAA-H-8083-25

121, DSP

8604. (Refer to Figures 51 and 52.) What is the total time from starting to the alternate through completing the approach for Operating Conditions L-2?

A—36 minutes.
B—55 minutes.
C—40 minutes.

40	*minutes*	*En route*
+15	*minutes*	*Holding*
55	*minutes*	*Total*

(PLT012) — FAA-H-8083-25

121, DSP

8605. (Refer to Figures 51 and 52.) What is the total time from starting to the alternate through completing the approach for Operating Conditions L-3?

A—1 hour.
B—1 hour 15 minutes.
C—1 hour 24 minutes.

60	*minutes*	*En route*
+15	*minutes*	*Holding*
75	*minutes*	*Total (1 hour 15 minutes)*

(PLT012) — FAA-H-8083-25

121, DSP

8606. (Refer to Figures 51 and 52.) What is the total time from starting to the alternate through completing the approach for Operating Conditions L-4?

A—35 minutes.
B—19 minutes.
C—20 minutes.

20	*minutes*	*En route*
+15	*minutes*	*Holding*
35	*minutes*	*Total*

(PLT012) — FAA-H-8083-25

Answers

8665 [B]	8666 [C]	8667 [C]	8603 [B]	8604 [B]	8605 [B]
8606 [A]					

121, DSP
8607. (Refer to Figures 51 and 52.) What is the total time from starting to the alternate through completing the approach for Operating Conditions L-5?

A—1 hour 3 minutes.
B—48 minutes.
C—55 minutes.

48	*minutes*	*En route*
+15	*minutes*	*Holding*
63	*minutes*	*Total (1 hour 3 minutes)*

(PLT012) — FAA-H-8083-25

121, DSP
8608. (Refer to Figures 51 and 52.) What is the approximate landing weight for Operating Conditions L-1?

A—78,850 pounds.
B—80,300 pounds.
C—81,600 pounds.

85,000	*lbs*	*Weight (start to alt)*
– 3,400	*lbs*	*Fuel burn*
81,600	*lbs*	*Landing weight*

(PLT008) — FAA-H-8083-25

121, DSP
8609. (Refer to Figures 51 and 52.) What is the approximate landing weight for Operating Conditions L-2?

A—65,200 pounds.
B—65,800 pounds.
C—69,600 pounds.

70,000	*lbs*	*Weight (start to alt)*
– 4,200	*lbs*	*Fuel burn*
65,800	*lbs*	*Landing weight*

(PLT008) — FAA-H-8083-25

121, DSP
8610. (Refer to Figures 51 and 52.) What is the approximate landing weight for Operating Conditions L-3?

A—80,300 pounds.
B—85,400 pounds.
C—77,700 pounds.

86,000	*lbs*	*Weight (start to alt)*
– 5,700	*lbs*	*Fuel burn*
80,300	*lbs*	*Landing weight*

(PLT008) — FAA-H-8083-25

121, DSP
8611. (Refer to Figures 51 and 52.) What is the approximate landing weight for Operating Conditions L-4?

A—73,200 pounds.
B—74,190 pounds.
C—73,500 pounds.

76,000	*lbs*	*Weight (start to alt)*
– 2,800	*lbs*	*Fuel burn*
73,200	*lbs*	*Landing weight*

(PLT008) — FAA-H-8083-25

121, DSP
8612. (Refer to Figures 51 and 52.) What is the approximate landing weight for Operating Conditions L-5?

A—78,600 pounds.
B—77,000 pounds.
C—76,300 pounds.

82,000	*lbs*	*Weight (start to alt)*
– 4,700	*lbs*	*Fuel burn*
77,300	*lbs*	*Landing weight*

(PLT008) — FAA-H-8083-25

Answers

8607 [A]	8608 [C]	8609 [B]	8610 [A]	8611 [A]	8612 [B]

Typical Flight Logs

Flight logs are used to accurately plan the time and fuel required for a flight. In the following paragraphs we describe all the steps required to complete a flight log.

1. Determine the magnetic courses for each leg and determine the leg distances.
2. Apply variations to the winds aloft.
3. Determine the temperature in relation to ISA.
4. Determine Mach number and convert to TAS.
5. Compute ground speed.
6. Calculate and record the time for each leg.
7. Compute fuel flow.
8. Compute total fuel.
9. Determine the reserve fuel.
10. Compute fuel burn to alternate.
11. Add the en route, reserve, alternate, and missed approach fuel to find the total fuel required for the flight.

Computation of Temperature at Cruise Altitude

Temperature is often expressed as a deviation from ISA which is the standard day temperature (i.e., ISA -2°). This temperature can be computed by the following procedure:

1. Compute ISA by multiplying the altitude in thousands of feet times -2° and then adding 15°. For example: ISA at 27,000 feet = 27 x (-2°) +15 = -39°
2. Apply the deviation from ISA. ISA -2° at 27,000 feet = (-39°) + (-2°) = -41°

Computation of True Airspeed Using Mach Number

True Airspeed (TAS) can be computed from Mach number and Outside Air Temperature (OAT).

Using the CX-2 computer, select "Plan Mach#" from the menu, then enter the OAT and the Mach number at the appropriate prompts.

Using an E6-B computer, follow these steps:

1. In the small window labeled "Airspeed Correction" or "True Airspeed," align the arrow labeled "Mach Number" with the OAT on the scale adjacent the window.
2. Find the Mach number on the inner of the two main scales and then read the TAS opposite it on the outer scale.

Note: Some "CR"-type mechanical computers have a window in which a Mach Index is aligned with a Mach number inside the window. Don't use this scale. It is designed to use Indicated Temperature and will give an inaccurate TAS when OAT is used.

See the instruction manual of your individual computer for more detailed instructions.

Specific Range

Specific range is the term used to describe the rate of fuel burn per nautical air mile flown. It is calculated by using TAS and fuel flow only. Wind has no effect on specific range. To calculate specific range in nautical air miles per 1,000 pounds, use the formula:

NM/1,000 = TAS x 1,000 ÷ PPH.

TAS should be calculated from the Mach number as in the paragraph above. PPH can be taken directly from the flight log.

121, 135, DSP

9546. (Refer to Figures 94, 95, and 96.) What is the ETE from Chicago Midway Airport to Greater Buffalo Intl?

A—2 hours 12 minutes.
B—2 hours 15 minutes.
C—2 hours 18 minutes.

To answer this question, complete the flight log in FAA Figure 94, using the information given in the problem:

1. *Change the winds aloft of FL190 at GIJ (Gipper) from true to magnetic, using the variation from the flight log (FAA Figure 94).* Note: *The variation changes with each leg. Winds at GIJ at FL190 are 230° true at 51 knots.*

   230° True
   + 01° West variation
   231° Magnetic

2. *Find the distance flown for the leg. In this case it is 19 nautical miles, since the L/O point is on the GIJ 270° radial at 19 NM. Distances for subsequent legs must be determined by referring to the enroute charts (FAA Figure 96).*
3. *Find the ground speed for the leg from L/O to GIJ using a flight computer:*

   Wind direction 231° (calculated in Step 1).
   Wind speed 51 knots (given on the flight log).
   Course 090° (L/O point is on the GIJ 270° radial).
   TAS 160 knots (given in the flight plan in FAA Figure 94).

   Calculated GS is 196.4 knots.
4. *Compute the time en route for the leg from L/O to GIJ:*

   Distance 19 NM (determined in Step 2).
   Ground speed 196.4 (calculated in Step 3).

   Calculated leg time is 5 minutes, 48 seconds.
5. *Since the next question asks for the fuel required for the flight, calculate the fuel burned on this leg:*

   ETE is :5:48 (calculated in Step 4).
   Fuel flow is 610 PPH (in note at the bottom of the flight log).

   Fuel burned is 59.0 pounds.
6. *Repeat steps for the subsequent legs to fill in the flight log:*

FROM	***TO***	***CRS***	***TAS***	***GS***	***NM***	***ETE***	***FUEL***
MDW	*L/O*	—	—	—	—	*:19:00**	*327.0**
L/O	*GIJ*	*090*	*160*	*196*	*19*	*:05:48*	*59.0*
GIJ	*CRL*	*085*	*160*	*200*	*129*	*:38:58*	*392.8*
CRL	*YXU*	*065*	*160*	*219*	*118*	*:32:20*	*328.7*
YXU	*T/D*	*101*	*160*	*215*	*80*	*:22:18*	*236.7*
T/D	*BUF*	—	—	—	*30*	*:14:00**	*121.5**
						2:12:04	*1,455.7*

**(given)*

(PLT012) — FAA-H-8083-15

121, 135, DSP

9547. (Refer to Figures 94, 95, and 96.) What are the fuel requirements from Chicago Midway Airport to Greater Buffalo Intl?

A—2,224 pounds.
B—1,987 pounds.
C—1,454 pounds.

Fuel from Question 9546: 1,455.7
Fuel for alternate to ROC: 236.0
Fuel for reserve: + 532.5
2,224.2 pounds

(PLT012) — FAA-H-8083-15

Answers

9546 [A] 9547 [A]

121, 135, DSP

9548. (Refer to Figures 94, 95, and 96.) What TAS should be maintained to arrive over CRL VORTAC 42 minutes after level-off?

A—166 knots.
B—168 knots.
C—171 knots.

1. *Determine the required ground speed using a flight calculator:*

   Distance flown is 148 NM (distance from L/O to GIJ to CRL).
   Time is 42 minutes (given in the problem).

   Calculated GS is 211.4 knots.

2. *Determine the required TAS:*

   Wind direction is 231° Magnetic (see Steps 1 and 2 in Question 9546).
   Wind speed is 51 knots (given in flight log).
   Course is 087° (average of the two legs).
   GS is 211.4 (calculated in Step 1).

   TAS is 172.8.

(PLT012) — FAA-H-8083-15

121, 135, DSP

9556. (Refer to Figures 98, 99, 100, and 102.) What is the ETE from DFW Intl to IAH?

A—1 hour 2 minutes.
B—1 hour 4 minutes.
C—1 hour 6 minutes.

To answer this question, complete the flight log in FAA Figure 98, using the information given in the problem:

1. *Change the winds aloft at 15,000 from true to magnetic, using the variation from the Airport/Facility Directory. Winds at BILEE are 230° True at 42 knots.*

   230° True
   – 08° East variation
   222° Magnetic

2. *Find the distance flown for the leg. In this case it is 80 NM. The sum of the distances from DFW to BILEE is 107 NM, and 27 NM of that is used by the climb. Distances for subsequent legs must be determined by referring to the enroute charts and the CUGAR Four Arrival (FAA Figures 100 and 102).*

3. *Find the ground speed for the leg from L/O to BILEE using a flight computer:*

   Wind direction 222° (calculated in Step 1).
   Wind speed 42 knots (given on the flight log).
   Course 154° (V369 is on the DFW 154° radial).
   TAS 248 knots (given in the flight plan on FAA Figure 98).

   Calculated GS is 229.2 knots.

4. *Compute the time en route for the leg from L/O to BILEE:*

   Distance 80 NM (determined in Step 2).
   Ground speed 229.2 (calculated in Step 3).

   Calculated leg time is 20 minutes, 57 seconds.

5. *Since the next question asks for the fuel required for the flight, calculate the fuel burned on this leg:*

   ETE is :20:57 (calculated in Step 4).
   Fuel flow is 850 PPH (in note at the bottom of the flight log).

   Fuel burned is 296.7 pounds.

6. *Repeat steps for the subsequent legs to fill in the flight log:*

FROM	*TO*	*CRS*	*TAS*	*GS*	*NM*	*ETE*	*FUEL*
DFW	*L/O*	—	—	—	*27*	*:12:00**	*231.0**
L/O	*BILEE*	*154*	*248*	*229.2*	*80*	*:20:57*	*296.7*
BILEE	*CUGAR*	*125*	*248*	*249.6*	*47*	*:11:18*	*160.1*
CUGAR	*T/D*	*125*	*248*	*249.6*	*21*	*:05:03*	*71.5*
T/D	*IAH*	—	—	—	*25*	*:14:00**	*132.0**
						1:03:18	*891.3*

**(given)*

(PLT012) — FAA-H-8083-15

121, 135, DSP

9557. (Refer to Figures 98, 99, 100, and 102.) What is the total fuel required from DFW Intl to IAH?

A—1,555 pounds.
B—1,863 pounds.
C—1,941 pounds.

In this question, the regulation-required fuel must be calculated using the summary box at the bottom of the flight plan.

1. *Compute the fuel required to fly the flight:*

   Fuel required is 891.3 pounds (see flight plan in Question 9556).

Answers

9548 [C] 9556 [B] 9557 [B]

2. *Compute the regulation-required reserve fuel (see Question 9556, Step 5):*

   Time required is 45 minutes (14 CFR §135.223).
   Fuel flow is 880 PPH (given at bottom of flight plan in FAA Figure 98).

   Reserve fuel is 660.0 pounds.

3. *Compute ETE to alternate (see Question 9556, Step 4):*

   GS is 194 knots (given in flight log).
   Distance 68 NM (given in flight log).

   ETE to alternate is 21 minutes, 2 seconds.

4. *Compute fuel burn to alternate (see Question 9556, Step 5):*

   Time to alternate is :21:02 (computed in previous step).
   Fuel flow is 880 PPH (given in flight log).

   Fuel burn to alternate is 308.5 pounds.

5. *Add the totals of Steps 1 through 4:*

FUEL	
891.3	*En route*
660.0	*Reserve*
+ 308.5	*Alternate*
1,859.8	*Total*

(PLT012) — FAA-H-8083-15, 14 CFR §91.167

121, 135, DSP

9558. (Refer to Figures 98, 99, 100, and 102.) Determine the TAS required to arrive at CUGAR, 31 minutes after level-off.

A—269 knots.
B—264 knots.
C—258 knots.

1. *Determine the required ground speed using a flight calculator:*

   Distance flown is 127 NM (80 + 47). Time is 31 minutes (stated in the problem).

   Calculated GS is 245.8 knots.

2. *Determine the required TAS:*

   Wind direction is 222° Magnetic (see Step 1 in Question 9556).
   Wind speed is 42 knots (given in flight log).
   Course is 154° (V369).
   GS is 245.8 knots (calculated in Step 1).

   TAS is 264.4 knots.

(PLT012) — FAA-H-8083-15

121, 135, DSP

9559. (Refer to Figures 98, 99, 100, and 102.) Determine the TAS required to arrive at CUGAR, 29 minutes after level-off.

A—285 knots.
B—290 knots.
C—295 knots.

1. *Determine the required ground speed using a flight calculator:*

   Distance flown is 127 NM (80 + 47).
   Time is 29 minutes (stated in the problem).

   Calculated GS is 262.8 knots.

2. *Determine the required TAS:*

   Wind direction is 222° Magnetic (see Step 1 in Question 9556).
   Wind speed is 42 knots (given in flight log).
   Course is 154° (V369).
   GS is 262.8 knots (calculated in Step 1).

   TAS is 281.2 knots (the closest answer is 285).

(PLT012) — FAA-H-8083-15

121, 135, DSP

9560. (Refer to Figures 103, 104, 105, and 106.) Estimate the total fuel required to be on the aircraft, prior to taxi at Tucson Intl. (Use 13°E for problem magnetic variation.)

A—2,223 pounds.
B—2,327 pounds.
C—2,447 pounds.

To answer this question, complete the flight log in FAA Figure 103, using the information given in the problem:

1. *Change the winds aloft at FL220 from true to magnetic, using the variation given in the question. Winds at GBN are 280° true at 46 knots.*

280°	*True*
– 13°	*East variation*
267°	*Magnetic*

2. *Find the distance flown for the leg. In this case it is 31 NM. The sum of the distances from Tucson to GBN is 104 NM (see the TUCSON 4 Departure), and 73 NM of that is used by the climb. Distances for subsequent legs must be determined by referring to the enroute charts and the DOWNE 3 Arrival (FAA Figures 105 and 106).*

Continued

Answers

9558 [B] 9559 [A] 9560 [B]

3. *Find the ground speed for the leg from L/O to GBN using a flight computer:*
   Wind direction 267° (calculated in Step 1).
   Wind speed 46 knots (given on the flight log).
   Course 289° (inbound on GBN 109 radial).
   TAS 233 knots (given in the flight plan on FAA Figure 103).

   Calculated GS is 189 knots.
4. *Compute the time en route for the leg from L/O to GBN:*
   Distance 31 NM (determined in Step 2).
   Ground speed 189 (calculated in Step 3).

   Calculated leg time is 9 minutes, 49 seconds.
5. *Calculate the fuel burned on this leg:*
   ETE is :09:49 (calculated in Step 4).
   Fuel flow is 676 PPH (in note at the bottom of the flight log).

   Fuel burned is 110.6 pounds.
6. *Repeat steps for the subsequent legs to fill in the flight log:*

FROM	***TO***	***CRS***	***TAS***	***GS***	***NM***	***ETE***	***FUEL***
TUS	*L/O*	—	—	—	*73*	*:25:00**	*350.0**
L/O	*GBN*	*289*	*233*	*189*	*31*	*:09:49*	*110.6*
GBN	*Int J104*	*298*	*233*	*192*	*76*	*:23:45*	*267.5*
Int J104	*PKE*	*277*	*233*	*188*	*47*	*:15:01*	*169.2*
PKE	*TNP*	*256*	*233*	*187*	*54*	*:17:17*	*194.7*
TNP	*T/D*	*250*	*233*	*188*	*81***	*:25:49*	*290.8*
T/D	*LAX*	—	—	—	*52*	*:18:00**	*170.0**
						2:14:41	*1,552.8*

**(given)*

**Note: *The DOWNE Three arrival is only used when LAX is landing on runways 6 or 7 and the total distance from TNP to LAX on the arrival is 176 NM. However, it is apparent from the answer choices and the approach charts that a straight-in landing on Runway 25L is assumed. This gives a distance of 133 from TNP to LAX. The distance from TNP to T/D was determined by taking the distance from TNP to PIONE Int to LAX (FAA Figure 106) and subtracting the 52 NM used for the descent. In this question, the regulation-required fuel must be calculated using the summary box at the bottom of the flight plan.*

7. *Compute the fuel required to fly the flight:*
   Fuel required is 1,552.8 pounds (see flight plan).
8. *Compute the required reserve fuel (see Step 5):*
   Time required is 45 minutes (§135.223).
   Fuel flow is 726 PPH (given at bottom of flight plan in FAA Figure 103).

   Reserve fuel is 544.2 pounds.
9. *Compute fuel burn to alternate (see Step 5):*
   Time to alternate is :19:00 (given in FAA Figure 103).
   Fuel flow is 726 PPH (given in flight log).

   Fuel burn to alternate is 229.9 pounds.
10. *Add the totals of Steps 7 through 9:*

FUEL	
1552.8	*En route*
544.5	*Reserve*
+ 229.9	*Alternate*
2,327.2	*Total*

(PLT012) — FAA-H-8083-15

121, 135, DSP

9561. (Refer to Figures 103, 104, 105, and 106.) Determine the ETE for the flight from Tucson Intl to Los Angeles Intl.

A—2 hours 10 minutes.
B—2 hours 15 minutes.
C—2 hours 19 minutes.

Refer to Question 9560, Steps 1-6.

(PLT012) — FAA-H-8083-15

121, 135, DSP

9578. (Refer to Figures 107, 115, 116, 117, 118, and 118C.) What is the ETE at .78 Mach?

A—1 hour 08 minutes.
B—1 hour 02 minutes.
C—1 hour 05 minutes.

To answer this question, complete the flight log in FAA Figure 115, using the information given in the problem:

1. *Change the winds aloft at FL270 from true to magnetic, using the variation from the Airport/Facility Directory entries for the LAX and PHX VORTACs. Winds at IPL are 300° True at 43 knots.*

300°	*True*
– 15°	*East variation*
285°	*Magnetic*

Answers

9561 [B] 9578 [A]

2. *Calculate the TAS for Mach .78:*

   ISA temperature at FL270 is -39°C
   ((27 x (-2°)) + 15°).
   ISA -2° = -41°C

   Using a flight calculator, determine that Mach .78 = 463.1 knots.

3. *Find the distance flown for the leg. In this case it is 50 NM. The remarks section of the flight plan notes that the L/O point is on the OCN 270° radial at 50 NM. The first leg of the IMPERIAL THREE Departure is from there to OCN. Even though only one line is used for the cruise portion of the IMPERIAL 3 departure, each leg of it should be calculated separately for best accuracy. The same is true of the ARLIN 9 Arrival into PHX.*

4. *Find the ground speed for the leg from L/O to OCN using a flight computer:*

   Wind direction 285° (calculated in Step 1).
   Wind speed 43 knots (given on the flight log).
   Course 090° (L/O point is on the OCN 270° radial).
   TAS 463.1 knots (calculated in Step 2).

   Calculated GS is 504.5 knots.

5. *Compute the time en route for the leg from L/O to OCN:*

   Distance 50 NM (determined in Step 3).
   Ground speed 504.5 (calculated in Step 4).

   Calculated leg time is 5 minutes, 57 seconds.

6. *Since the next question asks for the fuel required for the flight, calculate the fuel burned on this leg:*

   ETE is :5:57 (calculated in Step 5).
   Fuel flow is 9,600 PPH (in note at the bottom of the flight log).

   Fuel burned is 951.4 pounds.

7. *Repeat steps for the subsequent legs to fill in the flight log:*

FROM	***TO***	***CRS***	***TAS***	***GS***	***NM***	***ETE***	***FUEL***
LAX	*L/O*	*—*	*—*	*—*	*43*	*:19:00**	*4,510.0**
L/O	*OCN*	*090*	*463.1*	*504.5*	*50*	*:05:57*	*951.4*
OCN	*JLI*	*083*	*463.1*	*502.7*	*42*	*:05:01*	*802.1*
JLI	*KUMBA*	*115*	*463.1*	*505.4*	*35*	*:04:09*	*664.8*
KUMBA	*IPL*	*078*	*463.1*	*501.0*	*27*	*:03:14*	*517.4*
IPL	*BZA*	*074*	*463.1*	*499.4*	*46*	*:05:32*	*885.3*
BZA	*Mohak Int*	*075*	*463.1*	*499.0*	*32*	*:03:51*	*615.6*
Mohak	*T/P*	*067*	*463.1*	*494.7*	*15*	*:01:49*	*290.7*
T/P	*HYDRR*	*035*	*463.1*	*473.9*	*42*	*:05:19*	*850.7*
HYDRR	*ARLIN*	*076*	*463.1*	*499.0*	*19*	*:02:17*	*365.3*
ARLIN	*PHX*	*—*	*—*	*—*	*—*	*:12:00**	*1,140.0**
						1:08:09	*11,593.3*

**(given)*

(PLT012) — FAA-H-8083-15

121, 135, DSP

9579. (Refer to Figures 115, 116, 117, 118, and 118C.) What is the total fuel required at .78 Mach?

A—22,140 pounds.
B—22,556 pounds.
C—22,972 pounds.

In this question, the regulation-required fuel must be calculated using the summary box at the bottom of the flight plan:

1. *Compute the fuel required to fly the flight:*

   Fuel required is 11,593.3 pounds (see flight plan in Question 9578).

2. *Compute the required reserve fuel (see Question 9578, Step 6):*

   Time required is 45 minutes (§121.639).
   Fuel flow is 9,250 PPH (given at bottom of flight plan in FAA Figure 115).

   Reserve fuel is 6,937.5 pounds.

3. *Determine ETE to alternate:*

   ETE to alternate is 26 minutes (given in the flight log).

4. *Compute fuel burn to alternate (see Question 9578, Step 6):*

   Time to alternate is :26:00 (from Step 3).
   Fuel flow is 9,250 PPH (given in flight log).

   Fuel burn to alternate is 4,008.3 pounds.

Continued

Answers

9579 [B]

5. *Add the totals of Steps 1 through 4:*

FUEL	
11,593.3	*En route*
6,937.5	*Reserve*
+ 4,008.3	*Alternate*
22,539.1	*Total*

(PLT012) — FAA-H-8083-15

121, 135, DSP

9580. (Refer to Figures 115, 116, 117, 118, and 118C.) What is the specific range in nautical miles per 1,000 pounds of fuel from level-off to the ARLIN Intersection using .78 Mach?

A—48.8 NAM/1,000 pounds.
B—48.2 NAM/1,000 pounds.
C—47.9 NAM/1,000 pounds.

NM/1,000 is calculated by multiplying TAS times 1,000 and dividing the answer by the fuel flow.

TAS is 463.1 (calculated in Step 2 of Question 9578).
Fuel flow is 9,600 PPH (given in the flight log).

NM/1,000 = 48.2

(PLT015) — ANA

121, 135, DSP

9581. (Refer to Figures 107, 115, 116, 117, 118, and 118C.) What is the ETE at .80 Mach?

A—1 hour 02 minutes.
B—1 hour 04 minutes.
C—1 hour 07 minutes.

To answer this question, complete the flight log in FAA Figure 115, using the information given in the problem:

1. *Change the winds aloft at FL270 from true to magnetic, using the variation from the Airport/Facility Directory entries for the LAX and PHX VORTACs. Winds at IPL are 300° True at 43 knots.*

300°	*True*
– 15°	*East variation*
285°	*Magnetic*

2. *Calculate the TAS for Mach .80:*

   ISA temperature at FL270 is -39°C ((27 x (-2°)) + 15°).
   ISA -2°= -41°C

   Using a flight calculator, determine that Mach .80 = 475.0 knots.

3. *Find the distance flown for the leg. In this case it is 50 NM. The remarks section of the flight plan notes that the L/O point is on the OCN 270° radial at 50 NM. The first leg of the IMPERIAL THREE Departure is from there to OCN. Even though only one line is used for the cruise portion of the IMPERIAL 3 departure, each leg of it should be calculated separately for best accuracy. The same is true of the ARLIN 9 Arrival into PHX.*

4. *Find the ground speed for the leg from L/O to OCN using a flight computer:*

   Wind direction 285° (calculated in Step 1).
   Wind speed 43 knots (given on the flight log).
   Course 090° (L/O point is on the OCN 270° radial).
   TAS 475.0 knots (calculated in Step 2).

   Calculated GS is 516.4 knots.

5. *Compute the time en route for the leg from L/O to OCN:*

   Distance 50 NM (determined in Step 3).
   Ground speed 516.4 (calculated in Step 4).

   Calculated leg time is 5 minutes, 49 seconds.

6. *Since the next question asks for the fuel required for the flight, calculate the fuel burned on this leg:*

   ETE is :5:49 (calculated in Step 5).
   Fuel flow is 9,600 PPH (in note at the bottom of the flight log).

   Fuel burned is 930.7 pounds.

7. *Repeat steps for the subsequent legs to fill in the flight log:*

FROM	*TO*	*CRS*	*TAS*	*GS*	*NM*	*ETE*	*FUEL*
LAX	*L/O*	*—*	*—*	*—*	*43*	*:19:00**	*4,510.0**
L/O	*OCN*	*090*	*475*	*516.4*	*50*	*:05:49*	*930.7*
OCN	*JLI*	*083*	*475*	*514.6*	*42*	*:04:54*	*784.0*
JLI	*KUMBA*	*115*	*475*	*517.3*	*35*	*:04:04*	*650.7*
KUMBA	*IPL*	*078*	*475*	*512.9*	*27*	*:03:10*	*506.7*
IPL	*BZA*	*074*	*475*	*510.9*	*46*	*:05:24*	*864.3*
BZA	*Mohak Int*	*075*	*475*	*510.5*	*32*	*:03:46*	*601.8*
Mohak	*T/P*	*067*	*475*	*506.6*	*15*	*:01:47*	*285.3*
T/P	*HYDRR*	*035*	*475*	*485.8*	*42*	*:05:11*	*829.3*
HYDRR	*ARLIN*	*076*	*475*	*510.9*	*19*	*:02:14*	*357.3*
ARLIN	*PHX*	*—*	*—*	*—*	*—*	*:12:00**	*1,140.0**
						1:07:19	*11,460.1*

**(given)*

(PLT012) — FAA-H-8083-15

Answers

9580 [B]   9581 [C]

121, 135, DSP

9582. (Refer to Figures 115, 116, 117, 118, and 118C.) What is the total fuel required at .80 Mach?

A—22,836 pounds.
B—22,420 pounds.
C—22,556 pounds.

In this question, the regulation-required fuel must be calculated using the summary box at the bottom of the flight plan.

1. *Compute the fuel required to fly the flight:*

   Fuel required is 11,460.1 pounds (see flight plan in Question 9581).

2. *Compute the required reserve fuel (see Question 9581, Step 6):*

   Time required is 45 minutes (§121.639).
   Fuel flow is 9,250 PPH (given at bottom of flight plan in FAA Figure 115).

   Reserve fuel is 6,937.5 pounds.

3. *Determine ETE to alternate:*

   ETE to alternate is 26 minutes (given in the flight log).

4. *Compute fuel burn to alternate (see Question 9581, Step 6):*

   Time to alternate is :26:00 (from Step 3).
   Fuel flow is 9,250 PPH (given in flight log).

   Fuel burn to alternate is 4,008.3 pounds.

5. *Add the totals of Steps 1 through 4:*

FUEL	
11,460.1	*En route*
6,937.5	*Reserve*
+ 4,008.3	*Alternate*
22,405.9	*Total*

(PLT012) — FAA-H-8083-15

121, 135, DSP

9583. (Refer to Figures 115, 116, 117, and 118C.) What approximate indicated Mach should be maintained to arrive over the BZA VORTAC 6 minutes after passing IPL VORTAC?

A—.73 Mach.
B—.74 Mach.
C—.715 Mach.

1. *Determine the required GS:*

   Distance is 46 NM (FAA Figure 117).
   Time is 6 minutes (given in the question).

   GS = 460 knots

2. *Determine the required TAS:*

   Wind direction is 287° magnetic (300° True from flight log, less 13° east variation).
   Wind speed is 43 knots (given in flight log).
   Course is 074° (FAA Figure 117).
   GS is 460 knots (calculated in Step 1).

   Required TAS = 424.6 knots.

3. *Calculate the required Mach number:*

   TAS is 424.6 knots (calculated in Step 2).
   Temperature is -41°C (calculated in Question 9581, Step 2).

   Required Mach number is .715.

(PLT012) — FAA-H-8083-15

121, 135, DSP

9584. (Refer to Figures 107, 115, 116, 117, 118, and 118C.) What is the ETE at .82 Mach?

A—1 hour 05 minutes.
B—1 hour 07 minutes.
C—1 hour 03 minutes.

To answer this question, complete the flight log in FAA Figure 115, using the information given in the problem:

1. *Change the winds aloft at FL270 from true to magnetic, using the variation from the Airport/Facility Directory entries for the LAX and PHX VORTACs. Winds at IPL are 300° true at 43 knots.*

300°	*True*
– 15°	*East variation*
285°	*Magnetic*

2. *Calculate the TAS for Mach .82:*

   ISA temperature at FL270 is -39°C ((27 x (-2°)) + 15°).
   ISA -2° = -41°C

Continued

Answers

9582 [B] 9583 [C] 9584 [B]

Using a flight calculator, determine that Mach .82 = 486.9 knots.

3. *Find the distance flown for the leg. In this case it is 50 NM. The remarks section of the flight plan form notes that the L/O point is on the OCN 270° radial at 50 NM. The first leg of the IMPERIAL 3 Departure is from there to OCN. Even though only one line is used for the cruise portion of the IMPERIAL 3 departure, each leg of it should be calculated separately for best accuracy. The same is true of the ARLIN 9 Arrival into PHX.*
4. *Find the ground speed for the leg from L/O to OCN using a flight computer:*

    Wind direction 285° (calculated in Step 1).
    Wind speed 43 knots (given in the flight log).
    Course 090° (L/O point is on the OCN 270° radial).
    TAS 486.9 knots (calculated in Step 2).

    Calculated GS is 528.3 knots.
5. *Compute the time en route for the leg from L/O to OCN:*

    Distance 50 NM (determined in Step 3).
    Ground speed 528.3 (calculated in Step 4).

    Calculated leg time is 5 minutes, 41 seconds.
6. *Since the next question asks for the fuel required for the flight, calculate the fuel burned on this leg:*

    ETE is :5:41 (calculated in Step 5).
    Fuel flow is 9,600 PPH (in note at the bottom of the flight log).

    Fuel burned is 909.3 pounds.
7. *Repeat steps for the subsequent legs to fill in the flight log:*

FROM	***TO***	***CRS***	***TAS***	***GS***	***NM***	***ETE***	***FUEL***
LAX	*L/O*	—	—	—	*43*	*:19:00**	*4,510.0**
L/O	*OCN*	*090*	*486.9*	*528.3*	*50*	*:05:41*	*909.3*
OCN	*JLI*	*083*	*486.9*	*526.5*	*42*	*:04:47*	*765.3*
JLI	*KUMBA*	*115*	*486.9*	*529.2*	*35*	*:03:58*	*634.7*
KUMBA	*IPL*	*078*	*486.9*	*524.8*	*27*	*:03:05*	*493.3*
IPL	*BZA*	*074*	*486.9*	*522.4*	*46*	*:05:17*	*845.4*
BZA	*Mohak Int*	*075*	*486.9*	*522.4*	*32*	*:03:41*	*588.1*
Mohak	*T/P*	*067*	*486.9*	*518.5*	*15*	*:01:44*	*277.3*
T/P	*HYDRR*	*035*	*486.9*	*497.7*	*42*	*:05:04*	*810.7*
HYDRR	*ARLIN*	*076*	*486.9*	*522.8*	*19*	*:02:11*	*349.3*
ARLIN	*PHX*	—	—	—	—	*:12:00**	*1,140.0**
						1:06:28	*11,323.4*

**(given)*

(PLT012) — FAA-H-8083-15

121, 135, DSP

9585. (Refer to Figures 115, 116, 117, 118, and 118C.) What is the total fuel required at .82 Mach?

A—22,420 pounds.
B—22,284 pounds.
C—22,700 pounds.

In this question, the regulation-required fuel must be calculated using the summary box at the bottom of the flight plan.

1. *Compute the fuel required to fly the flight:*

    Fuel required is 11,323.4 pounds (see flight plan in Question 9584).
2. *Compute the required reserve fuel (see Question 9584, Step 6):*

    Time required is 45 minutes (§121.639).
    Fuel flow is 9,250 PPH (given at bottom of flight plan in FAA Figure 115).

    Reserve fuel is 6,937.5 pounds.
3. *Determine ETE to alternate:*

    ETE to alternate is 26 minutes (given in the flight log).
4. *Compute fuel burn to alternate (see Question 9584, Step 6):*

    Time to alternate is :26:00 (computed in previous step).
    Fuel flow is 9,250 PPH (given in flight log).

    Fuel burn to alternate is 4,008.3 pounds.
5. *Add the totals of Steps 1 through 4:*

FUEL	
11,323.4	*En route*
6,937.5	*Reserve*
+ 4,008.3	*Alternate*
22,269.2	*Total*

(PLT012) — FAA-H-8083-15

121, 135, DSP

9591. (Refer to Figures 119, 120, 121, and 122.) What is the ETE from BUF to ORD using .78 Mach?

A—1 hour 09 minutes.
B—1 hour 07 minutes.
C—1 hour 05 minutes.

To answer this question, complete the flight log in FAA Figure 119, using the information given in the problem:

1. *Change the winds aloft at FL310 from true to magnetic, using the variation from the remarks section*

Answers

9585 [B] 9591 [C]

of the flight plan. Winds at BUF are 330° True at 39 knots.

330° True
+ 08° West variation
338° Magnetic

2. *Calculate the TAS for Mach .78:*

   ISA temperature at FL310 is -47°C ((31 x (-2°)) + 15).
   ISA -6° = -53°C

   Using a flight calculator, determine that Mach .78 = 451.0 knots.

3. *Find the distance flown for the leg, in this case 40 NM. FAA Figure 121 shows that it is 110 NM from BUF to YXU via J547. The flight log indicates that 70 NM is needed for the climb.*
4. *Find the ground speed for the leg from L/O to YXU using a flight computer:*

   Wind direction 338° (calculated in Step 1).
   Wind speed 39 knots (given in the flight log).
   Course 282° (L/O point is on the BUF 282° radial).
   TAS 451.0 knots (calculated in Step 2).

   Calculated GS is 428.0 knots.

5. *Compute the time en route for the leg from L/O to YXU:*

   Distance 40 NM (determined in Step 3).
   Ground speed 428.0 (calculated in Step 4).

   Calculated leg time is 5 minutes, 36 seconds.

6. *Since the next question asks for the fuel required for the flight, calculate the fuel burned on this leg:*

   ETE is :5:36 (calculated in Step 5).
   Fuel flow is 9,300 PPH (in note at the bottom of the flight log).

   Fuel burned is 869.1 pounds.

7. *Repeat steps for the subsequent legs to fill in the flight log:*

FROM	***TO***	***CRS***	***TAS***	***GS***	***NM***	***ETE***	***FUEL***
BUF	*L/O*	*—*	*—*	*—*	*70*	*:16:00**	*4,960.0**
L/O	*YXU*	*282*	*451*	*428.0*	*40*	*:05:36*	*869.1*
YXU	*FNT*	*270*	*451*	*431.9*	*114*	*:15:50*	*2454.5*
FNT	*T/D*	*258*	*451*	*439.3*	*94*	*:12:50*	*1989.9*
T/D	*PMM*	*—*	*—*	*—*	*15*	*:02:00**	*216.7**
PMM	*ORD*	*—*	*—*	*—*	*89*	*:13:00**	*1,408.3**
						1:05:16	*11,898.5*

**(given)*

(PLT012) — FAA-H-8083-15

121, 135, DSP

9592. (Refer to Figures 119, 120, 121, and 122.) What is the total fuel required for the flight from BUF to ORD using .78 Mach?

A—19,033 pounds.
B—21,739 pounds.
C—22,189 pounds.

In this question, the regulation-required fuel must be calculated using the summary box at the bottom of the flight plan.

1. *Compute the fuel required to fly the flight:*

   Fuel required is 11,898.5 pounds (see flight plan in Question 9591).

2. *Compute the required reserve fuel (see Question 9591, Step 6):*

   Time required is 45 minutes (§121.639).
   Fuel flow is 9,550 PPH (given at bottom of flight plan in FAA Figure 119).

   Reserve fuel is 7,162.5 pounds.

3. *Determine ETE to alternate:*

   ETE to alternate is 17 minutes (given in the flight log).

4. *Compute fuel burn to alternate (see Question 9591, Step 6):*

   Time to alternate is :17:00 (from Step 3).
   Fuel flow is 9,550 PPH (given in flight log).

   Fuel burn to alternate is 2,705.8 pounds.

5. *Add the totals of Steps 1 through 4:*

FUEL	
11,898.5	*En route*
7,162.5	*Reserve*
+ 2,705.8	*Alternate*
21,766.8	*Total*

(PLT012) — FAA-H-8083-15

Answers

9592 [B]

121, 135, DSP

9593. (Refer to Figures 119, 120, 121, and 122.) What is the specific range in nautical air miles per 1,000 pounds of fuel from level-off to start of descent using .78 Mach?

A—48.9 NAM/1000.
B—52.5 NAM/1000.
C—55.9 NAM/1000.

NM/1,000 is calculated by multiplying TAS times 1,000 and dividing the answer by the fuel flow:

TAS is 451 (calculated in Question 9591, Step 2).
Fuel flow is 9,300 PPH (given in the flight log).

NM/1,000 = 48.5

(PLT015) — ANA

121, 135, DSP

9594. (Refer to Figures 119, 120, 121, and 122.) What is the ETE from BUF to ORD using .80 Mach?

A—1 hour 01 minutes.
B—1 hour 04 minutes.
C—1 hour 08 minutes.

To answer this question, complete the flight log in FAA Figure 119, using the information given in the problem:

1. *Change the winds aloft at FL310 from true to magnetic, using the variation from the remarks section of the flight plan. Winds at BUF are 330° True at 39 knots.*

   330° True
   + 08° West variation
   338° Magnetic

2. *Calculate the TAS for Mach .80:*

   ISA temperature at FL310 is -47°C ((31 x (-2°)) + 15°).
   ISA -6° = -53°C

   Using a flight calculator, determine that Mach .80 = 462.5 knots.

3. *Find the distance flown for the leg. In this case it is 40 NM. FAA Figure 121 shows that it is 110 NM from BUF to YXU via J547. The flight log indicates that 70 NM is needed for the climb.*

4. *Find the ground speed for the leg from L/O to YXU using a flight computer:*

   Wind direction 338° (calculated in Step 1).
   Wind speed 39 knots (given in the flight log).
   Course 282° (L/O point is on the BUF 282° radial).
   TAS 462.5 knots (calculated in Step 2).

   Calculated GS is 428.0 knots.

5. *Compute the time enroute for the leg from L/O to YXU:*

   Distance 40 NM (determined in Step 3).
   Ground speed 439.6 (calculated in Step 4).

   Calculated leg time is 5 minutes, 28 seconds.

6. *Since the next question asks for the fuel required for the flight, calculate the fuel burned on this leg:*

   ETE is :5:28 (calculated in Step 5).
   Fuel flow is 9,300 PPH (in note at the bottom of the flight log).

   Fuel burned is 846.2 pounds.

7. *Repeat steps for the subsequent legs to fill in the flight log:*

FROM	*TO*	*CRS*	*TAS*	*GS*	*NM*	*ETE*	*FUEL*
BUF	*L/O*	—	—	—	*70*	*:16:00**	*4,960.0**
L/O	*YXU*	*282*	*462.5*	*439.6*	*40*	*:05:28*	*846.2*
YXU	*FNT*	*270*	*462.5*	*443.5*	*114*	*:15:25*	*2,390.3*
FNT	*T/D*	*258*	*462.5*	*450.9*	*94*	*:12:30*	*1938.7*
T/D	*PMM*	—	—	—	*15*	*:02:00**	*216.7**
PMM	*ORD*	—	—	—	*89*	*:13:00**	*1,408.3**
						1:04:23	*11,760.2*

**(given)*

(PLT012) — FAA-H-8083-15

121, 135, DSP

9595. (Refer to Figures 119, 120, 121, and 122.) What is the total fuel required for the flight from BUF to ORD using .80 Mach?

A—19,388 pounds.
B—22,494 pounds.
C—21,644 pounds.

In this question, the regulation-required fuel must be calculated using the summary box at the bottom of the flight plan.

1. *Compute the fuel required to fly the flight:*

   Fuel required is 11,760.2 pounds (see flight plan in Question 9594).

2. *Compute the required reserve fuel (see Question 9594, Step 6):*

   Time required is 45 minutes (§121.639).
   Fuel flow is 9,550 PPH (given at bottom of flight plan in FAA Figure 119).

   Reserve fuel is 7,162.5 pounds.

3. *Determine ETE to alternate:*

   ETE to alternate is 17 minutes (given in the flight log).

Answers

9593 [A] 9594 [B] 9595 [C]

4. *Compute fuel burn to alternate (see Question 9594, Step 6):*

   Time to alternate is :17:00 (from Step 3).
   Fuel flow is 9,550 PPH (given in flight log).

   Fuel burn to alternate is 2,705.8 pounds.

5. *Add the totals of Steps 1 through 4:*

FUEL	
11,760.2	*En route*
7,162.5	*Reserve*
+ 2,705.8	*Alternate*
21,628.5	*Total*

(PLT012) — FAA-H-8083-15

121, 135, DSP

9597. (Refer to Figures 158, 159, 160, 160A, and 161.) The estimated time enroute from STL to LGA for N711JB is

A—1 hour 46 minutes.
B—1 hour 50 minutes.
C—1 hour 54 minutes.

To answer this question, complete the flight log in FAA Figure 158, using the information given in the problem:

1. *Change the winds aloft at FL370 from true to magnetic, using the variation from the remarks section of the flight plan. Notice that the variation changes with each leg. Winds at ROD are 350° True at 96 knots.*

350°	*True*
+ 01°	*West variation*
351°	*Magnetic*

2. *Find the distance flown for the leg. In this case it is 131 NM. This is the distance from BIB to SHB in FAA Figure 160. Although the flight log provides only one line for the leg from BIB to ROD, the turn at SHB makes it more accurate to compute the ETEs and fuel burns separately.*

3. *Find the ground speed for the leg from BID to SHB using a flight computer:*

   Wind direction 351° (calculated in Step 1).
   Wind speed 96 knots (given in the flight log).
   Course 067° (V394).
   TAS 506 knots (given in the flight plan in FAA Figure 158).

   Calculated GS is 474.1 knots.

4. *Compute the time en route for the leg from BIB to SHB:*

   Distance 131 NM (determined in Step 2).
   Ground speed 474.1 (calculated in Step 3).

   Calculated leg time is 16 minutes, 35 seconds.

5. *Since the next question asks for the fuel required for the flight, calculate the fuel burned on this leg:*

   ETE is :16:35 (calculated in Step 4).
   Fuel flow is 2,389 PPH (in note at the bottom of the flight log).

   Fuel burned is 660.1 pounds.

6. *Repeat steps for the subsequent legs to fill in the flight log:*

FROM	***TO***	***CRS***	***TAS***	***GS***	***NM***	***ETE***	***FUEL***
STL	*BIB*	*—*	*—*	*—*	*95*	*:16:00**	*987.0**
BIB	*SHB*	*067*	*506*	*474.1*	*131*	*:16:35*	*660.1*
SHB	*ROD*	*063*	*506*	*468.0*	*91*	*:11:40*	*464.5*
ROD	*DJB*	*059*	*506*	*456.5*	*107*	*:14:04*	*560.0*
DJB	*PSB*	*109*	*506*	*532.4*	*191*	*:21:32*	*857.1*
PSB	*MIP*	*093*	*506*	*505.2*	*61*	*:07:15*	*288.4*
MIP	*SBJ*	*117*	*506*	*540.7*	*92*	*:10:13*	*406.5*
SBJ	*LGA*	*—*	*—*	*—*	*52*	*:16:26**	*269.0**
						1:53:45	*4,492.6*

**(given)*

(PLT012) — FAA-H-8083-15

121, 135, DSP

9598. (Refer to Figures 158, 159, 160, 160A, and 161.) The required amount of fuel (in pounds) to be on N711JB at STL, prior to taxi, is

A—5,933 pounds.
B—6,408 pounds.
C—6,641 pounds

In this question, the regulation-required fuel must be calculated using the summary box at the bottom of the flight plan.

1. *Compute the fuel required to fly the flight:*

   Fuel required is 4,492.6 pounds (see flight plan in Question 9597).

2. *Compute the required reserve fuel (see Question 9597, Step 5):*

   Time required is 45 minutes (§135.223).
   Fuel flow is 1,898 PPH (given at bottom of flight plan in FAA Figure 158).

   Reserve fuel is 1,423.5 pounds.

Continued

Answers

9597 [C] 9598 [B]

3. *Determine ETE to alternate:*

   ETE to alternate is 15 minutes (given in the flight log).

4. *Compute fuel burn to alternate (see Question 9597, Step 5):*

   Time to alternate is :15:00 (from Step 3).
   Fuel flow is 1,898 PPH (given in flight log).

   Fuel burn to alternate is 474.5 pounds.

5. *Add the totals of Steps 1 through 4:*

FUEL	
4,492.6	*En route*
1,423.5	*Reserve*
+ 474.5	*Alternate*
6,390.6	*Total*

(PLT012) — FAA-H-8083-15

121, 135, DSP

9610. (Refer to Figures 168, 169, 169A, 171, and 172.) What is the ETE for PTZ 70 from Chicago Pal-Waukee Airport to Greater Buffalo Intl Airport?

A—2 hours 15 minutes.
B—2 hours 18 minutes.
C—2 hours 21 minutes.

To answer this question, complete the flight log in FAA Figure 168, using the information given in the problem:

1. *Change the winds aloft at FL190 from true to magnetic, using the variation from the remarks section of the flight plan. Notice that the variation changes with each leg. Winds at PMM are 020° True at 61 knots.*

020°	*True*
+ 01°	*West variation*
021°	*Magnetic*

2. *Find the distance flown for the leg. In this case it is 47 NM. This is noted in the remarks section of the flight plan.*

3. *Find the ground speed for the leg from L/O to PMM using a flight computer:*

   Wind direction 021° (calculated in Step 1).
   Wind speed 61 knots (given in the flight log).
   Course 081° (L/O point is the PMM 261° at 47 NM).
   TAS 247 knots (given in the flight plan in FAA Figure 168).

   Calculated GS is 210.0 knots.

4. *Compute the time en route for the leg from L/O to PMM:*

   Distance 47 NM (determined in Step 2).
   Ground speed 210.0 (calculated in Step 3).

   Calculated leg time is 13 minutes, 26 seconds.

5. *Since the next question asks for the fuel required for the flight, calculate the fuel burned on this leg:*

   ETE is :13:26 (calculated in Step 4).
   Fuel flow is 676 PPH (in note at the bottom of the flight log).

   Fuel burned is 151.3 pounds.

6. *Repeat steps for the subsequent legs to fill in the flight log:*

FROM	***TO***	***CRS***	***TAS***	***GS***	***NM***	***ETE***	***FUEL***
PWK	*L/O*	*—*	*—*	*—*	*49*	*:24:00**	*410.0**
L/O	*PMM*	*081*	*247*	*210.0*	*47*	*:13:26*	*151.3*
PMM	*FNT*	*076*	*247*	*205.4*	*109*	*:31:50*	*358.7*
FNT	*YXU*	*094*	*247*	*217.6*	*114*	*:31:26*	*354.1*
YXU	*T/D*	*101*	*247*	*222.2*	*70*	*:18:54*	*213.0*
T/D	*BUF*	*—*	*—*	*—*	*19*	*:19:00**	*163.0**
						2:18:36	*1,650.1*

**(given)*

(PLT012) — FAA-H-8083-15

121, 135, DSP

9611. (Refer to Figures 168, 169, 169A, 171, and 172.) What is the computed fuel usage for PTZ 70 from start of taxi at Chicago Pal-Waukee to landing at Greater Buffalo Intl?

A—1,642 pounds.
B—2,005 pounds.
C—2,550 pounds.

See *Steps 1 through 6 in Question 9610. Question 9611 asks only for the fuel burn, so it is not necessary to compute reserve or alternate fuel. (PLT012) — FAA-H-8083-15*

Answers

9610 [B] 9611 [A]

121, 135, DSP

9612. (Refer to Figures 168, 171, 172, and 173.) What TAS should PTZ 70 maintain to arrive at FNT 30 minutes after passing PMM?

A—255 knots.
B—265 knots.
C—260 knots.

1. Determine the required GS from PMM to FNT:

Distance is 109 NM (FAA Figure 171).
Time is 30 minutes (given in question).

Required GS is 218.0 knots.

2. Calculate the required TAS:

Wind direction is 023° magnetic (020 + 3° west variation).
Wind speed is 61 knots (given in flight log).
Course is 078° (inbound on the FNT 258° radial).
GS is 218.0 (calculated in Step 1).

Required TAS is 257.9.

(PLT012) — FAA-H-8083-15

121, 135, DSP

9626. (Refer to Figures 179, 180A, 181, 182, and 182A.) The time enroute from Newport News/Williamsburg Intl to Philadelphia Intl via the flight plan of EAB 90 is

A—1 hour 27 minutes.
B—1 hour 29 minutes.
C—1 hour 31 minutes.

To answer this question, complete the flight log in FAA Figure 179, using the information given in the problem:

1. Change the winds aloft at FL190 from true to magnetic. Variation is provided in the remarks section of the flight plan (FAA Figure 179). Winds at ORF are 300° True at 70 knots.

300° True
+ 10° West variation
310° Magnetic

2. Find the distance flown for the leg. In this case it is 42 NM, the distance from ORF to SAWED (FAA Figure 181).

3. Find the ground speed for the leg from ORF to SAWED using a flight computer:

Wind direction 310° (calculated in Step 1).

Wind speed 70 knots (given in the flight log).
Course 030° (J121 is the ORF 030° radial).
TAS 236 knots (given in the flight plan in FAA Figure 179).

Calculated GS is 213.6 knots.

4. Compute the time enroute for the leg from ORF to SAWED:

Distance 42 NM (determined in Step 2).
Ground speed 213.6 (calculated in Step 3).

Calculated leg time is 11 minutes, 48 seconds.

5. Since the next question asks for the fuel required for the flight, calculate the fuel burned on this leg:

ETE is :11:48 (calculated in Step 4).
Fuel flow is 689 PPH (in note at the bottom of the flight log).
Fuel burned is 133.5 pounds.

6. Repeat steps for the subsequent legs to fill in the flight log:

FROM	***TO***	***CRS***	***TAS***	***GS***	***NM***	***ETE***	***FUEL***
PHF	*ORF*	*—*	*—*	*—*	*40*	*:19:00**	*312.0**
ORF	*SAWED*	*030*	*236*	*213.6*	*42*	*:11:48*	*133.5*
SAWED	*SWL*	*039*	*236*	*224.2*	*37*	*:09:54*	*113.7*
SWL	*SIE*	*035*	*236*	*219.4*	*70*	*:19:09*	*219.9*
SIE	*VCN*	*354*	*236*	*180.6*	*28*	*:09:18*	*106.8*
VCN	*OOD*	*301*	*236*	*166.6*	*17*	*:06:07*	*70.3*
OOD	*PHL*	*—*	*—*	*—*	*30*	*:16:00**	*177.0**
						1:31:16	*1,133.2*

**(given)*

(PLT012) — FAA-H-8083-15

121, 135, DSP

9627. (Refer to Figures 179, 180, 181, 182, and 182A.) The planned fuel usage from Newport News/Williamsburg Intl to Philadelphia Intl for EAB 90 is

A—1,132 pounds.
B—1,107 pounds.
C—1,084 pounds.

See Steps 1 through 6 in Question 9626. Question 9627 asks only for the fuel burn, so it is not necessary to compute reserve or alternate fuel. (PLT012) — FAA-H-8083-15

Answers

9612 [C] 9626 [C] 9627 [A]

121, 135, DSP

9628. (Refer to Figures 179, 180, 181, 182, and 183.) The required fuel from Newport News/Williamsburg Intl to Philadelphia Intl for EAB 90 is

A—1,860 pounds.
B—1,908 pounds.
C—2,003 pounds.

In this question, the regulation-required fuel must be calculated using the summary box at the bottom of the flight plan.

1. *Compute the fuel required to fly the flight:*

   Fuel required is 1,133.2 pounds (see flight plan in Question 9626).

2. *Compute the required reserve fuel (see Question 9626, Step 5):*

   Time required is 45 minutes (§135.223).
   Fuel flow is 739 PPH (given at bottom of flight plan in FAA Figure 113).

   Reserve fuel is 554.3 pounds.

3. *Determine ETE to alternate:*

   ETE to alternate is 18 minutes (given in the flight log).

4. *Compute fuel burn to alternate (see Question 9626, Step 5):*

   Time to alternate is :18:00 (from Step 3).
   Fuel flow is 739 PPH (given in flight log).

   Fuel burn to alternate is 221.7 pounds.

5. *Add the totals of Steps 1 through 4:*

FUEL	
1,133.2	*En route*
554.3	*Reserve*
+ 221.7	*Alternate*
1,909.2	*Total*

(PLT012) — FAA-H-8083-15

121, 135, DSP

9635. (Refer to Figures 179, 180, 181, 182, and 183.) What "TAS" would EAB 90 need to maintain from SWL to SIE in an attempt to cut 3 minutes off of the flight plan (SWL-SIE) ETE?

A—276.
B—280.
C—284.

1. *Determine the required GS from SWL to SIE:*

   Distance is 70 NM (FAA Figure 181).
   Time is 16 minutes, 9 seconds (The time in the flight log in Question 9626, minus 3 minutes).

   Required GS is 260.1 knots.

2 *Calculate the required TAS:*

   Wind direction is 310° magnetic (300 plus 10° west variation).
   Wind speed is 70 knots (given in flight log).
   Course is 035° (J121).
   GS is 260.1 (calculated in Step 1).

   Required TAS is 275.1.

(PLT012) — FAA-H-8083-15

121, 135, DSP

9642. (Refer to Figures 190, 191, 192, 193, 193A, 194, 195, and 195A.) The estimated time enroute from MSP to DEN for PIL 10 is

A—1 hour 54 minutes.
B—1 hour 57 minutes.
C—2 hours 00 minutes.

To answer this question, complete the flight log in FAA Figure 190, using the information given in the problem:

1. *Change the winds aloft at FL430 from true to magnetic, using the variation from the remarks section of the flight plan. Winds at FSD are 290° True at 89 knots.*

290°	*True*
– 09°	*East variation*
281°	*Magnetic*

2. *The TAS is 456 knots from the flight plan.*
3. *Find the distance flown for the leg. In this case it is 90 NM. The flight log shows the L/O point as the FSD 048° radial at 90 NM.*

Answers

9628 [B]	9635 [A]	9642 [A]

4. *Find the ground speed for the leg from L/O to FSD using a flight computer:*

   Wind direction 281° (calculated in Step 1).
   Wind speed 89 knots (given in the flight log).
   Course 228° (L/O point is on the FSD 048° radial).
   TAS 456 knots (see Step 2).

   Calculated GS is 396.9 knots.

5. *Compute the time en route for the leg from L/O to FSD:*

   Distance 90 NM (determined in Step 3).
   Ground speed 396.9 (calculated in Step 4).

   Calculated leg time is 13 minutes, 36 seconds.

6. *Since the next question asks for the fuel required for the flight, calculate the fuel burned on this leg:*

   ETE is :13:36 (calculated in Step 5).
   Fuel flow is 9,026 PPH (in note at the bottom of the flight log).

   Fuel burned is 2,045.9 pounds.

7. *Repeat steps for the subsequent legs to fill in the flight log:*

FROM	***TO***	***CRS***	***TAS***	***GS***	***NM***	***ETE***	***FUEL***
MSP	*L/O*	—	—	—	*90*	*:19:00**	*4,170.0**
L/O	*FSD*	*228*	*456*	*396.9*	*90*	*:13:36*	*2,045.9*
FSD	*OBH*	*198*	*456*	*436.5*	*153*	*:21:02*	*3,163.6*
OBH	*LBF*	*249*	*456*	*390.1*	*110*	*:16:55*	*2,545.2*
LBF	*MODES*	*250*	*456*	*389.3*	*104*	*:16:02*	*2,411.3*
MODES	*AMWAY*	*218*	*456*	*423.5*	*11*	*:01:34*	*234.5*
AMWAY	*DEN*	—	—	—	*97*	*:25:00**	*3,107.0**
						1:53:09	*17,677.5*

**(given)*

(PLT012) — FAA-H-8083-15

121, 135, DSP

9643. (Refer to Figures 190, 191, 192, 193, 193A, 194, 195 and 195A.) The required fuel that should be onboard PIL 10 at MSP is

A—28,053 pounds.
B—29,057 pounds.
C—29,960 pounds.

In this question, the regulation-required fuel must be calculated using the summary box at the bottom of the flight plan.

1. *Compute the fuel required to fly the flight:*

   Fuel required is 17,677.5 pounds (see flight plan in Question 9643).

2. *Compute the required reserve fuel (see Question 9643, Step 6):*

   Time required is 45 minutes (§121.639).
   Fuel flow is 7,688 PPH (given at bottom of flight plan in FAA Figure 190).

   Reserve fuel is 5,766.0 pounds.

3. *Determine ETE to alternate:*

   ETE to alternate is 36 minutes (given in the flight log).

4. *Compute fuel burn to alternate (see Question 9642, Step 6):*

   Time to alternate is :36:00 (from Step 3).
   Fuel flow is 7,688 PPH (given in flight log).

   Fuel burn to alternate is 4,612.8 pounds.

5. *Add the totals of Steps 1 through 4:*

FUEL	
17,677.5	*En route*
5,766.0	*Reserve*
+ 4,612.8	*Alternate*
28,056.3	*Total*

(PLT012) — FAA-H-8083-15

121, 135, DSP

9656. (Refer to Figures 185A, 202, 203, 203A, 204, 205A, and 206.) For PTL 55 to be dispatched on this flight plan (LAS-SFO), how much fuel is required to be onboard at the start of taxi?

A—27,800 pounds.
B—28,317 pounds
C—29,450 pounds.

To answer this question, complete the flight log in FAA Figure 202, using the information given in the problem:

1. *Change the winds aloft at FL390 from true to magnetic, using the variation from the Airport/Facility Directory entries for the LAS and SFO VORTACs. Winds at BTY are 340° True at 53 knots.*

   340° True
   – 15° East variation
   325° Magnetic

2. *Determine the TAS:*

   The TAS is 460 Knots (given in the flight plan).

3. *Find the distance flown for the leg. In this case it is 41 NM. FAA Figure 203 shows a distance of 106 NM from LAS to BTY on the OASIS 8 Departure. The flight log shows 65 NM is required for the climb.*

Continued

Answers

9643 [A]   9656 [B]

4. *Find the ground speed for the leg from L/O to BTY using a flight computer:*

   Wind direction 325° (calculated in Step 1).
   Wind speed 53 knots (given in the flight log).
   Course 306° (inbound on the BTY 125° radial).
   TAS 460 knots (Step 2).

   Calculated GS is 409.6 knots.
5. *Compute the time en route for the leg from L/O to BTY:*

   Distance 41 NM (determined in Step 3).
   Ground speed 409.6 (calculated in Step 4).

   Calculated leg time is 6 minutes.
6. *Calculate the fuel burned on this leg:*

   ETE is :06:00 (calculated in Step 5).
   Fuel flow is 12,000 PPH (in note at the bottom of the flight log).

   Fuel burned is 1,201.3 pounds.
7. *Repeat steps for the subsequent legs to fill in the flight log:*

FROM	***TO***	***CRS***	***TAS***	***GS***	***NM***	***ETE***	***FUEL***
LAS	*L/O*	—	—	—	*65*	*:21:00**	*5,600.0**
L/O	*BTY*	*306*	*460*	*409.6*	*41*	*:06:00*	*1,201.3*
BTY	*OAL*	*310*	*460*	*408.4*	*87*	*:12:47*	*2,556.3*
OAL	*GROAN*	*245*	*460*	*447.8*	*170*	*:22:47*	*4,970.4*
GROAN	*SFO*	—	—	—	*75*	*:26:00**	*4,556.7**
						1:28:34	*18,884.7*

**(given)*

In this question, the regulation-required fuel must be calculated using the summary box at the bottom of the flight plan.

8. *Compute the fuel required to fly the flight:*

   Fuel required is 18,828.0 pounds (see flight plan).
9. *Compute the required reserve fuel (see Step 5):*

   Time required is 45 minutes (§121.639).
   Fuel flow is 11,000 PPH (given at bottom of flight plan in FAA Figure 202).

   Reserve fuel is 8,250 pounds.
10. *Determine ETE to alternate:*

    ETE to alternate is 9 minutes.
11. *Compute fuel burn to alternate (see Step 5):*

    Time to alternate is :09:00 (from Step 10).
    Fuel flow is 11,000 PPH (given in flight log).

    Fuel burn to alternate is 1,650 pounds.
12. *Add the totals of Steps 7 through 11:*

FUEL	
18,884.7	*En route*
8,250.0	*Reserve*
+ 1,650.0	*Alternate*
28,784.7	*Total*

(PLT012) — FAA-H-8083-15

121, 135, DSP

9657. (Refer to Figures 185A, 202, 203, 203A, 204, 205A, and 206.) The ETE on this flight (PTL 55 LAS-SFO) is

A—1 hour 25 minutes.
B—1 hour 27 minutes.
C—1 hour 29 minutes.

See Steps 1 through 7 in Question 9656.

(PLT012) — FAA-H-8083-15

121, 135, DSP

9664. (Refer to Figure 202.) In block 3 of the flight plan, there is the following entry: B/B747/R. What does the prefix "B" indicate?

A—Foreign air carrier (Brazil).
B—TCAS and heavy.
C—DME and transponder but no altitude encoding capability.

Prefixes in the aircraft type box can be T/ (TCAS equipped), H/ (heavy aircraft), or B/ (heavy aircraft and TCAS equipped). (PLT053) — AIM ¶5-1-8

121, 135, DSP

9665. (Refer to Figure 202.) In block 3 of the flight plan, there is the following entry: B/B747/R. The "/R" will change to "/I" in February 1999. What will the "/I" indicate?

A—That the flight plan contains an RNAV route.
B—RNAV/Transponder/altitude encoding capability.
C—RNAV/TCAS/Transponder/altitude encoding capability.

The suffix /I means the aircraft is equipped with LORAN, VOR/DME, or INS (RNAV), and transponder with Mode C (altitude encoding capability). (PLT053) — AIM ¶5-1-8

Answers

9657 [C] 9664 [B] 9665 [B]

121, 135, DSP

9676. (Refer to Figure 214.) In block 3 of the flight plan, the G following MD90/ indicates the aircraft is equipped with

A—GPS/GNSS that has oceanic, en route, and terminal capability.
B—Traffic Alert and Collision Avoidance System (TCAS) with /R capability.
C—Electronic Flight Instrument System (EFIS).

The /G indicates Global Positioning System (GPS)/ Global Navigation Satellite System (GNSS) oceanic, en route, and terminal capability. (PLT053) — AIM ¶5-1-8

121, 135, DSP

9677. (Refer to Figures 182, 214, 216, 216A, 217, and 218.) The time enroute between BDL and PHL for TNA 90 is

A—54 minutes.
B—52 minutes.
C—50 minutes.

To answer this question, complete the flight log in FAA Figure 214, using the information given in the problem:

1. *Change the winds aloft at FL330 from true to magnetic, using the variation from the Airport/Facility Directory entries for the BDL and PHL VORTACs. Winds at SHERL are 340° True at 55 knots.*

   340° True
   + 14° West variation
   354° Magnetic

2. *Determine the TAS:*

   The TAS is 440 knots (given in the flight plan).

3. *Find the distance flown for the leg. In this case it is 30 NM. FAA Figure 216 shows a distance of 30 NM from YODER to CCC on the COASTAL 1 Departure. A separate leg should be computed for each segment between turns.*

4. *Find the ground speed for the leg from YODER to CCC using a flight computer:*

   Wind direction 354° (calculated in Step 1).
   Wind speed 55 knots (given in the flight log).
   Course 237° (FAA Figure 216).
   TAS 440 knots (Step 2).

   Calculated GS is 462.2 knots.

5. *Compute the time en route for the leg from YODER to CCC:*

   Distance 30 NM (determined in Step 3).
   Ground speed 462.2 (calculated in Step 4).

   Calculated leg time is 3 minutes, 54 seconds.

6. *Calculate the fuel burned on this leg:*

   ETE is :03:54 (calculated in Step 5).
   Fuel flow is 6,150 PPH (in note at the bottom of the flight log).

   Fuel burned is 399.2 pounds.

7. *Repeat steps for the subsequent legs to fill in the flight log:*

FROM	*TO*	*CRS*	*TAS*	*GS*	*NM*	*ETE*	*FUEL*
BDL	*YODER*	—	—	—	*45*	*:15:00**	*2,560.0**
YODER	*CCC*	*237*	*440*	*462.2*	*30*	*:03:54*	*399.2*
CCC	*SHERL*	*213*	*440*	*481.4*	*43*	*:05:22*	*549.4*
SHERL	*MANTA*	*236*	*440*	*461.3*	*29*	*:03:46*	*386.6*
MANTA	*BRIGS*	*239*	*440*	*458.6*	*36*	*:04:43*	*482.8*
BRIGS	*VCN*	*281*	*440*	*417.3*	*38*	*:05:28*	*560.1*
VCN	*PHL*	—	—	—	*46*	*:14:00**	*1190.0**
						:52:13	*6,128.1*

**(given)*

(PLT012) — FAA-H-8083-15

121, 135, DSP

9678. (Refer to Figures 214, 216, 216A, 217, and 218.) The total fuel required to be onboard TNA 90 before starting to taxi at BDL is

A—11,979 pounds.
B—11,735 pounds.
C—11,851 pounds

In this question, the regulation-required fuel must be calculated using the summary box at the bottom of the flight plan.

1. *Compute the fuel required to fly the flight:*

   Fuel required is 6,128.1 pounds (see flight plan).

2. *Compute the required reserve fuel (see Step 5):*

   Time required is 45 minutes (§121.639).
   Fuel flow is 5,900 PPH (given at bottom of flight plan in FAA Figure 202).

   Reserve fuel is 4,425 pounds.

3. *Determine ETE to alternate:*

   ETE to alternate is 9 minutes.

Continued

Answers

9676 [A] 9677 [B] 9678 [B]

4. Compute fuel burn to alternate (see Step 5):

Time to alternate is :12:00 (computed in previous step).

Fuel flow is 5,900 PPH (given in flight log).

Fuel burn to alternate is 1,180 pounds.

5. Add the totals of Steps 1 through 4:

FUEL	
6,128.1	*En route*
4,425.0	*Reserve*
+ 1,180.0	*Alternate*
11,733.1	*Total*

(PLT012) — FAA-H-8083-15

121, 135, DSP

9679. (Refer to Figures 214, 216, 216A, 217, and 218.) The estimated fuel usage between BDL and PHL for TNA 90 is

A—10,555 pounds.
B—10,799 pounds.
C—6,130 pounds.

See Steps 1 through 7 of Question 9677. Since the question asks only for the fuel usage, it is not necessary to compute the reserve or alternate fuel requirements. (PLT012) — FAA-H-8083-15

121, 135, DSP

8795. (Refer to Figure 103.) What CAS should be used to maintain the fixed TAS at the proposed altitude?

A—157 knots.
B—167 knots.
C—172 knots.

The required CAS (Calibrated Airspeed) can be computed given TAS, pressure altitude, and OAT. The filed TAS from the flight plan in FAA Figure 103 is 233 knots. The OAT is -31°C, and the pressure altitude is 22,000 feet. Using a flight computer, the CAS is 167 knots. (PLT012) — FAA-H-8083-15

121, 135, DSP

8835. (Refer to Figures 115, 116, and 117.) Due to traffic, LAX Center radar vectored PTL 130 to TRM, then cleared the flight to PHX via J169 BLH, direct to Arlin Intersection. What approximate indicated Mach should be maintained to arrive over the BLH VORTAC 8 minutes after passing TRM VORTAC?

A—.84 Mach.
B—.82 Mach.
C—.86 Mach.

1. Compute the ground speed:

Distance between TRM VORTAC and BLH VORTAC is 70 NM (from FAA Figure 117).
Time enroute is 8 minutes (given in question).

Therefore, ground speed is 525 knots.

2. Compute the required true airspeed:

Wind is 285° (300° – 15° variation) at 43 knots (from FAA Figure 115).
Magnetic course is 078° (from FAA Figure 117).
Ground speed is 525 knots (from Step 1).

Therefore, required TAS is 487.1 knots.

3. Compute required Mach number:

TAS is 487.1 knots (from Step 2).
OAT is -41°C (ISA at FL270 is -2° = -41°C, from FAA Figure 115).

Therefore, required Mach number is .82.

(PLT012) — FAA-H-8083-15

RTC

9566. (Refer to Figures 108, 109, 110, and 112.) What is the ETE from DFW to landing at IAH?

A—1 hour 55 minutes.
B—1 hour 59 minutes.
C—2 hours 03 minutes.

To answer this question, complete the flight log in FAA Figure 108, using the information given in the problem:

1. Change the winds aloft at 7,000 from true to magnetic, using the variation from the Airport/Facility Directory. Winds at BILEE are 220° True at 36 knots.

220°	*True*
– 08°	*East variation*
212°	*Magnetic*

2. Find the distance flown for the leg. In this case it is 84 NM. The sum of the distances from DFW to BILEE is 107 NM, and 23 NM of that is used by the climb. Distances for subsequent legs must be determined

---

Answers

9679 [C]	8795 [B]	8835 [B]	9566 [B]

by referring to the enroute charts and the CUGAR Four Arrival (FAA Figures 110 and 112).

3. *Find the ground speed for the leg from L/O to BILEE using a flight computer:*

   Wind direction 212° (calculated in Step 1).
   Wind speed 36 knots (given on the flight log).
   True course 154° (V369 is on the DFW 154° radial).
   TAS 115 knots (given in the flight plan on FAA Figure 108).

   Calculated GS is 91.8 knots.

4. *Compute the time en route for the leg from L/O to BILEE:*

   Distance 84 NM (determined in Step 2).
   Ground speed 91.8 (calculated in Step 3).

   Calculated leg time is 54 minutes, 54 seconds.

5. *Since the next question asks for the fuel required for the flight, calculate the fuel burned on this leg:*

   ETE is :54:54 (calculated in Step 4).
   Fuel flow is 165 PPH (in note at the bottom of the flight log).

   Fuel burned is 151.0 pounds.

6. *Repeat steps for the subsequent legs to fill in the flight log:*

FROM	***TO***	***CRS***	***TAS***	***GS***	***NM***	***ETE***	***FUEL***
DFW	*L/O*	*—*	*—*	*—*	*23*	*:14:00**	*123.0**
L/O	*BILEE*	*154*	*115*	*91.8*	*84*	*:54:54*	*151.0*
BILEE	*T/P**	*154*	*115*	*91.8*	*34*	*:22:13*	*61.1*
*T/P***	*CUGAR*	*125*	*115*	*107.4*	*13*	*:07:16*	*20.0*
CUGAR	*T/D*	*125*	*115*	*107.4*	*9*	*:05:02*	*13.8*
T/D	*IAH*	*—*	*—*	*—*	*37*	*:16:00**	*140.0**
						1:59:25	*508.9*

**(given)*

**Note: *Split the leg from BILEE to CUGAR into two legs at 34 NM SE of BILEE. The 29° turn will affect the GS and the ETE and leg fuel will change as well.*

(PLT012) — FAA-H-8083-15

RTC

9567. (Refer to Figures 108, 109, 110, and 112.) What is the total fuel required for the flight from DFW to IAH?

A—693 pounds.
B—595 pounds.
C—638 pounds.

In this question, the regulation-required fuel must be calculated using the summary box at the bottom of the flight plan.

1. *Compute the fuel required to fly the flight:*

   Fuel required is 508.9 pounds (see flight plan in Question 9566).

2. *Compute the required reserve fuel (see Question 9566, Step 5):*

   Time required is 30 minutes (§135.223).
   Fuel flow is 172 PPH (given at bottom of flight plan in FAA Figure 108).

   Reserve fuel is 86 pounds.

3. *Determine ETE to alternate:*

   ETE to alternate is 15 minutes (given in the flight log).

4. *Compute fuel burn to alternate (see Question 9566, Step 5):*

   Time to alternate is :15:00 (from previous step).
   Fuel flow is 172 PPH (given in flight log).

   Fuel burn to alternate is 43.0 pounds.

5. *Add any additional fuel required for a missed approach at the destination. In this case add 55 pounds (given at the bottom of the flight log).*

6. *Add the totals of Steps 1 through 5.*

FUEL	
508.9	*En route*
86.0	*Reserve*
+ 98.0	*Alternate (43.0 + 55)*
692.9	*Total*

(PLT012) — FAA-H-8083-15

Answers

9567 [A]

RTC

9568. (Refer to Figures 108, 109, 110, and 112.) What TAS should be maintained to arrive at Cugar Four Arrival initial point 1 hour 6 minutes after level-off?

A—140 knots.
B—143 knots.
C—146 knots.

Compute the distance and enroute time from the flight log used in Question 9567. Using the time and distance, compute the average ground speed at 115 TAS. The difference between the ground speed 115 knots and the required ground speed to travel the computed distance in 1 hour and 6 minutes will equal the change in TAS needed:

FROM	TO	GS	DISTANCE	TIME
L/O	BILEE	91.8	84	:54:54
BILEE	X	91.8	34	:22:13
x	CUGAR	107.4	13	:07:16
			131	1:24:23

131 / 1:06:00 = 119.1 (required)
131 / 1:24:23 = – 93.1 (at 115 TAS)
26
+ 115.0 (TAS)
141 (required)

(PLT012) — FAA-H-8083-15

RTC

9574. (Refer to Figures 113 and 114.) What is the ETE for the IFR helicopter flight from Baker Airport to LAX?

A—1 hour 32 minutes.
B—1 hour 35 minutes.
C—1 hour 38 minutes.

To answer this question, complete the flight log in FAA Figure 113, using the information given in the problem:

1. *Change the winds aloft at 12,000 from true to magnetic, using the variation from the Airport/Facility Directory entries for Baker and LAX. Winds at DAG are 290° True at 36 knots.*

   290° True
   – 15° East variation
   275° Magnetic

2. *Find the distance flown for the leg. In this case it is 15 NM, given in the flight log.*
3. *Find the ground speed for the leg from L/O to DAG using a flight computer:*

   Wind direction 275° (calculated in Step 1).
   Wind speed 36 knots (given on the flight log).
   Course 211° (V394).
   TAS 110 knots (given in the flight plan on FAA Figure 113).

   Calculated GS is 89.4 knots.
4. *Compute the time en route for the leg from L/O to DAG:*

   Distance 15 NM (determined in Step 2).
   Ground speed 89.4 (calculated in Step 3).

   Calculated leg time is 10 minutes, 4 seconds.
5. *Since the next question asks for the fuel required for the flight, calculate the fuel burned on this leg:*

   ETE is :10:04 (calculated in Step 4).
   Fuel flow is 1,045 PPH (in note at the bottom of the flight log).

   Fuel burned is 175.4 pounds.
6. *Repeat steps for the subsequent legs to fill in the flight log:*

FROM	TO	CRS	TAS	GS	NM	ETE	FUEL
O02	L/O	—	—	—	8	:10:00*	250.0*
L/O	DAG	211	110	89.4	15	:10:04	175.4
DAG	POM	214	110	87.9	80	:54:35	950.6
POM	PIRRO	225	110	83.3	5	:03:36	62.7
PIRRO	LAX	—	—	—	29	:17:00*	348.0*
						1:35:15	1,786.7

**(given)*

(PLT012) — FAA-H-8083-15

RTC

9575. (Refer to Figures 113 and 114.) What is the total fuel required for the IFR helicopter flight from Baker Airport to LAX, with LGB as an alternate?

A—2,625 pounds.
B—2,536 pounds.
C—2,335 pounds.

In this question, the regulation-required fuel must be calculated using the summary box at the bottom of the flight plan.

1. *Compute the fuel required to fly the flight:*

   Fuel required is 1,786.7 pounds (see flight plan in Question 9574).

Answers

9568 [A] 9574 [B] 9575 [A]

2. *Compute the required reserve fuel (see Question 9574, Step 5):*

   Time required is 30 minutes (§135.223).
   Fuel flow is 1,095 PPH (given at bottom of flight plan in FAA Figure 113).

   Reserve fuel is 547.5 pounds.

3. *Determine ETE to alternate:*

   ETE to alternate is 11 minutes (given in the flight log).

4. *Compute fuel burn to alternate (see Question 9574, Step 5):*

   Time to alternate is :11:00 (from Step 3).
   Fuel flow is 1,095 PPH (given in flight log).

   Fuel burn to alternate is 200.7 pounds.

5. *Add any additional fuel required for a missed approach at the destination. In this case add 89 pounds (given at the bottom of the flight log).*

6. *Add the totals of Steps 1 through 5:*

FUEL	
1,786.7	*En route*
547.5	*Reserve*
+ 289.7	*Alternate (200.7 + 89)*
2,623.9	*Total*

(PLT012) — FAA-H-8083-15

RTC

9621. (Refer to Figures 174, 175, 176, 177, 177A, and 178.) The estimated time enroute from LWS to HQM, for Sea Hawk 1 is

A—2 hours 31 minutes.
B—2 hours 34 minutes.
C—2 hours 37 minutes.

To answer this question, complete the flight log in FAA Figure 174, using the information given in the problem:

1. *Change the winds aloft at 12,000 feet from true to magnetic, using the variation from the Airport/Facility Directory entries for the LWS and HQM VORTACs. Winds at ALW are 340° True at 40 knots.*

   340° True
   – 20° East variation
   320° Magnetic

2. *Calculate the TAS for existing temperature:*

   ISA temperature at 12,000 feet is -9°C ((12 x (-2°)) + 15°).
   ISA -8°= -17°C

   Using a flight calculator, determine that a CAS of 132 knots gives a TAS of 155.6. Note that the temperature and altitude changes many times in flight log.

3. *Find the distance flown for the leg. In this case it is 43 NM. FAA Figure 175 shows a distance of 43 NM from CLOVA Int. to ALW. The flight log shows the first cruise leg to be from MQG to ALW, but this an apparent misprint. The mileage shown in the climb leg agrees with the SID distance to CLOVA (FAA Figure 177A). The time and fuel calculations only work out if CLOVA is substituted for MQG in the flight log.*

4. *Find the ground speed for the leg from CLOVA to ALW using a flight computer:*

   Wind direction 320° (calculated in Step 1).
   Wind speed 40 knots (given on the flight log).
   Course 234° (V520 is the MQG 234 radial).
   TAS 155.6 knots (calculated in Step 2).

   Calculated GS is 147.6 knots.

5. *Compute the time en route for the leg from CLOVA to ALW:*

   Distance 43 NM (determined in Step 3).
   Ground speed 147.6 (calculated in Step 4).

   Calculated leg time is 17 minutes, 29 seconds.

6. *Since the next question asks for the fuel required for the flight, calculate the fuel burned on this leg:*

   ETE is :17:29 (calculated in Step 5).
   Fuel flow is 525 PPH (in note at the bottom of the flight log).

   Fuel burned is 152.9 pounds.

7. *Repeat steps for the subsequent legs to fill in the flight log:*

FROM	***TO***	***CRS***	***TAS***	***GS***	***NM***	***ETE***	***FUEL***
LWS	*CLOVA*	—	—	—	*15*	*:15:00**	*191.0**
CLOVA	*ALW*	*234*	*155.6*	*147.6*	*43*	*:17:29*	*152.9*
ALW	*PSC*	*267*	*157.1*	*129.9*	*36*	*:16:39*	*145.6*
PSC	*YKM*	*268*	*157.1*	*126.4*	*58*	*:27:33*	*241.0*
YKM	*SEA*	*283*	*158.3*	*129.4*	*93***	*:43:08*	*377.4*
SEA	*ULESS*	*228*	*139.9*	*134.0*	*64*	*:28:40*	*250.8*
ULESS	*HQ*	—	—	—	—	*:02:00**	*16.0**
HQ	*HQM*	—	—	—	—	*:04:00**	*32.0**
						2:34:29	*1,406.7*

**(given)*

***Caution: the 78 NM shown in the box on V4 is the distance from SEA to the compulsory reporting point TITON. It is another 15 NM from TITON to YKM.*

(PLT012) — FAA-H-8083-15

Answers

9621 [B]

RTC

9622. (Refer to Figures 174, 175, 176, 177, 177A, and 178.) The fuel that is required to be on the aircraft prior to taxi for the flight plan of Sea Hawk 1 is

A—1,656 pounds.
B—1,788 pounds.
C—1,828 pounds.

In this question, the regulation-required fuel must be calculated using the summary box at the bottom of the flight plan.

1. *Compute the fuel required to fly the flight:*

   Fuel required is 1,406.1 pounds (see flight plan in Question 9621).

2. *Compute the required reserve fuel (see Question 9621, Step 5):*

   Time required is 30 minutes (§135.223).
   Fuel flow is 499 PPH (given at bottom of flight plan in FAA Figure 113).

   Reserve fuel is 249.5 pounds.

3. *Determine ETE to alternate:*

   GS is 158 knots and distance is 42 NM (both given in the flight log).

   ETE to alternate is 15 minutes, 57 seconds.

4. *Compute fuel burn to alternate (see Question 9621, Step 5):*

   Time to alternate is :15:57 (from Step 3).
   Fuel flow is 499 PPH (given in flight log).

   Fuel burn to alternate is 132.6 pounds.

5. *Add any additional fuel required for a missed approach at the destination. In this case add 40 pounds (given at the bottom of the flight log).*

6. *Add the totals of Steps 1 through 5:*

FUEL	
1406.7	*En route*
249.5	*Reserve*
+ 172.6	*Alternate (132.6 + 40)*
1,828.8	*Total*

(PLT012) — FAA-H-8083-15

RTC

9671. (Refer to Figures 208, 210, 211, and 212.) The estimated time enroute from EYW to DAB (descent time is 9 minutes and 8 seconds) for Sling 2 is

A—2 hours 7 minutes.
B—2 hours 9 minutes.
C—2 hours 11 minutes.

To answer this question, complete the flight log in FAA Figure 208, using the information given in the problem:

1. *Change the winds aloft at 8,000 feet from true to magnetic, using the variation from the Airport/Facility Directory entries for the EYW and DAB VORTACs. Winds at MIA are 220° True at 23 knots.*

220°	*True*
+ 01°	*West variation*
221°	*Magnetic*

2. *Calculate the TAS for existing temperature:*

   ISA temperature at 9,000 feet is -3°C ((9 x (-2°)) + 15°).
   ISA +10° = 7°C. Notice the wind and temperature change on each leg.

   Using a flight calculator, determine that a CAS of 120 knots gives a TAS of 139.7.

3. *Find the distance flown for the leg. In this case it is 82 NM. FAA Figure 210 shows a distance of 82 NM from TIGAR to MIA.*

4. *Find the ground speed for the leg from TIGAR to MIA using a flight computer:*

   Wind direction 221° (calculated in Step 1).
   Wind speed 23 knots (given on the flight log).
   Course 041° (V157).
   TAS 139.7 knots (calculated in Step 2).

   Calculated GS is 162.7 knots.

5. *Compute the time en route for the leg from TIGAR to MIA:*

   Distance 82 NM (determined in Step 3).
   Ground speed 139.7 (calculated in Step 4).

   Calculated leg time is 30 minutes, 14 seconds.

6. *Since the next question asks for the fuel required for the flight, calculate the fuel burned on this leg:*

   ETE is :30:14 (calculated in Step 5).
   Fuel flow is 7,405 PPH (in note at the bottom of the flight log).

   Fuel burned is 372.9 pounds.

7. *Repeat steps for the subsequent legs to fill in the flight log. Calculate the MIA to PHK leg in two segments, to account for the turn in the airway.*

Answers

9622 [C] 9671 [A]

FROM	*TO*	*CRS*	*TAS*	*GS*	*NM*	*ETE*	*FUEL*
EYW	*TIGAR*	*—*	*—*	*—*	*28*	*:14:00**	*205.4**
TIGAR	*MIA*	*041*	*139.7*	*162.7*	*82*	*:30:14*	*372.9*
MIA	*T/P*	*337*	*139.7*	*153.0*	*27*	*:10:35*	*130.6*
T/P	*PHK*	*255*	*139.7*	*156.7*	*25*	*:09:34*	*118.1*
PHK	*MLB*	*003*	*137.5*	*154.4*	*79*	*:30:42*	*378.5*
MLB	*SMYRA*	*342*	*137.5*	*153.9*	*57*	*:22:13*	*274.0*
SMYRA	*DAB*	*—*	*—*	*—*	*16*	*:09:08**	*139.3**
						2:06:26	*1,618.8*

**(given)*

(PLT012) — FAA-H-8083-15

RTC

9672. (Refer to Figures 208, 210, 211, and 212.) The estimated fuel usage from EYW to DAB for Sling 2 is

A—2,273 pounds.
B—2,223 pounds.
C—1,623 pounds.

Refer to Steps 1 through 7 in Question 9671. Since the question only asks for fuel usage, it is not necessary to compute reserve or alternate fuel. (PLT012) — FAA-H-8083-15

RTC

9673. (Refer to Figures 208, 210, 211, and 212.) The total fuel requirement for this IFR flight from EYW to DAB, for Sling 2 is

A—1,623 pounds.
B—2,226 pounds.
C—2,277 pounds.

In this question, the regulation-required fuel must be calculated using the summary box at the bottom of the flight plan.

1. *Compute the fuel required to fly the flight:*

   Fuel required is 1,618.8 pounds (see flight plan in Question 9671).

2. *Compute the required reserve fuel (see Question 9671, Step 5):*

   Time required is 30 minutes (§135.223).
   Fuel flow is 705 PPH (given at bottom of flight plan in FAA Figure 113).

   Reserve fuel is 352.5 pounds.

3. *Determine ETE to alternate:*

   ETE to alternate is 21 minutes.

4. *Compute fuel burn to alternate (see Question 9671, Step 5):*

   Time to alternate is :21:00 (from Step 3).
   Fuel flow is 705 PPH (given in flight log).

   Fuel burn to alternate is 246.7 pounds.

5. *Add any additional fuel required for a missed approach at the destination. In this case add 51 pounds (given at the bottom of the flight log).*

6. *Add the totals of Steps 1 through 5:*

FUEL	
1,618.8	*En route*
352.5	*Reserve*
+ 297.7	*Alternate (246.7 + 51)*
2,269.0	*Total*

(PLT012) — FAA-H-8083-15

RTC

9607. (Refer to Figures 162, 163, 163A, 164, 165, 166A and 167.) What is the ETE for the IFR helicopter flight from Tucson/Ryan to Albuquerque/Double Eagle II?

A—2 hours 14 minutes.
B—2 hours 16 minutes.
C—2 hours 18 minutes.

To answer this question, complete the flight log in FAA Figure 162, using the information given in the problem:

1. *Change the winds aloft at 11,000 feet from true to magnetic, using the variation from the Airport/Facility Directory entries for the TUS and ABQ VORTACs. Winds at CIE are 240° True at 31 knots.*

   240° True
   – 12° East variation
   228° Magnetic

2. *Calculate the TAS for existing temperature:*

   ISA temperature at 11,000 feet is -7°C ((11 x (-2°)) + 15°).
   ISA -6° = -13°C

   Using a flight calculator, determine that a CAS of 125 knots gives a TAS of 145.7. Note that the temperature changes three times in flight log.

3. *Find the distance flown for the leg. In this case it is 38 NM. FAA Figure 164 shows a distance of 38 NM from MESCA Int. to CIE.*

Continued

Answers

9672 [C]	9673 [C]	9607 [B]

4. *Find the ground speed for the leg from MESCA to CIE using a flight computer:*

    Wind direction 228° (calculated in Step 1).
    Wind speed 31 knots (given in the flight log).
    Course 065° (MESCA is on the CIE 245° radial).
    TAS 145.7 knots (calculated in Step 2).

    Calculated GS is 175.1 knots.

5. *Compute the time en route for the leg from MESCA to CIE:*

    Distance 38 NM (determined in Step 3).
    Ground speed 175.1 (calculated in Step 4).

    Calculated leg time is 13 minutes, 1 second.

6. *Since the next question asks for the fuel required for the flight, calculate the fuel burned on this leg:*

    ETE is :13:01 (calculated in Step 5).
    Fuel flow is 523 PPH (in note at the bottom of the flight log).

    Fuel burned is 113.5 pounds.

7. *Repeat steps for the subsequent legs to fill in the flight log:*

FROM	***TO***	***CRS***	***TAS***	***GS***	***NM***	***ETE***	***FUEL***
RYN	*MESCA*	—	—	—	*38*	*:17:00**	*180.0**
MESCA	*CIE*	*065*	*145.7*	*175.1*	*38*	*:13:01*	*113.5*
CIE	*SSO*	*048*	*145.7*	*176.7*	*29*	*:09:51*	*85.8*
SSO	*DMN*	*077*	*147.1*	*172.4*	*84*	*:29:14*	*254.9*
DMN	*TCS*	*003*	*147.1*	*160.9*	*62*	*:23:07*	*201.5*
TCS	*T/P***	*015*	*147.9*	*179.9*	*39*	*:13:00*	*113.4*
*T/P***	*ONM*	*356*	*147.9*	*174.9*	*29*	*:09:57*	*86.7*
ONM	*ABQ*	*347*	*147.9*	*171.2*	*42*	*:14:43*	*128.3*
ABQ	*AEG*	—	—	—	*6*	*:06:00**	*49.0**
						2:15:53	*1,213.1*

**(given)*

**Note: *The leg from TCS to ONM should be split in two (39 NM from TCS) due to the turn in the airway. This results in more accurate time and fuel burn numbers.*

(PLT012) — FAA-H-8083-15

RTC

9608. (Refer to Figures 162, 163, 164, 165, and 167.) What is the total estimated fuel required for this IFR helicopter flight from RYN to AEG with ABQ as an alternate?

A—1,462 pounds.
B—1,503 pounds.
C—1,543 pounds.

In this question, the regulation-required fuel must be calculated using the summary box at the bottom of the flight plan.

1. *Compute the fuel required to fly the flight:*

    Fuel required is 1,213.1 pounds (see flight plan in Question 9607).

2. *Compute the required reserve fuel (see Step 5 in Question 9607):*

    Time required is 30 minutes (§135.223).
    Fuel flow is 1,898 PPH (given at bottom of flight plan in FAA Figure 113).

    Reserve fuel is 497 pounds.

3. *Determine ETE to alternate:*

    ETE to alternate is 5 minutes (given in the flight log).

4. *Compute fuel burn to alternate (see Step 5 in Question 9607):*

    Time to alternate is :05:00 (computed in previous step).
    Fuel flow is 497 PPH (given in flight log).

    Fuel burn to alternate is 41.4 pounds.

5. *Add any additional fuel required for a missed approach at the destination. In this case add 40 pounds (given at the bottom of the flight log).*

6. *Add the totals of Steps 1 through 5:*

FUEL	
1,213.1	*En route*
248.5	*Reserve*
+ 81.4	*Alternate (41.4 + 40)*
1,543.0	*Total*

(PLT012) — FAA-H-8083-15

Answers

9608 [C]

RTC

9639. (Refer to Figures 184, 186, 187, and 188A.) What is the ETE for Hoss 1 from Las Vegas to Provo?

A—1 hour 31 minutes.
B—1 hour 33 minutes.
C—1 hour 35 minutes.

To answer this question, complete the flight log in FAA Figure 184, using the information given in the problem:

1. *Change the winds aloft at 15,000 feet from true to magnetic, using the variation from the Airport/Facility Directory entries for the LAS and PVU VORTACs. Winds at MMM are 210° True at 71 knots.*

   210° True
   – 15° East variation
   195° Magnetic

2. *Calculate the TAS for existing temperature:*

   ISA temperature at 15,000 feet is -15°C ((15 x (-2°)) + 15°).
   ISA -10° = -25°C

   Using a flight calculator, determine that a CAS of 135 knots gives a TAS of 166.1.

3. *Find the distance flown for the leg. In this case it is 43 NM. FAA Figure 186 shows a distance of 43 NM from ACLAM Int. to MMM.*

4. *Find the ground speed for the leg from ACLAM to MMM using a flight computer:*

   Wind direction 195° (calculated in Step 1).
   Wind speed 71 knots (given on the flight log).
   Course 360° (V8 is the MMM 180° radial).
   TAS 166.1 knots (calculated in Step 2).

   Calculated GS is 233.7 knots.

5. *Compute the time en route for the leg from ACLAM to MMM:*

   Distance 43 NM (determined in Step 3).
   Ground speed 233.7 (calculated in Step 4).

   Calculated leg time is 11 minutes, 2 seconds.

6. *Since the next question asks for the fuel required for the flight, calculate the fuel burned on this leg:*

   ETE is :11:02 (calculated in Step 5).
   Fuel flow is 496 PPH (in note at the bottom of the flight log).

   Fuel burned is 91.3 pounds.

7. *Repeat steps for the subsequent legs to fill in the flight log. Although the leg from MMM to REEKA is shown on one line on the flight log, each segment should be calculated separately to get the most accurate time and fuel.*

FROM	*TO*	*CRS*	*TAS*	*GS*	*NM*	*ETE*	*FUEL*
LAS	*ACLAM*	—	—	—	*31*	*:15:00**	*152.0**
ACLAM	*MMM*	*360*	*166.1*	*233.7*	*43*	*:11:02*	*91.3*
MMM	*MLF*	*016*	*166.1*	*237.1*	*113*	*:28:36*	*236.4*
MLF	*DTA*	*007*	*166.1*	*236.1*	*61*	*:15:13*	*125.8*
DTA	*REEKA*	*008*	*166.1*	*236.4*	*50*	*:12:42*	*105.0*
REEKA	*PVU*	—	—	—	*19*	*:10:00**	*87.0**
						1:32:50	*797.5*

**(given)*

(PLT012) — FAA-H-8083-15

RTC

9640. (Refer to Figures 184, 186, 187, 188, and 188A.) What is the minimum fuel required under 14 CFR Part 135 for this IMC helicopter flight from LAS to PVU? The visibility is forecast to be 15 SM over the entire route.

A—1,304 pounds.
B—985 pounds.
C—1,224 pounds.

In this question, the regulation-required fuel must be calculated using the summary box at the bottom of the flight plan.

1. *Compute the fuel required to fly the flight:*

   Fuel required is 797.5 pounds (see flight plan in Question 9639).

2. *Compute the required reserve fuel (see Question 9639, Step 5):*

   Time required is 30 minutes (§135.223).
   Fuel flow is 480 PPH (given at bottom of flight plan in FAA Figure 113).

   Reserve fuel is 240.0 pounds.

3. *Determine ETE to alternate:*

   ETE to alternate is 17 minutes, 20 seconds.

4. *Compute fuel burn to alternate (see Question 9639, Step 5):*

   Time to alternate is :17:20 (from Step 3).
   Fuel flow is 480 PPH (given in flight log).

   Fuel burn to alternate is 138.7 pounds.

5. *Add any additional fuel required for a missed approach at the destination. In this case add 40 pounds (given at the bottom of the flight log).*

6. *Add the totals of Steps 1 through 5:*

FUEL	
7,97.5	*En route*
240.0	*Reserve*
+ 178.7	*Alternate (138.7 + 40)*
1,216.2	*Total*

(PLT012) — FAA-H-8083-15

Answers

9639 [B]   9640 [C]

RTC

9652. (Refer to Figures 197, 199, and 200.) What is the ETE for the IFR helicopter flight from Eagle County Regional to Salt Lake City Intl? (PUC to FFU should read "14,000" for altitude. Use PUC magnetic variation for entire problem.)

A—1 hour 28 minutes.
B—1 hour 35 minutes.
C—1 hour 31 minutes.

To answer this question, complete the flight log in FAA Figure 197, using the information given in the problem:

1. *Change the winds aloft at 14,000 feet from true to magnetic, using the variation for PUC given in the remarks of the flight plan. Winds at JNC are 100° true at 41 knots.*

   100° True
   – 12° East variation
   088° Magnetic

2. *Calculate the TAS for existing temperature:*

   ISA temperature at 14,000 feet is -13°C ((14 x (-2°)) + 15°).
   ISA +20° = 7°C. Notice the wind and temperature change at FFU.

   Using a flight calculator, determine that a CAS of 139 Kts gives a TAS of 178.1.

3. *Find the distance flown for the leg. In this case it is 91 NM. FAA Figure 199 shows a distance of 91 NM from DBL to JNC.*

4. *Find the ground speed for the leg from DBL to JNC using a flight computer:*

   Wind direction 088° (calculated in Step 1).
   Wind speed 41 knots (given in the flight log).
   Course 242° (V134).
   TAS 178.1 knots (calculated in Step 2).

   Calculated GS is 214.5 knots.

5. *Compute the time en route for the leg from DBL to JNC:*

   Distance 91 NM (determined in Step 3).
   Ground speed 214.5 (calculated in Step 4).

   Calculated leg time is 25 minutes, 28 seconds.

6. *Since the next question asks for the fuel required for the flight, calculate the fuel burned on this leg:*

   ETE is :25:28 (calculated in Step 5).
   Fuel flow is 495 PPH (in note at the bottom of the flight log).

   Fuel burned is 210.0 pounds.

7. *Repeat steps for the subsequent legs to fill in the flight log:*

FROM	*TO*	*CRS*	*TAS*	*GS*	*NM*	*ETE*	*FUEL*
EGE	*DBL*	*—*	*—*	*—*	*13*	*:10:00**	*101.0**
DBL	*JNC*	*242*	*178.1*	*214.5*	*91*	*:25:28*	*210.0*
JNC	*PUC*	*275*	*178.1*	*218.5*	*97*	*:26:38*	*219.7*
PUC	*FFU*	*291*	*176.2*	*210.1*	*68*	*:19:25*	*160.2*
FFU	*JAURN*	*341*	*176.2*	*193.5*	*12*	*:03:43*	*30.7*
JAURN	*SLC*	*—*	*—*	*—*	*18*	*:10:00**	*92.0**
						1:35:14	*813.6*

**(given)*

(PLT012) — FAA-H-8083-15

RTC

9653. (Refer to Figures 197, 199, and 200.) What is the total estimated fuel required for this IFR helicopter flight from EGE to SLC with OGD as an alternate?

A—1,152 pounds.
B—1,119 pounds.
C—1,049 pounds.

In this question, the regulation-required fuel must be calculated using the summary box at the bottom of the flight plan.

1. *Compute the fuel required to fly the flight:*

   Fuel required is 813.6 pounds (see flight plan in Question 9652).

2. *Compute the required reserve fuel (see Question 9652, Step 5):*

   Time required is 30 minutes (§135.223).
   Fuel flow is 469 PPH (given at bottom of flight plan in FAA Figure 113).

   Reserve fuel is 234.5 pounds.

3. *Determine ETE to alternate:*

   ETE to alternate is 9 minutes.

4. *Compute fuel burn to alternate (see Question 9652, Step 5):*

   Time to alternate is :09:00 (from Step 3).
   Fuel flow is 469 PPH (given in flight log).

   Fuel burn to alternate is 70.3 pounds.

5. *Add any additional fuel required for a missed approach at the destination. In this case add 33 pounds (given at the bottom of the flight log).*

6. *Add the totals of Steps 1 through 5:*

FUEL	
813.6	*En route*
234.5	*Reserve*
+ 103.3	*Alternate (70.3 + 33)*
1,151.4	*Total*

(PLT012) — FAA-H-8083-15

Answers

9652 [B] 9653 [A]

Chapter 5
Weight and Balance

Center of Gravity Computation

The first step in the solution of any weight and balance problem is the calculation of the total weight of the aircraft (gross weight) and the total moment. All weight and balance problems on the ATP-121 test use a moment index rather than the actual moment. The moment index is the actual moment divided by 1,000. Questions 8697 through 8711 require the calculation of the total weight and moment index for a Boeing 727-type aircraft. To determine the total weight and moment index, a separate weight and moment must be calculated for the Basic Operating Weight, the passenger loads in the forward and aft passenger compartments, the cargo loads in the forward and aft cargo compartments, and the fuel loads in fuel tanks 1, 2, and 3. The following example references Question 8697.

Basic Operating Weight (BOW) is defined as the empty weight of the aircraft plus the weight of the required crew, their baggage and other standard items such as meals and potable water. The BOW and the Basic Operating Index (Moment/1,000) are the same for all questions. The BOW is 105,500 pounds and the Basic Operating Index is 92,837. *See* FAA Figure 79.

The number of passengers is stated for each question. For example, Question 8697 refers to Load Condition WT-1. *(See* FAA Figure 76.) Load Condition WT-1 states that there are 18 passengers in the forward compartment and 95 passengers in the aft compartment. The weight of the passengers can be determined by use of the Passenger Loading Table in the upper left-hand corner of FAA Figure 80. Since neither 18 passengers for the forward compartment nor 95 passengers for the aft compartment is listed in the table, the weight must be calculated by multiplying the number of passengers times the average weight per passenger. A quick examination of the table reveals that the average passenger weight is 170 pounds. The weights are:

FWD Comp = 18 x 170 lbs = 3,060 lbs

AFT Comp = 95 x 170 lbs = 16,150 lbs

The Moment Index (MOM/1,000) is calculated by using the formula:

Weight x Arm/1,000 = MOM/1,000

The arms for the passenger compartments are listed at the top of each of the compartment loading tables after the words, “Forward Compartment Centroid” and “Aft Compartment Centroid.” The arm for the forward compartment is 582.0 inches, and the aft compartment arm is 1028.0 inches. The easiest way to apply the 1,000 reduction factor is to move the decimal on the arm three places to the left (i.e., 582.0"/1,000 = .582). In the example used, the Moment/1,000 for the forward and aft passengers compartments (rounded to the nearest whole number) are:

FWD Comp Moment/1,000 = 3,060 x .582 = 1,781

AFT Comp Moment/1,000 = 16,150 x 1.028 = 16,602

The weights for the forward and aft cargo holds are stated for each question. For example, Load Condition WT-1 states that there is 1,500 pounds in the forward hold and 2,500 pounds in the aft hold. The Moment/1,000 can be determined from the tables in the upper right-hand corner of FAA Figure 80. For example, the Moment/1,000 for 1,500 pounds in the forward cargo hold is determined by adding the Moment/1,000 for 1,000 pounds (680) and the Moment/1,000 for 500 pounds (340). If necessary, the Moment/1,000 can also be determined by multiplying weight times arm (divided by 1,000). The Moment/1,000 for the cargo holds are:

FWD Hold = 1,020

AFT Hold = 2,915

Continued

Fuel tanks 1 and 3 are the wing tanks and are always loaded with the same weight of fuel. They will always have the Moment/1,000 as well. The number 2 tank is the center fuselage tank and will often have a fuel weight different from tanks 1 and 3. It will always have a different Moment/1,000. For example, Load Condition WT-1 states that the fuel load in tanks 1 and 3 is 10,500 pounds each and that the load in tank 2 is 28,000 pounds. The Moment/1,000 for each tank is determined from the table in the bottom portion of FAA Figure 80. The Moment/1,000 can be calculated, if necessary, by multiplying weight times arm (divided by 1,000). Notice that the arm varies with the fuel load in each tank. The Moment/1,000 for each tank is:

Tank 1 Moment/1,000 = 10,451

Tank 3 Moment/1,000 = 10,451

Tank 2 Moment/1,000 = 25,589

The total weight and total Moment/1,000 is the sum of all the items discussed above. The Total Weight and Moment/1,000 for Load Condition WT-1 is:

	Weight	**Moment/1,000**
BOW	105,500	92,837
18 PAX FWD	3,060	1,781
95 PAX AFT	16,150	16,602
FWD Cargo	1,500	1,020
AFT Cargo	2,500	2,915
Fuel Tank 1	10,500	10,451
Fuel Tank 3	10,500	10,451
Fuel Tank 2	+ 28,000	+ 25,589
Total	177,710	161,646

The Center of Gravity (CG) in inches aft of the Datum line can be determined by using the formula:

CG = Total Moment / Total Weight

Since these questions use a Moment Index instead of Moment, it is necessary to modify this formula by multiplying the (Total Moment/Total Weight) by the reduction factor (1,000). The formula then becomes:

CG = (Total Moment Index / Total Weight) x 1,000

Using the weight and Moment/1,000 we calculated above:

CG = (161,646/177,710) x 1,000 = 909.6 inches

The Center of Gravity of a properly loaded airplane must always fall somewhere along the **Mean Aerodynamic Chord (MAC)**. The CG is often expressed as a percent of MAC. If the CG was at the Leading Edge of MAC (LEMAC), it would be at 0% of MAC. If it were at the Trailing Edge of MAC (TEMAC), it would be at 100% of MAC. *See* Figure 5-1. The CG's percent of MAC is calculated by:

1. Determine the CG in inches aft of LEMAC by subtracting the distance Datum to LEMAC from the CG in inches aft of Datum. The distance from Datum to LEMAC is given in FAA Figure 79 as 860.5 inches. This is used for all calculations of percent of MAC for the 727. The CG in inches aft of Datum is calculated in the previous paragraph. Using those numbers:

   CG (inches aft of LEMAC) = 909.6" – 860.5" = 49.1 inches

2. Determine the CG in percent of MAC by dividing the CG in inches aft of LEMAC by the length of MAC. The length of MAC is distance in inches from LEMAC to TEMAC. It is given in FAA Figure 79 and is 180.9 inches. The formula is:

   CG (% of MAC) = (CG in inches aft of LEMAC ÷ MAC) x 100%

   Using the numbers from above:

   CG (% of MAC) = (49.1" ÷ 180.9") x 100% = 27.1%

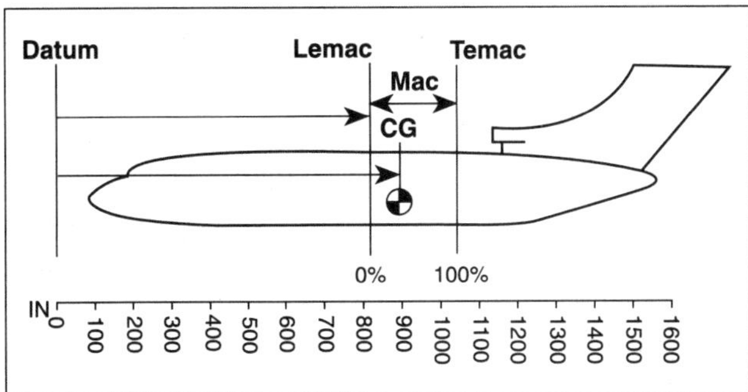

Figure 5-1. Center of gravity terms

121, DSP

8697. (Refer to Figures 76, 79, and 80.) What is the CG in percent of MAC for Loading Conditions WT 1?

A—26.0 percent MAC.
B—27.1 percent MAC.
C—27.9 percent MAC.

Weight and Moment/1,000 for Load Condition WT-1 is:

	Weight	***Moment/1,000***
BOW	*105,500*	*92,837*
18 PAX FWD	*3,060*	*1,781*
95 PAX AFT	*16,150*	*16,602*
FWD Cargo	*1,500*	*1,020*
AFT Cargo	*2,500*	*2,915*
Fuel Tank 1	*10,500*	*10,451*
Fuel Tank 3	*10,500*	*10,451*
Fuel Tank 2	*+ 28,000*	*+ 25,589*
Total	*177,710*	*161,646*

(Total Moment/1,000)/Total Weight x 1,000 = CG inches
(161,646/177,710) x 1,000 = 909.6 inches

(CG inches – LEMAC)/MAC = % MAC
(909.6" – 860.5")/180.9 = 27.1% MAC

(PLT121) — FAA-H-8083-1

121, DSP

8698. (Refer to Figures 76, 79, and 80.) What is the CG in inches aft of datum for Loading Conditions WT-2?

A—908.8 inches.
B—909.6 inches.
C—910.7 inches.

	Weight	***Moment/1,000***
BOW	*105,500*	*92,837*
23 PAX FWD	*3,910*	*2,276*
112 PAX AFT	*19,040*	*19,573*
FWD Cargo	*2,500*	*1,700*
AFT Cargo	*3,500*	*4,081*
Fuel Tank 1	*11,000*	*10,957*
Fuel Tank 3	*11,000*	*10,957*
Fuel Tank 2	*+ 27,000*	*+ 24,678*
Total	*183,450*	*167,059*

(Total Moment/1,000)/Total Weight x 1,000 = CG inches
(167,059/183,450) x 1,000 = 910.7 inches

(PLT121) — FAA-H-8083-1

121, DSP

8699. (Refer to Figures 76, 79, and 80.) What is the CG in percent of MAC for Loading Conditions WT 3?

A—27.8 percent MAC.
B—28.9 percent MAC.
C—29.1 percent MAC.

	Weight	***Moment/1,000***
BOW	*105,500*	*92,837*
12 PAX FWD	*2,040*	*1,187*
75 PAX AFT	*12,750*	*13,107*
FWD Cargo	*3,500*	*2,380*
AFT Cargo	*4,200*	*4,897*
Fuel Tank 1	*12,000*	*11,970*
Fuel Tank 3	*12,000*	*11,970*
Fuel Tank 2	*+ 24,250*	*+ 22,172*
Total	*176,240*	*160,520*

Continued

Answers

8697 [B] 8698 [C] 8699 [A]

(Total Moment/1,000)/Total Weight x 1,000 = CG inches
(176,240/160,520) x 1,000 = 910.8 inches

(CG inches – LEMAC)/MAC = % MAC
(910.8 – 860.5)/180.9 = 27.8%

(PLT121) — FAA-H-8083-1

121, DSP
8700. (Refer to Figures 76, 79, and 80.) What is the CG in inches aft of datum for Loading Conditions WT-4?

A—908.4 inches.
B—909.0 inches.
C—909.5 inches.

	Weight	***Moment/1,000***
BOW	*105,500*	*92,837*
28 PAX FWD	*4,760*	*2,770*
122 PAX AF	*20,740*	*21,321*
FWD Cargo	*850*	*578*
AFT Cargo	*1,500*	*1,749*
Fuel Tank 1	*10,000*	*9,947*
Fuel Tank 3	*10,000*	*9,947*
Fuel Tank 2	*+ 26,200*	*+ 23,952*
Total	*179,550*	*163,101*

(Total Moment/1,000)/Total Weight x 1,000 = CG inches
(163,101/179,550) x 1,000 = 908.4 inches

(PLT121) — FAA-H-8083-1

121, DSP
8701. (Refer to Figures 76, 79, and 80.) What is the CG in percent of MAC for Loading Conditions WT 5?

A—25.6 percent MAC.
B—26.7 percent MAC.
C—27.2 percent MAC.

	Weight	***Moment/1,000***
BOW	*105,500*	*92,837*
26 PAX FWD	*4,420*	*2,572*
103 PAX AFT	*17,510*	*18,000*
FWD Cargo	*1,400*	*952*
AFT Cargo	*2,200*	*2,565*
Fuel Tank 1	*11,500*	*11,463*
Fuel Tank 3	*11,500*	*11,463*
Fuel Tank 2	*+ 25,200*	*+ 23,038*
Total	*179,230*	*162,890*

(Total Moment/1,000)/Total Weight x 1,000 = CG inches
(162,890/179,230) x 1,000 = 908.8 inches

(CG inches – LEMAC)/MAC = % MAC
(908.8 – 860.5)/180.9 = 26.7%

(PLT121) — FAA-H-8083-1

121, DSP
8702. (Refer to Figures 77, 79, and 80.) What is the gross weight index for Loading Conditions WT-6?

A—181,340.5 index.
B—156,545.0 index.
C—165,991.5 index.

	Weight	***Moment/1,000***
BOW	*105,500*	*92,837.0*
10 PAX FWD	*1,700*	*989.4*
132 PAX AFT	*22,440*	*23,068.3*
FWD Cargo	*5,000*	*3,400.0*
AFT Cargo	*6,000*	*6,966.0*
Fuel Tank 1	*9,500*	*9,442.0*
Fuel Tank 3	*9,500*	*9,442.0*
Fuel Tank 2	*+ 21,700*	*+ 19,846.8*
Total	*181,340*	*165,991.5*

(PLT021) — FAA-H-8083-1

121, DSP
8703. (Refer to Figures 77, 79, and 80.) What is the CG in percent of MAC for Loading Conditions WT 7?

A—21.6 percent MAC.
B—22.9 percent MAC.
C—24.0 percent MAC.

	Weight	***Moment/1,000***
BOW	*105,500*	*92,837*
27 PAX FWD	*4,590*	*2,671*
83 PAX AFT	*14,110*	*14,505*
FWD Cargo	*4,500*	*3,060*
AFT Cargo	*5,500*	*6,413*
Fuel Tank 1	*9,000*	*8,937*
Fuel Tank 3	*9,000*	*8,937*
Fuel Tank 2	*+ 19,800*	*+ 18,115*
Total	*172,000*	*155,475*

(Total Moment/1,000)/Total Weight x 1,000 = CG inches
(155,475/172,000) x 1,000 = 903.9 inches

(CG inches – LEMAC)/MAC = % MAC
(903.9 – 860.5)/180.9 = 24.0%

(PLT121) — FAA-H-8083-1

Answers

8700 [A] 8701 [B] 8702 [C] 8703 [C]

121, DSP

8704. (Refer to Figures 77, 79, and 80.) What is the CG in percent of MAC for Loading Conditions WT 8?

A—29.4 percent MAC.
B—30.0 percent MAC.
C—31.3 percent MAC.

	Weight	***Moment/1,000***
BOW	*105,500*	*92,837*
6 PAX FWD	*1,020*	*594*
98 PAX AFT	*16,660*	*17,126*
FWD Cargo	*1,300*	*884*
AFT Cargo	*3,300*	*3,848*
Fuel Tank 1	*12,000*	*11,970*
Fuel Tank 3	*12,000*	*11,970*
Fuel Tank 2	*+ 12,000*	*+ 10,993*
Total	*163,780*	*150,222*

(Total Moment/1,000)/Total Weight x 1,000 = CG inches
(150,222/163,780) x 1,000 = 917.2 inches

(CG inches – LEMAC)/MAC = % MAC
(917.2 – 860.5)/180.9 = 31.4%

(PLT121) — FAA-H-8083-1

121, DSP

8705. (Refer to Figures 77, 79, and 80.) What is the gross weight index for Loading Conditions WT-9?

A—169,755.2 index.
B—158,797.9 index.
C—186,565.5 index.

	Weight	***Moment/1,000***
BOW	*105,500*	*92,837.0*
29 PAX FWD	*4,930*	*2,869.3*
133 PAX AFT	*22,610*	*23,243.1*
FWD Cargo	*975*	*663.0*
AFT Cargo	*1,250*	*1,457.5*
Fuel Tank 1	*11,000*	*10,957.0*
Fuel Tank 3	*11,000*	*10,957.0*
Fuel Tank 2	*+ 29,300*	*+ 26,771.4*
Total	*186,565*	*169,755.3*

(PLT021) — FAA-H-8083-1

121, DSP

8706. (Refer to Figures 77, 79, and 80.) What is the CG in percent of MAC for Loading Conditions WT 10?

A—27.0 percent MAC.
B—27.8 percent MAC.
C—28.0 percent MAC.

	Weight	***Moment/1,000***
BOW	*105,500*	*92,837*
21 PAX FWD	*3,570*	*2,078*
127 PAX AFT	*21,590*	*22,195*
FWD Cargo	*2,300*	*1,564*
AFT Cargo	*2,400*	*2,798*
Fuel Tank 1	*10,500*	*10,451*
Fuel Tank 3	*10,500*	*10,451*
Fuel Tank 2	*+ 22,700*	*+ 20,759*
Total	*179,060*	*163,133*

(Total Moment/1,000)/Total Weight x 1,000 = CG inches
(163,133/179,060) x 1,000 = 911.1 inches

(CG inches – LEMAC)/MAC = % MAC
(911.1 – 860.5)/180.9 = 27.97%

(PLT121) — FAA-H-8083-1

121, DSP

8707. (Refer to Figures 78, 79, and 80.) What is the CG in percent of MAC for Loading Conditions WT 11?

A—26.8 percent MAC.
B—27.5 percent MAC.
C—28.6 percent MAC.

	Weight	***Moment/1,000***
BOW	*105,500*	*92,837*
11 PAX FWD	*1,870*	*1,088*
99 PAX AFT	*16,830*	*17,301*
FWD Cargo	*3,100*	*2,108*
AFT Cargo	*5,500*	*6,413*
Fuel Tank 1	*8,500*	*8,433*
Fuel Tank 3	*8,500*	*8,433*
Fuel Tank 2	*+ 19,600*	*+ 17,932*
Total	*169,400*	*154,545*

(Total Moment/1,000)/Total Weight x 1,000 = CG inches
(154,545/169,400) x 1,000 = 912.3 inches

(CG inches – LEMAC)/MAC = % MAC
(912.3 – 860.5)/180.9 = 28.6%

(PLT121) — FAA-H-8083-1

Answers

8704 [C] 8705 [A] 8706 [C] 8707 [C]

121, DSP

8708. (Refer to Figures 78, 79, and 80.) What is the CG in percent of MAC for Loading Conditions WT 12?

A—25.8 percent MAC.
B—26.3 percent MAC.
C—27.5 percent MAC.

	Weight	Moment/1,000
BOW	105,500	92,837
28 PAX FWD	4,760	2,770
105 PAX AFT	17,850	18,350
FWD Cargo	4,200	2,856
AFT Cargo	4,400	5,130
Fuel Tank 1	11,500	11,463
Fuel Tank 3	11,500	11,463
Fuel Tank 2	+ 27,800	+ 25,406
Total	187,510	170,275

(Total Moment/1,000)/Total Weight x 1,000 = CG inches
(170,275/187,510) x 1,000 = 908.1 inches

(CG inches – LEMAC)/MAC = % MAC
(908.1 – 860.5)/180.9 = 26.3%

(PLT121) — FAA-H-8083-1

121, DSP

8709. (Refer to Figures 78, 79, and 80.) What is the CG in percent of MAC for Loading Conditions WT 13?

A—28.6 percent MAC.
B—29.4 percent MAC.
C—30.1 percent MAC.

	Weight	Moment/1,000
BOW	105,500	92,837
22 PAX FWD	3,740	2,177
76 PAX AFT	12,920	13,282
FWD Cargo	1,600	1,088
AFT Cargo	5,700	6,646
Fuel Tank 1	12,000	11,970
Fuel Tank 3	12,000	11,970
Fuel Tank 2	+ 29,100	+ 26,589
Total	182,560	166,559

(Total Moment/1,000)/Total Weight x 1,000 = CG inches
(166,559/182,560) x 1,000 = 912.4 inches

(CG inches – LEMAC)/MAC = % MAC
(912.4 – 860.5)/180.9 = 28.7%

(PLT021) — FAA-H-8083-1

121, DSP

8710. (Refer to Figures 78, 79, and 80.) What is the CG in percent of MAC for Loading Conditions WT 14?

A—30.1 percent MAC.
B—29.5 percent MAC.
C—31.5 percent MAC.

	Weight	Moment/1,000
BOW	105,500	92,837
17 PAX FWD	2,890	1,682
124 PAX AFT	21,080	21,670
FWD Cargo	3,800	2,584
AFT Cargo	4,800	5,597
Fuel Tank 1	11,000	10,957
Fuel Tank 3	11,000	10,957
Fuel Tank 2	+ 25,400	+ 23,221
Total	185,470	169,505

(Total Moment/1,000)/Total Weight x 1,000 = CG inches
(169,505/185,470) x 1,000 = 913.9 inches

(CG inches – LEMAC)/MAC = % MAC
(913.9 – 860.5)/180.9 = 29.5%

(PLT021) — FAA-H-8083-1

121, DSP

8711. (Refer to Figures 78, 79, and 80.) What is the CG in Percent of MAC for Loading Conditions WT 15?

A—32.8 percent MAC.
B—31.5 percent MAC.
C—29.5 percent MAC.

	Weight	Moment/1,000
BOW	105,500	92,837
3 PAX FWD	510	297
130 PAX AFT	22,100	22,719
FWD Cargo	1,800	1,224
AFT Cargo	3,800	4,431
Fuel Tank 1	10,500	10,451
Fuel Tank 3	10,500	10,451
Fuel Tank 2	+ 21,900	+ 20,030
Total	176,610	162,440

(Total Moment/1,000)/Total Weight x 1,000 = CG inches
(162,440/176,610) x 1,000 = 919.8 inches

(CG inches – LEMAC)/MAC = % MAC
(919.8 – 860.5)/180.9 = 32.8%

(PLT021) — FAA-H-8083-1

Answers

8708 [B] 8709 [A] 8710 [B] 8711 [A]

Stabilizer Trim Setting

The correct horizontal stabilizer trim setting is very critical for proper takeoff performance of jet aircraft. The main determinants are the CG location and possibly the flap setting. Some aircraft, such as the DC-9, have their stabilizer trim indicators calibrated in percent of MAC, so it is necessary to calculate the CG to know the trim setting. Other aircraft (such as the B-737 and B-727) have their trim indicators marked off in units of nose up trim. In such cases it is necessary to refer to the trim table to determine the proper setting for a given CG. *See* FAA Figure 55.

The **Stab Trim Setting** Table at the bottom left side of FAA Figure 55 is used to determine the takeoff trim setting for a B-737. CG location in percent of MAC is used to determine the setting. For example, if the CG is at 8.0% of MAC, the stab trim setting is 7-3/4 units ANU (Airplane Nose Up).

The Stab Trim Setting Table at the left side of FAA Figure 83 is used to determine the takeoff trim setting for a B-727. Flap setting and CG location in percent of MAC are used to determine the setting. For example, if the CG is at 28% of MAC and the flaps are set at 15°, the stab trim setting is 4-1/2 units ANU.

121, DSP

8588. (Refer to Figures 45, 46, and 47.) What is the STAB TRIM setting for Operating Conditions A-1?

A—29 percent MAC.
B—32 percent MAC.
C—36 percent MAC.

Using FAA Figure 45, calculate the CG in inches and % of MAC:

CG (inches aft of LEMAC) = 590.2" – 549.13" = 41.07"
CG (% of MAC) = 41.07"/141.5" = 29.0%

(PLT010) — FAA-H-8083-25

121, DSP

8589. (Refer to Figures 45, 46, and 47.) What is the STAB TRIM setting for Operating Conditions A-2?

A—26 percent MAC.
B—20 percent MAC.
C—22 percent MAC.

Using FAA Figure 45, convert CG from index arm to inches aft of datum:

CG (inches aft of Datum) = 580.0" – 3.1" = 576.9"
CG (inches aft of LEMAC) = 576.9" – 549.13" = 27.77"
CG (% of MAC) = 27.77"/141.5" = 19.6%

(PLT010) — FAA-H-8083-25

121, DSP

8590. (Refer to Figures 45, 46, and 47.) What is the STAB TRIM setting for Operating Conditions A-3?

A—18 percent MAC.
B—20 percent MAC.
C—22 percent MAC.

Using FAA Figure 45 calculate the CG in inches and % of MAC:

CG (inches aft of LEMAC) = 580.3" – 549.13" = 31.17"
CG (% of MAC) = 31.17"/141.5" = 22.0%

(PLT010) — FAA-H-8083-25

121, DSP

8591. (Refer to Figures 45, 46, and 47.) What is the STAB TRIM setting for Operating Conditions A-4?

A—26 percent MAC.
B—22 percent MAC.
C—18 percent MAC.

Using FAA Figure 45, convert CG from index arm to inches aft of datum:

CG (inches aft of Datum) = 580.0" + 5.9" = 585.9"
CG (inches aft of LEMAC) = 585.9" – 549.13" = 36.77"
CG (% of MAC) = 36.77"/141.5" = 26.0%

(PLT010) — FAA-H-8083-25

Answers

8588 [A] 8589 [B] 8590 [C] 8591 [A]

121, DSP

8592. (Refer to Figures 45, 46, and 47.) What is the STAB TRIM setting for Operating Conditions A-5?

A—26 percent MAC.
B—30 percent MAC.
C—32 percent MAC.

Using FAA Figure 45, calculate the CG in inches and % of MAC:

CG (inches aft of LEMAC) = 594.4" – 549.13" = 45.27"
CG (% of MAC) = 45.27"/141.5" = 32.0%

(PLT010) — FAA-H-8083-25

121, DSP

8623. (Refer to Figures 53 and 55.) What is the STAB TRIM setting for Operating Conditions R-1?

A—8 ANU.
B—7-5/8 ANU.
C—7-3/4 ANU.

CG (inches aft of LEMAC) = 635.7" – 625.0" = 10.7"
CG (% of MAC) = 10.7"/134.0" = 8.0%
Stab Trim = 7-3/4 ANU

(PLT010) — FAA-H-8083-25

121, DSP

8624. (Refer to Figures 53 and 55.) What is the STAB TRIM setting for Operating Conditions R-2?

A—5-3/4 ANU.
B—7 ANU.
C—6-3/4 ANU.

CG (inches aft of LEMAC) = 643.8" – 625.0" = 18.8"
CG (% of MAC) = 18.8"/134.0" = 14.0%
Stab Trim = 6-3/4 ANU

(PLT010) — FAA-H-8083-25

121, DSP

8625. (Refer to Figures 53 and 55.) What is the STAB TRIM setting for Operating Conditions R-3?

A—3 ANU.
B—4-1/2 ANU.
C—5 ANU.

CG (inches aft of LEMAC) = 665.2" – 625.0" = 40.2"
CG (% of MAC) = 40.2"/134.0" = 30.0%
Stab Trim = 3 ANU

(PLT010) — FAA-H-8083-25

121, DSP

8626. (Refer to Figures 53 and 55.) What is the STAB TRIM setting for Operating Conditions R-4?

A—4-1/4 ANU.
B—4-1/2 ANU.
C—5 ANU.

CG (inches aft of LEMAC) = 657.2" – 625.0" = 32.2"
CG (% of MAC) = 32.2"/134.0" = 24.0%
Stab Trim = 4-1/2 ANU

(PLT010) — FAA-H-8083-25

121, DSP

8627. (Refer to Figures 53 and 55.) What is the STAB TRIM setting for Operating Conditions R-5?

A—6-3/4 ANU.
B—8 ANU.
C—7-1/2 ANU.

CG (inches aft of LEMAC) = 638.4" – 625.0" = 13.4"
CG (% of MAC) = 13.4"/134.0" = 10.0%
Stab Trim = 7-1/2 ANU

(PLT010) — FAA-H-8083-25

121, DSP

8722. (Refer to Figures 81 and 83.) What is the STAB TRIM setting for Operating Conditions G-1?

A—4 ANU.
B—4-1/2 ANU.
C—4-3/4 ANU.

CG (inches aft of LEMAC) = 911.2" – 860.5" = 50.7"
CG (% of MAC) = 50.7"/180.9" = 28.0%
Flaps = 15°
Stab Trim = 4-1/2 ANU

(PLT010) — FAA-H-8083-25

121, DSP

8723. (Refer to Figures 81 and 83.) What is the STAB TRIM setting for Operating Conditions G-2?

A—6-1/2 ANU.
B—7-1/4 ANU.
C—5-3/4 ANU.

CG (inches aft of LEMAC) = 882.2" – 860.5" = 21.7"
CG (% of MAC) = 21.7"/180.9" = 12.0%
Flaps = 5°
Stab Trim = 6-1/2 ANU

(PLT010) — FAA-H-8083-25

Answers

8592 [C]	8623 [C]	8624 [C]	8625 [A]	8626 [B]	8627 [C]
8722 [B]	8723 [A]				

121, DSP

8724. (Refer to Figures 81 and 83.) What is the STAB TRIM setting for Operating Conditions G-3?

A—3-3/4 ANU.
B—4 ANU.
C—4-1/4 ANU.

CG (inches aft of LEMAC) = 914.8" – 860.5" = 54.3"
CG (% of MAC) = 54.3"/180.9" = 30.0%
Flaps = 25°
Stab Trim = 4-1/4 ANU

(PLT010) — FAA-H-8083-25

121, DSP

8725. (Refer to Figures 81 and 83.) What is the STAB TRIM setting for Operating Conditions G-4?

A—2-3/4 ANU.
B—4 ANU.
C—2-1/2 ANU.

CG (inches aft of LEMAC) = 932.9" – 860.5" = 72.4"
CG (% of MAC) = 72.4"/180.9" = 40.0%
Flaps = 15°
Stab Trim = 2-1/2 ANU

(PLT010) — FAA-H-8083-25

121, DSP

8726. (Refer to Figures 81 and 83.) What is the STAB TRIM setting for Operating Conditions G-5?

A—3-1/4 ANU.
B—2-3/4 ANU.
C—2-1/2 ANU.

CG (inches aft of LEMAC) = 925.6" – 860.5" = 65.1"
CG (% of MAC) = 65.1"/180.9" = 36.0%
Flaps = 5°
Stab Trim = 2-3/4 ANU

(PLT010) — FAA-H-8083-25

Changing Loading Conditions

Anytime weight is either added to or subtracted from a loaded airplane, both the gross weight and the center of gravity location will change. The solution of such a problem is really a simplified loading problem. Instead of calculating a weight and moment for every section of the aircraft, it is only necessary to compute the original weight and moment and then the effect the change in weight had. Often in these problems the original CG is expressed in percent of MAC and it is necessary to convert this to an arm for the entire aircraft. The following example references Question 8578.

It is sometimes necessary to convert a CG position expressed in percent of MAC to the CG in inches aft of Datum. This is just the reverse of the process described above. This is done in two steps.

1. Convert the CG in percent of MAC to CG in inches aft of LEMAC. This is done by using the formula:

   CG (inches aft of LEMAC) = (CG % of MAC ÷ 100%) x MAC.

   Load Condition WS-1 (FAA Figure 44) gives a CG of 22.5% and a length of MAC of 141.5 inches. The formula is:

   CG (inches aft of LEMAC) = (22.5% ÷ 100%) x 141.5" = 31.84 inches.

2. Add the CG in inches aft of LEMAC to the Distance from Datum to LEMAC. In FAA Figure 44, LEMAC is 549.13 inches aft of Datum.

   CG (inches aft of Datum) = 549.13" + 31.84" = 580.97 inches.

Use Question 8578 and Conditions WS-1 (FAA Figure 44) for this example. Use the original weight and CG to calculate the original Moment/1,000. Next use the weight change and station to determine the Moment/1,000 change.

Continued

Answers

8724 [C] 8725 [C] 8726 [B]

	Weight	Moment/1,000
Original Weight	90,000	52,287.08
Weight Change	– 2,500	– 880.25
New Weight	87,500	51,406.83

Note: A reduction in weight results in a reduction in Moment/1,000. An increase in weight results in an increase in Moment/1,000.

Determine the new CG:

CG = (51,406.83 ÷ 87,500) × 1,000 = 587.51 inches

Convert CG to percent of MAC:

CG (inches aft of LEMAC) = 587.51" – 549.13" = 38.38"

CG (% of MAC) = (38.38" ÷ 141.5") × 100% = 27.1%

When a portion of an aircraft's load is shifted from one location to another, the CG of the loaded aircraft will change as well. Also, the CG will follow the weight. That is, if weight is shifted rearward, the CG will move rearward as well; and if weight is shifted forward, the CG will move forward. To calculate the effect of a weight shift on CG position, three numbers must be known: the weight shifted, the distance the weight was moved, and the total weight of the aircraft. The formula used is:

$$\text{Change in CG} = \frac{\text{Weight Shifted x Distance Shifted}}{\text{Total Weight}}$$

Question 8573 asks what the effect on CG is if weight is shifted from the forward to aft cargo compartment under Load Condition WS-1 (*See* FAA Figure 44). Load Condition WS-1 gives the total weight as 90,000 pounds and the weight shifted as 2,500 pounds. The distance shifted is the difference between the forward compartment centroid (352.1 inches) and the aft compartment centroid (724.9 inches), which is 372.8 inches (724.9 – 352.1).

Note: These centroids are distances aft of the Datum line. The index arms are distances from the CG Index and will be discussed in a later example. Notice however, that the difference between the two index arms is also 372.8 inches (144.9 – (-227.9) = 372.8). The solution is:

$$\text{Change in CG} = \frac{\text{2,500 lbs x 372.8"}}{\text{90,000 lbs}} = 10.4"$$

If weight is shifted forward, the CG will move forward as well. This is expressed by writing the distance shifted as a negative number. If weight is shifted from the aft to the forward cargo compartment, the distance shifted is -372.8 inches. For example, Question 8574 asks about such a shift of 1,800 pounds with an aircraft total weight of 85,000 pounds. The formula is:

$$\text{Change in CG} = \frac{\text{1,800 lbs x (-372.8)}}{\text{85,000 lbs}} = -7.89"$$

Questions 8573 and 8576 require an answer in percent of MAC. The change in CG can be converted to a percent of MAC by using the formula:

Change in CG (% of MAC) = (Change in CG/MAC) × 100%

Questions 8577 through 8581 express CG as an **Index Arm**. Index Arm is the distance, in inches, from an index set at a point close to the normal CG location. A positive Index Arm is a point aft of the index and a negative Index Arm is a point forward of the index. For example, FAA Figure 44 shows LEMAC as having an Index Arm of -30.87 inches or 30.87 inches forward of the index. The index point for all questions on this test is 580.0 inches. CG in Index Arm is calculated by the formula:

CG (Index Arm) = CG (inches aft of Datum) – 580 inches

Using the data from Load Condition WS-5, the formula is:

CG (Index Arm) = 585.21" – 580" = +5.21 inches

Questions 8429 through 8433 require calculation of the maximum weight that can be carried on a pallet. The limiting factor is the amount of weight that the aircraft floor can support per square foot. The calculation refers to FAA legends and involves the following steps:

1. Determine the area covered by the pallet. This is done by multiplying the pallet width by length. Since the pallet dimensions are in inches and the floor load limit is expressed in square feet, it is necessary to convert the pallet area from square inches to square feet. This is done by dividing by 144 (there are 144 square inches in a square foot). The formula is:

   Pallet Area (square feet) = (Width x Length) ÷ 144

   Using the example of Question 8429 which is a pallet 76" x 76", the area covered is:

   Pallet Area = (76" x 76") ÷ 144 = 40.11 square feet

2. Determine the floor load limit by multiplying the pallet area in square feet times the floor load limit per square foot. Again using the example of Question 8429, if the floor load limit is 186 lbs/sq ft:

   Floor Load Limit = 40.11 sq ft x 186 lbs/sq ft = 7,460.7 pounds

3. Determine the cargo weight which can be placed on the pallet by subtracting the weight of the pallet and tie-down devices. Since the floor has to support the pallet and tiedown weight, this reduces the total cargo which can be placed on the pallet. Once again, using the example of Question 8429, where the pallet weighs 93 pounds and the tiedown devices weigh 39 pounds (132 pounds total):

   Allowable Weight = 7,460.7 lbs – 132 lbs = 7,328.7 pounds

121, DSP

8578. (Refer to Figure 44.) What is the new CG if the weight is removed from the forward compartment under Loading Conditions WS 1?

A—27.1 percent MAC.
B—26.8 percent MAC.
C—30.0 percent MAC.

1. Calculate original CG in inches aft of datum:

CG (inches aft of LEMAC) = (22.5% / 100%) × 141.5" = 31.84"

CG (inches aft of Datum) = 549.13" + 31.84" = 580.97"

2. Use the original weight and the CG to calculate the original Moment/1,000. Next use the weight change and station to determine the Moment/1,000 change:

	Weight	***Moment/1,000***
Original Weight	*90,000*	*52,287.08*
Weight Change	*– 2,500*	*– 880.25*
New Weight	*87,500*	*51,406.83*

3. Determine the new CG:

CG = (51,406.83/87,500) × 1,000 = 587.51"

4. Convert CG to percent of MAC:

CG (inches aft of LEMAC) = 587.51" – 549.13" = 38.38"

CG (% of MAC) = (38.38/141.5) = 27.1%

(PLT021) — FAA-H-8083-1

Answers

8578 [A]

121, DSP

8579. (Refer to Figure 44.) Where is the new CG if the weight is added to the aft compartment under Loading Conditions WS 2?

A—+17.06 index arm.
B—+14.82 index arm.
C—+12.13 index arm.

1. *Calculate original CG in inches aft of datum:*

   CG (inches aft of LEMAC) = (28.4% / 100%) × 141.5" = 40.19"

   CG (inches aft of Datum) = 549.13" + 40.19" = 589.32"

2. *Use the original weight and the CG to calculate the original Moment/1,000. Next use the weight change and station to determine the Moment/1,000 change:*

	Weight	***Moment/1,000***
Original Weight	*85,000*	*50,091.87*
Weight Change	*+ 1,800*	*+ 1,304.82*
New Weight	*86,800*	*51,396.69*

3. *Determine the new CG:*

   CG = (51,396.69/86,800) x 1,000 = 592.13"

4. *Convert to Index Arm (0 Index Arm = 580.0"):*

   CG (Index Arm) = 592.13" – 580" = +12.13"

(PLT021) — FAA-H-8083-1

121, DSP

8580. (Refer to Figure 44.) What is the new CG if the weight is added to the forward compartment under Loading Conditions WS 3?

A—11.4 percent MAC.
B—14.3 percent MAC.
C—14.5 percent MAC.

1. *Calculate original CG in inches aft of datum:*

   CG (inches aft of LEMAC) = (19.8% / 100%) × 141.5" = 28.02"

   CG (inches aft of Datum) = 549.13" + 28.02" = 577.15"

2. *Use the original weight and the CG to calculate the original Moment/1,000. Next use the weight change and station to determine the Moment/1,000 change:*

	Weight	***Moment/1,000***
Original Weight	*84,500*	*48,768.92*
Weight Change	*+ 3,000*	*+ 1,056.30*
New Weight	*87,500*	*49,825.22*

3. *Determine the new CG:*

   CG = (49,825.22/87,500) x 1,000 = 569.43

4. *Convert CG to percent of MAC:*

   CG (inches aft of LEMAC) = 569.43" – 549.13" = 20.3"

   CG (% of MAC) = (20.3"/141.5") x 100% = 14.3%

(PLT021) — FAA-H-8083-1

121, DSP

8581. (Refer to Figure 44.) Where is the new CG if the weight is removed from the aft compartment under Loading Conditions WS 4?

A—+15.53 index arm.
B—+8.50 index arm.
C—-93.51 index arm.

1. *Calculate original CG in inches aft of datum:*

   CG (inches aft of LEMAC) = (30.3% / 100%) x 141.5" = 42.87"

   CG (inches aft of Datum) = 549.13" + 42.87" = 592.00"

2. *Use the original weight and the CG to calculate the original Moment/1,000. Next use the weight change and station to determine the Moment/1,000 change:*

	Weight	***Moment/1,000***
Original Weight	*81,700*	*48,366.40*
Weight Change	*– 2,100*	*– 1,522.29*
New Weight	*79,600*	*46,844.11*

3. *Determine the new CG:*

   CG = (46,844.11/79,600) x 1,000 = 588.49"

4. *Convert to Index Arm:*

   CG (Index Arm) = 588.49" – 580" = +8.50"

(PLT021) — FAA-H-8083-1

121, DSP

8582. (Refer to Figure 44.) What is the new CG if the weight is removed from the forward compartment under Loading Conditions WS 5?

A—31.9 percent MAC.
B—19.1 percent MAC.
C—35.2 percent MAC.

Answers

8579 [C] 8580 [B] 8581 [B] 8582 [A]

1. *Calculate original CG in inches aft of datum:*

   CG (inches aft of LEMAC) = (25.5% / 100%) x 141.5" = 36.08"
   CG (inches aft of Datum) = 549.13" + 36.08" = 585.21"

2. *Use the original weight and the CG to calculate the original Moment/1,000. Next use the weight change and station to determine the Moment/1,000 change:*

	Weight	*Moment/1,000*
Original Weight	*88,300*	*51,674.04*
Weight Change	*– 3,300*	*– 1,161.93*
New Weight	*85,000*	*50,512.11*

3. *Determine the new CG:*

   CG = (50,512.11 ÷ 85,000) × 1,000 = 594.26"

4. *Convert CG to percent of MAC:*

   CG (inches aft of LEMAC) = 594.26" – 549.13" = 45.13"
   CG(% of MAC)=(45.13"/141.5") x 100% = 31.9%

(PLT021) — FAA-H-8083-1

121, DSP
8573. (Refer to Figure 44.) What is the new CG if the weight is shifted from the forward to the aft compartment under Loading Conditions WS 1?

A—15.2 percent MAC.
B—29.8 percent MAC.
C—30.0 percent MAC.

Change in CG = (2,500 lbs x 372.8)/90,000 lbs = +10.4"
Change in CG (% of MAC) = (10.4"/141.5") × 100% = 7.35%
New CG = 22.5% + 7.35% = 29.85%

(PLT021) — FAA-H-8083-1

121, DSP
8574. (Refer to Figure 44.) What is the new CG if the weight is shifted from the aft to the forward compartment under Loading Conditions WS 2?

A—26.1 percent MAC.
B—20.5 percent MAC.
C—22.8 percent MAC.

Change in CG = (1,800 lbs × (-372.8))/85,000 lbs = -7.89"
Change in CG (% of MAC) = (-7.89"/141.5") × 100% = -5.6%
New CG = 28.4% – 5.6% = 22.8%

(PLT021) — FAA-H-8083-1

121, DSP
8575. (Refer to Figure 44.) What is the new CG if the weight is shifted from the forward to the aft compartment under Loading Conditions WS 3?

A—29.2 percent MAC.
B—33.0 percent MAC.
C—28.6 percent MAC.

Change in CG = (3,000 lbs x 372.8")/84,500 lbs = 13.24"
Change in CG (% of MAC) = (13.24"/141.5") x 100% = 9.4%
New CG = 19.8% + 9.4% = 29.2%

(PLT021) — FAA-H-8083-1

121, DSP
8576. (Refer to Figure 44.) What is the new CG if the weight is shifted from the aft to the forward compartment under Loading Conditions WS 4?

A—37.0 percent MAC.
B—23.5 percent MAC.
C—24.1 percent MAC.

Change in CG = (2,100 lbs × (-372.8")/81,700 lbs = -9.58"
Change in CG (% of MAC) = (-9.58"/141.5") x 100% = -6.8%
New CG = 30.3% – 6.8% = 23.5%.

(PLT021) — FAA-H-8083-1

121, DSP
8577. (Refer to Figure 44.) Where is the new CG if the listed weight is shifted from the forward to the aft compartment under Loading Conditions WS 5?

A—+19.15 index arm.
B—+13.93 index arm.
C—-97.92 index arm.

CG (inches aft of LEMAC) = (25.5%/100%) x 141.5 = 36.08"
CG (inches aft of Datum) = 549.13" + 36.08" = 585.21"
CG (Index Arm) = 585.21" – 580" = +5.21"
Change in CG = (3,300 lbs x 372.8")/88,300 lbs = 13.93"
New CG (Index Arm) = +5.21" + 13.93" = 19.14"

(PLT021) — FAA-H-8083-1

Answers

8573 [B]	8574 [C]	8575 [A]	8576 [B]	8577 [A]

121, 135, DSP

8431. What is the maximum allowable weight that may be carried on a pallet which has the dimensions of 36 x 48 inches?

Floor load limit .... 169 lbs/sq ft
Pallet weight ....47 lbs
Tiedown devices ....33 lbs

A—1,948.0 pounds.
B—1,995.0 pounds.
C—1,981.0 pounds.

Pallet area = (36 x 48)/144 = 12 sq ft
Floor load limit = 12 sq ft x 169 lbs/sq ft = 2,028 lbs
Allowable weight = 2,028 lbs – 80 lbs = 1,948 lbs

(PLT121) — FAA-H-8083-1

121, 135, DSP

8432. What is the maximum allowable weight that may be carried on a pallet which has the dimensions of 76 x 74 inches?

Floor load limit .... 176 lbs/sq ft
Pallet weight ....77 lbs
Tiedown devices ....29 lbs

A—6,767.8 pounds.
B—6,873.7 pounds.
C—6,796.8 pounds.

Floor load limit – 176 lbs/sq ft
Pallet weight – 77 lbs
Tiedown devices – 29 lbs
Pallet area = (76" x 74")/144 = 39.1 sq ft
Floor load limit = 39.1 sq ft x 176 lbs/sq ft = 6,873.8 lbs
Allowable weight = 6,873.8 lbs – 106 lbs = 6,767.8 lbs

(PLT121) — FAA-H-8083-1

121, 135, DSP

8433. What is the maximum allowable weight that may be carried on a pallet which has the dimensions of 81 x 83 inches?

Floor load limit .... 180 lbs/sq ft
Pallet weight ....82 lbs
Tiedown devices ....31 lbs

A—8,403.7 pounds.
B—8,321.8 pounds.
C—8,290.8 pounds.

Floor load limit – 180 lbs/sq ft
Pallet weight – 82 lbs
Tiedown devices – 31 lbs
Pallet area = (81" x 83")/144 = 46.7 sq ft
Floor load limit = 46.7 sq ft x 180 lbs/sq ft = 8,403.8 lbs
Allowable weight = 8,403.8 lbs – 113 lbs = 8,290.8 lbs

(PLT121) — FAA-H-8083-1

Beech 1900 Weight and Balance

Note: By definition, "Basic Empty Weight" does not include crew weight, so you must include crew in the calculation. By definition, "Basic Operating Weight" includes crew weight so you do not include crew in the calculation.

135

8049. The weight and CG of an aircraft used in 135 operations must have been calculated from those values established by actual weighing of the aircraft within what period of time?

A—Multiengine aircraft, preceding 36 calendar months.
B—Multiengine and single-engine aircraft, preceding 36 calendar months.
C—Multiengine aircraft, last 36 calendar months; single-engine, last 24 calendar months.

No person may operate a multi-engine aircraft unless the current empty weight and center of gravity are calculated from values established by actual weighing of the aircraft within the preceding 36 calendar months. (PLT454) — 14 CFR §135.185

Answers

8431 [A]	8432 [A]	8433 [C]	8049 [A]

135

8067. What are the empty weight and balance currency requirements for aircraft used in 135 operations?

A—The empty weight and CG of multiengine and single-engine aircraft must have been calculated from an actual weighing within the previous 36 calendar months.
B—The empty weight and CG must have been calculated from an actual weighing within the previous 24 calendar months unless the original Airworthiness Certificate was issued within the previous 36 calendar months.
C—The empty weight and CG of multiengine aircraft must have been calculated from an actual weighing within the previous 36 calendar months.

No person may operate a multi-engine aircraft unless the current empty weight and center of gravity are calculated from values established by actual weighing of the aircraft within the preceding 36 calendar months. (PLT454) — 14 CFR §135.185

135, DSP

8434. (Refer to Figures 3, 6, 8, 9, 10, and 11.) What is the CG in inches from datum under Loading Conditions BE-1?

A—Station 290.3.
B—Station 285.8.
C—Station 291.8.

	Weight	***Moment/100***
Basic Empty Wt.	*9,226*	*25,823*
Crew	*360*	*464*
Row 1	*350*	*700*
Row 2	*260*	*598*
Row 3	*200*	*520*
Row 4	*340*	*986*
Row 5	*120*	*384*
Row 6	*400*	*1,400*
Row 7	*120*	*456*
Row 8	*250*	*1,025*
Row 9	*—*	*—*
Baggage		
Nose	*60*	*39*
FWD Cabin	*250*	*409*
Aft (FWD Sec)	*500*	*2,418*
Aft (Aft Sec)	*—*	*—*
Fuel (Jet B @ 6.6)	*+ 2,442*	*+ 7,299*
Total	*14,878*	*42,521*

CG = (42,521 ÷ 14,878) x 100 = 285.8"

(PLT021) — FAA-H-8083-1

135, DSP

8435. (Refer to Figures 3, 6, 8, 9, 10, and 11.) What is the CG in inches from datum under Loading Conditions BE-2?

A—Station 295.2.
B—Station 292.9.
C—Station 293.0.

	Weight	***Moment/100***
Basic Empty Wt.	*9,226*	*25,823*
Crew	*340*	*439*
Row 1	*300*	*600*
Row 2	*250*	*575*
Row 3	*190*	*494*
Row 4	*170*	*493*
Row 5	*190*	*608*
Row 6	*340*	*1,190*
Row 7	*190*	*722*
Row 8	*—*	*—*
Row 9	*—*	*—*
Baggage		
Nose	*—*	*—*
FWD Cabin	*100*	*164*
Aft (FWD Sec)	*200*	*967*
Aft (Aft Sec)	*600*	*3,198*
Fuel (Jet A @ 6.8)	*+ 2,652*	*+ 7,924*
Total	*14,748*	*43,197*

CG = (43,197 ÷ 14,748) x 100 = 292.9"

(PLT021) — FAA-H-8083-1

Answers

8067 [C] 8434 [B] 8435 [B]

135, DSP

8436. (Refer to Figures 3, 6, 8, 9, 10, and 11.) What is the CG in inches from datum under Loading Conditions BE-3?

A—Station 288.2.
B—Station 285.8.
C—Station 290.4.

	Weight	*Moment/100*
Basic Empty Wt.	*9,226*	*25,823*
Crew	*350*	*452*
Row 1	*120*	*240*
Row 2	*340*	*782*
Row 3	*350*	*910*
Row 4	*300*	*870*
Row 5	*170*	*544*
Row 6	—	—
Row 7	—	—
Row 8	—	—
Row 9	—	—
Baggage		
Nose	*80*	*52*
FWD Cabin	*120*	*197*
Aft (FWD Sec)	*250*	*1,209*
Aft (Aft Sec)	*500*	*2,665*
Fuel (Jet B @ 6.7)	*+ 2,680*	*+ 8,007*
Total	*14,486*	*41,751*

CG = (41,751 ÷ 14,486) x 100 = 288.2"

(PLT021) — FAA-H-8083-1

135, DSP

8437. (Refer to Figures 3, 6, 8, 9, 10, and 11.) What is the CG in inches from datum under Loading Conditions BE-4?

A—Station 297.4.
B—Station 299.6.
C—Station 297.7.

	Weight	*Moment/100*
Basic Empty Wt.	*9,226*	*25,823*
Crew	*340*	*439*
Row 1	—	—
Row 2	*370*	*851*
Row 3	*400*	*1,040*
Row 4	*290*	*841*
Row 5	*200*	*640*
Row 6	*170*	*595*
Row 7	*210*	*798*
Row 8	*190*	*779*
Row 9	*420*	*1,848*
(continued)	***Weight***	***Moment/100***
Baggage		
Nose	—	—
FWD Cabin	—	—
Aft (FWD Sec)	*800*	*3,868*
Aft (Aft Sec)	—	—
Fuel (Jet A @ 6.8)	*+ 1,972*	*+ 5,912*
Total	*14,588*	*43,434*

CG = (43,434 ÷ 14,588) x 100 = 297.7"

(PLT021) — FAA-H-8083-1

135, DSP

8438. (Refer to Figures 3, 6, 8, 9, 10, and 11.) What is the CG in inches from datum under Loading Conditions BE-5?

A—Station 288.9.
B—Station 290.5.
C—Station 289.1.

	Weight	*Moment/100*
Basic Empty Wt.	*9,226*	*25,823*
Crew	*360*	*464*
Row 1	—	—
Row 2	—	—
Row 3	*170*	*442*
Row 4	*200*	*580*
Row 5	*290*	*928*
Row 6	*400*	*1,400*
Row 7	*370*	*1,406*
Row 8	*340*	*1,394*
Row 9	*430*	*1,892*
Baggage		
Nose	*100*	*66*
FWD Cabin	*200*	*327*
Aft (FWD Sec)	—	—
Aft (Aft Sec)	—	—
Fuel (Jet B @ 6.5)	*+ 2,210*	*+ 6,610*
Total	*14,296*	*41,332*

CG = (41,332 ÷ 14,296) x 100 = 289.1"

(PLT021) — FAA-H-8083-1

Answers

8436 [A] 8437 [C] 8438 [C]

135, DSP

8439. (Refer to Figures 3, 6, 8, 9, 10, and 11.) What is the CG shift if the passengers in row 1 are moved to seats in row 9 under Loading Conditions BE-1?

A—1.5 inches aft.
B—5.6 inches aft.
C—6.2 inches aft.

	Weight	***Moment/100***
Basic Empty Wt.	*9,226*	*25,823*
Crew	*360*	*464*
Row 1	*350*	*700*
Row 2	*260*	*598*
Row 3	*200*	*520*
Row 4	*340*	*986*
Row 5	*120*	*384*
Row 6	*400*	*1,400*
Row 7	*120*	*456*
Row 8	*250*	*1,025*
Row 9	*—*	*—*
Baggage		
Nose	*60*	*39*
FWD Cabin	*250*	*409*
Aft (FWD Sec)	*500*	*2,418*
Aft (Aft Sec)	*—*	*—*
Fuel (Jet B @ 6.6)	*+ 2,442*	*+ 7,299*
Total	*14,878*	*42,521*

Change in CG = 350 lbs x 240" ÷ 14,878 lbs = 5.6" aft

(PLT021) — FAA-H-8083-1

135, DSP

8440. (Refer to Figures 3, 6, 8, 9, 10, and 11.) What is the CG shift if the passengers in row 1 are moved to row 8, and the passengers in row 2 are moved to row 9 under Loading Conditions BE-2?

A—9.2 inches aft.
B—5.7 inches aft.
C—7.8 inches aft.

	Weight	***Moment/100***
Basic Empty Wt.	*9,226*	*25,823*
Crew	*340*	*439*
Row 1	*300*	*600*
Row 2	*250*	*575*
Row 3	*190*	*494*
Row 4	*170*	*493*
Row 5	*190*	*608*
Row 6	*340*	*1,190*
Row 7	*190*	*722*
Row 8	*—*	*—*
Row 9	*—*	*—*
(continued)		
Baggage		
Nose	*—*	*—*
FWD Cabin	*100*	*164*
Aft (FWD Sec)	*200*	*967*
Aft (Aft Sec)	*600*	*3,198*
Fuel (Jet A @ 6.6)	*+ 2,652*	*+ 7,924*
Total	*14,748*	*43,197*

Change in CG = 550 lbs x 210" ÷ 14,748 lbs = 7.8" aft

(PLT021) — FAA-H-8083-1

135, DSP

8441. (Refer to Figures 3, 6, 8, 9, 10, and 11.) What is the CG shift if four passengers weighing 170 pounds each are added; two to seats in row 6 and two to seats in row 7 under Loading Conditions BE-3?

A—3.5 inches aft.
B—2.2 inches forward.
C—1.8 inches aft.

Use the following steps:

1. Compute the CG position prior to changes:

	Weight	***Moment/100***
Basic Empty Wt.	*9,226*	*25,823*
Crew	*350*	*452*
Row 1	*120*	*240*
Row 2	*340*	*782*
Row 3	*350*	*910*
Row 4	*300*	*870*
Row 5	*170*	*544*
Row 6	*—*	*—*
Row 7	*—*	*—*
Row 8	*—*	*—*
Row 9	*—*	*—*
Baggage		
Nose	*80*	*52*
FWD Cabin	*120*	*197*
Aft (FWD Sec)	*250*	*1,209*
Aft (Aft Sec)	*500*	*2,665*
Fuel (Jet B @ 6.7)	*+ 2,680*	*+ 8,007*
Total	*14,486*	*41,751*

CG = (41,751 ÷ 14,486) x 100 = 288.2"

Continued

Answers

8439 [B] 8440 [C] 8441 [A]

2. *Calculate the weight additions:*

	Weight	***Moment/100***
Original Wt.	*14,486*	*41,751*
2 pax in row 6	*+ 340*	*+ 1,190*
2 pax in row 7	*+ 340*	*+ 1,292*
New Total	*15,166*	*44,233*

3. *CG = (44,233 ÷ 15,166) × 100 = 291.7"*
4. *The CG moved from station 288.2 to station 291.7, a movement of 3.5 inches aft.*

(PLT021) — FAA-H-8083-1

135, DSP

8442. (Refer to Figures 3, 6, 8, 9, 10, and 11.) What is the CG shift if all passengers in rows 2 and 4 are deplaned under Loading Conditions BE-4?

A—2.5 inches aft.
B—2.5 inches forward.
C—2.0 inches aft.

Use the following steps:

1. *Compute the CG position prior to changes:*

	Weight	***Moment/100***
Basic Empty Wt.	*9,226*	*25,823*
Crew	*340*	*439*
Row 1	—	—
Row 2	*370*	*851*
Row 3	*400*	*1,040*
Row 4	*290*	*841*
Row 5	*200*	*640*
Row 6	*170*	*595*
Row 7	*210*	*798*
Row 8	*190*	*779*
Row 9	*420*	*1,848*
Baggage		
Nose	—	—
FWD Cabin	—	—
Aft (FWD Sec)	*800*	*3,868*
Aft (Aft Sec)	—	—
Fuel (Jet A @ 6.8)	*+1,972*	*+ 5,912*
Total	*14,588*	*43,434*

CG = (43,434 ÷ 14,588) x 100 = 297.7"

2. *Calculate the weight reductions:*

	Weight	***Moment/100***
Original Wt.	*14,588*	*43,434*
2 pax in row 2	*– 370*	*– 851*
2 pax in row 4	*– 290*	*– 841*
New Total	*13,928*	*41,742*

3. *Calculate the new CG:*
   CG = (41,742 ÷ 13,928) × 100 = 299.7"
4. *The CG moved from station 297.7 to station 299.7, a movement of 2.0 inches aft.*

(PLT021) — FAA-H-8083-1

135, DSP

8443. (Refer to Figures 3, 6, 8, 9, 10, and 11.) What is the CG shift if the passengers in row 8 are moved to row 2, and the passengers in row 7 are moved to row 1 under Loading Conditions BE-5?

A—1.0 inches forward.
B—8.9 inches forward.
C—6.5 inches forward.

	Weight	***Moment/100***
Basic Empty Wt.	*9,226*	*25,823*
Crew	*360*	*464*
Row 1	—	—
Row 2	—	—
Row 3	*170*	*442*
Row 4	*200*	*580*
Row 5	*290*	*928*
Row 6	*400*	*1,400*
Row 7	*370*	*1,406*
Row 8	*340*	*1,394*
Row 9	*430*	*1,892*
Baggage		
Nose	*100*	*66*
FWD Cabin	*200*	*327*
Aft (FWD Sec)	—	—
Aft (Aft Sec)	—	—
Fuel (Jet B @ 6.5)	*+ 2,210*	*+ 6,610*
Total	*14,296*	*41,332*

Change in CG = 710 lbs x 180" ÷ 14,296 lbs = 8.9" forward

(PLT021) — FAA-H-8083-1

Answers

8442 [C]   8443 [B]

135, DSP

8444. (Refer to Figures 4, 7, 9, 10, and 11.) What is the CG in inches from datum under Loading Conditions BE-6?

A—Station 300.5.
B—Station 296.5.
C—Station 300.8.

	Weight	*Moment/100*
Basic Operating Wt.	*9,005*	*25,934*
Sec A	*500*	*1,125*
Sec B	*500*	*1,275*
Sec C	*550*	*1,567.5*
Sec D	*550*	*1,732.5*
Sec E	*600*	*2,070*
Sec F	*600*	*2,250*
Sec G	*450*	*1,822.5*
Sec H	—	—
Sec J	*350*	*1,627.5*
Sec K	—	—
Sec L	—	—
Fuel (Jet B @ 6.5)	*+2,210*	*+ 6,610*
Total	*15,315*	*46,014*

CG = (46,014 ÷ 15,315) x 100 = 300.5"

(PLT021) — FAA-H-8083-1

135, DSP

8445. (Refer to Figures 4, 7, 9, 10, and 11.) What is the CG in inches from datum under Loading Conditions BE-7?

A—Station 296.0.
B—Station 297.8.
C—Station 299.9.

	Weight	*Moment/100*
Basic Operating Wt.	*9,005*	*25,934*
Sec A	—	—
Sec B	*400*	*1,020*
Sec C	*450*	*1,282.5*
Sec D	*600*	*1,890*
Sec E	*600*	*2,070*
Sec F	*600*	*2,250*
Sec G	*500*	*2,025*
Sec H	—	—
Sec J	—	—
(continued)	***Weight***	***Moment/100***
Sec K	—	—
Sec L	—	—
Fuel	*+ 2,442*	*+ 7,299*
Total	*14,597*	*43,770.5*

CG = (43,770.5 ÷ 14,597) x 100 = 299.9"

(PLT021) — FAA-H-8083-1

135, DSP

8446. (Refer to Figures 4, 7, 9, 10, and 11.) What is the CG in inches from datum under Loading Conditions BE-8?

A—Station 297.4.
B—Station 298.1.
C—Station 302.0.

	Weight	*Moment/100*
Basic Operating Wt.	*9,005*	*25,934*
Sec A	*600*	*1,350*
Sec B	*200*	*510*
Sec C	*400*	*1,140*
Sec D	*400*	*1,260*
Sec E	*200*	*690*
Sec F	*200*	*750*
Sec G	*200*	*810*
Sec H	*200*	*870*
Sec J	*300*	*1,395*
Sec K	*250*	*1,248.75*
Sec L	*100*	*533*
Fuel	*+ 2,652*	*+ 7,924*
Total	*14,707*	*44,414.75*

CG = (44,414.75 ÷ 14,707) x 100 = 302.0"

(PLT021) — FAA-H-8083-1

Answers

8444 [A] 8445 [C] 8446 [C]

135, DSP

8447. (Refer to Figures 4, 7, 9, 10, and 11.) What is the CG in inches from datum under Loading Conditions BE-9?

A—Station 296.7.
B—Station 297.1.
C—Station 301.2.

	Weight	*Moment/100*
Basic Operating Wt.	*9,005*	*25,934*
Sec A	*600*	*1,350*
Sec B	*600*	*1,530*
Sec C	*600*	*1,710*
Sec D	*600*	*1,890*
Sec E	*550*	*1,897.5*
Sec F	*350*	*1,312.5*
Sec G	*250*	*1,012.5*
Sec H	*250*	*1,087.5*
Sec J	*150*	*697.5*
Sec K	*200*	*999*
Sec L	*100*	*533*
Fuel (Jet A @ 6.8)	*+ 1,972*	*+ 5,912*
Total	*15,227*	*45,865.5*

CG = (45,865.5 ÷ 15,227) x 100 = 301.2"

(PLT021) — FAA-H-8083-1

135, DSP

8448. (Refer to Figures 4, 7, 9, 10, and 11.) What is the CG in inches from datum under Loading Conditions BE-10?

A—Station 298.4.
B—Station 298.1.
C—Station 293.9.

	Weight	*Moment/100*
Basic Operating Wt.	*9,005*	*25,934*
Sec A	*350*	*787.5*
Sec B	*450*	*1,147.5*
Sec C	*450*	*1,282.5*
Sec D	*550*	*1,732.5*
Sec E	*550*	*1,897.5*
Sec F	*600*	*2,250*
Sec G	*600*	*2,430*
Sec H	—	—
Sec J	—	—
Sec K	—	—
Sec L	—	—
Fuel (Jet B @ 6.7)	*+ 2,680*	*+ 8,007*
Total	*15,235*	*45,468.5*

CG = (45,468.5 ÷ 15,235) x 100 = 298.4"

(PLT021) — FAA-H-8083-1

135, DSP

8449. (Refer to Figures 4, 7, 9, 10, and 11.) What is the CG shift if 300 pounds of cargo in section A is moved to section H under Loading Conditions BE-6?

A—4.1 inches aft.
B—3.5 inches aft.
C—4.0 inches aft.

	Weight	*Moment/100*
Basic Operating Wt.	*9,005*	*25,934*
Sec A	*500*	*1,125*
Sec B	*500*	*1,275*
Sec C	*550*	*1,567.5*
Sec D	*550*	*1,732.5*
Sec E	*600*	*2,070*
Sec F	*600*	*2,250*
Sec G	*450*	*1,822.5*
Sec H	—	—
Sec J	*350*	*1,627.5*
Sec K	—	—
Sec L	—	—
Fuel (Jet B @ 6.5)	*+ 2,210*	*+ 6,610*
Total	*15,315*	*46,014*

Change in CG = 300 lbs x 210" ÷ 15,315 lbs = 4.1" aft

(PLT021) — FAA-H-8083-1

135, DSP

8450. (Refer to Figures 4, 7, 9, 10, and 11.) What is the CG shift if the cargo in section F is moved to section A, and 200 pounds of the cargo in section G is added to the cargo in section B, under Loading Conditions BE-7?

A—7.5 inches forward.
B—8.0 inches forward.
C—8.2 inches forward.

	Weight	*Moment/100*
Basic Operating Wt.	*9,005*	*25,934*
Sec A	—	—
Sec B	*400*	*1,020*
Sec C	*450*	*1,282.5*
Sec D	*600*	*1,890*
Sec E	*600*	*2,070*
Sec F	*600*	*2,250*
Sec G	*500*	*2,025*
Sec H	—	—
Sec J	—	—

Answers

8447 [C] 8448 [A] 8449 [A] 8450 [C]

(continued)	***Weight***	***Moment/100***
Sec K	—	—
Sec L	—	—
Fuel	*+ 2,442*	*+ 7,299*
Total	*14,597*	*43,770.5*

CG change moving weight from F to A is:
600 x 150 ÷ 14,597 = 6.2 inches forward
CG change moving weight from G to B is:
200 x 150 ÷ 14,597 = 2.1 inches forward
Total change:
6.2 inches + 2.1 inches = 8.3 inches forward

(PLT021) — FAA-H-8083-1

135, DSP

8451. (Refer to Figures 4, 7, 9, 10, and 11.) What is the CG if all cargo in sections A, B, J, K, and L are off-loaded under Loading Conditions BE-8?

A—Station 292.7.
B—Station 297.0.
C—Station 294.6.

Calculate the weight and CG accounting for the off-loaded items:

	Weight	***Moment/100***
Basic Operating Wt.	*9,005*	*25,934*
Sec A	—	—
Sec B	—	—
Sec C	*400*	*1,140*
Sec D	*400*	*1,260*
Sec E	*200*	*690*
Sec F	*200*	*750*
Sec G	*200*	*810*
Sec H	*200*	*870*
Sec J	—	—
Sec K	—	—
Sec L	—	—
Fuel	*+ 2,652*	*+ 7,924*
Total	*13,257*	*39,378*

CG = (39,378 ÷ 13,257) x 100 = 297.0"

(PLT021) — FAA-H-8083-1

135, DSP

8452. (Refer to Figures 4, 7, 9, 10, and 11.) What is the CG if cargo is loaded to bring sections F, G, and H to maximum capacity under Loading Conditions BE-9?

A—Station 307.5.
B—Station 305.4.
C—Station 303.5.

Use the following steps:

1. *Calculate the weight and CG prior to weight changes:*

	Weight	***Moment/100***
Basic Operating Wt.	*9,005*	*25,934*
Sec A	*600*	*1,350*
Sec B	*600*	*1,530*
Sec C	*600*	*1,710*
Sec D	*600*	*1,890*
Sec E	*550*	*1,897.5*
Sec F	*350*	*1,312.5*
Sec G	*250*	*1,012.5*
Sec H	*250*	*1,087.5*
Sec J	*150*	*697.5*
Sec K	*200*	*999*
Sec L	*100*	*533*
Fuel (Jet A @ 6.8)	*+ 1,972*	*+ 5,912*
Total	*15,227*	*45,865.5*

CG = (45,865.5 ÷ 15,227) x 100 = 301.2"

2. *250 pounds can be added to Section F, 350 pounds to Section G and 350 pounds to Section H. Apply these weight additions to the previously calculated weight and Moment/100, then calculate the new CG:*

	Weight	***Moment/100***
Original Wt.	*15,227*	*45,865.5*
Sec F	*250*	*937.5*
Sec G	*350*	*1,417.5*
Sec H	*+ 350*	*+1,522.5*
New Wt.	*16,177*	*49,743*

CG = (49,743 ÷ 16,177) x 100 = 307.5"

(PLT021) — FAA-H-8083-1

Answers

8451 **[B]** 8452 [A]

135, DSP

8453. (Refer to Figures 4, 7, 9, 10, and 11.) What is the CG shift if the cargo in section G is moved to section J under Loading Conditions BE-10?

A—2.7 inches aft.
B—2.4 inches aft.
C—3.2 inches aft.

	Weight	**Moment/100**
Basic Operating Wt.	*9,005*	*25,934*
Sec A	*350*	*787.5*
Sec B	*450*	*1,147.5*
Sec C	*450*	*1,282.5*
Sec D	*550*	*1,732.5*
Sec E	*550*	*1,897.5*
Sec F	*600*	*2,250*
Sec G	*600*	*2,430*
Sec H	—	—
Sec J	—	—
Sec K	—	—
Sec L	—	—
Fuel (Jet B @ 6.7)	*+ 2,680*	*+ 8,007*
Total	*15,235*	*45,468.5*

Change in CG = 600 lbs x 60" ÷ 15,235 lbs = 2.4" aft

(PLT021) — FAA-H-8083-1

135, DSP

8454. (Refer to Figures 5, 7, 9, and 11.) What limit is exceeded under Operating Conditions BE-11?

A—ZFW limit is exceeded.
B—Aft CG limit is exceeded at takeoff weight.
C—Aft CG limit is exceeded at landing weight.

Use the following steps:

1. *Calculate the Zero Fuel Weight (ZWF) using Operating Conditions BE-11:*

	Weight	**Moment/100**
Basic Empty Wt.	*9,225*	*25,820*
Crew	*340*	*439*
Pax & Bags	*+ 4,200*	*+ 15,025*
Zero Fuel Wt.	*13,765*	*41,284*

2. *Determine the Takeoff Weight:*

	Weight	**Moment/100**
Zero Fuel Wt.	*13,765*	*41,284*
T/O Fuel (340 gal)	*+ 2,312*	*+ 6,915*
Takeoff Wt.	*16,077*	*48,199*

3. *Calculate the Takeoff CG:*

   CG = (48,199 ÷ 16,077) x 100 = 299.8

4. *Determine the Landing Weight:*

	Weight	**Moment/100**
Zero Fuel Wt.	*13,765*	*41,284*
Land Fuel (100 gal)	*+ 680*	*+ 2,068*
Land Wt.	*14,445*	*43,352*

5. *Calculate the Landing Weight CG. Refer to CG = (43,352 ÷ 14,445) x 100 = 300.1*

This exceeds the aft CG limit (300.0) at landing.

(PLT021) — FAA-H-8083-1

135, DSP

8455. (Refer to Figures 5, 7, 9, and 11.) What limit(s) is(are) exceeded under Operating Conditions BE-12?

A—ZFW limit is exceeded.
B—Landing aft CG limit is exceeded.
C—ZFW and maximum takeoff weight limits are exceeded.

Use the following steps:

1. *Calculate the Zero Fuel Weight using Operating Conditions BE-12:*

	Weight	**Moment/100**
Basic Empty Wt.	*9,100*	*24,990*
Crew	*380*	*490*
Pax & Bags	*+ 4,530*	*+ 16,480*
Zero Fuel Wt.	*14,010*	*41,960*

2. *Determine the Takeoff Weight:*

	Weight	**Moment/100**
Zero Fuel Wt.	*14,010*	*41,960*
T/O Fuel (300 gal)	*+ 2,040*	*+ 6,112*
Takeoff Wt.	*16,050*	*48,072*

3. *Calculate the Takeoff CG:*

   CG = (48,172 ÷ 16,050) x 100 = 299.5

4. *Determine the Landing Weight:*

	Weight	**Moment/100**
Zero Fuel Wt.	*14,010*	*41,960*
Land Fuel (160 gal)	*+ 1,088*	*+ 3,303*
Land Wt.	*15,098*	*45,263*

5. *Calculate the Landing Weight CG. Refer to CG = (45,263 ÷ 15,098) x 100 = 299.8*

This exceeds the maximum Zero Fuel Weight.

(PLT021) — FAA-H-8083-1

Answers

8453 [B] 8454 [C] 8455 [A]

135, DSP

8456. (Refer to Figures 5, 7, 9, and 11.) What limit, if any, is exceeded under Operating Conditions BE-13?

A—Takeoff forward CG limit is exceeded.
B—No limit is exceeded.
C—Landing aft CG limit is exceeded.

Use the following steps:

1. *Calculate the Zero Fuel Weight using Operating Conditions BE-13:*

	Weight	***Moment/100***
Basic Empty Wt.	*9,000*	*24,710*
Crew	*360*	*464*
Pax & Bags	*+ 4,630*	*+ 16,743*
Zero Fuel Wt.	*13,990*	*41,917*

2. *Determine the Takeoff Weight:*

	Weight	***Moment/100***
Zero Fuel Wt.	*13,990*	*41,917*
T/O Fuel (330 gal)	*+ 2,244*	*+ 6,713*
Takeoff Wt.	*16,234*	*48,630*

3. *Calculate the Takeoff CG:*

   CG = (48,630 ÷ 16,234) x 100 = 299.6

4. *Determine the Landing Weight:*

	Weight	***Moment/100***
Zero Fuel Wt.	*13,990*	*41,917*
Land Fuel (140 gal)	*+ 952*	*+ 2,893*
Land Wt.	*14,942*	*44,810*

5. *Calculate the Landing Weight CG. Refer to CG = (44,810 ÷ 14,942) x 100 = 299.9*

No limits are exceeded.

(PLT021) — FAA-H-8083-1

135, DSP

8457. (Refer to Figures 5, 7, 9, and 11.) What limit(s) is(are) exceeded under Operating Conditions BE-14?

A—Maximum ZFW limit is exceeded.
B—Takeoff forward CG limit is exceeded.
C—Maximum landing weight and landing forward CG limits are exceeded.

Use the following steps:

1. *Calculate the Zero Fuel Weight using Operating Conditions BE-14:*

	Weight	***Moment/100***
Basic Empty Wt.	*8,910*	*24,570*
Crew	*400*	*516*
Pax & Bags	*+ 4,690*	*+ 13,724*
Zero Fuel Wt.	*14,000*	*38,810*

2. *Determine the Takeoff Weight:*

	Weight	***Moment/100***
Zero Fuel Wt.	*14,000*	*38,810*
T/O Fuel (290 gal)	*+ 1,972*	*+ 5,912*
Takeoff Wt.	*15,972*	*44,722*

3. *Calculate the Takeoff CG:*

   CG = (44,722 ÷ 15,972) x 100 = 280.0

The forward takeoff CG limit is exceeded.

(PLT021) — FAA-H-8083-1

Answers

8456 [B] 8457 [B]

135, DSP

8458. (Refer to Figures 5, 7, 9, and 11.) What limit(s) is(are) exceeded under Operating Conditions BE-15?

A—Maximum takeoff weight limit is exceeded.
B—Maximum ZFW and takeoff forward CG limits are exceeded.
C—Maximum takeoff weight and takeoff forward CG limits are exceeded.

Use the following steps:

1. Calculate the Zero Fuel Weight using Operating Conditions BE-15:

	Weight	*Moment/100*
Basic Empty Wt.	*9,150*	*25,240*
Crew	*370*	*477*
Pax & Bags	*+ 4,500*	*+ 13,561*
Zero Fuel Wt.	*14,020*	*39,278*

2. Determine the Takeoff Weight:

	Weight	*Moment/100*
Zero Fuel Wt.	*14,020*	*39,278*
T/O Fuel (380 gal)	*+ 2,584*	*+ 7,722*
Takeoff Wt.	*16,604*	*47,000*

3. Calculate the Takeoff CG:

CG = (47,000 ÷ 16.604) x 100 = 283.1

This exceeds the maximum takeoff gross weight of 16,600.

(PLT021) — FAA-H-8083-1

Helicopter Weight and Balance

RTC

8419. What is the result of loading a helicopter so that the CG is aft of the rearward limit?

A—Insufficient aft cyclic control to decelerate properly during an approach.
B—Inability of the pilot to recognize this dangerous condition when hovering in a strong headwind.
C—Insufficient forward cyclic control to fly in the upper allowable airspeed range.

If the center of gravity is too far aft of the mast, the helicopter hangs with the nose tilted up. If flight is attempted in this condition, the pilot may find it impossible to fly in the upper allowable airspeed range due to insufficient forward cyclic displacement to maintain a nose low attitude. (PLT240) — FAA-H-8083-21

RTC

8513. (Refer to Figures 29, 31, 32, and 33.) Where is the longitudinal CG located under Operating Conditions BL-1?

A—Station 214.3.
B—Station 235.6.
C—Station 237.8.

Condition	***BL-1***	***Weight***	***Moment***
Empty		*9,387.5*	*2,327,105*
Crew		*340.0*	*39,780*
Pax Row	*1*	*700.0*	*109,830*
	2	*830.0*	*154,546*
	3	*800.0*	*172,320*
	4	*—*	*—*
Baggage	*Center*	*500.0*	*148,500*
	L & R	*200.0*	*59,040*
Fuel		*+ 2,040.0*	*+ 475,400*
Total		*14,797.5*	*3,486,521*

Total Moment ÷ Total Weight = Longitudinal CG

$$\frac{3,486,521}{14,797.5} = 235.6$$

(PLT021) — FAA-H-8083-1

Answers

8458 [A] 8419 [C] 8513 [B]

RTC

8514. (Refer to Figures 29, 31, 32, and 33.) Where is the longitudinal CG located under Operating Conditions BL-2?

A—Station 237.6.
B—Station 238.5.
C—Station 262.3.

Condition	*BL-2*	*Weight*	*Moment*
Empty		*9387.5*	*2,327,105*
Crew		*400.0*	*46,800*
Pax Row	*1*	*620.0*	*97,278*
	2	*700.0*	*130,340*
	3	*680.0*	*146,472*
	4	*400.0*	*97,840*
Baggage	*Center*	*550.0*	*163,350*
	L & R	*250.0*	*73,800*
Fuel		*+ 1,625.0*	*+ 389,400*
Total		*14,612.5*	*3,472,385*

Total Moment ÷ Total Weight = Longitudinal CG

$$\frac{3{,}472{,}385}{14{,}612.5} = 237.6$$

(PLT021) — FAA-H-8083-1

RTC

8515. (Refer to Figures 29, 31, 32, and 33.) Where is the longitudinal CG located under Operating Conditions BL-3?

A—Station 223.4.
B—Station 239.0.
C—Station 240.3.

Condition	*BL-3*	*Weight*	*Moment*
Empty		*9,387.5*	*2,327,105*
Crew		*360.0*	*42,120*
Pax Row	*1*	—	—
	2	*750.0*	*139,650*
	3	*810.0*	*174,474*
	4	*650.0*	*158,990*
Baggage	*Center*	*300.0*	*89,100*
	L & R	—	—
Fuel		*+ 2,448.0*	*+ 583,900*
Total		*14,705.5*	*3,515,339*

Total Moment ÷ Total Weight = Longitudinal CG

$$\frac{3{,}515{,}339}{14{,}705.5} = 239.05$$

(PLT021) — FAA-H-8083-1

RTC

8516. (Refer to Figures 29, 31, 32, and 33.) Where is the longitudinal CG located under Operating Conditions BL-4?

A—Station 238.1.
B—Station 220.4.
C—Station 236.5.

Condition	*BL-4*	*Weight*	*Moment*
Empty		*9,387.5*	*2,327,105*
Crew		*380.0*	*44,460*
Pax Row	*1*	*180.0*	*28,242*
	2	*800.0*	*148,960*
	3	*720.0*	*155,088*
	4	*200.0*	*48,920*
Baggage	*Center*	*200.0*	*59,400*
	L & R	*100.0*	*29,520*
Fuel		*+ 2,600.0*	*+ 627,400*
Total		*14,567.5*	*3,469,095*

Total Moment ÷ Total Weight = Longitudinal CG

$$\frac{3{,}469{,}095}{14{,}567.5} = 238.14$$

(PLT021) — FAA-H-8083-1

RTC

8517. (Refer to Figures 29, 31, 32, and 33.) Where is the longitudinal CG located under Operating Conditions BL-5?

A—Station 232.0.
B—Station 235.4.
C—Station 234.9.

Condition	*BL-5*	*Weight*	*Moment*
Empty		*9,387.5*	*2,327,105*
Crew		*370.0*	*43,290*
Pax Row	*1*	*680.0*	*106,692*
	2	*950.0*	*176,890*
	3	*850.0*	*183,090*
	4	*500.0*	*122,300*
Baggage	*Center*	*450.0*	*133,650*
	L & R	—	—
Fuel		*+ 1,768.0*	*+ 420,000*
Total		*14,955.5*	*3,513,017*

Total Moment ÷ Total Weight = Longitudinal CG

$$\frac{3{,}513{,}017}{14{,}955.5} = 234.9$$

(PLT021) — FAA-H-8083-1

Answers

8514 [A] 8515 [B] 8516 [A] 8517 [C]

Helicopter Weight and Balance: CG Shifts

These questions require a re-computation of CG based on a shift of weight only, i.e., CG will change but total weight does not change. AC 91-23A, Chapter 5 gives us a formula for working this type of problem.

$$\frac{\text{Weight Shifted}}{\text{Total Weight}} = \frac{\text{Change of CG}}{\text{Distance of Shift}}$$

These problems may also be worked with a flight computer as shown in AC 91-23A, Chapter 5 in the following manner:

1. Set Weight Shifted (mile scale) over Total Weight (minute scale).
2. Find the Change in CG on the mile scale over the distance shifted on the minute scale.

Question 8518 is solved using both methods.

RTC

8518. (Refer to Figures 29, 31, 32, and 33.) What is the CG shift if all passengers in row 1 are moved to row 4 under Operating Conditions BL-1?

A—5.0 inches aft.
B—4.1 inches aft.
C—0.19 inch aft.

Using FAA Figure 29:
Weight shifted = 700 lbs, Total weight = 14,797.5
The distance shifted is the difference between Row 4 (Station 244.6) and Row 1 (Station 156.9):

$$244.6 - 156.9 = 87.7$$

To find the CG shift:

$$\frac{700}{14{,}797.5} = \frac{\text{CG Shift}}{87.7}$$

The shift from Row 1 to Row 4 is 4.15 aft.

or:

On the E6-B, set 700 (miles scale) over 14,797.5 (round it out: 14,800) on the minutes scale. Find 87.7 on the minutes scale and read 4.15 above it.

(PLT021) — FAA-H-8083-1

RTC

8519. (Refer to Figures 29, 31, 32, and 33.) What is the CG shift if one passenger weighing 150 pounds in row 2 is moved to row 4 under Operating Conditions BL-2?

A—0.1 inch aft.
B—0.6 inch aft.
C—1.1 inches aft.

$$\frac{\text{Weight Shifted (150)}}{\text{Total Weight (14,612.5)}} = \frac{\text{CG Shift}}{\text{Distance of Shift (58.4)}}$$

= .60 aft

(PLT021) — FAA-H-8083-1

RTC

8520. (Refer to Figures 29, 31, 32, and 33.) What is the CG shift if all passengers in row 4 are moved to row 1 under Operating Conditions BL-3?

A—3.7 inches forward.
B—0.4 inch forward.
C—3.9 inches forward.

$$\frac{\text{Weight Shifted (650)}}{\text{Total Weight (14,705.5)}} = \frac{\text{CG Shift}}{\text{Distance of Shift (87.7)}}$$

= -3.88 FWD

(PLT021) — FAA-H-8083-1

Answers

8518 [B] 8519 [B] 8520 [C]

RTC
8521. (Refer to Figures 29, 31, 32, and 33.) What is the CG shift if the passengers in row 1 are moved to row 4 under Operating Conditions BL-4?

A—1.1 inches aft.
B—1.6 inches aft.
C—0.2 inch aft.

$$\frac{\textit{Weight Shifted (180)}}{\textit{Total Weight (14,567.5)}} = \frac{\textit{CG Shift}}{\textit{Distance of Shift (87.7)}}$$

= 1.08 aft

(PLT021) — FAA-H-8083-1

RTC
8522. (Refer to Figures 29, 31, 32, and 33.) What is the CG shift if one passenger, weighing 100 pounds, seated in row 1 is moved to row 3 under Operating Conditions BL-5?

A—1.0 inch aft.
B—0.4 inch aft.
C—1.3 inches aft.

$$\frac{\textit{Weight Shifted (100)}}{\textit{Total Weight (14,955.5)}} = \frac{\textit{CG Shift}}{\textit{Distance of Shift (58.5)}}$$

= .39 aft

(PLT021) — FAA-H-8083-1

Helicopter Weight and Balance: Load Limits

In these questions, it will be necessary to compute both a takeoff and a landing weight and balance. Since the stations (CG) for fuel vary with weight, the most simple method of solving these problems is to compute the zero fuel weight for the given conditions, then perform a separate weight and balance for takeoff and landing. Some moments are given; others are not and therefore must be computed. Also, the fuel is stated in gallons, not pounds, which can be converted using the Jet A Table (FAA Figure 33).

RTC
8523. (Refer to Figures 30, 32, 33, and 35.) What limits are exceeded under Loading Conditions BL-6?

A—Aft CG limits are exceeded at takeoff and landing.
B—Takeoff aft CG and landing forward CG limits are exceeded.
C—Maximum takeoff weight and takeoff aft CG limits are exceeded.

Condition BL-6	***Weight***	***Moment***
Empty/basic	*10,225*	*2,556,250*
Crew	*340*	*39,780*
Passengers	*3,280*	*672,250*
Baggage Center	*+ 700*	*+ 207,900*
Zero Fuel wt.	*14,545*	*3,476,180*

Takeoff	***Weight***	***Moment***
Zero fuel wt.	*14,545*	*3,476,180*
Fuel 435 gal.	*+ 2,958*	*+ 719,900*
Total	*17,503*	*4,196,080*

Landing	***Weight***	***Moment***
Zero fuel wt.	*14,545*	*3,476,180*
Fuel 80 gal.	*+ 544*	*+ 125,600*
Total	*15,089*	*3,601,780*

CG = 4,196,080 ÷ 17,503 = 239.73
CG = 3,601,780 ÷ 15,089 = 238.7

Checking the longitudinal CG envelope (FAA Figure 35), we find that at takeoff, the aircraft is both over maximum gross weight and out of aft CG. (PLT021) — FAA-H-8083-1

RTC
8524. (Refer to Figures 30, 32, 33, and 35.) What limit, if any, is exceeded under Loading Conditions BL-7?

A—No limit is exceeded.
B—Forward CG limit is exceeded at landing only.
C—Forward CG limit is exceeded at takeoff and landing.

Condition BL-7	***Weight***	***Moment***
Empty/basic	*9,450*	*2,323,600*
Crew	*380*	*44,460*
Passengers	*2,880*	*541,860*
Baggage (center)	*+ 600*	*+ 178,200*
Zero Fuel	*13,310*	*3,088,120*

Note: *For this problem, it is easier to compute the landing data by subtracting the weight and moment of the fuel used from the takeoff data.*

Continued

Answers

8521 [A] 8522 [B] 8523 [C] 8524 [B]

Takeoff	***Weight***	***Moment***
Zero fuel wt.	*13,310*	*3,088,120*
Fuel 290 gal.	*+ 1,972*	*+ 457,900*
Total	*15,282*	*3,546,020*

Landing	***Weight***	***Moment***
Takeoff	*15,282*	*3,546,020*
Fuel 190 gal.	*– 1,292*	*– 315,900*
Total	*13,990*	*3,230,120*

CG = 3,546,020 ÷ 15,282 = 232.04
CG = 3,230,120 ÷ 13,990 = 230.89

Checking the longitudinal CG envelope (FAA Figure 35), we find that at landing only, the forward CG limit is exceeded. (PLT021) — FAA-H-8083-1

RTC
8525. (Refer to Figures 30, 32, 33, and 35.) What limit, if any, is exceeded under Loading Conditions BL-8?

A—No limit is exceeded.
B—Forward CG limit is exceeded at landing only.
C—Forward CG limit is exceeded at takeoff and landing.

Condition BL-8	***Weight***	***Moment***
Empty/basic	*9,000*	*2,202,050*
Crew	*410*	*47,970*
Passengers	*3,150*	*642,580*
Bags center	*+ 300*	*+ 89,100*
Zero Fuel	*12,860*	*2,981,700*

Note: *For this problem, it is easier to compute the landing data by subtracting the weight and moment of the fuel used from the takeoff data.*

Takeoff	***Weight***	***Moment***
Zero fuel wt.	*12,860*	*2,981,700*
Fuel 220 gal.	*+ 1,496*	*+ 369,400*
Total	*14,356*	*3,351,100*

Landing	***Weight***	***Moment***
Takeoff	*14,356*	*3,351,100*
Fuel 190 gal.	*– 1,292*	*– 315,900*
Total	*13,064*	*3,035,200*

CG = 3,351,100 ÷ 14,356 = 233.43
CG = 3,035,200 ÷ 13,064 = 232.33

Checking the longitudinal CG envelope (FAA Figure 35), we find that no limits are exceeded. (PLT021) — FAA-H-8083-1

RTC
8526. (Refer to Figures 30, 32, 33, and 35.) What limit, if any, is exceeded under Loading Conditions BL-9?

A—No limit is exceeded.
B—Aft CG limit is exceeded at takeoff only.
C—Aft CG limit is exceeded at takeoff and landing.

Condition BL-9	***Weight***	***Moment***
Empty/basic	*9,510*	*2,349,990*
Crew	*360*	*42,120*
Passengers	*2,040*	*473,220*
Bags center	*+ 550*	*+ 163,350*
Zero Fuel	*12,460*	*3,028,680*

Takeoff	***Weight***	***Moment***
Zero fuel wt.	*12,460*	*3,028,680*
Fuel 435 gal.	*+ 2,958*	*+ 719,900*
Total	*15,418*	*3,748,580*

Landing	***Weight***	***Moment***
Zero fuel wt.	*12,460*	*3,028,680*
Fuel 110 gal.	*+ 748*	*+ 170,900*
Total	*13,208*	*3,199,580*

CG = 3,748,580 ÷ 15,418 = 243.13
CG = 3,199,580 ÷ 13,208 = 242.24

Checking the longitudinal CG envelope (FAA Figure 35), we find that aft CG is exceeded at takeoff only. (PLT021) — FAA-H-8083-1

RTC
8527. (Refer to Figures 30, 32, 33, and 35.) What limit, if any, is exceeded under Loading Conditions BL-10?

A—No limit is exceeded.
B—Aft CG limit is exceeded at takeoff.
C—Forward CG limit is exceeded at landing.

Condition BL-10	***Weight***	***Moment***
Empty/basic	*9,375*	*2,329,680*
Crew	*400*	*46,800*
Passengers	*2,400*	*456,070*
Baggage (center)	*+ 650*	*+ 193,050*
Zero Fuel	*12,825*	*3,025,600*

Note: *For this problem, it is easier to compute the landing data by subtracting the weight and moment of the fuel used from the takeoff data.*

Answers

8525 [A]　　8526 [B]　　8527 [A]

Takeoff	***Weight***	***Moment***
Zero fuel wt.	*12,825*	*3,025,600*
Fuel 380 gal.	*+ 2,584*	*+ 620,200*
Total	*15,409*	*3,645,800*

Landing	***Weight***	***Moment***
Takeoff	*15,409*	*3,645,800*
Fuel 330 gal.	*– 2,244*	*– 529,600*
Total	*13,165*	*3,116,200*

CG = 3,645,800 ÷ 15,409 = 236.6
CG = 3,116,200 ÷ 13,165 = 236.7

Checking the longitudinal CG envelope (FAA Figure 35), we find that no limits are exceeded. (PLT021) — FAA-H-8083-1

Helicopter Weight and Balance: Lateral CG

These questions are answered by using the formula given in AC 91-23A.

1. For shifted weight:

$$\frac{\text{Weight Shifted (WS)}}{\text{Total Weight (TW)}} = \frac{\text{CG Shift (CS)}}{\text{Distance Shifted (DS)}}$$

2. For added/removed weight (WA or WR):

$$\frac{\text{(WA or WR)}}{\text{New Total Weight (NTW)}} = \frac{\text{CG Shift (CS)}}{\text{Distance shifted (DS)}}$$

Refer to answers to Questions 8523 through 8527 for total weights.

RTC

8528. (Refer to Figures 30, 31, 32, 33, and 34.) Given Loading Conditions BL-6, what is the effect on lateral CG if the outside passengers from each row on the left side are deplaned? Deplaned passenger weights are 170 pounds each.

A—CG shifts 1.5 inches right, out of limits.
B—CG shifts 1.4 inches right, within limits.
C—CG shifts 1.6 inches left, out of limits.

1. *Total weight for BL-6 = 17,503 (from Question 8523)*
2. *Weight removed = 170 x 4 = 680*
3. *New total weight = 17,503 – 680 = 16,823*
4. *Distance shifted = the average arm of the four outboard seats (34 + 35.4 + 35.4 + 39.4) ÷ 4 = 36.05.*

$$\frac{WR}{NTW} = \frac{CS}{DS}$$

(Very close to 1.5 inches.)

Since the deplaned passengers were all on the left, CG shift is to the right. Referring to FAA Figure 34, at our new weight of 16,823, the lateral CG is out of limits to the right. (PLT021) — FAA-H-8083-1

RTC

8529. (Refer to Figures 30, 31, 32, 33, and 34.) Given Loading Conditions BL-7, what is the effect on lateral CG if additional passengers, each weighing 200 pounds, are seated, one in each outside right seat of rows 1, 2, 3, and 4?

A—CG shifts 1.5 inches left, out of limits.
B—CG shifts 0.2 inch right, within limits.
C—CG shifts 1.8 inches right, out of limits.

1. *Total weight for BL-7 = 15,282 (from Question 8524)*
2. *Weight added = 800*
3. *New total weight = 15,282 + 800 = 16,082*
4. *Distance shifted = the average arm of the four outboard seats (34 + 35.4 + 35.4 + 39.4) ÷ 4 = 36.05.*

$$\frac{WA}{NTW} = \frac{CS}{DS}$$

CG shifts right 1.79 inches, out of limits.

(PLT021) — FAA-H-8083-1

Answers

8528 [A] 8529 [C]

RTC

8530. (Refer to Figures 30, 31, 32, 33, and 34.) Given Loading Conditions BL-8, what is the effect on lateral CG if a passenger weighing 200 pounds is added to the outer left seat of row 1, and a passenger weighing 220 pounds is added to the outer left seat of row 4?

A—CG shifts 1.5 inches left, out of limits.
B—CG shifts 1.2 inches left, within limits.
C—CG shifts 1.0 inch left, within limits.

1. *Total weight for BL-8 = 14,356 (from Question 8525)*
2. *Weight added = 420*
3. *New total weight = 14,356 + 420 = 15,776*
4. *Distance shifted = the average arm of Row 1 and Row 4 (34 + 39.4) ÷ 2 = 36.7*

$$\frac{WA}{NTW} = \frac{CS}{DS}$$

CG shifts left 1.04 inches, within limits.

(PLT021) — FAA-H-8083-1

RTC

8531. (Refer to Figures 30, 31, 32, 33, and 34.) Given Loading Conditions BL-9, what is the effect on lateral CG if passengers, each weighing 160 pounds, are added to the outer left seats of rows 1 and 2; and passengers, each weighing 180 pounds, are added to the outer right seats of rows 3 and 4?

A—CG shifts 0.14 inch left.
B—CG shifts 0.15 inch right.
C—CG does not shift.

It is obvious without doing the math, considering the weights and arms involved, that answer B is the only possible answer. However:

1. *Total weight for BL-9 = 15,418 (from Question 8526)*
2. *Weight added = 680*
3. *New total weight = 15,418 + 680 = 16,098*
4. *Distance shifted = the difference between the average left and right arms. If we consider the left side as negative, then distance shifted = [-(34 + 35.4) ÷ 2] + (35.4 + 39.4) ÷ 2 = -34.7 + 37.4 = + 2.7 inches.*

$$\frac{WA}{NTW} = \frac{CS}{DS}$$

CG shifts .11 inches right.

(PLT021) — FAA-H-8083-1

RTC

8532. (Refer to Figures 30, 31, 32, 33, and 34.) Given Loading Conditions BL-10, what is the effect on lateral CG if a passenger, weighing 240 pounds, is shifted from the outer right seat of row 4 to the outer left seat of row 1?

A—CG shifts 1.1 inches left, within limits.
B—CG shifts 1.5 inches left, out of limits.
C—CG shifts 1.7 inches left, out of limits.

1. *Total weight for BL-10 = 15,409 (from Question 8756)*
2. *Weight Shifted = 240*
3. *Distance shifted = the distance between Row 1 and Row 4 (34 + 39.4) = 73.4*

$$\frac{WS}{NTW} = \frac{CS}{DS}$$

CG shifts left 1.14 inches, within limits.

(PLT021) — FAA-H-8083-1

Answers

8530 [C] 8531 [B] 8532 [A]

Floor Loading Limits

In addition to ensuring that an aircraft is loaded within its weight and balance limits, it is important to make sure that the floor of a cargo compartment is not overloaded. The load limit of a floor is stated in pounds per square foot. The questions on the test require you to determine the maximum load that can be placed on a pallet of certain dimensions.

For example: what is the maximum weight that may be carried on a pallet which has the dimensions of 37 x 39 inches, when the floor load limit is 115 pounds per square foot, the pallet weight is 37 pounds, and the weight of the tiedown devices is 21 pounds?

The first step is to determine the area of the floor (in square feet) covered by the pallet. This is done by multiplying the given dimensions (which calculates the area in square inches) and dividing by 144 (which converts the area to square feet):

37 inches x 39 inches ÷ 144 square inches = 10.02 square feet.

The next step is to determine the total weight that the floor under the pallet can support, by multiplying the area times the floor load limit given in the question:

10.02 square feet x 115 pounds per square foot = 1,152.39 pounds.

The final step is to determine the maximum weight which can be placed on the pallet by subtracting the weight of the pallet and the tiedown devices from the total load limit:

1,152.39 pounds – 58 pounds = 1,094.39 pounds.

The weight on the pallet must be equal to or less than this number (1,094.39, in this example). If it is more than this number, the combination of cargo, pallet, and tiedown weight would exceed the floor load limit. A review of the test questions reveals that the closest answer choice is always equal to or slightly less than the floor limit. All the calculations in this section were performed with a calculator carrying all digits to the right of the decimal point forward for the next step of the problem. The explanations show only two places to the right of the decimal.

A variation of the pallet loading problem is to determine the minimum floor load limit (in pounds per square foot) required to carry a particular loaded pallet. For example: what is the minimum floor load limit to carry a pallet of cargo with a pallet dimension of 78.9 inches x 98.7 inches, and a combination weight of pallet, cargo, and tiedown devices of 9,896.5 pounds?

The first step is to determine the floor area, multiplying the dimensions and dividing by 144 (78.9 x 98.7 ÷ 144 = 54.08 square feet). The second step is to determine the minimum required floor limit by dividing the total weight of the pallet, cargo, and tiedowns by the pallet area (9,896.5 ÷ 54.08 = 183.00 pounds). The correct answer must be at or *above* this weight (183.00 pounds, in this example).

ALL
8769. What is the maximum allowable weight that may be carried on a pallet which has the dimensions of 33.5 x 48.5 inches?

Floor load limit — 76 lb/sq ft
Pallet weight — 44 lb
Tiedown devices — 27 lb

A—857.4 pounds.
B—830.4 pounds.
C—786.5 pounds.

1. *Determine the area.*

   33.5 x 48.5 ÷ 144 = 11.28 square feet.
2. *Determine the floor load limit.*

   11.28 x 76 = 857.51 pounds.
3. *Subtract the weight of the pallet and tiedown devices.*

   857.51 – 71 = 786.51 pounds.

(PLT121) — FAA-H-8083-1

ALL
8770. What is the maximum allowable weight that may be carried on a pallet which has the dimensions of 36.5 x 48.5 inches?

Floor load limit — 112 lb/sq ft
Pallet weight — 45 lb
Tiedown devices — 29 lb

A—1,331.8 pounds.
B—1,302.8 pounds.
C—1,347.8 pounds.

1. *Determine the area.*

   36.5 x 48.5 ÷ 144 = 12.29 square feet.
2. *Determine the floor load limit.*

   12.29 x 112 = 1,376.86 pounds.
3. *Subtract the weight of the pallet and tiedown devices.*

   1,376.86 – 74 = 1,302.86 pounds.

(PLT121) — FAA-H-8083-1

ALL
8771. What is the maximum allowable weight that may be carried on a pallet which has the dimensions of 42.6 x 48.7 inches?

Floor load limit — 121 lb/sq ft
Pallet weight — 47 lb
Tiedown devices — 33 lb

A—1,710.2 pounds.
B—1,663.2 pounds.
C—1,696.2 pounds.

1. *Determine the area.*

   42.6 x 48.7 ÷ 144 = 14.41 square feet.
2. *Determine the floor load limit.*

   14.41 x 121 = 1,743.25 pounds.
3. *Subtract the weight of the pallet and tiedown devices.*

   1,743.25 – 80 = 1,663.25 pounds.

(PLT121) — FAA-H-8083-1

ALL
8772. What is the maximum allowable weight that may be carried on a pallet which has the dimensions of 24.6 x 68.7 inches?

Floor load limit — 85 lb/sq ft
Pallet weight — 44 lb
Tiedown devices — 29 lb

A—924.5 pounds.
B—968.6 pounds.
C—953.6 pounds.

1. *Determine the area.*

   24.6 x 68.7 ÷ 144 = 11.74 square feet.
2. *Determine the floor load limit.*

   11.74 x 85 = 997.58 pounds.
3. *Subtract the weight of the pallet and tiedown devices.*

   997.58 – 73 = 924.58 pounds.

(PLT121) — FAA-H-8083-1

Answers

8769 [C]　8770 [B]　8771 [B]　8772 [A]

ALL

8773. What is the maximum allowable weight that may be carried on a pallet which has the dimensions of 34.6 x 46.4 inches?

Floor load limit — 88 lb/sq ft
Pallet weight — 41 lb
Tiedown devices — 26 lb

A—914.1 pounds.
B—940.1 pounds.
C—981.1 pounds.

1. *Determine the area.*

   34.6 x 46.4 ÷ 144 = 11.15 square feet.

2. *Determine the floor load limit.*

   11.15 x 88 = 981.10 pounds.

3. *Subtract the weight of the pallet and tiedown devices.*

   981.10 – 67 = 914.10 pounds.

(PLT121) — FAA-H-8083-1

ALL

8776. What is the maximum allowable weight that may be carried on a pallet which has the dimensions of 33.5 x 48.5 inches?

Floor load limit — 66 lb/sq ft
Pallet weight — 34 lb
Tiedown devices — 29 lb

A—744.6 pounds.
B—681.6 pounds.
C—663.0 pounds.

1. *Determine the area.*

   33.5 x 48.5 ÷ 144 = 11.28 square feet.

2. *Determine the floor load limit.*

   11.28 x 66 = 744.68 pounds.

3. *Subtract the weight of the pallet and tiedown devices.*

   744.68 – 63 = 681.68 pounds.

(PLT121) — FAA-H-8083-1

ALL

8777. What is the maximum allowable weight that may be carried on a pallet which has the dimensions of 36.5 x 48.5 inches?

Floor load limit — 107 lb/sq ft
Pallet weight — 37 lb
Tiedown devices — 33 lb

A—1,295.3 pounds.
B—1,212.3 pounds.
C—1,245.3 pounds.

1. *Determine the area.*

   36.5 x 48.5 ÷ 144 = 12.29 square feet.

2. *Determine the floor load limit.*

   12.29 x 107 = 1,315.39 pounds.

3. *Subtract the weight of the pallet and tiedown devices.*

   1,315.39 – 70 = 1,245.39 pounds.

(PLT121) — FAA-H-8083-1

ALL

8778. What is the maximum allowable weight that may be carried on a pallet which has the dimensions of 42.6 x 48.7 inches?

Floor load limit — 117 lb/sq ft
Pallet weight — 43 lb
Tiedown devices — 31 lb

A—1,611.6 pounds.
B—1,654.6 pounds.
C—1,601.6 pounds.

1. *Determine the area.*

   42.6 x 48.7 ÷ 144 = 14.41 square feet.

2. *Determine the floor load limit.*

   14.41 x 117 = 1,685.63 pounds.

3. *Subtract the weight of the pallet and tiedown devices.*

   1,685.63 – 74 = 1,611.63 pounds.

(PLT121) — FAA-H-8083-1

Answers

8773 [A] 8776 [B] 8777 [C] 8778 [A]

ALL

8779. What is the maximum allowable weight that may be carried on a pallet which has the dimensions of 24.6 x 68.7 inches?

Floor load limit — 79 lb/sq ft
Pallet weight — 43 lb
Tiedown devices — 27 lb

A—884.1 pounds.
B—857.1 pounds.
C—841.1 pounds.

1. *Determine the area.*

   24.6 x 68.7 ÷ 144 = 11.74 square feet.

2. *Determine the floor load limit.*

   11.74 x 79 = 927.16 pounds.

3. *Subtract the weight of the pallet and tiedown devices.*

   927.16 – 70 = 857.16 pounds.

(PLT121) — FAA-H-8083-1

ALL

8781. What is the maximum allowable weight that may be carried on a pallet which has the dimensions of 143 x 125.2 inches?

Floor load limit — 209 lb/sq ft
Pallet weight — 197 lb
Tiedown devices — 66 lb

A—25,984.9 pounds.
B—25,787.9 pounds.
C—25,721.9 pounds.

1. *Determine the area.*

   143 x 125.2 ÷ 144 = 124.33 square feet.

2. *Determine the floor load limit.*

   124.33 x 209 = 25,985.09 pounds.

3. *Subtract the weight of the pallet and tiedown devices.*

   25,985.09 – 263 = 25,722.09 pounds.

(PLT121) — FAA-H-8083-1

ALL

8787. What is the maximum allowable weight that may be carried on a pallet which has the dimensions of 138.5 x 97.6 inches?

Floor load limit — 235 lb/sq ft
Pallet weight — 219 lb
Tiedown devices — 71 lb

A—21,840.9 pounds.
B—21,769.9 pounds.
C—22,059.9 pounds.

1. *Determine the area.*

   138.5 x 97.6 ÷ 144 = 93.87 square feet.

2. *Determine the floor load limit.*

   93.87 x 235 = 22,059.97 pounds.

3. *Subtract the weight of the pallet and tiedown devices.*

   22,059.97 – 290 = 21,769.97 pounds.

(PLT121) — FAA-H-8083-1

ALL

8788. What is the maximum allowable weight that may be carried on a pallet which has the dimensions of 96.1 x 133.3 inches?

Floor load limit — 249 lb/sq ft
Pallet weight — 347 lb
Tiedown devices — 134 lb

A—21,669.8 pounds.
B—21,803.8 pounds.
C—22,120.8 pounds.

1. *Determine the area.*

   96.1 x 133.3 ÷ 144 = 88.96 square feet.

2. *Determine the floor load limit.*

   88.96 x 249 = 22,150.85 pounds.

3. *Subtract the weight of the pallet and tiedown devices.*

   22,150.85 – 481 = 21,669.85 pounds.

(PLT121) — FAA-H-8083-1

Answers

8779 [B] 8781 [C] 8787 [B] 8788 [A]

ALL

8789. What is the maximum allowable weight that may be carried on a pallet which has the dimensions of 87.7 x 116.8 inches?

Floor load limit — 175 lb/sq ft
Pallet weight — 137 lb
Tiedown devices — 49 lb

A—12,262.4 pounds.
B—12,448.4 pounds.
C—12,311.4 pounds.

1. *Determine the area.*

   87.7 x 116.8 ÷ 144 = 71.13 square feet.

2. *Determine the floor load limit.*

   71.13 x 175 = 12,448.52 pounds.

3. *Subtract the weight of the pallet and tiedown devices.*

   12,448.52 – 186 = 12,262.52 pounds.

(PLT121) — FAA-H-8083-1

ALL

8790. What is the maximum allowable weight that may be carried on a pallet which has the dimensions of 98.7 x 78.9 inches?

Floor load limit — 183 lb/sq ft
Pallet weight — 161 lb
Tiedown devices — 54 lb

A—9,896.5 pounds.
B—9,735.5 pounds.
C—9,681.5 pounds.

1. *Determine the area.*

   98.7 x 78.9 ÷ 144 = 54.08 square feet.

2. *Determine the floor load limit.*

   54.08 x 183 = 9,896.53 pounds.

3. *Subtract the weight of the pallet and tiedown devices.*

   9,896.53 – 215 = 9,681.53 pounds.

(PLT121) — FAA-H-8083-1

ALL

8791. What minimum floor load limit must an aircraft have to carry the following pallet of cargo?

Pallet size is 78.9 wide and 98.7 long
Pallet weight — 161 lb
Tiedown devices — 54 lb
Cargo weight — 9,681.5 lb

A—182 lb/sq ft.
B—180 lb/sq ft.
C—183 lb/sq ft.

1. *Determine the area.*

   78.9 x 98.7 ÷ 144 = 54.08 square feet.

2. *Determine the total weight.*

   9,681.5 + 54 + 161 = 9,896.5

3. *Determine the minimum floor load limit.*

   9,896.5 ÷ 54.08 = 183.00 lbs/sq ft.

(PLT121) — FAA-H-8083-1

ALL

8844. What is the minimum floor load limit that an aircraft must have to carry the following pallet of cargo?

Pallet dimensions are 39 x 37 inches.
Pallet weight — 37 lbs.
Tiedown devices — 21 lbs
Cargo weight — 1,094.3 lbs.

A—115 lbs/sq ft.
B—112 lbs/sq ft.
C—109 lbs/sq ft.

1. *Determine the area.*

   39 x 37 ÷ 144 = 10.02 sq ft.

2. *Determine the total weight.*

   1,094.3 + 21 + 37 = 1,152.3

3. *Determine the minimum floor load limit.*

   1,152.3 ÷ 10.02 = 114.99 lbs/sq ft.

(PLT121) — FAA-H-8083-1

Answers

8789 [A] 8790 [C] 8791 [C] 8844 [A]

ALL

8845. What is the minimum floor load limit that an aircraft must have to carry the following pallet of cargo?

Pallet dimensions are 37.5 x 35 inches.
Pallet weight — 34 lbs.
Tiedown devices — 23 lbs.
Cargo weight — 1,255.4 lbs.

A—152 lbs/sq ft.
B—148 lbs/sq ft.
C—144 lbs/sq ft.

1. *Determine the area.*

   37.5 x 35 ÷ 144 = 9.12 sq ft.
2. *Determine the total weight.*

   1,255.4 + 23 + 34 = 1,312.4
3. *Determine the minimum floor load limit.*

   1,312.4 ÷ 9.12 = 143.99 lbs/sq ft.

(PLT121) — FAA-H-8083-1

ALL

8846. What is the minimum floor load limit that an aircraft must have to carry the following pallet of cargo?

Pallet dimensions are 48.5 x 33.5 inches
Pallet weight — 44 lbs.
Tiedown devices — 27 lbs.
Cargo weight — 786.5 lbs.

A—79 lbs/sq ft.
B—76 lbs/sq ft.
C—73 lbs/sq ft.

1. *Determine the area.*

   48.5 x 33.5 ÷ 144 = 11.28 sq ft.
2. *Determine the total weight.*

   786.5 + 27 + 44 = 857.5
3. *Determine the minimum floor load limit.*

   857.5 ÷ 11.28 = 76.00 lbs/sq ft.

(PLT121) — FAA-H-8083-1

ALL

8847. What is the minimum floor load limit that an aircraft must have to carry the following pallet of cargo?

Pallet dimensions are 116.8 x 87.7 inches
Pallet weight — 137 lbs.
Tiedown devices — 49 lbs.
Cargo weight — 12,262.4 lbs.

A—172 lbs/sq ft.
B—176 lbs/sq ft.
C—179 lbs/sq ft.

1. *Determine the area.*

   116.8 x 87.7 ÷ 144 = 71.13 sq ft.
2. *Determine the total weight.*

   12,262.4 + 49 + 137 = 12,448.4
3. *Determine the minimum floor load limit.*

   12,448.4 ÷ 71.13 = 175.00 lbs/sq ft.

(PLT121) — FAA-H-8083-1

ALL

8848. What is the minimum floor load limit that an aircraft must have to carry the following pallet of cargo?

Pallet dimensions are 78.9 x 98.7 inches
Pallet weight — 161 lbs.
Tiedown devices — 54 lbs.
Cargo weight — 9,681.5 lbs.

A—180 lbs/sq ft.
B—186 lbs/sq ft.
C—183 lbs/sq ft.

1. *Determine the area.*

   78.9 x 98.7 ÷ 144 = 54.08 sq ft.
2. *Determine the total weight.*

   9,681.5 + 54 + 161 = 9,896.5
3. *Determine the minimum floor load limit.*

   9,896.5 ÷ 54.08 = 183.00 lbs/sq ft.

(PLT121) — FAA-H-8083-1

Answers

8845 [C] 8846 [B] 8847 [B] 8848 [C]

Chapter 6
Flight Operations

Airspace

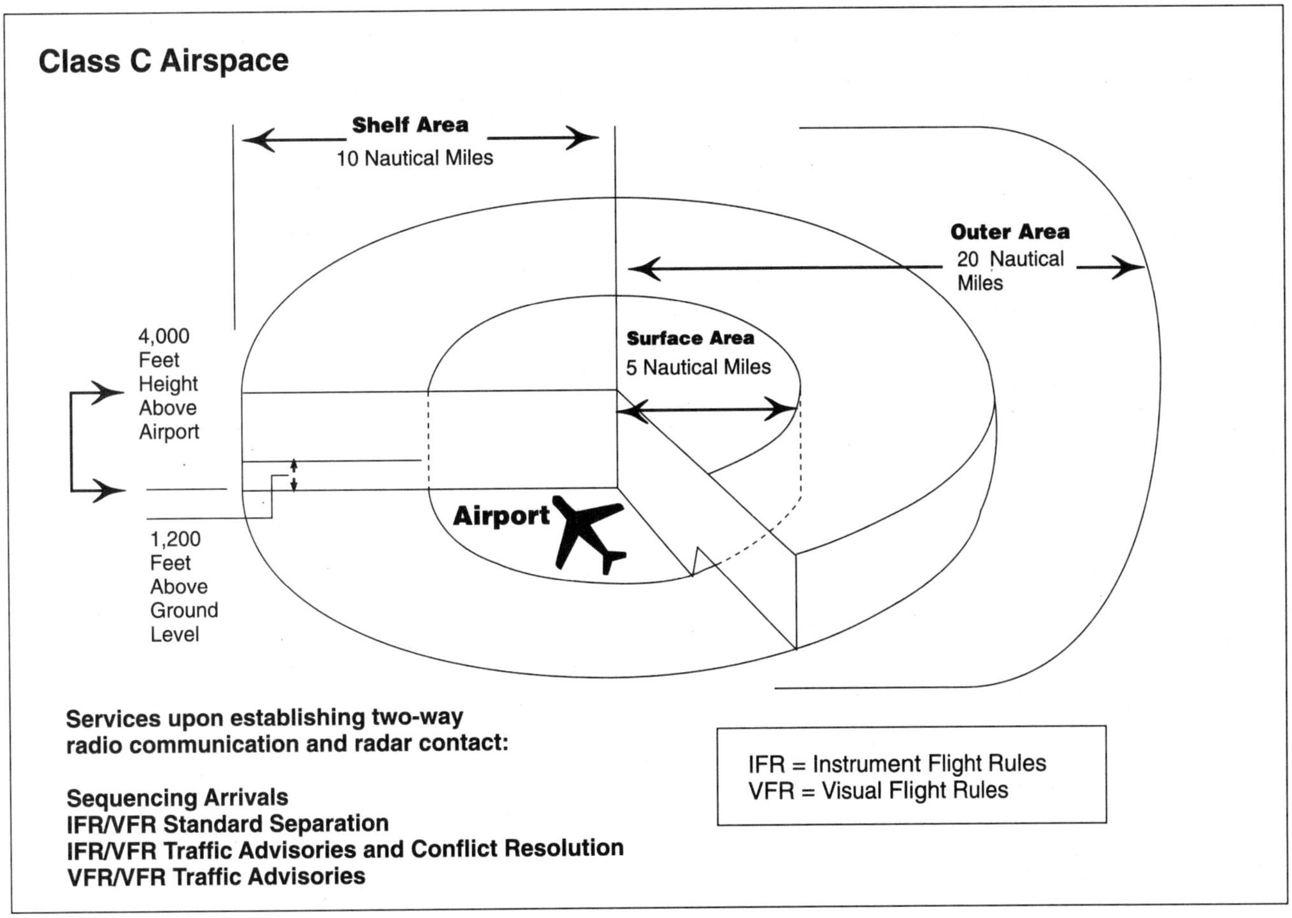

Class C Airspace
Shelf Area
10 Nautical Miles
Outer Area
20 Nautical Miles
Surface Area
5 Nautical Miles
4,000 Feet Height Above Airport
1,200 Feet Above Ground Level
Airport
Services upon establishing two-way radio communication and radar contact:
Sequencing Arrivals
IFR/VFR Standard Separation
IFR/VFR Traffic Advisories and Conflict Resolution
VFR/VFR Traffic Advisories
IFR = Instrument Flight Rules
VFR = Visual Flight Rules

Figure 6-1. Class C airspace

60,000 MSL (FL600)
18,000 MSL (FL180)
Class A
Class E
14,500 MSL
Class B
4,000 AGL
Class C
2,500 AGL
Class D
4 NM
1,200 AGL
700 AGL
Transition Area
Nontowered Airport
Class G
Class G
Class G
Class G
AGL: above ground level
FL: flight level
MSL: mean sea level

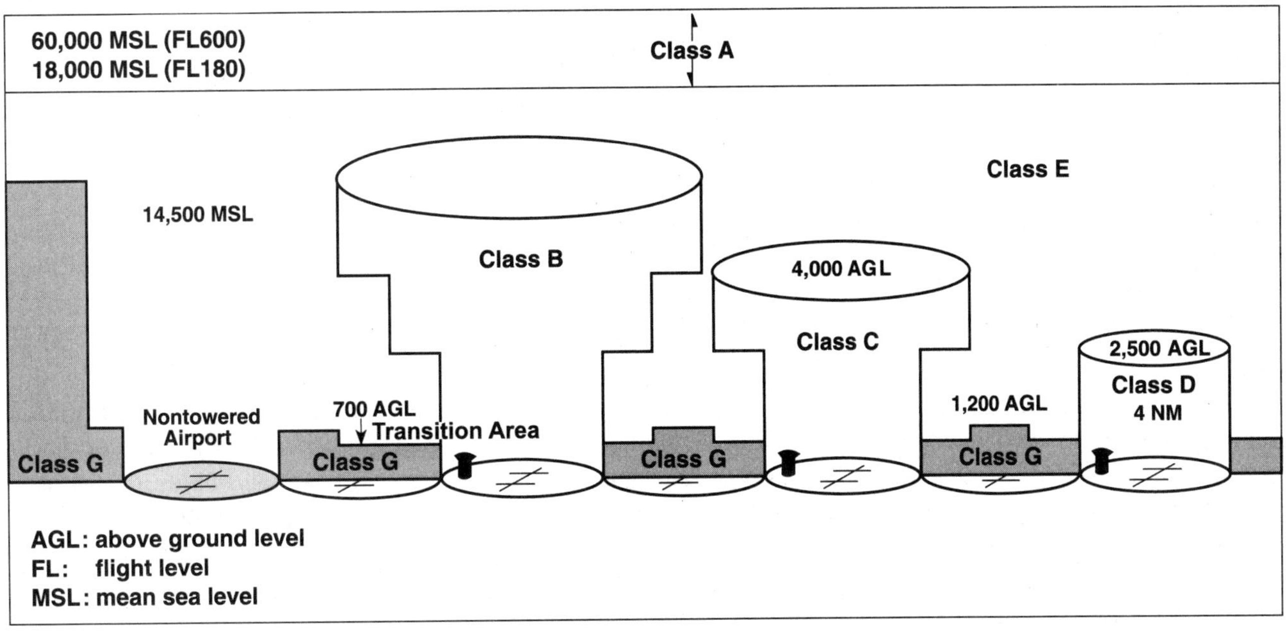

Figure 6-2. Airspace

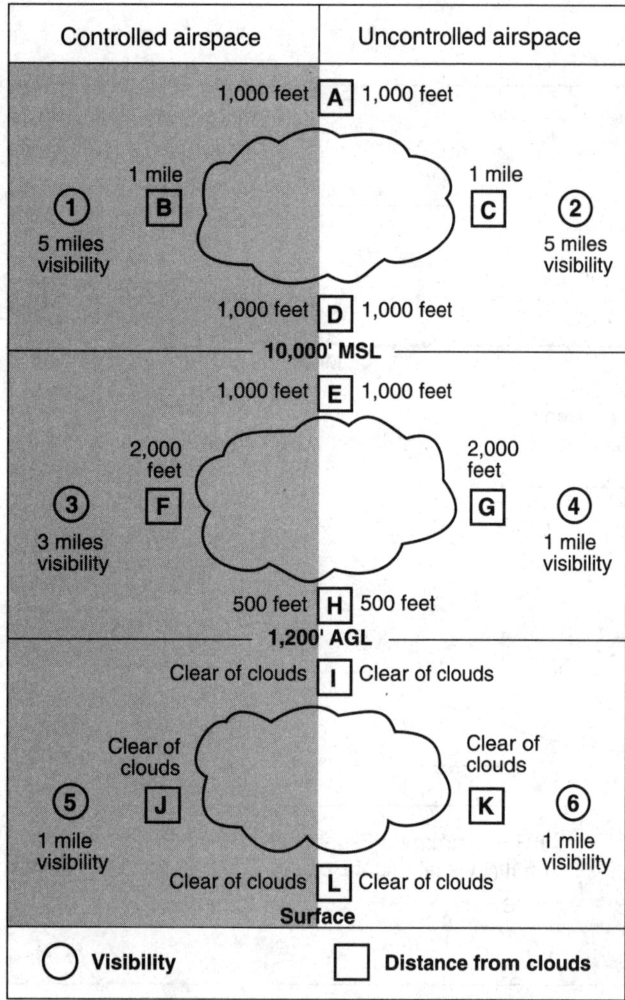

Figure 6-3. Minimum in-flight visibility and distance from clouds

A turbine-powered airplane or a large airplane must enter **Class D airspace** at an altitude of at least 1,500 feet AGL and maintain that altitude in the traffic pattern until a lower altitude is necessary for a safe landing. When taking off, the pilot of a turbine-powered airplane or a large airplane must climb as rapidly as practicable to an altitude of 1,500 feet AGL.

No person may operate an aircraft within **Class B airspace** unless a proper authorization from ATC has been received prior to entry. An IFR clearance is not necessarily required. Unless otherwise authorized by ATC, every person flying a large turbine-engine-powered airplane to or from the primary airport in Class B airspace must operate at or above the floor of Class B airspace.

All **Class C airspace** has the same dimensions with minor site variations. They are composed of two circles both centered on the primary airport. The surface area has a radius of 5 nautical miles and extends from the surface up to 4,000 feet above the airport. The shelf area has a radius of 10 nautical miles and extends vertically from 1,200 feet AGL up to 4,000 feet above the primary airport. In addition to the Class C airspace proper, there is an outer area with a radius of 20 nautical miles and vertical coverage from the lower limits of the radio/radar coverage up to the top of the approach control facility's delegated airspace.

The only equipment requirements for an aircraft to operate within Class C airspace are a two-way radio and a transponder. No specific pilot certification is required.

The following services are provided within Class C airspace:

- Sequencing of all arriving aircraft to the primary airport.
- Standard IFR separation between IFR aircraft.
- Between IFR and VFR aircraft — traffic advisories and conflict resolution so that radar targets do not touch, or 500 feet vertical separation.
- Between VFR aircraft, traffic advisories and as appropriate, safety alerts.

The same services are provided in the outer area when two-way radio and radar contact is established. There is no requirement for VFR participation in the outer area.

No one may operate an aircraft below 10,000 feet MSL at an indicated speed greater than 250 knots. No one may operate an aircraft within Class D airspace at an indicated airspeed of more than 200 knots. There is no special speed limit for operations within Class B airspace other than the 250-knot limit when below 10,000 feet MSL. When operating beneath the lateral limits of Class B airspace, the indicated airspeed cannot exceed 200 knots. If the minimum safe airspeed for any particular operation is greater than the maximum speed prescribed by 14 CFR §91.117, the aircraft may be operated at that minimum speed.

Warning Areas are so designated because they are located in international (and therefore uncontrolled) airspace and have invisible hazards to flight. The purpose of a **Military Operating Area (MOA)** is to separate IFR traffic from military training activities. Normally, ATC will not clear an IFR flight into an MOA if it is in use by the military. In an MOA, the individual pilots are responsible for collision avoidance. **VR** Military Training Routes which extend above 1,500 feet AGL, and **IR** Training Routes are depicted on IFR Enroute Low Altitude Charts.

When a flight is to penetrate an **Air Defense Identification Zone (ADIZ)**, it must be on either an IFR or a DVFR flight plan. The flight must penetrate the ADIZ within ±5 minutes of the flight plan estimate and within 10 miles when over land or within 20 miles when over water. These were formerly referred to as domestic and coastal ADIZs in the AIM.

A **VFR-On-Top** clearance is an IFR authorization to fly the cleared route at the VFR altitude of the pilot's choice. To request VFR-On-Top, the flight must be able to maintain the minimum VFR visibility and cloud clearances appropriate for the airspace and altitude. This may be done above, below or between the clouds, if any. While the pilot is expected to comply with all IFR rules, ATC will provide traffic advisories only. VFR-On-Top will not be authorized in Class A airspace. VFR weather minimums must be observed when operating under a VFR-On-Top clearance.

An air carrier flight may conduct day Over-the-Top operations below the minimum IFR altitude if the following are observed:

- The flight must be at least 1,000 feet above the top of a broken or overcast layer.
- The top of the clouds are generally uniform and level.
- The flight visibility is at least five miles.
- The base of any higher ceiling is at least 1,000 feet above the minimum IFR altitude.

OROCA is an off-route altitude which provides obstruction clearance with a 1,000-foot buffer in nonmountainous terrain areas, and a 2,000-foot buffer in designated mountainous areas within the U.S. **Minimum Vectoring Altitudes (MVAs)** are established for use by ATC when radar is exercised; MVA charts are prepared by air traffic facilities at locations where there are many different minimum IFR altitudes. **Minimum Safe/Sector Altitudes (MSA)** are published for emergency use on IAP charts; they are expressed in MSL and normally have a 25 NM radius; however, this radius may be expanded to 30 NM if necessary to encompass the airport landing surfaces.

ALL

8881. (Refer to Figure 127.) Which altitude is appropriate for circle 4 (top of Class G airspace)?

A—700 feet AGL.
B—1,200 feet AGL.
C—1,500 feet AGL.

Circle 4 corresponds to 1,200 feet, the floor of controlled airspace or the ceiling of Class G airspace in this area. (PLT040) — AIM ¶3-2-1

Answer (A) is incorrect because 700 feet AGL is the base of Class E airspace when used in conjunction with an instrument approach that has been prescribed, as in circle 6. Answer (C) is incorrect because 1,500 feet AGL is not an altitude which defines a certain airspace.

ALL

8882. (Refer to Figure 127.) Which altitude is normally appropriate for circle 5 (top of Class D airspace)?

A—1,000 feet AGL.
B—2,500 feet AGL.
C—3,000 feet AGL.

Circle 5 corresponds to the ceiling of Class D airspace which is normally at 2,500 feet AGL. There will be variations to suit special conditions. (PLT040) — AIM ¶3-2-5

Answer (A) is incorrect because 1,000 feet AGL is the normal traffic pattern altitude for piston aircraft, not the ceiling of Class D airspace. Answer (C) is incorrect because 3,000 feet AGL is not an altitude which defines a certain airspace.

ALL

8883. (Refer to Figure 127.) Which altitude is appropriate for circle 6 (top of Class G airspace)?

A—700 or 2,500 feet AGL.
B—500 or 2,000 feet AGL.
C—700 or 1,200 feet AGL.

Circle 6 corresponds to the floor of the Class G airspace, which could be 700 or 1,200 feet AGL, depending on whether an instrument approach is present. (PLT040) — AIM ¶3-2-1

ALL

8884. (Refer to Figure 127.) Which altitude is appropriate for circle 1 (top of Class E airspace)?

A—14,000 feet MSL.
B—14,500 feet MSL.
C—18,000 feet MSL.

Circle 1 corresponds to 18,000 feet MSL, the ceiling of the Class E airspace. (PLT040) — AIM ¶3-2-1

Answer (A) is incorrect because 14,000 feet MSL is not an altitude which defines a certain airspace. Answer (B) is incorrect because 14,500 MSL is the base of Class E airspace when it is not set lower.

ALL

8885. (Refer to Figure 127.) Which altitude is appropriate for circle 2 (top of Class C airspace)?

A—3,000 feet AGL.
B—4,000 feet AGL.
C—3,500 feet MSL.

Circle 2 indicates the ceiling of Class C airspace, generally 4,000 AGL. (PLT040) — AIM ¶3-2-4

Answers (A) and (C) are incorrect because neither 3,000 feet AGL nor 3,500 feet MSL are altitudes which define a certain airspace.

ALL

8886. (Refer to Figure 127.) Which altitude is appropriate for circle 3 (top of Class A airspace)?

A—FL 600.
B—FL 450.
C—FL 500.

Circle 3 indicates the upper limits of Class A airspace which is FL600. (PLT040) — AIM ¶3-2-1

Answers (B) and (C) are incorrect because neither FL450 nor FL500 are altitudes which define a certain airspace.

ALL

8888. (Refer to Figure 127.) What is the base of the Class A airspace?

A—12,000 feet AGL.
B—14,500 feet MSL.
C—FL 180.

The base of Class A airspace is 18,000 feet MSL. (PLT040) — AIM ¶3-2-1

Answer (A) is incorrect because 12,000 feet AGL is not an altitude which defines a certain airspace. Answer (B) is incorrect because 14,500 MSL is the base of Class E airspace when it is not set lower.

Answers

8881 [B] 8882 [B] 8883 [C] 8884 [C] 8885 [B] 8886 [A]
8888 [C]

ALL

9409. In what altitude structure is a transponder required when operating in controlled airspace?

A—Above 12,500 feet MSL, excluding the airspace at and below 2,500 feet AGL.
B—Above 10,000 feet MSL, excluding the airspace at and below 2,500 feet AGL.
C—Above 14,500 feet MSL, excluding the airspace at and below 2,500 feet AGL.

A transponder is required at and above 10,000 feet MSL and below the floor of Class A airspace, excluding the airspace at and below 2,500 feet AGL. (PLT429) — 14 CFR §91.215

Answer (A) is incorrect because 12,500 feet MSL was the old altitude above which a transponder was needed; it is now 10,000 feet MSL. Answer (C) is incorrect because 14,500 feet MSL is the base of Class E airspace when it is not set lower.

ALL

9424. Pilots should state their position on the airport when calling the tower for takeoff

A—from a runway intersection.
B—from a runway intersection, only at night.
C—from a runway intersection, only during instrument conditions.

Pilots should state their position on the airport when calling the tower for takeoff from a runway intersection. (PLT434) — AIM ¶4-3-10

Answers (B) and (C) are incorrect because this rule applies to all operations.

ALL

9780. When flying in the airspace underlying Class B airspace, the maximum speed authorized is

A—200 knots.
B—230 knots.
C—250 knots.

No person may operate an aircraft in the airspace underlying Class B airspace at a speed of more than 200 knots. (PLT161) — 14 CFR §91.117

ALL

8889. What restriction applies to a large, turbine-powered airplane operating to or from a primary airport in Class B airspace?

A—Must not exceed 200 knots within Class B airspace.
B—Must operate above the floor when within lateral limits of Class B airspace.
C—Must operate in accordance with IFR procedures regardless of weather conditions.

Unless otherwise authorized by ATC, each person operating a large turbine-engine-powered airplane to or from a primary airport in Class B airspace shall operate at or above the designated floors while within the lateral limits of the Class B airspace. (PLT161) — AIM ¶3-2-3

Answer (A) is incorrect because the speed limit within a Class B airspace is 250 knots for all aircraft. Answer (C) is incorrect because VFR is permitted in Class B airspace; i.e., an IFR clearance is not required as it is at FL180 and above.

ALL

8872. (Refer to Figure 126.) What is the normal radius from the airport of the outer area, B?

A—10 miles.
B—20 miles.
C—25 miles.

The normal radius of the outer area of Class C airspace is 20 NM. (PLT040) — AIM ¶3-2-4

Answer (A) is incorrect because 10 NM is the radius of the outer circle of Class C airspace. Answer (C) is incorrect because 25 NM does not pertain to any set radius of Class C airspace.

ALL

8873. (Refer to Figure 126.) What is the usual radius from the airport of the inner circle (now called surface area), C?

A—5 miles.
B—7 miles.
C—10 miles.

The usual radius from the airport of the surface area is 5 NM in Class C airspace. (PLT040) — AIM ¶3-2-4

Answer (B) is incorrect because 7 NM is not established as the radius for any portion of Class C airspace. Answer (C) is incorrect because 10 NM is the radius of the outer circle of Class C airspace.

Answers

9409 [B]	9424 [A]	9780 [A]	8889 [B]	8872 [B]	8873 [A]

ALL

8874. (Refer to Figure 126.) What is the radius from the airport of the outer circle (now called shelf area), A?

A—5 miles.
B—10 miles.
C—15 miles.

The radius of the shelf area is 10 NM in Class C airspace. (PLT040) — AIM ¶3-2-4

Answer (A) is incorrect because 5 NM is the radius of the surface area of Class C airspace. Answer (C) is incorrect because 15 NM is not established as the radius for any area of Class C airspace.

ALL

8875. (Refer to Figure 126.) Which altitude (box 2) is applicable to the base of the outer circle (now called shelf area)?

A—700 feet AGL.
B—1,000 feet AGL.
C—1,200 feet AGL.

The base of the shelf area is 1,200 feet AGL. (PLT040) — AIM ¶3-2-4

Answer (A) is incorrect because 700 feet AGL is not applicable to Class C airspace (it is the base of some Class E airspace). Answer (B) is incorrect because 1,000 feet AGL is not applicable to Class C airspace (it is the normal traffic pattern altitude for propeller airplanes).

ALL

8876. (Refer to Figure 126.) Which altitude (box 1) is applicable to the vertical extent of the inner and outer circles (now called surface and shelf areas)?

A—3,000 feet AGL.
B—3,000 feet above airport.
C—4,000 feet above airport.

The vertical extent of the surface and shelf areas is 4,000 feet above the airport. (PLT040) — AIM ¶3-2-4

Answers (A) and (B) are incorrect because 3,000 feet AGL does not define an airspace.

ALL

8877. What minimum aircraft equipment is required for operation within Class C airspace?

A—Two-way communications.
B—Two-way communications and transponder.
C—Transponder and DME.

An encoding transponder and two-way radio is required in order to operate within Class C airspace. (PLT040) — AIM ¶3-2-4

Answer (A) is incorrect because a Mode C transponder is also required. Answer (C) is incorrect because two-way communications are also required and DME is not required.

ALL

8878. What service is provided for aircraft operating within the outer area of Class C airspace?

A—The same as within Class C airspace when communications and radar contact is established.
B—Radar vectors to and from secondary airports within the outer area.
C—Basic radar service only when communications and radar contact is established.

The same services are provided for aircraft operating within the outer area, as within the Class C airspace, when two-way communication and radar contact are established. (PLT161) — AIM ¶3-2-4

Answer (B) is incorrect because providing radar vectors to and from secondary airports within the outer circle is not a mandated service of ATC. Answer (C) is incorrect because the same services are provided in the outer area as within Class C airspace, once two-way communications and radar contact are established.

ALL

8879. What services are provided for aircraft operating within Class C airspace?

A—Sequencing of arriving aircraft, separation of aircraft (except between VFR aircraft), and traffic advisories.
B—Sequencing of arriving aircraft (except VFR aircraft), separation between all aircraft, and traffic advisories.
C—Sequencing of all arriving aircraft, separation between all aircraft, and traffic advisories.

ATC services within an Class C airspace include:

1. *Sequencing of all arriving aircraft to the primary Class C airport,*
2. *Standard IFR separation between IFR aircraft,*
3. *Between IFR and VFR aircraft — traffic advisories and conflict resolution so that radar targets do not touch, or 500 feet vertical separation, and*
4. *Between VFR aircraft — traffic advisories and as appropriate, safety alerts.*

(PLT161) — AIM ¶3-2-4

Answer (B) is incorrect because the services in the Class C airspace provide sequencing of all aircraft to the primary/Class C airspace airport. Answer (C) is incorrect because the services in the Class C airspace do not provide separation between VFR aircraft, only traffic advisories and safety alerts.

Answers

8874 [B]	8875 [C]	8876 [C]	8877 [B]	8878 [A]	8879 [A]

ALL

8880. What pilot certification and aircraft equipment are required for operating in Class C airspace?

A—No specific certification but a two-way radio.
B—At least a Private Pilot Certificate and two-way radio.
C—At least a Private Pilot Certificate, two-way radio, and a TSO-C74b transponder.

No specific pilot certification is required for operation within Class C airspace. The aircraft must be equipped with a two-way radio for operations within Class C airspace. (PLT161) — AIM ¶3-2-4

Answers (B) and (C) are incorrect because there is no specific pilot certificate required, although two-way radio and transponder are required.

ALL

9399. What is the maximum indicated airspeed a turbine-powered aircraft may be operated below 10,000 feet MSL?

A—288 knots.
B—250 knots.
C—230 knots.

Unless otherwise authorized by the Administrator, no person may operate an aircraft below 10,000 feet MSL at an indicated airspeed of more than 250 knots (288 MPH). (PLT161) — 14 CFR §91.117

ALL

8890. Why are certain areas that start 3 nautical miles from the coastline of the U.S. and extend outward, classified as Warning Areas?

A—To inform pilots of participating aircraft to maintain extreme vigilance while conducting flight within the area.
B—To warn all aircraft pilots that flying within the area may be extremely hazardous to aircraft and occupants.
C—To warn pilots of nonparticipating aircraft of a potential danger within the area.

A Warning Area is airspace of defined dimensions, extending from three nautical miles outward from the coast of the United States, that contains activity that may be hazardous to nonparticipating aircraft. The purpose of such warning areas is to warn nonparticipating pilots of the potential danger. A warning area may be located over domestic or international waters or both. (PLT161) — AIM ¶3-4-4

ALL

8891. What is the purpose of MOAs?

A—To protect military aircraft operations from civil aircraft.
B—To separate military training activities from IFR traffic.
C—To separate military training activities from both IFR and VFR traffic.

Military Operations Areas (MOAs) consist of airspace of defined vertical and lateral limits established for the purpose of separating certain military training activities from IFR traffic. (PLT161) — AIM ¶3-4-5

Answer (A) is incorrect because MOAs are to separate (not protect) military training activities from IFR traffic. Answer (C) is incorrect because MOAs are established for the purpose of separating IFR traffic from military training activities.

ALL

8892. Who is responsible for collision avoidance in an MOA?

A—Military controllers.
B—ATC controllers.
C—Each pilot.

Pilots operating under VFR should exercise extreme caution while flying within an MOA when military activity is being conducted. (PLT162) — AIM ¶3-4-5

ALL

9049. Which aeronautical chart depicts Military Training Routes (MTR) above 1,500 feet?

A—IFR Low Altitude En Route Chart.
B—IFR High Altitude En Route Chart.
C—IFR Planning Chart.

The IFR Enroute Low Altitude Chart depicts all Military Training Routes (MTR) that accommodate operations above 1,500 feet AGL. (PLT100) — AIM ¶3-5-2

Answer (B) is incorrect because IFR High Altitude Enroute Charts do not depict MTRs. Answer (C) is incorrect because VFR Planning Charts depict MTRs.

Answers

8880 [A]	9399 [B]	8890 [C]	8891 [B]	8892 [C]	9049 [A]

ALL

9100. What is the maximum acceptable position tolerance for penetrating a domestic ADIZ over land?

A—Plus or minus 10 miles; plus or minus 10 minutes.
B—Plus or minus 20 miles; plus or minus 5 minutes.
C—Plus or minus 10 miles; plus or minus 5 minutes.

The aircraft position tolerances over land in a domestic ADIZ is within ±5 minutes from the estimated time over a reporting point or point of penetration and within 10 NM from the centerline of an intended track over an estimated reporting point or penetration point. (PLT161) — AIM ¶5-6-1

Answer (A) is incorrect because penetration of an ADIZ within 10 minutes is not an acceptable tolerance for either over water or land. Answer (B) is incorrect because the maximum acceptable tolerance for penetrating over water (a coastal ADIZ) is within 20 NM of the intended track and within 5 minutes of the estimated penetration time.

ALL

9741. What is the maximum acceptable position tolerance for penetrating a domestic ADIZ over water?

A—Plus or minus 10 miles; plus or minus 10 minutes.
B—Plus or minus 10 miles; plus or minus 5 minutes.
C—Plus or minus 20 miles; plus or minus 5 minutes.

The aircraft position tolerances over water in a domestic ADIZ is plus or minus five minutes from the estimated time over a reporting point or point of penetration and within 20 NM from the centerline of the intended track over an estimated reporting point or point of penetration. (PLT161) — AIM ¶5-6-1

ALL

9046. Under what conditions may a pilot on an IFR flight plan comply with authorization to maintain "VFR on Top"?

A—Maintain IFR flight plan but comply with visual flight rules while in VFR conditions.
B—Maintain VFR altitudes, cloud clearances, and comply with applicable instrument flight rules.
C—Maintain IFR altitudes, VFR cloud clearances, and comply with applicable instrument flight rules.

When operating in VFR conditions with an ATC authorization to "maintain VFR-On-Top" pilots on IFR flight plans must:

1. *Fly an appropriate VFR altitude,*
2. *Comply with VFR visibility and distance from cloud criteria, and*
3. *Comply with instrument flight rules that are applicable to the flight.*

(PLT370) — AIM ¶5-5-13

Answer (A) is incorrect because not only will a pilot remain on the IFR flight plan and comply with VFR altitudes, visibility, and cloud clearances, he/she must also comply with applicable IFR rules, e.g., position reporting, minimum IFR altitudes. Answer (C) is incorrect because, while operating on a "VFR-On-Top" clearance, a pilot must maintain VFR altitudes.

ALL

9047. What cloud clearance must be complied with when authorized to maintain "VFR on Top"?

A—May maintain VFR clearance above, below, or between layers.
B—Must maintain VFR clearance above or below.
C—May maintain VFR clearance above or below, but not between layers.

ATC authorization to "maintain VFR-On-Top" is not intended to restrict pilots so that they must operate only above an obscuring meteorological formation. Instead, it permits operations above, below, between or in areas where there is no meteorological obstruction. (PLT370) — AIM ¶5-5-13

ALL

9048. In what airspace will ATC not authorize "VFR on Top"?

A—Class C airspace.
B—Class B airspace.
C—Class A airspace.

ATC will not authorize VFR or VFR-On-Top operations in Class A airspace. (PLT161) — AIM ¶5-5-13

ALL

9093. What separation or service by ATC is afforded pilots authorized "VFR on Top"?

A—The same afforded all IFR flights.
B—3 miles horizontally instead of 5.
C—Traffic advisories only.

Answers

9100 [C]	9741 [C]	9046 [B]	9047 [A]	9048 [C]	9093 [C]

Pilots operating VFR-On-Top may receive traffic information from ATC on other pertinent IFR or VFR aircraft. (PLT172) — AIM ¶5-5-13

Answer (A) is incorrect because separation will be provided for all IFR flights except those operating with a VFR-On-Top clearance. In that case, only traffic advisories may be provided. Answer (B) is incorrect because, when radar is employed for separation of aircraft at the same altitude, a minimum of 3 miles separation is provided between airplanes operating within 40 miles of the radar antenna site, and 5 miles between aircraft operating beyond 40 miles from the antenna site.

ALL

9018. A minimum instrument altitude for enroute operations off of published airways which provides obstruction clearance of 1,000 feet in nonmountainous terrain areas and 2,000 feet in designated mountainous areas within the United States is called

A—Minimum Obstruction Clearance Altitude (MOCA).
B—Off-Route Obstruction Clearance Altitude (OROCA).
C—Minimum Safe/Sector Altitude (MSA).

OROCA is an off-route altitude which provides obstruction clearance with a 1,000-foot buffer in nonmountainous terrain areas, and a 2,000-foot buffer in designated mountainous areas within the U.S. (PLT162) — AIM ¶4-4-9

Answer (A) is incorrect because MOCAs provide the lowest published altitude in effect between radio fixes on VOR airways, off-airway routes, or route segments which meets obstacle clearance requirements for the entire route segment and which ensures acceptable navigational signal coverage only within 25 SM (22 NM) of a VOR. Answer (C) is incorrect because MSAs are published for emergency use on IAP charts; they are expressed in feet above mean sea level and normally have a 25 NM radius; however, this radius may be expanded to 30 NM if necessary to encompass the airport landing surfaces.

ALL

8893. What is the required flight visibility and distance from clouds if you are operating in Class E airspace at 9,500 feet with a VFR-on-Top clearance during daylight hours?

A—3 statute miles, 1,000 feet above, 500 feet below, and 2,000 feet horizontal.
B—5 statute miles, 500 feet above, 1,000 feet below, and 2,000 feet horizontal.
C—3 statute miles, 500 feet above, 1,000 feet below, and 2,000 feet horizontal.

A pilot receiving authorization for VFR-On-Top must comply with VFR visibility, distance from cloud criteria, and minimum IFR altitudes. When operating at more than 1,200 feet AGL but less than 10,000 feet MSL, pilots are required to maintain flight visibility of 3 statute miles and a distance from clouds of 1,000 feet above, 500 feet below, and 2,000 feet horizontal. (PLT163) — 14 CFR §91.155

Answer (B) is incorrect because the visibility requirement is 3 miles (not 5 miles) and the distances from clouds above and below are reversed. They should be 1,000 feet above and 500 feet below. Answer (C) is incorrect because the distances from clouds above and below are reversed. They should be 1,000 feet above and 500 feet below.

ALL

8894. (Refer to Figure 128.) What is the minimum in-flight visibility and distance from clouds required for a VFR-on-Top flight at 9,500 feet MSL (above 1,200 feet AGL) during daylight hours for the circle 3 area?

A—2,000 feet; (E) 1,000 feet; (F) 2,000 feet; (H) 500 feet.
B—5 miles; (E) 1,000 feet; (F) 2,000 feet; (H) 500 feet.
C—3 miles; (E) 1,000 feet; (F) 2,000 feet; (H) 500 feet.

When operating at more than 1,200 feet AGL but less than 10,000 feet MSL during the day in controlled airspace, pilots are required to maintain a flight visibility of 3 statute miles and a distance of 1,000 feet above, 500 feet below, and 2,000 feet horizontally from clouds. (PLT163) — 14 CFR §91.155

ALL

8895. (Refer to Figure 128.) A flight is to be conducted in VFR-on-Top conditions at 12,500 feet MSL (above 1,200 feet AGL). What is the in flight visibility and distance from clouds required for operations during daylight hours for the circle 1 area?

A—5 miles; (A) 1,000 feet; (B) 2,000 feet; (D) 500 feet.
B—5 miles; (A) 1,000 feet; (B) 1 mile; (D) 1,000 feet.
C—3 miles; (A) 1,000 feet; (B) 2,000 feet; (D) 1,000 feet.

A pilot on an IFR flight plan requesting and receiving authorization to operate VFR-On-Top must comply with instrument flight rules as well as VFR visibilities and distances from clouds. When operating at more than 1,200 feet AGL and at or above 10,000 feet MSL, pilots are required to maintain flight visibility of 5 statute miles and distances of 1,000 feet above, 1,000 feet below, and 1 mile horizontally from clouds. (PLT163) — 14 CFR §91.155

Answer (A) is incorrect because the distance from cloud requirements listed are for below 10,000 feet MSL. Answer (C) is incorrect because the visibility requirement is 5 miles and the horizontal separation from clouds requirement is 1 mile.

Answers

9018 [B]	8893 [A]	8894 [C]	8895 [B]

ALL

8896. (Refer to Figure 128.) What is the minimum in-flight visibility and distance from clouds required in VFR conditions above clouds at 13,500 feet MSL (above 1,200 feet AGL) during daylight hours for the circle 2 area?

A—5 miles; (A) 1,000 feet; (C) 2,000 feet; (D) 500 feet.
B—3 miles; (A) 1,000 feet; (C) 1 mile; (D) 1,000 feet.
C—5 miles; (A) 1,000 feet; (C) 1 mile; (D) 1,000 feet.

The minimum flight visibility in Class G airspace during daylight hours in VFR conditions above clouds at 13,500 feet MSL (above 1,200 feet AGL) is 5 statute miles. The distance from clouds required is 1,000 feet below, 1,000 feet above, and 1 mile horizontally. (PLT163) — 14 CFR §91.155

Answer (A) is incorrect because 1,000 feet above, 2,000 feet horizontal, and 500 feet below are the minimum cloud distances for VFR in Class G airspace above 1,200 feet AGL and below 10,000 feet MSL. Answer (B) is incorrect because visibility minimum is 5 miles.

ALL

8897. (Refer to Figure 128.) What in-flight visibility and distance from clouds is required for a flight at 8,500 feet MSL (above 1,200 feet AGL) in VFR conditions during daylight hours for the circle 4 area?

A—1 mile; (E) 1,000 feet; (G) 2,000 feet; (H) 500 feet.
B—3 miles; (E) 1,000 feet; (G) 2,000 feet; (H) 500 feet.
C—5 miles; (E) 1,000 feet; (G) 1 mile; (H) 1,000 feet.

At 8,500 feet MSL (above 1,200 feet AGL) in Class G airspace in VFR conditions during daylight hours the visibility is 1 statute mile and distance from clouds required is 500 feet below, 1,000 feet above, and 2,000 feet horizontally. (PLT163) — 14 CFR §91.155

Answer (B) is incorrect because these are the VFR weather minimums for Class E airspace below 10,000 feet MSL and above 1,200 feet AGL. Answer (C) is incorrect because these are the VFR weather minimums for Class E and Class G airspace above 10,000 feet MSL.

ALL

8898. (Refer to Figure 128.) What is the minimum in-flight visibility and distance from clouds required for an airplane operating less than 1,200 feet AGL during daylight hours in the circle 6 area?

A—3 miles; (I) 1,000 feet; (K) 2,000 feet; (L) 500 feet.
B—1 mile; (I) clear of clouds; (K) clear of clouds; (L) clear of clouds.
C—1 mile; (I) 500 feet; (K) 1,000 feet; (L) 500 feet.

When operating outside Class E airspace at less than 1,200 feet AGL during daylight hours, pilots are required to maintain flight visibility of 1 statute mile and operate clear of clouds. (PLT163) — 14 CFR §91.155

Answer (A) is incorrect because 3 miles visibility, 1,000 feet above, 500 feet below, and 2,000 feet horizontal are the VFR weather minimums for a flight in airspace above 1,200 feet AGL and below 10,000 feet MSL at night. Answer (C) is incorrect because no such combination of requirements exists in any airspace.

ALL

8899. (Refer to Figure 128.) What is the minimum in-flight visibility and distance from clouds required for an airplane operating less than 1,200 feet AGL under special VFR during daylight hours in the circle 5 area?

A—1 mile; (I) 2,000 feet; (J) 2,000 feet; (L) 500 feet.
B—3 miles; (I) clear of clouds; (J) clear of clouds; (L) 500 feet.
C—1 mile; (I) clear of clouds; (J) clear of clouds; (L) clear of clouds.

When operating under special VFR during daylight hours, pilots are required to maintain flight visibility of 1 statute mile and operate clear of clouds. (PLT163) — 14 CFR §91.157

Answers (A) and (B) are incorrect because special VFR permits operation just clear of clouds and with a minimum visibility of 1 mile.

ALL

8900. What is the minimum flight visibility and distance from clouds for flight at 10,500 feet, in Class E airspace, with a VFR-on-Top clearance during daylight hours?

A—3 statute miles, 1,000 feet above, 500 feet below, and 2,000 feet horizontal.
B—5 statute miles, 1,000 feet above, 1,000 feet below, and 1 mile horizontal.
C—5 statute miles, 1,000 feet above, 500 feet below, and 1 mile horizontal.

A pilot on an IFR flight plan requesting and receiving authorization to operate VFR-On-Top must comply with instrument flight rules as well as VFR visibilities and distances from clouds. When operating at more than 1,200 feet AGL and at or above 10,000 feet MSL pilots are required to maintain flight visibility of 5 statute miles and distances of 1,000 feet above, 1,000 feet below, and 1 mile horizontally from clouds. (PLT163) — 14 CFR §91.155

Answer (A) is incorrect because it presents the VFR weather minimums for below 10,000 feet MSL. Answer (C) is incorrect because the vertical separation from clouds is 1,000 feet both above and below.

Answers

8896 [C]	8897 [A]	8898 [B]	8899 [C]	8900 [B]

121, DSP

8253. Which in-flight conditions are required by a supplemental air carrier to conduct a day, over-the-top flight below the specified IFR minimum en route altitude?

A—The flight must remain clear of clouds by at least 1,000 feet vertically and 1,000 feet horizontally and have at least 3 miles flight visibility.
B—The flight must be conducted at least 1,000 feet above an overcast or broken cloud layer, any higher broken/overcast cloud cover is a minimum of 1,000 feet above the IFR MEA, and have at least 5 miles flight visibility.
C—The height of any higher overcast or broken layer must be at least 500 feet above the IFR MEA.

A person may conduct day Over-the-Top operations in an airplane at flight altitudes lower than the minimum enroute IFR altitudes if—

1. *The operation is conducted at least 1,000 feet above the top of lower broken or overcast cloud cover;*
2. *The top of the lower cloud cover is generally uniform and level;*
3. *Flight visibility is at least 5 miles; and*
4. *The base of any higher broken or overcast cloud cover is generally uniform and level, and is at least 1,000 feet above the minimum enroute IFR altitude for that route segment.*

(PLT468) — 14 CFR §121.657

Answer (A) is incorrect because the flight must remain at least 1,000 feet above the cloud layer with a flight visibility of at least 5 miles. Answer (C) is incorrect because the height of any higher ceiling must be at least 1,000 feet above the IFR MEA.

121, 135, DSP

9395. At what minimum altitude is a turbine-engine-powered, or large airplane, required to enter Class D airspace?

A—1,500 feet AGL.
B—2,000 feet AGL.
C—2,500 feet AGL.

When operating to an airport with an operating control tower, each pilot of a turbine-powered airplane or a large airplane shall, unless otherwise required by the applicable distance from cloud criteria, enter Class D airspace at an altitude of at least 1,500 feet above the surface of the airport. (PLT161) — 14 CFR §91.129

121, 135, DSP

9401. A pilot of a turbine-powered airplane should climb as rapidly as practicable after taking off to what altitude?

A—1,000 feet AGL.
B—1,500 feet AGL.
C—5,000 feet AGL.

When taking off from an airport with an operating control tower, each pilot of a turbine-powered airplane shall climb to an altitude of 1,500 feet above the surface as rapidly as practicable. (PLT459) — 14 CFR §91.129

121, 135, DSP

9396. What is the maximum indicated airspeed a reciprocating-engine-powered airplane may be operated within Class B airspace?

A—180 knots.
B—230 knots.
C—250 knots.

Unless otherwise authorized by the Administrator, no person may operate an aircraft below 10,000 feet MSL at an indicated airspeed of more than 250 knots (288 MPH). There is no specific speed restriction which applies to operation within Class B airspace. (PLT161) — 14 CFR §91.117

Answer (A) is incorrect because 180 knots is the old maximum airspeed for turbine-powered aircraft while operating within Class D airspace (it is now 200 knots). Answer (B) is incorrect because 230 knots is the maximum authorized holding speed for all civil turbojet aircraft while operating from the minimum holding altitude to 14,000 feet. It is not an airspeed limitation in Class B airspace.

121, 135, DSP

8887. The maximum indicated airspeed that an aircraft may be flown in Class B airspace, after departing the primary airport, while at 1,700 feet AGL and 3.5 nautical miles from the airport is

A—200 knots.
B—230 knots.
C—250 knots.

Unless otherwise authorized by the Administrator, no person may operate an aircraft below 10,000 feet MSL at an indicated airspeed of more than 250 knots (288 MPH). (PLT161) — 14 CFR §91.117

Answers

8253 [B] 9395 [A] 9401 [B] 9396 [C] 8887 [C]

121, 135, DSP

9397. At what maximum indicated airspeed can a B-727 operate within Class B airspace without special ATC authorization?

A—230 knots.
B—250 knots.
C—275 knots.

Unless otherwise authorized by the Administrator, no person may operate an aircraft below 10,000 feet MSL at an indicated airspeed of more than 250 knots (288 MPH). There is no specific speed restriction which applies to operation within Class B airspace. (PLT161) — 14 CFR §91.117

Answer (A) is incorrect because 230 knots is not an airspeed limitation in Class B airspace. Answer (C) is incorrect because 275 knots is not an established maximum speed for any type of operation.

121, 135, DSP

9398. At what maximum indicated airspeed may a reciprocating-engine-powered airplane be operated within Class D airspace?

A—156 knots.
B—180 knots.
C—200 knots.

Unless otherwise authorized or required by ATC, no person may operate an aircraft within Class D airspace at an indicated airspeed of more than 200 knots. (PLT161) — 14 CFR §91.117

Answer (A) is incorrect because 156 knots was the old maximum authorized airspeed for reciprocating aircraft in Class D airspace (it is now 200 knots). Answer (B) is incorrect because 180 knots was the old maximum authorized airspeed for turbine-powered aircraft in Class D airspace (it is now 200 knots).

121, 135, DSP

9400. At what maximum indicated airspeed can a reciprocating-engine airplane operate in the airspace underlying Class B airspace?

A—180 knots.
B—200 knots.
C—230 knots.

No person may operate an aircraft in the airspace underlying Class B airspace at an indicated airspeed of more than 200 knots (230 MPH). (PLT161) — 14 CFR §91.117

Answer (A) is incorrect because 180 knots was the old published maximum airspeed for turbine-powered aircraft in Class D airspace (it is now 200 knots). Answer (C) is incorrect because the limitation is 200 knots or 230 MPH, not 230 knots.

Answers

9397 [B] 9398 [C] 9400 [B]

NOTAMs (NOtices To AirMen)

Notices to Airmen (NOTAMs) provide the most current information available. They provide time-critical information on airports and changes that affect the national airspace system and are of concern to instrument flight rule (IFR) operations. NOTAM information is classified into four categories: NOTAM (D) or distant, Flight Data Center (FDC) NOTAMs, pointer NOTAMs, and military NOTAMs.

NOTAM-Ds are attached to hourly weather reports and are available at flight service stations (AFSS/FSS). FDC NOTAMs are issued by the National Flight Data Center and contain regulatory information, such as temporary flight restrictions or an amendment to instrument approach procedures.

Pointer NOTAMs highlight or point out another NOTAM, such as an FDC or NOTAM (D). This type of NOTAM will assist pilots in cross-referencing important information that may not be found under an airport or NAVAID identifier. Military NOTAMs pertain to U.S. Air Force, Army, Marine, and Navy NAVAIDs/airports that are part of the NAS.

NOTAM-Ds and FDC NOTAMs are contained in the Notices to Airmen publication, which is issued every 28 days. Prior to any flight, pilots should check for any NOTAMs that could affect their intended flight.

ALL

9086. What are FDC NOTAMs?

A—Conditions of facilities en route that may cause delays.

B—Time critical aeronautical information of a temporary nature from distant centers.

C—Regulatory amendments to published IAPs and charts not yet available in normally published charts.

FDC NOTAMs contain such things as amendments to published IAPs and other current aeronautical charts and other information which is considered regulatory in nature. (PLT323) — AIM ¶5-1-3

Answer (A) is incorrect because NOTAM (D) contains information on navigational facilities en route that may cause delays. Answer (B) is incorrect because time critical aeronautical information of a temporary nature from distant centers will be included in a NOTAM (D) ("distant").

ALL

9087. What type information is disseminated by NOTAM (D)s?

A—Status of navigation aids, ILSs, radar service available, and other information essential to planning.

B—Airport or primary runway closings, runway and taxiway conditions, and airport lighting aids outages.

C—Temporary flight restrictions, changes in status in navigational aids, and updates on equipment such as VASI.

NOTAM (D) information is disseminated for all navigational facilities that are part of the national airspace system, all IFR airports with approved instrument approaches, and those VFR airports annotated with the NOTAM service symbol (§) in the Airport/Facility Directory. NOTAM (D) information could affect a pilot's decision to make a flight. It includes such information as airport or primary runway closures, changes in the status of navigational aids, ILS's, radar service availability, and other information essential to planned en route, terminal or landing operations. (PLT323) — AIM ¶5-1-3

Answer (B) is incorrect because NOTAM (L)s ("local") contain information on runway closings, runway and taxiway conditions, and airport lighting aids outages that do not affect instrument approach criteria. Answer (C) is incorrect because temporary flight restrictions are normally disseminated in FDC NOTAMs.

ALL

9089. How often are NOTAMs broadcast to pilots on a scheduled basis?

A—15 minutes before and 15 minutes after the hour.

B—Between weather broadcasts on the hour.

C—Hourly, appended to the weather broadcast.

NOTAM (D) information is appended to the hourly weather reports via the Service A (ATC/FSS) telecommunications system. (PLT323) — AIM ¶5-1-3

Answer (A) is incorrect because SIGMETs and AIRMETs are broadcast by FSS's and/or by the Hazardous In-flight Weather Advisory Service (HIWAS) to pilots 15 minutes before and 15 minutes after the hour during the valid period. Answer (B) is incorrect because NOTAM (D)s are appended to the hourly weather broadcast, not a separate broadcast between weather reports.

Answers

9086 [C] 9087 [A] 9089 [C]

Items on the Flight Plan

An IFR flight plan should be filed at least 30 minutes prior to the departure time, and pilots should request their IFR clearance no more than 10 minutes prior to taxi.

In a composite flight plan, one portion of the flight is IFR and the other is VFR. Both the VFR and IFR boxes of the flight plan form should be checked and the route defined in the route of flight box as with any other flight plan. The flight plan should also note where the switch from one type of clearance to the other is planned. If the first part of the flight is IFR, the pilot should cancel with ATC and open the VFR portion with the nearest Flight Service Station by radio. If the first portion is VFR, the pilot should close the VFR portion with the nearest Flight Service Station and request the IFR clearance at least five minutes prior to the IFR portion of the flight.

If the flight is to be flown on established airways, the route should be defined using the airways or jet routes with transitions. Intermediate VORs and fixes on an airway need not be listed. If filing for an off-airway direct route, list all the radio fixes over which the flight will pass. Pilots of appropriately equipped aircraft may file for random RNAV routes. The following rules must be observed:

- Radar monitoring by ATC must be available along the entire proposed route.
- Plan the random route portion to begin and end over appropriate departure and arrival transition fixes or navigation aids appropriate for the altitude structure used for the flight. Use of DPs and STARs, where available, is recommended.
- Define the random route by waypoints. Use degree-distance fixes based on navigational aids appropriate for the altitude structure used. Above FL390 latitude/longitude fixes may be used to define the route.
- List at least one waypoint for each Air Route Traffic Control Center through which the flight will pass. The waypoint must be within 200 NM of the preceding Center's boundary.

A pilot may file a flight plan to an airport containing a special or privately-owned instrument approach procedure only upon approval of the owner.

Air ambulance flights and air carrier flights responding to medical emergencies will receive expedited handling by ATC when necessary. When appropriate, the word "Lifeguard" should be entered in the remarks section of the flight plan. It should also be used in the flight's radio call sign as in, "Lifeguard Delta Thirty-Seven."

ALL

9031. What is the suggested time interval for filing and requesting an IFR flight plan?

A—File at least 30 minutes prior to departure and request the clearance not more than 10 minutes prior to taxi.

B—File at least 30 minutes prior to departure and request the clearance at least 10 minutes prior to taxi.

C—File at least 1 hour prior to departure and request the clearance at least 10 minutes prior to taxi.

Pilots should file IFR flight plans at least 30 minutes prior to the estimated time of departure to preclude possible delay in receiving a departure clearance from ATC. Pilots should call clearance delivery or ground control for their IFR clearance not more than 10 minutes before the proposed taxi time. (PLT224) — AIM ¶5-1-8, 5-2-1

ALL

9028. When a composite flight plan indicates IFR for the first portion of the flight, what is the procedure for the transition?

A—The IFR portion is automatically canceled and the VFR portion is automatically activated when the pilot reports VFR conditions.

B—The pilot should advise ATC to cancel the IFR portion and contact the nearest FSS to activate the VFR portion.

C—The pilot should advise ATC to cancel the IFR portion and activate the VFR portion.

When a flight plan indicates IFR for the first portion of flight and VFR for the latter portion, the pilot will normally be cleared to the point at which the change is proposed. Once the pilot has reported over the clearance limit and does not desire further IFR clearance, he/she should

Answers

9031 [A] 9028 [B]

advise ATC to cancel the IFR portion of the flight plan. The pilot should then contact the FSS to activate the VFR portion of the flight plan. (PLT433) — AIM ¶5-1-7

Answer (A) is incorrect because a pilot can operate on an IFR flight plan in VFR conditions without the IFR portion being automatically canceled. Answer (C) is incorrect because the pilot should contact the nearest FSS, not ATC, to activate the VFR portion.

ALL

9029. Which IFR fix(es) should be entered on a composite flight plan?

A—All compulsory reporting points en route.
B—The VORs that define the IFR portion of the flight.
C—The fix where the IFR portion is to be terminated.

The IFR clearance limit should be listed on a composite flight plan. (PLT433) — AIM ¶5-1-7

Answer (A) is incorrect because compulsory reporting points are not listed on an IFR flight plan unless they define a point of transition, direct route segments or the clearance limit fix. Also, there are no compulsory reporting points for a VFR flight. Answer (B) is incorrect because IFR fixes can be defined as intersections, waypoints, and DME distance, along with VORs.

ALL

9030. When a composite flight plan indicates VFR for the first portion of the flight, what is the procedure for the transition?

A—The VFR portion is automatically canceled and the IFR portion is automatically activated when the pilot reports IFR conditions.
B—The pilot should advise ATC to cancel VFR and activate the IFR portion of the flight.
C—The pilot should close the VFR portion with the nearest FSS and request the IFR clearance at least 5 minutes prior to IFR.

If VFR flight is conducted for the first portion of a composite flight plan, the pilot should report the departure time to the FSS with which he/she filed. He/she should close the VFR portion and request ATC clearance from the FSS nearest the point at which the change from VFR to IFR is proposed. (PLT433) — AIM ¶5-1-7

Answer (A) is incorrect because VFR flight plans are never automatically closed; it is the pilot's responsibility to close a VFR flight plan. An IFR clearance must be requested and received before entering IFR conditions in controlled airspace. It is not automatically activated when the pilot reports IFR conditions. Answer (B) is incorrect because the pilot should cancel a VFR flight plan with the nearest FSS, not ATC.

ALL

9032. How should the route of flight be defined on an IFR flight plan?

A—A simplified route via airways or jet routes with transitions.
B—A route via airways or jet routes with VORs and fixes used.
C—A route via airways or jet routes with only the compulsory reporting points.

Pilots are requested to file via airways or jet routes established for use at the altitude or flight level planned. If the flight is to be conducted via designated airways or jet routes, describe the route by indicating the type and number designators of the airway(s) or jet route(s) requested. If more than one airway or jet route is to be used, clearly indicate points of transition. (PLT224) — AIM ¶5-1-8

Answer (B) is incorrect because, to simplify the route, all VORs and fixes are not used to define a route on an IFR flight plan. Answer (C) is incorrect because compulsory reporting points might not define the transitions between airways or jet routes.

ALL

9033. How should an off-airway direct flight be defined on an IFR flight plan?

A—The initial fix, the true course, and the final fix.
B—All radio fixes over which the flight will pass.
C—The initial fix, all radio fixes which the pilot wishes to be compulsory reporting points, and the final fix.

Any portions of the route which will not be flown on the radials or courses of established airways or routes, such as direct route flights, must be clearly defined by indicating the radio fixes over which the flight will pass. (PLT225) — AIM ¶5-1-8

Answer (A) is incorrect because true course is not an item that is reported on an IFR flight plan. The initial fix and the final fix are listed as radio fixes that define the start and finish points of a flight. Answer (C) is incorrect because initial and final fixes are required to define random RNAV (not direct flight) routes. All radio fixes that define the route of a direct flight automatically become compulsory reporting points, not just those the pilot chooses.

Answers

9029 [C]	9030 [C]	9032 [A]	9033 [B]

ALL

9026. How are random RNAV routes below FL 390 defined on the IFR flight plan?

A—Define route waypoints using degree-distance fixes based on appropriate navigational aids for the route and altitude.
B—List the initial and final fix with at least one waypoint each 200 NM.
C—Begin and end over appropriate arrival and departure transition fixes or navigation aids for the altitude being flown, define the random route waypoints by using degree-distance fixes based on navigation aids appropriate for the altitude being flown.

Pilots of aircraft equipped with operational area navigation equipment may file for random RNAV routes throughout the national airspace system, where radar monitoring by ATC is available, in accordance with the following:

1. *File airport-to-airport flight plans prior to departure.*
2. *File the appropriate RNAV capability suffix in the flight plan.*
3. *Plan the random route portion of the flight plan to begin and end over appropriate arrival and departure fixes.*
4. *Define the random route by waypoints. File route description waypoints by using degree/distance fixes based on navigation aids which are appropriate to the altitude.*
5. *File a minimum of one route description waypoint for each ARTCC through whose area the random route will be flown. These waypoints must be located within 200 NM of the preceding center's boundary.*

(PLT224) — AIM ¶5-1-8

Answer (A) is incorrect because RNAV routes defined on an IFR flight plan must also begin and end over an established radio fix. Answer (B) is incorrect because RNAV waypoints have no established distance requirement. A minimum of one waypoint must be filed for each ARTCC through which the route is planned, and this must be located within 200 NM of the preceding center's boundary.

ALL

9027. What is one limitation when filing a random RNAV route on an IFR flight plan?

A—The waypoints must be located within 200 NM of each other.
B—The entire route must be within radar environment.
C—The waypoints may only be defined by degree-distance fixes based on appropriate navigational aids.

Random RNAV routes can only be approved in a radar environment. Aircraft operating at or above FL390 may file waypoints based on latitude/longitude fixes, under some circumstances. (PLT225) — AIM ¶5-1-8

Answer (A) is incorrect because VOR/VORTAC facilities must be within 200 NM of each other when operating above FL450 to define a direct route. Answer (C) is incorrect because random RNAV waypoints may be defined by degree-distance fixes based on appropriate navigational aids, of latitude/longitude coordinate navigation, independent of VOR/TACAN references, operating at and above FL390 in the conterminous U.S.

ALL

9040. Under what condition may a pilot file an IFR flight plan containing a special or privately owned IAP?

A—Upon approval of ATC.
B—Upon approval of the owner.
C—Upon signing a waiver of responsibility.

Pilots planning flights to locations served by special IAPs should obtain advance approval from the owner of the procedure. Approval by the owner is necessary because special procedures are for the exclusive use of the single interest unless otherwise authorized by the owner. Controllers assume a pilot has obtained approval and is aware of any details of the procedure if he/she files an IFR flight plan to that airport. (PLT083) — AIM ¶5-4-7

Answer (A) is incorrect because ATC is not required to question pilots to determine whether they have the owner's permission to use the procedure. Answer (C) is incorrect because a pilot is responsible for the safe operation of the airplane. To sign a waiver of responsibility is contrary to a pilot's duty.

ALL

9053. To assure expeditious handling of a civilian air ambulance flight, the word "LIFEGUARD" should be entered in which section of the flight plan?

A—Aircraft type/special equipment block.
B—Pilot's name and address block.
C—Remarks block.

When expeditious handling is necessary because of a medical emergency, add the word "LIFEGUARD" in the remarks section of the flight plan. (PLT225) — AIM ¶4-2-4

Answer (A) is incorrect because only the airplane's designator or manufacturer's name and the transponder DME and/or RNAV equipment code is entered in the aircraft type/special equipment block. Answer (B) is incorrect because the complete name, address, and telephone number of the pilot-in-command are entered in the pilot's name and address block. Sufficient information is listed here to identify home base, airport, or operator. This information would be essential in the event of a search and rescue operation.

Answers

9026 [C] 9027 [B] 9040 [B] 9053 [C]

Alternate Airport Planning

An airport may not be available for alternate use if the airport NAVAID is unmonitored, is GPS-based, or if it does not have weather-reporting capabilities.

For an airport to be used as an alternate, the forecast weather at that airport must meet certain qualifications at the estimated time of arrival. Standard alternate minimums for a precision approach are a 600-foot ceiling and 2 SM visibility. For a nonprecision approach, the minimums are an 800-foot ceiling and 2 SM visibility. Standard alternate minimums apply unless higher alternate minimums are listed for an airport.

Alternate Airport for Destination—Domestic Air Carriers: Unless the weather at the destination meets certain criteria, an alternate must be listed in the dispatch release (and flight plan) for each destination airport. If the weather at the first listed alternate is marginal (as defined by the operations specifications) at least one additional alternate must be listed.

Alternate Airport for Destination—Flag Carriers: An alternate airport must be listed in the dispatch release (and flight plan) for all flag air carrier flights longer than 6 hours. An alternate is not required for a flag air carrier flight if it is scheduled for less than 6 hours and the weather forecast for the destination meets certain criteria. For the period from 1 hour before to 1 hour after the estimated time of arrival:

- The ceiling must be forecast to be at least 1,500 feet above the lowest minimums or 2,000 feet, whichever is higher; and
- The visibility must be forecast to be 3 miles, or 2 miles greater than the lowest applicable visibility minimum, whichever is greater.

Alternate Airport for Destination—Supplemental Air Carriers and Commercial Operators: Except for certain operations, a supplemental air carrier or commercial operator must always list an alternate airport regardless of existing or forecast weather conditions.

An airport cannot be listed as an alternate in the dispatch or flight release unless the appropriate weather reports and forecasts indicate that the weather conditions will be at or above the alternate weather minimums specified in the certificate holder's operations specifications for that airport, when the flight arrives. Alternate weather minimums are for planning purposes only and do not apply to actual operations. If an air carrier flight actually diverts to an alternate airport, the crew may use the actual weather minimums shown on the IAP (Instrument Approach Procedure) Chart for that airport.

If the weather conditions at the departure airport are below landing minimums in the airline's operations specifications, a departure alternate must be listed in the dispatch or the flight release. Weather at alternate airports must meet the conditions for alternates in the operations specifications. The maximum distance to the departure alternate for a two-engine airplane cannot be more than 1 hour from the departure airport in still air with one engine operating. The distance to the departure alternate for an airplane with three or more engines cannot be more than 2 hours from the departure airport in still air with one engine inoperative.

ALL

9394-1. When proceeding to the alternate airport, which minimums apply?

A—The IFR alternate minimums section in front of the NOAA IAP book.
B—2000-3 for at least 1 hour before until 1 hour after the ETA.
C—The actual minimums shown on the IAP chart for the airport.

When the approach procedure being used provides for and requires the use of a DH or MDA, the authorized decision height or authorized minimum descent altitude is the DH or MDA prescribed by the approach procedure, the DH or MDA prescribed for the pilot-in-command, or the DH or MDA for which the aircraft is equipped, whichever is highest.

Note: *The alternate airport minimums are used only during preflight planning to determine the suitability of an airport as an IFR alternate. They impose no additional restrictions should a flight actually divert to the filed alternate. (PLT421) — 14 CFR §91.175*

Answer (A) is incorrect because the alternate minimums listed in the NOAA IAP (National Oceanic and Atmospheric Administration Instrument Approach Procedure) book refer to the ceiling and visibility requirements for that airport in order to file it as an alternate, not the ceiling and visibility required to execute an instrument approach. Answer (B) is incorrect because 2000-3 minimums apply to the destination airport. If your destination airport has a forecast ceiling of at least 2,000 feet and a visibility of at least 3 miles, an alternate airport need not be filed in the flight plan.

ALL

9394-2. An airport may not be qualified for alternate use if

A—the airport has AWOS-3 weather reporting.
B—the airport is located next to a restricted or prohibited area.
C—the NAVAIDs used for the final approach are unmonitored.

Not all airports can be used as an alternate. An airport may not be qualified for alternate use if the airport NAVAID is unmonitored, is GPS-based, or if it does not have weather reporting capability. (PLT379) — FAA-H-8261-1

Answer (A) is incorrect because an airport can qualify for alternate use if it has any weather reporting capability. Answer (B) is incorrect because an airport can qualify for alternate use even if it is located in a restricted or prohibited area.

ALL

9770. When planning to use RNAV equipment with GPS input for an instrument approach at a destination airport, any required alternate airport must have an available instrument approach procedure that does not

A—require the use of GPS except when the RNAV system has a WAAS input.
B—require the use of GPS except when the RNAV system has an IRU input.
C—require the use of GPS except when dual, independent GPS receivers are installed.

Aircraft using GPS navigation equipment under IFR for domestic en route, terminal operations, and certain IAPs must be equipped with an approved and operational alternate means of navigation appropriate to the flight. However, a required alternate airport may be selected if it uses an RNAV system with WAAS equipment. (PLT420) — FAA-H-8083-15

121, DSP

8247. When the forecast weather conditions for a destination and alternate airport are considered marginal for a domestic air carrier's operation, what specific action should the dispatcher or pilot in command take?

A—List an airport where the forecast weather is not marginal as the alternate.
B—Add 1 additional hour of fuel based on cruise power settings for the airplane in use.
C—List at least one additional alternate airport.

When weather conditions forecast for the destination and first alternate airport are marginal, at least one additional alternate must be designated. (PLT379) — 14 CFR §121.619

121, DSP

8256. Which dispatch requirement applies to a flag air carrier that is scheduled for a 7-hour IFR flight?

A—No alternate airport is required if the forecast weather at the ETA at the destination airport is at least 1,500 feet and 3 miles.
B—An alternate airport is not required if the ceiling will be at least 1,500 feet above the lowest circling MDA.
C—An alternate airport is required.

All flag air carrier flights over 6 hours require an alternate airport. (PLT379) — 14 CFR §121.621

Answers (A) and (B) are incorrect because whenever the scheduled flight exceeds 6 hours, a flag air carrier must list an alternate regardless of the weather.

Answers

9394-1 [C]	9394-2 [C]	9770 [A]	8247 [C]	8256 [C]

121, DSP

8262. An alternate airport is not required to dispatch a flag air carrier airplane for a flight of less than 6 hours when the visibility for at least 1 hour before and 1 hour after the ETA at the destination airport is forecast to be

A—2 miles or greater.
B—at least 3 miles, or 2 miles more than the lowest applicable minimum.
C—3 miles.

An alternate airport need not be listed if the destination weather, from an hour before to an hour after the ETA, is forecast to have the required ceiling criteria and the visibility is forecast to be at least 3 miles, or 2 miles more than the lowest visibility minimums, whichever is greater, for the instrument approach procedures to be used at the destination airport. (PLT379) — 14 CFR §121.621

121, DSP

8251. When is a supplemental air carrier, operating under IFR, required to list an alternate airport for each destination airport within the 48 contiguous United States?

A—When the forecast weather indicates the ceiling will be less than 1,000 feet and visibility less than 2 miles at the estimated time of arrival.
B—On all flights, an alternate is required regardless of existing or forecast weather conditions at the destination.
C—When the flight is scheduled for more than 6 hours en route.

A supplemental air carrier must declare an alternate airport for all IFR operations. (PLT379) — 14 CFR §121.623

121, DSP

8254. Prior to listing an airport as an alternate airport in the dispatch or flight release, weather reports and forecasts must indicate that weather conditions will be at or above authorized minimums at that airport

A—for a period 1 hour before or after the ETA.
B—during the entire flight.
C—when the flight arrives.

No person may list an airport as an alternate airport in the dispatch release or flight release unless the appropriate weather reports or forecasts, or any combination thereof, indicate that the weather conditions will be at or above the alternate weather minimums specified in the certificate holder's operations specifications for that airport when the flight arrives. (PLT380) — 14 CFR §121.625

121, DSP

8255. The minimum weather conditions that must exist for an airport to be listed as an alternate in the dispatch release for a domestic air carrier flight are

A—those listed in the NOAA IAP charts for the alternate airport, at the time the flight is expected to arrive.
B—those specified in the certificate holder's Operations Specifications for that airport, when the flight arrives.
C—those listed in the NOAA IAP charts for the alternate airport, from 1 hour before or after the ETA for that flight.

No person may list an airport as an alternate airport in the dispatch release or flight release unless the appropriate weather reports or forecasts, or any combination thereof, indicate that the weather conditions will be at or above the alternate weather minimums specified in the certificate holder's operations specifications for that airport when the flight arrives. (PLT380) — 14 CFR §121.625

Answers (A) and (C) are incorrect because, although the alternate minimums in IAP charts may coincide with the air carrier's operation specifications, it is the operations specifications that determine alternate weather minimums.

121, DSP

8248. An alternate airport for departure is required

A—if weather conditions are below authorized landing minimums at the departure airport.
B—when the weather forecast at the ETD is for landing minimums only at the departure airport.
C—when destination weather is marginal VFR (ceiling less than 3,000 feet and visibility less than 5 SM).

If the weather conditions at the airport of takeoff are below the landing minimums in the certificate holder's operations specifications for that airport, no person may dispatch or release an aircraft from that airport unless the dispatch or flight release specifies an alternate airport located within the following distances from the airport of takeoff.

1. *Aircraft having two engines: Not more than 1 hour from the departure airport at normal cruising speed in still air with one engine inoperative.*
2. *Aircraft having three or more engines: Not more than 2 hours from the departure airport at normal cruising speed in still air with one engine inoperative.*

(PLT379) — 14 CFR §121.617

Answers

8262 [B]	8251 [B]	8254 [C]	8255 [B]	8248 [A]

121, DSP

8249. What is the maximum distance that a departure alternate airport may be from the departure airport for a two-engine airplane?

A—1 hour at normal cruise speed in still air with both engines operating.
B—1 hour at normal cruise speed in still air with one engine operating.
C—2 hours at normal cruise speed in still air with one engine operating.

If the weather conditions at the airport of takeoff are below the landing minimums in the certificate holder's operations specifications for that airport, no person may dispatch or release an aircraft from that airport unless the dispatch or flight release specifies an alternate airport located within the following distances from the airport of takeoff.

1. *Aircraft having two engines: Not more than 1 hour from the departure airport at normal cruising speed in still air with one engine inoperative.*
2. *Aircraft having three or more engines: Not more than 2 hours from the departure airport at normal cruising speed in still air with one engine inoperative.*

(PLT379) — 14 CFR §121.617

Answer (A) is incorrect because the maximum distance is determined with one engine operating. Answer (C) is incorrect because 2 hours is the limit for airplanes with three or more engines with one engine inoperative.

121, DSP

8250. If a four-engine air carrier airplane is dispatched from an airport that is below landing minimums, what is the maximum distance that a departure alternate airport may be located from the departure airport?

A—Not more than 2 hours at cruise speed with one engine inoperative.
B—Not more than 2 hours at normal cruise speed in still air with one engine inoperative.
C—Not more than 1 hour at normal cruise speed in still air with one engine inoperative.

If the weather conditions at the airport of takeoff are below the landing minimums in the certificate holder's operations specifications for that airport, no person may dispatch or release an aircraft from that airport unless the dispatch or flight release specifies an alternate airport located within the following distances from the airport of takeoff.

1. *Aircraft having two engines: Not more than 1 hour from the departure airport at normal cruising speed in still air with one engine inoperative.*
2. *Aircraft having three or more engines: Not more than 2 hours from the departure airport at normal cruising speed in still air with one engine inoperative.*

(PLT396) — 14 CFR §121.617

121, DSP

8252. When a departure alternate is required for a three-engine air carrier flight, it must be located at a distance not greater than

A—2 hours from the departure airport at normal cruising speed in still air with one engine not functioning.
B—1 hour from the departure airport at normal cruising speed in still air with one engine inoperative.
C—2 hours from the departure airport at normal cruising speed in still air.

If the weather conditions at the airport of takeoff are below the landing minimums in the certificate holder's operations specifications for that airport, no person may dispatch or release an aircraft from that airport unless the dispatch or flight release specifies an alternate airport located within the following distances from the airport of takeoff.

1. *Aircraft having two engines: Not more than 1 hour from the departure airport at normal cruising speed in still air with one engine inoperative.*
2. *Aircraft having three or more engines: Not more than 2 hours from the departure airport at normal cruising speed in still air with one engine inoperative.*

(PLT379) — 14 CFR §121.617

Answer (B) is incorrect because 1 hour is correct for a two-engine airplane. Answer (C) is incorrect because it does not contain the words "with one engine inoperative."

Answers

8249 [B]   8250 [B]   8252 [A]

ATC Clearances

No one may operate an aircraft in Class A, B, C, D or E airspace under Instrument Flight Rules (IFR) unless he/she has filed an IFR flight plan and received an appropriate ATC clearance. No flight plan or clearance is required for IFR operations in Class G airspace.

IFR clearances always contain:

- A clearance limit (usually the destination);
- Route of flight;
- Altitude assignment; and
- Departure instructions (could be a DP).

The words "cleared as filed" replace only the route of flight portion of a normal clearance. The controller will still state the destination airport, the enroute altitude (or initial altitude and expected final altitude) and DP if appropriate. If a STAR is filed on the flight plan, it is considered part of the enroute portion of the flight plan and is included in the term "cleared as filed."

When an ATC clearance has been received, you may not deviate from it (except in an emergency) unless an amended clearance is received. If you are uncertain of the meaning of an ATC clearance or the clearance appears to be contrary to a regulation, you should immediately request a clarification. When you receive a clearance you should always read back altitude assignments, altitude restrictions, and vectors. A Departure Procedure (DP) may contain these elements but they need not be included in the readback unless the ATC controller specifically states them.

At airports with pretaxi clearance delivery, a pilot should call for the clearance 10 minutes prior to the desired taxi time. After receiving clearance on the clearance delivery frequency, the pilot should call ground control for taxi when ready.

Occasionally, an aircraft with an IFR release will be held on the ground for traffic management reasons. The traffic may be too heavy or weather may be causing ATC delays. If this happens to an aircraft waiting for takeoff, it will be given a hold for release instruction.

When ATC can anticipate long delays for IFR aircraft, they will establish gate hold procedures. The idea is to hold aircraft at the gate rather than cause congestion and unnecessary fuel burn on the taxiways while waiting for an IFR release. Ground control will instruct aircraft when to start engines. ATC expects that turbine-powered aircraft will be ready for takeoff as soon as they reach the runway after having been released from gate hold.

When departing uncontrolled airports, IFR flights will often receive a void time with their clearance. The void time is a usually a 30-minute window of time during which the aircraft must takeoff for its IFR clearance to be valid. If unable to comply with the void time, a pilot must receive another clearance with an amended void time.

The flight plan of an airborne IFR aircraft may only be canceled when the aircraft is in VFR weather conditions and outside of Class A airspace.

ALL

9374. A pilot is operating in Class G airspace. If existing weather conditions are below those for VFR flight, an IFR flight plan must be filed and an ATC clearance received prior to

A—takeoff if weather conditions are below IFR minimums.
B—entering controlled airspace.
C—entering IFR weather conditions.

No person may operate an aircraft in Class A, B, C, D or E airspace under IFR unless an IFR flight plan has been filed and an appropriate ATC clearance has been received. (PLT162) — 14 CFR §91.173

Answers (A) and (C) are incorrect because an IFR flight plan and an ATC clearance are not required to fly in IMC (instrument meteorological conditions) in Class G airspace.

ALL

9006. What minimum information does an abbreviated departure clearance "cleared as filed" include?

A—Clearance limit and en route altitude.
B—Clearance limit, transponder code, and DP, if appropriate.
C—Destination airport, en route altitude, transponder code, and DP, if appropriate.

The following apply to "cleared as filed" clearances:

1. *The clearance as issued will include the destination airport filed in the flight plan. "Cleared to (destination) as filed."*
2. *The controller will state the DP name and number.*
3. *STARs, when filed in a flight plan, are considered a part of the filed route of flight and will not normally be stated in an initial clearance.*
4. *An enroute altitude will be stated in the clearance or the pilot will be advised to expect an assigned or filed altitude within a given time frame or at a certain point after departure. This may be done verbally in the departure clearance or stated in the DP.*

(PLT370) — AIM ¶5-2-5

Answer (A) is incorrect because a clearance limit may be a fix, point, or location. An abbreviated clearance will be a clearance to the destination airport. In some cases, a clearance is issued to a fix (limit) from which another clearance limit will be issued. DPs are stated in all IFR departure clearances when appropriate. Answer (B) is incorrect because a clearance will state the destination airport's name, not a clearance limit.

ALL

9439. An ATC "instruction"

A—is the same as an ATC "clearance."
B—is a directive issued by ATC for the purpose of requiring a pilot to take a specific action.
C—must be "read back" in full to the controller and confirmed before becoming effective.

Instructions are directives issued by air traffic control for the purpose of requiring a pilot to take specific actions; e.g., "Turn left heading two five zero," "Go around," "Clear the runway." (PLT370) — Pilot/Controller Glossary

Answer (A) is incorrect because an ATC clearance is not the same as an ATC instruction. Answer (C) is incorrect because an ATC instruction does not have to be read back in full to the controller and confirmed before becoming effective.

ALL

9402. What action should a pilot take when a clearance is received from ATC that appears to be contrary to a regulation?

A—Read the clearance back in its entirety.
B—Request a clarification from ATC.
C—Do not accept the clearance.

If a pilot is uncertain of the meaning of an ATC clearance, he/she shall immediately request clarification from ATC. (PLT444) — 14 CFR §91.123

Answer (A) is incorrect because reading the clearance back in its entirety does not inform ATC of the possible conflict to a regulation. A pilot should actively seek clarification if there is any doubt. Answer (C) is incorrect because not accepting a clearance is not the proper procedure to use when, in a pilot's opinion, it would conflict with a regulation. First, a pilot should receive a clarification from ATC, then ask for an amended clearance, if necessary.

Answers

9374 [B]	9006 [C]	9439 [B]	9402 [B]

ALL

9045. What is the pilot's responsibility for clearance or instruction readback?

A—Except for SIDs, read back altitude assignments, altitude restrictions, and vectors.
B—If the clearance or instruction is understood, an acknowledgment is sufficient.
C—Read back the entire clearance or instruction to confirm the message is understood.

Pilots of airborne aircraft should read back those parts of ATC clearances and instructions containing altitude assignments or vectors. Altitudes contained in charted procedures such as DPs, instrument approaches, etc., should not be read back unless they are specifically stated by the controller. (PLT370) — AIM ¶4-4-7

Answer (B) is incorrect because the best way to know that the clearance or instruction is understood is to read back the "numbers" as a double-check between the pilot and ATC. This reduces the kinds of communication errors that occur when a number is either misheard or is incorrect. Answer (C) is incorrect because the pilot's responsibility is to read back the clearances and instructions containing altitude assignments, altitude restrictions, and vectors, not the entire clearance or instruction.

ALL

9008. What is the normal procedure for IFR departures at locations with pretaxi clearance programs?

A—Pilots request IFR clearance when ready to taxi. The pilot will receive taxi instruction with clearance.
B—Pilots request IFR clearance when ready to taxi. Pilots will receive taxi clearance, then receive IFR clearance while taxiing or on runup.
C—Pilots request IFR clearance 10 minutes or less prior to taxi, then request taxi clearance from ground control.

When operating at airports with pretaxi clearance delivery, participating pilots should call clearance delivery or ground control not more than 10 minutes before taxi. When the IFR clearance is received on clearance delivery frequency, pilots should call ground control when ready to taxi. (PLT370) — AIM ¶5-2-1

Answers (A) and (B) are incorrect because the pilot will first be given the IFR clearance, then the taxi instruction or clearance.

ALL

9009. What is the purpose of the term "hold for release" when included in an IFR clearance?

A—A procedure for delaying departure for traffic volume, weather, or need to issue further instructions.
B—When an IFR clearance is received by telephone, the pilot will have time to prepare for takeoff prior to being released.
C—Gate hold procedures are in effect and the pilot receives an estimate of the time the flight will be released.

ATC may issue "hold for release" instructions in a clearance to delay an aircraft's departure for traffic management reasons (i.e., weather, traffic volume, etc.). (PLT370) — AIM ¶5-2-6

Answer (B) is incorrect because, when a pilot receives an IFR clearance via telephone, it is normally because he/she is departing from an uncontrolled airport. In this case, ATC would issue a clearance void time, not a hold for release. Answer (C) is incorrect because gate hold procedures are in effect whenever departure delays exceed (or are expected to exceed) 15 minutes. This procedure is not a way for ATC to delay an airplane's departure.

ALL

9056. What action should the pilot take when "gate hold" procedures are in effect?

A—Contact ground control prior to starting engines for sequencing.
B—Taxi into position and hold prior to requesting clearance.
C—Start engines, perform pretakeoff check, and request clearance prior to leaving the parking area.

When gate hold procedures are in effect, pilots should contact ground control or clearance delivery prior to starting engines, because departure delays are expected to exceed 15 minutes. (PLT434) — AIM ¶4-3-15

Answer (B) is incorrect because taxi into position means that the pilot is on the active runway and ready for takeoff. This is not a position where ATC would issue an IFR clearance. Answer (C) is incorrect because pilots should contact ground control for sequencing before starting engines.

Answers

9045 [A]   9008 [C]   9009 [A]   9056 [A]

ALL

9057. What special consideration is given for turbine-powered aircraft when "gate hold" procedures are in effect?

A—They are given preference for departure over other aircraft.
B—They are expected to be ready for takeoff when they reach the runway or warmup block.
C—They are expected to be ready for takeoff prior to taxi and will receive takeoff clearance prior to taxi.

Even with gate holds in effect, the tower controller will consider that pilots of turbine-powered aircraft are ready for takeoff when they reach the runway or warm up block unless advised otherwise. (PLT149) — AIM ¶4-3-15

Answer (A) is incorrect because, when gate hold procedures are in effect, sequencing of all airplanes is based on the initial call-up to ground control or clearance delivery. Answer (C) is incorrect because a pilot of any airplane should be ready to taxi prior to requesting taxi, and takeoff clearance is received prior to takeoff.

ALL

9007. Under what condition does a pilot receive a "void time" specified in the clearance?

A—On an uncontrolled airport.
B—When "gate hold" procedures are in effect.
C—If the clearance is received prior to starting engines.

If operating from an airport not served by a control tower, the pilot may receive a clearance containing a provision that if the flight has not departed by a specific time, the clearance is void. (PLT370) — AIM ¶5-2-6

Answer (B) is incorrect because gate hold procedures are in effect whenever departure delays exceed or are anticipated to exceed 15 minutes. Answer (C) is incorrect because clearances can be issued before starting the airplane's engine(s).

ALL

9005. Under what condition may a pilot cancel an IFR flight plan prior to completing the flight?

A—Anytime it appears the clearance will cause a deviation from FARs.
B—Anytime within controlled airspace by contacting ARTCC.
C—Only if in VFR conditions in other than Class A airspace.

An IFR flight plan may be canceled anytime the flight is operating in VFR conditions outside Class A airspace. (PLT224) — AIM ¶5-1-14

Answer (A) is incorrect because anytime a clearance appears to deviate from a regulation, the pilot should request clarification from ATC and an amended clearance. Answer (B) is incorrect because all aircraft in Class A airspace (above FL180) or when operating in IMC in Class B, C, D or E must be operating under an IFR flight plan.

ALL

9737. (Refer to Runway Incursion Figure.) You have requested taxi instructions for takeoff using Runway 16. The controller issues the following taxi instructions: "N123, Taxi to runway 16." Where are you required to stop in order to be in compliance with the controller's instructions?

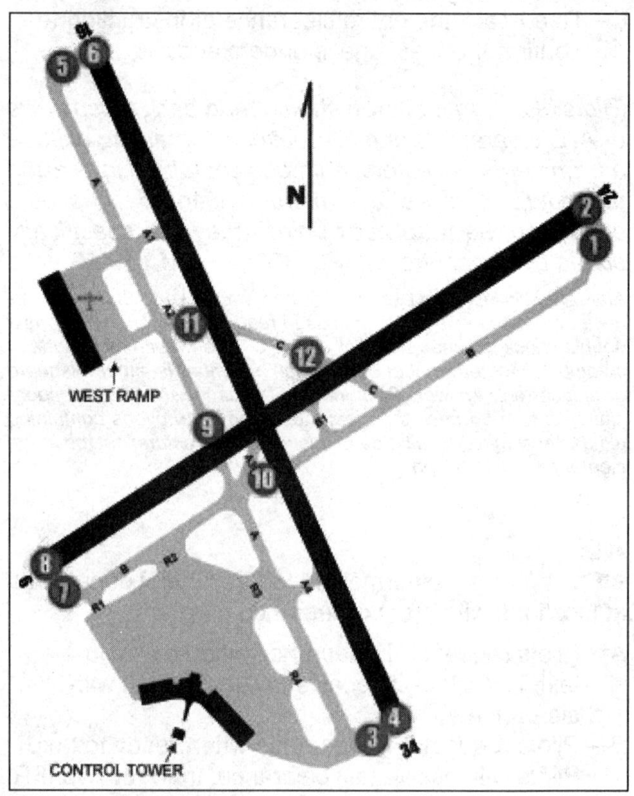

Runway Incursion

A—5 (Five).
B—6 (Six).
C—9 (Nine).

When ATC clears an aircraft to "taxi to" an assigned takeoff runway, the absence of holding instructions does not authorize the aircraft to "cross" all runways which the taxi route intersects except the assigned takeoff runway. A clearance must be obtained prior to crossing any runway. It does not include authorization to "taxi onto" or "cross" the assigned takeoff runway at any point. You should taxi and hold short of runway 16, which is position 5. (PLT141) — AIM 4-3-18

Answer (B) is incorrect because "taxi to" does not authorize the aircraft to "taxi onto" the assigned takeoff runway. Answer (C) is incorrect because the airplane should taxi the most direct route to the assigned runway unless instructed otherwise; position 9 would not be encountered for the airplane at the west ramp to taxi to runway 16.

Answers

9057 [B]	9007 [A]	9005 [C]	9737 [A]

ALL

9788. As you call for taxi instructions, the key words to understand are

A—cleared to runway.
B—hold short of or "cross."
C—taxi to and "expedite."

When issuing taxi instructions to any point other than an assigned takeoff runway, ATC will specify the point to taxi to, issue taxi instructions, and state any hold short instructions or runway crossing clearances if the taxi route will cross a runway. (PLT149) — AIM ¶4-3-18

Answer (A) is incorrect because ATC will not use the word "cleared" in conjunction with authorization for aircraft to taxi. Answer (C) is incorrect because "expedite" is not as critical as being told to hold short or cross.

ALL

9789. You received these ATC taxi instructions: "Taxi to Runway 30 via Lima and hold short of Runway 25L". Your airplane is on the ramp by the terminal and NWS on the east side of the airport. Your taxi route

A—requires crossing of Runway 25L at Lima.
B—involves transiting HS 4.
C—requires crossing Runway 34R en route to the assigned runway.

When assigned a takeoff runway, ATC will first specify the runway, issue taxi instructions, and state any hold short instructions or runway crossing clearances if the taxi route will cross a runway. (PLT149) — AIM ¶4-3-18

ALL

9790. As you rolled out long on Runway 30 after landing at Long Beach (LGB) (figures 241 and 242), you slowed and turned left on very wide pavement and now see Taxiway D signs on both sides of your pavement. You notice your heading is about 250°. Tower is urging you to turn left on D, cross 16R/34L, then taxi to G and hold short of Runway 30. You now know you

A—exited onto Runway 25R and transited HS 2.
B—exited onto Taxiway G.
C—exited at Taxiway J and transited HS 4.

The very wide pavement and taxiway D signs on both sides of the pavement indicate you are now on runway 25R in the HS1 and HS2 region. (PLT149) — AIM ¶4-3-18

Answers

9788 [B] 9789 [A] 9790 [A]

Takeoff Procedures

Unless otherwise authorized by the FAA, an air carrier flight may not takeoff unless the weather meets the prescribed takeoff minimums for that airport. If takeoff minimums are not published for the airport, the following visibility is required for takeoff:

- For aircraft having two engines or less: 1 statute mile visibility.
- For aircraft having three or more engines: 1/2 statute mile visibility.

If an air carrier flight is going to takeoff from an airport that is not listed in its operations specifications, the pilot must observe the takeoff weather minimums published for that airport. If no takeoff weather minimums are published for that airport, then the pilot must be sure that the ceiling and visibility meet a sliding scale requirement of 800-2 or 900-1-1/2 or 1,000-1.

121, 135, DSP

9370. When takeoff minimums are not prescribed for a civil airport, what are the takeoff minimums under IFR for a three-engine airplane?

A—1 SM.
B—1/2 SM.
C—300 feet and 1/2 SM.

If takeoff minimums are not prescribed under 14 CFR Part 97, the takeoff minimums under IFR for aircraft having more than two engines are 1/2 statute mile visibility. (PLT421) — 14 CFR §91.175

Answer (A) is incorrect because 1 SM visibility is for aircraft, other than helicopters, having two engines or less. Answer (C) is incorrect because minimum ceilings are not specified for takeoff minimums.

121, DSP

8257. An airport is not listed in a domestic Air Carrier's Operations Specifications and does not have the prescribed takeoff weather minimums. What are the minimum weather conditions required for takeoff?

A—800-2.
B—900-1.
C—1000-1/2.

When departing from an alternate airport within the United States which is not listed in the operations specifications, an air carrier must use the 14 CFR Part 97 takeoff minimums. When no takeoff minimums are specified, the air carrier must use a ceiling and visibility of 800-2, 900-1-1/2, or 1,000-1. (PLT398) — 14 CFR §121.637

Answer (B) is incorrect because a 900-foot ceiling requires 1.5 miles visibility. Answer (C) is incorrect because a 1,000-foot ceiling requires 1 mile visibility.

121, DSP

8261. The weather conditions that meet the minimum requirements for a flag air carrier to take off from an alternate airport that is not listed in the Operations Specifications are

A—800-2, 900-1-1/2, or 1000-1.
B—800-1/2, 900-1, or 1000-2.
C—800-1, 900-2, or 1000-3.

When departing from an alternate airport within the United States which is not listed in the operations specifications, an air carrier must use the 14 CFR Part 97 takeoff minimums. When no takeoff minimums are specified, the air carrier must use a ceiling and visibility of 800-2, 900-1-1/2, or 1,000-1. (PLT380) — 14 CFR §121.637

121, DSP

8263. The minimum weather conditions that must exist for a domestic air carrier flight to take off from an airport that is not listed in the Air Carrier's Operations Specifications (takeoff minimums are not prescribed for that airport) is

A—800-2, 1,100-1, or 900-1-1/2.
B—1,000-1, 900-1-1/4, or 800-2.
C—1,000-1, 900-1-1/2, or 800-2.

When departing from an alternate airport within the United States which is not listed in the operations specifications, an air carrier must use the 14 CFR Part 97 takeoff minimums. When no takeoff minimums are specified, the air carrier must use a ceiling and visibility of 800-2, 900-1-1/2, or 1,000-1. (PLT459) — 14 CFR §121.637

Answers

9370 [B] 8257 [A] 8261 [A] 8263 [C]

121, DSP

8264. When an alternate airport outside the United States has no prescribed takeoff minimums and is not listed in a Flag Air Carrier's Operations Specifications, the minimum weather conditions that will meet the requirements for takeoff is

A—800-1-1/2.
B—600-2.
C—900-1-1/2.

When departing from an alternate airport outside the United States which is not listed in the operations specifications, an air carrier must use the takeoff minimums approved by the government of the country in which the airport is located. When no takeoff minimums are specified, the air carrier must use a ceiling and visibility of 800-2, 900-1-1/2, or 1,000-1. (PLT380) — 14 CFR §121.637

Instrument Approaches

This section is limited to rules and procedures common to most, or all approaches, or procedures that may be used in connection with published instrument approaches.

Contact and visual approaches are both IFR authorizations to proceed to an airport visually. A **visual approach** may be authorized by ATC to reduce pilot or controller workload and to expedite traffic by shortening flight paths to the airport. The weather must be VFR and the pilot must report either the airport or the preceding aircraft in sight. Either the pilot or ATC may initiate a visual approach. A **contact approach** may be initiated only by the pilot. The weather need not be VFR but the aircraft must be clear of the clouds, have at least 1 mile visibility and be able to proceed to the landing airport visually.

When an airport has ILS or MLS approaches to parallel runways at least 4,300 feet apart, ATC may conduct approaches to both runways simultaneously. The pilots will be informed if simultaneous approaches are in progress. To ensure safe separation between aircraft, radar monitoring is provided on the tower frequency. A pilot must report any malfunctioning aircraft receivers if he/she has been informed that simultaneous approaches are in progress.

Occasionally, a pilot will be asked to fly an instrument approach to a runway and then fly a visual "sidestep" maneuver to land on a parallel runway. This sidestep maneuver should be executed as soon as possible after the runway environment is in sight.

If a pilot is being radar vectored when an approach clearance is received, he/she must maintain the last assigned altitude until the aircraft is established on a segment of a published route or approach procedure unless a different altitude is assigned by ATC. If a flight is being radar vectored to the final approach course and intercepts a published portion of the course, the pilot may not descend to the published altitudes until cleared for the approach. If a flight has not been cleared for approach while on a radar vector and it becomes apparent that the current vector will take it across the final approach course, the pilot should advise ATC of the situation. Do not turn to intercept the approach course unless cleared to do so.

Unless ATC issues a clearance otherwise, no pilot may make a procedure turn on an instrument approach if any of the following apply:

- The flight is radar vectored to the final approach course or fix
- The flight makes a timed approach from a holding fix
- The approach procedure specifies "No PT"

When the approach procedure involves a procedure turn, a maximum speed of not greater than 200 KIAS should be observed from first overheading the course reversal IAF through the procedure turn maneuver, to ensure containment with the obstruction clearance area.

Answers

8264 [C]

Except for Category II and III approaches, if RVR minimums for takeoff or landing are prescribed in an instrument approach procedure, but the RVR is not reported for the runway intended, the ground visibilities may be substituted. These may be found in FAA Legend 7.

A pilot may not continue an approach past the final approach fix or on to the final approach segment unless the latest weather report for the airport indicates that the visibility is equal to, or greater than, the visibility required for the approach procedure. If a pilot has begun the final approach segment and then receives a report of below minimum conditions, he/she may continue the approach to the DH or MDA.

To descend below the published DH or MDA on an instrument approach, one of the following must be distinctly visible and identifiable to the pilot:

- Approach light system, except that the pilot may not descend below 100 feet above the touchdown zone elevation using the approach lights as a reference unless the red terminating bars or red side row bars are also distinctly visible and identifiable.
- Threshold
- Threshold markings
- Threshold lights
- Runway end identifier lights
- Visual approach slope indicator
- Touchdown zone or touchdown zone markings
- Touchdown zone lights
- Runway or runway markings
- Runway lights

A pilot must initiate a missed approach from an ILS upon arrival at the DH on the glide slope if none of the required visual references is distinctly visible. If visual contact is lost anytime after descending below the DH but before touchdown, the pilot must start a missed approach.

If a pilot loses visual reference while circling to land from an instrument approach, he/she should follow the missed approach procedure published for the approach used. The pilot should make an initial climbing turn toward the landing runway to establish the aircraft on the missed approach course.

An **Airport Surveillance Radar (ASR)** approach is one in which an ATC radar controller provides directional guidance and distance to the runway information to the pilot. The only airborne equipment required is an operating radio receiver. The controller will tell the pilot when the aircraft is at the missed approach point and give missed approach instructions as required. If the pilot desires to execute a missed approach prior to the missed approach point, he/she should inform the controller, who will then issue missed approach instructions.

Precision Runway Monitor (PRM) is a high update-rate radar surveillance system, certified to provide simultaneous independent approaches to closely-spaced parallel runways.

If there is penetration of the obstacle identification surfaces (OIS), the published approach visibility can be no lower than 3/4 SM.

ALL

9091. What is the difference between a visual and a contact approach?

A—A visual approach is an IFR authorization while a contact approach is a VFR authorization.

B—A visual approach is initiated by ATC while a contact approach is initiated by the pilot.

C—Both are the same but classified according to the party initiating the approach.

Visual approaches are initiated by ATC to reduce pilot/controller workload and expedite traffic. Pilots operating on IFR flight plans may request a contact approach if the appropriate weather conditions exist. (PLT170) — AIM ¶5-4-22, 5-4-24

Answer (A) is incorrect because both a visual and contact approach are initiated from an IFR flight plan. Answer (C) is incorrect because a visual approach is one in which the pilot has a preceding aircraft or the airport in sight and can maintain basic VFR weather minimums. A contact approach is used by a pilot in lieu of conducting a standard or special instrument approach to an airport.

Answers

9091 [B]

ALL

8953. When simultaneous approaches are in progress, how does each pilot receive radar advisories?

A—On tower frequency.
B—On approach control frequency.
C—One pilot on tower frequency and the other on approach control frequency.

Whenever simultaneous approaches are in progress, radar advisories will be provided on the tower frequency. (PLT420) — AIM ¶5-4-15

Answer (B) is incorrect because pilots will be advised to monitor the tower (not approach control) frequency to receive radar advisories and instructions. Answer (C) is incorrect because both pilots would receive radar advisories on the tower frequency.

ALL

8955. When simultaneous ILS approaches are in progress, which of the following should approach control be advised of immediately?

A—Any inoperative or malfunctioning aircraft receivers.
B—If a simultaneous ILS approach is desired.
C—If radar monitoring is desired to confirm lateral separation.

When advised that simultaneous ILS approaches are in progress, pilots shall advise approach control immediately of malfunctioning or inoperative receivers or if simultaneous approach is not desired. (PLT170) — AIM ¶5-4-15

Answer (B) is incorrect because simultaneous approaches are issued at any time according to ATC needs, and it is not the responsibility of the pilot to request such an approach. Answer (C) is incorrect because radar monitoring is always provided during simultaneous approaches.

ALL

8954. When cleared to execute a published side-step maneuver, at what point is the pilot expected to commence this maneuver?

A—At the published DH.
B—At the MDA published or a circling approach.
C—As soon as possible after the runway environment is in sight.

Pilots are expected to execute the side-step maneuver as soon as possible after the runway or runway environment is in sight. (PLT083) — AIM ¶5-4-19

Answers (A) and (B) are incorrect because the side-step maneuver can only be performed and should be performed as soon as possible after the runway or runway environment is in sight.

ALL

9438. When cleared to execute a published side-step maneuver for a specific approach and landing on the parallel runway, at what point is the pilot expected to commence this maneuver?

A—At the published minimum altitude for a circling approach.
B—As soon as possible after the runway or runway environment is in sight.
C—At the localizer MDA minimums and when the runway is in sight.

Pilots are expected to execute the side-step maneuver as soon as possible after the runway or runway environment is in sight. (PLT170) — AIM 5-4-19

Answers (A) and (C) are incorrect because the maneuver should be started as soon as the runway environment is in sight not at a DH or MDA of an approach.

ALL

9037. While being vectored to the final approach course of an IFR approach, when may the pilot descend to published altitudes?

A—Anytime the flight is on a published leg of an approach chart.
B—When the flight is within the 10-mile ring of a published approach.
C—Only when approach control clears the flight for the approach.

When operating on an unpublished route or while being radar vectored, the pilot, when approach clearance is received, in addition to complying with the minimum altitudes for IFR operations, shall maintain the last assigned altitude unless a different altitude is assigned by ATC, or until the aircraft is established on a segment of published route or IAP. This implies that even if a radar vector should happen to put a flight on a published route, the pilot may not descend until cleared for the approach. (PLT420) — AIM ¶5-4-7

Answer (A) is incorrect because you may only descend if cleared. Answer (B) is incorrect because the 10-mile ring has nothing to do with descent clearance.

Answers

8953 [A]	8955 [A]	8954 [C]	9438 [B]	9037 [C]

ALL

9383. What action should be taken when a pilot is "cleared for approach" while being radar vectored on an unpublished route?

A—Descend to minimum vector altitude.
B—Remain at last assigned altitude until established on a published route segment.
C—Descend to initial approach fix altitude.

When operating on an unpublished route or while being radar vectored, the pilot, when an approach clearance is received, shall maintain the last altitude assigned until the aircraft is established on a segment of a published route or instrument approach procedure, unless a different altitude is assigned by ATC. (PLT421) — 14 CFR §91.175

Answer (A) is incorrect because a pilot should maintain the last altitude assigned by ATC and should use the minimum vector altitude only during lost communication procedures. Answer (C) is incorrect because a pilot should maintain the last altitude assigned by ATC.

ALL

9385. What altitude is a pilot authorized to fly when cleared for an ILS approach? The pilot

A—may begin a descent to the procedure turn altitude.
B—must maintain the last assigned altitude until established on a published route or segment of the approach with published altitudes.
C—may descend from the assigned altitude only when established on the final approach course.

When operating on an unpublished route or while being radar vectored, the pilot, when an approach clearance is received, shall maintain the last altitude assigned until the aircraft is established on a segment of a published route or instrument approach procedure, unless a different altitude is assigned by ATC. (PLT421) — 14 CFR §91.175

Answer (A) is incorrect because descent to the procedure turn altitude can be commenced only when you are established on that route segment or instrument approach. Answer (C) is incorrect because the pilot does not have to be established on the final approach course to descend from the last assigned altitude, if established on a published route segment with a specified lower minimum altitude than the last assigned altitude.

ALL

9036. What action(s) should a pilot take if vectored across the final approach course during an IFR approach?

A—Continue on the last heading issued until otherwise instructed.
B—Contact approach control, and advise that the flight is crossing the final approach course.
C—Turn onto final, and broadcast in the blind that the flight has proceeded on final.

Aircraft will normally be informed when it is necessary to vector across the final approach course for spacing or other reasons. If approach course crossing is imminent and the pilot has not been informed that he will be vectored across the final approach course, he should query the controller. (PLT420) — AIM ¶5-4-3

Answer (A) is incorrect because the pilot should maintain last heading issued, but should also advise approach control that the flight is crossing the final approach course. Answer (C) is incorrect because a pilot should broadcast in the blind that the flight has turned onto final when operating VFR at an uncontrolled airport.

ALL

9369. If being radar vectored to the final approach course of a published instrument approach that specifies "NO PT," the pilot should

A—advise ATC that a procedure turn will not be executed.
B—not execute the procedure turn unless specifically cleared to do so by ATC.
C—execute a holding-pattern type procedure turn.

In the case of a radar vector to a final approach course or fix, a timed approach from a holding fix, or an approach for which the approach procedure specifies "NoPT," no pilot may make a procedure turn unless cleared to do so by ATC. (PLT420) — 14 CFR §91.175

Answer (A) is incorrect because a procedure turn is not authorized or expected to be executed for this instrument approach; therefore, advising ATC of your intention to omit a procedure turn is not necessary. Answer (C) is incorrect because if the published instrument approach specifies "NoPT," you should follow the published procedure rather than automatically reverting to a holding-pattern-type procedure turn.

Answers

9383 [B]	9385 [B]	9036 [B]	9369 [B]

ALL

9021. When the approach procedure involves a procedure turn the maximum speed that should be observed from first overheading the course reversal IAF through the procedure turn is

A—180 knots IAS.
B—200 knots TAS.
C—200 knots IAS.

When the approach procedure involves a procedure turn, a maximum speed of not greater than 200 knots (IAS) should be observed from first overheading the course reversal IAF through the procedure turn maneuver, to ensure containment within the obstruction clearance area. (PLT420) — AIM ¶5-4-9

ALL

9391. What minimum ground visibility may be used instead of a prescribed visibility criteria of RVR 16 when that RVR value is not reported?

A—1/4 SM.
B—3/4 SM.
C—3/8 SM.

RVR minimum may be converted to ground visibility using FAA Legend 7. (PLT420) — 14 CFR §91.175

ALL

9392. The prescribed visibility criteria of RVR 32 for the runway of intended operation is not reported. What minimum ground visibility may be used instead of the RVR value?

A—3/8 SM.
B—5/8 SM.
C—3/4 SM.

RVR minimum may be converted to ground visibility using FAA Legend 7. (PLT420) — 14 CFR §91.175

ALL

9393. The visibility criteria for a particular instrument approach procedure is RVR 40. What minimum ground visibility may be substituted for the RVR value?

A—5/8 SM.
B—3/4 SM.
C—7/8 SM.

RVR minimum may be converted to ground visibility using FAA Legend 7. (PLT420) — 14 CFR §91.175

ALL

9384. Under which condition, if any, may a pilot descend below DH or MDA when using the ALSF-1 approach light system as the primary visual reference for the intended runway?

A—Under no condition can the approach light system serve as a necessary visual reference for descent below DH or MDA.
B—Descent to the intended runway is authorized as long as any portion of the approach light system can be seen.
C—The approach light system can be used as a visual reference, except that descent below 100 feet above TDZE requires that the red light bars be visible and identifiable.

A pilot may descend below the MDA or DH using the approach light system as the sole visual reference. However, the pilot may not descend below 100 feet above touchdown zone elevation (TDZE) using the approach lights as a reference unless the red terminating bars or the red side row bars are also distinctly visible and identifiable. (PLT420) — 14 CFR §91.175

Answer (A) is incorrect because approach lighting systems can be used as a reference below the DH or MDA up to 100 feet above the TDZE, at which point the red terminating bars must be in sight. Answer (B) is incorrect because the approach lighting system can only be used to within 100 feet of the TDZE, at which point the red side row bars must be in sight.

ALL

9368. When must the pilot initiate a missed approach procedure from an ILS approach?

A—At the DA/DH when the runway is not clearly visible.
B—When the time has expired after reaching the DA/DH and the runway environment is not clearly visible.
C—At the DA/DH, if the visual references for the intended runway are not distinctly visible or anytime thereafter that visual reference is lost.

A pilot must initiate a missed approach procedure from an ILS approach at the DA/DH, if the required visual references for intended runway are not distinctly visible or anytime thereafter if visual reference is lost. (PLT420) — 14 CFR §91.175

Answer (A) is incorrect because the runway itself does not have to be visible at the DA/DH to continue with the approach; a pilot may use the required visual references. Answer (B) is incorrect because as soon as the DA/DH is reached on an ILS approach, regardless of the elapsed time, a missed approach procedure should be executed if visual references are not obtained, or any time thereafter that visual reference is lost.

Answers

9021 [C]	9391 [A]	9392 [B]	9393 [B]	9384 [C]	9368 [C]

ALL

9382. Assuming that all ILS components are operating and the required visual references are not acquired, the missed approach should be initiated upon

A—arrival at the DH on the glide slope.
B—arrival at the visual descent point.
C—expiration of the time listed on the approach chart for missed approach.

A pilot must initiate a missed approach procedure from an ILS approach at the DH, if the required visual references for intended runway are not distinctly visible or any time thereafter if visual reference is lost. (PLT356) — 14 CFR §91.175

Answer (B) is incorrect because a visual descent point is a point in which an aircraft operating visually can descend from a specified altitude to the runway and land. If on an ILS approach and no visual flight is encountered, a missed approach should be executed at the DH. Answer (C) is incorrect because time listed on the approach chart is used only if the glide slope were to fail. If the aircraft reaches the DH prior to the time listed on the chart, the pilot should execute a missed approach.

ALL

9041. When may a pilot execute a missed approach during an ASR approach?

A—Anytime at the pilot's discretion.
B—Only at the MAP.
C—Only when advised by the controller.

Controllers will terminate guidance on an ASR approach and instruct the pilot to execute a missed approach unless at the MAP, the pilot has the runway or airport in sight. Also, if at any time during an ASR approach the controller considers that safe guidance for the remainder of the approach cannot be provided, he will terminate the approach and instruct the pilot to execute a missed approach. A missed approach will also be effected upon pilot request. (PLT420) — AIM ¶5-4-10

Answer (B) is incorrect because the controller will instruct the pilot to execute a missed approach at the MAP or anytime during the approach that the controller considers that safe guidance cannot be provided. Answer (C) is incorrect because a missed approach will be effected upon pilot request.

ALL

9090-1. If visual reference is lost while circling to land from an instrument approach, what action(s) should the pilot take?

A—Make a climbing turn toward the landing runway until established on the missed approach course.
B—Turn toward the landing runway maintaining MDA, and if visual reference is not regained, perform missed approach.
C—Make a climbing turn toward the VOR/NDB, and request further instructions.

If visual reference is lost while circling to land from an instrument approach, the missed approach specified for that particular procedure must be followed. To become established on the prescribed missed approach course, the pilot should make an initial climbing turn toward the landing runway and continue the turn until he is established on the missed approach course. (PLT170) — AIM ¶5-4-21

Answer (B) is incorrect because while turning toward the runway, a climbing turn should be established. Answer (C) is incorrect because a pilot should make a climbing turn toward the runway to ensure obstacle clearance while becoming established on the missed approach course.

ALL

9090-2. Precision Runway Monitoring (PRM) is

A—an airborne RADAR system for monitoring approaches to two runways.
B—a RADAR system for monitoring approaches to closely spaced parallel runways.
C—a high update rate RADAR system for monitoring multiple aircraft ILS approaches to a single runway.

Precision Runway Monitoring (PRM) is a high update-rate radar surveillance system, certified to provide simultaneous independent approaches to closely spaced parallel runways. (PLT172) — FAA-H-8261-1

Answer (A) is incorrect because PRM is not an airborne radar system; it is ground based. Answer (C) is incorrect because PRM monitors simultaneous approaches to two closely spaced parallel runways.

Answers

9382 [A] 9041 [A] 9090-1 [A] 9090-2 [B]

ALL

9760. Precision runway monitoring requires

A—pilot responsibility to monitor 2 simultaneous radios.
B—pilot responsibility to monitor 2 ILS receivers.
C—detailed performance during the "decision region": 1/3 dot localizer and 1/2 dot glideslope.

The aircraft flying the ILS/PRM or LDA/PRM approach must have the capability of enabling the pilots to listen to two communications frequencies simultaneously. (PLT172) — AIM ¶5-4-16

ALL

9090-3. How can the pilot determine, for an ILS runway equipped with MALSR, that there may be penetration of the obstacle identification surfaces (OIS), and care should be taken in the visual segment to avoid any obstacles?

A—The runway has a visual approach slope indicator (VASI).
B—The published visibility for the ILS is no lower than 3/4 SM.
C—The approach chart has a visual descent point (VDP) published.

The visibility published on an approach chart is dependent on many variables, including the height above touchdown for straight-in approaches, or height above airport elevation for circling approaches. Other factors include the approach light system coverage, and type of approach procedure, such as precision, nonprecision, circling or straight-in. Another factor determining the minimum visibility is the penetration of the 34:1 and 20:1 surfaces. These surfaces are inclined planes that begin 200 feet out from the runway and extend outward to 10,000 feet. If there is a penetration of the 34:1 surface, the published visibility can be no lower than 3/4 SM. If there is penetration of the 20:1 surface, the published visibility can be no lower than 1 SM with a note prohibiting approaches to the affected runway at night (both straight-in and circling). Pilots should be aware of these penetrating obstacles when entering the visual and/or circling segments of an approach and take adequate precautions to avoid them. (PLT170) — FAA-H-8261-1

Answers (A) and (C) are incorrect because a VASI or VDP are not indicators to a possible penetration of the OIS.

ALL

9738. To conduct a localizer performance with vertical guidance (LPV) RNAV (GPS) approach, the aircraft must be furnished with

A—a GPS/WAAS receiver approved for an LPV approach by the AFM supplement.
B—a GPS (TSO-129) receiver certified for IFR operations.
C—an IFR approach-certified system with required navigation performance (RNP) of 0.5.

"LPV" is the acronym for localizer performance with vertical guidance. LPV identifies the APV minimums with electronic lateral and vertical guidance. The lateral guidance is equivalent to localizer, and the protected area is considerably smaller than the protected area for the present LNAV and LNAV/VNAV lateral protection. Aircraft can fly this minima line with a statement in the Aircraft Flight Manual that the installed equipment supports LPV approaches. This includes Class 3 and 4 TSO-C146 WAAS equipment, and future LAAS equipment. (PLT354) — AIM ¶5-4-5

ALL

9744. Pilots are not authorized to fly a published RNAV or RNP procedure unless it is retrievable by the procedure name from

A—the aircraft navigation database, or manually loaded with each individual waypoint in the correct sequence.
B—the aircraft navigation database, or manually loaded with each individual waypoint and verified by the pilot(s).
C—the aircraft navigation database.

Pilots are not authorized to fly a published RNAV or RNP procedure (instrument approach, departure, or arrival procedure) unless it is retrievable by the procedure name from the aircraft navigation database and conforms to the charted procedure. (PLT354) — AIM ¶5-5-16

Answers

9760 [A]	9090-3 [B]	9738 [A]	9744 [C]

ALL

9773. Pilots are responsible for knowing

A—if they can conduct an RNP approach with an arc at a designated airspeed.

B—if the RNP missed approach is normal or reduced.

C—if the RNP registration is complete.

Some RNP approaches have a curved path, also called a radius-to-a-fix (RF) leg. Since not all aircraft have the capability to fly these arcs, pilots are responsible for knowing whether or not they can conduct an RNP approach with an arc. (PLT300) — AIM ¶5-4-18

121, DSP

8279. Under what conditions may an air carrier pilot continue an instrument approach to the DH, after receiving a weather report indicating that less than minimum published landing conditions exist at the airport?

A—If the instrument approach is conducted in a radar environment.

B—When the weather report is received as the pilot passes the FAF.

C—When the weather report is received after the pilot has begun the final approach segment of the instrument approach.

If a pilot has begun the final approach segment of an instrument approach procedure with the reported weather at or above landing minimums and later receives a report indicating below minimum conditions, he may continue the approach to DH or MDA. The pilot may land from that approach if he discovers that the visibility is at least that required by the approach, he/she has the required visual references in sight and a normal descent and landing can be made. (PLT420) — 14 CFR §121.651

Landing

Except for emergencies, the landing priority of aircraft arriving at a tower controlled airport is on "first-come, first-served" basis. When landing at a tower controlled airport, an aircraft should exit the runway at the first suitable taxiway and remain on the tower frequency until instructed to do otherwise. The aircraft should not turn onto any other taxiway unless a clearance to do so has been received.

If a flight is making an IFR approach at an uncontrolled airport, radar service will be terminated when the aircraft lands or when the controller tells the pilot to change to advisory frequency. After changing to the advisory frequency, the pilot should broadcast his/her intentions and continually update position reports. The advisory frequency will be an FSS frequency, or if there is no FSS on the field, a UNICOM frequency.

ATC furnishes pilots' braking action reports using the terms "good," "fair," "poor" and "nil." If you give a braking action report to ATC, you should use the same terminology.

ALL

9092. Except during an emergency, when can a pilot expect landing priority?

A—When cleared for an IFR approach.

B—When piloting a large, heavy aircraft.

C—In turn, on a first-come, first-serve basis.

Air Traffic Control towers handle all aircraft, regardless of the type of flight plan, on a "first-come, first-served" basis. (PLT170) — AIM ¶5-4-25

Answer (A) is incorrect because a clearance for an IFR approach does not mean landing priority will be given over other traffic. Answer (B) is incorrect because a large, heavy aircraft will be sequenced for landing on a first-come, first-served basis, with no special priority over other traffic.

Answers

9773 [A] 8279 [C] 9092 [C]

ALL

9044. What action is expected of an aircraft upon landing at a controlled airport?

A—Continue taxiing in the landing direction until advised by the tower to switch to ground control frequency.
B—Exit the runway at the nearest suitable taxiway and remain on tower frequency until instructed otherwise.
C—Exit the runway at the nearest suitable taxiway and switch to ground control upon crossing the taxiway holding lines.

After landing, unless otherwise instructed by the control tower, continue to taxi in the landing direction, proceed to the nearest suitable taxiway and exit the runway without delay. Do not turn on another runway or make a 180° turn to taxi back on an active runway or change to ground control frequency while on the active runway without authorization from the tower. A pilot who has just landed should not change from the tower frequency to the ground control frequency until he is directed to do so by the controller. (PLT434) — AIM ¶4-3-20

Answer (A) is incorrect because upon landing, the pilot should exit the runway at the nearest suitable taxiway to clear the runway for other traffic. Answer (C) is incorrect because while the crossing of the taxiway hold lines indicates clearing of the active runway, a pilot should not switch to ground control until directed to do so by the controller. Switching without permission may be confusing to ATC.

ALL

9038. When is radar service terminated while vectored for an IFR approach at an uncontrolled airport?

A—Only upon landing or advised to change to advisory frequency.
B—When aligned on the final approach course.
C—When cleared for the approach.

Whether aircraft are vectored to the appropriate final approach course or provide their own navigation on published routes to it, radar service is automatically terminated when the landing is completed or when instructed to change to advisory frequency at uncontrolled airports, whichever occurs first. (PLT420) — AIM ¶5-4-3

Answer (B) is incorrect because when established on the final approach course, radar separation will be maintained and the pilot is expected to complete the approach utilizing the approach aid designated in the clearance (ILS, VOR, etc.). Answer (C) is incorrect because when cleared for the approach, approach control will continue to maintain radar separation and the pilot is expected to complete the approach utilizing the approach aid designated in the clearance (ILS, VOR, etc.).

ALL

9039. When cleared for an IFR approach to an uncontrolled airport with no FSS, what precaution should the pilot take after being advised to change to advisory frequency?

A—Monitor ATC for traffic advisories as well as UNICOM.
B—Broadcast position and intentions on the Common Traffic Advisory Frequency and monitor the frequency.
C—Wait until visual contact is made with the airport and then broadcast position and intentions to land on UNICOM.

When making an IFR approach to an airport not served by a tower or FSS, after the ATC controller advises, "CHANGE TO ADVISORY FREQUENCY APPROVED" you should broadcast your intentions, including the type of approach being executed, your position, and when you are over the outer marker or final approach fix. Continue to monitor the appropriate frequency (UNICOM, etc.) for reports from other pilots. (PLT170) — AIM ¶5-4-4

Answer (A) is incorrect because after ATC advises the pilot to change to advisory frequency, ATC will no longer be able to provide traffic advisories. Answer (C) is incorrect because a pilot should always broadcast intentions and continually update position reports on UNICOM, not wait until visual contact is made with airport.

ALL

9055. How should a pilot describe braking action?

A—00 percent, 50 percent, 75 percent, or 100 percent.
B—Zero-zero, fifty-fifty, or normal.
C—Nil, poor, fair, or good.

Pilots should describe the quality of braking action by using the terms "good," "fair," "poor," and "nil." (PLT144) — AIM ¶4-3-8

Answers

9044 [B]	9038 [A]	9039 [B]	9055 [C]

Communications

The "Sterile Cockpit" Rule: Regulations say only those duties required for the safe operation of the aircraft are allowed during critical phases of flight. Critical phases of flight are defined as climb and descent when below 10,000 feet, taxi, takeoff, and landing. Excluded from the definition of critical phase of flight are any operations at or above 10,000 feet and cruise flight below 10,000 feet. Activities which are prohibited during critical phases of flight include filling out logs, ordering galley supplies, making passenger announcements or pointing out sights of interest. Activities such as eating meals or engaging in nonessential conversations are also prohibited.

The following should be reported without ATC request:

- Vacating a previously assigned altitude for a newly assigned one.
- An altitude change when operating under a VFR-On-Top clearance.
- When unable to climb or descend at a rate of at least 500 feet per minute.
- When an approach has been missed.
- A change in cruising true airspeed of 10 knots or 5%, whichever is greater.
- The time and altitude (or Flight Level) upon reaching a holding fix or clearance limit.
- When leaving an assigned holding fix or point.
- The malfunction of navigation, approach or communication equipment.
- Any information pertaining to the safety of flight.

In addition to the reports listed above, when not in radar contact a pilot must report:

- When over designated compulsory reporting points.
- When leaving the final approach fix inbound on an instrument approach.
- When it becomes apparent that an estimate of arrival time over a fix is in error by more than 3 minutes.

Occasionally an ATC controller will query a pilot about the aircraft's altitude or course. For example, a controller says "Verify 9000," meaning he/she wants confirmation that the aircraft is at 9,000 feet altitude. If the aircraft is not at that altitude, the pilot should reply, "Negative, maintaining 8,000 as assigned." No climb or descent should be started unless specifically assigned by the controller.

Pilots should notify controllers on initial contact that they have received the ATIS broadcast by repeating the alphabetical code used appended to the broadcast. For example, "Information Sierra received."

ALL

8854. What report should the pilot make at a clearance limit?

A—Time and altitude/flight level arriving or leaving.
B—Time, altitude/flight level, and expected holding speed.
C—Time, altitude/flight level, expected holding speed, and inbound leg length.

Pilots should report to ATC the time and altitude/flight level at which the aircraft reaches the clearance limit, and report when leaving the clearance limit. (PLT171) — AIM ¶5-3-2

Answer (B) is incorrect because ATC does not need the expected holding speed reported since it will be below the maximum holding airspeed. For all aircraft between MHA and 6,000 feet MSL, holding speed is 200 KIAS; for all aircraft between 6,001 and 14,000 feet MSL, holding speed is 230 KIAS; for all aircraft 14,001 feet MSL and above, holding speed is 265 KIAS. For turbojet airplanes, the maximum holding airspeed is 230 knots IAS from minimum holding altitude to 14,000 feet. Answer (C) is incorrect because inbound leg lengths are set by time or DME distance. At or below 14,000 feet MSL there is a 1-minute inbound leg. Above 14,000 feet MSL the inbound leg is 1-1/2 minutes.

ALL

9014. Where are position reports required on an IFR flight on airways or routes?

A—Over all designated compulsory reporting points.
B—Only where specifically requested by ARTCC.
C—When requested to change altitude or advise of weather conditions.

A position report is required by all flights regardless of altitude over each designated compulsory reporting point along the route being flown.

Note: *When the controller states "radar contact," this requirement is removed. However, the question states nothing about being in "radar contact." (PLT421) — AIM ¶5-3-2*

Answer (B) is incorrect because the "on request" reporting point is indicated on enroute charts by an open triangle. Reports passing an "on request" reporting point are only necessary when requested by ARTCC. Answer (C) is incorrect because pilots in IFR are expected to report weather conditions which have not been forecast, or hazardous conditions which have been forecast.

ALL

9015. Which reports are required when operating IFR in radar environment?

A—Position reports, vacating an altitude, unable to climb 500 ft/min, and time and altitude reaching a holding fix or point to which cleared.
B—Position reports, vacating an altitude, unable to climb 500 ft/min, time and altitude reaching a holding fix or point to which cleared, and a change in average true airspeed exceeding 5 percent or 10 knots.
C—Vacating an altitude, unable to climb 500 ft/min, time and altitude reaching a holding fix or point to which cleared, a change in average true airspeed exceeding 5 percent or 10 knots, and leaving any assigned holding fix or point.

The following reports should be made to ATC or FSS facilities without specific ATC request:

1. *Vacating any previously assigned altitude.*
2. *Making an altitude change when VFR-On-Top.*
3. *Unable to climb or descend at least 500 feet per minute.*
4. *Making a missed approach.*
5. *Changing true airspeed from flight plan by 5% or 10 knots (whichever is greater).*
6. *Time and altitude of reaching a clearance holding fix or point.*
7. *Leaving any holding fix.*

(PLT171) — AIM ¶5-3-3

Answers (A) and (B) are incorrect because position reports are not required in a radar environment.

ALL

9016. Which reports are always required when on an IFR approach not in radar contact?

A—Leaving FAF inbound or outer marker inbound and missed approach.
B—Leaving FAF inbound, leaving outer marker inbound or outbound, and missed approach.
C—Leaving FAF inbound, leaving outer marker inbound or outbound, procedure turn outbound and inbound, and visual contact with the runway.

The following reports should be made when not in radar contact:

1. *When over designated compulsory reporting points*
2. *When leaving the final approach fix inbound*

Continued

Answers

8854 [A] 9014 [A] 9015 [C] 9016 [A]

3. When it becomes apparent that an ETA is in error by more than 3 minutes.

(PLT171) — AIM ¶5-3-3

Answer (B) is incorrect because a pilot is required to report leaving the outer marker inbound on final approach. Answer (C) is incorrect because a pilot is not required to report leaving the outer marker outbound, the execution of a procedure turn, and/or visual contact with the runway.

ALL

9013. What action should a pilot take if asked by ARTCC to "VERIFY 9,000" and the flight is actually maintaining 8,000?

A—Immediately climb to 9,000.
B—Report climbing to 9,000.
C—Report maintaining 8,000.

At times controllers will ask pilots to verify that they are at a particular altitude. Pilots should confirm that they are at the altitude stated. If this is not the case, they should inform the controller of the actual altitude being maintained. Pilots should not take action to change their actual altitude to the altitude stated in the controller's verification request unless the controller specifically authorizes a change. (PLT171) — AIM ¶5-3-1

Answers (A) and (B) are incorrect because pilots should not take action to change their actual altitude to the altitude stated in the controller's verification request unless the controller specifically authorizes a change.

ALL

9022. Pilots should notify controllers on initial contact that they have received the ATIS broadcast by

A—stating "Have Numbers".
B—stating "Have Weather".
C—repeating the alphabetical code word appended to the broadcast.

Pilots should notify controllers on initial contact that they have received the ATIS broadcast by repeating the alphabetical code word appended to the broadcast. For example, "Information Sierra received." (PLT196) — AIM ¶4-1-13

121, DSP

8297. Below what altitude, except when in cruise flight, are non-safety related cockpit activities by flight crewmembers prohibited?

A—10,000 feet.
B—14,500 feet.
C—FL 180.

No certificate holder shall require, nor may any flight crewmember perform, any duties during a critical phase of flight except those duties required for the safe operation of the aircraft. For purposes of this section, critical phases of flight include all ground operations involving taxi, takeoff and landing, and all other flight operations conducted below 10,000 feet, except cruise flight. (PLT430) — 14 CFR §121.542

121, DSP

8298. With regard to flight crewmember duties, which of the following operations are considered to be in the "critical phase of flight"?

A—Taxi, takeoff, landing, and all other operations conducted below 10,000 feet MSL, including cruise flight.
B—Descent, approach, landing, and taxi operations, irrespective of altitudes MSL.
C—Taxi, takeoff, landing, and all other operations conducted below 10,000 feet, excluding cruise flight.

No certificate holder shall require, nor may any flight crewmember perform, any duties during a critical phase of flight except those duties required for the safe operation of the aircraft. For purposes of this section, critical phases of flight include all ground operations involving taxi, takeoff and landing, and all other flight operations conducted below 10,000 feet, except cruise flight. (PLT029) — 14 CFR §121.542

Answer (A) is incorrect because critical phase of flight includes all operations (except cruise flight) below 10,000 feet. Answer (B) is incorrect because 14,500 feet is the base of Class E airspace (if not set lower).

Answers

9013 [C]	9022 [C]	8297 [A]	8298 [C]

Speed Adjustments

ATC controllers often issue speed adjustments to radar controlled aircraft to achieve or maintain the desired separation. The following minimum speeds are usually observed:

- Turbine-powered aircraft below 10,000 feet: 210 knots.
- Turbine-powered aircraft departing an airport: 230 knots.

If an ATC controller assigns a speed which is too fast or too slow for the operating limitations of the aircraft under the existing circumstances, the pilot should advise ATC of the speed that will be used. The controller will then issue instructions based on that speed.

Because of the great differences in speed and operating characteristics of helicopters and airplanes, they are usually assigned different routing. Occasionally, larger/faster helicopters are integrated with fixed-wing aircraft. These situations could occur on IFR flights, routes that avoid noise-sensitive areas, or when the helicopter is assigned runways or taxiways to avoid downwash in congested areas.

ALL

9094. When a speed adjustment is necessary to maintain separation, what minimum speed may ATC request of a turbine-powered aircraft operating below 10,000 feet?

A—200 knots.
B—210 knots.
C—250 knots.

When a speed adjustment is necessary to maintain separation, the minimum airspeed for a turbine-powered aircraft operated below 10,000 feet is 210 knots. (PLT161) — AIM ¶4-4-12

Answer (A) is incorrect because 200 knots is the maximum airspeed of any airplane operating within Class C or D airspace, a VFR-designated corridor through Class B airspace, or in airspace underlying Class B airspace. Answer (C) is incorrect because 250 knots is the maximum airspeed of any airplane operating below 10,000 feet MSL.

ALL

9095. When a speed adjustment is necessary to maintain separation, what minimum speed may ATC request of a turbine-powered aircraft departing an airport?

A—188 knots.
B—210 knots.
C—230 knots.

When a speed adjustment is necessary to maintain separation, the minimum airspeed for a turbine-powered aircraft on departure is 230 knots. (PLT161) — AIM ¶4-4-12

Answer (A) is incorrect because 188 knots is not an applicable airspeed for any ATC operation. All airspeeds used by ATC/regulations are expressed in 10-knot increments. Answer (B) is incorrect because it is the minimum airspeed that ATC can request of a turbine-powered airplane operating below 10,000 feet, excluding departing airplanes.

ALL

9096. If ATC requests a speed adjustment that is not within the operating limits of the aircraft, what action must the pilot take?

A—Maintain an airspeed within the operating limitations as close to the requested speed as possible.
B—Attempt to use the requested speed as long as possible, then request a reasonable airspeed from ATC.
C—Advise ATC of the airspeed that will be used.

The pilots retain the prerogative of rejecting the application of speed adjustment by ATC if the minimum safe airspeed for any particular operation is greater than the speed adjustment. In such cases, the pilots are expected to advise ATC of the speed that will be used. (PLT172) — AIM ¶4-4-12

Answer (A) is incorrect because while a pilot should maintain at least the minimum safe airspeed for any particular operation, a pilot is expected to advise ATC of the airspeed being used when it differs from ATC's requested speed adjustment. Answer (B) is incorrect because a pilot who uses an airspeed that is not within the operating limits of the airplane is not only in violation of regulations, but is also risking the safety of all on board the airplane. A pilot must operate the airplane in a safe manner and advise ATC of the airspeed that will be used.

Answers

9094 [B] 9095 [C] 9096 [C]

RTC

9042. Under what situations are faster/larger helicopters integrated with fixed-wing aircraft?

A—IFR flights, noise avoidance routes, and use of runways or taxiways.
B—Use of taxiways, sequencing for takeoff and landing, and use of the same traffic patterns.
C—Use of taxiways, sequencing for takeoff and landing, and use of the same loading ramps.

There will be situations where faster/larger helicopters may be integrated with fixed-wing aircraft. These include IFR flights, avoidance of noise-sensitive areas, or use of runway/taxiways to minimize the hazardous effects of rotor downwash in congested areas. (PLT434) — AIM ¶4-3-17

Holding

Holding may be necessary when ATC is unable to clear a flight to its destination. VORs, nondirectional beacons, airway intersections, and DME fixes may all be used as holding points. Flying a holding pattern involves two turns and two straight-and-level legs as shown in Figure 6-4.

At and below 14,000 feet MSL (no wind), the aircraft flies the specified course inbound to the fix, turns to the right 180°, flies a parallel course outbound for 1 minute, again turns 180° to the right, and flies 1 minute inbound to the fix. Above 14,000 feet MSL, the inbound leg length is 1-1/2 minutes. If a nonstandard pattern is to be flown, ATC will specify left turns.

When 3 minutes or less from the holding fix, the pilot is expected to start a speed reduction so as to cross the fix at or below the maximum holding airspeed. For all aircraft between MHA (minimum holding altitude) and 6,000 feet MSL, holding speed is 200 KIAS. For all aircraft between 6,001 and 14,000 feet MSL, holding speed is 230 KIAS. For all aircraft 14,001 feet MSL and above, holding speed is 265 KIAS. Exceptions to these speeds will be indicated by an icon.

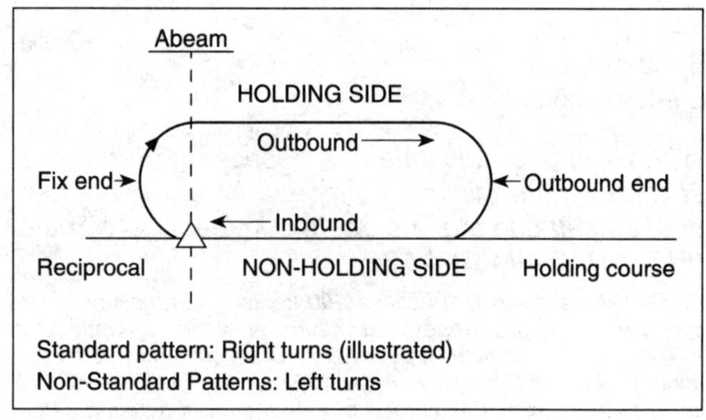

Figure 6-4

The aircraft is in a holding pattern as of the initial time of arrival over the fix, and that time should be reported to ATC. The initial outbound leg is flown for 1 minute at or below 14,000 feet MSL. Subsequently, timing of the outbound leg should be adjusted as necessary to arrive at the proper inbound leg length. Timing of the outbound leg begins over or abeam the fix, whichever occurs later. If the abeam position cannot be determined, start timing when the turn to outbound is completed. The same entry and holding procedures apply to DME holding, except distance in nautical miles are used to establish leg length.

The FAA has three recommended methods for entering a holding pattern, as shown in Figure 6-5. An aircraft approaching from within sector (A) would fly a parallel entry by turning left to parallel the outbound course, making another left turn to remain in protected airspace, and returning to the holding fix. Aircraft approaching from sector (B) would fly a teardrop entry, by flying outbound on a track of 30° or less to the holding course, and then making a right turn to intercept the holding course inbound to the fix. Those approaching from within sector (C) would fly a direct entry by turning right to fly the pattern.

Answers

9042 [A]

If the holding pattern is charted, the controller may omit all holding instructions, except the holding direction and the statement "as published." Pilots are expected to hold in the pattern depicted even if it means crossing the clearance limit. If the holding pattern to be used is not depicted on charts, ATC will issue general holding instructions. The holding clearance will include the following information: direction of holding from the fix in terms of the eight cardinal compass points; holding fix; radial, course, bearing, airway, or route on which the aircraft is to hold; leg length in miles if DME or RNAV is to be used; direction of turn if left turns are to be made; time to expect further clearance and any pertinent additional delay information.

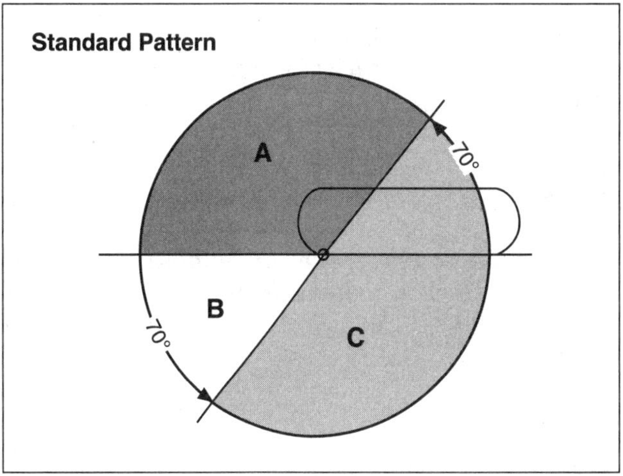

Figure 6-5

ALL

8853. What action should a pilot take if within 3 minutes of a clearance limit and further clearance has not been received?

A—Assume lost communications and continue as planned.
B—Plan to hold at cruising speed until further clearance is received.
C—Start a speed reduction to holding speed in preparation for holding.

When an aircraft is 3 minutes or less from a clearance limit and a clearance beyond the fix has not been received, the pilot is expected to start a speed reduction so that he will cross the fix, initially, at or below the maximum holding airspeed. (PLT296) — AIM ¶5-3-7

Answer (A) is incorrect because if two-way communications are lost, the pilot is required to hold at the clearance limit in a standard pattern on the course that was used to approach the fix. If an expected further clearance time was received, plan on leaving the fix at that time. If none was given and the fix is an IAF, plan your arrival as close as possible to the estimated time of arrival. Answer (B) is incorrect because cruising speed may be greater than maximum holding speed.

ALL

8855. The maximum speed a propeller-driven airplane may hold at is

A—265 knots.
B—230 knots.
C—156 knots.

For all aircraft between MHA and 6,000 feet MSL, holding speed is 200 KIAS. For all aircraft between 6,001 and 14,000 feet MSL, holding speed is 230 KIAS. For all aircraft 14,000 feet MSL and above, holding speed is 265 KIAS. Exceptions to these speeds will be indicated by an icon. Since this question does not specify what altitude the airplane is holding at, both answers (A) and (B) are correct. Choosing either of these will result in a correct response. (PLT296) — AIM ¶5-3-7

ALL

8856. Maximum holding speed for a turbojet airplane above 14,000 feet is

A—210 knots.
B—230 knots.
C—265 knots.

For all aircraft between MHA (minimum holding altitude) and 6,000 feet MSL, holding speed is 200 KIAS. For all aircraft between 6,001 and 14,000 feet MSL, holding speed is 230 KIAS. For all aircraft 14,000 feet MSL and above, holding speed is 265 KIAS. Exceptions to these speeds will be indicated by an icon. (PLT296) — AIM ¶5-3-7

Answers

8853 [C]	8855 [A] or [B]	8856 [C]

ALL

8857. Maximum holding speed for a civil turbojet aircraft at a joint use airport (civil/Navy) between 7,000 and 14,000 feet is

A—265 knots.
B—230 knots.
C—200 knots.

The following are exceptions to the maximum holding airspeeds: Holding patterns at Navy fields only 230 KIAS maximum, unless otherwise depicted. (PLT296) — AIM ¶5-3-7

ALL

9418. What is the maximum holding speed for a civil turbojet holding at a civil airport at 15,000 feet MSL, unless a higher speed is required due to turbulence or icing and ATC is notified?

A—265 knots.
B—230 knots.
C—250 knots.

For all aircraft between MHA (minimum holding altitude) and 6,000 feet MSL, holding speed is 200 KIAS. For all aircraft between 6,001 and 14,000 feet MSL, holding speed is 230 KIAS. For all aircraft 14,000 feet MSL and above, holding speed is 265 KIAS. Exceptions to these speeds will be indicated by an icon. (PLT296) — AIM ¶5-3-7

ALL

9419. Civil aircraft holding at an altitude of 14,000 feet at a military or joint civil/military use airports should expect to operate at which holding pattern airspeed?

A—250 knots.
B—260 knots.
C—230 knots.

Aircraft holding at military or joint civil/military use airports should expect to operate at a maximum holding pattern airspeed of 230 knots up to and including 14,000 feet. (PLT296) — AIM ¶5-3-7

ALL

8858. When using a flight director system, what rate of turn or bank angle should a pilot observe during turns in a holding pattern?

A—3° per second or 25° bank, whichever is less.
B—3° per second or 30° bank, whichever is less.
C—1-1/2° per second or 25° bank, whichever is less.

When making turns in the holding pattern, use whichever of the following requires the least angle of bank:

1. 3° per second;

2. 30° bank angle; or

3. 25° bank provided a flight director system is used. (PLT047) — AIM ¶5-3-7

ALL

8859. When holding at an NDB, at what point should the timing begin for the second leg outbound?

A—Abeam the holding fix or when the wings are level after completing the turn to the outbound heading, whichever occurs first.
B—At the end of a 1-minute standard rate turn after station passage.
C—When abeam the holding fix.

Outbound leg timing begins over or abeam the holding fix, whichever occurs later. If the abeam position cannot be determined, start timing when the turn to outbound is complete. (PLT296) — AIM ¶5-3-7

Answer (A) is incorrect because the pilot should start the timing when the turn is complete, only when a position abeam the fix cannot be determined. Answer (B) is incorrect because abeam the fix is preferable and should be used rather than at the completion of a standard rate turn, especially if turn completion occurs before coming abeam the fix.

ALL

8860. When entering a holding pattern above 14,000 feet, the initial outbound leg should not exceed

A—1 minute.
B—1-1/2 minutes.
C—1-1/2 minutes or 10 NM, whichever is less.

Inbound leg time should not exceed 1 minute when holding at or below 14,000 feet, or 1-1/2 minutes when holding above 14,000 feet. The outbound leg should be flown for 1 minute or 1-1/2 minutes as appropriate on the first leg and then adjusted on subsequent legs to get the correct time on the inbound leg. (PLT296) — AIM ¶5-3-7

Answer (A) is incorrect because an initial outbound leg of 1 minute should be used only when below 14,000 feet. Answer (C) is incorrect because a DME distance is issued only by the specified controller for aircraft equipped with DME capability. A DME distance is not required unless specified by the controller.

Answers

8857 [B] 9418 [A] 9419 [C] 8858 [A] 8859 [C] 8860 [B]

ALL

8861. (Refer to Figure 123.) You receive this ATC clearance:

"...HOLD EAST OF THE ABC VORTAC ON THE ZERO NINER ZERO RADIAL, LEFT TURNS..."

What is the recommended procedure to enter the holding pattern?

A—Parallel only.
B—Direct only.
C—Teardrop only.

Determine the holding pattern by placing your pencil on the holding fix and dragging it on the holding radial given by ATC, then returning back to the fix. Then draw the pattern from the fix with turns in the direction specified. Holding east on the 090° radial with left turns means you will be south of R-090.

The entry procedure is based on the aircraft's heading. To determine which entry procedure to use, draw a line at a 70° angle from the holding fix, and cutting the outbound leg at about one-third its length. With a heading of 055°, we are in the middle-size piece of pie, so a parallel entry would be used. See *the figure below. (PLT296) — AIM ¶5-3-7*

Answer (B) is incorrect because a direct entry would be appropriate if you were coming in on R-340 to R-160. Answer (C) is incorrect because a teardrop entry would be appropriate if you were coming in from R-270 to R-340.

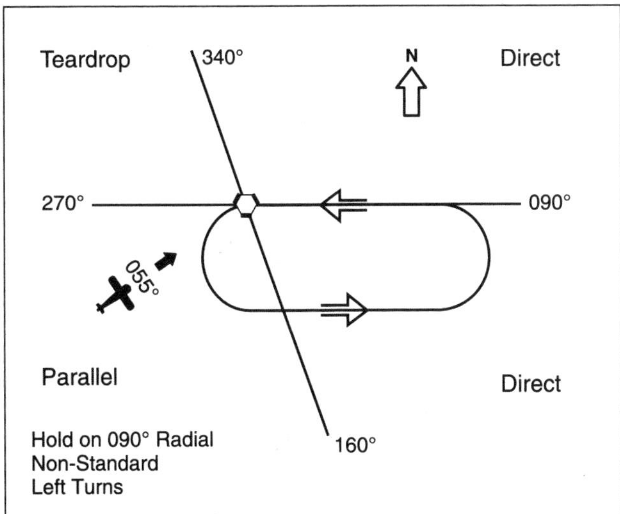

Question 8861

ALL

8862. (Refer to Figure 123.) You receive this ATC clearance:

"...CLEARED TO THE ABC VORTAC. HOLD SOUTH ON THE ONE EIGHT ZERO RADIAL..."

What is the recommended procedure to enter the holding pattern?

A—Teardrop only.
B—Direct only.
C—Parallel only.

Determine the holding pattern by placing your pencil on the holding fix and dragging it on the holding radial given by ATC, then returning back to the fix. Then draw the pattern from the fix with turns in the direction specified. Holding south on the 180° radial with right turns means you will be east of R-180.

The entry procedure is based on the aircraft's heading. To determine which entry procedure to use, draw a line at a 70° angle from the holding fix, and cutting the outbound leg at about one-third its length. With a heading of 055°, we are in the largest piece of pie, so a direct entry would be used. See *the figure below. (PLT087) — AIM ¶5-3-7*

Answer (A) is incorrect because a teardrop entry would be appropriate only from R-290 to R-360. Answer (C) is incorrect because a parallel entry would only be appropriate from R-360 to R-110.

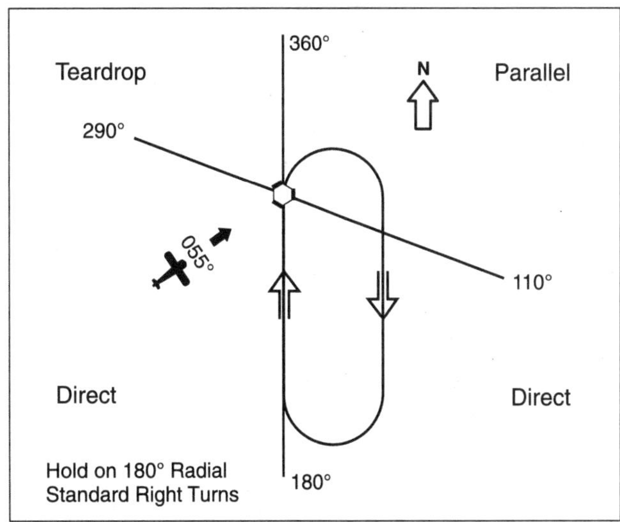

Question 8862

Answers

8861 [A]   8862 [B]

ALL

8863. (Refer to Figure 123.) You receive this ATC clearance:

"...CLEARED TO THE XYZ VORTAC. HOLD NORTH ON THE THREE SIX ZERO RADIAL, LEFT TURNS..."

What is the recommended procedure to enter the holding pattern?

A—Parallel only.
B—Direct only.
C—Teardrop only.

Determine the holding pattern by placing your pencil on the holding fix and dragging it on the holding radial given by ATC, then returning back to the fix. Then draw the pattern from the fix with turns in the direction specified. Holding north on the 360° radial with left turns means you will be east of R-090.

The entry procedure is based on the aircraft's heading. To determine which entry procedure to use, draw a line at a 70° angle from the holding fix, and cutting the outbound leg at about one-third its length. With a heading of 055°, we are in the smallest piece of pie, so a teardrop entry would be used. See the figure below. *(PLT296) — AIM ¶5-3-7*

Answer (A) is incorrect because a parallel entry would be appropriate only from R-070 to R-180. Answer (B) is incorrect because a direct entry would only be appropriate from R-250 to R-070.

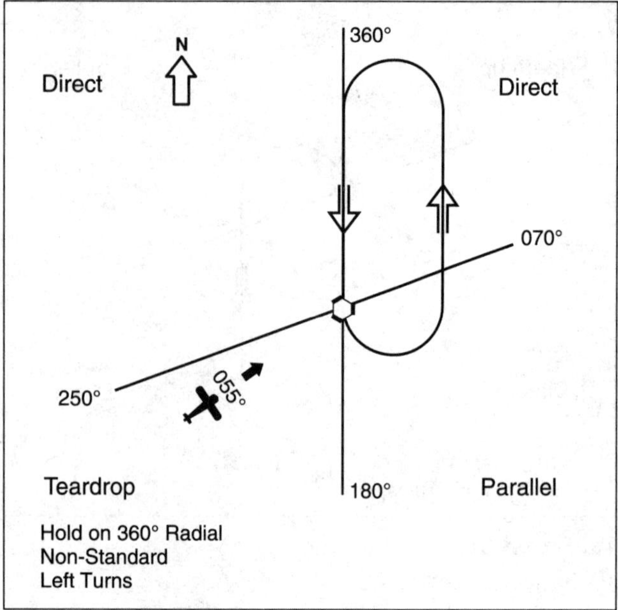

Question 8863

ALL

8864. (Refer to Figure 123.) You receive this ATC clearance:

"...CLEARED TO THE ABC VORTAC. HOLD WEST ON THE TWO SEVEN ZERO RADIAL..."

What is the recommended procedure to enter the holding pattern?

A—Parallel only.
B—Direct only.
C—Teardrop only.

Determine the holding pattern by placing your pencil on the holding fix and dragging it on the holding radial given by ATC, then returning back to the fix. Then draw the pattern from the fix with turns in the direction specified. Holding west on the 270° radial with right turns means you will be south of R-090.

The entry procedure is based on the aircraft's heading. To determine which entry procedure to use, draw a line at a 70° angle from the holding fix, and cutting the outbound leg at about one-third its length. With a heading of 055°, we are in the largest piece of pie, so a direct entry would be used. See the figure below. *(PLT296) — AIM ¶5-3-7*

Answer (A) is incorrect because a parallel entry would be appropriate only from R-090 to R-200. Answer (C) is incorrect because a teardrop entry would only be appropriate from R-020 to R-090.

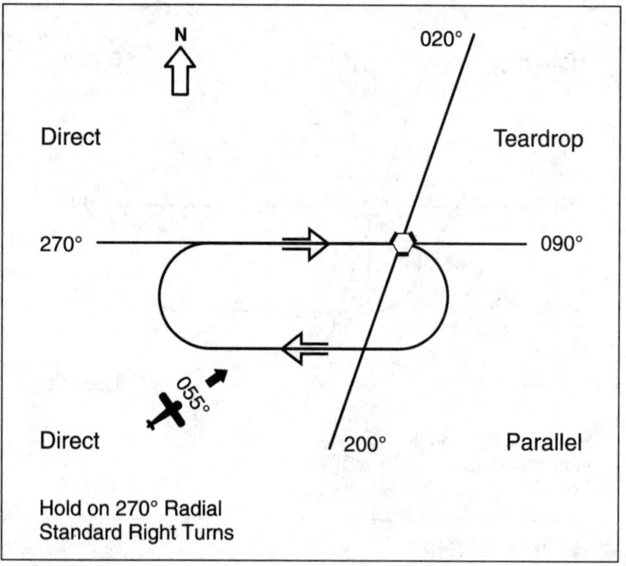

Question 8864

Answers

8863 [C] 8864 [B]

ALL

8865. (Refer to Figure 124.) A pilot receives this ATC clearance:

"...CLEARED TO THE ABC VORTAC. HOLD WEST ON THE TWO SEVEN ZERO RADIAL..."

What is the recommended procedure to enter the holding pattern?

A—Parallel or teardrop.
B—Parallel only.
C—Direct only.

Determine the holding pattern by placing your pencil on the holding fix and dragging it on the holding radial given by ATC, then returning back to the fix. Then draw the pattern from the fix with turns in the direction specified. Holding west on the 270° radial with right turns means you will be south of R-090.

The entry procedure is based on the aircraft's heading. To determine which entry procedure to use, draw a line at a 70° angle from the holding fix, and cutting the outbound leg at about one-third its length. With a heading of 155°, we are in the largest piece of pie, so a direct entry would be used. See *the figure below. (PLT296) — AIM ¶5-3-7*

Answer (A) is incorrect because the parallel or teardrop entries are alternatives only when approaching on R-090. Answer (B) is incorrect because a parallel entry would only be appropriate when approaching from R-090 to R-200.

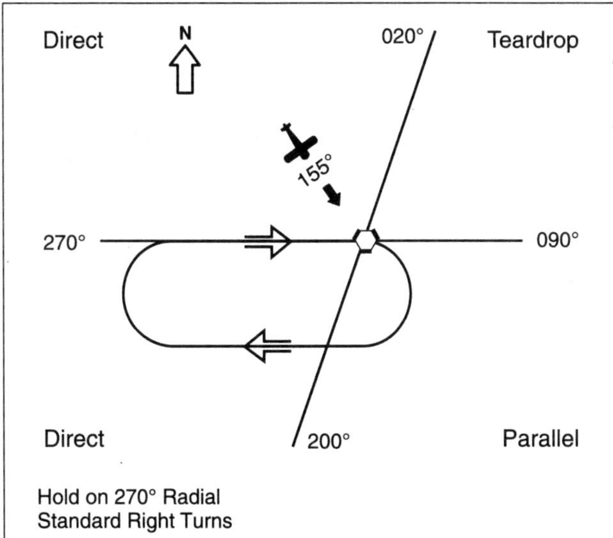

Question 8865

ALL

8866. (Refer to Figure 124.) A pilot receives this ATC clearance:

"...CLEARED TO THE XYZ VORTAC. HOLD NORTH ON THE THREE SIX ZERO RADIAL, LEFT TURNS..."

What is the recommended procedure to enter the holding pattern?

A—Teardrop only.
B—Parallel only.
C—Direct.

Determine the holding pattern by placing your pencil on the holding fix and dragging it on the holding radial given by ATC, then returning back to the fix. Then draw the pattern from the fix with turns in the direction specified. Holding north on the 360° radial with left turns means you will be east of R-360.

The entry procedure is based on the aircraft's heading. To determine which entry procedure to use, draw a line at a 70° angle from the holding fix, and cutting the outbound leg at about one-third its length. With a heading of 155°, we are in the largest piece of pie, so a direct entry would be used. See *the figure below. (PLT296) — AIM ¶5-3-7*

Answer (A) is incorrect because a teardrop entry would be appropriate only from R-180 to R-250. Answer (B) is incorrect because, if you were approaching on R-070 to R-180, you would make a parallel entry.

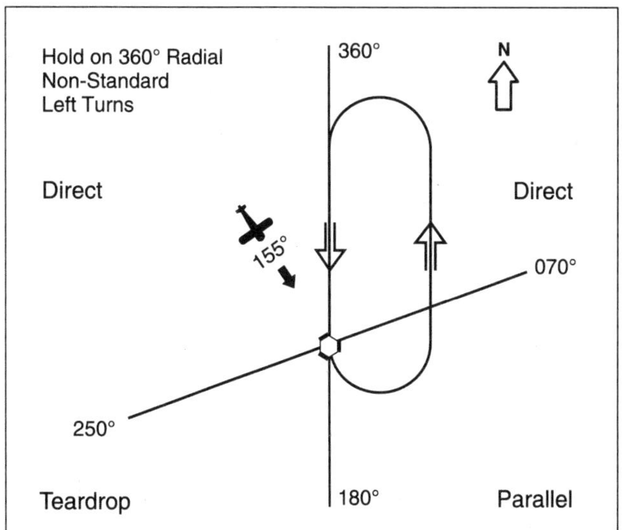

Question 8866

Answers

8865 [C]
8866 [C]

ALL

8867. (Refer to Figure 124.) A pilot receives this ATC clearance:

"...CLEARED TO THE ABC VORTAC. HOLD SOUTH ON THE ONE EIGHT ZERO RADIAL..."

What is the recommended procedure to enter the holding pattern?

A—Teardrop only.
B—Parallel only.
C—Direct only.

Determine the holding pattern by placing your pencil on the holding fix and dragging it on the holding radial given by ATC, then returning back to the fix. Then draw the pattern from the fix with turns in the direction specified. Holding south on the 180° radial with right turns means you will be east of R-360.

The entry procedure is based on the aircraft's heading. To determine which entry procedure to use, draw a line at a 70° angle from the holding fix, and cutting the outbound leg at about one-third its length. With a heading of 155°, we are in the smallest piece of pie, so a teardrop entry would be used. See the following figure. (PLT087) — AIM ¶5-3-7

Answer (B) is incorrect because a parallel entry would be appropriate only from R-360 to R-110. Answer (C) is incorrect because a direct entry would only be appropriate from R-110 to R-290.

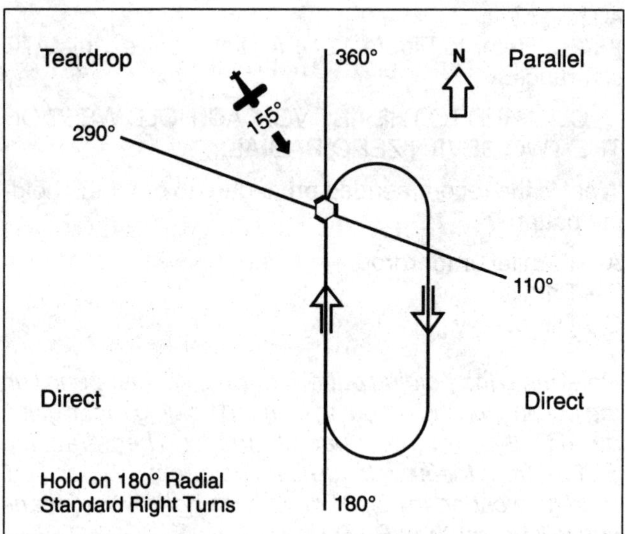

Question 8867

Answers

8867 [A]

Charts

The pilot-in-command must ensure that the appropriate aeronautical charts are on board the aircraft for each flight.

There are a number of questions that require reference to a segment of the Airport/Facility Directory. The legend for this publication is available in the FAA Legends 13 through 19.

Most of the questions concerning interpretation of Approach Charts, DPs and STARs can be answered by referring to the appropriate legend. These legends are available during the test in FAA Legend 40.

There are a few questions that require you to interpret the symbology on Enroute Charts. Unlike the other charts, no legend is available in the test book.

Departure Procedures (DPs) are depicted in one of two basic forms. Pilot Navigation (Pilot NAV) DPs are established where the pilot is primarily responsible for navigation on the DP route. Vector DPs are established where ATC will provide radar navigational guidance to an assigned route or fix. A vector DP will often include procedures to be followed in the event of a two-way communication radio failure.

Standard Terminal Arrival Routes (STARs) are ATC-coded IFR arrival routes established for certain airports. STARs purpose is to simplify clearance delivery procedures. ATC will assign a STAR to a civil aircraft whenever they deem it appropriate.

The Jet Route system consists of jet routes established from 18,000 feet MSL to FL450 inclusive.

The GPS Approach Overlay Program permits pilots to use GPS avionics under IFR for flying existing instrument approach procedures, except localizer (LOC), localizer directional aid (LDA), and simplified directional facility (SDF) procedures. Aircraft navigating by GPS are considered to be RNAV aircraft. Therefore, the appropriate equipment suffix must be included in the ATC flight plan. The word "or" in the approach title indicates that approach is in Phase III of the GPS Overlay Program. This allows the approach to be flown without reference of any kind to the ground-based NAVAIDs associated with the approach. When using GPS for the approach at the destination airport, the alternate must be an approach other than a GPS.

ALL

9012. In what way are SIDs depicted in plan view?

A—"Vectors" provided for navigational guidance or "Pilot NAV" with courses the pilot is responsible to follow.
B—"Vectors" and "Pilot NAV" for pilots to use at their discretion.
C—Combined textual and graphic form which are mandatory routes and instructions.

Pilot navigation (Pilot NAV) DPs (previously called SIDs) are established where the pilot is primarily responsible for navigation on the DP route. Vector DPs are established where ATC will provide radar navigational guidance to a filed/assigned route or to a fix depicted on the DP. (PLT201) — AIM ¶5-2-8

Answer (B) is incorrect because DPs are departure procedures and must be followed as depicted (not at the pilot's discretion). If a pilot does not wish to use a DP, then he/she must notify ATC. Answer (C) is incorrect because a NOS DP does not list the textual description in the plan view. The plan view depicts a DP as either a "pilot nav" or "vector" to signify if navigation is provided by the pilot or by radar vectors.

ALL

9034. What is the primary purpose of a STAR?

A—Provide separation between IFR and VFR traffic.
B—Simplify clearance delivery procedures.
C—Decrease traffic congestion at certain airports.

A STAR is an ATC-coded IFR arrival route established for application to arriving IFR aircraft destined for certain airports. Its purpose is to simplify clearance delivery procedures. (PLT170) — AIM ¶5-4-1

Answer (A) is incorrect because separation between IFR and VFR traffic is provided by Stage III radar service, not a STAR. Answer (C) is incorrect because controlled airspace, e.g., Class B, Class C, can be used to decrease traffic congestion at some airports by allowing ATC to regulate traffic flow and volume.

Answers

9012 [A] 9034 [B]

ALL

9035. When does ATC issue a STAR?

A—Only when ATC deems it appropriate.
B—Only to high priority flights.
C—Only upon request of the pilot.

Pilots of IFR civil aircraft destined to locations for which STARs have been published may be issued a clearance containing a STAR whenever ATC deems it appropriate. (PLT170) — AIM ¶5-4-1

Answer (B) is incorrect because any type of IFR flight can be issued a STAR. High priority flights will normally be handled in an expeditious manner by ATC. Answer (C) is incorrect because a STAR is a clearance delivery procedure that is issued by ATC. A pilot has the responsibility to accept or refuse that clearance. A pilot can list a STAR in the flight plan, but ATC will issue one only if appropriate.

ALL

9549. (Refer to Figures 94, 95, and 96.) What action should be taken by the pilot, if communications are lost, while IMC, after takeoff on RWY 13L at Chicago Midway Airport?

A—Return and land immediately at Chicago Midway Airport.
B—Complete initially assigned turn south of DPA R-096, maintain 3,000 feet or lower if assigned. Then 10 minutes after departure, climb to FL 190, direct to GIJ, then flight plan route.
C—Complete initially assigned turn within 4 DME of Midway and maintain 3,000 feet or lower, if assigned. Then 10 minutes after departure, climb to FL 190, direct to GIJ, and then flight plan route.

If a two-way radio communications failure occurs while in IFR conditions, the pilot of a flight should comply with the route and altitude listed.
Route:

1. *By the route assigned in the last IFR clearance.*
2. *If being radar vectored, by the direct route from the point of failure to the fix, route, or airway specified in the vector clearance.*
3. *In the absence of an assigned route, by the route that ATC has advised may be expected in a further clearance.*
4. *If there is no assigned or expected route, by the route in the flight plan.*

The pilot should maintain the highest of the following altitudes for the route segment flown:

1. *The altitude or flight level in the last clearance.*
2. *The minimum altitude for IFR operations (i.e., MEA).*
3. *The altitude or flight level that ATC has advised may be expected in a further clearance.*

(PLT052) — 14 CFR §91.185

Answer (A) is incorrect because the pilot is in IFR conditions, and therefore must continue the flight according to the route and altitude assigned in the last clearance. Answer (B) is incorrect because these are the actions for a non-DME equipped aircraft, and the question deals with an aircraft which has DME.

ALL

9613. (Refer to Figures 168, 169, and 169A.) What action should be taken by the pilot if communications are lost after departure from RWY 16 at PWK if VMC?

A—Continue the flight under VMC and land as soon as practicable.
B—Climb to 3,000 feet; after 3 minutes, turn direct to PMM and climb to FL 190.
C—Start right turn within 1 mile of the departure end of RWY, remain east of ORD VOR/DME R-345, and maintain 3,000 feet; 3 minutes after departure, turn direct to PMM, and climb to FL 190.

If a radio failure occurs in VFR conditions, or if VFR conditions are encountered after the failure, each pilot shall continue the flight under VFR and land as soon as practicable. (PLT078) — 14 CFR §91.185

ALL

9616. (Refer to Figure 169A.) What action should be taken by the PIC of PTZ 70 if the communication radios fail after takeoff from RWY 16 at PWK while in IMC conditions?

A—Climb to 3,000 feet on RWY heading; after 3 minutes, turn direct to PMM and climb to FL 190.
B—Start right turn within 1 mile of the departure end of RWY 16 and remain east of the 345 radial of the ORD VOR/DME while climbing to 3,000 feet; after 3 minutes, turn direct to PMM and climb to FL 190.
C—Set 7600 in Mode 3 of the transponder, turn direct to Northbrook (the IAF), climb to 2,700 feet, and fly the ILS RWY 16 to land at PWK.

If a two-way radio communications failure occurs while in IFR conditions, the pilot of a flight should comply with the route and altitude listed.
Route:

1. *By the route assigned in the last IFR clearance.*
2. *If being radar vectored, by the direct route from the point of failure to the fix, route, or airway specified in the vector clearance.*

Answers

9035 [A] 9549 [C] 9613 [A] 9616 [B]

3. *In the absence of an assigned route, by the route that ATC has advised may be expected in a further clearance.*
4. *If there is no assigned or expected route, by the route in the flight plan.*

The pilot should maintain the highest of the following altitudes for the route segment flown:

1. *The altitude or flight level in the last clearance.*
2. *The minimum altitude for IFR operations (i.e., MEA).*
3. *The altitude or flight level that ATC has advised may be expected in a further clearance.*

(PLT391) — 14 CFR §91.185

ALL

9550. (Refer to Figure 97.) In the profile view of the RNAV or GPS RWY 32 approach to Buffalo Intl, between CYUGA and the MAP, the following appears: ∠ 2 91°. What is it?

A—The required pitch attitude change at CYUGA, to ensure arriving at 1,220 feet and 1.5 miles at the same time.
B—It indicates that 2.91° below level is recommended on the attitude indicator.
C—The Final Approach Angle for Vertical Path Computers.

The symbol in the profile section of a nonprecision approach is used to designate the Final Approach Path Angle for Vertical Path Computers (RNAV Descent). (PLT083) — Instrument Approach Procedures

Answer (A) is incorrect because it is the vertical path from CYUGA to the runway, not the vertical path from CYUGA to arrive at 1,220 feet and 1.5 miles at the same time. Answer (B) is incorrect because it must indicate 2.91° below the level flight attitude indication on the attitude indicator which may or may not be 2.91° below the horizon on the attitude indicator.

ALL

9551. (Refer to Figure 97A.) Greater Buffalo Intl was closed upon N60JB's arrival and was not expected to be open for 4 hours, for snow removal. N60JB received clearance to ROC. Upon arrival at ROC, the flight was cleared for the ILS RWY 28. To fly the ILS RWY 28 at ROC, the aircraft must have the following navigation equipment:

A—Radar and VOR/DME.
B—VOR/DME and ADF.
C—Radar and VOR/ILS/DME.

The remarks listed in the bottom left-hand corner of the instrument approach procedure (FAA Figure 97) indicates ADF is required for this approach. (PLT083) — Instrument Approach Procedures

ALL

9552. (Refer to Figure 97.) How can the FAF on the RNAV RWY 32 approach at BUF be identified?

A—The RNAV receiver will indicate 175.1° and 2.5 DME miles from BUF VORTAC.
B—The RNAV receiver will indicate a change from TO to FROM and 0 deflection of the course needle.
C—Two flashes/second on the OM beacon light.

The FAF (CYUGA waypoint) is indicated by a Maltese Cross in the profile view of the chart. When passing CYUGA, the RNAV receiver will either indicate station passage by changing from TO to FROM, or by changing the active waypoint to the missed approach point. The RNAV displays navigational information to the pilot with reference to the active waypoint and not to the NAVAIDs that define the waypoint. (PLT083) — Instrument Approach Procedures

Answer (A) is incorrect because the waypoint would be set as the BUF 175.1° at 2.5 DME; thus at the waypoint, the indication would change from TO to FROM without any right or left deflection of the course line. Answer (C) is incorrect because the approach chart does not indicate that there is a marker beacon at the FAF.

ALL

9553. (Refer to Figure 97.) Which of the following will define the position of the RNAV MAP for Greater Buffalo Intl?

A—116.4 BUF 286.9°, -3.5 NM.
B—42°56.44'N - 78°38.48'W.
C—42°56.26'N - 78°43.57'W.

The MAP waypoint identification box lists the coordinates as N42°56.29' W78°43.54', with a frequency of 116.4. The identifier is BUF, and the radial-distance (facility to waypoint) is 286.9° – 3.5NM. (PLT083) — Instrument Approach Procedures

Answers (B) and (C) are incorrect because they are the wrong coordinates for the MAP waypoint.

Answers

9550 [C]	9551 [B]	9552 [B]	9553 [A]

ALL

9554. (Refer to Figure 97.) What is the procedure for initiating the missed approach on the RNAV RWY 32 approach at BUF?

A—Climbing left turn, select GANIS Waypoint and establish a direct course, climbing to 2,800 feet.
B—Select and maintain R-302 of BUF VORTAC climbing to 2,800 feet.
C—Establish and maintain R-286.9 of BUF VORTAC climbing to 2,800 feet.

The missed approach procedure is in the upper left-hand corner of the profile view. It specifies, "Missed approach climbing left turn to 2800 direct GANIS WPT and hold." (PLT083) — Instrument Approach Procedures

Answers (B) and (C) are incorrect because the course guidance is in reference to the RNAV or GPS waypoint labeled GANIS, not the VORTAC.

ALL

9555. (Refer to Figures 97A, 97B, and 97C.) N60JB desired to list ROC as an alternate for BUF. The active RWY at ROC was expected to be RWY 28. What weather forecast was required at Greater Rochester Intl, for N60JB to list it as an alternate?

A—Nonprecision approach 800-2, precision approach 800-2.
B—Nonprecision approach 800-2, precision approach 600-2.
C—Nonprecision approach 800-2 1/4, precision approach 600-2.

The A in the triangle in the bottom left-hand corner of the approach chart indicates that Rochester has non-standard alternate minimums. FAA Figure 97C lists all the approaches at Rochester which have nonstandard alternate minimums. Assuming that N60JB is in approach category A or B, the minimums for the ILS 28 approach are 800-2. Since no nonstandard minimums are listed for a nonprecision approach to runway 28 in a category A or B aircraft, we can assume that standard alternate minimums (800-2) apply. The standard alternate minimums are listed at the top of FAA Figure 97B. (PLT083) — Instrument Approach Procedures

Answers (B) and (C) are incorrect because the lowest minimums for a precision approach in any category aircraft is 800-2.

ALL

9562. (Refer to Legends 43 and 43A.) The filed flight plan, for N91JB, block 13, indicates that BUR is the alternate, for LAX. The RWYs at LAX are closed and expected to remain closed for 2 hours when N91JB arrives. N91JB requests 4,000 feet, Tower Enroute Control (TEC) with radar vectors to BUR. If radar vectors are not available, what route can be expected from LAX to BUR?

A—Direct SMO, VNY, BUR.
B—LAX LAX316 SILEX .
C—Direct SMO, UR, SILEX BUR.

There is no listing in the Legend for a TEC flight from LAX to BUR. The first listing under Coast TRACON is for flights from other Los Angeles basin airports to BUR via the LAX VORTAC. The remainder of the route from LAX is the best answer for the information given. (PLT073) — Airport/Facility Directory

ALL

9565. (Refer to Legends 43 and 43A and Figure 103.) The RWYs at LAX are closed and expected to remain closed for 2 hours when N91JB arrives. N91JB requests 4,000 feet, Tower Enroute Control (TEC) with radar vectors to BUR. What altitude can N91JB expect based upon the type aircraft?

A—4,000 feet.
B—5,000 feet.
C—6,000 feet.

Aircraft types (J, M, P, and Q) are given for each city pair and should be used with the route of flight filed. M = Turbo Props/Special (cruise speed 190 knots or greater). Not all city pairs are depicted; however, geographic areas connected by tower enroute control are shown. The closest paring given is the Coast Tracon to Burbank (BUR). N91JB has filed the Downe Three Arrival, which terminates with vectors to final approach course for runways 6 and 7 at Los Angeles International (LAX). This indicates that LAX is landing to the East, therefore the CSTJ2 route should be used. The altitude for CSTJ2 is 5,000 feet (JM50PQ40). (PLT073) — Airport/Facility Directory

Answers

9554 [A] 9555 [A] 9562 [B] 9565 [B]

ALL

9563. (Refer to Figures 103 and 104.) If communications are lost soon after takeoff on RWY 11R at Tucson Intl, what altitude restrictions apply, in IMC conditions?

A—Fly assigned heading for vectors to intercept appropriate transition, maintain 17,000 feet to GBN, then climb to assigned altitude.
B—Fly assigned heading for vectors to intercept the Gila Bend transition; climb to 17,000 feet or lower assigned altitude; climb to FL 220, 10 minutes after departure.
C—Fly assigned heading for vectors to intercept the Gila Bend transition; climb to 17,000 feet; 10 minutes after departure, climb to FL 220.

If a two-way radio communications failure occurs while in IFR conditions, the pilot of a flight should comply with the route and altitude listed.
Route:

1. *By the route assigned in the last IFR clearance.*
2. *If being radar vectored, by the direct route from the point of failure to the fix, route, or airway specified in the vector clearance.*
3. *In the absence of an assigned route, by the route that ATC has advised may be expected in a further clearance.*
4. *If there is no assigned or expected route, by the route in the flight plan.*

The pilot should maintain the highest of the following altitudes for the route segment flown:

1. *The altitude or flight level in the last clearance.*
2. *The minimum altitude for IFR operations (i.e., MEA).*
3. *The altitude or flight level that ATC has advised may be expected in a further clearance.*

See *FAA Figure 104, DP Route Description. (PLT052) — 14 CFR §91.185*

Answers (A) and (C) are incorrect because the statement "or assigned lower altitude" omits an important altitude restriction. The pilot would be required to maintain the lower assigned altitude until 10 minutes after the departure before climbing to the filed flight level.

ALL

9564. (Refer to Figure 104.) What are the takeoff minimums for RWY 11R at Tucson Intl that apply to N91JB?

A—1 SM.
B—800/1.
C—4,000/3.

Reference the DP plan view, first note (FAA Figure 104): Rwys 3, 11L/R require a ceiling of 4,000 and 3 miles visibility or you must be able to meet the minimum climb requirements of 250 feet per NM to 6,500 feet. For this question, do not assume the aircraft can meet the required climb gradient. (PLT052) — Departure Procedure (DP) Chart

ALL

9569. (Refer to Figures 110 and 112.) In addition to VOR and DME, what other avionics equipment is required to be operational, at takeoff, to fly the VOR/DME RWY 32R approach at IAH?

A—Altitude alerting system.
B—Standby VOR and DME receivers.
C—VHF communications and transponder equipment.

The charts and the Airport/Facility Directory indicate Houston Intercontinental Airport is a Class II TCA. TCAs have been redesignated as Class B airspace. To operate in Class B airspace, an aircraft must be equipped with a VOR, two-way radio, and a transponder. (PLT083) — 14 CFR §91.131

ALL

9571. (Refer to Figures 111 and 112.) Which approach lighting is available for RWY 32R?

A—MALSR with RAIL.
B—HIRL.
C—TDZ and CL.

The A_5 inside the circle with the dot indicates that runway 32R has MALSR approach lights with RAIL. See *FAA Legend 5. (PLT083) — Instrument Approach Procedures*

Answer (B) is incorrect because HIRL indicates High Intensity Runway Lights. Answer (C) is incorrect because TDZ and CL lighting is not available for RWY 32R, and it is in-runway, not approach, lighting.

ALL

9573. (Refer to Figure 112.) At what point must the missed approach be initiated on the VOR/DME RWY 32R approach at IAH, if still IMC?

A—Anytime after the FAF.
B—IAH 1.3 DME.
C—IAH 1 DME.

The profile view of the approach chart shows the change from the procedure track to the missed approach track at the IAH 1.0 DME fix. (PLT083) — Instrument Approach Procedures

Answer (A) is incorrect because the question asks what is the latest point at which you must initiate the missed approach. Answer (B) is incorrect because it is the VDP, not the MAP.

Answers

9563 [B]	9564 [C]	9569 [C]	9571 [A]	9573 [C]

ALL

9587. (Refer to Figure 118C.) What instrument approach light system or RWY lighting system is available for the LOC BC RWY 26L approach at Phoenix Sky Harbor Intl?

A—HIRL and REIL.
B—MALS and REIL.
C—SALS and ODALS.

The Airport/Facility Directory indicates that runway 26L has HIRL and REIL. (PLT078) — Airport/Facility Directory

ALL

9588. (Refer to Figure 118A.) The touchdown zone elevation of the LOC BC RWY 26L approach at Phoenix Sky Harbor Intl is

A—1,123 feet.
B—1,130 feet.
C—1,640 feet.

The notation "TDZE 1130" indicates the touchdown zone elevation on runway 26L is 1,130 feet MSL. See FAA Legend 4. (PLT049) — Instrument Approach Procedures

ALL

9590. (Refer to Figure 118A.) What is the HAT a Category B aircraft may descend to if the pilot has identified HADEN INT on the LOC BC RWY 26L approach at Phoenix Sky Harbor Intl?

A—510 feet.
B—667 feet.
C—670 feet.

The Height Above Touchdown (HAT) for a Category B airplane with HADEN DME is 510 feet. See FAA Legend 7. (PLT083) — Instrument Approach Procedures

Answer (B) is incorrect because this is the HAT for the Category B circling approach without HADEN INT. Answer (C) is incorrect because this is the HAT for straight-in approach RWY 26L without HADEN.

ALL

9596. (Refer to Figure 121, upper panel.) On the airway J220 (BUF R-158) SE of Buffalo, the MAA is 39,000 feet. What is the MAA on J547 between BUF and PMM (lower panel)?

A—60,000 feet.
B—43,000 feet.
C—45,000 feet.

The Maximum Authorized Altitude is the maximum usable altitude or flight level on an airway or jet route for which an MEA is published. FL450 is the upper limit of all jet routes. (PLT055) — Pilot/Controller Glossary

Answer (A) is incorrect because this is the upper limit of Class A airspace, but the jet route system stops above 45,000 feet. Answer (B) is incorrect because the upper limit of the jet route system includes FL450 unless marked otherwise.

ALL

9599. (Refer to Figure 161.) To receive the DME information from the facility labeled "DME Chan 22" at La Guardia requires that

A—N711JB be equipped with a UHF NAV radio, which is tuned to channel 22.
B—a military TACAN tuned to channel 22.
C—the VHF NAV radio be tuned to the ILS (108.5) frequency.

VOR/ILS frequencies are paired with TACAN channels so that the VHF and UHF facilities can be tuned together. FAA Legend 21 indicates that DME channel 22 is paired with LOC frequency 108.5. (PLT202) — FAA-H-8083-15

ALL

9600. (Refer to Figure 161A.) The La Guardia weather goes below minimums and New York Approach Control issues a clearance to N711JB, via radar vectors, to ASALT Intersection. As N711JB is approaching ASALT, Approach Control clears the aircraft to fly the VOR RWY 13L/13R approach. What is the distance from ASALT Intersection to RWY 13L?

A—12.3 NM.
B—12.4 NM.
C—13.3 NM.

The plan view indicates that the distance from ASALT Int to CRI is 6.0 NM. The distance from CRI to the lead-in lights is 1.7 NM. The distance from the lead-in lights to runway 13L is 4.7 NM. (PLT083) — Instrument Approach Procedures

Answer (A) is incorrect because the RWY 13R and radar required distance have been used. Answer (C) is incorrect because the radar required distance has been used rather than the correct distance to the first lead-in light cluster.

Answers

9587 [A]	9588 [B]	9590 [A]	9596 [C]	9599 [C]	9600 [B]

ALL

9601. (Refer to Figure 161A.) The La Guardia weather goes below minimums and New York Approach Control issues a clearance to N711JB, via radar vectors, to ASALT Intersection. What is the lowest altitude that Approach Control may clear N711JB to cross ASALT Intersection?

A—3,000 feet.
B—2,500 feet.
C—2,000 feet.

The profile view shows a mandatory altitude of 3,000 feet at ASALT unless advised by ATC, then 2,000 feet is the minimum. (PLT049) — Instrument Approach Procedures

ALL

9602. (Refer to Figure 161A.) For landing on RWY 31L at JFK, how much RWY is available?

A—11,248 feet.
B—11,966 feet.
C—14,572 feet.

The runway 31L landing distance of 11,248 feet is listed in the upper right-hand corner of the airport diagram. (PLT083) — Instrument Approach Procedures

ALL

9603. (Refer to Figure 161A.) What must be operational for N711JB to execute the VOR RWY 13L/13R approach to JFK?

A—Radar and DME.
B—LDIN and VOR.
C—Lead-in Light System, VOR and Radar.

VOR is required, as indicated by the approach title. The plan view indicates that radar is required. The notes section at the bottom of the chart shows that the lead-in light system must be operational to execute the procedure. (PLT083) — Instrument Approach Procedures

ALL

9604. (Refer to Figure 161A.) The distance from Canarsie (CRI) to RWY 13R at JFK is

A—5.4 NM.
B—6.3 NM.
C—7.3 NM.

The plan view indicates that the distance from CRI to the lead-in lights is 1.7 NM and the distance from the lead-in lights to runway 13R is 3.7 NM. (PLT083) — Instrument Approach Procedures

ALL

9614. (Refer to Figure 169A.) The PIC of PTZ 70 will use 25° of bank during the turn after departing RWY 16 at PWK. What is the maximum TAS that the aircraft may maintain during the turn and remain east of the ORD VOR/DME R-345 under a no wind condition?

A—160 knots.
B—162 knots.
C—164 knots.

The table at the top of the plan view shows the maximum TAS for various bank angles. (PLT052) — Instrument Approach Procedures

ALL

9615. (Refer to Figure 169A.) To remain east of the ORD VOR/DME R-345, while flying the PAL-WAUKEE TWO DEPARTURE, requires a turn radius of

A—over 5,000 feet.
B—5,000 feet.
C—less than 5,000 feet.

The note in the bottom left corner of the plan view states that a turn radius of less than 5,000 feet is required. (PLT052) — Instrument Approach Procedures

Answers

9601 [C]	9602 [A]	9603 [C]	9604 [A]	9614 [B]	9615 [C]

ALL

9617. (Refer to Figure 172A.) The airport diagram of Greater Buffalo Intl Airport has a symbol (appears to be a triangle balanced on top of another triangle) located close to the end of RWYs 14 and 32. What do these symbols indicate?

A—Helicopter landing areas.
B—That special takeoff and landing minimums apply to RWYs 14 and 32.
C—RWY Radar Reflectors.

The double triangle symbol in the airport diagram indicates runway radar reflectors. See *FAA Legend 4. (PLT083) — Instrument Approach Procedures*

ALL

9619. (Refer to Figure 173A.) During the approach (ILS RWY 10 at SYR) while maintaining an on glide slope indication with a groundspeed of 110 knots, what was the approximate rate of descent for PTZ 70?

A—475 feet per minute.
B—585 feet per minute.
C—690 feet per minute.

The profile view indicates that the final approach angle is 3.0°. The table in FAA Legend 9 shows that a 555 fpm descent rate is required for 105 knots, and 635 fpm is required for 120 knots. 585 is the only answer choice to fall within these limits. (PLT049) — FAA-H-8083-15, Chapter 7

ALL

9620. (Refer to Figure 171, top panel.) The facility (Kankakee) that is located 9 miles NE of Chicago Midway or 27 miles SSE of Northbrook (OBK) is a/an

A—Aeronautical Radio Inc. (AIRINC) transmitter.
B—Automated Weather Observing System (AWOS-ASOS) with frequency.
C—Flight Service, Remote Communications Outlet.

This is the symbol for a Remote Communication Outlet (RCO). The remoted FSS and frequency are shown. (PLT058) — Enroute Low Altitude Chart Legend

ALL

9623. (Refer to Figure 175.) Four airways (V298, V25, V448 and V204) near YKM have a series of dots that overlay the airway. What do these dots indicate?

A—That the airways penetrate a Prohibited and Restricted Airspace.
B—That 2 miles either side of the airway, where shaded, is a Controlled Firing Area.
C—That the airways penetrate a Military Operations Area (MOA) and a special clearance must be received from ATC.

A series of dots overlaying an airway segment indicates that it penetrates restricted or prohibited airspace. (PLT058) — Enroute Low Altitude Chart Legend

ALL

9629. (Refer to Figure 182A.) EAB 90 is a "CAT B" aircraft and received a clearance to fly the LOC RWY 09R approach, to circle to land RWY 27R. The Baldn fix was received. What are the minimums?

A—540-1.
B—600-1.
C—680-1.

The note in the plan view shows category B circling minimums of 600-1 if the BALDN fix is received. (PLT083) — Instrument Approach Procedures

ALL

9630. (Refer to Figure 182A.) EAB 90 is a "CAT B" aircraft and received a clearance to fly the LOC RWY 09R, to land RWY 09R. The Baldn fix was received. What are the minimums?

A—520/24.
B—600/24.
C—680/24.

The note in the plan view shows category A and B LOC 9R minimums of 520-24 when the BALDN fix is received. (PLT083) — Instrument Approach Procedures

Answers

9617 [C]	9619 [B]	9620 [C]	9623 [A]	9629 [B]	9630 [A]

ALL

9631. (Refer to Figure 182A.) EAB 90 is a "CAT B" aircraft and received a clearance to fly the LOC RWY 09R to sidestep and land RWY 09L. The Baldn fix was received. What are the minimums?

A—680-1.
B—520/24.
C—600-1.

The note in the plan view shows category B sidestep 9L minimums are 600-1 when BALDN is received. (PLT083) — Instrument Approach Procedures

Answer (A) is incorrect because it is the sidestep minimums without BALDN fix. Answer (B) is incorrect because it is the LOC minimums with the BALDN fix.

ALL

9641. (Refer to Figure 186.) The NAVAID box at Mormon Mesa (MMM) has a black square in the upper left corner. What does this indicate?

A—That Hazardous Inflight Weather Advisory Service is available.
B—That the National Observatory transmits a time signal on the VOR frequency.
C—That the facility has a Transcribed Weather Broadcast (TWEB) service on the frequency.

A black square in the upper left-hand corner of a NAVAID box indicates HIWAS (Hazardous Inflight Weather Advisory Service) is available. (PLT058) — Enroute Low Altitude Chart Legend

Answer (B) is incorrect because the National Observatory does not transmit time signals on VOR frequencies. Answer (C) is incorrect because the TWEB is indicated by the T inside the circle.

ALL

9644. (Refer to Figure 192.) On the airway J10 between OBH and LBF, the MAA is 41,000 feet. What is the MAA on J197 between FSD and OBH?

A—43,000 feet.
B—45,000 feet.
C—60,000 feet.

The Maximum Authorized Altitude is the maximum usable altitude or flight level on an airway or jet route which has a published MEA. FL450 is the upper limit of all jet routes. (PLT100) — Pilot/Controller Glossary

Answer (A) is incorrect because the upper limit of the jet route system includes FL450 unless marked otherwise. Answer (C) is incorrect because this is the upper limit of Class A airspace, but the jet route system stops above 45,000 feet.

ALL

9645. (Refer to Figures 193, 193A, 194, 195, 195A, 196, and 196A.) While being radar vectored for the ILS/DME RWY 35R, Denver Approach Control tells PIL 10 to contact the tower, without giving the frequency. What frequency should PIL 10 use for tower?

A—121.85.
B—132.35.
C—124.3.

The approach chart, NOAA airport diagram, and the Airport/Facility Directory all list 124.3 as the tower frequency for runway 35R. (PLT049) — Instrument Approach Procedures

Answer (A) is incorrect because this is the ground control frequency for runway 8-26. Answer (B) is incorrect because this is the Denver Tower frequency for runway 17R-35L.

ALL

9647. (Refer to Figures 193, 193A, and 194.) The entry points for the (NORTHEAST GATE) LANDR ONE and SAYGE ONE arrivals are approximately

A—11 NM apart.
B—12 NM apart.
C—13 NM apart.

The Arrival Overview in the upper left-hand corner of FAA Figure 194 states that entry points for parallel arrivals are approximately 12 NM apart. (PLT080) — Instrument Approach Procedures

ALL

9648. (Refer to Figures 195, 195A, 196, and 196A.) When PIL 10 becomes visual, at 3.8 NM from the end of Runway 35R, if the aircraft is on glide slope and on course, what should the pilot see for a Visual Glideslope Indicator?

A—Two white and two red lights on the left side of the runway, in a row.
B—One white and one red light on the left or right side of the runway, in a row.
C—Two red and two white lights, in a row, on the right side of the runway.

FAA Figure 196A indicates Runway 35R has PAPI(P4R), which means the visual glide slope indicators are 4 identical light units placed on the right side of the runway. See FAA Legend 17. (PLT147) — Airport/Facility Directory

Answers

9631 [C]	9641 [A]	9644 [B]	9645 [C]	9647 [B]	9648 [C]

ALL

9649. (Refer to Figures 195, 195A, 196, and 196A.) All of the runways at Denver Intl have what type of Visual Glideslope Indicators?

A—PVASI.
B—PAPI.
C—APAP.

The airport diagrams display the P inside the circle at both end of all runways. FAA Legend 6 indicates this symbol means the runway has Precision Approach Path Indicator (PAPI) approach lighting system. (PLT147) — Instrument Approach Procedures

ALL

9654. (Refer to Figure 198A.) The highest terrain shown in the planview section of the LOC-B approach to Eagle County Regional is

A—11,275 feet.
B—11,573 feet.
C—12,354 feet.

The highest terrain is indicated by a large dot in the plan view. In this case, the large dot is in the upper left corner above Denver Center. (PLT083) — Instrument Approach Procedures

ALL

9655. (Refer to Figures 201 and 201A.) What type of weather information would normally be expected to be available from the Weather Data Source at Ogden-Hinckley?

A—Cloud height, weather, obstructions to vision, temperature, dewpoint, altimeter, surface winds, and any pertinent remarks.
B—Cloud bases/tops, obstructions to vision, altimeter, winds, precipitation, and the intensity of the precipitation.
C—Cloud height, obstructions to vision, temperature, dewpoint, altimeter, wind data, and density altitude.

The Airport/Facility Directory indicates the weather source for Ogden-Hinckley is LAWRS (Limited Aviation Weather Reporting Station). Observers report cloud height, weather, obstructions to vision, temperature and dew point (in most cases), surface wind, altimeter, and pertinent remarks. See *FAA Legend 18. (PLT078) — Airport/Facility Directory*

Answer (B) is incorrect because it does not report cloud bases or tops, nor precipitation and intensity. Answer (C) is incorrect because it does not report wind data and density altitude.

ALL

9658. (Refer to Figure 185A.) The maximum gross weight that an L1011 can be operated on RWY 07R/25L at McCarran Intl is

A—521,000 pounds.
B—633,000 pounds.
C—620,000 pounds.

The runway weight limit for a dual tandem gear airplane on runway 07R-25L is listed as 633,000 pounds. However, the remarks section states the maximum weight for an L1011 is 521,000 pounds. (PLT078) — Airport/Facility Directory

ALL

9659. (Refer to Figures 185 and 185A.) The threshold of RWY 07L at McCarran Intl is displaced

A—874 feet, due to a pole.
B—2,133 feet, due to a hangar.
C—1,659 feet, due to a pole.

The Airport/Facility Directory listing for runway 07L states the threshold is displaced 2,133 feet for a hangar. (PLT078) — Airport/Facility Directory

ALL

9660. (Refer to Figures 203 and 203A.) PTL 55 will be unable to cross the Oasis Intersection at 9,500 feet. What should the crew do?

A—Enter holding on R-211 LAS at 15 DME, right-hand turns, advise Departure Control, climb to 9,500 prior to Oasis.
B—Advise Las Vegas Departure Control and request radar vectors.
C—Continue the climb on LAS R-211 to 9,500 feet, then turn right to 260°.

The note on the plan view states that aircraft on the Beatty transition which are unable to cross OASIS at 9,500 feet should continue the climb on the LAS R-211 until reaching 9,500 feet and then turn right to 260°. (PLT052) — Instrument Approach Procedures

Answers

9649 [B]	9654 [C]	9655 [A]	9658 [A]	9659 [B]	9660 [C]

ALL

9661. (Refer to Figures 205 and 206.) What is the maximum weight that PTL 55 may weigh for landing at San Francisco Intl (SFO)?

A—710,000 pounds.
B—715,000 pounds.
C—720,000 pounds.

The remarks section of the Airport/Facility Directory states that the maximum weight for a B-747 is 710,000 pounds. (PLT078) — Airport/Facility Directory

ALL

9662. (Refer to Figures 202 and 206.) PTL 55 received the following clearance from Bay Approach Control. PTL 55 is cleared ILS RWY 19L at SFO, sidestep to RWY 19R. 1.3 times the V_{SO} speed, of PTL 55, is 165 knots. What is the lowest minimum descent altitude (MDA) and the lowest visibility that PTL 55 may accomplish the sidestep?

A—340-1
B—340-1-1/2.
C—340-2.

The ILS RWY 19L shows the sidestep minimums are 340-2 for category D. (PLT083) — Instrument Approach Procedures

ALL

9666. (Refer to Figures 205 and 206A.) At San Francisco Intl (SFO), the runway hold position signs are

A—all on the left-hand side of the taxiways.
B—all on the right-hand side of the taxiways.
C—on either side of the taxiways.

The note on the airport diagram states that several runway hold position signs are on the right rather than the left side of the taxiways. (PLT083) — Instrument Approach Procedures

ALL

9667. (Refer to Figures 207 and 207A.) Due to weather PTL 55 was unable to land at SFO. PTL 55 was given radar vectors to COMMO Intersection and clearance to fly the ILS RWY 11 approach at Oakland Intl. What frequencies will PTL 55 use for Oakland Tower and Oakland Ground Control?

A—118.3 and 121.75.
B—127.2 and 121.75.
C—127.2 and 121.9.

The frequency listing on the plan view of the approach chart lists the runway 11-29 tower frequency as 127.2 and the ground control frequency as 121.75. The Airport/ Facility Directory lists these frequencies for the south complex, which is runway 11-29. (PLT083) — Instrument Approach Procedures

ALL

9670. (Refer to Figure 210.) The route between FIS (near Key West) and MTH, which is labeled B646, is an example of a

A—LF/MF Airway.
B—LF/MF Oceanic Route.
C—Military Training Route.

The brown color and airway designation (Blue 646) indicate it is a LF/MF airway. A solid narrow line indicates that it is an Oceanic route. A domestic LF/MF airway would be represented by a broader, shaded line. (PLT058) — Enroute Low Altitude Chart Legend

ALL

9674. (Refer to Figures 210 and 211.) The Miami Flight Service Station has

A—Hazardous Inflight Weather Advisory Service (HIWAS).
B—Remote Communications Outlet (RCO) northeast of MIA which operates on 122.3.
C—Transcribed Weather Broadcast.

A black square in the upper left-hand corner of a NAVAID box indicates HIWAS is available. (PLT058) — Enroute Low Altitude Chart Legend

Answer (B) is incorrect because the RCO northeast of MIA operates on 126.7. Answer (C) is incorrect because the TWEB symbol (the T inside a circle) is not shown in the communication box.

Answers

9661 [A]	9662 [C]	9666 [C]	9667 [B]	9670 [B]	9674 [A]

ALL

9675. (Refer to Figure 210.) The Miami ARTCC remote site located near Pahokee has a discrete VHF frequency of

A—123.45.
B—133.55.
C—135.35.

The Miami ARTCC remote site communication box is the light blue scalloped box southwest of the Pahokee VORTAC. (PLT058) — Enroute Low Altitude Chart Legend

ALL

9681. (Refer to Figures 214 and 182A.) TNA 90 is a "CAT C" aircraft and has received a clearance to fly the ILS 9R approach and sidestep to RWY 9L at PHL. What are the minimums?

A—520/40.
B—600/1 1/2.
C—680/1 3/4.

The minimums for the side-step approach are listed in the plan view. We are not told we are receiving BALDN INT, therefore we cannot assume that we are. (PLT083) — Instrument Approach Procedures

Answer (A) is incorrect because these are the minimums for the Category C S-LOC 9R with the BALDN INT. Answer (B) is incorrect because these are the minimums for the sidestep 9L with the BALDN INT.

ALL

9682. (Refer to Figures 214 and 182A.) TNA 90 is a "CAT C" aircraft and has received clearance to fly the LOC RWY 09R and circle to land RWY 27R. Baldn fix is received. What are the minimums?

A—640/2.
B—600/1 1/2.
C—680/1 3/4.

The minimums for the circling approach are listed in the plan view. (PLT083) — Instrument Approach Procedures

Answer (A) is incorrect because they are the minimums for the circling approach with the BALDN fix for category D aircraft. Answer (C) is incorrect because they are the minimums without the BALDN INT.

ALL

9683. (Refer to Figures 214 and 182A.) TNA 90 is a "CAT C" aircraft and has received a clearance to fly the LOC RWY 09R; Baldn fix is received and TNA 90 is cleared to land 09R. What are the minimums?

A—520/40.
B—680/60.
C—600/1-1/2.

The minimums for the LOC approach are listed in the plan view. (PLT083) — Instrument Approach Procedures

Answer (B) is incorrect because they are the minimums without the BALDN INT. Answer (C) is incorrect because they are the minimums for the circling approach with the BALDN fix.

ALL

9686. (Refer to Figure 215A.) The airport diagram of Bradley Intl Airport has a symbol (appears to be a triangle balanced on top of another triangle) located close to the approach end of RWY 19. What does this symbol indicate?

A—Runway Radar Reflectors.
B—Practice hover area for the Army National Guard helicopters.
C—Two course lights, back to back, which flash beams of light along the course of an airway.

The double triangle symbol in the airport diagram stands for runway radar reflectors. See FAA Legend 4. (PLT083) — Instrument Approach Procedures

ALL

9688. (Refer to Legends 42 and 42B and Figure 214.) The filed flight plan for TNA 90 indicates, if it becomes necessary to divert to the alternate, that tower enroute (TEC), radar vectors, and 3,000 feet are requested to ACY. If radar vectors are not available, what route can be expected from PHL to ACY?

A—Direct SAVVY Intersection, V166 OOD, V184 ACY.
B—Direct WILJR Intersection, VCN, V184 ACY.
C—OOD VCN V184 ACY.

Philadelphia is listed second-to-last in the left column, and Atlantic City is the second entry in the right column. In this case, the route is OOD VCN V184 ACY with the highest altitude listed as 3,000 feet. (PLT073) — Instrument Approach Procedures

Answers

9675 [B]	9681 [C]	9682 [B]	9683 [A]	9686 [A]	9688 [C]

ALL

9689. (Refer to Legends 42 and 42B and Figure 214.) The filed flight plan for TNA 90 indicates, if it becomes necessary to divert to the alternate, that tower enroute (TEC), radar vectors and 3,000 feet are requested to ACY. What is the maximum altitude that TNA 90 may be cleared to under TEC?

A—2,000 feet.
B—3,000 feet.
C—4,000 feet.

Philadelphia is listed second-to-last in the left column, and Atlantic City is the second entry in the right column. In this case, the route is OOD VCN V184 ACY with the highest altitude listed as 3,000 feet. (PLT073) — Instrument Approach Procedures

ALL

9691. The GPS Approach Overlay Program permits pilots to use GPS avionics when IFR for flying existing instrument approach procedures, except

A—LOC, LDA and ADF.
B—LDA, TAC and SDF.
C—SDF, LOC and LDA.

The GPS Approach Overlay Program permits pilots to use GPS avionics under IFR for flying existing instrument approach procedures, except localizer (LOC), localizer directional aid (LDA), and simplified directional facility (SDF) procedures. (PLT354) — AIM ¶1-1-19

ALL

9692. Aircraft navigating by GPS are considered, on the flight plan, to be

A—RNAV equipped.
B—Astrotracker equipped.
C—FMS/EFIS equipped.

Aircraft navigating by GPS are considered to be RNAV aircraft. Therefore, the appropriate equipment suffix must be included in the ATC flight plan. (PLT354) — Pilot/Controller Glossary

ALL

9693. The Instrument Approach Procedure Chart top margin identification is VOR or GPS RWY 25, AL-5672 (FAA), LUKACHUKAI, ARIZONA. In what phase of the approach overlay program is this GPS approach?

A—Phase I.
B—Phase III.
C—Phase II.

The word "or" in the approach title indicates that approach is in Phase III of the GPS Overlay Program. This allows the approach to be flown without reference of any kind to the ground-based NAVAIDs associated with the approach. (PLT354) — AIM ¶1-1-19

ALL

9694. The weather forecast requires an alternate for LUKACHUKAI (GPS RWY 25) ARIZONA. The alternate airport must have an approved instrument approach procedure, which is anticipated to be operational and available at the estimated time of arrival, other than

A—GPS or VOR.
B—ILS or GPS.
C—GPS.

When using GPS for the approach at the destination airport, the alternate must be an approach other than a GPS. (PLT354) — AIM ¶1-1-19

ALL

8793. (Refer to Figure 104.) Determine the DEP CON frequency for the TUS3.GBN SID after takeoff from RWY 11R at Tucson Intl.

A—125.1 MHz.
B—118.5 MHz.
C—128.5 MHz.

The A/FD segment indicates that the Departure Control Frequency for runway 11, departures on bearings 090° through 285° from the airport, is 125.1 Mhz. (PLT078) — Airport/Facilities Directory

Answer (B) is incorrect because 118.5 MHz is for departures from RWY 11 with a departure heading between 286° to 089°. Answer (C) is incorrect because 128.5 is the general approach/departure control frequency.

Answers

9689 [B]	9691 [C]	9692 [A]	9693 [B]	9694 [C]	8793 [A]

ALL

8794. (Refer to Figure 104.) Using an average groundspeed of 140 knots, what minimum indicated rate of climb must be maintained to meet the required climb rate (feet per NM) to 9,000 as specified on the SID?

A—349 ft/min.
B—560 ft/min.
C—584 ft/min.

The DP (previously called SID) requires a 250 foot per NM climb gradient to 9,000 feet. FAA Legend 10 indicates that at 140 knots, a rate of climb of 583 feet per minute is required to meet this climb gradient. (PLT052) — DP Chart

ALL

8796. (Refer to Figure 104.) How can the pilot receive the latest NOTAMs for the TUS LAX flight?

A—Monitor ATIS on 123.8 MHz.
B—Contact the FSS on 122.2 MHz.
C—Request ADCUS on any FSS or Tower frequency.

NOTAMs can be received through the FSS located on the airport. The standard FSS frequency is 122.2 MHz. (PLT078) — Airport/Facility Directory

ALL

8797. (Refer to Figure 104.) What distance is available for takeoff on RWY 11R at Tucson Intl?

A—7,000 feet.
B—9,129 feet.
C—10,994 feet.

Runway 11R-29L is 9,129 feet long. The displaced threshold does not reduce takeoff distance. (PLT078) — Airport/Facility Directory

Answer (A) is incorrect because 7,000 feet is the length of RWY 03-21. Answer (C) is incorrect because 10,994 feet is the length of RWY 11L-29R.

ALL

8798. (Refer to Figure 104.) What effect on the takeoff run can be expected on RWY 11R at Tucson Intl?

A—Takeoff length shortened to 6,986 feet by displaced threshold.
B—Takeoff run shortened by 0.6 percent runway slope to the SE.
C—Takeoff run will be lengthened by the 0.6 percent upslope of the runway.

There is a 0.6% upslope to the southeast on runway 11R-29L. (PLT052) — Airport/Facility Directory

Answer (A) is incorrect because a displaced threshold will shorten the usable runway but has no effect on an airplane's required takeoff distance. Answer (B) is incorrect because the takeoff run is lengthened due to the 0.6% upslope to the southeast.

ALL

8782. (Refer to Figures 99 and 101.) Which frequency should be selected to check airport conditions and weather prior to departure at DFW Intl?

A—117.0 MHz.
B—134.9 MHz.
C—135.5 MHz.

The departure ATIS for DFW airport is listed as 135.5 Mhz. (PLT078) — Airport/Facility Directory

Answer (A) is incorrect because 117.0 MHz is the listed ATIS frequency for arriving airplanes. Answer (B) is incorrect because 134.9 MHz is also a listed ATIS frequency for arriving airplanes.

ALL

8783. (Refer to Figures 99, 100, and 101.) The frequency change from departure control to ARTCC after departing DFW Intl for IAH is

A—135.5 to 126.0 MHz.
B—118.55 to 127.95 MHz.
C—127.75 to 127.95 MHz.

The A/FD lists 127.75 MHz as the southbound Departure Control frequency for DFW. That would be most appropriate for V369. The "postage stamp" symbol just above and to the right of KILLR INT in FAA Figure 100 indicates that the Fort Worth Center frequency in that area is 127.95 Mhz. (PLT078) — Airport/Facility Directory

Answer (A) is incorrect because 135.5 MHz is the ATIS frequency for departing airplanes and 126.0 MHz is the ARTCC discrete frequency for northwest of DFW. Answer (B) is incorrect because 118.55 MHz is the departure control frequency for eastbound traffic from DFW.

Answers

8794 [C]	8796 [B]	8797 [B]	8798 [C]	8782 [C]	8783 [C]

ALL

8784. (Refer to Figure 100.) Where is the VOR changeover point on V369 between DFW Intl and TNV?

A—Ft. Worth/Houston ARTCC boundary.
B—81 NM from DFW Intl.
C—TORNN Int.

The distance on V369 between DFW and TNV is 162 NM. Since there is no VOR changeover symbol on the chart, the changeover point is halfway (81 NM from DFW). (PLT058) — Enroute Low Altitude Chart Legend

Answer (A) is incorrect because the Ft. Worth/Houston ARTCC boundary would indicate a radio communication frequency change, not a navigational frequency change. Answer (C) is incorrect because TORNN intersection is 85 NM from DFW VORTAC. Even though a DME distance fix is shown for TORNN, there is no special VOR changeover indicator that designates TORNN as a COP, thus the midway point (81 NM) should be used.

ALL

8785. (Refer to Figure 100 or 101.) What is the magnetic variation at both DFW Intl and IAH?

A—08 E.
B—0.
C—08 W.

The Magnetic Variation for the DFW VOR is 8° East. (PLT078) — Airport/Facility Directory

ALL

8786. (Refer to Figures 100 and 102.) How should the pilot identify the position to leave V369 for the Cugar Four Arrival?

A—Intercept R-305 of IAH.
B—21 DME miles from TNV.
C—141 DME miles from DFW.

The CUGAR4 STAR begins at BILEE INT but the initial course is inbound on the TNV 334° radial which is the same as V369. This is the course to be flown until intercepting the IAH 305° radial. (PLT080) — Enroute Low Altitude Chart Legend

Answer (B) is incorrect because there is no DME fix for 21 DME miles from TNV VORTAC on either the STAR or the IFR Enroute Chart. Answer (C) is incorrect because a pilot should have changed over to TNV VORTAC, and there is no 141 DME fix from DFW on V369.

ALL

8824. (Refer to Figure 114.) The changeover point on V394 between DAG VORTAC and POM VORTAC is

A—halfway.
B—38 DME miles from DAG VORTAC.
C—64 DME miles from DAG VORTAC.

There is a changeover point marked on the enroute chart at 64 NM from DAG VORTAC and 16 NM from POM VORTAC. (PLT058) — Enroute Low Altitude Chart

Answer (A) is incorrect because when the changeover point is not located at the midway point, aeronautical charts will depict the location and give mileage to the radio aids. Answer (B) is incorrect because 38 DME miles from DAG VORTAC is APLES INT (not the change-over point).

ALL

8825. (Refer to Figure 114.) The minimum crossing altitude at APLES INT southwest bound on V394 is

A—7,500 feet.
B—9,100 feet.
C—11,500 feet.

There is a Minimum Crossing Altitude (MCA) of 9,100 feet at APLES INT when southwest bound on V394. (PLT058) — Enroute Low Altitude Chart Legend

Answer (A) is incorrect because 7,500 feet is the minimum enroute altitude (MEA) from DAG VORTAC to APLES INT on V394, not the MCA at APLES INT. Answer (C) is incorrect because 11,500 feet is the minimum enroute altitude after APLES INT, not the MCA at APLES INT.

ALL

8826. (Refer to Figure 114, lower panel.) What is the minimum enroute altitude on V210, when crossing the POM VORTAC southwest bound and continuing on the same airway?

A—10,700 feet.
B—10,300 feet.
C—5,300 feet.

The Minimum Enroute Altitude (MEA) approaching POM VORTAC southwest bound is 5,300 feet. Absent an MCA, this is the minimum crossing altitude at the VORTAC. There is no published MCA for V210 Southwest bound crossing POM. (PLT058) — Enroute Low Altitude Chart Legend

Answer (A) is incorrect because 10,700 feet is the MEA from MEANT INT to CALBE INT when southwest bound on V210. Answer (B) is incorrect because 10,300 feet is the MCA on V210 northeast bound.

Answers

8784 [B]	8785 [A]	8786 [A]	8824 [C]	8825 [B]	8826 [C]

ALL

8810. (Refer to Figures 110 and 112.) How should the pilot identify the position to leave V369 for the Cugar Four Arrival?

A—Intercept R-305 of IAH.
B—21 DME miles from TNV.
C—141 DME miles from DFW.

The CUGAR4 STAR begins at BILEE INT but the initial course is inbound on the TNV 334° radial which is the same as V369. This is the course to be flown until intercepting the IAH 305° radial. (PLT080) — Enroute Chart

Answer (B) is incorrect because the pilot should leave V369 when he/she intercepts R-305 of IAH, not 21 DME miles from TNV. Answer (C) is incorrect because the pilot should have switched to TNV VORTAC for V369 course guidance before BILEE INT and should not be using DFW VORTAC for any navigational guidance.

ALL

8811. (Refer to Figure 112.) What action should the pilot take if communications were lost during the Cugar Four Arrival, after turning on the 305 radial of IAH?

A—Proceed direct to IAH VORTAC, then outbound on the IAH R-125 for a procedure turn for final approach.
B—From BANTY INT, proceed to the IAF on the IAH R-290, then continue on the IAH 10 DME Arc to final approach.
C—Proceed direct to IAH VORTAC, then to either IAF on the IAH 10 DME Arc to final approach.

In the event of a two-way radio failure, the pilot should fly the cleared route to the clearance limit. If the clearance limit is not a fix from which an approach begins, he/she should leave the clearance limit (if no EFC has been received) upon arrival at the limit and proceed to a fix from which an approach begins. Then, commence descent and approach as close as possible to the estimated time of arrival. (PLT208) — 14 CFR §91.185(c)

Answer (A) is incorrect because there is no procedure turn indicated for the VOR/DME RWY 32R approach; thus, no procedure turn can be made to reverse course. Answer (B) is incorrect because the pilot would proceed to the IAH VORTAC, then to either IAF on the IAH 10 DME arc to final approach.

ALL

8812. (Refer to Figure 112.) The Cugar Four Arrival ends

A—at BANTY INT.
B—at IAH VORTAC.
C—when cleared to land.

The text portion of the STAR says to expect radar vectors to final after BANTY INT. (PLT080) — STAR Chart

Answer (B) is incorrect because the arrival ends at BANTY INT. Answer (C) is incorrect because a clearance to land will be given during the instrument approach procedure, which is after the end of the Cugar Four Arrival.

ALL

8816. (Refer to Figure 112.) What effect on approach minimums, if any, does an inoperative MALSR have for an aircraft with an approach speed of 120 knots at IAH?

A—None.
B—Increases RVR to 5,000 feet.
C—Increases RVR to 6,000 feet.

A 120-knot approach speed makes this a Category B aircraft for approach minimum purposes. When MALSR is inoperative on a VOR/DME approach, 1/2 mile must be added to the required visibility. The published Category B minimums are 2,400 RVR (1/2 mile). The higher minimums are therefore 1 mile or 5,000 RVR. (PLT082) — FAA Legend 11

Answer (A) is incorrect because, with an inoperative MALSR, there is no change to the MDA, but an increase of 1/2 mile to the required visibility. Answer (C) is incorrect because an airplane with an approach speed of 141-165 knots (Category D) or greater would have the visibility increased to RVR of 6,000 feet.

Answers

8810 [A] 8811 [C] 8812 [A] 8816 [B]

ALL

8817. (Refer to Figure 112.) When is the earliest time the pilot may initiate a descent from 460 feet MSL to land at IAH?

A—Anytime after GALES INT if the runway environment is visible.
B—Only after the IAH 1.3 DME if the runway environment is visible.
C—Only after the IAH 1 DME if the runway environment is visible.

The "v" symbol in the profile view at the 1.3 DME fix is a Visual Descent Point (VDP). If able to receive the VDP, a pilot may not begin descent below the MDA until reaching the VDP. (PLT083) — FAA Legend 3

Answer (A) is incorrect because GALES INT is the final approach fix (FAF) which indicates the point at which a pilot may initiate a descent to the MDA. Answer (C) is incorrect because, at the IAH, 1 DME is the MAP which must be initiated if the runway environment is not in sight. The earliest time the pilot may initiate a descent from the MDA of 460 feet at IAH is at the VDP.

ALL

8818. (Refer to Figure 112.) How should the pilot identify the MAP on the IAH VOR/DME RWY 32R?

A—After time has elapsed from FAF.
B—IAH 1.3 DME.
C—IAH 1 DME.

The IAH 1 DME fix is designated as the MAP. (PLT083) — IFR Approach Plate

Answer (A) is incorrect because timing from the FAF is used on non-precision approaches that are not VOR/DME procedures, or when the facility is on the airport and the facility is the MAP. Answer (B) is incorrect because IAH 1.3 DME is the VDP.

ALL

8849. (Refer to Figure 122.) What is the lowest altitude at which the glide slope may be intercepted when authorized by ATC?

A—2,500 feet.
B—3,000 feet.
C—4,000 feet.

The lowest altitude at which the glide slope may be intercepted when authorized by ATC is 2,500 feet as indicated by the number 2500 with a solid line below the altitude next to the FAF. (PLT083) — FAA-H-8083-15

Answer (B) is incorrect because 3,000 feet is the minimum altitude at GRETI, which is prior to the FAF unless ATC authorizes a descent to 2,500 feet. Answer (C) is incorrect because 4,000 feet is the minimum altitude at KITTS (not the FAF), unless ATC authorizes a descent to 2,500 feet.

ALL

8850. (Refer to Figure 122.) What would be the DME reading at the lowest altitude at which the glide slope may be intercepted when authorized by ATC?

A—12.4 miles.
B—9.4 miles.
C—7.7 miles.

The DME reading at the lowest altitude for glide slope intercept is at the FAF marker which indicates altitude of 2500 at 7.7 DME. See *FAA Figure 122 around Jockey LOM. (PLT083) — Instrument Approach Procedure*

Answer (A) is incorrect because 12.4 NM is the DME reading at KITTS. Answer (B) is incorrect because 9.4 NM is the DME reading at GRETI.

ALL

8851. (Refer to Figure 122.) At what altitude and indicated airspeed would you expect to cross PIVOT INT on the approach to ORD?

A—FL 200 and 300 KIAS.
B—10,000 feet and 250 KIAS.
C—12,000 feet and 200 KIAS.

You would expect to cross PIVOT intersection at 10,000 feet and 250 KIAS as indicated by the vertical navigation planning information on the upper left-hand corner of the STAR. (PLT083) — STAR Chart

Answer (A) is incorrect because turbojet arrivals should expect a clearance to cross PMM VORTAC at FL200. Answer (C) is incorrect because 12,000 feet and 200 KIAS is not an expected clearance for either a turbojet or turboprop aircraft using the Pullman Two Arrival to ORD.

ALL

8852. (Refer to Figure 122 and Legend 9.) What is the approximate rate of descent required (for planning purposes) to maintain the electronic glide slope at 120 KIAS with a reported headwind component of 15 knots?

A—555 ft/min.
B—635 ft/min.
C—650 ft/min.

The ILS RWY 32L approach into Chicago-O'Hare has a 3° glide slope as indicated within the profile section of the approach chart. Using FAA Legend 9 along with a ground speed of 105 (120 KIAS – 15K headwind component) we find a rate of descent of 555 feet per minute. (PLT083) — IFR Approach Procedure

Answer (B) is incorrect because a rate of descent of 635 fpm is appropriate for the indicated airspeed of 120 knots. Answer (C) is incorrect because a rate of descent of 650 fpm is appropriate for a 3.5° glide slope angle.

Answers

8817 [B]	8818 [C]	8849 [A]	8850 [C]	8851 [B]	8852 [A]

ALL

8799. (Refer to Figures 106 and 107.) Which approach control frequency is indicated for the TNP.DOWNE3 Arrival with LAX as the destination?

A—128.5 MHz.
B—124.9 MHz.
C—124.5 MHz.

FAA Figure 106 lists the approach frequency as 124.5 for the TNP.DOWNE3 arrival. (PLT080) — Airport/Facility Directory

Answers (A) and (B) are incorrect because the A/FD frequencies do not apply when using the TNP.DOWNE3 arrival.

ALL

8800. (Refer to Figures 106 and 107.) At what point does the flight enter the final approach phase of the ILS RWY 25L at LAX?

A—FUELR INT.
B—HUNDA INT.
C—Intercept of glide slope.

The "Lightning Bolt" symbol indicates that 3,500 feet is the glide slope interception altitude. An aircraft is considered to be on the final approach segment past this point when on an ILS approach. (PLT083) — Instrument Approach Procedure

Answer (A) is incorrect because FUELR INT is an initial approach fix (IAF) and does not mark the point at which the flight enters the final approach phase. Answer (B) is incorrect because HUNDA INT is a fix at which the pilot should intercept the glide slope, but the intersection itself does not mark the point at which the airplane enters the final approach phase.

ALL

8801. (Refer to Figures 106 and 107.) What is the DH for the ILS RWY 25L at LAX if the pilot has completed the initial Category II certification within the preceding 6 months, but has flown no CAT II approaches?

A—201 feet.
B—251 feet.
C—301 feet.

The IAP minimums section shows DHs of 251 and 201 MSL, which represent 150-foot and 100-foot DHs respectively. Regulations restrict a pilot who has completed initial CAT II qualification to the 150-foot DH until he/she has completed 3 approaches to that minimum within the previous 6 months. (PLT420) — 14 CFR §61.13

Answer (A) is incorrect because a DH of 201 feet is available once the initial Category II limitation is removed. Answer (C) is incorrect because a DH of 150 feet is the limitation placed upon an original issue for Category II operations.

ALL

8802. (Refer to Figures 106 and 107.) The radio altimeter indication for the DH at the inner marker on the ILS RWY 25L approach at LAX is

A—101.
B—111.
C—201.

The RA height at the IM is also the DH for the 100-foot minimums. At this point, the radio altimeter will indicate 111 feet. (PLT083) — Instrument Approach Procedure

Answer (A) is incorrect because 101 feet is the touchdown zone elevation (TDZE). Answer (C) is incorrect because 201 feet is the height above touchdown (HAT) at the inner marker.

ALL

8803. (Refer to Figures 106 and 107.) If the glide slope indication is lost upon passing HUNDA INT on the ILS RWY 25L approach at LAX, what action should the pilot take?

A—Continue the approach as an LOC and add 100 feet to the DH.
B—Immediately start the missed approach direct to INISH INT.
C—Continue to the MAP and execute the missed approach as indicated.

A CAT II approach cannot be flown without a glide slope indication so a missed approach should be executed. Obstacle clearance on a missed approach is predicated on the assumption that the abort is initiated at the MAP and not lower than the DH or MDA. When an early missed approach is executed, pilots should, unless otherwise authorized by ATC, fly the instrument approach procedure to the MAP at or above the DH or MDA before executing any turning maneuver. (PLT406) — AIM ¶5-5-5

Answer (A) is incorrect because there are no LOC minimums for this approach. A pilot must have a glide slope indication, or execute the missed approach procedure. Answer (B) is incorrect because protected obstacle clearance areas for missed approaches are made on the assumption that the abort is made at the MAP. No consideration is made for an abnormally early turn, unless otherwise directed by ATC.

Answers

8799 [C] 8800 [C] 8801 [B] 8802 [B] 8803 [C]

ALL

8804. (Refer to Figures 106 and 107.) What approach lights are available for the ILS RWY 25L approach at LAX?

A—ALSF-2 with sequenced flashing lights.
B—MALSR with a displayed threshold.
C—HIRL and TDZ/CL.

The "A" in the circle next to runway 25L on the Airport Diagram indicates that ALSF-2 approach lights are available. The dot at the top of the circle indicates that sequenced flashing lights are part of the system. See *FAA Legend 5. (PLT078) — Airport/Facility Directory*

Answer (B) is incorrect because RWY 25R has an out of service (see Remarks) MALSR and a displaced threshold. Answer (C) is incorrect because high intensity runway lights (HIRL), touchdown zone lights (TDZ), and centerline lights (CL) are runway lighting systems.

ALL

8805. (Refer to Figures 106 and 107.) How can DOWNE INT be identified?

A—ILAX 15 DME.
B—LAX 15 DME.
C—LAX R-249 and SLI R-327.

DOWNE INT can be identified by the SLI 327° radial crossing the localizer or by the I-LAX 15 DME. (PLT083) — Instrument Approach Procedure

Answer (B) is incorrect because LAX 15 DME defines a circle, not an intersection. The correct answer would be LAX R-068, 15 DME. Answer (C) is incorrect because DOWNE INT is located on the ILS localizer, not on a radial from the LAX VORTAC.

ALL

8806. (Refer to Figure 107.) How should the IFR flight plan be closed upon landing at LAX?

A—Contact Hawthorne FSS on 123.6 MHz.
B—Phone Hawthorne FSS on 644-1020.
C—LAX tower will close it automatically.

If operating IFR to an airport with a functioning control tower, the flight plan is automatically closed upon landing. (PLT224) — AIM ¶5-1-14

Answer (A) is incorrect because an IFR flight plan would be closed with a flight service station only if there is no operating tower. According to the A/FD in Figure 6-37, 123.6 MHz is not a frequency on which to contact Hawthorne FSS. Answer (B) is incorrect because an IFR flight plan is closed with a FSS only if there is no operating tower at the destination airport.

ALL

8950. (Refer to Figure 134.) What are the required weather minimums to execute the CONVERGING ILS RWY 9R approach procedure?

A—Ceiling 700 feet and 2-1/2 miles visibility.
B—At least 1,000 feet and 3 miles visibility.
C—Ceiling 800 feet and 2 miles visibility.

The required weather minimums are listed as 721–2-1/2 700 (700–2-1/2). The 721 refers to DH in MSL. The 2-1/2 refers to visibility in miles. The 700 refers to height of DH above touchdown zone. The minimums in parentheses apply only to military pilots. (PLT083) — FAA-H-8083-15

Answer (B) is incorrect because 1,000 feet and 3 miles visibility defines the minimums required for VFR flight in Class D or E airspace for an airport. Answer (C) is incorrect because 800 feet and 2 miles visibility with a nonprecision approach are alternate minimums.

ALL

8951. (Refer to Figure 134.) What is the final approach fix for the CONVERGING ILS RWY 9R approach procedure?

A—BWINE INT and 3,000 feet MSL.
B—KELEE INT.
C—1,800 feet MSL and glide slope interception.

The final approach fix is designated on government charts by the Maltese Cross symbol for nonprecision approaches and the lightning bolt symbol for precision approaches. The lightning bolt symbol on the descent profile portion of the chart indicates the FAF as 1,800 feet MSL. See *Legend 3. (PLT083) — Pilot/Controller Glossary*

Answer (A) is incorrect because by looking at the plan view of the chart (Figure 6-38), BWINE Intersection is labeled as an IAF or initial approach fix (not a final approach fix). Answer (B) is incorrect because KELEE Intersection only identifies the outer marker. Note that the lightning bolt arrow is depicted just prior to KELEE Intersection.

ALL

8952. (Refer to Figure 134.) What is the MINIMUM airborne equipment required to execute the CONVERGING ILS RWY 9R approach procedure?

A—Localizer and DME.
B—Localizer and glide slope.
C—Localizer only.

Neither the chart title nor the missed approach procedure specifies DME. The category minimums specify "NA" (not authorized) for localizer only and circling approaches. (PLT083) — AIM ¶5-5-4

Answers

8804 [A]	8805 [A]	8806 [C]	8950 [A]	8951 [C]	8952 [B]

ALL

8836. (Refer to Figure 118A.) Straight-in minimums for a Category B aircraft equipped with DME on the LOC BC RWY 26L approach are

A—1,800/1.
B—700/1.
C—1,640/1.

The straight in minimums for the LOC 26L approach are an MDA of 1,640 and visibility of 1 mile if the HADEN DME fix is identified. (PLT083) — Instrument Approach Procedure

Answer (A) is incorrect because 1,800/1 are the straight-in minimums without DME. Answer (B) is incorrect because 700/1 are military minimums without operating DME.

ALL

8837. (Refer to Figure 118A.) How is course reversal accomplished when outbound on the LOC BC RWY 26L approach at Phoenix Sky Harbor Intl?

A—Radar vector only.
B—Procedure turn beyond 10 NM.
C—Holding pattern entry beyond 10 NM.

The absence of the procedure turn barb in the plan view indicates that a procedure turn is not authorized for that procedure. (PLT083) — AIM ¶5-4-9

Answer (B) is incorrect because no procedure turn is depicted on the IAP chart, and when radar is used for vectoring, no pilot may make a procedure turn unless he/she requests and is issued a clearance by ATC. Answer (C) is incorrect because no holding pattern is depicted on the IAP chart and it may not be used as a course reversal.

ALL

8839. (Refer to Figure 118A.) Identify the final approach fix on the LOC BC RWY 26L approach at Phoenix Sky Harbor Intl.

A—Upon intercepting the glide slope beyond I PHX 5 DME.
B—When crossing I-PHX 5 DME at 3,000 feet.
C—When crossing the SRP VORTAC on the glide slope.

The "Maltese Cross" symbol indicates the final approach fix for the LOC 26L approach is at the I-PHX 5 DME fix. The minimum altitude for crossing that fix is 3,000 feet MSL. See FAA Legend 3. (PLT083) — Instrument Approach Procedure

Answers (A) and (C) are incorrect because an LOC approach is a nonprecision approach that does not provide glide slope information. On the back course a glide slope indication may be indicated, but it must be ignored.

121, DSP

8242. Assuring that appropriate aeronautical charts are aboard an aircraft is the responsibility of the

A—aircraft dispatcher.
B—flight navigator.
C—pilot-in-command.

The pilot-in-command shall ensure that appropriate aeronautical charts containing adequate information concerning navigation aids and instrument approach procedures are aboard the aircraft for each flight. (PLT444) — 14 CFR §121.549

Answer (A) is incorrect because the dispatcher may be hundreds of miles from the origination of the flight, e.g., in a central dispatch office. Answer (B) is incorrect because although a flight navigator may be assigned the task of carrying aeronautical charts, the pilot-in-command is responsible for ensuring that adequate charts are aboard the aircraft.

RTC

9572. (Refer to Figures 111 and 112.) While N131JB was flying the VOR/DME RWY 32R approach to Houston Intercontinental, approach control told them to contact copter control, at GALES. Approach control did not tell them what frequency, and did not respond when N131JB asked what frequency. What frequency should N131JB use for copter control?

A—118.1.
B—121.5.
C—135.15.

The frequency for copter control at Houston Intercontinental is found under the tower listing in the Airport/Facility Directory. (PLT078) — Airport/Facility Directory

Answers

8836 [C]	8837 [A]	8839 [B]	8242 [C]	9572 [C]

Chapter 7
Emergencies, Hazards, and Flight Physiology

Flight Emergencies and Hazards

The Pilot/Controller Glossary divides emergencies into two categories: **distress** and **urgency**. Distress is a condition of being threatened by serious and/or imminent danger and of requiring immediate assistance. Distress conditions include fire, mechanical failure or structural damage. An urgency condition is one of being concerned about safety and of requiring timely but not immediate assistance. At least an urgency condition exists the moment a pilot becomes doubtful about position, fuel endurance, weather or any other condition that could adversely affect the safety of flight. A pilot should declare an emergency when either an urgency or a distress condition exists.

When a distress or urgency condition exists, the pilot should set the radar beacon transponder to code 7700. If an aircraft is being hijacked or illegally interfered with, the pilot can alert ATC to that fact by setting the transponder to code 7500. If an aircraft has experienced a two-way communications radio failure, the pilot should set the transponder to code 7600. The pilot should also conform to the radio failure procedures of 14 CFR §91.185 (IFR operations: Two-way radio communications failure). In order to avoid false alarms, pilots should take care not to inadvertently switch through codes 7500, 7600 and 7700 when changing the transponder.

If a two-way radio failure occurs in VFR conditions, or if VFR conditions are encountered after the failure, the pilot must continue the flight under VFR and land as soon as practicable. If IFR conditions prevail, the pilot must follow the rules listed below for route, altitude and time to leave a clearance limit:

Route to be Flown

- The route assigned in the last ATC clearance received.
- If being radar vectored, fly by the direct route from the point of the radio failure to the fix, route or airway specified in the vector clearance.
- In the absence of an assigned route, fly by the route that ATC has advised may be expected in a further clearance.
- In the absence of an assigned route or expected further routing, fly by the route filed in the flight plan.

Altitude

Fly the highest of the following altitudes or flight levels for the route segment being flown:

- The altitude or flight level assigned in the last ATC clearance received.
- The minimum IFR altitude for the route segment being flown (MEA).
- The altitude or flight level that ATC has advised may be expected in a further clearance.

When to Leave a Clearance Limit

- When the clearance limit is a fix from which an approach begins, commence descent or descent and approach as close as possible to the expect further clearance (EFC) time if one has been received; or if one has not been received, as close as possible to the estimated time of arrival (ETA) as calculated from the filed or amended estimated time en route.
- If the clearance limit is not a fix from which an approach begins, leave the clearance limit at the expect further clearance (EFC) time if one has been received; or if none has been received, upon arrival over the clearance limit, and proceed to a fix from which an approach begins and commence descent or descent and approach as close as possible to the estimated time of arrival (ETA) as calculated from the filed or amended time en route.

Continued

A near midair collision is defined as an occurrence in which the possibility of a collision existed as the result of two aircraft coming within 500 feet or less of each other.

A minimum fuel advisory is used by a pilot to inform ATC that the fuel supply has reached a state where the pilot cannot accept any undue delay upon arrival at the destination. The minimum fuel advisory is not a declaration of an emergency, nor is it a request for priority. It does indicate that an emergency situation may develop if any undue delay occurs during the rest of the flight.

Some airports have a number of wind indicators located around the perimeter of the field as well as a center field windsock. When there is a significant difference in speed or direction between the center field windsock and one or more of the boundary wind indicators, the tower can report that a wind shear condition exists.

A safety alert will be issued to pilots being controlled by ATC in either of two circumstances. A controller will issue a safety alert when, in the controller's opinion, the aircraft's altitude will put it in unsafe proximity to the surface or an obstacle. A controller will also issue an alert if he/she becomes aware of another aircraft, not controlled by him/her, that will put both aircraft in an unsafe proximity to each other.

The **wake turbulence** developed by large aircraft can present a significant flight hazard to other aircraft that encounter them. The main component of wake turbulence is **wing-tip vortices.** These are twin vortices of air trailing behind an aircraft in flight. The **vortex** is a by-product of lift. The pressure under each wing is greater than the pressure above it and this induces a flow of air outward, upward and around the wing tip. This leaves two counter-rotating spirals of air trailing behind the aircraft. *See* Figure 7-1.

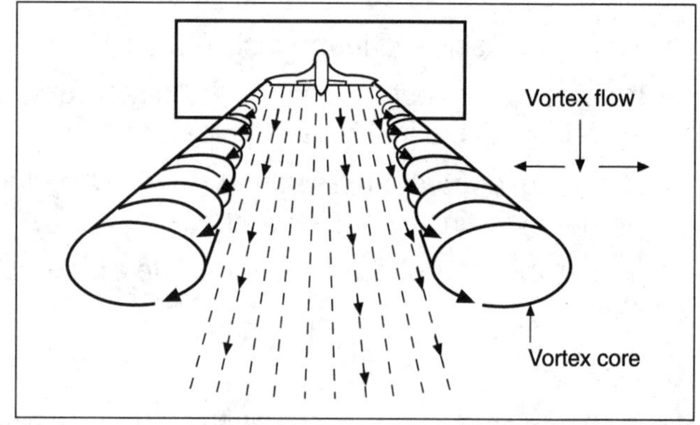

Figure 7-1

The characteristics of a vortex can be altered by changing the aircraft's configuration. The most intense vortices will be produced by an airplane that is heavy, flying slowly, and with the landing gear and flaps retracted.

The vortices generated by a large aircraft will slowly sink below its flight path and dissipate by the time they have descended about 1,000 feet. They will also tend to drift away from each other at a speed of about five knots. In a light crosswind, the upwind vortex will tend to stay over the same position on the ground while the downwind vortex will move away at about twice its normal rate. It is good wake turbulence avoidance technique to stay above and on the upwind side of the flight path of a preceding large airplane.

If the vortices reach the ground before dissipating, they will move away from each other as noted above. In a light crosswind, the upwind vortex can remain on the runway long after a large airplane has taken off or landed. The most hazardous situation is a light quartering tailwind, which not only keeps a vortex on the runway but also inhibits its dissipation.

If you plan to take off behind a large airplane, try to rotate prior to that airplane's point of rotation and climb out above and on the upwind side of the other airplane's flight path. If you plan to takeoff from a runway on which a large airplane has just landed, try to plan your lift-off point to be beyond the point where that aircraft touched down. *See* Figure 7-2.

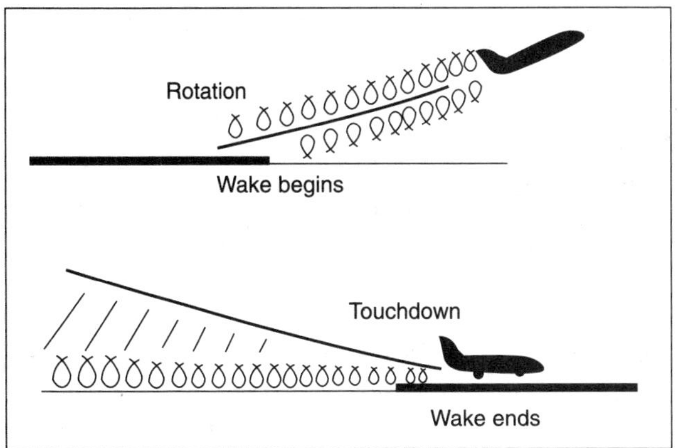

Figure 7-2

ALL

9097. What minimum condition is suggested for declaring an emergency?

A—Anytime the pilot is doubtful of a condition that could adversely affect flight safety.
B—When fuel endurance or weather will require an en route or landing priority.
C—When distress conditions such as fire, mechanical failure, or structural damage occurs.

An aircraft is in an emergency condition the moment the pilot becomes doubtful about position, fuel endurance, weather or any other condition that could adversely affect flight safety. (PLT394) — AIM ¶6-1-2

Answer (B) is incorrect because, if fuel endurance or weather will require an en route or landing priority, this is beyond an urgency situation and is now a distress condition. Answer (C) is incorrect because this is the step after an urgency condition, one that has the potential to become a distress condition.

ALL

9051. What is the hijack code?

A—7200.
B—7500.
C—7777.

Only nondiscrete transponder code 7500 will be decoded as the hijack code. (PLT497) — AIM ¶4-1-20

Answer (A) is incorrect because 7200 is a code that is used for normal operating procedures. Answer (C) is incorrect because code 7777 is reserved for military interceptor operations.

ALL

9052. Which range of codes should a pilot avoid switching through when changing transponder codes?

A—0000 through 1000.
B—7200 and 7500 series.
C—7500, 7600, and 7700 series.

When making routine transponder code changes, pilot should avoid inadvertent selection of codes 7500, 7600 or 7700 thereby causing momentary false alarms at automated ground facilities. (PLT497) — AIM ¶4-1-20

Answer (A) is incorrect because codes 0000 through 1000 are acceptable codes that may be used during normal operations. Answer (B) is incorrect because the 7200 series can be used during normal operations, while 7500 should be avoided except in the case of a hijack.

ALL

9362. After experiencing two-way radio communications failure en route, when should a pilot begin the descent for the instrument approach?

A—Upon arrival at any initial approach fix for the instrument approach procedure but not before the flight plan ETA as amended by ATC.
B—Upon arrival at the holding fix depicted on the instrument approach procedure at the corrected ETA, plus or minus 3 minutes.
C—At the primary initial approach fix for the instrument approach procedure at the ETA shown on the flight plan or the EFC time, whichever is later.

During a two-way radio communications failure, if the clearance limit is a fix from which an approach begins, commence descent or descent and approach as close as possible to the expect further clearance (EFC) time, if one has been received. If no EFC time has been received, commence descent and approach as close as possible to the estimated time of arrival as calculated from the filed or amended (with ATC) estimated time en route. (PLT391) — 14 CFR §91.185

Answer (B) is incorrect because an approach should begin at the initial approach fix, not at a holding fix, as close as possible to the ETA. Answer (C) is incorrect because an EFC time supersedes a flight plan ETA and should be used if one has been received (no matter if the EFC is sooner or later than the flight plan ETA).

Answers

9097 [A] 9051 [B] 9052 [C] 9362 [A]

ALL

9363. If a pilot is being radar vectored in IFR conditions and loses radio communications with ATC, what action should be taken?

A—Fly directly to the next point shown on the IFR flight plan and continue the flight.
B—Squawk 7700 and climb to VFR on Top.
C—Fly direct to a fix, route, or airway specified in the vector clearance.

If a two-way radio communication failure occurs while in IFR conditions the pilot should continue the flight by the following route:

1. *By the route assigned in the last ATC clearance;*
2. *If being radar vectored, by the direct route from the point of the radio failure to the fix, route or airway specified in the vector clearance;*
3. *In the absence of an assigned route, by the route that ATC has advised may be expected in a further clearance; or*
4. *In the absence of an assigned route or a route that ATC has advised may be expected in a further clearance, by the route filed in the flight plan.*

(PLT406) — 14 CFR §91.185

Answer (A) is incorrect because the route shown on the flight plan should be the last route to be used and only if an assigned route, vector, or expected route has not been received. Answer (B) is incorrect because a climb should only be initiated in order to establish the highest of either the assigned, MEA, or expected altitude. The squawk of 7700 is no longer correct.

ALL

9364. A pilot is flying in IFR weather conditions and has two-way radio communications failure. What altitude should be used?

A—Last assigned altitude, altitude ATC has advised to expect, or the MEA, whichever is highest.
B—An altitude that is at least 1,000 feet above the highest obstacle along the route.
C—A VFR altitude that is above the MEA for each leg.

A pilot should maintain the highest altitude or flight level of the following for each route segment:

1. *The altitude or flight level assigned in the last ATC clearance received;*
2. *The minimum altitude or flight level for IFR operations (MEA); or*
3. *The altitude that ATC has advised may be expected in a further clearance.*

(PLT391) — 14 CFR §91.185

Answer (B) is incorrect because 1,000 feet above the highest obstacle along the route is what a MOCA, (minimum obstruction clearance altitude) provides. Answer (C) is incorrect because VFR altitudes or regulations should never be used while flying in IFR weather conditions.

ALL

9365. A pilot is holding at an initial approach fix after having experienced two-way radio communications failure. When should that pilot begin descent for the instrument approach?

A—At the EFC time, if this is within plus or minus 3 minutes of the flight plan ETA as amended by ATC.
B—At flight plan ETA as amended by ATC.
C—At the EFC time as amended by ATC.

During a two-way radio communications failure, if the clearance limit is a fix from which an approach begins, commence descent or descent and approach as close as possible to the expect further clearance (EFC) time, if one has been received. If no EFC time has been received, commence descent and approach as close as possible to the estimated time of arrival as calculated from the filed or amended (with ATC) estimated time enroute. (PLT391) — 14 CFR §91.185

Answer (A) is incorrect because the approach should begin at the EFC time, regardless of whether it is close to the planned ETA or not; ETA is only used if an EFC has not been received. Answer (B) is incorrect because a pilot who is holding at an initial approach fix after having experienced a two-way radio communication failure without an EFC time should begin descent for the instrument approach so that the arrival will coincide as closely as possible with the ETA.

ALL

9389. What altitude and route should be used if the pilot is flying in IFR weather conditions and has two-way radio communications failure?

A—Continue on the route specified in the clearance and fly the highest of the following: the last assigned altitude, altitude ATC has informed the pilot to expect, or to the MEA.
B—Descend to MEA and, if clear of clouds, proceed to the nearest appropriate airport. If not clear of clouds, maintain the highest of the MEAs along the clearance route.
C—Fly the most direct route to the destination, maintaining the last assigned altitude or MEA, whichever is higher.

Answers

9363 [C] 9364 [A] 9365 [C] 9389 [A]

If a two-way radio communication failure occurs while in IFR conditions the pilot should continue the flight by the following route:

1. *By the route assigned in the last ATC clearance;*
2. *If being radar vectored, by the direct route from the point of the radio failure to the fix, route or airway specified in the vector clearance;*
3. *In the absence of an assigned route, by the route that ATC has advised may be expected in a further clearance; or*
4. *In the absence of an assigned route or a route that ATC has advised may be expected in a further clearance, by the route filed in the flight plan.*

A pilot should maintain the highest altitude or flight level of the following for each route segment:

1. *The altitude or flight level assigned in the last ATC clearance received;*
2. *The minimum altitude or flight level for IFR operations (MEA); or*
3. *The altitude that ATC has advised may be expected in a further clearance.*

(PLT391) — 14 CFR §91.185

Answer (B) is incorrect because the highest of either the MEA, expected altitude, or assigned altitude should be used. Answer (C) is incorrect because, if ATC advises you may expect an altitude after reaching a clearance limit, and it is higher than the published MEA or assigned altitude, the expected altitude should be used. The route to be used should be the one assigned by ATC, as specified in a vector clearance, by the route ATC has advised may be expected, or in the absence of all of these, the route as filed in the flight plan, not the most direct route.

ALL

9390. While in IFR conditions, a pilot experiences two-way radio communications failure. Which route should be flown in the absence of an ATC assigned route or a route ATC has advised to expect in a further clearance?

A—The most direct route to the filed alternate airport.
B—An off-airway route to the point of departure.
C—The route filed in the flight plan.

If a two-way radio communication failure occurs while in IFR conditions the pilot should continue the flight by the following route:

1. *By the route assigned in the last ATC clearance;*
2. *If being radar vectored, by the direct route from the point of the radio failure to the fix, route or airway specified in the vector clearance;*
3. *In the absence of an assigned route, by the route that ATC has advised may be expected in a further clearance; or*
4. *In the absence of an assigned route or a route that ATC has advised may be expected in a further clearance, by the route filed in the flight plan.*

(PLT391) — 14 CFR §91.185

ALL

9098. It is the responsibility of the pilot and crew to report a near midair collision as a result of proximity of at least

A—50 feet or less to another aircraft.
B—500 feet or less to another aircraft.
C—1,000 feet or less to another aircraft.

A near midair collision is defined as an incident associated with the operation on an aircraft in which a possibility of collision occurs as a result of proximity of less than 500 feet to another aircraft, or a report is received from a pilot or flight crewmember stating that a collision hazard existed between two or more aircraft. (PLT366) — AIM ¶7-6-3

ALL

9010. Under what condition should a pilot on IFR advise ATC of minimum fuel status?

A—When the fuel supply becomes less than that required for IFR.
B—If the remaining fuel suggests a need for traffic or landing priority.
C—If the remaining fuel precludes any undue delay.

A pilot should advise ATC of his/her minimum fuel status when the fuel supply has reached a state where, upon reaching destination, he/she cannot accept any undue delay. It indicates a possible future emergency, but does not declare one and does not get priority handling. (PLT318) — AIM ¶5-5-15

Answer (A) is incorrect because a pilot must ensure the minimum amount of fuel is on board the airplane for the planned IFR flight and alternatives, if needed, during the flight planning phase. Answer (B) is incorrect because, if the remaining fuel suggests a need for traffic or landing priority, the pilot should declare an emergency (not minimum fuel status), and report fuel remaining in minutes.

Answers

9390 [C]	9098 [B]	9010 [C]

ALL

9011. What does the term "minimum fuel" imply to ATC?

A—Traffic priority is needed to the destination airport.
B—Emergency handling is required to the nearest suitable airport.
C—Advisory that indicates an emergency situation is possible should an undue delay occur.

A pilot should advise ATC of his/her minimum fuel status when the fuel supply has reached a state where, upon reaching destination, he/she cannot accept any undue delay. It indicates a possible future emergency, but does not declare one and does not get priority handling. (PLT318) — AIM ¶5-5-15

Answer (A) is incorrect because a pilot should declare an emergency and report fuel remaining in minutes if traffic priority is needed to the destination airport. Answer (B) is incorrect because emergency handling to the nearest suitable airport would be required in a distress or urgent type of emergency.

ALL

9420. You should advise ATC of minimum fuel status when your fuel supply has reached a state where, upon reaching your destination, you cannot accept any undue delay.

A—This will ensure your priority handling by ATC.
B—ATC will consider this action as if you had declared an emergency.
C—If your remaining usable fuel supply suggests the need for traffic priority to ensure a safe landing, declare an emergency due to low fuel and report fuel remaining in minutes.

If the remaining usable fuel supply suggests the need for traffic priority to ensure a safe landing, you should declare an emergency due to low fuel and report fuel remaining in minutes. (PLT318) —AIM ¶5-5-15

Answer (A) is incorrect because minimum fuel advisory does not imply a need for a traffic priority. Answer (B) is incorrect because minimum fuel advisory is not an emergency situation, but merely an advisory that indicates an emergency situation is possible should any undue delay occur.

ALL

9054. What airport condition is reported by the tower when more than one wind condition at different positions on the airport is reported?

A—Light and variable.
B—Wind shear.
C—Frontal passage.

The Low-Level Wind Shear Alert System (LLWAS) is a computerized system which detects low level wind shear by continuously comparing the winds measured by sensors installed around the periphery of an airport with the wind measured at the center field location. When a significant difference exists, the tower controller will provide aircraft with an advisory of the situation which includes the center field wind plus the remote site location and wind. (PLT518) — AIM ¶4-3-7

Answer (A) is incorrect because "light and variable" is used to report the wind conditions when wind speed is less than 5 knots. Answer (C) is incorrect because a frontal passage is normally indicated by a change in wind direction, but it is usually not reported by the tower.

ALL

9050. Under what condition does ATC issue safety alerts?

A—When collision with another aircraft is imminent.
B—If the aircraft altitude is noted to be in close proximity to the surface or an obstacle.
C—When weather conditions are extreme and wind shear or large hail is in the vicinity.

A safety alert will be issued to pilots of aircraft being controlled by ATC if the controller is aware the aircraft is at an altitude which, in the controller's judgment, places the aircraft in unsafe proximity to terrain, obstructions or other aircraft. (PLT172) — AIM ¶4-1-16

Answer (A) is incorrect because a safety alert is issued to a pilot if ATC believes that his/her airplane is at an altitude which would place it in unsafe proximity to another airplane. Answer (C) is incorrect because, when weather conditions are extreme and wind shear or large hail is in the vicinity, a Convective SIGMET would be broadcast.

ALL

9119. Which flight conditions of a large jet airplane create the most severe flight hazard by generating wingtip vortices of the greatest strength?

A—Heavy, slow, gear and flaps up.
B—Heavy, slow, gear and flaps down.
C—Heavy, fast, gear and flaps down.

The greatest vortex strength occurs when the generating aircraft is heavy, clean (gear and flaps up) and slow. (PLT509) — AIM ¶7-3-3

Answers

9011 [C]	9420 [C]	9054 [B]	9050 [B]	9119 [A]

ALL

9120. Hazardous vortex turbulence that might be encountered behind large aircraft is created only when that aircraft is

A—developing lift.
B—operating at high airspeeds.
C—using high power settings.

Lift is generated by the creation of a pressure differential over the wing surface. The lowest pressure occurs over the upper wing surface and the highest pressure under the wing. This pressure differential triggers the roll up of the airflow aft of the wing resulting in swirling air masses trailing downstream of the wing tips. (PLT509) — AIM ¶7-3-2

Answer (B) is incorrect because hazardous vortex turbulence is created only when the aircraft is developing lift, which can be while operating at low or high airspeeds. A slow, heavy, and clean airplane will generate the most hazardous vortex turbulence. Answer (C) is incorrect because most takeoff rolls are at high power settings, but the generation of vortices does not occur until lift is produced. Landing approaches are also conducted at lower power settings; vortex turbulence is produced whenever an airplane is producing lift.

ALL

9121. Wingtip vortices created by large aircraft tend to

A—sink below the aircraft generating the turbulence.
B—rise from the surface to traffic pattern altitude.
C—accumulate and remain for a period of time at the point where the takeoff roll began.

Flight tests have shown that the vortices from large aircraft sink at a rate of several hundred feet per minute. They tend to level off at a distance about 900 feet below the flight path of the generating aircraft. (PLT509) — AIM ¶7-3-4

Answer (B) is incorrect because vortices created by large aircraft tend to sink from (not rise into) the traffic pattern altitude. Answer (C) is incorrect because wing-tip vortices are not generated until the aircraft's wings develop lift, so no wing-tip vortices are generated at the point where the takeoff roll begins.

ALL

9122. How does the wake turbulence vortex circulate around each wingtip?

A—Inward, upward, and around the wingtip.
B—Counterclockwise when viewed from behind the aircraft.
C—Outward, upward, and around the wingtip.

The vortex circulation is outward, upward and around the wing tips when viewed from either ahead or behind the aircraft. (PLT509) — AIM ¶7-3-4

ALL

9123. Which statement is true concerning the wake turbulence produced by a large transport aircraft?

A—Vortices can be avoided by flying 300 feet below and behind the flightpath of the generating aircraft.
B—The vortex characteristics of any given aircraft may be altered by extending the flaps or changing the speed.
C—Wake turbulence behind a propeller-driven aircraft is negligible because jet engine thrust is a necessary factor in the formation of vortices.

The strength of the vortex is governed by the weight, speed and shape of the wing of the generating aircraft. The vortex characteristics of a given aircraft can be changed by extension of flaps or other wing configuring devices as well as by a change in speed. (PLT509) — AIM ¶7-3-3

Answer (A) is incorrect because the vortices generated by large transport aircraft tend to sink below and behind, thus vortices can be expected by flying 300 feet below and behind the flight path of the generating aircraft. Answer (C) is incorrect because wake turbulence vortices are generated by lift produced by any airplane. It does not matter whether the airplane is powered by propeller or jet engines.

ALL

9124. What effect would a light crosswind have on the wingtip vortices generated by a large airplane that has just taken off?

A—The upwind vortex will tend to remain on the runway longer than the downwind vortex.
B—A crosswind will rapidly dissipate the strength of both vortices.
C—The downwind vortex will tend to remain on the runway longer than the upwind vortex.

A crosswind will decrease the lateral movement of the upwind vortex and increase the movement of the downwind vortex. Thus, a light wind of 3 to 7 knots could result in the upwind vortex remaining in the touchdown zone for a period of time and hasten the drift of the downwind vortex toward another runway. (PLT509) — AIM ¶7-3-4

Answer (B) is incorrect because a crosswind will hold the upwind vortex in the vicinity of the runway. Answer (C) is incorrect because the downwind vortex moves away at a faster rate than the upwind vortex.

Answers

9120 [A]	9121 [A]	9122 [C]	9123 [B]	9124 [A]

ALL

9125. To avoid the wingtip vortices of a departing jet airplane during takeoff, the pilot should

A—lift off at a point well past the jet airplane's flightpath.
B—climb above and stay upwind of the jet airplane's flightpath.
C—remain below the flightpath of the jet airplane.

When departing behind a large aircraft, note its rotation point and rotate prior to that point. During the climb, stay above and upwind of the large aircraft's climb path until turning clear of its wake. (PLT509) — AIM ¶7-3-4

Answer (A) is incorrect because, if you rotate beyond the jet's rotation point, you will have to fly up into the jet's vortices. Answer (C) is incorrect because the jet's vortices will sink. If you stay below the jet's flight path, you will fly in the area of the vortices.

ALL

9126. What wind condition prolongs the hazards of wake turbulence on a landing runway for the longest period of time?

A—Direct tailwind.
B—Light quartering tailwind.
C—Light quartering headwind.

A crosswind will decrease the lateral movement of the upwind vortex and increase the movement of the downwind vortex. Thus a light wind of 3 to 7 knots could result in the upwind vortex remaining in the touchdown zone for a period of time and hasten the drift of the downwind vortex toward another runway. Similarly, a tailwind condition can move the vortices of the preceding aircraft forward into the touchdown zone. The light, quartering tailwind requires maximum caution. (PLT509) — AIM ¶7-3-4

Answer (A) is incorrect because, even though a direct tailwind can move the vortices of a preceding aircraft forward into the touchdown zone, it is not as hazardous because both vortices would move to the sides (and not remain on the runway). Answer (C) is incorrect because a light quartering headwind would move the vortices toward the runway threshold, away from (not into) the touchdown zone on a landing runway.

ALL

9127. If you take off behind a heavy jet that has just landed, you should plan to lift off

A—prior to the point where the jet touched down.
B—beyond the point where the jet touched down.
C—at the point where the jet touched down and on the upwind edge of the runway.

When departing behind a large aircraft which has just landed, note the aircraft's touchdown point and rotate past that point on the runway. (PLT509) — AIM ¶7-3-6

Answer (A) is incorrect because a lift-off prior to the point where the jet touched down would force you to climb through the jet's vortices. Answer (C) is incorrect because lift-off should be planned beyond the point of touchdown to ensure that you avoid the vortices, and you should remain on the center of the runway during takeoff.

ALL

9715. To allow pilots of in-trail lighter aircraft to make flight path adjustments to avoid wake turbulence, pilots of heavy and large jet aircraft should fly

A—below the established glidepath and slightly to either side of the on-course centerline.
B—on the established glidepath and on the approach course centerline or runway centerline extended.
C—above the established glidepath and slightly downwind of the on-course centerline.

Pilots of aircraft that produce strong wake vortices should make every attempt to fly on the established glidepath, and as closely as possible to the approach course centerline or to the extended centerline of the runway of intended landing. (PLT509) — AIM ¶7-3-8

Answers

9125 [B]	9126 [B]	9127 [B]	9715 [B]

Flight Physiology

Even small amounts of alcohol have an adverse effect on reaction and judgment. This effect is magnified as altitude increases. No one may serve as a crewmember on a civil aircraft:

- Within 8 hours of the consumption of any alcoholic beverage.
- While having a blood alcohol level of .04% or higher.

Runway width illusion—A runway that is narrower than usual can create the illusion that the aircraft is higher than it really is. This can cause an unwary pilot to descend too low on approach. A wide runway creates an illusion of being too low on glide slope.

Featureless terrain illusion—An absence of ground feature, as when landing over water, darkened areas and terrain made featureless by snow can create the illusion that the aircraft is higher than it really is.

Autokinesis—In the dark, a static light will appear to move about when stared at for a period of time.

An effective scan pattern is necessary to ensure that a pilot will see other aircraft in time to avoid potential midair collisions. This means that 2/3 to 3/4 of a pilot's time should be spent scanning outside the aircraft. The best method would be to look outside for about 15 seconds and then inside for about 5 seconds. It is much easier to see an aircraft which is moving relative to the observer. Unfortunately, aircraft which present a collision hazard are usually on the horizon with little or no apparent horizontal or vertical movement. The image only grows larger as the threat aircraft gets closer. Special vigilance must be exercised for this type of situation. A pilot's most acute night vision is off-center in his/her peripheral vision. When looking for other aircraft at night, scan slowly to allow sufficient time for this off-center viewing.

All pilots who fly in instrument conditions or at night are subject to spatial disorientation. This occurs when body sensations are used to interpret flight attitudes, and there is no visual reference to the horizon. The only reliable way to overcome this disorientation is to rely entirely on the indications of the flight instruments. Some types of vertigo include:

The leans—An abrupt correction of a banked angle can create the illusion of banking in the opposite direction.

Coriolis illusion—An abrupt head movement during a constant rate turn can create the illusion of rotation in an entirely different axis. This illusion can be overwhelming and so rapid head movements in turns should be avoided.

Inversion illusion—An abrupt change from a climb to straight and level flight can create the illusion of tumbling backwards.

Somatogravic illusion—A rapid acceleration during takeoff can create the illusion of being in a nose up attitude.

Hypoxia is caused by insufficient oxygen reaching the brain. The most usual reason is the low partial pressure of oxygen encountered at altitude. Carbon monoxide poisoning is similar to hypoxia in that it causes too little oxygen to reach the brain. Carbon monoxide (usually from an exhaust leak) binds with the hemoglobin in the blood, preventing its usual oxygen-carrying function. The symptoms of both are similar and include dizziness, tingling of the hands, feet and legs, loss of higher thought processes, and unconsciousness. The sufferer may not notice or react to any of the symptoms due to his degraded mental faculties. Hyperventilation is caused by a reduction of carbon dioxide in the blood, usually due to rapid breathing in a stressful situation. The symptoms of hyperventilation are similar to hypoxia, but recovery is rapid once the rate of breathing is brought under control.

ALL

9354. A person may not act as a crewmember of a civil aircraft if alcoholic beverages have been consumed by that person within the preceding

A—8 hours.
B—12 hours.
C—24 hours.

No person may act or attempt to act as a crewmember of a civil aircraft within 8 hours after the consumption of any alcoholic beverage. (PLT409) — 14 CFR §91.17

ALL

9111. What is the effect of alcohol consumption on functions of the body?

A—Alcohol has an adverse effect, especially as altitude increases.
B—Small amounts of alcohol in the human system increase judgment and decision-making abilities.
C—Alcohol has little effect if followed by equal quantities of black coffee.

The adverse effect of alcohol is greatly multiplied when a person is exposed to altitude. Two drinks on the ground are equivalent to three or four at altitude. (PLT205) — AIM ¶8-1-1

Answer (B) is incorrect because even small amounts of alcohol impair judgment and decision-making abilities. Answer (C) is incorrect because there is no way to increase the body's metabolism of alcohol or to alleviate a hangover (including drinking black coffee).

ALL

9107. When making an approach to a narrower-than-usual runway, without VASI assistance, the pilot should be aware that the approach

A—altitude may be higher than it appears.
B—altitude may be lower than it appears.
C—may result in leveling off too high and landing hard.

An approach to a narrower-than-usual runway can create the illusion that the aircraft is higher than it actually is. (PLT280) — AIM ¶8-1-5

Answer (A) is incorrect because wider-than-usual runways may result in higher than desired approaches. Answer (C) is incorrect because leveling off too high is not affected by the runway width, but rather by the pilot's landing proficiency.

ALL

9109. In the dark, a stationary light will appear to move when stared at for a period of time. This illusion is known as

A—somatogravic illusion.
B—ground lighting illusion.
C—autokinesis.

In the dark, a stationary light will appear to move about when stared at for many seconds. This illusion is known as Autokinesis. (PLT280) — AIM ¶8-1-5

Answer (A) is incorrect because somatogravic illusion occurs with a rapid acceleration during takeoff, creating the illusion of being in a nose-up attitude. Answer (B) is incorrect because a ground lighting illusion refers to lights on a straight path such as a road being mistaken by a pilot for runway or approach lights.

ALL

9110. When making a landing over darkened or featureless terrain such as water or snow, a pilot should be aware of the possibility of illusion. The approach may appear to be too

A—high.
B—low.
C—shallow.

An absence of ground features, when landing over water, darkened areas and terrain made featureless by snow, can create the illusion that the aircraft is at a higher altitude than it actually is. (PLT280) — AIM ¶8-1-5

ALL

9108. The illusion of being in a noseup attitude which may occur during a rapid acceleration takeoff is known as

A—inversion illusion.
B—autokinesis.
C—somatogravic illusion.

A rapid acceleration during takeoff can create the illusion of being in a nose-up attitude. This is known as a Somatogravic Illusion. (PLT280) — AIM ¶8-1-5

Answer (A) is incorrect because the inversion illusion results from an abrupt change from climb to straight-and-level flight which can create an illusion of tumbling backwards. Answer (B) is incorrect because autokinesis refers to a stationary light appearing to move about when stared at for many seconds in the dark.

Answers

9354 [A] 9111 [A] 9107 [B] 9109 [C] 9110 [A] 9108 [C]

ALL

9114. What is the most effective way to use the eyes during night flight?

A—Look only at far away, dim lights.
B—Scan slowly to permit offcenter viewing.
C—Concentrate directly on each object for a few seconds.

One should scan slowly at night to permit off-center viewing of dim objects. (PLT099) — FAA-H-8083-3

Answer (A) is incorrect because pilots must look at their gauges and instruments, which are about 2 feet in front of them. Answer (C) is incorrect because peripheral (off-center) vision is more effective at night.

ALL

9116. Which observed target aircraft would be of most concern with respect to collision avoidance?

A—One which appears to be ahead and moving from left to right at high speed.
B—One which appears to be ahead and moving from right to left at slow speed.
C—One which appears to be ahead with no lateral or vertical movement and is increasing in size.

Any aircraft that appears to have no relative motion and stays in one scan quadrant is likely to be on a collision course. If a target shows no lateral or vertical motion, but increases in size, take evasive action. (PLT099) — AIM ¶8-1-8

Answers (A) and (B) are incorrect because an airplane which is ahead of you and moving from left to right, or from right to left, should pass in front of you.

ALL

9117. Scanning procedures for effective collision avoidance should constitute

A—looking outside for 15 seconds, then inside for 5 seconds, then repeat.
B—1 minute inside scanning, then 1 minute outside scanning, then repeat.
C—looking outside every 30 seconds except in radar contact when outside scanning is unnecessary.

Studies show that the time a pilot spends on visual tasks inside the cabin should represent no more than 1/4 to 1/3 of the scan time outside, or no more than 4 to 5 seconds on the instrument panel for every 16 seconds outside. (PLT099) — AIM ¶8-1-6

Answer (B) is incorrect because pilots should spend the majority of scan time outside the airplane when in VFR conditions. Answer (C) is incorrect because pilots should spend the majority of scan time outside the airplane, and outside scanning is necessary when in radar contact in VFR conditions.

ALL

9118. When using the Earth's horizon as a reference point to determine the relative position of other aircraft, most concern would be for aircraft

A—above the horizon and increasing in size.
B—on the horizon with little relative movement.
C—on the horizon and increasing in size.

Any aircraft that appears to have no relative motion and stays in one scan quadrant is likely to be on a collision course. If a target shows no lateral or vertical motion, but increases in size, take evasive action. (PLT099) — AIM ¶8-1-8

Answer (A) is incorrect because an airplane above the horizon is probably at a higher altitude. Answer (B) is incorrect because an airplane on the horizon without movement may be traveling in the same direction as you.

ALL

9112. A pilot is more subject to spatial disorientation when

A—ignoring or overcoming the sensations of muscles and inner ear.
B—eyes are moved often in the process of cross-checking the flight instruments.
C—body sensations are used to interpret flight attitudes.

When seated on an unstable moving platform at altitude with your vision cut off from the earth, horizon or other fixed reference, you are susceptible to misinterpreting certain body sensations caused by angular accelerations. (PLT334) — AIM ¶8-1-5

Answer (A) is incorrect because ignoring or overcoming the sensations of muscles and inner ear is a means of avoiding (not becoming subject to) spatial disorientation. Answer (B) is incorrect because rapid eye movements have little or no impact on spatial disorientation and vision reference to reliable flight instruments helps avoid spatial disorientation.

Answers

9114 [B]	9116 [C]	9117 [A]	9118 [C]	9112 [C]

ALL

9113. Which procedure is recommended to prevent or overcome spatial disorientation?

A— Reduce head and eye movement to the greatest possible extent.
B— Rely on the kinesthetic sense.
C— Rely entirely on the indications of the flight instruments.

The best method to prevent or overcome spatial disorientation is to rely entirely on the indications of the flight instruments. (PLT334) — AIM ¶8-1-5

Answer (A) is incorrect because head and eye movement have little effect on spatial disorientation. Answer (B) is incorrect because relying on the kinesthetic sense encourages (not prevents) spatial disorientation.

ALL

9115. While making prolonged constant rate turns under IFR conditions, an abrupt head movement can create the illusion of rotation on an entirely different axis. This is known as

A— autokinesis.
B— Coriolis illusion.
C— the leans.

An abrupt head movement while making a prolonged constant rate turn, can produce a strong sensation of rotation or movement in an entirely different axis. The phenomenon is known as Coriolis Illusion. (PLT280) — AIM ¶8-1-5

Answer (A) is incorrect because Autokinesis refers to a stationary light appearing to move about when stared at for many seconds in the dark. Answer (C) is incorrect because the "leans" refer to an abrupt correction of a banked attitude which can create the illusion of bank in the opposite direction.

ALL

9433. Haze can give the illusion that the aircraft is

A— closer to the runway than it actually is.
B— farther from the runway than it actually is.
C— the same distance from the runway as when there is no restriction to visibility.

Atmospheric haze can create an illusion of being at a greater distance from the runway than you actually are. (PLT280) — AIM ¶8-1-5

ALL

9434. Sudden penetration of fog can create the illusion of

A— pitching up.
B— pitching down.
C— leveling off.

Penetration of fog can create an illusion of pitching up. (PLT280) — AIM ¶8-1-5

ALL

9435. What illusion, if any, can rain on the windscreen create?

A— Does not cause illusions.
B— Lower than actual.
C— Higher than actual.

Rain on the windscreen can create an illusion of being at a higher altitude than you are. (PLT280) — AIM ¶8-1-5

ALL

9101. What is a symptom of carbon monoxide poisoning?

A— Rapid, shallow breathing.
B— Pain and cramping of the hands and feet.
C— Dizziness.

Carbon monoxide poisoning produces the same symptoms as hypoxia, which include dizziness. (PLT097) — AIM ¶8-1-4

Answer (A) is incorrect because rapid breathing can result in hyperventilation, but it is not a symptom of carbon monoxide poisoning. Answer (B) is incorrect because tingling in the extremities (not pain and cramping) is one symptom of hyperventilation (not carbon monoxide poisoning).

ALL

9102. Which would most likely result in hyperventilation?

A— A stressful situation causing anxiety.
B— The excessive consumption of alcohol.
C— An extremely slow rate of breathing and insufficient oxygen.

You are most likely to hyperventilate when under stress or at high altitudes. (PLT332) — AIM ¶8-1-3

Answer (B) is incorrect because excessive consumption of alcohol results in intoxication, not hyperventilation. Answer (C) is incorrect because a slow rate of breathing is the cure for hyperventilation, and insufficient oxygen is the cause of hypoxia, not hyperventilation.

Answers

9113 [C] 9115 [B] 9433 [B] 9434 [A] 9435 [C] 9101 [C]
9102 [A]

ALL

9103. What causes hypoxia?

A—Excessive carbon dioxide in the atmosphere.
B—An increase in nitrogen content of the air at high altitudes.
C—A decrease of oxygen partial pressure.

Low partial pressure of oxygen causes hypoxia. (PLT330) — AIM ¶8-1-2

Answer (A) is incorrect because the percentage of carbon dioxide and oxygen in the atmosphere remains constant with changes in altitude, but there is less pressure as you increase in altitude. Answer (B) is incorrect because relative nitrogen content also remains constant at high altitudes, but there is less pressure.

ALL

9104. Which is a common symptom of hyperventilation?

A—Tingling of the hands, legs, and feet.
B—Increased vision keenness.
C—Decreased breathing rate.

Symptoms of hyperventilation include dizziness, tingling of the extremities, sensation of body heat, rapid heart rate, blurring of vision, muscle spasm and, finally, unconsciousness. (PLT332) — AIM ¶8-1-3

Answer (B) is incorrect because hyperventilation distorts one's abilities. Answer (C) is incorrect because decreasing the breathing rate overcomes hyperventilation and is not a symptom of it.

ALL

9105. Loss of cabin pressure may result in hypoxia because as cabin altitude increases

A—the percentage of nitrogen in the air is increased.
B—the percentage of oxygen in the air is decreased.
C—oxygen partial pressure is decreased.

Low partial pressure of oxygen causes hypoxia. (PLT330) — AIM ¶8-1-2

Answers (A) and (B) are incorrect because the percentage of nitrogen, carbon dioxide and oxygen in the atmosphere remains constant with changes in altitude, but there is less pressure as you increase in altitude.

ALL

9106. Hypoxia is the result of which of these conditions?

A—Insufficient oxygen reaching the brain.
B—Excessive carbon dioxide in the bloodstream.
C—Limited oxygen reaching the heart muscles.

Hypoxia is a result of too little oxygen reaching the brain. (PLT330) — AIM ¶8-1-2

Answer (B) is incorrect because excessive carbon dioxide in the blood stream causes hyperventilation. Answer (C) is incorrect because it is the result of insufficient oxygen to the brain.

ALL

9778. An experienced pilot trying to meet a schedule

A—can expect the flight crew to alert them to problems or areas of concern.
B—will always err on the side of caution.
C—can fail to perceive operational pitfalls.

Although more experienced pilots are likely to make more automatic decisions, there are tendencies or operational pitfalls that come with the development of pilot experience. These are classic behavioral traps into which pilots have been known to fall. More experienced pilots (as a rule) try to complete a flight as planned, please passengers, and meet schedules. The desire to meet these goals can have an adverse effect on safety and contribute to an unrealistic assessment of piloting skills. (PLT104) — FAA-H-8083-25

ALL

9804. The crew monitoring function is essential,

A—particularly during high altitude cruise flight modes to prevent CAT issues.
B—particularly during approach and landing to prevent CFIT.
C—during RNAV departures in class B airspace.

Effective monitoring and cross-checking can be the last line of defense that prevents an accident because detecting an error or unsafe situation may break the chain of events leading to an accident. This monitoring function is always essential, and particularly so during approach and landing when controlled flight into terrain (CFIT) accidents are most common. (PLT104) — AC 120-71A

Answers

9103 [C]	9104 [A]	9105 [C]	9106 [A]	9778 [C]	9804 [B]

ALL

9805. CRM training refers to

A—the two components of flight safety and resource management, combined with mentor feedback.
B—the three components of initial indoctrination awareness, recurrent practice and feedback, and continual reinforcement.
C—the five components of initial indoctrination awareness, communication principles, recurrent practice and feedback, coordination drills, and continual reinforcement.

The critical components of effective crew resource management (CRM) training include initial indoctrination awareness, recurrent practice and feedback, and continual reinforcement. (PLT104) — AC120-51

ALL

9806. In evaluating error management, one

A—should recognize not all errors can be prevented.
B—may include error evaluation that should have been prevented.
C—must mark errors as disqualifying.

It is certainly desirable to prevent as many errors as possible, but since they cannot all be prevented, detection and recovery from errors should be addressed in training. Error management (error prevention, detection, and recovery) should be considered in the evaluation of pilots, as well as the fact that since not all errors can be prevented, it is important that errors be managed properly. (PLT104) — AC120-51

Answers

9805 [B] 9806 [A]

Chapter 8
Meteorology and Weather Services

The Atmosphere

The primary cause of all the Earth's weather is the variation in solar radiation received at the surface. When the surface is warmed by the sun, the air next to it is, in turn, heated and it expands. This creates a low pressure area where the air rises and, at altitude, expands outward. Air from regions of relatively high pressure descends and then moves away from the center of the high toward the lower pressure areas. On both a global and local scale, this movement of air sets off an immensely complex process that generates all the Earth's weather. *See* Figure 8-1.

Another major influence in the pattern of the weather is a phenomenon known as **Coriolis effect**. This is an apparent force, caused by the Earth's rotation, acting on any movement of air. If the Earth did not rotate, air would move directly from areas of high pressure to areas of low pressure. Coriolis force bends the track of the air over the ground to right in the northern hemisphere and to the left in the southern hemisphere. Viewed from above (as on a weather map) this makes air rotate clockwise around high pressure areas in the northern hemisphere and counterclockwise around lows. In the southern hemisphere, the rotation around highs and lows is just the opposite. In the northern hemisphere, the rotation of air around a low pressure area is called a cyclone and that around a high is called an anticyclone.

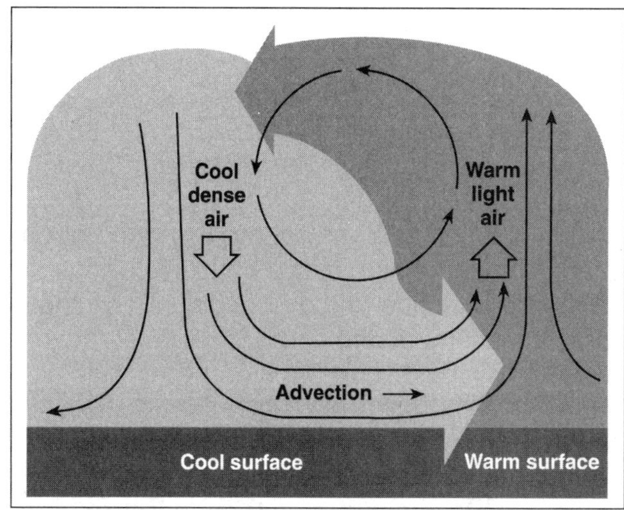

Figure 8-1. Circulation of air due to uneven surface heating

The strength of the Coriolis force is determined by wind speed and the latitude. Coriolis has the least effect at the equator and the most at the poles. It is also reduced in effect when wind speed decreases. Air moving near the Earth's surface is slowed by friction. This reduces the Coriolis force. However, the gradient pressure causing the air to move remains the same. The reduced Coriolis allows air to spiral out away from the center of a high and in toward the center of a low, and at an angle to winds aloft which are out of the friction level.

If the Earth did not rotate, air would move from the poles to the equator at the surface and from the equator to the poles at altitude. Because the Earth does rotate, Coriolis force and the pressure gradients tend to form three bands of prevailing winds in each hemisphere. Weather systems tend to move from east to west in the subtropical regions on the "trade winds." In the mid latitudes, the prevailing westerlies move weather systems from west to east. *See* Figure 8-2.

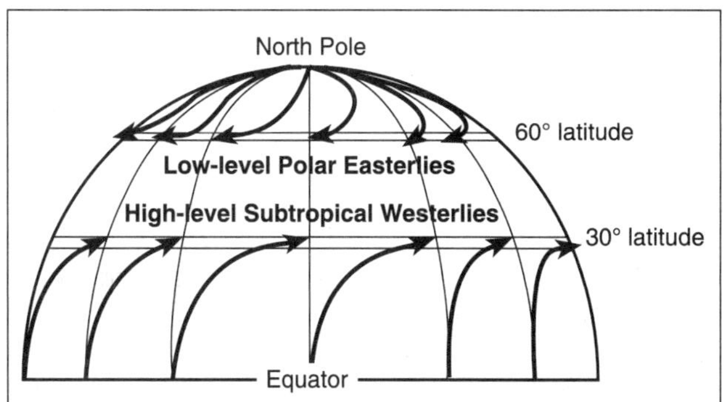

Figure 8-2. Global wind systems

All air carrier flights take place in the two lowest levels of the atmosphere. These are the **troposphere** and the **stratosphere**. The troposphere starts at the surface and extends vertically to roughly 35,000 feet. The thickness of the troposphere varies with latitude, being thicker over the equator than over the poles and with the season of the year (thicker in the summer than in the winter). The stratosphere extends from the top of the troposphere to about 26 to 29 miles altitude. *See* Figure 8-3. The main characteristic that distinguishes the troposphere from the stratosphere is the temperature lapse rate. In the troposphere, the temperature decreases with increasing altitude at an average rate of two degrees Celsius per one thousand feet of altitude. In the stratosphere, there is little or no change in temperature with altitude. In fact, in some regions the temperature increases with increasing altitude causing temperature inversions.

The thin boundary layer between the troposphere and the stratosphere is called the **tropopause**. The height of the tropopause is of great interest to the pilots of jet aircraft for two reasons. First, there is an abrupt change in the temperature lapse rate at the tropopause and that has a significant effect on jet engine performance. Second, maximum winds (the jet stream) and narrow zones of wind shear are found at the tropopause.

The **jet stream** is a few thousand feet thick and a few hundred miles wide. By arbitrary definition, it has wind speeds of fifty knots or greater. The highest wind speeds can be found on the polar side of the jet core. *See* Figure 8-4. There may be two or more jet streams in existence at one time. The jet stream is always found at a vertical break in the tropopause where the tropical and polar tropopauses meet. In addition to the high speed horizontal winds, the jet stream contains a circular rotation with rising air on the tropical side and descending air on the polar side. Because of the rising air, cirrus clouds will sometimes form on the equatorial side of the jet.

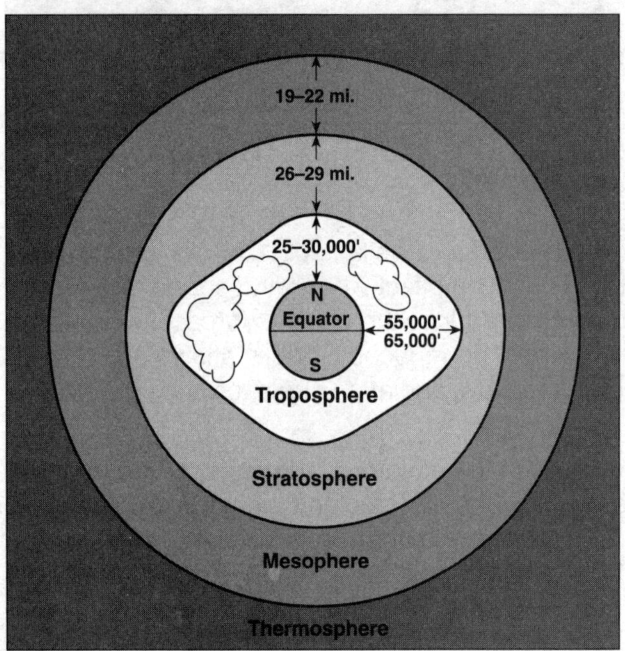

Figure 8-3. Levels of atmosphere

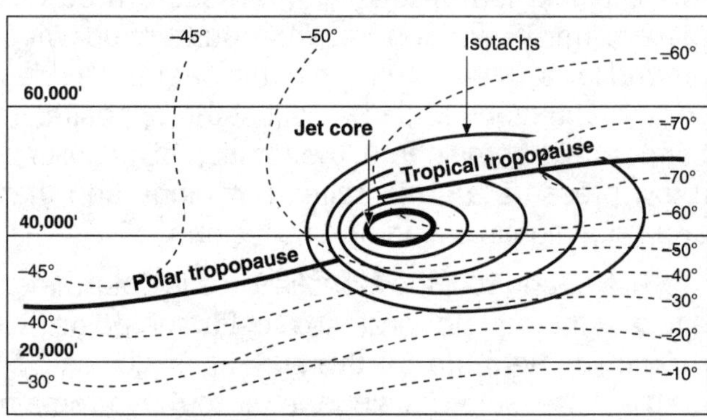

Figure 8-4. Cross-section of the jet stream

ALL
9152. What is the primary cause of all changes in the Earth's weather?

A—Variations of solar energy at the Earth's surface.
B—Changes in air pressure over the Earth's surface.
C—Movement of air masses from moist areas to dry areas.

Every physical process of weather is accompanied by or is the result of a heat exchange. Differences in solar energy create temperature variations. These temperature variations create forces that drive the atmosphere in its endless motion. (PLT510) — AC 00-6A, Chapter 2

Answer (B) is incorrect because changes in air pressure are due to temperature variations. Answer (C) is incorrect because movement of air masses is a result of varying temperatures and pressures.

Answers
9152 [A]

ALL

9160. Where is the usual location of a thermal low?

A—Over the arctic region.
B—Over the eye of a hurricane.
C—Over the surface of a dry, sunny region.

A dry, sunny region can become quite warm from intense surface heating, thus generating a surface low-pressure area. This is called a thermal low. (PLT511) — AC 00-6A, Chapter 4

Answer (A) is incorrect because thermal lows develop over dry, sunny regions, not in the arctic. Answer (B) is incorrect because the eye of a hurricane marks the center of a well-developed tropical cyclone.

ALL

9159. What is a feature of air movement in a high pressure area?

A—Ascending from the surface high to lower pressure at higher altitudes.
B—Descending to the surface and then outward.
C—Moving outward from the high at high altitudes and into the high at the surface.

Air in a high pressure system (in the northern hemisphere) tends to descend to the surface and then spiral out from the center of the high. (PLT173) — AC 00-6A, Chapter 4

ALL

9157. At lower levels of the atmosphere, friction causes the wind to flow across isobars into a low because the friction

A—decreases windspeed and Coriolis force.
B—decreases pressure gradient force.
C—creates air turbulence and raises atmospheric pressure.

Frictional force slows wind speed near the surface and Coriolis force is decreased. The stronger pressure gradient force turns the wind at an angle across the isobars toward lower pressure until the three forces (Coriolis, pressure gradient, and friction) are in balance. (PLT173) — AC 00-6A, Chapter 4

Answer (B) is incorrect because friction does not affect the pressure gradient force. Answer (C) is incorrect because of surface obstructions.

ALL

9176. At which location does Coriolis force have the least effect on wind direction?

A—At the poles.
B—Middle latitudes (30° to 60°).
C—At the Equator.

Coriolis force varies with latitude from zero at the Equator to a maximum at the poles. (PLT510) — AC 00-6A, Chapter 4

ALL

9177. How does Coriolis force affect wind direction in the Southern Hemisphere?

A—Causes clockwise rotation around a low.
B—Causes wind to flow out of a low toward a high.
C—Has exactly the same effect as in the Northern Hemisphere.

Coriolis force deflects air flow to the right causing winds above the friction level to flow parallel to the isobars. In the northern hemisphere, winds flow clockwise around high-pressure areas and counterclockwise around low-pressure areas. In the southern hemisphere, Coriolis force causes a counterclockwise flow around highs and a clockwise flow around lows. (PLT516) — AC 00-6A, Chapter 4

Answer (B) is incorrect because the wind flows from a high to a low (not a low to a high) in both the northern and southern hemispheres. Answer (C) is incorrect because the Coriolis force deflects air to the left in the southern hemisphere, which is the opposite effect from the northern hemisphere.

ALL

9178. Which weather condition is defined as an anticyclone?

A—Calm.
B—High pressure area.
C—COL.

The clockwise flow of air around a high-pressure area in the northern hemisphere is called an anticyclone. (PLT173) — AC 00-6A, Chapter 4

Answer (A) is incorrect because calm is defined as the absence of wind or of apparent motion of the air. Answer (C) is incorrect because COL is the neutral area between two highs or two lows. It is also the intersection of a trough and a ridge.

Answers

9160 [C]	9159 [B]	9157 [A]	9176 [C]	9177 [A]	9178 [B]

ALL
9156. Which area or areas of the Northern Hemisphere experience a generally east to west movement of weather systems?

A—Arctic only.
B—Arctic and subtropical.
C—Subtropical only.

Polar easterlies carry storms from east to west. The northeasterly trade winds carry tropical storms from east to west. The prevailing westerlies drive mid-latitude storms generally from west to east. (PLT510) — AC 00-6A, Chapter 4

ALL
9233. Summer thunderstorms in the arctic region will generally move

A—northeast to southwest in polar easterlies.
B—southwest to northeast with the jetstream flow.
C—directly north to south with the low-level polar airflow.

Arctic thundershowers, usually circumnavigable, move generally from northeast to southwest in the polar easterlies which is opposite from the general movement in mid-latitudes. (PLT495) — AC 00-6A, Chapter 14

ALL
9151. What is a characteristic of the troposphere?

A—It contains all the moisture of the atmosphere.
B—There is an overall decrease of temperature with an increase of altitude.
C—The average altitude of the top of the troposphere is about 6 miles.

The troposphere is the layer of atmosphere from the surface to an average altitude of 7 miles. It is characterized by an overall decrease of temperature with increasing altitude. (PLT203) — AC 00-6A, Chapter 1

Answer (A) is incorrect because moisture can be found in the stratosphere, as evidenced by some of the largest thunderstorms. Answer (C) is incorrect because the average altitude of the top of the troposphere is about 7 miles.

ALL
9240. What weather feature occurs at altitude levels near the tropopause?

A—Maximum winds and narrow wind shear zones.
B—Abrupt temperature increase above the tropopause.
C—Thin layers of cirrus (ice crystal) clouds at the tropopause level.

Maximum winds generally occur at levels near the tropopause. These strong winds create narrow zones of wind shear which often generate hazardous turbulence. (PLT203) — AC 00-6A, Chapter 13

Answer (B) is incorrect because temperature is fairly constant above the tropopause. Answer (C) is incorrect because thin layers of cirrus (ice crystal) clouds can develop at altitudes below the tropopause level and extend into the lower stratosphere.

ALL
9209. Which feature is associated with the tropopause?

A—Absence of wind and turbulence.
B—Absolute upper limit of cloud formation.
C—Abrupt change of temperature lapse rate.

An abrupt change in the temperature lapse rate characterizes the tropopause. (PLT203)—AC 00-6A, Chapter 13

Answer (A) is incorrect because the jet stream (wind) and clear air turbulence are found extensively in the tropopause. Answer (B) is incorrect because clouds can be present into the stratosphere, as in very large thunderstorms and cirrus clouds made up of ice crystals.

ALL
9168. Where is a common location for an inversion?

A—At the tropopause.
B—In the stratosphere.
C—At the base of cumulus clouds.

Inversions are common in the stratosphere. (PLT203) — AC 00-6A, Chapter 2

Answer (A) is incorrect because a common location for an inversion is in the stratosphere, not at the tropopause. Answer (C) is incorrect because the base of cumulus clouds is where the dew point lapse rate and the dry adiabatic lapse rate converge. It is not a common location for an inversion.

Answers

9156 [B] 9233 [A] 9151 [B] 9240 [A] 9209 [C] 9168 [B]

ALL

9241. Where are jetstreams normally located?

A—In areas of strong low pressure systems in the stratosphere.
B—In a break in the tropopause where intensified temperature gradients are located.
C—In a single continuous band, encircling the Earth, where there is a break between the equatorial and polar tropopause.

The jet stream is usually associated with a break in the tropopause where intensified temperature gradients are located. (PLT302) — AC 00-6A, Chapter 13

Answer (A) is incorrect because a jet stream is located in a break in the tropopause, not in the stratosphere. Answer (C) is incorrect because there may be more than one jet stream at any time; up to three at one time are not uncommon.

ALL

9779. The tropopause is generally found when the free air temperatures are

A—between -55°C and -65°C.
B—between -40°C and -55°C.
C—colder than -60°C.

In the absence of other information, the tropopause will generally have a temperature of between -55°C and -65°C. (PLT302) — AC 00-30

ALL

9229. Which type clouds may be associated with the jetstream?

A—Cumulonimbus cloud line where the jetstream crosses the cold front.
B—Cirrus clouds on the equatorial side of the jetstream.
C—Cirrostratus cloud band on the polar side and under the jetstream.

When high-level moisture is present, cirriform clouds form on the equatorial side of the jet stream. (PLT192) — AC 00-6A, Chapter 13

Answer (A) is incorrect because cirriform, not cumulonimbus, clouds are associated with the jet stream. Answer (C) is incorrect because cirriform clouds form on the equatorial side of the jet stream.

ALL

9238. Where do the maximum winds associated with the jetstream usually occur?

A—In the vicinity of breaks in the tropopause on the polar side of the jet core.
B—Below the jet core where a long straight stretch of the jetstream is located.
C—On the equatorial side of the jetstream where moisture has formed cirriform clouds.

Maximum winds in a jet stream occur near a break in the tropopause and on the polar side. (PLT302) — AC 00-6A, Chapter 13

Answer (B) is incorrect because in the jet stream, the maximum winds are found in, not below, the core. Answer (C) is incorrect because when moisture is available, cirriform clouds will form on the upward motion of air of the jet stream on the equatorial side. This will occur in the slower winds of the jet stream.

Answers

9241 [B] 9779 [A] 9229 [B] 9238 [A]

Weather Systems

When air masses of different temperature or moisture content collide, they force air aloft along the area where they meet. An elongated line of low pressure is referred to as a trough.

A **front** is defined as the boundary between two different air masses. The formation of a front is called frontogenesis. When a front dissipates, the area experiences frontolysis. All fronts lie in troughs. This means that winds flow around a front more or less parallel to the front, and in a counterclockwise direction. As an aircraft flies toward a front in the northern hemisphere, the pilot will notice a decreasing pressure and a wind from the left of the aircraft. After passing through the front, the pilot will note a wind shift to the right and increasing air pressure.

A front is usually the boundary between air masses of different temperatures. If cold air is displacing warm air, it is called a cold front. When warm air displaces cold air, it is a warm front. The speed of movement of the front is determined by the winds aloft. A cold front will move at about the speed of the wind component perpendicular to the front just above the friction level. It is harder for warm air to displace cold air and so warm fronts move at about half the speed of cold fronts under the same wind conditions.

A stationary front is one with little or no movement. Stationary fronts or slow moving cold fronts can form frontal waves and low pressure areas. A small disturbance can cause a bend in the frontal line that induces a counterclockwise flow of air around a deepening low pressure area. The wave forms into a warm front followed by a cold front. The cold front can then overtake the warm front and force the warm air between the two aloft. This is called an occluded front or an occlusion.

Most fronts mark the line between two air masses of different temperature. However, this is not always the case. Sometimes, air masses with virtually the same temperatures will form a front. The only difference between the two is the moisture content. The front formed in such conditions is called a dew point front or a dry line.

The surface position of a front often marks the line where an arctic and a tropical air mass meet at the surface. The jet stream is located in the area where these air masses meet at the altitude of the tropopause. There is often a rough correlation between the surface position of a front and the location of the jet stream. Generally speaking, the jet stream will lie to the north of the surface position of a front. As a frontal wave forms, the jet will move toward the center of the deepening low pressure area. If an occluded front forms, the jet stream will often cross the front near the point of the occlusion.

ALL

9165. What term describes an elongated area of low pressure?

A—Trough.
B—Ridge.
C—Hurricane or typhoon.

A trough is an elongated area of low pressure with the lowest pressure along a line marking maximum anti-cyclonic curvature. (PLT173) — AC 00-6A, Chapter 3

Answer (B) is incorrect because a ridge is an elongated area of high pressure. Answer (C) is incorrect because a hurricane or typhoon is a tropical cyclone (low) with highest sustained winds of 65 knots or greater.

ALL

9191. What is a feature of a stationary front?

A—The warm front surface moves about half the speed of the cold front surface.
B—Weather conditions are a combination of strong cold front and strong warm front weather.
C—Surface winds tend to flow parallel to the frontal zone.

The opposing forces exerted by adjacent air masses in a stationary front are such that the frontal surface between them shows little or no movement. In such cases, the surface winds tend to blow parallel to the frontal zone. (PLT511) — AC 00-6A, Chapter 8

Answer (A) is incorrect because the movement of a warm front surface in comparison to a cold front surface has nothing to do with a stationary front. Answer (B) is incorrect because weather conditions that are a combination of strong cold front and strong warm front weather are a feature of an occluded front.

Answers

9165 [A]   9191 [C]

ALL
9192. Which event usually occurs after an aircraft passes through a front into the colder air?

A—Temperature/dewpoint spread decreases.
B—Wind direction shifts to the left.
C—Atmospheric pressure increases.

A front lies in a pressure trough, and pressure generally is higher in the cold air. Thus, when crossing a front directly into colder air, the pressure will usually rise abruptly. (PLT511) — AC 00-6A, Chapter 8

Answer (A) is incorrect because the temperature/dew point spread usually differs across a front. But it might not decrease if you fly in to a cold, dry air mass. Answer (B) is incorrect because in the northern hemisphere the wind always shifts to the right due to the Coriolis force.

ALL
9213. What type weather change is to be expected in an area where frontolysis is reported?

A—The frontal weather is becoming stronger.
B—The front is dissipating.
C—The front is moving at a faster speed.

When the temperature and pressure differences across a front equalize, the front dissipates. This process is called frontolysis. (PLT511) — AC 00-6A, Chapter 8

Answers (A) and (C) are incorrect because frontal weather becoming stronger or faster are weather changes that are to be expected in an area where frontogenesis is reported.

ALL
9215. Which atmospheric factor causes rapid movement of surface fronts?

A—Upper winds blowing across the front.
B—Upper low located directly over the surface low.
C—The cold front overtaking and lifting the warm front.

Cold fronts move at about the speed of the wind component perpendicular to the front just above the frictional layer. (PLT511) — AC 00-6A, Chapter 8

Answer (B) is incorrect because an upper low located directly over a surface low would be a factor in how extensive the weather would be, not in how fast it would move. Answer (C) is incorrect because a cold front overtaking and lifting the warm front is a characteristic of an advancing cold front.

ALL
9216. In which meteorological conditions can frontal waves and low pressure areas form?

A—Warm fronts or occluded fronts.
B—Slow-moving cold fronts or stationary fronts.
C—Cold front occlusions.

Frontal waves and cyclones (areas of low pressure) usually form on slow moving cold fronts or on stationary fronts. (PLT511) — AC 00-6A, Chapter 8

Answer (A) is incorrect because occluded fronts are formed by frontal waves and areas of low pressure which cause a cold front to close together with a warm front. Frontal waves and low pressure areas normally form on slow-moving cold fronts. Answer (C) is incorrect because a cold front occlusion occurs when the air behind the cold front is colder than the air in advance of the warm front, lifting the warm front aloft.

ALL
9217. What weather difference is found on each side of a "dry line"?

A—Extreme temperature difference.
B—Dewpoint difference.
C—Stratus versus cumulus clouds.

A dew point front or "dry line" is formed when two air masses of similar density and temperature meet. Except for the moisture differences, there is little contrast across the front. (PLT511) — AC 00-6A, Chapter 8

Answer (A) is incorrect because except for moisture (not extreme temperature) difference, there is seldom any significant air mass contrast across the "dry line." Answer (C) is incorrect because the side with moisture may have clouds, while generally clear skies mark the dry side.

ALL
9227. Where is the normal location of the jetstream relative to surface lows and fronts?

A—The jetstream is located north of the surface systems.
B—The jetstream is located south of the low and warm front.
C—The jetstream is located over the low and crosses both the warm front and the cold front.

Development of a surface low is usually south of the jet stream. As the low deepens, it moves nearer the jet center. When a low occludes, the jet stream usually crosses the frontal system at the point of the occlusion. (PLT302) — AC 00-6A, Chapter 13

Answer (B) is incorrect because the jet stream is located to the north of the low and warm front. Answer (C) is incorrect because the jet stream crosses the occlusion of the warm and cold front at the point of occlusion.

Answers

9192 [C]	9213 [B]	9215 [A]	9216 [B]	9217 [B]	9227 [A]

ALL

9228. Which type frontal system is normally crossed by the jetstream?

A—Cold front and warm front.
B—Warm front.
C—Occluded front.

Development of a surface low is usually south of the jet stream. As the low deepens, it moves nearer the jet center. When a low occludes, the jet stream usually crosses the frontal system at the point of the occlusion. (PLT302) — AC 00-6A, Chapter 13

ALL

9776. A jet stream is a narrow, shallow, meandering river of maximum winds extending around the globe in a wavelike pattern with speeds of

A—50 knots or greater.
B—71 knots or greater.
C—100 knots or greater.

The concentrated winds, by arbitrary definition, must be 50 knots or greater to be classified as a jet stream. (PLT302) — AC 00-6

Stability and Instability of Air

When a parcel of air is forced to rise it expands because its pressure decreases. Air that is forced to descend is compressed. When the pressure and volume change, so does the temperature. When air expands, it cools and when it is compressed, it warms. This cooling or heating is referred to as being **adiabatic**, meaning that no heat was removed from or added to the air.

When unsaturated air is forced to rise or descend it cools or heats at a rate of about 3°C per 1,000 feet of altitude change. This called the dry adiabatic rate. The saturated adiabatic rate is normally much lower.

When moist air is forced upward, the temperature and the dew point converge on each other at a rate of about 2.5°C per 1,000 feet. At the altitude where the dew point lapse rate and the dry adiabatic rate meet, cloud bases will form. Once the condensation starts taking place the adiabatic rate slows considerably because the process of condensation releases latent heat into the air and partially offsets the expansional cooling.

Saturated air flowing downward will also warm at less than the dry adiabatic rate because vaporization of water droplets uses heat. Once the air is no longer saturated it will heat at the normal dry rate. An example of this is the "katabatic wind" which becomes warmer and dryer as it flows downslope.

The adiabatic rate should not be confused with the actual (ambient) lapse rate. The actual lapse rate is the rate at which the air temperature varies with altitude when air is not being forced to rise or descend. The actual lapse averages about 2°C per 1,000 feet, but that is highly variable. When a parcel of air is forced to rise, the adiabatic rate may be different than the ambient rate.

When a parcel of air becomes colder (and more dense) than the air around it, it will tend to sink back toward its original altitude. If the parcel becomes warmer than the surrounding air, it will tend to rise convectively even though the original lifting force may have disappeared. If this happens, the air is said to be unstable. When a parcel of air resists convective movement through it, it is said to be stable.

The best indication of the stability or instability of an air mass is the ambient temperature lapse rate. If the temperature drops rapidly as the altitude increases, the air is unstable. If the temperature remains unchanged or decreases only slightly as altitude is increased, the air mass is stable. If the temperature actually increases as altitude increases, a temperature inversion exists. This is the most stable of weather conditions.

Answers

9228 [C] 9776 [A]

ALL

9170. Which term applies when the temperature of the air changes by compression or expansion with no heat added or removed?

A—Katabatic.
B—Advection.
C—Adiabatic.

When air expands, it cools; and when compressed, it warms. These changes are adiabatic, meaning that no heat is removed from or added to the air. (PLT024) — AC 00-6A, Chapter 6

Answer (A) is incorrect because katabatic is a wind blowing down an incline caused by cold, heavier air spilling down the incline displacing warmer, less dense air. Answer (B) is incorrect because advection is the horizontal flow in a convective current, i.e., wind.

ALL

9186. Which process causes adiabatic cooling?

A—Expansion of air as it rises.
B—Movement of air over a colder surface.
C—Release of latent heat during the vaporization process.

When air expands, it cools; and when compressed, it warms. These changes are adiabatic, meaning that no heat is removed from or added to the air. (PLT024) — AC 00-6A, Chapter 6

Answer (B) is incorrect because adiabatic cooling means that no heat is removed from the air, as would be the case if the air was moved over a colder surface. Answer (C) is incorrect because adiabatic cooling is the process in which no heat is removed from or added to the air.

ALL

9158. Which type wind flows downslope becoming warmer and dryer?

A—Land breeze.
B—Valley wind.
C—Katabatic wind.

A katabatic wind is any wind blowing down an incline when the incline is influential in causing the wind. Any katabatic wind originates because cold, heavy air spills down sloping terrain displacing warmer, less dense air ahead of it. Air is heated and dried as it flows downslope. (PLT516) — AC 00-6A, Chapter 4

Answer (A) is incorrect because a land breeze is a wind that flows from the cooler land toward warmer water. Answer (B) is incorrect because a valley wind is wind flowing up out of a valley because colder, denser air settles downward and forces the warmer air near the ground up a mountain slope.

ALL

9171. What is the approximate rate unsaturated air will cool flowing upslope?

A—3°C per 1,000 feet.
B—2°C per 1,000 feet.
C—4°C per 1,000 feet.

Unsaturated air moving upward and downward cools and warms at about 3.0°C (5.4°F) per 1,000 feet. (PLT024) — AC 00-6A, Chapter 6

Answers (B) and (C) are incorrect because unsaturated air will cool flowing upslope at 3°C per 1,000 feet.

ALL

9182. What is the result when water vapor changes to the liquid state while being lifted in a thunderstorm?

A—Latent heat is released to the atmosphere.
B—Latent heat is transformed into pure energy.
C—Latent heat is absorbed from the surrounding air by the water droplet.

When water vapor condenses to liquid water or sublimates directly to ice, energy originally used in the evaporation reappears as heat and is released to the atmosphere. This energy is "latent heat." (PLT512) — AC 00-6A, Chapter 5

Answer (B) is incorrect because latent heat cannot create pure energy. Latent heat is returned to the surrounding atmosphere. Answer (C) is incorrect because this is the process of latent heat in vaporization, which is changing liquid water to vapor.

ALL

9185. What weather condition occurs at the altitude where the dewpoint lapse rate and the dry adiabatic lapse rate converge?

A—Cloud bases form.
B—Precipitation starts.
C—Stable air changes to unstable air.

Unsaturated air in a convective current cools at about 5.4°F (3°C) per 1,000 feet. The dew point decreases at about 1°F (5/9°C) per 1,000 feet. When the temperature and dew point converge, cloud bases will form. (PLT512) — AC 00-6A, Chapter 6

Answer (B) is incorrect because precipitation starts when precipitation particles have grown to a size and weight that the atmosphere can no longer suspend, and the particles fall as precipitation. Answer (C) is incorrect because air stability depends on the ambient or existing temperature lapse rate, not the convergence of the dew point lapse rate and the dry adiabatic lapse rate.

Answers

9170 [C] 9186 [A] 9158 [C] 9171 [A] 9182 [A] 9185 [A]

ALL

9187. When saturated air moves downhill, its temperature increases

A—at a faster rate than dry air because of the release of latent heat.
B—at a slower rate than dry air because vaporization uses heat.
C—at a slower rate than dry air because condensation releases heat.

The saturated adiabatic rate of heating is slower than the dry rate because vaporization uses heat. (PLT024) — AC 00-6A, Chapter 6

Answer (A) is incorrect because when saturated air moves downhill, its temperature increases at a slower rate than dry air because of the absorption of latent heat. Answer (C) is incorrect because as air moves downhill, its temperature increases at a slower rate than dry air because vaporization uses heat, not because of the release of heat through condensation.

ALL

9154. What feature is associated with a temperature inversion?

A—A stable layer of air.
B—An unstable layer of air.
C—Air mass thunderstorms.

A temperature inversion is defined as an increase in temperature with increasing altitude, or a negative temperature lapse rate. Stable air masses have a low or negative lapse rate. (PLT301) — AC 00-6A, Chapter 2

Answer (B) is incorrect because instability occurs when the temperature decreases (not increases as in a temperature inversion) with an increase in altitude, and the rising air continues to rise. Answer (C) is incorrect because air mass thunderstorms result from instability. They do not occur when there is a temperature inversion.

ALL

9184. What is indicated about an air mass if the temperature remains unchanged or decreases slightly as altitude is increased?

A—The air is unstable.
B—A temperature inversion exists.
C—The air is stable.

A mass of air in which the temperature decreases rapidly with height favors instability. Air tends to be stable if the temperature changes little or not at all with altitude. (PLT512) — AC 00-6A, Chapter 6

Answer (A) is incorrect because unstable air would have a uniform decrease (approaching 3°C/1,000 feet) in temperature with an increase in altitude. Answer (B) is incorrect because in a temperature inversion, the temperature increases with increases in altitude.

ALL

9188. Which condition is present when a local parcel of air is stable?

A—The parcel of air resists convection.
B—The parcel of air cannot be forced uphill.
C—As the parcel of air moves upward, its temperature becomes warmer than the surrounding air.

A parcel of air which resists convection when forced upward is called stable. (PLT173) — AC 00-6A, Chapter 6

Answer (B) is incorrect because stable air can be forced uphill to form a mountain wave. Answer (C) is incorrect because rising air, warmer than the surrounding air, describes unstable air.

ALL

9195. How can the stability of the atmosphere be determined?

A—Ambient temperature lapse rate.
B—Atmospheric pressure at various levels.
C—Surface temperature/dewpoint spread.

A mass of air in which the temperature decreases rapidly with height favors instability. Air tends to be stable if the temperature changes little or not at all with altitude. The rate of temperature decrease with altitude is referred to as the temperature lapse rate. (PLT173) — AC 00-6A, Chapter 6

Answer (B) is incorrect because the difference between ambient temperature and adiabatic lapse rate, not atmospheric pressure at various levels, determines stability. Answer (C) is incorrect because the surface temperature/dew point spread is used to indicate probability of fog, not atmospheric stability.

Answers

9187 [B]	9154 [A]	9184 [C]	9188 [A]	9195 [A]

Fog and Rain

Fog is a surface-based cloud that always forms in stable air conditions. The three main types are radiation fog, advection fog and upslope fog.

Radiation fog occurs when there is a surface-based temperature inversion. On a clear, relatively calm night the surface rapidly cools by radiating heat into space. This in turn cools the air within a few hundred feet of the surface and leaves warmer air aloft. If the temperature drops to the dew point, fog will form. Since the minimum temperature during the day occurs just after sunrise, this type of fog often forms then. This fog will dissipate when the air warms up enough to raise the temperature above the dew point again. However, if the inversion persists, visibility can remain limited due to lingering fog and haze. Wind or any significant movement of air will disperse both radiation fog and haze.

Advection fog and **upslope fog** both require wind to form. Advection fog forms when warm moist air flows over a colder surface. The temperature of the air drops to the dew point and fog forms. This commonly occurs over bodies of water such as lakes or oceans. The fog can drift over land on the leeward (downwind) side of the body of water lowering visibility at nearby airports. If the wind increases to over about 15 knots, the fog will tend to lift into low stratus clouds.

Upslope fog forms when moist, stable air is gradually moved over higher ground by the wind. As the air rises, it cools adiabatically and fog forms. This type of fog is common in mountainous areas.

All clouds are composed of tiny droplets of water (or ice crystals). As these drops of water collide with each other, they form larger drops until they precipitate out as rain. As a general rule, clouds need to be at least 4,000 feet thick to produce precipitation reported as light or greater intensity.

ALL

9153. What characterizes a ground-based inversion?

A—Convection currents at the surface.
B—Cold temperatures.
C—Poor visibility.

Inversions can occur in warm and cold temperatures in stable air, and usually trap particles in the air causing poor visibility. (PLT301) — AC 00-6A, Chapter 2

Answer (A) is incorrect because convective currents at the surface do not occur when there is a ground-based inversion. Answer (B) is incorrect because when the temperature is cold, it is difficult for the earth to radiate enough heat to become colder than the overlying air.

ALL

9155. When does minimum temperature normally occur during a 24-hour period?

A—After sunrise.
B—About 1 hour before sunrise.
C—At midnight.

At night, solar radiation ceases, but terrestrial radiation continues and cools the surface. Cooling continues after sunrise until solar radiation again exceeds terrestrial radiation. Minimum temperature usually occurs after sunrise, sometimes as much as 1 hour after. (PLT512) — AC 00-6A, Chapter 2

Answer (B) is incorrect because the minimum temperature normally occurs after sunrise. Answer (C) is incorrect because the minimum temperature normally occurs after sunrise, not at midnight.

ALL

9169. What condition produces the most frequent type of ground- or surface-based temperature inversion?

A—The movement of colder air under warm air or the movement of warm air over cold air.
B—Widespread sinking of air within a thick layer aloft resulting in heating by compression.
C—Terrestrial radiation on a clear, relatively calm night.

An inversion often develops near the ground on clear, cool nights when the wind is light. The ground radiates and cools much faster than the overlying air. Air in contact with the ground becomes cold while the temperature a few hundred feet above changes very little. Thus, temperature increases with height. (PLT301) — AC 00-6A, Chapter 2

Answer (A) is incorrect because the movement of colder air under warm air is what happens when a cold front is advancing, and the movement of warm air over cold air is the process of an advancing warm front. Answer (B) is incorrect because widespread sinking of air describes compressional or adiabatic heating.

Answers

9153 [C] 9155 [A] 9169 [C]

ALL
9208. How are haze layers cleared or dispersed?

A—By convective mixing in cool night air.
B—By wind or the movement of air.
C—By evaporation similar to the clearing of fog.

Haze or smoke must be dispersed by movement of air. (PLT510) — AC 00-6A, Chapter 12

Answer (A) is incorrect because convective mixing would be caused by heating during the day, not by the cool night air. Answer (C) is incorrect because haze must be dispersed by movement of air, it cannot evaporate in a similar manner to the clearing of fog.

ALL
9206. When advection fog has developed, what may tend to dissipate or lift the fog into low stratus clouds?

A—Temperature inversion.
B—Wind stronger than 15 knots.
C—Surface radiation.

Wind much stronger than 15 knots lifts the fog into a layer of low stratus or stratocumulus. (PLT226) — AC 00-6A, Chapter 12

Answer (A) is incorrect because winds stronger than 15 knots tend to dissipate or lift advection fog, not a temperature inversion. Answer (C) is incorrect because surface temperature radiation is a factor in radiation fog.

ALL
9207. Which conditions are necessary for the formation of upslope fog?

A—Moist, stable air being moved over gradually rising ground by a wind.
B—A clear sky, little or no wind, and 100 percent relative humidity.
C—Rain falling through stratus clouds and a 10- to 25-knot wind moving the precipitation up the slope.

Upslope fog forms as a result of moist, stable air being cooled adiabatically as it moves up sloping terrain. (PLT226) — AC 00-6A, Chapter 12

Answer (B) is incorrect because these are conditions necessary for the formation of radiation fog which usually occurs at night. Answer (C) is incorrect because upslope fog is formed by moist air being moved gradually over rising ground, not by rain falling through stratus clouds and not by a wind blowing the precipitation up the slope.

ALL
9194. Which condition produces weather on the lee side of a large lake?

A—Warm air flowing over a colder lake may produce fog.
B—Cold air flowing over a warmer lake may produce advection fog.
C—Warm air flowing over a cool lake may produce rain showers.

When warm air flows over a colder lake, the air may become saturated by evaporation from the water while also becoming cooler in the low levels by contact with the cool water. Fog often becomes extensive and dense to the lee (downwind) side of the lake. (PLT226) — AC 00-6A, Chapter 5

Answer (B) is incorrect because cold air flowing over a warmer lake may produce rain showers, not advection fog, on the lee side of the lake. Answer (C) is incorrect because warm air flowing over a cool lake may produce fog, not rain showers, on the lee side of the lake.

ALL
9193. What minimum thickness of cloud layer is indicated if precipitation is reported as light or greater intensity?

A—4,000 feet thick.
B—2,000 feet thick.
C—A thickness which allows the cloud tops to be higher than the freezing level.

When arriving at or departing from a terminal reporting precipitation of light or greater intensity, expect clouds to be more than 4,000 feet thick. (PLT192) — AC 00-6A, Chapter 5

Answer (B) is incorrect because to produce significant precipitation (light or greater intensity), clouds are normally at least 4,000 feet thick. Answer (C) is incorrect because a cloud thickness resulting in cloud tops above the freezing level means ice droplets and super-cooled water will develop.

Answers

9208 [B] 9206 [B] 9207 [A] 9194 [A] 9193 [A]

Thunderstorms

Thunderstorms are always generated in very unstable conditions. Warm, moist air is forced upward either by heating from below or by frontal lifting, and becomes unstable. When the rising air cools to its dew point, a cumulus cloud forms. This "cumulus stage" is the first of three in a thunderstorm's life. It is characterized by a continuous updraft as the cloud builds. As the raindrops and ice pellets in the cloud grow larger, their weight begins to overpower the lifting force of the updrafts. As the drops fall through the cloud, they cool the air making it more dense than in the surrounding updrafts. This process causes downdrafts to form within the cloud.

When the **downdrafts** become strong enough to allow the first precipitation to reach the surface, the mature stage of the thunderstorm has begun. Eventually, the downdrafts cut off the updrafts and the storm loses the source of warm air that is its driving force. When the storm is characterized predominantly by downdrafts, it is in the dissipating stage.

Air mass thunderstorms are associated with local surface heating. On a clear, sunny day, local hot spots form that are capable of making the air over them unstable enough to generate a thunderstorm. Because the downdrafts in an air mass thunderstorm shut off the updrafts fairly quickly, this type of storm is relatively short-lived.

Steady-state thunderstorms are usually associated with weather systems. Fronts, converging winds and troughs aloft force upward motion. In a steady-state storm the precipitation falls outside the updraft allowing the storm to continue without abating for several hours.

The most violent type of steady-state thunderstorms are those generated by cold fronts or by squall lines. A **squall line** is a non-frontal instability line that often forms ahead of a fast moving cold front. Thunderstorms generated under these conditions are the most likely to develop cumulonimbus mamma clouds, funnel clouds and tornadoes. A severe thunderstorm is one which has surface winds of 50 knots or more, and/or has hail 3/4-inch or more in diameter.

Pressure usually falls rapidly with the approach of a thunderstorm, then rises sharply with onset of the first gust and arrival of the cold downdraft and heavy rain showers. As the storm passes on, the pressure returns to normal.

Even though thunderstorms are cumulus clouds formed in unstable air they can sometimes penetrate overlying bands of stratiform clouds. These are known as "**embedded thunderstorms**." Because these thunderstorms are obscured by other clouds and it is impossible for a pilot to visually detour around them, they present a particular hazard to IFR flight.

When they can, most pilots prefer to visually avoid thunderstorms by flying around them or, if they can maintain a high enough altitude, by flying over the storm. If you are going to fly over the top of a thunderstorm, a good rule of thumb to follow is that the cloud should be overflown by at least 1,000 feet for each 10 knots of wind speed. Radar is a very useful tool in thunderstorm avoidance, especially at night or in IFR weather. The radar displays an area of precipitation size rain drops as a bright spot on the screen. Since thunderstorms often contain large water drops, they usually show up on the radar screen. A dark area on the screen is one in which no precipitation drops are detected. Areas of clouds may or may not be displayed depending on the size of the drops that make up the clouds. *See* Figure 8-5.

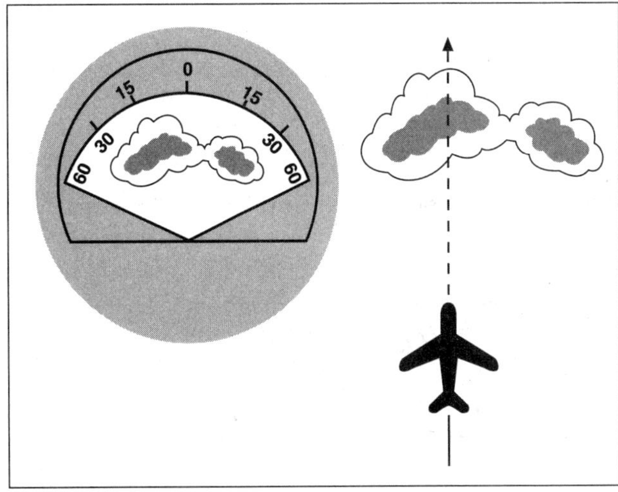

Figure 8-5. Radar display of a thunderstorm

ALL

9196. Which weather phenomenon signals the beginning of the mature stage of a thunderstorm?

A—The appearance of an anvil top.
B—The start of rain at the surface.
C—Growth rate of the cloud is at its maximum.

Precipitation beginning to fall from the cloud base is the indication that a downdraft has developed and a thunderstorm cell has entered the mature stage. (PLT495) — AC 00-6A, Chapter 11

Answer (A) is incorrect because the appearance of an anvil top occurs during the dissipating stage. Answer (C) is incorrect because the growth rate of a thunderstorm is at its greatest during the cumulus stage.

ALL

9197. During the life cycle of a thunderstorm, which stage is characterized predominately by downdrafts?

A—Cumulus.
B—Dissipating.
C—Mature.

Downdrafts characterize the dissipating stage of the thunderstorm. (PLT495) — AC 00-6A, Chapter 11

Answer (A) is incorrect because the cumulus stage is the building stage characterized by updrafts. Answer (C) is incorrect because the mature stage has both updrafts and downdrafts, which create strong wind shears.

ALL

9198. What feature is normally associated with the cumulus stage of a thunderstorm?

A—Beginning of rain at the surface.
B—Frequent lightning.
C—Continuous updraft.

The key feature of the cumulus stage is a continuous updraft. (PLT495) — AC 00-6A, Chapter 11

Answer (A) is incorrect because the beginning of rain at a surface marks the beginning of the mature stage. Answer (B) is incorrect because frequent lightning occurs after the downdrafts have developed and produce the static electricity which causes lighting.

ALL

9203. Why are downdrafts in a mature thunderstorm hazardous?

A—Downdrafts are kept cool by cold rain which tends to accelerate the downward velocity.
B—Downdrafts converge toward a central location under the storm after striking the surface.
C—Downdrafts become warmer than the surrounding air and reverse into an updraft before reaching the surface.

Precipitation beginning to fall from the cloud base is the indication that a downdraft has developed and a thunderstorm cell has entered the mature stage. Cold rain in the downdraft retards compressional heating, and the downdraft remains cooler than surrounding air. Therefore, its downward speed is accelerated and may exceed 2,500 feet per minute. (PLT495) — AC 00-6A, Chapter 11

Answer (B) is incorrect because after striking the ground the downdrafts will move away from the storm's center. Answer (C) is incorrect because downdrafts remain colder than the surrounding air and accelerate downward into an updraft.

ALL

9200. Where do squall lines most often develop?

A—In an occluded front.
B—Ahead of a cold front.
C—Behind a stationary front.

A squall line is a non-frontal, narrow band of active thunderstorms. Often it develops ahead of a cold front in moist, unstable air. (PLT475) — AC 00-6A, Chapter 11

Answer (A) is incorrect because squall lines most often develop ahead of a cold front, not in an occluded front. Answer (C) is incorrect because squall lines most often develop ahead of a cold front, not behind a stationary front.

ALL

9204. What is a difference between an air mass thunderstorm and a steady-state thunderstorm?

A—Air mass thunderstorms produce precipitation which falls outside of the updraft.
B—Air mass thunderstorm downdrafts and precipitation retard and reverse the updrafts.
C—Steady-state thunderstorms are associated with local surface heating.

Air mass thunderstorms most often result from surface heating. When the storm reaches the mature stage, rain falls through or immediately beside the updraft. Falling precipitation induces frictional drag, retards the updraft and reverses it to a downdraft. Such a self-destructive cell usually has a life cycle of 20 minutes to 1-1/2 hours. In a steady-state thunderstorm, the precipitation falls outside the downdraft and so the cell can last several hours. (PLT495) — AC 00-6A, Chapter 11

Answer (A) is incorrect because steady-state, not air mass, thunderstorms produce precipitation which falls outside the updraft. Answer (C) is incorrect because air mass, not steady-state, thunderstorms are associated with local surface heating.

Answers

9196 [B]	9197 [B]	9198 [C]	9203 [A]	9200 [B]	9204 [B]

ALL

9205. Which type storms are most likely to produce funnel clouds or tornadoes?

A—Air mass thunderstorms.
B—Cold front or squall line thunderstorms.
C—Storms associated with icing and supercooled water.

Tornadoes occur with isolated thunderstorms at times, but much more frequently, they form with steady-state thunderstorms associated with cold fronts or squall lines. (PLT495) — AC 00-6A, Chapter 11

Answer (A) is incorrect because even though air mass thunderstorms may produce funnel clouds or tornadoes, they are most likely to occur with steady-state thunderstorms. Answer (C) is incorrect because all thunderstorms that have updrafts and carry water above the freezing level can produce icing and supercooled water. But thunderstorms associated with cold fronts and squall lines are most likely to produce funnel clouds or tornadoes.

ALL

9210. Which type cloud is associated with violent turbulence and a tendency toward the production of funnel clouds?

A—Cumulonimbus mamma.
B—Standing lenticular.
C—Stratocumulus.

Frequently, cumulonimbus mamma clouds occur in connection with violent thunderstorms and tornadoes. (PLT501) — AC 00-6A, Chapter 11

Answer (B) is incorrect because standing lenticular clouds mark mountain waves that are the product of stable air flowing over an obstruction. Answer (C) is incorrect because stratocumulus sometimes form from the breaking up of stratus or the spreading out of cumulus, and they are associated with some turbulence and possible icing at subfreezing levels.

ALL

9214. Which weather condition is an example of a nonfrontal instability band?

A—Squall line.
B—Advective fog.
C—Frontogenesis.

An instability line is a narrow, nonfrontal line or band of convective activity. If the activity is fully developed thunderstorms, the line is a squall line. (PLT511) — AC 00-6A, Chapter 8

Answer (B) is incorrect because advective fog occurs when warm moist air moves over a cool surface. It forms in stable air. Answer (C) is incorrect because frontogenesis is the generation of a frontal zone.

ALL

9706. A severe thunderstorm is one in which the surface wind is

A—58 MPH or greater and/or surface hail is 3/4 inch or more in diameter.
B—50 knots or greater and/or surface hail is 1/2 inch or more in diameter.
C—45 knots or greater and/or surface hail is 1 inch or more in diameter.

A severe thunderstorm is one which has surface winds of 50 knots (58 MPH) or more, and/or has hail 3/4 inch or more in diameter. (PLT495) — AC 00-45

ALL

9708. A squall is a sudden increase of at least 16 knots in average wind speed to a sustained speed of

A—24 knots or more for at least 1 minute.
B—22 knots or more for at least 1 minute.
C—20 knots or more for at least 1 minute.

A squall (SQ) means there has been a sudden increase in wind speed of at least 16 knots to a speed of 22 knots or more, and it lasted at least one minute. (PLT475) — AC 00-45

ALL

9202. Atmospheric pressure changes due to a thunderstorm will be at the lowest value

A—during the downdraft and heavy rain showers.
B—when the thunderstorm is approaching.
C—immediately after the rain showers have stopped.

Pressure usually falls rapidly with the approach of a thunderstorm, then rises sharply with the onset of the first gust and the arrival of the cold downdraft and heavy rain showers, falling back to normal as the storm moves on. (PLT495) — AC 00-6A, Chapter 11

Answer (A) is incorrect because during the downdraft and heavy rain showers the pressure rises sharply. Answer (C) is incorrect because immediately after the rain showers have stopped the pressure will return to normal.

Answers

9205 [B]	9210 [A]	9214 [A]	9706 [A]	9708 [B]	9202 [B]

ALL

9189. Convective clouds which penetrate a stratus layer can produce which threat to instrument flight?

A—Freezing rain.
B—Clear air turbulence.
C—Embedded thunderstorms.

A layer of stratiform clouds may sometimes form in a mildly stable layer while a few convective clouds penetrate the layer thus merging stratiform with cumuliform. Under the right conditions, the cumuliform clouds can become thunderstorms which are completely obscured by the surrounding stratus clouds. (PLT192) — AC 00-6A, Chapter 6

Answer (A) is incorrect because the formation of freezing rain is dependent on rain falling through colder air. Convective clouds that penetrate a stratus layer may or may not produce precipitation. Answer (B) is incorrect because clear air turbulence is turbulence encountered in air where no clouds are present.

ALL

9199. What is indicated by the term "embedded thunderstorms"?

A—Severe thunderstorms are embedded in a squall line.
B—Thunderstorms are predicted to develop in a stable air mass.
C—Thunderstorms are obscured by other types of clouds.

A layer of stratiform clouds may sometimes form in a mildly stable layer while a few convective clouds penetrate the layer, thus merging stratiform with cumuliform. Under the right conditions, the cumuliform clouds can become thunderstorms which are completely obscured by the surrounding stratus clouds. (PLT495) — AC 00-6A, Chapter 6

Answer (A) is incorrect because a squall line consists of severe thunderstorms which can always be seen. Answer (B) is incorrect because thunderstorms do not occur in stable air masses.

ALL

9211. A clear area in a line of thunderstorm echoes on a radar scope indicates

A—the absence of clouds in the area.
B—an area of no convective turbulence.
C—an area where precipitation drops are not detected.

Airborne weather radar detects only precipitation size raindrops and hail. Absence of a radar return does indicate an area free of clouds or turbulence. (PLT495) — AC 00-6A, Chapter 11

Answer (A) is incorrect because weather radar detects only precipitation drops, not clouds. Answer (B) is incorrect because convective turbulence, which would not be detected by radar, could be found under cumulus clouds. Radar does not detect turbulence.

ALL

9212. When flying over the top of a severe thunderstorm, the cloud should be overflown by at least

A—1,000 feet for each 10 knots windspeed.
B—2,500 feet.
C—500 feet above any moderate to severe turbulence layer.

When flying over the top of a severe thunderstorm, clear the top by 1,000 feet for each 10 knots of wind at the cloud top. (PLT495) — AC 00-6A, Chapter 11

Answer (B) is incorrect because the cloud should be overflown by at least 1,000 feet for each 10 knots wind speed, which would normally be greater than 2,500 feet with a severe thunderstorm. Answer (C) is incorrect because the exact location of a turbulence layer will not usually be known.

Answers

9189 [C]	9199 [C]	9211 [C]	9212 [A]

Wind Shear

Normally we think of changes in wind speed or direction as having an effect only on an aircraft's ground speed and track. However, when there is a very rapid shift in wind speed or direction there is a noticeable change in the aircraft's indicated airspeed as well.

In a situation where there is a sudden increase in headwind (or decrease in tailwind) the aircraft's momentum keeps it moving through space at the same ground speed as before. This means that the aircraft will be moving through the air faster than before and there will be an increase in its indicated airspeed. The aircraft will react to this increase by pitching up and by tending to climb (or descend more slowly). When there is a sudden increase in a tailwind (or decrease in the headwind), just the opposite occurs. There will be a loss of indicated airspeed accompanied by a tendency to pitch down and descend.

Wind shear is defined as any rapid change in wind direction or velocity. Often, there is little or no turbulence associated with wind shear. Severe wind shear is defined as a rapid change in wind direction or velocity causing airspeed changes greater than 15 knots or vertical speed changes greater than 500 feet per minute.

Wind shear may be associated with either a wind shift or a wind speed gradient at any level in the atmosphere. Three common generators of wind shear conditions are thunderstorms, temperature inversions and jet stream winds. Thunderstorms generate a very significant wind shear hazard for two reasons. The shear from thunderstorms is usually encountered close to the ground where there is little time or altitude to recover. The magnitude of the shear is often very severe, especially in situations involving microbursts, which we will discuss shortly. Wind shear can be encountered on all sides and directly under the thunderstorm cell. Often, in a low altitude temperature inversion the winds are very light but just above the inversion layer the wind is much stronger. When an aircraft either climbs or descends through the top of the inversion it can encounter significant wind shear because of the change in wind speed. A jet stream is a narrow "river" of wind where the speed can change a great deal over a very short distance. This is the very definition of wind shear.

Microbursts are a very localized, but very dangerous, wind shear condition. They can occur anywhere that convective weather conditions exist. This includes rain showers, virga and thunderstorms. It is believed that about five percent of thunderstorms produce a microburst.

A microburst is a very narrow downdraft of very high speed wind. The downdraft is typically a few hundred to 3,000 feet across with vertical speeds up to 6,000 feet per minute. When the downdraft approaches the surface, the wind flows outward from the core in all directions. Not only are these outflow winds very strong (up to 45 knots) but their effect is doubled when an aircraft flies through the shear. For example, a 45 knot headwind approaching the microburst will be a 45 knot tailwind flying out the other side—a change of 90 knots. This is usually a short-lived phenomena, seldom lasting more than 15 minutes from the time the burst strikes the ground until it dissipates.

An aircraft approaching a microburst will first experience an increasing headwind as it encounters the outflow. The increasing headwind shear causes the indicated airspeed to increase and gives the aircraft a tendency to pitch up and climb. This increase in performance without an increase in power might induce an unwary pilot into reducing power to maintain airspeed and flight path. As the aircraft flies into the core of the microburst the headwind shifts to a downdraft. The sudden loss of headwind will cause indicated airspeed to drop and cause the aircraft to pitch down and descend. The strong downdraft increases the tendency to descend and the aircraft can quickly get into the situation of having low airspeed and a very high rate of descent. As the aircraft flies out the backside of the microburst, it encounters an increasing tailwind shear that further reduces indicated airspeed and performance.

Continued

There are some wind shear conditions that exceed the performance capability of typical air carrier aircraft. For this reason it is imperative that pilots avoid situations where severe wind shear is either reported or is likely to exist. At this time only a couple of airports in the United States have experimental Doppler radar units capable of detecting wind shear. Many airports have the less sophisticated Low-Level Wind Shear Alert System (LLWAS), which is used to alert pilots to the possibility of wind shear on or near the airport. This system consists of wind sensors located around the perimeter of the airport as well as a center field wind sensor. When there is a significant difference in speed or direction between any of these sensors and the center field sensor, the tower will broadcast the difference. A typical tower transmission would be:

> "SOUTH BOUNDARY WIND ONE SIX ZERO AT TWO FIVE, WEST BOUNDARY WIND TWO FOUR ZERO AT THREE FIVE."

The greatest danger from a wind shear encounter at low altitude is that the aircraft will pick up such a high rate of descent that the pilots will be unable to stop it before hitting the ground. The technique to be used during a wind shear encounter essentially involves trading airspeed for altitude. The exact procedures vary from one aircraft to another but if an aircraft encounters severe wind shear, the pilot should maintain or increase the pitch attitude, increase power to the maximum available and accept lower than normal airspeed indications. If this does not arrest the descent, the pilot should continue to pitch up until the descent does stop or until "stick shaker" is encountered.

ALL

9139. Which is a definition of "severe wind shear"?

A—Any rapid change of horizontal wind shear in excess of 25 knots; vertical shear excepted.

B—Any rapid change in wind direction or velocity which causes airspeed changes greater than 15 knots or vertical speed changes greater than 500 ft/min.

C—Any change of airspeed greater than 20 knots which is sustained for more than 20 seconds or vertical speed changes in excess of 100 ft/min.

Severe wind shear is defined as any rapid change in wind direction or velocity which causes airspeed changes greater than 15 knots or vertical speed changes greater than 500 feet per minute. (PLT518) — AC 00-54

Answer (A) is incorrect because a severe wind shear can be caused to both horizontal and vertical shears. Answer (C) is incorrect because a severe wind shear causes airspeed changes greater than 15 knots or vertical speed changes greater than 500 fpm.

ALL

9220. In comparison to an approach in a moderate headwind, which is an indication of a possible wind shear due to a decreasing headwind when descending on the glide slope?

A—Less power is required.

B—Higher pitch attitude is required.

C—Lower descent rate is required.

When a headwind shears to calm or a tailwind, the aircraft tends to lose airspeed, get low, and pitch nose down. The aircraft will require more power and a higher pitch attitude to stay on glide slope. (PLT518) — AC 00-54

Answer (A) is incorrect because as airspeed decreases, more power is required. Answer (C) is incorrect because as the headwind decreases, ground speed will increase, requiring a higher descent rate.

ALL

9133. Which INITIAL cockpit indications should a pilot be aware of when a headwind shears to a calm wind?

A—Indicated airspeed decreases, aircraft pitches up, and altitude decreases.

B—Indicated airspeed increases, aircraft pitches down, and altitude increases.

C—Indicated airspeed decreases, aircraft pitches down, and altitude decreases.

With a headwind shearing to a calm wind there is a loss of lift as airspeed decreases, the aircraft pitches down, and the aircraft drops below glide slope (altitude decreases). Responding promptly by adding power and pitching up, a pilot may overshoot the glide slope and airspeed target but then recover. (PLT518) — AC 00-54

Answer (A) is incorrect because the aircraft will pitch down due to the relatively small angle of attack used during the headwind and the sudden decrease in the airflow over the wing when the wind shears to calm. Answer (B) is incorrect because less power is required to maintain an indicated airspeed in a headwind than in calm air because of ram air; thus, a shear from a headwind to calm would be indicated by a decrease in airspeed and a decrease in altitude.

Answers

9139 [B] 9220 [B] 9133 [C]

ALL

9134. Which condition would INITIALLY cause the indicated airspeed and pitch to increase and the sink rate to decrease?

A—Sudden decrease in a headwind component.
B—Tailwind which suddenly increases in velocity.
C—Sudden increase in a headwind component.

An increase in headwind component (which could also be caused by a tailwind shearing to calm) causes airspeed and pitch to increase, sink rate to decrease. (PLT518) — AC 00-54

Answer (A) is incorrect because a sudden decrease in a headwind component would decrease aircraft performance and would be indicated by a decrease in airspeed, pitch, and altitude. Answer (B) is incorrect because an increase in tailwind velocity would decrease performance and be indicated by a decrease in airspeed, pitch, and altitude.

ALL

9135. Which INITIAL cockpit indications should a pilot be aware of when a constant tailwind shears to a calm wind?

A—Altitude increases; pitch and indicated airspeed decrease.
B—Altitude, pitch, and indicated airspeed decrease.
C—Altitude, pitch, and indicated airspeed increase.

When a tailwind on final shears to calm (or headwind), descent rate decreases. The closest answer suggests altitude decreases, which is still true when one considers the ground speed decreases in this situation. Indicated airspeed and pitch increase. An overshoot can result from insufficient power reduction. (PLT518) — AC 00-54

Answer (A) is incorrect because pitch and indicated airspeed also increase. Answer (B) is incorrect because altitude, pitch, and indicated airspeed decrease when a headwind (not tailwind) shears to a calm wind.

ALL

9137. Which wind-shear condition results in a loss of airspeed?

A—Decreasing headwind or tailwind.
B—Decreasing headwind and increasing tailwind.
C—Increasing headwind and decreasing tailwind.

Decreasing headwind by itself or with a shear to a tailwind will result in loss of indicated airspeed. (PLT518) — AC 00-54

Answer (A) is incorrect because in a decreasing tailwind condition, airspeed initially increases. Answer (C) is incorrect because an increasing headwind and a decreasing tailwind both initially increase airspeed.

ALL

9138. Which wind-shear condition results in an increase in airspeed?

A—Increasing tailwind and decreasing headwind.
B—Increasing tailwind and headwind.
C—Decreasing tailwind and increasing headwind.

A headwind increasing against the pitot and airframe will result in an airspeed increase. (PLT518) — AC 00-54

Answers (A) and (B) are incorrect because when a headwind shears to a tailwind, increasing tailwind component and decreasing headwind component, the reduction of the ram air pressure on the pitot tube causes an initial reduction of indicated airspeed. The reduced headwind component will also cause a pitch down moment and a decrease in altitude.

ALL

9141. Which airplane performance characteristics should be recognized during takeoff when encountering a tailwind shear that increases in intensity?

A—Loss of, or diminished, airspeed performance.
B—Decreased takeoff distance.
C—Increased climb performance immediately after takeoff.

When a tailwind is encountered at liftoff, airspeed will decrease. The pilot must overcome the instinct to lower pitch attitude to recover airspeed or the aircraft may sink beyond recovery limits. Use all available performance by commanding a higher-than-normal pitch attitude and accepting the lower airspeed. (PLT518) — AC 00-54

Answer (B) is incorrect because as a tailwind shear increases, takeoff distance is increased because more power or distance is required to attain lift-off speed. Answer (C) is incorrect because as a tailwind shear increases during climb-out, the climb performance will decrease.

ALL

9142. Thrust is being managed to maintain desired indicated airspeed and the glide slope is being flown. Which characteristics should be observed when a tailwind shears to a constant headwind?

A—PITCH ATTITUDE: Increases. VERTICAL SPEED: Increases. INDICATED AIRSPEED: Decreases, then increases to approach speed.
B—PITCH ATTITUDE: Increases. VERTICAL SPEED: Decreases. INDICATED AIRSPEED: Increases, then decreases.
C—PITCH ATTITUDE: Decreases. VERTICAL SPEED: Decreases. INDICATED AIRSPEED: Decreases, then increases to approach speed.

Answers

9134 [C] 9135 [C] 9137 [B] 9138 [C] 9141 [A] 9142 [B]

Shearing to a headwind will create an increased airspeed condition, causing a pitch up with a vertical speed decrease. (PLT518) — AC 00-54

Answer (A) is incorrect because indicated airspeed will initially increase. Answer (C) is incorrect because pitch and indicated airspeed will both initially increase.

ALL

9166. What is an important characteristic of wind shear?

A—It is primarily associated with the lateral vortices generated by thunderstorms.
B—It usually exists only in the vicinity of thunderstorms, but may be found near a strong temperature inversion.
C—It may be associated with either a wind shift or a windspeed gradient at any level in the atmosphere.

Wind shear may be associated with either a wind shift or a wind speed gradient at any level in the atmosphere. (PLT518) — AC 00-6A, Chapter 9

Answer (A) is incorrect because wind shear can be vertical (as well as lateral) in thunderstorm clouds between the updrafts and downdrafts, as well as in other areas such as frontal zones and low-level temperature inversions. Answer (B) is incorrect because wind shear can be encountered in areas other than thunderstorms; e.g., within a frontal zone, in and near the jet stream, low level inversions.

ALL

9201. Where can the maximum hazard zone caused by wind shear associated with a thunderstorm be found?

A—In front of the thunderstorm cell (anvil side) and on the southwest side of the cell.
B—Ahead of the roll cloud or gust front and directly under the anvil cloud.
C—On all sides and directly under the thunderstorm cell.

Wind shear can be found on all sides of a thunderstorm cell and in the downdraft directly under the cell. (PLT495) — AC 00-54

Answer (A) is incorrect because the wind shear associated with a thunderstorm is on all sides and directly under the cell, not just in the front and on the southwest side. Answer (B) is incorrect because the wind shear associated with a thunderstorm is on all sides and directly under the cell. A roll cloud is not present on all thunderstorms, and when present it marks the eddies of the shear zone between the downdraft and surrounding air.

ALL

9225. Which is a necessary condition for the occurrence of a low-level temperature inversion wind shear?

A—The temperature differential between the cold and warm layers must be at least 10°C.
B—A calm or light wind near the surface and a relatively strong wind just above the inversion.
C—A wind direction difference of at least 30° between the wind near the surface and the wind just above the inversion.

When taking off or landing in calm wind under clear skies within a few hours before or after sunrise, a pilot should be prepared for a temperature inversion near the ground. A shear zone in the inversion is relatively certain if the wind at 2,000 to 4,000 is 25 knots or more. (PLT501) — AC 00-6A, Chapter 9

Answer (A) is incorrect because magnitude of temperature differential in the inversion is not important; the wind shear is caused by the variation in wind speed. Answer (C) is incorrect because surface wind and a relatively strong wind just above the inversion, not a wind direction difference of at least 30°, are needed to form a low-level temperature inversion wind shear. The wind shear is caused by wind speed variation, not variation in wind direction.

ALL

9701. The horizontal wind shear, critical for turbulence (moderate or greater) per 150 miles is

A—18 knots or less.
B—greater than 18 knots.
C—not a factor, only vertical shear is a factor.

Horizontal wind shear can be determined from the spacing of isotachs. The horizontal wind shear critical for turbulence (moderate or greater) is greater than 18 knots per 150 miles. 150 nautical miles is equal to 2-1/2 degrees latitude. (PLT263) — AC 00-45

ALL

9130. What is the expected duration of an individual microburst?

A—Two minutes with maximum winds lasting approximately 1 minute.
B—One microburst may continue for as long as 2 to 4 hours.
C—Seldom longer than 15 minutes from the time the burst strikes the ground until dissipation.

Wind speeds intensify for about 5 minutes after a microburst initially contacts the ground. An encounter during the initial stage of microburst development may not be considered significant, but an airplane following

Answers

9166 [C]	9201 [C]	9225 [B]	9701 [B]	9130 [C]

may experience an airspeed change two to three times greater. Microbursts typically dissipate within 10 to 20 minutes after ground contact. (PLT317) — AC 00-54

ALL

9131. Maximum downdrafts in a microburst encounter may be as strong as

A—8,000 ft/min.
B—7,000 ft/min.
C—6,000 ft/min.

The downdrafts can be as strong as 6,000 feet per minute. Horizontal winds near the surface can be as strong as 45 knots resulting in a 90-knot shear across the microburst. (PLT317) — AIM ¶7-1-26

ALL

9132. An aircraft that encounters a headwind of 40 knots, within a microburst, may expect a total shear across the microburst of

A—40 knots.
B—80 knots.
C—90 knots.

With a headwind of 40 knots, the pilot may expect a total shear of 80 knots across the microburst. (PLT317) — AIM ¶7-1-26

Answers (A) and (C) are incorrect because the total shear is the total headwind to tailwind change of a traversing airplane, thus a 40-knot headwind would shear 80 knots to a 40-knot tailwind.

ALL

9140. Doppler wind measurements indicate that the windspeed change a pilot may expect when flying through the peak intensity of a microburst is approximately

A—15 knots.
B—25 knots.
C—45 knots.

The downdrafts can be as strong as 6,000 feet per minute. Horizontal winds near the surface can be as strong as 45 knots resulting in a 90-knot shear across the microburst. (PLT317) — AIM ¶7-1-26

ALL

9143. Maximum downdrafts in a microburst encounter may be as strong as

A—8,000 ft/min.
B—7,000 ft/min.
C—6,000 ft/min.

The downdrafts can be as strong as 6,000 feet per minute. Horizontal winds near the surface can be as strong as 45 knots resulting in a 90-knot shear across the microburst. (PLT317) — AIM ¶7-1-26

ALL

9144. An aircraft that encounters a headwind of 45 knots, within a microburst, may expect a total shear across the microburst of

A—40 knots.
B—80 knots.
C—90 knots.

With a headwind of 45 knots, the pilot may expect a total shear of 90 knots across the microburst. (PLT317) — AIM ¶7-1-26

Answer (A) is incorrect because the total shear is the total headwind to tailwind change of a traversing airplane, thus a 45-knot headwind would shear 90 knots, not 40 knots, to a 45-knot tailwind. Answer (B) is incorrect because the total shear is the total headwind to tailwind change of a traversing airplane, thus a 45-knot headwind would shear 90 knots, not 80 knots, to a 45-knot tailwind.

ALL

9145. (Refer to Figure 144.) If involved in a microburst encounter, in which aircraft positions will the most severe downdraft occur?

A—4 and 5.
B—2 and 3.
C—3 and 4.

An airplane flying through the microburst as depicted in FAA Figure 144 would encounter increasing performance in position 1, followed by a decreasing headwind in position 2. At position 3 the aircraft would encounter the strong downdraft followed by a strong tailwind at position 4. Position 5 represents the situation just before ground contact. (PLT317) — AIM ¶7-1-26

Answer (A) is incorrect because position 5 has significantly less downdraft even though it has considerably more tailwind. Answer (B) is incorrect because position 2 has not as significant a downdraft as 3 and 4, but it contains a significant headwind even though it is decreasing.

Answers

9131 [C]	9132 [B]	9140 [C]	9143 [C]	9144 [C]	9145 [C]

ALL

9146. (Refer to Figure 144.) When penetrating a microburst, which aircraft will experience an increase in performance without a change in pitch or power?

A—3.
B—2.
C—1.

An airplane flying through the microburst as depicted in FAA Figure 144 would encounter increasing performance in position 1, followed by a decreasing headwind in position 2. At position 3 the aircraft would encounter the strong downdraft followed by a strong tailwind at position 4. Position 5 represents the situation just before ground contact. (PLT317) — AIM ¶7-1-26

Answer (A) is incorrect because position 3 indicates where the most severe downdraft occurs, which results in a decrease in performance. Answer (B) is incorrect because position 2 does not have as significant a headwind component as position 1, and thus performance is less than at position 1.

ALL

9147. (Refer to Figure 144.) What effect will a microburst encounter have upon the aircraft in position 3?

A—Decreasing headwind.
B—Increasing tailwind.
C—Strong downdraft.

An airplane flying through the microburst as depicted in FAA Figure 144 would encounter increasing performance in position 1, followed by a decreasing headwind in position 2. At position 3 the aircraft would encounter the strong downdraft followed by a strong tailwind at position 4. Position 5 represents the situation just before ground contact. (PLT317) — AIM ¶7-1-26

Answer (A) is incorrect because at position 2, not 3, the airplane encounters decreasing headwind. Answer (B) is incorrect because at position 5 the airplane encounters an increasing tailwind and it may result in an extreme situation as pictured, i.e., just before impact.

ALL

9148. (Refer to Figure 144.) What effect will a microburst encounter have upon the aircraft in position 4?

A—Strong tailwind.
B—Strong updraft.
C—Significant performance increase.

An airplane flying through the microburst as depicted in FAA Figure 144 would encounter increasing performance in position 1, followed by a decreasing headwind in position 2. At position 3 the aircraft would encounter the strong downdraft followed by a strong tailwind at position 4. Position 5 represents the situation just before ground contact. (PLT317) — AIM ¶7-1-26

Answer (B) is incorrect because updrafts will occur in thunderstorms, not in microbursts from thunderstorms. Answer (C) is incorrect because the significant increase in performance occurs at position 1 where the headwind component is the greatest.

ALL

9149. (Refer to Figure 144.) How will the aircraft in position 4 be affected by a microburst encounter?

A—Performance increasing with a tailwind and updraft.
B—Performance decreasing with a tailwind and downdraft.
C—Performance decreasing with a headwind and downdraft.

An airplane flying through the microburst as depicted in FAA Figure 144 would encounter increasing performance in position 1, followed by a decreasing headwind in position 2. At position 3 the aircraft would encounter the strong downdraft followed by a strong tailwind at position 4. Position 5 represents the situation just before ground contact. (PLT317) — AIM ¶7-1-26

Answer (A) is incorrect because performance will decrease with a tailwind, and thunderstorms (not microbursts) will have updrafts. Answer (C) is incorrect because the airplane at position 2 indicates where performance will decrease due to a headwind and downdraft.

ALL

9150. What is the expected duration of an individual microburst?

A—Two minutes with maximum winds lasting approximately 1 minute.
B—One microburst may continue for as long as 2 to 4 hours.
C—Seldom longer than 15 minutes from the time the burst strikes the ground until dissipation.

An individual microburst will seldom last longer than 15 minutes from the time it strikes the ground until dissipation. The horizontal winds continue to increase during the first 5 minutes with the maximum intensity winds lasting approximately 2 to 4 minutes. (PLT317) — AIM ¶7-1-26

Answer (A) is incorrect because microbursts last 15, not 2 minutes, and maximum winds last 2 to 4 minutes, not 1 minute. Answer (B) is incorrect because the maximum winds last 2 to 4 minutes, not 2 to 4 hours, and the microburst is usually limited to about 15 minutes.

Answers

9146 [C] 9147 [C] 9148 [A] 9149 [B] 9150 [C]

ALL

9167. What information from the control tower is indicated by the following transmission?

"SOUTH BOUNDARY WIND ONE SIX ZERO AT TWO FIVE, WEST BOUNDARY WIND TWO FOUR ZERO AT THREE FIVE."

A—A downburst is located at the center of the airport.
B—Wake turbulence exists on the west side of the active runway.
C—There is a possibility of wind shear over or near the airport.

The Low-Level Wind Shear Alert System (LLWAS) is a computerized system which detects the presence of a possible hazardous low-level wind shear by continuously comparing the winds measured by sensors installed around the periphery on an airport with the wind measured at the center field location. If the difference between the center field wind sensor and peripheral sensor becomes excessive, a thunderstorm or thunderstorm gust front wind shear is probable. When this condition exists, the tower controller will provide arrival and departure aircraft with an advisory of the situation which includes the center field wind plus the remote location and wind. The broadcast quoted in the question is an example of this type of advisory. (PLT044) — AIM ¶4-3-7

Answer (A) is incorrect because a downburst is a vertical movement of air which is not measured by the LLWAS until it has horizontal movement. Also the wind direction is toward the center of the airport, not away from it. Answer (B) is incorrect because wake turbulence does not produce wind. It is generated by an aircraft that is producing lift, which could be on either side of the active runway.

ALL

9136. What is the recommended technique to counter the loss of airspeed and resultant lift from wind shear?

A—Lower the pitch attitude and regain lost airspeed.
B—Avoid overstressing the aircraft, "pitch to airspeed," and apply maximum power.
C—Maintain, or increase, pitch attitude and accept the lower-than-normal airspeed indications.

Pitch attitude must be maintained or increased even when lower-than-normal airspeed indications are required. (PLT518) — AC 00-54

Answer (A) is incorrect because lowering the pitch attitude to regain lost airspeed is a result of past training emphasis on airspeed control, not recovering from a wind shear. Answer (B) is incorrect because the recommended technique to recover from a wind shear is to maintain or increase pitch attitude and not "pitch to airspeed," which may decrease pitch to regain lost airspeed.

Frost and Ice

No person may dispatch or release an aircraft, continue to operate en route, or land when in the opinion of the pilot-in-command or aircraft dispatcher, icing conditions are expected or met that might adversely affect the safety of the flight. No person may takeoff when frost, snow or ice is adhering to the wings, control surfaces or propellers of the aircraft.

Deicing is a procedure in which frost, ice, or snow is removed from the aircraft in order to provide clean surfaces. Anti-icing is a process that provides some protection against the formation of frost or ice for a limited period of time.

The equipment most commonly used for **deicing** and **anti-icing** airplanes on the ground is the truck-mounted mobile deicer/anti-icer. The two basic types of fluids used are Type 1 (unthickened) fluids and Type 2 (thickened) fluids. Type 1 fluids have a minimum 80% glycol content and a relatively low viscosity, except at very low temperatures. The viscosity of Type 1 fluids depends only on temperature. The holdover time is relatively short for Type 1 fluids. Type 2 fluids have a significantly higher holdover time. Type 2 fluids have a minimum glycol content of 50% with 45% to 50% water plus thickeners and inhibitors. Water decreases the freeze point. The freeze point should be no greater than 20°F below ambient or surface temperature, whichever is less.

There is a one-step process and a two-step process for deicing and anti-icing. The one-step process uses heated fluid to remove snow, ice and frost. The primary advantage of this process is that it is quick and uncomplicated. However, where large deposits of snow or ice must be flushed off, fluid usage will be greater than with the two-step process. The two-step process consists of separate deicing and anti-

Answers

9167 [C]   9136 [C]

icing steps. A diluted fluid, usually heated, is used to deice and a more concentrated fluid (either 100% or diluted, depending on the weather), usually cold, is used to anti-ice. Type 1 or 2 fluids can be used for both steps, or Type 1 for step 1 and Type 2 for step 2.

Two precautions to observe when using this equipment are:

1. Do not spray deice/anti-ice fluid at or into pitot inlets, TAT probes, or static ports; and
2. Apply deice/anti-ice fluid on pressure relief doors, lower door sills, and bottom edges of doors prior to closing for flight.

Icing

For ice to form, there must be moisture present in the air and the air must be cooled to a temperature of 0°C (32°F) or less. Aerodynamic cooling can lower the temperature of an airfoil to 0°C even though the ambient temperature is a few degrees warmer.

Ice is identified as clear, rime, or mixed. *Rime ice* forms if the droplets are small and freeze immediately when contacting the aircraft surface. This type of ice usually forms on areas such as the leading edges of wings or struts. It has a somewhat rough looking appearance and is a milky white color. *Clear ice* is usually formed from larger water droplets or freezing rain that can spread over a surface. This is the most dangerous type of ice since it is clear, hard to see, and can change the shape of the airfoil. *Mixed ice* is a mixture of clear ice and rime ice. It has the bad characteristics of both types and can form rapidly.

There are two kinds of icing that are significant to aviation: structural icing and induction icing. *Structural icing* refers to the accumulation of ice on the exterior of the aircraft; *induction icing* affects the powerplant operation. Structural icing occurs on an aircraft whenever supercooled droplets of water make contact with any part of the aircraft that is also at a temperature below freezing.

One inflight condition necessary for structural icing is visible moisture (clouds or raindrops). *Freezing rain* always occurs in a temperature inversion. As the rain falls through air that is below freezing, its temperature begins to fall below freezing yet it does not freeze solid—i.e., freezing rain. The process requires the temperature of the rain to be above freezing before it becomes supercooled. Eventually, the water drops will freeze into ice pellets. Any encounter with ice pellets in flight indicates that there is freezing rain at a higher altitude.

Aircraft structural ice will most likely have the highest accumulation in freezing rain; therefore, an operational consideration if you fly into rain which freezes on impact is that temperatures are above freezing at some higher altitude.

Hazards of Structural Icing

The most hazardous aspect of structural icing is its aerodynamic effects. Ice can alter the shape of an airfoil. This can cause control problems, change the angle of attack at which the aircraft stalls, and cause the aircraft to stall at a significantly higher airspeed. Ice can reduce the amount of lift that an airfoil will produce and increase the amount of drag by several times. It can partially block or limit control surfaces, which will limit or make control movements ineffective. If the extra weight caused by ice accumulation is too great, the aircraft might not be able to become airborne, and if in flight, might not be able to maintain altitude.

For this reason, regulations prohibit takeoff when snow, ice, or frost is adhering to wings, propellers, or control surfaces of an aircraft. Yet another hazard of structural icing is the possible uncommanded and uncontrolled roll phenomenon referred to as "roll upset," which is associated with severe inflight icing. Therefore, pilots flying airplanes certificated for flight in known icing conditions should be aware that severe icing is a condition that is outside of the airplane's certificated icing envelope.

Structural icing can also cause tailplane (empennage) stall. The tail can collect ice faster than the wing and because it is not visible to the pilot inflight, the situation could go undetected. A tailplane stall occurs when, same as with the wing, the critical angle of attack is exceeded. Since the horizontal stabilizer counters the natural nose-down tendency caused by the center of lift of the main wing, the airplane will react by pitching down, sometimes uncontrollably, when the tailplane is stalled. Application of flaps can aggravate or initiate the stall.

Because of this, the pilot should use caution when applying flaps during an approach if there is the possibility of icing on the tailplane. Ice buildup will cause the airplane to require more power to maintain cruise airspeed. Ice on the tailplane can cause diminished nose-up pitch control and heavy elevator forces, and the aircraft may buffet if flaps are applied. Ice on the rudder or ailerons can cause control oscillations or vibrations.

For an airplane to be approved for flight into icing conditions, the airplane must be equipped with systems that will adequately protect various components. Not all airplanes with these components are approved for flight into known icing; check your POH to know if your airplane has been certificated to operate in known icing conditions.

Frost Formation

Frost is described as ice deposits formed by sublimation on a surface when the temperature of the collecting surface is at or below the dew point of the adjacent air and the dew point is below freezing. Frost causes early airflow separation on an airfoil resulting in a loss of lift. Therefore, all frost should be removed from the lifting surfaces of an airplane before flight or it may prevent the airplane from becoming airborne.

Snow always forms in colder than freezing temperatures by the process of sublimation. This is when water goes straight from its vapor state into ice without ever being a liquid. Wet snow occurs when it falls to altitudes with above freezing temperatures and begins to melt.

Test data indicate that ice, snow, or frost formations having a thickness and surface roughness similar to medium or course sandpaper on the leading edge and upper surface of a wing can reduce wing lift by as much as 30% and increase drag by 40%.

ALL
9440. Which is an effect of ice, snow, or frost formation on an airplane?

A—Increased stall speed.
B—Increased pitchdown tendencies.
C—Increased angle of attack for stalls.

Aircraft with ice, snow, or frost on the wings may experience increased stall speed, decreased angle of attack for stalls, and increased pitchup tendencies. (PLT493) — AC 20-117

ALL
9449. Clear ice generally forms in outside temperature ranges of

A—-15 to -25°C.
B—0 to -10°C.
C—colder than -25°C.

Temperatures close to the freezing point, large amounts of liquid water, high aircraft velocities, and large droplets are conducive to the formation of clear ice. (PLT493) — AC 91-51

Answers

9440 [A]   9449 [B]

ALL

9451. Test data indicate that ice, snow, or frost having a thickness and roughness similar to medium or coarse sandpaper on the leading edge and upper surface of a wing can

A—reduce lift by as much as 40 percent and increase drag by 30 percent.
B—increase drag and reduce lift by as much as 40 percent.
C—reduce lift by as much as 30 percent and increase drag by 40 percent.

Test data indicate that ice, snow, or frost formations having a thickness and surface roughness similar to medium or coarse sandpaper on the leading edge and upper surface of a wing can reduce wing lift by as much as 30 percent and increase drag by 40 percent. (PLT128) — AC 120-58

ALL

9695. The adverse effects of ice, snow, or frost on aircraft performance and flight characteristics include decreased lift and

A—increased thrust.
B—a decreased stall speed.
C—an increased stall speed.

Ice, frost, or snow on an aircraft can cause decreased lift, increased stall speed, and loss of thrust. (PLT493) — AC 120-58

ALL

9441. Which is a disadvantage of the one-step over the two-step process when deicing/anti-icing an airplane?

A—It is more complicated.
B—The holding time is increased.
C—More fluid is used with the one-step method when large deposits of ice and snow must be flushed off airplane surfaces.

Use the two-stage process to remove ice deposits with hot water or a mix of FPD (Freezing Point Depressant) and water. This reduces the amount of fluid required. (PLT108) — AC 120-58

Answer (A) is incorrect because the one-step process is less complicated. Answer (B) is incorrect because one of the advantages of the one-step process is increased holding time.

ALL

9442. The purpose of diluting ethylene glycol deicing fluid with water in non-precipitation conditions is to

A—raise the eutectic point.
B—decrease the freeze point.
C—increase the minimum freezing point (onset of crystallization).

Pure ethylene glycol will freeze at warmer temperatures than aqueous solutions of ethylene glycol. (PLT108) — AC 20-117

Answer (A) is incorrect because diluting ethylene glycol lowers the eutectic point. Answer (C) is incorrect because diluting ethylene glycol decreases the minimum freezing point (onset of crystallization).

ALL

9443. Which procedure increases holding time when deicing/anti-icing an airplane using a two-step process?

A—Heated Type 1 fluid followed by cold Type 2 fluid.
B—Cold Type 2 fluid followed by hot Type 2 fluid.
C—Heated Type 1 or 2 fluid followed by cold Type 1 fluid.

Type 2 fluid is applied cold to increase its thickness and increase holding time. (PLT108) — AC 120-58

Answer (B) is incorrect because cold Type 2 would not be an effective deicer. Answer (C) is incorrect because step 2 should be Type 2.

ALL

9444. Which of the following will decrease the holding time during anti-icing using a two-step process?

A—Apply heated Type 2 fluid.
B—Decrease the water content.
C—Increase the viscosity of Type 1 fluid.

Heating fluids increases their deicing effectiveness; however, in the anti-icing process, unheated fluids are more effective. (PLT108) — AC 120-58

Answer (B) is incorrect because decreasing the water content will increase the holding time. Answer (C) is incorrect because increasing the viscosity of Type 1 fluid will increase the holding time.

Answers

9451 [C]	9695 [C]	9441 [C]	9442 [B]	9443 [A]	9444 [A]

ALL

9752. Pilots should check for ice accumulation prior to flight by

A—using a flashlight to reflect off a white wing.
B—using ice detection lights.
C—feeling the control surface, especially the leading edges.

Early ice detection is critical and is particularly difficult during night flight. Use a flashlight to check for ice accumulation on the wings. (PLT493) — AC 91-51

Answer (B) is incorrect because ice detection lights are used inflight to monitor wing accumulation at night. Answer (C) is incorrect because not all control surfaces are within physical reach, particularly the tailplane.

ALL

9753. When icing is detected, particularly while operating an aircraft without deicing equipment, the pilot should

A—fly to an area with liquid precipitation.
B—fly to a lower altitude.
C—leave the area of precipitation or go to an altitude where the temperature is above freezing.

When icing is detected, a pilot should do one of two things, particularly if the aircraft is not equipped with deicing equipment: leave the area of precipitation or go to an altitude where the temperature is above freezing. This "warmer" altitude may not always be a lower altitude. Proper preflight action includes obtaining information on the freezing level and the above-freezing levels in precipitation areas. (PLT493) — FAA-H-8083-15

Answer (A) is incorrect because pilots should leave the area of any precipitation; even liquid precip can result in an icing scenario. Answer (B) is incorrect because a "warmer" altitude may not always be a lower altitude.

ALL

9754. Tailplane icing can be detected by

A—a slow and steady decrease in altitude.
B—flaps failing to operate.
C—a sudden change in elevator force or uncommanded nose-down pitch.

Since the tailplane is ordinarily thinner than the wing, it is a more efficient collector of ice. It is important the pilot be alert to the possibility of a tailplane stall, particularly on approach and landing. Any of the following symptoms, occurring singly or in combination, may be a warning of tailplane icing: elevator control pulsing, oscillations or vibrations; abnormal nose-down trim change; reduction or loss of elevator effectiveness; sudden change in elevator force; sudden uncommanded nose-down pitch. (PLT493) — FAA-H-8083-15

Answer (A) is incorrect because tailplane ice or a taiplane stall typically results in a rapid change in pitch. Answer (B) is incorrect because flaps may operate even with tailplane icing, further aggravating or initiating a stall.

ALL

9755. If tailplane icing or a tailplane stall is detected, the pilot should

A—lower the flaps to decrease airspeed.
B—decrease power to V_{FE}.
C—retract flaps and increase power.

If a tailplane stall is suspected, the pilot should immediately retract flaps to the previous setting and apply appropriate nose-up elevator pressure; increase airspeed appropriately for the reduced flap extension setting; apply sufficient power for aircraft configuration and conditions; make nose-down pitch changes slowly; and if a pneumatic deicing system is used, operate the system several times in an attempt to clear the tailplane of ice. (PLT493) — FAA-H-8083-15

Answer (A) is incorrect because flaps should not be used if tailplane ice is suspected. Answer (B) is incorrect because power should be increased if tailplane ice is suspected.

ALL

9756. If icing is suspected on an airplane equipped with deicing equipment, the pilot should

A—first confirm ice with the ice light prior to deploying the pneumatic boots.
B—operate the pneumatic deicing system several times to clear the ice.
C—operate the pneumatic deicing system once to allow time for the ice removal.

Pneumatic boots are one method capable of removing ice from an aircraft surface. This system is commonly used on smaller aircraft and usually provides ice removal for the wing and tail section by inflating a rubber boot. (PLT493) — FAA-H-8083-15

Answer (A) is incorrect because the deicing system should be used as soon as icing is suspected. Answer (C) is incorrect because the pneumatic boots should be inflated/deflated several times to try to remove the ice.

Answers

9752 [A]	9753 [C]	9754 [C]	9755 [C]	9756 [B]

ALL

9757. The first place ice is likely to form on an aircraft is the

A—wings.
B—tailplane.
C—windshield.

Small and/or narrow objects are the best collectors of droplets and ice up most rapidly. This is why a small protuberance within sight of the pilot can be used as an "ice evidence probe." It is generally one of the first parts of the airplane on which an appreciable amount of ice forms. An aircraft's tailplane is a better collector than its wings, because the tailplane presents a thinner surface to the airstream. (PLT493) — FAA-H-8083-15

Answer (A) is incorrect because the wings are thicker than the tailplane so not as likely to first build up with ice. Answer (C) is incorrect because the windshield does not protrude into the airstream as much as the control surfaces so less likely than the tailplane to develop ice.

ALL

9445. What should the deice/anti-ice fluid temperature be during the last step of a two-phase process?

A—Hot.
B—Warm.
C—Cold.

The two-step procedure involves both deicing and anti-icing. Deicing is accomplished with hot water or a hot mixture of FPD and water. The ambient weather conditions and the type of accumulation to be removed from the aircraft must be considered when determining which type of deicing fluid to use. The second (anti-icing) step involves applying a mixture of SAE or ISO Type 2 and water to the critical surfaces of the aircraft. (PLT108) — AC 120-58

Answers (A) and (B) are incorrect because heated fluids are used during the first step of a two-phase process.

ALL

9446. What is the minimum glycol content of Type 1 deicing/anti-icing fluid?

A—30 percent.
B—50 percent.
C—80 percent.

SAE and ISO Type 1 fluids in the concentrated form contain a minimum of 80 percent glycols and are considered "unthickened" because of their relatively low viscosity. (PLT108) — AC 120-58

ALL

9447. What is the minimum glycol content of Type 2 deicing/anti-icing fluid?

A—30 percent.
B—50 percent.
C—80 percent.

SAE and ISO Type 2 fluids contain a minimum of 50 percent glycols and are considered "thickened" because of added thickening agents that enable the fluid to be deposited in a thicker film and to remain on the aircraft surfaces until the time for takeoff. (PLT108) — AC 120-58

ALL

9448. Anti-icing fluid should provide freezing point protection to

A—-20°F ambient temperature.
B—+32°F outside temperature or below.
C—a freezing point no greater than 20°F below the ambient or airplane surface temperature.

In any case the freezing point of residual fluids (water, FPD fluids or mixtures) should not be greater than 20°F below ambient or surface temperature, whichever is less. (PLT108) — AC 20-117

Answers (A) and (B) are incorrect because anti-icing fluid should protect from icing to a freezing point no greater than 20°F below the ambient or aircraft surface temperature.

ALL

9450. Freezing Point Depressant (FPD) fluids used for deicing

A—provide ice protection during flight.
B—are intended to provide ice protection on the ground only.
C—on the ground, cause no performance degradation during takeoff.

FPD fluids are used to aid the ground deicing process and provide a protective film to delay formations of frost, snow, or other ice. (PLT108) — AC 120-58

Answer (A) is incorrect because FPD does not provide inflight protection. Answer (C) is incorrect because some large aircraft experience performance degradation and may require weight or other compensation.

Answers

9757 [B]	9445 [C]	9446 [C]	9447 [B]	9448 [C]	9450 [B]

ALL

9452. Snow on top of deicing or anti-icing fluids

A—need not be considered as adhering to the aircraft.
B—must be considered as adhering to the aircraft.
C—must be considered as adhering to the aircraft, but a safe takeoff can be made as it will blow off.

FPD fluids are highly soluble in water; however, ice is slow to absorb FPD or to melt when in contact with it. If frost, ice, or snow is adhering to an aircraft surface, the formation may be melted by repeated application of proper quantities of FPD fluid. This process can be significantly accelerated by thermal energy from heated fluids. As the ice melts, the FPD mixes with the water thereby diluting the FPD. As dilution occurs, the resulting mixture may begin to run off. If all the ice is not melted, additional applications of FPD become necessary until the fluid penetrates to the aircraft surface. When all the ice has melted, the remaining liquid residue is a mixture of water and FPD. The resulting film could freeze (begin to crystallize) with only a slight temperature decrease. (PLT108) — AC 120-58

Answer (A) is incorrect because snow (ice) needs to be considered as adhering to the aircraft (see explanation). Answer (C) is incorrect because snow may not necessarily blow off during takeoff.

ALL

9453. Freezing Point Depressant (FPD) fluids are highly soluble in water; however,

A—ice is slow to absorb it but fast to melt when in contact with FPD.
B—ice absorbs it very fast but is slow to melt when in contact with it.
C—ice is slow to absorb it, and to melt when in contact with it.

FPD fluids are highly soluble in water; however, ice is slow to absorb FPD or to melt when in contact with it. (PLT108) — AC 120-58

Answer (A) is incorrect because FPD fluids are slow to melt. Answer (B) is incorrect because FPD fluids are slow to absorb.

ALL

9454. Freezing Point Depressant (FPD) fluid residue on engine fan or compressor blades

A—can increase performance and cause stalls or surges.
B—could cause FDP vapors to enter the aircraft but would have no affect on engine thrust or power.
C—can reduce engine performance and cause surging and/or compressor stalls.

Fluid residue on engine fan or compressor blades can reduce engine performance or cause stall or surge. (PLT108) — AC 120-58

Answer (A) is incorrect because fluid residue would cause a decrease in performance. Answer (B) is incorrect because fluid residue would have an affect on engine thrust or power.

ALL

9698. The practice developed and accepted by the North American air carrier industry using traditional North American fluids is to ensure that the freeze point of the remaining film is below ambient temperature by at least

A—10°F.
B—20°F.
C—20°C.

As it is applied, deicing fluid is often diluted by melted snow and ice. It is standard practice to ensure that the remaining film of diluted fluid has a freeze point at least 20°F below the ambient temperature. (PLT108) — AC 120-58

ALL

9700. What is the effect of Freezing Point Depressant (FPD) fluid residue on engine fan or compressor blades?

A—could cause FPD vapors to enter the aircraft but would have no affect on engine thrust or power.
B—It can increase performance and cause stalls or surges.
C—It can reduce engine performance and cause surging and/or compressor stalls.

Fluid residue on engine fan or compressor blades can reduce engine performance or cause stall or surge. In addition, this could increase the possibility of, or the quantity of, glycol vapors entering the aircraft through the engine bleed air system. (PLT108) — AC 120-58

Answers

9452 [B]	9453 [C]	9454 [C]	9698 [B]	9700 [C]

ALL

9183. What is a feature of supercooled water?

A—The water drop sublimates to an ice particle upon impact.
B—The unstable water drop freezes upon striking an exposed object.
C—The temperature of the water drop remains at 0°C until it impacts a part of the airframe, then clear ice accumulates.

Rain or drizzle is always formed in temperatures which are above freezing. Rain falling through colder air may become supercooled, freezing on impact as freezing rain. (PLT512) — AC 00-6A, Chapter 5

Answer (A) is incorrect because sublimation is the process of changing water vapor to ice crystals (not liquid water to ice). Answer (C) is incorrect because supercooled water temperature is below 0°C.

ALL

9221. What condition is necessary for the formation of structural icing in flight?

A—Supercooled water drops.
B—Water vapor.
C—Visible water.

For structural icing to form, the aircraft must be flying through visible moisture and the temperature where the moisture strikes the aircraft must 0°C or colder. Note that the moisture does not need to be supercooled. (PLT274) — AC 00-6A, Chapter 10

Answer (A) is incorrect because supercooled water drops increase the rate of icing, but are not a condition necessary for the formation of structural icing. Answer (B) is incorrect because water must be visible, not in a gaseous (vapor) state.

ALL

9224. Which type of icing is associated with the smallest size of water droplet similar to that found in low-level stratus clouds?

A—Clear ice.
B—Frost ice.
C—Rime ice.

Rime ice forms when drops are small, such as those in stratified clouds or light drizzle. (PLT274) — AC 00-6A, Chapter 10

Answer (A) is incorrect because clear ice forms when drops are large, not small, as found in rain or cumuliform clouds. Answer (B) is incorrect because frost is not a structural icing condition found in flight; it happens to airplanes parked on the ground as well.

ALL

9161. Freezing rain encountered during climb is normally evidence that

A—a climb can be made to a higher altitude without encountering more than light icing.
B—a layer of warmer air exists above.
C—ice pellets at higher altitudes have changed to rain in the warmer air below.

Rain or drizzle is always formed in temperatures which are above freezing. Rain falling through colder air may become supercooled, freezing on impact as freezing rain. (PLT512) — AC 00-6A, Chapter 5

Answer (A) is incorrect because freezing rain only means that a layer of warmer air exists above; it does not indicate the amount of icing that may be encountered during a climb. Answer (C) is incorrect because freezing rain is formed by rain falling through colder air, not from ice pellets melting through warmer air.

ALL

9223. Which type precipitation is an indication that supercooled water is present?

A—Wet snow.
B—Freezing rain.
C—Ice pellets.

Rain or drizzle is always formed in temperatures which are above freezing. Rain falling through colder air may become supercooled, freezing on impact as freezing rain. (PLT344) — AC 00-6A, Chapter 10

Answer (A) is incorrect because wet snow is an indication that temperature is above freezing at the present level. Answer (C) is incorrect because ice pellets indicate that water has frozen, not that is has become supercooled.

ALL

9180. What condition is indicated when ice pellets are encountered during flight?

A—Thunderstorms at higher levels.
B—Freezing rain at higher levels.
C—Snow at higher levels.

Rain or drizzle is always formed in temperatures which are above freezing. Rain falling through colder air may become supercooled, freezing on impact as freezing rain. As it continues to fall in the freezing temperature, it will form into ice pellets. (PLT493) — AC 00-6A, Chapter 5

Answer (A) is incorrect because ice pellets always indicate freezing rain, not thunderstorms, at higher altitudes. Answer (C) is incorrect because freezing rain, not snow, is indicated at higher altitude when ice pellets are encountered.

Answers

9183 [B] 9221 [C] 9224 [C] 9161 [B] 9223 [B] 9180 [B]

ALL

9774. Which of the following weather conditions are conducive to inflight icing?

A—Droplets that splash or splatter on impact at temperatures below -10°C ambient temperature.
B—Droplets that impact at temperatures below -20°C ambient air temperature.
C—Droplets that splash or splatter on impact at temperatures below 10°C and above 1°C ambient temperature.

The following weather conditions may be conducive to severe in-flight icing: visible rain at temperatures below 0°C ambient air temperature; droplets that splash or splatter on impact at temperatures below 0°C ambient air temperature. (PLT274) — AC 91-51

ALL

9775. Which of the following weather conditions are conducive to inflight icing?

A—Visible rain with temperatures below 0°C.
B—Visible rain with temperatures below 10°C.
C—Visible moisture with temperatures below 5°C.

The following weather conditions may be conducive to severe in-flight icing: visible rain at temperatures below 0°C ambient air temperature; droplets that splash or splatter on impact at temperatures below 0°C ambient air temperature. (PLT274) — AC 91-51

ALL

9162. What temperature condition is indicated if precipitation in the form of wet snow occurs during flight?

A—The temperature is above freezing at flight altitude.
B—The temperature is above freezing at higher altitudes.
C—There is an inversion with colder air below.

Snowflakes are formed by sublimation in below-freezing temperatures. If the snow falls into an area of above-freezing temperatures it will start to melt, become wet snow and eventually turn into rain. (PLT493) — AC 00-6A, Chapter 5

Answer (B) is incorrect because wet snow indicates above-freezing temperature at flight level, not at higher altitudes. The temperature was below freezing at the altitudes where the snow formed. Answer (C) is incorrect because wet snow indicates falling snow that has begun to melt due to above-freezing temperature at flight level. An inversion may or may not be the cause of the warmer air.

ALL

9179. Which conditions result in the formation of frost?

A—The temperature of the collecting surface is at or below freezing and small droplets of moisture are falling.
B—Dew collects on the surface and then freezes because the surface temperature is lower than the air temperature.
C—Temperature of the collecting surface is below the dewpoint and the dewpoint is also below freezing.

Frost forms when both the temperature and the dew point of the collecting surface are below freezing. When this occurs, water vapor sublimates directly into frost. This condition most often occurs on clear nights with little or no wind. (PLT493) — AC 00-6A, Chapter 5

Answer (A) is incorrect because moisture that falls on a collecting surface that is at or below freezing will form ice. Answer (B) is incorrect because frozen dew is hard and transparent, while frost is the sublimation of vapor into ice, and is white and opaque.

ALL

9748. The following weather condition may be conducive to severe in-flight icing:

A—visible rain at temperatures below 0°C ambient air temperature.
B—visible moisture at temperatures below 5°C ambient temperature.
C—visible rain at temperatures below 10°C ambient temperature.

Visible rain at temperatures below 0°C ambient air temperature are conditions conducive to severe inflight icing. (PLT274) — Airworthiness Directives

ALL

9181. When will frost most likely form on aircraft surfaces?

A—On clear nights with stable air and light winds.
B—On overcast nights with freezing drizzle precipitation.
C—On clear nights with convective action and a small temperature/dewpoint spread.

Frost forms when both the temperature and the dew point of the collecting surface are below freezing. When this occurs, water vapor sublimates directly into frost. This condition most often occurs on clear nights with little or no wind. (PLT493) — AC 00-6A, Chapter 10

Answer (B) is incorrect because freezing drizzle would produce ice on the aircraft surfaces, not frost. Answer (C) is incorrect because stable air is required. Convective action requires unstable conditions.

Answers

9774 [A]	9775 [A]	9162 [A]	9179 [C]	9748 [A]	9181 [A]

ALL

9736. During an IFR cross-country flight you picked up rime icing which you estimate is 1/2" thick on the leading edge of the wings. You are now below the clouds at 2000 feet AGL and are approaching your destination airport under VFR. Visibility under the clouds is more than 10 miles, winds at the destination airport are 8 knots right down the runway, and the surface temperature is 3 degrees Celsius. You decide to:

A—use a faster than normal approach and landing speed.
B—approach and land at your normal speed since the ice is not thick enough to have any noticeable effect.
C—fly your approach slower than normal to lessen the "wind chill" effect and break up the ice.

Ice will accumulate unevenly on the airplane. It will add weight and drag, and decrease thrust and lift. With ice accumulations, landing approaches should be made with a minimum wing flap setting and with an added margin of airspeed. Sudden and large configuration and airspeed changes should be avoided. (PLT274) — FAA-H-8083-3

Answer (B) is incorrect because ice having a thickness similar to sandpaper on the leading edge and upper surface of a wing can reduce wing lift by as much as 30% and increase drag by 40%. Answer (C) is incorrect because ice will increase drag, requiring additional lift (airspeed); "wind chill" effect cannot be relied upon to melt/remove the ice that has already accumulated; flying slower than normal increases the possibility of a stall due to the decreased lift.

121, DSP

8258. The pilot in command of an airplane en route determines that icing conditions can be expected that might adversely affect safety of the flight. Which action is appropriate?

A—The pilot in command may continue to the original destination airport, after climbing to a higher altitude.
B—The pilot in command shall not continue flight into the icing conditions.
C—The flight may continue to the original destination airport, provided all anti-icing and deicing equipment is operational and is used.

No person may dispatch or release an aircraft, continue to operate an aircraft en route, or land an aircraft when, in the opinion of the pilot-in-command or aircraft dispatcher, icing conditions are expected or met that might adversely affect the safety of flight. (PLT379) — 14 CFR §121.629

121, DSP

8265. What action is required prior to takeoff if snow is adhering to the wings of an air carrier airplane?

A—Sweep off as much snow as possible and the residue must be polished smooth.
B—Assure that the snow is removed from the airplane.
C—Add 15 knots to the normal VR speed as the snow will blow off.

No person may take off in an aircraft when frost, snow, or ice is adhering to the wings, control surfaces, or propellers of the aircraft. (PLT493) — 14 CFR §121.629

Answer (A) is incorrect because all of the snow must be removed prior to takeoff. Answer (C) is incorrect because there is no V-speed adjustment authorized to compensate for snow on the wings; it must be removed prior to takeoff.

135

9696. A pretakeoff contamination check for snow, ice or frost is required by 14 CFR Part 135. This check is required to

A—be made within 2 minutes of starting the takeoff roll.
B—be completed within 5 minutes prior to beginning the taxi to the runway.
C—see that the aircraft is clean, therefore, a safe takeoff can be made during the next 5 minutes.

A pre-takeoff inspection for ice, snow, or frost must be completed no more than 5 minutes prior to takeoff anytime conditions require it. (PLT108) — 14 CFR §135.227

135

9697. Deicing procedures and equipment developed for large transport airplanes

A—will not be appropriate for the smaller aircraft, used under 14 CFR Part 135.
B—will be appropriate for all of the smaller aircraft, used under 14 CFR Part 135.
C—may not be appropriate for some of the smaller aircraft, used under 14 CFR Part 135.

Deicing procedures developed for large transport aircraft may not be appropriate for smaller, slower aircraft. (PLT108) — AC 120-58

Answers

9736 [A]	8258 [B]	8265 [B]	9696 [C]	9697 [C]

Turbulence

Light chop causes slight, rapid and somewhat erratic bumpiness without appreciable changes in altitude or attitude. Light turbulence causes momentary slight erratic changes in altitude and/or attitude. Light chop causes rapid bumps or jolts without appreciable changes in aircraft altitude or attitude. Moderate turbulence is similar to light turbulence, but of greater intensity. Changes in altitude or attitude occur but the aircraft remains in positive control at all times. It usually causes variations in indicated airspeed. Severe turbulence causes large, abrupt changes in altitude or attitude. It usually causes large variations in indicated airspeed. The aircraft may be momentarily out of control. In extreme turbulence the aircraft is violently tossed about and is practically impossible to control. Extreme turbulence may cause structural damage.

Turbulence that occurs less than 1/3 of the time should be reported as occasional. Turbulence that occurs 1/3 to 2/3 of the time is intermittent. Turbulence that occurs more than 2/3 of the time is continuous. High altitude turbulence (normally above 15,000 feet MSL) not associated with cumuliform cloudiness should be reported as **CAT (Clear Air Turbulence)**.

Strong winds across mountain crests can cause turbulence for 100 or more miles downwind of the mountains and to altitudes as high as 5,000 feet above the tropopause. If there is enough moisture in the air, a mountain wave can be marked by standing lenticular clouds. These clouds mark the crest of each wave. Under the right conditions, several lenticulars can form one above another. A rotor current forms below the crest of a mountain wave. This is sometimes marked by a rotor cloud which will be the lowest of a group of stationary clouds. *See* Figure 8-6.

The jet stream is a common source of CAT. The strong winds and steep wind gradients will almost always produce some turbulence. The most likely place to find turbulence is on the polar side of the stream in an upper trough. The strongest turbulence will be found in a curving jet stream associated with such a trough. If you encounter turbulence in the jet stream and you have a direct headwind or tailwind you should change course or altitude. With the wind parallel to your heading, you are likely to remain in the jet and the turbulence for a considerable distance. If you approach a jet stream from the polar side the temperature will drop. When you approach it from the tropical side, the temperature rises. Recall that there is a downdraft on the polar side and an updraft on the tropical side. Therefore, to avoid jet stream turbulence descend if the temperature is falling and climb if the temperature is rising as you approach the stream.

Fronts often have turbulence due to the wind shift associated with a sharp pressure trough. Try to cross the front at right angles to minimize the time you are exposed to this turbulence.

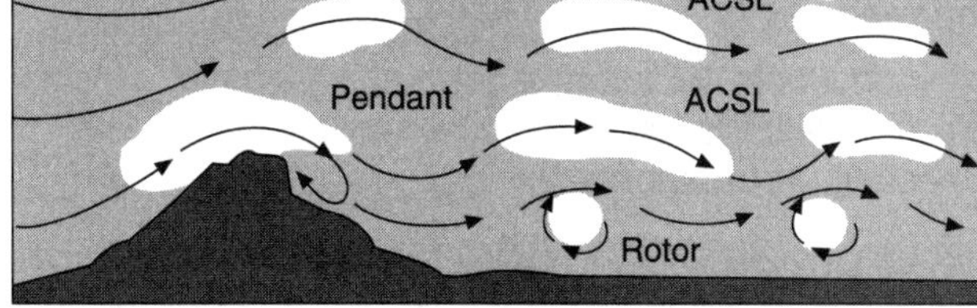

Figure 8-6. Mountain wave and associated clouds

ALL

9262. What type turbulence should be reported when it causes slight, rapid, and somewhat rhythmic bumpiness without appreciable changes in attitude or altitude, less than one-third of the time?

A—Occasional light chop.
B—Moderate turbulence.
C—Moderate chop.

This description meets the criteria for occasional light chop. (PLT501) — AC 00-45

Answer (B) is incorrect because moderate turbulence causes a change in the aircraft's attitude and/or altitude. Answer (C) is incorrect because moderate chop causes rapid, not rhythmic, bumps or jolts, which are not "slight."

Answers

9262 [A]

ALL

9263. What type turbulence should be reported when it causes changes in altitude and/or attitude more than two-thirds of the time, with the aircraft remaining in positive control at all times?

A—Continuous severe chop.
B—Continuous moderate turbulence.
C—Intermittent moderate turbulence.

This description meets the criteria for continuous moderate turbulence. (PLT501) — AC 00-45

Answer (A) is incorrect because severe chop is not a turbulence reporting term. Answer (C) is incorrect because intermittent means that turbulence is occurring from one-third to two-thirds of the time.

ALL

9264. What type turbulence should be reported when it momentarily causes slight, erratic changes in altitude and/or attitude, one-third to two-thirds of the time?

A—Occasional light chop.
B—Moderate chop.
C—Intermittent light turbulence.

This description meets the criteria for intermittent light turbulence. (PLT501) — AC 00-45

Answer (A) is incorrect because light chop does not cause any appreciable changes in altitude and/or attitude, and occasional is less than one-third of the time. Answer (B) is incorrect because moderate chop does not cause any appreciable changes in altitude and/or attitude.

ALL

9235. Turbulence encountered above 15,000 feet AGL, not associated with cloud formations, should be reported as

A—convective turbulence.
B—high altitude turbulence.
C—clear air turbulence.

High-level turbulence (normally above 15,000 feet AGL) not associated with cumuliform cloudiness, including thunderstorms, should be reported as CAT (Clear Air Turbulence). (PLT501) — AC 00-45

Answer (A) is incorrect because convective turbulence is normally associated with cumuliform clouds and is reported as turbulence. Answer (B) is incorrect because turbulence above 15,000 feet AGL, not associated with clouds, is termed clear air turbulence, not high altitude turbulence.

ALL

9190. Which type clouds are indicative of very strong turbulence?

A—Nimbostratus.
B—Standing lenticular.
C—Cirrocumulus.

Standing lenticular clouds form at the crests of waves created by barriers to the wind flow. Though the clouds do not move, they are indicative of strong winds and possible turbulence. (PLT192) — AC 00-6A, Chapter 7

Answer (A) is incorrect because nimbostratus is a gray or dark massive cloud layer, diffused by continuous rain, snow, or ice pellets. The stratus feature indicates very little turbulence. Answer (C) is incorrect because cirrocumulus are thin clouds appearing as small white flakes or patches of cotton. Their presence indicates some turbulence and possible icing.

ALL

9226. What is the lowest cloud in the stationary group associated with a mountain wave?

A—Rotor cloud.
B—Standing lenticular.
C—Low stratus.

Mountain waves can generate standing lenticular clouds and rotor clouds. Of these, the rotor cloud is likely to be the lowest. (PLT501) — AC 00-6A, Chapter 9

Answer (B) is incorrect because the standing lenticular clouds mark the crest, or the top, of each standing wave. Answer (C) is incorrect because low stratus clouds are not associated with a mountain wave.

ALL

9232. Clear air turbulence (CAT) associated with a mountain wave may extend as far as

A—1,000 miles or more downstream of the mountain.
B—5,000 feet above the tropopause.
C—100 miles or more upwind of the mountain.

Mountain wave CAT may extend from the mountain crests to as high as 5,000 feet above the tropopause, and can range 100 miles or more downwind from the mountains. (PLT501) — AC 00-6A, Chapter 13

Answer (A) is incorrect because mountain wave CAT can range 100, not 1,000, miles or more downstream of the mountain. Answer (C) is incorrect because mountain wave CAT is downwind of the mountain.

Answers

9263 [B]	9264 [C]	9235 [C]	9190 [B]	9226 [A]	9232 [B]

ALL

9777. Clear air turbulence associated with a jet stream is

A—most commonly found in temperatures between -40 and -50 degrees C.
B—most commonly found in the vicinity of the tropopause.
C—similar to that associated with a tropical maritime front.

Maximum winds generally occur at levels near the tropopause. These strong winds create narrow zones of wind shear which often generate hazardous turbulence.

Answer (A) is incorrect because temperatures are not a primary factor for CAT. Answer (C) is incorrect because CAT is associated with the tropopause, not a tropical maritime front. (PLT501) — AC 00-6

ALL

9237. What is a likely location of clear air turbulences?

A—In an upper trough on the polar side of a jetstream.
B—Near a ridge aloft on the equatorial side of a high pressure flow.
C—Downstream of the equatorial side of a jetstream.

A likely location of CAT is in an upper trough on the cold (polar) side of the jet stream. (PLT302) — AC 00-6A, Chapter 13

Answer (B) is incorrect because CAT is likely to occur on the polar side of the jet stream in an upper trough. Answer (C) is incorrect because CAT is likely on the polar, not equatorial, side of a jet stream.

ALL

9239. Which type jetstream can be expected to cause the greater turbulence?

A—A straight jetstream associated with a high pressure ridge.
B—A jetstream associated with a wide isotherm spacing.
C—A curving jetstream associated with a deep low pressure trough.

A frequent CAT location is along the jet stream where it curves north and northeast of a rapidly deepening surface low. (PLT302) — AC 00-6A, Chapter 13

Answer (A) is incorrect because greater turbulence is expected in a curved, not a straight jet stream. Answer (B) is incorrect because greater turbulence is more pronounced when isotherm spacing is narrow.

ALL

9230. Which action is recommended if jetstream turbulence is encountered with a direct headwind or tailwind?

A—Increase airspeed to get out of the area quickly.
B—Change course to fly on the polar side of the jetstream.
C—Change altitude or course to avoid a possible elongated turbulent area.

If jet stream turbulence is encountered with direct tailwinds or headwinds, a change of flight level or course should be initiated since these turbulent areas are elongated with the wind, and are shallow and narrow. (PLT263) — AC 00-30, Appendix 1

Answer (A) is incorrect because an increase in airspeed may overstress the airplane in turbulent conditions. Normally, a reduction in airspeed is required for turbulent air penetration. Answer (B) is incorrect because CAT is normally on the polar side of the jet stream, so you would be flying into more turbulent weather.

ALL

9231. Which action is recommended regarding an altitude change to get out of jetstream turbulence?

A—Descend if ambient temperature is falling.
B—Descend if ambient temperature is rising.
C—Maintain altitude if ambient temperature is not changing.

If you want to traverse an area of CAT more quickly, watch the temperature gauge for a minute or two. If the temperature is rising—climb; if the temperature is falling—descend. Application of these rules will prevent you from following the sloping tropopause and staying in the turbulent area. If the temperature remains constant, the flight is probably close to the level of the core, so either climb or descend as is convenient. (PLT263) — AC 00-30, Appendix 1

Answer (B) is incorrect because to get out of jet stream turbulence with a rising ambient temperature you would climb, not descend. Answer (C) is incorrect because you would need to make an altitude change due to jet stream turbulence, and there should be a temperature change due to a sloping tropopause.

Answers

9777 [B]	9237 [A]	9239 [C]	9230 [C]	9231 [A]

ALL

9219. What action is recommended when encountering turbulence due to a wind shift associated with a sharp pressure trough?

A—Establish a straight course across the storm area.
B—Climb or descend to a smoother level.
C—Increase speed to get out of the trough as soon as possible.

If turbulence is encountered in an abrupt wind shift associated with a sharp pressure storm area, establish a straight course across the storm area rather than parallel to it. A change in flight level is not likely to alleviate the bumpiness. (PLT501) — AC 00-30, Appendix 1

Answer (B) is incorrect because there is no indication to identify in which direction the turbulence is stronger. A change in altitude will normally remove the aircraft from the turbulent zone. Answer (C) is incorrect because speed should be decreased to the recommended airspeed for rough air. This will avoid overstressing the airplane.

Arctic and Tropical Weather Hazards

"Whiteout" is a visibility restricting phenomenon that occurs in the Arctic when a layer of cloudiness of uniform thickness overlies a snow or ice covered surface. Parallel rays of the sun are broken up and diffused when passing through the cloud layer so that they strike the snow surface from many angles. The diffused light then reflects back and forth between the clouds and the snow eliminating all shadows. The result is a loss of depth perception that makes takeoff or landing on snow-covered surfaces very dangerous.

"Tropical Cyclone" is the term for any low that originates over tropical oceans. Tropical cyclones are classified according to their intensity based on average one minute wind speeds. These classifications are:

Tropical Depression—highest sustained winds up to 34 knots.

Tropical Storm—highest sustained winds of 35 knots through 64 knots.

Hurricane or Typhoon—highest sustained winds of 65 knots or more.

The movement of hurricanes is erratic and very difficult to predict with any degree of precision. As a general rule, hurricanes in the northern hemisphere tend to move to the northwest while they are in the lower latitudes and under the influence of the trade winds. Once they move far enough north to come under the influence of the prevailing westerlies of the mid-latitudes their track tends to curve back to the northeast.

ALL

9234. Which arctic flying hazard is caused when a cloud layer of uniform thickness overlies a snow or ice covered surface?

A—Ice fog.
B—Whiteout.
C—Blowing snow.

"Whiteout" is a visibility restricting phenomenon that occurs in the Arctic when a layer of cloudiness of uniform thickness overlies a snow or ice covered surface. The result is a loss of depth perception. (PLT512) — AC 00-6A, Chapter 14

Answer (A) is incorrect because ice fog forms in moist air during extremely cold conditions. It is not formed by a cloud layer overlying a snow-covered surface. Answer (C) is incorrect because blowing snow is snow that is blown by light or greater winds, causing decreased visibility. It is not formed by a cloud layer overlying a snow-covered surface.

ALL

9259. Which weather condition is present when the tropical storm is upgraded to a hurricane?

A—Highest windspeed, 100 knots or more.
B—A clear area or hurricane eye has formed.
C—Sustained winds of 65 knots or more.

Tropical cyclone international classifications are:

1. *Tropical depression — highest sustained winds up to 34 knots;*
2. *Tropical storm — highest sustained winds of 35 through 64 knots;*
3. *Hurricane or typhoon — highest sustained winds of 65 knots or more.*

(PLT511) — AC 00-6A, Chapter 15

Answer (A) is incorrect because tropical cyclones are classified based on the sustained winds, not the highest wind speed. Answer (B) is incorrect because a clear area, or eye, usually forms in the tropical storm stage and continues through the hurricane stage.

Answers

9219 [A] 9234 [B] 9259 [C]

ALL

9260. What is the general direction of movement of a hurricane located in the Caribbean or Gulf of Mexico region?

A—Northwesterly curving to northeasterly.
B—Westerly, until encountering land, then easterly.
C—Counterclockwise over open water, then dissipating outward over land.

Hurricanes located in the Caribbean or Gulf of Mexico move northwesterly in the lower latitudes curving to northeasterly in the higher latitudes. (PLT068) — AC 00-6A, Chapter 15

Answer (B) is incorrect because a hurricane will curve easterly because of prevailing winds, not because of land. Answer (C) is incorrect because the windflow in the hurricane is counterclockwise, not the general movement of the hurricane itself.

Aviation Routine Weather Report (METAR)

Weather reports (METAR) and forecasts (TAF) follow the format shown in Figure 8-7.

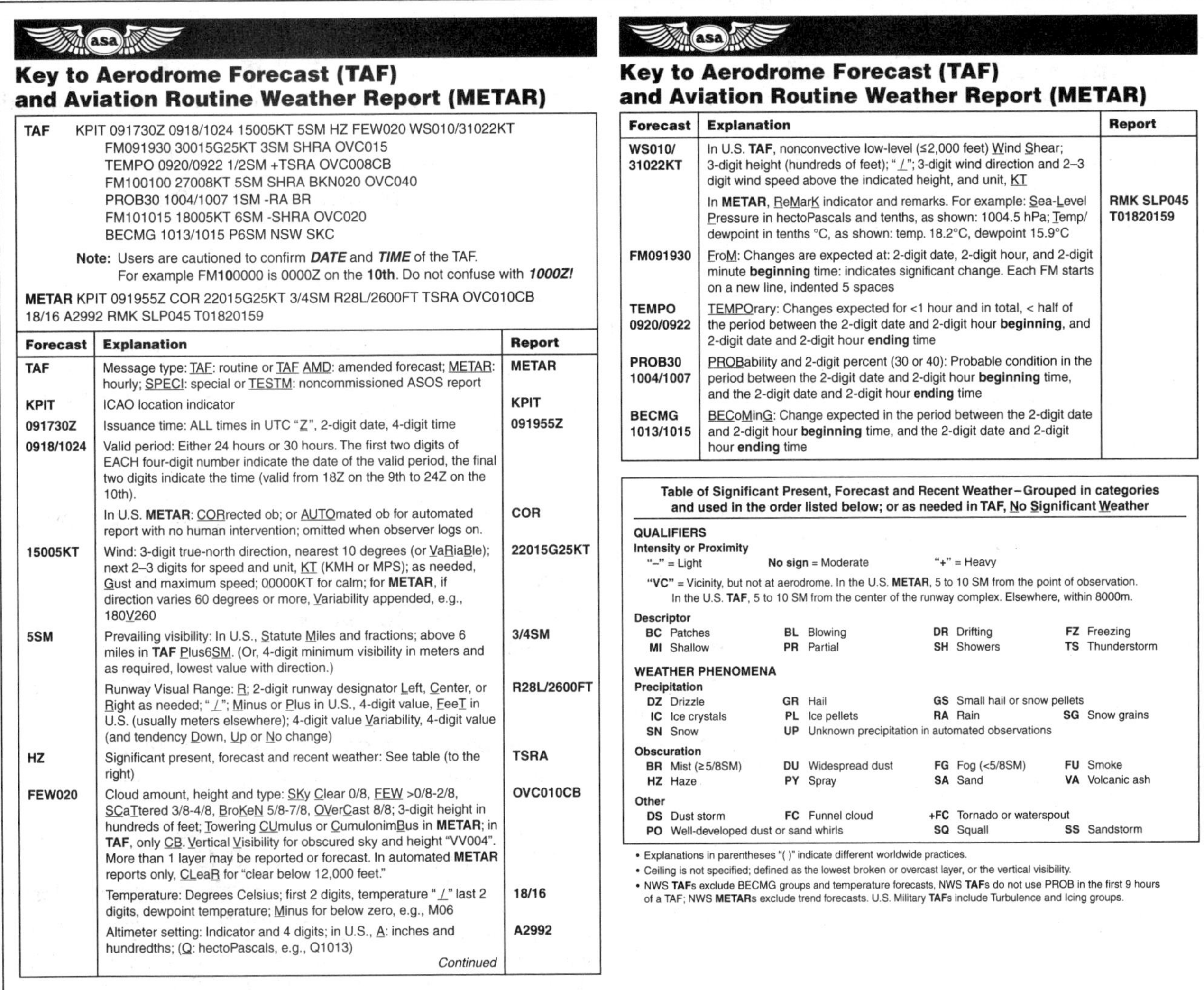

Key to Aerodrome Forecast (TAF) and Aviation Routine Weather Report (METAR)

TAF KPIT 091730Z 0918/1024 15005KT 5SM HZ FEW020 WS010/31022KT
FM091930 30015G25KT 3SM SHRA OVC015
TEMPO 0920/0922 1/2SM +TSRA OVC008CB
FM100100 27008KT 5SM SHRA BKN020 OVC040
PROB30 1004/1007 1SM -RA BR
FM101015 18005KT 6SM -SHRA OVC020
BECMG 1013/1015 P6SM NSW SKC

Note: Users are cautioned to confirm ***DATE*** and ***TIME*** of the TAF. For example FM**10**0000 is 0000Z on the **10th**. Do not confuse with ***1000Z!***

METAR KPIT 091955Z COR 22015G25KT 3/4SM R28L/2600FT TSRA OVC010CB 18/16 A2992 RMK SLP045 T01820159

Forecast	Explanation	Report
TAF	Message type: TAF: routine or TAF AMD: amended forecast; METAR: hourly; SPECI: special or TESTM: noncommissioned ASOS report	**METAR**
KPIT	ICAO location indicator	**KPIT**
091730Z	Issuance time: ALL times in UTC "Z", 2-digit date, 4-digit time	**091955Z**
0918/1024	Valid period: Either 24 hours or 30 hours. The first two digits of EACH four-digit number indicate the date of the valid period, the final two digits indicate the time (valid from 18Z on the 9th to 24Z on the 10th).	
	In U.S. **METAR**: CORrected ob; or AUTOmated ob for automated report with no human intervention; omitted when observer logs on.	**COR**
15005KT	Wind: 3-digit true-north direction, nearest 10 degrees (or VaRiaBle); next 2–3 digits for speed and unit, KT (KMH or MPS); as needed, Gust and maximum speed; 00000KT for calm; for **METAR**, if direction varies 60 degrees or more, Variability appended, e.g., 180V260	**22015G25KT**
5SM	Prevailing visibility: In U.S., Statute Miles and fractions; above 6 miles in **TAF** Plus6SM. (Or, 4-digit minimum visibility in meters and as required, lowest value with direction.)	**3/4SM**
	Runway Visual Range: R; 2-digit runway designator Left, Center, or Right as needed; "/"; Minus or Plus in U.S., 4-digit value, FeeT in U.S. (usually meters elsewhere); 4-digit value Variability, 4-digit value (and tendency Down, Up or No change)	**R28L/2600FT**
HZ	Significant present, forecast and recent weather: See table (to the right)	**TSRA**
FEW020	Cloud amount, height and type: SKy Clear 0/8, FEW >0/8-2/8, SCaTtered 3/8-4/8, BroKeN 5/8-7/8, OVerCast 8/8; 3-digit height in hundreds of feet; Towering CUmulus or CumulonimBus in **METAR**; in **TAF**, only CB. Vertical Visibility for obscured sky and height "VV004". More than 1 layer may be reported or forecast. In automated **METAR** reports only, CLeaR for "clear below 12,000 feet."	**OVC010CB**
	Temperature: Degrees Celsius; first 2 digits, temperature "/" last 2 digits, dewpoint temperature; Minus for below zero, e.g., M06	**18/16**
	Altimeter setting: Indicator and 4 digits; in U.S., A: inches and hundredths; (Q: hectoPascals, e.g., Q1013) *Continued*	**A2992**

Key to Aerodrome Forecast (TAF) and Aviation Routine Weather Report (METAR)

Forecast	Explanation	Report
WS010/ 31022KT	In U.S. **TAF**, nonconvective low-level (≤2,000 feet) Wind Shear; 3-digit height (hundreds of feet); "/"; 3-digit wind direction and 2–3 digit wind speed above the indicated height, and unit, KT	
	In **METAR**, ReMarK indicator and remarks. For example: Sea-Level Pressure in hectoPascals and tenths, as shown: 1004.5 hPa; Temp/dewpoint in tenths °C, as shown: temp. 18.2°C, dewpoint 15.9°C	**RMK SLP045 T01820159**
FM091930	FroM: Changes are expected at: 2-digit date, 2-digit hour, and 2-digit minute **beginning** time: indicates significant change. Each FM starts on a new line, indented 5 spaces	
TEMPO 0920/0922	TEMPOrary: Changes expected for <1 hour and in total, < half of the period between the 2-digit date and 2-digit hour **beginning**, and 2-digit date and 2-digit hour **ending** time	
PROB30 1004/1007	PROBability and 2-digit percent (30 or 40): Probable condition in the period between the 2-digit date and 2-digit hour **beginning** time, and the 2-digit date and 2-digit hour **ending** time	
BECMG 1013/1015	BECoMinG: Change expected in the period between the 2-digit date and 2-digit hour **beginning** time, and the 2-digit date and 2-digit hour **ending** time	

Table of Significant Present, Forecast and Recent Weather – Grouped in categories and used in the order listed below; or as needed in TAF, No Significant Weather

QUALIFIERS

Intensity or Proximity

"–" = Light **No sign** = Moderate "+" = Heavy

"VC" = Vicinity, but not at aerodrome. In the U.S. **METAR**, 5 to 10 SM from the point of observation. In the U.S. **TAF**, 5 to 10 SM from the center of the runway complex. Elsewhere, within 8000m.

Descriptor

BC	Patches	**BL**	Blowing	**DR**	Drifting	**FZ**	Freezing
MI	Shallow	**PR**	Partial	**SH**	Showers	**TS**	Thunderstorm

WEATHER PHENOMENA

Precipitation

DZ	Drizzle	**GR**	Hail	**GS**	Small hail or snow pellets		
IC	Ice crystals	**PL**	Ice pellets	**RA**	Rain	**SG**	Snow grains
SN	Snow	**UP**	Unknown precipitation in automated observations				

Obscuration

BR	Mist (≥5/8SM)	**DU**	Widespread dust	**FG**	Fog (<5/8SM)	**FU**	Smoke
HZ	Haze	**PY**	Spray	**SA**	Sand	**VA**	Volcanic ash

Other

DS	Dust storm	**FC**	Funnel cloud	**+FC**	Tornado or waterspout		
PO	Well-developed dust or sand whirls			**SQ**	Squall	**SS**	Sandstorm

- Explanations in parentheses "( )" indicate different worldwide practices.
- Ceiling is not specified; defined as the lowest broken or overcast layer, or the vertical visibility.
- NWS **TAF**s exclude BECMG groups and temperature forecasts, NWS **TAF**s do not use PROB in the first 9 hours of a TAF; NWS **METAR**s exclude trend forecasts. U.S. Military **TAF**s include Turbulence and Icing groups.

Figure 8-7. TAF/METAR weather card

Answers

9260 [A]

ALL

9266. (Refer to Figure 145.) What was the local Central Standard Time of the Aviation Routine Weather Report at Austin (KAUS)?

A—11:53 a.m.
B—5:53 p.m.
C—10:53 p.m.

"131753Z" indicates that this METAR report is for the thirteenth day of the month, at 1753 Coordinated Universal Time (UTC). Central Standard Time is UTC minus 6 hours (1753 – 6 = 1153). (PLT059) — AC 00-45

ALL

9267. (Refer to Figure 145.) What type of report is listed for Lubbock (KLBB) at 1818Z?

A—An Aviation selected special weather report.
B—A special report concerning very low station pressure.
C—A Special METAR weather observation, concerning significant weather changes.

The designation "SPECI" means that this is a special weather observation. (PLT059) — AC 00-45

ALL

9268. (Refer to Figure 146.) What method was used to obtain the METAR at Tyler (KTYR) at 1753Z?

A—Automated Surface Observing System (ASOS), having a precipitation discriminator.
B—Automatic Meteorological Observing Station (AMOS), with a precipitation discriminator.
C—Automated Weather Observing System (AWOS), without a precipitation discriminator.

The word "AUTO" after the date-time group indicates that this is an automated weather report. The "A02" in the remarks section indicate that it was made by an unattended Automated Surface Observing System (ASOS). The ASOS will automatically report precipitation amounts. (PLT059) — AC 00-45

ALL

9269. (Refer to Figure 145.) What condition is reported at Childress (KCDS)?

A—Light rain showers.
B—Heavy rain showers began 42 minutes after the hour.
C—The ceiling is solid overcast at an estimated 1,800 feet above sea level.

"-SHRA" indicates an observation of light rain showers. (PLT059) — AC 00-45

Answer (B) is incorrect because the remark "RERAB42" means an observed recent weather event was that rain began at 42 minutes past the previous hour. Answer (C) is incorrect because the abbreviation "OVC180" indicated the base of an overcast layer is at 18,000 feet above the station.

ALL

9270. (Refer to Figure 145.) What condition is reported at Dallas (KDAL)?

A—The tops of the overcast is 10,000 feet.
B—Temperature/dewpoint spread is 8°F.
C—Altimeter setting is 30.07.

"A3007" indicates an altimeter setting of 30.07" Hg. (PLT059) — AC 00-45

Answer (A) is incorrect because "OVC100" indicates the base of the overcast layer is 10,000 feet above the station. Answer (B) is incorrect because the reported temperature/dew point are given in degrees Celsius, not Fahrenheit.

ALL

9272. SPECI KGLS 131802Z 10012G21KT 060V140 2SM +SHRA SCT005 BKN035 OVC050CB 24/23 A2980 RMK RAB57 WS TKO RW09L WSHFT 58 FROPA.

This SPECI report at Galveston (KGLS) indicates which condition?

A—Wind steady at 100° magnetic at 12 knots, gusts to 21.
B—Precipitation started at 57 after the hour.
C—5,000 feet overcast with towering cumulus.

The remarks "RAB57" indicates rain began at 57 minutes past the hour. (PLT059) — AC 00-45

Answer (A) is incorrect because "10012G21KT 060V140" indicates the wind was from 100° at 12 knots with gusts to 21 knots; however, the wind direction was variable between 60-140°, and wind direction in written reports and forecasts are referenced to true north. Answer (C) is incorrect because "OVC050CB" indicates there was an overcast cumulonimbus cloud at 5,000 feet. Towering cumulus is abbreviated "TCU."

Answers

9266 [A]	9267 [A]	9268 [A]	9269 [A]	9270 [C]	9272 [B]

ALL

9273. (Refer to Figure 145.) What weather improvement was reported at Lubbock (KLBB) between 1750 and 1818 UTC?

A—The wind shift and frontal passage at 1812Z.
B—The vertical visibility improved by 2,000 feet.
C—The temperature and dew point spread improved.

At 1750Z, the vertical visibility was 1,000 feet (VV010). At 1818Z, the vertical visibility had improved to 3,000 feet (VV030). (PLT059) — AC 00-45

Answer (A) is incorrect because even though the description of the events is correct, it is not an improvement since the wind speed increased and the visibility decreased. Answer (C) is incorrect because the temperature and dew point actually converged. At 1750Z, the spread was 4°C (03/M01), but by 1818Z, there was a 0° spread (M01/M01).

ALL

9274. METAR KMAF 131756Z 02020KT 12SM BKN025 OVC250 27/18 A3009 RMK RAE44.

Which weather condition is indicated by this METAR report at Midland (KMAF)?

A—Rain of unknown intensity ended 16 minutes before the hour.
B—The ceiling was at 25,000 feet MSL.
C—Wind was 020° magnetic at 20 knots.

"RAE44" indicates the rain ended 44 minutes past the hour, which is the same thing as 16 minutes before the hour. (PLT059) — AC 00-45

Answer (B) is incorrect because a ceiling is the lowest broken or overcast layer. "BKN025" indicates a broken layer (not a ceiling) at 2,500 feet. The base of the higher overcast layer is at 25,000 feet above the station (not sea level). Answer (C) is incorrect because the winds are 020° true at 20 knots.

ALL

9275. METAR KSPS 131757Z 09014KT 6SM -RA SCT025 OVC090 24/22 A3005.
SPECI KSPS 131820Z 01025KT 3SM +RA FC OVC015 22/21 A3000.

Which change took place at Wichita Falls (KSPS) between 1757 and 1820 UTC?

A—The rain became lighter.
B—Atmospheric pressure increased.
C—A funnel cloud was observed.

"FC" in the 1820Z report indicates that a funnel cloud/tornado/waterspout was observed. (PLT059) — AC 00-45

Answer (A) is incorrect because the rain increased in intensity. At 1757Z, the abbreviation "-RA" indicated light rain; at 1820Z, "+RA" indicates heavy rain. Answer (B) is incorrect because the atmospheric pressure is decreased. The altimeter went from 30.05" Hg (A3005) at 1757Z, to 30.00" Hg (A3000) at 1820Z.

ALL

9276. (Refer to Figure 146.) What was the ceiling at Walnut Ridge (KARG)?

A—1,000 feet AGL.
B—2,400 feet AGL.
C—1,000 feet MSL.

"OVC010" indicates an overcast layer with bases at 1,000 feet AGL. (PLT059) — AC 00-45

Answer (B) is incorrect because 2,400 feet is the RVR for runway 28 at Walnut Ridge airport. Answer (C) is incorrect because the sky condition is reported in feet above the ground.

ALL

9277. METAR KHRO 131753Z 09007KT 7SM FEW020 BKN040 30/27 A3001.
SPECI KHRO 131815Z 13017G26KT 3SM +TSRA SCT020 BKN045TCU 29/24 A2983 RMK
RAB12 WS TKO LDG RW14R FRQ LTGICCG VC.

What change has taken place between 1753 and 1815 UTC at Harrison (KHRO)?

A—The ceiling lowered and cumulonimbus clouds developed.
B—Thundershowers began at 12 minutes past the hour.
C—Visibility reduced to IFR conditions.

At 1815Z, the station was reporting heavy thunderstorms and rain (+TSRA). "RAB12" indicates that the rain began at 12 minutes past the hour. (PLT059) — AC 00-45

Answers

9273 [B]	9274 [A]	9275 [C]	9276 [A]	9277 [B]

ALL

9704. Data that may be added (manual weather augmentation) to the Automated Weather Observing System (AWOS) report is limited to

A—the precipitation accumulation report, an automated variable visibility, and wind direction remark.
B—thunderstorms (intensity and direction), precipitation (type and intensity), and obstructions to visibility (dependent on the visibility being 3 miles or less).
C—density Altitude, NOTAMs, and reported slant range visibility.

In addition to the information automatically included in an AWOS report, information can be manually added. The remarks are limited to thunderstorms (type and intensity), and obstructions to vision when the visibility is 3 SM or less. Augmentation is identified in the observation as "observer weather." (PLT515) — AIM ¶7-1-12

ALL

9716. The prevailing visibility in the following METAR is

METAR KFSM 131756Z AUTO 00000KT M1/4SM R25/0600V1000FT -RA FG VV004 06/05
A2989 RMK AO2 $

A—less than 1/4 statute mile.
B—measured 1/4 statute mile.
C—a mean (average) of 1/4 statute mile.

The prevailing visibility is less than 1/4 statute miles, indicated by the "M1/4SM." (PLT059) — AIM ¶7-1-12

ALL

9717. The symbol ($) at the end of the following METAR indicates that

METAR KFSM 131756Z AUTO 00000KT M1/4SM R25/0600V1000FT -RA FG VV004 06/05
A2989 RMK AO2 $

A—the latest information is transmitted over a discrete VHF frequency at KFSM.
B—the latest information is broadcast on the voice portion of a local navaid at KFSM.
C—maintenance is needed on the system.

The dollar symbol ($) indicates the system may need maintenance. (PLT059) — AIM ¶7-1-12

ALL

9718. The VV001 in the following METAR indicates

METAR KFSM 131756Z AUTO 00000KT M1/4SM R25/0600V1000FT -RA FG VV001
A2989 RMK AO2 VIS 3/4 RWY19 CHINO RWY19 $

A—an observer reported the vertical visibility as 100 feet.
B—a 100 foot indefinite ceiling.
C—the variability value is 100 feet.

The height into an indefinite ceiling is preceded by "VV" and followed by three digits indicating the vertical visibility in hundreds of feet. This layer indicates total obscuration. The indefinite ceiling is 100 feet. (PLT059) — AIM ¶7-1-12

ALL

9242. METAR KFSO 031053Z VRB02KT 7SM MIFG SKC 15/14 A3012 RMK SLP993 6//// T01500139 56012

In the above METAR, the SLP993 6//// indicates

A—sea-level pressure 999.3 hectopascals which in the last 6 hours has dropped .4 hectopascals.
B—sea-level pressure 999.3 hectopascals and an indeterminable amount of precipitation has occurred over the last 3 hours.
C—sea-level pressure 999.3 hectopascals and in the last 6 hours that four-tenths of an inch of precipitation has fallen.

The "SLP993 6////" in the Remarks section of the METAR indicates sea-level pressure 999.3 hectopascals and an indeterminable amount of precipitation has occurred over the last 3 hours. (PLT059) — AC 00-45

Answers

9704 [B] 9716 [A] 9717 [C] 9718 [B] 9242 [B]

The Weather Depiction Chart

The Weather Depiction Chart is put together from surface aviation (SA) reports to give a broad view of the weather conditions at the time of the observations. The chart shows the actual sky cover, visibility restrictions, and type of precipitation at the reporting stations. In addition, the chart groups stations that are reporting VFR, Marginal VFR or IFR weather conditions.

Stations that report a ceiling of less than 1,000 feet or a visibility of less than 3 miles are classified as IFR and included in a hatched area surrounded by a smooth line. Stations that report ceilings of 1,000 to 3,000 feet or visibilities of 3 to 5 miles are MVFR (marginal VFR) and are included in a non-hatched area surrounded by a smooth line. Stations that have a ceiling greater than 3,000 feet and visibilities greater than 5 miles are VFR and are not outlined. This chart says nothing about the weather between reporting stations.

ALL

9265. What conditions are indicated on a Weather Depiction Chart?

A—Actual sky cover, visibility restrictions, and type of precipitation at reporting stations.
B—Forecast ceilings and visibilities over a large geographic area.
C—Actual en route weather conditions between reporting stations.

The Weather Depiction Chart is computer-generated from surface aviation (SA) reports to give a broad overview of observed flying weather. It depicts actual sky cover, visibility, and the type of precipitation at reporting stations. Areas of IFR or marginal VFR are established by the station observations only, not by en route observations. (PLT289) — AC 00-45

Answer (B) is incorrect because the weather depiction chart is not a forecast, but an overview of observed conditions at a specified time. Answer (C) is incorrect because the weather depiction chart indicates actual weather conditions at a reporting station only as of a specified time.

ALL

9299. What is indicated on the Weather Depiction Chart by a continuous smooth line enclosing a hatched geographic area?

A—The entire area has ceilings less than 1,000 feet and/or visibility less than 3 miles.
B—More than 50 percent of the area enclosed by the smooth line is predicted to have IFR conditions.
C—Reporting stations within the enclosed area are all showing IFR conditions at the time of the report.

Areas in which all stations are reporting IFR condition are enclosed in a continuous smooth line and the area is hatched. (PLT075) — AC 00-45

Answers (A) and (B) are incorrect because the weather depiction chart reflects only surface aviation reports which are made only at reporting stations.

ALL

9297. (Refer to Figure 150.) The IFR conditions in the vicinity of Lakes Superior, Huron and Michigan were caused by

A—overcast sky and haze.
B—convective action during the front's passage.
C—obscured skies.

A station model that has an X over the station circle means that the sky coverage is obscured or partially obscured, the = symbol means mist. (PLT075) — AC-00-45

Answer (A) is incorrect because the station model depicts fog (not haze). Answer (B) is incorrect because fog is indicated (not thunderstorm activity).

Answers

9265 [A] 9299 [C] 9297 [C]

ALL

9298. (Refer to Figure 150.) The Weather Depiction Chart indicates that the coastal sections of Texas and Louisiana are reporting

A—all ceilings at or above 20,000 feet with visibilities of 20 miles or more.
B—marginal VFR conditions due to broken ceilings of 3,200 feet.
C—VFR conditions with scattered clouds at 3,200 feet.

While the Texas coast is clear, the Louisiana coast shows scattered clouds at 3,200 to the west, and overcast MVR at 1,600 toward the east. Since the area is not contoured, the conditions are VFR. There is no indication of higher cirriform along the coast. (PLT075) — AC 00-45

Answer (A) is incorrect because a station with the visibility data missing means only that visibility is greater than 6 (not 20) miles and most of the coast is reporting clear skies (not ceilings at or above 20,000 feet). Answer (B) is incorrect because the station in southeast Louisiana is reporting overcast skies with a ceiling of 1,600 (not 3,200) feet.

The Terminal Aerodrome Forecast (TAF)

TAFs use the same code used in the METAR weather reports. (*See* Figure 8-7.)

ALL

9244. Which primary source contains information regarding the expected weather at the destination airport, at the ETA?

A—Low-Level Prog Chart.
B—Radar Summary and Weather Depiction Charts.
C—Terminal Aerodrome Forecast.

A Terminal Aerodrome Forecast (TAF) is a concise statement of the expected meteorological conditions at an airport during a specified period (usually 24 hours). (PLT288) — AC 00-45

Answer (A) is incorrect because a Low-Level Prog Chart is a forecast of significant weather for the United States, not a forecast for a specific destination. Answer (B) is incorrect because Weather Depiction Charts and Radar Summary Charts are national weather maps of observed weather at a specific time; they do not provide specific information about a particular destination.

ALL

9245. Weather conditions expected to occur in the vicinity of the airport, but not at the airport, are denoted by the letters "VC." When VC appears in a Terminal Aerodrome Forecast, it covers a geographical area of

A—a 5 to 10 statute mile radius from the airport.
B—a 5-mile radius of the center of a runway complex.
C—10 miles of the station originating the forecast.

Proximity applies to weather conditions expected to occur in the vicinity of the airport (between a 5 to 10 mile radius of the airport), but not at the airport itself. It is denoted by the letters "VC." (PLT288) — AC 00-45

ALL

9248. What weather is predicted by the term VCTS in a Terminal Aerodrome Forecast?

A—Thunderstorms are expected in the vicinity.
B—Thunderstorms may occur over the station and within 50 miles of the station.
C—Thunderstorms are expected between 5 and 25 miles of the runway complex.

"VC" applies to weather conditions expected to occur in the vicinity of the airport (between a 5 to 10 statute mile radius of the airport), but not at the airport itself. "TS" denotes thunderstorms. Therefore, "VCTS" in a Terminal Aerodrome Forecast indicates thunderstorms are expected between a 5 to 10 mile radius of the airport, but not at the airport itself. (PLT288) — AC 00-45

ALL

9246. Which are the only cloud types forecast in the Terminal Aerodrome Forecast?

A—Altocumulus
B—Cumulonimbus
C—Stratocumulus

If cumulonimbus clouds are expected at the airport, the contraction "CB" is appended to the cloud layer which represents the base of the cumulonimbus cloud(s). Cumulonimbus clouds are the only cloud type forecast in TAFs. (PLT288) — AC 00-45

Answers

9298 [C]	9244 [C]	9245 [A]	9248 [A]	9246 [B]

ALL

9278. A PROB40 (PROBability) HHhh group in an International Terminal Aerodrome Forecast (TAF) indicates the probability of

A—thunderstorms or other precipitation.
B—precipitation or low visibility.
C—thunderstorms or high wind.

A PROB40 (PROBability) HHhh group in a TAF indicates the probability of occurrence of thunderstorms or other precipitation events. (PLT288) — AC 00-45

ALL

9279. (Refer to Figure 147.) At which time is IFR weather first predicted at Lubbock (KLBB)?

A—0100Z.
B—2100Z.
C—0400Z.

"PROB40 0103 06025G35KT 1/8SM +SHRASNPE OVC003" indicates there is a 40% probability between 0100 and 0300Z that the visibility will drop to 1/8 SM in heavy rain showers, snow and ice pellets, and the ceiling will be 300 feet overcast. Consequently, the first prediction of IFR would be at 0100Z. (PLT072) — AC 00-45

ALL

9280. (Refer to Figure 147.) What type conditions can be expected for a flight scheduled to land at San Angelo (KSJT) at 1500Z?

A—Chance of 1 statute mile visibility and cumulonimbus clouds.
B—IFR conditions due to low visibility, rain, and mist.
C—IFR conditions due to low ceilings, rain, and fog.

The last line of the forecast includes 1500Z. "FM0900 17020G34KT 2SM RA BR OVC010CB" means that from 0900Z on, the forecast is for winds from 170° at 20 knots with gusts to 34 knots. The visibility will be two statute miles. There will be a 1,000 foot overcast ceiling, with cumulonimbus clouds along with rain (RA) and mist (BR). (PLT072) — AC 00-45

ALL

9707. On the constant pressure analysis chart, satellite and aircraft observations are used in the analysis, over areas of sparse data. A satellite observation plotted using

A—a station circle at the cloud top location.
B—a square at the cloud top location.
C—a star at the cloud top location.

A star identifies satellite wind estimates made from cloud types. (PLT283) — AC 00-45

Answer (A) is incorrect because a station circle is used for the weather stations which send up radiosondes. Answer (B) is incorrect because a square at the aircraft location is used to signify an aircraft report.

ALL

9709. A calm wind that is forecast, in the International Terminal Aerodrome Forecast (TAF) is encoded as

A—VRB00KT.
B—00000KT.
C—00003KT.

A forecast of "00000KT" (calm) on a TAF means winds are expected at 3 knots or less. (PLT288) — AC 00-45

ALL

9710. In the International Terminal Aerodrome Forecast (TAF), a variable wind direction is noted by "VRB" where the three digit direction usually appears. A calm wind appears in the TAF as

A—00003KT.
B—VRB00KT.
C—00000KT.

Calm wind forecasts in a TAF are entered as "00000KT." (PLT288) — AC 00-45

ALL

9711. On the constant pressure analysis chart, aircraft and satellite observations are used in the analysis over areas of sparse data. An aircraft observation is plotted using

A—a station circle at the aircraft location.
B—a square at the aircraft location.
C—a star at the aircraft location.

A square is used to signify an aircraft report. (PLT283) — AC 00-45

Answer (A) is incorrect because a station circle is used for the weather stations which send up radiosondes. Answer (C) is incorrect because a star identifies satellite wind estimates made from cloud tops.

Answers

9278 [A] 9279 [A] 9280 [B] 9707 [C] 9709 [B] 9710 [C]
9711 [B]

Enroute Forecasts

The **Area Forecast (FA)** is the single source reference that contains information regarding frontal movement, turbulence and icing conditions for a specific region.

Winds and temperatures aloft are forecast for various stations around the country. Wind directions are always relative to true north and the speed is in knots. Temperatures, in degrees Celsius, are forecast for all altitudes except for 3,000 feet. At altitudes where the wind or temperature is not forecast, a blank space is used to signify the omission. At 30,000 feet and above the minus sign is deleted from the temperature to save space.

When winds are light and variable the notation 9900 is used. When wind speeds exceed 99 knots, fifty is added to the wind direction and only the last two digits of the wind speed is printed. For example, an FB (previously FD) forecast of "731960" at FL390 is 230° true (73 – 50 = 23) at 119 knots with a temperature of -60°C. When winds exceed 199 knots they are coded as 199 knots. For example, winds from 280° at 205 knots are coded as 7899.

The temperature in the tropopause (36,000 feet and above) is approximately -56°C. ISA at sea level is 15°C and decreases at a rate of 2°/1,000 feet up to 36,000 feet MSL.

Forecast winds and temperatures aloft for international flights may be obtained by consulting wind and temperature aloft charts prepared by a Regional Area Forecast Center (RAFC).

ALL

9243. What is the single source reference that contains information regarding volcanic eruption, turbulence, and icing conditions for a specific region?

A—Weather Depiction Chart.
B—In-Flight Weather Advisories.
C—Area Forecast.

In-Flight Weather Advisories are forecast to advise en route aircraft of development of potentially hazardous weather. The advisories are in three types, Convective SIGMET (WST), SIGMET (WS), and AIRMET (WA). All In-Flight Advisories use the same location identifiers (either VORs, airports, or well-known geographic areas) to describe the hazardous weather areas (PLT294) — AC 00-45

Answer (A) is incorrect because a Weather Depiction Chart is a computer analysis of surface aviation reports that gives a broad overview of observed weather at the time of the chart. Answer (C) is incorrect because an Area Forecast is a forecast of general weather conditions over an area the size of several states, and does not include hazardous weather.

ALL

9281. Constant Pressure Analysis Charts contain contours, isotherms and some contain isotachs. The contours depict

A—ridges, lows, troughs and highs aloft.
B—highs, lows, troughs, and ridges on the surface.
C—highs, lows, troughs, and ridges corrected to MSL.

Heights of the specified pressure for each station are analyzed through the use of solid lines called contours. This contour analysis gives the charts a height pattern. The contours depict highs, lows, troughs, and ridges, and all heights are given as pressure altitude. (PLT283) — AC 00-45

ALL

9283. Vertical wind shear can be determined by comparing winds on vertically adjacent constant pressure charts. The vertical wind shear that is critical for probability of turbulence is

A—4 knots or greater per 1,000 feet.
B—6 knots or more per 1,000 feet.
C—greater than 8 knots per 1,000 feet.

Vertical wind shear can be identified by comparing winds on vertically adjacent constant pressure charts. The vertical shear critical for probable turbulence is 6 knots per 1,000 feet. (PLT518) — AC 00-45

Answers

9243 [B] 9281 [A] 9283 [B]

ALL

9287. (Refer to Figure 149.) What approximate wind direction, speed, and temperature (relative to ISA) are expected for a flight over OKC at FL 370?

A—265° true; 27 knots; ISA +1°C.
B—260° true; 27 knots; ISA +6°C.
C—260° magnetic; 27 knots; ISA +10°C.

The Winds Aloft forecast (FD) for OKC at 34,000 feet shows a wind of 250° at 27 knots and a temperature of -43°C. At 39,000 feet the winds are 270° at 27 knots with a temperature of -54°C. Interpolation of these forecasts yields a forecast for 37,000 feet of winds from 260° at 27 knots with a temperature of -50°C. The ISA temperature at 37,000 feet is -56°C. All wind directions on a FD are relative to true north. (PLT076) — AC 00-45

Answer (A) is incorrect because the direction is rounded to the nearest 10° and temperature is ISA +6°C. Answer (C) is incorrect because winds are in degrees true, not magnetic.

ALL

9288. (Refer to Figure 149.) What approximate wind direction, speed, and temperature (relative to ISA) are expected for a flight over TUS at FL 270?

A—347° magnetic; 5 knots; ISA -10°C.
B—350° true; 5 knots; ISA +5°C.
C—010° true; 5 knots; ISA +13°C.

The FD forecast of TUS at 24,000 feet is 050° at 5 knots with a temperature of -17°C. At 30,000 feet, the winds are 330° at 5 knots with a temperature of -33°C. Interpolation of these forecasts yields a forecast for 27,000 feet of winds from 010° at 5 knots with a temperature of -25°C. ISA temperature at 27,000 feet is -39°C. All wind directions on a FD are referenced to true north. (PLT076) — AC 00-45

Answer (A) is incorrect because wind direction is degrees true (not magnetic). Answer (B) is incorrect because wind direction is 010° true, not 350° true.

ALL

9289. (Refer to Figure 149.) What will be the wind and temperature trend for an SAT ELP TUS flight at 16,000 feet?

A—Temperature decrease slightly.
B—Windspeed decrease.
C—Wind direction shift from southwest to east.

SAT — Winds 196° at 8 knots, temperature -2°C.
ELP — Winds 026° at 14 knots, temperature -1°C.
TUS — Winds 080° at 11 knots, temperature 0°C.

(PLT076) — AC 00-45

Answer (A) is incorrect because temperatures increase slightly. Answer (B) is incorrect because wind speed increases between SAT and ELP, but decreases from ILP to TUS.

ALL

9290. (Refer to Figure 149.) What will be the wind and temperature trend for an STL MEM MSY flight at FL 330?

A—Windspeed decrease.
B—Wind shift from west to north.
C—Temperature increase 5°C.

Interpolation of the FD forecasts for 30,000 and 34,000 feet yield a forecast at 33,000 feet of:

STL — Wind 260° at 56 knots, temperature -42°C.
MEM — Wind 260° at 20 knots, temperature -41°C.
MSY — Wind light and variable (9900), temperature -41°C.

(PLT076) — AC 00-45

Answer (B) is incorrect because the wind direction remains westerly and becomes light and variable (not shifting to the north). Answer (C) is incorrect because temperature increases 1°C.

ALL

9291. (Refer to Figure 149.) What will be the wind and temperature trend for a DEN ICT OKC flight at 11,000 feet?

A—Temperature decrease.
B—Windspeed increase slightly.
C—Wind shift from calm to a westerly direction.

Interpolation of the FD forecasts for 9,000 and 12,000 feet yield a forecast at 11,000 feet of:

DEN — Wind light and variable (9900), temperature +6°C.
ICT — Wind 060° at 2 knots, temperature +5°C.
OKC — Wind 110° at 2 knots, temperature +6°C.

(PLT076) — AC 00-45

Answer (A) is incorrect because the temperature increases at OKC to a temperature higher than DEN. Answer (C) is incorrect because the wind shifts from calm to an easterly direction.

Answers

9287 [B]	9288 [C]	9289 [C]	9290 [A]	9291 [B]

ALL

9292. (Refer to Figure 149.) What will be the wind and temperature trend for a DSM LIT SHV flight at 12,000 feet?

A—Windspeed decrease.
B—Temperature decrease.
C—Wind direction shift from northwest to southeast.

The FB forecasts for 12,000 feet are:

DSM — Wind 300° at 22 knots, temperature 0°C.
LIT — Wind 280° at 8 knots, temperature +6°C.
SHV — Wind 210° at 6 knots, temperature +6°C.

(PLT076) — AC 00-45

Answer (B) is incorrect because the temperature increases en route. Answer (C) is incorrect because the wind direction shifts from northwest to southwest.

ALL

9293. (Refer to Figure 149.) What is the forecast temperature at ATL for the 3,000-foot level?

A—+6°C.
B—+6°F.
C—Not reported.

Temperatures are not forecast on the FB at the 3,000 feet level when within 2,500 of the surface. (PLT076) — AC 00-45

Answer (A) is incorrect because no temperature is reported at the 3,000-foot level or for a level within 2,500 feet of station elevation. Answer (B) is incorrect because no temperature is reported at the 3,000-foot level, and temperatures are depicted in degrees Celsius, not Fahrenheit. The 06 indicates wind velocity of 6 knots.

ALL

9294. (Refer to Figure 149.) What approximate wind direction, speed, and temperature (relative to ISA) are expected for a flight over MKC at FL 260?

A—260° true; 43 knots; ISA +10°C.
B—260° true; 45 knots; ISA -10°C.
C—260° magnetic; 42 knots; ISA +9°C.

The FB forecast for MKC at 24,000 feet is winds of 260° at 38 knots with a temperature of -21°C. At 30,000 feet, the winds are forecast to be 260° at 50 knots with a temperature of -36°C. Interpolation for 26,000 feet yields a forecast of 260° at 42 knots with a temperature of -26°C. ISA temperature for 26,000 feet is -36°C. All wind directions on a FD are referenced to true north. (PLT076) — AC 00-45

Answer (B) is incorrect because forecast temperature is warmer than ISA. Answer (C) is incorrect because wind direction is degrees true, not magnetic.

ALL

9295. What wind direction and speed aloft are forecast by this WINDS AND TEMPERATURE ALOFT FORECAST (FD) for FL390 — "750649"?

A—350° at 64 knots.
B—250° at 106 knots.
C—150° at 6 knots.

For FB forecasts of wind speeds from 100 knots through 199 knots, subtract 50 from the wind direction code and add 100 to the speed code. The forecast of 750649 decodes as a wind of 250° (75 – 50 = 25) at 106 knots (100 + 06 = 106) with a temperature of –49°C (temperatures above FL240 are always negative). (PLT076) — AC 00-45

Answer (A) is incorrect because direction is the first two, not three, digits, and 50 must be subtracted from the first two digits. Speed is the second group of digits, not the fourth and fifth digits. Answer (C) is incorrect because 50 must be subtracted from the first two digits and 100 added to the second two digits, because the wind speed is forecast to be greater than 100 knots.

ALL

9296. What wind direction and speed aloft are forecast by this WINDS AND TEMPERATURE ALOFT FORECAST (FD) for FL390 — "731960"?

A—230° at 119 knots.
B—131° at 96 knots.
C—073° at 196 knots.

FB forecasts of wind speeds from 100 knots through 199 knots have 50 added to the wind direction code and 100 subtracted from the speed. The forecast of 731960 decodes as a wind of 230° at 119 knot with a temperature of -60°C. (PLT076) — AC 00-45

Answer (B) is incorrect because coded directions with wind speed over 100 knots range from 51 through 86. The direction is 230° (not 131°) at 119 knots (not 96 knots). Answer (C) is incorrect because 50 must be subtracted from the first two digits and 100 added to the second two digits. The last two digits are the temperature.

Answers

9292 [A] 9293 [C] 9294 [A] 9295 [B] 9296 [A]

ALL

9251. Forecast winds and temperatures aloft for an international flight may be obtained by consulting

A—Area Forecasts published by the departure location host country.
B—The current International Weather Depiction Chart appropriate to the route.
C—Wind and Temperature Aloft Charts prepared by the U.S. National Centers of Environmental Prediction (NCEP).

Computer-generated forecast charts of winds and temperatures aloft are available for international flights at specified levels. The U.S. National Centers for Environmental Prediction (NCEP), near Washington D.C., prepares and supplies to users charts of forecast winds, temperatures, and significant weather. (PLT284) — AC 00-45

Answer (A) is incorrect because Area Forecasts are forecasts of general weather conditions over an area of several states and do not contain forecasts of the winds and temperatures aloft. Answer (B) is incorrect because the International Weather Depiction Chart indicates current weather and does not forecast winds and temperatures aloft.

ALL

9255. A station is forecasting wind and temperature aloft to be 280° at 205 knots; temperature -51°C at FL390. How would this data be encoded in the FD?

A—780051.
B—789951.
C—280051.

FB forecasts of wind speeds from 100 knots through 199 knots have 50 added to the wind direction code and 100 subtracted from the speed code. Winds over 200 knots are coded as 199 knots. A wind of 280° at 205 knots with a temperature of -51°C is coded as "789951." (PLT076) — AC 00-45

Answer (A) is incorrect because it indicates a wind at 280° at 100 knots. The minus sign is to be omitted above 24,000 feet MSL. Answer (C) is incorrect because if the wind is 0 knots, the direction and wind group is coded "9900."

Surface Analysis and Constant Pressure Charts

The Surface Analysis Chart shows pressure patterns, fronts and information on individual reporting stations. The pressure patterns are shown by lines called **isobars**. The isobars on a surface weather map represent lines of equal pressure reduced to sea level.

Constant pressure charts are similar in many ways to the surface analysis chart in that they show the pressure patterns and some weather conditions for reporting stations. These charts show conditions at one of five pressure levels from 850 millibars to 200 millibars. These pressure levels correspond roughly with altitudes from 5,000 feet MSL to 39,000 feet MSL. The chart is for a pressure level rather than an altitude. The altitude (in meters) of the pressure level is shown by height contours. In addition to the height contour lines, constant pressure charts can contain lines of equal temperature (isotherms) and lines of equal wind speed (isotachs). Since these are both dotted lines, be careful not to get them confused when looking at a chart. Six items of information are shown on the charts for reporting stations. These are the wind, temperature, temperature/dew point spread, height of the pressure level and the change of the height level over the previous 12 hours.

These charts can be used to locate the jet stream and its associated turbulence and wind shear. When there is a large change in wind speed over a short distance, as indicated by closely spaced isotachs, the probability of turbulence is greatly increased. Since the jet stream is associated with discontinuities in the temperature lapse rate at breaks in the tropopause, closely spaced isotherms indicate the possibility of turbulence or wind shear.

Charts can be used together to get a three dimensional view of the weather. For example, lows usually slope to the west with increasing height. If a low stops moving, it will extend almost vertically. This type of low is typical of a slow moving storm that may cause extensive and persistent cloudiness, precipitation, and generally adverse flying weather.

Answers

9251 [C] 9255 [B]

ALL

9175. Isobars on a surface weather chart represent lines of equal pressure

A—at the surface.
B—reduced to sea level.
C—at a given atmospheric pressure altitude.

Sea level pressures are plotted on a surface weather chart and lines are drawn connecting lines of equal pressure. These lines of equal pressure are called isobars. (PLT287) — AC 00-45

Answer (A) is incorrect because the isobars are depicted at the sea level pressure pattern at 4 MB intervals, not just at the surface. Answer (C) is incorrect because the isobars are reduced to sea level pressure, not any given atmospheric pressure altitude.

ALL

9218. Under what conditions would clear air turbulence (CAT) most likely be encountered?

A—When constant pressure charts show 20-knot isotachs less than 150 NM apart.
B—When constant pressure charts show 60-knot isotachs less than 20 NM apart.
C—When a sharp trough is moving at a speed less than 20 knots.

Clear Air Turbulence (CAT) is likely in areas where the vertical wind shear exceeds 6 knots per 1,000 feet or horizontal shear exceeds 40 knots per 150 miles. (PLT501) — AC 00-6A, Chapter 13

Answer (B) is incorrect because when constant pressure charts show 20-knot isotachs less than 60 NM, CAT is most likely to be encountered, and "60-knot isotachs" do not exist. Answer (C) is incorrect because CAT can be expected upwind of the base of a deep upper trough, not because a sharp trough is moving.

ALL

9236. A strong wind shear can be expected

A—on the low pressure side of a 100-knot jetstream core.
B—where the horizontal wind shear is 15 knots, in a distance equal to 2.5° longitude.
C—if the 5°C isotherms are spaced 100 NM or closer together.

Jet streams stronger than 110 knots are apt to have significant turbulence in them in the sloping tropopause above the core, in the jet stream front below the core and on the low-pressure side of the core. (PLT518) — AC 00-30, Appendix 1

ALL

9310. (Refer to Figure 154.) What is the height of the 300-millibar level at the low pressure center in Canada?

A—9,120 meters MSL.
B—18,000 meters MSL.
C—11,850 meters MSL.

The "912" just below the "L" symbol indicates an altitude of 9,120 meters. (PLT043) — AC 00-45

Answer (B) is incorrect because the height of the 300-mb level of the low-pressure center is 9,120 meters MSL. Answer (C) is incorrect because it is the 200 MB (not 300 MB) pressure chart. The "185" below the "L" indicates the height of the 200 MB low-pressure center to be 18,500 meters MSL.

ALL

9311. (Refer to Figures 153, 154, and 155.) Interpret the path of the jetstream.

A—Southern California, Nevada, Utah, Nebraska/Kansas, and then southeastward.
B—Oregon, Idaho, Wyoming, Nebraska, Iowa, and across the Great Lakes.
C—The Alaska area, across Canada to Montana, South Dakota, then across the Great Lakes area.

The strongest winds shown on the 300 MB and 200 MB charts indicate the jet stream flows from Alaska, across Canada to Montana, etc. Hatching indicates winds of 70 to 110 knots. (PLT042) — AC 00-45

Answer (A) is incorrect because the isotachs in southern California area are between 30 to 50 knots Answer (B) is incorrect because Oregon has isotachs between 10 to 30 knots.

ALL

9312. (Refer to Figure 153.) What type weather system is approaching the California Coast from the west?

A—LOW.
B—HIGH.
C—Cold front.

The counterclockwise flow around the system indicates that it is a low-pressure area. (PLT043) — AC 00-45

Answer (B) is incorrect because the height contours would increase for a high. Answer (C) is incorrect because pressure systems, not fronts, are depicted on a Constant Pressure Chart.

Answers

9175 [B]	9218 [A]	9236 [A]	9310 [A]	9311 [C]	9312 [A]

ALL

9313. (Refer to Figures 153, 154, and 155.) What type weather is inferred by the almost vertical extent of the LOW in Canada?

A—A rapid-moving system with little chance of developing cloudiness, precipitation, and adverse flying conditions.
B—A slow-moving storm which may cause extensive and persistent cloudiness, precipitation, and generally adverse flying weather.
C—A rapid-moving storm, leaning to west with altitude, which encourages line squalls ahead of the system with a potential of severe weather.

An old, non-developing low-pressure system tilts little with height. The low becomes almost vertical and is clearly evident on both surface and upper air maps. Upper winds encircle the surface low rather than blow across it. Thus, the storm moves very slowly and usually causes extensive and persistent cloudiness, precipitation, and generally adverse flying weather. (PLT042) — AC 00-45

Answer (A) is incorrect because the wind aloft is encircling the low which means it is a slow moving system. Answer (C) is incorrect because the low is leaning very little, and the winds are encircling the low, which means it is a slow moving system.

ALL

9314. (Refer to Figures 153 through 155.) What is the approximate temperature for a flight from southern California to central Kansas at FL 350?

A—-16°C.
B—-39°C.
C—-41°C.

Temperatures at the 300 MB level are in the -30°C range, and the temperatures at the 200 MB level are in the -52°C range. (PLT043) — AC 00-45

Answer (A) is incorrect because -16°C is the approximate temperature at FL230 (not FL350) from southern CA to central KS. Answer (B) is incorrect because -39°C is the approximate temperature at FL340 (not FL350).

ALL

9315. (Refer to Figures 153 through 155.) Determine the approximate wind direction and velocity at FL 240 over the station in central Oklahoma.

A—280° at 10 knots.
B—320° at 10 knots.
C—330° at 13 knots.

The 500 MB chart shows a wind of WNW at 10 knots. The 300 MB chart shows a north wind at 10 knots. (PLT043) — AC 00-45

Answer (A) is incorrect because 280° at 10 knots is the wind direction and speed at the 500 MB pressure level (FL180). Answer (C) is incorrect because 330° at 13 knots is the interpolation between the 200 and 300 MB Charts, not the 300 and 500 MB Charts.

ALL

9316. (Refer to Figures 153 through 155.) What is the relative moisture content of the air mass approaching the California coast?

A—Dry.
B—Moist enough for condensation.
C—Very wet with high potential for clouds and precipitation.

The 500 MB chart shows either dry air or high temperature/dewpoint spreads for stations at the leading edge of the air mass. (PLT043) — AC 00-45

Answer (B) is incorrect because there is very little moisture, as depicted by the wide temperature/dewpoint spread. Answer (C) is incorrect because the air mass is dry, not wet.

Answers

9313 [B] 9314 [C] 9315 [B] 9316 [A]

Prognostic Charts

A prognostic chart depicts weather conditions that are forecast to exist at a specific time in the future shown on the chart. The **Low-Level Significant Prog Chart** forecasts weather conditions from the surface to the 400 millibar level (about 24,000 feet). The **High-Level Significant Weather Prog Chart** forecasts conditions from 25,000 feet to 63,000 feet. This encompasses FL250 to FL600.

The upper two panels of the Low-Level Significant Weather Prognostic Chart are the 12-hour and the 24-hour Significant Weather Prog Charts. These two charts forecast areas of MVFR and IFR weather as well as areas of moderate or greater turbulence. In each case, the turbulence is forecast to be of moderate intensity as shown by the inverted "V" symbol. The underlined number next to the turbulence symbol indicates that the turbulence goes up to that altitude. If the turbulence started at an altitude other than the surface, a number would also appear below the line.

The lower two panels are the 12-hour and the 24-hour surface prog charts. These forecast frontal positions and areas of precipitation.

An area surrounded by a dotted and dashed line has showery precipitation. An area surrounded by a continuous line has either continuous or intermittent precipitation. If the area is shaded, the precipitation will cover more than half the area. The type of precipitation expected is shown by the symbols used within the area.

ALL

9253. For international flights, a U.S. High-Level Significant Weather Prognostic Chart is prepared for use

A—at any altitude above 29,000.
B—between 25,000 feet and 60,000 feet pressure altitude.
C—between FL180 and FL600.

The High-Level Significant Weather Prog encompasses airspace from 25,000 feet to 60,000 feet pressure altitude over the conterminous U.S., Mexico, Central America, portions of South America, the western Atlantic, and eastern Pacific. (PLT286) — AC 00-45

ALL

9254. The U.S. Low-Level Significant Weather Prognostic Chart depicts weather conditions

A—that are forecast to exist at a specific time shown on the chart.
B—as they existed at the time the chart was prepared.
C—that are forecast to exist 6 hours after the chart was prepared.

The Low-Level Prognostic Chart depicts weather conditions that are forecast to exist at a specific time shown on the chart. (PLT286) — AC 00-45

Answer (B) is incorrect because prognostic charts forecast conditions, not report observed conditions (as the Weather Depiction Chart does). Answer (C) is incorrect because Low-Level Prognostic Charts forecast conditions 12 and 24 hours (not 6 hours) after the time of issuance.

ALL

9304. A prognostic chart depicts the conditions

A—existing at the surface during the past 6 hours.
B—which presently exist from the 1,000-millibar through the 700-millibar level.
C—forecast to exist at a specific time in the future.

The Low-Level Prog is a four-panel chart. The two lower panels are 12- and 24-hour surface progs. The two upper panels are 12- and 24-hour progs of significant weather from the surface to 400 millibars (24,000 feet). The charts show conditions as they are forecast to be at the valid time of the chart. (PLT286) — AC 00-45

Answer (A) is incorrect because prognostic charts relate to the future, not the past. Answer (B) is incorrect because Low-Level Prognostic Charts are issued for the surface and 24,000 feet, which is 400 MB. 1,000 MB is very close to the surface; i.e., sea level is 1,013 MB. 700 MB is approximately 10,000 feet MSL.

Answers

9253 [B] 9254 [A] 9304 [C]

ALL

9300. (Refer to Figure 151.) The 12-Hour Significant Weather Prognostic Chart indicates that West Virginia will likely experience

A—continuous or showery precipitation covering half or more of the area.
B—thunderstorms and rain showers covering half or more of the area.
C—continuous rain covering less than half of the area.

West Virginia is included in the shaded area enclosed by a dash-dot line. The symbols over Oklahoma indicate rain and rain showers are present in this area. More than half of the area is indicated by the shading. (PLT068) — AC00-45

Answer (B) is incorrect because thunderstorms would be indicated by an R with an arrow on its front leg. Answer (C) is incorrect because absence of shading indicates less than half area coverage. Continuous rain would be outlined in a solid (not dash-dot) line.

ALL

9301. (Refer to Figure 151.) The 12-Hour Significant Weather Prognostic Chart indicates that eastern Kentucky and eastern Tennessee can expect probable ceilings

A—less than 1,000 feet and/or visibility less than 3 miles.
B—less than 1,000 feet and/or visibility less than 3 miles, and moderate turbulence below 10,000 feet MSL.
C—less than 1,000 feet and/or visibility less than 3 miles, and moderate turbulence above 10,000 feet MSL.

Use panels A and B (12-hour prog). Note the legend near the center of the chart which explains methods of depiction. (PLT068) — AC 00-45

Answers (B) and (C) are incorrect because no turbulence is forecast for eastern Kentucky and eastern Tennessee on the 12-hour significant weather prog. Turbulence is indicated by a dashed line such as the one which includes parts of western Kentucky and Tennessee.

ALL

9302. (Refer to Figure 151.) The chart symbols over southern California on the 12-Hour Significant Weather Prognostic Chart indicate

A—expected top of moderate turbulent layer to be 12,000 feet MSL.
B—expected base of moderate turbulent layer to be 12,000 feet MSL.
C—light turbulence expected above 12,000 feet MSL.

Use panel A (12-hour prog). Forecast areas of moderate or greater turbulence are enclosed by long-dashed lines. The symbol indicates moderate turbulence. Figures below and above a short line show expected base and top of the turbulent layer in hundreds of feet MSL. Absence of a figure below the line indicates turbulence from the surface upward. No figure above this line indicates turbulence extending above the upper limit of the chart. (PLT068) — AC 00-45

Answer (B) is incorrect because 12,000 feet MSL is the top of the turbulence, not the base. Answer (C) is incorrect because the symbol shows moderate (not light) turbulence and it is below (not above) 12,000 feet MSL.

ALL

9303. (Refer to Figure 151.) A planned low-altitude flight from central Oklahoma to western Tennessee at 1200Z is likely to encounter

A—continuous or intermittent rain or rain showers, moderate turbulence, and freezing temperatures below 8,000 feet.
B—continuous or showery rain over half or more of the area, moderate turbulence, and freezing temperatures above 10,000 feet.
C—showery precipitation covering less than half the area, no turbulence below 18,000 feet, and freezing temperatures above 12,000 feet.

Use panels C and D (24-hour prog). These are valid at 1200Z. The symbol indicates showery precipitation (i.e., rain showers) embedded in an area of continuous rain covering half or more of the area. The moderate turbulence is indicated, and is found on chart C. The symbol drawn at 4,000-foot intervals indicates the freezing level above mean sea level, and is found on chart C. In this case, the freezing level for the planned route of flight is between 8,000 feet and 12,000 feet, or at approximately 10,000 feet. (PLT063) — AC 00-45

Answer (A) is incorrect because the route of flight is south of the 8,000-foot freezing level. Answer (C) is incorrect because the shading indicates more than 50% rain coverage, the turbulence is up to 18,000 feet, and temperatures are probably freezing above 10,000 feet (not 12,000 feet).

Answers

9300 [A] 9301 [A] 9302 [A] 9303 [B]

Reports and Forecasts of Hazardous Weather

The Radar Summary Chart graphically displays a collection of radar reports. It shows the types of precipitation echoes and indicates their intensity, trend, tops and bases. Shaded areas are those with significant radar returns. Areas surrounded by one line are classified as having weak to moderate echoes. Areas enclosed with two lines have strong to very strong echoes and areas inside three lines have intense to extreme echoes. The direction of an individual cell is indicated by an arrow and its speed of movement is shown by a number near the point of the arrow. Line or area movement is shown with an arrow using the "feathers" associated with wind on other charts. The wind is not shown on this chart.

A **Convective Outlook (AC)** describes the prospects for general thunderstorm activity during the following 24 hours. Areas in which there is a high, moderate or slight risk of severe thunderstorms are included as well as areas where thunderstorms may approach severe limits.

The **Severe Weather Outlook Chart** is a preliminary 24-hour outlook for thunderstorms presented in two panels. A line with an arrowhead delineates an area of probable general thunderstorm activity. An area labeled APCHG indicates probable general thunderstorm activity may approach severe intensity. "Approaching" means winds of 35 to 50 knots or hail 1/2 to 3/4 of an inch in diameter.

AIRMETs and **SIGMETs** are issued to alert pilots to potentially hazardous weather not adequately forecast in the current Area Forecast (FA). They are appended to the current FA and are broadcast by the FSS upon issue and at H+15 and H+45 while they are in effect. ARTCC facilities will announce that a SIGMET is in effect and the pilot can then contact the nearest FSS for the details.

AIRMET forecast:

- Moderate icing
- Moderate turbulence
- Sustained winds of 30 knots or more at the surface
- Widespread areas of ceilings less than 1,000 feet or visibilities of less than 3 miles
- Extensive mountain obscurement

SIGMET forecast:

- Severe and extreme turbulence
- Severe icing
- Widespread dust storms, sandstorms or volcanic ash lowering visibility to below three miles

Convective SIGMETs cover the following:

- Tornadoes
- Lines of thunderstorms
- Embedded thunderstorms
- Thunderstorm areas greater than or equal to intensity level 4
- Hail greater than 3/4 of an inch in diameter

Convective SIGMETs are each valid for one hour and are removed at H+40. They are reissued as necessary. On an hourly basis, an outlook is made up for each of the WST regions. This outlook covers the prospects for 2 to 6 hours.

Telephone Information Briefing Service (TIBS) is provided by automated flight service stations (AFSS). It is a continuous recording of meteorological and aeronautical information, available by telephone by calling 1-800-WX-BRIEF. Each AFSS provides at least four route and/or area briefings. In addition, airspace procedures and special announcements (if applicable) concerning aviation interests may also be available. Depending on user demand, other items may be provided, such as METAR observations, terminal aerodrome forecasts, wind/temperatures aloft forecasts, etc. TIBS is not intended to substitute for specialist-provided preflight briefings. It is, however, recommended for use as a preliminary briefing, and often will be valuable in helping you to make a "go or no go" decision.

ALL

9306. (Refer to Figure 152.) What weather conditions are depicted in the area indicated by arrow A on the Radar Summary Chart?

A—Moderate to strong echoes; echo tops 30,000 feet MSL; line movement toward the northwest.

B—Weak to moderate echoes; average echo bases 30,000 feet MSL; cell movement toward the southeast; rain showers with thunder.

C—Strong to very strong echoes; echo tops 30,000 feet MSL; thunderstorms and rain showers.

Area A indicates an echo intensity level of 3 to 4, strong to very strong. Echo heights are displayed in hundreds of feet MSL and are approximations. Tops are entered above a short line while any available bases are entered below. Echo tops in area A are at 30,000 feet. Thunderstorms and rain showers are indicated by the contraction "TRW." (PLT063) — AC 00-45

Answer (A) is incorrect because an area within the first contour is level 1 or 2, or weak to moderate echo intensity. Strong echo intensity is a level 3, which is indicated within the second contour. The line movement is to the northeast, not northwest, as indicated by the shaft and barb combination located in southern Nebraska. Answer (B) is incorrect because weak to moderate echoes are located in the first contour. The "300" over the line indicates the tops, not bases, at 30,000 feet MSL.

ALL

9307. (Refer to Figure 152.) What weather conditions are depicted in the area indicated by arrow D on the Radar Summary Chart?

A—Echo tops 4,100 feet MSL; strong to very strong echoes within the smallest contour; area movement toward the northeast at 50 knots.

B—Intense to extreme echoes within the smallest contour; echo tops 29,000 feet MSL; cell movement toward the northeast at 50 knots.

C—Strong to very strong echoes within the smallest contour; echo bases 29,000 feet MSL; cell in northeast Nebraska moving northeast at 50 knots.

Area D indicates an echo intensity level of 5 to 6, intense to extreme. Echo heights are displayed in hundreds of feet MSL and are approximations. Tops are entered above a short line while any available bases are entered below. Echo tops in area D are at 29,000 feet. Individual cell movement is indicated by an arrow with the speed in knots entered as a number and the arrow shaft pointing in the direction of movement. In area D an individual cell is moving northeast at 50 knots. (PLT063) — AC 00-45

Answer (A) is incorrect because the 410 is for another area of echoes to the northeast of the D arrow, with tops of 41,000 feet MSL (not 4,100). Answer (C) is incorrect because the tops, not bases, are 29,000 feet MSL and the echoes are intense to extreme, not strong to very strong.

ALL

9308. (Refer to Figure 152.) What weather conditions are depicted in the area indicated by arrow C on the Radar Summary Chart?

A—Average echo bases 2,800 feet MSL; thundershowers; intense to extreme echo intensity.

B—Cell movement toward the northwest at 20 knots; intense echoes; echo bases 28,000 feet MSL.

C—Area movement toward the northeast at 20 knots; strong to very strong echoes; echo tops 28,000 feet MSL.

Area C indicates an echo intensity level of 3 to 4, strong to very strong. Echo heights are displayed in hundreds of feet MSL and are approximations. Tops are entered above a short line while any available bases are entered below. Echo tops in area C are at 28,000 feet. Line or area movement is indicated by a shaft and barb combination with the shaft indicating the direction and the barbs the speed. A whole barb is 10 knots, a half barb is 5 knots, and a pennant is 50 knots. In area C, the area movement is toward the northeast at 20 knots. (PLT063) — AC 00-45

Answer (A) is incorrect because 280 indicates tops of 28,000 feet MSL (not bases of 2,800 feet MSL). Answer (B) is incorrect because point C would be in the third (not second) level of contour if the echoes were intense.

Answers

9306 [C] 9307 [B] 9308 [C]

ALL

9309. (Refer to Figure 152.) What weather conditions are depicted in the area indicated by arrow B on the Radar Summary Chart?

A—Weak echoes; heavy rain showers; area movement toward the southeast.

B—Weak to moderate echoes; rain showers increasing in intensity.

C—Strong echoes; moderate rain showers; no cell movement.

Area B indicates an echo intensity of 1 to 2, weak to moderate. The contraction "RW+" means rain showers, the "plus" indicating that the intensity is increasing or there is a new echo. (PLT063) — AC 00-45

Answer (A) is incorrect because only intensity of precipitation (increasing or decreasing), is depicted on a Radar Summary Chart. The area of movement is to the northeast, as indicated by the shaft and barb combination located in southern Nebraska. Answer (C) is incorrect because strong echoes are level 3, which is indicated by the second contour line on the Radar Summary Chart. Precipitation is indicated by increasing or decreasing.

ALL

9305. What information is provided by a Convective Outlook (AC)?

A—It describes areas of probable severe icing and severe or extreme turbulence during the next 24 hours.

B—It provides prospects of both general and severe thunderstorm activity during the following 24 hours.

C—It indicates areas of probable convective turbulence and the extent of instability in the upper atmosphere (above 500 mb).

A Convective Outlook (AC) describes the prospects for general and severe thunderstorm activity during the following 24 hours. Use the outlook primarily for planning (or canceling) flights later in the day. (PLT514) — AC 00-45

Answer (A) is incorrect because severe icing and severe or extreme turbulence are the subjects of SIGMETs. Answer (C) is incorrect because it describes a 500-mb Constant Pressure Analysis Chart.

ALL

9252. How will an area of thunderstorm activity, that may grow to severe intensity, be indicated on the Severe Weather Outlook Chart?

A—SLGT within cross-hatched areas.

B—APCHG within any area.

C—SVR within any area.

On a Severe Weather Outlook Chart, an area labeled "APCHG" indicates that probable general thunderstorm activity may approach severe intensity. (PLT518) — AC 00-45

Answer (A) is incorrect because a cross-hatched area identifies a tornado watch area. Answer (C) is incorrect because the term SVR is not used on the Severe Weather Outlook Chart.

ALL

9578. When are severe weather watch bulletins (WW) issued?

A—Every 12 hours as required.

B—Every 24 hours as required.

C—Unscheduled and issued as required.

A severe weather watch bulletin (WW) defines areas of possible severe thunderstorms or tornado activity. They are unscheduled and are issued as required. (PLT316) — AC 00-45

ALL

9705. The Hazardous Inflight Weather Advisory Service (HIWAS) is a continuous broadcast over selected VORs of

A—SIGMETs, CONVECTIVE SIGMETs, AIRMETs, Severe Weather Forecast Alerts (AWW), and Center Weather Advisories (CWA).

B—SIGMETs, CONVECTIVE SIGMETs, AIRMETs, Wind Shear Advisories, and Severe Weather Forecast Alerts (AWW).

C—Wind Shear Advisories, Radar Weather Reports, SIGMETs, CONVECTIVE SIGMETs, AIRMETs, and Center Weather Advisories (CWA).

HIWAS broadcasts include SIGMETs, Convective SIGMETs, AIRMETs, Severe Weather Forecast Alerts, and Center Weather Advisories. (PLT515) — AC 00-45

Answers

9309 [B]	9305 [B]	9252 [B]	9758 [C]	9705 [A]

ALL

9256. At what time are current AIRMETs broadcast in their entirety by the Hazardous Inflight Weather Advisory Service (HIWAS)?

A—15 and 45 minutes after the hour during the first hour after issuance, and upon receipt.
B—Every 15 minutes until the AIRMET is canceled.
C—There is a continuous broadcast over selected VORs of Inflight Weather Advisories.

The Hazardous Inflight Weather Advisory Service (HIWAS) is a continuous broadcast service over selected VORs of In-Flight Weather Advisories; i.e. SIGMETs, CONVECTIVE SIGMETs, AIRMETs, Severe Weather Forecast Alerts (AWW), and Center Weather Advisories (CWA). (PLT515) — AC 00-45

ALL

9257. If a SIGMET alert is announced, how can information contained in the SIGMET be obtained?

A—ATC will announce the hazard and advise when information will be provided in the FSS broadcast.
B—By contacting a weather watch station.
C—By contacting the nearest AFSS.

SIGMETs, CWAs, and AIRMETs are broadcast by FSS's upon receipt and at 30-minute intervals at H+15 and H+45 for the first hour after issuance. Thereafter, a summarized alert notice will be broadcast at H+15 and H+45 during the valid period of the advisories. If a pilot has not previously received the SIGMET, etc., he/she should call the nearest FSS. (PLT290) — AC 00-45

Answer (A) is incorrect because ATC does not advise when to listen to an FSS broadcast, rather they tell you to contact FSS. Answer (B) is incorrect because the pilot may monitor an FSS broadcast or contact any FSS. "Weather watch station" is a nonexistent term.

ALL

9249. If squalls are reported at the destination airport, what wind conditions existed at the time?

A—Sudden increases in windspeed of at least 15 knots, to a sustained wind speed of 20 knots, lasting for at least 1 minute.
B—A sudden increase in wind speed of at least 16 knots, the speed rising to 22 knots or more for 1 minute or longer.
C—Rapid variation in wind direction of at least 20° and changes in speed of at least 10 knots between peaks and lulls.

A squall (SQ) is a sudden increase in wind speed of at least 16 knots, the speed rising to 22 knots or more, and lasting at least 1 minute. (PLT475) — AC 00-45

ALL

9284. (Refer to Figure 148.) Which system in the Convective SIGMET listing has the potential of producing the most severe storm?

A—The storms in Texas and Oklahoma.
B—The storms in Colorado, Kansas, and Oklahoma.
C—The isolated storm 50 miles northeast of Memphis (MEM).

Convective SIGMET 44C forecasts level 5 thunderstorms 50 miles northeast of MEM. (PLT067) — AC 00-45

ALL

9285. (Refer to Figure 148.) What time period is covered by the outlook section of the Convective SIGMET?

A—24 hours after the valid time.
B—2 to 6 hours after the valid time.
C—No more than 2 hours after the valid time.

Convective SIGMETs are valid for 6 hours, with the last 4 hours being the outlook. (PLT067) — AC 00-45

Answer (A) is incorrect because the outlook is for a period of 2 to 6 hours, not 24, after the valid time. Answer (C) is incorrect because the outlook can cover up to 6 hours after the valid time.

ALL

9286. Which type weather conditions are covered in the Convective SIGMET?

A—Embedded thunderstorms, lines of thunderstorms, and thunderstorms with 3/4-inch hail or tornadoes.
B—Cumulonimbus clouds with tops above the tropopause and thunderstorms with 1/2-inch hail or funnel clouds.
C—Any thunderstorm with a severity level of VIP 2 or more.

Convective SIGMET forecast:

1. *Severe thunderstorms which have either surface winds greater than 50 knots, hail equal to or greater than 3/4 inches in diameter, or tornadoes;*
2. *Embedded thunderstorms;*
3. *Lines of thunderstorms; or*
4. *Thunderstorms equal to or greater than VIP level 4 affecting 40% or more of an area at least 3,000 square miles.*

(PLT290) — AC 00-45

Answer (B) is incorrect because cumulonimbus clouds with tops above the tropopause is not a weather condition covered in a Convective SIGMET. Answer (C) is incorrect because thunderstorms must be at least VIP level 4.

Answers

9256 [C]	9257 [C]	9249 [B]	9284 [C]	9285 [B]	9286 [A]

ALL

9747. The Telephone Information Briefing Service (TIBS) recordings are provided by selected Automated Flight Service Stations and

A—are updated on the hour.
B—are designed to replace the standard briefing given by a flight service specialist.
C—contain area briefings encompassing a 50 NM radius.

TIBS provides continuous telephone recordings of meteorological and aeronautical information, specifically area and route briefings, as well as airspace procedures and special announcements, if applicable. It is designed to be a preliminary briefing tool and is not intended to replace a standard briefing from a flight service specialist. TIBS is available 24 hours a day by calling 1-800-WX-BRIEF and is updated when conditions change. As a minimum, area briefings encompass a 50 NM radius. (PLT515) — AIM ¶7-1-8

PIREPs

A pilot weather report (PIREP) is often the most timely source of information about such weather conditions as icing and multiple cloud layers. While area forecasts and freezing level charts can give the pilot a good idea of the potential for icing, only a PIREP can let the pilot know what is happening currently. A typical PIREP appended to an SA is:

FTW UA /OV DFW 18005/TM1803/FL095/TP PA 30/SK 036 OVC 060/070 OVC 075/OVC ABV

The translation is:

FTW / UA — PIREP from reporting station FTW.

OV DFW 18005 — location is the DFW 180° radial at 5 miles.

TM 1803 — time of the report is 1803.

FL095 — altitude is 9,500 feet.

TP PA 30 — Type of aircraft is a PA 30.

SK 036 OVC 060/070 OVC 075/OVC ABV — Sky condition. The base of an overcast layer is at 3,600 feet with top at 6,000 feet. A second overcast layer has its base at 7,000 feet and its top is 7,500 feet. There is another overcast layer above the aircraft's altitude of 9,500 feet.

ALL

9247. What sources reflect the most accurate information on current and forecast icing conditions?

A—Low-Level Sig Weather Prog Chart, RADATs, and the Area Forecast.
B—PIREPs, Area Forecast, and the Freezing Level Chart.
C—PIREPs, AIRMETs, and SIGMETs.

PIREPs are the only source for actual icing conditions; AIRMETs and SIGMETs advise of forecast moderate and severe icing conditions, respectively. (PLT294) — AC 00-45

Answer (A) is incorrect because Low-Level Sig Weather Prog Charts do not forecast icing conditions but do forecast freezing levels. Answer (B) is incorrect because although the Freezing Level Panel of the Composite Moisture Stability Chart gives you the lowest observed freezing level, it does not indicate the presence of clouds or precipitation, which must be present for icing to occur.

ALL

9250. Which type of weather can only be directly observed during flight and then reported in a PIREP?

A—Turbulence and structural icing.
B—Jetstream-type winds and icing.
C—Level of the tropopause and turbulence.

Aircraft in flight are the only means of directly observing cloud tops, icing and turbulence. (PLT061) — AC 00-45

Answer (B) is incorrect because a pilot would not be able to determine from observation if jet stream type winds or other CAT were encountered. Answer (C) is incorrect because the level of the tropopause is determined by radiosondes released by ground weather observing stations. It is not a type of weather that can be directly observed by a pilot during flight.

Answers

9747 [C] 9247 [C] 9250 [A]

ALL

9271. (Refer to Figure 145.) The peak wind at KAMA was reported to be from 320° true at 39 knots,

A—which occurred at 1743Z.
B—with gusts to 43 knots.
C—with .43 of an inch liquid precipitation since the last report.

"PK WND 32039/43" indicates that the peak wind was reported to be from 320° true at 39 knots, which occurred at 43 minutes past the hour. (PLT059) — AC 00-45

ALL

9713. KFTW UA/OV DFW/TM 1645/FL100/TP PA30/ SK SCT031-TOP043/BKN060-TOP085/OVC097-TOPUNKN/WX FV00SM RA/TA 07.

This pilot report to Fort Worth (KFTW) indicates

A—the aircraft is in light rain.
B—that the top of the ceiling is 4,300 feet.
C—the ceiling at KDFW is 6,000 feet.

"BKN060-TOP085" indicates the ceiling (defined as the lowest broken or overcast layer aloft) is broken at 6,000 feet with tops at 8,500 feet. (PLT061) — AC 00-45

Answer (A) is incorrect because "WX FV00SM RA" indicates the flight visibility is 0 statute miles due to moderate rain. Answer (B) is incorrect because the scattered layer has tops at 4,300 feet, but this does not constitute a ceiling.

Answers

9271 [A] 9713 [C]

Cross-Reference A
Question Number and Page Number

The following list of the numbered questions included in this ASA Test Prep is given in sequential order; however, as a result of our ongoing review of FAA test question databases, some question numbers may have been removed due to changes in the database. **All currently existing questions are accounted for in this list.** For more information about the questions included in ASA Test Preps, please read Pages v–vi in the front matter for this book.

Question Number	Page Number
8685	4–51
8686	4–51
8687	4–39
8688	4–39
8689	4–39
8690	4–40
8691	4–40
8692	4–38
8693	4–39
8694	4–39
8695	4–39
8696	4–39
8697	5–5
8698	5–5
8699	5–5
8700	5–6
8701	5–6
8702	5–6
8703	5–6
8704	5–7
8705	5–7
8706	5–7
8707	5–7
8708	5–8
8709	5–8
8710	5–8
8711	5–8
8712	4–19
8713	4–19
8714	4–20
8715	4–20
8716	4–20
8717	4–15
8718	4–15
8719	4–15
8720	4–15
8721	4–16
8722	5–10
8723	5–10
8724	5–11
8725	5–11
8726	5–11
8727	4–43
8728	4–44
8729	4–44
8730	4–44
8731	4–44
8732	4–44
8733	4–44
8734	4–45
8735	4–45
8736	4–45
8737	4–46
8738	4–46
8739	4–46
8740	4–46
8741	4–47
8742	4–35
8743	4–35
8744	4–35

Question Number	Page Number
8745	4–35
8746	4–36
8747	4–36
8748	4–36
8749	4–36
8750	4–36
8751	4–36
8752	4–36
8753	4–37
8754	4–37
8755	4–37
8756	4–37
8757	4–37
8758	4–37
8759	4–37
8760	4–38
8761	4–38
8762	4–38
8763	4–38
8764	4–38
8765	4–38
8766	4–38
8767	1–31
8768	1–32
8769	5–34
8770	5–34
8771	5–34
8772	5–34
8773	5–35
8774	4–11
8775	4–12
8776	5–35
8777	5–35
8778	5–35
8779	5–36
8780	4–12
8781	5–36
8782	6–62
8783	6–62
8784	6–63
8785	6–63
8786	6–63
8787	5–36
8788	5–36
8789	5–37
8790	5–37
8791	5–37
8792	1–72
8793	6–61
8794	6–62
8795	4–104
8796	6–62
8797	6–62
8798	6–62
8799	6–66
8800	6–66
8801	6–66
8802	6–66
8803	6–66
8804	6–67

Question Number	Page Number
8805	6–67
8806	6–67
8807	1–82
8808	1–82
8809	1–83
8810	6–64
8811	6–64
8812	6–64
8813	1–83
8814	1–83
8815	1–83
8816	6–64
8817	6–65
8818	6–65
8819	1–83
8820	1–84
8821	1–84
8822	4–79
8823	4–79
8824	6–63
8825	6–63
8826	6–63
8827	1–84
8828	1–84
8829	1–84
8830	1–84
8831	1–72
8832	1–72
8833	1–72
8834	1–85
8835	4–104
8836	6–68
8837	6–68
8838	1–85
8839	6–68
8840	1–85
8841	1–85
8842	1–73
8843	1–85
8844	5–37
8845	5–38
8846	5–38
8847	5–38
8848	5–38
8849	6–65
8850	6–65
8851	6–65
8852	6–65
8853	6–43
8854	6–39
8855	6–43
8856	6–43
8857	6–44
8858	6–44
8859	6–44
8860	6–44
8861	6–45
8862	6–45
8863	6–46
8864	6–46

Question Number	Page Number
8865	6–47
8866	6–47
8867	6–48
8868	2–27
8869	2–27
8870	2–28
8871	2–28
8872	6–7
8873	6–7
8874	6–8
8875	6–8
8876	6–8
8877	6–8
8878	6–8
8879	6–8
8880	6–9
8881	6–6
8882	6–6
8883	6–6
8884	6–6
8885	6–6
8886	6–6
8887	6–13
8888	6–6
8889	6–7
8890	6–9
8891	6–9
8892	6–9
8893	6–11
8894	6–11
8895	6–11
8896	6–12
8897	6–12
8898	6–12
8899	6–12
8900	6–12
8901	2–44
8902	2–44
8903	2–45
8904	2–45
8905	2–43
8906	2–43
8907	2–45
8908	2–52
8909	2–54
8910	2–54
8911	2–52
8912	2–52
8913	2–52
8914	2–47
8915	2–47
8916	2–53
8917	2–53
8918	2–53
8919	2–53
8920	2–53
8921	2–52
8922	2–45
8923	2–45
8924	2–45

Question Number	Page Number
9184	8–12
9185	8–11
9186	8–11
9187	8–12
9188	8–12
9189	8–18
9190	8–36
9191	8–8
9192	8–9
9193	8–14
9194	8–14
9195	8–12
9196	8–16
9197	8–16
9198	8–16
9199	8–18
9200	8–16
9201	8–22
9202	8–17
9203	8–16
9204	8–16
9205	8–17
9206	8–14
9207	8–14
9208	8–14
9209	8–6
9210	8–17
9211	8–18
9212	8–18
9213	8–9
9214	8–17
9215	8–9
9216	8–9
9217	8–9
9218	8–50
9219	8–38
9220	8–20
9221	8–32
9222	2–8
9223	8–32
9224	8–32
9225	8–22
9226	8–36
9227	8–9
9228	8–10
9229	8–7
9230	8–37
9231	8–37
9232	8–36
9233	8–6
9234	8–38
9235	8–36
9236	8–50
9237	8–37
9238	8–7
9239	8–37
9240	8–6
9241	8–7
9242	8–42
9243	8–46
9244	8–44
9245	8–44
9246	8–44
9247	8–58
9248	8–44
9249	8–57
9250	8–58
9251	8–49
9252	8–56
9253	8–52
9254	8–52
9255	8–49
9256	8–57
9257	8–57
9258	2–14
9259	8–38
9260	8–39
9261	2–14
9262	8–35
9263	8–36
9264	8–36
9265	8–43
9266	8–40
9267	8–40
9268	8–40
9269	8–40
9270	8–40
9271	8–59
9272	8–40
9273	8–41
9274	8–41
9275	8–41
9276	8–41
9277	8–41
9278	8–45
9279	8–45
9280	8–45
9281	8–46
9283	8–46
9284	8–57
9285	8–57
9286	8–57
9287	8–47
9288	8–47
9289	8–47
9290	8–47
9291	8–47
9292	8–48
9293	8–48
9294	8–48
9295	8–48
9296	8–48
9297	8–43
9298	8–44
9299	8–43
9300	8–53
9301	8–53
9302	8–53
9303	8–53
9304	8–52
9305	8–56
9306	8–55
9307	8–55
9308	8–55
9309	8–56
9310	8–50
9311	8–50
9312	8–50
9313	8–51
9314	8–51
9315	8–51
9316	8–51
9317	4–11
9318	3–26
9319	4–11
9320	4–42
9321	4–41
9322	4–31
9323	4–31
9324	4–11
9325	1–5
9326	1–30
9327	4–11
9328	1–4
9329	1–4
9330	1–5
9331	1–5
9332	1–5
9333	1–6
9334	1–11
9335	1–6
9336	1–86
9337	1–86
9338	1–87
9339	1–10
9340	1–6
9341	1–87
9342	1–11
9343	1–6
9344	1–11
9345	1–11
9346	1–12
9347	1–12
9348	1–12
9349	1–5
9350	1–4
9351	1–5
9352	2–29
9353	2–29
9354	7–12
9355	4–49
9356	2–11
9357	2–11
9358	4–49
9359	4–49
9360	4–50
9361	4–50
9362	7–5
9363	7–6
9364	7–6
9365	7–6
9366	1–87
9367	1–87
9368	6–33
9369	6–32
9370	6–28
9371	1–88
9372	1–88
9373	1–88
9374	6–24
9375	2–17
9376	2–18
9377	2–18
9378	2–51
9379	1–54
9380	2–3
9381	2–3
9382	6–34
9383	6–32
9384	6–33
9385	6–32
9386	2–4
9387	2–4
9388	1–54
9389	7–6
9390	7–7
9391	6–33
9392	6–33
9393	6–33
9394-1	6–20
9394-2	6–20
9395	6–13
9396	6–13
9397	6–14
9398	6–14
9399	6–9
9400	6–14
9401	6–13
9402	6–24
9403	2–34
9404	2–18
9405	2–18
9406	2–18
9407	2–3
9408	2–19
9409	6–7
9410	2–11
9411	2–36
9412	2–36
9413	2–37
9414	1–88
9415	1–88
9416	2–50
9416-1	2–48
9416-2	2–48
9417	2–49
9418	6–44
9419	6–44
9420	7–8
9421	2–43

Cross-Reference B
Learning Statement Code and Question Number

The expression "learning statement," as used in FAA airman testing, refers to measurable statements about the knowledge a student should be able to demonstrate following a certain segment of training. When you take the applicable airman knowledge test required for an airman pilot certificate or rating, you will receive an Airman Knowledge Test Report. The test report will list the learning statement codes for questions you have answered incorrectly. Match the codes given on your test report to the ones in the official FAA Learning Statement Codes (listed in this cross-reference). Use Cross-Reference A in this book to find the page number for the question numbers listed in the table beginning on the next page.

Your instructor is required to provide instruction on each of the areas of deficiency listed on your Airman Knowledge Test Report (as LSCs), and give you an endorsement for this instruction. The Airman Knowledge Test Report must be presented to the examiner conducting your practical test. During the oral portion of the practical test, the examiner is required to evaluate the noted areas of deficiency.

The FAA's learning statement codes are a hierarchical sequence of classification codes that places a knowledge item in a unique category, which can then be used for reference to source textbooks and study material. The LSCs are assigned to all FAA test questions in order to categorize them for placement on a given Knowledge Exam. This classification code system uses the following hierarchy (which is further detailed in the cross-reference table):

- *Topic*—this is the overall subject matter topic code, the highest classification of overall subject matter a knowledge test item was developed to assess (for example, "Aerodynamics").
- *Content*—the secondary level subject matter code (for example, "Airspeed").
- *Specific*—the basic hierarchical classification code the subject matter for a knowledge test item (for example, "Thrust").

If you received a code on your Airman Test Report that is not listed in this cross-reference, email ASA at **cfi@asa2fly.com**. We will provide the definition so you can review that subject area.

The FAA appreciates testing experience feedback. You can contact the branch responsible for the FAA Knowledge Exams directly at:

Federal Aviation Administration
AFS-630, Airman Testing Standards Branch
PO Box 25082
Oklahoma City, OK 73125
Email: AFS630comments@faa.gov

Learning Statement Code	*FAA Reference*	***Subject Description** (or **Topic** 〉 **Content** 〉 **Specific** classification)* ***Question Numbers***
PLT002	FAA-H-8083-25	Aircraft Performance 〉 Computations 〉 Airspeeds *8563, 8564, 8565, 8566, 8567, 8696*
PLT004	FAA-H-8083-25	Aircraft Performance 〉 Charts 〉 Climb; Engine Out Performance *8382, 8474, 8475, 8476, 8477, 8478, 8553, 8554, 8555, 8556, 8557, 8558, 8559, 8560, 8561, 8562, 8593, 8594, 8595, 8596, 8597, 8598, 8599, 8600, 8601, 8602, 8628, 8629, 8630, 8631, 8632, 8635, 8636, 8637, 8682, 8683, 8684, 8685, 8686, 8740, 8741*
PLT006	FAA- H-8083-3	Calculate aircraft performance—glide *8381*
PLT007	FAA-H-8083-25	Aircraft Performance 〉 Charts 〉 Holding; Landing; Takeoff *8613, 8615, 8616, 8617, 8638, 8639, 8640, 8641, 8642-1, 8668, 8669, 8670, 8671, 8672, 8687, 8688, 8689, 8690, 8691, 8712, 8713, 8714, 8715, 8716, 8727, 8728, 8729, 8730, 8731, 8763*
PLT008	FAA-H-8083-25	Aircraft Performance 〉 Charts 〉 Landing *8117, 8118, 8119, 8120, 8121, 8122, 8123, 8124, 8125, 8126, 8129, 8504, 8505, 8506, 8507, 8508, 8509, 8510, 8511, 8512, 8608, 8609, 8610, 8611, 8612, 8692, 8742, 8743, 8745, 8746, 8747, 8748, 8749, 8750, 8751, 8752, 8753, 8754, 8755, 8756, 8757, 8758, 8759, 8760, 8761, 8762, 8764, 8765, 8766*
PLT009	FAA-H-8083-21	Aircraft Performance 〉 Charts 〉 Power Check; Turbine Engine *8533, 8534, 8535, 8536, 8537*
PLT010	FAA-H-8083-25	Aircraft Performance 〉 Charts 〉 Takeoff *8588, 8589, 8590, 8591, 8592, 8623, 8624, 8625, 8626, 8627, 8722, 8723, 8724, 8725, 8726*
PLT011	FAA-H-8083-25	Aircraft Performance 〉 Charts 〉 Takeoff *8115, 8116, 8464, 8465, 8466, 8467, 8468, 8469, 8470, 8471, 8472, 8473, 8548, 8549, 8550, 8551, 8552, 8568, 8569, 8570, 8571, 8572, 8583, 8584, 8585, 8586, 8587, 8614, 8618, 8619, 8620, 8642-4, 8717, 8718, 9076, 9797*
PLT012	FAA-H-8083-15 FAA-H-8083-25	Aircraft Performance 〉 Computations 〉 ETE; Fuel; Mach; Preflight Planning Aircraft Performance 〉 Charts Alternate; Cruise Flight Operations 〉 Normal Procedures 〉 Flight Plan *8479, 8480, 8481, 8482, 8483, 8489, 8490, 8491, 8492, 8493, 8494, 8495, 8496, 8497, 8498, 8499, 8500, 8603, 8604, 8605, 8606, 8607, 8621, 8622, 8633, 8634, 8643, 8644, 8645, 8646, 8647, 8648, 8649, 8650, 8651, 8652, 8658, 8659, 8660, 8661, 8662, 8663, 8664, 8665, 8666, 8667, 8673, 8674, 8675, 8676, 8677, 8693, 8694, 8695, 8719, 8720, 8721, 8732, 8733, 8734, 8735, 8736, 8737, 8738, 8739, 8795, 8822, 8823, 8835, 9546, 9547, 9548, 9556, 9557, 9558, 9559, 9560, 9561, 9566, 9567, 9568, 9574, 9575, 9578, 9579, 9581, 9582, 9583, 9584, 9585, 9591, 9592, 9594, 9595, 9597, 9598, 9607, 9608, 9610, 9611, 9612, 9621, 9622, 9626, 9627, 9628, 9635, 9639, 9640, 9642, 9643, 9652, 9653, 9656, 9657, 9671, 9672, 9673, 9677, 9678, 9679*
PLT015	ANA FAA-H-8083-25	Aircraft Performance 〉 Computations 〉 Specific Range: NAM/1000# Fuel Aircraft Performance 〉 Limitations 〉 Best Range Flight Operations 〉 Cruise 〉 Range *8397, 9077, 9078, 9580, 9593*

Learning Statement Code	*FAA Reference*	***Subject Description** (or **Topic** 〉 **Content** 〉 **Specific** classification)* ***Question Numbers***
PLT016	FAA-H-8083-1	Aircraft Performance 〉 Computations 〉 Fuel Dump *8678, 8679, 8680, 8681*
PLT018	FAA-H-8083-25	Aerodynamics 〉 Load Factor 〉 Angle of Bank *8354*
PLT020	FAA-H-8083-25	Aircraft Performance 〉 Charts 〉 Turbulent Air Penetration *8653, 8654, 8655, 8656, 8657*
PLT021	FAA-H-8083-1 FAA-H-8083-25	Aircraft Performance 〉 Charts 〉 Climb; Landing Weight and Balance 〉 Aircraft Loading 〉 Weight/Moment Indexes Weight and Balance 〉 Center of Gravity 〉 Computations; Shifting Weight *8434, 8435, 8436, 8437, 8438, 8439, 8440, 8441, 8442, 8443, 8444, 8445, 8446, 8447, 8448, 8449, 8450, 8451, 8452, 8453, 8454, 8455, 8456, 8457, 8458, 8513, 8514, 8515, 8516, 8517, 8518, 8519, 8520, 8521, 8522, 8523, 8524, 8525, 8526, 8527, 8528, 8529, 8530, 8531, 8532, 8573, 8574, 8575, 8576, 8577, 8578, 8579, 8580, 8581, 8582, 8702, 8705, 8709, 8710, 8711*
PLT023	AC 00-6	Weather 〉 Meteorology 〉 Pressure *9164, 9172, 9173*
PLT024	AC 00-6	Weather 〉 Meteorology 〉 Stability *9170, 9171, 9186, 9187*
PLT029	14 CFR 121 FAA-H-8261-1	Flight Operations 〉 Normal Procedures 〉 Sterile Cockpit Regulations 〉 14 CFR Part 121 〉 Flight Crewmember Duties *8106, 8298*
PLT032	FAA-H-8083-25	Aerodynamics 〉 Airspeed 〉 Mach *8387*
PLT034	14 CFR 121	Regulations 〉 14 CFR Part 121 〉 Takeoff Minimums *8134*
PLT040	14 CFR 71 AIM	Airspace 〉 Controlled 〉 Class A; Class C Airspace 〉 Uncontrolled 〉 Class G *8872, 8873, 8874, 8875, 8876, 8877, 8881, 8882, 8883, 8884, 8885, 8886, 8888*
PLT042	AC 00-45	Weather 〉 Aeronautical Weather Reports 〉 Constant Pressure Analysis Charts *9311, 9313*
PLT043	AC 00-45	Weather 〉 Aeronautical Weather Reports 〉 Constant Pressure Analysis Charts *9310, 9312, 9314, 9315, 9316*
PLT044	AIM	Air Traffic Control Procedures 〉 Departure 〉 Speed Adjustments; Takeoff Air Traffic Control Procedures 〉 En Route 〉 Speed Adjustments Air Traffic Control Procedures 〉 Ground 〉 Ground Hold Delays Airport Operations 〉 Communications 〉 Exiting the Runway after Landing *9167, 9388*
PLT045	FAA-H-8083-25	Aircraft Performance 〉 Charts 〉 Descent *8501, 8502, 8503, 9749*
PLT047	AIM	Instrument Procedures 〉 En Route Procedures 〉 Holding *8858*

Learning Statement Code	*FAA Reference*	*Subject Description (or Topic ⟩ Content ⟩ Specific classification) Question Numbers*
PLT048	FAA-H-8083-21	Interpret Hovering Ceiling Chart *8538, 8539, 8540, 8541, 8542, 8543, 8544, 8545, 8546, 8547*
PLT049	AIM FAA-H-8083-15 U.S. Terminal Procedures	Instrument Procedures ⟩ Approach Procedures ⟩ ILS; ILS Function; Final Approach Segment; Localizer (LOC) Publications ⟩ Aeronautical Charts ⟩ IAP Publications ⟩ U.S. Terminal Procedures ⟩ IAP *8972, 9588, 9601, 9619, 9645*
PLT052	U.S. Terminal Procedures	Instrument Procedures ⟩ Instrument Departures ⟩ SID *8794, 8798, 9549, 9563, 9564, 9614, 9615, 9660*
PLT053	AIM	Flight Operations ⟩ Preflight ⟩ Flight Plan *9664, 9665, 9676*
PLT055	AIM IFR Enroute High Altitude Chart	Air Traffic Control Procedures ⟩ En Route ⟩ Airways and Route Systems Publications ⟩ Aeronautical Charts ⟩ IFR En Route *9596*
PLT058	FAA-H-8083-15 FAA-H-8261-1 IFR Enroute Low Altitude Chart	Instrument Procedures ⟩ En Route ⟩ Chart Interpretation Publications ⟩ Aeronautical Charts ⟩ IFR En Route *8784, 8824, 8825, 8826, 9620, 9623, 9641, 9670, 9674, 9675*
PLT059	AC 00-45	Weather ⟩ Aeronautical Weather Reports ⟩ Aviation Routine Weather Reports (METAR); Aviation Selected Special Report (SPECI); Aviation Weather Reports *9242, 9266, 9267, 9268, 9269, 9270, 9271, 9272, 9273, 9274, 9275, 9276, 9277, 9716, 9717, 9718*
PLT061	AC 00-45	Weather ⟩ Aeronautical Weather Reports ⟩ Aviation Routine Weather Reports (METAR) *9250, 9713*
PLT063	AC 00-45	Weather ⟩ Aeronautical Weather Reports ⟩ Radar Summary Charts *9303, 9306, 9307, 9308, 9309*
PLT065	FAA-H-8083-25	Aircraft Performance ⟩ Charts ⟩ Engine Out Performance *8484, 8485, 8486, 8487, 8488*
PLT067	AC 00-45	Weather ⟩ Aeronautical Weather Forecasts ⟩ SIGMETS *9284, 9285*
PLT068	AC 00-45	Weather ⟩ Aeronautical Weather Reports ⟩ Significant Weather Prognostic Charts *9260, 9300, 9301, 9302*
PLT069	FAA-H-8083-25	Aircraft Performance ⟩ Charts ⟩ Takeoff *8642-3*
PLT072	AC 00-45	Weather ⟩ Aeronautical Weather Forecasts ⟩ TAF *9279, 9280*
PLT073	A/FD	Instrument Procedures ⟩ Flight Planning ⟩ Tower Enroute Control (TEC) *9562, 9565, 9688, 9689*

Learning Statement Code	*FAA Reference*	***Subject Description** (or **Topic** 〉 **Content** 〉 **Specific** classification)* ***Question Numbers***
PLT075	AC 00-45	Weather 〉 Aeronautical Weather Reports 〉 Weather Depiction Charts *9297, 9298, 9299*
PLT076	AC 00-45	Weather 〉 Aeronautical Weather Forecasts 〉 Winds/Temperatures Aloft Forecasts *9255, 9287, 9288, 9289, 9290, 9291, 9292, 9293, 9294, 9295, 9296*
PLT078	A/FD U.S. Terminal Procedures	Publications 〉 Airport Facility Directory 〉 Communications; Runway; Runway Lighting *8782, 8783, 8785, 8793, 8796, 8797, 8804, 9572, 9587, 9613, 9625, 9637, 9655, 9658, 9659, 9661, 9782*
PLT080	AIM U.S. Terminal Procedures	Instrument Procedures 〉 Air Traffic Control 〉 Navigation Procedures Publications 〉 Aeronautical Charts 〉 STAR *8786, 8799, 8810, 8812, 9647*
PLT082	U.S. Terminal Procedures	Regulations 〉 14 CFR Parts 121/135 〉 Flight Planning *8816, 8842, 9618*
PLT083	AIM FAA-H-8083-15 U.S. Terminal Procedures	Air Traffic Control Procedures 〉 Arrival 〉 Instrument Approach Procedures; Missed Approach Instrument 〉 Procedures 〉 Approach Procedures Final Approach Segment Navigation 〉 Radio 〉 DME Publications 〉 Aeronautical Charts 〉 IAP Publications 〉 U.S. Terminal Procedures 〉 IAP *8800, 8802, 8805, 8817, 8818, 8836, 8837, 8839, 8849, 8850, 8851, 8852, 8950, 8951, 8952, 8954, 9040, 9550, 9551, 9552, 9553, 9554, 9555, 9569, 9571, 9573, 9586, 9589, 9590, 9600, 9602, 9603, 9604, 9617, 9629, 9630, 9631, 9654, 9662, 9666, 9667, 9681, 9682, 9683, 9686*
PLT085	FAA-H-8083-25	Aircraft Performance 〉 Charts 〉 Takeoff *8642-2*
PLT087	FAA-H-8083-15	Instrument Procedures 〉 En Route Procedures 〉 Holding *8862, 8867*
PLT091	FAA-H-8083-15	Navigation 〉 Radio 〉 Radio Magnetic Indicator (RMI) *8868, 8869, 8870, 8871*
PLT097	AIM	Human Factors 〉 Aeromedical Factors 〉 Physiological *9101*
PLT099	AIM	Human Factors 〉 Aeromedical Factors 〉 Flight Illusions; Physiological *9114, 9116, 9117, 9118*
PLT100	AIM IFR Enroute Low Altitude Chart	Recall aeronautical charts—IFR En Route Low Altitude *9049, 9644*
PLT104	FAA-H-8083-25	Aeronautical Decision Making (ADM) 〉 Judgment 〉 Automatic Decisions *9778, 9804, 9805, 9806*
PLT108	AC 120-58 AC 135-17	Airport Operations 〉 Ground De-icing 〉 De-icing Procedures; Fluid Types *9441, 9442, 9443, 9444, 9445, 9446, 9447, 9448, 9450, 9452, 9453, 9454, 9696, 9697, 9698, 9700*

Learning Statement Code	*FAA Reference*	***Subject Description** (or **Topic** 〉 **Content** 〉 **Specific** classification)* ***Question Numbers***
PLT112	AIM	Recall aircraft controls—proper use/techniques *9043*
PLT121	FAA-H-8083-1	Weight and Balance 〉 Aircraft Loading 〉 Limitations *8431, 8432, 8433, 8697, 8698, 8699, 8700, 8701, 8703, 8704, 8706, 8707, 8708, 8769, 8770, 8771, 8772, 8773, 8776, 8777, 8778, 8779, 8781, 8787, 8788, 8789, 8790, 8791, 8844, 8845, 8846, 8847, 8848*
PLT124	FAA-H-8083-25	Aircraft Performance 〉 Atmospheric Effects 〉 Instrumentation Error *8374, 8404, 8405, 8417, 8418, 9767*
PLT127	FAA-8083-3	Aircraft Systems 〉 Powerplant 〉 Density Altitude *9059, 9061, 9062, 9063*
PLT128	AC 120-58	Publications 〉 Advisory Circulars 〉 Clean Aircraft Concept *9080, 9451*
PLT129	FAA-H-8083-25	Aircraft Performance 〉 Charts 〉 Runway *9083*
PLT130	FAA-H-8083-25	Recall aircraft performance—fuel *9071*
PLT131	FAA-H-8083-25	Aerodynamics 〉 Principles of Flight 〉 Ground Effect *8375, 8379*
PLT132	FAA-H-8083-25	Aircraft Systems 〉 Flight Instruments 〉 Airspeed Indicator *8364, 8744, 9321*
PLT134	ANA	Recall aircraft performance—takeoff *9075, 9801, 9802*
PLT139	14 CFR 121 ANA	Aerodynamics 〉 Stall/Spins 〉 Stall Warning Devices Regulations 〉 14 CFR Part 121 Subpart K 〉 Navigation Equipment *8069, 8070, 8071, 8154*
PLT140	AIM	Recall airport operations—LAHSO *9731, 9732, 9733, 9734*
PLT141	AIM	Airport Operations 〉 Lighting 〉 In-Runway Lighting Airport Operations 〉 Marking/Signs 〉 Runway Hold Position Sign; Runway Hold Short/Edge Markings; Runway Markings; Runway Signs; Taxi Way Markings/Signs *8903, 8904, 8905, 8906, 8907, 8922, 8923, 8924, 8925, 8926, 8927, 8928, 8929, 8930, 8931, 8932, 9416, 9416-1, 9416-2, 9417, 9421, 9422, 9423, 9423-1, 9436, 9437, 9735, 9735-1, 9735-2, 9735-3, 9737, 9764, 9772, 9785, 9786, 9798, 9799*
PLT143	A/FD	Publications 〉 Airport Facility Directory 〉 Public Protection *9636, 9668, 9669, 9690*
PLT144	AIM FAA-H-8083-3	Airport Operations 〉 Runway Conditions 〉 Braking Action *8133, 8933, 8934, 8935, 8936, 8937, 8938, 8939, 9055*
PLT145	AIM	Airport Operations 〉 Lighting 〉 Runway; Runway Edge Light Systems *8914, 8915*

Learning Statement Code	*FAA Reference*	*Subject Description (or Topic 〉 Content 〉 Specific classification)* *Question Numbers*
PLT147	AIM U.S. Terminal Procedures	Airport Operations 〉 Lighting 〉 PAPI; Visual Approach Slope Indicator Instrument Procedures 〉 Approach Procedures 〉 VASI *8908, 8909, 8910, 8911, 8912, 8913, 8916, 8917, 8918, 8919, 8920, 8921, 9378, 9648, 9649*
PLT148	AIM A/FD	Airport Operations 〉 Lighting 〉 Runway; Approach Lights *8901, 8902*
PLT149	AIM	Air Traffic Control Procedures 〉 Communications 〉 Gate Hold Procedures Instrument Procedures 〉 Air Traffic Control 〉 Communications *9057, 9783, 9784, 9787, 9788, 9789, 9790*
PLT161	14 CFR 91 AIM	Airspace 〉 Controlled 〉 Class A; Class B; Class C Airspace 〉 Special Use 〉 Warning Areas Publications 〉 Aeronautical Charts Airspace Regulations 〉 14 CFR Part 91 〉 Flight Rules *8878, 8879, 8880, 8887, 8889, 8890, 8891, 9048, 9094, 9095, 9100, 9395, 9396, 9397, 9398, 9399, 9400, 9741, 9780*
PLT162	14 CFR 91 AIM	Airspace 〉 Controlled 〉 Class C Airspace 〉 Special Use 〉 MOA Flight Operations 〉 Collision Avoidance 〉 UA/UAV Regulations 〉 14 CFR Part 91 〉 Flight Rules *8892, 9018, 9374, 9381*
PLT163	14 CFR 91 14 CFR 121	Regulations 〉 14 CFR Part 91 〉 Enroute Regulations 〉 14 CFR Part 121 〉 Flight Rules *8114, 8893, 8894, 8895, 8896, 8897, 8898, 8899, 8900*
PLT166	AC 00-6	Weather 〉 Meteorology 〉 Pressure *9099, 9163, 9174*
PLT168	FAA-H-8083-25	Recall angle of attack—characteristics/forces/principles *8378*
PLT169	FAA-H-8083-25	Recall antitorque system—components/functions *8459, 8460, 8461, 8462, 8463*
PLT170	FAA-H-8083-3 FAA-H-8261-1	Aircraft Systems 〉 Landing Gear 〉 Brakes Flight Operations 〉 Approach 〉 Rate of Descent *8402, 8403, 8955, 8969, 9034, 9035, 9039, 9074, 9084, 9090-1, 9090-3, 9091, 9092, 9438, 9791, 9792, 9793*
PLT171	AIM	Air Traffic Control Procedures 〉 Communications 〉 Radio Procedures Flight Operations 〉 Normal Procedures 〉 Bird/Wildlife Reports Instrument Procedures 〉 Air Traffic Control 〉 Communications *8854, 9013, 9015, 9016*
PLT172	AIM	Air Traffic Control Procedures 〉 Arrival 〉 Approach Control; Uncontrolled Field Air Traffic Control Procedures 〉 Communications 〉 ATC Attitude Alerts Air Traffic Control Procedures 〉 En Route 〉 Traffic Separation Airspace 〉 Controlled 〉 Class C Instrument Procedures 〉 Communications 〉 IFR Flight Plans *9050, 9090-2, 9093, 9096, 9760*

Learning Statement Code	*FAA Reference*	***Subject Description** (or **Topic** 〉 **Content** 〉 **Specific** classification)* ***Question Numbers***
PLT173	AC 00-6	Weather 〉 Meteorology 〉 Pressure; Stability *9157, 9159, 9165, 9178, 9188, 9195*
PLT192	AC 00-6	Weather 〉 Meteorology 〉 Clouds *9189, 9190, 9193, 9229*
PLT195	AIM	Flight Operations 〉 Collision Avoidance 〉 Traffic Alert/Collision Avoidance System *9425, 9426, 9427, 9428*
PLT196	AIM	Recall communications—ATIS broadcasts *9022*
PLT197	FAA-H-8083-21	Aerodynamics 〉 Principles of Flight 〉 Forces Acting on Rotary Wing *8420*
PLT201	FAA-H-8261-1	Instrument Procedures 〉 Departure 〉 Departure Procedures Charts *9012*
PLT202	FAA-H-8083-15	Instrument Procedures 〉 En Route 〉 Instrument Interpretation *9023, 9024, 9570, 9599*
PLT203	AC 00-6	Weather 〉 Meteorology 〉 Troposphere *9151, 9168, 9209, 9240*
PLT205	AIM	Human Factors 〉 Aeromedical Factors 〉 Alcohol *9111*
PLT208	14 CFR 91 14 CFR 121	Instrument Procedures 〉 Terminal Area Operations 〉 Published Procedures Regulations 〉 14 CFR Part 121 〉 Crew Equipment/Publications/Checklists *8360, 8369, 8406, 8811*
PLT213	FAA-H-8083-25	Aerodynamics 〉 Stability/Control 〉 Static *8366, 8372*
PLT214	FAA-H-8083-3 FAA-H-8083-25	Aerodynamics 〉 Flight Controls 〉 Normal Flight Aerodynamics 〉 Principles of Flight 〉 Forces Acting on Aircraft; Lift *8368, 8388, 8389, 8390, 8391, 8392, 8393, 8394, 8395*
PLT223	FAA-H-8083-3	Aerodynamics 〉 Airspeed 〉 V_y Aerodynamics 〉 Principles of Flight 〉 Drag *8051, 8241, 8357, 8359, 8370*
PLT224	AIM	Flight Operations 〉 Normal Procedures 〉 Flight Plan Instrument Procedures 〉 Approach Procedures 〉 Approach Criteria *8806, 9005, 9026, 9031, 9032*
PLT225	AIM	Flight Operations 〉 Preflight 〉 Flight Plan Instrument Procedures 〉 Flight Planning 〉 Flight Plan *9027, 9033, 9053*
PLT226	AC 00-6	Weather 〉 Meteorology 〉 Fog *9194, 9206, 9207*
PLT234	FAA-H-8083-15	Recall forces acting on aircraft—3 axis intersect *9740*

Learning Statement Code	*FAA Reference*	*Subject Description (or Topic 〉 Content 〉 Specific classification) Question Numbers*
PLT236	FAA-H-8083-25	Recall forces acting on aircraft—airfoil/center of pressure/mean camber line *8365, 8367, 8373, 8376*
PLT237	ANA	Aerodynamics 〉 Load Factor 〉 Lift *8421*
PLT240	FAA-H-8083-25	Weight and Balance 〉 Center of Gravity 〉 Stability *8380, 8419*
PLT242	FAA-H-8083-3 FAA-H-8083-25	Aerodynamics 〉 Airspeed 〉 General Aerodynamics 〉 Principles of Flight 〉 Forces Acting on Wing; Lift *8377, 8383*
PLT244	FAA-H-8083-25	Recall forces acting on aircraft—stability/controllability *9079*
PLT245	FAA-H-8083-25	Aerodynamics 〉 Load Factor 〉 Lift *8344*
PLT248	FAA-H-8083-3 FAA-H-8083-25	Aerodynamics 〉 Principles of Flight 〉 Forces Acting on Aircraft *8345, 8352, 8353, 8422, 8423*
PLT263	AC 00-6 AC 00-45	Weather 〉 Hazardous 〉 Turbulence Weather 〉 Meteorology 〉 Icing; Moisture *9230, 9231, 9701*
PLT266	ANA FAA-H-8083-25	Aerodynamics 〉 Performance 〉 Normal Flight Aerodynamics 〉 Principles of Flight 〉 Lift *8341, 8356, 8384, 8385, 8386, 9759, 9766, 9771, 9803*
PLT268	FAA-H-8083-25	Aerodynamics 〉 Performance 〉 Normal Flight *8409*
PLT274	AC 00-6	Weather 〉 Meteorology 〉 Icing *9221, 9224, 9736, 9748, 9774, 9775*
PLT276	FAA-H-8083-15	Instrument Procedures 〉 Approach Procedures 〉 ILS *8968, 8971, 8973, 8984, 8985, 8986, 8987, 8988, 8989*
PLT277	14 CFR 91 FAA-H-8083-15	Navigation 〉 Radio 〉 ILS Regulations 〉 14 CFR Part 91 〉 Instrument Flight Rules *8959, 8960, 8962, 8970*
PLT279	14 CFR 121 AIM	Navigation 〉 Inertial 〉 System/Components Navigation 〉 Radio 〉 RNAV Regulations 〉 14 CFR Part 121 Subpart K 〉 Navigation Equipment *9025*
PLT280	AIM	Human Factors 〉 Aeromedical Factors 〉 Flight Illusions; Spatial Disorientation *9107, 9108, 9109, 9110, 9115, 9433, 9434, 9435*
PLT282	14 CFR 121	Regulations 〉 14 CFR Part 121 〉 Dispatch/Redispatch *8004, 8005, 8011, 8019, 8068, 8093*
PLT283	AC 00-45	Weather 〉 Aeronautical Weather Reports 〉 Constant Pressure Analysis Charts *9281, 9707, 9711*

Learning Statement Code	*FAA Reference*	***Subject Description** (or **Topic** 〉 **Content** 〉 **Specific** classification)* ***Question Numbers***
PLT284	AC 00-06	Recall information on a Forecast Winds and Temperatures Aloft (FD) *9251*
PLT286	AC 00-45	Weather 〉 Aeronautical Weather Forecasts 〉 Significant Weather Prognostic Charts *9253, 9254, 9304*
PLT287	AC 00-45	Weather 〉 Aeronautical Weather Reports 〉 Surface Analysis Charts *9175*
PLT288	AC 00-45	Weather 〉 Aeronautical Weather Forecasts 〉 Aviation Weather Forecasts; TAF *9244, 9245, 9246, 9248, 9278, 9709, 9710*
PLT289	AC 00-45	Weather 〉 Aeronautical Weather Reports 〉 Weather Depiction Charts *9265*
PLT290	AC 00-45	Weather 〉 Aeronautical Weather Forecasts 〉 Sigmets Weather 〉 Aeronautical Weather Reports 〉 HIWAS *9257, 9286*
PLT294	AC 00-45 AIM	Weather 〉 Aeronautical Weather Forecasts 〉 En Route Flight Advisory Service (EFAS); Inflight Aviation Weather Advisories *9243, 9247*
PLT296	AIM FAA-H-8083-15	Instrument Procedures 〉 Approach Procedures 〉 Holding; MAP Instrument Procedures 〉 En Route Procedures 〉 Holding Navigation Radio ADF/NDB *8853, 8855, 8856, 8857, 8859, 8860, 8861, 8863, 8864, 8865, 8866, 9418, 9419*
PLT300	AIM	Recall instrument/navigation system checks/inspections—limits/tuning/identifying/logging *9019, 9020, 9773*
PLT301	AC 00-6	Weather 〉 Meteorology 〉 Stability; Temperature *9153, 9154, 9169*
PLT302	AC 00-6	Weather 〉 High Altitude 〉 Clear Air Turbulence *9227, 9228, 9237, 9238, 9239, 9241, 9776, 9779*
PLT303	ANA FAA-H-8083-3 FAA-H-8083-25	Recall L/D ratio *8346, 8398, 8399, 8400, 8401*
PLT309	FAA-H-8083-25	Aerodynamics 〉 Load Factor 〉 Angle of Bank *8396*
PLT310	FAA-H-8083-25	Aerodynamics 〉 Load Factor 〉 Lift *8347, 8355*
PLT314	FAA-H-8083-3	Recall longitudinal axis—aerodynamics/center of gravity/direction of motion *8362*
PLT316	AC 00-45	Weather 〉 Hazardous 〉 Thunderstorms *9758*
PLT317	AC 00-54 AIM	Weather 〉 Hazardous 〉 Microburst *9130, 9131, 9132, 9140, 9143, 9144, 9145, 9146, 9147, 9148, 9149, 9150*

Learning Statement Code	***FAA Reference***	***Subject Description** (or **Topic** 〉 **Content** 〉 **Specific** classification) **Question Numbers***
PLT318	AIM	Instrument Procedures 〉 Air Traffic Control 〉 Fuel Status Publications 〉 AIM 〉 Low Fuel *9010, 9011, 9420*
PLT322	14 CFR 121	Regulations 〉 14 CFR Part 121 Subpart K 〉 Navigation Equipment *8145, 8147*
PLT323	14 CFR 121 AIM	Flight operations 〉 Normal Procedures 〉 NOTAMS Regulations 〉 14 CFR Part 121 〉 Crew Equipment/Publications/Checklists *8283, 9086, 9087, 9089*
PLT325	49 CFR 830	Recall operations manual—transportation of prisoner *8132, 8136*
PLT330	AIM FAA-H-8083-25	Human Factors 〉 Aeromedical Factors 〉 Physiological *9103, 9105, 9106*
PLT332	AIM	Human Factors 〉 Aeromedical Factors 〉 Physiological *9102, 9104*
PLT334	AIM	Human Factors 〉 Aeromedical Factors 〉 Flight Illusions *9112, 9113*
PLT337	AC 91-43	Aircraft Systems 〉 De-icing/Anti-icing 〉 Airspeed Indications Aircraft Indicator 〉 Pitot/Static 〉 Airspeed Indicator *9081, 9082, 9222*
PLT343	14 CFR 1 FAA-H-8083-25	Aircraft Systems 〉 Powerplant 〉 Density Altitude; Operation *9064, 9065, 9066, 9067, 9072, 9073*
PLT344	AC 00-6	Weather 〉 Meteorology 〉 Icing *9223*
PLT346	AC 65-15	Aerodynamics 〉 Flight Characteristics 〉 Normal Flight Aircraft Systems 〉 Flight Controls/Primary 〉 Ailerons; Primary *8324, 8325, 8326, 8337, 8342, 8343*
PLT347	ANA FAA-H-8083-3	Recall principles of flight—critical engine *8361, 9085*
PLT348	FAA-H-8083-3	Aerodynamics 〉 Principles of Flight 〉 Forces Acting on Wing *8349, 8350, 8351*
PLT354	AIM U.S. Terminal Procedures	Flight Operations 〉 Normal Procedures 〉 Flight Plan Instrument Procedures 〉 Approach Procedures 〉 GPS Instrument Procedures 〉 Departure 〉 GPS Publications 〉 U.S. Terminal Procedures 〉 IAP Navigation 〉 Radio 〉 GPS *9429, 9430, 9431, 9432, 9691, 9692, 9693, 9694, 9722, 9723, 9725, 9726, 9727, 9728, 9729, 9729-1, 9730, 9738, 9739, 9743, 9744, 9794, 9795, 9796*
PLT355	FAA-H-8083-15	Navigation 〉 Radio 〉 HSI; ILS *8990, 8991, 8992, 8993, 8994, 8995, 8996, 8997, 8998, 8999, 9000, 9001, 9002, 9003, 9004*

Learning Statement Code	*FAA Reference*	***Subject Description** (or **Topic** 〉 **Content** 〉 **Specific** classification)* ***Question Numbers***
PLT356	AIM	Instrument Procedures 〉 Approach Procedures 〉 Category (CAT) 11; CAT 111A Navigation 〉 Radio 〉 ILS *8956, 8957, 8958, 8963, 8966, 8967, 8975, 9380, 9382, 9403, 9412*
PLT358	AIM	Navigation 〉 Radio 〉 ILS *8961*
PLT361	FAA-H-8083-15	Navigation 〉 Radio 〉 SDF *8965*
PLT365	FAA-H-8083-25	Recall reciprocating engine—components/operating principles/characteristics *9068, 9069*
PLT366	49 CFR 830	Regulations 〉 NTSB Part 830 〉 Definitions; Reports/Reporting *8233, 8236, 8246, 8317, 8318, 8322, 8323, 9098*
PLT367	14 CFR 91 14 CFR 121 14 CFR 135	Regulations 〉 14 CFR Part 91 〉 Limitations Regulations 〉 14 CFR Part 121 〉 Aircraft Equipment Regulations 〉 14 CFR Part 135 〉 Aircraft Equipment *8061, 8062, 9355, 9358, 9359, 9360, 9361*
PLT370	AIM	Instrument Procedures 〉 Departure 〉 Clearances Instrument Procedures 〉 En Route Procedures 〉 Holding *9006, 9007, 9008, 9009, 9045, 9046, 9047, 9439*
PLT373	14 CFR 121	Regulations 〉 14 CFR Part 121 〉 Flight Crewmember Duties *9745, 9807*
PLT374	14 CFR 121 14 CFR 135	Recall regulations—aircraft owner/operator responsibilities *8006, 8199, 8200*
PLT375	14 CFR 135	Regulations 〉 14 CFR Part 135 Subpart B 〉 Records Keeping *8012*
PLT379	14 CFR 121	Regulations 〉 14 CFR Part 121 〉 Flight Release *8063, 8064, 8086, 8247, 8248, 8249, 8251, 8252, 8256, 8258, 8262, 9394-2*
PLT380	14 CFR 121	Regulations 〉 14 CFR Part 121 〉 Alternate/Weather/Fuel/Requirements *8087, 8254, 8255, 8261, 8264*
PLT382	14 CFR 121 U.S. Terminal Procedures	Regulations 〉 14 CFR Part 121 〉 Landing Minimums Airport Operations 〉 Lighting 〉 MALSR *9721*
PLT383	14 CFR 121	Regulations 〉 14 CFR Part 121 〉 Icing Conditions *9379*
PLT384	14 CFR 121	Regulations 〉 14 CFR Part 121 〉 Seat Belts/Cabin Announcements *8027, 8029, 8225*
PLT385	14 CFR 121	Regulations 〉 14 CFR Part 121 〉 Cargo *8032, 8038, 8039, 8040, 8041, 8042, 8138, 8139, 8175, 8832*
PLT388	14 CFR 91 14 CFR 121	Regulations 〉 14 CFR Part 91 〉 Equipment/Instrument/Certificate Rating; Limitations Regulations 〉 14 CFR Part 121 Subpart K 〉 FDR *8047, 8141, 8142, 8143, 8833, 9356, 9357, 9410*

Learning Statement Code	*FAA Reference*	*Subject Description (or Topic 〉 Content 〉 Specific classification) Question Numbers*
PLT389	14 CFR 119 AIM	Navigation 〉 Radio 〉 GPS Regulations 〉 14 CFR Part 119 〉 Definitions *8003, 8192, 8193, 8196, 8197, 8201, 8202, 8430, 8767, 8768, 9724*
PLT390	14 CFR 121 AIM	Instrument Procedures 〉 En Route Procedures 〉 Clearance Limits Regulations 〉 14 CFR Part 121 〉 Communications *8135*
PLT391	14 CFR 91	Regulations 〉 14 CFR Part 91 〉 Instrument Flight Rules *9362, 9364, 9365, 9389, 9390, 9616*
PLT392	14 CFR 135	Regulations 〉 14 CFR Part 135 〉 Operator/Control/Manual(s) *8010*
PLT393	14 CFR 91	Regulations 〉 14 CFR Part 91 〉 Flight Rules *9352, 9353*
PLT394	14 CFR 121 AIM	Recall regulations—declaration of an emergency *8239, 9097*
PLT395	14 CFR 1 14 CFR 119	Regulations 〉 14 CFR Part 1 〉 General Definitions Regulations 〉 14 CFR Part 119 〉 Definitions *8319, 8320, 8429, 9324, 9325, 9327*
PLT396	14 CFR 121	Regulations 〉 14 CFR Part 121 〉 Takeoff Minimums *8250*
PLT398	14 CFR 121	Flight Operations 〉 Normal Procedures 〉 Flight Plan Regulations 〉 14 CFR Part 121 Subpart E 〉 ETOPS *8257, 8259, 8260, 8266, 8267, 8280, 8284, 9746-2*
PLT400	14 CFR 121	Regulations 〉 14 CFR Part 121 〉 Dispatch/Redispatch *8007, 8226, 8286, 8292, 8296*
PLT402	14 CFR 121	Regulations 〉 14 CFR Part 121 〉 Emergency Equipment/Survival *8171*
PLT403	14 CFR 121	Regulations 〉 14 CFR Part 121 〉 Emergency Authority/Actions/Reports *8240, 8245*
PLT404	14 CFR 121	Regulations 〉 14 CFR Part 121 〉 Emergency Equipment/Survival Regulations 〉 14 CFR Part 121 Subpart K 〉 Emergency Equipment *8058, 8059, 8060, 8144, 8157, 8159, 8163, 8164, 8165, 8167, 8168, 8170, 8172, 8177, 8834*
PLT405	14 CFR 91 14 CFR 121	Regulations 〉 14 CFR Part 91 〉 Equipment/Instrument/Certificate Requirement Regulations 〉 14 CFR Part 121 〉 Crew Equipment/Publications/Checklists; Dispatch/Redispatch *8045, 8046, 8053, 8054, 8140, 8146, 8194, 8235, 8808, 9407, 9414, 9415*
PLT406	14 CFR 91	Regulations 〉 14 CFR Part 91 〉 Equipment/Instrument/Certificate Requirement; Flight Rules *8237, 8803, 9363, 9386, 9387*

Learning Statement Code	*FAA Reference*	***Subject Description*** *(or* ***Topic*** 〉 ***Content*** 〉 ***Specific*** *classification)* ***Question Numbers***
PLT407	14 CFR 61 14 CFR 121	Regulations 〉 14 CFR Part 61 〉 Limitations Regulations 〉 14 CFR Part 121 〉 Landing Minimums; Line-Oriented Simulator Training Course; Training/Currency *8034, 8103, 8108, 8109, 8110, 8111, 8205, 8207, 8215, 8216, 8217, 8218, 8820, 8821, 8827, 8829, 9346, 9347, 9618, 9632, 9720*
PLT408	14 CFR 121	Regulations 〉 14 CFR Part 121 〉 Emergency Equipment/Survival *8176*
PLT409	14 CFR 61 14 CFR 121	Regulations 〉 14 CFR Part 61 〉 Limitations Regulations 〉 14 CFR Part 121 〉 Flight Time/Duty/Rest/Requirements *8002, 8104, 8189, 8211, 8219, 8220, 8221, 8222, 8223, 8224, 8227, 8228, 8229, 8814, 8815, 9342, 9354, 9714*
PLT412	14 CFR 121	Regulations 〉 14 CFR Part 121 〉 Flight Release *8293, 8294, 8295*
PLT413	14 CFR 121	Regulations 〉 14 CFR Part 121 〉 Alternate/Weather/Fuel/Requirements *8088, 8089, 8268, 8269, 8270, 8271, 8272, 8273, 8274, 8275, 8276, 8277*
PLT416	49 CFR 830	Regulations 〉 NTSB Part 830 〉 Reports/Reporting *8321*
PLT417	14 CFR 121	Regulations 〉 14 CFR Part 121 Subpart K 〉 Emergency Equipment *8166, 8169*
PLT420	14 CFR 91	Regulations 〉 14 CFR Part 91 〉 Instrument Flight Rules *8092, 8279, 8801, 8953, 9021, 9036, 9037, 9038, 9041, 9348, 9368, 9369, 9384, 9391, 9392, 9393, 9411, 9413, 9742, 9770*
PLT421	14 CFR 91 AIM	Recall regulations—instrument flight rules *9014, 9370, 9383, 9385, 9394-1*
PLT424	14 CFR 135	Regulations 〉 14 CFR Part 135 〉 Aircraft Equipment; Flight Operations Regulations 〉 14 CFR Part 135 Subpart B 〉 Flight/Crewmember Duties *8013, 8014, 8015, 8016, 8017, 8037, 8102*
PLT425	14 CFR 135	Regulations 〉 14 CFR Part 135 Subpart B 〉 Records Keeping *9746, 9746-1, 9761, 9762*
PLT426	14 CFR 121	Recall regulations—maintenance requirements *8278*
PLT427	14 CFR 61	Regulations 〉 14 CFR Part 61 〉 Limitations *9333, 9335, 9340, 9343, 9349*
PLT428	14 CFR 135	Regulations 〉 14 CFR Part 135 〉 MEL/CDL; Operator/Control/Manual(s)/ Operation Specs *8052, 8807*
PLT429	14 CFR 121	Regulations 〉 14 CFR Part 121 Subpart K 〉 Navigation Equipment *8149, 8152, 8195, 8203, 9408, 9409*
PLT430	14 CFR 91	Regulations 〉 14 CFR Part 91 〉 Instrument Flight Rules *8297, 9366, 9367, 9373*

Learning Statement Code	*FAA Reference*	*Subject Description (or Topic 〉 Content 〉 Specific classification) Question Numbers*
PLT432	14 CFR 1	Regulations 〉 14 CFR Part 1 〉 General Definitions *9326*
PLT433	AIM	Recall regulations—operational flight plan requirements *9028, 9029, 9030*
PLT434	AIM	Airport Operations 〉 Taxiing 〉 Airport Taxi Modes; Taxiing After Landing Airport Operations 〉 Tower Controlled 〉 ATC Procedures *9042, 9044, 9056, 9424*
PLT436	14 CFR 121	Regulations 〉 14 CFR Part 121 〉 Crew Equipment/Publications/Checklists; Enroute *8198*
PLT437	14 CFR 135	Regulations 〉 14 CFR Part 135 〉 Aircraft Equipment; Performance Requirements *8050, 8078, 8079, 8838, 8840, 8841*
PLT438	14 CFR 121	Regulations 〉 14 CFR Part 121 〉 Supplemental; Oxygen for Sustenance: Turbine Engine Regulations 〉 14 CFR Part 121 Subpart K 〉 Emergency Equipment *8020, 8021, 8022, 8023, 8024, 8025, 8028, 8030, 8031, 8055, 8056, 8072, 8073, 8074, 8080, 8081, 8156, 8173, 8174, 8180, 8181, 8182, 8183, 8184, 8185, 8186, 8187, 9576, 9577, 9605, 9606, 9638, 9650, 9651*
PLT440	14 CFR 121	Regulations 〉 14 CFR Part 121 Subpart M 〉 Flight Engineer Requirements *8008, 8009, 8026, 8033, 8043, 8113, 8155, 8188, 8190, 8212, 8213*
PLT442	14 CFR 61 14 CFR 121 14 CFR 135	Regulations 〉 14 CFR Part 61 〉 Instrument Currency Regulations 〉 14 CFR Part 121 〉 Recent Experience; Training Currency Regulations 〉 14 CFR Part 135 〉 Crew Requirements *8095, 8096, 8097, 8098, 8099, 8100, 8101, 8105, 8208, 8209, 8210, 8809, 8830, 9329, 9334, 9336, 9337, 9338, 9339, 9341, 9344, 9345*
PLT443	14 CFR 61 14 CFR 121	Regulations 〉 14 CFR Part 61 〉 Type Rating Regulations 〉 14 CFR Part 121 〉 Flight Time/Duty/Rest/Requirements Regulations 〉 14 CFR Part 121 Subpart M 〉 Flight Engineer Requirements *8035, 8036, 8044, 8082, 8083, 8107, 8112, 8191, 8289, 9328, 9350, 9633, 9634, 9646, 9663, 9680, 9684, 9685, 9687*
PLT444	14 CFR 121	Regulations 〉 14 CFR Part 121 〉 Crew Equipment/Publications/Checklists; Emergency Authority/Actions/ Reports; Operational Control/ Flight Release *8018, 8234, 8242, 8243, 8244, 8281, 8282, 8285, 8819, 9402*
PLT447	14 CFR 61	Regulations 〉 14 CFR Part 61 〉 Limitations *9351*
PLT449	14 CFR 121	Regulations 〉 14 CFR Part 121 〉 Training/Currency *8214*
PLT450	14 CFR 63 14 CFR 121	Regulations 〉 14 CFR Part 63 〉 Experience Requirements Regulations 〉 14 CFR Part 121 〉 Flight Time/Duty/Rest Requirements Regulations 〉 14 CFR Part 121 Subpart P 〉 Dispatcher Duty Limitations *8211, 8230, 8231, 8238*

Learning Statement Code	***FAA Reference***	***Subject Description*** *(or **Topic** 〉 **Content** 〉 **Specific** classification)* ***Question Numbers***
PLT452	14 CFR 121	Recall regulations — re-dispatch *8232*
PLT453	14 CFR 121	Regulations 〉 14 CFR Part 121 〉 Records Keeping *8287, 8288*
PLT454	14 CFR 135	Regulations 〉 14 CFR Part 135 〉 Aircraft Equipment *8001, 8049, 8067*
PLT455	14 CFR 121	Regulations 〉 14 CFR Part 121 〉 Dispatch/Redispatch *8290, 8291*
PLT456	14 CFR 121	Regulations 〉 14 CFR Part 121 〉 Landing Minimums *8094, 8127, 8128, 8130, 8831*
PLT459	14 CFR 91 14 CFR 121	Regulations 〉 14 CFR Part 91 〉 Instrument Flight Rules Regulations 〉 14 CFR Part 121 〉 Takeoff Minimums *8057, 8065, 8066, 8085, 8090, 8091, 8158, 8263, 8358, 8363, 8843, 9371, 9372, 9401*
PLT460	14 CFR 121 14 CFR 135	Regulations 〉 14 CFR Part 121 〉 Training/Currency Regulations 〉 14 CFR Part 121/135 〉 Training *8204, 8828, 9330, 9331*
PLT462	14 CFR 121 14 CFR 135	Regulations 〉 14 CFR Part 121 Subpart K 〉 Emergency Equipment Regulations 〉 14 CFR Part 121/135 〉 Aircraft Equipment *8048, 8160, 8161, 8162, 8178, 8179, 8792*
PLT463	14 CFR 61 14 CFR 121	Regulations 〉 14 CFR Part 61 〉 Limitations Regulations 〉 14 CFR Part 121 〉 Passenger/Flight Events/Disturbances *8813, 9332*
PLT464	14 CFR 135	Regulations 〉 14 CFR 135 〉 Aircraft Equipment Regulations 〉 14 CFR Part 135 Subpart B 〉 Flight/Crewmember Duties *8075, 8076, 8077*
PLT465	14 CFR 121	Regulations 〉 14 CFR Part 121 〉 Seat Belts/Cabin Announcements *8153*
PLT466	FAA-H-8083-3	Weight and Balance 〉 Center of Gravity 〉 Limitations *8371, 8775, 8780, 9317, 9318, 9319, 9320, 9322, 9323*
PLT468	14 CFR 121	Recall regulations — Visual Meteorological Conditions (VMC) *8253*
PLT469	14 CFR 121	Regulations 〉 14 CFR Part 121 Subpart K 〉 Navigation Equipment *8148, 8150, 8151*
PLT470	ANA	Aerodynamics 〉 Principles of Flight 〉 Helicopter Rotary Wings *8407, 8408, 8410, 8411, 8412, 8413, 8424, 8425, 8426, 8427, 8428, 9781*
PLT472	FAA-H-8083-21	Aircraft Systems 〉 Rotor 〉 Vibrations *8414, 8415, 8416, 9800*
PLT473	ANA FAA-H-8083-25	Recall secondary flight controls — types / purpose / functionality *8327, 8328, 8329, 8330, 8331, 8332, 8333, 8334, 8336, 8338, 8339, 8340*

Learning Statement Code	*FAA Reference*	*Subject Description (or Topic 〉 Content 〉 Specific classification) Question Numbers*
PLT475	AC 00-6	Weather 〉 Aeronautical Weather Reports 〉 Aviation Routine Weather Reports (METAR) *9200, 9249, 9708*
PLT477	FAA-H-8083-25	Aerodynamics 〉 Load Factor 〉 Stall Speed *8348, 9808*
PLT493	14 CFR 121 AC 00-6 AC 20-117 AC 135-17	Aerodynamics 〉 Principles of Flight 〉 Hazards Regulations 〉 14 CFR Part 121 〉 Icing Conditions Weather 〉 Meteorology 〉 Frost *8084, 8265, 9162, 9179, 9180, 9181, 9440, 9449, 9695, 9752, 9753, 9754, 9755, 9756, 9757*
PLT495	AC 00-6	Weather 〉 Meteorology 〉 Arctic Flying; Thunderstorms *9196, 9197, 9198, 9199, 9201, 9202, 9203, 9204, 9205, 9211, 9212, 9233, 9706*
PLT497	AIM	Aircraft Systems 〉 Avionics 〉 Transponder Operation *9051, 9052*
PLT498	49 CFR 830	Recall Transportation Security Regulations *8131, 8137, 9763*
PLT499	FAA-H-8083-25	Aircraft Systems 〉 Powerplant 〉 Compressor Stalls *8974, 9058, 9060, 9768*
PLT500	FAA-H-8083-3	Aircraft Systems 〉 Powerplant 〉 Limitations *9070*
PLT501	AC 00-6 AC 00-30 AIM	Weather 〉 Aeronautical Weather Reports 〉 PIREPS Weather 〉 Hazardous 〉 Turbulence Weather 〉 Meteorology 〉 Turbulence *9128, 9129, 9210, 9218, 9219, 9225, 9226, 9232, 9235, 9262, 9263, 9264, 9777*
PLT506	14 CFR 1	Regulations 〉 14 CFR Part 1 〉 General Definitions *8774*
PLT508	14 CFR 91	Regulations 〉 14 CFR Part 91 〉 Equipment/Instrument/Certificate Requirement *9375, 9376, 9377, 9404, 9405, 9406*
PLT509	AIM	Airport Operations 〉 Wake Turbulence 〉 Turbulence Factors *9119, 9120, 9121, 9122, 9123, 9124, 9125, 9126, 9127, 9715*
PLT510	AC 00-6	Weather 〉 Meteorology 〉 Haze; Temperature; Wind *9152, 9156, 9176, 9208*
PLT511	AC 00-6	Weather 〉 Meteorology 〉 Air Masses; Fronts; Thunderstorms *9160, 9191, 9192, 9213, 9214, 9215, 9216, 9217, 9259*
PLT512	AC 00-6	Weather 〉 Meteorology 〉 Moisture; Stability *9155, 9161, 9182, 9183, 9184, 9185, 9234*
PLT514	AC 00-6	Weather 〉 Meteorology 〉 Pressure *9305*
PLT515	AIM	Weather 〉 Aeronautical Weather Reports 〉 Data Dissemination; HIWAS *9256, 9258, 9261, 9702, 9704, 9705, 9712-1, 9712-2, 9747*

Learning Statement Code	*FAA Reference*	*Subject Description (or Topic 〉 Content 〉 Specific classification) Question Numbers*
PLT516	AC 00-6	Weather 〉 Meteorology 〉 Wind *9158, 9177*
PLT518	AC 00-6 AC 00-54	Weather 〉 Hazardous 〉 Wind Shear *9054, 9133, 9134, 9135, 9136, 9137, 9138, 9139, 9141, 9142, 9166, 9220, 9236, 9252, 9283*
PLT524	FAA-H-8083-6	Navigation 〉 Radio 〉 Electronic Displays *8206, 9750, 9751, 9769*

Notes

COMPUTER TESTING SUPPLEMENT FOR AIRLINE TRANSPORT PILOT AND AIRCRAFT DISPATCHER

2005

U.S. DEPARTMENT OF TRANSPORTATION
FEDERAL AVIATION ADMINISTRATION
Flight Standards Service

PREFACE

This computer testing supplement is designed by the Flight Standards Service of the Federal Aviation Administration (FAA) for use by computer testing designees (CTDs) and testing centers in the administration of airman knowledge tests in the following knowledge areas:

Airline Transport Pilot (FAR 121) Airplane (ATP)
Airline Transport Pilot (FAR 135) Airplane (ATA)
Airline Transport Pilot (FAR 135) Added Rating—Airplane (ARA)
Airline Transport Pilot (FAR 135) Helicopter (ATH)
Airline Transport Pilot (FAR 135) Added Rating—Helicopter (ARH)
Aircraft Dispatcher

FAA-CT-8080-7C supercedes FAA-CT-8080-7B dated 1998.

Comments regarding this supplement should be sent to:

U.S. Department of Transportation
Federal Aviation Administration
Flight Standards Service
Airman Testing Standards Branch, AFS-630
P.O. Box 25082
Oklahoma City, OK 73125

CONTENTS

APPENDIX 1

APPENDIX 2

CONTENTS—Continued

CONTENTS—Continued

CONTENTS—Continued

CONTENTS—Continued

CONTENTS—Continued

APPENDIX 1

94118

GENERAL INFO

ABBREVIATIONS

ADF .... Automatic Direction Finder
ALS .... Approach Light System
ALSF .... Approach Light System with Sequenced Flashing Lights
APP CON .... Approach Control
ARR .... Arrival
ASR/PAR .... Published Radar Minimums at this Airport
ATIS .... Automatic Terminal Information Service
AWOS .... Automated Weather Observing System
AZ .... Azimuth
BC .... Back Course
C .... Circling
CAT .... Category
CCW .... Counter Clockwise
Chan .... Channel
CLNC DEL .... Clearance Delivery
CTAF .... Common Traffic Advisory Frequency
CW .... Clockwise
DH .... Decision Heights
DME .... Distance Measuring Equipment
DR .... Dead Reckoning
ELEV .... Elevation
FAF .... Final Approach Fix
FM .... Fan Marker
GPI .... Ground Point of Interception
GPS .... Global Positioning System
GS .... Glide Slope
HAA .... Height Above Airport
HAL .... Height Above Landing
HAT .... Height Above Touchdown
HIRL .... High Intensity Runway Lights
IAF .... Initial Approach Fix
ICAO .... International Civil Aviation Organization
IM .... Inner Marker
Intcp .... Intercept
INT .... Intersection
LDA .... Localizer Type Directional Aid
Ldg .... Landing
LDIN .... Lead in Light System
LIRL .... Low Intensity Runway Lights
LOC .... Localizer
LR .... Lead Radial. Provides at least 2 NM (Copter 1 NM) of lead to assist in turning onto the intermediate/final course
MALS .... Medium Intensity Approach Light System
MALSR .... Medium Intensity Approach Light Systems with RAIL
MAP .... Missed Approach Point
MDA .... Minimum Descent Altitude
MIRL .... Medium Intensity Runway Lights
MLS .... Microwave Landing System
MM .... Middle Marker
NA .... Not Authorized
NDB .... Non-directional Radio Beacon
NM .... Nautical Miles
NoPT .... No Procedure Turn Required (Procedure Turn shall not be executed without ATC clearance)
ODALS .... Omnidirectional Approach Light System
OM .... Outer Marker
R .... Radial
RA .... Radio Altimeter setting height
Radar Required .... Radar vectoring required for this approach
RAIL .... Runway Alignment Indicator Lights
RBn .... Radio Beacon
RCLS .... Runway Centerline Light System
REIL .... Runway End Identifier Lights
RNAV .... Area Navigation
RPI .... Runway Point of Intercept(ion)
RRL .... Runway Remaining Lights
Runway Touchdown Zone..... First 3000' of Runway
Rwy .... Runway
RVR .... Runway Visual Range
S .... Straight-in
SALS .... Short Approach Light System
SSALR .... Simplified Short Approach Light System with RAIL
SDF .... Simplified Directional Facility
TA .... Transition Altitude
TAC .... TACAN
TCH .... Threshold Crossing Height (height in feet Above Ground Level)
TDZ .... Touchdown Zone
TDZE .... Touchdown Zone Elevation
TDZ/CL .... Touchdown Zone and Runway Centerline Lighting
TDZL .... Touchdown Zone Lights
TLv .... Transition Level
VASI .... Visual Approach Slope Indicator
VDP .... Visual Descent Point
WPT .... Waypoint (RNAV)
X .... Radar Only Frequency

PILOT CONTROLLED AIRPORT LIGHTING SYSTEMS

Available pilot controlled lighting (PCL) systems are indicated as follows:

1. Approach lighting systems that bear a system identification are symbolized using negative symbology, e.g., Ⓐ1, Ⓥ, ✪
2. Approach lighting systems that do not bear a system identification are indicated with a negative " ● " beside the name.

A star (*) indicates non-standard PCL, consult Directory/Supplement, e.g., ●*

To activate lights use frequency indicated in the communication section of the chart with a ● or the appropriate lighting system identification e.g., UNICOM 122.8 ●, Ⓐ1, Ⓥ

KEY MIKE	FUNCTION
7 times within 5 seconds	Highest intensity available
5 times within 5 seconds	Medium or lower intensity (Lower REIL or REIL-off)
3 times within 5 seconds	Lowest intensity available (Lower REIL or REIL-off)

LEGEND 1.—General Information and Abbreviations.

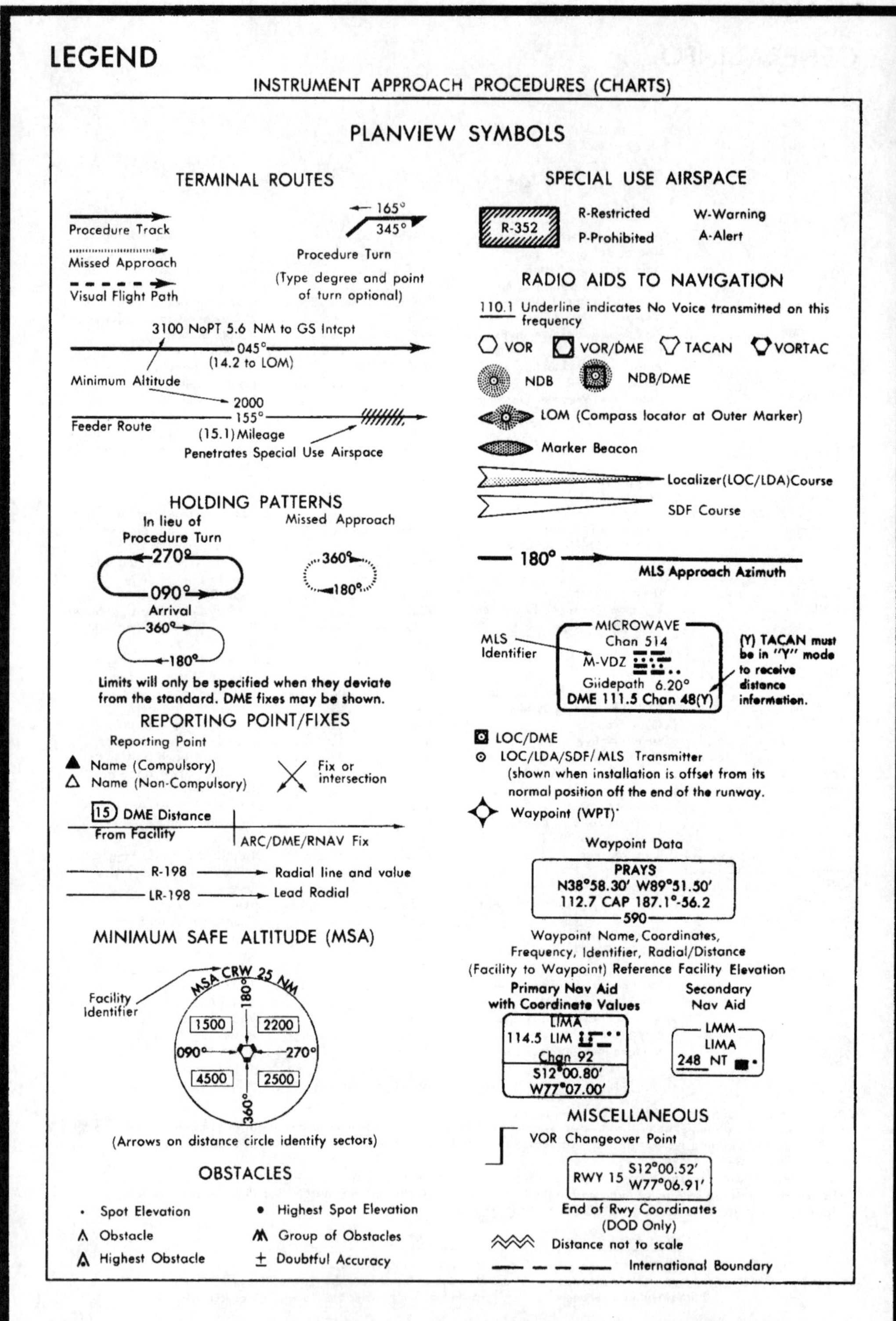

LEGEND 2.—Planview Symbols.

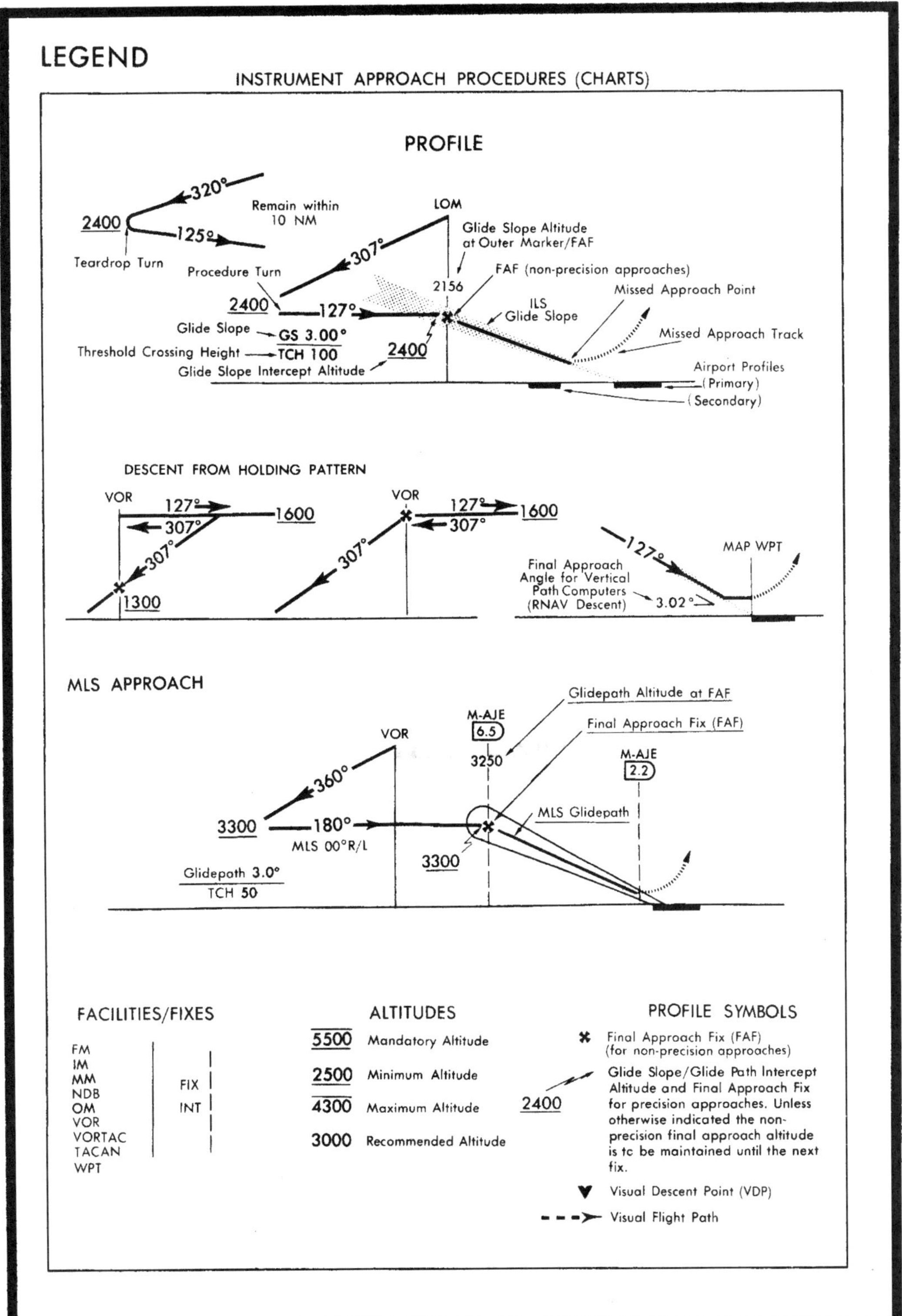

LEGEND 3.—Profile.

94286

LEGEND

INSTRUMENT APPROACH PROCEDURES (CHARTS)

AIRPORT DIAGRAM/ AIRPORT SKETCH

Runways

Hard Surface | Other Than Hard Surface | Stopways, Taxiways, Parking Areas | Displaced Threshold

Closed Runway | Closed Taxiway | Under Construction | Metal Surface | Runway Centerline Lighting

ARRESTING GEAR: Specific arresting gear systems; e.g., BAK-12, MA-1A etc., shown on airport diagrams, not applicable to Civil Pilots. Military Pilots Refer to Appropriate DOD Publications.

uni-directional | bi-directional | Jet Barrier

REFERENCE FEATURES

Buildings ....

Tanks ....

Obstruction ....

Airport Beacon # ....

Runway Radar Reflectors ....

Control Tower # ....

When Control Tower and Rotating Beacon are co-located, Beacon symbol will be used and further identified as TWR.

Runway length depicted is the physical length of the runway (end-to-end, including displaced thresholds if any) but excluding areas designated as overruns or stopways. Where a displaced threshold is shown and/or part of the runway is otherwise not available for landing, an annotation is added to indicate the landing length of the runway; e.g., RWY 13 ldg 5000'.

Runway Weight Bearing Capacity is shown as a codified expression. Refer to the appropriate Supplement/Directory for applicable codes, e.g., RWY 14-32 S75, T185, ST175, TT325

Helicopter Alighting Areas

Negative Symbols used to identify Copter Procedures landing point............

Runway TDZ elevation .... TDZE 123

Runway Slope .... 0.3% DOWN / 0.8% UP
(shown when runway slope exceeds 0.3%)

NOTE:
Runway Slope measured to midpoint on runways 8000 feet or longer.

U.S. Navy Optical Landing System (OLS) "OLS" location is shown because of its height of approximately 7 feet and proximity to edge of runway may create an obstruction for some types of aircraft.

Approach light symbols are shown in the Flight Information Handbook.

Airport diagram scales are variable.

True/magnetic North orientation may vary from diagram to diagram.

Coordinate values are shown in 1 or ½ minute increments. They are further broken down into 6 second ticks, within each 1 minute increment.

Positional accuracy within ±600 feet unless otherwise noted on the chart.

NOTE:
All new and revised airport diagrams are shown referenced to the World Geodetic System (W G S) (noted on appropriate diagram), and may not be compatible with local coordinates published in FLIP. (Foreign Only)

Runway Slope | FIELD ELEV 174 | Rwy 2 ldg 8000'
BAK-12 | 0.7% UP | 20 | 2 Runway Identification
9000 X 200 | 023.2° | 1000 X 200
Runway End Elevation — ELEV 164 | Runway Dimensions (in feet) | Runway Heading (Magnetic) | Overrun/Stopway Dimensions (in feet)

SCOPE

Airport diagrams are specifically designed to assist in the movement of ground traffic at locations with complex runway/taxiway configurations and provide information for updating Computer Based Navigation Systems (I.E., INS, GPS) aboard aircraft. Airport diagrams are not intended to be used for approach and landing or departure operations. For revisions to Airport Diagrams: Consult FAA Order 7910.4B.

LEGEND 4.—Airport Diagram/Airport Sketch.

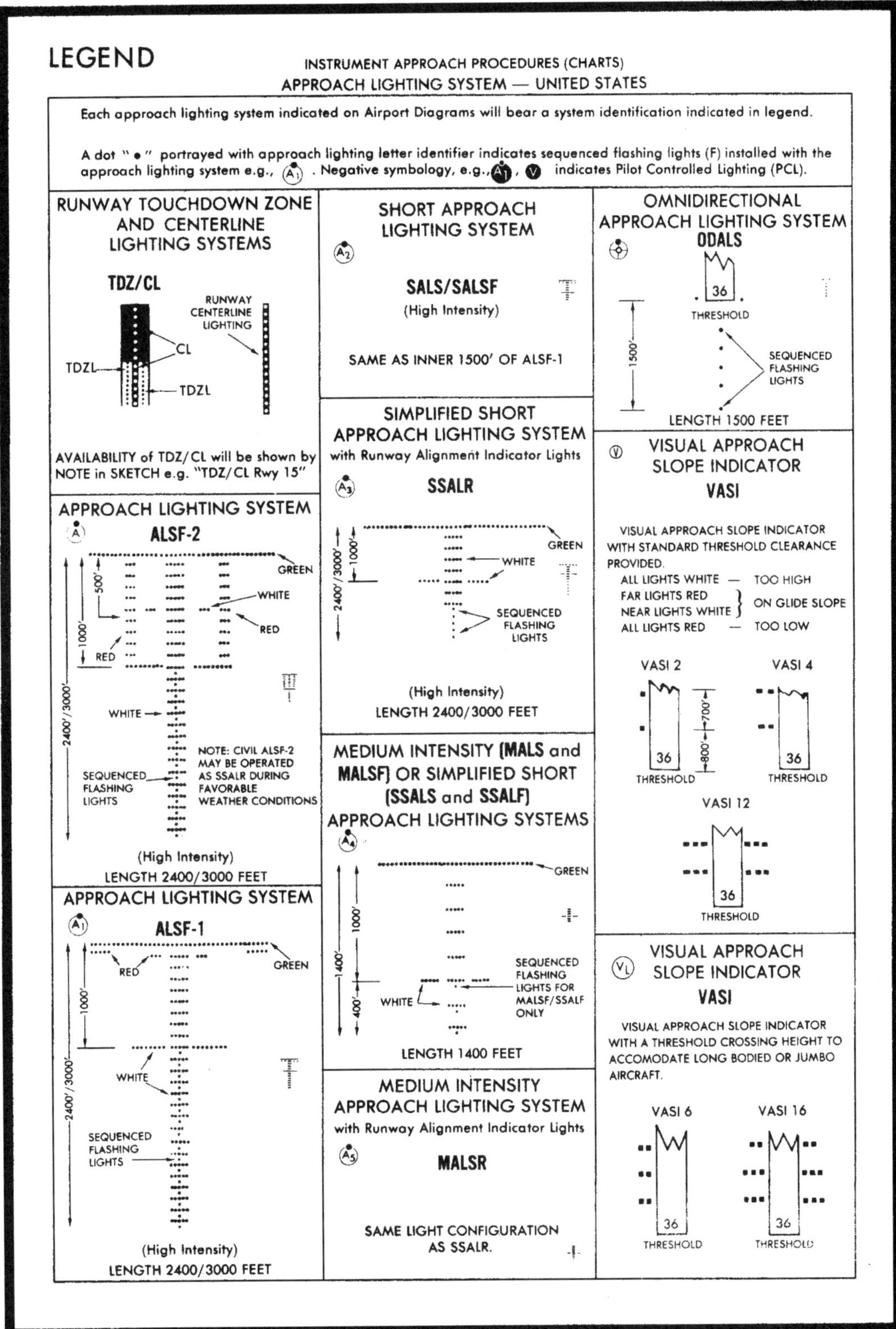

LEGEND 5.—Approach Lighting System—United States.

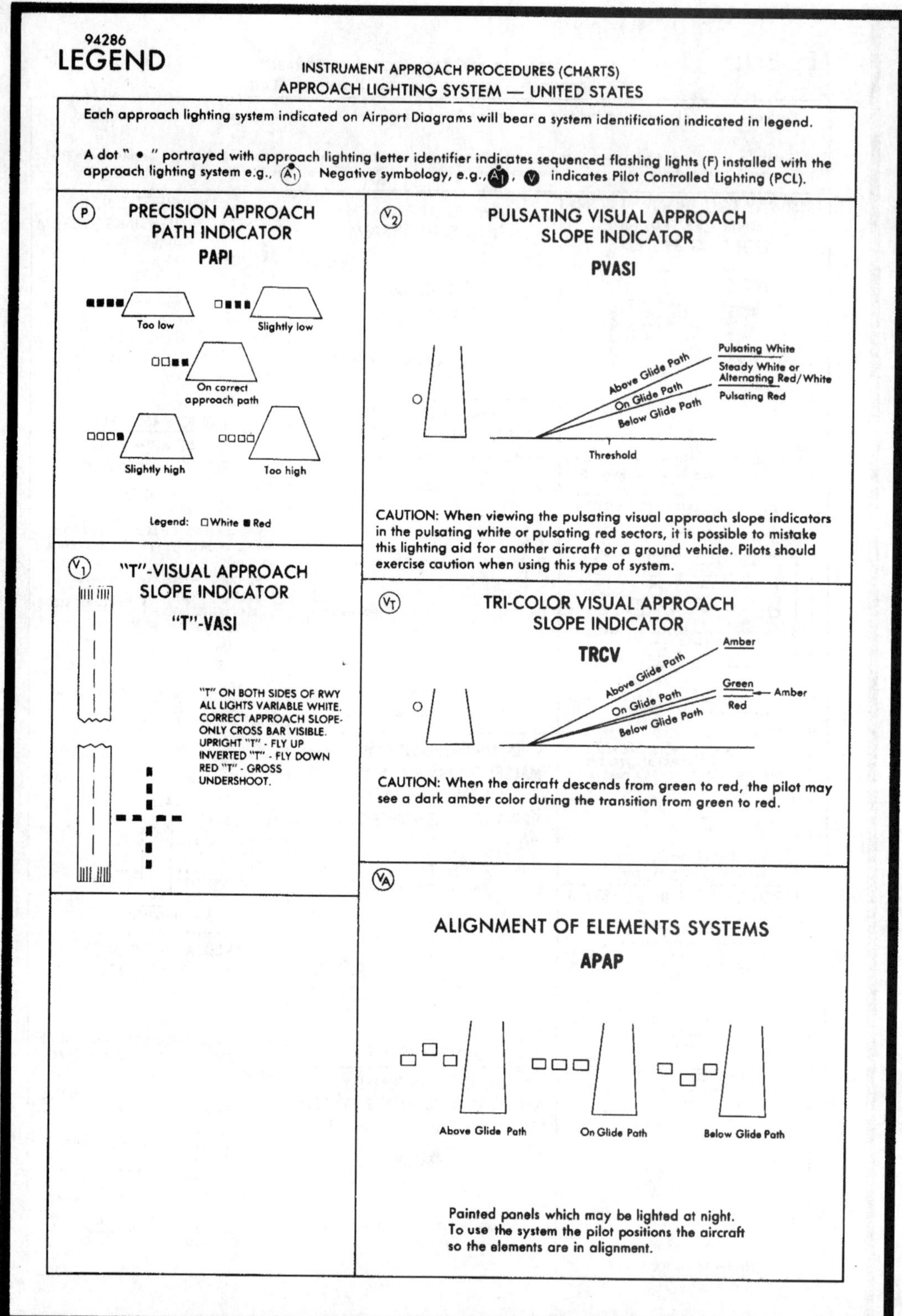

LEGEND 6.—Approach Lighting System—United States.

TERMS/LANDING MINIMA DATA

IFR LANDING MINIMA

Landing minima are established for six aircraft approach categories (ABCDE and COPTER). In the absence of COPTER MINIMA, helicopters may use the CAT A minimums of other procedures. The standard format for portrayal of landing minima is as follows:

AIRCRAFT APPROACH CATEGORIES

Speeds are based on 1.3 times the stall speed in the landing configuration of maximum gross landing weight. An aircraft shall fit in only one category. If it is necessary to maneuver at speeds in excess of the upper limit of a speed range for a category, the minimums for the next higher category should be used. For example, an aircraft which falls in Category A, but is circling to land at a speed in excess of 91 knots, should use the approach Category B minimums when circling to land. See following category limits:

MANEUVERING TABLE

Approach Category	A	B	C	D	E
Speed (Knots)	0-90	91-120	121-140	141-165	Abv 165

RVR/Meteorological Visibility Comparable Values

The following table shall be used for converting RVR to meteorological visibility when RVR is not reported for the runway of intended operation. Adjustment of landing minima may be required — see Inoperative Components Table.

RVR (feet)	Visibility (statute miles)	RVR (feet)	Visibility (statute miles)
1600	¼	4000	¾
2000	⅜	4500	⅞
2400	½	5000	1
3200	⅝	6000	1¼

LANDING MINIMA FORMAT

In this example airport elevation is 1179, and runway touchdown zone elevation is 1152.

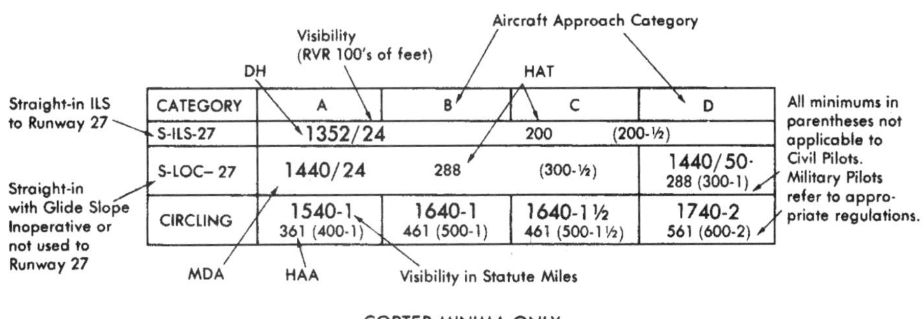

COPTER MINIMA ONLY

CATEGORY	COPTER
H-176°	680-½ 363 (400-½)

Copter Approach Direction

Height of MDA/DH Above Landing Area (HAL)

No circling minimums are provided

LEGEND 7.—IFR Landing Minima.

TERMS/LANDING MINIMA DATA

91262

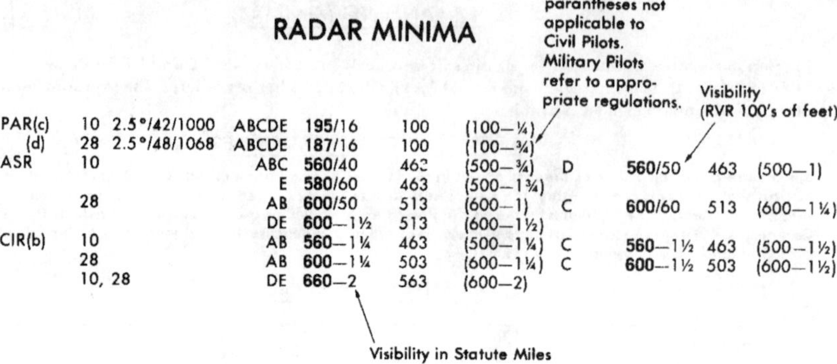

Radar Minima:

1. Minima shown are the lowest permitted by established criteria. Pilots should consult applicable directives for their category of aircraft.
2. The circling MDA and weather minima to be used are those for the runway to which the final approach is flown - not the landing runway. In the above RADAR MINIMA example, a category C aircraft flying a radar approach to runway 10, circling to land on runway 28, must use an MDA of 560 feet with weather minima of 500-1½.

▲ Alternate Minimums not standard. Civil users refer to tabulation. USA/USN/USAF pilots refer to appropriate regulations.

▲ NA Alternate minimums are Not Authorized due to unmonitored facility or absence of weather reporting service.

▼ Take-off Minimums not standard and/or Departure Procedures are published. Refer to tabulation.

EXPLANATION OF TERMS

The United States Standard for Terminal Instrument Procedures (TERPS) is the approved criteria for formulating instrument approach procedures.

LEGEND 8.—Radar Minima.

INSTRUMENT APPROACH PROCEDURE CHARTS
RATE OF DESCENT TABLE
(ft. per min.)

A rate of descent table is provided for use in planning and executing precision descents under known or approximate ground speed conditions. It will be especially useful for approaches when the localizer only is used for course guidance. A best speed, power, attitude combination can be programmed which will result in a stable glide rate and attitude favorable for executing a landing if minimums exist upon breakout. Care should always be exercised so that the minimum descent altitude and missed approach point are not exceeded.

ANGLE OF DESCENT (degrees and tenths)	GROUND SPEED (knots)										
	30	45	60	75	90	105	120	135	150	165	180
2.0	105	160	210	265	320	370	425	475	530	585	635
2.5	130	200	265	330	395	465	530	595	665	730	795
3.0	160	240	320	395	480	555	635	715	795	875	955
3.5	185	280	370	465	555	650	740	835	925	1020	1110
4.0	210	315	425	530	635	740	845	955	1060	1165	1270
4.5	240	355	475	595	715	835	955	1075	1190	1310	1430
5.0	265	395	530	660	795	925	1060	1190	1325	1455	1590
5.5	290	435	580	730	875	1020	1165	1310	1455	1600	1745
6.0	315	475	635	795	955	1110	1270	1430	1590	1745	1905
6.5	345	515	690	860	1030	1205	1375	1550	1720	1890	2065
7.0	370	555	740	925	1110	1295	1480	1665	1850	2035	2220
7.5	395	595	795	990	1190	1390	1585	1785	1985	2180	2380
8.0	425	635	845	1055	1270	1480	1690	1905	2115	2325	2540
8.5	450	675	900	1120	1345	1570	1795	2020	2245	2470	2695
9.0	475	715	950	1190	1425	1665	1900	2140	2375	2615	2855
9.5	500	750	1005	1255	1505	1755	2005	2255	2510	2760	3010
10.0	530	790	1055	1320	1585	1845	2110	2375	2640	2900	3165
10.5	555	830	1105	1385	1660	1940	2215	2490	2770	3045	3320
11.0	580	870	1160	1450	1740	2030	2320	2610	2900	3190	3480
11.5	605	910	1210	1515	1820	2120	2425	2725	3030	3335	3635
12.0	630	945	1260	1575	1890	2205	2520	2835	3150	3465	3780

LEGEND 9.—Rate-of-Descent Table.

INSTRUMENT TAKEOFF PROCEDURE CHARTS

RATE OF CLIMB TABLE

(ft. per min.)

A rate of climb table is provided for use in planning and executing takeoff procedures under known or approximate ground speed conditions.

REQUIRED CLIMB RATE (ft. per NM)	GROUND SPEED (KNOTS)						
	30	60	80	90	100	120	140
200	100	200	267	300	333	400	467
250	125	250	333	375	417	500	583
300	150	300	400	450	500	600	700
350	175	350	467	525	583	700	816
400	200	400	533	600	667	800	933
450	225	450	600	675	750	900	1050
500	250	500	667	750	833	1000	1167
550	275	550	733	825	917	1100	1283
600	300	600	800	900	1000	1200	1400
650	325	650	867	975	1083	1300	1516
700	350	700	933	1050	1167	1400	1633

REQUIRED CLIMB RATE (ft. per NM)	GROUND SPEED (KNOTS)					
	150	180	210	240	270	300
200	500	600	700	800	900	1000
250	625	750	875	1000	1125	1250
300	750	900	1050	1200	1350	1500
350	875	1050	1225	1400	1575	1750
400	1000	1200	1400	1600	1700	2000
450	1125	1350	1575	1800	2025	2250
500	1250	1500	1750	2000	2250	2500
550	1375	1650	1925	2200	2475	2750
600	1500	1800	2100	2400	2700	3000
650	1625	1950	2275	2600	2925	3250
700	1750	2100	2450	2800	3150	3500

LEGEND 10.—Rate-of-Climb Table.

INOPERATIVE COMPONENTS OR VISUAL AIDS TABLE

Landing minimums published on instrument approach procedure charts are based upon full operation of all components and visual aids associated with the particular instrument approach chart being used. Higher minimums are required with inoperative components or visual aids as indicated below. If more than one component is inoperative, each minimum is raised to the highest minimum required by any single component that is inoperative. ILS glide slope inoperative minimums are published on instrument approach charts as localizer minimums. This table may be amended by notes on the approach chart. Such notes apply only to the particular approach category(ies) as stated. See legend page for description of components indicated below.

(1) ILS, MLS, and PAR

Inoperative Component or Aid	Approach Category	Increase Visibility
ALSF 1 & 2, MALSR, & SSALR	ABCD	1/4 mile

(2) ILS with visibility minimum of 1,800 RVR.

ALSF 1 & 2, MALSR, &SSALR	ABCD	To 4000 RVR
TDZI RCLS	ABCD	To 2400 RVR
RVR	ABCD	To 1/2 mile

(3) VOR, VOR/DME, VORTAC, VOR (TAC), VOR/DME (TAC), LOC, LOC/DME, LDA, LDA/DME, SDF, SDF/DME, RNAV, and ASR

Inoperative Visual Aid	Approach Category	Increase Visibility
ALSF 1 & 2, MALSR, & SSALR	ABCD	1/2 mile
SSALS, MALS, & ODALS	ABC	1/4 mile

(4) NDB

ALSF 1 & 2, MALSR & SSALR	C	1/2 mile
	ABD	1/4 mile
MALS, SSALS, ODALS	ABC	1/4 mile

LEGEND 11.—Inoperative Components or Visual Aids Table.

DIRECTORY LEGEND

ABBREVIATIONS

The following abbreviations are those commonly used within this directory. Other abbreviations may be found in the Legend and are not duplicated below:

AAS	airport advisory service	ldg	landing
acft	aircraft	med	medium
apch	approach	NFCT	non-federal control tower
arpt	airport	ngt	night
avbl	available	NSTD	nonstandard
bcn	beacon	ntc	notice
blo	below	opr	operate
byd	beyond	ops	operates operation
clsd	closed	ovrn	overrun
ctc	contact	p-line	power line
dalgt	daylight	PPR	prior permission required
dsplc	displace	req	request
dsplcd	displaced	rqr	requires
durn	duration	rgt tfc	right traffic
emerg	emergency	rwy	runway
extd	extend, extended	svc	service
fld	field	tmpry	temporary, temporarily
FSS	Flight Service Station	tkf	takeoff
ints	intensity	tfc	traffic
lgtd	lighted	thld	threshold
lgts	lights	twr	tower

LEGEND 12.—Abbreviations.

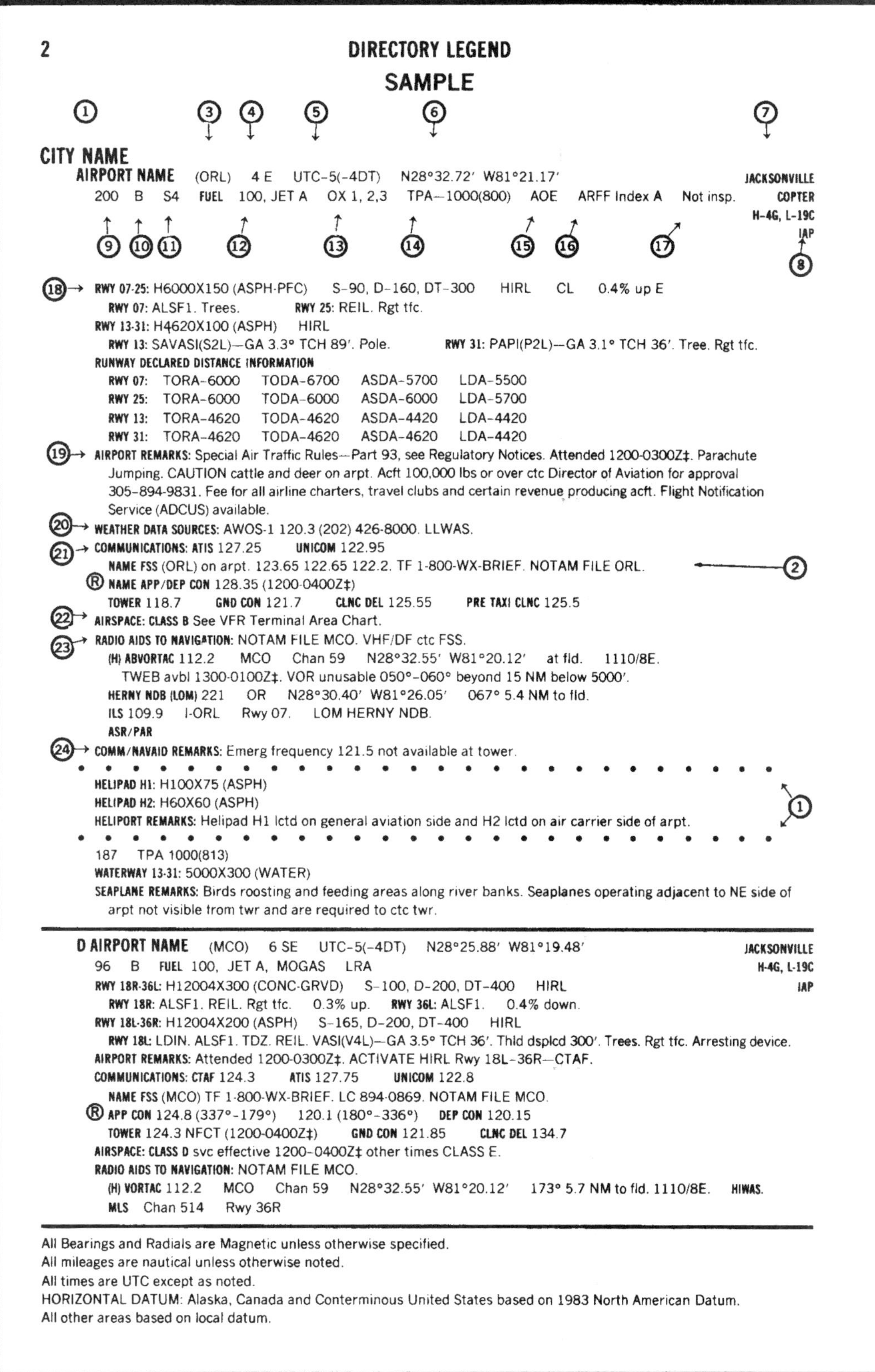

2 DIRECTORY LEGEND

SAMPLE

(1) (3) (4) (5) (6) (7)

CITY NAME

AIRPORT NAME (ORL) 4 E UTC-5(-4DT) N28°32.72′ W81°21.17′ JACKSONVILLE

200 B S4 FUEL 100, JET A OX 1, 2,3 TPA—1000(800) AOE ARFF Index A Not insp. COPTER

H-4G, L-19C

IAP

(9) (10) (11) (12) (13) (14) (15) (16) (17) (8)

(18)→ RWY 07-25: H6000X150 (ASPH-PFC) S-90, D-160, DT-300 HIRL CL 0.4% up E

RWY 07: ALSF1. Trees. RWY 25: REIL. Rgt tfc.

RWY 13-31: H4620X100 (ASPH) HIRL

RWY 13: SAVASI(S2L)—GA 3.3° TCH 89′. Pole. RWY 31: PAPI(P2L)—GA 3.1° TCH 36′. Tree. Rgt tfc.

RUNWAY DECLARED DISTANCE INFORMATION

RWY 07: TORA-6000 TODA-6700 ASDA-5700 LDA-5500

RWY 25: TORA-6000 TODA-6000 ASDA-6000 LDA-5700

RWY 13: TORA-4620 TODA-4620 ASDA-4420 LDA-4420

RWY 31: TORA-4620 TODA-4620 ASDA-4620 LDA-4420

(19)→ AIRPORT REMARKS: Special Air Traffic Rules—Part 93, see Regulatory Notices. Attended 1200-0300Z‡. Parachute Jumping. CAUTION cattle and deer on arpt. Acft 100,000 lbs or over ctc Director of Aviation for approval 305-894-9831. Fee for all airline charters, travel clubs and certain revenue producing acft. Flight Notification Service (ADCUS) available.

(20)→ WEATHER DATA SOURCES: AWOS-1 120.3 (202) 426-8000. LLWAS.

(21)→ COMMUNICATIONS: ATIS 127.25 UNICOM 122.95

NAME FSS (ORL) on arpt. 123.65 122.65 122.2. TF 1-800-WX-BRIEF. NOTAM FILE ORL. ←(2)

Ⓡ NAME APP/DEP CON 128.35 (1200-0400Z‡)

TOWER 118.7 GND CON 121.7 CLNC DEL 125.55 PRE TAXI CLNC 125.5

(22)→ AIRSPACE: CLASS B See VFR Terminal Area Chart.

(23)→ RADIO AIDS TO NAVIGATION: NOTAM FILE MCO. VHF/DF ctc FSS.

(H) ABVORTAC 112.2 MCO Chan 59 N28°32.55′ W81°20.12′ at fld. 1110/8E.

TWEB avbl 1300-0100Z‡. VOR unusable 050°-060° beyond 15 NM below 5000′.

HERNY NDB (LOM) 221 OR N28°30.40′ W81°26.05′ 067° 5.4 NM to fld.

ILS 109.9 I-ORL Rwy 07. LOM HERNY NDB.

ASR/PAR

(24)→ COMM/NAVAID REMARKS: Emerg frequency 121.5 not available at tower.

• •

HELIPAD H1: H100X75 (ASPH)

HELIPAD H2: H60X60 (ASPH) (1)

HELIPORT REMARKS: Helipad H1 lctd on general aviation side and H2 lctd on air carrier side of arpt.

• •

187 TPA 1000(813)

WATERWAY 13-31: 5000X300 (WATER)

SEAPLANE REMARKS: Birds roosting and feeding areas along river banks. Seaplanes operating adjacent to NE side of arpt not visible from twr and are required to ctc twr.

D AIRPORT NAME (MCO) 6 SE UTC-5(-4DT) N28°25.88′ W81°19.48′ JACKSONVILLE

96 B FUEL 100, JET A, MOGAS LRA H-4G, L-19C

RWY 18R-36L: H12004X300 (CONC-GRVD) S-100, D-200, DT-400 HIRL IAP

RWY 18R: ALSF1. REIL. Rgt tfc. 0.3% up. RWY 36L: ALSF1. 0.4% down.

RWY 18L-36R: H12004X200 (ASPH) S-165, D-200, DT-400 HIRL

RWY 18L: LDIN. ALSF1. TDZ. REIL. VASI(V4L)—GA 3.5° TCH 36′. Thld dsplcd 300′. Trees. Rgt tfc. Arresting device.

AIRPORT REMARKS: Attended 1200-0300Z‡. ACTIVATE HIRL Rwy 18L-36R—CTAF.

COMMUNICATIONS: CTAF 124.3 ATIS 127.75 UNICOM 122.8

NAME FSS (MCO) TF 1-800-WX-BRIEF. LC 894-0869. NOTAM FILE MCO.

Ⓡ APP CON 124.8 (337°-179°) 120.1 (180°-336°) DEP CON 120.15

TOWER 124.3 NFCT (1200-0400Z‡) GND CON 121.85 CLNC DEL 134.7

AIRSPACE: CLASS D svc effective 1200-0400Z‡ other times CLASS E.

RADIO AIDS TO NAVIGATION: NOTAM FILE MCO.

(H) VORTAC 112.2 MCO Chan 59 N28°32.55′ W81°20.12′ 173° 5.7 NM to fld. 1110/8E. HIWAS.

MLS Chan 514 Rwy 36R

All Bearings and Radials are Magnetic unless otherwise specified.
All mileages are nautical unless otherwise noted.
All times are UTC except as noted.
HORIZONTAL DATUM: Alaska, Canada and Conterminous United States based on 1983 North American Datum.
All other areas based on local datum.

LEGEND 13.—Directory Legend Sample.

DIRECTORY LEGEND 3

LEGEND

This Directory is an alphabetical listing of data on record with the FAA on all airports that are open to the public, associated terminal control facilities, air route traffic control centers and radio aids to navigation within the conterminous United States, Puerto Rico and the Virgin Islands. Airports are listed alphabetically by associated city name and cross referenced by airport name. Facilities associated with an airport, but with a different name, are listed individually under their own name, as well as under the airport with which they are associated.

The listing of an airport in this directory merely indicates the airport operator's willingness to accommodate transient aircraft, and does not represent that the facility conforms with any Federal or local standards, or that it has been approved for use on the part of the general public.

The information on obstructions is taken from reports submitted to the FAA. It has not been verified in all cases. Pilots are cautioned that objects not indicated in this tabulation (or on charts) may exist which can create a hazard to flight operation.

Detailed specifics concerning services and facilities tabulated within this directory are contained in Airman's Information Manual, Basic Flight Information and ATC Procedures.

The legend items that follow explain in detail the contents of this Directory and are keyed to the circled numbers on the sample on the preceding page.

① CITY/AIRPORT NAME

Airports and facilities in this directory are listed alphabetically by associated city and state. Where the city name is different from the airport name the city name will appear on the line above the airport name. Airports with the same associated city name will be listed alphabetically by airport name and will be separated by a dashed rule line. All others will be separated by a solid rule line. (Designated Helipads and Seaplane Landing Areas (Water) associated with a land airport will be separated by a dotted line.)

② NOTAM SERVICE

All public use landing areas are provided NOTAM "D" (distant dissemination) and NOTAM "L" (local dissemination) service. Airport NOTAM file identifier is shown following the associated FSS data for individual airports, e.g. "NOTAM FILE IAD". See AIM, Basic Flight Information and ATC Procedures for detailed description of NOTAM's.

③ LOCATION IDENTIFIER

A three or four character code assigned to airports. These identifiers are used by ATC in lieu of the airport name in flight plans, flight strips and other written records and computer operations.

④ AIRPORT LOCATION

Airport location is expressed as distance and direction from the center of the associated city in nautical miles and cardinal points, i.e., 4 NE.

⑤ TIME CONVERSION

Hours of operation of all facilities are expressed in Coordinated Universal Time (UTC) and shown as "Z" time. The directory indicates the number of hours to be subtracted from UTC to obtain local standard time and local daylight saving time UTC−5(−4DT). The symbol ‡ indicates that during periods of Daylight Saving Time effective hours will be one hour earlier than shown. In those areas where daylight saving time is not observed that (−4DT) and ‡ will not be shown. All states observe daylight savings time except Arizona and that portion of Indiana in the Eastern Time Zone and Puerto Rico and the Virgin Islands.

⑥ GEOGRAPHIC POSITION OF AIRPORT

Positions are shown in degrees, minutes and hundredths of a minute.

⑦ CHARTS

The Sectional Chart and Low and High Altitude Enroute Chart and panel on which the airport or facility is located. Helicopter Chart locations will be indicated as, i.e., COPTER.

⑧ INSTRUMENT APPROACH PROCEDURES

IAP indicates an airport for which a prescribed (Public Use) FAA Instrument Approach Procedure has been published.

⑨ ELEVATION

Elevation is given in feet above mean sea level and is the highest point on the landing surface. When elevation is sea level it will be indicated as (00). When elevation is below sea level a minus (-) sign will precede the figure.

⑩ ROTATING LIGHT BEACON

B indicates rotating beacon is available. Rotating beacons operate dusk to dawn unless otherwise indicated in AIRPORT REMARKS.

⑪ SERVICING

S1: Minor airframe repairs.
S2: Minor airframe and minor powerplant repairs.
S3: Major airframe and minor powerplant repairs.
S4: Major airframe and major powerplant repairs.

LEGEND 14.—Directory Legend.

4

DIRECTORY LEGEND

⑫ FUEL

CODE	FUEL
80	Grade 80 gasoline (Red)
100	Grade 100 gasoline (Green)
100LL	100LL gasoline (low lead) (Blue)
115	Grade 115 gasoline
A	Jet A—Kerosene freeze point-40° C.
A1	Jet A-1—Kerosene freeze point-50°C.
A1 +	Jet A-1—Kerosene with icing inhibitor, freeze point-50° C.
B	Jet B—Wide-cut turbine fuel, freeze point-50° C.
B +	Jet B—Wide-cut turbine fuel with icing inhibitor, freeze point-50° C.
MOGAS	Automobile gasoline which is to be used as aircraft fuel.

NOTE: Automobile Gasoline. Certain automobile gasoline may be used in specific aircraft engines if a FAA supplemental type cetificate has been obtained. Automobile gasoline which is to be used in aircraft engines will be identified as "MOGAS", however, the grade/type and other octane rating will not be published.

Data shown on fuel availability represents the most recent information the publisher has been able to acquire. Because of a variety of factors, the fuel listed may not always be obtainable by transient civil pilots. Confirmation of availability of fuel should be made directly with fuel dispensers at locations where refueling is planned.

⑬ OXYGEN

OX 1 High Pressure
OX 2 Low Pressure
OX 3 High Pressure—Replacement Bottles
OX 4 Low Pressure—Replacement Bottles

⑭ TRAFFIC PATTERN ALTITUDE

Traffic Pattern Altitude (TPA)—The first figure shown is TPA above mean sea level. The second figure in parentheses is TPA above airport elevation.

⑮ AIRPORT OF ENTRY, LANDING RIGHTS, AND CUSTOMS USER FEE AIRPORTS

U.S. CUSTOMS USER FEE AIRPORT—Private Aircraft operators are frequently required to pay the costs associated with customs processing.

AOE—Airport of Entry—A customs Airport of Entry where permission from U.S. Customs is not required, however, at least one hour advance notice of arrival must be furnished.

LRA—Landing Rights Airport—Application for permission to land must be submitted in advance to U.S. Customs. At least one hour advance notice of arrival must be furnished.

NOTE: Advance notice of arrival at both an AOE and LRA airport may be included in the flight plan when filed in Canada or Mexico, where Flight Notification Service (ADCUS) is available the airport remark will indicate this service. This notice will also be treated as an application for permission to land in the case of an LRA. Although advance notice of arrival may be relayed to Customs through Mexico, Canadian, and U.S. Communications facilities by flight plan, the aircraft operator is solely responsible for insuring that Customs receives the notification. (See Customs, Immigration and Naturalization, Public Health and Agriculture Department requirements in the International Flight Information Manual for further details.)

⑯ CERTIFICATED AIRPORT (FAR 139)

Airports serving Department of Transportation certified carriers and certified under FAR, Part 139, are indicated by the ARFF index; i.e., ARFF Index A, which relates to the availability of crash, fire, rescue equipment.

FAR-PART 139 CERTIFICATED AIRPORTS

INDICES AND AIRCRAFT RESCUE AND FIRE FIGHTING EQUIPMENT REQUIREMENTS

Airport Index	*Required No. Vehicles*	*Aircraft Length*	*Scheduled Departures*	*Agent + Water for Foam*
A	1	$<90'$	≥ 1	500#DC or HALON 1211 or 450#DC + 100 gal H_2O
B	1 or 2	$\geq 90'$, $<126'$	≥ 5	Index A + 1500 gal H_2O
		$\geq 126'$, $<159'$	<5	
C	2 or 3	$\geq 126'$, $<159'$	≥ 5	Index A + 3000 gal H_2O
		$\geq 159'$, $<200'$	<5	
D	3	$\geq 159'$, $<200'$	≥ 5	Index A + 4000 gal H_2O
		$>200'$	<5	
E	3	$\geq 200'$	≥ 5	Index A + 6000 gal H_2O

$>$ Greater Than; $<$ Less Than; $\geq$ Equal or Greater Than; $\leq$ Equal or Less Than; H_2O–Water; DC–Dry Chemical.

LEGEND 15.—Directory Legend.

DIRECTORY LEGEND

NOTE: The listing of ARFF index does not necessarily assure coverage for non-air carrier operations or at other than prescribed times for air carrier. ARFF Index Ltd.—indicates ARFF coverage may or may not be available, for information contact airport manager prior to flight.

⑰ FAA INSPECTION

All airports not inspected by FAA will be identified by the note: Not insp. This indicates that the airport information has been provided by the owner or operator of the field.

⑱ RUNWAY DATA

Runway information is shown on two lines. That information common to the entire runway is shown on the first line while information concerning the runway ends are shown on the second or following line. Lengthy information will be placed in the Airport Remarks.

Runway direction, surface, length, width, weight bearing capacity, lighting, gradient and appropriate remarks are shown for each runway. Direction, length, width, lighting and remarks are shown for sealanes. The full dimensions of helipads are shown, i.e., 50X150.

RUNWAY SURFACE AND LENGTH

Runway lengths prefixed by the letter "H" indicate that the runways are hard surfaced (concrete, asphalt). If the runway length is not prefixed, the surface is sod, clay, etc. The runway surface composition is indicated in parentheses after runway length as follows:

(AFSC)—Aggregate friction seal coat
(ASPH)—Asphalt
(CONC)—Concrete
(DIRT)—Dirt
(GRVD)—Grooved
(GRVL)—Gravel, or cinders
(PFC)—Porous friction courses
(PSP)—Pierced steel plank
(RFSC)—Rubberized friction seal coat
(TURF)—Turf
(TRTD)—Treated
(WC)—Wire combed

RUNWAY WEIGHT BEARING CAPACITY

Runway strength data shown in this publication is derived from available information and is a realistic estimate of capability at an average level of activity. It is not intended as a maximum allowable weight or as an operating limitation. Many airport pavements are capable of supporting limited operations with gross weights of 25-50% in excess of the published figures. Permissible operating weights, insofar as runway strengths are concerned, are a matter of agreement between the owner and user. When desiring to operate into any airport at weights in excess of those published in the publication, users should contact the airport management for permission. Add 000 to figure following S, D, DT, DDT, AUW, etc., for gross weight capacity:

S—Single-wheel type landing gear. (DC-3), (C-47), (F-15), etc.
D—Dual-wheel type landing gear. (DC-6), etc.
T—Twin-wheel type landing gear. (DC-6), (C-9A), etc.
ST—Single-tandem type landing gear. (C-130).
SBTT—Single-belly twin tandem landing gear (KC-10).
DT—Dual-tandem type landing gear, (707), etc.
TT—Twin-tandem type (includes quadricycle) landing gear (707), (B-52), (C-135), etc.
TRT—Triple-tandem landing gear, (C-17)
DDT—Double dual-tandem landing gear. (E4A/747).
TDT—Twin delta-tandem landing gear. (C-5, Concorde).
AUW—All up weight. Maximum weight bearing capacity for any aircraft irrespective of landing gear configuration.
SWL—Single Wheel Loading. (This includes information submitted in terms of Equivalent Single Wheel Loading (ESWL) and Single Isolated Wheel Loading). SWL figures are shown in thousands of pounds with the last three figures being omitted.
PSI—Pounds per square inch. PSI is the actual figure expressing maximum pounds per square inch runway will support, e.g., (SWL 000/PSI 535).

Quadricycle and dual-tandem are considered virtually equal for runway weight bearing consideration, as are single-tandem and dual-wheel.

Omission of weight bearing capacity indicates information unknown.

RUNWAY LIGHTING

Lights are in operation sunset to sunrise. Lighting available by prior arrangement only or operating part of the night only and/or pilot controlled and with specific operating hours are indicated under airport remarks. Since obstructions are usually lighted, obstruction lighting is not included in this code. Unlighted obstructions on or surrounding an airport will be noted in airport remarks. Runway lights nonstandard (NSTD) are systems for which the light fixtures are not FAA approved L-800 series: color, intensity, or spacing does not meet FAA standards. Nonstandard runway lights, VASI, or any other system not listed below will be shown in airport remarks.

Temporary, emergency or limited runway edge lighting such as flares, smudge pots, lanterns or portable runway lights will also be shown in airport remarks.

Types of lighting are shown with the runway or runway end they serve.

NSTD—Light system fails to meet FAA standards.
LIRL—Low Intensity Runway Lights
MIRL—Medium Intensity Runway Lights
HIRL—High Intensity Runway Lights
RAIL—Runway Alignment Indicator Lights
REIL—Runway End Identifier Lights
CL—Centerline Lights
TDZ—Touchdown Zone Lights
ODALS—Omni Directional Approach Lighting System.
AF OVRN—Air Force Overrun 1000′ Standard Approach Lighting System.
LDIN—Lead-In Lighting System.
MALS—Medium Intensity Approach Lighting System.
MALSF—Medium Intensity Approach Lighting System with Sequenced Flashing Lights.

LEGEND 16.—Directory Legend.

6 **DIRECTORY LEGEND**

MALSR—Medium Intensity Approach Lighting System with Runway Alignment Indicator Lights.

SALS—Short Approach Lighting System.

SALSF—Short Approach Lighting System with Sequenced Flashing Lights.

SSALS—Simplified Short Approach Lighting System.

SSALF—Simplified Short Approach Lighting System with Sequenced Flashing Lights.

SSALR—Simplified Short Approach Lighting System with Runway Alignment Indicator Lights.

ALSAF—High Intensity Approach Lighting System with Sequenced Flashing Lights

ALSF1—High Intensity Approach Lighting System with Sequenced Flashing Lights, Category I, Configuration.

ALSF2—High Intensity Approach Lighting System with Sequenced Flashing Lights, Category II, Configuration.

VASI—Visual Approach Slope Indicator System.

NOTE: Civil ALSF-2 may be operated as SSALR during favorable weather conditions.

VISUAL GLIDESLOPE INDICATORS

APAP—A system of panels, which may or may not be lighted, used for alignment of approach path.
- PNIL APAP on left side of runway
- PNIR APAP on right side of runway

PAPI—Precision Approach Path Indicator
- P2L 2-identical light units placed on left side of runway
- P2R 2-identical light units placed on right side of runway
- P4L 4-identical light units placed on left side of runway
- P4R 4-identical light units placed on right side of runway

PVASI—Pulsating/steady burning visual approach slope indicator, normally a single light unit projecting two colors.
- PSIL- PVASI on left side of runway
- PSIR- PVASI on right side of runway

SAVASI—Simplified Abbreviated Visual Approach Slope Indicator
- S2L 2-box SAVASI on left side of runway
- S2R 2-box SAVASI on right side of runway

TRCV—Tri-color visual approach slope indicator, normally a single light unit projecting three colors.
- TRIL TRCV on left side of runway
- TRIR TRCV on right side of runway

VASI—Visual Approach Slope Indicator
- V2L 2-box VASI on left side of runway
- V2R 2-box VASI on right side of runway
- V4L 4-box VASI on left side of runway
- V4R 4-box VASI on right side of runway
- V6L 6-box VASI on left side of runway
- V6R 6-box VASI on right side of runway
- V12 12-box VASI on both sides of runway
- V16 16-box VASI on both sides of runway

NOTE: Approach slope angle and threshold crossing height will be shown when available; i.e., -GA 3.5° TCH 37'.

PILOT CONTROL OF AIRPORT LIGHTING

Key Mike	Function
7 times within 5 seconds	Highest intensity available
5 times within 5 seconds	Medium or lower intensity (Lower REIL or REIL-Off)
3 times within 5 seconds	Lowest intensity available (Lower REIL or REIL-Off)

Available systems will be indicated in the Airport Remarks, as follows:

ACTIVATE MALSR Rwy 07, HIRL Rwy 07-25-122.8 (or CTAF).
or
ACTIVATE MIRL Rwy 18-36-122.8 (or CTAF).
or
ACTIVATE VASI and REIL, Rwy 07-122.8 (or CTAF).

Where the airport is not served by an instrument approach procedure and/or has an independent type system of different specification installed by the airport sponsor, descriptions of the type lights, method of control, and operating frequency will be explained in clear text. See AIM, "Basic Flight Information and ATC Procedures," for detailed description of pilot control of airport lighting.

RUNWAY SLOPE

Runway slope will be shown only when it is 0.3 percent or more. On runways less than 8000 feet: When available the direction of the slope upward will be indicated, ie., 0.3% up NW. On runways 8000 feet or greater: When available the slope will be shown on the runway end line, ie., RWY 13: 0.3% up., RWY 21: Pole. Rgt tfc. 0.4% down.

RUNWAY END DATA

Lighting systems such as VASI, MALSR, REIL; obstructions; displaced thresholds will be shown on the specific runway end. "Rgt tfc"—Right traffic indicates right turns should be made on landing and takeoff for specified runway end.

LEGEND 17.—Directory Legend Visual Glide Slope Indicators.

DIRECTORY LEGEND

RUNWAY DECLARED DISTANCE INFORMATION

TORA—Take-off Run Available
TODA—Take-off Distance Available
ASDA—Accelerate-Stop Distance Available
LDA—Landing Distance Available

(19) AIRPORT REMARKS

Landing Fee indicates landing charges for private or non-revenue producing aircraft, in addition, fees may be charged for planes that remain over a couple of hours and buy no services, or at major airline terminals for all aircraft.
Remarks—Data is confined to operational items affecting the status and usability of the airport.
Parachute Jumping.—See "PARACHUTE" tabulation for details.
Unless otherwise stated, remarks including runway ends refer to the runway's approach end.

(20) WEATHER DATA SOURCES

ASOS—Automated Surface Observing System. Reports the same as an AWOS-3 plus precipitation identification and intensity, and freezing rain occurrence (future enhancement).
AWOS—Automated Weather Observing System

AWOS-A—reports altimeter setting.
AWOS-1—reports altimeter setting, wind data and usually temperature, dewpoint and density altitude.
AWOS-2—reports the same as AWOS-1 plus visibility.
AWOS-3—reports the same as AWOS-1 plus visibility and cloud/ceiling data.
See AIM, Basic Flight Information and ATC Procedures for detailed description of AWOS.

HIWAS—See RADIO AIDS TO NAVIGATION
LAWRS—Limited Aviation Weather Reporting Station where observers report cloud height, weather, obstructions to vision, temperature and dewpoint (in most cases), surface wind, altimeter and pertinent remarks.
LLWAS—indicates a Low Level Wind Shear Alert System consisting of a center field and several field perimeter anemometers.
SAWRS—identifies airports that have a Supplemental Aviation Weather Reporting Station available to pilots for current weather information.
SWSL—Supplemental Weather Service Location providing current local weather information via radio and telephone.

(21) COMMUNICATIONS

Communications will be listed in sequence in the order shown below:
Common Traffic Advisory Frequency (CTAF), Automatic Terminal Information Service (ATIS) and Aeronautical Advisory Stations (UNICOM) along with their frequency is shown, where available, on the line following the heading "COMMUNICATIONS." When the CTAF and UNICOM is the same frequency, the frequency will be shown as CTAF/UNICOM freq.
Flight Service Station (FSS) information. The associated FSS will be shown followed by the identifier and information concerning availability of telephone service, e.g., Direct Line (DL), Local Call (LC-384-2341), Toll free call, dial (TF 800-852-7036 or TF 1-800-227-7160), Long Distance (LD 202-426-8800 or LD 1-202-555-1212) etc. The airport NOTAM file identifier will be shown as "NOTAM FILE IAD." Where the FSS is located on the field it will be indicated as "on arpt" following the identifier. Frequencies available will follow. The FSS telephone number will follow along with any significant operational information. FSS's whose name is not the same as the airport on which located will also be listed in the normal alphabetical name listing for the state in which located. Remote Communications Outlet (RCO) providing service to the airport followed by the frequency and name of the Controlling FSS.
FSS's provide information on airport conditions, radio aids and other facilities, and process flight plans. Local Airport Advisory Service is provided on the CTAF by FSS's located at non-tower airports or airports where the tower is not in operation.
(See AIM, Par. 157/158 Traffic Advisory Practices at airports where a tower is not in operation or AC 90 - 42C.)
Aviation weather briefing service is provided by FSS specialists. Flight and weather briefing services are also available by calling the telephone numbers listed.
Remote Communications Outlet (RCO)—An unmanned air/ground communications facility, remotely controlled and providing UHF or VHF communications capability to extend the service range of an FSS.
Civil Communications Frequencies—Civil communications frequencies used in the FSS air/ground system are now operated simplex on 122.0, 122.2, 122.3, 122.4, 122.6, 123.6; emergency 121.5; plus receive-only on 122.05, 122.1, 122.15, and 123.6.

a. 122.0 is assigned as the Enroute Flight Advisory Service channel at selected FSS's.
b. 122.2 is assigned to most FSS's as a common enroute simplex service.
c. 123.6 is assigned as the airport advisory channel at non-tower FSS locations, however, it is still in commission at some FSS's collocated with towers to provide part time Local Airport Advisory Service.
d. 122.1 is the primary receive-only frequency at VOR's. 122.05, 122.15 and 123.6 are assigned at selected VOR's meeting certain criteria.
e. Some FSS's are assigned 50 kHz channels for simplex operation in the 122-123 MHz band (e.g. 122.35). Pilots using the FSS A/G system should refer to this directory or appropriate charts to determine frequencies available at the FSS or remoted facility through which they wish to communicate.

Part time FSS hours of operation are shown in remarks under facility name.

Emergency frequency 121.5 is available at all Flight Service Stations, Towers, Approach Control and RADAR facilities, unless indicated as not available.
Frequencies published followed by the letter "T" or "R", indicate that the facility will only transmit or receive respectively on that frequency. All radio aids to navigation frequencies are transmit only.

LEGEND 18.—Directory Legend.

8

DIRECTORY LEGEND

TERMINAL SERVICES

CTAF—A program designed to get all vehicles and aircraft at uncontrolled airports on a common frequency.
ATIS—A continuous broadcast of recorded non-control information in selected areas of high activity.
UNICOM—A non-government air/ground radio communications facility utilized to provide general airport advisory service.
APP CON —Approach Control. The symbol Ⓡ indicates radar approach control.
TOWER—Control tower
GND CON—Ground Control
DEP CON—Departure Control. The symbol Ⓡ indicates radar departure control.
CLNC DEL—Clearance Delivery.
PRE TAXI CLNC—Pre taxi clearance
VFR ADVSY SVC—VFR Advisory Service. Service provided by Non-Radar Approach Control.
Advisory Service for VFR aircraft (upon a workload basis) ctc APP CON.
TOWER, APP CON and DEP CON RADIO CALL will be the same as the airport name unless indicated otherwise.

㉒ AIRSPACE

CLASS C—CLASS C service provided
CLASS B—Radar Sequencing and Separation Service for all aircraft in CLASS B airspace
TRSA—Radar Sequencing and Separation Service for participating VFR Aircraft within a Terminal Radar Service Area

㉓ RADIO AIDS TO NAVIGATION

The Airport Facility Directory lists by facility name all Radio Aids to Navigation, except Military TACANS, that appear on National Ocean Service Visual or IFR Aeronautical Charts and those upon which the FAA has approved an Instrument Approach Procedure. All VOR, VORTAC ILS and MLS equipment in the National Airspace System has an automatic monitoring and shutdown feature in the event of malfunction. Unmonitored, as used in this publication for any navigational aid, means that FSS or tower personnel cannot observe the malfunction or shutdown signal. The NAVAID NOTAM file identifier will be shown as "NOTAM FILE IAD" and will be listed on the Radio Aids to Navigation line. When two or more NAVAIDS are listed and the NOTAM file identifier is different than shown on the Radio Aids to Navigation line, then it will be shown with the NAVAID listing. NOTAM file identifiers for ILS's and their components (e.g., NDB (LOM) are the same as the identifiers for the associated airports and are not repeated. Hazardous Inflight Weather Advisory Service (HIWAS) will be shown where this service is broadcast over selected VOR's.

NAVAID information is tabulated as indicated in the following sample:

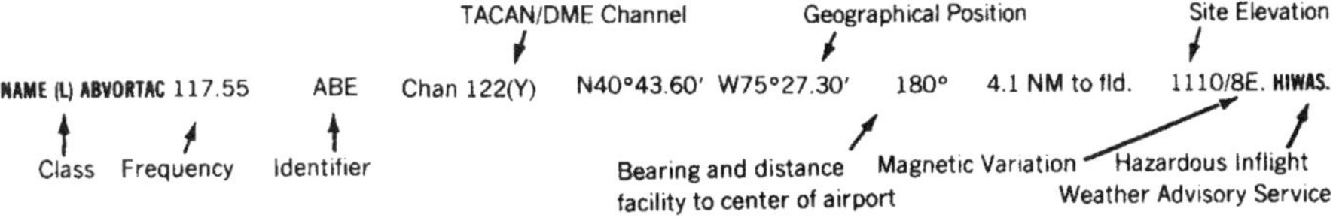

VOR unusable 020°-060° beyond 26 NM below 3500′

Restriction within the normal altitude/range of the navigational aid (See primary alphabetical listing for restrictions on VORTAC and VOR/DME).

Note: Those DME channel numbers with a (Y) suffix require TACAN to be placed in the "Y" mode to receive distance information.

HIWAS—Hazardous Inflight Weather Advisory Service is a continuous broadcast of inflight weather advisories including summarized SIGMETs, convective SIGMETs, AIRMETs and urgent PIREPs. HIWAS is presently broadcast over selected VOR's and will be implemented throughout the conterminous U.S.

ASR/PAR—Indicates that Surveillance (ASR) or Precision (PAR) radar instrument approach minimums are published in U.S. Government Instrument Approach Procedures.

RADIO CLASS DESIGNATIONS

VOR/DME/TACAN Standard Service Volume (SSV) Classifications

SSV Class	Altitudes	Distance (NM)
(T) Terminal	1000′ to 12,000′	25
(L) Low Altitude	1000′ to 18,000′	40
(H) High Altitude	1000′ to 14,500′	40
	14,500′ to 18,000′	100
	18,000′ to 45,000′	130
	45,000′ to 60,000′	100

NOTE: Additionally, (H) facilities provide (L) and (T) service volume and (L) facilities provide (T) service. Altitudes are with respect to the station's site elevation. Coverage is not available in a cone of airspace directly above the facility.

LEGEND 19.—Directory Legend.

DIRECTORY LEGEND 9

The term VOR is, operationally, a general term covering the VHF omnidirectional bearing type of facility without regard to the fact that the power, the frequency protected service volume, the equipment configuration, and operational requirements may vary between facilities at different locations.

AB ______ Automatic Weather Broadcast
DF ______ Direction Finding Service.
DME ______ UHF standard (TACAN compatible) distance measuring equipment.
DME(Y) ______ UHF standard (TACAN compatible) distance measuring equipment that require TACAN to be placed in the "Y" mode to receive DME.
H ______ Non-directional radio beacon (homing), power 50 watts to less than 2,000 watts (50 NM at all altitudes).
HH ______ Non-directional radio beacon (homing), power 2,000 watts or more (75 NM at all altitudes).
H-SAB ______ Non-directional radio beacons providing automatic transcribed weather service.
ILS ______ Instrument Landing System (voice, where available, on localizer channel).
ISMLS ______ Interim Standard Microwave Landing System.
LDA ______ Localizer Directional Aid.
LMM ______ Compass locator station when installed at middle marker site (15 NM at all altitudes).
LOM ______ Compass locator station when installed at outer marker site (15 NM at all altitudes).
MH ______ Non-directional radio beacon (homing) power less than 50 watts (25 NM at all altitudes).
MLS ______ Microwave Landing System
S ______ Simultaneous range homing signal and/or voice.
SABH ______ Non-directional radio beacon not authorized for IFR or ATC. Provides automatic weather broadcasts.
SDF ______ Simplified Direction Facility.
TACAN ______ UHF navigational facility-omnidirectional course and distance information.
VOR ______ VHF navigational facility-omnidirectional course only.
VOR/DME ______ Collocated VOR navigational facility and UHF standard distance measuring equipment.
VORTAC ______ Collocated VOR and TACAN navigational facilities.
W ______ Without voice on radio facility frequency.
Z ______ VHF station location marker at a LF radio facility.

FREQUENCY PAIRING PLAN AND MLS CHANNELING

MLS CHANNEL	VHF FREQUENCY	TACAN CHANNEL	MLS CHANNEL	VHF FREQUENCY	TACAN CHANNEL	MLS CHANNEL	VHF FREQUENCY	TACAN CHANNEL
500	108.10	18X	568	109.45	31Y	634	114.05	87Y
502	108.30	20X	570	109.55	32Y	636	114.15	88Y
504	108.50	22X	572	109.65	33Y	638	114.25	89Y
506	108.70	24X	574	109.75	34Y	640	114.35	90Y
508	108.90	26X	576	109.85	35Y	642	114.45	91Y
510	109.10	28X	578	109.95	36Y	644	114.55	92Y
512	109.30	30X	580	110.05	37Y	646	114.65	93Y
514	109.50	32X	582	110.15	38Y	648	114.75	94Y
516	109.70	34X	584	110.25	39Y	650	114.85	95Y
518	109.90	36X	586	110.35	40Y	652	114.95	96Y
520	110.10	38X	588	110.45	41Y	654	115.05	97Y
522	110.30	40X	590	110.55	42Y	656	115.15	98Y
524	110.50	42X	592	110.65	43Y	658	115.25	99Y
526	110.70	44X	594	110.75	44Y	660	115.35	100Y
528	110.90	46X	596	110.85	45Y	662	115.45	101Y
530	111.10	48X	598	110.95	46Y	664	115.55	102Y
532	111.30	50X	600	111.05	47Y	666	115.65	103Y
534	111.50	52X	602	111.15	48Y	668	115.75	104Y
536	111.70	54X	604	111.25	49Y	670	115.85	105Y
538	111.90	56X	606	111.35	50Y	672	115.95	106Y
540	108.05	17Y	608	111.45	51Y	674	116.05	107Y
542	108.15	18Y	610	111.55	52Y	676	116.15	108Y
544	108.25	19Y	612	111.65	53Y	678	116.25	109Y
546	108.35	20Y	614	111.75	54Y	680	116.35	110Y
548	108.45	21Y	616	111.85	55Y	682	116.45	111Y
550	108.55	22Y	618	111.95	56Y	684	116.55	112Y
552	108.65	23Y	620	113.35	80Y	686	116.65	113Y
554	108.75	24Y	622	113.45	81Y	688	116.75	114Y
556	108.85	25Y	624	113.55	82Y	690	116.85	115Y
558	108.95	26Y	626	113.65	83Y	692	116.95	116Y
560	109.05	27Y	628	113.75	84Y	694	117.05	117Y
562	109.15	28Y	630	113.85	85Y	696	117.15	118Y
564	109.25	29Y	632	113.95	86Y	698	117.25	119Y
566	109.35	30Y						

Legend 20.—Frequency Pairing Plan and MLS Channeling.

DIRECTORY LEGEND

FREQUENCY PAIRING PLAN AND MLS CHANNELING

The following is a list of paired VOR/ILS VHF frequencies with TACAN channels and MLS channels.

TACAN CHANNEL	VHF FREQUENCY	MLS CHANNEL	TACAN CHANNEL	VHF FREQUENCY	MLS CHANNEL	TACAN CHANNEL	VHF FREQUENCY	MLS CHANNEL
17X	108.00	-	50Y	111.35	606	94X	114.70	-
17Y	108.05	540	51X	111.40	-	94Y	114.75	648
18X	108.10	500	51Y	111.45	608	95X	114.80	-
18Y	108.15	542	52X	111.50	534	95Y	114.85	650
19X	108.20	-	52Y	111.55	610	96X	114.90	-
19Y	108.25	544	53X	111.60	-	96Y	114.95	652
20X	108.30	502	53Y	111.65	612	97X	115.00	-
20Y	108.35	546	54X	111.70	536	97Y	115.05	654
21X	108.40	-	54Y	111.75	614	98X	115.10	-
21Y	108.45	548	55X	111.80	-	98Y	115.15	656
22X	108.50	504	55Y	111.85	616	99X	115.20	-
22Y	108.55	550	56X	111.90	538	99Y	115.25	658
23X	108.60	-	56Y	111.95	618	100X	115.30	-660
23Y	108.65	552	57X	112.00	-	100Y	115.35	-
24X	108.70	506	57Y	112.05	-	101X	115.40	662
24Y	108.75	554	58X	112.10	-	101Y	115.45	-
25X	108.80	-	58Y	112.15	-	102X	115.50	664
25Y	108.85	556	59X	112.20	-	102Y	115.55	-
26X	108.90	508	59Y	112.25	-	103X	115.60	666
26Y	108.95	558	70X	112.30	-	103Y	115.65	-
27X	109.00	-	70Y	112.35	-	104X	115.70	668
27Y	109.05	560	71X	112.40	-	104Y	115.75	-
28X	109.10	510	71Y	112.45	-	105X	115.80	670
28Y	109.15	562	72X	112.50	-	105Y	115.85	-
29X	109.20	-	72Y	112.55	-	106X	115.90	672
29Y	109.25	564	73X	112.60	-	106Y	115.95	-
30X	109.30	512	73Y	112.65	-	107X	116.00	674
30Y	109.35	566	74X	112.70	-	107Y	116.05	-
31X	109.40	-	74Y	112.75	-	108X	116.10	676
31Y	109.45	568	75X	112.80	-	108Y	116.15	-
32X	109.50	514	75Y	112.85	-	109X	116.20	678
32Y	109.55	570	76X	112.90	-	109Y	116.25	-
33X	109.60	-	76Y	112.95	-	110X	116.30	680
33Y	109.65	572	77X	113.00	-	110Y	116.35	-
34X	109.70	516	77Y	113.05	-	111X	116.40	682
34Y	109.75	574	78X	113.10	-	111Y	116.45	-
35X	109.80	-	78Y	113.15	-	112X	116.50	684
35Y	109.85	576	79X	113.20	-	112Y	116.55	-
36X	109.90	518	79Y	113.25	-	113X	116.60	686
36Y	109.95	578	80X	113.30	-	113Y	116.65	-
37X	110.00	-	80Y	113.35	620	114X	116.70	688
37Y	110.05	580	81X	113.40	-	114Y	116.75	-
38X	110.10	520	81Y	113.45	622	115X	116.80	690
38Y	110.15	582	82X	113.50	-	115Y	116.85	-
39X	110.20	-	82Y	113.55	624	116X	116.90	692
39Y	110.25	584	83X	113.60	-	116Y	116.95	-
40X	110.30	522	83Y	113.65	626	117X	117.00	694
40Y	110.35	586	84X	113.70	-	117Y	117.05	-
41X	110.40	-	84Y	113.75	628	118X	117.10	696
41Y	110.45	588	85X	113.80	-	118Y	117.15	-
42X	110.50	524	85Y	113.85	630	119X	117.20	698
42Y	110.55	590	86X	113.90	-	119Y	117.25	-
43X	110.60	-	86Y	113.95	632	120X	117.30	-
43Y	110.65	592	87X	114.00	-	120Y	117.35	-
44X	110.70	526	87Y	114.05	634	121X	117.40	-
44Y	110.75	594	88X	114.10	-	121Y	117.45	-
45X	110.80	-	88Y	114.15	636	122X	117.50	-
45Y	110.85	596	89X	114.20	-	122Y	117.55	-
46X	110.90	528	89Y	114.25	638	123X	117.60	-
46Y	110.95	598	90X	114.30	-	123Y	117.65	-
47X	111.00	-	90Y	114.35	640	124X	117.70	-
47Y	111.05	600	91X	114.40	-	124Y	117.75	-
48X	111.10	530	91Y	114.45	642	125X	117.80	-
48Y	111.15	602	92X	114.50	-	125Y	117.85	-
49X	111.20	-	92Y	114.55	644	126X	117.90	-
49Y	111.25	604	93X	114.60	-	126Y	117.95	
50X	111.30	532	93Y	114.65	646			

(23) **COMM/NAVAID REMARKS:**

Pertinent remarks concerning communications and NAVAIDS.

LEGEND 21.—Frequency Pairing Plan and MLS Channeling.

AIRPORT LORAN TD CORRECTION TABLE

The following LORAN - C time difference (TD) table contains the TD correction values for each airport with a published LORAN RNAV instrument approach procedure. This TD correction value must be entered into the LORAN airborne receiver prior to beginning the approach. TD values from this table should be transferred to the TD correction box shown in the plan view of the LORAN RNAV approach for the destination airport.

Pilots are advised to check LORAN (LRN) NOTAM's to obtain the status of the LORAN chain or group repetition interval (GRI) and NOTAM's for the LORAN monitor location identifier (MLID) at their destination and alternate airport.

	AIRPORT											
CITY	ST	NAME	LID	MLID	GRI	TRI	V	W	X	Y	Z
BURLINGTON	VT	BURLINGTON INTL	BTV	BTV	9960	MWX		10.0	09.8		
COLUMBUS	OH	OHIO STATE UNIV.	OSU	OSU	9960	MYZ				08.6	11.6
NEW ORLEANS	LA	LAKEFRONT	NEW	NEW	7980	MWX		11.5	11.2		
ORLANDO	FL	ORLANDO EXEC.	ORL	ORL	7980	MYZ				11.3	11.7
PORTLAND	OR	PORTLAND INTL.	PDX	PDX	9940	MWX		11.7	09.4		
VENICE	LA	CHEVRON	8LA5	NEW	7980	MWX		11.5	11.2		

LEGEND 22.—Airport Loran TD Correction Table.

EXCERPT FROM CFR 49 PART 175

PART 175—CARRIAGE BY AIRCRAFT

Subpart A—General Information and Regulations

Sec.
175.1 Purpose and scope.
175.3 Unacceptable hazardous materials shipments.
175.5 Applicability.
175.10 Exceptions.
175.20 Compliance.
175.30 Accepting shipments.
175.33 Notification of pilot-in-command.
175.35 Shipping papers aboard aircraft.
175.40 Keeping and replacement of labels.
175.45 Reporting hazardous materials dents.

Subpart B—Loading, Unloading and Handling

175.75 Quantity limitations aboard aircraft.
175.78 Stowage compatibility of cargo.
175.79 Orientation of cargo.
175.85 Cargo location.
175.90 Damaged shipments.

Subpart C—Specific Regulations Applicable According to Classification of Material

175.305 Self-propelled vehicles.
175.310 Transportation of flammable liquid fuel in small, passenger-carrying aircraft.
175.320 Cargo-only aircraft; only means of transportation.
175.630 Special requirements for poisons.

LEGEND 23.—Excerpt from CFR 49 Part 175.

EXCERPT FROM CFR 49 PART 175

Sec.
175.640 Special requirements for other regulated materials.
175.700 Special requirements for radioactive materials.
175.710 Special requirements for fissile Class III radioactive materials.

AUTHORITY: 49 U.S.C. 1803, 1804, 1808; 49 CFR 1.53(e), unless otherwise noted.

SOURCE: Amdt. 175-1, 41 FR 16106, Apr. 15, 1976, unless otherwise noted.

NOTE: Nomenclature changes to Part 175 appear at 43 FR 48645, Oct. 19, 1978 (Amdt. 175-6).

Subpart A—General Information and Regulations

§ 175.1 Purpose and scope.

This part prescribes requirements, in addition to those contained in Parts 171, 172 and 173 of this subchapter, to be observed by aircraft operators with respect to the transportation of hazardous materials aboard (including attached to or suspended from) civil aircraft.

§ 175.3 Unacceptable hazardous materials shipments.

A shipment of hazardous materials that is not prepared for shipment in accordance with Parts 172 and 173 of this subchapter may not be accepted for transportation or transported aboard an aircraft.

§ 175.5 Applicability.

This part contains regulations pertaining to the acceptance of hazardous materials for transportation, and the loading and transportation of hazardous materials, in any civil aircraft in the United States and in civil aircraft of United States registry anywhere in air commerce, except aircraft of United States registry under lease to and operated solely by foreign nationals outside the United States.

§ 175.10 Exceptions.

(a) This subchapter does not apply to—

(1) Aviation fuel and oil in tanks that are in compliance with the installation provisions of 14 CFR, Chapter 1.

(2) Aircraft parts, equipment, and supplies (other than fuel) carried by an aircraft operator if authorized or required aboard his aircraft for their operation including:

(i) Fire extinguishers;

(ii) Cylinders containing compressed gases;

(iii) Aerosol dispensers;

(iv) Distilled spirits;

(v) Hydraulic accumulators;

(vi) Non-spillable batteries;

(vii) First-aid kits;

(viii) Signaling devices;

(ix) Tires; and

(x) Items of replacement therefor, except that batteries, aerosol dispensers, and signaling devices must be packed in strong outside containers, and tires must be deflated to a pressure not greater than 100 p.s.i.g.

(3) Hazardous materials loaded and carried in hoppers or tanks of aircraft certificated for use in aerial seeding, dusting, spraying, fertilizing, crop improvement, or pest control, to be dispensed during such an operation.

(4) Medicinal and toilet articles carried by a crewmember or passenger in his baggage (including carry-on baggage) when:

(i) The total capacity of all the containers used by a crewmember or passenger does not exceed 75 ounces (net weight ounces and fluid ounces);

(ii) The capacity of each container other than an aerosol container does not exceed 16 fluid ounces or 1 pound of material.

(5) Small-arms ammunition for personal use carried by a crewmember or passenger in his baggage (excluding carry-on baggage) if securely packed in fiber, wood, or metal boxes.

(6) Prior to May 3, 1981, radioactive materials which meet the requirements of § 173.391(a), (b), or (c) of this subchapter in effect on May 3, 1979.

(7) Oxygen, or any hazardous material used for the generation of oxygen, carried for medical use by a passenger in accordance with 14 CFR 121.574 or 135.114.

(8) Human beings and animals with an implanted medical device, such as a heart pacemaker, that contains radioactive material or with radio-pharmaceuticals that have been injected or ingested.

LEGEND 24.—Excerpt from CFR 49 Part 175.

EXCERPT FROM CFR 49 PART 175

Chapter I—Research and Special Programs Administration **§ 175.30**

(9) Smoke grenades, flares, or similar devices carried only for use during a sport parachute jumping activity.

(10) Personal smoking materials intended for use by any individual when carried on his person except lighters with flammable liquid reservoirs and containers containing lighter fluid for use in refilling lighters.

(11) Smoke grenades, flares, and pyrotechnic devices affixed to aircraft carrying no person other than a required flight crewmember during any flight conducted at and as a part of a scheduled air show or exhibition of aeronautical skill. The affixed installation accommodating the smoke grenades, flares, or pyrotechnic devices on the aircraft must be approved by the FAA for its intended use.

(12) Hazardous materials which are loaded and carried on or in cargo-only aircraft and which are to be dispensed or expended during flight for weather control, forest preservation and protection, or avalanche control purposes when the following requirements are met:

(i) Operations may not be conducted over densely populated areas, in a congested airway, or near any airport where air carrier passenger operations are conducted.

(ii) Each operator shall prepare and keep current a manual containing operational guidelines and handling procedures, for the use and guidance of flight, maintenance, and ground personnel concerned in the dispensing or expending of hazardous materials. The manual must be approved by the FAA District Office having jurisdiction over the operator's certificate, if any, or the FAA Regional Office in the region where the operator is located. Each operation must be conducted in accordance with the manual.

(iii) No person other than a required flight crewmember, FAA inspector, or person necessary for handling or dispensing the hazardous material may be carried on the aircraft.

(iv) The operator of the aircraft must have advance permission from the owner of any airport to be used for the dispensing or expending operation.

(v) When dynamite and blasting caps are carried for avalanche control flights, the explosives must be handled, and, at all times be, under the control of a blaster who is licensed under a state or local authority identified in writing to the FAA district office having jurisdiction over the operator's certificate, if any, or the FAA regional office in the region where the operator is located.

(49 U.S.C. 1803, 1804, 1806, 1808; 49 CFR 1.53 and App. A to Part 1)

[Amdt. 175-1, 41 FR 16106, Apr. 15, 1976, as amended by Amdt. 175-1A, 41 FR 40686, Sept. 20, 1976]

NOTE: For amendments to § 175.10 see the List of CFR Sections Affected appearing in the Finding Aids section of this volume.

§ 175.20 Compliance.

Unless the regulations in this subchapter specifically provide that another person must perform a duty, each operator shall comply with all the regulations in Parts 102, 171, 172, and 175 of this subchapter and shall thoroughly instruct his employees in relation thereto. (See 14 CFR 121.135, 121.401, 121.433a, 135.27 and 135.140.)

§ 175.30 Accepting shipments.

(a) No person may accept a hazardous material for transportation aboard an aircraft unless the hazardous material is—

(1) Authorized, and is within the quantity limitations specified for carriage aboard aircraft according to § 172.101 of this subchapter;

(2) Described and certified on a shipping paper prepared in duplicate in accordance with Subpart C of Part 172 of this subchapter. The originating aircraft operator must retain one copy of each shipping paper for 90 days;

(3) Labeled and marked, or placarded (when required), in accordance with Subparts D, E and F of Part 172 of this subchapter; and

(4) Labeled with a "CARGO AIRCRAFT ONLY" label (see § 172.448 of this subchapter) if the material as presented is not permitted aboard passenger-carrying aircraft.

(b) Except as provided in paragraph (c) of this section, no person may carry any hazardous material aboard an aircraft unless, prior to placing the material aboard the aircraft, the operator of the aircraft has inspected the package, or the outside container pre-

LEGEND 25.—Excerpt from CFR 49 Part 175.

EXCERPT FROM CFR 49 PART 175

pared in accordance with § 173.25 of this subchapter which contains the material, and has determined that—it has no holes, leakage, or other indication that its integrity has been compromised, and for radioactive materials that the package seal has not been broken.

(c) The requirements of paragraph (b) of this section do not apply to ORM-D materials packed in a freight container and offered for transportation by one consignor.

[Amdt. 175-1, 41 FR 16106, Apr. 15, 1976, as amended by Amdt. 175-1A, 41 FR 40686, Sept. 20, 1976; Amdt. 175-1B, 41 FR 57072, Dec. 30, 1976]

§ 175.33 Notification of pilot-in-command.

When materials subject to the provisions of this subchapter are carried in an aircraft, the operator of the aircraft shall give the pilot-in-command the following information in writing before takeoff:

(a) The information required by §§ 172-202 and 172.203 of this subchapter;

(b) The location of the hazardous material in the aircraft; and

(c) The results of the inspection required by § 175.30(b).

[Amdt. 175-1A, 41 FR 40686, Sept. 20, 1976]

§ 175.35 Shipping papers aboard aircraft.

(a) A copy of the shipping papers required by § 175.30(a)(2) must accompany the shipment it covers during transportation aboard an aircraft.

(b) The documents required by paragraph (a) of this section and § 175.33 may be combined into one document if it is given to the pilot-in-command before departure of the aircraft.

§ 175.40 Keeping and replacement of labels.

(a) Aircraft operators who engage in the transportation of hazardous materials must keep an adequate supply of the labels specified in Subpart E of Part 172 of this subchapter, on hand at each location where shipments are loaded aboard aircraft.

(b) Lost or detached labels for packages of hazardous materials must be replaced in accordance with the information provided on the shipping papers.

§ 175.45 Reporting hazardous materials incidents.

(a) Each operator that transports hazardous materials shall report to the nearest Air Carrier District Office (ACDO), Flight Standards District Office (FSDO), General Aviation District Office (GADO) or other FAA facility, except that in place of reporting to the nearest of those facilities a certificate holder under 14 CFR Part 121, 127, or 135 may report to the FAA District Office holding the carrier's operating certificate and charged with overall inspection of its operations, by telephone at the earliest practicable moment after each incident that occurs during the course of transportation (including loading, unloading or temporary storage) in which as a direct result of any hazardous material—

(1) A person is killed;

(2) A person receives injuries requiring his or her hospitalization;

(3) Estimated carrier or other property damage, or both, exceeds $50,000;

(4) Fire, breakage, or spillage or suspected radioactive contamination occurs involving shipment of radioactive materials (see § 175.700(b));

(5) Fire, breakage, spillage, or suspected contamination occurs involving shipment of etiologic agents. In addition to the report required by paragraph (a) of this section, a report on an incident involving etiologic agents should be telephoned directly to the Director, Center for Disease Control, U.S. Public Health, Atlanta, Georgia, area code 404-633-5313; or

(6) A situation exists of such a nature that, in the judgment of the carrier, it should be reported to the Department even though it does not meet the criteria of paragraph (b)(1), (2), or (3) of this section, e.g., a continuing danger to life exists at the scene of the incident.

(7) If the operator conforms to the provisions of this section, the carrier requirements of § 171.15 except § 171.15(c) of this subchapter shall be deemed to have been satisfied.

(b) The following information shall be furnished in each report:

LEGEND 26.—Excerpt from CFR 49 Part 175.

EXCERPT FROM CFR 49 PART 175

Chapter I—Research and Special Programs Administration **§ 175.85**

(1) Name of reporting person;

(2) Name and address of carrier represented by reporter;

(3) Phone number where reporter can be contacted;

(4) Date, time, and location of incident;

(5) The extent of the injuries, if any; and

(6) Classification, name and quantity of hazardous material involvement and whether a continuing danger to life exists at the scene.

(c) Each operator who transports hazardous materials shall report in writing, in duplicate, on DOT Form F 5800.1 within 15 days of the date of discovery, each incident that occurs during the course of transportation (including loading, unloading, or temporary storage) in which, as a direct result to hazardous materials, any of the circumstances set forth in paragraph (a) of this section occurs or there has been an unintentional release of hazardous materials from a package. Each operator making a report under this section shall send that report to the Materials Transportation Bureau, Office of Hazardous Materials Regulation, Department of Transportation, Washington, D.C. 20590, with a separate copy to the FAA facility indicated in paragraph (a) of this section.

[Amdt. 175-1, 41 FR 16106, Apr. 15, 1976, as amended by Amdt. 175-1A, 41 FR 40686, Sept. 20, 1976]

Subpart B—Loading, Unloading and Handling

§ 175.75 Quantity limitations aboard aircraft.

(a) Except as provided in § 175.85(b), no person may carry on an aircraft—

(1) A hazardous material except as permitted in Part 172 of this subchapter;

(2) More than 50 pounds net weight of hazardous material (and in addition thereto, 150 pounds net weight of nonflammable compressed gas) permitted to be carried aboard passenger-carrying aircraft—

(i) In an inaccessible cargo compartment,

(ii) In any freight container within an accessible cargo compartment, or

(iii) In any accessible cargo compartment in a cargo-only aircraft in a manner that makes it inaccessible unless in a freight container;

(3) Packages containing radioactive materials when their combined transport indices exceed 50.

(b) No limitation applies to the number of packages of ORM material aboard an aircraft.

[Amdt. 175-1A, 41 FR 40686, Sept. 20, 1976]

§ 175.78 Stowage compatibility of cargo.

(a) No person may stow a package of a corrosive material on an aircraft next to or in a position that will allow contact with a package of flammable solids, oxidizing materials, or organic peroxides.

(b) No person may stow a package labeled BLASTING AGENT on an aircraft next to, or in a position that will allow contact with a package of special fireworks or railway torpedoes.

[Amdt. 175-1, 41 FR 16106, Apr. 15, 1976, as amended by Amdt. 175-8, 44 FR 31184, May 31, 1979]

§ 175.79 Orientation of cargo.

(a) A package containing hazardous materials marked "THIS SIDE UP", "THIS END UP", or with arrows to indicate the proper orientation of the package, must be stored, loaded abroad an aircraft in accordance with such markings, and secured in a manner that will prevent any movement which would change the orientation of the package.

(b) A package containing liquid hazardous materials not marked as indicated in paragraph (a) of this section must be stored and loaded with closures up.

§ 175.85 Cargo location.

(a) No person may carry a hazardous material subject to the requirements of this subchapter in the cabin of a passenger-carrying aircraft.

(b) Each person carrying materials acceptable only for cargo-only aircraft shall carry those materials in a location accessible to a crewmember during flight. However, when materials acceptable for cargo-only or pas-

LEGEND 27.—Excerpt from CFR 49 Part 175.

EXCERPT FROM CFR 49 PART 175

senger carrying aircraft are carried on a small, single pilot, cargo-only aircraft being used where other means of transportation are impracticable or not available, they may be carried without quantity limitation as specified in § 175.75 in a location that is not accessible to the pilot subject to the following conditions.

(1) No person other than the pilot, an FAA inspector, the shipper or consignee of the material or a representative of the shipper or consignee so designated in writing, or a person necessary for handling the material may be carried on the aircraft.

(2) The pilot must be provided with written instructions on characteristics and proper handling of the material.

(3) Whenever a change of pilots occurs while the material is on board, the new pilot must be briefed under a hand-to-hand signature service provided by the operator of the aircraft.

(c) No person may load magnetized material (which might cause an erroneous magnetic compass reading) on an aircraft, in the vicinity of a magnetic compass, or compass master unit, that is a part of the instrument equipment of the aircraft, in a manner that affects its operation. If this requirement cannot be met, a special aircraft swing and compass calibration may be made. No person loading magnetized materials may obscure the warning labels.

(d) No person may carry materials subject to the requirements of this subchapter in an aircraft unless they are suitably safeguarded to prevent their becoming a hazard by shifting. For packages bearing "RADIOACTIVE YELLOW-II" or "RADIOACTIVE YELLOW-III" labels, such safeguarding must prevent movement that would permit the package to be closer to a space that is occupied by a person or an animal than is permitted by § 175.700.

(e) No person may carry a material subject to the requirements of this subchapter that is acceptable for carriage in a passenger-carrying aircraft (other than magnetized materials) unless it is located in the aircraft in a place that is inaccessible to persons other than crew-members.

[Amdt. 175-1, 41 FR 16106, Apr. 15, 1976, as amended by Amdt. 175-1A, 41 FR 40686, Sept. 20, 1976]

§ 175.90 Damaged shipments.

Except as provided for in § 175.700, the operator of an aircraft shall remove from the aircraft any package subject to this subchapter that appears to be damaged or leaking. No person shall place or transport a package that is damaged or appears to be damaged or leaking aboard an aircraft subject to this Part.

[Amdt. 175-1, 41 FR 16106, Apr. 15, 1976, as amended by Amdt 175-1A, 41 FR 40686, Sept. 20, 1976]

Subpart C—Specific Regulations Applicable According to Classification of Material

§ 175.305 Self-propelled vehicles.

(a) Self-propelled vehicles are exempt from the drainage requirements of § 173.120 of this subchapter when carried in aircraft designed or modified for vehicle ferry operations and when all of the following conditions are met:

(1) Authorization for this type operation has been given by the appropriate authority in the government of the country in which the aircraft is registered;

(2) Each vehicle is secured in an upright position;

(3) Each fuel tank is filled in a manner and only to a degree that will preclude spillage of fuel during loading, unloading, and transportation; and

(4) Ventilation rates to be maintained in the vehicle storage compartment have been approved by the appropriate authority in the government of the country in which the aircraft is registered.

§ 175.310 Transportation of flammable liquid fuel in small, passenger-carrying aircraft.

A small aircraft or helicopter operated entirely within the State of Alaska or into a remote area elsewhere in the United States may carry, in other than scheduled passenger operations, not more than 20 gallons of flammable liquid fuel, if—

LEGEND 28.—Excerpt from CFR 49 Part 175.

EXCERPT FROM CFR 49 PART 175

Chapter I—Research and Special Programs Administration **§ 175.320**

(a) Transportation by air is the only practical means of providing suitable fuel;

(b) The flight is necessary to meet the needs of a passenger;

(c) The fuel is carried in metal containers that are either—

(1) DOT Specification 2A containers of not more than 5 gallons capacity, each packed inside a DOT Specification 12B fiberboard box or each packed inside a DOT Specification 15A, 15B, 15C, 16A, 19A or 19B wooden box, or in the case of a small aircraft in Alaska, each packed inside a wooden box of at least one-half inch thickness;

(2) Airtight, leakproof, inside containers of not more than 10 gallons capacity and of at least 28-gauge metal, each packed inside a DOT Specification 15A, 15B, 15C, 16A, 19A, or 19B wooden box or, in the case of a small aircraft in Alaska, each packed inside a wooden box of at least one-half inch thickness;

(3) DOT Specification 17E containers of not more than 5 gallons capacity; or

(4) Fuel tanks attached to flammable liquid fuel powered equipment under the following conditions:

(i) Each piece of equipment is secured in an upright position;

(ii) Each fuel tank is filled in a manner that will preclude spillage of fuel during loading, unloading, and transportation; and

(iii) Ventilation rates which are maintained in the compartment in which the equipment is carried have been approved by the FAA district office responsible for inspection and surveillance of the aircraft on which the equipment is carried.

(d) In the case of a helicopter, the fuel is carried on external cargo racks;

(e) The area or compartment in which the fuel is loaded is ventilated so as to prevent the accumulation of fumes;

(f) Before each flight, the pilot-in-command—

(1) Informs each passenger of the location of the fuel and the hazards involved; and

(2) Prohibits smoking, lighting matches, the carrying of any lighted cigar, pipe, cigarette or flame, and the use of anything that might cause an open flame or spark, while loading or unloading or in flight; and

(g) Fuel is transferred to the fuel tanks only while the aircraft is on the surface.

[Amdt. 175-1, 41 FR 16106, Apr. 15, 1976, as amended by Amdt. 175-1A, 41 FR 40686, Sept. 20, 1976]

§ 175.320 Cargo-only aircraft; only means of transportation.

(a) Notwithstanding § 172.101, when means of transportation other than air are impracticable or not available, hazardous materials listed in the following table may be carried on a cargo-only aircraft subject to the conditions stated in the table and in paragraph (b) of this section and, when appropriate, paragraph (c) of this section:

Material description	Class	Conditions
Electric blasting caps (more than 1,000).	Class A explosives	Permitted only when no other cargo is aboard the aircraft. However, if the electric blasting caps are packed in an IME 22 container (see 49 CFR 171.7(d)(9)), they may be transported in the same aircraft with materials that are not classed as hazardous materials.
Electric blasting caps (1,000 or less).	Class C explosives	Permitted only when no other cargo is aboard the aircraft. However, if the electric blasting caps are packed in a DOT MC 201 container (49 CFR 178.318) or an IME 22 container (see 49 CFR 171.7(d)(9)), they may be transported in the same aircraft with materials other than class A or class B explosives.
Gasoline	Flammable liquid	Permitted in metal drums having rated capacities of 55 gal. or less. May not be transported in the same aircraft with materials classed as class A, B, or C explosives, blasting agents, corrosive materials or oxidizing materials. Permitted in installed tanks each having a capacity of more than 110 gal. Subject to the conditions specified in para. (c) of this section.

LEGEND 29.—Excerpt from CFR 49 Part 175.

EXCERPT FROM CFR 49 PART 175

§ 175.320 **Title 49—Transportation**

Material description	Class	Conditions
High explosives	Class A explosives	Limited to explosives to be used for blasting. Permitted only when no other cargo is aboard the aircraft or when being transported in the same aircraft with an authorized shipment of any 1 or more of the following materials to be used for blasting: Ammonium nitrate-fuel oil mixtures Blasting agent, n.o.s. Cordeau detonant fuse. Propellant explosive (solid) class B (water gels only). Propellant explosive (liquid) class B (water gels only).
Oil n.o.s.; petroleum oil or petroleum oil, n.o.s.	Flammable liquid	Permitted in metal drums having rated capacities of 55 gal. or less. May not be transported in the same aircraft with materials classed as class A, B, or C explosives, blasting agents, corrosive materials, or oxidizing materials. Permitted in installed tanks each having a capacity of more than 110 gal. subject to the conditions specified in para. (c) of this section.
Combustible liquid, n.o.s	Combustible liquid	Permitted in installed tanks each having a capacity of more than 110 gal subject to the conditions specified in par. (c) of this section.

(b) The following conditions apply to the carriage of hazardous materials performed under the authority of this section:

(1) No person other than a required flight crewmember, an FAA inspector, the shipper or consignee of the material or a representative of the shipper or consignee so designated in writing, or a person necessary for handling the material may be carried on the aircraft.

(2) The operator of the aircraft must have advance permission from the owner or operator of each manned airport where the material is to be loaded or unloaded or where the aircraft is to land while the material is on board. When the destination is changed after departure because of weather or other unforeseen circumstances, permission from the owner or operator of the alternate airport should be obtained as soon as practicable before landing.

(3) At any airport where the airport owner or operator or authorized representative thereof has designated a location for loading or unloading the material concerned, the material may not be loaded or unloaded at any other location.

(4) If the material concerned can create destructive forces or have lethal or injurious effects over an appreciable area as a result of an accident involving the aircraft or the material, the loading and unloading of the aircraft and its operation in takeoff, en route, and in landing must be conducted at a safe distance from heavily populated areas and from any place of human abode or assembly.

(5) If the aircraft is being operated by a holder of a certificate issued under 14 CFR Part 121, Part 127, or Part 135, operations must be conducted in accordance with conditions and limitations specified in the certificate holder's operations specifications or operations manual accepted by the FAA. If the aircraft is being operated under 14 CFR Part 91, operations must be conducted in accordance with an operations plan accepted and acknowledged in writing by the operator's FAA District Office.

(6) Each pilot of the aircraft must be provided written instructions stating the conditions and limitations of the operation being conducted and the name of the airport official[s] granting the advance permission required by the first sentence of paragraph (b)(2) of this section.

(7) The aircraft and the loading arrangement to be used must be approved for safe carriage of the particular materials concerned by the FAA District Office holding the operator's certificate and charged with overall inspection of its operations, or the appropriate FAA District Office serving the place where the material is to be loaded.

LEGEND 30.—Excerpt from CFR 49 Part 175.

EXCERPT FROM CFR 49 PART 175

Chapter I—Research and Special Programs Administration **§ 175.700**

(8) When Class A explosives are carried under the authority of this section, the operator of the aircraft shall obtain route approval from the FAA inspector in the operator's FAA District Office.

(9) During loading and unloading, no person may smoke, carry a lighted cigarette, cigar, or pipe, or operate any device capable of causing an open flame or spark within 50 feet of the aircraft.

(c) The following additional conditions apply to the carriage of flammable liquids and combustible liquids in tanks each having a capacity of more than 110 gallons under the authority of this section:

(1) The tanks and their associated piping and equipment and the installation thereof must have been approved for the material to be transported by the appropriate FAA Regional Office.

(2) In the case of an aircraft being operated by a certificate holder, the operator shall list the aircraft and the approval information in its operating specifications. If the aircraft is being operated by other than a certificate holder, a copy of the FAA Regional Office approval required by this section must be carried on the aircraft.

(3) The crew of the aircraft must be thoroughly briefed on the operation of the particular bulk tank system being used.

(4) During loading and unloading and thereafter until any remaining fumes within the aircraft are dissipated:

(i) Only those electrically operated bulk tank shutoff valves that have been approved under a supplemental type certificate may be electrically operated.

(ii) No engine or electrical equipment, avionic equipment, or auxiliary power units may be operated, except position lights in the steady position and equipment required by approved loading or unloading procedures, as set forth in the operator's operations manual, or for operators that are not certificate holders, as set forth in a written statement.

(iii) No person may fill a container, other than an approved bulk tank, with a flammable or combustible liquid or discharge a flammable or combustible liquid from a container, other than an approved bulk tank, while that container is inside or within 50 feet of the aircraft.

(iv) When filling an approved bulk tank by hose from inside the aircraft, the doors and hatches must be fully open to insure proper ventilation.

(v) Static ground wires must be connected between the storage tank or fueler and the aircraft, and between the aircraft and a positive ground device.

[Amdt. 175-1, 41 FR 16106, Apr. 15, 1976, as amended by Amdt. 175-1A, 41 FR 40686, Sept. 20, 1976]

NOTE: For amendments to § 175.320 see the List of CFR Sections Affected appearing in the Finding Aids section of this volume.

§ 175.630 Special requirements for poisons.

(a) No person may transport a package bearing a POISON label aboard an aircraft in the same cargo compartment with material which is marked as or known to be food stuff, feed, or any other edible material intended for consumption by humans or animals.

(b) No person may operate an aircraft that has been used to transport any package bearing a POISON label unless, upon removal of such package, the area in the aircraft in which it was carried is visually inspected for evidence of leakage, spillage, or other contamination. All contamination discovered must be either isolated or removed from the aircraft. The operation of an aircraft contaminated with such poisons is considered to be the carriage of poisonous materials under paragraph (a) of this section.

§ 175.640 Special requirements for other regulated materials.

Asbestos must be loaded, handled, and unloaded, and any asbestos contamination of aircraft removed, in a manner that will minimize occupational exposure to airborne asbestos particles released incident to transportation. (See § 173.1090 of this subchapter.)

[Amdt. 175-7, 43 FR 56668, Dec. 4, 1978]

§ 175.700 Special requirements for radioactive materials.

(a) No person may place any package of radioactive materials bearing

LEGEND 31.—Excerpt from CFR 49 Part 175.

EXCERPT FROM CFR 49 PART 175

§ 175.700 **Title 49—Transportation**

"RADIOACTIVE YELLOW-II" or "RADIOACTIVE YELLOW-III" labels in an aircraft closer than the distances shown in the following table to a space (or dividing partition between spaces) which may be continuously occupied by people, or shipments of animals, or closer than the distances shown in the following table to any package containing undeveloped film (if so marked). If more than one of these packages is present, the distance shall be computed from the following table on the basis of the total transport index numbers shown on labels of the individual packages in the aircraft:

Total transport index	Minimum separation distances in feet to the nearest undeveloped film for various times of transit					Minimum distance in feet to area of persons, or minimum distance in feet from dividing partition of cargo compartment
	Up to 2 hr	2–4 hr	4–8 hr	8–12 hr	Over 12 hr	
None	0	0	0	0	0	0
0.1 to 1.0	1	2	3	4	5	1
1.1 to 5.0	3	4	6	8	11	2
5.1 to 10.0	4	6	9	11	15	3
10.1 to 20.0	5	8	12	16	22	4
20.1 to 30.0	7	10	15	20	29	5
30.1 to 40.0	8	11	17	22	33	6
40.1 to 50.0	9	12	19	24	36	7

(b) In addition to the reporting requirements of § 175.45, the carrier must also notify the shipper at the earliest practicable moment following any incident in which there has been breakage, spillage, or suspected radioactive contamination involving radioactive materials shipments. Aircraft in which radioactive materials have been spilled may not be again placed in service or routinely occupied until the radiation does rate at any accessible surface is less than 0.5 millirem per hour and there is no significant removable radioactive surface contamination (see § 173.397 of this subchapter). In these instances, the package or materials should be segregated as far as practicable from personnel contact. If radiological advice or assistance is needed, the U.S. Energy Research and Development Administration should also be notified. In case of obvious leakage, or if it appears likely that the inside container may have been damaged, care should be taken to avoid inhalation, ingestion, or contact with the radioactive materials. Any loose radioactive materials should be left in a segregated area pending disposal instructions from qualified persons.

(c) No person may carry aboard a passenger-carrying aircraft any package of radioactive material which contains a large quantity (large radioactive source) of radioactivity (as defined in § 173.389(b) of this subchapter), except as specifically approved by the Director, Office of Hazardous Materials Regulation, Materials Transportation Bureau, Department of Transportation.

(d) Except as provided in this paragraph, no person may carry aboard a passenger-carrying aircraft any radioactive material other than a radioactive material intended for use in, or incident to, research or medical diagnosis or treatment. Prior to May 3, 1981, this prohibition does not apply to materials which meet the requirements of § 173.391(a), (b), or (c) of this subchapter in effect on May 3, 1979.

(49 U.S.C. 1803, 1804, 1806, 1808; 49 CFR 1.53 and App. A to Part 1)

[Amdt. 175-1, 41 FR 16106, Apr. 15, 1976, as amended by Amdt. 175-4, 42 FR 22367, May 3, 1977]

NOTE: For amendments to § 175.700 see the List of CFR Sections Affected appearing in the Finding Aids section of this volume.

LEGEND 32.—Excerpt from CFR 49 Part 175.

EXCERPT FROM CFR 49 PART 175

Chapter I—Research and Special Programs Administration **§ 175.710**

§ 175.710 **Special requirements for fissile Class III radioactive materials.**

(a) No person may carry aboard any aircraft any package of fissile Class III radioactive material (as defined in § 173.389(a)(3) of this subchapter), except as follows:

(1) On a cargo-only aircraft which has been assigned for the sole use of the consignor for the specific shipment of fissile radioactive material. Instructions for such sole use must be provided for in special arrangements between the consignor and carrier, with instructions to that effect issued with shipping papers; or

(2) On any aircraft on which there is no other package of radioactive materials required to bear one of the RADIOACTIVE labels described in §§ 172.436, 172.438, and 172.440 of this subchapter. Specific arrangements must be effected between the shipper and carriers, with instructions to that effect issued with the shipping papers; or

(3) In accordance with any other procedure specifically approved by the Director, Office of Hazardous Materials Regulation, Materials Transportation Bureau.

[Amdt. 175-1, 41 FR 16106, Apr. 15, 1976, as amended by Amdt. 175-6, 43 FR 48645, Oct. 19, 1978]

LEGEND 33.—Excerpt from CFR 49 Part 175.

EXCERPT FROM CFR 49 PART 172

§172.101 Hazardous Materials Table (cont'd)

(1)	(2)	(3)	(4)	(5)		(6)		(7)		
□/ W/ A	Hazardous materials descriptions and proper shipping names	Hazard class	Label(s) required (if not excepted)	Packaging		Maximum net quantity in one package		Water shipments		
				(a) Exceptions	(b) Specific require-ments	(a) Passenger carrying aircraft or railcar	(b) Cargo only aircraft	(a) Cargo vessel	(b) Pas-senger vessel	(c) Other requirements
	Accumulator, pressurized (pneumatic or hydraulic), containing nonflammable gas	Nonflamma-ble gas	Nonflamma-ble gas	173.306		No limit	No limit	1,2	1,2	
	Acetal	Flammable liquid	Flammable liquid	173.118	173.119	1 quart	10 gallons	1,3	4	
	Acetaldehyde (ethyl aldehyde)	Flammable liquid	Flammable liquid	None	173.119	Forbid-den	10 gallons	1,3	5	
A	Acetaldehyde ammonia	ORM-A	None	173.505	173.510	No limit	No limit			
□	Acetic acid (aqueous solution)	Corrosive material	Corrosive	173.244	173.245	1 quart	10 gallons	1,2	1,2	Stow separate from nitric acid or oxidiz-ing materials.
	Acetic acid, glacial	Corrosive material	Corrosive	173.244	173.245	1 quart	10 gallons	1,2	1,2	Stow separate from nitric acid or oxidiz-ing materials. Segregation same as for flammable liquids
	Acetic anhydride	Corrosive material	Corrosive	173.244	173.245	1 quart	1 gallon	1,2	1,2	
	Acetone	Flammable liquid	Flammable liquid	173.118	173.119	1 quart	10 gallons	1,3	4	
	Acetone cyanohydrin	Poison B	Poison	None	173.346	Forbid-den	55 gallons	1	5	Shade from radiant heat. Stow away from corrosive materials.
	Acetone oil	Flammable liquid	Flammable liquid	173.118	173.119	1 quart	10 gallons	1,2	1	
	Acetonitrile	Flammable liquid	Flammable liquid	173.118	173.119	1 quart	10 gallons	1	4	Shade from radiant heat.
	Acetyl benzoyl peroxide, solid	Forbidden								
	Acetyl benzoyl peroxide solution, not over 40% peroxide	Organic peroxide	Organic peroxide	None	173.222	Forbid-den	1 quart	1,2	1	
	Acetyl bromide	Corrosive material	Corrosive	173.244	173.247	1 quart	1 gallon	1	1	Keep dry. Glass carboys not permitted on passenger vessels.

LEGEND 34.—Hazardous Materials Table (CFR 49 Part 172).

EXCERPT FROM CFR 49 PART 172

§172.101 Hazardous Materials Table (cont'd)

(1) □/W/A	(2) Hazardous materials descriptions and proper shipping names	(3) Hazard class	(4) Label(s) required (if not excepted)	(5) Packaging (a) Exceptions	(5) Packaging (b) Specific require-ments	(6) Maximum net quantity in one package (a) Passenger carrying aircraft or railcar	(6) Maximum net quantity in one package (b) Cargo only aircraft	(7) Water shipments (a) Cargo vessel	(7) Water shipments (b) Pas-senger vessel	(7) Water shipments (c) Other requirements
	Acetyl chloride	Flammable liquid	Flammable liquid	173.244	173.247	1 quart	1 gallon	1	1	Stow away from alcohols. Keep cool and dry. Separate longitudinally by an intervening complete compartment or hold from explosives.
	Acetylene	Flammable gas	Flammable gas	None	173.303	Forbid-den	300 pounds	1	1	Shade from radiant heat.
A	Acetylene tetrabromide	ORM-A	None	173.505	173.510	10 gallons	55 gallons			
	Acetyl iodide	Corrosive material	Corrosive	173.244	173.247	1 quart	1 gallon	1	1	Keep dry. Glass carboys not permitted on passenger vessels.
	Acetyl peroxide solution, not over 25% peroxide	Organic peroxide	Organic peroxide	173.153	173.222	Forbid-den	1 quart	1,2	1	
	Acid butyl phosphate	Corrosive material	Corrosive	173.244	173.245	1 quart	5 gallons	1,2	1,2	Glass carboys in hampers not permitted under deck.
	Acid carboy empty. See Carboy, empty									
□	Acid, liquid, n.o.s.	Corrosive material	Corrosive	173.244	173.245	1 quart	5 pints	1	4	Keep cool.
□	Acid, sludge	Corrosive material	Corrosive	None	173.248	Forbid-den	1 quart	1,2	1	
	Acrolein, inhibited	Flammable liquid	Flammable liquid and Poison	None	173.122	Forbid-den	1 quart	1,2	5	Keep cool. Stow away from living quar-ters.
	Acrylic acid	Corrosive material	Corrosive	173.244	173.245	1 quart	5 pints	1	1	
	Acrylonitrile	Flammable liquid	Flammable liquid and Poison	None	173.119	Forbid-den	1 quart	1,2	5	Keep cool.

LEGEND 35.—Hazardous Materials Table (CFR 49 Part 172) (Cont'd).

EXCERPT FROM CFR 49 PART 172

§172.101 Title 49—Transportation

§172.101 Hazardous Materials Table (cont'd)

(1)	(2)	(3)	(4)	(5)		(6)		(7)		
				Packaging		Maximum net quantity in one package		Water shipments		
□/ W/ A	Hazardous materials descriptions and proper shipping names	Hazard class	Label(s) required (if not excepted)	(a) Exceptions	(b) Specific require-ments	(a) Passenger carrying aircraft or railcar	(b) Cargo only aircraft	(a) Cargo vessel	(b) Pas-senger vessel	(c) Other requirements
	Alkyl aluminum halides. See Pyrophoric liquid, n.o.s.									
A	Allethrin	ORM-A	None	173.505	173.510	No limit	No limit			
	Allyl alcohol	Flammable liquid	Flammable liquid and Poison	173.118	173.119	1 quart	10 gallons	1,2	1	
	Allyl bromide	Flammable liquid	Flammable liquid	173.118	173.119	Forbid-den	10 gallons	1,2	1	
	Allyl chloride	Flammable liquid	Flammable liquid	None	173.119	Forbid-den	10 gallons	1,3	5	
	Allyl chlorocarbonate	Flammable liquid	Flammable liquid	None	173.288	Forbid-den	5 pints	1	5	Keep dry. Separate longitudinally by an intervening complete hold or compart-ment from explosives. Segregation same as for corrosive materials.
	Allyl chloroformate. See Allyl chlorocarbonate									
	Allyl trichlorosilane	Corrosive material	Corrosive	None	173.280	Forbid-den	10 gallons	1	1	Keep dry.
	Aluminum alkyls. See Pyrophoric liquid n.o.s.									
	Aluminum bromide, anhydrous	Corrosive material	Corrosive	173.244	173.245b	25 pounds	100 pounds	1,2	1,2	Keep dry.
	Aluminum dross, wet or hot. See Sec. 173.173									
	Aluminum hydride	Flammable solid	Flammable solid and Dangerous when wet	None	173.206	Forbid-den	25 pounds	1,2	5	Segregation same as for flammable solid labeled Dangerous When Wet.

LEGEND 36.—Hazardous Materials Table (CFR 49 Part 172) (Cont'd).

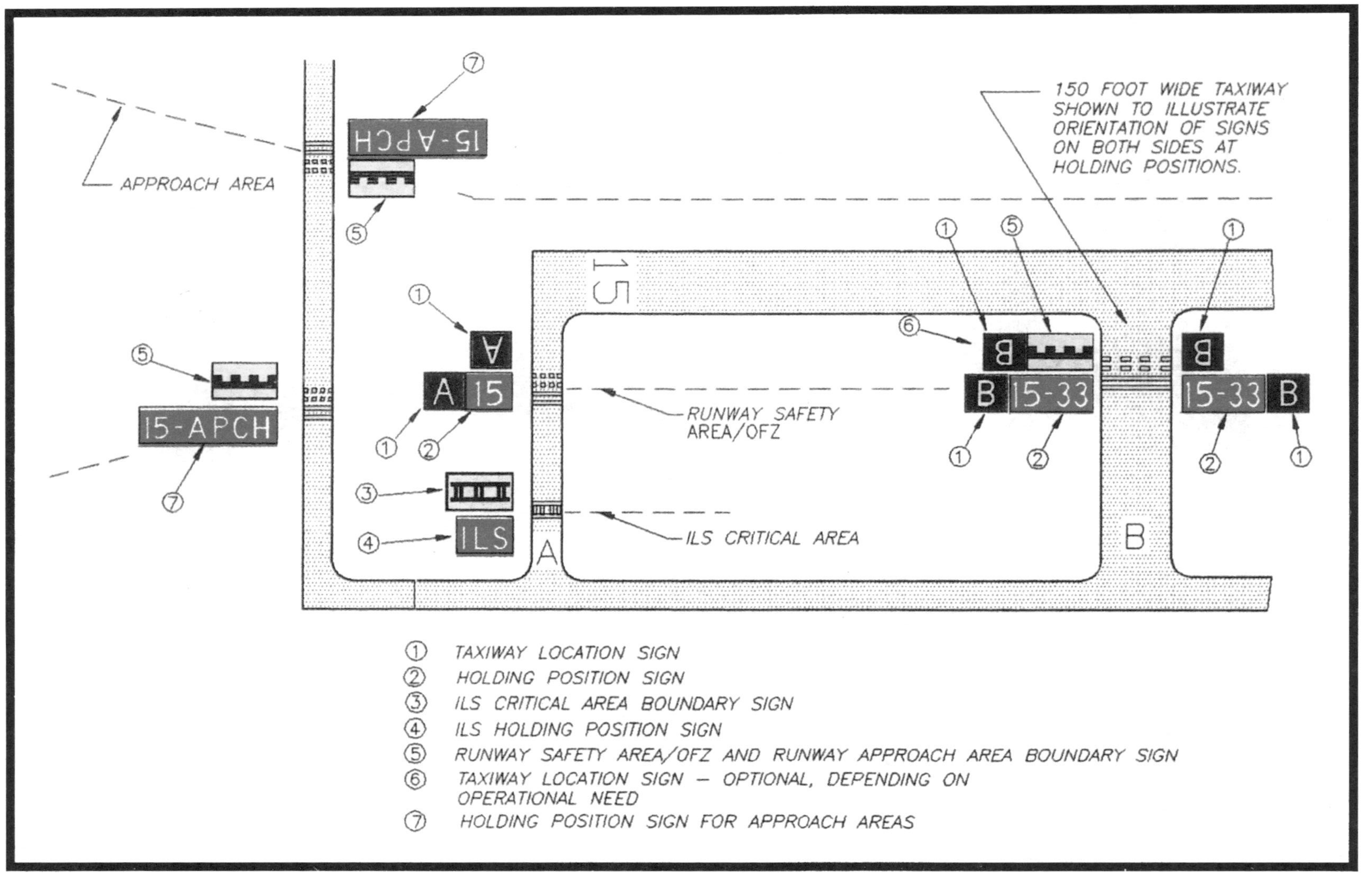

LEGEND 37.—Application Examples for Holding Position Signs.

Airspace Reclassification at a Glance

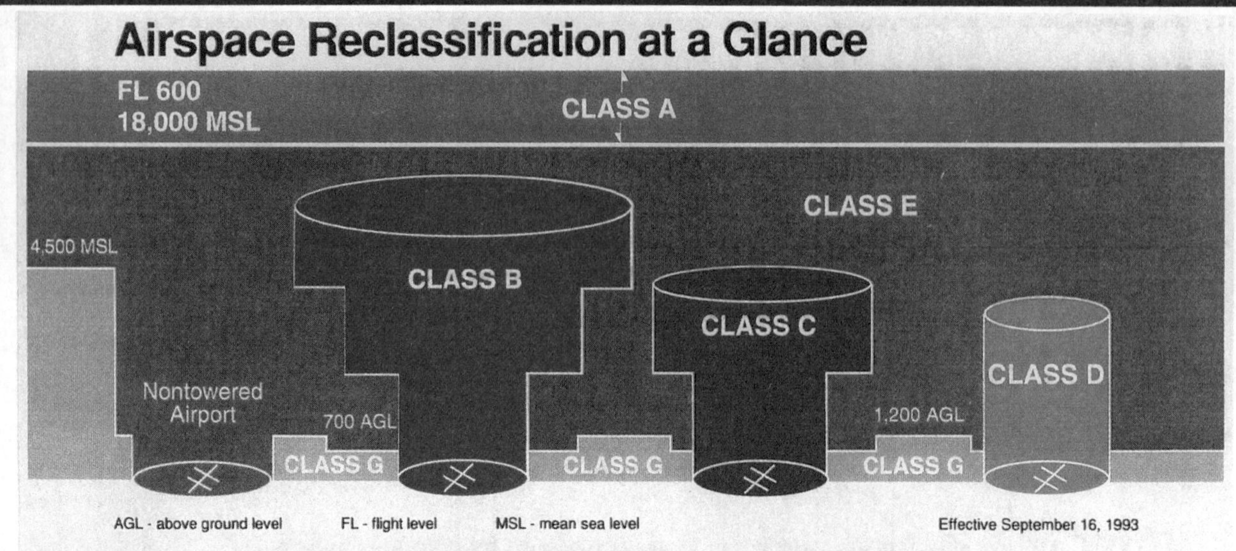

AGL - above ground level FL - flight level MSL - mean sea level Effective September 16, 1993

And an Easy-to-Read Chart

Airspace Features	Class A	Class B	Class C	Class D	Class E	Class G
Former Airspace Equivalent	Positive Control Area (PCA)	Terminal Control Area (TCA)	Airport Radar Service Area (ARSA)	Airport Traffic Area (ATA) and Control Zone (CZ)	General Controlled Airspace	Uncontrolled Airspace
Operations Permitted	IFR	IFR and VFR	IFR and VFR	IFR and VFR	IFR and VFR	IFR and VFR
Entry Requirements	ATC clearance	ATC clearance	ATC clearance for IFR. All require radio contact.	ATC clearance for IFR. All require radio contact.	ATC clearance for IFR. All IFR require radio contact.	None
Minimum Pilot Qualifications	Instrument Rating	Private or student certificate	Student certificate	Student certificate	Student certificate	Student certificate
Two-way Radio Communications	Yes	Yes	Yes	Yes	Yes for IFR	No
VFR Minimum Visibility	N/A	3 statute miles	3 statute miles	3 statute miles	[1]3 statute miles	[2]1 statute mile
VFR Minimum Distance from Clouds	N/A	Clear of clouds	500' below, 1,000' above, and 2,000' horizontal	500' below, 1,000' above, and 2,000' horizontal	[1]500' below, 1,000' above, and 2,000' horizontal	Clear of clouds
Aircraft Separation	All	All	IFR, SVFR, and runway operations	IFR, SVFR, and runway operations	IFR and SVFR	None
Conflict Resolution	N/A	N/A	Between IFR and VFR ops	No	No	No
Traffic Advisories	N/A	N/A	Yes	Workload permitting	Workload permitting	Workload permitting
Safety Advisories	Yes	Yes	Yes	Yes	Yes	Yes
Differs from ICAO	No	[3]Yes	[3,4]Yes	[4]Yes for VFR	No	[5]Yes for VFR
Changes the Existing Rule	No	[6]Yes for VFR	No	[7,8,9]Yes	No	No

[1] Different visibility minima and distance from cloud requirements exist for operations above 10,000 feet MSL

[2] Different visibility minima and distance from cloud requirements exist for night operations above 10,000 feet MSL, and operations below 1,200 feet AGL

[3] ICAO does not have speed restrictions in this class - U.S. will retain the 250 KIAS rule

[4] ICAO requires an ATC clearance for VFR

[5] ICAO requires 3 statute miles visibility

[6] Reduces the cloud clearance distance from standard to clear of clouds

[7] Generally, the upper limits of the Control Zone have been lowered from 14,500 MSL to 2,500 feet AGL

[8] Generally, the upper limits of the Airport Traffic Area has been lowered from 2,999 feet AGL to 2,500 feet AGL

[9] The requirement for two-way communications for Airport Traffic Areas has been retained

LEGEND 38.—Airspace Reclassification.

94062

GENERAL INFO

GENERAL INFORMATION

This publication includes Instrument Approach Procedures (IAPs), Standard Instrument Departures (SIDs), Standard Terminal Arrivals (STARs) and Profile Descent Procedures for use by both civil and military aviation and is issued every 56 days.

STANDARD TERMINAL ARRIVAL AND STANDARD INSTRUMENT DEPARTURES

The use of the associated codified STAR/SID and transition identifiers are requested of users when filing flight plans via teletype and are required for users filing flight plans via computer interface. It must be noted that when filing a STAR/SID with a transition, the first three coded characters of the STAR and the last three coded characters of the SID are replaced by the transition code. Examples: ACTON SIX ARRIVAL, file (AQN.AQN6); ACTON SIX ARRIVAL EDNAS TRANSITION, file (EDNAS.AQN6). FREEHOLD THREE DEPARTURE, file (FREH3. RBV), FREEHOLD THREE DEPARTURE, ELWOOD CITY TRANSITION, file (FREH3.EWC).

PROFILE DESCENT PROCEDURAL NOTE

A profile descent is an uninterrupted descent (except where level flight is required for speed adjustment, e.g., 250 knots at 10,000 feet MSL) from cruising altitude/level to interception of a glide slope or to a minimum altitude specified for the initial or intermediate approach segment of a non-precision instrument approach. The profile descent normally terminates at the approach gate or where the glide slope or other appropriate minimum altitude is intercepted.

Profile descent clearances are subject to traffic conditions and may be altered by ATC if necessary. Acceptance, by the pilot, of a profile descent clearance; i.e., "cleared for Runway 28 profile descent," requires the pilot to adhere to all depicted procedures on the profile descent chart.

After a profile descent has been issued and accepted:

(1) Any subsequent ATC revision of altitude or route cancels the remaining portion of the charted profile descent procedure. ATC will then assign necessary altitude, route, and speed clearances.

(2) Any subsequent revision of depicted speed restriction voids all charted speed restrictions. Charted route and altitude restrictions are not affected by revision to depicted speed restrictions. If the pilot cannot comply with charted route and/or altitude restrictions because of revised speed, he is expected to so advise ATC.

THE PROFILE DESCENT CLEARANCES DOES NOT CONSTITUTE CLEARANCE TO FLY AN INSTRUMENT APPROACH PROCEDURE (IAP). The last "maintain altitude" specified in the PROFILE DESCENT procedure constitutes that the last ATC assigned altitude and the pilot must maintain such altitude until he is cleared for an approach unless another altitude is assigned by ATC.

PILOTS SHOULD REVIEW RUNWAY PROFILE DESCENT CHARTS BEFORE FLIGHT INTO AIRPORTS WITH CHARTED PROCEDURES.

MISCELLANEOUS

★ Indicates control tower or ATIS operates non-continuously.
# Indicates control tower temporarily closed UFN.
Distances in nautical miles (except visibility in statute miles and Runway Visual Range in hundreds of feet). Runway Dimensions in feet. Elevations in feet. Mean Sea Level (MSL). Ceilings in feet above airport elevation. Radials/bearings/headings/courses are magnetic. Horizontal Datum: Unless otherwise noted on the chart, all coordinates are referenced to North American Datum 1983 (NAD 83), which for charting purposes is considered equivalent to World Geodetic System 1984 (WGS84).

LEGEND 39.—General Information on SIDs, STARs, and PROFILE DESCENTS.

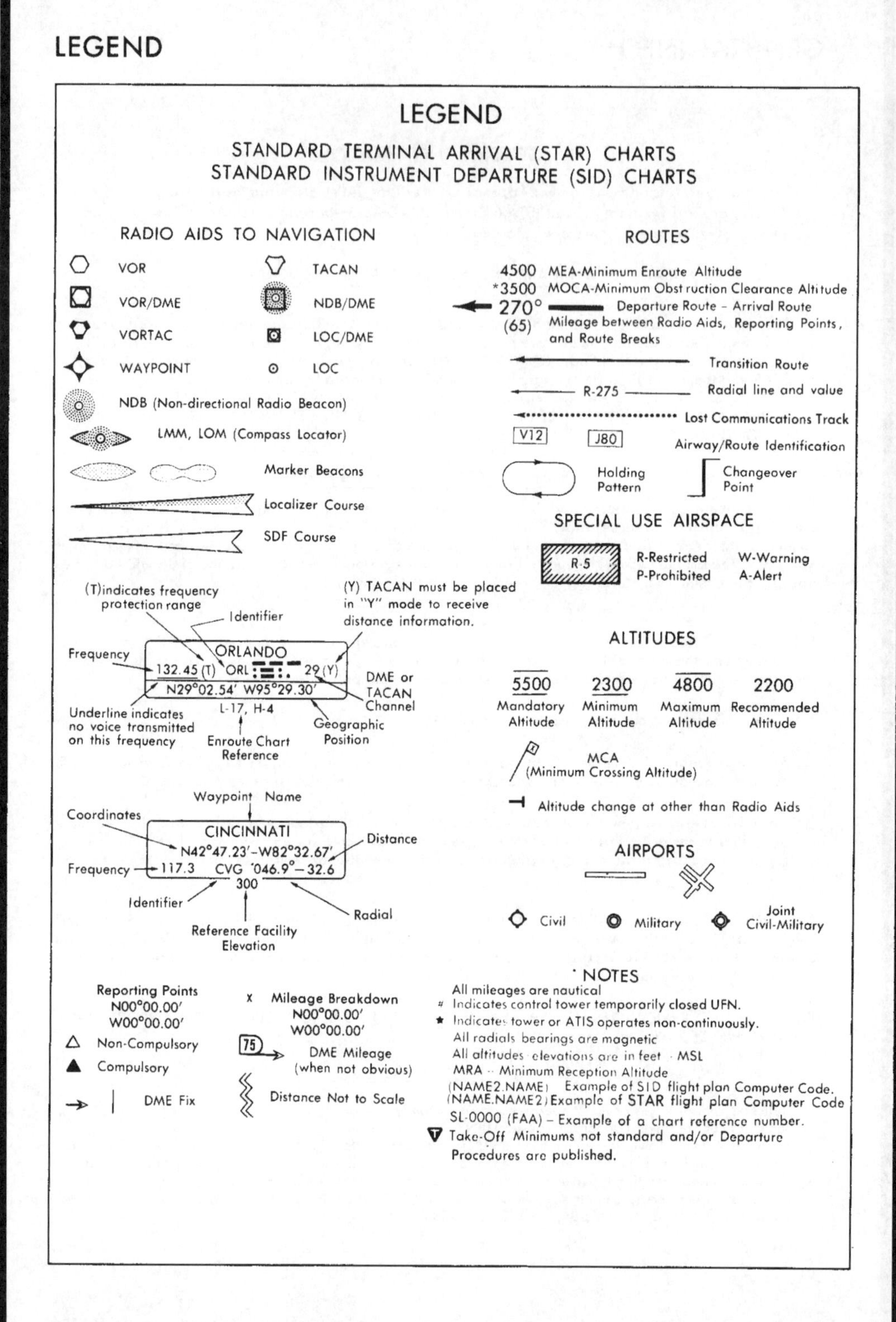

LEGEND 40.—SIDs and STARs.

LEGEND

LEGEND

PROFILE DESCENT PROCEDURES

RADIO AIDS TO NAVIGATION

VOR

VORTAC

NDB (Non-Directional Radiobeacon)

LOM (Compass Locator)

Marker Beacon

Localizer Course

NAME
000.0 NAM 00
DME or TACAN Channel

Underline indicates no voice transmitted on this frequency

R-117 Radial line and value

Reporting Point
N00°00.00′
W00°00.00′
△ Non-Compulsory
▲ Compulsory

DME fix

15 DME Mileage (when not obvious)

x Mileage Breakdown
N00°00.00′
W00°00.00′

Changeover Point

ROUTES

Non-Radar Route

2900 MEA
169°
(69) Mileage

Radar Route
Headings are approximate)

270°

Transition Route

15000 MEA
214°
(28) Mileage

Altitude change at other than Radio Aids

(65) Mileage between Radio Aids, Reporting Points and Route Breaks

V25 J54 Airway/Route identification

Holding Pattern

4200 MEA-Minimum Enroute Altitude

* 3600 MOCA – Minimum Obstruction Clearance Altitude

RENO,
(RNO.MOD4) – Computer Code

AIR TRAFFIC CLEARANCE

Cross at or above ***13,000′***.
Descend and maintain ***11,000′***
Turn left ***350°***. Vector to final.

All radials/bearings are magnetic
All mileages are nautical
All altitudes in feet – MSL

LEGEND 41.—Profile Descent Procedures.

TOWER ENROUTE CONTROL (TEC)

Within the national airspace system it is possible for a pilot to fly IFR from one point to another without leaving approach control airspace. This is referred to as "tower enroute" which allows flight beneath the enroute structure. The tower enroute concept has been expanded (where practical) by reallocating airspace vertically/geographically to allow flight planning between city pairs while remaining within approach control airspace. Pilots are encouraged to solicit tower enroute information from FSS's and to use the route descriptions provided in this directory when filing flight plans. Other airways which appear to be more direct between two points may take the aircraft out of approach control airspace thereby resulting in additional delays or other complications. All published TEC routes are designed to avoid enroute airspace and the majority are within radar coverage. Additional routes and other changes will appear in forthcoming editions as necessary. The acronym "TEC" should be included in the remarks section of the flight plan. This will advise ATC that the pilot intends to remain within approach control airspace for the entire flight. The following items should be noted before using the graphics and route descriptions:

1. The graphic is not to be used for navigation nor detailed flight planning. Not all city pairs are depicted. It is intended to show general geographic areas connected by tower enroute control. Pilots should refer to route descriptions for specific flight planning.
2. The route description contains four columns of information; i.e., the approach control area (listed alphabetically) within which the departure airport is located (check appropriate flight information publications), the specific route (airway, radial, etc.), the highest altitude allowed for the route, and the destination airport (listed alphabetically).
3. The word "DIRECT" will appear as the route when radar vectors will be used or no airway exists. Also, this indicates that a Standard Instrument Departure (SID) or Standard Terminal Arrival Route (STAR) may be applied by ATC.
4. When a NAVAID or intersection identifier appears with no airway immediately preceding or following the identifier, the routing is understood to be DIRECT to or from that point unless otherwise cleared by ATC.
5. Routes beginning or ending with an airway indicate that the airway essentially overflies the airport or radar vectors will be applied.
6. Where more than one route is listed to the same destination, the pilot may select which route is desired. Unless otherwise stated, all routes may be flown in either direction.
7. Routes are effective only during each respective terminal facility's normal operating hours. Pilots are cautioned to check NOTAMS to ensure appropriate terminal facilities will be operating for the planned flight time.
8. All identifiers used for NAVAIDS, airports, and intersections are official identifiers.
9. Altitudes are listed in thousands of feet. ATC may require altitude changes to maintain flight within approach control airspace. ATC will provide radar monitoring and, if necessary, course guidance if the highest altitude assigned by ATC is below the Minimum Enroute Altitude (MEA).
10. Although all airports are not listed under the destination column, IFR flight may be planned to satellite airports in proximity to major airports via the same routing.
11. Flight plans should be filed with a Flight Service Station (FSS).

TOWER ENROUTE CONTROL CITY PAIRS

(1) Single Engine only.
(2) Props less than 210 KT IAS.
(3) Props less than 250 KT IAS.
(4) Jets and Props greater than 210 KT IAS.

Boston—NO SATS = BED/LWM/BVY/AYE/FIT/B09/6B6/2B2
SO SATS = BOS/OWD/NZW/1B9/3B2

Approach Control Area (Including Satellites)	Route	Highest Altitude	Destination
Albany	V14 V428 V29	6000	Binghamton
	V130	7000	Bradley
	V14	10,000	Buffalo
	V14 V428	8000	Elmira
	V14 V428	8000	Ithaca
	V2	10,000	Rochester
	V14 BEEPS	10,000	Rochester
	V2	10,000	Utica/Rome
Allentown	FJC V149 LHY	8000	Albany
	ETX LHY	8000	Albany
	FJC ARD V276 DIXIE V229	5000 (only)	Atlantic City
	V93 LRP	8000	Baltimore
	ETX V162 DUMMR V93 LRP	6000	Baltimore
	V39 LRP	8000 (only)	Baltimore
	FJC BWZ	5000 (only)	Caldwell, NJ
	(2) ETX V30 SBJ	5000 (only)	Farmingdale, NY
	(2) FJC V6 SBJ	5000 (only)	Farmingdale, NY
	ETX V162 HAR	8000	Harrisburg
	ETX ETX004 WEISS	4000 (only)	Hazleton
	ETX V39	4000	Lancaster
	(2) ETX V30 SBJ	5000 (only)	Newark

LEGEND 42.—Tower Enroute Control (NE).

TOWER ENROUTE CONTROL

Approach Control Area (Including Satellites)		Route	Highest Altitude	Destination
New York /Kennedy	....	SAX V249 SBJ V30 ETX (Non jet/Non turboprop)	8000	Allentown
	....	DIXIE V229 ACY (Props only)	6000	Atlantic City
	....	DIXIE V1 HOWIE (Jets only)	8000	Atlantic City
	....	DIXIE V1 V308 OTT (Props only)	6000	Andrews AFB
	....	DIXIE V16 ENO V268 SWANN (Props only)	6000	Baltimore
	....	COL	2000	Belmar
	....	BDR MAD V475 V188 TMU	9000	Block Island
	....	BDR V229 HFD V3 WOONS	9000	Boston
	....	BDR V229 HFD HFD053 DREEM	9000	Boston (North)
	....	BDR BDR014 JUDDS V419 BRISS	9000	Bradley
	....	BDR BDR014 JUDDS V419 BRISS (Jets only)	10000	Bradley
	....	BDR	3000	Bridgeport
	....	SAX V249 SBJ V30 ETX V162 HAR (Non jet/Non turboprop)	8000	Capital City
	....	DIXIE V1 LEEAH V268 BAL BAL291 KROLL AML (Non-pressurized aircraft only)	6000	Dulles
	....	BDR MAD MAD126 MONDI	9000	Groton
	....	R/V CCC 232 CCC HTO	3000	Hampton
	....	BDR V229 HFD	9000	Hartford
	....	BDR V229 HFD V167 PVD V151 GAILS	9000	Hyannis
	....	R/V ILS 6 LOC (Text Info)	3000	Islip
	....	R/V CCC232 CCC	3000	Islip
	....	Direct	2000	LaGuardia
	....	SAX V249 SBJ V30 ETX V162 V93 LRP (Props only)	8000	Lancaster
	....	DIXIE V16 CYN	6000	McGuire
	....	BDR MAD V475 V188 TMU V374 MVY	9000	Martha's Vineyard
	....	BDR MAD	3000	Meriden Markham
	....	DIXIE V16 VCN (Props only)	6000	Millville
	....	BDR MAD V475 V188 TMU V374 MVY	9000	Nantucket
	....	COL V232 SBJ	3000	Newark
	....	BDR MAD V475 V188 TMU V374 MINNK	9000	New Bedford
	....	DIXIE V1 (Props only)	6000	Norfolk
	....	DIXIE V276 ARD	4000	N. Philadelphia
	....	DIXIE V16 CYN V312 OOD (Props only)	6000	Philadelphia
	....	DIXIE V16 CYN V312 OOD (Jets only)	8000	Philadelphia
	....	BDR MAD V475 V188 TMU (210 kts +)	9000	Providence
	....	BDR MAD V475 V188 TMU	9000	Quonset
	....	SAX V249 SBJ V30 ETX V39 FLOAT (Non jet/Non turboprop only)	8000	Reading
	....	DIXIE V16 (Props only)	6000	Richmond
	....	DIXIE V1 (Props only)	6000	Salisbury
	....	DIXIE V1 V308 OTT (Props only)	6000	Washington
	....	DPK V483 CMK	2000	Westchester Co
	....	BDR MAD V475 V188 TMU	9000	Westerly
	....	DIXIE V229 PANZE V44 SIE (Props only)	6000	Wildwood
	....	DIXIE V1 HOWIE (Jets only)	8000	Wildwood
	....	BDR MAD V1 GRAYM	9000	Worcester
New York/ LaGuardia	....	SAX V249 SBJ V30 ETX	8000	Allentown
	....	DIXIE V229 ACY (Props only)	6000	Atlantic City
	....	DIXIE V1 HOWIE (Jets only)	8000	Atlantic City
	....	ABBYS V403 GLOMO V408 V93 BAL (Props only)	7000	Andrews AFB
	....	ABBYS V403 BELAY V378 BAL (Props only)	7000	Baltimore
	....	JFK COL	6000	Belmar
	....	BDR MAD V475 V188 TMU	9000	Block Island
	....	BDR V229 HFD V3 WOONS	9000	Boston
	....	BDR V229 HFD HFD053 DREEM	9000	Boston (North)
	....	BDR BDR014 JUDDS V419 BRISS (Props only)	9000	Bradley
	....	BDR BDR014 JUDDS V419 BRISS (Jets only)	10000	Bradley
	....	BDR 248 CCC285 PUGGS V229 BDR	5000	Bridgeport
	....	R/V BDR248 BDR. . .(Helicopter Route)	5000	Bridgeport (Points NE)
	....	SAX V249 SBJ V30 ETX V162 HAR	8000	Capital City
	....	SAX V249 SBJ V30 ETX V162 V93 V143 ROBRT AML (Props only)	8000	Dulles

LEGEND 42A—Tower Enroute Control Continued.

346 **TOWER ENROUTE CONTROL**

Approach Control Area (Including Satellites)	Route	Highest Altitude	Destination
	ABBYS V403 GLOMO V408 V93 BAL (Props only)	7000	Washington
	DIXIE V229 PANZE V44 SIE (Props only)	6000	Wildwood
	DIXIE V1 HOWIE (Jets only)	8000	Wildwood
	CMK V3 HFD V1 GRAYM	9000	Worcester
Norfolk	CCV CCV345 PXT175 PXT	5000	Patuxent River
	HPW V260 RIC (West-bound only)	9000	Richmond
	CCV V1 SBY	5000	Salisbury
	CCV V139 SWL (Northeast-bound only)	5000	Snow Hill
	HCM HCM330 SVILL	7000	Washington
Patuxent	SWL V139	5000	Atlantic City
	PXT V16 V44	5000	Atlantic City
	SBY V1 V44	5000	Atlantic City
	SBY332 BAL130	4000	Baltimore
	PXT V93	5000	Baltimore
	SBY V29 ENO	5000	Dover AFB
	PXT V16 ENO	5000	Dover AFB
	PXT V16	5000	Dover AFB
	SBY VI ATR	5000	Dover AFB
	PXT V213 V286 FLUKY	6000	Dulles
	COLIN V33 HCM	6000	Newport News
	SBY V1 CCV	6000	Norfolk
	SWL V139 CCV	6000	Norfolk
	WHINO V33 V286 STEIN	5000	Norfolk
	PXT V213 ENO V29 DQO	5000	Philadelphia
	SBY V29 DQO	5000	Philadelphia
	PXT V16	6000	Richmond
	SBY V1 JAMIE HCM	6000	Richmond
	COLIN V33 HCM	6000	Richmond
	PXT V31 OTT (No Overflight of D.C. Area)	4000	Washington
	SBY CHURK OTT (No Overflight of D.C. Area)	4000	Washington
Pease	RAYMY LWM	8000	Boston
	EXALT V139 V141 GAILS	10000	Hyannis
	V106 GDM V14 ORW V16 CCC	10000	Islip
	V106 GDM V14 ORW V16 DPK	10000	Kennedy
	EXALT V139 BURDY	10000	Providence
Philadelphia	RV FJC180 FJC	4000	Allentown
	OOD VCN V184 ACY	3000	Atlantic City
	MXE V378 BAL	6000	Baltimore
	DQO V166 V378 BAL	6000	Baltimore
	OOD V157 ENO	4000	Dover AFB
	DQO V29 ENO	4000	Dover AFB
	MXE V408 ROBRT AML	8000	Dulles
	MXE V184 MXE283027 V469 HAR	6000	Harrisburg
	PNE PNE090 ARD126 V16 DIXIE (Direct) (Single Engine only)	5000	Kennedy
	PNE PNE090 ARD126 V16 V276 ZIGGI (Direct) (No Single Engine)	5000	Kennedy
	RBV V123 PROUD	7000	LaGuardia
	MXE MXE295 HABER LRP137 LRP	4000	Lancaster
	ARD V214 METRO (Non Turbojets only)	5000	Newark
	RBV V213 WARRD (Turbojets only)	7000	Newark
	MXE MXE334 HUMEL	4000	Reading
	ARD V214 METRO	5000	Teterboro
	MXE V408 VINNY V93 BAL	8000	Washington
	DQO V166 V93 BAL	8000	Washington
	RV FJC180 FJC BWZ SAX V39 BREZY	5000	Westchester Co.
	RV FJC180 FJC V149 RITTY	5000	Wilkes Barre/Scranton
Pittsburgh	BSV (Westbound only)	8000	Akron-Canton
	V37 (Southbound only)	8000	Clarksburg
	EWC V37 (Northbound only)	8000	Erie

LEGEND 42B.—Tower Enroute Control Continued.

TOWER ENROUTE CONTROL (TEC) 235

Within the national airspace system it is possible for a pilot to fly IFR from one point to another without leaving approach control airspace. This is referred to as "Tower Enroute" which allows flight beneath the enroute structure. The tower enroute concept has been expanded (where practical) by reallocating airspace vertically/geographically to allow flight planning between city pairs while remaining within approach control airspace. Pilots are encouraged to use the TEC route descriptions provided in the Southwest U.S. Airport/Facility Directory when filing flight plans. Other airways which appear to be more direct between two points may take the aircraft out of approach control airspace thereby resulting in additional delays or other complications. All published TEC routes are designed to avoid enroute airspace and the majority are within radar coverage. The following items should be noted before using the graphics and route descriptions.

1. The graphic is not to be used for navigation nor detailed flight planning. Not all city pairs are depicted. It is intended to show geographic areas connected by tower enroute control. Pilots should refer to route descriptions for specific flight planning.
2. The route description contains four colums of information after approach control area listed in the heading, where the departure airport is located; i.e., the airport/airports of intended landing using FAA three letter/letter-two number identifiers, the coded route number (this should be used when filing the flight plan and will be used by ATC in lieu of reading out the full route description), the specific route (airway, radial, etc.), the altitude allowed for type of aircraft and the routes.
3. The word "DIRECT" will appear as the route when radar vectors will be used or no airway exists. Also this indicates that a Standard Instrument Departure (SID) or Standard Terminal Arrival (STAR) may be applied by ATC.
4. When a NAVAID or intersection identifier appears with no airway immediately preceding or following the identifier, the routing is understood to be DIRECT to or from that point unless otherwise cleared by ATC or radials are listed (See item 5).
5. Routes beginning and ending with an airway indicate that the airway essentially overflies the airport or radar vectors will be applied.
6. Where more than one route is listed to the same destination, ensure you file correct route for type of aircraft which are denoted after the route in the altitude column using J,M,P, or Q. These are listed after item 10 under Aircraft Classification.
7. Although all airports are not listed under the destination column, IFR flight may be planned to satellite airports in the proximity to major airports via the same routing.
8. Los Angeles International Airport (LAX) and four other airports (ONT-SAN-TOA-SNA) have two options due to winds and these affect the traffic flows and runways in use. To indicate the difference the following symbols are used after the airport: Runway Number, W for west indicating normal conditions, E for East and N for North indicating other than normal operation. If nothing follows the airport use this route on either West, East or North plan. Other destinations have different arrivals due to LAX being East and they have the notation "(LAXE)." Torrance Airport is also unique in that the airport is split between Los Angeles and Coast TRACON, for Runway 11 departures use Coast TRACON routings and for Runway 29 departures use LAX TRACON routings.
9. When filing flight plans, the coded route identifier i.e. SANJ2, VTUJ4, POMJ3 may be used in lieu of the route of flight.
10. Aircraft types i.e. J, M, P, and Q are listed at the beginning of the altitude and should be used with the route of flight filed. (See Aircraft Classification below). The altitudes shown are to be used for the route. This allows for separation of various arrival routes, departure routes, and overflights to, from, and over all airports in the Southern California area.

LEGENDS

AIRCRAFT CLASSIFICATION

(J) = Jet powered
(M) = Turbo Props/Special (cruise speed 190 knots or greater)
(P) = Non-jet (cruise speed 190 knots or greater)
(Q) = Non-jet (cruise speed 189 knots or less)

LEGEND 43.—Tower Enroute Control (SW).

236 **TOWER ENROUTE CONTROL**

BURBANK TRACON
FROM: BUR VNY WHP

TO:	ROUTE ID	ROUTE	ALTITUDE
FUL LGB SLI TOA (RWY 29)	BURJ1	V186 V394 SLI	MPQ50
LAX	BURJ2	V186 PURMS	JMPQ40
LAX (LAXE)	BURJ3	VNY SMO	JM50PQ40
TOA (RWY 11)	BURJ4	VNY VNY095 DARTS SMO	JMPQ40
SMO	BURJ5	V186 DARTS	JMPQ30
CCB CNO EMT HMT L12 L65 L66 L67 F70 ONT POC RAL RIR RIV SBD	BURJ6	V186 PDZ	JM70PQ50
CRQ NFG NKX L39 L32	BURJ7	V186 V363 V23 OCN	JM70PQ50
MYF NRS NZY SAN SDM SEE	BURJ8	V186 V363 V23 MZB	PQ50
MYF NRS NZY SAN SDM SEE	BURJ9	V186 POM164 V208 MZB320 MZB	JM70
OXR CMA	BURJ10	VNY	JMPQ40
SBA	BURJ11	FIM V186 V27 KWANG	JMPQ60
SNA	BURJ12	V186 V363 V8 SLI	JMPQ50
SAN (SANE)	BURJ13	V186 V363 V23 V165 SARGS	PQ50
SAN (SANE)	BURJ14	V186 POM164 V25 V165 SARGS	JM70
NZJ NTK	BURJ15	V186 V363 V23 DAMPS	JM70PQ50
AVX	BURJ16	V186 V363 KRAUZ SXC	JM70PQ50
HHR	BURJ17	V186 ELMOO	JMPQ40
LGB	BURJ18	V186 V363 V23 SLI	J70

COAST TRACON
FROM: FUL LGB SLI SNA TOA (RWY 11) NTK NZJ

TO:	ROUTE ID	ROUTE	ALTITUDE
BUR VNY WHP	CSTJ1	SLI V23 LAX LAX316 SILEX	JM60PQ40
BUR VNY WHP (LAXE)	CSTJ2	SLI SLI333 V186 VNY	JM50PQ40
CMA OXR (LAXE)	CSTJ3	SLI SLI333 V186 FIM	JM50PQ40
LAX	CSTJ4	SLI	JM70PQ40
LAX (LAXE)	CSTJ5	SLI V8 TANDY	JM50PQ40
SMO	CSTJ6	SLI V23 LAX LAX046 ELMOO	JM70PQ40
SMO (LAXE)	CSTJ7	SLI SLI333 V186 DARTS	JM50PQ40
CCB EMT POC	CSTJ8	SLI V8 V363 POM	JMPQ50
CNO HMT L12 L65 L66 L67 F70 ONT RAL RIR RIV SBD	CSTJ9	SLI V8 PDZ (SNA RWY 19 ONLY)	JM60
CNO HMT L12 L65 L66 L67 F70 ONT RAL RIR RIV SBD	CSTJ10	SLI V8 PDZ	JMPQ50
CRQ L39 NFG NKX L32	CSTJ12	V25 V208 OCN	JM70
MYF NRS NZY SAN SDM SEE	CSTJ14	V25 V208 MZB320 MZB	J110M90
OXR CMA	CSTJ15	SLI V23 LAX VNY	M50PQ40
OXR CMA	CSTJ16	SXC SXC295 VTU160 VTU	J80
SBA	CSTJ17	SLI V23 LAX VTU KWANG	PQ40
SBA (LAXE)	CSTJ18	SLI SLI333 V186 V27 KWANG	M50PQ40
SBA	CSTJ19	SXC SXC295 VTU160 VTU KWANG	JM80
SAN (SANE)	CSTJ21	V25 V165 SARGS	J110M90
HHR	CSTJ26	SLI SLI340 WELLZ LOC	JM70PQ40

FROM: SNA NTK NZJ and when SNAN FUL LGB SLI TOA RWY 11

TO:	ROUTE ID	ROUTE	ALTITUDE
CRQ L39 NFG NKX L32	CSTJ11	V23 OCN	PQ50
MYF NRS NZY SAN SDM SEE	CSTJ13	V23 MZB	PQ50
SAN (SANE)	CSTJ20	V23 V165 SARGS	PQ50

FROM: AVX (DEPARTURES ONLY)

TO:	ROUTE ID	ROUTE	ALTITUDE
CRQ L39 NFG NKX L32	CSTJ22	SXC V208 OCN	JMPQ50
MYF NRS NZY SAN SDM SEE	CSTJ23	SXC V208 MZB320 MZB	J110M90
MYF NRS NZY SAN SDM SEE (SANE)	CSTJ24	SXC V208 OCN V165 SARGS	PQ50
MYF NRS NZY SAN SDM SEE	CSTJ25	SXC V208 OCN V23 MZB	PQ50

LEGEND 43A.—Tower Enroute Control Continued.

APPENDIX 2

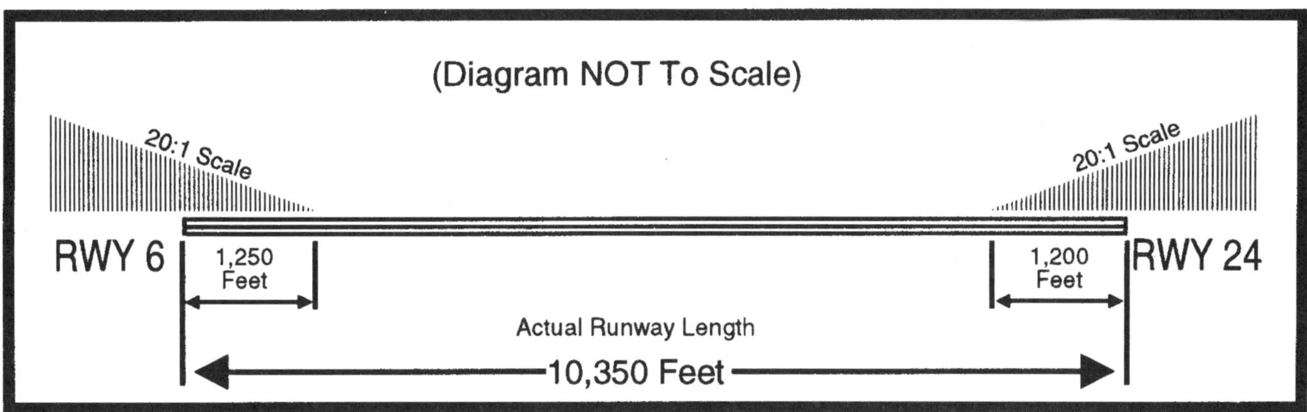

FIGURE 1.—Runway Diagram.

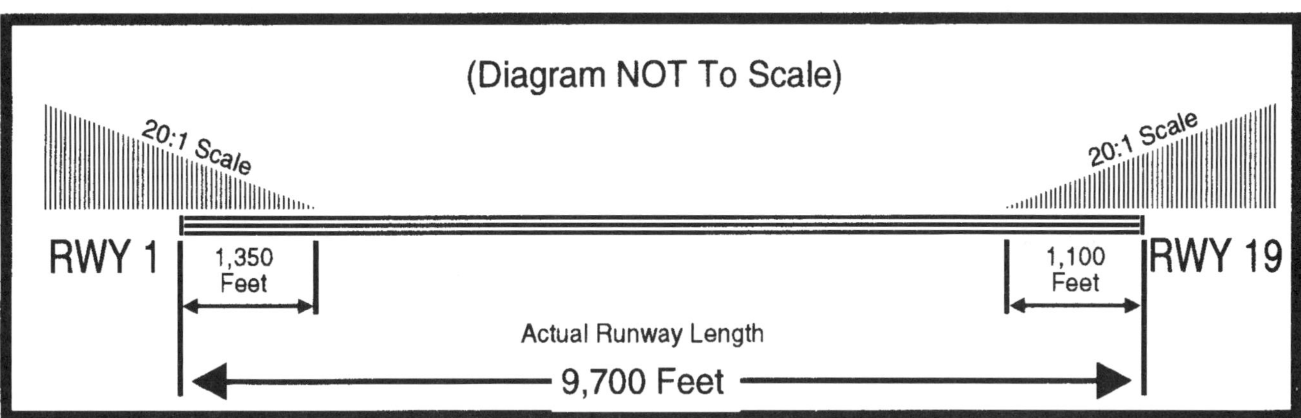

FIGURE 2.—Runway Diagram.

LOADING CONDITIONS	BE-1	BE-2	BE-3	BE-4	BE-5
CREW	360	340	350	340	360
PASSENGERS ROW 1	350	300	120	-	-
ROW 2	260	250	340	370	-
ROW 3	200	190	350	400	170
ROW 4	340	170	300	290	200
ROW 5	120	190	170	200	290
ROW 6	400	340	-	170	400
ROW 7	120	190	-	210	370
ROW 8	250	-	-	190	340
ROW 9	-	-	-	420	430
BAGGAGE NOSE	60	-	80	-	100
FWD CABIN	250	100	120	-	200
AFT (FWD SEC)	500	200	250	800	-
AFT (AFT SEC)	-	600	500	-	-
FUEL GAL	370	390	400	290	340
TYPE	JET B	JET A	JET B	JET A	JET B
TEMP	+5 °C	+15 °C	-15 °C	+10 °C	+25 °C

FIGURE 3.—Beech 1900 – Loading Passenger Configuration.

LOADING CONDITIONS	BE-6	BE-7	BE-8	BE-9	BE-10
CREW	360	340	350	370	420
CARGO SECTION A	500	-	600	600	350
B	500	400	200	600	450
C	550	450	400	600	450
D	550	600	400	600	550
E	600	600	200	550	550
F	600	600	200	350	600
G	450	500	200	250	600
H	-	-	200	250	-
J	350	-	300	150	-
K	-	-	250	200	-
L	-	-	100	100	-
FUEL GAL	340	370	390	290	400
TYPE	JET B	JET B	JET A	JET A	JET B
TEMP	+25 °C	+5 °C	+15 °C	+10 °C	-15 °C
BASIC OPERATING WEIGHT - 9,005 POUNDS, 25,934 MOM/100					

FIGURE 4.—Beech 1900 – Loading Cargo Configuration.

OPERATING CONDITIONS	BE-11	BE-12	BE-13	BE-14	BE-15
BASIC EMPTY WT WEIGHT MOM/100	 9,225 25,820	 9,100 24,990	 9,000 24,710	 8,910 24,570	 9,150 25,240
CREW WEIGHT	340	380	360	400	370
PASS AND BAG WEIGHT MOM/100	 4,200 15,025	 4,530 16,480	 4,630 16,743	 4,690 13,724	 4,500 13,561
FUEL (6.8 LB/GAL) RAMP LOAD-GAL USED START AND TAXI REMAIN AT LDG	 360 20 100	 320 20 160	 340 10 140	 310 20 100	 410 30 120

FIGURE 5.—Beech 1900 – Loading Limitations.

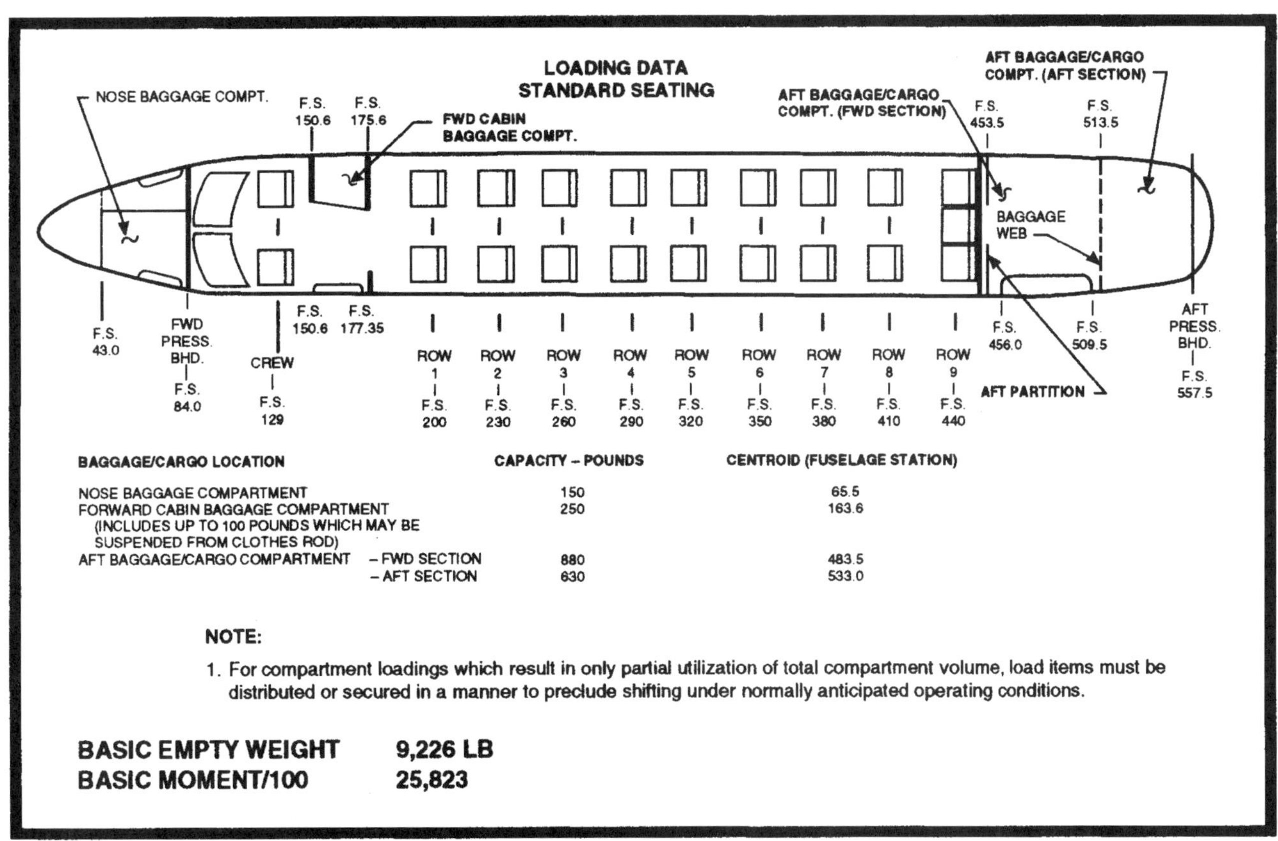

BAGGAGE/CARGO LOCATION	CAPACITY – POUNDS	CENTROID (FUSELAGE STATION)
NOSE BAGGAGE COMPARTMENT	150	65.5
FORWARD CABIN BAGGAGE COMPARTMENT (INCLUDES UP TO 100 POUNDS WHICH MAY BE SUSPENDED FROM CLOTHES ROD)	250	163.6
AFT BAGGAGE/CARGO COMPARTMENT – FWD SECTION	880	483.5
– AFT SECTION	630	533.0

NOTE:

1. For compartment loadings which result in only partial utilization of total compartment volume, load items must be distributed or secured in a manner to preclude shifting under normally anticipated operating conditions.

BASIC EMPTY WEIGHT 9,226 LB
BASIC MOMENT/100 25,823

FIGURE 6.—Airplane – Loading Data.

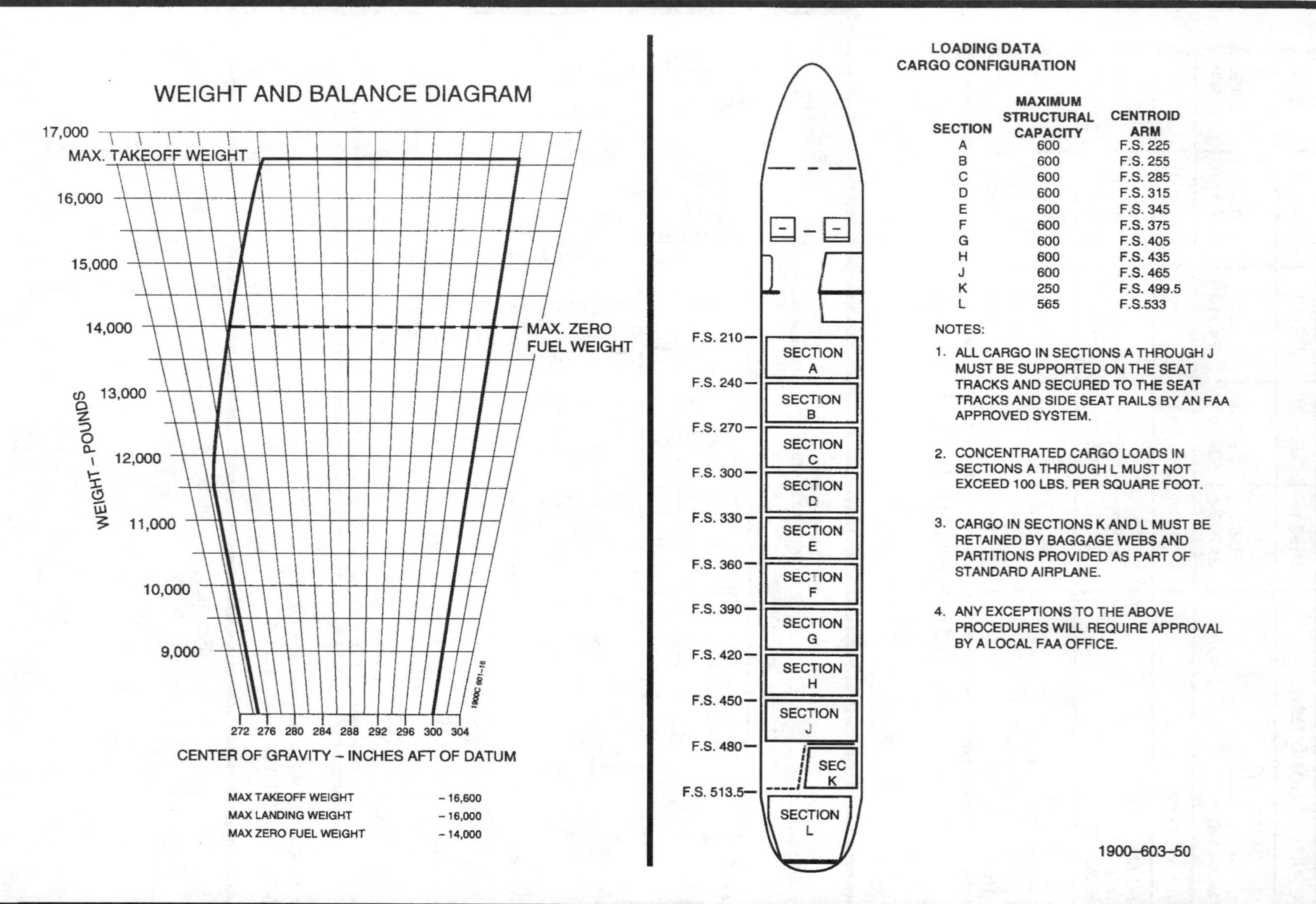

FIGURE 7.—Beech 1900 – CG Envelope and Cargo Loading Data.

USEFUL LOAD WEIGHTS AND MOMENTS

BAGGAGE

WEIGHT	NOSE BAGGAGE COMPART-MENT F.S. 65.5	FORWARD CABIN BAGGAGE COMPART-MENT F.S. 163.6	AFT BAGGAGE/ CARGO COMPART-MENT (FORWARD SECTION) F.S. 483.5	AFT BAGGAGE/ CARGO COMPART-MENT (AFT SECTION) F.S. 533.0
	MOMENT/100			
10	7	16	48	53
20	13	33	97	107
30	20	49	145	160
40	26	65	193	213
50	33	82	242	266
60	39	98	290	320
70	46	115	338	373
80	52	131	387	426
90	59	147	435	480
100	66	164	484	533
150	98	245	725	800
200		327	967	1066
250		409	1209	1332
300			1450	1599
350			1692	1866
400			1934	2132
450			2176	2398
500			2418	2665
550			2659	2932
600			2901	3198
630			3046	3358
650			3143	
700			3384	
750			3626	
800			3868	
850			4110	
880			4255	

FIGURE 8.—Airplane – Weights and Moments – Baggage.

USEFUL LOAD WEIGHTS AND MOMENTS

OCCUPANTS

WEIGHT	CREW	CABIN SEATS								
	F.S. 129	F.S. 200	F.S. 230	F.S. 260	F.S. 290	F.S. 320	F.S. 350	F.S. 380	F.S. 410	F.S. 440
	MOMENT/100									
80	103	160	184	208	232	256	280	304	328	352
90	116	180	207	234	261	288	315	342	369	396
100	129	200	230	260	290	320	350	380	410	440
110	142	220	253	286	319	352	385	418	451	484
120	155	240	276	312	348	384	420	456	492	528
130	168	260	299	338	377	416	455	494	533	572
140	181	280	322	364	406	448	490	532	574	616
150	194	300	345	390	435	480	525	570	615	660
160	206	320	368	416	464	512	560	608	656	704
170	219	340	391	442	493	544	595	646	697	748
180	232	360	414	468	522	576	630	684	738	792
190	245	380	437	494	551	608	665	722	779	836
200	258	400	460	520	580	640	700	760	820	880
210	271	420	483	546	609	672	735	798	861	924
220	284	440	506	572	638	704	770	836	902	968
230	297	460	529	598	667	736	805	874	943	1012
240	310	480	552	624	696	768	840	912	984	1056
250	323	500	575	650	725	800	875	950	1025	1100

Note: Weights reflected in above table represent weight per seat.

FIGURE 9.—Beech 1900 – Weights and Moments – Occupants.

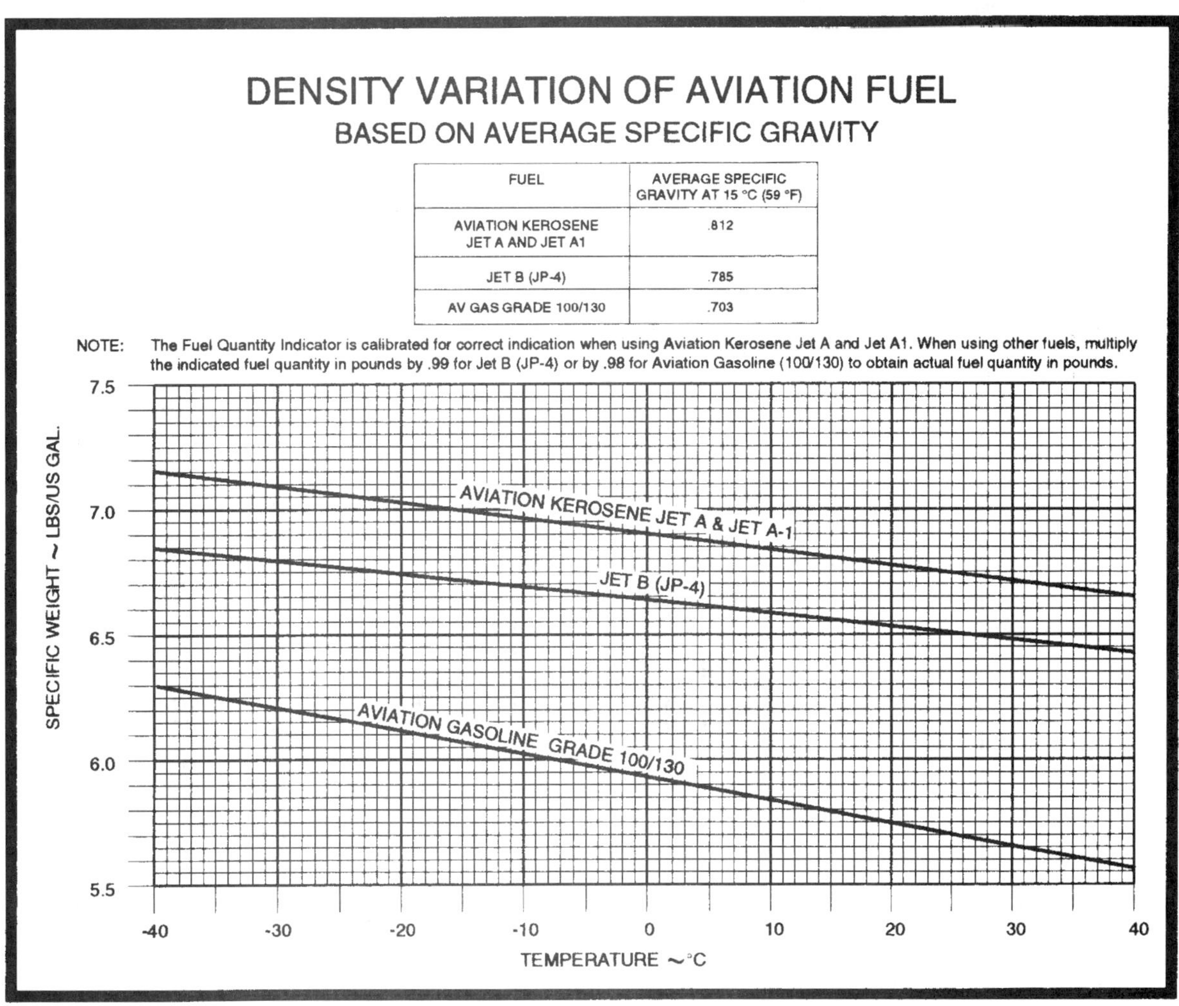

FUEL	AVERAGE SPECIFIC GRAVITY AT 15 °C (59 °F)
AVIATION KEROSENE JET A AND JET A1	.812
JET B (JP-4)	.785
AV GAS GRADE 100/130	.703

FIGURE 10.—Density Variation of Aviation Fuel.

USEFUL LOAD WEIGHTS AND MOMENTS

USABLE FUEL

GALLONS	6.5 LB/GAL WEIGHT	6.5 LB/GAL MOMENT/100	6.6 LB/GAL WEIGHT	6.6 LB/GAL MOMENT/100	6.7 LB/GAL WEIGHT	6.7 LB/GAL MOMENT/100	6.8 LB/GAL WEIGHT	6.8 LB/GAL MOMENT/100
10	65	197	66	200	67	203	68	206
20	130	394	132	401	134	407	136	413
30	195	592	198	601	201	610	204	619
40	260	789	264	802	268	814	272	826
50	325	987	330	1002	335	1018	340	1033
60	390	1185	396	1203	402	1222	408	1240
70	455	1383	462	1404	469	1426	476	1447
80	520	1581	528	1605	536	1630	544	1654
90	585	1779	594	1806	603	1834	612	1861
100	650	1977	660	2007	670	2038	680	2068
110	715	2175	726	2208	737	2242	748	2275
120	780	2372	792	2409	804	2445	816	2482
130	845	2569	858	2608	871	2648	884	2687
140	910	2765	924	2808	938	2850	952	2893
150	975	2962	990	3007	1005	3053	1020	3099
160	1040	3157	1056	3205	1072	3254	1088	3303
170	1105	3351	1122	3403	1139	3454	1156	3506
180	1170	3545	1188	3600	1206	3654	1224	3709
190	1235	3739	1254	3797	1273	3854	1292	3912
200	1300	3932	1320	3992	1340	4053	1360	4113
210	1365	4124	1386	4187	1407	4250	1428	4314
220	1430	4315	1452	4382	1474	4448	1496	4514
230	1495	4507	1518	4576	1541	4646	1564	4715
240	1560	4698	1584	4770	1608	4843	1632	4915
250	1625	4889	1650	4964	1675	5040	1700	5115
260	1690	5080	1716	5158	1742	5236	1768	5315
270	1755	5271	1782	5352	1809	5433	1836	5514
280	1820	5462	1848	5546	1876	5630	1904	5714
290	1885	5651	1914	5738	1943	5825	1972	5912
300	1950	5842	1980	5932	2010	6022	2040	6112
310	2015	6032	2046	6125	2077	6218	2108	6311
320	2080	6225	2112	6321	2144	6416	2176	6512
330	2145	6417	2178	6516	2211	6615	2244	6713
340	2210	6610	2244	6711	2278	6813	2312	6915
350	2275	6802	2310	6907	2345	7011	2380	7116
360	2340	6995	2376	7103	2412	7210	2448	7318
370	2405	7188	2442	7299	2479	7409	2516	7520
380	2470	7381	2508	7495	2546	7609	2584	7722
390	2535	7575	2574	7691	2613	7808	2652	7924
400	2600	7768	2640	7888	2680	8007	2720	8127
410	2665	7962	2706	8085	2747	8207	2788	8330
420	2730	8156	2772	8282	2814	8407	2856	8532
425	2763	8259	2805	8386	2848	8513	2890	8640

FIGURE 11.—Beech 1900 – Weights and Moments – Usable Fuel.

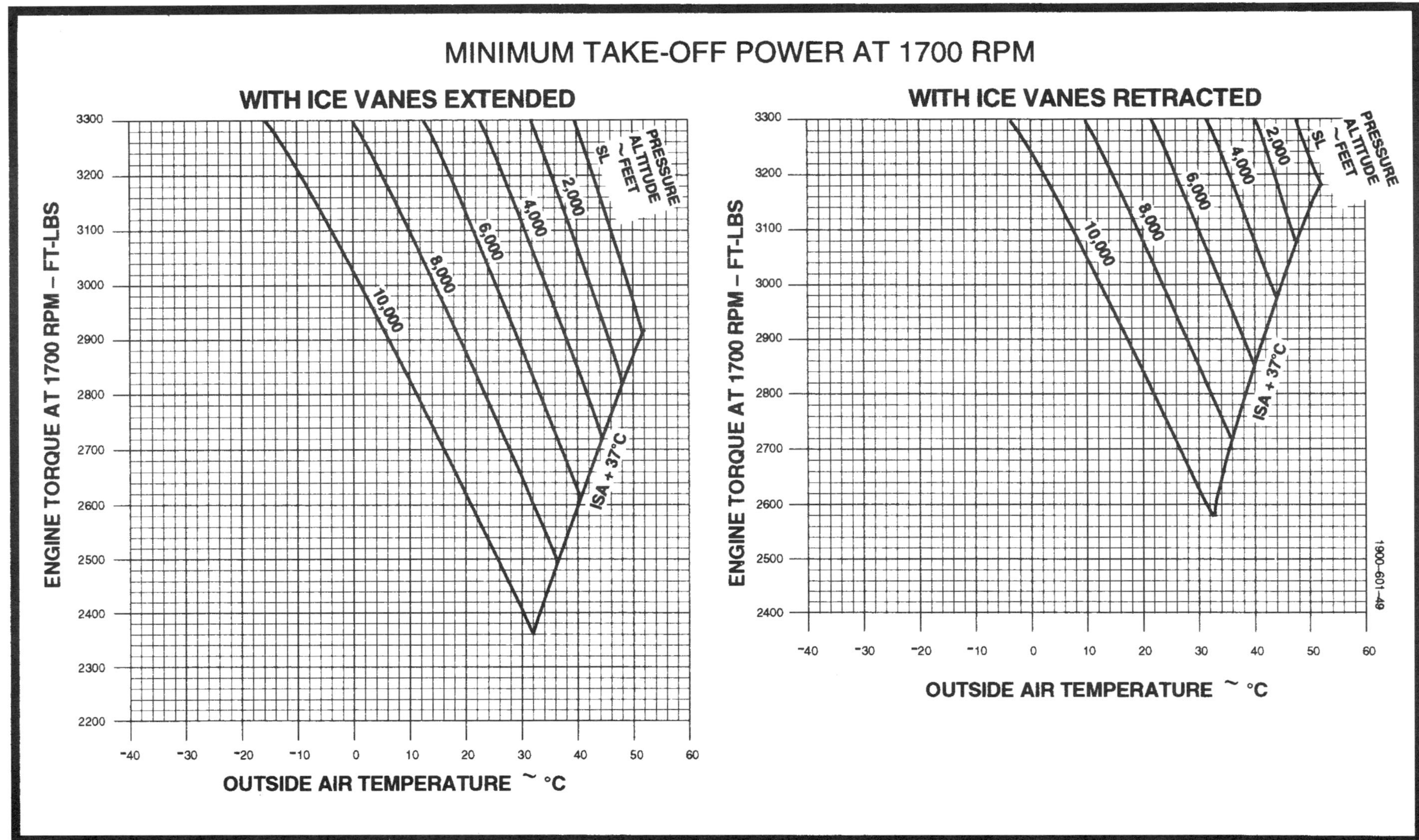

FIGURE 12.—Minimum Takeoff Power at 1700 RPM.

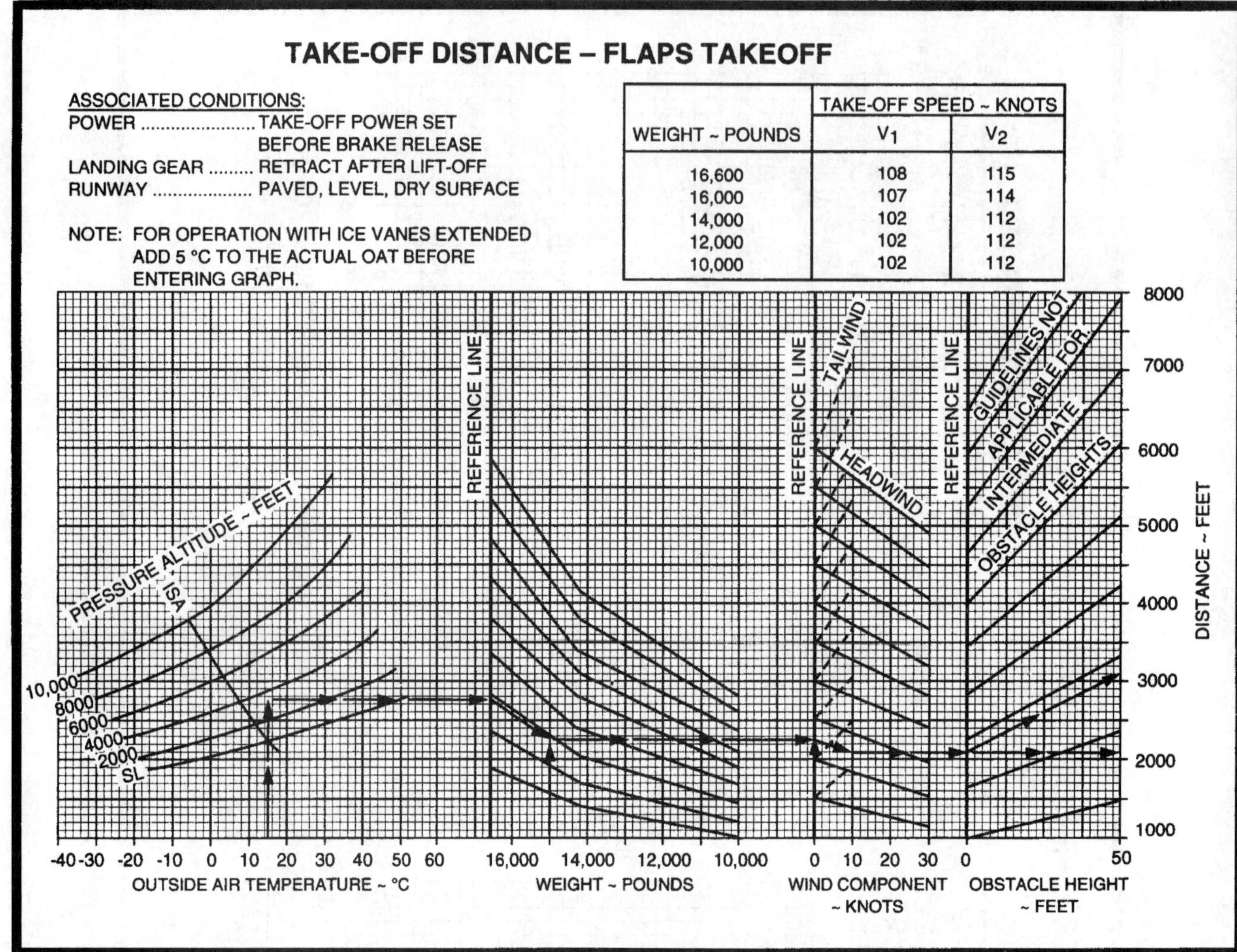

FIGURE 13.—Takeoff Distance – Flaps Takeoff.

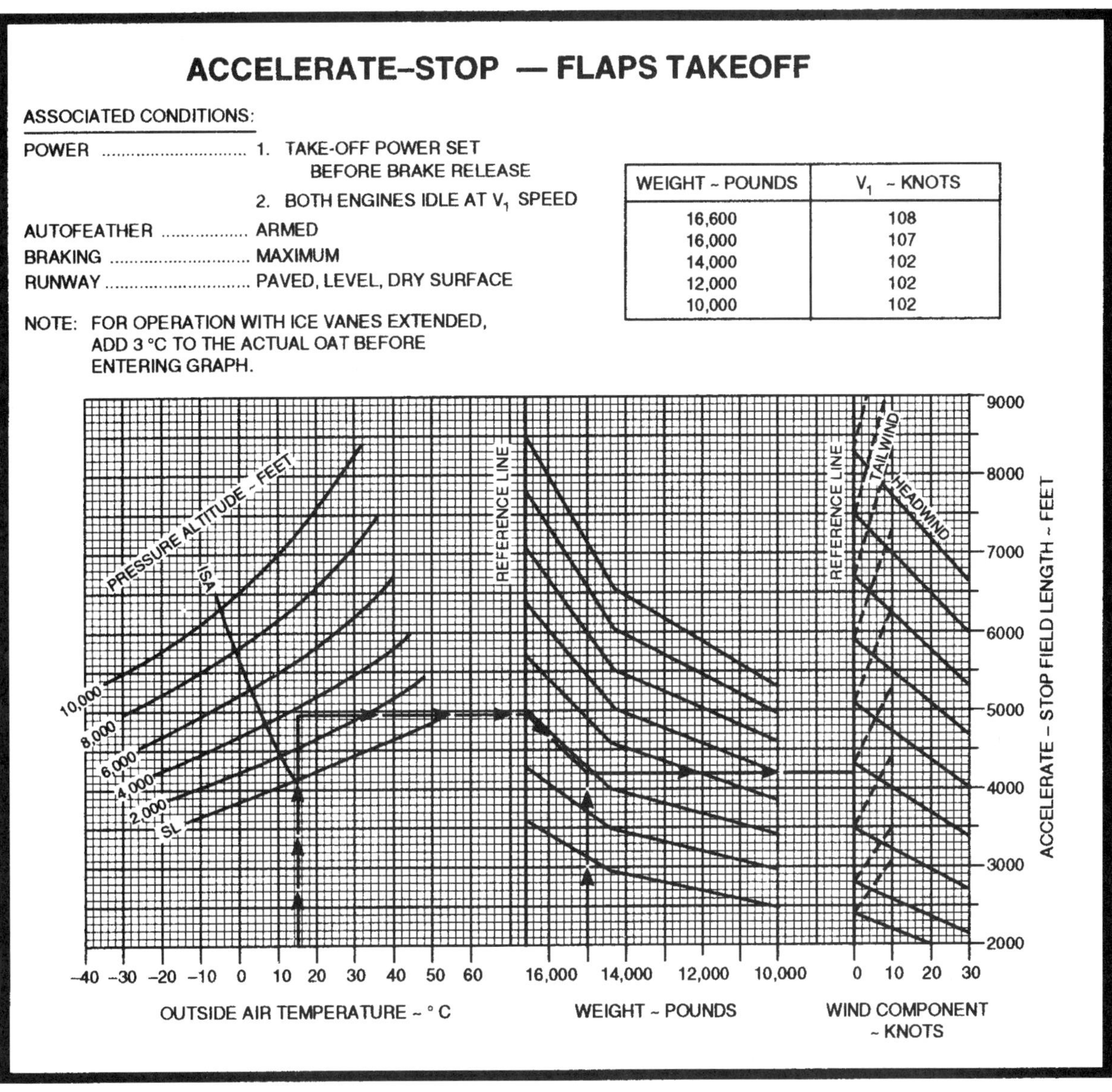

FIGURE 14.—Accelerate-Stop – Flaps Takeoff.

OPERATING CONDITIONS	BE-21	BE-22	BE-23	BE-24	BE-25
OAT AT TAKEOFF	+10 °C	0 °C	+20 °C	+25 °C	−10 °C
OAT AT CRUISE	−20 °C	−25 °C	ISA	0 °C	−40 °C
AIRPORT PRESS ALTITUDE	2,000	1,000	3,000	4,000	5,000
CRUISE ALTITUDE	16,000	18,000	20,000	14,000	22,000
INITIAL CLIMB WEIGHT	16,600	14,000	15,000	16,000	14,000
ICE VANES	RETRACT	EXTEND	RETRACT	RETRACT	EXTEND

FIGURE 15.— Beech 1900 – Climb.

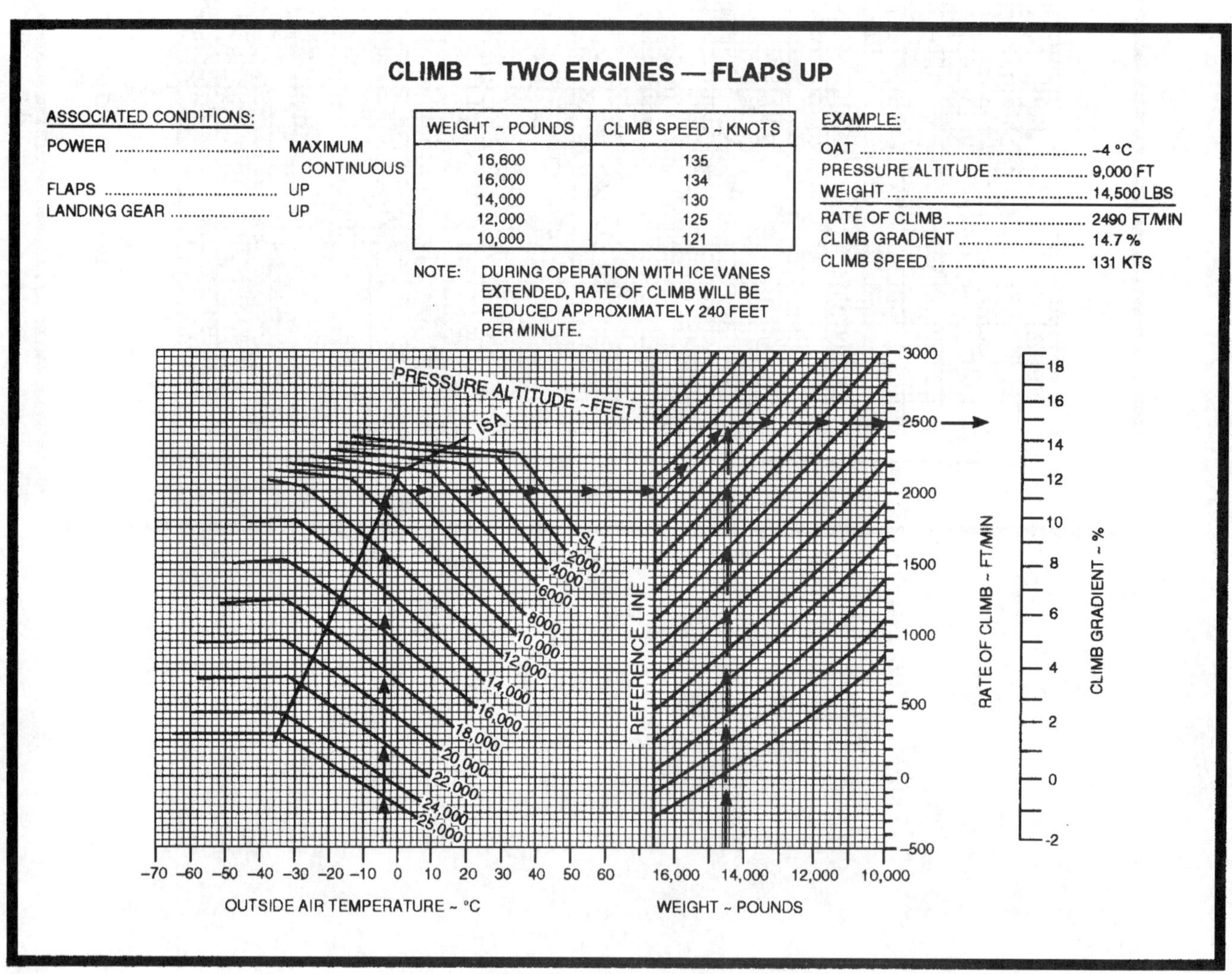

FIGURE 16.—Climb – Two Engines – Flaps Up.

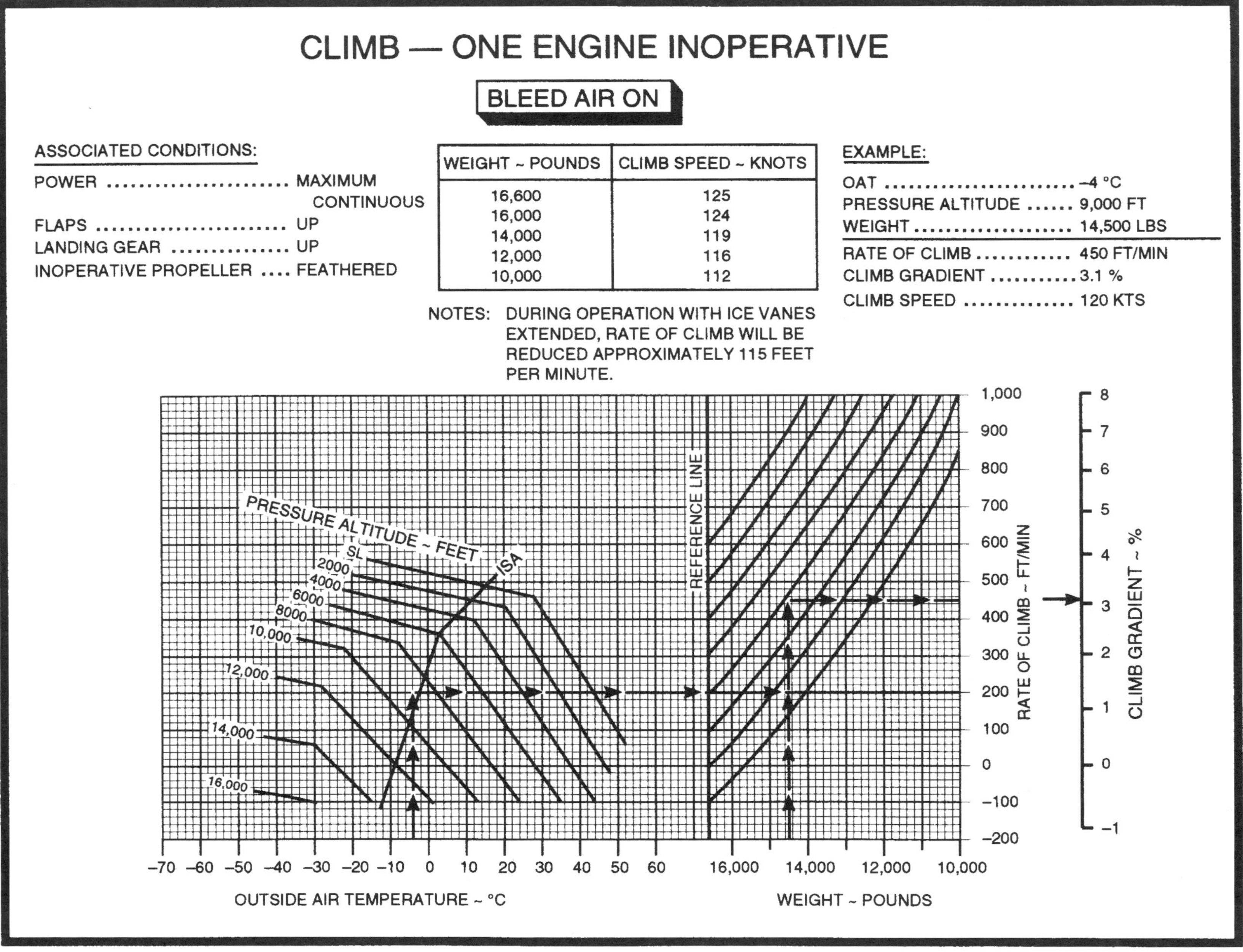

FIGURE 17.—Climb – One Engine Inoperative.

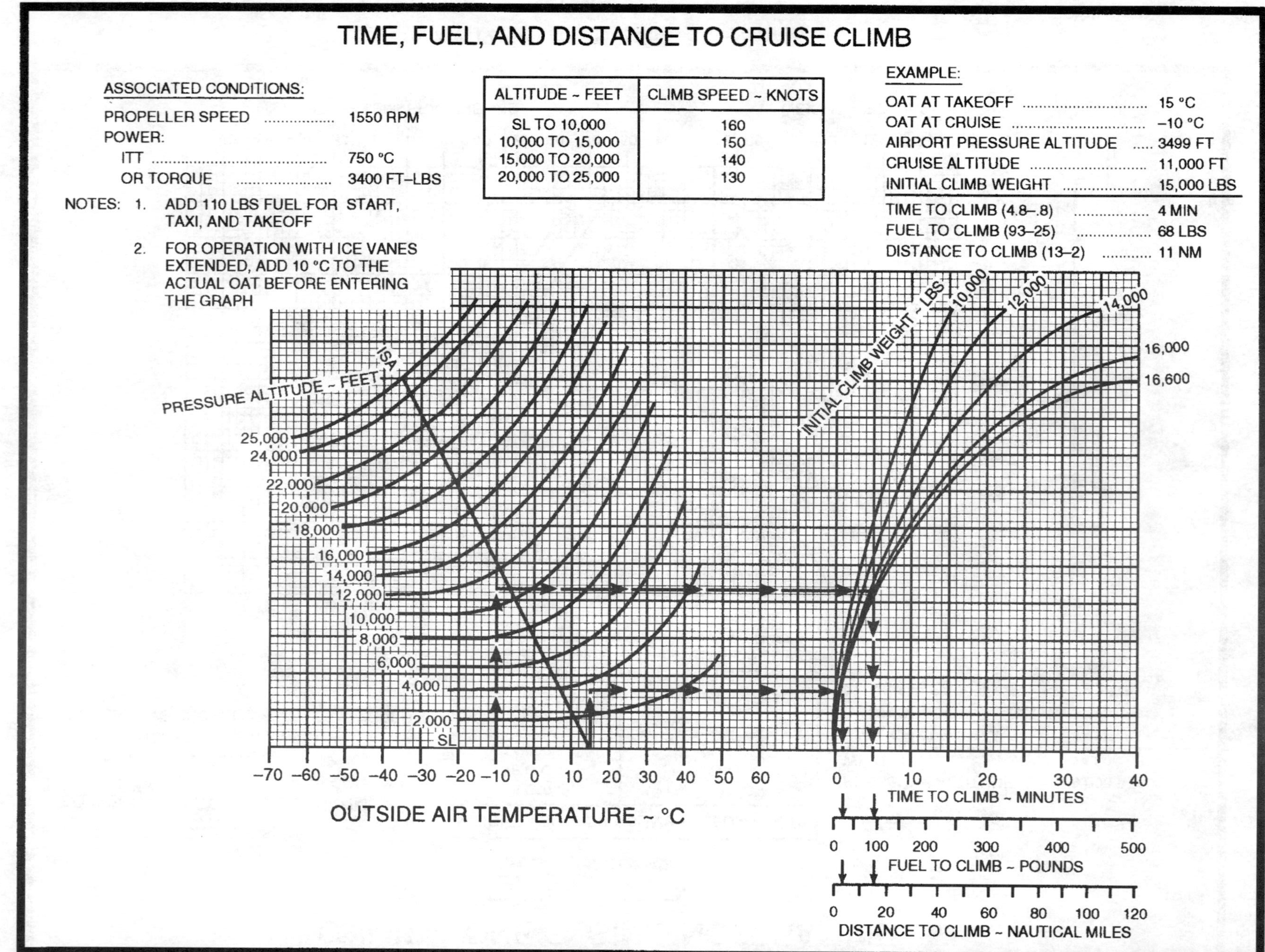

FIGURE 18.—Time, Fuel, and Distance to Cruise Climb.

OPERATING CONDITIONS	BE-26	BE-27	BE-28	BE-29	BE-30
OAT AT MEA	–8 °C	+30 °C	+5 °C	+18 °C	+22 °C
WEIGHT	15,500	16,600	16,000	16,300	14,500
ROUTE SEGMENT MEA	6,000	5,500	9,000	7,000	9,500
BLEED AIR	ON	ON	OFF	ON	OFF

FIGURE 19.—Beech 1900 – Service Ceiling.

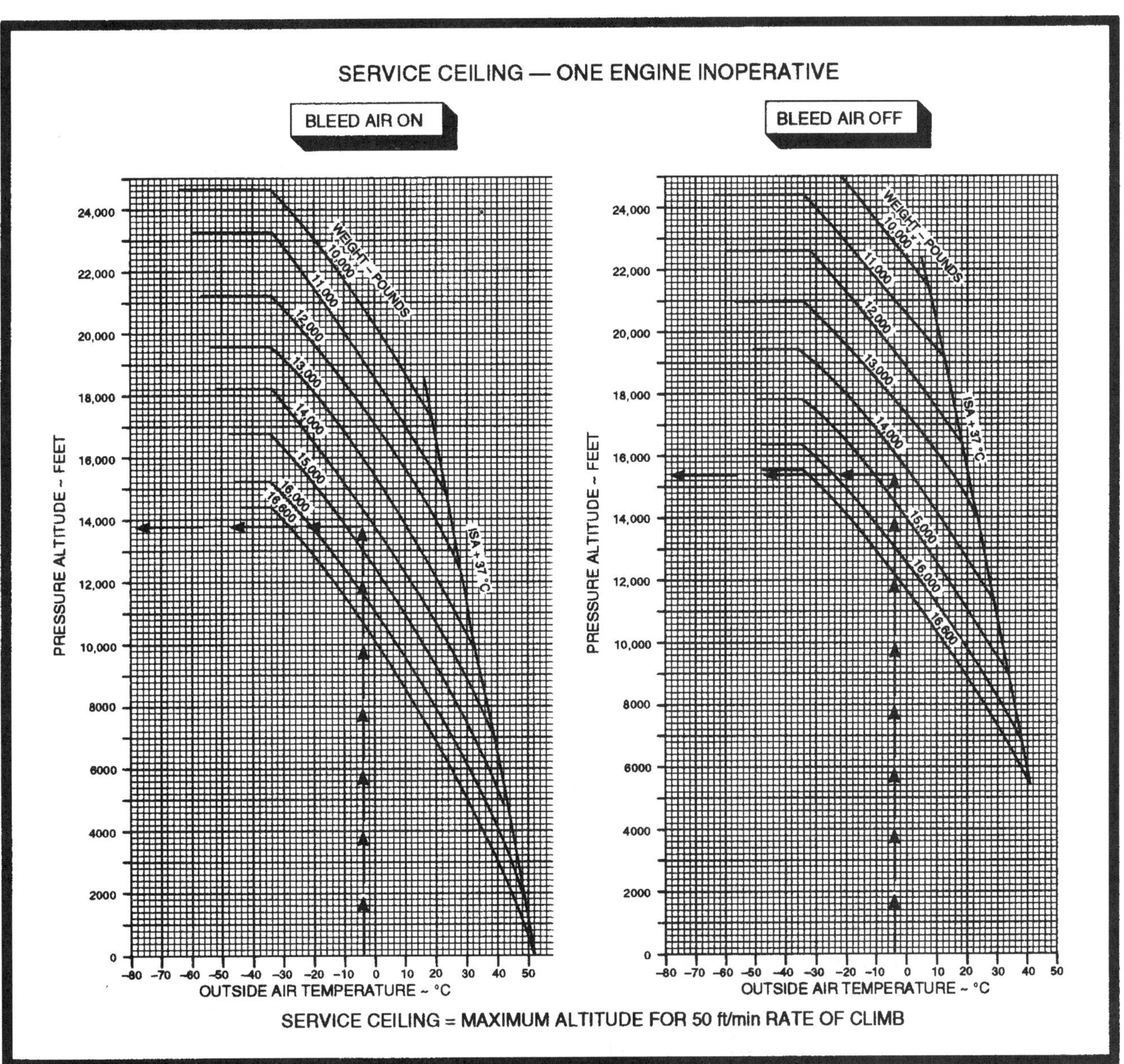

FIGURE 20.—Service Ceiling – One Engine Inoperative.

OPERATING CONDITIONS	BE-31	BE-32	BE-33	BE-34	BE-35
WEIGHT	15,000	14,000	13,000	16,000	11,000
PRESSURE ALTITUDE	22,000	17,000	20,000	23,000	14,000
TEMPERATURE (OAT)	–19 °C	–19 °C	–35 °C	–31 °C	–3 °C
TRUE COURSE	110	270	185	020	305
WIND	180/30	020/35	135/45	340/25	040/50
CRUISE DISTANCE	280	320	400	230	300

FIGURE 21.—Beech 1900 – Cruise.

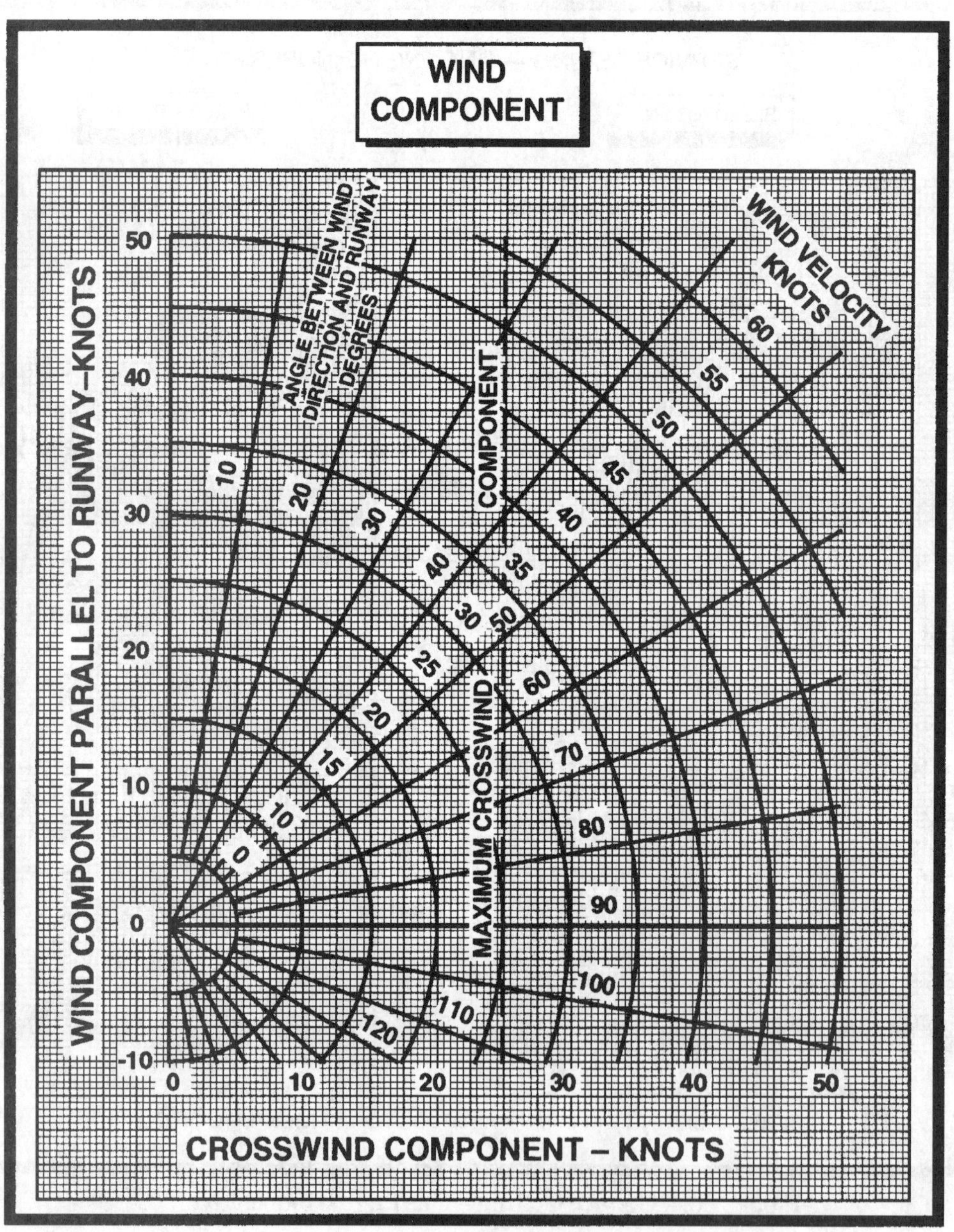

FIGURE 22.—Wind Component Chart.

RECOMMENDED CRUISE POWER

1550 RPM

ISA +10 °C

WEIGHT			16,000 POUNDS					14,000 POUNDS					12,000 POUNDS					10,000 POUNDS				
PRESSURE ALTITUDE	IOAT	OAT	TORQUE PER ENG	FUEL FLOW PER ENG	TOTAL FUEL FLOW	IAS	TAS	TORQUE PER ENG	FUEL FLOW PER ENG	TOTAL FUEL FLOW	IAS	TAS	TORQUE PER ENG	FUEL FLOW PER ENG	TOTAL FUEL FLOW	IAS	TAS	TORQUE PER ENG	FUEL FLOW PER ENG	TOTAL FUEL FLOW	IAS	TAS
FEET	°C	°C	FT-LBS	LBS/HR	LBS/HR	KTS	KTS	FT-LBS	LBS/HR	LBS/HR	KTS	KTS	FT-LBS	LBS/HR	LBS/HR	KTS	KTS	FT-LBS	LBS/HR	LBS/HR	KTS	KTS
SL	30	25	3294	577	1154	232	239	3301	577	1154	235	241	3307	577	1154	237	243	3312	577	1154	238	245
2000	26	21	3191	551	1102	227	240	3198	551	1102	230	243	3204	552	1104	232	245	3209	552	1104	233	247
4000	22	17	3092	527	1054	222	242	3100	528	1056	224	244	3106	528	1056	227	247	3111	528	1056	228	249
6000	19	13	2992	504	1008	216	243	3000	505	1010	219	246	3006	505	1010	222	249	3012	505	1010	224	251
8000	15	9	2886	481	962	211	244	2896	482	964	214	247	2903	482	964	216	250	2909	482	964	219	253
10,000	11	5	2778	458	916	205	244	2789	458	916	208	248	2797	459	918	211	252	2804	459	918	213	254
12,000	7	1	2636	432	864	198	243	2648	433	866	202	248	2657	433	866	205	252	2664	434	868	207	255
14,000	3	-3	2495	408	816	190	241	2508	409	818	195	247	2518	409	818	198	251	2525	409	818	201	255
16,000	-1	-7	2352	384	768	182	239	2367	385	770	188	246	2378	385	770	192	251	2386	386	772	195	255
18,000	-6	-11	2208	361	722	174	235	2226	362	724	180	243	2239	363	726	185	250	2248	363	726	188	254
20,000	-10	-15	2063	338	676	164	229	2085	340	680	172	240	2100	341	682	177	248	2111	341	682	181	253
22,000	-14	-19	1911	316	632	153	221	1939	317	634	163	235	1957	319	638	169	245	1969	319	638	174	252
24,000	-19	-23	1749	292	584	137	206	1790	295	590	152	229	1812	297	594	161	241	1827	298	596	167	249
25,000	-21	-25	1649	279	558	122	187	1714	284	568	147	224	1739	286	572	156	238	1756	287	574	163	248

FIGURE 23.—Recommended Cruise Power – ISA + 10 °C.

RECOMMENDED CRUISE POWER

1550 RPM

ISA

WEIGHT			16,000 POUNDS					14,000 POUNDS					12,000 POUNDS					10,000 POUNDS				
PRESSURE ALTITUDE	IOAT	OAT	TORQUE PER ENG	FUEL FLOW PER ENG	TOTAL FUEL FLOW	IAS	TAS	TORQUE PER ENG	FUEL FLOW PER ENG	TOTAL FUEL FLOW	IAS	TAS	TORQUE PER ENG	FUEL FLOW PER ENG	TOTAL FUEL FLOW	IAS	TAS	TORQUE PER ENG	FUEL FLOW PER ENG	TOTAL FUEL FLOW	IAS	TAS
FEET	°C	°C	FT-LBS	LBS/HR	LBS/HR	KTS	KTS	FT-LBS	LBS/HR	LBS/HR	KTS	KTS	FT-LBS	LBS/HR	LBS/HR	KTS	KTS	FT-LBS	LBS/HR	LBS/HR	KTS	KTS
SL	20	15	3400	586	1172	237	239	3400	585	1170	239	241	3400	585	1170	241	243	3400	585	1170	242	244
2000	17	11	3400	573	1146	234	244	3400	573	1146	236	246	3400	572	1144	238	248	3400	572	1144	240	249
4000	13	7	3400	560	1120	232	248	3400	559	1118	234	250	3400	559	1118	236	252	3400	559	1118	237	254
6000	9	3	3397	548	1096	229	252	3400	548	1096	231	255	3400	547	1094	233	257	3400	547	1094	235	259
8000	5	-1	3253	521	1042	223	253	3260	522	1044	225	256	3265	522	1044	228	258	3270	522	1044	229	260
10,000	1	-5	3092	494	988	216	252	3100	494	988	219	256	3107	495	990	221	258	3112	495	990	223	261
12,000	-3	-9	2929	466	932	208	251	2937	467	934	212	255	2945	467	934	214	258	2950	467	934	217	261
14,000	-7	-13	2772	440	880	201	250	2781	441	882	205	255	2789	441	882	208	258	2795	442	884	210	261
16,000	-11	-17	2606	414	828	193	248	2618	414	828	197	253	2626	415	830	201	258	2633	415	830	203	261
18,000	-15	-21	2435	288	776	184	244	2449	389	778	189	251	2459	389	778	193	256	2467	390	780	196	260
20,000	-19	-25	2263	363	726	175	239	2282	364	728	181	248	2294	365	730	186	254	2302	365	730	189	259
22,000	-24	-29	2094	338	676	164	233	2118	340	680	172	244	2133	341	682	178	251	2144	342	684	182	257
24,000	-28	-33	1931	315	630	152	223	1960	317	634	163	238	1979	318	636	169	248	1991	319	638	174	255
25,000	-30	-35	1846	303	606	145	216	1880	305	610	157	235	1901	307	614	165	246	1915	308	616	170	253

FIGURE 24.—Recommended Cruise Power – ISA.

RECOMMENDED CRUISE POWER

1550 RPM

ISA -10 °C

WEIGHT			16,000 POUNDS					14,000 POUNDS					12,000 POUNDS					10,000 POUNDS				
PRESSURE ALTITUDE	IOAT	OAT	TORQUE PER ENG	FUEL FLOW PER ENG	TOTAL FUEL FLOW	IAS	TAS	TORQUE PER ENG	FUEL FLOW PER ENG	TOTAL FUEL FLOW	IAS	TAS	TORQUE PER ENG	FUEL FLOW PER ENG	TOTAL FUEL FLOW	IAS	TAS	TORQUE PER ENG	FUEL FLOW PER ENG	TOTAL FUEL FLOW	IAS	TAS
FEET	°C	°C	FT-LBS	LBS/HR	LBS/HR	KTS	KTS	FT-LBS	LBS/HR	LBS/HR	KTS	KTS	FT-LBS	LBS/HR	LBS/HR	KTS	KTS	FT-LBS	LBS/HR	LBS/HR	KTS	KTS
SL	10	5	3400	582	1164	238	237	3400	582	1164	240	239	3400	581	1162	242	240	3400	581	1162	243	242
2000	6	1	3400	569	1138	236	241	3400	569	1138	238	243	3400	568	1136	240	245	3400	568	1136	241	246
4000	3	-3	3400	558	1116	233	245	3400	557	1114	236	248	3400	557	1114	237	249	3400	557	1114	239	251
6000	-1	-7	3400	548	1096	231	250	3400	547	1094	233	252	3400	547	1094	235	254	3400	546	1092	236	256
8000	-5	-11	3400	538	1076	228	254	3400	538	1076	231	257	3400	538	1076	232	259	3400	537	1074	234	261
10,000	-9	-15	3400	530	1060	226	259	3400	530	1060	228	262	3400	530	1060	230	264	3400	529	1058	232	266
12,000	-13	-19	3200	499	998	218	258	3208	500	1000	221	261	3215	500	1000	223	264	3220	501	1002	225	266
14,000	-17	-23	3010	470	940	210	256	3019	471	942	213	260	3026	471	942	216	263	3032	472	944	218	266
16,000	-21	-27	2823	442	884	202	254	2833	442	884	205	258	2841	443	886	209	262	2848	443	886	211	265
18,000	-25	-31	2641	414	828	193	251	2652	415	830	198	256	2661	416	832	201	261	2668	416	832	204	264
20,000	-29	-35	2456	387	774	184	247	2471	388	776	189	254	2481	389	778	193	259	2489	390	780	196	263
22,000	-33	-39	2277	361	722	174	242	2296	363	726	181	250	2308	363	726	185	256	2318	364	728	189	261
24,000	-37	-43	2105	336	672	163	234	2128	338	676	172	246	2144	339	678	177	254	2155	340	680	181	260
25,000	-40	-45	2017	324	648	157	230	2044	326	652	167	243	2061	327	654	173	252	2073	328	656	177	258

FIGURE 25.—Recommended Cruise Power – ISA –10 °C.

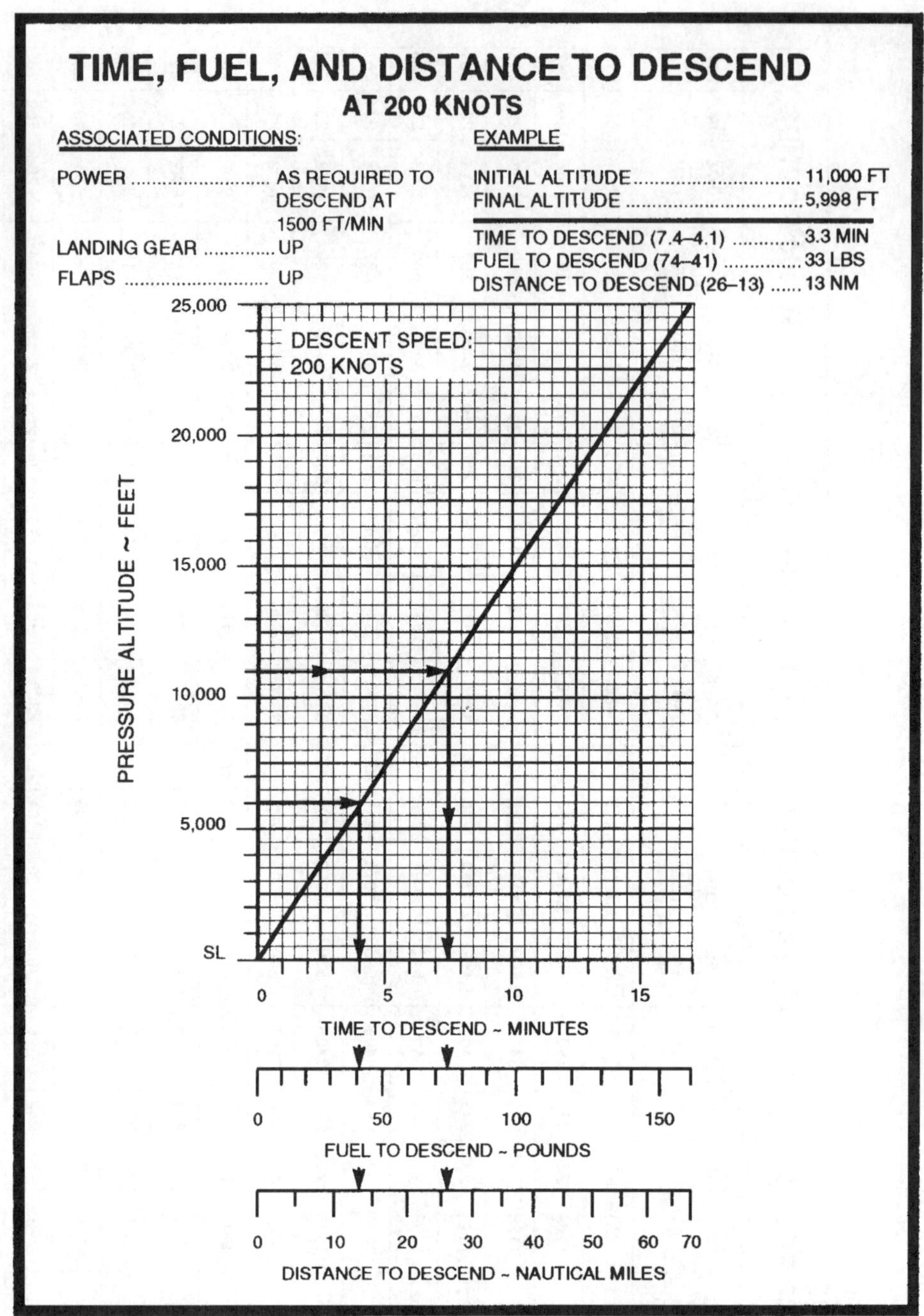

FIGURE 26.—Time, Fuel, and Distance to Descend.

OPERATING CONDITIONS	B–36	B–37	B–38	B–39	B–40
PRESSURE ALTITUDE	SL	1,000	2,000	4,000	5,000
TEMPERATURE (OAT)	+30 °C	+16 °C	0 °C	+20 °C	ISA
WEIGHT	16,000	14,500	13,500	15,000	12,500
WIND COMPONENT (KTS)	20 HW	10 TW	15 HW	5 TW	25 HW
RUNWAY LENGTH (FT)	4,000	4,500	3,800	5,000	4,000

FIGURE 27.—Beech 1900 – Landing.

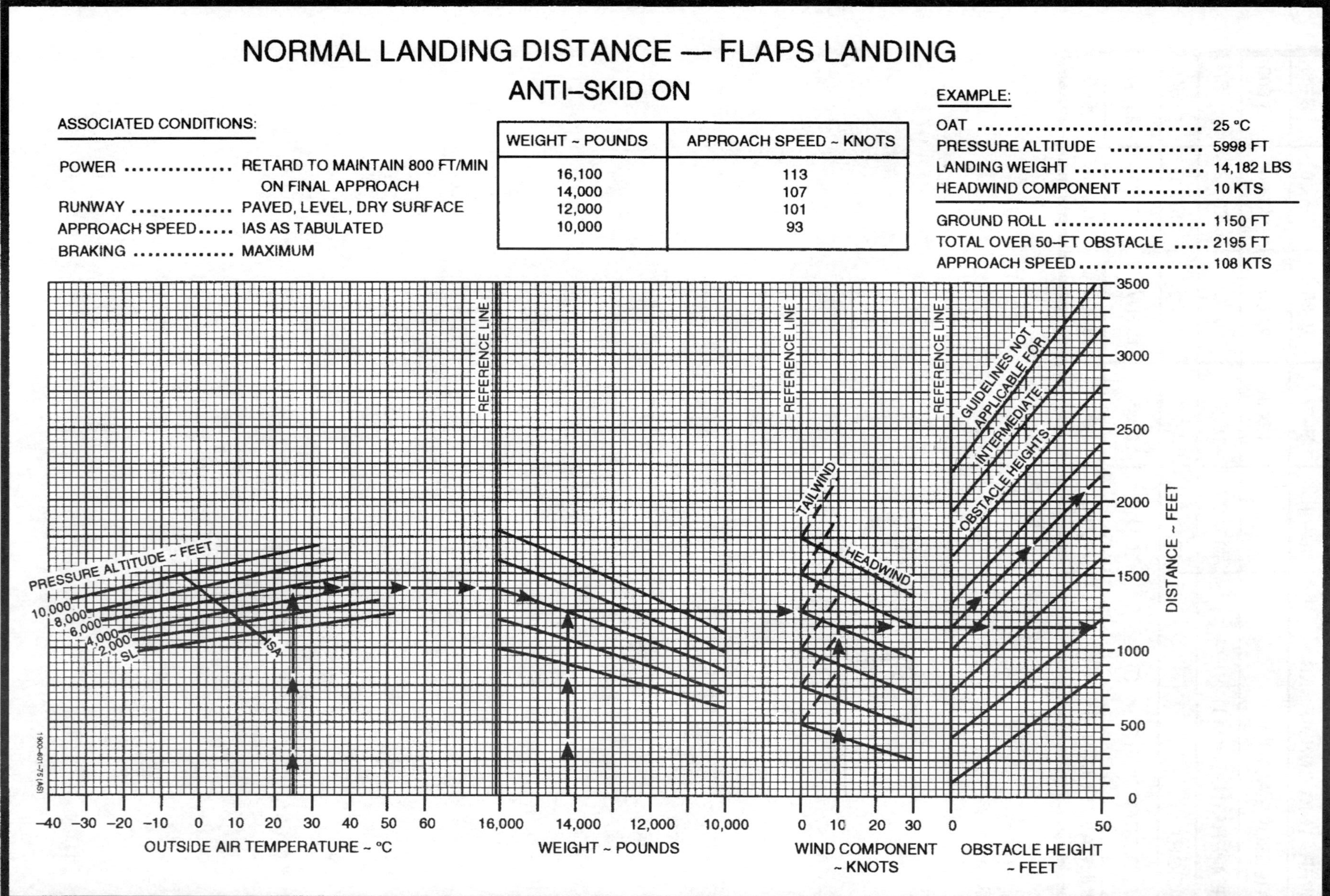

FIGURE 28.—Normal Landing Distance – Flaps Landing.

OPERATING CONDITIONS	BL-1	BL-2	BL-3	BL-4	BL-5
CREW WEIGHT	340	400	360	380	370
PASSENGER WT ROW 1	700	620	—	180	680
ROW 2	830	700	750	800	950
ROW 3	800	680	810	720	850
ROW 4	—	400	650	200	500
BAGGAGE CENTER	500	550	300	200	450
LEFT AND RIGHT	200	250	—	100	—
FUEL GALLONS	300	250	360	400	260
TYPE	JET A	JET B	JET A	JET B	JET A

FIGURE 29.—Bell 214 ST – Loading.

LOADING CONDITIONS	BL-6	BL-7	BL-8	BL-9	BL-10
BASIC WEIGHT	10,225	9,450	9,000	9,510	9,375
BASIC MOM/100	25562.5	23236.0	22020.5	23499.9	23296.8
CREW WEIGHT	340	380	410	360	400
PASSENGER WEIGHT	3,280	2,880	3,150	2,040	2,400
PASSENGER MOM/100	6722.5	5418.6	6425.8	4732.2	4560.7
BAGGAGE (CENTER)	700	600	300	550	650
FUEL LOAD (6.8 LB/GAL)	435	290	220	435	380
TRIP FUEL BURN (GAL)	355	190	190	325	330
LATERAL CG IS ON LONGITUDINAL AXIS					

FIGURE 30.—Bell 214 ST – Weight Shift and Limits.

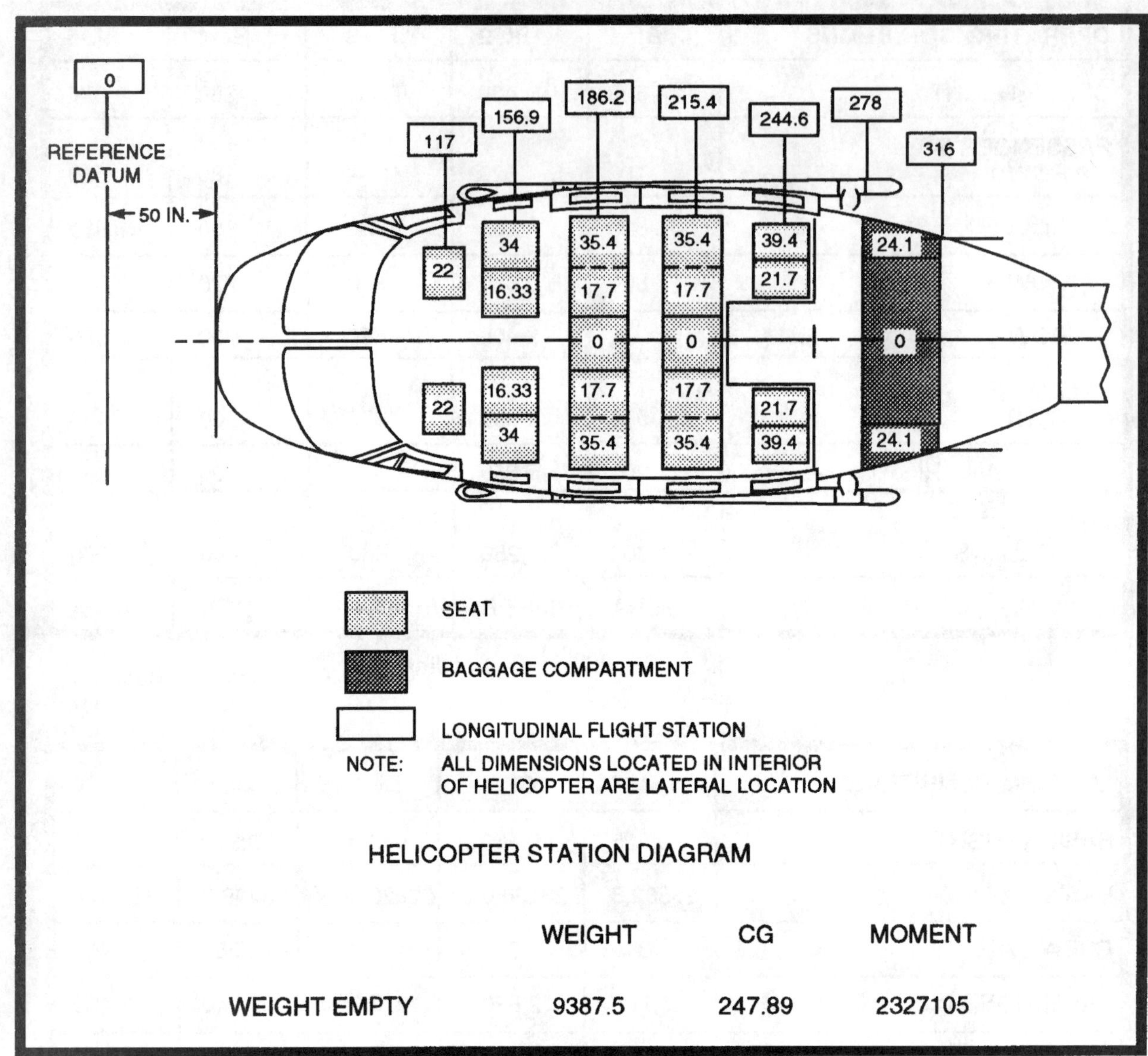

	WEIGHT	CG	MOMENT
WEIGHT EMPTY	9387.5	247.89	2327105

FIGURE 31.—Helicopter – Loading Data.

CREW AND PASSENGER TABLE OF MOMENTS (IN-LB)

WEIGHT LBS	CREW SEATS F.S. 117	AIRLINE PASSENGEER SEATS FIRST ROW (FOUR PASSENGER) SEATS F.S. 156.9	SECOND ROW (FIVE PASSENGER) SEATS F.S. 186.2	THIRD ROW (FIVE PASSENGER) SEATS F.S. 215.4	FOURTH ROW (FOUR PASSENGER) SEATS F.S. 244.6
100	11700	15690	18620	21540	24460
110	12870	17259	20482	23694	26906
120	14040	18828	22344	25848	29352
130	15210	20397	24206	28002	31798
140	16380	21966	26068	30156	34244
150	17550	23535	27930	32310	36690
160	18720	25104	29792	34464	39136
170	19890	26673	31654	36618	41582
180	21060	28242	33516	38772	44028
190	22230	29811	35378	40926	46474
200	23400	31380	37240	43080	48920
210	24570	32949	39102	45234	51366
220	25740	34518	40964	47388	53812

BAGGAGE COMPARTMENT LOADING TABLE (IN-LB ÷ 100)

BAGGAGE WEIGHT LBS	LEFT AND RIGHT BAGGAGE COMPARTMENT STA. 278.0 TO 316.0 F.S. 295.2	CENTER BAGGAGE COMPARTMENT STA. 278.0 TO 316.0 F.S. 297.0
50	147.6	148.5
100	295.2	297.0
150	442.8	445.5
200	590.4	594.0
250	738.0	742.5
300	885.6	891.0
350	1033.2	1039.5
400	1180.8	1188.0
450	1328.4	1336.5
500	1476.0	1485.0
530	1564.6	1574.1
550		1633.5
600		1782.0
650		1930.5
700		2079.0
740		2197.8

FIGURE 32.—Helicopter – Weights and Moments – Crew, Passengers, and Baggage.

USABLE FUEL LOADING TABLE (ENGLISH)

JET A, JET A-1, JP-5 (6.8 LBS/GAL)							
U.S. GAL	WEIGHT LBS	C.G.	MOMENT IN. LB. ÷ 100	U.S. GAL	WEIGHT LBS	C.G.	MOMENT IN. LB. ÷ 100
10	68	244.3	166	220	1496	246.9	3694
20	136	244.3	332	230	1564	244.3	3820
30	204	244.4	499	240	1632	241.8	3947
**37.1	252	244.4	616	250	1700	239.6	4073
40	272	242.8	660	260	1768	237.6	4200
50	340	237.8	808	270	1836	235.6	4326
60	408	234.5	957	280	1904	233.9	4453
70	476	232.1	1105	290	1972	232.2	4579
80	544	230.9	1256	**291.4	1982	232.0	4597
90	612	229.2	1403	300	2040	233.1	4754
*99.7	678	228.2	1546	310	2108	234.0	4934
*109.2	743	228.2	1695	320	2176	235.1	5115
110	748	228.5	1709	330	2244	236.0	5296
120	816	231.7	1890	340	2312	236.9	5477
130	884	234.4	2072	350	2380	237.7	5658
140	952	236.7	2253	360	2448	238.5	5839
150	1020	238.6	2434	370	2516	239.3	6021
160	1088	240.4	2615	380	2584	240.0	6202
170	1156	242.0	2798	390	2652	240.7	6383
180	1224	243.3	2978	400	2720	241.3	6564
190	1292	244.5	3159	410	2788	241.9	6745
200	1360	245.6	3340	420	2856	242.5	6927
210	1428	246.6	3521	430	2924	243.1	7108
*218.4	1484	247.3	3673	435.0	2958	243.4	7199

JET B, JP-4 (6.5 LBS/GAL)							
U.S. GAL	WEIGHT LBS	C.G.	MOMENT IN. LB. ÷ 100	U.S. GAL	WEIGHT LBS	C.G.	MOMENT IN. LB. ÷ 100
10	65	244.3	159	220	1430	246.9	3531
20	130	244.3	318	230	1495	244.3	3652
30	195	244.5	477	240	1560	241.8	3772
**37.1	241	244.4	589	250	1625	239.6	3894
40	260	242.8	631	260	1690	237.6	4015
50	325	237.8	773	270	1755	235.6	4135
60	390	234.5	915	280	1820	233.9	4257
70	455	232.1	1056	290	1885	232.2	4377
80	520	230.9	1201	**291.4	1894	232.0	4394
90	585	229.2	1341	300	1950	233.1	4545
*99.7	648	228.2	1479	310	2015	234.0	4715
*109.2	710	228.2	1620	320	2080	235.1	4890
110	715	228.5	1634	330	2145	236.0	5062
120	780	231.7	1807	340	2210	236.9	5235
130	845	234.4	1981	350	2275	237.7	5408
140	910	236.7	2154	360	2340	238.5	5581
150	975	238.6	2326	370	2405	239.3	5755
160	1040	240.4	2500	380	2470	240.5	5928
170	1105	242.0	2674	390	2535	240.7	6102
180	1170	243.3	2847	400	2600	241.3	6274
190	1235	244.5	3020	410	2665	241.9	6447
200	1300	245.6	3193	420	2730	242.5	6620
210	1365	246.6	3366	430	2795	243.1	6795
*218.4	1420	247.3	3512	435	2827.5	243.4	6882

* Extreme limits of fuel C.G.
** Point of C.G. direction change.

Weights given are nominal weights at 15 °C.

FIGURE 33.—Helicopter – Weights and Moments – Usable Fuel.

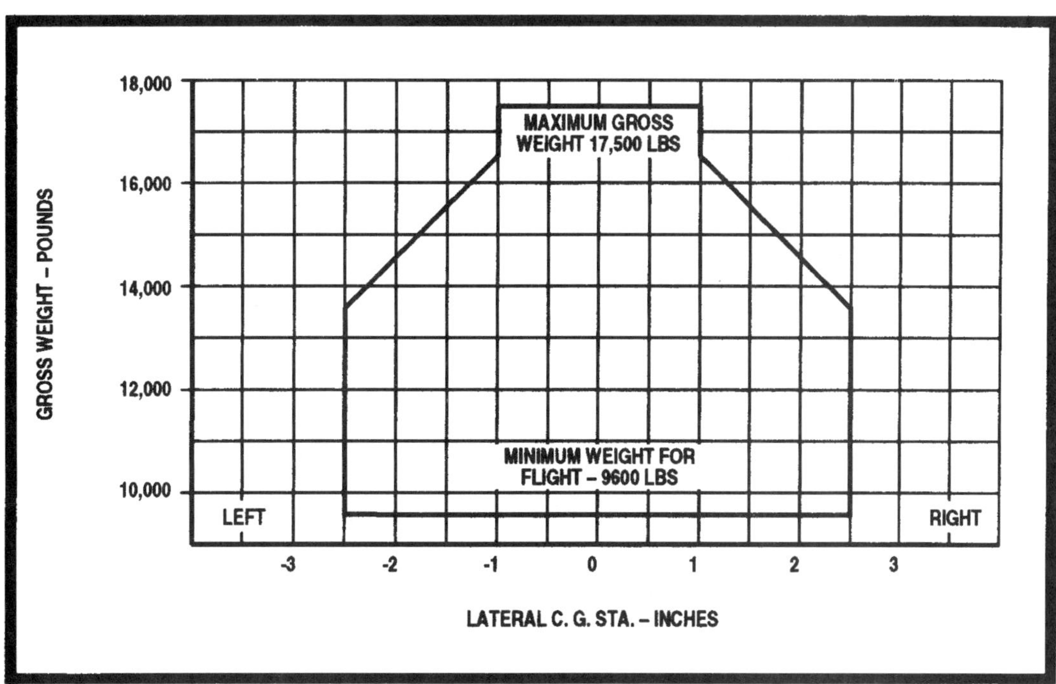

FIGURE 34.—Helicopter – Lateral CG Envelope.

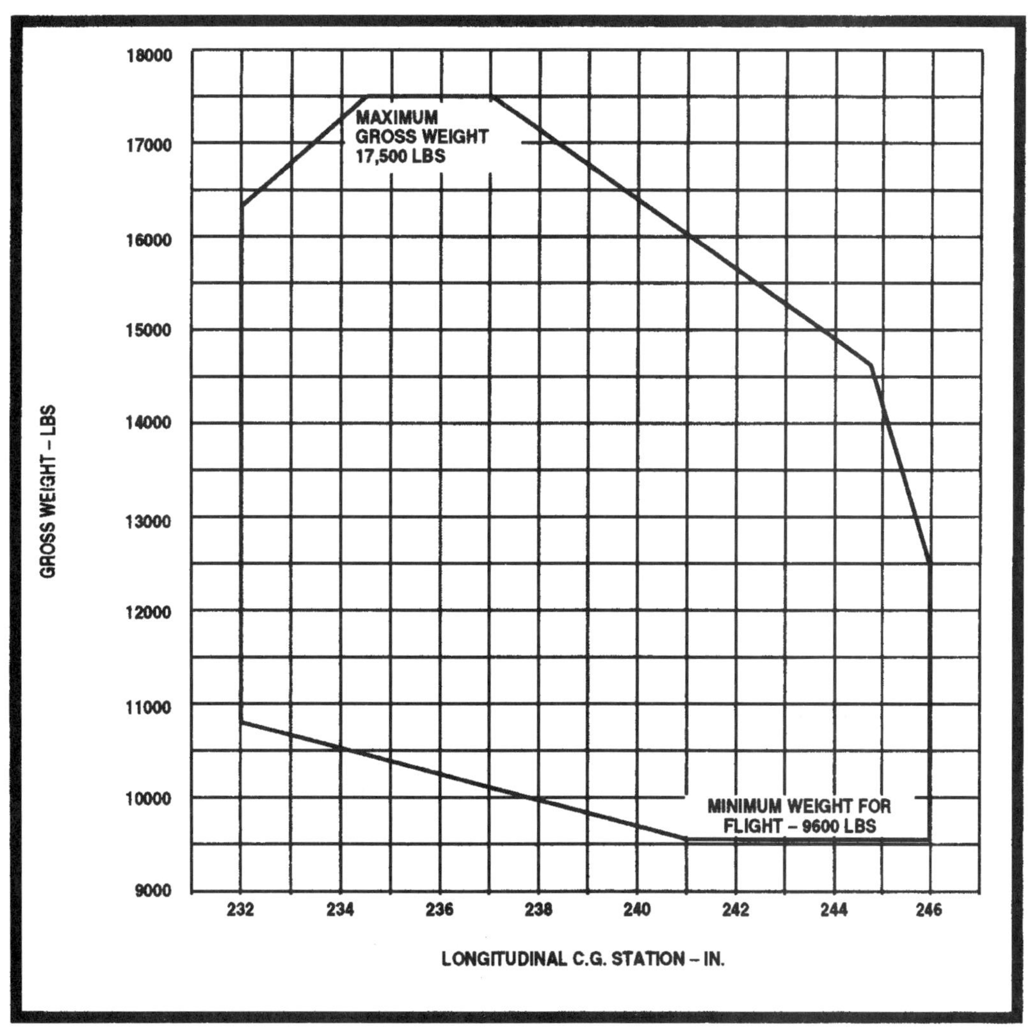

FIGURE 35.—Helicopter – Longitudinal CG Envelope.

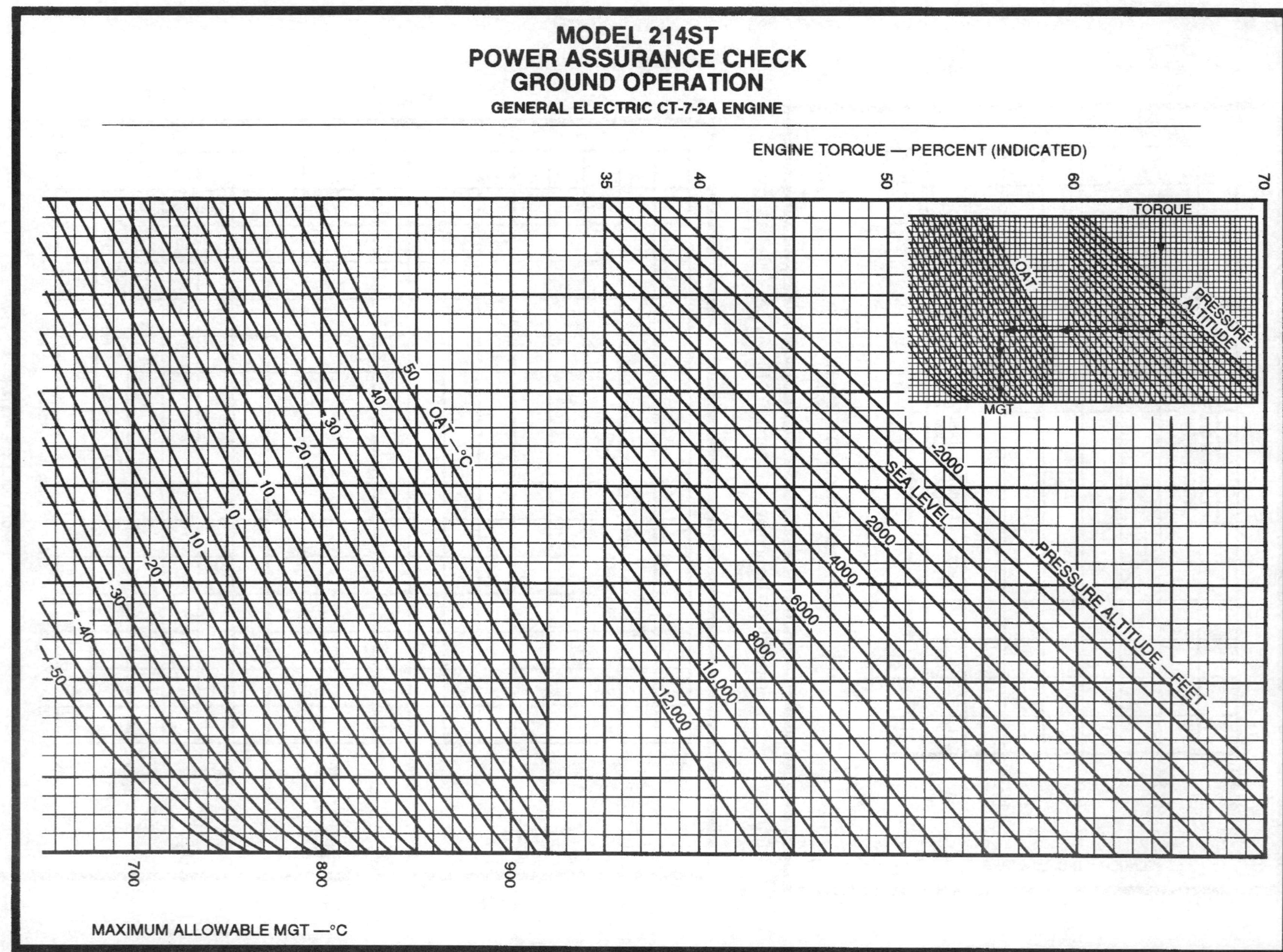

FIGURE 36.—Bell 214 – Power Assurance Check.

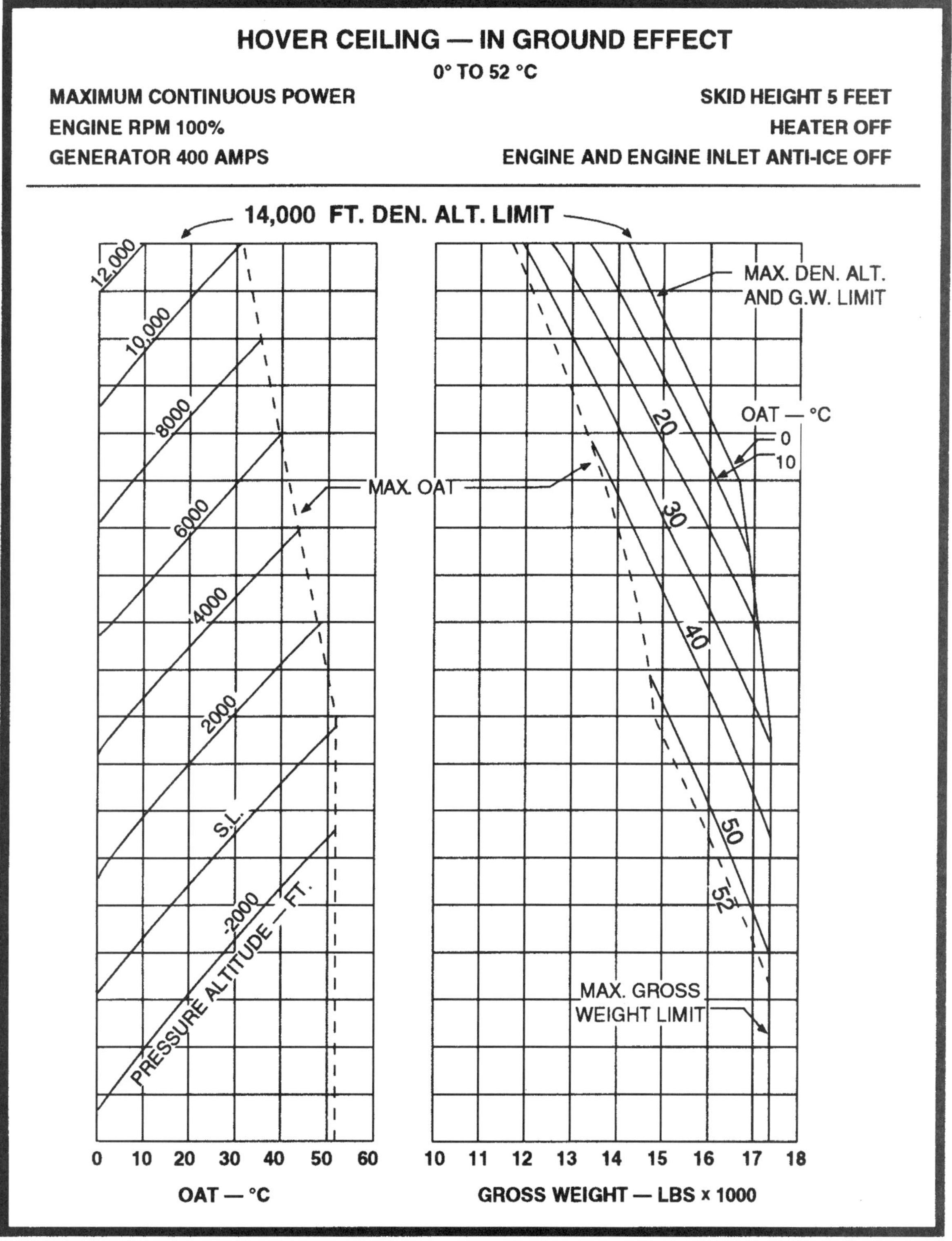

FIGURE 37.—Hovering Ceiling – In Ground Effect.

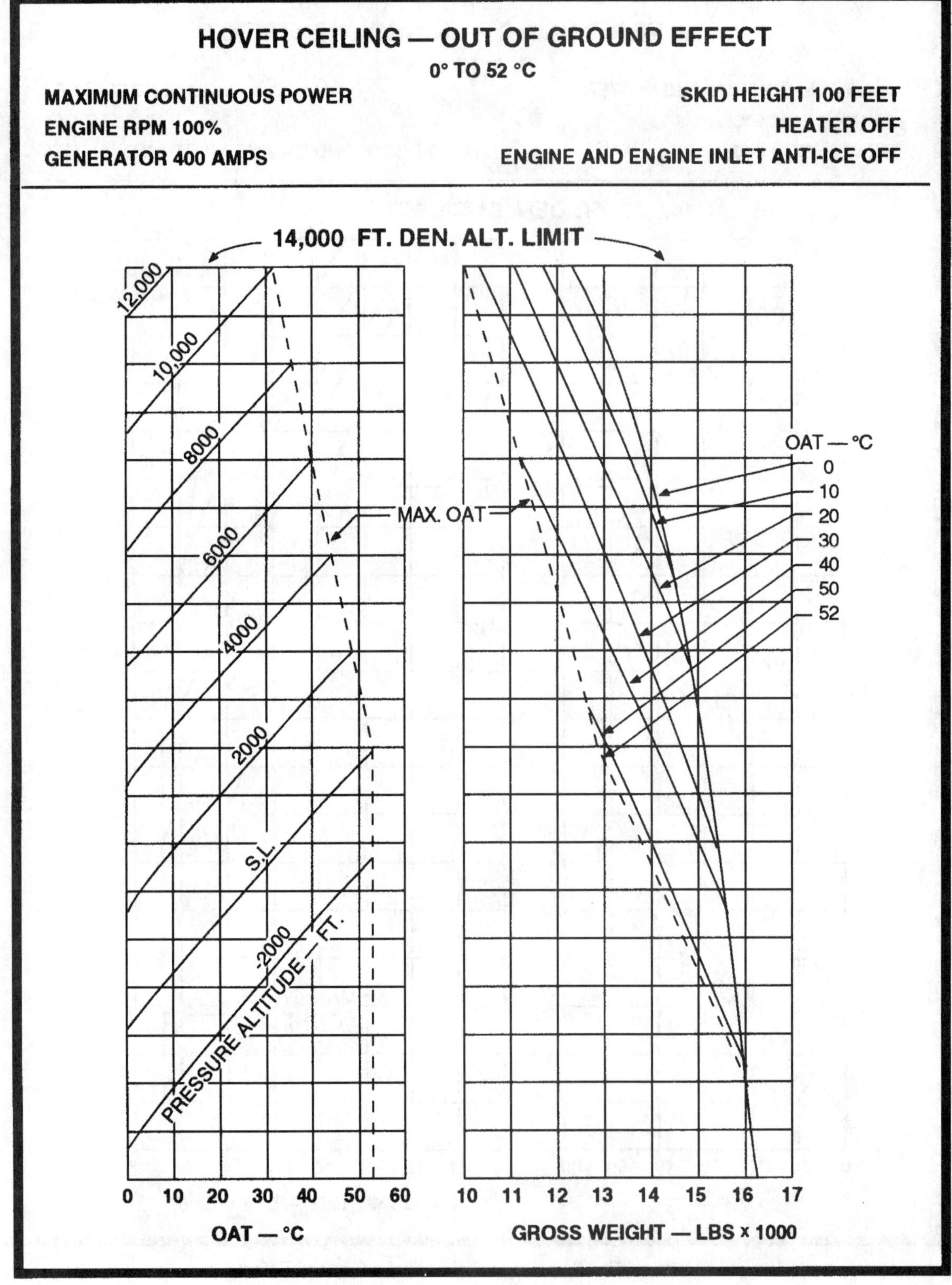

FIGURE 38.—Hovering Ceiling – Out of Ground Effect.

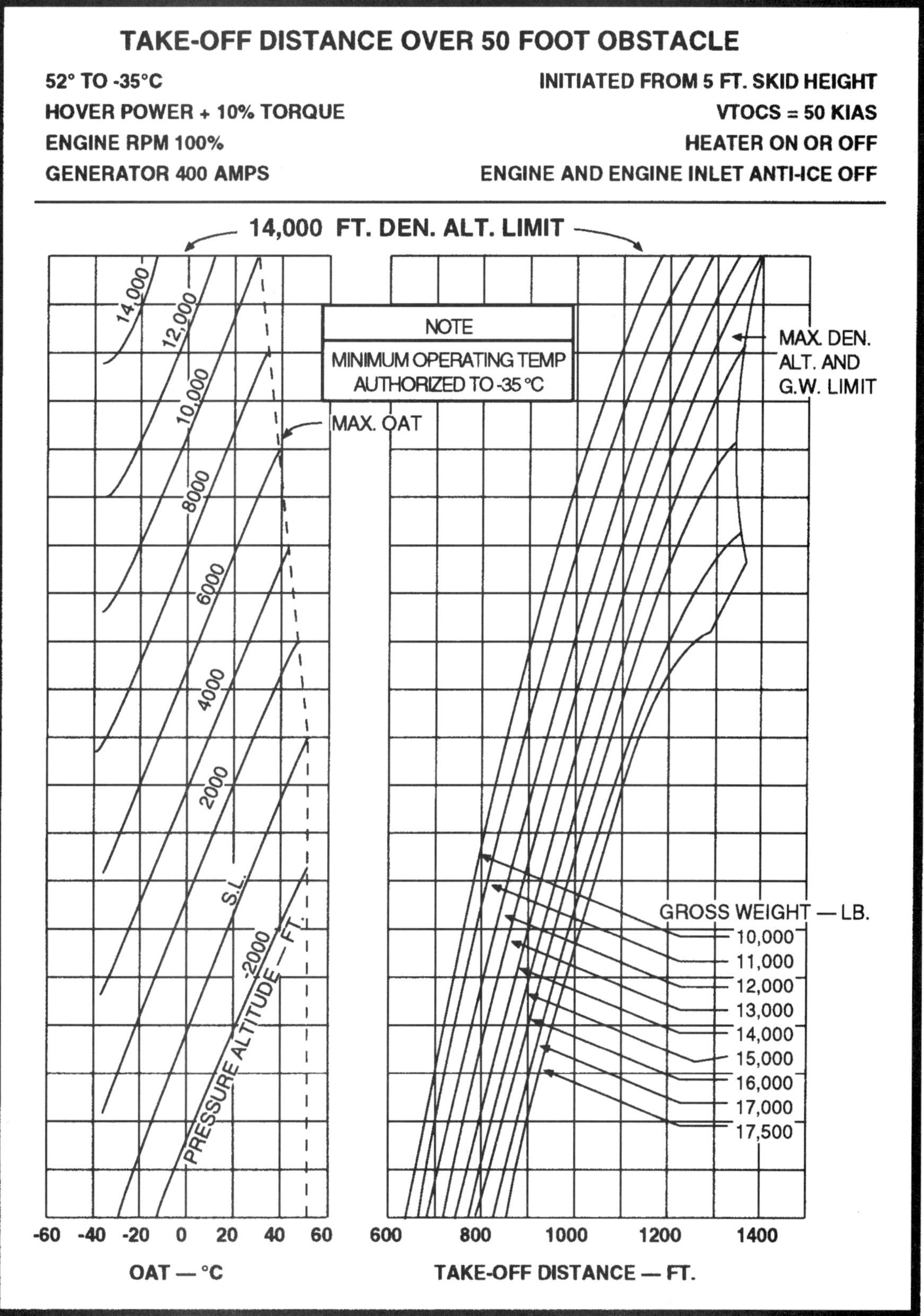

FIGURE 39.—Takeoff Distance Over 50-Foot Obstacle.

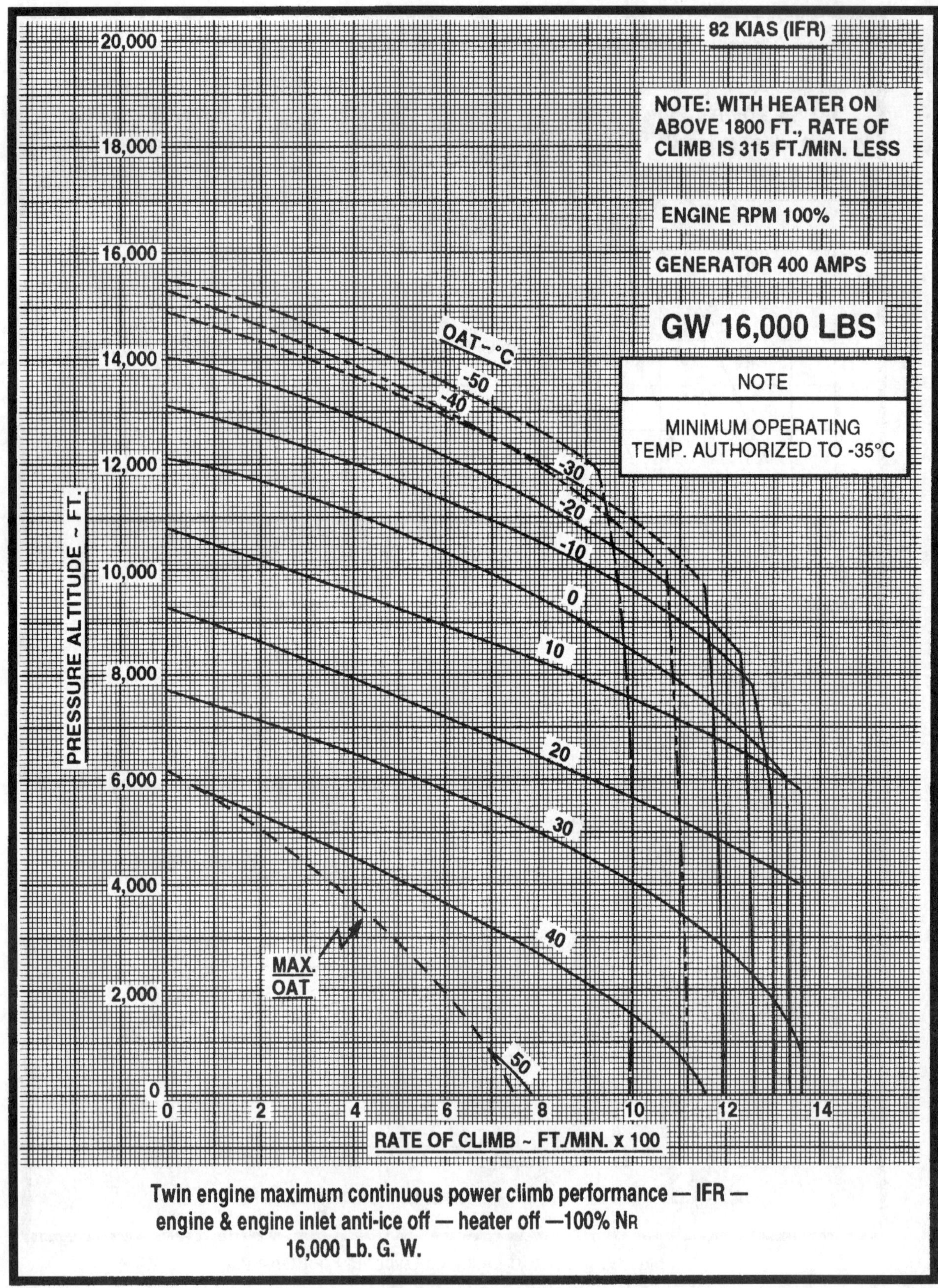

FIGURE 40.—Twin-Engine Climb Performance.

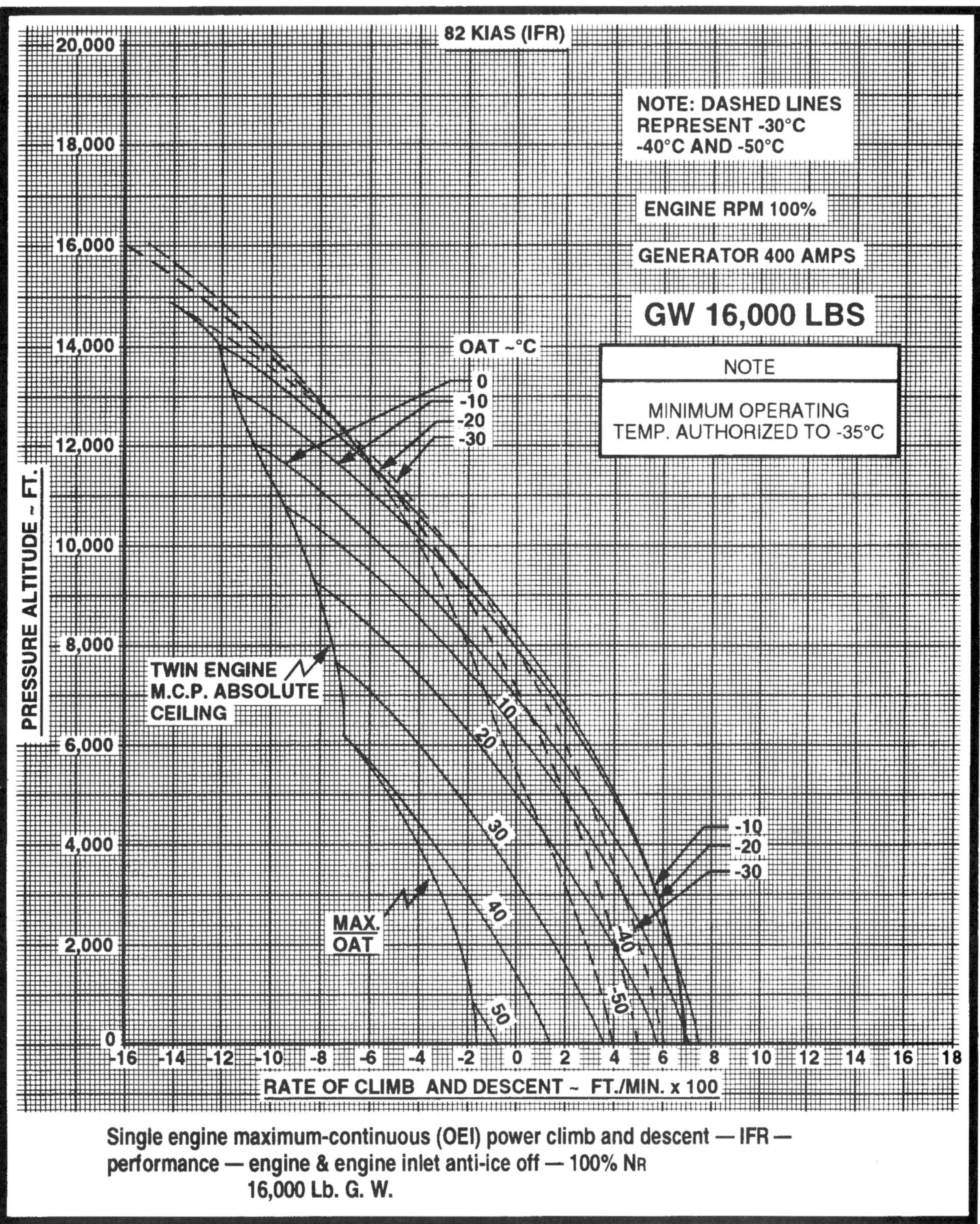

FIGURE 41.—Single-Engine Climb Performance.

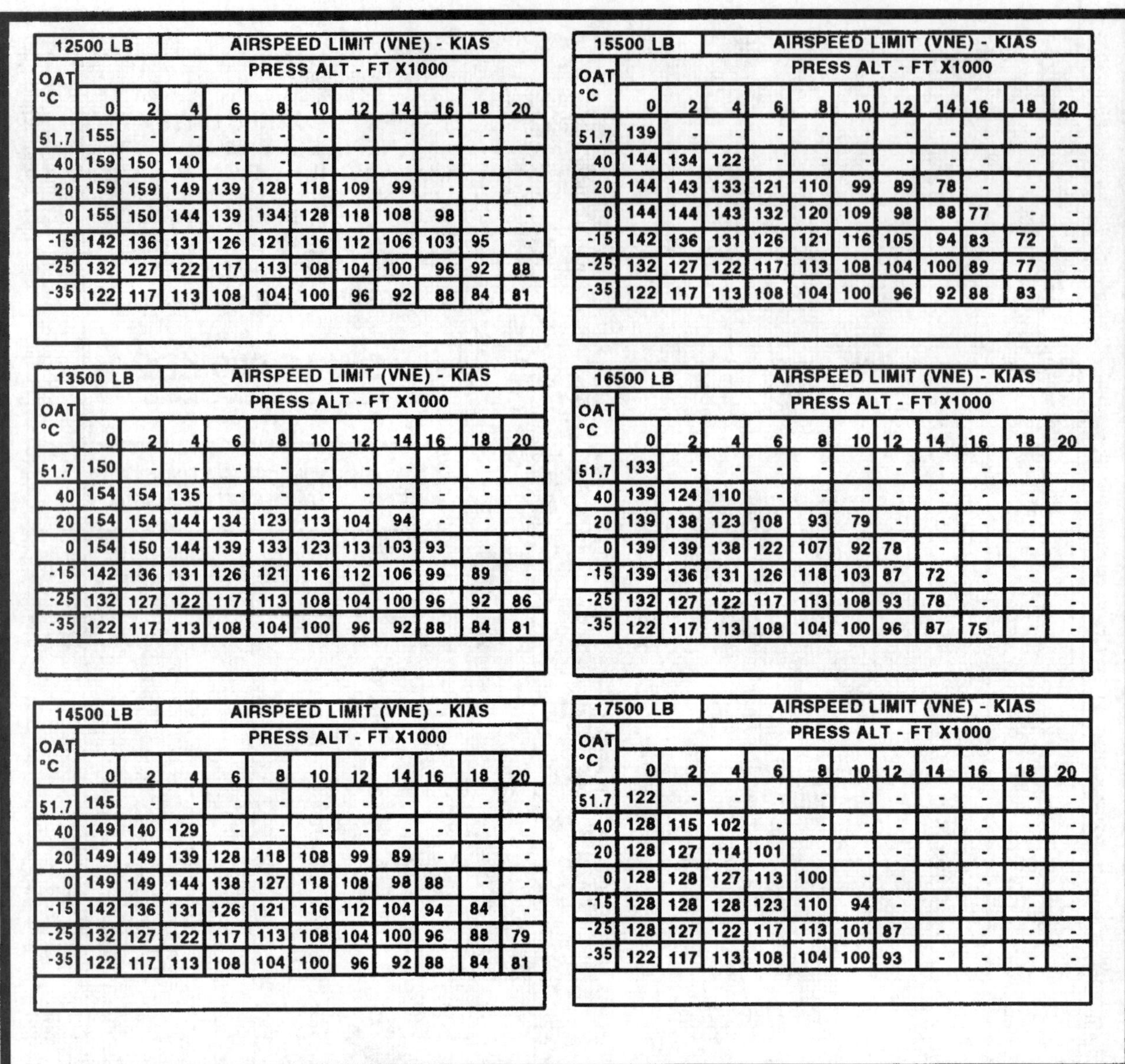

12500 LB	AIRSPEED LIMIT (VNE) - KIAS										
OAT °C	PRESS ALT - FT X1000										
	0	2	4	6	8	10	12	14	16	18	20
51.7	155	-	-	-	-	-	-	-	-	-	-
40	159	150	140	-	-	-	-	-	-	-	-
20	159	159	149	139	128	118	109	99	-	-	-
0	155	150	144	139	134	128	118	108	98	-	-
-15	142	136	131	126	121	116	112	106	103	95	-
-25	132	127	122	117	113	108	104	100	96	92	88
-35	122	117	113	108	104	100	96	92	88	84	81

15500 LB	AIRSPEED LIMIT (VNE) - KIAS										
OAT °C	PRESS ALT - FT X1000										
	0	2	4	6	8	10	12	14	16	18	20
51.7	139	-	-	-	-	-	-	-	-	-	-
40	144	134	122	-	-	-	-	-	-	-	-
20	144	143	133	121	110	99	89	78	-	-	-
0	144	144	143	132	120	109	98	88	77	-	-
-15	142	136	131	126	121	116	105	94	83	72	-
-25	132	127	122	117	113	108	104	100	89	77	-
-35	122	117	113	108	104	100	96	92	88	83	-

13500 LB	AIRSPEED LIMIT (VNE) - KIAS										
OAT °C	PRESS ALT - FT X1000										
	0	2	4	6	8	10	12	14	16	18	20
51.7	150	-	-	-	-	-	-	-	-	-	-
40	154	154	135	-	-	-	-	-	-	-	-
20	154	154	144	134	123	113	104	94	-	-	-
0	154	150	144	139	133	123	113	103	93	-	-
-15	142	136	131	126	121	116	112	106	99	89	-
-25	132	127	122	117	113	108	104	100	96	92	86
-35	122	117	113	108	104	100	96	92	88	84	81

16500 LB	AIRSPEED LIMIT (VNE) - KIAS										
OAT °C	PRESS ALT - FT X1000										
	0	2	4	6	8	10	12	14	16	18	20
51.7	133	-	-	-	-	-	-	-	-	-	-
40	139	124	110	-	-	-	-	-	-	-	-
20	139	138	123	108	93	79	-	-	-	-	-
0	139	139	138	122	107	92	78	-	-	-	-
-15	139	136	131	126	118	103	87	72	-	-	-
-25	132	127	122	117	113	108	93	78	-	-	-
-35	122	117	113	108	104	100	96	87	75	-	-

14500 LB	AIRSPEED LIMIT (VNE) - KIAS										
OAT °C	PRESS ALT - FT X1000										
	0	2	4	6	8	10	12	14	16	18	20
51.7	145	-	-	-	-	-	-	-	-	-	-
40	149	140	129	-	-	-	-	-	-	-	-
20	149	149	139	128	118	108	99	89	-	-	-
0	149	149	144	138	127	118	108	98	88	-	-
-15	142	136	131	126	121	116	112	104	94	84	-
-25	132	127	122	117	113	108	104	100	96	88	79
-35	122	117	113	108	104	100	96	92	88	84	81

17500 LB	AIRSPEED LIMIT (VNE) - KIAS										
OAT °C	PRESS ALT - FT X1000										
	0	2	4	6	8	10	12	14	16	18	20
51.7	122	-	-	-	-	-	-	-	-	-	-
40	128	115	102	-	-	-	-	-	-	-	-
20	128	127	114	101	-	-	-	-	-	-	-
0	128	128	127	113	100	-	-	-	-	-	-
-15	128	128	128	123	110	94	-	-	-	-	-
-25	128	127	122	117	113	101	87	-	-	-	-
-35	122	117	113	108	104	100	93	-	-	-	-

FIGURE 42.—Airspeed Limit.

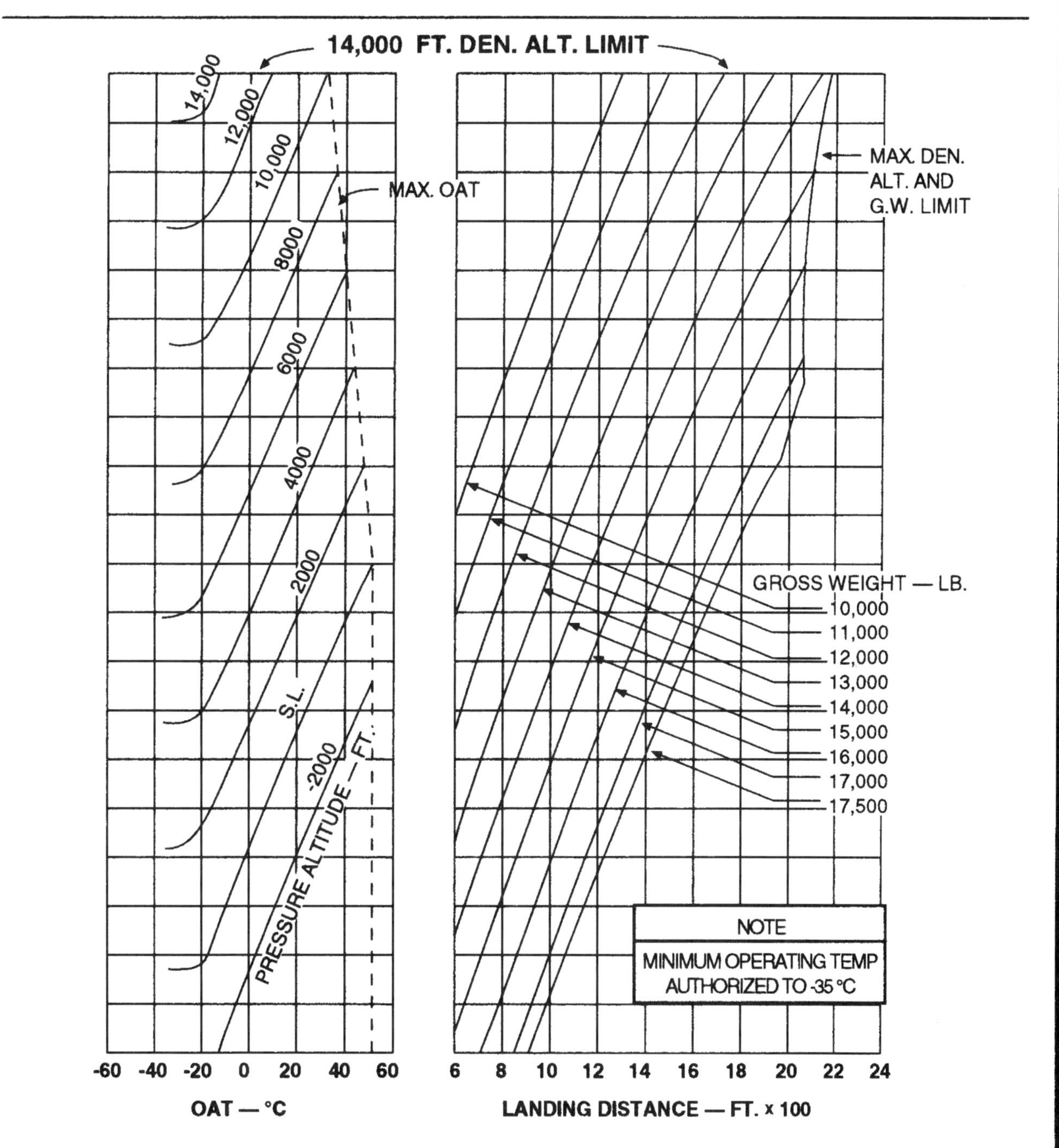

FIGURE 43.—Single-Engine Landing Distance Over 50-Foot Obstacle.

LOADING CONDITIONS	WS-1	WS-2	WS-3	WS-4	WS-5
LOADED WEIGHT	90,000	85,000	84,500	81,700	88,300
LOADED CG (% MAC)	22.5%	28.4%	19.8%	30.3%	25.5%
WEIGHT CHANGE (POUNDS)	2,500	1,800	3,000	2,100	3,300

FWD COMPT CENTROID – STA 352.1 AND –227.9 INDEX ARM
AFT COMPT CENTROID – STA 724.9 AND +144.9 INDEX ARM
MAC – 141.5 INCHES, LEMAC – STA 549.13, AND –30.87 INDEX ARM

FIGURE 44.—DC-9 – Weight Shift.

OPERATING CONDITIONS	A-1	A-2	A-3	A-4	A-5
FIELD ELEVATION	2,500	600	4,200	5,100	2,100
ALTIMETER SETTING	29.40"	30.50"	1020mb	29.35"	1035mb
AMBIENT TEMPERATURE	+10 °F	+80 °F	0 °C	+30 °F	+20 °C
WEIGHT (X1000)	75	85	90	80	65
FLAP POSITION	20°	20°	20°	20°	20°
RUNWAY SLOPE %	+1%	–1.5%	0	+1.5%	–2%
WIND COMPONENT	10 HW	10 TW	15 HW	5 TW	20 HW
ICE PROTECTION	BOTH	NONE	BOTH	ENGINE	NONE
CG STATION	590.2	—	580.3	—	594.4
CG INDEX ARM	—	–3.1	—	+5.9	—

INDEX ARM REF – STA 580.0, LEMAC – STA 549.13, AND –30.87 INDEX, MAC 141.5
CG % MAC = STAB TRIM SETTING

FIGURE 45.—DC-9 – Takeoff.

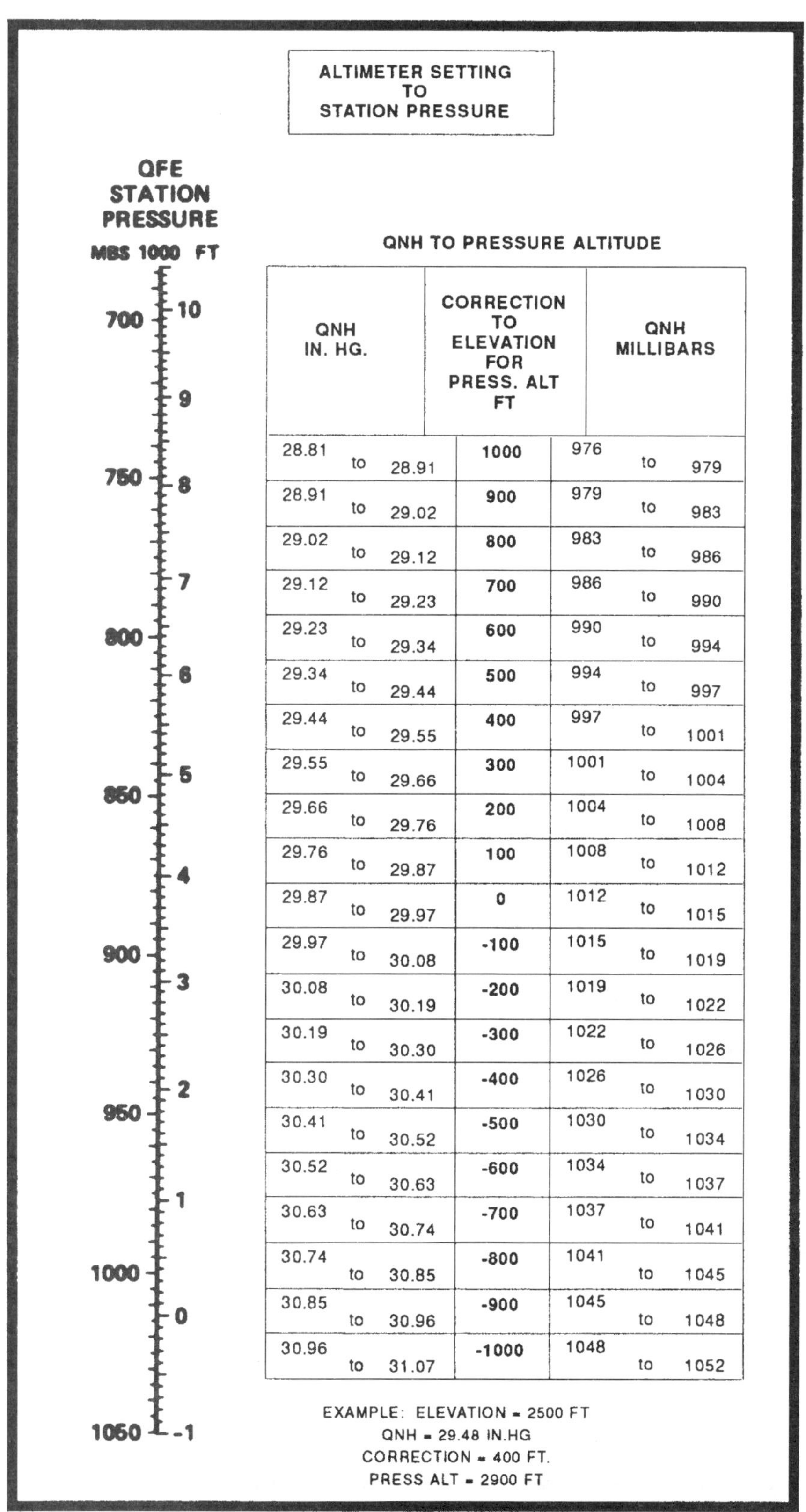

QNH IN. HG.	CORRECTION TO ELEVATION FOR PRESS. ALT FT	QNH MILLIBARS
28.81 to 28.91	1000	976 to 979
28.91 to 29.02	900	979 to 983
29.02 to 29.12	800	983 to 986
29.12 to 29.23	700	986 to 990
29.23 to 29.34	600	990 to 994
29.34 to 29.44	500	994 to 997
29.44 to 29.55	400	997 to 1001
29.55 to 29.66	300	1001 to 1004
29.66 to 29.76	200	1004 to 1008
29.76 to 29.87	100	1008 to 1012
29.87 to 29.97	0	1012 to 1015
29.97 to 30.08	-100	1015 to 1019
30.08 to 30.19	-200	1019 to 1022
30.19 to 30.30	-300	1022 to 1026
30.30 to 30.41	-400	1026 to 1030
30.41 to 30.52	-500	1030 to 1034
30.52 to 30.63	-600	1034 to 1037
30.63 to 30.74	-700	1037 to 1041
30.74 to 30.85	-800	1041 to 1045
30.85 to 30.96	-900	1045 to 1048
30.96 to 31.07	-1000	1048 to 1052

FIGURE 46.—Altimeter Setting to Pressure Altitude.

MODEL DC–9
TAKEOFF SPEEDS
JT8D–1 ENGINES

TAKEOFF SPEED – 20 ° FLAPS EITHER NO ICE PROTECTION OR ENGINE ICE PROTECTION ONLY								
TAKEOFF WEIGHT (1000 LB)	60	65	70	75	80	85	90	95
V_1 (KNOTS, IAS)	104.0	110.0	115.0	120.5	125.0	129.5	133.5	136.0
V_R (KNOTS, IAS)	106.5	112.5	118.0	123.5	129.0	134.0	139.0	143.5
V_2 (KNOTS, IAS)	117.0	121.5	126.5	130.5	135.0	139.0	143.0	147.0

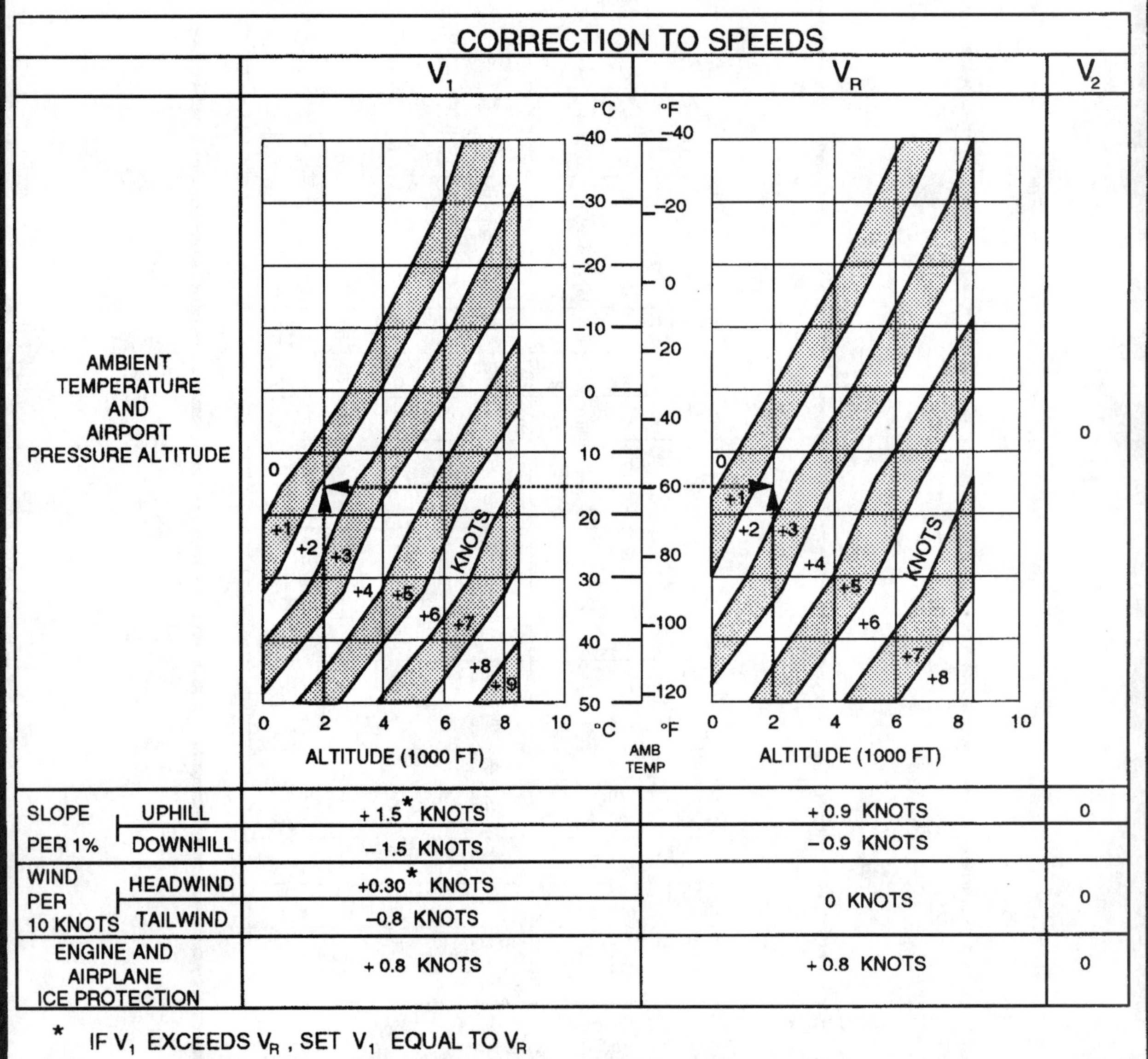

CORRECTION TO SPEEDS				
		V_1	V_R	V_2
AMBIENT TEMPERATURE AND AIRPORT PRESSURE ALTITUDE		(chart)	(chart)	0
SLOPE PER 1%	UPHILL	+ 1.5* KNOTS	+ 0.9 KNOTS	0
	DOWNHILL	– 1.5 KNOTS	– 0.9 KNOTS	
WIND PER 10 KNOTS	HEADWIND	+0.30* KNOTS	0 KNOTS	0
	TAILWIND	–0.8 KNOTS		
ENGINE AND AIRPLANE ICE PROTECTION		+ 0.8 KNOTS	+ 0.8 KNOTS	0

* IF V_1 EXCEEDS V_R, SET V_1 EQUAL TO V_R

FIGURE 47.—DC-9 – Takeoff Speeds.

OPERATING CONDITIONS	W-1	W-2	W-3	W-4	W-5
CLIMB SCHEDULE	LR	HS	LR	HS	HS
INITIAL WEIGHT (X1000)	84	86	78	88	92
CRUISE PRESS ALTITUDE	34,000	28,000	32,000	22,000	24,000
ISA TEMPERATURE	ISA	ISA	ISA	ISA	ISA
AVG WIND COMP (KTS)	20 HW	30 HW	10 TW	20 TW	40 HW

FIGURE 48.—DC-9 – En Route Climb.

TIME, FUEL, AND DISTANCE TO CLIMB
JT8D-1 ENGINES - NORMAL BLEED
DC-9 SERIES 10 - HIGH SPEED CLIMB SCHEDULE
CLIMB AT 320 KNOTS IAS TO 23500 FT ALTITUDE THEN CLIMB AT M .74

INITIAL WEIGHT = 86000. POUNDS				INITIAL WEIGHT = 90000. POUNDS			
PRES. ALT. FEET	TIME MIN.	FUEL BURNED LB.	DIST. N. MI.	PRES. ALT. FEET	TIME MIN.	FUEL BURNED LB.	DIST. N. MI.
0.	0.	0.	0.	0.	0.	0.	0.
2000.	0.5	133.	2.8	2000.	0.6	140.	3.0
4000.	1.1	267.	5.9	4000.	1.1	282.	6.3
6000.	1.7	403.	9.3	6000.	1.8	426.	9.8
8000.	2.3	541.	13.0	8000.	2.5	573.	13.8
10000.	3.0	684.	17.2	10000.	3.2	724.	18.2
12000.	3.8	830.	21.3	12000.	4.0	879.	23.1
14000.	4.6	982.	27.0	14000.	4.8	1041.	28.6
16000.	5.5	1141.	32.9	16000.	5.8	1211.	34.9
18000.	6.4	1309.	39.6	18000.	6.9	1390.	42.1
20000.	7.6	1489.	47.4	20000.	8.0	1583.	50.4
22000.	8.8	1684.	56.6	22000.	9.4	1793.	60.3
23500.	9.9	1845.	64.7	23500.	10.6	1968.	69.1
23500.	9.9	1845.	64.7	23500.	10.6	1968.	69.1
24000.	10.2	1886.	66.8	24000.	10.9	2013.	71.5
26000.	11.4	2052.	75.9	26000.	12.3	2196.	81.5
28000.	12.8	2225.	85.8	28000.	13.8	2389.	92.6
30000.	14.3	2410.	97.1	30000.	15.5	2598.	105.4
32000.	16.2	2613.	110.3	32000.	17.6	2833.	120.6
34000.	18.4	2844.	126.3	34000.	20.3	3110.	139.8
36000.	21.4	3136.	147.8	36000.	24.3	3494.	168.0

INITIAL WEIGHT = 88000. POUNDS				INITIAL WEIGHT = 92000. POUNDS			
0.	0.	0.	0.	0.	0.	0.	0.
2000.	0.5	136.	2.9	2000.	0.6	144.	3.1
4000.	1.1	274.	6.1	4000.	1.2	290.	6.4
6000.	1.7	414.	9.6	6000.	1.8	438.	10.1
8000.	2.4	557.	13.4	8000.	2.5	589.	14.2
10000.	3.1	703.	17.7	10000.	3.3	744.	18.7
12000.	3.9	855.	22.5	12000.	4.1	905.	23.8
14000.	4.7	1012.	27.8	14000.	5.0	1072.	29.5
16000.	5.6	1176.	33.9	16000.	6.0	1247.	36.0
18000.	6.6	1349.	40.8	18000.	7.1	1432.	43.4
20000.	7.8	1535.	48.9	20000.	8.3	1631.	52.0
22000.	9.1	1738.	58.4	22000.	9.7	1850.	62.3
23500.	10.3	1906.	66.9	23500.	11.0	2032.	71.5
23500.	10.3	1906.	66.9	23500.	11.0	2032.	71.5
24000.	10.6	1949.	69.1	24000.	11.3	2079.	73.9
26000.	11.9	2123.	78.6	26000.	12.7	2272.	84.4
28000.	13.3	2306.	89.1	28000.	14.3	2476.	96.2
30000.	14.9	2502.	101.2	30000.	16.2	2693.	109.8
32000.	16.9	2720.	115.3	32000.	18.4	2951.	126.2
34000.	19.3	2973.	132.8	34000.	21.4	3258.	147.4
36000.	22.7	3304.	157.2	36000.	26.1	3713.	181.0

FIGURE 49.—High-Speed Climb Schedule.

TIME, FUEL, AND DISTANCE TO CLIMB
JT8D-1 ENGINES - NORMAL BLEED
DC-9 SERIES 10 - LONG RANGE CLIMB SCHEDULE
CLIMB AT 290 KNOTS IAS TO 26860 FT ALTITUDE THEN CLIMB AT M .72

INITIAL WEIGHT = 78000. POUNDS

PRES. ALT. FEET	TIME MIN.	FUEL BURNED LB.	DIST. N. MI.
0.	0.	0.	0.
2000.	0.5	113.	2.2
4000.	0.9	227.	4.6
6000.	1.5	342.	7.3
8000.	2.0	457.	10.2
10000.	2.6	574.	13.3
12000.	3.2	693.	16.8
14000.	3.9	815.	20.7
16000.	4.6	941.	25.0
18000.	5.4	1070.	29.9
20000.	6.3	1205.	35.4
22000.	7.2	1347.	41.7
24000.	8.3	1498.	49.0
26000.	9.5	1661.	57.6
26860.	10.1	1736.	61.8
26860.	10.1	1736.	61.8
28000.	10.7	1813.	66.2
30000.	11.9	1953.	74.6
32000.	13.3	2102.	84.2
34000.	14.9	2267.	95.4
36000.	16.9	2456.	109.2

INITIAL WEIGHT = 82000. POUNDS

PRES. ALT. FEET	TIME MIN.	FUEL BURNED LB.	DIST. N. MI.
0.	0.	0.	0.
2000.	0.5	120.	2.4
4000.	1.0	241.	4.9
6000.	1.5	363.	7.7
8000.	2.1	486.	10.8
10000.	2.7	610.	14.2
12000.	3.4	737.	17.9
14000.	4.1	868.	22.1
16000.	4.9	1002.	26.7
18000.	5.7	1141.	31.9
20000.	6.7	1286.	37.9
22000.	7.7	1439.	44.6
24000.	8.9	1602.	52.5
26000.	10.2	1780.	61.9
26860.	10.9	1863.	66.5
26860.	10.9	1863.	66.5
28000.	11.6	1948.	71.4
30000.	12.9	2104.	80.8
32000.	14.4	2274.	91.7
34000.	16.3	2464.	104.6
36000.	18.7	2693.	121.3

INITIAL WEIGHT = 80000. POUNDS

PRES. ALT. FEET	TIME MIN.	FUEL BURNED LB.	DIST. N. MI.
0.	0.	0.	0.
2000.	0.5	117.	2.3
4000.	1.0	234.	4.8
6000.	1.5	352.	7.5
8000.	2.1	471.	10.5
10000.	2.7	592.	13.7
12000.	3.3	715.	17.4
14000.	4.0	841.	21.4
16000.	4.7	971.	25.9
18000.	5.6	1105.	30.9
20000.	6.5	1245.	36.6
22000.	7.5	1392.	43.2
24000.	8.6	1549.	50.7
26000.	9.9	1719.	59.7
26860.	10.5	1798.	64.1
26860.	10.5	1798.	64.1
28000.	11.1	1879.	68.7
30000.	12.4	2027.	77.7
32000.	13.8	2186.	87.8
34000.	15.6	2362.	99.8
36000.	17.7	2570.	114.9

INITIAL WEIGHT = 84000. POUNDS

PRES. ALT. FEET	TIME MIN.	FUEL BURNED LB.	DIST. N. MI.
0.	0.	0.	0.
2000.	0.5	124.	2.4
4000.	1.0	248.	5.1
6000.	1.6	374.	8.0
8000.	2.2	500.	11.1
10000.	2.8	629.	14.6
12000.	3.5	760.	18.5
14000.	4.2	894.	22.8
16000.	5.1	1033.	27.6
18000.	5.9	1177.	33.0
20000.	6.9	1327.	39.1
22000.	8.0	1486.	46.2
24000.	9.2	1656.	54.4
26000.	10.6	1841.	64.1
26860.	11.3	1928.	69.0
26860.	11.3	1928.	69.0
28000.	12.0	2018.	74.1
30000.	13.4	2183.	84.1
32000.	15.0	2364.	95.7
34000.	17.1	2570.	109.7
36000.	19.7	2826.	128.3

FIGURE 50.—Long-Range Climb Schedule.

OPERATING CONDITIONS	L-1	L-2	L-3	L-4	L-5
WEIGHT (START TO ALT)	85,000	70,000	86,000	76,000	82,000
DISTANCE (NAM)	110	190	330	50	240
WIND COMPONENT (KTS)	15 HW	40 TW	50 HW	20 TW	45 HW
HOLDING TIME AT ALT (MIN)	15	15	15	15	15

FIGURE 51.—DC-9 – Alternate Planning.

ALTERNATE PLANNING CHART

DIST. - NAM	20	30	40	50	60	70	80	90	100	110	120	130	140
OPTM. ALT.	2000	3000	4000	5000	6000	7000	8000	9000	10000	11000	12000	13000	14000
TIME:	:16	:17	:19	:20	:22	:23	:25	:26	:28	:29	:30	:32	:33
FUEL	2500	2600	2700	2800	2900	3000	3100	3200	3300	3400	3500	3600	3700
TAS	275	280	283	286	289	292	296	300	303	306	309	312	315
DIST. NAM	150	160	170	180	190	200	210	220	230	240	250	260	270
OPTM. ALT.	15000	16000	17000	18000	19000	20000	21000	22000	23000	24000	25000	26000	27000
TIME:	:35	:36	:38	:39	:40	:42	:43	:45	:46	:48	:49	:50	:52
FUEL	3800	3900	4000	4100	4200	4300	4400	4500	4600	4700	4800	4900	5000
TAS	319	323	326	330	334	338	341	345	349	353	357	361	365
DIST. - NAM	280	290	300	310	320	330	340	350	360	370	380	390	400
OPTM. ALT.	27000	28000	28000	29000	29000	30000	30000	31000	31000	31000	31000	31000	31000
TIME:	:53	:55	:56	:58	:59	1:00	1:02	1:03	1:04	1:05	1:07	1:08	1:10
FUEL	5150	5250	5350	5450	5600	5700	5800	5900	6050	6150	6250	6350	6500
TAS	368	372	376	380	385	388	392	397	397	397	397	397	397

NOTES:

1. Fuel includes 1/2 climb distance en route credit, fuel to cruise remaining distance at LRC schedule, 15 minutes holding at alternate, and 800 lbs. for descent.
2. Time includes 1/2 climb distance credit, time to cruise distance shown at LRC schedule and 8 minutes for descent. 15 minutes holding is not included in time.

FIGURE 52.—DC-9 – Alternate Planning Chart.

OPERATING CONDITIONS	R-1	R-2	R-3	R-4	R-5
FIELD ELEVATION	100	4,000	950	2,000	50
ALTIMETER SETTING	29.50"	1032 mb	29.40"	1017 mb	30.15"
TEMPERATURE (OAT)	+50 °F	−15 °C	+59 °F	0 °C	+95 °F
WEIGHT (X1000)	90	110	100	85	95
FLAP POSITION	15°	5°	5°	1°	1°
WIND COMPONENT (KTS)	5 HW	5 TW	20 HW	10 TW	7 HW
RUNWAY SLOPE %	1% UP	1% DN	1% UP	2% DN	1.5% UP
AIR CONDITIONING	ON	ON	OFF	ON	OFF
ENGINE ANTI-ICE	OFF	ON	OFF	ON	OFF
CG STATION	635.7	643.8	665.2	657.2	638.4
LEMAC STA 625.0, MAC 134.0					

FIGURE 53.—B-737 – Takeoff.

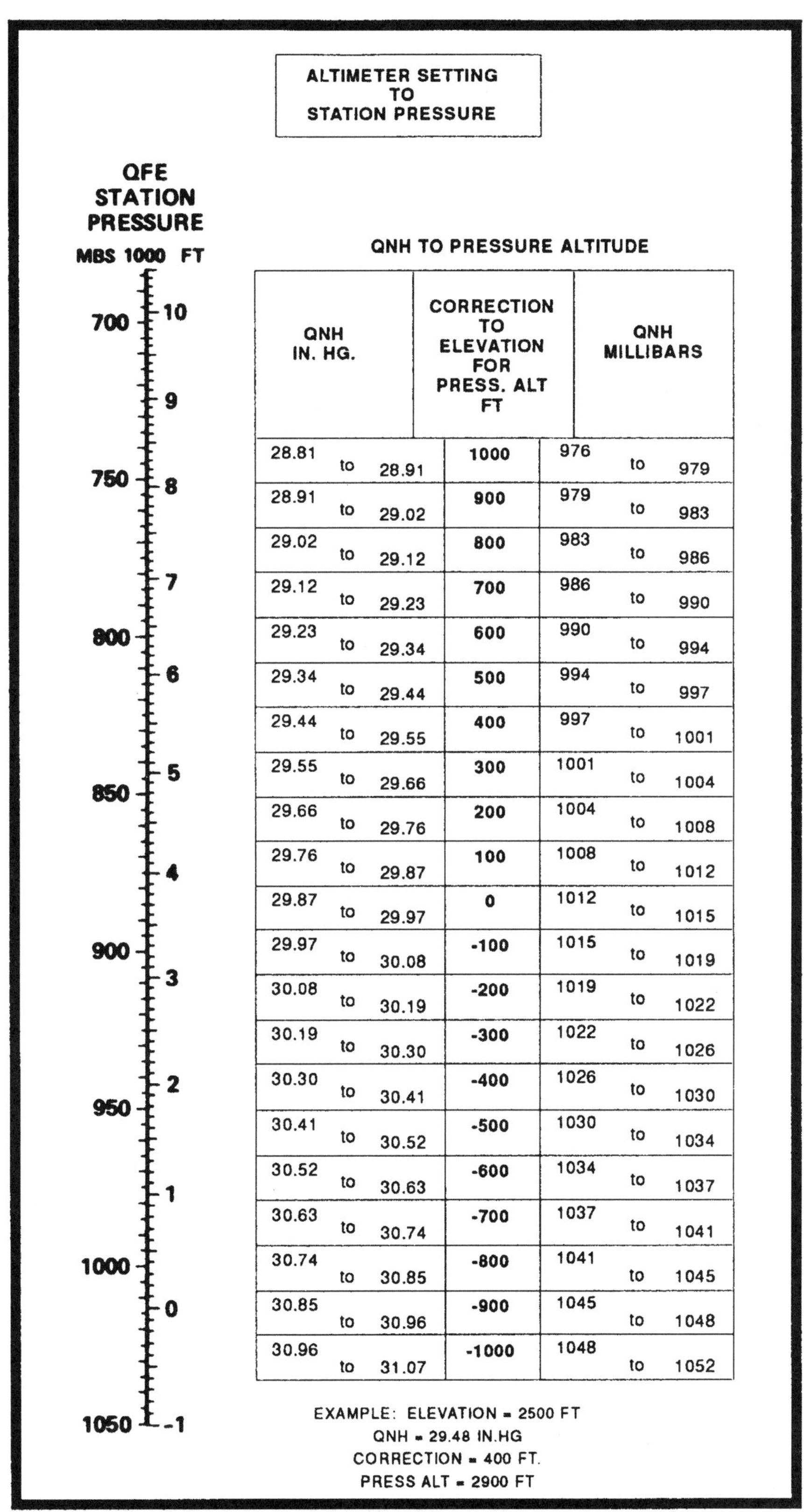

QNH IN. HG.			CORRECTION TO ELEVATION FOR PRESS. ALT FT	QNH MILLIBARS		
28.81	to	28.91	1000	976	to	979
28.91	to	29.02	900	979	to	983
29.02	to	29.12	800	983	to	986
29.12	to	29.23	700	986	to	990
29.23	to	29.34	600	990	to	994
29.34	to	29.44	500	994	to	997
29.44	to	29.55	400	997	to	1001
29.55	to	29.66	300	1001	to	1004
29.66	to	29.76	200	1004	to	1008
29.76	to	29.87	100	1008	to	1012
29.87	to	29.97	0	1012	to	1015
29.97	to	30.08	-100	1015	to	1019
30.08	to	30.19	-200	1019	to	1022
30.19	to	30.30	-300	1022	to	1026
30.30	to	30.41	-400	1026	to	1030
30.41	to	30.52	-500	1030	to	1034
30.52	to	30.63	-600	1034	to	1037
30.63	to	30.74	-700	1037	to	1041
30.74	to	30.85	-800	1041	to	1045
30.85	to	30.96	-900	1045	to	1048
30.96	to	31.07	-1000	1048	to	1052

EXAMPLE: ELEVATION = 2500 FT
QNH = 29.48 IN.HG
CORRECTION = 400 FT.
PRESS ALT = 2900 FT

FIGURE 54.—Altimeter Setting to Pressure Altitude.

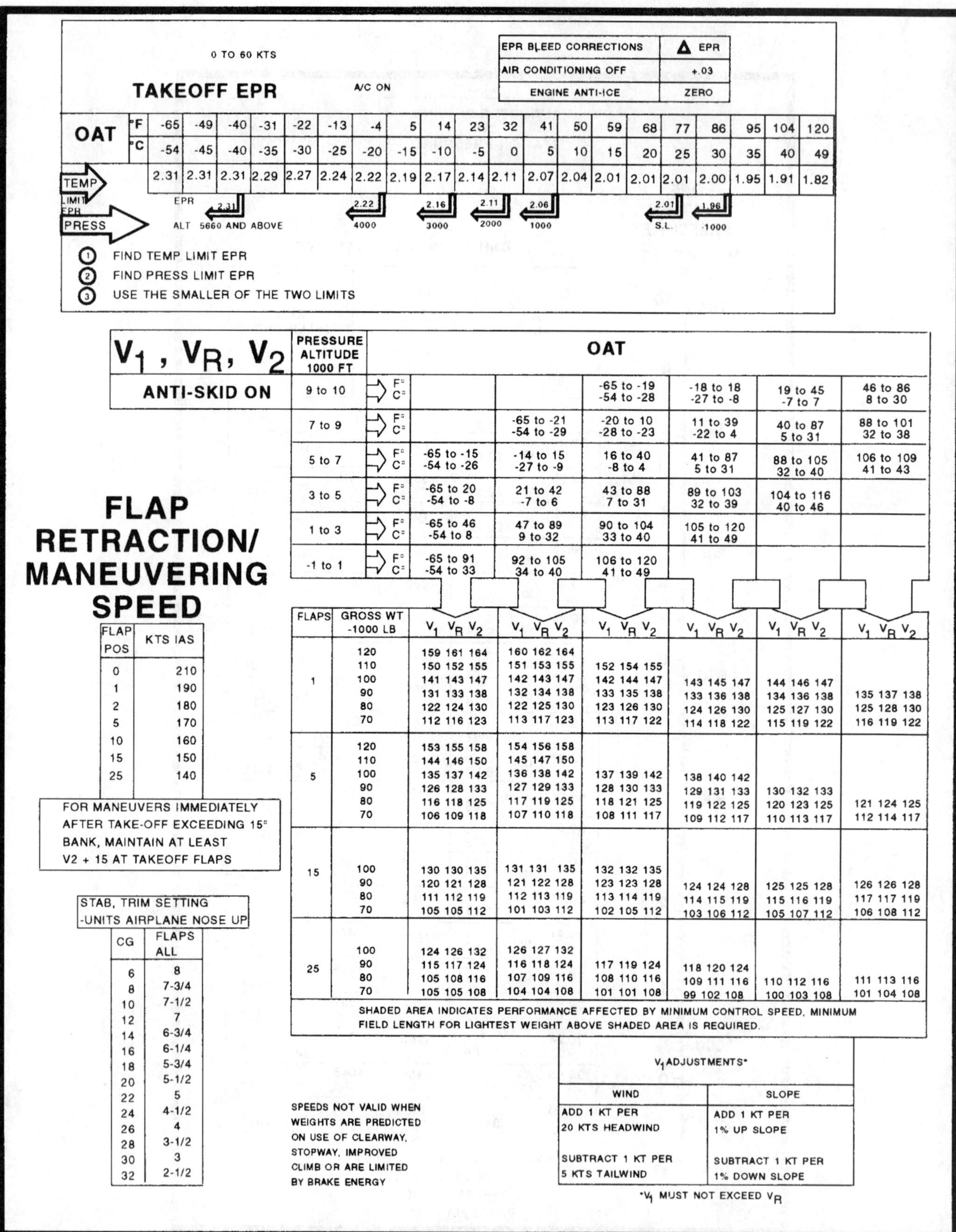

TAKEOFF EPR

0 TO 60 KTS

A/C ON

EPR BLEED CORRECTIONS	Δ EPR
AIR CONDITIONING OFF	+.03
ENGINE ANTI-ICE	ZERO

OAT °F	-65	-49	-40	-31	-22	-13	-4	5	14	23	32	41	50	59	68	77	86	95	104	120
OAT °C	-54	-45	-40	-35	-30	-25	-20	-15	-10	-5	0	5	10	15	20	25	30	35	40	49
TEMP LIMIT EPR	2.31	2.31	2.31	2.29	2.27	2.24	2.22	2.19	2.17	2.14	2.11	2.07	2.04	2.01	2.01	2.01	2.00	1.95	1.91	1.82

PRESS LIMIT EPR	2.31	2.22	2.16	2.11	2.06	2.01	1.96
ALT	5660 AND ABOVE	4000	3000	2000	1000	S.L.	-1000

1. FIND TEMP LIMIT EPR
2. FIND PRESS LIMIT EPR
3. USE THE SMALLER OF THE TWO LIMITS

V_1, V_R, V_2

ANTI-SKID ON

PRESSURE ALTITUDE 1000 FT	OAT					
9 to 10 (F°/C°)			-65 to -19 / -54 to -28	-18 to 18 / -27 to -8	19 to 45 / -7 to 7	46 to 86 / 8 to 30
7 to 9 (F°/C°)		-65 to -21 / -54 to -29	-20 to 10 / -28 to -23	11 to 39 / -22 to 4	40 to 87 / 5 to 31	88 to 101 / 32 to 38
5 to 7 (F°/C°)	-65 to -15 / -54 to -26	-14 to 15 / -27 to -9	16 to 40 / -8 to 4	41 to 87 / 5 to 31	88 to 105 / 32 to 40	106 to 109 / 41 to 43
3 to 5 (F°/C°)	-65 to 20 / -54 to -8	21 to 42 / -7 to 6	43 to 88 / 7 to 31	89 to 103 / 32 to 39	104 to 116 / 40 to 46	
1 to 3 (F°/C°)	-65 to 46 / -54 to 8	47 to 89 / 9 to 32	90 to 104 / 33 to 40	105 to 120 / 41 to 49		
-1 to 1 (F°/C°)	-65 to 91 / -54 to 33	92 to 105 / 34 to 40	106 to 120 / 41 to 49			

FLAPS	GROSS WT -1000 LB	V_1 V_R V_2	V_1 V_R V_2	V_1 V_R V_2	V_1 V_R V_2	V_1 V_R V_2	V_1 V_R V_2
1	120	159 161 164	160 162 164				
	110	150 152 155	151 153 155	152 154 155			
	100	141 143 147	142 143 147	142 144 147	143 145 147	144 146 147	
	90	131 133 138	132 134 138	133 135 138	133 136 138	134 136 138	135 137 138
	80	122 124 130	122 125 130	123 126 130	124 126 130	125 127 130	125 128 130
	70	112 116 123	113 117 123	113 117 122	114 118 122	115 119 122	116 119 122
5	120	153 155 158	154 156 158				
	110	144 146 150	145 147 150				
	100	135 137 142	136 138 142	137 139 142	138 140 142		
	90	126 128 133	127 129 133	128 130 133	129 131 133	130 132 133	
	80	116 118 125	117 119 125	118 121 125	119 122 125	120 123 125	121 124 125
	70	106 109 118	107 110 118	108 111 117	109 112 117	110 113 117	112 114 117
15	100	130 130 135	131 131 135	132 132 135			
	90	120 121 128	121 122 128	123 123 128	124 124 128	125 125 128	126 126 128
	80	111 112 119	112 113 119	113 114 119	114 115 119	115 116 119	117 117 119
	70	105 105 112	101 103 112	102 105 112	103 106 112	105 107 112	106 108 112
25	100	124 126 132	126 127 132				
	90	115 117 124	116 118 124	117 119 124	118 120 124		
	80	105 108 116	107 109 116	108 110 116	109 111 116	110 112 116	111 113 116
	70	105 105 108	104 104 108	101 101 108	99 102 108	100 103 108	101 104 108

SHADED AREA INDICATES PERFORMANCE AFFECTED BY MINIMUM CONTROL SPEED. MINIMUM FIELD LENGTH FOR LIGHTEST WEIGHT ABOVE SHADED AREA IS REQUIRED.

FLAP RETRACTION/ MANEUVERING SPEED

FLAP POS	KTS IAS
0	210
1	190
2	180
5	170
10	160
15	150
25	140

FOR MANEUVERS IMMEDIATELY AFTER TAKE-OFF EXCEEDING 15° BANK, MAINTAIN AT LEAST V2 + 15 AT TAKEOFF FLAPS

STAB, TRIM SETTING -UNITS AIRPLANE NOSE UP

CG	FLAPS ALL
6	8
8	7-3/4
10	7-1/2
12	7
14	6-3/4
16	6-1/4
18	5-3/4
20	5-1/2
22	5
24	4-1/2
26	4
28	3-1/2
30	3
32	2-1/2

SPEEDS NOT VALID WHEN WEIGHTS ARE PREDICTED ON USE OF CLEARWAY, STOPWAY, IMPROVED CLIMB OR ARE LIMITED BY BRAKE ENERGY

V_1 ADJUSTMENTS*

WIND	SLOPE
ADD 1 KT PER 20 KTS HEADWIND	ADD 1 KT PER 1% UP SLOPE
SUBTRACT 1 KT PER 5 KTS TAILWIND	SUBTRACT 1 KT PER 1% DOWN SLOPE

*V_1 MUST NOT EXCEED V_R

FIGURE 55.—B-737 – Takeoff Performance.

OPERATING CONDITIONS	V-1	V-2	V-3	V-4	V-5
BRK REL WEIGHT (X1000)	110	95	85	105	75
CRUISE PRESS ALT	33,000	27,000	35,000	22,000	31,000
AIRPORT ELEVATION	2,000	3,000	2,000	4,000	2,000
ISA TEMPERATURE	+10°	ISA	ISA	+10°	+10°
AVG WIND COMP (KTS)	20 HW	20 TW	30 HW	10 TW	40 HW

FIGURE 56.—B-737 – En Route Climb.

EN ROUTE CLIMB 280/.70 ISA

PRESSURE ALTITUDE -FT	UNITS MIN/LB NM/KNOTS	BRAKE RELEASE WEIGHT - LB										
		120000	115000	110000	105000	100000	95000	90000	85000	80000	75000	65000
37000	TIME/FUEL DIST./TAS		41/5700 251/387	32/4700 192/384	27/4100 162/382	24/3700 140/380	21/3400 124/379	19/3100 111/378	17/2800 100/377	16/2500 90/376	14/2300 82/375	12/1900 67/374
36000	TIME/FUEL DIST./TAS	41/ 5900 246/386	33/4900 194/383	28/4300 164/381	25/3900 143/379	22/3500 127/378	20/3200 114/377	18/2900 103/376	16/2700 93/375	15/2500 84/374	14/2300 77/374	11/1900 63/373
35000	TIME/FUEL DIST./TAS	33/5100 197/382	29/4500 168/380	25/4100 147/378	23/3700 131/377	21/3400 117/376	19/3100 106/375	17/2800 96/374	16/2600 87/373	14/2400 80/373	13/2200 73/372	11/1800 60/371
34000	TIME/FUEL DIST./TAS	29/4700 171/379	26/4300 150/377	23/3900 134/376	21/3500 120/375	19/3200 109/374	18/3000 99/373	16/2700 90/372	15/2500 82/372	14/2300 75/371	12/2100 69/371	10/1800 57/370
33000	TIME/FUEL DIST./TAS	27/4400 153/376	24/4000 137/375	22/3700 123/374	20/3400 112/373	18/3100 102/372	17/2900 93/371	15/2700 85/370	14/2500 78/370	13/2300 71/369	12/2100 65/369	10/1700 54/368
32000	TIME/FUEL DIST./TAS	25/4200 139/374	23/3900 126/372	21/3600 114/371	19/3300 104/370	17/3000 95/370	16/2800 87/369	15/2600 80/368	14/2400 74/368	12/2200 67/367	11/2000 62/367	10/1700 51/366
31000	TIME/FUEL DIST./TAS	23/4000 128/371	21/3700 117/370	19/3400 107/369	18/3200 98/368	16/2900 90/367	15/2700 82/367	14/2500 76/366	13/2300 70/366	12/2100 64/365	11/2000 59/365	9/1700 49/364
30000	TIME/FUEL DIST./TAS	22/3900 119/368	20/3600 109/367	18/3300 100/366	17/3100 92/365	16/2800 84/365	15/2600 78/364	13/2400 72/364	12/2300 66/363	11/2100 61/363	11/1900 56/363	9/1600 47/362
29000	TIME/FUEL DIST./TAS	21/3700 111/365	19/3400 102/364	18/3200 93/363	16/3000 86/363	15/2700 79/362	14/2500 73/362	13/2400 68/361	12/2200 62/361	11/2000 57/361	10/1900 53/360	9/1600 44/360
28000	TIME/FUEL DIST./TAS	19/3600 103/362	18/3300 95/361	17/3100 88/360	15/2900 81/360	14/2700 75/359	13/2500 69/359	12/2300 64/359	11/2100 59/358	11/2000 54/358	10/1800 50/358	8/1500 42/357
27000	TIME/FUEL DIST./TAS	19/3400 96/358	17/3200 89/358	16/3000 82/357	15/2800 76/357	14/2600 71/356	13/2400 65/356	12/2200 60/356	11/2100 56/356	10/1900 52/355	9/1800 47/355	8/1500 40/355
26000	TIME/FUEL DIST./TAS	17/3300 88/354	16/3000 82/354	15/2800 76/353	14/2600 70/353	13/2500 65/352	12/2300 60/352	11/2100 56/352	10/2000 52/352	10/1800 48/351	9/1700 44/351	7/1400 37/351
25000	TIME/FUEL DIST./TAS	16/3100 81/350	15/2900 75/350	14/2700 70/349	13/2500 65/349	12/2400 60/349	11/2200 56/348	11/2000 52/348	10/1900 48/348	9/1800 45/348	8/1600 41/348	7/1400 35/347
24000	TIME/FUEL DIST./TAS	15/3000 75/346	14/2800 69/346	13/2600 65/345	12/2400 60/345	12/2300 56/345	11/2100 52/345	10/2000 48/345	9/1800 45/344	9/1700 41/344	8/1600 38/344	7/1300 32/344
23000	TIME/FUEL DIST./TAS	14/2800 69/342	13/2700 64/342	13/2500 60/342	12/2300 56/342	11/2200 52/342	10/2000 48/341	9/1900 45/341	9/1800 41/341	8/1600 38/341	8/1500 35/341	6/1300 30/341
22000	TIME/FUEL DIST./TAS	14/2700 63/339	13/2500 59/339	12/2400 55/338	11/2200 51/338	10/2100 48/338	10/1900 45/338	9/1800 41/338	8/1700 38/338	8/1600 36/338	7/1400 33/338	6/1200 28/337
6000	TIME/FUEL DIST./TAS	4/1000 9/295	4/1000 9/295	4/900 8/295	4/800 8/295	3/800 7/295	3/700 7/295	3/700 6/295	3/700 6/295	3/600 5/295	2/600 5/295	2/500 4/295
1500	TIME/FUEL	2/600	2/600	2/500	2/500	2/500	2/400	2/400	2/400	1/400	1/300	1/300

FUEL ADJUSTMENT FOR HIGH ELEVATION AIRPORTS	AIRPORT ELEVATION	2000	4000	6000	8000	10000	12000
EFFECT ON TIME AND DISTANCE IS NEGLIGIBLE	FUEL ADJUSTMENT	-100	-200	-400	-500	-600	-700

FIGURE 57.—En Route Climb 280/.70 ISA.

EN ROUTE CLIMB 280/.70 ISA +10 °C

PRESSURE ALTITUDE -FT	UNITS MIM/LB NM/KNOTS	BRAKE RELEASE WEIGHT - LB										
		120000	115000	110000	105000	100000	95000	90000	85000	80000	75000	65000
37000	TIME/FUEL DIST./TAS			42/5700 263/395	34/4700 206/391	29/4100 174/389	25/3700 151/388	23/3300 133/386	20/3000 119/385	18/2700 107/384	16/2500 96/384	13/2100 78/382
36000	TIME/FUEL DIST./TAS		43/5900 266/394	35/5000 211/391	30/4400 179/389	26/3900 156/387	23/3500 138/385	21/3200 123/384	19/2900 111/383	17/2700 100/383	16/2400 90/382	13/2000 74/381
35000	TIME/FUEL DIST./TAS	45/6200 275/394	36/5300 219/390	31/4600 186/388	27/4100 162/386	24/3700 143/385	22/3400 128/384	20/3100 115/383	10/2800 104/382	16/2600 94/381	15/2400 85/380	12/2000 70/379
34000	TIME/FUEL DIST./TAS	38/5600 228/390	32/4900 193/387	28/4400 168/386	25/3900 149/384	23/3600 133/383	21/3300 120/382	19/3000 108/381	17/2700 98/380	16/2500 89/379	14/2300 81/379	12/1900 67/378
33000	TIME/FUEL DIST./TAS	34/5100 200/387	30/4600 174/385	26/4100 154/383	24/3800 138/382	22/3400 124/381	20/3100 113/380	18/2900 102/379	16/2600 93/378	15/2400 85/378	14/2200 77/377	11/1900 64/376
32000	TIME/FUEL DIST./TAS	31/4800 180/384	28/4400 160/382	25/4000 143/381	23/3600 129/379	21/3300 116/378	19/3000 106/378	17/2800 96/377	16/2600 88/376	14/2400 80/376	13/2200 73/375	11/1800 61/374
31000	TIME/FUEL DIST./TAS	29/4600 165/381	26/4200 147/379	23/3800 133/378	21/3500 120/377	20/3200 109/376	18/2900 100/375	16/2700 91/375	15/2500 83/374	14/2300 76/374	13/2100 70/373	11/1800 58/372
30000	TIME/FUEL DIST./TAS	27/4400 152/378	24/4000 137/376	22/3700 124/375	20/3400 113/374	19/3100 103/374	17/2900 94/373	16/2600 86/372	14/2400 79/372	13/2200 72/371	12/2100 66/371	10/1700 55/370
29000	TIME/FUEL DIST./TAS	25/4200 141/375	23/3800 128/374	21/3500 116/373	19/3200 106/372	18/3000 97/371	16/2800 89/370	15/2600 82/370	14/2400 75/369	13/2200 69/369	12/2000 63/369	10/1700 52/368
28000	TIME/FUEL DIST./TAS	24/4000 131/371	22/3700 119/370	20/3400 109/369	18/3100 100/369	17/2900 91/368	16/2700 84/368	14/2500 77/367	13/2300 71/367	12/2100 65/366	11/1900 60/366	9/1600 50/365
27000	TIME/FUEL DIST./TAS	22/3800 121/368	21/3500 111/367	19/3300 102/366	18/3000 93/366	16/2800 86/365	15/2600 79/364	14/2400 73/364	13/2200 67/364	12/2000 61/363	11/1900 56/363	9/1600 47/363
26000	TIME/FUEL DIST./TAS	21/3600 110/363	19/3400 101/362	18/3100 93/362	16/2900 86/361	15/2700 79/361	14/2500 73/360	13/2300 67/360	12/2100 62/360	11/2000 57/359	10/1800 52/359	9/1500 44/359
25000	TIME/FUEL DIST./TAS	19/3400 101/358	18/3200 93/358	17/3000 85/357	15/2800 79/357	14/2600 73/357	13/2400 67/356	12/2200 62/356	11/2000 57/356	10/1900 53/356	10/1700 48/355	8/1500 41/355
24000	TIME/FUEL DIST./TAS	18/3300 92/354	17/3000 85/354	16/2800 78/353	15/2600 72/353	13/2400 67/353	12/2300 62/352	12/2100 57/352	11/1900 53/352	10/1800 49/352	9/1700 45/352	8/1400 38/351
23000	TIME/FUEL DIST./TAS	17/3100 84/350	16/2900 78/350	15/2900 72/350	14/2500 67/349	13/2300 62/349	12/2200 57/349	11/2000 53/349	10/1900 49/348	9/1700 45/348	9/1600 42/348	7/1300 35/348
22000	TIME/FUEL DIST./TAS	16/3000 77/346	15/2800 71/346	14/2600 66/346	13/2400 61/346	12/2200 57/345	11/2100 53/345	10/1900 49/345	10/1800 45/345	9/1700 42/345	8/1500 38/345	7/1300 32/344
6000	TIME/FUEL DIST./TAS	5/1100 10/301	4/1000 10/301	4/900 9/301	4/900 9/301	4/800 8/301	3/800 8/301	3/700 7/301	3/700 7/301	3/600 6/301	3/600 6/301	2/500 5/301
1500	TIME/FUEL	3/600	2/600	2/500	2/500	2/500	2/500	2/400	2/400	2/400	1/300	1/300

FUEL ADJUSTMENT FOR HIGH ELEVATION AIRPORTS	AIRPORT ELEVATION	2000	4000	6000	8000	10000	12000
EFFECT ON TIME AND DISTANCE IS NEGLIGIBLE	FUEL ADJUSTMENT	-100	-300	-400	-500	-600	-800

FIGURE 58.—En Route Climb 280/.70 ISA +10 °C.

OPERATING CONDITIONS	T-1	T-2	T-3	T-4	T-5
TOTAL AIR TEMP (TAT)	+10 °C	0 °C	–15 °C	–30 °C	+15 °C
ALTITUDE	10,000	5,000	25,000	35,000	18,000
ENGINE ANTI-ICE	ON	ON	ON	ON	OFF
WING ANTI-ICE	OFF	2 ON	2 ON	1 ON	OFF
AIR CONDITIONING	ON	OFF	ON	ON	OFF

FIGURE 59.—B-737 – Climb and Cruise Power.

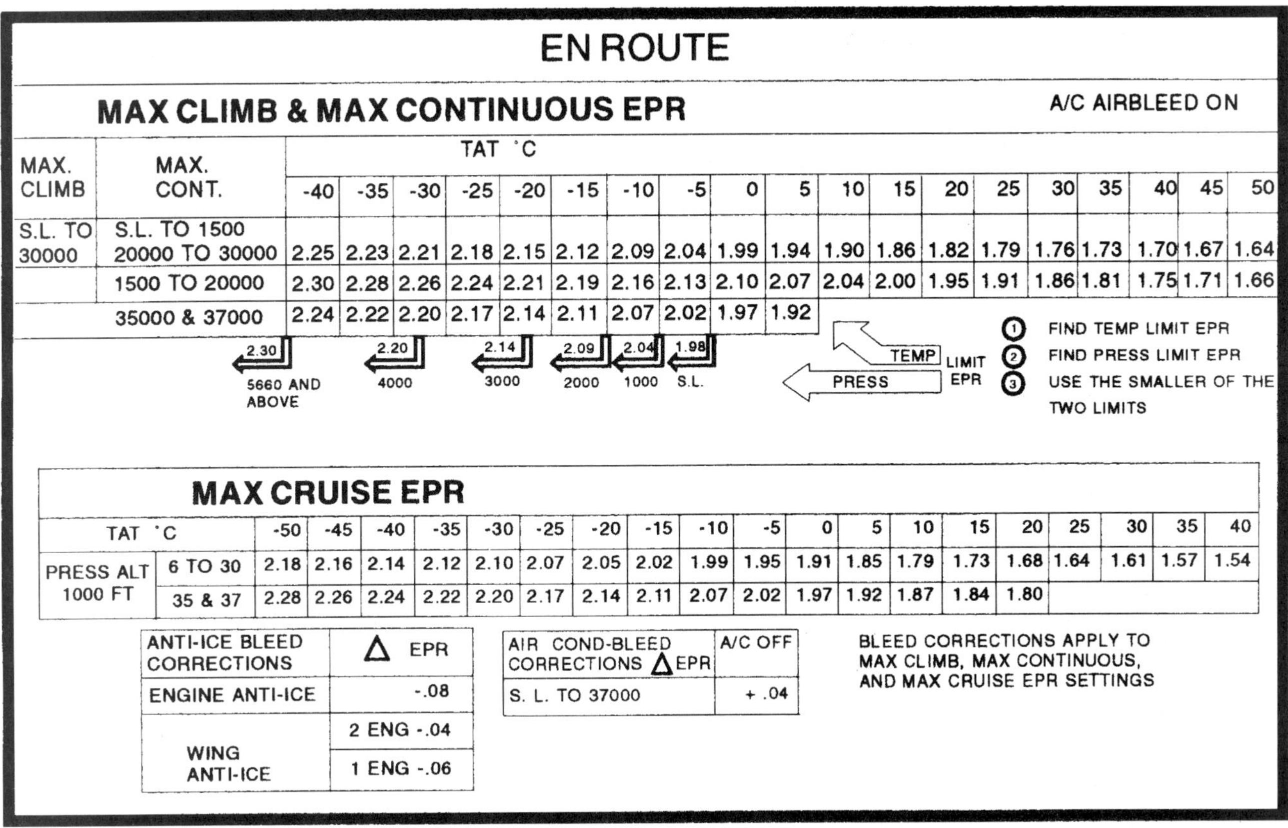

EN ROUTE

MAX CLIMB & MAX CONTINUOUS EPR

A/C AIRBLEED ON

MAX. CLIMB	MAX. CONT.	TAT °C -40	-35	-30	-25	-20	-15	-10	-5	0	5	10	15	20	25	30	35	40	45	50
S.L. TO 30000	S.L. TO 1500 20000 TO 30000	2.25	2.23	2.21	2.18	2.15	2.12	2.09	2.04	1.99	1.94	1.90	1.86	1.82	1.79	1.76	1.73	1.70	1.67	1.64
	1500 TO 20000	2.30	2.28	2.26	2.24	2.21	2.19	2.16	2.13	2.10	2.07	2.04	2.00	1.95	1.91	1.86	1.81	1.75	1.71	1.66
	35000 & 37000	2.24	2.22	2.20	2.17	2.14	2.11	2.07	2.02	1.97	1.92									

MAX CRUISE EPR

TAT °C		-50	-45	-40	-35	-30	-25	-20	-15	-10	-5	0	5	10	15	20	25	30	35	40
PRESS ALT 1000 FT	6 TO 30	2.18	2.16	2.14	2.12	2.10	2.07	2.05	2.02	1.99	1.95	1.91	1.85	1.79	1.73	1.68	1.64	1.61	1.57	1.54
	35 & 37	2.28	2.26	2.24	2.22	2.20	2.17	2.14	2.11	2.07	2.02	1.97	1.92	1.87	1.84	1.80				

ANTI-ICE BLEED CORRECTIONS		Δ EPR
ENGINE ANTI-ICE		-.08
WING ANTI-ICE	2 ENG	-.04
	1 ENG	-.06

AIR COND-BLEED CORRECTIONS Δ EPR	A/C OFF
S. L. TO 37000	+ .04

BLEED CORRECTIONS APPLY TO MAX CLIMB, MAX CONTINUOUS, AND MAX CRUISE EPR SETTINGS

FIGURE 60.—B-737 – Climb and Cruise Power.

OPERATING CONDITIONS	X-1	X-2	X-3	X-4	X-5
DISTANCE (NM)	2,000	2,400	1,800	2,800	1,200
WIND COMPONENT (KTS)	50 TW	50 HW	20 HW	50 TW	30 HW
CRUISE PRESS ALTITUDE	27,000	35,000	20,000	29,000	37,000
ISA TEMPERATURE	+10°	ISA	+20°	–10°	+10°
LANDING WEIGHT (X1000)	70	75	75	65	90

FIGURE 61.—Flight Planning at .78 Mach Cruise.

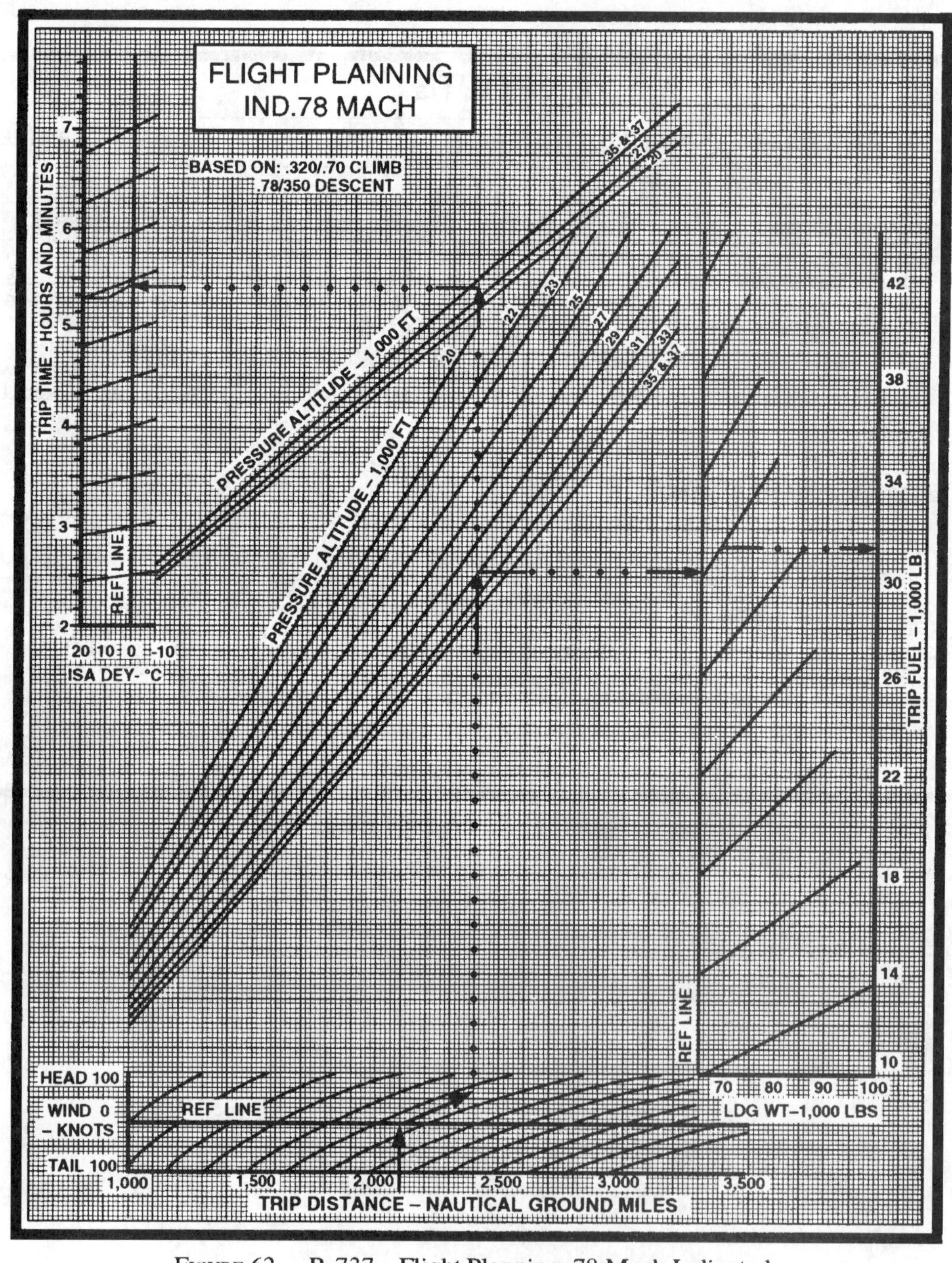

FIGURE 62.—B-737 – Flight Planning .78 Mach Indicated.

OPERATING CONDITIONS	Q–1	Q–2	Q–3	Q–4	Q–5
WEIGHT (X1000)	110	70	90	80	100
PRESSURE ALTITUDE	30,000	25,000	35,000	20,000	10,000
TOTAL AIR TEMP (TAT)	–8 °C	–23 °C	–16 °C	+4 °C	–6 °C

FIGURE 63.—B-737– Turbulent Air RPM.

TURBULENT AIR PENETRATION

TARGET SPEED IAS/MACH	PRESS ALT -1000 FT	GROSS WEIGHT - 1000 LB / APPROXIMATE POWER SETTING -%N1 RPM 70	80	90	100	110	ISA TAT -°C	% N1 ADJUSTMENT PER 10 °C VARIATION FROM TABLE TAT COLDER - WARMER +
280/.70	35	77.1	79.0	81.0	83.4		-36	1.6
	30	77.2	78.2	79.4	81.1	82.4	-23	1.6
	25	76.7	77.5	78.3	79.2	80.1	-13	1.5
	20	74.7	75.4	76.1	77.0	77.9	-6	1.4
	15	72.7	73.5	74.2	74.8	75.7	1	1.2
	10	70.5	71.3	72.1	72.9	73.9	9	1.3

FIGURE 64.—B-737 – Turbulent Air Penetration.

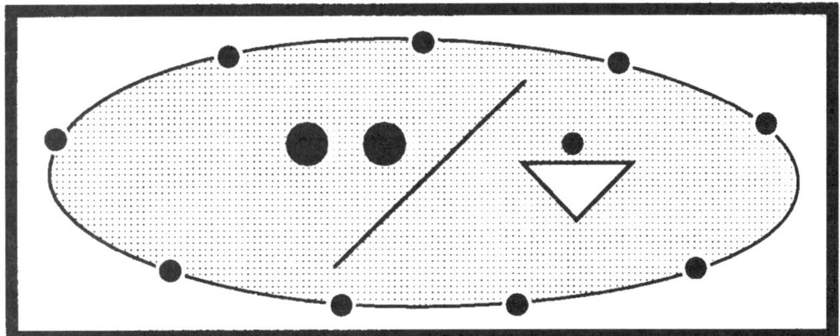

FIGURE 65.—Symbol Used on U.S. Low Level Significant Weather Prog Chart.

OPERATING CONDITIONS	Z-1	Z-2	Z-3	Z-4	Z-5
DISTANCE (NM)	340	650	900	290	400
AVG WIND COMP (KTS)	25 TW	45 HW	35 TW	25 HW	60 HW

FIGURE 66.—Flight Planning at .74 Mach Cruise.

ABBREVIATED FLIGHT PLANNING
.280/.70 CLIMB
.74/320/340 DESCENT
250 KTS CRUISE BELOW 10000 FT.
320 KTS CRUISE 10000 THRU 23000 FT.
.74 MACH CRUISE 24000 FT. AND ABOVE

DIST. N. MI.	REC. ALT.	TAS KTS	AIR TIME MINS.	FUEL LBS.
50	6000-7000	279	16	1800
60	6000-7000	279	18	1950
260	26000-27000	447	44	4600
270	26000-27000	447	45	4750
280	27000-28000	445	47	4850
290	28000-29000	443	48	4950
300	28000-29000	443	49	5100
310	28000-29000	443	51	5200
320	29000-31000	441	52	5300
330	29000-31000	441	53	5400
340	31000-33000	438	55	5550
350	31000-33000	438	56	5650
400	33000-35000	433	62	6250
450	33000-35000	433	69	6850
500	33000-35000	433	76	7500
550	33000-35000	433	82	8100
600	33000-35000	433	89	8700
650	33000-35000	433	96	9300
700	33000-35000	433	102	9900
750	33000-35000	433	109	10500
800	33000-35000	433	115	11100
850	33000-35000	433	122	11700
900	33000-35000	433	129	12300
950	33000-35000	433	135	12900
1000	33000-35000	433	142	13500

TIME AND FUEL CORRECTION FOR WIND
Δ TIME = TIME X WIND COMPONENT ÷ TAS
Δ FUEL = FUEL X WIND COMPONENT ÷ TAS

EXAMPLE: DIST. = 250
STILL AIR TIME = 43 MIN.
STILL AIR FUEL = 4500 LBS.
WIND COMPONENT = 20 KTS.

Δ TIME = 43 X 20 ÷ 449 = MIN.
Δ FUEL = 4500 X 20 = 449 = 200 LBS.

ADD Δ TIME AND Δ FUEL FOR THE HEADWIND; SUBTRACT FOR TAILWIND

FIGURE 67.—Abbreviated Flight Planning.

OPERATING CONDITIONS	O-1	O-2	O-3	O-4	O-5
ALTITUDE	31,000	23,000	17,000	8,000	4,000
WEIGHT (X1000)	102	93	104	113	109
ENGINES OPERATING	2	2	2	2	2
HOLDING TIME (MIN)	20	40	35	15	25

FIGURE 68.—B-737 – Holding.

HOLDING

EPR
IAS KNOTS
FF PER ENGINE LB/HR

FLIGHT LEVEL	GROSS WEIGHT 1000 LB										
	115	110	105	100	95	90	85	80	75	70	65
	2.13	2.07	2.01	1.95	1.90	1.85	1.80	1.76	1.71	1.67	1.64
350	234	228	223	217	211	210	210	210	210	210	210
	2830	2810	2630	2460	2290	2180	2070	1960	1870	1780	1700
	1.86	1.82	1.79	1.75	1.71	1.67	1.64	1.60	1.57	1.54	1.51
300	231	226	220	215	210	210	210	210	210	210	210
	2740	2600	2470	2370	2250	2140	2050	1960	1880	1790	1720
	1.69	1.66	1.63	1.60	1.57	1.54	1.51	1.48	1.45	1.43	1.41
250	229	224	218	213	210	210	210	210	210	210	210
	2710	2610	2490	2370	2260	2180	2080	1980	1920	1840	1780
	1.56	1.53	1.50	1.48	1.45	1.43	1.40	1.38	1.36	1.34	1.32
200	227	222	217	211	210	210	210	210	210	210	210
	2716	2590	2490	2390	2310	2230	2130	2060	2000	1920	1860
	1.45	1.43	1.40	1.38	1.36	1.34	1.32	1.31	1.29	1.27	1.26
150	226	221	216	210	210	210	210	210	210	210	210
	2790	2680	2570	2470	2380	2290	2220	2140	2070	2000	1990
	1.36	1.34	1.33	1.31	1.29	1.28	1.26	1.25	1.24	1.22	1.21
100	225	220	215	210	210	210	210	210	210	210	210
	2860	2780	2670	2560	2470	2390	2310	2240	2170	2100	2030
	1.29	1.28	1.27	1.25	1.24	1.23	1.21	1.20	1.19	1.18	1.17
050	224	219	214	210	210	210	210	210	210	210	210
	2960	2870	2770	2670	2580	2500	2420	2350	2290	2230	2150
	1.25	1.24	1.23	1.22	1.21	1.20	1.19	1.18	1.17	1.16	1.15
015	224	219	214	210	210	210	210	210	210	210	210
	3050	2950	2850	2790	2670	2590	2510	2430	2370	2300	2240

FIGURE 69.—B-737 – Holding Performance Chart.

INITIAL FUEL WEIGHT 1000 LB	ENDING FUEL WEIGHT - 1000 LB															
	10	14	18	22	26	30	34	38	42	46	50	54	58	62	64	70
70	28	27	25	23	22	20	18	17	15	13	12	10	8	5	3	0
66	26	25	23	21	20	18	16	15	13	12	10	8	5	3	0	
62	23	23	20	18	17	15	13	11	10	8	7	5	3	0		
58	21	20	18	16	15	13	11	10	8	6	5	3	0			
54	18	16	15	13	12	10	8	7	5	3	2	0				
50	16	15	13	12	10	8	7	5	3	2	0					
46	15	13	12	10	8	7	5	3	2	0						
42	13	12	10	8	7	5	3	2	0		FUEL DUMP TIME					
38	12	10	8	7	5	3	2	0								
34	10	8	7	5	3	2	0									
30	8	7	5	3	2	0										
26	7	5	3	2	0											
22	5	3	2	0												
18	3	2	0						FUEL JETTISON TIME-MINUTES							
14	2	0														
10	0															

FIGURE 70.—Fuel Dump Time.

OPERATING CONDITIONS	D-1	D-2	D-3	D-4	D-5
WT AT ENG FAIL (X1000)	100	110	90	80	120
ENGINE ANTI-ICE	ON	OFF	ON	ON	ON
WING ANTI-ICE	OFF	OFF	ON	ON	OFF
ISA TEMPERATURE	ISA	+10°	−10°	−10°	+20°
AIR CONDITIONING	OFF	OFF	OFF	OFF	OFF

FIGURE 71. —B-737– Drift-Down.

1 ENGINE INOP

ENGINE A/I OFF

GROSS WEIGHT 1000 LB		OPTIMUM DRIFTDOWN SPEED KIAS	ISA DEV °C			
AT ENGINE FAILURE	AT LEVEL OFF (APPROX)		-10	0	10	20
			APPROX GROSS LEVEL OFF PRESS ALT FT			
80	77	184	27900	26800	25400	22800
90	86	195	25000	23800	21700	20000
100	96	206	22000	20500	20000	18500
110	105	216	20000	19100	17500	15400
120	114	224	18200	16600	14700	12200

ENGINE A/I ON

GROSS WEIGHT 1000 LB		OPTIMUM DRIFTDOWN SPEED KIAS	ISA DEV °C			
AT ENGINE FAILURE	AT LEVEL OFF (APPROX)		-10	0	10	20
			APPROX GROSS LEVEL OFF PRESS ALT FT			
80	77	184	25500	24600	22800	20000
90	86	195	23000	21400	20000	19400
100	96	206	20000	19400	18700	15600
110	105	216	18100	16600	14700	12200
120	114	224	15500	13800	11800	8800

ENGINE AND WING A/I ON

GROSS WEIGHT 1000 LB		OPTIMUM DRIFTDOWN SPEED KIAS	ISA DEV °C			
AT ENGINE FAILURE	AT LEVEL OFF (APPROX)		-10	0	10	20
			APPROX GROSS LEVEL OFF PRESS ALT FT			
80	77	184	24400	23400	21400	20000
90	86	195	21600	20100	19800	18000
100	96	206	19600	18000	16400	14200
110	105	216	16800	15100	13300	10700
120	114	224	14000	12200	10300	7200

NOTE:

WHEN ENGINE BLEED FOR AIR CONDITIONING IS OFF BELOW 17,000 FT., INCREASE LEVEL-OFF ALTITUDE BY 800 FT.

FIGURE 72.—Drift-Down Performance Chart.

OPERATING CONDITIONS	L-1	L-2	L-3	L-4	L-5
TEMPERATURE	+15 °C TAT	+27 °F OAT	−8 °C OAT	−10 °C TAT	+55 °F OAT
PRESSURE ALTITUDE	500	3,100	2,500	2,100	1,200
AIR CONDITIONING	OFF	ON	ON	ON	ON
WING ANTI-ICE	OFF	2 ON	1 ON	2 ON	OFF
WEIGHT (X1000)	100	95	90	105	85
FLAP SETTING	30°	25°	15°	40°	30°
RUNWAY ASSIGNED	35	04	27	34	09
SURFACE WIND	300/20	350/15	310/20	030/10	130/15

FIGURE 73.—B-737 – Landing.

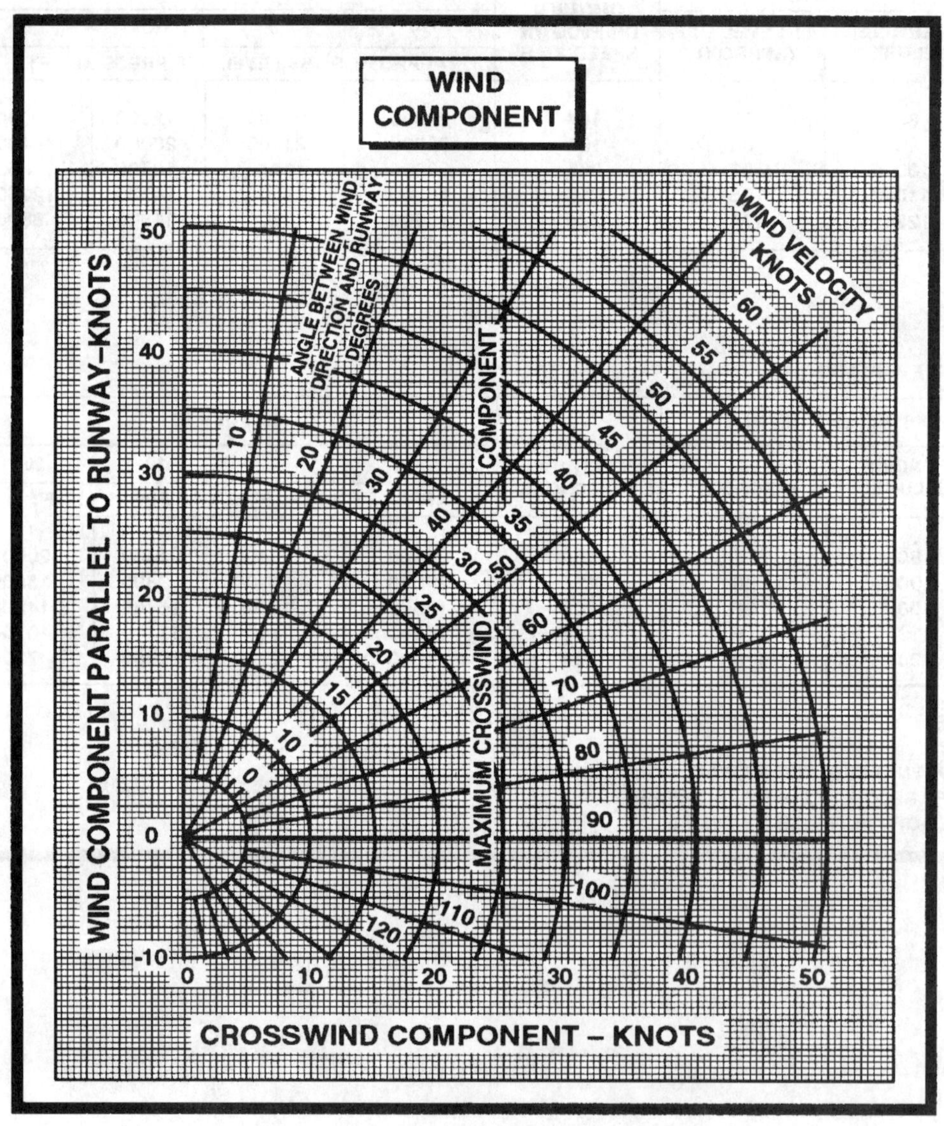

FIGURE 74.—Wind Component Chart.

LANDING

GO AROUND EPR
A/C BLEED ON

EPR BLEED CORRECTIONS	1 ENG	2 ENG
AIR CONDITIONING OFF	+.03	+.03
ENGINE ANTI-ICE ON	ZERO	ZERO
WING ANTI-ICE ON (FLT)	-.06	-.04

REPORTED OAT °F	-61	-52	-43	-35	-25	-17	-10	0	10	18	27	38	47	55	64	73	83	91	100	119
REPORTED OAT °C	-52	-47	-42	-37	-32	-27	-23	-18	-13	-8	-3	3	8	13	18	23	28	33	38	48
TAT °C	-50	-45	-40	-35	-30	-25	-20	-15	-10	-5	0	5	10	15	20	25	30	35	40	50
TEMP LIMIT EPR	2.30	2.30	2.30	2.29	2.26	2.24	2.21	2.19	2.16	2.13	2.10	2.07	2.04	2.00	1.98	1.98	1.98	1.94	1.90	1.80

PRESS LIMIT EPR	5660 AND ABOVE	4000	3000	2000	1000	S.L.	-1000
EPR	2.30	2.20	2.15	2.09	2.04	1.98	1.93

1. FIND TEMP LIMIT EPR
2. FIND PRESS LIMIT EPR
3. USE THE SMALLER OF THE TWO LIMITS

FLAP EXTENSION/ MANEUVERING SPEED

NORMAL MANEUVER AND FLAP EXTENSION SPEEDS

FLAP POS	NORMAL MANEUVER	SELECT FLAP
0	210	1
1	190	5
5	170	10/15
10	160	15
15	150	25/30/40
25	140	30/40
30	VREF	—
40	VREF	—

LANDING SPEED

GROSS WT 1000 LB	REFERENCE SPEED AT FLAP POSITION 40	30	25	15
110	138	142	153	158
105	134	138	149	154
100	130	135	144	150
95	127	131	140	145
90	123	127	136	141
85	119	124	132	136
80	115	120	127	132
75	113	116	123	127
70	109	112	119	123

ADD WIND FACTOR OF: 1/2 HEADWIND COMPONENT + GUST (MAX: 20 KTS)

FIGURE 75.—B-737 – Landing Performance Chart.

LOADING CONDITIONS	WT-1	WT-2	WT-3	WT-4	WT-5
PASSENGERS					
FORWARD COMPT	18	23	12	28	26
AFT COMPT	95	112	75	122	103
CARGO					
FORWARD HOLD	1,500	2,500	3,500	850	1,400
AFT HOLD	2,500	3,500	4,200	1,500	2,200
FUEL					
TANKS 1 AND 3 (EACH)	10,500	11,000	FULL	10,000	11,500
TANK 2	28,000	27,000	24,250	26,200	25,200

FIGURE 76.—B-727 – Loading.

LOADING CONDITIONS	WT-6	WT-7	WT-8	WT-9	WT-10
PASSENGERS					
FORWARD COMPT	10	27	6	29	21
AFT COMPT	132	83	98	133	127
CARGO					
FORWARD HOLD	5,000	4,500	1,300	975	2,300
AFT HOLD	6,000	5,500	3,300	1,250	2,400
FUEL					
TANKS 1 AND 3 (EACH)	9,500	9,000	FULL	11,000	10,500
TANK 2	21,700	19,800	12,000	29,300	22,700

FIGURE 77.—B-727 – Loading.

LOADING CONDITIONS	WT-11	WT-12	WT-13	WT-14	WT-15
PASSENGERS					
FORWARD COMPT	11	28	22	17	3
AFT COMPT	99	105	76	124	130
CARGO					
FORWARD HOLD	3,100	4,200	1,600	3,800	1,800
AFT HOLD	5,500	4,400	5,700	4,800	3,800
FUEL					
TANKS 1 AND 3 (EACH)	8,500	11,500	12,000	11,000	10,500
TANK 2	19,600	27,800	29,100	25,400	21,900

FIGURE 78.—B-727 – Loading.

AIRPLANE DATUM CONSTANTS

MAC	180.9 inches
L.E. of MAC	860.5 inches
Basic Operating Index	$\frac{92,837.0}{1,000}$

OPERATING LIMITATIONS

Maximum Takeoff Slope	±2%
Maximum Takeoff / Landing Crosswind Component	32 knots
Maximum Takeoff / Landing Tailwind Component	12 knots

WEIGHT LIMITATIONS

Basic Operating Weight	105,500 pounds
Maximum Zero Fuel Weight	138,500 pounds
Maximum Taxi Weight	185,700 pounds
Maximum Takeoff Weight (Brake Release)	184,700 pounds
Maximum In-flight Weight (Flaps 30)	155,500 pounds
(Flaps 40)	144,000 pounds
Maximum Landing Weight (Flaps 30)	155,000 pounds
(Flaps 40)	143,000 pounds

FIGURE 79.—B-727 – Table of Weights and Limits.

PASSENGER LOADING TABLE

Number of Pass.	Weight Lbs.	Moment 1000
Forward Compartment Centroid-582.0		
5	850	495
10	1,700	989
15	2,550	1,484
20	3,400	1,979
25	4,250	2,473
29	4,930	2,869
AFT Compartment Centroid-1028.0		
10	1,700	1,748
20	3,400	3,495
30	5,100	5,243
40	6,800	6,990
50	8,500	8,738
60	10,200	10,486
70	11,900	12,233
80	13,600	13,980
90	15,300	15,728
100	17,000	17,476
110	18,700	19,223
120	20,400	20,971
133	22,610	23,243

CARGO LOADING TABLE

	Moment 1000	
	Forward Hold	Aft Hold
Weight Lbs.	Arm 680.0	Arm 1166.0
6,000		6,966
5,000	3,400	5,830
4,000	2,720	4,664
3,000	2,040	3,498
2,000	1,360	2,332
1,000	680	1,166
900	612	1,049
800	544	933
700	476	816
600	408	700
500	340	583
400	272	466
300	204	350
200	136	233
100	68	117

NOTE: These computations are to be used for testing purposes only.

FUEL LOADING TABLE

TANKS 1 & 3 (EACH)

Weight Lbs.	Arm	Moment 1000
8,500	992.1	8,433
9,000	993.0	8,937
9,500	993.9	9,442
10,000	994.7	9,947
10,500	995.4	10,451
11,000	996.1	10,957
11,500	996.8	11,463
12,000	997.5	11,970
FULL CAPACITY		

**Note: Computations for Tank 2 weights for 12,500 lbs. to 18,000 lbs. have been purposely omitted.

TANKS 2 (3 CELL)

Weight Lbs.	Arm	Moment 1000	Weight Lbs.	Arm	Moment 1000
8,500	917.5	7,799	22,500	914.5	20,576
9,000	917.2	8,255	23,000	914.5	21,034
9,500	917.0	8,711	23,500	914.4	21,488
10,000	916.8	9,168	24,000	914.3	21,943
10,500	916.6	9,624	24,500	914.3	22,400
11,000	916.5	10,082	25,000	914.2	22,855
11,500	916.3	10,537	25,500	914.2	23,312
12,000	916.1	10,993	26,000	914.1	23,767
**(See note at lower left)			26,500	914.1	24,244
			27,000	914.0	24,678
18,500	915.1	16,929	27,500	913.9	25,132
19,000	915.0	17,385	28,000	913.9	25,589
19,500	914.9	17,841	28,500	913.8	26,043
20,000	914.9	18,298	29,000	913.7	26,497
20,500	914.8	18,753	29,500	913.7	26,954
21,000	914.7	19,209	30,000	913.6	27,408
21,500	914.6	19,664			
22,000	914.6	20,121	**FULL CAPACITY**		

FIGURE 80.—Loading Tables.

OPERATING CONDITIONS	G-1	G-2	G-3	G-4	G-5
FIELD ELEVATION FT	1,050	2,000	4,350	3,050	2,150
ALTIMETER SETTING	29.36"	1016 mb	30.10"	1010 mb	29.54"
TEMPERATURE	+23 °F	+10 °C	+68 °F	–5 °C	+5 °F
AIR COND ENGS 1 AND 3	OFF	ON	ON	ON	ON
ANTI-ICE ENG 2	ON	OFF	OFF	ON	ON
GROSS WEIGHT (X1000)	140	190	180	160	120
6TH STAGE BLEED	OFF	ON	ON	OFF	OFF
FLAP POSITION	15°	5°	25°	15°	5°
CG STATION	911.2	882.2	914.8	932.9	925.6
LEMAC – STA 860.5, MAC 180.9"					

FIGURE 81.—B-727 – Takeoff.

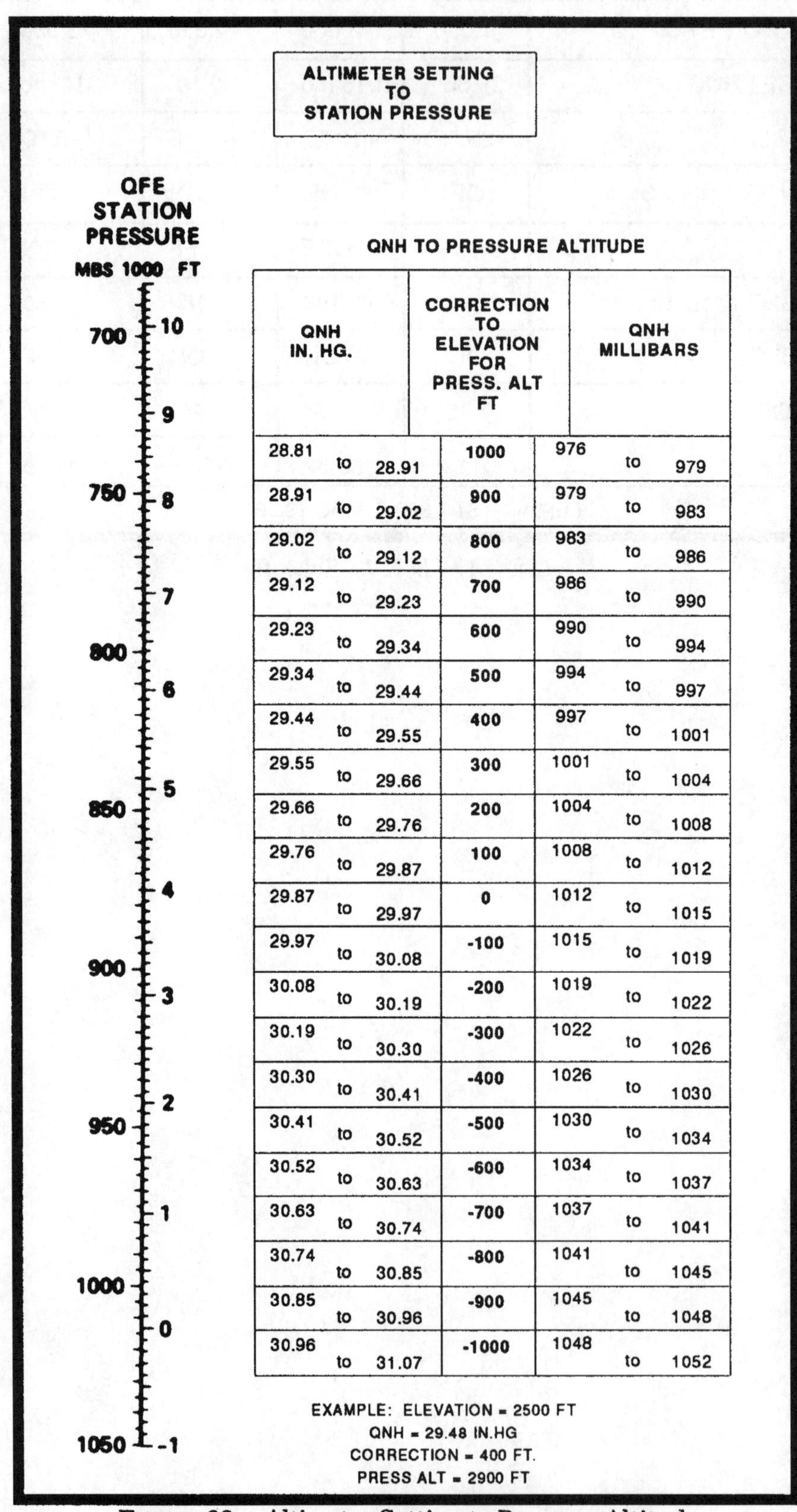

QNH TO PRESSURE ALTITUDE

QNH IN. HG.	CORRECTION TO ELEVATION FOR PRESS. ALT FT	QNH MILLIBARS
28.81 to 28.91	1000	976 to 979
28.91 to 29.02	900	979 to 983
29.02 to 29.12	800	983 to 986
29.12 to 29.23	700	986 to 990
29.23 to 29.34	600	990 to 994
29.34 to 29.44	500	994 to 997
29.44 to 29.55	400	997 to 1001
29.55 to 29.66	300	1001 to 1004
29.66 to 29.76	200	1004 to 1008
29.76 to 29.87	100	1008 to 1012
29.87 to 29.97	0	1012 to 1015
29.97 to 30.08	-100	1015 to 1019
30.08 to 30.19	-200	1019 to 1022
30.19 to 30.30	-300	1022 to 1026
30.30 to 30.41	-400	1026 to 1030
30.41 to 30.52	-500	1030 to 1034
30.52 to 30.63	-600	1034 to 1037
30.63 to 30.74	-700	1037 to 1041
30.74 to 30.85	-800	1041 to 1045
30.85 to 30.96	-900	1045 to 1048
30.96 to 31.07	-1000	1048 to 1052

FIGURE 82.—Altimeter Setting to Pressure Altitude.

TAKEOFF EPR, SPEEDS AND STAB TRIM SETTING

MAX TAKEOFF EPR

0 - 60 KNOTS ENG 1 & 3 AIRBLEED ON
ENG 2 NO AIRBLEED

PRESS ALT FT	OAT °F / °C	-67 TO -9 / -55 TO -23	-4 / -20	5 / -15	14 / -10	23 / -5	32 / 0	41 / 5	50 / 10	59 / 15	68 / 20	77 / 25	86 / 30	95 / 35	104 / 40	113 / 45	120 / 49
-1000	1 & 3	2.04	2.04	2.04	2.04	2.04	2.04	2.04	2.04	2.04	2.04	2.04	2.04	2.03	1.99	1.94	1.91
	2	2.06	2.06	2.06	2.06	2.06	2.06	2.06	2.06	2.06	2.06	2.06	2.06	2.05	2.00	1.96	1.92
S.L.	1 & 3	2.10	2.10	2.10	2.10	2.10	2.10	2.10	2.10	2.10	2.10	2.10	2.08	2.03	1.99	1.94	1.91
	2	2.11	2.11	2.11	2.11	2.11	2.11	2.11	2.11	2.11	2.11	2.11	2.10	2.05	2.00	1.96	1.92
1000	1 & 3	2.15	2.15	2.15	2.15	2.15	2.15	2.15	2.13	2.12	2.12	2.11	2.08	2.03	1.99	1.94	1.91
	2	2.16	2.16	2.16	2.16	2.16	2.16	2.16	2.15	2.13	2.13	2.12	2.10	2.05	2.00	1.96	1.92
2000	1 & 3	2.21	2.21	2.21	2.21	2.21	2.20	2.17	2.14	2.14	2.14	2.11	2.08	2.03	1.99	1.94	1.91
	2	2.22	2.22	2.22	2.22	2.22	2.21	2.18	2.16	2.16	2.15	2.12	2.10	2.05	2.00	1.96	1.92
3000	1 & 3	2.26	2.26	2.26	2.25	2.23	2.20	2.17	2.14	2.14	2.14	2.11	2.08	2.03	1.99	1.94	1.91
	2	2.28	2.28	2.28	2.27	2.24	2.21	2.18	2.16	2.16	2.15	2.12	2.10	2.05	2.00	1.96	1.92
3856 & ABOVE	1 & 3	2.31	2.29	2.27	2.25	2.23	2.20	2.17	2.14	2.14	2.14	2.11	2.08	2.03	1.99	1.94	1.91
	2	2.32	2.31	2.29	2.27	2.24	2.21	2.18	2.16	2.16	2.15	2.12	2.10	2.05	2.00	1.96	1.92

EPR BLEED CORRECTIONS	ENG 1 & 3	ENG 2
AIR CONDITIONING	OFF +.04	-
ENGINE ANTI-ICE ON	-	-.03

REDUCE ENG 2 EPR BY .05 WITH 6TH STAGE BLEED ON (IF INSTALLED) FOR 10 °C (50 °F) OAT & WARMER

V_1, V_R, V_2

ANTI-SKID OPERATIVE

PRESSURE ALT - 1000 FT		OAT			
9 TO 11	°F / °C	(ABOVE CERTIFIED ALTITUDE)		-65 TO 25 / -54 TO -4	26 TO 87 / -3 TO 31
7 TO 9	°F / °C		-65 TO 9 / -54 TO -13	10 TO 75 / -12 TO 24	76 TO 104 / 25 TO 40
5 TO 7	°F / °C	-65 TO -10 / -54 TO -23	-8 TO 42 / -22 TO 5	43 TO 97 / 6 TO 36	98 TO 111 / 37 TO 44
3 TO 5	°F / °C	-65 TO 32 / -54 TO 0	33 TO 90 / 1 TO 32	91 TO 113 / 33 TO 45	114 TO 120 / 46 TO 49
1 TO 3	°F / °C	-65 TO 83 / -54 TO 28	84 TO 106 / 29 TO 41	107 TO 120 / 42 TO 49	
-1 TO 1	°F / °C	-65 TO 99 / -54 TO 37	100 TO 120 / 38 TO 49		

FLAPS	GROSS WEIGHT 1000 LB	$V_1 = V_R$	V_2	$V_1 = V_R$	V_2	$V_1 = V_R$	V_2	$V_1 = V_R$	V_2
5	210	165	175	166	175				
	200	160	171	162	171				
	190	155	167	157	167	158	167		
	180	150	163	152	163	154	163		
	170	144	159	147	159	149	159	150	158
	160	140	154	141	153	143	153	145	153
	150	135	149	136	149	138	149	140	148
	140	129	145	130	145	132	144	134	144
	130	124	140	125	139	126	138	128	138
	120	119	135	120	134	120	134	121	133
15	210	156	166	157	166				
	200	151	162	153	162				
	190	146	158	148	158	149	158		
	180	141	154	143	154	145	154		
	170	136	150	138	150	140	150	141	149
	160	132	146	133	145	135	145	137	145
	150	127	141	128	141	130	141	132	140
	140	122	137	123	137	124	136	126	136
	130	117	133	118	132	118	131	120	131
	120	112	128	113	127	113	127	115	126
20	210	151	161	152	161				
	200	146	157	148	157				
	190	141	153	143	153	144	153		
	180	136	150	138	150	140	149		
	170	132	146	133	146	135	145	136	145
	160	128	142	129	141	131	141	133	141
	150	123	137	124	137	126	136	128	136
	140	118	133	119	133	120	132	122	132
	130	113	129	114	128	114	127	116	127
	120	109	124	109	123	109	123	111	122
25	210	146	157	147	157				
	200	141	153	143	153				
	190	137	149	138	149	139	149		
	180	132	145	134	145	136	145		
	170	127	141	129	141	131	141	132	140
	160	123	137	124	137	126	137	128	136
	150	119	133	120	133	122	133	124	132
	140	114	129	115	129	116	128	118	128
	130	109	125	110	124	110	124	112	123
	120	105	120	106	120	106	119	108	118

STAB TRIM SETTING

CG	FLAPS 5	FLAPS 15 / 20	FLAPS 25
	UNITS AIRPLANE NOSE UP		
10	6 3/4	7 1/2	8 1/4
12	6 1/2	7 1/4	8
14	6 1/4	7	7 3/4
16	6	6 3/4	7 1/2
18	5 3/4	6 1/2	7
20	5 1/2	6	6 1/2
22	5	5 3/4	6 1/4
24	4 3/4	5 1/4	5 3/4
26	4 1/2	4 3/4	5 1/4
28	4	4 1/2	4 3/4
30	3 3/4	4	4 1/4
32	3 1/2	3 3/4	4
34	3 1/4	3 1/4	3 1/2
36	2 3/4	3	3
38	2 1/2	2 1/2	2 1/2
40	2 1/2	2 1/2	2 1/2
42	2 1/2	2 1/2	2 1/2

FLAP RETRACTION/ MANEUVERING SPEEDS

GROSS WEIGHT LB	FLAP POSITION 15	5	2	0
154500 & BELOW	150	160	190	200
154501 TO 176000	160	170	200	210
176001 TO 191000	170	180	210	220
ABOVE 191000	180	190	225	235

FOR MANEUVERS IMMEDIATELY AFTER TAKEOFF EXCEEDING 15° BANK MAINTAIN AT LEAST V_2 +10 AT TAKEOFF FLAPS

FIGURE 83.—Takeoff Performance.

OPERATING CONDITIONS	H-1	H-2	H-3	H-4	H-5
ALTITUDE	24,000	17,000	8,000	18,000	22,000
WEIGHT (X1000)	195	185	155	135	175
ENGINES OPERATING	3	3	3	3	3
HOLDING TIME (MIN)	15	30	45	25	35

FIGURE 84.—B-727 – Holding.

EPR IAS - KTS FF PER ENG - LB/HR	HOLDING				B-727				
PRESSURE ALTITUDE FT	GROSS WEIGHT - 1000 LB								
	200	190	180	170	160	150	140	130	120
25000	1.85 268 3600	1.81 261 3400	1.77 253 3210	1.73 246 3030	1.69 238 2860	1.64 230 2680	1.60 222 2510	1.55 213 2340	1.51 205 2180
20000	1.69 265 3630	1.66 258 3450	1.62 251 3280	1.59 244 3110	1.55 236 2940	1.51 228 2770	1.48 220 2600	1.44 212 2440	1.40 204 2270
15000	1.56 263 3670	1.53 256 3500	1.50 249 3340	1.47 242 3170	1.44 235 3000	1.41 227 2850	1.38 219 2680	1.35 211 2520	1.32 203 2350
10000	1.45 262 3800	1.43 255 3640	1.40 248 3460	1.38 241 3310	1.35 234 3140	1.33 226 2970	1.30 218 2810	1.28 210 2640	1.25 202 2480
5000	1.36 260 3890	1.34 254 3720	1.32 247 3550	1.30 240 3380	1.28 233 3220	1.26 225 3060	1.24 218 2890	1.22 210 2730	1.20 201 2560

FIGURE 85.—B-727 – Holding Performance Chart.

OPERATING CONDITIONS	S–1	S–2	S–3	S–4	S–5
FLIGHT LEVEL	370	350	410	390	330
LANDING WEIGHT (X1000)	130	150	135	155	125
DESCENT TYPE	.80M/ 250	.80M/ 280/250	.80M/ 320/250	.80M/ 350/250	.80M/ 320/250

FIGURE 86.—Descent Performance.

.80M/250 KIAS

FLIGHT LEVEL	TIME MIN	FUEL LB	DISTANCE NAM AT LANDING WEIGHTS		
			120,000 LB	140,000 LB	160,000 LB
410	27	1610	133	137	138
390	27	1600	130	134	136
370	26	1570	123	128	129
350	25	1540	116	120	122
330	24	1510	110	113	115
310	23	1480	103	107	108
290	22	1450	97	100	101
270	21	1420	90	93	95
250	20	1390	84	87	88
230	19	1360	78	80	81
210	18	1320	72	74	75
190	17	1280	66	68	68
170	16	1240	60	62	62
150	14	1190	54	56	56
100	11	1050	39	40	40
050	8	870	24	24	24
015	5	700	12	12	12

.80M/280/250 KIAS

FLIGHT LEVEL	TIME MIN	FUEL LB	DISTANCE NAM AT LANDING WEIGHTS		
			120,000 LB	140,000 LB	160,000 LB
410	25	1550	123	129	132
390	24	1540	121	127	130
370	24	1520	115	121	125
350	23	1500	111	117	120
330	23	1480	106	111	115
310	22	1450	100	105	108
290	21	1430	94	99	102
270	20	1400	88	93	95
250	19	1370	83	87	89
230	18	1350	77	81	83
210	17	1310	72	75	76
190	16	1280	66	69	70
170	15	1240	61	63	64
150	14	1200	55	57	58
100	12	1080	42	42	42
050	8	870	24	24	24
015	5	700	12	12	12

.80M/320/250 KIAS

FLIGHT LEVEL	TIME MIN	FUEL LB	DISTANCE NAM AT LANDING WEIGHTS		
			120,000 LB	140,000 LB	160,000 LB
410	22	1490	113	120	123
390	22	1480	111	117	121
370	21	1460	105	112	116
350	21	1440	101	107	111
330	20	1420	96	103	107
310	20	1400	92	98	102
290	19	1390	89	94	98
270	19	1370	85	90	94
250	18	1350	80	85	88
230	17	1330	75	79	82
210	17	1300	71	74	77
190	16	1270	66	69	71
170	15	1240	61	64	65
150	14	1210	56	59	60
100	12	1110	45	46	46
050	8	870	24	24	24
015	5	700	12	12	12

.80M/350/250 KIAS

FLIGHT LEVEL	TIME MIN	FUEL LB	DISTANCE NAM AT LANDING WEIGHTS		
			120,000 LB	140,000 LB	160,000 LB
410	21	1440	106	112	116
390	21	1430	103	110	114
370	20	1420	99	106	110
350	20	1400	95	101	106
330	19	1390	91	98	102
310	19	1380	88	94	98
290	18	1360	85	90	95
270	18	1350	82	87	91
250	17	1330	78	83	87
230	17	1310	74	78	81
210	16	1290	70	74	76
190	16	1270	65	69	71
170	15	1240	61	64	66
150	14	1210	57	60	61
100	13	1130	47	48	49
050	8	870	24	24	24
015	5	700	12	12	12

NOTE: FUEL FOR A STRAIGHT-IN APPROACH IS INCLUDED

FIGURE 87.—Descent Performance Chart.

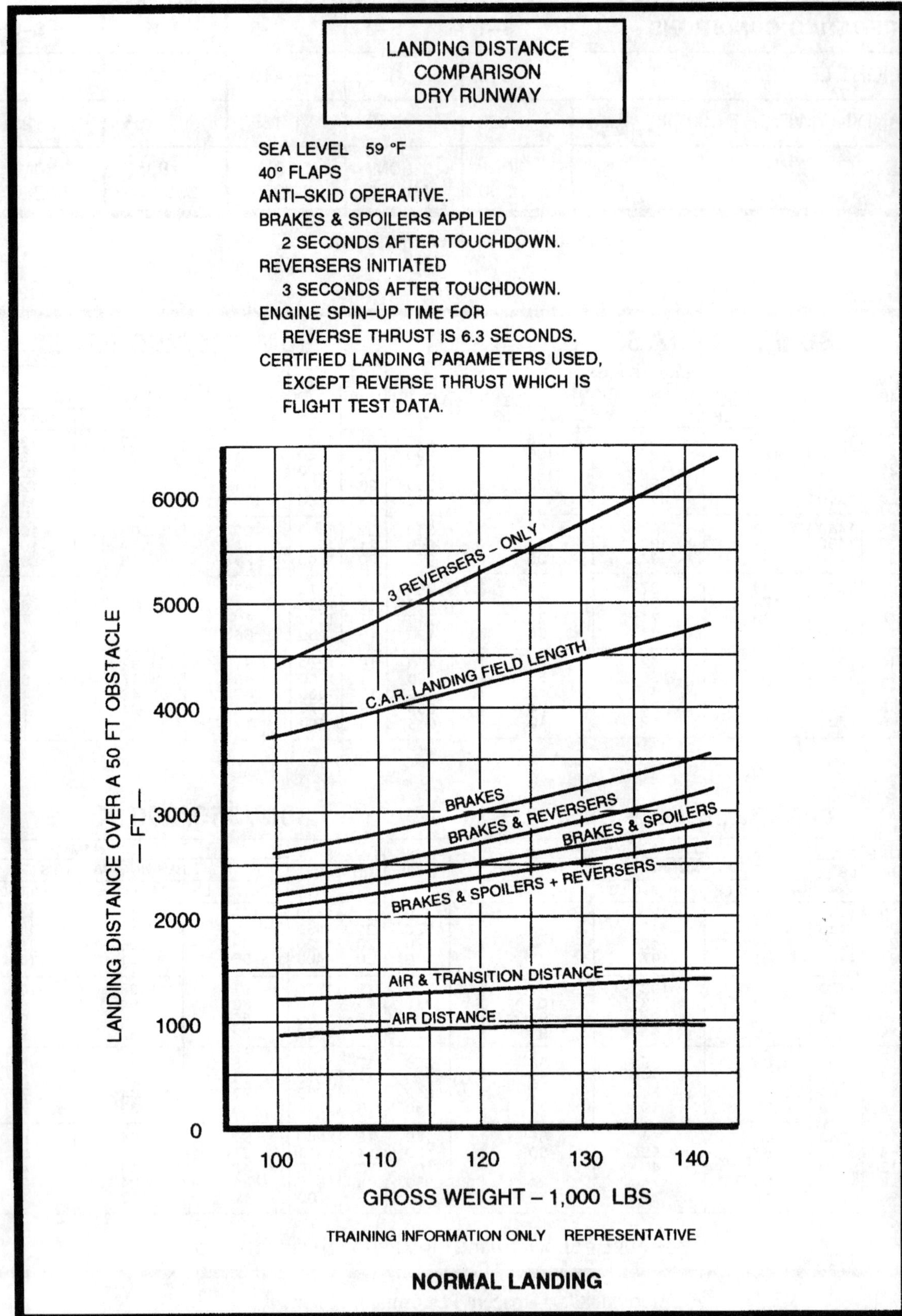

FIGURE 88.—B-727 – Normal Landing – Dry Runway.

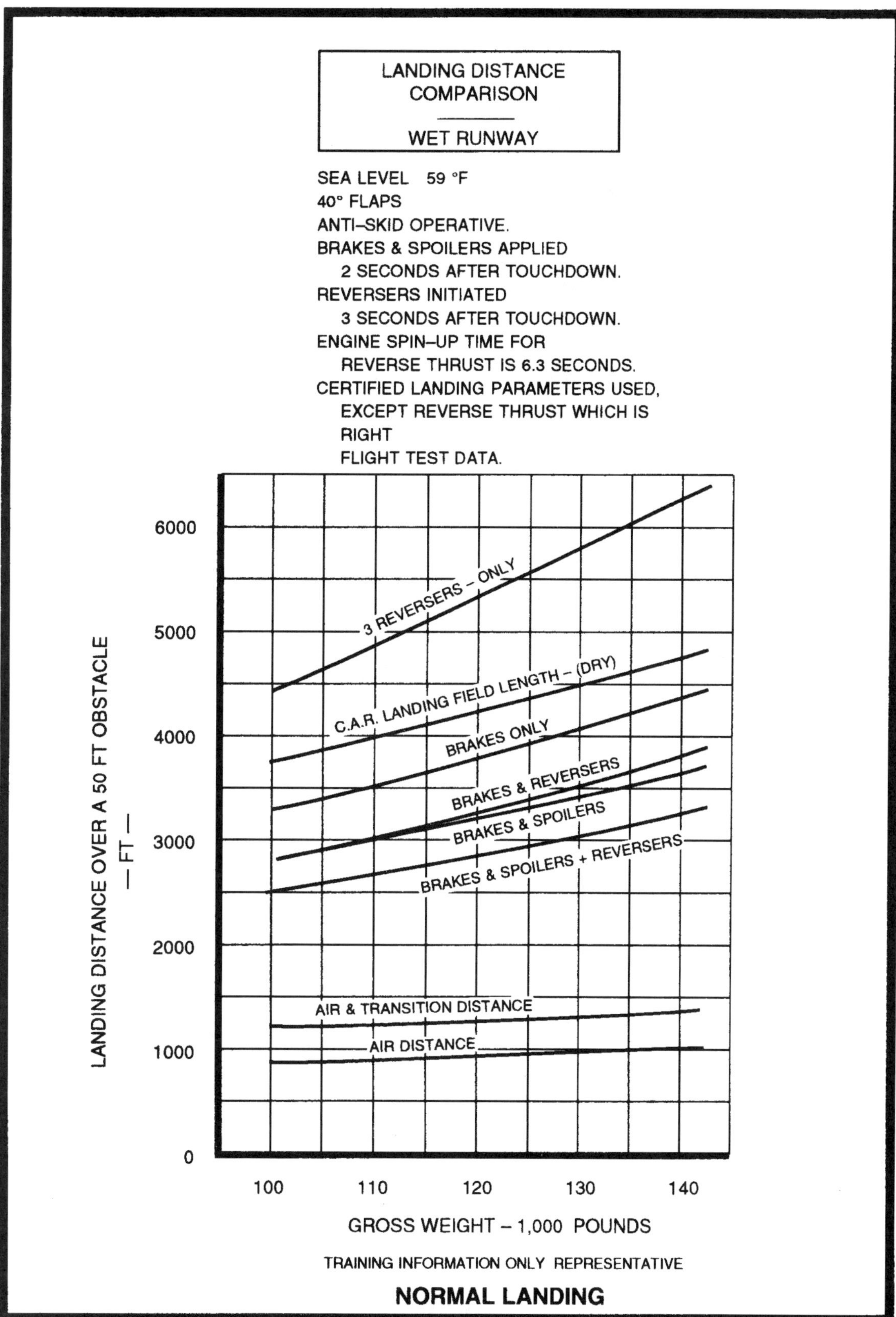

FIGURE 89.—B-727 – Normal Landing – Wet Runway.

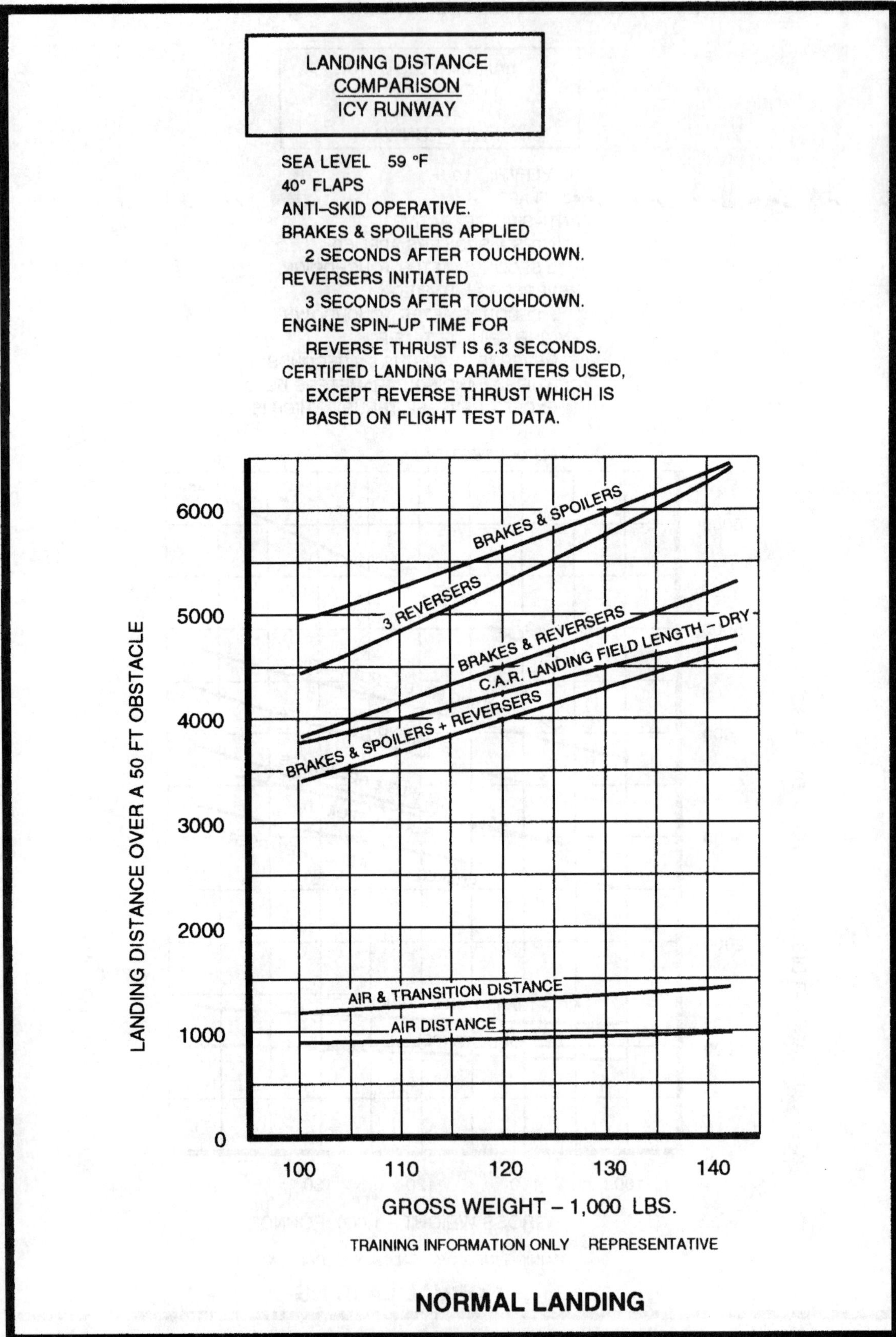

FIGURE 90.—B-727 – Normal Landing – Icy Runway.

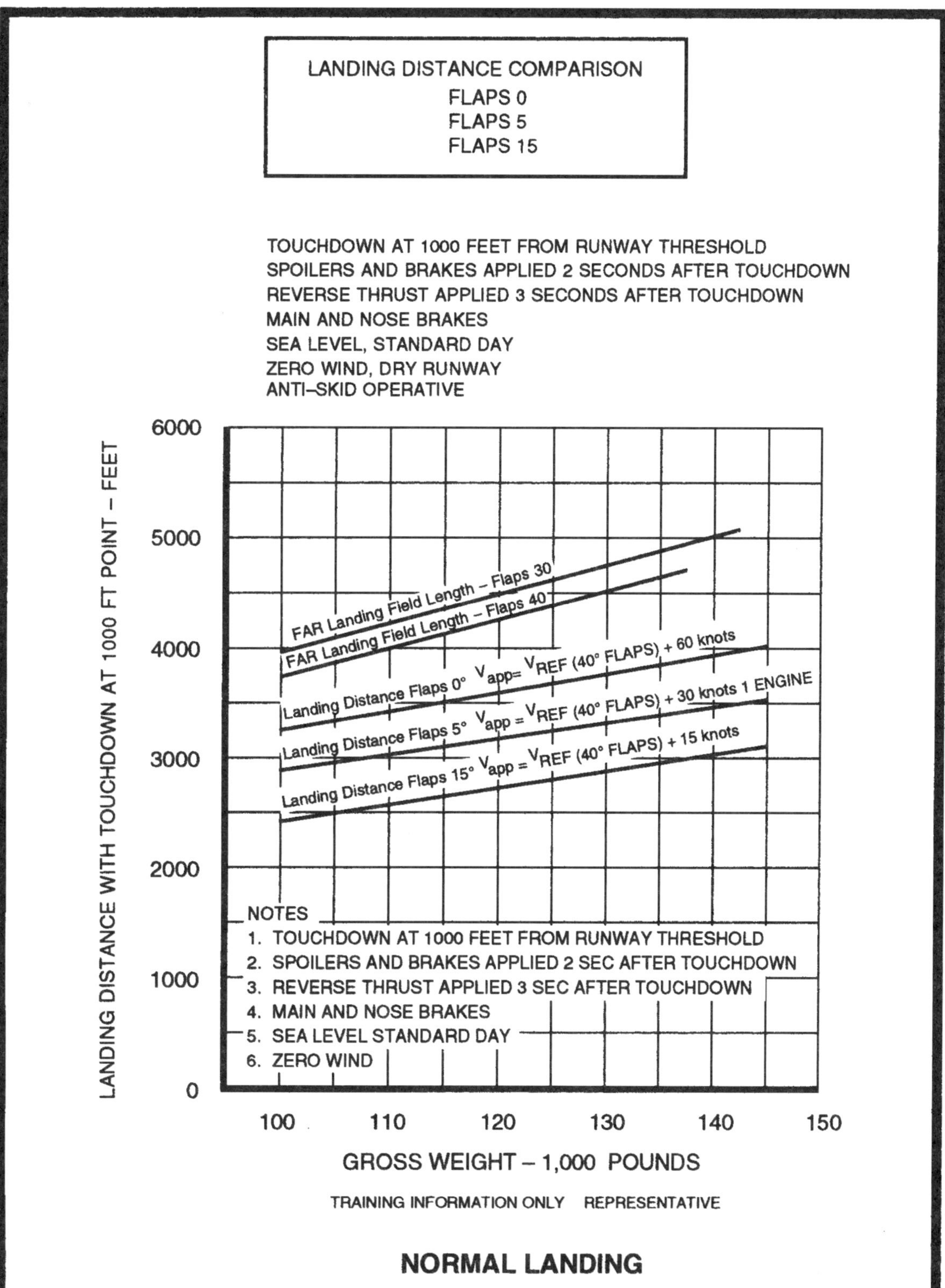

FIGURE 91.—B-727 – Normal Landing Distance Comparison.

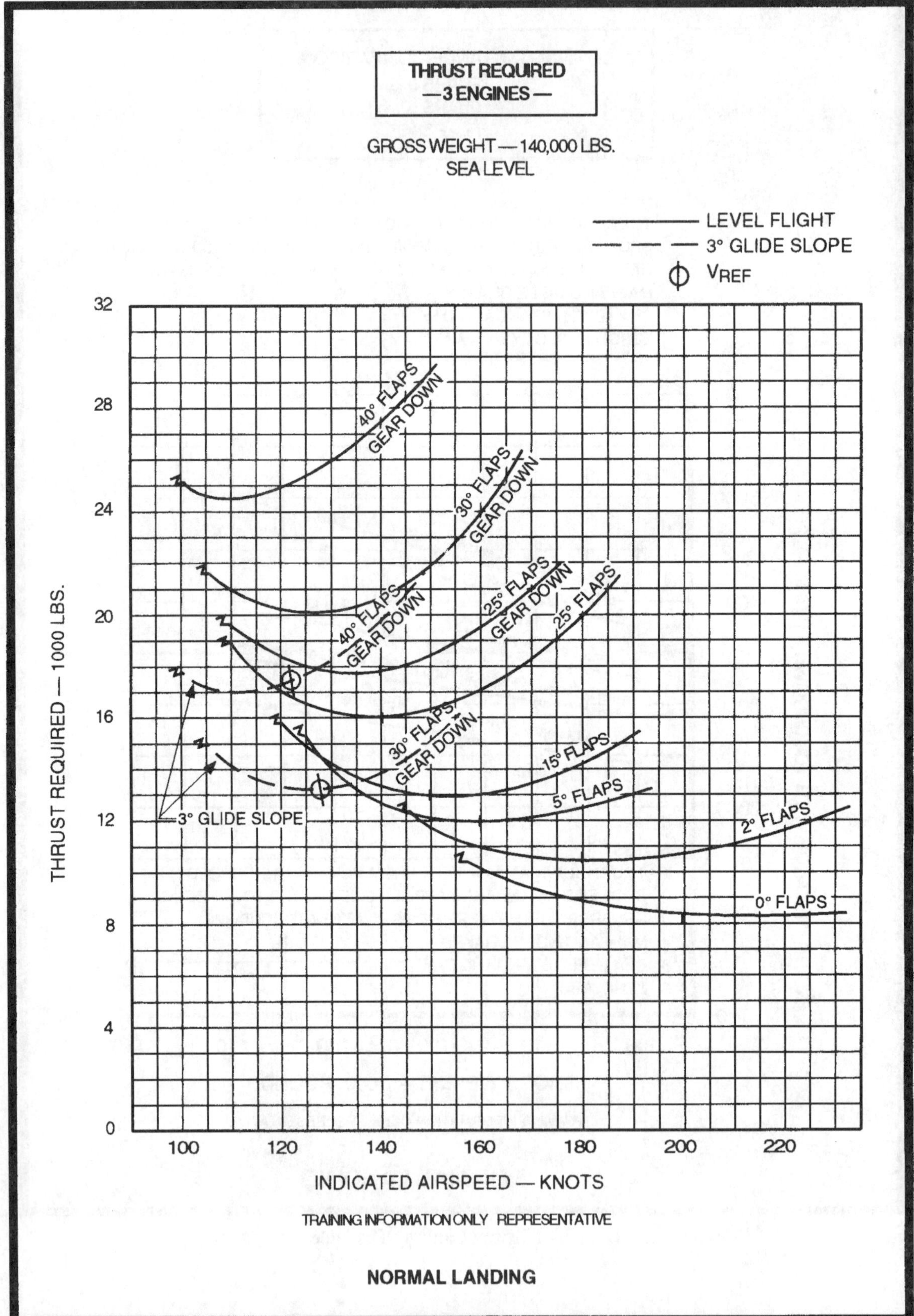

FIGURE 92.—B-727 – Landing Thrust – 140,000 Pounds.

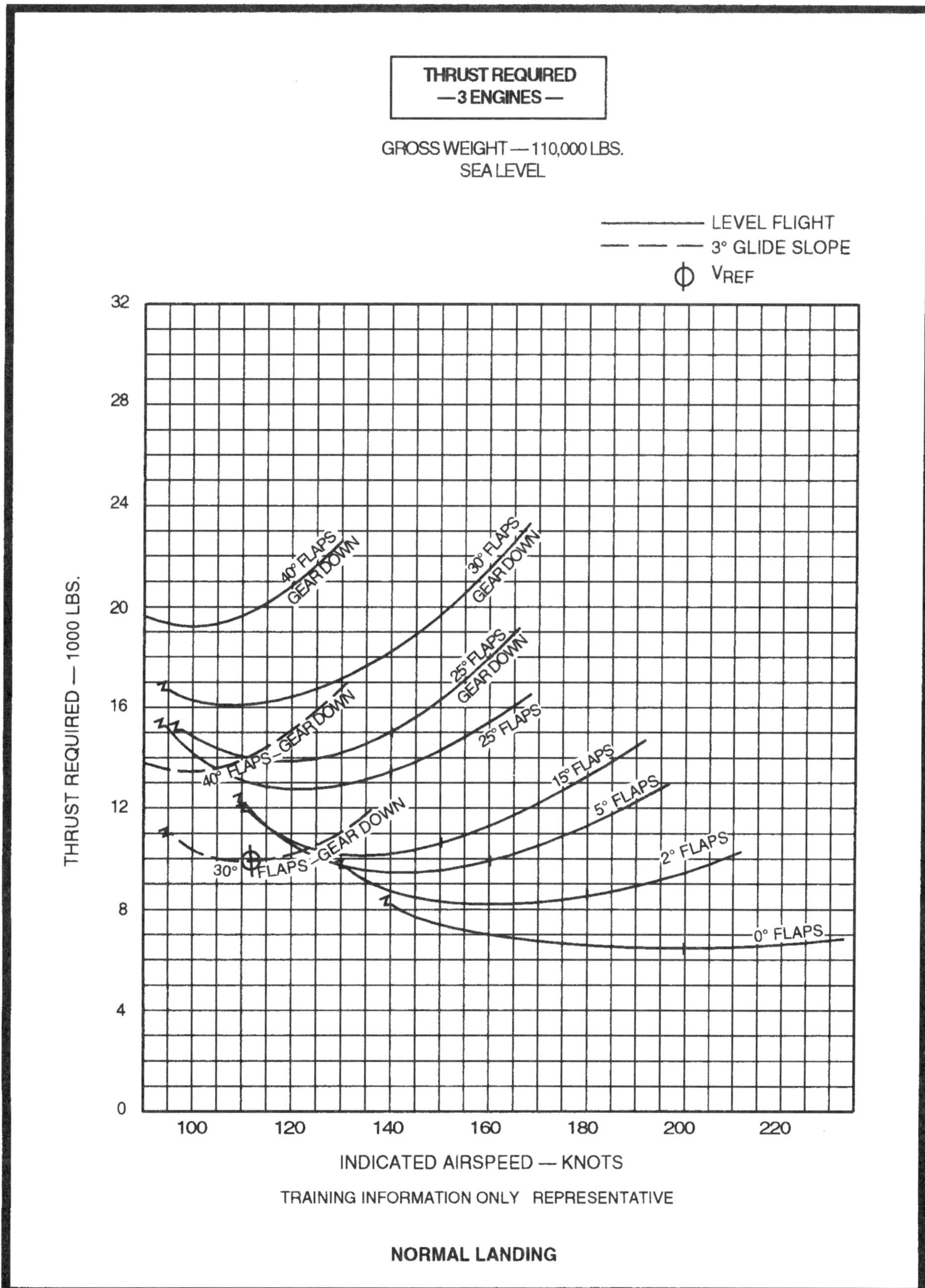

FIGURE 93.—B-727 – Landing Thrust – 110,000 Pounds.

Form Approved: OMB No. 2120-0034

U.S. DEPARTMENT OF TRANSPORTATION FEDERAL AVIATION ADMINISTRATION **FLIGHT PLAN**	(FAA USE ONLY) ☐ PILOT BRIEFING ☐ VNR ☐ STOPOVER	TIME STARTED	SPECIALIST INITIALS

1. TYPE	2. AIRCRAFT IDENTIFICATION	3. AIRCRAFT TYPE/ SPECIAL EQUIPMENT	4. TRUE AIRSPEED	5. DEPARTURE POINT	6. DEPARTURE TIME PROPOSED (Z)	ACTUAL (Z)	7. CRUISING ALTITUDE
VFR / X IFR / DVFR	N60JB	C208/R	160 KTS	MDW Chicago Midway			FL190

8. ROUTE OF FLIGHT
Midway Four Dep. GIJ, J554 CRL, J586 YXU, J547 BUF

9. DESTINATION (Name of airport and city)	10. EST. TIME ENROUTE HOURS	MINUTES	11. REMARKS
BUF Greater Buffalo Int'l. Buffalo			L/O = Level off. PPH = Pounds Per Hour L/O R-270/19 GIJ Variation: GIJ 1W, CRL 3W, YXU 6W, BUF 8W.

12. FUEL ON BOARD HOURS	MINUTES	13. ALTERNATE AIRPORT(S)	14. PILOT'S NAME, ADDRESS & TELEPHONE NUMBER & AIRCRAFT HOME BASE / 17. DESTINATION CONTACT/TELEPHONE (OPTIONAL)	15. NUMBER ABOARD
3	20	ROC Rochester		2

16. COLOR OF AIRCRAFT	
Brown/Tan	CIVIL AIRCRAFT PILOTS. FAR Part 91 requires you file an IFR flight plan to operate under instrument flight rules in controlled airspace. Failure to file could result in a civil penalty not to exceed $1,000 for each violation (Section 901 of the Federal Aviation Act of 1958, as amended). Filing of a VFR flight plan is recommended as a good operating practice. See also Part 99 for requirements concerning DVFR flight plans.

FAA Form 7233-1 (8-82) CLOSE VFR FLIGHT PLAN WITH ________________ FSS ON ARRIVAL

FLIGHT LOG

CHECK POINTS FROM	TO	ROUTE ALTITUDE	COURSE	WIND TEMP	SPEED-KTS TAS	GS	DIST NM	TIME LEG	TOT	FUEL LEG	TOT
MDW	L/O GIJ R-270/19	MDW 4 Climb					49	:19:00		327*	
R-270/19 GIJ	GIJ	Direct FL190		230/51 ISA							
GIJ	CRL	J554 FL190									
CRL	YXU	J586 FL190		240/59 ISA							
YXU	BUF R-282/30	J547 FL190		250/62 ISA							
BUF R-282/30	BUF	Descent & Approach					30	:14:00		121.5	
BUF	ROC	V2 7000			150		44	:20:00			

OTHER DATA: * Includes Taxi Fuel
NOTE: Use 610 PPH Total Fuel Flow From L/O To Start Of Descent.
Use 710 PPH Total Fuel Flow For Reserve And Alternate Requirements.
A Missed Approach Requires 81# of Fuel.

TIME and FUEL: As required by FARs.

TIME	FUEL (LB)	
		EN ROUTE
		RESERVE
		ALTERNATE
		TOTAL

FIGURE 94.—Flight Plan/Flight Log.

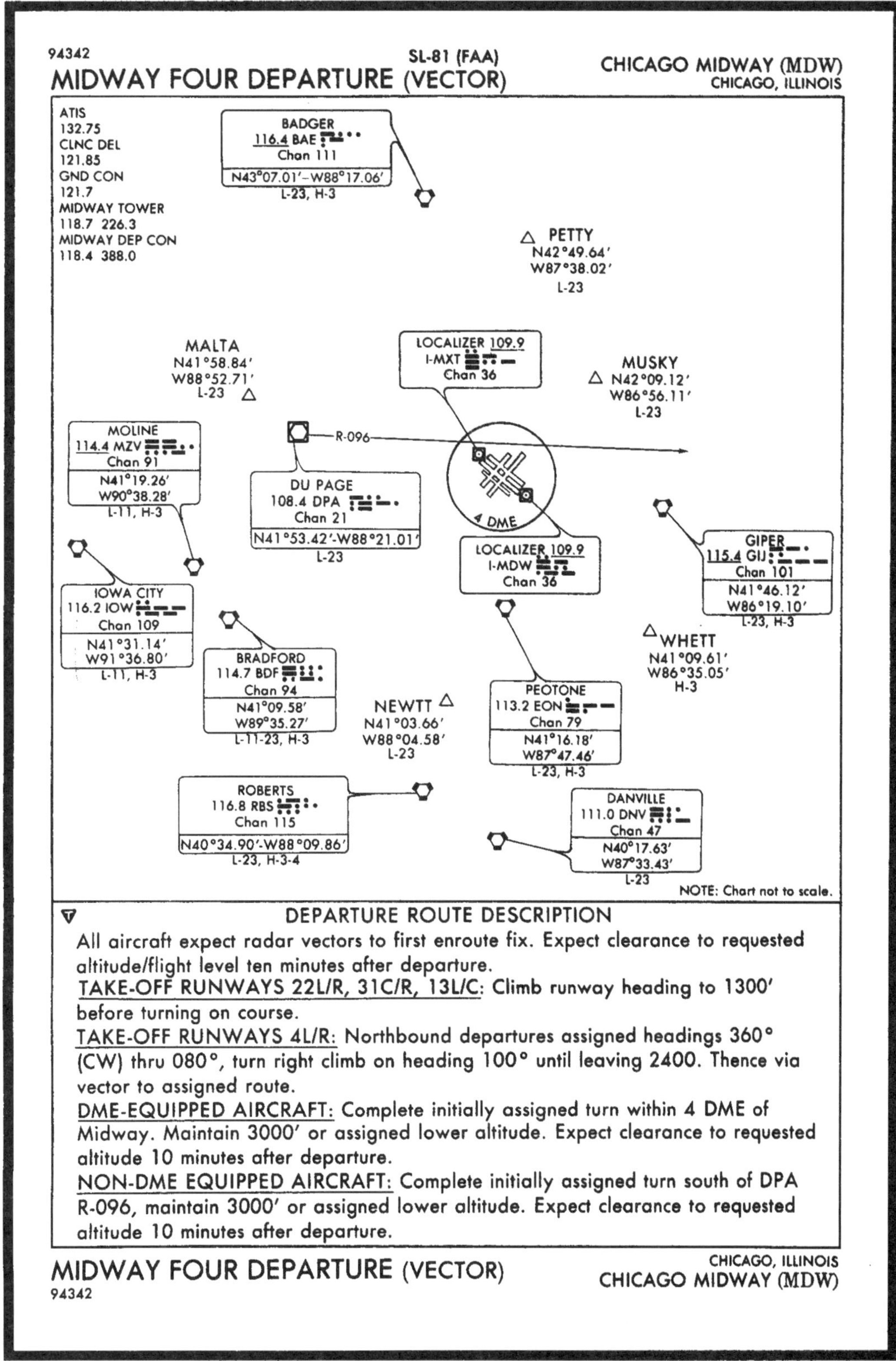

DEPARTURE ROUTE DESCRIPTION

All aircraft expect radar vectors to first enroute fix. Expect clearance to requested altitude/flight level ten minutes after departure.

TAKE-OFF RUNWAYS 22L/R, 31C/R, 13L/C: Climb runway heading to 1300' before turning on course.

TAKE-OFF RUNWAYS 4L/R: Northbound departures assigned headings 360° (CW) thru 080°, turn right climb on heading 100° until leaving 2400. Thence via vector to assigned route.

DME-EQUIPPED AIRCRAFT: Complete initially assigned turn within 4 DME of Midway. Maintain 3000' or assigned lower altitude. Expect clearance to requested altitude 10 minutes after departure.

NON-DME EQUIPPED AIRCRAFT: Complete initially assigned turn south of DPA R-096, maintain 3000' or assigned lower altitude. Expect clearance to requested altitude 10 minutes after departure.

MIDWAY FOUR DEPARTURE (VECTOR) CHICAGO, ILLINOIS CHICAGO MIDWAY (MDW)
94342

FIGURE 95.—Midway Four Departure (Vector).

THIS PAGE INTENTIONALLY LEFT BLANK

16 ILLINOIS

CHAMPAIGN (URBANA)
UNIVERSITY OF ILLINOIS-WILLARD (CMI) 5 SW UTC–6(–5DT) CHICAGO
N40°02.36′ W88°16.68′ H-4H, L-23A
754 B S4 **FUEL** 100LL, JET A1 + OX 1 ARFF Index A IAP
RWY 14R-32L: H8100X150 (CONC-GRVD) S-100, D-180, DT-260 HIRL
RWY 14R: VASI(V4L)—GA 3.0° TCH 31′. **RWY 32L:** MALSR. VASI(V4L)—GA 3.0° TCH 54′.
RWY 04L-22R: H6500X150 (CONC-GRVD) S-100, D-180, DT-260 MIRL
RWY 04L: VASI(V4L)—GA 3.0° TCH 45′. **RWY 22R:** VASI(V4L)—GA 3.0° TCH 41′. Tree.
RWY 18-36: H5299X150 (CONC) S-40, D-50, DT-90 MIRL
RWY 36: VASI(V4L)—GA 3.0° TCH 40′. Tree.
AIRPORT REMARKS: Attended continuously. Rwy 18-36 CLOSED 0600-1200Z‡ indefinitely. PPR for unscheduled air carrier operations with more than 30 passenger seats between 0400-1200Z‡, call arpt manager 217-244-8604. Rwy 04R-22L and Rwy 14L-32R VFR day only, restricted to authorized Flight Schools only. When twr clsd HIRL Rwy 14R-32L preset low ints, to increase ints and ACTIVATE MIRL Rwys 04L-22R MALSR Rwy 32L—CTAF. Itinerant parking on southeast ramp only. Taxiway D not available for air carrier ops with more than 30 passenger seats. NOTE: See Land and Hold Short Operations Section.
COMMUNICATIONS: CTAF 120.4 **ATIS** 124.85 **UNICOM** 122.95
ST LOUIS FSS (STL) TF 1-800-WX-BRIEF. NOTAM FILE CMI.
CHAMPAIGN RCO 122.1R 110.0T (KANKAKEE FSS)
CHAMPAIGN (URBANA) RCO 122.45 (ST LOUIS FSS)
® **CHAMPAIGN APP/DEP CON** 132.85 (134°-312°) 121.35 (313°-133°) 118.25 (1200-0600Z‡)
CHICAGO CENTER APP/DEP CON 121.35 (0600-1200Z‡)
CHAMPAIGN TOWER 120.4 (1200-0600Z‡) **GND CON** 121.8 **CLNC DEL** 128.75
AIRSPACE: CLASS C svc 1200-0600Z‡ ctc **APP CON** other times CLASS G.
RADIO AIDS TO NAVIGATION: NOTAM FILE CMI.
CHAMPAIGN (L) VORTAC 110.0 CMI Chan 37 N40°02.07′ W88°16.56′ at fld. 750/3E.
VEALS NDB (LOM) 407 CM N39°57.97′ W88°10.95′ 315°6.2 NM to fld.
ILS 109.1 I-CMI Rwy 32L. LOM VEALS NDB. ILS unmonitored when twr clsd.
ASR

CHICAGO
CHICAGO MIDWAY (MDW) 9SW UTC–6(–5DT) N41°47.16′ W87°45.15′ CHICAGO
620 B S4 **FUEL** 100LL, JET A1 + OX 2, 4 AOE ARFF Index C COPTER
RWY 13C-31C: H6522X150 (CONC-GRVD) S-95, D-165, DT-250 HIRL H-3H, L-23A, A
RWY 13C: ALSF1. PAPI (P4L)—GA 3.0° TCH 47′. Thld dsplcd 462′. Pole. IAP
RWY 31C: LDIN. REIL. VASI(V4L)—GA 3.0° TCH 52′. Thld dsplcd 696′. Tree.
RWY 04R-22L: H6446X150 (CONC-ASPH-GRVD) S-95, D-165, DT-250 HIRL
RWY 04R: REIL. VASI(V4R)—GA 3.4° TCH 64. Thld dsplcd 518′. Building.
RWY 22L: REIL. VASI(V4R)—GA 3.0° TCH 53′. Thld dsplcd 634′. Pole.
RWY 04L-22R: H5509X150 (ASPH) S-30, D-40 MIRL
RWY 04L: VASI(V4R). Thld dsplcd 758′. Tree. **RWY 22R:** VASI(V4L). Building.
RWY 13L-31R: H5412X150 (ASPH) S-30, D-40 MIRL
RWY 13L: Thld dsplcd 753′. Tree. **RWY 31R:** Pole.
Rwy 13R-31L: H3859X60 (CONC) S-12.5 MIRL
RWY 13R: Pole. **RWY 31L:** Tree.
AIRPORT REMARKS: Attended continuously. Landing fee. Arpt CLOSED to solo student training. Birds on and in vicinity of arpt. Noise abatement procedures: All departures are requested to expedite climb through 1500 ft MSL 0400-1200Z‡. Rwys 13L-31R and 04R-22R not avbl for air carrier ops with more than 30 passenger seats. Rwy 13C PAPI and RVR out of svc indefinitely. Flight Notification Service (ADCUS) available.
WEATHER DATA SOURCES: LAWRS.
COMMUNICATIONS: ATIS 132.75 **UNICOM** 122.95
KANKAKEE FSS (IKK) TF 1-800-WX-BRIEF. NOTAM FILE MDW.
® **APP/DEP CON** 118.4 126.05
MIDWAY TOWER 118.7 135.2 (helicopter ops) **GND CON** 121.7 **CLNC DEL** 121.85 **PRE TAXI CLNC** 121.85
AIRSPACE: CLASS C svc continuous ctc **MIDWAY RADAR** 119.45
RADIO AIDS TO NAVIGATION: NOTAM FILE IKK.
CHICAGO HEIGHTS (L) VORTAC 114.2 CGT Chan 89 N41°30.60′ W87°34.29′ 332° 18.5 NM to fld. 630/2E.
ERMIN NDB (MHW/LOM) 332 HK N41°43.14′ W87°50.19′ 044° 5.5 NM to fld. NOTAM FILE MDW.
KEDZI NDB (MHW/LOM) 248 MX N41°44.49′ W87°41.38′ 315° 3.9 NM to fld. NOTAM FILE MDW.
ILS/DME 109.9 I-MDW Chan 36 Rwy 13C.
ILS 111.5 I-HKH Rwy 04R. LOM ERMIN NDB.
ILS/DME 109.9 I-MXT Chan 36 Rwy 31C. LOM KEDZI NDB.
MLS Chan 660 Rwy 22L. MLS unusable 246°-262° byd 10NM blo 3500′; unusable clockwise byd 262°; elevation unusable clockwise beyond 226° blo 2.0°; elevation unusable counterclockwise byd 222° blo 2.0°. Disregard guidance signals found clockwise byd 314°. Disregard guidance signals found counterclockwise byd 184°.

FIGURE 95A.—Excerpt (MDW).

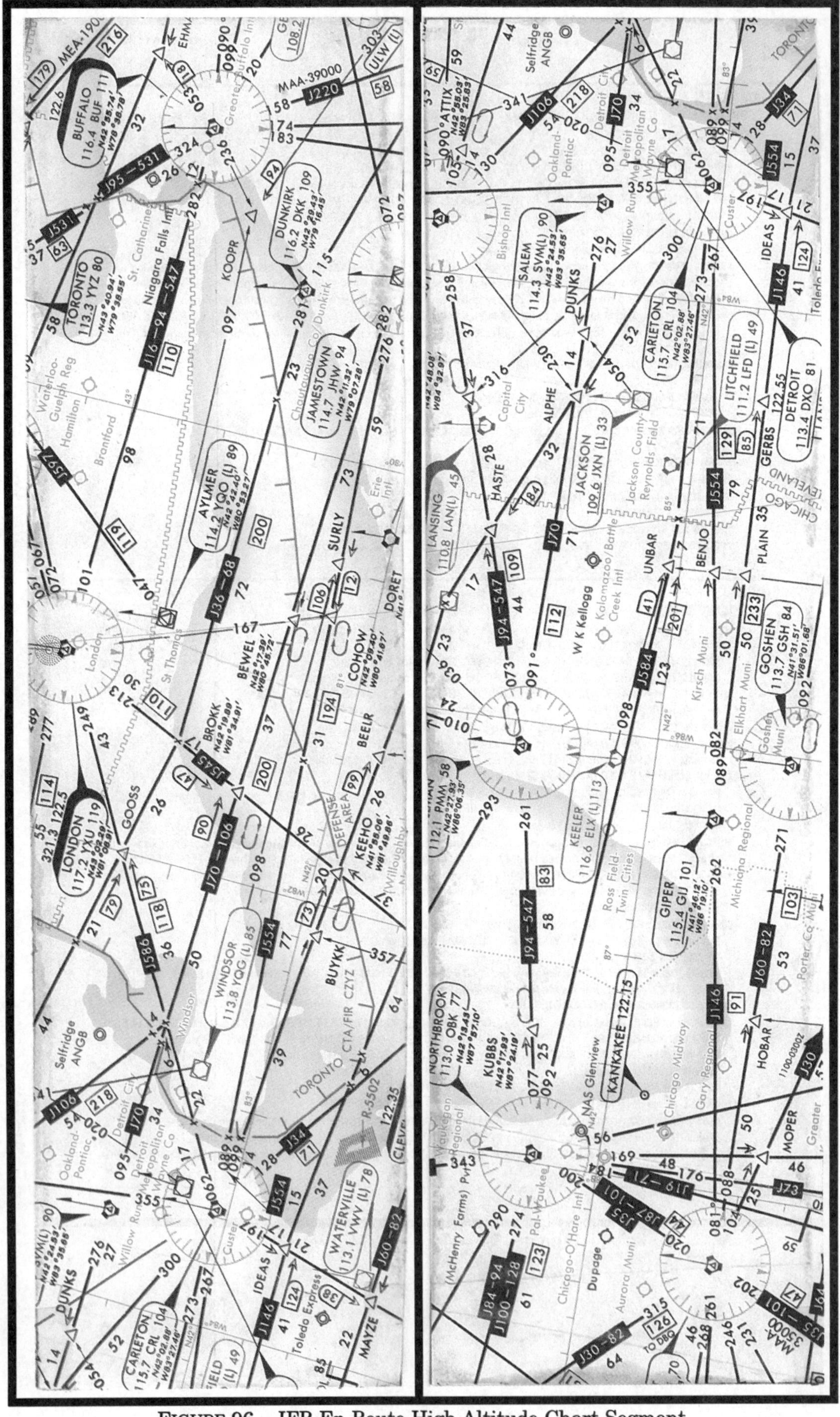

FIGURE 96.—IFR En Route High Altitude Chart Segment.

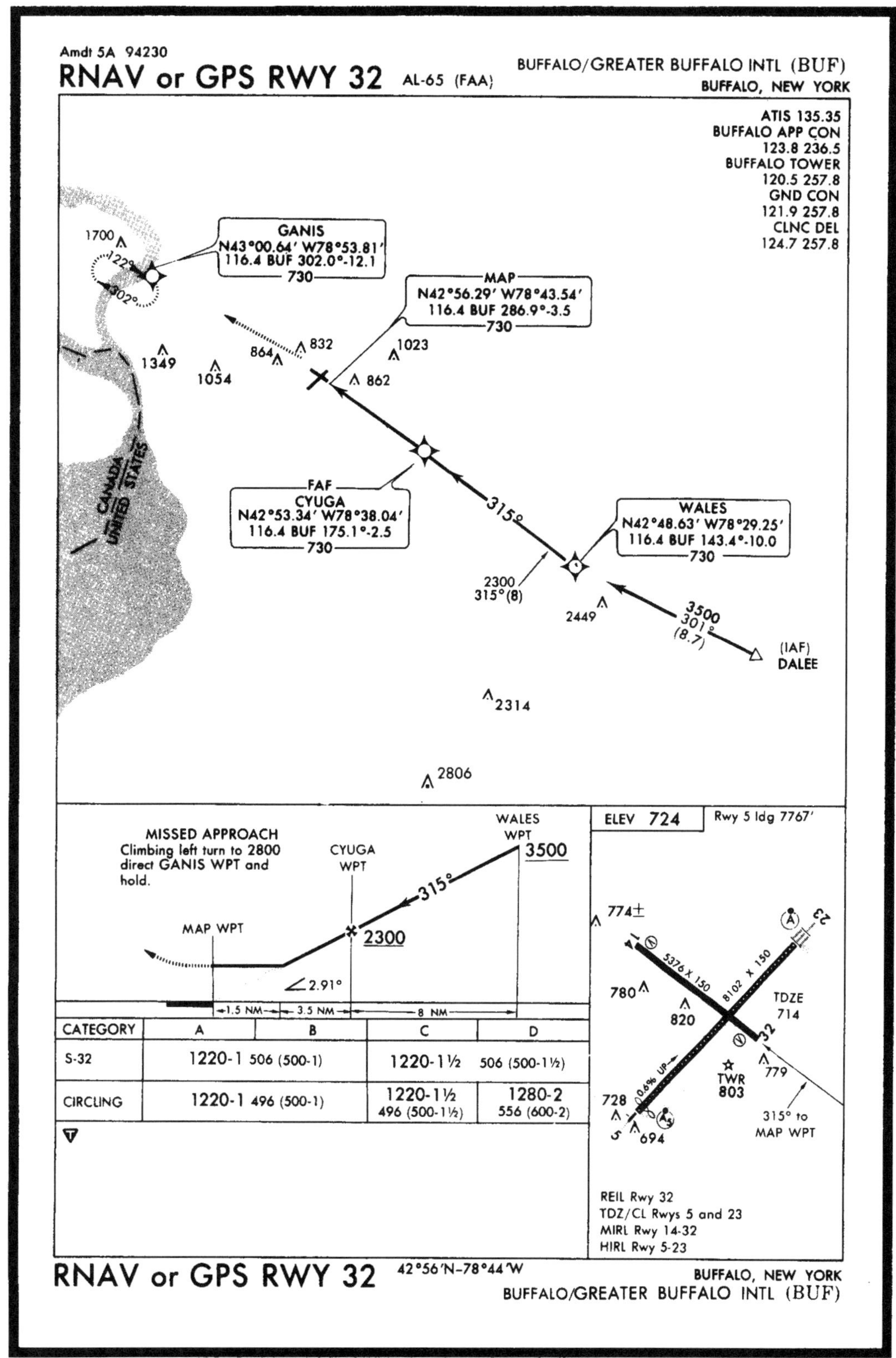

FIGURE 97.—RNAV or GPS RWY 32 (BUF).

THIS PAGE INTENTIONALLY LEFT BLANK

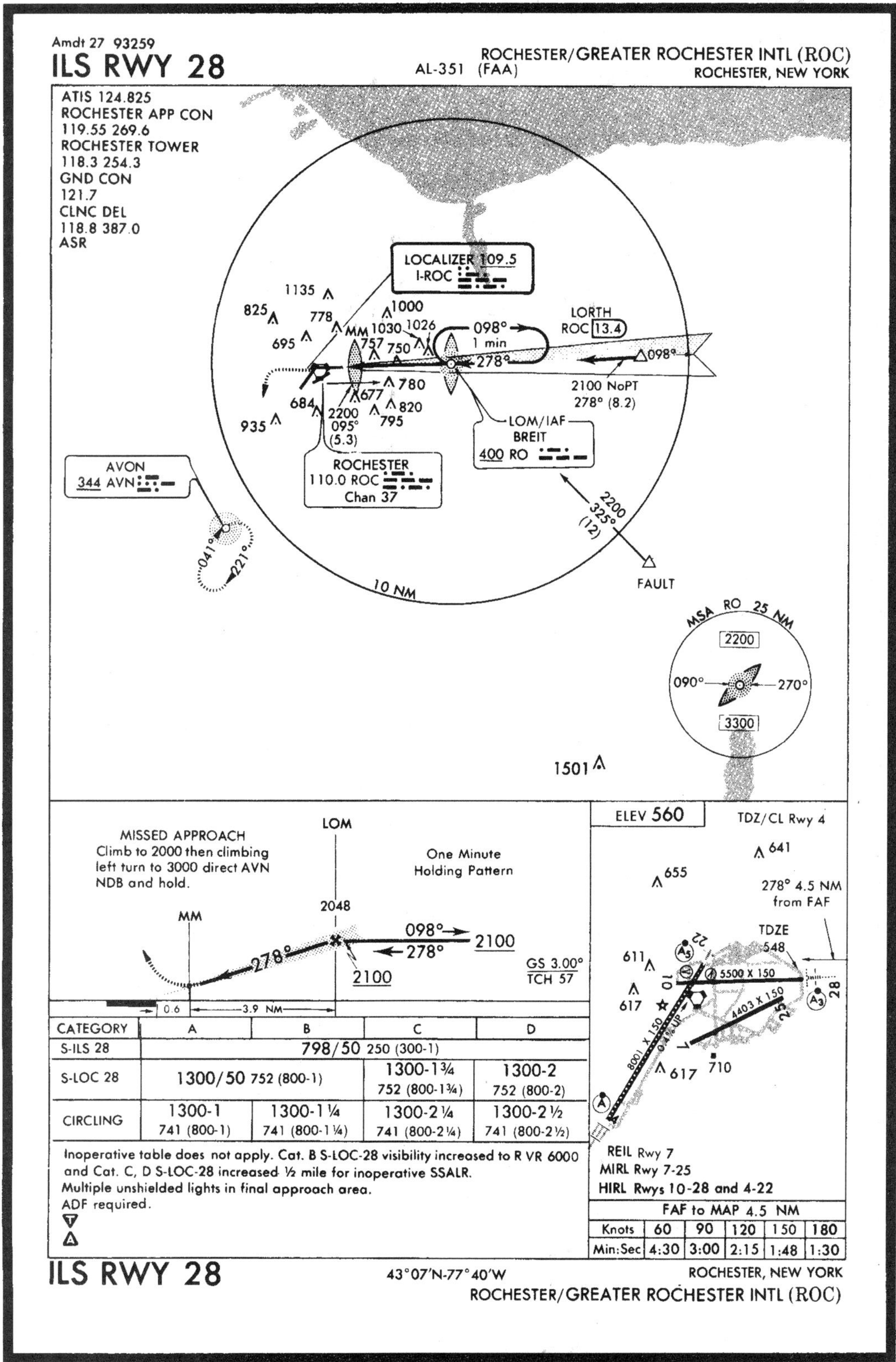

FIGURE 97A.—ILS RWY 28 (ROC).

 ALTERNATE MINS 

94342

INSTRUMENT APPROACH PROCEDURE CHARTS

IFR ALTERNATE MINIMUMS

(NOT APPLICABLE TO USA/USN/USAF)

Standard alternate minimums for non precision approaches are 800-2 (NDB, VOR, LOC, TACAN, LDA, VORTAC, VOR/DME or ASR); for precision approaches 600-2 (ILS or PAR). Airports within this geographical area that require alternate minimums other than standard or alternate minimums with restrictions are listed below. NA - means alternate minimums are not authorized due to unmonitored facility or absence of weather reporting service. Civil pilots see FAR 91. USA/USN/USAF pilots refer to appropriate regulations.

NAME — ALTERNATE MINIMUMS

ALBANY, NY

ALBANY COUNTY .... ILS Rwy 1[1]
ILS Rwy 19[1]
VOR/DME or GPS Rwy 1[1]
VOR Rwy 1[2]
VOR or GPS Rwy 19[1]
VOR or GPS Rwy 28[1]

[1]Category D, 800-2½.
[2]Category C, 800-2¼; Category D, 800-2½.

ALLENTOWN, PA

LEHIGH VALLEY INTL .... ILS Rwy 13
ILS, Categories A,B,C, 700-2; Category D, 700-2¼. LOC, Category D, 800-2¼.

ALTOONA, PA

ALTOONA-BLAIR COUNTY .... ILS Rwy 20[1]
VOR or GPS-A[2]

[1]Categories A,B,C, 900-2½, Category D, 1100-3.
[2]Category D, 1100-3.

BRADFORD, PA

BRADFORD REGIONAL .... VOR/DME Rwy 14
NA when BFD FSS closed.

CORTLAND, NY

CORTLAND COUNTY-CHASE FIELD .... VOR or GPS-A
Categories A,B, 1100-2,Categories C,D, 1100-3.

DUBOIS, PA

DUBOIS-JEFFERSON COUNTY ..... ILS Rwy 25
LOC, NA.

NAME — ALTERNATE MINIMUMS

ELMIRA, NY

ELMIRA/CORNING REGIONAL .... ILS Rwy 6[12]
ILS Rwy 24,1200-3
NDB or GPS Rwy 24,1200-3

[1]Categories A,B, 1200-2; Categories C,D, 1200-3.
[2]NA when control tower closed.

ERIE, PA

ERIE INTL .... ILS Rwy 6[1]
ILS Rwy 24[1]
NDB Rwy 6
NDB Rwy 24
RADAR-1

NA when control tower closed.
[1]ILS, 700-2.

FARMINGDALE, NY

REPUBLIC .... ILS Rwy 14[1]
NDB or GPS Rwy 1[2]

[1]NA when control tower closed.
[2]NA when control zone not effective.

HARRISBURG, PA

CAPITAL CITY .... ILS Rwy 8
Categories A,B, 900-2; Categories C,D, 900-2¾.
NA when control tower closed.

HARRISBURG INTL .... ILS Rwy 13[1]
ILS Rwy 31[1]
VOR or GPS Rwy 31[2]

[1]ILS, Categories C,D, 700-2. LOC, NA.
[2]Categories A,B, 900-2, Category C, 900-2¾, Category D, 900-3.

NE-2

 ALTERNATE MINS 

94342

FIGURE 97B.—IFR Alternate Minimums.

 ALTERNATE MINS 

94342

NAME	ALTERNATE MINIMUMS

PHILADELPHIA, PA(CON'T)

PHILADELPHIA INTL .... **ILS Rwy 9L[1]**
ILS Rwy 9R[2]
ILS Rwy 17[3]
ILS Rwy 27L[2]
ILS Rwy 27R[2]
NDB or GPS Rwy 27L#
RNAV or GPS Rwy 17*

[1]ILS, Category D, 700-2.
[2]ILS, 700-2.
[3]ILS, Categories A,B,C, 700-2; Category D, 700-2¼. LOC, Category D,800-2¼.
#Category C, 800-2¼; Category D, 800-2½.
*Category D, 800-2¼.

PHILIPSBURG, PA

MID-STATE .... **ILS Rwy 16[1]**
NDB Rwy 16[2]
VOR Rwy 24[3]

[1]ILS, Category C, 700-2; Category D, 700-2¼. LOC, Category D, 800-2¼.
[2]Category D, 800-2¼.
[3]Categories A,B, 900-2; Category C, 900-2¼; Category D, 900-2½.

PITTSBURGH, PA

PITTSBURGH INTL .... **ILS Rwy 10L[1]**
ILS Rwy 10R[1]
ILS Rwy 28L[1]
ILS Rwy 28R[1]
ILS Rwy 32[1]
VOR or TACAN Rwy 28L/C[2]

[1]ILS, Category E, 700-2¼. LOC, Category E, 800-2¼.
[2]Category E, 800-2¼.

POUGHKEEPSIE, NY

DUTCHESS COUNTY .... **ILS Rwy 6**

ILS, Categories B,C,D, 700-2.

READING, PA

READING REGIONAL/CARL A. SPAATZ FIELD .... **ILS Rwy 36[1]#**
NDB Rwy 36[2]#
RNAV or GPS Rwy 13[2]*
RNAV or GPS Rwy 18[3]*

[1]ILS, Categories A,B,C, 700-2; Category D, 800-2½. LOC, Category D, 800-2½.
[2]Category D, 800-2½.
[3]Category C, 800-2¼; Category D, 800-2½.
#NA when control tower closed.
*NA when control zone not in effect.

NAME	ALTERNATE MINIMUMS

REEDSVILLE, PA

MIFFLIN COUNTY .... **LOC Rwy 6**

NA when airport unattended.
Category D, 1500-3.

ROCHESTER, NY

GREATER ROCHESTER INTL .... **ILS Rwy 4[1]**
ILS Rwy 22[1]
ILS Rwy 28[2]
NDB or GPS Rwy 28[3]
RADAR-1#
VOR/DME or GPS Rwy 4#
VOR Rwy 4#

[1]ILS, Category D, 700-2¼. LOC, Category D, 800-2¼.
[2]Categories A,B, 800-2; Category C, 800-2¼; Category D, 800-2½.
[3]Category C, 800-2¼; Category D, 800-2½.
#Category D, 800-2¼.

SARANAC LAKE, NY

ADIRONDACK REGIONAL .... **VOR/DME or GPS Rwy 5[1]**
VOR or GPS Rwy 9[2,3]

[1]NA except Categories A,B, 1200-2; Categories C,D, 1200-3, for operators with approved weather reporting service.
[2]Category A, 1000-2; Category B, 1100-2; Categories C,D, 1100-3.
[3]NA except for operators with approved weather reporting service.

STATE COLLEGE, PA

UNIVERSITY PARK .... **ILS Rwy 24[1,2]**
VOR/DME RNAV or GPS Rwy 6[1]
VOR or GPS-B,1300-3

[1]Category D, 900-2¾.
[2]NA when airport unattended.

UTICA, NY

ONEIDA COUNTY .... **NDB Rwy 33**

Category D, 800-2¼.

WATERTOWN, NY

WATERTOWN INTL .... **ILS Rwy 7**

LOC, NA.

WESTHAMPTON BEACH, NY

THE FRANCIS S. GABRESKI .... **ILS Rwy 24**
NDB Rwy 24

NA when control zone not in effect.

NE-2

 ALTERNATE MINS 

94342

FIGURE 97C.—IFR Alternate Minimums.

Form Approved: OMB No. 2120-0034

U.S. DEPARTMENT OF TRANSPORTATION
FEDERAL AVIATION ADMINISTRATION

FLIGHT PLAN

(FAA USE ONLY) ☐ PILOT BRIEFING ☐ VNR | TIME STARTED | SPECIALIST INITIALS
☐ STOPOVER

1. TYPE	2. AIRCRAFT IDENTIFICATION	3. AIRCRAFT TYPE/ SPECIAL EQUIPMENT	4. TRUE AIRSPEED	5. DEPARTURE POINT	6. DEPARTURE TIME PROPOSED (Z)	6. DEPARTURE TIME ACTUAL (Z)	7. CRUISING ALTITUDE
VFR / X IFR / DVFR	N55JB	BE90/A	248 KTS	DFW Dallas Ft. Worth			15,000

8. ROUTE OF FLIGHT
DFW V369 BILEE, CUGAR 4 IAH

9. DESTINATION (Name of airport and city)
IAH
Houston Intercontinental
Houston

10. EST. TIME ENROUTE — HOURS | MINUTES

11. REMARKS L/O = Level off. PPH = Pounds Per Hour

12. FUEL ON BOARD — HOURS | MINUTES

13. ALTERNATE AIRPORT(S)
BPT
Beaumont-Port Arthur
Jefferson County

14. PILOT'S NAME, ADDRESS & TELEPHONE NUMBER & AIRCRAFT HOME BASE

15. NUMBER ABOARD
4

17. DESTINATION CONTACT/TELEPHONE (OPTIONAL)

16. COLOR OF AIRCRAFT
BLUE/YELLOW

CIVIL AIRCRAFT PILOTS. FAR Part 91 requires you file an IFR flight plan to operate under instrument flight rules in controlled airspace. Failure to file could result in a civil penalty not to exceed $1,000 for each violation (Section 901 of the Federal Aviation Act of 1958, as amended). Filing of a VFR flight plan is recommended as a good operating practice. See also Part 99 for requirements concerning DVFR flight plans.

FAA Form 7233-1 (8-82) CLOSE VFR FLIGHT PLAN WITH ____________ FSS ON ARRIVAL

FLIGHT LOG

CHECK POINTS FROM	CHECK POINTS TO	ROUTE ALTITUDE	COURSE	WIND TEMP	SPEED-KTS TAS	SPEED-KTS GS	DIST NM	TIME LEG	TIME TOT	FUEL LEG	FUEL TOT
DFW	L/O	V369 Climb					27		:12:00		231*
L/O	Bilee	V369 15,000		230/42 ISA							
Bilee	Cugar	Cugar 4 15,000									
Cugar	Start Descent	Cugar 4 15,000		230/42 ISA							
Start Descent	IAH	Descent & Approach					25	:14:00		132	
IAH	BPT	Vectors 3000				194	68				

OTHER DATA: * Includes Taxi Fuel
NOTE: Use 850 PPH Total Fuel Flow From L/O To Start Of Descent.
Use 880 PPH Total Fuel Flow For Reserve And Alternate Requirements.
A Missed Approach Requires 82# of Fuel.

TIME and FUEL: As required by FARs.

TIME	FUEL (LB)	
		EN ROUTE
		RESERVE
		ALTERNATE
		TOTAL

FIGURE 98.—Flight Plan/Flight Log.

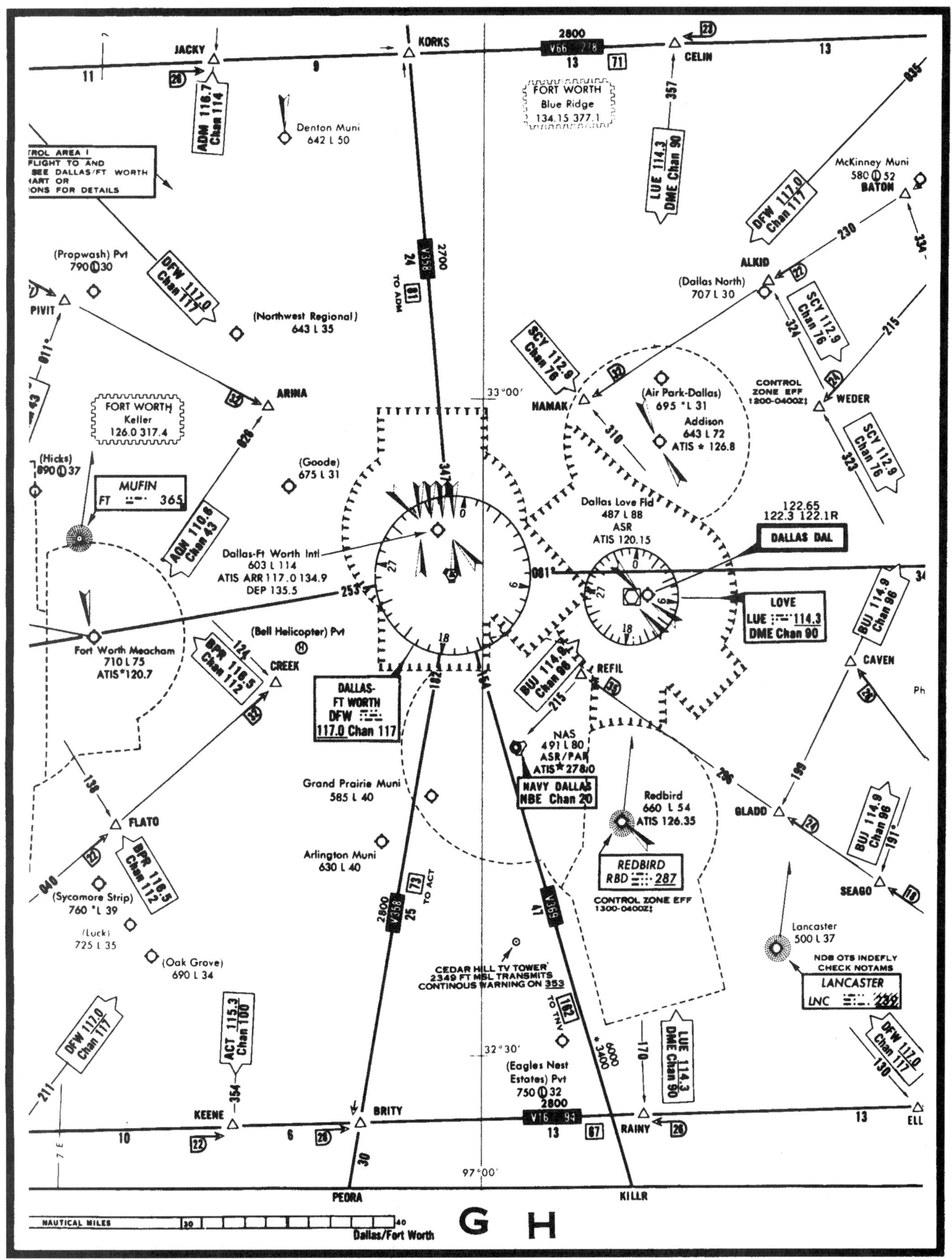

FIGURE 99.—IFR Area Chart Segment.

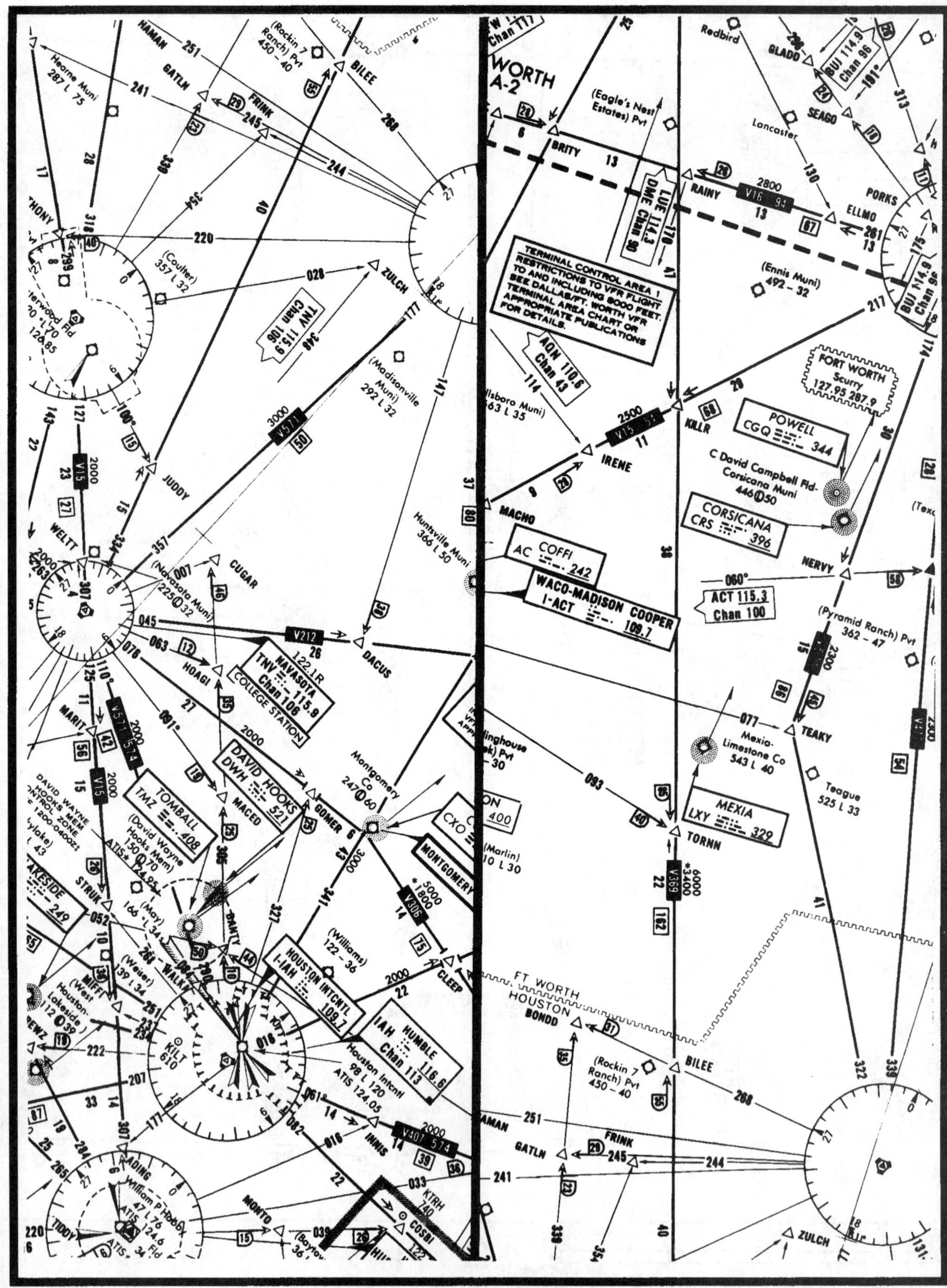

FIGURE 100.—IFR En Route Low Altitude Chart Segment.

TEXAS

§ DALLAS-FORT WORTH INTL (DFW) 12 NW UTC-6(-5DT) 32°53'47"N 97°02'28"W **DALLAS-FT. WORTH H-2K, 4F, 5B, L-13C, A IAP**

603 B FUEL 100LL, JET A OX 1,3 LRA CFR Index E

RWY 17L-35R: H11,388X150 (CONC-GRVD) S-120, D-200, DT-600, DDT-850 HIRL, CL
RWY 17L: ALSF2. TDZ. **RWY 35R:** MALSR. TDZ.
RWY 17R-35L: H11,388X200 (CONC-GRVD) S-120, D-200, DT-600, DDT-850 HIRL, CL
RWY 17R: SSALR TDZ. **RWY 35L:** TDZ. VASI(V6L).
RWY 18L-36R: H11,387X200 (CONC-GRVD) S-120, D-200, DT-600, DDT-850 HIRL, CL
RWY 18L: SSALR.TDZ. **RWY 36R:** TDZ. VASI(V6L).
RWY 18R-36L: H11,388X150 (CONC-GRVD) S-120, D-200, DT-600, DDT-850 HIRL, CL
RWY 18R: ALSF2. TDZ **RWY 36L:** MALSR. TDZ.
RWY 13L-31R: H9000X200 (CONC-GRVD) S-120, D-200, DT-600, DDT-850 HIRL, CL .5% up NW.
RWY 13L: TDZ. VASI(V6L)—Upper GA 3.25°TCH 93'. Lower GA 3.0°TCH 47'.
RWY 31 R: MALSR. TDZ.
RWY 13R-31L: H9300X150 (CONC-GRVD) S-120, D-220, DT-600, DDT-850 HIRL, CL
RWY 13 R: MALSR. TDZ. **RWY 31L:** TDZ.
RWY 18S-36S: H4000X100 (CONC)

AIRPORT REMARKS: Attended continuously. Prior Permission Required from arpt ops for General Aviation acft to proceed to airline terminal gate except to General Aviation Facility. Rwy 18S-36S located on taxiway G, 4000' long 100' wide restricted to prop acft 12,500 lbs. & below and stol acft daylight VFR plus IFR departures. Prior permission required from the primary tenant airlines to operate within central terminal area. CAUTION: proper minimum clearance may not be maintained within the central terminal area. Landing fee. Clearways 500x1000 each end Rwy 17L-35R, Rwy 17R-35L, Rwy 18L-36R and Rwy 18R-36L. Flight Notification Service (ADCUS)available.

WEATHER DATA SOURCES: LLWAS.

COMMUNICATIONS: ATIS 117.0 134.9 (ARR) 135.5 (DEP) **UNICOM** 122.95
FORT WORTH FSS (FTW) LC 624-8471, Toll free call, dial 1-800-WX-BRIEF. NOTAM FILE DFW
® **REGIONAL APP CON** 119.05(E) 119.4(E) 125.8(W) 132.1(W)
REGIONAL TOWER 126.55 (E) 124.15 (W) **GND CON** 121.65 133.15(E) 121.8 (W) **CLNC DEL** 128.25 127.5
® **REGIONAL DEP CON** 118.55 (E) 124.25 (WEST) 127.75 (NORTH-SOUTH)
TCA Group I: See VFR Terminal Area chart.

RADIO AIDS TO NAVIGATION: NOTAM FILE DFW.
(H) VORTACW 117.0 DFW Chan 117 32°51'57"N97°01'40"W at fld. 560/08E.
VOR Portion unusable 045°-050° all altitudes and distances, 350-100° beyond 30 NM below 2100'.
ISSUE NDB (LOM) 233 PK 32°47'35"N97°01'49"W 353° 5.1 NM to fld.
JIFFY NDB (LOM) 219 FL 32°59'45"N97°01'46"W 173° 5.1 NM to fld.
ILS/DME 109.5 I-LWN Chan 32 Rwy 13R
ILS/DME 109.1 I-FLQ Chan 28 Rwy 17L LOM JIFFY NDB
ILS 111.5 I-JHZ Rwy 17R LOM JIFFY NDB
ILS 111.3 I-CIX Rwy 18L
ILS/DME 111.9 I-VYN Chan 56 Rwy 18R
ILS 110.9 I-RRA Rwy 31R
ILS/DME 109.1 I-PKQ Chan 28 Rwy 35R LOM ISSUE NDB
ILS/DME 111.9 I-BXN Chan 56 Rwy 36L

---

§ HOUSTON INTERCONTINENTAL (IAH) 15N UTC-6(-5DT) 29°58'49"N 95°20'22"W **HOUSTON H-5B, L-17B IAP**

98 B S4 FUEL 100LL, JET A OX2 LRA CFR Index D

RWY 14L-32R: H1200X150 (CONC-GRVD) S-100, D-200, DT-400, DDT-778 HIRL, CL
RWY 14L: MALSR. VASI(V4L)—GA 3.0°TCH 54'. **RWY 32R:** MALSR.
RWY 09-27: H10000X150 (ASPH-GRVD) S-75, D-191, DT-400, DDT-850 HIRL,CL
RWY 09: MALSR. TDZ. PAPI(P4L)—GA 3.0°TCH 63'.
RWY 27: ALSF2. TDZ. PAPI(P4L)—GA 3.0°TCH 63'.
RWY 08-26: H9401X150 (CONC-GRVD) S-120, D-155, DT-265 HIRL, CL
RWY 08: MALSR. TDZ. **RWY 26:** ALSF2. TDZ. VASI(V4L)—GA 3.0°TCH 53'.
RWY 14R-32L: H6038X100 (ASPH-GRVD) S-30, D-60, DT-60 MIRL
RWY 14R: VASI(V4L)—GA 3.0°TCH 40'. Road. **RWY 32L:** VASI(V4L)—GA 3.0°TCH 45'.

AIRPORT REMARKS: Attended continuously. CAUTION: Birds on and in vicinity of arpt. CAUTION—Approach end of rwy 26 bright lgts approximately one mile from thld and 900' South of centerline. Caution—Deer on and in vicinity of arpt. Rwy 14R-32L CLOSED to acft over 140,000 lbs gross weight. Landing Fee. Flight Notification Service (ADCUS) available.

WEATHER DATA SOURCES: LLWAS

COMMUNICATIONS: ATIS 124.05 **UNICOM** 122.95
MONTGOMERY COUNTY FSS (CXO) Toll free call, dial 1-800-WX-BRIEF. NOTAM FILE IAH.
® **APP CON** 124.35 (West) 127.25 (North and East)
TOWER 118.1 (135.15 copter control) **GND CON** 121.7 **CLNC DEL** 128.1 (135.15 copter control)
® **DEP CON** 123.8 (West) 119.7 (North and East)
TCA Group II: VFR Terminal Area chart.

RADIO AIDS TO NAVIGATION: NOTAM FILE IAH.
HUMBLE (H) VORTACW 116.6 IAH Chan 113 29°57'24"N95°20'44"W at fld. 90/08E. **HIWAS.**
MARBE NDB (LOM) 379 HS 30°04'29"N 95°24'45"W 146° 5.9 NM to fld.
NIXIN NDB (LOM) 326 JY 29°59'36"N 95°12'54"W 257° 6.5 NM to fld.
ILS/DME 109.7 I-JYV Chan 34 Rwy 26 LOM NIXIN NDB
ILS 111.9 I-HSQ Rwy 14L LOM MARBE NDB
ILS/DME 109.7 I-IAH Chan 34 Rwy 08
ILS/DME 110.9 I-UYO Chan 34 Rwy 09
ILS 111.9 I-CDG Rwy 32R

FIGURE 101.—Airport/Facility Directory Excerpts.

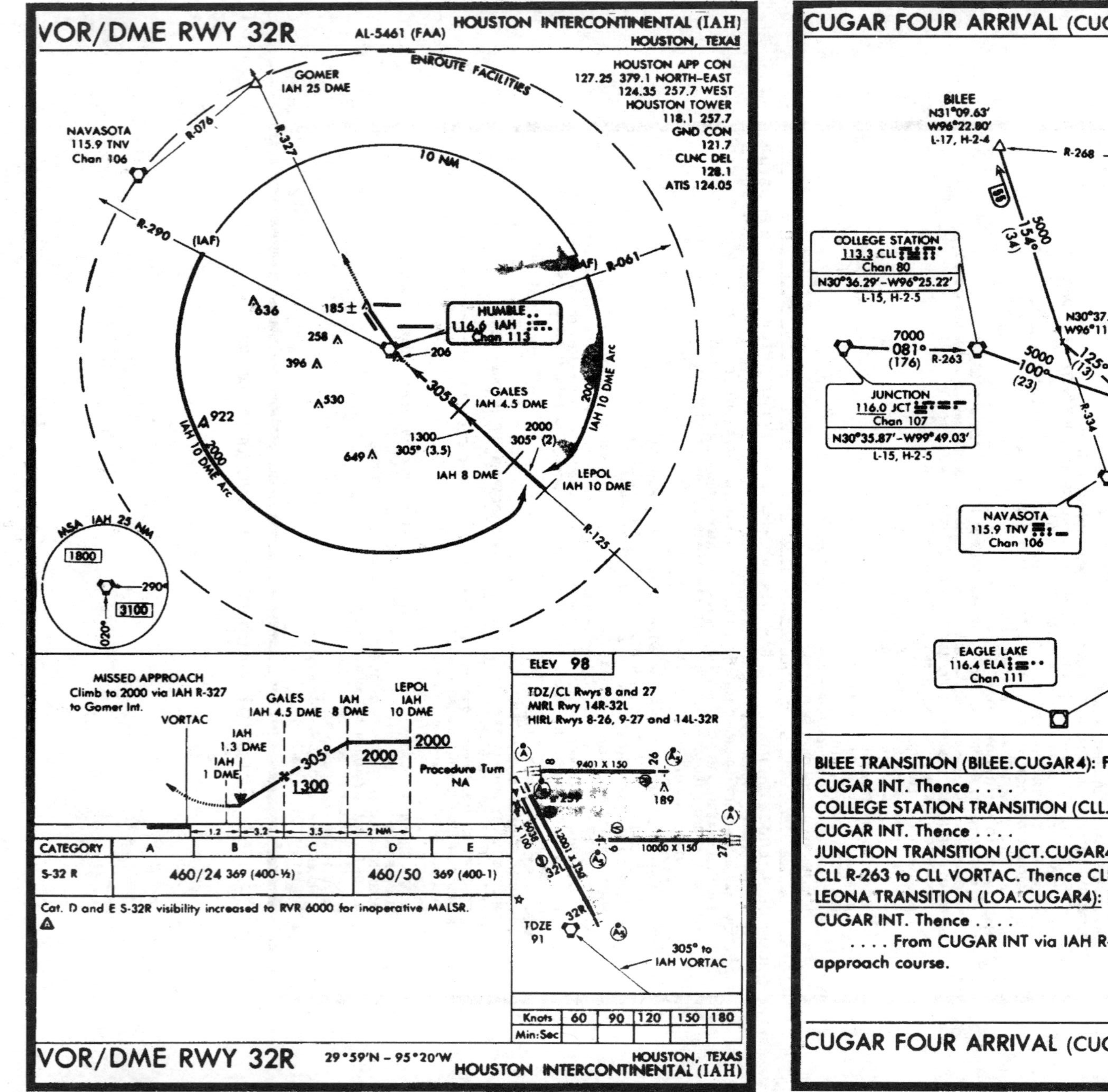

FIGURE 102.—VOR/DME RWY 32R (IAH)/Cugar Four Arrival (Cugar.Cugar4).

Form Approved: OMB No. 2120-0026

U.S. DEPARTMENT OF TRANSPORTATION FEDERAL AVIATION ADMINISTRATION **FLIGHT PLAN**	(FAA USE ONLY) ☐ PILOT BRIEFING ☐ VNR ☐ STOPOVER	TIME STARTED	SPECIALIST INITIALS

1. TYPE	2. AIRCRAFT IDENTIFICATION	3. AIRCRAFT TYPE/ SPECIAL EQUIPMENT	4. TRUE AIRSPEED	5. DEPARTURE POINT	6. DEPARTURE TIME PROPOSED (Z)	ACTUAL (Z)	7. CRUISING ALTITUDE
VFR / X IFR / DVFR	N91JB	BE1900/A	233 KTS	TUS TUCSON			FL220

8. ROUTE OF FLIGHT
TUS TUS3.GBN, J104TNP, TNP.DOWNE 3 LAX

9. DESTINATION (Name of airport and city)	10. EST. TIME ENROUTE HOURS	MINUTES	11. REMARKS
LAX LOS ANGELES INT'L Los Angeles			L/O = Level Off PPH = Pounds Per Hour TEC = Tower Enroute Control This flight is operating under FAR 135.

12. FUEL ON BOARD HOURS	MINUTES	13. ALTERNATE AIRPORT(S)	14. PILOT'S NAME, ADDRESS & TELEPHONE NUMBER & AIRCRAFT HOME BASE	15. NUMBER ABOARD
		BUR Burbank-Glendale-Pasadena	17. DESTINATION CONTACT/TELEPHONE (OPTIONAL)	18

16. COLOR OF AIRCRAFT Maroon/White

CIVIL AIRCRAFT PILOTS. FAR Part 91 requires you file an IFR flight plan to operate under instrument flight rules in controlled airspace. Failure to file could result in a civil penalty not to exceed $1,000 for each violation (Section 901 of the Federal Aviation Act of 1958, as amended). Filing of a VFR flight plan is recommended as a good operating practice. See also Part 99 for requirements concerning DVFR flight plans.

FAA Form 7233-1 (8-82) CLOSE VFR FLIGHT PLAN WITH ________________ FSS ON ARRIVAL

FLIGHT LOG

CHECK POINTS FROM	TO	ROUTE ALTITUDE	COURSE	WIND TEMP	SPEED-KTS TAS	GS	DIST NM	TIME LEG	TOT	FUEL LEG	TOT
TUS	L/O	TUS3.GBN Climb					73		:25:00		350*
L/O	GBN	TUS3.GBN FL220		280/46 ISA-3							
GBN	INT. J104	J104 FL220									
INT J104	PKE			280/46 ISA-3							
PKE	TNP										
TNP	Start Descent										
Start Descent	Downe 3 LAX	Descent & Approach					52	:18:00		170	
LAX	BUR	TEC 3000					31	:19:00			

OTHER DATA: * Includes Taxi Fuel
NOTE: Use 676 PPH Total Fuel Flow From L/O To Start Of Descent.
Use 726 PPH Total Fuel Flow For Reserve And Alternate Requirements.
A Missed Approach Requires 120# of Fuel.

TIME and FUEL: As required by FARs.

TIME	FUEL (LB)	
		EN ROUTE
		RESERVE
		ALTERNATE
		TOTAL

FIGURE 103.—Flight Plan/Flight Log.

THIS PAGE INTENTIONALLY LEFT BLANK

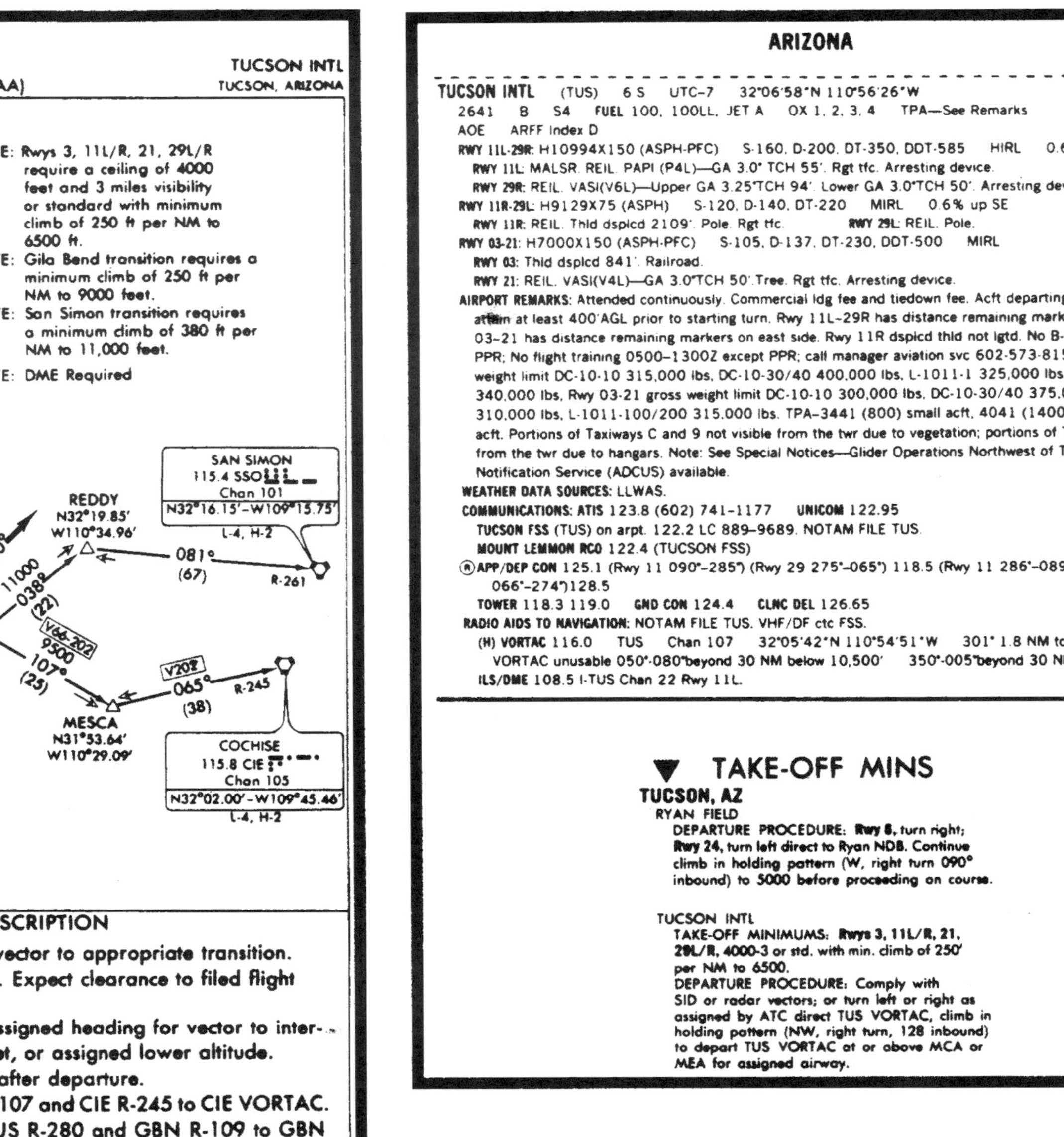

DEPARTURE ROUTE DESCRIPTION

TAKE-OFF RUNWAY 3: Fly heading 030° for vector to appropriate transition. Maintain 17,000 feet or assigned lower altitude. Expect clearance to filed flight level 10 minutes after departure.

TAKE-OFF RUNWAYS 11L/R, 21, 29L/R: Fly assigned heading for vector to intercept appropriate transition. Maintain 17,000 feet, or assigned lower altitude. Expect clearance to filed flight level 10 minutes after departure.

COCHISE TRANSITION (TUS3.CIE): Via TUS R-107 and CIE R-245 to CIE VORTAC.

GILA BEND TRANSITION (TUS3.GBN): Via TUS R-280 and GBN R-109 to GBN VORTAC.

SAN SIMON TRANSITION (TUS3.SSO): Via TUS R-038 and SSO R-261 to SSO VORTAC.

TOTEC TRANSITION (TUS3.TOTEC): Via TUS R-308 to TOTEC INT.

TUCSON THREE DEPARTURE
(PILOT NAV) (TUS3.TUS)

TUCSON, ARIZONA
TUCSON INTL

ARIZONA

TUCSON INTL (TUS) 6 S UTC-7 32°06'58"N 110°56'26"W **PHOENIX**
2641 B S4 **FUEL** 100, 100LL, JET A OX 1, 2, 3, 4 TPA—See Remarks **H-2H, L-4F**
AOE ARFF Index D **IAP**

RWY 11L-29R: H10994X150 (ASPH-PFC) S-160, D-200, DT-350, DDT-585 HIRL 0.6% up SE
RWY 11L: MALSR. REIL. PAPI (P4L)—GA 3.0° TCH 55'. Rgt tfc. Arresting device.
RWY 29R: REIL. VASI(V6L)—Upper GA 3.25°TCH 94'. Lower GA 3.0°TCH 50'. Arresting device.
RWY 11R-29L: H9129X75 (ASPH) S-120, D-140, DT-220 MIRL 0.6% up SE
RWY 11R: REIL. Thld dsplcd 2109'. Pole. Rgt tfc. **RWY 29L:** REIL. Pole.
RWY 03-21: H7000X150 (ASPH-PFC) S-105, D-137, DT-230, DDT-500 MIRL
RWY 03: Thld dsplcd 841'. Railroad.
RWY 21: REIL. VASI(V4L)—GA 3.0°TCH 50'. Tree. Rgt tfc. Arresting device.

AIRPORT REMARKS: Attended continuously. Commercial ldg fee and tiedown fee. Acft departing Rwy 11R reqd to attain at least 400'AGL prior to starting turn. Rwy 11L-29R has distance remaining markers on both sides. Rwy 03-21 has distance remaining markers on east side. Rwy 11R dsplcd thld not lgtd. No B-747 training except PPR; No flight training 0500-1300Z except PPR; call manager aviation svc 602-573-8152. Rwy 11L-29R gross weight limit DC-10-10 315,000 lbs, DC-10-30/40 400,000 lbs, L-1011-1 325,000 lbs, L-1011-100/200 340,000 lbs, Rwy 03-21 gross weight limit DC-10-10 300,000 lbs, DC-10-30/40 375,000 lbs, L-1011-01 310,000 lbs, L-1011-100/200 315,000 lbs. TPA-3441 (800) small acft, 4041 (1400) large/heavy turbojet acft. Portions of Taxiways C and 9 not visible from the twr due to vegetation; portions of Taxiway 2 not visible from the twr due to hangars. Note: See Special Notices—Glider Operations Northwest of Tucson, Arizona. Flight Notification Service (ADCUS) available.

WEATHER DATA SOURCES: LLWAS.

COMMUNICATIONS: ATIS 123.8 (602) 741-1177 **UNICOM** 122.95
TUCSON FSS (TUS) on arpt. 122.2 LC 889-9689. NOTAM FILE TUS.
MOUNT LEMMON RCO 122.4 (TUCSON FSS)
Ⓡ **APP/DEP CON** 125.1 (Rwy 11 090°-285°) (Rwy 29 275°-065°) 118.5 (Rwy 11 286°-089°) (Rwy 29 066°-274°)128.5
TOWER 118.3 119.0 **GND CON** 124.4 **CLNC DEL** 126.65

RADIO AIDS TO NAVIGATION: NOTAM FILE TUS. VHF/DF ctc FSS.
(H) **VORTAC** 116.0 TUS Chan 107 32°05'42"N 110°54'51"W 301° 1.8 NM to fld. 2670/12E.
VORTAC unusable 050°-080°beyond 30 NM below 10,500' 350°-005°beyond 30 NM below 11,200'
ILS/DME 108.5 I-TUS Chan 22 Rwy 11L.

▼ TAKE-OFF MINS

TUCSON, AZ

RYAN FIELD
DEPARTURE PROCEDURE: **Rwy 6**, turn right; **Rwy 24**, turn left direct to Ryan NDB. Continue climb in holding pattern (W, right turn 090° inbound) to 5000 before proceeding on course.

TUCSON INTL
TAKE-OFF MINIMUMS: **Rwys 3, 11L/R, 21, 29L/R**, 4000-3 or std. with min. climb of 250' per NM to 6500.
DEPARTURE PROCEDURE: Comply with SID or radar vectors; or turn left or right as assigned by ATC direct TUS VORTAC, climb in holding pattern (NW, right turn, 128 inbound) to depart TUS VORTAC at or above MCA or MEA for assigned airway.

FIGURE 104.—Tucson Three Departure (Pilot Nav) (TUS3.TUS).

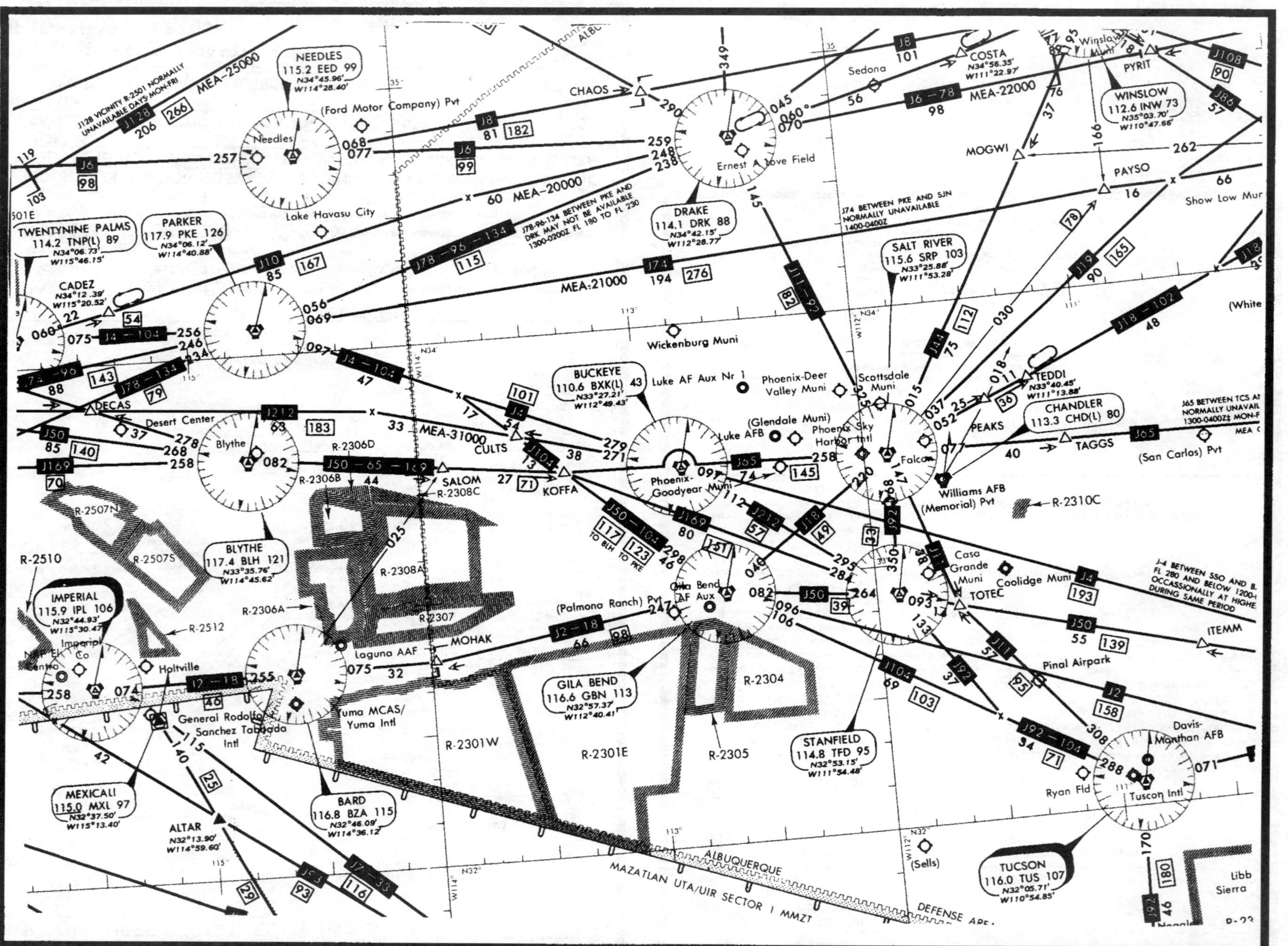

FIGURE 105.—IFR En Route High Altitude Chart Segment.

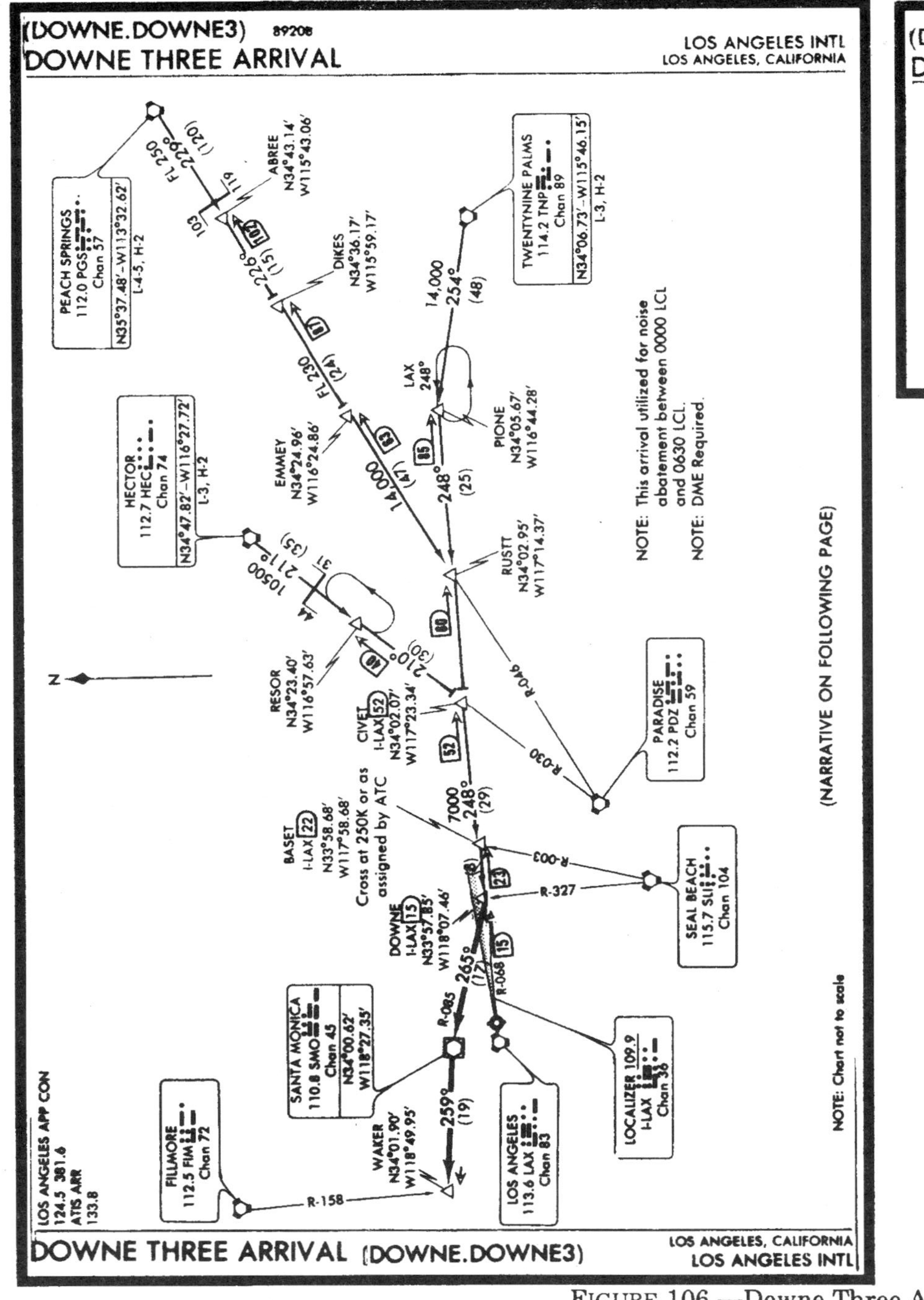

(DOWNE.DOWNE3) 89208

LOS ANGELES INTL
LOS ANGELES, CALIFORNIA

DOWNE THREE ARRIVAL

ARRIVAL DESCRIPTION

HECTOR TRANSITION (HEC.DOWNE3): From over HEC VORTAC via HEC R-211 and PDZ R-030 to CIVET INT, then LAX R-068 to DOWNE INT. Thence . . . .
PEACH SPRINGS TRANSITION (PGS.DOWNE3): From over PGS VORTAC via PGS R-229 and PDZ R-046 to RUSTT INT, then LAX R-068 to DOWNE INT. Thence . . . .
TWENTYNINE PALMS TRANSITION (TNP.DOWNE3): From over TNP VORTAC via TNP R-254 to PIONE DME, then LAX R-068 to DOWNE INT. Thence . . . .
. . . . From DOWNE INT via SMO R-085 to SMO VOR/DME, then via SMO R-259 to WAKER INT, expect vector to final approach course for runways 6 and 7.

FIGURE 106.—Downe Three Arrival (Downe.Downe3).

THIS PAGE INTENTIONALLY LEFT BLANK

Amdt 2 90347 (CAT II)
ILS RWY 25L AL-237 (FAA)
LOS ANGELES INTL (LAX)
LOS ANGELES, CALIFORNIA

ATIS ARR 133.8
DEP 135.65
LOS ANGELES APP CON
124.5 381.6
LOS ANGELES TOWER
N 133.9 239.3
S 120.95 379.1
GND CON
N 121.65 327.0
S121.75 327.0
CLNC DEL
121.4 327.0

MISSED APPROACH
Climb to 520 then climbing left turn to 2000 via heading 220° and LAX R-192 then climb to 3000 to INISH INT/LAX 12 DME.

CATEGORY	A	B	C	D
S-ILS 25L	251/16 150 (RA 163)			
S-ILS 25L	201/12 100 (RA 111)			

Simultaneous approaches authorized with Runway 24L/R.

ELEV 126
Rwy 25R ldg 11134'
Rwy 6R ldg 9964'

TDZ/CL Rwys 6R, 24R and 25L
HIRL all rwys

CATEGORY II ILS-SPECIAL AIRCREW
& AIRCRAFT CERTIFICATION REQUIRED

ILS RWY 25L (CAT II) 33°57'N - 118°24'W
LOS ANGELES, CALIFORNIA
LOS ANGELES INTL (LAX)

CALIFORNIA

LOS ANGELES

LOS ANGELES INTL (LAX) 9 SW UTC-8(-7DT) 33°56'33"N 118°24'26"W LOS ANGELES COPTER H-2G, L-3B, A IAP
126 B S4 FUEL 100, 100LL, JET A OX 4 LRA ARFF Index E
RWY 07L-25R: H12091X150 (CONC-GRVD) S-175, D-225, DT-400, DDT-900 HIRL CL
RWY 07L: MALSR. VASI(V4L)—GA 2.5° TCH 59'. Building. Rgt tfc.
RWY 25R: MALSR. Thld dsplcd 957'. Railroad.
RWY 07R-25L: H11096X200 (CONC-GRVD) S-175, D-225, DT-400, DDT-900 HIRL CL
RWY 07R: MALSR. VASI(V4L)—GA 3.0° TCH 56'. Pole. Rgt tfc. RWY 25L: ALSF2. TDZ. Railroad
RWY 06R-24L: H10285X150 (CONC-GRVD) S-175, D-225, DT-400, DDT-900 HIRL CL
RWY 06R: MALSR. TDZ. VASI(V6L)—Upper GA 3.25° TCH 92'. Lower GA 3.0° TCH 60'. Thld dsplcd 321'. Pole.
RWY 24L: MALSR. Rgt tfc.
RWY 06L-24R: H8925X150 (CONC-GRVD) S-175, D-225, DT-400, DDT-900 HIRL CL
RWY 06L: MALSR. VASI(V6L)—Upper GA 3.25° TCH 94'. Lower GA 3.0°TCH 54'. Pole.
RWY 24R: ALSF2. TDZ. Sign. Rgt tfc.
AIRPORT REMARKS: Attended continuously. Turbulence may be deflected upward from the blast fence 180' E of Rwy 25R. CAUTION: Impaired wing clearance may exist on taxiway J between 30K and 19K when taxilane K occupied; 165' centerline to centerline Twr will advise. Rwy 07L-25R hold lines have been relocated N on Taxiways 28J, 30J, 32J, 36J and 42J. Practice instrument approaches and touch and go landings are prohibited. Taxiway Tango between taxilanes 32 and 35 north of terminal one is restricted to B-767 or smaller acft. Taxiways 2L, 8L, 11F, 20G, 30L, 32L, 36L, 42L, T33, and 32S will not accommodate A B747-200. Noise sensitive arpt. On westerly this no turns before crossing shoreline. Over-ocean apchs utilized 0800-1430Z‡. Rwy 24R ALSF2 operates as SSALR till weather goes below VFR. Rwy 25L ALSF2 operates as SSALR until weather goes below VFR. Rwy 07L-25R FAA strength evaluation DC-10-10 540,000 lbs, DC-10-30 600,000 lbs, L-1011 600,000 lbs, B747-SP 630,000 lbs. Rwy 06L-24R FAA strength evaluation DC-10-10 430,000 lbs, DC-10-30 510,000 lbs, L-1011 510,000 lbs. Rwy 07R-25L FAA strength evaluation DC-10-10 400,000 lbs, DC-10-30 600,000 lbs, L-1011 600,000 lbs, B747-SP 710,000 lbs. Rwy 06R-24L FAA strength evaluation DC-10-10 340,000 lbs, DC-10-30 480,000 lbs, L-1011 480,000 lbs. Dual wheel acft up to 200,000 lbs, dual tandem wheel up to 350,000 lbs and double dual tanden wheel up to 834,000 lbs and DC10-10 to 430,000 lbs, DC10-30 to 578,000 lbs and L-1011 to 466,000 lbs and A-300 to 366,000 lbs regularly opr on all rwys. A 700'X500' clearway has been established at west end of Rwy 24L. Numerous birds on and in vicinity of airport. Rwy 25R MALSR out of service indefinitely. Flight Notification Service (ADCUS) available.
WEATHER DATA SOURCES: LLWAS.
COMMUNICATIONS: ATIS ARRIVAL 133.8 ATIS DEP 135.65 (213) 646-2297 UNICOM 122.95
HAWTHORNE FSS (HHR) TF 1-800-WX-BRIEF. NOTAM FILE LAX.
Ⓡ APP CON 128.5 (045°-089°), 124.9 (090°-224°), 124.5 (225°-044°)
TOWER 133.9 (N. complex), 120.95 (S. complex), 119.8 120.35 (helicopters).
GND CON 121.75 (S. complex), 121.65 (N. complex) CLNC DEL 121.4
Ⓡ DEP CON 125.2 (249°-044°) 124.3 (045°-248°)
TCA: See VFR Terminal Area chart.
RADIO AIDS TO NAVIGATION: NOTAM FILE LAX.
(H) VORTAC 113.6 LAX Chan 83 33°55'59"N 118°25'52"W 050° 1.3 NM to fld. 180/15E
VOR unusable 270-280° 17-35 NM below 8000' 280-300° 10-20 NM below 8000'
ROMEN NDB (LOM) 278 OS 33°57'54"N 118°16'37"W 244° 6.6 NM to fld.
ILS/DME 108.5 I-OSS Chan 22 Rwy 24R LOM ROMEN NDB
ILS/DME 108.5 I-HQB Chan 22 Rwy 24L LOM ROMEN NDB
ILS/DME 109.9 I-LAX Chan 36 Rwy 25L
ILS/DME 109.9 I-CFN Chan 36 Rwy 25R
ILS/DME 111.7 I-UWU Chan 54 Rwy 06L
ILS/DME 111.7 I-GPE Chan 54 Rwy 06R
ILS/DME 111.1 I-IAS Chan 48 Rwy 07L
ILS/DME 111.1 I-MKZ Chan 48 Rwy 07R

FIGURE 107.—ILS RWY 25L (CAT II) – LAX.

THIS PAGE INTENTIONALLY LEFT BLANK

Form Approved: OMB No. 2120-0034

U.S. DEPARTMENT OF TRANSPORTATION FEDERAL AVIATION ADMINISTRATION **FLIGHT PLAN**	(FAA USE ONLY) ☐ PILOT BRIEFING ☐ VNR ☐ STOPOVER	TIME STARTED	SPECIALIST INITIALS

1. TYPE	2. AIRCRAFT IDENTIFICATION	3. AIRCRAFT TYPE/ SPECIAL EQUIPMENT	4. TRUE AIRSPEED	5. DEPARTURE POINT	6. DEPARTURE TIME PROPOSED (Z)	ACTUAL (Z)	7. CRUISING ALTITUDE
VFR							
X IFR	N131JB	BH206/A	115 KTS	DFW Dallas Ft. Worth			7,000
DVFR							

8. ROUTE OF FLIGHT
DFW V369 BILEE, CUGAR 4 IAH

9. DESTINATION (Name of airport and city)	10. EST. TIME ENROUTE HOURS	MINUTES	11. REMARKS
IAH Houston Intercontinental Houston			L/O = Level off. PPH = Pounds Per Hour

12. FUEL ON BOARD HOURS	MINUTES	13. ALTERNATE AIRPORT(S)	14. PILOT'S NAME, ADDRESS & TELEPHONE NUMBER & AIRCRAFT HOME BASE	15. NUMBER ABOARD
		HOU William P. Hobby	17. DESTINATION CONTACT/TELEPHONE (OPTIONAL)	4

16. COLOR OF AIRCRAFT BLUE/YELLOW

CIVIL AIRCRAFT PILOTS. FAR Part 91 requires you file an IFR flight plan to operate under instrument flight rules in controlled airspace. Failure to file could result in a civil penalty not to exceed $1,000 for each violation (Section 901 of the Federal Aviation Act of 1958, as amended). Filing of a VFR flight plan is recommended as a good operating practice. See also Part 99 for requirements concerning DVFR flight plans.

FAA Form 7233-1 (8-82) CLOSE VFR FLIGHT PLAN WITH ________________ FSS ON ARRIVAL

FLIGHT LOG

CHECK POINTS FROM	TO	ROUTE / ALTITUDE	COURSE	WIND / TEMP	SPEED-KTS TAS	GS	DIST NM	TIME LEG	TOT	FUEL LEG	TOT
DFW	L/O	V369 / Climb					23		:14:00		123*
L/O	Bilee	V369 / 7,000		220/36 / ISA							
Bilee	Cugar	Cugar 4 / 7,000									
Cugar	Start Descent	Cugar 4 / 7,000		220/36 / ISA							
Start Descent	IAH	Descent & / Approach					37	:16:00		140	
IAH	HOU	Direct / 3000					22	:15:00			

OTHER DATA: * Includes Taxi Fuel
NOTE: Use 165 PPH Total Fuel Flow From L/O To Start Of Descent.
Use 172 PPH Total Fuel Flow For Reserve And Alternate Requirements.
A Missed Approach Requires 55# of Fuel.

TIME and FUEL: As required by FARs.

TIME	FUEL (LB)	
		EN ROUTE
		RESERVE
		ALTERNATE
		TOTAL

FIGURE 108.—Flight Plan/Flight Log.

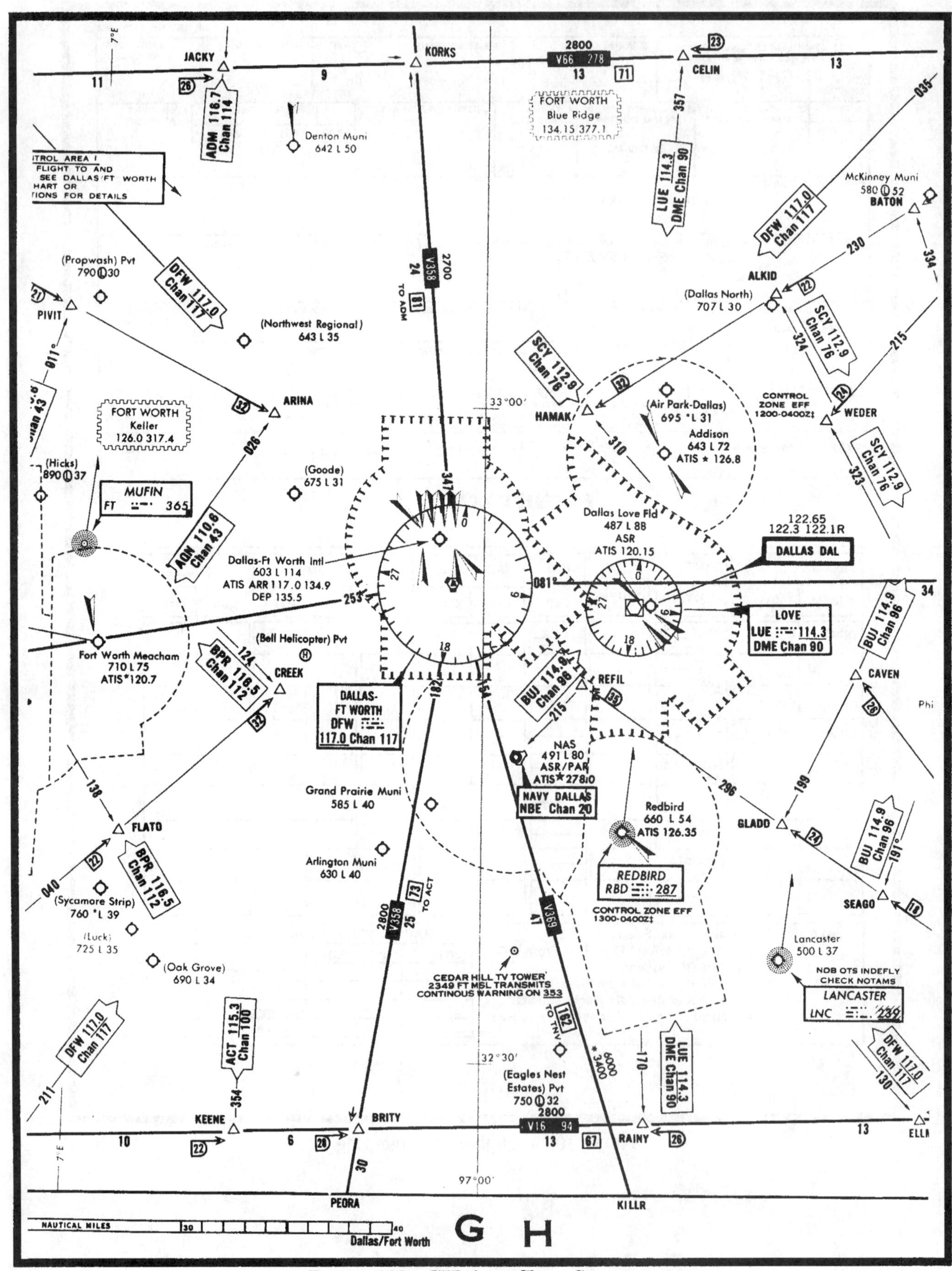

FIGURE 109.—IFR Area Chart Segment.

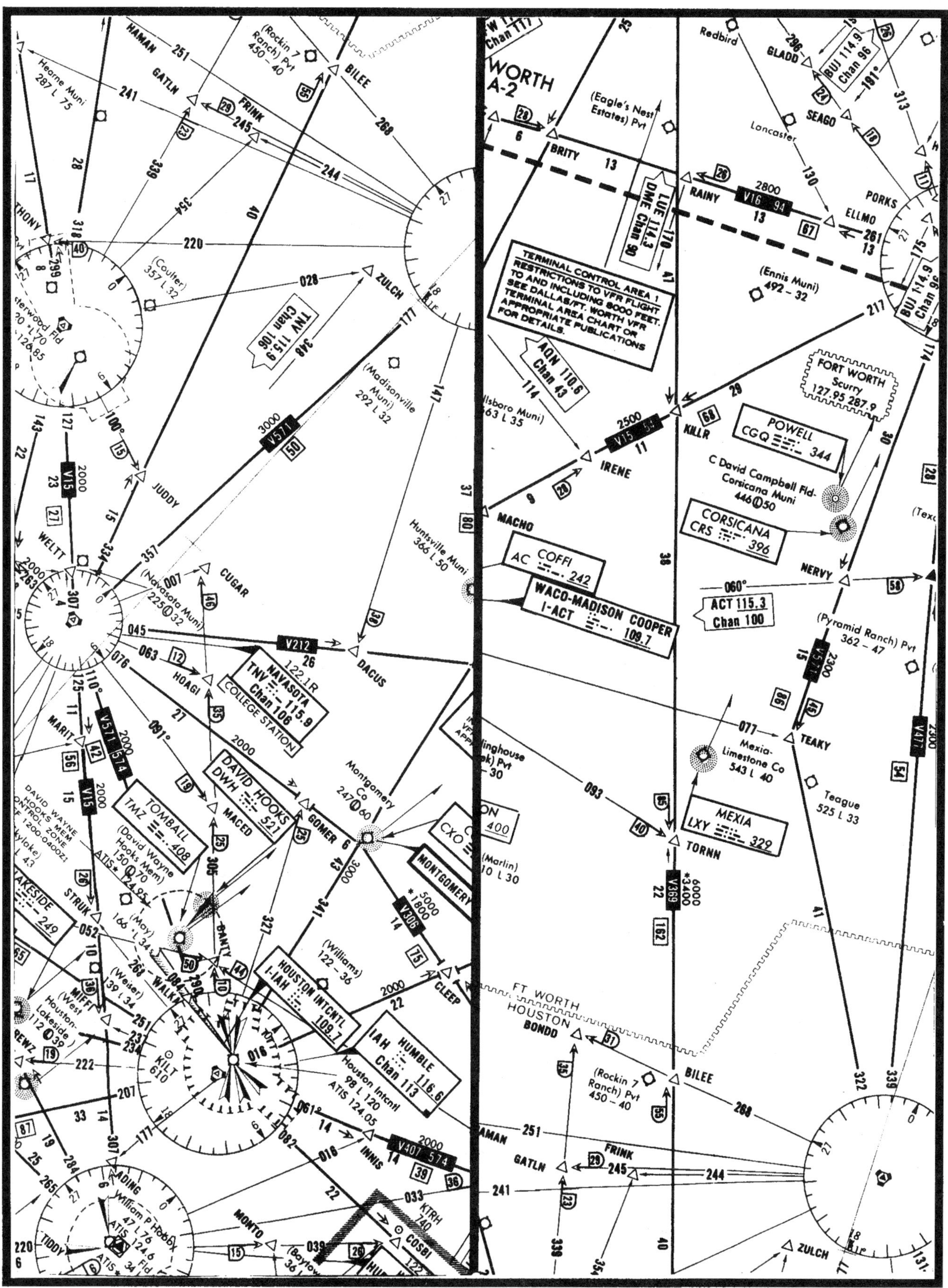

FIGURE 110.—IFR En Route Low Altitude Chart Segment.

TEXAS

§ DALLAS-FORT WORTH INTL (DFW) 12 NW UTC-6(-5DT) 32°53'47"N 97°02'28"W **DALLAS-FT. WORTH H-2K, 4F, 5B, L-13C, A IAP**

603 B **FUEL** 100LL, JET A OX 1,3 LRA CFR Index E

RWY 17L-35R: H11,388X150 (CONC-GRVD) S-120, D-200, DT-600, DDT-850 HIRL, CL
RWY 17L: ALSF2. TDZ. **RWY 35R:** MALSR. TDZ.

RWY 17R-35L: H11,388X200 (CONC-GRVD) S-120, D-200, DT-600, DDT-850 HIRL, CL
RWY 17R: SSALR TDZ. **RWY 35L:** TDZ. VASI(V6L).

RWY 18L-36R: H11,387X200 (CONC-GRVD) S-120, D-200, DT-600, DDT-850 HIRL, CL
RWY 18L: SSALR.TDZ. **RWY 36R:** TDZ. VASI(V6L).

RWY 18R-36L: H11,388X150 (CONC-GRVD) S-120, D-200, DT-600, DDT-850 HIRL, CL
RWY 18R: ALSF2. TDZ **RWY 36L:** MALSR. TDZ.

RWY 13L-31R: H9000X200 (CONC-GRVD) S-120, D-200, DT-600, DDT-850 HIRL, CL .5% up NW.
RWY 13L: TDZ. VASI(V6L)—Upper GA 3.25°TCH 93'. Lower GA 3.0°TCH 47'.
RWY 31 R: MALSR. TDZ.

RWY 13R-31L: H9300X150 (CONC-GRVD) S-120, D-220, DT-600, DDT-850 HIRL, CL
RWY 13 R: MALSR. TDZ. **RWY 31L:** TDZ.

RWY 18S-36S: H4000X100 (CONC)

AIRPORT REMARKS: Attended continuously. Prior Permission Required from arpt ops for General Aviation acft to proceed to airline terminal gate except to General Aviation Facility. Rwy 18S-36S located on taxiway G, 4000' long 100' wide restricted to prop acft 12,500 lbs. & below and stol acft daylight VFR plus IFR departures. Prior permission required from the primary tenant airlines to operate within central terminal area, CAUTION: proper minimum clearance may not be maintained within the central terminal area. Landing fee. Clearways 500x1000 each end Rwy 17L-35R, Rwy 17R-35L, Rwy 18L-36R and Rwy 18R-36L. Flight Notification Service (ADCUS) available.

WEATHER DATA SOURCES: LLWAS.

COMMUNICATIONS: ATIS 117.0 134.9 (ARR) 135.5 (DEP) **UNICOM** 122.95
FORT WORTH FSS (FTW) LC 624-8471, Toll free call, dial 1-800-WX-BRIEF. NOTAM FILE DFW
® **REGIONAL APP CON** 119.05(E) 119.4(E) 125.8(W) 132.1(W)
REGIONAL TOWER 126.55 (E) 124.15 (W) **GND CON** 121.65 133.15(E) 121.8 (W) **CLNC DEL** 128.25 127.5
® **REGIONAL DEP CON** 118.55 (E) 124.25 (WEST) 127.75 (NORTH-SOUTH)
TCA Group I: See VFR Terminal Area chart.

RADIO AIDS TO NAVIGATION: NOTAM FILE DFW.
(H) **VORTACW** 117.0 DFW Chan 117 32°51'57"N97°01'40"W at fld. 560/08E.
VOR Portion unusable 045°-050° all altitudes and distances, 350-100° beyond 30 NM below 2100'.
ISSUE NDB (LOM) 233 PK 32°47'35"N97°01'49"W 353° 5.1 NM to fld.
JIFFY NDB (LOM) 219 FL 32°59'45"N97°01'46"W 173° 5.1 NM to fld.
ILS/DME 109.5 I-LWN Chan 32 Rwy 13R
ILS/DME 109.1 I-FLQ Chan 28 Rwy 17L LOM JIFFY NDB
ILS 111.5 I-JHZ Rwy 17R LOM JIFFY NDB
ILS 111.3 I-CIX Rwy 18L
ILS/DME 111.9 I-VYN Chan 56 Rwy 18R
ILS 110.9 I-RRA Rwy 31R
ILS/DME 109.1 I-PKQ Chan 28 Rwy 35R LOM ISSUE NDB
ILS/DME 111.9 I-BXN Chan 56 Rwy 36L

§ HOUSTON INTERCONTINENTAL (IAH) 15N UTC-6(-5DT) 29°58'49"N 95°20'22"W **HOUSTON H-5B, L-17B IAP**

98 B S4 FUEL 100LL, JET A OX2 LRA CFR Index D

RWY 14L-32R: H1200X150 (CONC-GRVD) S-100, D-200, DT-400, DDT-778 HIRL, CL
RWY 14L: MALSR. VASI(V4L)—GA 3.0°TCH 54'. **RWY 32R:** MALSR.

RWY 09-27: H10000X150 (ASPH-GRVD) S-75, D-191, DT-400, DDT-850 HIRL,CL
RWY 09: MALSR. TDZ. PAPI(P4L)—GA 3.0°TCH 63'.
RWY 27: ALSF2. TDZ. PAPI(P4L)—GA 3.0°TCH 63'.

RWY 08-26: H9401X150 (CONC-GRVD) S-120, D-155, DT-265 HIRL, CL
RWY 08: MALSR. TDZ. **RWY 26:** ALSF2. TDZ. VASI(V4L)—GA 3.0°TCH 53'.

RWY 14R-32L: H6038X100 (ASPH-GRVD) S-30, D-60, DT-60 MIRL
RWY 14R: VASI(V4L)—GA 3.0°TCH 40'. Road. **RWY 32L:** VASI(V4L)—GA 3.0°TCH 45'.

AIRPORT REMARKS: Attended continuously. CAUTION: Birds on and in vicinity of arpt. CAUTION—Approach end of rwy 26 bright lgts approximately one mile from thld and 900' South of centerline. Caution—Deer on and in vicinity of arpt. Rwy 14R-32L CLOSED to acft over 140,000 lbs gross weight. Landing Fee. Flight Notification Service (ADCUS) available.

WEATHER DATA SOURCES: LLWAS

COMMUNICATIONS: ATIS 124.05 **UNICOM** 122.95
MONTGOMERY COUNTY FSS (CXO) Toll free call, dial 1-800-WX-BRIEF. NOTAM FILE IAH.
® **APP CON** 124.35 (West) 127.25 (North and East)
TOWER 118.1 (135.15 copter control) **GND CON** 121.7 **CLNC DEL** 128.1 (135.15 copter control)
® **DEP CON** 123.8 (West) 119.7 (North and East)
TCA Group II: VFR Terminal Area chart.

RADIO AIDS TO NAVIGATION: NOTAM FILE IAH.
HUMBLE (H) VORTACW 116.6 IAH Chan 113 29°57'24"N95°20'44"W at fld. 90/08E. **HIWAS.**
MARBE NDB (LOM) 379 HS 30°04'29"N 95°24'45"W 146° 5.9 NM to fld.
NIXIN NDB (LOM) 326 JY 29°59'36"N 95°12'54"W 257° 6.5 NM to fld.
ILS/DME 109.7 I-JYV Chan 34 Rwy 26 LOM NIXIN NDB
ILS 111.9 I-HSQ Rwy 14L LOM MARBE NDB
ILS/DME 109.7 I-IAH Chan 34 Rwy 08
ILS/DME 110.9 I-UYO Chan 34 Rwy 09
ILS 111.9 I-CDG Rwy 32R

FIGURE 111.—Airport/Facility Directory Excerpts.

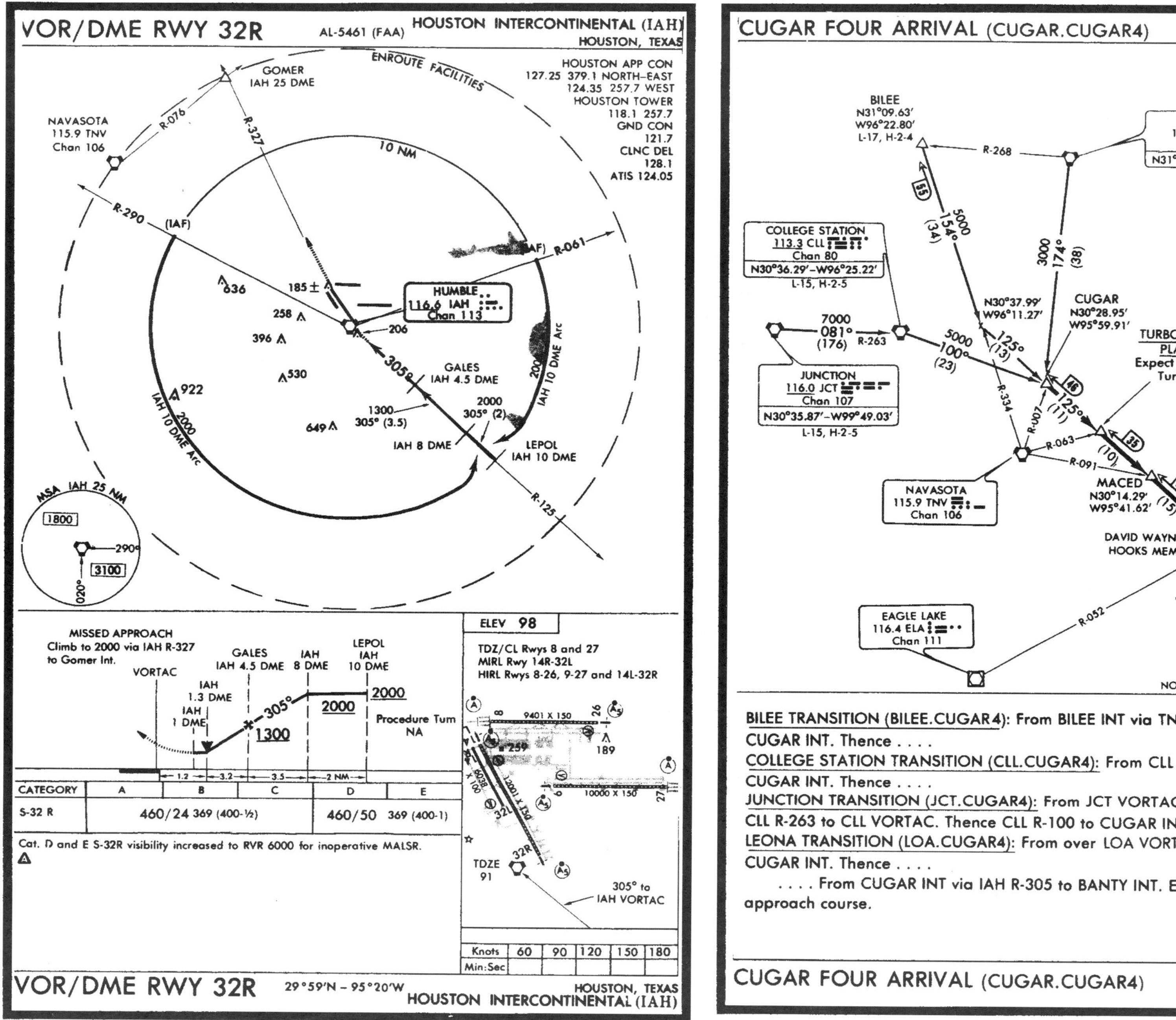

FIGURE 112.—VOR / DME RWY 32R (IAH) / Cugar Four Arrival (Cugar.Cugar4).

Form Approved: OMB No. 2120-0034

U.S. DEPARTMENT OF TRANSPORTATION FEDERAL AVIATION ADMINISTRATION **FLIGHT PLAN**	(FAA USE ONLY) ☐ PILOT BRIEFING ☐ VNR ☐ STOPOVER	TIME STARTED	SPECIALIST INITIALS

1. TYPE	2. AIRCRAFT IDENTIFICATION	3. AIRCRAFT TYPE/ SPECIAL EQUIPMENT	4. TRUE AIRSPEED	5. DEPARTURE POINT	6. DEPARTURE TIME PROPOSED (Z)	ACTUAL (Z)	7. CRUISING ALTITUDE
VFR / X IFR / DVFR	N60BJ	BH214/A	110 KTS	ØO2 Baker			12,000

8. ROUTE OF FLIGHT
Hdg 270 degrees to V394, V394 POM, V210 LAX

9. DESTINATION (Name of airport and city)	10. EST. TIME ENROUTE HOURS	MINUTES	11. REMARKS
LAX Los Angeles Int'l Los Angeles			L/O = Level Off PPH = Pounds Per Hour TEC = Tower to Tower

12. FUEL ON BOARD HOURS	MINUTES	13. ALTERNATE AIRPORT(S)	14. PILOT'S NAME, ADDRESS & TELEPHONE NUMBER & AIRCRAFT HOME BASE / 17. DESTINATION CONTACT/TELEPHONE (OPTIONAL)	15. NUMBER ABOARD
		LGB Long Beach		15

16. COLOR OF AIRCRAFT
Brown/White

CIVIL AIRCRAFT PILOTS. FAR Part 91 requires you file an IFR flight plan to operate under instrument flight rules in controlled airspace. Failure to file could result in a civil penalty not to exceed $1,000 for each violation (Section 901 of the Federal Aviation Act of 1958, as amended). Filing of a VFR flight plan is recommended as a good operating practice. See also Part 99 for requirements concerning DVFR flight plans.

FAA Form 7233-1 (8-82) CLOSE VFR FLIGHT PLAN WITH ________________ FSS ON ARRIVAL

FLIGHT LOG

CHECK POINTS FROM	TO	ROUTE ALTITUDE	COURSE	WIND TEMP	SPEED-KTS TAS	GS	DIST NM	TIME LEG	TOT	FUEL LEG	TOT
ØO2	V394	HDG 270 Climb					8		:10:00		250*
Join V394	DAG	V394 12000		290/36 ISA-2			15				
DAG	POM										
POM	PIRRO Int.	V210 12000									
PIRRO Int.	LAX	Descent & Approach					29	:17:00		348	
LAX	LGB	TEC 3000			110	120	22	:11:00			

OTHER DATA: * Includes Taxi Fuel
NOTE: Use 1045 PPH Total Fuel Flow From L/O To Start Of Descent.
Use 1095 PPH Total Fuel Flow For Reserve / Alternate Requirements
A missed approach requires 89# of fuel.

TIME and FUEL: As required by FARs.

TIME	FUEL (LB)	
		EN ROUTE
		RESERVE
		ALTERNATE
		TOTAL

FIGURE 113.—Flight Plan/Flight Log.

LONG BEACH (DAUGHERTY FLD) (LGB) 3 NE UTC–8(–7DT) N33°49.06′ W118°09.10′ **LOS ANGELES**
57 B S4 **FUEL** 100LL, JET A OX 1, 2, 3, 4 LRA ARFF Index C **COPTER**
RWY 12-30: H10000X200 (ASPH-GRVD) S-30 +, D-200, DT-300 HIRL 0.4% up NW **H-2B, L-3B, A**
RWY 12: VASI(V4L)—GA 3.0° TCH 47′. Thld dsplcd 1340′. Railroad. **IAP**
RWY 30: MALSR. PAPI(P4L) —GA 3.0° TCH 70′. Thld dsplcd 1990′. Tree.
RWY 07L-25R: H6192X150 (ASPH-PFC) S-30, D-70, DT-110 MIRL 0.3% up W
RWY 07L: Thld dsplcd 1305′. Railroad.
RWY 25R: REIL. VASI(V4L)—GA 4.0° TCH 57′. Thld dsplcd 531′. Road. Rgt tfc.
RWY 07R-25L: H5420X150 (ASPH) S-30, D-75 HIRL 0.4% up W
RWY 07R: Tower. Rgt tfc. **RWY 25L:** REIL. VASI(V4L)—GA 4.0° TCH 58′. Trees.
RWY 16R-34L: H4470X75 (ASPH) S-12.5
RWY 16R: VASI(V4L)—GA 4.0° TCH 36′. Thld dsplcd 310′. Fence. Rgt tfc. **RWY 34L:** Road.
RWY 16L-34R: H4267X75 (ASPH) S-12.5
RWY 16L: Thld dsplcd 415′. Fence. **RWY 34R:** Thld dsplcd 292′. Road. Rgt tfc.
AIRPORT REMARKS: Attended continuously. All rwys CLOSED 0600–1500Z‡ except Rwy 12-30. Flocks of seagulls on and in vicinity of arpt especially during rain. Unlighted twr 152′ AGL 2500′ W and 500′ S of Rwy 07 thld. 255′ AGL obstruction 1200′ S of Rwy 07 thld. Broken pavement on NE Police helipad between perimeter road and Twy F. Prior notification requested 24 hours in advance for all acft over 75,000 pounds certificated maximum gross weight and civilian Non-Stage III jets and all military jets, ctc Noise Abatement 310-429-6647 Mon–Fri 1500–0100Z‡. Noise abatement information on 122.85. Noise limits (single event noise exposure level), Rwy 25 tkf 92.0 DB-ldg 88.0 DB; Rwy 07 tkf 88.0 DB-ldg 92.0 DB; Rwys 12 and 30 tkf 102.5 DB-ldg 101.5 DB except 0600–1500Z‡ tkf 79.0-ldg 79.0 DB. Touch and go, stop and go, low apch only permitted 1500–0300Z‡ weekdays and 1600–2300Z‡ weekends and holidays only on Rwy 07L-25R and Rwy 07R-25L unless weather conditions require twr to direct such operations to Rwy 16R-34L and Rwy 16L-34R. Rwy 12-30 arpt manager limits gross weight to 300,000 lbs dual tandom wheel except DC-10 series 30/40 and MD11 limited to 378,000 lbs. No twy access to Rwy 07L W of Twy D, 4897′ remaining on Rwy 07L from Twy D, Twy A clsd W of compass rose. Taxiway K east of Taxiway C clsd to acft with a wingspan greater than 117′. Engine run-ups other than preflight are limited to hours of 1500–0500Z‡ weekdays and 1700–0500Z‡ weekends and holidays. Rwy 07R-25L limited to acft with a maximum wing span of 90′. ACTIVATE MALSR Rwy 30 when tower clsd—CTAF. Rwy 12-30 HIRL lighted during hours tower clsd. NOTE: See SPECIAL NOTICE—Land and Hold Short Operations.
COMMUNICATIONS: CTAF 119.4 **ATIS** 127.75 (310) 595-8564 **UNICOM** 122.95
HAWTHORNE FSS (HHR) TF 1-800-WX-BRIEF. NOTAM FILE LGB.
Ⓡ **SOCAL APP CON** 124.65
Ⓡ **SOCAL DEP CON** 127.2
LONG BEACH TOWER 119.4 (Rwy 30 apch, Rwy 12 dep) 120.5 (Rwy 12 apch, Rwy 30 dep) (1415–0745Z‡)
GND CON 133.0 **CLNC DEL** 118.15
AIRSPACE: CLASS D svc effective 1415–0745Z‡ other times CLASS G.
RADIO AIDS TO NAVIGATION: NOTAM FILE HHR.
SEAL BEACH (L) VORTACW 115.7 SLI Chan 104 N33°47.00′ W118°03.29′ 278° 5.3 NM to fld. 20/15E.
HIWAS.
BECCA NDB (LOM) 233 LG N33°45.40′ W118°04.64′ 301° 5.2 NM to fld.
ILS 110.3 I-LGB Rwy 30. LOM BECCA NDB. Unmonitored when twr clsd. MM unmonitored.
• •
HELIPAD H1: H20X20 (ASPH-CONC)
HELIPAD H2: H20X20 (ASPH-CONC)
HELIPAD H3: H20X20 (ASPH-CONC)
HELIPORT REMARKS: Training helipads H1, H2 and H3 located N of Rwy 12-30 midfield between Taxiways G and K.

LONNIE POOL FLD/WEAVERVILLE (See WEAVERVILLE)

LOS ALAMITOS AAF (ARMED FORCES RESERVE CENTER): **LOS ANGELES**
AIRSPACE: CLASS D svc effective Sat–Mon 1600–0000Z‡, Tue–Fri 1500–0600Z‡ other times CLASS G. **L-3B, A**

FIGURE 113A.—Data from Southwest U.S. Airport/Facility Directory.

AVENAL N35°38.82′ W119°58.72′ NOTAM FILE HHR. **LOS ANGELES**
(H) VORTAC 117.1 AVE Chan 118 080° 14.4 NM to Lost Hills-Kern Co. 710/16E. **H-2A, L-2E, 3A**
RCO 122.1R 117.1T (BAKERSFIELD FSS)

BAKER (Ø02) 2 NW UTC–8(–7DT) N35°17.13′ W116°04.95′ **LOS ANGELES**
922 B TPA—1922(1000) **L-3C, 5B**
RWY 15-33: H3157X50 (ASPH) MIRL
RWY 33: P-line. Rgt tfc.
AIRPORT REMARKS: Unattended. Mountain ½ mile W of arpt. Unlit towers and unmarked powerlines across apch path Rwy 33.
COMMUNICATIONS: CTAF 122.9
RIVERSIDE FSS (RAL) TF 1-800-WX-BRIEF. NOTAM FILE RAL.
RADIO AIDS TO NAVIGATION: NOTAM FILE DAG.
DAGGETT (L) VORTAC 113.2 DAG Chan 79 N34°57.75′ W116°34.69′ 036° 31.1 NM to fld. 1760/15E.
HIWAS.

BAKERSFIELD N35°26.02′ W119°03.41′ **LOS ANGELES**
FSS (BFL) at Meadows Fld. 123.65 122.45 122.2. LD 805-399-1787. **L-3B, 5A**

BAKERSFIELD

BAKERSFIELD MUNI (L45) 3 S UTC–8(–7DT) N35°19.49′ W118°59.75′ **LOS ANGELES**
376 B S4 FUEL 80, 100LL TPA—1176(800) **L-3B, 5A**
RWY 16-34: H4000X75 (ASPH) S-20 MIRL **IAP**
RWY 16: Road. Rgt tfc. RWY 34: PAPI(P2L)—GA 4.0° TCH 54′. P-line.
AIRPORT REMARKS: Attended 1500-0100Z‡. 100′ pole line ½ mile south of arpt.
COMMUNICATIONS: CTAF/UNICOM 122.8
BAKERSFIELD FSS (BFL) LC 399-1787 NOTAM FILE BFL.
Ⓡ BAKERSFIELD APP/DEP CON 126.45 (1400-0700Z‡)
Ⓡ L.A. CENTER APP/DEP CON 127.1 (0700-1400Z‡)
RADIO AIDS TO NAVIGATION: NOTAM FILE BFL.
SHAFTER (H) VORTACW 115.4 EHF Chan 101 N35°29.07′ W119°05.84′ 138° 10.8 NM to fld. 550/14E.
HIWAS.

- - - - - - - - - - - - - - - - - - - -

MEADOWS FLD (BFL) 3 NW UTC–8(–7DT) N35°26.02′ W119°03.41′ **LOS ANGELES**
507 B S4 FUEL 80, 100, 100LL, JET A ARFF Index B **H-2B, L-2E, 3B, 5A**
RWY 12L-30R: H10857X150 (ASPH-GRVD) S-110, D-200, DT-500, DDT-850 HIRL 0.3% up NW **IAP**
RWY 12L: VASI(V4L)—GA 3.0° TCH 52′.
RWY 30R: MALSR. PAPI(P4L)—GA 3.0° TCH 64′. Thld dsplcd 3428′. P-line. Rgt tfc.
RWY 12R-30L: H3700X75 (ASPH) S-18 MIRL
RWY 12R: Rgt tfc. RWY 30L: VASI(NSTD)—GA 3.0°. Tree.
AIRPORT REMARKS: Attended 1330-0700Z‡, fee for call out service other hours. Rwy 12L 16′ pump 525′ from thld 550′ left. Distance remaining at the 2000′ mark on Rwy 30R is actually 2850′. Noise sensitive areas S and E of arpt recommended turbojet training hours weekdays 1600-0600Z‡, weekends 2000-0600Z‡ no more than ten practice approaches per hour. Rwy 30L NSTD VASI single light source visibility 1 mile, red blo glide path. When twr clsd ACTIVATE PAPI Rwy 30R—CTAF. For MIRL Rwy 12R-30L and taxiway lgts when tower clsd ctc FSS—CTAF.
COMMUNICATIONS: CTAF 118.1 ATIS 118.6 (805) 399-9425 UNICOM 122.95
BAKERSFIELD FSS (BFL) on arpt. 123.65 122.45 122.2. LC 399-1787. NOTAM FILE BFL.
BAKERSFIELD APP CON 118.9 (N) 118.8 (S) (1400-0700Z‡)
BAKERSFIELD DEP CON 126.45 (N,S) (1400-0700Z‡)
Ⓡ L.A. CENTER APP/DEP CON 127.1 (0700-1400Z‡)
BAKERSFIELD TOWER 118.1 (1400-0700Z‡) GND CON 121.7
AIRSPACE: CLASS D svc effective 1400-0700Z‡ other times CLASS E.
RADIO AIDS TO NAVIGATION: NOTAM FILE BFL.
SHAFTER (H) VORTACW 115.4 EHF Chan 101 N35°29.07′ W119°05.84′ 133° 3.6 NM to fld. 550/14E.
HIWAS.
NILEY NDB (LOM) 385 BF N35°21.65′ W118°58.12′ 301° 6.1 NM to fld.
ILS/DME 111.9 I-BFL Chan 56 Rwy 30R. LOM NILEY NDB. ILS unmonitored when twr clsd.

- - - - - - - - - - - - - - - - - - - -

FIGURE 113B.—Data from Southwest U.S. Airport/Facility Directory.

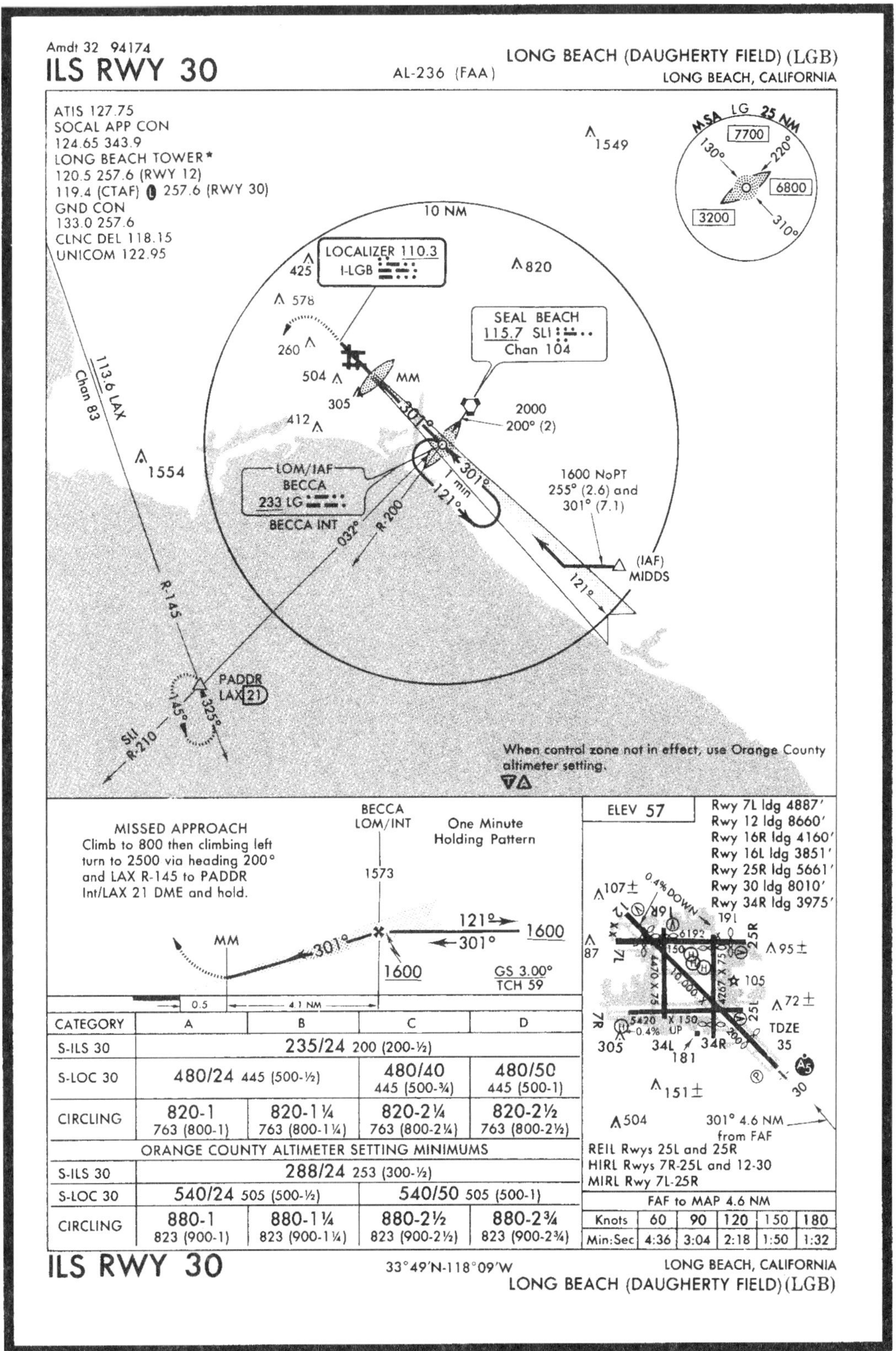

FIGURE 113C.—ILS RWY 30 (LGB).

THIS PAGE INTENTIONALLY LEFT BLANK

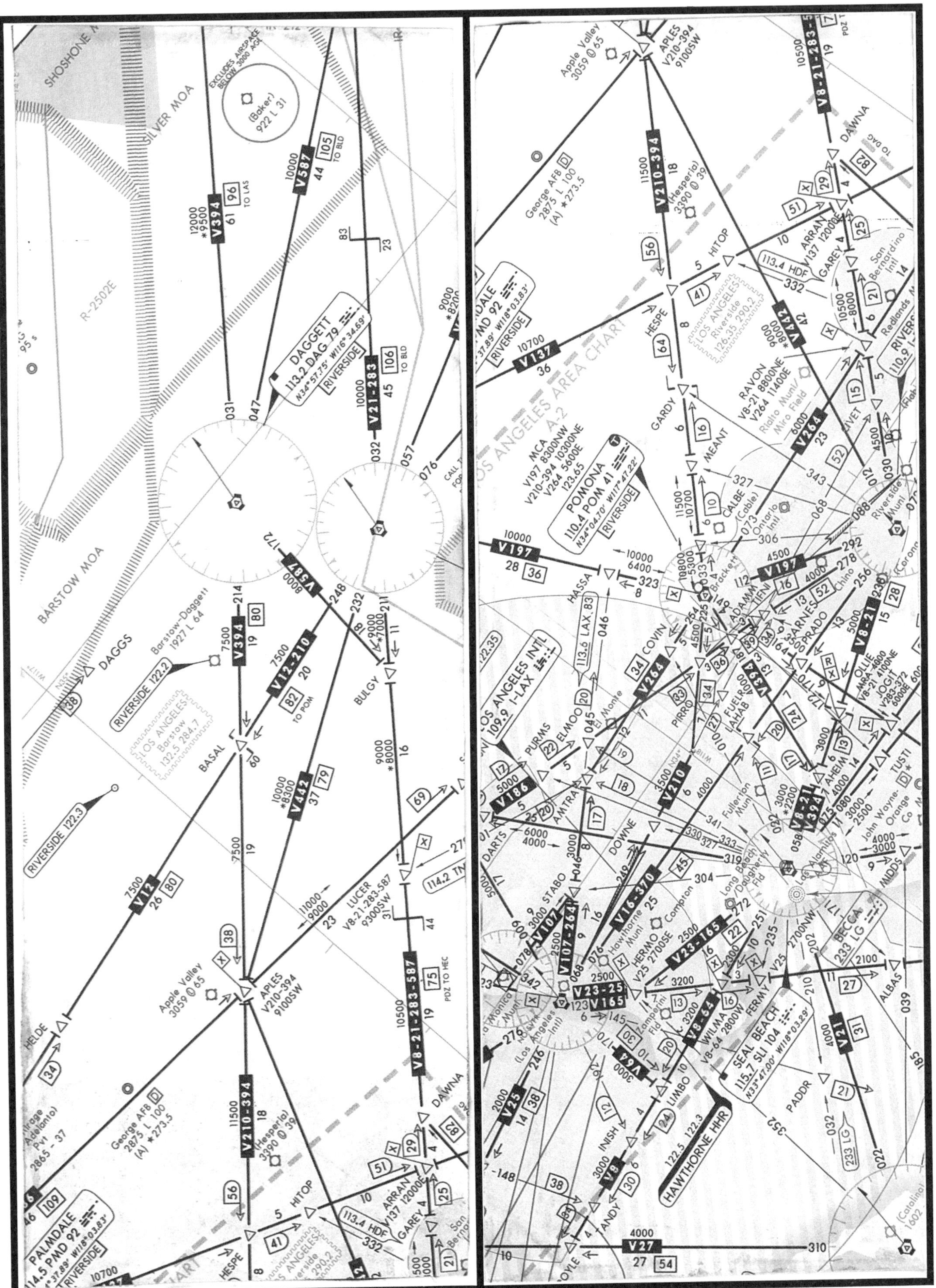

FIGURE 114.—En Route Low Altitude Chart Segment.

Form Approved: OMB No. 2120-0034

U.S. DEPARTMENT OF TRANSPORTATION FEDERAL AVIATION ADMINISTRATION **FLIGHT PLAN**	(FAA USE ONLY) ☐ PILOT BRIEFING ☐ VNR ☐ STOPOVER	TIME STARTED	SPECIALIST INITIALS

1. TYPE	2. AIRCRAFT IDENTIFICATION	3. AIRCRAFT TYPE/ SPECIAL EQUIPMENT	4. TRUE AIRSPEED	5. DEPARTURE POINT	6. DEPARTURE TIME PROPOSED (Z)	ACTUAL (Z)	7. CRUISING ALTITUDE
VFR / X IFR / DVFR	PTL 130	B727/R	*** KTS	LAX			FL270

8. ROUTE OF FLIGHT
LAX INP3.IPL, J2 MOHAK, ARLIN 9 PHX

9. DESTINATION (Name of airport and city)	10. EST. TIME ENROUTE HOURS	MINUTES	11. REMARKS
PHX PHOENIX SKY HARBOR PHOENIX			L/O = Level Off PPH = Pounds Per Hour ** L/O at OCN R-270/50 *** MACH .78

12. FUEL ON BOARD HOURS	MINUTES	13. ALTERNATE AIRPORT(S)	14. PILOT'S NAME, ADDRESS & TELEPHONE NUMBER & AIRCRAFT HOME BASE	15. NUMBER ABOARD
		TUS TUCSON INT'L	17. DESTINATION CONTACT/TELEPHONE (OPTIONAL)	83

16. COLOR OF AIRCRAFT	
RED/BLACK	CIVIL AIRCRAFT PILOTS. FAR Part 91 requires you file an IFR flight plan to operate under instrument flight rules in controlled airspace. Failure to file could result in a civil penalty not to exceed $1,000 for each violation (Section 901 of the Federal Aviation Act of 1958, as amended). Filing of a VFR flight plan is recommended as a good operating practice. See also Part 99 for requirements concerning DVFR flight plans.

FAA Form 7233-1 (8-82)

CLOSE VFR FLIGHT PLAN WITH ________________ FSS ON ARRIVAL

FLIGHT LOG

CHECK POINTS		ROUTE	COURSE	WIND	SPEED-KTS		DIST	TIME		FUEL	
FROM	TO	ALTITUDE		TEMP	TAS	GS	NM	LEG	TOT	LEG	TOT
LAX	L/O**	IPL3.IPL Climb					43		:19:00		4510*
L/O	IPL	IPL3.IPL FL270		300/43 ISA-2							
IPL	BZA	J2 FL270									
BZA	Mohak Int	J2 FL270		300/43 ISA-2							
Mohak	Arlin Int	Arlin 9 FL270									
Arlin	PHX	Radar Vec DES/APP						:12:00		1140	
PHX	TUS	Radar V FL190					97	:26:00			

OTHER DATA: * Includes Taxi Fuel
NOTE: Use 9600 PPH Total Fuel Flow From L/O To Start Of Descent.
Use 9250 PPH Total Fuel Flow For Reserve And Alternate Requirements.
A Missed Approach Requires 416# of Fuel.

TIME and FUEL: As required by FARs.

TIME	FUEL (LB)	
		EN ROUTE
		RESERVE
		ALTERNATE
		TOTAL

FIGURE 115.—Flight Plan/Flight Log.

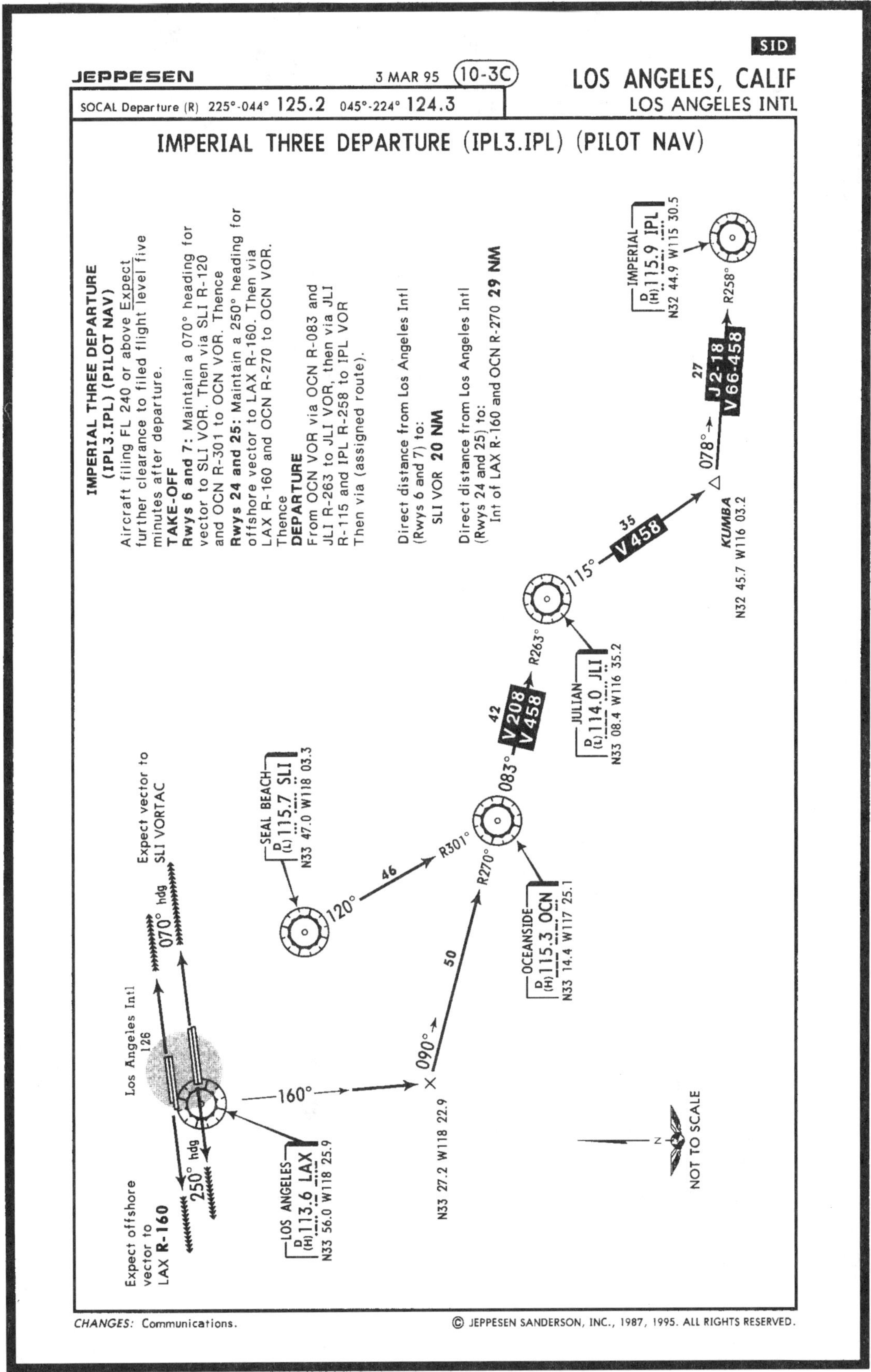

FIGURE 116.—Imperial Three Departure (IPL3.IPL) (PILOT NAV).

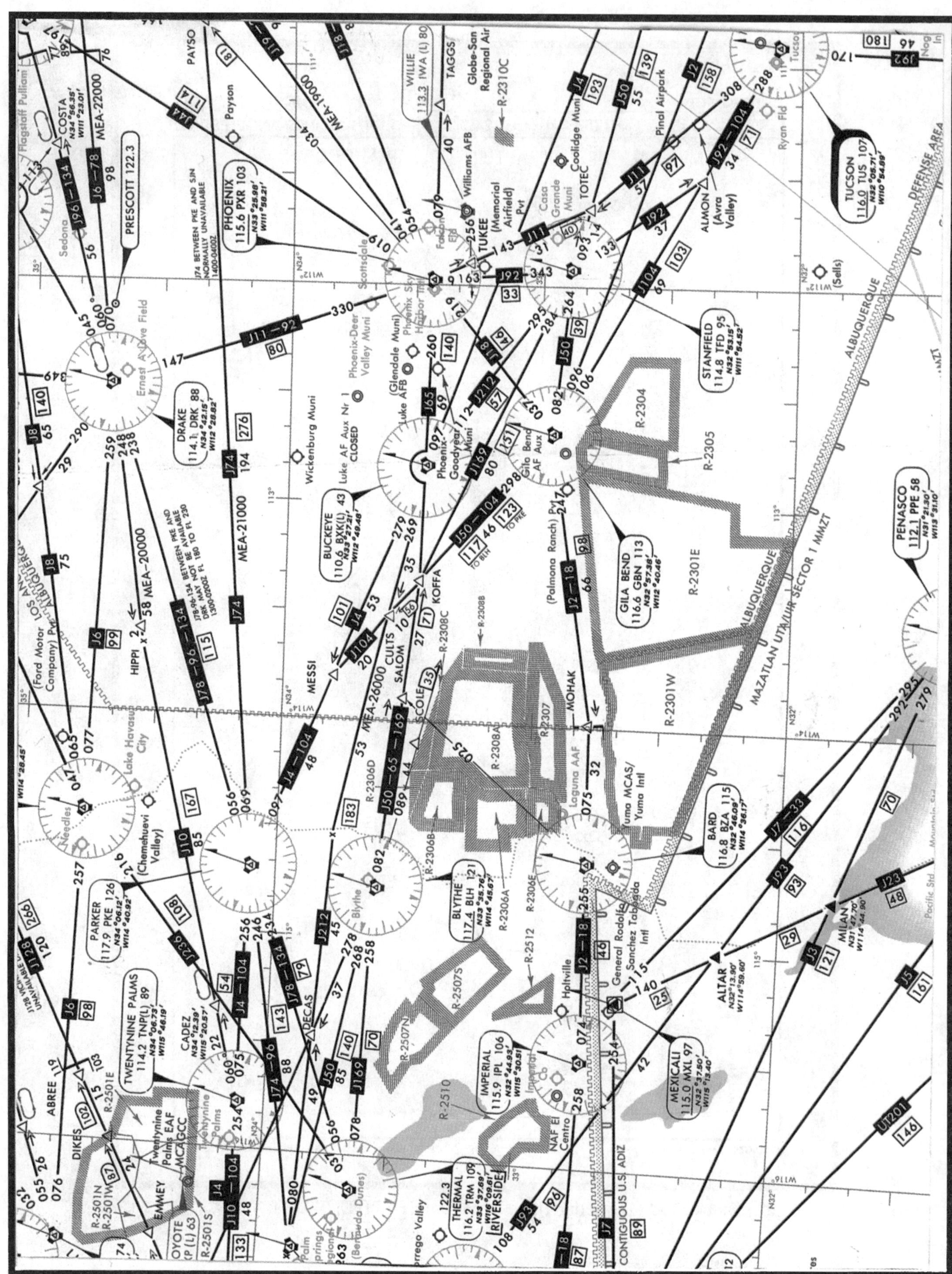

FIGURE 117.—IFR En Route High Altitude Chart Segment.

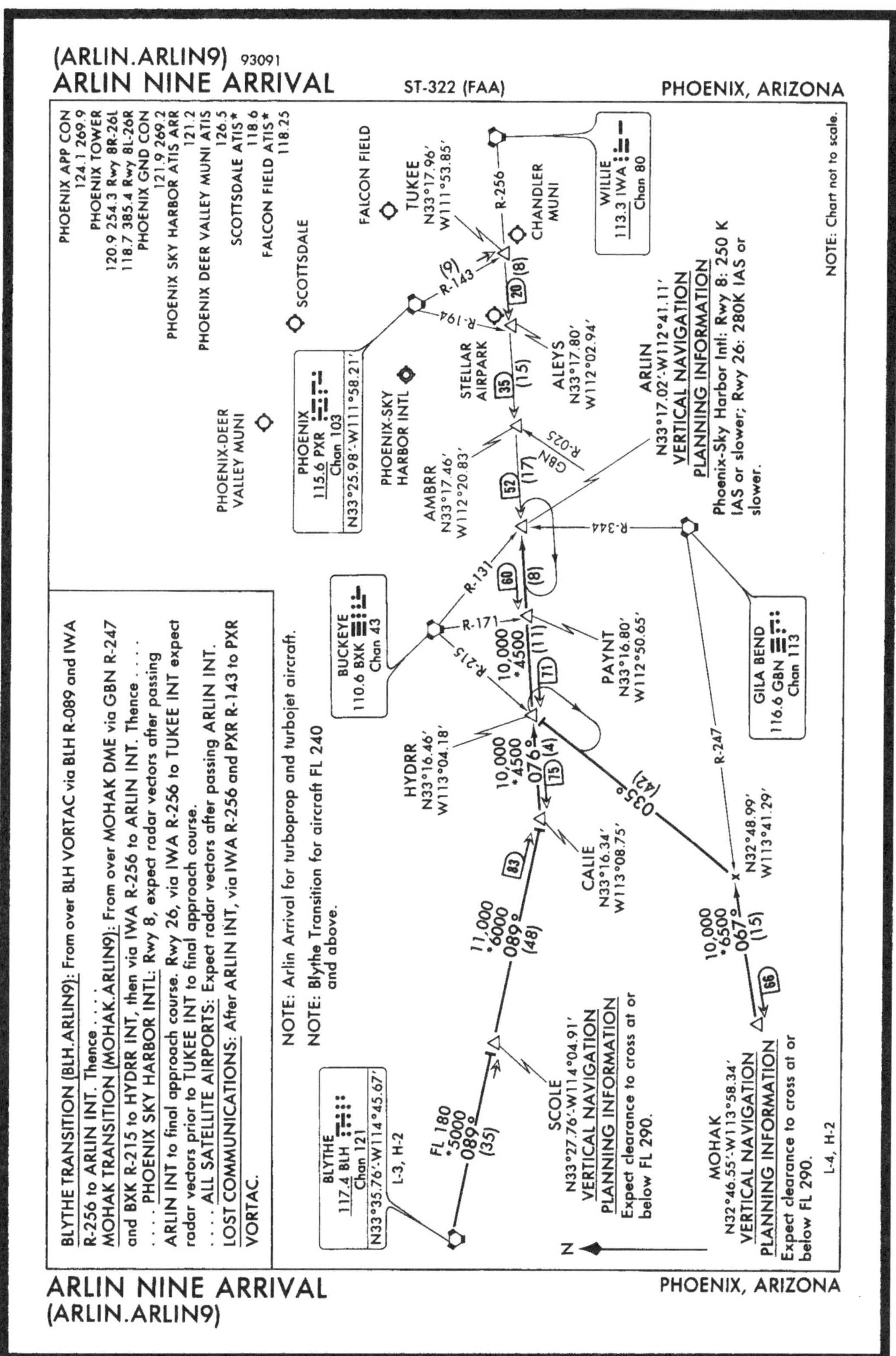

FIGURE 118.—ARLIN NINE ARRIVAL (ARLIN.ARLIN9).

THIS PAGE INTENTIONALLY LEFT BLANK

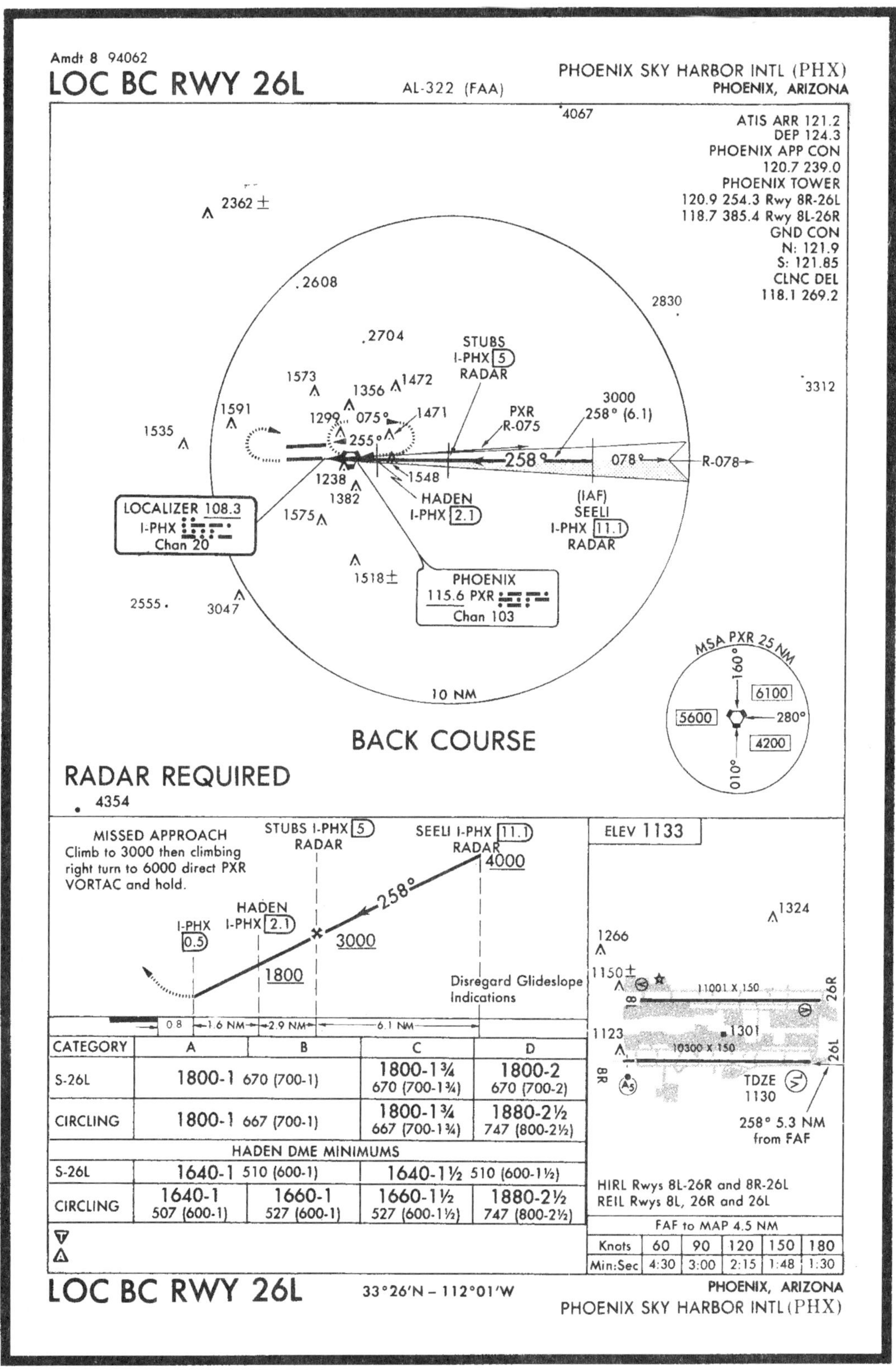

CATEGORY	A	B	C	D
S-26L	1800-1 670 (700-1)		1800-1¾ 670 (700-1¾)	1800-2 670 (700-2)
CIRCLING	1800-1 667 (700-1)		1800-1¾ 667 (700-1¾)	1880-2½ 747 (800-2½)
HADEN DME MINIMUMS				
S-26L	1640-1 510 (600-1)		1640-1½ 510 (600-1½)	
CIRCLING	1640-1 507 (600-1)	1660-1 527 (600-1)	1660-1½ 527 (600-1½)	1880-2½ 747 (800-2½)

FAF to MAP 4.5 NM

Knots	60	90	120	150	180
Min:Sec	4:30	3:00	2:15	1:48	1:30

FIGURE 118A.—LOC BC RWY 26L (PHX).

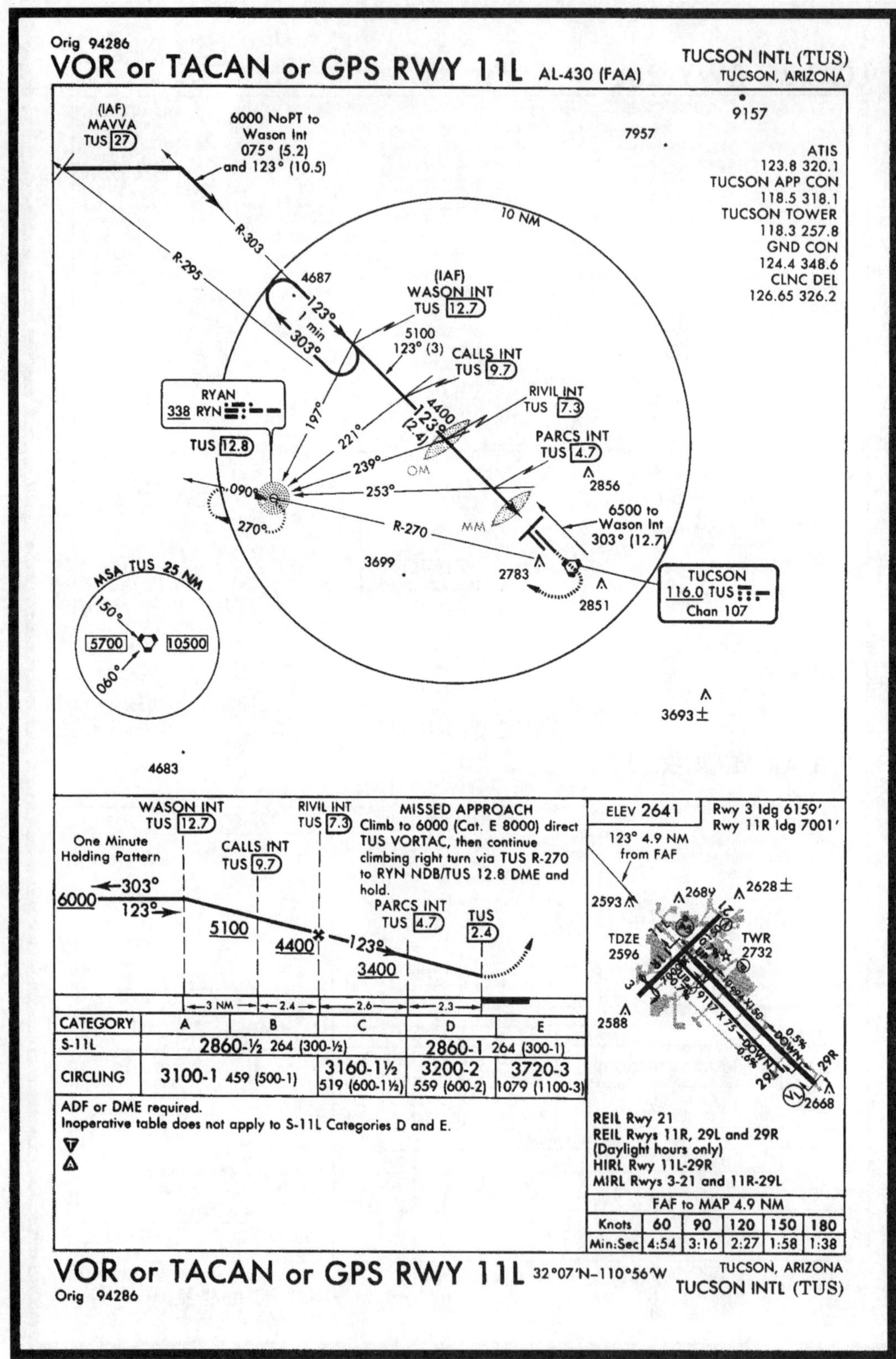

CATEGORY	A	B	C	D	E
S-11L	2860-½ 264 (300-½)			2860-1 264 (300-1)	
CIRCLING	3100-1 459 (500-1)		3160-1½ 519 (600-1½)	3200-2 559 (600-2)	3720-3 1079 (1100-3)

Knots	60	90	120	150	180
Min:Sec	4:54	3:16	2:27	1:58	1:38

FIGURE 118B.—VOR or TACAN or GPS RWY 11L (TUS).

PHOENIX

PHOENIX-DEER VALLEY MUNI (DVT) 15 N UTC–7 N33°41.30′ W112°04.93′ **PHOENIX**
1476 B S4 **FUEL** 80, 100LL, JET A OX 1, 3 TPA—See Remarks **H-2C, L-4E**
RWY 07R-25L: H8200X100 (ASPH) S-40, D-50, DT-80 MIRL **IAP**
RWY 07R: REIL. VASI(V2L)—GA 3.0° Thld dsplcd 900′. Rgt tfc.
RWY 25L: REIL. VASI(V2L)—GA 3.0° Thld dsplcd 920′.
RWY 07L-25R: H4500X75 (ASPH) S-20 MIRL
RWY 07L: REIL. PVASI(PSIL)—GA 3.0° TCH 40′. **RWY 25R:** REIL. PVASI(PSIL)—GA 3.5° TCH 47′. Hill. Rgt tfc.
AIRPORT REMARKS: Attended 1300–0400Z. Fuel avbl only during hours 1400–0300Z 7 days. Lgtd hills NE, E, SE and W. Hot air balloon ops fall, winter, and spring months and ultralight opr South and West of arpt. Rwy 07L-25R is designated training rwy. Aerobatic practice area approximately 8½ miles northwest of the Deer Valley Arpt from the surface to 6000′ MSL. Parallel taxiway north and close proximity to Rwy 07L-25R. Rwy 07R VASI and REIL, Rwy 25L VASI and REIL, Rwy 07L PVASI and Rwy 25R PVASI on when twr clsd. Fee for all charters; travel clubs and certain revenue producing acft. TPA-2501(1025) single engine and 3001(1525) multi engine.
COMMUNICATIONS: CTAF 118.4 **ATIS** 126.5 **UNICOM** 122.95
PRESCOTT FSS (PRC) TF 1-800-992-7433. NOTAM FILE DVT.
PHOENIX RCO 122.6 122.2 (PRESCOTT FSS)
® **PHOENIX APP/DEP CON** 120.7
DEER VALLEY TOWER 118.4 (Rwy 07R-25L) 120.2 (Rwy 07L-25R) (1300–0400Z) **GND CON** 121.8
CLNC DEL 119.5
AIRSPACE: CLASS D svc effective 1300–0400Z other times CLASS G.
RADIO AIDS TO NAVIGATION: NOTAM FILE PRC.
PHOENIX (H) VORTACW 115.6 PXR Chan 103 N33°25.98′ W111°58.21′ 328° 16.3 NM to fld. 1180/12E.
HIWAS.
SCOTTSDALE NDB (MHW) 224 SDL N33°37.75′ W111°54.47′ 279° 9.4 NM to fld. NOTAM FILE SDL.
Unmonitored when twr closed.
COMM/NAVAID REMARKS: Emerg frequency 121.5 not available at twr.

PHOENIX SKY HARBOR INTL (PHX) 3 E UTC–7 N33°26.17′ W112°00.57′ **PHOENIX**
1133 B S4 **FUEL** 100LL, JET A OX 1, 2, 3, 4 TPA—See Remarks **H-2C, L-4E**
LRA ARFF Index D **IAP**
RWY 08L-26R: H11001X150 (ASPH-GRVD) S-30, D-170, DT-280, DDT-620 HIRL
RWY 08L: REIL. VASI(V4L)—GA 3.0° TCH 55′. Building.
RWY 26R: REIL. VASI(V4L)—GA 3.0° TCH 60′. Road. Rgt tfc.
RWY 08R-26L: H10300X150 (ASPH-GRVD) S-30, D-200, DT-400, DDT-620 HIRL
RWY 08R: MALSR. Pole. Rgt tfc.
RWY 26L: REIL. VASI(V6L)—Upper GA 3.25° TCH 90′. Lower GA 3.0° TCH 53′. Antenna.
AIRPORT REMARKS: Attended continuously. Training by civil turbojet acft prohibited except PPR. TPA—2133(1000) lgt acft and non-turbo jets; 2633(1500) heavy acft and turbojets. Unless advised by ATC all turbine acft and acft 12,500 lbs and over remain at or above 3000′ MSL until established on final. Fly base leg at least 5 mile from arpt. Overnight parking fee. Fee for all charters; travel clubs and certain revenue producing aircraft. Taxiway A-6 limited to 68,000 GWT. Rwy 08L-26R FAA strength evaluation DC-10-10 505,000 pounds, DC-10-30/40 500,000 pounds, L-1011-1 450,000 pounds, aircraft up to DDTW 620,000 pounds, DC-10-10 505,000 pounds, DC-10-30/40 540,000 pounds, L-1011-1 450,000 pounds regularly operate on rwy. Rwy 08R-26L gross weight limit DC-10-10 430,000 pounds, DC-10-30/40 540,000 pounds, L-1011-1 430,000 pounds. Flight Notification Service (ADCUS) available.
WEATHER DATA SOURCES: ASOS (602) 231-8557. LLWAS.
COMMUNICATIONS: ATIS ARR 121.2 DEP 124.3 (602) 244-0963 **UNICOM** 122.95
PRESCOTT FSS (PRC) TF 1-800-992-7433. NOTAM FILE PHX.
RCO 122.6 122.2 (PRESCOTT FSS)
® **APP/DEP CON** 126.8 (259°-309°) 124.9 (053°-146°) 124.1(147°-258° above 5500′) 123.7 (147°-258° 5500′ and below) 120.7 120.4 (Rwy 08L 275°-290° blo 6000′, Rwy 26R 030°-080°) (310°-052° 5500′ and below) 119.2 (310°-052° above 5500′)
TOWER 118.7 (Rwy 08L-26R) 120.9 (Rwy 08R-26L) **GND CON** 121.9 (North) 121.85 (South) **CLNC DEL** 118.1
AIRSPACE: CLASS B See VFR Terminal Area Chart.
RADIO AIDS TO NAVIGATION: NOTAM FILE PRC.
PHOENIX (H) VORTACW 115.6 PXR Chan 103 N33°25.98′ W111°58.21′ 263° 2.0 NM to fld. 1180/12E.
HIWAS.
ILS 111.75 I-PZZ Rwy 26R (LOC only).
ILS/DME 108.3 I-PHX Chan 20 Rwy 08R. GS unusable below 1280′. LOC back course unusable beyond 20° south of course.

FIGURE 118C.—Excerpt from Airport/Facilities Directory.

Form Approved: OMB No. 2120-0034

U.S. DEPARTMENT OF TRANSPORTATION FEDERAL AVIATION ADMINISTRATION **FLIGHT PLAN**	(FAA USE ONLY) ☐ PILOT BRIEFING ☐ VNR ☐ STOPOVER	TIME STARTED	SPECIALIST INITIALS

1. TYPE	2. AIRCRAFT IDENTIFICATION	3. AIRCRAFT TYPE/ SPECIAL EQUIPMENT	4. TRUE AIRSPEED	5. DEPARTURE POINT	6. DEPARTURE TIME PROPOSED (Z)	ACTUAL (Z)	7. CRUISING ALTITUDE
VFR / X IFR / DVFR	N130JB	B727/A	** KTS	BUF Greater Buffalo Int'l			FL310

8. ROUTE OF FLIGHT
Buffalo One Dep. J547 FNT, FNT.PMM 2 ORD

9. DESTINATION (Name of airport and city)	10. EST. TIME ENROUTE HOURS	MINUTES	11. REMARKS
ORD Chicago-Ohare Int'l Chicago			L/O = Level Off PPH = Pounds Per Hour ** MACH .78 Variation: BUF 8W, FNT 3W, ORD 2E ATC cleared N130JB to maintain FL310 until PMM R-073/15 cross PMM at FL200, cross Pivot at 10,000 feet.

12. FUEL ON BOARD HOURS	MINUTES	13. ALTERNATE AIRPORT(S)	14. PILOT'S NAME, ADDRESS & TELEPHONE NUMBER & AIRCRAFT HOME BASE / 17. DESTINATION CONTACT/TELEPHONE (OPTIONAL)	15. NUMBER ABOARD
		RFD Greater Rockford Rockford, Ill		101

16. COLOR OF AIRCRAFT
RED/WHITE/BLUE

CIVIL AIRCRAFT PILOTS. FAR Part 91 requires you file an IFR flight plan to operate under instrument flight rules in controlled airspace. Failure to file could result in a civil penalty not to exceed $1,000 for each violation (Section 901 of the Federal Aviation Act of 1958, as amended). Filing of a VFR flight plan is recommended as a good operating practice. See also Part 99 for requirements concerning DVFR flight plans.

FAA Form 7233-1 (8-82) CLOSE VFR FLIGHT PLAN WITH ____________ FSS ON ARRIVAL

FLIGHT LOG

CHECK POINTS FROM	TO	ROUTE ALTITUDE	COURSE	WIND TEMP	SPEED-KTS TAS	GS	DIST NM	TIME LEG	TOT	FUEL LEG	TOT
BUF	L/O	Buffalo 1 Climb					70		:16:00		4960*
L/O	YXU	J547 FL310		330/39 ISA-6							
YXU	FNT										
FNT	R-073/15 PMM	FNT.PMM2		330/39 ISA-6							
R-073/15 PMM	PMM	FNT.PMM2 Descent	253				15	:02:00		216.7	
PMM	ORD	FNT.PMM2 Descent & Approach	261/216				89	:13:00		1408.3	
ORD	RFD	Radar V 10,000					97	:17:00			

OTHER DATA: * Includes Taxi Fuel
NOTE: Use 9300 PPH Total Fuel Flow From L/O To Start Of Descent.
Use 9550 PPH Total Fuel Flow For Reserve And Alternate Requirements.
A Missed Approach Requires 450# of Fuel.

TIME and FUEL: As required by FARs.

TIME	FUEL (LB)	
		EN ROUTE
		RESERVE
		ALTERNATE
		TOTAL

FIGURE 119.—Flight Plan/Flight Log.

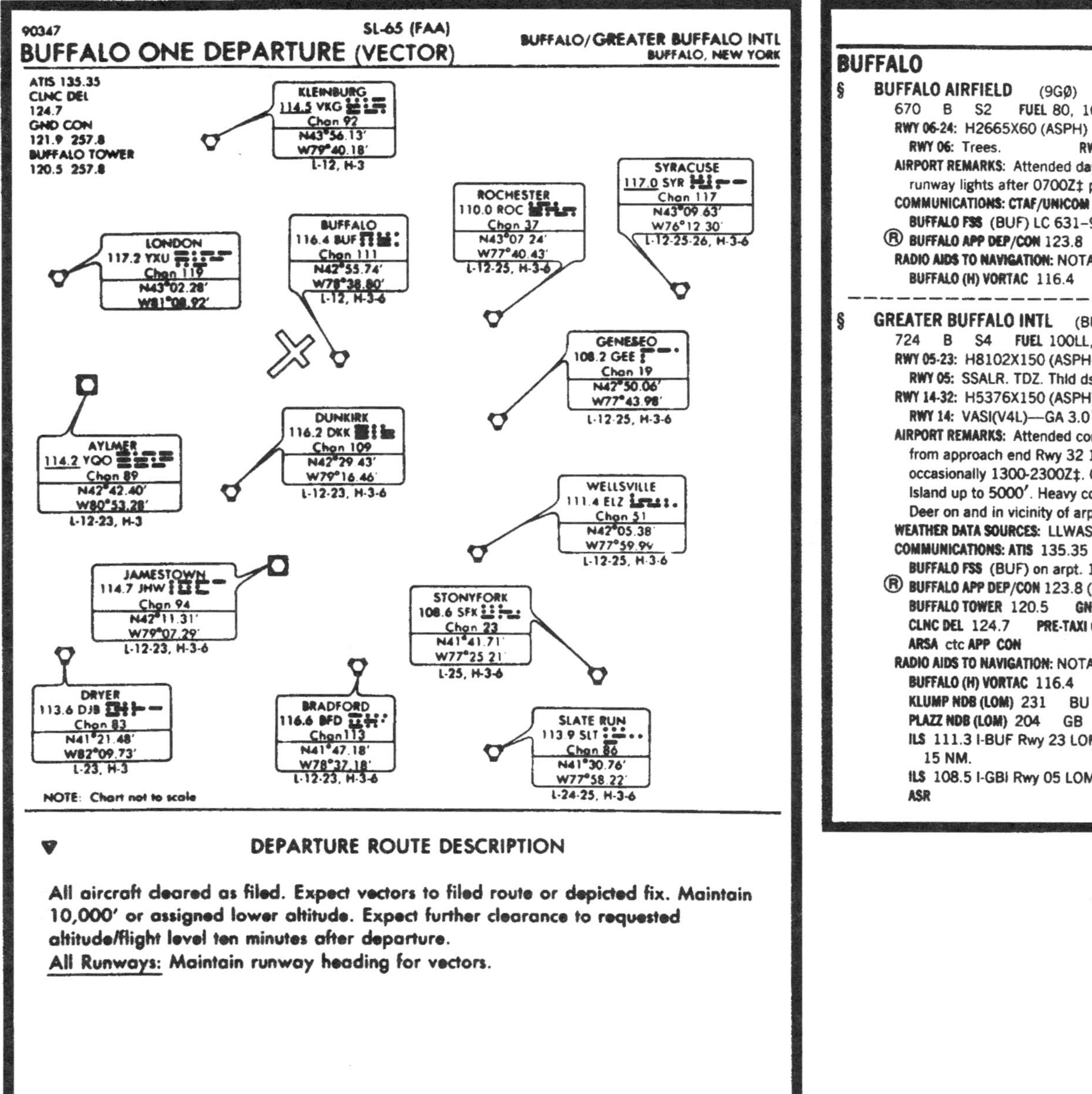

NEW YORK

BUFFALO

§ **BUFFALO AIRFIELD** (9GØ) 6.1 SE UTC–5(–4DT) 42°51'40"N 78°43'00"W — DETROIT L-12H IAP
670 B S2 **FUEL** 80, 100LL
RWY 06-24: H2665X60 (ASPH) S-8 MIRL
RWY 06: Trees. **RWY 24:** Tree.
AIRPORT REMARKS: Attended daylight hours. CAUTION: Ultralight activity on arpt. Airport lgts opr dusk-0700Z‡. For runway lights after 0700Z‡ phone 716-668-4900.
COMMUNICATIONS: CTAF/UNICOM 122.8
BUFFALO FSS (BUF) LC 631-9830. NOTAM FILE BUF.
Ⓡ **BUFFALO APP DEP/CON** 123.8
RADIO AIDS TO NAVIGATION: NOTAM FILE BUF.
BUFFALO (H) VORTAC 116.4 BUF Chan 111 42°55'44"N 78°38'48"W 225°4.9 NM to fld. 730/08W.

---

§ **GREATER BUFFALO INTL** (BUF) 5.2 E UTC–5(–4DT) 42°56'26"N 78°43'57"W — DETROIT H-3C, 6I, L-12H IAP
724 B S4 **FUEL** 100LL, JET A OX 1, 2, 3, 4 LRA CFR Index D
RWY 05-23: H8102X150 (ASPH-GRVD) S–75, D–195, DT–450 HIRL CL 0.6% up NE
RWY 05: SSALR. TDZ. Thld dsplcd 335'. Pole. **RWY 23:** SSALR. TDZ.
RWY 14-32: H5376X150 (ASPH) S-75, D-100, DT-160 MIRL 0.3% up SE.
RWY 14: VASI(V4L)—GA 3.0°TCH 55'. Tree. **RWY 32:** REIL. VASI(V4L)—GA 3.0°TCH 55'. Fence.
AIRPORT REMARKS: Attended continuously. Landing fee. CAUTION-Jet engine test stand located approximately 3600' from approach end Rwy 32 1400' south center line . Jet exhaust may reach altitude 100' AGL. Ops conducted occasionally 1300-2300Z‡. CAUTION: Numerous types of birds may be encountered in holding pattern over Grand Island up to 5000'. Heavy concentration of Gulls, Blackbirds, and Starlings up to 5000' on and in vicinity of arpt. Deer on and in vicinity of arpt. Flight Notification Service (ADCUS) available.
WEATHER DATA SOURCES: LLWAS.
COMMUNICATIONS: ATIS 135.35 **UNICOM** 122.95
BUFFALO FSS (BUF) on arpt. 122.6 122.2 122.1R 116.4T DL NOTAM FILE BUF.
Ⓡ **BUFFALO APP DEP/CON** 123.8 (055°-194°) 126.5 (195°-279°) 126.15 (280°-054°)
BUFFALO TOWER 120.5 **GND CON** 121.9
CLNC DEL 124.7 **PRE-TAXI CLNC** 124.7
ARSA ctc **APP CON**
RADIO AIDS TO NAVIGATION: NOTAM FILE BUF.
BUFFALO (H) VORTAC 116.4 BUF Chan 111 42°55'44"N 78°38'48"W 288°3.5 NM to fld. 730/08W.
KLUMP NDB (LOM) 231 BU 43°00'01"N 78°39'04"W 233° 4.4 NM to fld.
PLAZZ NDB (LOM) 204 GB 42°52'26"N 78°49'00"W 053° 4.8 NM to fld.
ILS 111.3 I-BUF Rwy 23 LOM KLUMP NDB. Inner marker out of svc indefinitely. Back course unusable beyond 15 NM.
ILS 108.5 I-GBI Rwy 05 LOM PLAZZ NDB.
ASR

FIGURE 120.—Buffalo One Departure (Vector).

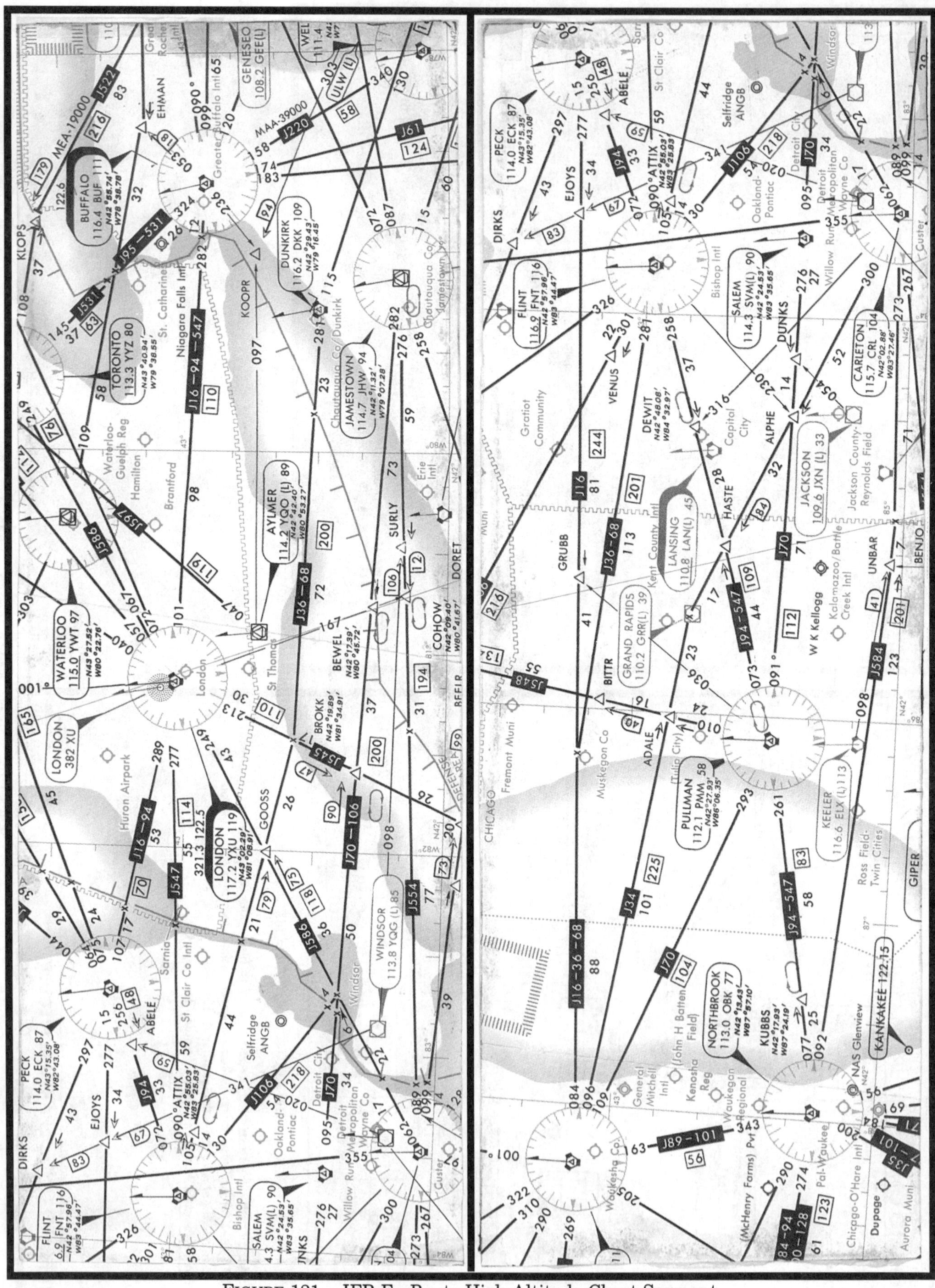

FIGURE 121.—IFR En Route High Altitude Chart Segment.

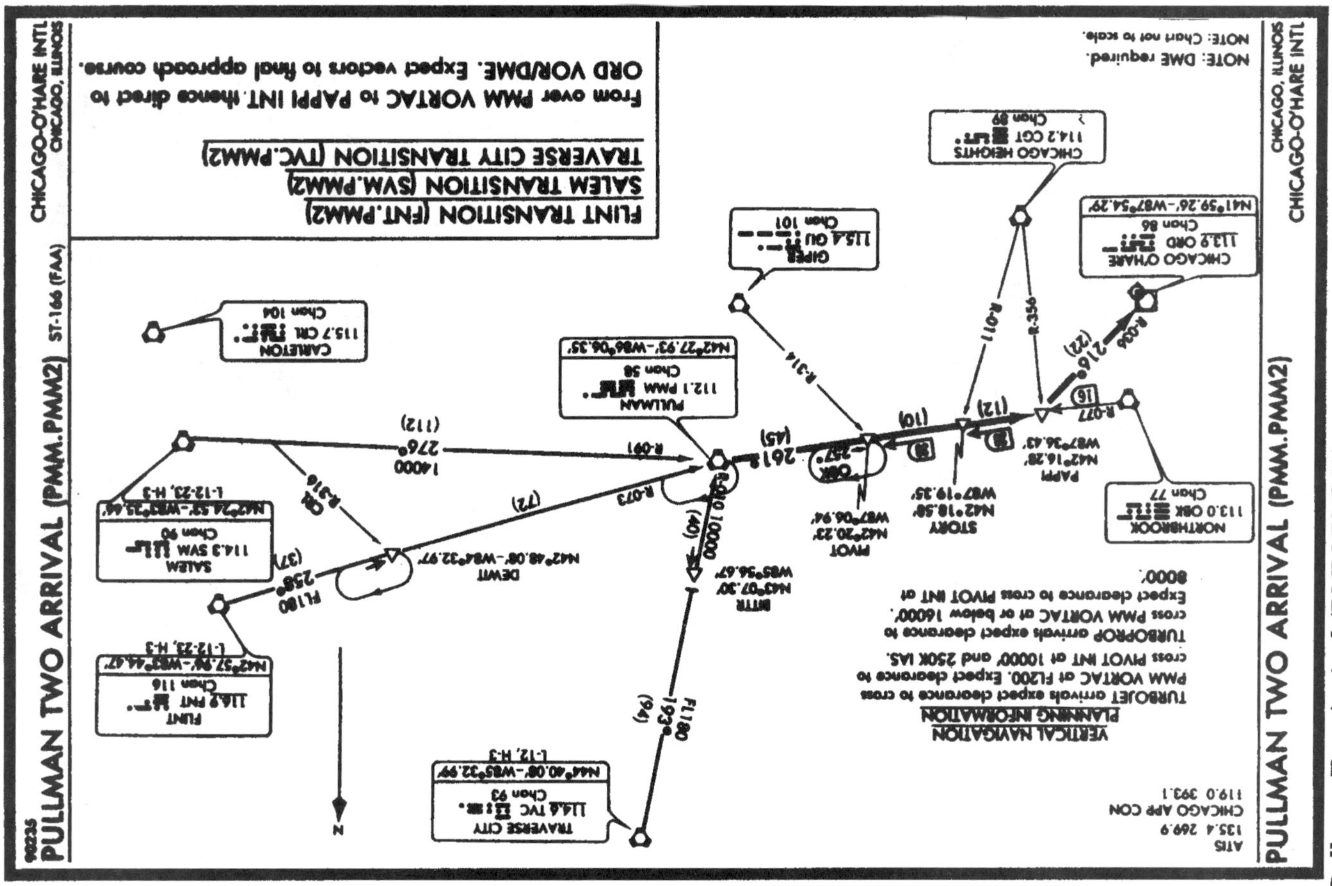

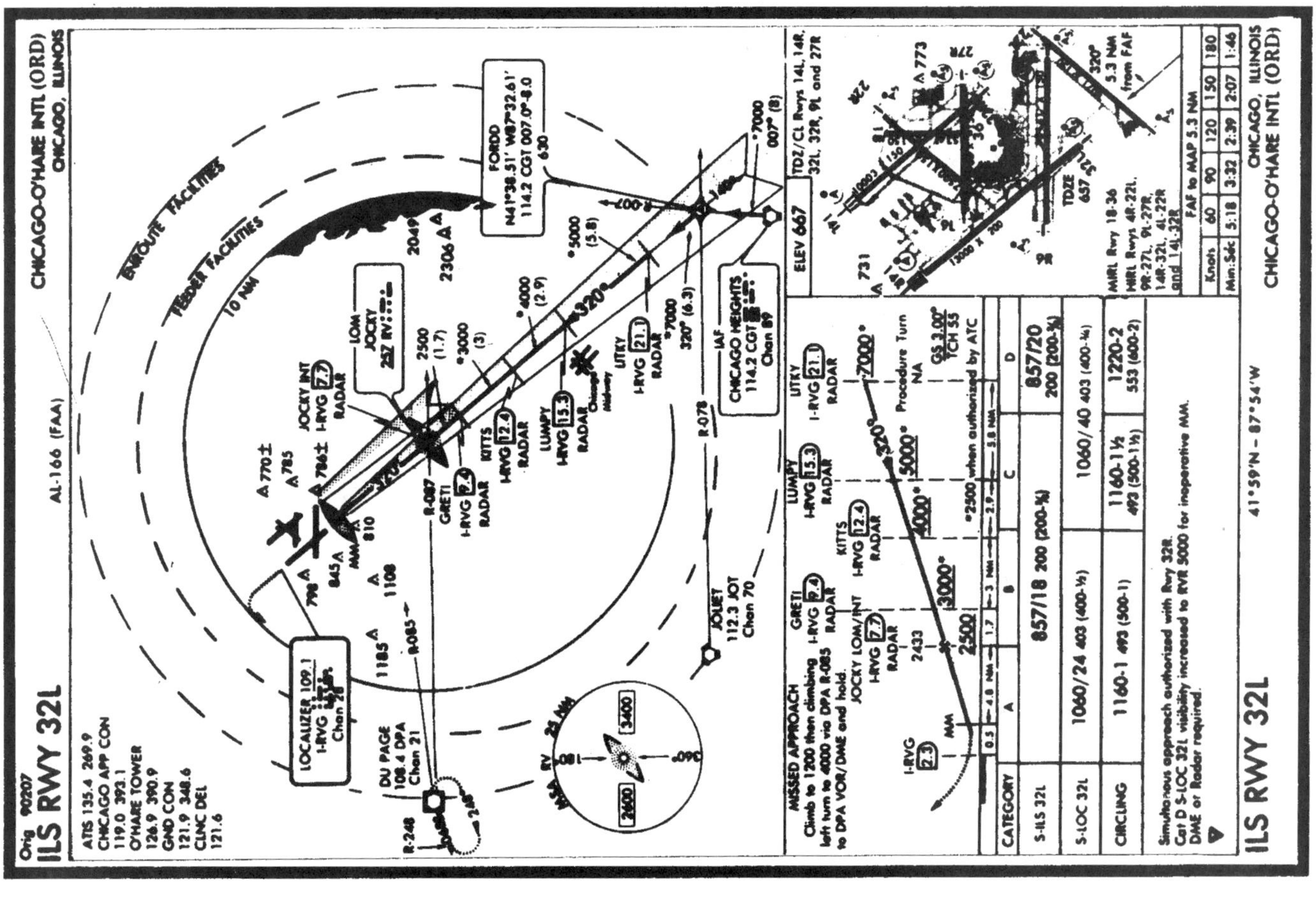

FIGURE 122.—ILS RWY 32L (ORD) / Pullman Two Arrival (PMM.PMM2).

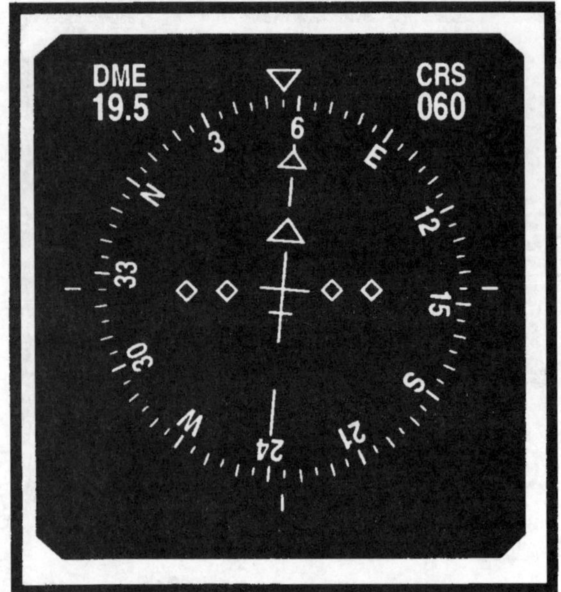

FIGURE 123.—Aircraft Course and DME Indicator.

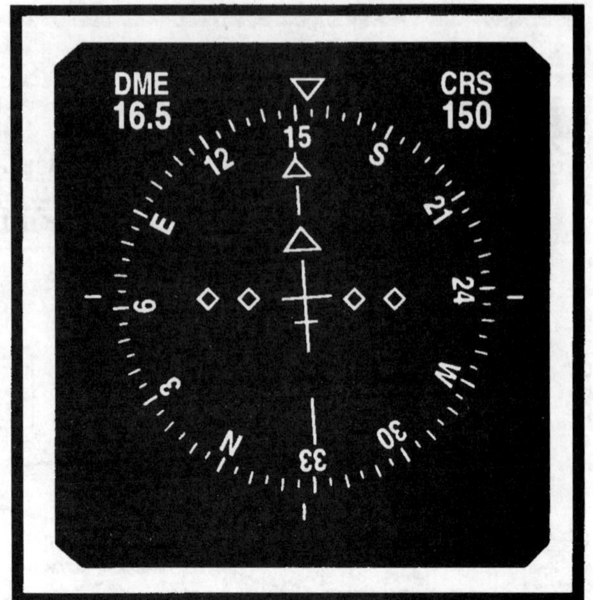

FIGURE 124.—Aircraft Course and DME Indicator.

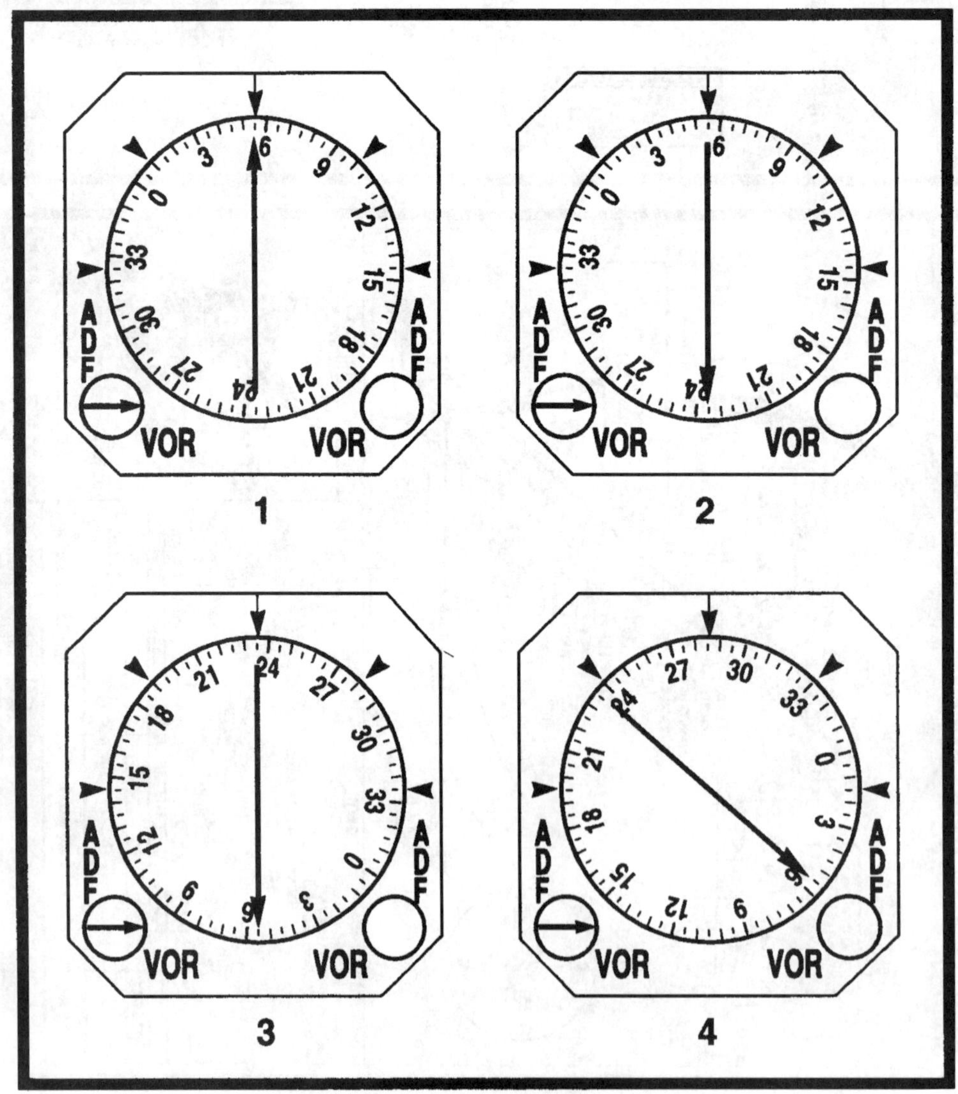

FIGURE 125.—RMI Illustrations.

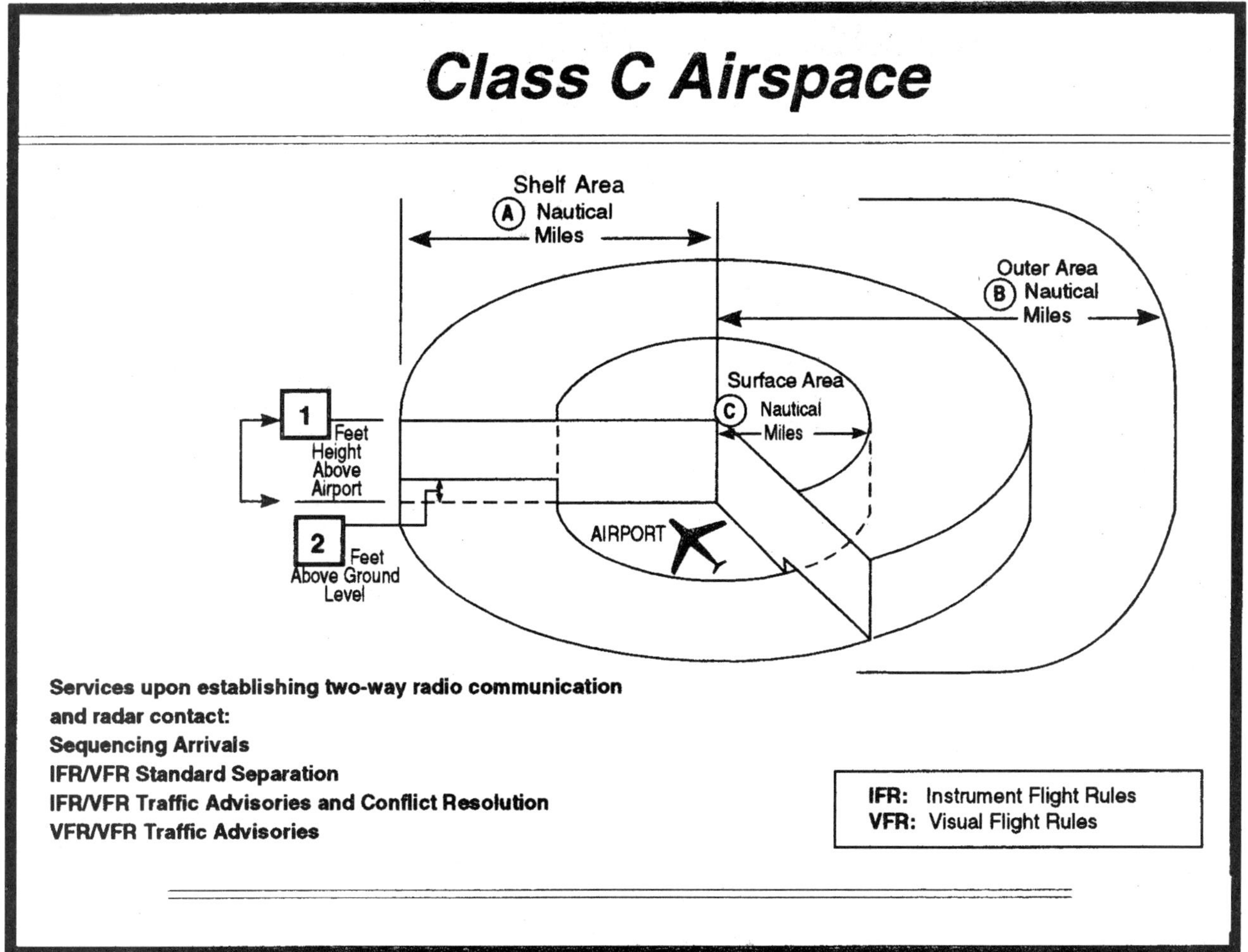

FIGURE 126.—Class C Airspace.

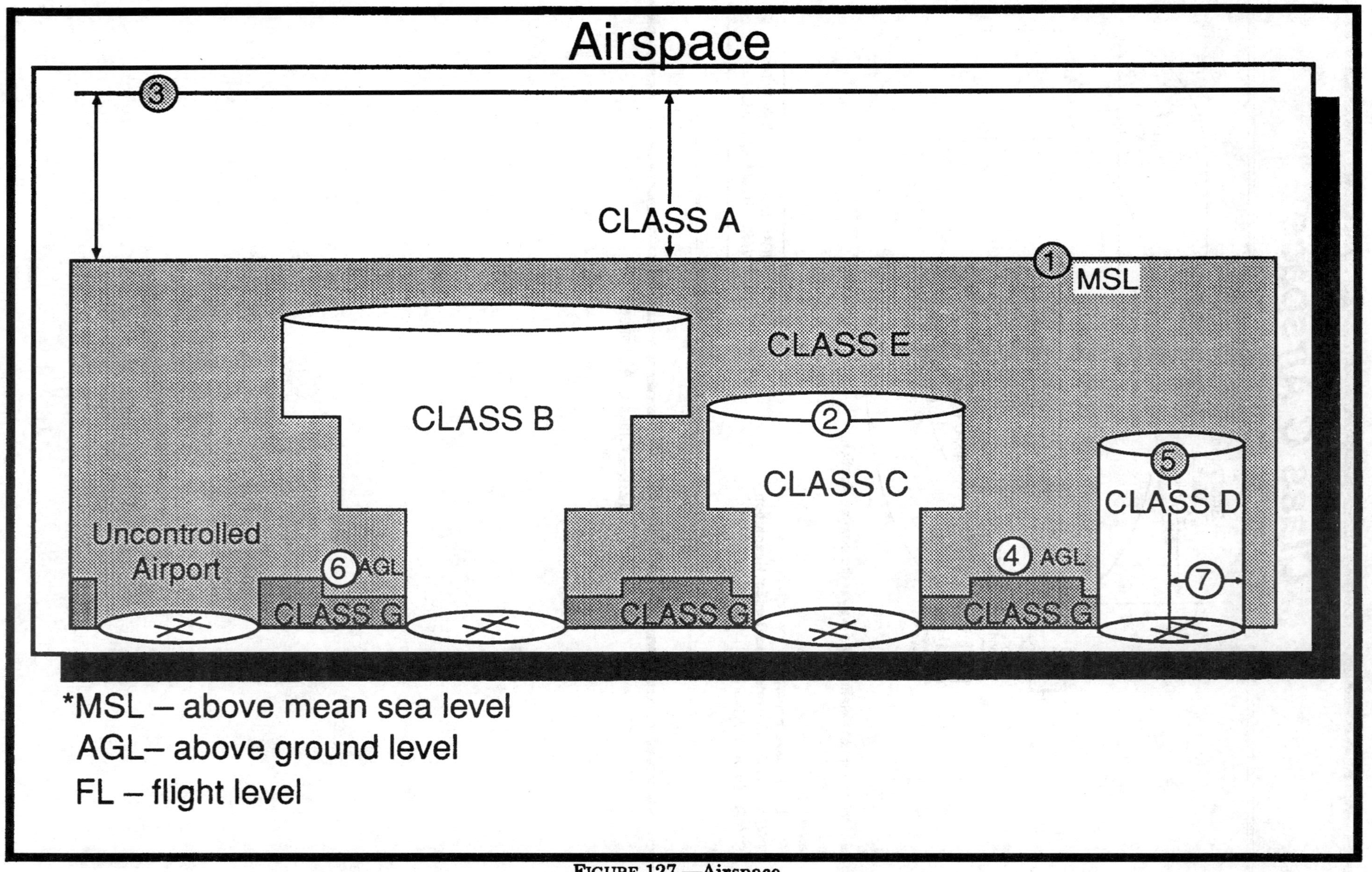

FIGURE 127.—Airspace.

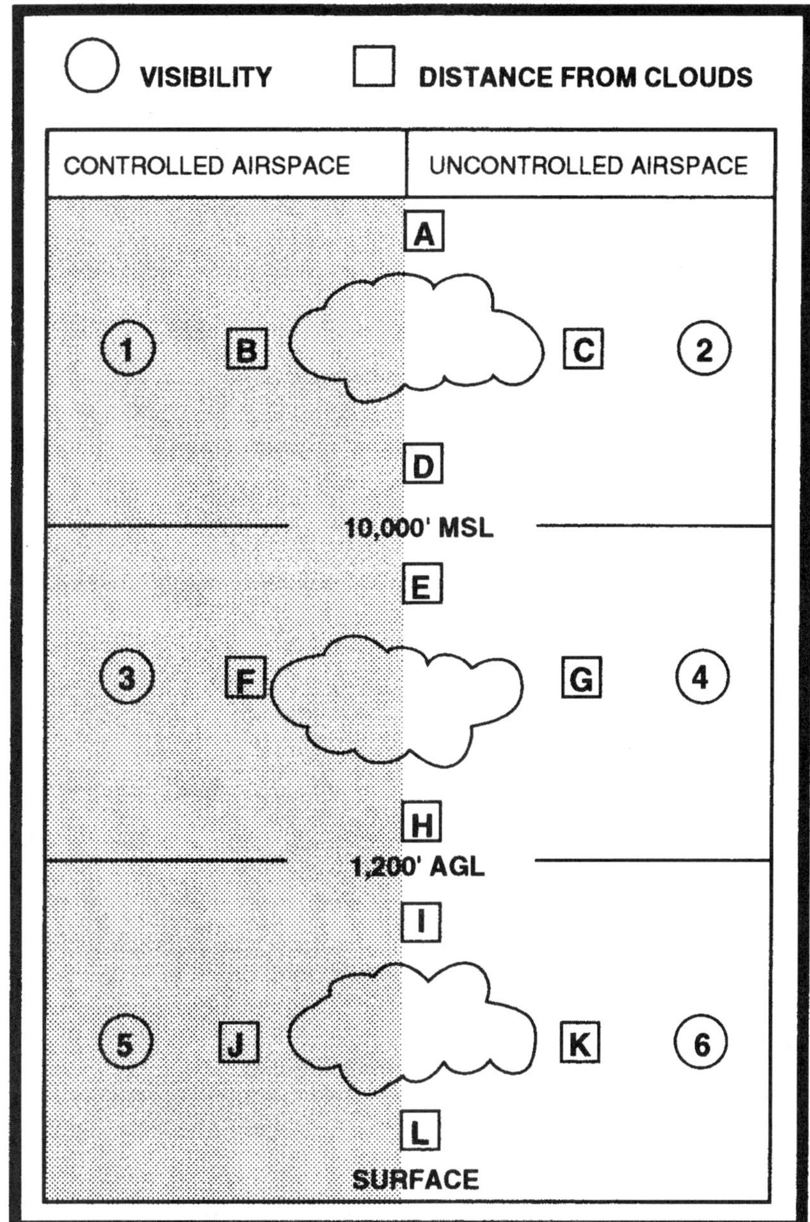

FIGURE 128.—Minimum In-Flight Visibility and Distance From Clouds.

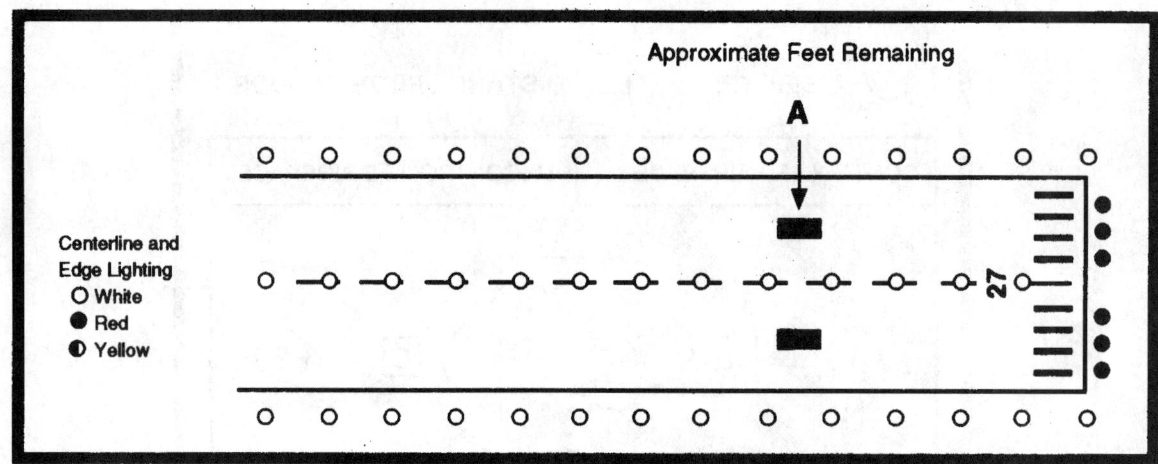

FIGURE 129.—FAA Nonprecision Approach Runway Markings and Lighting.

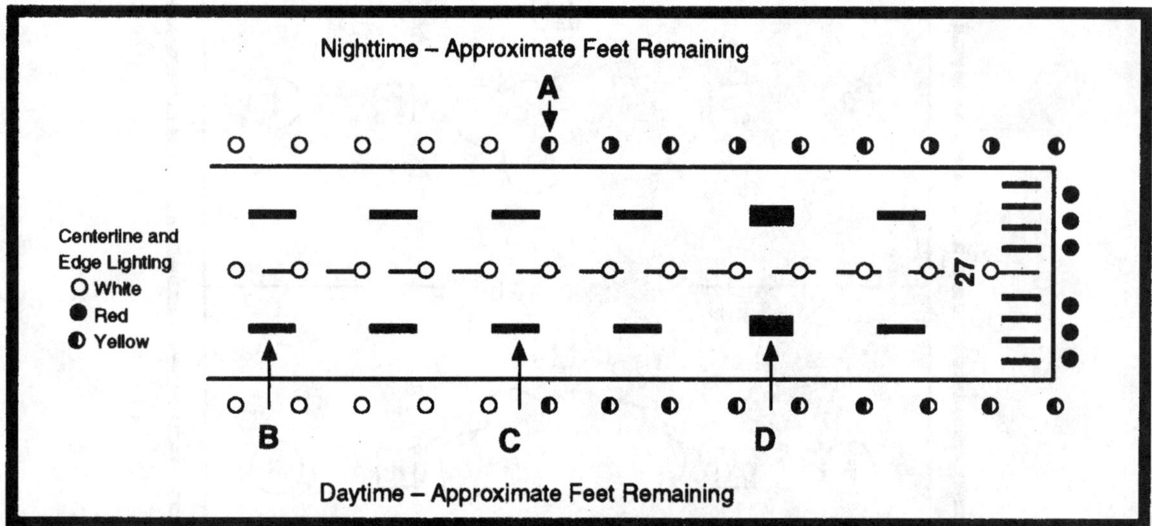

FIGURE 130.—ICAO Nonprecision Approach Runway Markings and Lighting.

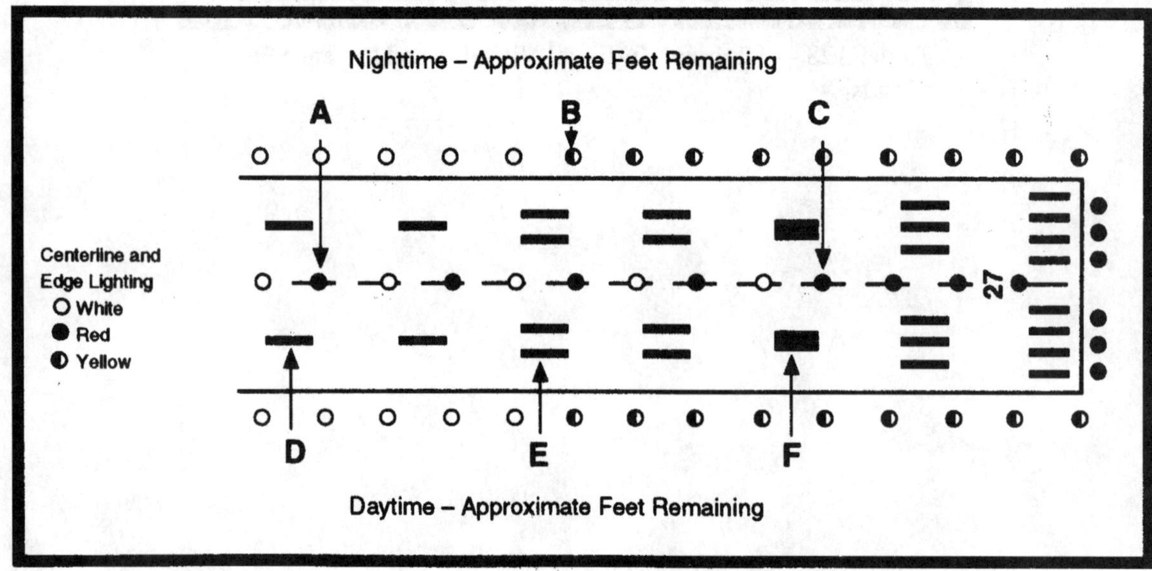

FIGURE 131.—FAA ICAO Precision Approach Runway Markings and Lighting.

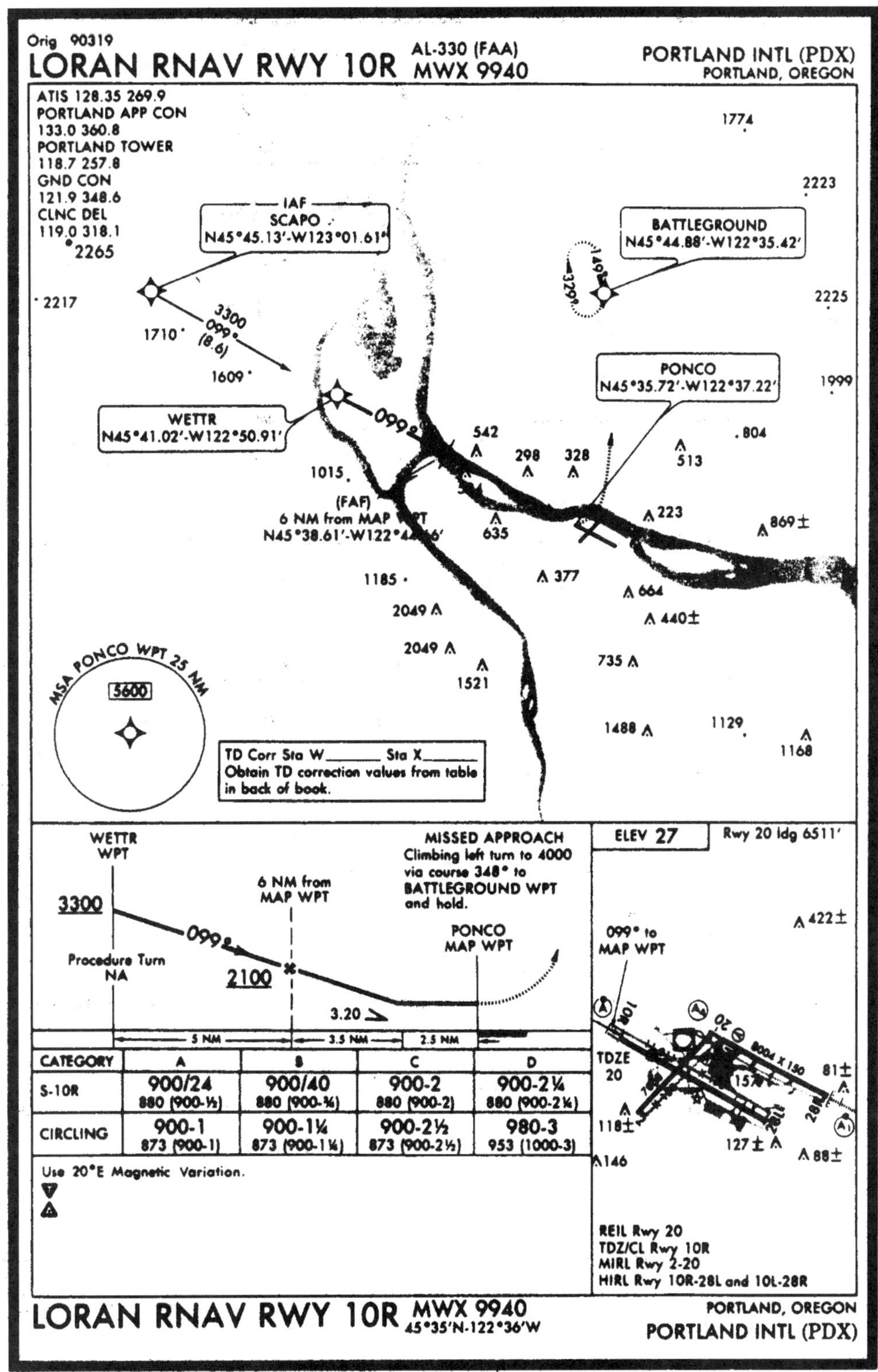

CATEGORY	A	B	C	D
S-10R	900/24 880 (900-½)	900/40 880 (900-¾)	900-2 880 (900-2)	900-2¼ 880 (900-2¼)
CIRCLING	900-1 873 (900-1)	900-1¼ 873 (900-1¼)	900-2½ 873 (900-2½)	980-3 953 (1000-3)

FIGURE 132.—LORAN RNAV RWY 10R – MWX 9940 – (PDX).

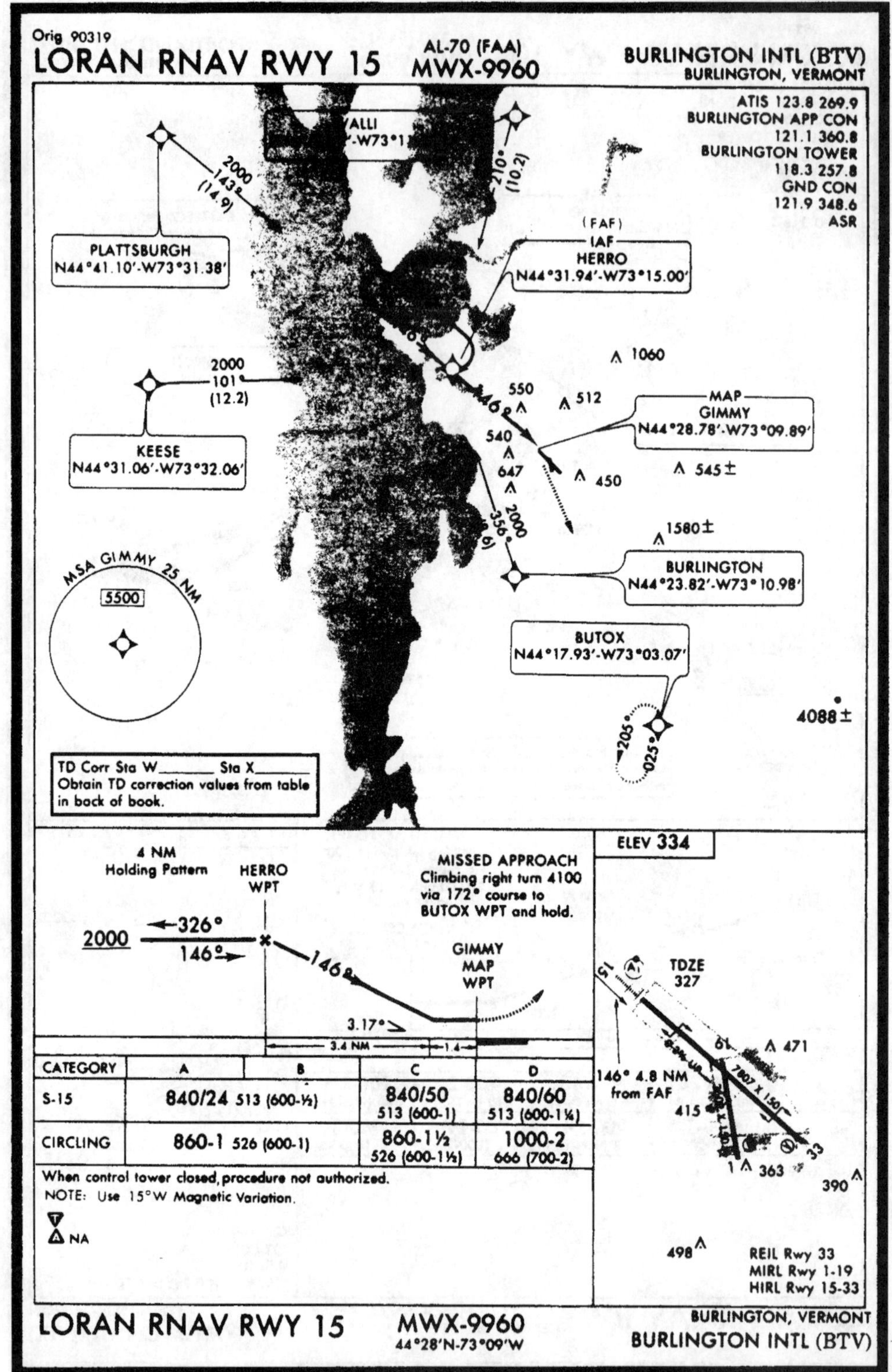

CATEGORY	A	B	C	D
S-15	840/24 513 (600-½)		840/50 513 (600-1)	840/60 513 (600-1¼)
CIRCLING	860-1 526 (600-1)		860-1½ 526 (600-1½)	1000-2 666 (700-2)

FIGURE 133.—LORAN RNAV RWY 15 – MWX-9960 – (BTV).

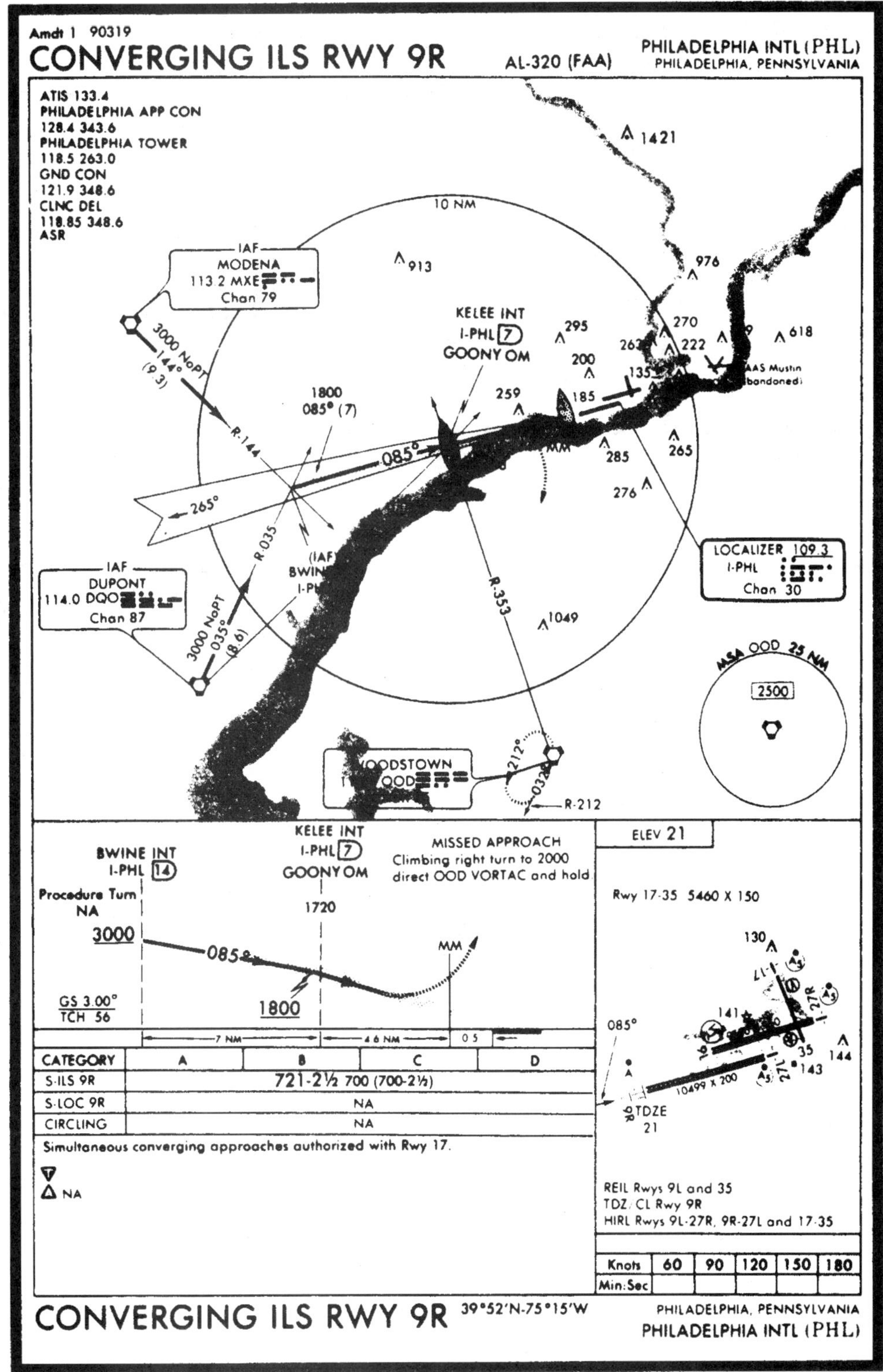

FIGURE 134.—Converging ILS RWY 9R (PHL).

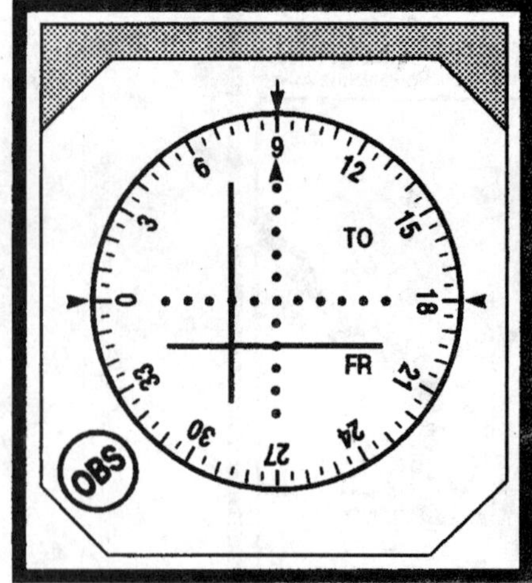

FIGURE 135.—OBS, ILS, and GS Displacement.

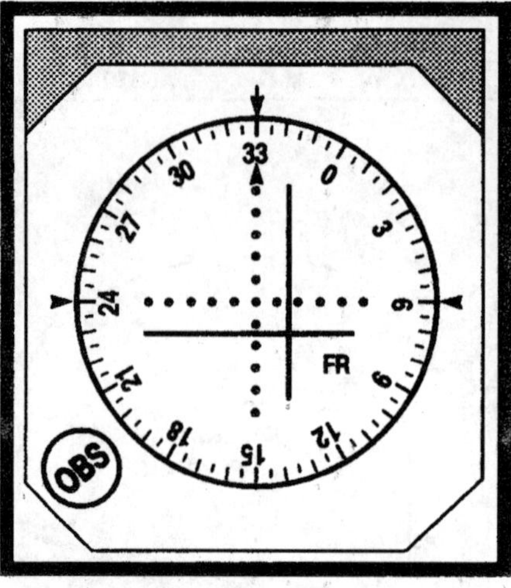

FIGURE 136.—OBS, ILS, and GS Displacement.

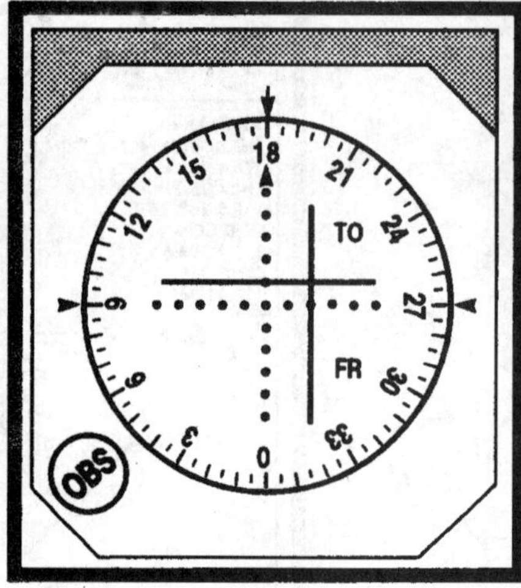

FIGURE 137.—OBS, ILS, and GS Displacement.

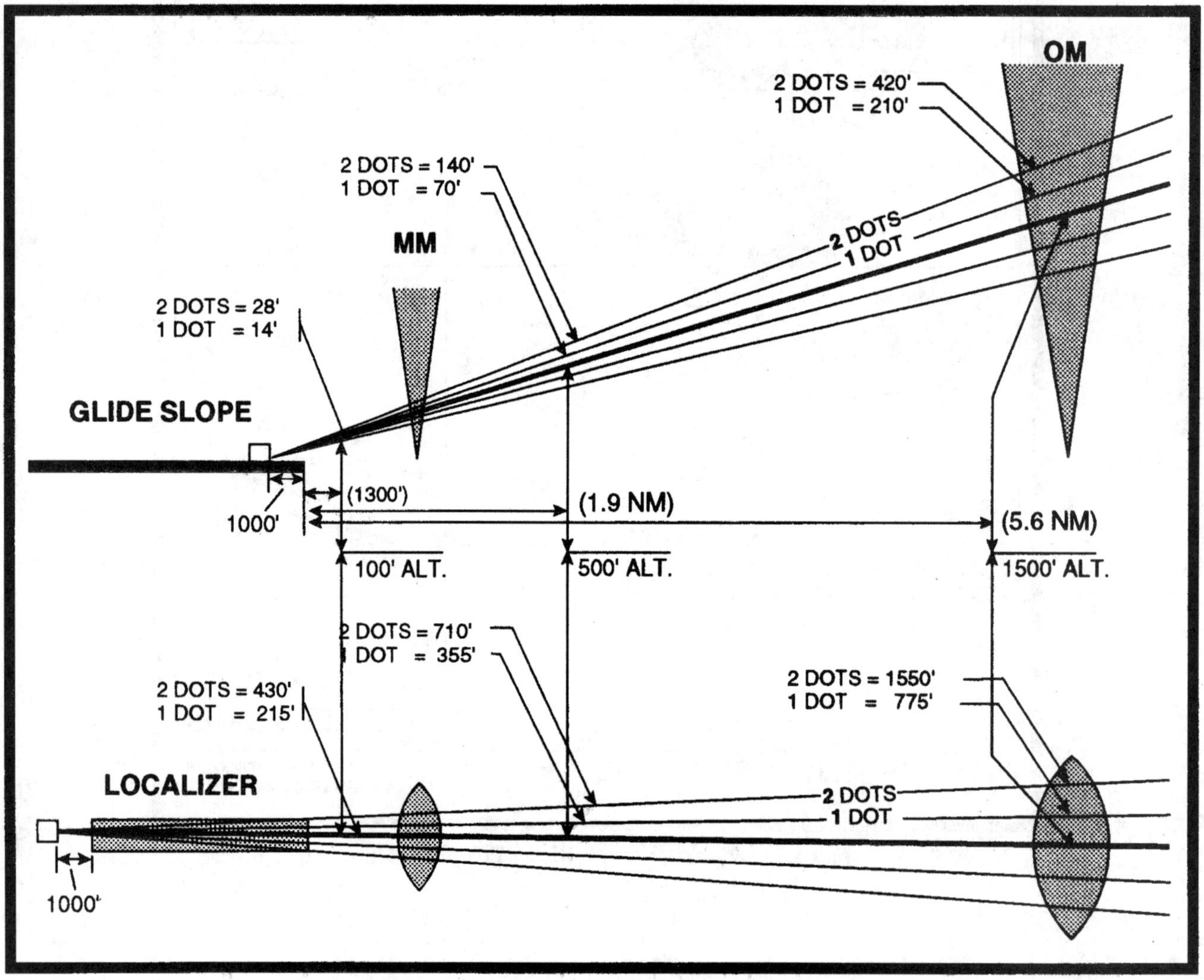

FIGURE 138.—Glide Slope and Localizer Illustration.

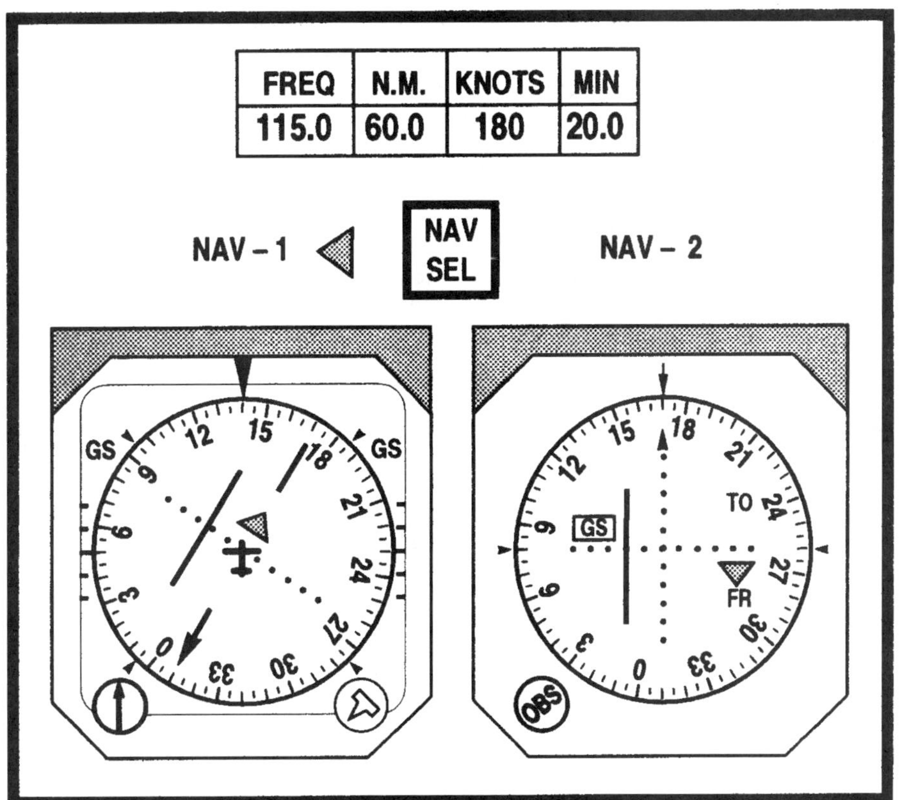

FIGURE 139.—No. 1 and No. 2 NAV Presentation.

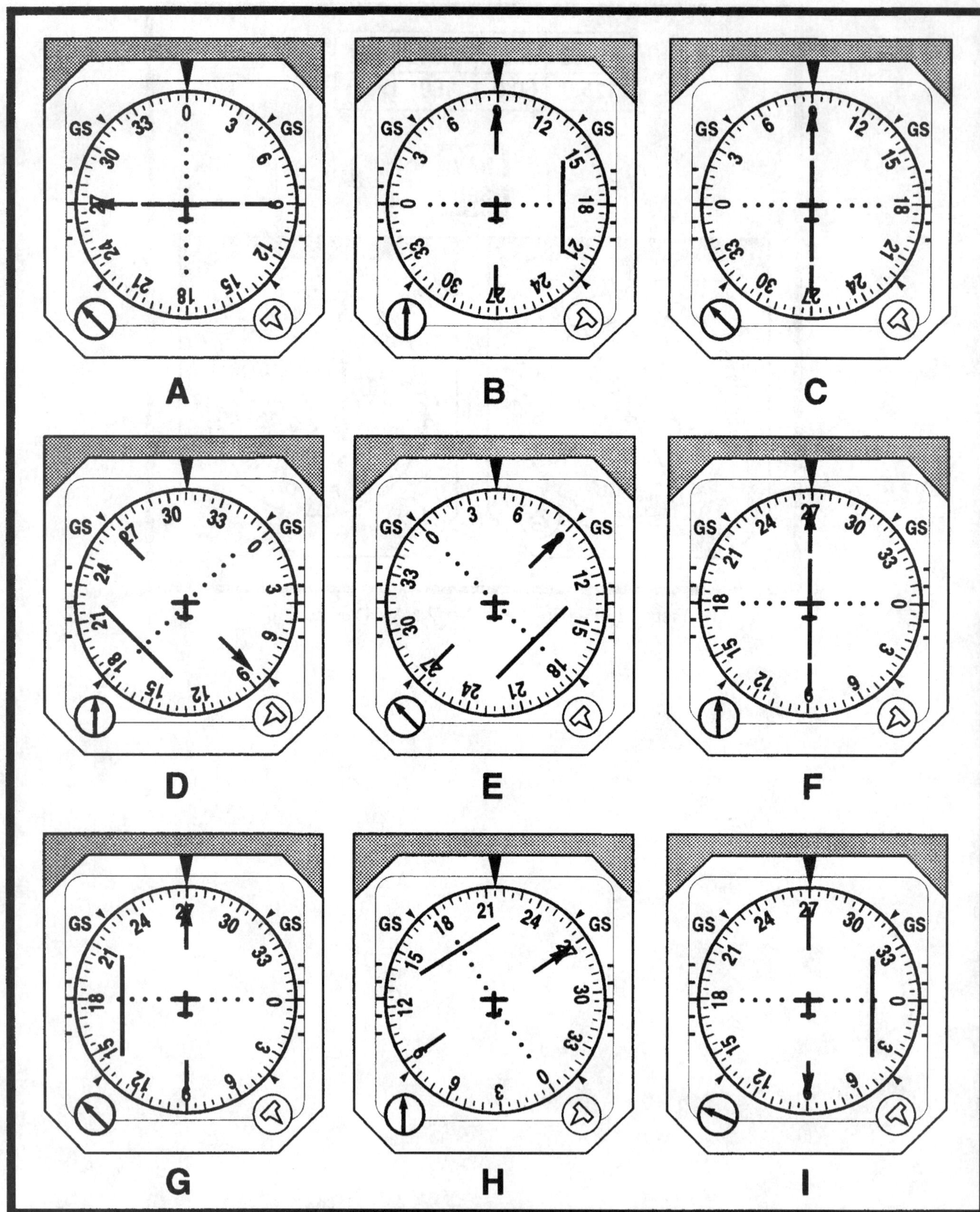

FIGURE 140.—HSI Presentation.

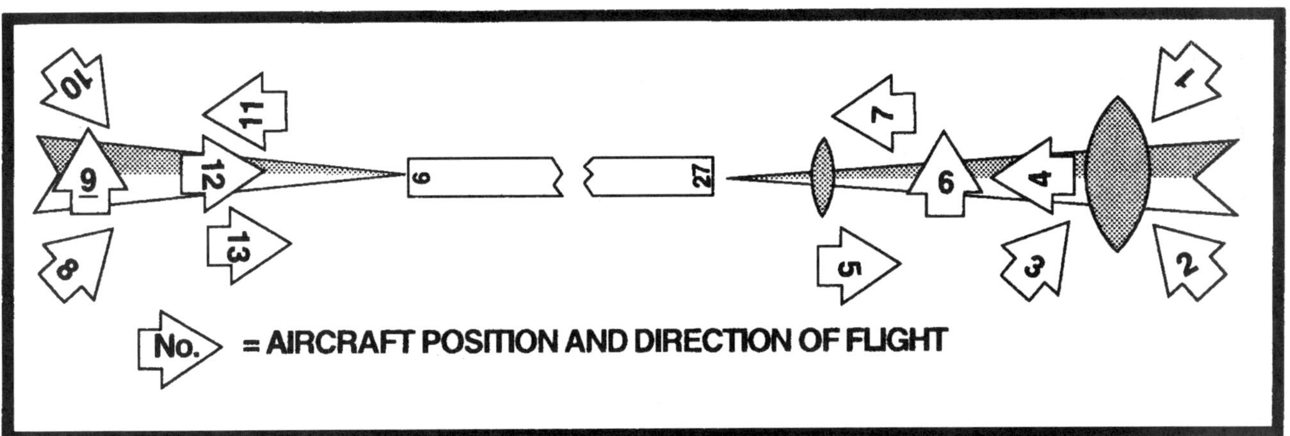

FIGURE 141.—Aircraft Position and Direction of Flight.

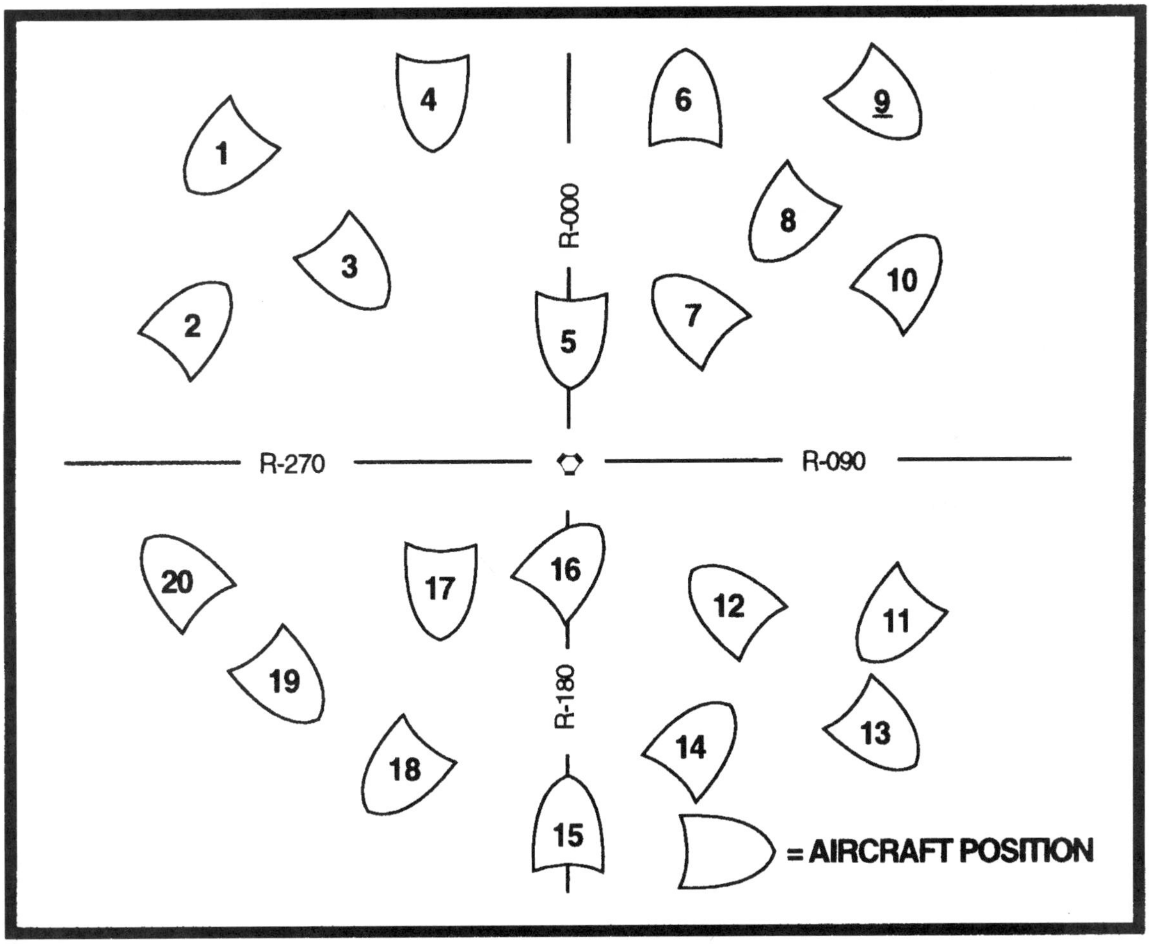

R-000
R-270
R-090
R-180
= AIRCRAFT POSITION

FIGURE 142.—Aircraft Position.

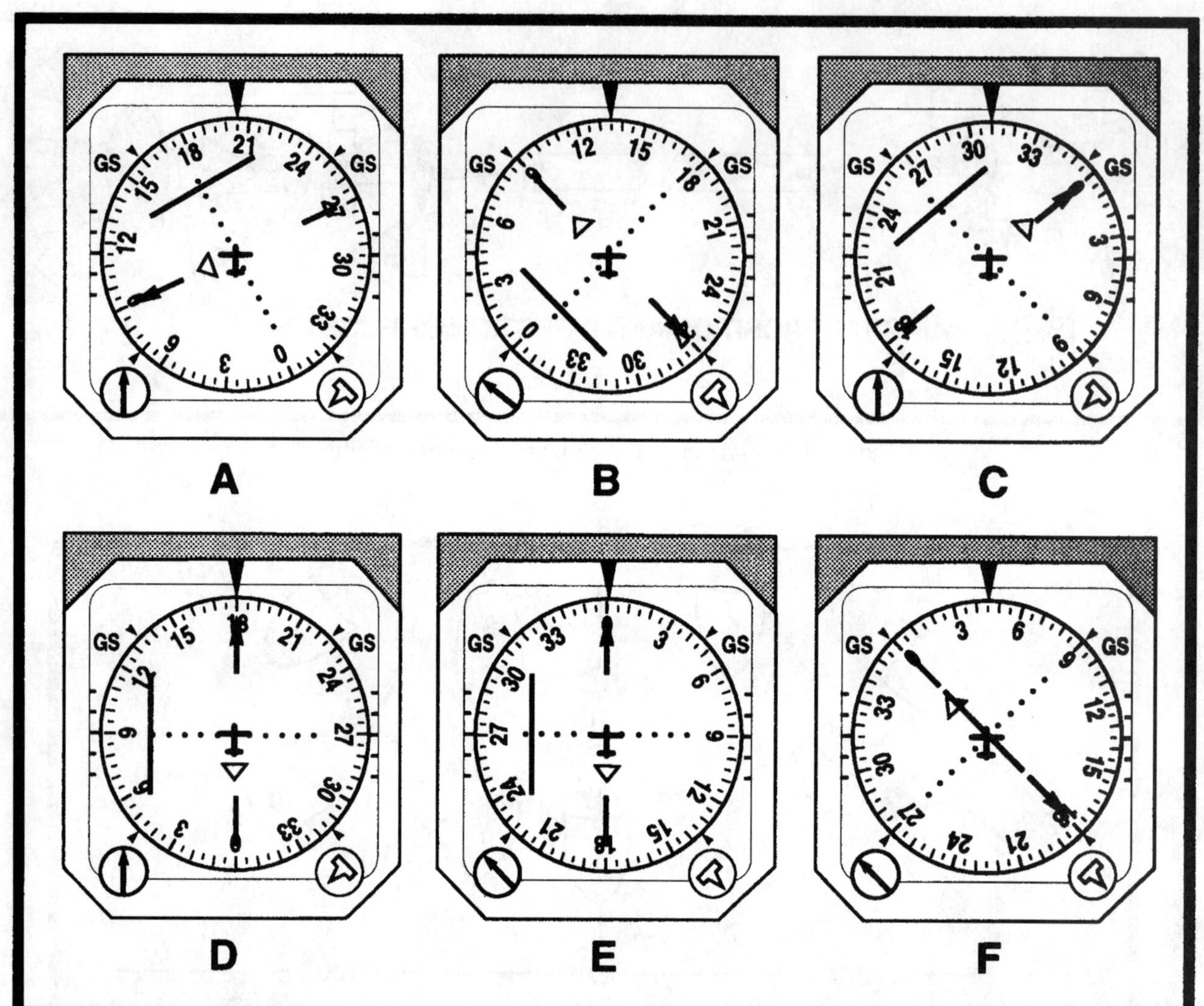

FIGURE 143.—HSI Presentation.

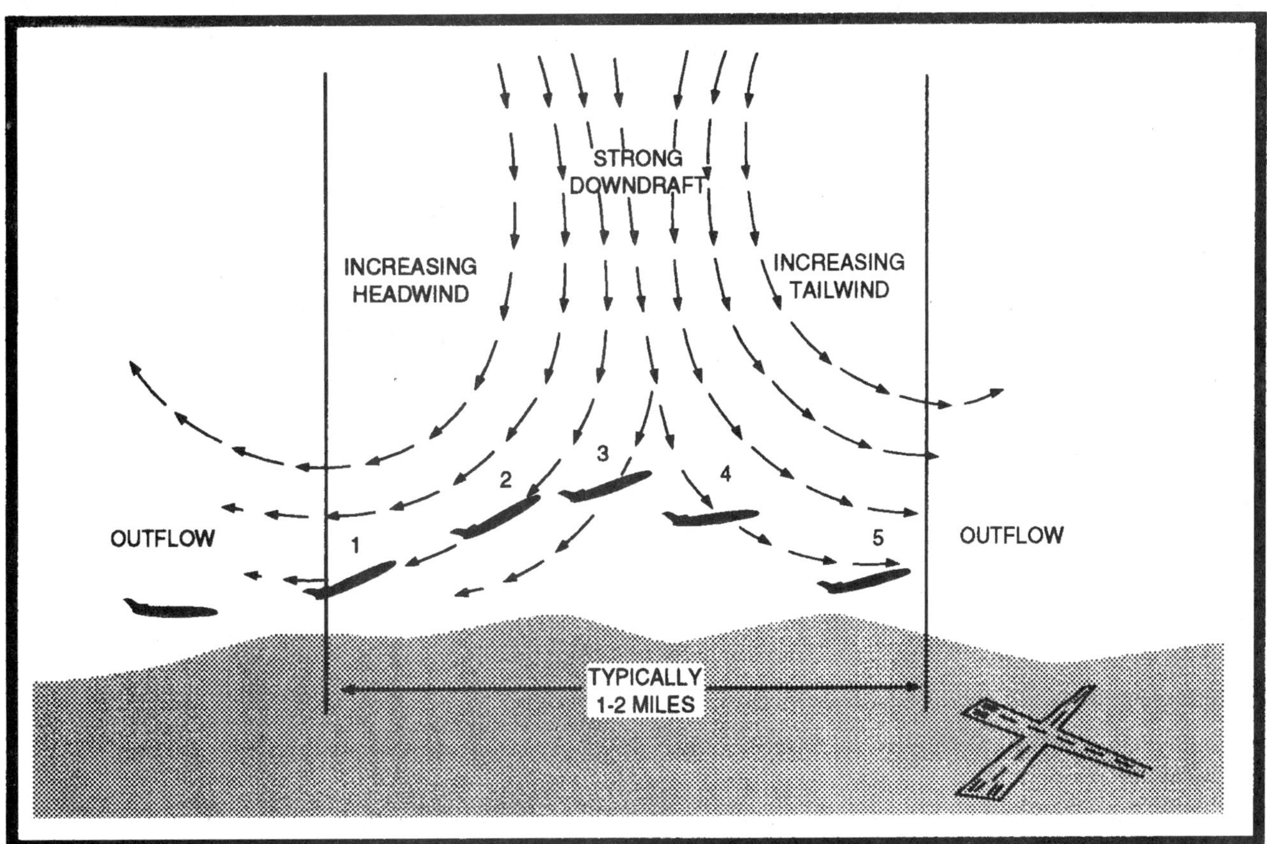

FIGURE 144.—Microburst Section Chart.

AVIATION ROUTINE WEATHER REPORTS (METAR)

TX

METAR KAMA 131755Z 33025G35KT 3/4SM IC OVC003 M02/M01 A2952 RMK PK WND 32039/43 WSHFT 1735 PRESFR P0003.

METAR KAUS 131753Z 19011G17KT 8SM SCT040 BKN250 31/21 A3006 RMK SLPNO.

METAR KBPT 131755Z 17004KT 7SM FEW001 SCT030 BKN250 34/23 A2979 RMK VIS E 2.

METAR KBRO 131755Z 14015KT 6SM HZ SCT034 OVC250 34/30 A2985 RMK PRESRR.

METAR KCDS 131758Z 11013KT 7SM -SHRA OVC180 23/21 A3012 RMK RAB42 VIRGA SW.

METAR KCLL 131749Z 21011KT 7SM SCT003 BKN025 OVC100 34/21 A3008 RMK BKN025 V OVC.

METAR KCOT 131749Z 13010KT 10SM SCT040 SCT200 31/21 A3002 RMK RAE24.

METAR KCRP 131753Z 16016KT 10SM SCT028 BKN250 32/24 A3003.

METAR KDAL 131755Z 16005KT 7SM SCT023 OVC100 30/22 A3007.

METAR KDFW 131800Z 17007KT 10SM SCT035 OVC120 29/20 A3008.

METAR KDHT 131756Z 04014KT 15SM BKN025 22/15 A3026.

METAR KDRT 131756Z 12012KT 10SM FEW006 SCT020 BKN100 OVC250 29/22 A3000 RMK CONS LTG DSTN ESE TS SE MOVG NW VIRGA W.

METAR KELP 131755Z 09007KT 60SM VCBLDU FEW070 SCT170 BKN210 29/13 A3015.

METAR KFTW 131750Z 18007KT 7SM SCT025 OVC100 29/20 A3008.

FTW 131815Z UA /OV DFW/TM 1803/FL095/TP PA30/SK 036 OVC 060/075 OVC/RM TOPS UNKN.

METAR KGGG 131745Z 15008KT 15SM SKC 32/21 A3011.

METAR KGLS 131750Z VRB04KT 6SM VCSH SCT041 BKN093 26/22 A2995.

SPECI KGLS 131802Z 10012G21KT 060V140 2SM +SHRA SCT005 BKN035 OVC050CB 24/23 A2980 RMK RAB57 WSHFT 58 FROPA

METAR KHOU 131752Z 15008KT 7SM SCT030 OVC250 31/27 A3008.

METAR KHRL 131753Z 14015KT 8SM SKC 30/25 A3010.

METAR KIAH 131755Z VRB03KT 1/4SM R33L/1200FT BCFG VV007 27/26 A3005.

METAR KINK 131755Z 04027G36KT 2SM BLSA PO OVC015TCU 24/13 A2985.

METAR KLBB 131750Z 06029G43KT 1SM BLSNDU SQ VV010 03/M01 A2949.

LBB 131808Z UUA /OV LBB /TM 1800 /FL UNKN /TP B737 /TB MDT /RM LLWS -17 KT SFC-010 DURC RWY 36 LBB.

SPECI KLBB 131818Z 35031G40KT 1/2SM FZDZ VV030 M01/M01 A2946 RMK WSHFT 12 FROPA.

LBB 131821Z UUA /OV LBB/TM1817/FL011/TP B727/SK UNKN OVC/TA -06 /TB MOD/IC MDT CLR.

METAR KLFK 131756Z 24007KT 7SM BKN100 33/19 A3008.

METAR KMAF 131756Z 02020KT 12SM BKN025 OVC250 27/18 A3009 RMK RAE44.

METAR KMFE 131756Z 13015KT 7SM BKN125 33/19 A2998.

METAR KMRF 131752Z 09012G20KT 60SM SKC 28/14 A3000.

MRF 131801Z UUA/OV MRF/TM1758/FL450/TP B767/TB MDT CAT.

FIGURE 145.—Aviation Routine Weather Reports (METAR).

AVIATION ROUTINE WEATHER REPORTS (METAR)

TX

METAR KABI 131755Z AUTO 21016G24KT 180V240 1SM R11/P6000FT -RA BR BKN015 OVC025 19/15 A2990 RMK AO2 PK WND 20035/25 WSHFT 1715 VIS 3/4V1 1/2 VIS 3/4 RWY11 RAB07 CIG 013V 017 CIG 014 RWY11 PRESFR SLP125 P0003 60009 T01940154 10196 20172 58033 TSNO $.

METAR KMWL 131756Z 13011KT 10SM BKN011 OVC050 25/23 A3006.

METAR KPSX 131755Z 20010KT 7SM SCT018 OVC200 31/24 A3007.

METAR KPVW 131750Z 05006KT 10SM SCT012 OVC030 30/20 A3011 RMK RAE47.

METAR KSAT 131756Z 15016KT 7SM SCT028 OVC250 30/20 A3005.

SAT 131756Z UA /OV SAT/TM 1739Z/ FL UNKN/TP UNKN/SK OVC 040.

METAR KSJT 131755Z 22012KT 7SM BKN018 OVC070 25/23 A3002.

METAR KSPS 131757Z 09014KT 6SM -RA SCT025 OVC090 24/22 A3005.

SPECI KSPS 131820Z 01025KT 2SM +RA OVC015TCU 22/21 A3000 RMK DSNT TORNADO B15 N MOV E.

SPS 131820Z UA/OV SPS/TM 1818/FL090/TP C402/SK OVC 075.

METAR KTPL 131751Z 17015KT 15SM SCT015 SCT100 OVC250 31/20 A3007.

METAR KTYR 131753Z AUTO 26029G41KT 2SM +TSRA BKN008 OVC020 31/24 A3001 RMK A02 TSB44 RAB46.

METAR KVCT 131755Z 17013KT 7SM SCT030 OVC250 30/24 A3005.

AR

METAR KARG 131753Z AUTO 22015G25KT 3/4SM R28/2400FT +RA OVC010 29/28 A2985 RMK AO2.

METAR KELD 131755Z 06005G10KT 3SM FU BKN050 OVC100 30/21 A3010.

METAR KFSM 131756Z 00000KT 5SM SKC 30/20 A2982.

FSM 131830Z UA/OV HRO-FSM/TM 1825/FL290/TP B737/SK SCT 290.

METAR KFYV 131755Z 170018G32KT 2SM +TSRA SQ SCT030 BKN060OVC100CB 28/21 A2978 RMK RAB47.

FYV 131801Z UA/OV 1 E DAK/TM 1755Z/FL 001/TP CV440/RM WS LND RWY16 FYV.

METAR KHOT 131751Z 34006KT 18SM SCT040 OVC150 32/18 A3010.

METAR KHRO 131753Z 09007KT 7SM FEW020 BKN040CB 30/27 A3001.

SPECI KHRO 131815Z 13017G26KT 2SM +TSRA SCT020 BKN045TCU 29/24A2983 RMK RAB12 FRQ LTGICCG VC PRESFR.

HRO 131830Z UUA/OV 6 S HRO/TM 1825Z/FL 001/TP DC6/RM WS TKO RWY 18.

METAR KLIT 131754Z 07004KT 10SM SCT030 BKN250 34/29 A3007.

METAR KPBF 131753Z 29007KT 5SM SCT040 BKN100 35/19 A3008.

METAR KTXK 131753Z 25003KT 7SM SCT100 BKN200 33/19 A3010.

FIGURE 146.—Aviation Routine Weather Reports (METAR).

INTERNATIONAL TERMINAL AERODROME FORECASTS (TAF)

TX

TAF
KALI 031745Z 031818 14015KT 6SM HZ BKN012
FM2000 15015G25KT P6SM BKN030 WS009/02045KT
FM2200 16011G21KT 4SM SCT040 BKN250 TEMPO 2301 3SM TSGS BKN020
FM0100 13015KT 5SM SCT015
FM0700 12008KT 5SM BKN008 BECMG 0912 3SM BKN015

TAF
KAMA 031745Z 031818 05012KT 5SM RA BR BKN010 BKN080 TEMPO1803 03015KT 2SM +TSRA OVC010
FM0400 03015KT 3SM BKN020 OVC080 TEMPO 0410 2SM +TSRA OVC010
FM1100 03012KT 5SM RA BR OVC010 BECMG 1618 1/2SM RA FG OVC008

TAF
KAUS 031745Z COR 031818 17010KT P6SM BKN025 OVC100
FM2100 15008KT 4SM BKN030 OVC100 TEMPO 2223 1SM TSPE OVC010
FM0100 16005KT 5SM BKN014 TEMPO 0809 1SM +TSRA BKN014 BECMG 1214 3SM TSRA BKN020
FM1500 17008KT 5SM SCT050

TAF
KCRP 031745Z 031818 15015G20KT P6SM SCT020 BKN250
FM2300 16015G25KT 4SM SCT030 BKN250 TEMPO 0001 TSRA
FM0100 16015KT 2SM BKN015 BECMG 0911 5SM SCT030

TAF
KDAL 031745Z 031818 00000KT P6SM SCT030 BKN100
FM2200 17007KT 5SM BR BKN030 OVC100 PROB40 0002 2SM TSRA OVC010
FM0200 09005KT 4SM -RA BKN020 PROB30 0407 3SM TSRA
FM0700 07004KT 1/2SM FG OVC002 BECMG 0912 3SM TSRA SCT040

TAF AMD
KDRT 031745Z 031818 14010KT P6SM OVC014
FM1900 VBR05KT 5SM BKN020 OVC100 TEMPO 2021 2SM +TSRA
FM2300 14012KT 5SM HZ BKN030 BKN100 PROB40 0205 3SM TSRA BKN020
FM0500 27006KT 6SM BR SCT035 BKN080 TEMPO 0709 2SM FU BR BKN020
FM1000 00000KT 4SM OVC030 BECMG 1416 3SM TSRA OVC020

TAF
KELP 031745Z 031818 08012KT P6SM SCT070 SCT100
FM2000 13010KT 6SM SCT070 BKN120 TEMPO 2223 15026G35KT 3SM BLSA BKN050
FM0600 07012KT 5SM BKN070 PROB40 0709 2SM -TSRA BKN025
FM1200 07020G34KT 1SM +TSRA BKN020CB WS008/25040KT

TAF
KHOU 031745Z 031818 18010KT 6SM HZ SCT020
FM2100 18015KT 4SM HZ SCT035 SCT250
FM0100 19010KT 3SM HZ SCT 250
FM0700 20005KT 1SM BR FU BKN005 OVC025
FM1300 13007KT 4SM HZ BKN040

TAF
KIAH 031745Z 031818 18010KT 5SM HZ SCT020
FM2000 16008KT 4SM HZ SCT015 SCT250
FM0500 17012KT 1SM BR FU BKN008 OVC020
FM1000 00000KT 1/4SM -RA FG BKN010 OVC031
FM1400 14005KT 5SM BKN004 OVC080 BECMG 1618 NSW

TAF
KINK 031745Z 031818 10010KT P6SM SCT020 SCT100
FM2100 08013KT 3SM DZ BKN025 BKN080 PROB40 0002 06026G35KT1SM +TSRAGR
FM0400 05019KT 2SM DU BKN020 OVC050 PROB40 0709 1SM +TSRA FEW002 OVC010
FM0900 02004KT 1/2SM RA FG SCT025 BKN045 OVC100CB
FM1400 34035G45KT 2SM SS SKC

FIGURE 147.—International Terminal Aerodrome Forecasts (TAF).

INTERNATIONAL TERMINAL AERODROME FORECASTS (TAF), CONTINUED

TAF
KLBB 031745Z 031818 06012KT 3SM -TSRA SCT010 OVC020
FM2100 04015KT 5SM BR BKN020 OVC060 PROB40 0103 06025G35KT 1/8SM +SHRASNPE OVC003
FM0400 05018KT 3SM -RA BR OVC010 PROB30 0608 07020KT 1SM +TSRA
FM0900 00000KT 1/4SM -RA FG VV002
FM1300 01005G12KT 1SM FZRA
FM1600 VBR04KT 1/8SM FG VV001

TAF
KSAT 031745Z 031818 17010KT 6SM HZ BKN016 OVC030
FM2000 17015KT P6SM BKN025
FM2200 19012KT 4SM FU BKN030 OVC250 PROB40 0104 07020G30KT 3SM TSRA BKN020
FM0500 12015KT 3SM SG BKN010 BKN035 PROB40 0709 05015G23KT 1SM +TSRA OVC010
FM1000 35008G16KT 4SM BLSN OVC020

TAF
KSJT 031745Z 031818 12012KT 6SM HZ BKN016
FM2000 17018KT 4SM BR BKN025
FM2200 14020G28KT 3SM GS BKN030 OVC250 PROB40 0103 16025G32KT 1SM +TSRA OVC008CB
FM0900 17020G34KT 2SM RA BR OVC010CB

TAF
KSPS 031745Z 031818 07012KT 4SM -RA FG SCT030 BKN080 TEMPO 0203 09022G30KT 1SM FZDZ OVC020
FM0900 05015KT 2SM BR SCT001 BKN005 OVC010 SNRA WS090/09035KT

FIGURE 147.—International Terminal Aerodrome Forecasts (TAF), Continued.

CONVECTIVE SIGMET

MKCC WST Ø31755
CONVECTIVE SIGMET 42C
VALID UNTIL 1955Z
TX OK
FROM 5W MLC–PEQ–SJT–5W MLC
AREA SCT EMBDD TSTMS MOVG LTL. TOPS 3ØØ.

CONVECTIVE SIGMET 43C
VALID UNTIL 1955Z
CO KS OK
FROM AKO–OSW–3ØWNW OKC–AKO
AREA SCT TSTMS OCNLY EMBDD MOVG FROM 322Ø. TOPS 38Ø.

CONVECTIVE SIGMET 44C
VALID UNTIL 1955Z
5ØNE MEM
ISOLD INSTD LVL5 TSTM DIAM 1Ø MOVG FROM 2625. TOP ABV 45Ø.

OUTLOOK VALID UNTIL 2355Z
TSTMS OVR TX AND SE OK WL MOV SEWD 15 KTS.
TSTMS OVER CO, KS, AND N OK WL CONT MOVG SEWD 2Ø KTS.
TSTM OVR TN WL CONT MOVG EWD 25 KTS.

FIGURE 148.—Convective Sigmet.

WINDS AND TEMPERATURES ALOFT FORECASTS

DATA BASED ON Ø312ØØZ
VALID Ø4ØØØØZ FOR USE 18ØØ–Ø3ØØZ. TEMPS NEG ABV 24ØØØ

FT	3ØØØ	6ØØØ	9ØØØ	12ØØØ	18ØØØ	24ØØØ	3ØØØØ	34ØØØ	39ØØØ
ABI		13Ø6+16	16Ø7+11	18Ø7+Ø6	21Ø8-Ø7	22Ø8-18	24Ø833	25Ø942	3ØØ753
ABO			Ø81Ø+14	Ø511+Ø8	3415-Ø6	322Ø-18	312333	312543	3Ø2554
AMA		Ø614	Ø814+1Ø	Ø7Ø9+Ø5	321Ø-Ø7	2914-19	281934	282243	292554
ATL	Ø9Ø6	99ØØ+17	99ØØ+12	Ø2Ø5+Ø7	35Ø7-Ø7	33Ø5-19	29Ø534	28Ø543	99ØØ54
BNA	99ØØ	99ØØ+17	32Ø5+12	31Ø9+Ø7	3Ø18-Ø7	2918-19	272134	262444	262855
BRO	151Ø	1614+2Ø	1611+14	17Ø8+Ø8	99ØØ-Ø7	99ØØ-19	99ØØ34	99ØØ43	99ØØ55
DAL	Ø91Ø	17Ø6+17	2ØØ9+11	2Ø11+Ø6	2Ø15-Ø8	2214-19	231333	241342	271153
DEN			99ØØ+Ø9	99ØØ+Ø4	3Ø2Ø-1Ø	3Ø29-21	3Ø3636	3Ø4145	294756
DSM	3615	3315+Ø7	3118+Ø4	3Ø22+ØØ	2835-12	2748-24	276438	277348	277957
ELP		Ø61Ø	Ø614+13	Ø615+Ø8	Ø113-Ø5	3614-17	361433	361442	251354
GCK		Ø611+11	Ø8Ø9+Ø8	99ØØ+Ø3	2817-Ø9	2823-2Ø	273135	273644	284155
HLC		Ø4Ø9+Ø9	Ø4Ø5+Ø7	31Ø6+Ø2	2822-1Ø	273Ø-21	273936	274545	275256
HOU	Ø9Ø9	16Ø7+19	16Ø6+13	16Ø6+Ø7	16Ø5-Ø8	99ØØ-2Ø	99ØØ34	99ØØ43	99ØØ54
ICT	Ø516	Ø613+12	Ø6Ø7+Ø8	99ØØ+Ø4	2718-Ø9	2626-2Ø	263635	264144	274655
IND	3611	32Ø7+12	2912+Ø8	2818+Ø3	2733-Ø9	2643-21	265635	265944	256255
INK		Ø6Ø9+16	Ø7Ø9+12	Ø6Ø8+Ø7	Ø1Ø7-Ø6	36Ø7-18	35Ø833	34Ø842	35Ø855
JAN	3612	3613+18	3611+13	36Ø9+Ø7	Ø1Ø5-Ø8	99ØØ-19	99ØØ34	99ØØ43	23Ø854
LIT	Ø31Ø	36Ø8+16	32Ø6+11	28Ø8+Ø6	2517-Ø8	2518-19	252Ø34	252243	262454
LOU	Ø1Ø5	99ØØ+15	29Ø8+1Ø	2913+Ø5	2825-Ø8	2731-2Ø	263834	264143	254454
MEM	Ø1Ø9	Ø1Ø8+17	34Ø8+12	311Ø+Ø6	2916-Ø7	2717-19	261934	262144	262555
MKC	Ø316	Ø211+11	34Ø9+Ø7	3Ø13+Ø3	2728-1Ø	2638-21	265Ø36	265645	276356
MSY	Ø315	Ø216+19	Ø315+13	Ø414+Ø7	Ø51Ø-Ø8	Ø6Ø5-2Ø	99ØØ34	99ØØ43	21Ø854
OKC	Ø715	Ø81Ø+14	11Ø6+1Ø	99ØØ+Ø5	2414-Ø8	2419-19	252534	252743	272754
SAT	11Ø7	1713+18	1813+13	1911+Ø7	2ØØ6-Ø7	19Ø6-19	18Ø734	17Ø743	99ØØ54
SGF	Ø414	Ø41Ø+14	36Ø5+Ø9	29Ø8+Ø4	2624-Ø9	2632-2Ø	254135	264444	264655
SHV	Ø5Ø9	99ØØ+18	99ØØ+12	21Ø6+Ø6	2Ø12-Ø8	21Ø9-19	22Ø734	24Ø743	26Ø754
STL	Ø314	Ø11Ø+12	321Ø+Ø8	2915+Ø3	273Ø-Ø9	2741-21	265435	265744	266Ø55
TUS		Ø8Ø7+23	Ø814+16	Ø814+1Ø	Ø81Ø-Ø5	Ø5Ø5-17	33Ø533	31Ø842	29Ø954

FIGURE 149.—Winds and Temperatures Aloft Forecast.

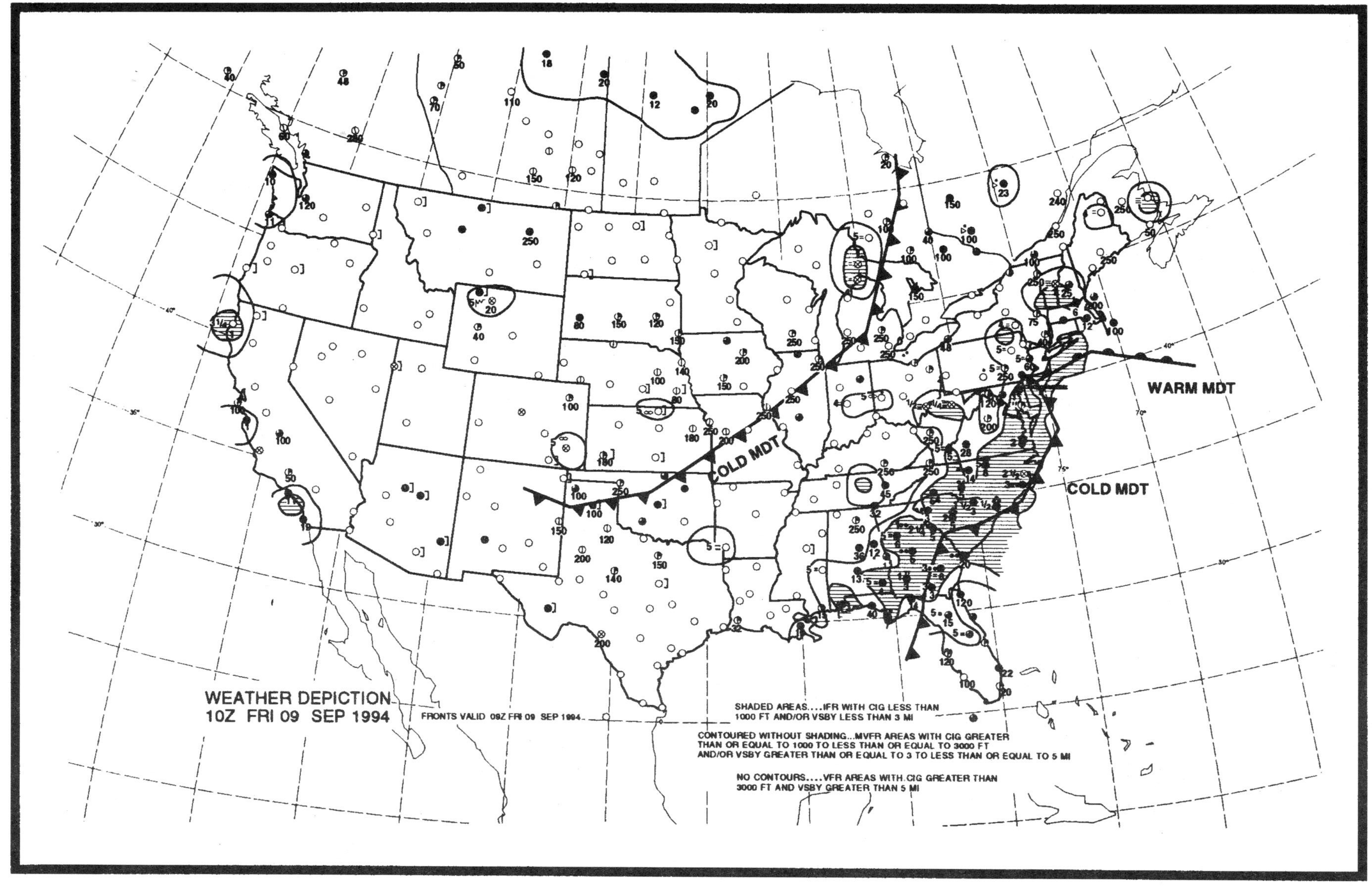

FIGURE 150.—Weather Depiction Chart.

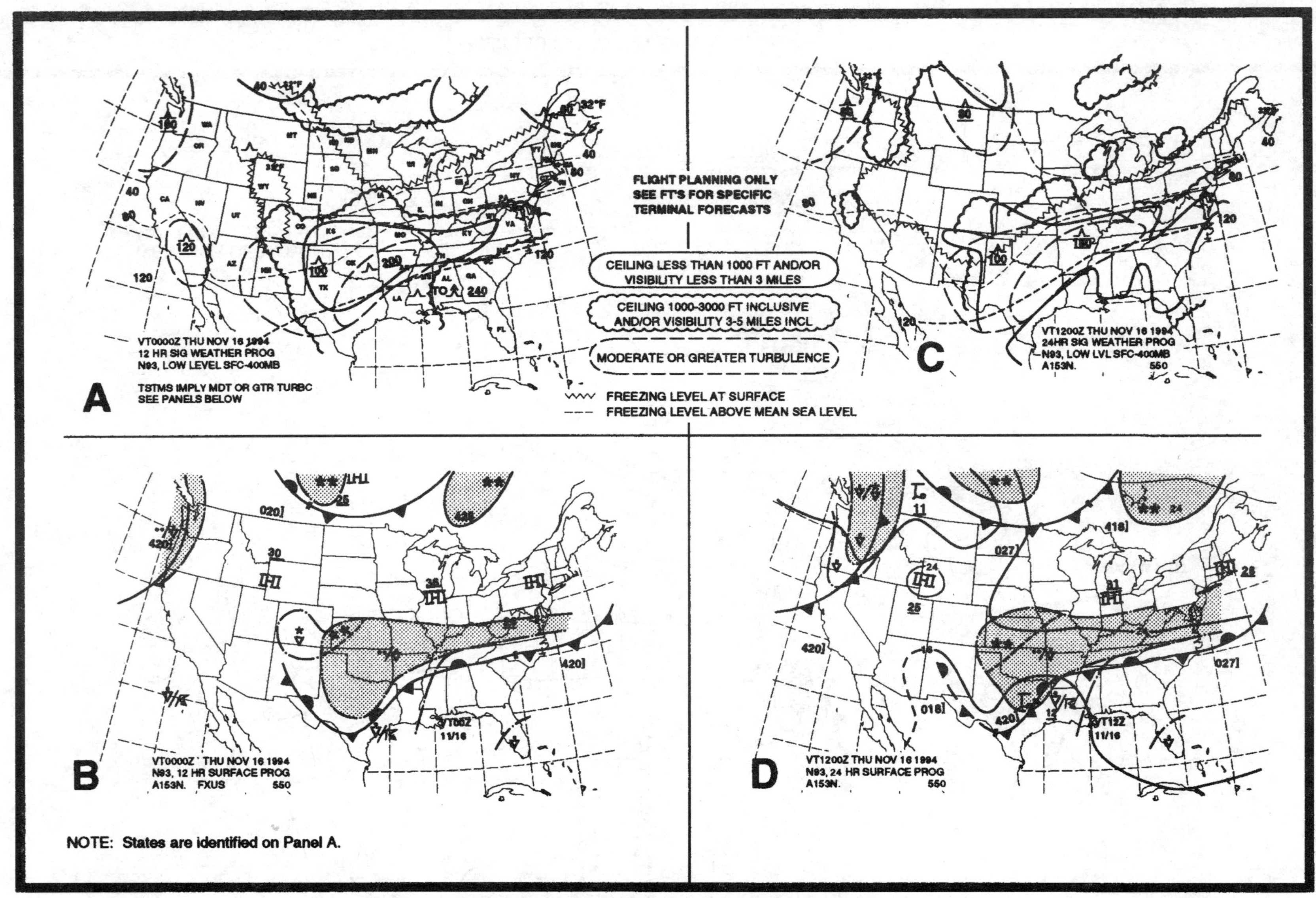

FIGURE 151.—U.S. Low-Level Significant Prog Chart.

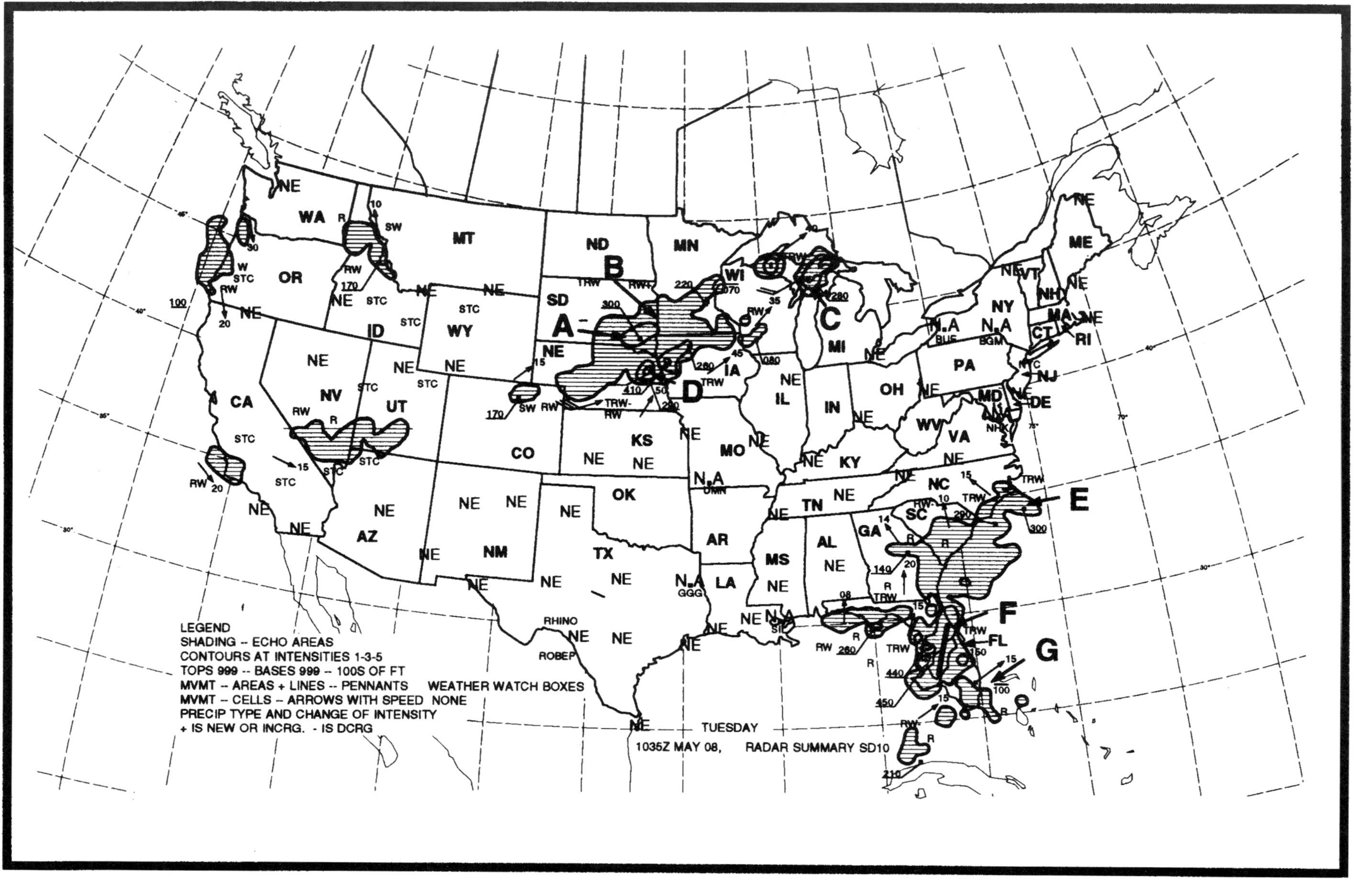

FIGURE 152.—Radar Summary Chart.

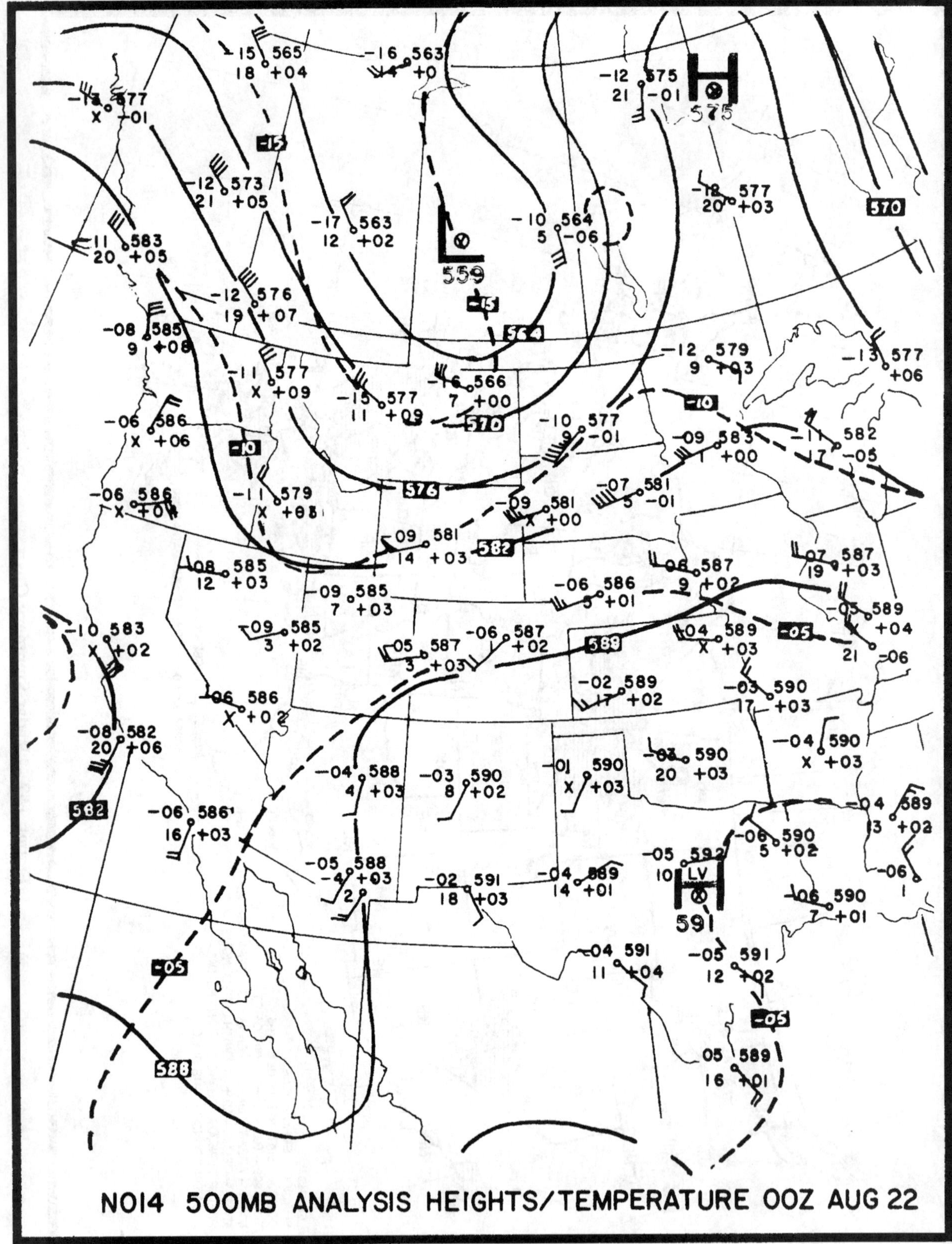

FIGURE 153.—500 MB Analysis Heights/Temperature Chart.

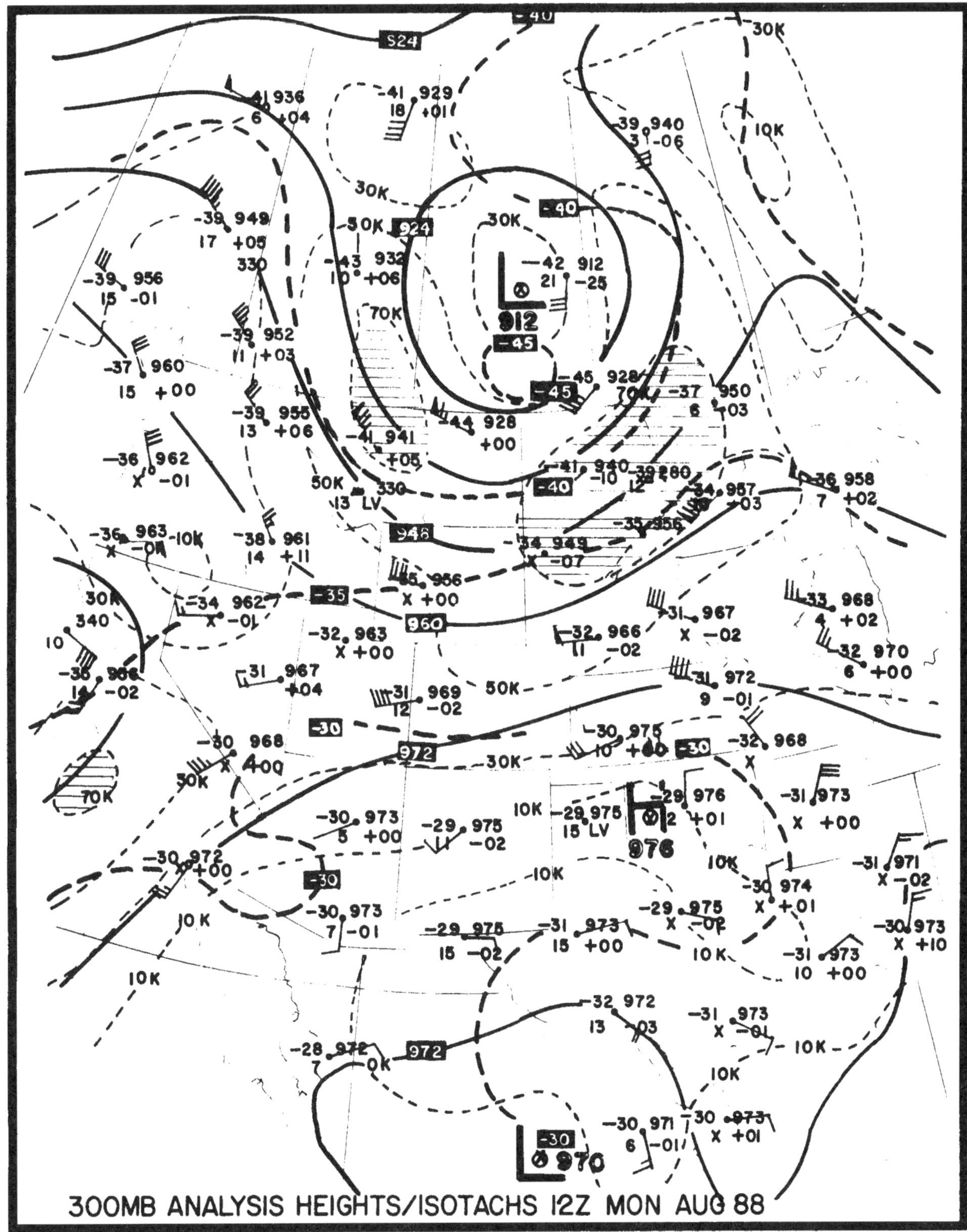

FIGURE 154.—300 MB Analysis Heights/Isotachs Chart.

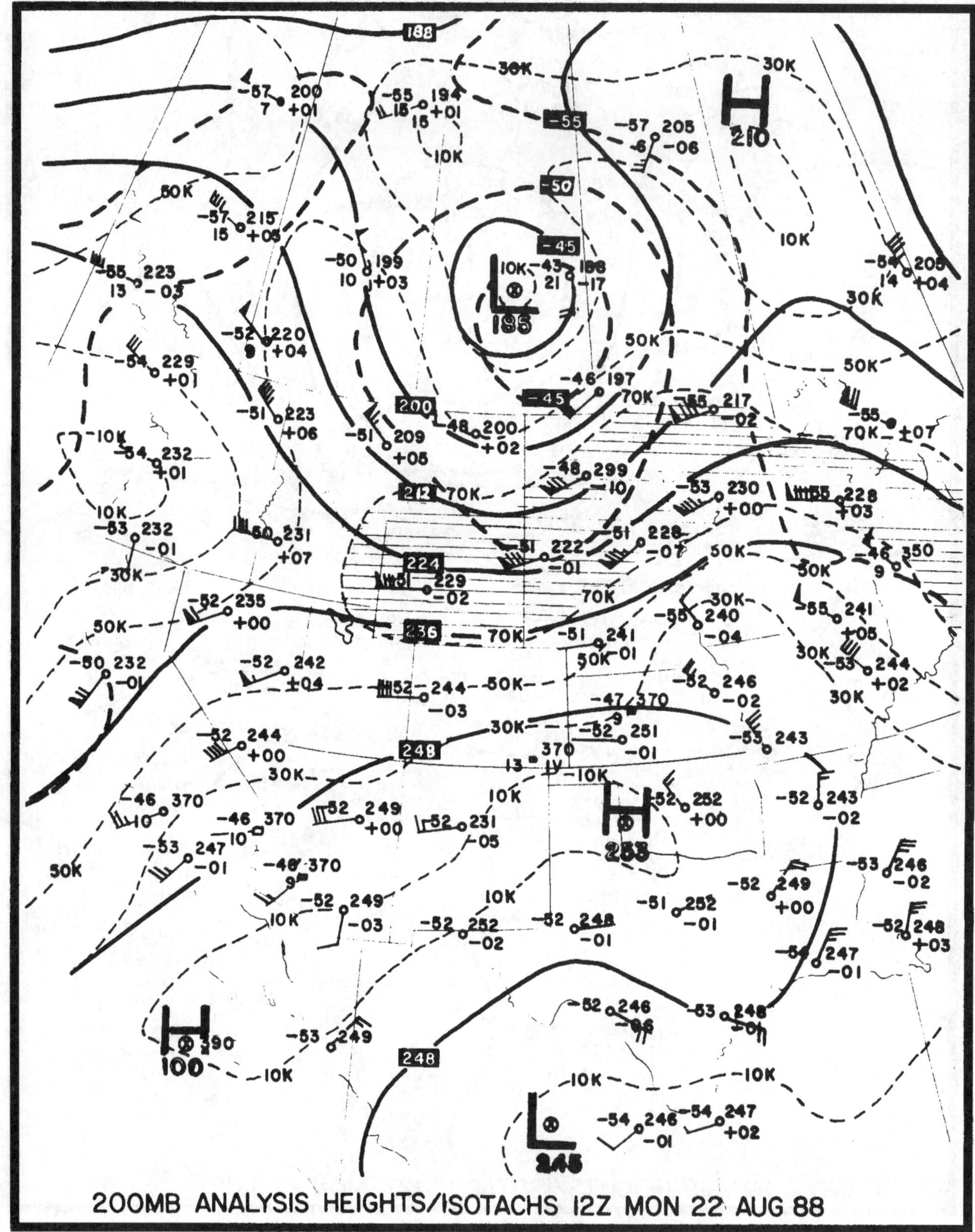

FIGURE 155.—200 MB Analysis Heights/Isotachs Chart.

FIGURE 156.—Airport Sign.

FIGURE 157.—Airport Sign.

Form Approved: OMB No. 2120-0034

U.S. DEPARTMENT OF TRANSPORTATION FEDERAL AVIATION ADMINISTRATION **FLIGHT PLAN**	(FAA USE ONLY) ☐ PILOT BRIEFING ☐ VNR ☐ STOPOVER	TIME STARTED	SPECIALIST INITIALS

1. TYPE	2. AIRCRAFT IDENTIFICATION	3. AIRCRAFT TYPE/ SPECIAL EQUIPMENT	4. TRUE AIRSPEED	5. DEPARTURE POINT	6. DEPARTURE TIME PROPOSED (Z)	ACTUAL (Z)	7. CRUISING ALTITUDE
VFR X IFR DVFR	N711JB	G1159/A	506 KTS	STL ST LOUIS, MO			FL370

8. ROUTE OF FLIGHT
STL, GATWAY2.ROD, J29 DJB, J60 PSB, PSB.MIP8, LGA

9. DESTINATION (Name of airport and city)	10. EST. TIME ENROUTE HOURS	MINUTES	11. REMARKS
LGA LA GAURDIA NEW YORK, NY			L/O = LEVEL OFF PPH = POUNDS PER HOUR TEC = TOWER ENROUTE CONTROL VARIATION: BIB 1E, ROD 1W, DJB 5W, PSB 8W, MIP 11W, SBJ 11W, LGA 12W

12. FUEL ON BOARD HOURS	MINUTES	13. ALTERNATE AIRPORT(S)	14. PILOT'S NAME, ADDRESS & TELEPHONE NUMBER & AIRCRAFT HOME BASE / 17. DESTINATION CONTACT/TELEPHONE (OPTIONAL)	15. NUMBER ABOARD
4	00	JFK NEW YORK, NY		12

16. COLOR OF AIRCRAFT
BLACK/RED

CIVIL AIRCRAFT PILOTS. FAR Part 91 requires you file an IFR flight plan to operate under instrument flight rules in controlled airspace. Failure to file could result in a civil penalty not to exceed $1,000 for each violation (Section 901 of the Federal Aviation Act of 1958, as amended). Filing of a VFR flight plan is recommended as a good operating practice. See also Part 99 for requirements concerning DVFR flight plans.

FAA Form 7233-1 (8-82) CLOSE VFR FLIGHT PLAN WITH ________ FSS ON ARRIVAL

FLIGHT LOG

CHECK POINTS FROM	TO	ROUTE ALTITUDE	COURSE	WIND TEMP	SPEED-KTS TAS	GS	DIST NM	TIME LEG	TOT	FUEL LEG	TOT
STL	BIB	GATWAY 2.ROD CLIMB					95		:16:00		987*
BIB	ROD	GATWAY2.ROD FL370		350/96 ISA-1							
ROD	DJB	J29 FL370									
DJB	PSB	J60 FL370									
PSB	MIP	PSB.MIP8 FL370									
MIP	SBJ										
SBJ	LGA	DESCENT					52	:16:26		269	
LGA	JFK	TEC 4000			260			:15:00			

OTHER DATA: * Includes Taxi Fuel
NOTE: Use 2389 PPH Total Fuel Flow From L/O To Start Of Descent.
Use 1898 PPH Total Fuel Flow For Reserve And Alternate Requirements.

A Missed Approach Requires 233# of Fuel.

TIME and FUEL: As required by FARs.

TIME	FUEL (LB)	
		EN ROUTE
		RESERVE
		ALTERNATE
		TOTAL

FIGURE 158.—Flight Plan/Flight Log.

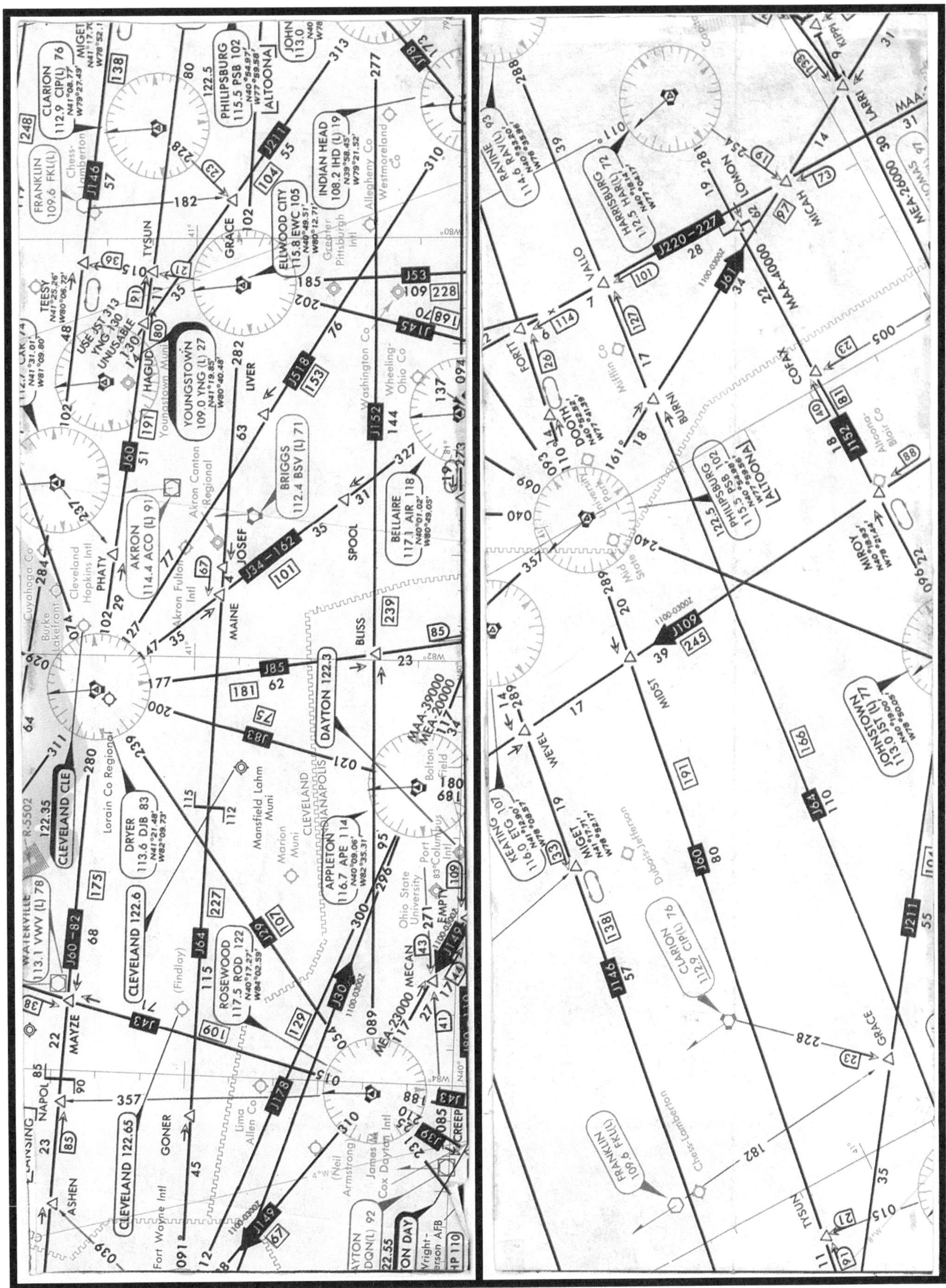

FIGURE 159.—High Altitude Airways.

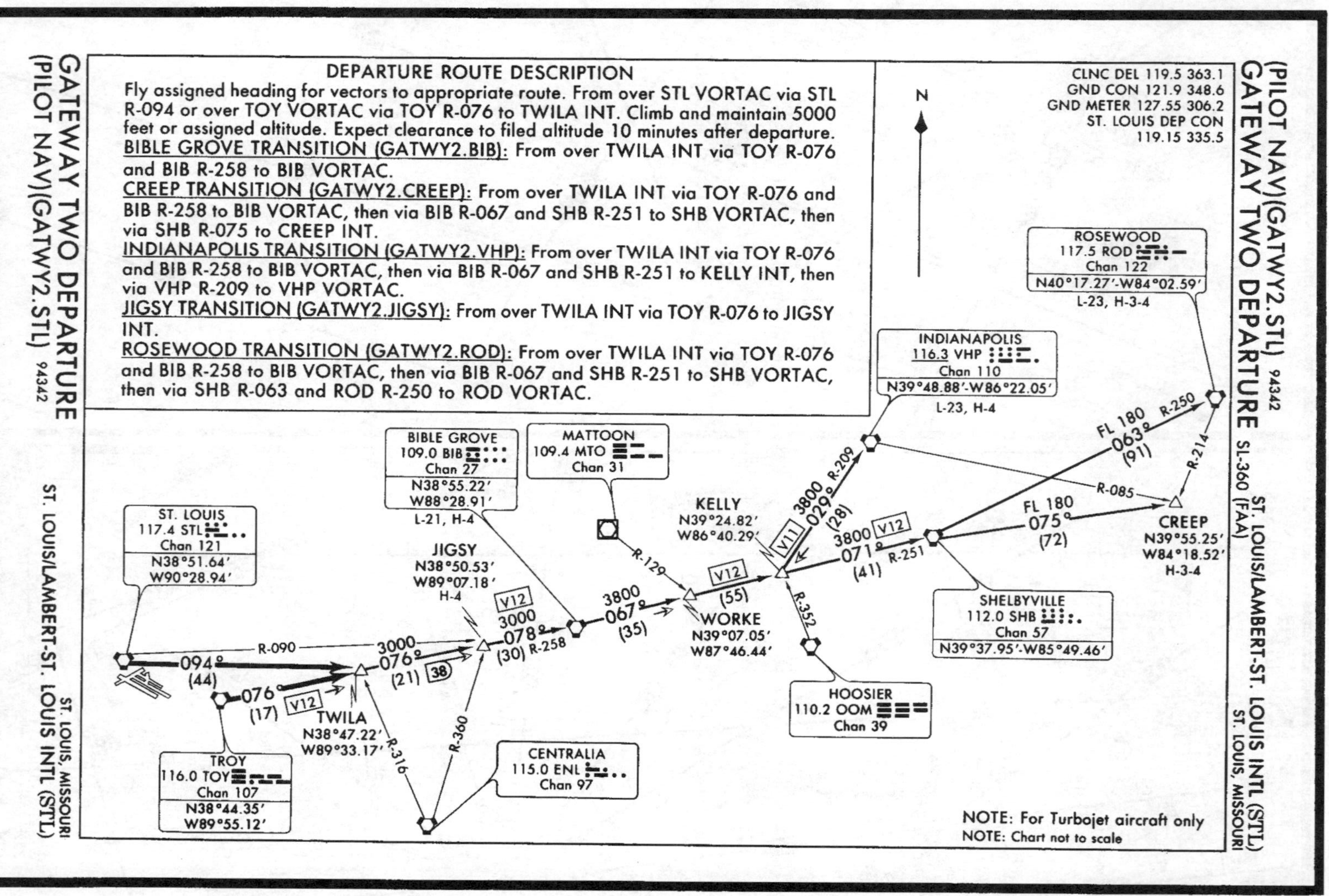

FIGURE 160.—GATEWAY TWO DEPARTURE (STL).

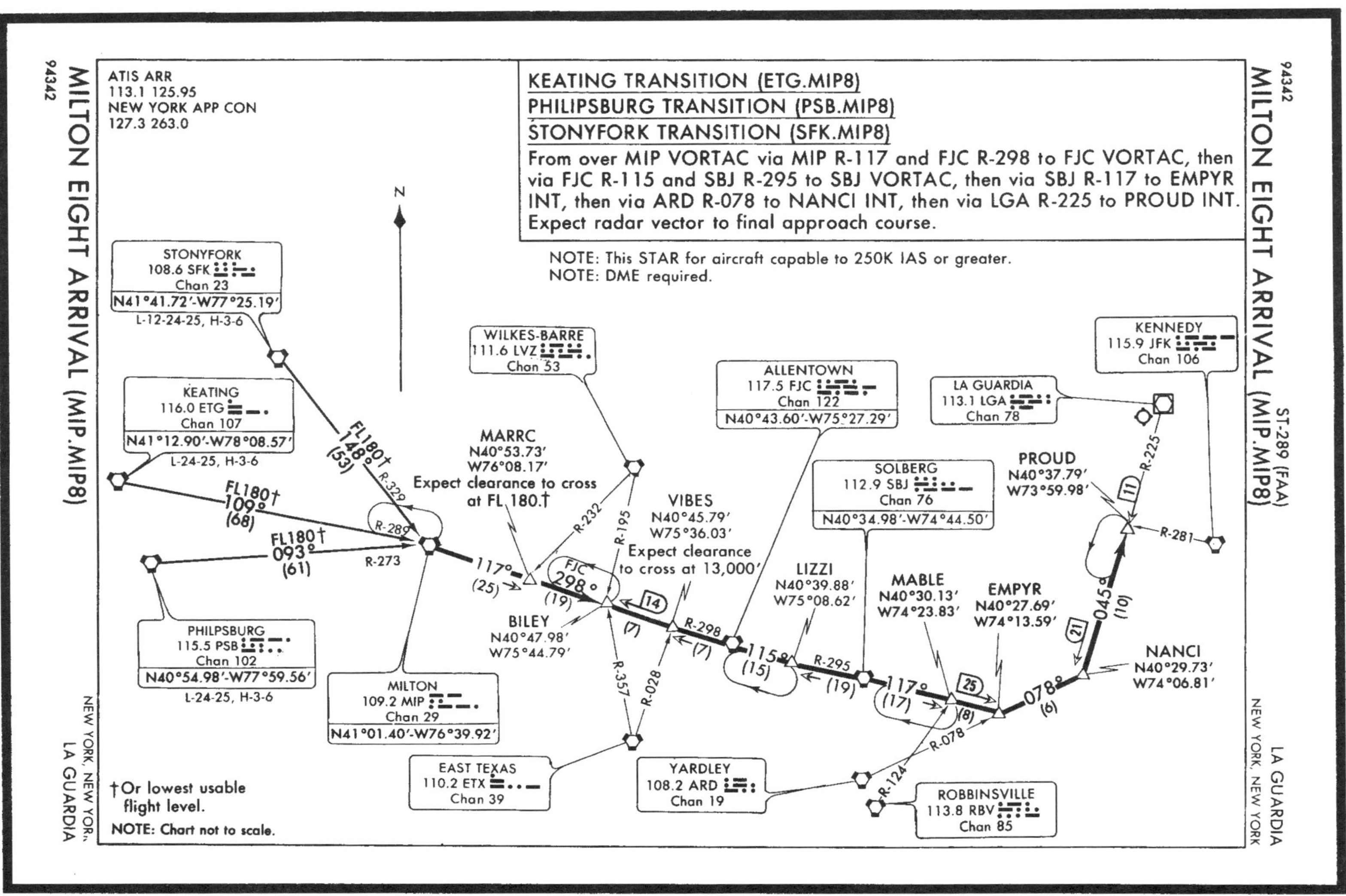

FIGURE 160A.—MILTON EIGHT ARRIVAL (MIP.MIP8).

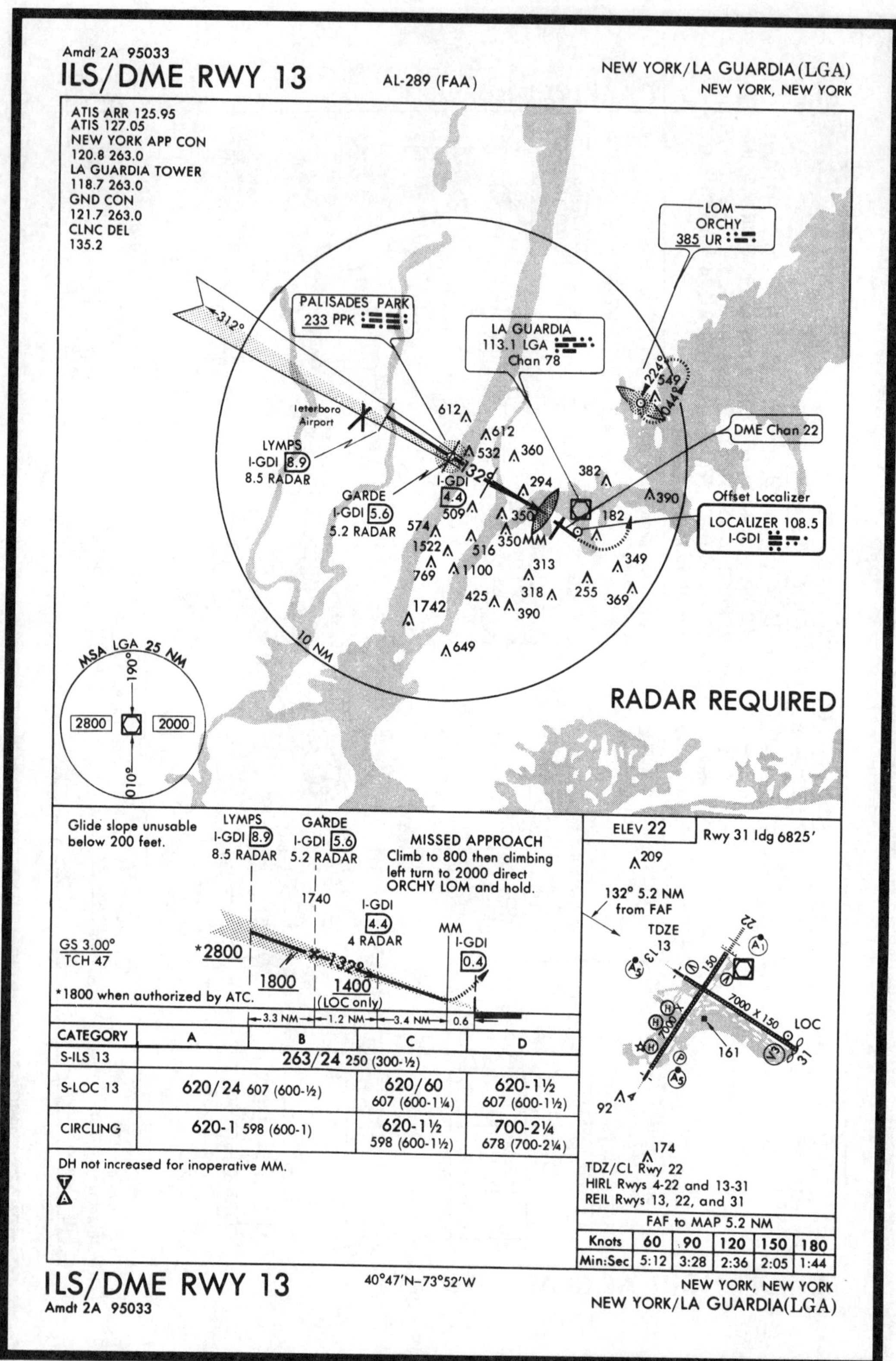

CATEGORY	A	B	C	D
S-ILS 13	263/24 250 (300-½)			
S-LOC 13	620/24 607 (600-½)		620/60 607 (600-1¼)	620-1½ 607 (600-1½)
CIRCLING	620-1 598 (600-1)		620-1½ 598 (600-1½)	700-2¼ 678 (700-2¼)

FAF to MAP 5.2 NM					
Knots	60	90	120	150	180
Min:Sec	5:12	3:28	2:36	2:05	1:44

FIGURE 161.—ILS/DME RWY 13 (LGA).

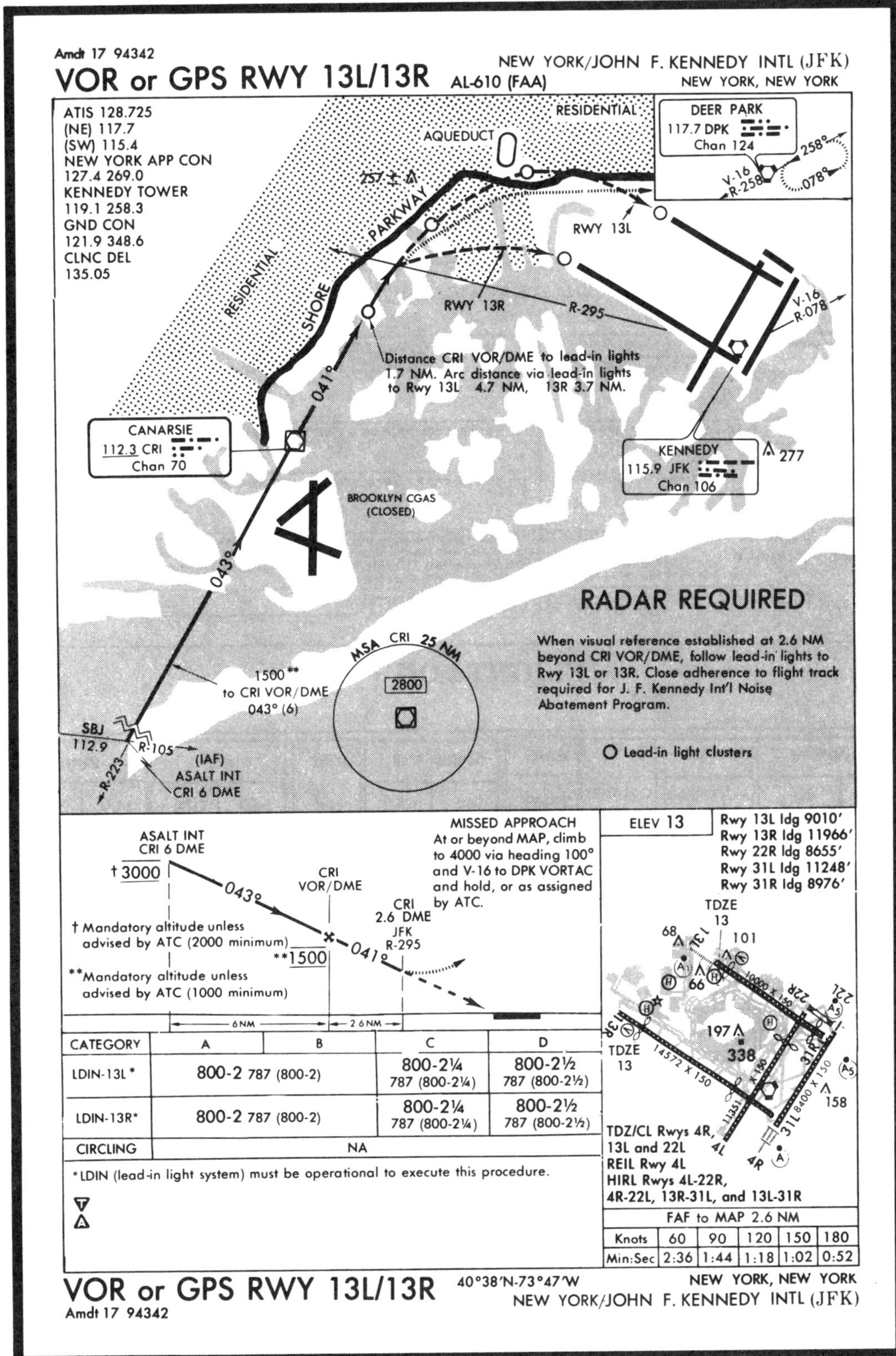

CATEGORY	A	B	C	D
LDIN-13L*	800-2 787 (800-2)		800-2¼ 787 (800-2¼)	800-2½ 787 (800-2½)
LDIN-13R*	800-2 787 (800-2)		800-2¼ 787 (800-2¼)	800-2½ 787 (800-2½)
CIRCLING	NA			

*LDIN (lead-in light system) must be operational to execute this procedure.

Knots	60	90	120	150	180
Min:Sec	2:36	1:44	1:18	1:02	0:52

FIGURE 161A.—VOR or GPS RWY 13L/13R (JFK).

Form Approved: OMB No. 2120-0034

U.S. DEPARTMENT OF TRANSPORTATION FEDERAL AVIATION ADMINISTRATION **FLIGHT PLAN**	(FAA USE ONLY) ☐ PILOT BRIEFING ☐ VNR ☐ STOPOVER	TIME STARTED	SPECIALIST INITIALS

1. TYPE	2. AIRCRAFT IDENTIFICATION	3. AIRCRAFT TYPE/ SPECIAL EQUIPMENT	4. TRUE AIRSPEED	5. DEPARTURE POINT	6. DEPARTURE TIME PROPOSED (Z)	ACTUAL (Z)	7. CRUISING ALTITUDE
VFR / X IFR / DVFR	CHIEF 4	BH230/A	** KTS	RYN TUCSON/RYAN FIELD			11000

8. ROUTE OF FLIGHT
TUS, V202 SSO, V94 DMN, V110 TCS, V19ABQ, AEG.

9. DESTINATION (Name of airport and city)	10. EST. TIME ENROUTE HOURS	MINUTES	11. REMARKS
AEG ALBUQUERQUE/DOUBLE EAGLE II			L/O = LEVEL OFF PPH = POUNDS PER HOUR **CAS 125 ISA -6 TO +2

12. FUEL ON BOARD HOURS	MINUTES	13. ALTERNATE AIRPORT(S)	14. PILOT'S NAME, ADDRESS & TELEPHONE NUMBER & AIRCRAFT HOME BASE / 17. DESTINATION CONTACT/TELEPHONE (OPTIONAL)	15. NUMBER ABOARD
		ABQ ALBUQUERQUE INT'L		9

16. COLOR OF AIRCRAFT	
ORANGE/BLACK	CIVIL AIRCRAFT PILOTS. FAR Part 91 requires you file an IFR flight plan to operate under instrument flight rules in controlled airspace. Failure to file could result in a civil penalty not to exceed $1,000 for each violation (Section 901 of the Federal Aviation Act of 1958, as amended). Filing of a VFR flight plan is recommended as a good operating practice. See also Part 99 for requirements concerning DVFR flight plans.

FAA Form 7233-1 (8-82) CLOSE VFR FLIGHT PLAN WITH ____________ FSS ON ARRIVAL

FLIGHT LOG

CHECK POINTS FROM	TO	ROUTE ALTITUDE	COURSE	WIND TEMP	SPEED-KTS TAS	GS	DIST NM	TIME LEG	TOT	FUEL LEG	TOT
RYN	MESCA						38		:17:00		180*
MESCA	CIE	V202 11000		240/31 ISA-6							
CIE	SSO										
SSO	DMN	V94 11000		250/27 ISA-1							
DMN	TCS	V110 11000									
TCS	ONM	V19 11000		220/33 ISA+2							
ONM	ABQ										
ABQ	AEG	DIRECT DESCENT					6	:06:00		49.0	
AEG	ABQ	TWR-TWR 8000					11	:05:00			

OTHER DATA: * Includes Taxi Fuel
NOTE: Use 523 PPH Total Fuel Flow From L/O To Start Of Descent.
Use 497 PPH Total Fuel Flow For Reserve And Alternate Requirements.

A Missed Approach Requires 40# of Fuel.

TIME and FUEL: As required by FARs.

TIME	FUEL (LB)	
		EN ROUTE
		RESERVE
		ALTERNATE
		TOTAL

FIGURE 162.—Flight Plan/Flight Log.

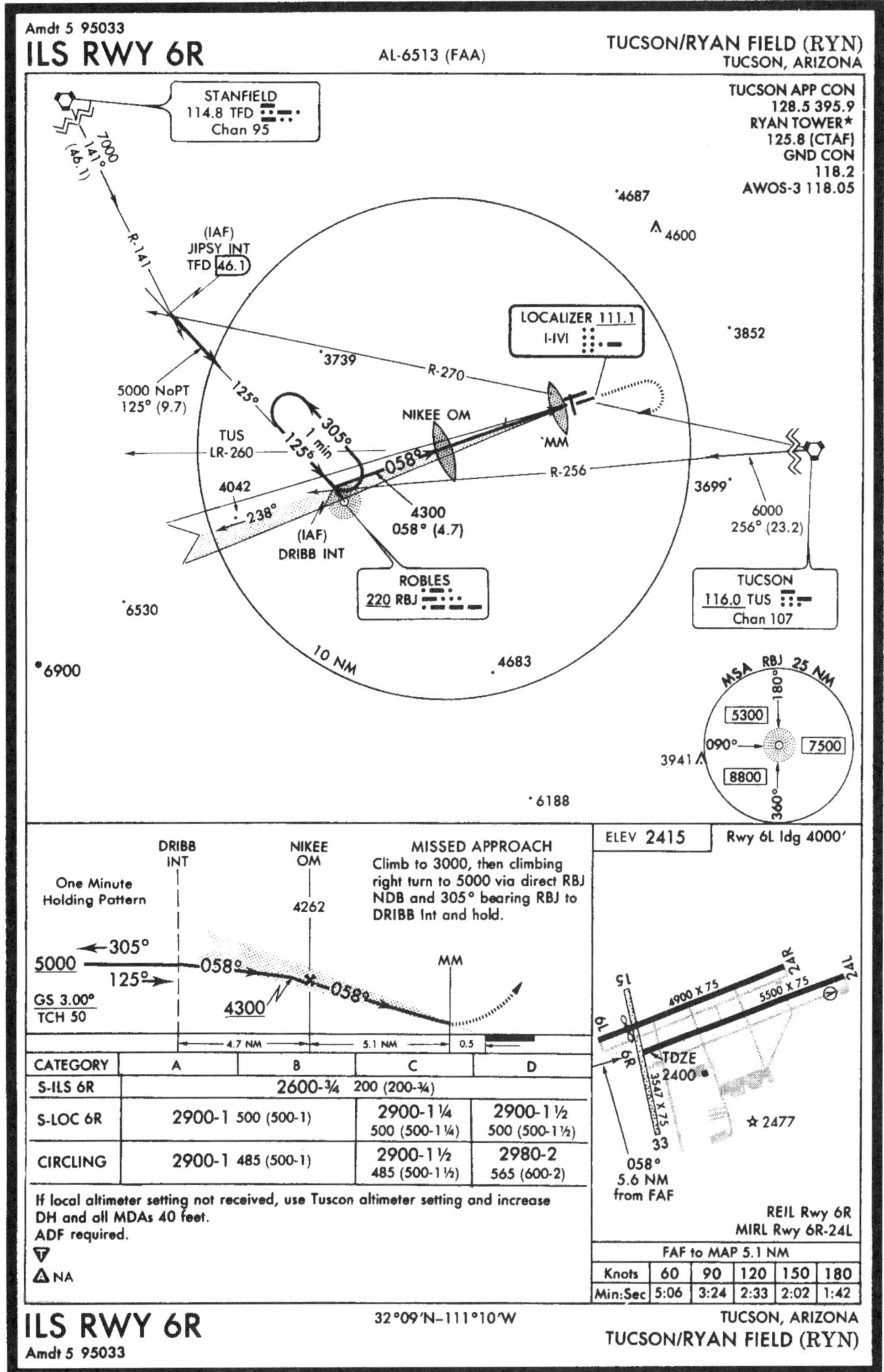

FIGURE 163.—ILS RWY 6R (RYN).

TOYEI SCHOOL (See GANADO)

TUBA CITY (T03) 5W UTC–7 N36°05.57′ W111°22.96′ **LAS VEGAS**
4513 B **H-2C, L-4E,5C**
RWY 15-33: H6230X75 (ASPH) S-12.5 MIRL
RWY 15: PAPI(P2L)—GA 3.0° TCH 40′. **RWY 33:** PAPI(P2L)—GA 3.0° TCH 40′.
AIRPORT REMARKS: Unattended. Daylight operations only 1300–0100Z. Ngt operations not authorized. MIRL Rwy 15–33 out of svc indefinitely. Livestock on airport.
COMMUNICATIONS: CTAF 122.9
PRESCOTT FSS (PRC) TF 1-800-WX-BRIEF. NOTAM FILE PRC.
RCO 122.05R 113.5T (PRESCOTT FSS)
RADIO AIDS TO NAVIGATION: NOTAM FILE PRC.
(H) VORTAC 113.5 TBC Chan 82 N36°07.28′ W111°16.18′ 238° 5.8 NM to fld. 4960/15E.

TUCSON

AVRA VALLEY (E14) 13 NW UTC–7 N32°24.56′ W111°13.11′ **PHOENIX**
2031 B S3 **FUEL** 100LL, JET A **H-2C, L-4F**
RWY 12-30: H6901X100 (ASPH)
RWY 30: Road. Rgt tfc.
RWY 03-21: H4201X75 (ASPH) MIRL
RWY 03: VASI(V2L)-GA 3.0° TCH 43′. Thld dsplcd 295′. Road. Rgt tfc.
RWY 21: VASI(V2L)-GA 3.0° TCH 31′. Tree.
AIRPORT REMARKS: Attended 1400–0100Z. Parachute Jumping. Ditch apch end Rwy 21. Aerobatic activities 2–10 miles south of arpt, surface 5000′ MSL dalgt hours indefinitely. Extensive parachute training high and low levels all hours NW quadrant of arpt. ACTIVATE MIRL Rwy 03–21, VASI Rwy 03 and Rwy 21—CTAF. Note: See Special Notices—Glider Operations Northwest of Tucson, Arizona.
COMMUNICATIONS: CTAF/UNICOM 123.0
PRESCOTT FSS (PRC) TF 1-800-WX-BRIEF. NOTAM FILE PRC.
RADIO AIDS TO NAVIGATION: NOTAM FILE PRC.
TUCSON (H) VORTACW 116.0 TUS Chan 107 N32°05.71′ W110°54.89′ 309° 24.3 NM to fld. 2670/12E.
HIWAS.

- -

CASCABEL AIR PARK (05A) 35 N UTC–7 N32°18.01′ W110°21.91′ **PHOENIX**
3374
RWY 02-20: 2750X60 (DIRT)
RWY 20: Road.
AIRPORT REMARKS: Unattended. Rwy 20 10′ brush within primary surface. Rwy 02 25′ power lines 1/2 mile south of rwy. – 15′ down slope beginning at end of Rwy 02.
COMMUNICATIONS: CTAF 122.9
PRESCOTT FSS (PRC) TF 1-800 WX-BRIEF. NOTAM FILE PRC.

- -

RYAN FLD (RYN) 10 SW UTC–7 N32°08.53′ W111°10.46′ **PHOENIX**
2415 B S4 **FUEL** 80, 100LL TPA—See Remarks **H-2C, L-4F**
RWY 06R-24L: H5500X75 (ASPH) S-12.5, D-30 MIRL **IAP**
RWY 06R: REIL. Rgt tfc. **RWY 24L:** VASI(V4L)—GA 3.0° TCH 26′.
RWY 06L-24R: H4900X75 (ASPH) S-12.5, D-30
RWY 06L: Thld dsplcd 900′. Pole. **RWY 24R:** Tree. Rgt tfc.
RWY 15-33: 3547X75 (DIRT)
RWY 33: Tree.
AIRPORT REMARKS: Attended 1300–0100Z. Self svc fuel avbl 1300–0400Z. Rwy 06L–24R CLOSED 0100–1300Z. Rwy 06R preferential rwy up to 10 knot tailwind. Rwy 06L–24R paved shoulders 30′ wide both sides. TPA–3215(800), 3415(1000) when twr closed. Note: See Special Notices—Glider Operations Northwest of Tucson, Arizona.
WEATHER DATA SOURCES: AWOS-3 118.05 (602) 578-0269.
COMMUNICATIONS: CTAF 125.8
PRESCOTT FSS (PRC) TF 1-800-WX-BRIEF. NOTAM FILE PRC.
® **TUCSON APP/DEP CON** 128.5
TOWER 125.8 NFCT (Apr–Sep 1300–0300Z, Oct–Mar 1300–0100Z) **GND CON** 118.2
AIRSPACE: CLASS D svc Apr–Sep 1300–0300Z, Oct–Mar 1300–0100Z other times CLASS E.
RADIO AIDS TO NAVIGATION: NOTAM FILE PRC.
TUCSON (H) VORTACW 116.0 TUS Chan 107 N32°05.71′ W110°54.89′ 270° 13.5 NM to fld. 2670/12E.
HIWAS.
NDB (HW-SAB) 338 RYN N32°08.30′ W111°09.69′ at fld. TWEB avbl 1200–0500Z.
ILS 111.1 I-IVI Rwy 06R. Unmonitored.

- -

FIGURE 163A.—Excerpt from Airport/Facilities Directory.

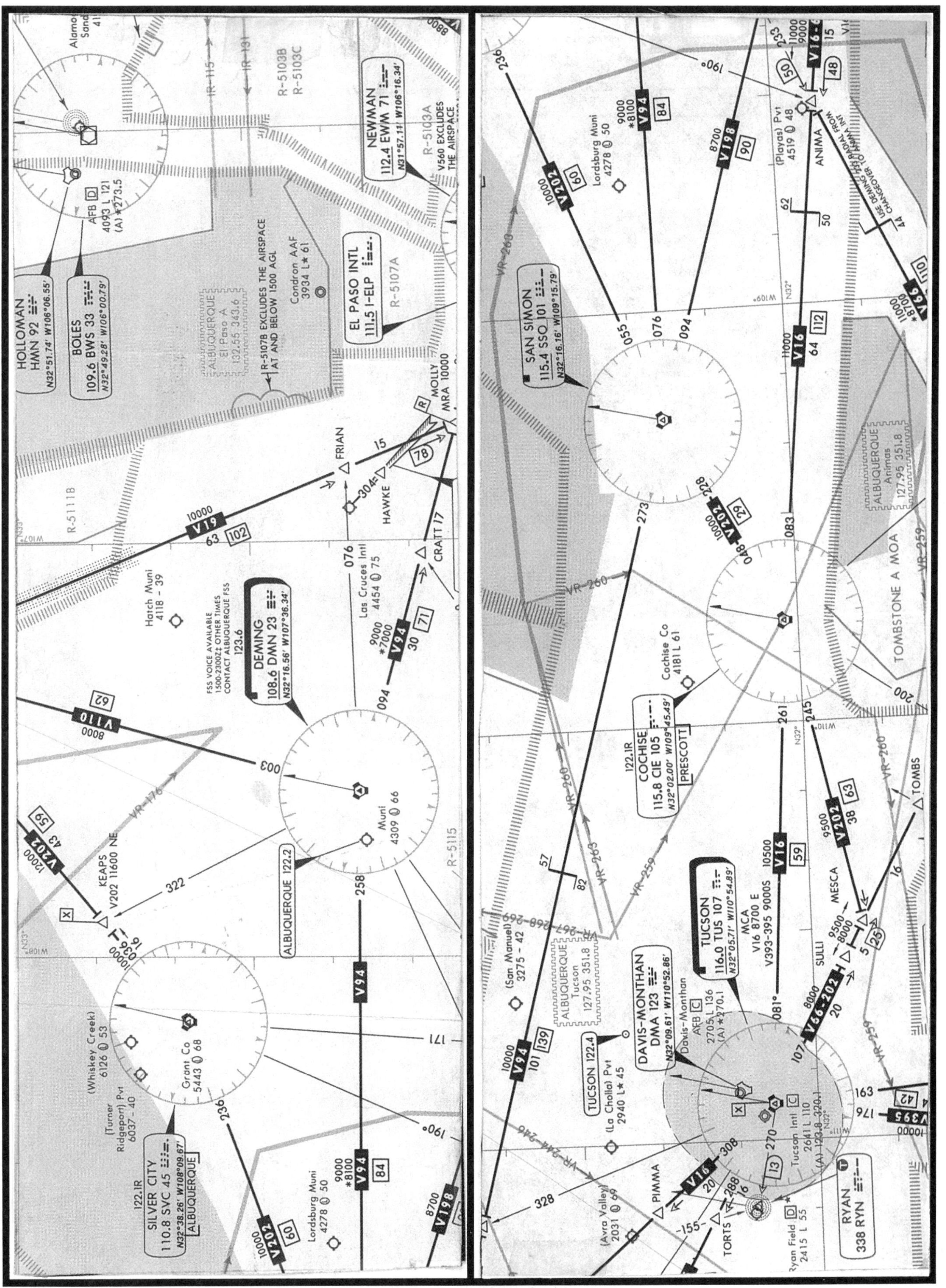

FIGURE 164.—Low Altitude Airways.

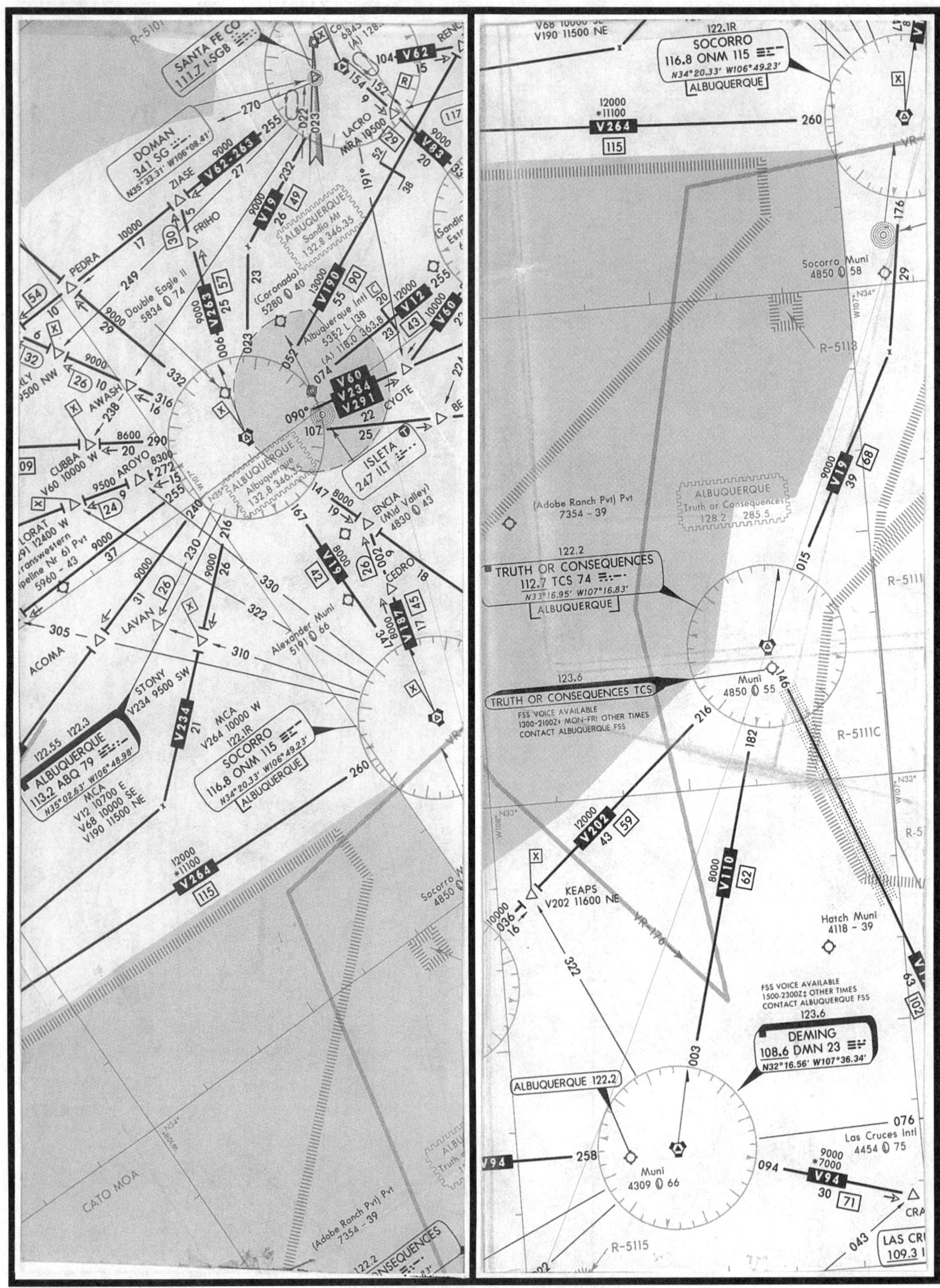

FIGURE 165.—Low Altitude Airways.

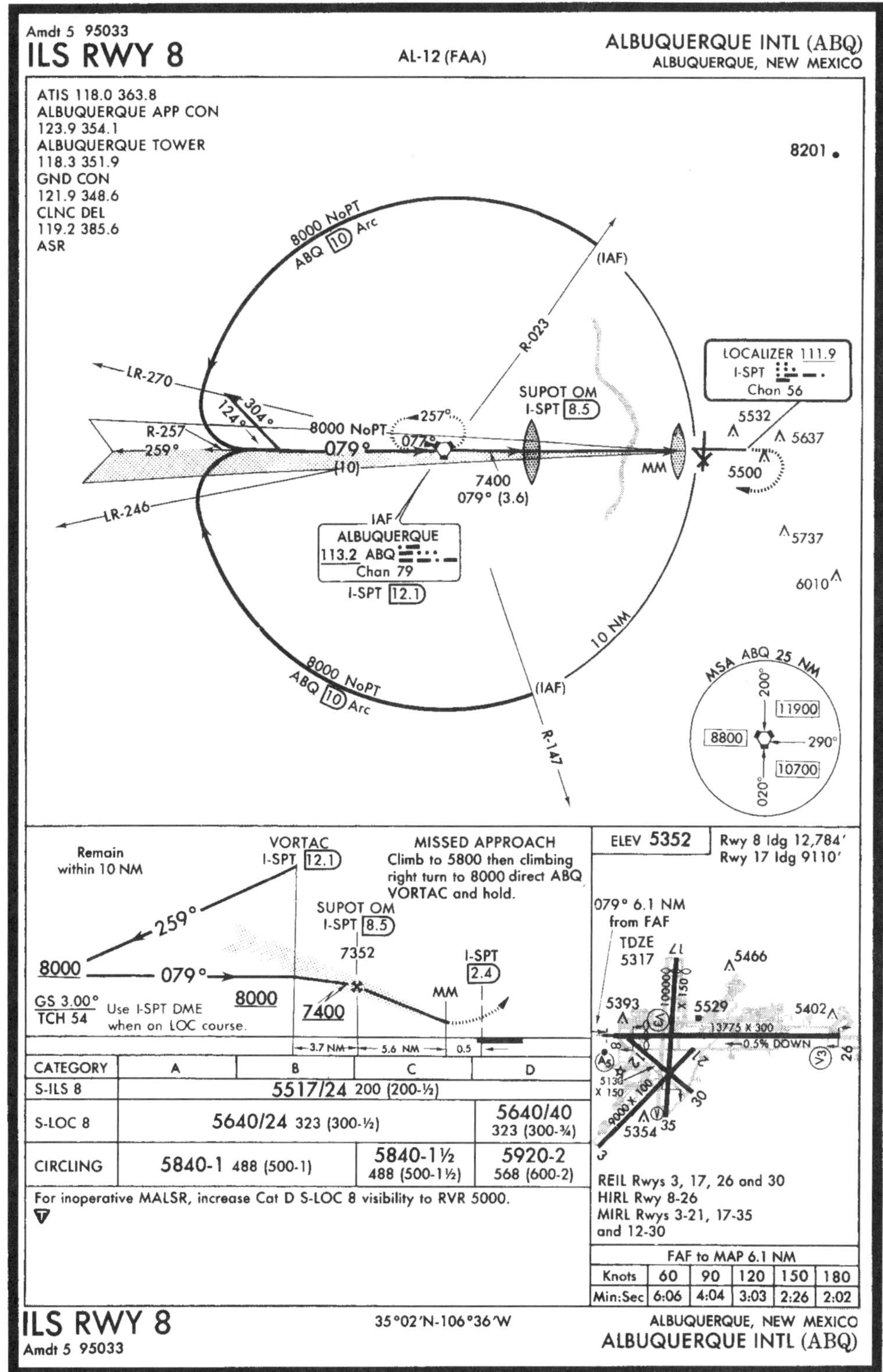

CATEGORY	A	B	C	D
S-ILS 8	5517/24 200 (200-½)			
S-LOC 8	5640/24 323 (300-½)			5640/40 323 (300-¾)
CIRCLING	5840-1 488 (500-1)		5840-1½ 488 (500-1½)	5920-2 568 (600-2)

FAF to MAP 6.1 NM					
Knots	60	90	120	150	180
Min:Sec	6:06	4:04	3:03	2:26	2:02

FIGURE 166.—ILS RWY 8 (ABQ).

ALBUQUERQUE

ALBUQUERQUE INTL (ABQ) 3 SE UTC–7(–6DT) N35°02.45′ W106°36.52′ **ALBUQUERQUE**

5352 B S4 **FUEL** 100LL, JET A, A1, A1 + OX 1, 2, 3, 4 LRA ARFF Index C **H-2D, L-4G, 6E**

RWY 08-26: H13775X300 (ASPH-CONC-GRVD) S-100, D-210, DT-360 HIRL 0.3% up E **IAP**

RWY 08: MALSR. VASI(V6L)—GA 3.0° TCH 54′. Thld dsplcd 991′. Rgt tfc. Arresting device.

RWY 26: REIL. VASI(V6L)—Upper GA 3.25° TCH 77′. Lower GA 3.0° TCH 47′. Arresting device.

RWY 17-35: H10000X150 (ASPH-CONC-GRVD) S-100, D-210, DT-360 MIRL

RWY 17: REIL. VASI(V4L)—GA 3.0° TCH 53′. Thld dsplcd 890′. Road. Rgt tfc.

RWY 35: VASI(V4L)—GA 3.0° TCH 55′. Arresting device.

RWY 03-21: H9000X100 (ASPH) S-45, D-65 MIRL

RWY 03: REIL. Rgt tfc.

RWY 12-30: H5130X150 (ASPH) S-45, D-65 MIRL

RWY 12: Rgt tfc. **RWY 30:** REIL.

RUNWAY DECLARED DISTANCE INFORMATION

RWY 03:	TORA-9000	TODA-9000	ASDA-9000	LDA-9000
RWY 21:	TORA-9000	TODA-9000	ASDA-9000	LDA-9000
RWY 08:	TORA-13775	TODA-13775	ASDA-13775	LDA-12784
RWY 26:	TORA-13775	TODA-13775	ASDA-13775	LDA-13775
RWY 12:	TORA-5130	TODA-5130	ASDA-5130	LDA-5130
RWY 30:	TORA-5130	TODA-5130	ASDA-5130	LDA-5130
RWY 17:	TORA-10000	TODA-10000	ASDA-10000	LDA-9110
RWY 35:	TORA-10000	TODA-10000	ASDA-10000	LDA-10000

AIRPORT REMARKS: Attended continuously. Bird hazard Oct–Dec, and Mar–May. Heavy student copter traffic, control firing area S of arpt. Fighter acft depart S only, no military depart on Rwy 35. Rwy 03-21 SW 200′ CLOSED to acft weighing over 12,500 pounds. Ramp W of Rwy 17-35 and N of Rwy 08-26 CLOSED to helicopters. Rwy 08-26 and Rwy 17-35 grooved 130′ wide. Takeoff Rwy 03 prohibited except for emergency conditions on fld. Takeoff Rwy 35 requires prior coordination with twr. Twy H closed between Twy G and Rwy 17-35 indefinitely. Twy F S of freight ramp closed to acft over 65,000 pounds. Twy F between Twy F1 and Twy C restricted to maximum wing span 108′ B727 or smaller acft. Portions of Twy D N of Twy D-3 not visible from twr. Arresting cables at Rwy 26 thld. Recessed arresting cables at Rwy 08 and Rwy 35 thld. Flight Notification Service (ADCUS) available. NOTE: See Land and Hold Short Operations Section.

WEATHER DATA SOURCES: LLWAS.

COMMUNICATIONS: ATIS 118.0 (505) 856-4928 **UNICOM** 122.95

ALBUQUERQUE FSS (ABQ) on arpt. 122.55 122.3 TF 1-800-WX-BRIEF. LC 505-243-7831. NOTAM FILE ABQ.

® **APP CON** 124.4 (on or N of V12 and W of SANDIA MTNS) 134.8 (S of V12 and W of Manzano Mtns) 123.9 (S of V12 and E of Manzano Mtns) 127.4 (on or N of V12 and E of Sandia Mtns) 126.3

® **DEP CON** 127.4 (on or N of V12 and E of Sandia Mtns) 124.4 (on or N of V12 and W of Sandia Mtns) 123.9 (S of V12 and E of Manzano Mtns) 134.8 (S of V12 and W of Manzano Mtns)

TOWER 118.3 120.3 **GND CON** 121.9 **CLNC DEL** 119.2

AIRSPACE: CLASS C svc ctc **APP CON**

RADIO AIDS TO NAVIGATION: NOTAM FILE ABQ.

(H) VORTACW 113.2 ABQ Chan 79 N35°02.63′ W106°48.98′ 078° 10.2 NM to fld. 5740/13E. **HIWAS.**

ISLETA NDB (HW) 247 ILT N34°59.22′ W106°37.22′ 359° 3.3 NM to fld.

ILS/DME 111.9 I-SPT Chan 56 Rwy 08.

ASR

CORONADO (4AC) 6 NE UTC–7(–6DT) N35°11.75′ W106°34.40′ **ALBUQUERQUE**

5280 B S4 **FUEL** 100LL OX 3 **L-4G, 6E**

RWY 17-35: H4010X60 (ASPH) S-22, D-28 LIRL (NSTD)

RWY 17: Thld dsplcd 200′. Hill. Rgt tfc. **RWY 35:** Thld dsplcd 200′. Trees.

RWY 03-21: H3500X40 (ASPH) S-22, D-28

RWY 03: Building.

AIRPORT REMARKS: Attended continuously. Rising terrain East of airport. Rwy 03 rgt tfc for ultralight operations below 300′ and E of Rwy 17-35. Rwy 03-21 cracked and heavily weeded. ACTIVATE LIRL Rwy 17-35—CTAF.

COMMUNICATIONS: CTAF/UNICOM 122.8

ALBUQUERQUE FSS (ABQ) LC 243-7831 NOTAM FILE ABQ.

RADIO AIDS TO NAVIGATION: NOTAM FILE ABQ.

ALBUQUERQUE (H) VORTACW 113.2 ABQ Chan 79 N35°02.63′ W106°48.98′ 040° 15 NM to fld. 5740/13E. **HIWAS.**

FIGURE 166A.—Excerpt from Airport/Facilities Directory.

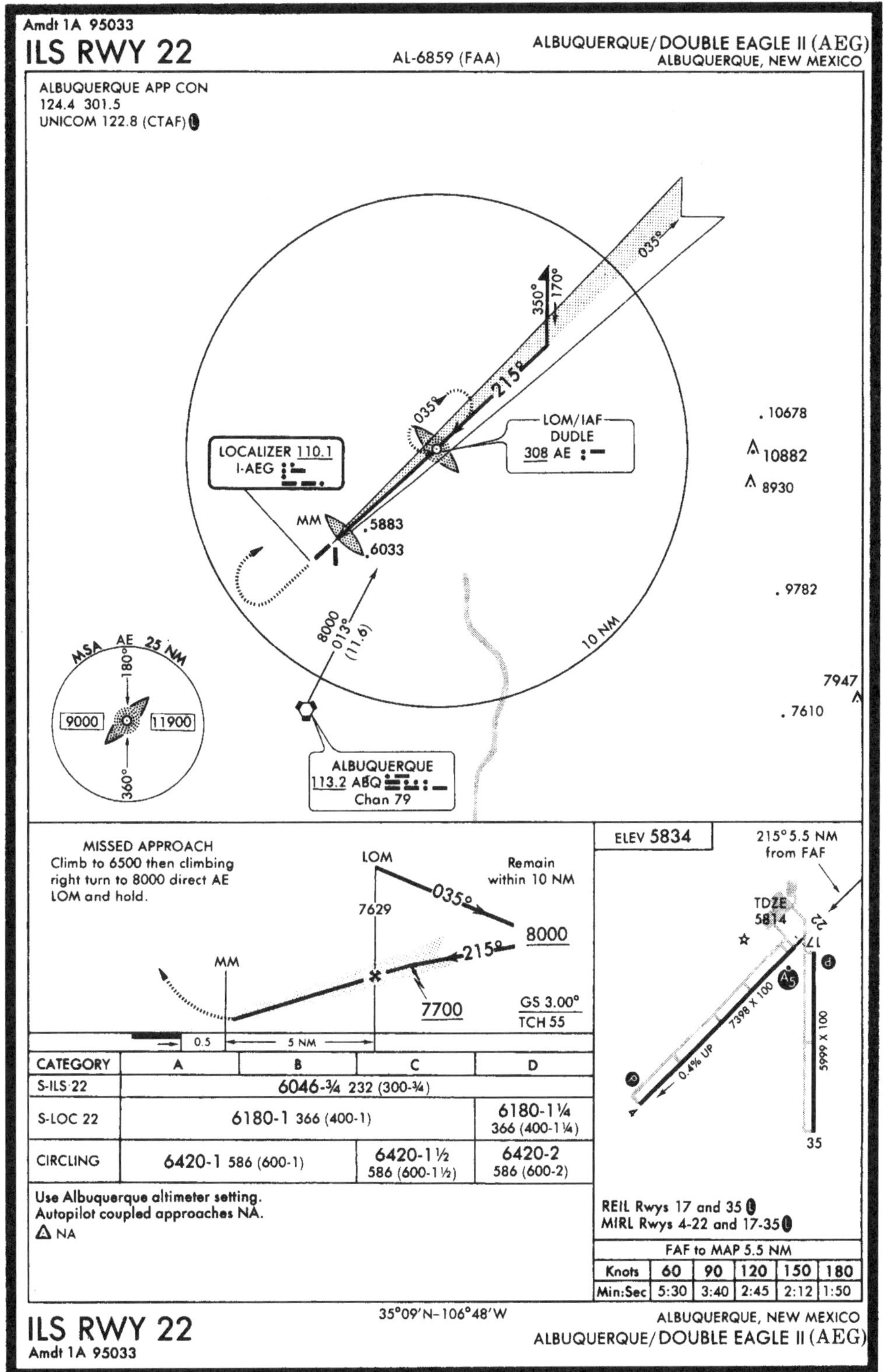

CATEGORY	A	B	C	D
S-ILS 22	6046-¾ 232 (300-¾)			
S-LOC 22	6180-1 366 (400-1)			6180-1¼ 366 (400-1¼)
CIRCLING	6420-1 586 (600-1)		6420-1½ 586 (600-1½)	6420-2 586 (600-2)

FAF to MAP 5.5 NM					
Knots	60	90	120	150	180
Min:Sec	5:30	3:40	2:45	2:12	1:50

FIGURE 167.—ILS RWY 22 (AEG).

DOUBLE EAGLE II (AEG) 7 NW UTC–7(–6DT) N35°08.71′ W106°47.71′ ALBUQUERQUE
5834 B S4 FUEL 100LL, JET A OX 3 H-2D, L-4G, 6E
RWY 04-22: H7398X100 (ASPH) S-30 MIRL 0.4% up SW IAP
RWY 04: PAPI(P4L). RWY 22: MALSR. Rgt tfc.
RWY 17-35: H5999X100 (ASPH) S-30 MIRL
RWY 17: REIL. PAPI(P4L). RWY 35: REIL. Rgt tfc.
AIRPORT REMARKS: Attended continuously. Fixed-base operator and arpt services 505-842-7007 or 505-836-7717. ACTIVATE MIRL Rwy 04-22 and Rwy 17-35, REIL Rwy 17 and Rwy 35 and MALSR Rwy 22 and PAPI Rwy 04 and Rwy 17—CTAF.
COMMUNICATIONS: CTAF/UNICOM 122.8
ALBUQUERQUE FSS (ABQ) LC 243-7831. NOTAM FILE ABQ
® ALBUQUERQUE APP/DEP CON 124.4
RADIO AIDS TO NAVIGATION: NOTAM FILE ABQ.
ALBUQUERQUE (H) VORTACW 113.2 ABQ Chan 79 N35°02.63′ W106°48.98′ 357° 6.2 NM to fld. 5740/13E. HIWAS.
DUDLE NDB (LOM) 308 AE N35°13.04′ W106°42.77′ 212° 5.9 NM to fld.
ILS 110.1 I-AEG Rwy 22 LOM DUDLE NDB. ILS unmonitored.

ALEXANDER MUNI (See BELEN)

ANTON CHICO N35°06.70′ W105°02.40′ NOTAM FILE ABQ. ALBUQUERQUE
(H) VORTAC 117.8 ACH Chan 125 105° 22.3 NM to Santa Rosa Muni. 5450/12E. H-2D, L-4G, 6E
RCO 122.1R 117.8T (ALBUQUERQUE FSS)

ANGEL FIRE (AXX) 1 N UTC–7(–6DT) N36°25.24′ W105°17.40′ DENVER
8382 S4 FUEL 100LL, JET A H-2D, L-6E
RWY 17-35: H8900X100 (ASPH) S-22 0.6% up S
RWY 17: Ground. RWY 35: Road.
AIRPORT REMARKS: Attended dalgt hours. Airport located in mountain valley, rising terrain in all directions. Aerobatics will be conducted adjacent to and east of arpt 14,500′ and below. Ramp asph surfaces deteriorated with numerous cracks and soft spots. Dirt berm located approximately 1700′ up NW from rwy end 17 250 ft E of E NW edge.
COMMUNICATIONS: CTAF/UNICOM 122.8
ALBUQUERQUE FSS (ABQ) TF 1-800-WX-BRIEF. NOTAM FILE ABQ.
RADIO AIDS TO NAVIGATION: NOTAM FILE ABQ.
TAOS (L) VORTAC 117.6 TAS Chan 123 N36°36.53′ W105°54.38′ 098° 31.9 NM to fld. 7860/13E.

APACHE CREEK

JEWETT MESA (Q13) 10 N UTC–7(–6DT) N34°00.20′ W108°40.69′ ALBUQUERQUE
7681
RWY 06-24: 5200X40 (DIRT)
RWY 06: Pole. RWY 24: Fence.
AIRPORT REMARKS: Unattended. Arpt open May–Sep; other times CLOSED. Livestock on runway. Rwy 06-24 recommend visual inspection before using, infrequent maintenance. Rwy 06-24 heavily weeded with large rocks on rwy edges +2′ rocks 45′ from rwy centerline.
COMMUNICATIONS: CTAF 122.9
ALBUQUERQUE FSS (ABQ) TF 1-800-WX-BRIEF. NOTAM FILE ABQ.

ARTESIA MUNI (ATS) 3 W UTC–7(–6DT) N32°51.15′ W104°28.06′ ALBUQUERQUE
3548 B S4 FUEL 100LL, JET A1 H-2D, 5A, L-4H
RWY 03-21: H6300X150 (ASPH-PFC) S-40, D-57 MIRL 0.4% up SW IAP
RWY 03: P-line. RWY 21: PVASI(PSIL)—GA 3.0° TCH 25′. Road.
RWY 12-30: H5399X150 (ASPH-PFC) S-40, D-57 MIRL 0.5% up NW
RWY 12: Fence.
AIRPORT REMARKS: Attended 1400–0100Z‡. Fuel on call after hours 505-748-9053, 746-4196, 457-2399/2268, fee charged. MIRL Rwy 03-21 and Rwy 12-30 preset at low ints dusk–0800Z‡, ACTIVATE higher ints—CTAF. Rwys 03-21 and 12-30 PFC center 75′ width only.
WEATHER DATA SOURCES: AWOS-3 126.725 (505) 748-2103.
COMMUNICATIONS: CTAF/UNICOM 122.8
ALBUQUERQUE FSS (ABQ) TF 1-800-WX-BRIEF. NOTAM FILE ABQ.
ROSWELL APP/DEP CON 119.6 (1300–0400Z‡) ® ALBUQUERQUE CENTER APP/DEP CON 132.65 (0400–1300Z‡)
RADIO AIDS TO NAVIGATION: NOTAM FILE ROW.
CHISUM (H) VORTACW 116.1 CME Chan 108 N33°20.25′ W104°37.28′ 153° 30.1 NM to fld. 3770/12E. HIWAS.
NDB (MHW) 414 ATS N32°51.16′ W104°27.70′ at fld. NOTAM FILE ABQ.

FIGURE 167A.—Excerpt from Airport/Facilities Directory.

Form Approved: OMB No. 2120-0034

U.S. DEPARTMENT OF TRANSPORTATION FEDERAL AVIATION ADMINISTRATION **FLIGHT PLAN**	(FAA USE ONLY) ☐ PILOT BRIEFING ☐ VNR ☐ STOPOVER	TIME STARTED	SPECIALIST INITIALS

1. TYPE	2. AIRCRAFT IDENTIFICATION	3. AIRCRAFT TYPE/ SPECIAL EQUIPMENT	4. TRUE AIRSPEED	5. DEPARTURE POINT	6. DEPARTURE TIME PROPOSED (Z)	ACTUAL (Z)	7. CRUISING ALTITUDE
VFR				KPWK			
X IFR	PTZ 70	BE 1900/R	247 KTS	CHICAGO/ PAL-WAUKEE			FL190
DVFR							

8. ROUTE OF FLIGHT
PAL-WAUKEE TWO DEPARTURE, PMM J547 BUF

9. DESTINATION (Name of airport and city)	10. EST. TIME ENROUTE HOURS	MINUTES	11. REMARKS
BUF GREATER BUFFALO INT'L BUFFALO			L/O = LEVEL OFF PPH = POUNDS PER HOUR L/O PMM R-261/47 VARIATION: PWK 1W, FNT 3W, BUF 8W

12. FUEL ON BOARD HOURS	MINUTES	13. ALTERNATE AIRPORT(S)	14. PILOT'S NAME, ADDRESS & TELEPHONE NUMBER & AIRCRAFT HOME BASE / 17. DESTINATION CONTACT/TELEPHONE (OPTIONAL)	15. NUMBER ABOARD
3	35	SYR SYRACUSE HANCOCK INT'L		13

16. COLOR OF AIRCRAFT
WHITE/BLACK

CIVIL AIRCRAFT PILOTS. FAR Part 91 requires you file an IFR flight plan to operate under instrument flight rules in controlled airspace. Failure to file could result in a civil penalty not to exceed $1,000 for each violation (Section 901 of the Federal Aviation Act of 1958, as amended). Filing of a VFR flight plan is recommended as a good operating practice. See also Part 99 for requirements concerning DVFR flight plans.

FAA Form 7233-1 (8-82) CLOSE VFR FLIGHT PLAN WITH ______________ FSS ON ARRIVAL

FLIGHT LOG

CHECK POINTS FROM	TO	ROUTE ALTITUDE	COURSE	WIND TEMP	SPEED-KTS TAS	GS	DIST NM	TIME LEG	TOT	FUEL LEG	TOT
PWK	L/O	VECTORS CLIMB					49		:24:00		410*
L/O	PMM	J547 FL190		020/61 ISA							
PMM	FNT										
FNT	YXU										
YXU	BUF R-282/40										
BUF R-282/40	BUF	J547 DESCENT					40	:19:00		163	
BUF	SYR	VECTORS 4000					112	:30:00			

OTHER DATA: * Includes Taxi Fuel
NOTE: Use 676 PPH Total Fuel Flow From L/O To Start Of Descent.
Use 726 PPH Total Fuel Flow for Reserve And Alternate Requirements.

A Missed Approach Requires 76# of Fuel.

TIME and FUEL: As required by FARs.

TIME	FUEL (LB)	
		EN ROUTE
		RESERVE
		ALTERNATE
		TOTAL

FIGURE 168.—Flight Plan/Flight Log.

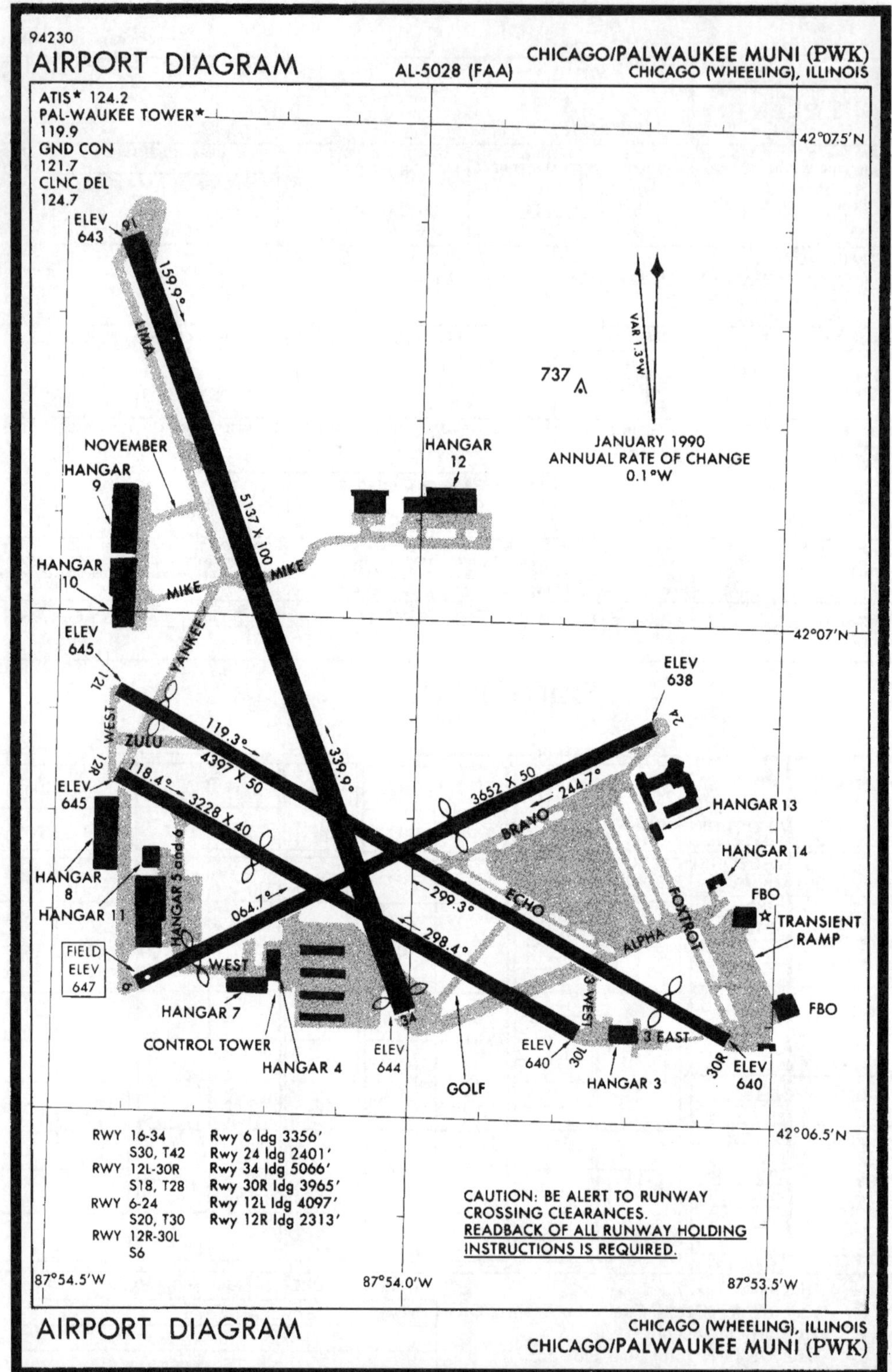

FIGURE 169.—AIRPORT DIAGRAM (PWK).

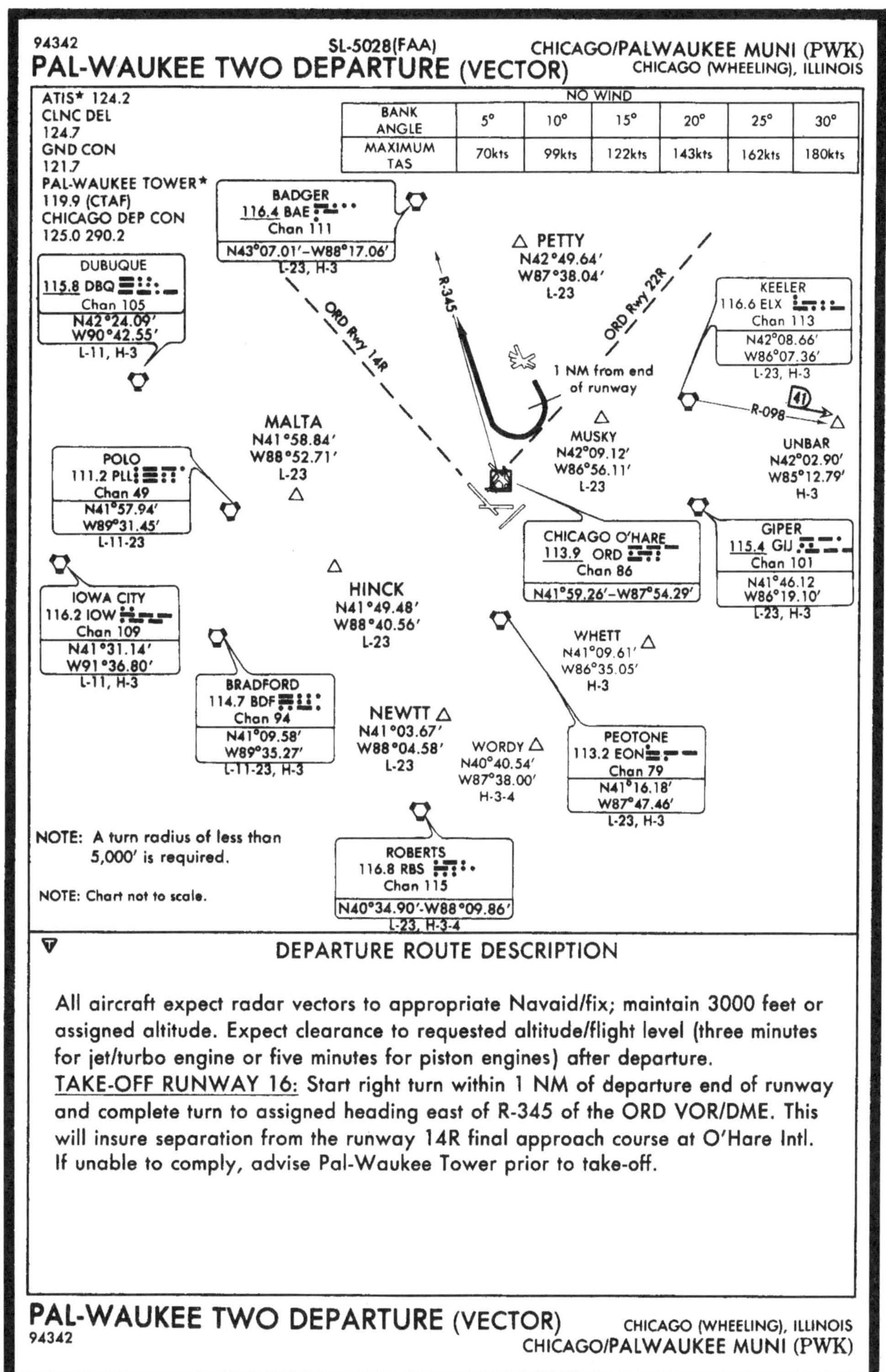

FIGURE 169A.—PAL-WAUKEE TWO DEPARTURE (VECTOR) (PWK).

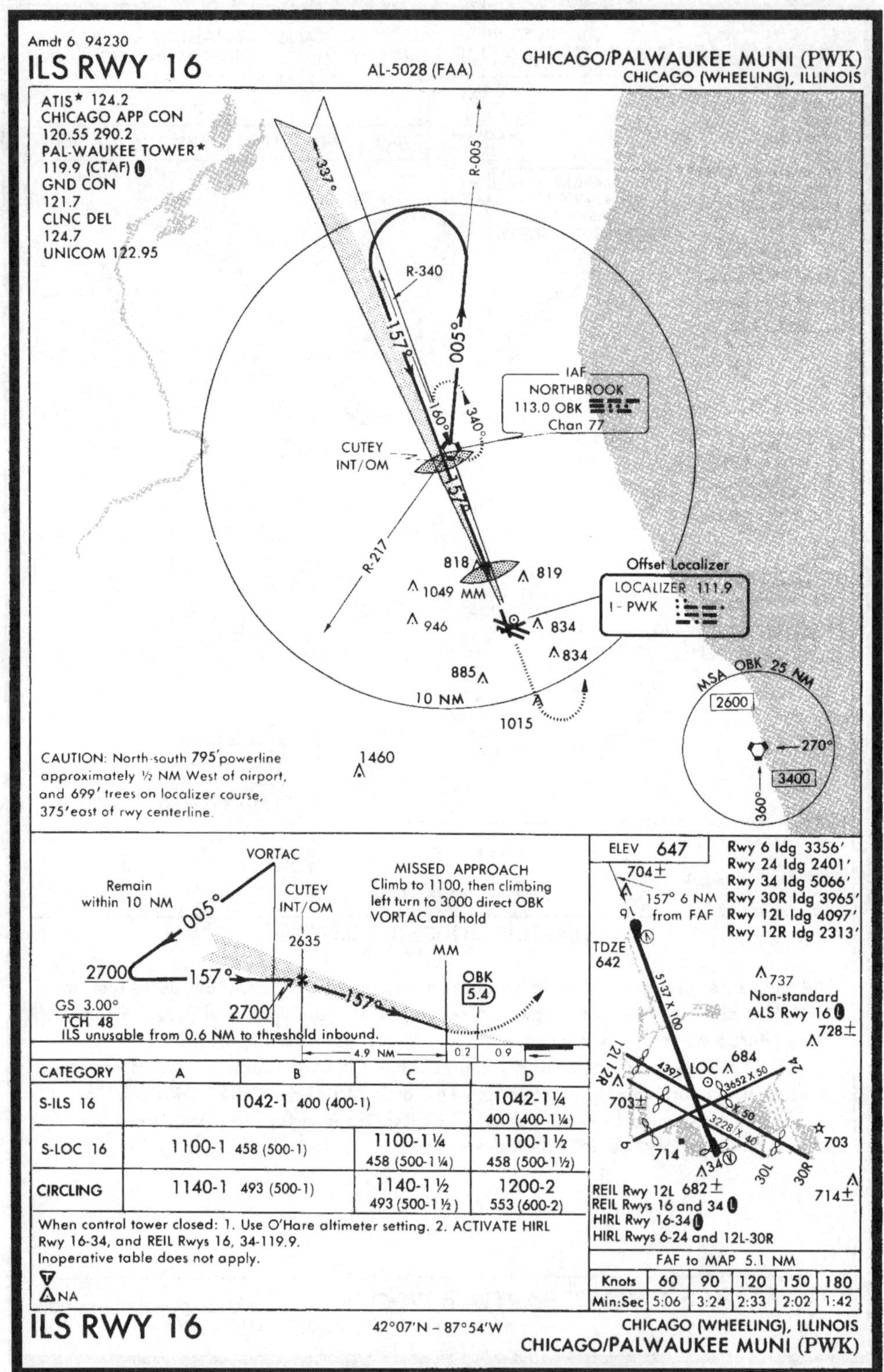

CATEGORY	A	B	C	D
S-ILS 16	1042-1 400 (400-1)			1042-1¼ 400 (400-1¼)
S-LOC 16	1100-1 458 (500-1)		1100-1¼ 458 (500-1¼)	1100-1½ 458 (500-1½)
CIRCLING	1140-1 493 (500-1)		1140-1½ 493 (500-1½)	1200-2 553 (600-2)

FAF to MAP 5.1 NM

Knots	60	90	120	150	180
Min:Sec	5:06	3:24	2:33	2:02	1:42

FIGURE 170.—ILS RWY 16 (PWK).

TAKE-OFF MINS

94286

CHICAGO, IL

CHICAGO MIDWAY

TAKE-OFF MINIMUMS: **Rwys 13R, 31L,** 300-1. **Rwy 13L,** 300-1 or std. with min. climb of 300' per NM to 900. **Rwy 31C,** 300-1 or std. with min. climb of 330' per NM to 900. **Rwy 31R,** 300-1 or std. with min. climb of 225' per NM to 900. **Rwy 22L,** 300-1 or std. with min. climb of 400' per NM to 900. **Rwy 22R,** 300-1 or std. with min. climb of 340' per NM to 900.
DEPARTURE PROCEDURE: **Rwys 4L, 4R,** Northbound Departures (360° CW 080°), climbing right turn to 2400 heading 100° before proceeding on course. **Rwys 22L, 22R, 31C, 31R, 31L, 13R, 13L, 13C,** climb runway heading to 1300' before turning.

CHICAGO-O'HARE INTL

TAKE-OFF MINIMUMS: **Rwy 22R,** 300-1. **Rwy 32L,** straight out or right turn, std.; left turn 1000-3 or std. with a min. climb of 240' per NM to 1800. **Rwy 18,** NA. **Rwy 36,** 500-1.

LANSING MUNI

DEPARTURE PROCEDURE: **Rwy 9,** 300-1. **Rwy 36,** 400-1.

CHICAGO/ROMEOVILLE, IL

LEWIS UNIVERSITY

DEPARTURE PROCEDURE: **Rwy 6,** climb on heading 065° to 1200 before proceeding on course.

CHICAGO/WAUKEGAN, IL

WAUKEGAN REGIONAL

TAKE-OFF MINIMUMS: **Rwy 14,** 300-1.

CHICAGO (WHEELING), IL

PALWAUKEE MUNI

TAKE-OFF MINIMUMS: **Rwys 6, 12L/R, 24, 30L/R, 34,** 300-1.

CLINTONVILLE, WI

CLINTONVILLE MUNI

DEPARTURE PROCEDURE: **Rwys 4, 9,** climb on runway heading to 2000 before turning on course.

DE KALB, IL

DE KALB TAYLOR MUNI

TAKE-OFF MINIMUMS: **Rwys 9, 27,** 300-1.

DECATUR, IL

DECATUR

DEPARTURE PROCEDURE: Northbound Departures; **Rwy 36,** left turn, climb to 3000 via DEC R-340 before proceeding North. **Rwy 30,** right turn, climb to 3000 via DEC R-340 before proceeding North. **Rwy 18,** climb runway heading to 1200 before turning North. **Rwys 6, 12, 24,** climb runway heading to 1600 before turning North.

DELAVAN, WI

LAKE LAWN

TAKE-OFF MINIMUMS: **Rwys 18, 36,** 300-1.

DIXON, IL

DIXON MUNI-CHARLES R. WALGREEN FIELD

TAKE-OFF MINIMUMS: **Rwys 26, 30,** 300-1.

EAU CLAIRE, WI

CHIPPEWA VALLEY REGIONAL

TAKE-OFF MINIMUMS: **Rwy 14,** 500-1.
DEPARTURE PROCEDURE: **Rwys 14, 22,** climb runway heading to 2500 before turning southbound.

EFFINGHAM, IL

EFFINGHAM COUNTY MEMORIAL

TAKE-OFF MINIMUMS: **Rwy 1,** 500-1.
DEPARTURE PROCEDURE: **Rwy 29,** climb runway heading to 2100 before turning right.

FAIRFIELD, IL

FAIRFIELD MUNI

TAKE-OFF MINIMUMS: **Rwy 9,** 400-1.
DEPARTURE PROCEDURE: **Rwy 36,** climb runway heading to 2100 before turning right. **Rwy 18,** climb runway heading to 2100 before turning left. **Rwy 27,** climb runway heading to 1500 before turning eastbound. **Rwy 9,** climb to 2100 on heading 120° before proceeding eastbound or northbound.

FLORA, IL

FLORA MUNI

DEPARTURE PROCEDURE: **Rwys 3, 33,** climb runway heading to 1100' before turning left. **Rwy 21,** climb runway heading to 1100 before turning right.

FOND DU LAC, WI

FOND DU LAC COUNTY

DEPARTURE PROCEDURE: **Rwy 9,** climb runway heading to 2000 before turning North. **Rwy 36,** climb runway heading to 2000 before turning East.

FRANKFORT, IL

FRANKFORT

TAKE-OFF MINIMUMS: **Rwy 27,** 300-1.
DEPARTURE PROCEDURE: **Rwy 9,** climb runway heading to 1200 before turning northbound.

GRANTSBURG, WI

GRANTSBURG MUNI

TAKE-OFF MINIMUMS: **Rwy 23,** 300-1.

GRAYSLAKE, IL

CAMPBELL

TAKE-OFF MINIMUMS: **Rwy 24,** 300-1.
DEPARTURE PROCEDURE: **Rwy 9,** climb runway heading to 1200 before turning.

4286 EC-3

TAKE-OFF MINS

FIGURE 170A.—TAKE-OFF MINS.

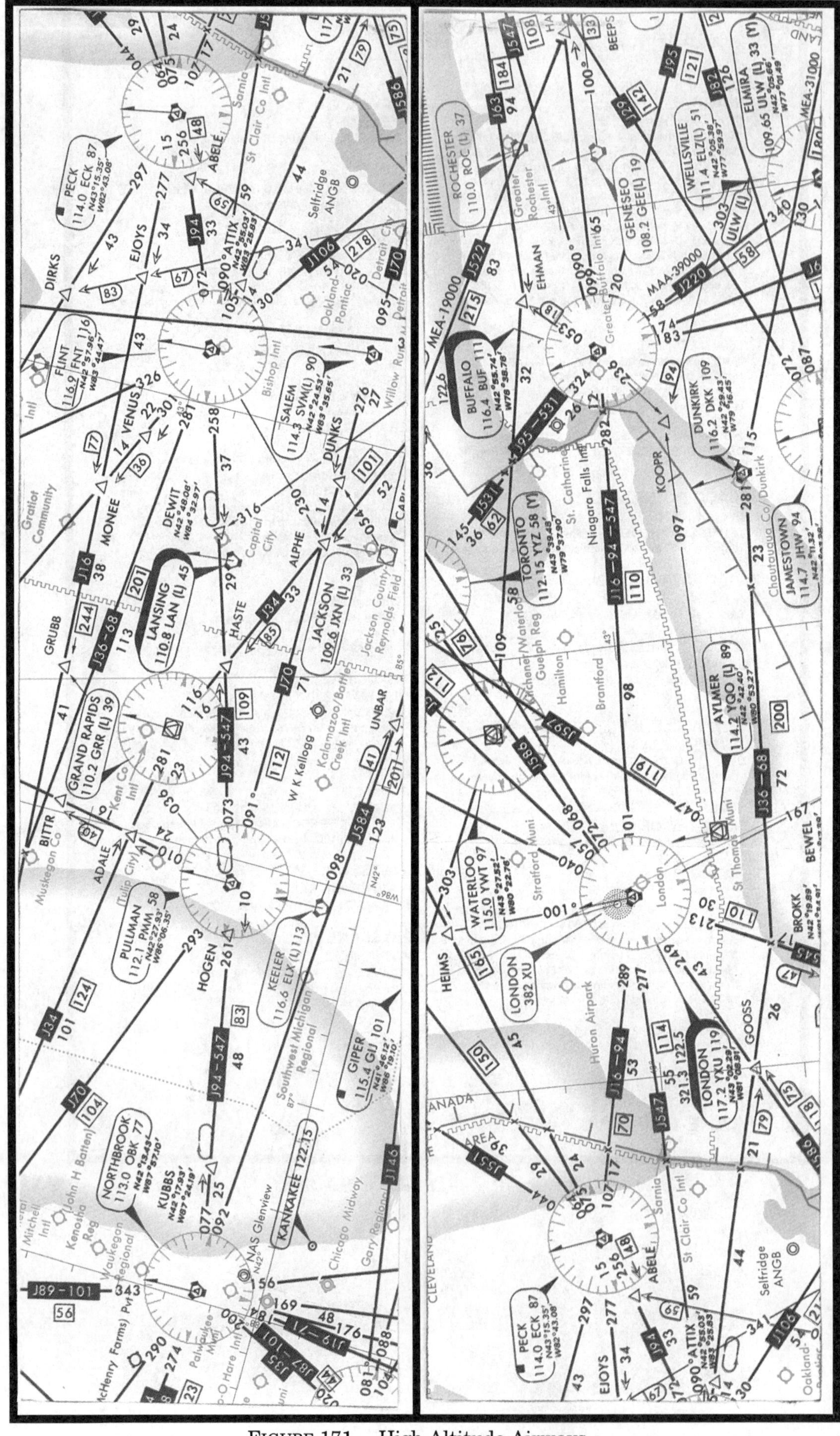

FIGURE 171.—High Altitude Airways.

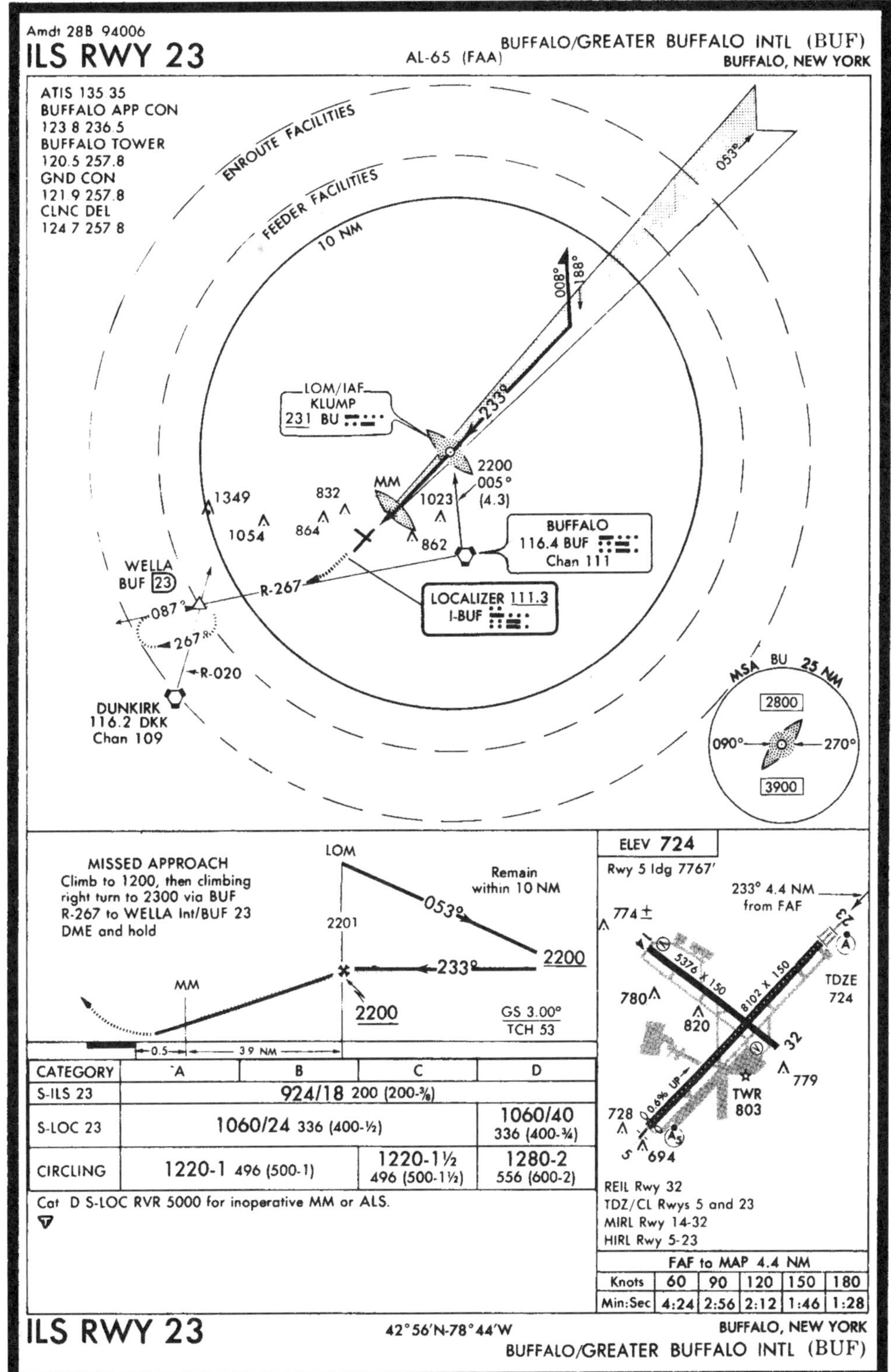

CATEGORY	A	B	C	D
S-ILS 23	924/18 200 (200-¾)			
S-LOC 23	1060/24 336 (400-½)			1060/40 336 (400-¾)
CIRCLING	1220-1 496 (500-1)		1220-1½ 496 (500-1½)	1280-2 556 (600-2)

FAF to MAP 4.4 NM					
Knots	60	90	120	150	180
Min:Sec	4:24	2:56	2:12	1:46	1:28

FIGURE 172.—ILS RWY 23 (BUF).

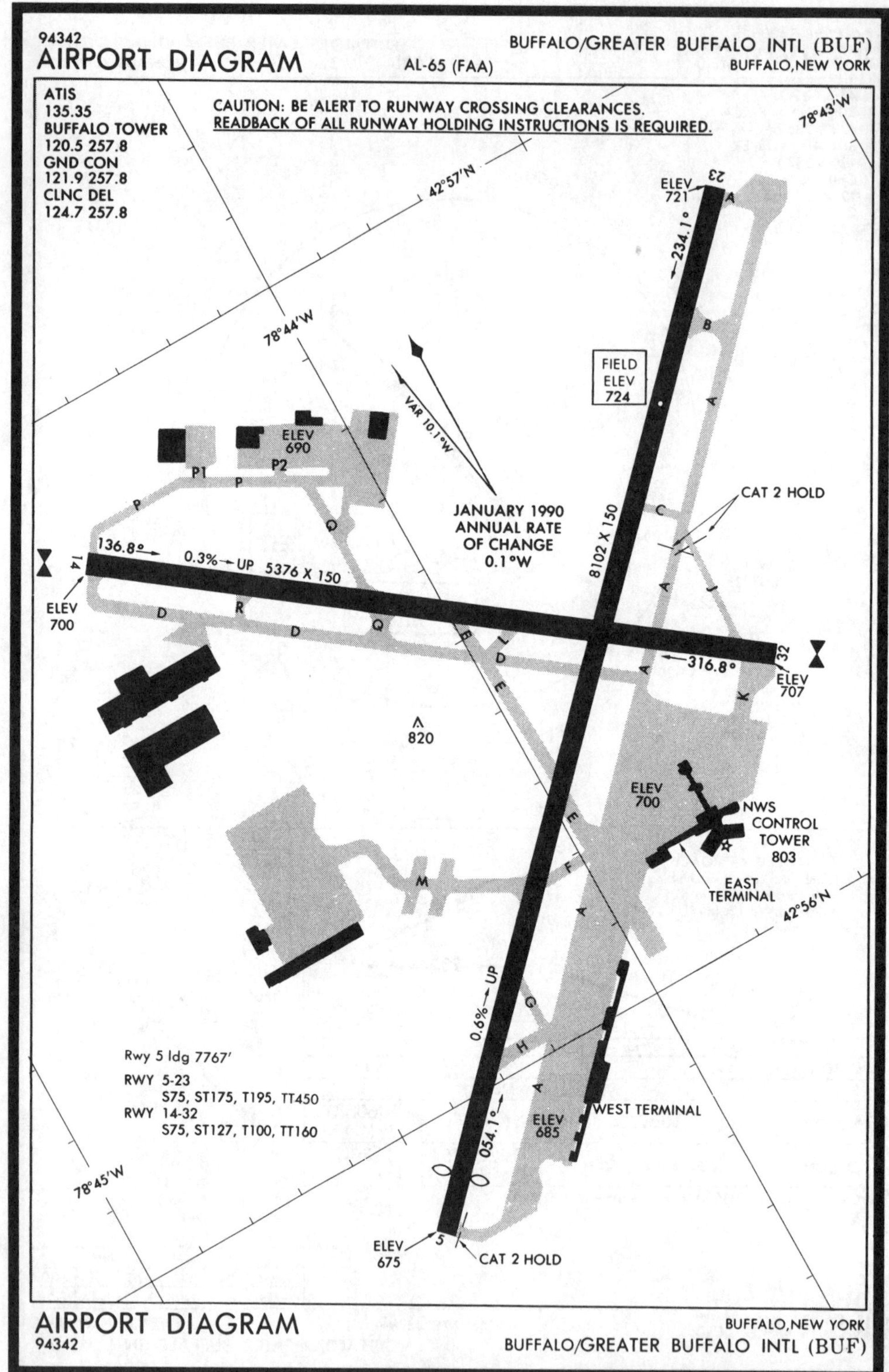

FIGURE 172A.—AIRPORT DIAGRAM (BUF).

ALTERNATE MINS

95033

INSTRUMENT APPROACH PROCEDURE CHARTS

IFR ALTERNATE MINIMUMS

(NOT APPLICABLE TO USA/USN/USAF)

Standard alternate minimums for non precision approaches are 800-2 (NDB, VOR, LOC, TACAN, LDA, VORTAC, VOR/DME or ASR); for precision approaches 600-2 (ILS or PAR). Airports within this geographical area that require alternate minimums other than standard or alternate minimums with restrictions are listed below. NA - means alternate minimums are not authorized due to unmonitored facility or absence of weather reporting service. Civil pilots see FAR 91. USA/USN/USAF pilots refer to appropriate regulations.

NAME — ALTERNATE MINIMUMS

ALBANY, NY

ALBANY COUNTY .... **ILS Rwy 1[1]**
ILS Rwy 19[1]
VOR/DME or GPS Rwy 1[1]
VOR Rwy 1[2]
VOR or GPS Rwy 19[1]
VOR or GPS Rwy 28[1]

[1]Category D, 800-2½.
[2]Category C, 800-2¼; Category D, 800-2½.

ALLENTOWN, PA

LEHIGH VALLEY INTL .... **ILS Rwy 13**

ILS, Categories A,B,C, 700-2; Category D, 700-2¼. LOC, Category D, 800-2¼.

ALTOONA, PA

ALTOONA-BLAIR COUNTY .... **ILS Rwy 20[1]**
VOR or GPS-A[2]

[1]Categories A,B,C, 900-2½, Category D, 1100-3.
[2]Category D, 1100-3.

BRADFORD, PA

BRADFORD REGIONAL .... **VOR/DME or GPS Rwy 14**

NA when BFD FSS closed.

CORTLAND, NY

CORTLAND COUNTY-CHASE FIELD .... **VOR or GPS-A**

Categories A,B, 1100-2,Categories C,D, 1100-3.

DUBOIS, PA

DUBOIS-JEFFERSON COUNTY ..... **ILS Rwy 25**

LOC, NA.

NAME — ALTERNATE MINIMUMS

ELMIRA, NY

ELMIRA/CORNING REGIONAL .... **ILS Rwy 6[12]**
ILS Rwy 24,1200-3
NDB or GPS Rwy 24,1200-3

[1]Categories A,B, 1200-2; Categories C,D, 1200-3.
[2]NA when control tower closed.

ERIE, PA

ERIE INTL .... **ILS Rwy 6[1]**
ILS Rwy 24[1]
NDB Rwy 6
NDB Rwy 24
RADAR-1

NA when control tower closed.
[1]ILS, 700-2.

FARMINGDALE, NY

REPUBLIC .... **ILS Rwy 14[1]**
NDB or GPS Rwy 1[2]

[1]NA when control tower closed.
[2]NA when control zone not effective.

HARRISBURG, PA

CAPITAL CITY .... **ILS Rwy 8**

Categories A,B, 900-2; Categories C,D, 900-2¾.
NA when control tower closed.

HARRISBURG INTL .... **ILS Rwy 13[1]**
ILS Rwy 31[1]
VOR or GPS Rwy 31[2]

[1]ILS, Categories C,D, 700-2. LOC, NA.
[2]Categories A,B, 900-2, Category C, 900-2¾, Category D, 900-3.

NE-2

ALTERNATE MINS

95033

FIGURE 173.—IFR ALTERNATE MINIMUMS.

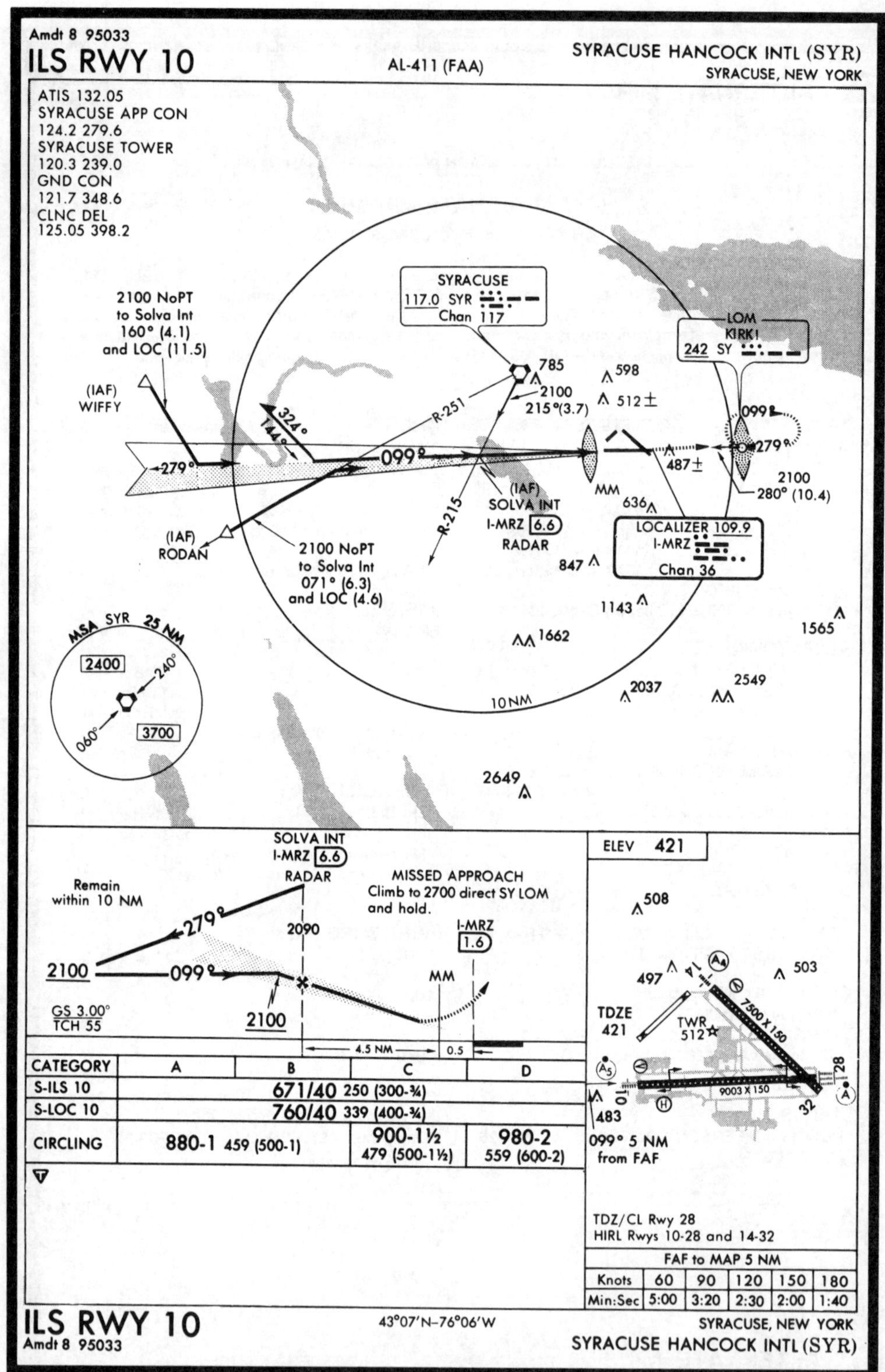

CATEGORY	A	B	C	D
S-ILS 10	671/40 250 (300-¾)			
S-LOC 10	760/40 339 (400-¾)			
CIRCLING	880-1 459 (500-1)		900-1½ 479 (500-1½)	980-2 559 (600-2)

FAF to MAP 5 NM					
Knots	60	90	120	150	180
Min:Sec	5:00	3:20	2:30	2:00	1:40

FIGURE 173A.—ILS RWY 10 (SYR).

Form Approved: OMB No. 2120-0034

U.S. DEPARTMENT OF TRANSPORTATION FEDERAL AVIATION ADMINISTRATION **FLIGHT PLAN**	(FAA USE ONLY) ☐ PILOT BRIEFING ☐ VNR ☐ STOPOVER	TIME STARTED	SPECIALIST INITIALS

1. TYPE	2. AIRCRAFT IDENTIFICATION	3. AIRCRAFT TYPE/ SPECIAL EQUIPMENT	4. TRUE AIRSPEED	5. DEPARTURE POINT	6. DEPARTURE TIME PROPOSED (Z)	ACTUAL (Z)	7. CRUISING ALTITUDE
VFR / X IFR / DVFR	SEA HAWK 1	BH230/R	** KTS	LWS LEWISTON-ZEZ PERCE CO.			*** 12000

8. ROUTE OF FLIGHT
POTOR2.CLOVA, V520PSC, V204YKM, V4 SEA, V27 ULESS, HQ, HQM

9. DESTINATION (Name of airport and city)	10. EST. TIME ENROUTE HOURS	MINUTES	11. REMARKS
HQM BOWERMAN HOQUIAM, WA			L/O = LEVEL OFF PPH = POUNDS PER HOUR **CAS 132 ISA - 8 TO ±0 *** AFTER SEA DESCEND TO 4000 FEET

12. FUEL ON BOARD HOURS	MINUTES	13. ALTERNATE AIRPORT(S)	14. PILOT'S NAME, ADDRESS & TELEPHONE NUMBER & AIRCRAFT HOME BASE / 17. DESTINATION CONTACT/TELEPHONE (OPTIONAL)	15. NUMBER ABOARD
		OLM OLYMPIA, WA		8

16. COLOR OF AIRCRAFT YELLOW/BLACK

CIVIL AIRCRAFT PILOTS. FAR Part 91 requires you file an IFR flight plan to operate under instrument flight rules in controlled airspace. Failure to file could result in a civil penalty not to exceed $1,000 for each violation (Section 901 of the Federal Aviation Act of 1958, as amended). Filing of a VFR flight plan is recommended as a good operating practice. See also Part 99 for requirements concerning DVFR flight plans.

FAA Form 7233-1 (8-82) CLOSE VFR FLIGHT PLAN WITH ____________ FSS ON ARRIVAL

FLIGHT LOG

CHECK POINTS FROM	TO	ROUTE / ALTITUDE	COURSE	WIND / TEMP	SPEED-KTS TAS	GS	DIST NM	TIME LEG	TOT	FUEL LEG	TOT
LWS	MQG	POTOR2.CLOVA / CLIMB					15		:15:00		191*
MQG	ALW	V520 / 12000		340/40 / ISA-8							
ALW	PSC	V520 / 12000		340/40 / ISA-3							
PSC	YKM	V204 / 12000		320/35 / ISA-3							
YKM	SEA	V4 / 12000		300/29 / ISA+1							
SEA	ULESS	V27 / 4000		280/7 / ISA							
ULESS	HQ	DIRECT / DESCENT						:02:00		16.0	
HQ	HQM	DIRECT / DESCENT						:04:00		32.0	
HQM	OLM	V204 / 5000		270/16 / ISA		158	42				

OTHER DATA: * Includes Taxi Fuel
NOTE: Use 525 PPH Total Fuel Flow From L/O To Start Of Descent.
Use 499 PPH Total Fuel Flow For Reserve And Alternate Requirements.

A Missed Approach Requires 40# of Fuel.

TIME and FUEL: As required by FARs.

TIME	FUEL (LB)	
		EN ROUTE
		RESERVE
		ALTERNATE
		TOTAL

FIGURE 174.—Flight Plan/Flight Log.

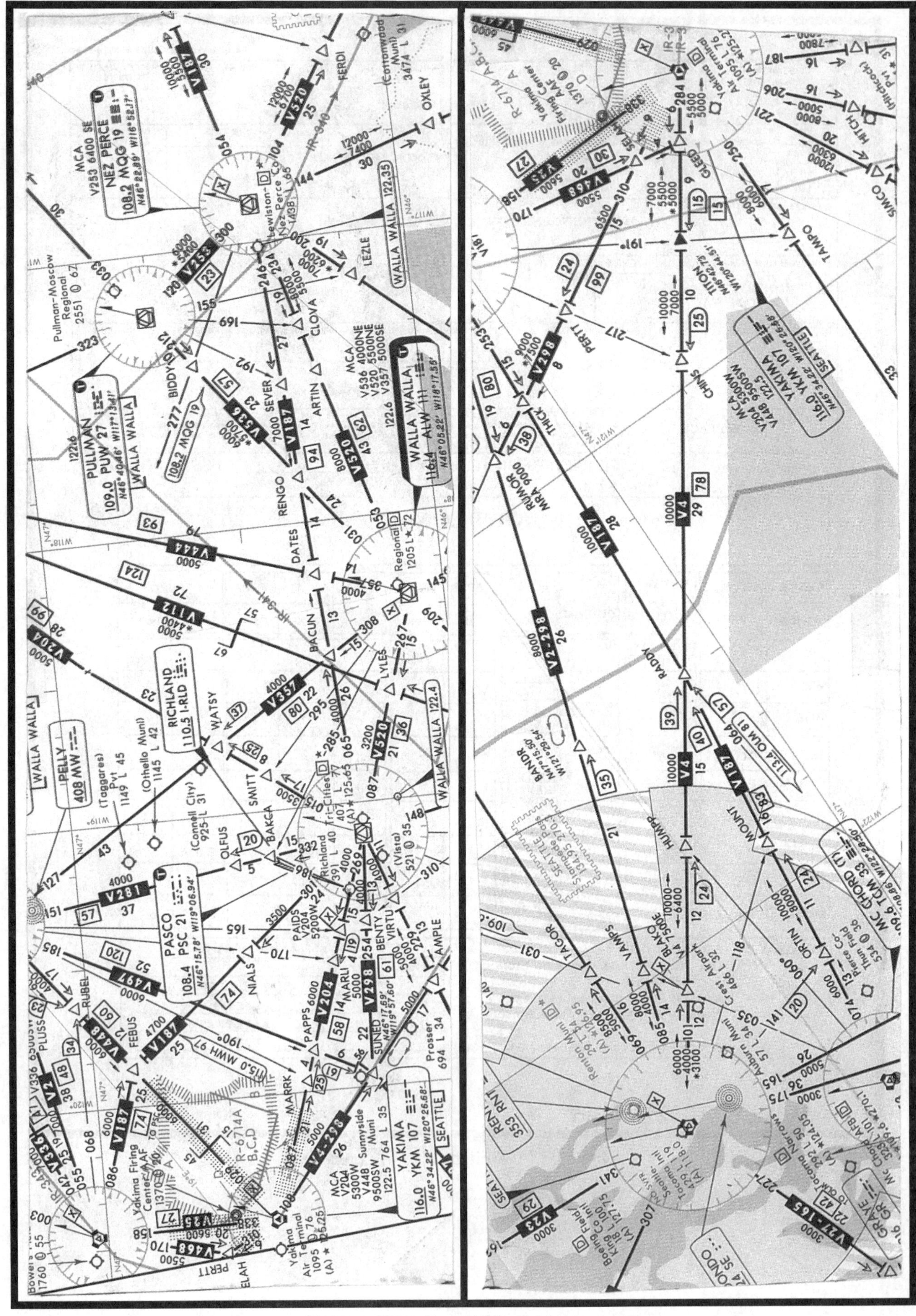

FIGURE 175.—Low Altitude Airways.

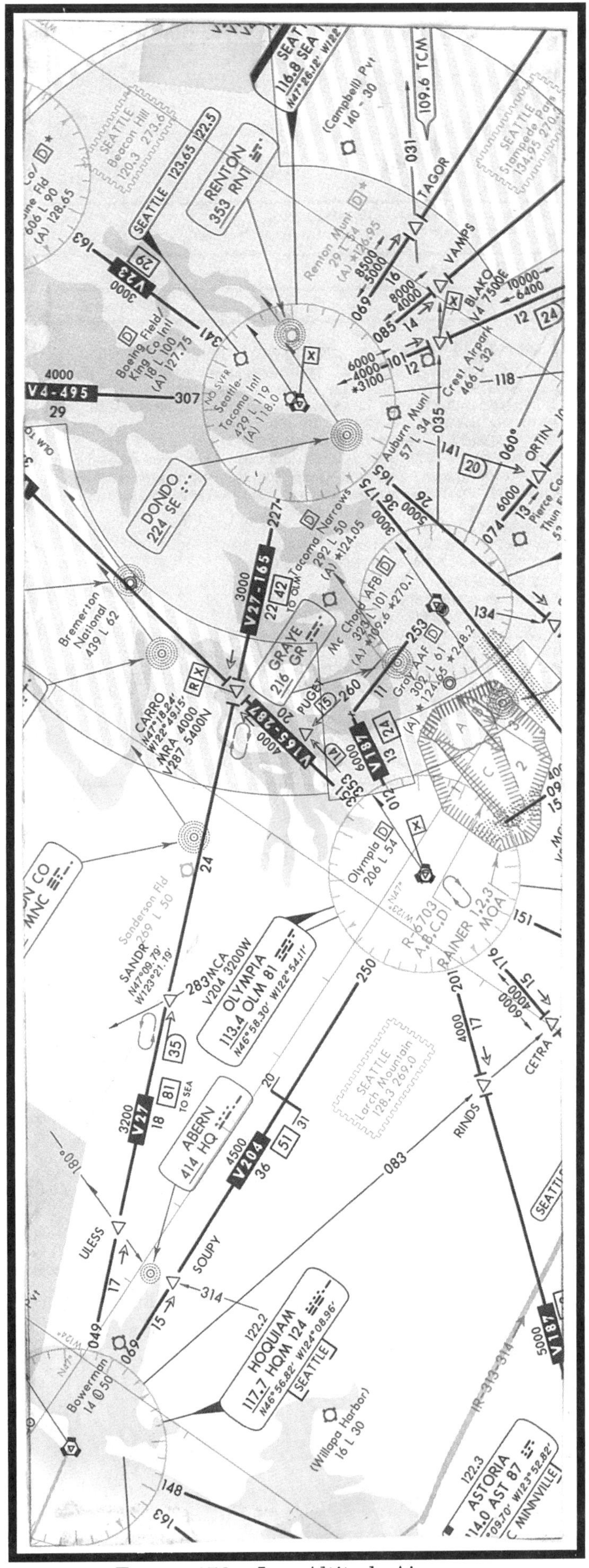

FIGURE 176.—Low Altitude Airways.

26 **IDAHO**

LEENY N47°44.57′ W116°57.66′. NOTAM FILE COE. **GREAT FALLS**
NDB (LOM) 347 CO 053° 6.0 NM to Coeur D'Alene Air Terminal.

LEE WILLIAMS MEM (See MIDVALE)

LEMHI CO (See SALMON)

LEWISTON
LEWISTON-NEZ PERCE CO (LWS) 2 S UTC−8(−7DT) N46°22.47′ W117°00.92′ **SEATTLE**
1438 B S4 **FUEL** 100, 100LL, JET A TPA—See Remarks ARFF Index A **H-1B, L-9A**
RWY 08-26: H6512X150 (ASPH-PFC) S-150, D-180, DT-400 HIRL **IAP**
RWY 08: REIL. VASI(V4R)—GA 3.0° TCH 45′. Antenna. Rgt tfc. **RWY 26:** MALSR. Tree.
RWY 11-29: H5001X100 (ASPH) S-70, D-94, DT-150 MIRL
RWY 11: REIL. Rgt tfc. **RWY 29:** VASI(V4R)—GA 3.0° TCH 47′.
AIRPORT REMARKS: Attended 1330–0500Z‡. CLOSED to unscheduled air carrier ops with more than 30 passenger seats 1500–0100Z‡ except PPR call arpt manager 208-746-7962 other times call station number 4 208-743-0172. TPA—turbine powered heavy acft 3000 (1562) all others 2500 (1062). When twr clsd ACTIVATE MALSR Rwy 26, REIL Rwy 08 and Rwy 11—CTAF.
WEATHER DATA SOURCES: LAWRS.
COMMUNICATIONS: CTAF 119.4 **UNICOM** 122.95
BOISE FSS (BOI) TF 1-800-WX-BRIEF. NOTAM FILE LWS.
RCO 122.35 (BOISE FSS)
SEATTLE CENTER APP/DEP CON 120.05
TOWER 119.4 (1400–0600Z‡) **GND CON** 121.9
AIRSPACE: CLASS D svc effective 1400–0600Z‡ other times CLASS G.
RADIO AIDS TO NAVIGATION: NOTAM FILE LWS.
NEZ PERCE (L) VORW/DME 108.2 MQG Chan 19 N46°22.89′ W116°52.17′ 246° 6.1 NM to fld. 1720/20E.
ILS 109.7 I-LWS Rwy 26. ILS unmonitored when tower closed.

WASHINGTON 105

HOQUIAM
BOWERMAN (HQM) 2 W UTC−8(−7DT) N46°58.27′ W123°56.19′ **SEATTLE**
14 B S4 **FUEL** 80, 100LL, JET A1+ LRA **L-1C**
RWY 06-24: H4999X150 (ASPH) S-30, D-40, DT-80 HIRL **IAP**
RWY 06: MALSR. REIL. VASI(V4L)—GA 3.0° TCH 52′. Antenna. Rgt tfc.
RWY 24: VASI(V4L)—GA 3.0° TCH 50′. Sign.
AIRPORT REMARKS: Attended 1600–0200Z‡. Fuel avbl between 0100–1700Z‡, call 533-6655, call-out fee required. CAUTION—Flocks of waterfowl on and in vicinity of arpt. Service road south of rwy in primary surface. Ultralights prohibited without written permission from arpt manager. ACTIVATE HIRL Rwy 06-24 and REIL Rwy 06—CTAF.
COMMUNICATIONS: CTAF/UNICOM 122.7
SEATTLE FSS (SEA) TF 1-800-WX-BRIEF. NOTAM FILE HQM.
RCO 122.2 (SEATTLE FSS)
AIRSPACE: CLASS E svc effective 1400–0600Z‡ other times CLASS G.
RADIO AIDS TO NAVIGATION: NOTAM FILE HQM. VHF/DF ctc FSS.
HOQUIAM (H) VORTACW 117.7 HQM Chan 124 N46°56.82′ W124°08.96′ 062° 8.9 NM to fld. 10/19E. **HIWAS.**
ABERN NDB (LOM) 414 HQ N46°59.26′ W123°47.86′ 241° 5.8 NM to fld. Unmonitored 0600–1400Z‡. Out of service indefinitely.
ILS/DME 108.7 I-HQM Chan 24 Rwy 24 LOM ABERN NDB. LOM out of service indefinitely. LOC/LOM/DME unmonitored 0600–1400Z‡.

FIGURE 177.—Excerpt from Airport/Facilities Directory.

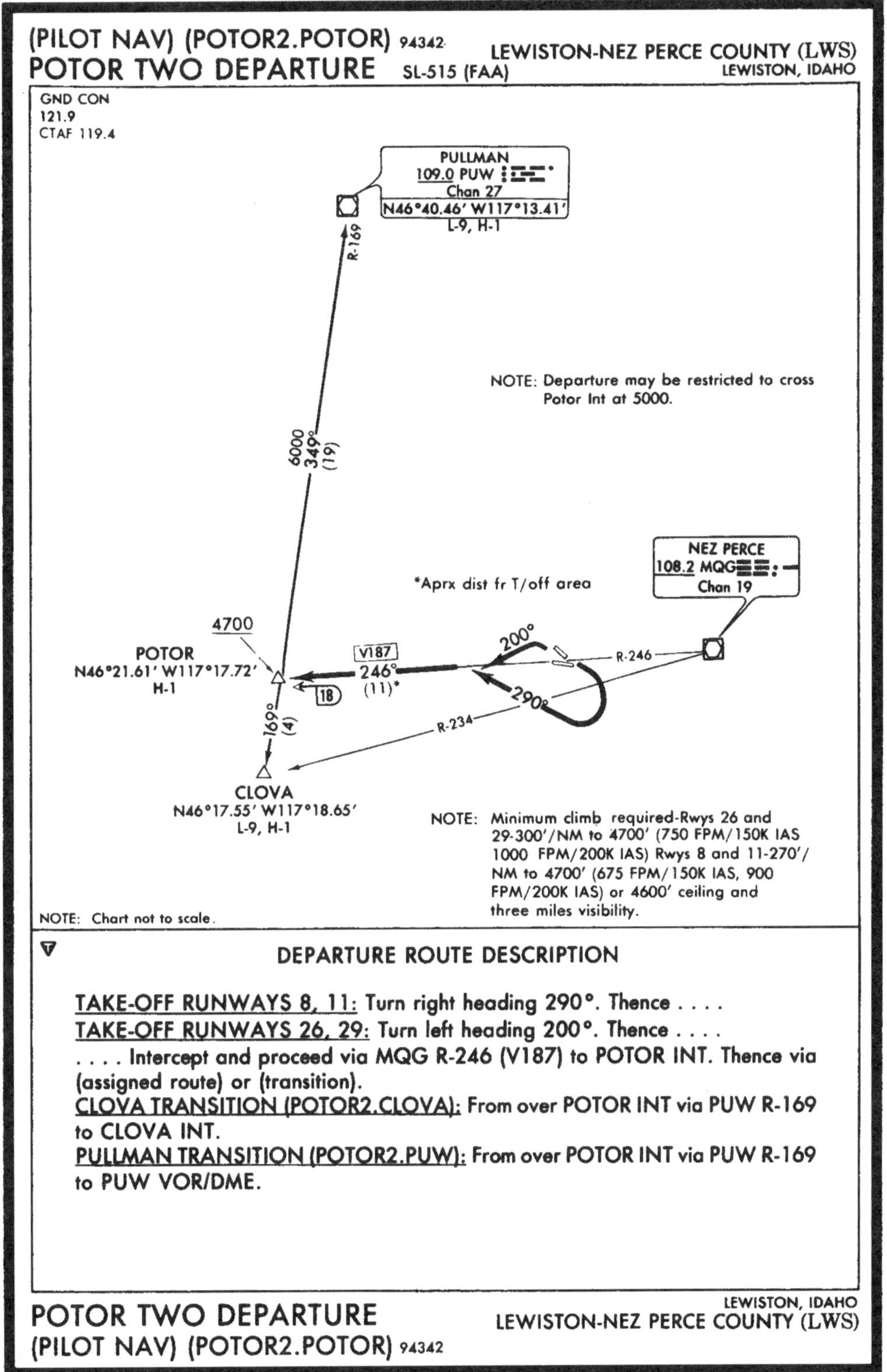

FIGURE 177A.—POTOR TWO DEPARTURE (LWS).

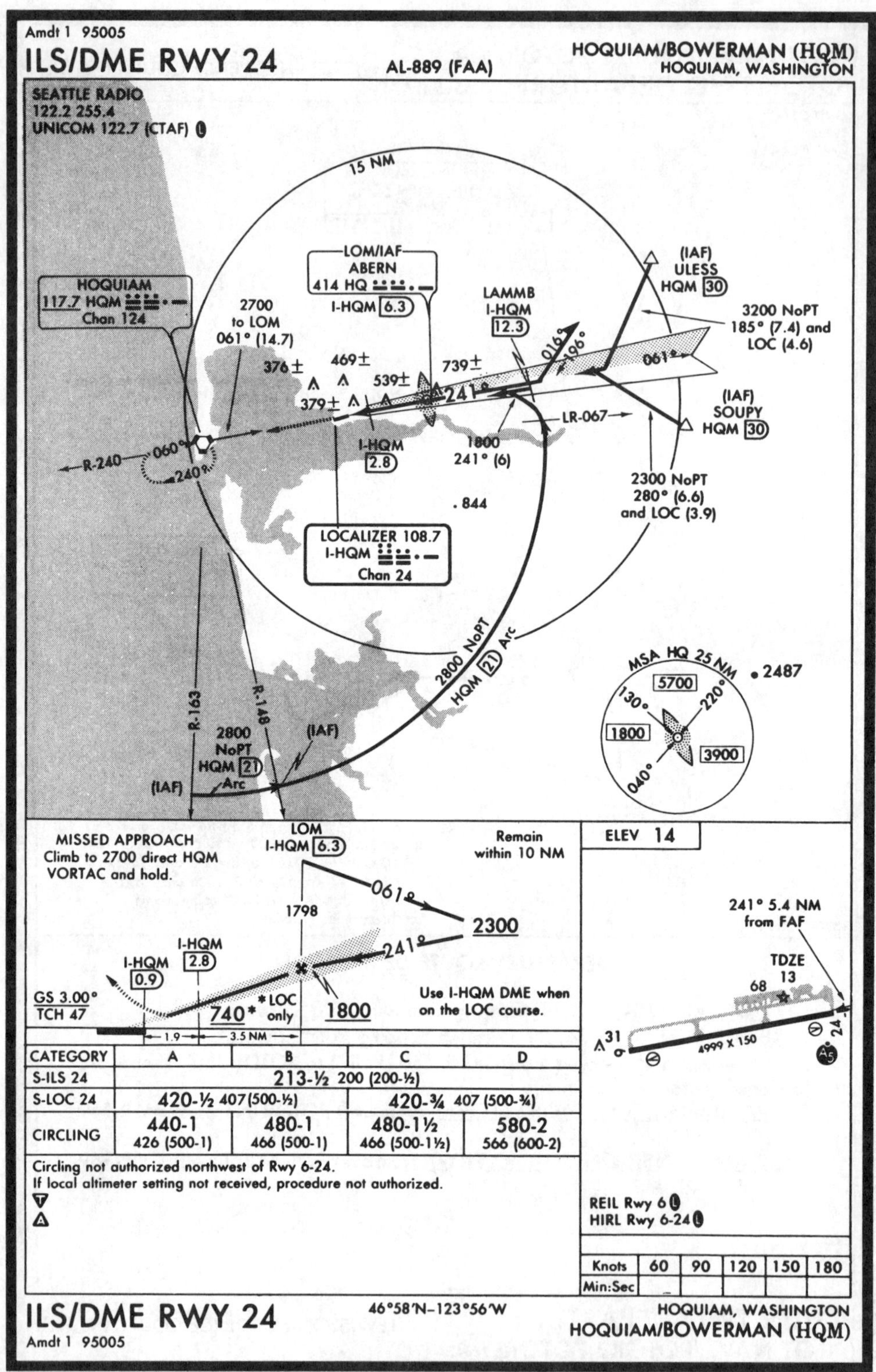

CATEGORY	A	B	C	D
S-ILS 24	213-½ 200 (200-½)			
S-LOC 24	420-½ 407 (500-½)		420-¾ 407 (500-¾)	
CIRCLING	440-1 426 (500-1)	480-1 466 (500-1)	480-1½ 466 (500-1½)	580-2 566 (600-2)

FIGURE 178.—ILS/DME RWY 24 (HQM).

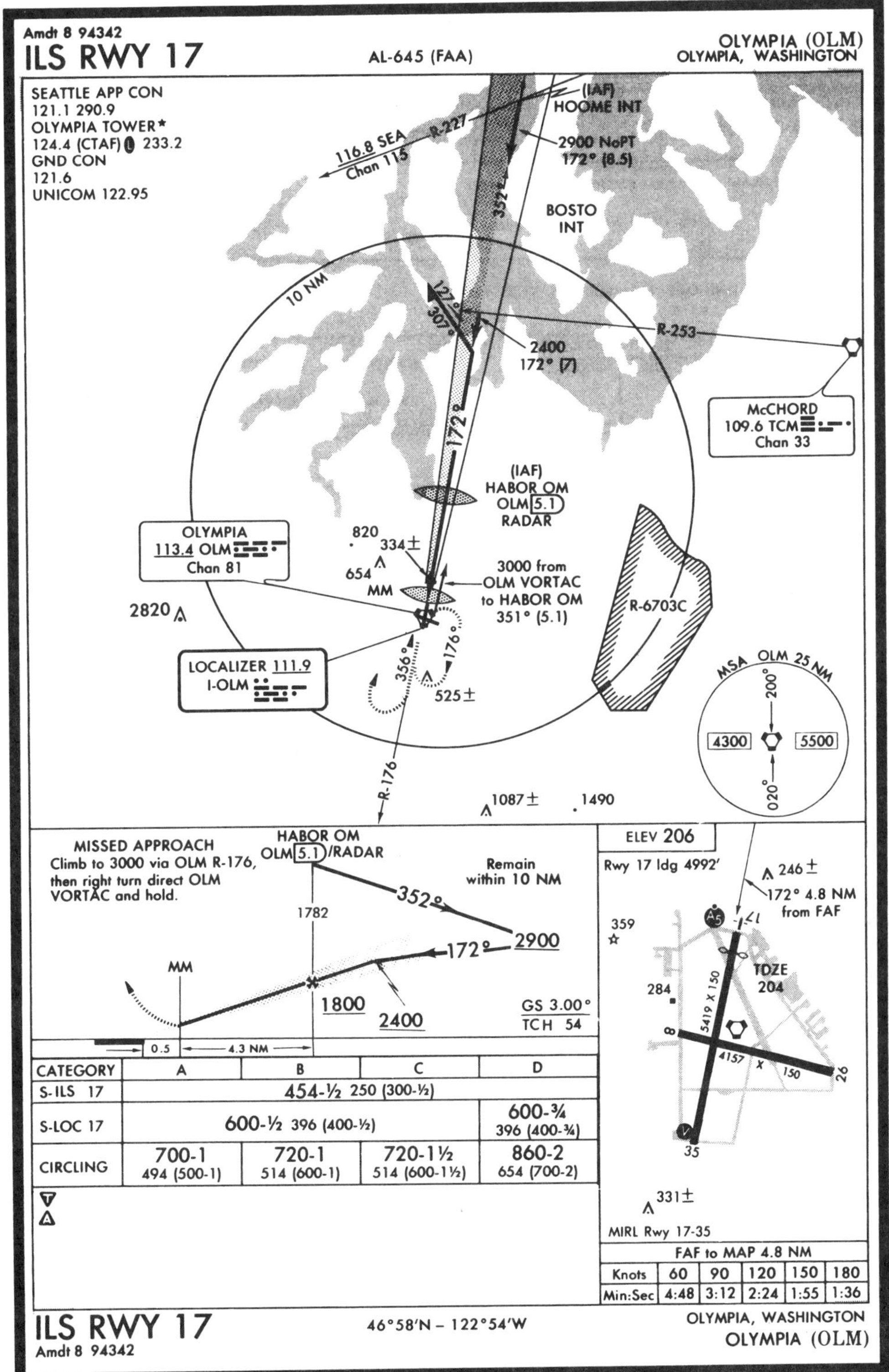

FIGURE 178A.—ILS RWY 17 (OLM).

Form Approved: OMB No. 2120-0034

U.S. DEPARTMENT OF TRANSPORTATION FEDERAL AVIATION ADMINISTRATION **FLIGHT PLAN**	(FAA USE ONLY) ☐ PILOT BRIEFING ☐ VNR ☐ STOPOVER	TIME STARTED	SPECIALIST INITIALS

1. TYPE	2. AIRCRAFT IDENTIFICATION	3. AIRCRAFT TYPE/ SPECIAL EQUIPMENT	4. TRUE AIRSPEED	5. DEPARTURE POINT	6. DEPARTURE TIME PROPOSED (Z)	ACTUAL (Z)	7. CRUISING ALTITUDE
VFR / X IFR / DVFR	EAB 90	BE1900/A	236 KTS	KPHF NEWPORT NEWS, VA			FL190

8. ROUTE OF FLIGHT
HENRY ONE ORF, J121 SIE, SIE. VCN5 PHL

9. DESTINATION (Name of airport and city)	10. EST. TIME ENROUTE HOURS	MINUTES	11. REMARKS
KPHL PHILADELPHIA INT'L PHILADELPHIA			VARIATION: PHF 7°W, PHL 10°W. TEC = TOWER ENROUTE CONTROL PPH = Pounds Per Hour

12. FUEL ON BOARD HOURS	MINUTES	13. ALTERNATE AIRPORT(S)	14. PILOT'S NAME, ADDRESS & TELEPHONE NUMBER & AIRCRAFT HOME BASE / 17. DESTINATION CONTACT/TELEPHONE (OPTIONAL)	15. NUMBER ABOARD
2	45	KACY ATLANTIC CITY INT'L		13

16. COLOR OF AIRCRAFT	
BLUE/RED	CIVIL AIRCRAFT PILOTS. FAR Part 91 requires you file an IFR flight plan to operate under instrument flight rules in controlled airspace. Failure to file could result in a civil penalty not to exceed $1,000 for each violation (Section 901 of the Federal Aviation Act of 1958, as amended). Filing of a VFR flight plan is recommended as a good operating practice. See also Part 99 for requirements concerning DVFR flight plans.

FAA Form 7233-1 (8-82) CLOSE VFR FLIGHT PLAN WITH ______________ FSS ON ARRIVAL

FLIGHT LOG

CHECK POINTS FROM	TO	ROUTE ALTITUDE	COURSE	WIND TEMP	SPEED-KTS TAS	GS	DIST NM	TIME LEG	TOT	FUEL LEG	TOT
PHF	ORF	VECTORS CLIMB					40		:19:00		312*
ORF	SAWED	J/21 FL190		300/70 ISA+5							
SAWED	SWL										
SWL	SIE										
SIE	VCN										
VCN	OOD										
OOD	PHL	DESCENT & APPROACH					30	:16:00		177	
PHL	ACY	TEC 3000					46	:18:00			

OTHER DATA: * Includes Taxi Fuel
NOTE: Use 689 PPH Total Fuel Flow From L/O To Start Of Descent.
Use 739 PPH Total Fuel Flow For Reserve And Alternate Requirements.
A Missed Approach Requires 95# of Fuel.

TIME and FUEL: As required by FARs.

TIME	FUEL (LB)	
		EN ROUTE
		RESERVE
		ALTERNATE
		TOTAL

FIGURE 179.—Flight Plan/Flight Log.

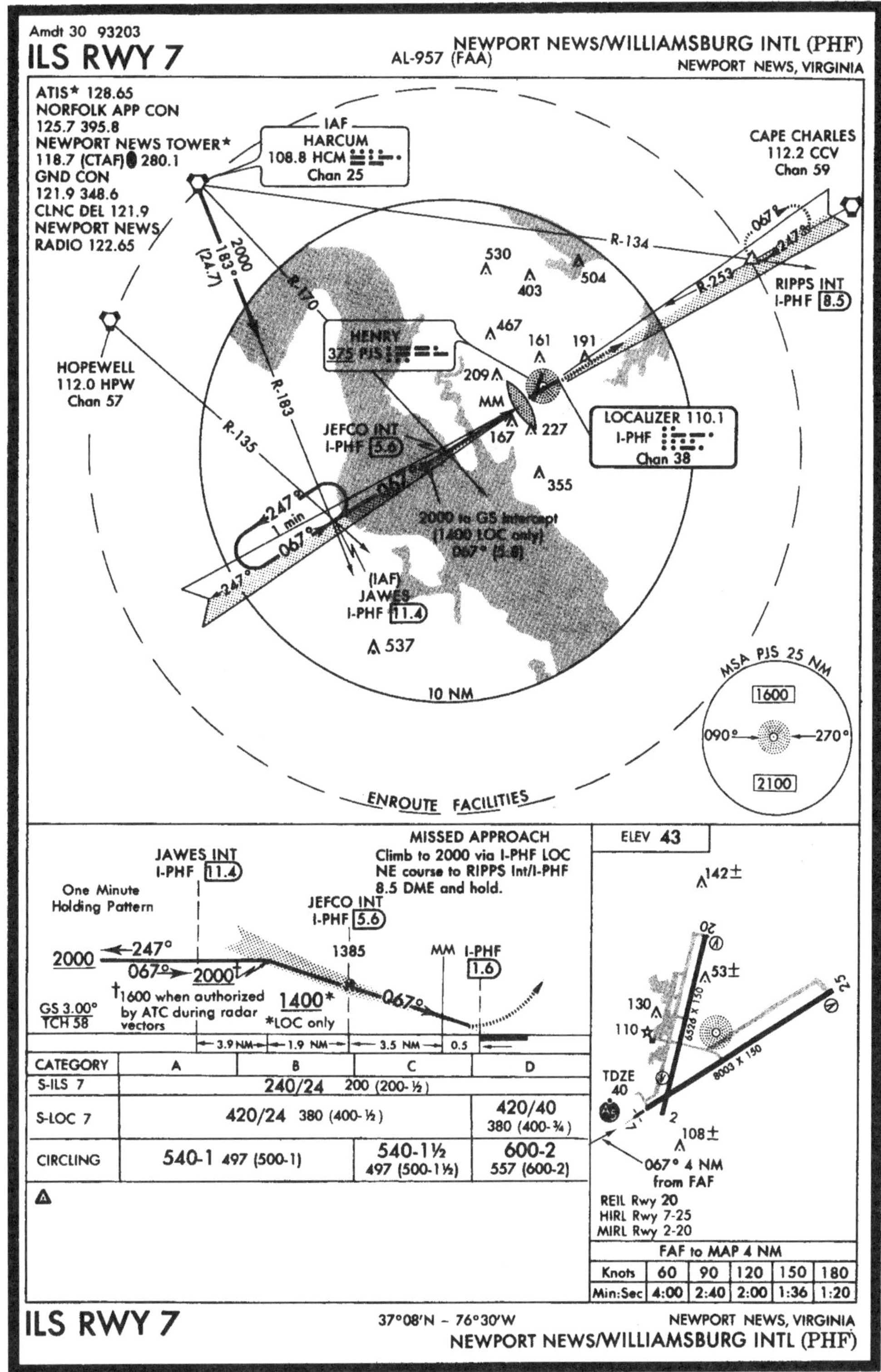

CATEGORY	A	B	C	D
S-ILS 7	240/24 200 (200-½)			
S-LOC 7	420/24 380 (400-½)			420/40 380 (400-¾)
CIRCLING	540-1 497 (500-1)		540-1½ 497 (500-1½)	600-2 557 (600-2)

FAF to MAP 4 NM					
Knots	60	90	120	150	180
Min:Sec	4:00	2:40	2:00	1:36	1:20

FIGURE 180.—ILS RWY 7 (PHF).

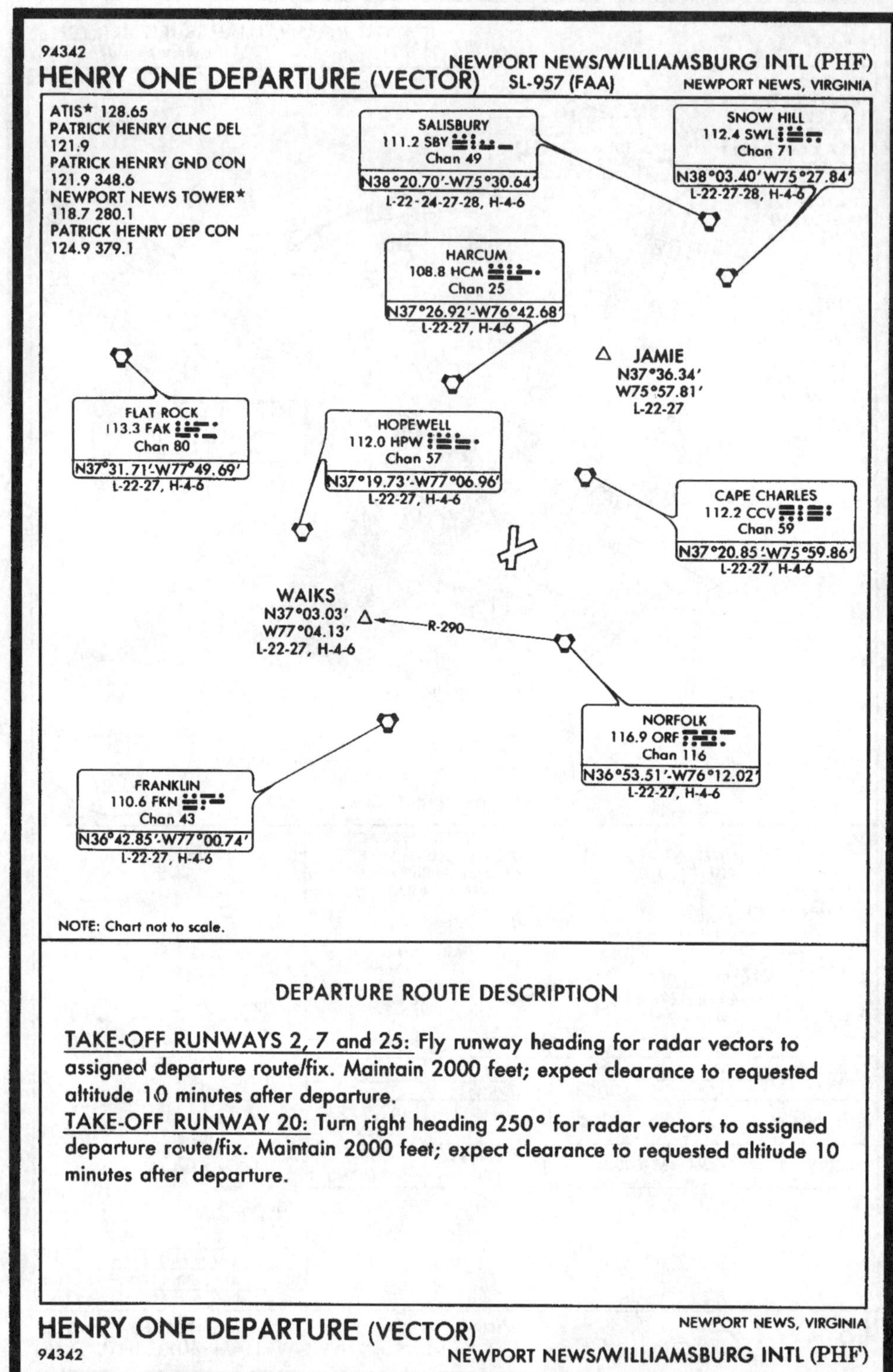

FIGURE 180A.—HENRY ONE DEPARTURE (VECTOR) (PHF).

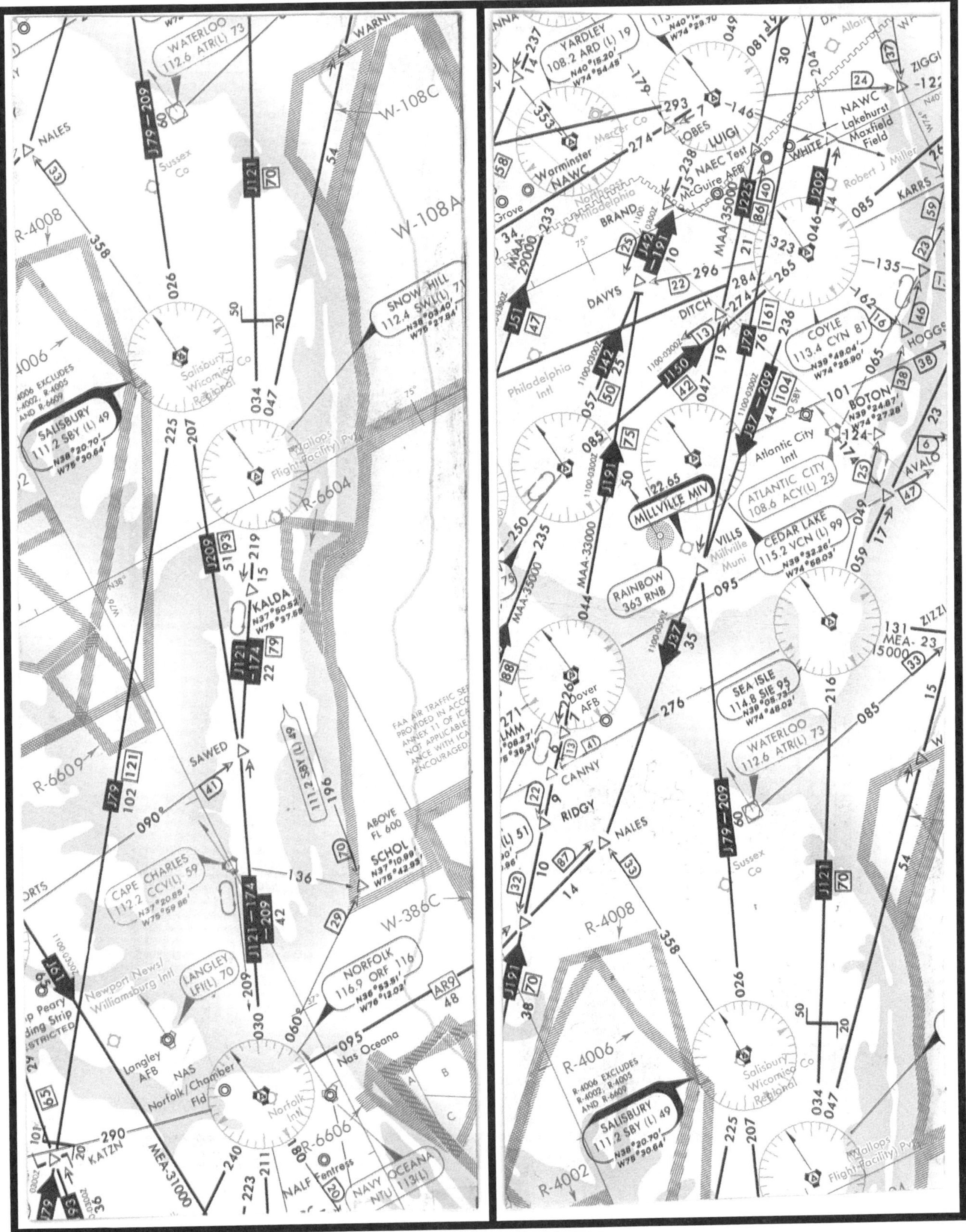

FIGURE 181.—High Altitude Airways.

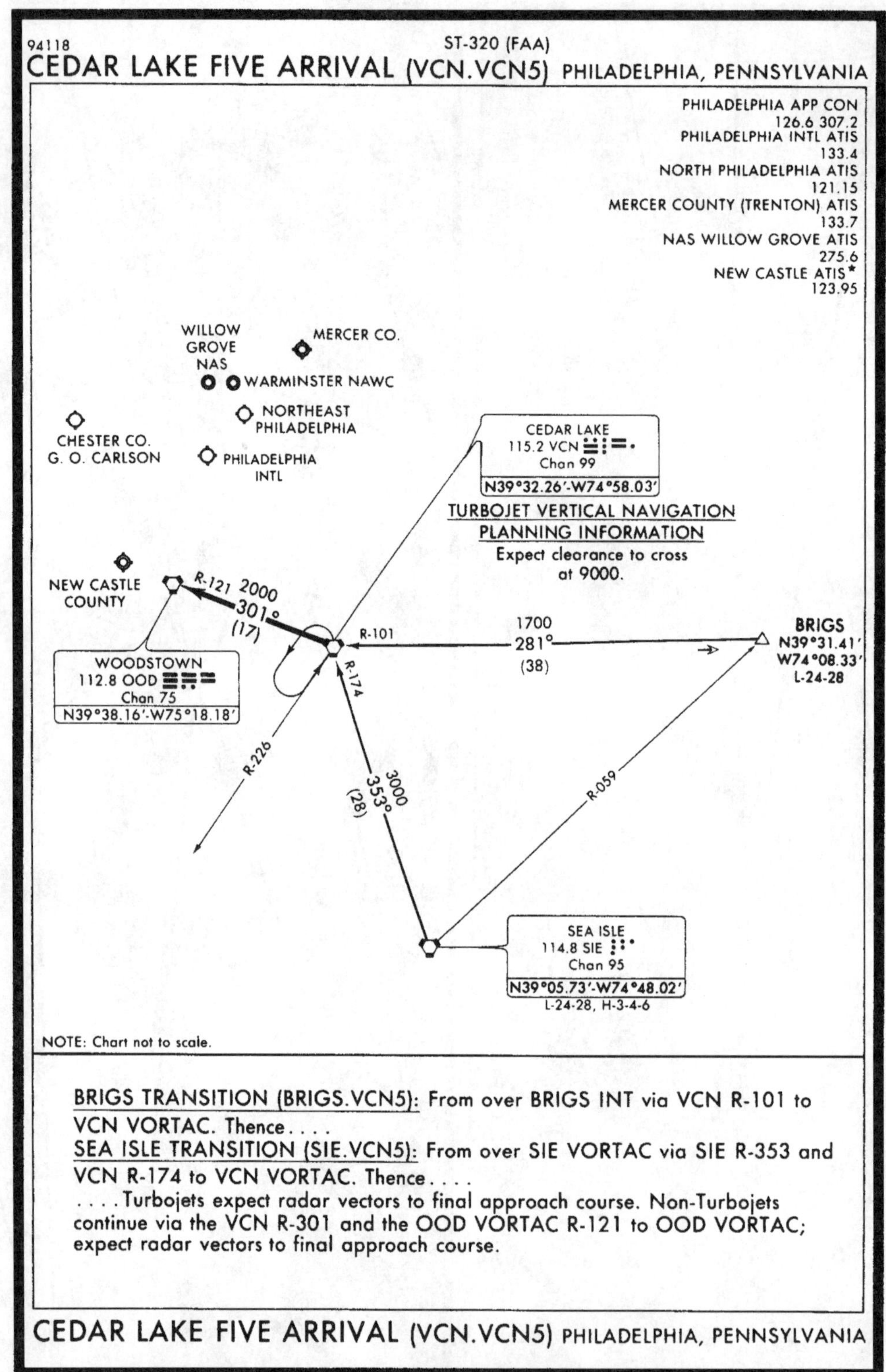

FIGURE 182.—CEDAR LAKE FIVE ARRIVAL (VCN.VCN5).

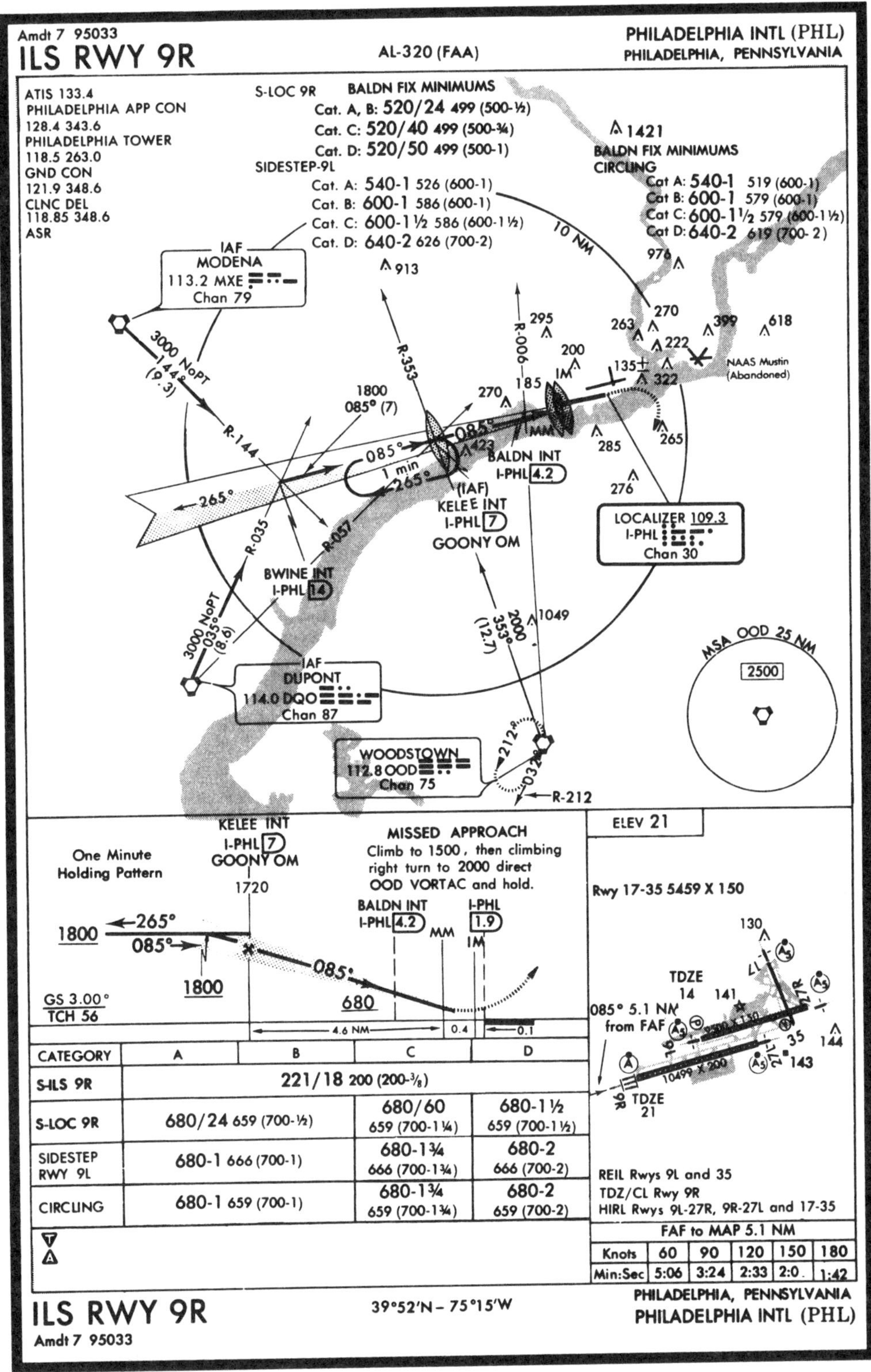

CATEGORY	A	B	C	D
S-ILS 9R	221/18 200 (200-³⁄₈)			
S-LOC 9R	680/24 659 (700-½)		680/60 659 (700-1¼)	680-1½ 659 (700-1½)
SIDESTEP RWY 9L	680-1 666 (700-1)		680-1¾ 666 (700-1¾)	680-2 666 (700-2)
CIRCLING	680-1 659 (700-1)		680-1¾ 659 (700-1¾)	680-2 659 (700-2)

FAF to MAP 5.1 NM

Knots	60	90	120	150	180
Min:Sec	5:06	3:24	2:33	2:0	1:42

FIGURE 182A.—ILS RWY 9R (PHL).

A ALTERNATE MINS 
94342

INSTRUMENT APPROACH PROCEDURE CHARTS

 IFR ALTERNATE MINIMUMS

(NOT APPLICABLE TO USA/USN/USAF)

Standard alternate minimums for non precision approaches are 800-2 (NDB, VOR, LOC, TACAN, LDA, VORTAC, VOR/DME or ASR); for precision approaches 600-2 (ILS or PAR). Airports within this geographical area that require alternate minimums other than standard or alternate minimums with restrictions are listed below. NA - means alternate minimums are not authorized due to unmonitored facility or absence of weather reporting service. Civil pilots see FAR 91. USA/USN/USAF pilots refer to appropriate regulations.

NAME — ALTERNATE MINIMUMS

ATLANTIC CITY, NJ

ATLANTIC CITY INTL .... **ILS Rwy 13**[1]
RADAR-1[2]
VOR/DME or GPS Rwy 22[3]
VOR or GPS Rwy 4[3]
VOR or GPS Rwy 13[3]
VOR or GPS Rwy 31[3]

[1]ILS, Category D, 700-2; Category E, 700-2½. LOC, Category E, 800-2½.
[2]Category D, 700-2; Category E, 800-2½.
[3]Category E, 800-2½.

BALTIMORE, MD

BALTIMORE-WASHINGTON INTL .... **VOR or GPS Rwy 10,1000-3**

MARTIN STATE .... **ILS Rwy 33**[1]
VOR/DME or TACAN 1 Rwy 15[2]

[1]ILS, Category D, 700-2.
[2]Categories A,B, 900-2; Categories C,D, 900-2¾.

BECKLEY, WV

RALEIGH COUNTY MEMORIAL ..... **ILS Rwy 19**[1]
VOR or GPS Rwy 19[2]

[1]ILS, Categories B,C,D, 700-2. LOC, NA.
[2]Category D, 800-2¼.

BLUEFIELD, WV

MERCER COUNTY .... **ILS Rwy 23**[1]
VOR/DME or GPS Rwy 23
VOR Rwy 23

NA when FSS is closed.
[1]ILS, Categories C,D, 700-2.

CHARLESTON, WV

YEAGER .... **ILS Rwy 5, 700-2**
ILS Rwy 23, 700-2
VOR/DME RNAV or GPS Rwy 33[1]
VOR or GPS-A[1]

[1]Category D, 800-2¼.

NAME — ALTERNATE MINIMUMS

CHARLOTTESVILLE, VA

CHARLOTTESVILLE-ALBEMARLE .. **ILS Rwy 3**[1]
NDB Rwy 3[2]

NA when control tower closed.
[1]ILS, Category D, 900-2¾. LOC, NA.
[2]Category D, 900-2¾.

CLARKSBURG, WV

BENEDUM .... **ILS Rwy 21**[1]
VOR or GPS Rwy 3[2]

NA when control tower is closed, except for operators with approved weather reporting service.
[1]Categories A,B, 800-2; Category C, 900-2½; Category D, 900-2¾.
[2]Category C, 900-2½; Category D, 900-2¾.

DANVILLE, VA

DANVILLE REGIONAL .... **RNAV Rwy 20**

NA when control zone not in effective.

ELKINS, WV

ELKINS-RANDOLPH COUNTY JENNINGS-RANDOLPH FIELD .... **LDA-C**[1]
VOR/DME-B[2]

NA at night.
[1]Categories A,B, 1200-2; Categories C,D, 1500-3.
[2]Categories A,B, 1500-2; Categories C,D, 1500-3.

HAGERSTOWN, MD

WASHINGTON COUNTY REGIONAL..................................... **ILS Rwy 27**[1]
VOR or GPS Rwy 9[2]

[1]NA when control zone not in effect.
[2]NA when control zone not in effect except for operators with approved weather reporting service.

NE-3

 ALTERNATE MINS A
94342

FIGURE 183.—IFR ALTERNATE MINIMUMS.

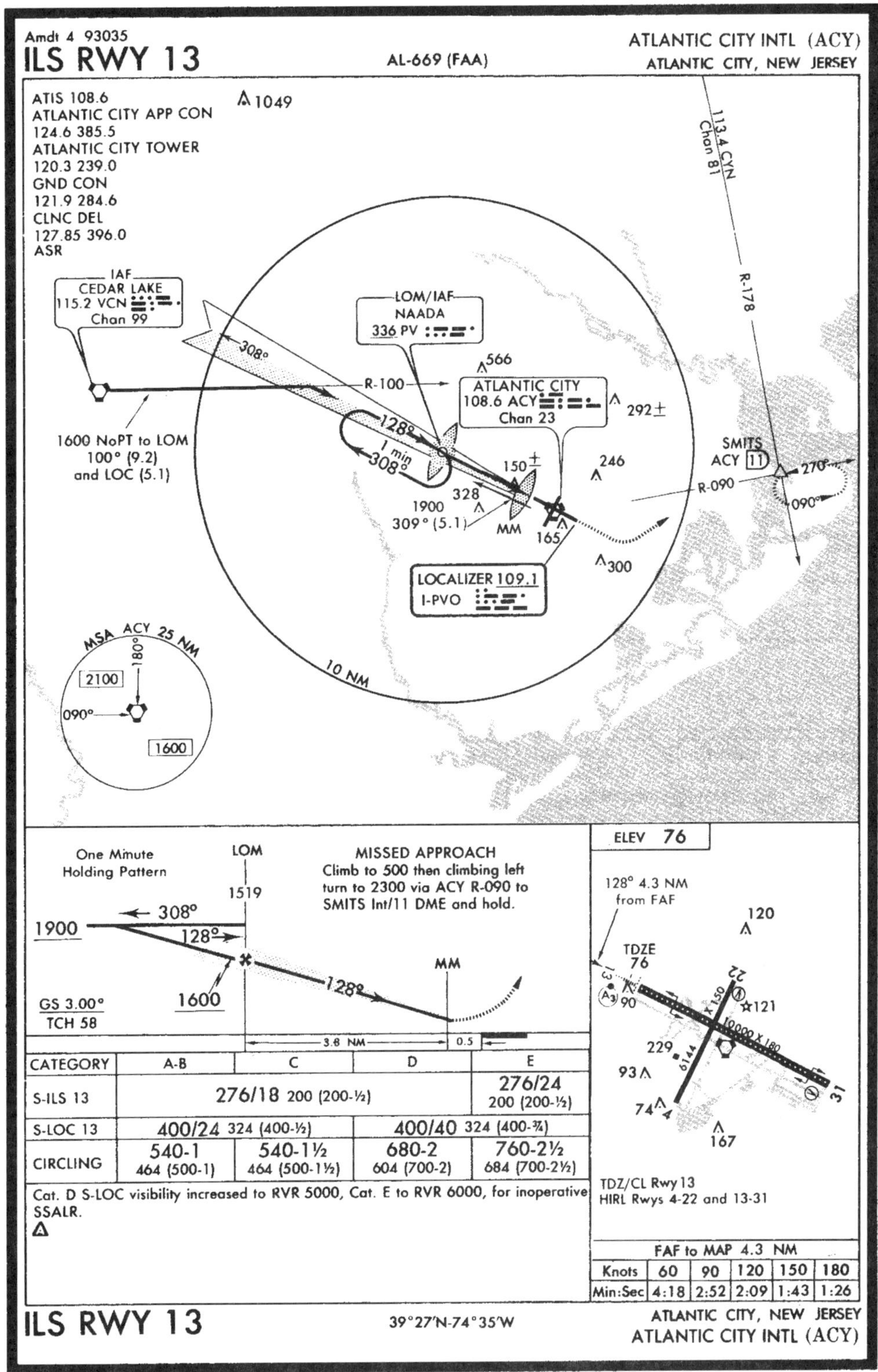

FIGURE 183A.—ILS RWY 13 (ACY).

Form Approved: OMB No. 2120-0034

U.S. DEPARTMENT OF TRANSPORTATION FEDERAL AVIATION ADMINISTRATION **FLIGHT PLAN**	(FAA USE ONLY) ☐ PILOT BRIEFING ☐ VNR ☐ STOPOVER	TIME STARTED	SPECIALIST INITIALS

1. TYPE	2. AIRCRAFT IDENTIFICATION	3. AIRCRAFT TYPE/ SPECIAL EQUIPMENT	4. TRUE AIRSPEED	5. DEPARTURE POINT	6. DEPARTURE TIME PROPOSED (Z)	ACTUAL (Z)	7. CRUISING ALTITUDE
VFR / X IFR / DVFR	HOSS 1	A109K2/A	** KTS	LAS			15000

8. ROUTE OF FLIGHT LAS, ACLAM, V8 MMM, V21 REEKA, PVU

9. DESTINATION (Name of airport and city)	10. EST. TIME ENROUTE HOURS	MINUTES	11. REMARKS
PVU PROVO MUNI PROVO, UTAH			L/O = LEVEL OFF PPH = POUNDS PER HOUR **CAS 135, ISA -10

12. FUEL ON BOARD HOURS	MINUTES	13. ALTERNATE AIRPORT(S)	14. PILOT'S NAME, ADDRESS & TELEPHONE NUMBER & AIRCRAFT HOME BASE / 17. DESTINATION CONTACT/TELEPHONE (OPTIONAL)	15. NUMBER ABOARD
		SLC SALT LAKE CITY		8

16. COLOR OF AIRCRAFT	
GREEN/GOLD	CIVIL AIRCRAFT PILOTS. FAR Part 91 requires you file an IFR flight plan to operate under instrument flight rules in controlled airspace. Failure to file could result in a civil penalty not to exceed $1,000 for each violation (Section 901 of the Federal Aviation Act of 1958, as amended). Filing of a VFR flight plan is recommended as a good operating practice. See also Part 99 for requirements concerning DVFR flight plans.

FAA Form 7233-1 (8-82) CLOSE VFR FLIGHT PLAN WITH ______________ FSS ON ARRIVAL

FLIGHT LOG

CHECK POINTS FROM	TO	ROUTE ALTITUDE	COURSE	WIND TEMP	SPEED-KTS TAS	GS	DIST NM	TIME LEG	TOT	FUEL LEG	TOT
LAS	ACLAM	DIRECT CLIMB					31		:15:00		152*
ACLAM	MMM	V-8 15000		210/71 ISA-10							
MMM	REEKA	V-21 15000									
REEKA	PVU	DESCENT & APPROACH					19	:10:00		87	
PVU	SLC	DIRECT 7000					41	:17:20			

OTHER DATA: NOTE:
* Includes Taxi Fuel
Use 496 PPH Total Fuel Flow From L/O To Start Of Descent.
Use 480 PPH Total Fuel Flow For Reserve And Alternate Requirements.

A Missed Approach Requires 40# of Fuel.

TIME and FUEL: As required by FARs.

TIME	FUEL (LB)	
		EN ROUTE
		RESERVE
		ALTERNATE
		TOTAL

FIGURE 184.—Flight Plan/Flight Log.

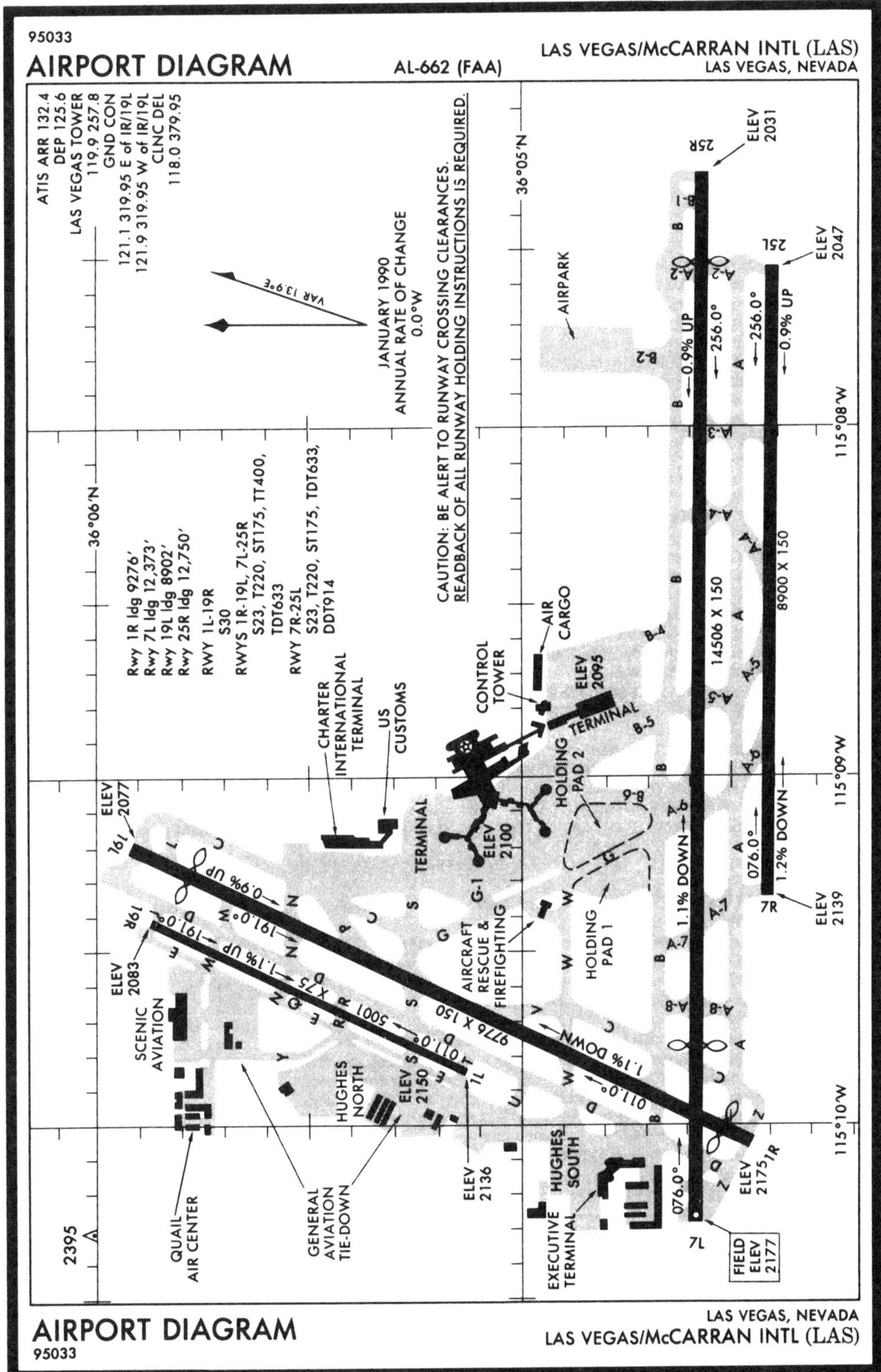

FIGURE 185.—AIRPORT DIAGRAM.

McCARRAN INTL (LAS) 5 S UTC–8(–7DT) N36°04.82′ W115°09.02′ **LAS VEGAS**
2177 B S4 **FUEL** 100, 100LL, JET A1 + OX 1, 2, 3 LRA ARFF Index D **H-2B, L-5B**
RWY 07L-25R: H14506X150 (ASPH-PFC) S-23, D-220, DT-633 HIRL 1.0% up W **IAP**
RWY 07L: VASI(V6L)—Upper GA 3.25° TCH 94′. Lower GA 3.0° TCH 47′. Thld dsplcd 2133′. Hangar.
RWY 25R: MALSR. Thld dsplcd 1400′.
RWY 01R-19L: H9776X150 (ASPH-PFC) S-23, D-220, DT-633 MIRL 1.0% up S
RWY 01R: VASI(V4L)—GA 3.0° TCH 50′. Thld dsplcd 500′. Railroad. Rgt tfc.
RWY 19L: VASI(V6L)—Upper GA 3.25° TCH 66′. Lower GA 3.0° TCH 35′. Thld dsplcd 874′. Pole.
RWY 07R-25L: H8900X150 (ASPH-PFC) S-23, D-220, DT-633, DDT-914 HIRL
RWY 07R: REIL. Pole. **RWY 25L:** MALSF.
RWY 01L-19R: H5001X75 (ASPH) S-30 MIRL 1.1% up S
RWY 01L: REIL. VASI(V4L)—GA 3.0° TCH 35′. Antenna.
RWY 19R: REIL. VASI(V4L)—GA 3.0° TCH 60′. Pole. Rgt tfc.
AIRPORT REMARKS: Attended continuously. Rwy 19R CLOSED arrival, Rwy 01L CLOSED departure Mon–Fri 1500–2300Z‡. Extensive glider/soaring operations weekends and holidays. Sunrise to sunset, LAS 187020, altitudes up to but not including FL180. Gliders remain clear of the CLASS B airspace but otherwise operate within the entire SW quadrant of the CLASS B airspace Veil. Lgtd crane 950′ AGL 4 miles N of arpt. Rotating bcn not visible 115°–240° NE to SW from McCarran Twr. Acft may experience reflection of sun from glass pyramid located NW of arpt. Reflection may occur at various altitudes, headings and distances from arpt. Rwy 07R-25L DDT GWT 521,000 lbs for L-1011, 620,000 lbs for DC-10, 633,000 lbs for MD-11. PAEW between Rwy 01R-19L and Twy D north of Twy N. PAEW west of Rwy 01L-19R. PAEW west of Twy D. Twy E clsd between Twy Q and Twy R. Twy N, Twy S and Twy T clsd between Rwy 01L-19R and Twy D. Twy Y acft be alert keep nosewheel on centerline and acft with wing span greater than 70′ prohibited north of New Quail Gate. Twy D clsd to B747 and clsd to all acft with wingspan 171′ or greater north of Rwy 07L-25R. All non-standard rwy operations PPR from Department of Aviation. Turbojet operations not permitted Rwy 01R-19L and Rwy 01L-19R between 0400–1600Z‡. Exceptions will be made due to weather. Rwy 07L VASI out of svc indefinitely. Rwy 25 MALSR out of svc indefinitely. Tiedown fee. Flight Notification Service (ADCUS) available. NOTE: See Land and Hold Short Operations Section.
WEATHER DATA SOURCES: LLWAS.
COMMUNICATIONS: ATIS 132.4 (ARR) 125.6 (DEP) **UNICOM** 122.95
RENO FSS (RNO) TF 1-800-WX-BRIEF. NOTAM FILE LAS.
Ⓡ **LAS VEGAS APP CON** 127.15
Ⓡ **LAS VEGAS DEP CON** 133.95 (North) 125.9 (South)
LAS VEGAS TOWER 119.9 **GND CON** 121.9 (West of Rwy 01R-19L) 121.1 (East of Rwy 01R-19L) **CLNC DEL** 118.0
AIRSPACE: CLASS B See VFR Terminal Area Chart.
RADIO AIDS TO NAVIGATION: NOTAM FILE LAS.
LAS VEGAS (H) VORTACW 116.9 LAS Chan 116 N36°04.78′ W115°09.59′ at fld. 2140/15E.
ILS 110.3 I-LAS Rwy 25R.
ILS 111.75 I-RLE Rwy 25L. Loc unusable byd 19° South of course.

NORTH LAS VEGAS AIR TERMINAL (VGT) 3 NW UTC–8(–7DT) **LAS VEGAS**
N36°12.75′ W115°11.82′ **H-2B, L-5B**
2207 B S4 **FUEL** 100LL, JET A OX 2 TPA—3007(800)
RWY 07-25: H5005X75 (ASPH) S-30 MIRL
RWY 07: PAPI(P4L)—GA 3.0° TCH 37′. Pole. **RWY 25:** PAPI(P4L)—GA 3.0° TCH 36′.
RWY 12-30: H5000X75 (ASPH) S-30 MIRL
RWY 12: PAPI(P4L)—GA 3.0° TCH 25′.
RWY 30: MIRL. PAPI(P4L)—GA 3.0° TCH 45′. Thld dsplcd 290′. P-line.
AIRPORT REMARKS: Attended 1400–0630Z‡. Rwy 30 PAPI OTS indef. When twr clsd ACTIVATE MIRL Rwy 07-25 and Rwy 12-30—CTAF. NOTE: See Land and Hold Short Operations Section.
COMMUNICATIONS: CTAF 125.7 **ATIS** 118.05 (1400–0400Z‡) **UNICOM** 122.95
RENO FSS (RNO) TF 1-800-WX-BRIEF. NOTAM FILE RNO.
TOWER 125.7 (1400–0400Z‡) **GND CON** 121.7
AIRSPACE: CLASS D svc effective 1400–0400Z‡ other times CLASS G.
RADIO AIDS TO NAVIGATION: NOTAM FILE LAS.
LAS VEGAS (H) VORTACW 116.9 LAS Chan 116 N36°04.78′ W115°09.59′ 332° 8.2 NM to fld. 2140/15E.

LIDA JUNCTION (See GOLDFIELD)

LINCOLN CO (See PANACA)

FIGURE 185A.—Excerpt from Airport/Facilities Directory.

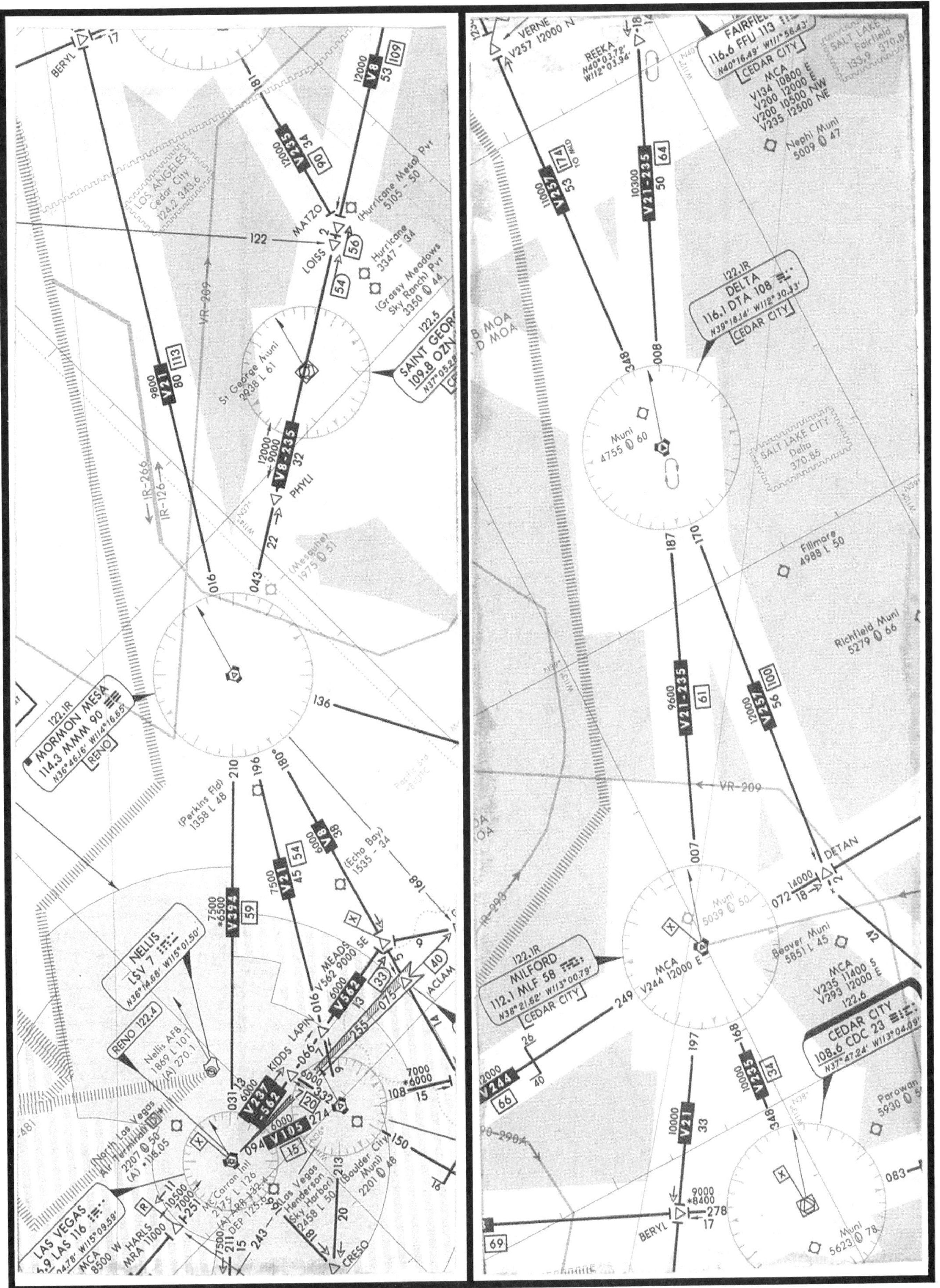

FIGURE 186.—Low Altitude Airways.

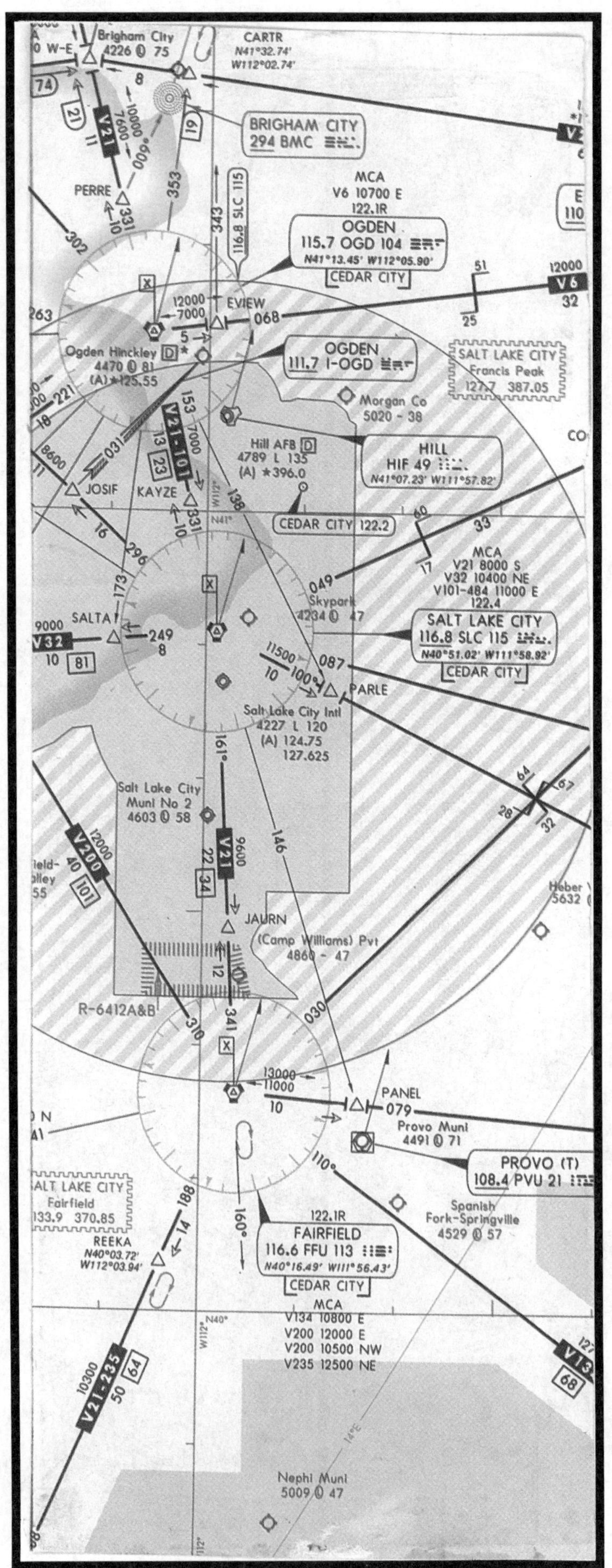

FIGURE 187.—Low Altitude Airways.

PROVO MUNI (PVU) 2 SW UTC–7(–6DT) N40°12.94′ W111°43.29′ SALT LAKE CITY
4491 B S4 FUEL 100, JET A OX 2 H-2C, L-8E, 7D, 5C
RWY 13-31: H7091X150 (ASPH-PFC) S-65, D-85, DT-140 MIRL IAP
RWY 13: MALSF. VASI(V2L)—GA 3.0°TCH 40′. Thld dsplcd 438′. Road.
RWY 31: VASI(V2L)—GA 3.0°TCH 56′. Thld dsplcd 290′. Road.
RWY 18-36: H6937X150 (ASPH) S-50, D-70, DT-110 MIRL
RWY 18: VASI(V2L)—GA 3.0°TCH 55′. Thld dsplcd 373′. Road.
RWY 36: VASI(V2L)—GA 3.0°TCH 55′. Thld dsplcd 334′. Road.
RWY 06-24: H5596X150 (ASPH) S-50, D-70, DT-110
RWY 06: Thld dsplcd 611′. Brush. RWY 24: Thld dsplcd 598′. Road.
AIRPORT REMARKS: Attended 1400–0200Z‡. ACTIVATE ALS Rwy 13, MIRL and VASI Rwys 13-31 and 18-36 122.8.
WEATHER DATA SOURCES: AWOS-3 135.175 (801) 373-9782.
COMMUNICATIONS: CTAF/UNICOM 122.8
CEDAR CITY FSS (CDC) TF 1-800-WX-BRIEF. NOTAM FILE PVU.
Ⓡ SALT LAKE CITY APP CON 124.3
Ⓡ SALT LAKE CITY DEP CON 118.85
AIRSPACE: CLASS E svc effective 1400–0200Z‡ other times CLASS G.
RADIO AIDS TO NAVIGATION: NOTAM FILE PVU.
(T) VORW/DME 108.4 PVU Chan 21 N40°12.90′ W111°43.28′ at fld. 4490/15E.
Unusable 330°–170°beyond 10 NM below 13,000′
ILS/DME 110.3 I-PVU Chan 40 Rwy 13. LOC unusable inside threshold. ILS unmonitored 0200–1400Z‡.
• •
HELIPAD H1: H40X40 (CONC)
HELIPAD H2: H40X40 (CONC)

RICHFIELD MUNI (RIF) 1 SW UTC–7(–6DT) N38°44.50′ W112°05.71′ LAS VEGAS
5279 B FUEL 100, JET A H-2C, L-5C
RWY 01-19: H6645X75 (ASPH) S-19 MIRL
RWY 01: Rgt tfc.
AIRPORT REMARKS: Attended Mon–Fri 1530–0000Z‡. For fuel after hours call 801-896-8918/7258. ACTIVATE MIRL Rwy 01-19—CTAF.
COMMUNICATIONS: CTAF/UNICOM 122.8
CEDAR CITY FSS (CDC) TF 1-800-WX-BRIEF. NOTAM FILE CDC.
RCO 122.5 (CEDAR CITY FSS)
RADIO AIDS TO NAVIGATION: NOTAM FILE CDC.
DELTA (H) VORTAC 116.1 DTA Chan 108 N39°18.14′ W112°30.33′ 134° 38.7 NM to fld. 4600/16E.

ROOSEVELT MUNI (74V) 3 SW UTC–7(–6DT) N40°16.70′ W110°03.08′ SALT LAKE CITY
5172 B FUEL 100, JET A, MOGAS H-2C, L-8E, 5C
RWY 07-25: H6500X75 (ASPH) S-12 MIRL 1.0% up W IAP
RWY 07: VASI(V2L)—GA 3.0° TCH 34′. RWY 25: VASI(V2L)—GA 3.0° TCH 27′.
AIRPORT REMARKS: Attended on call. For svc call 801-722-4741. ACTIVATE MIRL and VASI Rwy 07-25—CTAF.
COMMUNICATIONS: CTAF/UNICOM 122.8
CEDAR CITY FSS (CDC) TF 1-800-WX-BRIEF. NOTAM FILE CDC.
MYTON RCO 122.1R 112.7T (CEDAR CITY FSS)
RADIO AIDS TO NAVIGATION: NOTAM FILE CDC.
MYTON (H) VORTAC 112.7 MTU Chan 74 N40°08.70′ W110°07.66′ 010° 8.7 NM to fld. 5332/14E.

ST GEORGE MUNI (SGU) 1 W UTC–7(–6DT) N37°05.48′ W113°35.58′ LAS VEGAS
2938 B S4 FUEL 100, 100LL, JET A, MOGAS OX 2 ARFF Index Ltd. H-2B, L-5B
RWY 16-34: H6101X100 (ASPH-PFC) S-26 MIRL 1.1% up N IAP
RWY 16: VASI(V2R)—GA 4.0° TCH 44′. Road. RWY 34: REIL. VASI(V2L)—GA 3.0° TCH 43′.
AIRPORT REMARKS: Attended 1330–0230Z‡. CLOSED to Air Carrier ops with more than 30 passenger seat except PPR. Call arpt manager 801-634-5800. ACTIVATE REIL Rwy 34—CTAF.
WEATHER DATA SOURCES: AWOS-3 135.075 (801) 634-0940.
COMMUNICATIONS: CTAF/UNICOM 122.8
CEDAR CITY FSS (CDC) TF 1-800-WX-BRIEF. NOTAM FILE SGU.
RCO 122.5 (CEDAR CITY FSS)
RADIO AIDS TO NAVIGATION: NOTAM FILE CDC.
(T) VORW/DME 109.8 OZN Chan 35 N37°05.28′ W113°35.51′ at fld. 2898/15E.
VOR/DME unusable:
210°–235° beyond 15 NM below 8500′
235°–270° beyond 15 NM below 9700′
270°–350° all altitudes and distances;
350°–020° beyond 10 NM below 14000′.

FIGURE 188.—Excerpt from Airport/Facilities Directory.

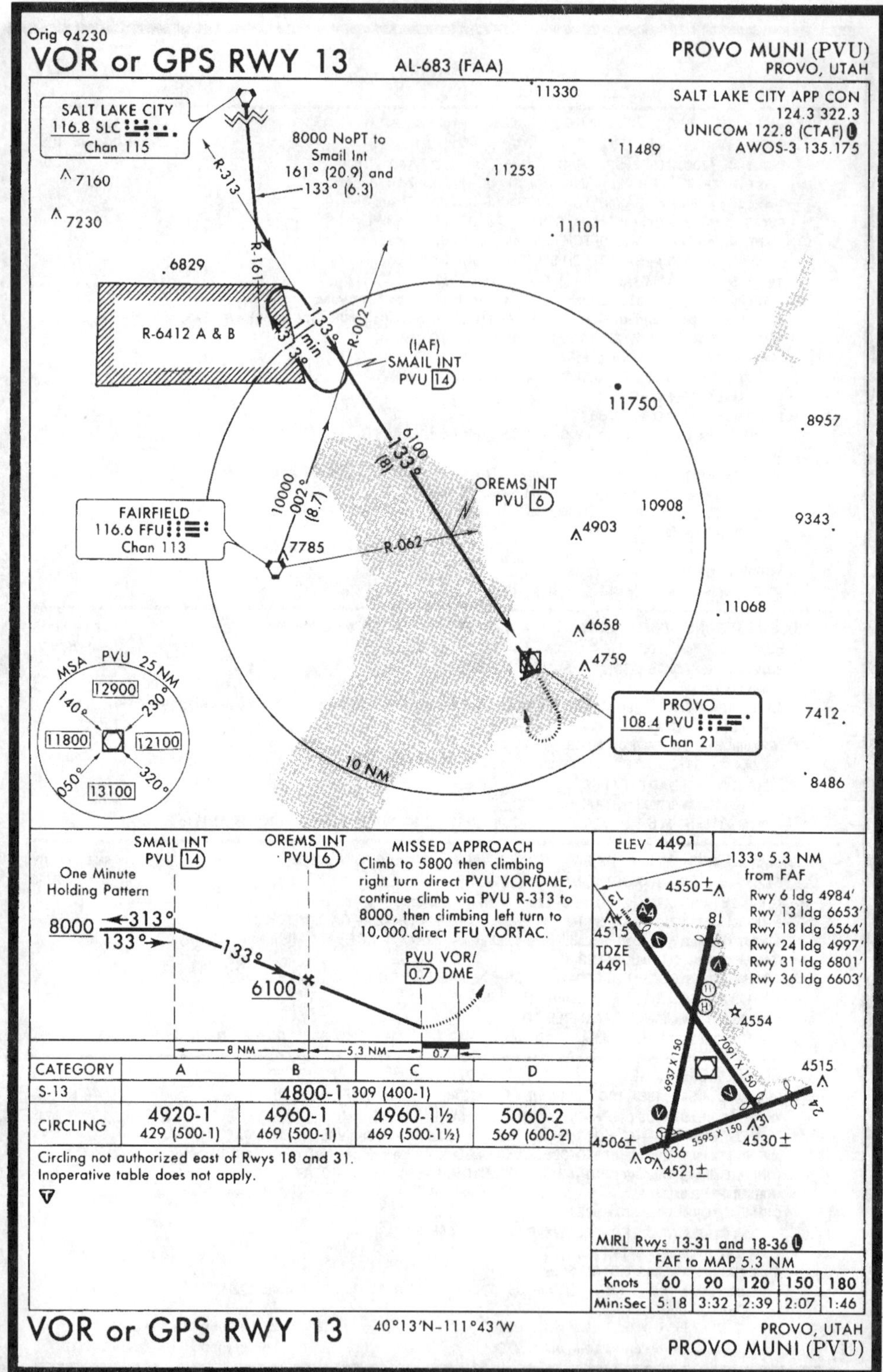

CATEGORY	A	B	C	D
S-13	4800-1 309 (400-1)			
CIRCLING	4920-1 429 (500-1)	4960-1 469 (500-1)	4960-1½ 469 (500-1½)	5060-2 569 (600-2)

Knots	60	90	120	150	180
Min:Sec	5:18	3:32	2:39	2:07	1:46

FIGURE 188A.—VOR or GPS RWY 13 (PVU).

SALINA-GUNNISON (44U) 5 NE UTC–7(–6DT) N39°01.75′ W111°50.30′ **LAS VEGAS**
5159 B **L-5C**
RWY 02-20: H3815X60 (ASPH) S-6 MIRL
AIRPORT REMARKS: Unattended. ACTIVATE MIRL Rwy 02-20—CTAF.
COMMUNICATIONS: CTAF 122.9
CEDAR CITY FSS (CDC) TF 1-800-WX-BRIEF. NOTAM FILE CDC.
RADIO AIDS TO NAVIGATION: NOTAM FILE CDC.
DELTA (H) VORTAC 116.1 DTA Chan 108 N39°18.14′ W112°30.33′ 102° 35.2 NM to fld. 4600/16E.

SALT LAKE CITY

SALT LAKE CITY INTL (SLC) 3 W UTC–7(–6DT) N40°47.21′ W111°58.13′ **SALT LAKE CITY**
4227 B S4 **FUEL** 100, 100LL, JET A1 OX 1, 2, 3, 4 LRA ARFF Index D **H-1C, L-7D**
RWY 16-34: H12003X150 (ASPH-PFC) S-60+, D-200+, DT-350 HIRL CL **IAP**
RWY 16: ALSF2. TDZ. REIL. PAPI(P4L). **RWY 34:** ALSF2. TDZ. REIL. PAPI(P4L).
RWY 17-35: H9596X150 (ASPH-PFC) S-60 +, D-170, DT-320 HIRL
RWY 17: MALSR. PAPI(P4R)—GA 3.0° TCH 55′. **RWY 35:** PAPI(P4L)—GA 3.0° TCH 75′.
RWY 14-32: H4758X150 (ASPH-PFC) S-60 +, D-170, DT-320 MIRL
RWY 14: Thld dsplcd 202′. **RWY 32:** Thld dsplcd 479′. Road.
AIRPORT REMARKS: Attended continuously. CAUTION: Flocks of birds on and in vicinity of arpt. Preferential rwys, use Rwy 34 and Rwy 35 when wind and temperature permit. Rwy 14-32 GWT strengths for S, D, DT apply to center 75′ only. 180° turns by acft over 12,500 lbs, prohibited on all rwys and taxiways. Rwy 17-35 tfc on Twy K not visible from twr. Due to the high volume of tfc at SLC arpt during the following time periods: 1800-1845Z‡, 2200-2230Z‡, 0145-0230Z‡ and 0300-0330Z‡ local departures and arrivals are discouraged. Greater than normal delays can be expected during these time periods. Rwy 16R-34L under construction. Flight Notification Service (ADCUS) available. NOTE: See Land and Hold Short Operations Section. NOTE: See Special Notice—Runway Under Construction.
WEATHER DATA SOURCES: LLWAS.
COMMUNICATIONS: ATIS 127.625 124.75 (801) 539-2581 **UNICOM** 122.95
CEDAR CITY FSS (CDC) TF 1-800-WX-BRIEF. NOTAM FILE SLC.
RCO 122.4 (CEDAR CITY FSS)
Ⓡ **APP/DEP CON** 121.1 (North of 41° latitude blo 8000′) 124.3 (105°-249°) 124.9 (297°-005° N of 41° N latitude above 8000′) 135.5 (250°-296° and 006°-104°) 125.7
TOWER 118.3 (E of Rwy 17-35) 119.05 (W of Rwy 16-34) 127.3 **GND CON** 121.9 **CLNC DEL** 127.3
PRE-TAXI CLNC 127.3
AIRSPACE: CLASS B See VFR Terminal Area Chart. Ctc **APP CON** 134.35 (West) all other quadrants 120.9.
RADIO AIDS TO NAVIGATION: NOTAM FILE SLC.
(H) VORTACW 116.8 SLC Chan 115 N40°51.02′ W111°58.92′ 155° 3.9 NM to fld. 4220/16E.
Unusable:
100°-140°beyond 30 NM below 13,000′
200°-230°beyond 25 NM below 10,800′
280°-290°beyond 30 NM below 8,100′
290°-315°beyond 35 NM below 8600′.
315°-330°beyond 20 NM below 8600′.
330°-345°beyond 24 NM below 7000′.
360°-070°beyond 20 NM below 11,200′
KERNN NDB (LOM) 338 SL N40°40.87′ W111°57.78′ 343° 6.3 NM to fld.
ILS/DME 110.7 I-MOY Chan 44 Rwy 16.
ILS 109.5 I-SLC Rwy 34 LOM KERNN NDB.
ILS/DME 111.5 I-BNT Chan 52 Rwy 17.
ILS 110.1 I-UTJ Rwy 35 LOM KERNN NOB. Localizer unusable byd 25° west of rwy centerline.
ASR

• •

HELIPAD H1: H100X75 (ASPH)
HELIPAD H2: H60X60 (ASPH)
HELIPAD H3: H60X60 (ASPH)
HELIPAD H4: H60X60 (ASPH)
HELIPAD H5: H60X60 (ASPH)
HELIPAD H6: H60X60 (ASPH)
HELIPORT REMARKS: Helipads H1 through H4 located on general aviation side of arpt and Helipads H5 and H6 located on air carrier side of arpt.

FIGURE 189.—Excerpt from Airport/Facilities Directory.

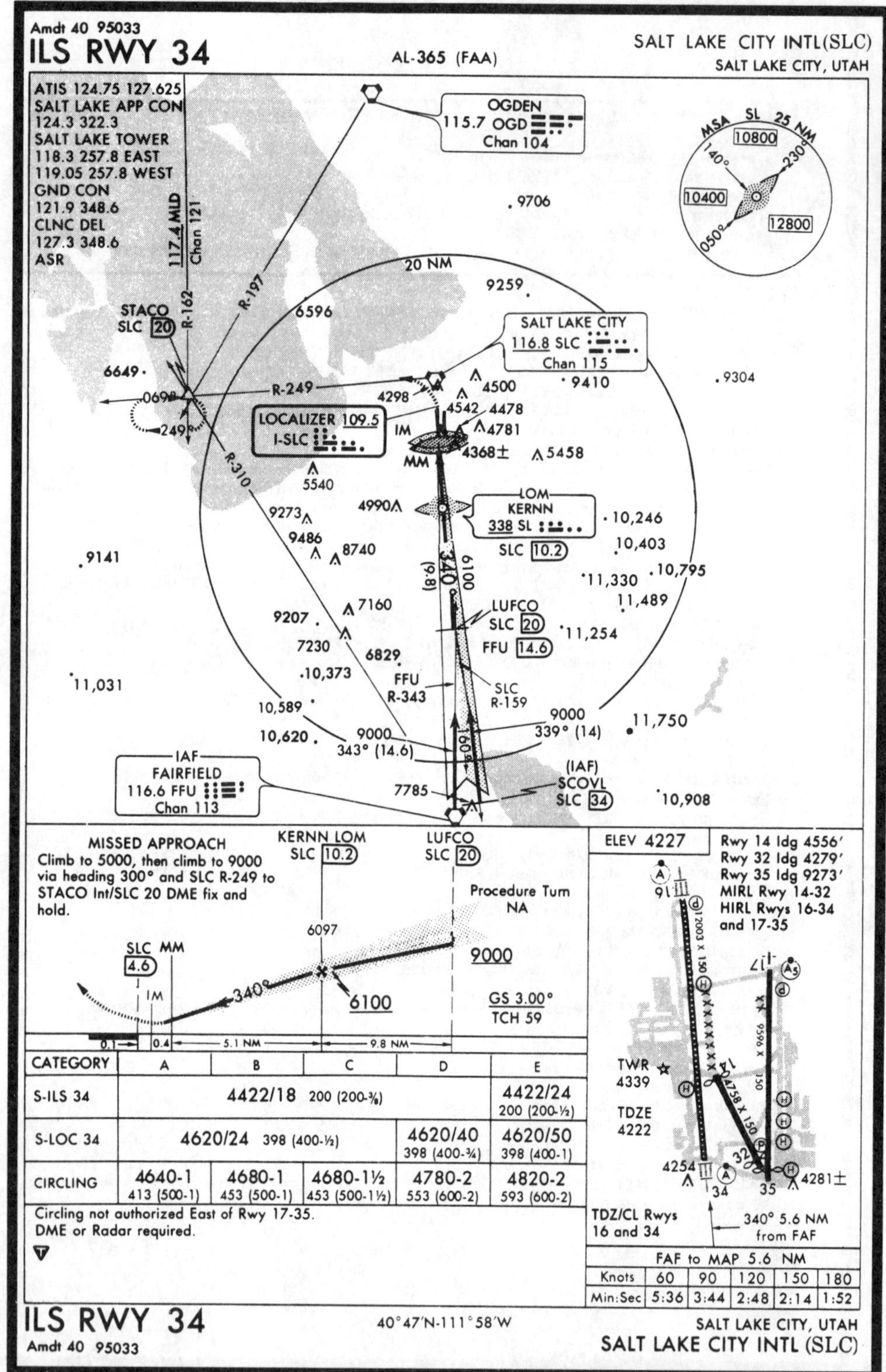

FIGURE 189A.—ILS RWY 34 (SLC).

Form Approved: OMB No. 2120-0034

U.S. DEPARTMENT OF TRANSPORTATION FEDERAL AVIATION ADMINISTRATION **FLIGHT PLAN**	(FAA USE ONLY) ☐ PILOT BRIEFING ☐ VNR ☐ STOPOVER	TIME STARTED	SPECIALIST INITIALS

1. TYPE	2. AIRCRAFT IDENTIFICATION	3. AIRCRAFT TYPE/ SPECIAL EQUIPMENT	4. TRUE AIRSPEED	5. DEPARTURE POINT	6. DEPARTURE TIME PROPOSED (Z)	ACTUAL (Z)	7. CRUISING ALTITUDE
VFR / X IFR / DVFR	PIL 10	B767/G	456 KTS	MSP			FL430

8. ROUTE OF FLIGHT
MINNEAPOLIS FOUR DEPARTURE FSD, J197 OBH, J10 LBF, SAYGE.SAYGE1

9. DESTINATION (Name of airport and city)	10. EST. TIME ENROUTE HOURS	MINUTES	11. REMARKS
DEN			L/O = LEVEL OFF PPH = POUNDS PER HOUR L/O FSD R-048/90 VARIATION: FSD 9E, LBF 10E, MSP 3E

12. FUEL ON BOARD HOURS	MINUTES	13. ALTERNATE AIRPORT(S)	14. PILOT'S NAME, ADDRESS & TELEPHONE NUMBER & AIRCRAFT HOME BASE / 17. DESTINATION CONTACT/TELEPHONE (OPTIONAL)	15. NUMBER ABOARD
		ABQ ALBUQUERQUE		190

16. COLOR OF AIRCRAFT SILVER/RED

CIVIL AIRCRAFT PILOTS. FAR Part 91 requires you file an IFR flight plan to operate under instrument flight rules in controlled airspace. Failure to file could result in a civil penalty not to exceed $1,000 for each violation (Section 901 of the Federal Aviation Act of 1958, as amended). Filing of a VFR flight plan is recommended as a good operating practice. See also Part 99 for requirements concerning DVFR flight plans.

FAA Form 7233-1 (8-82) CLOSE VFR FLIGHT PLAN WITH ______________ FSS ON ARRIVAL

FLIGHT LOG

CHECK POINTS FROM	TO	ROUTE / ALTITUDE	COURSE	WIND / TEMP	SPEED-KTS TAS	GS	DIST NM	TIME LEG	TOT	FUEL LEG	TOT
MSP	FSD R-048/90	VECTORS / CLIMB					90		:19:00		4170*
FSD R-048/90	FSD	DIRECT / FL430		290/89 / ISA-6							
FSD	OBH	J197 / FL430									
OBH	LBF	J10 / FL410		300/83 / ISA-5							
LBF	MODES	SAUGE.SAUGE.1 / FL410									
MODES	AMWAY	SAUGE.SAUGE.1 / FL410									
AMWAY	DEN	DESCENT & APPROACH					97	:25:00		3107	
DEN	ABQ	VECTORS / FL410						:36:00			

OTHER DATA:* Includes Taxi Fuel
NOTE: Use 9026 PPH Total Fuel Flow From L/O To Start Of Descent.
Use 7688 PPH Total Fuel Flow For Reserve And Alternate Requirements.

A Missed Approach Requires 1050# of Fuel.

TIME and FUEL: As required by FARs.

TIME	FUEL (LB)	
		EN ROUTE
		RESERVE
		ALTERNATE
		TOTAL

FIGURE 190.—Flight Plan/Flight Log.

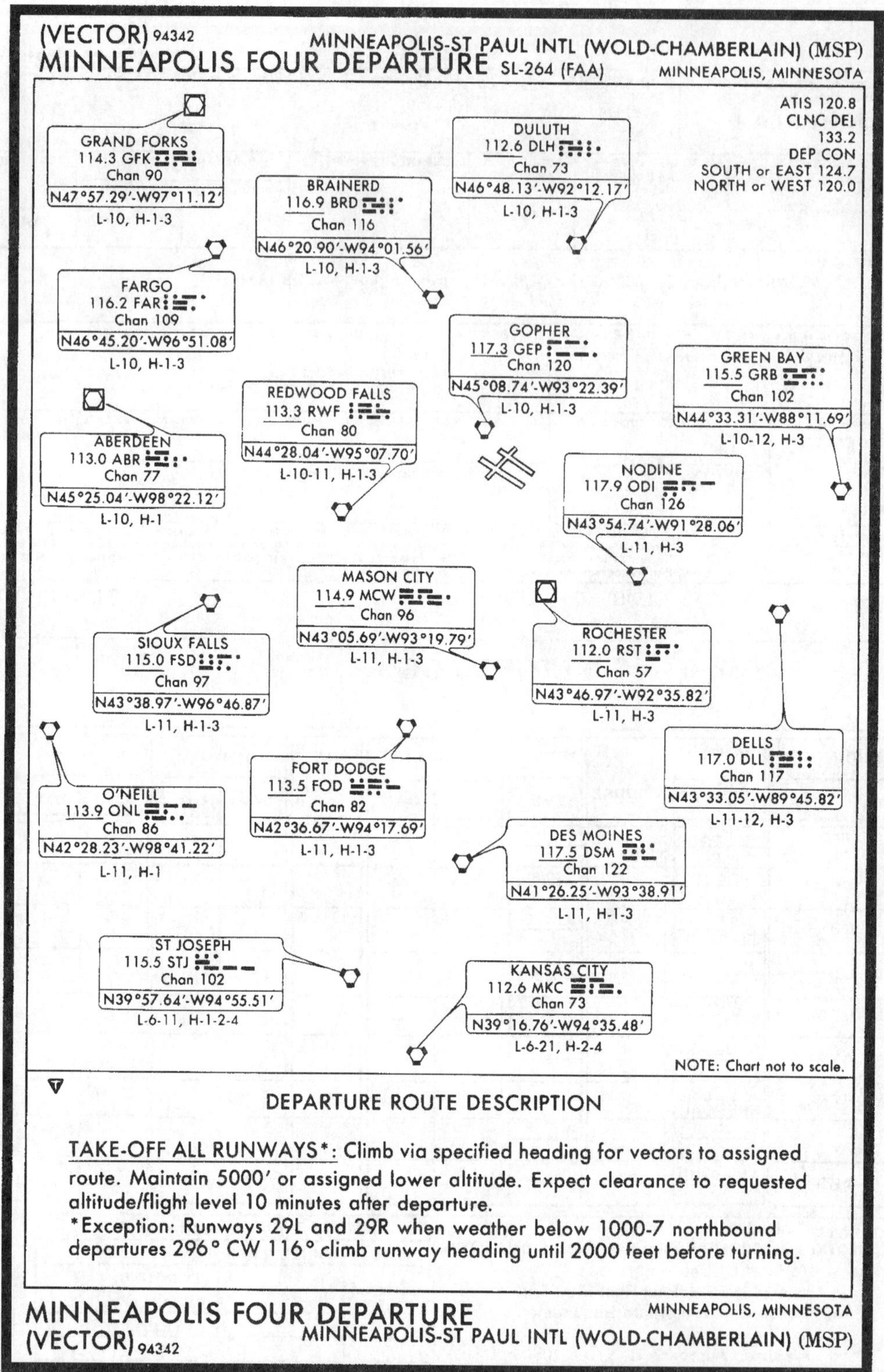

FIGURE 191.—MINNEAPOLIS FOUR DEPARTURE (MSP).

FLYING CLOUD (FCM) 11 SW UTC–6(–5DT) N44°49.63′ W93°27.43′ TWIN CITIES
906 B S4 **FUEL** 100, 100LL, JET A OX 3, 4 TPA–1906(1000) L–10G, A
RWY 09R-27L: H3909X75 (ASPH) S–30 HIRL IAP
RWY 09R: MALSR. VASI(V4L)—GA 3.0° TCH 41′ Rgt tfc.
RWY 27L: REIL. VASI(V4L)—GA 3.0° TCH 45′. Thld dsplcd 200′. Road.
RWY 09L-27R: H3599X75 (ASPH) S–30
RWY 27R: Road. Rgt tfc.
RWY 18-36: H2691X75 (ASPH) S–23 MIRL
RWY 18: VASI(V4L)—GA 3.0° TCH 37′. Barn. **RWY 36:** REIL. VASI(V4L)—GA 3.0° TCH 31′. Road.
AIRPORT REMARKS: Attended 1300–0400Z‡. Rwy 09L–27R CLOSED when twr clsd. Arpt CLOSED to jet acft not meeting FAR 36, jet training and jet acft over 20,000 lbs. Deer and waterfowl on and in vicinity of arpt. Rwy 09R and Rwy 27R rgt tfc during twr hours only. When twr clsd ACTIVATE VASI Rwy 09R, VASI Rwy 18, MALSR Rwy 09R, HIRL Rwy 09R–27L and MIRL Rwy 18–36—118.1.
WEATHER DATA SOURCES: LAWRS
COMMUNICATIONS: CTAF 118.1 **ATIS** 124.9 (612) 944–2970 **UNICOM** 122.95
PRINCETON FSS (PNM) TF 1–800–WX–BRIEF. NOTAM FILE FCM
® **MINNEAPOLIS APP/DEP CON** 125.0
MINNEAPOLIS CLNC DEL 121.7 (When twr closed)
TOWER 118.1 125.2 (Apr–Oct 1300–0400Z‡, Nov–Mar 1300–0300Z‡) **GND CON** 121.7 **CLNC DEL** 121.7
AIRSPACE: CLASS D svc effective Apr–Oct 1300–0400Z‡ Nov–Mar 1300–0300Z‡ other times CLASS G.
RADIO AIDS TO NAVIGATION: NOTAM FILE FCM.
(L) **ABVORW/DME** 111.8 FCM Chan 55 N44°49.54′ W93°27.41′ at fld. 900/6E.
Route forecast only on TWEB 0400–1100Z‡.
ILS 109.7 I–FCM RWY 09R. LOC unusable byd 30 degrees either side of centerline. GS unusable byd 5 degrees left of course.

MINNEAPOLIS-ST PAUL INTL (WOLD-CHAMBERLAIN) (MSP) 6 SW UTC–6(–5DT) TWIN CITIES
N44°53.05′ W93°12.90′ H–1E, 3G, L–10G, A
841 B S4 **FUEL** 100, JET A, A1 + OX 1, 2, 3, 4 LRA ARFF Index E IAP
RWY 11R-29L: H10000X200 (ASPH–CONC–GRVD) S–65, D–85, DT–145 HIRL CL 0.3% up W
RWY 11R: MALSR. PAPI(P4L)—GA 3.0° TCH 65′. Tree.
RWY 29L: ALSF1. TDZ. PAPI(P4L)—GA 3.0° TCH 73′. Pole.
RWY 04-22: H8256X150 (CONC–GRVD) S–65, D–85, DT–145 HIRL
RWY 04: SSALR. PAPI(P4L)—GA 3.0° TCH 76′.
RWY 22: MALSR. PAPI(P4L)—GA 3.0° TCH 42′. Thld dsplcd 988′. Fence.
RWY 11L-29R: H8200X150 (ASPH–CONC–GRVD) S–100, D–125, DT–210 HIRL
RWY 11L: MALSR. PAPI(P4L)—GA 3.0° TCH 75′. Tree. 0.3% down.
RWY 29R: REIL. PAPI(P4L)—GA 3.0° TCH 73′.
AIRPORT REMARKS: Attended continuously. Birds on and in vicinity of arpt. Training prohibited. Only Initial departure and full stop termination training flights permitted. PPR for noise abatement procedures —call 612–726–9411. No stage 1 noise Category Civil acft. Landing fee. Flight Notification Service (ADCUS) available. NOTE: See Land and Hold Short Operations Section.
WEATHER DATA SOURCES: LLWAS.
COMMUNICATIONS: ATIS 135.35 (612) 726–9240. 120.8 (TCA ARR INFO) **UNICOM** 122.95
PRINCETON FSS (PNM) TF 1–800–WX–BRIEF. NOTAM FILE MSP.
RCO 122.55 122.3 (PRINCETON FSS) **RCO** 122.1R 115.3T (PRINCETON FSS)
® **APP CON** 119.3 (N or E of arrival rwy) 126.95 (S or W of arrival rwy and Rwys 04, 11R and 29L)
TOWER 126.7 (Rwys 11R–29L and 04–22) 123.95 (Rwy 11L–29R) **GND CON** 121.9 (S) 121.8 (N) **CLNC DEL** 133.2
® **DEP CON** 127.925 (N or E of arrival rwy) 124.7 (S or W of arrival rwy)
AIRSPACE: CLASS B: See VFR Terminal Area Chart.
RADIO AIDS TO NAVIGATION: NOTAM FILE MSP.
(H) **VORTAC** 115.3 MSP Chan 100 N44°52.92′ W93°13.99′ at fld. 850/3E.
VOR portion unusable below 3000′, beyond 20 NM below 4000′, 205°–235°/265°–025° all distances and altitudes, 235°–265° below 7000′.
NARCO NDB (MH-SAB/LOM) 266 MS N44°49.55′ W93°05.48′ 299° 6.3 NM to fld.
Route forecast only on TWEB 0400–1100Z‡.
VAGEY NDB (LOM) 338 AP N44°49.45′ W93°18.36′ 042° 5.3 NM to fld. Unmonitored.
ILS 109.9 I–INN Rwy 29R
ILS/DME 110.3 I–MSP Chan 40 Rwy 29L LOM NARCO NDB.
ILS 109.3 I–APL Rwy 04 LOM VAGEY NDB. Glide slope unusable for coupled approaches below 1085.
ILS/DME 110.3 I–HKZ Chan 40 Rwy 11R.
ILS 110.7 I–PJL Rwy 11L.
ILS 110.5 I–SIJ Rwy 22.

MOBERG AIR BASE SPB (See BEMIDJI)

FIGURE 191A.—Excerpt from Airport/Facilities Directory.

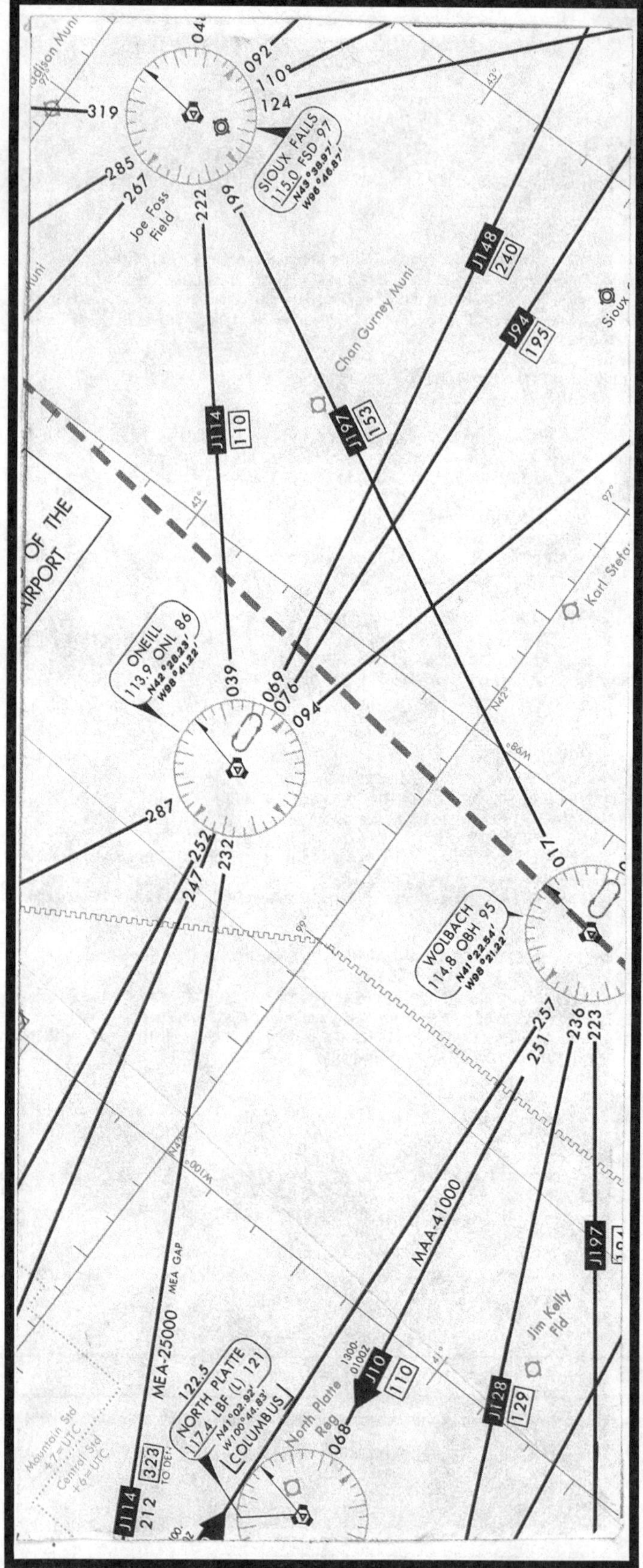

FIGURE 192.—High Altitude Airways.

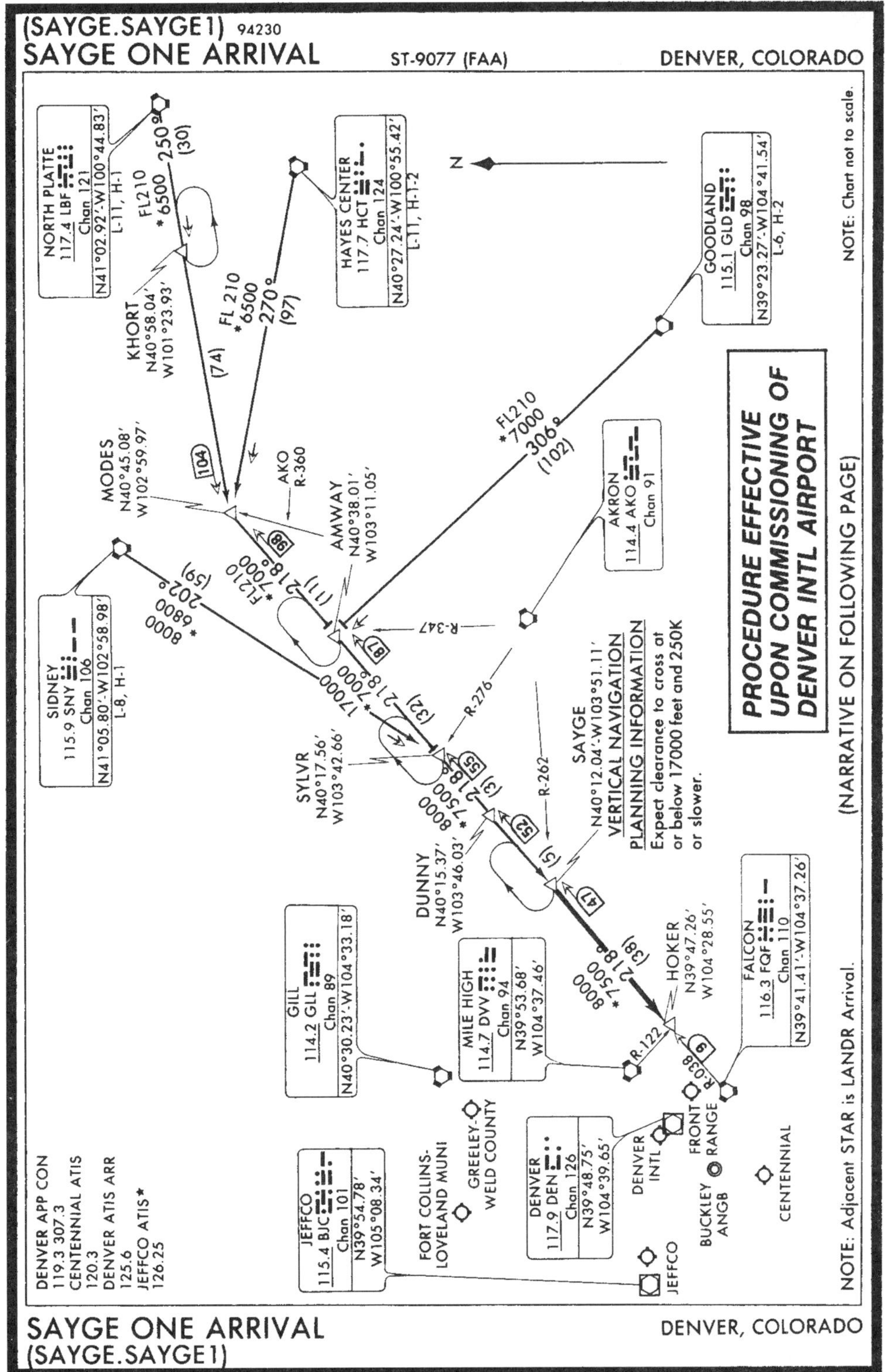

FIGURE 193.—SAYGE ONE ARRIVAL (SAYGE.SAYGE1).

(SAYGE.SAYGE1) 93315

SAYGE ONE ARRIVAL ST-9077 (FAA) DENVER, COLORADO

ARRIVAL DESCRIPTION

GOODLAND TRANSITION (GLD.SAYGE1): From over GLD VORTAC via GLD R-306 and FQF R-038 to SAYGE INT. Thence. . . .
HAYES CENTER TRANSITION (HCT.SAYGE1): From over HCT VORTAC via HCT R-270 and FQF R-038 to SAYGE INT. Thence. . . .
NORTH PLATTE TRANSITION (LBF.SAYGE1): From over LBF VORTAC via LBF R-250 and FQF R-038 to SAYGE INT. Thence. . . .
SIDNEY TRANSITION (SNY.SAYGE1): From over SNY VORTAC via SNY R-202 and FQF R-038 to SAYGE INT. Thence. . . .
. . . .From over SAYGE INT via FQF R-038 to HOKER INT. Expect radar vectors to the final approach course at or before HOKER INT.

PROCEDURE EFFECTIVE UPON COMMISSIONING OF DENVER INTL AIRPORT

SAYGE ONE ARRIVAL DENVER, COLORADO
(SAYGE.SAYGE1)

FIGURE 193A.—SAYGE ONE ARRIVAL (SAYGE.SAYGE1).

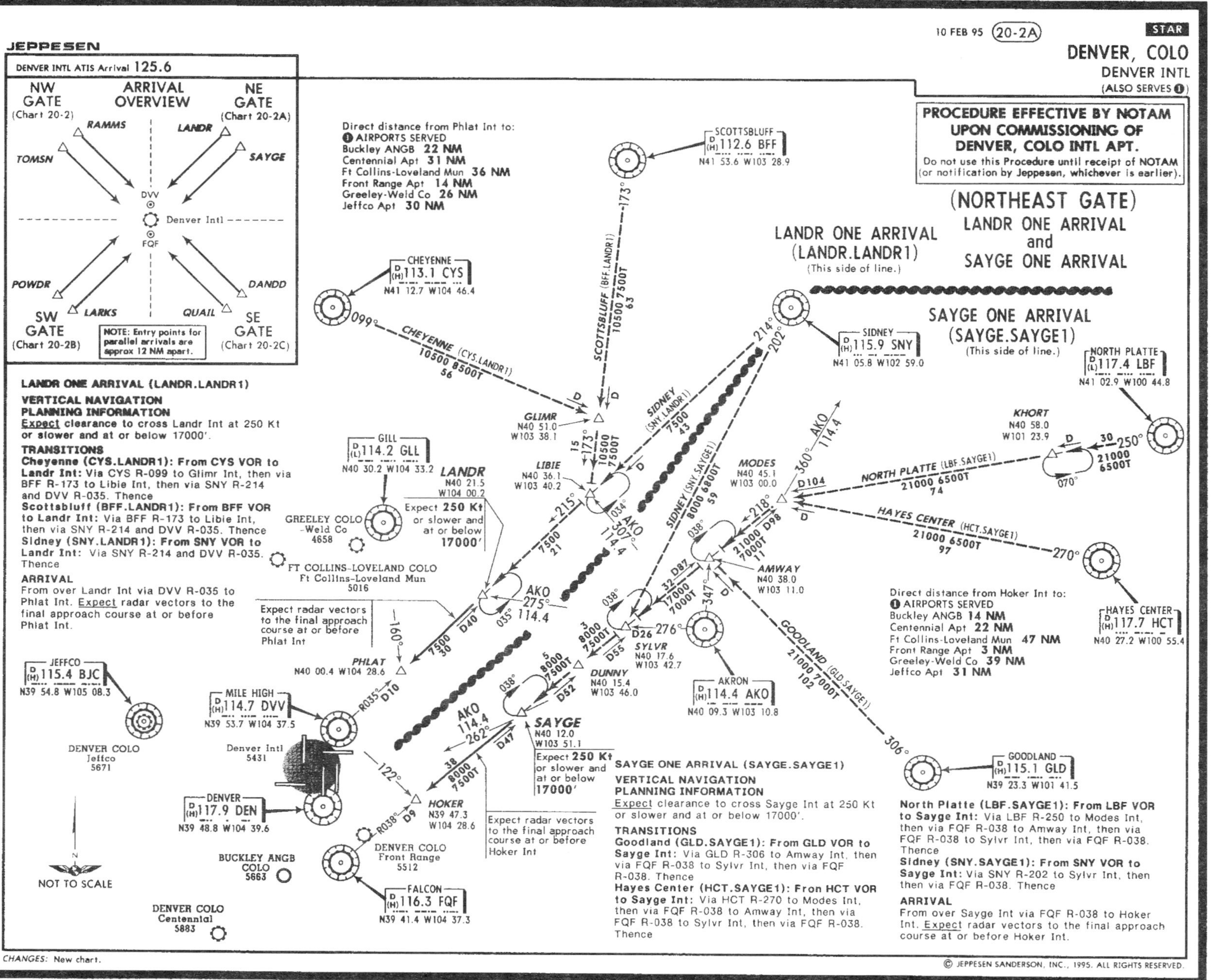

FIGURE 194.—Landr One Arrival/Sayge One Arrival.

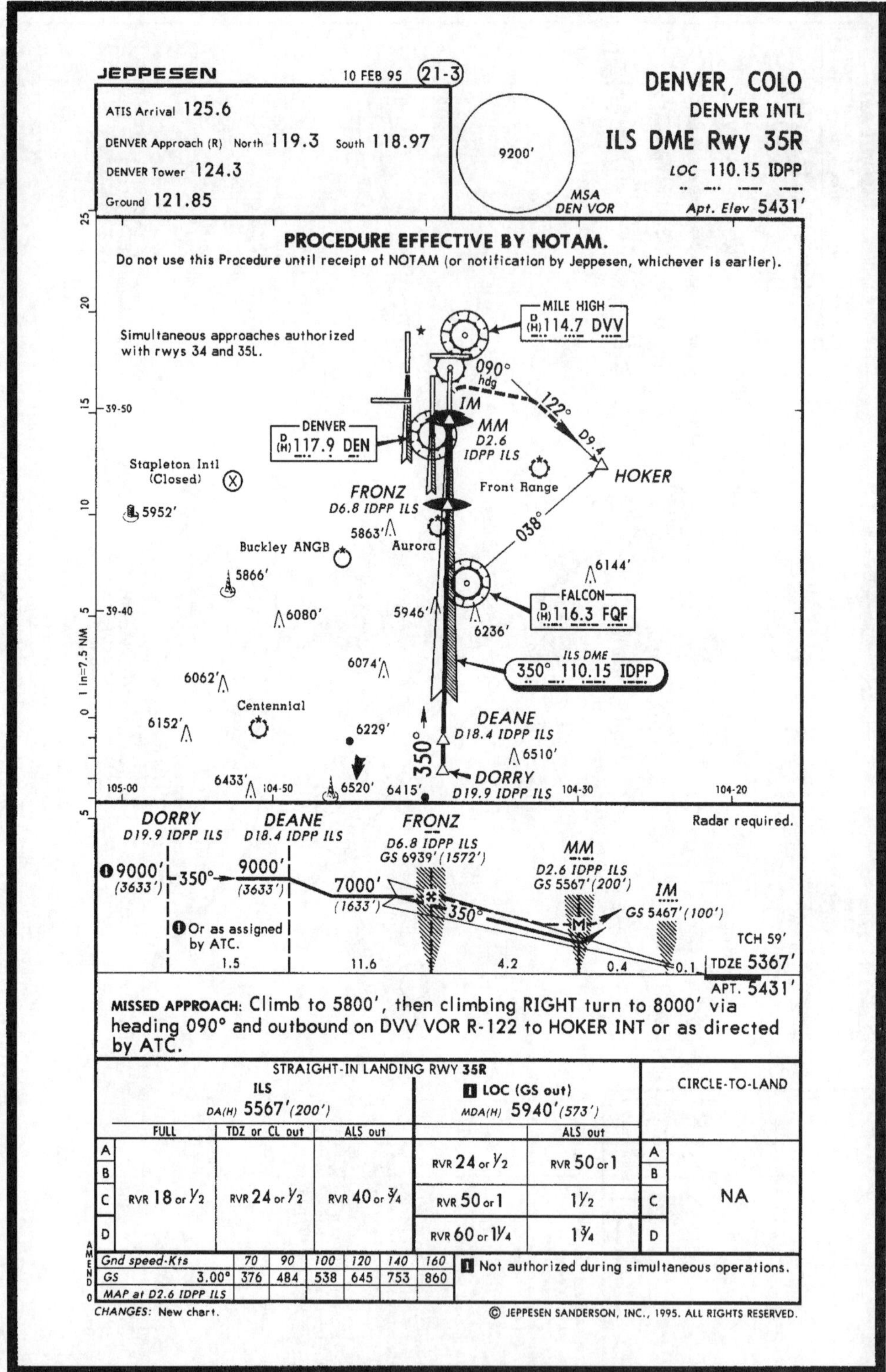

FIGURE 195.—ILS DME RWY 35R.

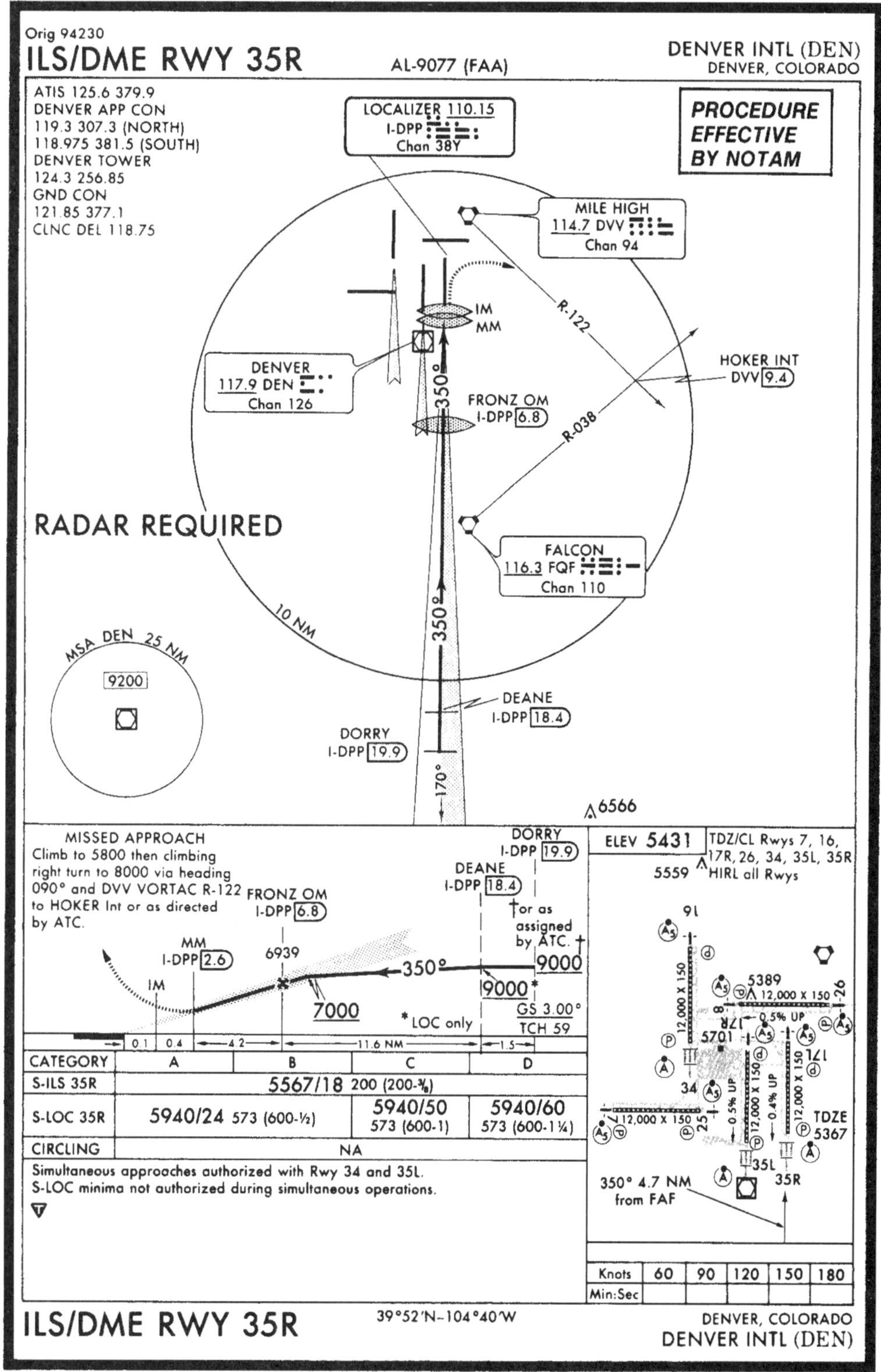

CATEGORY	A	B	C	D
S-ILS 35R	5567/18 200 (200-¾)			
S-LOC 35R	5940/24 573 (600-½)		5940/50 573 (600-1)	5940/60 573 (600-1¼)
CIRCLING	NA			

Simultaneous approaches authorized with Rwy 34 and 35L.
S-LOC minima not authorized during simultaneous operations.

Knots	60	90	120	150	180
Min:Sec					

FIGURE 195A.—ILS/DME RWY 35R (DEN).

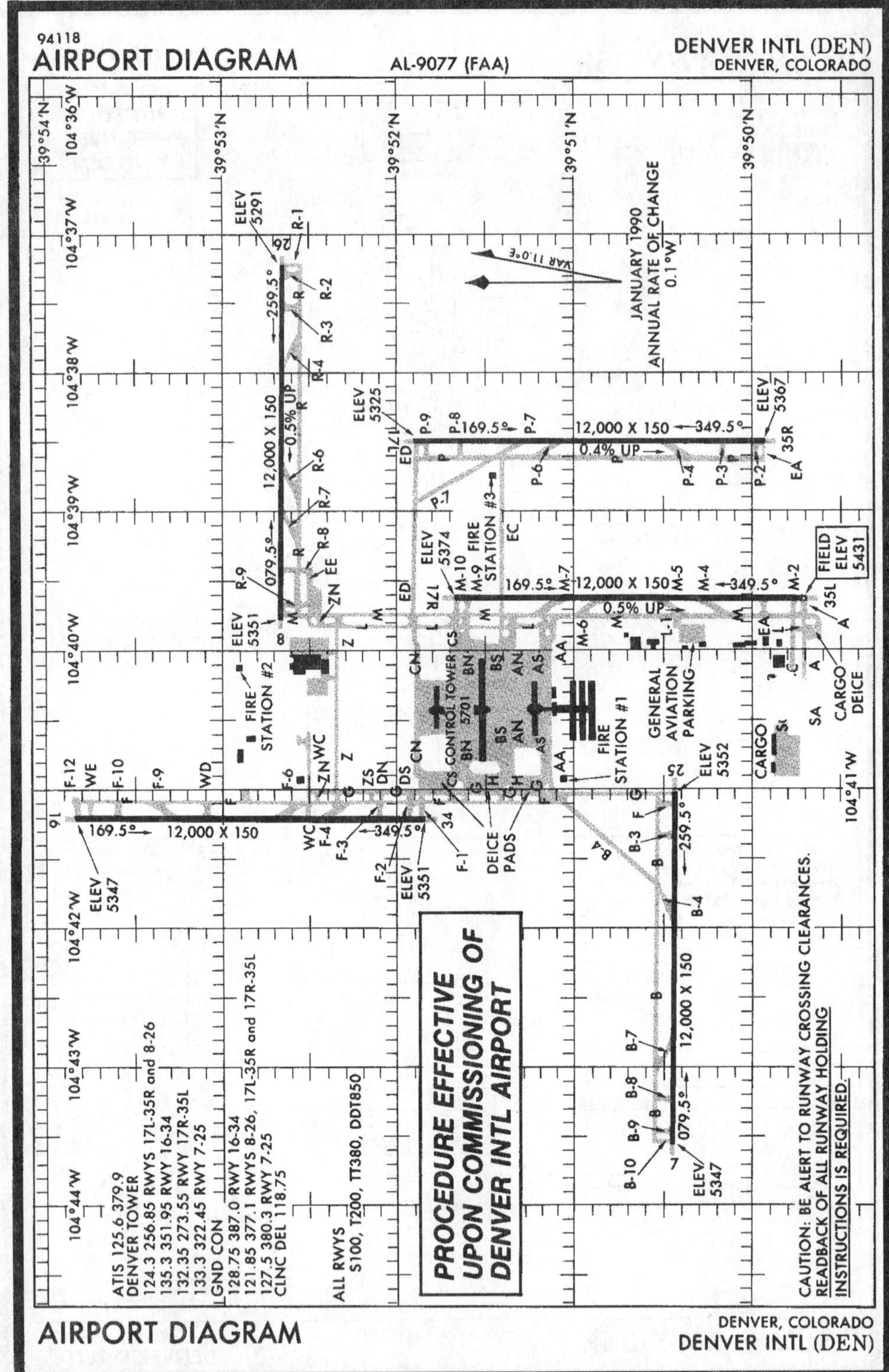

FIGURE 196.—AIRPORT DIAGRAM.

184 **SPECIAL NOTICES**

DENVER INTL (DEN) 16 NE UTC-7(-6DT) N39°51.51′ W104°40.02′ **DENVER**
5431 B S4 **FUEL** 100, 100LL, MOGAS OX 1, 3 **H-2D, L-6E, 8G, A**
RWY 07-25: H12000X150 (CONC-GRVD) S-100, D-200, DT-380, DDT-850 HIRL CL **IAP**
RWY 07: MALSR. TDZ. PAPI(P4R)—GA 3.0° TCH 55.′ **RWY 25:** MALSR. PAPI(P4L)—GA 3.0° TCH 55′.
RWY 08-26: H12000X150 (CONC-GRVD) S-100, D-200, DT-380, DDT-850 HIRL CL
RWY 08: MALSR. PAPI(P4L)—GA 3.0° TCH 55′. **RWY 26:** MALSR. TDZ. PAPI(P4L)—GA 3.0° TCH 55′.
RWY 16-34: H12000X150 (CONC-GRVD) S-100, D-200, DT-380, DDT-850 HIRL CL
RWY 16: MALSR. TDZ. PAPI(P4L)—GA 3.0° TCH 55.′ **RWY 34:** ALSF2. TDZ. PAPI(P4L)—GA 3.0° TCH 55′.
RWY 17R-35L: H12000X150 (CONC-GRVD) S-100, D-200, DT-380, DDT-850 HIRL CL
RWY 17R: MALSR. TDZ. PAPI(P4L)—GA 3.0° TCH 55′. **RWY 35L:** ALSF2. TDZ. PAPI(P4R)—GA 3.0° TCH 55′.
RWY 17L-35R: H12000X150 (CONC-GRVD) S-100, D-200, DT-380, DDT-850 HIRL CL
RWY 17L: MALSR. PAPI(P4L)—GA 3.0° TCH 55′. **RWY 35R:** ALSF2. TDZ. PAPI(P4R)—GA 3.0° TCH 55′.
AIRPORT REMARKS: Attended continuously. Overhead walk-way on South side of concourse 'A' provides 44 ft high tail and 117 ft wide wing span clearance. Insufficient twy corner fillet pavement in the SE corner of the Twy M/M2 intersection for acft with wingspan over 107 ft. Noise abatement: Stage III or quieter acft only allowed to depart Rwy 25. Ldg fee.
WEATHER DATA SOURCES: ASOS (303)342-0838. LLWAS.
COMMUNICATIONS: ATIS 125.6 (Arr) (303) 342-0819 134.025 (Dep) (303) 342-0820 **UNICOM** 122.95
FSS (DEN) TF 1-800-WX-BRIEF. NOTAM FILE DEN.
Ⓡ **APP CON** 119.3 (North) 118.975 (South) **FINAL CON** 120.8
TOWER 135.3 (Rwy 16-34) 133.3 (Rwy 07-25) 132.35 (Rwy 17R-35L) 124.3 (Rwys 08-26 and 17L-35R)
GND CON 128.75 (Rwy 16-34) 127.5 (Rwy 07-25) 121.85 (Rwys 08-26, 17L-35R and 17R-35L) **CLNC DEL** 118.75
Ⓡ **DEP CON** 128.25 (East/South) 127.05 (North) 126.1 (West/South)
AIRSPACE: CLASS B See VFR Terminal Area Chart.
RADIO AIDS TO NAVIGATION: NOTAM FILE DEN.
(H) VORW/DME 117.9 DEN Chan 126 N39°48.75′ W104°39.65′ 343° 2.8 NM to fld. 5440/11E.

ILS/DME	111.1	I-LTT	Chan 48	Rwy 16.
ILS/DME	111.1	I-OUF	Chan 48	Rwy 34.
ILS/DME	108.9	I-FUI	Chan 26	Rwy 08.
ILS/DME	108.9	I-JOY	Chan 26	Rwy 26.
ILS/DME	108.5	I-ACX	Chan 22	Rwy 17R.
ILS/DME	108.5	I-AQD	Chan 22	Rwy 35L.
ILS/DME	110.15	I-BXP	Chan 38(Y)	Rwy 17L.
ILS/DME	110.15	I-DPP	Chan 38(Y)	Rwy 35R.
ILS/DME	111.55	I-DZG	Chan 52(Y)	Rwy 07.
ILS/DME	111.55	I-ERP	Chan 52(Y)	Rwy 25.

COMM/NAVAID REMARKS: Emerg frequency 121.5 not avbl at twr.

SATELLITE AIRPORT COMMUNICATIONS

DENVER, CENTENNIAL (APA)
DENVER APP/DEP CON 126.375
DENVER, FRONT RANGE (FTG)
DENVER APP/DEP CON 128.45
DENVER, JEFFCO (BJC)
DENVER APP/DEP CON 128.45
ERIE, TRI-COUNTY (48V)
DENVER APP/DEP CON 128.45
FORT COLLINS, DOWNTOWN FORT COLLINS AIRPARK (3V5)
DENVER APP/DEP CON 134.85
FORT COLLINS-LOVELAND MUNI (FNL)
DENVER APP/DEP CON 134.85 **CLNC DEL** 120.25
GREELEY-WELD COUNTY (GXY)
DENVER APP/DEP CON 134.85 **CLNC DEL** 126.65
LONGMONT, VANCE BRAND (2V2)
DENVER APP/DEP CON 128.45

FIGURE 196A.—Excerpt from Airport/Facilities Directory.

Form Approved: OMB No. 2120-0034

U.S. DEPARTMENT OF TRANSPORTATION FEDERAL AVIATION ADMINISTRATION **FLIGHT PLAN**	(FAA USE ONLY) ☐ PILOT BRIEFING ☐ VNR ☐ STOPOVER	TIME STARTED	SPECIALIST INITIALS

1. TYPE	2. AIRCRAFT IDENTIFICATION	3. AIRCRAFT TYPE/ SPECIAL EQUIPMENT	4. TRUE AIRSPEED	5. DEPARTURE POINT	6. DEPARTURE TIME		7. CRUISING ALTITUDE
VFR X IFR DVFR	HOSS 2	A109K2/A	** KTS	EGE EAGLE CO. REGIONAL	PROPOSED (Z)	ACTUAL (Z)	14000

8. ROUTE OF FLIGHT
DBL, VI34 FFU, V21 JAURN, SL, SLC

9. DESTINATION (Name of airport and city)	10. EST. TIME ENROUTE		11. REMARKS
SLC SALT LAKE CITY INT'L	HOURS	MINUTES	L/O = LEVEL OFF PPH = POUNDS PER HOUR **CAS 139 ISA +20 TO +14 VARIATION: PUC 14E.

12. FUEL ON BOARD		13. ALTERNATE AIRPORT(S)	14. PILOT'S NAME, ADDRESS & TELEPHONE NUMBER & AIRCRAFT HOME BASE 17. DESTINATION CONTACT/TELEPHONE (OPTIONAL)	15. NUMBER ABOARD
HOURS	MINUTES	OGD OGDEN HINCKLEY		8

16. COLOR OF AIRCRAFT	CIVIL AIRCRAFT PILOTS. FAR Part 91 requires you file an IFR flight plan to operate under instrument flight rules in controlled airspace. Failure to file could result in a civil penalty not to exceed $1,000 for each violation (Section 901 of the Federal Aviation Act of 1958, as amended). Filing of a VFR flight plan is recommended as a good operating practice. See also Part 99 for requirements concerning DVFR flight plans.
ORANGE/BLUE	

FAA Form 7233-1 (8-82) CLOSE VFR FLIGHT PLAN WITH ________ FSS ON ARRIVAL

FLIGHT LOG

CHECK POINTS		ROUTE	COURSE	WIND	SPEED-KTS		DIST	TIME		FUEL	
FROM	TO	ALTITUDE		TEMP	TAS	GS	NM	LEG	TOT	LEG	TOT
EGE	DBL						13		:10:00		101*
DBL	JNC	V-134 14000		100/41 ISA+20							
JNC	PUC	V134 14000									
PUC	FFU	V134 1400		120/34 ISA+14							
FFU	JAURN	V21 14000									
JAURN	SLC	DESCENT & APPROACH					18	:10:00		92	
SLC	OGD					160	24	:09:00			

OTHER DATA: * Includes Taxi Fuel
NOTE: Use 495 PPH Total Fuel Flow From L/O To Start Of Descent.
Use 469 PPH Total Fuel Flow For Reserve And Alternate Requirements.

A Missed Approach Requires 33# of Fuel.

TIME and FUEL: As required by FARs.

TIME	FUEL (LB)	
		EN ROUTE
		RESERVE
		ALTERNATE
		TOTAL

FIGURE 197.—Flight Plan/Flight Log.

EAGLE CO REGIONAL (EGE) 4 W UTC–7(–6DT) N39°38.55′ W106°55.06′ **DENVER**
6535 B S4 **FUEL** 100, 100LL, JET A1, JET A1 + OX 1, 3 ARFF Index C **H-2C, L-5D, 6E, 8F**
RWY 07-25: H8000X150 (ASPH) S-60, D-115 MIRL **IAP**
RWY 07: REIL. Tree. Rgt tfc. **RWY 25:** MALSR. REIL. PAPI(P4L)—GA 3.0° TCH 45′.
AIRPORT REMARKS: Attended 1400–0200Z‡. CLOSED to unscheduled air carrier operations with more than 30 passenger seats except PPR call arpt manager 303-524-9490. High unmarked terrain all quadrants. Ngt ops discouraged to pilots unfamiliar with arpt. Rwy 07 mountain top 10:1 clearance 12000′ from thld 1500′ left of rwy centerline extended. Recommend all acft departing Rwy 25 initiate a left turn as soon as altitude and safety permit to avoid high terrain. Extensive military helicopter training operations surface to 1000′ AGL within 25 NM radius Eagle County Arpt 1330–0500Z‡. Wildlife in vicinity of arpt. No snow removal at nights. Rwy 25 PAPI only visible to 6° left of centerline due to terrain. After 0300Z‡ ACTIVATE MALSR Rwy 25 MIRL Rwy 07-25, PAPI Rwy 25 and REIL Rwy 07 and Rwy 25—CTAF.
WEATHER DATA SOURCES: AWOS-3 135.575 (303) 524-7386. Frequency 135.575 out of svc 1400–0200Z‡.
COMMUNICATIONS: **CTAF** 118.2 **UNICOM** 122.95
DENVER FSS (DEN) TF 1-800-WX-BRIEF. NOTAM FILE EGE.
RCO 122.2 (DENVER FSS)
DENVER CENTER APP/DEP CON 134.5
TOWER 118.2 NFCT (1400–0200Z‡) VFR only. **GND CON** 121.8
AIRSPACE: **CLASS D** svc effective 1400–0200Z‡ other times CLASS E.
RADIO AIDS TO NAVIGATION: NOTAM FILE DEN.
SNOW (L) VORW/DME 109.2 SXW Chan 29 N39°37.77′ W106°59.47′ 065° 3.5 NM to fld. 8060/12E. Unmonitored.
ILS/DME 110.1 I-EGE Chan 38 Rwy 25 (LOC only). LOC/DME unmonitored.

EASTON (VALLEY VIEW) (See GREELEY)

ECKERT/ORCHARD CITY

DOCTORS MESA (E00) 3 W UTC–7(–6DT) N38°51.17′ W108°01.04′ **DENVER**
5600 **FUEL** 100, MOGAS Not insp.
RWY 08-26: 6750X110 (DIRT-TURF)
RWY 08: Hill. **RWY 26:** Tree.
AIRPORT REMARKS: Attended continuously. Powerlines across middle of rwy 2352′ from Rwy 26 end perpendicular to centerline, marked with 3 red powerline marker balls. Vegetation/grass on and in vicinity of rwy. +200′ mountains 3 miles west of arpt. Wildlife on and in vicinity of arpt. Rwy 26 has end reflectors. Takeoffs to the east and landings to the west preferred, SW winds predominant. Rwy 08-26 soft when wet.
COMMUNICATIONS: **CTAF/UNICOM** 122.8
DENVER FSS (DEN) TF 1-800-WX-BRIEF. NOTAM FILE DEN.

ELLICOTT

COLORADO SPRINGS EAST (CO50) 3 NW UTC–7(–6DT) N38°52.47′ W104°24.60′ **DENVER**
6145 S2 **FUEL** 100LL **H-2D, L-6E**
RWY 17-35: H5000X60 (ASPH) RWY LGTS (NSTD)
RWY 35: Rgt tfc.
RWY 08-26: 3440X60 (GRVL)
RWY 08: Fence.
AIRPORT REMARKS: Attended 1500–0000Z‡. P-line runs perpendicular to Rwy 08 1500′ from rwy end. Rwy 17-35 asph broken and loose in areas. Rwy 08-26 rough at intersection of Rwy 17-35. 4′ fence 125′ from centerline both sides of Rwy 08-26 W of Rwy 17-35. Rwy 17-35 lights on E side of rwy only. For NSTD rwy lights call 719-683-2701. Fee for commercial acft ctc arpt manager 719-683-2701.
COMMUNICATIONS: **CTAF** 122.9
DENVER FSS (DEN) TF 1-800-WX-BRIEF. NOTAM FILE DEN.
RADIO AIDS TO NAVIGATION: NOTAM FILE COS.
COLORADO SPRINGS (L) VORTACW 112.5 COS Chan 72 N38°56.67′ W104°38.01′ 099° 11.3 NM to fld. 6930/13E.

FIGURE 198.—Excerpt from Airport/Facilities Directory.

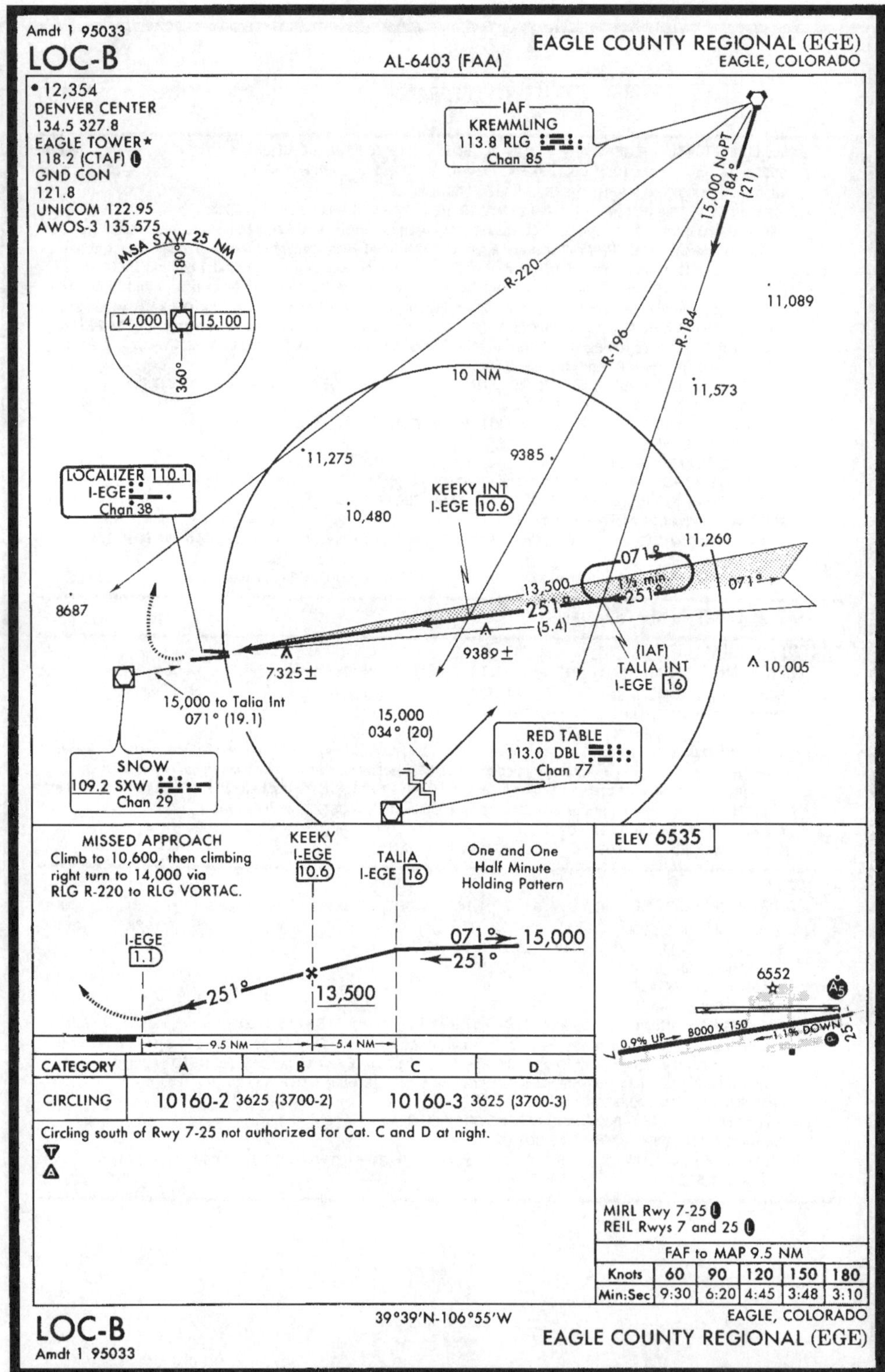

FIGURE 198A.—LOC-B (EGE).

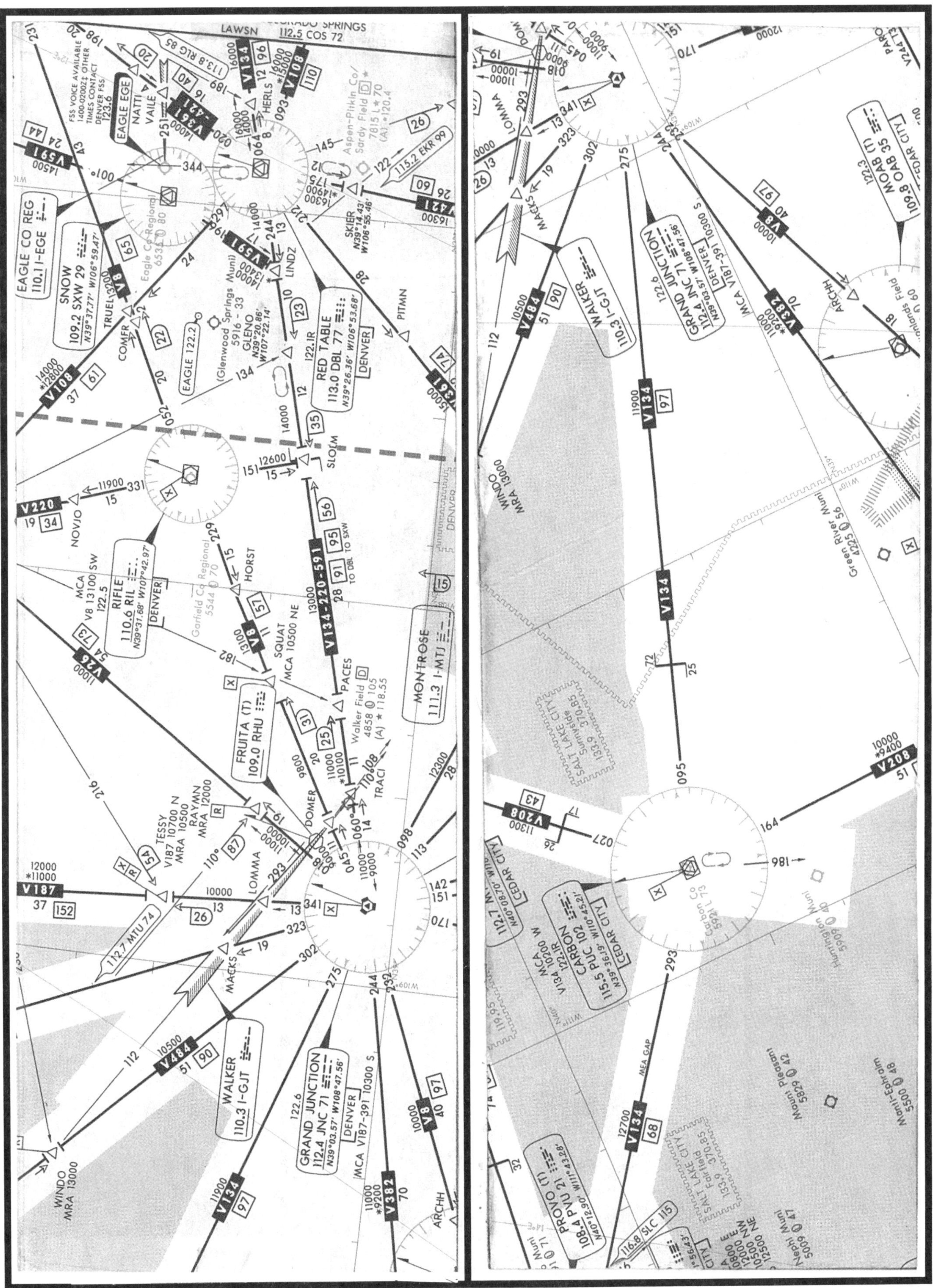

FIGURE 199.—Low Altitude Airways.

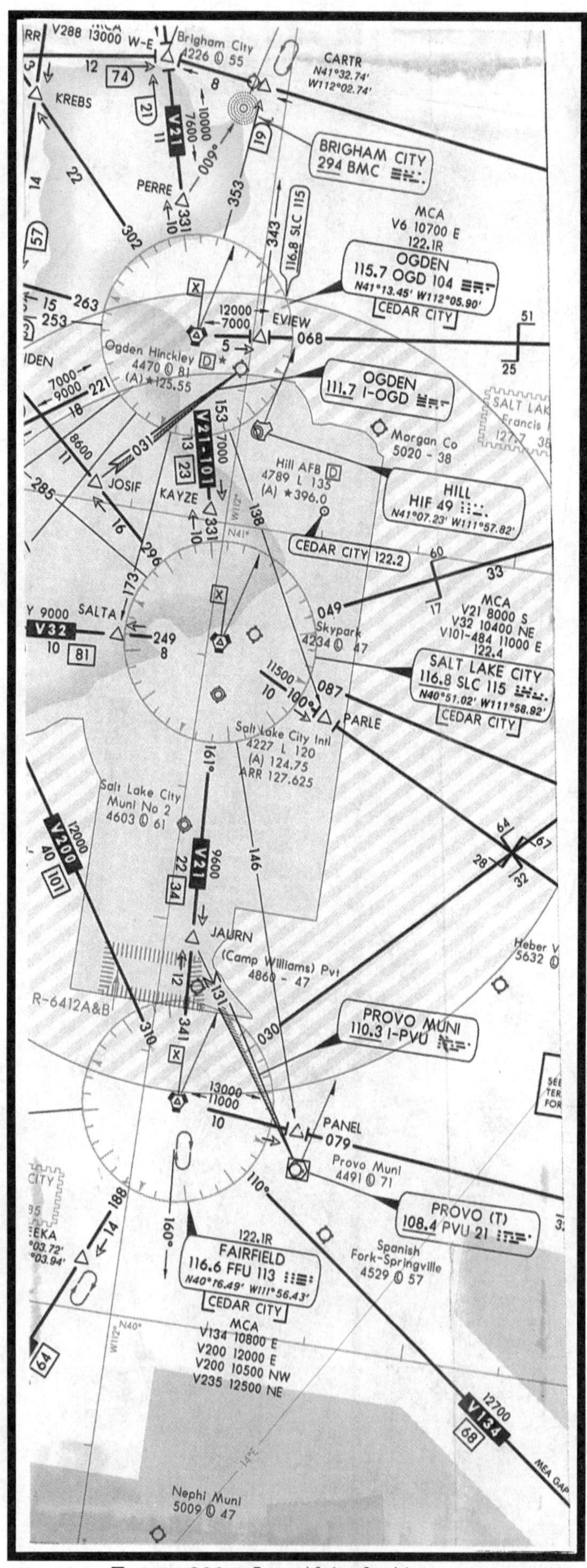

FIGURE 200.—Low Altitude Airways.

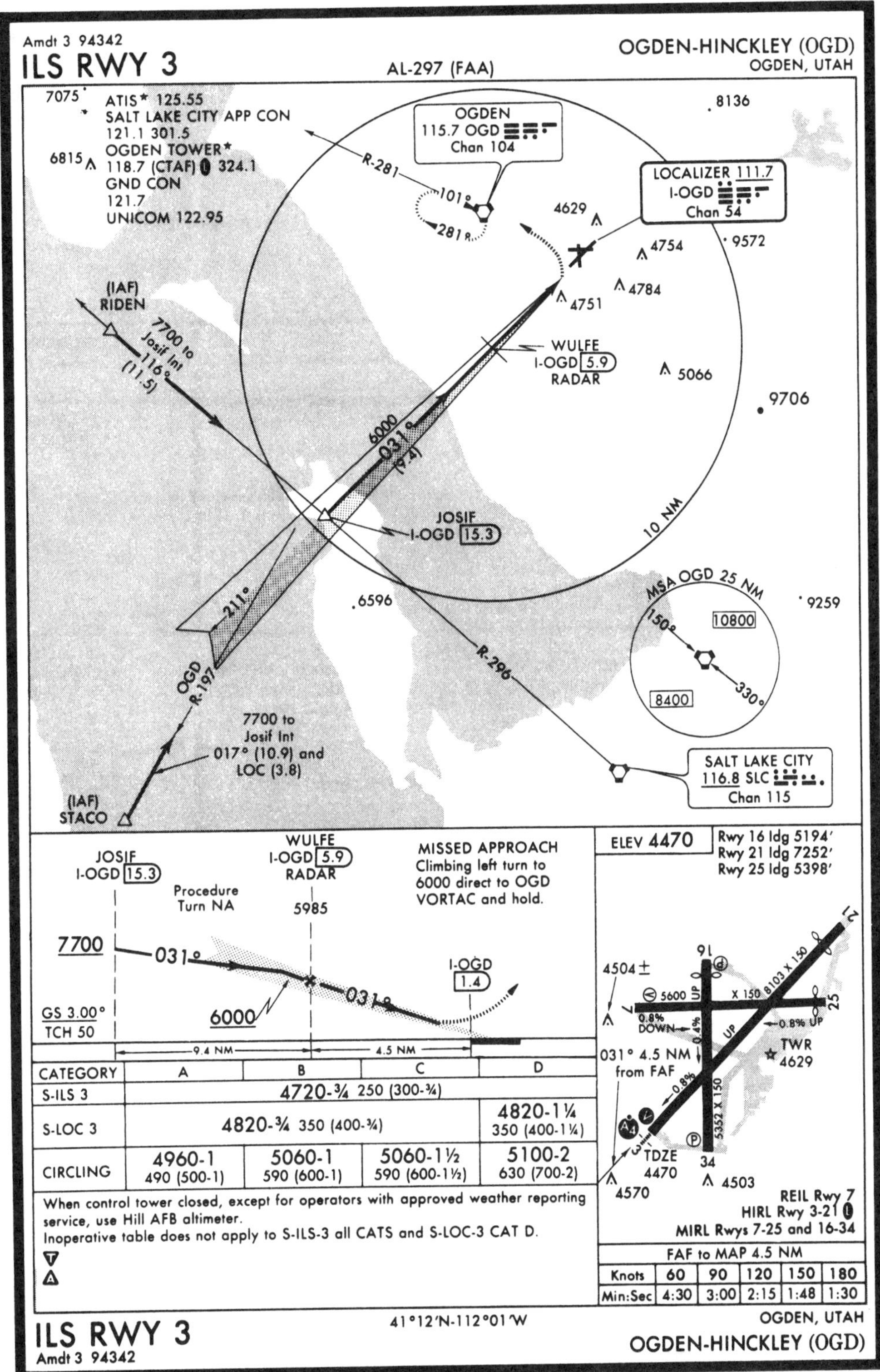

CATEGORY	A	B	C	D
S-ILS 3	4720-¾ 250 (300-¾)			
S-LOC 3	4820-¾ 350 (400-¾)			4820-1¼ 350 (400-1¼)
CIRCLING	4960-1 490 (500-1)	5060-1 590 (600-1)	5060-1½ 590 (600-1½)	5100-2 630 (700-2)

When control tower closed, except for operators with approved weather reporting service, use Hill AFB altimeter.
Inoperative table does not apply to S-ILS-3 all CATS and S-LOC-3 CAT D.

FAF to MAP 4.5 NM					
Knots	60	90	120	150	180
Min:Sec	4:30	3:00	2:15	1:48	1:30

ILS RWY 3
Amdt 3 94342
OGDEN, UTAH
OGDEN-HINCKLEY (OGD)

FIGURE 201.—ILS RWY 3 (OGD).

MOUNT PLEASANT (43U) 2 SW UTC–7(–6DT) N39°31.48′ W111°28.51′ LAS VEGAS
5829 B L-5C
RWY 02-20: H4260X60 (ASPH) MIRL
RWY 20: Road.
AIRPORT REMARKS: Unattended. Rwy 02-20 marked by stripes. For runway lights key 122.8 7 times.
COMMUNICATIONS: CTAF 122.9
CEDAR CITY FSS (CDC) TF 1-800-WX-BRIEF. NOTAM FILE CDC.
RADIO AIDS TO NAVIGATION: NOTAM FILE CDC.
DELTA (H) VORTAC 116.1 DTA Chan 108 N39°18.14′ W112°30.33′ 058° 49.7 NM to fld. 4600/16E.

MYTON N40°08.70′ W110°07.66′ NOTAM FILE CDC. SALT LAKE CITY
(H) VORTAC 112.7 MTU Chan 74 270° 12.0 NM to Duchesne Muni. 5332/14E. H-2C, L-5C, 8E
RCO 122.1R 112.7T (CEDAR CITY FSS)

NEPHI MUNI (U14) 3 NW UTC–7(–6DT) N39°44.33′ W111°52.30′ LAS VEGAS
5009 B S4 **FUEL** 100LL, JET A L-5C, 7D, 8E
RWY 16-34: H4700X75 (ASPH) S-21 MIRL
RWY 16: Thld dsplcd 200′. **RWY 34:** Thld dsplcd 400′.
AIRPORT REMARKS: Attended continuously. Rwy 16–34 cracking and loose chips on apron and rwy. Rwy 16 thld relocated 200′ for ngt operations, Rwy 34 thld relocated 400′ for ngt operations, 4100′ of rwy avbl for ngt operations. ACTIVATE MIRL Rwy 16-34—CTAF.
COMMUNICATIONS: CTAF/UNICOM 122.8
CEDAR CITY FSS (CDC) TF 1-800-WX-BRIEF. NOTAM FILE CDC.
RADIO AIDS TO NAVIGATION: NOTAM FILE PVU.
PROVO (T) VORW/DME 108.4 PVU Chan 21 N40°12.90′ W111°43.28′ 179° 29.4 NM to fld. 4490/15E.

OGDEN-HINCKLEY (OGD) 3 SW UTC–7(–6DT) N41°11.76′ W112°00.73′ SALT LAKE CITY
4470 B S4 **FUEL** 80, 100, JET A1 + OX 1, 2 TPA—5215(745) ARFF Index Ltd. H-1C, L-7D
RWY 03-21: H8103X150 (ASPH-PFC) S-75, D-100, DT-170 HIRL 0.8% up SW IAP
RWY 03: MALS. VASI(V2L). Trees. **RWY 21:** Thld dsplcd 851′. Signs. Rgt tfc.
RWY 07-25: H5600X150 (ASPH) S-20, D-50, DT-70 MIRL 0.3% up W
RWY 07: REIL. VASI(V4L)—GA 3.5° TCH 50′. Tree. **RWY 25:** Thld dsplcd 202′. Road. Rgt tfc.
RWY 16-34: H5352X150 (ASPH) S-50, D-75, DT-120 MIRL 0.4% up S
RWY 16: PAPI(P2L)—GA 3.0° TCH 40′. Thld dsplcd 158′. Ditch. Rgt tfc.
RWY 34: PAPI(P2L)—GA 3.0° TCH 40′. Sign.
AIRPORT REMARKS: Attended continuously. Parachute jumping on arpt between Rwys 21 and 25. Rwy 07-25 CLOSED indefinitely. Flocks of birds on and in vicinity of arpt. No multiple approaches. No practice approaches—full stop ldgs only from 0500-1400Z‡. CLOSED to air carrier ops with more than 30 passenger seats except PPR call arpt manager or twr 801-629-8251/625-5569. Be alert parking lot lgts off the apch end of Rwy 34 can be confused for rwy lgts. Acft exceeding S-50, D-75 and DT-120 use Taxiway C and conc apron except PPR call arpt manager 801-629-8251/625-5569. Air carriers use Rwy 03-21 and Taxiway C only. No snow removal after twr closes. When twr clsd ACTIVATE HIRL Rwy 03-21 and taxiway lights—CTAF. MIRL Rwys 07-25 and 16-34 and REIL Rwy 07 not avbl when twr closed. NOTE: See Land and Hold Short Operations Section.
WEATHER DATA SOURCE: LAWRS.
COMMUNICATIONS: CTAF 118.7 **ATIS** 125.55 (1400-0500Z‡) **UNICOM** 122.95
CEDAR CITY FSS (CDC) TF 1-800-WX-BRIEF. NOTAM FILE OGD.
RCO 122.1R 115.7T (CEDAR CITY FSS)
Ⓡ **SALT LAKE CITY APP/DEP CON** 121.1
TOWER 118.7 (1400-0500Z‡) **GND CON** 121.7
AIRSPACE: CLASS D svc effective 1400-0500Z‡ other times CLASS G.
RADIO AIDS TO NAVIGATION: NOTAM FILE OGD.
(L) VORTAC 115.7 OGD Chan 104 N41°13.45′ W112°05.90′ 099° 4.3 NM to fld. 4220/14E.
VORTAC unusable 010°-130° beyond 25 NM below 11,300′ 350°-010° beyond 38 NM below 11,000′
ILS/DME 111.7 I-OGD Chan 54 Rwy 03. ILS/DME unmonitored when twr clsd.
COMM/NAVAID REMARKS: Emerg frequency 121.5 not avbl at twr.

FIGURE 201A.—Excerpt from Airport/Facilities Directory.

Form Approved: OMB No. 2120-0034

U.S. DEPARTMENT OF TRANSPORTATION FEDERAL AVIATION ADMINISTRATION **FLIGHT PLAN**	(FAA USE ONLY) ☐ PILOT BRIEFING ☐ VNR ☐ STOPOVER	TIME STARTED	SPECIALIST INITIALS

1. TYPE	2. AIRCRAFT IDENTIFICATION	3. AIRCRAFT TYPE/ SPECIAL EQUIPMENT	4. TRUE AIRSPEED	5. DEPARTURE POINT	6. DEPARTURE TIME PROPOSED (Z)	ACTUAL (Z)	7. CRUISING ALTITUDE
VFR				LAS			
X IFR	PTL 55	B/B747/R	460 KTS	LAS VEGAS			FL390
DVFR							

8. ROUTE OF FLIGHT
LAS OASIS8.BTY, J92 OAL, OAL.LOCKE SFO

9. DESTINATION (Name of airport and city)	10. EST. TIME ENROUTE HOURS	MINUTES	11. REMARKS
SFO SAN FRANCISCO INT'L.			L/O = LEVEL OFF PPH = POUNDS PER HOUR

12. FUEL ON BOARD HOURS	MINUTES	13. ALTERNATE AIRPORT(S)	14. PILOT'S NAME, ADDRESS & TELEPHONE NUMBER & AIRCRAFT HOME BASE / 17. DESTINATION CONTACT/TELEPHONE (OPTIONAL)	15. NUMBER ABOARD
4	30	OAK METROPOLITAN OAKLAND INT'L		339

16. COLOR OF AIRCRAFT	
WHITE/GREEN	CIVIL AIRCRAFT PILOTS. FAR Part 91 requires you file an IFR flight plan to operate under instrument flight rules in controlled airspace. Failure to file could result in a civil penalty not to exceed $1,000 for each violation (Section 901 of the Federal Aviation Act of 1958, as amended). Filing of a VFR flight plan is recommended as a good operating practice. See also Part 99 for requirements concerning DVFR flight plans.

FAA Form 7233-1 (8-82) CLOSE VFR FLIGHT PLAN WITH ________________ FSS ON ARRIVAL

FLIGHT LOG

CHECK POINTS FROM	TO	ROUTE ALTITUDE	COURSE	WIND TEMP	SPEED-KTS TAS	GS	DIST NM	TIME LEG	TOT	FUEL LEG	TOT
LAS	L/O	OASIS8.BTY CLIMB					65		:21:00		5600*
L/O	BTY	BTY R-126 FL390		340/53 ISA+4							
BTY	OAL	J92 FL390									
OAL	GROAN	OAL LOCKE.1 FL390									
GROAN	SFO	DESCENT					75	:26:00		4500	
SFO	OAK	VECTORS 3000					29	:09:00			

OTHER DATA: * Includes Taxi Fuel
NOTE: Use 12,000 PPH Total Fuel Flow From L/O To Start Of Descent.
Use 11,000 PPH Total Fuel Flow For Reserve/Alternate Requirements.

A Missed Approach Requires 1133# of Fuel

TIME and FUEL: As required by FARs.

TIME	FUEL (LB)	
		EN ROUTE
		RESERVE
		ALTERNATE
		TOTAL

FIGURE 202.—Flight Plan/Flight Log.

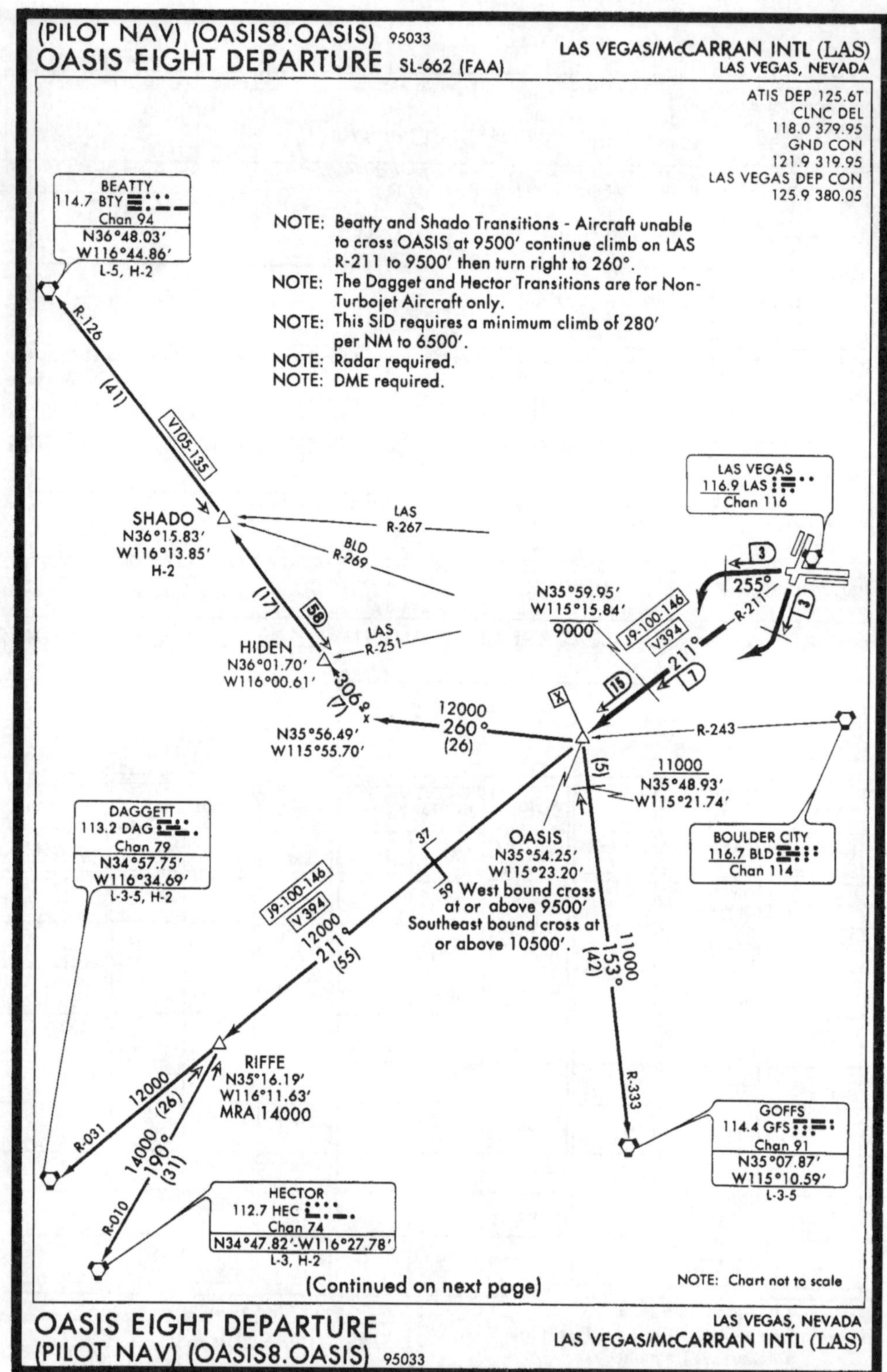

FIGURE 203.—OASIS EIGHT DEPARTURE (LAS).

(PILOT NAV) (OASIS8.OASIS) 95033
OASIS EIGHT DEPARTURE SL-662 (FAA)

LAS VEGAS/McCARRAN INTL (LAS)
LAS VEGAS, NEVADA

DEPARTURE ROUTE DESCRIPTION

TAKE-OFF RUNWAYS 19L/R: Climb on runway heading until reaching 3 DME, then turn right to intercept and proceed via the LAS R-211, thence. . . .
TAKE-OFF RUNWAYS 25L/R: Fly heading 255° until reaching 3 DME, then turn left to intercept and proceed via the LAS R-211, thence. . . .
. . . . cross the LAS R-211 7 DME at or below 9000′, then climb via LAS R-211 to OASIS INT, then via (transition) or (assigned route).
ALL RUNWAYS: Aircraft filing 17000 or above expect filed altitude/flight level 10 minutes after departure.
BEATTY TRANSITION (OASIS8.BTY): From over OASIS INT via heading 260° to intercept BTY R-126 to BTY VORTAC.
DAGGETT TRANSITION (OASIS8.DAG): From over OASIS INT via LAS R-211 and DAG R-031 to DAG VORTAC.
GOFFS TRANSITION (OASIS8.GFS): From over OASIS INT via GFS R-333 to GFS VORTAC.
HECTOR TRANSITION (OASIS8.HEC): From over OASIS INT via LAS R-211 and DAG R-031 to RIFFE INT, then via HEC R-010 to HEC VORTAC.
SHADO TRANSITION (OASIS8.SHADO): From over OASIS INT via heading 260° to intercept BTY R-126 to SHADO INT.

OASIS EIGHT DEPARTURE
(PILOT NAV) (OASIS8.OASIS) 95033

LAS VEGAS, NEVADA
LAS VEGAS/McCARRAN INTL (LAS)

FIGURE 203A.—OASIS EIGHT DEPARTURE (DEPARTURE ROUTE DESCRIPTIONS).

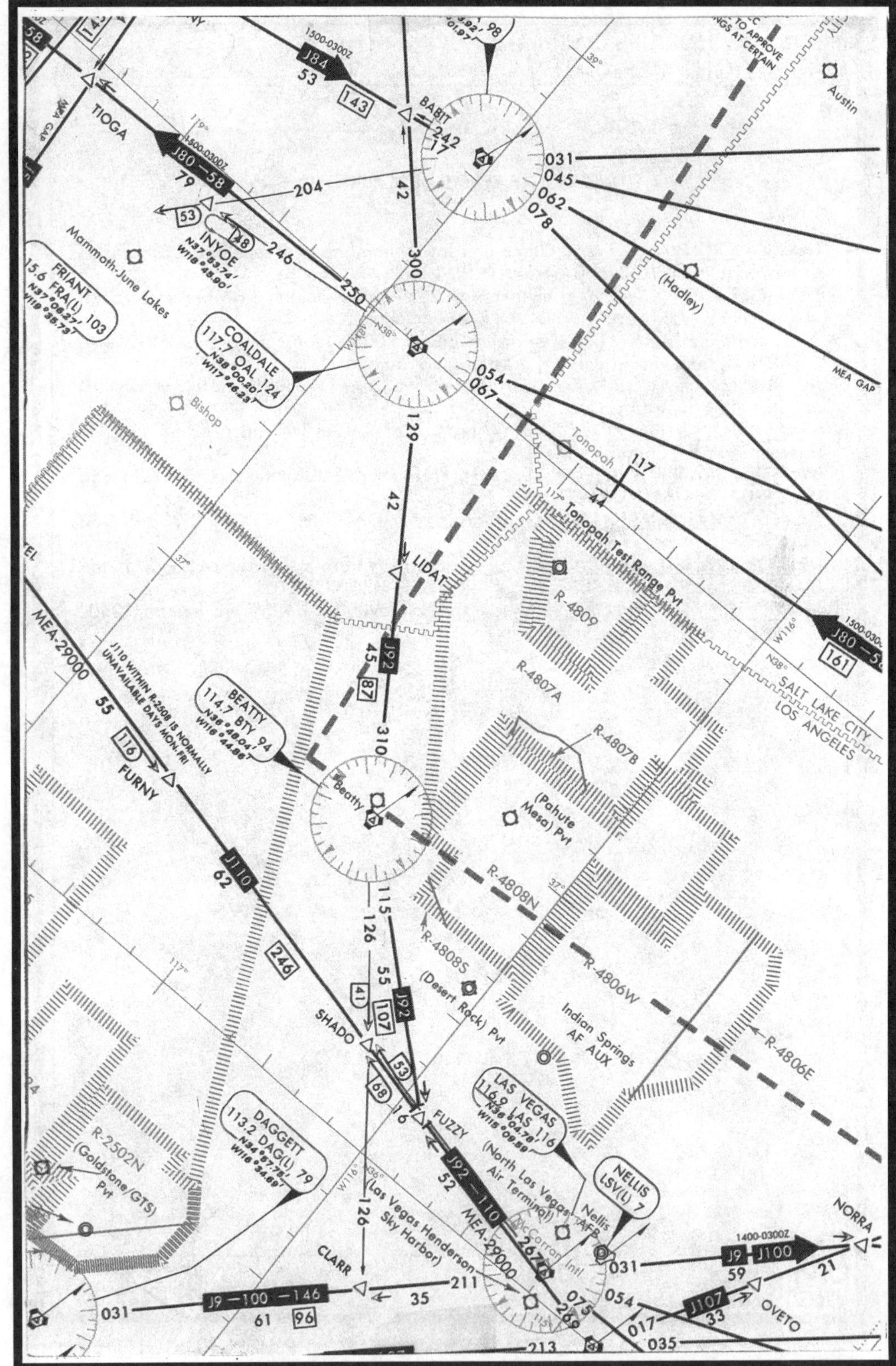

FIGURE 204.—High Altitude Airways.

SAN FRANCISCO INTL (SFO) 8 SE UTC–8(–7DT) N37°37.14′ W122°22.49′ **SAN FRANCISCO**
11 B S4 **FUEL** 100, 100LL OX 1, 2, 3, 4 ARFF Index E **H-2A, L-2F, A**
RWY 10L-28R: H11870X200 (ASPH-GRVD) S-60, D-200, DT-355, DDT-710 HIRL CL **IAP**
RWY 10L: REIL. VASI(V6L)—Upper GA 3.25° TCH 109′, Lower GA 3.0° TCH 69′. Transmission twr.
RWY 28R: ALSF2. TDZ. PAPI(P4L)—GA 3.0° TCH 51′. Rgt tfc.
RWY 10R-28L: H10600X200 (ASPH-GRVD) S-60, D-200, DT-355, DDT-710 HIRL CL
RWY 10R: VASI(V6L)—Upper GA 3.25° TCH 101′, Lower GA 3.0° TCH 60′. Transmission twr. Rgt tfc.
RWY 28L: SSALR.
RWY 01R-19L: H8901X200 (ASPH-GRVD) S-60, D-195, DT-325, DDT-710 HIRL CL
RWY 01R: REIL. Thld dsplcd 492′. Blast fence. **RWY 19L:** SSALS. TDZ.
RWY 01L-19R: H7001X200 (ASPH) S-60, D-170, DT-270, DDT-710 HIRL
RWY 01L: REIL. Trees. **RWY 19R:** VASI(V6L)—Upper GA 3.25° TCH 79′, Lower GA 3.0° TCH 47′.
AIRPORT REMARKS: Attended continuously. Rwy 19L SALSF are only 1100′ long with only one flasher on the last light station. Flocks of birds feeding along shoreline adjacent to arpt; on occasions fly across various parts of arpt. Noise sensitive arpt. For noise abatement procedures ctc arpt noise office Monday-Friday 1600-0100Z‡ by calling 415-876-2220. Ldg fee. Rubber accumulated on first 3000 feet of Rwys 28L-28R. No grooving exists at arpt rwy intersections. Rwy 01R-19L is grooved full length except area between Rwys 28L and 28R and 535′ from Taxiway Charlie north. Rwy 10L-28R grooved full length except from Taxiway Tango to Rwy 10L thld. Rwy 10R-28L grooved full length except from east edge of Rwy 01R-19L to Taxiway Kilo. Rwy 01L-19R grooved full length except from south edge of Taxiway Foxtrot to north edge of Rwy 10L-28R. Widebody acft restricted on Taxiway M west of Taxiway A. Several rwy hold position signs are on the right rather than the left side of the taxiways. Rwys 01L-19R, 01R-19L, 10L-28R and 10R-28L gross weight limit DC-10-10 430,000 pounds, DC-10-30 555,000 pounds, L-1011-100 450,000 pounds, L-1011-200 466,000 pounds, B-747 710,000 pounds. 747-400's shall taxi at a speed of less than 10 miles per hour on all non-restricted taxiways on the terminal side of the intersecting rwys. Movement speed of not more than 5 miles per hour is required when two 747-400's pass or overtake each other on parallel taxiways A and B. 747-400 are restricted from using Twy E to or from Twy B. Airline pilots shall strictly follow the painted nose-gear lines and no oversteering adjustment is permitted. Acft with wingspan of 140-156′ must be under tow with wing walkers on Twy R southwest of the fix-base operator, acft with wingspan exceeding 156′ are prohibited. B747 and larger acft are prohibited from using Twy A between Twy S and the United Airline Freedom area. Twy M clsd west of Gate 16 to acft exceeding a wingspan of 125′. Flight Notification Service (ADCUS) available. NOTE: See Land and Hold Short Operations Section.
WEATHER DATA SOURCES: AWOS-1 118.05 (San Bruno Hill). LLWAS.
COMMUNICATIONS: ATIS (ARR) 118.85 113.7 108.9 (415) 877-3585 (DEP) 135.45 (415) 877-8422/8423
UNICOM 122.95
OAKLAND FSS (OAK) TF 1-800-WX-BRIEF. NOTAM FILE SFO.
Ⓡ **BAY APP CON** 134.5 132.55 135.65
Ⓡ **BAY DEP CON** 135.1 (SE-W) 120.9 (NW-E)
TOWER 120.5 **GND CON** 121.8 (Gates 53-90 W side) 124.25 (Gates 1-52 E side) **CLNC DEL** 118.2
PRE TAXI CLNC 118.2
AIRSPACE: CLASS B See VFR Terminal Area Chart.
RADIO AIDS TO NAVIGATION: NOTAM FILE SFO.
(L) VORW/DME 115.8 SFO Chan 105 N37°37.17′ W122°22.43′ at fld. 10/17E.
VOR/DME unusable:
035-055°beyond 15 NM below 6500′ 190-260°beyond 10 NM below 4500′
025-065°beyond 30 NM 260-295°beyond 35 NM below 3000′
150-190°beyond 25 NM below 4500′ 295-330°beyond 20 NM below 4000′
BRIJJ NDB (LOM) 379 SF N37°34.33′ W122°15.59′ 280° 6.2 NM to fld.
Unusable 160°-195° byd 6 NM all altitudes.
ILS/DME 109.55 I-SFO Chan 32Y Rwy 28L. LOM BRIJJ NDB. LOM unusable 160°-195° byd 6 NM all altitudes.
ILS/DME 111.7 I-GWQ Chan 54 Rwy 28R. LOM BRIJJ NDB. LOM unusable 160°-195° byd 6 NM all altitudes.
ILS/DME 108.9 I-SIA Chan 26 Rwy 19L.
LDA/DME 110.75 Chan 44(Y) Rwy 28R.
COMM/NAVID REMARKS: ILS Rwy 19L-pilots be alert for momentary LOC course excursions due to large acft opr in vicinity of LOC antenna. ATIS frequency 108.9 avbl when SFO VOR out of service 415-877-3585.

SAN JACINTO N33°47.70′ W116°59.96′ NOTAM FILE RAL. **LOS ANGELES**
NDB (MHW) 227 SJY 184° 3.8 NM to Hemet-Ryan. NDB unmonitored. **L-3C**

FIGURE 205.—Excerpt from Airport/Facilities Directory.

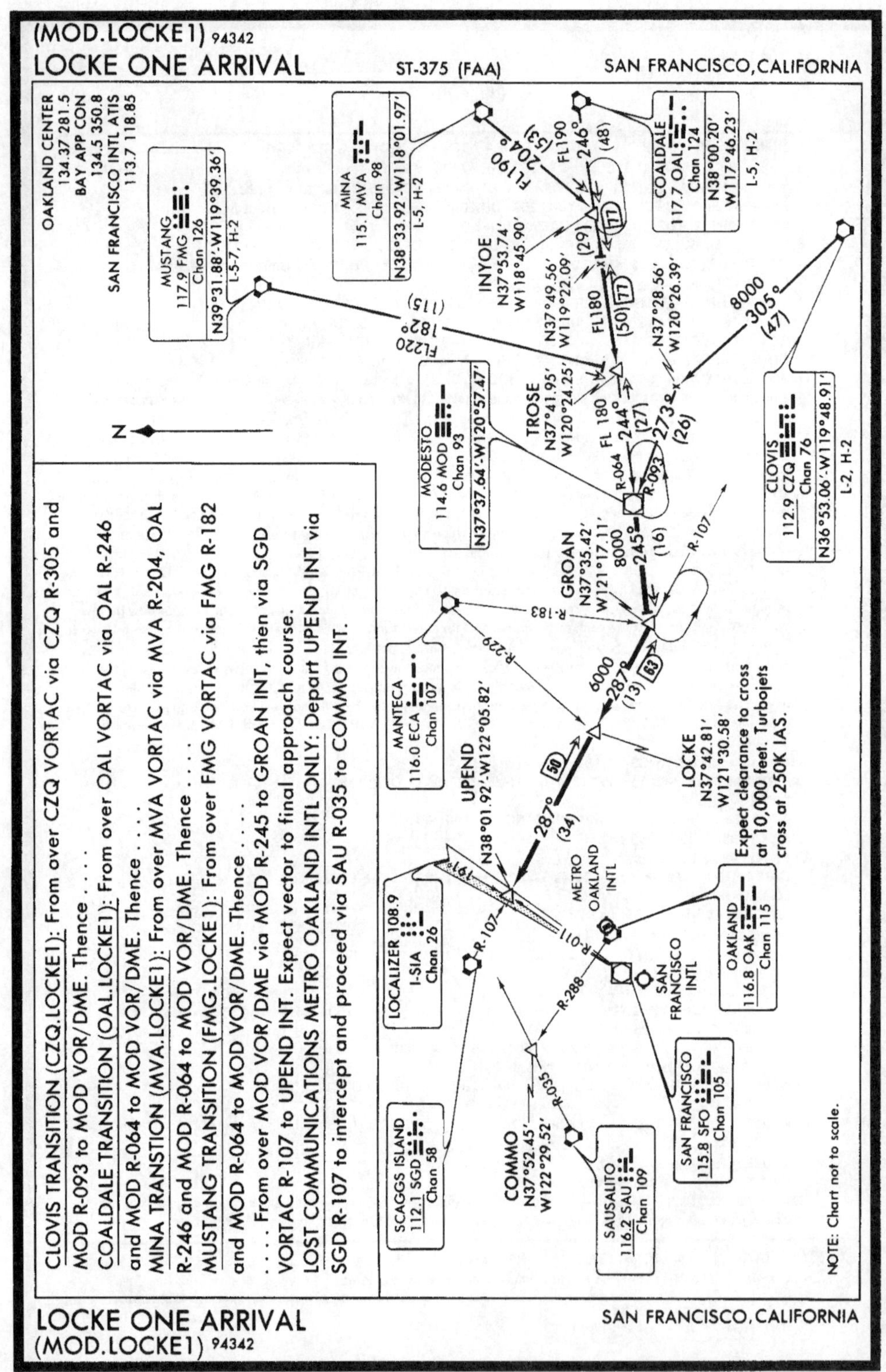

FIGURE 205A.—LOCKE ONE ARRIVAL (MOD.LOCKE1).

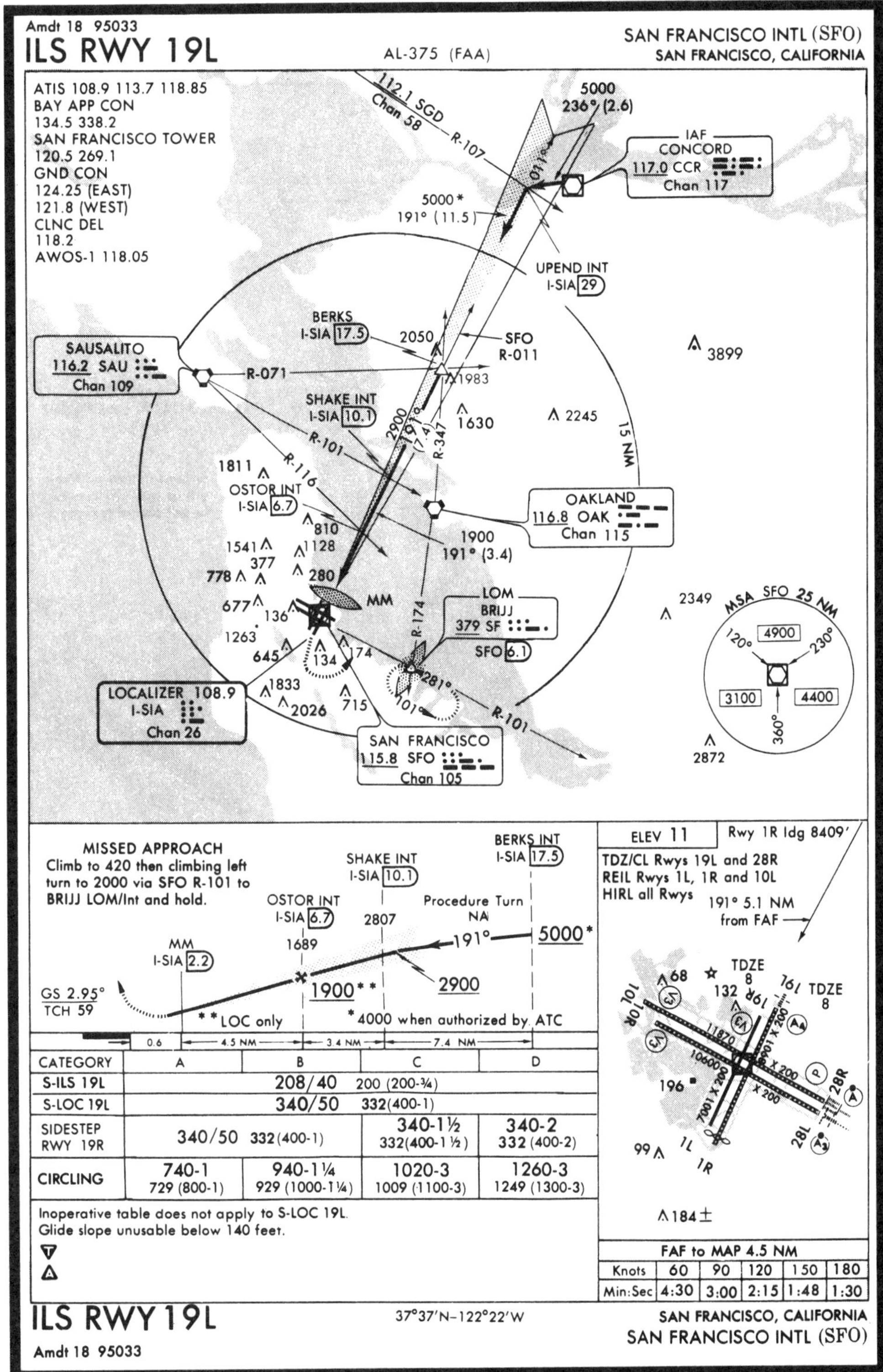

FIGURE 206.—ILS RWY 19L (SFO).

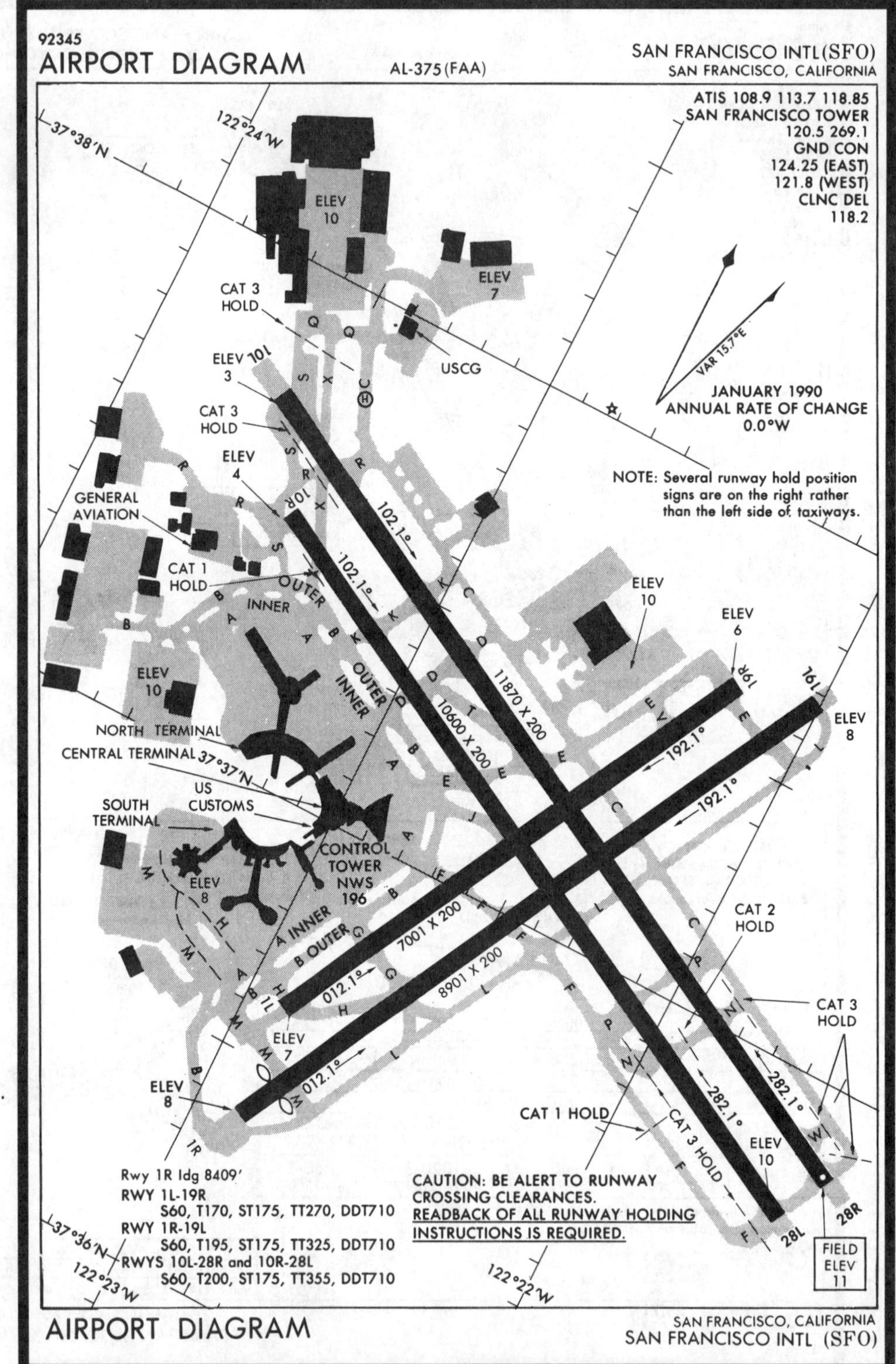

FIGURE 206A.—AIRPORT DIAGRAM.

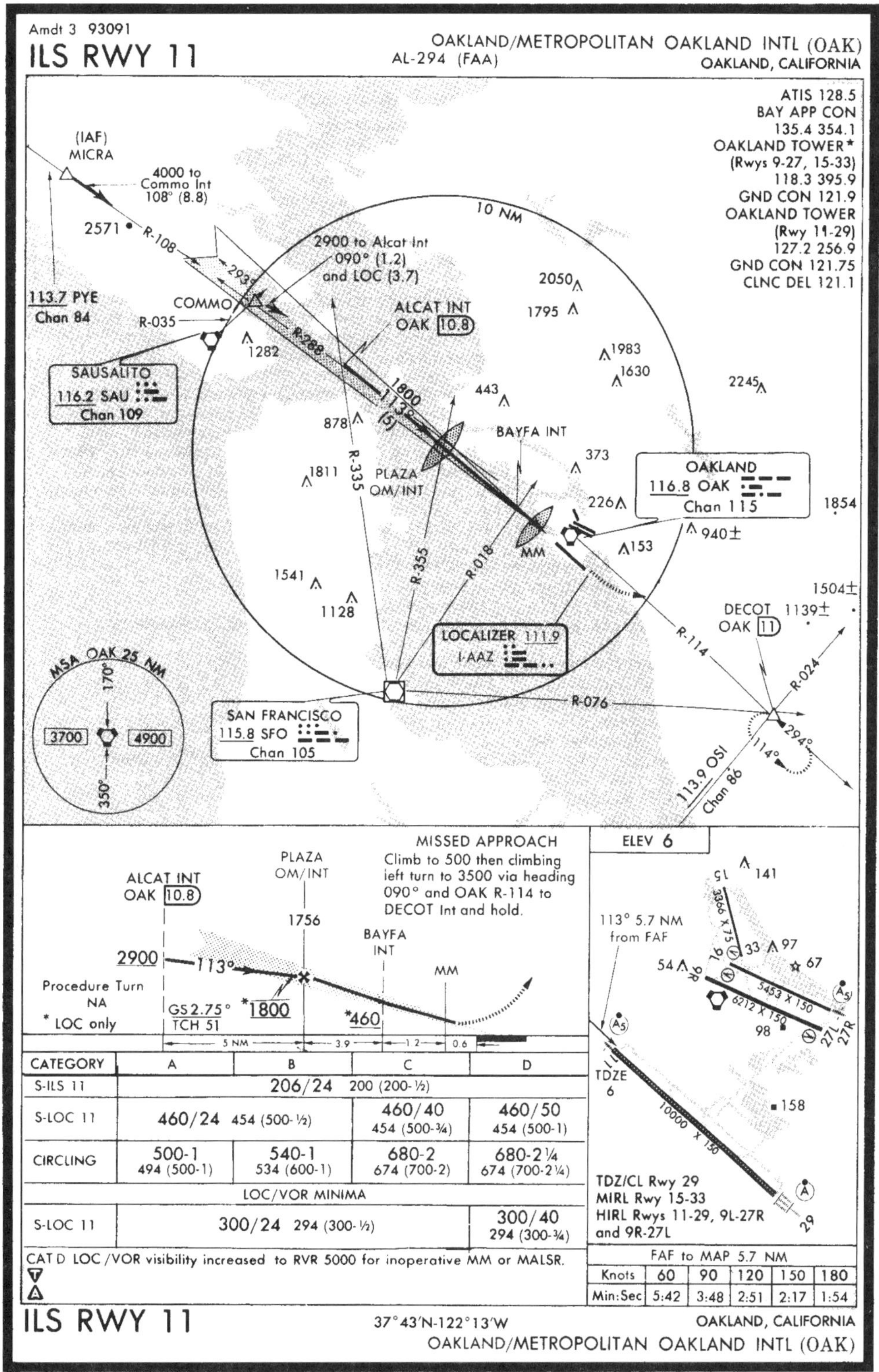

CATEGORY	A	B	C	D
S-ILS 11	206/24 200 (200-½)			
S-LOC 11	460/24 454 (500-½)		460/40 454 (500-¾)	460/50 454 (500-1)
CIRCLING	500-1 494 (500-1)	540-1 534 (600-1)	680-2 674 (700-2)	680-2¼ 674 (700-2¼)
LOC/VOR MINIMA				
S-LOC 11	300/24 294 (300-½)			300/40 294 (300-¾)

FAF to MAP 5.7 NM					
Knots	60	90	120	150	180
Min:Sec	5:42	3:48	2:51	2:17	1:54

FIGURE 207.—ILS RWY 11 (OAK).

OAKLAND

METROPOLITAN OAKLAND INTL (OAK) 4 S UTC–8(–7DT) N37°43.28′ W122°13.24′ **SAN FRANCISCO**
06 B S4 **FUEL** 100LL, JET A OX 1, 2, 3, 4 TPA—See Remarks LRA ARFF Index D **H-2A, L-2F, A**
RWY 11-29: H10000X150 (ASPH-PFC) S-200, D-200, DT-400, DDT-900 HIRL CL **IAP**
RWY 11: MALSR. Rgt tfc. **RWY 29:** ALSF2. TDZ.
RWY 09R-27L: H6212X150 (ASPH-PFC) S-75, D-200, DT-400, DDT-800 HIRL
RWY 09R: VASI(V4L)—GA 3.0° TCH 46′. Tree. **RWY 27L:** VASI(V4L)—GA 3.0° TCH 55′.
RWY 09L-27R: H5453X150 (ASPH) S-75, D-115, DT-180 HIRL
RWY 09L: VASI(V4L)—GA 3.0° TCH 38′. **RWY 27R:** MALSR. Building. Rgt tfc.
RWY 15-33: H3366X75 (ASPH) S-12.5, D-65, DT-100 MIRL
RWY 33: Rgt tfc.
AIRPORT REMARKS: Attended continuously. Fee Rwy 11-29 and tiedown Birds on and in vicinity of arpt. Rwy 09L-27R and Rwy 15-33 CLOSED to air carrier acft, except air carrier acft may use Rwy 09L and 27R for taxiing. Rwy 09L-27R and Rwy 09R-27L CLOSED to 4 engine wide body acft except Rwy 09R-27L operations avbl PPR call operations supervisor 510-577-4067. All turbo-jet/fan acft, all 4-engine acft and turbo-prop acft with certificated gross weight over 12,500 pounds are prohibited from tkf Rwys 27R/27L or ldg Rwy 09L and Rwy 09R. Preferential rwy use program in effect 0600-1400Z‡: All acft preferred north fld arrive Rwys 27R/27L or Rwy 33; all acft preferred north fld dep Rwys 09R/09L or Rwy 15. If these rwys unacceptable for safety or ATC instructions then Rwy 11-29 must be used. Prohibitions not applicable in emerg or whenever Rwy 11-29 is closed due to maintenance, construction or safety. For noise abatement information ctc noise abatement office at 510-577-4276. 400′ blast pad Rwy 29 and 500′ blast pad Rwy 11. Rwy 29 and Rwy 27L distance remaining signs left side. Acft with experimental or limited certification having over 1,000 horsepower or 4,000 pounds are restricted to Rwy 11-29. Rwy 09R-27L FAA gross weight strength DC 10-10 350,000 pounds, DC 10-30 450,000 pounds, L-1011 350,000 pounds. Rwy 11-29 FAA gross weight strength DC 10-10 600,000 pounds, DC 10-30 700,000 pounds, L-1011 600,000 pounds. TPA—Rwy 27L 606(600), TPA—Rwy 27R 1006(1000). Rwy 29 centerline lgts 6500′. Flight Notification Service (ADCUS) available.
COMMUNICATIONS: ATIS 128.5 (510) 635-5850 (N and S Complex) **UNICOM** 122.95
OAKLAND FSS (OAK) on arpt. 122.5 122.2. TF 1-800-WX-BRIEF. NOTAM FILE OAK.
® **BAY APP CON** 135.65 133.95 (South) 135.4 134.5 (East) 135.1 (West) 127.0 (North) 120.9 (Northwest) 120.1 (Southeast)
® **BAY DEP CON** 135.4 (East) 135.1 (West) 127.0 (North) 120.9 (Northwest)
OAKLAND TOWER 118.3 (N Complex) 127.2 (S Complex) 124.9
GND CON 121.75 (S Complex) 121.9 (N Complex) **CLNC DEL** 121.1
AIRSPACE: CLASS C svc ctc **APP CON**
RADIO AIDS TO NAVIGATION: NOTAM FILE OAK.
OAKLAND (H) VORTACW 116.8 OAK Chan 115 N37°43.55′ W122°13.42′ at fld. 10/17E. **HIWAS.**
RORAY NDB (LMM) 341 AK N37°43.28′ W122°11.65′ 253° 1.3 NM to fld.
ILS 108.7 I-INB Rwy 29
ILS 111.9 I-AAZ Rwy 11
ILS 109.9 I-OAK Rwy 27R LMM RORAY NDB.

OAKLAND N37°43.56′ W122°13.42′ NOTAM FILE OAK. **SAN FRANCISCO**
(H) VORTACW 116.8 OAK Chan 115 at Metropolitan Oakland Intl. 10/17E. **HIWAS.** **H-2A, L-2F, A**
VOR unusable: 307°-323° byd 10 NM blo 5,000′ 307°-323° byd 17 NM blo 12,500′
DME unusable:
307°-323° byd 30 NM blo 1,500′ 040°-065° byd 30 NM blo 4,100′
350°-030° byd 20 NM blo 3,500′
FSS (OAK) at Metropolitan Oakland Intl. 122.5 122.2 TF 1-800-WX-BRIEF.

OCEANO CO (L52) 1 W UTC–8(–7DT) N35°06.08′ W120°37.33′ **LOS ANGELES**
14 B S4 **FUEL** 100LL TPA—1000(986)
RWY 11-29: H2325X50 (ASPH) S-12.5 MIRL
RWY 11: P-line. Rgt tfc. **RWY 29:** Pole.
AIRPORT REMARKS: Attended 1600-0100Z‡. Arpt unattended Christmas day. For fuel after hours call 805-481-6100. Ultralight activity on and in vicinity of arpt. Recurring flocks of waterfowl on and in vicinity of arpt. Be alert for kites flown along beach 1/2 mile west of rwy. Unsurfaced areas soft and unusable. Taxilanes very narrow near buildings and parked acft. Extremely noise sensitive arpt and community, for tkf Rwy 29 pilots are requested to maintain rwy heading until crossing the shoreline. ACTIVATE MIRL Rwy 11-29—CTAF.
COMMUNICATIONS: CTAF/UNICOM 122.7
HAWTHORNE FSS (HHR) TF 1-800-WX-BRIEF. NOTAM FILE HHR.

OCEAN RIDGE (See GUALALA)

OCEANSIDE N33°14.44′ W117°25.06′ NOTAM FILE CRQ. **LOS ANGELES**
(H) VORTAC 115.3 OCN Chan 100 097° 3.6 NM to Oceanside Muni. 90/15E. **H-2B, L-3C**
VOR unusable 260°-265° byd 20NM.

FIGURE 207A.—Excerpt from Airport/Facilities Directory.

Form Approved: OMB No. 2120-0034

U.S. DEPARTMENT OF TRANSPORTATION FEDERAL AVIATION ADMINISTRATION **FLIGHT PLAN**	(FAA USE ONLY) ☐ PILOT BRIEFING ☐ STOPOVER	☐ VNR	TIME STARTED	SPECIALIST INITIALS

1. TYPE	2. AIRCRAFT IDENTIFICATION	3. AIRCRAFT TYPE/ SPECIAL EQUIPMENT	4. TRUE AIRSPEED	5. DEPARTURE POINT	6. DEPARTURE TIME PROPOSED (Z)	ACTUAL (Z)	7. CRUISING ALTITUDE
VFR ☐ / X IFR / DVFR ☐	SLING 2	S76/A	** KTS	EYW KEY WEST INT'L			9000

8. ROUTE OF FLIGHT: TIGAR, V157 MIA, V51 PHK, V437 MLB, V3 SMYRA, DA, DAB

9. DESTINATION (Name of airport and city)	10. EST. TIME ENROUTE HOURS	MINUTES	11. REMARKS
DAB DAYTONA BEACH			PPH = POUNDS PER HOUR **CAS 120 TEMP ISA+10 TO ISA+1

12. FUEL ON BOARD HOURS	MINUTES	13. ALTERNATE AIRPORT(S)	14. PILOT'S NAME, ADDRESS & TELEPHONE NUMBER & AIRCRAFT HOME BASE / 17. DESTINATION CONTACT/TELEPHONE (OPTIONAL)	15. NUMBER ABOARD
		SGJ ST. AUGUSTINE		6

16. COLOR OF AIRCRAFT: GREY/RED

CIVIL AIRCRAFT PILOTS. FAR Part 91 requires you file an IFR flight plan to operate under instrument flight rules in controlled airspace. Failure to file could result in a civil penalty not to exceed $1,000 for each violation (Section 901 of the Federal Aviation Act of 1958, as amended). Filing of a VFR flight plan is recommended as a good operating practice. See also Part 99 for requirements concerning DVFR flight plans.

FAA Form 7233-1 (8-82)   CLOSE VFR FLIGHT PLAN WITH ________________ FSS ON ARRIVAL

FLIGHT LOG

CHECK POINTS FROM	TO	ROUTE / ALTITUDE	COURSE	WIND / TEMP	SPEED-KTS TAS	GS	DIST NM	TIME LEG	TOT	FUEL LEG	TOT
EYW	TIGAR	DIRECT / CLIMB					28		:14:00		205.4*
TIGAR	MIA	V157 / 9000		220/23 / ISA+10							
MIA	PHK	V51		200/19 / ISA+10							
PHK	MLB	V437		180/17 / ISA+1							
MLB	SMYRA	V3		190/19 / ISA+1							
SMYRA	DAB	DESCENT / APPROACH					16	:09:08		139.3	
DAB	SGJ	DIRECT / 4000					48	:21:00			

OTHER DATA: * Includes Taxi Fuel

NOTE: Use 740 PPH Total Fuel Flow From L/O To Start Of Descent.
Use 705 PPH Total Fuel Flow For Reserve And Alternate Requirements.

A Missed Approach Requires 51# of Fuel.

TIME and FUEL: As required by FARs.

TIME	FUEL (LB)	
		EN ROUTE
		RESERVE
		ALTERNATE
		TOTAL

FIGURE 208.—Flight Plan/Flight Log.

KEYSTONE HEIGHTS

KEYSTONE AIRPARK (42J) 3 N UTC-5(-4DT) N29°50.66' W82°03.01' **JACKSONVILLE**
196 B S4 **FUEL** 100LL, JET A TPA 1196 (1000) **H-5E, L-18H, 19B**
RWY 04-22: H5025X100 (ASPH) S-40, D-80 MIRL **IAP**
RWY 04: PAPI(P2L)—GA 3.0° TCH 40'. Trees. **RWY 22:** PAPI(P2L)—GA 3.0° TCH 40'. Trees.
RWY 10-28: H4900X75 (ASHP) S-30, D-60
RWY 10: Trees. **RWY 28:** Trees.
RWY 16-34: H4400X150 (ASPH) S-25
RWY 16: Thld dsplcd 200'. Trees. **RWY 34:** Trees.
AIRPORT REMARKS: Attended 1300-2300Z‡. Parachute Jumping. CAUTION: Ultralgt activity on and in vicinity of arpt. CAUTION—Animals on and in vicinity of arpt. ACTIVATE MIRL Rwy 04-22—CTAF. Rwy 16-34 cracked with loose grvl and weeds growing thru cracks.
COMMUNICATIONS: CTAF/UNICOM 122.7
GAINESVILLE FSS (GNV) TF 1-800-WX-BRIEF. NOTAM FILE GNV.
® **JACKSONVILLE APP/DEP CON** 123.8
RADIO AIDS TO NAVIGATION: NOTAM FILE GNV.
GAINESVILLE (L) VORTAC 116.2 GNV Chan 109 N29°34.33' W82°21.76' 044° 23.1 NM to fld. 60/01E.
HIWAS.

KEY WEST N24°35.15' W81°48.03' NOTAM FILE EYW. **MIAMI**
(H) VORTAC 113.5 EYW Chan 82 128° 2.8 NM to Key West Intl. 10/01E. **HIWAS.** **H-5E, L-19D**
RCO 123.65 122.2 122.1R 113.5T (MIAMI FSS)
FSS freqs 123.65 and 122.2 unusable 330°-015° byd 20 NM blo 1500'.

KEY WEST INTL (EYW) 2 E UTC-5(-4DT) N24°33.37' W81°45.57' **MIAMI**
4 B S4 **FUEL** 100, JET A AOE ARFF Index A **L-19D**
RWY 09-27: H4800X100 (ASPH-GRVD) S-40, D-95, DT-130 MIRL **IAP**
RWY 09: REIL. VASI(V4L)—GA 3.0° TCH 34'. Tree. **RWY 27:** REIL. VASI(V4L)—GA 3.0° TCH 34'. Tree.
AIRPORT REMARKS: Attended 1300-2300Z‡. CAUTION: Numerous flocks of birds on and in the vicinity of airport. CAUTION—Restricted area R-2916 14 NM NE of arpt has strobe-lgtd and marked balloon and cable to 14,000 ft. Noise Sensitive Area: all jet acft use NBAA close in noise abatement procedures. PPR for unscheduled air carrier operations with more than 30 passenger seats 0430-1045Z‡; Call arpt manager 305-296-7223. ACTIVATE MIRL Rwy 09-27, VASI/REIL Rwys 09-27—CTAF. Intensive military jet tfc S and E of arpt; acft entering arpt tfc area from SE through W. Enter arpt tfc area blo 2000'; refer to MIAMI VFR Terminal Area Chart for suggested VFR flyway routes. Flight Notification Service (ADCUS) available.
COMMUNICATIONS: CTAF 118.2 **UNICOM** 122.95
MIAMI FSS (MIA) TF 1-800-WX-BRIEF. NOTAM FILE EYW.
RCO 123.65 122.2 122.1R 113.5T (MIAMI FSS)
® **NAVY KEY WEST APP/DEP CON** 124.45 (1200-0500Z‡) ® **MIAMI CENTER APP/DEP CON** 132.2 (0500-1200Z‡)
TOWER 118.2 (1100-0300Z‡) **GND CON** 121.9 **CLNC DEL** 121.9
AIRSPACE: CLASS D svc effective 1100-0300Z‡ other times CLASS G.
RADIO AIDS TO NAVIGATION: NOTAM FILE EYW. VHF/DF ctc MIAMI FSS.
(H) VORTAC 113.5 EYW Chan 82 N24°35.15' W81°48.03' 127° 2.9 NM to fld. 10/01E. **HIWAS.**
FISH HOOK NDB (H) 332 FIS N24°32.90' W81°47.18' 073° 1.5 NM to fld.
ASR
COMM/NAVAID REMARKS: FSS freqs 123.65 and 122.2 unusable 330°-015° beyond 20 NM below 1500'.

KEY WEST NAS **AIRSPACE: CLASS D** svc effective 1100-0300Z‡ other times CLASS G.

KISSIMMEE MUNI (See ORLANDO)

KNIGHT N27°54.50' W82°27.26' NOTAM FILE PIE. **MIAMI**
NDB (MHW) 270 TPF at Peter O'Knight. NDB unusable byd 20NM. **L-19B**

KOBRA N30°51.19' W86°32.20' NOTAM FILE CEW. **NEW ORLEANS**
NDB (LOM) 201 CE 170° 4.5 NM to Bob Sikes.

LA BELLE N26°49.69' W81°23.49' NOTAM FILE MIA **MIAMI**
(L) VORTAC 110.4 LBV Chan 41 205° 5.2 NM to La Belle Muni. 30/01E. **H-5E, L-19C**
RCO 122.1R 110.4T (MIAMI FSS)

FIGURE 209.—Excerpt from Airport/Facilities Directory.

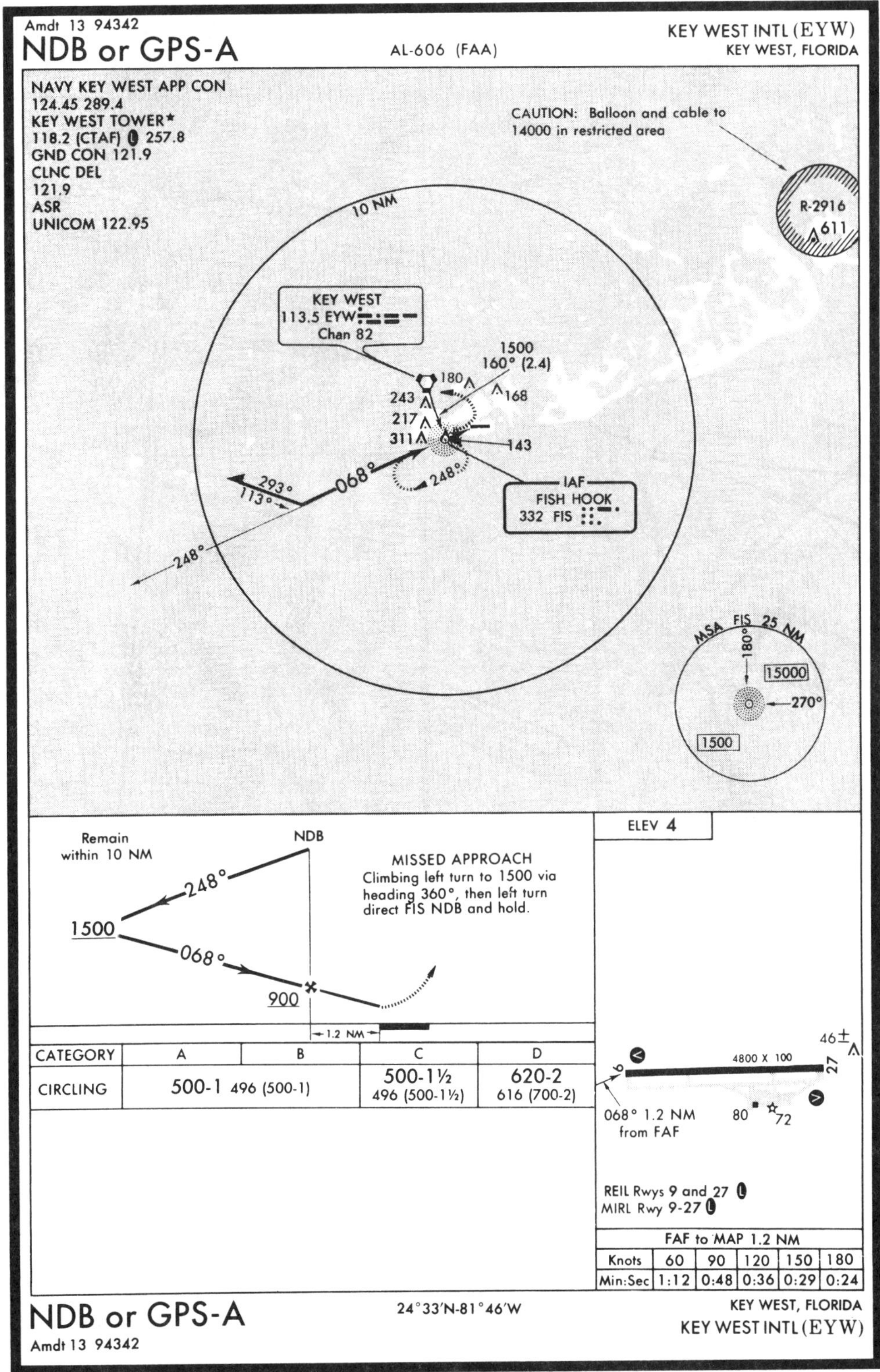

CATEGORY	A	B	C	D
CIRCLING	500-1 496 (500-1)		500-1½ 496 (500-1½)	620-2 616 (700-2)

FAF to MAP 1.2 NM					
Knots	60	90	120	150	180
Min:Sec	1:12	0:48	0:36	0:29	0:24

FIGURE 209A.—NDB or GPS-A (EYW).

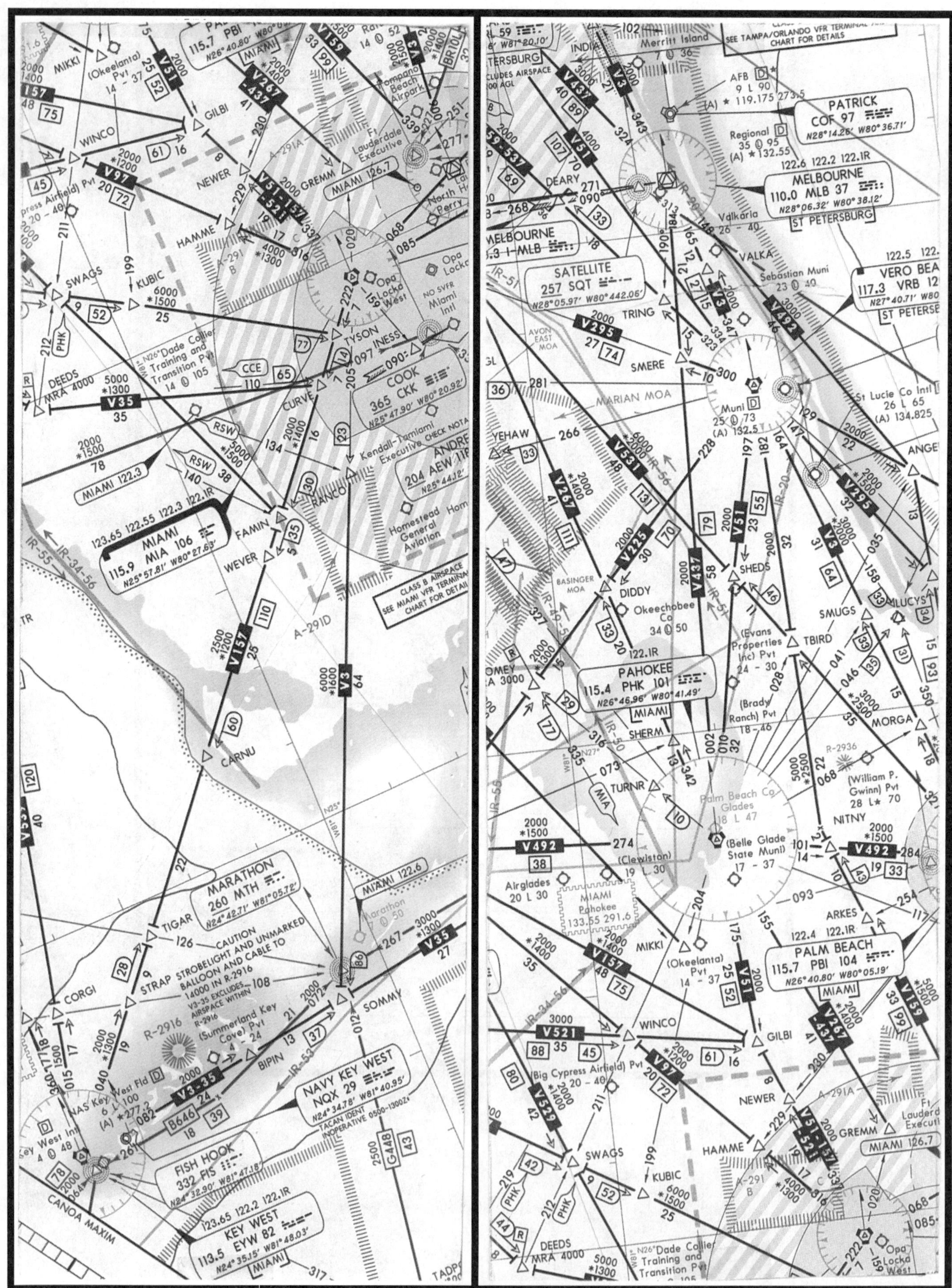

FIGURE 210.—Low Altitude Airways.

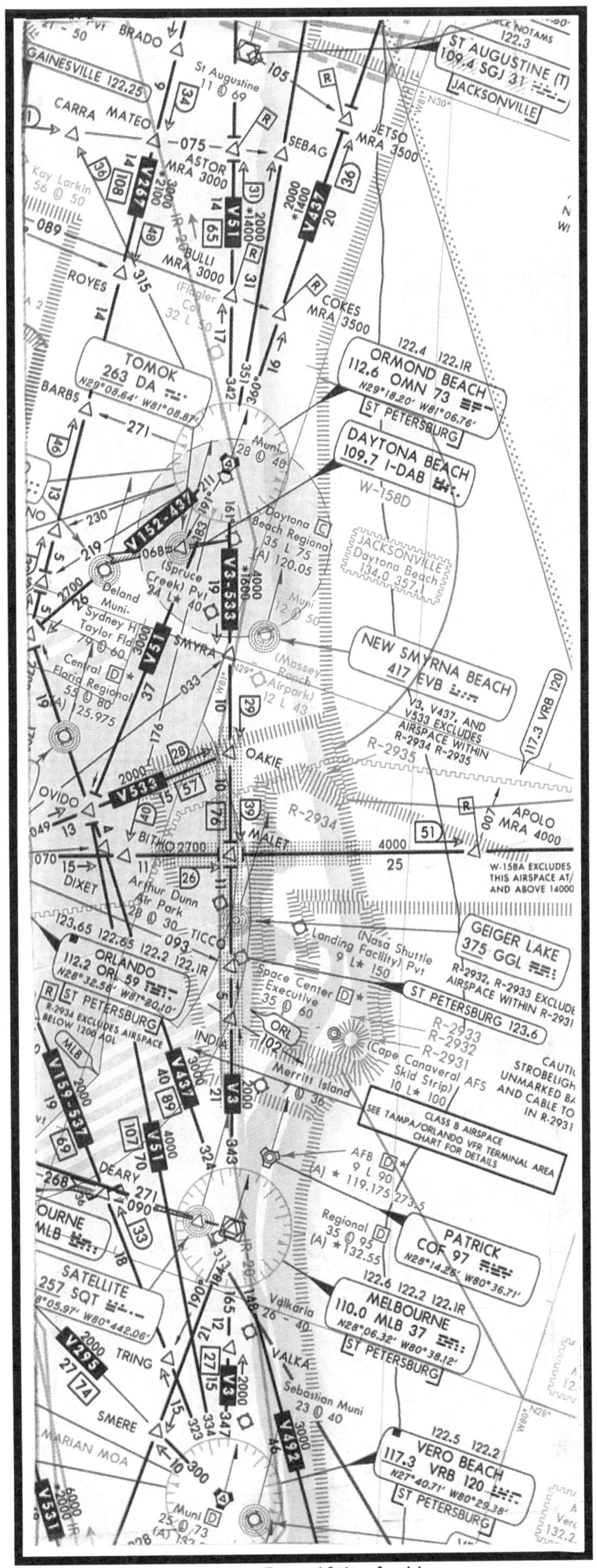

FIGURE 211.—Low Altitude Airways.

CROSS CITY (CTY) 1 E UTC−5(−4DT) N29°38.07′ W83°06.34′ — JACKSONVILLE H-5D, L-18G, 19B IAP
42 B **FUEL** 100LL
RWY 13-31: H5001X150 (ASPH) S-13 MIRL
RWY 13: Trees. **RWY 31:** Trees.
RWY 04-22: H4050X150 (ASPH) S-13 MIRL
RWY 04: Trees. **RWY 22:** Trees.
AIRPORT REMARKS: Attended continuously. ACTIVATE MIRL Rwy 13-31 and 04-22—CTAF.
COMMUNICATIONS: CTAF/UNICOM 122.8
GAINESVILLE FSS (GNV) TF 1-800-WX-BRIEF. NOTAM FILE GNV.
RCO 122.1R 112.0T (GAINESVILLE FSS)
® **JAX CENTER APP/DEP CON** 127.8
RADIO AIDS TO NAVIGATION: NOTAM FILE GNV.
(L) VORTAC 112.0 CTY Chan 57 N29°35.94′ W83°02.92′ 308° 3.7 NM to fld. 30/02W.

CRYSTAL RIVER (X31) 3 SE UTC−5(−4DT) N28°52.07′ W82°34.47′ — JACKSONVILLE L-19B IAP
10 B S2 **FUEL** 100LL, JET A
RWY 09-27: H4297X60 (ASPH) LIRL
RWY 09: PAPI(P2R)—GA 3.0° TCH 38′. P-line. **RWY 27:** REIL. Trees.
RWY 18-36: 3020X57 (TURF)
RWY 18: Thld dsplcd 517′. Building. **RWY 36:** Thld dsplcd 748′. Road.
AIRPORT REMARKS: Attended 1200–0000Z‡. Aerobatic activity along N side of Rwy 09-27 and 4 NM SW of arpt. Ctc unicom for tfc info and ST PETERSBURG FSS for specific times. Rwy 18-36 dsplcd thld marked with green pipes. Rwy 18-36 marked with white pipes every 200′. Rwy 18-36 acft parked 45′ from rwy W edge. ACTIVATE LIRL Rwy 09-27—CTAF. Glider ops within 25 NM.
COMMUNICATIONS: CTAF/UNICOM 122.7
ST PETERSBURG FSS (PIE) TF 1-800-WX-BRIEF. NOTAM FILE PIE.
® **JAX CENTER APP/DEP CON** 135.75
RADIO AIDS TO NAVIGATION: NOTAM FILE OCF.
OCALA (L) VORTAC 113.7 OCF Chan 84 N29°10.65′ W82°13.58′ 225° 26 NM to fld. 80/00E.

CYPRESS N26°09.21′ W81°46.69′ NOTAM FILE APF. — MIAMI H-5E, L-19C
(T) VORW/DME 108.6 CCE Chan 23 at Naples Muni. 10/00E.

DADE-COLLIER TRAINING AND TRANSITION (See MIAMI)

DAVIE N26°04.34′ W80°14.69′ — MIAMI L-19C, A
RCO 126.7 (MIAMI FSS)

DAYTONA BEACH INTL (DAB) 3 SW UTC−5(−4DT) N29°10.80′ W81°03.48′ — JACKSONVILLE H-5E, L-18H, 19B IAP
35 B S4 **FUEL** 100LL, JET A OX 2 TPA—See Remarks ARFF Index C
RWY 07L-25R: H10500X150 (ASPH-GRVD) S-75, D-140, DT-220 HIRL
RWY 07L: MALSR. Thld dsplcd 700′.
RWY 25R: REIL. VASI(V6L)—Upper GA 3.25° TCH 95.3′. Lower GA 2.75° TCH 53.4′. Rgt tfc.
RWY 16-34: H6000X150 (ASPH-GRVD) S-75, D-150, DT-260 MIRL
RWY 16: REIL. PAPI(P4L)—GA 3.0° TCH 45′. Road. Rgt tfc.
RWY 34: REIL. PAPI(P4L)—GA 3.0° TCH 45′. Trees.
RWY 07R-25L: H3197X100 (ASPH) S-30 MIRL
RWY 07R: PAPI(P2L)—GA 2.86° TCH 40′. Trees. Rgt tfc. **RWY 25L:** PAPI(P2L)—GA 2.86° TCH 32′. Ground.
AIRPORT REMARKS: Attended continuously. Heavy migratory bird activity on and in vicinity of arpt. TPA—835(800) lgt acft; 1235(1200) high performance acft. E end of Twy S is non-movement area. NOTE: See Land and Hold Short Operations Section.
WEATHER DATA SOURCES: LLWAS.
COMMUNICATIONS: ATIS 120.05 **UNICOM** 122.95
ST PETERSBURG FSS (PIE) TF 1-800-WX-BRIEF. NOTAM FILE DAB.
® **APP CON** 135.57 (9000′ and above) 118.85 (N 4000′-8500′) 127.07 (S 4000′-8500′) 125.8 (N 3500′ and blo) 125.35 (S 3500′ and blo)
TOWER 120.7 118.1 **GND CON** 121.9 **CLNC DEL** 119.3
® **DEP CON** 123.9
AIRSPACE: CLASS C svc continuous ctc **APP CON**
RADIO AIDS TO NAVIGATION: NOTAM FILE PIE.
ORMOND BEACH (H) VORTAC 112.6 OMN Chan 73 N29°18.20′ W81°06.76′ 159° 7.9 NM to fld. 20/00E.
TOMOK NDB (LOM) 263 DA N29°08.66′ W81°08.87′ 069° 5.2 NM to fld.
ILS 109.7 I-DAB Rwy 07L. LOM TOMOK NDB.
ASR

FIGURE 212.—Excerpt from Airport/Facilities Directory.

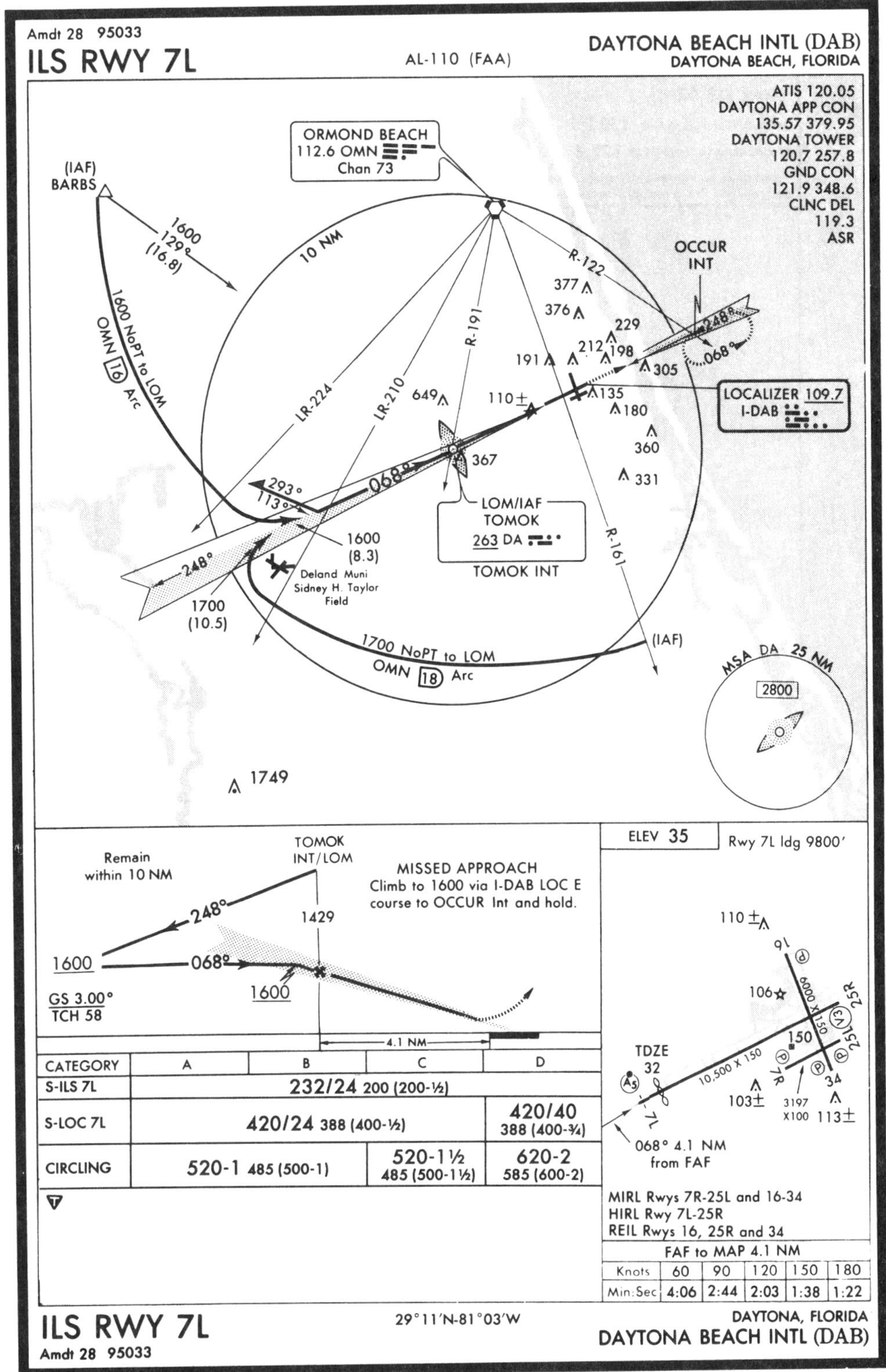

CATEGORY	A	B	C	D
S-ILS 7L	232/24 200 (200-½)			
S-LOC 7L	420/24 388 (400-½)			420/40 388 (400-¾)
CIRCLING	520-1 485 (500-1)		520-1½ 485 (500-1½)	620-2 585 (600-2)

FAF to MAP 4.1 NM					
Knots	60	90	120	150	180
Min:Sec	4:06	2:44	2:03	1:38	1:22

FIGURE 212A.—ILS RWY 7L (DAB).

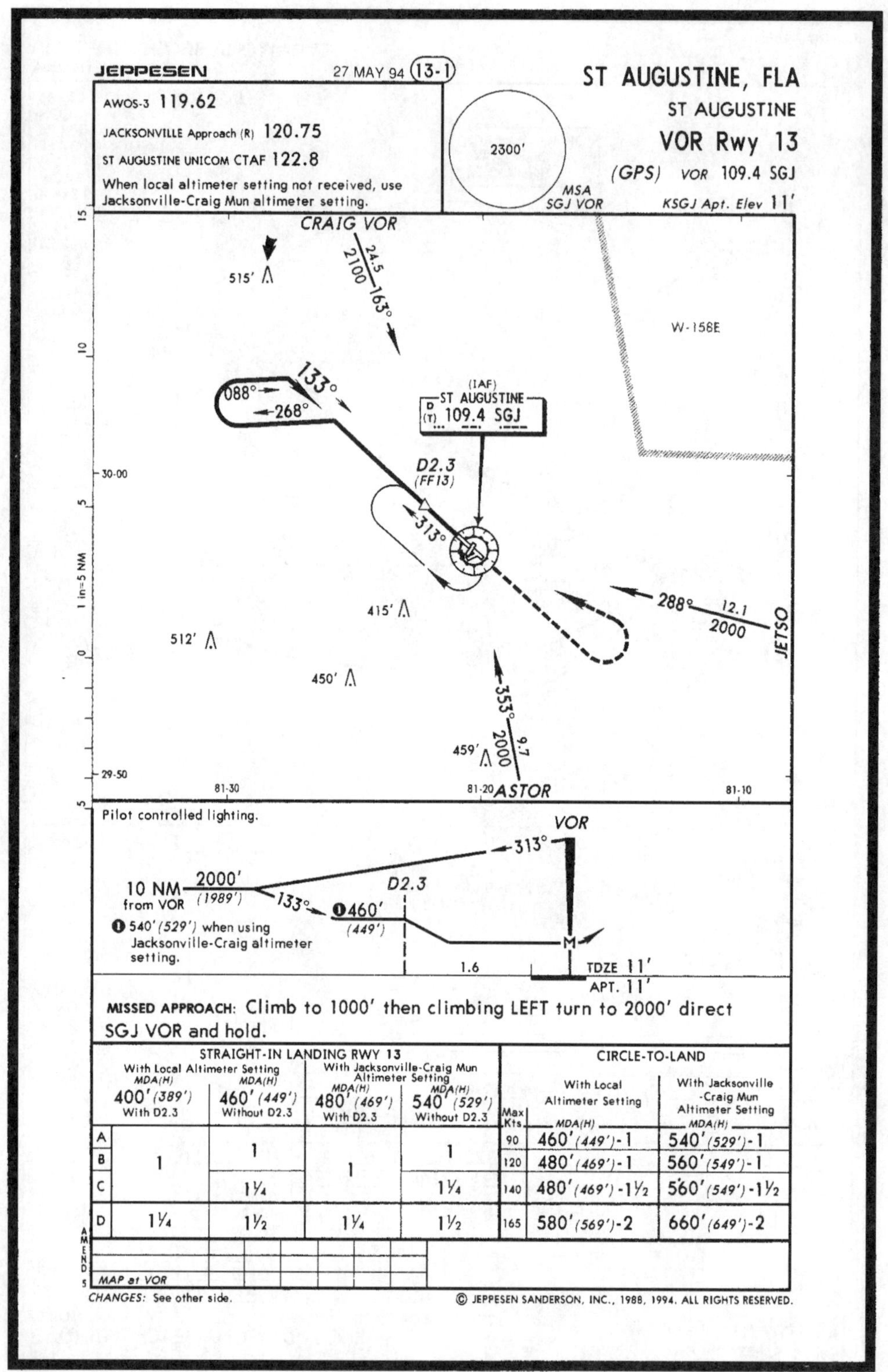

FIGURE 213.—VOR RWY 13.

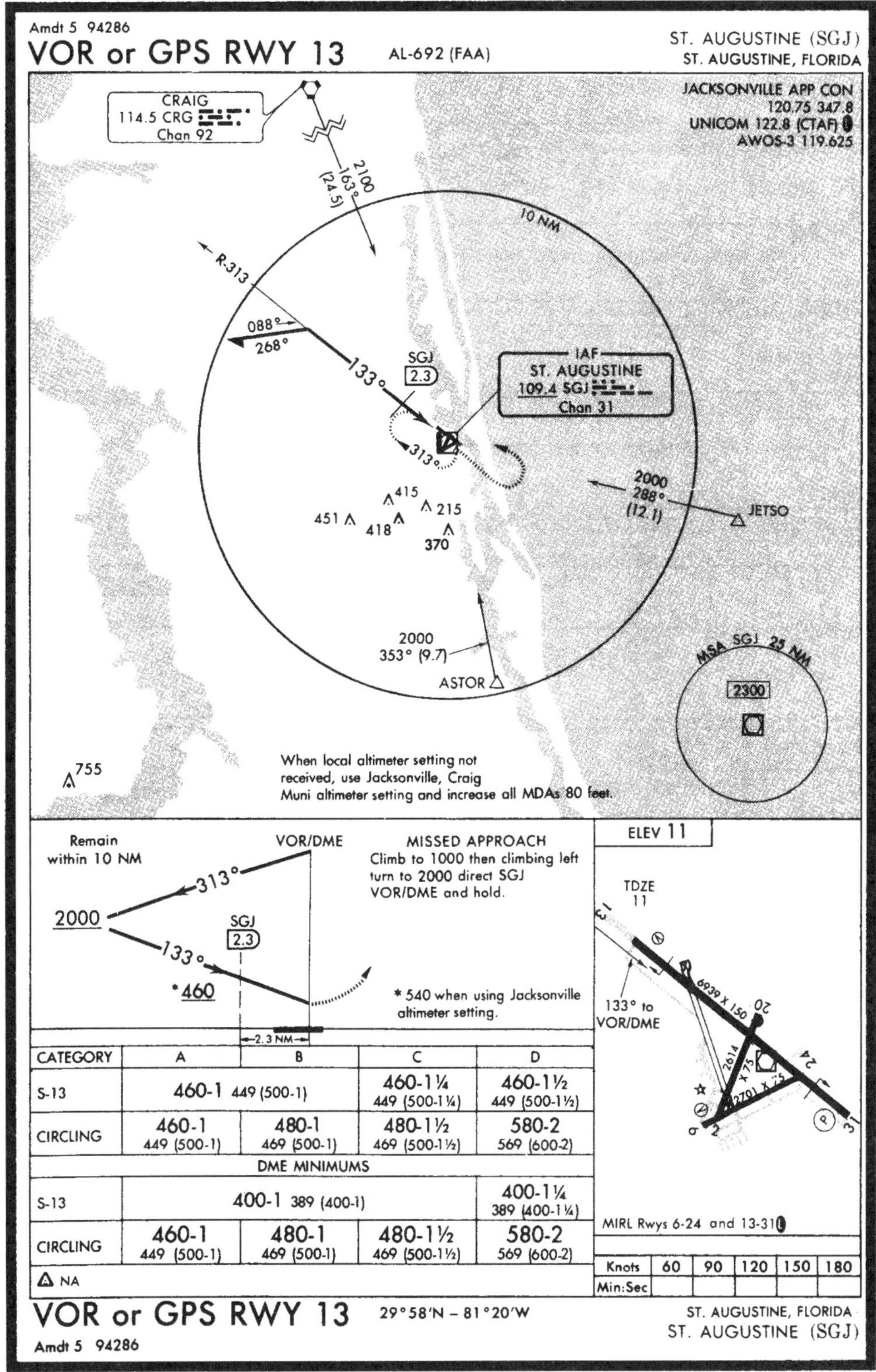

CATEGORY	A	B	C	D
S-13	460-1 449 (500-1)		460-1¼ 449 (500-1¼)	460-1½ 449 (500-1½)
CIRCLING	460-1 449 (500-1)	480-1 469 (500-1)	480-1½ 469 (500-1½)	580-2 569 (600-2)
DME MINIMUMS				
S-13	400-1 389 (400-1)			400-1¼ 389 (400-1¼)
CIRCLING	460-1 449 (500-1)	480-1 469 (500-1)	480-1½ 469 (500-1½)	580-2 569 (600-2)

△ NA

FIGURE 213A.—VOR or GPS RWY 13 (SGJ).

Form Approved: OMB No. 2120-0034

U.S. DEPARTMENT OF TRANSPORTATION FEDERAL AVIATION ADMINISTRATION **FLIGHT PLAN**	(FAA USE ONLY) ☐ PILOT BRIEFING ☐ VNR ☐ STOPOVER	TIME STARTED	SPECIALIST INITIALS

1. TYPE	2. AIRCRAFT IDENTIFICATION	3. AIRCRAFT TYPE/ SPECIAL EQUIPMENT	4. TRUE AIRSPEED	5. DEPARTURE POINT	6. DEPARTURE TIME PROPOSED (Z)	ACTUAL (Z)	7. CRUISING ALTITUDE
VFR X IFR DVFR	TNA 90	MD90/G	440 KTS	KBDL Bradley Int'l			FL330

8. ROUTE OF FLIGHT
CSTL.SHERL, J121 BRIGS, BRIGS.VCN 5 PHL

9. DESTINATION (Name of airport and city)	10. EST. TIME ENROUTE HOURS	MINUTES	11. REMARKS
KPHL PHILADELPHIA INT'L PHILADELPHIA			L/O = Level Off PPH = Pounds Per Hour TEC = Tower to Tower Variation: BDL 14W, PHL 10W

12. FUEL ON BOARD HOURS	MINUTES	13. ALTERNATE AIRPORT(S)	14. PILOT'S NAME, ADDRESS & TELEPHONE NUMBER & AIRCRAFT HOME BASE / 17. DESTINATION CONTACT/TELEPHONE (OPTIONAL)	15. NUMBER ABOARD
2	20	KACY ATLANTIC CITY INT'L		99

16. COLOR OF AIRCRAFT	
BLACK/RED	CIVIL AIRCRAFT PILOTS. FAR Part 91 requires you file an IFR flight plan to operate under instrument flight rules in controlled airspace. Failure to file could result in a civil penalty not to exceed $1,000 for each violation (Section 901 of the Federal Aviation Act of 1958, as amended). Filing of a VFR flight plan is recommended as a good operating practice. See also Part 99 for requirements concerning DVFR flight plans.

FAA Form 7233-1 (8-82) CLOSE VFR FLIGHT PLAN WITH ______________ FSS ON ARRIVAL

FLIGHT LOG

CHECK POINTS		ROUTE		WIND	SPEED-KTS		DIST	TIME		FUEL	
FROM	TO	ALTITUDE	COURSE	TEMP	TAS	GS	NM	LEG	TOT	LEG	TOT
BDL	YODER INTER	CSTL1.SHERL CLIMB					45		:15:00		2560*
Yoder Inter	SHERL INTER	CSTL1.SHERL FL330		340/55 ISA							
Sherl Inter	BRIGS INTER	J121 FL330									
Brigs Inter	VCN	BRIGS.VCN5 FL300									
VCN	PHL	BRIGS.VCN5 DESCENT & APPROACH					46	:14:00		1190	
PHL	ACY	TEC 3000					44	:12:00			

OTHER DATA: * Includes Taxi Fuel
NOTE: Use 6150 PPH Total Fuel Flow From L/O To Start Of Descent.
Use 5900 PPH Total Fuel Flow For Reserve And Alternate Requirements.

A Missed Approach Requires 244# of Fuel.

TIME and FUEL: As required by FARs.

TIME	FUEL (LB)	
		EN ROUTE
		RESERVE
		ALTERNATE
		TOTAL

FIGURE 214.—Flight Plan/Flight Log.

20 **CONNECTICUT**

WINDSOR LOCKS

BRADLEY INTL (BDL) 3 W UTC-5(-4DT) N41°56.33′ W72°40.99′ **NEW YORK**

174 B S4 **FUEL** 100LL, JET A OX 1, 2, 3, 4 TPA—See Remarks **H-3J, 6J, L-25C, 28I**

LRA ARFF Index D **IAP**

RWY 06-24: H9502X200 (ASPH-GRVD) S-200, D-200, DT-350,DDT-710 HIRL CL

RWY 06: ALSF2 TDZ. **RWY 24:** MALSR. VASI(V4L)—GA 3.0°TCH 56′. Trees.

RWY 15-33: H6846X150 (ASPH-GRVD) S-200, D-200, DT-350 HIRL

RWY 15: REIL. VASI(V4L)—GA 3.5°TCH 59′. Trees. **RWY 33:** MALSF. VASI(V4R)—GA 3.0°TCH 59′. Trees.

RWY 01-19: H5145X100 (ASPH) S-60, D-190, DT-328 MIRL

RWY 01: Building. **RWY 19:** Trees.

AIRPORT REMARKS: Attended continuously. Numerous birds frequently on or in vicinity or arpt. TPA—1174(1000) light acft, 1874(1700) heavy acft. Landing fee for business, corporate and revenue producing aircraft. Flight Notification Service (ADCUS) available. NOTE: See Land and Hold Short Operations Section.

WEATHER DATA SOURCES: LLWAS.

COMMUNICATIONS: ATIS 118.15 (203-627-3423) **UNICOM** 122.95

BRIDGEPORT FSS (BDR) TF 1-800-WX-BRIEF. NOTAM FILE BDL.

WINDSOR LOCKS RCO 122.3 (BRIDGEPORT FSS)

® **BRADLEY APP CON** 125.8 (within 20 miles)

® **BRADLEY DEP CON** 127.8 (South) 125.35 (North and West) 123.95 (Northeast)

TOWER 120.3 **GND CON** 121.9 **CLNC DEL** 121.75

AIRSPACE: CLASS C svc continuous ctc **APP CON**

RADIO AIDS TO NAVIGATION: NOTAM FILE BDL.

(T) **VORTACW** 109.0 BDL Chan 27 N41°56.45′ W72°41.32′ at fld. 160/14W.

VOR portion unusable:

093°-103° byd 24 NM blo 5000′ 140°-170° byd 15 NM blo 6000′

104°-139° byd 10 NM blo 6000′ 260°-290° byd 15 NM blo 6000′

DME unusable 250°-290° byd 18 NN blo 6000′.

CHUPP NDB (LOM) 388 BD N41°52.64′ W72°45.98′ 059° 5.2 NM to fld.

ILS/DME 111.1 I-BDL Chan 48 Rwy 06. LOM CHUPP NDB.

ILS/DME 108.55 I-IKX Chan 22Y Rwy 33.

ILS/DME 111.1 I-MYQ Chan 48 Rwy 24.

T **TAKE-OFF MINS** 

94286

INSTRUMENT APPROACH PROCEDURE CHARTS

T IFR TAKE-OFF MINIMUMS AND DEPARTURE PROCEDURES

Civil Airports and Selected Military Airports

CIVIL USERS: FAR 91 prescribes take-off rules and establishes take-off minimums for certain operators as follows: (1) Aircraft having two engines or less - one statute mile. (2) Aircraft having more than two engines - one-half statute mile. Airports with IFR take-off minimums other than standard are listed below. Departure procedures and/or ceiling visibility minimums are established to assist all pilots conducting IFR flight in avoiding obstacles during climb to the minimum enroute altitude. Take-off minimums and departures apply to all runways unless otherwise specified. Altitudes, unless otherwise indicated, are minimum altitudes in feet MSL.

MILITARY USERS: Special IFR departures not published as Standard Instrument Departure (SIDS) and civil take-off minima are included below and are established to assist pilots in obstacle avoidance. Refer to appropriate service directives for take-off minimums.

WINDSOR LOCKS, CT

BRADLEY INTL

TAKE-OFF MINIMUMS: **Rwy 15,** 300-1 or std. with a min. climb of 350′ per NM to 300. **Rwy 33,** 700-1 or std. with a min. climb of 300′ per NM to 1000.

DEPARTURE PROCEDURE: **Rwy 1,** climb to 1000 via runway heading before turning westbound.

FIGURE 215.—Excerpts from Airport/Facilities Directory.

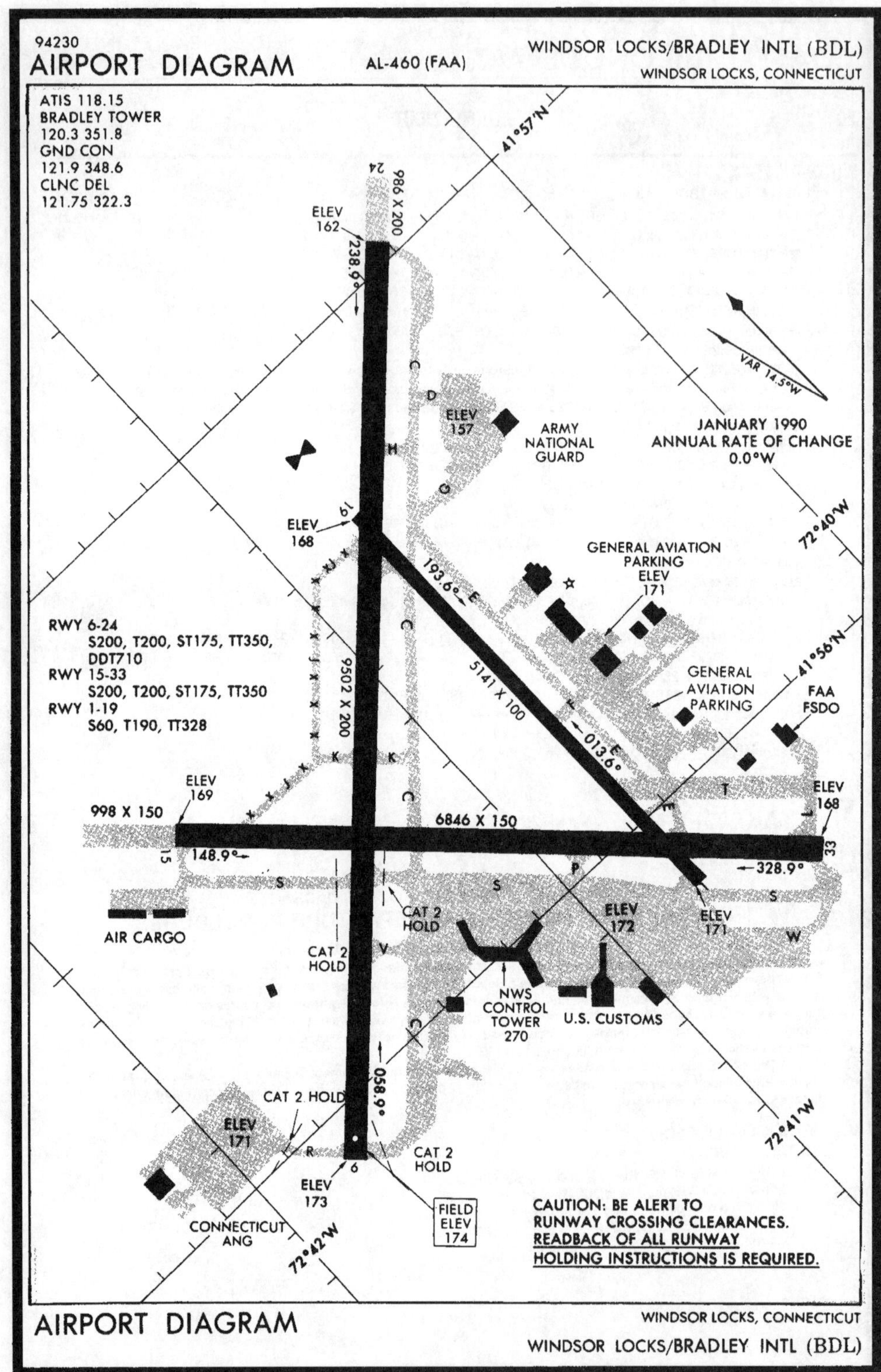

FIGURE 215A.—AIRORT DIAGRAM.

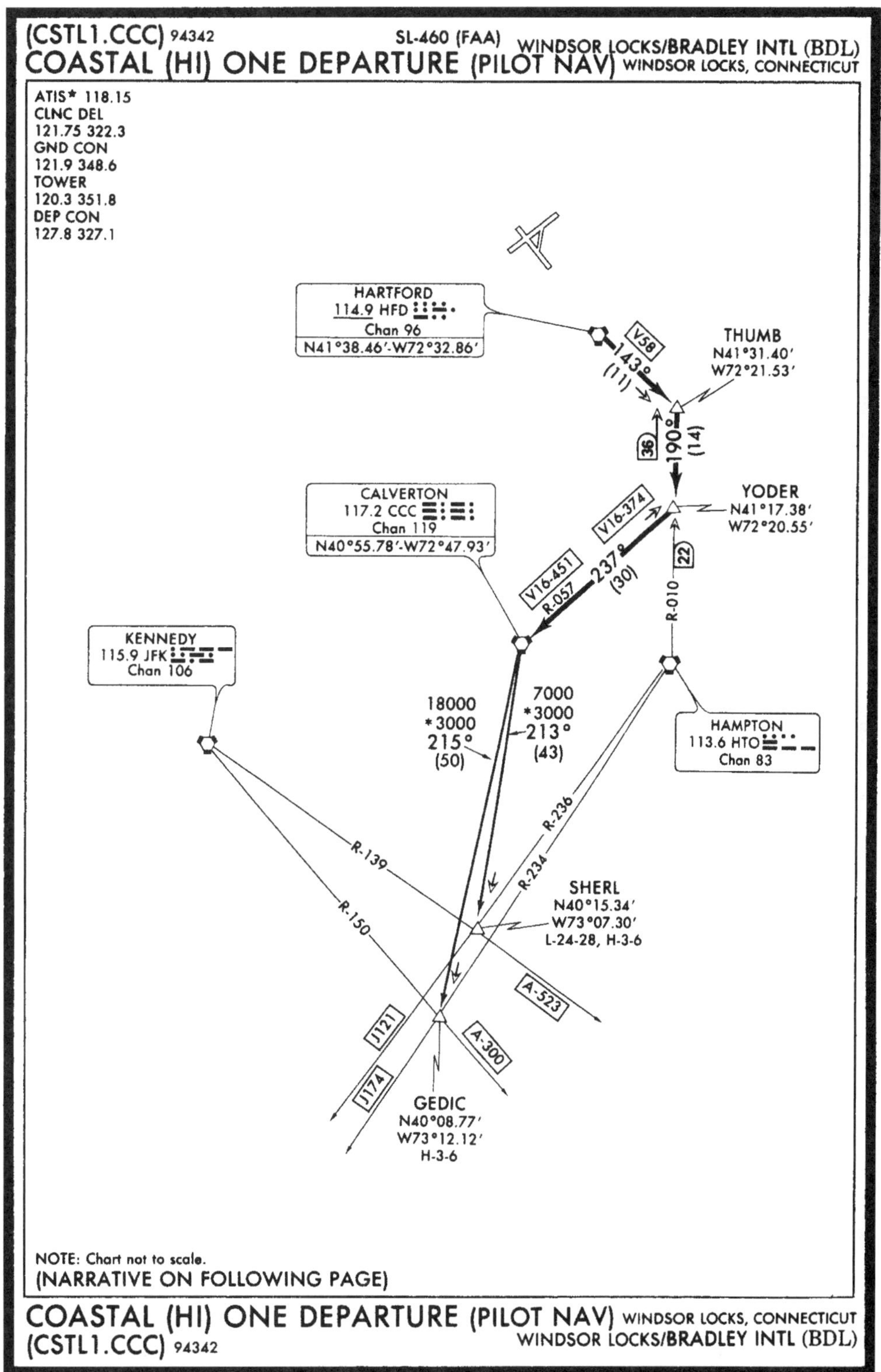

FIGURE 216.—COASTAL (HI) DEPARTURE (BDL).

(CSTL1.CCC) 94342 SL-460 (FAA) WINDSOR LOCKS/BRADLEY INTL (BDL)
COASTAL (HI) ONE DEPARTURE (PILOT NAV) WINDSOR LOCKS, CONNECTICUT

DEPARTURE ROUTE DESCRIPTION

TAKE-OFF RWY 6: Turn right heading 075° or as assigned for radar vectors to HFD VORTAC.
TAKE-OFF ALL OTHER RUNWAYS: Fly runway heading or as assigned for radar vectors to HFD VORTAC. Maintain 4000 feet or assigned altitude. Expect clearance to requested flight level ten (10) minutes after departure.
. . . . From over HFD VORTAC proceed via the HFD R-143 to THUMB INT, then proceed via the HTO R-010 to YODER INT, then via the CCC R-057 to CCC VORTAC. Then via (transition) or (assigned route).
GEDIC TRANSITION (CSTL1.GEDIC): From over CCC VORTAC via CCC R-215 to GEDIC INT.
SHERL TRANSITION (CSTL1.SHERL): From over CCC VORTAC via CCC R-213 to SHERL INT.

COASTAL (HI) ONE DEPARTURE (PILOT NAV) WINDSOR LOCKS, CONNECTICUT
(CSTL1.CCC) 94342 WINDSOR LOCKS/BRADLEY INTL (BDL)

FIGURE 216A.—Departure Route Description.

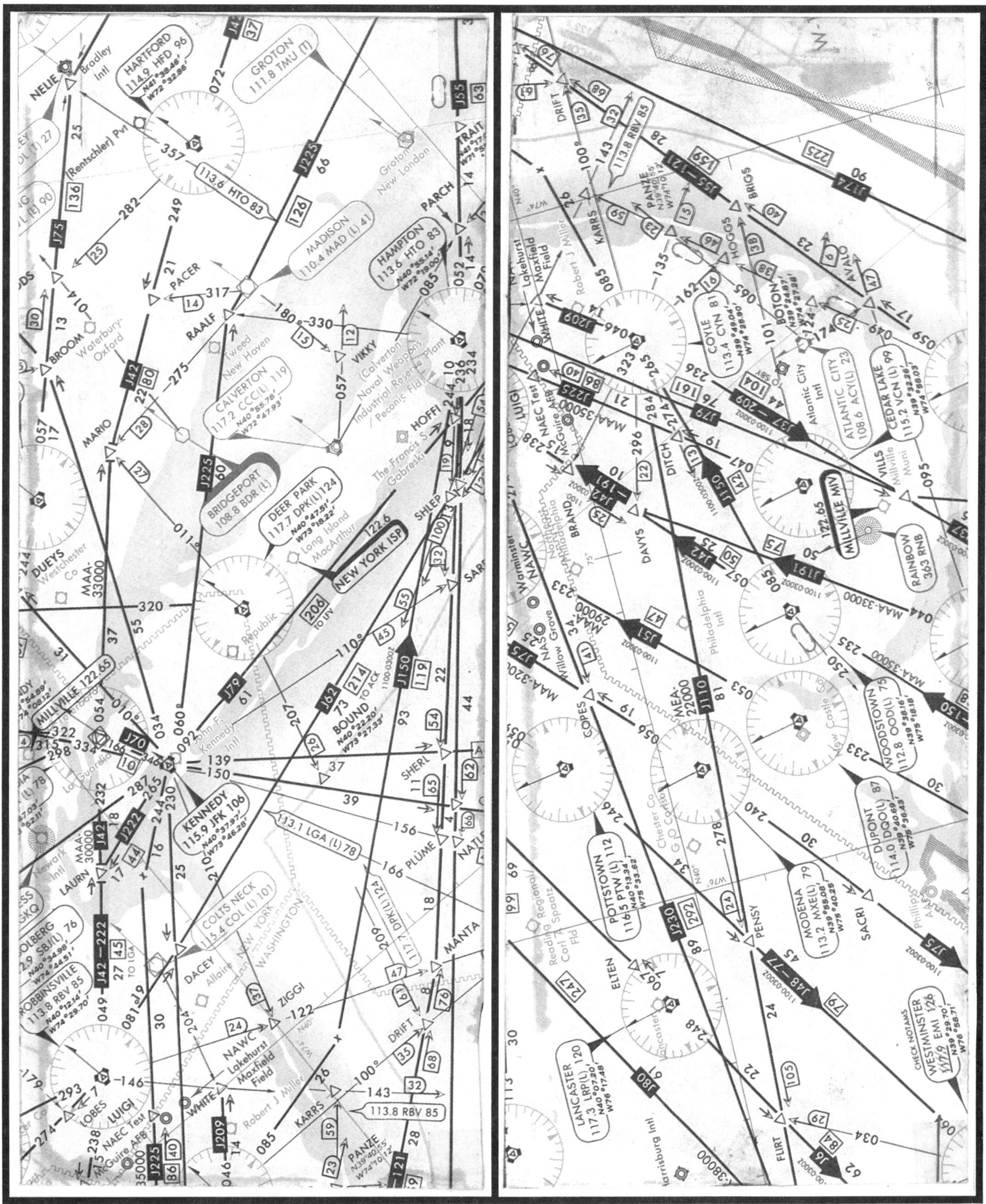

FIGURE 217.—High Altitude Airways.

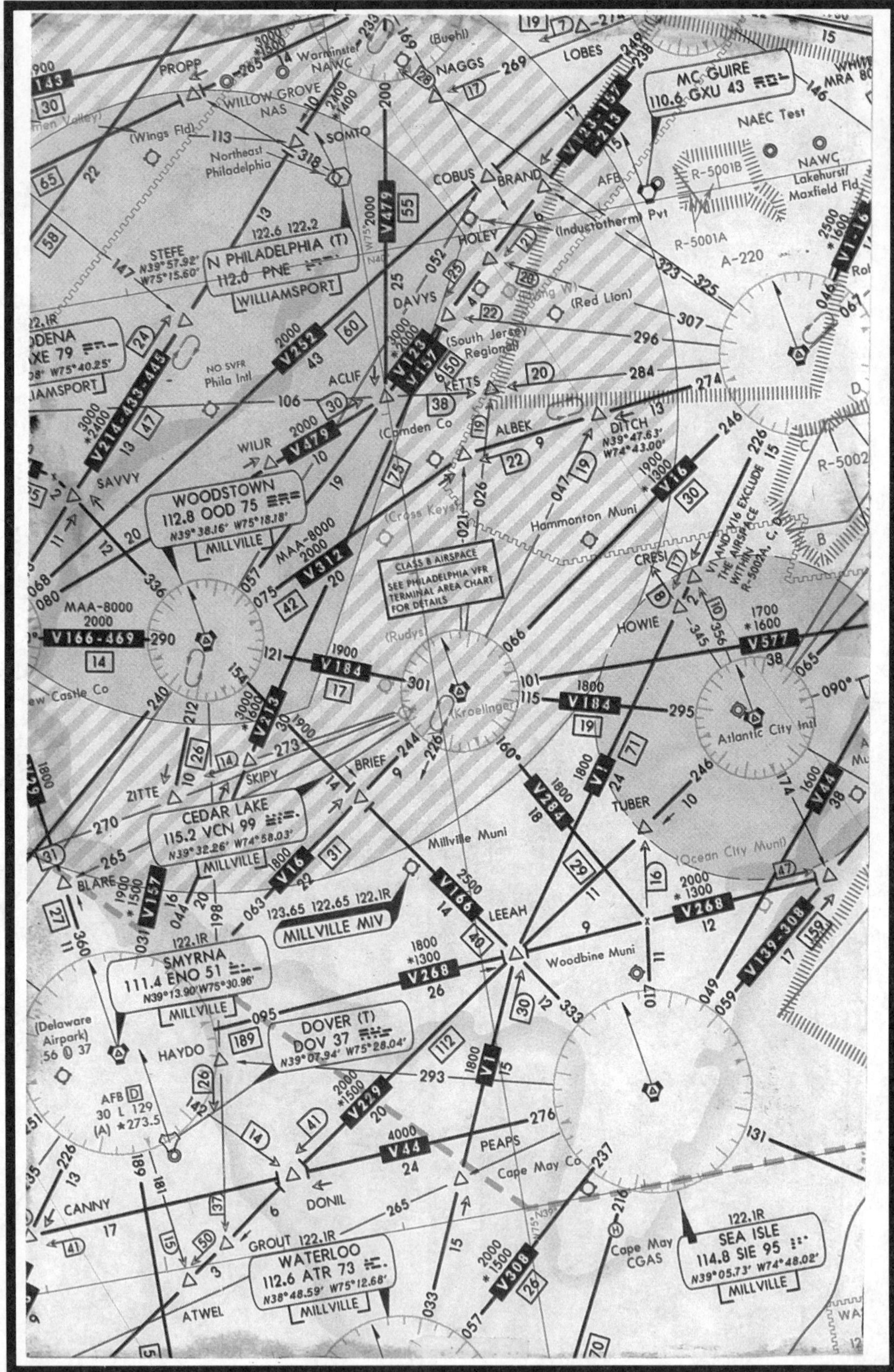

FIGURE 218.—Low Altitude Airways.

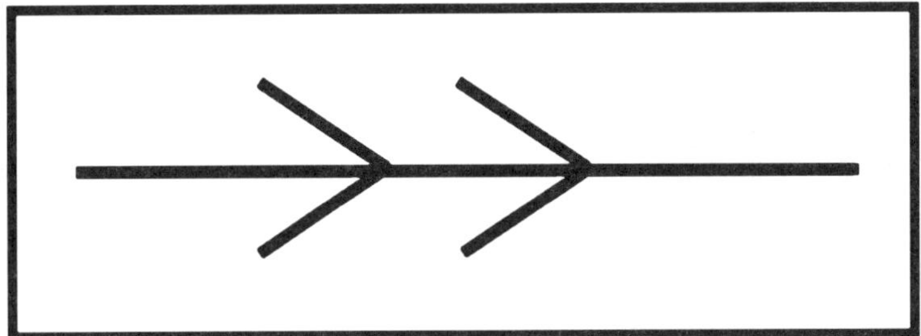

FIGURE 219.—Chart and Navigation Symbol.

FIGURE 220.—Chart and Navigation Symbol.

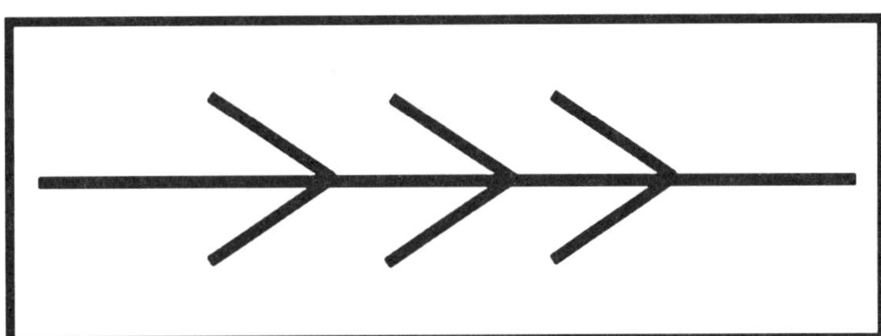

FIGURE 221.—Chart and Navigation Symbol.

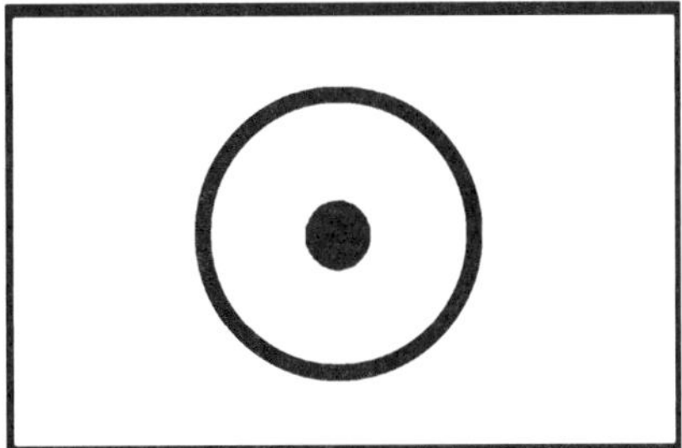

FIGURE 222.—Chart and Navigation Symbol.

FIGURE 223.—Holding Position Markings.

FIGURE 224.—ILS Critical Area Markings.

FIGURE 225.—No Entry.

FIGURE 226.—Outbound Destination.

FIGURE 227.—Taxiway End Marker.

FIGURE 228.—TWY-RWY Hold Position.

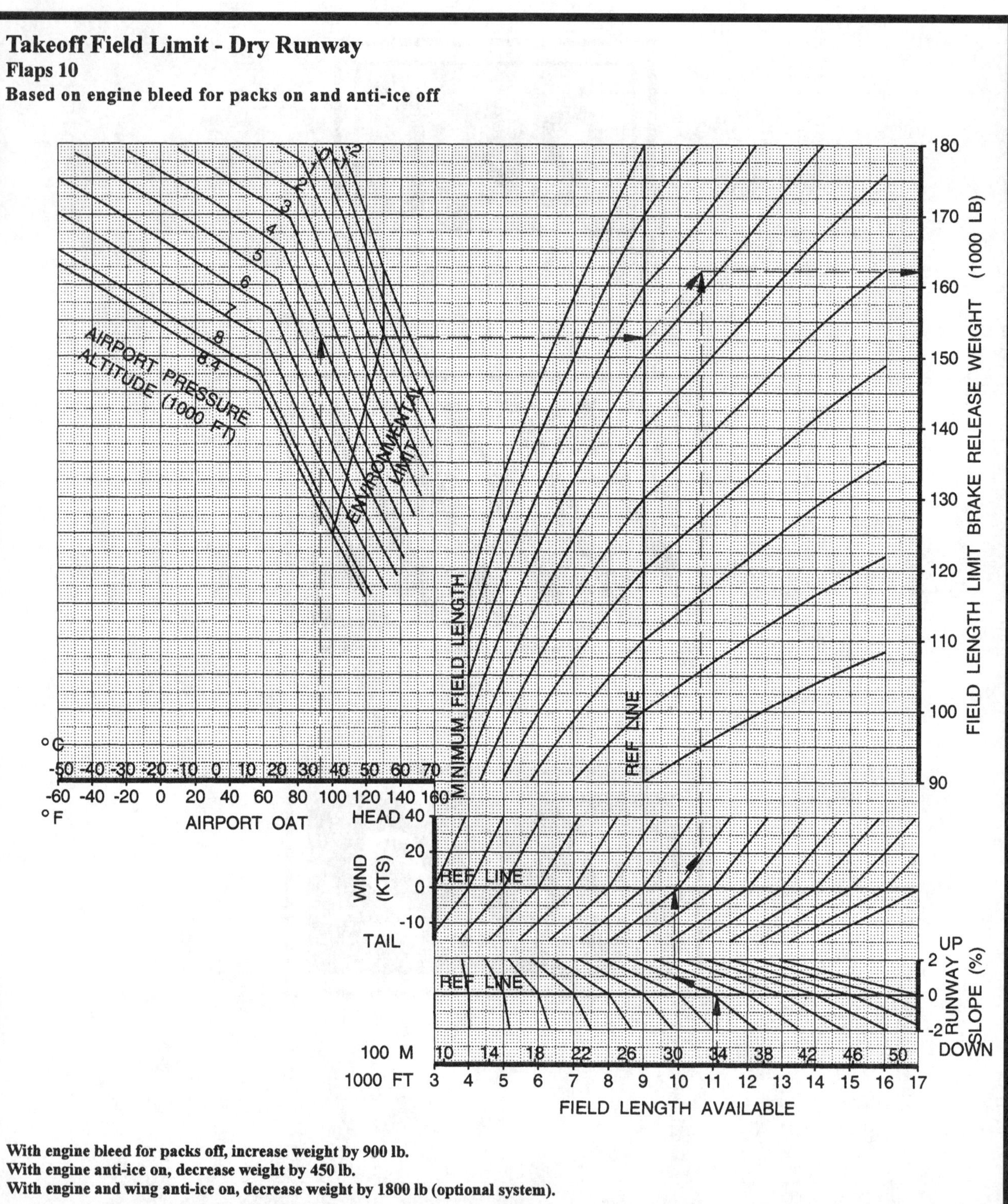

FIGURE 229.—Takeoff Field Limit—Dry Runway.

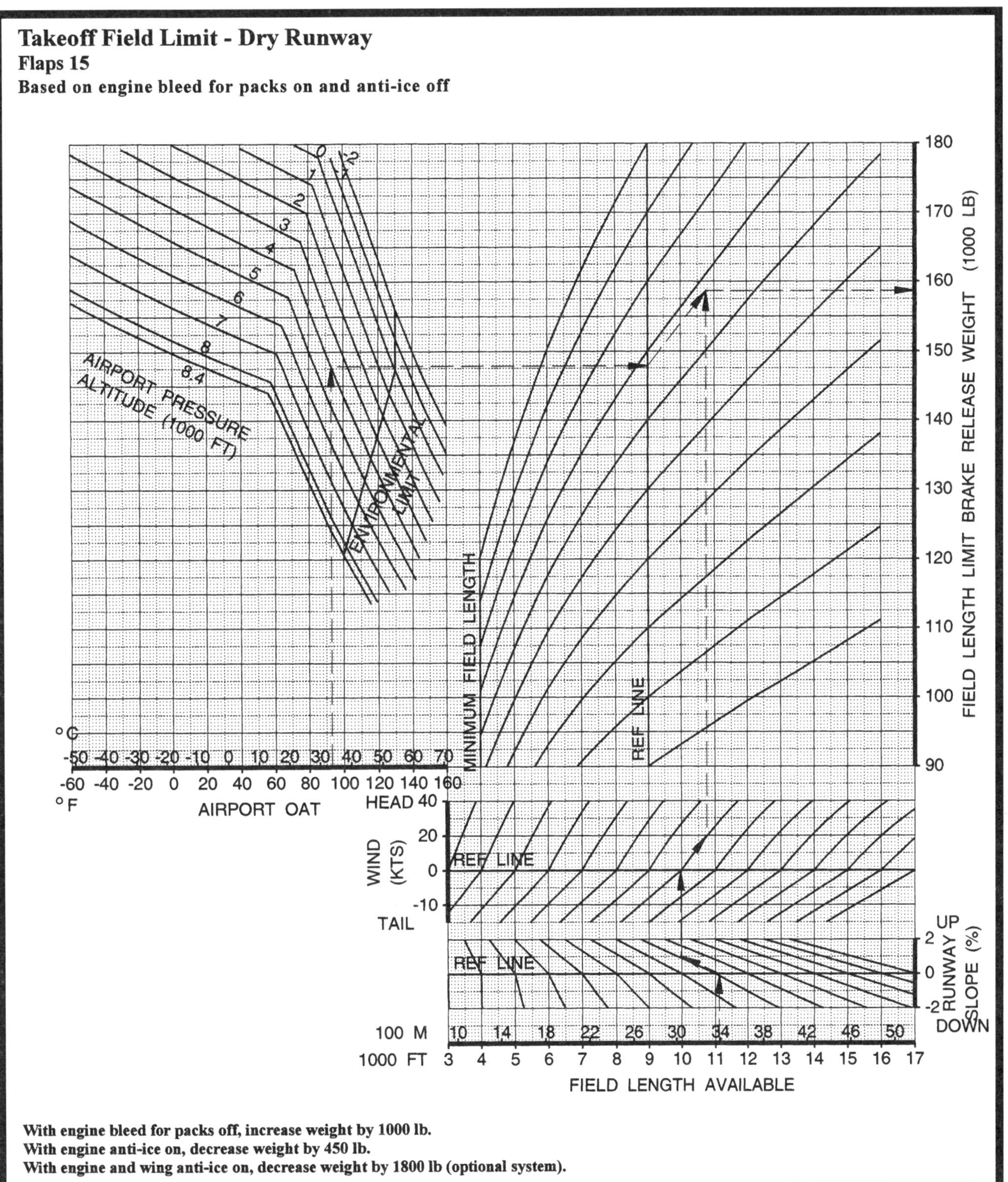

FIGURE 230.—Takeoff Field Limit—Dry Runway.

Takeoff Climb Limit

Flaps 15

Based on engine bleed for packs on and anti-ice off

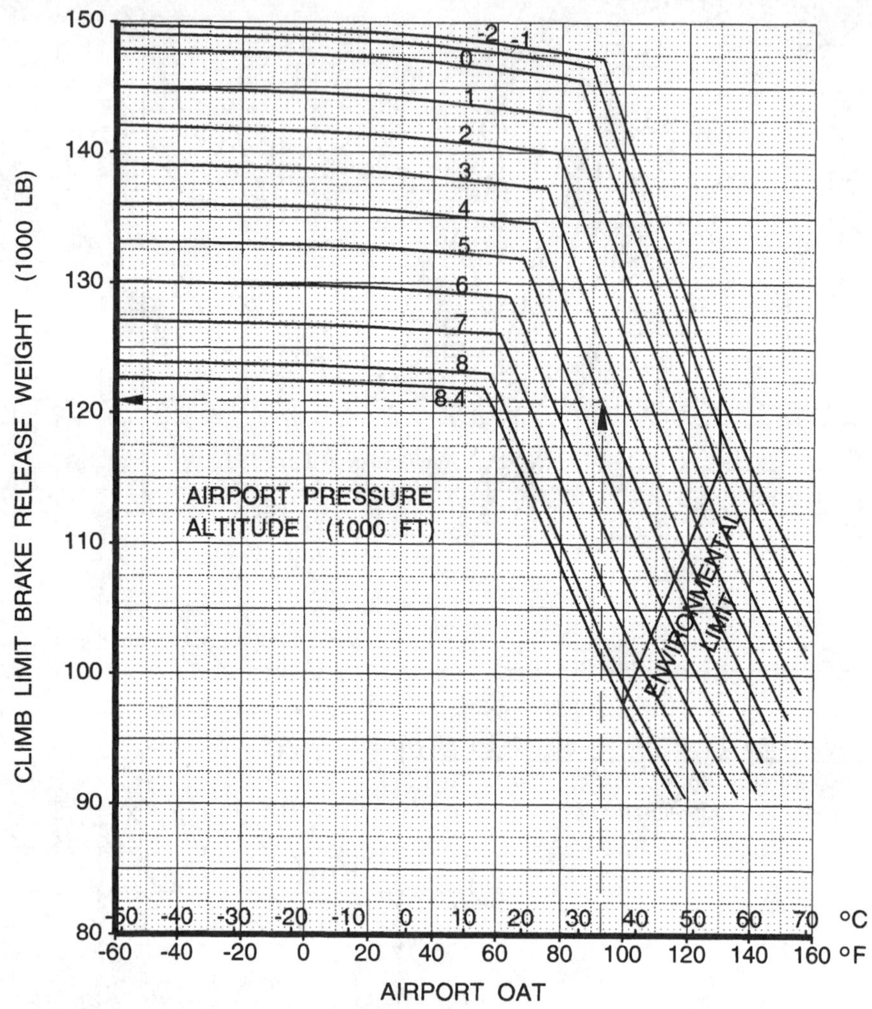

With engine bleed for packs off, increase weight by 2400 lb.
With engine anti-ice on, decrease weight by 400 lb.
With engine and wing anti-ice on, decrease weight by 2300 lb (optional system).

FIGURE 231.—Takeoff Climb Limit.

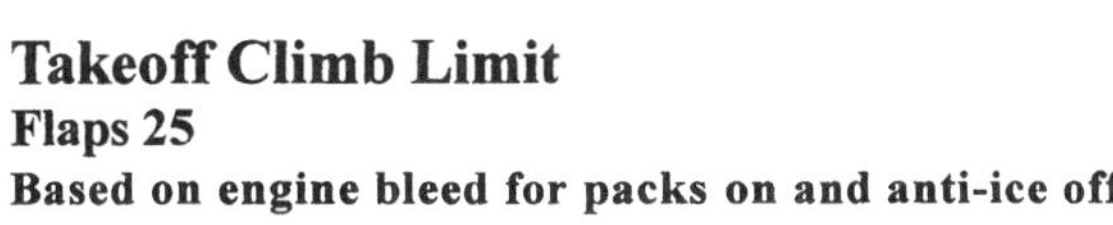

Takeoff Climb Limit
Flaps 25
Based on engine bleed for packs on and anti-ice off

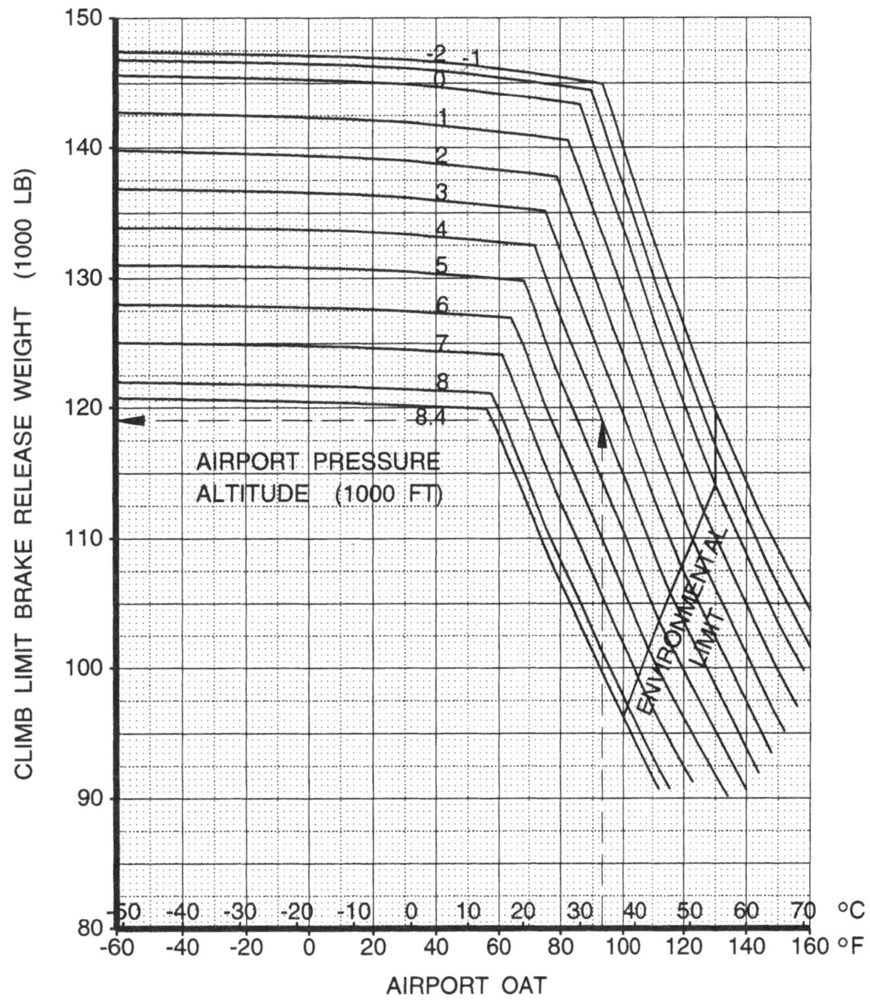

With engine bleed for packs off, increase weight by 2300 lb.
With engine anti-ice on, decrease weight by 400 lb.
With engine and wing anti-ice on, decrease weight by 2300 lb (optional system).

FIGURE 232.—Takeoff Climb Limit.

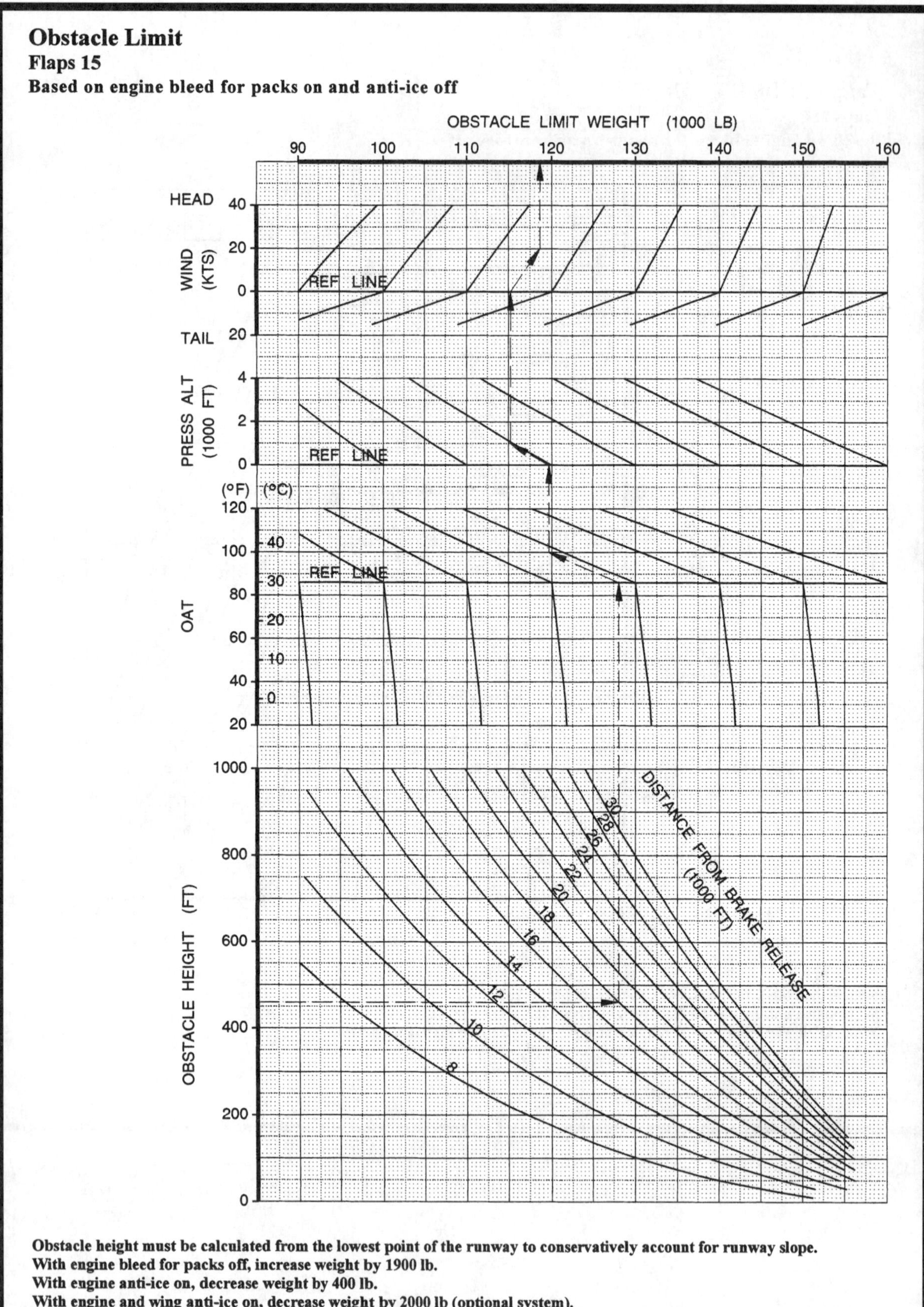

FIGURE 233.—Obstacle Limit.

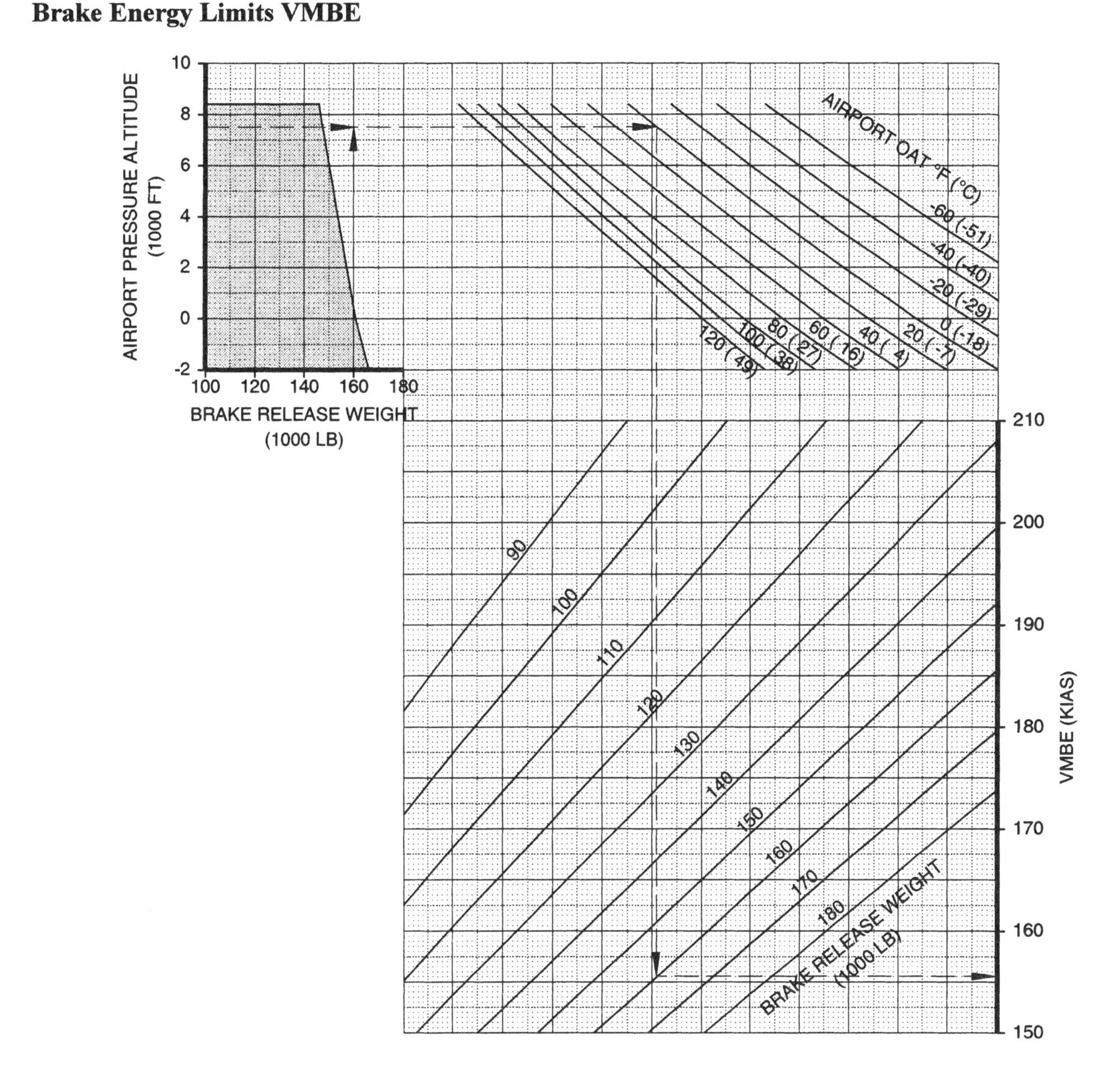

Check VMBE when outside shaded area or when operating with tailwind.
Increase VMBE by 2 knots per 1% uphill runway slope. Decrease VMBE by 3 knots per 1% downhill runway slope.
Increase VMBE by 4 knots per 10 knots headwind. Decrease VMBE by 18 knots per 10 knots tailwind.

Normal takeoff:
Decrease brake release weight by 1100 lb for each knot V1 exceeds VMBE.
Determine normal V1, VR, V2 speeds for lower brake release weight.

Improved climb takeoff:
Decrease climb weight improvement by 550 lb for each knot V1 exceeds VMBE.
Determine V1, VR, V2 speed increments for the lower climb weight improvement.

FIGURE 234.—Brake Energy Limits VMBE.

ADVISORY INFORMATION

Slush/Standing Water Takeoff

No Reverse Thrust

Weight Adjustments (1000 LB)

DRY FIELD/OBSTACLE LIMIT WEIGHT (1000 LB)	SLUSH/STANDING WATER DEPTH								
	0.12 INCHES (3 mm)			0.25 INCHES (6 mm)			0.50 INCHES (13 mm)		
	PRESSURE ALTITUDE (FT)			PRESSURE ALTITUDE (FT)			PRESSURE ALTITUDE (FT)		
	S.L.	5000	10000	S.L.	5000	10000	S.L.	5000	10000
180	-26.4	-34.9	-43.4	-30.8	-39.3	-47.8	-39.8	-48.3	-56.8
170	-23.4	-31.9	-40.4	-27.0	-35.5	-44.0	-34.4	-42.9	-51.4
160	-20.7	-29.2	-37.7	-23.6	-32.1	-40.6	-29.6	-38.1	-46.6
150	-18.3	-26.8	-35.3	-20.6	-29.1	-37.6	-25.5	-34.0	-42.5
140	-16.2	-24.7	-33.2	-18.0	-26.5	-35.0	-22.0	-30.5	-39.0
130	-14.4	-22.9	-31.4	-15.9	-24.4	-32.9	-19.3	-27.8	-36.3
120	-12.9	-21.4	-29.9	-14.2	-22.7	-31.2	-17.2	-25.7	-34.2
110	-11.6	-20.1	-28.6	-12.8	-21.3	-29.8	-15.6	-24.1	-32.6
100	-10.4	-18.9	-27.4	-11.4	-19.9	-28.4	-14.1	-22.6	-31.1
90	-9.1	-17.6	-26.1	-10.1	-18.6	-27.1	-12.5	-21.0	-29.5

V1(MCG) Limit Weight (1000 LB)

ADJUSTED FIELD LENGTH (FT)	SLUSH/STANDING WATER DEPTH								
	0.12 INCHES (3 mm)			0.25 INCHES (6 mm)			0.50 INCHES (13 mm)		
	PRESSURE ALTITUDE (FT)			PRESSURE ALTITUDE (FT)			PRESSURE ALTITUDE (FT)		
	S.L.	5000	10000	S.L.	5000	10000	S.L.	5000	10000
5800							86.8		
6200				78.7			105.6		
6600	81.1			99.7			125.0	86.8	
7000	103.5			120.8	78.7		145.2	105.6	
7400	125.9	81.1		142.1	99.7		166.1	125.0	86.8
7800	148.3	103.5		163.6	120.8	78.7	188.0	145.2	105.6
8200	170.6	125.9	81.1	185.2	142.1	99.7		166.1	125.0
8600	192.9	148.3	103.5		163.6	120.8		188.0	145.2
9000		170.6	125.9		185.2	142.1			166.1
9400		192.9	148.3			163.6			188.0
9800			170.6			185.2			
10200			192.9						

1. **Enter Weight Adjustment table with slush/standing water depth and dry field/obstacle limit weight to obtain slush/standing water weight adjustment.**
2. **Adjust field length available by -150 ft/+140 ft for every 5°C above/below 4°C.**
3. **Find V1(MCG) limit weight for adjusted field length and pressure altitude.**
4. **Max allowable slush/standing water limited weight is lesser of weights from 1 and 3.**

V1 Adjustment (KIAS)

WEIGHT (1000 LB)	SLUSH/STANDING WATER DEPTH								
	0.12 INCHES (3 mm)			0.25 INCHES (6 mm)			0.50 INCHES (13 mm)		
	PRESSURE ALTITUDE (FT)			PRESSURE ALTITUDE (FT)			PRESSURE ALTITUDE (FT)		
	S.L.	5000	10000	S.L.	5000	10000	S.L.	5000	10000
180	-21	-16	-11	-12	-7	-2	0	0	0
170	-22	-17	-12	-14	-9	-4	0	0	0
160	-23	-18	-13	-17	-12	-7	-3	0	0
150	-24	-19	-14	-19	-14	-9	-7	-2	0
140	-25	-20	-15	-21	-16	-11	-10	-5	0
130	-26	-21	-16	-22	-17	-12	-14	-9	-4
120	-27	-22	-17	-24	-19	-14	-17	-12	-7
110	-28	-23	-18	-25	-20	-15	-20	-15	-10
100	-29	-24	-19	-27	-22	-17	-22	-17	-12
90	-29	-24	-19	-28	-23	-18	-25	-20	-15

1. **Obtain V1, VR and V2 for the actual weight using the Dry Runway Takeoff Speeds table.**
2. **If V1(MCG) limited, set V1 = V1(MCG). If not V1(MCG) limited, enter V1 Adjustment table with the actual weight to obtain V1 speed adjustment. If adjusted V1 is less than V1(MCG), set V1 = V1(MCG).**

FIGURE 235.—Slush/Standing Water Takeoff.

ADVISORY INFORMATION

Slush/Standing Water Takeoff

Maximum Reverse Thrust
Weight Adjustments (1000 LB)

DRY FIELD/OBSTACLE LIMIT WEIGHT (1000 LB)	SLUSH/STANDING WATER DEPTH								
	0.12 INCHES (3 mm)			0.25 INCHES (6 mm)			0.50 INCHES (13 mm)		
	PRESSURE ALTITUDE (FT)			PRESSURE ALTITUDE (FT)			PRESSURE ALTITUDE (FT)		
	S.L.	5000	10000	S.L.	5000	10000	S.L.	5000	10000
180	-21.9	-27.4	-32.9	-26.4	-31.9	-37.4	-37.5	-43.0	-48.5
170	-19.3	-24.8	-30.3	-22.8	-28.3	-33.8	-31.1	-36.6	-42.1
160	-17.0	-22.5	-28.0	-19.7	-25.2	-30.7	-25.8	-31.3	-36.8
150	-15.0	-20.5	-26.0	-17.2	-22.7	-28.2	-21.7	-27.2	-32.7
140	-13.3	-18.8	-24.3	-15.1	-20.6	-26.1	-18.8	-24.3	-29.8
130	-11.9	-17.4	-22.9	-13.4	-18.9	-24.4	-16.6	-22.1	-27.6
120	-10.5	-16.0	-21.5	-11.7	-17.2	-22.7	-14.4	-19.9	-25.4
110	-9.1	-14.6	-20.1	-10.0	-15.5	-21.0	-12.2	-17.7	-23.2
100	-7.6	-13.1	-18.6	-8.2	-13.7	-19.2	-10.0	-15.5	-21.0
90	-6.2	-11.7	-17.2	-6.5	-12.0	-17.5	-7.8	-13.3	-18.8

V1(MCG) Limit Weight (1000 LB)

ADJUSTED FIELD LENGTH (FT)	SLUSH/STANDING WATER DEPTH								
	0.12 INCHES (3 mm)			0.25 INCHES (6 mm)			0.50 INCHES (13 mm)		
	PRESSURE ALTITUDE (FT)			PRESSURE ALTITUDE (FT)			PRESSURE ALTITUDE (FT)		
	S.L.	5000	10000	S.L.	5000	10000	S.L.	5000	10000
4600							74.3		
5000	75.8			82.9			93.3		
5400	94.0			100.9			111.9		
5800	112.6			119.4	73.9		130.1	83.9	
6200	131.5	84.9		138.2	91.9		147.9	102.7	
6600	150.8	103.2		157.4	110.1		165.4	121.1	74.3
7000	170.6	122.0	75.8	177.0	128.7	82.9	182.6	139.1	93.3
7400	190.9	141.1	94.0	197.0	147.7	100.9	199.5	156.7	111.9
7800		160.7	112.6		167.1	119.4		174.0	130.1
8200		180.7	131.5		186.9	138.2		191.0	147.9
8600			150.8			157.4			165.4
9000			170.6			177.0			182.6
9400			190.9			197.0			199.5

1. **Enter Weight Adjustment table with slush/standing water depth and dry field/obstacle limit weight to obtain slush/standing water weight adjustment.**
2. **Adjust field length available by -120 ft/+110 ft for every 5°C above/below 4°C.**
3. **Find V1(MCG) limit weight for adjusted field length and pressure altitude.**
4. **Max allowable slush/standing water limited weight is lesser of weights from 1 and 3.**

V1 Adjustment (KIAS)

WEIGHT (1000 LB)	SLUSH/STANDING WATER DEPTH								
	0.12 INCHES (3 mm)			0.25 INCHES (6 mm)			0.50 INCHES (13 mm)		
	PRESSURE ALTITUDE (FT)			PRESSURE ALTITUDE (FT)			PRESSURE ALTITUDE (FT)		
	S.L.	5000	10000	S.L.	5000	10000	S.L.	5000	10000
180	-15	-12	-10	-8	-5	-3	-3	0	0
170	-16	-13	-11	-10	-7	-5	-3	-1	0
160	-17	-15	-12	-12	-10	-7	-4	-2	0
150	-18	-16	-13	-14	-11	-9	-6	-3	-1
140	-19	-16	-14	-15	-13	-10	-8	-5	-3
130	-20	-17	-15	-17	-14	-12	-10	-7	-5
120	-20	-18	-15	-18	-16	-13	-12	-10	-7
110	-21	-19	-16	-19	-17	-14	-15	-12	-10
100	-23	-20	-18	-21	-18	-16	-17	-14	-12
90	-24	-21	-19	-22	-20	-17	-19	-17	-14

1. **Obtain V1, VR and V2 for the actual weight using the Dry Runway Takeoff Speeds table.**
2. **If V1(MCG) limited, set V1 = V1(MCG). If not V1(MCG) limited, enter V1 Adjustment table with the actual weight to obtain V1 speed adjustment. If adjusted V1 is less than V1(MCG), set V1 = V1(MCG).**

FIGURE 236.—Slush/Standing Water Takeoff.

Takeoff Speeds - Dry Runway

Flaps 10, 15 and 25
V1, VR, V2 for Max Takeoff Thrust

WEIGHT (1000 LB)	FLAPS 10			FLAPS 15			FLAPS 25		
	V1	VR	V2	V1	VR	V2	V1	VR	V2
170	138	140	145	136	136	141			
160	134	135	141	132	132	138	131	131	136
150	129	131	137	128	128	135	126	126	133
140	124	126	133	123	123	131	121	122	129
130	118	121	129	117	118	127	116	117	125
120	112	115	124	111	113	122	109	111	121
110	106	109	119	105	107	117	103	106	116
100	99	103	114	98	101	113	97	100	111
90	92	97	109	91	95	107	90	94	106

Check V1(MCG).

V1, VR, V2 Adjustments*

TEMP		V1						VR						V2					
		PRESSURE ALTITUDE (1000 FT)						PRESSURE ALTITUDE (1000 FT)						PRESSURE ALTITUDE (1000 FT)					
°F	°C	-2	0	2	4	6	8	-2	0	2	4	6	8	-2	0	2	4	6	8
140	60	5	6	7	9			3	4	5	6			-2	-2	-2	-3		
120	49	3	4	5	7	8	10	2	3	4	5	6	6	-1	-1	-2	-2	-3	-3
100	38	1	2	3	5	6	8	1	1	2	3	4	5	0	-1	-1	-2	-2	-3
80	27	0	0	1	3	5	6	0	0	1	2	3	4	0	0	0	-1	-1	-2
60	16	0	0	1	2	3	4	0	0	1	1	2	3	0	0	0	0	-1	-1
-60	-51	0	0	1	2	3	3	0	0	1	1	2	3	0	0	0	0	-1	-1

Slope and Wind V1 Adjustments*

WEIGHT (1000 LB)	SLOPE (%)					WIND (KTS)							
	-2	-1	0	1	2	-15	-10	-5	0	10	20	30	40
170	-3	-1	0	1	1	-1	-1	-1	0	0	1	1	1
160	-3	-1	0	1	2	-1	-1	-1	0	0	1	1	1
150	-3	-1	0	1	2	-1	-1	-1	0	0	1	1	1
140	-2	-1	0	1	2	-2	-1	-1	0	0	1	1	1
130	-2	-1	0	1	2	-2	-1	-1	0	0	1	1	1
120	-2	-1	0	1	2	-2	-1	-1	0	0	1	1	1
110	-2	-1	0	1	2	-2	-1	-1	0	0	1	1	2
100	-2	-1	0	1	2	-2	-1	-1	0	0	1	2	2
90	-1	-1	0	1	1	-2	-1	-1	0	0	1	2	2

Clearway and Stopway V1 Adjustments*

NORMAL V1 (KIAS)	CLEARWAY MINUS STOPWAY (FT)								
	800	600	400	200	0	-200	-400	-600	-800
140	-3	-3	-3	-2	0	2	2	2	2
120	-3	-3	-3	-2	0	2	2	2	2
100	-3	-3	-2	-1	0	1	1	1	1

***V1 not to exceed VR.**

Max Allowable Clearway for V1 Adjustment

FIELD LENGTH (FT)	4000	6000	8000	10000	12000	14000
MAX ALLOWABLE CLEARWAY (FT)	450	650	850	1000	1450	1550

V1(MCG)

Max Takeoff Thrust

TEMP		PRESSURE ALTITUDE (FT)						
°F	°C	-2000	0	2000	4000	6000	8000	10000
160	71	102						
140	60	102	99	97	96			
120	49	104	102	98	96	94	92	90
100	38	110	107	103	100	96	92	90
80	27	112	111	109	105	101	97	93
60	16	112	112	109	107	104	101	97
-60	-51	113	113	110	108	105	102	100

FIGURE 237.—Takeoff Speeds—Dry Runway.

Takeoff Speeds - Wet Runway
Flaps 10, 15 and 25
V1, VR, V2 for Max Takeoff Thrust

WEIGHT (1000 LB)	FLAPS 10			FLAPS 15			FLAPS 25		
	V1	VR	V2	V1	VR	V2	V1	VR	V2
170	133	139	145	133	136	141			
160	128	135	141	128	132	138	126	131	136
150	123	131	137	122	128	135	121	126	133
140	117	126	133	117	123	131	115	122	129
130	111	121	129	111	118	127	109	117	125
120	105	115	124	104	113	122	103	111	121
110	99	109	119	98	107	117	97	106	116
100	92	103	114	92	101	112	91	100	111
90	86	97	109	85	95	107	84	94	106

Check V1(MCG).

V1, VR, V2 Adjustments*

TEMP		V1						VR						V2					
		PRESSURE ALTITUDE (1000 FT)						PRESSURE ALTITUDE (1000 FT)						PRESSURE ALTITUDE (1000 FT)					
°F	°C	-2	0	2	4	6	8	-2	0	2	4	6	8	-2	0	2	4	6	8
140	60	6	7	9	10			3	4	5	6			-2	-2	-2	-3		
120	49	4	4	6	8	9	11	2	3	4	4	5	6	-1	-1	-2	-2	-3	-3
100	38	1	2	3	5	7	9	1	1	2	3	4	5	0	-1	-1	-2	-2	-2
80	27	0	0	1	3	4	6	0	0	1	2	3	4	0	0	0	-1	-1	-2
60	16	0	0	1	2	3	4	0	0	1	1	2	3	0	0	0	0	-1	-1
-60	-51	0	0	1	2	3	4	0	0	1	1	2	3	0	0	0	0	-1	-1

Slope and Wind V1 Adjustments*

WEIGHT (1000 LB)	SLOPE (%)					WIND (KTS)							
	-2	-1	0	1	2	-15	-10	-5	0	10	20	30	40
170	-4	-2	0	2	4	-3	-2	-1	0	1	1	2	3
160	-4	-2	0	2	4	-3	-2	-1	0	1	1	2	3
150	-4	-2	0	2	4	-3	-2	-1	0	1	1	2	3
140	-4	-2	0	2	3	-4	-2	-1	0	1	1	2	3
130	-3	-1	0	1	3	-4	-3	-1	0	1	2	2	3
120	-3	-1	0	1	3	-4	-3	-1	0	1	2	2	3
110	-2	-1	0	1	2	-4	-3	-1	0	1	2	2	3
100	-2	-1	0	1	2	-4	-3	-1	0	1	2	2	3
90	-2	-1	0	1	2	-4	-3	-1	0	1	2	3	3

Stopway V1 Adjustments*

NORMAL V1 (KIAS)	STOPWAY (FT)				
	0	200	400	600	800
160	0	1	2	2	3
140	0	1	2	2	3
120	0	1	2	3	4
100	0	1	2	3	4

Use of clearway not allowed on wet runways.
***V1 not to exceed VR.**

V1(MCG)
Max Takeoff Thrust

TEMP		PRESSURE ALTITUDE (FT)						
°F	°C	-2000	0	2000	4000	6000	8000	10000
160	71	102						
140	60	102	99	97	96			
120	49	104	102	98	96	94	92	90
100	38	110	107	103	100	96	92	90
80	27	112	111	109	105	101	97	93
60	16	112	112	109	107	104	101	97
-60	-51	113	113	110	108	105	102	100

FIGURE 238.—Takeoff Speeds—Wet Runway.

Takeoff %N1

Based on engine bleeds for packs on, engine and wing anti-ice on or off

OAT (°F)	AIRPORT PRESSURE ALTITUDE (FT)												
	-2000	-1000	0	1000	2000	3000	4000	5000	6000	7000	8000	9000	10000
170	87.6	88.0	88.9	89.4	89.8	90.4	91.0	91.7	92.4	92.9	93.4	93.5	93.6
160	88.5	89.0	89.3	89.2	89.1	89.7	90.3	91.0	91.7	92.2	92.6	92.8	92.9
150	89.4	89.9	90.3	90.2	90.1	90.1	90.0	90.3	91.0	91.4	91.9	92.0	92.1
140	90.3	90.8	91.2	91.2	91.1	91.1	91.0	91.1	91.2	91.0	91.2	91.3	91.4
130	91.1	91.7	92.1	92.1	92.0	92.0	92.0	92.0	92.0	91.9	91.8	91.4	90.9
120	92.0	92.6	93.0	93.0	93.0	92.9	92.9	92.9	92.9	92.8	92.7	92.4	92.0
110	92.9	93.5	93.9	93.9	93.8	93.8	93.8	93.7	93.7	93.6	93.6	93.4	93.1
100	93.8	94.3	94.8	94.7	94.7	94.7	94.6	94.6	94.5	94.4	94.4	94.3	94.2
90	94.2	95.3	95.7	95.7	95.7	95.6	95.6	95.5	95.4	95.4	95.3	95.2	95.2
80	93.3	94.5	95.6	96.1	96.5	96.5	96.4	96.4	96.3	96.2	96.2	96.1	96.1
70	92.5	93.7	94.8	95.3	95.8	96.4	97.1	97.4	97.3	97.2	97.1	97.1	97.0
60	91.6	92.8	93.9	94.4	95.0	95.6	96.2	96.9	97.6	98.3	98.5	98.4	98.3
50	90.8	92.0	93.0	93.6	94.1	94.7	95.3	96.0	96.7	97.5	98.2	99.1	100.0
40	89.9	91.1	92.2	92.7	93.2	93.8	94.4	95.1	95.8	96.6	97.4	98.3	99.2
30	89.1	90.2	91.3	91.8	92.3	92.9	93.6	94.2	94.9	95.7	96.5	97.4	98.3
20	88.2	89.3	90.4	90.9	91.4	92.0	92.7	93.4	94.0	94.8	95.6	96.6	97.5
10	87.3	88.4	89.5	90.0	90.5	91.1	91.7	92.4	93.1	93.9	94.7	95.7	96.6
0	86.4	87.5	88.6	89.1	89.6	90.2	90.8	91.5	92.2	93.0	93.8	94.8	95.8
-10	85.5	86.6	87.6	88.1	88.6	89.3	89.9	90.6	91.3	92.1	92.9	94.0	94.9
-20	84.6	85.7	86.7	87.2	87.7	88.3	89.0	89.7	90.4	91.2	92.0	93.1	94.0
-30	83.6	84.7	85.7	86.2	86.7	87.4	88.0	88.7	89.4	90.2	91.1	92.2	93.1
-40	82.7	83.8	84.8	85.3	85.8	86.4	87.0	87.8	88.5	89.3	90.1	91.2	92.2
-50	81.7	82.8	83.8	84.3	84.8	85.4	86.1	86.8	87.5	88.3	89.2	90.3	91.3
-60	80.8	81.8	82.8	83.3	83.8	84.4	85.1	85.8	86.5	87.3	88.2	89.4	90.3

%N1 Adjustments for Engine Bleeds

BLEED CONFIGURATION	PRESSURE ALTITUDE (FT)												
	-2000	-1000	0	1000	2000	3000	4000	5000	6000	7000	8000	9000	10000
PACKS OFF	0.7	0.7	0.7	0.7	0.7	0.7	0.8	0.8	0.8	0.8	0.8	0.9	1.0

FIGURE 239.—Takeoff % N1.

Stab Trim Setting

Max Takeoff Thrust

Flaps 1 and 5

WEIGHT (1000 LB)	C.G. (%MAC)								
	9	11	13	16	20	24	28	30	33
160-180	8 1/2	8 1/2	8 1/2	7 3/4	6 3/4	6	5 1/4	4 3/4	4 1/4
140	8 1/2	8 1/2	8	7 1/4	6 1/2	5 1/2	4 3/4	4 1/2	3 3/4
120	8 1/2	8	7 1/2	6 1/2	5 3/4	5	4 1/4	4	3 1/4
80-100	6 3/4	6 1/2	6	5 1/2	5	4 1/4	3 1/2	3 1/4	2 3/4

Flaps 10, 15 and 25

WEIGHT (1000 LB)	C.G. (%MAC)								
	9	11	13	16	20	24	28	30	33
160-180	8 1/2	8 1/2	8 1/2	7 1/4	6 1/2	5 1/2	4 1/2	4 1/4	3 1/2
140	8 1/2	8 1/2	7 3/4	6 3/4	6	5	4 1/4	3 3/4	3
120	8 1/2	7 3/4	7	6	5 1/4	4 1/2	3 3/4	3 1/4	2 3/4
80-100	6 1/4	6	5 1/2	5	4 1/2	3 3/4	3	2 3/4	2 3/4

FIGURE 240.—Stab Trim Setting.

Table of Contents

COMPUTER TESTING SUPPLEMENT
FOR
AIRLINE TRANSPORT PILOT
AND
AIRCRAFT DISPATCHER

ADDENDUM A
JULY 2011

U.S. DEPARTMENT OF TRANSPORTATION
FEDERAL AVIATION ADMINISTRATION
Flight Standards Service

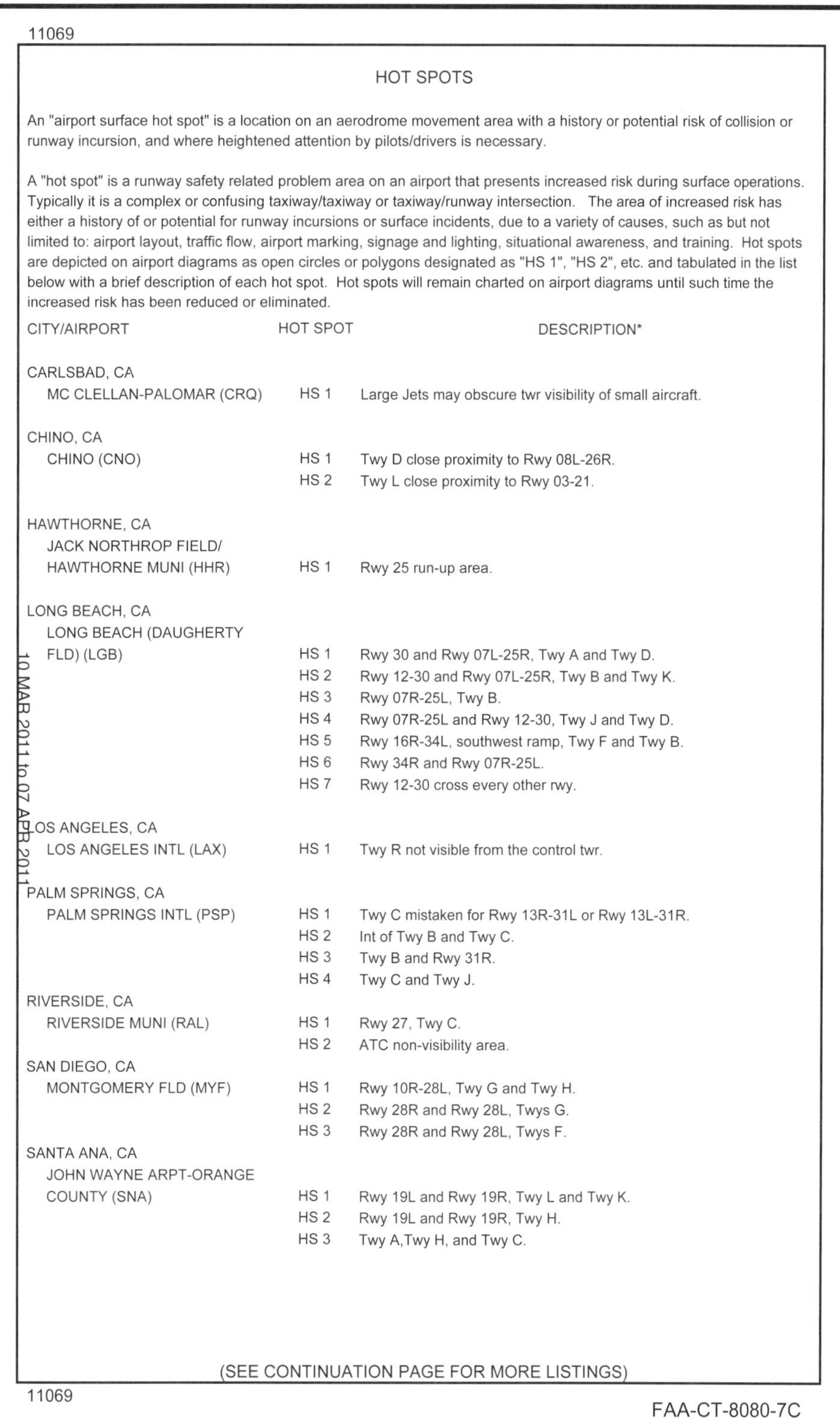

11069

HOT SPOTS

An "airport surface hot spot" is a location on an aerodrome movement area with a history or potential risk of collision or runway incursion, and where heightened attention by pilots/drivers is necessary.

A "hot spot" is a runway safety related problem area on an airport that presents increased risk during surface operations. Typically it is a complex or confusing taxiway/taxiway or taxiway/runway intersection. The area of increased risk has either a history of or potential for runway incursions or surface incidents, due to a variety of causes, such as but not limited to: airport layout, traffic flow, airport marking, signage and lighting, situational awareness, and training. Hot spots are depicted on airport diagrams as open circles or polygons designated as "HS 1", "HS 2", etc. and tabulated in the list below with a brief description of each hot spot. Hot spots will remain charted on airport diagrams until such time the increased risk has been reduced or eliminated.

CITY/AIRPORT	HOT SPOT	DESCRIPTION*
CARLSBAD, CA		
MC CLELLAN-PALOMAR (CRQ)	HS 1	Large Jets may obscure twr visibility of small aircraft.
CHINO, CA		
CHINO (CNO)	HS 1	Twy D close proximity to Rwy 08L-26R.
	HS 2	Twy L close proximity to Rwy 03-21.
HAWTHORNE, CA		
JACK NORTHROP FIELD/ HAWTHORNE MUNI (HHR)	HS 1	Rwy 25 run-up area.
LONG BEACH, CA		
LONG BEACH (DAUGHERTY FLD) (LGB)	HS 1	Rwy 30 and Rwy 07L-25R, Twy A and Twy D.
	HS 2	Rwy 12-30 and Rwy 07L-25R, Twy B and Twy K.
	HS 3	Rwy 07R-25L, Twy B.
	HS 4	Rwy 07R-25L and Rwy 12-30, Twy J and Twy D.
	HS 5	Rwy 16R-34L, southwest ramp, Twy F and Twy B.
	HS 6	Rwy 34R and Rwy 07R-25L.
	HS 7	Rwy 12-30 cross every other rwy.
LOS ANGELES, CA		
LOS ANGELES INTL (LAX)	HS 1	Twy R not visible from the control twr.
PALM SPRINGS, CA		
PALM SPRINGS INTL (PSP)	HS 1	Twy C mistaken for Rwy 13R-31L or Rwy 13L-31R.
	HS 2	Int of Twy B and Twy C.
	HS 3	Twy B and Rwy 31R.
	HS 4	Twy C and Twy J.
RIVERSIDE, CA		
RIVERSIDE MUNI (RAL)	HS 1	Rwy 27, Twy C.
	HS 2	ATC non-visibility area.
SAN DIEGO, CA		
MONTGOMERY FLD (MYF)	HS 1	Rwy 10R-28L, Twy G and Twy H.
	HS 2	Rwy 28R and Rwy 28L, Twys G.
	HS 3	Rwy 28R and Rwy 28L, Twys F.
SANTA ANA, CA		
JOHN WAYNE ARPT-ORANGE COUNTY (SNA)	HS 1	Rwy 19L and Rwy 19R, Twy L and Twy K.
	HS 2	Rwy 19L and Rwy 19R, Twy H.
	HS 3	Twy A,Twy H, and Twy C.

10 MAR 2011 to 07 APR 2011

(SEE CONTINUATION PAGE FOR MORE LISTINGS)

11069

FAA-CT-8080-7C

Figure 241—Hot Spots.

11069

HOT SPOTS

(CONTINUED)

CITY/AIRPORT	HOT SPOT	DESCRIPTION*
SANTA BARBARA, CA		
SANTA BARBARA MUNI (SBA)	HS 1	Rwy 07-25, Twy C.
	HS 2	Rwy 15L and Rwy 15R, Twy C, wide pavement.
	HS 3	Rwy 15L-33R, Rwy 15R-33L, Rwy 07-25. Rwy 15L-33R and Rwy 15R-33L utilized for taxi.
	HS 4	Rwy 25, Twy H and Twy J.
SANTA MARIA, CA		
CAPTAIN G. ALLAN HANCOCK FLD (SMX)	HS 1	Twy A, Twy C, and Twy D.
	HS 2	Rwy 20 and Twy A.
	HS 3	Rwy 12 and Twy B.
VICTORVILLE, CA		
SOUTHERN CALIFORNIA LOGISTICS (VCV)	HS 1	Wrong rwy departure risk.

10 MAR 2011 to 07 APR 2011

*See appropriate A/FD, Alaska or Pacific Supplement HOT SPOT table for additional information.

11069 FAA-CT-8080-7C

Figure 241—Hot Spots, continued.

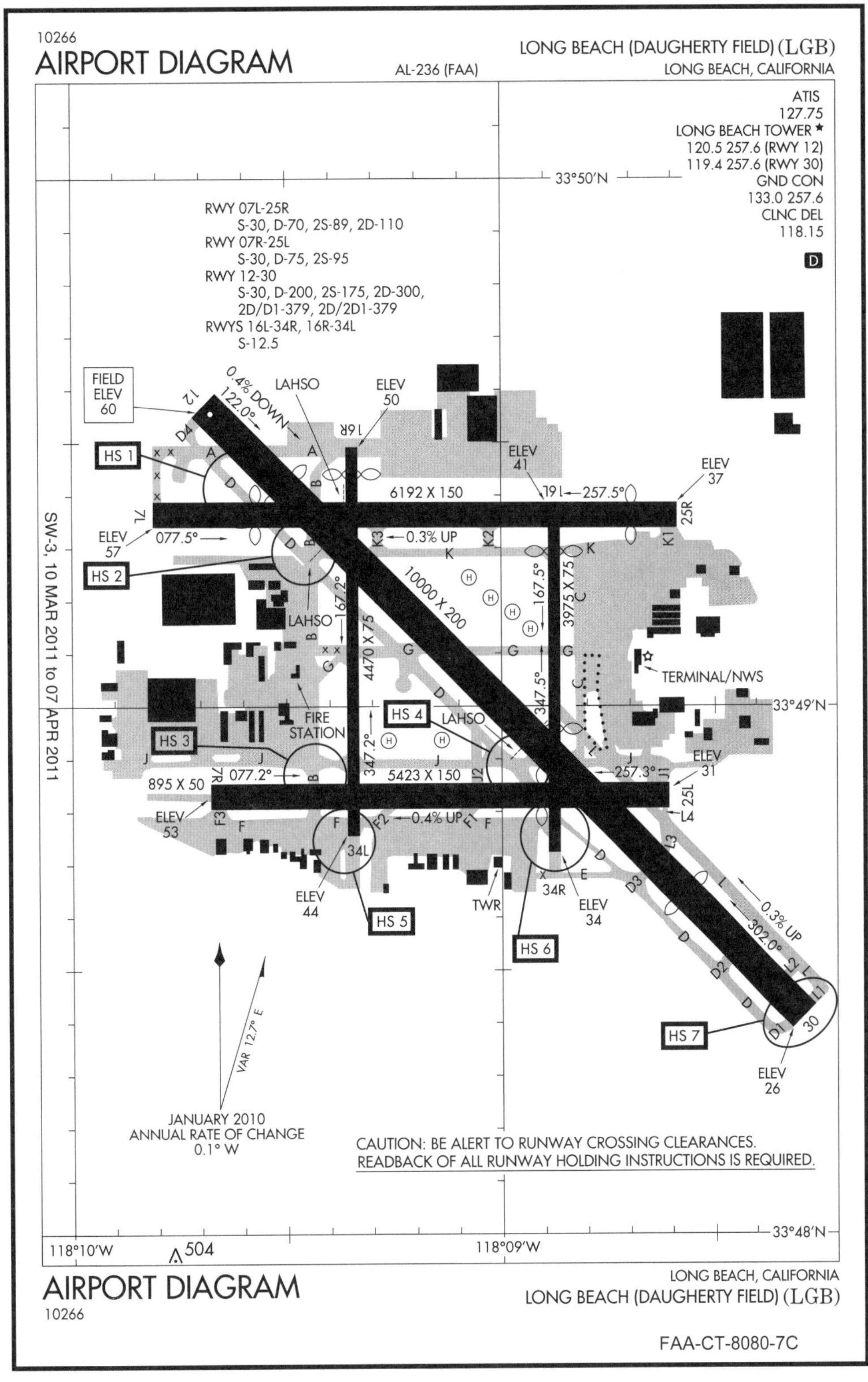

Figure 242—Airport Diagram.

11069

10 MAR 2011 to 07 APR 2011

HOT SPOTS

An "airport surface hot spot" is a location on an aerodrome movement area with a history or potential risk of collision or runway incursion, and where heightened attention by pilots/drivers is necessary.

A "hot spot" is a runway safety related problem area on an airport that presents increased risk during surface operations. Typically it is a complex or confusing taxiway/taxiway or taxiway/runway intersection. The area of increased risk has either a history of or potential for runway incursions or surface incidents, due to a variety of causes, such as but not limited to: airport layout, traffic flow, airport marking, signage and lighting, situational awareness, and training. Hot spots are depicted on airport diagrams as open circles or polygons designated as "HS 1", "HS 2", etc. and tabulated in the list below with a brief description of each hot spot. Hot spots will remain charted on airport diagrams until such time the increased risk has been reduced or eliminated.

CITY/AIRPORT	HOT SPOT	DESCRIPTION*
DAYTONA BEACH, FL		
DAYTONA BEACH INTL (DAB)	HS 1	Int of Twy W and Twy S.
FORT LAUDERDALE, FL		
FORT LAUDERDALE-HOLLYWOOD INTL (FLL)	HS 1	Twy E at Rwy 09L-27R.
	HS 2	Twy D at Rwy 09L-27R.
	HS 3	Twy Q at Rwy 09L-27R.
	HS 4	Twy E int departure for Rwy 27L.
	HS 5	Twy departure risk. Twy B instead of Rwy 09L.
HOLLYWOOD, FL		
NORTH PERRY (HWO)	HS 1	Southbound on Twy D for Rwy 27R departures.
	HS 2	The hold line for Rwy 36L is also the hold line for Rwy 09R.
	HS 3	Aircraft taxiing on Twy L westbound to depart on Rwy 18R-36L.
MIAMI, FL		
MIAMI INTL (MIA)	HS 1	Short twy risk.
	HS 2	Short twy risk.
	HS 3	Rwy 27 and Rwy 30 wrong rwy departure risk.
	HS 4	Short twy between rwys.
MIAMI, FL		
OPA-LOCKA EXECUTIVE (OPF)	HS 1	Surface painted LOCATION and DIRECTION signs ONLY.
ORLANDO, FL		
ORLANDO SANFORD INTL (SFB)	HS 1	Twy C is beyond the Rwy 09C APCH hold sign and marking.
	HS 2	Hold line for Rwy 09R on Twy R northbound is adjacent to Twy S.
STUART, FL		
WITHAM FIELD (SUA)	HS 1	Intersecting rwys, wrong rwy departure risk.
	HS 2	Rwy 12 and Twy A1.

*See appropriate A/FD, Alaska or Pacific Supplement HOT SPOT table for additional information.

11069

10 MAR 2011 to 07 APR 2011

FAA-CT-8080-7C

Figure 243—Hot Spots.

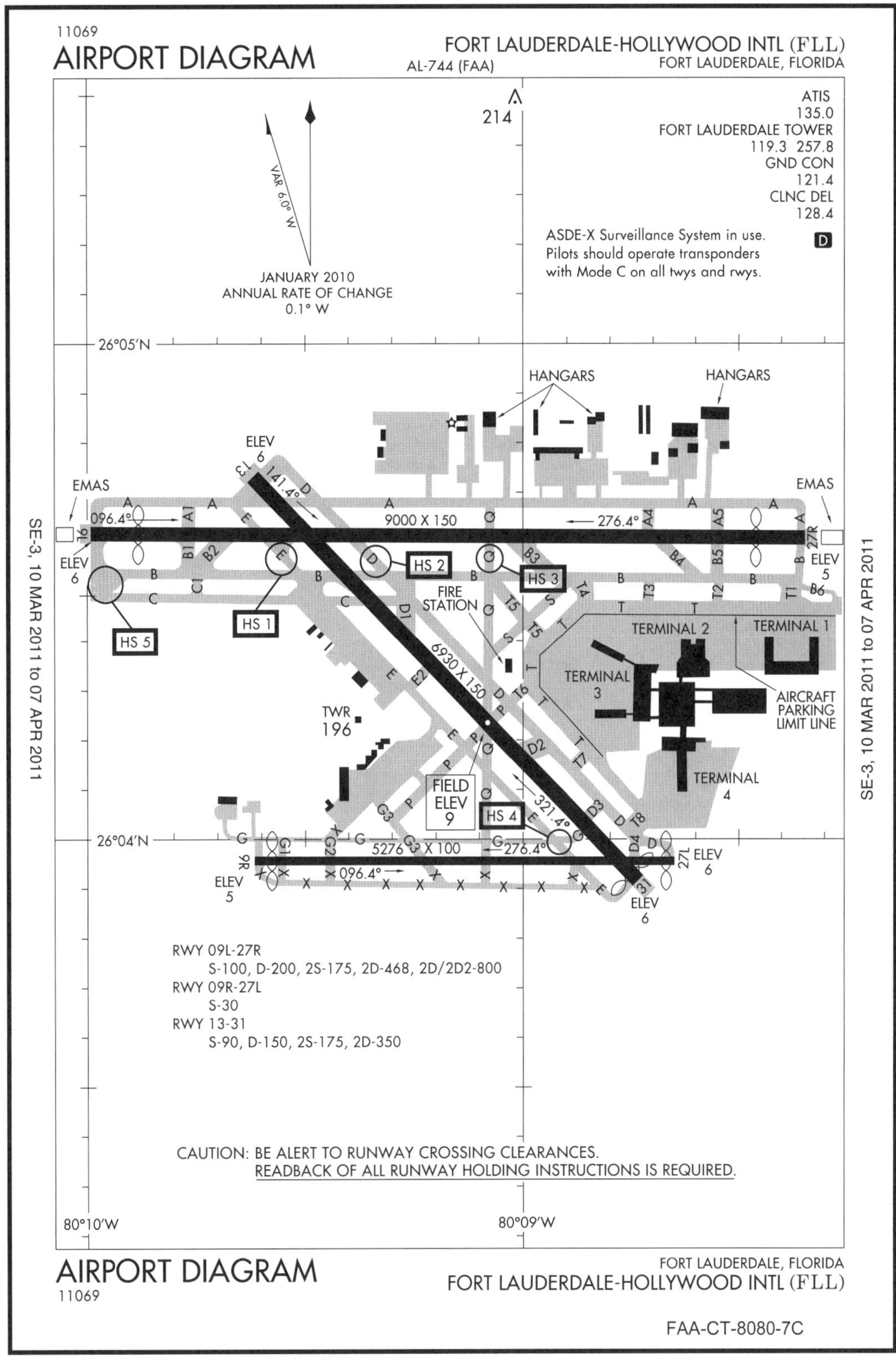

Figure 244—Airport Diagram.

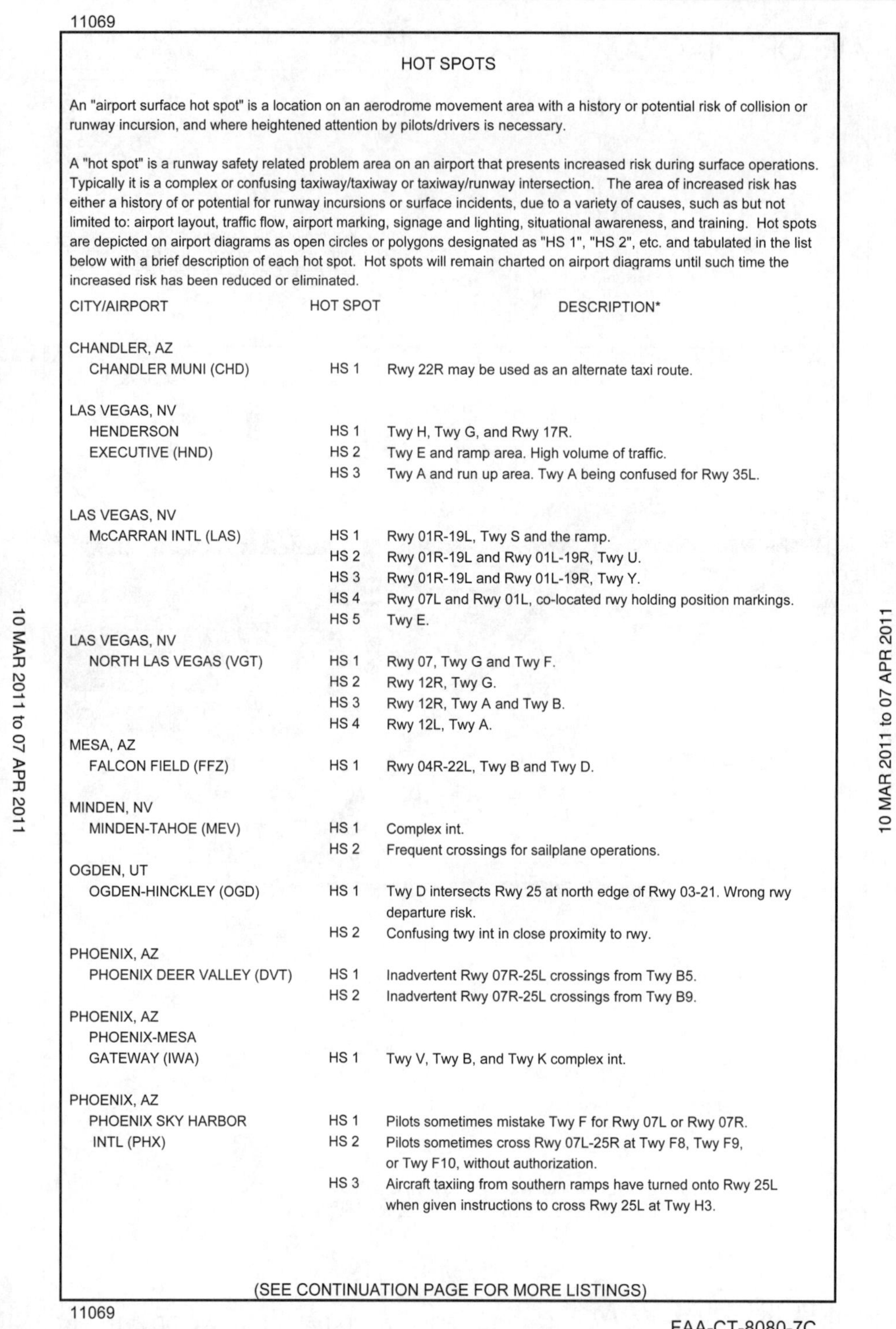

11069

10 MAR 2011 to 07 APR 2011

HOT SPOTS

An "airport surface hot spot" is a location on an aerodrome movement area with a history or potential risk of collision or runway incursion, and where heightened attention by pilots/drivers is necessary.

A "hot spot" is a runway safety related problem area on an airport that presents increased risk during surface operations. Typically it is a complex or confusing taxiway/taxiway or taxiway/runway intersection. The area of increased risk has either a history of or potential for runway incursions or surface incidents, due to a variety of causes, such as but not limited to: airport layout, traffic flow, airport marking, signage and lighting, situational awareness, and training. Hot spots are depicted on airport diagrams as open circles or polygons designated as "HS 1", "HS 2", etc. and tabulated in the list below with a brief description of each hot spot. Hot spots will remain charted on airport diagrams until such time the increased risk has been reduced or eliminated.

CITY/AIRPORT	HOT SPOT	DESCRIPTION*
CHANDLER, AZ		
CHANDLER MUNI (CHD)	HS 1	Rwy 22R may be used as an alternate taxi route.
LAS VEGAS, NV		
HENDERSON	HS 1	Twy H, Twy G, and Rwy 17R.
EXECUTIVE (HND)	HS 2	Twy E and ramp area. High volume of traffic.
	HS 3	Twy A and run up area. Twy A being confused for Rwy 35L.
LAS VEGAS, NV		
McCARRAN INTL (LAS)	HS 1	Rwy 01R-19L, Twy S and the ramp.
	HS 2	Rwy 01R-19L and Rwy 01L-19R, Twy U.
	HS 3	Rwy 01R-19L and Rwy 01L-19R, Twy Y.
	HS 4	Rwy 07L and Rwy 01L, co-located rwy holding position markings.
	HS 5	Twy E.
LAS VEGAS, NV		
NORTH LAS VEGAS (VGT)	HS 1	Rwy 07, Twy G and Twy F.
	HS 2	Rwy 12R, Twy G.
	HS 3	Rwy 12R, Twy A and Twy B.
	HS 4	Rwy 12L, Twy A.
MESA, AZ		
FALCON FIELD (FFZ)	HS 1	Rwy 04R-22L, Twy B and Twy D.
MINDEN, NV		
MINDEN-TAHOE (MEV)	HS 1	Complex int.
	HS 2	Frequent crossings for sailplane operations.
OGDEN, UT		
OGDEN-HINCKLEY (OGD)	HS 1	Twy D intersects Rwy 25 at north edge of Rwy 03-21. Wrong rwy departure risk.
	HS 2	Confusing twy int in close proximity to rwy.
PHOENIX, AZ		
PHOENIX DEER VALLEY (DVT)	HS 1	Inadvertent Rwy 07R-25L crossings from Twy B5.
	HS 2	Inadvertent Rwy 07R-25L crossings from Twy B9.
PHOENIX, AZ		
PHOENIX-MESA GATEWAY (IWA)	HS 1	Twy V, Twy B, and Twy K complex int.
PHOENIX, AZ		
PHOENIX SKY HARBOR	HS 1	Pilots sometimes mistake Twy F for Rwy 07L or Rwy 07R.
INTL (PHX)	HS 2	Pilots sometimes cross Rwy 07L-25R at Twy F8, Twy F9, or Twy F10, without authorization.
	HS 3	Aircraft taxiing from southern ramps have turned onto Rwy 25L when given instructions to cross Rwy 25L at Twy H3.

(SEE CONTINUATION PAGE FOR MORE LISTINGS)

11069

10 MAR 2011 to 07 APR 2011

FAA-CT-8080-7C

Figure 245—Hot Spots.

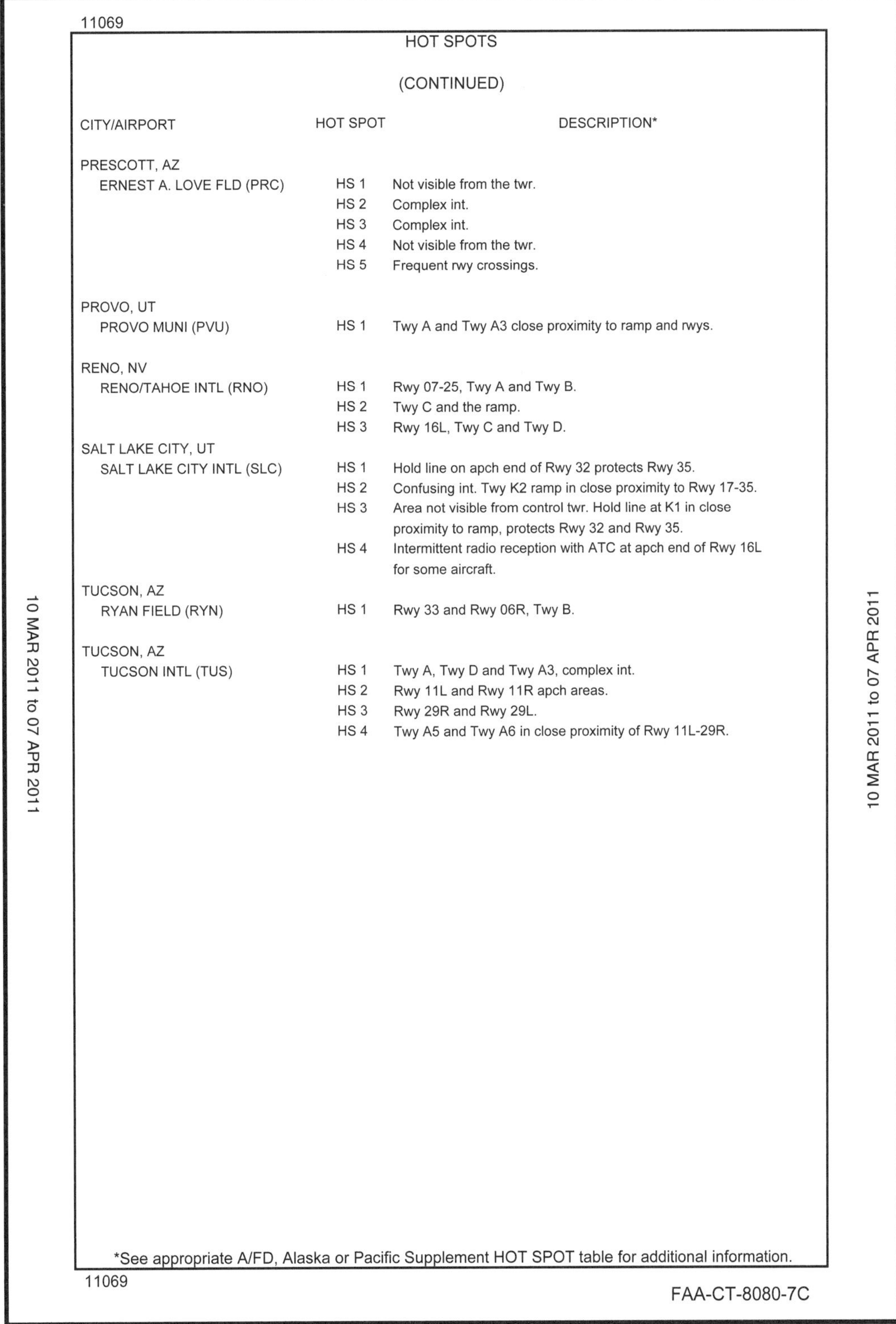

11069

HOT SPOTS

(CONTINUED)

CITY/AIRPORT	HOT SPOT	DESCRIPTION*
PRESCOTT, AZ		
ERNEST A. LOVE FLD (PRC)	HS 1	Not visible from the twr.
	HS 2	Complex int.
	HS 3	Complex int.
	HS 4	Not visible from the twr.
	HS 5	Frequent rwy crossings.
PROVO, UT		
PROVO MUNI (PVU)	HS 1	Twy A and Twy A3 close proximity to ramp and rwys.
RENO, NV		
RENO/TAHOE INTL (RNO)	HS 1	Rwy 07-25, Twy A and Twy B.
	HS 2	Twy C and the ramp.
	HS 3	Rwy 16L, Twy C and Twy D.
SALT LAKE CITY, UT		
SALT LAKE CITY INTL (SLC)	HS 1	Hold line on apch end of Rwy 32 protects Rwy 35.
	HS 2	Confusing int. Twy K2 ramp in close proximity to Rwy 17-35.
	HS 3	Area not visible from control twr. Hold line at K1 in close proximity to ramp, protects Rwy 32 and Rwy 35.
	HS 4	Intermittent radio reception with ATC at apch end of Rwy 16L for some aircraft.
TUCSON, AZ		
RYAN FIELD (RYN)	HS 1	Rwy 33 and Rwy 06R, Twy B.
TUCSON, AZ		
TUCSON INTL (TUS)	HS 1	Twy A, Twy D and Twy A3, complex int.
	HS 2	Rwy 11L and Rwy 11R apch areas.
	HS 3	Rwy 29R and Rwy 29L.
	HS 4	Twy A5 and Twy A6 in close proximity of Rwy 11L-29R.

10 MAR 2011 to 07 APR 2011

*See appropriate A/FD, Alaska or Pacific Supplement HOT SPOT table for additional information.

11069

FAA-CT-8080-7C

Figure 245—Hot Spots, continued.

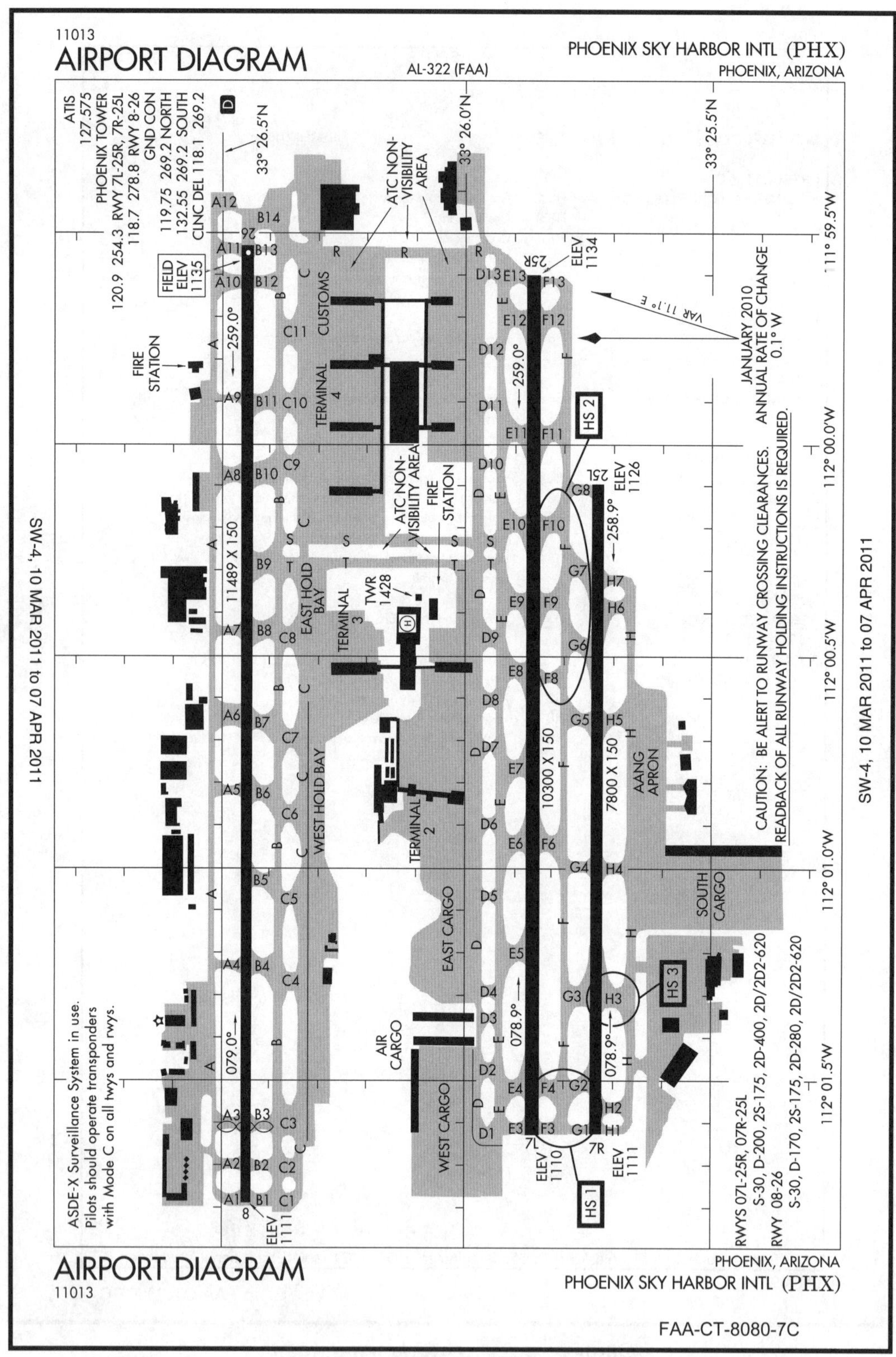

Figure 246—Airport Diagram.

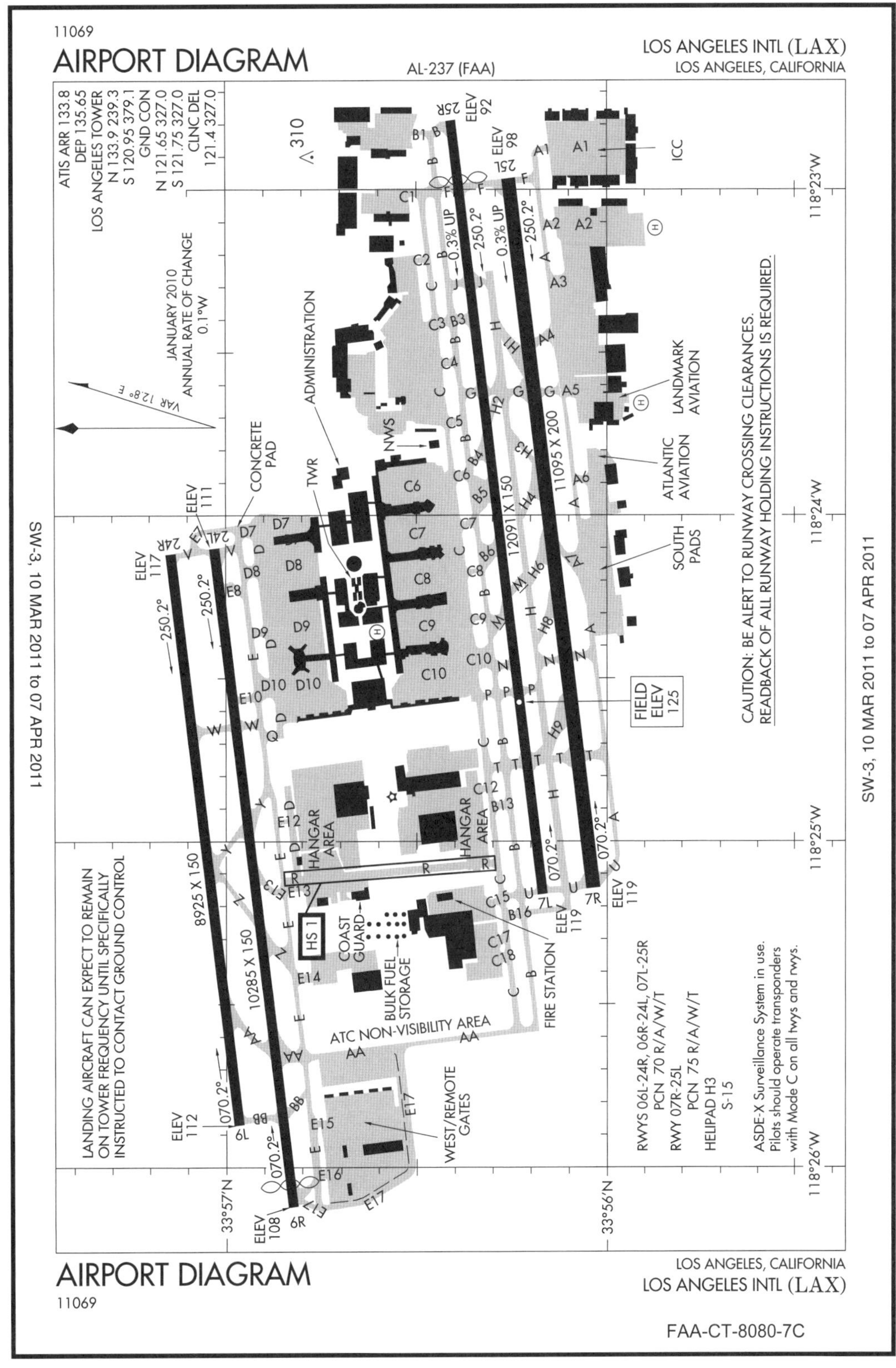

Figure 247—Airport Diagram.

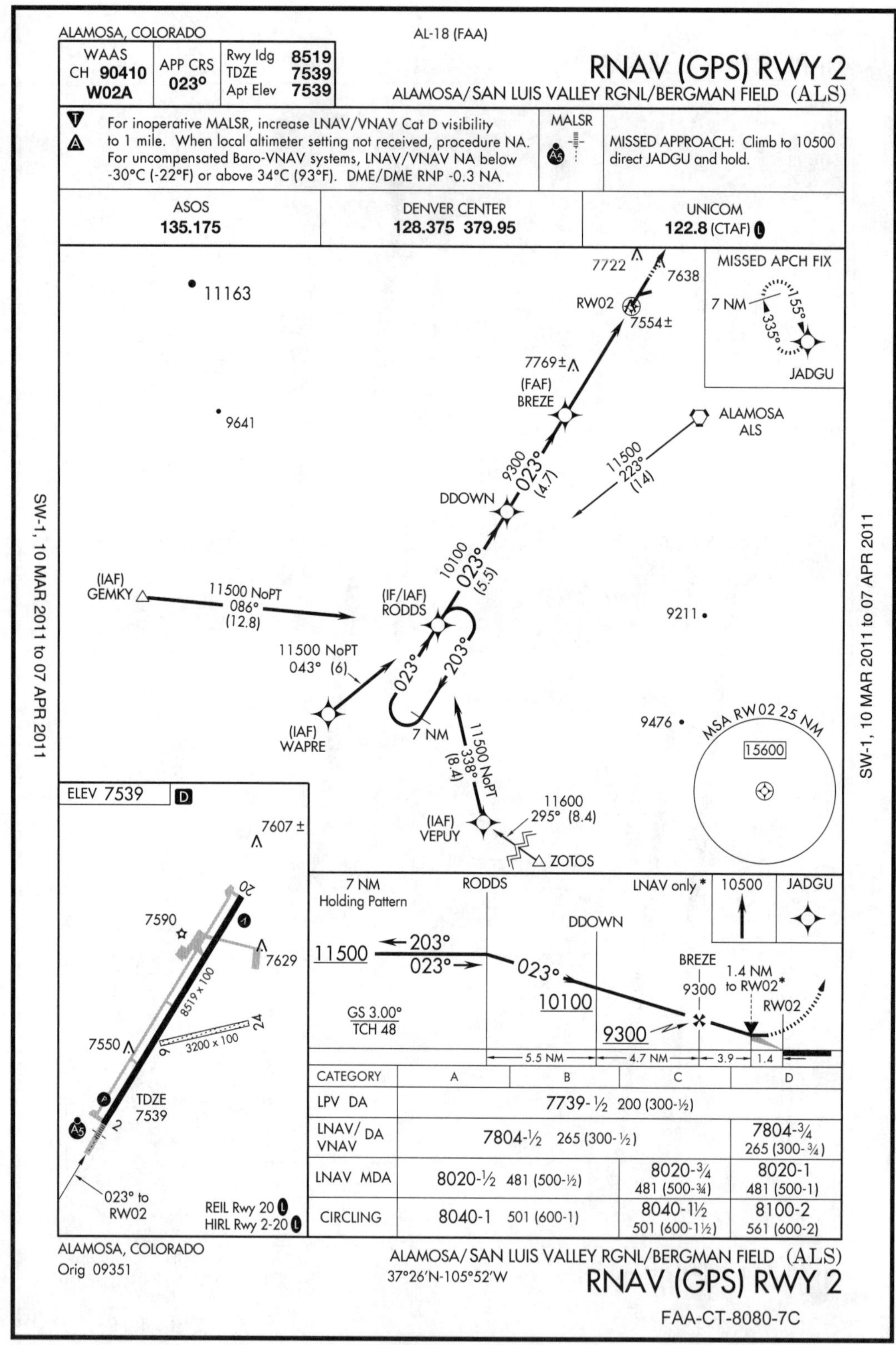

CATEGORY	A	B	C	D
LPV DA	7739-½ 200 (300-½)			
LNAV/VNAV DA	7804-½ 265 (300-½)			7804-¾ 265 (300-¾)
LNAV MDA	8020-½ 481 (500-½)		8020-¾ 481 (500-¾)	8020-1 481 (500-1)
CIRCLING	8040-1 501 (600-1)		8040-1½ 501 (600-1½)	8100-2 561 (600-2)

Figure 248—Airport Diagram.

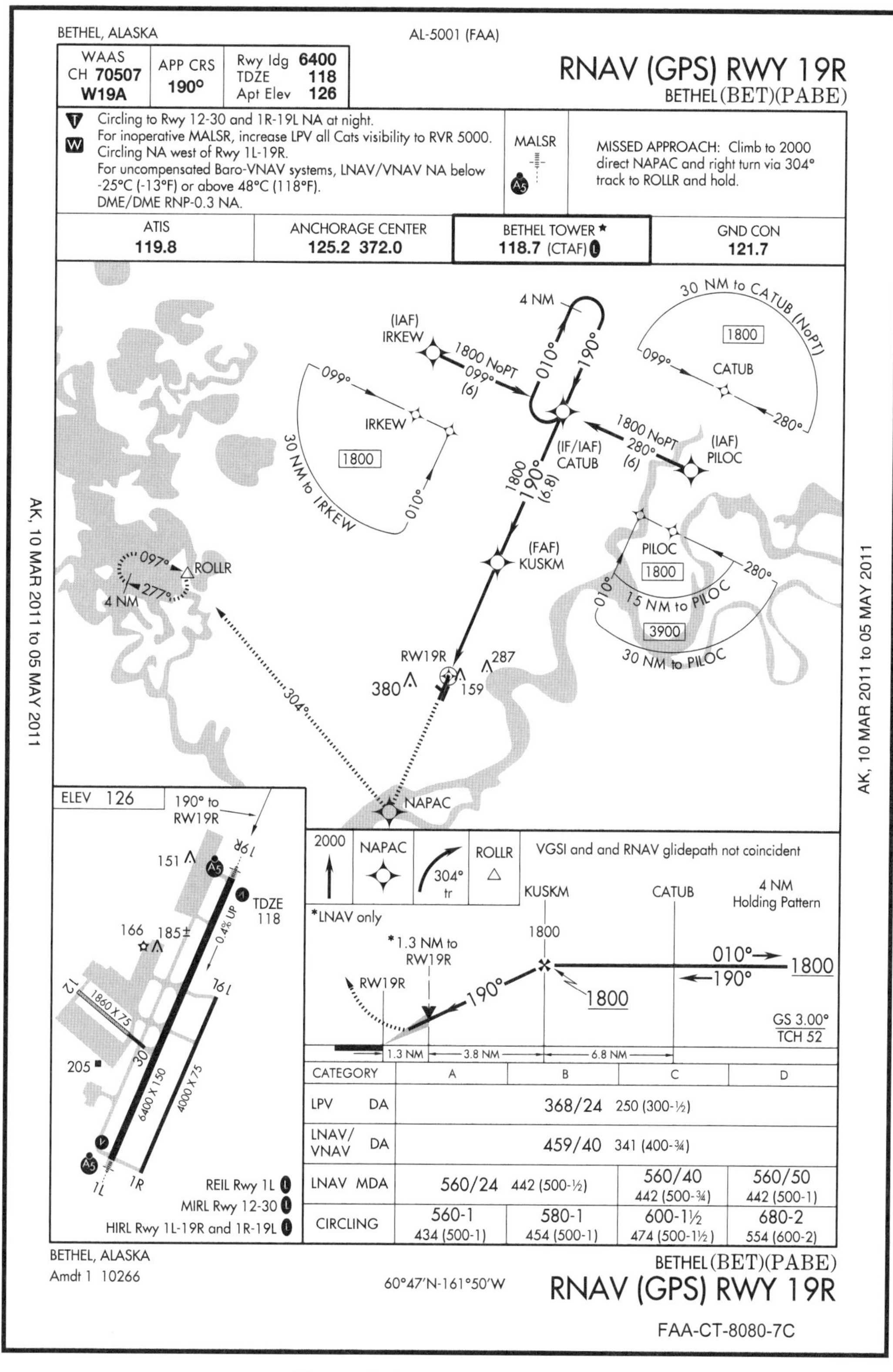

CATEGORY	A	B	C	D
LPV DA	368/24 250 (300-½)			
LNAV/VNAV DA	459/40 341 (400-¾)			
LNAV MDA	560/24 442 (500-½)		560/40 442 (500-¾)	560/50 442 (500-1)
CIRCLING	560-1 434 (500-1)	580-1 454 (500-1)	600-1½ 474 (500-1½)	680-2 554 (600-2)

Figure 249—Airport Diagram.

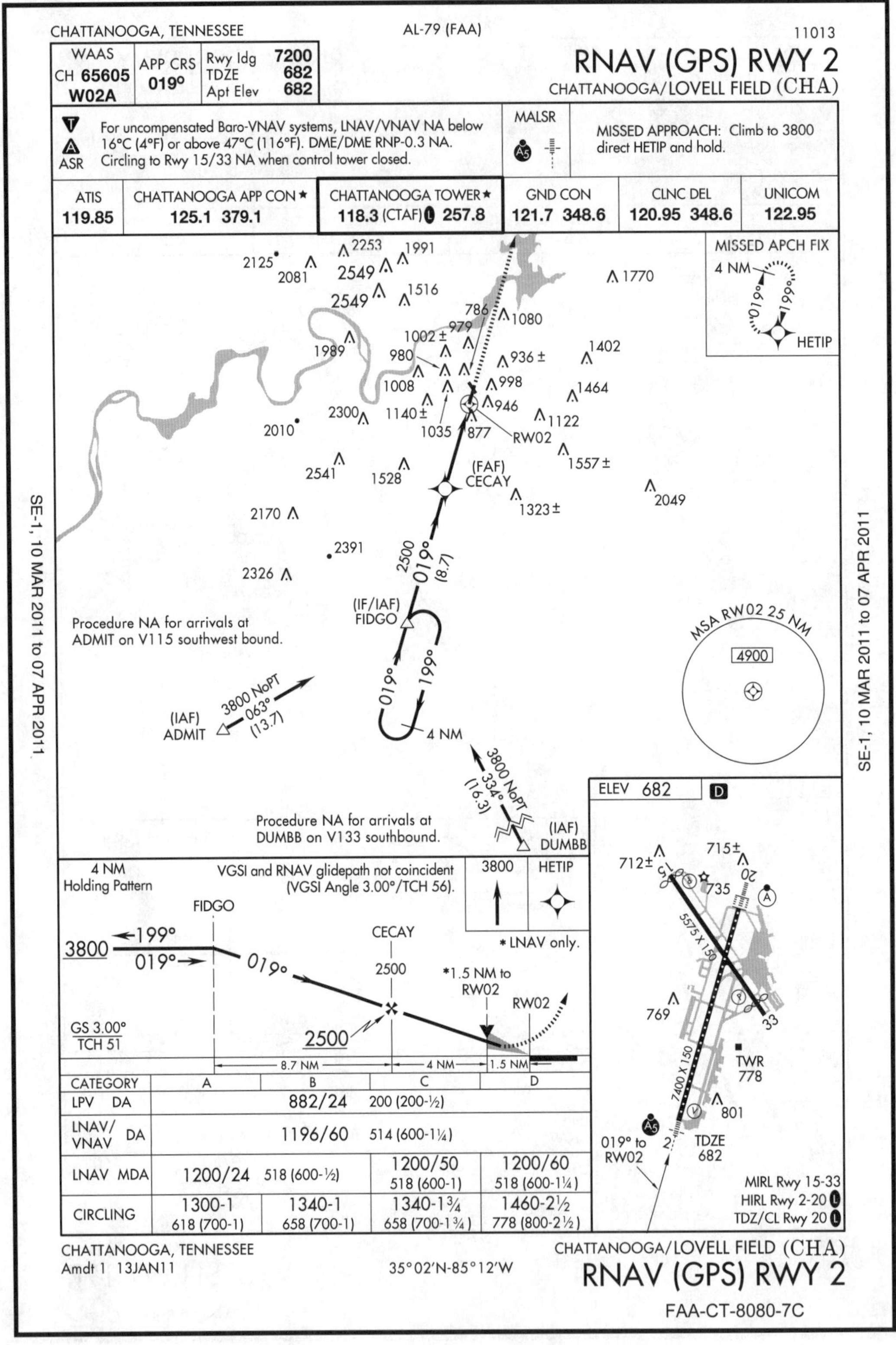

Figure 250—Airport Diagram.

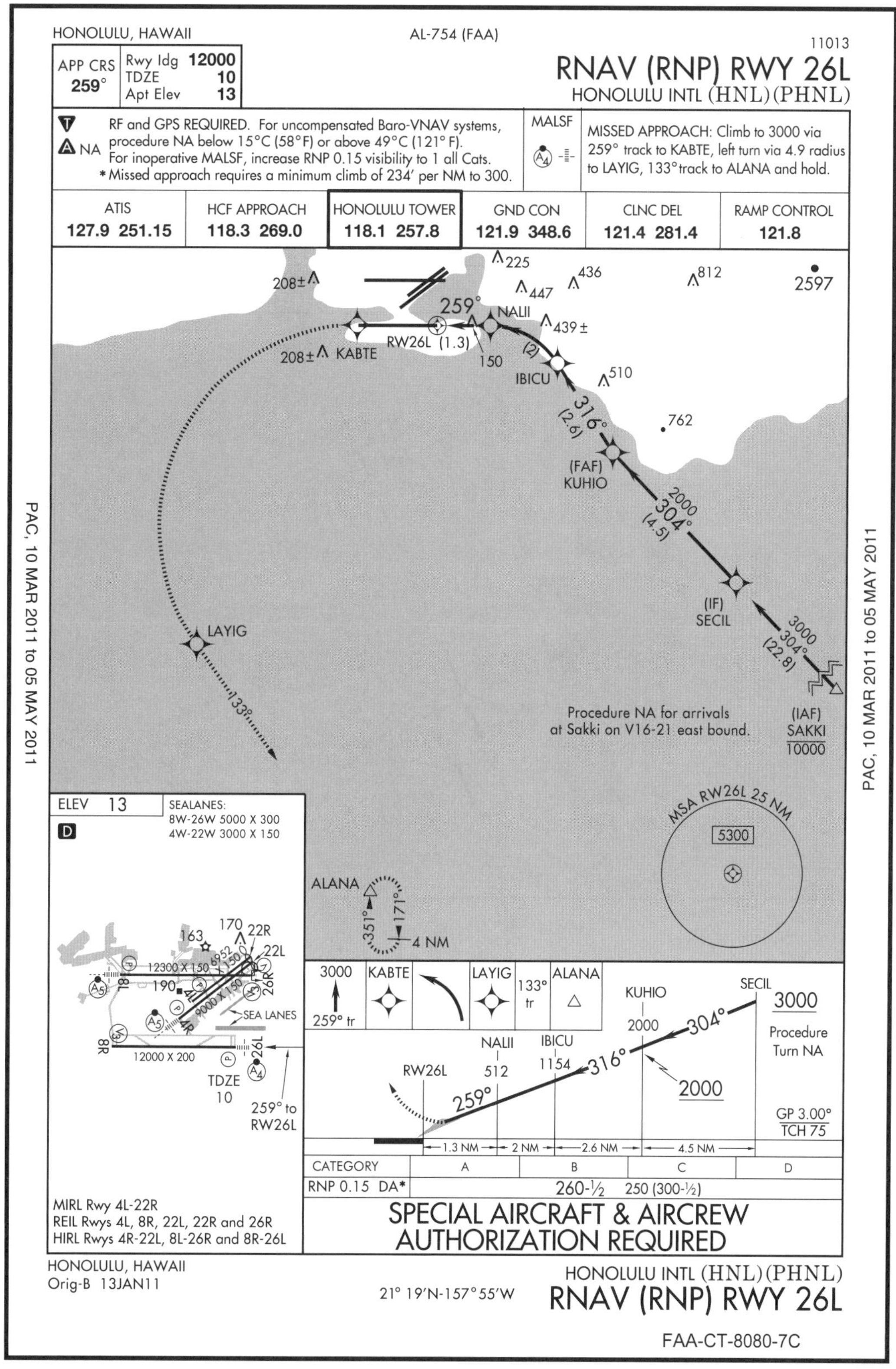

Figure 251—Airport Diagram.

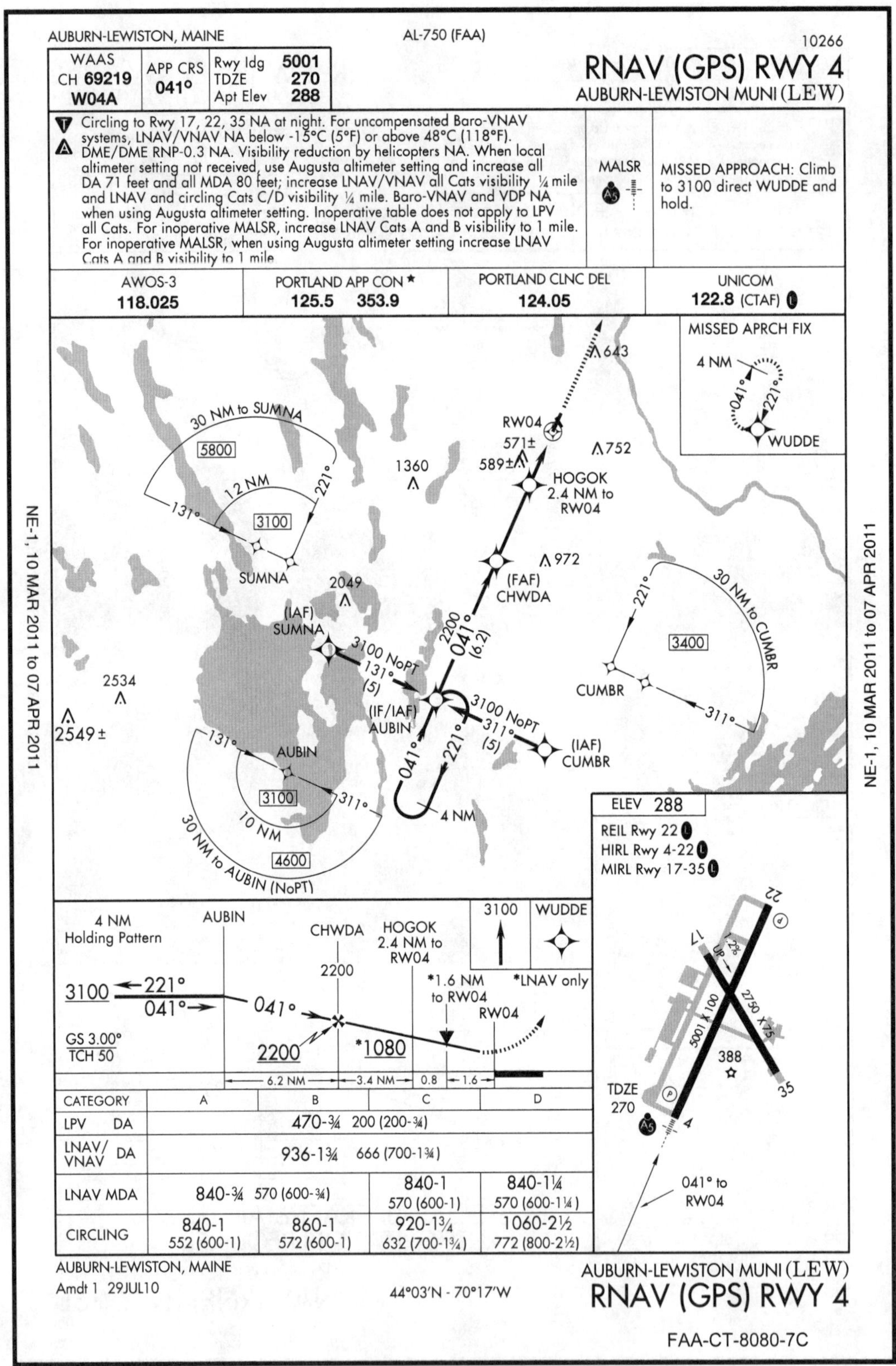

CATEGORY	A	B	C	D
LPV DA	470-¾ 200 (200-¾)			
LNAV/VNAV DA	936-1¾ 666 (700-1¾)			
LNAV MDA	840-¾ 570 (600-¾)		840-1 570 (600-1)	840-1¼ 570 (600-1¼)
CIRCLING	840-1 552 (600-1)	860-1 572 (600-1)	920-1¾ 632 (700-1¾)	1060-2½ 772 (800-2½)

Figure 252—Airport Diagram.

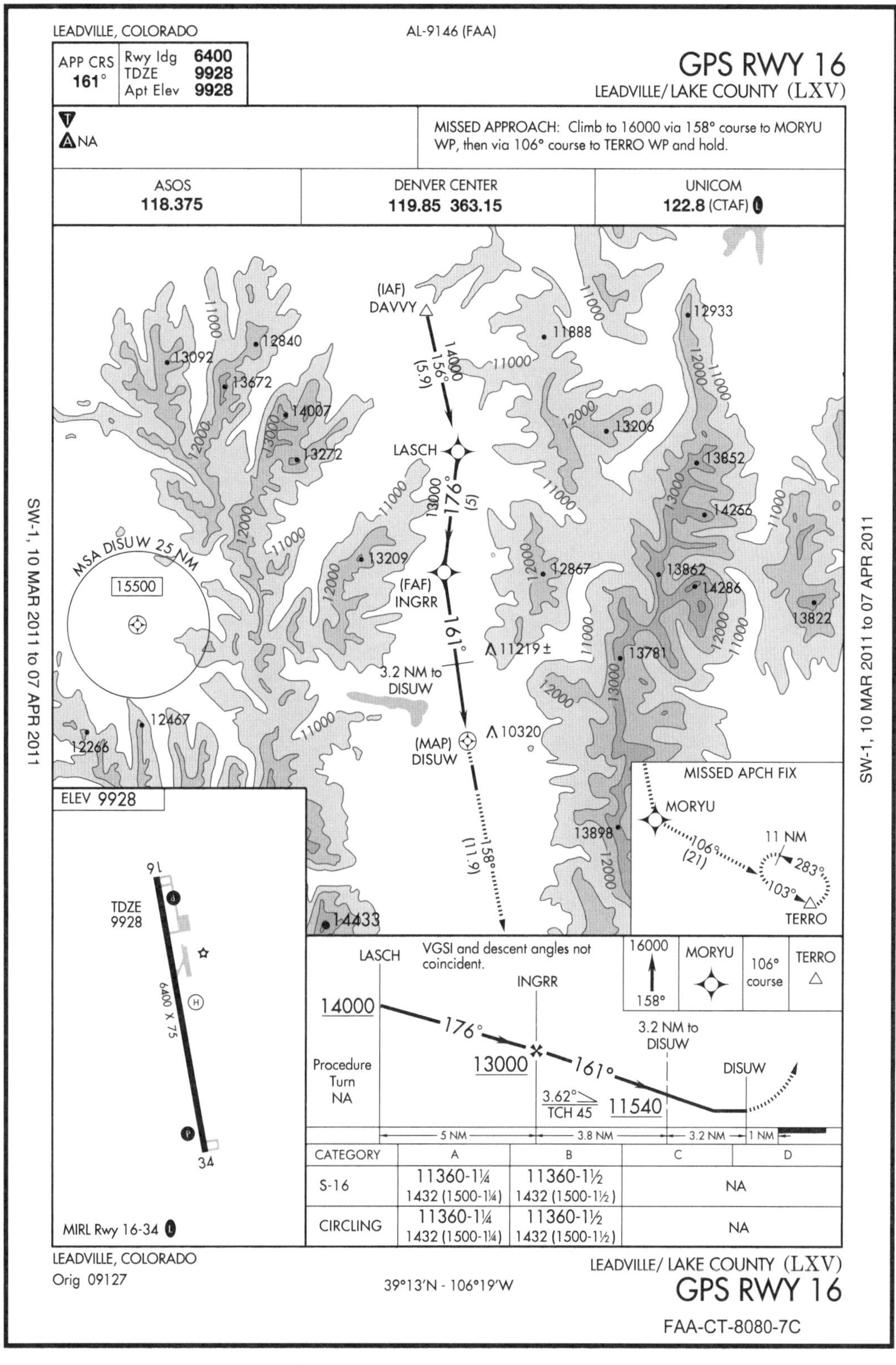

Figure 253—Airport Diagram.

COMPUTER TESTING SUPPLEMENT
FOR
AIRLINE TRANSPORT PILOT
AND
AIRCRAFT DISPATCHER

ADDENDUM B

ADDENDUM B
MAY 2012

U.S. DEPARTMENT OF TRANSPORTATION
FEDERAL AVIATION ADMINISTRATION
Flight Standards Service

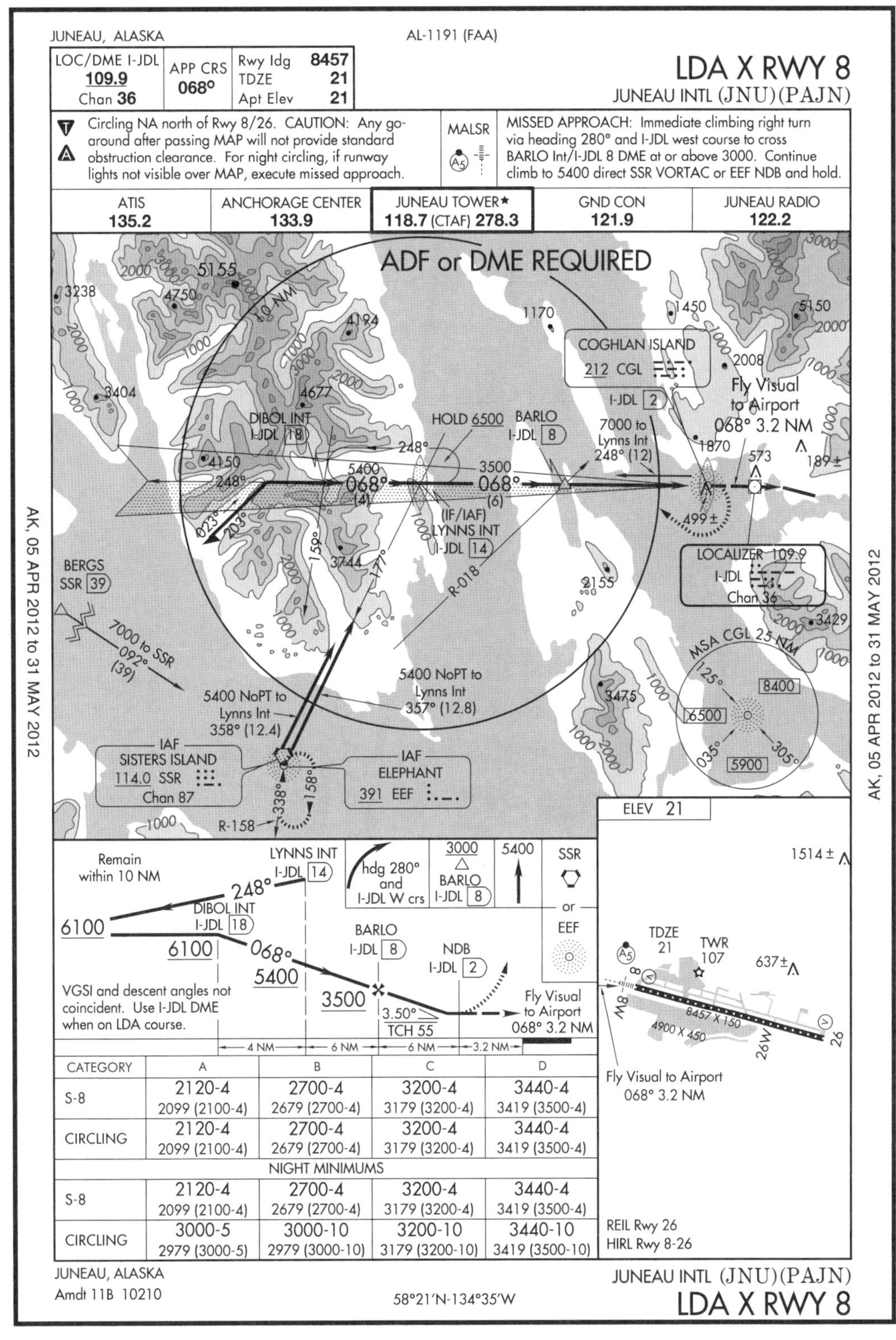

CATEGORY	A	B	C	D
S-8	2120-4 2099 (2100-4)	2700-4 2679 (2700-4)	3200-4 3179 (3200-4)	3440-4 3419 (3500-4)
CIRCLING	2120-4 2099 (2100-4)	2700-4 2679 (2700-4)	3200-4 3179 (3200-4)	3440-4 3419 (3500-4)
NIGHT MINIMUMS				
S-8	2120-4 2099 (2100-4)	2700-4 2679 (2700-4)	3200-4 3179 (3200-4)	3440-4 3419 (3500-4)
CIRCLING	3000-5 2979 (3000-5)	3000-10 2979 (3000-10)	3200-10 3179 (3200-10)	3440-10 3419 (3500-10)

Figure 254

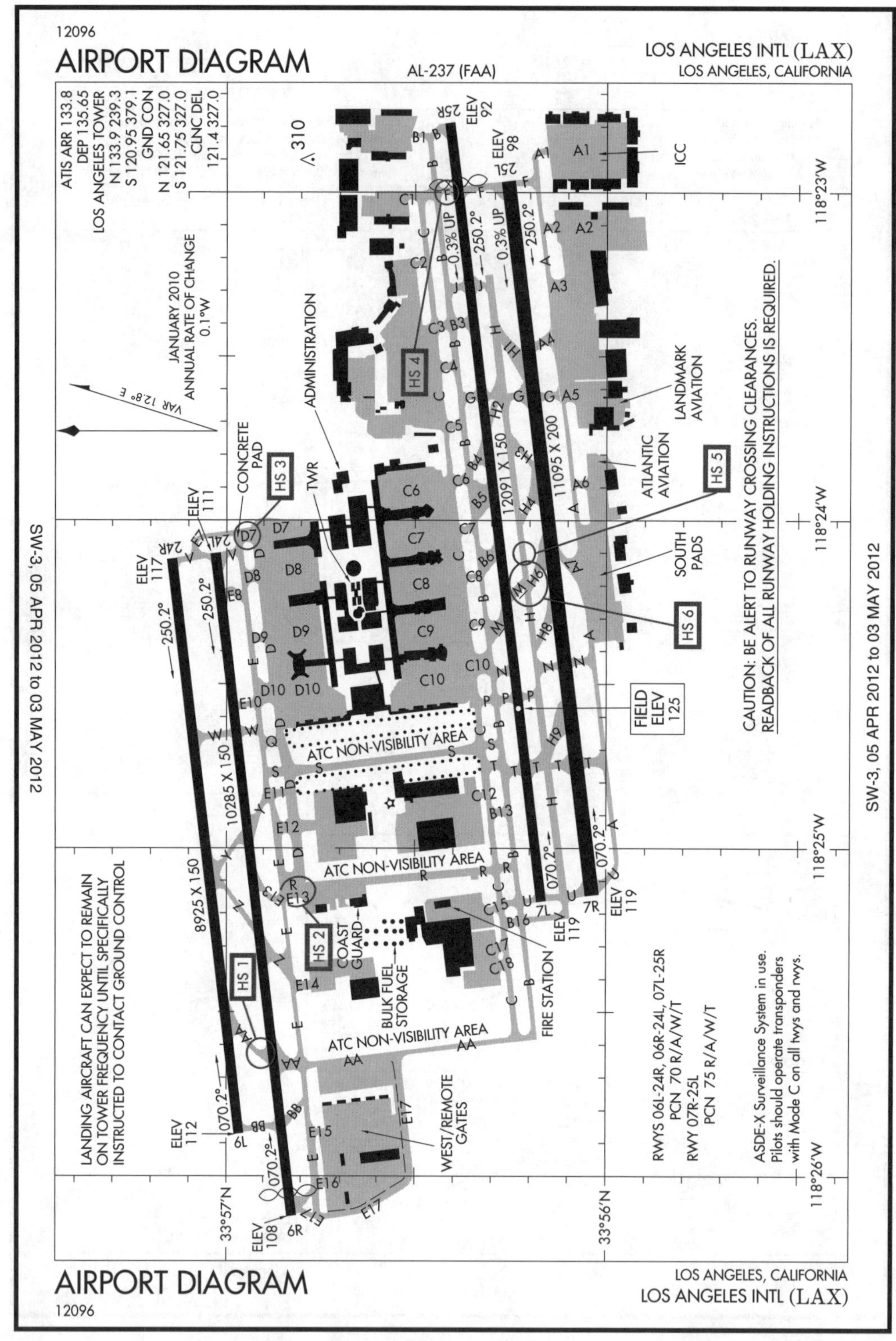

12096
AIRPORT DIAGRAM
AL-237 (FAA)
LOS ANGELES INTL (LAX)
LOS ANGELES, CALIFORNIA
ATIS ARR 133.8
DEP 135.65
LOS ANGELES TOWER
N 133.9 239.3
S 120.95 379.1
GND CON
N 121.65 327.0
S 121.75 327.0
CLNC DEL
121.4 327.0
JANUARY 2010
ANNUAL RATE OF CHANGE
0.1°W
VAR 12.8° E
SW-3, 05 APR 2012 to 03 MAY 2012
ADMINISTRATION
CONCRETE PAD
TWR
ICC
LANDMARK AVIATION
ATLANTIC AVIATION
SOUTH PADS
12091 X 150
11095 X 200
10285 X 150
8925 X 150
FIELD ELEV 125
ATC NON-VISIBILITY AREA
COAST GUARD
BULK FUEL STORAGE
FIRE STATION
WEST/REMOTE GATES
CAUTION: BE ALERT TO RUNWAY CROSSING CLEARANCES.
READBACK OF ALL RUNWAY HOLDING INSTRUCTIONS IS REQUIRED.
LANDING AIRCRAFT CAN EXPECT TO REMAIN ON TOWER FREQUENCY UNTIL SPECIFICALLY INSTRUCTED TO CONTACT GROUND CONTROL
RWYS 06L-24R, 06R-24L, 07L-25R
PCN 70 R/A/W/T
RWY 07R-25L
PCN 75 R/A/W/T
ASDE-X Surveillance System in use. Pilots should operate transponders with Mode C on all twys and rwys.
33°57'N
33°56'N
118°23'W
118°24'W
118°25'W
118°26'W

Figure 255

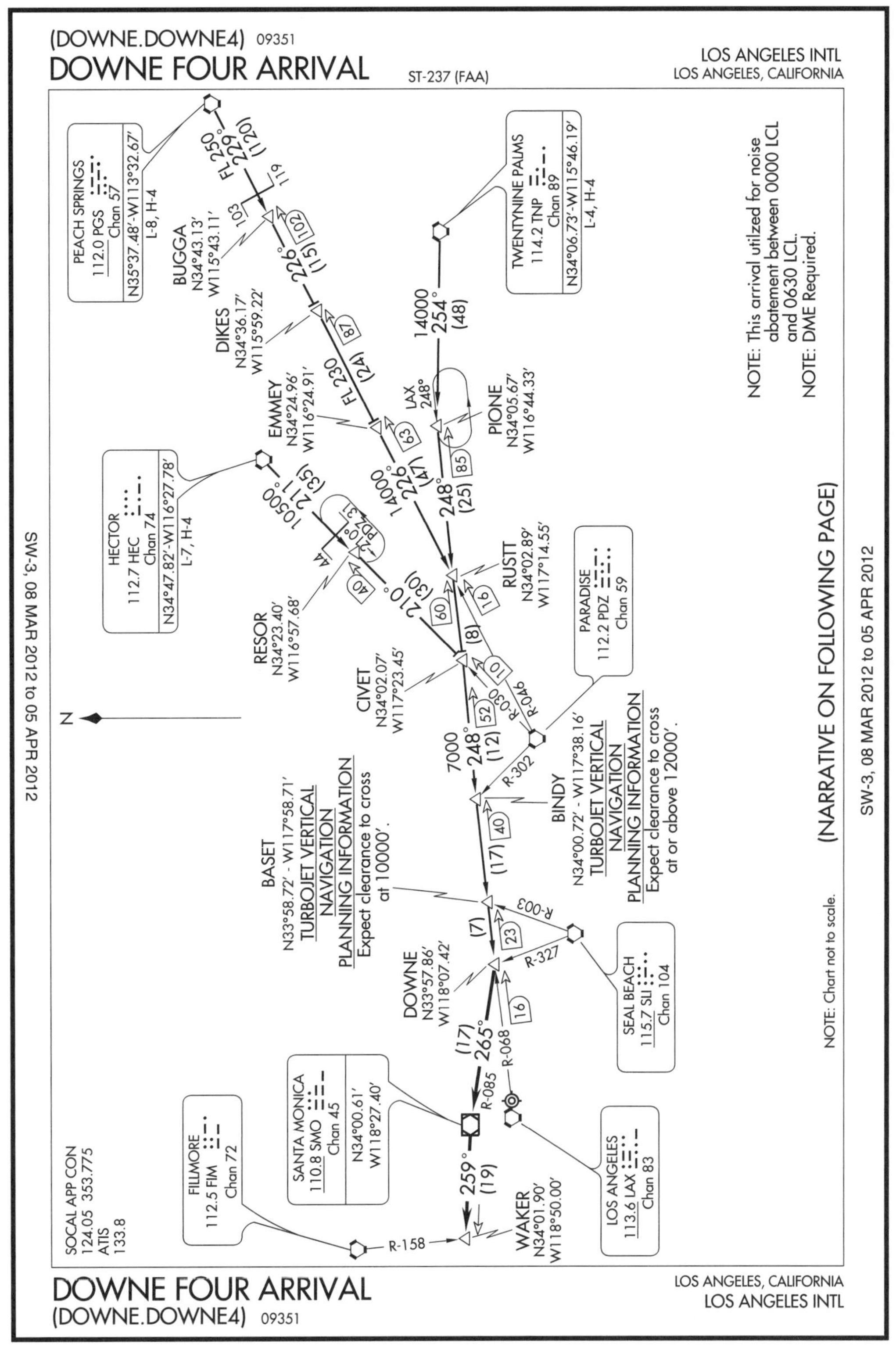
(DOWNE.DOWNE4) 09351
DOWNE FOUR ARRIVAL
ST-237 (FAA)
LOS ANGELES INTL
LOS ANGELES, CALIFORNIA
SW-3, 08 MAR 2012 to 05 APR 2012
PEACH SPRINGS
112.0 PGS
Chan 57
N35°37.48'-W113°32.67'
L-8, H-4
TWENTYNINE PALMS
114.2 TNP
Chan 89
N34°06.73'-W115°46.19'
L-4, H-4
NOTE: This arrival utilized for noise abatement between 0000 LCL and 0630 LCL.
NOTE: DME Required.
BUGGA
N34°43.13'
W115°43.11'
DIKES
N34°36.17'
W115°59.22'
EMMEY
N34°24.96'
W116°24.91'
PIONE
N34°05.67'
W116°44.33'
HECTOR
112.7 HEC
Chan 74
N34°47.82'-W116°27.78'
L-7, H-4
RESOR
N34°23.40'
W116°57.68'
RUSTT
N34°02.89'
W117°14.55'
PARADISE
112.2 PDZ
Chan 59
CIVET
N34°02.07'
W117°23.45'
BINDY
N34°00.72' - W117°38.16'
TURBOJET VERTICAL NAVIGATION PLANNING INFORMATION
Expect clearance to cross at or above 12000'.
BASET
N33°58.72' - W117°58.71'
TURBOJET VERTICAL NAVIGATION PLANNING INFORMATION
Expect clearance to cross at 10000'.
(NARRATIVE ON FOLLOWING PAGE)
DOWNE
N33°57.86'
W118°07.42'
SEAL BEACH
115.7 SLI
Chan 104
NOTE: Chart not to scale.
SANTA MONICA
110.8 SMO
Chan 45
N34°00.61'
W118°27.40'
FILLMORE
112.5 FIM
Chan 72
SOCAL APP CON
124.05 353.775
ATIS
133.8
LOS ANGELES
113.6 LAX
Chan 83
WAKER
N34°01.90'
W118°50.00'
DOWNE FOUR ARRIVAL
(DOWNE.DOWNE4) 09351
LOS ANGELES, CALIFORNIA
LOS ANGELES INTL

Figure 255A

(DOWNE.DOWNE4) 02276

DOWNE FOUR ARRIVAL ST-237 (FAA)

LOS ANGELES INTL
LOS ANGELES, CALIFORNIA

SW-3, 08 MAR 2012 to 05 APR 2012

ARRIVAL DESCRIPTION

HECTOR TRANSITION (HEC.DOWNE4): From over HEC VORTAC via HEC R-211 and PDZ R-030 to CIVET INT, then LAX R-068 to DOWNE INT. Thence....
PEACH SPRINGS TRANSITION (PGS.DOWNE4): From over PGS VORTAC via PGS R-229 and PDZ R-046 to RUSTT INT, then LAX R-068 to DOWNE INT. Thence....
TWENTYNINE PALMS TRANSITION (TNP.DOWNE4): From over TNP VORTAC via TNP R-254 to PIONE DME, then LAX R-068 to DOWNE INT.Thence....
....From DOWNE INT via SMO R-085 to SMO VOR/DME, then via SMO R-259 to WAKER INT, expect vector to final approach course for runways 6 and 7.

SW-3, 08 MAR 2012 to 05 APR 2012

DOWNE FOUR ARRIVAL
(DOWNE.DOWNE4) 02276

LOS ANGELES, CALIFORNIA
LOS ANGELES INTL

Figure 255B

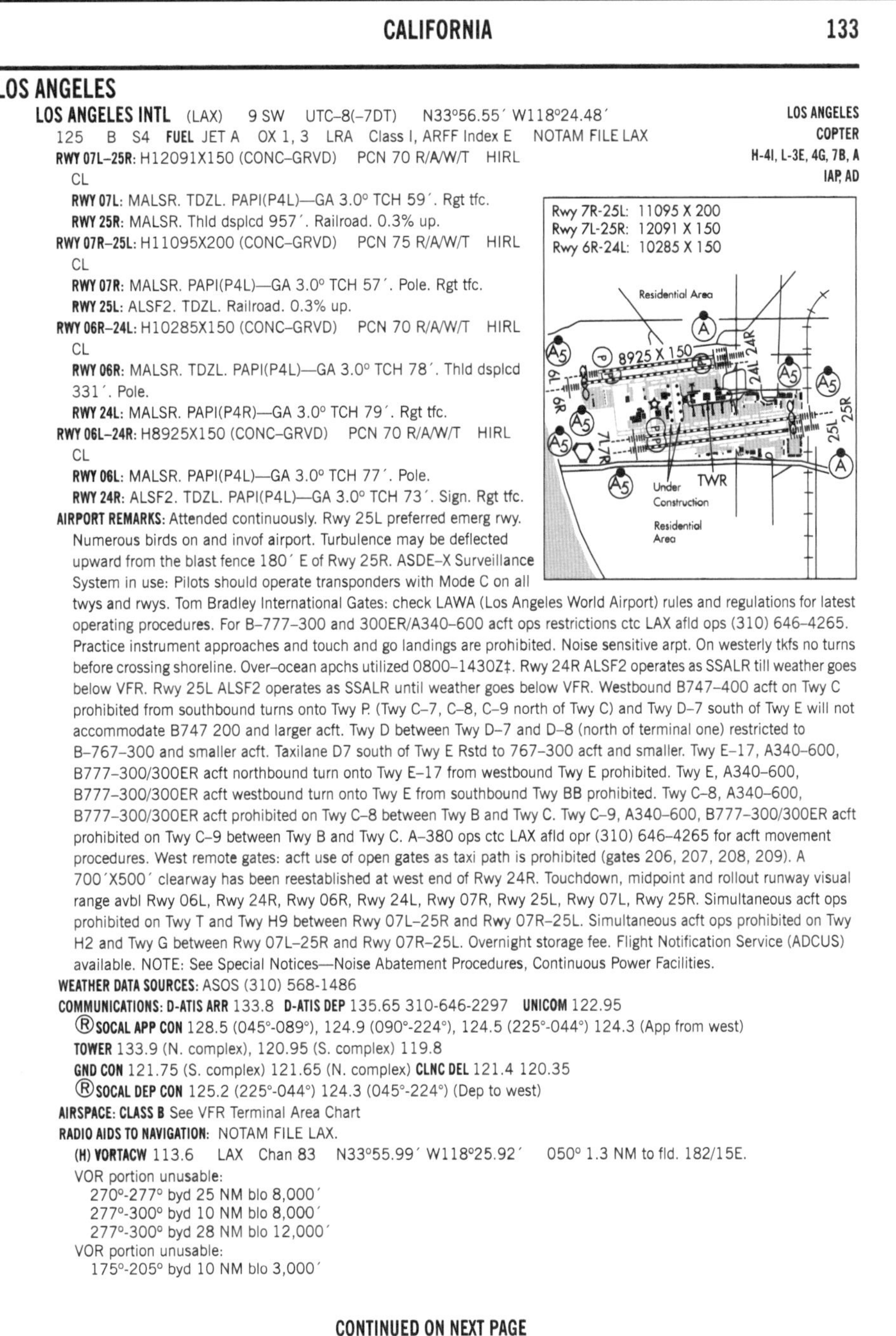

CALIFORNIA 133

LOS ANGELES

LOS ANGELES INTL (LAX) 9 SW UTC–8(–7DT) N33°56.55′ W118°24.48′ **LOS ANGELES**

125 B S4 **FUEL** JET A OX 1, 3 LRA Class I, ARFF Index E NOTAM FILE LAX **COPTER**

RWY 07L–25R: H12091X150 (CONC–GRVD) PCN 70 R/A/W/T HIRL CL **H–4I, L–3E, 4G, 7B, A IAP, AD**

RWY 07L: MALSR. TDZL. PAPI(P4L)—GA 3.0° TCH 59′. Rgt tfc.

RWY 25R: MALSR. Thld dsplcd 957′. Railroad. 0.3% up.

RWY 07R–25L: H11095X200 (CONC–GRVD) PCN 75 R/A/W/T HIRL CL

RWY 07R: MALSR. PAPI(P4L)—GA 3.0° TCH 57′. Pole. Rgt tfc.

RWY 25L: ALSF2. TDZL. Railroad. 0.3% up.

RWY 06R–24L: H10285X150 (CONC–GRVD) PCN 70 R/A/W/T HIRL CL

RWY 06R: MALSR. TDZL. PAPI(P4L)—GA 3.0° TCH 78′. Thld dsplcd 331′. Pole.

RWY 24L: MALSR. PAPI(P4R)—GA 3.0° TCH 79′. Rgt tfc.

RWY 06L–24R: H8925X150 (CONC–GRVD) PCN 70 R/A/W/T HIRL CL

RWY 06L: MALSR. PAPI(P4L)—GA 3.0° TCH 77′. Pole.

RWY 24R: ALSF2. TDZL. PAPI(P4L)—GA 3.0° TCH 73′. Sign. Rgt tfc.

AIRPORT REMARKS: Attended continuously. Rwy 25L preferred emerg rwy. Numerous birds on and invof airport. Turbulence may be deflected upward from the blast fence 180′ E of Rwy 25R. ASDE–X Surveillance System in use: Pilots should operate transponders with Mode C on all twys and rwys. Tom Bradley International Gates: check LAWA (Los Angeles World Airport) rules and regulations for latest operating procedures. For B–777–300 and 300ER/A340–600 acft ops restrictions ctc LAX afld ops (310) 646–4265. Practice instrument approaches and touch and go landings are prohibited. Noise sensitive arpt. On westerly tkfs no turns before crossing shoreline. Over–ocean apchs utilized 0800–1430Z‡. Rwy 24R ALSF2 operates as SSALR till weather goes below VFR. Rwy 25L ALSF2 operates as SSALR until weather goes below VFR. Westbound B747–400 acft on Twy C prohibited from southbound turns onto Twy P. (Twy C–7, C–8, C–9 north of Twy C) and Twy D–7 south of Twy E will not accommodate B747 200 and larger acft. Twy D between Twy D–7 and D–8 (north of terminal one) restricted to B–767–300 and smaller acft. Taxilane D7 south of Twy E Rstd to 767–300 acft and smaller. Twy E–17, A340–600, B777–300/300ER acft northbound turn onto Twy E–17 from westbound Twy E prohibited. Twy E, A340–600, B777–300/300ER acft westbound turn onto Twy E from southbound Twy BB prohibited. Twy C–8, A340–600, B777–300/300ER acft prohibited on Twy C–8 between Twy B and Twy C. Twy C–9, A340–600, B777–300/300ER acft prohibited on Twy C–9 between Twy B and Twy C. A–380 ops ctc LAX afld opr (310) 646–4265 for acft movement procedures. West remote gates: acft use of open gates as taxi path is prohibited (gates 206, 207, 208, 209). A 700′X500′ clearway has been reestablished at west end of Rwy 24R. Touchdown, midpoint and rollout runway visual range avbl Rwy 06L, Rwy 24R, Rwy 06R, Rwy 24L, Rwy 07R, Rwy 25L, Rwy 07L, Rwy 25R. Simultaneous acft ops prohibited on Twy T and Twy H9 between Rwy 07L–25R and Rwy 07R–25L. Simultaneous acft ops prohibited on Twy H2 and Twy G between Rwy 07L–25R and Rwy 07R–25L. Overnight storage fee. Flight Notification Service (ADCUS) available. NOTE: See Special Notices—Noise Abatement Procedures, Continuous Power Facilities.

WEATHER DATA SOURCES: ASOS (310) 568-1486

COMMUNICATIONS: D–ATIS ARR 133.8 **D–ATIS DEP** 135.65 310-646-2297 **UNICOM** 122.95

®**SOCAL APP CON** 128.5 (045°-089°), 124.9 (090°-224°), 124.5 (225°-044°) 124.3 (App from west)

TOWER 133.9 (N. complex), 120.95 (S. complex) 119.8

GND CON 121.75 (S. complex) 121.65 (N. complex) **CLNC DEL** 121.4 120.35

®**SOCAL DEP CON** 125.2 (225°-044°) 124.3 (045°-224°) (Dep to west)

AIRSPACE: CLASS B See VFR Terminal Area Chart

RADIO AIDS TO NAVIGATION: NOTAM FILE LAX.

(H) VORTACW 113.6 LAX Chan 83 N33°55.99′ W118°25.92′ 050° 1.3 NM to fld. 182/15E.

VOR portion unusable:
270°-277° byd 25 NM blo 8,000′
277°-300° byd 10 NM blo 8,000′
277°-300° byd 28 NM blo 12,000′

VOR portion unusable:
175°-205° byd 10 NM blo 3,000′

CONTINUED ON NEXT PAGE

Figure 256

CONTINUED FROM PRECEDING PAGE

ILS/DME 108.5 I-UWU Chan 22 Rwy 06L. Class IE. DME also serves Rwy 24R.
ILS/DME 111.7 I-GPE Chan 54 Rwy 06R. Class IE. MM OTS indef. DME also serves Rwy 24L.
ILS/DME 111.1 I-IAS Chan 48 Rwy 07L. Class ID. MM OTS indef. Glideslope unusable byd 5° right of localizer course. DME also serves Rwy 25R.
ILS/DME 109.9 I-MKZ Chan 36 Rwy 07R. Class IT. GS unuseable 5° left and 4° right of course. Coupled approaches not applicable below 264 ' MSL. DME also serves Rwy 25L.
ILS/DME 111.7 I-HQB Chan 54 Rwy 24L. Class IE. DME also serves Rwy 06R.
ILS/DME 108.5 I-OSS Chan 22 Rwy 24R. Class IIIE. DME also serves Rwy 06L
ILS/DME 109.9 I-LAX Chan 36 Rwy 25L. Class IIIE.
ILS/DME 111.1 I-CFN Chan 48 Rwy 25R. Class IE. DME also serves Rwy 07L.

WHITEMAN (WHP) 1 E UTC–8(–7DT) N34°15.56′ W118°24.81′ **LOS ANGELES**
1003 B S4 **FUEL** 100LL, JET A OX 1, 3 TPA—2003(1000) NOTAM FILE WHP **COPTER**
RWY 12–30: H4120X75 (ASPH) S–12.5 MIRL 1.0% up NW **L–3E, 4G, 7B, A**
RWY 12: REIL. PAPI(P2R)—GA 3.8° TCH 40′. Thld dsplcd 729′. P-line. **IAP, AD**
RWY 30: REIL. PAPI(P2L)—GA 3.8° TCH 40′. Thld dsplcd 478′. P-line. Rgt tfc. 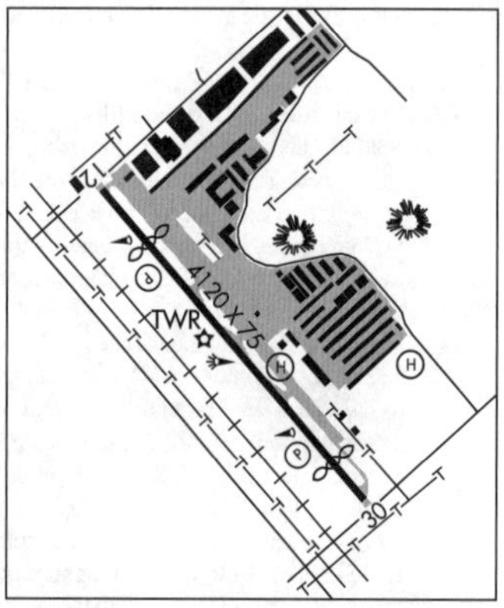
RUNWAY DECLARED DISTANCE INFORMATION
RWY 12: TORA–3442 TODA–4120 ASDA–3910 LDA–3181
RWY 30: TORA–3191 TODA–4120 ASDA–3940 LDA–3462
AIRPORT REMARKS: Attended continuously. Birds on and invof arpt. Helicopter ops 2500′ MSL (1500′ AGL) and below. Arpt CLOSED to helicopter training/pattern opr 0400–1600Z‡. Dirt infield areas. Helicopters advised to use care to prevent blasting dirt and debris onto movement areas.
WEATHER DATA SOURCES: AWOS-3PT 132.1 (818) 899-9820.
COMMUNICATIONS: CTAF 135.0 **ATIS** 132.1 818-899-9820
UNICOM 122.95
®**SOCAL APP/DEP CON** 120.4 134.2 (VNY 280°-BUR 050°) 134.2 (VNY 160°-VNY 280°)
TOWER 135.0 (1600-0400Z‡) **GND CON** 125.0
CLNC DEL For clnc del when ATCT clsd call Socal App 800-448-3724.
AIRSPACE: CLASS D svc 1600-0400Z‡ other times CLASS G
RADIO AIDS TO NAVIGATION: NOTAM FILE VNY.
VAN NUYS (L) VORW/DME 113.1 VNY Chan 78 N34°13.41′ W118°29.50′ 046° 4.4 NM to fld. 812/15E.
VOR/DME unusable:
010°-030° byd 20 NM blo 6,700′
030°-050° byd 25 NM blo 8,600′
330°-350° byd 25 NM blo 5,500′
350°-010° byd 15 NM blo 6,100′
DME unusable:
094°-096° byd 35 NM blo 5,000′
PACOIMA NDB (MHW) 370 PAI N34°15.58′ W118°24.80′ at fld. NOTAM FILE HHR. VFR only.
COMM/NAV/WEATHER REMARKS: Whiteman arpt altimeter setting not avbl.

LOS BANOS

LOS BANOS MUNI (LSN) 1 W UTC–8(–7DT) N37°03.83′ W120°52.19′ **SAN FRANCISCO**
121 B S2 **FUEL** 100LL, JET A TPA—921(800) NOTAM FILE RIU **L–3B**
RWY 14–32: H3801X75 (ASPH) S–23 MIRL **IAP**
RWY 14: REIL. PAPI(P2L)—GA 3.0° TCH 30′. Tree. Rgt tfc.
RWY 32: REIL. PAPI(P2L)—GA 3.0° TCH 38′. Tree.
AIRPORT REMARKS: Unattended. For cash fuel after hours call 209–827–7070. 24 hour automated fuel avbl with major credit card. Avoid overflight of houses south of arpt. No departures over housing areas to east of arpt. MIRL Rwy 14–32 preset low intensity until 0800Z‡. To increase intensity and ACTIVATE MIRL Rwy 14–32, REIL Rwy 14 and Rwy 32, and PAPI Rwy 14 and Rwy 32—CTAF.
WEATHER DATA SOURCES: AWOS-3 118.675 (209) 827-7084.
COMMUNICATIONS: CTAF/UNICOM 122.8
PANOCHE RCO 122.1R 112.6T (FRESNO RADIO)
®**NORCAL APP/DEP CON** 120.95
RADIO AIDS TO NAVIGATION: NOTAM FILE RIU.
PANOCHE (L) VORTAC 112.6 PXN Chan 73 N36°42.93′ W120°46.72′ 332° 21.3 NM to fld. 2060/16E.
VOR unusable:
230°-280° byd 7NM blo 9,000′

Figure 257

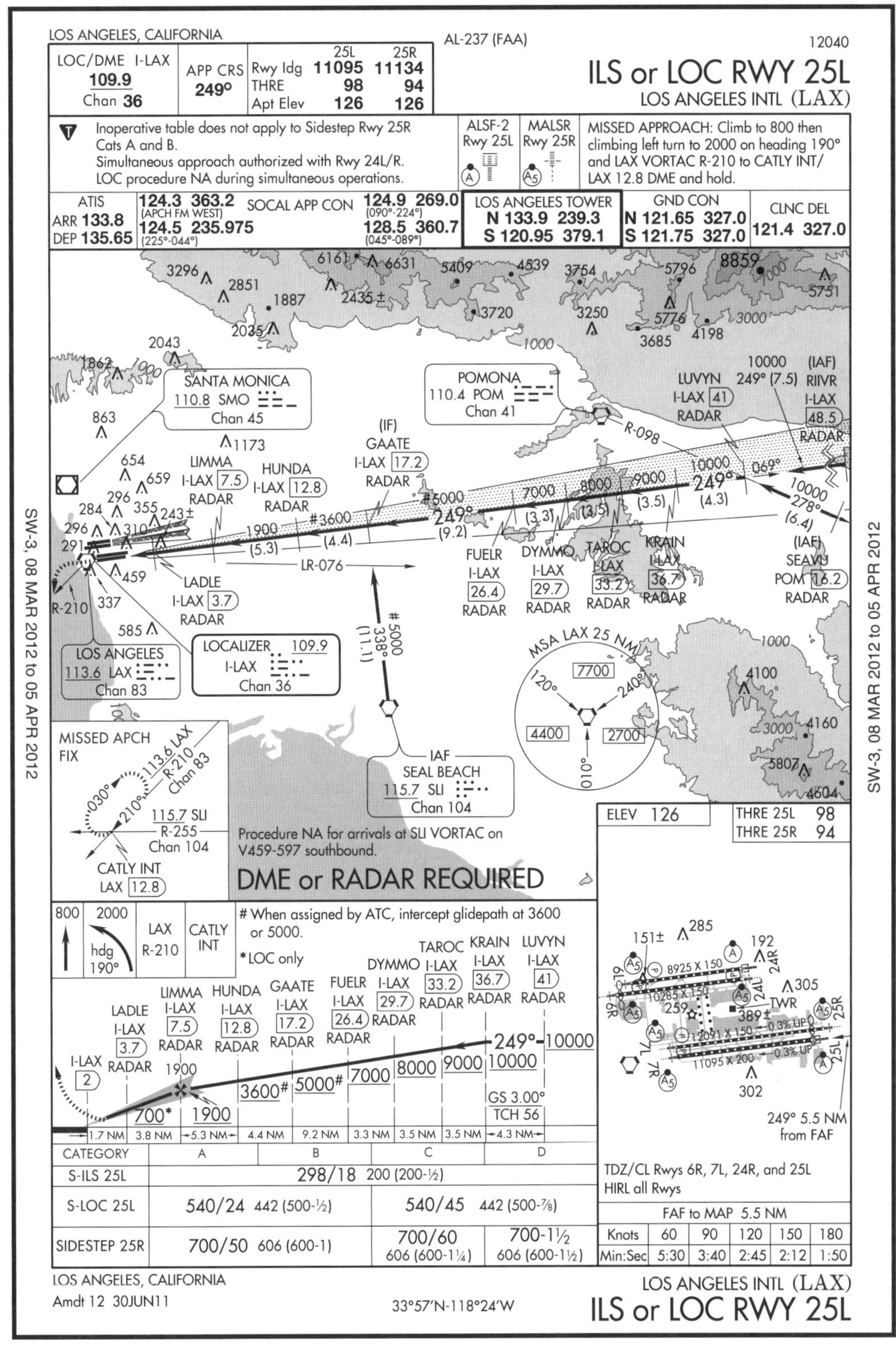

CATEGORY	A	B	C	D
S-ILS 25L	298/18 200 (200-½)			
S-LOC 25L	540/24 442 (500-½)		540/45 442 (500-⅞)	
SIDESTEP 25R	700/50 606 (600-1)		700/60 606 (600-1¼)	700-1½ 606 (600-1½)

FAF to MAP 5.5 NM					
Knots	60	90	120	150	180
Min:Sec	5:30	3:40	2:45	2:12	1:50

LOS ANGELES, CALIFORNIA
Amdt 12 30JUN11
33°57'N-118°24'W
LOS ANGELES INTL (LAX)
ILS or LOC RWY 25L

Figure 257A

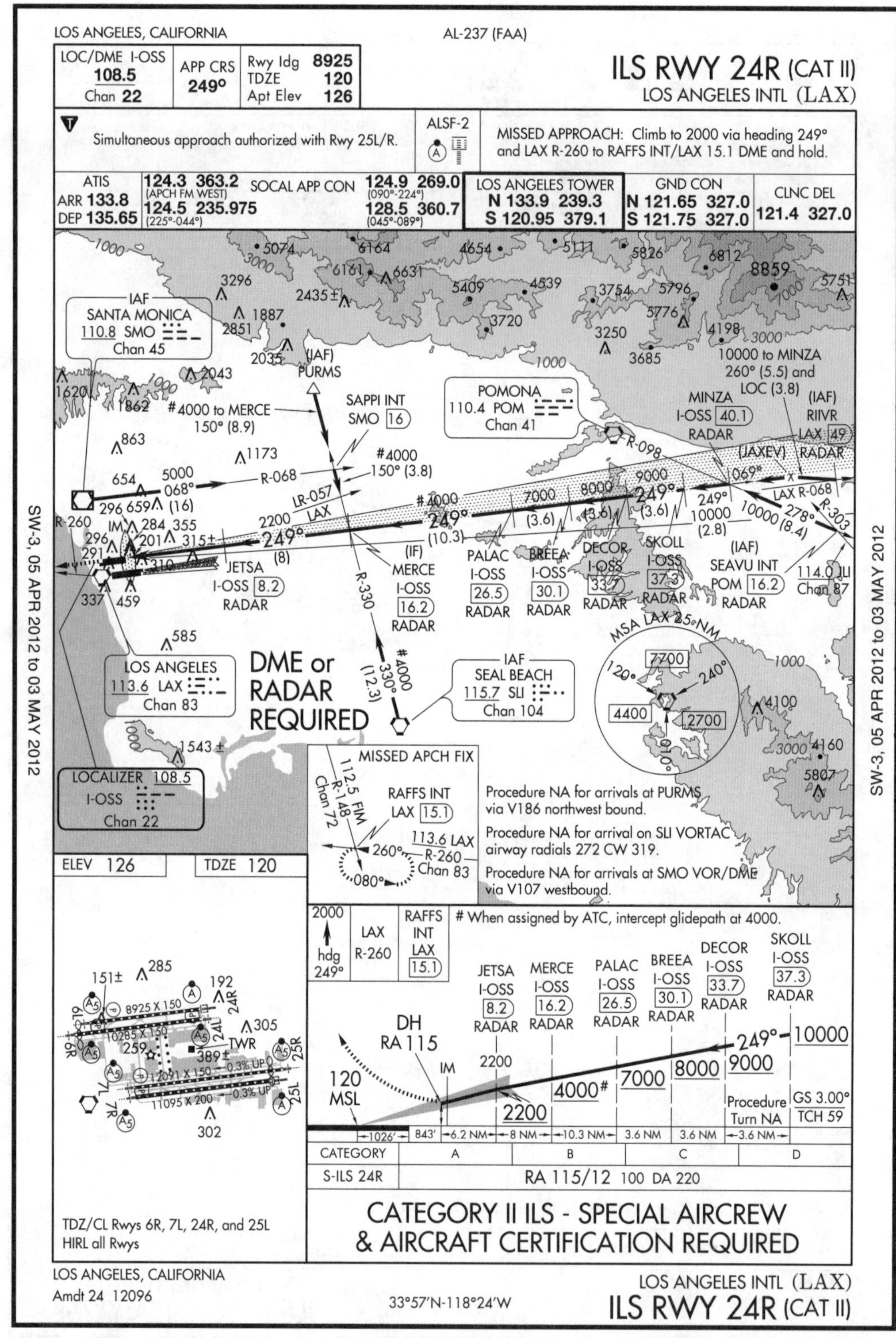

LOS ANGELES, CALIFORNIA
AL-237 (FAA)
LOC/DME I-OSS 108.5 Chan 22
APP CRS 249°
Rwy ldg 8925
TDZE 120
Apt Elev 126
ILS RWY 24R (CAT II)
LOS ANGELES INTL (LAX)
Simultaneous approach authorized with Rwy 25L/R.
ALSF-2
MISSED APPROACH: Climb to 2000 via heading 249° and LAX R-260 to RAFFS INT/LAX 15.1 DME and hold.
ATIS ARR 133.8 DEP 135.65
124.3 363.2 (APCH FM WEST) 124.5 235.975 (225°-044°)
SOCAL APP CON 124.9 269.0 (090°-224°) 128.5 360.7 (045°-089°)
LOS ANGELES TOWER N 133.9 239.3 S 120.95 379.1
GND CON N 121.65 327.0 S 121.75 327.0
CLNC DEL 121.4 327.0
IAF SANTA MONICA 110.8 SMO Chan 45
(IAF) PURMS
SAPPI INT SMO 16
#4000 to MERCE 150° (8.9)
#4000 150° (3.8)
POMONA 110.4 POM Chan 41
10000 to MINZA 260° (5.5) and LOC (3.8)
MINZA I-OSS 40.1 RADAR
(IAF) RIIVR LAX 49 RADAR
JAXEV
JETSA I-OSS 8.2 RADAR
(IF) MERCE I-OSS 16.2 RADAR
PALAC I-OSS 26.5 RADAR
BREEA I-OSS 30.1 RADAR
DECOR I-OSS 33.7 RADAR
SKOLL I-OSS 37.3 RADAR
(IAF) SEAVU INT POM 16.2 RADAR
114.0 JLI Chan 87
MSA LAX 25 NM
LOS ANGELES 113.6 LAX Chan 83
DME or RADAR REQUIRED
IAF SEAL BEACH 115.7 SLI Chan 104
LOCALIZER 108.5 I-OSS Chan 22
MISSED APCH FIX
RAFFS INT LAX 15.1
Procedure NA for arrivals at PURMS via V186 northwest bound.
Procedure NA for arrival on SLI VORTAC airway radials 272 CW 319.
Procedure NA for arrivals at SMO VOR/DME via V107 westbound.
ELEV 126
TDZE 120
When assigned by ATC, intercept glidepath at 4000.
DH RA 115
120 MSL
GS 3.00° TCH 59
Procedure Turn NA
CATEGORY A B C D
S-ILS 24R RA 115/12 100 DA 220
CATEGORY II ILS - SPECIAL AIRCREW & AIRCRAFT CERTIFICATION REQUIRED
TDZ/CL Rwys 6R, 7L, 24R, and 25L
HIRL all Rwys
LOS ANGELES, CALIFORNIA
Amdt 24 12096
33°57'N-118°24'W
LOS ANGELES INTL (LAX)
ILS RWY 24R (CAT II)
SW-3, 05 APR 2012 to 03 MAY 2012

Figure 257B

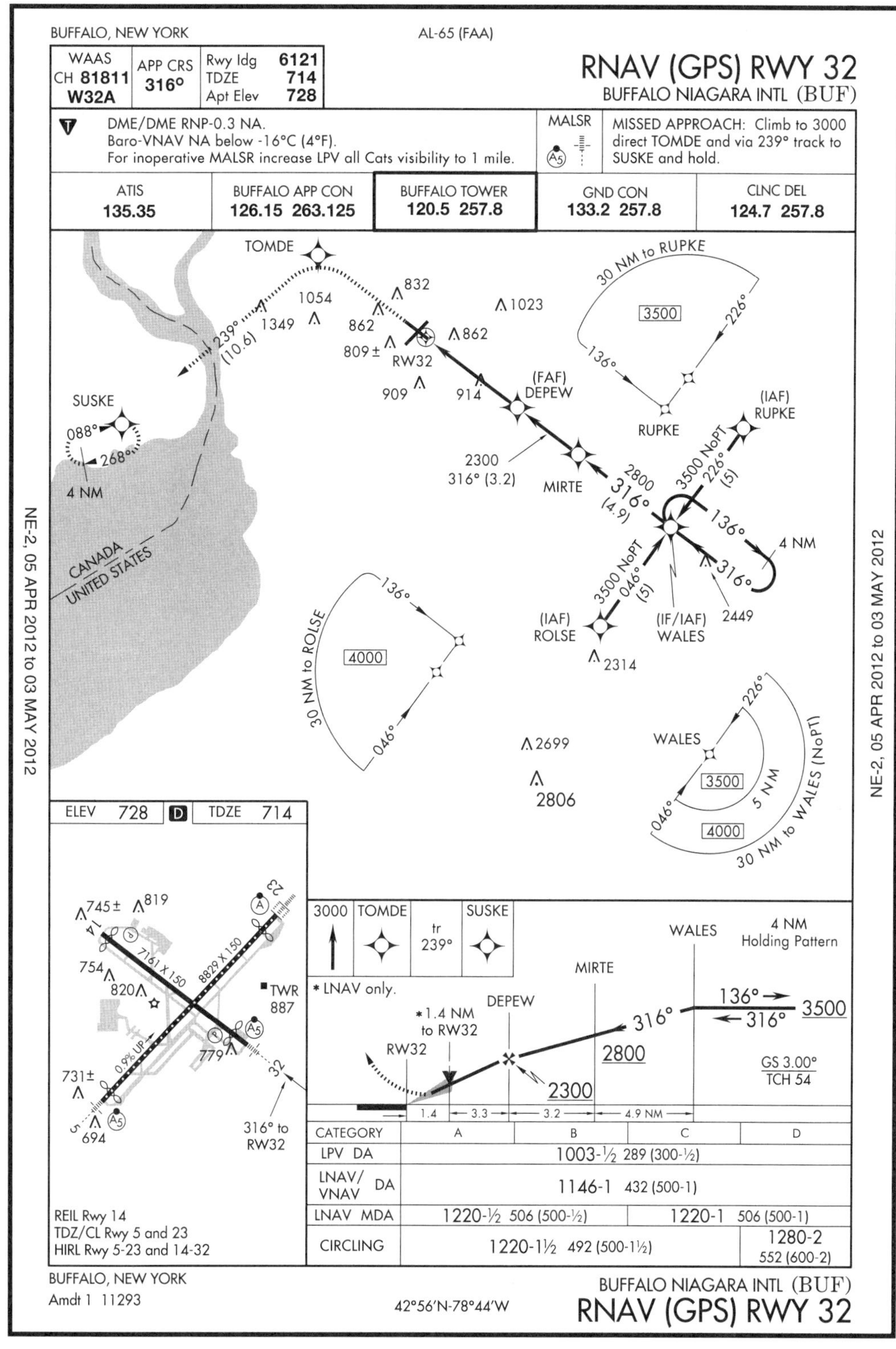

CATEGORY	A	B	C	D
LPV DA	1003-½ 289 (300-½)			
LNAV/VNAV DA	1146-1 432 (500-1)			
LNAV MDA	1220-½ 506 (500-½)		1220-1 506 (500-1)	
CIRCLING	1220-1½ 492 (500-1½)			1280-2 552 (600-2)

Figure 258

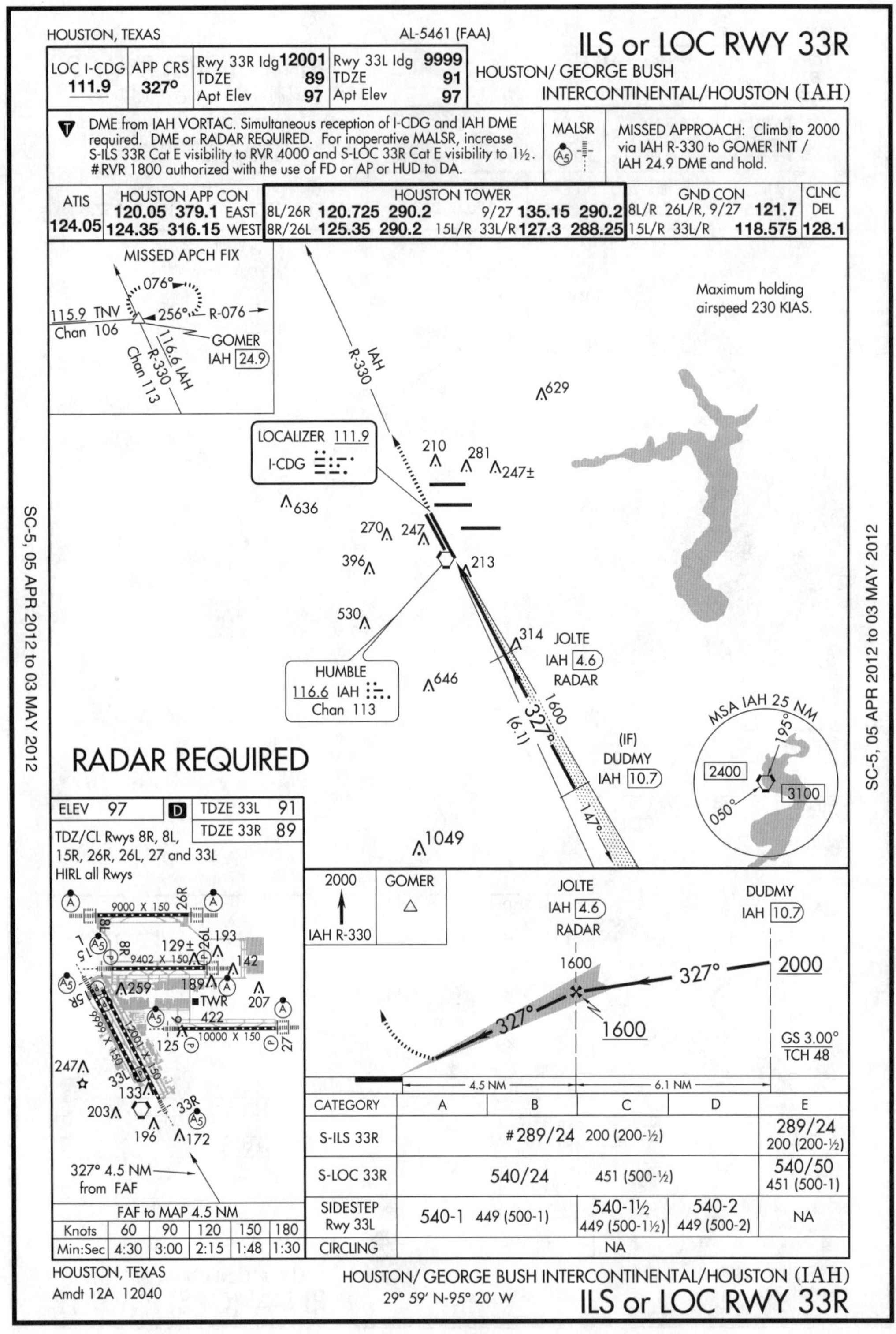

FAF to MAP 4.5 NM					
Knots	60	90	120	150	180
Min:Sec	4:30	3:00	2:15	1:48	1:30

CATEGORY	A	B	C	D	E
S-ILS 33R	#289/24 200 (200-½)				289/24 200 (200-½)
S-LOC 33R	540/24 451 (500-½)				540/50 451 (500-1)
SIDESTEP Rwy 33L	540-1 449 (500-1)		540-1½ 449 (500-1½)	540-2 449 (500-2)	NA
CIRCLING	NA				

Figure 259

TAKE-OFF MINIMUMS AND (OBSTACLE) DEPARTURE PROCEDURES

11349

CROCKETT, TX

HOUSTON COUNTY (DKR)
ORIG 11349 (FAA)

TAKE-OFF MINIMUMS: **Rwy 2,** 400-2 or std. w/min. climb of 280' per NM to 800. **Rwy 20,** 300-1½ or std. w/ min. climb of 459'per NM to 700.

NOTE: **Rwy 2,** multiple trees beginning 57' from DER, 61' right of centerline, up to 50' AGL/399' MSL. Multiple trees and terrain beginning 27' from DER, 109' left of centerline, up to 50' AGL/409' MSL. Tower 1.5 NM from DER, 2864' left of centerline 233' AGL/ 623' MSL. **Rwy 20,** multiple towers beginning 4567' from DER, 1025' right of centerline, up to 200' AGL/ 529' MSL.

EAGLE LAKE, TX

EAGLE LAKE

TAKE-OFF MINIMUMS: **Rwy 17,** 200-1, or std. with a min. climb of 420' per NM to 500.

NOTE: **Rwy 17,** tower 3068' from departure end of runway, 793' left of centerline, 192' AGL/317' MSL.

GALVESTON, TX

SCHOLES INTL AT GALVESTON (GLS)
AMDT 4 08157 (FAA)

TAKE-OFF MINIMUMS: **Rwy 17,** 300-1 or std. w/ min. climb of 502' per NM to 300.

NOTE: **Rwy 13,** bush 381' from departure end of runway, 533' left of centerline, 15' AGL/20' MSL. Fence 201' from departure end of runway, 490' left of centerline, 6' AGL/ 11' MSL. Tree 343' from departure end of runway, 468' right of centerline, 12' AGL/17' MSL. **Rwy 17,** building 3057' from departure end of runway, 339' left of centerline, 123' AGL/130' MSL. Multiple poles beginning 2034' from departure end of runway, 87' right of centerline, up to 60' AGL/70' MSL. Multiple transmission towers beginning 636' from departure end of runway, 551' right of centerline, up to 55' AGL/60' MSL. Tree 460' from departure end of runway, 316' right of centerline, 22' AGL/29' MSL. **Rwy 31,** multiple cranes beginning 4341' from departure end of runway, 1017' left of centerline, up to 131' AGL/ 131' MSL. **Rwy 35,** tree 730' from departure end of runway, 501' right of centerline, 27' AGL/32' MSL.

GIDDINGS, TX

GIDDINGS-LEE COUNTY (GYB)
ORIG 10210 (FAA)

NOTE: **Rwy 17,** numerous trees beginning 720' from DER, 58' right of centerline, up to 50' AGL/479' MSL. Numerous trees beginning 754' from DER, 340' left of centerline, up to 50' AGL/479' MSL. **Rwy 35,** numerous trees beginning 613' from DER, 272' right of centerline, up to 50' AGL/539' MSL. Numerous trees beginning 558' from DER, 265' left of centerline, up to 50' AGL/559' MSL. Vehicle on road 516' from DER, 246' left of centerline, 15' AGL/514' MSL.

HOUSTON, TX

DAN JONES INTL (T51)
ORIG 11321 (FAA)

TAKE-OFF MINIMUMS: **Rwy 17,** NA - numerous trees. **Rwy 35,** NA - numerous trees.

HOUSTON, TX (CON'T)

DAVID WAYNE HOOKS MEMORIAL (DWH)
AMDT 3 08157 (FAA)

TAKE-OFF MINIMUMS: **Rwys 17L, 35R,** NA-Environmental. **Waterways 17, 35,** NA - air traffic.

NOTE: **Rwy 17R,** multiple trees beginning 708' from departure end of runway, 68' left of centerline, up to 71' AGL/220' MSL. Multiple hangars beginning 433' from departure end of runway, 515' left of centerline, up to 37' AGL/182' MSL. DME antenna 653' from departure end of runway, 256' left of centerline, 13' AGL/162' MSL. Multiple trees and pole beginning 85' from departure end of runway, 294' right of centerline, up to 45' AGL/ 189' MSL. **Rwy 35L,** multiple trees and poles beginning 144' from departure end of runway, 32' left of centerline, up to 79' AGL/238' MSL. Multiple hangars and buildings beginning 85' from departure end of runway, 9' left of centerline, up to 53' AGL/202' MSL. Multiple trees, towers and pole beginning 100' from departure end of runway, 124' right of centerline, up to 93' AGL/247' MSL. Vehicle and road 315' from departure end of runway, on centerline 15' AGL/166' MSL. Building 894' from departure end of runway, 231' right of centerline, 23' AGL/173' MSL.

ELLINGTON FIELD (EFD)
AMDT 2 08157 (FAA)

NOTE: **Rwy 17R,** pole 1489' from departure end of runway, 817' right of centerline, 40' AGL/74' MSL. **Rwy 22,** antenna on building 1998' from departure end of runway, 598' right of centerline, 54' AGL/83' MSL. Obstruction light on glide slope 327' from departure end of runway, 543' left of centerline, 39' AGL/68' MSL. **Rwy 35R,** tree 1597' from departure end of runway, 32' left of centerline, 33' AGL/80' MSL. **Rwy 35L,** multiple trees beginning 1118' from departure end of runway, 679' right of centerline, up to 37' AGL/ 77' MSL. Crane 2352' from departure end of runway, 1024' left of centerline, 37' AGL/ 97' MSL.

GEORGE BUSH INTERCONTINENTAL/ HOUSTON (IAH)
AMDT 2 08157 (FAA)

NOTE: **Rwy 8L,** tree 2866' from departure end of runway, 921' left of centerline, 107' AGL/201' MSL. Multiple trees beginning 2750' from departure end of runway, 106' right of centerline, up to 80' AGL/174' MSL. **Rwy 15L,** multiple trees 2638' from departure end of runway, 758' right of centerline, up to 76' AGL/160' MSL. **Rwy 15R,** tower 1431' from departure end of runway, 591' left of centerline, 48' AGL/133' MSL. Antenna on glideslope 1469' from departure end of runway, 621' left of centerline, 49' AGL/133' MSL. **Rwy 26R,** pole 950' from departure end of runway, 660' right of centerline, 40' AGL/129' MSL. **Rwy 33R,** tree 2868' from departure end of runway, 1027' right of centerline, 73' AGL/172' MSL.

HOUSTON EXECUTIVE (TME)

DEPARTURE PROCEDURE: **Rwy 36,** Climb heading 355° to 700 before turning east.

NOTE: **Rwy 36,** power poles from left to right beginning 703' from departure end of runway, 623' left to 685' right of centerline, up to 32' AGL/196' MSL.

Figure 260

L4

TAKE-OFF MINIMUMS AND (OBSTACLE) DEPARTURE PROCEDURES

11349

HOUSTON, TX (CON'T)

HOUSTON-SOUTHWEST (AXH)
AMDT 5 08157 (FAA)

DEPARTURE PROCEDURE: **Rwy 9,** climb heading 089° to 2000 before turning left. **Rwy 27,** climb heading 269° to 2200 before turning right.

NOTE: **Rwy 9,** multiple hangars beginning 239' from departure end of runway, 360' right of centerline, up to 42' AGL/106' MSL. Multiple trees beginning 501' from departure end of runway, 355' right of centerline, up to 43' AGL/111' MSL. Multiple hangars beginning 119' from departure end of runway, 498' left of centerline, up to 41' AGL/105' MSL. Pole 332' from departure end of runway, 299' left of centerline, 43' AGL/97' MSL. Antenna 1172' from departure end of runway, 658' left of centerline, 51' AGL/115' MSL. Multiple trees beginning 558' from departure end of runway, 68' left of centerline, up to 58' AGL/122' MSL. **Rwy 27,** multiple trees beginning 1050' from departure end of runway, 40' left of centerline, up to 71' AGL/140' MSL. Vehicle and road 99' from departure end of runway, 291' right of centerline, 15' AGL/83' MSL. Multiple trees beginning 873' from departure end of runway, 514' right of centerline, up to 59' AGL/130' MSL. Multiple transmission poles beginning 1304' from departure end of runway, 131' right of centerline, up to 41' AGL/110' MSL.

LONE STAR EXECUTIVE (CXO)
AMDT 3 10266 (FAA)

NOTE: **Rwy 1,** trees beginning 194' from DER, 130' right of centerline, up to 100' AGL/374' MSL. Trees beginning 817' from DER, 15' left of centerline, up to 100' AGL/359' MSL. **Rwy 14,** trees and obstruction light on DME beginning 399' from DER, 80' right of centerline, up to 100' AGL/329' MSL. Trees beginning 640' from DER, 408' left of centerline, up to 100' AGL/329' MSL. **Rwy 19,** trees beginning 68' from DER, 64' right of centerline, up to 100' AGL/344' MSL. Trees beginning 1' from DER, 159' left of centerline, up to 100' AGL/339' MSL. **Rwy 32,** trees beginning 1785' from DER, 973' right of centerline, up to 100' AGL/339' MSL. Trees and vehicles on road beginning 603' from DER, 458' left of centerline, up to 100' AGL/354' MSL.

PEARLAND RGNL

DEPARTURE PROCEDURE: **Rwy 14,** climb heading 139° to 1600 before proceeding south through southwest. **Rwy 32,** climb heading 319° to 900 before proceeding on course.

NOTE: **Rwy 14,** multiple trees beginning 199' from departure end of runway, 226' right of centerline, up to 66' AGL/100' MSL. Vehicle on road 398' from departure end of runway, 405' left of centerline, 9' AGL/55' MSL. Trees 1287' from departure end of runway, 453' left of centerline, up to 56' AGL/90' MSL. **Rwy 32,** multiple trees beginning 690' from departure end of runway, 81' left of centerline, up to 79' AGL/128' MSL. Multiple poles beginning 745' from departure end of runway, 24' left of centerline, up to 40' AGL/80' MSL. Multiple trees and poles beginning 29' from departure end of runway, 11' right of centerline, up to 64' AGL/104' MSL. Building 237' from departure end of runway, 520' right of centerline, 32' AGL/72' MSL.

HOUSTON, TX (CON'T)

SUGAR LAND RGNL (SGR)
AMDT 7 08157 (FAA)

DEPARTURE PROCEDURE: **Rwy 17,** climb heading 170° to 1500 before turning eastbound. **Rwy 35,** climb heading 350° to 1100 before turning southbound.

NOTE: **Rwy 17,** multiple poles beginning 436' from departure end of runway, 172' right of centerline, up to 44' AGL/124' MSL. Railroad 110' from departure end of runway, 10' left of centerline, 23' AGL/104' MSL. Multiple poles beginning 135' from departure end of runway, 270' left of centerline, up to 44' AGL/ 111' MSL. **Rwy 35,** vehicle and road 65' from departure end of runway, 2' right of centerline, 15' AGL/ 96' MSL. Multiple trees beginning 37' from departure end of runway, 275' right of centerline, up to 81' AGL/164' MSL. DME antenna 380' from departure end of runway, 253' right of centerline, 24' AGL/100' MSL. Multiple trees beginning 83' from departure end of runway, 65' left of centerline, up to 81' AGL/155' MSL.

WEISER AIR PARK (EYQ)
AMDT 2 08157 (FAA)

TAKE-OFF MINIMUMS: **Rwy 9,** 200-1 or std. w/ min. climb of 399' per NM to 400.

NOTE: **Rwy 9,** tank 4127' from departure end of runway, 1455' left of centerline, 147' AGL/282' MSL. **Rwy 27,** railroad 462' from departure end of runway, 555' left of centerline, 23' AGL/165' MSL. Vehicle and road 650' from departure end of runway, 7' left of centerline, 17' AGL/159' MSL.

WEST HOUSTON (IWS)
AMDT 3 09295 (FAA)

NOTE: **Rwy 15,** vehicles on roadway beginning abeam DER, left and right of centerline, up to 15' AGL/124' MSL. Building 177' from DER, 398' left of centerline, 18' AGL/ 126' MSL. Trees beginning 178' from DER, 289' right of centerline, up to 100' AGL/209' MSL. **Rwy 33,** building 265' from DER, 364' left of centerline, 33' AGL/143' MSL. Trees beginning 2706' from DER, 700' left of centerline, up to 100' AGL/214' MSL. Trees beginning 3159' from DER, 747' right of centerline, up to 100' AGL/216' MSL.

05 APR 2012 to 03 MAY 2012

05 APR 2012 to 03 MAY 2012

11349

TAKE-OFF MINIMUMS AND (OBSTACLE) DEPARTURE PROCEDURES

SC-5

L4

Figure 261

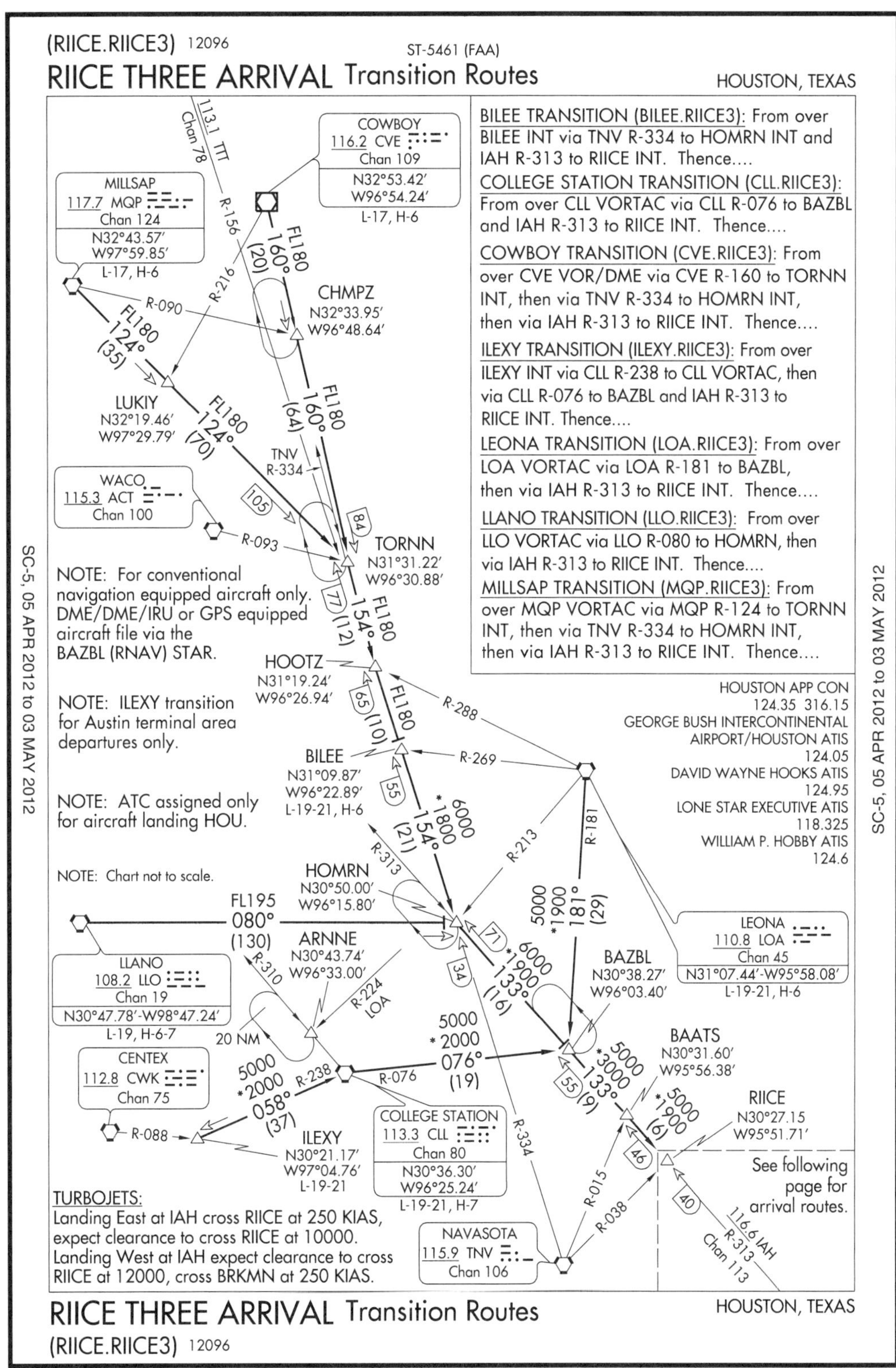
(RIICE.RIICE3) 12096
ST-5461 (FAA)
RIICE THREE ARRIVAL Transition Routes
HOUSTON, TEXAS
BILEE TRANSITION (BILEE.RIICE3): From over BILEE INT via TNV R-334 to HOMRN INT and IAH R-313 to RIICE INT. Thence....
COLLEGE STATION TRANSITION (CLL.RIICE3): From over CLL VORTAC via CLL R-076 to BAZBL and IAH R-313 to RIICE INT. Thence....
COWBOY TRANSITION (CVE.RIICE3): From over CVE VOR/DME via CVE R-160 to TORNN INT, then via TNV R-334 to HOMRN INT, then via IAH R-313 to RIICE INT. Thence....
ILEXY TRANSITION (ILEXY.RIICE3): From over ILEXY INT via CLL R-238 to CLL VORTAC, then via CLL R-076 to BAZBL and IAH R-313 to RIICE INT. Thence....
LEONA TRANSITION (LOA.RIICE3): From over LOA VORTAC via LOA R-181 to BAZBL, then via IAH R-313 to RIICE INT. Thence....
LLANO TRANSITION (LLO.RIICE3): From over LLO VORTAC via LLO R-080 to HOMRN, then via IAH R-313 to RIICE INT. Thence....
MILLSAP TRANSITION (MQP.RIICE3): From over MQP VORTAC via MQP R-124 to TORNN INT, then via TNV R-334 to HOMRN INT, then via IAH R-313 to RIICE INT. Thence....
HOUSTON APP CON 124.35 316.15
GEORGE BUSH INTERCONTINENTAL AIRPORT/HOUSTON ATIS 124.05
DAVID WAYNE HOOKS ATIS 124.95
LONE STAR EXECUTIVE ATIS 118.325
WILLIAM P. HOBBY ATIS 124.6
COWBOY 116.2 CVE Chan 109 N32°53.42' W96°54.24' L-17, H-6
113.1 TTT Chan 78
MILLSAP 117.7 MQP Chan 124 N32°43.57' W97°59.85' L-17, H-6
CHMPZ N32°33.95' W96°48.64'
LUKIY N32°19.46' W97°29.79'
WACO 115.3 ACT Chan 100
TORNN N31°31.22' W96°30.88'
HOOTZ N31°19.24' W96°26.94'
BILEE N31°09.87' W96°22.89' L-19-21, H-6
HOMRN N30°50.00' W96°15.80'
ARNNE N30°43.74' W96°33.00'
BAZBL N30°38.27' W96°03.40'
BAATS N30°31.60' W95°56.38'
RIICE N30°27.15 W95°51.71'
LEONA 110.8 LOA Chan 45 N31°07.44'-W95°58.08' L-19-21, H-6
LLANO 108.2 LLO Chan 19 N30°47.78'-W98°47.24' L-19, H-6-7
CENTEX 112.8 CWK Chan 75
ILEXY N30°21.17' W97°04.76' L-19-21
COLLEGE STATION 113.3 CLL Chan 80 N30°36.30' W96°25.24' L-19-21, H-7
NAVASOTA 115.9 TNV Chan 106
NOTE: For conventional navigation equipped aircraft only. DME/DME/IRU or GPS equipped aircraft file via the BAZBL (RNAV) STAR.
NOTE: ILEXY transition for Austin terminal area departures only.
NOTE: ATC assigned only for aircraft landing HOU.
NOTE: Chart not to scale.
See following page for arrival routes.
TURBOJETS:
Landing East at IAH cross RIICE at 250 KIAS, expect clearance to cross RIICE at 10000.
Landing West at IAH expect clearance to cross RIICE at 12000, cross BRKMN at 250 KIAS.
SC-5, 05 APR 2012 to 03 MAY 2012
RIICE THREE ARRIVAL Transition Routes
HOUSTON, TEXAS
(RIICE.RIICE3) 12096

Figure 262

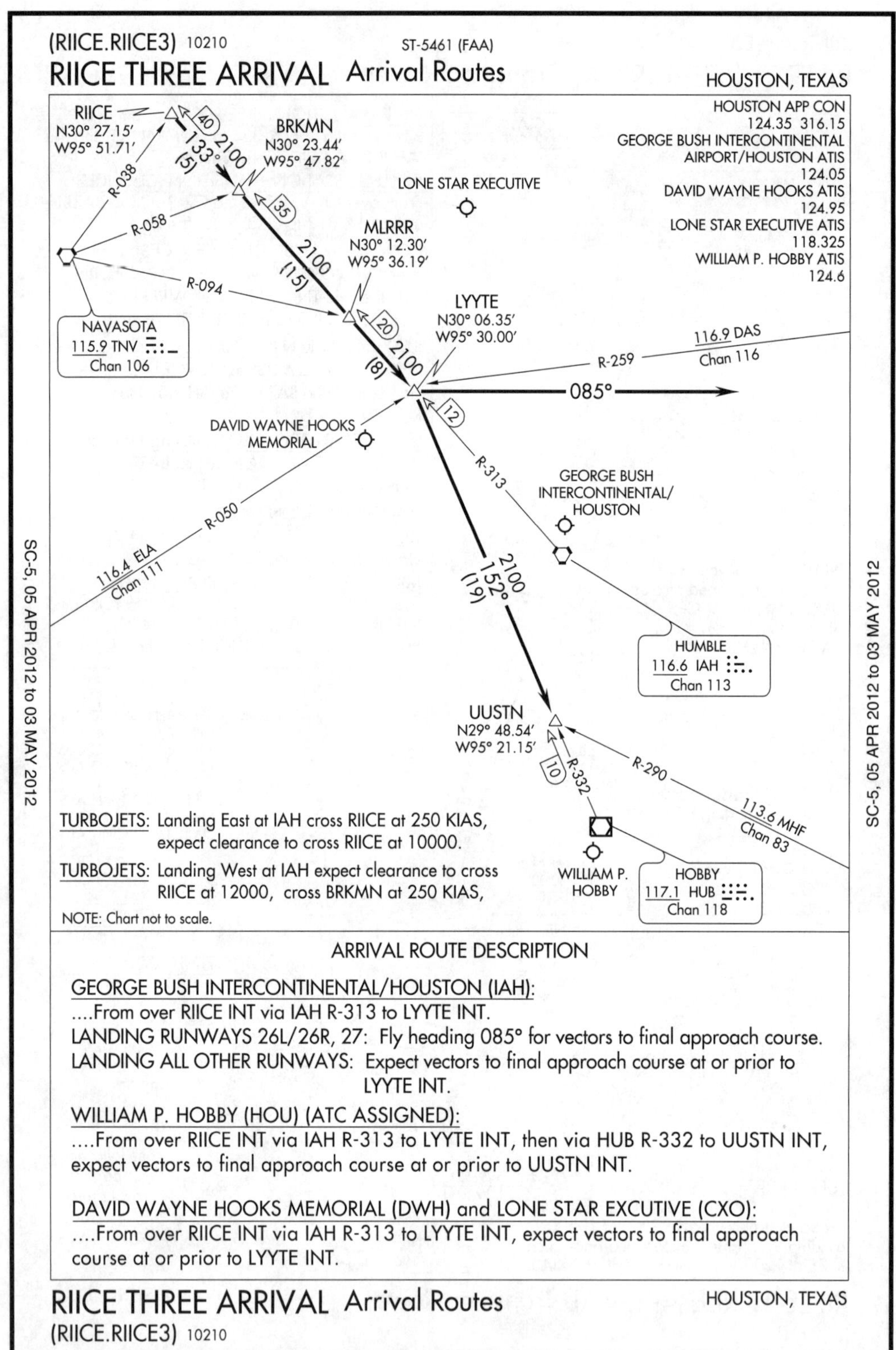

(RIICE.RIICE3) 10210
ST-5461 (FAA)
RIICE THREE ARRIVAL Arrival Routes
HOUSTON, TEXAS
HOUSTON APP CON
124.35 316.15
GEORGE BUSH INTERCONTINENTAL AIRPORT/HOUSTON ATIS
124.05
DAVID WAYNE HOOKS ATIS
124.95
LONE STAR EXECUTIVE ATIS
118.325
WILLIAM P. HOBBY ATIS
124.6
RIICE
N30° 27.15′
W95° 51.71′
BRKMN
N30° 23.44′
W95° 47.82′
133°
2100
(5)
40
35
LONE STAR EXECUTIVE
MLRRR
N30° 12.30′
W95° 36.19′
2100
(15)
20
LYYTE
N30° 06.35′
W95° 30.00′
2100
(8)
12
R-038
R-058
R-094
NAVASOTA
115.9 TNV
Chan 106
R-259
116.9 DAS
Chan 116
085°
DAVID WAYNE HOOKS MEMORIAL
R-313
GEORGE BUSH INTERCONTINENTAL/ HOUSTON
R-050
116.4 ELA
Chan 111
2100
152°
(19)
HUMBLE
116.6 IAH
Chan 113
UUSTN
N29° 48.54′
W95° 21.15′
10
R-332
R-290
113.6 MHF
Chan 83
WILLIAM P. HOBBY
HOBBY
117.1 HUB
Chan 118
SC-5, 05 APR 2012 to 03 MAY 2012
SC-5, 05 APR 2012 to 03 MAY 2012
TURBOJETS: Landing East at IAH cross RIICE at 250 KIAS, expect clearance to cross RIICE at 10000.
TURBOJETS: Landing West at IAH expect clearance to cross RIICE at 12000, cross BRKMN at 250 KIAS,
NOTE: Chart not to scale.
ARRIVAL ROUTE DESCRIPTION
GEORGE BUSH INTERCONTINENTAL/HOUSTON (IAH):
....From over RIICE INT via IAH R-313 to LYYTE INT.
LANDING RUNWAYS 26L/26R, 27: Fly heading 085° for vectors to final approach course.
LANDING ALL OTHER RUNWAYS: Expect vectors to final approach course at or prior to LYYTE INT.
WILLIAM P. HOBBY (HOU) (ATC ASSIGNED):
....From over RIICE INT via IAH R-313 to LYYTE INT, then via HUB R-332 to UUSTN INT, expect vectors to final approach course at or prior to UUSTN INT.
DAVID WAYNE HOOKS MEMORIAL (DWH) and LONE STAR EXCUTIVE (CXO):
....From over RIICE INT via IAH R-313 to LYYTE INT, expect vectors to final approach course at or prior to LYYTE INT.
RIICE THREE ARRIVAL Arrival Routes
HOUSTON, TEXAS
(RIICE.RIICE3) 10210

Figure 263

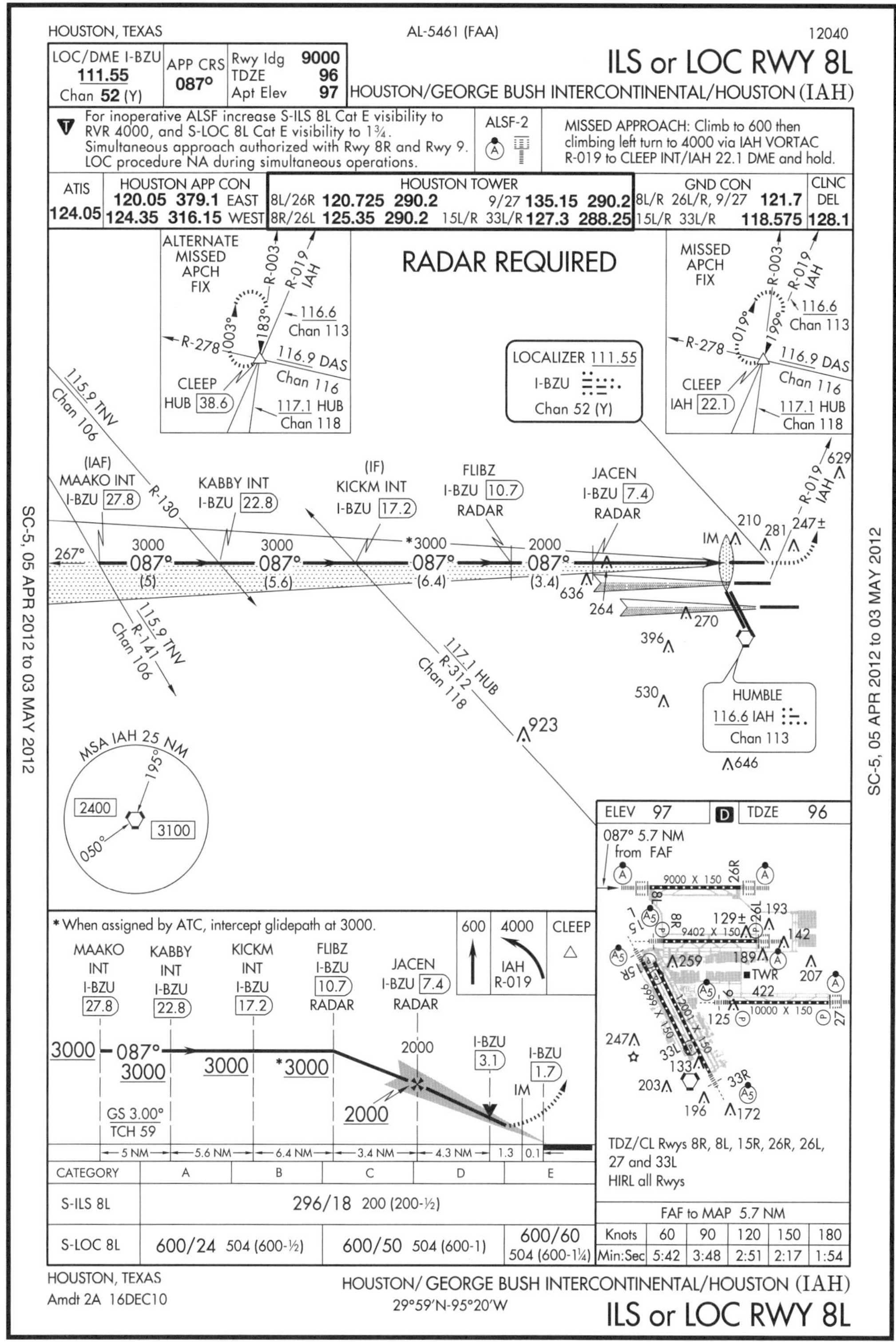
HOUSTON, TEXAS
AL-5461 (FAA)
12040
ILS or LOC RWY 8L
HOUSTON/GEORGE BUSH INTERCONTINENTAL/HOUSTON (IAH)
LOC/DME I-BZU 111.55 Chan 52 (Y)
APP CRS 087°
Rwy ldg 9000
TDZE 96
Apt Elev 97
For inoperative ALSF increase S-ILS 8L Cat E visibility to RVR 4000, and S-LOC 8L Cat E visibility to 1¾. Simultaneous approach authorized with Rwy 8R and Rwy 9. LOC procedure NA during simultaneous operations.
ALSF-2
MISSED APPROACH: Climb to 600 then climbing left turn to 4000 via IAH VORTAC R-019 to CLEEP INT/IAH 22.1 DME and hold.
ATIS 124.05
HOUSTON APP CON 120.05 379.1 EAST 124.35 316.15 WEST
HOUSTON TOWER 8L/26R 120.725 290.2 9/27 135.15 290.2 8R/26L 125.35 290.2 15L/R 33L/R 127.3 288.25
GND CON 8L/R 26L/R, 9/27 121.7 15L/R 33L/R 118.575
CLNC DEL 128.1
RADAR REQUIRED
ALTERNATE MISSED APCH FIX
MISSED APCH FIX
LOCALIZER 111.55 I-BZU Chan 52 (Y)
HUMBLE 116.6 IAH Chan 113
MSA IAH 25 NM
SC-5, 05 APR 2012 to 03 MAY 2012
* When assigned by ATC, intercept glidepath at 3000.
GS 3.00° TCH 59
CATEGORY A B C D E
S-ILS 8L 296/18 200 (200-½)
S-LOC 8L 600/24 504 (600-½) 600/50 504 (600-1) 600/60 504 (600-1¼)
ELEV 97 TDZE 96
087° 5.7 NM from FAF
TDZ/CL Rwys 8R, 8L, 15R, 26R, 26L, 27 and 33L
HIRL all Rwys
FAF to MAP 5.7 NM
Knots 60 90 120 150 180
Min:Sec 5:42 3:48 2:51 2:17 1:54
HOUSTON, TEXAS
Amdt 2A 16DEC10
HOUSTON/ GEORGE BUSH INTERCONTINENTAL/HOUSTON (IAH)
29°59'N-95°20'W
ILS or LOC RWY 8L

Figure 264

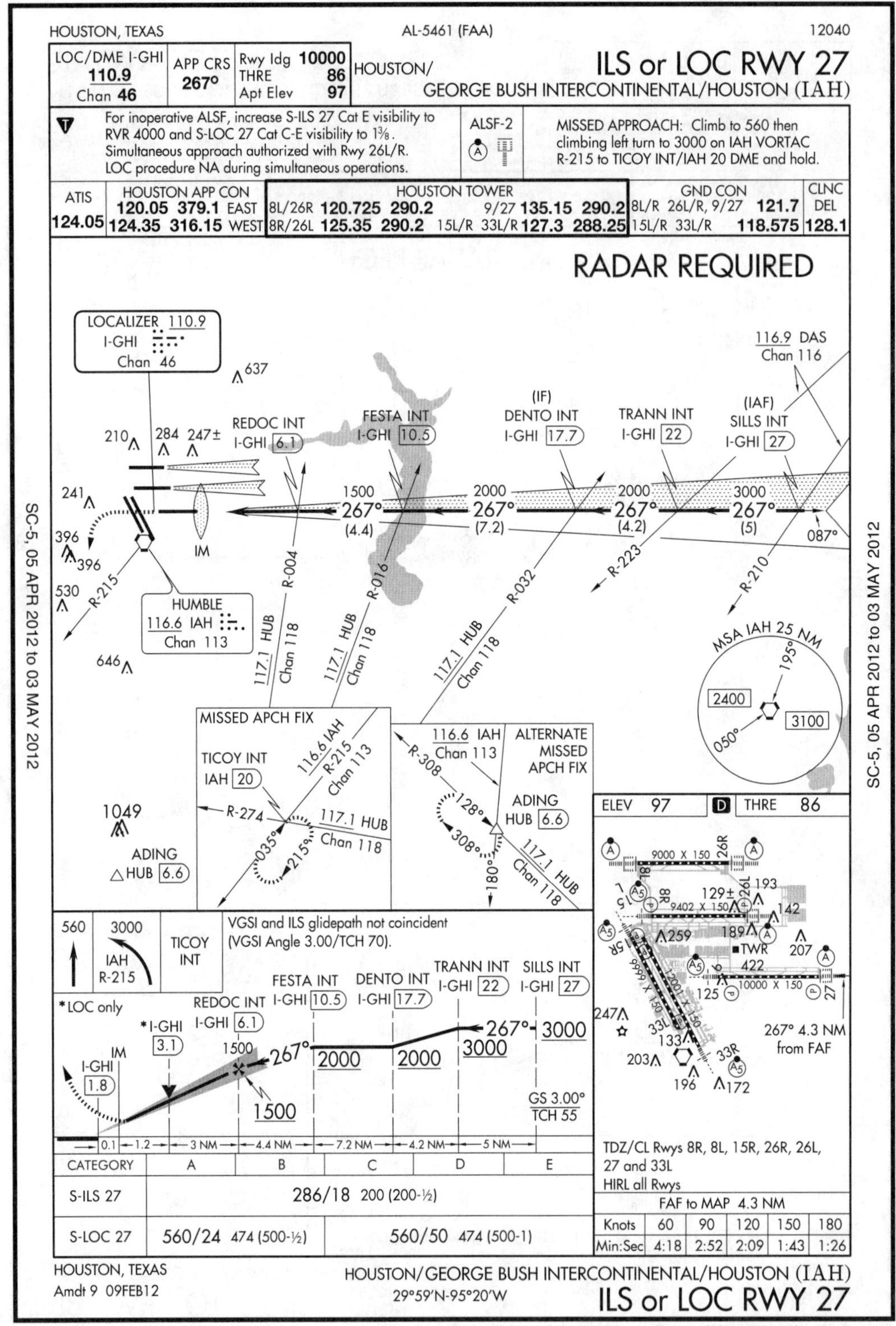

CATEGORY	A	B	C	D	E
S-ILS 27	286/18 200 (200-½)				
S-LOC 27	560/24 474 (500-½)		560/50 474 (500-1)		

FAF to MAP 4.3 NM					
Knots	60	90	120	150	180
Min:Sec	4:18	2:52	2:09	1:43	1:26

Figure 265

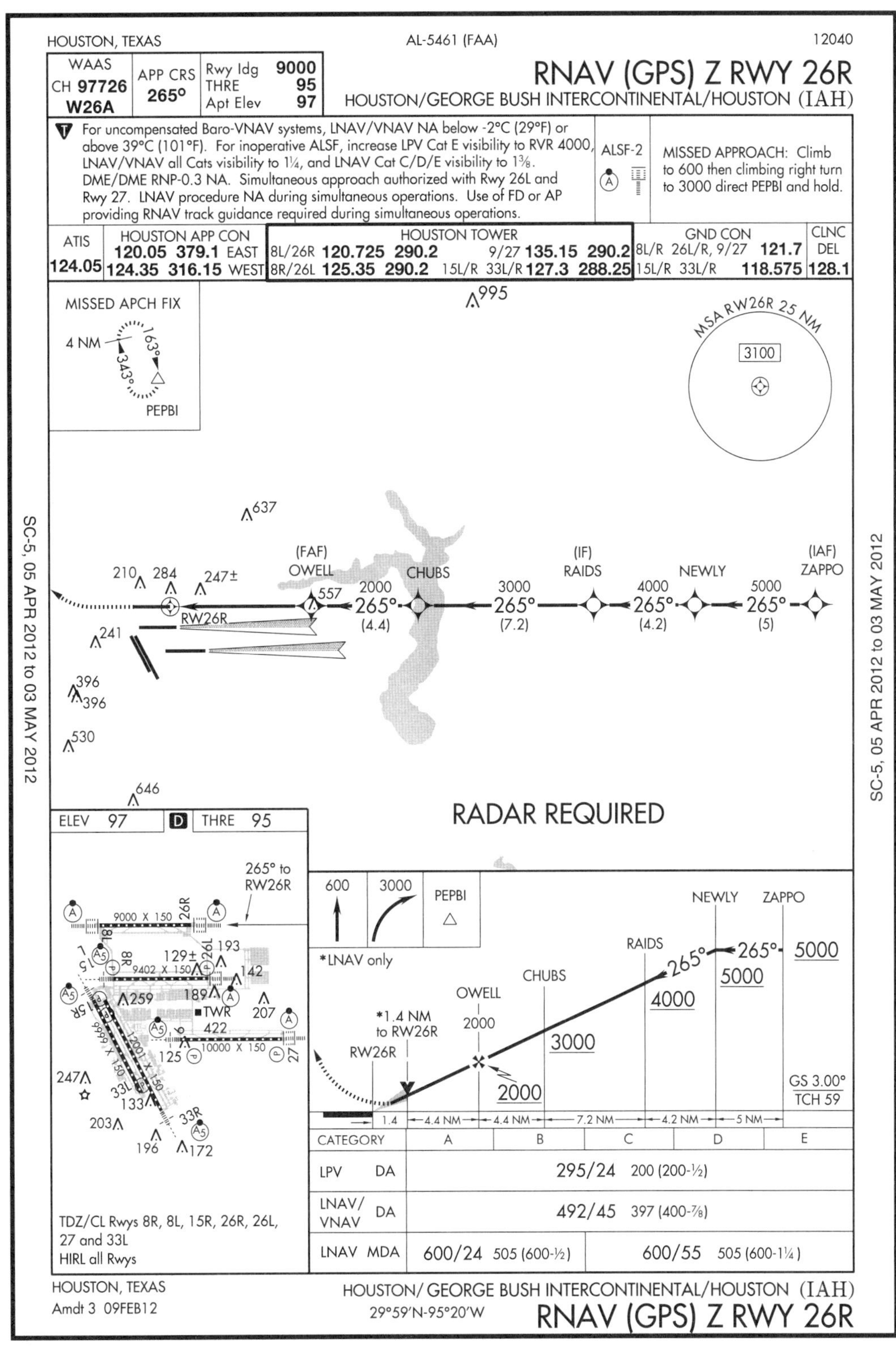
HOUSTON, TEXAS
AL-5461 (FAA)
12040
WAAS CH 97726 W26A
APP CRS 265°
Rwy ldg 9000
THRE 95
Apt Elev 97
RNAV (GPS) Z RWY 26R
HOUSTON/GEORGE BUSH INTERCONTINENTAL/HOUSTON (IAH)
For uncompensated Baro-VNAV systems, LNAV/VNAV NA below -2°C (29°F) or above 39°C (101°F). For inoperative ALSF, increase LPV Cat E visibility to RVR 4000, LNAV/VNAV all Cats visibility to 1¼, and LNAV Cat C/D/E visibility to 1⅜. DME/DME RNP-0.3 NA. Simultaneous approach authorized with Rwy 26L and Rwy 27. LNAV procedure NA during simultaneous operations. Use of FD or AP providing RNAV track guidance required during simultaneous operations.
ALSF-2
MISSED APPROACH: Climb to 600 then climbing right turn to 3000 direct PEPBI and hold.
ATIS 124.05
HOUSTON APP CON 120.05 379.1 EAST 124.35 316.15 WEST
HOUSTON TOWER 8L/26R 120.725 290.2 9/27 135.15 290.2 8R/26L 125.35 290.2 15L/R 33L/R 127.3 288.25
GND CON 8L/R 26L/R, 9/27 121.7 15L/R 33L/R 118.575
CLNC DEL 128.1
MISSED APCH FIX
4 NM
163°
343°
PEPBI
MSA RW26R 25 NM
3100
(FAF) OWELL
CHUBS
(IF) RAIDS
NEWLY
(IAF) ZAPPO
2000
3000
4000
5000
265° (4.4)
265° (7.2)
265° (4.2)
265° (5)
RW26R
RADAR REQUIRED
SC-5, 05 APR 2012 to 03 MAY 2012
ELEV 97
THRE 95
265° to RW26R
9000 X 150
9402 X 150
10000 X 150
12001 X 150
9999 X 150
TWR
TDZ/CL Rwys 8R, 8L, 15R, 26R, 26L, 27 and 33L
HIRL all Rwys
600
3000
PEPBI
*LNAV only
*1.4 NM to RW26R
GS 3.00°
TCH 59
1.4
4.4 NM
4.4 NM
7.2 NM
4.2 NM
5 NM
CATEGORY A B C D E
LPV DA 295/24 200 (200-½)
LNAV/VNAV DA 492/45 397 (400-⅞)
LNAV MDA 600/24 505 (600-½) 600/55 505 (600-1¼)
HOUSTON, TEXAS
Amdt 3 09FEB12
29°59'N-95°20'W
HOUSTON/GEORGE BUSH INTERCONTINENTAL/HOUSTON (IAH)
RNAV (GPS) Z RWY 26R

Figure 266

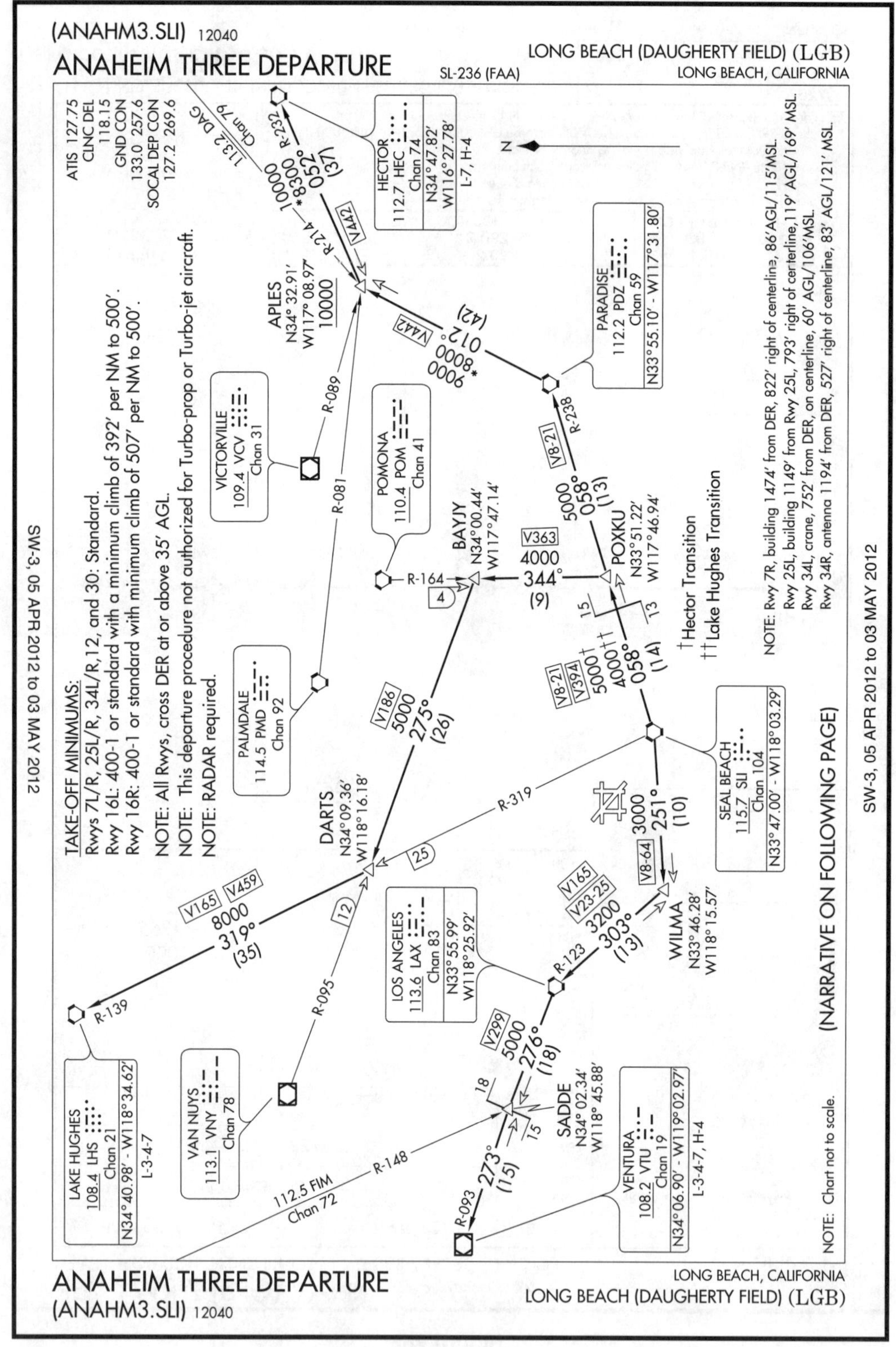

(ANAHM3.SLI) 12040
ANAHEIM THREE DEPARTURE
SL-236 (FAA)
LONG BEACH (DAUGHERTY FIELD) (LGB)
LONG BEACH, CALIFORNIA
ATIS 127.75
CLNC DEL 118.15
GND CON 133.0 257.6
SOCAL DEP CON 127.2 269.6
TAKE-OFF MINIMUMS:
Rwys 7L/R, 25L/R, 34L/R,12, and 30: Standard.
Rwy 16L: 400-1 or standard with a minimum climb of 392' per NM to 500'.
Rwy 16R: 400-1 or standard with minimum climb of 507' per NM to 500'.
NOTE: All Rwys, cross DER at or above 35' AGL.
NOTE: This departure procedure not authorized for Turbo-prop or Turbo-jet aircraft.
NOTE: RADAR required.
LAKE HUGHES 108.4 LHS Chan 21 N34°40.98' - W118°34.62' L-3-4-7
VAN NUYS 113.1 VNY Chan 78
PALMDALE 114.5 PMD Chan 92
VICTORVILLE 109.4 VCV Chan 31
APLES N34° 32.91' W117° 08.97' 10000
113.2 DAG Chan 79
10000 *8300 052° (37) R-232 R-214 V442
HECTOR 112.7 HEC Chan 74 N34°47.82' W116°27.78' L-7, H-4
DARTS N34°09.36' W118°16.18'
V165 V459 8000 319° (35) R-139
R-095 12 R-081 R-089
V186 5000 275° (26)
POMONA 110.4 POM Chan 41
R-164 4
BAYJY N34°00.44' W117°47.14'
V363 4000 344° (9)
9000 *8000 012° (42) V442
LOS ANGELES 113.6 LAX Chan 83 N33°55.99' W118°25.92'
112.5 FIM Chan 72 R-148
R-093 273° (15) 18 15
V299 5000 276° (18)
SADDE N34° 02.34' W118° 45.88'
R-123 V165 V23-25 3200 303° (13)
R-319 25
V8-21 V394 5000† 4000†† 058° (14) 15 13
POXKU N33°51.22' W117°46.94'
5000 058° (13) V8-21 R-238
PARADISE 112.2 PDZ Chan 59 N33°55.10' - W117°31.80'
N
VENTURA 108.2 VTU Chan 19 N34°06.90' - W119°02.97' L-3-4-7, H-4
V8-64 3000 251° (10)
WILMA N33°46.28' W118°15.57'
SEAL BEACH 115.7 SLI Chan 104 N33°47.00' - W118°03.29'
†Hector Transition
††Lake Hughes Transition
NOTE: Rwy 7R, building 1474' from DER, 822' right of centerline, 86'AGL/115'MSL.
Rwy 25L, building 1149' from Rwy 25L, 793' right of centerline,119' AGL/169' MSL.
Rwy 34L, crane, 752' from DER, on centerline, 60' AGL/106'MSL.
Rwy 34R, antenna 1194' from DER, 527' right of centerline, 83' AGL/121' MSL.
NOTE: Chart not to scale.
(NARRATIVE ON FOLLOWING PAGE)
SW-3, 05 APR 2012 to 03 MAY 2012

Figure 267

(ANAHM3.SLI) 08045

ANAHEIM THREE DEPARTURE

SL-236 (FAA)

LONG BEACH (DAUGHERTY FIELD) (LGB)
LONG BEACH, CALIFORNIA

SW-3, 05 APR 2012 to 03 MAY 2012

DEPARTURE ROUTE DESCRIPTION

HECTOR or LAKE HUGHES TRANSITION: Climb runway heading to 800′ then fly assigned heading for radar vectors to SLI VORTAC. Thence. . . .

VENTURA TRANSITION: Climb runway heading to 800′ then fly assigned heading for radar vectors to LAX VORTAC. Thence. . . .

. . . .via (transition) or (assigned route). Maintain assigned altitude. Expect clearance to filed altitude 10 minutes after departure.

HECTOR TRANSITION (ANAHM3.HEC): From over SLI VORTAC via SLI R-058 and PDZ R-238 to PDZ VORTAC, then via PDZ R-012 and HEC R-232 to HEC VORTAC.

LAKE HUGHES TRANSITION (ANAHM3.LHS): From over SLI VORTAC via SLI R-058 and PDZ R-238 to POXKU INT, then via POM R-164 to BAYJY INT, then via VNY R-095 to DARTS INT. Thence via SLI R-319 and LHS R-139 to LHS VORTAC.

VENTURA TRANSITION (ANAHM3.VTU): From over SLI VORTAC via SLI R-251 to WILMA INT, then via LAX R-123 to LAX VORTAC, then via LAX R-276 and VTU R-093 to VTU VOR/DME.

SW-3, 05 APR 2012 to 03 MAY 2012

ANAHEIM THREE DEPARTURE
(ANAHM3.SLI) 08045
LONG BEACH, CALIFORNIA
LONG BEACH (DAUGHERTY FIELD) (LGB)

Figure 268

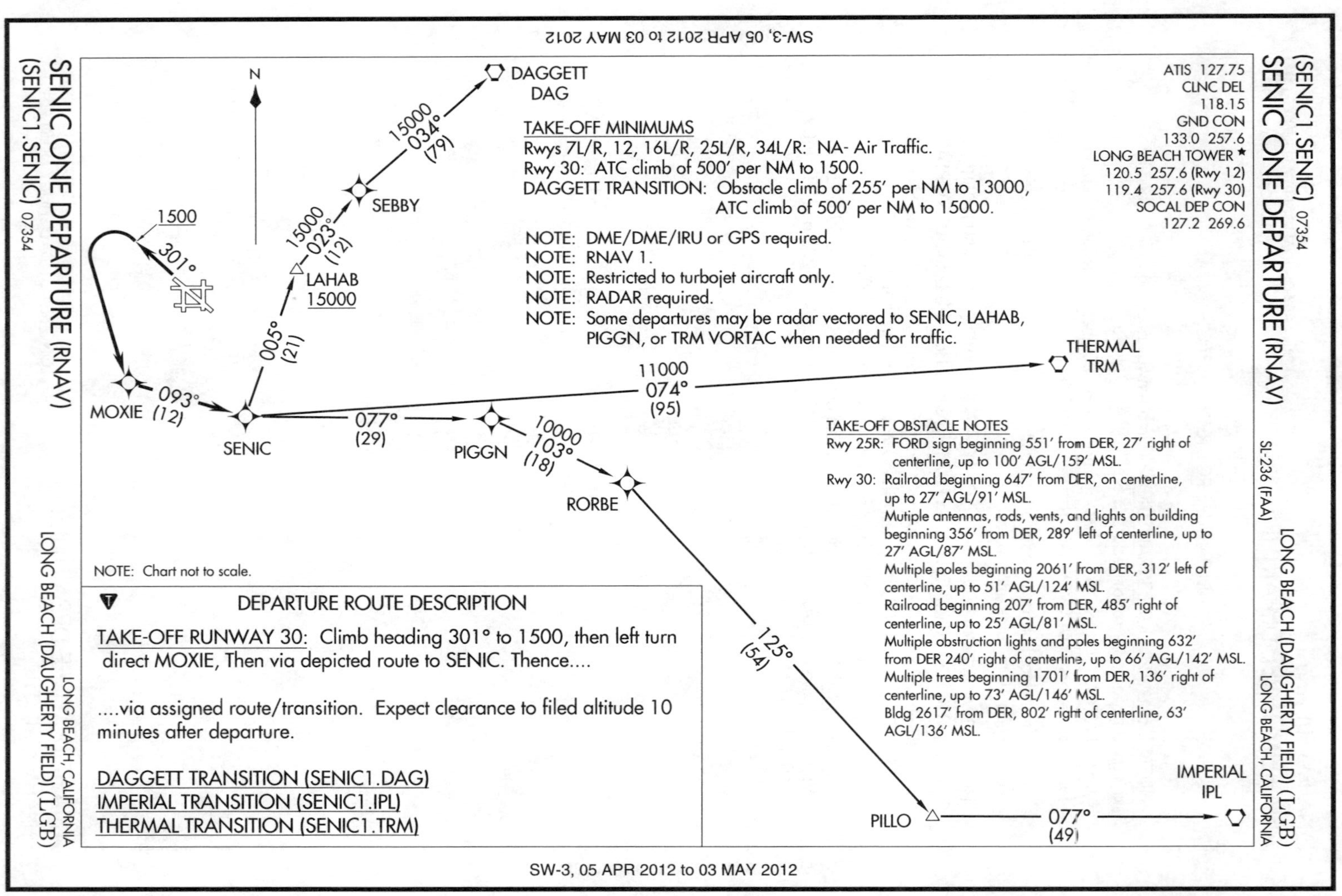

(SENIC1.SENIC) 07354
SENIC ONE DEPARTURE (RNAV)
SL-236 (FAA)
LONG BEACH (DAUGHERTY FIELD) (LGB)
LONG BEACH, CALIFORNIA
ATIS 127.75
CLNC DEL 118.15
GND CON 133.0 257.6
LONG BEACH TOWER ★ 120.5 257.6 (Rwy 12) 119.4 257.6 (Rwy 30)
SOCAL DEP CON 127.2 269.6
SW-3, 05 APR 2012 to 03 MAY 2012
DAGGETT DAG
TAKE-OFF MINIMUMS
Rwys 7L/R, 12, 16L/R, 25L/R, 34L/R: NA- Air Traffic.
Rwy 30: ATC climb of 500' per NM to 1500.
DAGGETT TRANSITION: Obstacle climb of 255' per NM to 13000, ATC climb of 500' per NM to 15000.
NOTE: DME/DME/IRU or GPS required.
NOTE: RNAV 1.
NOTE: Restricted to turbojet aircraft only.
NOTE: RADAR required.
NOTE: Some departures may be radar vectored to SENIC, LAHAB, PIGGN, or TRM VORTAC when needed for traffic.
THERMAL TRM
TAKE-OFF OBSTACLE NOTES
Rwy 25R: FORD sign beginning 551' from DER, 27' right of centerline, up to 100' AGL/153' MSL.
Rwy 30: Railroad beginning 647' from DER, on centerline, up to 27' AGL/91' MSL.
Mutiple antennas, rods, vents, and lights on building beginning 356' from DER, 289' left of centerline, up to 27' AGL/87' MSL.
Multiple poles beginning 2061' from DER, 312' left of centerline, up to 51' AGL/124' MSL.
Railroad beginning 207' from DER, 485' right of centerline, up to 25' AGL/81' MSL.
Multiple obstruction lights and poles beginning 632' from DER 240' right of centerline, up to 66' AGL/142' MSL.
Multiple trees beginning 1701' from DER, 136' right of centerline, up to 73' AGL/146' MSL.
Bldg 2617' from DER, 802' right of centerline, 63' AGL/136' MSL.
IMPERIAL IPL
077° (49)
PILLO
125° (54)
11000 074° (95)
10000 103° (18)
RORBE
PIGGN
077° (29)
15000 034° (79)
SEBBY
15000 023° (12)
LAHAB 15000
005° (21)
SENIC
N
1500
301°
093° (12)
MOXIE
NOTE: Chart not to scale.
DEPARTURE ROUTE DESCRIPTION
TAKE-OFF RUNWAY 30: Climb heading 301° to 1500, then left turn direct MOXIE, Then via depicted route to SENIC. Thence....
....via assigned route/transition. Expect clearance to filed altitude 10 minutes after departure.
DAGGETT TRANSITION (SENIC1.DAG)
IMPERIAL TRANSITION (SENIC1.IPL)
THERMAL TRANSITION (SENIC1.TRM)
SW-3, 05 APR 2012 to 03 MAY 2012
SENIC ONE DEPARTURE (RNAV)
(SENIC1.SENIC) 07354
LONG BEACH, CALIFORNIA
LONG BEACH (DAUGHERTY FIELD) (LGB)

Figure 269

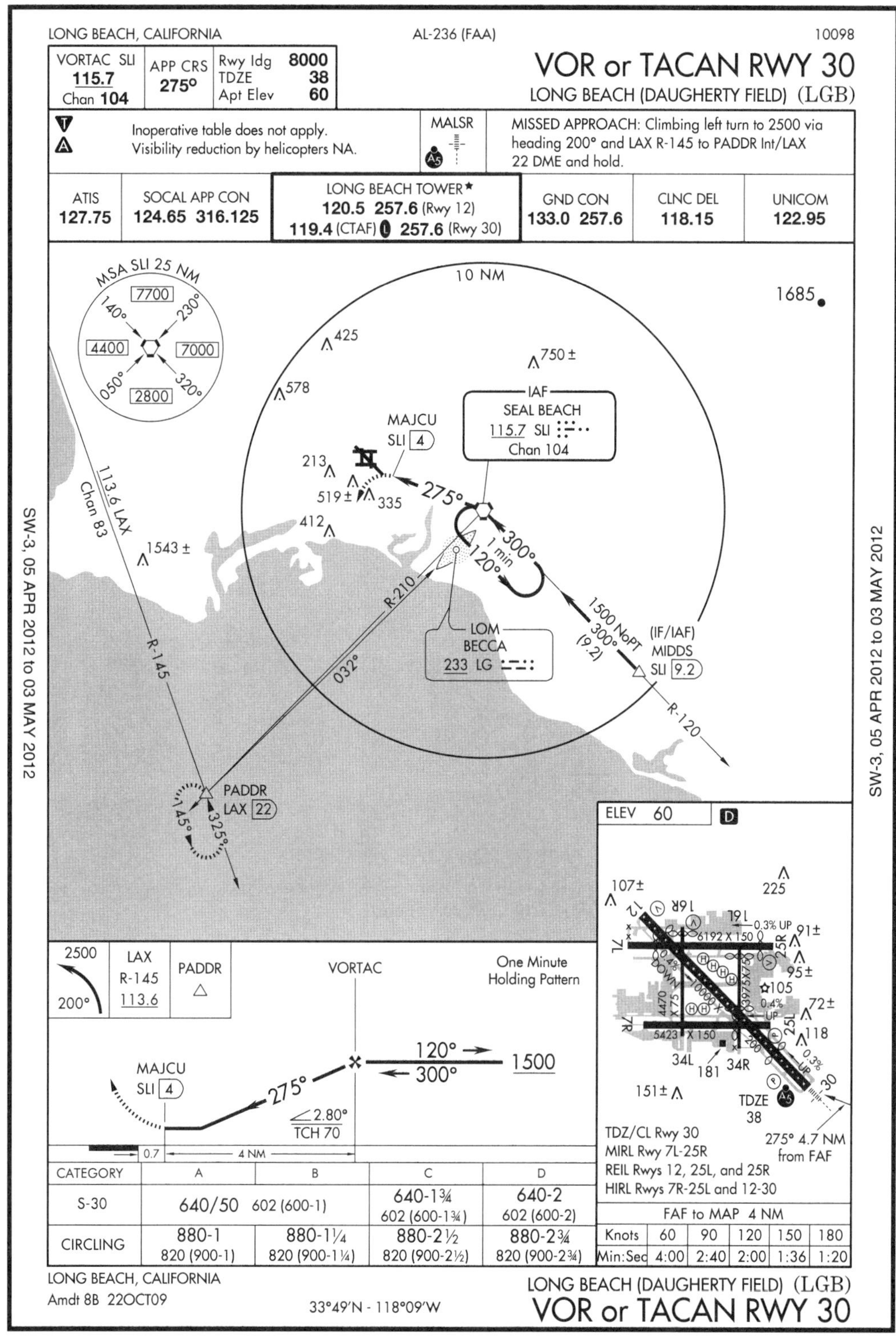

CATEGORY	A	B	C	D
S-30	640/50	602 (600-1)	640-1¾ 602 (600-1¾)	640-2 602 (600-2)
CIRCLING	880-1 820 (900-1)	880-1¼ 820 (900-1¼)	880-2½ 820 (900-2½)	880-2¾ 820 (900-2¾)

FAF to MAP 4 NM

Knots	60	90	120	150	180
Min:Sec	4:00	2:40	2:00	1:36	1:20

Figure 270

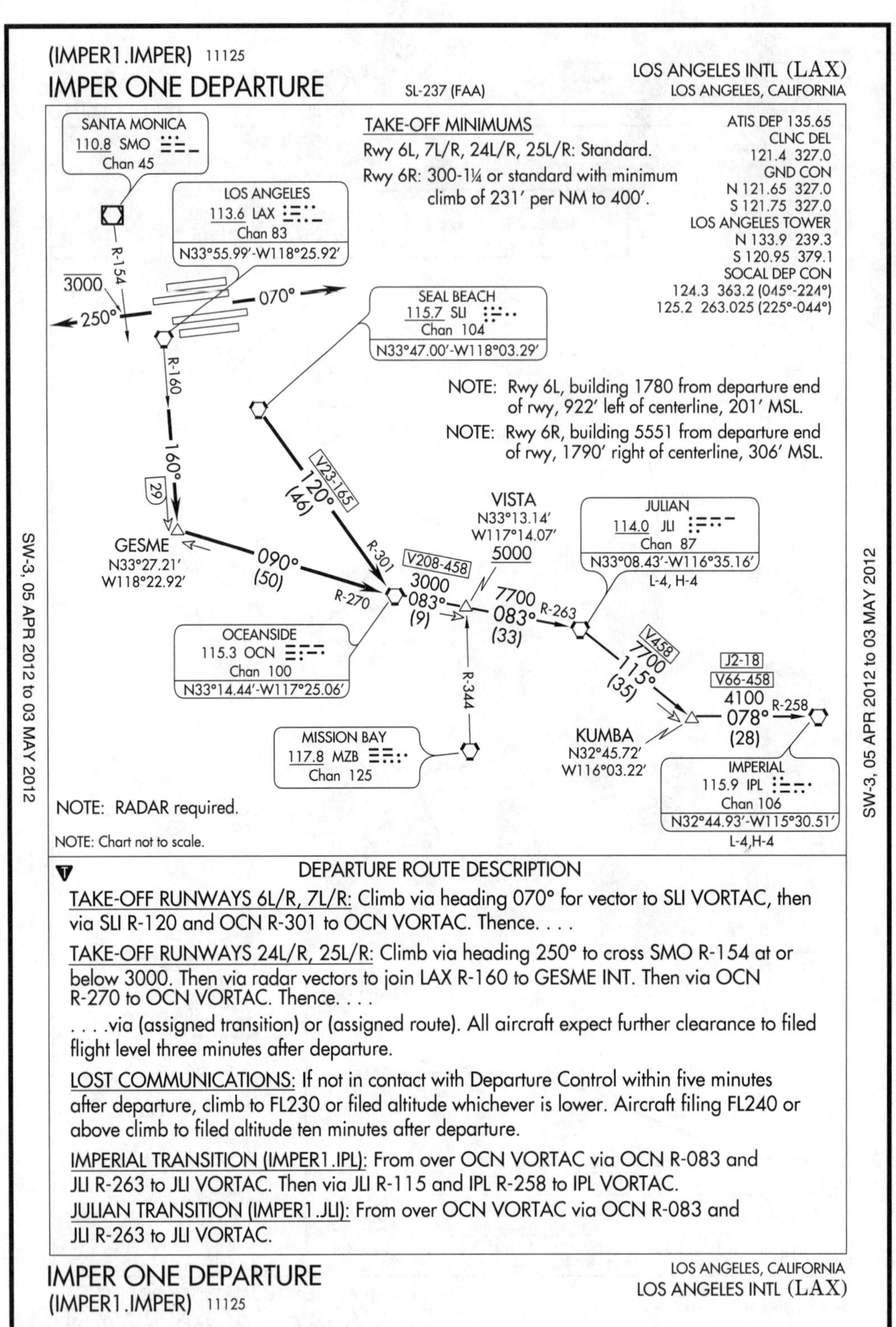

(IMPER1.IMPER) 11125
IMPER ONE DEPARTURE
SL-237 (FAA)
LOS ANGELES INTL (LAX)
LOS ANGELES, CALIFORNIA
TAKE-OFF MINIMUMS
Rwy 6L, 7L/R, 24L/R, 25L/R: Standard.
Rwy 6R: 300-1¼ or standard with minimum climb of 231′ per NM to 400′.
ATIS DEP 135.65
CLNC DEL 121.4 327.0
GND CON N 121.65 327.0 S 121.75 327.0
LOS ANGELES TOWER N 133.9 239.3 S 120.95 379.1
SOCAL DEP CON 124.3 363.2 (045°-224°) 125.2 263.025 (225°-044°)
SANTA MONICA 110.8 SMO Chan 45
LOS ANGELES 113.6 LAX Chan 83 N33°55.99′-W118°25.92′
SEAL BEACH 115.7 SLI Chan 104 N33°47.00′-W118°03.29′
NOTE: Rwy 6L, building 1780 from departure end of rwy, 922′ left of centerline, 201′ MSL.
NOTE: Rwy 6R, building 5551 from departure end of rwy, 1790′ right of centerline, 306′ MSL.
3000
R-154
250°
070°
R-160
160°
29
V23-165
120°
(46)
GESME N33°27.21′ W118°22.92′
090° (50)
R-270
R-301
V208-458
3000
083° (9)
VISTA N33°13.14′ W117°14.07′ 5000
7700
083° (33)
R-263
JULIAN 114.0 JLI Chan 87 N33°08.43′-W116°35.16′ L-4, H-4
OCEANSIDE 115.3 OCN Chan 100 N33°14.44′-W117°25.06′
R-344
V458
7700
115° (35)
J2-18
V66-458
4100
078° (28)
R-258
KUMBA N32°45.72′ W116°03.22′
MISSION BAY 117.8 MZB Chan 125
IMPERIAL 115.9 IPL Chan 106 N32°44.93′-W115°30.51′ L-4,H-4
NOTE: RADAR required.
NOTE: Chart not to scale.
SW-3, 05 APR 2012 to 03 MAY 2012
DEPARTURE ROUTE DESCRIPTION
TAKE-OFF RUNWAYS 6L/R, 7L/R: Climb via heading 070° for vector to SLI VORTAC, then via SLI R-120 and OCN R-301 to OCN VORTAC. Thence. . . .
TAKE-OFF RUNWAYS 24L/R, 25L/R: Climb via heading 250° to cross SMO R-154 at or below 3000. Then via radar vectors to join LAX R-160 to GESME INT. Then via OCN R-270 to OCN VORTAC. Thence. . . .
. . . .via (assigned transition) or (assigned route). All aircraft expect further clearance to filed flight level three minutes after departure.
LOST COMMUNICATIONS: If not in contact with Departure Control within five minutes after departure, climb to FL230 or filed altitude whichever is lower. Aircraft filing FL240 or above climb to filed altitude ten minutes after departure.
IMPERIAL TRANSITION (IMPER1.IPL): From over OCN VORTAC via OCN R-083 and JLI R-263 to JLI VORTAC. Then via JLI R-115 and IPL R-258 to IPL VORTAC.
JULIAN TRANSITION (IMPER1.JLI): From over OCN VORTAC via OCN R-083 and JLI R-263 to JLI VORTAC.
IMPER ONE DEPARTURE
(IMPER1.IMPER) 11125
LOS ANGELES, CALIFORNIA
LOS ANGELES INTL (LAX)

Figure 271

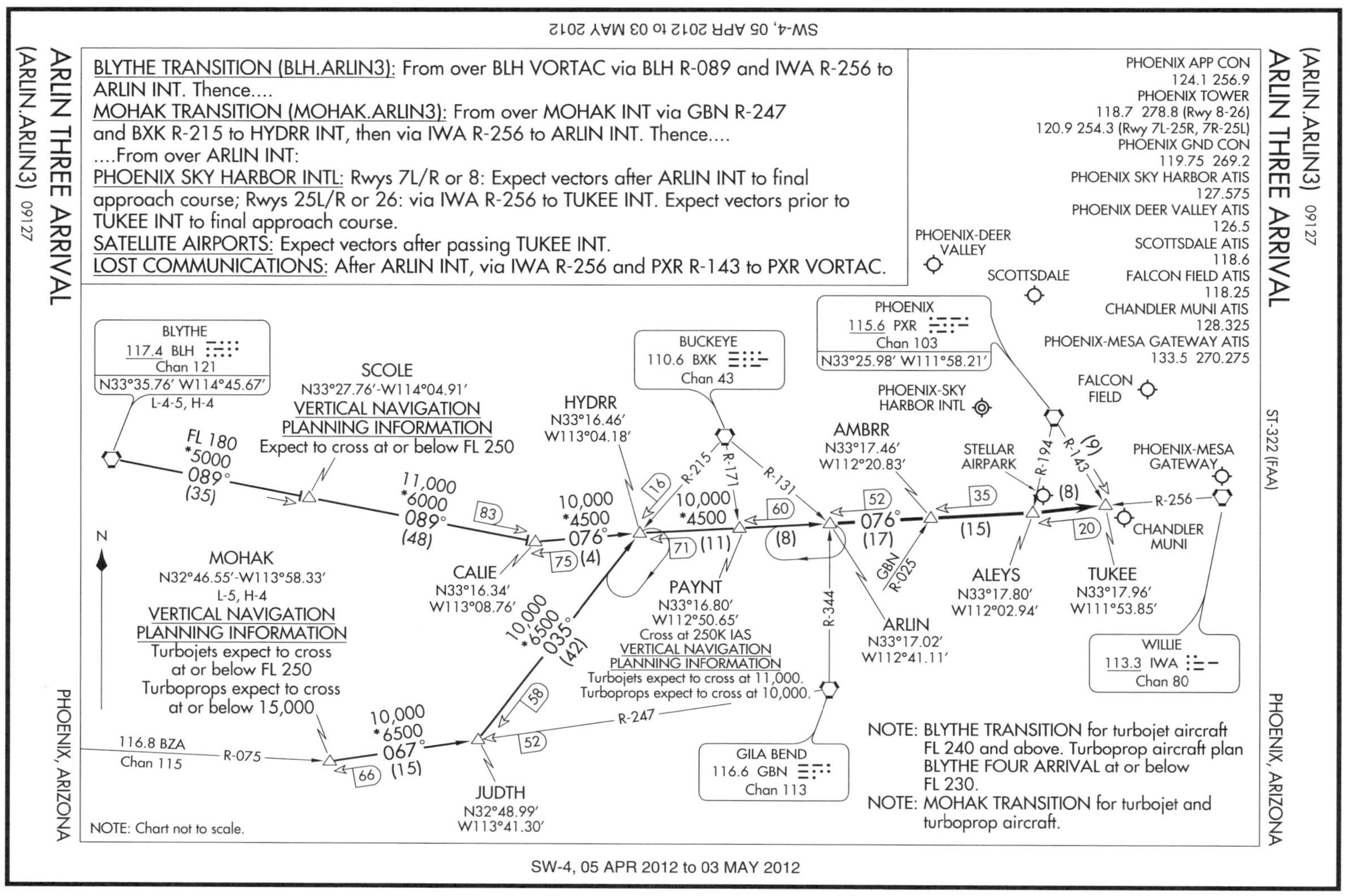
(ARLIN.ARLIN3) 09127
ARLIN THREE ARRIVAL
ST-322 (FAA)
PHOENIX, ARIZONA
SW-4, 05 APR 2012 to 03 MAY 2012
BLYTHE TRANSITION (BLH.ARLIN3): From over BLH VORTAC via BLH R-089 and IWA R-256 to ARLIN INT. Thence....
MOHAK TRANSITION (MOHAK.ARLIN3): From over MOHAK INT via GBN R-247 and BXK R-215 to HYDRR INT, then via IWA R-256 to ARLIN INT. Thence....
....From over ARLIN INT:
PHOENIX SKY HARBOR INTL: Rwys 7L/R or 8: Expect vectors after ARLIN INT to final approach course; Rwys 25L/R or 26: via IWA R-256 to TUKEE INT. Expect vectors prior to TUKEE INT to final approach course.
SATELLITE AIRPORTS: Expect vectors after passing TUKEE INT.
LOST COMMUNICATIONS: After ARLIN INT, via IWA R-256 and PXR R-143 to PXR VORTAC.
PHOENIX APP CON
124.1 256.9
PHOENIX TOWER
118.7 278.8 (Rwy 8-26)
120.9 254.3 (Rwy 7L-25R, 7R-25L)
PHOENIX GND CON
119.75 269.2
PHOENIX SKY HARBOR ATIS
127.575
PHOENIX DEER VALLEY ATIS
126.5
SCOTTSDALE ATIS
118.6
FALCON FIELD ATIS
118.25
CHANDLER MUNI ATIS
128.325
PHOENIX-MESA GATEWAY ATIS
133.5 270.275
PHOENIX-DEER VALLEY
SCOTTSDALE
PHOENIX 115.6 PXR Chan 103 N33°25.98′ W111°58.21′
PHOENIX-SKY HARBOR INTL
FALCON FIELD
BLYTHE 117.4 BLH Chan 121 N33°35.76′ W114°45.67′ L-4-5, H-4
BUCKEYE 110.6 BXK Chan 43
SCOLE N33°27.76′-W114°04.91′
VERTICAL NAVIGATION PLANNING INFORMATION
Expect to cross at or below FL 250
HYDRR N33°16.46′ W113°04.18′
AMBRR N33°17.46′ W112°20.83′
STELLAR AIRPARK
PHOENIX-MESA GATEWAY
R-256
CHANDLER MUNI
FL 180 *5000 089° (35)
11,000 *6000 089° (48)
10,000 *4500 076° (4)
10,000 *4500 (11)
076° (17)
(8)
(15)
(9)
R-215
R-171
R-131
R-194
R-143
GBN R-025
R-344
CALIE N33°16.34′ W113°08.76′
PAYNT N33°16.80′ W112°50.65′
Cross at 250K IAS
VERTICAL NAVIGATION PLANNING INFORMATION
Turbojets expect to cross at 11,000.
Turboprops expect to cross at 10,000.
ARLIN N33°17.02′ W112°41.11′
ALEYS N33°17.80′ W112°02.94′
TUKEE N33°17.96′ W111°53.85′
WILLIE 113.3 IWA Chan 80
MOHAK N32°46.55′-W113°58.33′ L-5, H-4
VERTICAL NAVIGATION PLANNING INFORMATION
Turbojets expect to cross at or below FL 250
Turboprops expect to cross at or below 15,000
10,000 *6500 035° (42)
10,000 *6500 067° (15)
116.8 BZA Chan 115
R-075
R-247
JUDTH N32°48.99′ W113°41.30′
GILA BEND 116.6 GBN Chan 113
NOTE: BLYTHE TRANSITION for turbojet aircraft FL 240 and above. Turboprop aircraft plan BLYTHE FOUR ARRIVAL at or below FL 230.
NOTE: MOHAK TRANSITION for turbojet and turboprop aircraft.
NOTE: Chart not to scale.
ARLIN THREE ARRIVAL
(ARLIN.ARLIN3) 09127
PHOENIX, ARIZONA
SW-4, 05 APR 2012 to 03 MAY 2012

Figure 272

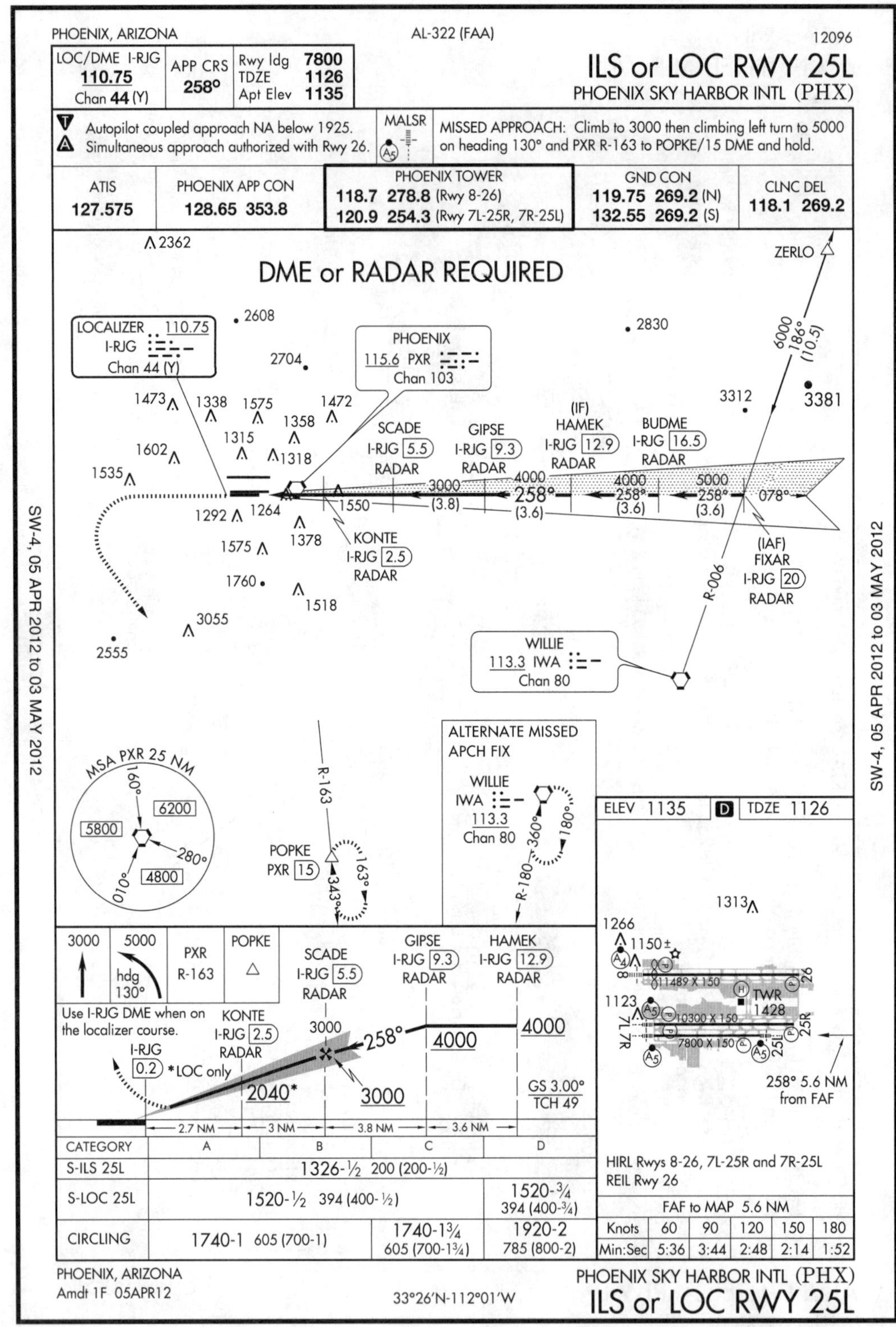

CATEGORY	A	B	C	D
S-ILS 25L	1326-½ 200 (200-½)			
S-LOC 25L	1520-½ 394 (400-½)			1520-¾ 394 (400-¾)
CIRCLING	1740-1 605 (700-1)		1740-1¾ 605 (700-1¾)	1920-2 785 (800-2)

Knots	60	90	120	150	180
Min:Sec	5:36	3:44	2:48	2:14	1:52

Figure 273

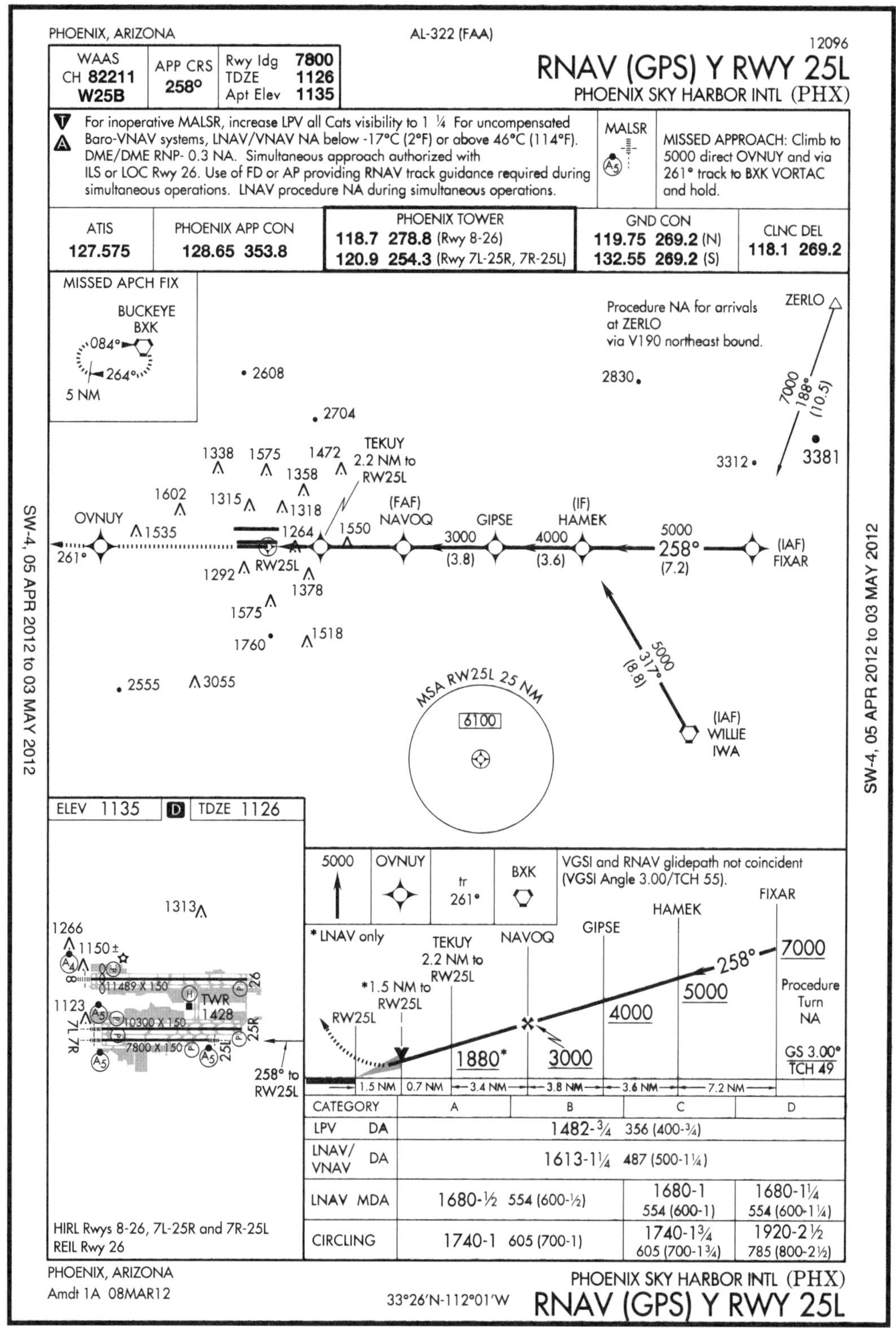

CATEGORY	A	B	C	D
LPV DA	1482-¾ 356 (400-¾)			
LNAV/VNAV DA	1613-1¼ 487 (500-1¼)			
LNAV MDA	1680-½ 554 (600-½)		1680-1 554 (600-1)	1680-1¼ 554 (600-1¼)
CIRCLING	1740-1 605 (700-1)		1740-1¾ 605 (700-1¾)	1920-2½ 785 (800-2½)

Figure 274

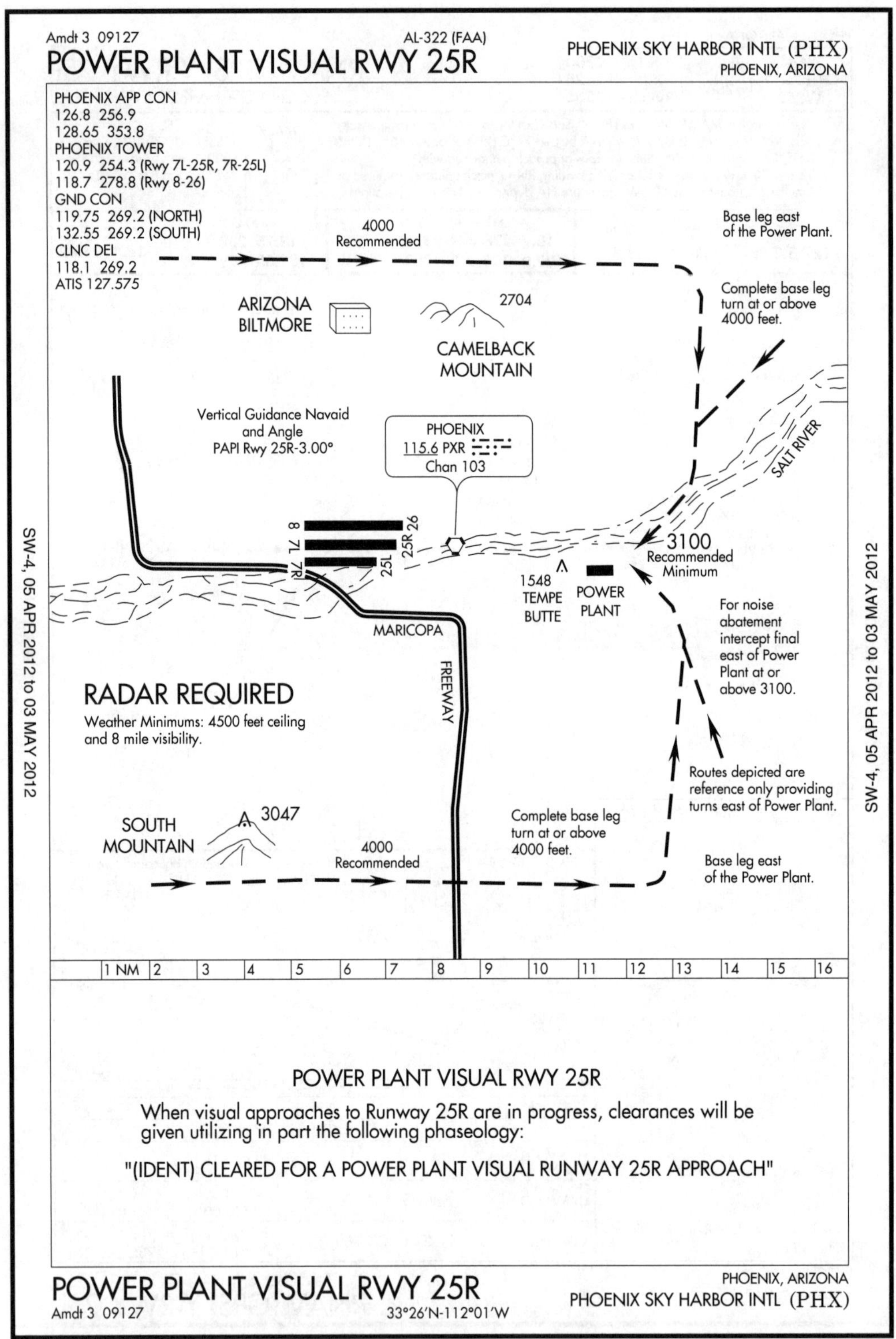

Amdt 3 09127
AL-322 (FAA)
POWER PLANT VISUAL RWY 25R
PHOENIX SKY HARBOR INTL (PHX)
PHOENIX, ARIZONA
PHOENIX APP CON
126.8 256.9
128.65 353.8
PHOENIX TOWER
120.9 254.3 (Rwy 7L-25R, 7R-25L)
118.7 278.8 (Rwy 8-26)
GND CON
119.75 269.2 (NORTH)
132.55 269.2 (SOUTH)
CLNC DEL
118.1 269.2
ATIS 127.575
4000
Recommended
Base leg east of the Power Plant.
Complete base leg turn at or above 4000 feet.
ARIZONA BILTMORE
2704
CAMELBACK MOUNTAIN
Vertical Guidance Navaid and Angle
PAPI Rwy 25R-3.00°
PHOENIX
115.6 PXR
Chan 103
SALT RIVER
8
7L
7R
26
25R
25L
3100
Recommended Minimum
1548
TEMPE BUTTE
POWER PLANT
MARICOPA
FREEWAY
For noise abatement intercept final east of Power Plant at or above 3100.
RADAR REQUIRED
Weather Minimums: 4500 feet ceiling and 8 mile visibility.
Routes depicted are reference only providing turns east of Power Plant.
SOUTH MOUNTAIN
3047
4000
Recommended
Complete base leg turn at or above 4000 feet.
Base leg east of the Power Plant.
1 NM 2 3 4 5 6 7 8 9 10 11 12 13 14 15 16
SW-4, 05 APR 2012 to 03 MAY 2012
SW-4, 05 APR 2012 to 03 MAY 2012
POWER PLANT VISUAL RWY 25R
When visual approaches to Runway 25R are in progress, clearances will be given utilizing in part the following phaseology:
"(IDENT) CLEARED FOR A POWER PLANT VISUAL RUNWAY 25R APPROACH"
POWER PLANT VISUAL RWY 25R
Amdt 3 09127
33°26'N-112°01'W
PHOENIX, ARIZONA
PHOENIX SKY HARBOR INTL (PHX)

Figure 275

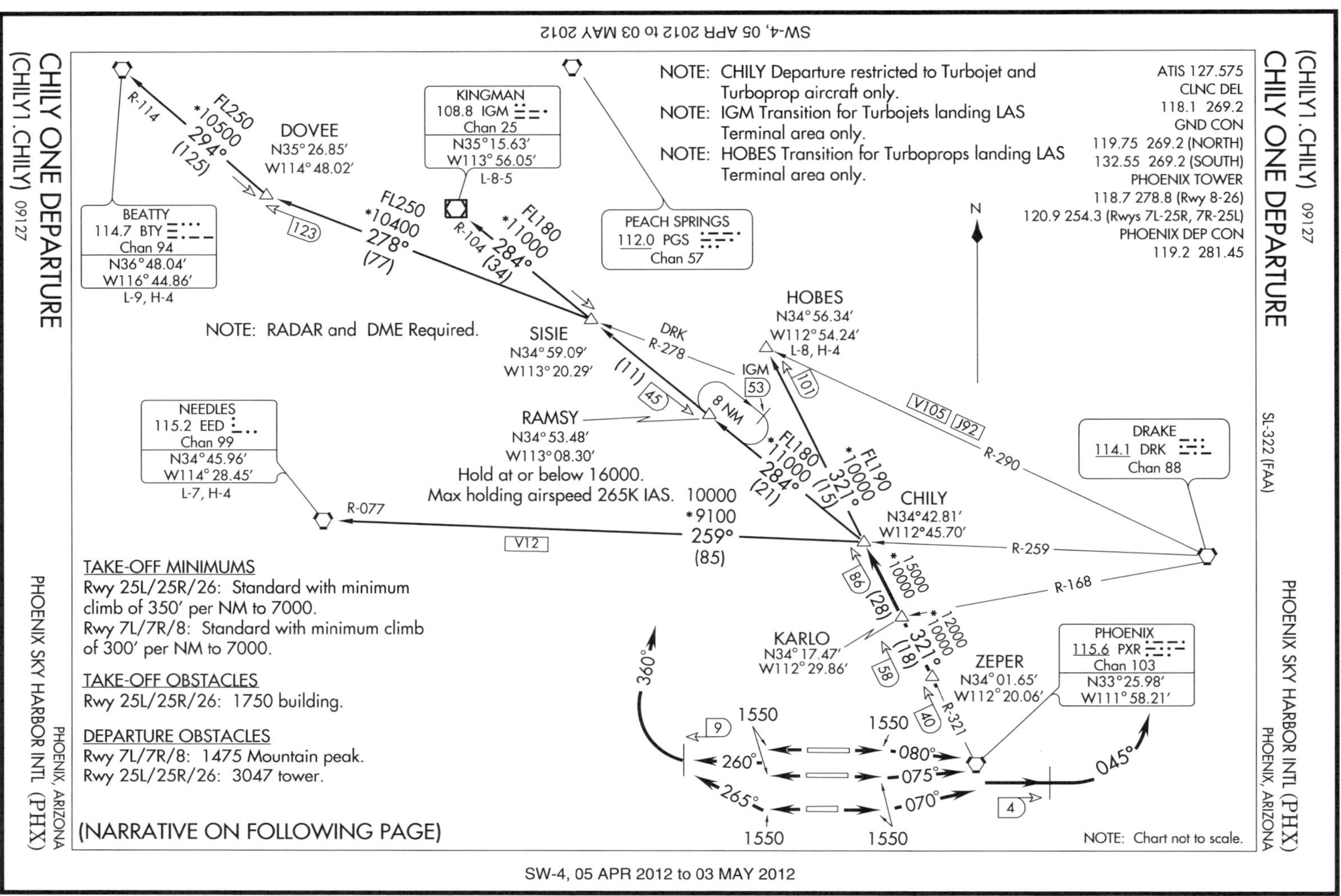
(CHILY1.CHILY) 09127
CHILY ONE DEPARTURE
SL-322 (FAA)
PHOENIX SKY HARBOR INTL (PHX)
PHOENIX, ARIZONA
ATIS 127.575
CLNC DEL
118.1 269.2
GND CON
119.75 269.2 (NORTH)
132.55 269.2 (SOUTH)
PHOENIX TOWER
118.7 278.8 (Rwy 8-26)
120.9 254.3 (Rwys 7L-25R, 7R-25L)
PHOENIX DEP CON
119.2 281.45
NOTE: CHILY Departure restricted to Turbojet and Turboprop aircraft only.
NOTE: IGM Transition for Turbojets landing LAS Terminal area only.
NOTE: HOBES Transition for Turboprops landing LAS Terminal area only.
N
NOTE: RADAR and DME Required.
BEATTY 114.7 BTY Chan 94 N36°48.04' W116°44.86' L-9, H-4
KINGMAN 108.8 IGM Chan 25 N35°15.63' W113°56.05' L-8-5
PEACH SPRINGS 112.0 PGS Chan 57
NEEDLES 115.2 EED Chan 99 N34°45.96' W114°28.45' L-7, H-4
DRAKE 114.1 DRK Chan 88
PHOENIX 115.6 PXR Chan 103 N33°25.98' W111°58.21'
DOVEE N35°26.85' W114°48.02'
SISIE N34°59.09' W113°20.29'
HOBES N34°56.34' W112°54.24' L-8, H-4
RAMSY N34°53.48' W113°08.30'
Hold at or below 16000.
Max holding airspeed 265K IAS.
CHILY N34°42.81' W112°45.70'
KARLO N34°17.47' W112°29.86'
ZEPER N34°01.65' W112°20.06'
FL250 *10500 294° (125)
FL250 *10400 278° (77)
FL180 *11000 284° (34)
FL180 *11000 284° (21)
FL190 *10000 321° (15)
10000 *9100 259° (85)
15000 *10000
(28)
12000 *10000 321° (18)
8 NM
(11)
R-114 R-104 R-278 R-290 R-259 R-168 R-077 R-321
DRK IGM
V105 J92 V12
360° 260° 265° 080° 075° 070° 045°
1550
TAKE-OFF MINIMUMS
Rwy 25L/25R/26: Standard with minimum climb of 350' per NM to 7000.
Rwy 7L/7R/8: Standard with minimum climb of 300' per NM to 7000.
TAKE-OFF OBSTACLES
Rwy 25L/25R/26: 1750 building.
DEPARTURE OBSTACLES
Rwy 7L/7R/8: 1475 Mountain peak.
Rwy 25L/25R/26: 3047 tower.
(NARRATIVE ON FOLLOWING PAGE)
NOTE: Chart not to scale.
SW-4, 05 APR 2012 to 03 MAY 2012

Figure 276

(CHILY1.CHILY) 02052

CHILY ONE DEPARTURE

SL-322 (FAA)

PHOENIX SKY HARBOR (PHX)
PHOENIX, ARIZONA

SW-4, 05 APR 2012 to 03 MAY 2012

DEPARTURE ROUTE DESCRIPTION

TAKE-OFF RUNWAY 7L: Climb runway heading to 1550, then climbing left turn heading 075°, at 4 DME east of PXR VORTAC, climbing left turn heading 045°. Thence....
TAKE-OFF RUNWAY 7R: Climb runway heading to 1550, then climbing left turn heading 070°, at 4 DME east of PXR VORTAC, climbing left turn heading 045°. Thence....
TAKE-OFF RUNWAY 8: Climb runway heading to 1550, then climbing right turn heading 080°, at 4 DME east of PXR VORTAC, climbing left turn heading 045°. Thence....
TAKE-OFF RUNWAY 25L: Climb runway heading to 1550, then climbing right turn heading 265°, at 9 DME west of PXR VORTAC, climbing right turn heading 360°. Thence....
TAKE-OFF RUNWAY 25R: Climb runway heading to 1550, then climbing right turn heading 260°, at 9 DME west of PXR VORTAC, climbing right turn heading 360°. Thence....
TAKE-OFF RUNWAY 26: Climb runway heading to 1550, then climbing right turn heading 260°, at 9 DME west of PXR VORTAC, climbing right turn heading 360°. Thence....

....maintain 7000. Expect radar vectors to PXR R-321 to ZEPER INT then CHILY INT. Then via (transition). Expect filed altitude 3 minutes after departure.

BEATTY TRANSITION (CHILY1.BTY): From over CHILY INT via IGM R-104 to SISIE INT, then via DRK R-278 to DOVEE INT, then via BTY R-114 to BTY VORTAC.
HOBES TRANSITION (CHILY1.HOBES): From over CHILY INT via PXR R-321 to HOBES INT.
KINGMAN TRANSITION (CHILY1.IGM): From over CHILY INT via IGM R-104 to IGM VOR/DME.
NEEDLES TRANSITION (CHILY1.EED): From over CHILY INT via DRK R-259 and EED R-077 to EED VORTAC.

SW-4, 05 APR 2012 to 03 MAY 2012

CHILY ONE DEPARTURE
(CHILY1.CHILY) 02052

PHOENIX, ARIZONA
PHOENIX SKY HARBOR (PHX)

Figure 277

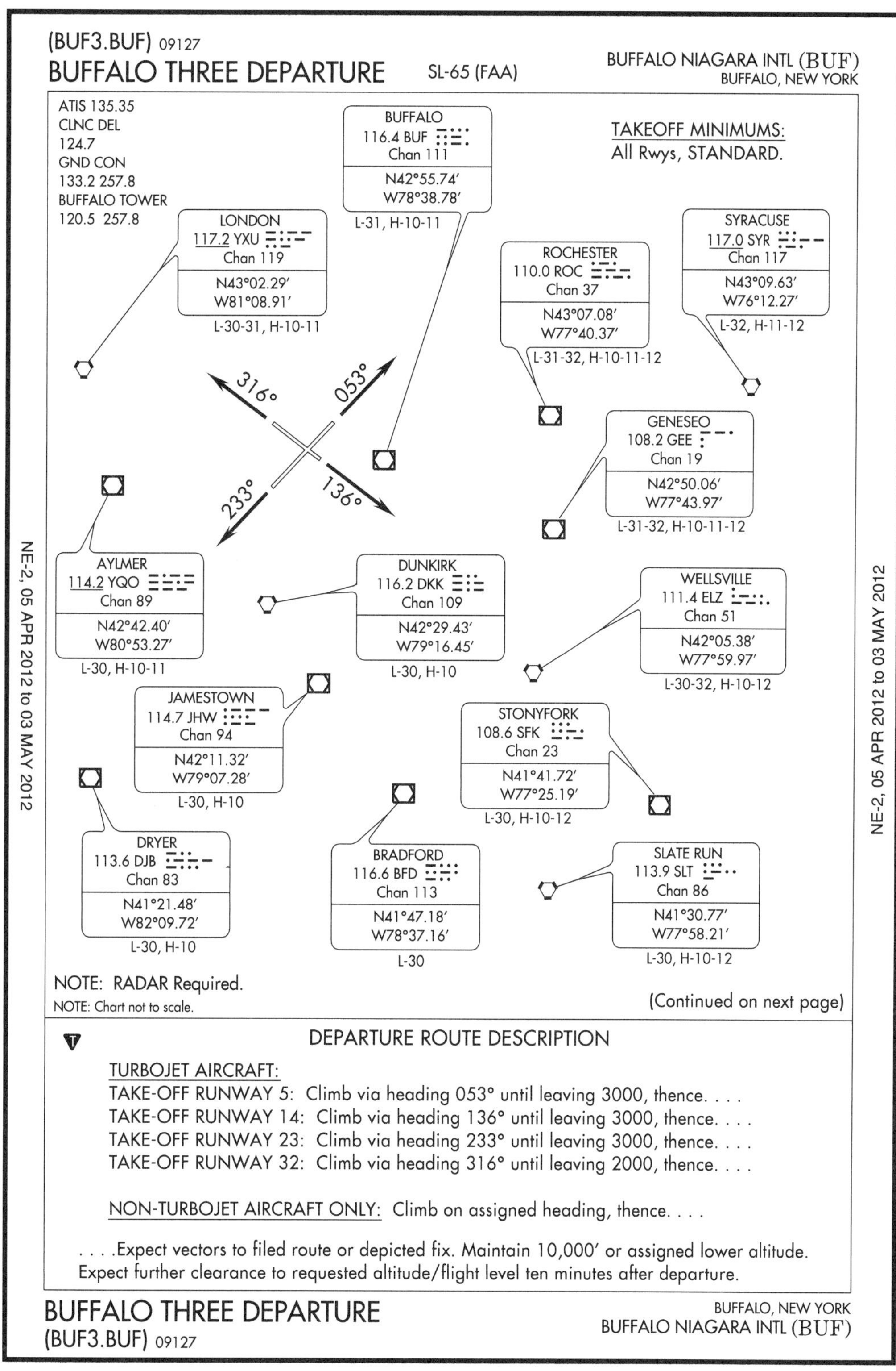

(BUF3.BUF) 09127
BUFFALO THREE DEPARTURE
SL-65 (FAA)
BUFFALO NIAGARA INTL (BUF)
BUFFALO, NEW YORK
ATIS 135.35
CLNC DEL
124.7
GND CON
133.2 257.8
BUFFALO TOWER
120.5 257.8
BUFFALO
116.4 BUF
Chan 111
N42°55.74′
W78°38.78′
L-31, H-10-11
TAKEOFF MINIMUMS:
All Rwys, STANDARD.
LONDON
117.2 YXU
Chan 119
N43°02.29′
W81°08.91′
L-30-31, H-10-11
ROCHESTER
110.0 ROC
Chan 37
N43°07.08′
W77°40.37′
L-31-32, H-10-11-12
SYRACUSE
117.0 SYR
Chan 117
N43°09.63′
W76°12.27′
L-32, H-11-12
316°
053°
233°
136°
GENESEO
108.2 GEE
Chan 19
N42°50.06′
W77°43.97′
L-31-32, H-10-11-12
AYLMER
114.2 YQO
Chan 89
N42°42.40′
W80°53.27′
L-30, H-10-11
DUNKIRK
116.2 DKK
Chan 109
N42°29.43′
W79°16.45′
L-30, H-10
WELLSVILLE
111.4 ELZ
Chan 51
N42°05.38′
W77°59.97′
L-30-32, H-10-12
JAMESTOWN
114.7 JHW
Chan 94
N42°11.32′
W79°07.28′
L-30, H-10
STONYFORK
108.6 SFK
Chan 23
N41°41.72′
W77°25.19′
L-30, H-10-12
DRYER
113.6 DJB
Chan 83
N41°21.48′
W82°09.72′
L-30, H-10
BRADFORD
116.6 BFD
Chan 113
N41°47.18′
W78°37.16′
L-30
SLATE RUN
113.9 SLT
Chan 86
N41°30.77′
W77°58.21′
L-30, H-10-12
NE-2, 05 APR 2012 to 03 MAY 2012
NOTE: RADAR Required.
NOTE: Chart not to scale.
(Continued on next page)
DEPARTURE ROUTE DESCRIPTION
TURBOJET AIRCRAFT:
TAKE-OFF RUNWAY 5: Climb via heading 053° until leaving 3000, thence. . . .
TAKE-OFF RUNWAY 14: Climb via heading 136° until leaving 3000, thence. . . .
TAKE-OFF RUNWAY 23: Climb via heading 233° until leaving 3000, thence. . . .
TAKE-OFF RUNWAY 32: Climb via heading 316° until leaving 2000, thence. . . .
NON-TURBOJET AIRCRAFT ONLY: Climb on assigned heading, thence. . . .
. . . .Expect vectors to filed route or depicted fix. Maintain 10,000′ or assigned lower altitude.
Expect further clearance to requested altitude/flight level ten minutes after departure.
BUFFALO THREE DEPARTURE
(BUF3.BUF) 09127
BUFFALO, NEW YORK
BUFFALO NIAGARA INTL (BUF)

Figure 278

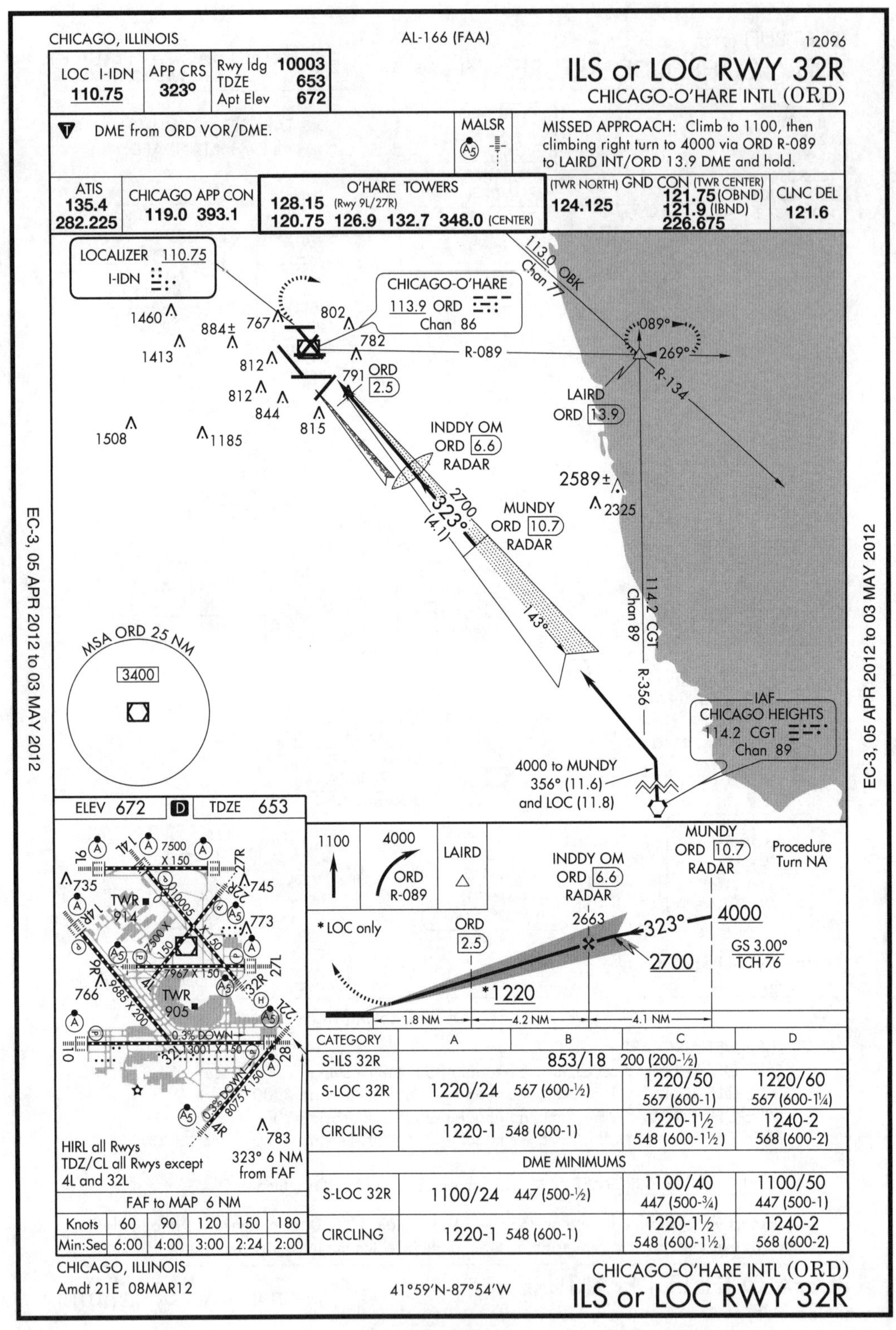

CATEGORY	A	B	C	D
S-ILS 32R	853/18 200 (200-½)			
S-LOC 32R	1220/24 567 (600-½)		1220/50 567 (600-1)	1220/60 567 (600-1¼)
CIRCLING	1220-1 548 (600-1)		1220-1½ 548 (600-1½)	1240-2 568 (600-2)
DME MINIMUMS				
S-LOC 32R	1100/24 447 (500-½)		1100/40 447 (500-¾)	1100/50 447 (500-1)
CIRCLING	1220-1 548 (600-1)		1220-1½ 548 (600-1½)	1240-2 568 (600-2)

Figure 279

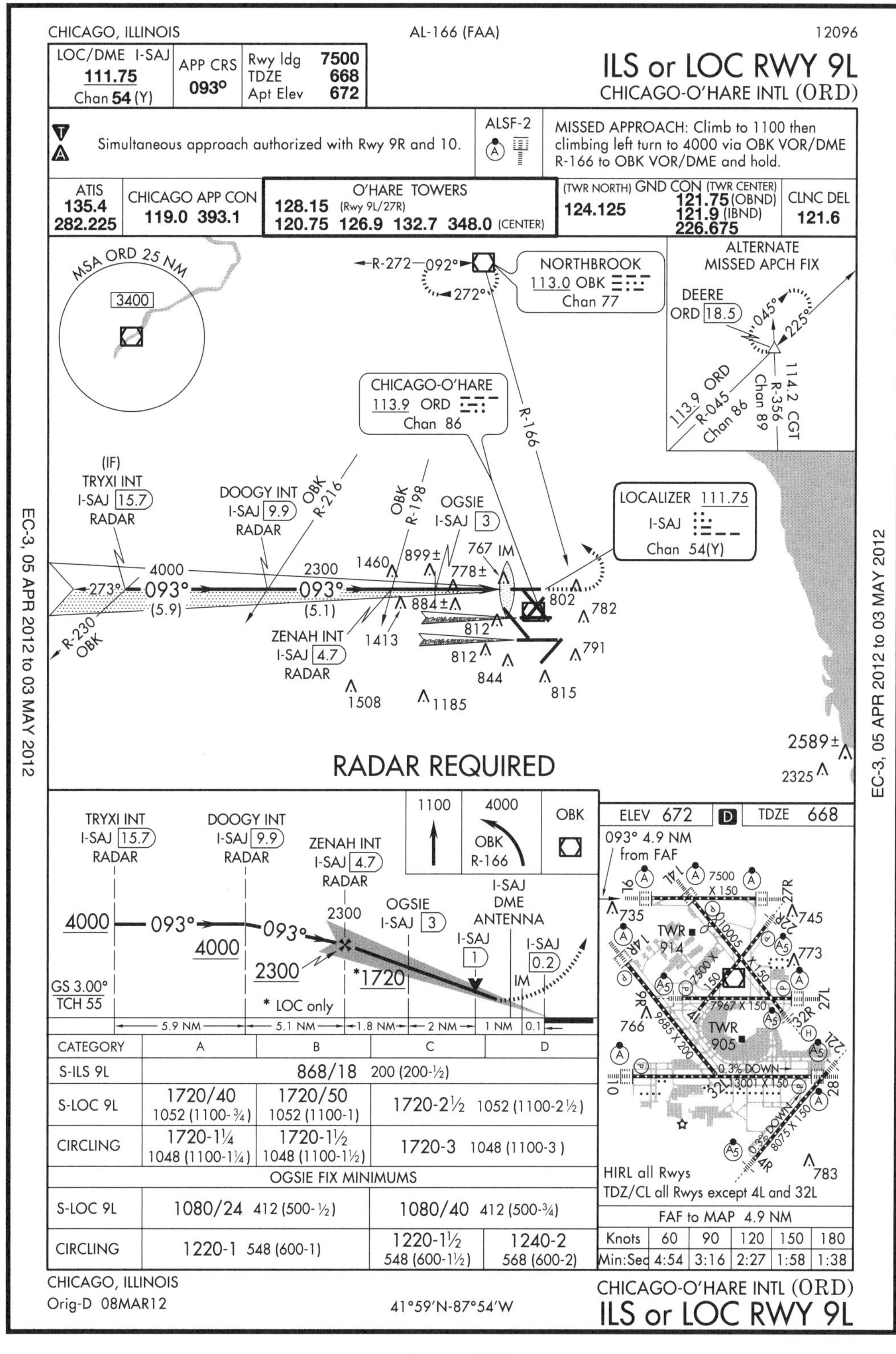

CATEGORY	A	B	C	D
S-ILS 9L	868/18 200 (200-½)			
S-LOC 9L	1720/40 1052 (1100-¾)	1720/50 1052 (1100-1)	1720-2½ 1052 (1100-2½)	
CIRCLING	1720-1¼ 1048 (1100-1¼)	1720-1½ 1048 (1100-1½)	1720-3 1048 (1100-3)	
OGSIE FIX MINIMUMS				
S-LOC 9L	1080/24 412 (500-½)		1080/40 412 (500-¾)	
CIRCLING	1220-1 548 (600-1)		1220-1½ 548 (600-1½)	1240-2 568 (600-2)

FAF to MAP 4.9 NM

Knots	60	90	120	150	180
Min:Sec	4:54	3:16	2:27	1:58	1:38

CHICAGO, ILLINOIS Orig-D 08MAR12 41°59′N-87°54′W CHICAGO-O'HARE INTL (ORD) ILS or LOC RWY 9L

Figure 280

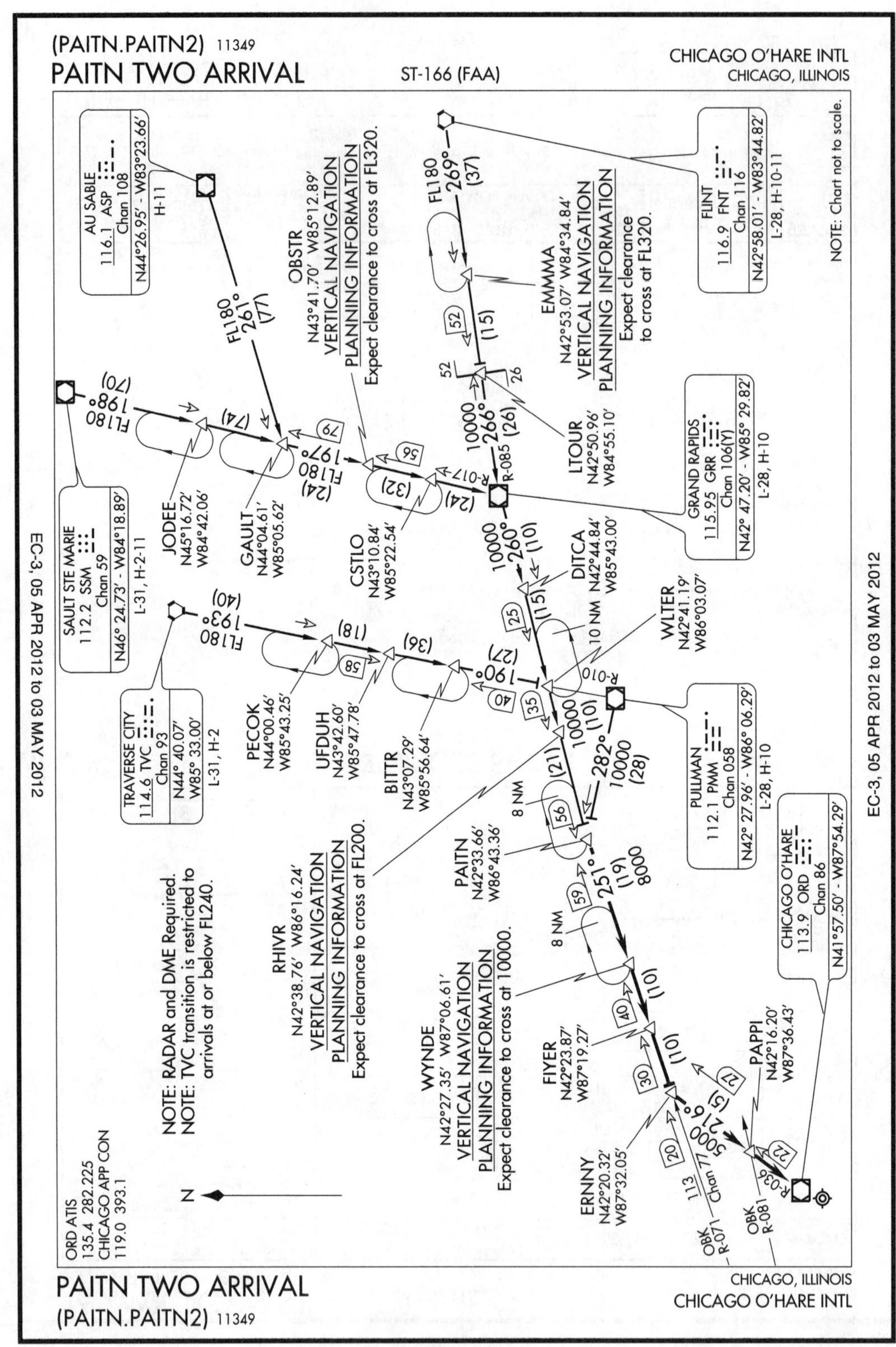
(PAITN.PAITN2) 11349
PAITN TWO ARRIVAL
ST-166 (FAA)
CHICAGO O'HARE INTL
CHICAGO, ILLINOIS
AU SABLE
116.1 ASP
Chan 108
N44°26.95' - W83°23.66'
H-11
FL180
261°
(77)
OBSTR
N43°41.70' W85°12.89'
VERTICAL NAVIGATION
PLANNING INFORMATION
Expect clearance to cross at FL320.
FL180
269°
(37)
EMMMA
N42°53.07' W84°34.84'
VERTICAL NAVIGATION
PLANNING INFORMATION
Expect clearance to cross at FL320.
FLINT
116.9 FNT
Chan 116
N42°58.01' - W83°44.82'
L-28, H-10-11
NOTE: Chart not to scale.
FL180
198°
(70)
JODEE
N45°16.72'
W84°42.06'
GAULT
N44°04.61'
W85°05.62'
FL180
197°
(24)
CSTLO
N43°10.84'
W85°22.54'
10000
266°
(26)
LTOUR
N42°50.96'
W84°55.10'
GRAND RAPIDS
115.95 GRR
Chan 106(Y)
N42° 47.20' - W85° 29.82'
L-28, H-10
SAULT STE MARIE
112.2 SSM
Chan 59
N46° 24.73' - W84°18.89'
L-31, H-2-11
10000
260°
(10)
DITCA
N42°44.84'
W85°43.00'
WLTER
N42°41.19'
W86°03.07'
FL180
193°
(40)
PECOK
N44°00.46'
W85°43.25'
UFDUH
N43°42.60'
W85°47.78'
BITTR
N43°07.29'
W85°56.64'
190°
TRAVERSE CITY
114.6 TVC
Chan 93
N44° 40.07'
W85° 33.00'
L-31, H-2
10000
282°
10000
(28)
PULLMAN
112.1 PMM
Chan 058
N42° 27.96' - W86° 06.29'
L-28, H-10
EC-3, 05 APR 2012 to 03 MAY 2012
NOTE: RADAR and DME Required.
NOTE: TVC transition is restricted to arrivals at or below FL240.
RHIVR
N42°38.76' W86°16.24'
VERTICAL NAVIGATION
PLANNING INFORMATION
Expect clearance to cross at FL200.
PAITN
N42°33.66'
W86°43.36'
251°
(19)
8000
8 NM
WYNDE
N42°27.35' W87°06.61'
VERTICAL NAVIGATION
PLANNING INFORMATION
Expect clearance to cross at 10000.
FIYER
N42°23.87'
W87°19.27'
CHICAGO O'HARE
113.9 ORD
Chan 86
N41°57.50' - W87°54.29'
ERNNY
N42°20.32'
W87°32.05'
PAPPI
N42°16.20'
W87°36.43'
5000
216°
ORD ATIS
135.4 282.225
CHICAGO APP CON
119.0 393.1
PAITN TWO ARRIVAL
(PAITN.PAITN2) 11349
CHICAGO, ILLINOIS
CHICAGO O'HARE INTL

Figure 281

(PAITN.PAITN2) 11013

PAITN TWO ARRIVAL ST-166 (FAA)

CHICAGO O'HARE INTL
CHICAGO, ILLINOIS

ARRIVAL ROUTE DESCRIPTION

AU SABLE TRANSITION (ASP.PAITN2): From over ASP VOR/DME via ASP R-261 to GAULT then via GRR R-017 to GRR VOR/DME then via GRR R-260 to PAITN. Thence....
FLINT TRANSITION (FNT.PAITN2): From over FNT VORTAC via FNT R-269 to LTOUR and GRR R-085 to GRR VOR/DME then via GRR R-260 to PAITN. Thence....
GRAND RAPIDS TRANSITION (GRR.PAITN2): From over GRR VOR/DME via GRR R-260 to PAITN. Thence....
PULLMAN TRANSITION (PMM.PAITN2): From over PMM VOR/DME via PMM R-282 to PAITN. Thence....
SAULT STE MARIE TRANSITION (SSM.PAITN2): From over SSM VOR/DME via SSM R-198 to GAULT then via GRR R-017 to GRR VOR/DME then via GRR R-260 to PAITN. Thence....
TRAVERSE CITY TRANSITION (TVC.PAITN2): From over TVC VORTAC via TVC R-193 to BITTR then via PMM R-010 to WLTER then via GRR R-260 to PAITN. Thence....

....From over PAITN via OBK VOR/DME R-071 to WYNDE/OBK 40 DME, then via OBK VOR/DME R-071 to FIYER/OBK 30 DME, then via OBK VOR/DME R-071 to ERNNY/OBK 20 DME, then via ORD VOR/DME R-036 to PAPPI/ORD 22 DME, then via ORD VOR/DME R-036 to ORD VOR/DME. Expect radar vectors to final approach course.

EC-3, 05 APR 2012 to 03 MAY 2012

EC-3, 05 APR 2012 to 03 MAY 2012

PAITN TWO ARRIVAL
(PAITN.PAITN2) 11013

CHICAGO, ILLINOIS
CHICAGO O'HARE INTL

Figure 282

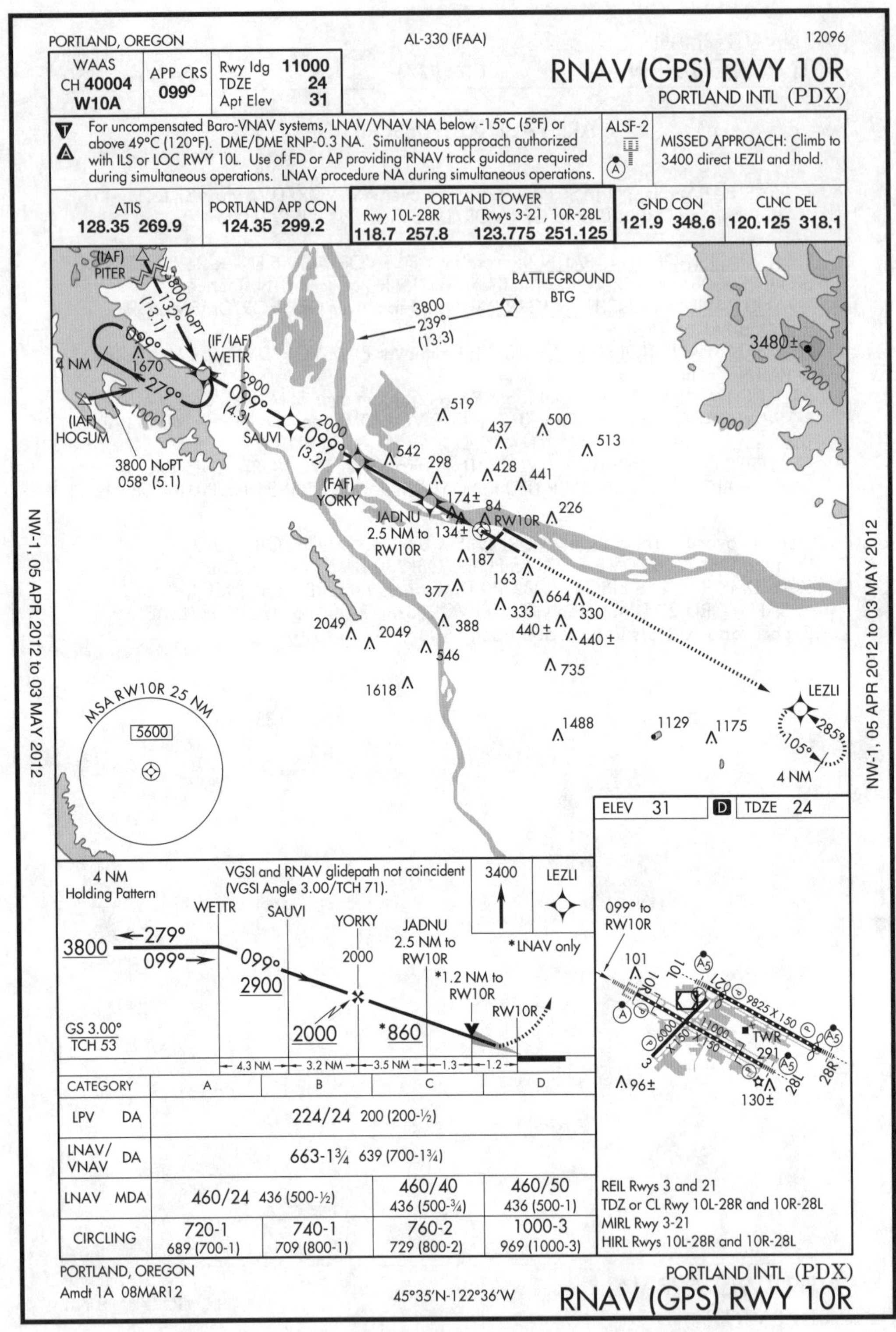

CATEGORY	A	B	C	D
LPV DA	224/24 200 (200-½)			
LNAV/VNAV DA	663-1¾ 639 (700-1¾)			
LNAV MDA	460/24 436 (500-½)		460/40 436 (500-¾)	460/50 436 (500-1)
CIRCLING	720-1 689 (700-1)	740-1 709 (800-1)	760-2 729 (800-2)	1000-3 969 (1000-3)

Figure 283

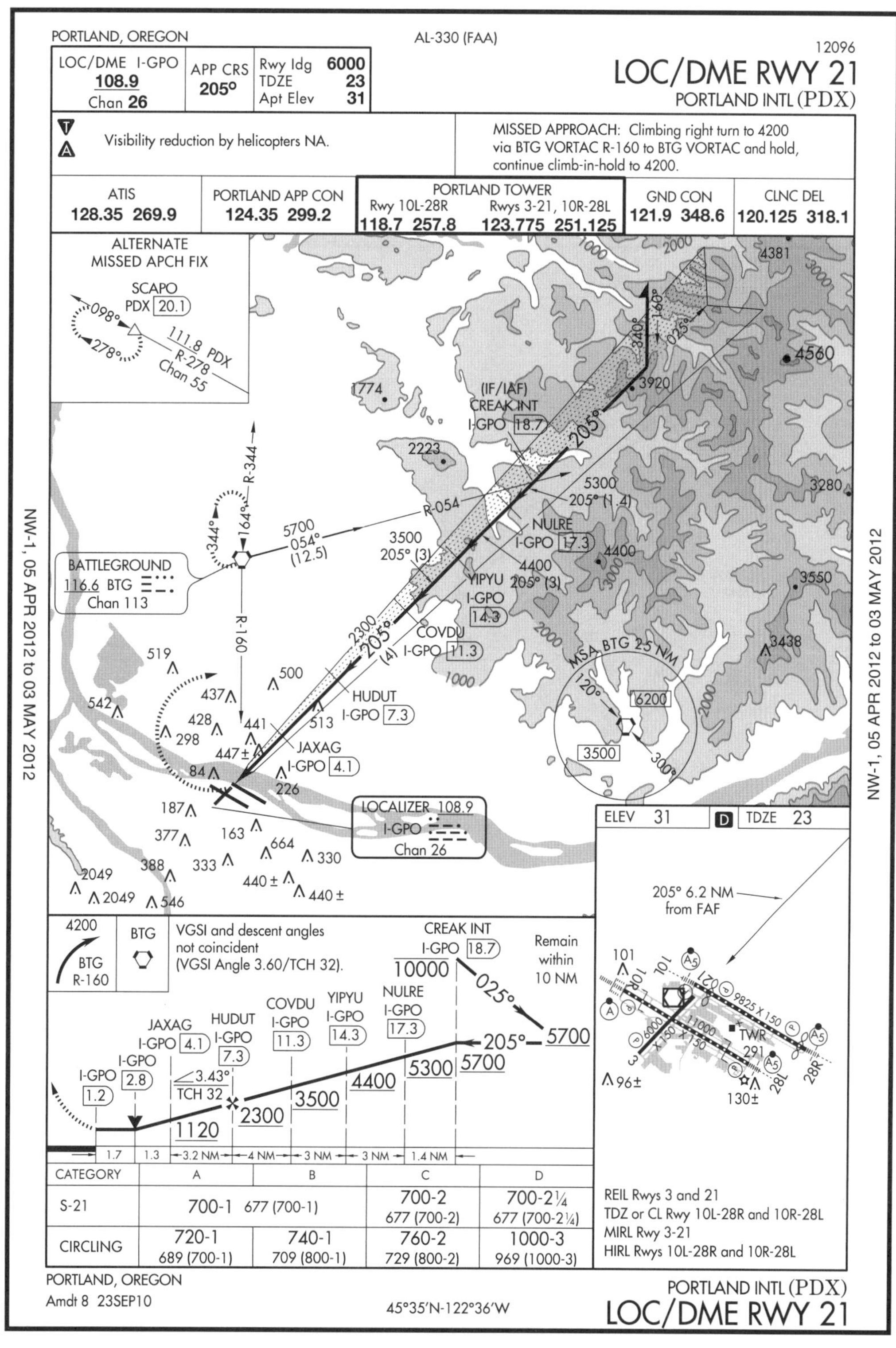
PORTLAND, OREGON
AL-330 (FAA)
12096
LOC/DME RWY 21
PORTLAND INTL (PDX)
LOC/DME I-GPO 108.9 Chan 26
APP CRS 205°
Rwy ldg 6000
TDZE 23
Apt Elev 31
Visibility reduction by helicopters NA.
MISSED APPROACH: Climbing right turn to 4200 via BTG VORTAC R-160 to BTG VORTAC and hold, continue climb-in-hold to 4200.
ATIS 128.35 269.9
PORTLAND APP CON 124.35 299.2
PORTLAND TOWER Rwy 10L-28R 118.7 257.8 Rwys 3-21, 10R-28L 123.775 251.125
GND CON 121.9 348.6
CLNC DEL 120.125 318.1
ALTERNATE MISSED APCH FIX
SCAPO PDX 20.1
111.8 PDX R-278 Chan 55
(IF/IAF) CREAK INT I-GPO 18.7
BATTLEGROUND 116.6 BTG Chan 113
NULRE I-GPO 17.3
YIPYU I-GPO 14.3
COVDU I-GPO 11.3
HUDUT I-GPO 7.3
JAXAG I-GPO 4.1
LOCALIZER 108.9 I-GPO Chan 26
MSA BTG 25 NM
ELEV 31
TDZE 23
205° 6.2 NM from FAF
VGSI and descent angles not coincident (VGSI Angle 3.60/TCH 32).
Remain within 10 NM
REIL Rwys 3 and 21
TDZ or CL Rwy 10L-28R and 10R-28L
MIRL Rwy 3-21
HIRL Rwys 10L-28R and 10R-28L
CATEGORY A B C D
S-21 700-1 677 (700-1) 700-2 677 (700-2) 700-2¼ 677 (700-2¼)
CIRCLING 720-1 689 (700-1) 740-1 709 (800-1) 760-2 729 (800-2) 1000-3 969 (1000-3)
PORTLAND, OREGON
Amdt 8 23SEP10
45°35′N-122°36′W
PORTLAND INTL (PDX)
LOC/DME RWY 21
NW-1, 05 APR 2012 to 03 MAY 2012

Figure 284

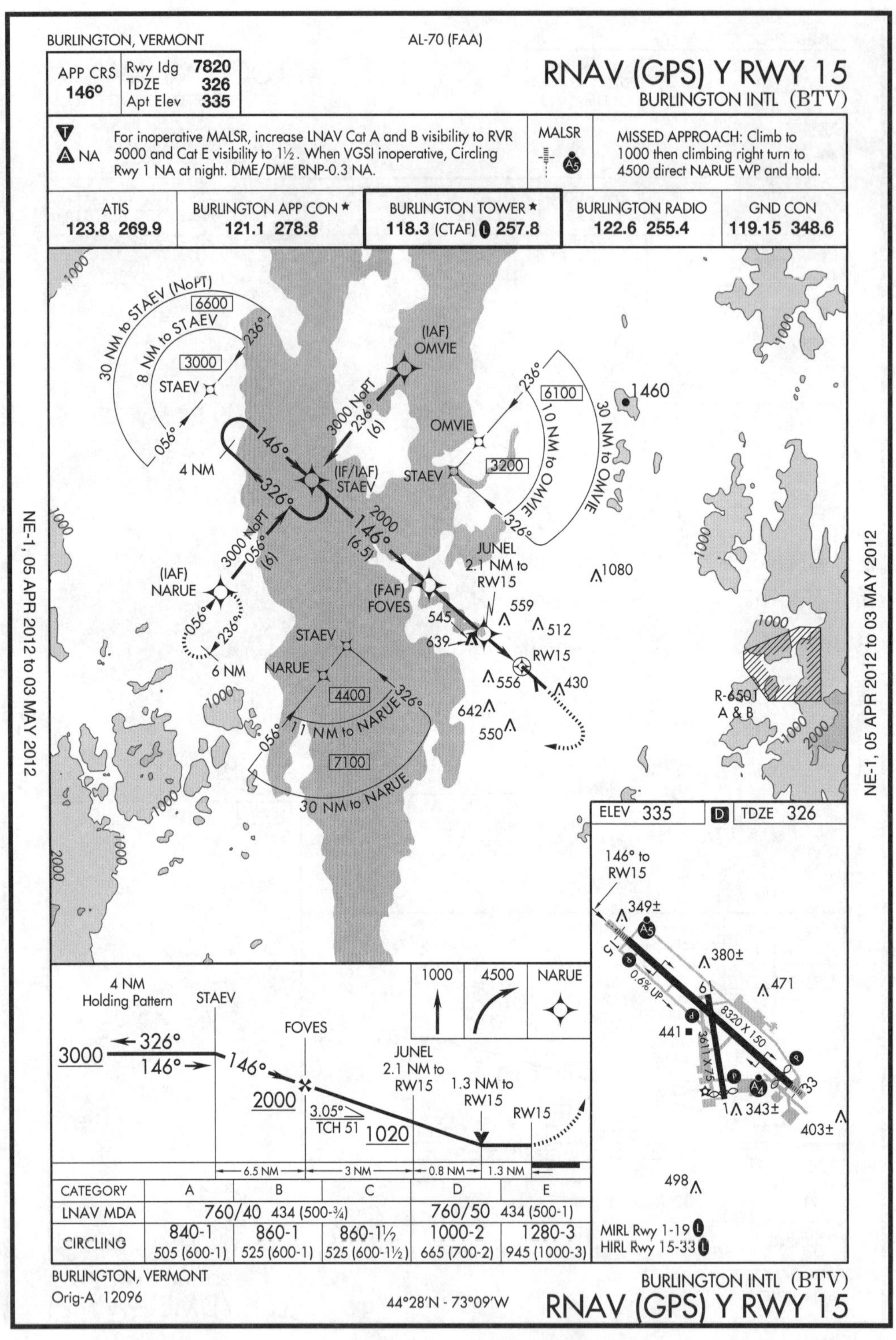

CATEGORY	A	B	C	D	E
LNAV MDA	760/40 434 (500-¾)			760/50 434 (500-1)	
CIRCLING	840-1 505 (600-1)	860-1 525 (600-1)	860-1½ 525 (600-1½)	1000-2 665 (700-2)	1280-3 945 (1000-3)

Figure 285

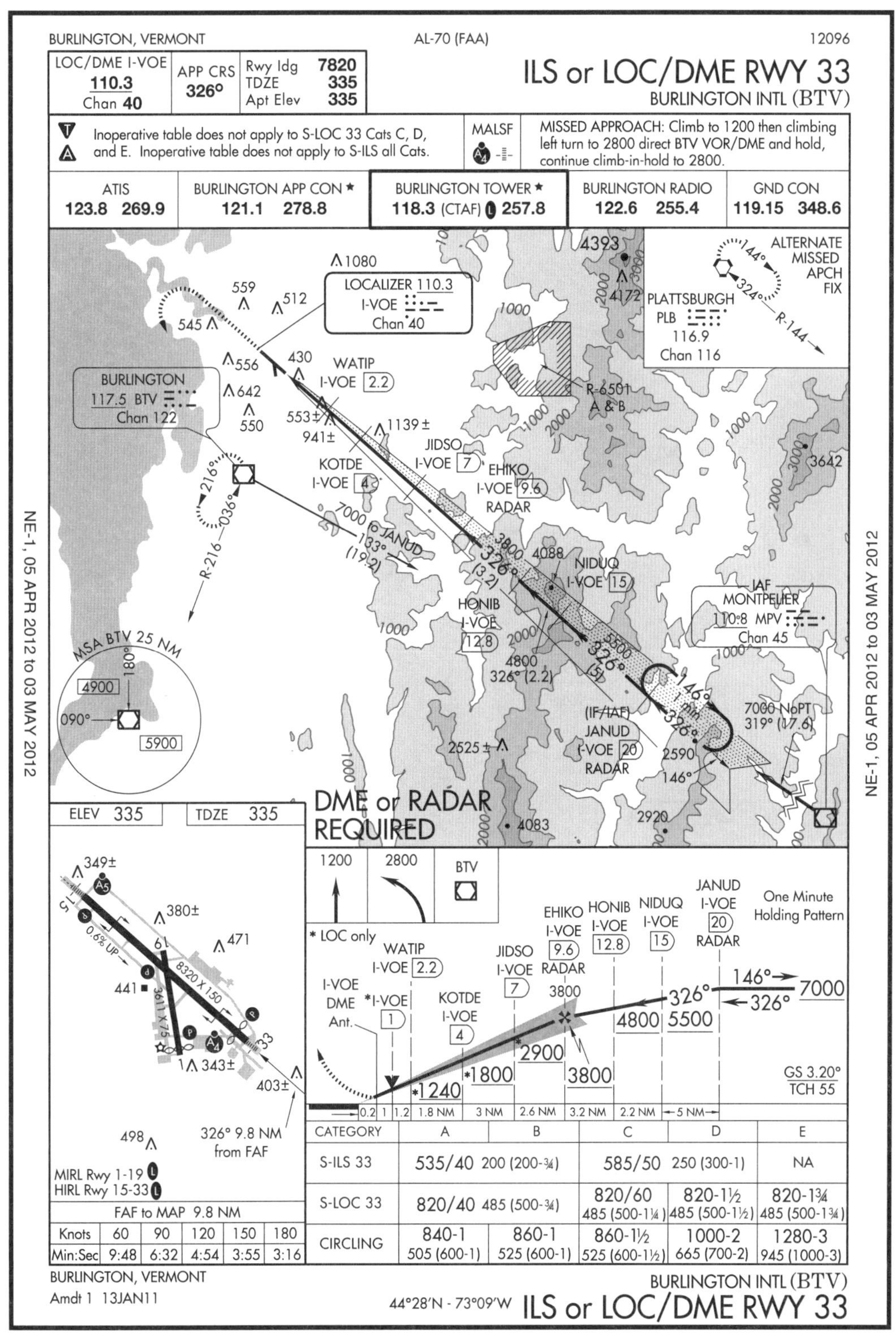

FAF to MAP 9.8 NM					
Knots	60	90	120	150	180
Min:Sec	9:48	6:32	4:54	3:55	3:16

CATEGORY	A	B	C	D	E
S-ILS 33	535/40 200 (200-¾)		585/50 250 (300-1)		NA
S-LOC 33	820/40 485 (500-¾)		820/60 485 (500-1¼)	820-1½ 485 (500-1½)	820-1¾ 485 (500-1¾)
CIRCLING	840-1 505 (600-1)	860-1 525 (600-1)	860-1½ 525 (600-1½)	1000-2 665 (700-2)	1280-3 945 (1000-3)

Figure 286

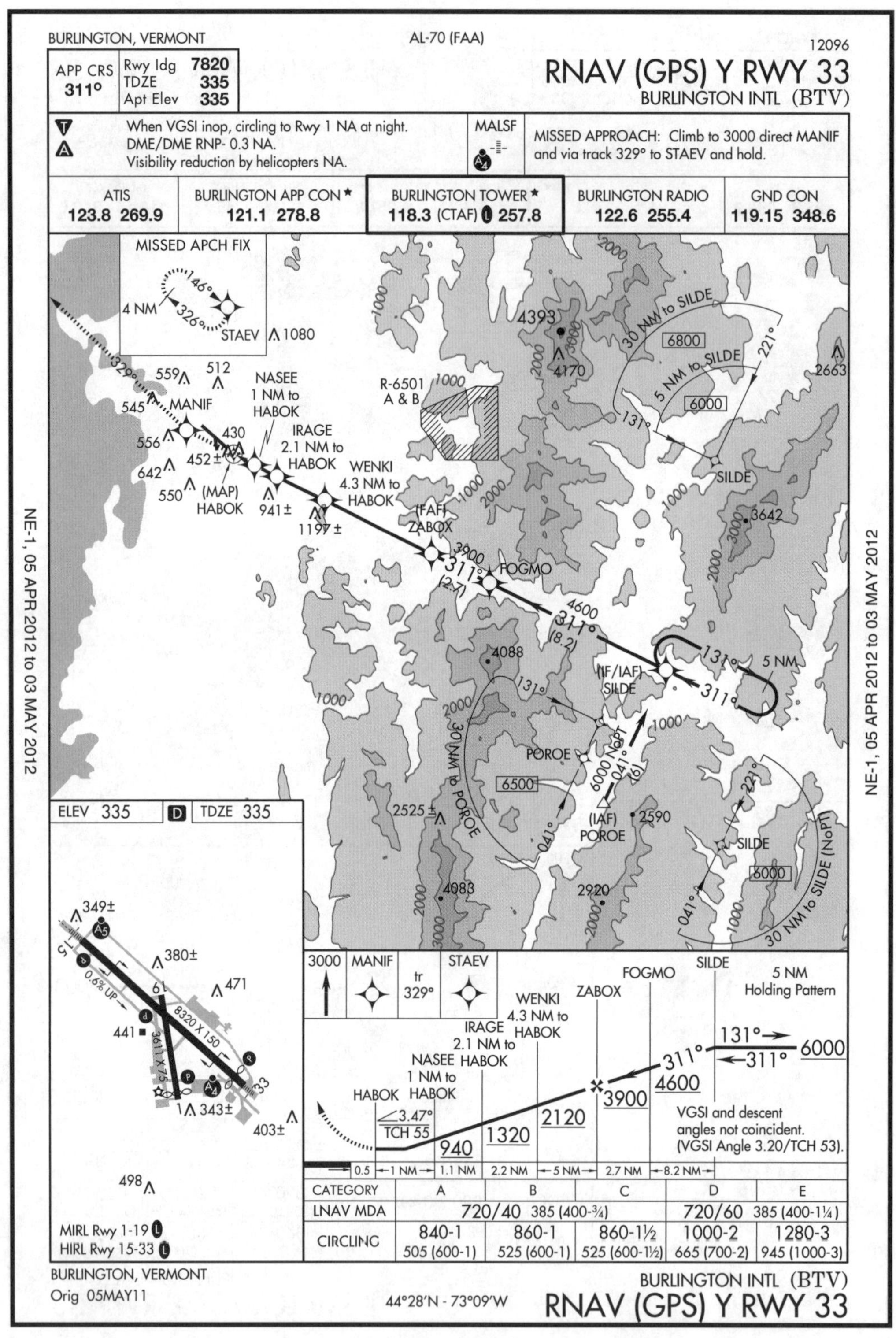

CATEGORY	A	B	C	D	E
LNAV MDA	720/40 385 (400-¾)			720/60 385 (400-1¼)	
CIRCLING	840-1 505 (600-1)	860-1 525 (600-1)	860-1½ 525 (600-1½)	1000-2 665 (700-2)	1280-3 945 (1000-3)

Figure 287

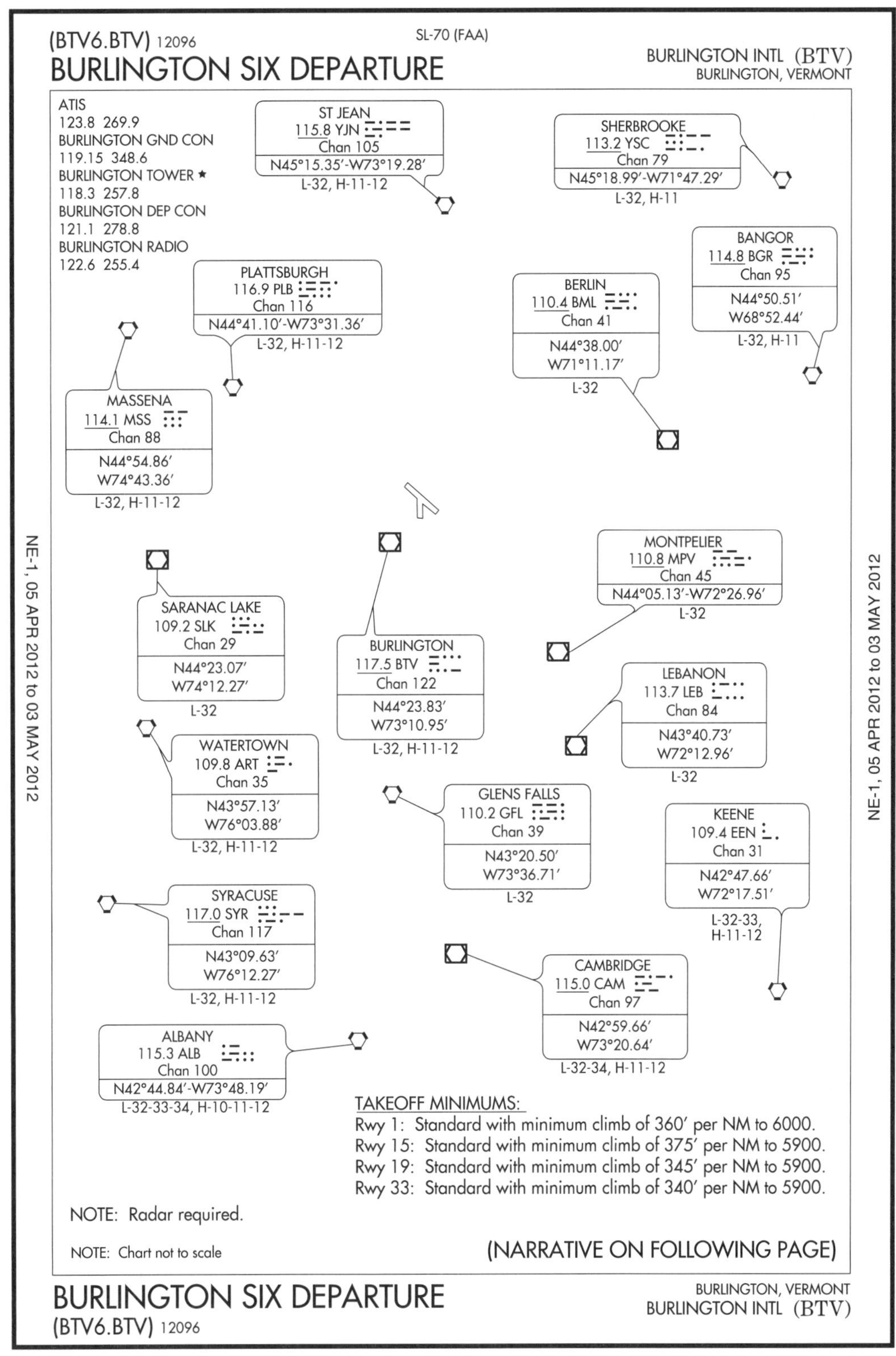
(BTV6.BTV) 12096
SL-70 (FAA)
BURLINGTON SIX DEPARTURE
BURLINGTON INTL (BTV)
BURLINGTON, VERMONT
ATIS
123.8 269.9
BURLINGTON GND CON
119.15 348.6
BURLINGTON TOWER ★
118.3 257.8
BURLINGTON DEP CON
121.1 278.8
BURLINGTON RADIO
122.6 255.4
ST JEAN
115.8 YJN
Chan 105
N45°15.35'-W73°19.28'
L-32, H-11-12
SHERBROOKE
113.2 YSC
Chan 79
N45°18.99'-W71°47.29'
L-32, H-11
BANGOR
114.8 BGR
Chan 95
N44°50.51'
W68°52.44'
L-32, H-11
PLATTSBURGH
116.9 PLB
Chan 116
N44°41.10'-W73°31.36'
L-32, H-11-12
BERLIN
110.4 BML
Chan 41
N44°38.00'
W71°11.17'
L-32
MASSENA
114.1 MSS
Chan 88
N44°54.86'
W74°43.36'
L-32, H-11-12
MONTPELIER
110.8 MPV
Chan 45
N44°05.13'-W72°26.96'
L-32
SARANAC LAKE
109.2 SLK
Chan 29
N44°23.07'
W74°12.27'
L-32
BURLINGTON
117.5 BTV
Chan 122
N44°23.83'
W73°10.95'
L-32, H-11-12
LEBANON
113.7 LEB
Chan 84
N43°40.73'
W72°12.96'
L-32
WATERTOWN
109.8 ART
Chan 35
N43°57.13'
W76°03.88'
L-32, H-11-12
GLENS FALLS
110.2 GFL
Chan 39
N43°20.50'
W73°36.71'
L-32
KEENE
109.4 EEN
Chan 31
N42°47.66'
W72°17.51'
L-32-33,
H-11-12
SYRACUSE
117.0 SYR
Chan 117
N43°09.63'
W76°12.27'
L-32, H-11-12
CAMBRIDGE
115.0 CAM
Chan 97
N42°59.66'
W73°20.64'
L-32-34, H-11-12
ALBANY
115.3 ALB
Chan 100
N42°44.84'-W73°48.19'
L-32-33-34, H-10-11-12
TAKEOFF MINIMUMS:
Rwy 1: Standard with minimum climb of 360' per NM to 6000.
Rwy 15: Standard with minimum climb of 375' per NM to 5900.
Rwy 19: Standard with minimum climb of 345' per NM to 5900.
Rwy 33: Standard with minimum climb of 340' per NM to 5900.
NOTE: Radar required.
NOTE: Chart not to scale
(NARRATIVE ON FOLLOWING PAGE)
BURLINGTON SIX DEPARTURE
(BTV6.BTV) 12096
BURLINGTON, VERMONT
BURLINGTON INTL (BTV)
NE-1, 05 APR 2012 to 03 MAY 2012
NE-1, 05 APR 2012 to 03 MAY 2012

Figure 288

(BTV6.BTV) 10210 SL-70 (FAA)

BURLINGTON SIX DEPARTURE

BURLINGTON INTL (BTV)
BURLINGTON, VERMONT

DEPARTURE ROUTE DESCRIPTION

TAKEOFF RUNWAYS 1, 15, 19, 33: Climb on assigned heading for vectors to filed navaid, fix, or airway to 10000 or assigned lower altitude. Expect filed altitude ten minutes after departure.

TAKEOFF OBSTACLE NOTES:

Rwy 1: Trees beginning 1396′ from DER, 216′ right of centerline, up to 64′ AGL/384′ MSL.
Trees 1694′ from DER, 200′ left of centerline, up to 80′ AGL/380′ MSL.

Rwy 15: Bush 318′ from DER, 292′ left of centerline, up to 23′ AGL/343′ MSL.
Trees beginning 1418′ from DER, 358′ right of centerline, up to 27′ AGL/387′ MSL.
Hopper and trees beginning 1801′ from DER, 377′ left of centerline, up to 63′ AGL/403′ MSL.
Building 3453′ from DER, 1145′ left of centerline, 110′ AGL/430′ MSL.

Rwy 19: Trees beginning 168′ from DER, 24′ right of centerline, up to 56′ AGL/436′ MSL.
Trees beginning 172′ from DER, 184′ left of centerline, up to 93′ AGL/413′ MSL.

Rwy 33: Pole and trees beginning 971′ from DER, 755′ left of centerline, up to 97′ AGL/357′ MSL.
Trees beginning 1091′ from DER, 590′ right of centerline, up to 34′ AGL/334′ MSL.

NE-1, 05 APR 2012 to 03 MAY 2012

NE-1, 05 APR 2012 to 03 MAY 2012

BURLINGTON SIX DEPARTURE
(BTV6.BTV) 10210
BURLINGTON, VERMONT
BURLINGTON INTL (BTV)

Figure 289

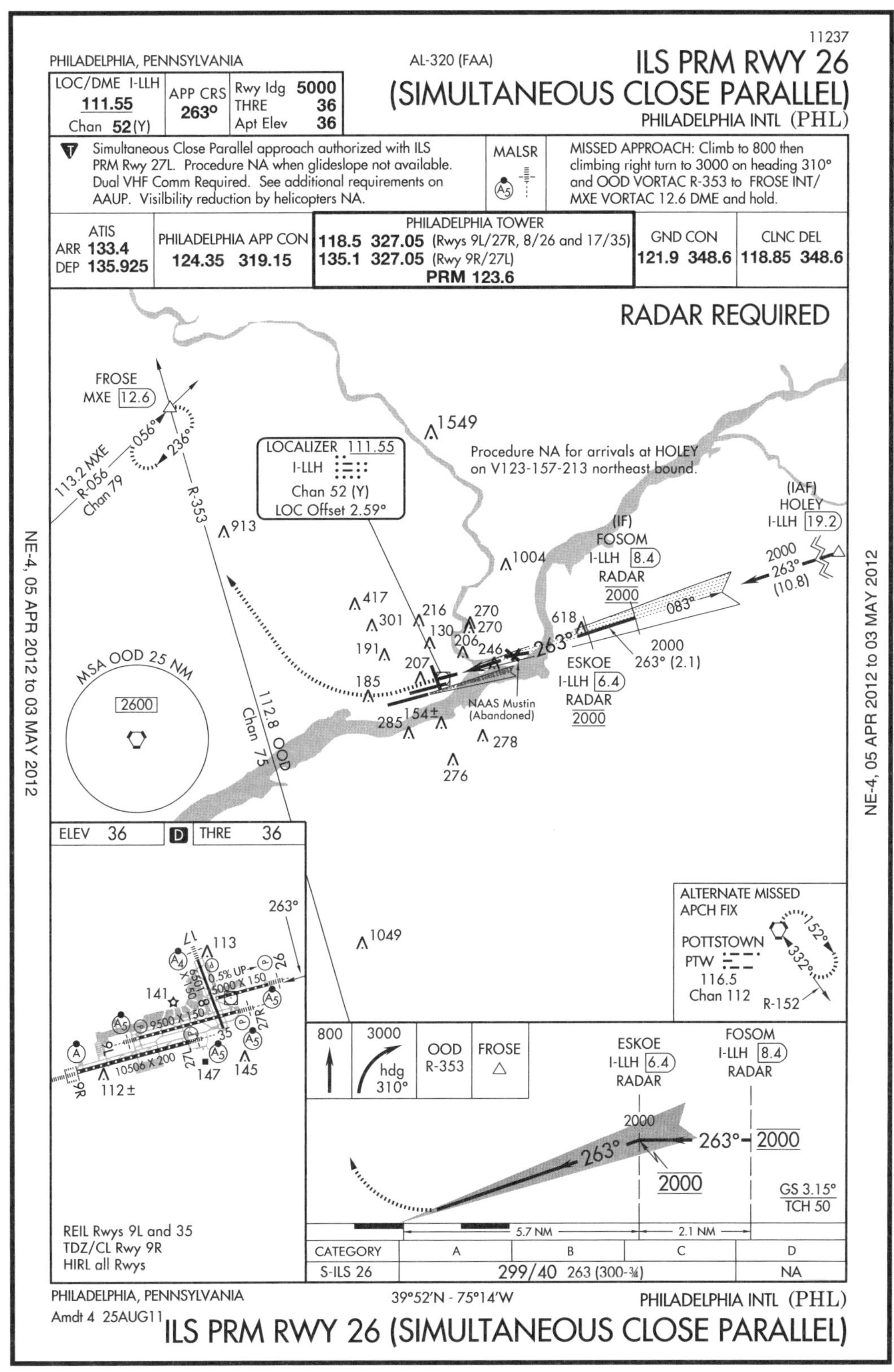

CATEGORY	A	B	C	D
S-ILS 26	299/40 263 (300-¾)			NA

Figure 290

ILS PRM RWY 26 Amdt 4 11237
(SIMULTANEOUS CLOSE PARALLEL) AL-320 (FAA)

PHILADELPHIA INTL (PHL)
PHILADELPHIA, PENNSYLVANIA

ATTENTION ALL USERS OF ILS PRECISION RUNWAY MONITOR (PRM)

NE-4, 05 APR 2012 to 03 MAY 2012

Condensed Briefing Point:
*When instructed, immediately switch to the tower frequency and select the monitor frequency audio.

1. **ATIS.** When the ATIS broadcast advises that simultaneous ILS/PRM and LDA/PRM approaches are in progress, pilots should brief to fly the ILS/PRM 26 approach. If later advised to expect an ILS 26 approach, the ILS/PRM 26 chart may be used after completing the following briefing items:
   (a) Minimums and missed approach procedures are unchanged.
   (b) Monitor frequency no longer required.

2. **Dual VHF Communication required.** To avoid blocked transmissions, each runway will have two frequencies, a primary and a monitor frequency. The tower controller will transmit on both frequencies. The monitor controller's transmissions, if needed, will override both frequencies. Pilots will ONLY transmit on the tower controller's frequency, but will listen to both frequencies. Select the monitor frequency audio only when instructed by ATC to contact the tower. The volume levels should be set about the same on both radios so that the pilots will be able to hear transmissions on at least one frequency if the other is blocked.

3. **ALL "Breakouts"** are to be hand flown to assure that the manuever is accomplished in the shortest amount of time. Pilots, when directed by ATC to break off an approach, must assume that an aircraft is blundering toward their course and a breakout must be initiated immediately.

   (a) ATC Directed "Breakouts": ATC directed breakouts will consist of a turn and a climb or descent. Pilots must always initiate the breakout in response to an air traffic controller instruction. Controllers will give a descending breakout only when there are no other reasonable options available, but in no case will the descent be below minimum vectoring altitude (MVA) which provides at least 1000 feet required obstruction clearance. The MVA in the final approach segment is 1800 feet at Philadelphia Intl Airport.

   (b) Phraseology - "TRAFFIC ALERT": If an aircraft enters the "NO TRANSGRESSION ZONE" (NTZ), the controller will breakout the threatened aircraft on the adjacent approach. The phraseology for the breakout will be:

   "TRAFFIC ALERT, (aircraft call sign) TURN (left/right) IMMEDIATELY, HEADING (degrees), CLIMB/DESCEND AND MAINTAIN (altitude)".

4. **ILS Navigation** Decending on ILS glideslope ensures complying with any charted crossing restrictions.

Special pilot training required. Pilots who are unable to participate will be afforded appropriate arrival services as operational conditions permit and must notify the controlling ARTCC as soon as practical, but at least 100 miles from destination.

NE-4, 05 APR 2012 to 03 MAY 2012

(SIMULTANEOUS CLOSE PARALLEL) 39° 52'N-75° 14'W
ILS PRM RWY 26 Amdt 4 11237

PHILADELPHIA, PENNSYLVANIA
PHILADELPHIA INTL (PHL)

Figure 291

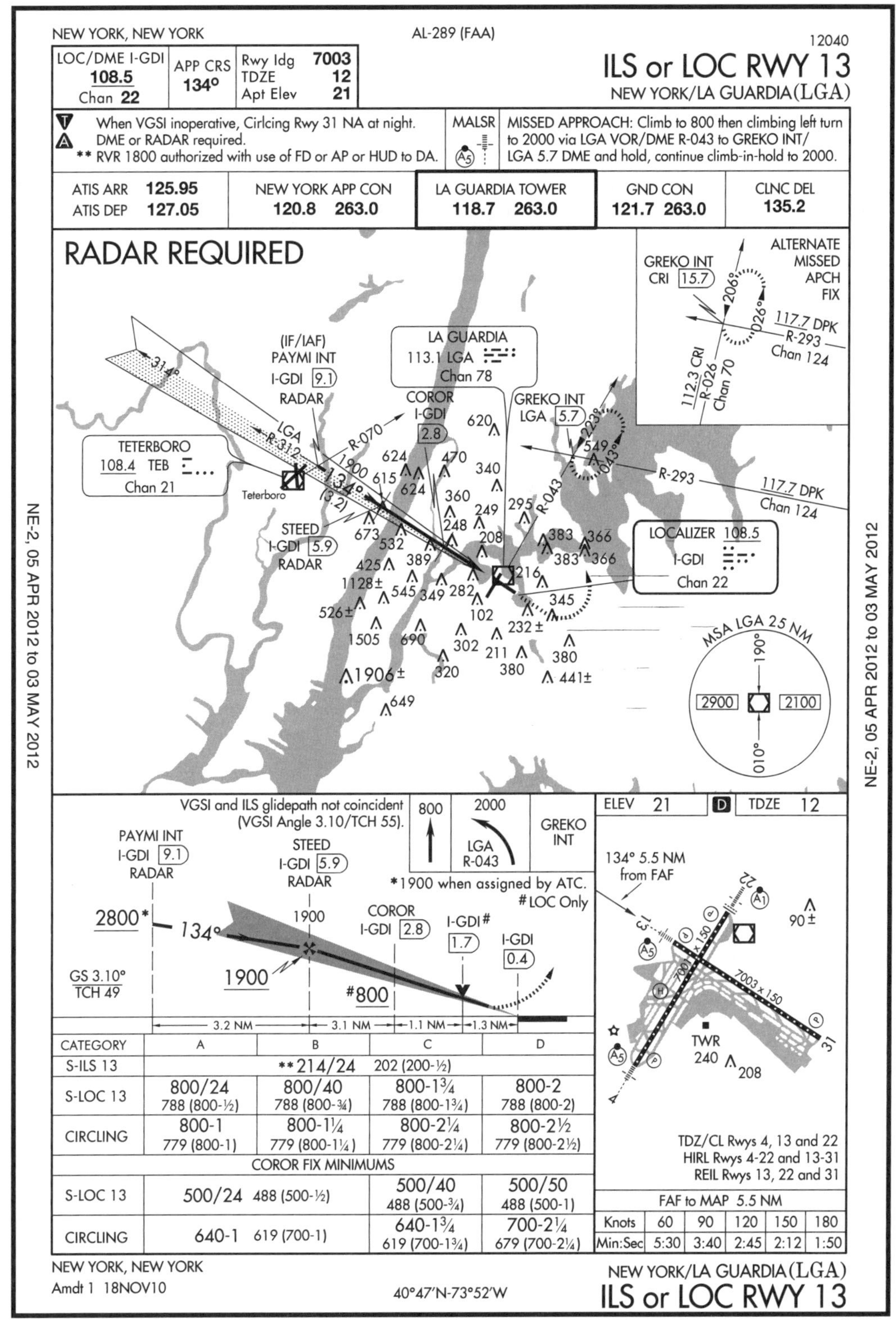

CATEGORY	A	B	C	D
S-ILS 13	**214/24 202 (200-½)			
S-LOC 13	800/24 788 (800-½)	800/40 788 (800-¾)	800-1¾ 788 (800-1¾)	800-2 788 (800-2)
CIRCLING	800-1 779 (800-1)	800-1¼ 779 (800-1¼)	800-2¼ 779 (800-2¼)	800-2½ 779 (800-2½)
COROR FIX MINIMUMS				
S-LOC 13	500/24 488 (500-½)		500/40 488 (500-¾)	500/50 488 (500-1)
CIRCLING	640-1 619 (700-1)		640-1¾ 619 (700-1¾)	700-2¼ 679 (700-2¼)

FAF to MAP 5.5 NM

Knots	60	90	120	150	180
Min:Sec	5:30	3:40	2:45	2:12	1:50

Figure 292

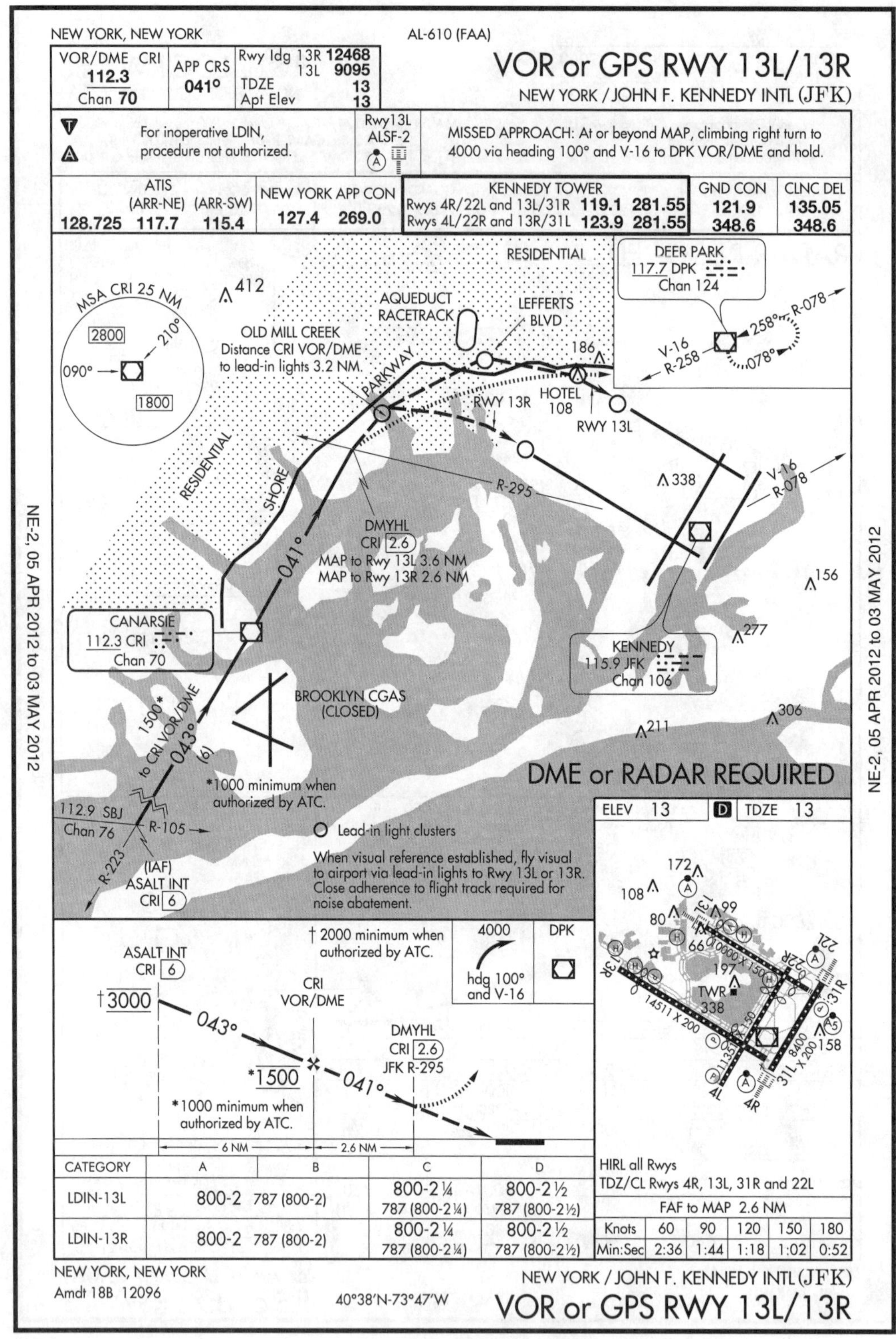

CATEGORY	A	B	C	D
LDIN-13L	800-2 787 (800-2)		800-2 ¼ 787 (800-2 ¼)	800-2 ½ 787 (800-2 ½)
LDIN-13R	800-2 787 (800-2)		800-2 ¼ 787 (800-2 ¼)	800-2 ½ 787 (800-2 ½)

HIRL all Rwys
TDZ/CL Rwys 4R, 13L, 31R and 22L

FAF to MAP 2.6 NM					
Knots	60	90	120	150	180
Min:Sec	2:36	1:44	1:18	1:02	0:52

NEW YORK, NEW YORK
Amdt 18B 12096
40°38'N-73°47'W
NEW YORK / JOHN F. KENNEDY INTL (JFK)
VOR or GPS RWY 13L/13R

NE-2, 05 APR 2012 to 03 MAY 2012

Figure 293

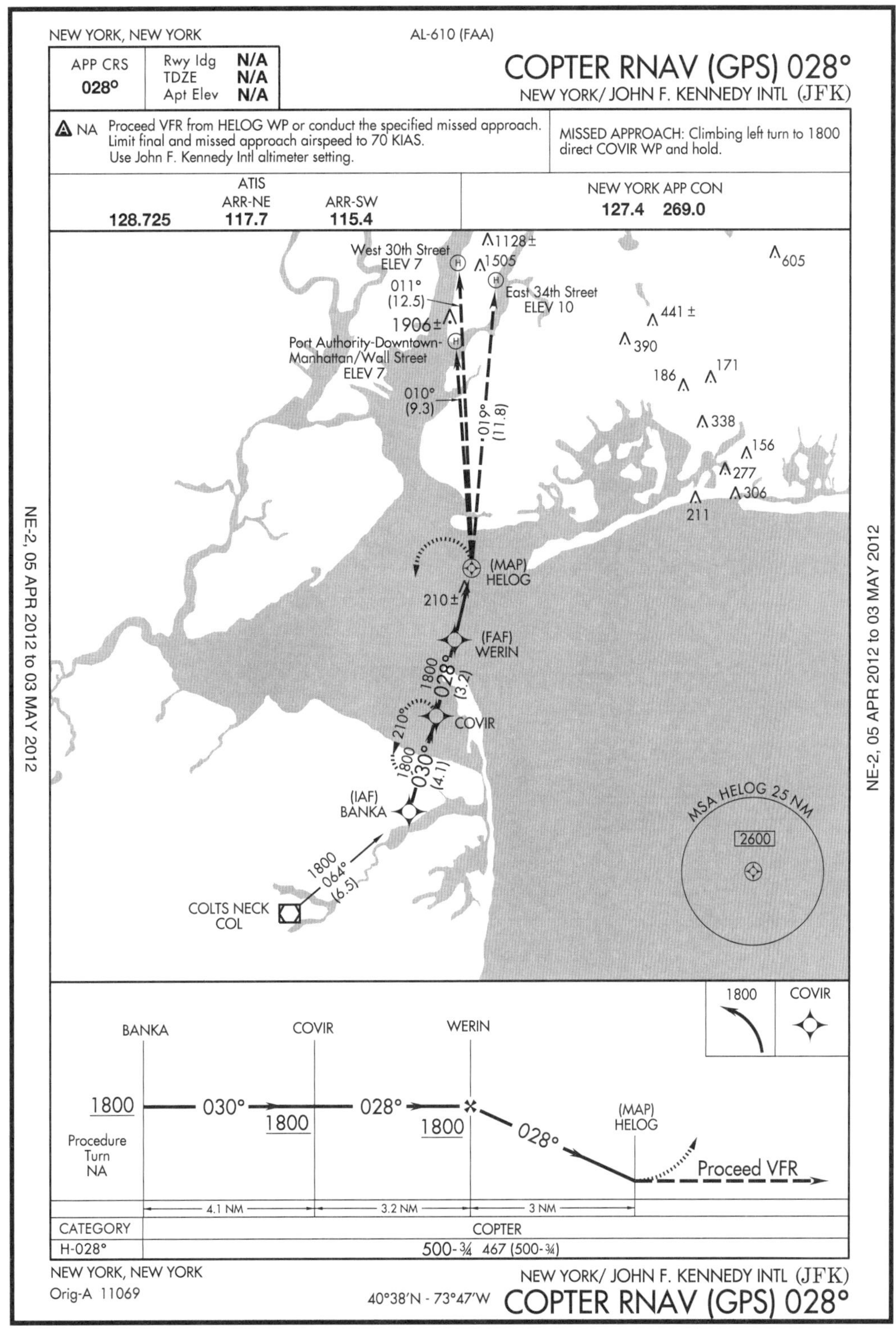
NEW YORK, NEW YORK
AL-610 (FAA)
APP CRS 028°
Rwy ldg N/A
TDZE N/A
Apt Elev N/A
COPTER RNAV (GPS) 028°
NEW YORK/ JOHN F. KENNEDY INTL (JFK)
NA Proceed VFR from HELOG WP or conduct the specified missed approach. Limit final and missed approach airspeed to 70 KIAS. Use John F. Kennedy Intl altimeter setting.
MISSED APPROACH: Climbing left turn to 1800 direct COVIR WP and hold.
ATIS 128.725 ARR-NE 117.7 ARR-SW 115.4
NEW YORK APP CON 127.4 269.0
West 30th Street ELEV 7
East 34th Street ELEV 10
Port Authority-Downtown-Manhattan/Wall Street ELEV 7
011° (12.5)
010° (9.3)
019° (11.8)
1906±
1128±
1505
605
441 ±
390
186
171
338
156
277
306
211
(MAP) HELOG
210±
(FAF) WERIN
1800 028° (3.2)
COVIR
210°
1800 030° (4.1)
(IAF) BANKA
1800 064° (6.5)
COLTS NECK COL
MSA HELOG 25 NM
2600
NE-2, 05 APR 2012 to 03 MAY 2012
1800
COVIR
BANKA
WERIN
1800
030°
028°
028°
Procedure Turn NA
(MAP) HELOG
Proceed VFR
4.1 NM
3.2 NM
3 NM
CATEGORY COPTER
H-028° 500-¾ 467 (500-¾)
Orig-A 11069
40°38'N - 73°47'W

Figure 294

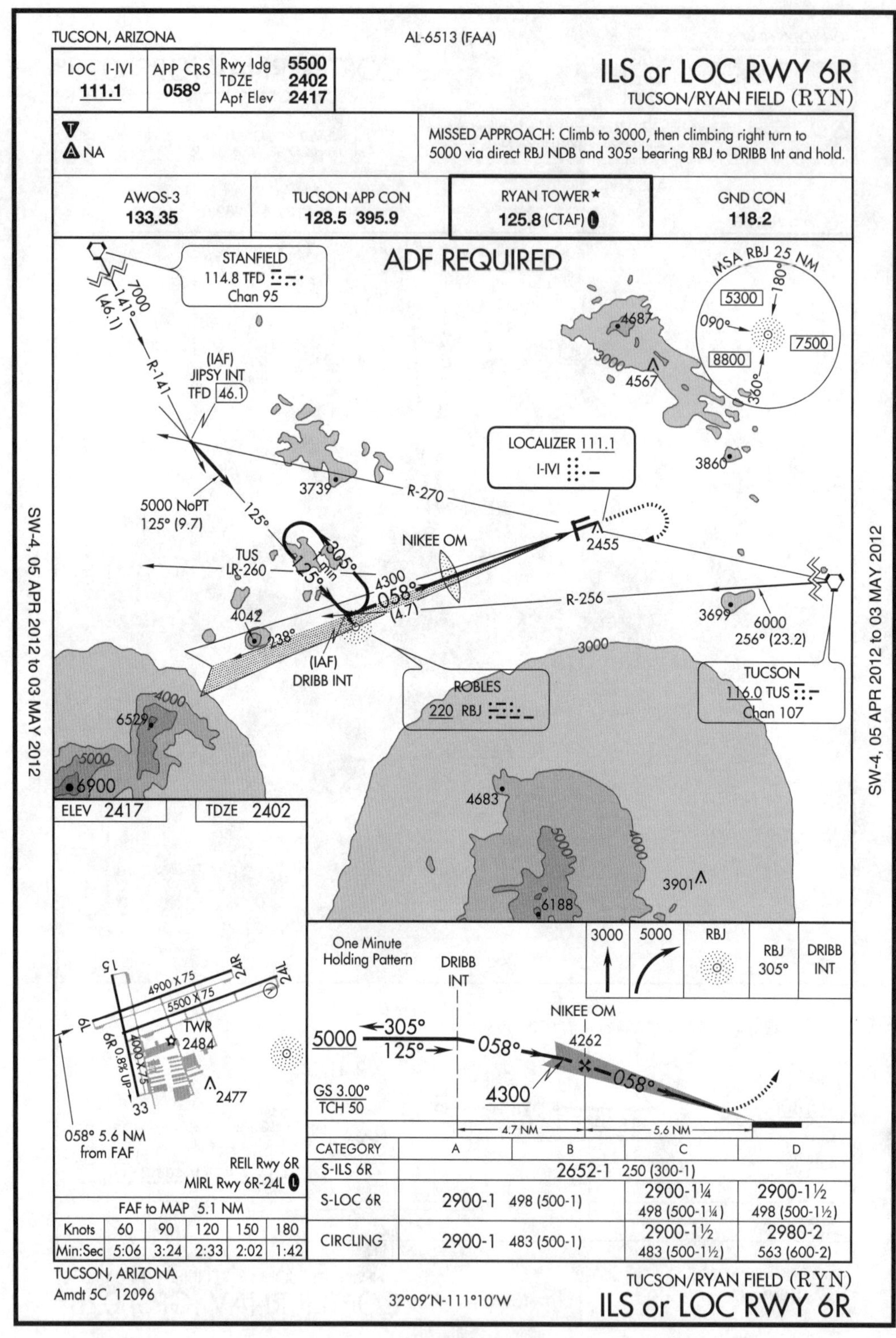

TUCSON, ARIZONA
AL-6513 (FAA)
LOC I-IVI 111.1
APP CRS 058°
Rwy ldg 5500
TDZE 2402
Apt Elev 2417
ILS or LOC RWY 6R
TUCSON/RYAN FIELD (RYN)
NA
MISSED APPROACH: Climb to 3000, then climbing right turn to 5000 via direct RBJ NDB and 305° bearing RBJ to DRIBB Int and hold.
AWOS-3 133.35
TUCSON APP CON 128.5 395.9
RYAN TOWER★ 125.8 (CTAF)
GND CON 118.2
ADF REQUIRED
STANFIELD 114.8 TFD Chan 95
7000 141° (46.1)
R-141
(IAF) JIPSY INT TFD 46.1
MSA RBJ 25 NM
5300
7500
8800
180°
090°
360°
4687
4567
3000
3860
LOCALIZER 111.1 I-IVI
3739
R-270
5000 NoPT 125° (9.7)
125°
305°
1 Min
NIKEE OM
2455
TUS LR-260
4300
058°
(4.7)
R-256
3699
6000 256° (23.2)
4042
238°
(IAF) DRIBB INT
ROBLES 220 RBJ
TUCSON 116.0 TUS Chan 107
4000
6529
5000
6900
4683
3901
6188
ELEV 2417
TDZE 2402
4900 X 75
5500 X 75
4000 X 75
TWR 2484
2477
6R 0.8% UP
24R
24L
15
33
058° 5.6 NM from FAF
REIL Rwy 6R
MIRL Rwy 6R-24L
FAF to MAP 5.1 NM
Knots 60 90 120 150 180
Min:Sec 5:06 3:24 2:33 2:02 1:42
One Minute Holding Pattern
DRIBB INT
3000
5000
RBJ
RBJ 305°
DRIBB INT
NIKEE OM
4262
5000
305°
125°
058°
GS 3.00° TCH 50
4300
4.7 NM
5.6 NM
CATEGORY A B C D
S-ILS 6R 2652-1 250 (300-1)
S-LOC 6R 2900-1 498 (500-1) 2900-1¼ 498 (500-1¼) 2900-1½ 498 (500-1½)
CIRCLING 2900-1 483 (500-1) 2900-1½ 483 (500-1½) 2980-2 563 (600-2)
SW-4, 05 APR 2012 to 03 MAY 2012
TUCSON, ARIZONA
Amdt 5C 12096
32°09'N-111°10'W
TUCSON/RYAN FIELD (RYN)
ILS or LOC RWY 6R

Figure 295

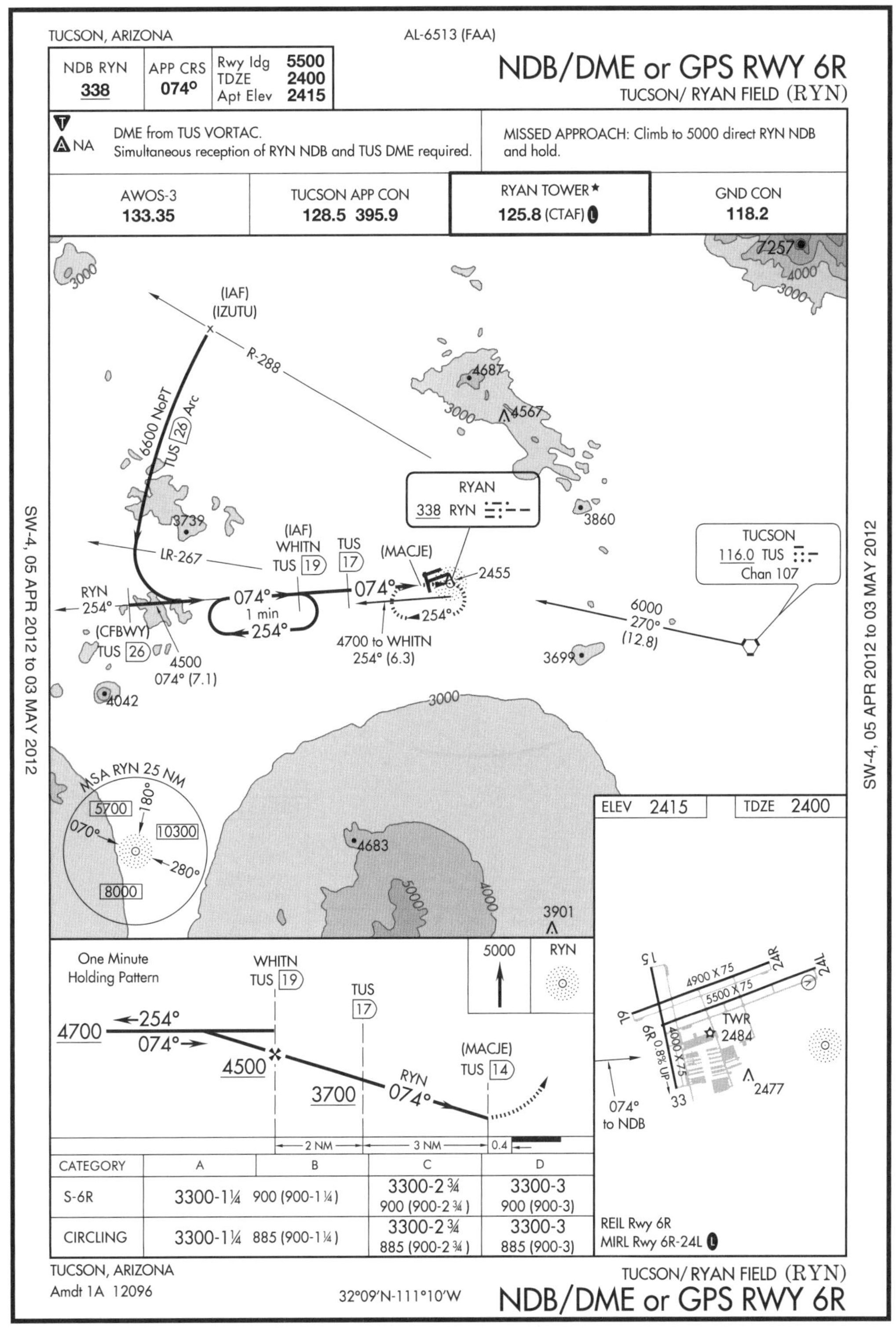
TUCSON, ARIZONA
AL-6513 (FAA)
NDB RYN 338
APP CRS 074°
Rwy ldg 5500
TDZE 2400
Apt Elev 2415
NDB/DME or GPS RWY 6R
TUCSON/ RYAN FIELD (RYN)
DME from TUS VORTAC.
Simultaneous reception of RYN NDB and TUS DME required.
MISSED APPROACH: Climb to 5000 direct RYN NDB and hold.
AWOS-3 133.35
TUCSON APP CON 128.5 395.9
RYAN TOWER★ 125.8 (CTAF)
GND CON 118.2
(IAF) (IZUTU)
R-288
6600 NoPT
TUS 26 Arc
RYAN 338 RYN
(IAF) WHITN TUS 19
TUS 17
(MACJE)
LR-267
RYN 254°
074°
1 min
254°
(CFBWY) TUS 26
4500 074° (7.1)
4700 to WHITN 254° (6.3)
TUCSON 116.0 TUS Chan 107
6000 270° (12.8)
MSA RYN 25 NM
5700
10300
8000
ELEV 2415
TDZE 2400
One Minute Holding Pattern
4700
4500
3700
5000
4900 X 75
5500 X 75
4000 X 75
TWR 2484
074° to NDB
CATEGORY A B C D
S-6R 3300-1¼ 900 (900-1¼) 3300-2¾ 900 (900-2¾) 3300-3 900 (900-3)
CIRCLING 3300-1¼ 885 (900-1¼) 3300-2¾ 885 (900-2¾) 3300-3 885 (900-3)
REIL Rwy 6R
MIRL Rwy 6R-24L
SW-4, 05 APR 2012 to 03 MAY 2012
TUCSON, ARIZONA
Amdt 1A 12096
32°09'N-111°10'W
TUCSON/ RYAN FIELD (RYN)
NDB/DME or GPS RWY 6R

Figure 296

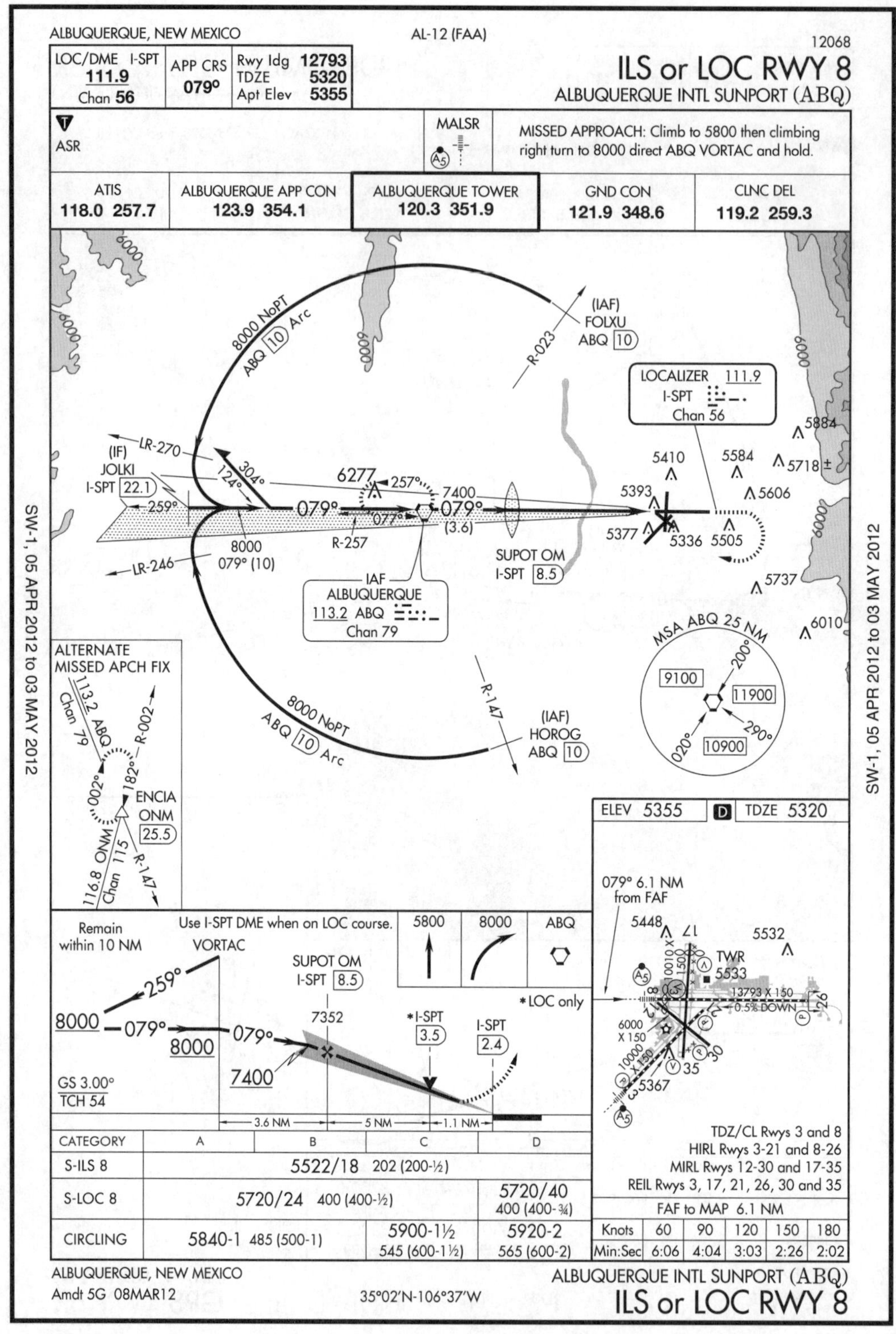

CATEGORY	A	B	C	D
S-ILS 8	5522/18 202 (200-½)			
S-LOC 8	5720/24 400 (400-½)			5720/40 400 (400-¾)
CIRCLING	5840-1 485 (500-1)		5900-1½ 545 (600-1½)	5920-2 565 (600-2)

FAF to MAP 6.1 NM

Knots	60	90	120	150	180
Min:Sec	6:06	4:04	3:03	2:26	2:02

Figure 297

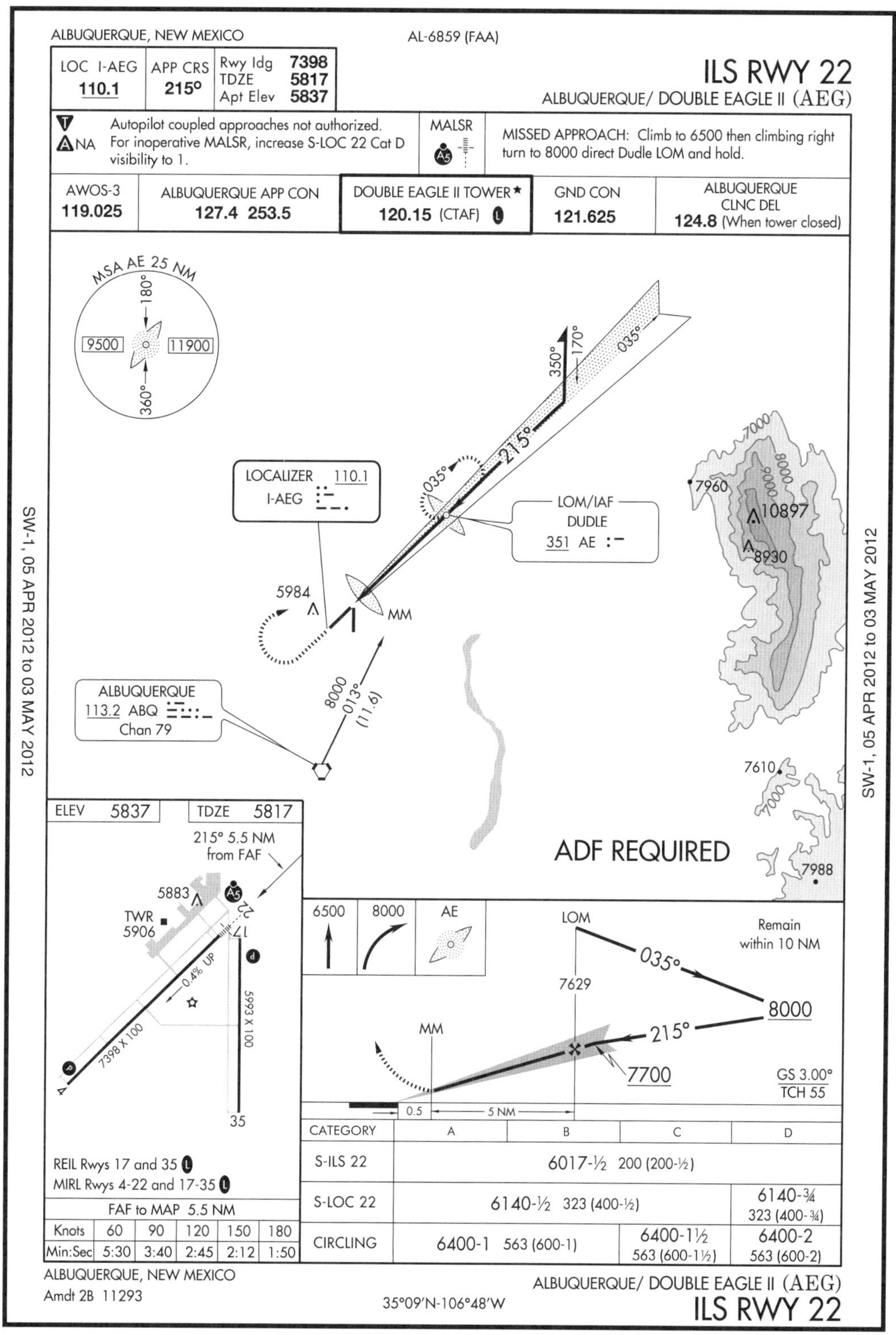
ALBUQUERQUE, NEW MEXICO
AL-6859 (FAA)
LOC I-AEG 110.1
APP CRS 215°
Rwy ldg 7398
TDZE 5817
Apt Elev 5837
ILS RWY 22
ALBUQUERQUE/ DOUBLE EAGLE II (AEG)
Autopilot coupled approaches not authorized.
NA For inoperative MALSR, increase S-LOC 22 Cat D visibility to 1.
MALSR
MISSED APPROACH: Climb to 6500 then climbing right turn to 8000 direct Dudle LOM and hold.
AWOS-3 119.025
ALBUQUERQUE APP CON 127.4 253.5
DOUBLE EAGLE II TOWER ★ 120.15 (CTAF)
GND CON 121.625
ALBUQUERQUE CLNC DEL 124.8 (When tower closed)
MSA AE 25 NM
9500
11900
LOCALIZER 110.1
I-AEG
LOM/IAF
DUDLE
351 AE
ALBUQUERQUE
113.2 ABQ
Chan 79
ADF REQUIRED
ELEV 5837
TDZE 5817
215° 5.5 NM from FAF
5883
TWR 5906
7398 X 100
5993 X 100
REIL Rwys 17 and 35
MIRL Rwys 4-22 and 17-35
FAF to MAP 5.5 NM
Knots 60 90 120 150 180
Min:Sec 5:30 3:40 2:45 2:12 1:50
6500
8000
AE
LOM
Remain within 10 NM
035°
8000
7629
215°
7700
GS 3.00°
TCH 55
5 NM
CATEGORY A B C D
S-ILS 22 6017-½ 200 (200-½)
S-LOC 22 6140-½ 323 (400-½) 6140-¾ 323 (400-¾)
CIRCLING 6400-1 563 (600-1) 6400-1½ 563 (600-1½) 6400-2 563 (600-2)
SW-1, 05 APR 2012 to 03 MAY 2012
ALBUQUERQUE, NEW MEXICO
Amdt 2B 11293
35°09'N-106°48'W
ALBUQUERQUE/ DOUBLE EAGLE II (AEG)
ILS RWY 22

Figure 298

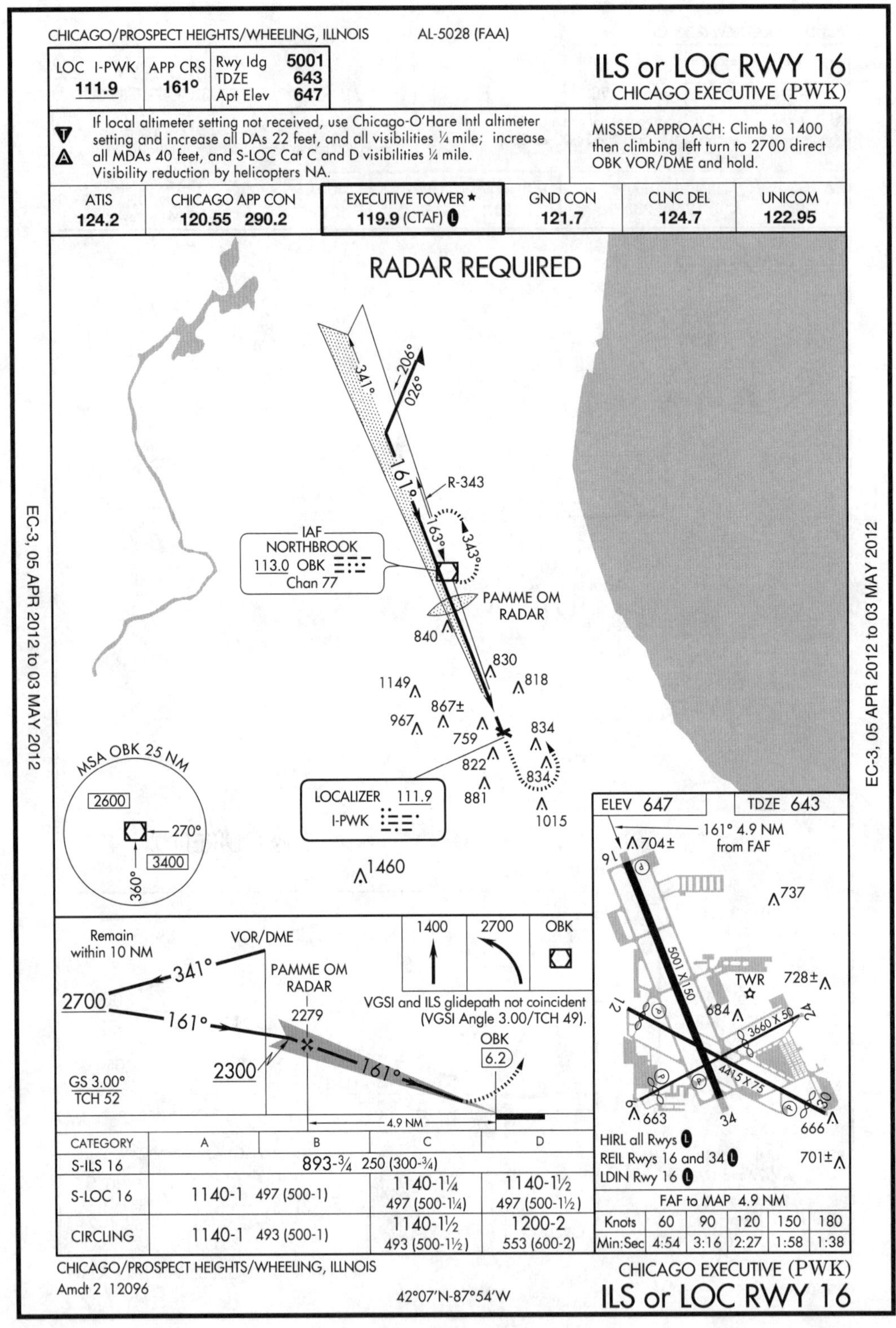

CATEGORY	A	B	C	D
S-ILS 16	893-¾ 250 (300-¾)			
S-LOC 16	1140-1 497 (500-1)		1140-1¼ 497 (500-1¼)	1140-1½ 497 (500-1½)
CIRCLING	1140-1 493 (500-1)		1140-1½ 493 (500-1½)	1200-2 553 (600-2)

FAF to MAP 4.9 NM

Knots	60	90	120	150	180
Min:Sec	4:54	3:16	2:27	1:58	1:38

Figure 299

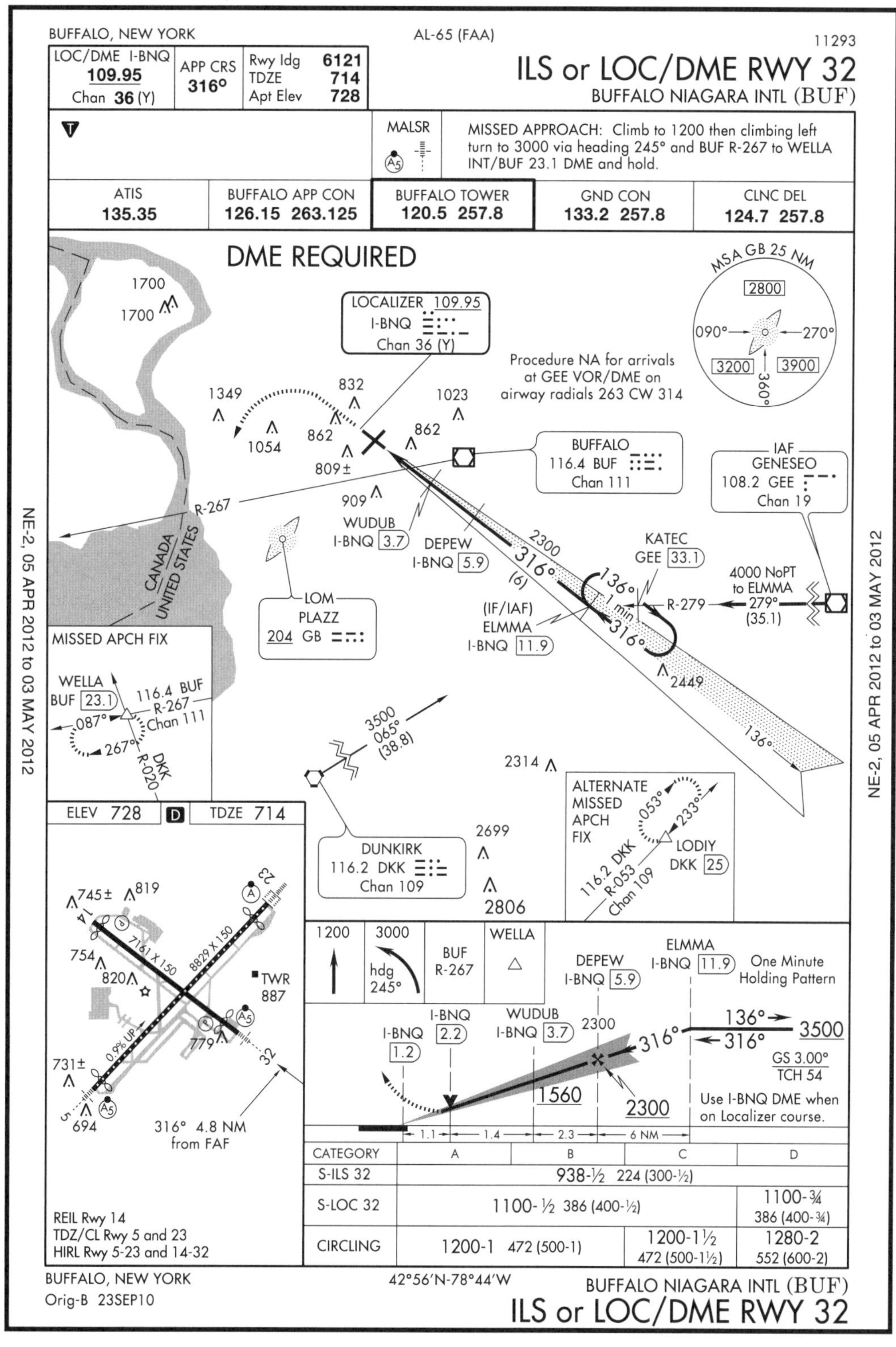

CATEGORY	A	B	C	D
S-ILS 32	938-½ 224 (300-½)			
S-LOC 32	1100-½ 386 (400-½)			1100-¾ 386 (400-¾)
CIRCLING	1200-1 472 (500-1)		1200-1½ 472 (500-1½)	1280-2 552 (600-2)

Figure 300

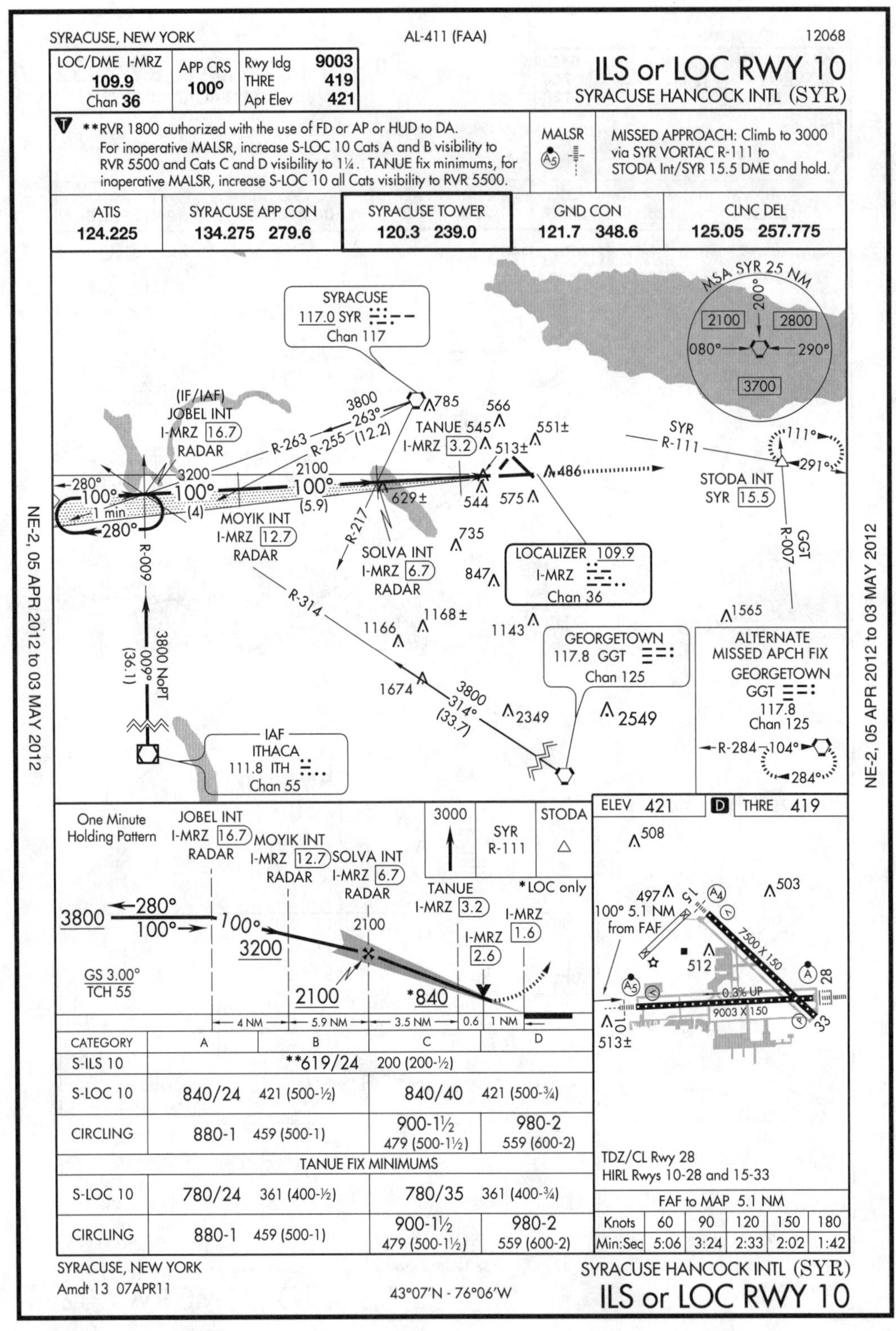

CATEGORY	A	B	C	D
S-ILS 10	**619/24 200 (200-½)			
S-LOC 10	840/24 421 (500-½)		840/40 421 (500-¾)	
CIRCLING	880-1 459 (500-1)		900-1½ 479 (500-1½)	980-2 559 (600-2)
TANUE FIX MINIMUMS				
S-LOC 10	780/24 361 (400-½)		780/35 361 (400-¾)	
CIRCLING	880-1 459 (500-1)		900-1½ 479 (500-1½)	980-2 559 (600-2)

FAF to MAP 5.1 NM

Knots	60	90	120	150	180
Min:Sec	5:06	3:24	2:33	2:02	1:42

SYRACUSE, NEW YORK
Amdt 13 07APR11
43°07'N - 76°06'W
SYRACUSE HANCOCK INTL (SYR)
ILS or LOC RWY 10

Figure 301

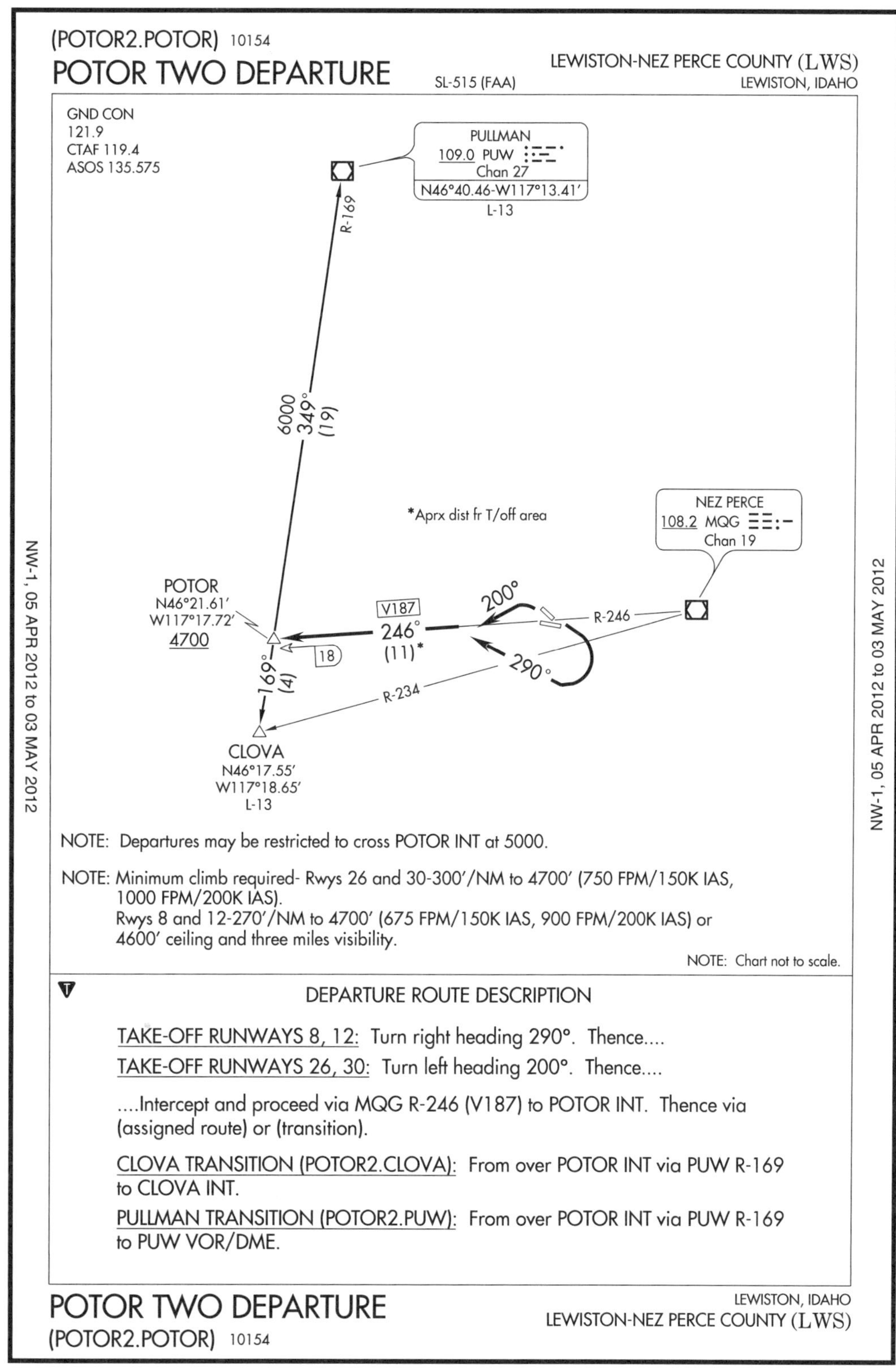

(POTOR2.POTOR) 10154
POTOR TWO DEPARTURE
SL-515 (FAA)
LEWISTON-NEZ PERCE COUNTY (LWS)
LEWISTON, IDAHO
GND CON
121.9
CTAF 119.4
ASOS 135.575
PULLMAN
109.0 PUW
Chan 27
N46°40.46-W117°13.41′
L-13
R-169
6000
349°
(19)
*Aprx dist fr T/off area
NEZ PERCE
108.2 MQG
Chan 19
POTOR
N46°21.61′
W117°17.72′
4700
V187
246°
(11)*
200°
R-246
290°
18
169°
(4)
R-234
CLOVA
N46°17.55′
W117°18.65′
L-13
NW-1, 05 APR 2012 to 03 MAY 2012
NOTE: Departures may be restricted to cross POTOR INT at 5000.
NOTE: Minimum climb required- Rwys 26 and 30-300′/NM to 4700′ (750 FPM/150K IAS, 1000 FPM/200K IAS).
Rwys 8 and 12-270′/NM to 4700′ (675 FPM/150K IAS, 900 FPM/200K IAS) or 4600′ ceiling and three miles visibility.
NOTE: Chart not to scale.
DEPARTURE ROUTE DESCRIPTION
TAKE-OFF RUNWAYS 8, 12: Turn right heading 290°. Thence....
TAKE-OFF RUNWAYS 26, 30: Turn left heading 200°. Thence....
....Intercept and proceed via MQG R-246 (V187) to POTOR INT. Thence via (assigned route) or (transition).
CLOVA TRANSITION (POTOR2.CLOVA): From over POTOR INT via PUW R-169 to CLOVA INT.
PULLMAN TRANSITION (POTOR2.PUW): From over POTOR INT via PUW R-169 to PUW VOR/DME.
POTOR TWO DEPARTURE
(POTOR2.POTOR) 10154
LEWISTON, IDAHO
LEWISTON-NEZ PERCE COUNTY (LWS)

Figure 302

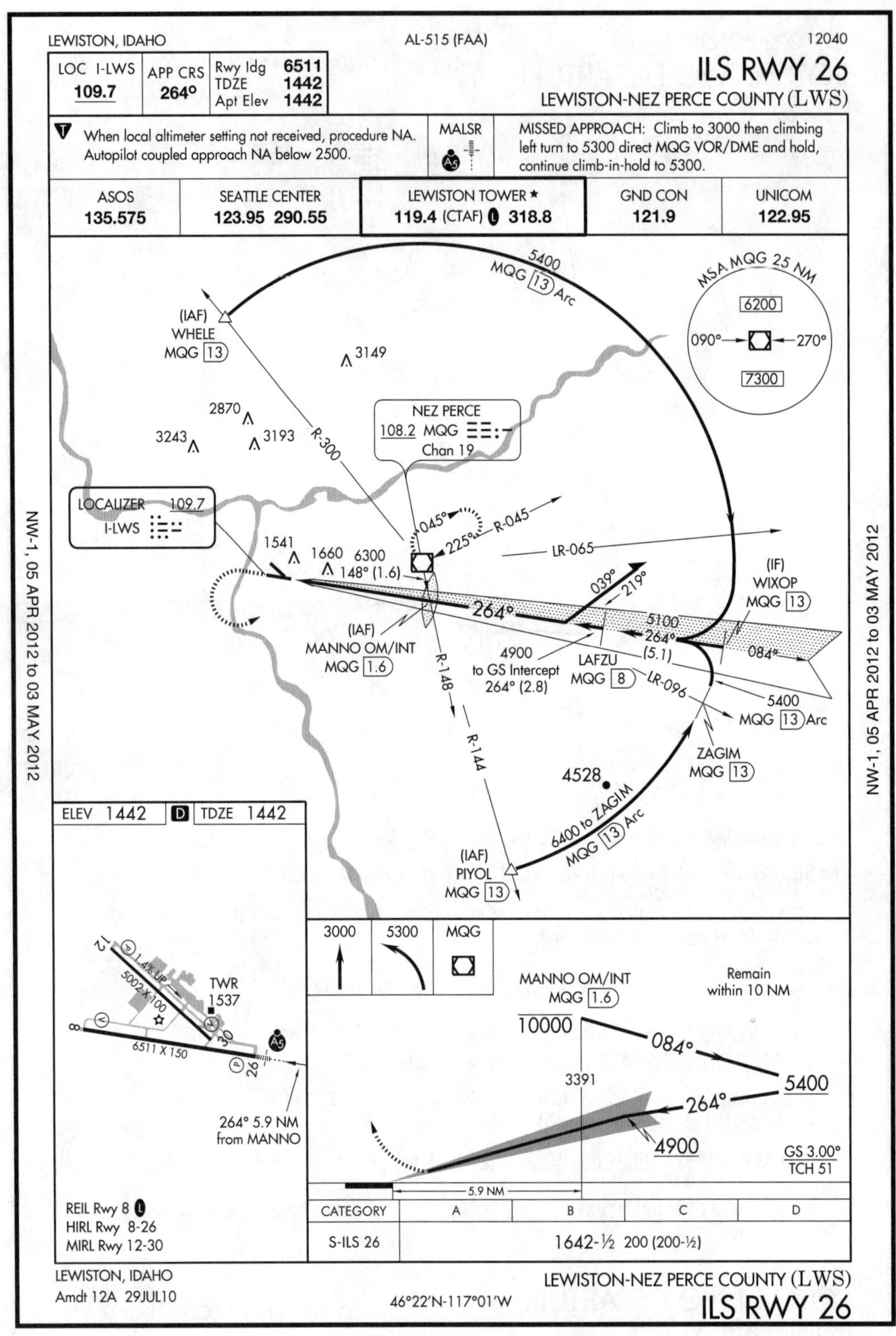

LEWISTON, IDAHO
AL-515 (FAA)
12040
ILS RWY 26
LEWISTON-NEZ PERCE COUNTY (LWS)
LOC I-LWS 109.7
APP CRS 264°
Rwy ldg 6511
TDZE 1442
Apt Elev 1442
When local altimeter setting not received, procedure NA. Autopilot coupled approach NA below 2500.
MALSR
MISSED APPROACH: Climb to 3000 then climbing left turn to 5300 direct MQG VOR/DME and hold, continue climb-in-hold to 5300.
ASOS 135.575
SEATTLE CENTER 123.95 290.55
LEWISTON TOWER ★ 119.4 (CTAF) 318.8
GND CON 121.9
UNICOM 122.95
5400 MQG 13 Arc
MSA MQG 25 NM
6200
090°
270°
7300
(IAF) WHELE MQG 13
3149
2870
3243
3193
R-300
NEZ PERCE 108.2 MQG Chan 19
LOCALIZER 109.7 I-LWS
045°
225°
R-045
1541
1660
6300
148° (1.6)
LR-065
039°
219°
(IF) WIXOP MQG 13
264°
5100
264°
(5.1)
084°
(IAF) MANNO OM/INT MQG 1.6
4900 to GS Intercept 264° (2.8)
LAFZU MQG 8
LR-096
5400 MQG 13 Arc
R-148
R-144
ZAGIM MQG 13
4528
6400 to ZAGIM MQG 13 Arc
(IAF) PIYOL MQG 13
NW-1, 05 APR 2012 to 03 MAY 2012
ELEV 1442
TDZE 1442
12
1.4% UP
5002 X 100
TWR 1537
30
8
6511 X 150
26
264° 5.9 NM from MANNO
REIL Rwy 8
HIRL Rwy 8-26
MIRL Rwy 12-30
3000
5300
MQG
MANNO OM/INT MQG 1.6
Remain within 10 NM
10000
084°
5400
3391
264°
4900
GS 3.00° TCH 51
5.9 NM
CATEGORY A B C D
S-ILS 26 1642-½ 200 (200-½)
LEWISTON, IDAHO
Amdt 12A 29JUL10
46°22'N-117°01'W
LEWISTON-NEZ PERCE COUNTY (LWS)
ILS RWY 26

Figure 303

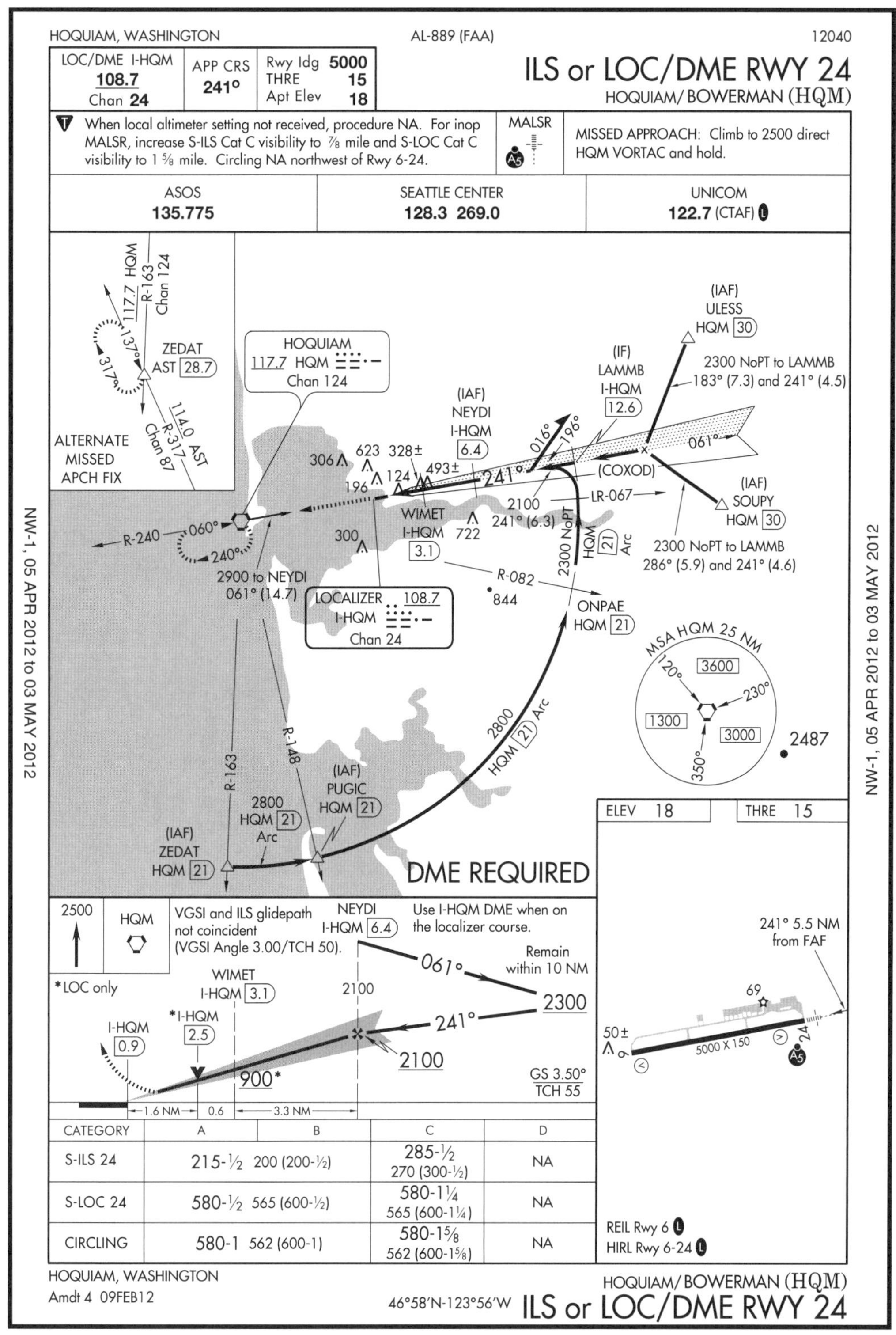

CATEGORY	A	B	C	D
S-ILS 24	215-½ 200 (200-½)		285-½ 270 (300-½)	NA
S-LOC 24	580-½ 565 (600-½)		580-1¼ 565 (600-1¼)	NA
CIRCLING	580-1 562 (600-1)		580-1⅝ 562 (600-1⅝)	NA

Figure 304

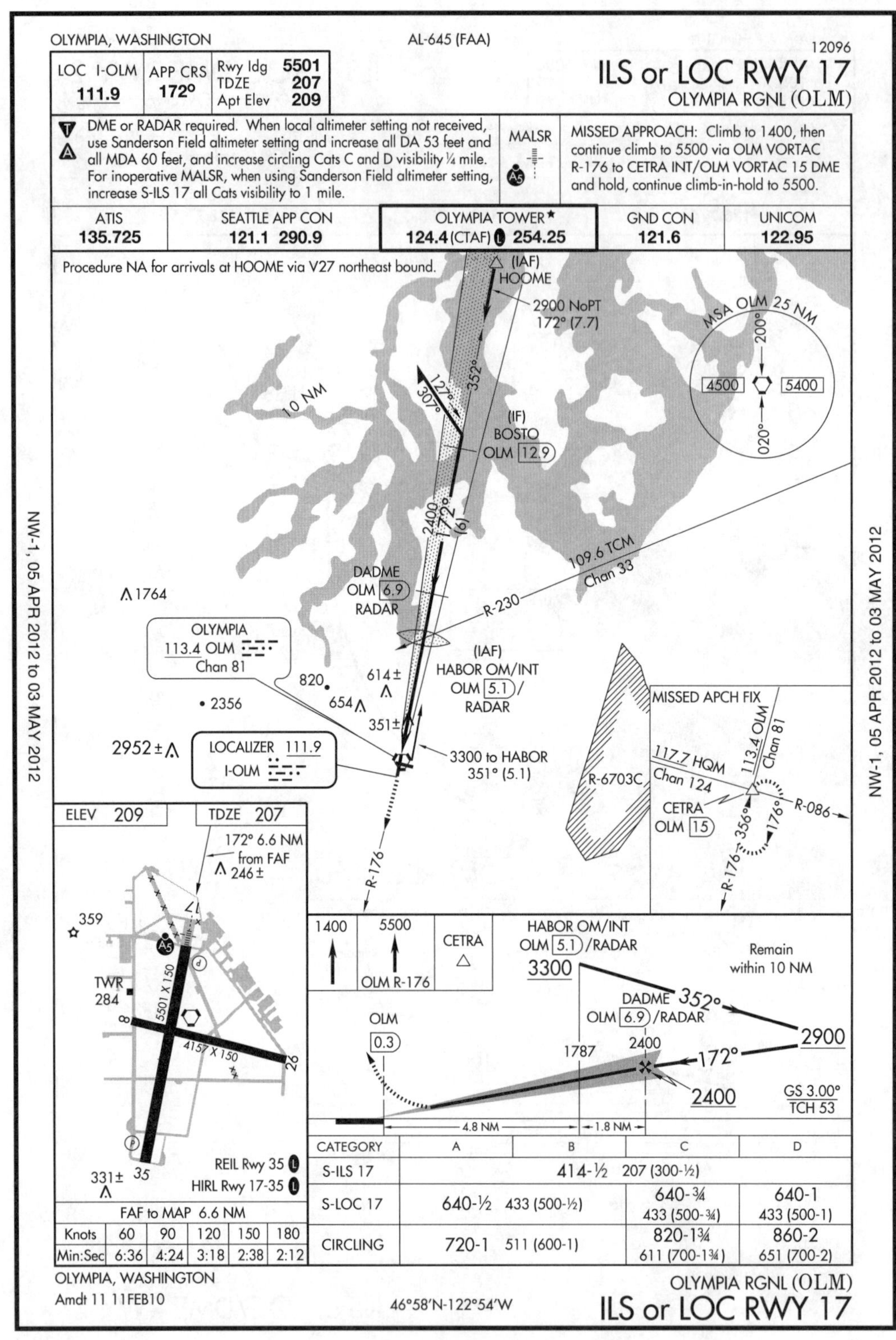
OLYMPIA, WASHINGTON
AL-645 (FAA)
12096
ILS or LOC RWY 17
OLYMPIA RGNL (OLM)
LOC I-OLM 111.9
APP CRS 172°
Rwy ldg 5501
TDZE 207
Apt Elev 209
DME or RADAR required. When local altimeter setting not received, use Sanderson Field altimeter setting and increase all DA 53 feet and all MDA 60 feet, and increase circling Cats C and D visibility ¼ mile. For inoperative MALSR, when using Sanderson Field altimeter setting, increase S-ILS 17 all Cats visibility to 1 mile.
MALSR
MISSED APPROACH: Climb to 1400, then continue climb to 5500 via OLM VORTAC R-176 to CETRA INT/OLM VORTAC 15 DME and hold, continue climb-in-hold to 5500.
ATIS 135.725
SEATTLE APP CON 121.1 290.9
OLYMPIA TOWER★ 124.4 (CTAF) 254.25
GND CON 121.6
UNICOM 122.95
Procedure NA for arrivals at HOOME via V27 northeast bound.
(IAF) HOOME
2900 NoPT 172° (7.7)
MSA OLM 25 NM
200°
4500
5400
020°
10 NM
127°
307°
352°
(IF) BOSTO OLM 12.9
2400 172° (6)
109.6 TCM Chan 33
R-230
DADME OLM 6.9 RADAR
1764
OLYMPIA 113.4 OLM Chan 81
(IAF) HABOR OM/INT OLM 5.1 / RADAR
820
614±
654
2356
351±
2952±
LOCALIZER 111.9 I-OLM
3300 to HABOR 351° (5.1)
MISSED APCH FIX
117.7 HQM Chan 124
113.4 OLM Chan 81
R-6703C
CETRA OLM 15
R-086
356°
176°
R-176
ELEV 209
TDZE 207
172° 6.6 NM from FAF
246±
359
TWR 284
5501 X 150
4157 X 150
8
26
35
331±
REIL Rwy 35
HIRL Rwy 17-35
FAF to MAP 6.6 NM
Knots 60 90 120 150 180
Min:Sec 6:36 4:24 3:18 2:38 2:12
1400
5500
OLM R-176
CETRA
HABOR OM/INT OLM 5.1 /RADAR
3300
Remain within 10 NM
DADME 352°
OLM 6.9 /RADAR
OLM 0.3
1787
2400
172°
2900
2400
GS 3.00° TCH 53
4.8 NM
1.8 NM
CATEGORY A B C D
S-ILS 17 414-½ 207 (300-½)
S-LOC 17 640-½ 433 (500-½) 640-¾ 433 (500-¾) 640-1 433 (500-1)
CIRCLING 720-1 511 (600-1) 820-1¾ 611 (700-1¾) 860-2 651 (700-2)
OLYMPIA, WASHINGTON
Amdt 11 11FEB10
46°58'N-122°54'W
OLYMPIA RGNL (OLM)
ILS or LOC RWY 17
NW-1, 05 APR 2012 to 03 MAY 2012

Figure 305

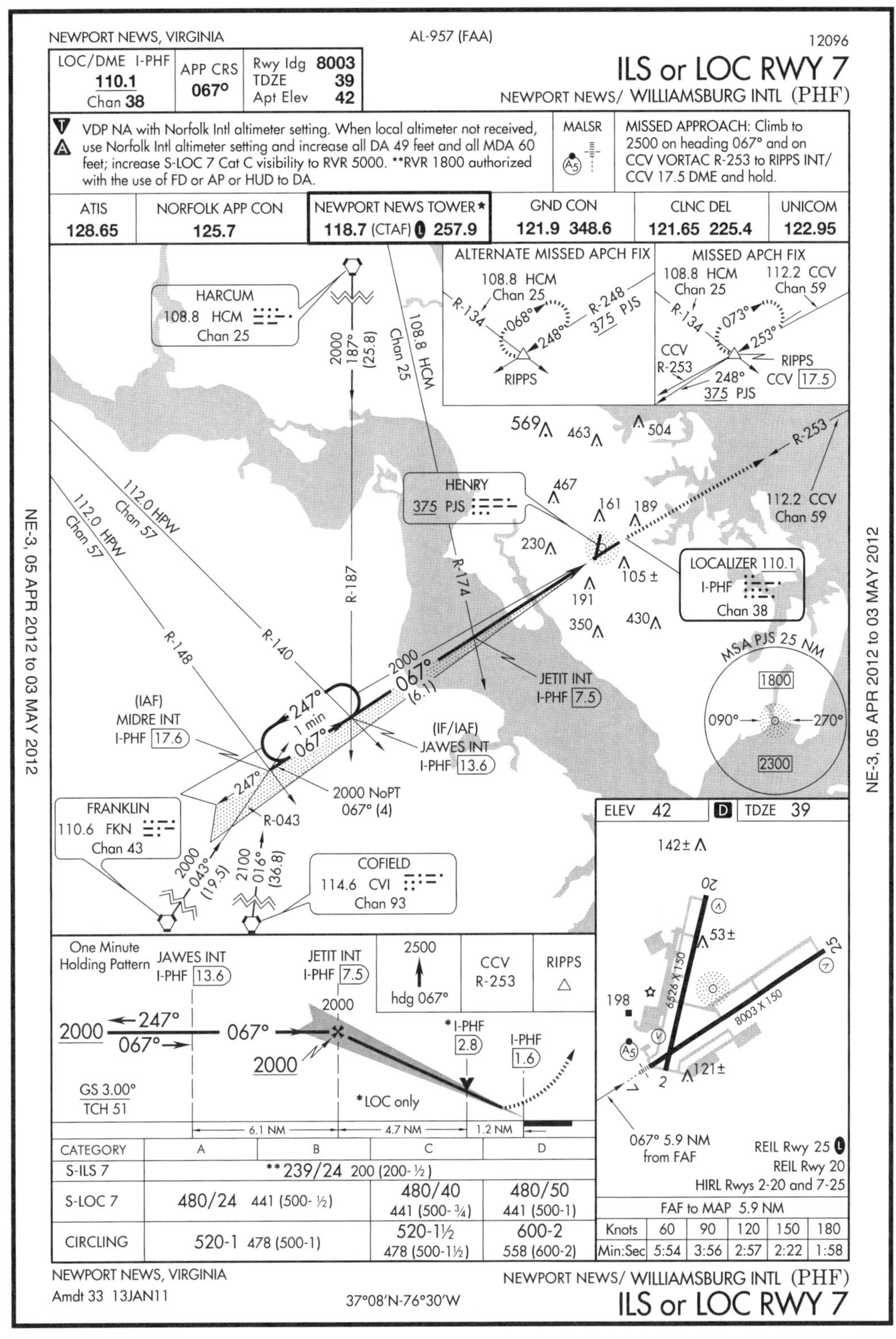

CATEGORY	A	B	C	D
S-ILS 7	**239/24 200 (200-½)			
S-LOC 7	480/24 441 (500-½)		480/40 441 (500-¾)	480/50 441 (500-1)
CIRCLING	520-1 478 (500-1)		520-1½ 478 (500-1½)	600-2 558 (600-2)

Knots	60	90	120	150	180
Min:Sec	5:54	3:56	2:57	2:22	1:58

Figure 306

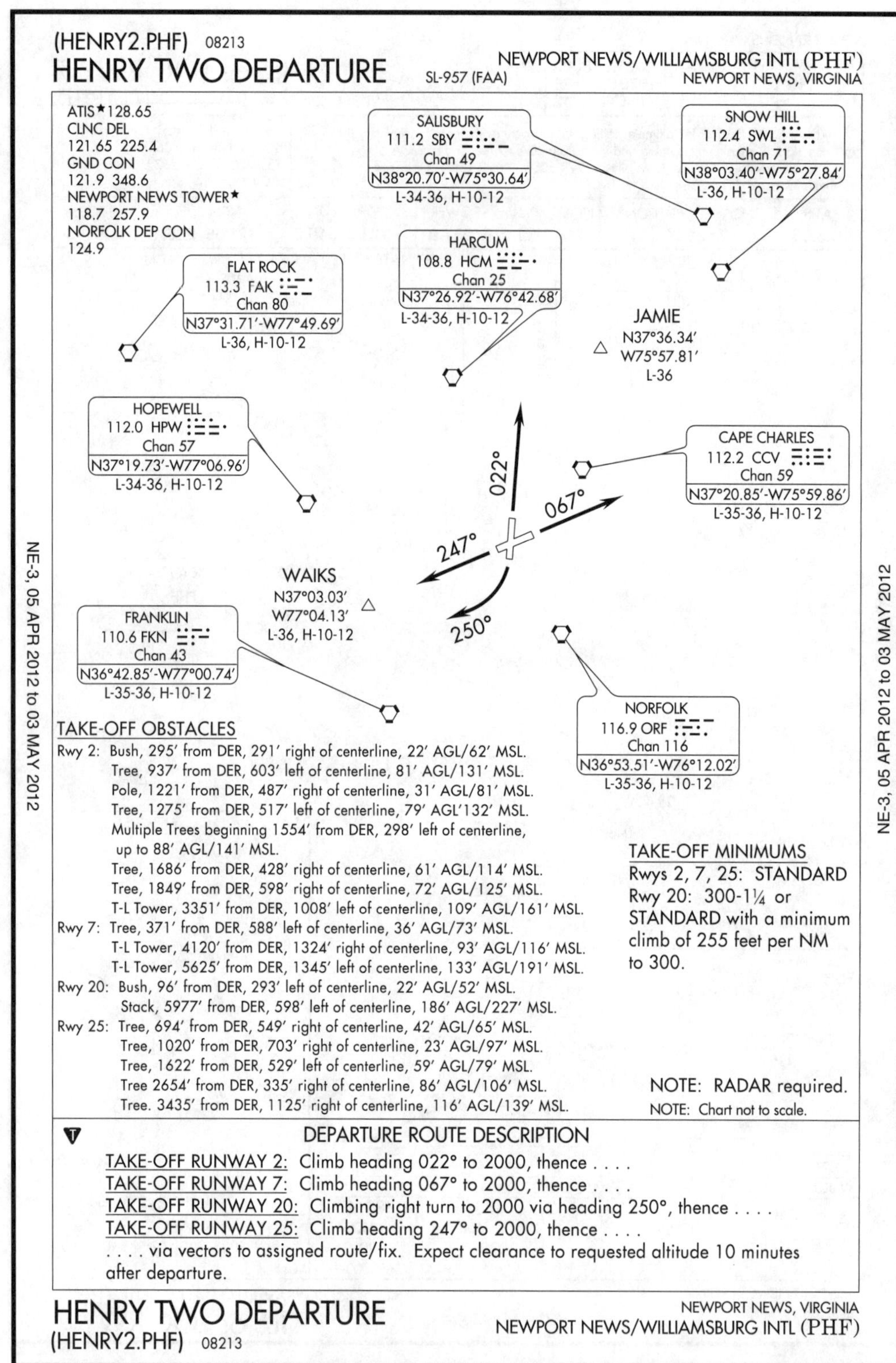
(HENRY2.PHF) 08213
HENRY TWO DEPARTURE
SL-957 (FAA)
NEWPORT NEWS/WILLIAMSBURG INTL (PHF)
NEWPORT NEWS, VIRGINIA
ATIS★128.65
CLNC DEL
121.65 225.4
GND CON
121.9 348.6
NEWPORT NEWS TOWER★
118.7 257.9
NORFOLK DEP CON
124.9
SALISBURY
111.2 SBY
Chan 49
N38°20.70'-W75°30.64'
L-34-36, H-10-12
SNOW HILL
112.4 SWL
Chan 71
N38°03.40'-W75°27.84'
L-36, H-10-12
HARCUM
108.8 HCM
Chan 25
N37°26.92'-W76°42.68'
L-34-36, H-10-12
FLAT ROCK
113.3 FAK
Chan 80
N37°31.71'-W77°49.69'
L-36, H-10-12
JAMIE
N37°36.34'
W75°57.81'
L-36
HOPEWELL
112.0 HPW
Chan 57
N37°19.73'-W77°06.96'
L-34-36, H-10-12
022°
067°
247°
250°
CAPE CHARLES
112.2 CCV
Chan 59
N37°20.85'-W75°59.86'
L-35-36, H-10-12
WAIKS
N37°03.03'
W77°04.13'
L-36, H-10-12
FRANKLIN
110.6 FKN
Chan 43
N36°42.85'-W77°00.74'
L-35-36, H-10-12
NORFOLK
116.9 ORF
Chan 116
N36°53.51'-W76°12.02'
L-35-36, H-10-12
NE-3, 05 APR 2012 to 03 MAY 2012
NE-3, 05 APR 2012 to 03 MAY 2012
TAKE-OFF OBSTACLES
Rwy 2: Bush, 295' from DER, 291' right of centerline, 22' AGL/62' MSL.
Tree, 937' from DER, 603' left of centerline, 81' AGL/131' MSL.
Pole, 1221' from DER, 487' right of centerline, 31' AGL/81' MSL.
Tree, 1275' from DER, 517' left of centerline, 79' AGL'132' MSL.
Multiple Trees beginning 1554' from DER, 298' left of centerline, up to 88' AGL/141' MSL.
Tree, 1686' from DER, 428' right of centerline, 61' AGL/114' MSL.
Tree, 1849' from DER, 598' right of centerline, 72' AGL/125' MSL.
T-L Tower, 3351' from DER, 1008' left of centerline, 109' AGL/161' MSL.
Rwy 7: Tree, 371' from DER, 588' left of centerline, 36' AGL/73' MSL.
T-L Tower, 4120' from DER, 1324' right of centerline, 93' AGL/116' MSL.
T-L Tower, 5625' from DER, 1345' left of centerline, 133' AGL/191' MSL.
Rwy 20: Bush, 96' from DER, 293' left of centerline, 22' AGL/52' MSL.
Stack, 5977' from DER, 598' left of centerline, 186' AGL/227' MSL.
Rwy 25: Tree, 694' from DER, 549' right of centerline, 42' AGL/65' MSL.
Tree, 1020' from DER, 703' right of centerline, 23' AGL/97' MSL.
Tree, 1622' from DER, 529' left of centerline, 59' AGL/79' MSL.
Tree 2654' from DER, 335' right of centerline, 86' AGL/106' MSL.
Tree. 3435' from DER, 1125' right of centerline, 116' AGL/139' MSL.
TAKE-OFF MINIMUMS
Rwys 2, 7, 25: STANDARD
Rwy 20: 300-1¼ or STANDARD with a minimum climb of 255 feet per NM to 300.
NOTE: RADAR required.
NOTE: Chart not to scale.
DEPARTURE ROUTE DESCRIPTION
TAKE-OFF RUNWAY 2: Climb heading 022° to 2000, thence . . . .
TAKE-OFF RUNWAY 7: Climb heading 067° to 2000, thence . . . .
TAKE-OFF RUNWAY 20: Climbing right turn to 2000 via heading 250°, thence . . . .
TAKE-OFF RUNWAY 25: Climb heading 247° to 2000, thence . . . .
. . . . via vectors to assigned route/fix. Expect clearance to requested altitude 10 minutes after departure.
HENRY TWO DEPARTURE
(HENRY2.PHF) 08213
NEWPORT NEWS, VIRGINIA
NEWPORT NEWS/WILLIAMSBURG INTL (PHF)

Figure 307

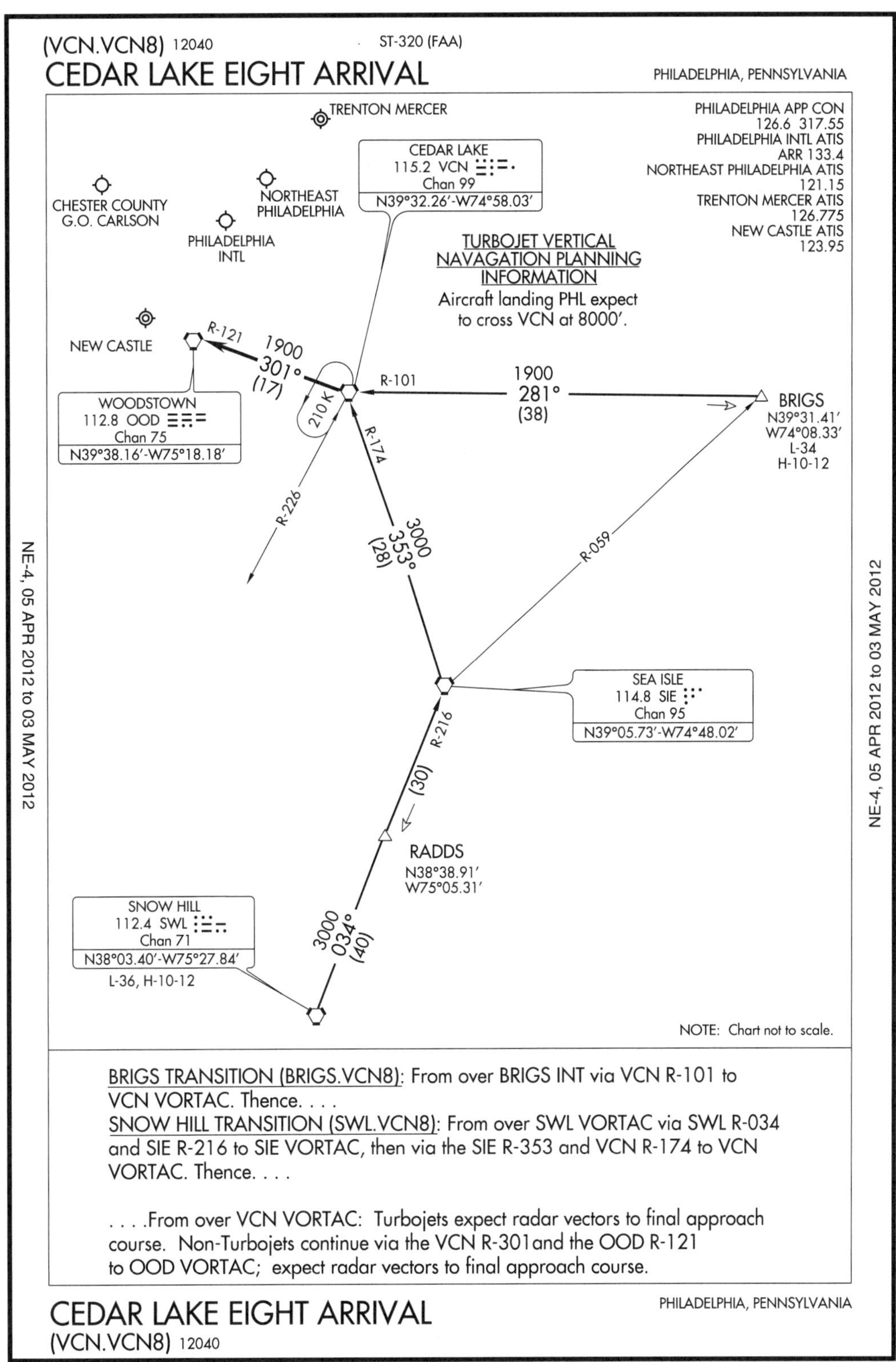

(VCN.VCN8) 12040
ST-320 (FAA)
CEDAR LAKE EIGHT ARRIVAL
PHILADELPHIA, PENNSYLVANIA
TRENTON MERCER
PHILADELPHIA APP CON 126.6 317.55
PHILADELPHIA INTL ATIS ARR 133.4
NORTHEAST PHILADELPHIA ATIS 121.15
TRENTON MERCER ATIS 126.775
NEW CASTLE ATIS 123.95
CEDAR LAKE 115.2 VCN Chan 99 N39°32.26′-W74°58.03′
CHESTER COUNTY G.O. CARLSON
NORTHEAST PHILADELPHIA
PHILADELPHIA INTL
TURBOJET VERTICAL NAVAGATION PLANNING INFORMATION
Aircraft landing PHL expect to cross VCN at 8000′.
NEW CASTLE
R-121
1900 301° (17)
210 K
R-101
1900 281° (38)
BRIGS N39°31.41′ W74°08.33′ L-34 H-10-12
WOODSTOWN 112.8 OOD Chan 75 N39°38.16′-W75°18.18′
R-174
R-226
3000 353° (28)
R-059
SEA ISLE 114.8 SIE Chan 95 N39°05.73′-W74°48.02′
R-216
(30)
RADDS N38°38.91′ W75°05.31′
SNOW HILL 112.4 SWL Chan 71 N38°03.40′-W75°27.84′
L-36, H-10-12
3000 034° (40)
NOTE: Chart not to scale.
BRIGS TRANSITION (BRIGS.VCN8): From over BRIGS INT via VCN R-101 to VCN VORTAC. Thence. . . .
SNOW HILL TRANSITION (SWL.VCN8): From over SWL VORTAC via SWL R-034 and SIE R-216 to SIE VORTAC, then via the SIE R-353 and VCN R-174 to VCN VORTAC. Thence. . . .
. . . .From over VCN VORTAC: Turbojets expect radar vectors to final approach course. Non-Turbojets continue via the VCN R-301and the OOD R-121 to OOD VORTAC; expect radar vectors to final approach course.
CEDAR LAKE EIGHT ARRIVAL
PHILADELPHIA, PENNSYLVANIA
(VCN.VCN8) 12040
NE-4, 05 APR 2012 to 03 MAY 2012
NE-4, 05 APR 2012 to 03 MAY 2012

Figure 308

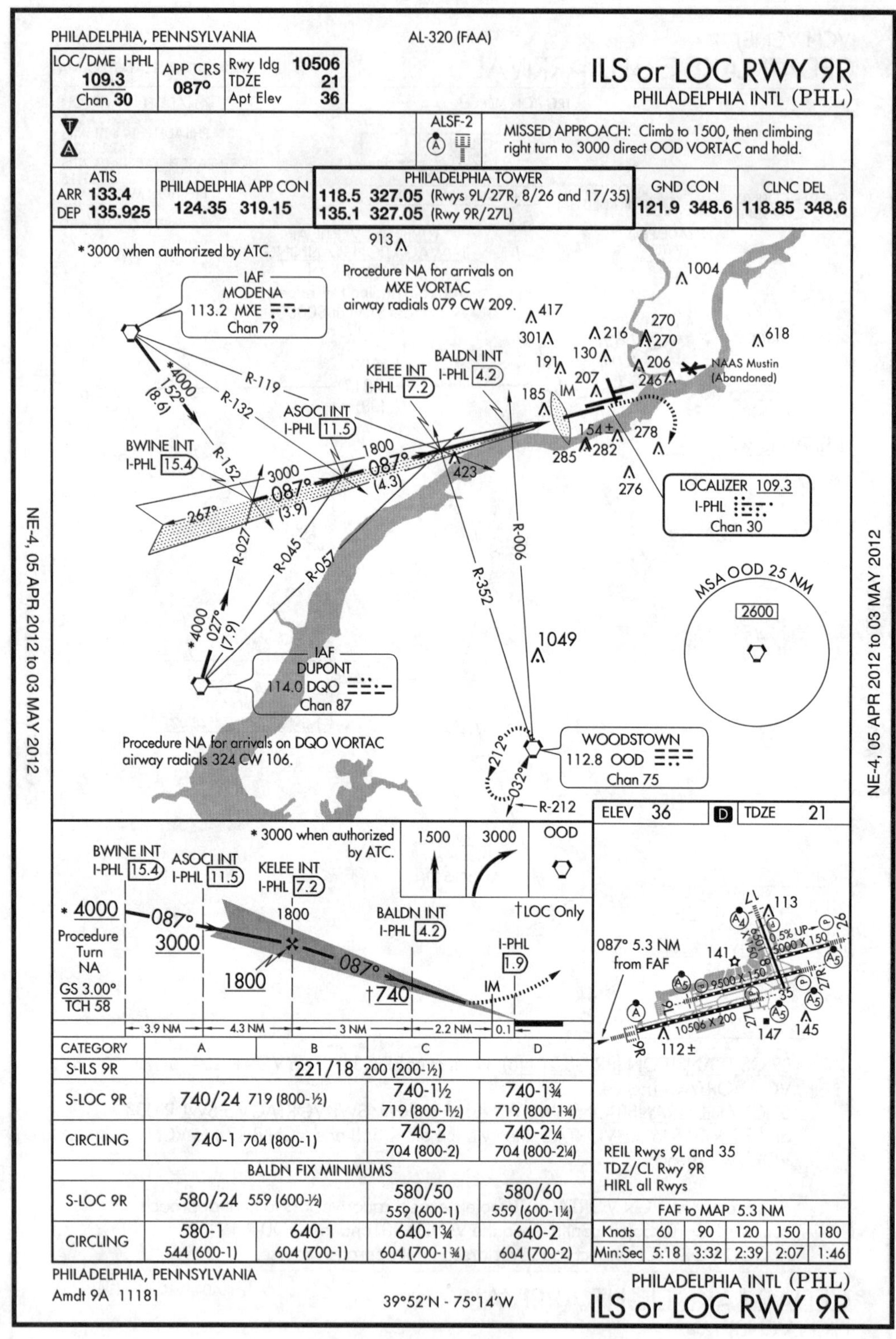
PHILADELPHIA, PENNSYLVANIA
AL-320 (FAA)
LOC/DME I-PHL 109.3 Chan 30
APP CRS 087°
Rwy ldg 10506
TDZE 21
Apt Elev 36
ILS or LOC RWY 9R
PHILADELPHIA INTL (PHL)
ALSF-2
MISSED APPROACH: Climb to 1500, then climbing right turn to 3000 direct OOD VORTAC and hold.
ATIS ARR 133.4 DEP 135.925
PHILADELPHIA APP CON 124.35 319.15
PHILADELPHIA TOWER
118.5 327.05 (Rwys 9L/27R, 8/26 and 17/35)
135.1 327.05 (Rwy 9R/27L)
GND CON 121.9 348.6
CLNC DEL 118.85 348.6
*3000 when authorized by ATC.
IAF MODENA 113.2 MXE Chan 79
Procedure NA for arrivals on MXE VORTAC airway radials 079 CW 209.
BALDN INT I-PHL 4.2
KELEE INT I-PHL 7.2
ASOCI INT I-PHL 11.5
BWINE INT I-PHL 15.4
NAAS Mustin (Abandoned)
LOCALIZER 109.3 I-PHL Chan 30
MSA OOD 25 NM
2600
IAF DUPONT 114.0 DQO Chan 87
Procedure NA for arrivals on DQO VORTAC airway radials 324 CW 106.
WOODSTOWN 112.8 OOD Chan 75
NE-4, 05 APR 2012 to 03 MAY 2012
ELEV 36
TDZE 21
*3000 when authorized by ATC.
1500
3000
OOD
*4000
Procedure Turn NA
GS 3.00°
TCH 58
†LOC Only
I-PHL 1.9
087° 5.3 NM from FAF
3.9 NM
4.3 NM
3 NM
2.2 NM
0.1
CATEGORY A B C D
S-ILS 9R 221/18 200 (200-½)
S-LOC 9R 740/24 719 (800-½) | 740-1½ 719 (800-1½) | 740-1¾ 719 (800-1¾)
CIRCLING 740-1 704 (800-1) | 740-2 704 (800-2) | 740-2¼ 704 (800-2¼)
BALDN FIX MINIMUMS
S-LOC 9R 580/24 559 (600-½) | 580/50 559 (600-1) | 580/60 559 (600-1¼)
CIRCLING 580-1 544 (600-1) | 640-1 604 (700-1) | 640-1¾ 604 (700-1¾) | 640-2 604 (700-2)
REIL Rwys 9L and 35
TDZ/CL Rwy 9R
HIRL all Rwys
FAF to MAP 5.3 NM
Knots 60 90 120 150 180
Min:Sec 5:18 3:32 2:39 2:07 1:46
PHILADELPHIA, PENNSYLVANIA
Amdt 9A 11181
39°52'N - 75°14'W
PHILADELPHIA INTL (PHL)
ILS or LOC RWY 9R

Figure 309

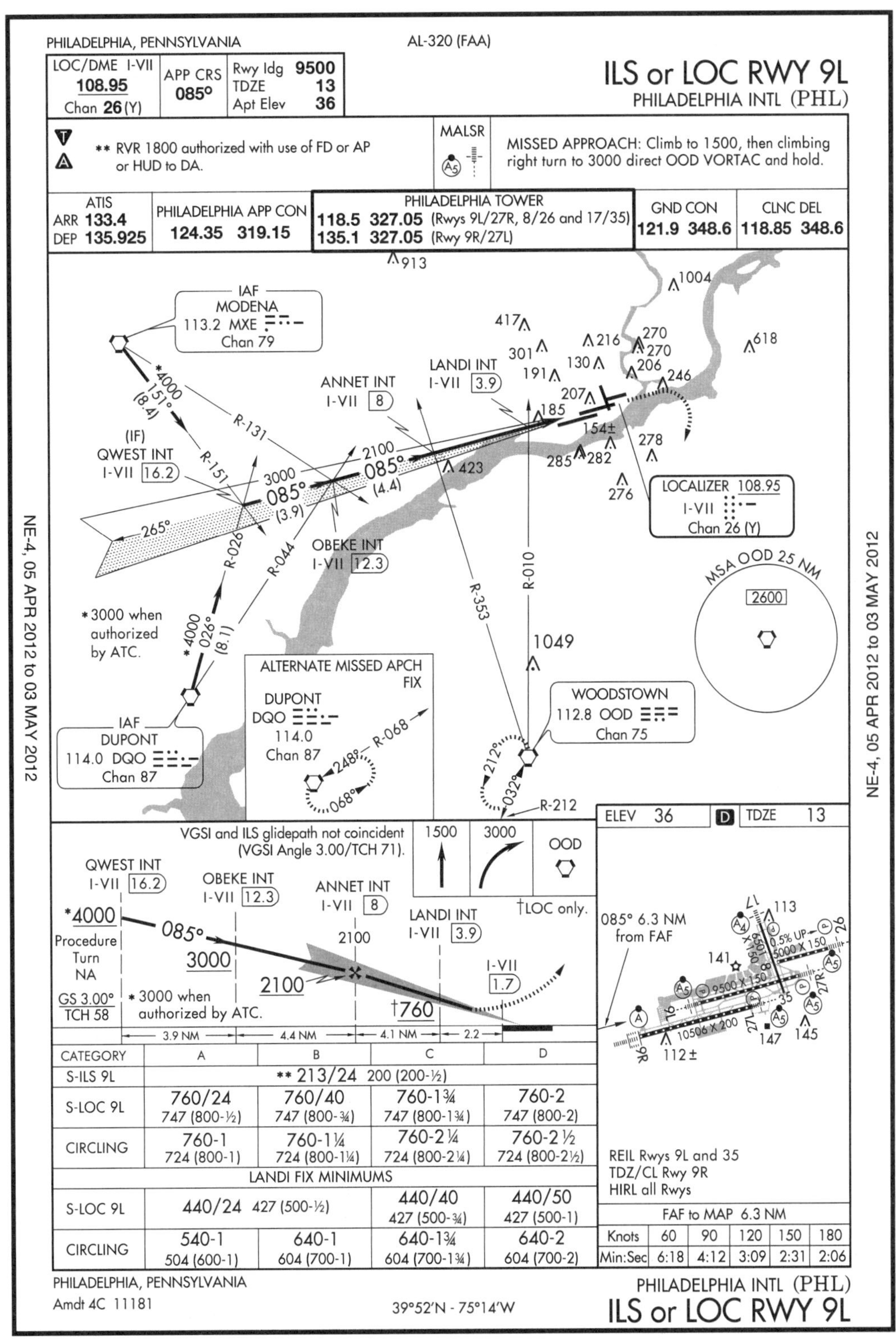
PHILADELPHIA, PENNSYLVANIA
AL-320 (FAA)
ILS or LOC RWY 9L
PHILADELPHIA INTL (PHL)
LOC/DME I-VII 108.95 Chan 26 (Y)
APP CRS 085°
Rwy ldg 9500
TDZE 13
Apt Elev 36
** RVR 1800 authorized with use of FD or AP or HUD to DA.
MALSR
MISSED APPROACH: Climb to 1500, then climbing right turn to 3000 direct OOD VORTAC and hold.
ATIS ARR 133.4 DEP 135.925
PHILADELPHIA APP CON 124.35 319.15
PHILADELPHIA TOWER 118.5 327.05 (Rwys 9L/27R, 8/26 and 17/35) 135.1 327.05 (Rwy 9R/27L)
GND CON 121.9 348.6
CLNC DEL 118.85 348.6
IAF MODENA 113.2 MXE Chan 79
(IF) QWEST INT I-VII 16.2
ANNET INT I-VII 8
LANDI INT I-VII 3.9
OBEKE INT I-VII 12.3
LOCALIZER 108.95 I-VII Chan 26 (Y)
MSA OOD 25 NM
2600
*3000 when authorized by ATC.
ALTERNATE MISSED APCH FIX
DUPONT DQO 114.0 Chan 87
IAF DUPONT 114.0 DQO Chan 87
WOODSTOWN 112.8 OOD Chan 75
VGSI and ILS glidepath not coincident (VGSI Angle 3.00/TCH 71).
ELEV 36
TDZE 13
Procedure Turn NA
GS 3.00° TCH 58
†LOC only.
085° 6.3 NM from FAF
CATEGORY A B C D
S-ILS 9L ** 213/24 200 (200-½)
S-LOC 9L 760/24 747 (800-½) | 760/40 747 (800-¾) | 760-1¾ 747 (800-1¾) | 760-2 747 (800-2)
CIRCLING 760-1 724 (800-1) | 760-1¼ 724 (800-1¼) | 760-2¼ 724 (800-2¼) | 760-2½ 724 (800-2½)
LANDI FIX MINIMUMS
S-LOC 9L 440/24 427 (500-½) | 440/40 427 (500-¾) | 440/50 427 (500-1)
CIRCLING 540-1 504 (600-1) | 640-1 604 (700-1) | 640-1¾ 604 (700-1¾) | 640-2 604 (700-2)
REIL Rwys 9L and 35
TDZ/CL Rwy 9R
HIRL all Rwys
FAF to MAP 6.3 NM
Knots 60 90 120 150 180
Min:Sec 6:18 4:12 3:09 2:31 2:06
NE-4, 05 APR 2012 to 03 MAY 2012
PHILADELPHIA, PENNSYLVANIA
Amdt 4C 11181
39°52'N - 75°14'W
PHILADELPHIA INTL (PHL)
ILS or LOC RWY 9L

Figure 310

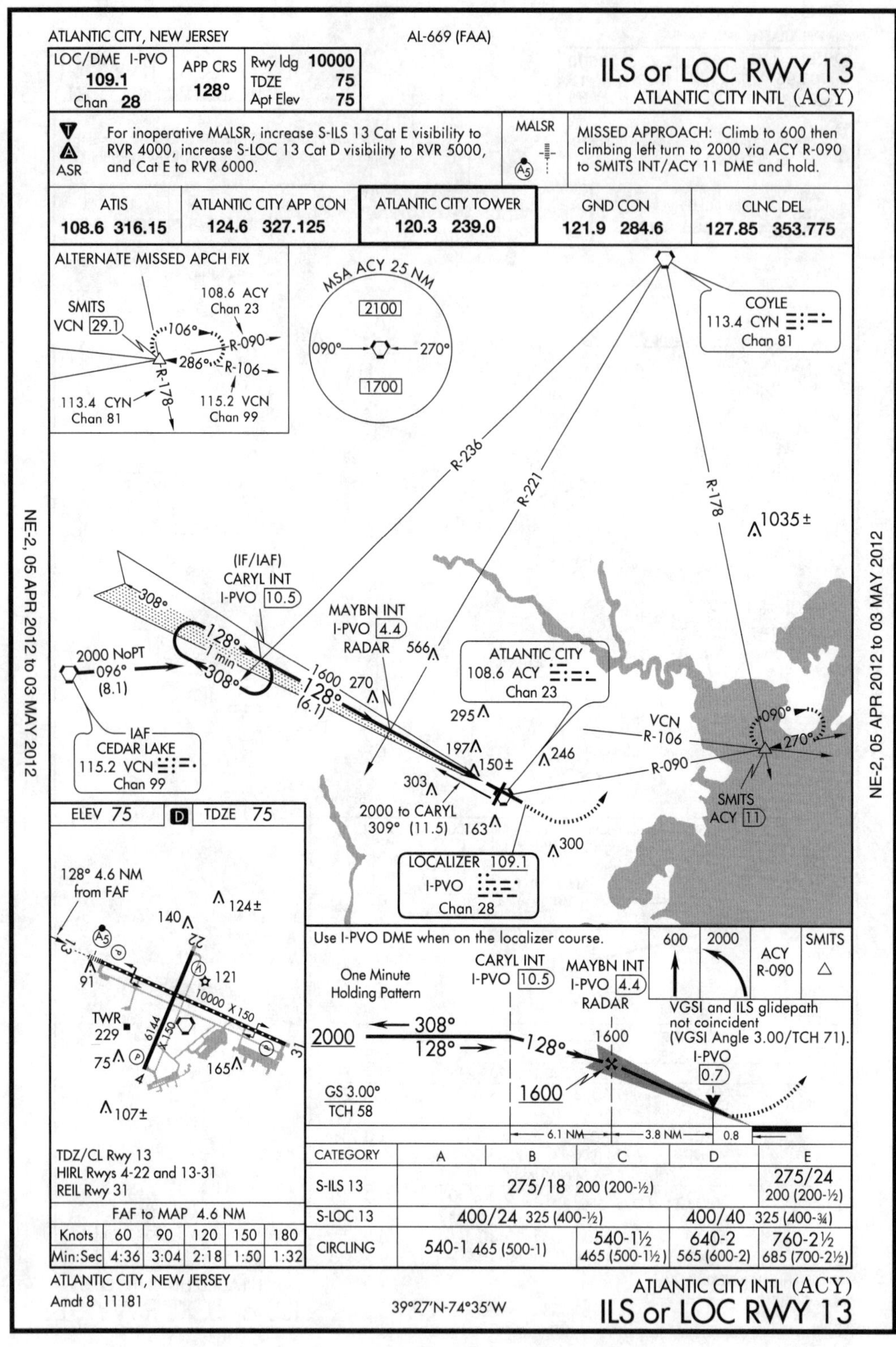

FAF to MAP 4.6 NM					
Knots	60	90	120	150	180
Min:Sec	4:36	3:04	2:18	1:50	1:32

CATEGORY	A	B	C	D	E
S-ILS 13	275/18 200 (200-½)				275/24 200 (200-½)
S-LOC 13	400/24 325 (400-½)			400/40 325 (400-¾)	
CIRCLING	540-1 465 (500-1)		540-1½ 465 (500-1½)	640-2 565 (600-2)	760-2½ 685 (700-2½)

NE-2, 05 APR 2012 to 03 MAY 2012

ATLANTIC CITY, NEW JERSEY
Amdt 8 11181
39°27'N-74°35'W
ATLANTIC CITY INTL (ACY)
ILS or LOC RWY 13

Figure 311

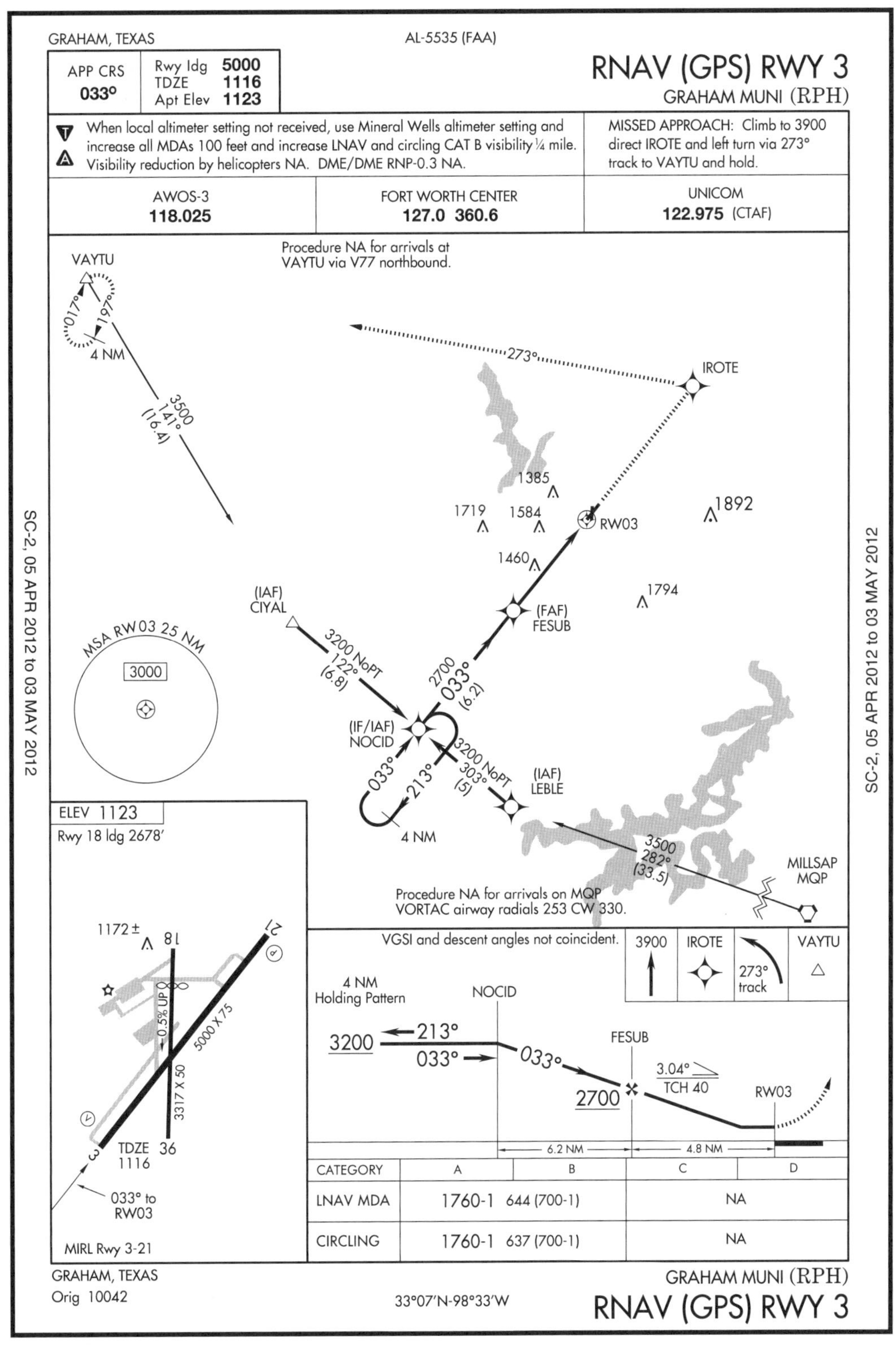

CATEGORY	A	B	C	D
LNAV MDA	1760-1 644 (700-1)		NA	
CIRCLING	1760-1 637 (700-1)		NA	

Figure 312

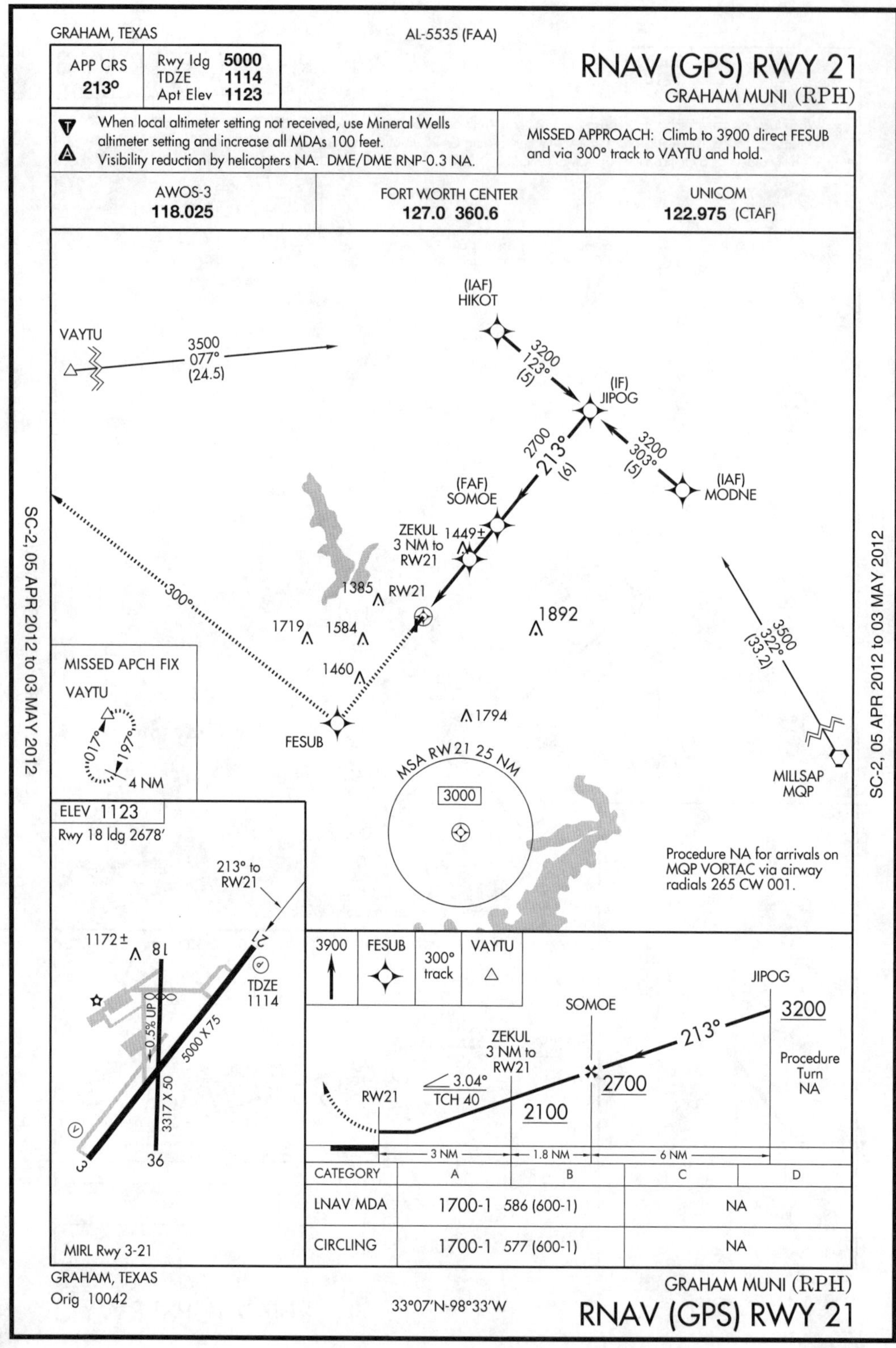

CATEGORY	A	B	C	D
LNAV MDA	1700-1 586 (600-1)		NA	
CIRCLING	1700-1 577 (600-1)		NA	

Figure 313

L7

TAKE-OFF MINIMUMS AND (OBSTACLE) DEPARTURE PROCEDURES

12096

05 APR 2012 to 03 MAY 2012

FORT WORTH, TX (CON'T)

FORT WORTH SPINKS

TAKE-OFF MINIMUMS: **Rwys 17L,35R**, NA. (Environmental)

DEPARTURE PROCEDURE: **Rwy 17R** climb heading 173° to 1200 before turning right.

NOTE: **Rwy 17R,** tree 4909' from departure end of runway, 1556' left of centerline, 60' AGL/830' MSL.

FORT WORTH NAS JRB (CARSWELL FLD)(KNFW)

FORT WORTH, TX 10014

Rwy 17, Cross DER at or above 6' AGL/656' MSL.

TAKE-OFF OBSTACLES: **Rwy 17,** rising terrain up to 670' MSL, 200'-600' from DER, 500'-560' right of centerline.

GAINESVILLE, TX

GAINESVILLE MUNI (GLE)

ORIG 09127 (FAA)

NOTE: **Rwy 17,** trees and poles beginning 1' from DER, 472' right and left of centerline, up to 25' AGL/819' MSL. **Rwy 30,** taxiways beginning 651' from DER, crossing centerline left to right 859' MSL. Trees and terrain beginning 2' from DER, 14' left and right of centerline, up to 64' AGL/890' MSL. **Rwy 35,** terrain, trees, poles, road, and vehicle beginning 149' from DER, 51' left of centerline, up to 95' AGL/940' MSL. Terrain and poles beginning 13' from DER, 85' right of centerline, up to 37' AGL/882' MSL.

GILMER, TX

FOX STEPHENS FIELD-GILMER MUNI (JXI)

ORIG 11293 (FAA)

DEPARTURE PROCEDURE: **Rwy 18,** climb heading 177° to 1000 before turning left.

NOTE: **Rwy 18,** trees beginning abeam the DER left and right of centerline, up to 100' AGL/500' MSL. **Rwy 36,** trees beginning abeam the DER left and right of centerline, up to 50' AGL/505' MSL.

GLADEWATER, TX

GLADEWATER MUNI (07F)

AMDT 1 11153 (FAA)

TAKE-OFF MINIMUMS: **Rwy 17,** 300-1¾ or std. w/min. climb of 285' per NM to 600. **Rwy 32,** 300-1. **Rwy 35,** Std. w/min. climb of 280' per NM to 1300 or 1100-2 ½ for climb in visual conditions.

DEPARTURE PROCEDURE: **Rwy 32,** climb heading 320° to 1100 before turning right. **Rwy 35,** for climb in visual conditions cross Gladewater Municipal Airport at or above 1200 before proceeding on course.

NOTE: **Rwy 14,** vehicles on roadway beginning 450' from DER, left and right of centerline, up to 17' AGL/311' MSL. Trees beginning 770' from DER, left and right of centerline, up to 100' AGL/394' MSL. Power lines 3524' from DER, left to right of centerline, 150' AGL/420' MSL. **Rwy 17,** vehicles on roadway beginning 212' from DER, left and right of centerline, up to 17' AGL/311' MSL. Trees beginning 624' from DER, left and right of centerline, up to 100' AGL/509' MSL. Power lines 1807' from DER, left to right of centerline, 150' AGL/439' MSL. **Rwy 32,** trees beginning 12' from DER, left and right of centerline, up to 100' AGL/429' MSL. **Rwy 35,** trees beginning 47' from DER, left and right of centerline, up to 100' AGL/429' MSL. Power lines 1.4 NM from DER, 844' right of centerline, 75' AGL/520' MSL.

GRAFORD, TX

POSSUM KINGDOM (F35)

ORIG-A 10154 (FAA)

TAKE-OFF MINIMUMS: **Rwy 20,** 400-2½ or std. w/ a min. climb of 212' per NM to 1500 or alternatively, with standard takeoff minimums and a normal 200' per NM climb gradient, takeoff must occur no later than 1600' prior to DER.

DEPARTURE PROCEDURE: **Rwy 20,** climb heading 204° to 1500 before turning left.

NOTE: **Rwy 2,** trees beginning 31' from DER, 22' left of centerline, up to 100' AGL/1099' MSL. Trees beginning 1023' from DER, 114' right of centerline, up to 100' AGL/1129' MSL. **Rwy 20,** vehicle on roadway 116' from DER, 498' right of centerline, 15' AGL/1024' MSL. Trees beginning 494' from DER, 126' right of centerline, up to 100' AGL/1109' MSL. Trees beginning 977' from DER, 115' left of centerline, up to 100' AGL/1109' MSL. Trees beginning 2.29 miles from DER, 1679' left of centerline, up to 100' AGL/1329' MSL.

GRAHAM, TX

GRAHAM MUNI

DEPARTURE PROCEDURE: **Rwys 17, 21,** climb runway heading to 2000 before proceeding on course.

NOTE: **Rwy 17,** light pole 21' from departure end of runway, 195' left of centerline, 30' AGL/1141' MSL. Light pole 86' from departure end of runway, 381' left of centerline, 50' AGL/1168' MSL.

GRANBURY, TX

GRANBURY RGNL (GDJ)

AMDT 2 11125 (FAA)

TAKE-OFF MINIMUMS: **Rwy 14,** 300-1.

DEPARTURE PROCEDURE: **Rwy 14,** climb heading 144° to 1700 before turning right.

NOTE: **Rwy 14,** vehicles on road beginning 1020' from DER, on centerline, 15' AGL/814' MSL. Trees and power poles beginning at DER, 75' right of centerline, up to 100' AGL/879' MSL. Trees, power poles, light poles and vehicles on road beginning at DER, 251' left of centerline, up to 100' AGL/899' MSL. **Rwy 32,** train on railroad tracks, transmission poles and tree beginning 339' from DER, 107' right of centerline, 76' AGL/845' MSL. Trees, vehicles on road and bush beginning 14' from DER, 198' left of centerline, up to 46' AGL/815' MSL.

GRAND PRAIRIE, TX

GRAND PRAIRIE MUNI (GPM)

AMDT 4 09295 (FAA)

DEPARTURE PROCEDURE: **Rwy 17,** climbing right turn to 2000 via heading 200° and TTT R-180 to NINAE/ TTT 24 DME before proceeding on course. DME Required. **Rwy 35,** climb heading 356° to 1400 before turning south.

NOTE: **Rwy 17,** antenna 190' from DER, 456' right of centerline, 26' AGL/615' MSL. Road, multiple poles and signs beginning 570' from DER, 410' right of centerline, up to 31' AGL/620' MSL. Tree 1506' from DER, 517' right of centerline, 37' AGL/617' MSL. **Rwy 35,** tree 837' from DER, 204' left of centerline, up to 100' AGL/665' MSL. Pole 2687' from DER, 122' left of centerline, up to 75' AGL/653' MSL.

05 APR 2012 to 03 MAY 2012

12096

TAKE-OFF MINIMUMS AND (OBSTACLE) DEPARTURE PROCEDURES

SC-2

L7

Figure 314

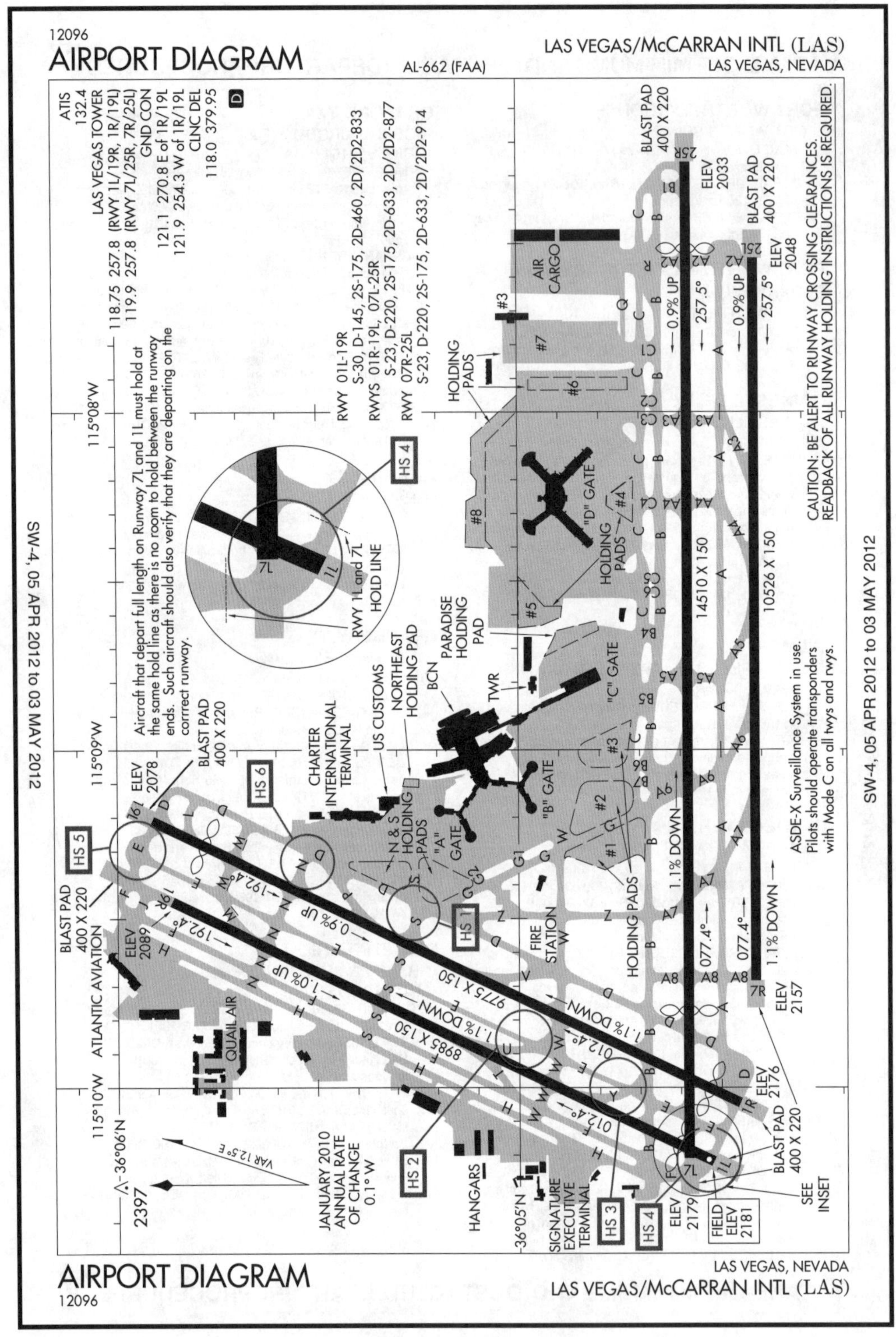
12096
AIRPORT DIAGRAM
AL-662 (FAA)
LAS VEGAS/McCARRAN INTL (LAS)
LAS VEGAS, NEVADA
ATIS 132.4
LAS VEGAS TOWER
118.75 257.8 (RWY 1L/19R, 1R/19L)
119.9 257.8 (RWY 7L/25R, 7R/25L)
GND CON
121.1 270.8 E of 1R/19L
121.9 254.3 W of 1R/19L
CLNC DEL
118.0 379.95
Aircraft that depart full length on Runway 7L and 1L must hold at the same hold line as there is no room to hold between the runway ends. Such aircraft should also verify that they are departing on the correct runway.
RWY 01L-19R
S-30, D-145, 2S-175, 2D-460, 2D/2D2-833
RWYS 01R-19L, 07L-25R
S-23, D-220, 2S-175, 2D-633, 2D/2D2-877
RWY 07R-25L
S-23, D-220, 2S-175, 2D-633, 2D/2D2-914
RWY 1L and 7L HOLD LINE
CAUTION: BE ALERT TO RUNWAY CROSSING CLEARANCES. READBACK OF ALL RUNWAY HOLDING INSTRUCTIONS IS REQUIRED.
ASDE-X Surveillance System in use. Pilots should operate transponders with Mode C on all twys and rwys.
SW-4, 05 APR 2012 to 03 MAY 2012
JANUARY 2010 ANNUAL RATE OF CHANGE 0.1° W
VAR 12.5° E
14510 X 150
10526 X 150
9775 X 150
8985 X 150
BLAST PAD 400 X 220
ELEV 2033
ELEV 2048
ELEV 2078
ELEV 2089
ELEV 2157
ELEV 2176
ELEV 2179
FIELD ELEV 2181
SEE INSET
AIR CARGO
HOLDING PADS
PARADISE HOLDING PAD
NORTHEAST HOLDING PAD
US CUSTOMS
CHARTER INTERNATIONAL TERMINAL
N & S HOLDING PADS
"A" GATE
"B" GATE
"C" GATE
"D" GATE
TWR
BCN
FIRE STATION
ATLANTIC AVIATION
QUAIL AIR
HANGARS
SIGNATURE EXECUTIVE TERMINAL
HS 1
HS 2
HS 3
HS 4
HS 5
HS 6
115°08'W
115°09'W
115°10'W
36°06'N
36°05'N

Figure 315

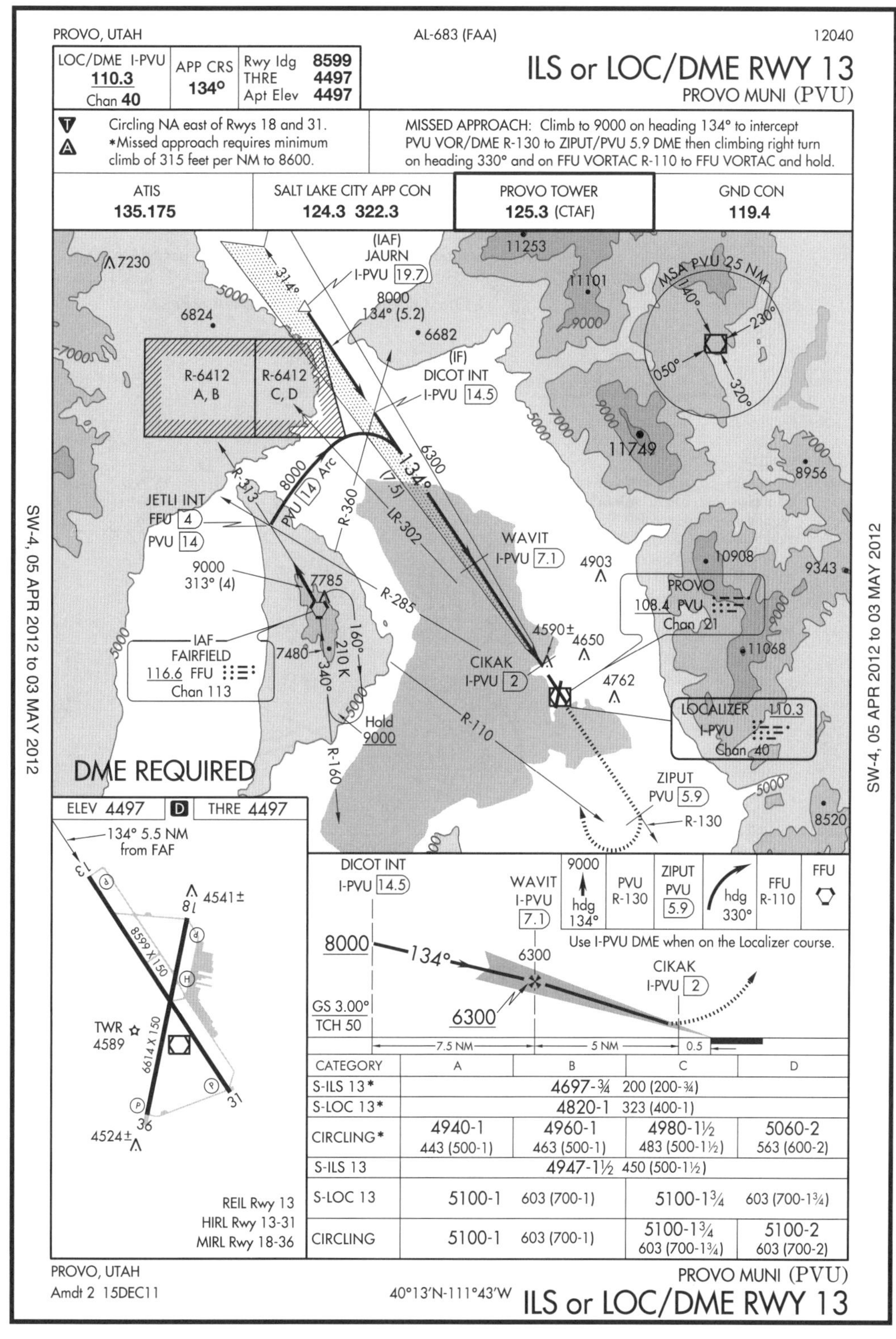

PROVO, UTAH
AL-683 (FAA)
12040
LOC/DME I-PVU 110.3 Chan 40
APP CRS 134°
Rwy ldg 8599
THRE 4497
Apt Elev 4497
ILS or LOC/DME RWY 13
PROVO MUNI (PVU)
Circling NA east of Rwys 18 and 31.
*Missed approach requires minimum climb of 315 feet per NM to 8600.
MISSED APPROACH: Climb to 9000 on heading 134° to intercept PVU VOR/DME R-130 to ZIPUT/PVU 5.9 DME then climbing right turn on heading 330° and on FFU VORTAC R-110 to FFU VORTAC and hold.
ATIS 135.175
SALT LAKE CITY APP CON 124.3 322.3
PROVO TOWER 125.3 (CTAF)
GND CON 119.4
(IAF) JAURN I-PVU 19.7
8000 134° (5.2)
(IF) DICOT INT I-PVU 14.5
R-6412 A, B
R-6412 C, D
MSA PVU 25 NM
JETLI INT FFU 4 PVU 14
9000 313° (4)
PVU 14 Arc 8000
WAVIT I-PVU 7.1
CIKAK I-PVU 2
PROVO 108.4 PVU Chan 21
IAF FAIRFIELD 116.6 FFU Chan 113
Hold 9000
LOCALIZER 110.3 I-PVU Chan 40
ZIPUT PVU 5.9
R-130
R-110
R-160
R-285
R-313
R-360
LR-302
DME REQUIRED
ELEV 4497
THRE 4497
134° 5.5 NM from FAF
8599 X 150
6614 X 150
TWR 4589
REIL Rwy 13
HIRL Rwy 13-31
MIRL Rwy 18-36
9000 hdg 134°
PVU R-130
ZIPUT PVU 5.9
hdg 330°
FFU R-110
FFU
Use I-PVU DME when on the Localizer course.
8000
134°
6300
GS 3.00°
TCH 50
7.5 NM
5 NM
0.5
CATEGORY A B C D
S-ILS 13* 4697-¾ 200 (200-¾)
S-LOC 13* 4820-1 323 (400-1)
CIRCLING* 4940-1 443 (500-1) | 4960-1 463 (500-1) | 4980-1½ 483 (500-1½) | 5060-2 563 (600-2)
S-ILS 13 4947-1½ 450 (500-1½)
S-LOC 13 5100-1 603 (700-1) | 5100-1¾ 603 (700-1¾)
CIRCLING 5100-1 603 (700-1) | 5100-1¾ 603 (700-1¾) | 5100-2 603 (700-2)
SW-4, 05 APR 2012 to 03 MAY 2012
PROVO, UTAH
Amdt 2 15DEC11
40°13'N-111°43'W
PROVO MUNI (PVU)
ILS or LOC/DME RWY 13

Figure 316

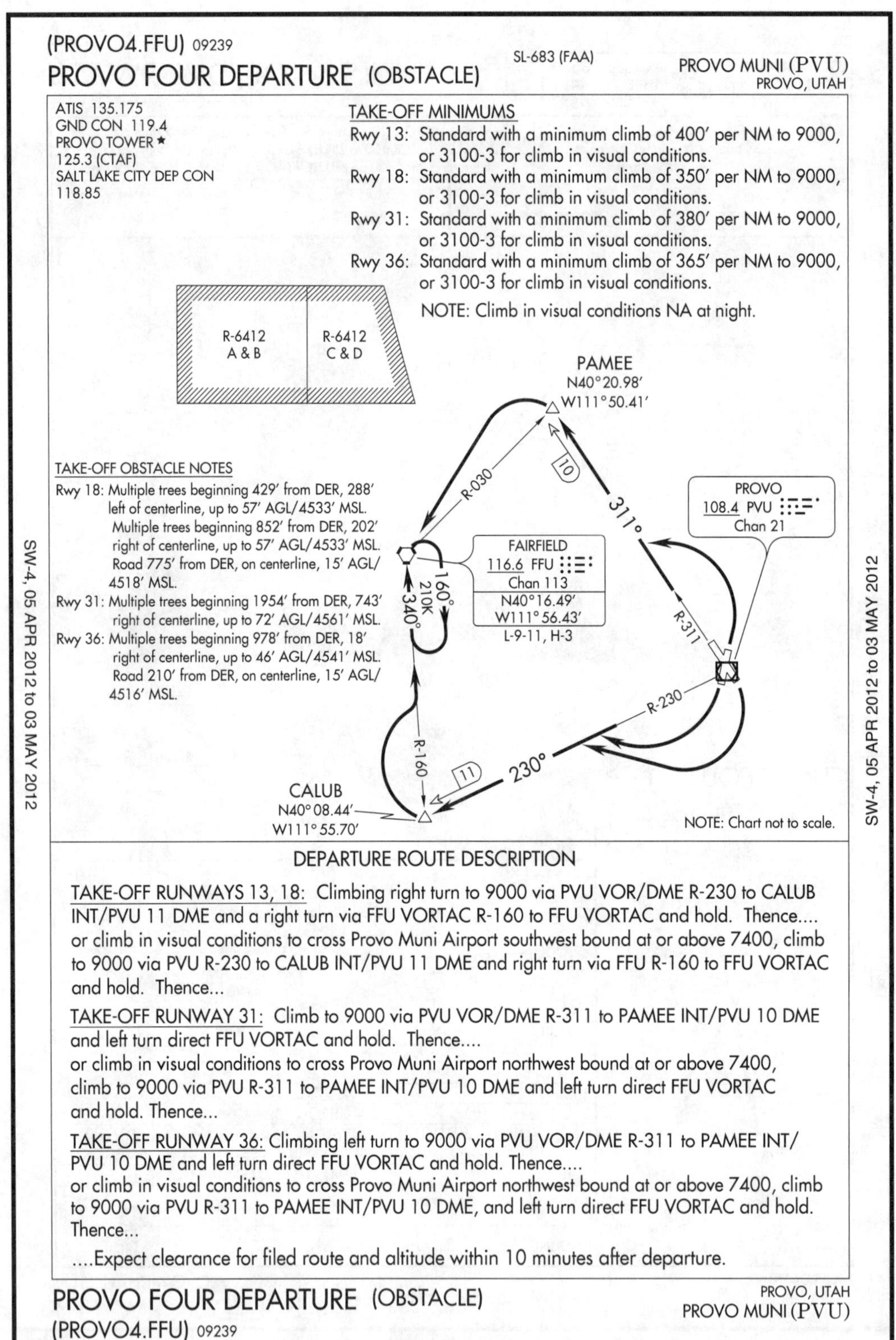
(PROVO4.FFU) 09239

SL-683 (FAA)

PROVO FOUR DEPARTURE (OBSTACLE)

PROVO MUNI (PVU)
PROVO, UTAH

ATIS 135.175
GND CON 119.4
PROVO TOWER ★
125.3 (CTAF)
SALT LAKE CITY DEP CON
118.85

TAKE-OFF MINIMUMS

Rwy 13: Standard with a minimum climb of 400′ per NM to 9000, or 3100-3 for climb in visual conditions.
Rwy 18: Standard with a minimum climb of 350′ per NM to 9000, or 3100-3 for climb in visual conditions.
Rwy 31: Standard with a minimum climb of 380′ per NM to 9000, or 3100-3 for climb in visual conditions.
Rwy 36: Standard with a minimum climb of 365′ per NM to 9000, or 3100-3 for climb in visual conditions.
NOTE: Climb in visual conditions NA at night.

TAKE-OFF OBSTACLE NOTES

Rwy 18: Multiple trees beginning 429′ from DER, 288′ left of centerline, up to 57′ AGL/4533′ MSL. Multiple trees beginning 852′ from DER, 202′ right of centerline, up to 57′ AGL/4533′ MSL. Road 775′ from DER, on centerline, 15′ AGL/4518′ MSL.
Rwy 31: Multiple trees beginning 1954′ from DER, 743′ right of centerline, up to 72′ AGL/4561′ MSL.
Rwy 36: Multiple trees beginning 978′ from DER, 18′ right of centerline, up to 46′ AGL/4541′ MSL. Road 210′ from DER, on centerline, 15′ AGL/4516′ MSL.

DEPARTURE ROUTE DESCRIPTION

TAKE-OFF RUNWAYS 13, 18: Climbing right turn to 9000 via PVU VOR/DME R-230 to CALUB INT/PVU 11 DME and a right turn via FFU VORTAC R-160 to FFU VORTAC and hold. Thence....
or climb in visual conditions to cross Provo Muni Airport southwest bound at or above 7400, climb to 9000 via PVU R-230 to CALUB INT/PVU 11 DME and right turn via FFU R-160 to FFU VORTAC and hold. Thence...

TAKE-OFF RUNWAY 31: Climb to 9000 via PVU VOR/DME R-311 to PAMEE INT/PVU 10 DME and left turn direct FFU VORTAC and hold. Thence....
or climb in visual conditions to cross Provo Muni Airport northwest bound at or above 7400, climb to 9000 via PVU R-311 to PAMEE INT/PVU 10 DME and left turn direct FFU VORTAC and hold. Thence...

TAKE-OFF RUNWAY 36: Climbing left turn to 9000 via PVU VOR/DME R-311 to PAMEE INT/PVU 10 DME and left turn direct FFU VORTAC and hold. Thence....
or climb in visual conditions to cross Provo Muni Airport northwest bound at or above 7400, climb to 9000 via PVU R-311 to PAMEE INT/PVU 10 DME, and left turn direct FFU VORTAC and hold. Thence...

....Expect clearance for filed route and altitude within 10 minutes after departure.

PROVO FOUR DEPARTURE (OBSTACLE)
(PROVO4.FFU) 09239
PROVO, UTAH
PROVO MUNI (PVU)

Figure 317

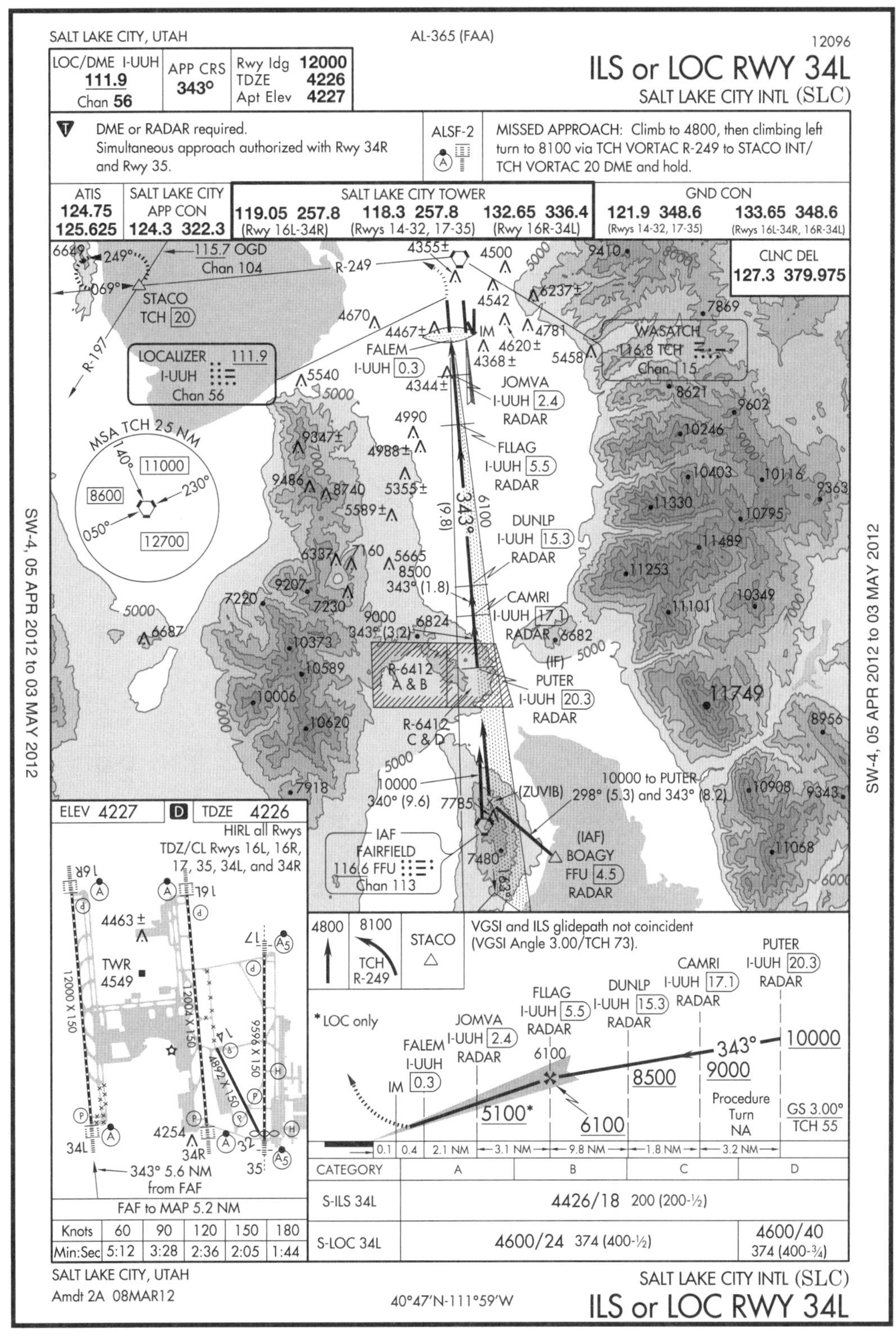

CATEGORY	A	B	C	D
S-ILS 34L	4426/18 200 (200-½)			
S-LOC 34L	4600/24 374 (400-½)			4600/40 374 (400-¾)

Knots	60	90	120	150	180
Min:Sec	5:12	3:28	2:36	2:05	1:44

Figure 318

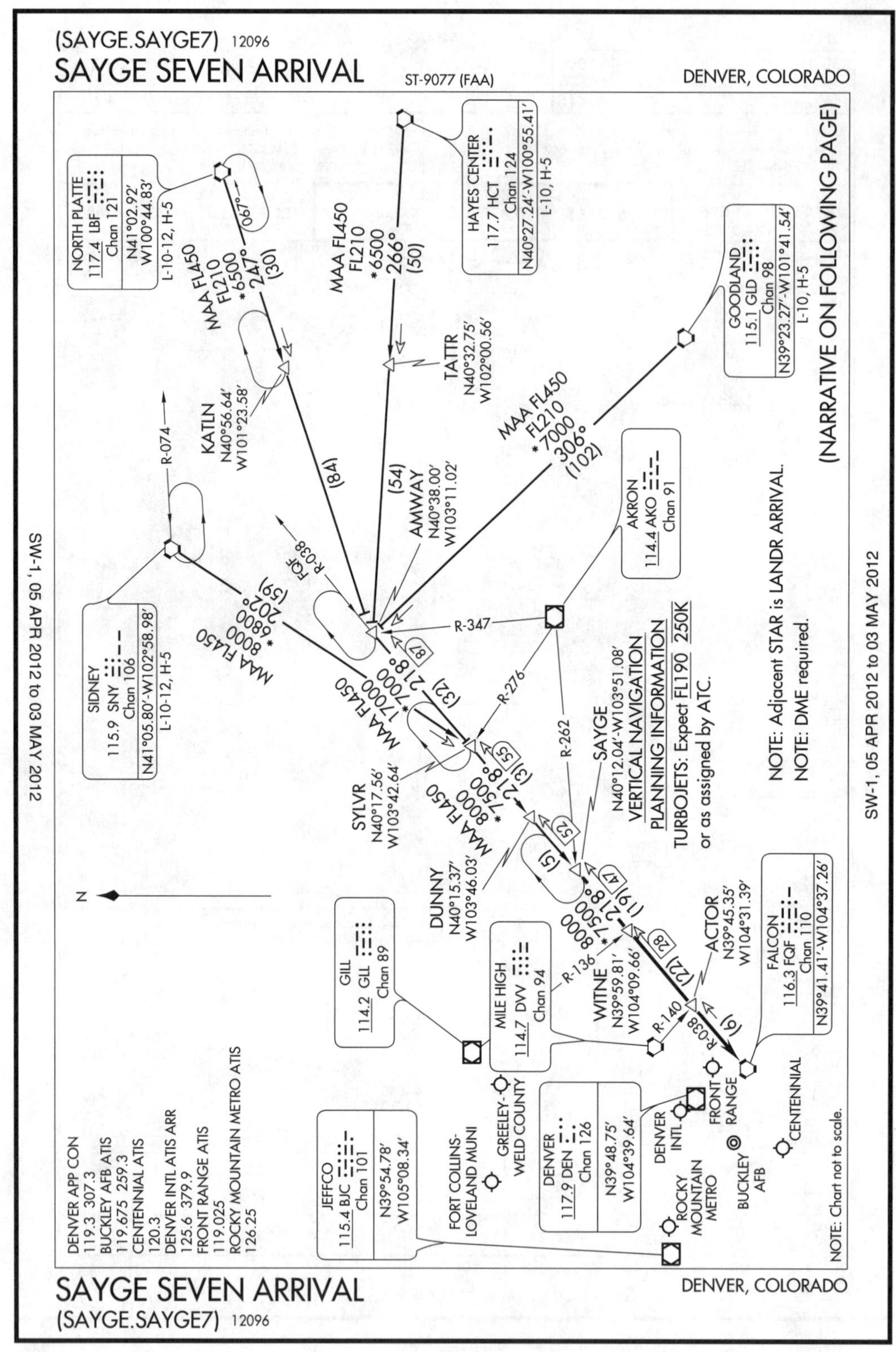
(SAYGE.SAYGE7) 12096
SAYGE SEVEN ARRIVAL
ST-9077 (FAA)
DENVER, COLORADO
NORTH PLATTE
117.4 LBF
Chan 121
N41°02.92'
W100°44.83'
L-10-12, H-5
HAYES CENTER
117.7 HCT
Chan 124
N40°27.24'-W100°55.41'
L-10, H-5
GOODLAND
115.1 GLD
Chan 98
N39°23.27'-W101°41.54'
L-10, H-5
(NARRATIVE ON FOLLOWING PAGE)
MAA FL450
FL210
*6500
247°
(30)
067°
266°
(50)
KATLN
N40°56.64'
W101°23.58'
TATTR
N40°32.75'
W102°00.56'
R-074
(84)
(54)
AMWAY
N40°38.00'
W103°11.02'
MAA FL450
FL210
*7000
306°
(102)
AKRON
114.4 AKO
Chan 91
SIDNEY
115.9 SNY
Chan 106
N41°05.80'-W102°58.98'
L-10-12, H-5
MAA FL450
*8000
6800
202°
(59)
R-038
FQF
R-347
MAA FL450
17000
*7000
218°
(32)
87
R-276
R-262
55
(31)
218°
*7500
8000
MAA FL450
SYLVR
N40°17.56'
W103°42.64'
SAYGE
N40°12.04'-W103°51.08'
VERTICAL NAVIGATION
PLANNING INFORMATION
TURBOJETS: Expect FL190 250K
or as assigned by ATC.
NOTE: Adjacent STAR is LANDR ARRIVAL.
NOTE: DME required.
SW-1, 05 APR 2012 to 03 MAY 2012
DUNNY
N40°15.37'
W103°46.03'
45
(5)
47
(19)
218°
*7500
8000
28
(22)
R-136
R-140
R-038
(6)
WITNE
N39°59.81'
W104°09.66'
ACTOR
N39°45.35'
W104°31.39'
FALCON
116.3 FQF
Chan 110
N39°41.41'-W104°37.26'
GILL
114.2 GLL
Chan 89
MILE HIGH
114.7 DVV
Chan 94
N
DENVER
117.9 DEN
Chan 126
N39°48.75'
W104°39.64'
JEFFCO
115.4 BJC
Chan 101
N39°54.78'
W105°08.34'
FORT COLLINS-LOVELAND MUNI
GREELEY-WELD COUNTY
DENVER INTL
FRONT RANGE
BUCKLEY AFB
CENTENNIAL
ROCKY MOUNTAIN METRO
DENVER APP CON
119.3 307.3
BUCKLEY AFB ATIS
119.675 259.3
CENTENNIAL ATIS
120.3
DENVER INTL ATIS ARR
125.6 379.9
FRONT RANGE ATIS
119.025
ROCKY MOUNTAIN METRO ATIS
126.25
NOTE: Chart not to scale.
SW-1, 05 APR 2012 to 03 MAY 2012
SAYGE SEVEN ARRIVAL
(SAYGE.SAYGE7) 12096
DENVER, COLORADO

Figure 319

(SAYGE.SAYGE7) 12096

SAYGE SEVEN ARRIVAL

ST-9077 (FAA) DENVER, COLORADO

ARRIVAL ROUTE DESCRIPTION

GOODLAND TRANSITION (GLD.SAYGE7): From over GLD VORTAC via GLD R-306 and FQF R-038 to SAYGE INT. Thence....

HAYES CENTER TRANSITION (HCT.SAYGE7): From over HCT VORTAC via HCT R-266 and FQF R-038 to SAYGE INT. Thence....

NORTH PLATTE TRANSITION (LBF.SAYGE7): From over LBF VORTAC via LBF R-247 and FQF R-038 to SAYGE INT. Thence....

SIDNEY TRANSITION (SNY.SAYGE7): From over SNY VORTAC via SNY R-202 and FQF R-038 to SAYGE INT. Thence....

....From over SAYGE INT via FQF R-038 to FQF VORTAC. Expect RADAR vectors to the final approach course at or before FQF VORTAC.

SW-1, 05 APR 2012 to 03 MAY 2012

SW-1, 05 APR 2012 to 03 MAY 2012

SAYGE SEVEN ARRIVAL DENVER, COLORADO

(SAYGE.SAYGE7) 12096

Figure 320

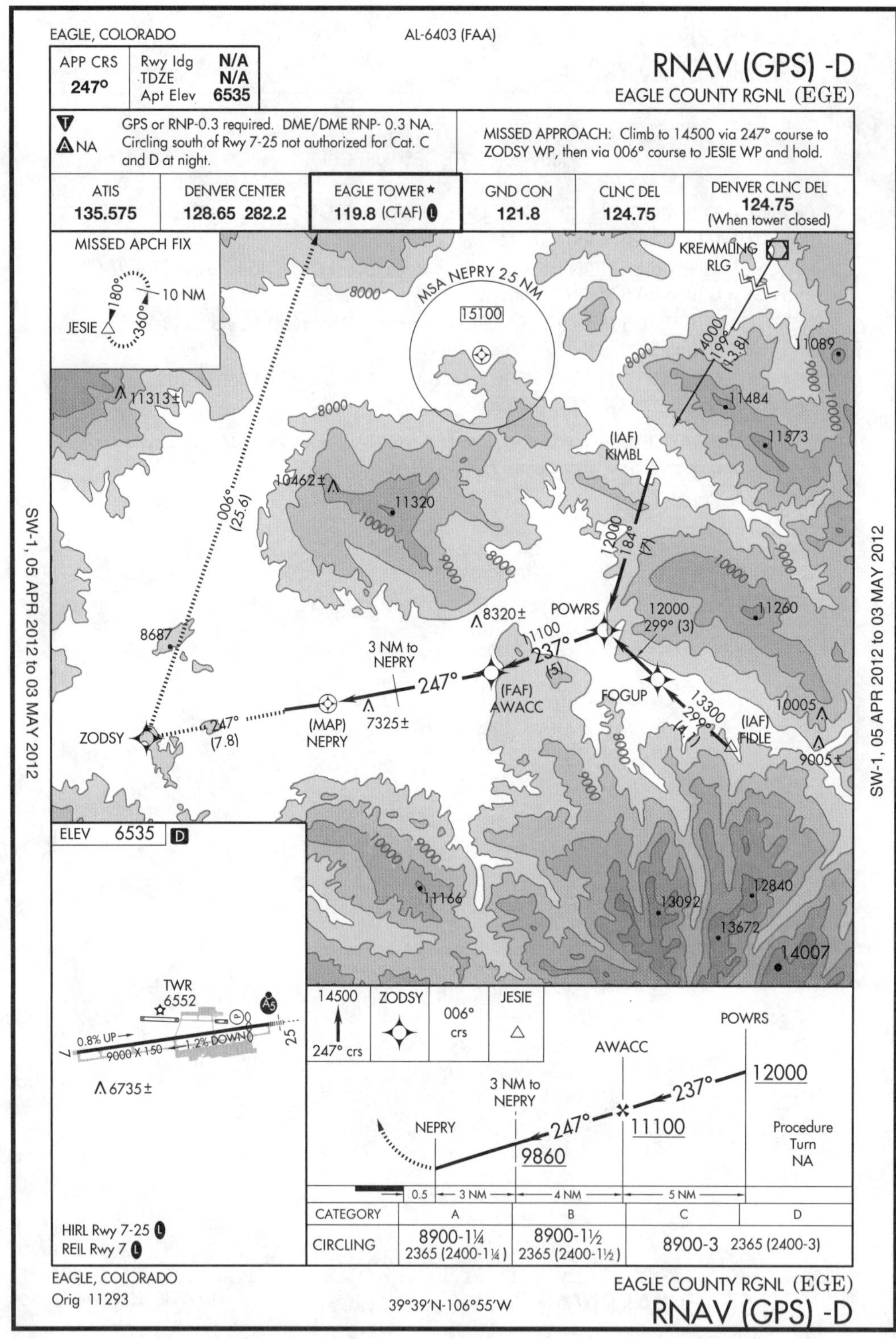

CATEGORY	A	B	C	D
CIRCLING	8900-1¼ 2365 (2400-1¼)	8900-1½ 2365 (2400-1½)	8900-3 2365 (2400-3)	

Figure 321

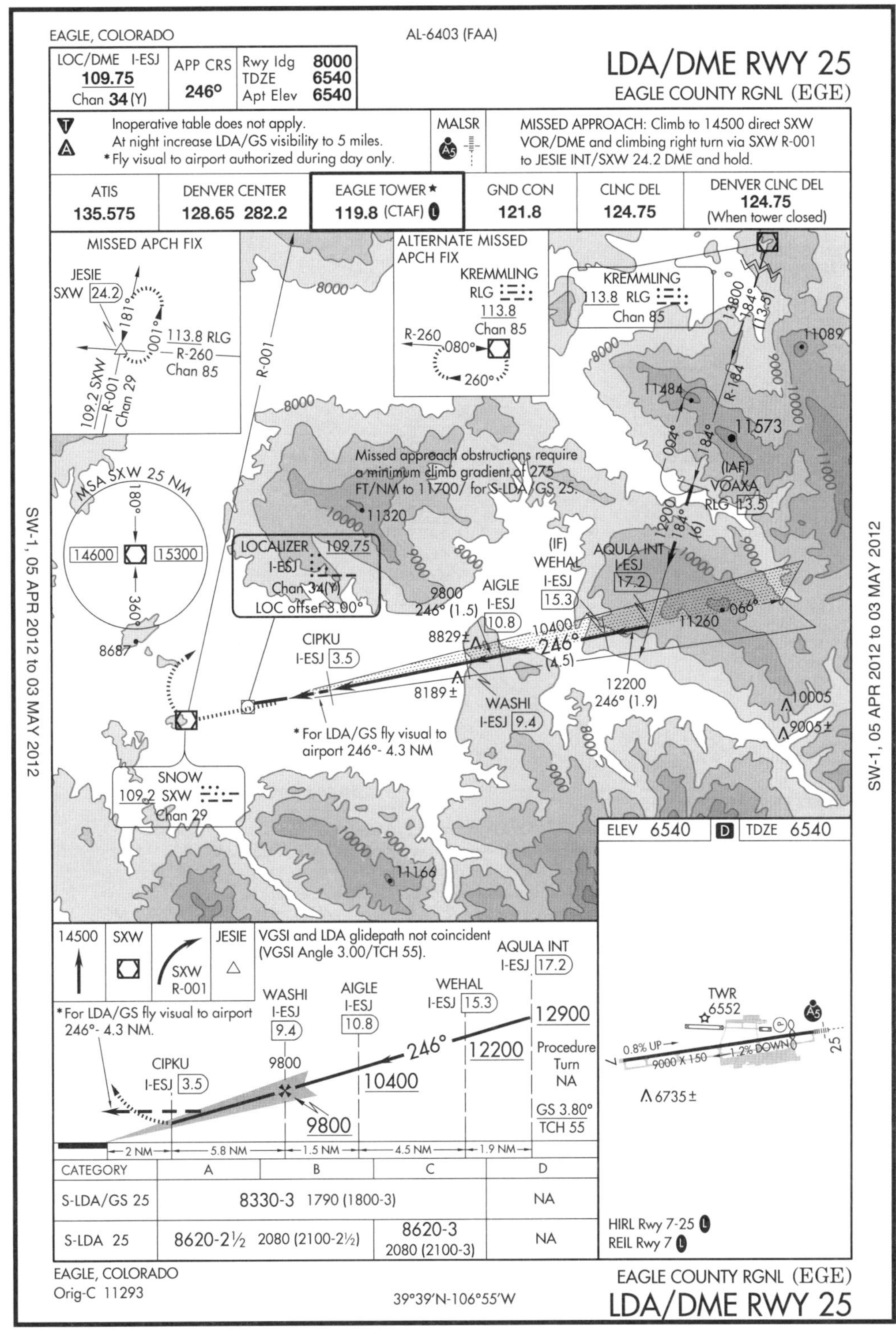

CATEGORY	A	B	C	D
S-LDA/GS 25	8330-3 1790 (1800-3)			NA
S-LDA 25	8620-2½ 2080 (2100-2½)		8620-3 2080 (2100-3)	NA

Figure 322

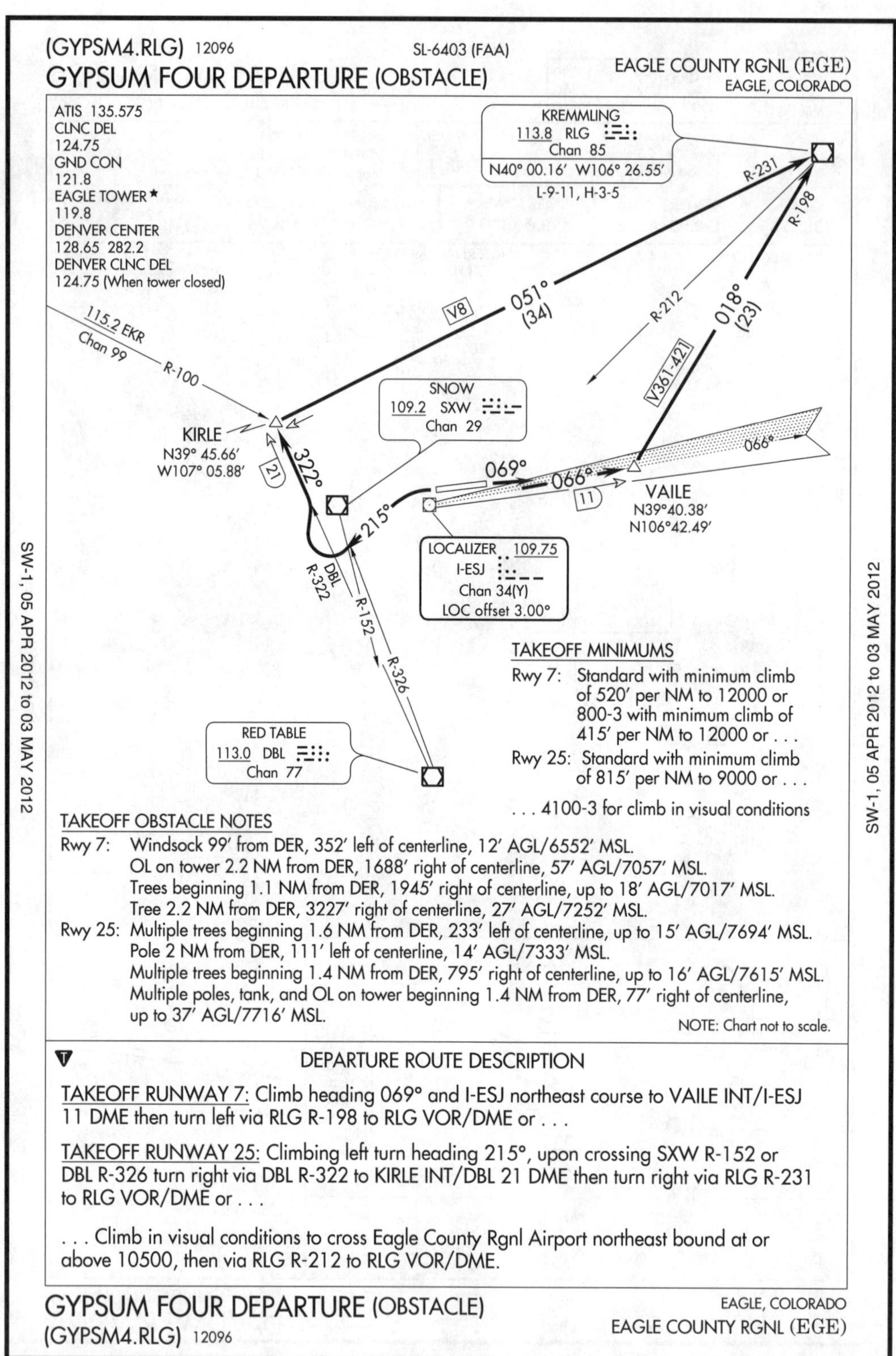

(GYPSM4.RLG) 12096
SL-6403 (FAA)
GYPSUM FOUR DEPARTURE (OBSTACLE)
EAGLE COUNTY RGNL (EGE)
EAGLE, COLORADO
ATIS 135.575
CLNC DEL 124.75
GND CON 121.8
EAGLE TOWER ★ 119.8
DENVER CENTER 128.65 282.2
DENVER CLNC DEL 124.75 (When tower closed)
KREMMLING
113.8 RLG
Chan 85
N40° 00.16′ W106° 26.55′
L-9-11, H-3-5
R-231
R-198
R-212
051° (34)
V8
018° (23)
V361-421
115.2 EKR
Chan 99
R-100
KIRLE
N39° 45.66′
W107° 05.88′
21
322°
SNOW
109.2 SXW
Chan 29
069°
066°
11
VAILE
N39°40.38′
N106°42.49′
215°
LOCALIZER 109.75
I-ESJ
Chan 34(Y)
LOC offset 3.00°
DBL R-322
R-152
R-326
RED TABLE
113.0 DBL
Chan 77
SW-1, 05 APR 2012 to 03 MAY 2012
TAKEOFF MINIMUMS
Rwy 7: Standard with minimum climb of 520′ per NM to 12000 or 800-3 with minimum climb of 415′ per NM to 12000 or . . .
Rwy 25: Standard with minimum climb of 815′ per NM to 9000 or . . .
. . . 4100-3 for climb in visual conditions
TAKEOFF OBSTACLE NOTES
Rwy 7: Windsock 99′ from DER, 352′ left of centerline, 12′ AGL/6552′ MSL.
OL on tower 2.2 NM from DER, 1688′ right of centerline, 57′ AGL/7057′ MSL.
Trees beginning 1.1 NM from DER, 1945′ right of centerline, up to 18′ AGL/7017′ MSL.
Tree 2.2 NM from DER, 3227′ right of centerline, 27′ AGL/7252′ MSL.
Rwy 25: Multiple trees beginning 1.6 NM from DER, 233′ left of centerline, up to 15′ AGL/7694′ MSL.
Pole 2 NM from DER, 111′ left of centerline, 14′ AGL/7333′ MSL.
Multiple trees beginning 1.4 NM from DER, 795′ right of centerline, up to 16′ AGL/7615′ MSL.
Multiple poles, tank, and OL on tower beginning 1.4 NM from DER, 77′ right of centerline, up to 37′ AGL/7716′ MSL.
NOTE: Chart not to scale.
DEPARTURE ROUTE DESCRIPTION
TAKEOFF RUNWAY 7: Climb heading 069° and I-ESJ northeast course to VAILE INT/I-ESJ 11 DME then turn left via RLG R-198 to RLG VOR/DME or . . .
TAKEOFF RUNWAY 25: Climbing left turn heading 215°, upon crossing SXW R-152 or DBL R-326 turn right via DBL R-322 to KIRLE INT/DBL 21 DME then turn right via RLG R-231 to RLG VOR/DME or . . .
. . . Climb in visual conditions to cross Eagle County Rgnl Airport northeast bound at or above 10500, then via RLG R-212 to RLG VOR/DME.
GYPSUM FOUR DEPARTURE (OBSTACLE)
(GYPSM4.RLG) 12096
EAGLE, COLORADO
EAGLE COUNTY RGNL (EGE)

Figure 323

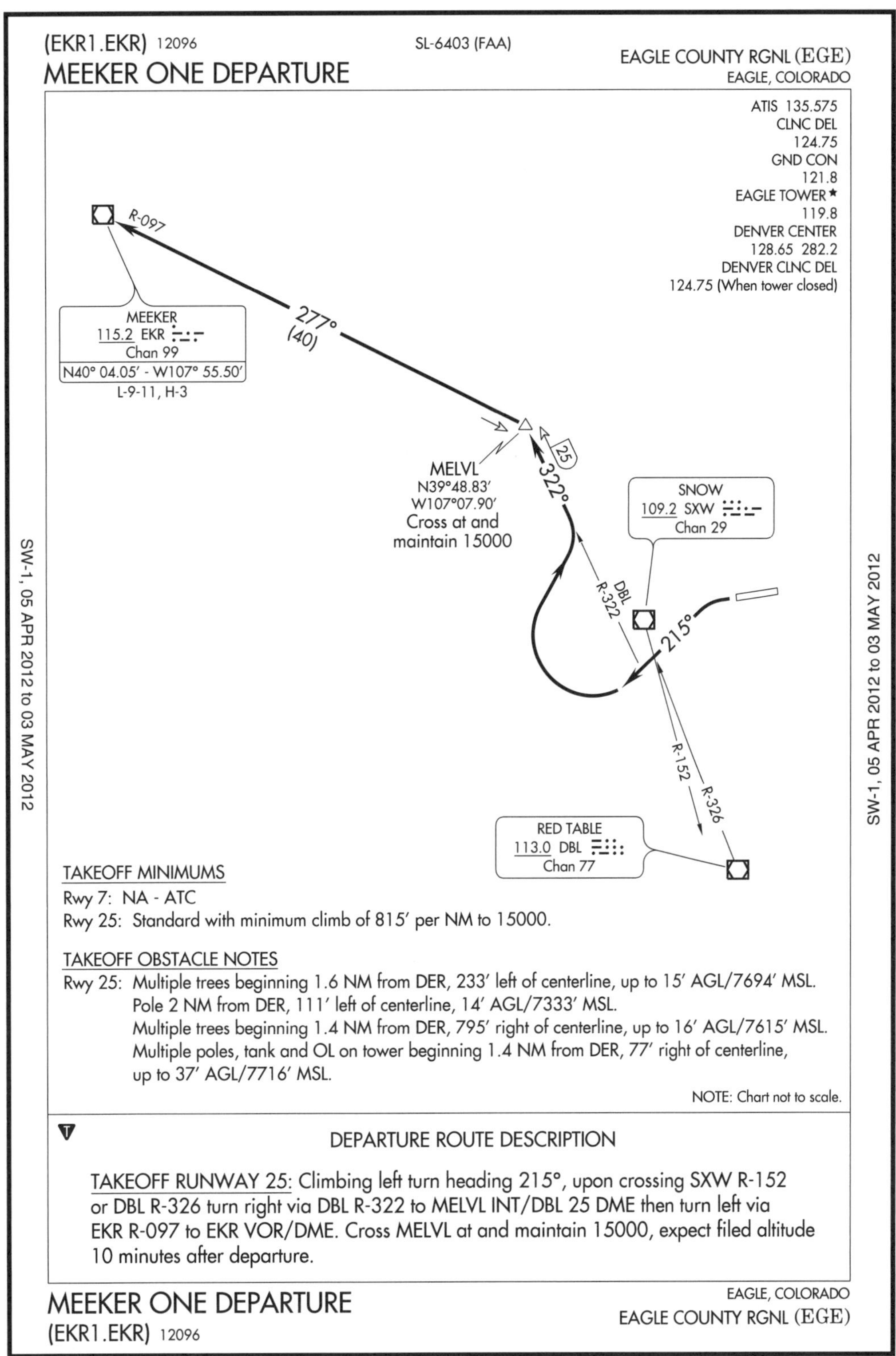
(EKR1.EKR) 12096
MEEKER ONE DEPARTURE
SL-6403 (FAA)
EAGLE COUNTY RGNL (EGE)
EAGLE, COLORADO
ATIS 135.575
CLNC DEL
124.75
GND CON
121.8
EAGLE TOWER★
119.8
DENVER CENTER
128.65 282.2
DENVER CLNC DEL
124.75 (When tower closed)
R-097
MEEKER
115.2 EKR
Chan 99
N40° 04.05′ - W107° 55.50′
L-9-11, H-3
277°
(40)
MELVL
N39°48.83′
W107°07.90′
Cross at and
maintain 15000
25
322°
SNOW
109.2 SXW
Chan 29
DBL
R-322
215°
R-152
R-326
RED TABLE
113.0 DBL
Chan 77
SW-1, 05 APR 2012 to 03 MAY 2012
TAKEOFF MINIMUMS
Rwy 7: NA - ATC
Rwy 25: Standard with minimum climb of 815′ per NM to 15000.
TAKEOFF OBSTACLE NOTES
Rwy 25: Multiple trees beginning 1.6 NM from DER, 233′ left of centerline, up to 15′ AGL/7694′ MSL.
Pole 2 NM from DER, 111′ left of centerline, 14′ AGL/7333′ MSL.
Multiple trees beginning 1.4 NM from DER, 795′ right of centerline, up to 16′ AGL/7615′ MSL.
Multiple poles, tank and OL on tower beginning 1.4 NM from DER, 77′ right of centerline, up to 37′ AGL/7716′ MSL.
NOTE: Chart not to scale.
DEPARTURE ROUTE DESCRIPTION
TAKEOFF RUNWAY 25: Climbing left turn heading 215°, upon crossing SXW R-152 or DBL R-326 turn right via DBL R-322 to MELVL INT/DBL 25 DME then turn left via EKR R-097 to EKR VOR/DME. Cross MELVL at and maintain 15000, expect filed altitude 10 minutes after departure.
MEEKER ONE DEPARTURE
(EKR1.EKR) 12096
EAGLE, COLORADO
EAGLE COUNTY RGNL (EGE)

Figure 324

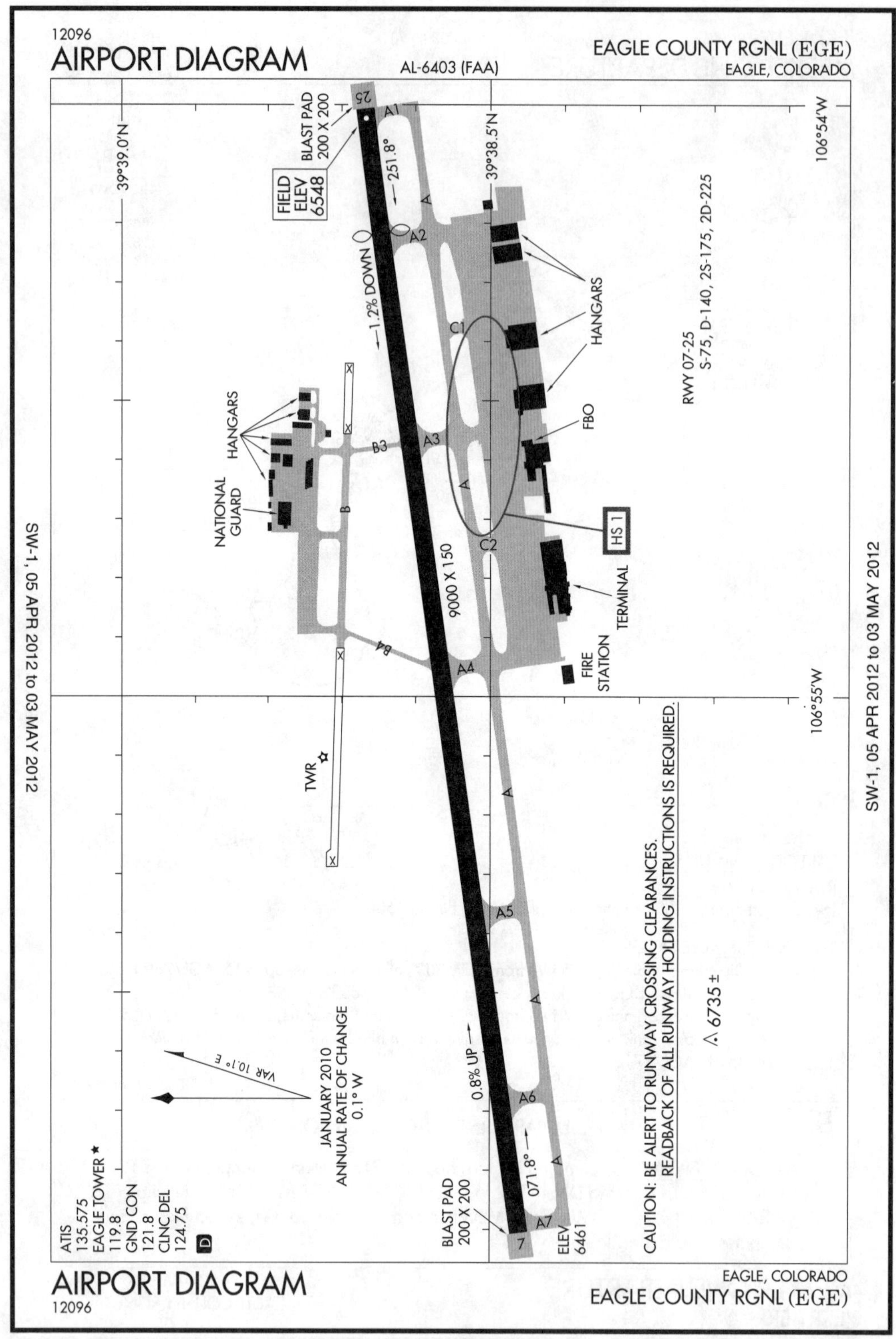
12096
AIRPORT DIAGRAM
AL-6403 (FAA)
EAGLE COUNTY RGNL (EGE)
EAGLE, COLORADO
FIELD ELEV 6548
BLAST PAD 200 X 200
25
A1
251.8°
39°39.0'N
39°38.5'N
106°54'W
106°55'W
1.2% DOWN
A2
A
HANGARS
C1
RWY 07-25
S-75, D-140, 2S-175, 2D-225
HANGARS
B3
A3
FBO
NATIONAL GUARD
B
HS 1
C2
9000 X 150
TERMINAL
B4
A4
FIRE STATION
TWR
CAUTION: BE ALERT TO RUNWAY CROSSING CLEARANCES.
READBACK OF ALL RUNWAY HOLDING INSTRUCTIONS IS REQUIRED.
A5
6735 ±
VAR 10.1° E
JANUARY 2010
ANNUAL RATE OF CHANGE
0.1° W
0.8% UP
A6
071.8°
BLAST PAD 200 X 200
A7
7
ELEV 6461
ATIS
135.575
EAGLE TOWER
119.8
GND CON
121.8
CLNC DEL
124.75
D
SW-1, 05 APR 2012 to 03 MAY 2012
SW-1, 05 APR 2012 to 03 MAY 2012
AIRPORT DIAGRAM
12096
EAGLE, COLORADO
EAGLE COUNTY RGNL (EGE)

Figure 325

ALTERNATE MINS
12096
M2

NAME	ALTERNATE MINIMUMS

DENVER, CO
CENTENNIAL (APA) ...... **ILS or LOC Rwy 35R**[1]
NDB Rwy 35R[2]
RNAV (GPS) Rwy 28[34]
RNAV (GPS) Rwy 35R[14]
[1]Categories A,B, 900-2; Category C, 900-2½, Category D, 900-2¾.
[2]Categories A,B, 1000-2; Categories C,D, 1000-3.
[3]Category D, 800-2¼.
[4]NA when local weather not available.

ROCKY MOUNTAIN METROPOLITAN (BJC) .... **ILS or LOC Y Rwy 29R**[123]
ILS or LOC Z Rwy 29R[23]
RNAV (GPS) Rwy 29L[3]
RNAV (GPS) Rwy 29R[3]
VOR/DME Rwy 29L/R[2]
[1]ILS, Categories A, B, C, D, 700-2.
[2]NA when control tower closed.
[3]NA when local weather not available.

EAGLE, CO
EAGLE COUNTY RGNL (EGE) .... **LDA/DME Rwy 25**
Categories A,B, 2100-2; Category C, 2100-3.
NA when control tower closed.
NA when local weather not available.

FARMINGTON, NM
FOUR CORNERS RGNL (FMN) .... **ILS or LOC Rwy 25**[12]
RNAV (GPS) Rwy 5[3]
RNAV (GPS) Rwy 7[3]
RNAV (GPS) Rwy 23[4]
RNAV (GPS) Rwy 25[3]
[1]NA when control tower closed.
[2]ILS, Categories B,C,D, 700-2.
[3]NA when local weather not available.
[4]Category D, 800-2¼.

FORT COLLINS/LOVELAND, CO
FORT COLLINS-LOVELAND MUNI (FNL) .... **RNAV (GPS) Rwy 15**
RNAV (GPS) Rwy 33
VOR/DME-A
NA when local weather not available.

GALLUP, NM
GALLUP MUNI (GUP) .... **RNAV (GPS) Rwy 6**[12]
RNAV (GPS) Rwy 24[3]
VOR Rwy 6[4]
[1]Categories A, B, 900-2; Category C, 900-2½; Category D, 900-3.
[2]NA when local weather not available.
[3]Category D, 900-3.
[4]Category C, 800-2¼; Category D, 900-3.

NAME	ALTERNATE MINIMUMS

GRAND JUNCTION, CO
GRAND JUNCTION RGNL (GJT) .... **ILS or LOC Rwy 11**[12]
LDA/DME Rwy 29[3]
RNAV (GPS) Y Rwy 11[3]
[1]ILS, Category D, 700-2¼.
[2]NA when local weather not available.
[3]Category D, 800-2¼.

GREELEY, CO
GREELEY-WELD COUNTY (GXY) .... **ILS or LOC Rwy 34**
RNAV (GPS) Rwy 16
RNAV (GPS) Rwy 27
RNAV (GPS) Rwy 34
VOR-A
NA when local weather not available.

GUNNISON, CO
GUNNISON-CRESTED BUTTE RGNL (GUC) .... **ILS or LOC Rwy 6**[1]
RNAV (RNP) Rwy 6, 800-2¼
VOR or GPS-A[23]
[1]ILS,LOC, Categories A, B, C, 1600-3.
[2]Categories A,B,C, 1700-3;Cat D, 2300-3.
[3]NA when local altimeter setting not available except for operators with approved weather reporting service.

HAYDEN, CO
YAMPA VALLEY (HDN) .... **ILS or LOC/DME Y Rwy 10**[12]
RNAV (GPS) Y Rwy 10[12]
RNAV (RNP) Z Rwy 10, 800-2¼[1]
VOR/DME-B[3]
[1]NA when local weather not availalbe.
[2]Categories A, B, 1200-2; Categories C, D, 1200-3.
[3]Categories A, B, 1300-2; Categories C, D, 1300-3.

HOBBS, NM
LEA COUNTY RGNL (HOB) .... **ILS or LOC Rwy 3**[1]
LOC/DME BC Rwy 21[2]
RNAV (GPS) Rwy 3[3]
RNAV (GPS) Rwy 21[2]
RNAV (GPS) Rwy 30[2]
VOR/DME or TACAN Rwy 21[2]
VOR or TACAN Rwy 3[2]
[1]NA when control tower closed.
[2]NA when control tower closed, except standard for operators with approved weather reporting service.
[3]NA when local weather not available.

05 APR 2012 to 03 MAY 2012

ALTERNATE MINS
12096
SW-1
M2

Figure 326

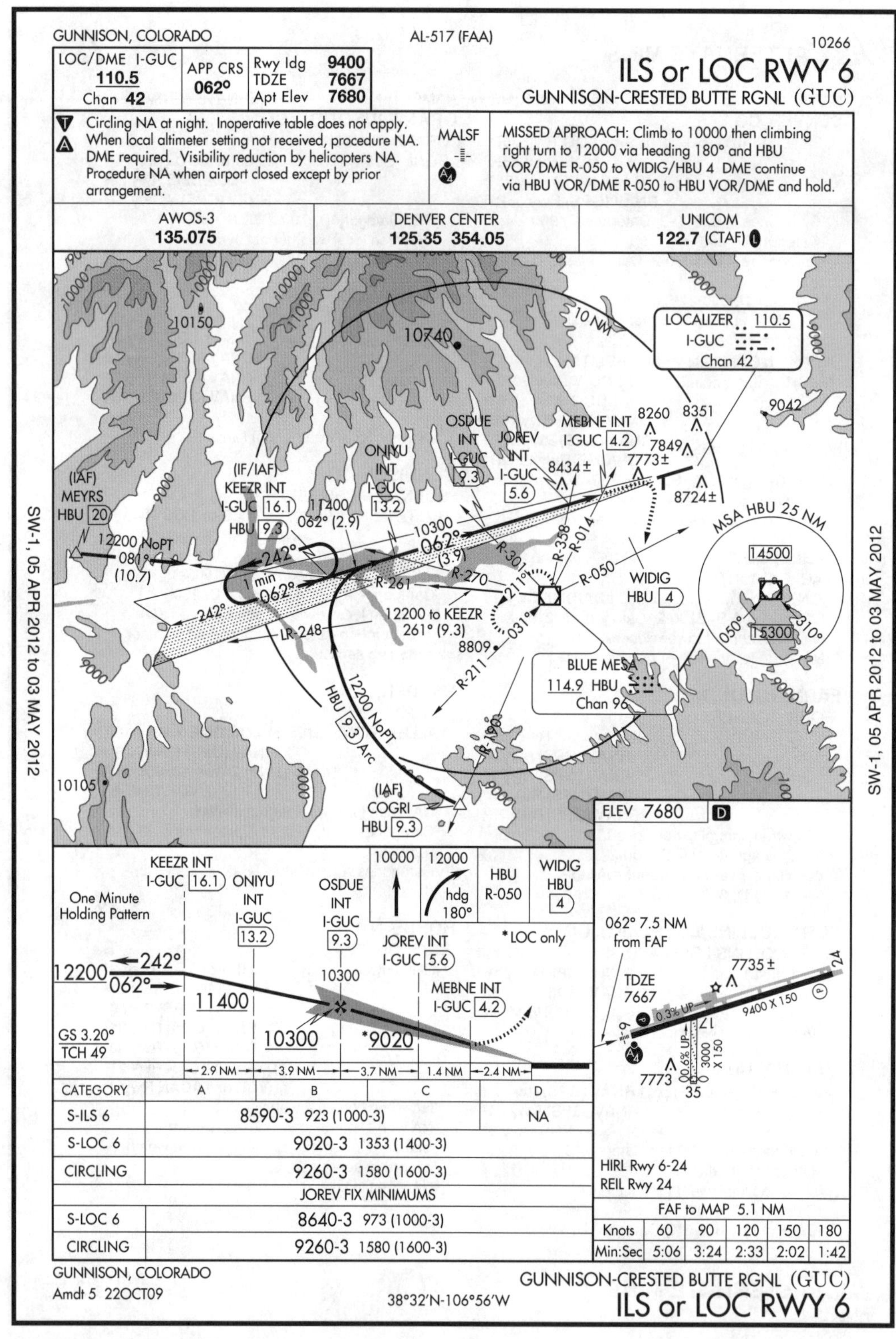
GUNNISON, COLORADO
AL-517 (FAA)
10266
LOC/DME I-GUC 110.5 Chan 42
APP CRS 062°
Rwy ldg 9400
TDZE 7667
Apt Elev 7680
ILS or LOC RWY 6
GUNNISON-CRESTED BUTTE RGNL (GUC)
Circling NA at night. Inoperative table does not apply.
When local altimeter setting not received, procedure NA. DME required. Visibility reduction by helicopters NA. Procedure NA when airport closed except by prior arrangement.
MALSF
MISSED APPROACH: Climb to 10000 then climbing right turn to 12000 via heading 180° and HBU VOR/DME R-050 to WIDIG/HBU 4 DME continue via HBU VOR/DME R-050 to HBU VOR/DME and hold.
AWOS-3 135.075
DENVER CENTER 125.35 354.05
UNICOM 122.7 (CTAF)
LOCALIZER 110.5 I-GUC Chan 42
10 NM
10740
10150
9042
MEBNE INT I-GUC 4.2
8260
8351
7849
7773±
8724±
OSDUE INT I-GUC 9.3
JOREV INT I-GUC 5.6
8434±
ONIYU INT I-GUC 13.2
(IF/IAF) KEEZR INT I-GUC 16.1 HBU 9.3
11400 062° (2.9)
10300 062° (3.9)
(IAF) MEYRS HBU 20
12200 NoPT 081° (10.7)
242°
1 min
062°
R-261
R-270
R-301
R-358
R-014
R-050
211°
031°
WIDIG HBU 4
MSA HBU 25 NM
14500
15300
050°
310°
12200 to KEEZR 261° (9.3)
8809
LR-248
R-211
BLUE MESA 114.9 HBU Chan 96
12200 NoPT HBU 9.3 Arc
R-196
(IAF) COGRI HBU 9.3
10105
SW-1, 05 APR 2012 to 03 MAY 2012
ELEV 7680
KEEZR INT I-GUC 16.1
ONIYU INT I-GUC 13.2
OSDUE INT I-GUC 9.3
10000
12000 hdg 180°
HBU R-050
WIDIG HBU 4
*LOC only
One Minute Holding Pattern
12200 242° 062°
11400
10300
JOREV INT I-GUC 5.6
MEBNE INT I-GUC 4.2
10300
*9020
GS 3.20° TCH 49
2.9 NM 3.9 NM 3.7 NM 1.4 NM 2.4 NM
062° 7.5 NM from FAF
TDZE 7667
7735±
0.3% UP
9400 X 150
0.6% UP
3000 X 150
7773
35
24
6
17
CATEGORY A B C D
S-ILS 6 8590-3 923 (1000-3) NA
S-LOC 6 9020-3 1353 (1400-3)
CIRCLING 9260-3 1580 (1600-3)
JOREV FIX MINIMUMS
S-LOC 6 8640-3 973 (1000-3)
CIRCLING 9260-3 1580 (1600-3)
HIRL Rwy 6-24
REIL Rwy 24
FAF to MAP 5.1 NM
Knots 60 90 120 150 180
Min:Sec 5:06 3:24 2:33 2:02 1:42
GUNNISON, COLORADO
Amdt 5 22OCT09
38°32'N-106°56'W
GUNNISON-CRESTED BUTTE RGNL (GUC)
ILS or LOC RWY 6

Figure 327

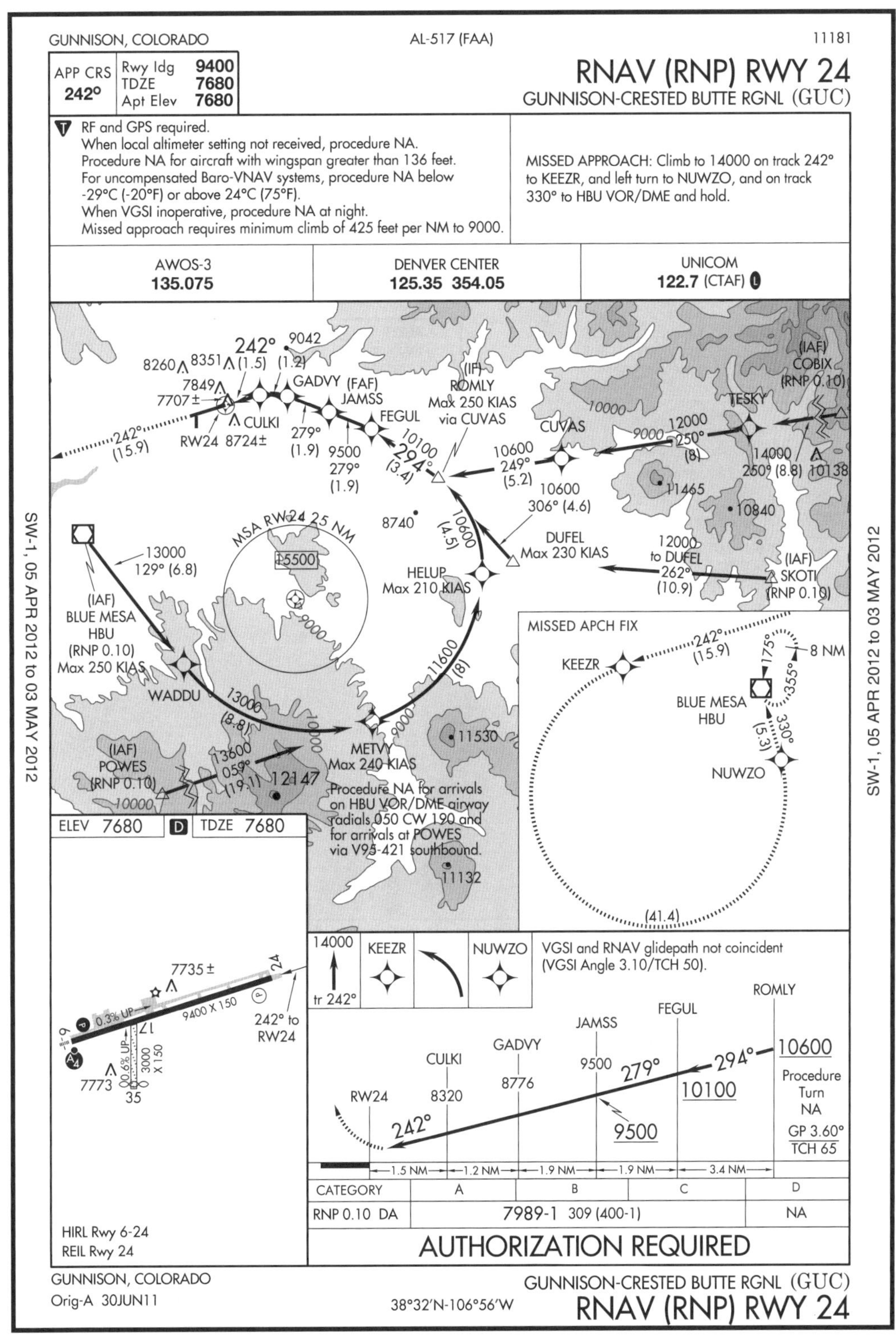
GUNNISON, COLORADO
AL-517 (FAA)
11181
APP CRS 242°
Rwy ldg 9400
TDZE 7680
Apt Elev 7680
RNAV (RNP) RWY 24
GUNNISON-CRESTED BUTTE RGNL (GUC)
RF and GPS required.
When local altimeter setting not received, procedure NA.
Procedure NA for aircraft with wingspan greater than 136 feet.
For uncompensated Baro-VNAV systems, procedure NA below -29°C (-20°F) or above 24°C (75°F).
When VGSI inoperative, procedure NA at night.
Missed approach requires minimum climb of 425 feet per NM to 9000.
MISSED APPROACH: Climb to 14000 on track 242° to KEEZR, and left turn to NUWZO, and on track 330° to HBU VOR/DME and hold.
AWOS-3 135.075
DENVER CENTER 125.35 354.05
UNICOM 122.7 (CTAF)
SW-1, 05 APR 2012 to 03 MAY 2012
(IAF) COBIX (RNP 0.10)
TESKY
(IF) ROMLY Max 250 KIAS via CUVAS
CUVAS
DUFEL Max 230 KIAS
(IAF) SKOTI (RNP 0.10)
HELUP Max 210 KIAS
(IAF) BLUE MESA HBU (RNP 0.10) Max 250 KIAS
WADDU
METVY Max 240 KIAS
(IAF) POWES (RNP 0.10)
GADVY (FAF)
JAMSS
FEGUL
CULKI
RW24 8724±
MSA RW24 25 NM
Procedure NA for arrivals on HBU VOR/DME airway radials 050 CW 190 and for arrivals at POWES via V95-421 southbound.
MISSED APCH FIX
KEEZR
BLUE MESA HBU
NUWZO
ELEV 7680
TDZE 7680
9400 X 150
3000 X 150
0.3% UP
0.6% UP
242° to RW24
HIRL Rwy 6-24
REIL Rwy 24
14000 tr 242°
VGSI and RNAV glidepath not coincident (VGSI Angle 3.10/TCH 50).
Procedure Turn NA
GP 3.60° TCH 65
CATEGORY A B C D
RNP 0.10 DA 7989-1 309 (400-1) NA
AUTHORIZATION REQUIRED
GUNNISON, COLORADO
Orig-A 30JUN11
38°32'N-106°56'W
GUNNISON-CRESTED BUTTE RGNL (GUC)
RNAV (RNP) RWY 24

Figure 328

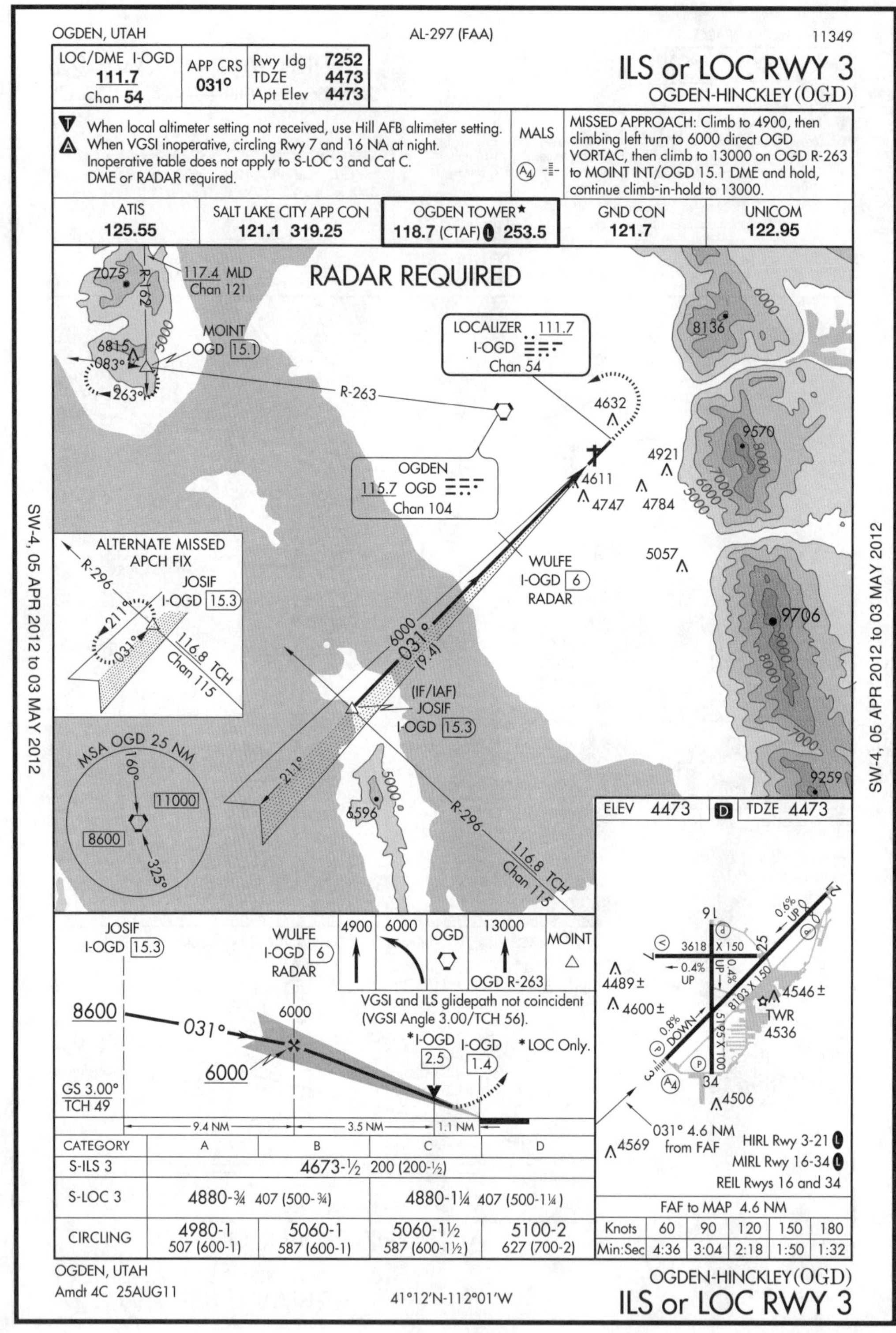

CATEGORY	A	B	C	D
S-ILS 3	4673-½ 200 (200-½)			
S-LOC 3	4880-¾ 407 (500-¾)		4880-1¼ 407 (500-1¼)	
CIRCLING	4980-1 507 (600-1)	5060-1 587 (600-1)	5060-1½ 587 (600-1½)	5100-2 627 (700-2)

FAF to MAP 4.6 NM

Knots	60	90	120	150	180
Min:Sec	4:36	3:04	2:18	1:50	1:32

Figure 329

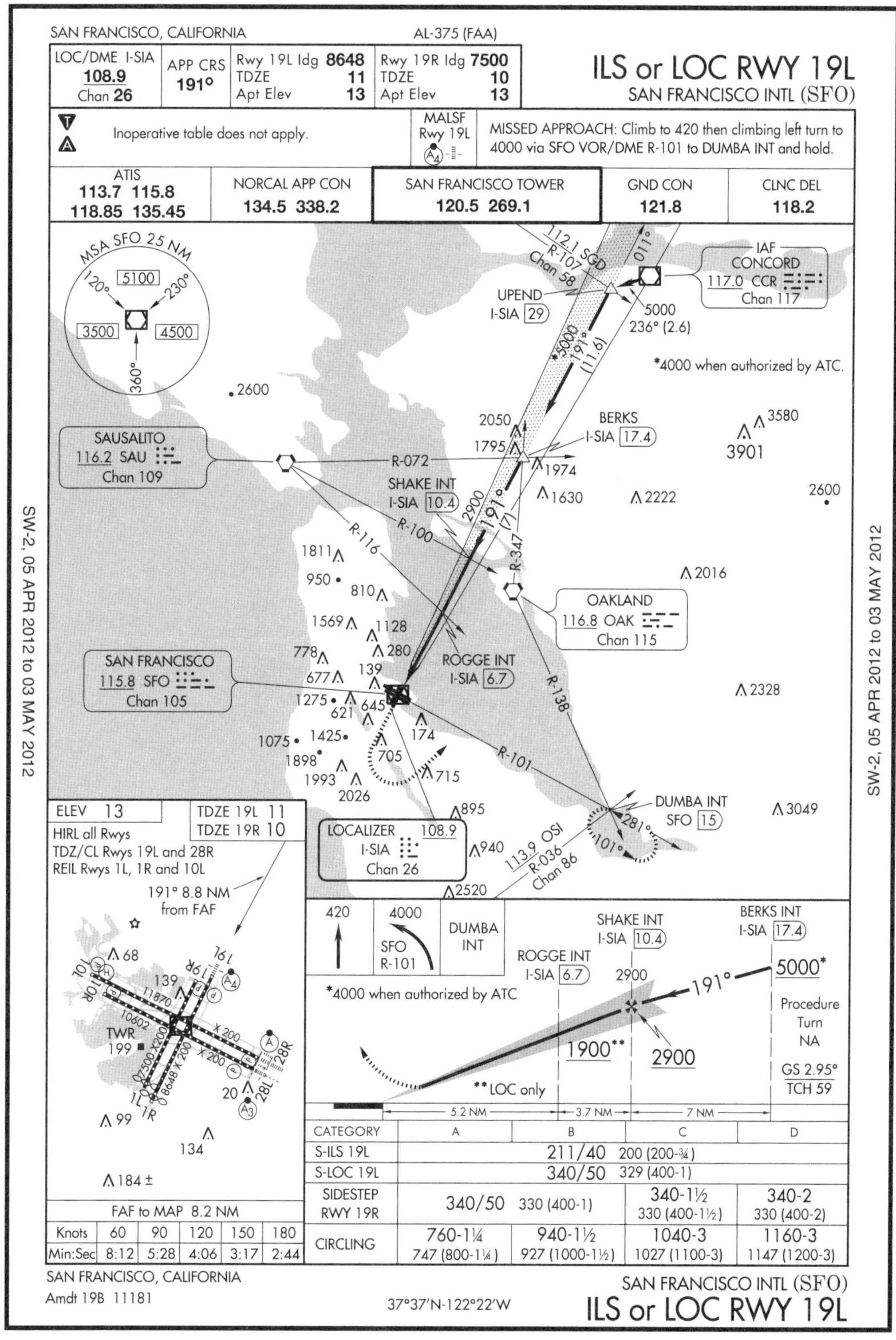

FAF to MAP 8.2 NM					
Knots	60	90	120	150	180
Min:Sec	8:12	5:28	4:06	3:17	2:44

CATEGORY	A	B	C	D
S-ILS 19L	211/40 200 (200-¾)			
S-LOC 19L	340/50 329 (400-1)			
SIDESTEP RWY 19R	340/50 330 (400-1)		340-1½ 330 (400-1½)	340-2 330 (400-2)
CIRCLING	760-1¼ 747 (800-1¼)	940-1½ 927 (1000-1½)	1040-3 1027 (1100-3)	1160-3 1147 (1200-3)

SAN FRANCISCO, CALIFORNIA
Amdt 19B 11181
37°37′N-122°22′W
SAN FRANCISCO INTL (SFO)
ILS or LOC RWY 19L

Figure 330

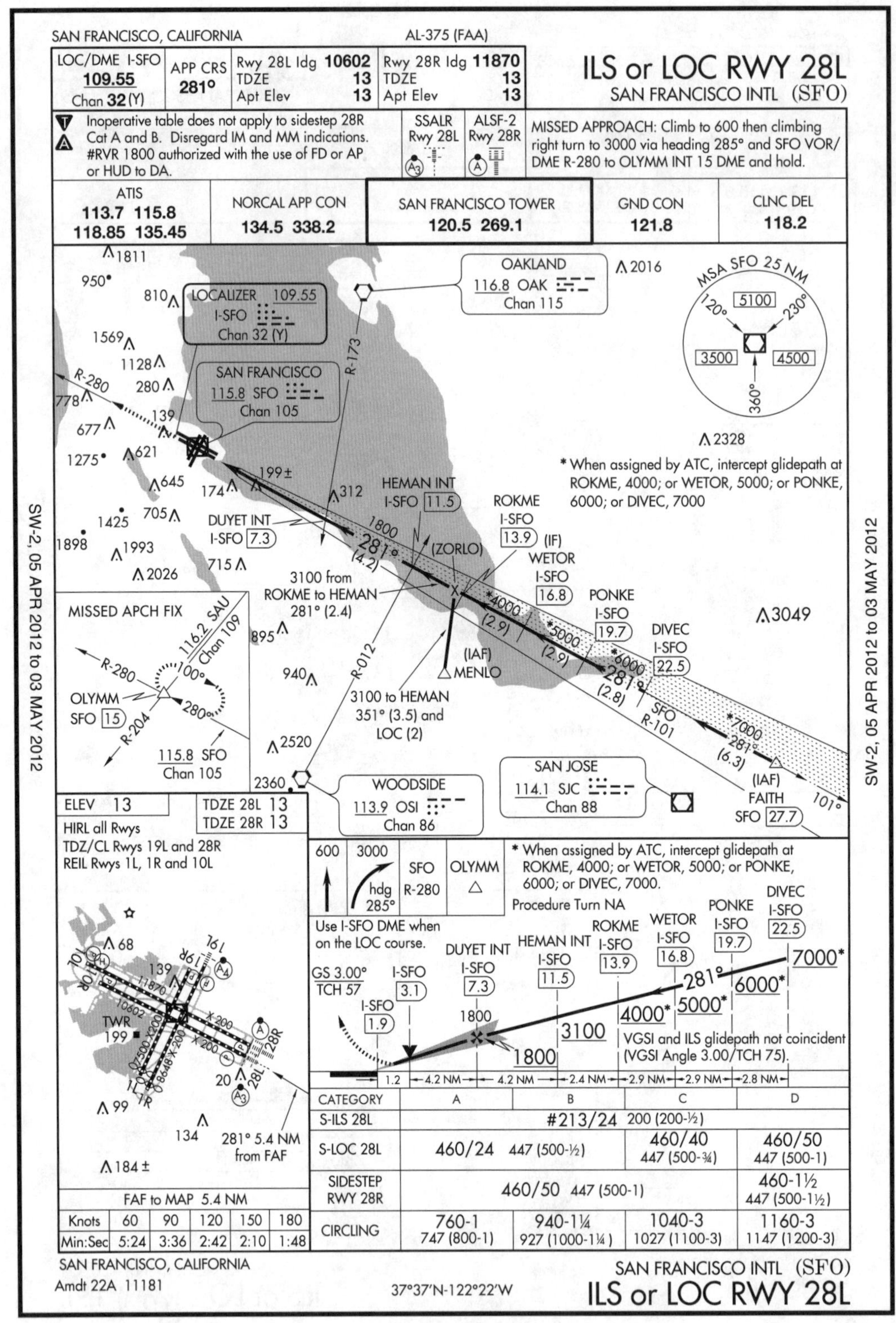

Knots	60	90	120	150	180
Min:Sec	5:24	3:36	2:42	2:10	1:48

CATEGORY	A	B	C	D
S-ILS 28L	#213/24 200 (200-½)			
S-LOC 28L	460/24 447 (500-½)		460/40 447 (500-¾)	460/50 447 (500-1)
SIDESTEP RWY 28R	460/50 447 (500-1)			460-1½ 447 (500-1½)
CIRCLING	760-1 747 (800-1)	940-1¼ 927 (1000-1¼)	1040-3 1027 (1100-3)	1160-3 1147 (1200-3)

Figure 331

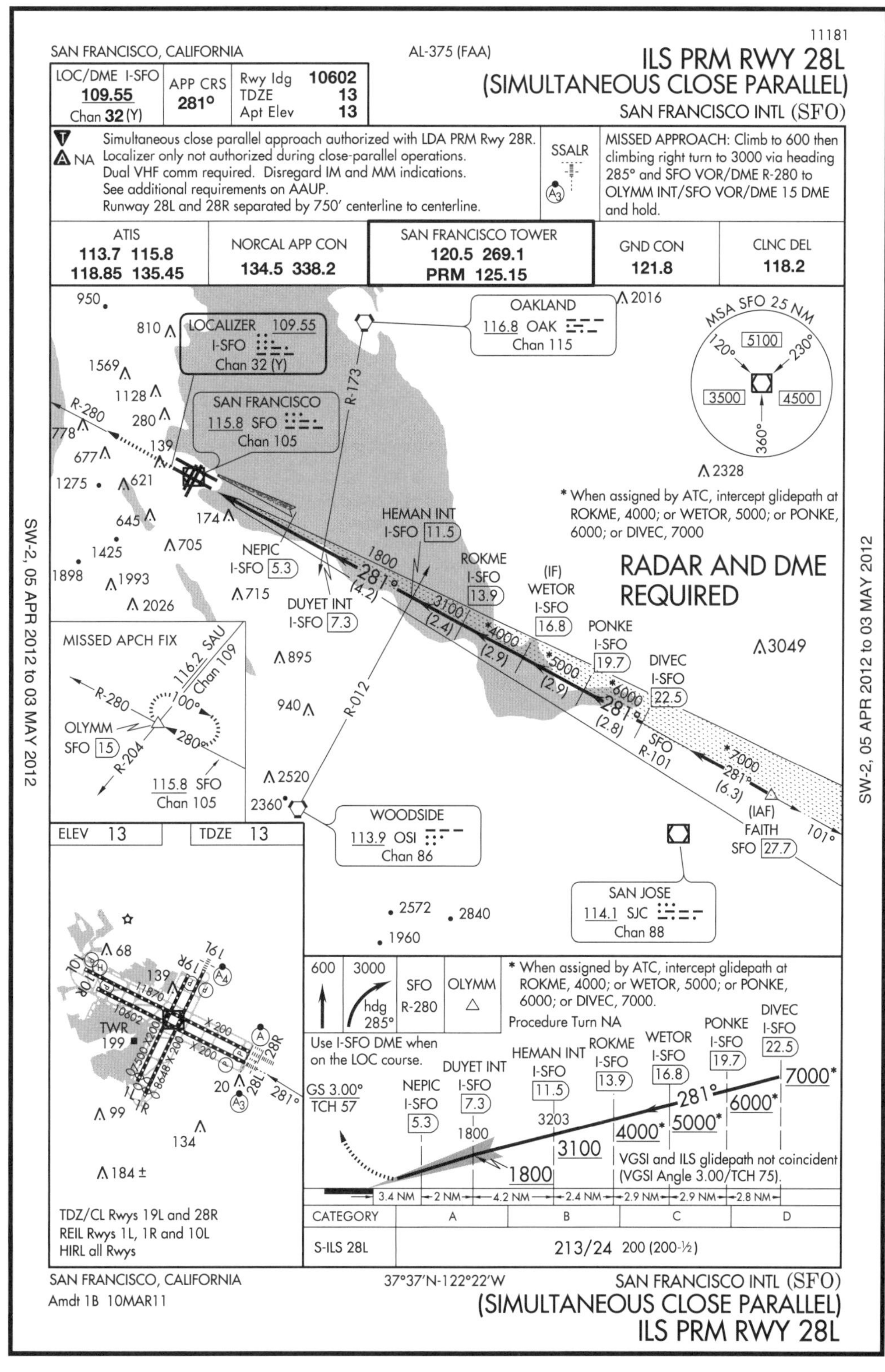

SAN FRANCISCO, CALIFORNIA
AL-375 (FAA)
11181
ILS PRM RWY 28L
(SIMULTANEOUS CLOSE PARALLEL)
SAN FRANCISCO INTL (SFO)
LOC/DME I-SFO 109.55 Chan 32 (Y)
APP CRS 281°
Rwy ldg 10602
TDZE 13
Apt Elev 13
Simultaneous close parallel approach authorized with LDA PRM Rwy 28R.
NA Localizer only not authorized during close-parallel operations.
Dual VHF comm required. Disregard IM and MM indications.
See additional requirements on AAUP.
Runway 28L and 28R separated by 750′ centerline to centerline.
SSALR
MISSED APPROACH: Climb to 600 then climbing right turn to 3000 via heading 285° and SFO VOR/DME R-280 to OLYMM INT/SFO VOR/DME 15 DME and hold.
ATIS 113.7 115.8 118.85 135.45
NORCAL APP CON 134.5 338.2
SAN FRANCISCO TOWER 120.5 269.1 PRM 125.15
GND CON 121.8
CLNC DEL 118.2
LOCALIZER 109.55 I-SFO Chan 32 (Y)
SAN FRANCISCO 115.8 SFO Chan 105
OAKLAND 116.8 OAK Chan 115
MSA SFO 25 NM
5100 3500 4500
* When assigned by ATC, intercept glidepath at ROKME, 4000; or WETOR, 5000; or PONKE, 6000; or DIVEC, 7000
RADAR AND DME REQUIRED
HEMAN INT I-SFO 11.5
NEPIC I-SFO 5.3
DUYET INT I-SFO 7.3
ROKME I-SFO 13.9
(IF) WETOR I-SFO 16.8
PONKE I-SFO 19.7
DIVEC I-SFO 22.5
(IAF) FAITH SFO 27.7
MISSED APCH FIX
OLYMM SFO 15
116.2 SAU Chan 109
115.8 SFO Chan 105
WOODSIDE 113.9 OSI Chan 86
SAN JOSE 114.1 SJC Chan 88
SW-2, 05 APR 2012 to 03 MAY 2012
ELEV 13
TDZE 13
TDZ/CL Rwys 19L and 28R
REIL Rwys 1L, 1R and 10L
HIRL all Rwys
600 3000 hdg 285° SFO R-280 OLYMM
* When assigned by ATC, intercept glidepath at ROKME, 4000; or WETOR, 5000; or PONKE, 6000; or DIVEC, 7000.
Procedure Turn NA
Use I-SFO DME when on the LOC course.
GS 3.00° TCH 57
VGSI and ILS glidepath not coincident (VGSI Angle 3.00/TCH 75).
3.4 NM 2 NM 4.2 NM 2.4 NM 2.9 NM 2.9 NM 2.8 NM
CATEGORY A B C D
S-ILS 28L 213/24 200 (200-½)
SAN FRANCISCO, CALIFORNIA
Amdt 1B 10MAR11
37°37′N-122°22′W
SAN FRANCISCO INTL (SFO)
(SIMULTANEOUS CLOSE PARALLEL)
ILS PRM RWY 28L

Figure 332

ILS PRM RWY 28L Amdt 1B 11069 AL-375 (FAA)
(SIMULTANEOUS CLOSE PARALLEL)

SAN FRANCISCO INTL (SFO)
SAN FRANCISCO, CALIFORNIA

SW-2, 05 APR 2012 to 03 MAY 2012

ATTENTION ALL USERS PAGE (AAUP)

Condensed Briefing Points:

- Listen to the PRM monitor frequency when communicating with NORCAL approach control (135.65), no later than LOC intercept.
- Expect to be switched to SFO Tower (120.5) at NEPIC (I-SFO 5.3 DME).
- PRM monitor frequency may be de-selected after determining that the aircraft is on the tower frequency.

1. **ATIS.** When the ATIS broadcast advises that simultaneous ILS/PRM and LDA/PRM approaches are in progress, pilots should brief to fly the ILS/PRM 28L approach. If later advised to expect an ILS 28L approach, the ILS/PRM 28L chart may be used after completing the following briefing items:

   (a) Minimums and missed approach procedures are unchanged.
   (b) Monitor frequency no longer required.
   (c) A different glideslope intercept altitude may be assigned when advised to expect the ILS 28L approach.

Simultaneous parallel approaches will only be offered/conducted when the weather is at least 2100 feet (ceiling) and 4 miles (visibility).

2. **Dual VHF Communication required.** To avoid blocked transmissions, each runway will have two frequencies, a primary and a PRM monitor frequency. The NORCAL approach controller will transmit on both frequencies. The PRM Monitor controller's transmissions, if needed, will override both frequencies. Pilots will ONLY transmit on the approach controller's frequency (135.65), but will listen to both frequencies. Select the PRM monitor frequency audio only when in contact with NORCAL approach control (135.65). The volume levels should be set about the same on both radios so that the pilots will be able to hear transmissions on at least one frequency if the other is blocked. The PRM monitor frequency may be de-selected passing NEPIC.

3. **ALL "Breakouts"** are to be hand flown to assure that the maneuver is accomplished in the shortest amount of time. Pilots, when directed by ATC to break off an approach, must assume that an aircraft is blundering toward their course and a breakout must be initiated immediately.

   (a) ATC Directed "Breakouts:" ATC directed breakouts will consist of a turn and a climb or descent. Pilots must always initiate the breakout in response to an air traffic controller instruction. Controllers will give a descending breakout only when there are no other reasonable options available, but in no case will the descent be below minimum vectoring altitude (MVA) which provides at least 1000 feet required obstruction clearance. The MVA in the final approach segment is 1600 feet at San Francisco International Airport.

   (b) Phraseology - "TRAFFIC ALERT:" If an aircraft enters the "NO TRANSGRESSION ZONE" (NTZ), the controller will breakout the threatened aircraft on the adjacent approach. The phraseology for the breakout will be:

   "TRAFFIC ALERT, (aircraft call sign) TURN (left/right) IMMEDIATELY, HEADING (degrees), CLIMB/DESCEND AND MAINTAIN (altitude)".

4. Descending on (not above) the ILS glideslope ensures complying with any charted crossing restrictions and assists traffic on the LDA PRM 28R approach to mitigate possible wake turbulence encounters without destabilizing the LDA approach and creating a go-around.

5. **LDA Traffic:** While conducting this ILS/PRM approach to Runway 28L, other aircraft may be conducting the offset LDA/PRM approach to Runway 28R. These aircraft will approach from the right-rear and will re-align with 28R after making visual contact with the ILS traffic.

Special pilot training required. Pilots who are unable to participate will be afforded appropriate arrival services as operational conditions permit and must notify the controlling ARTCC as soon as practical, but at least 100 miles from destination.

SW-2, 05 APR 2012 to 03 MAY 2012

(SIMULTANEOUS CLOSE PARALLEL)
ILS PRM RWY 28L Amdt 1B 11069 37°37'N-122°22'W

SAN FRANCISCO, CALIFORNIA
SAN FRANCISCO INTL (SFO)

Figure 333

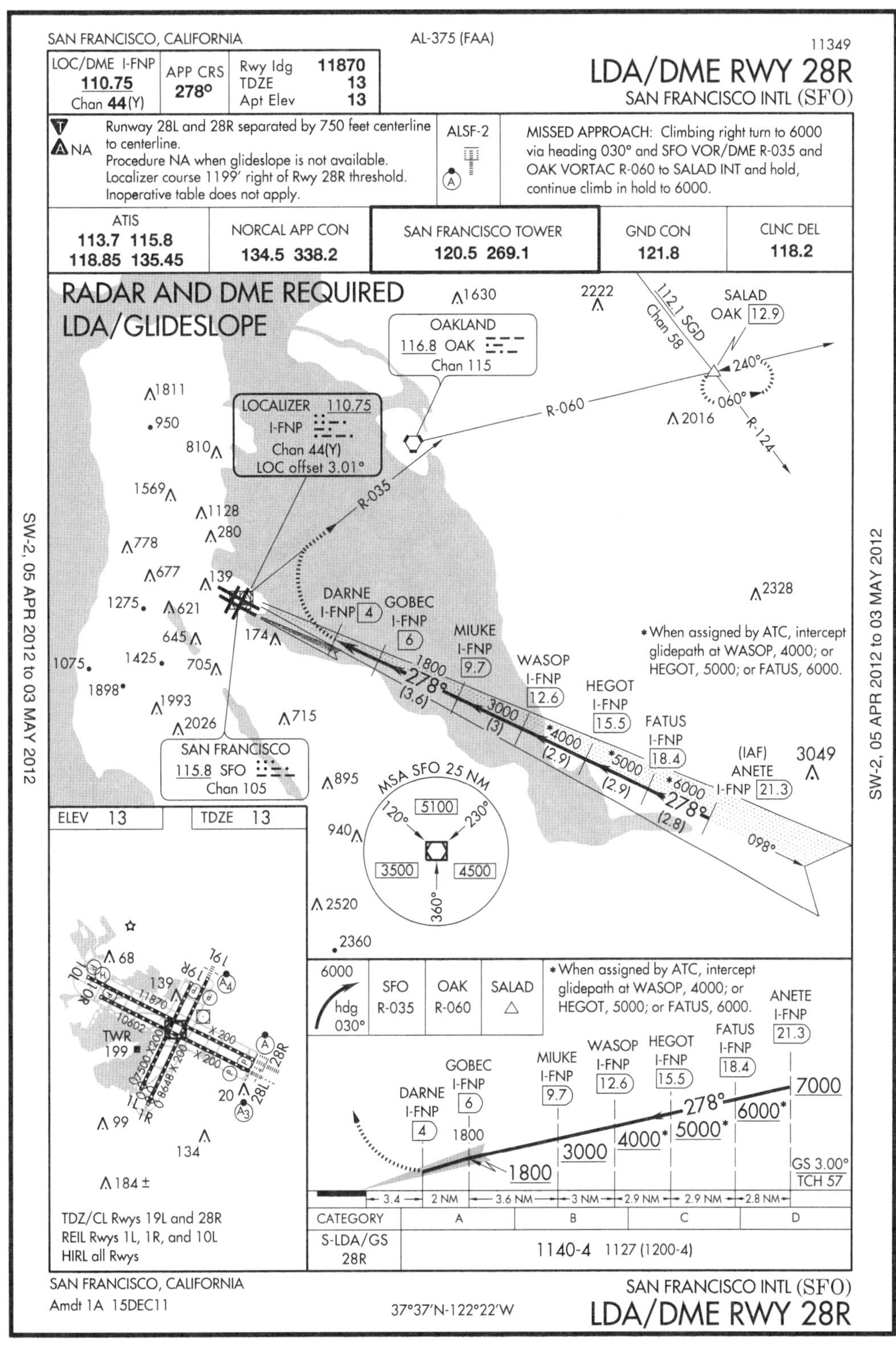
SAN FRANCISCO, CALIFORNIA
AL-375 (FAA)
11349
LDA/DME RWY 28R
SAN FRANCISCO INTL (SFO)
LOC/DME I-FNP 110.75 Chan 44(Y)
APP CRS 278°
Rwy ldg 11870
TDZE 13
Apt Elev 13
Runway 28L and 28R separated by 750 feet centerline to centerline.
Procedure NA when glideslope is not available.
Localizer course 1199' right of Rwy 28R threshold.
Inoperative table does not apply.
ALSF-2
MISSED APPROACH: Climbing right turn to 6000 via heading 030° and SFO VOR/DME R-035 and OAK VORTAC R-060 to SALAD INT and hold, continue climb in hold to 6000.
ATIS 113.7 115.8 118.85 135.45
NORCAL APP CON 134.5 338.2
SAN FRANCISCO TOWER 120.5 269.1
GND CON 121.8
CLNC DEL 118.2
RADAR AND DME REQUIRED
LDA/GLIDESLOPE
OAKLAND 116.8 OAK Chan 115
LOCALIZER 110.75 I-FNP Chan 44(Y) LOC offset 3.01°
SAN FRANCISCO 115.8 SFO Chan 105
112.1 SGD Chan 58
SALAD OAK 12.9
R-060
R-035
R-124
DARNE I-FNP 4
GOBEC I-FNP 6
MIUKE I-FNP 9.7
WASOP I-FNP 12.6
HEGOT I-FNP 15.5
FATUS I-FNP 18.4
(IAF) ANETE I-FNP 21.3
*When assigned by ATC, intercept glidepath at WASOP, 4000; or HEGOT, 5000; or FATUS, 6000.
MSA SFO 25 NM
5100
3500
4500
ELEV 13
TDZE 13
TDZ/CL Rwys 19L and 28R
REIL Rwys 1L, 1R, and 10L
HIRL all Rwys
6000 hdg 030°
SFO R-035
OAK R-060
SALAD
GS 3.00°
TCH 57
CATEGORY A B C D
S-LDA/GS 28R 1140-4 1127 (1200-4)
SAN FRANCISCO, CALIFORNIA
Amdt 1A 15DEC11
37°37'N-122°22'W
SAN FRANCISCO INTL (SFO)
LDA/DME RWY 28R
SW-2, 05 APR 2012 to 03 MAY 2012

Figure 334

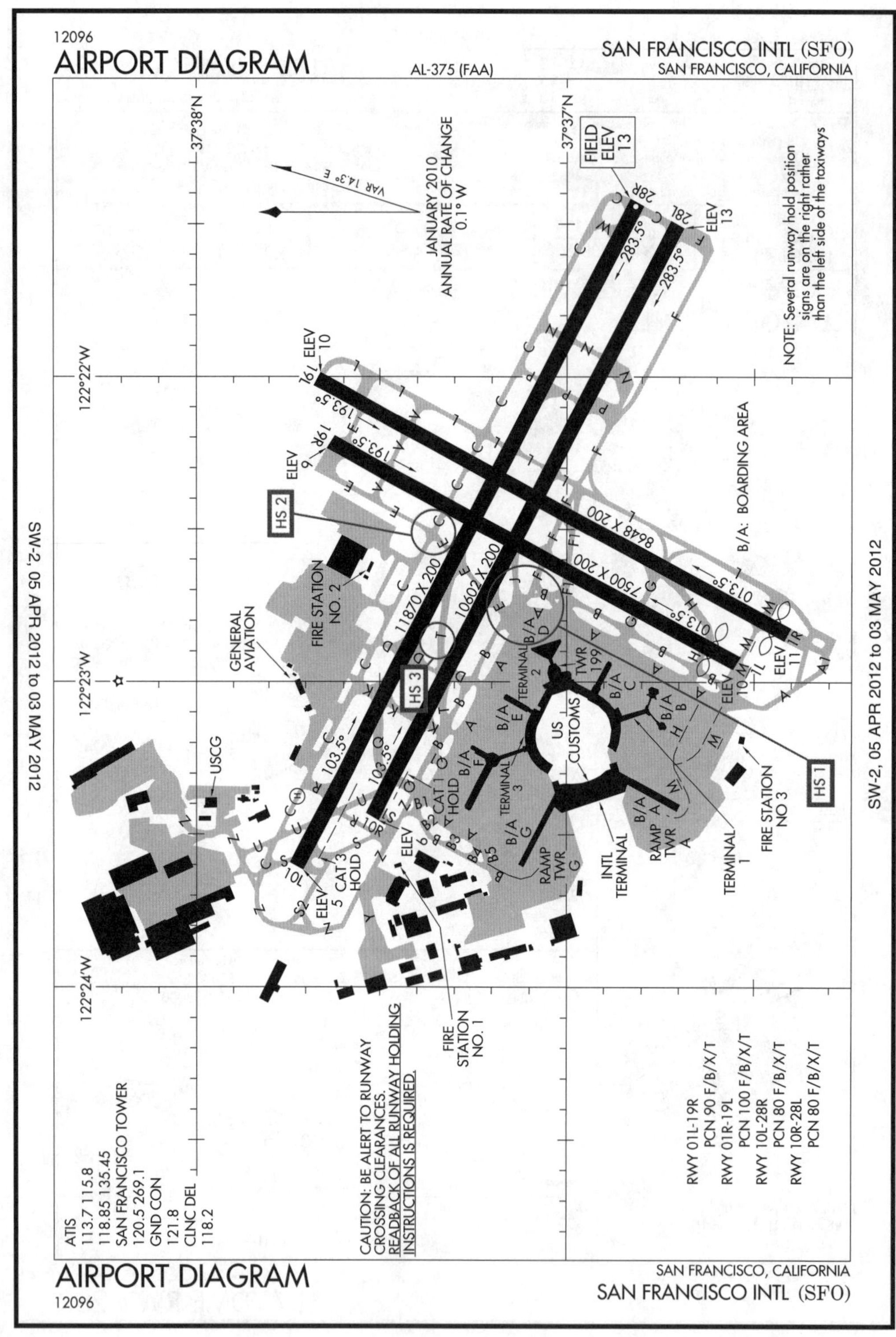
12096
AIRPORT DIAGRAM
AL-375 (FAA)
SAN FRANCISCO INTL (SFO)
SAN FRANCISCO, CALIFORNIA
SW-2, 05 APR 2012 to 03 MAY 2012
JANUARY 2010
ANNUAL RATE OF CHANGE
0.1° W
VAR 14.3° E
FIELD ELEV 13
NOTE: Several runway hold position signs are on the right rather than the left side of the taxiways
B/A: BOARDING AREA
ATIS
113.7 115.8
118.85 135.45
SAN FRANCISCO TOWER
120.5 269.1
GND CON
121.8
CLNC DEL
118.2
CAUTION: BE ALERT TO RUNWAY CROSSING CLEARANCES. READBACK OF ALL RUNWAY HOLDING INSTRUCTIONS IS REQUIRED.
RWY 01L-19R
PCN 90 F/B/X/T
RWY 01R-19L
PCN 100 F/B/X/T
RWY 10L-28R
PCN 80 F/B/X/T
RWY 10R-28L
PCN 80 F/B/X/T
11870 X 200
10602 X 200
8648 X 200
7500 X 200
FIRE STATION NO. 1
FIRE STATION NO. 2
FIRE STATION NO 3
GENERAL AVIATION
USCG
US CUSTOMS
INTL TERMINAL
TERMINAL 1
TERMINAL 2
TERMINAL 3
RAMP TWR A
RAMP TWR G
TWR 199
HS 1
HS 2
HS 3
CAT 3 HOLD
CAT 1 HOLD
AIRPORT DIAGRAM
12096
SAN FRANCISCO, CALIFORNIA
SAN FRANCISCO INTL (SFO)

Figure 335

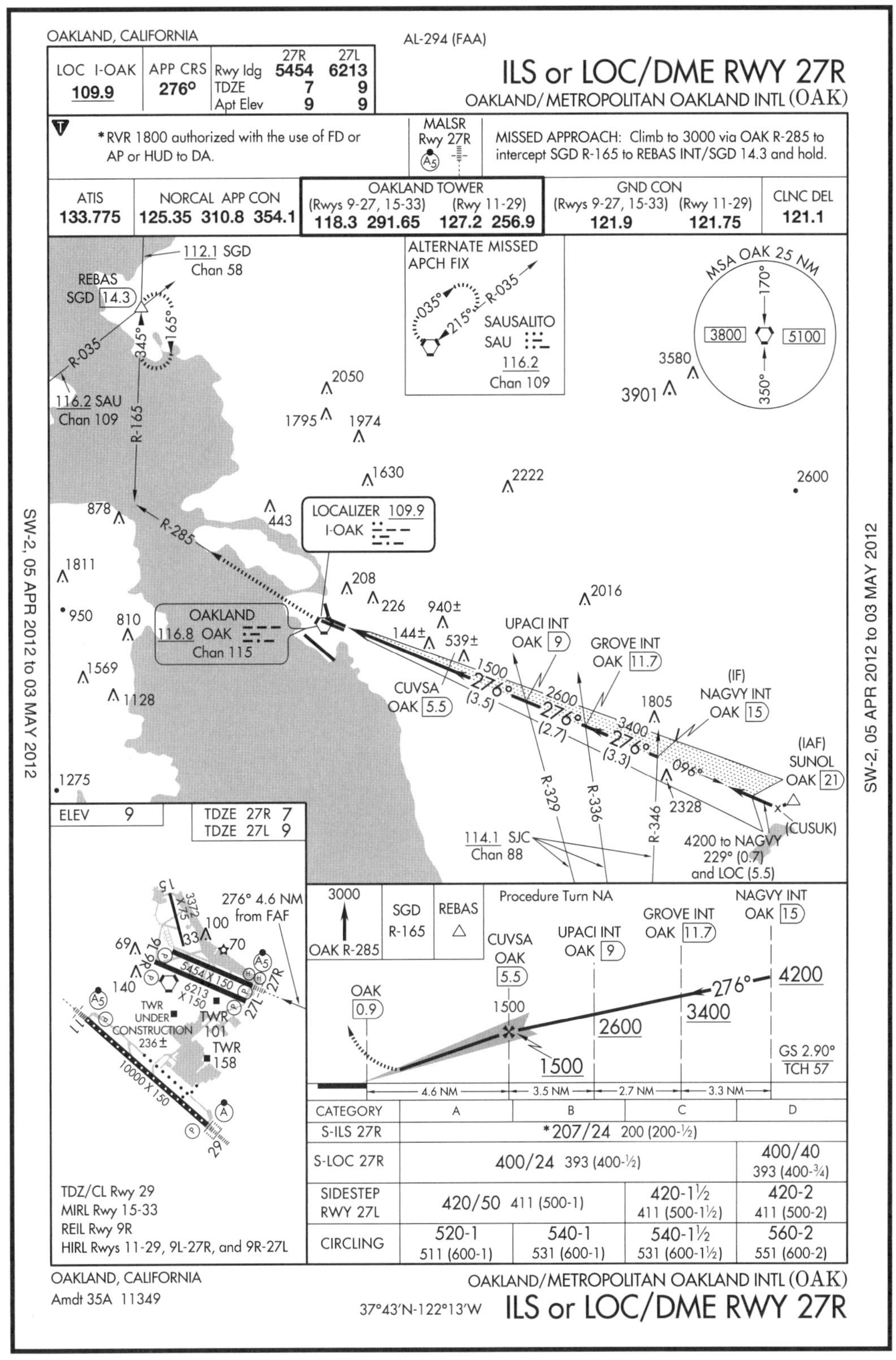

CATEGORY	A	B	C	D
S-ILS 27R	*207/24 200 (200-½)			
S-LOC 27R	400/24 393 (400-½)			400/40 393 (400-¾)
SIDESTEP RWY 27L	420/50 411 (500-1)		420-1½ 411 (500-1½)	420-2 411 (500-2)
CIRCLING	520-1 511 (600-1)	540-1 531 (600-1)	540-1½ 531 (600-1½)	560-2 551 (600-2)

Figure 336

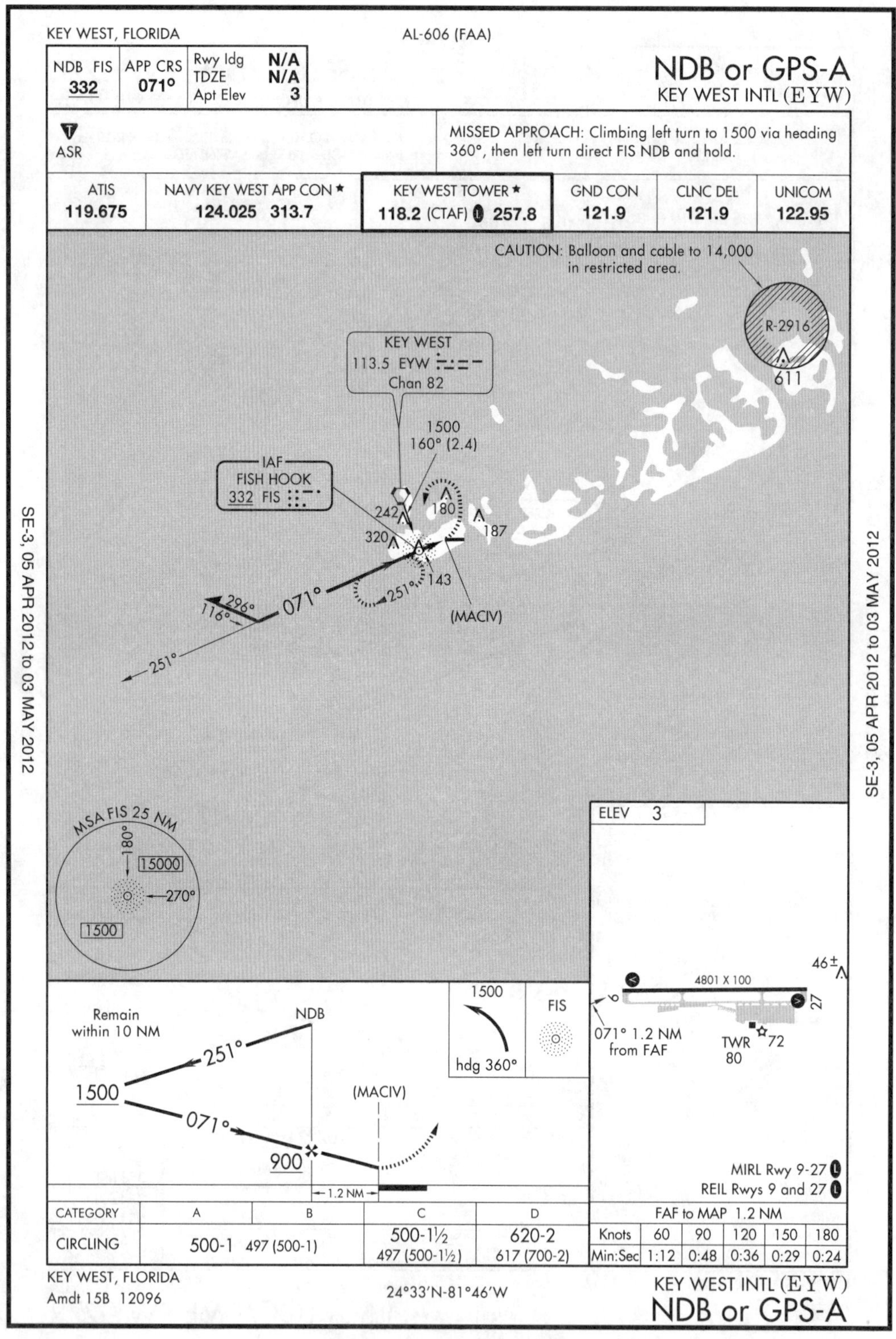
KEY WEST, FLORIDA
AL-606 (FAA)
NDB FIS 332
APP CRS 071°
Rwy ldg N/A
TDZE N/A
Apt Elev 3
NDB or GPS-A
KEY WEST INTL (EYW)
ASR
MISSED APPROACH: Climbing left turn to 1500 via heading 360°, then left turn direct FIS NDB and hold.
ATIS 119.675
NAVY KEY WEST APP CON ★ 124.025 313.7
KEY WEST TOWER ★ 118.2 (CTAF) 257.8
GND CON 121.9
CLNC DEL 121.9
UNICOM 122.95
CAUTION: Balloon and cable to 14,000 in restricted area.
R-2916
611
KEY WEST
113.5 EYW
Chan 82
1500
160° (2.4)
IAF
FISH HOOK
332 FIS
242
180
187
320
143
251°
071°
296°
116°
251°
(MACIV)
SE-3, 05 APR 2012 to 03 MAY 2012
MSA FIS 25 NM
180°
15000
270°
1500
ELEV 3
46±
4801 X 100
9
27
071° 1.2 NM from FAF
TWR 80
72
MIRL Rwy 9-27
REIL Rwys 9 and 27
Remain within 10 NM
NDB
251°
1500
071°
900
(MACIV)
1.2 NM
1500
hdg 360°
FIS
CATEGORY A B C D
CIRCLING 500-1 497 (500-1) 500-1½ 497 (500-1½) 620-2 617 (700-2)
FAF to MAP 1.2 NM
Knots 60 90 120 150 180
Min:Sec 1:12 0:48 0:36 0:29 0:24
KEY WEST, FLORIDA
Amdt 15B 12096
24°33'N-81°46'W
KEY WEST INTL (EYW)
NDB or GPS-A

Figure 337

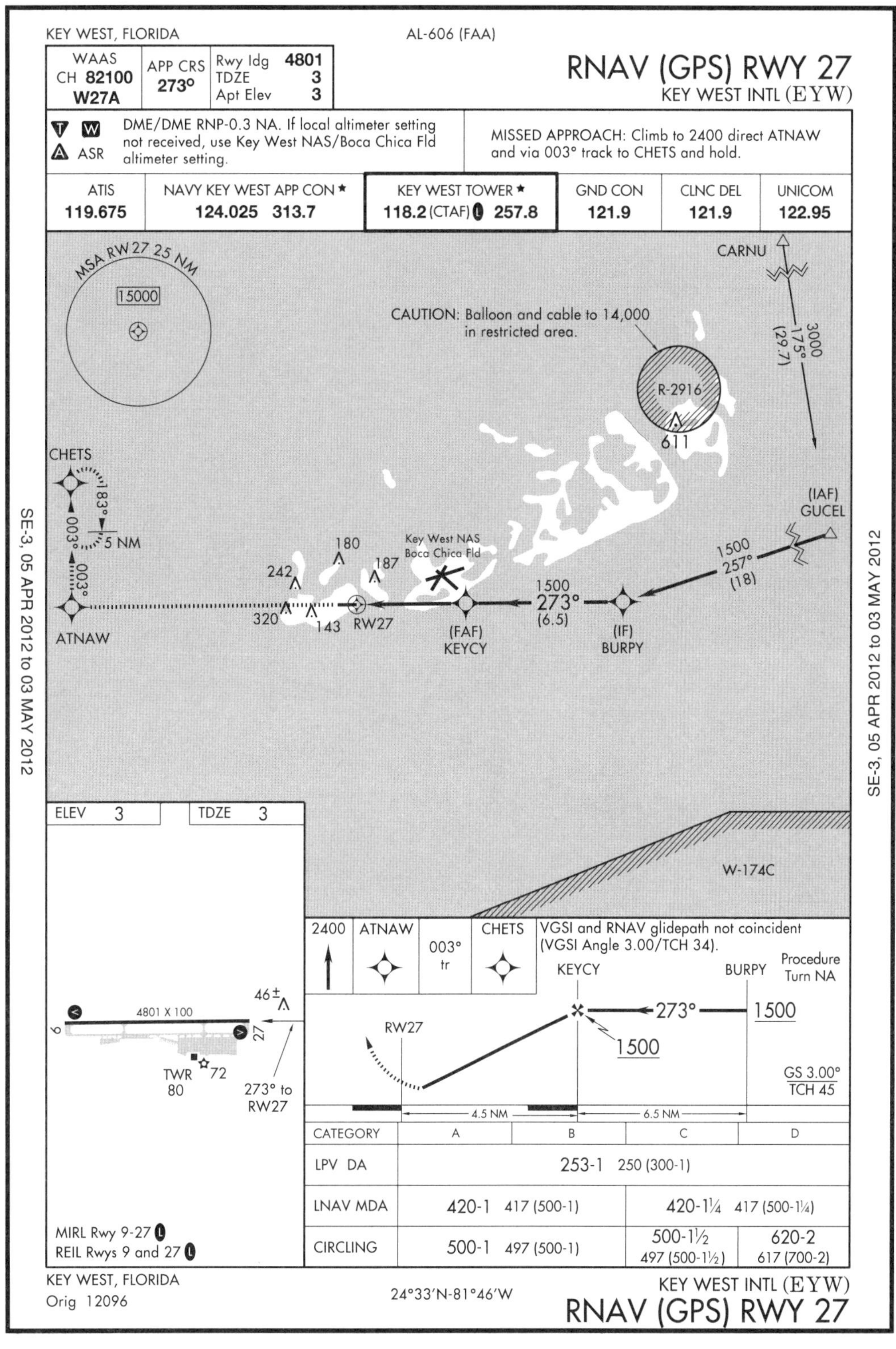

CATEGORY	A	B	C	D
LPV DA	253-1 250 (300-1)			
LNAV MDA	420-1 417 (500-1)		420-1¼ 417 (500-1¼)	
CIRCLING	500-1 497 (500-1)		500-1½ 497 (500-1½)	620-2 617 (700-2)

Figure 338

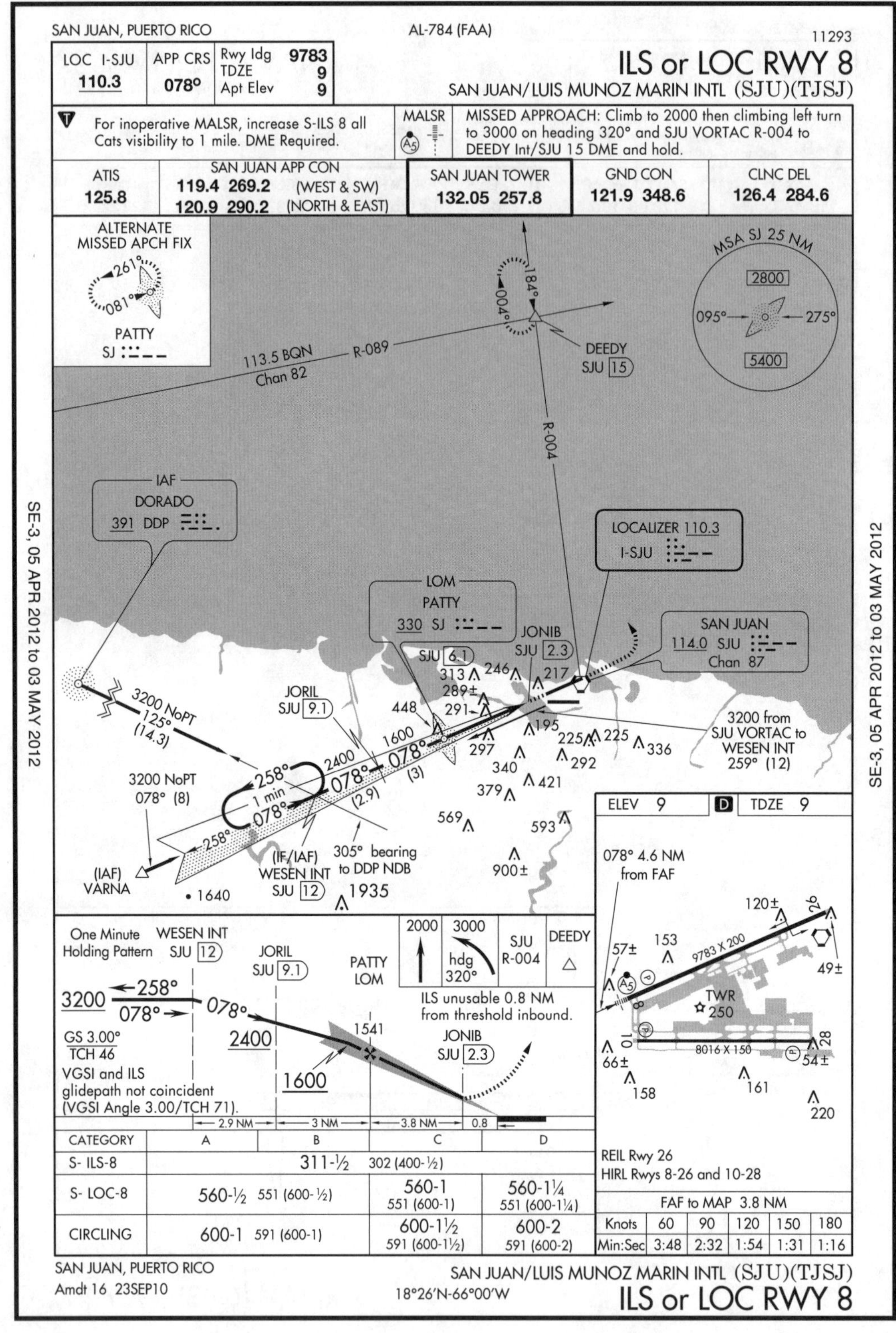

CATEGORY	A	B	C	D
S- ILS-8	311-½ 302 (400-½)			
S- LOC-8	560-½ 551 (600-½)		560-1 551 (600-1)	560-1¼ 551 (600-1¼)
CIRCLING	600-1 591 (600-1)		600-1½ 591 (600-1½)	600-2 591 (600-2)

FAF to MAP 3.8 NM					
Knots	60	90	120	150	180
Min:Sec	3:48	2:32	1:54	1:31	1:16

Figure 339

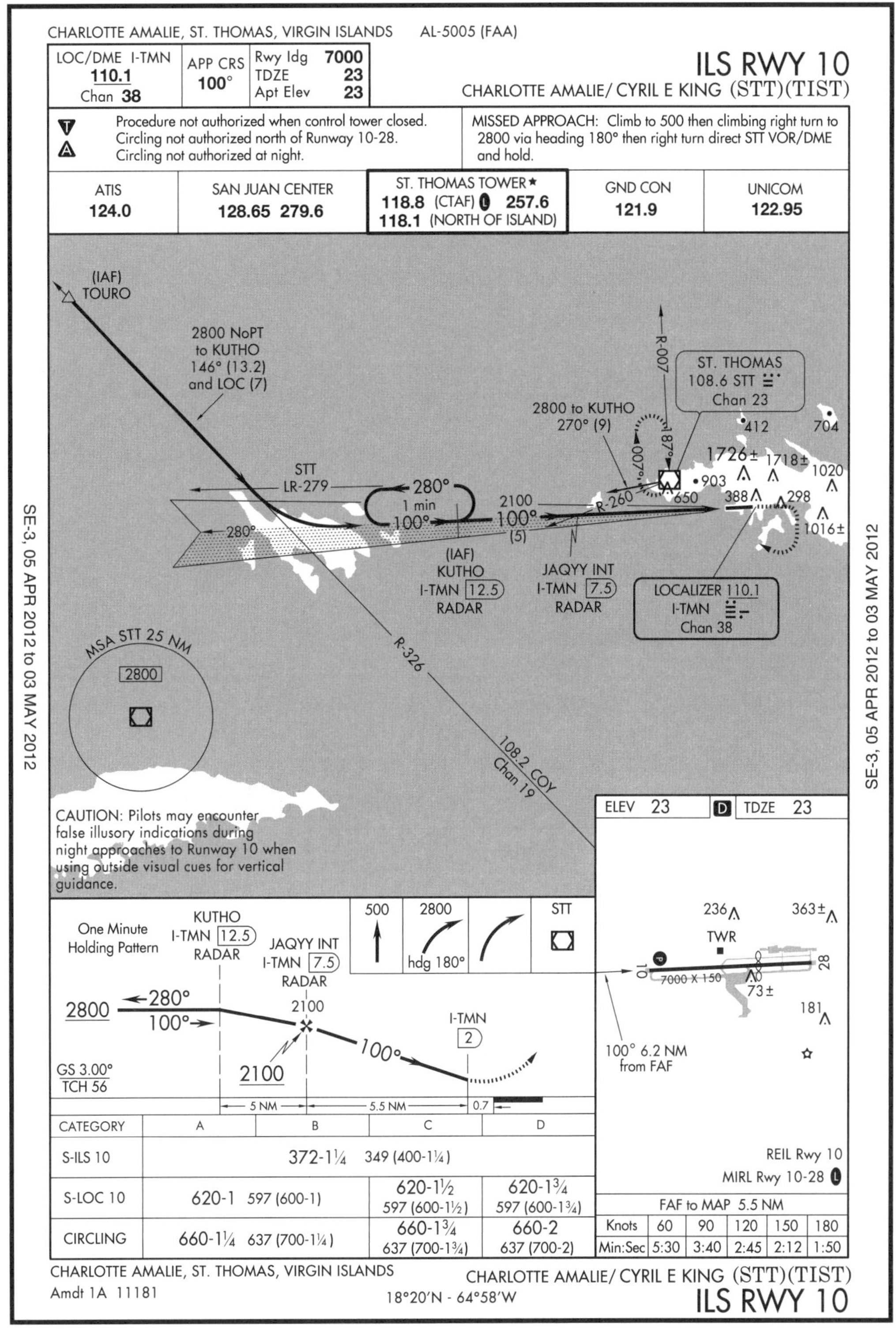

CATEGORY	A	B	C	D
S-ILS 10	372-1¼ 349 (400-1¼)			
S-LOC 10	620-1 597 (600-1)		620-1½ 597 (600-1½)	620-1¾ 597 (600-1¾)
CIRCLING	660-1¼ 637 (700-1¼)		660-1¾ 637 (700-1¾)	660-2 637 (700-2)

FAF to MAP 5.5 NM

Knots	60	90	120	150	180
Min:Sec	5:30	3:40	2:45	2:12	1:50

Figure 340

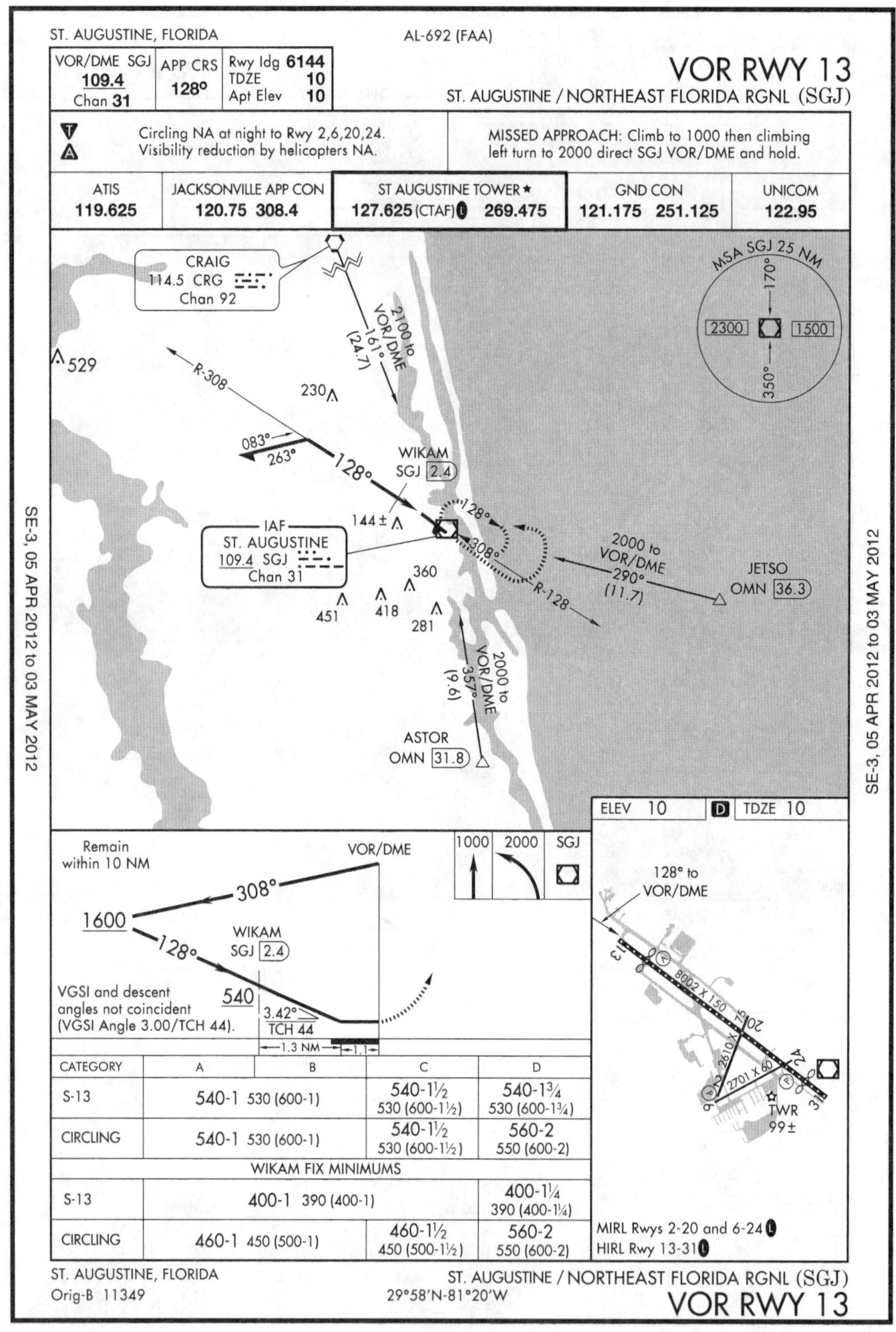

CATEGORY	A	B	C	D
S-13	540-1 530 (600-1)		540-1½ 530 (600-1½)	540-1¾ 530 (600-1¾)
CIRCLING	540-1 530 (600-1)		540-1½ 530 (600-1½)	560-2 550 (600-2)
WIKAM FIX MINIMUMS				
S-13	400-1 390 (400-1)			400-1¼ 390 (400-1¼)
CIRCLING	460-1 450 (500-1)		460-1½ 450 (500-1½)	560-2 550 (600-2)

Figure 341

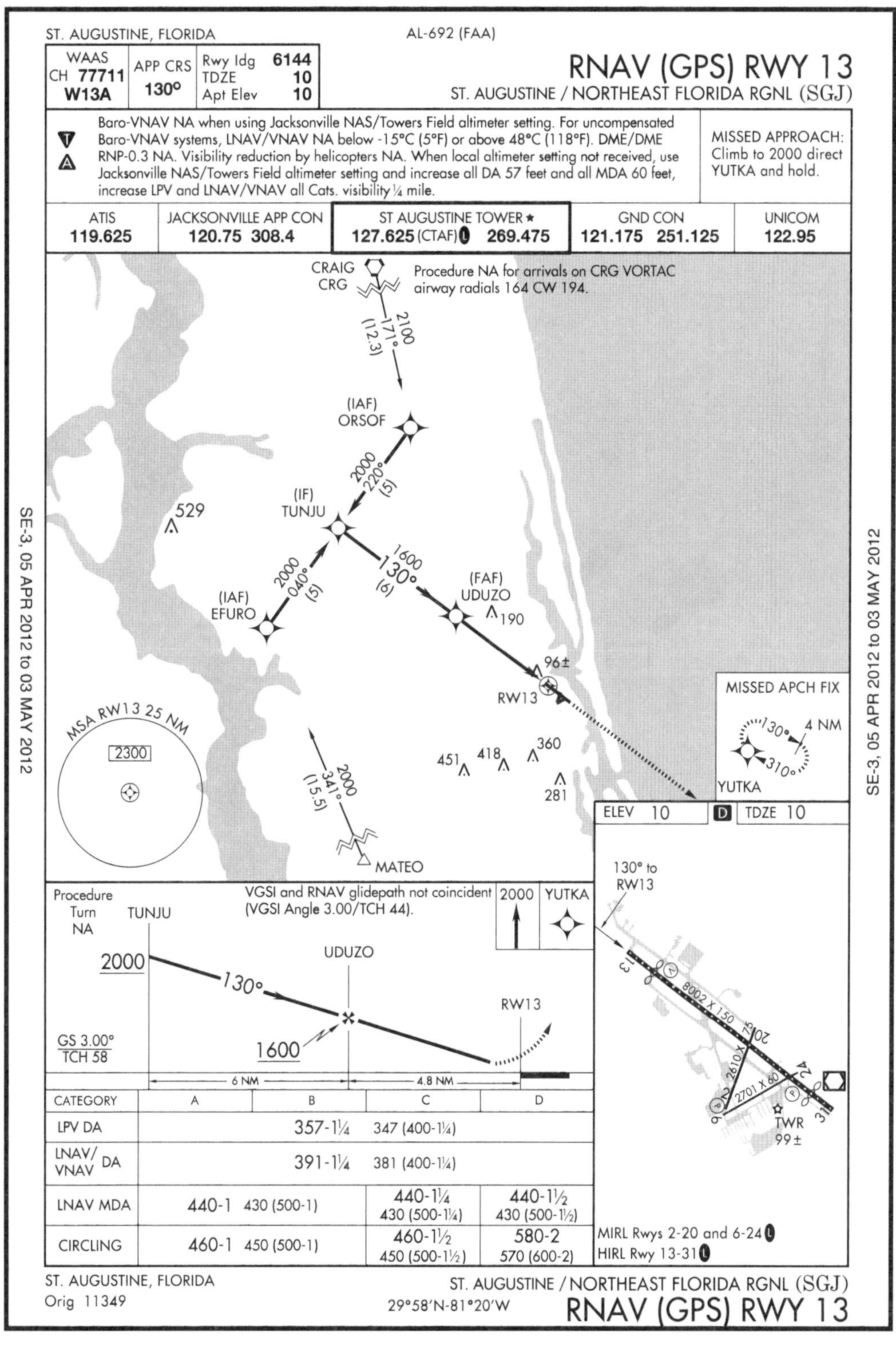

CATEGORY	A	B	C	D
LPV DA	357-1¼ 347 (400-1¼)			
LNAV/VNAV DA	391-1¼ 381 (400-1¼)			
LNAV MDA	440-1 430 (500-1)		440-1¼ 430 (500-1¼)	440-1½ 430 (500-1½)
CIRCLING	460-1 450 (500-1)		460-1½ 450 (500-1½)	580-2 570 (600-2)

Figure 342

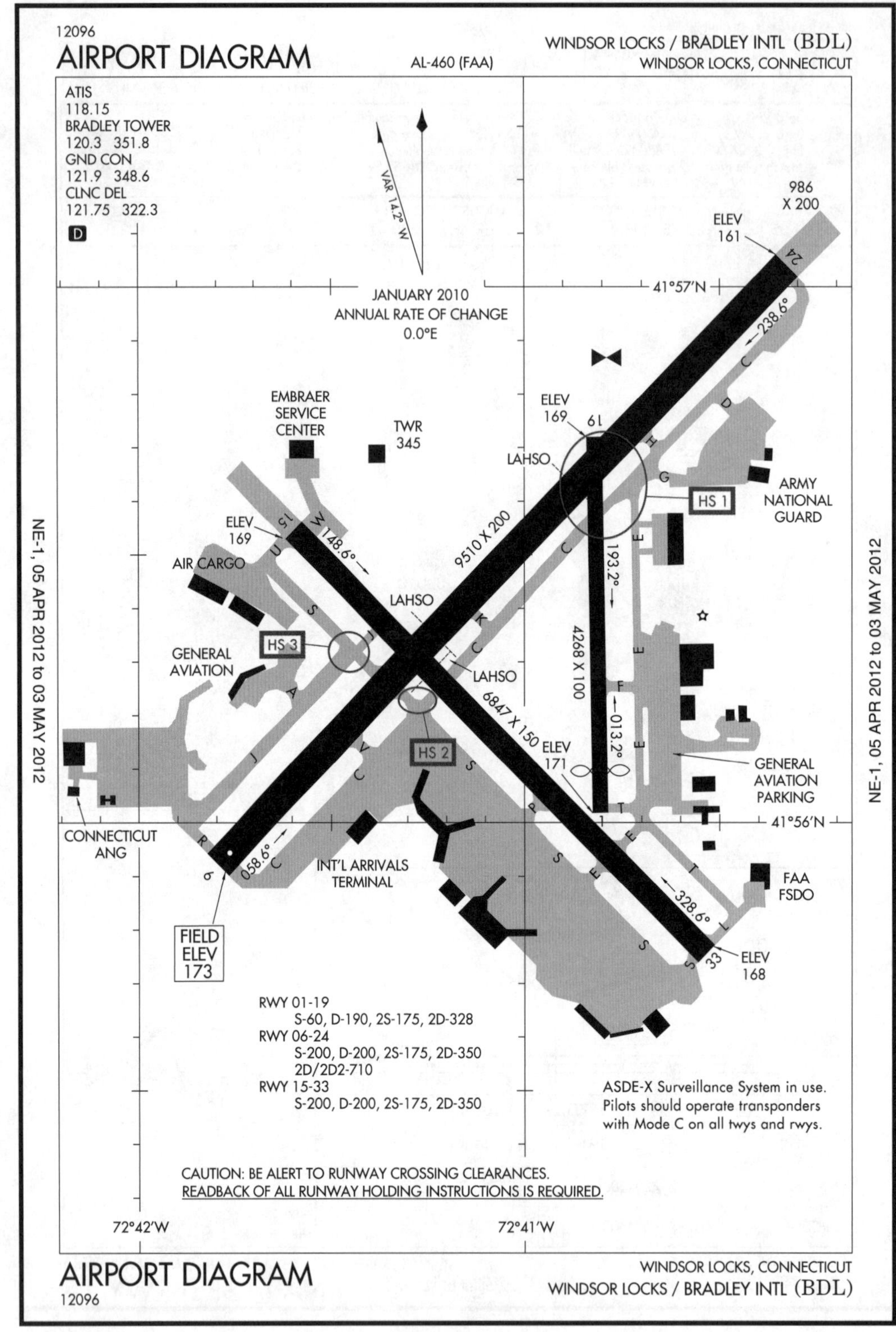
12096
AIRPORT DIAGRAM
AL-460 (FAA)
WINDSOR LOCKS / BRADLEY INTL (BDL)
WINDSOR LOCKS, CONNECTICUT
ATIS
118.15
BRADLEY TOWER
120.3 351.8
GND CON
121.9 348.6
CLNC DEL
121.75 322.3
VAR 14.2° W
JANUARY 2010
ANNUAL RATE OF CHANGE
0.0°E
41°57′N
986
X 200
ELEV
161
238.6°
EMBRAER
SERVICE
CENTER
TWR
345
ELEV
169
LAHSO
HS 1
ARMY
NATIONAL
GUARD
ELEV
169
148.6°
9510 X 200
193.2°
AIR CARGO
LAHSO
HS 3
GENERAL
AVIATION
LAHSO
4268 X 100
6847 X 150
013.2°
HS 2
ELEV
171
GENERAL
AVIATION
PARKING
41°56′N
CONNECTICUT
ANG
058.6°
INT'L ARRIVALS
TERMINAL
328.6°
FAA
FSDO
FIELD
ELEV
173
ELEV
168
RWY 01-19
S-60, D-190, 2S-175, 2D-328
RWY 06-24
S-200, D-200, 2S-175, 2D-350
2D/2D2-710
RWY 15-33
S-200, D-200, 2S-175, 2D-350
ASDE-X Surveillance System in use.
Pilots should operate transponders
with Mode C on all twys and rwys.
CAUTION: BE ALERT TO RUNWAY CROSSING CLEARANCES.
READBACK OF ALL RUNWAY HOLDING INSTRUCTIONS IS REQUIRED.
72°42′W
72°41′W
NE-1, 05 APR 2012 to 03 MAY 2012
AIRPORT DIAGRAM
12096
WINDSOR LOCKS, CONNECTICUT
WINDSOR LOCKS / BRADLEY INTL (BDL)

Figure 343

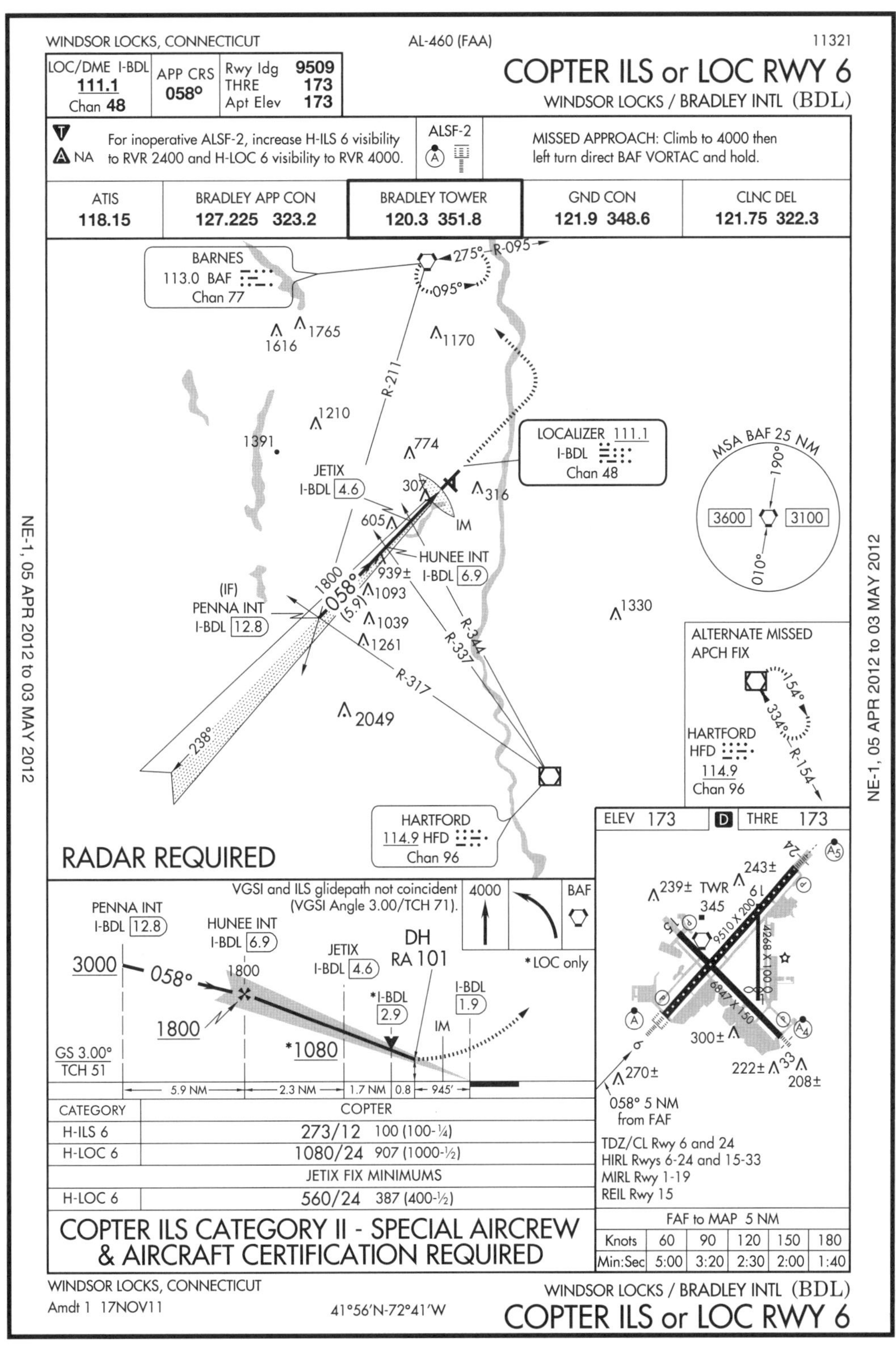

CATEGORY	COPTER
H-ILS 6	273/12 100 (100-¼)
H-LOC 6	1080/24 907 (1000-½)
	JETIX FIX MINIMUMS
H-LOC 6	560/24 387 (400-½)

COPTER ILS CATEGORY II - SPECIAL AIRCREW & AIRCRAFT CERTIFICATION REQUIRED

FAF to MAP 5 NM					
Knots	60	90	120	150	180
Min:Sec	5:00	3:20	2:30	2:00	1:40

Figure 344

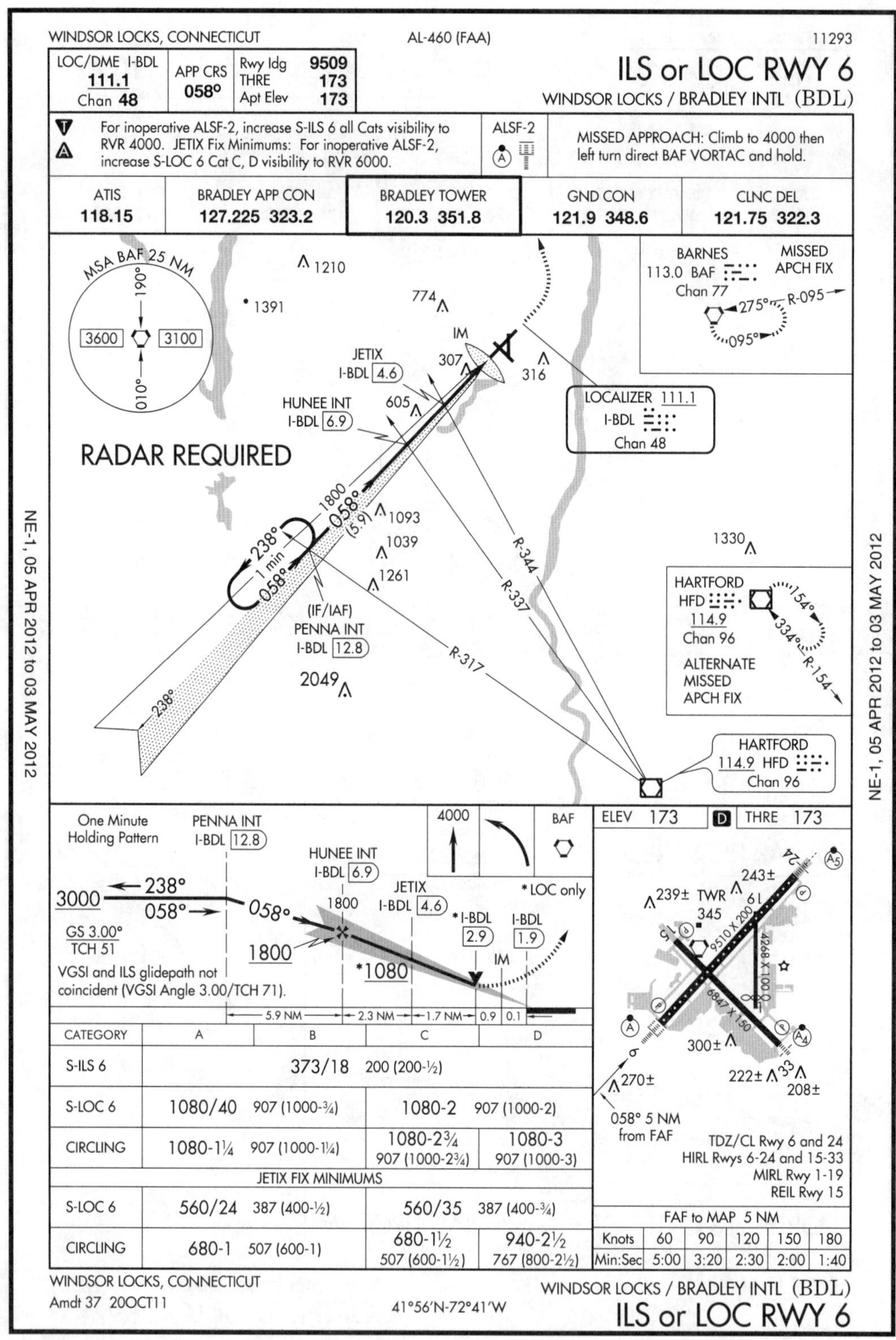

CATEGORY	A	B	C	D
S-ILS 6	373/18 200 (200-½)			
S-LOC 6	1080/40 907 (1000-¾)		1080-2 907 (1000-2)	
CIRCLING	1080-1¼ 907 (1000-1¼)		1080-2¾ 907 (1000-2¾)	1080-3 907 (1000-3)
JETIX FIX MINIMUMS				
S-LOC 6	560/24 387 (400-½)		560/35 387 (400-¾)	
CIRCLING	680-1 507 (600-1)		680-1½ 507 (600-1½)	940-2½ 767 (800-2½)

FAF to MAP 5 NM

Knots	60	90	120	150	180
Min:Sec	5:00	3:20	2:30	2:00	1:40

Figure 345

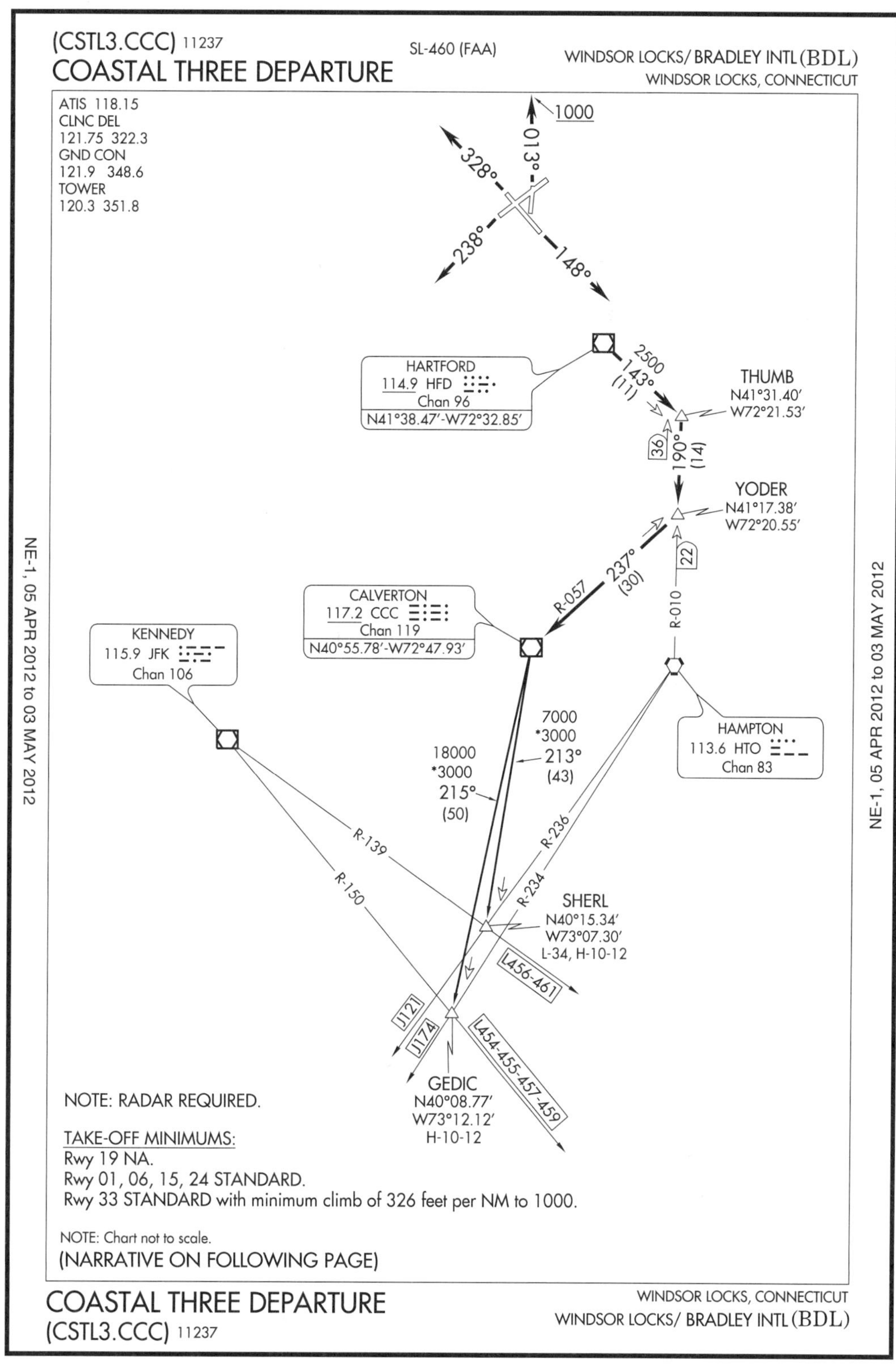
(CSTL3.CCC) 11237
COASTAL THREE DEPARTURE
SL-460 (FAA)
WINDSOR LOCKS/ BRADLEY INTL (BDL)
WINDSOR LOCKS, CONNECTICUT
ATIS 118.15
CLNC DEL
121.75 322.3
GND CON
121.9 348.6
TOWER
120.3 351.8
1000
013°
328°
238°
148°
HARTFORD
114.9 HFD
Chan 96
N41°38.47'-W72°32.85'
2500
143°
(11)
THUMB
N41°31.40'
W72°21.53'
36
190°
(14)
YODER
N41°17.38'
W72°20.55'
237°
(30)
R-057
22
R-010
CALVERTON
117.2 CCC
Chan 119
N40°55.78'-W72°47.93'
KENNEDY
115.9 JFK
Chan 106
HAMPTON
113.6 HTO
Chan 83
7000
*3000
213°
(43)
18000
*3000
215°
(50)
R-139
R-150
R-236
R-234
SHERL
N40°15.34'
W73°07.30'
L-34, H-10-12
L456-461
J121
J174
L454-455-457-459
GEDIC
N40°08.77'
W73°12.12'
H-10-12
NE-1, 05 APR 2012 to 03 MAY 2012
NOTE: RADAR REQUIRED.
TAKE-OFF MINIMUMS:
Rwy 19 NA.
Rwy 01, 06, 15, 24 STANDARD.
Rwy 33 STANDARD with minimum climb of 326 feet per NM to 1000.
NOTE: Chart not to scale.
(NARRATIVE ON FOLLOWING PAGE)
COASTAL THREE DEPARTURE
(CSTL3.CCC) 11237
WINDSOR LOCKS, CONNECTICUT
WINDSOR LOCKS/ BRADLEY INTL (BDL)

Figure 346

(CSTL3.CCC) 10154 SL-460 (FAA) WINDSOR LOCKS/ BRADLEY INTL (BDL)

COASTAL THREE DEPARTURE WINDSOR LOCKS, CONNECTICUT

DEPARTURE ROUTE DESCRIPTION

NOTE: INITIAL DEPARTURE HEADINGS ARE PREDICATED ON AVOIDING NOISE SENSITIVE AREAS. FLIGHT CREW AWARENESS AND COMPLIANCE IS IMPORTANT IN MINIMIZING NOISE IMPACTS ON SURROUNDING COMMUNITIES.

NOTE: APPROPRIATE DEPARTURE CONTROL FREQUENCY TO BE ASSIGNED BY ATC.

TAKE-OFF RWY 1: Climb heading 013° to 1000 or as assigned for radar vectors to HFD VOR/DME, thence . . .
TAKE-OFF RWY 6: Fly assigned heading for radar vectors to HFD VOR/DME, thence . . .
TAKE-OFF RWY 15: Climb heading 148° or as assigned for radar vectors to HFD VOR/DME, thence . . .
TAKE-OFF RWY 24: Climb heading 238° or as assigned for radar vectors to HFD VOR/DME, thence . . .
TAKE-OFF RWY 33: Climb heading 328° or as assigned for radar vectors to HFD VOR/DME, thence . . .
. . . . From over HFD VOR/DME proceed via HFD R-143 to THUMB INT, then proceed via HTO R-010 to YODER INT, then proceed via CCC R-057 to CCC VOR/DME. Then via (transition) or (assigned route). Maintain 4000 or assigned altitude. Expect clearance to requested flight level ten minutes after departure.

GEDIC TRANSITION (CSTL3.GEDIC): From over CCC VOR/DME via CCC R-215 to GEDIC.
SHERL TRANSITION (CSTL3.SHERL): From over CCC VOR/DME via CCC R-213 to SHERL.

TAKE-OFF OBSTACLE NOTES:

Rwy 1: Vehicle on road 342′ from DER, 564′ left of centerline, 15′ AGL/184′ MSL. Trees beginning 441′ from DER, 493′ left of centerline, up to 100′ AGL/269′ MSL. Trees beginning 1884′ from DER, 45′ right of centerline, up to 100′ AGL/299′ MSL.

Rwy 6: Trees beginning 21′ from DER, 464′ left of centerline, up to 100′ AGL/249′ MSL. Trees beginning 1956′ from DER, 921′ right of centerline, up to 100′ AGL/239′ MSL.

Rwy 15: Vehicle on roadway 531′ from DER, 606′ left of centerline, up to 15′ AGL/186′ MSL. Trees beginning 2341′ from DER, 767′ left of centerline, up to 100′ AGL/244′ MSL. Vehicle on roadway 429′ from DER, 572′ right of centerline, up to 15′ AGL/186′ MSL. Tree 1520′ from DER, 786′ right of centerline, up to 100′ AGL/259′ MSL.

Rwy 24: Trees beginning 3066′ from DER, 599′ left of centerline, up to 100′ AGL/269′ MSL. OL on fence 1239′ DER, 784′ left of centerline, up to 45′ AGL/215′ MSL. Trees beginning 2345′ from DER, 489′ right of centerline, up to 100′ AGL/299′ MSL.

Rwy 33: Trees beginning 1590′ from DER, 275′ left of centerline, up to 100′ AGL/256′ MSL. Tower 2.4 NM from DER, 3534′ left of centerline, 104′ AGL/774′ MSL. Trees beginning 1618′ from DER, 264′ right of centerline, up to 100′ AGL/263′ MSL.

NE-1, 05 APR 2012 to 03 MAY 2012

NE-1, 05 APR 2012 to 03 MAY 2012

COASTAL THREE DEPARTURE WINDSOR LOCKS, CONNECTICUT

(CSTL3.CCC) 10154 WINDSOR LOCKS/ BRADLEY INTL (BDL)

Figure 347

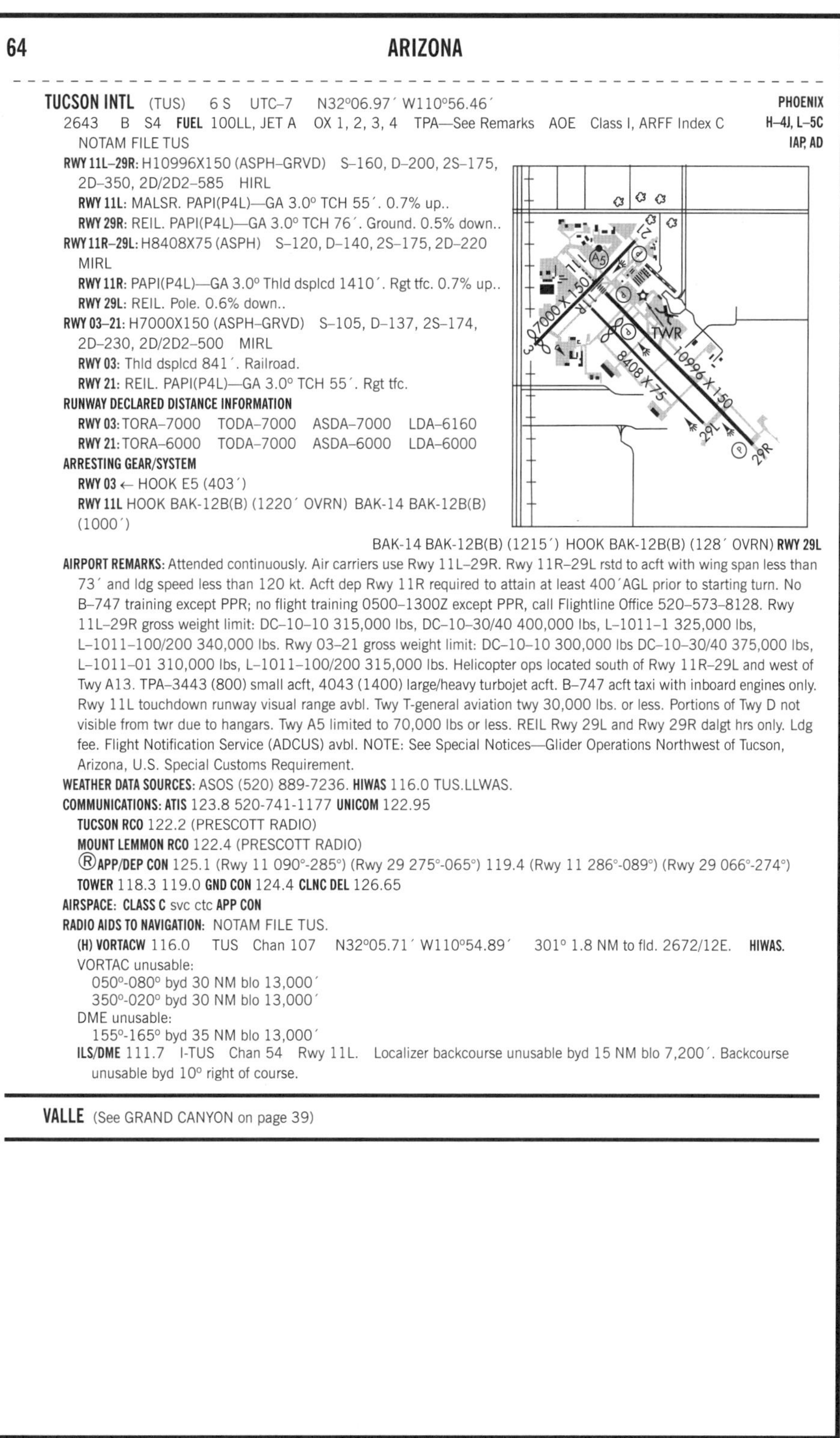

64 **ARIZONA**

TUCSON INTL (TUS) 6 S UTC–7 N32°06.97′ W110°56.46′ **PHOENIX**
2643 B S4 **FUEL** 100LL, JET A OX 1, 2, 3, 4 TPA—See Remarks AOE Class I, ARFF Index C **H–4J, L–5C**
NOTAM FILE TUS **IAP, AD**

RWY 11L–29R: H10996X150 (ASPH–GRVD) S–160, D–200, 2S–175, 2D–350, 2D/2D2–585 HIRL
RWY 11L: MALSR. PAPI(P4L)—GA 3.0° TCH 55′. 0.7% up..
RWY 29R: REIL. PAPI(P4L)—GA 3.0° TCH 76′. Ground. 0.5% down..
RWY 11R–29L: H8408X75 (ASPH) S–120, D–140, 2S–175, 2D–220 MIRL
RWY 11R: PAPI(P4L)—GA 3.0° Thld dsplcd 1410′. Rgt tfc. 0.7% up..
RWY 29L: REIL. Pole. 0.6% down..
RWY 03–21: H7000X150 (ASPH–GRVD) S–105, D–137, 2S–174, 2D–230, 2D/2D2–500 MIRL
RWY 03: Thld dsplcd 841′. Railroad.
RWY 21: REIL. PAPI(P4L)—GA 3.0° TCH 55′. Rgt tfc.
RUNWAY DECLARED DISTANCE INFORMATION
RWY 03: TORA–7000 TODA–7000 ASDA–7000 LDA–6160
RWY 21: TORA–6000 TODA–7000 ASDA–6000 LDA–6000
ARRESTING GEAR/SYSTEM
RWY 03 ← HOOK E5 (403′)
RWY 11L HOOK BAK-12B(B) (1220′ OVRN) BAK-14 BAK-12B(B) (1000′)
BAK-14 BAK-12B(B) (1215′) HOOK BAK-12B(B) (128′ OVRN) **RWY 29L**

AIRPORT REMARKS: Attended continuously. Air carriers use Rwy 11L–29R. Rwy 11R–29L rstd to acft with wing span less than 73′ and ldg speed less than 120 kt. Acft dep Rwy 11R required to attain at least 400′AGL prior to starting turn. No B–747 training except PPR; no flight training 0500–1300Z except PPR, call Flightline Office 520–573–8128. Rwy 11L–29R gross weight limit: DC–10–10 315,000 lbs, DC–10–30/40 400,000 lbs, L–1011–1 325,000 lbs, L–1011–100/200 340,000 lbs. Rwy 03–21 gross weight limit: DC–10–10 300,000 lbs DC–10–30/40 375,000 lbs, L–1011–01 310,000 lbs, L–1011–100/200 315,000 lbs. Helicopter ops located south of Rwy 11R–29L and west of Twy A13. TPA–3443 (800) small acft, 4043 (1400) large/heavy turbojet acft. B–747 acft taxi with inboard engines only. Rwy 11L touchdown runway visual range avbl. Twy T-general aviation twy 30,000 lbs. or less. Portions of Twy D not visible from twr due to hangars. Twy A5 limited to 70,000 lbs or less. REIL Rwy 29L and Rwy 29R dalgt hrs only. Ldg fee. Flight Notification Service (ADCUS) avbl. NOTE: See Special Notices—Glider Operations Northwest of Tucson, Arizona, U.S. Special Customs Requirement.
WEATHER DATA SOURCES: ASOS (520) 889-7236. **HIWAS** 116.0 TUS.LLWAS.
COMMUNICATIONS: ATIS 123.8 520-741-1177 **UNICOM** 122.95
TUCSON RCO 122.2 (PRESCOTT RADIO)
MOUNT LEMMON RCO 122.4 (PRESCOTT RADIO)
®**APP/DEP CON** 125.1 (Rwy 11 090°-285°) (Rwy 29 275°-065°) 119.4 (Rwy 11 286°-089°) (Rwy 29 066°-274°)
TOWER 118.3 119.0 **GND CON** 124.4 **CLNC DEL** 126.65
AIRSPACE: CLASS C svc ctc **APP CON**
RADIO AIDS TO NAVIGATION: NOTAM FILE TUS.
(H) VORTACW 116.0 TUS Chan 107 N32°05.71′ W110°54.89′ 301° 1.8 NM to fld. 2672/12E. **HIWAS.**
VORTAC unusable:
050°-080° byd 30 NM blo 13,000′
350°-020° byd 30 NM blo 13,000′
DME unusable:
155°-165° byd 35 NM blo 13,000′
ILS/DME 111.7 I-TUS Chan 54 Rwy 11L. Localizer backcourse unusable byd 15 NM blo 7,200′. Backcourse unusable byd 10° right of course.

VALLE (See GRAND CANYON on page 39)

Figure 348

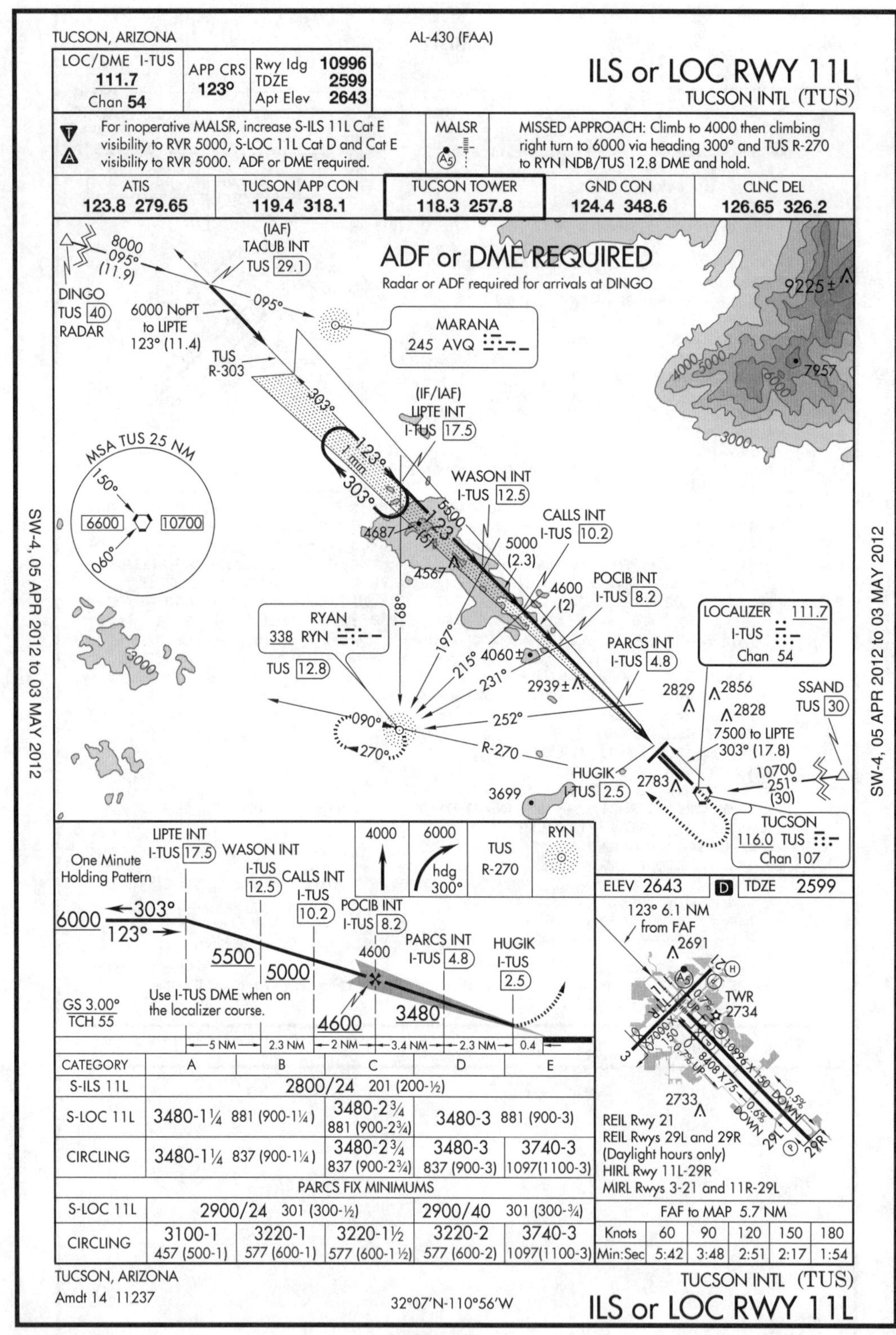

CATEGORY	A	B	C	D	E
S-ILS 11L	2800/24 201 (200-½)				
S-LOC 11L	3480-1¼ 881 (900-1¼)		3480-2¾ 881 (900-2¾)	3480-3 881 (900-3)	
CIRCLING	3480-1¼ 837 (900-1¼)		3480-2¾ 837 (900-2¾)	3480-3 837 (900-3)	3740-3 1097(1100-3)
PARCS FIX MINIMUMS					
S-LOC 11L	2900/24 301 (300-½)			2900/40 301 (300-¾)	
CIRCLING	3100-1 457 (500-1)	3220-1 577 (600-1)	3220-1½ 577 (600-1½)	3220-2 577 (600-2)	3740-3 1097(1100-3)

FAF to MAP 5.7 NM

Knots	60	90	120	150	180
Min:Sec	5:42	3:48	2:51	2:17	1:54

Figure 349

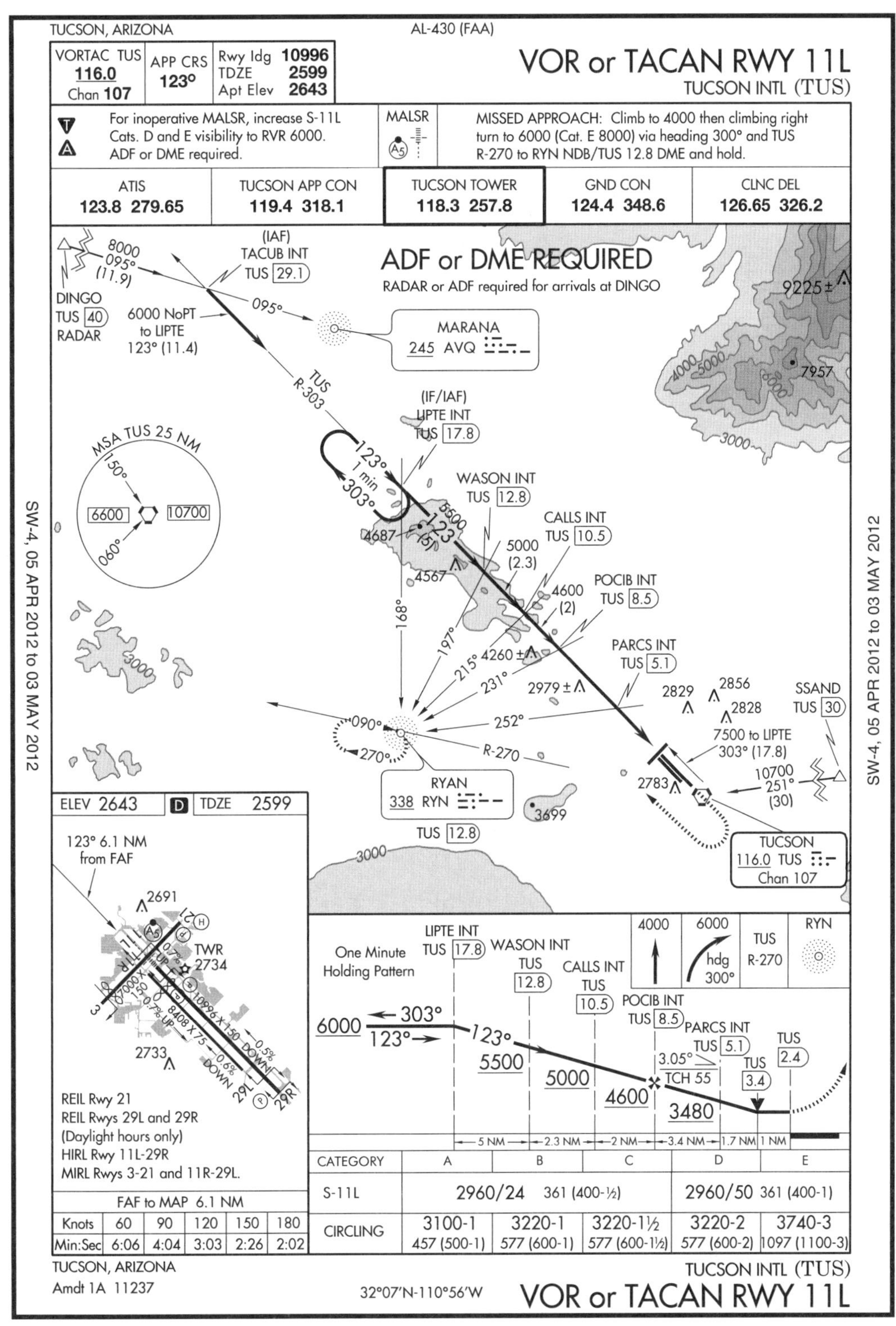

FAF to MAP 6.1 NM					
Knots	60	90	120	150	180
Min:Sec	6:06	4:04	3:03	2:26	2:02

CATEGORY	A	B	C	D	E
S-11L	2960/24 361 (400-½)			2960/50 361 (400-1)	
CIRCLING	3100-1 457 (500-1)	3220-1 577 (600-1)	3220-1½ 577 (600-1½)	3220-2 577 (600-2)	3740-3 1097 (1100-3)

Figure 350

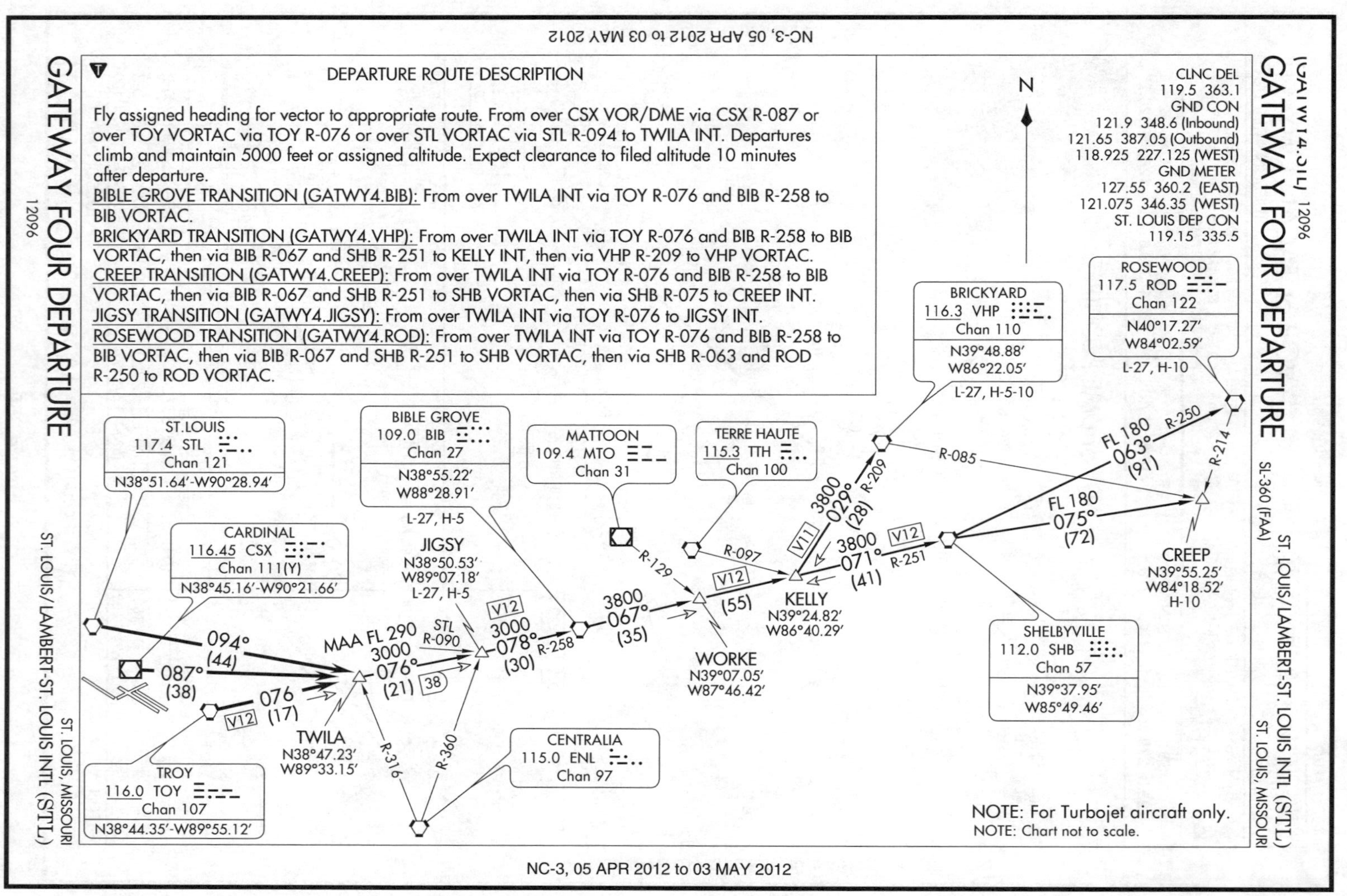
(GATWY4.STL) 12096
GATEWAY FOUR DEPARTURE
SL-360 (FAA)
ST. LOUIS/LAMBERT-ST. LOUIS INTL (STL)
ST. LOUIS, MISSOURI
NC-3, 05 APR 2012 to 03 MAY 2012
DEPARTURE ROUTE DESCRIPTION
Fly assigned heading for vector to appropriate route. From over CSX VOR/DME via CSX R-087 or over TOY VORTAC via TOY R-076 or over STL VORTAC via STL R-094 to TWILA INT. Departures climb and maintain 5000 feet or assigned altitude. Expect clearance to filed altitude 10 minutes after departure.
BIBLE GROVE TRANSITION (GATWY4.BIB): From over TWILA INT via TOY R-076 and BIB R-258 to BIB VORTAC.
BRICKYARD TRANSITION (GATWY4.VHP): From over TWILA INT via TOY R-076 and BIB R-258 to BIB VORTAC, then via BIB R-067 and SHB R-251 to KELLY INT, then via VHP R-209 to VHP VORTAC.
CREEP TRANSITION (GATWY4.CREEP): From over TWILA INT via TOY R-076 and BIB R-258 to BIB VORTAC, then via BIB R-067 and SHB R-251 to SHB VORTAC, then via SHB R-075 to CREEP INT.
JIGSY TRANSITION (GATWY4.JIGSY): From over TWILA INT via TOY R-076 to JIGSY INT.
ROSEWOOD TRANSITION (GATWY4.ROD): From over TWILA INT via TOY R-076 and BIB R-258 to BIB VORTAC, then via BIB R-067 and SHB R-251 to SHB VORTAC, then via SHB R-063 and ROD R-250 to ROD VORTAC.
N
CLNC DEL
119.5 363.1
GND CON
121.9 348.6 (Inbound)
121.65 387.05 (Outbound)
118.925 227.125 (WEST)
GND METER
127.55 360.2 (EAST)
121.075 346.35 (WEST)
ST. LOUIS DEP CON
119.15 335.5
ROSEWOOD
117.5 ROD
Chan 122
N40°17.27′
W84°02.59′
L-27, H-10
BRICKYARD
116.3 VHP
Chan 110
N39°48.88′
W86°22.05′
L-27, H-5-10
ST.LOUIS
117.4 STL
Chan 121
N38°51.64′-W90°28.94′
BIBLE GROVE
109.0 BIB
Chan 27
N38°55.22′
W88°28.91′
L-27, H-5
MATTOON
109.4 MTO
Chan 31
TERRE HAUTE
115.3 TTH
Chan 100
CARDINAL
116.45 CSX
Chan 111(Y)
N38°45.16′-W90°21.66′
JIGSY
N38°50.53′
W89°07.18′
L-27, H-5
KELLY
N39°24.82′
W86°40.29′
WORKE
N39°07.05′
W87°46.42′
CREEP
N39°55.25′
W84°18.52′
H-10
SHELBYVILLE
112.0 SHB
Chan 57
N39°37.95′
W85°49.46′
TWILA
N38°47.23′
W89°33.15′
CENTRALIA
115.0 ENL
Chan 97
TROY
116.0 TOY
Chan 107
N38°44.35′-W89°55.12′
FL 180
063°
(91)
R-250
R-214
R-085
FL 180
075°
(72)
3800
029°
(28)
R-209
V11
3800
071°
(41)
V12
R-251
R-097
V12
(55)
R-129
3800
067°
(35)
V12
3000
078°
(30)
R-258
STL
R-090
MAA FL 290
3000
076°
(21)
38
094°
(44)
087°
(38)
076
(17)
V12
R-316
R-360
NOTE: For Turbojet aircraft only.
NOTE: Chart not to scale.
NC-3, 05 APR 2012 to 03 MAY 2012
GATEWAY FOUR DEPARTURE
12096
ST. LOUIS, MISSOURI
ST. LOUIS/LAMBERT-ST. LOUIS INTL (STL)

Figure 351

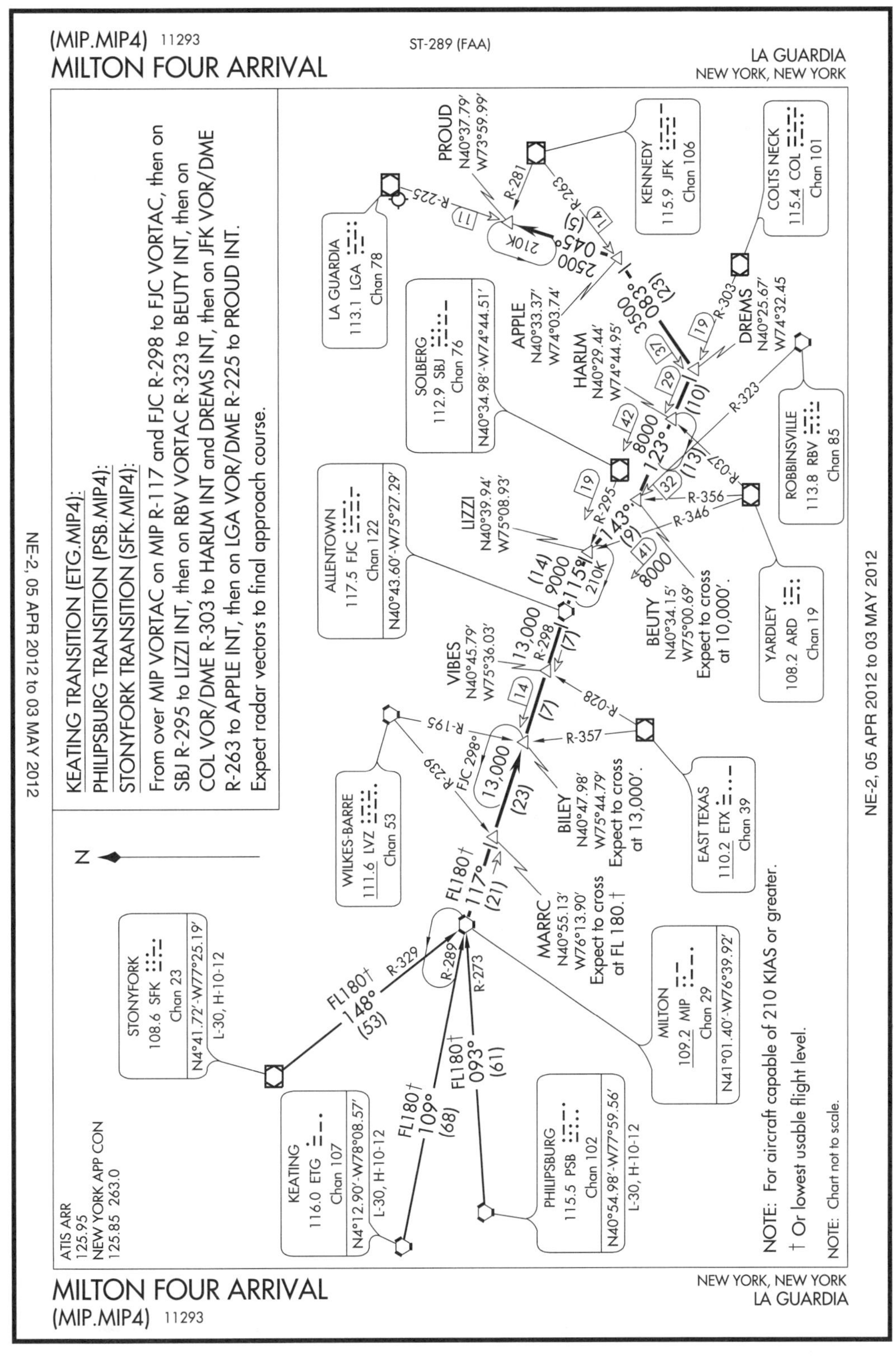
(MIP.MIP4) 11293
MILTON FOUR ARRIVAL
ST-289 (FAA)
LA GUARDIA
NEW YORK, NEW YORK
KEATING TRANSITION (ETG.MIP4):
PHILIPSBURG TRANSITION (PSB.MIP4):
STONYFORK TRANSITION (SFK.MIP4):
From over MIP VORTAC on MIP R-117 and FJC R-298 to FJC VORTAC, then on SBJ R-295 to LIZZI INT, then on RBV VORTAC R-323 to BEUTY INT, then on COL VOR/DME R-303 to HARLM INT and DREMS INT, then on JFK VOR/DME R-263 to APPLE INT, then on LGA VOR/DME R-225 to PROUD INT.
Expect radar vectors to final approach course.
NE-2, 05 APR 2012 to 03 MAY 2012
ATIS ARR
125.95
NEW YORK APP CON
125.85 263.0
KEATING
116.0 ETG
Chan 107
N4°12.90'-W78°08.57'
L-30, H-10-12
STONYFORK
108.6 SFK
Chan 23
N4°41.72'-W77°25.19'
L-30, H-10-12
PHILIPSBURG
115.5 PSB
Chan 102
N40°54.98'-W77°59.56'
L-30, H-10-12
MILTON
109.2 MIP
Chan 29
N41°01.40'-W76°39.92'
WILKES-BARRE
111.6 LVZ
Chan 53
EAST TEXAS
110.2 ETX
Chan 39
ALLENTOWN
117.5 FJC
Chan 122
N40°43.60'-W75°27.29'
SOLBERG
112.9 SBJ
Chan 76
N40°34.98'-W74°44.51'
LA GUARDIA
113.1 LGA
Chan 78
KENNEDY
115.9 JFK
Chan 106
COLTS NECK
115.4 COL
Chan 101
ROBBINSVILLE
113.8 RBV
Chan 85
YARDLEY
108.2 ARD
Chan 19
MARRC
N40°55.13'
W76°13.90'
Expect to cross at FL 180.†
BILEY
N40°47.98'
W75°44.79'
Expect to cross at 13,000'.
VIBES
N40°45.79'
W75°36.03'
LIZZI
N40°39.94'
W75°08.93'
BEUTY
N40°34.15'
W75°00.69'
Expect to cross at 10,000'.
HARLM
N40°29.44'
W74°44.95'
DREMS
N40°25.67'
W74°32.45
APPLE
N40°33.37'
W74°03.74'
PROUD
N40°37.79'
W73°59.99'
FL180† 148° (53)
FL180† 109° (68)
FL180† 093° (61)
FL180† 117° (21)
13,000 (23)
13,000 (7)
9000 115° (14)
210K
8000 143° (9)
8000 123° (13)
350° 083° (23)
2500 045° (5)
210K
NOTE: For aircraft capable of 210 KIAS or greater.
† Or lowest usable flight level.
NOTE: Chart not to scale.
MILTON FOUR ARRIVAL
(MIP.MIP4) 11293
NEW YORK, NEW YORK
LA GUARDIA

Figure 352

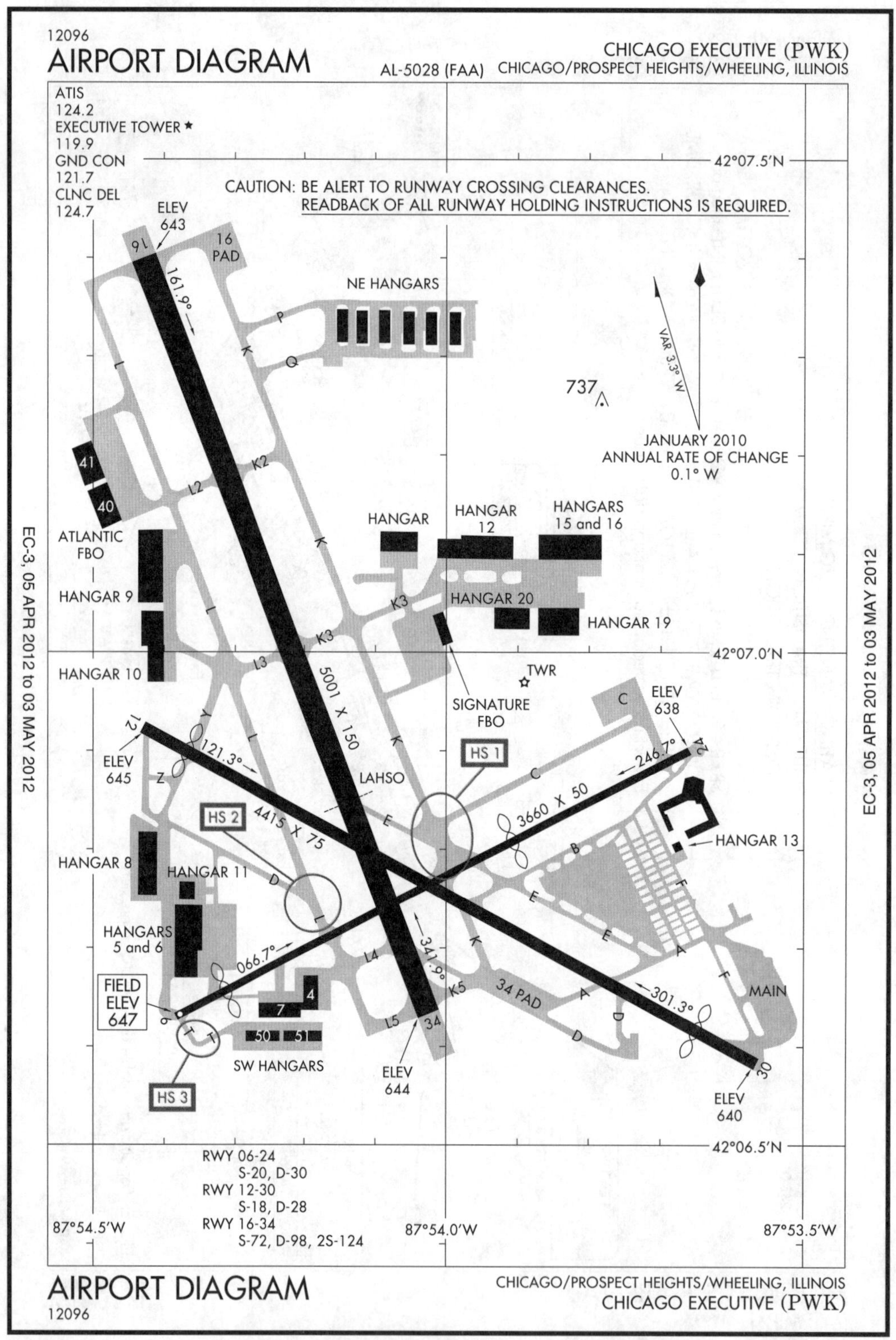
12096
AIRPORT DIAGRAM
AL-5028 (FAA)
CHICAGO EXECUTIVE (PWK)
CHICAGO/PROSPECT HEIGHTS/WHEELING, ILLINOIS
ATIS
124.2
EXECUTIVE TOWER ★
119.9
GND CON
121.7
CLNC DEL
124.7
CAUTION: BE ALERT TO RUNWAY CROSSING CLEARANCES.
READBACK OF ALL RUNWAY HOLDING INSTRUCTIONS IS REQUIRED.
42°07.5'N
ELEV
643
16 PAD
161.9°
NE HANGARS
VAR 3.3° W
737
JANUARY 2010
ANNUAL RATE OF CHANGE
0.1° W
HANGAR
HANGAR 12
HANGARS 15 and 16
ATLANTIC FBO
HANGAR 9
HANGAR 20
HANGAR 19
HANGAR 10
42°07.0'N
TWR
SIGNATURE FBO
ELEV
638
5001 X 150
ELEV
645
121.3°
246.7°
HS 1
LAHSO
3660 X 50
HS 2
4415 X 75
HANGAR 13
HANGAR 8
HANGAR 11
HANGARS
5 and 6
066.7°
341.9°
34 PAD
301.3°
MAIN
FIELD
ELEV
647
SW HANGARS
HS 3
ELEV
644
ELEV
640
42°06.5'N
RWY 06-24
S-20, D-30
RWY 12-30
S-18, D-28
RWY 16-34
S-72, D-98, 2S-124
87°54.5'W
87°54.0'W
87°53.5'W
EC-3, 05 APR 2012 to 03 MAY 2012
AIRPORT DIAGRAM
12096
CHICAGO/PROSPECT HEIGHTS/WHEELING, ILLINOIS
CHICAGO EXECUTIVE (PWK)

Figure 353

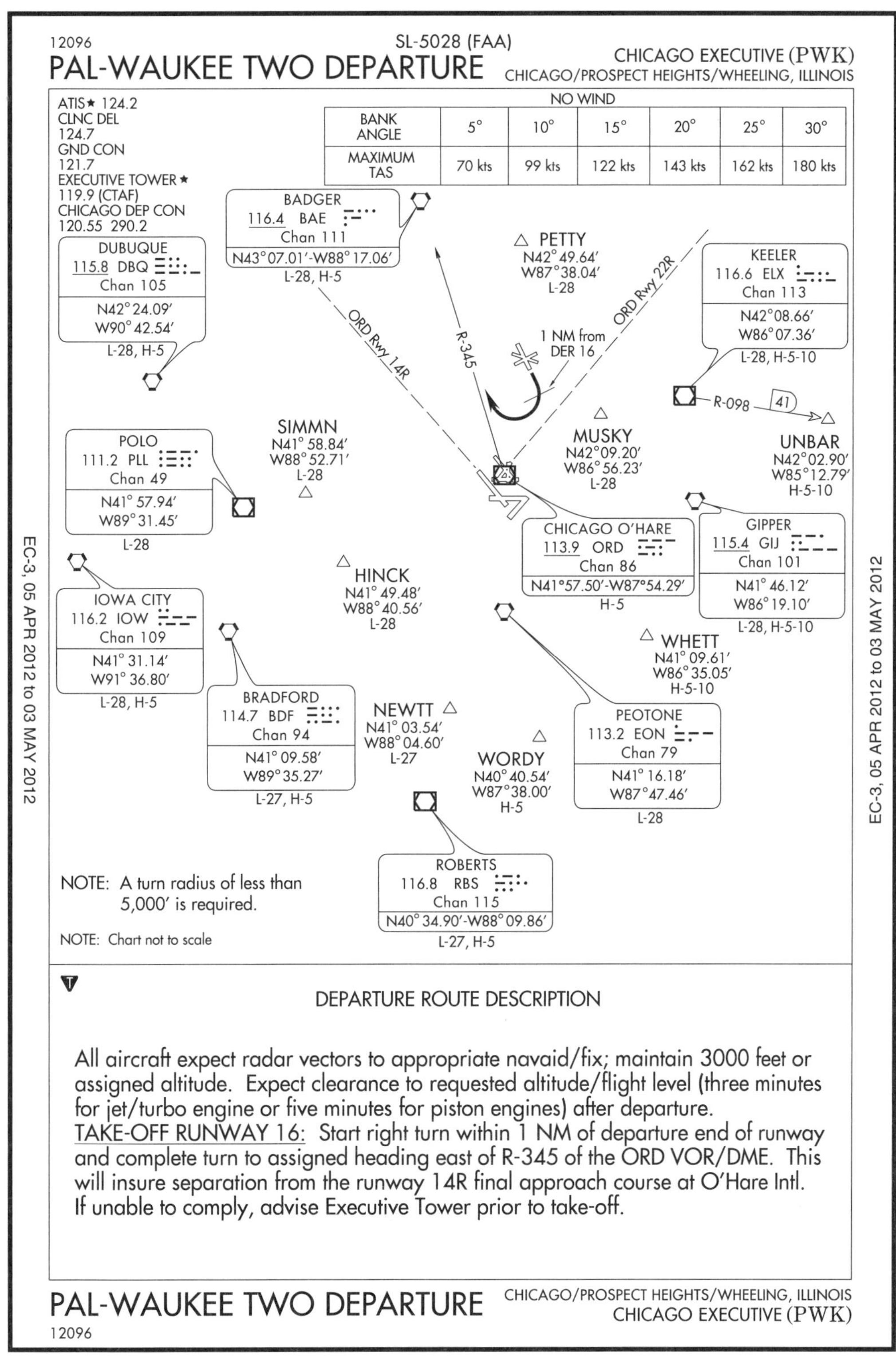
12096
SL-5028 (FAA)
PAL-WAUKEE TWO DEPARTURE
CHICAGO EXECUTIVE (PWK)
CHICAGO/PROSPECT HEIGHTS/WHEELING, ILLINOIS
ATIS★ 124.2
CLNC DEL 124.7
GND CON 121.7
EXECUTIVE TOWER★ 119.9 (CTAF)
CHICAGO DEP CON 120.55 290.2
NO WIND
BANK ANGLE | 5° | 10° | 15° | 20° | 25° | 30°
MAXIMUM TAS | 70 kts | 99 kts | 122 kts | 143 kts | 162 kts | 180 kts
BADGER 116.4 BAE Chan 111 N43°07.01'-W88°17.06' L-28, H-5
DUBUQUE 115.8 DBQ Chan 105 N42°24.09' W90°42.54' L-28, H-5
PETTY N42°49.64' W87°38.04' L-28
KEELER 116.6 ELX Chan 113 N42°08.66' W86°07.36' L-28, H-5-10
ORD Rwy 14R
R-345
ORD Rwy 22R
1 NM from DER 16
R-098
41
SIMMN N41°58.84' W88°52.71' L-28
MUSKY N42°09.20' W86°56.23' L-28
UNBAR N42°02.90' W85°12.79' H-5-10
POLO 111.2 PLL Chan 49 N41°57.94' W89°31.45' L-28
CHICAGO O'HARE 113.9 ORD Chan 86 N41°57.50'-W87°54.29' H-5
GIPPER 115.4 GIJ Chan 101 N41°46.12' W86°19.10' L-28, H-5-10
HINCK N41°49.48' W88°40.56' L-28
IOWA CITY 116.2 IOW Chan 109 N41°31.14' W91°36.80' L-28, H-5
WHETT N41°09.61' W86°35.05' H-5-10
BRADFORD 114.7 BDF Chan 94 N41°09.58' W89°35.27' L-27, H-5
NEWTT N41°03.54' W88°04.60' L-27
WORDY N40°40.54' W87°38.00' H-5
PEOTONE 113.2 EON Chan 79 N41°16.18' W87°47.46' L-28
ROBERTS 116.8 RBS Chan 115 N40°34.90'-W88°09.86' L-27, H-5
NOTE: A turn radius of less than 5,000' is required.
NOTE: Chart not to scale
EC-3, 05 APR 2012 to 03 MAY 2012
DEPARTURE ROUTE DESCRIPTION
All aircraft expect radar vectors to appropriate navaid/fix; maintain 3000 feet or assigned altitude. Expect clearance to requested altitude/flight level (three minutes for jet/turbo engine or five minutes for piston engines) after departure.
TAKE-OFF RUNWAY 16: Start right turn within 1 NM of departure end of runway and complete turn to assigned heading east of R-345 of the ORD VOR/DME. This will insure separation from the runway 14R final approach course at O'Hare Intl.
If unable to comply, advise Executive Tower prior to take-off.
PAL-WAUKEE TWO DEPARTURE
CHICAGO/PROSPECT HEIGHTS/WHEELING, ILLINOIS
CHICAGO EXECUTIVE (PWK)
12096

Figure 354

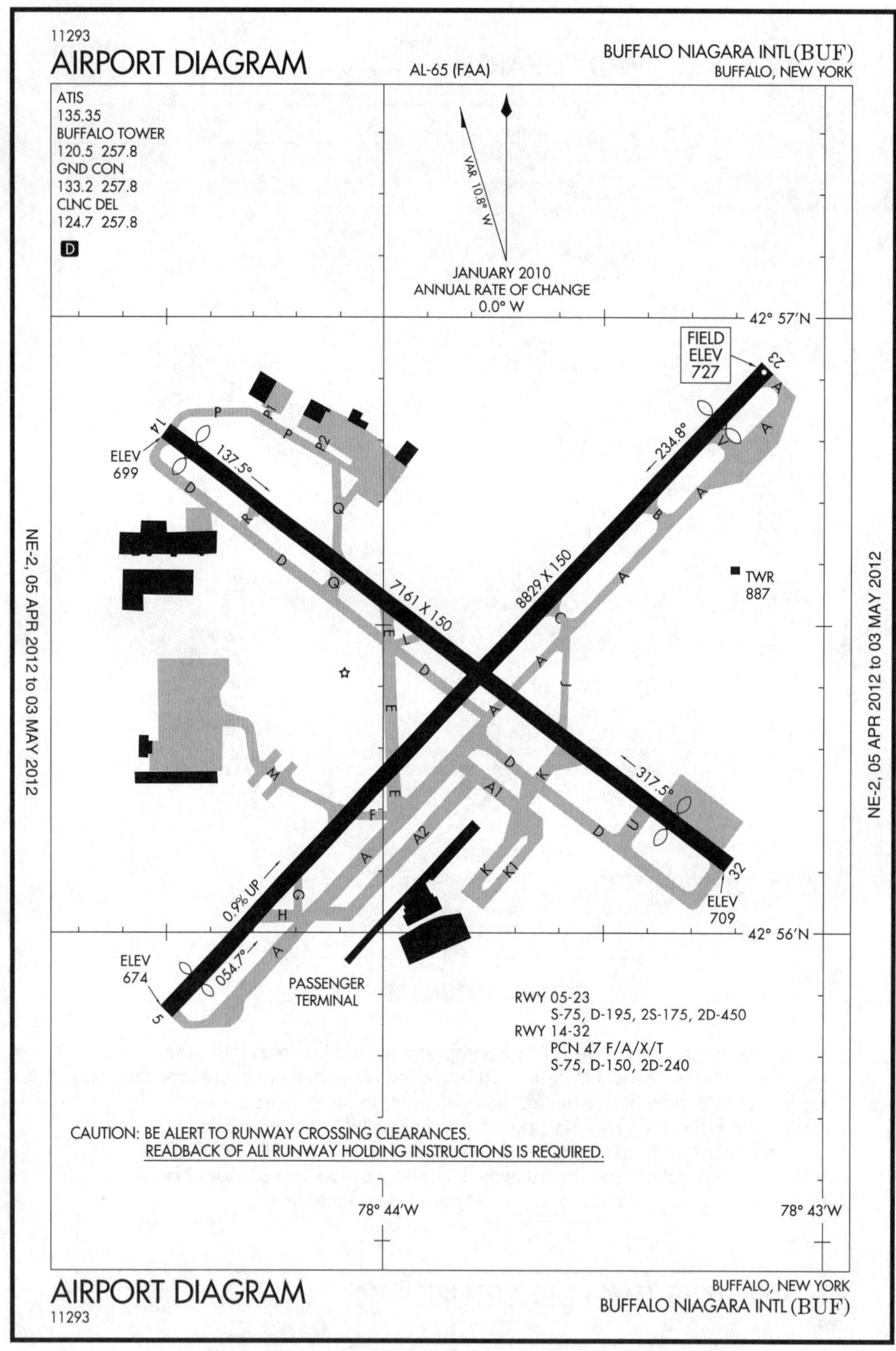
11293
AIRPORT DIAGRAM
AL-65 (FAA)
BUFFALO NIAGARA INTL (BUF)
BUFFALO, NEW YORK
ATIS
135.35
BUFFALO TOWER
120.5 257.8
GND CON
133.2 257.8
CLNC DEL
124.7 257.8
VAR 10.8° W
JANUARY 2010
ANNUAL RATE OF CHANGE
0.0° W
42° 57'N
FIELD ELEV 727
ELEV 699
137.5°
234.8°
8829 X 150
7161 X 150
TWR 887
317.5°
ELEV 709
42° 56'N
0.9% UP
054.7°
ELEV 674
PASSENGER TERMINAL
RWY 05-23
S-75, D-195, 2S-175, 2D-450
RWY 14-32
PCN 47 F/A/X/T
S-75, D-150, 2D-240
CAUTION: BE ALERT TO RUNWAY CROSSING CLEARANCES.
READBACK OF ALL RUNWAY HOLDING INSTRUCTIONS IS REQUIRED.
78° 44'W
78° 43'W
NE-2, 05 APR 2012 to 03 MAY 2012
AIRPORT DIAGRAM
11293
BUFFALO, NEW YORK
BUFFALO NIAGARA INTL (BUF)

Figure 355

BUFFALO NIAGARA INTL (BUF) 5 E UTC–5(–4DT) N42°56.43′ W78°43.84′ **DETROIT**
727 B S4 **FUEL** 100LL, JET A OX 1, 2, 3, 4 LRA Class I, ARFF Index D **H–10H, 11B, L–31E**
NOTAM FILE BUF **IAP, AD**
RWY 05–23: H8829X150 (ASPH–GRVD) S–75, D–195, 2S–175, 2D–450 HIRL CL
RWY 05: MALSR. TDZL. Thld dsplcd 535′. Bldg. 0.9% up.
RWY 23: ALSF2. TDZL. Thld dsplcd 725′. Tree.
RWY 14–32: H7161X150 (ASPH–GRVD) S–75, D–150, 2D–240 PCN 47 F/A/X/T HIRL
RWY 14: REIL. PAPI(P4L)—GA 3.0° TCH 45′. Thld dsplcd 320′. Tree.
RWY 32: MALSR. REIL. PAPI(P4L)—GA 3.0° TCH 54′. Thld dsplcd 720′. Sign. 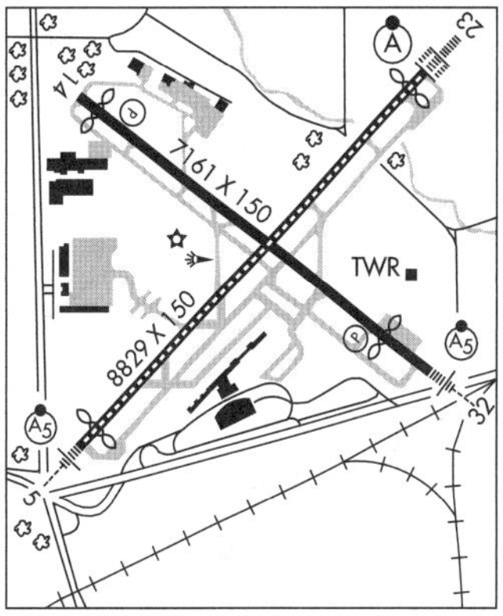
RUNWAY DECLARED DISTANCE INFORMATION
RWY 05: TORA–8827 TODA–8827 ASDA–8292 LDA–7757
RWY 14: TORA–7161 TODA–7161 ASDA–6441 LDA–6121
RWY 23: TORA–8827 TODA–8827 ASDA–8292 LDA–7567
RWY 32: TORA–7161 TODA–7161 ASDA–6841 LDA–6121
AIRPORT REMARKS: Attended continuously. Heavy concentration of gulls, blackbirds, and starlings up to 5000 ft on and invof arpt. Deer on and invof arpt. Twy K1 clsd 0200–1300Z‡ daily. Twy A SW runup area/holding bay marked design group 3 acft (generally B727 or smaller), unavbl design group 4 (includes but not limited to B757, DC8). For fixed–base operator svcs ctc 131.75; for cargo svcs ctc 122.95. Rwy 23 ALSF2 unmonitored. Ldg fee. Flight Notification Service (ADCUS) available.

7161 X 150
8829 X 150
TWR
A
A5
14
32
5
23

WEATHER DATA SOURCES: ASOS (716) 635–0532. WSP.
COMMUNICATIONS: D–ATIS 135.35
RCO 122.6 122.2 122.1R (BUFFALO RADIO)
Ⓡ **APP DEP/CON** 126.15 (053°–233°) 126.5 (234°–052°)
TOWER 120.5 **GND CON** 133.2 **CLNC DEL** 124.7 **PRE–TAXI CLNC** 124.7
AIRSPACE: CLASS C svc continuous, ctc **APP CON**
RADIO AIDS TO NAVIGATION: NOTAM FILE BUF.
(H) VOR/DME 116.4 BUF Chan 111 N42°55.74′ W78°38.78′ 288° 3.8 NM to fld. 730/08W.
VOR/DME unusable:
036°–261° blo 11,000′ 276°–305° blo 6000′
262°–275° blo 2300′
KLUMP NDB (LOM) 231 BU N43°00.02′ W78°39.05′ 233° 5.0 NM to fld.
PLAZZ NDB (LOM) 204 GB N42°52.43′ W78°48.99′ 053° 5.5 NM to fld.
ILS 111.3 I–BUF Rwy 23. Class IE. LOM KLUMP NDB. Glideslope unusable byd 5° rgt of course.
ILS 108.5 I–GBI Rwy 05. Class IA. LOM PLAZZ NDB.
ILS/DME 109.95 I–BNQ Chan 36(Y) Rwy 32.

CLARENCE AERODROME (D51) 5 NE UTC–5(–4DT) N43°04.00′ W78°40.99′ **DETROIT**
589 NOTAM FILE BUF
RWY 10–28: 2500X67 (TURF) LIRL
RWY 10: Fence. **RWY 28:** Trees.
AIRPORT REMARKS: Unattended. Ultralights on and invof arpt. Rwy 10–28 outlined with cones. ACTIVATE LIRL Rwy 10–28—122.7.
COMMUNICATIONS: CTAF/UNICOM 122.7

BUFFALO–LANCASTER RGNL (See LANCASTER)

BURRELLO–MECHANICVILLE (See MECHANICVILLE)

CALVERTON N40°55.78′ W72°47.93′ NOTAM FILE ISP. **NEW YORK**
(L) VORW/DME 117.2 CCC Chan 119 219° 7.2 NM to Brookhaven. 85/13W. **COPTER**
VOR portion unusable 280°–290° byd 25 NM. **H–10I, L–33B, 34I**

CAMBRIDGE N42°59.66′ W73°20.64′ NOTAM FILE BTV. **NEW YORK**
(L) VORW/DME 115.0 CAM Chan 97 159° 7.5 NM to Bennington State, Vt. 1490/14W. **H–11C, 12J, L–32G, 34J**
HIWAS.
DME unusable 050°–130° beyond 20 NM below 9000′.

NE, 05 APR 2012 to 31 MAY 2012

Figure 356

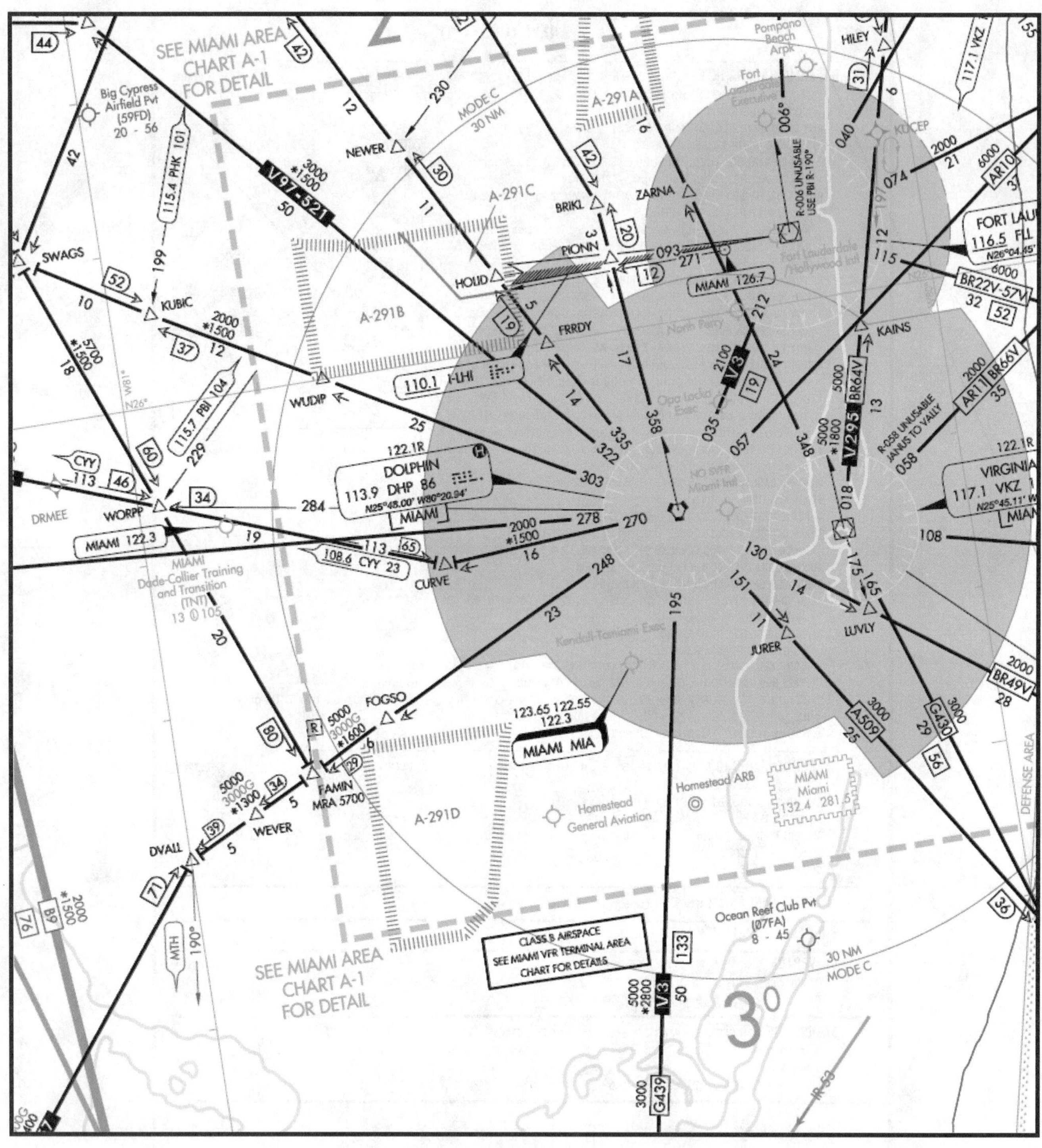

Figure 357

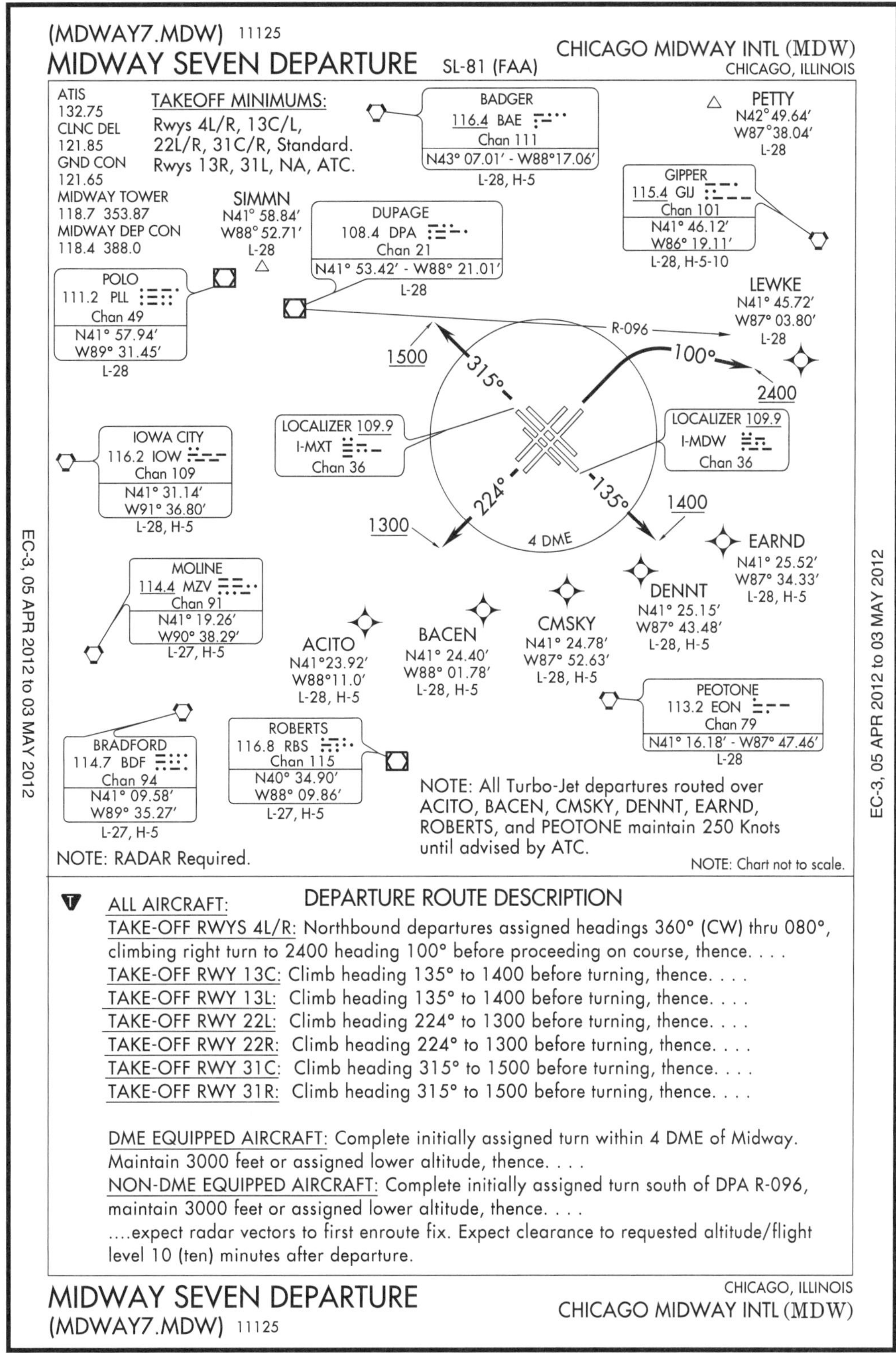
(MDWAY7.MDW) 11125
MIDWAY SEVEN DEPARTURE SL-81 (FAA)
CHICAGO MIDWAY INTL (MDW)
CHICAGO, ILLINOIS
ATIS 132.75
CLNC DEL 121.85
GND CON 121.65
MIDWAY TOWER 118.7 353.87
MIDWAY DEP CON 118.4 388.0
TAKEOFF MINIMUMS:
Rwys 4L/R, 13C/L, 22L/R, 31C/R, Standard.
Rwys 13R, 31L, NA, ATC.
BADGER 116.4 BAE Chan 111 N43° 07.01' - W88°17.06' L-28, H-5
PETTY N42°49.64' W87°38.04' L-28
SIMMN N41° 58.84' W88° 52.71' L-28
DUPAGE 108.4 DPA Chan 21 N41° 53.42' - W88° 21.01' L-28
GIPPER 115.4 GIJ Chan 101 N41° 46.12' W86° 19.11' L-28, H-5-10
POLO 111.2 PLL Chan 49 N41° 57.94' W89° 31.45' L-28
LEWKE N41° 45.72' W87° 03.80' L-28
R-096
1500
315°
100°
2400
LOCALIZER 109.9 I-MXT Chan 36
LOCALIZER 109.9 I-MDW Chan 36
IOWA CITY 116.2 IOW Chan 109 N41° 31.14' W91° 36.80' L-28, H-5
224°
135°
1300
1400
4 DME
EARND N41° 25.52' W87° 34.33' L-28, H-5
MOLINE 114.4 MZV Chan 91 N41° 19.26' W90° 38.29' L-27, H-5
DENNT N41° 25.15' W87° 43.48' L-28, H-5
CMSKY N41° 24.78' W87° 52.63' L-28, H-5
BACEN N41° 24.40' W88° 01.78' L-28, H-5
ACITO N41°23.92' W88°11.0' L-28, H-5
PEOTONE 113.2 EON Chan 79 N41° 16.18' - W87° 47.46' L-28
BRADFORD 114.7 BDF Chan 94 N41° 09.58' W89° 35.27' L-27, H-5
ROBERTS 116.8 RBS Chan 115 N40° 34.90' W88° 09.86' L-27, H-5
NOTE: All Turbo-Jet departures routed over ACITO, BACEN, CMSKY, DENNT, EARND, ROBERTS, and PEOTONE maintain 250 Knots until advised by ATC.
NOTE: RADAR Required.
NOTE: Chart not to scale.
EC-3, 05 APR 2012 to 03 MAY 2012
DEPARTURE ROUTE DESCRIPTION
ALL AIRCRAFT:
TAKE-OFF RWYS 4L/R: Northbound departures assigned headings 360° (CW) thru 080°, climbing right turn to 2400 heading 100° before proceeding on course, thence. . . .
TAKE-OFF RWY 13C: Climb heading 135° to 1400 before turning, thence. . . .
TAKE-OFF RWY 13L: Climb heading 135° to 1400 before turning, thence. . . .
TAKE-OFF RWY 22L: Climb heading 224° to 1300 before turning, thence. . . .
TAKE-OFF RWY 22R: Climb heading 224° to 1300 before turning, thence. . . .
TAKE-OFF RWY 31C: Climb heading 315° to 1500 before turning, thence. . . .
TAKE-OFF RWY 31R: Climb heading 315° to 1500 before turning, thence. . . .
DME EQUIPPED AIRCRAFT: Complete initially assigned turn within 4 DME of Midway. Maintain 3000 feet or assigned lower altitude, thence. . . .
NON-DME EQUIPPED AIRCRAFT: Complete initially assigned turn south of DPA R-096, maintain 3000 feet or assigned lower altitude, thence. . . .
....expect radar vectors to first enroute fix. Expect clearance to requested altitude/flight level 10 (ten) minutes after departure.
MIDWAY SEVEN DEPARTURE
(MDWAY7.MDW) 11125
CHICAGO, ILLINOIS
CHICAGO MIDWAY INTL (MDW)

Figure 358

(MDWAY7.MDW) 08325
MIDWAY SEVEN DEPARTURE SL-81 (FAA)
CHICAGO MIDWAY INTL (MDW)
CHICAGO, ILLINOIS

EC-3, 05 APR 2012 to 03 MAY 2012

TAKEOFF OBSTACLE NOTES:

NOTE: RWY 4L, Fence 18 feet from DER, 257 feet left of centerline, 12 feet AGL/616 feet MSL. Vehicle plus road 143 feet from DER, 163 feet left of centerline, 16 feet AGL/ 620 feet MSL. Bldg 251 feet from DER, 217 feet left of centerline, 26 feet AGL/630 feet MSL. Sign 1,912 feet from DER, 330 feet left of centerline, 88 feet AGL/692 feet MSL. Multiple Lt poles and trees beginning 375 feet from DER, 98 feet right of centerline, up to 75 feet AGL/679 feet MSL.

NOTE: RWY 4R, LOC 300 feet from DER, on centerline, 10 feet AGL/614 feet MSL. Lt pole and multiple trees beginning 40 feet from DER, 369 feet left of centerline, up to 75 feet AGL/679 feet MSL. Blast fence 277 feet from DER, 45 feet left of centerline, 9 feet AGL/613 feet MSL. Tower 3,983 feet from DER, 1,142 feet left of centerline, 109 feet AGL/708 feet MSL. Multiple lt poles and trees beginning 96 feet from DER, 21 feet right of centerline, up to 53 feet AGL/657 feet MSL. Train beginning 1,483 feet from DER, 570 feet right of centerline, 48 feet AGL/654 feet MSL.

NOTE: RWY 13C, LOC 248 feet from DER, on centerline, 8 feet AGL/619 feet MSL. Bldg 101 feet from DER, 254 feet left of centerline, 14 feet AGL/625 feet MSL. Trees beginning 288 feet from DER, 459 feet left of centerline, up to 76 feet AGL/680 feet MSL. Trees beginning 109 feet from DER, 402 feet right of centerline, up to 86 feet AGL/700 feet MSL.

NOTE: RWY 13L, Multiple poles and trees beginning 362 feet from DER, 215 feet left of centerline, up to 71 feet AGL/675 feet MSL. Trees beginning 1,136 feet from DER, 54 feet right of centerline, up to 76 feet AGL/680 feet MSL.

NOTE: RWY 22L, Multiple poles and trees beginning 74 feet from DER, 375 feet left of centerline, up to 70 feet AGL/689 feet MSL. Multiple poles and trees beginning 465 feet from DER, 49 feet right of centerline, up to 60 feet AGL/679 feet MSL. Tank 4,100 feet from DER, 161 feet right of centerline, 109 feet AGL/728 feet MSL.

NOTE: RWY 22R, Multiple poles and trees beginning 575 feet from DER, 168 feet left of centerline, up to 58 feet AGL/677 feet MSL. Tank 4,100 feet from DER, 161 feet left of centerline, 109 feet AGL/728 feet MSL. Fence 198 feet from DER, 3 feet right of centerline, 12 feet AGL/630 feet MSL. Trees beginning 183 feet from DER, 65 feet right of centerline, up to 72 feet AGL/686 feet MSL.

NOTE: RWY 31C, LOC 239 feet from DER, on centerline, 10 feet AGL/617 feet MSL. Trees beginning 452 feet from DER, 454 feet left of centerline, up to 63 feet AGL/667 feet MSL. Spire 2,207 feet from DER, 699 feet left of centerline, 78 feet AGL/684 feet MSL. Multiple poles and trees beginning 142 feet from DER, 28 feet right of centerline, up to 73 feet AGL/672 feet MSL. DME 183 feet from DER, 309 feet right of centerline, 17 feet AGL/624 feet MSL. Sign 1,528 feet from DER, 270 feet right of centerline, 52 feet AGL/652 feet MSL. Tank 5,576 feet from DER, 1,430 feet right of centerline, 162 feet AGL/756 feet MSL.

NOTE: RWY 31R, Multiple poles and trees beginning 379 feet from DER, 49 feet left of centerline, up to 65 feet AGL/664 feet MSL. Pole and trees beginning 70 feet from DER, 50 feet right of centerline, up to 68 feet AGL/667 feet MSL.

EC-3, 05 APR 2012 to 03 MAY 2012

MIDWAY SEVEN DEPARTURE
(MDWAY7.MDW) 08325
CHICAGO, ILLINOIS
CHICAGO MIDWAY INTL(MDW)

Figure 359

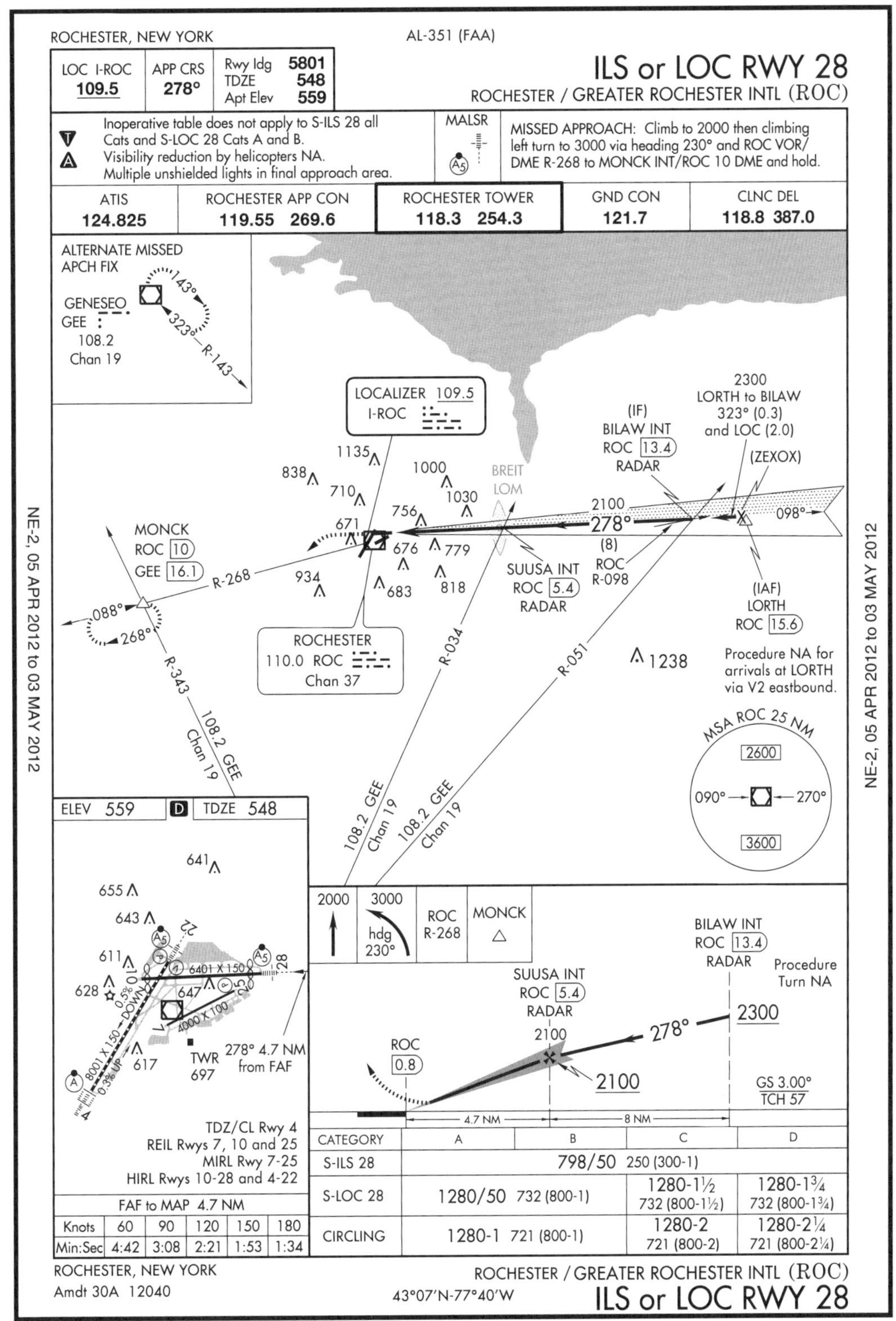

CATEGORY	A	B	C	D
S-ILS 28	798/50 250 (300-1)			
S-LOC 28	1280/50 732 (800-1)		1280-1½ 732 (800-1½)	1280-1¾ 732 (800-1¾)
CIRCLING	1280-1 721 (800-1)		1280-2 721 (800-2)	1280-2¼ 721 (800-2¼)

Figure 360

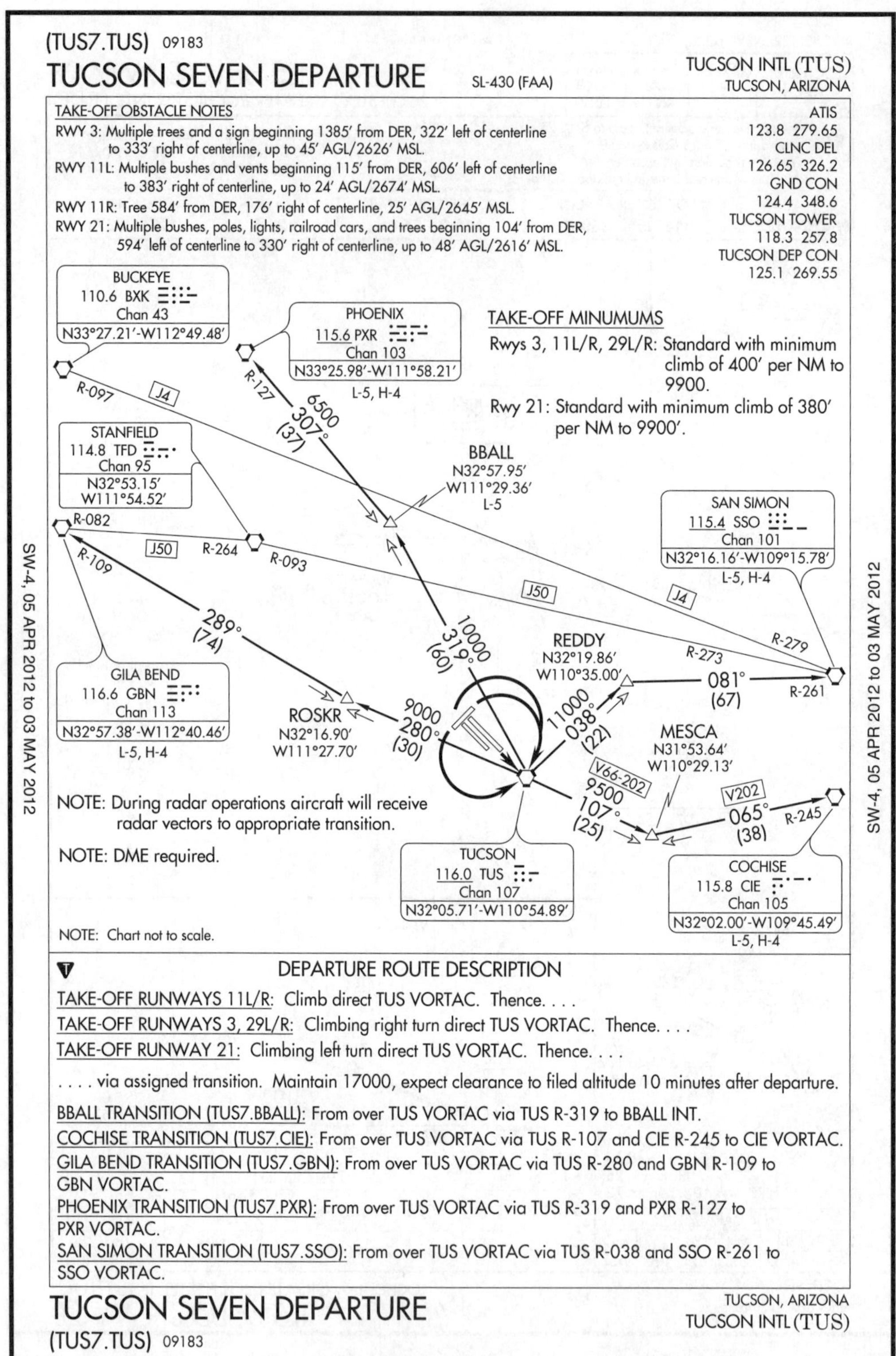
(TUS7.TUS) 09183
TUCSON SEVEN DEPARTURE
SL-430 (FAA)
TUCSON INTL (TUS)
TUCSON, ARIZONA
TAKE-OFF OBSTACLE NOTES
RWY 3: Multiple trees and a sign beginning 1385' from DER, 322' left of centerline to 333' right of centerline, up to 45' AGL/2626' MSL.
RWY 11L: Multiple bushes and vents beginning 115' from DER, 606' left of centerline to 383' right of centerline, up to 24' AGL/2674' MSL.
RWY 11R: Tree 584' from DER, 176' right of centerline, 25' AGL/2645' MSL.
RWY 21: Multiple bushes, poles, lights, railroad cars, and trees beginning 104' from DER, 594' left of centerline to 330' right of centerline, up to 48' AGL/2616' MSL.
ATIS
123.8 279.65
CLNC DEL
126.65 326.2
GND CON
124.4 348.6
TUCSON TOWER
118.3 257.8
TUCSON DEP CON
125.1 269.55
TAKE-OFF MINUMUMS
Rwys 3, 11L/R, 29L/R: Standard with minimum climb of 400' per NM to 9900.
Rwy 21: Standard with minimum climb of 380' per NM to 9900'.
BUCKEYE
110.6 BXK
Chan 43
N33°27.21'-W112°49.48'
PHOENIX
115.6 PXR
Chan 103
N33°25.98'-W111°58.21'
L-5, H-4
STANFIELD
114.8 TFD
Chan 95
N32°53.15'
W111°54.52'
BBALL
N32°57.95'
W111°29.36'
L-5
SAN SIMON
115.4 SSO
Chan 101
N32°16.16'-W109°15.78'
L-5, H-4
GILA BEND
116.6 GBN
Chan 113
N32°57.38'-W112°40.46'
L-5, H-4
ROSKR
N32°16.90'
W111°27.70'
REDDY
N32°19.86'
W110°35.00'
MESCA
N31°53.64'
W110°29.13'
TUCSON
116.0 TUS
Chan 107
N32°05.71'-W110°54.89'
COCHISE
115.8 CIE
Chan 105
N32°02.00'-W109°45.49'
L-5, H-4
R-097
J4
R-127
6500
307°
(37)
R-082
R-109
J50
R-264
R-093
289°
(74)
10000
319°
(60)
9000
280°
(30)
11000
038°
(22)
081°
(67)
R-273
R-279
R-261
V66-202
9500
107°
(25)
V202
065°
(38)
R-245
SW-4, 05 APR 2012 to 03 MAY 2012
NOTE: During radar operations aircraft will receive radar vectors to appropriate transition.
NOTE: DME required.
NOTE: Chart not to scale.
DEPARTURE ROUTE DESCRIPTION
TAKE-OFF RUNWAYS 11L/R: Climb direct TUS VORTAC. Thence. . . .
TAKE-OFF RUNWAYS 3, 29L/R: Climbing right turn direct TUS VORTAC. Thence. . . .
TAKE-OFF RUNWAY 21: Climbing left turn direct TUS VORTAC. Thence. . . .
. . . . via assigned transition. Maintain 17000, expect clearance to filed altitude 10 minutes after departure.
BBALL TRANSITION (TUS7.BBALL): From over TUS VORTAC via TUS R-319 to BBALL INT.
COCHISE TRANSITION (TUS7.CIE): From over TUS VORTAC via TUS R-107 and CIE R-245 to CIE VORTAC.
GILA BEND TRANSITION (TUS7.GBN): From over TUS VORTAC via TUS R-280 and GBN R-109 to GBN VORTAC.
PHOENIX TRANSITION (TUS7.PXR): From over TUS VORTAC via TUS R-319 and PXR R-127 to PXR VORTAC.
SAN SIMON TRANSITION (TUS7.SSO): From over TUS VORTAC via TUS R-038 and SSO R-261 to SSO VORTAC.
TUCSON SEVEN DEPARTURE
(TUS7.TUS) 09183
TUCSON, ARIZONA
TUCSON INTL (TUS)

Figure 361

L12

TAKE-OFF MINIMUMS AND (OBSTACLE) DEPARTURE PROCEDURES

12096

TONOPAH TEST RANGE (KTNX)

TONOPAH , NV. . . . . AMDT 1 12096

DEPARTURE PROCEDURE: **Rwy 14:** 1000-3 with minimum climb of 320 ft/NM to 10,700 or 2700-3 for Climb in Visual Conditions. Climb on a heading between 325° CW to 155° from departure end of runway or Climb in Visual Conditions to cross KZ-KTNX airport at or above 8100 MSL before proceeding on course. **Rwy 32:** 1000-3 with minimum climb of 260 ft/NM to 5900 or 2700-3 for Climb in Visual Conditions. Climb on a heading between 295° CW to 005° from departure end of runway or Climb in Visual Conditions to cross KZ-KTNX airport at or above 8100 MSL before proceeding on course.

TAKE-OFF OBSTACLES: **Rwy 14,** Terrain, 5582' MSL, 1204' from DER, 823' right of centerline. Terrain, 5565' MSL, 63' from DER, 517' right of centerline. Terrain, 5564' MSL, 46' from DER, 480' right of centerline. Terrain, 5561' MSL, 0' from DER, 353' right of centerline. Terrain, 5558' MSL, 62' from DER, 200' right of centerline. Terrain, 5561' MSL, 14' from DER, 292' right of centerline. Terrain, 5561' MSL, 0' from DER, 287' right of centerline. Terrain, 5559' MSL, 0' from DER, 222' right of centerline. Surveyed terrain, 5560' MSL, 215' from DER, 427' right of centerline. **Rwy 32,** Terrain, 5476' MSL, 0' from DER, 500' left of centerline. Terrain, 5476' MSL, 19' from DER, 465' left of centerline. Terrain, 5476' MSL, 110' from DER, 529' left of centerline.

TOOELE, UT

BOLINDER FIELD-TOOELE VALLEY

TAKE-OFF MINIMUMS: **Rwy 17,** std. with a min. climb of 490' per NM to 11000. **Rwy 35,** std. with a min. climb of 360' per NM to 9000.

DEPARTURE PROCEDURE: Use STACO DEPARTURE.

NOTE: **Rwy 17,** tree 794' from departure end of runway, 277' right of centerline, 35' AGL/4380' MSL. Tree 967' from departure end of runway, 432' right of centerline, 35' AGL/4394' MSL. Tree 1023' from departure end of runway, 313' right of centerline, 35' AGL/4395' MSL.

TUCSON, AZ

MARANA RGNL

TAKE-OFF MINIMUMS: **Rwys 3, 12,** N/A-Obstacles

DEPARTURE PROCEDURE: **Rwy 21,** climb to 6500 via heading 360° and TUS R-308 to TOTEC Int/TUS 57 DME, then as filed. **Rwy 30,** climb to 6500 via heading 303° intercept TUS R-308 above 3500, to TOTEC INT/TUS 57 DME, then as filed.

NOTE: **Rwy 21,** road 192' from departure end of runway, 527' left of centerline 15' AGL/2034' MSL.

RYAN FIELD (RYN)

AMDT 3 10210 (FAA)

TAKE-OFF MINIMUMS: **Rwys 6L, 15, 24R, 33,** NA, ATC.

DEPARTURE PROCEDURE: **Rwys 6R, 24L,** use ALMON DEPARTURE.

TUCSON, AZ (CON'T)

TUCSON INTL (TUS)

AMDT 4A 08241 (FAA)

TAKE-OFF MINIMUMS: **Rwy 3,** 300-1¾ or std. w/ min. climb of 228' per NM to 3000.

DEPARTURE PROCEDURE: **Rwys 3, 29L, 29R,** climbing right turn direct to TUS VORTAC. **Rwys 11L, 11R** climb via runway heading to 4000 then climbing left turn direct TUS VORTAC. **Rwy 21,** climbing left turn direct to TUS VORTAC. **All aircraft** continue climbing in holding pattern (NW, right turns, 128° inbound) to depart TUS VORTAC at or above 9000.

NOTE: **Rwy 3,** tower 9215' from departure end of runway, 1689' left of centerline, 246' AGL/2831' MSL.

VERNAL, UT

VERNAL RGNL

TAKE-OFF MINIMUMS: **Rwy 16,** 1500-2 or std with a min. climb of 250' per NM to 7000'. **Rwy 25,** 1500-2 or std. with a min. climb of 390' per NM to 7000. **Rwy 34,** 1600-2 pr std. with a min. climb of 330' per NM to 7000'.

DEPARTURE PROCEDURE: **Rwys 7,34,** turn right. **Rwys 16,25,** turn left. **All aircraft** climb direct VEL. Aircraft departing V391 S-bound climb on course. All others climb in holding pattern (SE, right turns, 322° inbound). Aircraft SW-bound V208 depart VEL at or above 8400', all others depart VEL at or above 9500'. Continue climb on course to MEA or assigned altitude.

WENDOVER, UT

WENDOVER

TAKE-OFF MINIMUMS: **Rwy 26,** standard with a min. climb of 300' per NM to 7000. **Rwy 30,** NA.

DEPARTURE PROCEDURE: **Rwys 8, 12, 26,** climbing left turn direct BVL VORTAC. Aircraft departing BVL VORTAC R-330 CW R-150 climb on course. All others continue climb in BVL VORTAC holding pattern (Hold NE right turns, 247° inbound) to cross at or above 7400, then climb on course.

WILLCOX, AZ

COCHISE COUNTY

DEPARTURE PROCEDURE: **Rwy 3,** turn right. **Rwy 21,** turn left. **All aircraft** climb direct CIE VORTAC.

WINDOW ROCK, AZ

WINDOW ROCK

TAKE-OFF MINIMUMS: **Rwy 2,** 700-2 or std. with a min. climb of 500' per NM to 8000. **Rwy 20,** 600-2 or std. with a min. climb of 260' per NM to 8200.

DEPARTURE PROCEDURE: **Rwy 2,** turn right. **Rwy 20,** turn left direct to GUP VORTAC before proceeding on course.

NOTE: **Rwy 2,** terrain 3832' from departure end of runway, 1025' right of centerline, 6926' MSL. Poles 5220' from departure end of runway, 245' right of centerline, 180' AGL/6922' MSL. Tower 7067' from departure end of runway, 3072' left of centerline, 71' AGL/7316' MSL. Terrain 7449' from departure end of runway, 1612' left of centerline, 6991' MSL. Terrain 8776' from departure end of runway, 1851' left of centerline, 7109' MSL. Tree 9665' from departure end of runway, 1326' right of centerline, 7340' MSL. Tree 11326' from departure end of runway, 355' left of centerline, 7351' MSL. **Rwy 20,** trees 1018' from departure end of runway, 620' left of centerline, 30' AGL/6768' MSL.

05 APR 2012 to 03 MAY 2012

05 APR 2012 to 03 MAY 2012

12096

TAKE-OFF MINIMUMS AND (OBSTACLE) DEPARTURE PROCEDURES

L12

SW-4

Figure 362

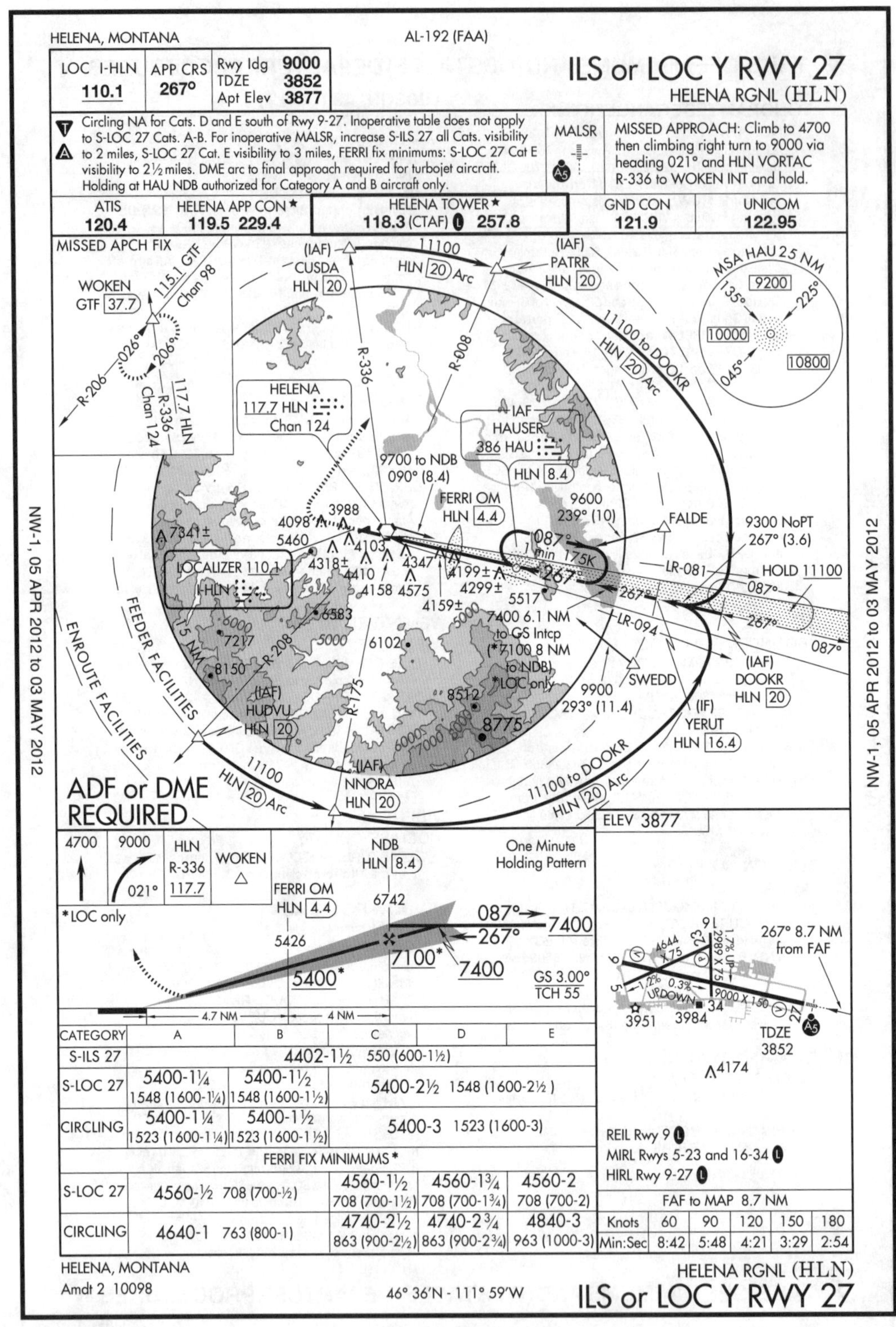
HELENA, MONTANA
AL-192 (FAA)
LOC I-HLN 110.1
APP CRS 267°
Rwy ldg 9000
TDZE 3852
Apt Elev 3877
ILS or LOC Y RWY 27
HELENA RGNL (HLN)
Circling NA for Cats. D and E south of Rwy 9-27. Inoperative table does not apply to S-LOC 27 Cats. A-B. For inoperative MALSR, increase S-ILS 27 all Cats. visibility to 2 miles, S-LOC 27 Cat. E visibility to 3 miles, FERRI fix minimums: S-LOC 27 Cat E visibility to 2½ miles. DME arc to final approach required for turbojet aircraft. Holding at HAU NDB authorized for Category A and B aircraft only.
MALSR
MISSED APPROACH: Climb to 4700 then climbing right turn to 9000 via heading 021° and HLN VORTAC R-336 to WOKEN INT and hold.
ATIS 120.4
HELENA APP CON ★ 119.5 229.4
HELENA TOWER ★ 118.3 (CTAF) 257.8
GND CON 121.9
UNICOM 122.95
MISSED APCH FIX
WOKEN GTF 37.7
115.1 GTF Chan 98
117.7 HLN Chan 124
R-206
R-336
(IAF) CUSDA HLN 20
(IAF) PATRR HLN 20
11100
HLN 20 Arc
11100 to DOOKR
MSA HAU 25 NM
9200
10000
10800
HELENA 117.7 HLN Chan 124
R-336
R-008
IAF HAUSER 386 HAU
HLN 8.4
9700 to NDB 090° (8.4)
FERRI OM HLN 4.4
9600 239° (10)
FALDE
9300 NoPT 267° (3.6)
HOLD 11100
LR-081
LR-094
087°
267°
1 Min 175K
(IAF) DOOKR HLN 20
(IF) YERUT HLN 16.4
SWEDD
9900 293° (11.4)
7400 6.1 NM to GS Intcp (*7100 8 NM to NDB) *LOC only
LOCALIZER 110.1 I-HLN
R-208
R-175
15 NM
FEEDER FACILITIES
ENROUTE FACILITIES
(IAF) HUDVU HLN 20
(IAF) NNORA HLN 20
11100 to DOOKR
ADF or DME REQUIRED
NW-1, 05 APR 2012 to 03 MAY 2012
ELEV 3877
4700
9000
021°
HLN R-336 117.7
WOKEN
*LOC only
NDB HLN 8.4
One Minute Holding Pattern
FERRI OM HLN 4.4
6742
5426
087°
267°
7400
7100*
5400*
GS 3.00° TCH 55
4.7 NM
4 NM
267° 8.7 NM from FAF
TDZE 3852
CATEGORY A B C D E
S-ILS 27 4402-1½ 550 (600-1½)
S-LOC 27 5400-1¼ 1548 (1600-1¼) | 5400-1½ 1548 (1600-1½) | 5400-2½ 1548 (1600-2½)
CIRCLING 5400-1¼ 1523 (1600-1¼) | 5400-1½ 1523 (1600-1½) | 5400-3 1523 (1600-3)
FERRI FIX MINIMUMS *
S-LOC 27 4560-½ 708 (700-½) | 4560-1½ 708 (700-1½) | 4560-1¾ 708 (700-1¾) | 4560-2 708 (700-2)
CIRCLING 4640-1 763 (800-1) | 4740-2½ 863 (900-2½) | 4740-2¾ 863 (900-2¾) | 4840-3 963 (1000-3)
REIL Rwy 9
MIRL Rwys 5-23 and 16-34
HIRL Rwy 9-27
FAF to MAP 8.7 NM
Knots 60 90 120 150 180
Min:Sec 8:42 5:48 4:21 3:29 2:54
HELENA, MONTANA
Amdt 2 10098
46° 36'N - 111° 59'W
HELENA RGNL (HLN)
ILS or LOC Y RWY 27

Figure 363

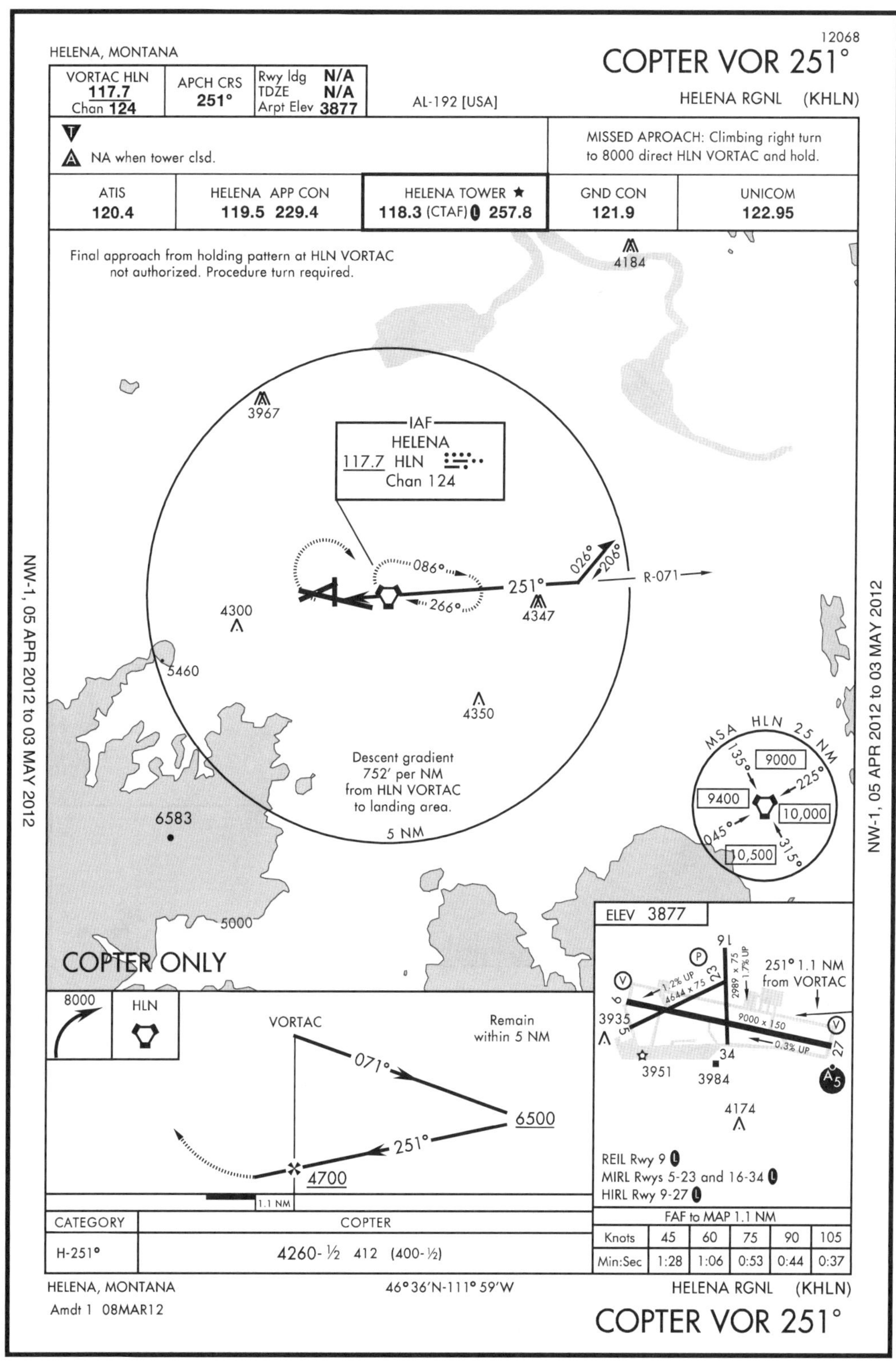
HELENA, MONTANA
12068
COPTER VOR 251°
VORTAC HLN 117.7 Chan 124
APCH CRS 251°
Rwy ldg N/A
TDZE N/A
Arpt Elev 3877
AL-192 [USA]
HELENA RGNL (KHLN)
NA when tower clsd.
MISSED APROACH: Climbing right turn to 8000 direct HLN VORTAC and hold.
ATIS 120.4
HELENA APP CON 119.5 229.4
HELENA TOWER ★ 118.3 (CTAF) 257.8
GND CON 121.9
UNICOM 122.95
Final approach from holding pattern at HLN VORTAC not authorized. Procedure turn required.
IAF HELENA 117.7 HLN Chan 124
086°
266°
251°
026°
206°
R-071
4184
3967
4300
4347
5460
4350
6583
5000
Descent gradient 752′ per NM from HLN VORTAC to landing area.
5 NM
MSA HLN 25 NM
135°
225°
045°
315°
9000
9400
10,000
10,500
COPTER ONLY
ELEV 3877
251° 1.1 NM from VORTAC
9000 x 150
4644 x 75
2989 x 75
1.2% UP
1.7% UP
0.3% UP
3935
3951
3984
4174
REIL Rwy 9
MIRL Rwys 5-23 and 16-34
HIRL Rwy 9-27
8000
HLN
VORTAC
Remain within 5 NM
071°
6500
251°
4700
1.1 NM
CATEGORY
COPTER
H-251°
4260-½ 412 (400-½)
FAF to MAP 1.1 NM
Knots 45 60 75 90 105
Min:Sec 1:28 1:06 0:53 0:44 0:37
HELENA, MONTANA
Amdt 1 08MAR12
46°36′N-111°59′W
HELENA RGNL (KHLN)
COPTER VOR 251°
NW-1, 05 APR 2012 to 03 MAY 2012

Figure 364

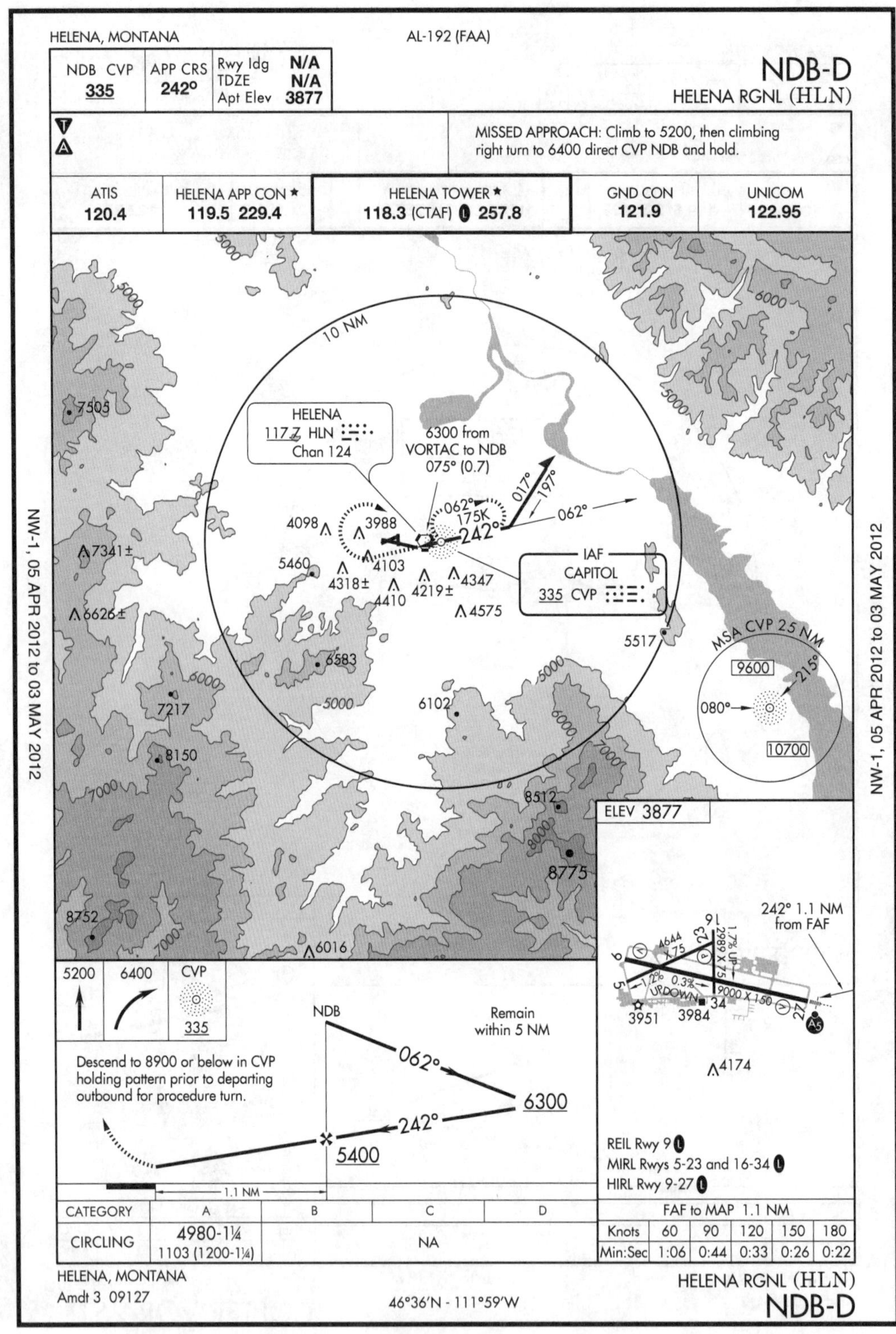
HELENA, MONTANA
AL-192 (FAA)
NDB CVP 335
APP CRS 242°
Rwy Idg N/A
TDZE N/A
Apt Elev 3877
NDB-D
HELENA RGNL (HLN)
MISSED APPROACH: Climb to 5200, then climbing right turn to 6400 direct CVP NDB and hold.
ATIS 120.4
HELENA APP CON ★ 119.5 229.4
HELENA TOWER ★ 118.3 (CTAF) 257.8
GND CON 121.9
UNICOM 122.95
HELENA 117.7 HLN Chan 124
6300 from VORTAC to NDB 075° (0.7)
IAF CAPITOL 335 CVP
MSA CVP 25 NM
9600
10700
NW-1, 05 APR 2012 to 03 MAY 2012
ELEV 3877
242° 1.1 NM from FAF
REIL Rwy 9
MIRL Rwys 5-23 and 16-34
HIRL Rwy 9-27
Descend to 8900 or below in CVP holding pattern prior to departing outbound for procedure turn.
Remain within 5 NM
CATEGORY A B C D
CIRCLING 4980-1¼ 1103 (1200-1¼) NA
FAF to MAP 1.1 NM
Knots 60 90 120 150 180
Min:Sec 1:06 0:44 0:33 0:26 0:22
HELENA, MONTANA
Amdt 3 09127
46°36'N - 111°59'W
HELENA RGNL (HLN)
NDB-D

Figure 365

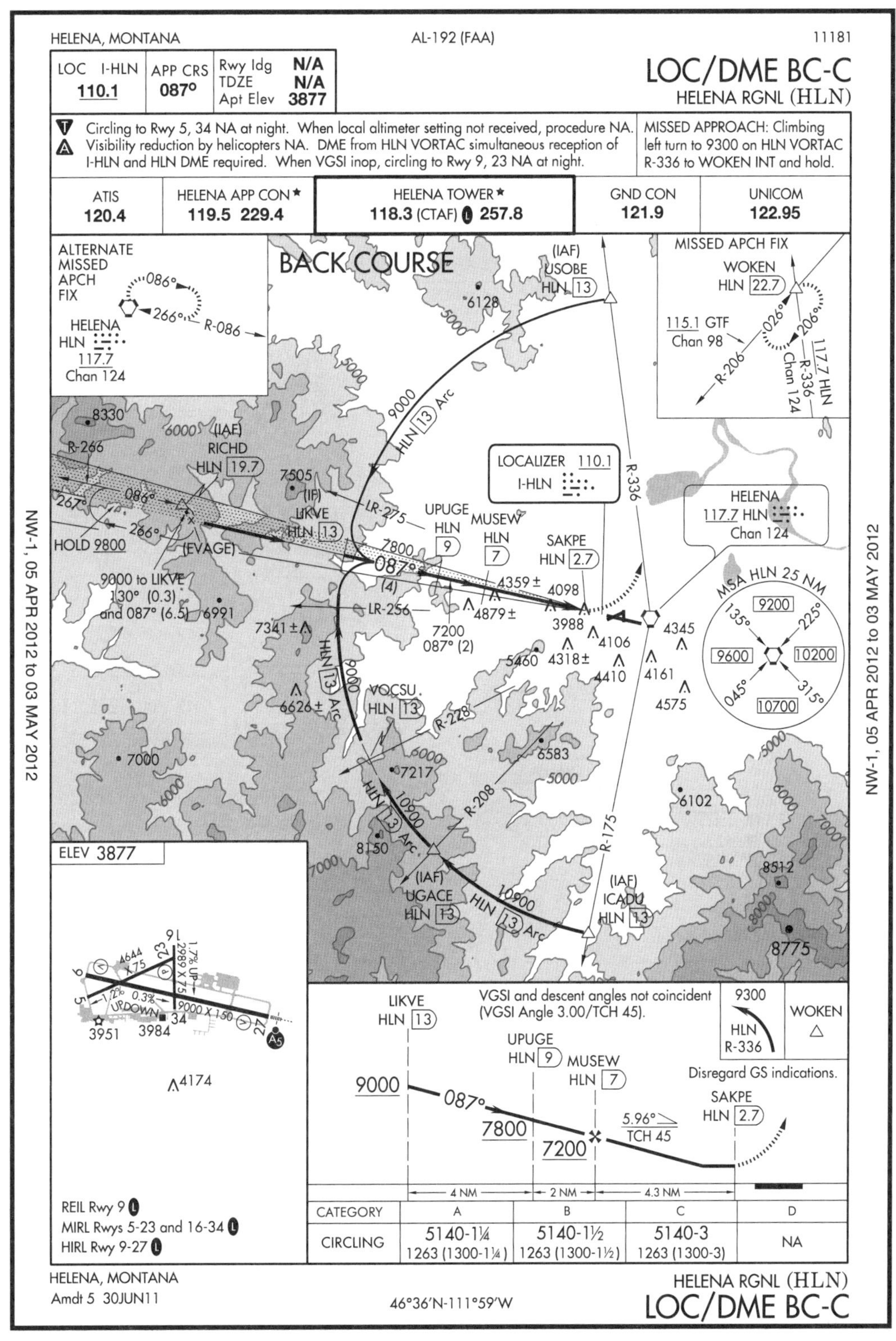
HELENA, MONTANA
AL-192 (FAA)
11181
LOC I-HLN 110.1
APP CRS 087°
Rwy ldg N/A
TDZE N/A
Apt Elev 3877
LOC/DME BC-C
HELENA RGNL (HLN)
Circling to Rwy 5, 34 NA at night. When local altimeter setting not received, procedure NA. Visibility reduction by helicopters NA. DME from HLN VORTAC simultaneous reception of I-HLN and HLN DME required. When VGSI inop, circling to Rwy 9, 23 NA at night.
MISSED APPROACH: Climbing left turn to 9300 on HLN VORTAC R-336 to WOKEN INT and hold.
ATIS 120.4
HELENA APP CON★ 119.5 229.4
HELENA TOWER★ 118.3 (CTAF) 257.8
GND CON 121.9
UNICOM 122.95
ALTERNATE MISSED APCH FIX
HELENA HLN 117.7 Chan 124
BACK COURSE
MISSED APCH FIX
WOKEN HLN 22.7
115.1 GTF Chan 98
LOCALIZER 110.1 I-HLN
HELENA 117.7 HLN Chan 124
MSA HLN 25 NM
ELEV 3877
REIL Rwy 9
MIRL Rwys 5-23 and 16-34
HIRL Rwy 9-27
VGSI and descent angles not coincident (VGSI Angle 3.00/TCH 45).
Disregard GS indications.
CATEGORY A B C D
CIRCLING 5140-1¼ 1263 (1300-1¼) 5140-1½ 1263 (1300-1½) 5140-3 1263 (1300-3) NA
HELENA, MONTANA
Amdt 5 30JUN11
46°36'N-111°59'W
HELENA RGNL (HLN)
LOC/DME BC-C
NW-1, 05 APR 2012 to 03 MAY 2012

Figure 366

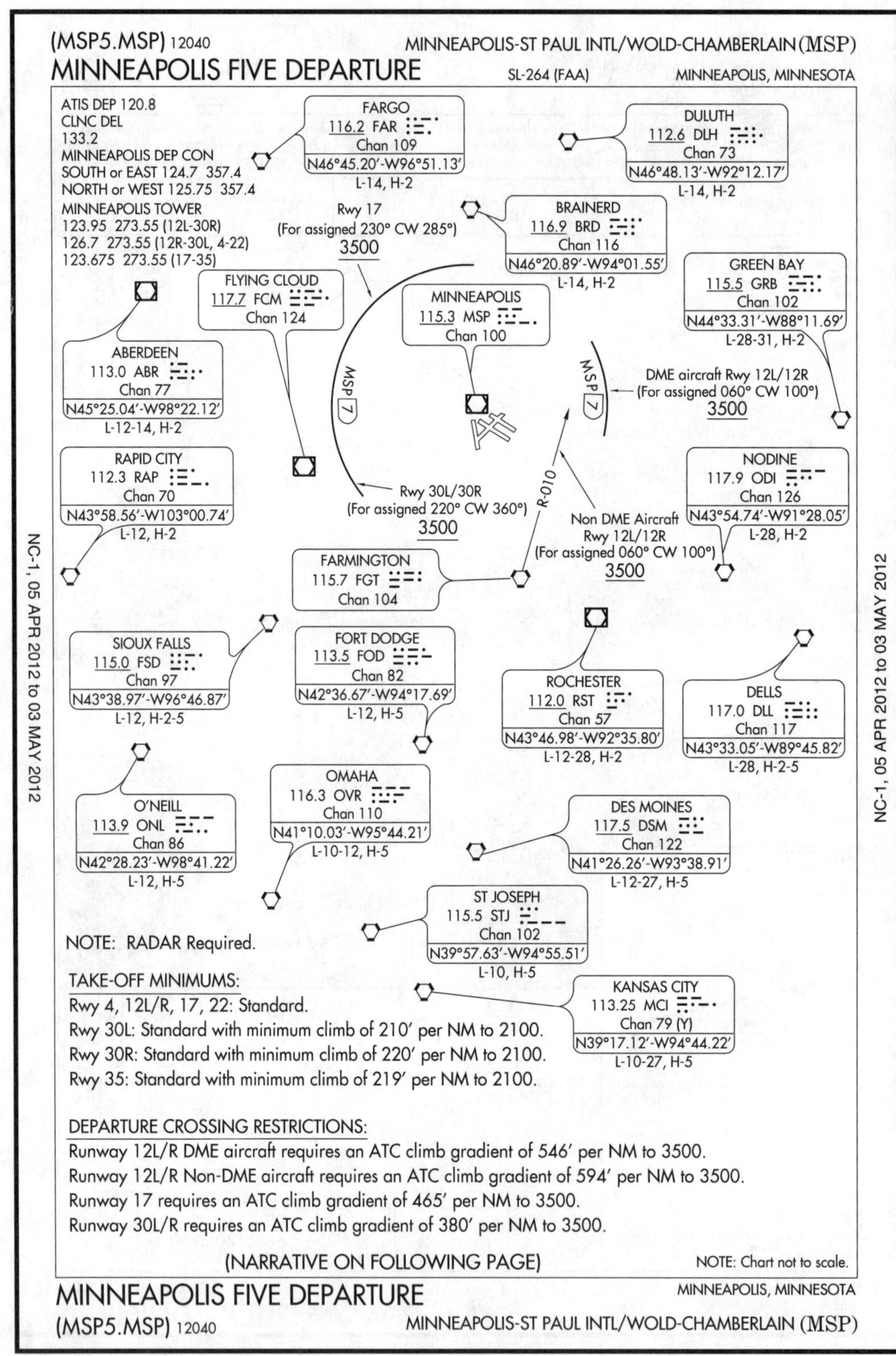

(MSP5.MSP) 12040
MINNEAPOLIS-ST PAUL INTL/WOLD-CHAMBERLAIN (MSP)
MINNEAPOLIS FIVE DEPARTURE
SL-264 (FAA)
MINNEAPOLIS, MINNESOTA
ATIS DEP 120.8
CLNC DEL
133.2
MINNEAPOLIS DEP CON
SOUTH or EAST 124.7 357.4
NORTH or WEST 125.75 357.4
MINNEAPOLIS TOWER
123.95 273.55 (12L-30R)
126.7 273.55 (12R-30L, 4-22)
123.675 273.55 (17-35)
FARGO 116.2 FAR Chan 109 N46°45.20'-W96°51.13' L-14, H-2
DULUTH 112.6 DLH Chan 73 N46°48.13'-W92°12.17' L-14, H-2
BRAINERD 116.9 BRD Chan 116 N46°20.89'-W94°01.55' L-14, H-2
Rwy 17 (For assigned 230° CW 285°) 3500
FLYING CLOUD 117.7 FCM Chan 124
MINNEAPOLIS 115.3 MSP Chan 100
GREEN BAY 115.5 GRB Chan 102 N44°33.31'-W88°11.69' L-28-31, H-2
ABERDEEN 113.0 ABR Chan 77 N45°25.04'-W98°22.12' L-12-14, H-2
MSP 7
DME aircraft Rwy 12L/12R (For assigned 060° CW 100°) 3500
RAPID CITY 112.3 RAP Chan 70 N43°58.56'-W103°00.74' L-12, H-2
Rwy 30L/30R (For assigned 220° CW 360°) 3500
R-010
Non DME Aircraft Rwy 12L/12R (For assigned 060° CW 100°) 3500
NODINE 117.9 ODI Chan 126 N43°54.74'-W91°28.05' L-28, H-2
FARMINGTON 115.7 FGT Chan 104
SIOUX FALLS 115.0 FSD Chan 97 N43°38.97'-W96°46.87' L-12, H-2-5
FORT DODGE 113.5 FOD Chan 82 N42°36.67'-W94°17.69' L-12, H-5
ROCHESTER 112.0 RST Chan 57 N43°46.98'-W92°35.80' L-12-28, H-2
DELLS 117.0 DLL Chan 117 N43°33.05'-W89°45.82' L-28, H-2-5
OMAHA 116.3 OVR Chan 110 N41°10.03'-W95°44.21' L-10-12, H-5
O'NEILL 113.9 ONL Chan 86 N42°28.23'-W98°41.22' L-12, H-5
DES MOINES 117.5 DSM Chan 122 N41°26.26'-W93°38.91' L-12-27, H-5
ST JOSEPH 115.5 STJ Chan 102 N39°57.63'-W94°55.51' L-10, H-5
KANSAS CITY 113.25 MCI Chan 79 (Y) N39°17.12'-W94°44.22' L-10-27, H-5
NOTE: RADAR Required.
TAKE-OFF MINIMUMS:
Rwy 4, 12L/R, 17, 22: Standard.
Rwy 30L: Standard with minimum climb of 210' per NM to 2100.
Rwy 30R: Standard with minimum climb of 220' per NM to 2100.
Rwy 35: Standard with minimum climb of 219' per NM to 2100.
DEPARTURE CROSSING RESTRICTIONS:
Runway 12L/R DME aircraft requires an ATC climb gradient of 546' per NM to 3500.
Runway 12L/R Non-DME aircraft requires an ATC climb gradient of 594' per NM to 3500.
Runway 17 requires an ATC climb gradient of 465' per NM to 3500.
Runway 30L/R requires an ATC climb gradient of 380' per NM to 3500.
(NARRATIVE ON FOLLOWING PAGE)
NOTE: Chart not to scale.
MINNEAPOLIS FIVE DEPARTURE
MINNEAPOLIS, MINNESOTA
(MSP5.MSP) 12040
MINNEAPOLIS-ST PAUL INTL/WOLD-CHAMBERLAIN (MSP)
NC-1, 05 APR 2012 to 03 MAY 2012
NC-1, 05 APR 2012 to 03 MAY 2012

Figure 367

(MSP5.MSP) 11349 MINNEAPOLIS-ST PAUL INTL/WOLD-CHAMBERLAIN (MSP)

MINNEAPOLIS FIVE DEPARTURE

SL-264 (FAA) MINNEAPOLIS, MINNESOTA

NC-1, 05 APR 2012 to 03 MAY 2012

DEPARTURE ROUTE DESCRIPTION

ALL RUNWAYS: Fly assigned heading for radar vectors to join filed/assigned route. Turbojet aircraft maintain 7000 or lower assigned altitude, all other aircraft maintain 5000 or lower assigned altitude. Expect clearance to assigned altitude/flight level 10 (ten) minutes after departure.

DME EQUIPPED AIRCRAFT RWY 12L/12R DEPARTURES: For assigned heading from 060° clockwise to 100°, cross MSP 7 DME at or above 3500, maintain assigned altitude. If unable to comply advise ATC as soon as possible prior to departure.

NON-DME EQUIPPED AIRCRAFT RWY 12L/12R DEPARTURES: For assigned headings from 060° clockwise to 100°, cross FGT R-010 at or above 3500, maintain assigned altitude. If unable to comply, advise ATC as soon as possible prior to departure.

TAKE-OFF RWY 17 DEPARTURES: For assigned headings from 230° clockwise to 285° cross MSP 7 DME at or above 3500, maintain assigned altitude. If unable to comply, advise ATC as soon as possible prior to departure.

TAKE-OFF RWYS 30L/30R DEPARTURES: For assigned headings from 220° clockwise to 360° cross MSP 7 DME at or above 3500, maintain assigned altitude. If unable to comply, advise ATC as soon as possible prior to departure.

TAKE-OFF OBSTACLE NOTES:

RWY 04: Multiple trees beginning 800′ from DER, 264′ left of centerline, up to 75′ AGL/921′ MSL.
Rod on building 2528′ from DER, 1175′ left of centerline, 78′ AGL/922′ MSL.
Fence 803′ from DER, 585′ left of centerline, 15′ AGL/860′ MSL.
Ant on OL building 456′ from DER, 319′ left of centerline, 13′ AGL/850′ MSL.
LT poles 1932′ from DER, 718′ left of centerline, 45′ AGL/885′ MSL.
Stack 4535′ from DER, 481′ left of centerline, 139′ AGL/949′ MSL.

RWY 12R: Multiple trees beginning 1477′ from DER, 407′ left of centerline, up to 86′ AGL/851′ MSL.
Multiple trees beginning 1426′ from DER, 124′ right of centerline, up to 111′ AGL/847′ MSL.
LT pole 1408′ from DER, 746′ right of centerline, 85′ AGL/843′ MSL.
Radar reflector 983′ from DER, 32′ left of centerline, 15′ AGL/829′ MSL.
Pipe on bldg, 826′ from DER, 576′ left of centerline, 10′ AGL/825′ MSL.
OL on LOC 766′ from DER, on centerline, 7′ AGL/821′ MSL.

RWY 17: Antenna 1272′ from DER, 562′ right of centerline, 57′ AGL/891′ MSL.
Pole 409′ from DER, 530′ right of centerline, 29′ AGL/866′ MSL.
Wind direction indicator on bldg 2619′ from DER, 881′ left of centerline, 97′ AGL/918′ MSL.
Bldg 2619′ from DER, 859′ left of centerline, 84′ AGL/905′ MSL.
LT 1176′ from DER, 291′ right of centerline, 11′ AGL/875′ MSL.
Tree 2619′ from DER, on centerline, 79′ AGL/900′ MSL.

RWY 22: Tree 2906′ from DER, 833′ right of centerline, 94′ AGL/934′ MSL.
Hopper 1717′ from DER, 456′ left of centerline, 48′ AGL/888′ MSL.

RWY 30L: Multiple trees beginning 1113′ from DER, 701′ left of centerline, up to 80′ AGL/919′ MSL.
Tree 1230′ from DER, 633′ right of centerline, 30′ AGL/877′ MSL.
Ground 28′ from DER, 490′ right of centerline, 0′ AGL/844′ MSL.

RWY 30R: Bldg 1056′ from DER, 198′ left of centerline, 13′ AGL/853′ MSL.
Multiple trees beginning 3010′ from DER, 334′ left of centerline, up to 94′ AGL/940′ MSL.
LT pole 1849′ from DER, 698′ right of centerline, 17′ AGL/871′ MSL.
Fence 1327′ from DER, 667′ right of centerline, 8′ AGL/857′ MSL.
Tree 3703′ from DER, 350′ right of centerline, 67′ AGL/914′ MSL.
Rod on pole 3143′ from DER, 47′ right of centerline, 38′ AGL/898′ MSL.

RWY 35: Tree 175′ from DER, 398′ right of centerline, 73′ AGL/883′ MSL.
Multiple trees beginning 1989′ from DER, 351′ left of centerline, up to 65′ AGL/902′ MSL.
Multiple buildings beginning 5.5 NM from DER, 1787′ left of centerline, up to 811′ AGL/1743′ MSL.

NC-1, 05 APR 2012 to 03 MAY 2012

MINNEAPOLIS FIVE DEPARTURE MINNEAPOLIS, MINNESOTA

(MSP5.MSP) 11349 MINNEAPOLIS-ST PAUL INTL/WOLD-CHAMBERLAIN (MSP)

Figure 368

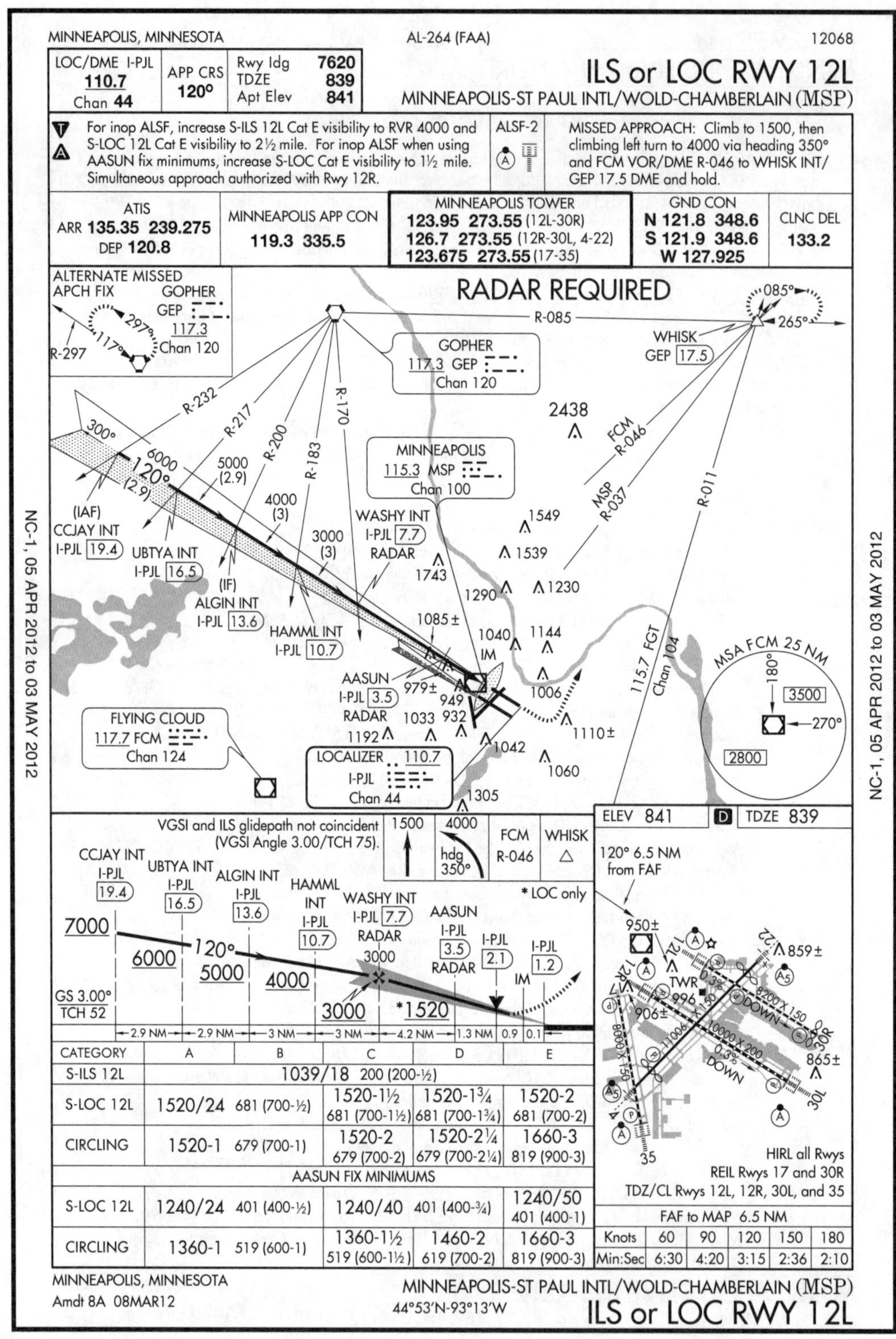

MINNEAPOLIS, MINNESOTA
AL-264 (FAA)
12068
LOC/DME I-PJL 110.7 Chan 44
APP CRS 120°
Rwy ldg 7620
TDZE 839
Apt Elev 841
ILS or LOC RWY 12L
MINNEAPOLIS-ST PAUL INTL/WOLD-CHAMBERLAIN (MSP)
For inop ALSF, increase S-ILS 12L Cat E visibility to RVR 4000 and S-LOC 12L Cat E visibility to 2½ mile. For inop ALSF when using AASUN fix minimums, increase S-LOC Cat E visibility to 1½ mile. Simultaneous approach authorized with Rwy 12R.
ALSF-2
MISSED APPROACH: Climb to 1500, then climbing left turn to 4000 via heading 350° and FCM VOR/DME R-046 to WHISK INT/GEP 17.5 DME and hold.
ATIS ARR 135.35 239.275 DEP 120.8
MINNEAPOLIS APP CON 119.3 335.5
MINNEAPOLIS TOWER 123.95 273.55 (12L-30R) 126.7 273.55 (12R-30L, 4-22) 123.675 273.55 (17-35)
GND CON N 121.8 348.6 S 121.9 348.6 W 127.925
CLNC DEL 133.2
ALTERNATE MISSED APCH FIX GOPHER GEP 117.3 Chan 120
RADAR REQUIRED
GOPHER 117.3 GEP Chan 120
MINNEAPOLIS 115.3 MSP Chan 100
FLYING CLOUD 117.7 FCM Chan 124
LOCALIZER 110.7 I-PJL Chan 44
WHISK GEP 17.5
(IAF) CCJAY INT I-PJL 19.4
UBTYA INT I-PJL 16.5
(IF) ALGIN INT I-PJL 13.6
HAMML INT I-PJL 10.7
WASHY INT I-PJL 7.7 RADAR
AASUN I-PJL 3.5 RADAR
MSA FCM 25 NM
ELEV 841
TDZE 839
120° 6.5 NM from FAF
VGSI and ILS glidepath not coincident (VGSI Angle 3.00/TCH 75).
GS 3.00° TCH 52
* LOC only
CATEGORY A B C D E
S-ILS 12L 1039/18 200 (200-½)
S-LOC 12L 1520/24 681 (700-½) 1520-1½ 681 (700-1½) 1520-1¾ 681 (700-1¾) 1520-2 681 (700-2)
CIRCLING 1520-1 679 (700-1) 1520-2 679 (700-2) 1520-2¼ 679 (700-2¼) 1660-3 819 (900-3)
AASUN FIX MINIMUMS
S-LOC 12L 1240/24 401 (400-½) 1240/40 401 (400-¾) 1240/50 401 (400-1)
CIRCLING 1360-1 519 (600-1) 1360-1½ 519 (600-1½) 1460-2 619 (700-2) 1660-3 819 (900-3)
HIRL all Rwys
REIL Rwys 17 and 30R
TDZ/CL Rwys 12L, 12R, 30L, and 35
FAF to MAP 6.5 NM
Knots 60 90 120 150 180
Min:Sec 6:30 4:20 3:15 2:36 2:10
MINNEAPOLIS, MINNESOTA
Amdt 8A 08MAR12
44°53'N-93°13'W
NC-1, 05 APR 2012 to 03 MAY 2012

Figure 369

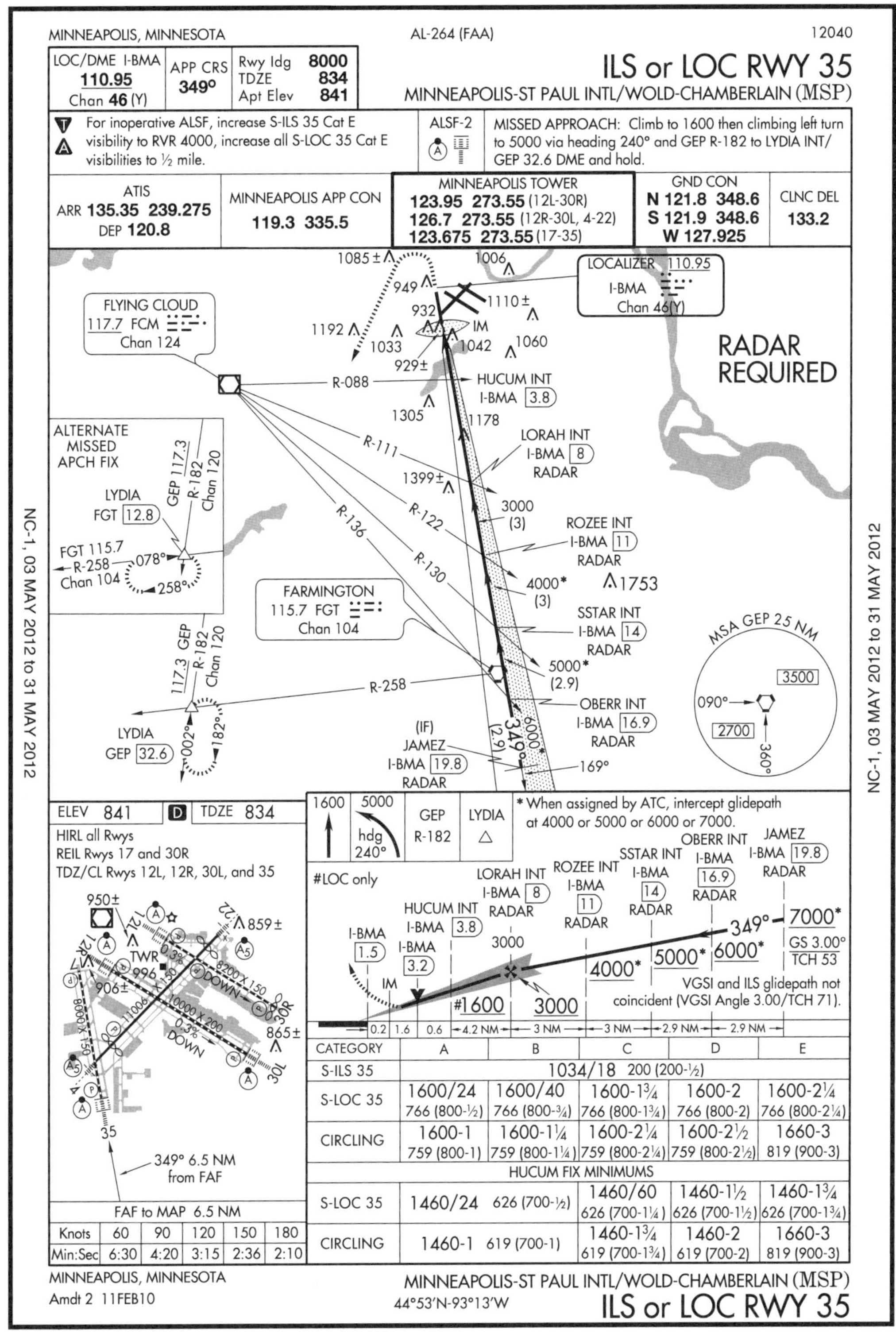

FAF to MAP 6.5 NM					
Knots	60	90	120	150	180
Min:Sec	6:30	4:20	3:15	2:36	2:10

CATEGORY	A	B	C	D	E
S-ILS 35	1034/18 200 (200-½)				
S-LOC 35	1600/24 766 (800-½)	1600/40 766 (800-¾)	1600-1¾ 766 (800-1¾)	1600-2 766 (800-2)	1600-2¼ 766 (800-2¼)
CIRCLING	1600-1 759 (800-1)	1600-1¼ 759 (800-1¼)	1600-2¼ 759 (800-2¼)	1600-2½ 759 (800-2½)	1660-3 819 (900-3)
HUCUM FIX MINIMUMS					
S-LOC 35	1460/24 626 (700-½)		1460/60 626 (700-1¼)	1460-1½ 626 (700-1½)	1460-1¾ 626 (700-1¾)
CIRCLING	1460-1 619 (700-1)		1460-1¾ 619 (700-1¾)	1460-2 619 (700-2)	1660-3 819 (900-3)

Figure 370

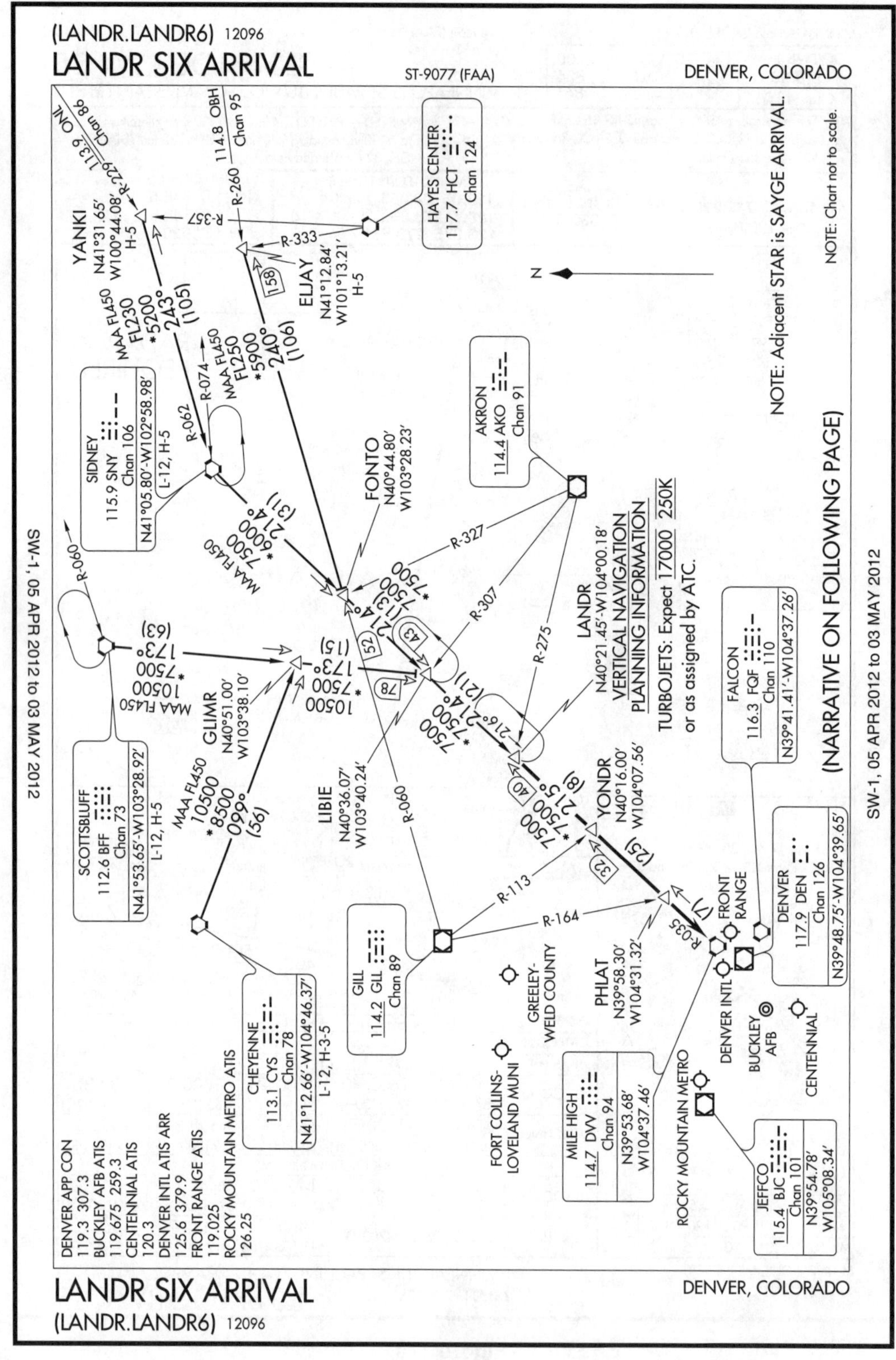
(LANDR.LANDR6) 12096
LANDR SIX ARRIVAL
ST-9077 (FAA)
DENVER, COLORADO
DENVER APP CON
119.3 307.3
BUCKLEY AFB ATIS
119.675 259.3
CENTENNIAL ATIS
120.3
DENVER INTL ATIS ARR
125.6 379.9
FRONT RANGE ATIS
119.025
ROCKY MOUNTAIN METRO ATIS
126.25
YANKI
N41°31.65′ W100°44.08′
H-5
ELJAY
N41°12.84′ W101°13.21′
H-5
HAYES CENTER
117.7 HCT
Chan 124
114.8 OBH
Chan 95
113.9 ONL
Chan 86
SIDNEY
115.9 SNY
Chan 106
N41°05.80′-W102°58.98′
L-12, H-5
SCOTTSBLUFF
112.6 BFF
Chan 73
N41°53.65′-W103°28.92′
L-12, H-5
CHEYENNE
113.1 CYS
Chan 78
N41°12.66′-W104°46.37′
L-12, H-3-5
FONTO
N40°44.80′ W103°28.23′
GLIMR
N40°51.00′ W103°38.10′
LIBIE
N40°36.07′ W103°40.24′
AKRON
114.4 AKO
Chan 91
GILL
114.2 GLL
Chan 89
LANDR
N40°21.45′-W104°00.18′
VERTICAL NAVIGATION PLANNING INFORMATION
TURBOJETS: Expect 17000 250K or as assigned by ATC.
YONDR
N40°16.00′ W104°07.56′
PHLAT
N39°58.30′ W104°31.32′
FALCON
116.3 FQF
Chan 110
N39°41.41′-W104°37.26′
DENVER
117.9 DEN
Chan 126
N39°48.75′-W104°39.65′
MILE HIGH
114.7 DVV
Chan 94
N39°53.68′ W104°37.46′
JEFFCO
115.4 BJC
Chan 101
N39°54.78′ W105°08.34′
FORT COLLINS-LOVELAND MUNI
GREELEY-WELD COUNTY
DENVER INTL
FRONT RANGE
BUCKLEY AFB
CENTENNIAL
ROCKY MOUNTAIN METRO
NOTE: Adjacent STAR is SAYGE ARRIVAL.
NOTE: Chart not to scale.
(NARRATIVE ON FOLLOWING PAGE)
SW-1, 05 APR 2012 to 03 MAY 2012
LANDR SIX ARRIVAL
(LANDR.LANDR6) 12096
DENVER, COLORADO

Figure 371

(LANDR.LANDR6) 12096

LANDR SIX ARRIVAL ST-9077 (FAA) DENVER, COLORADO

ARRIVAL ROUTE DESCRIPTION

CHEYENNE TRANSITION (CYS.LANDR6): From over CYS VORTAC via CYS R-099 and BFF R-173 to LIBIE INT; then via SNY R-214 and DVV R-035 to LANDR INT. Thence . . . .

ELJAY TRANSITION (ELJAY.LANDR6): From over ELJAY INT via GLL R-060 to FONTO INT, then via SNY R-214 and DVV R-035 to LANDR INT. Thence . . . .

SCOTTSBLUFF TRANSITION (BFF.LANDR6): From over BFF VORTAC via BFF R-173 to LIBIE INT; then via SNY R-214 and DVV R-035 to LANDR INT. Thence . . . .

SIDNEY TRANSITION (SNY.LANDR6): From over SNY VORTAC via SNY R-214 to FONTO INT, then via SNY R-214 and DVV R-035 to LANDR INT. Thence . . . .

YANKI TRANSITION (YANKI.LANDR6): From over YANKI INT via SNY R-062 to SNY VORTAC; then via SNY R-214 and DVV R-035 to LANDR INT. Thence . . . .

. . . . From over LANDR INT via DVV R-035 to DVV VORTAC. Expect RADAR vectors to the final approach course at or before DVV VORTAC.

SW-1, 05 APR 2012 to 03 MAY 2012

SW-1, 05 APR 2012 to 03 MAY 2012

LANDR SIX ARRIVAL DENVER, COLORADO

(LANDR.LANDR6) 12096

Figure 372

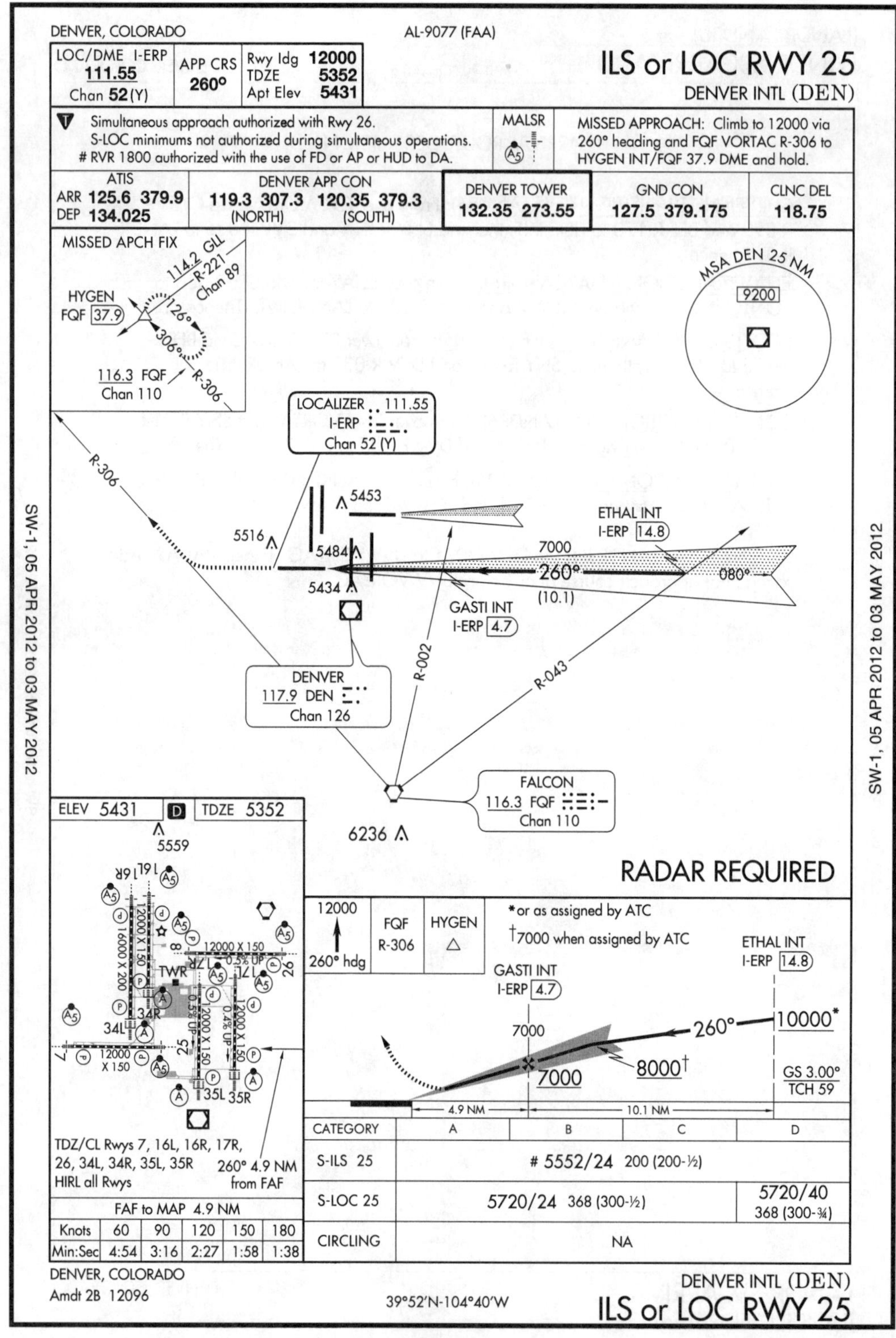

CATEGORY	A	B	C	D
S-ILS 25	# 5552/24 200 (200-½)			
S-LOC 25	5720/24 368 (300-½)			5720/40 368 (300-¾)
CIRCLING	NA			

FAF to MAP 4.9 NM

Knots	60	90	120	150	180
Min:Sec	4:54	3:16	2:27	1:58	1:38

Figure 373

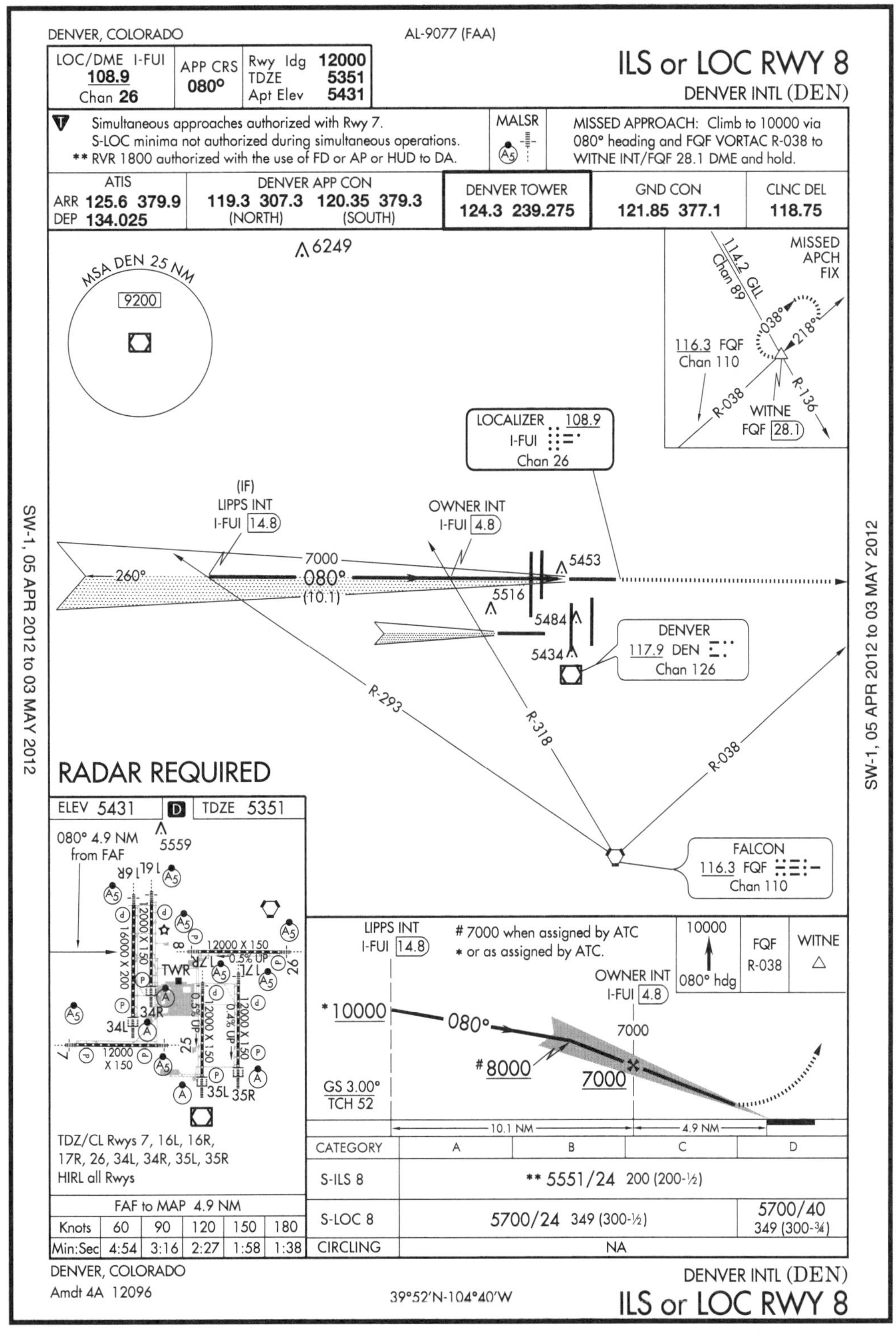

CATEGORY	A	B	C	D
S-ILS 8	** 5551/24 200 (200-½)			
S-LOC 8	5700/24 349 (300-½)			5700/40 349 (300-¾)
CIRCLING	NA			

FAF to MAP 4.9 NM					
Knots	60	90	120	150	180
Min:Sec	4:54	3:16	2:27	1:58	1:38

Figure 374

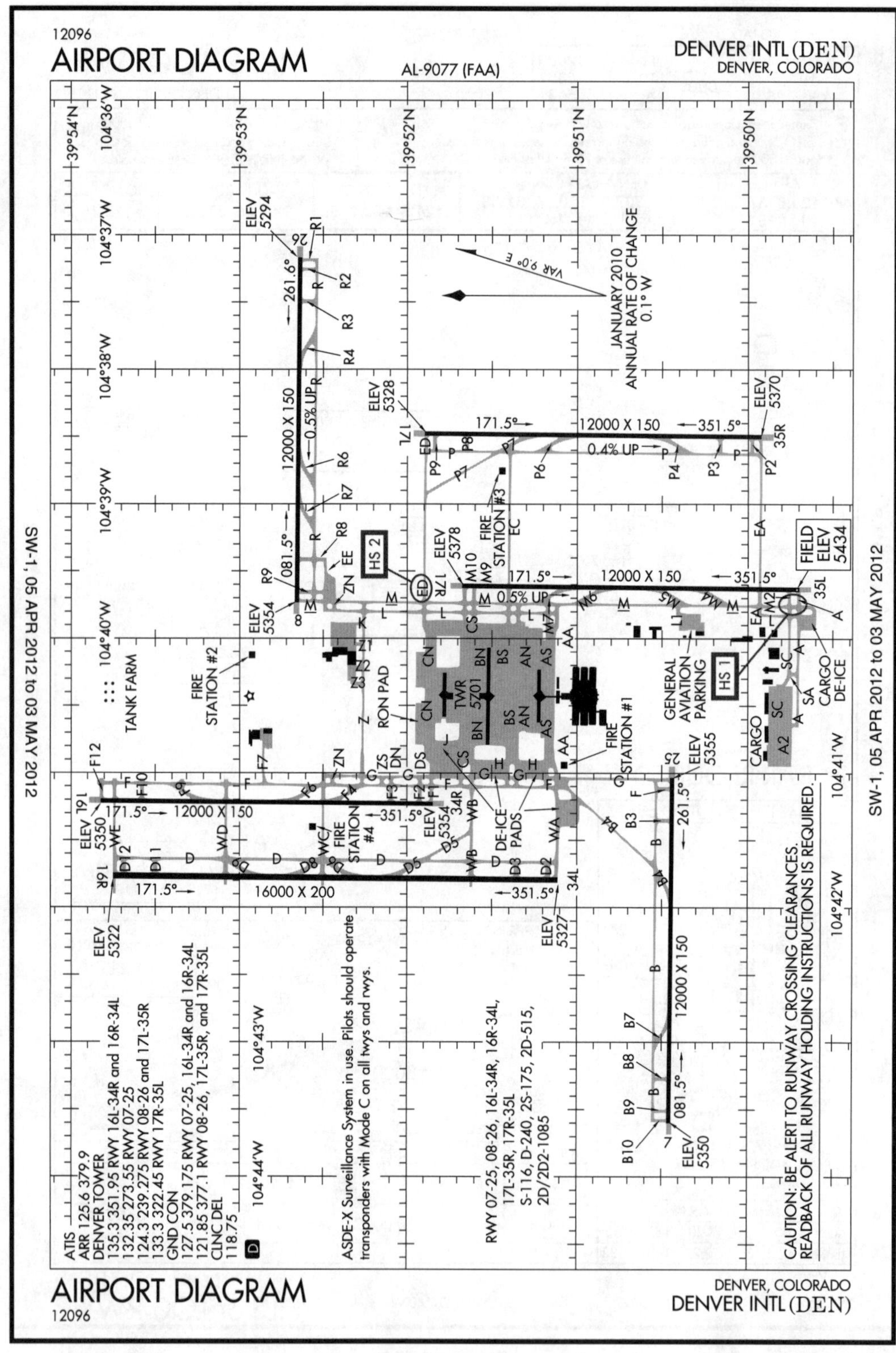

Figure 375

DENVER INTL (DEN) 16 NE UTC–7(–6DT) N39°51.70′ W104°40.39′ **DENVER**
5434 B S4 **FUEL** 100, 100LL, JET A, MOGAS OX 1, 3 Class I, ARFF Index E **H–3F, 5A, L–10F, A**
NOTAM FILE DEN **IAP, AD**

RWY 16R–34L: H16000X200 (CONC–GRVD) S–116, D–240, 2S–175, 2D–515, 2D/2D2–1085 PCN 92 R/B/W/T HIRL CL
RWY 16R: MALSR. TDZL. PAPI(P4R)—GA 3.0° TCH 55′.
RWY 34L: ALSF2. TDZL. PAPI(P4L)—GA 3.0° TCH 50′.
RWY 07–25: H12000X150 (CONC–GRVD) S–116, D–240, 2S–175, 2D–515, 2D/2D2–1085 PCN 92 R/B/W/T HIRL CL
RWY 07: MALSR. TDZL. PAPI(P4R)—GA 3.0° TCH 55′.
RWY 25: MALSR. PAPI(P4L)—GA 3.0° TCH 59′.
RWY 08–26: H12000X150 (CONC–GRVD) S–116, D–240, 2S–175, 2D–515, 2D/2D2–1085 PCN 92 R/B/W/T HIRL CL
RWY 08: MALSR. PAPI(P4L)—GA 3.0° TCH 52′.
RWY 26: MALSR. TDZL. PAPI(P4L)—GA 3.0° TCH 55′. 0.5% up.
RWY 16L–34R: H12000X150 (CONC–GRVD) S–116, D–240, 2S–175, 2D–515, 2D/2D2–1085 PCN 92 R/B/W/T HIRL CL
RWY 16L: MALSR. TDZL. PAPI(P4L)—GA 3.0° TCH 60′.
RWY 34R: ALSF2. TDZL. PAPI(P4L)—GA 3.0° TCH 59′.
RWY 17L–35R: H12000X150 (CONC–GRVD) S–116, D–240, 2S–175, 2D–515, 2D/2D2–1085 PCN 92 R/B/W/T HIRL CL
RWY 17L: MALSR. PAPI(P4L)—GA 3.0° TCH 55′. 0.4% up.
RWY 35R: ALSF2. TDZL. PAPI(P4R)—GA 3.0° TCH 59′.
RWY 17R–35L: H12000X150 (CONC–GRVD) S–116, D–240, 2S–175, 2D–515, 2D/2D2–1085 PCN 92 R/B/W/T HIRL CL
RWY 17R: MALSR. TDZL. PAPI(P4L)—GA 3.0° TCH 60′. 0.5% up.
RWY 35L: ALSF2. TDZL. PAPI(P4R)—GA 3.0° TCH 57′.

RUNWAY DECLARED DISTANCE INFORMATION
RWY 07: TORA–12000 TODA–12000 ASDA–12000 LDA–12000
RWY 08: TORA–12000 TODA–13000 ASDA–12000 LDA–12000
RWY 16L: TORA–12000 TODA–12000 ASDA–12000 LDA–12000
RWY 16R: TORA–16000 TODA–16000 ASDA–16000 LDA–16000
RWY 17L: TORA–12000 TODA–12000 ASDA–12000 LDA–12000
RWY 17R: TORA–12000 TODA–12000 ASDA–12000 LDA–12000
RWY 25: TORA–12000 TODA–13000 ASDA–12000 LDA–12000
RWY 26: TORA–12000 TODA–12000 ASDA–12000 LDA–12000
RWY 34L: TORA–16000 TODA–16000 ASDA–16000 LDA–16000
RWY 34R: TORA–12000 TODA–13000 ASDA–12000 LDA–12000
RWY 35L: TORA–12000 TODA–12000 ASDA–12000 LDA–12000
RWY 35R: TORA–12000 TODA–12000 ASDA–12000 LDA–12000

AIRPORT REMARKS: Attended continuously. Waterfowl and migratory bird activity invof arpt year round. ASDE–X Surveillance System in use: Pilots should opr transponders with Mode C on all twys and rwys. Arpt maintains clearways (500′ X 1000′). 1.25% slope) on departure Rwy 08, Rwy 26, and Rwy 34R. RVR Rwy 07 touchdown, rollout, RVR Rwy 25 touchdown, rollout, RVR Rwy 08 touchdown, rollout, RVR Rwy 26 touchdown, rollout, RVR Rwy 16L touchdown, midfield, rollout, RVR Rwy 34R touchdown, midfield, rollout, RVR Rwy 17L touchdown, midfield, rollout, RVR Rwy 35R touchdown, midfield, rollout, RVR Rwy 17R touchdown, midfield, rollout, RVR Rwy 35L touchdown, midfield, rollout. RVR Rwy 16R touchdown, midfield, rollout, RVR Rwy 34L touchdown, midfield, rollout. Overhead passenger bridge on South side of concourse ‘A′ provides 42 ft tail and 118 ft wingspan clearance when on twy centerline. Insufficient twy corner fillet pavement in the SE corner of the Twy M/M2 intersection for acft with wingspan over 107 ft. Informal rwy use program is in effect 24 hours a day. For additional noise abatement information contact airport management at 303–342–4200. Customs avbl with prior permission. Ldg fee. Flight Notification Service (ADCUS) avbl. NOTE: See Special Notices—Continuous Power Facilities.
WEATHER DATA SOURCES: ASOS (303) 342-0838 LLWAS-NE. TDWR.
COMMUNICATIONS: D-ATIS ARR 125.6 303-342-0819 **D-ATIS DEP** 134.025 303-342-0820 **UNICOM** 122.95
RCO 122.2 122.35 (DENVER RADIO)
RCO 123.65 (DENVER RADIO)
Ⓡ **APP CON** 119.3 124.95 (North) 120.35 126.55 (South) **FINAL CON** 120.8
TOWER 132.35 (Rwy 07-25) 135.3 (Rwy 16L-34R, Rwy 16R-34L) 133.3 (Rwy 17R-35L) 124.3 (Rwy 08-26 and 17L-35R)
GND CON 127.5 (Rwy 07-25, Rwy 16L-34R and Rwy 16R-34L) 121.85 (Rwys 08-26, 17L-35R and 17R-35L)
CLNC DEL 118.75
Ⓡ **DEP CON** 128.25 (East) 127.05 (North) 126.1 (West) 128.45 (South)
AIRSPACE: CLASS B See VFR Terminal Area Chart
RADIO AIDS TO NAVIGATION: NOTAM FILE DEN.

CONTINUED ON NEXT PAGE

Figure 376

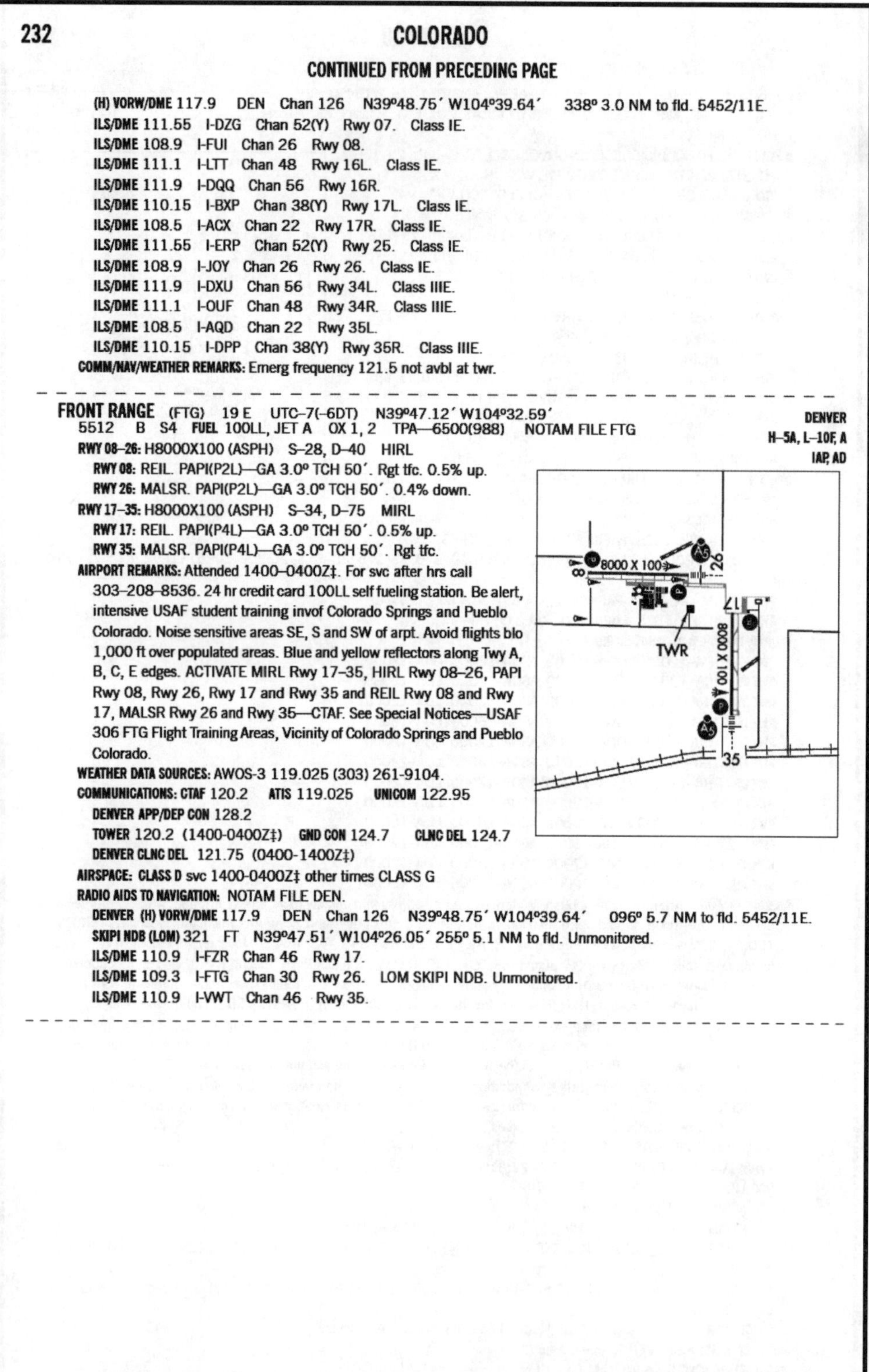

232 **COLORADO**

CONTINUED FROM PRECEDING PAGE

(H) VORW/DME 117.9 DEN Chan 126 N39°48.75′ W104°39.64′ 338° 3.0 NM to fld. 5452/11E.
ILS/DME 111.55 I-DZG Chan 52(Y) Rwy 07. Class IE.
ILS/DME 108.9 I-FUI Chan 26 Rwy 08.
ILS/DME 111.1 I-LTT Chan 48 Rwy 16L. Class IE.
ILS/DME 111.9 I-DQQ Chan 56 Rwy 16R.
ILS/DME 110.15 I-BXP Chan 38(Y) Rwy 17L. Class IE.
ILS/DME 108.5 I-ACX Chan 22 Rwy 17R. Class IE.
ILS/DME 111.55 I-ERP Chan 52(Y) Rwy 25. Class IE.
ILS/DME 108.9 I-JOY Chan 26 Rwy 26. Class IE.
ILS/DME 111.9 I-DXU Chan 56 Rwy 34L. Class IIIE.
ILS/DME 111.1 I-OUF Chan 48 Rwy 34R. Class IIIE.
ILS/DME 108.5 I-AQD Chan 22 Rwy 35L.
ILS/DME 110.15 I-DPP Chan 38(Y) Rwy 35R. Class IIIE.
COMM/NAV/WEATHER REMARKS: Emerg frequency 121.5 not avbl at twr.

FRONT RANGE (FTG) 19 E UTC–7(–6DT) N39°47.12′ W104°32.59′ **DENVER**
5512 B S4 **FUEL** 100LL, JET A OX 1, 2 TPA—6500(988) NOTAM FILE FTG **H–5A, L–10F, A**
IAP, AD
RWY 08–26: H8000X100 (ASPH) S–28, D–40 HIRL
RWY 08: REIL. PAPI(P2L)—GA 3.0° TCH 50′. Rgt tfc. 0.5% up.
RWY 26: MALSR. PAPI(P2L)—GA 3.0° TCH 50′. 0.4% down.
RWY 17–35: H8000X100 (ASPH) S–34, D–75 MIRL
RWY 17: REIL. PAPI(P4L)—GA 3.0° TCH 50′. 0.5% up.
RWY 35: MALSR. PAPI(P4L)—GA 3.0° TCH 50′. Rgt tfc.
AIRPORT REMARKS: Attended 1400–0400Z‡. For svc after hrs call 303–208–8536. 24 hr credit card 100LL self fueling station. Be alert, intensive USAF student training invof Colorado Springs and Pueblo Colorado. Noise sensitive areas SE, S and SW of arpt. Avoid flights blo 1,000 ft over populated areas. Blue and yellow reflectors along Twy A, B, C, E edges. ACTIVATE MIRL Rwy 17–35, HIRL Rwy 08–26, PAPI Rwy 08, Rwy 26, Rwy 17 and Rwy 35 and REIL Rwy 08 and Rwy 17, MALSR Rwy 26 and Rwy 35—CTAF. See Special Notices—USAF 306 FTG Flight Training Areas, Vicinity of Colorado Springs and Pueblo Colorado.
WEATHER DATA SOURCES: AWOS-3 119.025 (303) 261-9104.
COMMUNICATIONS: CTAF 120.2 **ATIS** 119.025 **UNICOM** 122.95
DENVER APP/DEP CON 128.2
TOWER 120.2 (1400-0400Z‡) **GND CON** 124.7 **CLNC DEL** 124.7
DENVER CLNC DEL 121.75 (0400-1400Z‡)
AIRSPACE: CLASS D svc 1400-0400Z‡ other times CLASS G
RADIO AIDS TO NAVIGATION: NOTAM FILE DEN.
DENVER (H) VORW/DME 117.9 DEN Chan 126 N39°48.75′ W104°39.64′ 096° 5.7 NM to fld. 5452/11E.
SKIPI NDB (LOM) 321 FT N39°47.51′ W104°26.05′ 255° 5.1 NM to fld. Unmonitored.
ILS/DME 110.9 I-FZR Chan 46 Rwy 17.
ILS/DME 109.3 I-FTG Chan 30 Rwy 26. LOM SKIPI NDB. Unmonitored.
ILS/DME 110.9 I-VWT Chan 46 Rwy 35.

Figure 377

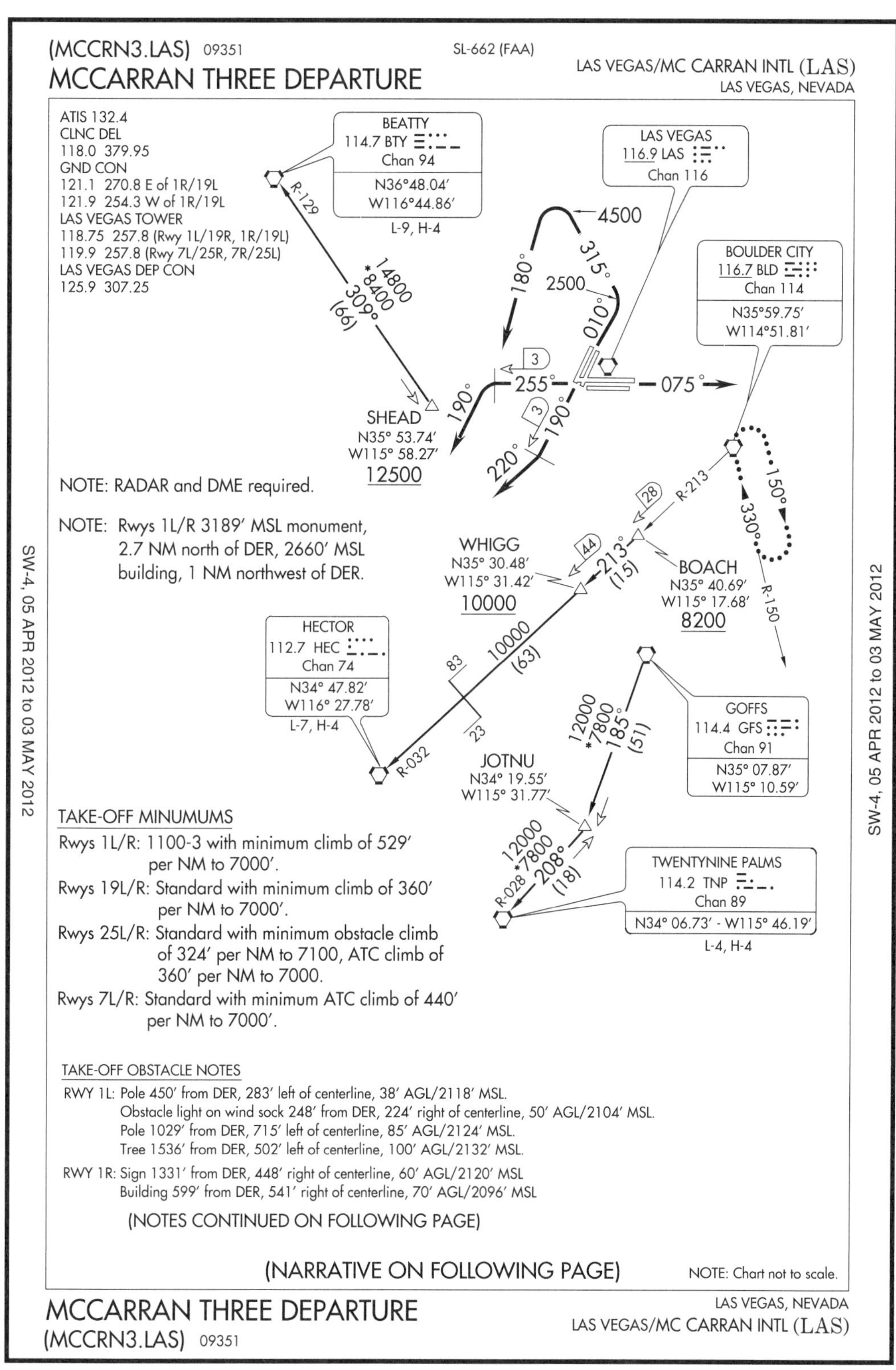

(MCCRN3.LAS) 09351
SL-662 (FAA)
MCCARRAN THREE DEPARTURE
LAS VEGAS/MC CARRAN INTL (LAS)
LAS VEGAS, NEVADA
ATIS 132.4
CLNC DEL
118.0 379.95
GND CON
121.1 270.8 E of 1R/19L
121.9 254.3 W of 1R/19L
LAS VEGAS TOWER
118.75 257.8 (Rwy 1L/19R, 1R/19L)
119.9 257.8 (Rwy 7L/25R, 7R/25L)
LAS VEGAS DEP CON
125.9 307.25
BEATTY
114.7 BTY
Chan 94
N36°48.04'
W116°44.86'
L-9, H-4
LAS VEGAS
116.9 LAS
Chan 116
BOULDER CITY
116.7 BLD
Chan 114
N35°59.75'
W114°51.81'
SHEAD
N35° 53.74'
W115° 58.27'
12500
NOTE: RADAR and DME required.
NOTE: Rwys 1L/R 3189' MSL monument, 2.7 NM north of DER, 2660' MSL building, 1 NM northwest of DER.
WHIGG
N35° 30.48'
W115° 31.42'
10000
BOACH
N35° 40.69'
W115° 17.68'
8200
HECTOR
112.7 HEC
Chan 74
N34° 47.82'
W116° 27.78'
L-7, H-4
GOFFS
114.4 GFS
Chan 91
N35° 07.87'
W115° 10.59'
JOTNU
N34° 19.55'
W115° 31.77'
TWENTYNINE PALMS
114.2 TNP
Chan 89
N34° 06.73' - W115° 46.19'
L-4, H-4
SW-4, 05 APR 2012 to 03 MAY 2012
TAKE-OFF MINUMUMS
Rwys 1L/R: 1100-3 with minimum climb of 529' per NM to 7000'.
Rwys 19L/R: Standard with minimum climb of 360' per NM to 7000'.
Rwys 25L/R: Standard with minimum obstacle climb of 324' per NM to 7100, ATC climb of 360' per NM to 7000.
Rwys 7L/R: Standard with minimum ATC climb of 440' per NM to 7000'.
TAKE-OFF OBSTACLE NOTES
RWY 1L: Pole 450' from DER, 283' left of centerline, 38' AGL/2118' MSL.
Obstacle light on wind sock 248' from DER, 224' right of centerline, 50' AGL/2104' MSL.
Pole 1029' from DER, 715' left of centerline, 85' AGL/2124' MSL.
Tree 1536' from DER, 502' left of centerline, 100' AGL/2132' MSL.
RWY 1R: Sign 1331' from DER, 448' right of centerline, 60' AGL/2120' MSL
Building 599' from DER, 541' right of centerline, 70' AGL/2096' MSL
(NOTES CONTINUED ON FOLLOWING PAGE)
(NARRATIVE ON FOLLOWING PAGE)
NOTE: Chart not to scale.
MCCARRAN THREE DEPARTURE
(MCCRN3.LAS) 09351
LAS VEGAS, NEVADA
LAS VEGAS/MC CARRAN INTL (LAS)

Figure 378

(MCCRN3.LAS) 09351 SL-662 (FAA)

MCCARRAN THREE DEPARTURE

LAS VEGAS/MC CARRAN INTL (LAS)
LAS VEGAS, NEVADA

DEPARTURE ROUTE DESCRIPTION

TAKE-OFF RUNWAYS 1L/R: Climb via heading 010° to 2500', then climbing left turn via heading 315° to 4500', then climbing left turn heading 180°, thence ....

TAKE-OFF RUNWAYS 7L/R: Climb via heading 075°, thence ....

TAKE-OFF RUNWAYS 19L/R: Climb via heading 190° until LAS VORTAC 3 DME, then right turn via heading 220°, thence ....

TAKE-OFF RUNWAYS 25L/R: Climb via heading 255° until LAS VORTAC 3 DME, then left turn via heading 190°, thence ....

....via radar vectors to transition or assigned route, maintain 7000', expect clearance to filed altitude 2 minutes after departure.

LOST COMMUNICATIONS: If no contact with ATC upon reaching 7000', proceed direct BLD VORTAC, then climb in BLD VORTAC holding pattern to the appropriate MEA for route of flight.

BEATTY TRANSITION (MCCRN3.BTY): From over SHEAD INT via BTY R-129 to BTY VORTAC.

HECTOR TRANSITION (MCCRN3.HEC): From over BOACH INT via BLD R-213 and HEC R-032 to HEC VORTAC.

TWENTY NINE PALMS TRANSITION (MCCRN3.TNP): From over GFS VORTAC via GFS R-185 to JOTNU INT, then via TNP R-028 to TNP VORTAC.

TAKE-OFF OBSTACLE NOTES (CONTINUED)

RWY 25R: Light pole 3115' from DER, 1033' right of centerline, 109' AGL/2301' MSL.
Light on pole 1.5 NM from DER, 2836' left of centerline, 124' AGL/2457' MSL.
Light pole 1.7 NM from DER, 2965' left of centerline, 139' AGL/2469' MSL.
Light on pole 1100' from DER, 508' left of centerline, 47' AGL/2226' MSL.
Building 1822' from DER, 652' left of centerline, 46' AGL/2238' MSL.
Building 2202' from DER, 596' left of centerline, 44' AGL/2246' MSL.
Rod on building 534' from DER, 369' left of centerline, 33' AGL/2202' MSL.
Road 678' from DER, 16' right of centerline, 35' AGL/2201' MSL.
Light on localizer antenna 533' from DER, 32' AGL/2195' MSL.

RWY 25L: Pole 2860' from DER, 813' left of centerline, 57' AGL/2236' MSL.
Sign 3672' from DER, 1302' left of centerline, 57' AGL/2256' MSL.
Antenna on building 1002' from DER, 251' left of centerline, 34' AGL/2183' MSL.
Pole 3677' from DER, 145' left of centerline, 67' AGL/2249' MSL.

RWY 7L: Tree 1257' from DER, 789' left of centerline, 85' AGL/2077' MSL.
Light pole 747' from DER, 441' right of centerline, 62' AGL/2057' MSL.
Tree 1007' from DER, 557' right of centerline, 70' AGL/2062' MSL.

RWY 7R: Light on wind sock 102' from DER, 300' right of centerline, 30' AGL/2051' MSL.

RWY 19L: Pole 1394' from DER, 533' right of centerline, 36' AGL/2236' MSL.
Sign 2181' from DER, 1062' right of centerline, 50' AGL/2256' MSL.
Rod on building 2921' from DER, 581' right of centerline, 50' AGL/2262' MSL.
Pole 2633' from DER, 319' right of centerline, 40' AGL/2246' MSL.

RWY 19R: Pole 1135' from DER, 619' right of centerline, 65' AGL/2249' MSL.
Pole 756' from DER, 618' right of centerline, 50' AGL/2231' MSL.
Sign 2182' from DER, 125' right of centerline, 50' AGL/2256' MSL.
Pole 1396' from DER, 403' left of centerline, 55' AGL/2236' MSL.
Rod on building 197' from DER, 441' right of centerline, 30' AGL/2202' MSL.
Rod on building 2922' from DER, 356' left of centerline, 50' AGL/2262' MSL.

SW-4, 05 APR 2012 to 03 MAY 2012

SW-4, 05 APR 2012 to 03 MAY 2012

MCCARRAN THREE DEPARTURE
(MCCRN3.LAS) 09351
LAS VEGAS, NEVADA
LAS VEGAS/MC CARRAN INTL (LAS)

Figure 379

184 CALIFORNIA

SAN FRANCISCO INTL (SFO) 8 SE UTC–8(–7DT) N37°37.14′ W122°22.49′ **SAN FRANCISCO**
13 B S4 **FUEL** 100, 100LL, JET A OX 1, 2, 3, 4 LRA Class I, ARFF Index E **H–3B, L–2F, 3B, A**
NOTAM FILE SFO **IAP, AD**

RWY 10L–28R: H11870X200 (ASPH–GRVD) PCN 80 F/B/X/T HIRL CL
RWY 10L: REIL. PAPI(P4L)—GA 3.0° TCH 80′. Tower.
RWY 28R: ALSF2. TDZL. PAPI(P4L)—GA 3.0° TCH 70′. Rgt tfc.
RWY 10R–28L: H10602X200 (ASPH–GRVD) PCN 80 F/B/X/T HIRL CL
RWY 10R: PAPI(P4L)—GA 3.0° TCH 75′. Tower. Rgt tfc.
RWY 28L: SSALR. PAPI(P4L)—GA 3.0° TCH 75′.
RWY 01R–19L: H8648X200 (ASPH–GRVD) PCN 100F/B/X/T HIRL CL
RWY 01R: REIL. Thld dsplcd 238′. Tree.
RWY 19L: MALSF. TDZL. PAPI(P4L)—GA 3.0° TCH 75′.
RWY 01L–19R: H7500X200 (ASPH–CONC–GRVD) PCN 90 F/B/X/T HIRL CL
RWY 01L: REIL. Thld dsplcd 491′.
RWY 19R: PAPI(P4L)—GA 3.0° TCH 73′.

Rwy 01L-19R: 7500 X 200

AIRPORT REMARKS: Attended continuously. PAEW AER 28L, Rwy 28R and Rwy 19L indef. Flocks of birds feeding along shoreline adjacent to arpt, on occasions fly across various parts of arpt. Due to obstructed vision, SFO twr is able to provide only limited arpt tfc control svc on Twy A between gates 88 and 89. Twr personnel are unable to determine whether this area is clear of traffic or obstructions. Rwy 10 preferred rwy between 0900–1400Z‡ weather and flight conditions permitting. Simultaneous ops in effect all rwys. Helicopter ldg area marked on Twy (C) west of Twy (R) opr for civil and military use. Noise sensitive arpt. For noise abatement procedures ctc arpt noise office Monday–Friday 1600–0100Z‡ by calling 650–821–5100. Airline pilots shall strictly follow the painted nose gear lines and no oversteering adjustment is permitted. No grooving exists at arpt rwy intersections. Rwy 01L–19R, 01R–19L, Rwy 10R–28L, Rwy 10L–28R grooved full length except at rwy intersections. B747, B777, A330, A340 or larger acft are restricted from using Twy A1 when B747–400, A340–600 or larger acft are holding short of Rwy 01R on Twy A. 747–400′s shall taxi at a speed of less than 10 miles per hour on all non–restricted taxiways on the terminal side of the intersecting rwys. All outbound Twy Y heavy aircraft with a wingspan of 171′ or greater under power prohibited from entering westbound Twy Z. Ramp clsd to acft with wingspan over 117′ at Terminal 1, gate C41 indef. Movement speed of not more than 5 miles per hour is required when two 747–400′s pass or overtake each other on parallel taxiways A and B. Rwy 19L MALSF has a NSTD length of 1115′ with 3 sequenced flashers. Ldg fee. Flight Notification Service (ADCUS) available. NOTE: See Special Notices—Intersection Departures During Period of Darkness, Expanded Charted Visual Flight Procedures. Continuous Power Facilities, Special Noise Abatement Procedures, Special Noise Abatement Procedures—Preferential Runways.

WEATHER DATA SOURCES: ASOS (650) 872-0246 LLWAS.

COMMUNICATIONS: D-ATIS 135.45 118.85 115.8 113.7 650-877-3585/8422 **UNICOM** 122.95
Ⓡ**NORCAL APP CON** 135.65 (S) 133.95
TOWER 120.5 **GND CON** 121.8 **CLNC DEL** 118.2 **PRE TAXI CLNC** 118.2
Ⓡ**NORCAL DEP CON** 135.1 (SE-W) 120.9 (NW-E)

AIRSPACE: CLASS B See VFR Terminal Area Chart

RADIO AIDS TO NAVIGATION: NOTAM FILE SFO.
(L) VORW/DME 115.8 SFO Chan 105 N37°37.17′ W122°22.43′ at fld. 13/17E.
VOR DME unusable:
025°-065° byd 30 NM blo 18,000′
035°-055° byd 12 NM blo 6,500′
150°-190° byd 25 NM blo 4,500′
190°-260° byd 10 NM blo 4,500′
260°-295° byd 35 NM blo 3,000′
295°-330° byd 20 NM blo 8,000′
BRIJJ NDB (LOM) 379 GW N37°34.33′ W122°15.59′ 282° 6.2 NM to fld. LOM unusable 160°-195° byd 6 NM.
ILS/DME 108.9 I-SIA Chan 26 Rwy 19L. Class IE. Ry 19L glideslope deviations are possible when critical areas are not required to be protected. Acft operating invof glideslope transmitter. Pilots should be alert for momentary localizer course excursions due to large aircraft operating in vicinity of localizer antenna.
ILS/DME 109.55 I-SFO Chan 32(Y) Rwy 28L. Class IE.
ILS/DME 111.7 I-GWQ Chan 54 Rwy 28R. Class IIIE. LOM BRIJJ NDB. LOM unusable 160°-195° byd 6 NM.
LDA/DME 110.75 I-FNP Chan 44(Y) Rwy 28R.

Figure 380

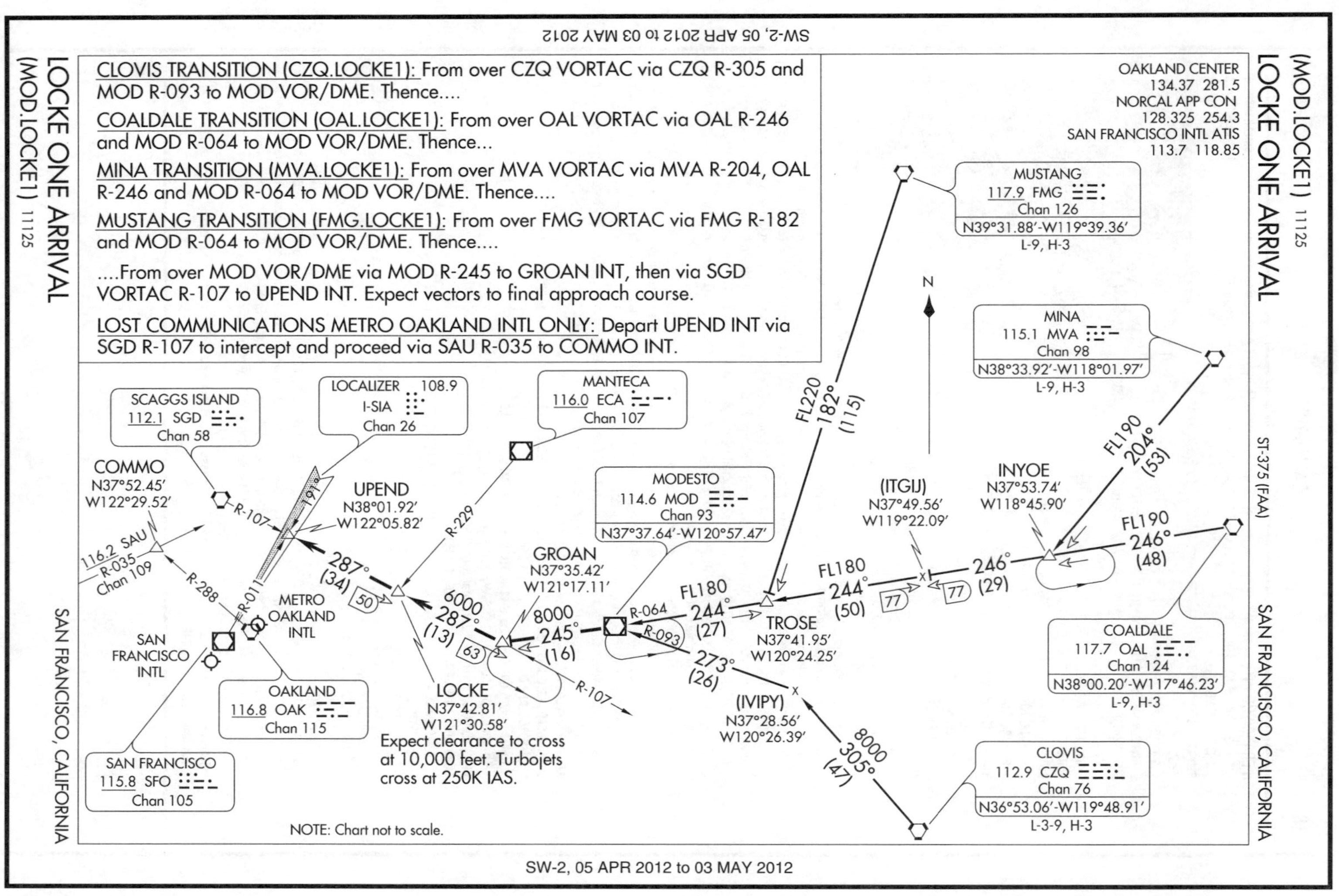
(MOD.LOCKE1) 11125
LOCKE ONE ARRIVAL
ST-375 (FAA)
SAN FRANCISCO, CALIFORNIA
CLOVIS TRANSITION (CZQ.LOCKE1): From over CZQ VORTAC via CZQ R-305 and MOD R-093 to MOD VOR/DME. Thence....
COALDALE TRANSITION (OAL.LOCKE1): From over OAL VORTAC via OAL R-246 and MOD R-064 to MOD VOR/DME. Thence...
MINA TRANSITION (MVA.LOCKE1): From over MVA VORTAC via MVA R-204, OAL R-246 and MOD R-064 to MOD VOR/DME. Thence....
MUSTANG TRANSITION (FMG.LOCKE1): From over FMG VORTAC via FMG R-182 and MOD R-064 to MOD VOR/DME. Thence....
....From over MOD VOR/DME via MOD R-245 to GROAN INT, then via SGD VORTAC R-107 to UPEND INT. Expect vectors to final approach course.
LOST COMMUNICATIONS METRO OAKLAND INTL ONLY: Depart UPEND INT via SGD R-107 to intercept and proceed via SAU R-035 to COMMO INT.
OAKLAND CENTER
134.37 281.5
NORCAL APP CON
128.325 254.3
SAN FRANCISCO INTL ATIS
113.7 118.85
MUSTANG
117.9 FMG
Chan 126
N39°31.88'-W119°39.36'
L-9, H-3
MINA
115.1 MVA
Chan 98
N38°33.92'-W118°01.97'
L-9, H-3
N
SCAGGS ISLAND
112.1 SGD
Chan 58
LOCALIZER 108.9
I-SIA
Chan 26
MANTECA
116.0 ECA
Chan 107
FL220
182°
(115)
FL190
204°
(53)
COMMO
N37°52.45'
W122°29.52'
UPEND
N38°01.92'
W122°05.82'
191°
R-107
R-229
MODESTO
114.6 MOD
Chan 93
N37°37.64'-W120°57.47'
(ITGIJ)
N37°49.56'
W119°22.09'
INYOE
N37°53.74'
W118°45.90'
FL190
246°
(48)
116.2 SAU
R-035
Chan 109
R-288
R-011
287°
(34)
50
METRO OAKLAND INTL
GROAN
N37°35.42'
W121°17.11'
FL180
244°
(50)
77
77
246°
(29)
6000
287°
(13)
63
8000
245°
(16)
R-064
R-093
FL180
244°
(27)
TROSE
N37°41.95'
W120°24.25'
SAN FRANCISCO INTL
273°
(26)
COALDALE
117.7 OAL
Chan 124
N38°00.20'-W117°46.23'
L-9, H-3
OAKLAND
116.8 OAK
Chan 115
LOCKE
N37°42.81'
W121°30.58'
Expect clearance to cross at 10,000 feet. Turbojets cross at 250K IAS.
R-107
(IVIPY)
N37°28.56'
W120°26.39'
8000
305°
(47)
SAN FRANCISCO
115.8 SFO
Chan 105
CLOVIS
112.9 CZQ
Chan 76
N36°53.06'-W119°48.91'
L-3-9, H-3
NOTE: Chart not to scale.
SW-2, 05 APR 2012 to 03 MAY 2012
SW-2, 05 APR 2012 to 03 MAY 2012
LOCKE ONE ARRIVAL
(MOD.LOCKE1) 11125
SAN FRANCISCO, CALIFORNIA

Figure 381

OAKLAND

METROPOLITAN OAKLAND INTL (OAK) 4 S UTC–8(–7DT) N37°43.28´ W122°13.24´ **SAN FRANCISCO**
9 B S4 **FUEL** 100LL, JET A OX 1, 2, 3, 4 TPA—See Remarks LRA Class I, ARFF Index D **H–3B, L–2F, 3B, A**
NOTAM FILE OAK **IAP, AD**

RWY 11–29: H10000X150 (ASPH–GRVD) PCN 71 F/A/W/T HIRL CL
RWY 11: MALSR. PAPI(P4L)—GA 2.75° TCH 65´. Rgt tfc.
RWY 29: ALSF2. TDZL. PAPI(P4L)—GA 3.0° TCH 71´.
RWY 09R–27L: H6213X150 (ASPH–GRVD) PCN 97 F/B/W/T HIRL
RWY 09R: REIL. PAPI(P4R)—GA 3.0° TCH 50´.
RWY 27L: PAPI(P4L)—GA 3.0° TCH 71´.
RWY 09L–27R: H5454X150 (ASPH–GRVD) PCN 69 F/C/W/T HIRL
RWY 09L: PAPI(P4R)—GA 3.0° TCH 49´.
RWY 27R: MALSR. PAPI(P4L)—GA 2.9° TCH 57´. Bldg. Rgt tfc.
RWY 15–33: H3372X75 (ASPH) S–12.5 MIRL
RWY 33: Rgt tfc.

AIRPORT REMARKS: Attended continuously. Rwy 15–33 CLOSED to air carrier acft. Birds on and invof arpt. Acft with experimental or limited certification having over 1,000 horsepower or 4,000 pounds are restricted to Rwy 11–29. 24 hr Noise abatement procedure–turbojet and turbofan powered acft, turborops over 17,000 lbs, four engine reciprocating powered acft, and surplus Military acft over 12,500 lbs should not depart Rwy 27L and Rwy 27R or land on Rwy 09L and Rwy 09R. For noise abatement information ctc noise abatement office at 510–563–6463. Intersection of Twy B, Twy W and Twy V not visible from twr. Twy K between Rwy 33 and Twy D and portions of Twy D not visible from twr. Twy A, Twy E, Twy G, Twy H between Rwy 27R and Twy C max acft weight 150,000 lbs. Twy G and Twy H between Rwy 27L and Rwy 27R, max acft weight 12,500 lbs. Twy P max acft weight 24,000 lbs single, 40,000 lbs dual. Twy C between Rwy 27R and Twy G and Twy B, Twy J, and Twy D max acft weight 9,000,000 lbs. Twy C between Twy G and Twy J max acft weight 25,000 lbs single, 175,000 lbs dual, 4,000,000 lbs tandem. Twy C between Twy J and Twy F max acft weight 25,000 lbs single, 150,000 lbs dual. 155,000 lbs tandem (dual tandem not authorized). Twy K between Twy D and intersection Twy F, Twy L, Twy K max acft weight 25,000 lbs single, 115,000 lbs dual, 140,000 lbs tandem. Twy K between Rwy 9R and intersection Twy F, Twy L, Twy K max acft weight 25,000 lbs single, 115,000 lbs dual, 140,000 lbs tandem. Twy K between Rwy 9R and intersection Twy F, Twy L, Twy K max acft weight 25,000 lbs single, 45,000 lbs dual, tandem not authorized. Preferential rwy use program in effect 0600–1400Z‡. North fld preferred arrival Rwy 27L, north fld preferred departure Rwys 09R or 27R. If these Rwys unacceptable for safety or twr instruction then Rwy 11–29 must be used. Noise prohibitions not applicable in emerg or whenever Rwy 11–29 is closed due to maintenance, safety, winds or weather. 400´ blast pad Rwy 29 and 500´ blast pad Rwy 11. Rwys 29, 27R and 27L distance remaining signs left side. TPA—Rwy 27L 606(597), TPA—Rwy 27R 1006(997). Ldg fee may apply for Rwy 11–29, rwy commercial ops and tiedown, ctc afld ops 510–563–3361. Flight Notification Service (ADCUS) avbl.

WEATHER DATA SOURCES: ASOS (510) 383-9514 **HIWAS** 116.8 OAK.

COMMUNICATIONS: D-ATIS 133.775 (510) 635-5850 (N and S Complex) **UNICOM** 122.95
OAKLAND RCO 122.2 122.5 (OAKLAND RADIO)
Ⓡ **NORCAL APP CON** 125.35 (East) 135.65 (South) 135.1 (West) 134.5 120.9
Ⓡ **NORCAL DEP CON** 135.1 (West) 120.9 (Northwest)
OAKLAND TOWER 118.3 (N Complex) 127.2 (S Complex) 124.9
GND CON 121.75 (S Complex) 121.9 (N Complex) **CLNC DEL** 121.1

AIRSPACE: CLASS C svc ctc **APP CON**

RADIO AIDS TO NAVIGATION: NOTAM FILE OAK.
OAKLAND (H) VORTACW 116.8 OAK Chan 115 N37°43.56´ W122°13.42´ at fld. 10/17E. **HIWAS.**
DME unusable:
335°-065° byd 30 NM blo 8,000´
ILS 111.9 I-AAZ Rwy 11. Class IE. Glideslope deviations are possible when critical areas are not required to be protected. Acft operating invof glideslope transmitter.
ILS 109.9 I-OAK Rwy 27R. Class IE.
ILS 108.7 I-INB Rwy 29. Class IIIE.

COMM/NAV/WEATHER REMARKS: Emerg frequency 121.5 not avbl at twr.

OAKLAND N37°43.56´ W122°13.42´ NOTAM FILE OAK. **SAN FRANCISCO**
(H) VORTACW 116.8 OAK Chan 115 at Metropolitan Oakland Intl. 10/17E. **HIWAS.** **H–3A, L–2F, 3B, A**
DME unusable:
335°-065° byd 30 NM blo 8,000´
RCO 122.2 122.5 (OAKLAND RADIO)
ASOS OAK N37°43.28´ W122°13.24´. (510) 383-9514.

OCEAN RIDGE (See GUALALA on page 113)

Figure 382

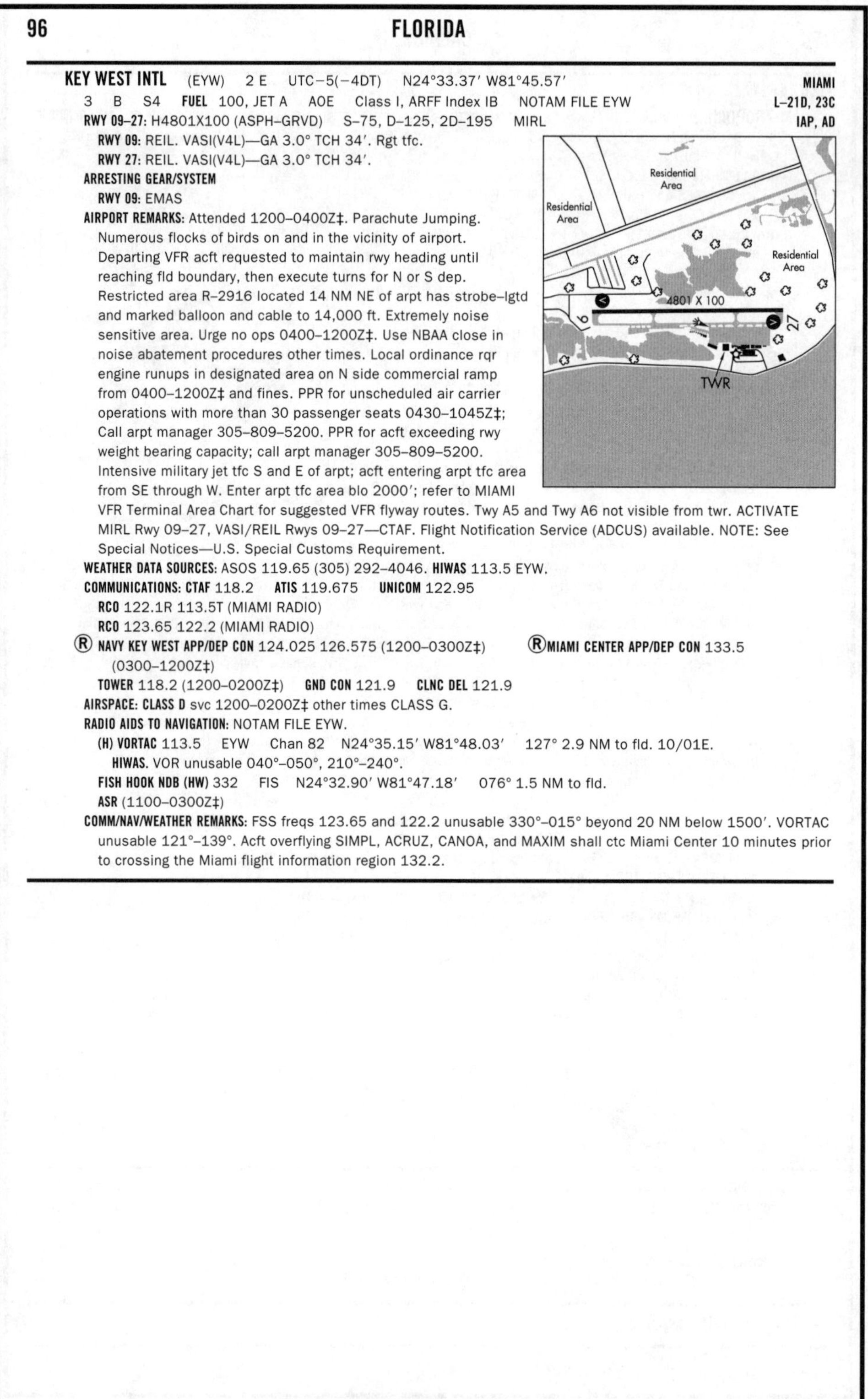

96 **FLORIDA**

KEY WEST INTL (EYW) 2 E UTC–5(–4DT) N24°33.37′ W81°45.57′ **MIAMI**
3 B S4 **FUEL** 100, JET A AOE Class I, ARFF Index IB NOTAM FILE EYW **L–21D, 23C**
RWY 09–27: H4801X100 (ASPH–GRVD) S–75, D–125, 2D–195 MIRL **IAP, AD**
RWY 09: REIL. VASI(V4L)—GA 3.0° TCH 34′. Rgt tfc.
RWY 27: REIL. VASI(V4L)—GA 3.0° TCH 34′.
ARRESTING GEAR/SYSTEM
RWY 09: EMAS
AIRPORT REMARKS: Attended 1200–0400Z‡. Parachute Jumping. Numerous flocks of birds on and in the vicinity of airport. Departing VFR acft requested to maintain rwy heading until reaching fld boundary, then execute turns for N or S dep. Restricted area R–2916 located 14 NM NE of arpt has strobe–lgtd and marked balloon and cable to 14,000 ft. Extremely noise sensitive area. Urge no ops 0400–1200Z‡. Use NBAA close in noise abatement procedures other times. Local ordinance rqr engine runups in designated area on N side commercial ramp from 0400–1200Z‡ and fines. PPR for unscheduled air carrier operations with more than 30 passenger seats 0430–1045Z‡; Call arpt manager 305–809–5200. PPR for acft exceeding rwy weight bearing capacity; call arpt manager 305–809–5200. Intensive military jet tfc S and E of arpt; acft entering arpt tfc area from SE through W. Enter arpt tfc area blo 2000′; refer to MIAMI VFR Terminal Area Chart for suggested VFR flyway routes. Twy A5 and Twy A6 not visible from twr. ACTIVATE MIRL Rwy 09–27, VASI/REIL Rwys 09–27—CTAF. Flight Notification Service (ADCUS) available. NOTE: See Special Notices—U.S. Special Customs Requirement.

WEATHER DATA SOURCES: ASOS 119.65 (305) 292–4046. **HIWAS** 113.5 EYW.
COMMUNICATIONS: CTAF 118.2 **ATIS** 119.675 **UNICOM** 122.95
RCO 122.1R 113.5T (MIAMI RADIO)
RCO 123.65 122.2 (MIAMI RADIO)
Ⓡ **NAVY KEY WEST APP/DEP CON** 124.025 126.575 (1200–0300Z‡) Ⓡ**MIAMI CENTER APP/DEP CON** 133.5 (0300–1200Z‡)
TOWER 118.2 (1200–0200Z‡) **GND CON** 121.9 **CLNC DEL** 121.9
AIRSPACE: CLASS D svc 1200–0200Z‡ other times CLASS G.
RADIO AIDS TO NAVIGATION: NOTAM FILE EYW.
(H) VORTAC 113.5 EYW Chan 82 N24°35.15′ W81°48.03′ 127° 2.9 NM to fld. 10/01E.
HIWAS. VOR unusable 040°–050°, 210°–240°.
FISH HOOK NDB (HW) 332 FIS N24°32.90′ W81°47.18′ 076° 1.5 NM to fld.
ASR (1100–0300Z‡)
COMM/NAV/WEATHER REMARKS: FSS freqs 123.65 and 122.2 unusable 330°–015° beyond 20 NM below 1500′. VORTAC unusable 121°–139°. Acft overflying SIMPL, ACRUZ, CANOA, and MAXIM shall ctc Miami Center 10 minutes prior to crossing the Miami flight information region 132.2.

SE, 05 APR 2012 to 31 MAY 2012

Figure 383

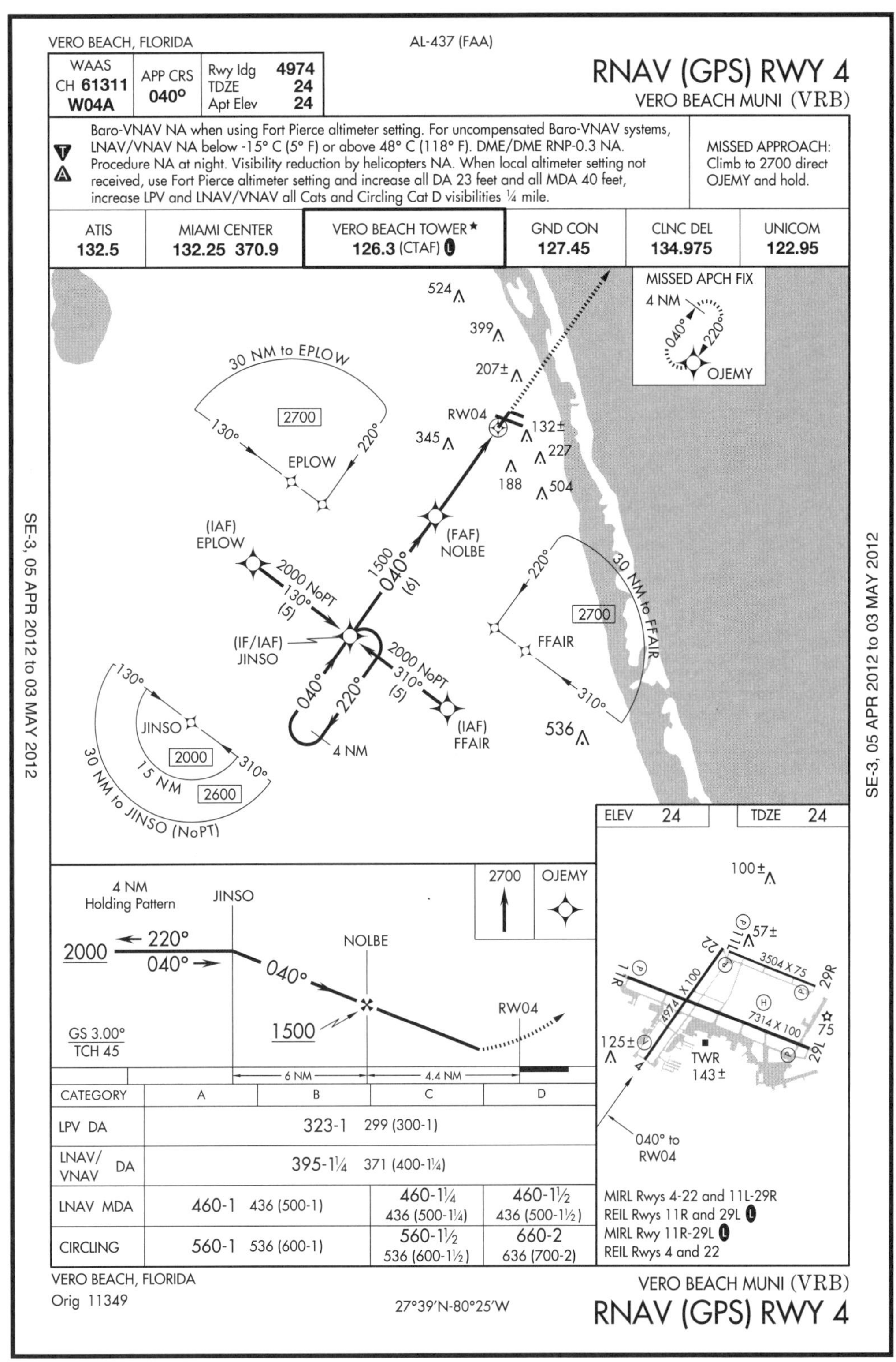

CATEGORY	A	B	C	D
LPV DA	323-1 299 (300-1)			
LNAV/VNAV DA	395-1¼ 371 (400-1¼)			
LNAV MDA	460-1 436 (500-1)		460-1¼ 436 (500-1¼)	460-1½ 436 (500-1½)
CIRCLING	560-1 536 (600-1)		560-1½ 536 (600-1½)	660-2 636 (700-2)

MIRL Rwys 4-22 and 11L-29R
REIL Rwys 11R and 29L
MIRL Rwy 11R-29L
REIL Rwys 4 and 22

Figure 384

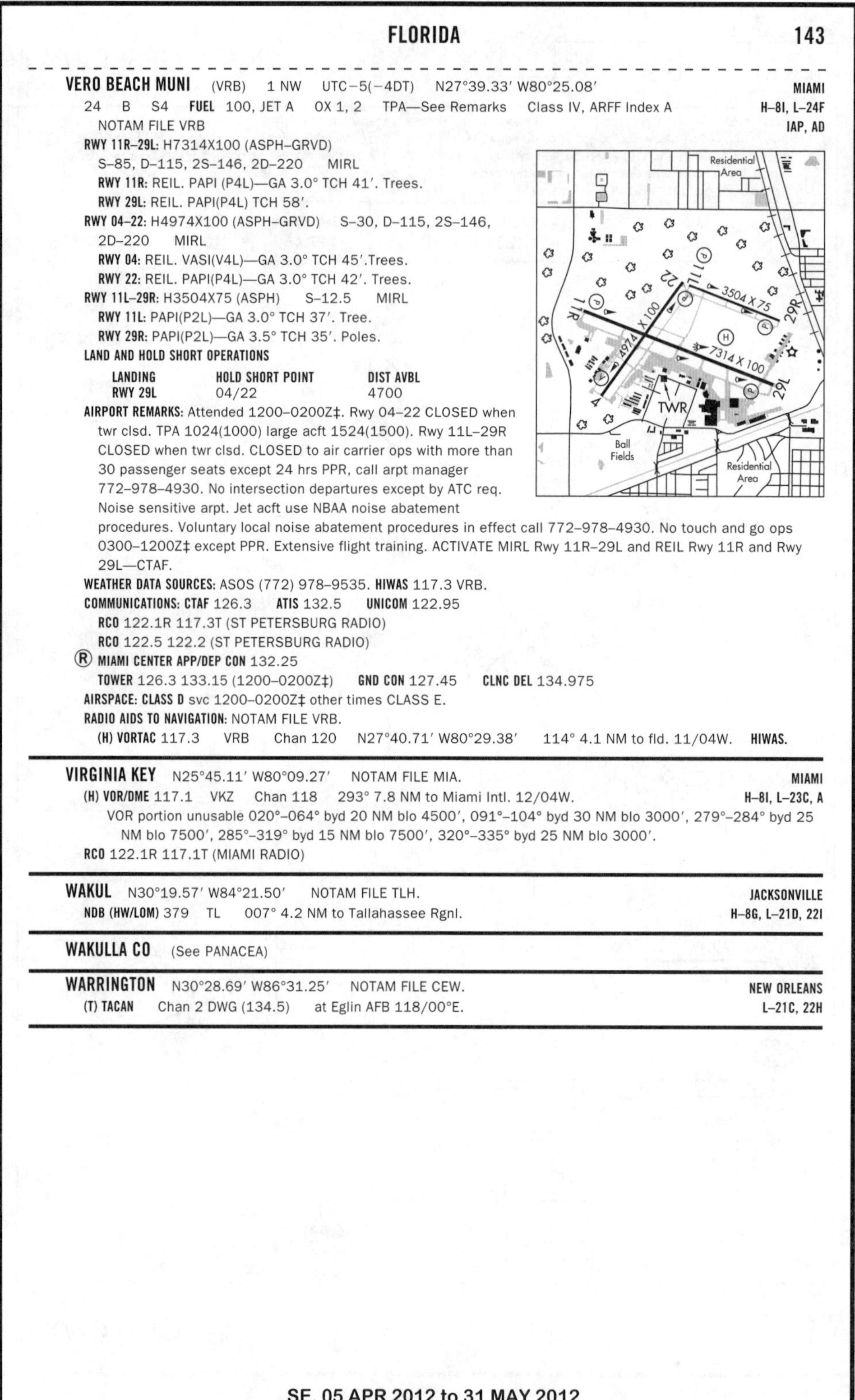

FLORIDA 143

VERO BEACH MUNI (VRB) 1 NW UTC–5(–4DT) N27°39.33′ W80°25.08′ **MIAMI**
24 B S4 **FUEL** 100, JET A OX 1, 2 TPA—See Remarks Class IV, ARFF Index A **H–8I, L–24F**
NOTAM FILE VRB **IAP, AD**

RWY 11R–29L: H7314X100 (ASPH–GRVD)
S–85, D–115, 2S–146, 2D–220 MIRL
RWY 11R: REIL. PAPI (P4L)—GA 3.0° TCH 41′. Trees.
RWY 29L: REIL. PAPI(P4L) TCH 58′.

RWY 04–22: H4974X100 (ASPH–GRVD) S–30, D–115, 2S–146, 2D–220 MIRL
RWY 04: REIL. VASI(V4L)—GA 3.0° TCH 45′.Trees.
RWY 22: REIL. PAPI(P4L)—GA 3.0° TCH 42′. Trees.

RWY 11L–29R: H3504X75 (ASPH) S–12.5 MIRL
RWY 11L: PAPI(P2L)—GA 3.0° TCH 37′. Tree.
RWY 29R: PAPI(P2L)—GA 3.5° TCH 35′. Poles.

LAND AND HOLD SHORT OPERATIONS

LANDING	HOLD SHORT POINT	DIST AVBL
RWY 29L	04/22	4700

AIRPORT REMARKS: Attended 1200–0200Z‡. Rwy 04–22 CLOSED when twr clsd. TPA 1024(1000) large acft 1524(1500). Rwy 11L–29R CLOSED when twr clsd. CLOSED to air carrier ops with more than 30 passenger seats except 24 hrs PPR, call arpt manager 772–978–4930. No intersection departures except by ATC req. Noise sensitive arpt. Jet acft use NBAA noise abatement procedures. Voluntary local noise abatement procedures in effect call 772–978–4930. No touch and go ops 0300–1200Z‡ except PPR. Extensive flight training. ACTIVATE MIRL Rwy 11R–29L and REIL Rwy 11R and Rwy 29L—CTAF.

WEATHER DATA SOURCES: ASOS (772) 978–9535. **HIWAS** 117.3 VRB.

COMMUNICATIONS: CTAF 126.3 **ATIS** 132.5 **UNICOM** 122.95
RCO 122.1R 117.3T (ST PETERSBURG RADIO)
RCO 122.5 122.2 (ST PETERSBURG RADIO)
Ⓡ **MIAMI CENTER APP/DEP CON** 132.25
TOWER 126.3 133.15 (1200–0200Z‡) **GND CON** 127.45 **CLNC DEL** 134.975

AIRSPACE: CLASS D svc 1200–0200Z‡ other times CLASS E.

RADIO AIDS TO NAVIGATION: NOTAM FILE VRB.
(H) VORTAC 117.3 VRB Chan 120 N27°40.71′ W80°29.38′ 114° 4.1 NM to fld. 11/04W. **HIWAS.**

VIRGINIA KEY N25°45.11′ W80°09.27′ NOTAM FILE MIA. **MIAMI**
(H) VOR/DME 117.1 VKZ Chan 118 293° 7.8 NM to Miami Intl. 12/04W. **H–8I, L–23C, A**
VOR portion unusable 020°–064° byd 20 NM blo 4500′, 091°–104° byd 30 NM blo 3000′, 279°–284° byd 25 NM blo 7500′, 285°–319° byd 15 NM blo 7500′, 320°–335° byd 25 NM blo 3000′.
RCO 122.1R 117.1T (MIAMI RADIO)

WAKUL N30°19.57′ W84°21.50′ NOTAM FILE TLH. **JACKSONVILLE**
NDB (HW/LOM) 379 TL 007° 4.2 NM to Tallahassee Rgnl. **H–8G, L–21D, 22I**

WAKULLA CO (See PANACEA)

WARRINGTON N30°28.69′ W86°31.25′ NOTAM FILE CEW. **NEW ORLEANS**
(T) TACAN Chan 2 DWG (134.5) at Eglin AFB 118/00°E. **L–21C, 22H**

SE, 05 APR 2012 to 31 MAY 2012

Figure 385

WINDSOR LOCKS

BRADLEY INTL (BDL) 3 W UTC–5(–4DT) N41°56.35′ W72°41.00′ **NEW YORK**
173 B S4 **FUEL** 100LL, JET A OX 1, 2, 3, 4 TPA—See Remarks **H–10I, 11D, 12K, L–33C, 34I**
LRA Class I, ARFF Index D NOTAM FILE BDL **IAP, AD**

RWY 06–24: H9510X200 (ASPH–GRVD) S–200, D–200, 2S–175, 2D–350, 2D/2D2–710 HIRL CL
RWY 06: ALSF2. TDZL. PAPI(P4L)—GA 3.0° TCH 71′. Trees.
RWY 24: MALSR. TDZL. PAPI(P4L)—GA 3.0° TCH 71′. Trees.
RWY 15–33: H6847X150 (ASPH–GRVD) S–200, D–200, 2S–175, 2D–350 HIRL
RWY 15: REIL. PAPI(P4L)—GA 3.5°TCH 61′. Trees.
RWY 33: MALSF. PAPI(P4R)—GA 3.0°TCH 72′. Trees.
RWY 01–19: H4268X100 (ASPH) S–60, D–190, 2S–175, 2D–328 MIRL
RWY 01: Thld dsplcd 475′. Acft. **RWY 19:** Trees.

LAND AND HOLD SHORT OPERATIONS

LANDING	HOLD SHORT POINT	DIST AVBL
RWY 06	01–19	6000
RWY 24	15–33	5850
RWY 33	06–24	4550

RUNWAY DECLARED DISTANCE INFORMATION

RWY 01:	TORA–4268	TODA–4268	ASDA–4268	
RWY 06:	TORA–9509	TODA–9509	ASDA–9509	LDA–9509
RWY 15:	TORA–6847	TODA–6847	ASDA–6847	LDA–6847
RWY 19:				LDA–4268
RWY 24:	TORA–9509	TODA–9509	ASDA–9509	LDA–9509
RWY 33:	TORA–6847	TODA–6847	ASDA–6847	LDA–6847

Rwy 1-19: 4268 X 100
TWR
9510 X 200
6847 X 150

AIRPORT REMARKS: Attended continuously. Numerous birds frequently on or invof arpt. No training flights; no practice apchs; no touch and go ldgs between: Mon–Sat 0400–1200Z‡ and Sun 0400–1700Z‡. Rwy 01–19 open for acft with wingspan less than 79′. Rwy 01 CLOSED for arrivals to all fixed wing acft. Rwy 19 CLOSED for departures to all fixed wing acft. Twy J clsd between S and R to acft with wing spans in excess of 171 ft. Air National Guard ramp PAEW barricaded adjacent northeast side. ASDE–X Surveilance System in Use. Pilots should operate transponders with Mode C on all twys and rwys. Rwy 33 touchdown RVR avbl. TPA—1873(1700) heavy acft. Rwy 06 visual glideslope indicator and glidepath not coincident. Rwy 24 visual glideslope indicator and glidepath not coincident. Rwy 33 visual glideslope indicator and glidepath not coincident. Ldg fee for business, corporate and revenue producing acft. Flight Notification Service (ADCUS) available. NOTE: See Special Notices–Land and Hold Short Lights.
WEATHER DATA SOURCES: ASOS (860) 627–9732. WSP.
COMMUNICATIONS: D–ATIS 118.15 (860–386–3570) **UNICOM** 122.95
WINDSOR LOCKS RCO 122.3 (BRIDGEPORT RADIO)
Ⓡ **BRADLEY APP CON** 123.95 (176°–240°) 125.35 (241°–060°) 127.8 (061°–175° and HFD area)
Ⓡ **BRADLEY DEP CON** 123.95 (176°–240°) 125.35 (241°–060°) 127.8 (061°–175° and HFD area)
TOWER 120.3 **GND CON** 121.9 **CLNC DEL** 121.75
AIRSPACE: CLASS C svc continuous ctc **APP CON**
RADIO AIDS TO NAVIGATION: NOTAM FILE BDL.
(T) VORTACW 109.0 BDL Chan 27 N41°56.46′ W72°41.32′ at fld. 160/14W.
ILS/DME 111.1 I–BDL Chan 48 Rwy 06. Class IIIE.
ILS/DME 108.55 I–IKX Chan 22(Y) Rwy 33. Class IE.
ILS/DME 111.1 I–MYQ Chan 48 Rwy 24. Class IT. DME unusable from .4 NM inbound to Rwy 24.

YALESVILLE HELIPORT (4C3) 2 N UTC–5(–4DT) N41°29.51′ W72°48.67′
65 B **FUEL** 100LL, JET A NOTAM FILE BDR
HELIPAD H1: H65X65 (CONC)
HELIPORT REMARKS: Attended 1400–2300Z‡. Pilots unfamiliar with heliport ctc 203–294–8800 prior to arrival for a briefing on current procedures. ACTIVATE rotating bcn—123.5
COMMUNICATIONS: CTAF/UNICOM 123.05

NE, 05 APR 2012 to 31 MAY 2012

Figure 386

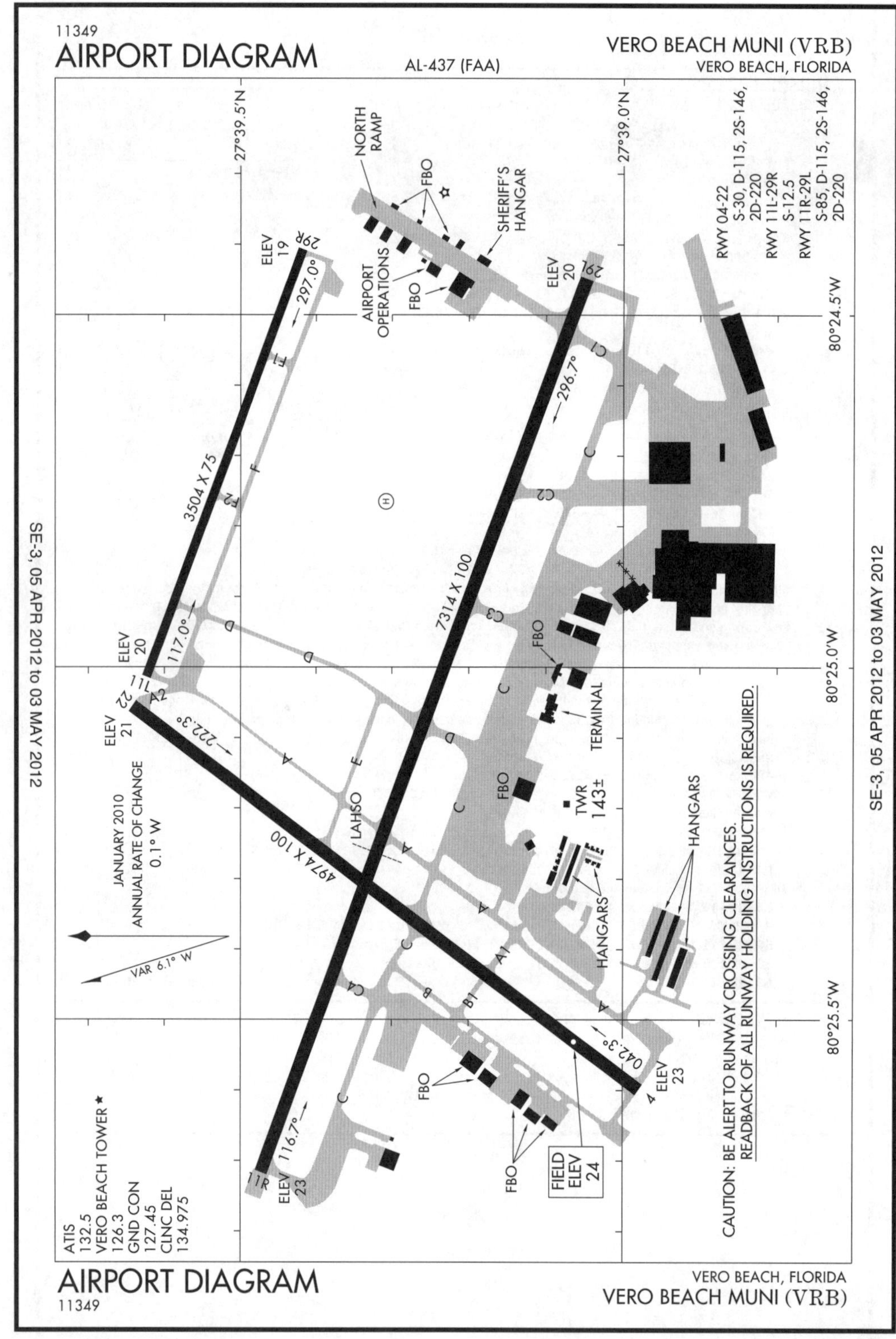

11349
AIRPORT DIAGRAM
AL-437 (FAA)
VERO BEACH MUNI (VRB)
VERO BEACH, FLORIDA
ATIS
132.5
VERO BEACH TOWER ★
126.3
GND CON
127.45
CLNC DEL
134.975
JANUARY 2010
ANNUAL RATE OF CHANGE
0.1° W
VAR 6.1° W
RWY 04-22
S-30, D-115, 2S-146,
2D-220
RWY 11L-29R
S-12.5
RWY 11R-29L
S-85, D-115, 2S-146,
2D-220
3504 X 75
7314 X 100
4974 X 100
NORTH RAMP
SHERIFF'S HANGAR
AIRPORT OPERATIONS
FBO
TERMINAL
HANGARS
TWR
143±
LAHSO
FIELD ELEV 24
CAUTION: BE ALERT TO RUNWAY CROSSING CLEARANCES.
READBACK OF ALL RUNWAY HOLDING INSTRUCTIONS IS REQUIRED.
SE-3, 05 APR 2012 to 03 MAY 2012
27°39.5'N
27°39.0'N
80°24.5'W
80°25.0'W
80°25.5'W
AIRPORT DIAGRAM
11349
VERO BEACH, FLORIDA
VERO BEACH MUNI (VRB)

Figure 387

ALTERNATE MINS

M4

12096

NAME ALTERNATE MINIMUMS

ROCHESTER, NY

GREATER ROCHESTER INTL (ROC) .... **ILS or LOC Rwy 4**[1]
ILS or LOC Rwy 22[1]
ILS or LOC Rwy 28[2]
RNAV (GPS) Rwy 4[3]
RNAV (GPS) Rwy 22[3]
RNAV (GPS) Rwy 25[3]
RNAV (GPS) Rwy 28[4]
VOR Rwy 4[3]
VOR/DME Rwy 4[3]

[1]ILS, Category D, 700-2¼. LOC, Category D, 800-2¼.
[2]ILS, Categories A,B,C, 800-2; Category D, 800-2¼. LOC, Category D, 800-2¼.
[3]Category D, 800-2¼.
[4]Category C, 800-2¼; Category D, 800-2½.

ROME, NY

GRIFFISS INTL (RME) .... **ILS or LOC Rwy 33**[12]
RNAV (GPS) Rwy 15[3]
RNAV (GPS) Rwy 33[3]

NA when local weather not available.
[1]NA when control tower closed.
[2]ILS, Categories A, B, 700-2; Category C, 800-2; Category D, 800-2½. LOC, Category D, 800-2½.
[3]Category D, 800-2½.

SARANAC LAKE, NY

ADIRONDACK RGNL (SLK) .... **VOR/DME Rwy 5**[1]
VOR or GPS Rwy 9[2]

[1]Category A, 1100-2; Category B, 1200-2; Categories C,D, 1200-3.
[2]Categories A,B, 1400-2; Categories C,D, 1400-3.

SHIRLEY, NY

BROOKHAVEN (HWV) .... **RNAV (GPS) Rwy 6**
RNAV (GPS) Rwy 15
RNAV (GPS) Y Rwy 24
RNAV (GPS) Z Rwy 24
RNAV (GPS) Rwy 33
VOR Rwy 6

NA when local weather not available.

SUSSEX, NJ

SUSSEX (FWN) .... **RNAV (GPS) Rwy 3**[1]
VOR-A[2]

NA when local weather not available.
[1]Categories A, B, 900-2; Category C, 900-2½.
[2]Categories A, B, 1400-2; Category C, 1400-3.

NAME ALTERNATE MINIMUMS

TETERBORO, NJ

TETERBORO (TEB) .... **ILS or LOC Rwy 6**[1]
ILS RWY 19[1]
RNAV (GPS) Y Rwy 6[3]
RNAV (RNP) Z Rwy 6, 800-2¼
VOR/DME-A[2]
VOR/DME-B[2]
VOR/DME Rwy 6[3]
VOR Rwy 24[4]

[1]ILS, Categories A,B, 800-2; Category C, 800-2¼; Category D, 900-2¾. LOC, Category C, 800-2¼; Category D, 900-2¾.
[2]Categories A,B, 1000-2; Categories C,D, 1000-3.
[3]Category C, 800-2¼; Category D, 900-2¾.
[4]Categories B,C,D, 1000-3.

TRENTON, NJ

TRENTON MERCER (TTN) .... **ILS Rwy 6**
NDB or GPS Rwy 6
VOR or GPS-A
VOR or GPS Rwy 24

NA when control tower closed.

WATERTOWN, NY

WATERTOWN INTL (ART) .... **ILS or LOC Rwy 7**[1]
RNAV (GPS) Rwy 7[2]
RNAV (GPS) Rwy 10[3]
RNAV (GPS) Rwy 28[3]
VOR Rwy 7[2]

[1]ILS, Categories A, B, C, 700-2; Category D 700-2¼. LOC, Category D, 800-2¼.
[2]Category D, 800-2¼.
[3]NA when local weather not available.

WELLSVILLE, NY

WELLSVILLE MUNI ARPT,TARANTINE FIELD (ELZ) .... **RNAV (GPS) Rwy 10**
RNAV (GPS) Rwy 28
VOR-A[1]

NA when local weather not available.
[1]Categories A,B, 1100-2; Categories C,D, 1100-3.

WESTHAMPTON BEACH, NY

FRANCIS S. GABRESKI (FOK) .... **ILS or LOC Rwy 24**[1]
RNAV (GPS) Rwy 24

NA when local weather not available.
[1]NA when control tower closed.
[2]NA when control tower closed.
[3]Category D, 800-2¼.

05 APR 2012 to 03 MAY 2012

05 APR 2012 to 03 MAY 2012

ALTERNATE MINS

12096

NE-2

M4

Figure 388